Calculus

Differentiation

1. $D_x u^n = n u^{n-1} D_x u$

2. $D_x(u+v) = D_x u + D_x v$

3. $D_x(uv) = u D_x v + v D_x u$

4. $D_x\left(\dfrac{u}{v}\right) = \dfrac{v D_x u - u D_x v}{v^2}$

5. $D_x \sin u = \cos u \, D_x u$

6. $D_x \cos u = -\sin u \, D_x u$

7. $D_x \tan u = \sec^2 u \, D_x u$

8. $D_x \cot u = -\csc^2 u \, D_x u$

9. $D_x \sec u = \sec u \tan u \, D_x u$

10. $D_x \csc u = -\csc u \cot u \, D_x u$

11. $D_x \sin^{-1} u = \dfrac{D_x u}{\sqrt{1-u^2}}$

12. $D_x \cos^{-1} u = \dfrac{-D_x u}{\sqrt{1-u^2}}$

13. $D_x \tan^{-1} u = \dfrac{D_x u}{1+u^2}$

14. $D_x \cot^{-1} u = \dfrac{-D_x u}{1+u^2}$

15. $D_x \sec^{-1} u = \dfrac{D_x u}{|u|\sqrt{u^2-1}}$

16. $D_x \csc^{-1} u = \dfrac{-D_x u}{|u|\sqrt{u^2-1}}$

17. $D_x \displaystyle\int_a^u f(t)\, dt = f(u) D_x u$

18. $D_x \ln u = \dfrac{D_x u}{u}$

19. $D_x e^u = e^u D_x u$

20. $D_x a^u = a^u \ln a \, D_x u$

21. $D_x \log_a u = \dfrac{D_x u}{u \ln a}$

22. $D_x \sinh u = \cosh u \, D_x u$

23. $D_x \cosh u = \sinh u \, D_x u$

24. $D_x \tanh u = \operatorname{sech}^2 u \, D_x u$

25. $D_x \coth u = -\operatorname{csch}^2 u \, D_x u$

26. $D_x \operatorname{sech} u = -\operatorname{sech} u \tanh u \, D_x u$

27. $D_x \operatorname{csch} u = -\operatorname{csch} u \coth u \, D_x u$

Basic Integration

1. $\displaystyle\int u^n\, dx = \dfrac{u^{n+1}}{n+1} + C \quad (n \neq -1)$

2. $\displaystyle\int \dfrac{du}{u} = \ln |u| + C$

3. $\displaystyle\int \sin u\, du = -\cos u + C$

4. $\displaystyle\int \cos u\, du = \sin u + C$

5. $\displaystyle\int \sec^2 u\, du = \tan u + C$

6. $\displaystyle\int \csc^2 u\, du = -\cot u + C$

7. $\displaystyle\int \sec u \tan u\, du = \sec u + C$

8. $\displaystyle\int \csc u \cot u\, du = -\csc u + C$

9. $\displaystyle\int \tan u\, du = -\ln |\cos u| + C$

10. $\displaystyle\int \cot u\, du = \ln |\sin u| + C$

11. $\displaystyle\int \sec u\, du = \ln |\sec u + \tan u| + C$

12. $\displaystyle\int \csc u\, du = \ln |\csc u - \cot u| + C$

13. $\displaystyle\int \sin^2 u\, du = \tfrac{1}{2}u - \tfrac{1}{4}\sin 2u + C$

14. $\displaystyle\int \cos^2 u\, du = \tfrac{1}{2}u + \tfrac{1}{4}\sin 2u + C$

15. $\displaystyle\int \dfrac{du}{\sqrt{a^2-u^2}} = \sin^{-1}\dfrac{u}{a} + C$

16. $\displaystyle\int \dfrac{du}{a^2+u^2} = \dfrac{1}{a}\tan^{-1}\dfrac{u}{a} + C$

17. $\displaystyle\int \dfrac{du}{u\sqrt{u^2-a^2}} = \dfrac{1}{a}\sec^{-1}\left|\dfrac{u}{a}\right| + C$

18. $\displaystyle\int \sinh u\, du = \cosh u + C$

19. $\displaystyle\int \cosh u\, du = \sinh u + C$

20. $\displaystyle\int \operatorname{sech}^2 u\, du = \tanh u + C$

21. $\displaystyle\int \operatorname{csch}^2 u\, du = -\coth u + C$

22. $\displaystyle\int \operatorname{sech} u \tanh u\, du = -\operatorname{sech} u + C$

23. $\displaystyle\int \operatorname{csch} u \coth u\, du = -\operatorname{csch} u + C$

24. $\displaystyle\int u\, dv = uv - \displaystyle\int v\, du + C$

25. $\displaystyle\int e^u\, du = e^u + C$

26. $\displaystyle\int a^u\, du = \dfrac{a^u}{\ln a} + C$

CALCULUS

with Analytic Geometry

Mustafa A. Munem   Macomb County Community College

David J. Foulis   University of Massachusetts

Worth Publishers, Inc.

Calculus
with Analytic Geometry

Printed in the United States of America

Library of Congress Catalog Card No. 770818

ISBN: 0-87901-087-8

First printing, February, 1978

Design by Malcolm Grear Designers

Composition by Santype International Limited

Art by Reproduction Drawings Limited

Printing by Rand McNally and Company

Worth Publishers, Inc.

444 Park Avenue South

New York, New York 10016

PREFACE

PURPOSE This textbook is intended for use in a standard college-level calculus course. It provides the necessary background in calculus and analytic geometry for students of mathematics, engineering, physics, chemistry, economics, or the life sciences.

PREREQUISITES Students who use this textbook should know the basic principles of algebra, geometry, and trigonometry covered in the usual precalculus mathematics courses. No previous study of analytic geometry is required; it is introduced as needed in the text.

OBJECTIVES The book was written with two main objectives in mind: First, that all explanations be sufficiently clear and accessible that students will have no difficulty in reading and learning from the book itself; second, that the facility gained from working through the book will enable students to apply the principles learned to the solution of practical problems. To accomplish these objectives, new topics are introduced in informal, everyday language and illustrated with simple and familiar examples. Formal definitions and technical theorems are introduced only after students have had an opportunity to understand the new concepts and to appreciate their usefulness. Some theorems are proved rigorously while others are made plausible by an appeal to geometric intuition. Some formal definitions, theorems, and proofs that are considered within the scope of the motivated student, but whose introduction might distract and exasperate such a student, are deferred to a separate subsection at the end of the section or to a separate section at the end of the chapter. This arrangement permits the individual instructor to introduce or to skip particular proofs without interrupting the flow of material critical to the chapter.

SPECIAL FEATURES The following features of the book are especially noteworthy:

1. The material is developed systematically through worked-out examples and geometric motivation, followed by clear definitions, step-by-step procedures, precise statements of theorems, and comprehensible proofs. The book contains nearly 950 worked-out examples and more than 850 figures and graphs. The step-by-step procedures assist students with im-

portant techniques that often cause difficulty (such as graph sketching, change of variable, and integration by parts).

2. Because so much of calculus is learned by working problems, special attention has been devoted to the problem sets at the end of each section and to the review problem sets at the end of each chapter. There are more than 6300 problems. These problems include applications to a wide variety of fields and are carefully graded so that the student can build confidence while progressing from simpler to more challenging problems. The odd-numbered problems are generally of the "drill" variety and strive for the level of understanding desired by most users. The worked-out examples in the text are similar to the odd-numbered problems, and answers to these problems are provided at the end of the book. Some of the even-numbered problems, especially those toward the ends of the problem sets, require considerable insight and should provide enough challenge to satisfy the most demanding instructor and the most highly motivated student. This arrangement of problems simplifies the task of making assignments.

The review problem sets at the end of each chapter highlight the essential material in the chapter. These additional problems will help students to gain confidence in their understanding of the material and will indicate to them those areas where additional study may be needed.

3. The power and elegance of calculus are amply demonstrated by the abundant applications, not only to geometry, engineering, physics, and chemistry, but also to economics, business, biology, ecology, sociology, and medicine.

4. The concepts and tools that are needed by students of engineering and science—derivatives, integrals, and differential equations—are developed early in the text.

5. Access to a hand calculator is *not* a prerequisite; however, if a calculator is available, many computational examples and problems can be worked and checked more easily.

6. Color is used for emphasis—to highlight definitions, theorems, properties, rules, step-by-step procedures, important statements, and parts of figures.

7. Pertinent formulas from algebra, geometry, trigonometry, and calculus are listed inside the front and back covers of the book for the student's convenience.

8. The entire manuscript has been class-tested a number of times over a two-year period by the authors and their colleagues.

CONTENTS   Although the table of contents gives an adequate indication of the order in which the material is presented, the following supplementary comments may be helpful to the potential user of the book.

Chapter 0 provides a review of basic precalculus mathematics, including inequalities, cartesian coordinates, trigonometry, and functions. Material on composition of functions and inverses of functions is placed at the end of Chapter 0 for ease of reference.

Calculus proper begins with Chapter 1 on limits and continuity of functions. Differentiation formulas for the trigonometric functions are introduced informally in Chapter 2 so as to be accessible to the student who does not take the whole calculus sequence, and for the convenience of engineering and physics students who need to see these formulas as early as possible. Instructors who desire an early *formal* treatment of the differentiation of trigonometric functions can cover Sections 1 and 2 of Chapter 8 immediately after Section 4 of Chapter 2 and then return to Section 5 of Chapter 2 with no loss of continuity. Linear approximation is introduced in Chapter 2 and is used to give a rigorous proof of the chain rule.

Chapter 4 gives a succinct, but reasonably complete account of the conic sections and their properties—more material on the conics (polar form and rotation of axes) can be found in Chapter 11.

Simple differential equations provide the main theme of Chapter 5. Finding the area under a curve is first treated as the problem of setting up and solving a separable differential equation. This approach not only shows how differential equations can arise in connection with the solution of practical problems, but it also provides a first informal glimpse of the definite integral and the fundamental theorem of calculus. The usual definition of the definite integral as a limit of Riemann sums is given in Chapter 6.

Throughout the book, the reader is continually encouraged to visualize analytic relationships in geometric form. In Chapters 14 and 15, direct geometric formulation of various problems is enhanced by the introduction of vectors. All concepts involving vectors are first introduced *geometrically;* thus the "analytic" treatment of vectors in terms of their scalar components is derived from simple geometric considerations. Chapter 14 treats only vectors in the plane so that the student can become comfortable with the more easily visualized planar configurations before studying vectors in three-dimensional space in Chapter 15.

Chapter 17 includes three sections (9, 10, and 11) on line integrals, surface integrals, Green's theorem, Gauss's divergence theorem, and Stokes' theorem. This material should be especially beneficial for those students who will not go on to study more advanced topics in calculus and analysis in subsequent courses.

PACE   The pace of the course, as well as the choice of which topics to cover and which to emphasize, will vary from school to school, depending on such factors as the requirements of the curriculum, the academic calendar, and the predilections of individual instructors. With students who are adequately prepared, the entire book can be covered in three semesters or in five quarters.

In general, Chapters 0 through 5 include enough material for a first-semester course, Chapters 6 through 13 are suitable for the second semester, and the remaining chapters can be covered in the third semester. Those instructors who might wish to postpone Chapters 12 and 13 until the third semester could replace them by Chapter 14.

The book is deliberately written for maximum flexibility; there are many ways to arrange the material coherently to conform to a wide variety of possible situations.

STUDY GUIDE   An accompanying *Study Guide* is available for students who need more drill or more assistance in any topic. The *Study Guide* is written in a semiprogrammed format; it conforms with the arrangement of topics in the book and contains a significant number of carefully graded fill-in statements and problems broken down into simple units. Study objectives and tests are also included for each chapter. Answers are given to all problems in the *Study Guide* in order to encourage self-study and self-testing at the student's own pace.

ACKNOWLEDGMENTS   We wish to thank the following individuals, who reviewed the manuscript and offered many helpful suggestions:

Professors Gerald L. Bradley, *Claremont Men's College*

Richard Dahlke, *University of Michigan,* Dearborn

Garret J. Etgen, *University of Houston*

Frank D. Farmer, *Arizona State University*

Brauch Fugate, *University of Kentucky*

Douglas W. Hall, *Michigan State University*

Franz X. Hiergeist, *West Virginia University*

Frank E. Higginbotham, *University of Puerto Rico*

Laurence D. Hoffman, *Claremont Men's College*

George W. Johnson, *University of South Carolina*

Kenneth Kalmanson, *Montclair State College*

Joseph F. Krebs, *Boston College*

Lynn C. Kurtz, *Arizona State University*

Stanley M. Lukawecki, *Clemson University*

George E. Mitchell, *University of Alabama*

Barbara Price, *Wayne State University*

Russell J. Rowlett, *University of Tennessee,* Knoxville

David Ryeburn, *Simon Fraser University*

Nevin Savage, *Arizona State University*

David A. Schedler, *Virginia Commonwealth University*

Jerry Silver, *Ohio State University*

Harold T. Slaby, *Wayne State University*

Gilbert Steiner, *Fairleigh Dickinson University,* Teaneck

Donald G. Stewart, *Arizona State University*

Neil A. Weiss, *Arizona State University*

We would also like to express our appreciation to Professors Donald Catlin, Thurlow Cook, Charles Randall, and Karen Zak of the University of Massachusetts for their special assistance.

We especially wish to thank the students who used preliminary versions of the book. Every teaching idea in this book is here because it worked in the classroom. Some ideas seemed good at first but were abandoned because they were not helpful to students. We are grateful to our students for reporting to us every frustration and for being willing to try new ideas.

Special thanks are due to Professor Steve Fasbinder of Wayne State University for reviewing the manuscript, reading page proofs, and solving many of the problems. We also are especially indebted to Hyla Gold Foulis for reviewing the manuscript, proofreading pages, and solving all the problems in the book. Finally, we would like to express our sincere gratitude to Paula Fasulo for her expert typing of the entire manuscript and to the staff of Worth Publishers for their constant help and encouragement.

Mustafa A. Munem
David J. Foulis

February, 1978

CONTENTS

0 PRECALCULUS REVIEW

This chapter provides the mathematical background required for understanding calculus. We begin by presenting some fundamental properties of the real number system but make no attempt to go deeply into this topic. We also develop some basic analytic geometry, set forth the notion of a function, and introduce the "algebra" of functions. The chapter includes a concise treatment of the trigonometric functions. Although much of this preliminary material may already be familiar to some readers, for proper comprehension of succeeding chapters, we advise that it be reviewed, at least briefly.

1 Real Numbers

Perhaps the most intuitively appealing approach to the *real number system* is to regard real numbers as corresponding to points along an infinite straight line. Suppose that a fixed point O, called the *origin*, and another fixed point U, called the *unit point*, are chosen on a straight line L (Figure 1). The distance between O and U is called the *unit distance*, and may be 1 inch, 1 centimeter, 1 mile, 1 parsec, or whatever unit of measure is desired. If the straight line L is horizontal, it is customary to place U to the right of O.

Each point P on line L is now assigned a *coordinate* x representing its *signed distance* from the origin O. Thus, $x = \pm d$, where d is the distance from O to P measured in terms of the given unit (Figure 2). The plus sign is used when P is to the right of O and the minus sign is used when P is to the left of O. The origin O is assigned the coordinate 0 and the unit point U is assigned the coordinate 1.

When each point on line L has been assigned a coordinate as described, the line is called a *number scale*, a *number line*, or a *coordinate axis*. Such a coordinate axis appears in Figure 3, where the numerical coordinates of a few selected points are shown explicitly. It is convenient to affix an arrowhead

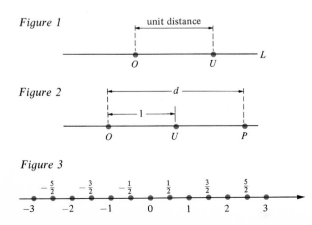

Figure 1

unit distance

O U L

Figure 2

d

1

O U P

Figure 3

$-\frac{5}{2}$ $-\frac{3}{2}$ $-\frac{1}{2}$ $\frac{1}{2}$ $\frac{3}{2}$ $\frac{5}{2}$

-3 -2 -1 0 1 2 3

to the coordinate axis to indicate the direction (to the *right* in Figure 3) in which the numerical coordinates are increasing.

It is not possible to show the coordinates of all the points on a number scale explicitly, since there are infinitely many such points and we soon run out of space, ink, and patience. Nevertheless, it is convenient to *imagine* all these coordinates displayed at once along the line. We call the set of *all* these coordinates the *set of real numbers*, and we denote this infinite set by the symbol $\mathbb{R}$.

Two real numbers x and y in $\mathbb{R}$ can be combined by means of the usual arithmetic operations to yield new real numbers $x + y, xy, x - y$, and (provided that $y \neq 0$) $x \div y$. (In algebra and calculus it is usually more convenient to write $\dfrac{x}{y}$ or x/y rather than $x \div y$.) It is also possible to compare x and y to see which is the larger. For instance, if y is larger than x, we write $y > x$ (or $x < y$). An assertion of the form $y > x$ (or $x < y$) is called an *inequality* and can be interpreted geometrically to mean that, on the number line, the point P whose coordinate is x lies to the left of the point Q whose coordinate is y (Figure 4). Intuitively, given two points on the number scale, either they are equal or one lies to the left of the other. This is expressed symbolically as follows.

Figure 4

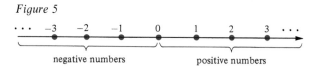

Trichotomy Principle

If x and y are any real numbers, then one and only one of the following conditions holds:

1 $x < y$,

2 $y < x$, or

3 $x = y$.

If we assume that $y = 0$ in the trichotomy principle, we see that one and only one of the following conditions is true:

1 $x < 0$, in which case we call x a *negative* real number;

2 $x > 0$, in which case we call x a *positive* real number; or

3 $x = 0$, in which case x is neither positive nor negative.

Figure 5

$$\cdots \quad -3 \quad -2 \quad -1 \quad 0 \quad 1 \quad 2 \quad 3 \quad \cdots$$

$$\underbrace{}_{\text{negative numbers}} \qquad \underbrace{}_{\text{positive numbers}}$$

On a horizontal number scale, the positive numbers are coordinates of points to the right of the origin and the negative numbers are coordinates of points to the left of the origin (Figure 5).

The following rules for handling inequalities should be familiar to the reader.

Rules for Manipulating Inequalities

Assume that a, b, c, and d are real numbers.

1 If $a < b$, then $a + c < b + c$.

2 If $a < b$ and $c < d$, then $a + c < b + d$.

3 If $a < b$ and $c > 0$, then $ac < bc$ and $\dfrac{a}{c} < \dfrac{b}{c}$.

4 If $a < b$ and $c < 0$, then $ac > bc$ and $\dfrac{a}{c} > \dfrac{b}{c}$.

5 If $a < b$ and $b < c$, then $a < c$.

Rule 5 is called the *transitive law.* The condition that $a < b$ and $b < c$ in the transitive law is often written simply $a < b < c$. According to Rule 3, an inequality can be multiplied or divided on both sides by a *positive* number; however, by Rule 4, multiplying or dividing both sides of an inequality by a *negative* number will cause the inequality to reverse.

EXAMPLE   Show that $-\frac{2}{9} < -\frac{13}{59}$.

SOLUTION
Notice that $13(9) = 117$, while $59(2) = 118$; hence, $13(9) < 59(2)$. Dividing both sides of this inequality by the positive number $59(9)$, we obtain

$$\frac{13(9)}{59(9)} < \frac{59(2)}{59(9)}.$$

Reducing fractions, we find that $\frac{13}{59} < \frac{2}{9}$. Multiplying both sides of this inequality by the negative number -1 causes the inequality to reverse and yields $-\frac{13}{59} > -\frac{2}{9}$, that is, $-\frac{2}{9} < -\frac{13}{59}$.

It is sometimes necessary to deal with statements which assert that certain inequalities *do not* hold. By the trichotomy principle, if the inequality $a < b$ does not hold, then either $a > b$ or $a = b$. In this case, we say that a is *greater than or equal to* b and we write $a \geq b$. By definition, $b \leq a$ means the same thing as $a \geq b$. If $b \leq a$ holds, we say that b is *less than or equal to a.*

In addition to the inequalities described previously, assertions of the form $a \leq b$ or $b \geq a$ are also called *inequalities,* in spite of the fact that they include the possibility of equality. An inequality of the form $a < b$ is called a *strict* inequality, whereas one of the form $a \leq b$ is called a *nonstrict* inequality. The rules for manipulating nonstrict inequalities are virtually the same as those for strict inequalities. Also, it is possible to combine strict and nonstrict inequalities; for example, $a < x \leq b$ means that $a < x$ and also that $x \leq b$.

Problem Set 1

1 Answer true or false. Justify the true assertions by citing suitable rules and give examples to show that the false assertions are indeed false.
   (a) If x is a positive number, then $5x$ is a positive number.
   (b) If $x < 3$ and $y > 3$, then $x < y$.
   (c) If $x \leq y$, then $-5x \leq -5y$.
   (d) If $x^2 \leq 9$, then $x \leq 3$.
   (e) If $x \geq 2$ and $y > x$, then $y > 0$.

2 Prove that if $x \neq 0$, then $x^2 > 0$. (*Hint:* By the trichotomy principle, if $x \neq 0$, then either $x > 0$ or $x < 0$. Consider the two cases $x > 0$ and $x < 0$ separately.)

3 Show that $\frac{3}{28} < \frac{25}{233}$.

4 Prove that if $0 < x < y$, then $1/x > 1/y$.

5 (a) Under what condition is $-x > 0$? Explain.
   (b) Under what condition is $-x < 0$? Explain.
   (c) Under what condition is $-x = 0$? Explain.

6 Assume that $a > 0$, $b > 0$, $c > 0$, and $d > 0$. Prove that if $a < b$ and $c < d$, then $ac < bd$.

7 Prove that if $0 < x < y$, then $x^2 < y^2$. (*Hint:* First multiply the inequality $x < y$ by x, then multiply it by y, and finish by using the transitive law.)

8 Prove that if $0 < x < y$, then $\sqrt{x} < \sqrt{y}$.

9 Which number is larger, $\sqrt{7}$ or 2.646? Justify your answer.

10 Prove that if $x < 0$, then $x^3 < 0$.

11 If $x^2 \geq 9$, is it true that $x \geq 3$? Explain.

12 Suppose that x and y are positive real numbers with $x < y$. Show that $-1/x < -1/y$. Illustrate with an example.

13 Explain why the product of two numbers is positive if and only if either both of the numbers are positive or both of the numbers are negative.

14 Prove that if $0 < x < 1$, then $x^3 < x$.

15 Prove that if $x > 0$, then $1/x > 0$.

16 Prove that if $x > 1$, then $0 < 1/x < 1$.

2   Solution of Inequalities, Intervals, and Absolute Value

If one is asked to "solve" an equation, it is usually clear what is wanted: namely, the value (or values) of the variable (or variables) that make the equation true. Similarly, to "solve" an inequality is to find all values of the variable (or variables) that make the inequality true. This set is called the *solution set* of the inequality.

EXAMPLE   Solve the inequality $x + 3 < 5x - 1$.

SOLUTION

$$x + 3 < 5x - 1$$

$$3 < 4x - 1 \qquad \text{adding } -x \text{ to both sides}$$

$$4 < 4x \qquad \text{adding } 1 \text{ to both sides}$$

$$1 < x \qquad \text{dividing both sides by 4}$$

Figure 1

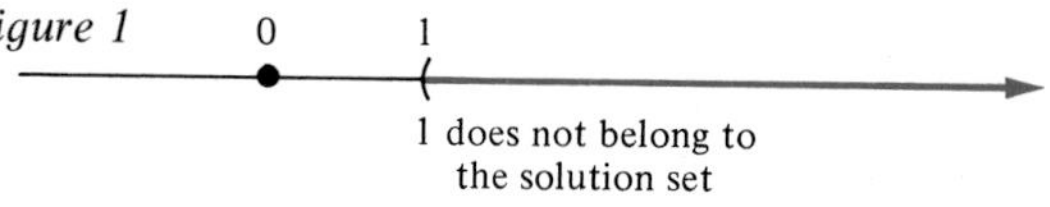

0   1
1 does not belong to the solution set

Hence, the solution is the set of all real numbers that are greater than 1 (Figure 1).

In Figure 1, notice that the solution set consists of one connected piece. Such sets, called *intervals,* often arise as the solution sets of inequalities. Intervals are classified as follows.

Bounded Intervals

Let a and b be real numbers with $a < b$.

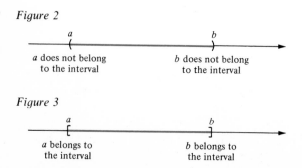

Figure 2

a   b
a does not belong to the interval   b does not belong to the interval

Figure 3

a   b
a belongs to the interval   b belongs to the interval

1 The *open interval* from a to b, denoted (a, b), is the set of all real numbers x such that $a < x < b$ (Figure 2). Notice that the *endpoints* a and b do not belong to the open interval (a, b).

2 The *closed interval* from a to b, denoted $[a, b]$, is the set of all real numbers x such that $a \leq x \leq b$ (Figure 3). Notice that the closed interval $[a, b]$ contains both of its *endpoints,* a and b.

Figure 4

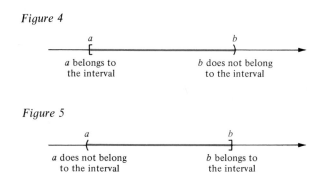

Figure 5

3 The *half-open interval on the right* from a to b, denoted $[a, b)$, is the set of all real numbers x such that $a \leq x < b$ (Figure 4). Here, the *left endpoint a* belongs to the interval, but the *right endpoint b* does not.

4 The *half-open interval on the left* from a to b, denoted $(a, b]$, is the set of all real numbers x such that $a < x \leq b$ (Figure 5). Here, the *right endpoint b* belongs to the interval, but the *left endpoint a* does not.

Unbounded Intervals

The unbounded intervals are written with the aid of the symbols $+\infty$ and $-\infty$, called *positive infinity* and *negative infinity*, respectively. Let a be a real number.

Figure 6

Figure 7

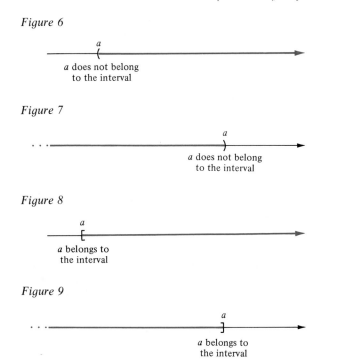

Figure 8

Figure 9

1 The *open interval* from a to $+\infty$, denoted $(a, +\infty)$, is the set of all real numbers x such that $x > a$ (Figure 6).

2 The *open interval* from $-\infty$ to a, denoted $(-\infty, a)$, is the set of all real numbers x such that $x < a$ (Figure 7).

3 The *closed interval* from a to $+\infty$. denoted $[a, +\infty)$, is the set of all real numbers x such that $x \geq a$ (Figure 8).

4 The *closed interval* from $-\infty$ to a, denoted $(-\infty, a]$, is the set of all real numbers x such that $x \leq a$ (Figure 9).

5 The interval notation $(-\infty, +\infty)$ is used to denote the set $\mathbb{R}$ of all real numbers.

It must be emphasized that $+\infty$ *and* $-\infty$ *are just convenient symbols and that they are not real numbers.* When these symbols are used in connection with unbounded intervals as described above, we usually write simply ∞ rather than $+\infty$. For instance, $(5, \infty)$ denotes the set of all real numbers that are greater than 5.

EXAMPLES   Solve the given inequality and show the solution set on the number line.

1 $x^2 + 3x + 2 \geq 0$

SOLUTION
Since $x^2 + 3x + 2 = (x + 1)(x + 2)$, the condition $x^2 + 3x + 2 \geq 0$ is equivalent to $(x + 1)(x + 2) \geq 0$. Here, equality holds exactly when $x = -1$ or $x = -2$. Also, the strict inequality $(x + 1)(x + 2) > 0$ holds if and only if $x + 1$ and $x + 2$ have the same algebraic sign.

   We first consider the situation in which both $x + 1$ and $x + 2$ are positive, that is, $x > -1$ and $x > -2$. Notice that, if $x > -1$, then $x > -2$ automatically holds; hence, both $x + 1$ and $x + 2$ will be positive precisely when $x > -1$.

Next, we consider the situation in which both $x + 1$ and $x + 2$ are negative, that is, $x < -1$ and $x < -2$. Notice that if $x < -2$, then $x < -1$ automatically holds; hence, both $x + 1$ and $x + 2$ will be negative precisely when $x < -2$.

The considerations above show that

$$x^2 + 3x + 2 \geq 0$$

holds precisely when $x \leq -2$ or $x \geq -1$. Therefore, the two intervals $(-\infty, -2]$ and $[-1, \infty)$ comprise the solution set (Figure 10).

Figure 10

2 $\dfrac{3x + 5}{x - 5} < 0$

SOLUTION
The inequality will hold if and only if the numerator $3x + 5$ and the denominator $x - 5$ have opposite algebraic signs. We first consider the situation in which $3x + 5$ is positive and $x - 5$ is negative, that is, $x > -\frac{5}{3}$ and $x < 5$. This situation obtains precisely when x belongs to the open interval $(-\frac{5}{3}, 5)$. Notice that the opposite situation, in which $3x + 5$ is negative and $x - 5$ is positive, cannot be obtained at all, for it would require that $x < -\frac{5}{3}$ and at the same time that $x > 5$. Therefore, the solution set of the inequality $(3x + 5)/(x - 5) < 0$ is the open interval $(-\frac{5}{3}, 5)$ (Figure 11).

Figure 11

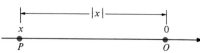

2.1 Absolute Value

The notion of absolute value plays an important role in analytic geometry and calculus, especially in expressions that involve the distance between two points on a number scale. The definition of absolute value is simple enough.

DEFINITION 1 **Absolute Value**

If x is a real number, then the *absolute value* of x, denoted $|x|$, is defined as follows:

$$|x| = \begin{cases} x & \text{if } x \geq 0 \\ -x & \text{if } x < 0. \end{cases}$$

For example, $|7| = 7$ because $7 \geq 0$, $|0| = 0$ because $0 \geq 0$, and $|-3| = -(-3) = 3$ because $-3 < 0$.

Notice that the absolute value of a real number is always nonnegative. Recalling the definition of the (principal) square root, we see that $|x| = \sqrt{x^2}$. Evidently, $|x|^2 = x^2$.

Geometrically, the absolute value of the real number x is the distance between the point P whose coordinate is x and the origin O, regardless of whether P is to the right or left of O (Figure 12). More generally, if P and Q are two points on the number scale with coordinates a and b, respectively, then the distance between P and Q is given by $|a - b|$ (Figure 13). For simplicity, we refer to this distance as "the distance between the two numbers a and b." For instance, the distance between 4 and 7 is $|4 - 7| = |-3| = 3$, the distance between -2 and 2 is $|-2 - 2| = |-4| = 4$, and so on.

Figure 12

Figure 13

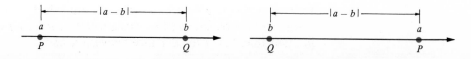

Many of the properties of absolute value can be established by considering all possible cases in which the quantities involved are positive, negative, or zero.

EXAMPLE   Show that $-|x| \le x \le |x|$ holds for every real number x.

SOLUTION
If $x \ge 0$, then $|x| = x$, so $x \le |x|$ holds. Also, if $x \ge 0$, then $-|x| = -x \le 0 \le x$, so $-|x| \le x$ holds. Thus, $-|x| \le x \le |x|$ holds when $x \ge 0$.
  On the other hand, if $x < 0$, then $|x| = -x$, so $-|x| = x < 0 < -x = |x|$; hence, $-|x| \le x \le |x|$ also holds when $x < 0$.

One of the most important properties of absolute value is the *triangle inequality* given in the following theorem.

THEOREM 1   **Triangle Inequality**
If a and b are real numbers, then $|a + b| \le |a| + |b|$.

PROOF
By the preceding example, $-|a| \le a \le |a|$ and $-|b| \le b \le |b|$. Adding these inequalities, we obtain

$$-(|a| + |b|) \le a + b \le (|a| + |b|).$$

If $a + b \ge 0$, then $|a + b| = a + b \le (|a| + |b|)$. On the other hand, if $a + b < 0$, then $|a + b| = -(a + b) \le (|a| + |b|)$. In either case, we have $|a + b| \le |a| + |b|$.

EXAMPLE   Use the triangle inequality to show that $|a - b| \le |a| + |b|$ holds for all real numbers a and b.

SOLUTION
Using the triangle inequality and the obvious fact that $|-b| = |b|$, we have

$$|a - b| = |a + (-b)| \le |a| + |-b| = |a| + |b|.$$

For future calculations involving absolute value, we present the following list of basic properties.

Properties of Absolute Value

Assume that x and y are real numbers. Then:

1 $-|x| \le x \le |x|$.

2 $|x + y| \le |x| + |y|$.

3 $|xy| = |x| \cdot |y|$.

4 $\left| \dfrac{x}{y} \right| = \dfrac{|x|}{|y|}$ if $y \ne 0$.

5 $|x| = |y|$ if and only if $x = \pm y$.

6 $|x| < y$ if and only if $-y < x < y$.

7 $|x| \ge y$ if and only if $x \ge y$ or $x \le -y$.

Properties 1 and 2 have already been established. Properties 3, 4, and 5 should be evident to the reader, and Properties 6 and 7 can be seen geometrically if one keeps it in mind that $|x|$ is the distance between x and 0.

EXAMPLES   Use Properties 1 through 7 to solve for x in each example.

1 $|x - 5| = |3x + 7|$

SOLUTION
By Property 5, the given equation is equivalent to

$$x - 5 = +(3x + 7) \quad \text{or} \quad x - 5 = -(3x + 7),$$

so $2x = -12$ or $4x = -2$. Therefore, $x = -6$ or $x = -\frac{1}{2}$.

2 $|3x - 2| < 4$

SOLUTION
By Property 6, the given inequality is equivalent to $-4 < 3x - 2 < 4$; that is, $-2 < 3x < 6$. The latter inequality is equivalent to $-\frac{2}{3} < x < 2$, so the solution set is $\left(-\frac{2}{3}, 2\right)$.

3 $|3x + 2| \geq 5$

SOLUTION
By Property 7, the given inequality is equivalent to $3x + 2 \geq 5$ or $3x + 2 \leq -5$, that is, $x \geq 1$ or $x \leq -\frac{7}{3}$. Hence, the solution set consists of the two intervals $\left(-\infty, -\frac{7}{3}\right]$ and $[1, \infty)$.

Problem Set 2

1 Following are solution sets for certain inequalities. Illustrate each solution set on a real line with suitable shading.
(a) $[-2, 3)$.
(b) All numbers x such that $-3 < x \leq 4$ and simultaneously $-6 \leq x < 2$.
(c) All numbers belonging to $[-2, 0]$ or to $\left[-\frac{1}{2}, 1\right]$ or to both of these intervals.
(d) All numbers belonging to $(0, \infty)$ or to $(-\infty, 0)$.
(e) All numbers x that belong to both intervals $(0, \infty)$ and $(3, \infty)$ simultaneously.
(f) All numbers x that belong to both intervals $(-\infty, -2]$ and $(-\infty, -5)$ simultaneously.

2 Use interval notation to represent each shaded set in Figure 14.

Figure 14

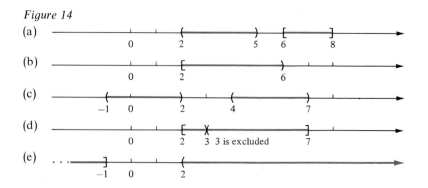

In problems 3 through 21, find all the real numbers that satisfy the inequality. Express the solution in interval notation and illustrate it on the number line.

3 $10x < 18 + 4x$

4 $\frac{9}{4} < \frac{5}{2} + \frac{2}{3}x$

5 $2 \leq 5 - 3x < 11$

6 $3 < 5x \leq 2x + 11$

7 $3 > -4 - 4x \geq -8$

8 $\frac{2}{x} - 4 < \frac{3}{x} - 8$

9 $\frac{3}{1 - x} \leq 1$

10 $\frac{5}{3 - x} \geq 2$

11 $x^2 > 9$

12 $x^2 \leq 4$

13 $x^2 - x - 2 < 0$

14 $2x^2 + 5x - 12 > 0$

15 $3x^2 - 13x \geq 10$ **16** $2x \geq 3x^2 - 16$ **17** $\dfrac{3 + x}{3 - x} \leq 1$ **18** $0 < \dfrac{x - 1}{2x - 1} < 2$

19 $2 \geq \dfrac{3x + 1}{x} > \dfrac{1}{x}$ **20** $\dfrac{1}{3x - 7} \leq \dfrac{4}{3 - 2x}$ **21** $\dfrac{x + 2}{x - 1} \leq \dfrac{x}{x + 4}$

22 Show that if $x \neq y$, then $x^2 + 2xy < 2x^2 + y^2$.

In problems 23 through 28, solve the equation for x.

23 $|x - 3| = 2$ **24** $|3x + 2| = 5$ **25** $|x - 5| = |3x - 1|$

26 $|x - 2| = |3 - 5x|$ **27** $|5x| = 3 - x$ **28** $|3x - 7| = x + 2$

In problems 29 through 35, find all real numbers that satisfy the inequality, express the solution in interval notation, and illustrate it on the number line.

29 $|2x - 5| < 1$ **30** $|4x - 6| \leq 3$ **31** $|3x + 5| > 2$ **32** $|9 - 2x| \geq |7x|$

33 $|3x + 5| \leq |2x + 1|$ **34** $|x - 2| \geq 4x + 1$ **35** $|5 - 1/x| \leq 2$

36 Suppose that a and b are real numbers with $a < b$. Show that every real number x that belongs to the interval $[a, b]$ can be expressed in the form $x = ta + (1 - t)b$, where t is a real number that belongs to the interval $[0, 1]$.

37 If $0 < k < 1$, solve the inequality $|(1/x) - 1| < k$ for x.

38 Prove that if x and y are real numbers:

(a) $|x| - |y| \leq |x - y|$ (b) $||x| - |y|| \leq |x - y|$

In problems 39 through 41, use the triangle inequality to show that the inequality holds under the given hypotheses.

39 If $|x - 2| < \frac{1}{2}$ and $|y - 2| < \frac{1}{3}$, then $|x - y| < \frac{5}{6}$.

40 If $|x + 2| < \frac{1}{2}$ and $|y + 2| < \frac{1}{3}$, then $|x - y| < \frac{5}{6}$.

41 If $|x - y| < \frac{1}{2}$ and $|x + 2| < \frac{1}{3}$, then $|y + 2| < \frac{5}{6}$.

42 A car can travel 220 miles on a full tank of gas. How many *full* tanks of gas would it need to travel at least 1,314 miles?

43 One of the dimensions of a rectangular floor is 4 meters and its area is less than 132 square meters. Letting x denote the other dimension of the floor:
(a) Find an inequality that x must satisfy.
(b) Solve this inequality.

44 A bank teller is entitled to a 2-week vacation for each of the first 5 years of employment. Thereafter, she is entitled to a 3-week vacation for each full year she works. How many full years must she work without taking a vacation to entitle her to at least a 30-week vacation?

3 The Cartesian Coordinate System

In Section 1 we have seen how a point P on a number line can be located by specifying a real number x called its *coordinate*. Similarly, it is possible to locate points in a plane by specifying *two* real numbers, also called *coordinates*. This is accomplished by establishing a suitable *coordinate system* in the plane so that points can be made to correspond to pairs of real numbers in a systematic way. We now describe the *cartesian coordinate system*, named in honor of the seventeenth-century French philosopher and mathematician, René Descartes.

The cartesian coordinate system is based on two perpendicular lines L_1 and L_2 in the plane (Figure 1). Ordinarily, the first line L_1 is taken to be horizontal and the second line L_2 is taken to be vertical. The point O where these two lines intersect is called the *origin*. Each of the lines L_1 and L_2 is made into a number scale as in Section 1 by choosing suitable unit points U_1 and U_2, respectively. It is traditional to choose U_1 to the *right* of the origin and U_2 *above* the origin, as in Figure 2, so that the positive direction on L_1 is to the right and the positive direction on L_2 is upward. We usually choose the *same* unit distance on L_1 and L_2 (although in some of our figures it will be convenient to use different units along the two lines). When the two lines L_1 and L_2 are made into number scales, they are called *coordinate axes.*

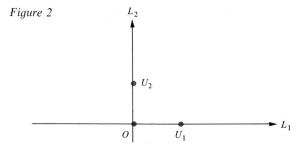

Figure 1

Figure 2

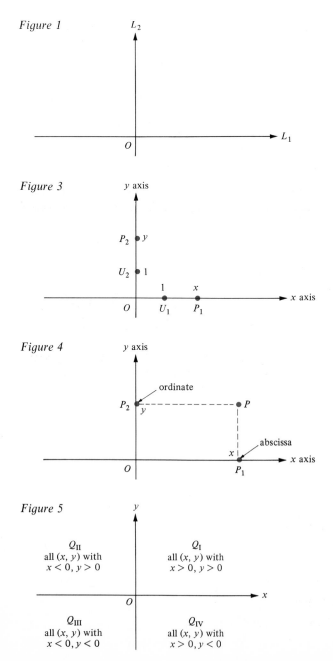

Figure 3

Figure 4

Figure 5

The horizontal axis is usually called the *x axis* and the coordinate of a general point P_1 on this axis is denoted by x. Similarly, the vertical coordinate axis is called the *y axis* and the coordinate of a general point P_2 on this axis is denoted by y (Figure 3).

Now, let P be any point in the plane and drop perpendiculars from P to the x and y axes (Figure 4). Let these perpendiculars intersect the x and y axes at points P_1 and P_2, respectively. The coordinate x of P_1 on the horizontal axis is called the *abscissa* of P, while the coordinate y of P_2 on the vertical axis is called the *ordinate* of P. The real numbers x and y are called the *coordinates* of the point P. Traditionally, these coordinates are written as an *ordered pair* (x, y) enclosed in parentheses, with the abscissa first and the ordinate second. (Unfortunately, this is the same symbolism that is used for an open interval; however, it is always clear from the context what is intended.)

A cartesian coordinate system establishes a one-to-one correspondence between points P in the plane and ordered pairs (x, y) of real numbers. If the geometric point P corresponds to the ordered pair (x, y) of coordinates, we write $P = (x, y)$. Thus, the set of all ordered pairs of real numbers is called the *cartesian plane*, the *xy plane*, or the *coordinate plane*, and an ordered pair (x, y) is referred to as a *point.*

The x and y axes divide the plane into four disjoint regions called *quadrants*, denoted by Q_I, Q_{II}, Q_{III}, and Q_{IV} and called the *first, second, third,* and *fourth* quadrants, respectively (Figure 5).

3.1 The Distance Formula

One of the attractive features of the cartesian coordinate system is the ease with which the distance d between two points P_1 and P_2 can be calculated in terms of their coordinates. We denote the line segment between P_1 and P_2 by $\overline{P_1P_2}$ and we use the notation $|\overline{P_1P_2}|$ for the length of this line segment, so that $d = |\overline{P_1P_2}|$. Then we have the following theorem.

THEOREM 1   **The Distance Formula**

If $P_1 = (x_1, y_1)$ and $P_2 = (x_2, y_2)$ are two points in the cartesian plane, then

$$|\overline{P_1P_2}| = \sqrt{(x_2 - x_1)^2 + (y_2 - y_1)^2}.$$

Figure 6

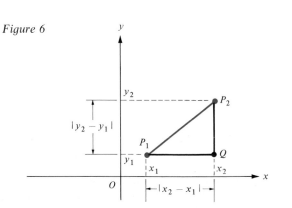

The distance formula is a simple consequence of the Pythagorean theorem, as can be seen in Figure 6. Notice that $\overline{P_1P_2}$ is the hypotenuse of the right triangle P_1QP_2, where $Q = (x_2, y_1)$, so

$$|\overline{P_1Q}| = |x_2 - x_1| \quad \text{and} \quad |\overline{QP_2}| = |y_2 - y_1|;$$

hence, by the Pythagorean theorem,

$$\begin{aligned}|\overline{P_1P_2}|^2 &= |x_2 - x_1|^2 + |y_2 - y_1|^2 \\ &= (x_2 - x_1)^2 + (y_2 - y_1)^2.\end{aligned}$$

Taking the square root on both sides of the latter equation, we obtain the distance formula in Theorem 1.

EXAMPLE   Find the distance $d = |\overline{P_1P_2}|$ if $P_1 = (-1, 2)$ and $P_2 = (3, -2)$. Plot P_1 and P_2.

Figure 7

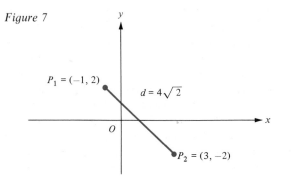

SOLUTION

$$\begin{aligned}d &= \sqrt{[3 - (-1)]^2 + (-2 - 2)^2} \\ &= \sqrt{4^2 + (-4)^2} = \sqrt{32} = 4\sqrt{2}.\end{aligned}$$

We use a wiggly equal sign, $\approx$, to mean "approximately equal to"; hence, in Figure 7 we have $d = 4\sqrt{2} \approx 5.657$ units.

Problem Set 3

In problems 1 and 2, plot the point M on the cartesian coordinate system and give the coordinates of the points N, R, and S such that:
(a) The line segment $\overline{MN}$ is perpendicular to the x axis and is bisected by it.
(b) The line segment $\overline{MR}$ is perpendicular to the y axis and is bisected by it.
(c) The line segment $\overline{MS}$ is bisected by the origin.

1 $M = (3, 2)$ **2** $M = (-4, -3)$

In problems 3 through 7, find the distance between the two given points.

3 $(-3, -4)$ and $(-5, -7)$ **4** $(-1, 7)$ and $(2, 11)$ **5** $(7, -1)$ and $(7, 3)$

6 $(0, 4)$ and $(-4, 0)$ **7** $(0, 0)$ and $(-8, -6)$

8 Check the derivation of the distance formula in the case where one point is in the third quadrant and the other point is in the second quadrant.

In problems 9 through 12, use the distance formula and the converse of the Pythagorean theorem to show that the triangle with the given vertices is a right triangle.

9 $(1, 1)$, $(5, 1)$, and $(5, 7)$

10 $(0, 0)$, $(-3, 3)$, and $(2, 2)$

11 $(-1, -2)$, $(3, -2)$, and $(-1, -7)$

12 $(-4, -4)$, $(0, 0)$, and $(5, -5)$

13 Show that the distance between the points (x_1, y_1) and (x_2, y_2) is the same as the distance between the point $(x_1 - x_2, y_1 - y_2)$ and the origin.

14 If P_1, P_2, and P_3 are three points in the plane, then P_2 lies on the line segment between P_1 and P_3 if and only if $|\overline{P_1 P_3}| = |\overline{P_1 P_2}| + |\overline{P_2 P_3}|$. Illustrate this geometric fact with diagrams.

In problems 15 through 17, determine whether P_2 lies on the line segment between P_1 and P_3 by checking whether $|\overline{P_1 P_3}| = |\overline{P_1 P_2}| + |\overline{P_2 P_3}|$. (See problem 14.)

15 $P_1 = (1, 2)$, $P_2 = (0, \frac{5}{2})$, $P_3 = (-1, 3)$

16 $P_1 = (-\frac{7}{2}, 0)$, $P_2 = (-1, 5)$, $P_3 = (2, 11)$

17 $P_1 = (2, 3)$, $P_2 = (3, -3)$, $P_3 = (-1, -1)$

In problems 18 and 19, use the distance formula to determine whether or not triangle ABC is isosceles.

18 $A = (-5, 1)$, $B = (-6, 5)$, $C = (-2, 4)$

19 $A = (6, -13)$, $B = (8, -2)$, $C = (21, -5)$

20 The point $P = (x, y)$ lies on the straight line passing through $P_1 = (-3, 5)$ and $P_2 = (-1, 2)$, and P satisfies $|\overline{PP_1}| = 4|\overline{P_1 P_2}|$. Use the distance formula to find the coordinates of P. (There are *two* solutions.)

4   Straight Lines and Their Slope

Straight lines in a plane have very simple equations relative to a cartesian coordinate system. These equations can be derived by using the idea of *slope*.

Consider the inclined line segment $\overline{AB}$ in Figure 1. The horizontal distance between A and B is called the *run*, and the vertical distance between A and B is called the *rise*. The ratio of rise to run is called the *slope* of the line segment and is traditionally denoted by the symbol m. Thus, by definition,

$$\text{slope of } \overline{AB} = m = \frac{\text{rise}}{\text{run}}.$$

Figure 1

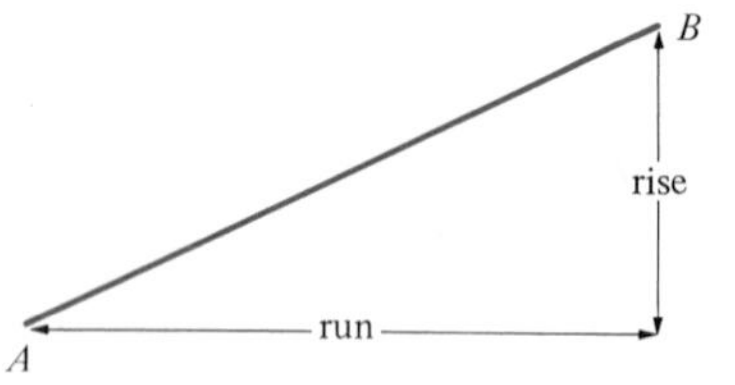

If the line segment $\overline{AB}$ is turned so that it becomes more nearly vertical, then the rise increases, the run decreases, and the slope $m = \text{rise/run}$ becomes very large. When the line segment becomes vertical, the slope $m = \text{rise/run}$ becomes undefined since the denominator is zero. In this case, we sometimes say that the slope is *infinite* and write $m = \infty$.

If the line segment $\overline{AB}$ is horizontal, its rise is zero, so its slope $m = \text{rise/run}$ is zero. If $\overline{AB}$ slants downward to the right as in Figure 2, its rise is considered negative; hence, its slope $m = \text{rise/run}$ is negative. (The run is always regarded as being nonnegative.)

Figure 2

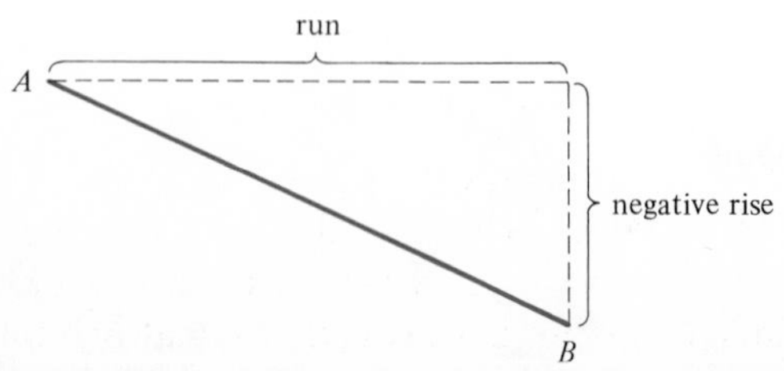

Figure 3

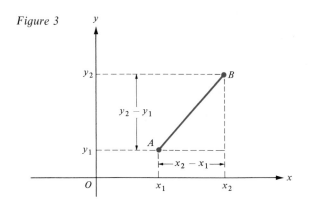

Now, set up a cartesian coordinate system and consider the line segment $\overline{AB}$, where $A = (x_1, y_1)$ and $B = (x_2, y_2)$ (Figure 3). Here, the rise is $y_2 - y_1$, the run is $x_2 - x_1$, and so the slope m is given by

$$m = \frac{y_2 - y_1}{x_2 - x_1}.$$

Of course, Figure 3 represents a special situation in which B lies above and to the right of A; however, the reader can check the other possible cases and see that the slope m of $\overline{AB}$ is always given by the preceding formula. Therefore, we have the following theorem.

THEOREM 1   The Slope Formula

Let $A = (x_1, y_1)$ and $B = (x_2, y_2)$ be any two points in the cartesian plane. Then, provided that $x_1 \neq x_2$, the slope m of the line segment $\overline{AB}$ is given by

$$m = \frac{y_2 - y_1}{x_2 - x_1}.$$

EXAMPLE   If $A = (8, -2)$ and $B = (3, 7)$, find the slope m of $\overline{AB}$.

SOLUTION

$$m = \frac{7 - (-2)}{3 - 8} = \frac{9}{-5} = -\frac{9}{5}.$$

Figure 4

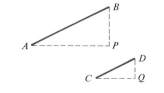

A consideration of the similar triangles in Figure 4 shows that two parallel line segments $\overline{AB}$ and $\overline{CD}$ have the same slope. Similarly, if two line segments $\overline{AB}$ and $\overline{CD}$ lie on the same infinite straight line as in Figure 5, they have the same slope. The common slope of *all* the segments of an infinite straight line L is called the *slope of L.*

From the fact that two parallel line *segments* have the same slope, it follows that two parallel straight lines have the same slope. Conversely, it is easy to see that two distinct straight lines having the same slope must be parallel, and we have the following theorem.

Figure 5

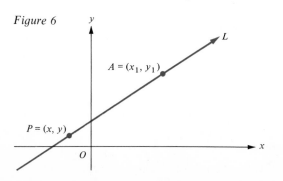

THEOREM 2   Parallelism Condition

Two distinct nonvertical straight lines are parallel if and only if they have the same slope.

Figure 6

Now, consider a nonvertical straight line L with slope m and containing a point $A = (x_1, y_1)$ (Figure 6). If $P = (x, y)$ is any other point on L, then, by Theorem 1, $m = (y - y_1)/(x - x_1)$; hence,

$$y - y_1 = m(x - x_1).$$

Notice that this equation holds even if $P = A$, when it simply reduces to $0 = 0$. In fact, we claim that it is the equation of the line L in the sense that not only do all points (x, y) on L satisfy this equation, but, conversely, any point (x, y) that satisfies this equation lies on line L. (The converse follows from Theorem 2.) The equation

$$y - y_1 = m(x - x_1)$$

is called the *point-slope form* for the equation of L.

EXAMPLE   Let L be the straight line of slope 5 containing the point $(3, 4)$. Write the equation of L in point-slope form, determine where L intersects the y axis, draw a diagram showing L and the coordinate axes, and decide whether or not the point $(4, 9)$ belongs to L.

Figure 7

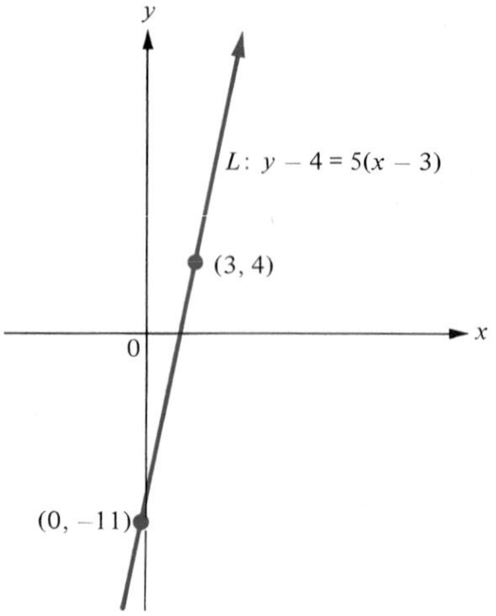

L: $y - 4 = 5(x - 3)$

$(3, 4)$

$(0, -11)$

SOLUTION
The point-slope form for the equation of the line L is $y - 4 = 5(x - 3)$. This equation can be rewritten as $y = 5x - 11$. If L intersects the y axis at the point $(0, b)$, then $b = 5(0) - 11 = -11$. Since both $(0, -11)$ and $(3, 4)$ belong to L, it is easy to draw L by drawing a straight line through these two points (Figure 7). If we put $x = 4$, $y = 9$ in the equation for L, we obtain $9 = 5(4) - 11$, which is true. Therefore, $(4, 9)$ does belong to L.

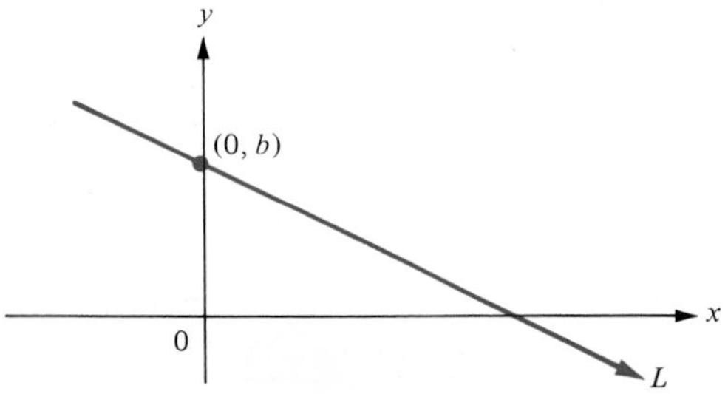

Now suppose that L is any nonvertical straight line with slope m. Since L is not parallel to the y axis, it must intersect it at some point $(0, b)$ (Figure 8). The ordinate b of this intersection point is called the *y intercept* of L. Since $(0, b)$ belongs to L, we can write the point-slope equation $y - b = m(x - 0)$ for L. The latter simplifies to

Figure 8

$(0, b)$

L

$$y = mx + b,$$

which is called the *slope-intercept form* of the equation for L.

EXAMPLE   In 1977, the Solar Electric Company showed a profit of $3.17 per share, and it expects this figure to increase by $0.24 per share per year. Counting the years so that 1977 corresponds to $x = 0$ and successive years correspond to $x = 1, 2, 3$, and so forth, find the equation $y = mx + b$ of the straight line which will enable the company to predict its profit y per share during future years. Draw a graph showing this line and find the predicted profit per share in 1985.

Figure 9

$y = 0.24x + 3.17$

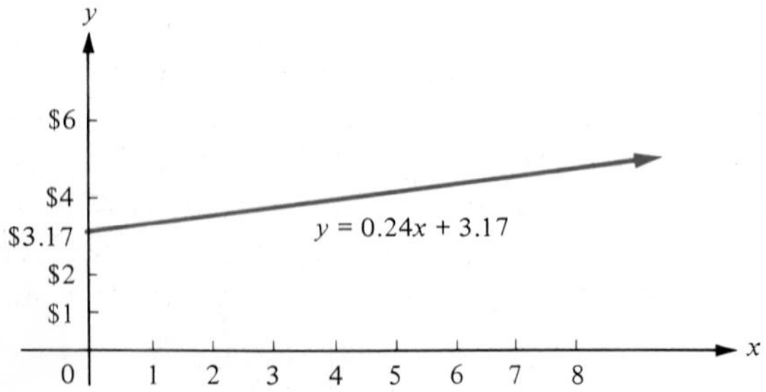

SOLUTION
When $x = 0$, $y = 3.17$; hence, $3.17 = m(0) + b$, and so $b = 3.17$. Thus, $y = mx + 3.17$. When x increases by 1, y increases by 0.24; hence, $m = 0.24$. The equation, therefore, is $y = 0.24x + 3.17$. In 1985, $x = 8$ and $y = (0.24)(8) + 3.17 = 5.09$. The predicted profit per share in 1985 is $5.09 (Figure 9).

Figure 10

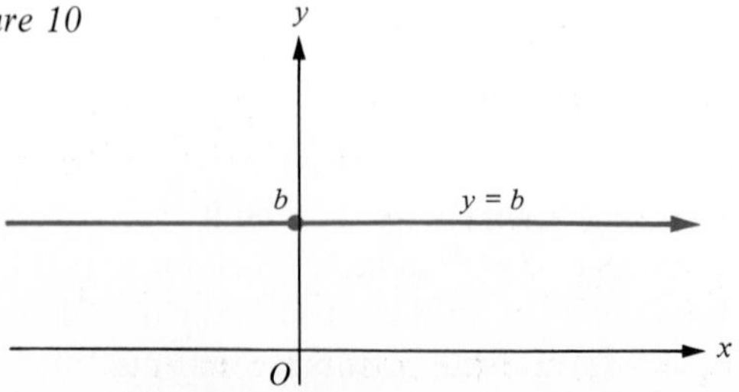

b $y = b$

A horizontal straight line has slope zero; hence, such a line has the equation $y = 0(x) + b$, or simply $y = b$, in slope-intercept form (Figure 10). The equation $y = b$ places no restriction whatsoever on the abscissa x of

Figure 11

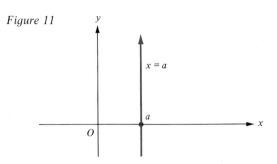

a point (x, y) on the horizontal line, but requires that all of the ordinates y have the same value b.

Of course, a vertical straight line has an undefined slope, so its equation cannot be written in slope-intercept form at all. However, since all points on a vertical line must have the same abscissa, the equation of such a line can be written as $x = a$, where a is the common value of all these abscissas (Figure 11).

The equation of any straight line can be put into the form

$$Ax + By + C = 0,$$

where A, B, and C are constants and not both A and B are zero. This is called the *general form* of the equation of a straight line. If $B \neq 0$, the equation $Ax + By + C = 0$ can be rewritten

$$y = -\frac{A}{B}x + -\frac{C}{B},$$

and therefore represents a straight line with slope $m = -A/B$ and y intercept $b = -C/B$. On the other hand, if $B = 0$, then $A \neq 0$ and the equation can be rewritten in the form $x = -C/A$, which represents a vertical straight line.

4.1 Perpendicular Lines

In Theorem 2 we have seen that two straight lines are parallel if and only if they have the same slope. The following theorem gives a condition for two straight lines to be perpendicular.

THEOREM 3   **Perpendicularity Condition**
Two nonvertical straight lines are perpendicular if and only if the slope of one of the lines is the negative of the reciprocal of the slope of the other line.

PROOF
Let the two lines be L_1 and L_2 and suppose that their slopes are m_1 and m_2, respectively. The condition that the slope of either one of the lines is the negative of the reciprocal of the slope of the other is equivalent to the condition that $m_1m_2 = -1$. (Why?) Neither the angle between the two lines nor the slopes of the lines are affected if the origin O of the coordinate system is placed at the point where the two lines intersect (Figure 12). The point $A = (1, m_1)$ belongs to L_1 and the point $B = (1, m_2)$ belongs to L_2. (Why?) By the Pythagorean theorem and its converse, angle AOB is a right angle if and only if

Figure 12

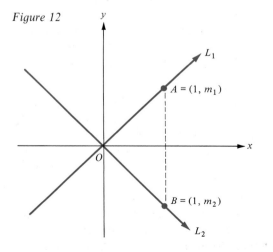

$$|\overline{AB}|^2 = |\overline{OA}|^2 + |\overline{OB}|^2,$$

that is, if and only if

$$(m_1 - m_2)^2 = [(1 - 0)^2 + (m_1 - 0)^2] + [(1 - 0)^2 + (m_2 - 0)^2].$$

This equation simplifies to

$$m_1^2 - 2m_1m_2 + m_2^2 = 1 + m_1^2 + 1 + m_2^2,$$

or $m_1m_2 = -1$, and the proof is complete.

EXAMPLE   Find:
(a) The equation of the line L_1 that contains the point $(-1, 2)$ and is parallel to the line L: $3x - y - 1 = 0$.

(b) The equation of the line L_2 that contains the point $(-1, 2)$ and is perpendicular to the line L: $3x - y - 1 = 0$.
Sketch the graphs of these lines.

Figure 13

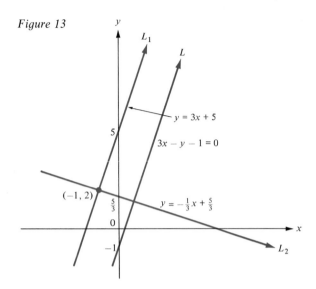

SOLUTION

(a) In slope-intercept form, the equation of L is $y = 3x - 1$; hence, L has slope $m = 3$. Since L_1 is to be parallel to L, the slope of L_1 must be $m_1 = m = 3$. Because L_1 contains $(-1, 2)$, its equation is $y - 2 = 3[x - (-1)]$ in point-slope form, or $y = 3x + 5$ in slope-intercept form.

(b) Since L_2 is to be perpendicular to L, it follows that the slope of L_2 must be $m_2 = -1/m = -\frac{1}{3}$. Therefore, because L_2 contains $(-1, 2)$, its equation in point-slope form is $y - 2 = -\frac{1}{3}[x - (-1)]$ or, in slope-intercept form, $y = -\frac{1}{3}x + \frac{5}{3}$ (Figure 13).

4.2 Intersecting Lines

If two different straight lines in the plane are not parallel, they will intersect in a single point. For instance, in Figure 13, the lines L_1 and L_2 meet in the point $(-1, 2)$. In order to find the point at which two nonparallel lines intersect, it is only necessary to solve the equations of the two lines simultaneously.

EXAMPLE   Find the coordinates of the point (x, y) at which the line whose equation is $3x + 2y + 8 = 0$ intersects the line whose equation is $y = \frac{1}{5}(2x - 1)$.

SOLUTION
Rewriting the second equation as $2x - 5y - 1 = 0$, we obtain the pair of simultaneous equations

$$\begin{cases} 3x + 2y + 8 = 0 \\ 2x - 5y - 1 = 0. \end{cases}$$

Multiplying the first equation by 2 and the second equation by 3 yields

$$\begin{cases} 6x + 4y + 16 = 0 \\ 6x - 15y - 3 = 0. \end{cases}$$

Subtracting the second equation from the first, we find that $19y + 19 = 0$, so $y = -1$. Substituting $y = -1$ into the original equation $3x + 2y + 8 = 0$, we obtain $3x + 6 = 0$, so $x = -2$. Hence, $(x, y) = (-2, -1)$.

Problem Set 4

In problems 1 through 6, find the slope of the straight line that contains the two given points.

1 $(6, 2)$ and $(3, 7)$ **2** $(3, -2)$ and $(5, -6)$ **3** $(14, 7)$ and $(2, 1)$

4 $(2, 2)$ and $(-4, -1)$ **5** $(-5, 3)$ and $(6, 8)$ **6** $(1, 3)$ and $(-1, -1)$

In problems 7 through 10, find the equation of the straight line with the slope m and containing the point (x_1, y_1). Draw the line.

7 $m = 2$, $(x_1, y_1) = (5, 4)$ **8** $m = -4$, $(x_1, y_1) = (6, 1)$ **9** $m = \frac{1}{4}$, $(x_1, y_1) = (3, 2)$ **10** $m = 0$, $(x_1, y_1) = (-5, -1)$

In problems 11 through 13, find the slope m of the line passing through the two given points, and find the equation of the line in point-slope form. Draw the line.

11 $(x_1, y_1) = (3, -5)$ and $(x_2, y_2) = (6, 8)$ **12** $(x_1, y_1) = (7, 11)$ and $(x_2, y_2) = (-1, 1)$

13 $(x_1, y_1) = (3, 2)$ and $(x_2, y_2) = (4, 8)$

14 Show that if $x_1 \neq x_2$, the equation of the straight line containing the two points (x_1, y_1) and (x_2, y_2) is

$$y = \frac{y_2 - y_1}{x_2 - x_1} x + \frac{x_2 y_1 - x_1 y_2}{x_2 - x_1}.$$

15 An advertising agency claims that a furniture store's sales revenue will increase by $20 per month for each additional dollar spent on advertising. The current average monthly sales revenue is $140,000, with an expenditure of $100 per month for advertising. Find the equation that relates the store's expected average monthly sales revenue y to the total expenditure x for advertising. Find the value of y if the store's management decides to spend $400 per month for advertising.

16 Prove that $\left(\dfrac{x_1 + x_2}{2}, \dfrac{y_1 + y_2}{2}\right)$ is the midpoint of the line segment between (x_1, y_1) and (x_2, y_2).

17 Use problem 16 to find the midpoint of the line segment between the given pair of points.
(a) $(8, 1)$ and $(7, 3)$ (b) $(9, 3)$ and $(-5, 7)$ (c) $(-1, 1)$ and $(5, 3)$ (d) $(1, -3)$ and $(5, 8)$

18 Show that, for each real number t, the point $(tx_1 + (1 - t)x_2, ty_1 + (1 - t)y_2)$ belongs to the straight line containing the distinct points (x_1, y_1) and (x_2, y_2). [Note that $t = 1$ gives (x_1, y_1), while $t = 0$ gives (x_2, y_2).]

19 The *x intercept* of a straight line in the cartesian plane is defined to be the abscissa of the point where the line cuts the x axis. Find the x intercept of the lines.
(a) $3x - 2y = 6$ (b) $y = 3x + 9$ (c) $y = \frac{2}{3}x - 1$ (d) $y = mx + b$, where $m \neq 0$

20 (a) Show that the equation of the straight line whose x intercept is $a \neq 0$ and whose y intercept is $b \neq 0$ can be written as $x/a + y/b = 1$. This equation is called the *intercept form* for the equation of the line.
(b) Write the equation of the straight line passing through the points $(3, 0)$ and $(0, 8)$ in the intercept form as in part (a).

In problems 21 through 24, suppose that m_1 and m_2 are the slopes of the distinct lines L_1 and L_2, respectively. Indicate whether the lines are (a) parallel, (b) perpendicular, or (c) neither parallel nor perpendicular. Then, supposing that L_1 contains the point $(3, 2)$ and that L_2 contains the point $(-2, 5)$, draw L_1 and L_2 on the same diagram.

21 $m_1 = \frac{2}{3}$ and $m_2 = \frac{4}{6}$ **22** $m_1 = \frac{2}{3}$ and $m_2 = \frac{3}{2}$ **23** $m_1 = \frac{3}{2}$ and $m_2 = -\frac{2}{3}$ **24** $m_1 = -1$ and $m_2 = 1$

In problems 25 through 27, find the point (x, y) at which the two given lines intersect. Illustrate graphically.

25 $\begin{cases} 3x - 2y = 1 \\ 2x + y = 0 \end{cases}$ **26** $\begin{cases} y = \frac{1}{6}x - \frac{2}{3} \\ y = -\frac{1}{6}x + \frac{2}{3} \end{cases}$ **27** $\begin{cases} y = x \\ y = \frac{57}{61}x \end{cases}$

28 If $m_1 \neq m_2$, show that the line $y = m_1 x + b_1$ intersects the line $y = m_2 x + b_2$ at the point

$$\left(\frac{b_2 - b_1}{m_1 - m_2}, \frac{m_1 b_2 - m_2 b_1}{m_1 - m_2}\right).$$

29 Show that the quadrilateral $ABCD$ is a parallelogram if $A = (-5, -2)$, $B = (1, -1)$, $C = (4, 4)$, and $D = (-2, 3)$. (*Hint*: Show that opposite sides have the same slope.)

30 Find the distance (measured perpendicularly) between the point $(-4, 3)$ and the line $y = 3x - 5$ by carrying out the following steps:

   (a) Find the equation of the line through $(-4, 3)$ which is perpendicular to the line $y = 3x - 5$.

   (b) Find the point (x_1, y_1) at which the line obtained in part (a) meets the line $y = 3x - 5$.

   (c) Use the distance formula to find the distance between $(-4, 3)$ and (x_1, y_1).

31 (a) Determine d so that the line containing $A = (d, 3)$ and $B = (-2, 1)$ is perpendicular to the line containing $C = (5, -2)$ and $D = (1, 4)$.

   (b) Determine k so that the line containing $E = (k, 3)$ and $B = (-2, 1)$ is parallel to the line containing $C = (5, -2)$ and $D = (1, 4)$.

32 Consider the quadrilateral $ABCD$ with $A = (3, 1)$, $B = (2, 4)$, $C = (7, 6)$, and $D = (8, 3)$.

   (a) Use the concept of slope to determine whether or not the diagonals $\overline{AC}$ and $\overline{BD}$ are perpendicular.

   (b) Is $ABCD$ a parallelogram? Is it a rectangle? Is it a rhombus? Is it a square? Justify your answer.

33 Show that the line $Ax + By + C = 0$ is perpendicular to the line $-Bx + Ay + D = 0$.

5   Functions and Their Graphs

The concept of a function is so fundamental in calculus that functions are almost literally the "name of the game." Indeed, many of the more sophisticated topics in calculus are studied in a subject called the *theory of functions*. Although the function concept is introduced and explored briefly in this section, it will be expanded upon throughout the book.

The general idea of a function is simple. Suppose that one variable quantity, say y, depends in a definite way on another variable quantity, say x. Then, to each particular value of x there is a unique corresponding value of y. Such a correspondence is called a *function*, and we say that *the variable y is a function of the variable x.*

For example, if x denotes the radius of a circle and y denotes the area of this circle, then y depends on x in a definite way, namely $y = \pi x^2$. Thus, we say that the area of a circle is a function of its radius.

Letters of the alphabet are often used to denote functions—f, g, and h as well as F, G, and H are favorites for this purpose. (Letters of the Greek alphabet are also used.) If f is a function, it is customary to write the value of y that corresponds to x as $f(x)$, read "f of x." For instance, if f is the function that gives the area of a circle in terms of its radius, then $f(x) = \pi x^2$. More generally, we have the following definition.

DEFINITION 1   **Function as a Rule or Correspondence**

A *function f* is a rule or correspondence that assigns one and only one value of a variable y to each value of a variable x. It is to be understood that the variable x, which is called the *independent variable*, can take on any value in a certain set of numbers called the *domain* of f. For each value of x in the domain of f, the corresponding value of y is denoted by $f(x)$, so that $y = f(x)$. The variable y is called the *dependent variable*, since its value depends on the value of x. The set of values assumed by y as x runs through the domain is called the *range* of f.

Usually, *but not always*, we use x for the independent variable and y for the dependent variable. An equation that gives y in terms of x determines a function f, and we say that the function f is *defined* by the equation (or *given* by the equation).

If a function f is to be defined by an equation, then (unless contrary provisions are made explicit) it is understood that the domain of f consists of those values of x for which the defining equation makes sense. Then, the range of f is automatically determined, since it consists of those values of y that correspond, by the defining equation, to some value of x in the domain.

DEFINITION 2   **Graph of a Function**

The *graph* of a function f is the set of all points (x, y) in the xy plane such that x is in the domain of f, y is in the range of f, and $y = f(x)$.

EXAMPLE   Sketch the graph of the function f defined by the equation $y = 2x^2$ with the constraint that $x > 0$.

Figure 1

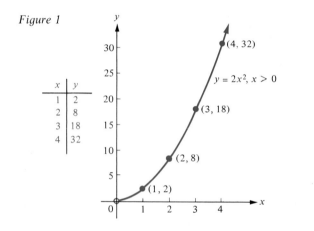

x	y
1	2
2	8
3	18
4	32

SOLUTION

We begin by assigning a few positive values to x and calculating the corresponding values of $y = f(x) = 2x^2$, as in the table in Figure 1. Here, $f(1) = 2(1)^2 = 2$, $f(2) = 2(2)^2 = 8$, $f(3) = 2(3)^2 = 18$, and $f(4) = 2(4)^2 = 32$; hence, the four points $(1, 2)$, $(2, 8)$, $(3, 18)$, and $(4, 32)$ belong to the graph. In Figure 1, we have connected these points by a smooth curve to obtain a sketch of the graph. Since the domain of f consists only of *positive* numbers (because of the constraint $x > 0$), the point $(0, 0)$ is excluded from the graph. This excluded point is indicated by a small open circle.

This procedure for graph sketching—plotting a few well-chosen points and connecting them by a smooth curve—does not *always* work, since it involves a guess about the shape of the graph between known points. If the function is fairly simple, it usually works reasonably well; however, more complicated functions require more sophisticated methods, which we shall study later.

In Definition 1, the requirement that a function f assign one and *only one* value of y to each value of x in its domain corresponds to the geometric condition that no two different points on the graph of f can have the same abscissa. Thus, the curve in Figure 2 *cannot be the graph of a function* because the two points P and Q have the same abscissa. The graph of a function cannot pass over or under itself.

The domain and the range of a function can be found easily from the graph of the function. Thus, *the domain of a function is the set of all abscissas of points on its graph* (Figure 3a), while *the range of a function is the set of all ordinates of points on its graph* (Figure 3b).

Figure 2

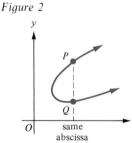

same abscissa

Figure 3

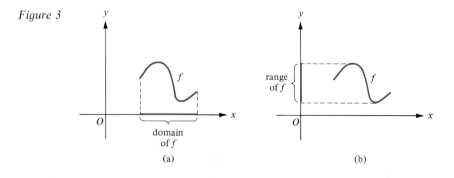

(a)   (b)

EXAMPLE   Let f be the function defined by the equation $y = \sqrt{x-1}$ with the constraint that $x \leq 2$. Sketch the graph of f and show the domain and the range of f by shading on the x and y axes, respectively.

Figure 4

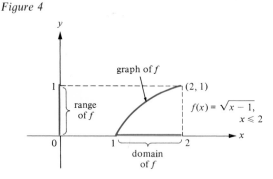

SOLUTION
Since $y = f(x) = \sqrt{x-1}$, we must have $x - 1 \geq 0$; that is, $x \geq 1$. We also have the constraint $x \leq 2$; hence, $1 \leq x \leq 2$. Choosing a few values of x between 1 and 2, we determine the corresponding values of $y = \sqrt{x-1}$. (A hand calculator or a table of square roots is useful here.) Plotting the corresponding points and connecting them by a smooth curve, we obtain the graph shown in Figure 4. Evidently, the domain of f is the interval $[1, 2]$ and the range of f is the interval $[0, 1]$.

The following examples illustrate the convention that, when a function f is defined by an equation and no constraints on the values of the variables are given, the domain of f consists of those values of x for which the defining equation makes sense.

EXAMPLES   Find the domain and the range of the function f defined by the equation and sketch the graph of the function.

1 $y = 3x + 1$

SOLUTION
Here, $y = f(x) = 3x + 1$ and the independent variable x can take on any value whatsoever, so the domain of f is the set $\mathbb{R}$ of all real numbers. Also, the dependent variable y can take on *any* value. In fact, suppose that we want y to take on a given value v; that is, we want $v = 3x + 1$. Solving the latter equation for x, we obtain $x = \frac{1}{3}(v - 1)$. Thus, when x has the value $\frac{1}{3}(v - 1)$, y will have the desired value v. Hence, the range of f is the set $\mathbb{R}$ (Figure 5).

Figure 5

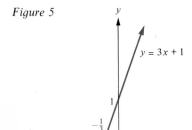

2 $y = \sqrt{4 - x}$

SOLUTION
Since $\sqrt{4 - x}$ is undefined except when $4 - x \geq 0$, that is, when $x \leq 4$, the domain of f is the interval $(-\infty, 4]$. Since $y = \sqrt{4 - x}$, it follows from the definition of the principal square root that $y \geq 0$. Notice that the dependent variable y can take on any nonnegative value v (just give x the value $4 - v^2$); hence, the range of f is the interval $[0, \infty)$ (Figure 6).

3 $y = |x|$

SOLUTION
The independent variable x can take on any value, so the domain is the set of real numbers $\mathbb{R}$. For $x < 0$, we have $y = -x$, while, for $x \geq 0$, we have $y = x$. The dependent variable y cannot be negative but can take

Figure 6

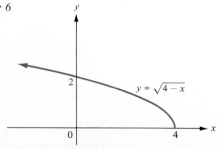

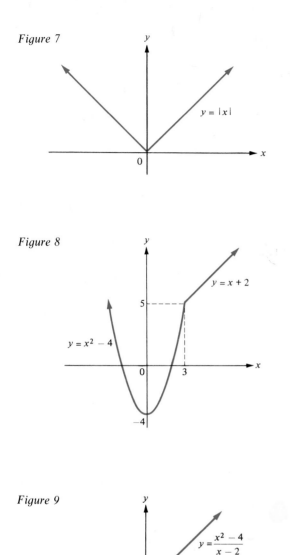

Figure 7

$y = |x|$

Figure 8

$y = x + 2$

$y = x^2 - 4$

Figure 9

$y = \dfrac{x^2 - 4}{x - 2}$

on any nonnegative value. Thus, the range of f is $[0, \infty)$ (Figure 7).

4 $y = \begin{cases} x + 2 & \text{if } x \geq 3 \\ x^2 - 4 & \text{if } x < 3 \end{cases}$

SOLUTION

The independent variable x can take on any value, so the domain is the set $\mathbb{R}$. In sketching the graph, we must consider separately the portion to the right of the vertical line $x = 3$ and the portion to the left of this line. To the right of $x = 3$, the graph is a portion of a straight line of slope 1 and contains the point $(3, 5)$. To the left of $x = 3$, the graph is part of a curve (called a *parabola*). These parts, *taken together*, form the graph of f (Figure 8). From this graph, we see that the dependent variable y can take on any value greater than or equal to -4; hence, the range of f is the interval $[-4, \infty)$.

5 $y = \dfrac{x^2 - 4}{x - 2}$

SOLUTION

The equation makes sense for all values of x *except for* $x = 2$ (which makes the denominator of the fraction zero); hence, the domain consists of the two intervals $(-\infty, 2)$ and $(2, \infty)$. Notice that

$$x^2 - 4 = (x + 2)(x - 2);$$

therefore, for $x \neq 2$,

$$\frac{x^2 - 4}{x - 2} = \frac{(x + 2)(x - 2)}{x - 2} = x + 2.$$

Hence, the condition $y = \dfrac{x^2 - 4}{x - 2}$ is equivalent to the condition $y = x + 2$, *provided that* $x \neq 2$. Therefore, the graph consists of all the points on the line $y = x + 2$ *except for the point* $(2, 4)$, which is excluded (Figure 9). Evidently, the range of f is all real numbers except for 4; that is, the range consists of the two intervals $(-\infty, 4)$ and $(4, \infty)$.

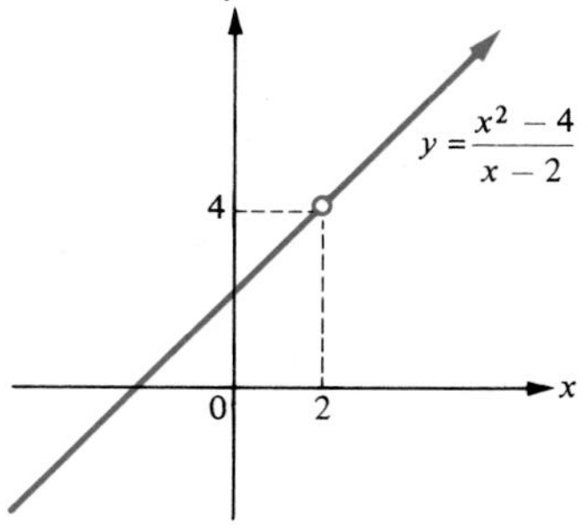

If f is a function and x is in the domain of f, then $f(x)$ is a number depending on the number x and is not the function f. Neither is the equation $y = f(x)$ the same as the function f. Sometimes, in the interest of brevity, people allow themselves to speak—incorrectly—of "the function $f(x)$" or "the function $y = f(x)$." Probably, there is no great harm in this practice, provided that the speakers (and their audiences) understand what is really meant. Although we shall avoid this practice whenever absolute precision is desired, we shall also indulge in it whenever it seems convenient.

EXAMPLES 1 Consider the function $f(x) = 1/(1 - x)$. Evaluate $f(-3)$, $f(-2)$, $f(-1)$, $f(0)$, $f(\pi)$, and $f(\frac{3}{2})$. What is $f(1)$? Sketch the graph of f and find the domain and the range of f.

Figure 10

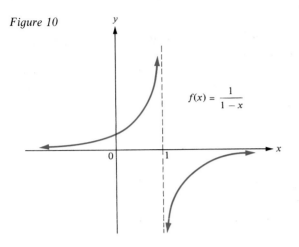

$f(x) = \dfrac{1}{1-x}$

SOLUTION
Here,

$$f(-3) = \frac{1}{1-(-3)} = \frac{1}{4}.$$

Similarly, $f(-2) = \frac{1}{3}$, $f(-1) = \frac{1}{2}$, $f(0) = 1$, $f(\pi) = 1/(1-\pi)$, and $f(\frac{3}{2}) = -2$. Since the denominator is zero when $x = 1$, $f(1)$ is undefined; that is, 1 is not in the domain of f. The domain of f consists of the two intervals $(-\infty, 1)$ and $(1, \infty)$. If x is near 1 but a little smaller than 1, then $1-x$ is small and positive, so $f(x) = 1/(1-x)$ is very large and positive. Similarly, if x is near 1 but a little larger than 1, then $f(x)$ is negative but very large in absolute value. Also, if x is very large in absolute value, then $f(x)$ is very small in absolute value. Plotting a few points and keeping these facts in mind, we can sketch the graph as in Figure 10.

From this graph, we can see that the range of f consists of all values of y except for $y = 0$; hence, the range consists of the two intervals $(-\infty, 0)$ and $(0, \infty)$.

2 Consider the function $g(x) = 4x + 7$. Calculate $\dfrac{g(3+h) - g(3)}{h}$ for $h \neq 0$, and "simplify" your answer.

SOLUTION

$$\frac{g(3+h) - g(3)}{h} = \frac{[4(3+h) + 7] - [4(3) + 7]}{h}$$

$$= \frac{12 + 4h + 7 - 12 - 7}{h} = \frac{4h}{h} = 4.$$

In writing equations involving functions, there is nothing essential about the choice of the letters x and y for the independent and dependent variables, respectively. Thus, we can write $y = f(t)$, $s = f(t)$, or even $x = f(y)$. For example, the volume V of a sphere is a function of its radius r, so $V = f(r)$ where $f(r) = \frac{4}{3}\pi r^3$.

5.1 Functions as Sets of Ordered Pairs

The idea of a function as a rule or correspondence is easy to understand and therefore will be used throughout this book. However, for some formal purposes this idea can be somewhat intangible, so mathematicians have developed an alternative definition, which is more concrete. Briefly, the idea is that the graph of a function determines the function uniquely; hence, it can be thought of as *being* the function. We have the following alternative definition.

DEFINITION 3   **Function as a Set of Ordered Pairs**
A *function* is a set of ordered pairs in which no two different ordered pairs have the same first member. If (x, y) is an ordered pair in this set, we say that y *corresponds* to x under the function.

EXAMPLE   According to Definition 3, which set is a function?
(a) The set of all ordered pairs (x, y) such that $xy^2 = 1$.
(b) The set of all ordered pairs (x, y) such that $xy = 1$.

SOLUTION

The set described in (a) cannot be a function since the two different ordered pairs $(1, 1)$ and $(1, -1)$ both belong to this set, and yet they have the same first member, 1. The set described in (b) is a function since, if (x, y_1) and (x, y_2) both belong to this set, then $xy_1 = 1 = xy_2$, so $y_1 = y_2$ and the two ordered pairs are the same.

Problem Set 5

In problems 1 through 14, find the domain and the range of the function defined by the given equation and sketch the graph of the function.

1 $y = -5x + 7$

2 $y = |3x|$

3 $y = |-2x|$

4 $y = -\sqrt{4 - x}$

5 $y = \sqrt{1 - x^2}$

6 $y = |2x - 3|$

7 $y = \dfrac{9x^2 - 4}{3x - .2}$

8 $y = \begin{cases} -1 & \text{if } x \le 2 \\ 1 & \text{if } x > 2 \end{cases}$

9 $y = \begin{cases} -3 & \text{if } x < -1 \\ -1 & \text{if } -1 \le x \le 1 \\ 2 & \text{if } x > 1 \end{cases}$

10 $y = \begin{cases} 6x + 7 & \text{if } x \le -2 \\ 4 - x & \text{if } x > -2 \end{cases}$

11 $y = \begin{cases} x^2 - 4 & \text{if } x < 3 \\ 2x - 1 & \text{if } x \ge 3 \end{cases}$

12 $y = \dfrac{x^3 - 4x^2}{x - 4}$

13 $y = \dfrac{(x^2 - 4x + 3)(x^2 - 4)}{(x^2 - 5x + 6)(x + 2)}$

14 $y = \dfrac{x^2 - 7x + 6}{x^2 + x - 6}$

15 Let f be the function defined by the equation $y = x + 1/x$.
(a) What is the domain of f?
(b) What is the range of f?
(c) Sketch the graph of f.
(d) Which of the following points lie on the graph of f? $(-1, -2)$, $(2, 1)$, $(1, 2)$, $(-2, -\frac{5}{2})$, $(3, \frac{7}{2})$

16 Let f be a function. Explain in your own words the distinction among f, $f(x)$, and $y = f(x)$.

17 Let f be the function defined by the equation $f(x) = x^2 - 3x - 4$. Sketch the graph of f, find the domain and the range of f, and evaluate the following:
(a) $f(1)$
(b) $f(2)$
(c) $f(-1)$
(d) $f(4)$
(e) $f(a)$
(f) $f(a + b)$
(g) $f(a - b)$
(h) $f(x_0)$

18 Let f be the function defined by the equation $f(x) = x^2$. Find $f(f(2))$. Find $f(f(3))$. What is $f(f(x))$?

19 Let f be the function defined by $f(x) = \dfrac{x - 2}{3x + 7}$. What is the domain of f? Evaluate:
(a) $f(\frac{1}{2})$
(b) $f(-\frac{1}{2})$
(c) $f(a/3)$
(d) $f(4/a)$
(e) $f(a + 2)$
(f) $f(a^2)$
(g) $[f(a)]^2$
(h) $f(x_0)$

20 A function f is called *additive* if the domain of f is $\mathbb{R}$ and $f(a + b) = f(a) + f(b)$ holds for all real numbers a and b.
(a) Give an example of an additive function.
(b) Give an example of a function that is not additive.
(c) Show that if f is an additive function, then $f(0) = 0$. (*Hint:* Put $a = b = 0$.)
(d) Show that an additive function f must satisfy $f(-x) = -f(x)$. [*Hint:* Put $a = x$, $b = -x$ and use part (c).]

21 If g is the function defined by the equation $g(x) = \sqrt{3x + 5}$, what is the domain of g? Evaluate:
(a) $g(-\frac{1}{3})$
(b) $g(\frac{4}{3})$
(c) $g(\frac{1}{3})$
(d) $g(-1)$
(e) $g(a^2)$
(f) $[g(a)]^2$
(g) $g(2x + 1)$
(h) $g(x_0 + h)$

22 Find the domain and the range of the function F defined by $F(x) = |x + 2|$. Also, evaluate:

(a) $F(-2)$
(b) $F(2)$
(c) $F(-3)$
(d) $[F(-3)]^2$
(e) $F(2) - F(-3)$
(f) $F(a^2)$

23 If G is the function defined by $G(t) = \dfrac{3t - 1}{1 + 2t}$, what is the domain of G? Evaluate:

(a) $G(2)$
(b) $G(-3)$
(c) $G(a)$
(d) $G(a^2)$
(e) $G(-a^2)$
(f) $G(2) - G(-3)$
(g) $G(x)$

24 Let h be the function defined by $h(x) = \begin{cases} |x|/x & \text{if } x \neq 0 \\ 1 & \text{if } x = 0. \end{cases}$ Evaluate:

(a) $h(-4)$
(b) $h(4)$
(c) $h(-1)$
(d) $h(1)$
(e) $h(x^2)$
(f) $h(-x^2)$
(g) $h(a + 1)$
(h) $h(a - 1)$
(i) $h(h(x))$

25 Let g be the function defined by $g(x) = x(x + 1)(x + 2)(x + 3)$.

(a) Evaluate $g(a + 1)$.
(b) Evaluate $g(a + 2)$.

(c) Show that for $a \neq -1$ and $a \neq -5$, $\dfrac{g(a + 1)}{a + 1} = \dfrac{g(a + 2)}{a + 5}$.

26 Which of the graphs in Figure 11 are graphs of functions?

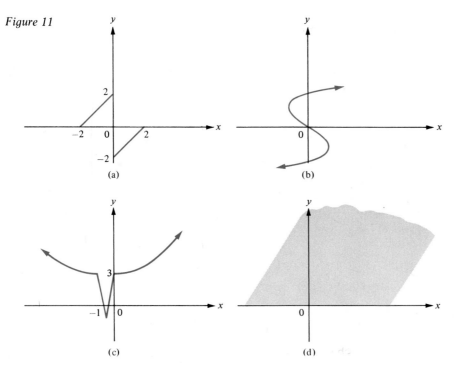

Figure 11

(a)

(b)

(c)

(d)

27 Is the radius r of a circle a function of its area A? If so, write an equation that defines this function.

28 A closed box with a square base x centimeters by x centimeters has a volume of 100 cubic centimeters. Express the total surface area A of the exterior of this box as a function of x. (Assume that the thickness of the walls of the box is negligible.)

29 The velocity v of a car is measured during a certain interval of time and found to vary according to the equation

$$v = \begin{cases} 60t & \text{for } 0 \leq t < 5 \\ 300 & \text{for } t \geq 5, \end{cases}$$

where v is measured in meters per minute and the elapsed time t is measured in seconds.

If V is the velocity measured in kilometers per hour and T is the elapsed time measured in minutes, find V as a function of T.

30 A baseball diamond is a square 90 feet on each side. A player is running from home plate to first base at the rate of 30 feet per second. Express the runner's distance s from second base as a function of the time t in seconds since he left home plate.

31 A rectangle is required to have an area of 25 square centimeters, but its dimensions may vary. If one side has length x, express the perimeter p as a function of x.

32 A rectangle with dimensions $2x$ by $2y$ is inscribed in a circle of radius 10. Express y and the area A of the rectangle as functions of x.

In problems 33 through 38, form the expression $\dfrac{f(x_0 + h) - f(x_0)}{h}$, $h \neq 0$, and "simplify."

33 $f(x) = 3$

34 $f(x) = 5x - 10$

35 $f(x) = -8x + 3$

36 $f(x) = \frac{1}{3}x^2$

37 $f(x) = 2/x$

38 $f(x) = \sqrt{1 + x}$

39 According to Definition 3, which of the following *is* a function?
(a) $y^2 = x^2$
(b) The set of all ordered pairs (x, y) with $x = y$.
(c) The set of all ordered pairs (x, y) with $y = x^2$.
(d) The set of all ordered pairs (x, y) with $x = y^2$.
(e) x^2
(f) y

40 For each equation, determine whether the set of all ordered pairs (x, y) that satisfy the equation is a function according to Definition 3.
(a) $x^2 y = 5$
(b) $y = 2/x^2$
(c) $x^2 + y^2 = 4$
(d) $y = \sqrt{16 - x^2}$
(e) $y^2 = x^3$
(f) $y = 5\sqrt[3]{x}$
(g) $3|x| + 2|y| = 6$
(h) $3|x| + 2y = 6$
(i) $y^2 = 1 + x^2$
(j) $y = |x|/x$

41 True or false: If f is a function, the graph of f consists of all points in the xy plane of the form $(x, f(x))$ as x runs through the domain of f.

42 State whether each of the following sets of ordered pairs is a function according to Definition 3.
(a) The set consisting of $(0, 1)$ and $(0, -1)$.
(b) The set consisting of $(1, 0)$ and $(-1, 0)$.
(c) The set consisting of all ordered pairs (x, y) such that x is a positive integer and y is the next larger positive integer.
(d) The set consisting of all ordered pairs (x, y) such that x and y are positive integers and y is an exact (integral) multiple of x.

6 Types of Functions

In this section we describe certain types or classes of functions that are considered in calculus. Among these are *even* functions, *odd* functions, *polynomial* functions, *rational* functions, *algebraic* functions, and *transcendental* functions.

6.1 Even and Odd Functions

Consider the functions f and g given by the equations $f(x) = x^2 - 4$ and $g(x) = x^3$ (Figure 1). In Figure 1a, notice that $f(-x) = f(x)$; hence, the graph of f is *symmetric about the y axis* in the sense that, if the point (x, y) belongs to the graph, so does the point $(-x, y)$. In Figure 1b, notice that $g(-x) = -g(x)$; hence, the graph of g is *symmetric about the origin* in the sense that, if the point (x, y) belongs to the graph, so does the point $(-x, -y)$.

Figure 1

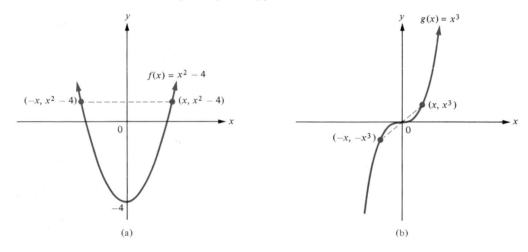

(a)   (b)

More generally, we make the following definition.

DEFINITION 1   **Even and Odd Functions**
(a) A function f is said to be *even* if, for every number x in the domain of f, $-x$ is also in the domain of f and $f(-x) = f(x)$.
(b) A function f is said to be *odd* if, for every number x in the domain of f, $-x$ is also in the domain of f and $f(-x) = -f(x)$.

Clearly, the even functions are precisely those functions whose graphs are symmetric about the vertical axis, while the odd functions are precisely those functions whose graphs are symmetric about the origin.

EXAMPLES **1** Show that each function is even.
(a) $g(x) = x^4$ (b) $f(t) = 2t^2 + 3|t|$

SOLUTION
(a) $g(-x) = (-x)^4 = x^4 = g(x)$, so g is an even function.
(b) $f(-t) = 2(-t)^2 + 3|-t| = 2t^2 + 3|t| = f(t)$, so f is an even function.

2 Show that each function is odd.
(a) $g(x) = x^5$ (b) $f(x) = x|x|$

SOLUTION
(a) $g(-x) = (-x)^5 = -x^5 = -g(x)$, so g is an odd function.
(b) $f(-x) = -x|-x| = -x|x| = -f(x)$, so f is an odd function.

There are many functions that are neither even nor odd. For example, the functions given by the equations $f(x) = 1 + x$ and $g(x) = \sqrt{x}$ are neither even nor odd; hence, neither of their graphs is symmetric about the y axis or the origin (Figure 2). However, if a function is found to be either even or odd, the job of sketching its graph becomes easier because of the symmetry involved.

Figure 2

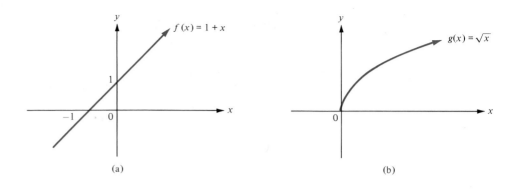

(a)   (b)

6.2 Polynomial Functions

A function f that is defined by an equation of the form

$$f(x) = a_0 + a_1 x + a_2 x^2 + \cdots + a_{n-1} x^{n-1} + a_n x^n,$$

where n is a nonnegative integer and the coefficients $a_0, a_1, a_2, \ldots, a_n$ are constant real numbers, is called a *polynomial function*. If $a_n \neq 0$, we say that this polynomial function has *degree n*. For instance, $f(x) = 7 + 5x - 3x^2 + 8x^3$ is a polynomial function of degree 3 with coefficients $a_0 = 7$, $a_1 = 5$, $a_2 = -3$, and $a_3 = 8$.

A polynomial function of the form $f(x) = a_0$ is called a *constant function*; its graph is a straight line of slope zero with y intercept a_0. If $a_0 \neq 0$, the polynomial function $f(x) = a_0$ has degree zero; however, the polynomial function given by $f(x) = 0$—that is, the constant function all of whose values are zero—is not assigned any degree.

Figure 3

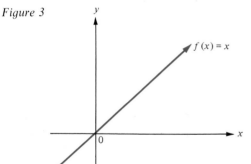

A polynomial function of the form $f(x) = a_0 + a_1 x$, where $a_1 \neq 0$, is called a *linear function* because its graph is a straight line with slope a_1 and y intercept a_0. Evidently, the linear functions are the same thing as the polynomial functions of degree 1. A particularly important linear function is the *identity function* defined by the equation $f(x) = x$. Of course, the graph of the identity function is a straight line of slope 1 passing through the origin (Figure 3).

6.3 Rational Functions and Algebraic Functions

The sum, difference, or product of two polynomials is again a polynomial, but the quotient of two polynomials is generally not a polynomial. For instance, $\dfrac{3x^2 - x + 1}{4x^5 - x^3 + 1}$ is not a polynomial. This observation motivates the following definition.

DEFINITION 2   **Rational Function**
A function f defined by an equation of the form $f(x) = p(x)/q(x)$, where p and q are polynomial functions and q is not the constant zero function, is called a *rational function*. (*Memory device:* When you see the word "*rational*," think of "*ratio*.")

The domain of the rational function defined by $f(x) = p(x)/q(x)$ consists of all values of x for which $q(x) \neq 0$. The graphs of rational functions, which can take on a variety of geometric shapes, will be discussed further in Chapter 1, Section 6 and again in Chapter 3, Section 3.

Note that polynomial functions themselves are special kinds of rational functions— just let the denominator $q(x)$ in Definition 2 be the constant function $q(x) = 1$. Some examples of rational functions are $f(x) = x/(x + 1)$, $g(x) = 2/x$, $h(x) = x^2 + 2x + 1$, $F(x) = x$, and $H(x) = (x^2 - 1)/(x - 1)$.

It is easy to see that if f and g are rational functions, then so are the functions $S, P, D,$ and Q defined by $S(x) = f(x) + g(x)$, $P(x) = f(x) \cdot g(x)$, $D(x) = f(x) - g(x)$, and $Q(x) = f(x)/g(x)$; that is, the sum, product, difference, or quotient of rational functions is again a rational function. However, by extracting roots of rational functions, it is possible to obtain functions that are no longer rational. A case in point is the function given by $h(x) = \sqrt{x^2}$, which is just the absolute value function, since $h(x) = \sqrt{x^2} = |x|$.

The class of rational functions is not sufficiently large to include many of the functions that we encounter in calculus, so we are led to the following definition.

DEFINITION 3   **Elementary Algebraic Function**
An *elementary algebraic function* is a function that can be formed by a finite number of algebraic operations (these operations being addition, subtraction, multiplication, division, and extraction of positive integral roots), starting with the identity function and constant functions.

Some examples of elementary algebraic functions are

$$f(x) = \sqrt{x^2}, \qquad g(x) = \frac{x}{\sqrt{x^2 + 5}}, \qquad F(x) = \frac{\sqrt[3]{x + 1} + 1}{\sqrt[5]{\sqrt{x^2 - 2} + 2}}.$$

It should be observed that any rational function is automatically an elementary algebraic function.

In advanced courses, a more inclusive class of functions, called *algebraic functions* (without the adjective "elementary"), is defined. Broadly speaking, these are the functions that are accessible by algebraic means. The remaining functions, those that are not algebraic, are called *transcendental functions*, since they transcend purely algebraic methods. For instance, the trigonometric functions, which we review in Section 7, are transcendental functions; so are the exponential, logarithmic, and hyperbolic functions, which we shall study in Chapter 9.

6.4 Discontinuous Functions

Many functions considered in elementary calculus have graphs that are "connected" in the sense that they consist of one continuous piece. Such functions, which are said to be *continuous*, are discussed in detail in Chapter 1, Section 4. In order to fully understand and appreciate the nature of continuous functions, it is sometimes useful to examine specific functions that are not continuous. One of the more interesting discontinuous functions is the *greatest integer function*, which, like the absolute value function, has its own special symbol.

DEFINITION 4   **Greatest Integer Function**
If x is a real number, the symbol $[\![x]\!]$ denotes the greatest integer not exceeding x; that is, $[\![x]\!]$ is the integer that is nearest to x but is less than or equal to x. The *greatest integer function* is the function f defined by $f(x) = [\![x]\!]$.

Notice that $[\![x]\!]$ is the unique integer satisfying the condition $[\![x]\!] \le x < [\![x]\!] + 1$. For instance, $[\![3.7]\!] = 3$, $[\![2 + \frac{9}{10}]\!] = 2$, $[\![3.234334]\!] = 3$, $[\![-2.7]\!] = -3$, $[\![-2.34334]\!] = -3$, $[\![-\frac{1}{2}]\!] = -1$, $[\![\sqrt{3}]\!] = 1$, $[\![2]\!] = 2$, and $[\![-2]\!] = -2$. A table of values of $[\![x]\!]$ for $-3 \le x < 4$ follows, and the corresponding graph of $f(x) = [\![x]\!]$ is shown in Figure 4.

Figure 4

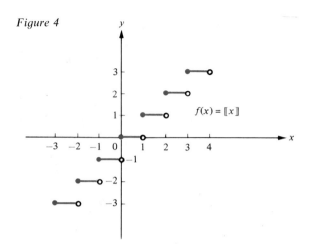

$$[\![x]\!] = \begin{cases} -3 & \text{for } -3 \le x < -2 \\ -2 & \text{for } -2 \le x < -1 \\ -1 & \text{for } -1 \le x < 0 \\ 0 & \text{for } 0 \le x < 1 \\ 1 & \text{for } 1 \le x < 2 \\ 2 & \text{for } 2 \le x < 3 \\ 3 & \text{for } 3 \le x < 4 \end{cases}$$

In Figure 4, we use small dots to emphasize that the left-hand endpoints of the horizontal line segments belong to the graph, and we use small open circles to indicate that the right-hand endpoints do not belong to the graph. The discontinuous nature of the greatest integer function is apparent from its graph.

Problem Set 6

In problems 1 through 9, decide whether the given function is even, odd, or neither.

1 $f(x) = x^4 + 3$

2 $g(x) = -x^4 + 2x^2 + 1$

3 $f(x) = x^4 + x$

4 $g(t) = t^2 + |t|$

5 $F(x) = 5x^3 + 7x$

6 $f(t) = -t^3 + 7t$

7 $h(x) = \sqrt{8x^3 + x}$

8 $f(y) = \dfrac{\sqrt{y^2 + 1}}{|y|}$

9 $f(x) = \dfrac{x + 1}{x^2 + 1}$

10 Discuss the symmetry of the graphs of the functions in problems 1 through 9.

In problems 11 through 18, decide whether the function is a polynomial function. If it is a polynomial function, indicate the degree (if any) and identify the coefficients.

11 $f(x) = 6x^2 - 3x - 8$

12 $f(x) = x^{-3} + 2x$

13 $g(x) = (x - 3)(x - 2) - x^3$

14 $f(x) = 2^{-1}$

15 $F(x) = \sqrt{2}\,x^4 - 5^{-1}x^3 + 20$

16 $f(x) = 210x^{117} - 11x - 40$

17 $g(x) = 0$

18 $h(x) = \sqrt[3]{x^3 - 6x^2 + 12x - 8}$

19 Explain in your own words the distinction between the constant function f defined by the equation $f(x) = 2$ and the real number 2.

20 Is the function f defined by the equation $f(x) = x^{-1} + \dfrac{x - 1}{x}$ a constant function?

In problems 21 through 24, determine a linear function f that satisfies the condition.

21 $f(2) = 5$ and $f(-3) = 7$

22 $f(2x + 3) = 2f(x) + 3$

23 $f(5x) = 5f(x)$

24 $f(x + 7) = f(x) + f(7)$

25 A number r is called a *root* (or a *zero*) of a function f if $f(r) = 0$. Prove that every linear function has a root.

26 Let f be a linear function. Prove that

$$f(tc + (1 - t)d) = tf(c) + (1 - t)f(d)$$

holds for every three numbers c, d, and t.

In problems 27 through 31, specify whether the algebraic function is a rational function. In each case, specify the domain of the function.

27 $f(x) = \dfrac{3x}{x-1}$

28 $g(x) = \dfrac{x+1}{\sqrt[3]{2x^2+5}}$

29 $f(x) = x^2 + 2x + 1$

30 $f(t) = \dfrac{t^2}{2t^3+5}$

31 $f(t) = \dfrac{6t^2}{\sqrt[5]{t+1}}$

32 Show that $f(x) = \dfrac{x}{1-x} - \dfrac{1}{1+x}$ is a rational function by rewriting it as a ratio of polynomial functions. What is the domain of f?

33 The *signum function* (abbreviated sgn) is defined by

$$\operatorname{sgn} x = \begin{cases} \dfrac{|x|}{x} & \text{if } x \neq 0 \\[2mm] 0 & \text{if } x = 0 \end{cases}$$

(a) Find sgn (-2), sgn (-3), sgn (0), sgn (2), sgn (3), and sgn (151).
(b) Prove that $|x| = x\operatorname{sgn} x$ is true for all values of x.
(c) Prove that $\operatorname{sgn}(ab) = (\operatorname{sgn} a)(\operatorname{sgn} b)$ is true for all values of a and b.
(d) Sketch the graph of the signum function.
(e) Find the domain and the range of the signum function.
(f) Sketch the graph of the function f defined by $f(x) = \operatorname{sgn}(x-1)$.
(g) Explain why the sgn function is discontinuous.

In problems 34 through 45, sketch the graph of the function and specify its domain and its range.

34 $f(x) = |x| + 1$

35 $f(x) = |3x| - 3x$

36 $f(x) = -|3x-2|$

37 $H(x) = |x+1| - |x|$

38 $h(x) = -3|x| + x$

39 $f(x) = [\![3x]\!]$

40 $h(x) = [\![x]\!] + x$

41 $f(x) = [\![\frac{1}{2}x]\!]$

42 $G(x) = [\![\,|x|\,]\!]$

43 $g(x) = |[\![x]\!]|$

44 $f(x) = \dfrac{x}{|x|} - \dfrac{|x|}{x}$

45 $f(x) = 3^{-1}$

46 Let f be any function with domain $\mathbb{R}$.

(a) Define a function g by the equation $g(x) = \dfrac{f(x) + f(-x)}{2}$. Prove that g is even.

(b) Define a function h by the equation $h(x) = \dfrac{f(x) - f(-x)}{2}$. Prove that h is odd.

(c) Prove that $f(x) = g(x) + h(x)$ holds for all x. Thus, conclude that any function with domain $\mathbb{R}$ is the sum of an even function and an odd function.

(d) Suppose that G is an even function with domain $\mathbb{R}$, that H is an odd function with domain $\mathbb{R}$, and that $f(x) = G(x) + H(x)$ holds for all x. Prove that $G(x) = g(x)$ and that $H(x) = h(x)$ for all values of x.

(e) Show that f is even if and only if $f(x) = g(x)$ holds for all x.

(f) Show that f is odd if and only if $f(x) = h(x)$ holds for all x.

(g) Is it possible for f to be both even and odd?

7 Trigonometric Functions

The six trigonometric functions, sine, cosine, tangent, secant, cosecant, and cotangent (abbreviated sin, cos, tan, sec, csc, and cot, respectively), are presumably already familiar to the reader, so we confine ourselves here to a brief review.

Certain fundamental formulas of calculus become much simpler if angles are measured in *radians* rather than in degrees. By definition, the number of radians in an angle θ (Figure 1) is the number of "radius units" contained in the arc s subtended by the central angle θ on a circle of radius r. That is,

Figure 1

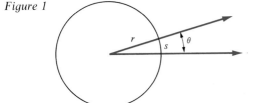

$$\theta \text{ (in radians)} = \frac{s}{r}.$$

Since the full circumference of the circle has arc length $s = 2\pi r$ and is subtended by a 360° angle, then $360° = 2\pi r/r$ radians; that is, $360° = 2\pi$ radians, or

$$\pi \text{ radians} = 180°.$$

$$\text{Hence, 1 radian} = (180/\pi)° \approx 57°18'.$$

Table 1 gives the corresponding degree and radian measures of certain special angles. Henceforth, *all angles will be measured in radians* unless otherwise explicitly indicated by the use of the symbol ° for degrees.

Table 1

Degree measure	30°	45°	60°	90°	120°	135°	150°	180°	270°	360°
Radian measure	$\dfrac{\pi}{6}$	$\dfrac{\pi}{4}$	$\dfrac{\pi}{3}$	$\dfrac{\pi}{2}$	$\dfrac{2\pi}{3}$	$\dfrac{3\pi}{4}$	$\dfrac{5\pi}{6}$	π	$\dfrac{3\pi}{2}$	2π

We can now define the six trigonometric functions.

DEFINITION 1   **Trigonometric Functions of Any Real Number t**

Let t be any real number and measure out an angle of $|t|$ radians, starting at the positive x axis and turning counterclockwise about the origin if $t \geq 0$ and clockwise about the origin if $t < 0$. Now construct a circle with radius 1 unit and center at the origin (Figure 2) and let (x, y) be the point where the terminal side of the angle meets this circle. The values at t of the six trigonometric functions are defined to be:

Figure 2

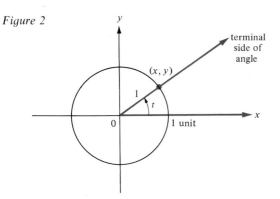

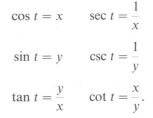

$$\cos t = x \qquad \sec t = \frac{1}{x}$$

$$\sin t = y \qquad \csc t = \frac{1}{y}$$

$$\tan t = \frac{y}{x} \qquad \cot t = \frac{x}{y}.$$

The domain of both the sine and cosine functions is $\mathbb{R}$. The domain of the remaining four functions is the set of values of t for which the denominator of the defining fraction is nonzero.

By using this definition and a little geometry, the values of the trigonometric functions at certain special values of t can be found. These values are shown in Table 2. (Dashes in the table indicate that the function is undefined at the corresponding value of t.)

Table 2

Value of t	0	$\dfrac{\pi}{6}$	$\dfrac{\pi}{4}$	$\dfrac{\pi}{3}$	$\dfrac{\pi}{2}$	π
Corresponding degrees	0°	30°	45°	60°	90°	180°
sin	0	$\dfrac{1}{2}$	$\dfrac{\sqrt{2}}{2}$	$\dfrac{\sqrt{3}}{2}$	1	0
cos	1	$\dfrac{\sqrt{3}}{2}$	$\dfrac{\sqrt{2}}{2}$	$\dfrac{1}{2}$	0	-1
tan	0	$\dfrac{\sqrt{3}}{3}$	1	$\sqrt{3}$	—	0
sec	1	$\dfrac{2\sqrt{3}}{3}$	$\sqrt{2}$	2	—	-1
csc	—	2	$\sqrt{2}$	$\dfrac{2\sqrt{3}}{3}$	1	—
cot	—	$\sqrt{3}$	1	$\dfrac{\sqrt{3}}{3}$	0	—

The six trigonometric functions satisfy certain identities that follow from standard arguments given in precalculus.

Standard Trigonometric Identities

The following identities hold for all real numbers s and t that belong to the domains of the functions involved:

1   $\tan t = \dfrac{\sin t}{\cos t}$

2   $\sec t = \dfrac{1}{\cos t}$

3   $\csc t = \dfrac{1}{\sin t}$

4   $\cot t = \dfrac{\cos t}{\sin t}$

5   $\sin t = \cos \left(\dfrac{\pi}{2} - t \right)$

6   $\cos t = \sin \left(\dfrac{\pi}{2} - t \right)$

7   $\sin (-t) = -\sin t$

8   $\cos (-t) = \cos t$

9   $\sin^2 t + \cos^2 t = 1$

10   $\tan^2 t + 1 = \sec^2 t$

11   $\cot^2 t + 1 = \csc^2 t$

12   $\sin (t + s) = \sin t \cos s + \sin s \cos t$ (addition formula for sine)

13   $\cos (t + s) = \cos t \cos s - \sin t \sin s$ (addition formula for cosine)

14 $\sin(t-s) = \sin t \cos s - \sin s \cos t$

15 $\cos(t-s) = \cos t \cos s + \sin t \sin s$

16 $\tan(t+s) = \dfrac{\tan t + \tan s}{1 - \tan t \tan s}$ (addition formula for tangent)

17 $\tan(t-s) = \dfrac{\tan t - \tan s}{1 + \tan t \tan s}$

18 $\sin 2t = 2 \sin t \cos t$ (double-angle formula for sine)

19 $\cos 2t = \cos^2 t - \sin^2 t$, or

$\cos 2t = 2 \cos^2 t - 1$, or $\Big\}$ (double-angle formulas for cosine)

$\cos 2t = 1 - 2 \sin^2 t$

20 $\sin^2\left(\dfrac{t}{2}\right) = \frac{1}{2}(1 - \cos t)$ (half-angle formula for sine)

21 $\cos^2\left(\dfrac{t}{2}\right) = \frac{1}{2}(1 + \cos t)$ (half-angle formula for cosine)

EXAMPLES Use the standard trigonometric identities to simplify the given expression.

1 $(\sin t + \cos t)^2 - \sin 2t$

SOLUTION

$$(\sin t + \cos t)^2 - \sin 2t = \sin^2 t + 2 \sin t \cos t + \cos^2 t - \sin 2t$$
$$= 1 + 2 \sin t \cos t - \sin 2t$$
$$= 1 + \sin 2t - \sin 2t$$
$$= 1.$$

2 $\sin(t + \pi)$

SOLUTION

$$\sin(t+\pi) = \sin t \cos \pi + \sin \pi \cos t$$
$$= (\sin t)(-1) + (0) \cos t$$
$$= -\sin t.$$

3 $\cos \dfrac{7\pi}{12}$

SOLUTION

$$\cos \frac{7\pi}{12} = \cos\left(\frac{\pi}{3} + \frac{\pi}{4}\right) = \cos \frac{\pi}{3} \cos \frac{\pi}{4} - \sin \frac{\pi}{3} \sin \frac{\pi}{4}$$
$$= \left(\frac{1}{2}\right)\left(\frac{\sqrt{2}}{2}\right) - \left(\frac{\sqrt{3}}{2}\right)\left(\frac{\sqrt{2}}{2}\right)$$
$$= \frac{\sqrt{2} - \sqrt{6}}{4}.$$

The graphs of the six trigonometric functions appear in Figure 3.

Notice the "wavelike" periodic appearance of these graphs. When the graphs are sketched over larger intervals, it is found that the geometric shapes shown in Figure 3 repeat themselves indefinitely. It is precisely the "wavelike" nature of the sine and cosine functions that makes them so useful in applied mathematics. Indeed, many natural phenomena, from electromagnetic waves to the ebb and flow of the tides, are periodic, and so these functions are indispensable in the construction of mathematical descriptions or models for such phenomena.

Figure 3

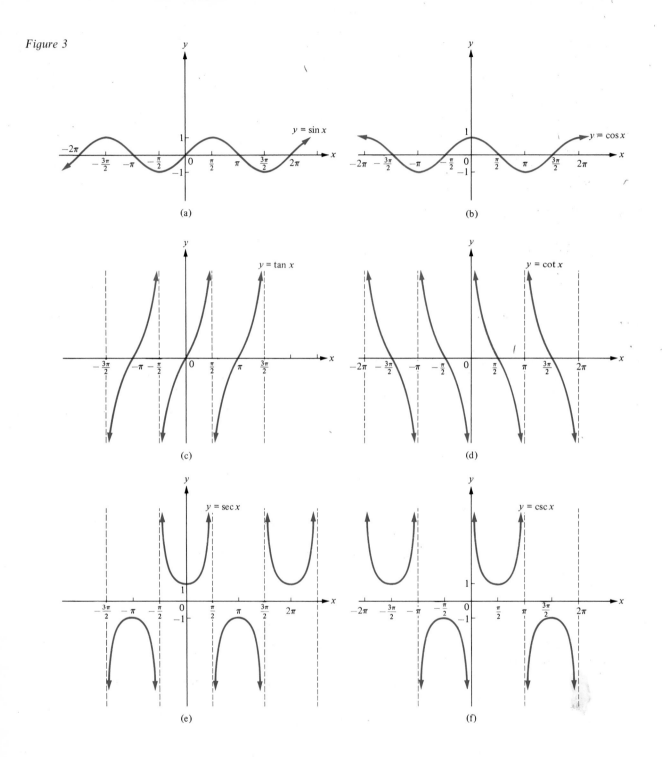

(a)

(b)

(c)

(d)

(e)

(f)

Problem Set 7

1 (a) Express the angle in radians that corresponds to the angle given in degrees.

　(i) $-45°$ (ii) $175°$ (iii) $-300°$

(b) Express the angle in degrees that corresponds to the angle given in radians.

　(i) $\dfrac{2\pi}{9}$ (ii) $-\dfrac{7\pi}{8}$ (iii) $\dfrac{43\pi}{6}$

2 (a) In Figure 4, show that:

    (i) $x = r \cos t$ (ii) $y = r \sin t$

    (iii) $r^2 = x^2 + y^2$ (iv) $\tan t = \dfrac{y}{x}$

Figure 4

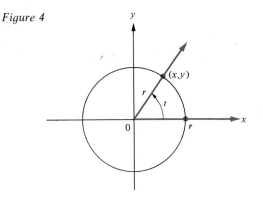

  (b) Find x and y for each value of r and t.

    (i) $r = 1,\ t = 0$

    (ii) $r = 6,\ t = -\dfrac{7\pi}{4}$ (iii) $r = 5,\ t = \dfrac{\pi}{6}$

3 If $\cos \theta = -\frac{4}{5}$ and the terminal side of θ lies in the third quadrant, find:

  (a) $\sin \theta$ (b) $\tan \theta$ (c) $\cot \theta$

  (d) $\sec \theta$ (e) $\csc \theta$

4 Use the fact that the trigonometric functions are periodic—that is, $T(t) = T(t + 2\pi k)$, where k is an integer—to determine each value.

  (a) $\sin \dfrac{43\pi}{4}$ (b) $\cos \dfrac{31\pi}{6}$ (c) $\tan \left(-\dfrac{22\pi}{3}\right)$

  (d) $\cot \left(-\dfrac{31\pi}{4}\right)$ (e) $\sec \dfrac{71\pi}{6}$ (f) $\csc \left(-\dfrac{91\pi}{3}\right)$

5 Use the standard identities to simplify each expression.

  (a) $(1 - \cos t)(1 + \cos t)$ (b) $2 \sin t \cos t \csc t$ (c) $\sec^2 t(\csc^2 t - 1)(\sin t + 1) - \csc t$

  (d) $\dfrac{1 + \cot^2 t}{\sec^2 t}$ (e) $\dfrac{\cos t - 1}{\sec t - 1}$

6 Show that:

  (a) $\cos \left(\dfrac{\pi}{2} - t\right) = \sin t$ (b) $\sin \left(\dfrac{\pi}{2} - t\right) = \cos t$ (c) $\tan (t + \pi) = \tan t$ (d) $\tan \left(t + \dfrac{\pi}{2}\right) = -\cot t$

7 (a) Use the fact that $\dfrac{5\pi}{6} + \dfrac{\pi}{4} = \dfrac{13\pi}{12}$ to find $\sin \dfrac{13\pi}{12}$ and $\cos \dfrac{13\pi}{12}$.

  (b) Find $\tan 195°$. (*Hint*: $195° = 150° + 45°$.)

8 Let t denote any number, put $s = t - 2\pi[\![t/2\pi]\!]$, and suppose that f is any one of the six trigonometric functions. Prove that $0 \le s < 2\pi$ and that $f(t) = f(s)$.

9 Simplify each expression.

  (a) $\dfrac{\sin^2 2t}{(1 + \cos 2t)^2} + 1$ (b) $\dfrac{\cos^4 t - \sin^4 t}{\sin 2t}$

  (c) $\cos^2 2t - \sin^2 t$ (d) $\tan t - \csc t(1 - 2 \cos^2 t) \sec t$

  (e) $\cos (s - t) \cos t - \sin (s - t) \sin t$

10 Prove the identity $\sin a \cos b = \frac{1}{2}[\sin (a + b) + \sin (a - b)]$. Then use this to prove the identity $\sin x + \sin y = 2[\sin (x + y)/2][\cos (x - y)/2]$.

11 If $\sin t = \frac{12}{13}$ and $\cos s = -\frac{4}{5}$, where $\pi/2 < t < \pi$ and $\pi/2 < s < \pi$, use the identities to evaluate each expression.

  (a) $\sin (s - t)$ (b) $\cos (s + t)$ (c) $\cot (s - t)$

12 Prove the identity $\cos 3t = 4 \cos^3 t - 3 \cos t$. Then use this to show that $\cos (\pi/9)$ is a solution of the equation $8x^3 - 6x - 1 = 0$.

13 Sketch the graph of each function over the interval $[-2\pi, 2\pi]$.

  (a) $f(x) = \sin 2x$ (b) $g(x) = 2 \cos x$

14 Show that the tangent, cotangent, and cosecant are odd functions. What about the secant function?

8   Algebra of Functions and Composition of Functions

In this section we see how functions can be combined in various ways so as to form new functions. In particular, we see how functions can be *added*, *subtracted*, *multiplied*, *divided*, and *composed*.

8.1   Algebra of Functions

If $f(x) = x^2 - 2$ and $g(x) = -\frac{1}{2}x + 1$, we can form the new function h defined by $h(x) = f(x) + g(x) = x^2 - \frac{1}{2}x - 1$ simply by adding $f(x)$ and $g(x)$. Naturally, we refer to the function h as the *sum* of the functions f and g and write $h = f + g$ (Figure 1). Notice that the graph of h is obtained from the graphs of f and g by adding corresponding ordinates; for instance, $h(-2) = f(-2) + g(-2)$.

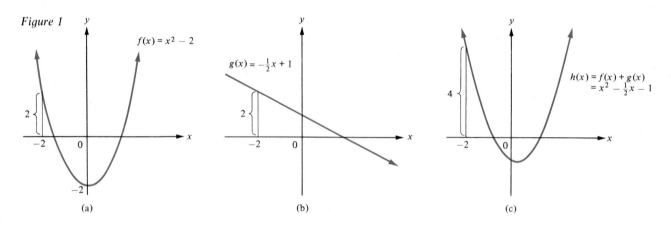

Figure 1

(a)   (b)   (c)

It should be clear that any two functions with overlapping domains can be added as shown. This idea is crucial in applied mathematics, since the functions that describe natural phenomena (for example, light waves or sound waves) often add when the phenomena are combined.

If functions can be added, it should also be possible to multiply, subtract, or divide them. The following definition shows exactly how this is done.

DEFINITION 1   **Sum, Difference, Product, and Quotient of Functions**
Let f and g be two functions with domains that overlap. We define functions $f + g$, $f - g$, $f \cdot g$, and f/g by the following equations:

$$(f + g)(x) = f(x) + g(x)$$
$$(f - g)(x) = f(x) - g(x)$$
$$(f \cdot g)(x) = f(x) \cdot g(x)$$
$$\left(\frac{f}{g}\right)(x) = \frac{f(x)}{g(x)}.$$

In each case, the domain of the defined function consists of all values of x common to the domains of f and g, except that in the fourth case the values of x for which $g(x) = 0$ are excluded.

Geometrically, the graph of the sum, difference, product, or quotient of f and g has at each point an ordinate that is the sum, difference, product, or quotient, respectively, of the ordinates of the graphs of f and g at the corresponding points. For sums or differences, this is usually relatively easy to visualize.

EXAMPLE   Let $f(x) = x^2 + 3$ and $g(x) = 2x - 1$. Evaluate: (a) $(f + g)(x)$ (b) $(f - g)(x)$ (c) $(f \cdot g)(x)$ (d) $\left(\dfrac{f}{g}\right)(x)$

SOLUTION
(a) $(f + g)(x) = f(x) + g(x) = (x^2 + 3) + (2x - 1) = x^2 + 2x + 2$
(b) $(f - g)(x) = f(x) - g(x) = (x^2 + 3) - (2x - 1) = x^2 - 2x + 4$
(c) $(f \cdot g)(x) = f(x) \cdot g(x) = (x^2 + 3)(2x - 1) = 2x^3 - x^2 + 6x - 3$
(d) $\left(\dfrac{f}{g}\right)(x) = \dfrac{f(x)}{g(x)} = \dfrac{x^2 + 3}{2x - 1}$

8.2 Composition of Functions

Given the equations

$$y = t^2 \quad \text{and} \quad t = 2x - 5,$$

we can solve for y in terms of x by substituting the second equation into the first one to obtain

$$y = (2x - 5)^2.$$

The equation $y = t^2$ defines a function $f(t) = t^2$, while the equation $t = 2x - 5$ defines a function $g(x) = 2x - 5$. The original equations can be rewritten as

$$y = f(t) \quad \text{and} \quad t = g(x),$$

and again we can solve for y in terms of x to obtain

$$y = f(g(x)).$$

To prevent a pileup of parentheses, we often replace the outside parentheses in the preceding equation by square brackets and write

$$y = f[g(x)].$$

The equation $y = f[g(x)]$ defines a new function h, so $h(x) = f[g(x)]$. The function h obtained by "chaining" f and g together in this way is called the *composition* of f and g and is written $h = f \circ g$. The notion of composition is made precise in the following definition.

DEFINITION 2   **Composition of Functions**

Let f and g be two functions satisfying the condition that at least one number in the range of g belongs to the domain of f. Then the composition of f and g, in symbols $f \circ g$, is the function defined by the equation

$$(f \circ g)(x) = f[g(x)].$$

Evidently, the domain of the composite function $f \circ g$ is the set of all values of x in the domain of g such that $g(x)$ belongs to the domain of f. The range of $f \circ g$ is just the set of all numbers of the form $f[g(x)]$ as x runs through the domain of $f \circ g$.

EXAMPLES   In Examples 1 through 5, let $f(x) = 3x - 1$, $g(x) = x^3$, and $p(x) = \frac{1}{3}(x + 1)$.

1 Evaluate $(f \circ g)(2)$ and $(g \circ f)(2)$.

SOLUTION

$$(f \circ g)(2) = f[g(2)] = f(2^3) = f(8) = 3(8) - 1 = 23$$
$$(g \circ f)(2) = g[f(2)] = g[(3)(2) - 1] = g(5) = 5^3 = 125.$$

2 Find $(f \circ g)(x)$ and $(g \circ f)(x)$.

SOLUTION

$$(f \circ g)(x) = f[g(x)] = f[x^3] = 3x^3 - 1$$
$$(g \circ f)(x) = g[f(x)] = g[3x - 1] = (3x - 1)^3 = 27x^3 - 27x^2 + 9x - 1.$$

3 Find $(f \circ p)(x)$ and $(p \circ f)(x)$.

SOLUTION

$$(f \circ p)(x) = f[p(x)] = f[\tfrac{1}{3}(x + 1)] = 3[\tfrac{1}{3}(x + 1)] - 1 = x$$
$$(p \circ f)(x) = p[f(x)] = p[3x - 1] = \tfrac{1}{3}(3x - 1 + 1) = x.$$

4 Find $(f \circ f)(x)$.

SOLUTION

$$(f \circ f)(x) = f[f(x)] = f[3x - 1] = 3(3x - 1) - 1 = 9x - 4.$$

5 Find $[f \circ (g + p)](x)$ and $[(f \circ g) + (f \circ p)](x)$.

SOLUTION

$$[f \circ (g + p)](x) = f[(g + p)(x)] = f[g(x) + p(x)] = f[x^3 + \tfrac{1}{3}(x + 1)]$$
$$= 3[x^3 + \tfrac{1}{3}(x + 1)] - 1 = 3x^3 + x$$
$$[(f \circ g) + (f \circ p)](x) = (f \circ g)(x) + (f \circ p)(x) = (3x^3 - 1) + x$$
$$= 3x^3 + x - 1.$$

Although the symbolism $f \circ g$ for the composition of f and g looks vaguely like some kind of a "product," it should not be confused with the actual product $f \cdot g$ of f and g. Whereas $f \cdot g = g \cdot f$, notice (in Example 2) that $f \circ g \neq g \circ f$. Whereas $f \cdot (g + p) = f \cdot g + f \cdot p$, notice (in Example 5) that $f \circ (g + p) \neq f \circ g + f \circ p$.
    Example 3 illustrates an interesting and important situation in which two functions f and p "undo each other" in the sense that $f \circ p$ and $p \circ f$ both give the identity function. We study this situation in more detail in Section 9.

Problem Set 8

In problems 1 through 4, use the graphs of the functions f and g to sketch the graph of $f + g$ by "adding ordinates."

1 $f(x) = 3x$ and $g(x) = 3$ **2** $f(x) = 2x^2$ and $g(x) = 5$

3 $f(x) = -2x$ and $g(x) = 1$ **4** $f(x) = x^3$ and $g(x) = -2x^2$

5 Let the functions f and g be defined by $f(x) = x^2 + 1$ for $-3 \leq x \leq \frac{3}{2}$ and $g(x) = 2x + 1$ for $-1 \leq x \leq 2$.
    (a) Find $(f + g)(1)$, $(f - g)(1)$, $(f \cdot g)(1)$, and $(f/g)(1)$.   (b) Find the domains of $f + g$, $f - g$, $f \cdot g$, and f/g.

In problems 6 through 10, find $f + g$, $f - g$, $f \cdot g$, and f/g. Also, find the domains of $f + g$, $f - g$, $f \cdot g$, and f/g.

6 $f(x) = 2x - 5$ and $g(x) = x^2 + 1$ **7** $f(x) = \sqrt{x}$ and $g(x) = x^2 + 4$ **8** $f(x) = 3x + 5$ and $g(x) = 7 - 4x$

9 $f(x) = \sqrt{x - 3}$ and $g(x) = 1/x$ **10** $f(x) = |x|$ and $g(x) = |x - 2|$

11 Suppose that f and g are even functions. Show that $f + g$, $f - g$, $f \cdot g$, and f/g are also even functions.

In problems 12 through 16, let $f(x) = \sin x$, $g(x) = x^2$, and $h(x) = \cos x$. Find a formula for the given function.

12 $f + g + h$ **13** $f \cdot (g - h)$ **14** $f \cdot f$

15 f/h **16** $2 \cdot f \cdot h$

17 Let f be the constant function defined by $f(x) = c$, where c is understood to be some fixed real number. Let g be the identity function, so that $g(x) = x$. Find:
(a) $f + g$ (b) $f - g$ (c) $f \cdot g$
(d) f/g (e) g/f

18 Let a, b, c, and d be constant numbers. Define functions f and g by $f(x) = ax + b$ and $g(x) = cx + d$. Find:
(a) $f + g$ (b) $f - g$ (c) $f \cdot g$ (d) f/g

19 Let f and g be defined by $f(x) = \sqrt{4 - x^2}$ and $g(x) = -\sqrt{4 - x^2}$. Sketch the graph.
(a) f (b) g (c) $f + g$
(d) $f - g$ (e) $f \cdot g$ (f) f/g

20 Let f be defined by $f(x) = x - 3$ and let g be defined by $g(x) = x^2 + 4$. Find:
(a) $(f \circ g)(4)$ (b) $(f \circ g)(2)$ (c) $(g \circ f)(4)$ (d) $(g \circ f)(2)$
(e) $(g \circ f)(5)$ (f) $(f \circ g)(5)$ (g) $(f \circ g)(x)$ (h) $(g \circ f)(x)$

21 Let $f(x) = \sin x$, $g(x) = x^2$, and $h(x) = \cos x$. Find a formula for the given function.
(a) $f \circ g$ (b) $g \circ f$ (c) $g \circ g$ (d) $g \circ (f + h)$
(e) $g \circ (f/h)$ (f) $(f/h) \circ (h/f)$ (g) $f \circ (g \circ h)$ (h) $(f \circ g) \circ h$

22 Let f, g, and h be defined by $f(x) = 4x$, $g(x) = x - 3$, and $h(x) = \sqrt{x}$. Express each of the following functions as a composition of functions chosen from f, g, and h.
(a) $F(x) = 4\sqrt{x}$ (b) $G(x) = \sqrt{x - 3}$ (c) $H(x) = 4x - 12$
(d) $J(x) = x - 6$ (e) $K(x) = \sqrt{4x}$

23 Let f be the function defined by $f(x) = 5x + 3$ and let g be the function defined by $g(x) = 3x + k$, where k is a constant number. Find a value of k so that $f \circ g$ and $g \circ f$ are the same function.

24 Let I be the identity function, so that $I(x) = x$ holds for all values of x. Show that if f is any function, then $I \circ f = f \circ I = f$.

25 If I denotes the identity function, then $I \circ I = I$. (Why?) Find another function f such that $f \circ f = I$. Explain why the function given by $f(x) = 1/x$ does *not* have the property that $f \circ f = I$.

26 A baseball diamond is a square 90 feet long on each side. A ball is hit down the third-base line at the rate of 50 feet per second. Let y denote the distance in feet of the ball from first base, let x denote its distance in feet from home plate, and let t denote the elapsed time in seconds since the ball was hit. Here y is a function of x, say $y = f(x)$, and x is a function of t, say $x = g(t)$. Find $f(x)$ and $g(t)$ explicitly. Find $(f \circ g)(t)$ explicitly. Explain why $y = (f \circ g)(t)$.

27 Let the functions f and g be given by $f(x) = -7x + 23$ and $g(x) = -\frac{1}{7}x + \frac{23}{7}$. Find $(f \circ g)(x)$ and $(g \circ f)(x)$.

28 A *fractional linear function* is defined to be a function of the form $f(x) = \dfrac{ax + b}{cx + d}$, where a, b, c, and d are constants and $ad \neq bc$. Is the composition of two fractional linear functions again a fractional linear function?

9 Inverse Functions

In Section 8 (Example 3, page 38), we saw that the two functions $f(x) = 3x - 1$ and $p(x) = \frac{1}{3}(x + 1)$ "undo each other" in the sense that whatever f "does to" x when we form the number $f(x)$, p will "undo" and give $p[f(x)] = x$, while whatever p "does to" x when we form the number $p(x)$, f will "undo" and give $f[p(x)] = x$. Two functions that "undo" each other in this way are said to be *inverses* of one another. Another example of functions that are inverses of one another is provided by

$$f(x) = \sqrt{x}$$
$$g(x) = x^2 \qquad \text{for } x \geq 0.$$

Indeed, for $x \geq 0$, we have

$$(f \circ g)(x) = f[g(x)] = \sqrt{x^2} = x$$
$$(g \circ f)(x) = g[f(x)] = (\sqrt{x})^2 = x.$$

These examples lead us to make the following definition.

DEFINITION 1 **Functions Inverse to Each Other**

Two functions f and g are said to be *inverses* of each other if the following four conditions are satisfied:

(i)   The range of g is contained in the domain of f.
(ii) For every number x in the domain of g, $(f \circ g)(x) = x$.
(iii) The range of f is contained in the domain of g.
(iv) For every number x in the domain of f, $(g \circ f)(x) = x$.

A function f for which such a function g exists is said to be *invertible*.

EXAMPLES Suppose that f and g are defined as shown. Prove that f and g are inverses of each other.

1 $f(x) = 3x$ and $g(x) = x/3$

SOLUTION
$\mathbb{R}$ is the range and also the domain of both f and g, so that conditions (i) and (iii) of Definition 1 hold. Also,

$$(f \circ g)(x) = f[g(x)] = f\left(\frac{x}{3}\right) = 3\left(\frac{x}{3}\right) = x$$

$$(g \circ f)(x) = g[f(x)] = g(3x) = \frac{3x}{3} = x;$$

hence, conditions (ii) and (iv) also hold, so f and g are inverses of each other.

2 $f(x) = x - 4$ for $-1 \leq x \leq 1$ and $g(x) = x + 4$ for $-5 \leq x \leq -3$

SOLUTION

The domain of f is $[-1, 1]$ and the range of g is the same interval $[-1, 1]$ (Figure 1); hence, condition (i) of Definition 1 holds. The domain of g is $[-5, -3]$ and the range of f is the same interval $[-5, -3]$; hence, condition (iii) of Definition 1 also holds. Since

$$(f \circ g)(x) = f[g(x)] = f[x + 4] = (x + 4) - 4 = x$$

for all x in $[-5, -3]$, condition (ii) holds. Also,

$$(g \circ f)(x) = g[f(x)] = g[x - 4] = (x - 4) + 4 = x$$

for all x in $[-1, 1]$, so that (iv) holds. Therefore, f and g are inverses of each other.

Figure 1

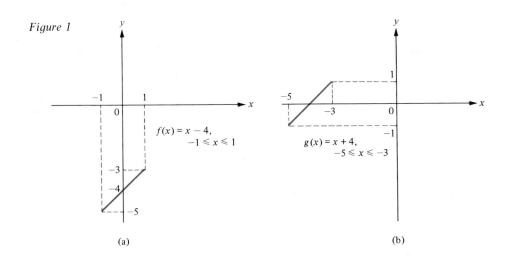

(a)

$f(x) = x - 4,$
$-1 \leqslant x \leqslant 1$

(b)

$g(x) = x + 4,$
$-5 \leqslant x \leqslant -3$

It follows from Definition 1 that, if f and g are inverses of each other, the domain of f must coincide with the range of g and the range of f must coincide with the domain of g.

Geometrically, it is easy to tell when two functions are inverses of each other. In fact, the functions f and g are inverses of each other precisely when the graph of g is the mirror image of the graph of f with respect to the straight line $y = x$ (Figure 2). Indeed, the mirror image of a point $P = (a, b)$ across the straight line $y = x$ is the point $Q = (b, a)$. But if f and g are inverses of each other and if $P = (a, b)$ belongs to the graph of f, then $b = f(a)$, so $g(b) = g[f(a)] = a$; that is, $Q = (b, a)$ belongs to the graph of g. On the other hand, it is easy to see that if f and g are inverses of each other and if $Q = (b, a)$ belongs to the graph of g, then $P = (a, b)$ belongs to the graph of f.

If the graph of f and the graph of g are mirror images of each other across the line $y = x$, then it is not difficult to show, conversely, that f and g are inverses of each other (Problem 17). From this geometric characterization of inverses we can draw two immediate conclusions:

Figure 2

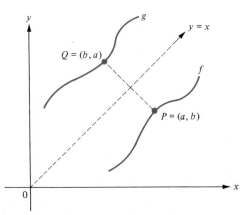

1 Not every function is invertible. Indeed, consider the function f whose graph appears

Figure 3

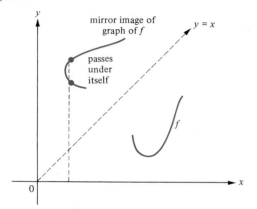

in Figure 3. The mirror image of the graph of f is not the graph of a *function* since part of this mirror image passes under itself. Hence, f cannot be invertible.

2 If a function f is invertible, there is *exactly one* function g such that f and g are inverses of each other; indeed, g must be the one and only function whose graph is the mirror image of the graph of f across the line $y = x$. It is traditional to call g the *inverse of f* and to write $g = f^{-1}$.

We formalize the preceding in the following definition.

DEFINITION 2   The Inverse of a Function

Suppose that f is an invertible function. Then we define *the inverse of the function f*, in symbols f^{-1}, to be the function whose graph is the mirror image of the graph of f across the straight line $y = x$ **(Figure 4)**. The function f^{-1} is read "f inverse."

Figure 4

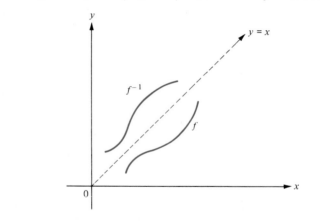

EXAMPLE   Let f be the function defined by the equation $f(x) = 2x + 1$. Sketch the graph of the inverse function f^{-1} and write a formula for $f^{-1}(x)$.

SOLUTION

Figure 5

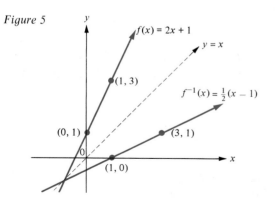

The graph of f is a straight line of slope 2 with y intercept 1. The mirror image of this straight line across the line $y = x$ is again a straight line. Since $(0, 1)$ and $(1, 3)$ belong to the graph of f, $(1, 0)$ and $(3, 1)$ belong to the mirror image, that is, to the graph of f^{-1}. Therefore, the graph of f^{-1} is a straight line of slope $\dfrac{1 - 0}{3 - 1} = \dfrac{1}{2}$ containing the point $(1, 0)$. From the point-slope form for the equation of a straight line, $y - y_1 = m(x - x_1)$, it follows that

$$y = \tfrac{1}{2}(x - 1) \qquad \text{so } f^{-1}(x) = \tfrac{1}{2}(x - 1) \quad \text{(Figure 5)}.$$

Notice that f^{-1}, the inverse of f, is not the same thing as $1/f$, *the reciprocal of f*. In the example above, $f^{-1}(x) = \tfrac{1}{2}(x - 1)$, while

$$\left(\frac{1}{f}\right)(x) = \frac{1}{f(x)} = \frac{1}{2x + 1}.$$

If f is an invertible function, then the graph of f^{-1} is the set of all ordered pairs (b, a) such that (a, b) belongs to the graph of f. This observation provides the basis of the following method for finding f^{-1}.

"Algebraic" Method for Finding f^{-1}

Step 1 Write an equation $y = f(x)$ which defines f.

Step 2 Solve the equation in Step 1 for x in terms of y to get $x = f^{-1}(y)$. This equation defines f^{-1}.

Step 3 (Optional) If you prefer having x as the independent variable in the equation for f^{-1}, or if you plan to sketch graphs of f and f^{-1} on the same diagram, then interchange x and y in the equation obtained in Step 2.

EXAMPLES   Use the algebraic method to find the inverse of the given function. Carry out the optional Step 3.

1 $f(x) = 2x + 1$

SOLUTION

$y = 2x + 1$, $2x = y - 1$, and $x = \frac{1}{2}(y - 1)$. Now interchange x and y in the latter equation to get $y = \frac{1}{2}(x - 1)$. Hence, f^{-1} is defined by the equation

$$f^{-1}(x) = \tfrac{1}{2}(x - 1).$$

(Graphs of f and f^{-1} appear in Figure 5.)

2 $f(x) = x^3 - 8$

SOLUTION

$y = x^3 - 8$, $x^3 = y + 8$, and $x = \sqrt[3]{y + 8}$. Now interchange x and y in the latter equation to get $y = \sqrt[3]{x + 8}$. Hence, f^{-1} is defined by the equation

$$f^{-1}(x) = \sqrt[3]{x + 8}.$$

If we are using certain variables to denote particular geometrical or physical quantities, it is best not to carry out the third step in the algebraic method, but simply to stop at the second step. For instance, the volume V of a sphere of radius r is given by the equation $V = \frac{4}{3}\pi r^3$. If we let f be the function defined by $f(r) = \frac{4}{3}\pi r^3$, then $V = f(r)$. Using the algebraic method, we must solve the equation $V = f(r)$ for r in terms of V to get $r = f^{-1}(V)$. Here we have $V = \frac{4}{3}\pi r^3$, $r^3 = 3V/4\pi$, $r = \sqrt[3]{3V/4\pi}$; hence, f^{-1} is the function defined by $f^{-1}(V) = \sqrt[3]{3V/4\pi}$. We do not bother to interchange the variables r and V in the equation $r = \sqrt[3]{3V/4\pi}$ since, if we were to do so, we would have to abandon our interpretation of V as the volume and r as the radius of a sphere.

Problem Set 9

In problems 1 through 6, the given function f has an inverse function f^{-1}. In each case, find f^{-1} by the algebraic method and then verify by calculation that $(f \circ f^{-1})(x) = x$ for all x in the domain of f^{-1} and that $(f^{-1} \circ f)(x) = x$ for all x in the domain of f.

1 $f(x) = 7x - 13$
2 $f(x) = mx$, $m \neq 0$
3 $f(x) = x^2 - 3$, $x \geq 0$

4 $f(x) = \dfrac{5}{x}$
5 $f(x) = \dfrac{2x - 3}{3x - 2}$
6 $f(x) = \dfrac{ax + b}{cx + d}$, $ad \neq bc$

7 Show that the functions f and g defined by $f(x) = \dfrac{3x - 7}{x + 1}$ and $g(x) = \dfrac{7 + x}{3 - x}$ are inverses of each other.

8 Prove that if m and b are constant numbers and $m \neq 0$, the linear functions f and g given by $f(x) = mx + b$ and $g(x) = \dfrac{1}{m}x - \dfrac{b}{m}$ are inverses of each other.

9 Let the function f be defined by $f(x) = \sqrt{4 - x^2}$ for $x \geq 0$. Show that f is its own inverse.

10 Let I be the identity function and let f be any function such that I and f are inverses of each other. What can you conclude about f?

11 Are the functions f and g defined by $f(x) = \dfrac{1 + \sqrt{x}}{1 - \sqrt{x}}$ for $x \geq 0$ and $g(x) = \left(\dfrac{x - 1}{x + 1}\right)^2$ for $x \neq \pm 1$ inverses of each other?

12 Suppose that f is an invertible function and that the numbers a and b belong to the domain of f. Prove that if $f(a) = f(b)$, then $a = b$.

13 Show geometrically that a function f is invertible if and only if no horizontal line intersects its graph in more than one place. (*Hint:* In taking mirror images across the line $y = x$, such a horizontal line would have a vertical mirror image.)

14 Which of the following functions is invertible?
(a) f consists of the four ordered pairs $(0, 1)$, $(1, 2)$, $(2, 1)$, and $(3, 2)$.
(b) f is defined by $f(x) = 3x + 5$.
(c) f is defined by $f(x) = x^2 + 1$.

15 Show that a function f is invertible if and only if it has the following property: If a and b are two different numbers in the domain of f, then $f(a)$ and $f(b)$ are always different numbers.

16 Decide whether each function f whose graph is shown in Figure 6 has an inverse function f^{-1}. If it does, sketch this inverse on the same xy coordinate system.

Figure 6

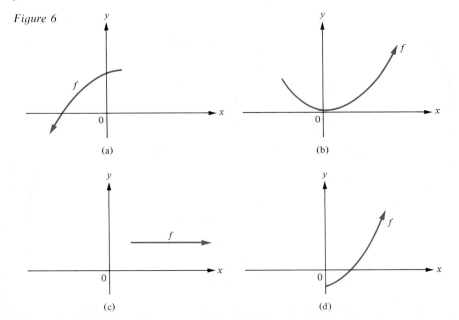

(a)   (b)

(c)   (d)

17 Suppose that f and g are functions such that the graph of f and the graph of g are mirror images of each other across the line $y = x$. Use Definition 1 directly and show that f and g are inverses of each other.

Review Problem Set

1 Assume that a and b are real numbers such that $a < b$. Which of the following are true and which are false?
(a) $a < 5b$ (b) $-3a < -3b$ (c) $5a < -(-5b)$
(d) $a + 5 < b + 5$ (e) $1/a > 1/b$ (f) $a - 7 > b - 7$

In problems 2 through 13, solve the inequality and illustrate the solution on the number line.

2 $x - 5 \leq 7$

3 $3x + 2 > 8$

4 $3x - 2 \geq 1 + 2x$

5 $\dfrac{2x - 6}{3} \geq 0$

6 $\dfrac{x - 1}{3} \geq 2 + \dfrac{x}{2}$

7 $x^2 + x - 20 < 0$

8 $x^2 - 6x - 7 \leq 0$

9 $2x^2 + 9 < 5$

10 $\dfrac{x - 2}{x + 3} > 0$

11 $\dfrac{2x - 1}{x - 6} < 0$

12 $\dfrac{5x - 1}{x - 2} \leq 1$

13 $\dfrac{x - 4}{x + 2} \geq 3$

14 State the conditions under which each inequality is true.
(a) $\dfrac{1}{x} < \dfrac{1}{10}$ (b) $\dfrac{1}{x} < \dfrac{1}{100}$ (c) $\dfrac{1}{x^2 - 1} < \dfrac{1}{1000}$
(d) $-x < x$ (e) $-x^2 < x^2$

15 Give two different examples of real numbers a, b, c, and d with $a < b$ and $c < d$ but $ac > bd$.

16 For positive real numbers a and b, show that $\dfrac{2ab}{a + b} \leq \sqrt{ab} \leq \dfrac{a + b}{2}$.

17 Use the absolute value notation to write an expression for the distance between the two given numbers on the number scale. Simplify your answer.
(a) -6 and 5 (b) $\dfrac{1}{x}$ and $\dfrac{1}{x + 1}$ (c) $\frac{3}{13}$ and $\frac{4}{17}$ (d) $\dfrac{x}{x + 1}$ and $\dfrac{x - 1}{x}$

In problems 18 through 23, solve the absolute value equation.

18 $|x + 1| = 3$

19 $|2x - 3| = 5$

20 $|x - 6| = -1$

21 $|x + 2| = 0$

22 $|x + \frac{5}{6}| = \frac{1}{6}$

23 $|x - \frac{4}{3}| = 0$

24 Prove or disprove the following: If x is a real number,
(a) $|x^2| = |x|^2$ (b) $|x^3| = |x|^3$

In problems 25 through 32, solve the absolute value inequality and indicate the solution on the number line.

25 $|2x + 5| \leq 6$

26 $|3x + 4| \leq 2$

27 $|1 - x| > |x|$

28 $|1 - 4x| \leq x$

29 $|7x - 6| > x$

30 $\dfrac{1}{|x - 1|} \geq 3$

31 $\dfrac{1}{|2x + 3|} \leq \dfrac{1}{4}$

32 $\dfrac{1}{|3x - 1|} \geq 5$

33 Show that the points $(12, 9)$, $(20, -6)$, $(5, -14)$, and $(-3, 1)$ are the vertices of a square.

34 Show that the points $(8, 1)$, $(-6, -7)$, and $(2, 7)$ are the vertices of an isosceles triangle.

35 Use the concept of the slope to show that the points $A = (-2, 1)$, $B = (2, 3)$, and $C = (10, 7)$ are collinear.

36 Use the distance formula and the concept of the slope to show that the points $(6, 1)$, $(5, 6)$, $(-4, 3)$, and $(-3, -2)$ are the vertices of a parallelogram.

In problems 37 through 41, find the equation of the line that satisfies the given condition.

37 The slope is -12 and the line contains the point $(1, 3)$.

38 The line contains the points $(5, 6)$ and $(-1, 2)$.

39 The slope is 0 and the line contains the point $(3, -7)$.

40 The line is parallel to the line $4x - 5y + 6 = 0$ and contains the point $(5, -3)$.

41 The line is perpendicular to the line $6x - 2y + 7 = 0$ and contains the point $(-8, 3)$.

In problems 42 through 47, find the domain of the given function and test the symmetry of its graph with respect to the y axis and with respect to the origin. Also, sketch the graph of the function.

42 $f(x) = \sqrt{x^2 - 4}$

43 $g(x) = (x^2 - 1)^{3/2}$

44 $f(x) = |x^2 - 9|$

45 $f(x) = |3x + 7|$

46 $g(x) = \dfrac{x}{x^2 + 1}$

47 $h(x) = \dfrac{|x|}{1 - x}$

48 Show that the mirror image across the line $y = x$ of the point (a, b) is the point (b, a).

49 Is the function f defined by $f(x) = \dfrac{x}{|x|} - \dfrac{|x|}{x}$ the zero function? Explain.

50 Let f be the function defined by $f(x) = \sqrt{\dfrac{x - 3}{x + 3}}$ and let $g(x)$ be the function defined by $g(x) = \dfrac{\sqrt{x - 3}}{\sqrt{x + 3}}$.

(a) Find the domains of f and of g.   (b) Are f and g algebraic functions?
(c) Is it true that $f = g$? Explain.

51 Let f be the function defined by $f(x) = 4x^2$. Compute:
(a) $f(-1)$ (b) $f(1)$ (c) $f(-3)$ (d) $f(3)$
(e) $f(4t)$ (f) $f(3x)$ (g) $f(x + h)$ (h) $f(x + h) - f(x)$
(i) $f(\sqrt{a})$ (j) $\sqrt{f(a)}$

52 For the function f, determine whether or not $[f(x)]^2 = f(x^2)$ holds for all values of x. If not, give values of x for which it is false.
(a) $f(x) = x$ (b) $f(x) = |x|$ (c) $f(x) = [\![x]\!]$

In problems 53 through 56, find the domain of the function, determine whether the function is even, odd, or neither, and sketch the graph.

53 $f(x) = |7x|$ **54** $g(x) = |x| - x$ **55** $h(x) = [\![2x]\!]$ **56** $F(x) = x^2 + 5$

In problems 57 through 62, use the standard trigonometric identities to simplify the given expression.

57 $\cos(x - y)\cos y - \sin(x - y)\sin y$

58 $\sin 2\theta \cos \theta + \cos 2\theta \sin \theta$

59 $\dfrac{\sec t + \tan t}{\cos t - \tan t - \sec t}$

60 $\dfrac{\sin x(\sec x - 1)\cos x}{\cos x - 1}$

61 $\dfrac{(\sin^2 t - \cos^2 t)^2}{\sin^4 t - \cos^4 t}$

62 $\dfrac{1}{1 + \sin t} + \dfrac{1}{1 - \sin t}$

63 Let f be the function defined by $f(x) = 5x^2 - 1$ and g be the function defined by $g(x) = 5x + 7$. Find each expression.
(a) $(f + g)(4)$ (b) $(f + 5g)(3)$ (c) $(f - 3g)(2)$ (d) $(f \cdot g)(2)$

(e) $\left(\dfrac{f}{g}\right)(3)$ (f) $(f + 2g)(x)$ (g) $\left(\dfrac{g}{2f}\right)(x)$ (h) $(f \cdot f)(x)$

(i) $\dfrac{f(x + h) - f(x)}{h}$ (j) $\dfrac{g(x + h) - g(x)}{h}$

64 Let f be the function defined by $f(x) = \dfrac{5x - 2}{x^2 - 4}$. Find constants A and B such that

$$f(x) = \frac{A}{x - 2} + \frac{B}{x + 2}.$$

In problems 65 through 67, find $f \circ g$ and $g \circ f$ if :

65 $f(x) = \sqrt{x}$ and $g(x) = \sqrt{x}$.

66 $f(x) = x^2 + 7$ and $g(x) = \sqrt{3 - x}$.

67 $f(x) = x^2 + 2x$ and $g(x) = 3x + 4$.

68 Suppose that f gives the area of a square as a function of the length of one of its diagonals. Express f as the composition of two other functions.

In problems 69 through 71, show that f and g are inverses of each other.

69 $f(x) = x^4$ for $x \geq 0$ and $g(x) = \sqrt[4]{x}$.

70 $f(x) = x^2 - 3x + 2$ for $x \leq \frac{3}{2}$ and $g(x) = \dfrac{3 - \sqrt{1 + 4x}}{2}$ for $x \geq -\frac{1}{4}$.

71 $f(x) = \dfrac{1}{1 - x}$ and $g(x) = \dfrac{x - 1}{x}$.

72 Let g be the function defined by $g(x) = |x + 2|$. Is g invertible? Why?

73 If f is invertible, show that f^{-1} is also invertible and $(f^{-1})^{-1} = f$.

74 Suppose that A, B, and C are constants with $A > 0$. Let f be the function defined by $f(x) = Ax^2 + Bx + C$ for $x \geq -B/2A$. Show that the inverse function f^{-1} is given by

$$f^{-1}(x) = \frac{-B + \sqrt{B^2 - 4AC + 4Ax}}{2A} \quad \text{for} \quad x \geq \frac{4AC - B^2}{4A}.$$

75 Use the "algebraic" method to find the inverse of each function.

(a) $f(x) = 7x - 19$ (b) $g(x) = -7x^3$ (c) $h(x) = \dfrac{13}{x}$

76 Assume that $f, g,$ and h are functions with domain $\mathbb{R}$. Prove that $(f + g) \circ h = f \circ h + g \circ h$. Is it true that $f \circ (g + h) = f \circ g + f \circ h$? Explain.

1 LIMITS AND CONTINUITY OF FUNCTIONS

The basic concept upon which calculus depends is the *limit* of a function. In this chapter we introduce the idea of a limit and use it to study the notion of *continuity* of a function. In subsequent chapters limits are used to formulate the definitions of *derivatives* and *integrals*. Calculus is the study of these extraordinary ideas—limits, continuity, derivatives, and integrals.

1 Limits and Continuity

The idea of a limit is easy to grasp intuitively. For instance, imagine a square metal plate that is expanding uniformly because it is being heated. If x is the length of an edge, the area of the plate is given by $A = x^2$. Evidently, if x comes closer and closer to 3 centimeters, the area A comes closer and closer to 9 square centimeters. We express this by saying that as x approaches 3, x^2 approaches 9 as a *limit*. In symbols, we write

$$\lim_{x \to 3} x^2 = 9,$$

where the notation "$x \to 3$" indicates that x comes closer and closer to 3 and "lim" stands for "the limit of."

EXAMPLE   If $f(x) = x^2$, show graphically that $\lim_{x \to 3} x^2 = 9$.

SOLUTION
From the graph in Figure 1, we can see clearly that as x approaches 3, the function values $f(x)$ approach 9 as a limit.

More generally, if f is a function and a is a number, we understand the notation

$$\lim_{x \to a} f(x) = L,$$

read "the limit of $f(x)$ as x approaches a is L," to mean that $f(x)$ comes closer and closer to the number L as x comes closer and closer to the number a. Although we present a more formal definition in Section 1.1, a working understanding of limits can be acquired by considering further examples and geometric illustrations.

Figure 1

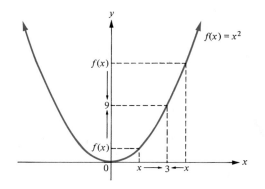

EXAMPLE   Determine $\lim\limits_{x \to 4} (5x + 7)$.

SOLUTION
As x comes closer and closer to 4, $5x$ comes closer and closer to 20, and $5x + 7$ comes closer and closer to 27. Therefore, $\lim\limits_{x \to 4} (5x + 7) = 27$.

Unfortunately, it is not always possible to determine the limit of a function by simple arithmetic considerations as in the preceding example. In the first place, the function values may jump around so erratically that they never settle down and approach a limit, in which case we say that the *limit does not exist*. (In Section 3 we give some examples in which limits do not exist; however, in the present section, our examples are chosen so that all required limits exist.) In the second place, the function may be so complicated that the limit, even though it exists, is not evident by superficial inspection.

For instance, let $f(x) = \dfrac{3x^2 - 4x - 4}{x - 2}$ and consider the problem of determining $\lim\limits_{x \to 2} f(x)$. Here, it is not immediately clear just how $f(x)$ behaves as x approaches 2; however, we can gain some insight into this behavior by calculating some values of $f(x)$ as x gets closer and closer to 2 but stays less than 2. These values are shown in the following table:

x	1	1.25	1.50	1.75	1.90	1.99	1.999
$f(x) = \dfrac{3x^2 - 4x - 4}{x - 2}$	5	5.75	6.50	7.25	7.70	7.97	7.997

Similarly, the following table shows some values of $f(x)$ as x gets closer and closer to 2 but stays greater than 2:

x	3	2.75	2.50	2.25	2.10	2.01	2.001
$f(x) = \dfrac{3x^2 - 4x - 4}{x - 2}$	11	10.25	9.50	8.75	8.30	8.03	8.003

Evidently, as x gets closer and closer to 2, $f(x)$ gets closer and closer to 8. We are tempted to conjecture that

$$\lim_{x \to 2} \frac{3x^2 - 4x - 4}{x - 2} = 8.$$

This conjecture can be verified by some elementary algebra, as in the following example.

EXAMPLE   Determine $\lim\limits_{x \to 2} \dfrac{3x^2 - 4x - 4}{x - 2}$.

SOLUTION

If $f(x) = \dfrac{3x^2 - 4x - 4}{x - 2}$, the domain of f consists of all real numbers x except for $x = 2$ (which makes the denominator zero). We are only concerned with the values of $f(x)$ as x *approaches* 2—what happens when x *reaches* 2 is not in question here. For $x \neq 2$,

$$f(x) = \frac{3x^2 - 4x - 4}{x - 2} = \frac{(3x + 2)(x - 2)}{x - 2} = 3x + 2.$$

Therefore, as x comes closer and closer to 2, $f(x)$ comes closer and closer to 8; that is,

$$\lim_{x \to 2} f(x) = \lim_{x \to 2} \frac{3x^2 - 4x - 4}{x - 2} = \lim_{x \to 2} (3x + 2) = 8.$$

The limit found in the preceding example can be illustrated geometrically by sketching a graph of the function $f(x) = \dfrac{3x^2 - 4x - 4}{x - 2}$ (Figure 2). Since $f(x) = 3x + 2$ holds for $x \neq 2$, this graph is a straight line with the point $(2, 8)$ excluded. Figure 2 makes it clear that the value of $f(x)$ can be made to come as close to 8 as we please simply by choosing x to be sufficiently close to 2 (but not equal to 2).

We see from Figure 2 that in finding the limit of $f(x)$ as x approaches a, it does not matter how f is defined *at* a (nor even whether it is defined there at all). The only thing that does matter is how f is defined for values of x *near* a. In fact, we can distinguish three possible cases as follows.

Figure 2

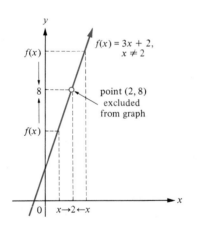

Suppose that $\lim\limits_{x \to a} f(x) = L$. Then, exactly one of these cases holds:

Case 1 f is defined at a and $f(a) = L$.

Case 2 f is not defined at a.

Case 3 f is defined at a and $f(a) \neq L$.

The three cases are illustrated by the following examples.

EXAMPLES   Determine $\lim\limits_{x \to 2} f(x)$ and sketch a graph of the function to illustrate the limit involved.

1 $f(x) = x + 2$

SOLUTION

Here, f is defined at 2 and $f(2) = 4$. As x approaches 2, $x + 2$ clearly approaches 4, so that $\lim\limits_{x \to 2} f(x) = 4$, the same as the value of the function f at $x = 4$ (Figure 3).

Figure 3

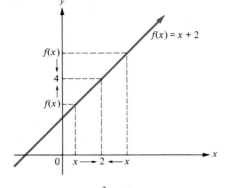

$$2\quad f(x) = \frac{x^2 - 4}{x - 2}$$

SOLUTION
Here, f is not defined at 2, but for $x \neq 2$,

$$f(x) = \frac{x^2 - 4}{x - 2} = \frac{(x - 2)(x + 2)}{x - 2} = x + 2;$$

hence, $\lim_{x \to 2} f(x) = \lim_{x \to 2} (x + 2) = 4$ (Figure 4).

$$3\quad f(x) = \begin{cases} x + 2 & \text{if } x \neq 2 \\ 6 & \text{if } x = 2 \end{cases}$$

SOLUTION
In calculating the desired limit, we let x approach 2 but make certain that $x \neq 2$. Then we have $\lim_{x \to 2} f(x) = \lim_{x \to 2} (x + 2) = 4$. Here, $f(2) = 6 \neq 4$, so that the limit of $f(x)$ as x approaches 2 is not the same as the value of the function at $x = 2$ (Figure 5).

In Examples 2 and 3, the function f "misbehaves" at $x = 2$, while in Example 1, the value $f(2)$ of f at 2 coincides with the limit value; that is, $\lim_{x \to 2} f(x) = f(2)$. In general, if $\lim_{x \to a} f(x) = f(a)$, we say that the function f is *continuous* at a. Thus, the functions graphed in Figures 4 and 5 are not continuous at 2, while the function graphed in Figure 3 is continuous at 2. The idea of continuity will be studied in more detail in Section 3.

Figure 4

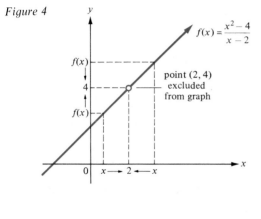

Figure 5

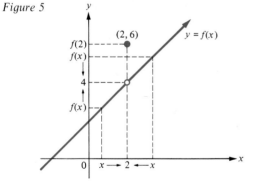

1.1 The Formal Definition of Limit

In the preceding discussion we have treated the idea of a limit intuitively; however, if one wishes to build calculus on a firm foundation, a more formal definition of limit is required. To say, "$f(x)$ comes closer and closer to L as x comes closer and closer to a" simply lacks precision. How close is $f(x)$ to L? How close is x to a?

In rigorous considerations involving limits, mathematicians customarily use the Greek letters ε and δ (called *epsilon* and *delta*, respectively) to denote positive real numbers that indicate how close $f(x)$ is to L and how close x is to a, respectively. Evidently, to say that $f(x)$ is close to L is equivalent to saying that $|f(x) - L|$ is small. Similarly, x is close to a when $|x - a|$ is small. Thus, to assert that $\lim_{x \to a} f(x) = L$ is to assert that, if we take any positive number ε, *no matter how small*, we can always find a sufficiently small positive number δ such that $|f(x) - L| < \varepsilon$ holds whenever $0 < |x - a| < \delta$. In most cases the value of δ will depend on the value of ε, and the smaller the ε we take, the smaller the δ that will be required.

In the above discussion, the condition $0 < |x - a|$ means that $x \neq a$ and reflects our statement that, in finding the limit of $f(x)$ as x approaches a, the value of $f(x)$ when $x = a$ does not matter. As Figure 6 shows, the condition $0 < |x - a| < \delta$ means that x lies in the open interval $(a - \delta,\ a + \delta)$, but $x \neq a$. Similarly, the condition $|f(x) - L| < \varepsilon$ means that $f(x)$ lies in the open interval $(L - \varepsilon,\ L + \varepsilon)$. Geometrically, $\lim_{x \to a} f(x) = L$ means that, for $x \neq a$, we can guarantee that $f(x)$ is in any given small open interval around L if we make sure that x is in a suitably small open interval around a.

Now, we summarize these considerations in a formal definition of limit.

Figure 6

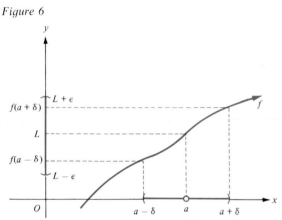

DEFINITION 1   Limit

Let f be a function defined on some open interval containing a, except possibly at the number a itself. The assertion $\lim_{x \to a} f(x) = L$ means that, for each positive number ε, there exists a positive number δ such that $|f(x) - L| < \varepsilon$ holds whenever $0 < |x - a| < \delta$.

EXAMPLES

1 Given $\varepsilon = 0.03$, determine a positive δ so that $|(3x + 7) - 1| < \varepsilon$ whenever $0 < |x - (-2)| < \delta$.

SOLUTION
We have

$$|(3x + 7) - 1| = |3x + 6| = |3(x + 2)| = 3|x + 2| \quad \text{and} \quad |x - (-2)| = |x + 2|;$$

hence, we must find a positive δ such that

$$3|x + 2| < 0.03 \quad \text{holds whenever} \quad 0 < |x + 2| < \delta.$$

The condition $3|x + 2| < 0.03$ is equivalent to $|x + 2| < 0.03/3 = 0.01$; hence, we must determine a positive δ such that

$$|x + 2| < 0.01 \quad \text{holds whenever} \quad 0 < |x + 2| < \delta.$$

Obviously, $\delta = 0.01$ works. So does any *smaller* positive value of δ.

2 Use Definition 1 to prove that $\lim_{x \to -2} (3x + 7) = 1$.

SOLUTION
Let $\varepsilon > 0$. We must find $\delta > 0$ such that $|(3x + 7) - 1| < \varepsilon$ holds whenever $0 < |x - (-2)| < \delta$, that is, whenever $0 < |x + 2| < \delta$. Just as in Example 1, we have

$$|(3x + 7) - 1| = 3|x + 2|.$$

Figure 7

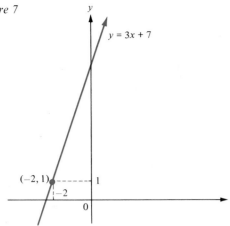

Therefore, the condition $|(3x + 7) - 1| < \varepsilon$ is equivalent to $3|x + 2| < \varepsilon$, that is, $|x + 2| < \varepsilon/3$. Hence, we must determine a positive δ such that

$$|x + 2| < \frac{\varepsilon}{3} \quad \text{holds whenever} \quad 0 < |x + 2| < \delta.$$

Obviously, $\delta = \varepsilon/3$ works. So does any *smaller* positive value of δ. This proves that

$$\lim_{x \to -2} (3x + 7) = 1$$

(Figure 7).

Problem Set 1

In problems 1 through 6, determine the limit and sketch a graph of each function to illustrate the limit involved.

1 $\lim\limits_{x \to 4} 3x$

2 $\lim\limits_{x \to 1} (3x - 6)$

3 $\lim\limits_{x \to -2} (2 - 3x)$

4 $\lim\limits_{x \to 5} \dfrac{2}{x}$

5 $\lim\limits_{x \to 1/2} |1 - 2x|$

6 $\lim\limits_{x \to 3} \dfrac{x^2 - 9}{x - 3}$

In problems 7 through 12, determine each limit. (Try to simplify by factoring and canceling if possible.)

7 $\lim\limits_{x \to 2} \dfrac{x^2 - 5x + 6}{x - 2}$

8 $\lim\limits_{t \to 0} \dfrac{t^2 + 2t + 1}{t + 5}$

9 $\lim\limits_{x \to 1} \dfrac{x^3 - 1}{x^2 - 1}$

10 $\lim\limits_{x \to -2} \dfrac{x^2 - 4}{x + 2}$

11 $\lim\limits_{x \to 1} \dfrac{x^2 - 2x + 1}{x - 1}$

12 $\lim\limits_{x \to 3} \dfrac{x^2 - 2x - 3}{x - 3}$

13 Use the graphs in problems 1 through 6 to decide which of the functions are continuous at the given value of x.

(a) $f(x) = 3x$ at $x = 4$

(b) $g(x) = 3x - 6$ at $x = 1$

(c) $F(x) = 2 - 3x$ at $x = -2$

(d) $f(x) = \dfrac{2}{x}$ at $x = 5$

(e) $h(x) = |1 - 2x|$ at $x = \frac{1}{2}$

(f) $f(x) = \dfrac{x^2 - 9}{x - 3}$ at $x = 3$

In problems 14 through 20, for the given ε, determine a positive δ so that $|f(x) - L| < \varepsilon$ holds whenever $0 < |x - a| < \delta$.

14 $f(x) = x + 3$, $L = 5$, $a = 2$, $\varepsilon = 0.01$, $\lim\limits_{x \to 2} (x + 3) = 5$

15 $f(x) = 4x - 1$, $L = 11$, $a = 3$, $\varepsilon = 0.01$, $\lim\limits_{x \to 3} (4x - 1) = 11$

16 $f(x) = 3 - 4x$, $L = 7$, $a = -1$, $\varepsilon = 0.02$, $\lim\limits_{x \to -1} (3 - 4x) = 7$

17 $f(x) = \dfrac{x^2 - 25}{x - 5}$, $L = 10$, $a = 5$, $\varepsilon = 0.01$, $\lim\limits_{x \to 5} \dfrac{x^2 - 25}{x - 5} = 10$

18 $f(x) = x - 1, L = 0, a = 1, \varepsilon = 0.1, \lim_{x \to 1} (x - 1) = 0$

19 $f(x) = \dfrac{x + 1}{2}, L = 3, a = 5, \varepsilon = 0.1, \lim_{x \to 5} \dfrac{x + 1}{2} = 3$

20 $f(x) = x^2, L = 4, a = 2, \varepsilon = 0.1, \lim_{x \to 2} x^2 = 4$

In problems 21 through 26, establish that each limit is correct by direct use of Definition 1. That is, for $\varepsilon > 0$, find $\delta > 0$ so that $|f(x) - L| < \varepsilon$ holds whenever $0 < |x - a| < \delta$.

21 $\lim_{x \to 4} (2x - 5) = 3$

22 $\lim_{x \to 0} (2 - 5x) = 2$

23 $\lim_{x \to 3} (4x - 1) = 11$

24 $\lim_{x \to 4} \dfrac{x^2 - 16}{x - 4} = 8$

25 $\lim_{x \to 3} a = a$, where a is constant

26 $\lim_{x \to 2} |x - 2| = 0$

2   Properties of Limits of Functions

Until now, we have evaluated limits of functions by intuition, by appealing to the graph of the function, by using elementary algebra, or by direct use of the definition of a limit in terms of ε and δ. In practical work, however, limits are often found by the use of certain properties, which we now state.

Basic Properties of Limits

Assume that $\lim_{x \to a} f(x) = L$ and that $\lim_{x \to a} g(x) = M$. Then:

1 $\lim_{x \to a} [f(x) + g(x)] = \lim_{x \to a} f(x) + \lim_{x \to a} g(x) = L + M$ and

$\lim_{x \to a} [f(x) - g(x)] = \lim_{x \to a} f(x) - \lim_{x \to a} g(x) = L - M.$

2 $\lim_{x \to a} [cf(x)] = c \lim_{x \to a} f(x) = cL$ (c is any constant).

3 $\lim_{x \to a} [f(x) \cdot g(x)] = \left[\lim_{x \to a} f(x) \right] \cdot \left[\lim_{x \to a} g(x) \right] = L \cdot M.$

4 If $\lim_{x \to a} g(x) = M \neq 0$, then $\lim_{x \to a} \dfrac{f(x)}{g(x)} = \dfrac{\lim_{x \to a} f(x)}{\lim_{x \to a} g(x)} = \dfrac{L}{M}.$

5 $\lim_{x \to a} [f(x)]^n = \left[\lim_{x \to a} f(x) \right]^n = L^n$ (n is any positive integer).

6 $\lim_{x \to a} \sqrt[n]{f(x)} = \sqrt[n]{\lim_{x \to a} f(x)} = \sqrt[n]{L}$ if $L > 0$ and n is a positive integer, or if $L \leq 0$ and n is an odd positive integer.

7 $\lim_{x \to a} |f(x)| = \left| \lim_{x \to a} f(x) \right| = |L|.$

8 $\lim\limits_{x \to a} c = c$ (c is any constant).

9 $\lim\limits_{x \to a} x = a$.

10 If h is a function such that $h(x) = f(x)$ holds for all values of x in some open interval containing a, except possibly for $x = a$, then

$$\lim_{x \to a} h(x) = \lim_{x \to a} f(x) = L.$$

Properties 1 and 3 can be extended (by mathematical induction) to any finite number of functions. All of the properties above should seem quite plausible to those readers who have managed an intuitive grasp of the idea of a limit. Properties 1 through 10 can be shown to follow deductively from the formal definition of a limit. However, we shall defer such proofs until Section 7, and simply accept these properties, for now, as being true.

EXAMPLES   In Examples 1 through 5, find each limit and indicate which of Properties 1 through 10 you use.

1 Assume that $\lim\limits_{x \to 3} f(x) = 9$ and that $\lim\limits_{x \to 3} g(x) = 4$. Find:

(a) $\lim\limits_{x \to 3} [f(x) + g(x)]$ (b) $\lim\limits_{x \to 3} [3f(x) - 2g(x)]$

(c) $\lim\limits_{x \to 3} \sqrt{f(x) \cdot g(x)}$ (d) $\lim\limits_{x \to 3} \left| \dfrac{f(x)}{g(x)} \right|$

SOLUTION

(a) $\lim\limits_{x \to 3} [f(x) + g(x)] = \lim\limits_{x \to 3} f(x) + \lim\limits_{x \to 3} g(x)$ (Property 1)

$$= 9 + 4 = 13.$$

(b) $\lim\limits_{x \to 3} [3f(x) - 2g(x)] = \lim\limits_{x \to 3} 3f(x) - \lim\limits_{x \to 3} 2g(x)$ (Property 1)

$$= 3 \lim_{x \to 3} f(x) - 2 \lim_{x \to 3} g(x) \quad \text{(Property 2)}$$

$$= 3(9) - 2(4) = 19.$$

(c) $\lim\limits_{x \to 3} \sqrt{f(x) \cdot g(x)} = \sqrt{\lim\limits_{x \to 3} [f(x) \cdot g(x)]}$ (Property 6)

$$= \sqrt{\left[\lim_{x \to 3} f(x) \right] \cdot \left[\lim_{x \to 3} g(x) \right]} \quad \text{(Property 3)}$$

$$= \sqrt{9(4)} = \sqrt{36} = 6.$$

(d) $\lim\limits_{x \to 3} \left| \dfrac{f(x)}{g(x)} \right| = \left| \lim\limits_{x \to 3} \dfrac{f(x)}{g(x)} \right|$ (Property 7)

$$= \left| \frac{\lim\limits_{x \to 3} f(x)}{\lim\limits_{x \to 3} g(x)} \right| \quad \text{(Property 4)}$$

$$= \left| \frac{9}{4} \right| = \frac{9}{4}.$$

2 $\lim\limits_{t \to 2} (4t^2 + 5t - 7)$

SOLUTION

$$
\begin{aligned}
\lim_{t \to 2} (4t^2 + 5t - 7) &= \lim_{t \to 2} 4t^2 + \lim_{t \to 2} 5t + \lim_{t \to 2} (-7) & \text{(Property 1)} \\
&= 4 \lim_{t \to 2} t^2 + 5 \lim_{t \to 2} t + \lim_{t \to 2} (-7) & \text{(Property 2)} \\
&= 4 \lim_{t \to 2} t^2 + 5(2) + \lim_{t \to 2} (-7) & \text{(Property 9)} \\
&= 4 \lim_{t \to 2} t^2 + 10 + (-7) & \text{(Property 8)} \\
&= 4 \left(\lim_{t \to 2} t \right)^2 + 3 & \text{(Property 5)} \\
&= 4(2)^2 + 3 = 19 & \text{(Property 9)}.
\end{aligned}
$$

3 $\lim\limits_{y \to 3} \sqrt[3]{\dfrac{y^2 + 5y + 3}{y^2 - 1}}$

SOLUTION
Proceeding as in Example 2, we have

$$\lim_{y \to 3} (y^2 + 5y + 3) = 3^2 + 5(3) + 3 = 27$$

$$\lim_{y \to 3} (y^2 - 1) = 3^2 - 1 = 8;$$

hence, by Property 4,

$$\lim_{y \to 3} \frac{y^2 + 5y + 3}{y^2 - 1} = \frac{27}{8}.$$

Therefore, using Property 6, we obtain

$$
\lim_{y \to 3} \sqrt[3]{\frac{y^2 + 5y + 3}{y^2 - 1}} = \sqrt[3]{\lim_{y \to 3} \frac{y^2 + 5y + 3}{y^2 - 1}}
$$

$$
= \sqrt[3]{\frac{27}{8}} = \frac{3}{2}.
$$

4 $\lim\limits_{x \to 7} \dfrac{x^2 - 49}{x - 7}$

SOLUTION

Property 4 is not applicable here since $\lim\limits_{x \to 7} (x - 7) = 0$. However, $\dfrac{x^2 - 49}{x - 7} = x + 7$ holds for all values of x except for $x = 7$; hence, by Property 10,

$$\lim_{x \to 7} \frac{x^2 - 49}{x - 7} = \lim_{x \to 7} (x + 7) = 7 + 7 = 14.$$

5 $\lim\limits_{x \to 0} \dfrac{\sqrt{4 + x} - 2}{x}$

SOLUTION
Again notice that Property 4 does not apply since the denominator approaches 0. Here we must resort to a clever trick—multiplying the numerator and denominator by $\sqrt{4 + x} + 2$ in order to rationalize the numerator. We have

$$\frac{\sqrt{4+x}-2}{x} = \frac{(\sqrt{4+x}-2)(\sqrt{4+x}+2)}{x(\sqrt{4+x}+2)}$$

$$= \frac{(\sqrt{4+x})^2 - 2^2}{x(\sqrt{4+x}+2)} = \frac{4+x-4}{x(\sqrt{4+x}+2)}$$

$$= \frac{x}{x(\sqrt{4+x}+2)} = \frac{1}{\sqrt{4+x}+2} \qquad \text{for } x \neq 0.$$

Thus, by Property 10,

$$\lim_{x\to 0}\frac{\sqrt{4+x}-2}{x} = \lim_{x\to 0}\frac{1}{\sqrt{4+x}+2} = \frac{\lim_{x\to 0} 1}{\lim_{x\to 0}(\sqrt{4+x}+2)}$$

$$= \frac{1}{\sqrt{4}+2} = \frac{1}{4}.$$

Problem Set 2

In problems 1 through 6, find each limit. Assume that $\lim_{x\to 2} f(x) = 4$ and $\lim_{x\to 2} g(x) = 3$.

1 $\lim_{x\to 2} [f(x) + g(x)]$

2 $\lim_{x\to 2} [f(x) - g(x)]$

3 $\lim_{x\to 2} [f(x) \cdot g(x)]$

4 $\lim_{x\to 2} \left[\dfrac{f(x)}{g(x)}\right]$

5 $\lim_{x\to 2} \sqrt{f(x) \cdot g(x)}$

6 $\lim_{x\to 2} \left|\dfrac{f(x)}{g(x)}\right|$

In problems 7 through 27, evaluate each limit. Indicate which of Properties 1 through 10 are used.

7 $\lim_{x\to -1} (5 - 3x - x^2)$

8 $\lim_{x\to 3} (5x^2 - 7x - 3)$

9 $\lim_{x\to 2} \dfrac{x^2 + x + 1}{x^2 + 2x}$

10 $\lim_{y\to -1} (3y^3 - 2y^2 + 5y - 1)$

11 $\lim_{x\to 5/2} \dfrac{4x^2 - 25}{2x - 5}$

12 $\lim_{t\to 2} \dfrac{2 - t^2}{4t}$

13 $\lim_{t\to 1/2} \dfrac{t^2 + 1}{1 + \sqrt{2t + 8}}$

14 $\lim_{x\to -2} \dfrac{x^3 - 5x}{x + 3}$

15 $\lim_{y\to 1} \sqrt[3]{\dfrac{27y^3 + 4y - 4}{y^{10} + 4y^2 + 3y}}$

16 $\lim_{x\to 1} \dfrac{\sqrt{4 - x^2}}{2 + x}$

17 $\lim_{z\to -1} \dfrac{z^2 + 4z + 3}{z^2 - 1}$

18 $\lim_{x\to 1} \sqrt{\dfrac{8x + 1}{x + 3}}$

19 $\lim_{x\to 8/3} \dfrac{9x^2 - 64}{3x - 8}$

20 $\lim_{t\to -7} \dfrac{t^2 - 49}{t + 7}$

21 $\lim_{x\to -3} \sqrt[3]{\dfrac{x - 4}{6x^2 + 2}}$

22 $\lim_{x\to -3} \left|\dfrac{x^2 + 4x + 3}{x + 3}\right|$

23 $\lim_{h\to 0} \dfrac{(3 + h)^2 - 9}{h}$

24 $\lim_{x\to 0} \dfrac{\sqrt{x + 2} - \sqrt{2}}{x}$

25 $\lim_{t\to 0} \left[\sqrt{1 + \dfrac{1}{|t|}} - \sqrt{\dfrac{1}{|t|}}\right]$

26 $\lim_{h\to 0} \dfrac{2 - \sqrt{4 - h}}{h}$

27 $\lim_{x\to 1} \dfrac{(1/\sqrt{x}) - 1}{1 - x}$

28 A point P is moving along the graph of f defined by the equation $f(x) = x^2$ toward the origin O. If the perpendicular bisector of the line segment $\overline{OP}$ intersects the y axis at a point D, what is the limit of D as P approaches the origin O?

3   Continuity—One-Sided Limits

We mentioned in Section 1 that if $\lim\limits_{x \to a} f(x) = f(a)$, the function f is said to be *continuous at a*. Let us now set this forth as an official definition.

DEFINITION 1   **Continuous Function**

We say that the function f is *continuous at the number a* if and only if the following conditions hold:

   (i) $f(a)$ is defined,

   (ii) $\lim\limits_{x \to a} f(x)$ exists, and

   (iii) $\lim\limits_{x \to a} f(x) = f(a)$.

EXAMPLES 1   Is the function f defined by $f(x) = x^2 - 3$ continuous at 0?

Figure 1

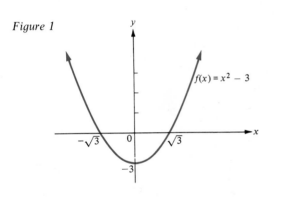

SOLUTION
Since $f(0) = 0^2 - 3 = -3$, $f(0)$ is defined. Also,

$$\lim_{x \to 0} f(x) = \lim_{x \to 0} (x^2 - 3) = \lim_{x \to 0} x^2 - \lim_{x \to 0} 3$$

$$= \left(\lim_{x \to 0} x \right)^2 - 3 = 0^2 - 3 = -3;$$

hence,

$$\lim_{x \to 0} f(x) = -3 = f(0).$$

Since conditions (i) through (iii) of Definition 1 hold, we conclude that f is continuous at 0 (Figure 1).

2 Show that the function f in Example 1 is actually continuous at every number a.

SOLUTION
Here, $f(a) = a^2 - 3$ is defined. Also, as in Example 1,

$$\lim_{x \to a} f(x) = \lim_{x \to a} (x^2 - 3) = a^2 - 3 = f(a);$$

hence, f is continuous at every number a.

If any of conditions (i) through (iii) of Definition 1 fails, we say that f is *discontinuous* at the number a.

EXAMPLE   Determine whether the function f defined by

$$f(x) = \begin{cases} \dfrac{2x^2 + 3x + 1}{x + 1} & \text{if } x \neq -1 \\ \\ 3 & \text{if } x = -1 \end{cases}$$

is continuous at the number -1.

Figure 2

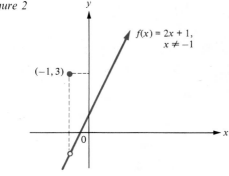

SOLUTION

For $x \neq -1$, we have

$$f(x) = \frac{2x^2 + 3x + 1}{x + 1} = \frac{(2x + 1)(x + 1)}{x + 1}$$

$$= 2x + 1,$$

so the straight line $y = 2x + 1$ serves as the graph of f everywhere except at $x = -1$, where a "hole" appears. The defined value of f at -1 produces a point $(-1, 3)$ on the graph of f, but above this "hole" (Figure 2). Since $f(x) = 2x + 1$ for all values of x except $x = -1$, Property 10 of Section 2 gives

$$\lim_{x \to -1} f(x) = \lim_{x \to -1} (2x + 1) = -1.$$

Here, $f(-1) = 3 \neq -1$, so we have

$$\lim_{x \to -1} f(x) \neq f(-1).$$

It follows that f is discontinuous at the number -1.

Another example of a discontinuous function is provided by

$$f(x) = \begin{cases} 3x - 2 & \text{if } x < 3 \\ 5 - x & \text{if } x \geq 3. \end{cases}$$

The graph of f (Figure 3) visually indicates a discontinuity at the number $x = 3$. In order to study this apparent discontinuity, it is convenient to introduce the idea of a *one-sided limit*, that is, a limit of $f(x)$ as x approaches 3 through values entirely on one side of 3. Figure 3 clearly shows that the numerical value $f(x)$ approaches 7 as x approaches 3 through values that are always less than 3. We write this fact symbolically as

$$\lim_{x \to 3^-} f(x) = 7,$$

Figure 3

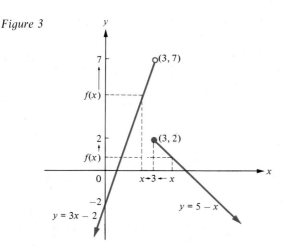

the condition that x approaches 3 *from the left* being denoted by $x \to 3^-$.

Likewise, the condition that x approaches 3 through values that are always greater than 3, so that x approaches 3 *from the right*, is denoted by $x \to 3^+$. From Figure 3 we see that

$$\lim_{x \to 3^+} f(x) = 2.$$

Since the two one-sided limits of $f(x)$ as $x \to 3^-$ and as $x \to 3^+$ are not the same, it follows that $\lim_{x \to 3} f(x)$ cannot exist. Therefore (as we have already guessed from our inspection of Figure 3), f is discontinuous at the number 3.

Although we give formal definitions of one-sided limits in Section 3.1, the idea of a limit from the right or from the left should be clear to the reader. Indeed, $\lim_{x \to a^-} f(x) = L$ simply means that we can make $|f(x) - L|$ as small as we please by taking x sufficiently close to a, but less than a. Likewise, $\lim\limits_{x \to a^+} f(x) = R$ means that we can make $|f(x) - R|$ as small as we please by taking x sufficiently close to a, but greater than a. If the two one-sided limits $\lim\limits_{x \to a^-} f(x)$ and $\lim\limits_{x \to a^+} f(x)$ exist and *have the same value*, it is clear that $\lim\limits_{x \to a} f(x)$ exists and that all three limits have the same value (Problem 29).

Conversely, if $\lim\limits_{x \to a} f(x)$ exists, the two one-sided limits $\lim\limits_{x \to a^-} f(x)$ and $\lim\limits_{x \to a^+} f(x)$ exist and all three limits have the same value (Problem 27). Consequently, if the two one-sided limits $\lim\limits_{x \to a^-} f(x)$ and $\lim\limits_{x \to a^+} f(x)$ exist, but have *different* values, it follows that $\lim\limits_{x \to a} f(x)$ *cannot exist*. (Why?) This method for showing that a certain limit does not exist has already been used for the function graphed in Figure 3.

It can be shown that one-sided limits satisfy appropriate analogs of Properties 1 through 10 of limits given in Section 2. For instance, suppose that $\lim\limits_{x \to a^+} f(x) = L$ and that $\lim\limits_{x \to a^+} g(x) = M$. Then we have

$$\lim_{x \to a^+} [f(x) + g(x)] = L + M,$$

$$\lim_{x \to a^+} [f(x) \cdot g(x)] = L \cdot M, \qquad \text{and so forth.}$$

EXAMPLES In each of the following examples, (a) sketch a graph of the function, (b) find the one-sided limits of the function as $x \to a^-$ and as $x \to a^+$, (c) determine the limit of the function as $x \to a$ (if this limit exists), and (d) decide whether the function is continuous at the number a.

1 $f(x) = \begin{cases} 2x + 1 & \text{if } x < 3 \\ 10 - x & \text{if } x \geq 3 \end{cases}$; $a = 3$

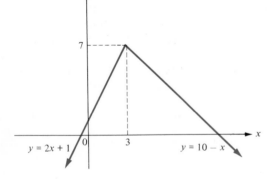

Figure 4

$y = 2x + 1$ $y = 10 - x$

SOLUTION

(a) A sketch of the graph of f is shown in Figure 4.

(b) $\lim\limits_{x \to 3^-} f(x) = \lim\limits_{x \to 3^-} (2x + 1) = 7$

$\lim\limits_{x \to 3^+} f(x) = \lim\limits_{x \to 3^+} (10 - x) = 7$

(c) Since both of the one-sided limits exist and have the same value 7, it follows that

$$\lim_{x \to 3} f(x) = 7.$$

(d) Here, $f(3) = 10 - 3 = 7$, so f is defined at 3. Also, $\lim\limits_{x \to 3} f(x)$ exists and $\lim\limits_{x \to 3} f(x) = 7 = f(3)$; hence, f is continuous at 3.

2 $f(x) = \begin{cases} |x - 2| & \text{if } x \neq 2 \\ 1 & \text{if } x = 2 \end{cases}$; $a = 2$

SOLUTION

(a) A sketch of the graph of f is shown in Figure 5.

Figure 5

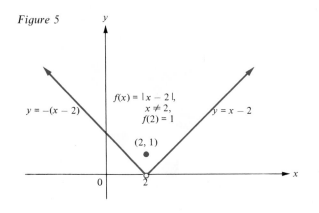

$f(x) = |x - 2|,$
$x \neq 2,$
$f(2) = 1$

$y = -(x - 2)$

$y = x - 2$

$(2, 1)$

(b) By Property 10 of Section 2,

$$\lim_{x \to 2} f(x) = \lim_{x \to 2} |x - 2|;$$

hence, by Property 7 of Section 2,

$$\lim_{x \to 2} f(x) = \left| \lim_{x \to 2} (x - 2) \right| = |0| = 0.$$

It follows that

$$\lim_{x \to 2^-} f(x) = \lim_{x \to 2^+} f(x) = \lim_{x \to 2} f(x) = 0.$$

(c) We have already seen in part (b) that $\lim_{x \to 2} f(x) = 0.$

(d) $f(2) = 1$, $\lim_{x \to 2} f(x) = 0 \neq 1$; hence, f is discontinuous at 2.

$$\textbf{3} \quad f(x) = \begin{vmatrix} 3 - x^2 & \text{if } x \leq 1 \\ 1 + x^2 & \text{if } x > 1 \end{vmatrix}; \, a = 1$$

Figure 6

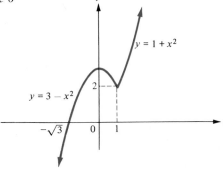

$y = 1 + x^2$

$y = 3 - x^2$

$-\sqrt{3}$

SOLUTION

(a) A sketch of the graph of f is shown in Figure 6.

(b) $\lim_{x \to 1^-} f(x) = \lim_{x \to 1^-} (3 - x^2) = 2$

   $\lim_{x \to 1^+} f(x) = \lim_{x \to 1^+} (1 + x^2) = 2$

(c) Since both of the one-sided limits exist and have the same value 2, it follows that $\lim_{x \to 1} f(x) = 2$.

(d) Here, $f(1) = 3 - 1^2 = 2$, so f is defined at 1. Also, $\lim_{x \to 1} f(x)$ exists and $\lim_{x \to 1} f(x) = 2 = f(1)$; hence, f is continuous at 1.

3.1 Formal Definitions of One-Sided Limits

The formal definitions of one-sided limits in terms of ε and δ are obtained by slightly modifying Definition 1 in Section 1.1.

DEFINITION 2   **Limit from the Right**
Let f be a function defined at least in some open interval (a, b). We say that *L is the limit of $f(x)$ as x approaches a from the right* and write $\lim_{x \to a^+} f(x) = L$ if, for each positive number ε, there exists a positive number δ such that $|f(x) - L| < \varepsilon$ holds whenever $0 < x - a < \delta$.

DEFINITION 3   **Limit from the Left**
Let f be a function defined at least in some open interval (c, a). We say that *L is the limit of $f(x)$ as x approaches a from the left* and write $\lim_{x \to a^-} f(x) = L$ if, for each positive number ε, there exists a positive number δ such that $|f(x) - L| < \varepsilon$ holds whenever $-\delta < x - a < 0$.

The connection between limits, limits from the right, and limits from the left is given by the following theorem, whose proof is left to the reader (Problems 28 and 30).

THEOREM 1   Limits and One-Sided Limits

The limit $\lim_{x \to a} f(x)$ exists and equals L if and only if both of the one-sided limits $\lim_{x \to a^-} f(x)$ and $\lim_{x \to a^+} f(x)$ exist and have the common value L.

Problem Set 3   odd 1-29

In problems 1 through 14, (a) sketch a graph of the given function, (b) find the one-sided limits of the given function as x approaches a from the right and from the left, (c) determine the limit of the function as x approaches a (if this limit exists), and (d) using the definition of continuity, decide whether the given function is continuous at the number a.

1 $f(x) = \begin{cases} 5 + x & \text{if } x \le 3 \\ 9 - x & \text{if } x > 3 \end{cases}$; $a = 3$

2 $F(x) = \begin{cases} -1 & \text{if } x < 0 \\ 0 & \text{if } x = 0 \\ 1 & \text{if } x > 0 \end{cases}$; $a = 0$

3 $g(x) = \begin{cases} 3 + x & \text{if } x \le 1 \\ 3 - x & \text{if } x > 1 \end{cases}$; $a = 1$

4 $G(x) = \begin{cases} 2x - 1 & \text{if } x < 1 \\ x^2 & \text{if } x \ge 1 \end{cases}$; $a = 1$

5 $H(x) = \begin{cases} \big||x - 5|\big| & \text{if } x \ne 5 \\ 2 & \text{if } x = 5 \end{cases}$; $a = 5$

6 $H(x) = \begin{cases} \dfrac{x - 2}{|x - 2|} & \text{if } x \ne 2 \\ 1 & \text{if } x = 2 \end{cases}$; $a = 2$

7 $f(x) = \begin{cases} 2 - x & \text{if } x > 1 \\ x^2 & \text{if } x \le 1 \end{cases}$; $a = 1$

8 $Q(x) = \begin{cases} \dfrac{1}{x - 2} & \text{if } x \ne 2 \\ 0 & \text{if } x = 2 \end{cases}$; $a = 2$

9 $F(x) = \begin{cases} \dfrac{x^2 - 9}{x - 3} & \text{if } x \ne 3 \\ 2 & \text{if } x = 3 \end{cases}$; $a = 3$

10 $R(x) = \begin{cases} 3 + x^2 & \text{if } x < -2 \\ 0 & \text{if } x = -2 \\ 11 - x^2 & \text{if } x > -2 \end{cases}$; $a = -2$

11 $S(x) = 5 + |6x - 3|$, $a = \frac{1}{2}$

12 $g(x) = [\![x]\!] + [\![5 - x]\!]$, $a = 4$

13 $f(x) = \dfrac{x^2 - 2x - 3}{x + 1}$, $a = -1$

14 $T(x) = [\![1 - x]\!] + [\![x - 1]\!]$, $a = 1$

In problems 15 through 21, sketch the graph of each function and determine the numbers a at which the function is continuous.

15 $f(x) = |2x|$

16 $f(x) = 3x + |3x|$

17 $F(x) = |x^5|$

18 $f(x) = 3x^2 - 4x + 5$

19 $G(x) = |x| - x$

20 $G(x) = [\![3x]\!]$

21 $H(x) = \dfrac{x}{|x|}$

22 Explain: If f is a rational function and if f is defined at the number a, then $\lim_{x \to a} f(x) = f(a)$.

23 (a) Explain why it often happens that $\lim_{x \to a} f(x)$ can be found simply by evaluating the function f at the number a to get $f(a)$.
(b) Give an example to show that $\lim_{x \to a} f(x) = f(a)$ can fail to be true.

24 Given a function g defined by the equation

$$g(x) = \begin{cases} \dfrac{x^3 - 1}{x - 1} & \text{if } x \neq 1 \\[2mm] a & \text{if } x = 1, \end{cases}$$

determine the value of a so that g will be continuous at 1.

25 Find $\lim_{x \to 2^+} \sqrt{x - 2}$ and explain why $\lim_{x \to 2} \sqrt{x - 2}$ does not exist.

26 Find $\lim_{x \to 0^-} \dfrac{|x|}{x}$ and explain why $\lim_{x \to 0} \dfrac{|x|}{x}$ does not exist.

27 Suppose that $\lim_{x \to a} f(x) = L$. Give an intuitive argument to explain why it should follow that $\lim_{x \to a^+} f(x) = L$.

28 Suppose that $\lim_{x \to a} f(x) = L$. Give a formal argument using the ε, δ definition of limit to prove that $\lim_{x \to a^+} f(x) = L$.

29 Suppose that $\lim_{x \to a^+} f(x) = \lim_{x \to a^-} f(x) = L$. Give an intuitive argument to explain why it should follow that $\lim_{x \to a} f(x) = L$.

30 Suppose that $\lim_{x \to a^+} f(x) = \lim_{x \to a^-} f(x) = L$. Give a formal argument using the ε, δ definition of limit to prove that $\lim_{x \to a} f(x) = L$.

31 State the analog of Property 3 of Section 2 for limits from the left.

32 State the analog of Property 10 of Section 2 for limits from the right.

4   Properties of Continuous Functions

Suppose that f and g are two functions that are continuous at the number a. Then $f(a)$ and $g(a)$ are both defined, so that $(f + g)(a) = f(a) + g(a)$ is defined. Moreover, by Property 1 of Section 2,

$$\lim_{x \to a} (f + g)(x) = \lim_{x \to a} f(x) + g(x) = \lim_{x \to a} f(x) + \lim_{x \to a} g(x)$$
$$= f(a) + g(a) = (f + g)(a).$$

We conclude that $f + g$ is continuous at a. Similar reasoning yields analogous results for the difference, product, quotient, and composition of f and g (see Section 7 for some of the details), and we have the following.

Basic Properties of Continuous Functions

1 If f and g are continuous at a, then so are $f + g$, $f - g$, and $f \cdot g$.

2 If f and g are continuous at a and $g(a) \neq 0$, then f/g is continuous at a.

3 If g is continuous at a and f is continuous at $g(a)$, then $f \circ g$ is continuous at a.

4 A polynomial function is continuous at every number.

5 A rational function is continuous at every number for which it is defined.

EXAMPLES   Use the basic properties of continuous functions to determine the numbers at which the given function is continuous. Sketch a graph of the function.

1 $f(x) = |x| + x$

Figure 1

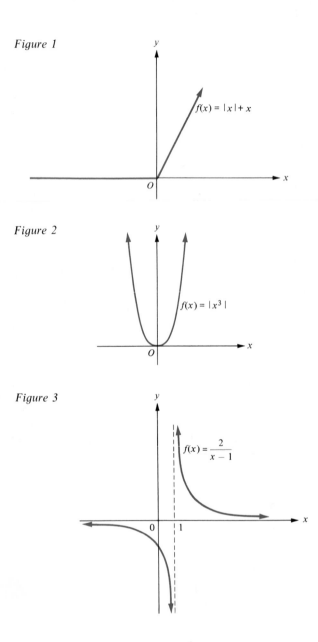

SOLUTION
Notice that $f = g + h$, where $g(x) = |x|$ and $h(x) = x$. By Property 4, the polynomial function h is continuous at every number. By Property 7 in Section 2,

$$\lim_{x \to a} g(x) = \lim_{x \to a} |x| = \left| \lim_{x \to a} x \right|$$
$$= |a| = g(a);$$

hence, g is continuous at every number a. It follows from Property 1 that $f = g + h$ is continuous at every number (Figure 1).

2 $f(x) = |x^3|$

Figure 2

SOLUTION
Here, $f = g \circ h$, where $h(x) = x^3$ and $g(x) = |x|$. In Example 1 we saw that the absolute value function g is continuous at every number, while the polynomial function h is also continuous at every number by Property 4. Therefore, by Property 3, $f = g \circ h$ is continuous at every number (Figure 2).

3 $f(x) = \dfrac{2}{x - 1}$

Figure 3

SOLUTION
Here, f is a rational function that is defined at every number x except for $x = 1$. By Property 5, f is continuous at every number x, except for $x = 1$ (Figure 3).

4.1 Continuity on an Interval

To say that a function f is *continuous on an open interval I* means, by definition, that f is continuous at every number in the interval I. For instance, the function $f(x) = \sqrt{9 - x^2}$ is continuous on the open interval $(-3, 3)$ (Figure 4).

Figure 4

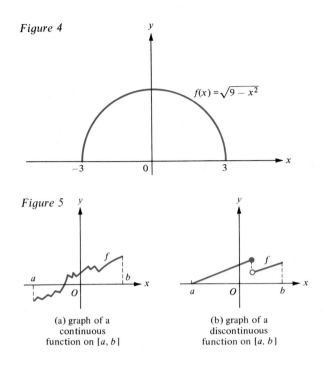

$f(x) = \sqrt{9 - x^2}$

Figure 5

(a) graph of a
continuous
function on $[a, b]$

(b) graph of a
discontinuous
function on $[a, b]$

Similarly, to say that a function f is *continuous on a closed interval* $[a, b]$ means, by definition, that f is continuous on the open interval (a, b) and that f satisfies the following "one-sided" continuity conditions at the endpoints a and b:

$$\lim_{x \to a^+} f(x) = f(a) \quad \text{and} \quad \lim_{x \to b^-} f(x) = f(b).$$

For instance, the function $f(x) = \sqrt{9 - x^2}$ is actually continuous on the closed interval $[-3, 3]$ (Figure 4).

Intuitively, the requirement that a function f be continuous on an (open or closed) interval I means that, if x is a number in I, then sufficiently small changes in x produce only small changes in the function value $f(x)$. The function values $f(x)$ of such a continuous function do not take any sudden "jumps" as x moves smoothly along the interval; hence, the graph of f is "connected" in the sense that it can be drawn without picking up one's pencil from the paper (Figure 5).

EXAMPLE   Determine whether or not the rational function $f(x) = \dfrac{x + 7}{x^2 - 36}$ is continuous on the following intervals: $(-\infty, -6)$, $[-6, 6]$, $(-6, 6)$, and $(6, \infty)$.

SOLUTION
Here, $f(x)$ is defined for all real numbers x except for the values $x = -6$ and $x = 6$ (which make the denominator zero). Thus, the rational function f is continuous on $(-\infty, -6)$, $(-6, 6)$, and $(6, \infty)$ by Property 5 of continuous functions. Since -6 and 6 do not belong to the domain of f, it follows that f is not continuous on the closed interval $[-6, 6]$.

Problem Set 4   odd 1~19

In problems 1 through 10, sketch the graph of the given function and determine the numbers a at which the function is continuous. Use the properties of continuous functions whenever they are relevant.

1 $f(x) = 2|x|$

2 $g(x) = |1 - x|$

3 $h(x) = x - 2|x|$

4 $F(x) = \dfrac{x + 1}{x - 1}$

5 $G(x) = \dfrac{1}{x}$

6 $f(x) = \left| \dfrac{1}{x} \right|$

7 $g(x) = \dfrac{x^2 - 4x + 3}{x - 1}$

8 $h(x) = \begin{cases} x^3 & \text{if } x \le 0 \\ x^2 & \text{if } x > 0 \end{cases}$

9 $F(x) = \dfrac{x^2 - 4x + 3}{x - 1} - \dfrac{x^2 - 2x - 3}{x + 1} + 1$

10 $G(x) = \dfrac{1}{|x| + 1}$

11 Let f and g be defined by the rules

$$f(x) = \begin{vmatrix} 3 - x & \text{if } x < 1 \\ 2 & \text{if } x \geq 1 \end{vmatrix} \quad \text{and} \quad g(x) = \begin{vmatrix} 2 & \text{if } x < 1 \\ 1 + x & \text{if } x \geq 1 \end{vmatrix}.$$

Which of the following functions are continuous at 1?
(a) $f + g$ (b) $f - g$ (c) $f \cdot g$

(d) $f \circ g$ (e) $\dfrac{f}{g}$

12 Suppose that f is continuous at the number a but that g is discontinuous at the number a. Is $f + g$ continuous or discontinuous at the number a? Illustrate by an example.

In problems 13 through 19, determine whether each function is continuous or discontinuous on each interval.

13 $f(x) = \sqrt{4 - x^2}$ on $[-2, 2]$, $[2, 3]$, $(-2, 2)$, and $(-1, 5)$

14 $g(x) = \dfrac{3}{x + 1}$ on $(-\infty, 1)$, $(-3, -1)$, $(-\infty, -1)$, $(-1, \infty)$, $[-1, \infty)$, and $[-2, 2]$

15 $F(x) = \dfrac{x + 6}{x^2 - 36}$ on $(-\infty, 6]$, $(-\infty, -4]$, $(-6, \infty)$, $[-6, 9]$, and $[-7, \infty)$

16 $f(x) = \dfrac{|x - 5|}{x - 5}$ on $[-1, 1]$, $(-1, 1)$, $(-5, \infty)$, $(-\infty, 5]$, and $[-8, 6]$

17 $G(x) = \dfrac{4x - 3}{16x^2 - 9}$ on $[-\frac{3}{4}, 0]$, $[-\frac{1}{2}, 0]$, $(-\frac{3}{4}, \infty)$, $[-2, \infty)$, and $(-1, -\frac{3}{4})$

18 $f(x) = \dfrac{5(2x + 1)}{4x^2 - 1}$ on $[-\frac{1}{2}, \frac{1}{2}]$, $(-\frac{1}{2}, \frac{1}{2})$, $(-\infty, 1)$, $[-\frac{1}{2}, 0]$, and $[\frac{1}{2}, \infty)$

19 $F(x) = \dfrac{3(2x - 3)}{4x^2 - 9}$ on $[-\frac{3}{2}, \frac{3}{2}]$, $(-\frac{3}{2}, \frac{3}{2})$, $(-\frac{3}{2}, \infty)$, $(\frac{3}{2}, \infty)$, and $(\frac{5}{2}, \infty)$

20 The weight of an object is given by

$$w(x) = \begin{cases} ax & \text{if } x \leq R \\ \dfrac{b}{x^2} & \text{if } x > R, \end{cases}$$

where x is the distance of the object from the center of the earth, R is the radius of the earth, and a and b are constants. What relation must exist between these constants if w is to be a continuous function? Assuming that w is continuous, sketch its graph.

21 If a hollow sphere of radius a is charged with one unit of static electricity, the field intensity E at a point P depends as follows on the distance x from the center of the sphere to P:

$$E(x) = \begin{cases} 0 & \text{if } 0 \leq x < a \\ \dfrac{1}{2a^2} & \text{if } x = a \\ \dfrac{1}{x^2} & \text{if } x > a. \end{cases}$$

(a) Sketch a graph of E.
(b) Discuss the continuity of E.

5   Limits Involving Infinity

In Section 1 the notion of the limit, $\lim\limits_{x \to a} f(x) = L$, of the function values $f(x)$ as x approaches the number a was introduced. A convenient extension of this idea is suggested by the behavior of the function f defined by $f(x) = 1/x^2$ near $x = 0$. (Of course, f is not defined at 0.) Certain values of $f(x)$ corresponding to smaller and smaller positive values of x are shown in the following table:

x	1	0.5	0.25	0.1	0.01	0.001	0.0001
$f(x) = 1/x^2$	1	4	16	100	10,000	1,000,000	100,000,000

For corresponding negative values of x the table would look the same, since f is an even function.

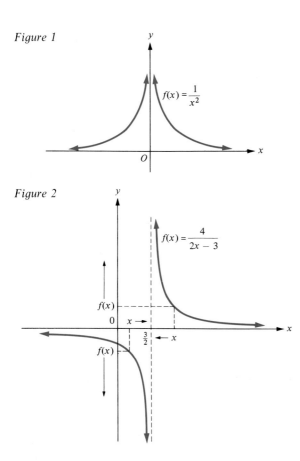

Figure 1

Figure 2

A glance at the table or at the graph of f (Figure 1) shows that the numerical values of $f(x)$ become large without bound as x gets closer and closer to 0. We express this symbolically by writing $\lim\limits_{x \to 0} 1/x^2 = +\infty$. (*Caution*: $+\infty$ is not a real number, so the symbolism $\lim\limits_{x \to 0} \dfrac{1}{x^2} = +\infty$ must be used with care!)

Now consider the rational function f defined by the equation $f(x) = \dfrac{4}{2x - 3}$. Notice that f is defined everywhere except at $x = \frac{3}{2}$. The graph of f (Figure 2) shows that as x approaches $\frac{3}{2}$ from the left, $f(x)$ decreases without bound. In fact, $f(x)$ can be made less than any preassigned number simply by taking x close enough to $\frac{3}{2}$ but less than $\frac{3}{2}$. This is expressed symbolically by writing

$$\lim_{x \to (3/2)^-} \frac{4}{2x - 3} = -\infty.$$

(Again, $-\infty$ is not a real number!)

Figure 2 also shows that as x approaches $\frac{3}{2}$ from the right, $f(x)$ increases without bound. In fact, $f(x)$ can be made larger than any preassigned number simply by taking x close enough to $\frac{3}{2}$ but greater than $\frac{3}{2}$. This is expressed symbolically by writing

$$\lim_{x \to (3/2)^+} \frac{4}{2x - 3} = +\infty.$$

In general, there are four possibilities for one-sided infinite limits:

1   The limit of $f(x)$ as x approaches a from the right is positive infinity, $\lim\limits_{x \to a^+} f(x) = +\infty$ (Figure 3a).

2   The limit of $f(x)$ as x approaches a from the left is positive infinity, $\lim\limits_{x \to a^-} f(x) = +\infty$ (Figure 3b).

Figure 3

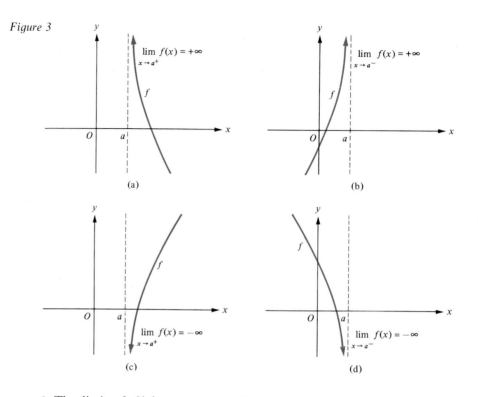

(a)

(b)

(c)

(d)

3 The limit of $f(x)$ as x approaches a from the right is negative infinity, $\lim\limits_{x \to a^+} f(x) = -\infty$ (Figure 3c).

4 The limit of $f(x)$ as x approaches a from the left is negative infinity, $\lim\limits_{x \to a^-} f(x) = -\infty$ (Figure 3d).

Although formal definitions of infinite limits are given in Section 5.2 for completeness, we usually determine such limits informally by inspection of the functions involved. (In a rigorous course in mathematical analysis, one would proceed much more carefully.)

In dealing with functions of the form $f(x) = p(x)/q(x)$, one should keep in mind that, if the denominator of a fraction is close to zero while the numerator is close to any number other than zero, the fraction will tend to have a large absolute value. More precisely,

$$\text{if } \lim_{x \to a} p(x) = L \neq 0 \quad \text{and} \quad \lim_{x \to a} q(x) = 0, \quad \text{then} \quad \lim_{x \to a} \left| \frac{p(x)}{q(x)} \right| = +\infty.$$

Naturally, the same thing holds for limits from the right or for limits from the left.

EXAMPLES   Evaluate (a) $\lim\limits_{x \to a^+} f(x)$, (b) $\lim\limits_{x \to a^-} f(x)$, and (c) $\lim\limits_{x \to a} f(x)$ for the given function.

1 $f(x) = \dfrac{2x^2 + 5x + 1}{x^2 - x - 6}; a = 3$

SOLUTION
Notice first that $\lim\limits_{x \to 3} (2x^2 + 5x + 1) = 34$ and $\lim\limits_{x \to 3} (x^2 - x - 6) = 0$; hence,

$$\lim_{x \to 3} \left| \frac{2x^2 + 5x + 1}{x^2 - x - 6} \right| = +\infty.$$

(a) As x approaches 3 from the right, so that $x > 3$, the numerator approaches 34; hence, the algebraic sign of the fraction is controlled by the algebraic sign of the denominator,

$$x^2 - x - 6 = (x - 3)(x + 2).$$

But if $x > 3$, then $x - 3 > 0$ and $x + 2 > 0$, so

$$x^2 - x - 6 > 0.$$

Therefore, for values of x near 3 but greater than 3, we have $f(x) > 0$. It follows that

$$\lim_{x \to 3^+} \frac{2x^2 + 5x + 1}{x^2 - x - 6} = +\infty.$$

(b) As x approaches 3 subject to the condition $x < 3$, the numerator approaches 34 while the denominator approaches zero. However, for $-2 < x < 3$, we have $x - 3 < 0$ and $x + 2 > 0$, so the denominator $x^2 - x - 6 = (x - 3)(x + 2)$ is negative. Thus, when x is close to 3 but less than 3, we have $f(x) < 0$. Therefore,

$$\lim_{x \to 3^-} \frac{2x^2 + 5x + 1}{x^2 - x - 6} = -\infty.$$

(c) From (a) and (b) we see that $\lim_{x \to a^+} f(x) = +\infty$ and $\lim_{x \to a^-} f(x) = -\infty$. It follows that $f(x)$ does not have a limit, finite or infinite, as $x \to 3$.

2 $f(x) = \dfrac{4x}{(x - 5)^2}; a = 5$

SOLUTION
The limit of the numerator is $\lim_{x \to 5} 4x = 20$. As x approaches 5, the denominator $(x - 5)^2$ approaches 0 through positive values; hence, for values of x near 5, the fraction is positive and very large. Therefore,

(a) $\lim_{x \to 5^+} f(x) = +\infty$ (b) $\lim_{x \to 5^-} f(x) = +\infty$ and   (c) $\lim_{x \to 5} f(x) = +\infty.$

5.1 Limits at Infinity

Consider the function $f(x) = 1/x$. As x becomes larger and larger without bound, the function values $f(x)$ approach 0 (Figure 4); symbolically,

$$\lim_{x \to +\infty} f(x) = 0.$$

Also, as x decreases without bound, the function values $f(x)$ approach 0; symbolically,

$$\lim_{x \to -\infty} f(x) = 0.$$

More generally, $\lim_{x \to +\infty} f(x) = L$ means that $|f(x) - L|$ can be made as small as we please by taking x to be sufficiently large. Similarly, $\lim_{x \to -\infty} f(x) = L$ means that

Figure 4

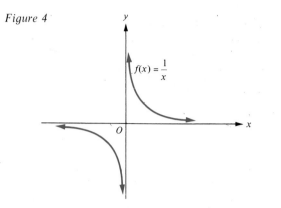

$|f(x) - L|$ can be made as small as we please by taking x to be negative with $|x|$ sufficiently large. (Formal definitions of limits at infinity can be found in Section 5.2.)

Naturally, we define $\lim\limits_{x \to +\infty} f(x) = +\infty$ to mean that $f(x)$ can be made as large as we please, provided that x is large enough. Similar definitions are understood to apply to such statements as $\lim\limits_{x \to +\infty} f(x) = -\infty$, $\lim\limits_{x \to -\infty} f(x) = -\infty$, and so forth (Problem 23).

In calculating limits at infinity, it is often useful to keep in mind that, for any positive integer p,

$$\lim_{x \to +\infty} \left(\frac{1}{x}\right)^p = \lim_{x \to +\infty} \frac{1}{x^p} = 0 \quad \text{and} \quad \lim_{x \to -\infty} \left(\frac{1}{x}\right)^p = \lim_{x \to -\infty} \frac{1}{x^p} = 0.$$

These facts are illustrated for odd values of p in Figure 5a and for even values of p in Figure 5b.

Figure 5

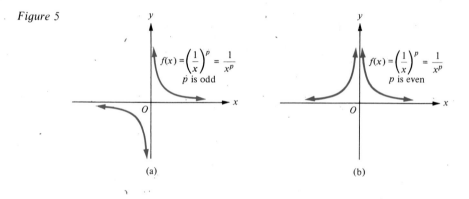

(a)   (b)

Also, in dealing with limits at infinity of rational functions, it is often helpful to divide numerator and denominator by the highest power of the independent variable appearing in the fraction. The following examples illustrate the techniques involved.

EXAMPLES   Evaluate the given limit.

1 $\lim\limits_{x \to +\infty} \dfrac{5x^2}{2x^2 - 3}$

SOLUTION
As x gets large, both the numerator and the denominator of the fraction $\dfrac{5x^2}{2x^2 - 3}$ get large, making it difficult to say offhand just what happens to the fraction. However, dividing both the numerator and the denominator by x^2, we have

$$\frac{5x^2}{2x^2 - 3} = \frac{\dfrac{5x^2}{x^2}}{\dfrac{2x^2}{x^2} - \dfrac{3}{x^2}} = \frac{5}{2 - \dfrac{3}{x^2}}.$$

As $x \to +\infty$, $1/x^2$ approaches 0 (Figure 5b), so $3/x^2$ approaches 0. Therefore,

$$\lim_{x \to +\infty} \frac{5x^2}{2x^2 - 3} = \lim_{x \to +\infty} \frac{5}{2 - \dfrac{3}{x^2}} = \frac{5}{2 - 0} = \frac{5}{2}.$$

2 $\displaystyle \lim_{x \to -\infty} \frac{x^3 + 1}{x^2 - 1}$

SOLUTION

$$\lim_{x \to -\infty} \frac{x^3 + 1}{x^2 - 1} = \lim_{x \to -\infty} \frac{\dfrac{x^3}{x^3} + \dfrac{1}{x^3}}{\dfrac{x^2}{x^3} - \dfrac{1}{x^3}} = \lim_{x \to -\infty} \frac{1 + \dfrac{1}{x^3}}{\dfrac{1}{x} - \dfrac{1}{x^3}}.$$

In the last fraction, as x approaches $-\infty$ the numerator approaches 1 while the denominator approaches 0. Hence, the absolute value of the fraction approaches $+\infty$. Whether the fraction itself approaches $+\infty$ or $-\infty$ depends on whether the denominator $1/x - 1/x^3$ is positive or negative when x is negative and has a large absolute value (that is, as $x \to -\infty$). However,

$$\frac{1}{x} - \frac{1}{x^3} = \frac{1}{x}\left(1 - \frac{1}{x^2}\right);$$

hence, if x is negative (so that $1/x$ is negative) and if $|x|$ is large (so that $1 - 1/x^2$ is close to 1), then $1/x - 1/x^3$ is negative. Consequently,

$$\lim_{x \to -\infty} \frac{x^3 + 1}{x^2 - 1} = \lim_{x \to -\infty} \frac{1 + \dfrac{1}{x^3}}{\dfrac{1}{x} - \dfrac{1}{x^3}} = -\infty.$$

3 $\displaystyle \lim_{x \to +\infty} \frac{5x}{\sqrt[3]{7x^3 + 3}}$

SOLUTION

Here the function is not rational; nevertheless, we try our trick of dividing numerator and denominator by some power of x. We divide by x itself, so that the numerator simplifies to 5. Thus,

$$\lim_{x \to +\infty} \frac{5x}{\sqrt[3]{7x^3 + 3}} = \lim_{x \to +\infty} \frac{5}{\dfrac{1}{x}\sqrt[3]{7x^3 + 3}} = \lim_{x \to +\infty} \frac{5}{\sqrt[3]{\dfrac{1}{x^3}(7x^3 + 3)}}$$

$$= \lim_{x \to +\infty} \frac{5}{\sqrt[3]{7 + \dfrac{3}{x^3}}} = \frac{5}{\sqrt[3]{7}}.$$

4 $\displaystyle \lim_{x \to -\infty} (7x^2 + 3x^3)$

SOLUTION

Factoring, we have $7x^2 + 3x^3 = x^2(7 + 3x)$. If x is negative and has a large absolute value, then x^2 is positive and large. Furthermore, if $x < -\frac{7}{3}$, then $7 + 3x$ is negative. It follows that

$$\lim_{x \to -\infty} (7x^2 + 3x^3) = \lim_{x \to -\infty} x^2(7 + 3x) = -\infty.$$

5.2 Formal Definitions of Limits Involving Infinity

Properties of limits involving infinity can be developed rigorously on the basis of the following formal definitions.

DEFINITION 1   **Infinite Limits from the Right**

Suppose that the function f is defined at least in an open interval (a, b). Then $\lim_{x \to a^+} f(x) = +\infty$ [respectively, $\lim_{x \to a^+} f(x) = -\infty$] means that for each positive number M there exists a positive number δ such that $f(x) > M$ [respectively, $f(x) < -M$] holds whenever $0 < x - a < \delta$.

DEFINITION 2   **Infinite Limits from the Left**

Suppose that the function f is defined at least in an open interval (c, a). Then $\lim_{x \to a^-} f(x) = +\infty$ [respectively, $\lim_{x \to a^-} f(x) = -\infty$] means that for each positive number M there exists a positive number δ such that $f(x) > M$ [respectively, $f(x) < -M$] holds whenever $-\delta < x - a < 0$.

DEFINITION 3   **Infinite Limits**

We take $\lim_{x \to a} f(x) = +\infty$ [respectively, $\lim_{x \to a} f(x) = -\infty$] to mean that $\lim_{x \to a^+} f(x) = +\infty$ and $\lim_{x \to a^-} f(x) = +\infty$ [respectively, $\lim_{x \to a^+} f(x) = -\infty$ and $\lim_{x \to a^-} f(x) = -\infty$].

DEFINITION 4   **Limits at Infinity**

Suppose that the function f is defined at least on an unbounded open interval (a, ∞) [respectively, $(-\infty, a)$]. We define $\lim_{x \to +\infty} f(x) = L$ [respectively, $\lim_{x \to -\infty} f(x) = L$] to mean that for each positive number ε there exists a positive number N such that $|f(x) - L| < \varepsilon$ holds whenever $x > N$ [respectively, $x < -N$].

Problem Set 5

In problems 1 through 22, evaluate the limit.

1 $\lim_{x \to 1^+} \dfrac{2x}{x - 1}$

2 $\lim_{x \to 2^-} \dfrac{x^2}{x - 2}$

3 $\lim_{x \to 0^+} \dfrac{\sqrt{4 + 3x^2}}{5x}$

4 $\lim_{x \to 3^+} \dfrac{x^2 + 5x + 1}{x^2 - 2x - 3}$

5 $\lim_{x \to 4^-} \dfrac{2x^2 + 3x - 2}{x^2 - 3x - 4}$

6 $\lim_{t \to 5^-} \dfrac{\sqrt{25 - t^2}}{t - 5}$

7 $\lim_{x \to 1^-} \dfrac{x^2 - 1}{|x^2 - 1|}$

8 $\lim_{x \to 2^-} \dfrac{[\![2 - x]\!]}{2 - x}$

9 $\lim_{x \to 2^-} \dfrac{x^2 + 1}{x - 2}$

10 $\lim_{z \to 2^+} \dfrac{z^2 + 1}{z - 2}$

11 $\lim_{x \to +\infty} \dfrac{1 + 6x}{-2 + x}$

12 $\lim_{x \to -\infty} \dfrac{2x^2 + x + 1}{-4x^2 + 5x + 10}$

13 $\lim_{x \to +\infty} \dfrac{5x^2 - 7x + 3}{8x^2 + 5x + 1}$

14 $\lim_{x \to -\infty} \dfrac{7x^3 + 3x + 1}{x^3 - 2x + 3}$

15 $\lim_{x \to -\infty} \dfrac{x^{100} + x^{99}}{x^{101} - x^{100}}$

16 $\lim_{x \to +\infty} \dfrac{x^{99} + x^{98}}{x^{100} - x^{99}}$

17 $\lim_{t \to +\infty} \dfrac{8t}{\sqrt[4]{3t^4 + 5}}$

18 $\lim_{x \to -\infty} \dfrac{6x^2}{\sqrt[3]{5x^6 - 1}}$

19 $\lim_{x \to +\infty} (5x^2 - 3x)$

20 $\lim_{x \to -\infty} \dfrac{x^3 - 5x^2}{3x}$

21 $\displaystyle\lim_{t \to -1^+} \left(\frac{3}{t+1} - \frac{5}{t^2-1} \right)$ **22** $\displaystyle\lim_{x \to +\infty} \frac{7x^3 - 15x^2}{13x}$

23 Write rigorous formal definitions, patterned after those given in the text, for:

    (a) $\displaystyle\lim_{x \to +\infty} f(x) = +\infty$ (b) $\displaystyle\lim_{x \to -\infty} f(x) = +\infty$

    (c) $\displaystyle\lim_{x \to +\infty} f(x) = -\infty$ (d) $\displaystyle\lim_{x \to -\infty} f(x) = -\infty$

24 Prove that $\displaystyle\lim_{t \to 0^+} f(1/t) = L$ holds if and only if $\displaystyle\lim_{x \to +\infty} f(x) = L$.

6   Horizontal and Vertical Asymptotes

Limits involving infinity are useful in graph sketching because they can be used to locate the *asymptotes* of the graphs. For instance, consider the graph of $f(x) = \dfrac{2x-6}{x-5}$ (Figure 1). Notice the way that the graph approaches the vertical straight line $x = 5$ just to the right of the line and also just to the left of the line; that is,

$$\lim_{x \to 5^+} f(x) = +\infty \quad \text{and} \quad \lim_{x \to 5^-} f(x) = -\infty.$$

Such a straight line is called a *vertical asymptote* of the graph. Similarly, the horizontal line $y = 2$ is called a *horizontal asymptote* of the graph, since

$$\lim_{x \to +\infty} f(x) = 2 \quad \text{and} \quad \lim_{x \to -\infty} f(x) = 2.$$

More precisely, we make the following definitions.

Figure 1

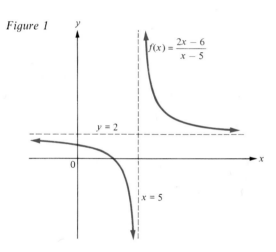

DEFINITION 1   **Vertical Asymptote**

The vertical straight line $x = a$ is called a *vertical asymptote* of the graph of the function f if at least one of the following four conditions holds:

    (i) $\displaystyle\lim_{x \to a^+} f(x) = +\infty.$

    (ii) $\displaystyle\lim_{x \to a^-} f(x) = +\infty.$

    (iii) $\displaystyle\lim_{x \to a^+} f(x) = -\infty.$

    (iv) $\displaystyle\lim_{x \to a^-} f(x) = -\infty.$

DEFINITION 2   **Horizontal Asymptote**

The horizontal straight line $y = b$ is called a *horizontal asymptote* of the graph of the function f if at least one of the following two conditions holds:

$$\lim_{x \to +\infty} f(x) = b \quad \text{or} \quad \lim_{x \to -\infty} f(x) = b.$$

Notice, for instance, in Figure 5 of Section 5 that the coordinate axes $x = 0$ and $y = 0$ are vertical and horizontal asymptotes of the graph of $y = (1/x)^p = 1/x^p$, p a positive integer.

Vertical asymptotes involve infinite limits, while horizontal asymptotes involve limits at infinity. Therefore, it is clear how to find horizontal asymptotes—simply calculate the appropriate limits as $x \to +\infty$ and as $x \to -\infty$. To spot potential vertical asymptotes $x = a$ of a function of the form f/g, simply look for values of a for which $g(a) = 0$. One has to be careful because there may be such values of a that do not give vertical asymptotes, nor does this procedure always yield all possible vertical asymptotes. (See Problem 11.)

EXAMPLES   Find the horizontal and vertical asymptotes of the graph of the function f and sketch this graph.

1 $f(x) = \dfrac{3x}{x - 1}$

SOLUTION
To find a horizontal asymptote, we calculate

$$\lim_{x \to +\infty} f(x) = \lim_{x \to +\infty} \frac{3x}{x - 1} = \lim_{x \to +\infty} \frac{3}{1 - 1/x} = 3;$$

thus, $y = 3$ is a horizontal asymptote. Note that we also have

$$\lim_{x \to -\infty} \frac{3x}{x - 1} = \lim_{x \to -\infty} \frac{3}{1 - 1/x} = 3.$$

Since $\displaystyle\lim_{x \to 1^+} \frac{3x}{x - 1} = +\infty$, it follows that $x = 1$ is a vertical asymptote. Note that we also have

$$\lim_{x \to 1^-} \frac{3x}{x - 1} = -\infty$$

(Figure 2).

Figure 2

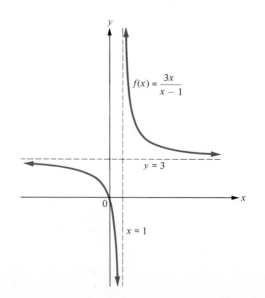

2 $f(x) = \dfrac{2x}{\sqrt{x^2 + 4}}$

SOLUTION

$$\lim_{x \to +\infty} f(x) = \lim_{x \to +\infty} \frac{2}{\dfrac{1}{x}\sqrt{x^2 + 4}} = \lim_{x \to +\infty} \frac{2}{\sqrt{\dfrac{x^2 + 4}{x^2}}} = \lim_{x \to +\infty} \frac{2}{\sqrt{1 + \dfrac{4}{x^2}}} = 2.$$

Thus, $y = 2$ is a horizontal asymptote. For negative values of x, we have $\sqrt{1/x^2} = 1/(-x)$, and so

$$\lim_{x \to -\infty} f(x) = \lim_{x \to -\infty} \frac{-2}{\dfrac{1}{-x}\sqrt{x^2 + 4}} = \lim_{x \to -\infty} \frac{-2}{\sqrt{\dfrac{x^2 + 4}{x^2}}}$$

$$= \lim_{x \to -\infty} \frac{-2}{\sqrt{1 + \dfrac{4}{x^2}}} = -2.$$

Hence, $y = -2$ is another horizontal asymptote. The graph (Figure 3) shows no vertical asymptotes. (Again note that the denominator $\sqrt{x^2 + 4}$ is never equal to 0.)

Figure 3

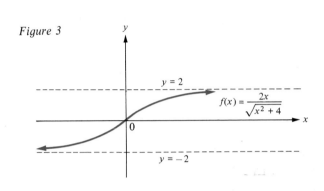

$y = 2$

$f(x) = \dfrac{2x}{\sqrt{x^2 + 4}}$

$y = -2$

3 $f(x) = \dfrac{2x^2 + 1}{2x^2 - 3x}$

SOLUTION
Since

$$\lim_{x \to +\infty} \frac{2x^2 + 1}{2x^2 - 3x} = \lim_{x \to +\infty} \frac{2 + 1/x^2}{2 - 3/x} = \frac{2}{2} = 1,$$

then $y = 1$ is the horizontal asymptote. $\left(\text{Note also that } \lim_{x \to -\infty} \dfrac{2x^2 + 1}{2x^2 - 3x} = 1.\right)$
The denominator factors as $2x^2 - 3x = x(2x - 3)$, so we suspect vertical asymptotes at $x = 0$ and at $x = \frac{3}{2}$. For small positive values of x, $2x - 3 < 0$, so

$$\frac{2x^2 + 1}{x(2x - 3)} < 0$$

and

$$\lim_{x \to 0^+} \frac{2x^2 + 1}{2x^2 - 3x} = \lim_{x \to 0^+} \frac{2x^2 + 1}{x(2x - 3)} = -\infty.$$

Similarly,

$$\lim_{x \to 0^-} \frac{2x^2 + 1}{2x^2 - 3x} = \lim_{x \to 0^-} \frac{2x^2 + 1}{x(2x - 3)} = +\infty,$$

$$\lim_{x \to (3/2)^+} \frac{2x^2 + 1}{2x^2 - 3x} = \lim_{x \to (3/2)^+} \frac{2x^2 + 1}{x(2x - 3)} = +\infty,$$

and

$$\lim_{x \to (3/2)^-} \frac{2x^2 + 1}{2x^2 - 3x} = \lim_{x \to (3/2)^-} \frac{2x^2 + 1}{x(2x - 3)} = -\infty,$$

confirming the existence of the suspected vertical asymptotes. The graph, showing the asymptotes, is sketched in Figure 4. Notice that the graph *crosses its own asymptote* at the point $\left(-\frac{1}{3}, 1\right)$.

Figure 4

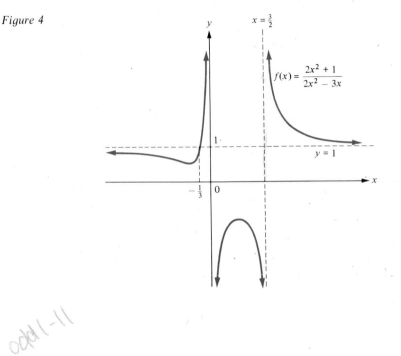

$$f(x) = \frac{2x^2 + 1}{2x^2 - 3x}$$

Problem Set 6

odd 1–11

In problems 1 through 10, find the horizontal and vertical asymptotes of the graph of each function and sketch the graph.

1 $f(x) = \dfrac{7x}{2x - 5}$

2 $F(x) = \dfrac{-2}{(x - 1)^2}$

3 $g(x) = \dfrac{1 - 2x}{3 + 5x}$

4 $G(x) = \dfrac{3x^2 + 1}{2x^2 - 7x}$

5 $f(x) = \dfrac{3x}{\sqrt{2x^2 + 1}}$

6 $f(x) = \sqrt{\dfrac{x}{x - 2}}$

7 $F(x) = \dfrac{-2x}{\sqrt{x^2 + 4}}$

8 $f(x) = \dfrac{x + 2}{\sqrt{1 - x}}$

9 $f(x) = \dfrac{x^2 - 1}{x}$

10 $f(x) = \dfrac{4x^2}{\sqrt{x^2 + 5x + 4}}$

11 (a) Show that even though $x = 0$ makes the denominator x of $\dfrac{\sqrt{1 + x} - 1}{x}$ zero, $x = 0$ is not a vertical asymptote of the function f defined by the equation $f(x) = \dfrac{\sqrt{1 + x} - 1}{x}$.

   (b) Show that even though $x = 0$ does not make the denominator $1 + x$ of $\dfrac{1 + 1/x}{1 + x}$ zero, nevertheless, $x = 0$ is a vertical asymptote of the function F defined by the equation $F(x) = \dfrac{1 + 1/x}{1 + x}$.

12 The amplitude $A(x)$ of a particle forced to oscillate in a resisting medium is given by the equation $A(x) = \dfrac{1}{\sqrt{(1 - x^2)^2 + kx^2}}$, where x is the ratio of the forcing frequency to the natural frequency of oscillation and k is a constant that measures the damping caused by the resisting medium.

(a) Sketch the graph of A for the values $k = 0$, $k = \frac{1}{2}$, and $k = 1$.

(b) Show that the graph of A has a vertical asymptote when $k = 0$, but that for $0 < k < 4$ the graph of A has no vertical asymptote.

13 Suppose that A, B, C, a, b, and c are constants so that $ax^2 + bx + c \neq 0$ for all values of x, with $b^2 - 4ac < 0$. Find all vertical and horizontal asymptotes of the graph of the function f defined by $f(x) = \dfrac{Ax^2 + Bx + C}{ax^2 + bx + c}$.

14 Let f be the function defined by $f(x) = (\sin x)/x$. Show that the graph of f crosses its asymptote infinitely many times.

7 Proofs of Basic Properties of Limits and of Continuous Functions

Although rigorous proofs of many of the theorems of calculus are best left to more advanced courses in analysis, it is perhaps not amiss to present a few typical proofs here so that the interested reader can gain some appreciation of the techniques involved. We begin by giving a proof of part of Property 1 in Section 2.

THEOREM 1   **Additivity of Limits**

If $\lim_{x \to a} f(x) = L$ and $\lim_{x \to a} g(x) = M$, then

$$\lim_{x \to a} [f(x) + g(x)] = \lim_{x \to a} f(x) + \lim_{x \to a} g(x) = L + M.$$

PROOF

According to the definition of limit (Section 1.1, p. 52), given a positive number ε, we must show that there exists a positive number δ such that

$$|[f(x) + g(x)] - [L + M]| < \varepsilon \quad \text{holds whenever} \quad 0 < |x - a| < \delta.$$

By the triangle inequality, we have

$$|[f(x) + g(x)] - [L + M]| = |[f(x) - L] + [g(x) - M]|$$
$$\leq |f(x) - L| + |g(x) - M|;$$

hence, it is enough to find a positive number δ such that

$$|f(x) - L| + |g(x) - M| < \varepsilon \quad \text{holds whenever} \quad 0 < |x - a| < \delta.$$

Now, in order to guarantee that

$$|f(x) - L| + |g(x) - M| < \varepsilon,$$

it certainly is enough to have

$$|f(x) - L| < \tfrac{1}{2}\varepsilon \quad \text{and} \quad |g(x) - M| < \tfrac{1}{2}\varepsilon.$$

Notice that $\frac{1}{2}\varepsilon$ is a positive number; hence, since $\lim_{x \to a} f(x) = L$, there exists (by definition of limit) a positive number δ_1 such that

$$|f(x) - L| < \tfrac{1}{2}\varepsilon \quad \text{holds whenever} \quad 0 < |x - a| < \delta_1.$$

Similarly, since $\lim\limits_{x \to a} g(x) = M$, there exists a positive number δ_2 such that

$$|g(x) - M| < \tfrac{1}{2}\varepsilon \quad \text{holds whenever} \quad 0 < |x - a| < \delta_2.$$

Let δ be the smaller of the two numbers δ_1 and δ_2 (or their common value if they are equal). Then

$$\delta \leq \delta_1 \quad \text{and} \quad \delta \leq \delta_2,$$

so that, if $0 < |x - a| < \delta$, both of the conditions

$$0 < |x - a| < \delta_1 \quad \text{and} \quad 0 < |x - a| < \delta_2$$

hold, and it follows that

$$|[f(x) + g(x)] - [L + M]| < \varepsilon,$$

as desired.

The following technical theorem can be quite helpful for establishing the remaining properties of limits.

THEOREM 2 **Boundedness Theorem**
If $\lim\limits_{x \to a} f(x)$ exists, then there exist positive numbers N and δ_1 such that $|f(x)| < N$ holds whenever $0 < |x - a| < \delta_1$.

PROOF

Suppose that $\lim\limits_{x \to a} f(x) = L$. Then, since 1 is a positive number, there exists a positive number δ_1 such that

$$|f(x) - L| < 1 \quad \text{holds whenever} \quad 0 < |x - a| < \delta_1.$$

(There is nothing special about our choice of the number 1 here—any positive number will work.) Now, let $N = 1 + |L|$. If $|f(x) - L| < 1$, then, adding $|L|$ to both sides, we have $|f(x) - L| + |L| < 1 + |L| = N$; therefore,

$$|f(x) - L| + |L| < N \quad \text{holds whenever} \quad 0 < |x - a| < \delta_1.$$

But, by the triangle inequality,

$$|f(x)| = |f(x) - L + L| \leq |f(x) - L| + |L|,$$

and it follows that

$$|f(x)| < N \quad \text{holds whenever} \quad 0 < |x - a| < \delta_1.$$

Theorem 2 is used in the course of the proof of the following theorem, which establishes Property 3 in Section 2.

THEOREM 3 **Multiplicativity of Limits**
If $\lim\limits_{x \to a} f(x) = L$ and $\lim\limits_{x \to a} g(x) = M$, then

$$\lim_{x \to a} [f(x) \cdot g(x)] = \left[\lim_{x \to a} f(x)\right] \cdot \left[\lim_{x \to a} g(x)\right] = LM.$$

PROOF

Let $\varepsilon > 0$ be given. We must show that there exists a positive number δ such that

$$|f(x)g(x) - LM| < \varepsilon \quad \text{holds whenever} \quad 0 < |x - a| < \delta.$$

Using some elementary algebra and the triangle inequality, we have

$$
\begin{aligned}
|f(x)g(x) - LM| &= |f(x)g(x) + 0 - LM| \\
&= |f(x)g(x) + [-f(x)M + f(x)M] - LM| \\
&= |f(x)g(x) - f(x)M + f(x)M - LM| \\
&= |f(x)[g(x) - M] + [f(x) - L]M| \\
&\leq |f(x)[g(x) - M]| + |[f(x) - L]M| \\
&= |f(x)||g(x) - M| + |f(x) - L||M|.
\end{aligned}
$$

Therefore, to guarantee that $|f(x)g(x) - LM| < \varepsilon$, it certainly is enough to have

(i) $|f(x)||g(x) - M| < \frac{1}{2}\varepsilon$ and (ii) $|f(x) - L||M| < \frac{1}{2}\varepsilon$.

In order to obtain the first condition, we plan to use Theorem 2 to make $|f(x)|$ smaller than a fixed positive number N, and then to make $|g(x) - M|$ less than $1/N$ times $\frac{1}{2}\varepsilon$. Thus, by Theorem 2, there exist positive numbers N and δ_1 such that

$$|f(x)| < N \quad \text{holds whenever} \quad 0 < |x - a| < \delta_1.$$

Also, since $\varepsilon/(2N)$ is a positive number and $\lim\limits_{x \to a} g(x) = M$, there exists a positive number δ_2 such that

$$|g(x) - M| < \frac{\varepsilon}{2N} \quad \text{holds whenever} \quad 0 < |x - a| < \delta_2.$$

If both inequalities $|f(x)| < N$ and $|g(x) - M| < \varepsilon/(2N)$ hold, then

$$|f(x)||g(x) - M| < N\left(\frac{\varepsilon}{2N}\right) = \frac{1}{2}\varepsilon$$

also holds; hence,

condition (i) holds if $0 < |x - a| < \delta_1$ and $0 < |x - a| < \delta_2$.

In order to obtain condition (ii), we plan to use the fact that $\lim\limits_{x \to a} f(x) = L$ to show that there exists a positive number δ_3 such that

$$|f(x) - L||M| < \frac{1}{2}\varepsilon \quad \text{holds whenever} \quad 0 < |x - a| < \delta_3.$$

If $M = 0$, the latter condition holds for any choice of δ_3, so we need only concern ourselves with the case in which $M \neq 0$. Then, since $\varepsilon/(2|M|)$ is a positive number and $\lim\limits_{x \to a} f(x) = L$, there exists a positive number δ_3 such that

$$|f(x) - L| < \frac{\varepsilon}{2|M|} \quad \text{holds whenever} \quad 0 < |x - a| < \delta_3.$$

Therefore,

condition (ii) holds if $0 < |x - a| < \delta_3$.

Now, to finish the proof, just let δ be the smallest of the three numbers δ_1, δ_2, and δ_3, so that, if $0 < |x - a| < \delta$, then

$$0 < |x - a| < \delta_1, \quad 0 < |x - a| < \delta_2, \quad 0 < |x - a| < \delta_3.$$

Consequently,

$$\text{conditions (i) and (ii)} \quad \text{hold whenever} \quad 0 < |x - a| < \delta.$$

It follows that

$$|f(x)g(x) - LM| < \varepsilon \quad \text{holds whenever} \quad 0 < |x - a| < \delta,$$

as desired.

Using the properties of limits and the definition of continuity (Section 3, p. 58), it is not difficult to establish the basic properties of continuous functions. The proof of the following theorem illustrates the general technique.

THEOREM 4   **Continuity of a Product**
If the functions f and g are both continuous at the number a, then the product function $f \cdot g$ is also continuous at a.

PROOF
Since f and g are both continuous at a, they are both defined at a; hence, $f \cdot g$ is defined at a and

$$(f \cdot g)(a) = f(a)g(a).$$

Because f and g are both continuous at a, $\lim_{x \to a} f(x)$ and $\lim_{x \to a} g(x)$ both exist and

$$\lim_{x \to a} f(x) = f(a) \quad \text{while} \quad \lim_{x \to a} g(x) = g(a).$$

Therefore, by Theorem 3, $\lim_{x \to a} [f(x) \cdot g(x)]$ exists and we have

$$\lim_{x \to a} (f \cdot g)(x) = \lim_{x \to a} [f(x)g(x)] = \left[\lim_{x \to a} f(x)\right]\left[\lim_{x \to a} g(x)\right]$$

$$= f(a)g(a) = (f \cdot g)(a).$$

It follows that $f \cdot g$ is continuous at a.

Problem Set 7

1 If c is a constant number and f is the constant function defined by $f(x) = c$, prove that $\lim_{x \to a} f(x) = c$ holds for each number a.

2 If f is the identity function, that is, if f is defined by $f(x) = x$, prove that $\lim_{x \to a} f(x) = a$ holds for each number a.

3 Combine problem 1 with Theorem 3 to prove that if c is a constant number and if the limit $\lim_{x \to a} g(x)$ exists, then $\lim_{x \to a} [cg(x)] = c \lim_{x \to a} g(x)$.

4 Prove that, if $\lim_{x \to a} f(x)$ exists, then $\lim_{x \to a} |f(x)| = \left|\lim_{x \to a} f(x)\right|$.

5 Theorem 1 provides a proof of part of Property 1 of limits in Section 2. Complete the proof of Property 1 by showing that, if $\lim_{x \to a} f(x) = L$ and $\lim_{x \to a} g(x) = M$, then $\lim_{x \to a} [f(x) - g(x)] = L - M$.

6 Assume that $\lim\limits_{x \to a} g(x) = M \neq 0$. Prove that there exist positive numbers N and δ_0 such that (i) $g(x) \neq 0$ holds for $0 < |x - a| < \delta_0$, and (ii) $|1/g(x)| < N$ holds for $0 < |x - a| < \delta_0$. [*Hint:* Select δ_0 such that $|g(x) - M| < |M|/2$ holds whenever $0 < |x - a| < \delta_0$. Put $N = 2/|M|$.]

7 Suppose that $\lim\limits_{x \to a} f(x)$, $\lim\limits_{x \to a} g(x)$, and $\lim\limits_{x \to a} h(x)$ all exist. Prove that

$$\lim\limits_{x \to a} [f(x) + g(x) + h(x)] = \lim\limits_{x \to a} f(x) + \lim\limits_{x \to a} g(x) + \lim\limits_{x \to a} h(x).$$

8 Assume that $\lim\limits_{x \to a} g(x) = M \neq 0$. Prove that $\lim\limits_{x \to a} 1/g(x) = 1/M$. [*Hints:* Select N and δ_0 as in problem 6. Notice that $|1/g(x) - 1/M|$ can be written as $\dfrac{1}{|g(x)|} \cdot \dfrac{1}{|M|} \cdot |g(x) - M|$. Given $\varepsilon > 0$, select δ_1 such that $|g(x) - M| < (|M|\varepsilon)/N$ holds whenever $0 < |x - a| < \delta_1$. Choose δ to be the smaller of the two numbers δ_0 and δ_1.]

9 Combine Theorem 3 and problem 8 and thus establish Property 4 of limits of functions in Section 2.

10 Assume that $\lim\limits_{x \to b} f(x) = f(b) = L$ and that $\lim\limits_{x \to a} g(x) = b$. Prove that $\lim\limits_{x \to a} (f \circ g)(x) = L$.

11 Using the result of problem 10, establish Property 3 of continuous functions in Section 4.

12 When we speak of *the* limit as x approaches a of $f(x)$ and use the notation $\lim\limits_{x \to a} f(x) = L$, we tacitly assume that there can be at most one number L such that, for each $\varepsilon > 0$, there exists $\delta > 0$ with the property that $|f(x) - L| < \varepsilon$ holds whenever $0 < |x - a| < \delta$. Prove that this tacit assumption is justified. (*Hint:* Suppose that there were two different values of L, say L_1 and L_2, satisfying the given condition, and let $\varepsilon = \frac{1}{2}|L_1 - L_2|$.)

Review Problem Set

In problems 1 through 6, determine a positive number δ for the given ε so that $|f(x) - L| < \varepsilon$ holds whenever $0 < |x - a| < \delta$. Draw a graph of f to illustrate the limit involved.

1 $f(x) = 2x - 7$, $a = -1$, $L = -9$, $\varepsilon = 0.01$

2 $f(x) = 1 - 5x$, $a = 3$, $L = -14$, $\varepsilon = 0.02$

3 $f(x) = 5x + 1$, $a = -2$, $L = -9$, $\varepsilon = 0.002$

4 $f(x) = \dfrac{4x^2 - 9}{2x - 3}$, $a = \frac{3}{2}$, $L = 6$, $\varepsilon = 0.001$

5 $f(x) = \dfrac{25x^2 - 1}{5x + 1}$, $a = -\frac{1}{5}$, $L = -2$, $\varepsilon = 0.01$

6 $f(x) = 2x - 7$, $a = -1$, $L = -9$, ε an arbitrarily small positive number

In problems 7 through 12, use the properties of limits (Section 2) to evaluate each limit, given that $\lim\limits_{x \to 3} f(x) = 12$ and $\lim\limits_{x \to 3} g(x) = 3$.

7 $\lim\limits_{x \to 3} [f(x) + g(x)]$

8 $\lim\limits_{x \to 3} [f(x) - g(x)]$

9 $\lim\limits_{x \to 3} \sqrt{f(x) \cdot g(x)}$

10 $\lim\limits_{x \to 3} \sqrt{\dfrac{f(x)}{g(x)}}$

11 $\lim\limits_{x \to 3} [f(x) - g(x)]^{3/2}$

12 $\lim\limits_{x \to 3} \dfrac{f(x) + g(x)}{f(x) - g(x)}$

In problems 13 through 24, use the properties of limits to evaluate each limit.

13 $\lim\limits_{t \to 5} (6t^2 + t - 4)$

14 $\lim\limits_{y \to 2} \dfrac{3y + 5}{4y^2 + 5y - 4}$

15 $\lim\limits_{t \to 1} \dfrac{1 - t^3}{1 - t^2}$

16 $\lim\limits_{z \to 5/2} \dfrac{4z^2 - 25}{2z - 5}$

17 $\lim\limits_{h \to 0} \dfrac{1}{h}\left(\dfrac{6 + h}{3 + 2h} - 2\right)$

18 $\lim\limits_{x \to 0} \dfrac{1}{x}\left[1 - \dfrac{1}{(x + 1)^2}\right]$

19 $\lim\limits_{t \to 1} \dfrac{\sqrt{4 - t^2}}{2 + t}$

20 $\lim\limits_{h \to -1} \dfrac{3 - \sqrt{h^2 + h + 9}}{h^3 + 1}$

21 $\lim\limits_{x \to 1} \dfrac{1 - x}{2 - \sqrt{x^2 + 3}}$

22 $\lim\limits_{t \to 0} \dfrac{\sqrt{6 + t} - \sqrt{6}}{t}$

23 $\lim\limits_{x \to 9} \dfrac{\sqrt{x} - 3}{x - 9}$

24 $\lim\limits_{t \to 0} \dfrac{\sqrt[3]{5 + t} - \sqrt[3]{5}}{t}$

In problems 25 through 28, evaluate each one-sided limit.

25 $\lim\limits_{t \to 3^-} \dfrac{t}{t^2 - 9}$

26 $\lim\limits_{y \to 2^+} \dfrac{\sqrt{y - 2}}{y^2 - 4}$

27 $\lim\limits_{x \to 0^+} \dfrac{x^2 - 3}{x^2 - x}$

28 $\lim\limits_{x \to 2^-} (3 + [\![2x - 4]\!])$

In problems 29 through 32, sketch the graph of the given function and find the indicated limit if it exists. If the limit does not exist, give the reason.

29 $f(x) = \begin{cases} 2x - 3 & \text{if } x \geq \frac{3}{2} \\ 6 - 4x & \text{if } x < \frac{3}{2} \end{cases}$; $\lim\limits_{x \to (3/2)^-} f(x)$, $\lim\limits_{x \to (3/2)^+} f(x)$, and $\lim\limits_{x \to 3/2} f(x)$

30 $h(x) = \begin{cases} x^2 + 2 & \text{if } x < 1 \\ 4 - x & \text{if } x \geq 1 \end{cases}$; $\lim\limits_{x \to 1^-} h(x)$, $\lim\limits_{x \to 1^+} h(x)$, and $\lim\limits_{x \to 1} h(x)$

31 $g(x) = \begin{cases} \dfrac{x^2 - 4}{x - 2} & \text{if } x \neq 2 \\ 1 & \text{if } x = 2 \end{cases}$; $\lim\limits_{x \to 2^-} g(x)$, $\lim\limits_{x \to 2^+} g(x)$, and $\lim\limits_{x \to 2} g(x)$

32 $f(x) = \begin{cases} \dfrac{5x - 5}{|x + 2|} & \text{if } x \neq -2 \\ 0 & \text{if } x = -2 \end{cases}$; $\lim\limits_{x \to -2^-} f(x)$, $\lim\limits_{x \to -2^+} f(x)$, and $\lim\limits_{x \to -2} f(x)$

In problems 33 through 38, evaluate the given limit if it exists.

33 $\lim\limits_{x \to +\infty} \dfrac{x^2 + 1}{5x + 3}$

34 $\lim\limits_{t \to +\infty} \dfrac{5t}{t^2 + 1}$

35 $\lim\limits_{y \to -\infty} \dfrac{4y^2 + y - 3}{(2y + 3)(3y + 4)}$

36 $\lim\limits_{h \to -\infty} \dfrac{h^2 - 3h}{\sqrt{5h^4 + 7h^2 + 3}}$

37 $\lim\limits_{t \to +\infty} \dfrac{3t^{-2} + 7t^{-3}}{7t^{-2} + 5t^{-3}}$

38 $\lim\limits_{x \to -\infty} \dfrac{\sqrt{7x^6 + 5x^4 + 7}}{x^2 + 2}$

39 Let $f(x) = 3x - 1$. We know that $\lim\limits_{x \to a} (3x - 1) = 3a - 1$.

(a) How close to a must you choose x so that $f(x)$ is within $\varepsilon > 0$ of $3a - 1$?

(b) If $\varepsilon = 0.01$, how close to a must you choose x? Would x in the interval $a - 0.1 < x < a + 0.1$ be close enough? Explain.

40 Show that if $\lim\limits_{x \to 0} \dfrac{f(x)}{x} = L$ and $b \neq 0$, then $\lim\limits_{x \to 0} \dfrac{f(bx)}{x} = bL$.

In problems 41 through 44, sketch the graph of the given function and indicate if the function is continuous at $x = a$.

41 $f(x) = \begin{cases} \dfrac{x^2 - 9}{x - 3} & \text{if } x \neq 3 \\ 6 & \text{if } x = 3 \end{cases}$; $a = 3$

42 $g(x) = \begin{cases} \dfrac{x^2 - 1}{x - 1} & \text{if } x \neq 1 \\ \frac{1}{2} & \text{if } x = 1 \end{cases}$; $a = 1$

43 $f(x) = \begin{cases} \sqrt{\dfrac{x - 1}{x^2 - 1}} & \text{if } x \neq 1 \\ \frac{1}{2}\sqrt{2} & \text{if } x = 1 \end{cases}$; $a = 1$

44 $h(x) = \begin{cases} \dfrac{2 - x}{2 - |x|} & \text{if } x \neq 2 \\ 1 & \text{if } x = 2 \end{cases}$; $a = 2$

45 Taxi fare is 60 cents plus 10 cents for each quarter mile or portion thereof. If we let $f(x)$ denote the fare for a ride of x miles, sketch the graph of f and indicate where it is discontinuous.

46 Assume that it takes 0.5 calorie of heat to raise the temperature of 1 gram of ice 1 degree Celsius, that it takes 80 calories to melt the ice at 0°C, and that it takes 1 calorie to raise the temperature of 1 gram of water 1 degree Celsius. Suppose that $-40 \leq x \leq 20$ and let $Q(x)$ be the number of calories of heat required to raise 1 gram of water from a temperature of -40°C to x°C. Sketch the graph of Q and indicate where Q is discontinuous.

47 Given a function f defined by the equation

$$f(x) = \begin{cases} -x & \text{if } x \leq 0 \\ x & \text{if } 0 < x \leq 1 \\ 2 - x & \text{if } 1 < x < 2 \\ 0 & \text{if } x \geq 2. \end{cases}$$

(a) Sketch the graph of f and discuss the continuity of f at the numbers 0, 1, and 2.
(b) Determine the constants A, B, C, D, and E so that

$$f(x) = Ax + B + C|x| + D|x - 1| + E|x - 2|.$$

48 Determine the values of the constants A and B so that the function f is continuous in the interval $(-\infty, \infty)$, and sketch the graph of the resulting function.

$$f(x) = \begin{cases} 3x & \text{if } x \leq 2 \\ Ax + B & \text{if } 2 < x < 5 \\ -6x & \text{if } x \geq 5. \end{cases}$$

49 Determine whether each function is continuous or discontinuous on each of the indicated intervals.

(a) $f(x) = \dfrac{3}{2x - 1}$; $[-1, 1]$, $[-\frac{1}{2}, \frac{1}{2}]$, $(-1, \frac{1}{2})$, $[\frac{1}{2}, \infty)$

(b) $g(x) = \begin{cases} 3x - 2 & \text{if } x < 1 \\ 2 - x & \text{if } 1 \leq x \leq 2 \end{cases}$; $(-\infty, 1)$, $(1, 2)$, $[1, 2]$

50 (a) Let f and g be functions defined by $f(x) = \dfrac{1}{x - 1}$ and $g(x) = \sqrt[3]{x}$. Determine all values of x for which each of the functions $f \circ g$ and $g \circ f$ is continuous.
(b) Same directions as part (a), but $g(x) = \sqrt{x}$.

In problems 51 and 52, find all horizontal and vertical asymptotes of the graph of each function and sketch the graph.

51 $f(x) = \dfrac{1}{x(x + 1)} - \dfrac{1}{x}$

52 $g(x) = \dfrac{x + 3}{2x + 1}$

53 Suppose that $\lim_{x \to a} f(x)$, $\lim_{x \to a} g(x)$, and $\lim_{x \to a} h(x)$ all exist. Prove that

$$\lim_{x \to a} [f(x)g(x)h(x)] = \left[\lim_{x \to a} f(x)\right]\left[\lim_{x \to a} g(x)\right]\left[\lim_{x \to a} h(x)\right].$$

54 Suppose that $\lim_{x \to a} f(x) = L \neq 0$. Prove that there exist positive numbers A, B, and δ such that $A < |f(x)| < B$ holds whenever $0 < |x - a| < \delta$.

55 If n is a positive integer and $\lim_{x \to a} f(x) = L$, prove that $\lim_{x \to a} [f(x)]^n = L^n$. (*Hint:* Use mathematical induction on n.)

56 Sketch the graph of the function f defined by $f(x) = [\![1/x]\!]$ for $x > 0$, and indicate where f is discontinuous.

2 THE DERIVATIVE

The limit concept, introduced in Chapter 1, will be used in this chapter to define a mathematical procedure called *differentiation*. A variety of problems that cannot be handled by strictly algebraic techniques—including problems involving the rate of change of a variable quantity—can be solved using this procedure. From a geometric point of view, such problems can be interpreted as questions involving a tangent line to the graph of a function. Useful computational rules for differentiation are also set forth.

1 Rates of Change and Slopes of Tangent Lines

In this section we use the concept of limit to solve two apparently unrelated problems; later, we shall see that they are really the same problem in disguise. The first problem is to find the rate of change of a variable quantity—for instance, the time rate of change of distance (speed). The second problem is to find the slope of the tangent line to the graph of a function at a given point.

1.1 Speed of an Automobile

An automobile is being driven along a straight road from city A to city B, perhaps at a variable rate of speed r. The distance d of the automobile from city A depends on the elapsed time t since the start of the journey (Figure 1). Suppose that the functions f and g give the distance d and the speed r, respectively, in terms of t, so that

$$d = f(t) \quad \text{and} \quad r = g(t).$$

For instance, if the rate of speed r is constant, say $r = 55$ miles per hour, we have the familiar formula

$$\text{distance} = \text{rate} \times \text{time} \quad \text{or} \quad d = rt = 55t;$$

hence, in this case, f and g are given by

$$f(t) = 55t \quad \text{and} \quad g(t) = 55.$$

Figure 1

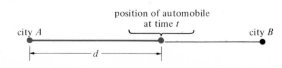

position of automobile
at time t

city A city B

$\longleftarrow\!\!-\!\!- d -\!\!-\!\!\longrightarrow$

In the more general case in which the automobile's speed is variable, the functions f and g are more complicated.

We now find a relationship between the distance function f and the speed function g. Choose and (temporarily) fix a value t of the time variable. Thus, at time t the automobile is $d = f(t)$ miles from city A and the speedometer reads $r = g(t)$ miles per hour. Now, let a short additional interval of time h elapse. At time $t + h$ the automobile is at a distance $f(t + h)$ miles from city A (Figure 2) and its speed is $g(t + h)$ miles per hour. Evidently, the automobile has gone $f(t + h) - f(t)$ miles during the time interval h; hence, its *average speed* (distance divided by time) during the time interval h is $\dfrac{f(t + h) - f(t)}{h}$ miles per hour.

Figure 2

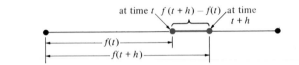

If the time interval h is sufficiently short, the speedometer reading $g(t)$ at the time t will not differ appreciably from the speedometer reading $g(t + h)$ at the slightly later time $t + h$. Furthermore, during this short interval of time, the speedometer readings should be approximately the same as the average rate of speed $\dfrac{f(t + h) - f(t)}{h}$. As the time interval h becomes shorter and shorter, these approximations should become more and more accurate, and the speed at the instant t should be given by

$$g(t) = \lim_{h \to 0} \frac{f(t + h) - f(t)}{h}.$$

This equation expresses the promised relationship between the functions f and g; in fact, it shows that we can calculate or derive the instantaneous speed $g(t)$ from the distance function f by evaluating a suitable limit.

1.2 Instantaneous Rate of Change in General

The above considerations concerning the rate of change of distance with respect to time can be generalized and thus made applicable to any variable quantities whatsoever. Indeed, let x and y denote variable quantities and suppose that y depends on x, so that $y = f(x)$, where f is a suitable function.

To find the rate of change of y per unit change in x, we naturally begin by considering a change in x, say from the value x_1 to the value x_2. Now, let

$$y_1 = f(x_1) \quad \text{and} \quad y_2 = f(x_2),$$

so that, as x changes from x_1 to x_2, y undergoes a corresponding change from y_1 to y_2.

It is traditional to denote the change in x by the symbol Δx (read "delta x"), so that

$$\Delta x = x_2 - x_1.$$

Similarly, the resulting change in y is denoted by the symbol Δy (read "delta y"), so that

$$\Delta y = y_2 - y_1 = f(x_2) - f(x_1).$$

The ratio of the change in y to the change in x that produced it is called the *average rate of change of y per unit change in x* (or *with respect to x*). More formally, we have the following definition.

DEFINITION 1   **Average Rate of Change**
The ratio

$$\frac{\Delta y}{\Delta x} = \frac{y_2 - y_1}{x_2 - x_1} = \frac{f(x_2) - f(x_1)}{x_2 - x_1}$$

is called the *average rate of change of y with respect to x as x changes from x_1 to x_2.*

Since $\Delta x = x_2 - x_1$, then $x_2 = x_1 + \Delta x$, and we can also write

$$\frac{\Delta y}{\Delta x} = \frac{f(x_1 + \Delta x) - f(x_1)}{\Delta x}.$$

If the average rate of change of y with respect to x approaches a limiting value as Δx approaches 0, it seems reasonable to refer to this limiting value as the *instantaneous rate of change of y with respect to x*; hence, we make the following definition.

DEFINITION 2   **Instantaneous Rate of Change**
If $y = f(x)$, we define the *instantaneous rate of change of y with respect to x at the instant when $x = x_1$* to be

$$\lim_{\Delta x \to 0} \frac{\Delta y}{\Delta x} = \lim_{\Delta x \to 0} \frac{f(x_1 + \Delta x) - f(x_1)}{\Delta x}.$$

EXAMPLE   A metal cube with an edge length x is expanding uniformly as a consequence of being heated. Find:
(a) The average rate of change of its volume with respect to edge length as x increases from 2 to 2.01 centimeters.
(b) The instantaneous rate of change of its volume with respect to edge length at the instant when $x = 2$ centimeters.

SOLUTION
Let y denote the volume of the cube, so that $y = x^3$ cubic centimeters.
(a) When $x = 2$ centimeters, $y = 2^3 = 8$ cubic centimeters. If x increases by $\Delta x = 0.01$ centimeter to 2.01 centimeters, y increases to $(2.01)^3$ cubic centimeters. Thus, a change in x of $\Delta x = 0.01$ centimeter produces a corresponding change in y of

$$\Delta y = (2.01)^3 - 8 = 0.120601 \text{ cubic centimeter.}$$

Hence, the average rate of change of y with respect to x over the interval Δx is given by

$$\frac{\Delta y}{\Delta x} = \frac{0.120601}{0.01} = 12.0601 \text{ cubic centimeters per centimeter of edge length.}$$

(b) More generally, if x changes by an amount Δx from 2 to $2 + \Delta x$ centimeters, then y changes by a corresponding amount.

$$\Delta y = (2 + \Delta x)^3 - 2^3 = 8 + 12\,\Delta x + 6(\Delta x)^2 + (\Delta x)^3 - 8$$
$$= 12\,\Delta x + 6(\Delta x)^2 + (\Delta x)^3 \text{ cubic centimeters.}$$

Therefore, the required instantaneous rate of change of y with respect to x is given by

$$\lim_{\Delta x \to 0} \frac{\Delta y}{\Delta x} = \lim_{\Delta x \to 0} \frac{12\,\Delta x + 6(\Delta x)^2 + (\Delta x)^3}{\Delta x} = \lim_{\Delta x \to 0} \left[12 + 6\,\Delta x + (\Delta x)^2\right]$$

$$= 12 \text{ cubic centimeters per centimeter of edge length.}$$

In calculating the rate of change of one variable with respect to another variable upon which it depends, it is not necessary to use the symbols y and x for the two variables. For instance, suppose that a particle P is moving along a straight line and is a distance s from its starting point A at time t (Figure 3). If $s = f(t)$, the speed of the particle at the instant when $t = t_1$ is given by

Figure 3   A $P \longrightarrow$

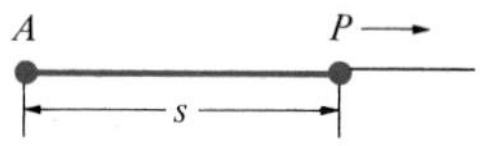

$$\lim_{\Delta t \to 0} \frac{\Delta s}{\Delta t} = \lim_{\Delta t \to 0} \frac{f(t_1 + \Delta t) - f(t_1)}{\Delta t}.$$

EXAMPLE   A particle is moving along a straight line in such a way that, at the end of t seconds, its distance s in meters from the starting point is given by $s = 3t^2 + t$. Find the speed of the particle at the instant when $t = 2$ seconds.

SOLUTION
Here, $s = f(t)$, where $f(t) = 3t^2 + t$. The speed when $t = 2$ is therefore given by

$$\lim_{\Delta t \to 0} \frac{f(2 + \Delta t) - f(2)}{\Delta t} = \lim_{\Delta t \to 0} \frac{[3(2 + \Delta t)^2 + (2 + \Delta t)] - [3(2)^2 + 2]}{\Delta t}$$

$$= \lim_{\Delta t \to 0} \frac{[12 + 12\,\Delta t + 3(\Delta t)^2 + 2 + \Delta t] - 14}{\Delta t}$$

$$= \lim_{\Delta t \to 0} \frac{3(\Delta t)^2 + 13\,\Delta t}{\Delta t}$$

$$= \lim_{\Delta t \to 0} (3\,\Delta t + 13) = 13 \text{ meters per second.}$$

1.3 Slope of the Tangent Line to a Graph at a Point

Suppose that $P = (x_1, y_1)$ is a point on the graph of a function f, so that $y_1 = f(x_1)$, and that we want to find the tangent line to the graph of f at P (Figure 4). Since this tangent line is the straight line that contains the point P and "best approximates" the graph of f near P, it is easy to sketch it roughly "by eye." However, suppose that we need to draw this tangent line accurately. Since a straight line in the plane is completely determined once we know its slope and one point P on it, we only need to find the slope m of the tangent line.

Figure 5 shows a point Q on the graph of f near the point P. A line segment such as $\overline{PQ}$ joining two points of a curve is called a *secant* and the straight line containing P and Q is called a *secant line* to the graph of f. The x coordinate of P is x_1 and, if the x coordinate of Q differs from the x coordinate of P by a small amount Δx, then the x coordinate of Q is $x_1 + \Delta x$.

Since Q lies on the graph of f, it follows that the y coordinate of Q is $f(x_1 + \Delta x)$. Again, the y coordinate of Q differs from the y coordinate of P by a small amount Δy, where

$$\Delta y = f(x_1 + \Delta x) - y_1 = f(x_1 + \Delta x) - f(x_1).$$

Thus,

$$Q = (x_1 + \Delta x, f(x_1 + \Delta x))$$
$$= (x_1 + \Delta x, y_1 + \Delta y).$$

By the slope formula, the slope of the secant $\overline{PQ}$ is

$$\frac{f(x_1 + \Delta x) - y_1}{(x_1 + \Delta x) - x_1} = \frac{f(x_1 + \Delta x) - f(x_1)}{\Delta x} = \frac{\Delta y}{\Delta x}.$$

Hence, the secant line also has slope $\dfrac{\Delta y}{\Delta x}$.

Now, if we allow Δx to approach 0, the point Q will move along the curve $y = f(x)$ and approach the point P; furthermore, the secant line will pivot about the point P and approach the tangent line. Thus, as Δx approaches 0, the slope $\Delta y / \Delta x$ of the secant line approaches the slope m of the tangent line; that is,

$$m = \lim_{\Delta x \to 0} \frac{\Delta y}{\Delta x} = \lim_{\Delta x \to 0} \frac{f(x_1 + \Delta x) - f(x_1)}{\Delta x}.$$

The preceding considerations lead us to the following formal definition of a tangent line to the graph of a function.

DEFINITION 3 **Tangent Line to a Graph**

Let f be a function defined at least in some open interval containing the number x_1 and let $y_1 = f(x_1)$. If the limit

$$m = \lim_{\Delta x \to 0} \frac{f(x_1 + \Delta x) - f(x_1)}{\Delta x}$$

exists, we say that the straight line in the xy plane containing the point (x_1, y_1) and having slope m is the *tangent line to the graph of f at (x_1, y_1)*.

EXAMPLES Find the slope m of the tangent line to the graph of the given function f at the indicated point P. Sketch a graph of f showing the tangent line at P.

1 $f(x) = x^2, P = (1, 1)$

Figure 4

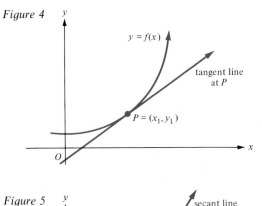

Figure 5

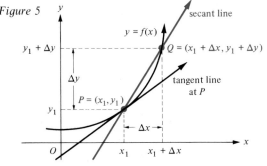

Figure 6

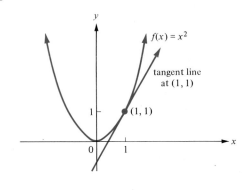

SOLUTION

$$m = \lim_{\Delta x \to 0} \frac{f(1 + \Delta x) - f(1)}{\Delta x}$$

$$= \lim_{\Delta x \to 0} \frac{(1 + \Delta x)^2 - 1^2}{\Delta x}$$

$$= \lim_{\Delta x \to 0} \frac{1 + 2\,\Delta x + (\Delta x)^2 - 1}{\Delta x}$$

$$= \lim_{\Delta x \to 0} (2 + \Delta x) = 2 \quad \text{(Figure 6)}.$$

2 $f(x) = \sqrt{x - 3}, \; P = (7, 2)$

SOLUTION

$$m = \lim_{\Delta x \to 0} \frac{f(7 + \Delta x) - f(7)}{\Delta x}$$

$$= \lim_{\Delta x \to 0} \frac{\sqrt{(7 + \Delta x) - 3} - \sqrt{7 - 3}}{\Delta x}$$

$$= \lim_{\Delta x \to 0} \frac{\sqrt{4 + \Delta x} - 2}{\Delta x}$$

$$= \lim_{\Delta x \to 0} \frac{(\sqrt{4 + \Delta x} - 2)(\sqrt{4 + \Delta x} + 2)}{\Delta x(\sqrt{4 + \Delta x} + 2)}$$

$$= \lim_{\Delta x \to 0} \frac{4 + \Delta x - 4}{\Delta x(\sqrt{4 + \Delta x} + 2)}$$

$$= \lim_{\Delta x \to 0} \frac{1}{\sqrt{4 + \Delta x} + 2} = \frac{1}{4} \quad \text{(Figure 7)}.$$

Figure 7

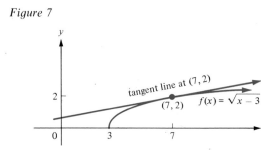

Naturally, if the function f is continuous at x_1 and

$$\lim_{\Delta x \to 0} \left| \frac{f(x_1 + \Delta x) - f(x_1)}{\Delta x} \right| = +\infty,$$

we say that *the vertical straight line $x = x_1$ is the tangent line to the graph of f at the point $(x_1, f(x_1))$.* For an example of a graph with a vertical tangent line, see Problem 18.

Problem Set 1

1 At a certain instant the speedometer of an automobile reads r miles per hour. During the next $\frac{1}{4}$ second the automobile travels 20 feet. Estimate r from this information.

2 Explain why the answer in problem 1 is only an estimate for r and need not be exactly the same as r.

In problems 3 and 4, assume that $y = f(x)$ as given. (a) Find the average rate of change of y with respect to x as x changes from x_1 to x_2. (b) Find the instantaneous rate of change of y with respect to x at the instant when $x = x_1$.

3 $y = f(x) = x^2 + x + 1$, $x_1 = 3$, $x_2 = 3.5$ $\qquad$ **4** $y = f(x) = \dfrac{4}{x}$, $x_1 = 5$, $x_2 = 6$

In problems 5 through 8, a particle is moving along a straight line according to the equation given, where s is the distance in feet of the particle from its starting point at the end of t seconds. Find (a) the average speed $\Delta s / \Delta t$ of the particle during the interval of time from $t = t_1$ to $t = t_2$ and (b) the instantaneous speed of the particle when $t = t_1$.

5 $s = 6t^2$, $t_1 = 2$, $t_2 = 3$ $\qquad\qquad$ **6** $s = 7t^3$, $t_1 = 1$, $t_2 = 2$

7 $s = t^2 + t$, $t_1 = 3$, $t_2 = 4$ $\qquad\qquad$ **8** $s = \sqrt{t}$, $t_1 = 9$, $t_2 = 16$

In problems 9 through 18, find the slope m of the tangent line to the graph of each function at the indicated point, sketch the graph, and show the tangent line at that point.

9 $f(x) = 2x - x^2$ at $(1, 1)$ $\qquad$ **10** $f(x) = (x - 2)^2$ at $(-2, 16)$ $\qquad$ **11** $f(x) = x^2 - 4x$ at $(3, -3)$

12 $f(x) = x^3$ at $(-1, -1)$ $\qquad$ **13** $f(x) = 3 + 2x - x^2$ at $(0, 3)$ $\qquad$ **14** $f(x) = \dfrac{1}{\sqrt{x}}$ at $(4, \frac{1}{2})$

15 $f(x) = \sqrt{x + 1}$ at $(3, 2)$ $\qquad$ **16** $f(x) = \sqrt{9 - 4x}$ at $(-4, 5)$

17 $f(x) = \dfrac{3}{x + 2}$ at $(1, 1)$ $\qquad$ **18** $f(x) = \sqrt[3]{x}$ at $(0, 0)$

19 An object falls from rest according to the equation $s = 16t^2$, where s is the number of feet through which the object falls during the first t seconds after being released. Find:
(a) The average speed during the first 5 seconds of fall.
(b) The instantaneous speed at the end of this 5-second interval.

20 A projectile is fired vertically upward and is s feet above the ground t seconds after being fired, where $s = 256t - 16t^2$. Find:
(a) The speed of the projectile 4 seconds after being fired.
(b) The time in seconds required for the projectile to reach its maximum height (at which point its speed is 0 feet per second).
(c) The maximum height to which the projectile ascends.

21 An equilateral triangle made of sheet metal is expanding because it is being heated. Its area A is given by $A = (\sqrt{3}/4)x^2$ square centimeters, where x is the length of one side in centimeters. Find the instantaneous rate of change of A with respect to x at the instant when $x = 10$ centimeters.

22 A spherical balloon of radius R inches has volume $V = \frac{4}{3}\pi R^3$ cubic inches. Find the instantaneous rate of change of V with respect to R at the moment when $R = 5$ inches.

23 The pressure P of a gas depends upon its volume V according to Boyle's law, $P = C/V$, where C is a constant. Suppose that $C = 2000$, that P is measured in pounds per square inch, and that V is measured in cubic inches. Find:
(a) The average rate of change of P with respect to V as V increases from 100 cubic inches to 125 cubic inches.
(b) The instantaneous rate of change of P with respect to V at the instant when $V = 100$ cubic inches.

24 Find the equation of the tangent line to the graph of the function $f(x) = 2x^2 - 5x + 1$ at the point $(2, -1)$.

25 Sketch the graph of the function f defined by $f(x) = \sqrt{x}$. Compute the value of the ratio $\dfrac{f(0 + h) - f(0)}{h}$ for positive numbers h. Identify these ratios as slopes of certain secant lines. What happens to these slopes as h approaches 0?

2   The Derivative of a Function

In Section 1.2 we found that if x and y are two variables related by an equation $y = f(x)$, then the instantaneous rate of change of y with respect to x when x has the value x_1 is given by

$$\lim_{\Delta x \to 0} \frac{\Delta y}{\Delta x} = \lim_{\Delta x \to 0} \frac{f(x_1 + \Delta x) - f(x_1)}{\Delta x}.$$

On the other hand, we found in Section 1.3 that the slope m of the tangent line to the graph of f at the point $(x_1, f(x_1))$ is also given by $m = \lim_{\Delta x \to 0} \frac{\Delta y}{\Delta x}$. Therefore, the problem of finding the rate of change of one variable with respect to another and the problem of finding the slope of the tangent line to a graph are both solved by calculating the same limit.

Limits of the form $\lim_{\Delta x \to 0} \Delta y / \Delta x$ arise so often in calculus that it is useful to introduce some special notation and terminology for them. A quotient of the form

$$\frac{\Delta y}{\Delta x} = \frac{f(x + \Delta x) - f(x)}{\Delta x}$$

is called a *difference quotient*. The limit as Δx approaches 0 of such a difference quotient defines a new function f', read "f *prime*," by the equation

$$f'(x) = \lim_{\Delta x \to 0} \frac{f(x + \Delta x) - f(x)}{\Delta x}.$$

Since the function f' is derived from the original function f, it is called the "*derivative*" of f. Thus, we have the following definition.

DEFINITION 1   **The Derivative**
Given a function f, the function f' defined by

$$f'(x) = \lim_{\Delta x \to 0} \frac{f(x + \Delta x) - f(x)}{\Delta x} = \lim_{\Delta x \to 0} \frac{\Delta y}{\Delta x}$$

is called the *derivative* of f.

In the definition it is understood that the domain of the derivative function f' is the set of all numbers x in the domain of f for which the limit of the difference quotient exists. In calculating this limit, one must be careful to treat x as a constant while letting Δx approach zero.

EXAMPLES   Find $f'(x)$ for the given function by direct use of Definition 1.

1 $f(x) = x^3$

SOLUTION

$$f'(x) = \lim_{\Delta x \to 0} \frac{f(x + \Delta x) - f(x)}{\Delta x} = \lim_{\Delta x \to 0} \frac{(x + \Delta x)^3 - x^3}{\Delta x}$$

$$= \lim_{\Delta x \to 0} \frac{x^3 + 3x^2 \Delta x + 3x(\Delta x)^2 + (\Delta x)^3 - x^3}{\Delta x}$$

$$= \lim_{\Delta x \to 0} [3x^2 + 3x \Delta x + (\Delta x)^2] = 3x^2.$$

2 $f(x) = \dfrac{1}{3x-2}$

SOLUTION

$$f'(x) = \lim_{\Delta x \to 0} \frac{f(x+\Delta x) - f(x)}{\Delta x} = \lim_{\Delta x \to 0} \frac{\dfrac{1}{3(x+\Delta x)-2} - \dfrac{1}{3x-2}}{\Delta x}$$

$$= \lim_{\Delta x \to 0} \frac{(3x-2) - [3(x+\Delta x)-2]}{[3(x+\Delta x)-2](3x-2)\,\Delta x}$$

$$= \lim_{\Delta x \to 0} \frac{-3}{[3(x+\Delta x)-2](3x-2)} = \frac{-3}{(3x-2)^2}.$$

2.1 The Derivative Notations

The derivative was invented independently by Isaac Newton and Gottfried Leibniz in the seventeenth century. Newton used the notation $\dot{s}$ to denote the time rate of change $\displaystyle\lim_{\Delta t \to 0} \frac{\Delta s}{\Delta t}$ of a variable quantity s, where $s = f(t)$. Thus, Newton wrote $\dot{s}$ for what we write as $f'(t)$, the value of the derivative f' at the time t. Newton's notation is still used in many physics textbooks.

Leibniz, on the other hand, realizing that the numerical value of a derivative is the limit of $\dfrac{\Delta y}{\Delta x}$, wrote this limit as $\dfrac{dy}{dx}$; that is,

$$\frac{dy}{dx} = \lim_{\Delta x \to 0} \frac{\Delta y}{\Delta x} = f'(x).$$

Henceforth, we make extensive use of the Leibniz notation; however, until the "differentials" dy and dx are given separate meanings (in Section 1 of Chapter 5), we do not regard $\dfrac{dy}{dx}$ as a fraction, only as a convenient symbol for the value of a derivative.

The notation f' for the derivative of the function f, which was introduced by Joseph Lagrange in the eighteenth century, is the preferred notation whenever precision and absolute clarity are demanded. Indeed, using the notation f', one can easily distinguish between the derivative f' (which is a function) and the numerical value $f'(x)$ of the derivative function at the number x.

The operation of finding the derivative f' of a function f [or of finding the value $f'(x)$] is called *differentiation*. Thus, the incomplete symbol $\dfrac{d}{dx}$ could be regarded as an instruction to differentiate whatever follows. For instance, the result of Example 1 above can be written in the Leibniz notation as $\dfrac{d}{dx}x^3 = 3x^2$.

Popular alternative notation for the symbol $\dfrac{d}{dx}$ is the simpler symbol D_x (or sometimes just D if the independent variable is understood), which is called the *differentiation operator*. Thus, if $y = x^3$, then $D_x y = D_x x^3 = 3x^2$.

EXAMPLE   Rewrite the result of Example 2 on page 93 in the Leibniz and operator notations.

SOLUTION

$$\frac{d}{dx}\left(\frac{1}{3x-2}\right) = D_x\left(\frac{1}{3x-2}\right) = \frac{-3}{(3x-2)^2}.$$

Preference for one notation over another is often just a matter of taste and convenience. In the remainder of this book, we use whichever symbolism seems appropriate to the problem at hand. Also, as the following example shows, we can use symbols other than y and x to denote the dependent and independent variables.

EXAMPLE   If $s = \frac{1}{2}gt^2$, where g is a constant, find $\dfrac{ds}{dt}$.

SOLUTION

$$\frac{ds}{dt} = \lim_{\Delta t \to 0} \frac{\frac{1}{2}g(t+\Delta t)^2 - \frac{1}{2}gt^2}{\Delta t} = \lim_{\Delta t \to 0} \left(gt + \tfrac{1}{2}g\,\Delta t\right) = gt.$$

Up to this point, we have taken care to distinguish between the derivative f' of a function f and the value $f'(x)$ of this derivative function at the number x. In practice, however, it is customary to use the word "derivative" to refer both to the derived function f' and to the value $f'(x)$ of this function at the number x. Henceforth, we follow this common practice, since one can usually tell what is intended from the context.

In summary, if $y = f(x)$, then the instantaneous rate of change of y with respect to x, or, what is the same thing, the slope of the tangent line to the graph of f at the point (x, y), is given by

$$\frac{dy}{dx} = D_x y = f'(x) = \lim_{\Delta x \to 0} \frac{f(x+\Delta x) - f(x)}{\Delta x} = \lim_{\Delta x \to 0} \frac{\Delta y}{\Delta x},$$

provided that the limit exists.

2.2 Differentiability and Continuity

Consider the function f defined by the equation

$$f(x) = \begin{cases} 5 - 2x & \text{if } x < 3 \\ 4x - 13 & \text{if } x \geq 3 \end{cases} \quad \text{(Figure 1)}.$$

Since $\lim_{x \to 3} f(x) = -1 = f(3)$, it follows that f is continuous at the number 3. However, if we form the difference quotient

$$\frac{f(3+\Delta x) - f(3)}{\Delta x} = \frac{f(3+\Delta x) + 1}{\Delta x}$$

and calculate its limits as Δx approaches zero both from the right and from the left, we obtain

$$\lim_{\Delta x \to 0^+} \frac{f(3+\Delta x) - f(3)}{\Delta x} = \lim_{\Delta x \to 0^+} \frac{[4(3+\Delta x) - 13] + 1}{\Delta x} = \lim_{\Delta x \to 0^+} \frac{4\,\Delta x}{\Delta x} = 4,$$

while

$$\lim_{\Delta x \to 0^-} \frac{f(3+\Delta x) - f(3)}{\Delta x} = \lim_{\Delta x \to 0^-} \frac{[5 - 2(3+\Delta x)] + 1}{\Delta x} = \lim_{\Delta x \to 0^-} \frac{-2\,\Delta x}{\Delta x} = -2.$$

Since the right and left limits of the difference quotient are not equal, it follows that the limit of the difference quotient cannot exist; that is, the derivative $f'(3)$ cannot exist. The nonexistence of the derivative of f at 3 might have been anticipated from the graph in Figure 1, since this graph *has no tangent line* at $(3, -1)$.

In general, we define the *derivative from the right* of a function f by

$$f'_{+}(x) = \lim_{\Delta x \to 0^{+}} \frac{f(x + \Delta x) - f(x)}{\Delta x}.$$

Figure 1

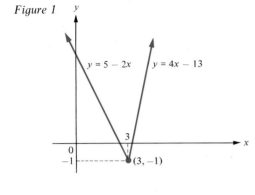

Similarly, the *derivative from the left* of f is defined by

$$f'_{-}(x) = \lim_{\Delta x \to 0^{-}} \frac{f(x + \Delta x) - f(x)}{\Delta x}.$$

Thus, for the function graphed in Figure 1, $f'_{+}(3) = 4$ and $f'_{-}(3) = -2$; hence, $f'(3)$ cannot exist. More generally, the derivative $f'(x)$ exists and has the value A if and only if both of the one-sided derivatives $f'_{+}(x)$ and $f'_{-}(x)$ exist and have the common value A.

EXAMPLE   Let the function f be defined by

$$f(x) = \begin{cases} x^2 & \text{if } x < 1 \\ 2x - 1 & \text{if } x \geq 1 \end{cases}$$

Figure 2

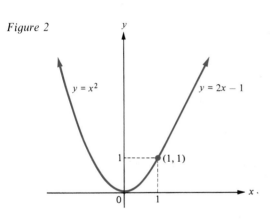

(Figure 2). Find the one-sided derivatives $f'_{+}(1)$ and $f'_{-}(1)$ and determine $f'(1)$, if it exists.

SOLUTION
Here,

$$f'_{+}(1) = \lim_{\Delta x \to 0^{+}} \frac{f(1 + \Delta x) - f(1)}{\Delta x}$$

$$= \lim_{\Delta x \to 0^{+}} \frac{[2(1 + \Delta x) - 1] - 1}{\Delta x}$$

$$= \lim_{\Delta x \to 0^{+}} \frac{2 \Delta x}{\Delta x} = 2.$$

Also,

$$f'_{-}(1) = \lim_{\Delta x \to 0^{-}} \frac{f(1 + \Delta x) - f(1)}{\Delta x} = \lim_{\Delta x \to 0^{-}} \frac{(1 + \Delta x)^2 - 1}{\Delta x} = \lim_{\Delta x \to 0^{-}} (2 + \Delta x) = 2.$$

Since $f'_{+}(1) = f'_{-}(1) = 2$, we conclude that $f'(1)$ exists and equals 2. This example shows that a function defined "piecewise" can have a derivative at the boundary number between the "pieces."

DEFINITION 2   **Differentiable Function**
A function f is said to be *differentiable* at the number x if f is defined at least on some open interval containing x and $f'(x)$ exists and is finite.

Evidently, f is differentiable at x if and only if both of the one-sided derivatives $f'_+(x)$ and $f'_-(x)$ exist and have the same finite value. A function f is said to be *differentiable on the open interval* (a,b) if it is differentiable at each number in this interval. If a function is differentiable at each number in its domain, it is called a *differentiable function.*

Geometrically, to say that a function f is differentiable at a number x is to say that the graph of f has a tangent line with slope $f'(x)$ at the point $(x, f(x))$. Obviously, if a graph has a tangent line at a point, it cannot have a discontinuity at that point. The following theorem confirms this analytically.

THEOREM 1   **Continuity of a Differentiable Function**
If a function f is differentiable at the number x, then it is continuous at x.

PROOF
Assume that f is differentiable at x. We show that f is continuous at x by showing that $\lim_{\Delta x \to 0} f(x + \Delta x) = f(x)$. Since the limit of a product is the product of the limits, we have

$$\lim_{\Delta x \to 0} [f(x + \Delta x) - f(x)] = \lim_{\Delta x \to 0} \left[\frac{f(x + \Delta x) - f(x)}{\Delta x} \Delta x \right]$$

$$= \left[\lim_{\Delta x \to 0} \frac{f(x + \Delta x) - f(x)}{x} \right] \left[\lim_{\Delta x \to 0} \Delta x \right]$$

$$= f'(x) \cdot 0 = 0.$$

Therefore, since the limit of a sum is the sum of the limits, we have

$$\lim_{\Delta x \to 0} f(x + \Delta x) = \lim_{\Delta x \to 0} [f(x + \Delta x) - f(x) + f(x)]$$

$$= \lim_{\Delta x \to 0} [f(x + \Delta x) - f(x)] + \lim_{\Delta x \to 0} f(x) = 0 + f(x) = f(x).$$

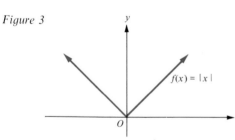

Figure 3

Although, as Theorem 1 shows, a differentiable function is automatically continuous, there are continuous functions that are not differentiable. The simplest example is the function f defined by $f(x) = |x|$ (Figure 3). Note that f is continuous at the number 0, but it is not differentiable at 0 since $f'_+(0) = 1$ and $f'_-(0) = -1$. A similar example is provided by the function graphed in Figure 1.

Problem Set 2

In problems 1 through 6, find $f'(x)$ directly from the definition of derivative.

1 $f(x) = x^2 + 4x$ **2** $f(x) = 2x^3 - 1$ **3** $f(x) = 2x^3 - 4x$

4 $f(x) = \dfrac{x^3}{2} + \dfrac{3}{2} x$ **5** $f(x) = \dfrac{2}{x}$ **6** $f(x) = \dfrac{-7}{x - 3}$

In problems 7 through 12, find the required derivative directly from the definition of derivative.

7 $s = \dfrac{3}{t - 1}$, $\dfrac{ds}{dt} = ?$

8 $s = \dfrac{t}{t + 1}$, $D_t s = ?$

9 $f(v) = \sqrt{v - 1}$, $f'(v) = ?$

10 $\dfrac{d}{du}\left(\sqrt{1 - 9u^2}\right) = ?$

11 $y = \dfrac{2}{x + 1}$, $D_x y = ?$

12 $h(t) = \dfrac{1}{\sqrt{t + 1}}$, $h'(t) = ?$

In problems 13 through 16, find $f'(x_1)$ for the given value of x_1 by direct calculation of
$$\lim_{\Delta x \to 0} \frac{f(x_1 + \Delta x) - f(x_1)}{\Delta x}.$$

13 $f(x) = 1 - 2x^2$; $x_1 = -1$

14 $f(x) = 7x^3$; $x_1 = -2$

15 $f(x) = \dfrac{7}{2x - 1}$; $x_1 = 3$

16 $f(x) = \dfrac{8}{7x - 1}$; $x_1 = -5$

In problems 17 through 19, find the value of the derivative for each function at the number indicated.

17 $f'(4)$ if $f(x) = \dfrac{1}{x - 1}$

18 $D_t s$ at $t = 3$ if $s = \sqrt{2t + 3}$

19 $\dfrac{dy}{dx}$ at $x = 2$ if $y = \dfrac{2}{2x + 1}$

20 Find $D_I P$ if $P = I^2 R$ and R is a constant.

21 Rewrite your answers to the odd-numbered problems 1 through 11 in both Leibniz and operator notation.

22 Given that $y = f(x)$, write the value of the derivative $f'(x)$ in as many different ways as possible.

In problems 23 and 24, find the indicated derivative for each function.

23 $D_t s$ if $s = 16t^2 + 30t + 10$

24 $\dfrac{du}{dv}$ if $u = 16v^2 + 30v + 10$

In problems 25 through 32, (a) sketch the graph of f, (b) determine whether f is continuous at the number x_1, and (c) determine whether f is differentiable at x_1 by finding $f'_+(x_1)$ and $f'_-(x_1)$.

25 $f(x) = x^2 - 2x$; $x_1 = 3$

26 $f(x) = \begin{cases} 2x + 9 & \text{if } x \le -1 \\ 5 - 2x & \text{if } x > -1 \end{cases}$; $x_1 = -1$

27 $f(x) = \begin{cases} 3x - 2 & \text{if } x \le 3 \\ 10 - x & \text{if } x > 3 \end{cases}$; $x_1 = 3$

28 $f(x) = \begin{cases} \sqrt{4 - x} & \text{if } x < 4 \\ (4 - x)^2 & \text{if } x \ge 4 \end{cases}$; $x_1 = 4$

29 $f(x) = \begin{cases} x^2 & \text{if } x \le 2 \\ 6 - x & \text{if } x > 2 \end{cases}$; $x_1 = 2$

30 $f(x) = \begin{cases} |x + 2| & \text{if } x \ge -2 \\ 0 & \text{if } x < -2 \end{cases}$; $x_1 = -2$

31 $f(x) = \begin{cases} (x - 1)^3 & \text{if } x > 0 \\ \frac{3}{2}x^2 + 3x - 1 & \text{if } x \le 0 \end{cases}$; $x_1 = 0$

32 $f(x) = 1 - |x - 3|$; $x_1 = 3$

33 (a) Explain, in your own words, why it is geometrically reasonable that a differentiable function must be continuous.
   (b) It is geometrically reasonable to believe that every continuous function is differentiable? Why or why not?

34 Find values of the constants a and b so that $f'(-1)$ exists, where

$$f(x) = \begin{cases} x^2 & \text{if } x < -1 \\ ax + b & \text{if } x \geq -1. \end{cases}$$

35 Sketch the graphs of each of the following functions and indicate where the functions are not differentiable.
   (a) $f(x) = |3x - 1|$ (b) $f(x) = |x^3 - 1|$

3   Basic Rules for Differentiation

In Section 2 we differentiated functions such as $f(x) = 1/(3x - 2)$ by direct use of the definition of a derivative (as a limit of a difference quotient). The direct calculation of derivatives in this way can be tedious, even for the relatively simple functions that we have considered. Relief from this tedium is forthcoming—there are general rules for differentiation that permit straightforward calculation of such derivatives. In this section we present rules for differentiating sums, products, and quotients of functions whose derivatives are already known. At first, we simply state these rules informally and illustrate their application in examples. Later, in Section 3.1, we state them precisely and give rigorous proofs.

RULE 1   **Constant Rule**
The derivative of a constant function is the zero function. In symbols, if c is a constant, then

$$D_x c = 0 \quad \text{or} \quad \frac{dc}{dx} = 0.$$

In using the constant rule, it is often convenient to deliberately confuse a constant number such as 7 and the constant function f which it determines according to the equation $f(x) = 7$. Thus, we often write "$D_x(7) = 0$" or "$\frac{d}{dx}(7) = 0$" instead of writing "f' is the function defined by the equation $f'(x) = 0$." Since *one can only differentiate functions—never numbers*—an expression of the form $D_x(7) = 0$ could only be interpreted sensibly as above. Thus, we have abbreviated the constant rule as $D_x c = 0$ or $\frac{dc}{dx} = 0$.

EXAMPLES 1 Let f be the constant function defined by the equation $f(x) = 5 + \pi$. Find f'.

SOLUTION
By the constant rule, f' is the constant function defined by the equation $f'(x) = 0$.

2 Find $D_t(5 + \sqrt{3})$.

SOLUTION
By the constant rule, $D_t(5 + \sqrt{3}) = 0$.

RULE 2   **Identity Rule**
The derivative of the identity function is the constant function 1. In symbols,

$$D_x x = 1 \quad \text{or} \quad \frac{dx}{dx} = 1.$$

EXAMPLE If f is the function defined by $f(x) = x$, find f'.

SOLUTION
By the identity rule, f' is the constant function defined by $f'(x) = 1$.

RULE 3 **Power Rule**
The derivative of a positive integral power of x is the exponent of x times x raised to the next lower power. In symbols, if n is a fixed positive integer, then

$$D_x x^n = nx^{n-1} \quad \text{or} \quad \frac{d}{dx} x^n = nx^{n-1}.$$

EXAMPLES 1 Differentiate the function $f(x) = x^7$.

SOLUTION
By the power rule, $f'(x) = 7x^{7-1} = 7x^6$.

2 If $u = t^{13}$, find $\dfrac{du}{dt}$.

SOLUTION
By the power rule, $\dfrac{du}{dt} = \dfrac{d}{dt} t^{13} = 13t^{12}$.

RULE 4 **Homogeneous Rule**
The derivative of a constant times a function is the constant times the derivative of the function. In symbols, if c is a constant and u is a differentiable function of x, then

$$D_x(cu) = cD_x u \quad \text{or} \quad \frac{d}{dx}(cu) = c\frac{du}{dx}.$$

EXAMPLES 1 Differentiate the function $f(x) = 5x^4$.

SOLUTION
Using both the homogeneous rule and the power rule, we have

$$f'(x) = D_x(5x^4) = 5D_x x^4 = 5(4x^3) = 20x^3.$$

2 Find $D_x(\tfrac{2}{3}x^7)$.

SOLUTION
By the homogeneous and power rules,

$$D_x\left(\frac{2}{3} x^7\right) = \frac{2}{3} D_x(x^7) = \frac{2}{3}(7x^6) = \frac{14}{3} x^6.$$

3 If c is a constant and n is a positive integer, find $\dfrac{d}{dx}(cx^n)$.

SOLUTION

$$\frac{d}{dx}(cx^n) = c\frac{d}{dx} x^n = c(nx^{n-1}) = ncx^{n-1}.$$

One important consequence of the homogeneous rule is that $D_x(-u) = -D_x u$. This follows from putting $c = -1$ in Rule 4.

Many functions encountered in practice are (or can be rewritten as) sums of simpler functions. For instance, the polynomial function $f(x) = 2x^2 + 5x - 1$ is a sum of $2x^2$, $5x$, and -1. Thus, one of the most useful differentiation rules is the following.

RULE 5   **Sum Rule**

The derivative of a sum is the sum of the derivatives. In symbols, if u and v are differentiable functions of x, then

$$D_x(u + v) = D_x u + D_x v \quad \text{or} \quad \frac{d}{dx}(u + v) = \frac{du}{dx} + \frac{dv}{dx}.$$

EXAMPLE   Find $D_x(3x^5 + 11x^8)$.

SOLUTION
We have

$$\begin{aligned}
D_x(3x^5 + 11x^8) &= D_x(3x^5) + D_x(11x^8) &&\text{(sum rule)}\\
&= 3D_x x^5 + 11D_x x^8 &&\text{(homogeneous rule)}\\
&= 3(5x^4) + 11(8x^7) &&\text{(power rule)}\\
&= 15x^4 + 88x^7.
\end{aligned}$$

If u, v, and w are three differentiable functions of x, then, by the sum rule,

$$\begin{aligned}
D_x(u + v + w) &= D_x[u + (v + w)]\\
&= D_x u + D_x(v + w)\\
&= D_x u + D_x v + D_x w.
\end{aligned}$$

More generally, *the derivative of the sum of any finite number of differentiable functions is the sum of their derivatives.* This rule is also called the *sum rule*.

Using the sum, homogeneous, power, identity, and constant rules, we can differentiate any polynomial function term by term. This is illustrated by the following examples.

EXAMPLES 1 Given $f(x) = 3x^{100} - 24x^3 + 7x^2 - x - 2$, find f'.

SOLUTION

$$\begin{aligned}
f'(x) &= D_x(3x^{100} - 24x^3 + 7x^2 - x - 2)\\
&= D_x(3x^{100}) + D_x(-24x^3) + D_x(7x^2) + D_x(-x) + D_x(-2)\\
&= 300x^{99} - 72x^2 + 14x - 1.
\end{aligned}$$

2 Find $\dfrac{dy}{du}$ if $y = \sqrt{2}\,u^5 - u^4/4 + 5u^3 + u + \pi^2$.

SOLUTION

$$\frac{dy}{du} = 5\sqrt{2}\,u^4 - \tfrac{1}{4}(4u^3) + 15u^2 + 1 + 0 = 5\sqrt{2}\,u^4 - u^3 + 15u^2 + 1.$$

The rule for differentiating the product of two functions is more complicated than the rule for differentiating their sum. Early in the development of calculus,

Leibniz found that, in general, *the derivative of a product is not the product of the derivatives* (see Problem 51). Leibniz did manage to find the correct rule, which is the following.

RULE 6 **Multiplication Rule, Product Rule, or Leibniz Rule**

The derivative of the product of two functions is the first function times the derivative of the second function plus the derivative of the first function times the second function. In symbols, if u and v are differentiable functions of x, then

$$D_x(uv) = u(D_x v) + (D_x u)v \quad \text{or} \quad \frac{d}{dx}(uv) = u\frac{dv}{dx} + \frac{du}{dx}v.$$

EXAMPLES 1 Find $D_x[(3x^2 + 1)(7x^3 + x)]$ by using the multiplication rule.

SOLUTION

$$\begin{aligned}
D_x[(3x^2 + 1)(7x^3 + x)] &= (3x^2 + 1)[D_x(7x^3 + x)] + [D_x(3x^2 + 1)](7x^3 + x) \\
&= (3x^2 + 1)(21x^2 + 1) + (6x)(7x^3 + x) \\
&= (63x^4 + 24x^2 + 1) + (42x^4 + 6x^2) \\
&= 105x^4 + 30x^2 + 1.
\end{aligned}$$

2 Suppose that f and g are differentiable functions at the number 2 and that $f(2) = 1$, $g(2) = 10$, $f'(2) = \frac{1}{2}$, and $g'(2) = 3$. If $h = f \cdot g$, find $h'(2)$.

SOLUTION
By the multiplication rule,

$$h'(x) = f(x) \cdot g'(x) + f'(x) \cdot g(x),$$

so that

$$h'(2) = f(2) \cdot g'(2) + f'(2) \cdot g(2)$$

or

$$h'(2) = (1)(3) + (\tfrac{1}{2})(10) = 8.$$

RULE 7 **Reciprocal Rule**

The derivative of the reciprocal of a function is the negative of the ratio of the derivative of the function to the square of the function. In symbols, if v is a differentiable function of x, then

$$D_x\!\left(\frac{1}{v}\right) = -\frac{D_x v}{v^2} \quad \text{or} \quad \frac{d}{dx}\left(\frac{1}{v}\right) = -\frac{dv/dx}{v^2}.$$

EXAMPLE Find $D_x(1/x)$.

SOLUTION
By the reciprocal rule,

$$D_x\!\left(\frac{1}{x}\right) = -\frac{D_x x}{x^2} = -\frac{1}{x^2}.$$

RULE 8 **Quotient Rule**

The derivative of a quotient of two functions is the denominator times the derivative of the numerator minus the numerator times the derivative of the denominator all divided by the square of the denominator. In symbols, if u and v are differentiable functions of x, then

$$D_x\left(\frac{u}{v}\right) = \frac{vD_x u - uD_x v}{v^2} \quad \text{or} \quad \frac{d}{dx}\left(\frac{u}{v}\right) = \frac{v\dfrac{du}{dx} - u\dfrac{dv}{dx}}{v^2}.$$

EXAMPLES 1 Find $D_x\left(\dfrac{x^2}{x^3 + 7}\right)$.

SOLUTION
By the quotient rule,

$$D_x\left(\frac{x^2}{x^3 + 7}\right) = \frac{(x^3 + 7)D_x(x^2) - x^2 D_x(x^3 + 7)}{(x^3 + 7)^2} = \frac{(x^3 + 7)(2x) - x^2(3x^2)}{(x^3 + 7)^2}$$

$$= \frac{14x - x^4}{(x^3 + 7)^2}.$$

2 Suppose that f and g are differentiable functions at the number 3 and that $f(3) = -2$, $g(3) = -5$, $f'(3) = 3$, and $g'(3) = 1$. If $h = f/g$, find the slope of the tangent line to the graph of h at the point $(3, \frac{2}{5})$.

SOLUTION
By the quotient rule,

$$h'(x) = \frac{g(x)f'(x) - f(x)g'(x)}{[g(\dot{x})]^2},$$

so the required slope is given by

$$h'(3) = \frac{g(3)f'(3) - f(3)g'(3)}{[g(3)]^2} = \frac{(-5)(3) - (-2)(1)}{(-5)^2} = -\frac{13}{25}.$$

Since a rational function is a ratio of polynomial functions and we can differentiate any polynomial function, we can use the quotient rule to differentiate any rational function. Example 1 above illustrates the technique.

Rule 3, the power rule, can be generalized to arbitrary integral powers as follows.

RULE 9   Power Rule for Integral Exponents
If n is a fixed integer, then

$$D_x x^n = nx^{n-1} \quad \text{or} \quad \frac{d}{dx}x^n = nx^{n-1}.$$

For $n = 1$, this formula gives $\dfrac{d}{dx}x = 1 \cdot x^0 = 1$, in conformity with the identity rule, except when $x = 0$, since 0^0 is undefined. It is traditional to overlook this slight difficulty and simply interpret 0^0 to be 1 *for purposes of this rule only*. Rule 9 permits us to use the same formula as in Rule 3, even when n is a negative integer or zero.

EXAMPLES 1 Given $f(x) = -2/x^3$, find $f'(x)$.

SOLUTION
Using the homogeneous rule and the power rule for integral exponents, we have

$$f'(x) = D_x\left(\frac{-2}{x^3}\right) = D_x(-2x^{-3}) = -2D_x(x^{-3}) = (-2)(-3x^{-3-1}) = 6x^{-4}.$$

2 Find $\dfrac{d}{dx}\left(x^2 + \dfrac{\sqrt{5}}{x^2} - \dfrac{\pi}{x^5}\right)$.

SOLUTION

$$\frac{d}{dx}\left(x^2 + \frac{\sqrt{5}}{x^2} - \frac{\pi}{x^5}\right) = \frac{d}{dx}(x^2 + \sqrt{5}\,x^{-2} - \pi x^{-5}) = 2x - 2\sqrt{5}\,x^{-3} + 5\pi x^{-6}.$$

In summary, we have the following basic rules for differentiation:

1 $\dfrac{dc}{dx} = 0$

2 $\dfrac{dx}{dx} = 1$

3 $\dfrac{d}{dx}x^n = nx^{n-1}$

4 $\dfrac{d}{dx}(cu) = c\dfrac{du}{dx}$

5 $\dfrac{d}{dx}(u + v) = \dfrac{du}{dx} + \dfrac{dv}{dx}$

6 $\dfrac{d}{dx}(uv) = u\dfrac{dv}{dx} + \dfrac{du}{dx}v$

7 $\dfrac{d}{dx}\left(\dfrac{1}{v}\right) = \dfrac{-\left(\dfrac{dv}{dx}\right)}{v^2}$

8 $\dfrac{d}{dx}\left(\dfrac{u}{v}\right) = \dfrac{v\dfrac{du}{dx} - u\dfrac{dv}{dx}}{v^2}$

In these rules, c denotes a constant, n is an integer, and u and v are differentiable functions of x.

3.1 Proofs of the Basic Differentiation Rules

We devote this section to the precise statements and rigorous proofs of the basic differentiation rules. It is convenient to prove the theorems in an order other than that in which the rules were stated.

THEOREM 1

Constant Rule

If c is a constant number and f is the constant function defined by $f(x) = c$, then f is differentiable at every number x and f' is the function defined by $f'(x) = 0$.

PROOF

$$f'(x) = \lim_{\Delta x \to 0} \frac{f(x + \Delta x) - f(x)}{\Delta x} = \lim_{\Delta x \to 0} \frac{c - c}{\Delta x} = \lim_{\Delta x \to 0} 0 = 0.$$

THEOREM 2

Identity Rule

If f is the function defined by $f(x) = x$, then f is differentiable at every number x and f' is the constant function defined by $f'(x) = 1$.

PROOF

$$f'(x) = \lim_{\Delta x \to 0} \frac{f(x + \Delta x) - f(x)}{\Delta x} = \lim_{\Delta x \to 0} \frac{x + \Delta x - x}{\Delta x} = \lim_{\Delta x \to 0} 1 = 1.$$

THEOREM 3

Sum Rule

Let f and g be functions both of which are differentiable at the number x_1 and let $h = f + g$. Then h is also differentiable at x_1 and

$$h'(x_1) = f'(x_1) + g'(x_1).$$

PROOF

$$h'(x_1) = \lim_{\Delta x \to 0} \frac{h(x_1 + \Delta x) - h(x_1)}{\Delta x}$$

$$= \lim_{\Delta x \to 0} \frac{[f(x_1 + \Delta x) + g(x_1 + \Delta x)] - [f(x_1) + g(x_1)]}{\Delta x}$$

$$= \lim_{\Delta x \to 0} \frac{f(x_1 + \Delta x) - f(x_1) + g(x_1 + \Delta x) - g(x_1)}{\Delta x}$$

$$= \lim_{\Delta x \to 0} \left[\frac{f(x_1 + \Delta x) - f(x_1)}{\Delta x} + \frac{g(x_1 + \Delta x) - g(x_1)}{\Delta x} \right]$$

$$= \lim_{\Delta x \to 0} \frac{f(x_1 + \Delta x) - f(x_1)}{\Delta x} + \lim_{\Delta x \to 0} \frac{g(x_1 + \Delta x) - g(x_1)}{\Delta x}$$

$$= f'(x_1) + g'(x_1).$$

THEOREM 4

Product Rule

Let f and g be functions both of which are differentiable at the number x_1, and let $h = f \cdot g$. Then h is also differentiable at x_1 and

$$h'(x_1) = f(x_1) \cdot g'(x_1) + f'(x_1) \cdot g(x_1).$$

PROOF

$$h'(x_1) = \lim_{\Delta x \to 0} \frac{h(x_1 + \Delta x) - h(x_1)}{\Delta x}$$

$$= \lim_{\Delta x \to 0} \frac{f(x_1 + \Delta x) \cdot g(x_1 + \Delta x) - f(x_1) \cdot g(x_1)}{\Delta x}.$$

We now use a curious but effective algebraic trick—the expression $f(x_1 + \Delta x) \cdot g(x_1)$ is subtracted from the numerator and then added back again (which, of course, leaves the value of the numerator unchanged). The result is

$$h'(x_1) = \lim_{\Delta x \to 0} \frac{f(x_1 + \Delta x) \cdot g(x_1 + \Delta x) - f(x_1 + \Delta x) \cdot g(x_1) + f(x_1 + \Delta x) \cdot g(x_1) - f(x_1) \cdot g(x_1)}{\Delta x}$$

$$= \lim_{\Delta x \to 0} \left[\frac{f(x_1 + \Delta x) \cdot g(x_1 + \Delta x) - f(x_1 + \Delta x) \cdot g(x_1)}{\Delta x} \right.$$

$$\left. + \frac{f(x_1 + \Delta x) \cdot g(x_1) - f(x_1) \cdot g(x_1)}{\Delta x} \right]$$

$$= \lim_{\Delta x \to 0} \left[f(x_1 + \Delta x) \frac{g(x_1 + \Delta x) - g(x_1)}{\Delta x} + \frac{f(x_1 + \Delta x) - f(x_1)}{\Delta x} g(x_1) \right]$$

$$= \left[\lim_{\Delta x \to 0} f(x_1 + \Delta x) \right] \cdot \left[\lim_{\Delta x \to 0} \frac{g(x_1 + \Delta x) - g(x_1)}{\Delta x} \right]$$

$$+ \left[\lim_{\Delta x \to 0} \frac{f(x_1 + \Delta x) - f(x_1)}{\Delta x} \right] \cdot \left[\lim_{\Delta x \to 0} g(x_1) \right]$$

$$= \left[\lim_{\Delta x \to 0} f(x_1 + \Delta x) \right] \cdot g'(x_1) + f'(x_1) \left[\lim_{\Delta x \to 0} g(x_1) \right].$$

Since f is differentiable at x_1, it is continuous at x_1 (Theorem 1, Section 2.2); hence,

$$\lim_{\Delta x \to 0} f(x_1 + \Delta x) = \lim_{x \to x_1} f(x) = f(x_1).$$

Also, since $g(x_1)$ is a constant,

$$\lim_{\Delta x \to 0} g(x_1) = g(x_1).$$

It follows that

$$h'(x_1) = f(x_1) \cdot g'(x_1) + f'(x_1) \cdot g(x_1).$$

THEOREM 5   **Homogeneous Rule**

Let g be a function that is differentiable at the number x_1 and let c be a constant. Let the function h be defined by $h(x) = cg(x)$. Then h is differentiable at x_1 and

$$h'(x_1) = cg'(x_1).$$

PROOF

Let f be the constant function defined by $f(x) = c$. By Theorem 1, $f'(x) = 0$. Evidently,

$$h(x) = cg(x) = f(x) \cdot g(x).$$

Therefore, by Theorem 4,

$$h'(x_1) = f(x_1) \cdot g'(x_1) + f'(x_1) \cdot g(x_1) = cg'(x_1) + 0 \cdot g(x_1) = cg'(x_1).$$

THEOREM 6   **Power Rule**

Let n be an integer greater than 1 and let f be the function defined by $f(x) = x^n$. Then f is differentiable at every number x and f' is the function defined by

$$f'(x) = nx^{n-1}.$$

PROOF

The proof proceeds by mathematical induction, starting with $n = 2$. For $n = 2$ we have, by the product and identity rules,

$$f'(x) = D_x(x^2) = D_x(x \cdot x) = x \cdot (D_x x) + (D_x x)x$$
$$= x + x = 2x = 2x^{2-1};$$

hence, the theorem holds when $n = 2$. Now, assuming that n is greater than 2 and that the theorem holds for exponents less than n, Theorems 4 and 2 imply that

$$f'(x) = D_x(x^n) = D_x(x^{n-1} \cdot x) = x^{n-1} \cdot (D_x x) + (D_x x^{n-1}) \cdot x$$
$$= x^{n-1} + [(n-1)x^{n-2}] \cdot x = x^{n-1} + (n-1)x^{n-1}$$
$$= nx^{n-1}.$$

THEOREM 7   **Reciprocal Rule**

Let g be a function which is differentiable at x_1 and suppose that $g(x_1) \neq 0$. Let h be the function defined by $h(x) = 1/g(x)$. Then h is differentiable at x_1 and

$$h'(x_1) = -\frac{g'(x_1)}{[g(x_1)]^2}.$$

PROOF

Since g is differentiable at x_1, it is defined in some open interval about x_1 and it is continuous at x_1. Therefore, for values of x close to x_1, the numerical values of $g(x)$ come close to $g(x_1)$. Since $g(x_1) \neq 0$, the numerical values $g(x)$ must differ from zero for values of x sufficiently close to x_1. This shows that $h(x) = 1/g(x)$ is defined at least in a small open interval around x_1. We have

$$h'(x_1) = \lim_{\Delta x \to 0} \frac{h(x_1 + \Delta x) - h(x_1)}{\Delta x} = \lim_{\Delta x \to 0} \frac{\dfrac{1}{g(x_1 + \Delta x)} - \dfrac{1}{g(x_1)}}{\Delta x}$$

$$= \lim_{\Delta x \to 0} \frac{1}{\Delta x} \left[\frac{g(x_1)}{g(x_1) \cdot g(x_1 + \Delta x)} - \frac{g(x_1 + \Delta x)}{g(x_1) \cdot g(x_1 + \Delta x)} \right]$$

$$= \lim_{\Delta x \to 0} \frac{1}{\Delta x} \cdot \frac{g(x_1) - g(x_1 + \Delta x)}{g(x_1) \cdot g(x_1 + \Delta x)}$$

$$= \lim_{\Delta x \to 0} (-1) \left[\frac{g(x_1 + \Delta x) - g(x_1)}{\Delta x} \cdot \frac{1}{g(x_1) \cdot g(x_1 + \Delta x)} \right]$$

$$= (-1) \left[\lim_{\Delta x \to 0} \frac{g(x_1 + \Delta x) - g(x_1)}{\Delta x} \right] \cdot \left[\lim_{\Delta x \to 0} \frac{1}{g(x_1) \cdot g(x_1 + \Delta x)} \right]$$

$$= (-1)g'(x_1) \cdot \frac{1}{g(x_1) \lim\limits_{\Delta x \to 0} g(x_1 + \Delta x)}.$$

Since g is continuous at x_1,

$$\lim_{\Delta x \to 0} g(x_1 + \Delta x) = \lim_{x \to x_1} g(x) = g(x_1).$$

It follows that

$$h'(x_1) = (-1)g'(x_1) \cdot \frac{1}{g(x_1) \cdot g(x_1)} = -\frac{g'(x_1)}{[g(x_1)]^2}.$$

THEOREM 8   **Quotient Rule**

Let f and g be functions both of which are differentiable at the number x_1 and suppose that $g(x_1) \neq 0$. Then, if $h = f/g$, it follows that h is differentiable at x_1 and

$$h'(x_1) = \frac{g(x_1) \cdot f'(x_1) - f(x_1) \cdot g'(x_1)}{[g(x_1)]^2}.$$

PROOF

Notice that $h = f \cdot (1/g)$; hence, by the product and reciprocal rules,

$$h'(x_1) = f(x_1) \cdot \frac{-g'(x_1)}{[g(x_1)]^2} + f'(x_1) \cdot \frac{1}{g(x_1)}$$

$$= \frac{-f(x_1) \cdot g'(x_1)}{[g(x_1)]^2} + \frac{g(x_1) \cdot f'(x_1)}{[g(x_1)]^2}$$

$$= \frac{g(x_1) \cdot f'(x_1) - f(x_1) \cdot g'(x_1)}{[g(x_1)]^2}.$$

THEOREM 9   **Power Rule for Integral Exponents**

If the function f is defined by $f(x) = x^n$, where n is any fixed integer, then f is differentiable and

$$f'(x) = nx^{n-1}.$$

Here we understand that:

(a) If $n \leq 0$, then x can be any number except 0.

(b) If $n = 1$, we interpret 0^0 as being the number 1 (for purposes of this theorem only).

PROOF

Theorem 6 takes care of the case $n \geq 2$, while Theorem 2 takes care of the case $n = 1$. For $n = 0$, $x^n = x^0 = 1$ (except for $x = 0$); hence, for $x \neq 0$ and $n = 0$, $f'(x) = D_x(x^n) = D_x(1) = 0 = 0 \cdot x^{-1} = 0 \cdot x^{0-1} = nx^{n-1}$, as desired. Finally, suppose that $n < 0$ and that $x \neq 0$. Note that $-n$ is a positive integer; hence, by what was already proved for positive exponents and by Theorem 7,

$$f'(x) = D_x(x^n) = D_x\left(\frac{1}{x^{-n}}\right) = -\frac{D_x(x^{-n})}{(x^{-n})^2} = -\frac{-nx^{-n-1}}{x^{-2n}}$$

$$= nx^{-n-1+2n} = nx^{n-1}.$$

Problem Set 3

In problems 1 through 31, differentiate each function by applying the basic rules for differentiation.

1 $f(x) = x^5 - 3x^3 + 1$

2 $f(x) = \frac{5}{6}x^6 - 9x^4$

3 $f(x) = \frac{x^{10}}{2} + \frac{x^5}{5} + 6$

4 $F(x) = \frac{x^4}{4} - \frac{x^3}{3} + 1$

5 $f(t) = t^8 - 2t^7 + 3t + 1$

6 $f(t) = 3t^2 + 7t + 17$

7 $F(x) = \frac{3}{x^2} + \frac{4}{x}$

8 $f(t) = \frac{1}{3t^3} - \frac{1}{2t^2} + 1$

9 $f(y) = \frac{5}{y^5} - \frac{25}{y}$

10 $f(u) = \frac{1}{u} - \frac{3}{u^3}$

11 $g(x) = 3x^{-2} - 7x^{-1} + 6$

12 $G(x) = \frac{1}{3}x^{-3} - \frac{1}{2}x^{-2} + 11$

13 $f(x) = \dfrac{2}{5x} - \dfrac{\sqrt{2}}{3x^2}$

14 $f(x) = \sqrt{3}(x^3 - x)$

15 $F(x) = x^2(3x^3 - 1)$

16 $f(x) = (x^2 + 1)(2x^3 + 5)$

17 $G(x) = (x^2 + 3x)(x^3 - 9x)$

18 $g(x) = (3x - x^2)(3x^3 - 4)$

19 $f(y) = (2y - 1)(4y^2 + 7)$

20 $f(t) = (6t^2 + 7)^2$

21 $f(x) = (x^3 - 8)\left(\dfrac{2}{x} - 1\right)$

22 $f(x) = \left(\dfrac{1}{x} + 3\right)\left(\dfrac{2}{x} + 7\right)$

23 $g(x) = \left(\dfrac{1}{x^2} + 3\right)\left(\dfrac{2}{x^3} + x\right)$

24 $g(u) = \left(u^2 + \dfrac{1}{u}\right)\left(u - \dfrac{1}{u^3}\right)$

25 $f(x) = \dfrac{2x + 7}{3x - 1}$

26 $f(x) = \dfrac{3x^2}{x - 2}$

27 $g(x) = \dfrac{2x^2 + x + 1}{x^2 - 3x + 2}$

28 $G(t) = \dfrac{t^3}{2t^4 + 5}$

29 $F(t) = \dfrac{3t^2 + 7}{t^2 - 1}$

30 $f(x) = \dfrac{x^2 - 19}{x^2 + 19}$

31 $f(x) = \left(\dfrac{3x + 1}{x + 2}\right)(x + 7)$

32 Suppose that f and g are differentiable functions. Define a function h by $h(x) = f(x) - g(x)$. Show that $h'(x) = f'(x) - g'(x)$.

33 Find $f'(2)$ in each case.

(a) $f(x) = \frac{1}{3}x^3 - 1$

(b) $f(x) = \dfrac{1}{x^3} - 1$

(c) $f(x) = (x^2 + 1)(1 - x)$

(d) $f(x) = \left(\dfrac{1}{x} + 2\right)\left(\dfrac{3}{x} - 1\right)$

(e) $f(x) = \dfrac{x}{x^2 + 2}$

(f) $f(x) = \dfrac{2x^2}{x + 7}$

34 Suppose that f, g, and h are differentiable functions. Let k be a function defined by $k(x) = f(x) \cdot g(x) \cdot h(x)$. Use the product rule to show that

$$k'(x) = f(x) \cdot g(x) \cdot h'(x) + f(x) \cdot g'(x) \cdot h(x) + f'(x) \cdot g(x) \cdot h(x).$$

35 Use the result of problem 34 to differentiate the following functions.

(a) $f(x) = (2x - 5)(x + 2)(x^2 - 1)$

(b) $f(x) = (1 - 3x)^2(2x + 5)$

(c) $f(x) = \left(\dfrac{1}{x^2} + 1\right)(3x - 1)(x^2 - 3x)$

(d) $f(x) = (2x^2 + 7)^3$

36 Let $f(t) = t^2 + t$ and $g(t) = t^2 - 1$. Compute $D_t[\frac{1}{2}f(t) - \frac{2}{3}g(t)]$.

37 Let f and g be differentiable functions at the number 1 and let $f(1) = 1$, $f'(1) = 2$, $g(1) = \frac{1}{2}$, and $g'(1) = -3$. Use the differentiation rules to find:

(a) $(f + g)'(1)$

(b) $(f - g)'(1)$

(c) $(2f + 3g)'(1)$

(d) $(fg)'(1)$

(e) $\left(\dfrac{f}{g}\right)'(1)$

(f) $\left(\dfrac{g}{f}\right)'(1)$

38 Suppose that f, g, and h are differentiable functions at the number 2 and let $f(2) = -2$, $f'(2) = 3$, $g(2) = -5$, $g'(2) = 1$, $h(2) = 2$, and $h'(2) = 4$. Use the differentiation rules to find:

(a) $(f + g + h)'(2)$

(b) $(2f - g + 3h)'(2)$

(c) $(fgh)'(2)$

(d) $\left(\dfrac{fg}{h}\right)'(2)$

39 Find the slope of the tangent line to the graph of the function f at the point whose x coordinate is 4.

(a) $f(x) = x^3 - 4x^2 - 1$

(b) $f(x) = \dfrac{3}{4x - 2}$

40 Determine the rate of change of volume with respect to the radius (a) of a sphere and (b) of a right circular cylinder with fixed height h.

41 Find the slope of the tangent line to the graph of $f(x) = \dfrac{x}{x^3 - 2}$ at the point $(1, -1)$.

42 For a thin lens of constant focal length p, the object distance x and the image distance y are related by the formula $1/x + 1/y = 1/p$.
(a) Solve for y in terms of x and p.
(b) Find the rate of change of y with respect to x.

43 An object is moving along a straight line in such a way that at the end of t seconds its distance s in feet from the starting point is given by $s = 8t + 2/t$, with $t > 0$. Find the speed of the object at the instant when $t = 2$ seconds.

44 The formula $D_x(x^n) = nx^{n-1}$, which works for integer values of n, suggests that perhaps $D_x(x^{1/2}) = \frac{1}{2}x^{(1/2)-1} = 1/(2x^{1/2})$; that is, $D_x(\sqrt{x}) = \dfrac{1}{2\sqrt{x}}$. Use the definition of derivative (as a limit of a difference quotient) to show that this is true for $x > 0$.

45 Criticize the following erroneous argument: We wish to compute the value of the derivative of $f(x) = 2x^2 + 3x - 1$ at $x = 2$. To this end, we put $x = 2$ and we have $f(2) = 2(2)^2 + 3(2) - 1 = 13$. But $D_x(13) = 0$, so $f'(2) = 0$.

46 Show that the reciprocal rule is a special case of the quotient rule when the numerator is the constant function $f(x) = 1$.

47 Let m be a given integer. If possible, find a constant c and an integer n such that $D_x(cx^n) = x^m$, at least for $x \neq 0$. For what value (or values) of m is this not possible?

48 Suppose that $a, b, c,$ and d are constant; that not both c and d are zero; and that the value of $\dfrac{ax + b}{cx + d}$ is independent of the value of x (provided $cx + d \neq 0$). Prove that $ad = bc$.

49 Use the product rule to prove that $D_x[f(x)]^2 = 2f(x) \cdot D_x[f(x)]$.

50 Use the product rule to prove that $D_x[f(x)]^3 = 3[f(x)]^2 \cdot D_x[f(x)]$.

51 Let $f(x) = x$ and $g(x) = 1$. Show that:
(a) The derivative of the product $f \cdot g$ is not the product of the derivatives of f and g.
(b) The derivative of the quotient f/g is not the quotient of the derivatives of f and g.

4   The Chain Rule

Suppose that $y = (x^2 + 5x)^3$ and that we wish to find dy/dx. One approach is to expand $(x^2 + 5x)^3$ and then differentiate the resulting polynomial. Thus,

$$y = (x^2 + 5x)^3 = x^6 + 15x^5 + 75x^4 + 125x^3,$$

so that

$$\frac{dy}{dx} = 6x^5 + 75x^4 + 300x^3 + 375x^2.$$

Another approach is to let $u = x^2 + 5x$, so that $y = u^3$, $dy/du = 3u^2$, and $du/dx = 2x + 5$. Thus,

$$\frac{dy}{dx} = \frac{dy}{du}\frac{du}{dx} = 3u^2(2x + 5) = 3(x^2 + 5x)^2(2x + 5) = 6x^5 + 75x^4 + 300x^3 + 375x^2.$$

The latter calculation produced the correct answer, but there is a catch to it! The expressions $\dfrac{dy}{du}$ and $\dfrac{du}{dx}$ are just symbols for derivatives in which the "numerators" and "denominators" have not yet been given any separate meanings, so we were not really justified in supposing that $\dfrac{dy}{dx} = \dfrac{dy}{du}\dfrac{du}{dx}$. In fact, the legitimacy of this calculation is guaranteed by one of the most important differentiation rules in calculus—the *chain rule*.

Although we give a precise statement and proof of the chain rule later (Theorem 3 in Section 7), we begin with the following informal version.

The Chain Rule

If y is a differentiable function of u and if u is a differentiable function of x, then y is a differentiable function of x and

$$\frac{dy}{dx} = \frac{dy}{du}\frac{du}{dx}.$$

The reader is asked to assume the truth of the chain rule for the time being and to become familiar with it before delving into its proof.

EXAMPLE   If $y = u^3$ and $u = 2x^2 + 3x - 1$, find dy/dx.

SOLUTION

$$\frac{dy}{dx} = \frac{dy}{du}\frac{du}{dx} = 3u^2(4x + 3) = 3(2x^2 + 3x - 1)^2(4x + 3).$$

Of course, the chain rule can be written in operator notation as

$$D_x y = (D_u y)(D_x u).$$

If we let $y = f(u)$, where u is a function of x, it becomes

$$D_x f(u) = f'(u)D_x u.$$

EXAMPLE   Use the fact that, if $f(u) = \sqrt{u}$, then $f'(u) = 1/(2\sqrt{u})$ (Problem 44 in Problem Set 3), and the chain rule to find $D_x \sqrt{x^2 + 1}$.

SOLUTION
If we put $f(u) = \sqrt{u}$ and $u = x^2 + 1$, then $f(u) = \sqrt{x^2 + 1}$. Therefore,

$$D_x \sqrt{x^2 + 1} = D_x f(u) = f'(u)D_x u = \frac{1}{2\sqrt{u}}(2x) = \frac{x}{\sqrt{x^2 + 1}}.$$

The chain rule is often used to calculate derivatives of the form $D_x u^n$, where u is a differentiable function of x and n is an integer. Thus, letting $f(u) = u^n$, so that $f'(u) = nu^{n-1}$, we obtain the important formula

$$D_x u^n = nu^{n-1}D_x u.$$

EXAMPLES 1 Find $D_x(x^2 + 5x)^{100}$.

SOLUTION
Here $u = x^2 + 5x$ and $n = 100$, so

$$D_x(x^2 + 5x)^{100} = 100(x^2 + 5x)^{99}D_x(x^2 + 5x) = 100(x^2 + 5x)^{99}(2x + 5).$$

2 If $F(x) = \dfrac{1}{(3x-1)^4}$, find $F'(x)$.

SOLUTION
Here $F(x) = (3x-1)^{-4}$, so

$$F'(x) = D_x F(x) = D_x(3x-1)^{-4} = (-4)(3x-1)^{-4-1}D_x(3x-1)$$
$$= (-4)(3x-1)^{-5}(3) = -12(3x-1)^{-5}.$$

3 Find $D_x\left(\dfrac{3x}{x^2+7}\right)^{10}$.

SOLUTION

$$D_x\left(\frac{3x}{x^2+7}\right)^{10} = 10\left(\frac{3x}{x^2+7}\right)^9 D_x\left(\frac{3x}{x^2+7}\right)$$

$$= 10\left(\frac{3x}{x^2+7}\right)^9\left[\frac{(x^2+7)(3) - (3x)(2x)}{(x^2+7)^2}\right]$$

$$= \frac{(3x)^9(210 - 30x^2)}{(x^2+7)^{11}}.$$

4 Find $g'(t)$ if $g(t) = (2t^2 - 5t + 1)^{-7}$.

SOLUTION

$$g'(t) = -7(2t^2 - 5t + 1)^{-8}(4t - 5) = \frac{35 - 28t}{(2t^2 - 5t + 1)^8}.$$

5 Find $D_x[(x^2 + 6x)^{10}(1 - 3x)^4]$.

SOLUTION

$$D_x[(x^2 + 6x)^{10}(1 - 3x)^4] = [D_x(x^2 + 6x)^{10}](1 - 3x)^4 + (x^2 + 6x)^{10}[D_x(1 - 3x)^4]$$
$$= [10(x^2 + 6x)^9(2x + 6)](1 - 3x)^4$$
$$+ (x^2 + 6x)^{10}[4(1 - 3x)^3(-3)]$$
$$= (x^2 + 6x)^9(1 - 3x)^3[10(2x + 6)(1 - 3x) - 12(x^2 + 6x)]$$
$$= (x^2 + 6x)^9(1 - 3x)^3(-72x^2 - 232x + 60).$$

In calculating derivatives it is sometimes necessary to use the chain rule repeatedly. For instance, if y is a function of v, v is a function of u, and u is a function of x, then $\dfrac{dy}{dx} = \dfrac{dy}{du}\dfrac{du}{dx}$ and $\dfrac{dy}{du} = \dfrac{dy}{dv}\dfrac{dv}{du}$, so that

$$\frac{dy}{dx} = \frac{dy}{dv}\frac{dv}{du}\frac{du}{dx}.$$

EXAMPLES 1   Let $y = (\sqrt{1 + x^2})^3$. Use the fact that $\dfrac{d}{du}\sqrt{u} = \dfrac{1}{2\sqrt{u}}$ and the chain rule to find $\dfrac{dy}{dx}$.

SOLUTION
Let $u = 1 + x^2$, $v = \sqrt{u}$, and $y = v^3$, so that $y = (\sqrt{u})^3 = (\sqrt{1 + x^2})^3$. Thus,

$$\frac{dy}{dx} = \frac{dy}{dv}\frac{dv}{du}\frac{du}{dx} = (3v^2)\left(\frac{1}{2\sqrt{u}}\right)(2x) = 3(\sqrt{u})^2\frac{x}{\sqrt{u}} = 3x\sqrt{u} = 3x\sqrt{1 + x^2}.$$

2 Find $D_x[1 + (1 + x^5)^6]^7$.

SOLUTION
Using the chain rule repeatedly, we have

$$D_x[1 + (1 + x^5)^6]^7 = 7[1 + (1 + x^5)^6]^6 D_x[1 + (1 + x^5)^6]$$
$$= 7[1 + (1 + x^5)^6]^6[6(1 + x^5)^5 D_x(1 + x^5)]$$
$$= 7[1 + (1 + x^5)^6]^6[6(1 + x^5)^5(5x^4)]$$
$$= 210x^4[1 + (1 + x^5)^6]^6(1 + x^5)^5.$$

The chain rule is actually a rule for differentiating the composition $f \circ g$ of two functions. To see this, let $y = f(u)$ and $u = g(x)$, so that

$$y = f(u) = f[g(x)] = (f \circ g)(x).$$

Therefore, by the chain rule,

$$\frac{dy}{dx} = \frac{dy}{du}\frac{du}{dx} = f'(u)g'(x) = f'[g(x)]g'(x).$$

Denoting the composition $f \circ g$ by h, we can write the chain rule as follows:

If $h = f \circ g$, then $h'(x) = (f \circ g)'(x) = f'[g(x)]g'(x).$

Here, of course, we are assuming that g is differentiable at the number x and that f is differentiable at the number $g(x)$.

EXAMPLES 1 Let $g(x) = \frac{1}{4}x^8 - \frac{2}{3}x^6 + x - \sqrt{2}$, $f(u) = u^4$. Find $(f \circ g)'(x)$.

SOLUTION
$f'(u) = 4u^3$ and $g'(x) = 2x^7 - 4x^5 + 1$; hence, by the chain rule,

$$(f \circ g)'(x) = f'[g(x)]g'(x) = 4[g(x)]^3 g'(x)$$
$$= 4(\frac{1}{4}x^8 - \frac{2}{3}x^6 + x - \sqrt{2})^3(2x^7 - 4x^5 + 1).$$

2 Let f and g be differentiable functions such that $g(7) = \frac{1}{4}$, $g'(7) = \frac{2}{3}$, and $f'(\frac{1}{4}) = 10$. If $h = f \circ g$, find $h'(7)$.

SOLUTION
By the chain rule,

$$h'(7) = (f \circ g)'(7) = f'[g(7)]g'(7) = f'(\frac{1}{4})g'(7) = (10)(\frac{2}{3}) = \frac{20}{3}.$$

If we have a composite function $f \circ g$ so that $(f \circ g)(x) = f[g(x)]$, let us call g the "inside function" and f the "outside function" in this composition (because of the positions they occupy in the expression $f[g(x)]$). Then we can state the chain rule in words as follows:

The derivative of the composite of two functions is the derivative of the outside function taken at the value of the inside function times the derivative of the inside function.

The reader should memorize these words and should see exactly how they correspond to the formal statement

$$(f \circ g)'(x) = f'[g(x)]g'(x).$$

EXAMPLES   Later (in Section 2 of Chapter 8) we show that if $f(x) = \sin x$, where x is an angle expressed in radians, then $f'(x) = \cos x$. In words, *the derivative of the sine function* *is the cosine function.* Assume this for now and calculate $h'(x)$ for the function h given in Examples 1 through 4 below.

1 $h(x) = \sin(x^3) = \sin x^3$

SOLUTION
Here the outside function is the sine function and the inside function is $g(x) = x^3$. The derivative of the outside function is the cosine function and the derivative of the inside function is $g'(x) = 3x^2$. Therefore, by the chain rule,

$$h'(x) = [\cos(x^3)][3x^2] = 3x^2 \cos(x^3).$$

2 $h(x) = (\sin x)^3 = \sin^3 x$

SOLUTION
Here the outside function is $f(u) = u^3$ and the inside function is the sine function. The derivative of the outside function is $f'(u) = 3u^2$ and the derivative of the inside function is the cosine function. Therefore, by the chain rule,

$$h'(x) = 3(\sin x)^2 \cos x = 3 \sin^2 x \cos x.$$

3 $h(x) = \cos x$

SOLUTION
Since $\cos x = \sin(\pi/2 - x)$, we begin by writing

$$h(x) = \sin\left(\frac{\pi}{2} - x\right).$$

Here, the outside function is the sine function, while the inside function is $g(x) = \pi/2 - x$. The derivative of the outside function is the cosine function and the derivative of the inside function is $g'(x) = -1$. Therefore, by the chain rule,

$$h'(x) = \left[\cos\left(\frac{\pi}{2} - x\right)\right](-1) = -\cos\left(\frac{\pi}{2} - x\right) = -\sin x.$$

We conclude that $D_x \cos x = -\sin x$; that is, *the derivative of the cosine function is the negative of the sine function.*

4 $h(x) = \cos^4(3x)$

SOLUTION

$$h'(x) = 4\cos^3(3x)[-\sin(3x)](3) = -12\cos^3 3x \sin 3x.$$

Problem Set 4

You may assume in this problem set that, for $x > 0$, $D_x\sqrt{x} = 1/(2\sqrt{x})$. In problems 1 through 4, find the required derivative by using the chain rule.

1 $y = \sqrt{u}$, $u = x^2 + x + 1$, find dy/dx.

2 $y = u^3 - 2u^{1/2}$, $u = x^2 + 2x$, find dy/dx.

3 $y = u^{-5}$, $u = x^4 + 1$, find dy/dx.

4 $y = u$, $u = (7 - x^2)(7 + x^2)^{-1}$, find $D_x y$.

In problems 5 through 32, find the derivative of each function with the aid of the chain rule.

5 $f(x) = (5 - 2x)^{10}$

6 $f(x) = (2x - 3)^8$

7 $f(y) = \dfrac{1}{(4y + 1)^5}$

8 $F(t) = (2t^4 - t + 1)^{-4}$

9 $f(u) = (u^3 + 2)^{15}$

10 $g(y) = (y^2 - 3y + 2)^7$

11 $F(x) = (x^5 - 2x^2 + x + 1)^{-7}$

12 $g(t) = (\sqrt{3}t^2 + t - \sqrt{11})^{-8}$

13 $g(x) = (3x^2 + 7)^2(5 - 3x)^3$

14 $G(t) = (5t^2 + 1)^2(3t^4 + 2)^4$

15 $f(x) = \left(3x + \dfrac{1}{x}\right)^2 (6x - 1)^5$

16 $f(t) = (3t - 1)^{-1}(2t + 5)^{-3}$

17 $g(y) = (7y + 3)^{-2}(2y - 1)^4$

18 $f(u) = \left(6u + \dfrac{1}{u}\right)^{-5} (2u - 2)^7$

19 $f(x) = \left(\dfrac{x^2 + x}{1 - 2x}\right)^4$

20 $f(t) = \left(\dfrac{1 + t^2}{1 - t^2}\right)^5$

21 $F(x) = \left(\dfrac{3x + 1}{x^2}\right)^3$

22 $G(x) = \left(\dfrac{7x + 1/x}{x^2 + 2x - 1}\right)^2$

23 $g(t) = \left(\dfrac{7t + 1/t}{t^3 + 2}\right)^7$

24 $f(x) = \left(\dfrac{16x}{x^2 - 7}\right)^{-3}$

25 $f(x) = \dfrac{1}{\sqrt{x}}$

26 $F(x) = \dfrac{1}{\sqrt{x^2 + 1}}$

27 $g(x) = \sqrt{x^2 + 2x - 1}$

28 $f(x) = \sqrt{\sqrt{x}} = x^{1/4}$

29 $f(t) = \sqrt{t^4 - t^2 + \sqrt{3}}$

30 $g(y) = \sqrt{y^3 - y + \sqrt{y}}$

31 $f(x) = (x + \sqrt{x})(x - 2\sqrt{x})$

32 $F(x) = (x - \sqrt{x})^4$

33 If $u = vw$, $v = \sqrt{t}$, $t = x^2 + 2$, $w = s^5$, and $s = x + 1$, find du/dx.

34 If $w = yz$, $y = \sqrt{u}$, $u = x^2 + 7$, $z = \dfrac{2t + 1}{t + 1}$, and $t = \sqrt{x}$, find $D_x w$.

35 If y depends on x and x depends on t, use the chain rule to show that the rate of change of y with respect to t is the product of the rate of change of y with respect to x and the rate of change of x with respect to t.

36 Let $f(u) = u^3$ and $g(x) = \sqrt[3]{x}$. Assuming that g is differentiable at each number except 0, use the fact that $(f \circ g)(x) = x$ and the chain rule to show that $D_x(\sqrt[3]{x}) = g'(x) = \frac{1}{3}\sqrt[3]{1/x^2}$.

37 Suppose that f and g are functions such that $g(2) = 123$, g is differentiable at the number 2, $g'(2) = 7$, f is differentiable at the number 123, and $f'(123) = \frac{1}{2}$. Let $h = f \circ g$. Find $h'(2)$.

38 Let f and g be functions such that $f(5) = -3$, $f'(5) = 10$, $f'(7) = 20$, $g(5) = 7$, $g'(5) = \frac{1}{4}$, and $g'(7) = \frac{2}{3}$. Find $(f \circ g)'(5)$.

39 Suppose that f is an even function, so that $f(x) = f(-x)$ holds for all values of x. Assuming that f is differentiable, show that f' is an odd function; that is, show that $f'(-x) = -f'(x)$ holds for all values of x. [*Hint:* Let $g(x) = -x$, and note that $f = f \circ g$ so that $f' = (f \circ g)'$. Apply the chain rule.]

40 Show that if f is a differentiable odd function, then f' is an even function.

41 Assume the following formulas, which give the derivatives of the six trigonometric functions for every value of x at which the respective functions are defined.

(i) $D_x \sin x = \cos x$

(ii) $D_x \cos x = -\sin x$

(iii) $D_x \tan x = \sec^2 x$

(iv) $D_x \cot x = -\csc^2 x$

(v) $D_x \sec x = \sec x \tan x$

(vi) $D_x \csc x = -\csc x \cot x$

Use this information to find the derivatives of each of the following functions.

(a) $f(x) = \sin 5x$

(b) $F(x) = \cos(8x - 1)$

(c) $g(t) = \tan(3t)$

(d) $F(x) = \cot(9x)$

(e) $f(t) = \sec(2t + 9)$

(f) $g(x) = \csc(15x - 2)$

(g) $f(\theta) = \sin^3 \theta$

(h) $f(\theta) = \sin \sqrt{\theta}$

(i) $g(x) = \sin^2 \dfrac{2\pi x}{360}$

(j) $g(\theta) = \sqrt{\cos \theta}$

(k) $f(\theta) = \sin^2 \theta + \cos^2 \theta$

(l) $f(\theta) = 2 \sin \theta \cos \theta$

42 Use the fact that $|u| = \sqrt{u^2}$ to find each of the following derivatives.

(a) $\dfrac{d}{dx}|x|$ (b) $\dfrac{d}{dx}|3x + 1|$ (c) $\dfrac{d}{dx}\left(\dfrac{x}{|x|}\right)$ (d) $D_x|x^3 + 2|$

43 Let g be a differentiable function, put $u = g(x)$, $f(u) = 1/u$. Use the chain rule together with the fact that $f'(u) = -1/u^2$ to give another proof of the reciprocal rule,

$$D_x\left[\frac{1}{g(x)}\right] = -\frac{g'(x)}{[g(x)]^2}.$$

44 The demand D for a certain product is related to its price P by the equation $D = 500/\sqrt{P-1}$. Find the instantaneous rate at which the demand is changing with respect to the price when the price is \$3.50.

45 A particle moves along a straight line and s denotes its distance in feet from the starting point after t seconds. Find ds/dt in each of the following cases.

(a) $s = \sqrt{t}(1 + t + t^2)$ (b) $s = \dfrac{\sqrt{t}}{1 + t + t^2}$

46 Explain the distinction between $D_x[f(7x + 3)]$ and $f'(7x + 3)$.

47 Find the slope of the tangent line to the graph of the function f defined by the equation $f(x) = 1/\sqrt{2x + 7}$ at the point $(1, \frac{1}{3})$.

48 Show that the chain rule can be expressed as $(f \circ g)' = (f' \circ g) \cdot g'$.

5   The Inverse Function Rule and the Rational Power Rule

In this section we give a rule, called the *inverse function rule,* that enables us to find the derivative of the inverse of a function, and we use this rule to find derivatives of the form $D_x x^r$, where r is a rational number.

If we solve the equation $y = x^3$ for x, we obtain $x = \sqrt[3]{y}$; in other words, the functions

$$f(x) = x^3 \quad \text{and} \quad g(y) = \sqrt[3]{y}$$

are inverses of one another (Section 9 of Chapter 0). Assuming that g is differentiable, let us find its derivative. Since $x = g(y)$, we want to find $dx/dy = g'(y)$. By the chain rule,

$$\frac{dx}{dy}\frac{dy}{dx} = \frac{dx}{dx} = 1;$$

hence,

$$\frac{dx}{dy} = \frac{1}{dy/dx}.$$

But, $y = x^3$, so $dy/dx = 3x^2 = 3(\sqrt[3]{y})^2 = 3y^{2/3}$. Therefore,

$$\frac{dx}{dy} = \frac{1}{3y^{2/3}},$$

provided that $y \neq 0$.

By generalizing the argument above, we obtain the inverse function rule. In Section 5.1 we shall state this rule precisely as a theorem; however, for now, we give the following informal statement.

The Inverse Function Rule Using Leibniz Notation

$$\text{If } \frac{dy}{dx} \neq 0, \quad \text{then} \quad \frac{dx}{dy} = \frac{1}{dy/dx}.$$

The following example illustrates the utility of the inverse function rule. In this example, we calculate the derivative of an inverse function by "brute force" and compare it with the same calculation using the inverse function rule.

EXAMPLE   Let $y = 3x^2 - 4x + 2$ for $x > \frac{2}{3}$. Find dx/dy when $y = 2$:
(a) By solving for x in terms of y and then differentiating.
(b) By using the inverse function rule.

SOLUTION
(a) Using the quadratic formula, we solve the equation $3x^2 - 4x + 2 - y = 0$ for x to obtain

$$x = \frac{4 \pm \sqrt{16 - 12(2 - y)}}{6} \quad \text{or} \quad x = \frac{2}{3} \pm \frac{1}{3}\sqrt{3y - 2}.$$

Since we require that $x > \frac{2}{3}$, we must use the plus sign in the latter equation, so that

$$x = \frac{2}{3} + \frac{1}{3}\sqrt{3y - 2}.$$

Recall that $\dfrac{d}{du}\sqrt{u} = \dfrac{1}{2\sqrt{u}}$ (Problem 44 in Problem Set 3); hence, by the chain rule,

$$\frac{dx}{dy} = \frac{1}{3} \cdot \frac{d}{dy}\sqrt{3y - 2} = \frac{1}{3} \cdot \frac{1}{2\sqrt{3y - 2}}\frac{d}{dy}(3y - 2) = \frac{1}{2\sqrt{3y - 2}}.$$

Therefore, when $y = 2$,

$$\frac{dx}{dy} = \frac{1}{2\sqrt{6 - 2}} = \frac{1}{4}.$$

(b) By the inverse function rule,

$$\frac{dx}{dy} = \frac{1}{dy/dx} = \frac{1}{\dfrac{d}{dx}(3x^2 - 4x + 2)} = \frac{1}{6x - 4}.$$

When $y = 2$, $x = \frac{4}{3}$ (why?), and so

$$\frac{dx}{dy} = \frac{1}{6(\frac{4}{3}) - 4} = \frac{1}{4}.$$

5.1 The Inverse Function Theorem

Now we give a precise statement of the inverse function rule in the form of a theorem. In order to understand this theorem, suppose that $y = f(x)$, where f is a differentiable and invertible function. Solving the equation $y = f(x)$ for x in terms of y, we obtain $x = g(y)$, where g is the inverse of f. Here, $dy/dx = f'(x)$ and, assuming that g is differentiable, $\dfrac{dx}{dy} = g'(y)$. If $f'(x) \neq 0$, the inverse function rule $\dfrac{dx}{dy} = \dfrac{1}{dy/dx}$ can be rewritten as $g'(y) = \dfrac{1}{f'(x)}$, or, since $x = g(y)$, $g'(y) = \dfrac{1}{f'[g(y)]}$.

In the applications of the inverse function rule, our main concern is to calculate the derivative g'. If the use of y as the independent variable in the expression $g'(y)$ is inconvenient, we can use the letter x rather than the letter y in the formula derived above. The result is $g'(x) = \dfrac{1}{f'[g(x)]}$. Our motivation is complete, and we can now state the theorem.

THEOREM 1   **Inverse Function Rule or Inverse Function Theorem**

Let f be a function whose domain is an open interval I, assume that f is differentiable on I, and suppose that $f'(c) \neq 0$ for every number c in I. Then f has an inverse g, g is differentiable, and

$$g'(x) = \frac{1}{f'[g(x)]}$$

holds for every number x in the domain of g.

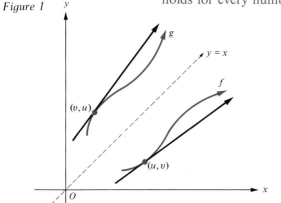

Figure 1

Not only does the inverse function theorem give us a formula for calculating the derivative g' of the inverse of f, it also guarantees the *existence* of g as well as its *differentiability*. For this reason its proof is too sophisticated to be given here. However, it is possible to gain some insight into the reason for the differentiability of g by considering Figure 1. Recall that the graph of g is the mirror image of the graph of f across the straight line $y = x$ (Section 9 of Chapter 0). By hypothesis, f is differentiable; hence, the graph of f has a tangent line at each point (u, v) with $v = f(u)$. Evidently, its mirror image—the graph of g—also has a tangent line at each point (v, u) with $u = g(v)$. But, to say that the graph of g has a tangent line at each point is to say that g is differentiable.

Recall from Section 9 of Chapter 0 that, if a function f has an inverse function g, we write $g = f^{-1}$, read "f inverse." Using the notation f^{-1}, we can rewrite the formula in the inverse function theorem as

$$(f^{-1})'(x) = \frac{1}{f'[f^{-1}(x)]}.$$

EXAMPLE   Let $f(x) = x^2$ for $x > 0$. Find $(f^{-1})'(x)$.

SOLUTION

Since f is defined and differentiable on the open interval $(0, \infty)$, and since $f'(x) = 2x \neq 0$ for all values of x in this interval, it follows from Theorem 1 that f is invertible, f^{-1} is differentiable, and

$$(f^{-1})'(x) = \frac{1}{f'[f^{-1}(x)]} = \frac{1}{2f^{-1}(x)}$$

holds for all x in the domain of f^{-1}. Since $f(x) = x^2$ for $x > 0$, it follows that

$f^{-1}(x) = \sqrt{x}$. Then $(f^{-1})'(x) = \dfrac{1}{2f^{-1}(x)}$ simply means that $D_x\sqrt{x} = \dfrac{1}{2\sqrt{x}}$.

5.2 The Power Rule for Rational Exponents

In the preceding example we used the inverse function theorem to find the derivative of the square root function. Generalizing this argument, we obtain the following theorem.

THEOREM 2   **Root Rule**

If n is a positive integer, then

$$D_x\sqrt[n]{x} = D_x x^{1/n} = \frac{1}{n}x^{(1/n)-1} = \frac{\sqrt[n]{x}}{nx}$$

holds for all values of x for which $\sqrt[n]{x}$ is defined, except for $x = 0$.

PROOF

Let $f(x) = x^n$, with the understanding that x is positive if n is even, but there is no restriction on x if n is odd. If we solve the equation $y = f(x)$ for x in terms of y, we obtain $x = \sqrt[n]{y}$; hence, the function $g(y) = \sqrt[n]{y}$ is the inverse of the function f. Using the letter x for the independent variable in the equation defining g, we find that $f^{-1}(x) = g(x) = \sqrt[n]{x}$. By Theorem 1, we have

$$D_x\sqrt[n]{x} = (f^{-1})'(x) = \frac{1}{f'[f^{-1}(x)]} = \frac{1}{f'(\sqrt[n]{x})} = \frac{1}{n(\sqrt[n]{x})^{n-1}}$$

$$= \frac{\sqrt[n]{x}}{n(\sqrt[n]{x})^n} = \frac{\sqrt[n]{x}}{nx} = \frac{1}{n}x^{(1/n)-1}, \qquad \text{provided that } x \neq 0.$$

EXAMPLES   **1** Find $D_x\sqrt[9]{x}$.

SOLUTION

Using Theorem 2, we have $D_x\sqrt[9]{x} = \dfrac{1}{9}x^{(1/9)-1} = \dfrac{1}{9}x^{-8/9} = \dfrac{\sqrt[9]{x}}{9x}$.

2 If $y = x^{1/8}$, find dy/dx.

SOLUTION

$$\frac{dy}{dx} = \frac{d}{dx}x^{1/8} = \frac{1}{8}x^{(1/8)-1} = \frac{1}{8}x^{-7/8}.$$

Using the root rule and the chain rule, we now establish the rule for differentiating rational powers of x. As the following theorem shows, this rule is formally the same as the rule for differentiating integral powers of x.

THEOREM 3   **Power Rule for Rational Exponents**

Let $r = m/n$ be a rational number, reduced to lowest terms so that n is a positive integer and the integers m and n have no common factors. Then

$$D_x x^r = rx^{r-1}$$

holds for all values of x for which $x^r = (x^{1/n})^m$ is defined, except possibly for $x = 0$. It also holds for $x = 0$, provided that n is odd and $m > n$.

PROOF
By the chain rule and Theorem 2, we have

$$D_x x^r = D_x(x^{1/n})^m = m(x^{1/n})^{m-1} D_x(x^{1/n}) = \left(mx^{(m-1)/n}\right)\left(\frac{1}{n} x^{(1/n)-1}\right)$$

$$= \frac{m}{n} x^{(m-1)/n+(1/n)-1} = \frac{m}{n} x^{(m/n)-1} = rx^{r-1},$$

provided that x^r is defined and $x \neq 0$.

To complete the proof, assume that $m > n$ and n is odd. Let $f(x) = x^r$, noting that $r > 1$. Thus, when $x = 0$, $rx^{r-1} = 0$, and we must prove that $f'(0) = 0$. We have

$$f'(0) = \lim_{\Delta x \to 0} \frac{f(0 + \Delta x) - f(0)}{\Delta x} = \lim_{\Delta x \to 0} \frac{[(\Delta x)^{1/n}]^m - 0}{\Delta x}$$

$$= \lim_{\Delta x \to 0} \frac{[(\Delta x)^{1/n}]^m}{[(\Delta x)^{1/n}]^n} = \lim_{\Delta x \to 0} [(\Delta x)^{1/n}]^{m-n} = 0.$$

If we combine the rational exponent rule with the chain rule, we obtain the following important result: If r is a rational number and u is a differentiable function of x, then

$$D_x u^r = ru^{r-1} D_x u \quad \text{or} \quad \frac{d}{dx} u^r = ru^{r-1} \frac{du}{dx}.$$

EXAMPLES 1 Find $D_x x^{3/2}$.

SOLUTION

$$D_x x^{3/2} = \tfrac{3}{2}x^{(3/2)-1} = \tfrac{3}{2}x^{1/2} = \tfrac{3}{2}\sqrt{x}.$$

2 If $y = \sqrt[3]{2x^2 - 3}$, find dy/dx.

SOLUTION

$$\frac{dy}{dx} = \frac{d}{dx} \sqrt[3]{2x^2 - 3} = \frac{d}{dx} (2x^2 - 3)^{1/3} = \frac{1}{3} (2x^2 - 3)^{(1/3)-1} \frac{d}{dx} (2x^2 - 3)$$

$$= \tfrac{1}{3}(2x^2 - 3)^{-2/3}(4x) = \tfrac{4}{3}x(2x^2 - 3)^{-2/3}.$$

3 If $f(x) = (1 - x)^{4/5}(1 + x^2)^{-2/3}$, find $f'(x)$.

SOLUTION
Using the product rule, the rational function rule, and the chain rule, we have

$$f'(x) = \tfrac{4}{5}(1 - x)^{-1/5}(-1)(1 + x^2)^{-2/3} + (1 - x)^{4/5}(-\tfrac{2}{3})(1 + x^2)^{-5/3}(2x)$$

$$= (1 - x)^{-1/5}(1 + x^2)^{-5/3}[-\tfrac{4}{5}(1 + x^2) - \tfrac{2}{3}(1 - x)(2x)]$$

$$= \tfrac{4}{15}(1 - x)^{-1/5}(1 + x^2)^{-5/3}(2x^2 - 5x - 3).$$

4 If u is a differentiable function of x, show that

$$D_x |u| = \frac{u}{|u|} D_x u \quad \text{holds for} \quad u \neq 0.$$

SOLUTION
Since $|u| = (u^2)^{1/2}$, then, for $u \neq 0$,

$$D_x |u| = D_x(u^2)^{1/2} = \tfrac{1}{2}(u^2)^{-1/2}D_x u^2 = \tfrac{1}{2}(u^2)^{-1/2}(2uD_x u)$$

$$= \frac{u}{(u^2)^{1/2}} D_x u = \frac{u}{|u|} D_x u.$$

5.3 Further Examples of the Inverse Function Theorem

Later in this book, in particular in Section 5 of Chapter 8 and in Section 3.2 of Chapter 9, we use the inverse function theorem to calculate certain derivatives. Here we give further examples of its use.

EXAMPLES **1** Let f be the function defined by $f(x) = x^3 + x + 1$.
(a) Show that f^{-1} exists.
(b) Find $(f^{-1})'(1)$.

SOLUTION
(a) Since $f'(x) = 3x^2 + 1 \neq 0$ for all real numbers x, it follows from Theorem 1 that f^{-1} exists.
(b) Evidently, $f(0) = 1$, so $f^{-1}(1) = f^{-1}[f(0)] = 0$. Hence, by Theorem 1,

$$(f^{-1})'(1) = \frac{1}{f'[f^{-1}(1)]} = \frac{1}{f'(0)} = \frac{1}{3(0)^2 + 1} = 1.$$

2 Suppose that the hypotheses of the inverse function theorem are satisfied by the function f, that $f(3) = 7$, and that $f'(3) = 2$. Find $(f^{-1})'(7)$.

SOLUTION
Since $f(3) = 7$, it follows that $f^{-1}(7) = 3$. (Why?) Hence,

$$(f^{-1})'(7) = \frac{1}{f'[f^{-1}(7)]} = \frac{1}{f'(3)} = \frac{1}{2}.$$

Problem Set 5

In problems 1 through 8, use the Leibniz notation and the inverse function rule to find the value of dx/dy when y has the value a.

1 $a = 1, y = x^5$

2 $a = 64, y = x^6, x > 0$

3 $a = 4, y = x^2 + 2x + 1, x > -1$

4 $a = -3, y = \dfrac{2x + 3}{x - 1}$

5 $a = -1, y = \dfrac{7x - 2}{2x - 7}$

6 $a = \frac{1}{2}, y = \sin x, -\dfrac{\pi}{2} < x < \dfrac{\pi}{2}$ $\left(\text{assume that } \dfrac{d}{dx} \sin x = \cos x\right)$

7 $a = \frac{2}{3}\sqrt{3}, y = \dfrac{x}{\sqrt{x^2 - 1}}, x > 1$

8 a is arbitrary, $y = mx + b, m \neq 0$

In problems 9 through 26, find the derivative of each function. (Use the root rule and the rational power rule together with the basic rules for differentiation.)

9 $f(x) = \sqrt[5]{x}$

10 $f(x) = \sqrt[7]{x^2}$

11 $g(x) = 36x^{-4/9}$

12 $f(x) = 21x^{5/7}$

13 $h(t) = (1 - t)^{-2/3}$

14 $f(x) = \dfrac{1}{\sqrt[5]{x^4}}$

15 $g(s) = \sqrt{\dfrac{9 - s^2}{9 + s^2}}$

16 $f(u) = \left(1 + \dfrac{2}{u}\right)^{3/4}$

17 $g(x) = x^{-1/2} + x^{-1/3} + x^{-1/4}$

18 $f(x) = \sqrt{x} + \sqrt[3]{x} + \sqrt[4]{x}$

19 $f(t) = \sqrt[5]{t^3} - \sqrt[4]{t}$

20 $g(y) = \sqrt{y^4 - y + \sqrt[3]{y}}$

21 $g(x) = \sqrt[10]{\dfrac{x}{x+1}}$

22 $f(x) = (x + \sqrt{x})(x - 2\sqrt{x})$

23 $h(x) = (1 + x)^{-3/4}(2x + 1)^{1/2}$

24 $f(t) = \dfrac{t}{\sqrt{36 - t^2}}$

25 $f(t) = \sqrt[4]{t} + 2\sqrt[5]{t} + 5$

26 $g(x) = \sqrt[3]{x}(1 + 2\sqrt{x})$

In problems 27 through 30, find the derivative of the given function. Assume the derivatives of the trigonometric functions as given in problem 41 of Problem Set 4.

27 $f(t) = \sqrt[5]{\sin t}$

28 $g(x) = \sqrt[7]{\cos 3x}$

29 $g(x) = \cos^{3/4} x$

30 $h(t) = \sin^{5/7}(4t - 1)$

31 Find $D_x(\sqrt[4]{x})$:
   (a) By writing $\sqrt[4]{x} = \sqrt{\sqrt{x}}$ and using the chain rule.
   (b) By the inverse function rule.
   (c) By writing $\sqrt[4]{x} = x^{1/4}$ and using the power rule for rational exponents.

32 In the last part of the proof of Theorem 3, (a) where did we use the assumption that n is odd, and (b) where did we use the assumption that $m > n$?

33 If $f(x) = x|x|$, find $f'(x)$. [To find $f'(0)$, use the definition of $f'(0)$ as a limit of a difference quotient.]

34 If f is the function defined by

$$f(x) = \begin{cases} x & \text{if } x < 1 \\ x^2 & \text{if } 1 \le x \le 9 \\ 27\sqrt{x} & \text{if } x > 9, \end{cases}$$

find $(f^{-1})'(x)$ if it exists.

In problems 35 through 40, use the given information and the inverse function theorem to find $(f^{-1})'(a)$. (You may assume that the hypotheses of the theorem are satisfied.)

35 $a = 7$, $f(3) = 7$, $f'(3) = 2$

36 $a = 2$, $f(2) = 5$, $f(5) = 2$, $f'(5) = 7$, $f'(2) = 6$

37 $a = -1$, $f(-1) = -2$, $f(4) = -1$, $f'(-1) = -3$, $f'(4) = \frac{1}{7}$

38 $a = \frac{1}{3}$, $f(\frac{1}{3}) = \frac{2}{3}$, $f(1) = \frac{1}{3}$, $f'(1) = \frac{2}{3}$, $f'(\frac{1}{3}) = 1$

39 $a = \dfrac{\sqrt{2}}{2}$, $f\left(\dfrac{\pi}{4}\right) = \dfrac{\sqrt{2}}{2}$, $f'\left(\dfrac{\pi}{4}\right) = \dfrac{\sqrt{2}}{2}$

40 $a = 0$, $f(0) = 0$, $f'(0) = 1$

41 Let f be the function defined by $f(x) = \dfrac{2x - 1}{3x - 2}$.

   (a) Show that $f^{-1} = f$.
   (b) Calculate $(f^{-1})'(x)$ by using part (a) and the quotient rule.
   (c) Calculate $(f^{-1})'(x)$ by using the inverse function theorem.

42 Let f be the function defined by $f(x) = \dfrac{x + 1}{2x - 3}$.

   (a) Show that $f^{-1}(x) = \dfrac{3x + 1}{2x - 1}$.
   (b) Calculate $(f^{-1})'(0)$ by using part (a) and the quotient rule.
   (c) Calculate $(f^{-1})'(0)$ by using the inverse function theorem.

43 Let f be the function defined by the equation $y = f(x) = 2x^2 - x + 1$, $x > \frac{1}{4}$.
   (a) Use the quadratic formula to solve the equation for x in terms of y.
   (b) Use part (a) to find an equation defining f^{-1}.
   (c) Calculate $(f^{-1})'(y)$ using part (b).
   (d) Find $(f^{-1})'(y)$ using the inverse function theorem.

44 Let f be the function defined by the equation $f(x) = x^3 - x^2 + 1$, $x > \frac{2}{3}$. Use the inverse function theorem to find $(f^{-1})'(5)$. [Note that $f(2) = 5$, so $f^{-1}(5) = 2$.]

45 Let f be the function defined by the equation $f(x) = \dfrac{x^3 - 1}{x^2 + 1}$ for $x > 0$. Since $f(1) = 0$, then $f^{-1}(0) = 1$. Use the inverse function rule to find $(f^{-1})'(0)$.

6   The Equations of Tangent and Normal Lines

Suppose that the function f is differentiable at x_1, so that $f'(x_1)$ is the slope of the tangent line to the graph of f at the point $(x_1, f(x_1))$. If $y_1 = f(x_1)$, the equation of this tangent line in point-slope form is

$$y - y_1 = f'(x_1)(x - x_1).$$

EXAMPLE   Find the equation of the tangent line to the graph of $f(x) = 4 - x^2$ at the point $(1, 3)$. Sketch the graph.

SOLUTION
Here, $f'(x) = -2x$, so that $f'(1) = -2$. The equation of the tangent line is

$$y - 3 = -2(x - 1) \quad \text{or} \quad y = -2x + 5 \quad \text{(Figure 1)}.$$

Figure 1   *Figure 2*

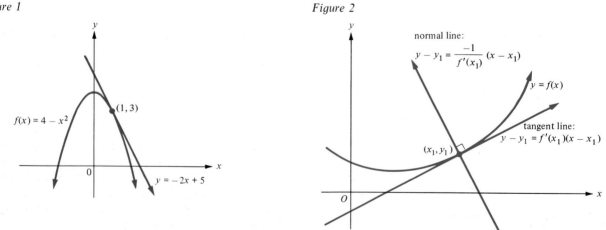

The *normal line* to the graph of f at the point (x_1, y_1) is defined to be the straight line through (x_1, y_1) that is perpendicular to the tangent line at (x_1, y_1) (Figure 2).

Since $f'(x_1)$ is the slope of the tangent line to the graph of f at (x_1, y_1), it follows from Theorem 3 in Section 4.1 of Chapter 0 that $-1/f'(x_1)$ is the slope of the normal line at (x_1, y_1). Consequently, the equation in point-slope form of the normal line is

$$y - y_1 = \frac{-1}{f'(x_1)}(x - x_1).$$

EXAMPLE   Find the equations of the tangent and normal lines to the graph of $f(x) = 1/x$ at the point $(\tfrac{1}{2}, 2)$. Illustrate graphically.

Figure 3

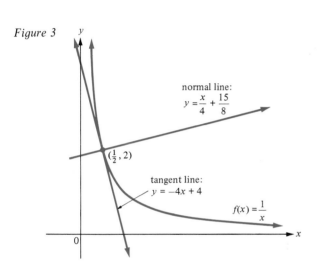

SOLUTION
Here, $f'(x) = -1/x^2$, so that $f'(\tfrac{1}{2}) = -1/(\tfrac{1}{2})^2 = -4$. Therefore, the equation of the tangent line is

$$y - 2 = -4(x - \tfrac{1}{2}) \quad \text{or} \quad y = -4x + 4.$$

Since the tangent line at $(\tfrac{1}{2}, 2)$ has slope -4, the normal line at this point has slope $-1/(-4) = \tfrac{1}{4}$. Hence, the equation of the normal line is

$$y - 2 = \frac{1}{4}\left(x - \frac{1}{2}\right) \quad \text{or} \quad y = \frac{x}{4} + \frac{15}{8}$$

(Figure 3).

It is sometimes useful to be able to find the point (or those points) $(x_1, f(x_1))$ on the graph of a differentiable function f at which the tangent line has a prescribed direction. If the slope of a straight line in this direction is m, it is only necessary to solve the equation $f'(x_1) = m$ for x_1 in order to find the desired value (or values) of x_1. This technique is illustrated by the following example.

EXAMPLE   If $f(x) = 2x^2 - x$, find the point on the graph of f where the tangent line is parallel to the straight line $3x - y - 4 = 0$, find the equation of the tangent line at this point, and sketch the graph.

Figure 4

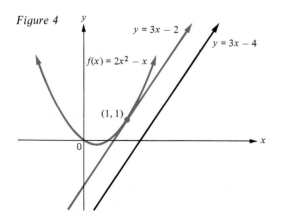

SOLUTION
Here, $f'(x) = 4x - 1$. The line $3x - y - 4 = 0$ has slope 3; hence, the x coordinate x_1 of the desired point must satisfy the equation $f'(x_1) = 3$; that is, $4x_1 - 1 = 3$. Solving the latter equation, we find that $x_1 = 1$; hence, the desired point on the graph of f is given by $(x_1, f(x_1)) = (1, f(1)) = (1, 1)$. The equation of the tangent line at $(1, 1)$ is

$$y - 1 = 3(x - 1) \quad \text{or} \quad y = 3x - 2$$

(Figure 4).

Problem Set 6

In problems 1 through 10, find the equations of the tangent and normal lines to the graph of the given function at the indicated point. Illustrate graphically in problems 1 through 5.

1 $f(x) = 2x^2 - 7$ at $(2, 1)$ **2** $F(x) = 5 + 2x - x^2$ at $(0, 5)$ **3** $g(x) = x^2 + x + 1$ at $(1, 3)$

4 $G(x) = \sqrt[3]{x} - 1$ at $(1, 0)$ **5** $h(x) = \sqrt[3]{x}$ at $(8, 2)$ **6** $H(x) = 3x^{2/3}$ at $(8, 12)$

7 $f(x) = \frac{3}{2}\sqrt{4 - x^2}$ at $(0, 3)$ **8** $F(x) = x^3 - 8x^2 + 9x + 20$ at $(4, -8)$

9 $g(x) = \dfrac{x^2 - 1}{x^2 + 1}$ at $(1, 0)$ **10** $G(x) = ax^2 + bx + c$ at $(0, c)$

11 Find the point where the normal line to $f(x) = 2/x$ at the point $(1, 2)$ crosses (a) the x axis and (b) the y axis.

12 Find the x and y intercepts of the tangent line to $y = 2\sqrt{x}$ at the point $(1, 2)$.

13 At what point on the curve $y = x^2 + 8$ is the slope of the tangent line 16? Write the equation of this tangent line.

14 Determine a value of the constant b so that the graph of $y = x^2 + bx + 17$ has a horizontal tangent at the point $(2, 21 + 2b)$.

15 At what point on the curve $y = 3x^2 + 5x + 6$ is the tangent line parallel to the x axis?

16 For what values of x is the tangent line to the curve $y = ax^3 + bx^2 + cx + d$ at the point (x, y) parallel to the x axis?

In problems 17 through 21, find a point on the graph of the given function where the tangent line (or normal line) satisfies the stipulated condition, and then write the equation of this tangent line (or normal line).

17 The tangent line to $f(x) = x - x^2$ is parallel to the line $x + y - 2 = 0$.

18 The tangent line to $f(x) = 2x^3 - x^2$ is parallel to the line $4x - y + 3 = 0$.

19 The normal line to $f(x) = x - 1/x$ is parallel to the line $x + 2y - 3 = 0$.

20 The normal line to $f(x) = \sqrt{4x - 3}$ is perpendicular to the line $3x - 2y + 3 = 0$.

21 The tangent line to $f(x) = 5 + x^2$ intersects the x axis at the point $(2, 0)$.

7   The Use of Derivatives to Approximate Function Values

The simple geometric fact that the tangent line to a curve is a good approximation to the curve near the point of tangency P (Figure 1) can be used to advantage in calculating the approximate values of functions.

Suppose, for instance, that we wish to find an approximate value for $\sqrt{1.02}$ with a minimum of calculation. We have $\sqrt{1} = 1$, so we know that $\sqrt{1.02}$ is a little larger than 1. To pin down its value more accurately, consider the graph of $y = \sqrt{x}$ near $x = 1$. Figure 2 shows the tangent line to this graph at $(1, 1)$. Since

$$\frac{dy}{dx} = \frac{d}{dx}\sqrt{x} = \frac{1}{2\sqrt{x}},$$

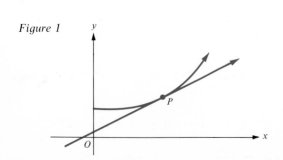

Figure 1

Figure 2

Figure 2

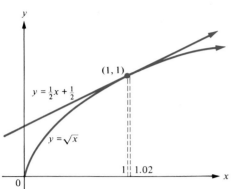

it follows that the slope of this tangent line is given by $m = 1/(2\sqrt{1}) = \frac{1}{2}$; hence, its equation in point-slope form is $y - 1 = \frac{1}{2}(x - 1)$. Solving the latter equation for y, we find that the equation of the tangent line is $y = \frac{1}{2}x + \frac{1}{2}$.

Since the graph of $y = \sqrt{x}$ and the graph of $y = \frac{1}{2}x + \frac{1}{2}$ are very nearly the same for values of x close to 1 (Figure 2), a good approximation to $\sqrt{1.02}$ should be provided by $\frac{1}{2}(1.02) + \frac{1}{2} = 1.01$. The correct value of $\sqrt{1.02}$ to six decimal places is 1.009950, so the approximation $\sqrt{1.02} \approx 1.01$ is quite accurate.

We now generalize the procedure above to make it applicable to a wide class of approximation problems. Let f be a function, suppose that x_1 is a number in the domain of f, and assume that the function value $y_1 = f(x_1)$ is known. Our problem is to estimate the function value $f(x)$ when x is a number close to x_1. Suppose that f is differentiable at x_1, so that the tangent line to the graph of f at (x_1, y_1) has a slope given by $m = f'(x_1)$; hence, its equation is

Figure 3

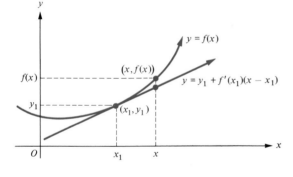

$$y - y_1 = f'(x_1)(x - x_1) \text{ or } y = y_1 + f'(x_1)(x - x_1)$$

(Figure 3).

Evidently, for values of x close to x_1, the height of the graph of f above $(x, 0)$ and the height of the tangent line above this point are approximately the same; that is, $f(x)$ is approximately the same as

$$y_1 + f'(x_1)(x - x_1).$$

This suggests the following procedure.

Linear Approximation Procedure

Suppose that the function f is differentiable at the number x_1 and that $y_1 = f(x_1)$ is known. Then, for values of x near x_1,

$$f(x) \approx y_1 + f'(x_1)(x - x_1);$$

that is,

$$f(x) \approx f(x_1) + f'(x_1)(x - x_1).$$

This procedure is called a *linear* approximation procedure because it is based on the use of a straight line (the tangent line).

EXAMPLE Use the linear approximation procedure to approximate 1/1.03. For comparison, also find the correct value to four decimal places.

SOLUTION
Let $f(x) = 1/x$, so $f'(x) = -1/x^2$. Using the linear approximation procedure with $x_1 = 1$, we obtain

$$\frac{1}{1.03} = f(1.03) \approx f(1) + f'(1)(1.03 - 1) = 1 + \frac{-1}{1^2}(1.03 - 1) = 0.97.$$

The correct value to four decimal places is 0.9709.

An approximation procedure is of limited use unless there is some means for specifying bounds on the error that might be involved. The error in the linear approximation procedure is given by

$$\text{error} = \text{true value} - \text{approximate value}$$
$$= f(x) - [f(x_1) + f'(x_1)(x - x_1)].$$

Notice that this error depends on x; hence, determines a function E by the equation

$$E(x) = f(x) - f(x_1) - f'(x_1)(x - x_1).$$

Here we do not concern ourselves with putting bounds on the numerical value $E(x)$ of the error—such matters are discussed in Section 5 of Chapter 12. However, we can now make a few simple remarks concerning the function E.

Figure 4

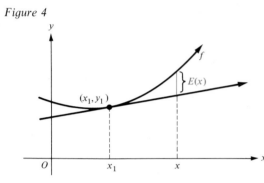

Geometrically, $E(x)$ is just the difference between the height above $(x, 0)$ of the graph of f and the height above $(x, 0)$ of the tangent line to this graph at (x_1, y_1) (Figure 4). Notice that

$$\lim_{x \to x_1} E(x) = \lim_{x \to x_1} [f(x) - f(x_1) - f'(x_1)(x - x_1)] = 0.$$

(Why?)

This equation expresses the fact that the error approaches 0 as x approaches x_1, a fact that is not really as significant as it might at first appear to be. Indeed, if, rather than using the tangent line to obtain a linear approximation for $f(x)$ (Figures 3 and 4), we had used *any other nonvertical straight line* containing the point (x_1, y_1), the error of approximation would still approach 0 as x approaches x_1 (Figure 5).

The really crucial feature of the error $E(x)$ is not merely that $E(x)$ approaches 0 as x approaches x_1, but that, as x *approaches* x_1, $E(x)$ approaches 0 so rapidly that the ratio $\dfrac{E(x)}{x - x_1}$ still approaches zero.

Figure 5

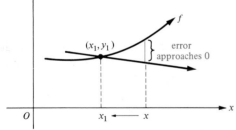

Notice that, as x approaches x_1, the denominator $x - x_1$ approaches 0, which might be expected to

make the ratio $\dfrac{E(x)}{x - x_1}$ become very large in absolute value; however, the numerator $E(x)$ approaches 0 so rapidly that the influence of the denominator is wiped out!

To see that $\lim\limits_{x \to x_1} \dfrac{E(x)}{x - x_1} = 0$, we calculate as follows:

$$\lim_{x \to x_1} \frac{E(x)}{x - x_1} = \lim_{x \to x_1} \frac{f(x) - f(x_1) - f'(x_1)(x - x_1)}{x - x_1}$$

$$= \lim_{x \to x_1} \left[\frac{f(x) - f(x_1)}{x - x_1} - f'(x_1) \right] = \lim_{x \to x_1} \left[\frac{f(x) - f(x_1)}{x - x_1} \right] - f'(x_1).$$

Now let $\Delta x = x - x_1$, so that $x = x_1 + \Delta x$ and the condition $x \to x_1$ is equivalent to the condition $\Delta x \to 0$. It follows that

$$\lim_{x \to x_1} \frac{E(x)}{x - x_1} = \lim_{\Delta x \to 0} \left[\frac{f(x_1 + \Delta x) - f(x_1)}{\Delta x} \right] - f'(x_1) = f'(x_1) - f'(x_1) = 0.$$

For future reference, we state the result just obtained in the form of the following theorem.

THEOREM 1   **Linear Approximation Theorem**
If the function f is differentiable at the number x_1, then there is a function ε with the same domain as f such that:

(i) $f(x) = f(x_1) + f'(x_1)(x - x_1) + \varepsilon(x)(x - x_1)$.

(ii) $\varepsilon(x_1) = 0$.

(iii) $\lim\limits_{x \to x_1} \varepsilon(x) = 0$.

PROOF
Let $E(x) = f(x) - f(x_1) - f'(x_1)(x - x_1)$. We have already seen that

$$\lim_{x \to x_1} \frac{E(x)}{x - x_1} = 0.$$

Define the function ε by the equation

$$\varepsilon(x) = \begin{cases} \dfrac{E(x)}{x - x_1} & \text{if } x \neq x_1 \text{ and } x \text{ is in the domain of } f \\ 0 & \text{if } x = x_1. \end{cases}$$

Then conditions (ii) and (iii) hold. Moreover, $\varepsilon(x)(x - x_1) = E(x)$ holds [even for $x = x_1$, since $E(x_1) = 0$]; hence,

$$\varepsilon(x)(x - x_1) = f(x) - f(x_1) - f'(x_1)(x - x_1);$$

that is,

$$f(x) = f(x_1) + f'(x_1)(x - x_1) + \varepsilon(x)(x - x_1).$$

Thus, (i) holds and the proof is complete.

In Theorem 1, notice that $E(x) = \varepsilon(x)(x - x_1)$. This formulation of $E(x)$ as a product emphasizes the fact that the error approaches 0 rapidly as x approaches x_1, since *both* of the factors $\varepsilon(x)$ and $(x - x_1)$ approach 0 as x approaches x_1.

EXAMPLE   Use the linear approximation theorem to write the function value

$$f(x) = 2x^3 + 5x^2 - x + 5$$

near $x_1 = 3$ as the sum of a linear term and an error term.

SOLUTION
Here $f'(x) = 6x^2 + 10x - 1$, so that $f'(3) = 83$. Since $f(3) = 101$, the linear approximation theorem yields

$$f(x) = f(3) + f'(3)(x - 3) + \varepsilon(x)(x - 3);$$

that is,

$$2x^3 + 5x^2 - x + 5 = 101 + 83(x - 3) + \varepsilon(x)(x - 3).$$

If $x \neq 3$, we can solve the latter equation for $\varepsilon(x)$ to get

$$\varepsilon(x) = \frac{2x^3 + 5x^2 - x + 5 - 101 - 83(x - 3)}{x - 3}$$

$$= \frac{2x^3 + 5x^2 - 84x + 153}{x - 3} = 2x^2 + 11x - 51$$

$$= (x - 3)(2x + 17).$$

Since $\varepsilon(3) = 0$, the equation $\varepsilon(x) = (x - 3)(2x + 17)$ holds even when $x = 3$. Thus, we have

$$\underbrace{2x^3 + 5x^2 - x + 5}_{f(x)} = \underbrace{101 + 83(x - 3)}_{\text{linear term}} + \underbrace{\overbrace{(x - 3)(2x + 17)}^{\varepsilon(x)}(x - 3)}_{\text{error term}},$$

where $\varepsilon(x)$ approaches 0 as x approaches 3.

The linear approximation theorem has a converse that provides yet another interpretation of the derivative (in addition to the "rate of change" and "slope of the tangent line" interpretations).

THEOREM 2   **Converse of the Linear Approximation Theorem**
Let f be a function defined at least on an open interval (a, b) containing the number x_1. Suppose that there is a function ε defined at least on the open interval (a, b) and that there are constants m and c such that

$$f(x) = mx + c + \varepsilon(x)(x - x_1) \qquad \text{for } a < x < b.$$

Then, if $\lim\limits_{x \to x_1} \varepsilon(x) = 0$, it follows that:

(i) f is differentiable at x_1.
(ii) $f'(x_1) = m$.
(iii) $c = f(x_1) - f'(x_1)x_1$.

The proof of Theorem 2 is straightforward and is left as an exercise (Problem 16).

7.1 A Proof of the Chain Rule

We have already mentioned that the chain rule is one of the most important of all the differentiation rules—the Leibniz notation $\dfrac{dy}{dx} = \dfrac{dy}{du}\dfrac{du}{dx}$ just makes a relatively deep analytical fact *appear* to be nothing but an algebraic triviality. In this section we state the chain rule precisely in the form of a theorem, and we use the linear approximation theorem and its converse to prove this theorem.

THEOREM 3   **The Chain Rule**
Let f and g be functions; suppose that g is differentiable at the number x_1 and that f is differentiable at the number $g(x_1)$. Then the composite function $f \circ g$ is differentiable at x_1 and

$$(f \circ g)'(x_1) = f'[g(x_1)]g'(x_1).$$

PROOF
For simplicity of calculation, put $u_1 = g(x_1)$, $A = f'(u_1)$, and $B = g'(x_1)$. We must prove that $(f \circ g)'(x_1) = AB$. Applying Theorem 1 to g near x_1, we obtain

$$g(x) = g(x_1) + g'(x_1)(x - x_1) + \varepsilon_1(x)(x - x_1);$$

that is,

$$g(x) - u_1 = [B + \varepsilon_1(x)](x - x_1), \qquad \text{where } \lim_{x \to x_1} \varepsilon_1(x) = 0 = \varepsilon_1(x_1).$$

Similarly, applying Theorem 1 to f near u_1, we obtain

$$f(u) = f(u_1) + f'(u_1)(u - u_1) + \varepsilon_2(u)(u - u_1);$$

that is,

$$f(u) = f[g(x_1)] + [A + \varepsilon_2(u)](u - u_1), \qquad \text{where } \lim_{u \to u_1} \varepsilon_2(u) = 0 = \varepsilon_2(u_1).$$

In the latter equation, we put $u = g(x)$ to get

$$f[g(x)] = f[g(x_1)] + \{A + \varepsilon_2[g(x)]\}[g(x) - u_1];$$

that is,

$$(f \circ g)(x) = (f \circ g)(x_1) + [A + (\varepsilon_2 \circ g)(x)][g(x) - u_1].$$

But, we saw above that $g(x) - u_1 = [B + \varepsilon_1(x)](x - x_1)$, so that

$$(f \circ g)(x) = (f \circ g)(x_1) + [A + (\varepsilon_2 \circ g)(x)][B + \varepsilon_1(x)](x - x_1).$$

The latter equation can be rewritten as

$$(f \circ g)(x) = ABx + c + \varepsilon(x)(x - x_1),$$

where

$$c = (f \circ g)(x_1) - ABx_1 \quad \text{and} \quad \varepsilon(x) = B(\varepsilon_2 \circ g)(x) + A\varepsilon_1(x) + [(\varepsilon_2 \circ g)(x)]\varepsilon_1(x).$$

If we can show that $\lim_{x \to x_1} \varepsilon(x) = 0$, we can apply Theorem 2 to the equation $(f \circ g)(x) = ABx + c + \varepsilon(x)(x - x_1)$ and conclude that $(f \circ g)'(x_1) = AB$, as desired. Since g is differentiable at the number x_1, then (by Theorem 1 of Section 2.2) g is continuous at x_1. Since $\lim_{u \to u_1} \varepsilon_2(u) = 0 = \varepsilon_2(u_1)$, then ε_2 is continuous at the number $u_1 = g(x_1)$. Hence, by Property 3 of continuous functions in Section 4 of Chapter 1 (see also Problems 10 and 11 in Chapter 1, Problem Set 7), $\varepsilon_2 \circ g$ is continuous at the number x_1. Therefore,

$$\lim_{x \to x_1} (\varepsilon_2 \circ g)(x) = (\varepsilon_2 \circ g)(x_1) = \varepsilon_2[g(x_1)] = \varepsilon_2(u_1) = 0.$$

Consequently,

$$\lim_{x \to x_1} \varepsilon(x) = \lim_{x \to x_1} \{B(\varepsilon_2 \circ g)(x) + A\varepsilon_1(x) + [(\varepsilon_2 \circ g)(x)]\varepsilon_1(x)\}$$
$$= B \lim_{x \to x_1} (\varepsilon_2 \circ g)(x) + A \lim_{x \to x_1} \varepsilon_1(x) + \left[\lim_{x \to x_1} (\varepsilon_2 \circ g)(x)\right]\left[\lim_{x \to x_1} \varepsilon_1(x)\right]$$
$$= B(0) + A(0) + (0)(0) = 0,$$

as desired.

Problem Set 7

In problems 1 through 8, use the linear approximation procedure

$$f(x) \approx f(x_1) + f'(x_1)(x - x_1)$$

to estimate each quantity.

1 $\sqrt{9.06}$

2 $(3.07)^3$

3 $\sqrt{36.1}$

4 $\sqrt{35.99}$

5 $\dfrac{1}{2.06}$

6 $\dfrac{1}{1.02}$

7 $x^2 + 2x - 3$ at $x = 1.07$

8 $x^2 + 2x - 3$ at $x = -3.02$

9 In the odd-numbered problems 1 through 7, calculate the true value of the quantity to several decimal places and compare it with your estimate.

In problems 10 through 15, represent the function near the given number x_1 as a linear function plus an error according to the linear approximation theorem.

10 $f(x) = x^2$ near $x_1 = 0$

11 $f(x) = x^3 - 3x - 2$ near $x_1 = 2$

12 $f(x) = 1 - 3x + 2x^2$ near $x_1 = -1$

13 $f(x) = 2x^3$ near $x_1 = 3$

14 $f(x) = \dfrac{2x}{x - 2}$ near $x_1 = 0$

15 $f(x) = \dfrac{5}{x}$ near $x_1 = 5$

16 Prove Theorem 2, the converse of the linear approximation theorem.

17 A *root* or a *zero* of a function f is defined to be a number a in the domain of f such that $f(a) = 0$. Suppose that a is a root of f and that the number b is near a as in Figure 6. If the graph of f has a tangent line at $(b, f(b))$, then, as Figure 6 shows, one might expect that the coordinate c of the point where this tangent line intersects the x axis is a better approximation to the root a than b is. Show that if $f'(b) \neq 0$, then $c = b - [f(b)/f'(b)]$.

Figure 6

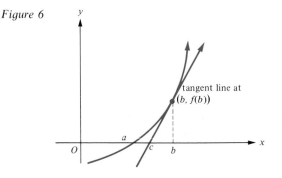

18 Problem 17 provides the basis of *Newton's method* for approximating the roots of functions. Given a first approximation b to a root of f,

$$c = b - [f(b)/f'(b)]$$

will usually be a better approximation to this root. Using $b = 2.1$ as a first approximation to a root of the function f defined by $f(x) = x^5 - 4x^3 - 2$ and Newton's method, find a better approximation to this root. Then, use this approximation and Newton's method again to find a still better approximation.

19 (a) Let $\varepsilon(x) = x^2 + x - 2$. Determine constants m and c so that

$$x^3 = mx + c + \varepsilon(x)(x - 1).$$

(b) Verify that $\lim\limits_{x \to 1} \varepsilon(x) = 0$.

(c) Use the converse of the linear approximation theorem to identify m as the value of a certain derivative at $x = 1$.

20 Use the linear approximation theorem to give another proof that if a function f is differentiable at x_1, it is continuous at x_1.

Review Problem Set

1 An object is thrown (vertically) upward with such a speed that at any time its distance s, measured in feet positively upward from the surface of the earth, is given by the equation $s = 200t - 16t^2$, where t is measured in seconds. Find:
   (a) The average speed during the third second of motion.
   (b) The instantaneous speed when $t = 2$ and when $t = 3$.
   (c) The highest point reached by the object.

2 The motion of a particle along the x axis is given by the equation $x = 2t - (t^2/2)$, where x is the coordinate of the particle at time t. Distance is in feet and time in seconds.
   (a) Determine whether the particle is moving in the positive or negative direction when $t = 0$.

(b) Find the instantaneous speed when $t = 1$.

(c) When does the particle change its direction of motion?

3 The side of a square metal plate is 20 inches when its temperature is 50°. If its temperature is raised to 75° in 10 minutes, each side expands 0.2 inch. Find:

(a) The average rate of change of the area of the plate per inch change in the length of a side.

(b) The average rate of change of the area per degree change in temperature.

4 The volume V of a certain quantity of gas varies with the pressure P according to the law $V = 100/P$. Find a general expression for the instantaneous rate of change of the volume per unit change in pressure.

In problems 5 through 8, find the equations of the tangent and normal lines to the graph of the given function at the indicated point.

5 $f(x) = x^2 - 4x + 2; (4, 2)$

6 $g(x) = \dfrac{4}{x + 1}; (2, \tfrac{4}{3})$

7 $f(x) = \tfrac{1}{3}x^4; (\tfrac{3}{2}, \tfrac{27}{16})$

8 $f(x) = 96x - \tfrac{1}{2}x^3; (0, 0)$

9 Show that the tangent line at the point $(1, 1)$ to the graph of $f(x) = x^k$, where k is a positive integer, intersects the y axis at a point that is $k - 1$ units away from the origin.

10 Indicate which of the functions in Figure 1 is (i) continuous on (a, b), (ii) differentiable on (a, b), and (iii) continuous and differentiable on (a, b).

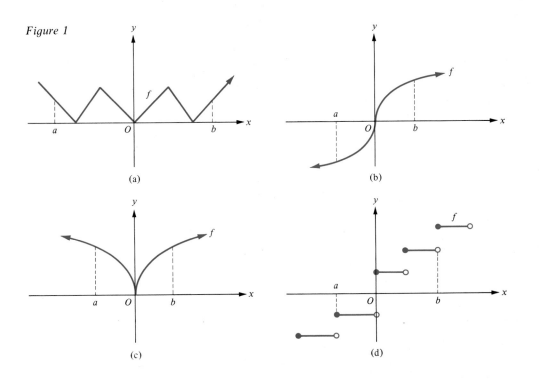

Figure 1

(a)

(b)

(c)

(d)

In problems 11 through 13: (a) Indicate whether the function f is continuous at $x = 2$. (b) Find $f'_-(a)$ and $f'_+(a)$. (c) Is f differentiable at $x = a$? (d) Sketch the graph.

11 $f(x) = \begin{vmatrix} 4 - 3x & \text{if } x \le 2 \\ x^2 - 6 & \text{if } x > 2 \end{vmatrix}; a = 2$

12 $f(x) = 2 + |x - 1|; a = 1$

13 $f(x) = \begin{vmatrix} 7 - x & \text{if } x \le -3 \\ 10 & \text{if } x > -3 \end{vmatrix}; a = -3$

14 Suppose that f is a differentiable function at $x = a$. Show that

$$f'(a) = \lim_{h \to 0} \frac{f(a + h) - f(a - h)}{2h}.$$

15 Let $f(x) = x^{2/3}$.
  (a) Write out the definition of $f'(x)$ in symbols.
  (b) With the aid of the definition of part (a), find

$$\lim_{\Delta x \to 0} \frac{(8 + \Delta x)^{2/3} - 4}{\Delta x}.$$

16 Let f be the function defined by the equation $f(x) = x^3$.
  (a) Find $[f'(x)]^2$ and $f'(x^2)$.
  (b) If $g(x) = f(x^2)$, compare $f'(x^2)$ and $g'(x)$.

17 Suppose that m and n are positive integers and $n > m$.
  (a) Does $D_x(x^n \cdot x^m) = D_x x^n \cdot D_x x^m$? Why?
  (b) Does $D_x \left[\dfrac{x^n}{x^m} \right] = \dfrac{D_x x^n}{D_x x^m}$, $x \neq 0$? Why?

18 Suppose that f is a differentiable function on an interval I such that $f(x + y) = f(x) + f(y)$. Then for x in I, $\dfrac{f(x + h) - f(x)}{h} = \dfrac{f(h)}{h}$, so that $f'(x) = f'(0)$ and f' is a constant function. What is the graph of f? Sketch the graph of f.

19 Let f be a function defined by the equation $f(x) = |x| + |x + 1|$.
  (a) Sketch the graph of f.
  (b) Determine the points at which f is not differentiable.

20 Suppose that the function g is continuous at 0, and f is a function defined by $f(x) = xg(x)$. Is f differentiable at 0? If so, find $f'(0)$ in terms of g.

21 Suppose that f and g are differentiable functions at 7 and that $f(7) = 10$, $f'(7) = 3$, $g(7) = 5$, and $g'(7) = -\frac{1}{30}$. Find:

  (a) $(f + g)'(7)$ (b) $(f - g)'(7)$ (c) $(fg)'(7)$ (d) $\left(\dfrac{f}{g}\right)'(7)$

  (e) $\left(\dfrac{f + 3g}{f}\right)'(7)$ (f) $(f + 2g)'(7)$ (g) $\left(\dfrac{f}{f + g}\right)'(7)$

22 Let f and g be differentiable functions such that $f(2) = 8$, $g(2) = 0$, $f'(0) = 12$, and $g'(2) = -1$. Let P be the function defined by $P(x) = (f \circ g)(x)$. Find $P'(2)$.

In problems 23 through 45, find the derivative of each function. In the even-numbered problems, you may assume the derivatives of the trigonometric functions as given in problem 41 of Problem Set 4.

23 $f(x) = \sqrt{x^2 + 12}$

24 $g(x) = \frac{1}{2} \sin 2x$

25 $f(x) = 1 + \sqrt{3x - 11}$

26 $h(t) = -\frac{3}{2} \cos 2t$

27 $f(x) = \sqrt{x^2 + \sqrt{1 + x^3}}$

28 $F(x) = \frac{5}{3} \sin (3x - 1)$

29 $g(x) = \sqrt[3]{\dfrac{2 + 3x^2}{3 - x^2}}$

30 $F(u) = \frac{1}{7} \tan 7u$

31 $h(t) = \dfrac{t^2 - 3t + 2}{2t^2 + 5}$

32 $G(x) = \cot (3x - 7)$

33 $k(u) = (u^2 + 7)^2 \sqrt{1 - u}$

34 $f(x) = \frac{1}{16} \sec (4x + 3)$

35 $g(z) = \sqrt{5 - z^2} \cdot (z^2 + 3)^4$

36 $h(x) = \sqrt{\sin x}$

37 $f(t) = \left(\dfrac{4t^2 - 3t + 2}{t^2 - 5t}\right)^{2/3}$

38 $F(t) = \sqrt[3]{\cos 5t}$

39 $f(y) = \dfrac{ay + b}{cy + d}$; a, b, c, and d are constants

40 $h(r) = \csc^{1/2} 5r$

41 $f(x) = x^2\sqrt{1 - 5x^3} \cdot (1 - 2x)^3$

42 $f(x) = \sec\left(\dfrac{\pi}{2} + 3x\right)$

43 $g(x) = \sqrt{x + \sqrt{x + \sqrt{x}}}$

44 $h(x) = \cot\left(\dfrac{x}{8} + \dfrac{\pi}{2}\right)$

45 $f(x) = \sqrt{\dfrac{ax + b}{cx + d}}$; a, b, c, and d are constants

46 Which of the following are examples of the chain rule?
  (a) $D_x y = D_x y \cdot D_y x$
  (b) $D_z y = D_z y \cdot D_t z$
  (c) $(f \circ g)'(x) = f'[g(x)]g'(x)$
  (d) $\dfrac{du}{dt} = \dfrac{du}{dy} \cdot \dfrac{dy}{dt}$

47 Let $y = \dfrac{1 - u}{1 + u}$ and $u = \dfrac{3 - x}{2 + x}$. Use the chain rule to find dy/dx.

48 The electric field intensity E on the axis of a uniformly charged ring at a point x units from the center of the ring is given by the formula $E = \dfrac{Qx}{(a^2 + x^2)^{3/2}}$, where a and Q are constants. Find the rate of change of field intensity with respect to distance along the x axis at a point on the axis 4 units from the center.

49 What is the slope of the line tangent to the graph of $f(x) = \sqrt{\dfrac{x}{x - 3}}$ at the point $(4, 2)$? Find the equation of the tangent line to the graph of f at the point $(4, 2)$.

50 Let f and g be differentiable functions and let a be a constant number. Find f' in terms of g' if
  (a) $f(x) = g[x + g(a)]$
  (b) $f(x) = g[x \cdot g(a)]$
  (c) $f(x) = g[x + g(x)]$
  (d) $f(x + 4) = g(x^2)$
  (e) $f(x) = g(a)(x - a)^7$

51 Use the linear approximation theorem to estimate $\sqrt[3]{65}$.

52 (a) Reduce the expression

$$\sqrt{x + h} - \sqrt{x} - \dfrac{h}{2\sqrt{x}} \quad \text{to} \quad \dfrac{-h^2}{2\sqrt{x}(\sqrt{x + h} + \sqrt{x})^2};$$

  thus, show that

$$\left|\sqrt{x + h} - \left(\sqrt{x} + \dfrac{h}{2\sqrt{x}}\right)\right| \le \dfrac{h^2}{8x\sqrt{x}} \quad \text{for } h > 0 \text{ and } x > 0.$$

  (b) Use part (a) to find a bound on the error involved in approximating $\sqrt{100.1}$ by 10.005.

53 Let $f(x) = |x| + |x - 1| + |x - 2| + |x - 3|$. What numbers are not in the domain of the derived function f'?

54 Let $f(x) = x - [\![x]\!]$. Find $f'_+(0)$.

55 Show that, if n is an odd positive integer, the lines tangent to the graph of the equation $y = x^n$ at the points $(1, 1)$ and $(-1, -1)$ are parallel.

56 Find a function f such that $\dfrac{d}{dx}(f^2) = f$.

57 If f is a differentiable function such that $f(x) \leq f(2)$ when $1 \leq x \leq 3$, show by the definition of the derivative that $f'(2) = 0$.

In problems 58 through 61, find a formula for $h'(x)$.

58 $h(x) = x^4 f(x)$ **59** $h(x) = f(x)[g(x)]^5$ **60** $h(x) = x^3 f(x) g(x)$ **61** $h(x) = \dfrac{[xf(x)]^3}{g(x)}$

62 Given the functions f and g satisfying the conditions $f' = g$ and $g' = -f$, show that

$$\frac{d}{dx}[f^2 + g^2] = 0.$$

63 Given functions f, g, and h such that $f' = g$, $g' = h$, and $h' = f$, show that

$$\frac{d}{dx}[f^3 + g^3 + h^3 - 3fgh] = 0.$$

64 A proof for the product rule

$$[f(x)g(x)]' = g(x)f'(x) + f(x)g'(x)$$

can be developed in a geometric fashion as follows. Let $g(x + \Delta x)$ and $f(x + \Delta x)$ represent adjacent sides of a rectangle (Figure 2). Set $\Delta z = g(x + \Delta x) - g(x)$ and $\Delta y = f(x + \Delta x) - f(x)$. Then, from Figure 2, we have

$$g(x + \Delta x)f(x + \Delta x) - f(x)g(x) = g(x)\,\Delta y + f(x)\,\Delta z + \Delta y\,\Delta z,$$

so that

$$\frac{g(x + \Delta x)f(x + \Delta x) - f(x)g(x)}{\Delta x}$$

Figure 2

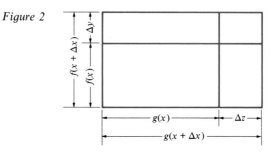

$$= g(x)\frac{\Delta y}{\Delta x} + f(x)\frac{\Delta z}{\Delta x} + \frac{\Delta y}{\Delta x}\,\Delta z.$$

Take the limits of both sides, using $\lim\limits_{\Delta x \to 0} \dfrac{\Delta y}{\Delta x} = f'(x)$

and $\lim\limits_{\Delta x \to 0} \dfrac{\Delta z}{\Delta x} = g'(x)$, and compute $\lim\limits_{\Delta x \to 0} \dfrac{\Delta y}{\Delta x}\,\Delta z$. Thus, prove that

$$[f(x)g(x)]' = g(x)f'(x) + f(x)g'(x).$$

65 Express the limit $\lim\limits_{x \to 2} \dfrac{(1 + x^2)^3 - 125}{x - 2}$ as a derivative, then evaluate the limit.

66 Criticize the fallacious argument:

$$x^2 = x \cdot x = \overbrace{x + x + x + \cdots + x}^{x \text{ times}}.$$

Differentiation with respect to x yields the equation

$$2x = \overbrace{1 + 1 + 1 + \cdots + 1}^{x \text{ times}}.$$

Thus, $2x = x$. Set $x = 1$; then $2 = 1$.

67 Is it true that if f is a differentiable function at 0, then the graph of $g(x) = f(x^3)$ has a horizontal tangent line at $(0, g(0))$? Explain.

68 Let f be the function defined by the equation $f(x) = |x|^3$. Find $f'(x)$.

69 Suppose that a function f satisfies $|f(x)| \leq |x|^n$, where $n > 1$. Show that f is differentiable at 0.

70 If g is the function inverse to the function f defined by $f(x) = x^3 - 75$, find $g'(-11)$.

71 Let f be the function defined by $f(x) = x^5 + 3x^3 - 1$. Find (a) $(f^{-1})'(-5)$ and (b) $(f^{-1})'(3)$.

72 The function f defined by the equation $y = f(x) = x^4 + 1$ has an inverse for $x > 0$. If g is the function inverse to f in this interval, so that $x = g(y)$, find:

(a) $\dfrac{dg}{dy}$ when $y = 2$ (b) $\dfrac{dg}{dy}$ when $y = 82$

73 If g is the inverse function of f where $f(x) = x^3 - 3x^2 + x$ for $x < 1 - (2/\sqrt{6})$, find $(g' \circ f)(x)$.

74 Let f and g be functions such that $(f \circ g)(x) = (g \circ f)(x) = x$. If $f(-1) = 2$, $f'(-1) = \frac{1}{3}$, and $f'(2) = -3$, find $g'(2)$.

75 Suppose that x units of a certain commodity are demanded when the manufacturing firm charges a price of $\$y$ per unit, so that the total revenue $\$R$ to the firm from the sale of the commodity is given by $R = xy$. The quantity E defined by $E = \dfrac{y}{x} \cdot \dfrac{dx}{dy}$ is called the *elasticity of demand* with respect to price. Show that the rate of change of total revenue with respect to price is given by $\dfrac{dR}{dy} = x(1 + E)$.

76 In problem 75 show that the elasticity of demand E is unaffected by a change in the units in which the commodity is measured (for instance, kilograms rather than pounds). Also show that it is unaffected by a change in monetary units (for instance, cents rather than dollars per unit of the commodity).

3
APPLICATIONS OF THE DERIVATIVE

We have seen in Chapter 2 that the derivative of a function can be interpreted as the slope of the tangent line to its graph. In the present chapter we exploit this fact and develop techniques for using derivatives as an aid in graph sketching. This chapter also includes applications of derivatives to problems in such diverse fields as geometry, engineering, physics, and economics.

1 The Intermediate Value Theorem and the Mean Value Theorem

We begin with two simple geometric observations about curves in the plane.

I If a continuous curve—that is, a curve consisting of one connected piece— begins on one side of a straight line l and ends on the other side of this line, it must intersect l in at least one point P (Figure 1).

II If A and B are the endpoints of a continuous curve, and if the curve has a tangent line at each of its intermediate points, then there is at least one intermediate point P at which the tangent line is parallel to the straight line through A and B (Figure 2).

We accept statements I and II as being true and use them for purposes of this section.

Figure 1

Figure 2

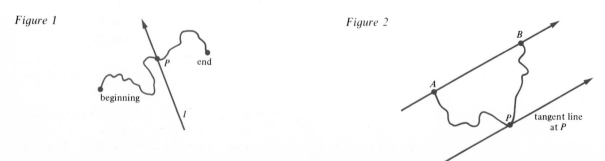

1.1 The Intermediate Value Theorem

As an indication of the usefulness of assertion I, consider the following problem: Show that there is a number c strictly between 0 and 1 such that $c^3 + c = 1$. (We are not obliged to *find* c, just to show that it *exists*.) The polynomial function f defined by the equation $f(x) = x^3 + x$ is continuous on the closed interval $[0, 1]$; also, $f(0) = 0$ and $f(1) = 2$. The portion of the graph of f between the points $(0,0)$ and $(1,2)$ is continuous, and the point $(0,0)$ is below the line $y = 1$ while the point $(1,2)$ is above this line (Figure 3). By assertion I, this graph must intersect the straight line $y = 1$ at some point, say at $(c, 1)$. Since $(c, 1)$ lies on the graph of f, $1 = f(c) = c^3 + c$, as desired.

More generally, we have the following theorem.

Figure 3

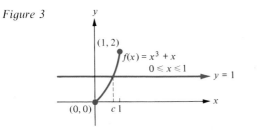

THEOREM 1   **Intermediate Value Theorem**

Let the function f be continuous on the closed interval $[a, b]$ and suppose that $f(a) \neq f(b)$. Then, if k is any number strictly between $f(a)$ and $f(b)$, there is at least one number c strictly between a and b such that $f(c) = k$. (In other words, *a continuous function takes on every value between any two of its values.*)

Figure 4

We make the intermediate value theorem plausible by appealing to assertion I as follows. Since f is continuous, its graph (Figure 4) consists of one connected piece. The points $(a, f(a))$ and $(b, f(b))$ lie on this graph, one below the horizontal line $y = k$ and the other above this line. By I, this graph must intersect the line $y = k$, say at the point (c, k). Hence, $k = f(c)$.

The intermediate value theorem is particularly useful for localizing the zeros (or roots) of a continuous function f. Recall that a number x is called a *zero* or a *root* of f if $f(x) = 0$. (Solutions of an equation are also called the roots of the equation.) Indeed, if we take $k = 0$ in Theorem 1, then it asserts the following: *If f is a continuous function on $[a, b]$, and if $f(a)$ and $f(b)$ have opposite signs, then there is a zero of f in the open interval (a, b)*; that is, there is a number c with $a < c < b$ and $f(c) = 0$.

EXAMPLE   Let f be the polynomial function defined by the equation $f(x) = x^5 - 2x^3 - 1$.
(a) Show that there is a root of f between 1 and 2.
(b) Show that, in fact, there is a root of f between 1.5 and 1.6.

SOLUTION
(a) $f(1) = -2$ and $f(2) = 15$ have opposite signs. Hence, since the polynomial function f is continuous, there is a number c with $1 < c < 2$ and $f(c) = c^5 - 2c^3 - 1 = 0$.
(b) Rounding off to two decimal places, $f(1.5) \approx -0.16$ and $f(1.6) \approx 1.29$; hence, f has a root between 1.5 and 1.6.

If we continue as in the example above, we can localize the desired root with more and more accuracy. Notice, however, that the intermediate value theorem itself merely assures us of the *existence* of a number c with a certain property—however, it does not tell us how to go about *finding* such a number.

1.2 The Mean Value Theorem

We now consider an example of assertion II. Let the function f be defined by $f(x) = \sqrt[3]{x}$. Consider the two points $A = (0,0)$ and $B = (8,2)$ on the graph of f (Figure 5). The straight line through A and B has slope $\frac{2-0}{8-0} = \frac{1}{4}$. By assertion II

Figure 5

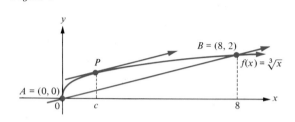

there is a point P along the graph of f between A and B at which the tangent line is parallel to the line through A and B. Let c be the abscissa of P. Then the slope of the tangent line at P is given by

$$f'(c) = \frac{1}{3(\sqrt[3]{c})^2}.$$

Since two straight lines that are parallel have the same slope, we have $\frac{1}{4} = \frac{1}{3(\sqrt[3]{c})^2}$, and so, $c = \left(\frac{2}{\sqrt{3}}\right)^3 = 1.5396\ldots.$

More generally, we have the following theorem.

THEOREM 2   **Mean Value Theorem**

If the function f is defined and continuous on the closed interval $[a,b]$ and is differentiable on the open interval (a, b), then there exists at least one number c with $a < c < b$ such that

$$f'(c) = \frac{f(b) - f(a)}{b - a}.$$

We make the mean value theorem plausible by appealing to assertion II. From the graph of f (Figure 6), we see that the slope of the straight line containing the points $A = (a, f(a))$ and $B = (b, f(b))$ on the graph of f is $\frac{f(b) - f(a)}{b - a}$. According to II, there is at least one point—say the point P (Figure 6)—on the graph of f between A and B at which the tangent line is parallel to the straight line containing A and B. Letting c denote the abscissa of P, we see that $a < c < b$ and the slope of the tangent line at P is $f'(c)$. If two straight lines are parallel, they have the same slope; hence, $f'(c) = \frac{f(b) - f(a)}{b - a}$.

Figure 6

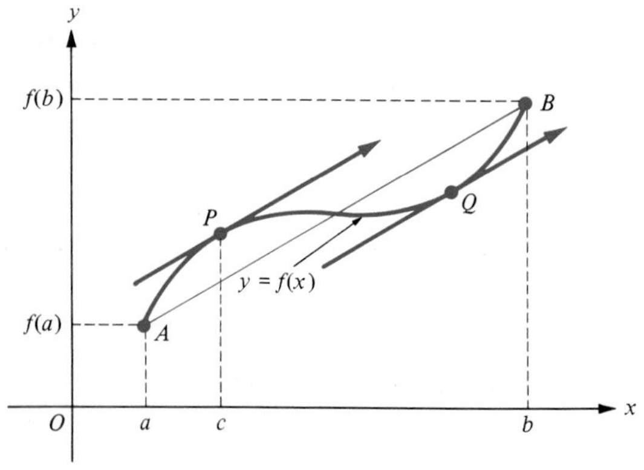

It is important to observe that there may be more than one value of c for which $f'(c) = \frac{f(b) - f(a)}{b - a}$. For instance, the abscissa of the point Q in Figure 6 would serve just as well as the chosen value of c.

Recall that a line segment joining two points of a curve is called a *secant* (Section 1.3 of Chapter 2). Therefore, the mean value theorem can be stated in words as follows: Given any secant to the graph of a differentiable function, one can always find a point on the graph between the endpoints of the secant at which the tangent line is parallel to the secant.

EXAMPLES   1 Let f be the function defined by $f(x) = x^2/6$.
   (a) Verify the hypotheses of the mean value theorem for the function f on the interval $[2, 6]$.

(b) Find a value of c on the interval $(2, 6)$ such that $f'(c) = \dfrac{f(6) - f(2)}{6 - 2}$.

(c) Interpret the result of part (b) geometrically and illustrate it on a graph.

SOLUTION

(a) Since f is a polynomial function, it is continuous on $[2, 6]$ and differentiable on $(2, 6)$.

(b) Here, $f'(x) = x/3$, $f(6) = 6$, and $f(2) = \frac{2}{3}$. Hence, we must solve the equation

$$\frac{c}{3} = \frac{6 - \frac{2}{3}}{6 - 2} = \frac{4}{3}.$$

Evidently, $c = 4$. Note that c belongs to the interval $(2, 6)$.

(c) The tangent line to the graph of $f(x) = x^2/6$ at the point $(4, f(4)) = (4, \frac{8}{3})$ is parallel to the secant between the points $(2, f(2)) = (2, \frac{2}{3})$ and $(6, f(6)) = (6, 6)$ (Figure 7).

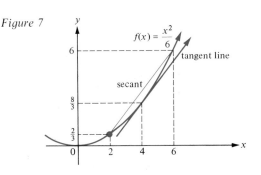

Figure 7

2 Let f be the function defined by $f(x) = x^3 + 2x^2 + 1$. By direct calculation, find a number c between 0 and 3 such that the tangent line to the graph of f at the point $(c, f(c))$ is parallel to the secant between the two points $(0, f(0))$ and $(3, f(3))$.

SOLUTION

We require $f'(c) = \dfrac{f(3) - f(0)}{3 - 0} = \dfrac{46 - 1}{3}$; that is,

$$3c^2 + 4c = 15 \quad \text{or} \quad 3c^2 + 4c - 15 = 0.$$

The solutions of this quadratic equation are $c = \frac{5}{3}$ and $c = -3$. Since we require that c belong to the interval $(0, 3)$, we must reject the solution $c = -3$. Therefore, the desired number is $c = \frac{5}{3}$.

Notice that the mean value theorem assures us of the *existence* of a solution to problems such as those in Examples 1 and 2 above, but it *does not tell us how to find such solutions*. Indeed, each problem has to be handled according to its own peculiarities.

In using the mean value theorem, one must make certain that all the hypotheses are satisfied. The following examples show that if these hypotheses do not hold, the conclusion of the mean value theorem need not hold.

EXAMPLES Show that at least one of the hypotheses of the mean value theorem fails to hold in the indicated interval for the given function, sketch the graph of the function, and observe that the conclusion of the mean value theorem also fails.

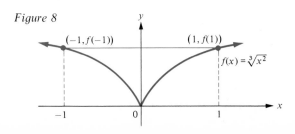

Figure 8

1 $f(x) = \sqrt[3]{x^2}$ in the interval $[-1, 1]$.

SOLUTION

The graph of f appears in Figure 8. The line containing the points $(-1, f(-1))$ and $(1, f(1))$ on the graph of f is parallel to the x axis, but the graph of f has no tangent line parallel to the x axis. This does not contradict the mean value theorem, since the hypothesis that f be differentiable on the open interval $(-1, 1)$ fails. Indeed, f is not differentiable at the number 0 in this interval.

Figure 9

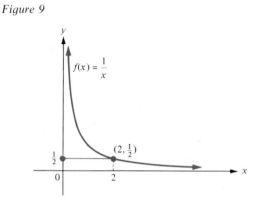

$$\textbf{2} \quad f(x) = \begin{cases} \dfrac{1}{x} & \text{if } x \neq 0 \\ \dfrac{1}{2} & \text{if } x = 0 \end{cases} \quad \text{in the interval } [0, 2].$$

SOLUTION

The graph appears in Figure 9. Here, f is differentiable on the open interval $(0, 2)$; however, it fails to be continuous on the closed interval $[0, 2]$; in fact,

$$\lim_{x \to 0^+} f(x) = +\infty \neq \frac{1}{2} = f(0).$$

The straight line through $(0, f(0))$ and $(2, f(2))$ is horizontal, yet the graph of f has no horizontal tangent line.

1.3 Rolle's Theorem

An interesting special case of the mean value theorem—the case in which $f(a) = f(b)$—is called *Rolle's theorem* in honor of the French mathematician Michel Rolle (1652–1719).

THEOREM 3   Rolle's Theorem

Let f be a continuous function on the closed interval $[a, b]$ such that f is differentiable on the open interval (a, b) and $f(a) = f(b)$. Then there is at least one number c in the open interval (a, b) such that $f'(c) = 0$.

Figure 10

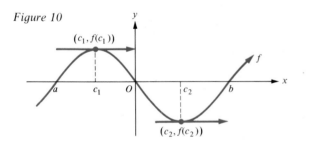

PROOF

Applying the mean value theorem to f, we conclude that there is at least one number c in the open interval (a, b) such that

$$f(b) - f(a) = (b - a)f'(c).$$

Since $f(b) = f(a)$, it follows that $f'(c) = 0$.

It should be noted that there may be more than one number c in (a, b) for which $f'(c) = 0$ (Figure 10).

EXAMPLES   Show that the hypotheses of Rolle's theorem are satisfied for the given function f on the interval $[a, b]$, find a value of c on the open interval (a, b) for which $f'(c) = 0$, and sketch the graph of the function.

1 $f(x) = 6x^2 - x^3$; $[a, b] = [0, 6]$

SOLUTION

Because f is a polynomial function, it is continuous and differentiable at every number. Clearly, $f(a) = f(0) = 0$ and $f(b) = f(6) = 0$, so $f(a) = f(b)$. Here, $f'(x) = 12x - 3x^2$ and we can solve the equation $f'(c) = 12c - 3c^2 = 0$ for c (between 0 and 6) to get $c = 4$ (Figure 11).

2 $f(x) = x^{4/3} - 3x^{1/3}$; $[a, b] = [0, 3]$

Figure 11

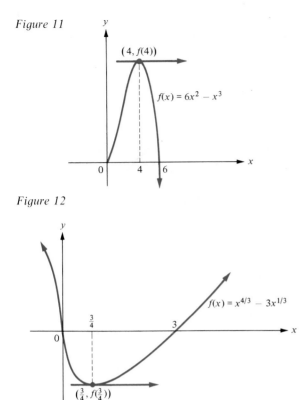

Figure 12

SOLUTION

Here, for $x \neq 0$, we have $f'(x) = \frac{4}{3}x^{1/3} - x^{-2/3} = \frac{1}{3}x^{-2/3}(4x - 3)$. Also,

$$f(x) = x^{4/3} - 3x^{1/3} = x^{1/3}(x - 3).$$

Evidently, f is continuous on $[0, 3]$ and differentiable on $(0, 3)$. Also, $f(0) = f(3) = 0$, so the hypotheses of Rolle's theorem are satisfied. (Notice that f is not differentiable at 0; in fact, the graph of f has a vertical tangent at the origin. However, differentiability of f at the endpoints of $[a, b]$ is not required in Rolle's theorem.) Solving the equation $f'(c) = \frac{1}{3}c^{-2/3}(4c - 3) = 0$ for c (between 0 and 3) we obtain $c = \frac{3}{4}$ (Figure 12).

Not only is Rolle's theorem a special case of the mean value theorem, but also it is possible to prove the mean value theorem on the basis of Rolle's theorem (Problem 31). In fact, in a rigorous course in analysis, Rolle's theorem is ordinarily proved first, and then the mean value theorem is obtained using Rolle's theorem.

Problem Set 1

In problems 1 through 4, use the intermediate value theorem to verify that each function f has a zero (that is, a root) on the indicated interval.

1 $f(x) = x^2 - 2$ between 1.4 and 1.5

2 $f(x) = x^3 + 3x - 6$ between 1 and 2

3 $f(x) = 2x^3 - x^2 + x - 3$ between 1 and 2

4 $f(x) = 2x^3 - x^2 + x - 3$ between 1.1 and 1.2

5 Let f be the function defined by the equation $f(x) = \dfrac{x + 1}{x^2 - 4}$. Notice that $f(0) = -\frac{1}{4}$ and $f(3) = \frac{4}{5}$ have opposite signs and yet there is no number c between 0 and 3 such that $f(c) = 0$. Explain why this does not contradict the intermediate value theorem.

6 Let f be a function that is continuous at every number but can take on only integral values. Use the intermediate value theorem to conclude that f must be a constant function.

In problems 7 through 12, verify the hypotheses of the mean value theorem for each function on the indicated interval $[a, b]$. Then find an explicit numerical value of c in the interval (a, b) such that $f(b) - f(a) = (b - a)f'(c)$.

7 $f(x) = 2x^3$, $[a, b] = [0, 2]$

8 $f(x) = \sqrt{x}$, $[a, b] = [1, 4]$

9 $f(x) = \dfrac{x - 1}{x + 1}$, $[a, b] = [0, 3]$

10 $f(x) = \sqrt{x + 1}$, $[a, b] = [3, 8]$

11 $f(x) = \sqrt{25 - x^2}$, $[a, b] = [-3, 4]$

12 $f(x) = \dfrac{x^2 - 2x - 3}{x + 4}$, $[a, b] = [-1, 3]$

In problems 13 through 16, find an explicit numerical value of c such that $a < c < b$ and the tangent line to the graph of each function f at $(c, f(c))$ is parallel to the secant between the points $(a, f(a))$ and $(b, f(b))$. Sketch the graph of f and show the tangent line and the secant.

13 $f(x) = x^2$, $a = 2$, $b = 4$

14 $f(x) = \sqrt{x}$, $a = 4$, $b = 9$

15 $f(x) = x^3$, $a = 1$, $b = 3$

16 $f(x) = \dfrac{1}{x-1}$, $a = 1.5$, $b = 1.6$

In problems 17 through 20, the conclusion of the mean value theorem fails for each function on the interval indicated. Sketch the graph of the function and determine which hypothesis of the mean value theorem fails to hold.

17 $f(x) = \sqrt{|x|}$, $[-1, 1]$

18 $g(x) = \dfrac{3}{x-2}$, $[1, 3]$

19 $f(x) = \begin{cases} x^2 + 1 & \text{if } x < 1 \\ 3 - x & \text{if } x \geq 1 \end{cases}$; $[0, 3]$

20 $G(x) = x - [\![x]\!]$, $[-1, 1]$

21 The functions whose graphs are shown in Figure 13 fail to satisfy the hypotheses of the mean value theorem on the interval from a to b. In each case, determine which hypothesis fails.

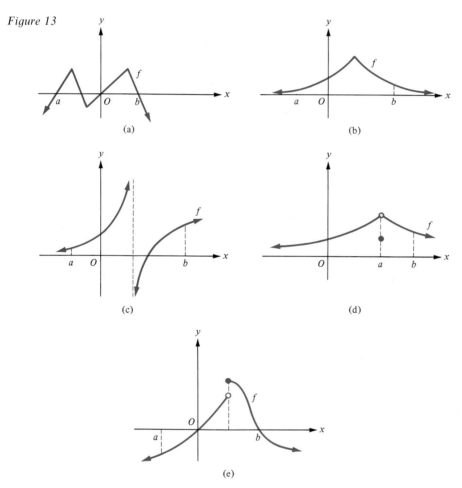

Figure 13

(a)

(b)

(c)

(d)

(e)

22 Let f be a differentiable function (and therefore continuous) at each number x in some open interval and suppose that $|f'(x)| \leq 1$ holds for each such x. Use the mean value theorem to show that $|f(b) - f(a)| \leq |b - a|$ for any two numbers a and b in the open interval.

23 Let x and y be variable quantities with $y = f(x)$, where f is a differentiable function. If x_1 and x_2 are two different values of x, show that there is a value x_0 between x_1 and x_2 such that the instantaneous rate of change of y with respect to x when $x = x_0$ is the same as the average rate of change of y with respect to x over the interval between x_1 and x_2.

SECTION 2 Higher-Order Derivatives   143

24 Explain why you cannot drive from city A to city B averaging 55 miles per hour unless, at some instant along the way, your speed is exactly 55 miles per hour. (See problem 23.)

In problems 25 through 29, verify the hypotheses of Rolle's theorem for each function on the indicated closed interval $[a, b]$ and then find a number c on the open interval (a, b) at which the derivative of the function is zero.

25 $f(x) = x^2 - 3x$, $[a, b] = [0, 3]$

26 $f(x) = x^2 - 5x + 6$, $[a, b] = [2, 3]$

27 $f(x) = x^3 - 3x^2 - x + 3$, $[a, b] = [-1, 3]$

28 $f(x) = \sqrt{x}\,(x^3 - 1)$, $[a, b] = [0, 1]$

29 $f(x) = x^{3/4} - 2x^{1/4}$, $[a, b] = [0, 4]$

30 Let f be the function defined by the equation $f(x) = x^5 + 2x^3 - 5x - 10$.
  (a) Use the intermediate value theorem to show that f has a zero between 1 and 2; that is, there exists at least one number c with $1 < c < 2$ such that

$$f(c) = c^5 + 2c^3 - 5c - 10 = 0.$$

  (b) Use Rolle's theorem to show that f cannot have two different zeros between 1 and 2. [*Hint:* If $1 < a < b < 2$ and $f(a) = f(b) = 0$, then f' would have a zero between 1 and 2.]

31 Let the function f be continuous on the closed interval $[a, b]$ and differentiable on the open interval (a, b). Define a function g on $[a, b]$ by the formula

$$g(x) = x[f(b) - f(a)] - (b - a)f(x) \text{ for } a \le x \le b.$$

Show that g satisfies the hypotheses of Rolle's theorem on the interval $[a, b]$. What conclusion can you draw?

32 Suppose that the function f is differentiable on the open interval (a, b) and that the one-sided limits $\lim_{x \to a^+} f(x)$ and $\lim_{x \to b^-} f(x)$ exist and are finite. Prove that there exists a number c in (a, b) such that $\lim_{x \to b^-} f(x) - \lim_{x \to a^+} f(x) = (b - a)f'(c)$.

33 Let f be a second-degree polynomial function. Given any two numbers a and b, find a formula for c strictly between a and b such that $f(b) - f(a) = (b - a)f'(c)$. [*Hint:* There are constants A, B, and C such that $f(x) = Ax^2 + Bx + C$.]

34 Let F and G be two functions both of which are continuous on $[a, b]$ and differentiable on (a, b). Define a function f on $[a, b]$ by the equation

$$f(x) = [G(a) - G(b)]F(x) - [F(a) - F(b)]G(x)$$

for $a \le x \le b$. Verify that f satisfies the hypotheses of Rolle's theorem and draw the conclusion that for some value of c strictly between a and b,

$$\frac{F'(c)}{G'(c)} = \frac{F(b) - F(a)}{G(b) - G(a)}$$

(provided that the denominators do not vanish).

2   Higher-Order Derivatives

The idea of a "second derivative" arises naturally in connection with the motion of a particle P along a linear scale (Figure 1). Call the scale the *s axis* and denote the variable coordinate of P by s, so that

Figure 1

$$s = f(t),$$

where f is a function determining the location of P

at time t. The equation $s = f(t)$ is called the *law of motion* or the *equation of motion* of the particle.

The *velocity* v of the particle P is defined to be the instantaneous rate of change of the coordinate s of P with respect to time. Thus,

$$v = \frac{ds}{dt}.$$

In physics, the instantaneous rate of change of velocity with respect to time is called the *acceleration* a of P; hence,

$$a = \frac{dv}{dt} = \frac{d}{dt}\,v = \frac{d}{dt}\left(\frac{ds}{dt}\right).$$

Therefore, the acceleration is the derivative of the derivative (or, as we say, the *second derivative*) of the coordinate s with respect to time. In operator notation

$$v = D_t s \quad \text{and} \quad a = D_t v = D_t(D_t s).$$

Notice that if distances along the s axis are given, say, in feet, and time is given, say, in seconds, then the velocity v is given in feet per second (ft/sec). Consequently, the acceleration a is given in feet per second per second (ft/sec²).

EXAMPLES **1** If $s = t + (2/t^2)$ for $t > 0$, with s in feet and t in seconds, find the values of v and a when $t = \frac{1}{2}$ second.

SOLUTION
Here,

$$v = \frac{ds}{dt} = \frac{d}{dt}\left(t + \frac{2}{t^2}\right) = 1 - \frac{4}{t^3}$$

and

$$a = \frac{dv}{dt} = \frac{d}{dt}\left(1 - \frac{4}{t^3}\right) = \frac{12}{t^4}.$$

Hence, when $t = \frac{1}{2}$, $v = -31$ ft/sec and $a = 192$ ft/sec².

2 If $s = 4\sqrt{1 + t^2} - t^2$ with s in centimeters and t in minutes, find the values of v and a in terms of t.

SOLUTION
Here,

$$v = D_t(4\sqrt{1 + t^2} - t^2) = \frac{4t}{\sqrt{1 + t^2}} - 2t \text{ cm/min}$$

and

$$a = D_t\left(\frac{4t}{\sqrt{1 + t^2}} - 2t\right) = \frac{4}{(1 + t^2)^{3/2}} - 2 \text{ cm/min}^2.$$

2.1 Derivatives of Order n

More generally, if f is any function that is differentiable on some open interval, then the derivative f' is again a function defined on this open interval and we can ask whether f' is differentiable on the interval. If it is, then its derivative $(f')'$ is written, for simplicity, as f'' (read "f double prime"). We call f'' the *second-order derivative* or simply the *second derivative* of the function f. For example, if a particle moves along a linear scale according to the law of motion $s = f(t)$, then v and a are written in the prime notation as

$$v = f'(t), \qquad a = D_t(f'(t)) = f''(t).$$

There is nothing to prevent us from successively taking derivatives of a function as many times as we please, provided that the derived functions remain differentiable at each stage. Thus, if f is a function and if f, f', and f'' are differentiable on an open interval, we can form the *third-order derivative*, or *third derivative*, $f''' = (f'')'$; if f''' is itself differentiable on the interval, we can form the *fourth-order derivative*, or *fourth derivative*, $f'''' = (f''')'$; and so forth. If f can be successively differentiated n times in this way, we say that f is n *times differentiable* and we write its *nth-order derivative*, or *nth derivative*, as $f\overbrace{'' \cdots ''}^{n}$ or, for simplicity, as $f^{(n)}$. (Here the parentheses around the n are to prevent its being confused with an exponent.) Thus,

$$f^{(1)} = f', \quad f^{(2)} = f'', \quad f^{(3)} = f''',$$

and so forth. Sometimes superscripts are written in small Roman numerals i, ii, iii, iv, v, vi, ... to denote derivatives of the corresponding orders. For instance,

$$f^{\text{iv}} = f'''' = f^{(4)} \quad \text{and} \quad f^{\text{v}} = f''''' = f^{(5)}.$$

EXAMPLE   Find all the higher-order derivatives of the polynomial function

$$f(x) = 15x^4 - 8x^3 + 3x^2 - 2x + 4.$$

SOLUTION

$$f'(x) = 60x^3 - 24x^2 + 6x - 2,$$
$$f''(x) = 180x^2 - 48x + 6,$$
$$f'''(x) = 360x - 48,$$
$$f''''(x) = f^{\text{iv}}(x) = f^{(4)}(x) = 360,$$
$$f'''''(x) = f^{\text{v}}(x) = f^{(5)}(x) = 0.$$

Since $f^{(4)}$ is a constant function, all subsequent derivatives are zero; that is,

$$f^{(n)}(x) = 0 \qquad \text{for } n \geq 5.$$

Just as with first derivatives, we often deliberately ignore the distinction between the nth-order derived function $f^{(n)}$ and the value of this function $f^{(n)}(x)$ at the number x, and both are referred to as "the nth derivative."

Operator notation for higher-order derivatives is self-explanatory; indeed, $D_x^n f(x)$ means $f^{(n)}(x)$. Corresponding Leibniz notation is motivated as follows: If $y = f(x)$, so that $\dfrac{dy}{dx} = D_x f(x) = f'(x)$, then the second derivative is given by

$$\frac{d\left(\dfrac{dy}{dx}\right)}{dx} = D_x^2 f(x) = f''(x).$$

The symbolism $d\left(\dfrac{dy}{dx}\right)\Big/ dx$ for the second derivative is cumbersome. Formal algebraic manipulation, as if actual fractions were involved, converts $\dfrac{d\left(\dfrac{dy}{dx}\right)}{dx}$ into $\dfrac{d^2 y}{(dx)^2}$. In practice, the parentheses in the "denominator" are usually dropped, and the second derivative is written as $\dfrac{d^2y}{dx^2}$. Similar notation is used for higher-order derivatives (assuming these exist) as shown in Table 1.

Table 1

$y = f(x)$	Prime	Operator	Leibniz
1. first derivative	$y' = f'(x)$	$D_x y = D_x f(x)$	$\dfrac{dy}{dx} = \dfrac{d}{dx} f(x)$
2. second derivative	$(y')' = y'' = f''(x)$	$D_x(D_x y) = D_x^2 y = D_x^2 f(x)$	$\dfrac{d^2y}{dx^2} = \dfrac{d^2}{dx^2} f(x)$
3. third derivative	$(y'')' = y''' = f'''(x)$	$D_x(D_x^2 y) = D_x^3 y = D_x^3 f(x)$	$\dfrac{d^3y}{dx^3} = \dfrac{d^3}{dx^3} f(x)$
$\vdots$ $\vdots$	$\vdots$	$\vdots$	$\vdots$
n. nth derivative	$(y^{(n-1)})' = y^{(n)} = f^{(n)}(x)$	$D_x(D_x^{n-1} y) = D_x^n y = D_x^n f(x)$	$\dfrac{d^n y}{dx^n} = \dfrac{d^n}{dx^n} f(x)$

EXAMPLES 1 If $y = 2x^2 + \dfrac{1}{x^2}$, find:

(a) $D_x y$ 　　　　　　　 (b) $D_x^2 y$ 　　　　　　　 (c) $D_x^3 y$

SOLUTION

(a) $D_x y = D_x\left(2x^2 + \dfrac{1}{x^2}\right) = 4x - \dfrac{2}{x^3}$

(b) $D_x^2 y = D_x\left(4x - \dfrac{2}{x^3}\right) = 4 + \dfrac{6}{x^4}$

(c) $D_x^3 y = D_x\left(4 + \dfrac{6}{x^4}\right) = -\dfrac{24}{x^5}$

2 Let $y = \sqrt{x}$. Find $d^n y/dx^n$ for all values of n.

SOLUTION

$y = \sqrt{x} = x^{1/2}$. Using direct calculation, we have

$$\frac{dy}{dx} = \frac{1}{2} x^{-1/2}$$

$$\frac{d^2y}{dx^2} = \frac{d}{dx}\left(\frac{1}{2} x^{-1/2}\right) = \frac{1}{2}\left(-\frac{1}{2}\right) x^{-3/2}$$

$$\frac{d^3y}{dx^3} = \frac{d}{dx}\left[\frac{1}{2}\left(-\frac{1}{2}\right) x^{-3/2}\right] = \frac{1}{2}\left(-\frac{1}{2}\right)\left(-\frac{3}{2}\right) x^{-5/2}$$

$$\frac{d^4y}{dx^4} = \frac{d}{dx}\left[\frac{1}{2}\left(-\frac{1}{2}\right)\left(-\frac{3}{2}\right) x^{-5/2}\right] = \frac{1}{2}\left(-\frac{1}{2}\right)\left(-\frac{3}{2}\right)\left(-\frac{5}{2}\right) x^{-7/2}.$$

The pattern is now emerging; in fact,

$$\frac{d^n y}{dx^n} = (-1)^{n+1} \frac{1 \cdot 3 \cdot 5 \cdot 7 \cdots (2n-3)}{2^n} x^{-(2n-1)/2}$$

apparently holds for $n \geq 2$. (Doubters are encouraged to prove this by mathematical induction on n.)

3 Let $f(x) = \dfrac{2x - 1}{3x + 2}$. Find:

(a) $f'(0)$ (b) $f''(1)$ (c) $f''(0)$

SOLUTION

$$f'(x) = \frac{(3x + 2)(2) - (2x - 1)(3)}{(3x + 2)^2} = \frac{7}{(3x + 2)^2}$$

and

$$f''(x) = D_x \left| \frac{7}{(3x + 2)^2} \right| = 7D_x[(3x + 2)^{-2}]$$

$$= -14(3x + 2)^{-3}D_x(3x + 2) = \frac{-42}{(3x + 2)^3}.$$

Thus,

(a) $f'(0) = \dfrac{7}{[3(0) + 2]^2} = \dfrac{7}{4}$

(b) $f''(1) = \dfrac{-42}{[3(1) + 2]^3} = \dfrac{-42}{125}$

(c) $f''(0) = \dfrac{-42}{[3(0) + 2]^3} = \dfrac{-42}{8} = -\dfrac{21}{4}$

The differentiation rules developed in Chapter 2 have generalizations for higher-order derivatives. (Most of these are proved by mathematical induction on n, the order of the derivative involved.) For instance, the addition rule and the homogeneous rule work for the nth-order derivative; that is,

$$D_x^n[f(x) + g(x)] = D_x^n f(x) + D_x^n g(x)$$

and

$$D_x^n[cf(x)] = cD_x^n f(x), \qquad c \text{ constant.}$$

The product (or Leibniz) rule for higher-order derivatives $(f \cdot g)^{(n)}$ is more complicated. For example, if $n = 2$, we have

$$(f \cdot g)'' = (f \cdot g' + f' \cdot g)' = (f \cdot g')' + (f' \cdot g)'$$
$$= f \cdot g'' + f' \cdot g' + f' \cdot g' + f'' \cdot g;$$

hence,

$$(f \cdot g)'' = f \cdot g'' + 2f' \cdot g' + f'' \cdot g.$$

Problem Set 2

In problems 1 through 6, a particle is moving along a linear scale according to the given law of motion $s = f(t)$. Find $v = ds/dt$ and $a = dv/dt$.

1 $s = t^3 + 2t^2$, s in feet, t in seconds

2 $s = (t^2 + 1)^{-1}$, s in centimeters, t in seconds

3 $s = 5t^2 + \sqrt{4t} - 3$, s in meters, t in seconds

4 $s = t\sqrt{t^2 + 4}$, s in miles, t in hours

5 $s = \frac{5}{2}t^{5/2} + \frac{2}{3}t^{3/2}$, s in kilometers, t in hours

6 $s = \frac{1}{2}gt^2 + v_0 t + s_0$, where g, v_0, and s_0 are constants

7 In problems 1, 3, and 5, find v and a at the instant when $t = 1$ unit.

8 Neglecting air resistance, a dropped body will fall through $s = 16t^2$ feet during a time interval of t seconds.
(a) Find the acceleration of this body.
(b) Find the velocity of this body after it has dropped through s feet.

In problems 9 through 24, find the first and second derivatives of the function defined by each equation.

9 $f(x) = 5x^3 + 4x + 2$

10 $g(x) = x^2(x^2 + 7)$

11 $f(t) = 7t^5 - 23t^2 + t + 9$

12 $F(x) = x^3(x + 2)^2$

13 $G(x) = (x^2 - 3)(x^4 + 3x^2 + 9)$

14 $f(u) = (u^2 + 1)^3$

15 $g(t) = t^3\sqrt{t} - 5t$

16 $f(x) = x - \dfrac{3}{x}$

17 $f(x) = x^2 - \dfrac{1}{x^3}$

18 $g(x) = \left(x + \dfrac{1}{x}\right)^2$

19 $f(u) = \dfrac{2u}{2 - u}$

20 $F(v) = \sqrt{v} + \dfrac{1}{\sqrt{v}}$

21 $f(t) = \sqrt{t^2 + 1}$

22 $g(y) = \sqrt{3y + 1}$

23 $F(r) = (1 - \sqrt{r})^2$

24 $h(x) = \dfrac{x}{\sqrt{x^2 + 1}}$

25 If f is the function defined by the equation $y = f(x) = 3x^2 + 2x$, evaluate and simplify the expression $x^2 y'' - 2xy' + 2y$.

26 Assume the Leibniz rule

$$(f \cdot g)'' = f \cdot g'' + 2f' \cdot g' + f'' \cdot g$$

for second derivatives.
(a) Derive the Leibniz rule

$$(f \cdot g)''' = f \cdot g''' + 3f' \cdot g'' + 3f'' \cdot g' + f''' \cdot g$$

for third derivatives.
(b) Assuming that $f(1) = -3$, $g(1) = \frac{1}{2}$, $f'(1) = -1$, $g'(1) = 4$, $f''(1) = 16$, and $g''(1) = \frac{2}{3}$, find $(f \cdot g)''(1)$.

27 Let g be a twice-differentiable function such that $g(2) = 3$, $g'(2) = \frac{1}{3}$, and $g''(2) = 5$. Define the function f by the equation $f(x) = x^4 g(x)$. Find the numerical value of $f''(2)$.

28 Let $y = (x + 2)^3(2x + 1)^4$. Find:

(a) $\dfrac{dy}{dx}$

(b) $\dfrac{d^2y}{dx^2}$

(c) $\dfrac{d^3y}{dx^3}$

29 Let $y = \sqrt{x^2 - 1}$. Find:

(a) $D_x y$

(b) $D_x^2 y$

(c) $D_x^3 y$

30 If $f(t) = (\sqrt{t + 1})^{-1}$, find $f^{(10)}(t)$.

31 Suppose that g is a twice-differentiable function such that $g(0) = -2$, $g'(0) = 3$, and $g''(0) = 5$. Define a function f by the equation $f(x) = \sqrt[3]{1 + g(x)}$. Find the numerical value of $f''(0)$.

32 Use the principle of mathematical induction to prove that

$$D_x^n(x^{100}) = \frac{100!}{(100 - n)!} x^{100 - n} \qquad \text{if } n \leq 100.$$

33 A particle is moving along a linear scale according to the given equation of motion, where s is the distance in feet from the origin at the end of t seconds. Find the time when the instantaneous acceleration is zero if:

(a) $s = t^3 - 6t^2 + 12t + 1$, $t \geq 0$ (b) $s = \sqrt{1 + t}$, $t \geq 0$ (c) $s = 5t + \dfrac{2}{t + 1}$, $t \geq 0$

34 Find formulas for $f'(x)$ and $f''(x)$ if:

(a) $f(x) = \begin{cases} x^2 & \text{if } x \le 1 \\ 2x - 1 & \text{if } x > 1 \end{cases}$

(b) $f(x) = \begin{cases} \dfrac{x^2}{2} & \text{if } x \ge 0 \\ -\dfrac{x^2}{2} & \text{if } x < 0 \end{cases}$

35 Show that $\dfrac{d^2 s}{dt^2} = v \dfrac{dv}{ds}$.

36 Use the principle of mathematical induction to prove the homogeneous rule $D_x^n(cf(x)) = cD_x^n(f(x))$.

37 Develop a formula for $D_x^n\left(\dfrac{1}{x}\right)$.

38 Find a formula for $D_x^n(\sqrt[3]{x})$, $x \ne 0$.

39 Given that $D_x \sin x = \cos x$ and $D_x \cos x = -\sin x$, find $D_x^{70} \sin x$.

40 Let f be a twice-differentiable function. A particle Q moves along the graph of f in such a way that the x coordinate of Q at time t is $x = g(t)$, where g is a twice-differentiable function. A particle P moves along the y axis in such a way that the y coordinate of P is always the same as the y coordinate of Q. Find a formula for the acceleration of P at time t.

41 Explain why the nth derivative of a rational function is again a rational function.

42 Let f and g be twice-differentiable functions. Suppose that $g(-1) = 27$, $g'(-1) = -1$, $g''(-1) = \frac{1}{4}$, $f(27) = 2$, $f'(27) = -4$, and $f''(27) = 1$. Find the numerical value of $(f \circ g)''(-1)$.

3 Geometric Properties of Graphs of Functions— Increasing and Decreasing Functions and Concavity of Graphs

The basic properties of functions and their graphs are usually discussed in precalculus courses; however, some of these properties require the use of calculus for their full understanding. This section is devoted to the study of certain features of functions and their graphs which can be apprehended geometrically and which, with the help of the mean value theorem, can be interpreted analytically.

3.1 Increasing and Decreasing Functions

The concept of an increasing, or decreasing, function can be motivated by consideration of the graphs of $f(x) = 2x + 3$ (Figure 1a) and $g(x) = -2x^3$ (Figure 1b). In Figure 1a, the function values $f(x) = 2x + 3$ increase (rise) as the values of x increase (vary from left to right); that is,

$$\text{if } x_1 < x_2, \quad \text{then} \quad f(x_1) < f(x_2).$$

Similarly, in Figure 1b, the function values $g(x) = -2x^3$ decrease (fall) as the values of x increase; that is,

$$\text{if } x_1 < x_2, \quad \text{then} \quad g(x_1) > g(x_2).$$

Figure 1

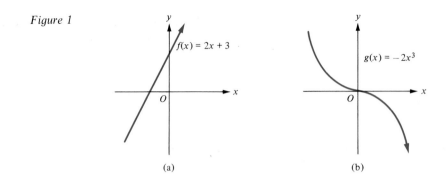

(a)   (b)

In general, we make the following definitions.

DEFINITION 1   **Increasing and Decreasing Functions**
A function f is said to be *increasing* (respectively, *decreasing*) on an interval I if f is defined on I and $f(a) < f(b)$ [respectively, $f(a) > f(b)$] holds whenever a and b are two numbers in I with $a < b$.

DEFINITION 2   **Monotone Function**
A function f is said to be *monotone* on an interval I if it is either increasing on I or decreasing on I.

Figure 2

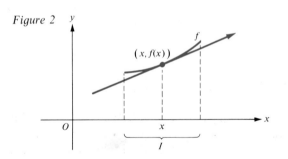

Now, suppose that f is a function with a positive derivative $f'(x)$ at each number x in some interval I. Then, for each value of x in I, the tangent line to the graph of f at the point $(x, f(x))$ is rising to the right since its slope $f'(x)$ is positive (Figure 2). Each such tangent line is a good approximation to the graph of f near the point of tangency, so it stands to reason that this graph must also be rising to the right. For similar reasons, a function with a negative derivative on an interval should be decreasing on this interval. In fact, we have the following theorem.

THEOREM 1   **Test for Increasing and Decreasing Functions**
Assume that the function f is defined and continuous on the interval I and that f is differentiable at every number in I, except possibly for the endpoints of I.

(i) If $f'(x) > 0$ for every number x in I, except possibly for the endpoints of I, then f is increasing on I.

(ii) If $f'(x) < 0$ for every number x in I, except possibly for the endpoints of I, then f is decreasing on I.

PROOF
(i) Suppose that $f'(x) > 0$ for every number x in I, except possibly for the endpoints of I, and let a and b be numbers in I with $a < b$. By the mean value theorem (Section 1.2), there exists a number c with $a < c < b$ such that

$$f(b) - f(a) = (b - a)f'(c).$$

Here, $b - a > 0$ and (by hypothesis) $f'(c) > 0$; hence,

$$f(b) - f(a) > 0; \quad \text{that is,} \quad f(a) < f(b).$$

(ii) The proof of (ii) is similar to the proof of (i), except that $f'(c) < 0$, so that $f(b) - f(a) < 0$; that is, $f(a) > f(b)$.

EXAMPLE Find the intervals where the function $f(x) = x^3 - 3x + 1$ is monotone (that is, either increasing or decreasing). Sketch the graph.

Figure 3

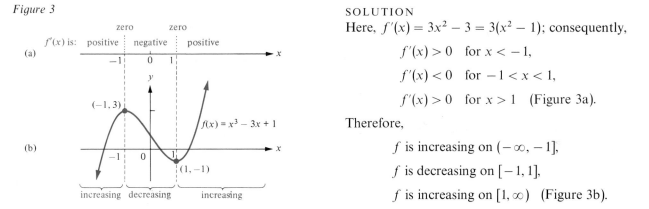

SOLUTION

Here, $f'(x) = 3x^2 - 3 = 3(x^2 - 1)$; consequently,

$$f'(x) > 0 \quad \text{for } x < -1,$$

$$f'(x) < 0 \quad \text{for } -1 < x < 1,$$

$$f'(x) > 0 \quad \text{for } x > 1 \quad \text{(Figure 3a)}.$$

Therefore,

f is increasing on $(-\infty, -1]$,

f is decreasing on $[-1, 1]$,

f is increasing on $[1, \infty)$ (Figure 3b).

Notice that a point on a continuous graph that separates a rising portion of the graph from a falling portion is either the "top of a hill" [for instance, the point $(-1, 3)$ in Figure 3b] or the "bottom of a valley" [for instance, the point $(1, -1)$ in Figure 3b]. The top of a hill is called a *relative maximum* point and the bottom of a valley is called a *relative minimum* point of the graph. Such points are studied in detail in Section 4, where formal definitions can be found.

3.2 Concavity of the Graph of a Function

We have seen in Section 3.1 that the algebraic sign of the *first* derivative of a function determines whether the graph is rising or falling. Here we see that the algebraic sign of the *second* derivative determines whether the graph is bending upward ("cup-shaped") or bending downward ("cap-shaped").

Figure 4a shows a cup-shaped graph. Notice that, as the point P on this graph moves to the right, the tangent line at P turns counterclockwise and its *slope increases*. We say that such a graph is *concave upward*. Similarly, in Figure 4b, the graph is cap-shaped and, as the point P moves to the right, the tangent line at P turns clockwise and its *slope decreases*. We say that such a graph is *concave downward*. These simple geometric considerations lead us to the following formal definition.

Figure 4

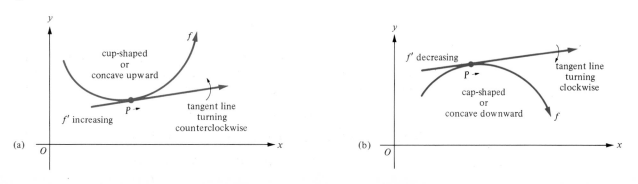

DEFINITION 3   **Concavity of a Graph**
Let the function f be differentiable on the open interval I. The graph of f is said to be *concave upward* (respectively, *concave downward*) on I if f' is an increasing function (respectively, a decreasing function) on I.

Using the test for increasing and decreasing functions (Theorem 1), we see that, if $(f')'$ has positive values on an open interval, then f' is increasing on that interval, so the graph of f is concave upward on the interval by Definition 3. Similarly, if $(f')'$ has negative values on an open interval, then f' is decreasing on the interval and the graph of f is concave downward on the interval. Thus, we have the following theorem.

THEOREM 2   **Test for Concavity of a Graph**
Let the function f be twice differentiable on the open interval I.

(i) If $f''(x) > 0$ for every number x in I, then the graph of f is concave upward on I.

(ii) If $f''(x) < 0$ for every number x in I, then the graph of f is concave downward on I.

EXAMPLE   Find the intervals where the graph of $f(x) = (x - 2)^2(x - 5)$ is concave upward or downward and sketch the graph.

Figure 5

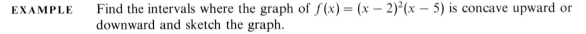

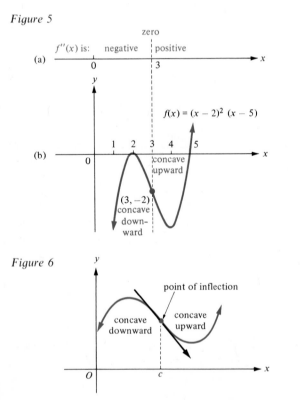

SOLUTION
Here,

$$f(x) = x^3 - 9x^2 + 24x - 20,$$

$$f'(x) = 3x^2 - 18x + 24, \quad \text{and}$$

$$f''(x) = 6x - 18 = 6(x - 3);$$

hence,

$$f''(x) < 0 \quad \text{for } x < 3 \quad \text{and}$$

$$f''(x) > 0 \quad \text{for } x > 3 \quad \text{(Figure 5a)}.$$

Therefore, the graph of f is concave downward on the interval $(-\infty, 3)$ and concave upward on the interval $(3, \infty)$ (Figure 5b).

A point on a smooth graph that separates a portion that is concave upward from a portion that is concave downward [for instance, the point $(3, -2)$ in Figure 5b] is called a *point of inflection*. At a point of inflection, a graph may "cut across its own tangent line" (Figure 6). The notion of a point of inflection is made more precise by the following definition.

Figure 6

point of inflection

concave downward

concave upward

DEFINITION 4   **Point of Inflection**
A point $(c, f(c))$ is called a *point of inflection* of the graph of a function f if the graph has a tangent line at this point, and if there is an open interval I containing the number c such that, for every pair of numbers a and b in I with $a < c < b$, $f''(a)$ and $f''(b)$ exist and have opposite algebraic signs.

3.3 Graph Sketching

A systematic approach to the problem of sketching the graph of a function f usually involves consideration of the following questions:

1 What is the domain of f?

2 At what numbers (if any) is f discontinuous?

3 Where does the graph of f intersect the coordinate axes?

4 Is f an even function? An odd function?

5 Does the graph of f have any asymptotes?

6 On what intervals is the graph of f rising? Falling?

7 On what intervals is the graph of f concave upward? Downward?

8 Where do relative maxima and minima occur?

9 Where do points of inflection occur?

The intervals on which the graph of a function f is rising or falling can often be found by carrying out the following simple procedure.

Procedure for Determining the Intervals on Which f' Is Positive or Negative

Step 1 Determine the numbers at which f' is undefined, or discontinuous, or takes the value 0, and arrange these numbers in increasing order: $x_1, x_2, x_3, \ldots, x_n$.

Step 2 On each open interval $(x_1, x_2), (x_2, x_3), \ldots, (x_{n-1}, x_n)$ the values $f'(x)$ will have constant algebraic sign. The same is true for an open interval contained in the domain of f' to the left of x_1 or to the right of x_n. To determine the sign for any one such interval, say (x_2, x_3), just select any number a in (x_2, x_3) and evaluate $f'(a)$: If $f'(a) > 0$, then $f'(x) > 0$ for all x in (x_2, x_3); and if $f'(a) < 0$, then $f'(x) < 0$ for all x in (x_2, x_3).

After carrying out this procedure, we can conclude by Theorem 1 that f is increasing (respectively, decreasing) on those intervals where the values of f' are positive (respectively, negative). Also, if f is defined and continuous on any *closed* interval, say $[x_2, x_3]$ for instance, and if f is increasing (respectively, decreasing) on (x_2, x_3), then f will be increasing (respectively, decreasing) on $[x_2, x_3]$ (Problem 36). Of course, similar remarks apply to half-closed intervals such as $[x_2, x_3)$ or $(x_2, x_3]$.

This procedure is justified by the intermediate value theorem in Section 1.1, which implies that a function that is continuous and has nonzero values on an interval cannot change algebraic sign on that interval. Naturally, the same procedure, with f' replaced by f'', can be used to determine the intervals on which the graph of f is concave upward or downward.

EXAMPLES For the given function f, (a) determine the intervals on which f is monotone, (b) determine the intervals on which the graph of f is concave upward or downward, (c) find all points of inflection of the graph of f, and (d) sketch the graph.

1 $f(x) = x + \dfrac{1}{x}$

SOLUTION

(a) Here, $f'(x) = 1 - 1/x^2$ and f' is defined and continuous at every number except 0. The zeros of f' occur at -1 and 1. Arranging the numbers at which f' is undefined or takes the value zero in increasing order, we obtain the follow-

Figure 7

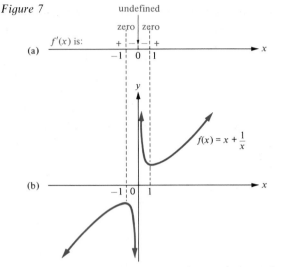

ing: $-1, 0, 1$. Therefore, f' has constant algebraic signs on each of the intervals

$$(-\infty, -1), \quad (-1, 0), \quad (0, 1), \quad \text{and} \quad (1, \infty).$$

To determine these signs, we select a number in each interval, say

$$-2, \quad -\tfrac{1}{2}, \quad \tfrac{1}{2}, \quad \text{and} \quad 2,$$

respectively. Evaluating f' at the selected numbers, we obtain

$$f'(-2) = \tfrac{3}{4} > 0, \quad f'(-\tfrac{1}{2}) = -3 < 0,$$
$$f'(\tfrac{1}{2}) = -3 < 0, \quad \text{and} \quad f'(2) = \tfrac{3}{4} > 0;$$

hence, the values of f' are positive on $(-\infty, -1)$, negative on $(-1, 0)$, negative on $(0, 1)$, and positive on $(1, \infty)$ (Figure 7a). It follows that f is increasing on the intervals $(-\infty, -1]$ and $[1, \infty)$, while it is decreasing on the intervals $[-1, 0)$ and $(0, 1]$.

(b) $f''(x) = 2/x^3$, so that

$$f''(x) < 0 \quad \text{for } x < 0,$$
$$f''(x) > 0 \quad \text{for } x > 0.$$

It follows that the graph of f is concave downward on the interval $(-\infty, 0)$ and concave upward on the interval $(0, \infty)$.

(c) Since the graph of f is concave downward to the left of the y axis and concave upward to the right of the y axis, the only possible point of inflection would be on the y axis. But, 0 is not in the domain of f, so the graph of f does not intersect the y axis. It follows that the graph of f has no point of inflection.

(d) Since f is an odd function, its graph is symmetric about the origin. By the methods of Section 6 of Chapter 1, the y axis is a vertical asymptote and there is no horizontal asymptote. Taking these facts into account, using the information in parts (a) and (b), and plotting a few points, we obtain the graph in Figure 7b.

2 $f(x) = x^2 - \dfrac{1}{x}$

SOLUTION

(a) Here, $f'(x) = 2x + (1/x^2)$. Solving the equation $f'(x) = 0$, we obtain the root $x = -2^{-1/3} \approx -0.8$. Also, $f'(x)$ is undefined for $x = 0$. Therefore, f' has constant algebraic signs on each of the intervals $(-\infty, -2^{-1/3})$, $(-2^{-1/3}, 0)$, and $(0, \infty)$. For instance, -1 belongs to $(-\infty, -2^{-1/3})$ and $f'(-1) = -1$, so $f'(x) < 0$ for x in $(-\infty, -2^{-1/3})$. Continuing in this way, we find that

$$f'(x) > 0 \text{ for } x \text{ in } (-2^{-1/3}, 0) \quad \text{and that} \quad f'(x) > 0 \text{ for } x \text{ in } (0, \infty).$$

Hence, f is decreasing on the interval $(-\infty, -2^{-1/3})$ and increasing on the intervals $(-2^{-1/3}, 0)$ and $(0, \infty)$.

(b) $f''(x) = 2 - 2x^{-3} = 2(1 - 1/x^3)$, so $f''(x)$ is undefined for $x = 0$ and $f''(x) = 0$ for $x = 1$. Therefore, f'' has constant algebraic signs on each of the intervals $(-\infty, 0)$, $(0, 1)$, and $(1, \infty)$. For instance, -1 belongs to $(-\infty, 0)$ and $f''(-1) = 4 > 0$, so $f''(x) > 0$ for x in $(-\infty, 0)$. Continuing in this way, we find that

$$f''(x) < 0 \text{ for } x \text{ in } (0, 1) \quad \text{and that} \quad f''(x) > 0 \text{ for } x \text{ in } (1, \infty).$$

It follows that the graph of f is concave upward on $(-\infty, 0)$, concave downward on $(0, 1)$, and concave upward again on $(1, \infty)$.

(c) Because $f''(x) < 0$ for x slightly to the left of 1 and $f''(x) > 0$ for x slightly to the right of 1, there *may be* a point of inflection at $(1, f(1))$. Consulting Definition 4, we find the additional requirement that the graph of f have a tangent line at $(1, f(1))$. Since $f'(1) = 3$, the graph does have a tangent line (of slope 3) at $(1, f(1))$; hence, $(1, f(1)) = (1, 0)$ is indeed a point of inflection of the graph of f.

(d) Solving the equation $f(x) = 0$, we obtain only the solution $x = 1$; hence, the graph of f intersects the x axis only at the point $(1, 0)$. The function f is neither even nor odd, so its graph will fail to be symmetric about the y axis and about the origin. The y axis is a vertical asymptote, and there is no horizontal asymptote. Taking these facts into account, using the information in parts (a), (b), and (c), and plotting a few points, we obtain the graph sketched in Figure 8.

Figure 8

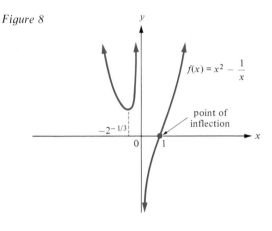

$f(x) = x^2 - \dfrac{1}{x}$

point of inflection

$-2^{-1/3}$

3 $f(x) = \sqrt{x} - \dfrac{1}{\sqrt{x}}$

SOLUTION

(a) Here, $f'(x) = \dfrac{1}{2\sqrt{x}} + \dfrac{1}{2(\sqrt{x})^3}$, and f' is defined, continuous, and positive on the entire domain $(0, \infty)$ of f. Thus, f is increasing on $(0, \infty)$.

(b) Since $f''(x) = \dfrac{-1}{4(\sqrt{x})^3} + \dfrac{-3}{4(\sqrt{x})^5}$, we see that $f''(x)$ is negative on the entire domain $(0, \infty)$ of f. Thus, the graph of f is concave downward on $(0, \infty)$.

(c) Since the graph is always concave downward, there can be no points of inflection.

(d) Solving the equation $f(x) = 0$, we obtain only the solution $x = 1$; hence, the graph intersects the x axis only at the point $(1, 0)$. The y axis is a vertical asymptote, and there is no horizontal asymptote. The graph is sketched in Figure 9.

Figure 9

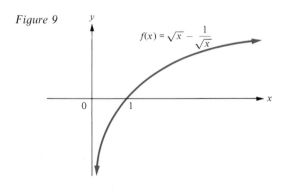

$f(x) = \sqrt{x} - \dfrac{1}{\sqrt{x}}$

4 $f(x) = x^{1/3}(x - 1)^{2/3}$

SOLUTION

(a) Here, $f(x) = [x(x-1)^2]^{1/3}$, so

$$f'(x) = \tfrac{1}{3}[x(x-1)^2]^{-2/3}[2x(x-1) + (x-1)^2]$$
$$= \tfrac{1}{3}x^{-2/3}(x-1)^{-1/3}(3x-1).$$

Evidently, f' is undefined at 0 and at 1. Also, $f'(x) = 0$ only for $x = \tfrac{1}{3}$. Therefore, f' has a constant algebraic sign on each of the intervals $(-\infty, 0)$, $(0, \tfrac{1}{3})$, $(\tfrac{1}{3}, 1)$, and $(1, \infty)$. Selecting numbers in each of these intervals and evaluating f' at each of the selected numbers, we find that f' is positive on $(-\infty, 0)$, $(0, \tfrac{1}{3})$, and $(1, \infty)$, while it is negative on $(\frac{1}{3}, 1)$. Therefore, f is increasing on $(-\infty, 0]$, $[0, \frac{1}{3}]$, and $[1, \infty)$, while f is decreasing on $[\frac{1}{3}, 1]$. Since f is increasing on the adjacent intervals $(-\infty, 0]$ and $[0, \frac{1}{3}]$, it is actually increasing on the entire interval $(-\infty, \frac{1}{3}]$.

(b) Differentiating f' and simplifying, we find that

$$f''(x) = -\tfrac{2}{9}x^{-5/3}(x-1)^{-4/3}.$$

Evidently, f'' is undefined at 0 and at 1, while the equation $f''(x) = 0$ has no solution. Here we find that

$$f''(x) > 0 \quad \text{for } x \text{ in } (-\infty, 0),$$
$$f''(x) < 0 \quad \text{for } x \text{ in } (0, 1),$$
$$f''(x) < 0 \quad \text{for } x \text{ in } (1, \infty);$$

hence, the graph of f is concave upward on the interval $(-\infty, 0)$ and concave downward on the intervals $(0, 1)$ and $(1, \infty)$.

(c) Since the graph is concave upward on the interval $(-\infty, 0)$ and concave downward on the interval $(0, 1)$, there is a possible point of inflection at the point $(0, f(0)) = (0, 0)$. We must see whether the graph of f has a tangent line at $(0, 0)$. Since f' is not defined at 0, the only possibility for a tangent line at $(0, 0)$ is a *vertical* tangent line.

According to the condition given in Chapter 2, Section 1, page 90, the graph of f has a vertical tangent line at $(0, 0)$ if f is continuous at 0 and $\displaystyle\lim_{\Delta x \to 0} \left| \frac{f(0 + \Delta x) - f(0)}{\Delta x} \right| = +\infty$. Evidently, f is continuous at 0. Also, we have

$$\lim_{\Delta x \to 0} \left| \frac{f(0 + \Delta x) - f(0)}{\Delta x} \right| = \lim_{\Delta x \to 0} \left| \frac{\Delta x^{1/3}(\Delta x - 1)^{2/3}}{\Delta x} \right|$$

$$= \lim_{\Delta x \to 0} \left| \frac{\Delta x - 1}{\Delta x} \right|^{2/3} = +\infty.$$

Therefore, the graph of f does have a vertical tangent line at $(0, 0)$. It follows that $(0, 0)$ is a point of inflection of the graph of f.

(d) The function f is defined and continuous on $(-\infty, \infty)$ and intersects the x axis at $x = 0$ and $x = 1$. The function is neither even nor odd and its graph has no horizontal or vertical asymptotes. The graph is sketched in Figure 10.

Figure 10

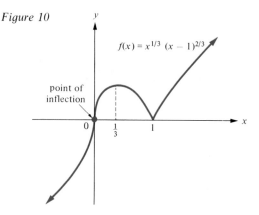

$f(x) = x^{1/3}\,(x-1)^{2/3}$

point of inflection

Problem Set 3

In problems 1 through 8, find the intervals where each function is monotone (that is, either increasing or decreasing). Sketch the graph.

1 $f(x) = x^3 - 12x + 11$

2 $g(x) = x^3 + x^2 - 5x$

3 $f(x) = x + \dfrac{3}{x^2}$

4 $h(x) = x^2 - \dfrac{3}{x^2}$

5 $f(x) = \sqrt{x} + \dfrac{4}{x}$

6 $g(x) = x^{7/5} - 8x^{3/5}$

7 $h(x) = \begin{cases} x^2 - 5 & \text{if } x < 4 \\ 10 - 3x & \text{if } x \geq 4 \end{cases}$

8 $f(x) = \begin{cases} \sqrt{25 - x^2} & \text{if } x \leq 4 \\ 7 - x & \text{if } x > 4 \end{cases}$

In problems 9 through 16, indicate the intervals where the graph of each function is concave upward or downward, sketch the graph, and locate the points of inflection.

9 $f(x) = 2x^3 - \frac{1}{2}x^2 - 7x + 2$

10 $g(x) = \frac{5}{3}x^3 - \frac{7}{2}x^2 - 6x + 4$

11 $f(x) = x^4 + 4x^3 + 6x^2 + 4x - 1$

12 $F(x) = 3x^4 + 8x^3 - 18x^2 + 12$

13 $h(x) = x - \frac{4}{x^2}$

14 $G(x) = \frac{2x}{x+2}$

15 $f(x) = \sqrt{x} - \frac{9}{x}$

16 $f(x) = \frac{4}{x+1}$

17 In Figure 11 consider the graphs of the given functions and the displayed interval $[a, e]$. In each case, the interval $[a, e]$ is broken up into four subintervals $[a, b]$, $[b, c]$, $[c, d]$, and $[d, e]$. Assume that the given functions are twice-differentiable on the interior of each subinterval. Determine on which of these subintervals the given function (i) is increasing, (ii) is decreasing, (iii) has a graph that is concave upward, and (iv) has a graph that is concave downward. Also, (v) find all inflection points.

Figure 11

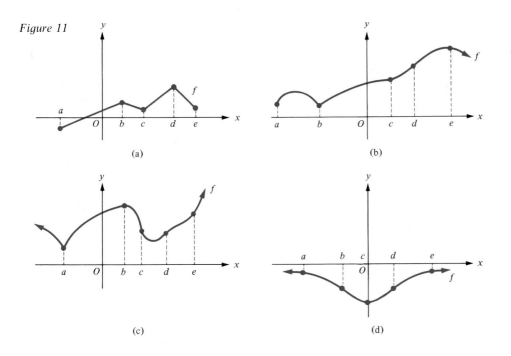

(a)

(b)

(c)

(d)

18 Sketch a graph of a continuous function f having the following properties: $f(-1) = 5$, $f(0) = 0$, $f(2) = 3$, $f(5) = 0$, $f'(x) < 0$ for $x < 0$, $f'(x) > 0$ for $0 < x < 2$, $f'(x) < 0$ for $x > 2$, $f''(x) > 0$ for $x < -1$, $f''(x) < 0$ for $-1 < x < 0$, and $f''(x) < 0$ for $x > 0$.

In problems 19 through 32, indicate (a) the intervals where f is increasing, (b) the intervals where f is decreasing, (c) the intervals where the graph of f is concave upward, (d) the intervals where the graph of f is concave downward, (e) the points of inflection, and (f) sketch the graph of f.

19 $f(x) = x^3 - 6x^2 + 9x + 1$

20 $f(x) = \frac{1}{3}x^3 + \frac{1}{2}x^2 - 2x + 1$

21 $f(x) = x^4 + 4x^2 - 16$

22 $f(x) = 8x - 2x^2 - x^4$

23 $f(x) = x^2(12 - x^2)$

24 $f(x) = x + \frac{1}{\sqrt{x}}$

25 $f(x) = 2x + \frac{2}{x}$

26 $f(x) = (x + 2)^{1/3}x^{-2/3}$

27 $f(x) = 3x^{2/3} - x^{5/3}$

28 $f(x) = x^{4/5}$

29 $f(x) = 1 + (x - 2)^{2/3}$

30 $f(x) = 1 + \left(\frac{x}{x-1}\right)^{1/3}$

31 $f(x) = \frac{2x}{x^2 + 1}$

32 $f(x) = \begin{cases} 10 - 3x & \text{if } x \geq 2 \\ x^2 & \text{if } x < 2 \end{cases}$

33 Assume that f is a continuous monotone function whose domain is an interval I. Explain why f must have an inverse f^{-1} and why the domain of f^{-1} must be an interval. (A geometric argument is acceptable, but an analytic argument is preferred.)

34 Suppose that f has a continuous derivative f' on an open interval I and that f is increasing on I. Show that $f'(x) \geq 0$ holds for all numbers x in I.

35 Show that, even if f is a function with a continuous second derivative, the condition $f''(c) = 0$ need not imply that $(c, f(c))$ is a point of inflection of the graph of f. [*Hint:* Consider $f(x) = x^4$.] On the other hand, show that, if f has a continuous second derivative and $(c, f(c))$ is a point of inflection of the graph of f, then $f''(c) = 0$.

36 Suppose that f is a continuous function on the closed interval $[a, b]$ and that f is monotone on the open interval (a, b). Show that f is monotone on $[a, b]$.

37 Let the function f be increasing on the interval I.
(a) Is $3f$ increasing on I? Why?
(b) Is $-3f$ increasing on I? Why?
(c) Is the function g defined by $g(x) = -1 + f(x)$ increasing on I? Why?

38 Let f and g be increasing functions on the interval I.
(a) Is $f + g$ necessarily increasing on I? Why?
(b) Is $f \cdot g$ necessarily increasing on I? Why?

39 If the function f is differentiable on the open interval I, some textbooks define the graph of f to be concave upward if f satisfies the following condition: For every pair of distinct numbers a and b in I, the point $(b, f(b))$ on the graph of f lies strictly above the tangent line to the graph of f at $(a, f(a))$. Show that this condition holds if and only if, for every pair of distinct numbers a and b in I, $f(b) > f(a) + f'(a)(b - a)$.

40 Assume that the function f is differentiable on the open interval I and that the graph of f is concave upward on I in the sense of Definition 3. If a and b are distinct numbers in I, prove that $f(b) > f(a) + f'(a)(b - a)$.

41 Some textbooks define the graph of the function f to be concave upward on the interval I provided that, for any two distinct numbers a and b in I, the portion of the graph between $(a, f(a))$ and $(b, f(b))$ lies below the secant between $(a, f(a))$ and $(b, f(b))$. Draw a diagram illustrating this condition.

42 Suppose that the function f is differentiable on the open interval I and that the graph of f is concave upward on I. Prove that, if a and b are two numbers in I, then $0 < t < 1$ implies that $f(ta + (1 - t)b) < tf(a) + (1 - t)f(b)$.

4   Relative Maximum and Minimum Values of Functions

In this section we apply the ideas developed in Section 3 to the problem of finding the relative maximum and minimum values of a function. Consider, for instance, the polynomial function

$$f(x) = x^3 - 3x^2 + 5.$$

A sketch of the graph of f (Figure 1) shows that the point $(0, 5)$ on this graph

Figure 1

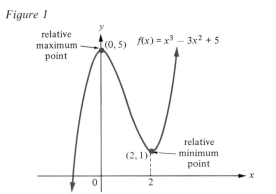

relative maximum point

$(0, 5)$ $f(x) = x^3 - 3x^2 + 5$

relative minimum point

$(2, 1)$

is higher than all its immediate neighboring points on the graph, although the graph does ultimately climb even higher. A point such as $(0, 5)$ is called a *relative maximum point* of the graph of f. Similarly, the point $(2, 1)$ is called a *relative minimum point* of the graph of f since it is lower than all its immediate neighboring points on the graph.

Our discussion suggests the following formal definitions.

DEFINITION 1   **Relative Maximum**

A function f is said to have a *relative maximum* (or a *local maximum*) at a number c if there is an open interval I containing c such that f is defined on I and $f(c) \geq f(x)$ holds for every number x in I.

DEFINITION 2   **Relative Minimum**

A function f is said to have a *relative minimum* (or a *local minimum*) at a number c if there is an open interval I containing c such that f is defined on I and $f(c) \leq f(x)$ holds for every number x in I.

Figure 2

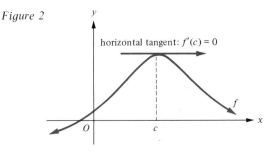

horizontal tangent: $f'(c) = 0$

f

If a function f has either a relative maximum or a relative minimum at a number c, we say that f has a *relative extremum* at c. It is geometrically clear that if a function f has a relative extremum at a number c and if the graph of f has a nonvertical tangent line at $(c, f(c))$, then this tangent line must be horizontal; that is, $f'(c) = 0$ (Figure 2). This observation suggests the following definition.

DEFINITION 3   **Critical Number**

We say that a number c is a *critical number* for the function f provided that f is defined at c and either f is not differentiable at c, or else $f'(c) = 0$.

We now give an analytic proof of the fact that a relative extremum can only occur at a critical number.

THEOREM 1   **Necessary Condition for Relative Extrema**

If the function f has a relative extremum at the number c, then c is a critical number for f.

PROOF

We consider the case in which f has a relative maximum at c, since the case of a relative minimum can be handled similarly. If f is not differentiable at c, then c is automatically a critical number for f by Definition 3, and we have nothing to prove. Thus, we can assume that f is differentiable at c, so that

$$f'(c) = \lim_{x \to c} \frac{f(x) - f(c)}{x - c}$$

exists. We must prove $f'(c) = 0$. If we prove that $f'(c)$ can be neither positive nor negative, then it must be zero and our argument will be complete. We prove that $f'(c)$ is not negative by showing that, if it were, then, contrary to hypothesis, f could not have a relative maximum at c. [That $f'(c)$ is not positive can be shown similarly.] Thus, suppose $f'(c)$ is negative. Since we can make $\dfrac{f(x) - f(c)}{x - c}$ as close as we please to the negative number $f'(c)$ by taking x sufficiently close to c (but not equal to c), there is a small open interval I containing c such that $\dfrac{f(x) - f(c)}{x - c}$ is negative if x is different from c and belongs to I.

If x belongs to I and $x < c$, then $x - c < 0$ and $\dfrac{f(x) - f(c)}{x - c} < 0$. If a fraction and its denominator are both negative, then the numerator must be positive; hence,

$$\text{if } x \text{ belongs to } I \text{ and } x < c, \quad \text{then} \quad f(x) > f(c).$$

The latter assertion contradicts the hypothesis that f has a relative maximum at c, since it says that the function values $f(x)$ are larger than $f(c)$ for values of x slightly to the left of c. This is the promised contradiction, and the proof is complete.

The numbers a, b, c, d, e, and k are critical numbers for the function f whose graph is sketched in Figure 3. Here, a, b, and k qualify as critical numbers for f because f is not differentiable at these numbers (although f does have one-sided derivatives at a and k). Notice that f does not have relative extrema at a or k, since no open interval about either of these numbers is contained in the domain of f. However, f does have a relative minimum at the critical number b (where f fails to be differentiable). Notice that f is differentiable at the remaining three critical numbers c, d, and e, so

Figure 3

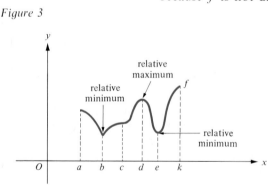

$$f'(c) = f'(d) = f'(e) = 0.$$

Clearly, f has a relative maximum at d and a relative minimum at e. However, in spite of the fact that the graph of f has a horizontal tangent at $(c, f(c))$, f *does not have a relative extremum at c.*

In order to find all relative extrema for a function f, one can begin by finding all the critical numbers for f. These critical numbers give *all possible candidates* for the numbers at which f has relative extrema; however, each critical number must be tested to see whether f really does have a relative extremum there. The simplest tests for relative extrema make use of the first or second derivatives of f.

4.1 First and Second Derivative Tests for Relative Extrema

Consider Figure 4a, in which the continuous function f has a positive first derivative on the interval (a, c) and a negative first derivative on the interval (c, b). By Theorem 1 in Section 3, f is increasing on $(a, c]$ and decreasing on $[c, b)$; hence, *f has a relative maximum at c.* Similarly, in Figure 4b, the continuous function f has a negative first derivative on (a, c) and a positive first derivative on (c, b); hence, *f has a relative minimum at c.*

The simple observations made above are summarized in the following theorem.

Figure 4

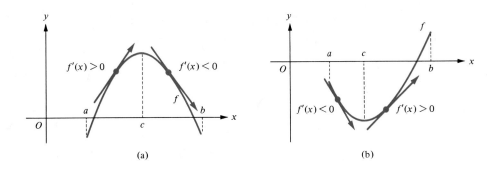

(a)   (b)

THEOREM 2

First Derivative Test for Relative Extrema

Let the function f be defined and continuous on the open interval (a,b), assume that the number c belongs to (a,b), and suppose that f is differentiable at every number in (a,b) except possibly at c.

(i)   If $f'(x) > 0$ for every number x in (a,c) and $f'(x) < 0$ for every number x in (c,b), then f has a relative maximum at c.

(ii) If $f'(x) < 0$ for every number x in (a,c) and $f'(x) > 0$ for every number x in (c,b), then f has a relative minimum at c.

In case (i) of Theorem 2, not only is there a relative maximum at c, but $f(c) \geq f(x)$ holds for *every* x in the interval (a,b) (Figure 4a). Similarly, in case (ii) of Theorem 2 (Figure 4b), $f(c) \leq f(x)$ holds for *every* x in the interval (a,b) (Problem 24).

EXAMPLE

If $f(x) = x^3 - 2x^2 + x + 1$, use the first derivative test to find all numbers at which f has a relative extremum and sketch the graph of f.

SOLUTION

Here, $f'(x) = 3x^2 - 4x + 1 = (x - 1)(3x - 1)$, so that the only critical numbers for f are the roots $x = 1$ and $x = \frac{1}{3}$ of the equation $f'(x) = 0$. Using the procedure for determining the intervals on which f' is positive or negative (Section 3.3), we find that

Figure 5

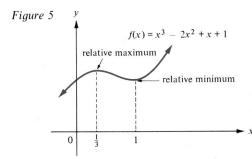

$$f'(x) > 0 \quad \text{for } x < \tfrac{1}{3};$$

$$f'(x) < 0 \quad \text{for } \tfrac{1}{3} < x < 1,$$

$$f'(x) > 0 \quad \text{for } x > 1.$$

Therefore, by the first derivative test, f has a relative maximum at $\frac{1}{3}$ and a relative minimum at 1. Using the second derivative to check the concavity of the graph (Section 3.2) and plotting a few points, we can sketch the graph of f (Figure 5).

The *second derivative test* for relative extrema is easily perceived (and remembered) geometrically. Figure 6a shows the graph of a function f and a critical number c such that $f''(c) > 0$. The condition $f''(c) > 0$ indicates that the graph of f is concave upward near the point $(c, f(c))$; hence, that f has a relative minimum at c. Similarly, Figure 6b shows a critical number c for f such that $f''(c) < 0$, so the graph of f is concave downward near $(c, f(c))$ and f has a relative maximum at c.

The second derivative test is stated and proved formally in the following theorem.

Figure 6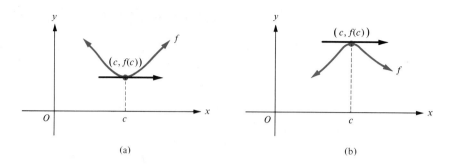

(a)   (b)

THEOREM 3 **Second Derivative Test for Relative Extrema**

Let the function f be differentiable on the open interval I and suppose that c is a number in I such that $f'(c) = 0$ and $f''(c)$ exists.

(i)   If $f''(c) > 0$, then f has a relative minimum at c.

(ii) If $f''(c) < 0$, then f has a relative maximum at c.

(iii) If $f''(c) = 0$, then the test is inconclusive.

PROOF

We prove (i) only, since the proof of (ii) is similar. [See Problem 25 for part (iii).] Thus, assume $f'(c) = 0$ and $f''(c) > 0$. By definition of $f''(c) = (f')'(c)$ and the fact that $f'(c) = 0$, we have

$$(f')'(c) = \lim_{x \to c} \frac{f'(x) - f'(c)}{x - c} = \lim_{x \to c} \frac{f'(x)}{x - c};$$

hence, we can make the ratio $\dfrac{f'(x)}{x - c}$ as close as we please to the positive number $f''(c)$ simply by taking x sufficiently close to c (but not equal to c). In particular, if x is close enough to c (but different from c), then $\dfrac{f'(x)}{x - c}$ will have to be positive. Therefore, there must be an open interval (a, b) containing c such that if x is different from c and belongs to (a, b), then $\dfrac{f'(x)}{x - c} > 0$. It follows that, if $a < x < c$, then $x - c < 0$ and $\dfrac{f'(x)}{x - c} > 0$, so that $f'(x) < 0$. Similarly, if $c < x < b$, we have $x - c > 0$ and $\dfrac{f'(x)}{x - c} > 0$, from which $f'(x) > 0$ follows. Thus, slightly to the left of c the derivative $f'(x)$ is negative, while slightly to the right of c it is positive. By the first derivative test (Theorem 2), we conclude that f has a relative minimum at c.

EXAMPLE   If $f(x) = x^3 - 6x^2 + 9x$, use the second derivative test to find all numbers at which f has a relative extremum and sketch the graph of f.

SOLUTION

Here, $f'(x) = 3x^2 - 12x + 9 = 3(x - 3)(x - 1)$ and $f''(x) = 6x - 12 = 6(x - 2)$. We set $f'(x) = 0$ and obtain the critical numbers $x = 1$ and $x = 3$. Since $f''(1) = -6 < 0$, we conclude that f has a relative maximum at 1. Similarly, since $f''(3) = 6 > 0$, it follows that f has a relative minimum at 3 (Figure 7).

Figure 7

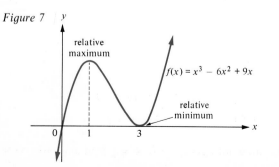

4.2 Procedure for Finding Relative Extrema

All relative extrema of a function f can be found systematically by carrying out the following procedure.

Procedure for Finding Relative Extrema of a Function f

Step 1   Find f'.

Step 2   Find the critical numbers for f; that is,
(a) Find all numbers c in the domain of f for which $f'(c)$ does not exist.
(b) Find all numbers c for which $f'(c) = 0$.

Step 3   Test each of the critical numbers to see whether it corresponds to a relative maximum, a relative minimum, or neither. Here, the first or second derivative tests can be used.

If f is an even or an odd function, we can begin by examining the nonnegative critical numbers, then use the symmetry of the graph of f to deal with the remaining ones.

EXAMPLES   Use the procedure above to find all numbers at which the given function f has a relative extremum. Sketch the graph of f.

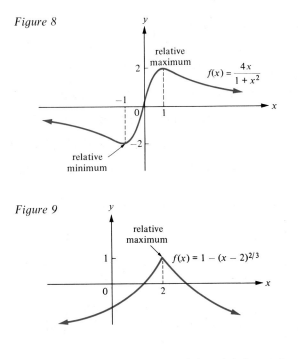

Figure 8

Figure 9

1 $f(x) = \dfrac{4x}{1 + x^2}$

SOLUTION
Using the quotient rule and simplifying, we find that $f'(x) = \dfrac{4(1 - x^2)}{(1 + x^2)^2}$, so the critical numbers are -1 and 1. If $-1 < x < 1$, then $f'(x) > 0$, while if $x > 1$, then $f'(x) < 0$. Therefore, by the first derivative test, f has a relative maximum at 1. Since f is an odd function, its graph is symmetric about the origin; hence, f has a relative minimum at -1 (Figure 8).

2 $f(x) = 1 - (x - 2)^{2/3}$

SOLUTION
Here, $f'(x) = -\frac{2}{3}(x - 2)^{-1/3}$, so f is differentiable at every number except 2. Also, for $x \neq 2$, $f'(x) \neq 0$; hence, 2 is the only critical number for f. If $x < 2$, then $f'(x) > 0$; if $x > 2$, then $f'(x) < 0$. Therefore, by the first derivative test, f has a relative maximum at 2 (Figure 9).

3 $f(x) = \frac{1}{12}(x^4 + 6x^3 - 18x^2)$

SOLUTION
We have

$$f'(x) = \tfrac{1}{12}(4x^3 + 18x^2 - 36x) = \tfrac{1}{6}x(2x^2 + 9x - 18)$$
$$= \tfrac{1}{6}x(2x - 3)(x + 6).$$

Thus, the roots of $f'(x) = 0$ are the critical numbers $x = 0$, $x = \frac{3}{2}$, and $x = -6$. Now,

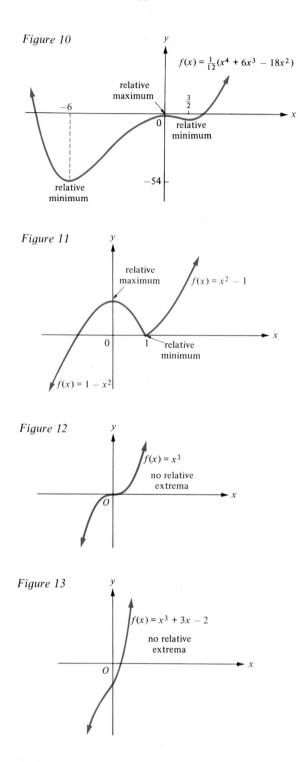

Figure 10

$f(x) = \frac{1}{12}(x^4 + 6x^3 - 18x^2)$

relative maximum

$\frac{3}{2}$

-6

0

relative minimum

-54

relative minimum

Figure 11

relative maximum

$f(x) = x^2 - 1$

0 1 relative minimum

$f(x) = 1 - x^2$

Figure 12

$f(x) = x^3$

no relative extrema

O

Figure 13

$f(x) = x^3 + 3x - 2$

no relative extrema

O

$$f''(x) = \tfrac{1}{12}(12x^2 + 36x - 36)$$
$$= x^2 + 3x - 3,$$

so that

$$f''(-6) = 15 > 0,$$
$$f''(0) = -3 < 0, \quad \text{and}$$
$$f''(\tfrac{3}{2}) = \tfrac{15}{4} > 0.$$

By the second derivative test, f has a relative minimum at -6, a relative maximum at 0, and a relative minimum at $\frac{3}{2}$ (Figure 10).

4 $f(x) = \begin{cases} x^2 - 1 & \text{if } x \geq 1 \\ 1 - x^2 & \text{if } x < 1 \end{cases}$

SOLUTION

Here,

$$f'(x) = \begin{cases} 2x & \text{if } x > 1 \\ -2x & \text{if } x < 1, \end{cases}$$

and $f'(1)$ is undefined. Thus, 1 is a critical number and, since $f'(0) = 0$, 0 is another critical number. For $x < 1$, $f''(x) = -2$; hence, $f''(0) = -2 < 0$, and it follows from the second derivative test that f has a relative maximum at 0. For $0 < x < 1$, $f'(x) = -2x < 0$, while for $x > 1$, $f'(x) = 2x > 0$; therefore, f has a relative minimum at 1 by the first derivative test (Figure 11).

5 $f(x) = x^3$

SOLUTION

Here, $f'(x) = 3x^2$, so 0 is the only critical number. However, for $x \neq 0$, $f'(x) = 3x^2 > 0$; hence, f is an increasing function and has no relative extrema whatsoever (Figure 12).

6 $f(x) = x^3 + 3x - 2$

SOLUTION

Here, $f'(x) = 3x^2 + 3$, so $f'(x) > 0$ for all values of x. Thus, there are no critical numbers for f and f has no relative extrema whatsoever (Figure 13).

Problem Set 4

In problems 1 through 10, (a) find the critical numbers for f and (b) use the first derivative test to see whether each of these critical numbers corresponds to a relative maximum, a relative minimum, or neither. Then, (c) sketch the graph of f.

1 $f(x) = 7 + 12x - 2x^2$　　　　**2** $f(x) = 2x^3 - 3x^2 - 4$　　　　**3** $f(x) = x^3 - x^2 - x - 1$

4 $f(x) = -x^3 + x^2 - x$ **5** $f(x) = x^4 - 4x$ **6** $f(x) = (x-1)^2(x-2)^2$

7 $f(x) = 2x^{1/2} - x$ **8** $f(x) = \dfrac{3}{x-2}$

9 $f(x) = \dfrac{x}{(x+1)^2}$ **10** $f(x) = \begin{cases} x^2 + 4 & \text{if } x \geq 1 \\ 8 - 3x & \text{if } x < 1 \end{cases}$

In problems 11 through 18, (a) find the critical numbers for f and (b) use the second derivative test on each of these critical numbers. Then, (c) sketch the graph of f.

11 $f(x) = x^2 - 5x + 4$ **12** $f(x) = x^3 + 3x^2 + 16$ **13** $f(x) = x^3 + 3x^2 - 3x$

14 $f(x) = 3x^4 - 4x^3 - 12x^2$ **15** $f(x) = (x-1)^{8/3} + (x-1)^2$ **16** $f(x) = \dfrac{5x}{x^2 + 7}$

17 $f(x) = x^2 + 5x^{-2}$ **18** $f(x) = \dfrac{1}{x^2 + x}$

In problems 19 and 20, find all numbers at which the given function has a relative extremum and sketch the graph.

19 $f(x) = \begin{cases} x^3 & \text{if } x \leq 1 \\ (x-2)^2 & \text{if } x > 1 \end{cases}$ **20** $f(x) = \begin{cases} 1 + \dfrac{1}{x^2} & \text{if } x \neq 0 \\ 0 & \text{if } x = 0 \end{cases}$

21 Find values of the constants a and b so that the function f defined by $f(x) = x^3 + ax + b$ has a relative minimum at the point $(1,3)$.

22 Find values of the constants p and q so that the function f defined by $f(x) = (p/x) + qx$ has a relative minimum at the point $(1,6)$.

23 Find values of the constants p and q so that the function g defined by $g(x) = px^{-1/2} + qx^{1/2}$ has a relative minimum at the point $(4,12)$.

24 In part (ii) of Theorem 2 in Section 4.1, prove that $f(c) \leq f(x)$ holds for all values of x in the interval (a,b).

25 Show that, if $f'(c) = 0$ and $f''(c) = 0$, as in part (iii) of Theorem 3 in Section 4.1, then the test is inconclusive. Do this by considering the three functions f, g, and h given by $f(x) = x^4$, $g(x) = x^3$, and $h(x) = -x^4$, and take $c = 0$.

5   Absolute Extrema

The techniques used in Section 4 to locate the relative extrema of functions are also useful for finding the *absolute extrema* of functions. The idea of an absolute extremum can be appreciated by considering the graph of the function f shown in Figure 1. Notice that f has relative maxima at $(p, f(p))$ and at $(r, f(r))$; however, the function value $f(r)$ is larger than the function value $f(p)$—in fact, it is larger than any other function value $f(x)$ for x in the closed interval $[a, b]$. Thus, we say that the function f *takes on an absolute maximum value $f(r)$ at r.*

In Figure 1, f has a relative minimum at $(q, f(q))$; however, the function value $f(b)$ is smaller than $f(q)$—in fact, it is smaller than any other function

Figure 1

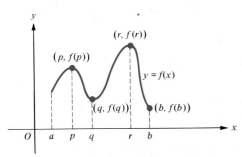

value $f(x)$ for x in the closed interval $[a,b]$. Thus, we say that the function f *takes on an absolute minimum value* $f(b)$ *at* b. Notice that, even though $f(b)$ is an absolute minimum value of f, it does not represent a relative minimum since the function f is not defined to the right of b.

The preceding discussion leads us to the following general definition.

DEFINITION 1   **Absolute Maximum and Minimum**

Suppose that the function f is defined (at least) on the interval I, and let c be a number in I. If $f(c) \geq f(x)$ [respectively, $f(c) \leq f(x)$] holds for all numbers x in I, then we say that, *on the interval I, the function f takes on its absolute maximum value* (respectively, its *absolute minimum value*) $f(c)$ *at the number* c.

If f takes on either an absolute maximum or an absolute minimum value at c, then we say that f takes on an *absolute extremum* at c. The question of whether a function has absolute extrema on an interval can often be settled by appealing to the following theorem.

THEOREM 1   **Existence of Absolute Extrema**

If a function f is defined and continuous on a closed interval $[a,b]$, then f takes on an absolute maximum value at some number in $[a,b]$ and f takes on an absolute minimum value at some number in $[a,b]$.

Geometrically, the property of continuous functions expressed by Theorem 1 is easy to accept since it asserts that a continuous curve drawn from point A to point B, as in Figure 2, has a highest point C and a lowest point D. An analytic proof of this important property of continuous functions can be found in more advanced textbooks.

If a function f takes on an absolute extremum on an interval I at a number c which is not an endpoint of I, then it is clear from the definitions involved that f has a relative extremum at c. This observation provides the basis for the following.

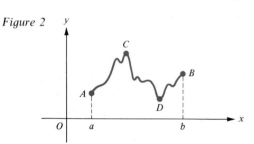

Figure 2

Procedure for Finding the Absolute Extrema of a Continuous Function on a Closed Interval

To find the absolute extrema of a continuous function f on a closed interval $[a,b]$, carry out the following steps:

Step 1   Find all critical numbers c for the function f on the open interval (a,b).

Step 2   Calculate the function values $f(c)$ for each of the numbers c obtained in Step 1.

Step 3   Calculate the function values $f(a)$ and $f(b)$ at the endpoints a and b of the interval.

Step 4   Conclude that the largest of all the numbers calculated in Steps 2 and 3 is the absolute maximum of f on $[a,b]$ and the smallest of these numbers is the absolute minimum of f on $[a,b]$.

If the function f is not continuous on the interval I, or if I is not a closed interval $[a,b]$, then perhaps the most effective method for finding the absolute extrema of f on I (when they exist) is to sketch the graph of f.

EXAMPLES   Find the absolute extrema of the given function f on the indicated interval and sketch the graph of f.

1 $f(x) = \sqrt{9 - x^2}$ on $[-3, 3]$

SOLUTION

Here, f is continuous on the closed interval $[-3, 3]$ and

$$f'(x) = \frac{-2x}{2\sqrt{9 - x^2}} = \frac{-x}{\sqrt{9 - x^2}}$$

Figure 3

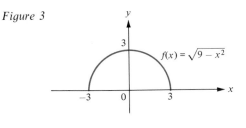

for x in the open interval $(-3, 3)$, so the only critical number for f on the interval $(-3, 3)$ is 0. Evaluating f at this critical number and at the endpoints of the interval, we obtain

$$f(-3) = 0, \quad f(0) = 3, \quad \text{and} \quad f(3) = 0.$$

Therefore, f takes on an absolute maximum value 3 at 0, and f takes on an absolute minimum value 0 at -3 and again at 3 (Figure 3).

2 $f(x) = \begin{cases} x^2 - 2x + 2 & \text{if } x \geq 0 \\ x^2 + 2x + 2 & \text{if } x < 0 \end{cases}$ on $[-\frac{1}{2}, \frac{3}{2}]$

SOLUTION

Here, f is continuous on the closed interval $[-\frac{1}{2}, \frac{3}{2}]$. Also,

$$f'(x) = \begin{cases} 2x - 2 & \text{if } 0 < x < \frac{3}{2} \\ 2x + 2 & \text{if } -\frac{1}{2} < x < 0, \end{cases}$$

so $f'(1) = 0$ and $f'(0)$ does not exist. Thus, we have the critical numbers 0 and 1 on $(-\frac{1}{2}, \frac{3}{2})$. Now, calculating the values of f at the endpoints and at the critical numbers, we obtain

Figure 4

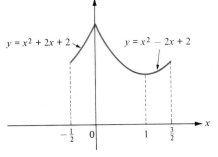

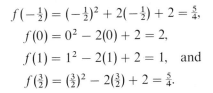

$$f(-\tfrac{1}{2}) = (-\tfrac{1}{2})^2 + 2(-\tfrac{1}{2}) + 2 = \tfrac{5}{4},$$
$$f(0) = 0^2 - 2(0) + 2 = 2,$$
$$f(1) = 1^2 - 2(1) + 2 = 1, \quad \text{and}$$
$$f(\tfrac{3}{2}) = (\tfrac{3}{2})^2 - 2(\tfrac{3}{2}) + 2 = \tfrac{5}{4}.$$

The largest of these function values is 2 and the smallest is 1; hence, f takes on the absolute maximum value 2 at 0 and the absolute minimum value 1 at 1 (Figure 4).

3 $f(x) = \dfrac{2 + x - x^2}{2 - x + x^2}$ on $[-2, 1]$

SOLUTION

Again, f is defined and continuous on the closed interval $[-2, 1]$. Using the quotient rule and simplifying, we have

$$f'(x) = \frac{(2 - x + x^2)(1 - 2x) - (2 + x - x^2)(-1 + 2x)}{(2 - x + x^2)^2} = \frac{4(1 - 2x)}{(2 - x + x^2)^2}.$$

The denominator $(2 - x + x^2)^2$ is nonzero for all values of x in $[-2, 1]$; hence, the only critical number in the interval $(-2, 1)$ is $\frac{1}{2}$. Now, calculating the values of f at the endpoints of $[-2, 1]$ and at the critical number $\frac{1}{2}$, we obtain

Figure 5

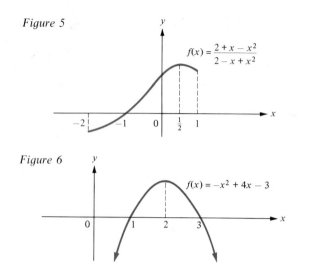

$$f(x) = \frac{2 + x - x^2}{2 - x + x^2}$$

Figure 6

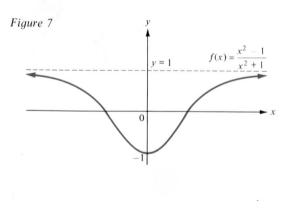

$$f(x) = -x^2 + 4x - 3$$

$f(-2) = -\frac{1}{2}$, $f(\frac{1}{2}) = \frac{9}{7}$, and $f(1) = 1$; hence, f takes on the absolute maximum value $\frac{9}{7}$ at $\frac{1}{2}$ and the absolute minimum value $-\frac{1}{2}$ at -2 (Figure 5).

4 $f(x) = -x^2 + 4x - 3$ on $(-\infty, \infty)$

SOLUTION
Here, $f'(x) = -2x + 4$ and $f''(x) = -2$, so $f'(x)$ is positive for x in $(-\infty, 2)$, $f'(x)$ is negative for x in $(2, \infty)$, and the graph of f is concave downward on $(-\infty, \infty)$ (Figure 6). Thus, f takes on an absolute maximum value 1 at 2, but f has no absolute minimum on the interval $(-\infty, \infty)$.

In Example 4, the interval $(-\infty, \infty)$ is the domain of the function f, and the absolute maximum of f on $(-\infty, \infty)$ is the largest of *all* function values of f. More generally, we say that a function f takes on the *absolute maximum value* (respectively, the *absolute minimum value*) $f(c)$ at the number c provided that c is a number in the domain of f and that $f(c) \geq f(x)$ [respectively, $f(c) \leq f(x)$] holds for every number x in the domain of f. Note that there is no reference here to any interval, it being understood that the entire domain of f is under consideration. In this connection, we sometimes even drop the word "absolute" and speak simply of the *maximum value* or the *minimum value* of the function f.

EXAMPLES Find the absolute extrema of the given function f and sketch the graph of f.

1 $f(x) = \dfrac{x^2 - 1}{x^2 + 1}$

SOLUTION
The domain of f is the interval $(-\infty, \infty)$. Using the quotient rule and simplifying, we have

$$f'(x) = \frac{4x}{(x^2 + 1)^2};$$

Figure 7

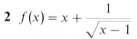

$$f(x) = \frac{x^2 - 1}{x^2 + 1}$$

hence, $f'(x) < 0$ for $x < 0$ and $f'(x) > 0$ for $x > 0$. It follows that f is decreasing on $(-\infty, 0]$ and increasing on $[0, \infty)$. Therefore, f has an absolute minimum of $f(0) = -1$ at 0. Since $\lim_{x \to +\infty} f(x) = 1$, it follows that the graph of f has the horizontal asymptote $y = 1$ (Figure 7). The function f has no absolute maximum, for although the values $f(x)$ can be made arbitrarily close to 1 by taking $|x|$ sufficiently large, these function values never *reach* 1.

2 $f(x) = x + \dfrac{1}{\sqrt{x - 1}}$

SOLUTION
The domain of f is the interval $(1, \infty)$. For $x > 1$,

$$f'(x) = 1 - \frac{1}{2\sqrt{(x - 1)^3}}.$$

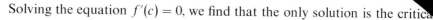

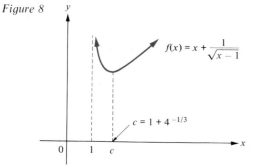

Figure 8

$f(x) = x + \dfrac{1}{\sqrt{x-1}}$

$c = 1 + 4^{-1/3}$

Solving the equation $f'(c) = 0$, we find that the only solution is the critica
$c = 1 + 4^{-1/3} \approx 1.63$. For $1 < x < c$, we
$f'(x) < 0$, while, for $x > c$, we have $f'(x) > 0$. The
fore, f is decreasing on $(1, c]$ and increasing on
$[c, \infty)$. It follows that f has an absolute minimum
value of $f(c) \approx 2.89$ at c. Notice that $\lim\limits_{x \to 1^+} f(x) =$
$+\infty$, so that the graph of f has the vertical
asymptote $x = 1$. Here, f has no absolute maximum,
since the values $f(x)$ can be made arbitrarily large
by taking $x > 1$ but close to 1 (or by letting x
approach $+\infty$) (Figure 8).

Problem Set 5

In problems 1 through 12, find the absolute extrema (if any) for the given function on the
given interval and sketch the graph.

1 $f(x) = -2x$ on $[-1, 2]$

2 $f(x) = 4x + 3$ on $[0, 2]$

3 $f(x) = (x + 1)^2$ on $[-2, 1]$

4 $f(x) = -x^2 + 5x - 4$ on $[0, 5]$

5 $f(x) = \sqrt{4 - x^2}$ on $[-2, 2]$

6 $f(x) = \sqrt{8 - 2x - x^2}$ on $[-3, 2]$

7 $f(x) = \begin{cases} x + 2 & \text{if } x < 1 \\ x^2 - 3x + 5 & \text{if } x \geq 1 \end{cases}$ on $[-6, 5]$

8 $f(x) = \begin{cases} 2x - 1 & \text{if } x \leq 2 \\ 2x^2 - 5 & \text{if } x > 2 \end{cases}$ on $[-3, 4]$

9 $f(x) = \begin{cases} \dfrac{1}{x + 1} & \text{if } x \neq -1 \\ 1 & \text{if } x = -1 \end{cases}$ on $[-2, 3]$

10 $f(x) = \begin{cases} |\,|x - 2| & \text{if } x \neq 2 \\ 4 & \text{if } x = 2 \end{cases}$ on $[1, 4]$

11 $f(x) = (x + 2)^{2/3}$ on $[-4, 3]$

12 $f(x) = 1 - (x - 2)^{2/3}$ on $[-5, 5]$

In problems 13 through 32 find (a) the intervals on which f is increasing or decreasing,
(b) the intervals on which the graph of f is concave upward or downward, (c) all numbers
where relative extrema of f occur, (d) the absolute extrema of f, and (e) the points of
inflection of the graph of f. Also, (f) sketch the graph of f.

13 $f(x) = 10 + 12x - 3x^2 - 2x^3$

14 $f(x) = x^3 + 2x^2 - 15x - 20$

15 $f(x) = x^4 - x^2 + 2$

16 $f(x) = x^5 - 3x^4$

17 $f(x) = 4x^5 - 20x^4$

18 $f(x) = x^2 + \dfrac{8}{x}$

19 $f(x) = \dfrac{3x}{x^2 + 9}$

20 $f(x) = \dfrac{x^2 + 4}{x^2 + 2}$

21 $f(x) = (3 + x)^2(1 - x)^2$

22 $f(x) = (1 + x)^2(3 + x)^3$

23 $f(x) = 2 + 4(x - 1)^{2/3}$

24 $f(x) = (3 + x)^{1/3}(1 - x)^{2/3}$

25 $f(x) = \dfrac{x + 1}{x^2 + 4x + 5}$

26 $f(x) = \dfrac{x + 2}{x^2 + 2x + 4}$

27 $f(x) = \dfrac{(1 - x)^3}{2 - 3x}$

28 $f(x) = \dfrac{4}{x} - \dfrac{6}{2 - x}$

29 $f(x) = (x + 1)^2 x^{1/3}$

30 $f(x) = \dfrac{x}{\sqrt{x^2 + 1}}$

31 $f(x) = \dfrac{x^2 + 1}{\sqrt{x^2 + 4}}$

32 $f(x) = \sqrt{\dfrac{9 - x}{9 + x}}$

33 Show that the absolute maximum value of the function f defined by

$$f(x) = \dfrac{1}{1 + |x|} + \dfrac{1}{1 + |x - 4|}$$

2. [*Hint:* Find the derivative on each of the intervals $(-\infty, 0)$,

onstants with $a > 0$. Find the absolute minimum value of the
$f(x) = ax^2 + bx + c.$

6   Maximum and Minimum—Applications to Geometry

In the following two sections we apply the methods of Section 5 to problems involving the maximum or minimum value of some quantity such as area, volume, power, time, profit, or cost. The present section is devoted to geometric problems, while the next section takes up problems arising in physics, engineering, business, and economics.

6.1   Geometric Applications

Before looking at some typical maximum–minimum problems involving geometry, we list a number of formulas for reference.

1 Plane area:
   (a) Square: $A = l^2$, $l =$ length of a side.
   (b) Rectangle: $A = lw$, $l =$ length, $w =$ width.
   (c) Circle: $A = \pi r^2$, $r =$ radius.
   (d) Triangle: $A = \frac{1}{2}bh$, $b =$ length of base, $h =$ height.
   (e) Trapezoid: $A = h\left(\dfrac{a+b}{2}\right)$, $h =$ height, $a =$ length of one base, $b =$ length of other base.

2 Perimeter:
   (a) Square: $p = 4l$.
   (b) Rectangle: $p = 2l + 2w$.
   (c) Circle: $p = 2\pi r$.

3 Surface area:
   (a) Closed rectangular box: $S = 2lw + 2lh + 2wh$, $l =$ length, $w =$ width, $h =$ height.
   (b) Right circular cylinder (open at top and bottom): $S = 2\pi rh$, $r =$ radius, $h =$ height.
   (c) Sphere: $S = 4\pi r^2$, $r =$ radius.
   (d) Right circular cone with open base: $S = \pi rl$, $h =$ height, $r =$ radius of base, $l =$ slant height $= \sqrt{r^2 + h^2}$.

4 Volume:
   (a) Rectangular box: $V = lwh$.
   (b) Right circular cylinder: $V = \pi r^2 h$.
   (c) Sphere: $V = \frac{4}{3}\pi r^3$.
   (d) Right circular cone: $V = \frac{1}{3}\pi r^2 h$.

EXAMPLES 1 Rodney has 1000 feet of fence with which he intends to enclose a rectangular exercise yard for his pet French poodle. What are the dimensions of the rectangular yard of maximum area?

SOLUTION

Figure 1

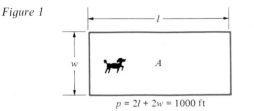

$$p = 2l + 2w = 1000 \text{ ft}$$

The quantity to be maximized is the area A of the yard. Here, $A = lw$, where l is the length of the yard and w is its width (Figure 1). (This diagram is so simple that it may seem ridiculous to bother drawing it. We do so anyway—it fixes ideas, displays the notation, and gives a start on the problem.) Now we can eliminate one of the two variables l or w from the formula $A = lw$, provided that we can find a suitable relationship between l and w. But there are 1000 feet of fence available, and the perimeter of the yard is given by $p = 2l + 2w$, so we have $2l + 2w = 1000$. Solving this equation for (say) l, we obtain $l = 500 - w$, which we substitute into the equation $A = lw$ to obtain

$$A = (500 - w)w = 500w - w^2.$$

We now have $A = f(w)$, where $f(w) = 500w - w^2$. Since the dimensions w and l of the yard cannot be negative, we have $w \geq 0$ and $l = 500 - w \geq 0$; that is, $0 \leq w \leq 500$. Our problem is to find the value of w that gives the absolute maximum of $f(w) = 500w - w^2$ on the closed interval $[0, 500]$. Here, $f'(w) = 500 - 2w$, so $w = 250$ gives the only critical number on the open interval $(0, 500)$. Evidently, $f'(w) > 0$ for $w < 250$ and $f'(w) < 0$ for $w > 250$; hence, f is increasing on $[0, 250]$ and decreasing on $[250, 500]$. Clearly, f takes on an absolute maximum value when $w = 250$ feet and $l = 500 - w = 500 - 250 = 250$ feet. (Notice that the area is largest when the yard has a square shape.)

2 Equal squares are cut off at each corner of a rectangular piece of cardboard 8 inches wide by 15 inches long and an open-topped box is formed by turning up the sides (Figure 2). Find the length x of the sides of the squares that must be cut off to produce a box of maximum volume.

SOLUTION

Figure 2

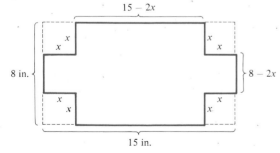

15 in.

Denote the volume of the open-topped box by V. From Figure 2, we see that the height of the box is x inches, the width is $8 - 2x$ inches, and the length is $15 - 2x$ inches. Thus,

$$V = x(8 - 2x)(15 - 2x) = 4x^3 - 46x^2 + 120x,$$
$$0 \leq x \leq 4.$$

Let f be the function defined by

$$f(x) = 4x^3 - 46x^2 + 120x,$$

so that

$$f'(x) = 12x^2 - 92x + 120 = (x - 6)(12x - 20).$$

The solutions of $f'(x) = 0$ are $x = \frac{5}{3}$ and $x = 6$; hence, f has only one critical number, $\frac{5}{3}$, on the interval $(0, 4)$. Since f is continuous on the closed interval $[0, 4]$ and $\frac{5}{3}$ is its only critical number on the open interval $(0, 4)$, the desired absolute maximum of f is the largest of the numbers $f(0) = 0$, $f(\frac{5}{3}) \approx 90.74$, and $f(4) = 0$. Therefore, the maximum volume (approximately 90.74 cubic inches) is achieved by cutting off squares of side length $\frac{5}{3}$ inches.

3 A cylindrical tin cup (with open top) is to hold 5 cubic inches. Find its dimensions if the amount of tin is to be minimum.

SOLUTION

Figure 3

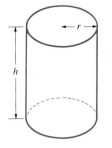

Let h be the height of the cup and let r be the radius of the circular base (Figure 3). Then, $5 = \pi r^2 h$, or, $h = 5/\pi r^2$. The lateral surface area of the cup is $2\pi rh$ and the area of the base is πr^2; hence, the total area S of tin required is given by

$$S = 2\pi rh + \pi r^2 = 2\pi r\left(\frac{5}{\pi r^2}\right) + \pi r^2 = \frac{10}{r} + \pi r^2, \qquad r > 0.$$

Let g be the function defined by $g(r) = (10/r) + \pi r^2$ for $r > 0$, so that

$$g'(r) = -\frac{10}{r^2} + 2\pi r \qquad \text{for } r > 0.$$

Setting $g'(r) = 0$, we obtain $2\pi r = 10/r^2$, or $r^3 = 10/(2\pi) = 5/\pi$. Thus, $r = \sqrt[3]{5/\pi}$ is the only critical number for g. Since

$$g'(r) < 0 \quad \text{for } 0 < r < \sqrt[3]{\frac{5}{\pi}} \quad \text{and} \quad g'(r) > 0 \quad \text{for } r > \sqrt[3]{\frac{5}{\pi}},$$

it follows that g is decreasing on $(0, \sqrt[3]{5/\pi}]$ and increasing on $[\sqrt[3]{5/\pi}, \infty)$. Consequently, g takes on an absolute minimum value $(S = g(\sqrt[3]{5/\pi}) \approx 12.85 \text{ in.}^2)$ when $r = \sqrt[3]{5/\pi} \approx 1.17$ in. and $h = 5/\pi r^2 = \sqrt[3]{5/\pi} \approx 1.17$ in.

Figure 4

4 A conical tent with no floor (Figure 4) is to have a capacity of 1000 cubic meters. Find the dimensions that minimize the amount of canvas required.

SOLUTION

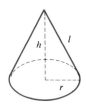

The lateral surface area of the cone is given by

$$S = \pi rl = \pi r\sqrt{r^2 + h^2},$$

while the volume is given by $1000 = \frac{1}{3}\pi r^2 h$. From the latter equation, $h = 3000/(\pi r^2)$; hence,

$$S = \pi r\sqrt{r^2 + \left(\frac{3000}{\pi r^2}\right)^2} = \sqrt{\pi^2 r^4 + \frac{3000^2}{r^2}}.$$

In order to minimize S, it is sufficient to minimize the quantity under the radical in the latter equation; hence, we define the function g by

$$g(r) = \pi^2 r^4 + \frac{3000^2}{r^2}, \qquad r > 0,$$

and seek the absolute minimum value of g. Here,

$$g'(r) = 4\pi^2 r^3 - \frac{2(3000)^2}{r^3}, \qquad r > 0.$$

Setting $g'(r) = 0$, we find only one positive critical number, namely,

$$r = \sqrt[6]{\frac{3000^2}{2\pi^2}} \approx 8.77 \text{ feet.}$$

Since

$$g'(r) < 0 \quad \text{for } r < \sqrt[6]{\frac{3000^2}{2\pi^2}} \quad \text{and} \quad g'(r) > 0 \quad \text{for } r > \sqrt[6]{\frac{3000^2}{2\pi^2}},$$

g takes on its absolute minimum value when

$$r = \sqrt[6]{\frac{3000^2}{2\pi^2}} \approx 8.77 \text{ meters} \quad \text{and} \quad h = \frac{3000}{\pi r^2} \approx 12.41 \text{ meters.}$$

5 Find the dimensions of the right circular cone of maximum volume V that can be inscribed in a sphere of radius a.

SOLUTION

Figure 5

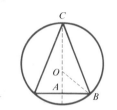

Figure 5 shows the vertical cross section through the center O of the sphere. Here the height h of the cone is $h = |\overline{AC}|$ and the radius r of the circular base of the cone is $r = |\overline{AB}|$. Also, the radius a of the sphere is $a = |\overline{OB}| = |\overline{OC}|$. Applying the Pythagorean theorem to right triangle OAB, we have

$$|\overline{OA}|^2 + |\overline{AB}|^2 = |\overline{OB}|^2;$$

that is,

$$(h - a)^2 + r^2 = a^2, \quad r^2 = a^2 - (h - a)^2 = 2ha - h^2.$$

Thus, the volume of the cone is

$$V = \tfrac{1}{3}\pi r^2 h = \tfrac{1}{3}\pi h(2ha - h^2) = \tfrac{2}{3}\pi a h^2 - \tfrac{1}{3}\pi h^3, \qquad 0 \le h \le 2a.$$

Therefore, $dV/dh = \tfrac{4}{3}\pi ah - \pi h^2$, and the critical numbers obtained by setting dV/dh equal to zero are $h = 0$ and $h = \tfrac{4}{3}a$. For $h = 0$, we have $V = 0$; for $h = \tfrac{4}{3}a$, we have $V = \tfrac{32}{81}\pi a^3$; for $h = 2a$, we have $V = 0$. Hence, the critical value $h = \tfrac{4}{3}a$ yields the maximum volume. When $h = \tfrac{4}{3}a$, $r^2 = 2ha - h^2 = \tfrac{8}{9}a^2$, so that $r = \tfrac{2}{3}a\sqrt{2}$.

6.2 General Procedure for Solving Maximum–Minimum Problems

We now set forth a step-by-step procedure for handling applied maximum–minimum problems—a procedure that is effective not only for geometric problems, but also for the physical and economic problems in the next section.

Procedure for Working Applied Maximum–Minimum Problems

Step 1   Attack the problem with determination! Just "have at it" and you will often find that the problem is not as formidable as it appears to be. Begin by reading the problem carefully (several times if necessary). Make certain that you understand just which quantity is to be maximized or minimized.

Step 2   Select a suitable symbol for the quantity to be maximized or minimized; for purposes of this discussion, call it Q. Determine the remaining quantities or variables upon which Q depends and select symbols for these variables. If it is feasible, draw a diagram and label the various parts with the corresponding symbols.

Step 3   Express the quantity Q whose extreme value is desired in terms of a formula involving the variables upon which it depends. If the formula involves more than one variable, use the conditions given in the statement of the problem to find relationships among these variables which can be used to eliminate all but one variable from the formula.

Step 4 We now have $Q = f(x)$, where (for purposes of this discussion) x denotes the single variable upon which Q was found to depend in Step 3, and f is the function determined by this dependence. If there are constraints on the quantity x imposed by the physical nature of the problem or by other practical considerations, specify these constraints explicitly. Apply the methods of Section 5 to find the desired absolute extremum of $f(x)$ subject to the imposed constraints on x.

In practice, the elimination of all but one of the variables upon which Q depends in Step 3 is often the trickiest part of the procedure. Sometimes it cannot be carried out at all because not enough relationships among these variables are given by the statement of the problem. In this case, the procedure given above fails, and more sophisticated methods must be used. (See Sections 10 and 11 of Chapter 16.)

After some experience solving maximum–minimum problems, one develops an intuition that suggests a somewhat cavalier treatment of the details in Step 4. In fact, one will often simply "set the first derivative equal to zero" for a solution, perhaps "checking with the second derivative" to see whether the solution represents a maximum or minimum. Such an informal solution is illustrated in the example given next. Of course, informal solutions should be viewed with skepticism since the desired extremum (if one exists at all) might occur at an endpoint, and the second derivative test (when it works at all) only indicates a *relative* maximum or minimum. The best advice here is to check all details whenever your conscience bothers you.

EXAMPLE   Give an *informal* solution to the following problem: Of all rectangles with area 25 square meters, find the one with the minimum perimeter p.

SOLUTION
Denoting the length of the rectangle by l and its width by w, we have $p = 2l + 2w$ and $lw = 25$; hence, $l = 25/w$ and $p = (50/w) + 2w$. Therefore, $dp/dw = -(50/w^2) + 2$. Setting $dp/dw = 0$ and solving for w, we obtain $w = \pm 5$. Since $w = -5$ is geometrically absurd, we have $w = 5$. Checking with the second derivative test, we find that

$$\frac{d^2p}{dw^2} = \frac{100}{w^3} = \frac{100}{5^3} > 0;$$

hence, $w = 5$ gives the desired minimum. (The reader is urged to show that the same result is obtained using rigorous methods.)

Problem Set 6

1 Find the dimensions of a rectangle of the smallest perimeter whose area is 100 square inches.

2 Find two positive real numbers whose sum is 20 and whose product is a maximum.

3 What are the dimensions of the rectangle of perimeter 48 inches that has the largest area?

4 Divide 40 into two parts in such a way that the sum of the squares of the parts is a minimum.

5 A rectangular area is to be enclosed by a fence 1500 feet long. Find the dimensions of the rectangle if the area is a maximum.

6 A rectangular field is to be adjacent to a river and is to have fencing on the three sides, the side on the river requiring no fencing. If 10,000 feet of fencing is available, find the dimensions of the field with largest area.

7 A farmer wishes to enclose a rectangular field by a fence and then divide it down the middle by another fence parallel to a side. What are the dimensions of the largest area that can be enclosed with 1800 feet of fencing?

8 Find the largest area of an isosceles triangle having a perimeter of 45 centimeters.

9 If three sides of a trapezoid are each 10 inches, how long must the fourth side be if the area is a maximum?

10 A cardboard box with a square base and no top is to be made from 400 square inches of cardboard. Find the dimensions of such a box with the largest volume.

11 A sheet of paper is to contain 18 square inches of print. The margins at the top and bottom are each 2 inches. Each margin at the side is to be 1 inch. What are the dimensions of the sheet of paper if its total area is to be a minimum?

12 Find the volume of the largest box that can be made from a piece of cardboard 8 centimeters square by cutting equal squares from the corners and turning up the sides.

13 A rectangle is placed inside a triangle with one side on the base of the triangle. If the base of the triangle is 10 inches and its height is 8 inches, find the largest possible area of the rectangle.

14 A wire 24 centimeters long is to be cut into two pieces. A circle is formed from one piece and a square from the other. How much wire should be used to form the square if the total area enclosed by the square and the circle is to be a minimum?

15 A conical tent is to have a volume of 3000 cubic feet. Find its dimensions if the amount of canvas is to be a minimum. (Disregard the floor.)

16 Find the dimensions of the cylinder of largest volume that can be inscribed in a sphere of radius 6 inches.

17 Find the dimensions of a cylindrical closed can of the largest volume if its surface area is 32π square centimeters.

18 Find the dimensions of the right circular cone of maximum volume having a slant height of 3 inches.

19 Find the dimensions of the right circular cone of minimum volume that can circumscribe a sphere of radius 4 inches.

20 Find the dimensions of the right circular cylinder of maximum volume that can be inscribed in a right circular cone with radius 6 meters and height 15 meters.

7 Maximum and Minimum—Applications to Physics, Engineering, Business, and Economics

7.1 Applications to Physics and Engineering

Many physical laws assert—or can be reformulated so as to assert—that physical motions or transformations take place in such a way that certain quantities are maximized or minimized. For instance, in optics the *principle of Fermat* asserts that light follows the path that minimizes the time of transit. The following example shows how Fermat's principle can be used to solve a problem in optics.

EXAMPLE   In Figure 1 a light beam starts at the point $(0, 1)$ on the y axis, strikes a horizontal mirror along the x axis at the point $(x, 0)$, and is reflected back up to the point $(4, 1)$. Use Fermat's principle to find the value of x. (Assume that light travels with a constant velocity c.)

Figure 1

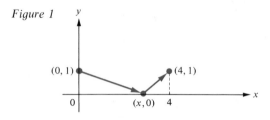

SOLUTION

The distance from $(0, 1)$ to $(x, 0)$ is given by

$$\sqrt{(x - 0)^2 + (0 - 1)^2} = \sqrt{x^2 + 1},$$

so the time required for the light beam to go from $(0, 1)$ to $(x, 0)$ is $\dfrac{\sqrt{x^2 + 1}}{c}$. Similarly, the time required for the light beam to go from $(x, 0)$ to $(4, 1)$ is given by

$$\frac{\sqrt{(4 - x)^2 + (1 - 0)^2}}{c} = \frac{\sqrt{x^2 - 8x + 17}}{c}.$$

Therefore, the total time of transit from point $(0, 1)$ to point $(4, 1)$ is given by

$$T = \frac{1}{c}\left(\sqrt{x^2 + 1} + \sqrt{x^2 - 8x + 17}\right).$$

Thus,

$$\frac{dT}{dx} = \frac{x}{c\sqrt{x^2 + 1}} + \frac{x - 4}{c\sqrt{x^2 - 8x + 17}}.$$

Setting dT/dx equal to zero and solving for x, we obtain

$$x\sqrt{x^2 - 8x + 17} = -(x - 4)\sqrt{x^2 + 1},$$
$$x^2(x^2 - 8x + 17) = (x - 4)^2(x^2 + 1),$$
$$x^4 - 8x^3 + 17x^2 = x^4 - 8x^3 + 17x^2 - 8x + 16,$$

so $8x = 16$ and $x = 2$. Thus, 2 is the only critical number for the continuous function expressing T in terms of x. Clearly $0 \le x \le 4$. When $x = 0$, $T = (1/c)(1 + \sqrt{17}) \approx 5.12/c$; when $x = 2$, $T = (1/c)(\sqrt{5} + \sqrt{5}) \approx 4.47/c$; and when $x = 4$, $T = (1/c)(\sqrt{17} + 1) \approx 5.12/c$. Therefore, T assumes its absolute minimum value when $x = 2$.

The following example is conceptually similar to the preceding one, but involves a different physical situation.

EXAMPLE   James lives on an island 6 miles away from a straight beach and his girl friend Jean lives 4 miles up the beach. James can row his boat 3 miles per hour and he can walk 5 miles per hour on the beach. Find the minimum time required for James to reach Jean's house from his island.

SOLUTION

Establish a coordinate system with the straight beach running along the x axis and with James's island at the point $(0, 6)$ on the y axis (Figure 2). Then Jean's house is located at the point $(4, 0)$ along the x axis. Suppose that James rows his boat from his island to the point $(x, 0)$ on the beach and then walks from $(x, 0)$ to Jean's house. Reasoning just as in the preceding example, but taking into account the different rates of speed for rowing and walking, we see that the time of transit is given by

$$T = \frac{\sqrt{(0-6)^2 + (x-0)^2}}{3} + \frac{\sqrt{(4-x)^2 + (0-0)^2}}{5};$$

that is,

$$T = \frac{\sqrt{x^2 + 36}}{3} + \frac{4-x}{5}, \qquad 0 \le x \le 4.$$

Thus, let f be the function defined by the equation

$$f(x) = \frac{\sqrt{x^2 + 36}}{3} + \frac{4-x}{5} \qquad \text{for } 0 \le x \le 4.$$

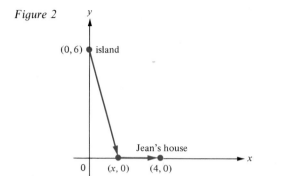

Figure 2

We must find the absolute minimum value of the continuous function f on the closed interval $[0, 4]$. Here $f'(x) = \frac{1}{3}x(x^2 + 36)^{-1/2} - \frac{1}{5}$. Solving the equation $f'(x) = 0$ for critical numbers, we obtain

$$5x = 3(x^2 + 36)^{1/2},$$

$$25x^2 = 9(x^2 + 36),$$

$$x = \pm\tfrac{9}{2}.$$

We must reject $x = -\frac{9}{2}$ because it is an extraneous root introduced by squaring, and we must reject $x = \frac{9}{2}$ because it does not belong to the interval $(0, 4)$. Thus, the function f has no critical numbers on the interval $(0, 4)$. Since $f(0) = 2.8$ while $f(4) = 2\sqrt{13}/3 \approx 2.4$, f takes on its minimum value when $x = 4$. Thus, for least time of transit, James should row directly to Jean's house. This requires $\frac{2}{3}\sqrt{13} \approx 2.4$ hours.

Additional examples showing how maximum–minimum problems arise in connection with physical situations follow.

EXAMPLES 1 Ship A is 65 miles due east of ship B and is sailing south at the rate of 15 miles per hour, while ship B is sailing east at the rate of 10 miles per hour. If the ships continue on their respective courses, find the minimum distance between them and when it occurs.

Figure 3

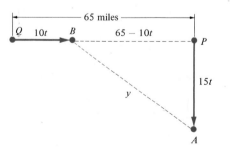

SOLUTION

In Figure 3, P shows the original position of ship A while Q shows the original position of ship B. After t hours have elapsed, B will have moved $10t$ miles while A will have moved $15t$ miles. By the Pythagorean theorem, the distance y between A and B at time t is given by

$$y = \sqrt{(15t)^2 + (65 - 10t)^2}$$
$$= \sqrt{325t^2 - 1300t + 4225}.$$

Clearly, y is minimum when the quantity

$$325t^2 - 1300t + 4225 = 325(t^2 - 4t + 13)$$

is minimum; that is, when the quantity $t^2 - 4t + 13$ is minimum. Let f be the function defined by $f(t) = t^2 - 4t + 13$. Since $f'(t) = 2t - 4$, we see that $t = 2$ gives the only critical number for f. Here, $f'(t) < 0$ for $0 < t < 2$ and $f'(t) > 0$ for $t > 2$; hence, f is decreasing on the interval $[0, 2]$ and increasing on the interval $[2, \infty)$. Consequently, f takes on its absolute minimum value when $t = 2$ hours. Therefore, the minimum distance between the ships occurs after 2 hours has elapsed and is given by

$$y = \sqrt{30^2 + 45^2} = \sqrt{2925} \approx 54.08 \text{ miles.}$$

2 Neglecting air resistance, the stream of water projected from a fire hose satisfies the equation

$$y = mx - 16(1 + m^2)\left(\frac{x}{v}\right)^2,$$

where m is the slope of the nozzle, v is the velocity of the stream at the nozzle in feet per second, and y is the height in feet of the stream x feet from the nozzle (Figure 4). Assume that v is a positive constant. Find (a) the value of x for which the height y of the stream is maximum for a fixed value of m, (b) the value of m for which the stream hits the ground at the greatest distance from the nozzle, and (c) the value of m for which the water reaches the greatest height on a vertical wall x feet from the nozzle.

Figure 4

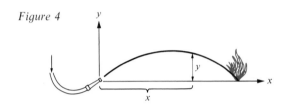

SOLUTION

(a) Here m and v are both constant and we seek the value of x that makes $y = mx - 16(1 + m^2)(x/v)^2$ a maximum. Since

$$\frac{dy}{dx} = m - \frac{32(1 + m^2)}{v^2}x \quad \text{and} \quad \frac{d^2y}{dx^2} = -\frac{32(1 + m^2)}{v^2} < 0,$$

the critical value $x = \dfrac{mv^2}{32(1 + m^2)}$ gives the desired maximum height; hence, the maximum value of y is $\dfrac{m^2 v^2}{64(1 + m^2)}$.

(b) For any given value of m, the stream hits the ground when $y = 0$; that is, when $mx = 16(1 + m^2)(x/v)^2$, $x > 0$. Solving for x, we obtain $x = \dfrac{mv^2}{16(1 + m^2)}$. [Comparing this with the result of part (a), we see that the stream reaches its maximum height halfway between the nozzle and the point where it hits the ground—all very reasonable.] Our problem here is to find the value of m that maximizes $\dfrac{mv^2}{16(1 + m^2)}$. To do this, set $D_m\left[\dfrac{mv^2}{16(1 + m^2)}\right] = 0$ and solve for the critical values m. Indeed,

$$D_m\left[\frac{mv^2}{16(1 + m^2)}\right] = \frac{v^2(1 - m^2)}{16(1 + m^2)^2},$$

so $m = \pm 1$ gives the critical values. We reject $m = -1$ on obvious physical grounds. The solution $m = 1$ would indicate that, to squirt the water a maximum distance, one holds the nozzle at a 45-degree angle. This seems so reasonable that only a dyed-in-the-wool skeptic would insist on completing the rigorous check to make certain that $m = 1$ gives an absolute maximum.

(c) Here, x and v are constants and y depends on the variable slope m according to $y = mx - 16(1 + m^2)(x/v)^2$. Thus,

$$\frac{dy}{dm} = x - 32\left(\frac{x}{v}\right)^2 m \quad \text{and} \quad \frac{d^2y}{dm^2} = -32\left(\frac{x}{v}\right)^2 < 0;$$

hence, the critical value $m = \dfrac{v^2}{32x}$ gives y the maximum value $\dfrac{v^2}{64} - \dfrac{16x^2}{v^2}$.

7.2 Applications to Business and Economics

In economics the term "marginal" is often used as a virtual synonym for "derivative of." For instance, if C is a cost function so that $C(x)$ is the cost of producing x units of a certain commodity, $C'(x)$ is called the *marginal cost* of producing x units and C' is called the *marginal cost function*. Thus, the marginal cost is the rate at which the total production cost changes per unit change in production.

In practical situations, x, the number of units produced, is usually a rather large number. Therefore, in comparison to x, the number 1 can be considered to be very small, so that

$$C'(x) = \lim_{\Delta x \to 0} \frac{C(x + \Delta x) - C(x)}{\Delta x} \approx \frac{C(x + 1) - C(x)}{1} = C(x + 1) - C(x).$$

Hence, when the number of units x is rather large, the marginal cost $C'(x)$ can be regarded as a good approximation to the cost $C(x + 1) - C(x)$ of producing one more unit.

EXAMPLE Solar Brush Co. finds that the total production cost for manufacturing x toothbrushes is given by $C(x) = \$(500 + 30\sqrt{x})$. If 5000 toothbrushes are manufactured, find the exact cost of manufacturing one more toothbrush and compare this with the marginal cost.

SOLUTION
The exact cost of manufacturing one more toothbrush would be

$$C(5001) - C(5000) = (500 + 30\sqrt{5001}) - (500 + 30\sqrt{5000})$$
$$= 30(\sqrt{5001} - \sqrt{5000}) \approx \$0.21212.$$

Since $C'(x) = 30/(2\sqrt{x}) = 15/\sqrt{x}$, then $C'(5000) = 15/\sqrt{5000} \approx \$0.21213$. Thus, the error made in using the marginal cost to estimate the true cost of manufacturing one more toothbrush is less than \$0.00002.

If $R(x)$ denotes the total revenue obtained when x units of a commodity are demanded, then the *marginal revenue* $R'(x)$ denotes the rate at which the revenue changes per unit change in demand. Again, for large values of x, the marginal revenue $R'(x)$ is a good approximation to the additional revenue $R(x + 1) - R(x)$ generated by one additional unit of demand.

Suppose that the total revenue takes on a maximum when x units are demanded. Then the marginal revenue $R'(x)$ must be zero. According to the interpretation of marginal revenue stated above, this would mean that when maximum revenue is generated by x units of demand, practically no additional revenue would be generated by one more unit of demand.

EXAMPLE   The total revenue for a certain kind of Swiss-made watch is given by the equation $R(x) = 2000x\sqrt{75 - x}$, $0 \leq x \leq 75$, where x denotes the demand in thousands of watches and the total revenue is in dollars. Find the maximum total revenue.

SOLUTION
The marginal revenue is given by

$$R'(x) = 2000\sqrt{75 - x} - \frac{1000x}{\sqrt{75 - x}}.$$

Setting $R'(x)$ equal to zero and solving for critical values, we obtain $x = 50$. Since R is continuous on the closed interval $[0, 75]$ and since $R(0) = 0$, $R(50) = 500{,}000$, $R(75) = 0$, we see that $x = 50$ corresponds to the maximum total revenue \$500,000. It is interesting to note that the additional revenue generated by a demand for 1000 more watches (that is, by one more unit of demand) is $-\$304.09$.

Miscellaneous additional examples follow.

EXAMPLES 1 A tire manufacturer charges \$24 per tire. The total cost of producing x tires per week is given by the equation $C(x) = 150 + 3.9x + 0.003x^2$ dollars.
(a) Find the marginal cost when $x = 1000$.
(b) Approximately how much does it cost the manufacturer to produce the 1001st tire?
(c) Exactly how much does it cost the manufacturer to produce the 1001st tire?
(d) Write the manufacturer's total profit P per week in terms of x.
(e) How many tires should the manufacturer make and sell per week to achieve maximum profit? What is the maximum profit?

SOLUTION
(a) $C'(x) = 3.9 + 0.006x$; $C'(1000) = \$9.90$.
(b) \$9.90.
(c) $C(1001) - C(1000) = 7059.903 - 7050.00 = \$9.903$.
(d) $P = 24x - C(x) = 20.1x - 150 - 0.003x^2$.
(e) $dP/dx = 20.1 - 0.006x$, so $x = 20.1/0.006 = 3350$ tires is a critical number. Since $d^2P/dx^2 = -0.006 < 0$, this critical number corresponds to a maximum. Thus, by manufacturing 3350 tires per week and selling them, the manufacturer will make a maximum profit of \$33,517.50 per week.

2 An automobile leasing agency rents cars to members of a teachers' credit union and discounts its total bill to these members by 2 percent for each rented car in excess of 12. For how many cars rented to the members would the total receipts to the agency be maximum?

SOLUTION
The total receipts to the agency is given by the equation

$$R(x) = \begin{cases} ax & \text{if } 0 \leq x \leq 12 \\ ax - [0.02(x - 12)]ax & \text{if } x > 12, \end{cases}$$

where x is the number of cars rented to the members and \$$a$ is the undiscounted rental fee per car. We have

$$R'(x) = \begin{cases} a & \text{if } 0 < x < 12 \\ 1.24a - 0.04ax & \text{if } 12 < x. \end{cases}$$

Therefore, the function R has critical numbers at 12 and at 31. Since $R'(x) > 0$

for $0 < x < 12$ and also for $12 < x < 31$, R is increasing on $[0, 12]$ as well as on $[12, 31]$; hence, R is increasing on $[0, 31]$. But $R'(x) < 0$ for $x > 31$, so R is decreasing on $[31, \infty)$. Hence, the critical number $x = 31$ gives the desired maximum.

3 A firm that manufactures body shirts for women estimates that the total cost $C(x)$ in dollars of making x body shirts is given by the equation

$$C(x) = 100 + 3x + \frac{x^2}{30}.$$

In any given week, the firm's total revenue $R(x)$ in dollars is given by the equation $R(x) = 25x + x^2/250$ where x is the number of body shirts sold.
(a) Assuming that the number x of body shirts sold per week is the same as the number manufactured, write an equation for the total weekly profit $P(x)$.
(b) Find the maximum weekly profit.

SOLUTION

(a) $P(x) = R(x) - C(x) = \left(25x + \dfrac{x^2}{250}\right) - \left(100 + 3x + \dfrac{x^2}{30}\right) = 22x - \dfrac{22}{750}x^2 - 100.$

(b) $P'(x) = 22 - \frac{44}{750}x$, so $x = 375$ is the only critical number. Since $P''(x) = -\frac{44}{750} < 0$, then $x = 375$ yields a maximum profit of \$4025 per week.

Problem Set 7

1 A cable television company has its master antenna located at point A on the bank of a straight river 1 mile wide. It is going to run a cable from A to a point P on the opposite bank of the river and then straight along the bank to a town T situated 3 miles downstream from A (Figure 5). It costs \$5 per foot to run the cable under the water and \$3 per foot to run the cable along the bank. What should be the distance from P to T in order to minimize the cost of the cable?

Figure 5

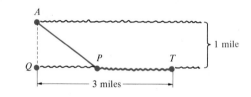

2 If a resistor of R ohms is connected across a battery of E volts whose internal resistance is r ohms, a current of I amperes will flow and generate P watts. Here P is given by the equation $P = I^2R$, where $I = \dfrac{E}{R + r}$. Find the resistance R if the power generated is to be maximum.

Figure 6

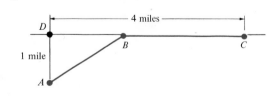

3 A vacationer runs out of gas in a trailer park. He is at point A, directly 1 mile from a point D on a paved road (Figure 6). He can reach a gas station at point C by walking through the woods in a straight line from A to a point B on the paved road at the rate of 3 miles per hour and then proceeding on the paved road at the rate of 5 miles per hour until he reaches C. If the distance from D to C is 4 miles along the paved road, how far should B be from D in order that he reach the gas station C in the shortest time?

4 If a ball is thrown upward (vertically) with a velocity of 36 feet per second, its height s after t seconds is given by the equation $s = 36t - 16t^2$. Find the time t at which the ball reaches its highest point.

5 Ship A is sailing north at the rate of 20 miles per hour, while ship B is sailing west at 30 miles per hour. At the start, ship B was 50 miles east of ship A. If they continue their respective courses, find the minimum distance between the ships.

6 Light travels with speed c in air and with (slower) speed v in water. Figure 7 shows the path by which a ray of light travels from a point A in the air to a point B on the surface of the water and then to a point W in the water. Here, α (the angle of incidence) is the angle between $\overline{AB}$ and the normal to the surface of the water, while β (the angle of refraction) is the angle between $\overline{BW}$ and this normal.

Figure 7

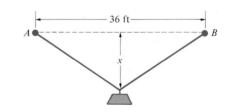

(a) Show that the total time required for the light to go from A to W is

$$T = \frac{\sqrt{a^2 + x^2}}{c} + \frac{\sqrt{b^2 + (k - x)^2}}{v},$$

where $a = |\overline{AD}|$, $x = |\overline{DB}|$, $b = |\overline{EW}|$, and $k = |\overline{DE}|$.

(b) Show that when x has the value that makes T minimum, then $\dfrac{\sin \alpha}{\sin \beta} = \dfrac{c}{v}$ (Snell's law of refraction).

7 A ball is thrown so that its height in feet is expressed by $y = mx - (m^2 + 1)x^2/800$, where m is the slope of the path of the ball at the origin and x is its distance from the origin at the horizontal level. Find the value of m for which the ball returns to the horizontal level at the maximum distance from the origin.

8 Suppose that n identical nickel–cadmium cells are to be arranged in series–parallel to furnish current to the motor of an experimental electric car. The motor has a resistance of R ohms, while each cell has an internal resistance of r ohms and an electromotive force of E volts. In the arrangement, x cells are to be connected in series so that the battery will have a net electromotive force of xE volts and an internal resistance of x^2r/n ohms. The current delivered to the motor is given by $I = \dfrac{xE}{R + (x^2r/n)}$. Solve for the value of x that will maximize the current I. (Although x represents a whole number here, you may treat it as a continuous variable.)

9 A sodium vapor lamp L is to be placed on top of a pole of height x meters to furnish illumination at a busy traffic intersection T (Figure 8). The foot P of the pole must be located 30 meters from T. If $r = |\overline{LT}|$ is the distance from the lamp to the point T and α is the angle PTL, then the intensity of illumination I at T will be proportional to the sine of α and inversely proportional to r^2; thus, $I = \dfrac{c \sin \alpha}{r^2}$, where c is a constant. Find the value of x that maximizes I.

Figure 8

10 A 4-ton weight is to be suspended from two identical cables fixed to points A and B (Figure 9). The distance between A and B is 36 feet, the perpendicular distance from the weight to the line $\overline{AB}$ is x feet, and the cable weighs 3 pounds per running foot. The resulting tension in the cable is given by

$$T = \frac{4000\sqrt{324 + x^2}}{x} + \frac{972}{x} + 3x \text{ pounds.}$$

Find the value of x that minimizes the tension.

Figure 9

11 The deflection of a beam 40 feet long supported at the ends and loaded at a point 30 feet from the left end is expressed by the equation $y = \dfrac{P}{3EI}\left(200x - \dfrac{x^3}{8}\right)$, where P, E, and I are constants; x is the distance in feet from the left end; and $0 < x < 30$. Find the maximum deflection.

12 The total cost C of manufacturing x toys is given by $C = 800 + 20x + 8000/x$. Find the production level x when the total cost is a minimum.

13 The total cost C of producing a certain commodity is given by the equation $C = 9000 - 240x + 4x^2$, where x is the number of units produced.
(a) Find the marginal cost when 40 units are produced.
(b) Find x when the total cost is a minimum.

14 If $C(x)$ is the total cost of manufacturing x units of a commodity, $C(x)/x$ is the average cost per unit of manufacturing the x units. Often it happens that the average cost per unit decreases as the number of units manufactured increases. Show that, when this is so, the marginal cost is always less than the average cost per unit.

15 A manufacturer produces x tons of a new alloy. The profit P in dollars earned by the manufacturer is expressed by the equation $P = 12{,}000x - 30x^2$. How many tons should be manufactured so as to maximize the total profit?

16 The total revenue R for selling a certain kind of leather belt is given by $R = 6x - 8\sqrt{x}$, where x is the demand for the leather belt and R is in dollars. Find the maximum total revenue.

17 A manufacturer of Christmas tree ornaments knows that the total cost C in dollars of making x thousand ornaments of a certain kind is given by $C = 600 + 60x$ and that the corresponding sales revenue R in dollars is given by $R = 300x - 4x^2$. Find the number of ornaments (in thousands) that will maximize the manufacturer's profit.

18 A cable television company plans to begin operations in a small town. It foresees that about 600 people will subscribe to the service if the price per subscriber is $5.00 per month. Experience shows that for each 5-cent increase in the individual subscription price per month, 4 of the original 600 people will decide not to subscribe. The cost to the company per month per subscription is estimated to be $3.50. (a) What price per month per subscription will bring in the greatest total revenue to the company? (b) What price will bring in the greatest profit to the company?

19 A mathematics department head observes that one secretary will work approximately 30 hours per week. However, if additional secretaries are hired, the resulting conversations reduce the effective number of work hours per week per secretary by $30(x - 1)^2/33$ hours, where x is the total number of secretaries hired. How many secretaries should be hired to turn out the most work?

20 In medicine it is often assumed that the *reaction* R to a dose of size x of a drug is given by an equation of the form $R = Ax^2(B - x)$, where A and B are certain positive constants. The *sensitivity* of the body to a dose of size x is defined to be the derivative dR/dx of the reaction with respect to the dose. (a) For what value of x is the reaction maximum? (b) For what value of x is the sensitivity dR/dx maximum?

21 One hundred animals belonging to an endangered species are placed in a protected game preserve. After t years the population p of these animals in the preserve is given by $p = 100 \dfrac{t^2 + 5t + 25}{t^2 + 25}$. After how many years is the population p a maximum?

8   Implicit Functions and Implicit Differentiation

An equation such as $y = 3x^2 - 5x + 12$ that is already solved for y in terms of x is said to give y *explicitly* as a function of x. On the other hand, an equation such as $xy + 1 = 2x - y$, which can be solved for y in terms of x but is not solved for y as it stands, is said to give y *implicitly* as a function of x. (In Section 8.2 we give a more precise definition of an implicit function.) In the present section we introduce a technique called *implicit differentiation* which allows us to calculate the derivative dy/dx of an implicitly given function without bothering to solve explicitly for y in terms of x.

8.1 Implicit Differentiation

The equation $2x + 3y = 1$ can be solved for y in terms of x to give $y = \frac{1}{3} - \frac{2}{3}x$, and so we have $dy/dx = -\frac{2}{3}$. The same result can be obtained directly from the original equation $2x + 3y = 1$ simply by differentiating both sides, term by term, to get $2 + 3(dy/dx) = 0$ and then solving for $dy/dx = -\frac{2}{3}$. The latter technique is called *implicit differentiation*. More generally, we have the following.

Procedure for Implicit Differentiation

Given an equation that determines y implicitly as a differentiable function of x, calculate dy/dx as follows:

Step 1 Differentiate both sides of the equation with respect to x; that is, apply d/dx to both sides of the equation, term by term. In so doing, keep in mind that y is regarded as a function of x and use the *chain rule* when necessary to differentiate expressions involving y.

Step 2 The result of Step 1 will be an equation involving not only x and y, but also dy/dx. Solve this equation for the desired derivative dy/dx.

When the procedure for implicit differentiation is executed, the result is often an equation that gives dy/dx in terms of *both* x and y. In this case, in order to calculate the numerical value of dy/dx, it is necessary to know not only the numerical value of x, but also the numerical value of y.

The procedure for implicit differentiation can only be used legitimately if it is known that the equation in question really does determine y implicitly as a differentiable function of x. However, in what follows we routinely apply the procedure and simply assume that this requirement is fulfilled.

EXAMPLES   Use implicit differentiation to solve each problem.

1 If $x^3 - 3x^2y^4 + 4y^3 = 6x + 1$, find dy/dx.

SOLUTION
The derivatives with respect to x of the individual terms in the equation are given by

$$\frac{d}{dx}(x^3) = 3x^2,$$

$$\frac{d}{dx}(3x^2y^4) = 3x^2\left[\frac{d}{dx}(y^4)\right] + 3\left[\frac{d}{dx}(x^2)\right]y^4 = 3x^2\left[4y^3\frac{dy}{dx}\right] + 3(2x)y^4,$$

$$\frac{d}{dx}(4y^3) = 12y^2\frac{dy}{dx},$$

$$\frac{d}{dx}(6x) = 6, \quad \text{and}$$

$$\frac{d}{dx}(1) = 0.$$

Therefore, term-by-term differentiation of the equation on both sides yields

$$3x^2 - 12x^2y^3\frac{dy}{dx} - 6xy^4 + 12y^2\frac{dy}{dx} = 6$$

or

$$(12x^2y^3 - 12y^2)\frac{dy}{dx} = 3x^2 - 6xy^4 - 6;$$

hence,

$$\frac{dy}{dx} = \frac{3x^2 - 6xy^4 - 6}{12x^2y^3 - 12y^2} = \frac{x^2 - 2xy^4 - 2}{4y^2(x^2y - 1)}.$$

2 Find the equations of the tangent and normal lines to the graph of the implicit function determined by $\sqrt{y} + \sqrt[3]{y} + \sqrt[4]{y} = 7xy$ at the point $(\frac{3}{7}, 1)$.

SOLUTION
Differentiating the equation $y^{1/2} + y^{1/3} + y^{1/4} = 7xy$ term by term with respect to x, we obtain

$$\tfrac{1}{2}y^{-1/2}\frac{dy}{dx} + \tfrac{1}{3}y^{-2/3}\frac{dy}{dx} + \tfrac{1}{4}y^{-3/4}\frac{dy}{dx} = 7y + 7x\frac{dy}{dx};$$

hence,

$$\frac{dy}{dx} = \frac{7y}{\tfrac{1}{2}y^{-1/2} + \tfrac{1}{3}y^{-2/3} + \tfrac{1}{4}y^{-3/4} - 7x}.$$

Therefore, when $x = \frac{3}{7}$ and $y = 1$, we have

$$\frac{dy}{dx} = \frac{7}{\tfrac{1}{2} + \tfrac{1}{3} + \tfrac{1}{4} - 7(\frac{3}{7})} = -\frac{84}{23}.$$

The slope of the tangent line at $(\frac{3}{7}, 1)$ is $-\frac{84}{23}$, the slope of the normal line is $\frac{23}{84}$, and the corresponding equations are

$$\text{tangent line:}\quad y - 1 = -\tfrac{84}{23}(x - \tfrac{3}{7}) \quad\text{or}\quad y = -\tfrac{84}{23}x + \tfrac{59}{23},$$
$$\text{normal line:}\quad y - 1 = \tfrac{23}{84}(x - \tfrac{3}{7}) \quad\text{or}\quad y = \tfrac{23}{84}x + \tfrac{173}{196}.$$

3 If $x^2 - 2y^2 = 4$, show that $D_x y = x/2y$ and that $D_x^2 y = -1/y^3$.

SOLUTION
Differentiating both sides of $x^2 - 2y^2 = 4$ with respect to x, we have $2x - 4yD_x y = 0$; hence, $D_x y = 2x/(4y) = x/(2y)$ as desired. Differentiating both sides of the latter equation with respect to x and using the quotient rule, we obtain

$$D_x^2 y = \frac{2yD_x x - xD_x(2y)}{(2y)^2} = \frac{2y - 2xD_x y}{4y^2} = \frac{y - xD_x y}{2y^2}.$$

Substituting $D_x y = x/(2y)$ into the latter equation and simplifying, we have

$$D_x^2 y = \frac{y - x[x/(2y)]}{2y^2} = \frac{2y^2 - x^2}{4y^3}.$$

Finally, since $x^2 - 2y^2 = 4$, we can rewrite the latter equation as

$$D_x^2 y = \frac{-4}{4y^3} = -\frac{1}{y^3}.$$

4 Given the equation $x^3 + xy + y^3 = 2$, consider x to be a differentiable function of y and find dx/dy.

SOLUTION

Applying d/dy to both sides of $x^3 + xy + y^3 = 2$, we obtain

$$3x^2 \frac{dx}{dy} + x + \frac{dx}{dy} y + 3y^2 = 0.$$

Solving the latter equation for dx/dy, we have

$$\frac{dx}{dy} = \frac{-x - 3y^2}{3x^2 + y}.$$

8.2 Remarks Concerning Implicit Functions

The idea of an implicit function is made more precise by the following definition.

DEFINITION 1   **Implicit Function**

A continuous function f, defined at least on an open interval, is said to be *implicit* in an equation involving the variables x and y provided that, when y is replaced by $f(x)$ in this equation, the resulting equation is true for all values of x in the domain of f.

EXAMPLES   If possible, find the functions implicit in the given equation by solving for y in terms of x.

1 $7 + x = y^2 - 3y$

SOLUTION

We have $y^2 - 3y - (7 + x) = 0$. Using the quadratic formula, we obtain $y = \dfrac{3 \pm \sqrt{4x + 37}}{2}$. The latter equation does not give y as an explicit function of x because of the ambiguous $\pm$ sign. However, the function f defined by $f(x) = \dfrac{3 + \sqrt{4x + 37}}{2}$ is implicit in the given equation. But so is the function h defined by $h(x) = \dfrac{3 - \sqrt{4x + 37}}{2}$.

2 $x^2 + y^2 + 1 = 0$

SOLUTION

This equation has no (real) solution, so it cannot define y implicitly as a function of x.

3 $\dfrac{2y - 3}{4y + 7} = 1 - x$

SOLUTION

Clearing the denominator, we have

$$2y - 3 = (4y + 7)(1 - x) = 4y - 4xy + 7 - 7x.$$

Thus,

$$2y - 4xy = 7x - 10 \quad \text{and so} \quad y = \frac{7x - 10}{2 - 4x}.$$

Thus, the function $f(x) = \dfrac{7x - 10}{2 - 4x}$ is implicit in the equation $\dfrac{2y - 3}{4y + 7} = 1 - x$.

4 $x = \dfrac{y^5}{40} + \dfrac{13}{600} y^3$

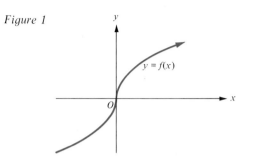

Figure 1

$y = f(x)$

SOLUTION

We must solve the equation explicitly for y in terms of x, if possible. Define the function g by the equation $g(y) = \dfrac{y^5}{40} + \dfrac{13}{600} y^3$. Since $g'(y) = \dfrac{5}{40} y^4 + \dfrac{13}{200} y^2$ is positive for $y \neq 0$, g is increasing on the interval $(-\infty, \infty)$; hence, g is invertible and, if we put $f = g^{-1}$, then f is implicit in the given equation. Here, however, although f is a perfectly definite function whose graph appears in Figure 1, it is not possible to find a tidy algebraic formula for $f(x)$.

As the examples above show, given an equation involving x and y, any one of the following can happen:

1 There are two or more functions implicit in the equation (Example 1).
2 There is no function implicit in the equation (Example 2).
3 There is only one function implicit in the equation (Examples 3 and 4).

Furthermore, even if there is a function f implicit in the equation, it may (Example 3) or may not (Example 4) be possible to find a tidy formula for $f(x)$ There is an important theorem in advanced calculus, called the *implicit function theorem*, which gives conditions that ensure the existence and differentiability of implicit functions. Roughly speaking, this theorem asserts that if the steps in the procedure for implicit differentiation given in Section 8.1 can be carried out, an implicit function actually exists and is differentiable.

Problem Set 8

In problems 1 through 20, find dy/dx using implicit differentiation.

1 $9x^2 + 4y^2 = 36$

2 $4xy^2 + 3x^2y = 2$

3 $x^2y - xy^2 + x^2 = 7$

4 $xy^2 + x^3 + y^3 = 5$

5 $x^2 - 3xy + y^2 = 3$

6 $xy^3 + 2y^3 = x^2 - 4y^2$

7 $x^{2/3} + y^{2/3} = 1$

8 $x^2 - \sqrt{xy} - y = 0$

9 $x^4y + \sqrt{xy} = 5$

10 $\sqrt{x} + \sqrt{y} = 9$

11 $x\sqrt{y} + y\sqrt{x} = 16$

12 $(4x - 1)^3 = 5y^3 + 2$

13 $x\sqrt{1 + y} + y\sqrt{1 + x} = 4$

14 $\sqrt[3]{xy} + 3y = 5\sqrt[3]{x}$

15 $\sqrt{x + y} + \sqrt{x - y} = 6$

16 $x^{1/n} + y^{1/n} = 1$

17 $\dfrac{x}{y} + \dfrac{y}{x} = 5$

18 $\dfrac{x}{x - y} + \dfrac{y}{x} = 4$

19 $\sqrt{\dfrac{y}{x}} + \sqrt{\dfrac{x}{y}} = 6$

20 $\sqrt[3]{y} + \sqrt[4]{y} + \sqrt[5]{y} = 4x$

In problems 21 through 24, find the equations of the tangent and normal lines to the graph of the implicit function determined by each of the following equations at the given point.

21 $x^2 + xy + 2y^2 = 28$ at $(2, 3)$.

22 $x^3 - 3xy^2 + y^3 = 1$ at $(2, -1)$.

23 $\sqrt{2x} + \sqrt{3y} = 5$ at $(2, 3)$.

24 $x^2 - 2\sqrt{xy} - y^2 = 52$ at $(8, 2)$.

25 Suppose that x and y satisfy the equation $x^2 + y^2 = 4$.

(a) Show that $\dfrac{dy}{dx} = -\dfrac{x}{y}$.

(b) Differentiate both sides of the equation in part (a) and conclude that $\dfrac{d^2 y}{dx^2} = \dfrac{x \dfrac{dy}{dx} - y}{y^2}$.

(c) Use parts (a) and (b) to show that $\dfrac{d^2 y}{dx^2} = -\dfrac{4}{y^3}$.

In problems 26 through 28, use the method in problem 25 to find d^2y/dx^2 in terms of x and/or y.

26 $x^4 + y^4 = 64$ **27** $x^3 + y^3 = 16$ **28** $\sqrt{x} + \sqrt{y} = 4$

In problems 29 through 32, consider x to be a function of y and find dx/dy by implicit differentiation.

29 $3x^2 + 5xy = 2$ **30** $x^2 y^2 = x^2 + y^2$

31 $x^2 = y^2 - y$ **32** $\sqrt{xy} + xy^4 = 5$

In problems 33 through 42, find all functions implicit in each equation by solving the equation for y in terms of x.

33 $5x - 4y = 6$ **34** $5x^2 - 4y^2 = 6$ **35** $xy + y^2 = x$

36 $3xy^2 + 4y + x = 0$ **37** $x = \dfrac{2y - 1}{3y + 1}$ **38** $x^2 - x^2 y + y^3 = y^2$

39 $x = y^4$ **40** $y^3(y^2 + 4) = x$

41 $y^3 - 3y^2 + 3y = 3(x + 1)$ **42** $\dfrac{x}{y} + \dfrac{y}{x} = 2$

43 Show that the function f defined by the equation $f(x) = \tfrac{3}{4}\sqrt{16 - x^2}$ is implicit in the equation $\dfrac{x^2}{16} + \dfrac{y^2}{9} = 1$. What does this imply about the relationship between the graph of f and the graph of the equation $\dfrac{x^2}{16} + \dfrac{y^2}{9} = 1$? (The *graph of an equation* is the set of all points in the xy plane whose coordinates satisfy the equation.)

44 Show that the function g defined by the equation $g(x) = -\tfrac{3}{4}\sqrt{16 - x^2}$ is implicit in the equation $\dfrac{x^2}{16} + \dfrac{y^2}{9} = 1$. What does this imply about the relationship between the graph of g and the graph of the equation $\dfrac{x^2}{16} + \dfrac{y^2}{9} = 1$? How is the graph of g related to the graph of the function f of problem 43?

45 Both of the functions f and g defined by $f(x) = \tfrac{3}{4}\sqrt{16 - x^2}$ and $g(x) = -\tfrac{3}{4}\sqrt{16 - x^2}$ are implicit in the equation $\dfrac{x^2}{16} + \dfrac{y^2}{9} = 1$.

(a) Calculate $f'(x)$ directly from $f(x) = \tfrac{3}{4}\sqrt{16 - x^2}$.

(b) Calculate $g'(x)$ directly from $g(x) = -\tfrac{3}{4}\sqrt{16 - x^2}$.

(c) Calculate dy/dx from the equation $\dfrac{x^2}{16} + \dfrac{y^2}{9} = 1$ by implicit differentiation.

(d) Show that the answer to part (c) is compatible with the answer to part (a).

(e) Show that the answer to part (c) is also compatible with the answer to part (b).

46 Interpret the answers to parts (a), (b), and (c) of problem 45 in terms of the slopes of tangent lines to the graphs of the function f, the function g, and the equation

$\dfrac{x^2}{16} + \dfrac{y^2}{9} = 1$, respectively. Thus, explain the "compatibility" found in the answers to parts (d) and (e) of problem 45.

47 Find the slope of the tangent line to the graph of the implicit function determined by the equation $\dfrac{x^2}{30} - \dfrac{y^2}{20} = 1$ at the point $(6, -2)$.
   (a) By implicit differentiation.
   (b) By solving the given equation to obtain y explicitly as a function f of x, then finding the value of $f'(x)$ when $x = 6$.

48 The volume of a cylinder is given by $V = \pi r^2 h$, where r is the radius of the base and h is the height. If r is changed while V is held constant, h will change accordingly. Use implicit differentiation to find $D_r h$.

49 The surface area of a right circular cone is given by $S = \pi r \sqrt{r^2 + h^2}$, where r is the radius of the base and h is the height. If h is changed while S is held constant, r will change accordingly. Use implicit differentiation to find dr/dh.

50 Find the slope of the tangent line to the graph of the equation $x = 5y^3 - 4y^5$ at the point $(1, 1)$.
   (a) By implicit differentiation.
   (b) By the inverse function rule.

9   Related Rates of Change

Let x and y be variable quantities related in such a way that they satisfy a certain equation—for instance, $x^2 + y^2 = 1$. Suppose that these quantities depend on the time t (that has elapsed since some fixed initial instant) according to equations $x = f(t)$ and $y = g(t)$. In this section we routinely assume that the variables are *differentiable* functions of time; hence, $dx/dt = f'(t)$ and $dy/dt = g'(t)$ give the instantaneous rates of change of x and y per unit of time.

If x and y are related according to the equation $x^2 + y^2 = 1$, it stands to reason that their rates of change dx/dt and dy/dt should also be related in some definite way. To find this relationship, we proceed much in the same way as in implicit differentiation, but this time we differentiate both sides of the equation $x^2 + y^2 = 1$ *with respect to t*. Since $\dfrac{d}{dt}(1) = 0$ and since (by the chain rule) $\dfrac{d}{dt}(x^2) = 2x\dfrac{dx}{dt}$ and $\dfrac{d}{dt}(y^2) = 2y\dfrac{dy}{dt}$, differentiation of

$$x^2 + y^2 = 1$$

on both sides yields

$$2x\frac{dx}{dt} + 2y\frac{dy}{dt} = 0, \qquad \text{so } x\frac{dx}{dt} + y\frac{dy}{dt} = 0.$$

The last equation gives the relationship between the rates of change dx/dt and dy/dt when $x^2 + y^2 = 1$.

The following examples show how this technique for finding the relationship between rates of change applies to problems in such areas as geometry, engineering, physics, business, and economics. Of course, if more than two variables are involved, the method is the same—we differentiate the equations relating the variables, on both sides, with respect to time t.

EXAMPLES   1   If the area of a circle is increasing at the constant rate of 4 square centimeters per second, at what rate is the radius increasing at the instant when the radius is 5 centimeters?

SOLUTION
Let r be the radius of the circle in centimeters, A the area of the circle in square centimeters, and t the time in seconds. The equation connecting A and r is $A = \pi r^2$. Taking the derivative on both sides with respect to t, we get

$$\frac{dA}{dt} = \frac{d}{dt}(\pi r^2); \qquad \text{that is, } \frac{dA}{dt} = 2\pi r \frac{dr}{dt}.$$

We are given that $dA/dt = 4$ cm^2/sec, so

$$4 = 2\pi r \frac{dr}{dt} \quad \text{and} \quad \frac{dr}{dt} = \frac{2}{\pi r} \text{ cm/sec.}$$

Thus, when $r = 5$ cm, $\dfrac{dr}{dt} = \dfrac{2}{5\pi}$ cm/sec ≈ 0.13 cm/sec.

2   The area A of a rectangle is decreasing at the constant rate of 9 square inches per second. At any instant, the length l of the rectangle is decreasing twice as fast as the width w. At a certain instant the rectangle is a 1-inch by 1-inch square. At this instant, how rapidly is the width decreasing?

SOLUTION
Here $A = lw$, so $D_t A = l(D_t w) + (D_t l)w$. Also, we are given that $D_t l = 2D_t w$, so

$$D_t A = l(D_t w) + (2D_t w)w = (l + 2w)D_t w \quad \text{or} \quad D_t w = \frac{D_t A}{l + 2w}.$$

Since $D_t A = -9$ square inches per second (the negative sign because the area is decreasing), $D_t w = -\dfrac{9}{l + 2w}$. At the instant when $l = w = 1$, $D_t w = -\frac{9}{3} = -3$ inches per second.

3   Two highways intersect at right angles (Figure 1). Car A on one highway is $\frac{1}{2}$ kilometer from the intersection and is moving toward it at the rate of 96 km/hr, while car B on the other highway is 1 kilometer from the intersection and moving toward it at the rate of 120 km/hr. At what rate is the distance between the two cars changing at this instant?

SOLUTION
In Figure 1, let

$x = $ the distance in kilometers of car B from the origin O at t hours,

$y = $ the distance in kilometers of car A from the origin O at t hours, and

$z = $ the distance in kilometers between the two cars at t hours.

Figure 1

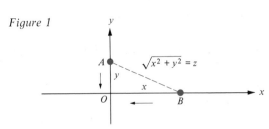

By the Pythagorean theorem, we have $z^2 = x^2 + y^2$. Differentiating both sides of this equation with respect to t, we obtain

$$2zD_t z = 2xD_t x + 2yD_t y,$$

and so

$$D_t z = \frac{xD_t x + yD_t y}{z}.$$

At the instant when $x = 1$, $y = \frac{1}{2}$, $D_t x = -120$, and $D_t y = -96$, we have

$$D_t z = \frac{1(-120) + \frac{1}{2}(-96)}{\sqrt{1^2 + (\frac{1}{2})^2}} \approx -150.26 \text{ km/hr.}$$

4 A man 6 feet tall is 12 feet away from the base of a lamppost that is 20 feet high, and he is walking directly away from the lamppost at the rate of 4.4 feet per second. At what rate is the length of his shadow increasing (Figure 2)?

Figure 2

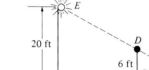

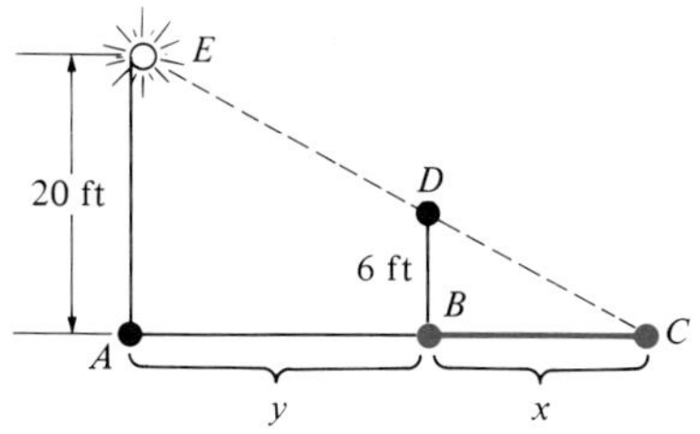

SOLUTION
Let

y = the distance in feet of the man from the post at t seconds and

x = the length in feet of the man's shadow at t seconds.

In Figure 2, triangle ACE is similar to triangle BCD; hence, $\dfrac{20}{y + x} = \dfrac{6}{x}$, so that $20x = 6y + 6x$, or $7x = 3y$. Thus, $7\dfrac{dx}{dt} = 3\dfrac{dy}{dt}$, and so, when $\dfrac{dy}{dt} = 4.4$, $\dfrac{dx}{dt} = \dfrac{3}{7}\dfrac{dy}{dt} = \dfrac{3}{7}(4.4) = \dfrac{66}{35}$ ft/sec.

5 Water is running out of a conical funnel at the rate of 3 cubic inches per second. The funnel has a radius of 2 inches and a height of 8 inches (Figure 3). How fast is the water level dropping when it is 3 inches from the top?

Figure 3

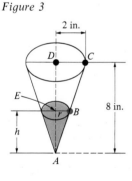

SOLUTION
Let

t = the time in seconds that has elapsed since the water started to run out of the funnel, and

V = the volume in cubic inches of the water in the funnel at t seconds.

In Figure 3, let $h = |\overline{AE}|$ and $r = |\overline{EB}|$. We must find dh/dt at the instant when $|\overline{DE}| = 3$, that is, when $h = 8 - |\overline{DE}| = 5$. Since triangle ADC is similar to triangle AEB,

$$\frac{|\overline{AD}|}{|\overline{AE}|} = \frac{|\overline{DC}|}{|\overline{EB}|}, \qquad \text{so that } r = |\overline{EB}| = |\overline{DC}|\frac{|\overline{AE}|}{|\overline{AD}|} = 2\left(\frac{h}{8}\right) = \frac{h}{4}.$$

At any time t, the volume V of water in the funnel can be expressed as the volume of a cone (Figure 3); hence,

$$V = \frac{1}{3}\pi r^2 h = \frac{1}{3}\pi\left(\frac{h}{4}\right)^2 h = \frac{\pi}{48}h^3.$$

Differentiating both sides of the latter equation with respect to t, we obtain

$$\frac{dV}{dt} = \frac{3\pi}{48}h^2\frac{dh}{dt}, \qquad \text{so that } \frac{dh}{dt} = \frac{16}{\pi h^2}\frac{dV}{dt}.$$

Since V is decreasing at the rate of 3 cubic inches per second, it follows that $dV/dt = -3$. Thus, when $h = 5$,

$$\frac{dh}{dt} = \frac{16}{\pi(25)}(-3) = -\frac{48}{25\pi} \approx -0.61 \text{ inch per second.}$$

6 The pressure P and the volume V of a sample of air that is undergoing adiabatic expansion are related by the equation $PV^{1.4} = C$, where C is a constant. At a certain instant the volume of such a sample is 4 cubic inches, the pressure is 4000 pounds per square inch, and the volume is increasing at the instantaneous rate of 2 cubic inches per second. At what rate is the pressure changing at this instant?

SOLUTION
Differentiation of both sides of the equation $PV^{1.4} = C$ with respect to t gives

$$P(1.4)V^{0.4}\frac{dV}{dt} + \frac{dP}{dt}V^{1.4} = 0, \quad \text{so that} \quad \frac{dP}{dt} = -\frac{1.4P}{V}\frac{dV}{dt}.$$

When $V = 4$, $P = 4000$, and $dV/dt = 2$, we have

$$\frac{dP}{dt} = -\frac{1.4(4000)}{4}(2) = -2800 \text{ pounds per square inch per second.}$$

7 The labor force y required by an industry to manufacture x units of a certain product is given by the equation $y = \frac{1}{2}\sqrt{x}$. Find the instantaneous rate at which the labor force should be increasing if, at present, there is a demand for 40,000 units of the product, but the demand is increasing at the constant rate of 10,000 units per year.

SOLUTION

$$\frac{dy}{dt} = \frac{1}{4\sqrt{x}}\frac{dx}{dt} = \frac{1}{4\sqrt{40,000}}(10,000)$$

$$= \frac{50}{4} = 12.5 \text{ workers per year.}$$

Problem Set 9

1 If the area of a circle is decreasing at the constant rate of 3 square centimeters per second, at what rate is the radius r decreasing at the instant when $r = 2$ cm?

2 A circular plate of metal expands when heated so that its radius increases at the constant rate of 0.02 inch per second. At what rate is the surface area (of one side) increasing when the radius is 4 inches?

3 A coin is cooled so that its radius is decreasing at the constant rate of 0.003 centimeter per second. At what rate is its total surface area—top and bottom, neglecting the area around the edge—decreasing at the instant when its radius is 1.02 centimeters?

4 The length of each side of a square is increasing at the constant rate of 2 inches per second. Find the rate of increase of the area and of the perimeter of the square at the instant when its side is 3 inches long.

5 Each side of a closed cubical box is decreasing at the rate of 10 centimeters per minute. (a) How fast is the volume of the box decreasing at the instant when its side length is 20 centimeters? (b) How fast is its total surface area (four sides plus top and bottom) decreasing at the instant when its side length is 20 centimeters?

6 Show that if each side of a cube is increasing at the constant rate of 0.1 inch per second, then, at the instant when the side length of the cube is 10 inches, the volume will be increasing at the rate of 30 cubic inches per second. Does this mean that during the next second exactly 30 cubic inches will be added to the volume? Explain.

7 At 2:00 P.M. a ship, which is steaming north at 20 miles per hour, is 4 miles west of a lighthouse. At the same time, a motor launch is proceeding south from a point 4 miles east of the lighthouse at 10 miles per hour. Find the rate of change of the distance between the ship and the launch at 3:30 P.M.

8 At 1:00 P.M., ship A is 100 miles north of ship B. Ship A is sailing south at 20 miles per hour, while ship B is sailing east at 15 miles per hour. How fast is the distance between the two ships changing at 7:00 P.M.?

9 A man 1.8 meters tall is 6 meters from the base of a lamppost that is 4 meters high. The man is walking directly away from the lamppost at the constant rate of 2 meters per second. (a) At what rate is his shadow lengthening? (b) At what rate is the tip of his shadow moving?

10 A woman 6 feet tall is walking toward a wall at the rate of 4 feet per second. Directly behind her and 40 feet from the wall is a spotlight 3 feet above ground level. How fast is the length of the woman's shadow on the wall changing when she is halfway from the spotlight to the wall? Is the shadow lengthening or shortening?

11 A ladder 4 meters long is leaning against the vertical wall of a house. If the base of the ladder is pulled horizontally away from the house at the rate of 0.7 meter per second, how fast is the top of the ladder sliding down the wall at the instant when it is 2 meters from the ground?

12 A boat is moored to a dock by a rope. A man on the dock begins pulling in the rope at the rate of 72 feet per minute. The man's hands (holding the rope) are 5 feet above the level of the bow of the boat (where the rope is fastened). How fast is the boat moving toward the dock at the instant when 13 feet of rope is out?

13 A spherical snowball is melting so that its volume is decreasing at the rate of 0.17 cubic meter per minute. Find the rate at which its radius is decreasing at the instant when the volume of the snowball is 0.4 cubic meter.

14 In a baseball game (which is played on a square-shaped diamond 90 feet on each side) between the Brewers and the Cubs, Joe (who is playing left field for the Brewers) hits the ball while at bat and runs toward first base at the rate of 20 feet per second. Just as Joe starts toward first base, his teammate Tony, who had taken a 10-foot lead off second base, starts running toward third base at the same rate of speed. How fast is the distance between Joe and Tony changing at the instant when Tony reaches third base?

15 The area A of a triangle is increasing at the rate of 0.5 square meter per minute while the length x of its base is decreasing at the rate of 0.25 meter per minute. How fast is the height y of the triangle changing at the instant when $x = 2$ meters and $A = 1$ square meter?

16 The volume of a soap bubble is increasing at the rate of 10 cubic centimeters per second. How fast is the surface area of the bubble increasing at the instant when its radius is 6 centimeters?

17 A spherical tank of radius R filled with liquid to a height h contains $V = \frac{1}{3}\pi h^2(3R - h)$ cubic units of liquid (Figure 4). A spherical gasoline tank has a radius of $R = 20$ feet. Gasoline is being pumped out of this tank at the rate of 200 gallons per minute. (One gallon is approximately 0.134 cubic foot.) At what rate is the level of gasoline in the tank dropping at the instant when $h = 5$ feet?

Figure 4

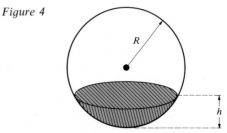

18 Water is being pumped at the rate of 1.5 cubic meters per minute into a swimming pool 20 meters long and 10 meters wide. The depth of the pool decreases uniformly from 7 meters at one end to 1 meter at the other end. How fast is the surface level of the water rising at the instant when its depth at the deepest end is 6 meters?

19 A horizontal eaves trough is 20 feet long and has a cross section in the shape of an isosceles triangle 8 inches across at the top and 10 inches deep. Because of a heavy rainstorm, the water in the trough is rising at the rate of $\frac{1}{2}$ inch per minute at the instant when it is 5 inches deep. How fast is the volume of water in the trough increasing at this instant?

20 Water is stored in a reservoir in the shape of a right circular cone with an open base facing upward. The radius of the base is 20 meters and the depth of the reservoir from top to vertex is 5 meters. The water in the reservoir evaporates at the rate of 0.00005 cubic meter per hour for each square meter of its surface exposed to the air. Show that, because of evaporation, the water level will drop at a uniform rate (independent of the depth of the water) and find this rate.

21 Sand is falling at the rate of 2 cubic feet per minute upon the tip of a sandpile which maintains the form of a right circular cone whose height is the same as the radius of the base. Find the rate at which the height of the pile is increasing at the instant when the pile is 6 feet high.

22 A cylindrical tank with vertical axis has a radius of 10 inches. The tank has a circular hole in its base with a 1-inch radius through which water pours out with a velocity of $v = 8\sqrt{h}$ feet per second, where h is the height in feet of the surface of the water above the base. How rapidly is the water level dropping at the instant when $h = 5$ feet?

23 A water tank has the shape of an inverted right circular cone with a radius of 5 meters at the top and a height of 12 meters. At the instant when the water in the tank is 6 meters deep, more water is being poured in at the rate of 10 cubic meters per minute. Find the rate at which the surface level of the water is rising at this instant.

24 A water tank has the shape of an inverted right circular cone with a radius of R units at the top and a height (from vertex to top) of H units. A small hole with cross-sectional area k square units allows water to leak out of the tank at the vertex with a velocity of $\sqrt{2gh}$ units per second, where the depth of the water in the tank is h units and where g is the acceleration of gravity in units per second per second. If water is being pumped into the tank at a uniform rate of c cubic units per second, find a formula for the rate at which the water level is changing.

25 If Q is the quantity of heat added to a unit mass of a substance and Θ is the corresponding rise in temperature, then $dQ/d\Theta = c$ is called the *specific heat* of the substance. The specific heat of copper is 0.093 calorie per degree C at 20°C. If 100 grams of copper at 20°C is absorbing 10 calories of heat per minute, what is the instantaneous rate of change of the temperature?

26 If the volume of one unit of mass of a substance at the temperature Θ is V, then $dV/d\Theta$ is called the *coefficient of cubical expansion* of the substance. The volume of 1 gram of water is given by $V = 1 + (8.38)(10^{-6})(\Theta - 4)^2$ cubic centimeters, where Θ is the temperature of the water in degrees C. Find the coefficient of cubical expansion of water at 10°C and find the rate of change of the volume of one gram of water at 10°C if its temperature is decreasing at the rate of 1.5°C per minute.

27 Boyle's law for the expansion of a gas held at constant temperature states that the pressure P and the volume V are related by the equation $PV = C$, where C is a constant. Suppose that a sample of 1000 cubic inches of gas is under a pressure of 150 pounds per square inch, but that the pressure is decreasing at the instantaneous rate of 5 pounds per square inch per second. Find the instantaneous rate of increase of the volume.

28 A boat sails parallel to a straight beach at a constant speed of 19.2 kilometers per hour, staying 6.4 kilometers offshore. How fast is it approaching a lighthouse on the shoreline at the instant it is exactly 8 kilometers from the lighthouse?

29 The price of apples in a certain marketing area is given by the equation $P = 2 + \dfrac{60}{30 + x}$, where x is the supply in thousands of bushels and P is the price per bushel in dollars. At what rate will the price per bushel be changing if the current supply of 10,000 bushels is decreasing at the rate of 200 bushels per day?

30 The demand D in thousands of boxes per week for a detergent is expressed by the equation $D = (1000/p) - 30$, where p is the price per box of detergent. The current price is $p = \$0.83$ per box, but inflation is increasing the price at the rate of $0.01 per month. Find the current instantaneous rate of change of the demand.

31 A certain microorganism has a spherical shape and a density of 1 gram per cubic centimeter. It grows by absorbing nutrients through its surface at the rate of $A/18{,}200$ grams per second, where A is its surface area. If its radius is now $\frac{1}{700}$ cm, what will be its radius after 2 hours?

Review Problem Set

1 Verify that the given function has a zero (that is, a root) on the given interval by using the intermediate value theorem.
(a) $f(x) = x^2 - 3$ between 1.7 and 1.8
(b) $f(x) = x^2 + x - 4$ between 1 and 2
(c) $f(x) = 2x^3 + x^2 - 7x + 3$ between 0 and 2

2 Which of the following graphs of functions (Figure 1) do not satisfy the hypotheses of the mean value theorem on the interval $[a, b]$? Indicate the hypothesis that does not hold.

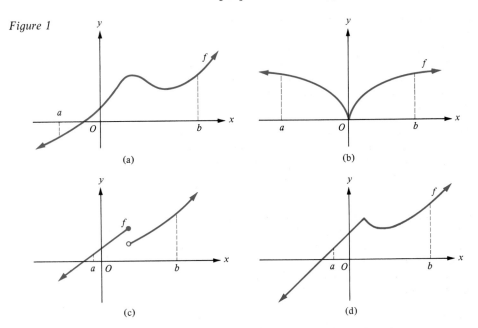

Figure 1

(a)

(b)

(c)

(d)

3 Find a suitable number c that satisfies the conclusion of the mean value theorem in each case. Be sure to verify the hypotheses of the theorem for the given function on the indicated interval first. Sketch the graph.

(a) $f(x) = \sqrt{x}; [1,4]$

(b) $g(x) = x^2 - 3x - 4; [-1,3]$

(c) $h(x) = x^3 - 2x^2 + 3x - 2; [0,2]$

(d) $f(x) = \dfrac{x-4}{x+4}; [0,4]$

(e) $f(x) = \begin{cases} \dfrac{3-x^2}{2} & \text{if } x \le 1 \\ \dfrac{1}{x} & \text{if } x > 1 \end{cases}; [0,2]$

4 Indicate which hypothesis of the mean value theorem fails to hold for the given function on the indicated interval. Sketch the graph of the function.

(a) $f(x) = \sqrt[3]{x}; [-8,8]$

(b) $g(x) = \dfrac{x+3}{2x+1}; [-1,2]$

(c) $f(x) = \dfrac{x^2 + 4x + 3}{x-1}; [0,2]$

(d) $h(x) = \begin{cases} -3x+1 & \text{if } x \le 1 \\ -3+x & \text{if } x > 1 \end{cases}; [-1,3]$

(e) $f(x) = \dfrac{1}{x}; [-1,1]$

5 Find a suitable number c that satisfies the conclusion of Rolle's theorem. Be sure to verify the hypotheses of Rolle's theorem for the given function on the indicated interval. Sketch the graph of the function.

(a) $f(x) = x^2 + 6x - 7; [-7,1]$

(b) $g(x) = x^3 - x; [0,1]$

(c) $h(x) = \sqrt{4 - x^2}; [-1,1]$

(d) $f(x) = 2x^3 - 27x^2 + 25x; [0,1]$

(e) $f(x) = 4x - x^3; [-2,0]$

6 Which of the following graphs of functions (Figure 2), for one reason or another, do not satisfy the hypotheses of Rolle's theorem on the interval $[a,b]$? Indicate the hypothesis that fails to hold.

Figure 2

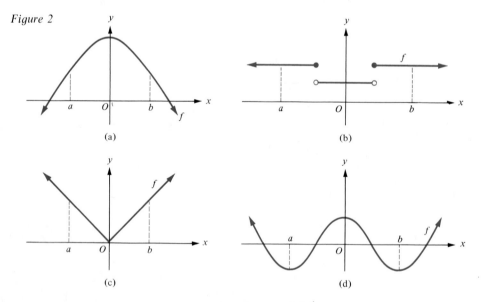

(a)

(b)

(c)

(d)

7 Use the mean value theorem to prove that $\sqrt{x} < \dfrac{x+1}{2}$ for $0 < x < 1$.

8 Suppose that f and f' are differentiable functions on the interval (a,b) and that $f(p) = f(q) = f(r) = 0$, for p, q, and r in (a,b) with $p \ne q \ne r$. Use the mean value theorem to prove that there is a number c in (a,b) such that $f''(c) = 0$.

9 Use Rolle's theorem to prove that if $f'(x) > 0$ for $a < x < b$, then there is at most one x for which $a < x < b$ and $f(x) = 0$.

10 Use Rolle's theorem to prove that the polynomial function f defined by

$$f(x) = x^3 - 3x + b$$

never has two roots in the interval $[0,1]$, no matter what b may be.

11 Suppose that f, g, and h are functions which are differentiable at the number -2, and that $f(-2) = 1$, $f'(-2) = -3$, $f''(-2) = -4$, $g(-2) = 4$, $g'(-2) = -\frac{1}{2}$, $g''(-2) = -3$, $h(-2) = 6$, $h'(-2) = -8$, and $h''(-2) = 7$. Find:

(a) $(fg)''(-2)$

(b) $(fh)''(-2)$

(c) $(f + g)''(-2)$

(d) $(g - h)''(-2)$

(e) $(fgh)''(-2)$

(f) $\left(\dfrac{f}{g}\right)''(-2)$

12 The *curvature* of the graph of the function f at the point $(x, f(x))$ is defined to be $k = \dfrac{f''(x)}{[1 + (f'(x))^2]^{3/2}}$. The equation of a semicircle of radius r with center at $(0,0)$ is $y = \sqrt{r^2 - x^2}$. Find the curvature of this semicircle at the point $(x, \sqrt{r^2 - x^2})$, where $-r < x < r$.

13 (a) If $(x - a)^2 + y^2 = b^2$, where a and b are constant numbers, show that
$$y D_x^2 y + 1 + (D_x y)^2 = 0.$$
(b) If $x^2 + y^2 = a^2$, where a is a constant number, show that $[1 + (y')^2]^3 = a^2(y'')^2$.

14 Suppose that g is a function defined by the rule $g(t) = \sqrt{1 - f(t)}$, where $f(-2) = -3$, $f'(-2) = 3$, and $f''(-2) = 5$. Find $g''(-2)$.

15 Find formulas for $f'(x)$ and $f''(x)$ if:

(a) $f(x) = |x|^3$

(b) $f(x) = \begin{cases} x^3 & \text{if } x \le 1 \\ 3x - 2 & \text{if } x > 1 \end{cases}$

16 Suppose that $xy + y = 2$. Show that $D_x^n y = \dfrac{2(-1)^n n!}{(x + 1)^{n+1}}$ by using the principle of mathematical induction.

17 Indicate the intervals where f is increasing or decreasing. Sketch the graph of the function.

(a) $f(x) = x^3 + 3x^2 - 2$

(b) $g(x) = x^3 + 6x^2 + 9x + 3$

(c) $g(x) = \sqrt[3]{x}(x - 4)^2$

(d) $h(x) = x^3 + \dfrac{4}{x}$

(e) $f(x) = \begin{cases} x + 1 & \text{if } x \le 0 \\ x^2 & \text{if } x > 0 \end{cases}$

(f) $g(x) = \begin{cases} & \text{if } x < 0 \\ (x - 1)^2 & \text{if } 0 \le x \le 2 \\ 1/x^3 & \text{if } x > 2 \end{cases}$

18 Indicate the intervals where the graph of f (Figure 3) is increasing or decreasing.

Figure 3

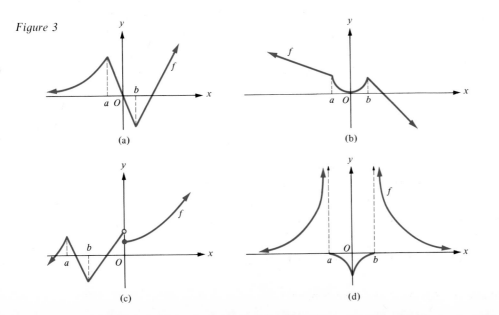

(a)

(b)

(c)

(d)

19 A particle is moving along a horizontal s axis according to the given equation of motion. Find the velocity and acceleration of the particle. Discuss and show its behavior along the s axis as t increases.
(a) $s = 6t^2 - 2t^3$ (b) $s = 8t^3 - 48t^2 + 72t$ (c) $s = 64t^2 - 16t^4$

20 A crew of x men unloads boxes of canned goods from a truck. The crew can unload y trucks per day, where y is expressed by the equation $y = \dfrac{x^2}{25}\left(3 - \dfrac{x}{6}\right)$ for $0 < x \le 18$.
Sketch the graph of y as a function of x, and discuss the effect of increasing the size of the crew on the number of trucks unloaded per day.

21 Let f be a differentiable function in an interval I, and let $f'(x_1) = f'(x_2) = f'(x_3) = 0$ where $x_1 < x_2 < x_3$ in I. Let a and b be numbers in I such that $a < x_1$ and $x_3 < b$. Sketch a graph of f in each of the following cases.
(a) $f(a) < f(x_1)$, $f(x_1) > f(x_2)$, $f(x_2) > f(x_3)$, and $f(x_3) < f(b)$
(b) $f(a) > f(x_1) > f(x_2) > f(x_3) > f(b)$

In problems 22 through 27, indicate the intervals where the graph of the given function is concave upward or concave downward.

22 $f(x) = x^3 - 8x$ **23** $g(x) = x^3 - 6x^2 + 9x + 1$

24 $f(x) = x^4 - 8x^3 + 64x + 8$ **25** $h(x) = \dfrac{3x}{x^2 - 9}$

26 $k(x) = \begin{cases} 2x^2 & \text{if } x < 0 \\ -2x^2 & \text{if } x \ge 0 \end{cases}$ **27** $f(x) = \begin{cases} -x^2 & \text{if } x \ge 1 \\ 3 - 7x + x^2 - x^3 & \text{if } x < 1 \end{cases}$

28 Show that the graph of f defined by $f(x) = ax^2 + bx + c$ is concave upward if $a > 0$ and concave downward if $a < 0$.

29 Find the point or points of inflection for each function. Sketch the graph of the function.
(a) $f(x) = -2x^3 + 4x^2 + 5$ (b) $g(x) = x^2(x^2 - 6)$ (c) $h(x) = 2x^3 + 4x^2 + 2x + 1$

(d) $k(x) = \frac{1}{3}(x^3 + 9x^2)$ (e) $f(x) = \dfrac{1}{1 + x^2}$

(f) $g(x) = \dfrac{2x}{(x + 3)^2}$ (g) $h(x) = \dfrac{x + 1}{x^2 + 1}$

30 Find the equation of the tangent line to the graph of the functions of problem 29 at each point of inflection.

31 Given the graph of the derived function f' in Figure 4, sketch a graph of the function f. Assume that $f(0) = 1$.

32 Let $f(x) = x^3 + ax^2 + bx + c$. Show that if $a^2 = 3b$, the graph of f has a horizontal tangent at its point of inflection.

33 Sketch a portion of the graph of the function f near the point indicated.
(a) $f(-2) = 24$, $f'(-2) = 8$, and $f''(-2) = -9$.
(b) $f(4) = 16$, $f'(4) = 22$, and $f''(4) = 18$.
(c) $f(0) = 8$, $f'(0) = -2$, and $f''(0) = -6$.

Figure 4

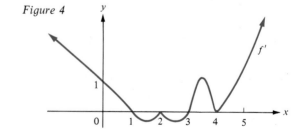

34 Let $f(x) = x^3 + px^2 + qx + r$ and suppose that there exists a number a such that $f(a) = 0$, $f'(a) = 0$, and $f''(a) = 0$. Show that $p = -3a$, $q = 3a^2$, and $r = -a^3$. Hence, show that $f(x) = (x - a)^3$.

In problems 35 through 42, (a) find the critical numbers of the function, (b) find all numbers at which f has a relative extremum, and (c) sketch the graph of the function.

35 $f(x) = 2x^3 - 9x^2 + 12x + 1$ **36** $g(x) = 2x^3 + 3x^2 - 12x - 2$ **37** $h(x) = x^4 - 6x^2 + 8x$

38 $k(x) = \dfrac{16}{x^2 + 4}$ 　　　　　**39** $g(x) = \dfrac{1}{x^2 - 16}$ 　　　　　**40** $g(x) = \sqrt{1 + x^2}$

41 $h(x) = \sqrt{2x^2 + 9}$ 　　　　　**42** $f(x) = \dfrac{10}{\sqrt{x + 2}}$

In problems 43 through 47, find the absolute extrema of the given function on the indicated interval. Also find the values of x at which the absolute extrema occur. Sketch the graph of the function.

43 $f(x) = x^3 - 6x^2 + 9x + 1; [0, 4]$ 　　　**44** $g(x) = x^3 - 2x + 1; [-1, 1]$ 　　　**45** $h(x) = (x - 1)^3; [-1, 2]$

46 $f(x) = x(x^2 + 2)^{-3/2}; [0, +\infty)$ 　　　**47** $h(x) = \dfrac{x + 1}{x - 1}; [-3, 3]$

In problems 48 through 53, find (a) the intervals where the function is increasing or decreasing, (b) the intervals where the graph of the function is concave upward or concave downward, (c) the maximum and minimum points, (d) the points of inflection, and (e) sketch the graph of the function.

48 $f(x) = x^4 + 4x$ 　　　　　**49** $g(x) = x^4 - 2x^3 + 1$ 　　　　　**50** $h(x) = x^3 + x - 1$

51 $k(x) = -x^3 + 2x + 5$ 　　　　**52** $f(x) = x^3 + x$ 　　　　**53** $g(x) = \dfrac{x + 1}{x^2 + 1}$

54 Suppose that f and g are continuous functions, both of which have relative minima at x_1. Given that $f(x_1) > 0$ and $g(x_1) > 0$, show that the product function fg has a relative minimum at x_1.

55 Let a, b, c, and d be constant numbers such that $ad \neq bc$ and $c \neq 0$. Show that the graph of the function f defined by the equation $f(x) = \dfrac{ax + b}{cx + d}$ has no points of inflection.

56 Suppose that the function f is defined by the equation $f(x) = ax^3 + bx^2 + cx + d$, where a, b, c, and d are constants with $a > 0$. Show that the graph of f has exactly one point of inflection, the point $\left(-\dfrac{b}{3a}, f\left(-\dfrac{b}{3a}\right)\right)$. Also show that f' is increasing when $x > -b/(3a)$.

57 If the period T and the length l of a conical pendulum are related by the equation $T = [\pi/(2\sqrt{2})](l^2 - r^2)^{1/4}$, where r is the radius of the path of the bob, sketch the graph of T as a function of l when $r = 5$.

58 Suppose that f and g are twice differentiable at $x = x_1$ and that f and g have points of inflection at $x = x_1$. Is it true that the product fg has a point of inflection at $x = x_1$? Explain.

59 Find the shortest distance from the point $(1, 0)$ to the graph of the function given by the equation $y = \sqrt{x^2 + 6x + 10}$.

60 Let f be the function defined by the equation

$$f(x) = (1 + x)^k - 1 - kx - \frac{k(k - 1)x^2}{2} \qquad \text{for } x > 0,$$

where k is a rational number and $k > 2$.
(a) Find $f'(x)$ and $f''(x)$.
(b) Show that f is an increasing function for $x > 0$.
(c) Show that f is concave upward for $x > 0$.
(d) Use part (b) to show that

$$(1 + x)^k > 1 + kx + \frac{k(k - 1)x^2}{2} \qquad \text{for } x > 0.$$

61 Let f be the function defined by the equation $f(x) = (1 + x)^{1/2} - 1 - \dfrac{x}{2}$.

    (a) Show that f is decreasing for $x > 0$ and increasing for $-1 < x < 0$.

    (b) Use part (a) to show that $\sqrt{1 + x} < 1 + x/2$ when $x > 0$.

62 A cylindrical can without a top is to be formed from a tin sheet of uniform thickness, and is to weigh $\frac{1}{4}$ pound. What is the relationship between its height and base radius if the volume of the can is to be maximum?

63 Find the ratio of height to radius of base of a closed cylinder of constant volume if the total surface area is minimum.

64 Find the dimensions of the smallest (in terms of area) piece of cardboard that will contain 50 square inches of printed matter, if it has 4-inch margins at the top and bottom and 2-inch margins on each side.

65 Find two positive numbers whose sum is 4 such that the sum of the square of the first number and the cube of the second number is minimum.

66 A newly independent country designs a flag that consists of a red rectangular region divided by a green stripe. The perimeter of the entire flag is 14 feet and the red part is to have an area of 9 square feet. What are the dimensions of the largest stripe?

67 Show that among all rectangles of given diagonal the square has the largest area.

68 The theory of probability tells us that the function f defined by the equation

$$f(p) = \frac{n!}{k!\,(n - k)!}\, p^k (1 - p)^{n-k}$$

is the probability of exactly k successes in n independent trials when the probability of success in each trial is p. Suppose that n and k are integers such that $n > 0$ and $0 \le k \le n$. Find the number p that maximizes f, $0 \le p \le 1$.

69 Find the rectangle of the largest area that can be inscribed in:

    (a) A circle of radius 5 inches.

    (b) A semicircle of radius r inches.

    (c) An isosceles triangle of base 10 inches and altitude 10 inches.

    (d) An isosceles trapezoid with bases 10 inches and 6 inches and altitude 8 inches.

70 A long rectangular sheet of tin is 8 inches wide. Find the depth of the V-shaped trough of maximum cross-sectional area that can be made by bending the plate along its central longitudinal axis.

71 Find the value of x, $0 \le x \le 5$ (Figure 5), that minimizes the given expression.

    (a) $|\overline{AP}| + |\overline{PB}|$ (b) $|\overline{AP}|^2 + |\overline{PB}|^2$

    (c) $|\overline{AP}|^2 - |\overline{PB}|^2$ (d) Area ACP + area BDP

72 A long sheet of paper is 8 inches wide. One corner of the paper is folded over (Figure 6). Find the value of x that gives the right triangle ABC the least possible area.

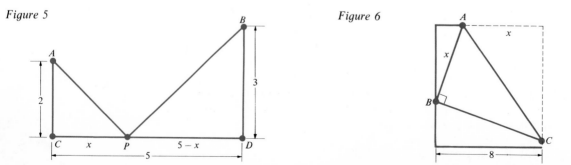

Figure 5

Figure 6

73 Find the largest volume of a right circular cone that can be inscribed in a sphere of radius a.

74 Find, among all right circular closed cylinders of fixed volume V, the one with the smallest surface area.

75 Ship A is anchored 3 miles directly from a point B off the shore of a lake (Figure 7). Opposite a point D, 5 miles farther along the shore, another ship E is anchored 9 miles directly from point D. A boat is to take some passengers from ship A to a point C on shore, then proceed to ship E. Find the shortest course of the boat.

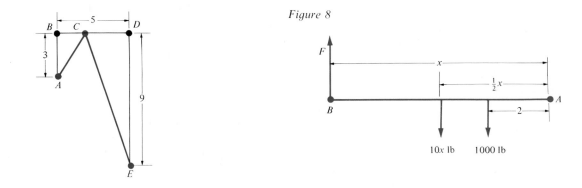

Figure 7

Figure 8

76 A weight of 1000 pounds hanging 2 feet from a point A at one end of a lever is to be lifted by an upward force F from the other end B. Suppose that the lever is to weigh 10 pounds per foot (Figure 8). Find the length of the lever if the force at end B is to be a minimum.

77 A sector ABC of central angle θ is cut out from a circular sheet of metal of radius a (Figure 9) and the remainder is used to form a conical shape. Find the maximum volume of such a cone.

Figure 9

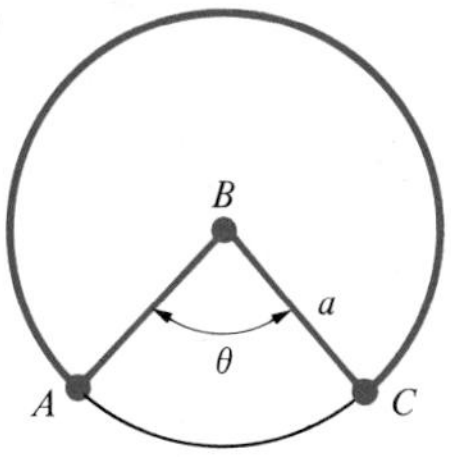

78 The manufacturer of a certain toy finds that, in order to sell x toys each week, she must price them at $\sqrt{5,000,000 - 2x^2}$ dollars each. How many toys per week will bring the largest total revenue?

79 Explain why you might expect that a business firm makes a maximum profit when its marginal total cost equals its marginal total revenue.

80 A garage owner finds that he can sell x tires per week at p dollars per tire where $p = \frac{1}{3}(375 - 5x)$ and his cost C in dollars of obtaining x tires per week to sell is expressed by $C = 500 + 15x + x^2/5$. Find the number of tires he must sell per week and the price he should charge per tire to maximize his profit.

In problems 81 through 84, use implicit differentiation to find $D_x y$ and $D_x^2 y$.

81 $3x^3 + 2y^3 = 1$

82 $x^4 + 4x^2y^2 = 25$

83 $5x^2 + 8y^3 = 16\sqrt{x + 1}$

84 $4x^3 - 5xy^2 + y^3 = 18$

In problems 85 through 88, find the equations of the tangent and normal line to the graph of the implicit function determined by the equation at the point indicated.

85 $y^2 + 2xy = 16$ at $(3, 2)$

86 $x^2 + 4xy + y^2 + 3 = 0$ at $(2, -1)$

87 $x^3 - xy^2 + y^3 = 8$ at $(2, 2)$

88 $y\sqrt{2x + 1} = 3y$ at $(4, 2)$

89 A 20-foot ladder leans against a vertical wall. The lower end is pulled along a horizontal floor at a constant rate of 4 feet per second and the top slides down the wall. Find the rate at which the top moves when:
(a) The lower end is 5 feet from the wall.
(b) The upper end is 16 feet above the floor.

90 Water is flowing through a hole in the bottom of a hemispherical bowl of radius 10 inches. At the instant the water is 6 inches deep, the rate of flow is 5 cubic inches per minute. Find the rate at which the surface of water is falling at this instant.

91 A highway crosses a railroad track at right angles. A car traveling at the constant rate of 40 miles per hour goes through the intersection 2 minutes before the engine of a train traveling at the constant rate of 36 miles per hour goes through the intersection. At what rate are the car and engine separating 10 minutes after the train goes through the intersection?

92 The height of a right circular cylinder is increasing at the rate of 3 inches per minute and the radius of the base is decreasing at the rate of 2 inches per minute. Find the rate at which the volume of the cylinder is changing when the height is 10 inches and the radius of the base is 4 inches.

93 A boy is going to lift a weight W by means of a rope mounted on a pulley (Figure 10) that is 36 feet above the ground. At the start, the weight is resting on the ground and the boy, standing directly under the pulley, grasps the rope 6 feet above the ground. If the boy, holding fast to the rope and keeping his hand 6 feet above the ground, walks away at a constant rate of 7 feet per second, how fast is the weight rising 2 seconds after he starts to walk?

Figure 10

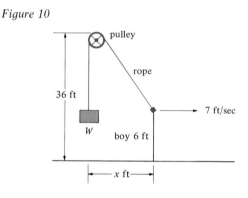

4 ANALYTIC GEOMETRY AND THE CONICS

The *conic sections* (*conics*, for short) are so called because they are curves obtained by sectioning or cutting circular cones by planes. These ubiquitous and graceful curves were well known to the ancients; however, their study is immensely enhanced by the use of analytic geometry and calculus. In this chapter we study the conics—the *circle*, the *ellipse*, the *parabola*, and the *hyperbola*—and we use the idea of translation of coordinate axes to simplify their equations.

By shining a flashlight onto a white wall, we can see examples of the conic sections. If the axis of the flashlight is perpendicular to the wall, then the illuminated region is *circular* (Figure 1a), while if the flashlight is tilted slightly upward, the illuminated region elongates and its boundary becomes an *ellipse* (Figure 1b). As the flashlight is tilted further, the ellipse becomes more and more elongated until it changes into a *parabola* (Figure 1c). Finally, if the flashlight is tilted still further, the edges of the parabola become straighter and it changes into a portion of a *hyperbola* (Figure 1d).

Figure 1

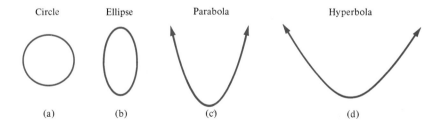

Circle   Ellipse   Parabola   Hyperbola

(a)   (b)   (c)   (d)

1 The Circle and Translation of Axes

In this section we derive the equation of a circle and we see how such an equation can be simplified by translation of axes.

1.1 The Circle

We begin by recalling (from precalculus mathematics) the definition of the graph of an equation.

DEFINITION 1   **Graph of an Equation**

The graph of an equation involving the variables x and y is the set of all points (x, y) in the xy plane, and only those points, whose coordinates x and y satisfy the equation.

If a curve in the xy plane is the graph of a certain equation, this equation is called an *equation of the curve*. To avoid cumbersome expressions, we occasionally write phrases such as "the circle $x^2 + y^2 = 25$" when we really mean "the circle whose equation is $x^2 + y^2 = 25$."

Of course, a *circle* of *radius* r with *center* at the point C in the xy plane is defined to be the set of all points P in the plane whose distance from C is r. Using the formula for the distance between two points $P = (x, y)$ and $C = (h, k)$, we see that $|\overline{PC}| = r$ if and only if

$$\sqrt{(x - h)^2 + (y - k)^2} = r;$$

that is,

$$(x - h)^2 + (y - k)^2 = r^2$$

(Figure 2). Therefore, we have the following theorem.

Figure 2

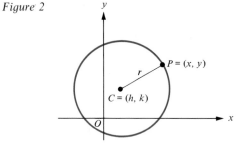

THEOREM 1   **Circle Equation**

Let $r > 0$ and let $C = (h, k)$. Then the graph of the equation

$$(x - h)^2 + (y - k)^2 = r^2$$

is the circle of radius r whose center is C.

The equation in Theorem 1 is called the *standard form* for the equation of a circle.

EXAMPLES **1** Find the center $C = (h, k)$ and the radius r of the circle

$$(x - 1)^2 + (y + 1)^2 = 9.$$

SOLUTION

The equation can be rewritten $(x - 1)^2 + [y - (-1)]^2 = 3^2$; hence, $C = (1, -1)$ and $r = 3$.

2 Find the equation of the circle (Figure 3) whose radius is 3 and whose center is $C = (-2, 3)$.

Figure 3

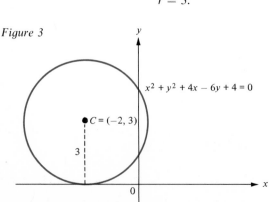

SOLUTION

By Theorem 1, the equation is

$$[x - (-2)]^2 + (y - 3)^2 = 3^2$$

or

$$(x + 2)^2 + (y - 3)^2 = 9.$$

Expanding the squares and simplifying, we can also write the equation in the form

$$x^2 + y^2 + 4x - 6y + 4 = 0.$$

The equation $x^2 + y^2 + 4x - 6y + 4 = 0$ obtained in Example 2 can be restored to standard form by "completing the squares." The work is arranged as follows:

$$x^2 + y^2 + 4x - 6y + 4 = 0,$$
$$x^2 + 4x \qquad + y^2 - 6y \qquad = -4,$$
$$x^2 + 4x + 4 + y^2 - 6y + 9 = -4 + 4 + 9,$$
$$(x + 2)^2 + (y - 3)^2 = 9.$$

Here, we added 4 to both sides of the equation to change $x^2 + 4x$ into the perfect square $x^2 + 4x + 4$, and we added 9 to both sides to change $y^2 - 6y$ into the perfect square $y^2 - 6y + 9$.

More generally, an expression of the form $x^2 \pm Bx$ becomes a perfect square if we add $(B/2)^2$ to obtain $x^2 \pm Bx + (B/2)^2 = (x \pm B/2)^2$.

EXAMPLES 1 Find the radius r and the center $C = (h, k)$ of the circle whose equation is $x^2 + y^2 + 2x + 8y - 8 = 0$.

SOLUTION
Completing the squares, we have

$$x^2 + 2x \qquad + y^2 + 8y \qquad = 8,$$
$$x^2 + 2x + \left(\frac{2}{2}\right)^2 + y^2 + 8y + \left(\frac{8}{2}\right)^2 = 8 + \left(\frac{2}{2}\right)^2 + \left(\frac{8}{2}\right)^2,$$
$$x^2 + 2x + 1 \quad + y^2 + 8y + 16 \quad = 8 + 1 + 16,$$
$$(x + 1)^2 + (y + 4)^2 = 25.$$

Thus, the center of the circle is $C = (-1, -4)$ and its radius is $r = \sqrt{25} = 5$ units.

2 Find the equation, in the form $x^2 + y^2 + Ax + By + C = 0$ and in the standard form, of the circle containing the three points $(-2, 5)$, $(1, 4)$, and $(-3, 6)$.

SOLUTION
We substitute the x and y coordinates of the three points into the equation $x^2 + y^2 + Ax + By + C = 0$ and thus obtain the three simultaneous equations

$$\begin{cases} -2A + 5B + C = -29 \\ A + 4B + C = -17 \\ -3A + 6B + C = -45. \end{cases}$$

Solving these simultaneous linear equations in the usual way, we get $A = -2$, $B = -18$, and $C = 57$. Thus, the equation of the circle is

$$x^2 + y^2 - 2x - 18y + 57 = 0.$$

Completing the squares, we have

$$x^2 - 2x + 1 + y^2 - 18y + 81 = -57 + 1 + 81,$$
$$(x - 1)^2 + (y - 9)^2 = 25.$$

Thus, the desired circle has radius 5 units and center $C = (1, 9)$.

As the following example shows, implicit differentiation can be used to find the slope of the tangent line to a circle at a specified point.

EXAMPLE Find the equations of the tangent and normal lines at the point $(-1, 2)$ on the circle $x^2 + y^2 - 6x - y - 9 = 0$.

SOLUTION

By implicit differentiation, we have $2x + 2y\dfrac{dy}{dx} - 6 - \dfrac{dy}{dx} = 0$, so that $\dfrac{dy}{dx} = \dfrac{6 - 2x}{2y - 1}$.

When $x = -1$ and $y = 2$, $\dfrac{dy}{dx} = \dfrac{8}{3}$. Therefore, the tangent line has slope $m = \frac{8}{3}$ and its equation is $y - 2 = \frac{8}{3}[x - (-1)]$, or $y = \frac{8}{3}x + \frac{14}{3}$. The normal line has slope $-1/m = -\frac{3}{8}$ and its equation is $y - 2 = -\frac{3}{8}[x - (-1)]$, or $y = -\frac{3}{8}x + \frac{13}{8}$.

1.2 Translation of Axes

The equation of a curve can often be simplified by changing to a more aptly chosen coordinate system. In practice, this is usually accomplished by choosing one or both of the new coordinate axes to coincide with an axis of symmetry of the curve. For instance, the equation of the curve C in Figure 4 would probably be simplified by switching from the xy coordinate system to the $\bar{x}\bar{y}$ coordinate system as shown, since C is symmetric about the $\bar{y}$ axis.

If two cartesian coordinate systems have corresponding axes that are parallel and have the same positive directions, then we say that these systems are obtained from one another by *translation*.

Figure 4

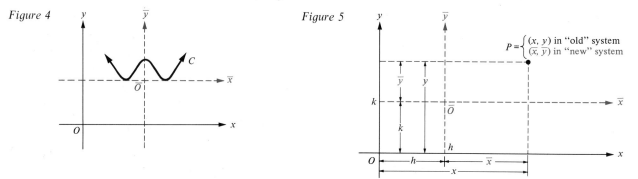

Figure 5

Figure 5 shows a translation of the "old" xy coordinate system to a "new" $\bar{x}\bar{y}$ system whose origin $\bar{O}$ has the "old" coordinates (h, k). Consider the point P in Figure 5 having old coordinates (x, y), but having new coordinates $(\bar{x}, \bar{y})$. Evidently, we have the following *rule for translation of cartesian coordinates*:

$$\begin{cases} x = h + \bar{x} \\ y = k + \bar{y} \end{cases} \quad \text{or} \quad \begin{cases} \bar{x} = x - h \\ \bar{y} = y - k. \end{cases}$$

EXAMPLE   Let the $\bar{x}\bar{y}$ axes be obtained from the xy axes by translation in such a way that the origin $\bar{O}$ of the "new" coordinate system has coordinates $(h, k) = (-3, 4)$ in the "old" coordinate system. Let P be the point whose old coordinates are $(x, y) = (2, 1)$. Find the coordinates $(\bar{x}, \bar{y})$ of P in the new coordinate system (Figure 6).

Figure 6

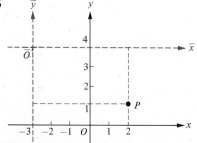

SOLUTION
According to the translation equations,

$$\bar{x} = x - h = 2 - (-3) = 5$$

and

$$\bar{y} = y - k = 1 - 4 = -3;$$

hence, the point P has coordinates $(\bar{x}, \bar{y}) = (5, -3)$ in the new coordinate system.

Figure 7

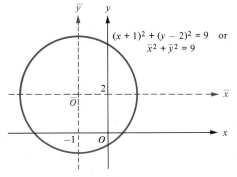

$(x + 1)^2 + (y - 2)^2 = 9$ or
$\bar{x}^2 + \bar{y}^2 = 9$

Notice that the equation of a curve in the plane depends not only on the set of points comprising this curve, but also on our choice of the coordinate system. For instance, the circle of radius 3 with center at $\bar{O}$ in Figure 7 has the equation

$$(x + 1)^2 + (y - 2)^2 = 9$$

with respect to the xy coordinate system; however, the very same circle has the equation

$$\bar{x}^2 + \bar{y}^2 = 9$$

with respect to the $\bar{x}\bar{y}$ coordinate system. This can be seen either by noticing that the center of the circle is the origin $\bar{O}$ in the $\bar{x}\bar{y}$ system, or by substituting $\bar{x} = x - h = x - (-1) = x + 1$ and $\bar{y} = y - k = y - 2$ from the translation equations into the equation $(x + 1)^2 + (y - 2)^2 = 9$. Notice that a translation of the coordinate axes does not change the position or the shape of a geometric curve in the plane—it only changes the *equation* of this curve.

EXAMPLE   Find a translation of axes that will reduce the equation $x^2 + y^2 + 4x - 4y = 0$ to the form $\bar{x}^2 + \bar{y}^2 = r^2$.

SOLUTION
The simplest procedure is to complete the squares so that the equation becomes

$$x^2 + 4x + 4 + y^2 - 4y + 4 = 8 \quad \text{or} \quad (x + 2)^2 + (y - 2)^2 = 8,$$

and then to put $\bar{x} = x + 2$ and $\bar{y} = y - 2$. The result is

$$\bar{x}^2 + \bar{y}^2 = 8 \quad \text{or} \quad \bar{x}^2 + \bar{y}^2 = (2\sqrt{2})^2.$$

An alternative solution is obtained by letting

$$x = \bar{x} + h \quad \text{and} \quad y = \bar{y} + k,$$

where the constants h and k are yet to be determined. Substitution of the latter into $x^2 + y^2 + 4x - 4y = 0$ and routine algebraic simplification yields

$$\bar{x}^2 + \bar{y}^2 + (2h + 4)\bar{x} + (2k - 4)\bar{y} = -h^2 - k^2 - 4h + 4k.$$

The terms involving $\bar{x}$ and $\bar{y}$ to the first power drop out when we put $h = -2$, $k = 2$, and the equation becomes $\bar{x}^2 + \bar{y}^2 = 8$.

Problem Set 1

In problems 1 through 6, find the equation of the circle in the xy plane that satisfies the conditions given:

1 Radius 3 and center at the point $(0, 2)$.

2 Radius 2 and center at the point $(-1, 4)$.

3 Radius 5 and center at the point $(3, 4)$.

4 Center at the point $(1, 6)$ and containing the point $(-2, 2)$.

5 Radius 4 and containing the points $(-3, 0)$ and $(5, 0)$.

6 The points $(3, 7)$ and $(-3, -1)$ are the endpoints of a diameter.

In problems 7 through 14, find the radius r and the coordinates (h, k) of the center of the circle for each equation and sketch a graph of the circle.

7 $(x + 1)^2 + (y - 2)^2 = 9$

8 $(x + 3)^2 + (y - 10)^2 = 100$

9 $x^2 + y^2 + 2x + 4y + 4 = 0$

10 $x^2 + y^2 - x - y - 1 = 0$

11 $4x^2 + 4y^2 + 8x - 4y + 1 = 0$ (*Hint:* Begin by dividing by 4.)

12 $3x^2 + 3y^2 - 6x + 9y = 27$

13 $4x^2 + 4y^2 + 4x - 4y + 1 = 0$

14 $4x^2 + 4y^2 + 12x + 20y + 25 = 0$

In problems 15 through 22, find the equation in the standard form of the circle or circles in the xy plane that satisfy the conditions given.

15 Containing the points $(-3, 1)$, $(7, 1)$, and $(-7, 5)$.

16 Containing the points $(1, 7)$, $(8, 6)$, and $(7, -1)$.

17 Radius $\sqrt{17}$, center on the x axis, and containing the point $(0, 1)$. (There are *two* circles in this case.)

18 Center on the line $x - 4y - 1 = 0$ and containing the points $(3, 7)$ and $(5, 5)$.

19 Tangent to the x axis and with center at the point $(1, -7)$.

20 Center on the line $x + 4 = 0$, radius 5, and tangent to the x axis. (There are *two* circles in this case.)

21 Containing the point $(3, 4)$, radius 2, and tangent to the line $y = 2$. (There are *two* circles in this case.)

22 Radius $\sqrt{10}$ and tangent to the line $3x + y = 6$ at the point $(3, -3)$. (There are *two* circles in this case.)

23 A "new" $\bar{x}\bar{y}$ coordinate system is obtained by translating the "old" xy coordinate system so that the origin $\bar{O}$ of the new system has old xy coordinates $(-1, 2)$. Find the new $\bar{x}\bar{y}$ coordinates of the points whose old xy coordinates are:

(a) $(0, 0)$ (b) $(-2, 1)$ (c) $(3, -3)$
(d) $(-3, -2)$ (e) $(5, 5)$ (f) $(6, 0)$

24 Describe a translation of coordinates that will reduce the equation

$$x^2 + y^2 + 4x - 2y + 1 = 0$$

of a circle in the old xy coordinate system to the simpler form $\bar{x}^2 + \bar{y}^2 = r^2$ in the new $\bar{x}\bar{y}$ system. Find the radius r and draw a graph showing the circle and the two coordinate systems.

25 A new $\bar{x}\bar{y}$ coordinate system is obtained by translating the old xy coordinate system so that the origin O of the old xy system has coordinates $(-3, 2)$ in the new $\bar{x}\bar{y}$ system. Find the old xy coordinates of the points whose new $\bar{x}\bar{y}$ coordinates are:

(a) $(0, 0)$ (b) $(3, 2)$ (c) $(-3, 4)$
(d) $(\sqrt{2}, -2)$ (e) $(0, -\pi)$ (f) $(-3, 2)$

In problems 26 through 28, find a translation of axes that reduces each equation to an equation of the form $\bar{x}^2 + \bar{y}^2 = r^2$. Sketch the graph, showing both the old and the new coordinate axes.

26 $x^2 + y^2 + 4x - 4y = 0$

27 $3x^2 + 3y^2 + 7x - 5y + 3 = 0$

28 $x^2 + y^2 - 8x - 9 = 0$

In problems 29 through 31, find the equations of the tangent and normal lines to each circle at the point indicated.

29 $(x - 3)^2 + (y + 5)^2 = 5$ at $(2, -3)$

30 $x^2 + y^2 = 169$ at $(5, -12)$

31 $x^2 + y^2 - 6x + 8y - 11 = 0$ at $(3, 2)$

32 Prove that the radius $\overline{CP}$ of a circle is perpendicular to the tangent line to the circle at the point P.

33 A point $P = (x, y)$ moves so that it is always twice as far from the point $(6, 0)$ as it is from the point $(0, 3)$.
   (a) Find the equation of the curve traced out by the point P.
   (b) Sketch a graph of this curve.

34 Generalize problem 33 by assuming that a point $P = (x, y)$ moves so that its distance from the point $(a, 0)$ is always c times its distance from the point $(0, b)$, where $a, b, c > 0$. Distinguish the cases (i) $0 < c < 1$, (ii) $c = 1$, and (iii) $c > 1$.

35 Discuss the symmetry of the circle $x^2 + y^2 = r^2$.

36 Find the length of the upper base of the trapezoid of the maximum area that can be cut from a semicircle of radius 6 as in Figure 8.

Figure 8

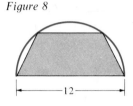

37 Salesperson A is allowed to sell encyclopedias anywhere within the circle $x^2 + y^2 = 100$, while salesperson B is allowed to sell them within the circle $(x - 20)^2 + y^2 = 144$. Do their sales territories overlap? Explain.

38 Find conditions on the constants $a, b, c, d, r,$ and R so that the circle

$$(x - a)^2 + (y - b)^2 = r^2$$

will intersect the circle $(x - c)^2 + (y - d)^2 = R^2$.

39 By expanding the squares and collecting the terms, show that the equation of the circle in the standard form $(x - h)^2 + (y - k)^2 = r^2$ can always be rewritten in the alternative form $x^2 + y^2 + Ax + By + C = 0$, where $A = -2h$, $B = -2k$, and $C = h^2 + k^2 - r^2$.

40 By completing the squares, show that the equation $x^2 + y^2 + Ax + By + C = 0$ represents a circle with center at the point $(h, k) = (-A/2, -B/2)$ and radius $r = \frac{1}{2}\sqrt{A^2 + B^2 - 4C}$, provided that $A^2 + B^2 > 4C$.

2 The Ellipse

Figure 1

In this section we discuss the ellipse, a type of conic that appears frequently in nature. For instance, a circular hoop, viewed from an angle, forms an ellipse (Figure 1), and an orbiting satellite moves in an elliptical path. We begin with the following definition.

DEFINITION 1 Ellipse

An *ellipse* is defined to be the set of all points P in the plane such that the sum of the distances from P to two fixed points F_1 and F_2 is constant. Here F_1 and F_2 are called the *focal points* or the *foci* of the ellipse. The midpoint C of the line segment $\overline{F_1 F_2}$ is called the *center* of the ellipse.

Figure 2

Figure 2 shows two fixed pins F_1 and F_2 and a loop of string of length l stretched tightly about them to the point P. Since $|\overline{PF_1}| + |\overline{PF_2}| + |\overline{F_1 F_2}| = l$, we have $|\overline{PF_1}| + |\overline{PF_2}| = l - |\overline{F_1 F_2}|$; hence, as P is moved about, $|\overline{PF_1}| + |\overline{PF_2}|$ always has the constant value $l - |\overline{F_1 F_2}|$. Thus, according to the definition, P lies on an ellipse having F_1 and F_2 as foci. If a pencil point P is inserted into the loop of string as in Figure 2 and moved about so as to keep the string tight, it traces out an ellipse. Figure 3 shows three successive positions P_1, P_2, and P_3 of P and the complete ellipse traced out by P.

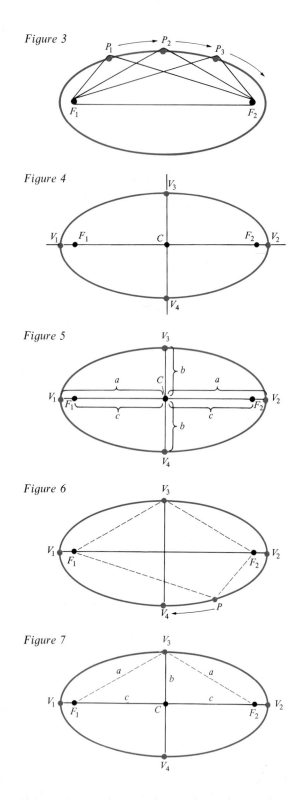

Figure 3

Figure 4

Figure 5

Figure 6

Figure 7

Evidently (Figure 4), an ellipse with foci F_1 and F_2 is symmetric about the straight line through F_1 and F_2. Let V_1 and V_2 be the points where this straight line intersects the ellipse. Notice that the center C of the ellipse bisects the line segment $\overline{V_1 V_2}$ as well as the line segment $\overline{F_1 F_2}$. The ellipse is also symmetric about the line through the center C perpendicular to $\overline{V_1 V_2}$. Let V_3 and V_4 be the points where this perpendicular intersects the ellipse. The four points V_1, V_2, V_3, and V_4 (Figure 4), where the two axes of symmetry intersect the ellipse, are called the *vertices* of the ellipse.

The line segment $\overline{V_1 V_2}$ between the two vertices that contains the two foci F_1 and F_2 is called the *major axis* of the ellipse, while the line segment $\overline{V_3 V_4}$ between the remaining two vertices is called the *minor axis* (Figure 5). Let $2a$ denote the length of the major axis $\overline{V_1 V_2}$, let $2b$ denote the length of the minor axis $\overline{V_3 V_4}$, and let $2c$ denote the distance between the two foci (Figure 5). The numbers a and b are called the *semimajor axis* and the *semiminor axis*, respectively.

Consider the ellipse in Figure 6 with semimajor and semiminor axes a and b, respectively, with foci F_1 and F_2, and with $|\overline{F_1 F_2}| = 2c$. If the point P moves along the ellipse, then, by definition,

$$|\overline{PF_1}| + |\overline{PF_2}|$$

remains the same. Therefore, the value of $|\overline{PF_1}| + |\overline{PF_2}|$ when P reaches V_1 is the same as the value of $|\overline{PF_1}| + |\overline{PF_2}|$ when P reaches V_3; that is,

$$|\overline{V_1 F_1}| + |\overline{V_1 F_2}| = |\overline{V_3 F_1}| + |\overline{V_3 F_2}|.$$

By symmetry $|\overline{V_3 F_1}| = |\overline{V_3 F_2}|$, so the equation above can be rewritten

$$|\overline{V_1 F_1}| + |\overline{V_1 F_2}| = 2|\overline{V_3 F_2}|.$$

But, by symmetry again, $|\overline{V_1 F_2}| = |\overline{V_2 F_1}|$; hence,

$$2|\overline{V_3 F_2}| = |\overline{V_1 F_1}| + |\overline{V_1 F_2}|$$
$$= |\overline{V_1 F_1}| + |\overline{V_2 F_1}| = |\overline{V_1 V_2}| = 2a,$$

from which $|\overline{V_3 F_2}| = a$ follows.

If we use the fact that $|\overline{V_3 F_2}| = a$ and consider the right triangle $V_3 C F_2$ in Figure 7, we can conclude that $a^2 = b^2 + c^2$. Since $c^2 > 0$, the latter equation shows that $a^2 > b^2$, so that $a > b$ and consequently $2a > 2b$; that is, *the major axis of an ellipse is always longer than its minor axis.*

If we place the ellipse of Figure 7 in the xy plane so that its center C is at the origin O and the foci F_1 and F_2 lie on the negative and positive portions of the x axis, respectively, then we can derive the equation of the ellipse as follows.

THEOREM 1   **Ellipse Equation**

The equation of the ellipse with foci at $F_1 = (-c,0)$ and $F_2 = (c,0)$ is

$$\frac{x^2}{a^2} + \frac{y^2}{b^2} = 1,$$

where a is the semimajor axis, b is the semiminor axis, and $a^2 = b^2 + c^2$.

Figure 8

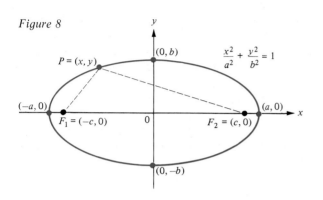

PROOF

We show that if $P = (x, y)$ is any point on the ellipse (Figure 8), then $x^2/a^2 + y^2/b^2 = 1$. The interested reader can complete the proof (Problem 38) by reversing the argument to show that if the equation $x^2/a^2 + y^2/b^2 = 1$ holds, then the point $P = (x, y)$ is on the ellipse. We have already seen (Figure 7) that $a^2 = b^2 + c^2$ and that when $P = (0, b)$,

$$|\overline{PF_1}| + |\overline{PF_2}| = 2a.$$

Therefore, from the definition of an ellipse, $|\overline{PF_1}| + |\overline{PF_2}| = 2a$ holds for every point $P = (x, y)$ on the ellipse.

By the distance formula, $|\overline{PF_1}| + |\overline{PF_2}| = 2a$ becomes

$$\sqrt{(x + c)^2 + y^2} + \sqrt{(x - c)^2 + y^2} = 2a,$$

so that

$$\sqrt{(x + c)^2 + y^2} = 2a - \sqrt{(x - c)^2 + y^2}.$$

Squaring both sides of the latter equation, we have

$$x^2 + 2cx + c^2 + y^2 = 4a^2 - 4a\sqrt{(x - c)^2 + y^2} + x^2 - 2cx + c^2 + y^2,$$

so that

$$4cx - 4a^2 = -4a\sqrt{(x - c)^2 + y^2} \quad \text{or} \quad cx - a^2 = -a\sqrt{(x - c)^2 + y^2}.$$

Squaring both sides of the latter equation, we obtain

$$c^2x^2 - 2a^2cx + a^4 = a^2(x^2 - 2cx + c^2 + y^2),$$

so that

$$a^4 - a^2c^2 = (a^2 - c^2)x^2 + a^2y^2 \quad \text{or} \quad a^2(a^2 - c^2) = (a^2 - c^2)x^2 + a^2y^2.$$

Since $a^2 = b^2 + c^2$, we have $a^2 - c^2 = b^2$ and the equation above can be rewritten as

$$a^2b^2 = b^2x^2 + a^2y^2.$$

If both sides of the latter equation are divided by a^2b^2, the result is

$$1 = \frac{x^2}{a^2} + \frac{y^2}{b^2}$$

as desired.

The equation $x^2/a^2 + y^2/b^2 = 1$, where $a > b$, is called the *standard form* for the equation of an ellipse.

EXAMPLES 1 Find the coordinates of the four vertices and of the two foci of the ellipse $4x^2 + 9y^2 = 36$ and sketch the graph.

SOLUTION

Divide both sides of the equation by 36 to obtain $x^2/9 + y^2/4 = 1$; that is,

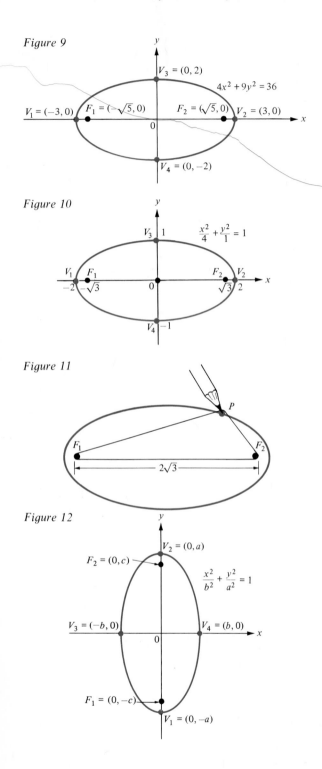

Figure 9

Figure 10

Figure 11

Figure 12

$x^2/a^2 + y^2/b^2 = 1$, with $a = 3$ and $b = 2$. By Theorem 1, this is the equation of an ellipse with vertices at $(-3,0)$, $(3,0)$, $(0,2)$, and $(0,-2)$ (Figure 9). Also, the foci are at $(-c,0)$ and $(c,0)$, where $c^2 = a^2 - b^2 = 9 - 4 = 5$; that is, $c = \sqrt{5}$. Thus, $F_1 = (-\sqrt{5},0)$ and $F_2 = (\sqrt{5},0)$.

2 Find the equation in the standard form of the ellipse with foci $F_1 = (-\sqrt{3},0)$, $F_2 = (\sqrt{3},0)$ and vertices $V_1 = (-2,0)$, $V_2 = (2,0)$. Also, find the coordinates of the remaining two vertices, V_3 and V_4, and sketch the graph.

SOLUTION

$c = \sqrt{3}$, $a = 2$; hence, $b = \sqrt{a^2 - c^2} = \sqrt{4 - 3} = 1$ and the equation is $x^2/4 + y^2/1 = 1$. Also, $V_3 = (0,1)$ and $V_4 = (0,-1)$. (Why?) The graph appears in Figure 10.

3 It is desired to construct the ellipse of Example 2 by driving two pins into a board at positions F_1 and F_2, $2\sqrt{3}$ units apart, and using the loop-of-string-and-pencil construction as in Figure 11. What should be the total length l of the loop?

SOLUTION

When P is at the upper vertex V_3 as in Figure 7,

$$l = a + a + c + c = 2a + 2c = (2)(2) + 2\sqrt{3}$$
$$= 4 + 2\sqrt{3} \approx 7.46 \text{ units.}$$

It is not difficult to derive the equation of an ellipse with center at the origin and with a vertical major axis (Figure 12). In this case, the ellipse has foci $F_1 = (0,-c)$ and $F_2 = (0,c)$ on the y axis and vertices $V_1 = (0,-a)$, $V_2 = (0,a)$, $V_3 = (-b,0)$, and $V_4 = (b,0)$. The semimajor axis is a and the semiminor axis is b. The equation can be derived as in Theorem 1, the argument being word for word the same except that the variables x and y interchange their roles (see Problem 39). Therefore, the equation is

$$\frac{x^2}{b^2} + \frac{y^2}{a^2} = 1, \qquad \text{where } a > b.$$

This equation is also called the *standard form* for the equation of the ellipse.

By using the translation equations $\bar{x} = x - h$ and $\bar{y} = y - k$, we easily find the equation of an ellipse whose axes of symmetry are parallel to the coordinate axes, but whose center is at (h,k) in the xy coordinate system. The resulting equation, depending on whether the major axis is horizontal or vertical, is, respectively,

$$\frac{(x-h)^2}{a^2} + \frac{(y-k)^2}{b^2} = 1 \quad \text{or} \quad \frac{(x-h)^2}{b^2} + \frac{(y-k)^2}{a^2} = 1,$$

where $a > b$.

EXAMPLES **1** Write the equation of the ellipse whose vertices are the points $(-5, 1)$, $(1, 1)$, $(-2, 3)$, and $(-2, -1)$; find the foci, and sketch the graph.

SOLUTION

The horizontal axis is the line segment from $(-5, 1)$ to $(1, 1)$, and its length is $|(-5) - 1| = 6$ units. The vertical axis is the line segment from $(-2, -1)$ to $(-2, 3)$, and its length is $|(-1) - 3| = 4$ units. Therefore, the ellipse has a horizontal major axis, $a = \frac{6}{2} = 3$, $b = \frac{4}{2} = 2$, and $c = \sqrt{a^2 - b^2} = \sqrt{5}$. Here, the center is at $(-2, 1)$ (why?), so $h = -2$ and $k = 1$. Consequently, the equation of the ellipse is $\dfrac{(x+2)^2}{9} + \dfrac{(y-1)^2}{4} = 1$. The foci are $c = \sqrt{5}$ units on either side of the center; hence, $F_1 = (-\sqrt{5} - 2, 1)$ and $F_2 = (\sqrt{5} - 2, 1)$ (Figure 13).

Figure 13

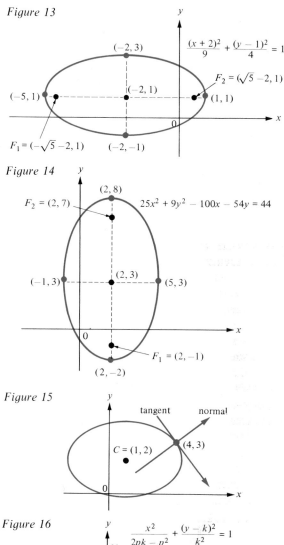

$$\frac{(x+2)^2}{9} + \frac{(y-1)^2}{4} = 1$$

2 Given the equation $25x^2 + 9y^2 - 100x - 54y = 44$ of an ellipse, find the coordinates of the center, the vertices, and the foci. Sketch the graph.

SOLUTION

Here $25(x^2 - 4x) + 9(y^2 - 6y) = 44$. Completing the squares in the latter equation, we have

$$25(x^2 - 4x + 4) + 9(y^2 - 6y + 9) = 44 + 25(4) + 9(9)$$

or

$$25(x - 2)^2 + 9(y - 3)^2 = 225$$

Dividing the latter equation by 225, we obtain $\dfrac{(x-2)^2}{9} + \dfrac{(y-3)^2}{25} = 1$. Thus, $(h, k) = (2, 3)$, $a = 5$, $b = 3$, $c = \sqrt{a^2 - b^2} = 4$, $F_1 = (2, -1)$, $F_2 = (2, 7)$, and the vertices are $(2, -2)$, $(2, 8)$, $(-1, 3)$, and $(5, 3)$ (Figure 14).

Figure 14

$25x^2 + 9y^2 - 100x - 54y = 44$

3 Find the equations of the tangent and normal lines to the ellipse $4x^2 - 8x + 9y^2 - 36y = 5$ at the point $(4, 3)$. Sketch the graph and show the tangent and normal lines.

SOLUTION

Using implicit differentiation we find that $8x - 8 + 18y\dfrac{dy}{dx} - 36\dfrac{dy}{dx} = 0$, so $\dfrac{dy}{dx} = -\dfrac{4(x-1)}{9(y-2)}$. When $x = 4$ and $y = 3$, we have $dy/dx = -\frac{4}{3}$. Thus, the equation of the tangent line is $y - 3 = -\frac{4}{3}(x - 4)$ and the equation of the normal line is $y - 3 = \frac{3}{4}(x - 4)$. By completing the squares after rewriting the equation in the form $4(x^2 - 2x) + 9(y^2 - 4y) = 5$, we obtain $\dfrac{(x-1)^2}{\frac{45}{4}} + \dfrac{(y-2)^2}{5} = 1$ (Figure 15).

Figure 15

Figure 16

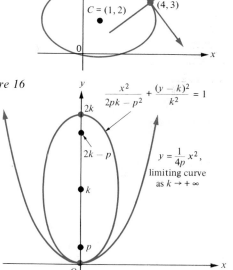

$$\frac{x^2}{2pk - p^2} + \frac{(y-k)^2}{k^2} = 1$$

$y = \dfrac{1}{4p}x^2$, limiting curve as $k \to +\infty$

Now, consider the ellipse in Figure 16 having vertical major axis, center at

$(0, k)$, foci at $(0, p)$ and $(0, 2k - p)$, and lower vertex at the origin O. If we hold the lower vertex fixed at O and hold the lower focus fixed at $(0, p)$, but allow the upper vertex $(0, 2k)$ to approach $+\infty$ along the y axis, then the ellipse approaches a limiting curve $y = (1/4p)x^2$ (Problem 50). The curve $y = (1/4p)x^2$ is no longer an ellipse, but it is still a conic section—namely, a *parabola*. We study parabolas in the next section.

Problem Set 2

In problems 1 through 8, find the coordinates of the vertices and foci of each ellipse and sketch its graph.

1 $\dfrac{x^2}{16} + \dfrac{y^2}{4} = 1$

2 $\dfrac{x^2}{9} + y^2 = 1$

3 $4x^2 + y^2 = 16$

4 $36x^2 + 9y^2 = 144$

5 $x^2 + 16y^2 = 16$

6 $16x^2 + 25y^2 = 400$

7 $9x^2 + 36y^2 = 4$

8 $x^2 + 4y^2 = 1$

In problems 9 through 14, find the equation of the ellipse that satisfies the conditions given.

9 Foci at $(-4, 0)$ and $(4, 0)$, vertices at $(-5, 0)$ and $(5, 0)$.

10 Foci at $(0, -2)$ and $(0, 2)$, vertices at $(0, -4)$ and $(0, 4)$.

11 Vertices at $(0, -8)$ and $(0, 8)$ and containing the point $(6, 0)$.

12 Vertices at $(0, -3)$ and $(0, 3)$ and containing the point $(\frac{2}{3}, 2\sqrt{2})$.

13 Center at the origin and containing the points $(4, 0)$ and $(3, 2)$.

14 Vertices at $(-2\sqrt{3}, 0)$, $(2\sqrt{3}, 0)$, $(0, -4)$, and $(0, 4)$.

In problems 15 through 22, find the coordinates of the center, the vertices, and the foci of each ellipse, and sketch its graph.

15 $\dfrac{(x-1)^2}{9} + \dfrac{(y+2)^2}{4} = 1$

16 $\dfrac{(x+2)^2}{16} + \dfrac{(y-1)^2}{4} = 1$

17 $4(x+3)^2 + y^2 = 36$

18 $25(x+1)^2 + 16(y-2)^2 = 400$

19 $x^2 + 2y^2 + 6x + 7 = 0$

20 $4x^2 + y^2 - 8x + 4y - 8 = 0$

21 $2x^2 + 5y^2 + 20x - 30y + 75 = 0$

22 $9x^2 + 4y^2 + 18x - 16y - 11 = 0$

In problems 23 through 26, use a suitable translation of axes $\bar{x} = x - h$ and $\bar{y} = y - k$ to reduce each equation to a "simpler" form (not involving first powers of the variables). Also, sketch the graph showing both the "old" xy and the "new" $\bar{x}\bar{y}$ coordinate systems.

23 $x^2 + 4y^2 + 2x - 8y + 1 = 0$

24 $9x^2 + y^2 - 18x + 2y + 9 = 0$

25 $6x^2 + 9y^2 - 24x - 54y + 51 = 0$

26 $9x^2 + 4y^2 - 18x + 16y - 11 = 0$

In problems 27 through 30, find the equation of the ellipse satisfying the conditions given.

27 Vertices at $(-2, -3)$, $(-2, 5)$, $(-7, 1)$, and $(3, 1)$.

28 Foci at $(1, 3)$ and $(5, 3)$ and major axis 10 units long.

29 Center at $(1, -2)$, major axis parallel to the y axis, major axis 6 units long, and minor axis 4 units long.

30 Ends of major axis at $(-3, 2)$, $(5, 2)$, and the length of the minor axis is 4 units.

In problems 31 through 34, find the equations of the tangent and normal lines to each ellipse at the point indicated.

31 $x^2 + 9y^2 = 225$ at $(9, 4)$ **32** $4x^2 + 9y^2 = 45$ at $(3, 1)$

33 $x^2 + 4y^2 - 2x + 8y = 35$ at $(3, 2)$ **34** $9x^2 + 25y^2 - 50y - 200 = 0$ at $(5, 1)$

35 Find the coordinates of the point where the normal line to the ellipse $x^2/a^2 + y^2/b^2 = 1$ at the point (x_0, y_0) intersects the x axis.

36 Suppose that the numbers x and y satisfy $x^2/a^2 + y^2/b^2 = 1$, where $a > b > 0$. Put $c = \sqrt{a^2 - b^2}$. Show analytically (without reference to graphs or geometry) that the following inequalities hold:

   (i) $c|x| < a^2$ (ii) $\sqrt{(x - c)^2 + y^2} < 2a$

37 The segment cut by an ellipse from a line containing a focus and perpendicular to the major axis is called a *latus rectum* of the ellipse.
   (a) Show that $2b^2/a$ is the length of a latus rectum of the ellipse whose equation is $b^2x^2 + a^2y^2 = a^2b^2$.
   (b) Find the length of a latus rectum of the ellipse whose equation is $9x^2 + 16y^2 = 144$.

38 Finish the proof of Theorem 1 by showing that if $x^2/a^2 + y^2/b^2 = 1$. where $a > b > 0$, then the point $P = (x, y)$ is on the ellipse with foci $F_1 = (-c, 0)$ and $F_2 = (c, 0)$, where $c = \sqrt{a^2 - b^2}$ and the semimajor axis is a.

39 Derive the equation of the ellipse in Figure 12 directly from Definition 1.

40 A point moves on the ellipse $x^2 + 4y^2 = 25$ in such a way that its abscissa is increasing at the constant rate of 8 units per second. How fast is the ordinate changing at the instant when it is equal to -2 and the abscissa is positive?

41 The point $P = (x, y)$ moves so that the sum of its distances from the two points $(3, 0)$ and $(-3, 0)$ is 8. Describe and write an equation for the curve traced out by the point.

42 A mathematician has accepted a position at a new university situated 6 miles from the straight shoreline of a large lake (Figure 17). The professor wishes to build a home that is half as far from the university as it is from the shore of the lake. The possible home-sites satisfying this condition lie along a curve. Describe this curve and find its equation with respect to a coordinate system having the shoreline as the x axis and the university at the point $(0, 6)$ on the y axis.

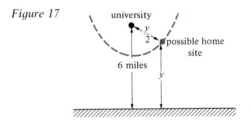

Figure 17

43 Find the maximum area of a rectangle that can be inscribed in the ellipse $x^2/a^2 + y^2/b^2 = 1$ with its sides parallel to the coordinate axes.

44 An isosceles triangle with vertex at the point $(0, 3)$ and base parallel to the x axis is to be inscribed in the ellipse $9x^2 + 16y^2 = 144$ in such a way that its area is maximum. What should the altitude of the triangle be?

45 How long a loop of string should be used to lay out an elliptical flower bed 20 feet wide and 60 feet long? How far apart should the two stakes (foci) be?

46 Except for minor perturbations, a satellite orbiting the earth moves in an ellipse with the center of the earth at one focus. Suppose that a satellite at perigee (nearest point to center of earth) is 400 kilometers from the surface of the earth and at apogee (farthest point to center of earth) is 600 kilometers from the surface of the earth. Assume that the earth is a sphere of radius 6371 kilometers. Find the semiminor axis b of the elliptical orbit.

47 An arch in the shape of the upper half of an ellipse with a horizontal major axis is to support a bridge over a river 100 feet wide. The center of the arch is to be 25 feet above the surface of the river. Find the equation in standard form for the ellipse.

48 Prove that the normal line to the ellipse in Figure 18 at the point $P = (x, y)$ bisects the angle $F_1 P F_2$. [*Hint:* The equation of the ellipse is $x^2/a^2 + y^2/b^2 = 1$. By implicit differentiation, the slope of the normal line at P is $m = a^2 y/(b^2 x)$. The slopes of $\overline{F_1 P}$ and $\overline{F_2 P}$, respectively, are $m_1 = \dfrac{y}{x + c}$ and $m_2 = \dfrac{y}{x - c}$, where $F_1 = (-c, 0)$, $F_2 = (c, 0)$, and $c = \sqrt{a^2 - b^2}$. Verify that $(m_1 + m_2)m^2 + 2(1 - m_1 m_2)m = m_1 + m_2$. Then show that the latter equation implies the desired result.]

49 A ray of light will be reflected from a curved mirror (Figure 19) in such a way that the angle α between the incident ray and the normal is equal to the angle α between the normal and the reflected ray. Use the result of problem 48 to show that a ray of light emanating from one focus of an elliptical mirror will be reflected through the other focus. (This is called the *reflecting property* of the ellipse.)

50 In Figure 16, show that, as $k \to +\infty$, the ellipse approaches the curve $y = (1/4p)x^2$ as a limit.

51 What happens to the ellipse in Figure 16 as $k \to p^+$?

Figure 18

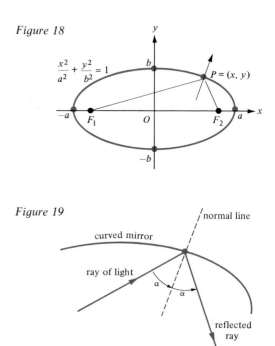

Figure 19

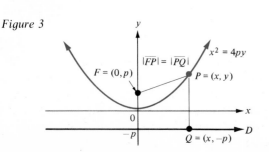

curved mirror

normal line

ray of light

α

α

reflected ray

3   The Parabola

At the end of Section 2 we saw that if one focus and a nearby vertex of an ellipse are held fixed while the opposite vertex is moved farther and farther away, the ellipse approaches a limiting curve called a parabola. Thus, although we give a precise geometric definition of this conic below, we can regard a parabola as an enormous ellipse with one vertex infinitely far away.

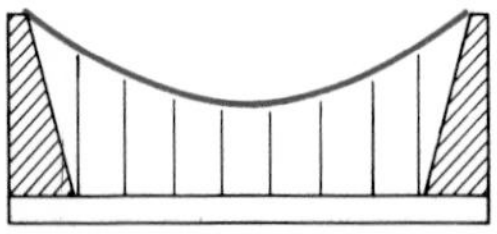

Parabolas often appear in nature. A rock thrown up at an angle travels along a parabolic arc (Figure 1), and the main cable in a suspension bridge forms an arc of a parabola (Figure 2).

We now give the precise definition promised above.

Figure 1

Figure 2

DEFINITION 1   Parabola

A *parabola* is the set of all points P in the plane such that the distance from P to a fixed point F called the *focus* is equal to the distance from P to a fixed straight line D called the *directrix*.

If the focus F of a parabola is placed on the y axis at the point $(0, p)$ and if the directrix D is placed parallel to the x axis and p units below it, the resulting parabola appears as in Figure 3. Its equation is derived in the following theorem.

Figure 3

THEOREM 1   **Parabola Equation**
The equation of the parabola with focus $F = (0, p)$ and with directrix $D: y = -p$ is $x^2 = 4py$, or $y = (1/4p)x^2$.

PROOF
Let $P = (x, y)$ be any point and let $Q = (x, -p)$ be the point at the foot of the perpendicular from P to the directrix D (Figure 3). The requirement for P to be on the parabola is $|\overline{FP}| = |\overline{PQ}|$; that is, $\sqrt{x^2 + (y - p)^2} = \sqrt{(y + p)^2}$. The latter equation is equivalent to $x^2 + (y - p)^2 = (y + p)^2$; that is, $x^2 + y^2 - 2py + p^2 = y^2 + 2py + p^2$, or $x^2 = 4py$.

The equation $x^2 = 4py$ [or $y = (1/4p)x^2$] is called the *standard form* for the equation of a parabola.

EXAMPLE   Write the equation of the parabola with focus $F = (0, \frac{1}{4})$ and with directrix $D: y = -\frac{1}{4}$ (Figure 4).

SOLUTION
Here $p = \frac{1}{4}$, so the equation is $x^2 = 4(\frac{1}{4})y$; that is, $y = x^2$.

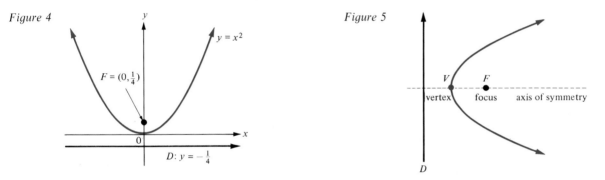

Figure 4

Figure 5

A parabola with focus F and directrix D is evidently symmetric about the straight line through F perpendicular to D (Figure 5). This straight line is called the *axis of symmetry* or simply the *axis* of the parabola. The *vertex* of the parabola is defined to be the point V where the parabola intersects its axis. Notice that the vertex of a parabola is located midway between the focus F and the directrix D. (Why?)

EXAMPLE   A parabola has its focus at $(0, \frac{1}{8})$ and directrix $D: y = -\frac{1}{8}$. Find its equation and sketch the graph. Also find its vertex V and its axis.

Figure 6

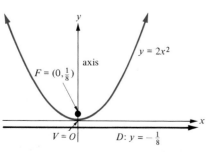

SOLUTION
The equation is $y = (1/4p)x^2$, where $p = \frac{1}{8}$; hence, the equation is $y = 2x^2$. Since the y axis is perpendicular to the directrix and passes through the focus, it is the axis of the parabola. The vertex V lies on the axis and is midway between the focus and the directrix, so $V = O = (0, 0)$ (Figure 6).

In Figure 7a–d, we see parabolas opening *upward, to the right, downward,* and *to the left,* respectively. In each case, the vertex is at the origin O and the distance from the vertex O to the focus F is p. The corresponding equations are shown in Figure 7 and can be determined by consideration of the symmetries involved.

Figure 7

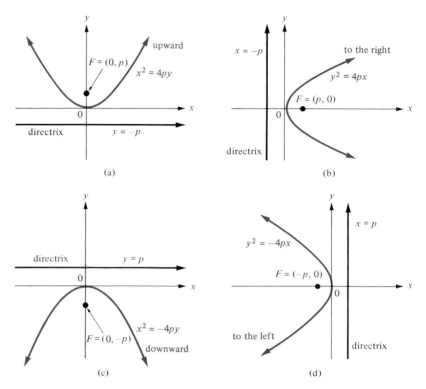

(a)

(b)

(c)

(d)

EXAMPLES   Find the coordinates of the focus and the equation of the directrix of the given parabola, determine its direction of opening, and sketch the graph.

1 $y^2 = -8x$

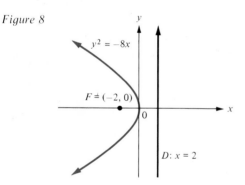

Figure 8

SOLUTION
The equation has the form $y^2 = -4px$ with $p = 2$; hence, it corresponds to Figure 7d. Therefore, the graph is a parabola opening to the left with focus given by

$$F = (-p, 0) = (-2, 0)$$

and directrix D: $x = p$; that is, D: $x = 2$ (Figure 8).

2 $x^2 = -16y$

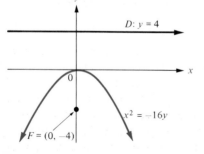

Figure 9

SOLUTION
The equation has the form $x^2 = -4py$ with $p = 4$; hence, it corresponds to Figure 7c. Therefore, the graph is a parabola opening downward with focus given by

$$F = (0, -p) = (0, -4)$$

and directrix D: $y = p$; that is, D: $y = 4$ (Figure 9).

The segment cut by a parabola on the straight line through the focus and perpendicular to its axis is called the *latus rectum* or the *focal chord* of the parabola. The length of the latus rectum can be found as in the following example.

EXAMPLE A parabola opens to the right, has its vertex at the origin, and contains the point $(3, 6)$. Find its equation, sketch the parabola, and find the length of its latus rectum.

Figure 10

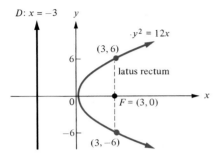

SOLUTION

The equation must have the form $y^2 = 4px$. Since the point $(3, 6)$ belongs to the graph, we put $x = 3$ and $y = 6$ in the equation to obtain $36 = 12p$ and we conclude that $p = 3$; thus, the equation of the parabola is $y^2 = 12x$ (Figure 10). The focus is given by $F = (p, 0) = (3, 0)$. The latus rectum lies along the line $x = 3$. Putting $x = 3$ in the equation $y^2 = 12x$ and solving for y, we obtain $y^2 = 36$, $y = \pm 6$. Thus, the points $(3, 6)$ and $(3, -6)$ are the endpoints of the latus rectum, and therefore its length is 12 units.

A parabola with its vertex at the origin which opens upward, to the right, downward, or to the left is said to be *in the standard position*. If a parabola is in the standard position, then its axis of symmetry is either horizontal or vertical. Conversely, if a parabola has an axis of symmetry that is either horizontal or vertical, then a translation of the coordinate axes to the vertex of the parabola will put it into standard position with respect to the "new" coordinate system.

Figure 11

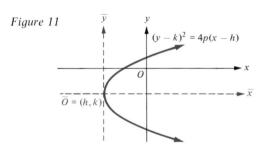

For instance, if the parabola in Figure 11 has its vertex at the point (h, k) with respect to the "old" xy coordinate system and opens to the right as shown, then its equation with respect to the "new" $\bar{x}\bar{y}$ coordinate system is $\bar{y}^2 = 4p\bar{x}$. Since $\bar{x} = x - h$ and $\bar{y} = y - k$, its equation with respect to the old xy coordinate system is $(y - k)^2 = 4p(x - h)$. Similar arguments can be made for parabolas opening upward, downward, or to the left, with vertices at the point (h, k).

EXAMPLES Find the coordinates of the vertex V and the focus F of the given parabola, determine the direction in which it opens, find the equation of its directrix, determine the length of its latus rectum, and sketch the graph.

1 $(y + 1)^2 = -12(x - 2)$

SOLUTION

The equation can be written as

$$(y - k)^2 = -4p(x - h)$$

with $p = 3$, $h = 2$, and $k = -1$; hence, the parabola opens to the left, $V = (2, -1)$, $F = (2 - 3, -1) = (-1, -1)$, and the directrix is $D: x = 5$. Since the x coordinate of the focus is -1, the latus rectum lies along the vertical line $x = -1$. Putting $x = -1$ in the equation of the parabola, we obtain $(y + 1)^2 = 36$, so that $y + 1 = \pm 6$; that is, $y = 5$ or $y = -7$. Therefore, the endpoints of the latus rectum are $(-1, 5)$ and $(-1, -7)$ and its length is 12 units (Figure 12).

Figure 12

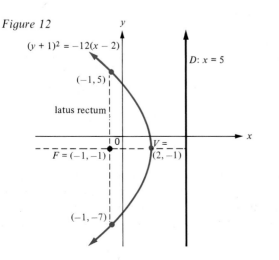

Figure 13

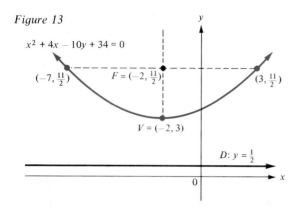

$x^2 + 4x - 10y + 34 = 0$

$(-7, \frac{11}{2})$ $F = (-2, \frac{11}{2})$ $(3, \frac{11}{2})$

$V = (-2, 3)$

$D: y = \frac{1}{2}$

Figure 14

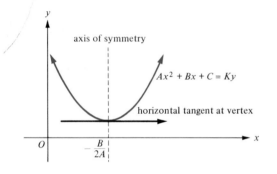

axis of symmetry

$Ax^2 + Bx + C = Ky$

horizontal tangent at vertex

$-\dfrac{B}{2A}$

2 $x^2 + 4x - 10y + 34 = 0$

SOLUTION

Completing the square, we obtain

$$x^2 + 4x + 4 - 10y + 34 = 4 \quad \text{or}$$
$$(x + 2)^2 = 10y - 30;$$

that is, $(x + 2)^2 = 10(y - 3)$. Thus, $p = \frac{10}{4} = \frac{5}{2}$, and the graph is a parabola opening upward with vertex $V = (-2, 3)$, focus $F = (-2, \frac{11}{2})$, and directrix D: $y = \frac{1}{2}$ (Figure 13). Here, the latus rectum lies along the horizontal line $y = \frac{11}{2}$. Putting $y = \frac{11}{2}$ in the equation of the parabola, we obtain $(x + 2)^2 = 10(\frac{11}{2} - 3) = 25$, so $x = 3$ or $x = -7$. Therefore, the endpoints of the latus rectum are $(-7, \frac{11}{2})$ and $(3, \frac{11}{2})$ and its length is 10 units.

Of course, one can find the slope dy/dx of the tangent line (and therefore the slope $\dfrac{-1}{dy/dx}$ of the normal line) to a parabola at a given point by differentiation as usual. (Implicit differentiation may be useful here.) This can be used for locating the vertex of a parabola. For instance, the parabola whose equation has the form $Ax^2 + Bx + C = Ky$, where A, B, C, and K are constants, has a vertical axis of symmetry (Figure 14). Its tangent line is horizontal only at its vertex. Here, $dy/dx = (1/K)(2Ax + B)$, so $dy/dx = 0$ when $x = -B/(2A)$. Thus, the vertex is the point $\left(-\dfrac{B}{2A}, \dfrac{1}{K}\left(\dfrac{B^2}{4A} - \dfrac{B^2}{2A} + C\right) \right)$.

EXAMPLE   Find the vertex of the parabola $3y = 2x^2 + 4x + 5$.

SOLUTION

This parabola has a vertical axis of symmetry, so we set dy/dx equal to zero; $dy/dx = \frac{1}{3}(4x + 4) = 0$ for $x = -1$. When $x = -1$, $y = \frac{1}{3}[2(-1)^2 + 4(-1) + 5] = 1$. Therefore, the vertex is at $(-1, 1)$.

The vertex of a parabola with horizontal axis of symmetry can be found similarly by setting $dx/dy = 0$.

One of the most important properties of a parabola is its so-called *reflecting property*: A ray of light emanating from the focus of a parabolic mirror is always reflected parallel to the axis (Figure 15). In order to demonstrate the reflecting property analytically, one must recall that a ray of light striking a curved mirror is reflected so that the angle between the incident ray and the normal to the mirror is the same as the angle between the normal and the reflected ray.

Thus, in Figure 16 it is desired to show that angle α is the same as angle β. By implicit differentiation of

Figure 15

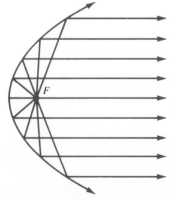

F

Figure 16

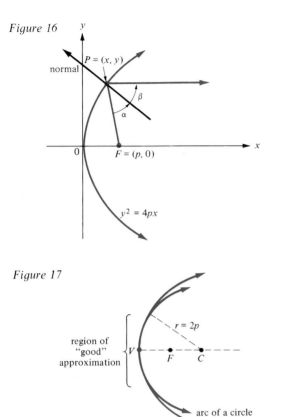

$y^2 = 4px$

Figure 17

region of "good" approximation

$r = 2p$

arc of a circle

parabola

the equation $4px = y^2$ of the parabola, $4p = 2y(dy/dx)$, $dy/dx = 2p/y$, so the slope of the normal is $m = -\dfrac{y}{2p}$. The slope of the segment $\overline{FP}$ is $m_1 = \dfrac{y}{x-p}$. It is easy to show that $\tan \alpha = \dfrac{m - m_1}{1 + mm_1}$ and that $\tan \beta = -m$ (Problem 45). Thus, the reflecting property follows if it can be shown that $\dfrac{m - m_1}{1 + mm_1} = -m$. The latter equation is easy to verify by using the relations $m = -\dfrac{y}{2p}$, $m_1 = \dfrac{y}{x-p}$, and $4px = y^2$ (Problem 46).

If an intense source of light such as a carbon arc or an incandescent filament is placed at the focus of a parabolic mirror, the light is reflected and projected in a parallel beam. The same principle is used in reverse in a reflecting telescope—parallel rays of light from a distant object are brought together at the focus of a parabolic mirror.

In practice it is very difficult to manufacture large parabolic mirrors, so it is often necessary to make do with mirrors whose cross section is a portion of a circle approximating the appropriate parabola (Figure 17). It can be shown (Problem 42) that the circle that "best approximates" the parabola near its vertex V has its center C located on the axis of the parabola twice as far from the vertex V as the focus F of the parabola and so has radius $r = 2|\overline{VF}| = 2p$.

Problem Set 3

In problems 1 through 6, find the coordinates of the vertex and the focus of the parabola. Also find the equation of the directrix and the length of the latus rectum (focal chord). Sketch the graph.

1 $y^2 = 4x$

2 $y^2 = -9x$

3 $x^2 - y = 0$

4 $x^2 - 4y = 0$

5 $x^2 + 9y = 0$

6 $3x^2 - 4y = 0$

7 Find the equation of the parabola whose focus is the point $(0, 3)$ and whose directrix is the line $y = -3$.

8 Find the vertex of the parabola $y = Ax^2 + Bx + C$, where A, B, and C are constants and $A \neq 0$.

In problems 9 through 16, find the coordinates of the vertex and the focus of the parabola. Also find the equation of the directrix and the length of the latus rectum. Sketch the graph.

9 $(y - 2)^2 = 8(x + 3)$

10 $(y + 1)^2 = -4(x - 1)$

11 $(x - 4)^2 = 12(y + 7)$

12 $(x + 1)^2 = -8y$

13 $y^2 - 8y - 6x - 2 = 0$

14 $2x^2 + 8x - 3y + 4 = 0$

15 $x^2 - 6x - 8y + 1 = 0$

16 $y^2 + 10y - x + 21 = 0$

In problems 17 through 20, find the equation of the parabola that satisfies the conditions given.

17 Focus at $(4, 2)$ and directrix $x = 6$.

18 Focus at $(3, -1)$ and directrix $y = 5$.

19 Vertex at $(-6, -5)$ and focus at $(2, -5)$.

20 Vertex at $(2, -3)$ and directrix $x = -8$.

21 Find the equation of the parabola whose axis is parallel to the x axis, whose vertex is at the point $(-\frac{1}{2}, -1)$, and which contains the point $(\frac{5}{8}, 2)$.

22 Find the equation of the parabola whose axis coincides with the y axis and which contains the points $(2, 3)$ and $(-1, -2)$.

In problems 23 through 25, reduce each equation to "simpler" form by a suitable translation of axes. Also, sketch the graph in the "old" as well as the "new" coordinate system.

23 $y^2 + 2y - 8x - 3 = 0$

24 $x^2 + 2x + 4y - 7 = 0$

25 $5y = x^2 + 4x + 19$

26 Let A, B, and C be constants with $A > 0$. Show that $y = Ax^2 + Bx + C$ is the equation of a parabola with a vertical axis of symmetry, opening upward. Find the coordinates of the vertex V and the focus F. Find p and the length of the latus rectum. Find conditions for the graph to intersect the x axis.

In problems 27 through 31, find the equations of the tangent and normal lines to each parabola at the points indicated.

27 $y^2 = 8x$ at $(2, -4)$

28 $2y^2 = 9x$ at $(2, -3)$

29 $x^2 = -12y$ at $(-6, -3)$

30 $x^2 + 8y + 4x - 20 = 0$ at $(1, \frac{15}{8})$

31 $y^2 - 2y + 10x - 44 = 0$ at $(\frac{9}{2}, 1)$

32 How can one tell immediately (without computation) that $3y - 2 = 4x^2 - 27x + 11$ is the equation of a parabola opening upward?

In problems 33 through 36, find the coordinates of the vertex of the parabola by differentiating its equation on both sides and then solving for a horizontal (or vertical) tangent.

33 $y = x^2 - 2x + 6$

34 $4x^2 + 24x + 39 - 3y = 0$

35 $y^2 - 10y = 4x - 21$

36 $3x = 14y - y^2 - 43$

37 Find the dimensions of the rectangle of largest area whose base is on the x axis and whose two upper vertices are on the parabola whose equation is $y = 12 - x^2$ (Figure 18).

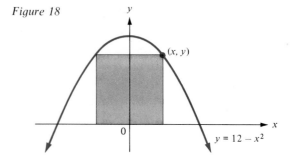

Figure 18

38 Show that, if p is the distance between the vertex and the focus of a parabola, then the latus rectum of the parabola has length $4p$.

39 A roadway 400 meters long is held up by a parabolic main cable (Figure 2). The main cable is 100 meters above the roadway at the ends and 4 meters above the roadway at the center. Vertical supporting cables run at 50-meter intervals along the roadway. Find the lengths of these vertical cables. (*Hint:* Set up an xy coordinate system with vertical y axis and having the vertex of the parabola 4 units above the origin.)

40 The surface of a roadway over a stone bridge follows a parabolic curve with the vertex in the middle of the bridge. The span of the bridge is 60 feet and the road surface is 1 foot higher in the middle than at the ends. How much higher than the ends is a point on the roadway 15 feet from an end?

41 For $a > 0$, let $(0, f(a))$ be the center of the circle tangent to the parabola whose equation is $4py = x^2$ at the points whose x coordinates are $-a$ and a (Figure 19). Find (a) a formula for $f(a)$ in terms of a and p and (b) $\lim_{a \to 0} f(a)$.

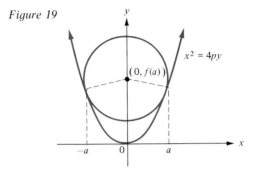

Figure 19

$x^2 = 4py$

$(0, f(a))$

42 Find the circle that "best fits" the parabola whose equation is $4py = x^2$ at and near its vertex by letting a approach 0 in Figure 19.

43 A point $P = (x, y)$ moves along the parabola whose equation is $y = 8 - \frac{1}{2}x^2$. When $P = (-2, 6)$, x is increasing at the rate of 2 units per second. How fast is the distance between P and the vertex changing at this instant?

44 For $a > 0$, let $(f(a), g(a))$ be the center of the circle that is tangent to the parabola whose equation is $y = (1/4p)x^2$, $p > 0$, at the point $(a, a^2/4p)$ and that contains the origin. Find:
(a) Formulas for $f(a)$ and $g(a)$.   (b) $\lim_{a \to 0} f(a)$ and $\lim_{a \to 0} g(a)$.

45 In Figure 16 show that $\tan \alpha = \dfrac{m - m_1}{1 + mm_1}$ and that $\tan \beta = -m$, where m is the slope of the normal line and m_1 is the slope of the segment $\overline{FP}$. [*Hint*: Use the formula

$$\tan (\theta - \theta_1) = \frac{\tan \theta - \tan \theta_1}{1 + \tan \theta \tan \theta_1}$$

from trigonometry.]

46 In Figure 16 verify that $\alpha = \beta$. (Use problem 45 and the relations $m = -\dfrac{y}{2p}$, $m_1 = \dfrac{y}{x - p}$, and $4px = y^2$.)

47 A calculus student says, "If you've seen one parabola, you've seen them all."
(a) Explain why this assertion is essentially true.
(b) Explain why a similar assertion cannot be made for ellipses.

4   The Hyperbola

A comet that enters the solar system with more than enough energy to escape the sun's gravitational pull traces out one branch of a conic called a *hyperbola* (Figure 1). Actually, a hyperbola has two "branches," each of which looks like the curve in Figure 1, but these branches open in opposite directions (Figure 2). Hyperbolas can be defined geometrically as follows.

Figure 1

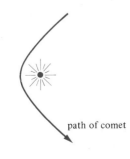

path of comet

Figure 2

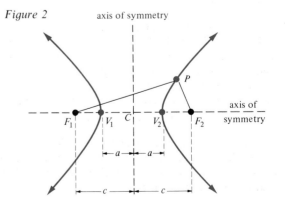

axis of symmetry

axis of symmetry

DEFINITION 1

The Hyperbola

A *hyperbola* is defined to be the set of all points P in the plane such that the absolute value of the difference of the distances from P to two fixed points F_1 and F_2 is a constant positive number k. The points F_1 and F_2 are called the *foci* of the hyperbola.

Figure 2 shows a hyperbola with foci F_1 and F_2. Evidently, the straight line through the two foci is an *axis of symmetry* for the hyperbola, and so is the perpendicular bisector of the line segment $\overline{F_1F_2}$. The point of intersection C of these two axes of symmetry is called the *center* of the hyperbola. The two points V_1 and V_2 where the two branches of the hyperbola intersect the axis of symmetry passing through the foci are called the *vertices* of the hyperbola.

The line segment $\overline{V_1V_2}$ between the two vertices (Figure 2) is called the *transverse axis*. We denote the length of the transverse axis by $2a$, and the distance between the two foci by $2c$. Thus, the distance $|\overline{CV_2}|$ from the center to a vertex is a, while the distance $|\overline{CF_2}|$ from the center to a focus is c. As the point P in Figure 2 moves along the right-hand branch toward V_2, the difference $|\overline{PF_1}| - |\overline{PF_2}|$ maintains, by definition, the constant value k; hence, when P reaches V_2, we have $|\overline{V_2F_1}| - |\overline{V_2F_2}| = k$. Since $|\overline{V_2F_1}| = c + a$ and $|\overline{V_2F_2}| = c - a$,

$$k = |\overline{V_2F_1}| - |\overline{V_2F_2}| = (c + a) - (c - a) = 2a.$$

Therefore, for any point P on the hyperbola, $\left\|\overline{PF_1}| - |\overline{PF_2}\right\| = 2a$.

The following theorem gives the equation of a hyperbola with a horizontal transverse axis whose center is at the origin.

THEOREM 1

Hyperbola Equation

The equation of the hyperbola with foci $F_1 = (-c, 0)$, $F_2 = (c, 0)$, and vertices $V_1 = (-a, 0)$, $V_2 = (a, 0)$ is $\dfrac{x^2}{a^2} - \dfrac{y^2}{b^2} = 1$, where $b = \sqrt{c^2 - a^2}$.

The proof of Theorem 1 is quite similar to the proof of Theorem 1 in Section 2; hence, it is left as an exercise (Problem 42). The equation

$$\frac{x^2}{a^2} - \frac{y^2}{b^2} = 1$$

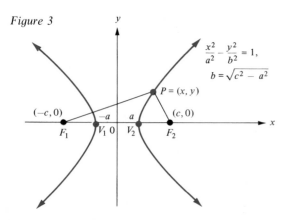

Figure 3

is called the *standard form* for the hyperbola (Figure 3).

Now, consider the hyperbola whose equation is $x^2/a^2 - y^2/b^2 = 1$ (Figure 3). We solve this equation for y in terms of x as follows: $y^2/b^2 = x^2/a^2 - 1$, so $y^2 = b^2[(x^2/a^2) - 1] = (b^2x^2/a^2)[1 - (a^2/x^2)]$; hence, $y = \pm(bx/a)\sqrt{1 - (a^2/x^2)}$, provided that $x \neq 0$. Since $\lim_{x \to \pm\infty} \sqrt{1 - (a^2/x^2)} = 1$, we see that, when x is large in absolute value, $y \approx \pm bx/a$. The two straight lines whose equations are

$$y = \frac{b}{a}x \quad \text{and} \quad y = -\frac{b}{a}x$$

are called the *asymptotes* of the hyperbola. They are good approximations to the hyperbola itself at suitably large distances from the origin.

Figure 4

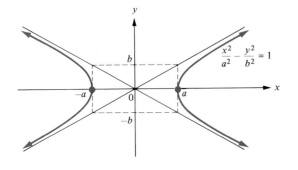

Although the asymptotes of the hyperbola are *not* part of the hyperbola itself, they are useful in sketching it. For instance, if we wish to sketch the graph of the equation $x^2/a^2 - y^2/b^2 = 1$, we begin by sketching the rectangle with height $2b$ and horizontal base $2a$ whose center is at the origin (Figure 4). The asymptotes are then drawn through the two diagonals of this rectangle. If one keeps in mind that the vertices of the hyperbola are located at the midpoints of the left and right sides of the rectangle, and that the hyperbola approaches the asymptotes as one moves out away from the vertices, then it becomes an easy matter to sketch the graph (Figure 4).

EXAMPLE   Find the coordinates of the foci and the vertices, and find the equations of the asymptotes of the hyperbola whose equation is $x^2/4 - y^2/1 = 1$. Also, sketch the graph.

SOLUTION
The equation has the form $x^2/a^2 - y^2/b^2 = 1$ with $a = 2$, $b = 1$. Hence, $c = \sqrt{a^2 + b^2} = \sqrt{5}$. Thus, the foci are $F_1 = (-\sqrt{5}, 0)$, $F_2 = (\sqrt{5}, 0)$, while the vertices are $V_1 = (-2, 0)$, $V_2 = (2, 0)$. The asymptotes are given by $y = \frac{1}{2}x$ and $y = -\frac{1}{2}x$ (Figure 5).

Suppose that we wish to find the equation of the hyperbola in Figure 6, which has a vertical transverse axis, center at the origin, vertices $V_1 = (0, -b)$ and

Figure 5   Figure 6

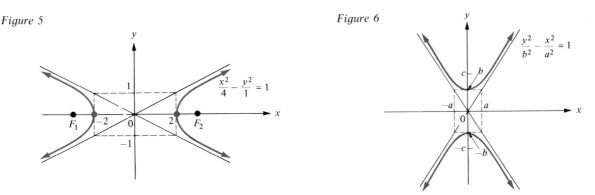

$V_2 = (0, b)$, and foci $F_1 = (0, -c)$ and $F_2 = (0, c)$. Using Theorem 1, but interchanging x with y and interchanging a with b, we obtain the equation

$$\frac{y^2}{b^2} - \frac{x^2}{a^2} = 1.$$

where $a = \sqrt{c^2 - b^2}$. This equation is also called the *standard form* for the hyperbola. The asymptotes are still given by $y = (b/a)x$ and $y = -(b/a)x$. (Why?)

Unlike the ellipse equations $x^2/a^2 + y^2/b^2 = 1$ or $x^2/b^2 + y^2/a^2 = 1$, there is no requirement that $a > b$ in the hyperbola equation $x^2/a^2 - y^2/b^2 = 1$ or in the hyperbola equation $y^2/b^2 - x^2/a^2 = 1$.

EXAMPLE   Which of the following hyperbolas have a horizontal transverse axis and which have a vertical transverse axis?

(a) $\dfrac{y^2}{16} - \dfrac{x^2}{9} = 1$ (b) $\dfrac{y^2}{9} - \dfrac{x^2}{4} = 1$ (c) $\dfrac{y^2}{4} - \dfrac{x^2}{4} = 1$

(d) $\dfrac{x^2}{9} - \dfrac{y^2}{4} = 1$ (e) $\dfrac{x^2}{4} - \dfrac{y^2}{9} = 1$ (f) $\dfrac{x^2}{4} - \dfrac{y^2}{4} = 1$

SOLUTION

Whether the hyperbola has a horizontal or a vertical transverse axis does not depend on the relative sizes of the denominators but on which term is subtracted from the other. Consequently, (a), (b), and (c) have vertical transverse axes, whereas (d), (e), and (f) have horizontal transverse axes.

A hyperbola whose transverse axis is either horizontal or vertical and whose center is at the point (h, k) will, of course, have the equation

$$\frac{(x - h)^2}{a^2} - \frac{(y - k)^2}{b^2} = 1 \quad \text{or} \quad \frac{(y - k)^2}{b^2} - \frac{(x - h)^2}{a^2} = 1,$$

respectively. In either case, the asymptotes have the equations

$$y - k = \frac{b}{a}(x - h) \quad \text{and} \quad y - k = -\frac{b}{a}(x - h),$$

and the distance from the center (h, k) to either focus is given by $c = \sqrt{a^2 + b^2}$.

EXAMPLES   1   Find the coordinates of the center, the foci, and the vertices of the hyperbola $y^2 - 4x^2 - 8x - 4y - 4 = 0$. Also find the equations of its asymptotes and sketch its graph.

Figure 7

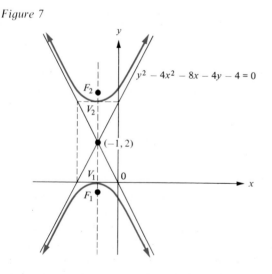

$y^2 - 4x^2 - 8x - 4y - 4 = 0$

SOLUTION

Completing the squares, we have

$y^2 - 4y + 4 - 4(x^2 + 2x + 1) = 4,$

or $(y - 2)^2 - 4(x + 1)^2 = 4$. Dividing by 4, we obtain

$$\frac{(y - 2)^2}{4} - \frac{(x + 1)^2}{1} = 1,$$

the equation of a hyperbola with center $(-1, 2)$ and with a vertical transverse axis. Since $a = 1$ and $b = 2$, the equations of the asymptotes are

$$y - 2 = \frac{2}{1}(x + 1) \quad \text{and} \quad y - 2 = \frac{-2}{1}(x + 1);$$

that is,

$$y = 2x + 4 \quad \text{and} \quad y = -2x.$$

Also, $c = \sqrt{a^2 + b^2} = \sqrt{5}$, so $F_1 = (-1, 2 - \sqrt{5})$, $F_2 = (-1, 2 + \sqrt{5})$, $V_1 = (-1, 0)$, and $V_2 = (-1, 4)$ (Figure 7).

2   Simplify the equation $4x^2 - 9y^2 - 24x - 90y - 225 = 0$ by using a suitable translation $x = \bar{x} + h$, $y = \bar{y} + k$. Sketch the graph, showing both the "old" xy and the "new" $\bar{x}\bar{y}$ coordinate axes.

SOLUTION
Completing the squares, we have

$$4(x^2 - 6x + 9) - 9(y^2 + 10y + 25) = 225 + 36 - 225$$

or

$$4(x - 3)^2 - 9(y + 5)^2 = 36.$$

Dividing by 36, we obtain

$$\frac{(x - 3)^2}{9} - \frac{(y + 5)^2}{4} = 1.$$

Thus, if we make the translation

$$\bar{x} = x - 3, \qquad \bar{y} = y + 5,$$

then the equation simplifies to $\bar{x}^2/9 - \bar{y}^2/4 = 1$ (Figure 8).

Figure 8

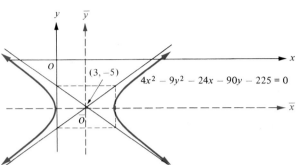

$4x^2 - 9y^2 - 24x - 90y - 225 = 0$

$(3, -5)$

3 The segment cut off by a hyperbola on a line perpendicular to the transverse axis and passing through a focus is called a *latus rectum*. Find a formula for the length of a latus rectum of the hyperbola $y^2/b^2 - x^2/a^2 = 1$.

SOLUTION
The hyperbola has a vertical transverse axis and the upper focus is at $(0, c)$, $c = \sqrt{a^2 + b^2}$. Putting $y = c$, we solve the equation of the hyperbola for x to obtain $x = \pm a^2/b$. The upper latus rectum is therefore the line segment from $(-a^2/b, c)$ to $(a^2/b, c)$; its length is $2a^2/b$.

4 Two microphones are located at the points $(-c, 0)$ and $(c, 0)$ on the x axis (Figure 9). An explosion occurs at an unknown point P to the right of the y axis. The sound of the explosion is detected by the microphone at $(c, 0)$ exactly T seconds before it is detected by the microphone at $(-c, 0)$. Assuming that sound travels in air at the constant speed of v feet per second, show that the point P must have been located on the right-hand branch of the hyperbola whose equation is $x^2/a^2 - y^2/b^2 = 1$, where

$$a = \frac{vT}{2} \quad \text{and} \quad b = \frac{\sqrt{4c^2 - v^2T^2}}{2}$$

Figure 9

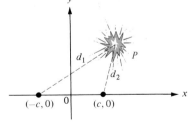

$(-c, 0)$ $(c, 0)$

SOLUTION
Let d_1 and d_2 denote the distances from P to $(-c, 0)$ and $(c, 0)$, respectively. The sound of the explosion reaches $(-c, 0)$ in d_1/v seconds and it reaches $(c, 0)$ in d_2/v seconds; hence, $d_1/v - d_2/v = T$, so $d_1 - d_2 = vT$. Putting $a = vT/2$, we notice that the condition $d_1 - d_2 = 2a$ requires (by Definition 1) that P belong to a hyperbola with foci $F_1 = (-c, 0)$ and $F_2 = (c, 0)$. By Theorem 1, the equation of this hyperbola is $x^2/a^2 - y^2/b^2 = 1$, where $a = vT/2$ and

$$b = \sqrt{c^2 - a^2} = \sqrt{c^2 - (vT/2)^2} = \frac{\sqrt{4c^2 - v^2T^2}}{2}.$$

Problem Set 4

In problems 1 through 8, find the coordinates of the vertices and the foci of each hyperbola. Also find the equations of the asymptotes and sketch the graph.

1 $\dfrac{x^2}{9} - \dfrac{y^2}{4} = 1$ **2** $\dfrac{x^2}{1} - \dfrac{y^2}{9} = 1$ **3** $\dfrac{y^2}{16} - \dfrac{x^2}{4} = 1$ **4** $\dfrac{y^2}{4} - \dfrac{x^2}{1} = 1$

5 $4x^2 - 16y^2 = 64$ **6** $49x^2 - 16y^2 = 196$ **7** $36y^2 - 10x^2 = 360$ **8** $y^2 - 4x^2 = 1$

In problems 9 through 11, find the equation of the hyperbola that satisfies the conditions given.

9 Vertices at $(-4,0)$ and $(4,0)$, foci at $(-6,0)$ and $(6,0)$. **10** Vertices at $(0,-\frac{1}{2})$ and $(0,\frac{1}{2})$, foci at $(0,-1)$ and $(0,1)$.

11 Vertices at $(-4,0)$ and $(4,0)$, the equations of the asymptotes are $y = -\frac{5}{4}x$ and $y = \frac{5}{4}x$.

12 Determine the values of a^2 and b^2 so that the graph of the equation $b^2x^2 - a^2y^2 = a^2b^2$ contains the pair of points
(a) $(2,5)$ and $(3,-10)$ (b) $(4,3)$ and $(-7,6)$

In problems 13 through 20, find the coordinates of the center, the vertices, and the foci of each hyperbola. Also find the equations of the asymptotes and sketch the graph.

13 $\dfrac{(x-1)^2}{9} - \dfrac{(y+2)^2}{4} = 1$ **14** $\dfrac{(x+3)^2}{1} - \dfrac{(y-1)^2}{9} = 1$

15 $\dfrac{(y+1)^2}{16} - \dfrac{(x+2)^2}{25} = 1$ **16** $4x^2 - y^2 - 8x + 2y + 7 = 0$

17 $x^2 - 4y^2 - 4x - 8y - 4 = 0$ **18** $16x^2 - 9y^2 + 180y = 612$

19 $9x^2 - 25y^2 + 72x - 100y + 269 = 0$ **20** $9x^2 - 16y^2 - 90x - 256y = 223$

21 In parts (a), (b), and (c), find the equation of the hyperbola that satisfies the conditions given.
(a) Foci at $(1,-1)$ and $(7,-1)$, length of transverse axis is 2.
(b) Vertices at $(-4,3)$ and $(0,3)$, foci at $(-\frac{9}{2},3)$ and $(\frac{1}{2},3)$.
(c) Center at $(2,3)$, one vertex at $(2,8)$, and one focus at $(2,-3)$.

22 (a) Show that the length of a latus rectum of a hyperbola whose equation is
$$b^2x^2 - a^2y^2 = a^2b^2 \text{ is } 2b^2/a.$$
(b) Find the length of a latus rectum of the hyperbola $x^2 - 8y^2 = 16$.

In problems 23 through 26, find the equations of the tangent and normal lines to each hyperbola at the point indicated.

23 $x^2 - y^2 = 9$ at $(-5,4)$ **24** $4y^2 - x^2 = 7$ at $(3,-2)$

25 $x^2 - 4x - y^2 - 2y = 0$ at $(0,0)$ **26** $9(x+1)^2 - 16(y-2)^2 = 144$ at $(3,2)$

27 Find the equation of the hyperbola whose asymptotes are the given lines and which contains the given point.
(a) $y = -2x$ and $y = 2x$, $(1,1)$ (b) $y = -2x + 3$ and $y = 2x + 1$, $(1,4)$
(c) $y = x + 5$ and $y = -x + 3$, $(2,4)$

In problems 28 through 31, reduce each hyperbola equation to an equation of a "simpler" form by a suitable translation of axes. Also, sketch the graph showing both the xy and $\bar{x}\bar{y}$ axes.

28 $3x^2 - y^2 + 12x + 8y = 7$ **29** $4x^2 - 25y^2 + 24x + 50y + 22 = 0$

30 $5y^2 - 9x^2 + 10y + 54x - 112 = 0$ **31** $x^2 - 4y^2 - 4x - 8y - 4 = 0$

32 A point moves so that it is equidistant from the point $(2,0)$ and the circle of radius 3 units with center at $(-2,0)$. Describe the path of the point.

33 A point moves on the hyperbola $4x^2 - 9y^2 = 27$ with its abscissa increasing at the constant rate of 6 units/sec. How fast is the ordinate changing at the point $(3, 1)$?

34 A point moves so that the product of the slopes of the line segments that join it to two given points is 9. Describe the path of the point.

35 Find the shortest (minimum) distance from the point $(3, 0)$ to the hyperbola $y^2 - x^2 = 18$.

36 A region in the plane is bounded by the line $x = 8$ and the hyperbola $x^2 - y^2 = 16$. Find the dimensions of the rectangle of maximum area which can be inscribed in this region.

37 Let m be the slope of the tangent line to the hyperbola $x^2/a^2 - y^2/b^2 = 1$ at the point (x_0, y_0), where $x_0 > a$ and $y_0 > 0$. Find $\lim\limits_{x_0 \to +\infty} m$ and relate your answer to the asymptote $y = (b/a)x$.

38 It can be shown that the graph of the equation $xy = 1$ is a hyperbola with center at the origin and with the x and y axes as asymptotes (Figure 10). Find the coordinates of the foci of this hyperbola. (*Hint:* The transverse axis makes a 45° angle with the x axis.)

Figure 10

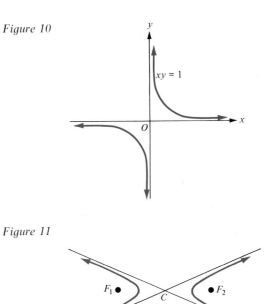

39 A hyperbola is said to be *equilateral* if its two asymptotes are perpendicular. Find the equation of an equilateral hyperbola with horizontal transverse axis and center at the origin. (Denote the distance from the center to a vertex by a.)

40 Sketch the graph of the hyperbola whose equation is

$$\frac{(y + b)^2}{b^2} - \frac{x^2}{2bp + p^2} = 1.$$

Show that, as $b \to +\infty$, the upper branch of this hyperbola approaches the parabola whose equation is $y = (1/4p)x^2$.

Figure 11

41 In Figure 11, hold the center C and the asymptotes fixed, but allow the foci F_1 and F_2 to move in toward the center. What happens to the hyperbola?

42 Give a proof of Theorem 1.

5 The Conic Sections

In the four preceding sections, we have derived the equations for the circle, the ellipse, the parabola, and the hyperbola from four special definitions—one for each type of curve. It is possible to give a unified geometric definition for these curves since they can all be obtained by cutting (or sectioning) a right circular cone of two nappes with a suitable plane (Figure 1).

Instead of going into the three-dimensional geometry required to prove that the conics can be obtained as in Figure 1, we give a different unified definition covering the ellipse, the parabola, and the hyperbola.

Figure 1

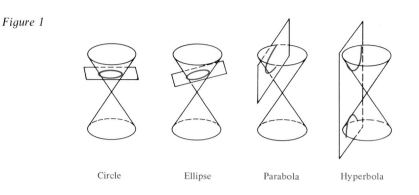

Circle   Ellipse   Parabola   Hyperbola

DEFINITION 1   **Conics in Terms of Focus, Directrix, and Eccentricity**

Figure 2

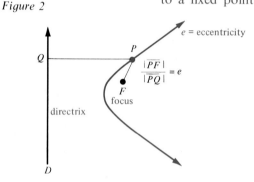

A *conic* is the set of all points P in the plane such that the distance from P to a fixed point F in the plane bears a constant ratio e to the distance from P to a fixed straight line D in the plane. The fixed point F, the fixed straight line D, and the constant number e are called the *focus*, the *directrix*, and the *eccentricity*, respectively, of the conic.

Thus, the point P belongs to the conic with focus F, directrix D, and eccentricity e, if and only if $\dfrac{|\overline{PF}|}{|\overline{PQ}|} = e$, where Q is the foot of the perpendicular from P to D (Figure 2).

EXAMPLE   Find the equation of the conic with eccentricity $e = 2$, whose focus is at the origin and whose directrix is given by $D\colon x = -3$.

SOLUTION

Figure 3

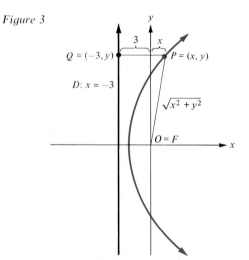

From Figure 3, the point $P = (x, y)$ belongs to the given conic if and only if $|\overline{PF}|/|\overline{PQ}| = 2$; that is, $\dfrac{\sqrt{x^2 + y^2}}{3 + x} = 2$, or $\sqrt{x^2 + y^2} = 6 + 2x$. Squaring both sides of the latter equation, we have

$$x^2 + y^2 = 36 + 24x + 4x^2$$

or

$$3x^2 + 24x - y^2 = -36.$$

Completing the square, we obtain

$$3(x^2 + 8x + 16) - y^2 = 3(16) - 36$$

or

$$3(x + 4)^2 - y^2 = 12;$$

that is, $\dfrac{(x + 4)^2}{4} - \dfrac{y^2}{12} = 1$. Therefore, the conic is a hyperbola with center at $(-4, 0)$.

Proceeding as in the example above, it can be shown that a conic, defined as in Definition 1, is an ellipse, a parabola, or a hyperbola (provided that its eccentricity is positive and that its focus does not lie on its directrix). In fact, we have the following theorem, whose proof is left as an exercise (Problems 22 and 24).

THEOREM 1   **Conic Theorem**

Suppose that a conic with focus F at the origin and directrix D: $x = -d$ has eccentricity e, where e and d are positive. Then, exactly one of the following holds:

Case i $e < 1$ and the conic is an ellipse with the equation

$$\frac{(x-c)^2}{a^2} + \frac{y^2}{b^2} = 1, \qquad \text{where } a = \frac{ed}{1-e^2},$$

$$b = \frac{ed}{\sqrt{1-e^2}}, \quad \text{and} \quad c = \sqrt{a^2 - b^2} = ae.$$

Case ii $e = 1$ and the conic is a parabola with the equation

$$4p(x + p) = y^2, \qquad \text{where } p = \frac{d}{2}.$$

Case iii $e > 1$ and the conic is a hyperbola with the equation

$$\frac{(x+c)^2}{a^2} - \frac{y^2}{b^2} = 1, \qquad \text{where } a = \frac{ed}{e^2-1},$$

$$b = \frac{ed}{\sqrt{e^2-1}}, \quad \text{and} \quad c = \sqrt{a^2 + b^2} = ae.$$

Consider Case i of Theorem 1, in which $e < 1$ and the conic is an ellipse with the equation

$$\frac{(x-c)^2}{a^2} + \frac{y^2}{b^2} = 1, \qquad \text{where } c = \sqrt{a^2 - b^2}.$$

This ellipse has its center at the point $C = (c, 0)$, a horizontal major axis, left focus $F_1 = (c - c, 0) = (0, 0)$, and right focus $F_2 = (c + c, 0) = (2c, 0)$ (Figure 4). Therefore, the given focus F really is the left focus F_1 of the ellipse. Since the ellipse is symmetric about its center C, it has a *second directrix* D_2 just as far to the right of C as the originally given directrix D was to the left of C; hence, the second directrix has the equation D_2: $x = 2c + d$. The second directrix D_2 is affiliated with the second focus F_2 in the same way that D is affiliated with $F = F_1$.

Consider Case ii of Theorem 1, in which $e = 1$ and the conic is a parabola with the equation $2d(x + d/2) = y^2$. This parabola has its vertex at $(-d/2, 0)$, and it opens to the right. Its focus is at $(-d/2 + d/2, 0) = (0, 0)$, and therefore coincides with F (Figure 5). The parabola has no second focus and no second directrix.

Consider Case iii of Theorem 1, in which $e > 1$ and the conic is a hyperbola with the equation

$$\frac{(x+c)^2}{a^2} - \frac{y^2}{b^2} = 1, \qquad \text{where } c = \sqrt{a^2 + b^2}.$$

This hyperbola has its center at the point $C = (-c, 0)$, a horizontal transverse axis, left focus $F_1 = (-c - c, 0) = (-2c, 0)$, and right focus $F_2 = (-c + c, 0) = (0, 0)$ (Figure 6). Therefore, the given focus F really is the right focus F_2 of the

Figure 4

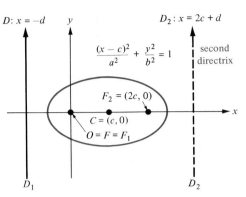

Figure 5

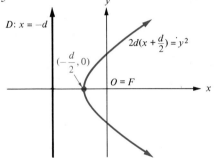

Figure 6

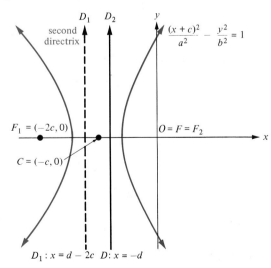

hyperbola. Since the hyperbola is symmetric about its center C, it has a *second directrix* D_1 affiliated with its left focus F_1 in the same way that the originally given directrix D is affiliated with the originally given focus $F = F_2$. Since D_1 is just as far to the left of the center C as D is to the right of C, the equation of D_1 is $D_1: x = d - 2c$.

EXAMPLES 1 Find the equation of the ellipse with focus F at the origin, directrix $D: x = -\frac{5}{2}$, and eccentricity $e = \frac{2}{3}$. Sketch the graph showing the second focus F_2 and the corresponding second directrix D_2.

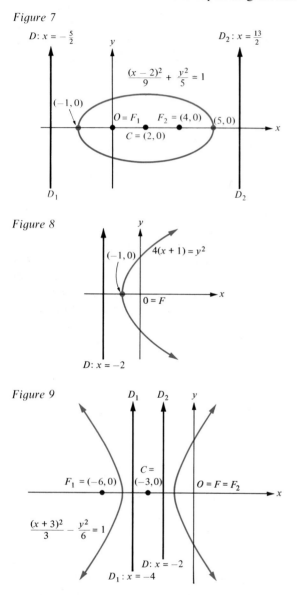

Figure 7

Figure 8

Figure 9

SOLUTION
Using Case i of Theorem 1 with $e = \frac{2}{3}$ and $d = \frac{5}{2}$, we have $a = ed/(1 - e^2) = 3$, $b = ed/\sqrt{1 - e^2} = \sqrt{5}$, $c = ae = 2$, and the equation is $(x - 2)^2/9 + y^2/5 = 1$. The focus $F_1 = F$ is at the origin, the second focus is $F_2 = (2c, 0) = (4, 0)$, and, since $2c + d = \frac{13}{2}$, the second directrix is $D_2: x = \frac{13}{2}$ (Figure 7).

2 Find the equation of the conic with eccentricity $e = 1$, directrix $D: x = -2$, and focus F at the origin. Sketch the graph.

SOLUTION
Using Case ii of Theorem 1 with $e = 1$ and $d = 2$, we see that the conic is a parabola with the equation $4(x + 1) = y^2$ (Figure 8). The focus is at the origin and the vertex is at the point $(-1, 0)$.

3 Find the equation of the hyperbola with focus F at the origin, directrix $D: x = -2$, and eccentricity $e = \sqrt{3}$. Sketch the graph showing the second focus F_1 and the corresponding second directrix D_1.

SOLUTION
Using Case iii of Theorem 1 with $e = \sqrt{3}$, $d = 2$, we have $a = ed/(e^2 - 1) = \sqrt{3}$, $b = ed/\sqrt{e^2 - 1} = \sqrt{6}$, $c = ae = 3$, and the equation is $(x + 3)^2/3 - y^2/6 = 1$. The center of this hyperbola is $C = (-3, 0)$, its left focus is $F_1 = (-2c, 0) = (-6, 0)$, and its right focus is $F_2 = F = O$ (Figure 9). Since $d - 2c = -4$, the second directrix, corresponding to the focus F_1, is $D_1: x = -4$.

Theorem 1 has a converse, since any ellipse, parabola, or hyperbola satisfies the focus–directrix–eccentricity definition of a conic (**Definition 1**). This is clear for the parabola because of its original definition; however, a simple calculation is required to find the directrices and the eccentricity of an ellipse or a hyperbola. Indeed, the equations in Case i of Theorem 1 give $a = ed/(1 - e^2)$ and $c = ae$, where $c = \sqrt{a^2 - b^2}$, and these equations can be solved for e and d in terms of a, b, and c as follows:

$$e = \frac{c}{a} \quad \text{and} \quad d = a\frac{1 - e^2}{e} = a\left(\frac{1}{e} - e\right) = a\left(\frac{a}{c} - \frac{c}{a}\right) = \frac{a^2}{c} - c = \frac{a^2 - c^2}{c} = \frac{b^2}{c}.$$

Similarly, the equations $a = ed/(e^2 - 1)$ and $c = ae$, where $c = \sqrt{a^2 + b^2}$, of Case iii can be solved for e and d to obtain $e = c/a$ and $d = b^2/c$ (Problem 29).

The preceding calculations provide a basis for the following assertions:

1 Any ellipse has two directrices D_1 and D_2 (affiliated with its two foci F_1 and F_2, respectively) and an eccentricity e with $0 < e < 1$ (Figure 10). If the ellipse has semimajor axis a and semiminor axis b, then the eccentricity is given by $e = c/a$, where $c = \sqrt{a^2 - b^2}$ is the distance from the center C to either focus. Each directrix is d units from its corresponding focus, where $d = b^2/c$, and the directrices are perpendicular to the major axis. The distance from the center C to either directrix is

$$c + d = c + \frac{b^2}{c} = \frac{c^2 + b^2}{c} = \frac{a^2}{c}$$

units.

Figure 10

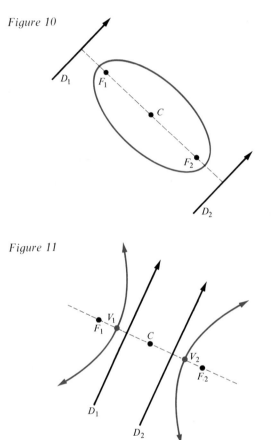

2 Any hyperbola has two directrices D_1 and D_2 (affiliated with its two foci F_1 and F_2, respectively) and an eccentricity e with $e > 1$ (Figure 11). If the distance from the center C of the hyperbola to either focus is c units, and if the vertices V_1 and V_2 are a units from the center, then the eccentricity is given by $e = c/a$, where $c = \sqrt{a^2 + b^2}$. Each directrix is d units from its corresponding focus, where $d = b^2/c$, and the directrices are perpendicular to the major axis. The distance from the center C to either directrix is

$$c - d = c - \frac{c^2 - a^2}{c} = \frac{a^2}{c}$$

units.

Figure 11

EXAMPLES In Examples 1 and 2, find the eccentricity and the equations of the directrices for the given conic. Sketch the graph.

Figure 12

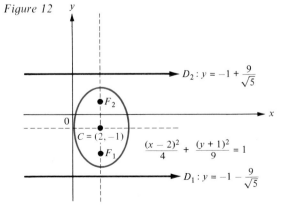

1 The ellipse $\dfrac{(x - 2)^2}{4} + \dfrac{(y + 1)^2}{9} = 1$.

SOLUTION

The major axis is vertical and the center is at $C = (2, -1)$. The semimajor axis is $a = 3$, the semiminor axis is $b = 2$, and the distance from the center to the foci is $c = \sqrt{9 - 4} = \sqrt{5}$. Hence, the eccentricity is $e = c/a = \sqrt{5}/3$, and the directrices are $a^2/c = 9/\sqrt{5}$ units from the center. The equations of the directrices are $D_1: y = -1 - (9/\sqrt{5})$ and $D_2: y = -1 + (9/\sqrt{5})$ (Figure 12).

Figure 13

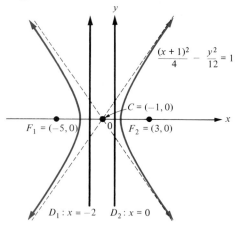

2   The hyperbola $\dfrac{(x+1)^2}{4} - \dfrac{y^2}{12} = 1$.

SOLUTION

The center is $C = (-1,0)$, the distance from the center to the vertices is $a = 2$, and the distance from the center to the foci is $c = \sqrt{4+12} = 4$. Hence, the eccentricity is $e = c/a = 2$, and the distance from the center to the directrices is $a^2/c = 1$. Thus, the equations of the directrices are $D_1: \; x = -1 - 1 = -2$ and $D_2: \; x = -1 + 1 = 0$ (Figure 13).

Problem Set 5

In problems 1 through 6, use Theorem 1 to find the equation of the conic whose focus F is at the origin, and whose eccentricity and directrix are given. Sketch the graph.

1 Eccentricity $e = \frac{2}{3}$, directrix $x = -\frac{5}{2}$.

2 Eccentricity $e = \frac{1}{2}$, directrix $x = -\frac{4}{5}$.

3 Eccentricity $e = 1$, directrix $x = -4$.

4 Eccentricity $e = 1$, directrix $x = -\frac{1}{3}$.

5 Eccentricity $e = 2$, directrix $x = -3$.

6 Eccentricity $e = \sqrt{5}$, directrix $x = -1$.

In problems 7 through 16, find the eccentricity and the equations of the directrices for each conic.

7 $\dfrac{x^2}{9} - \dfrac{y^2}{16} = 1$

8 $\dfrac{y^2}{4} - \dfrac{x^2}{2} = 1$

9 $2y^2 + 9x^2 = 18$

10 $\dfrac{(x+1)^2}{9} + \dfrac{(y+3)^2}{1} = 1$

11 $100x^2 + 36y^2 = 3600$

12 $(x-3)^2 - 4(y+2)^2 = 4$

13 $3x^2 - 5y^2 = 15$

14 $4x^2 - 3y^2 = 6(x+y)$

15 $(x+3)^2 = 24(y-2)$

16 $y^2 - 12y + 2x + 2 = 0$

17 Given the conic $x^2 + 12y^2 - 6x - 48y + 9 = 0$:
   (a) Find the coordinates of the center.
   (b) Find the coordinates of the vertices.
   (c) Find the equations of the directrices.
   (d) Find the eccentricity.
   (e) Sketch the graph.

18 Except for minor perturbations, the orbit of the earth is an ellipse with the sun at one focus. The least and greatest distances from the earth to the sun have a ratio $\frac{29}{30}$. Find the eccentricity of the elliptical orbit.

19 Suppose that the foci of a particular ellipse lie midway between the center and the vertices.
   (a) Find the eccentricity.
   (b) If the length of the major axis is $2a$, find the length of the minor axis and the distance from the center to the directrices.
   (c) Sketch the graph.

20 Show that the eccentricity e of the ellipse whose equation is $b^2x^2 + a^2y^2 = a^2b^2$, $a > b$, satisfies $b^2 = a^2(1 - e^2)$.

21 Show that the eccentricity e of the hyperbola whose equation is $b^2x^2 - a^2y^2 = a^2b^2$ satisfies $b^2 = a^2(e^2 - 1)$.

22 Let the focus F of a conic with eccentricity $e > 0$ be at the origin, and suppose that the directrix is D: $x = -d$, where $d > 0$. Using Definition 1 directly, prove that the equation of the conic is $(1 - e^2)x^2 - (2e^2 d)x + y^2 = e^2 d^2$. If $e = 1$, show that this equation becomes $y^2 = d^2 + (2d)x$. If $e \neq 1$, show that the equation can be rewritten as

$$(1 - e^2)\left(x - \frac{e^2 d}{1 - e^2}\right)^2 + y^2 = \frac{e^2 d^2}{1 - e^2}.$$

23 Find the equations of the directrices of the ellipse whose equation is

$$\frac{(x - h)^2}{b^2} + \frac{(y - k)^2}{a^2} = 1, \qquad \text{where } a > b > 0.$$

24 Using the results in problem 22, prove Theorem 1.

25 Let $a > b > 0$, and show that, if $b^2 > k > 0$, the ellipse having the equation

$$\frac{x^2}{a^2 - k} + \frac{y^2}{b^2 - k} = 1$$

has its foci at the points $(-\sqrt{a^2 - b^2}, 0)$ and $(\sqrt{a^2 - b^2}, 0)$. Find the equations of the directrices of this ellipse.

26 Sound travels with speed s in air and a bullet travels with speed b from a gun at $(-h, 0)$ to a target at $(h, 0)$ in the xy plane. At what points (x, y) can the boom of the gun and the ping of the bullet hitting the target be heard simultaneously?

27 If the two directrices of an ellipse are held fixed, and the eccentricity is decreased, what happens to the shape of the ellipse?

28 If the two directrices of a hyperbola are held fixed, and the eccentricity is increased, what happens to the shape of the hyperbola?

29 (a) Suppose that $a > 0$ and $b > 0$. Put $c = \sqrt{a^2 + b^2}$, $e = c/a$, and $d = b^2/c$. By direct calculation, show that $ed/(e^2 - 1) = a$ and that $ed/\sqrt{e^2 - 1} = b$.
  (b) Suppose that $e > 1$ and $d > 0$. Put $a = ed/(e^2 - 1)$, $b = ed/\sqrt{e^2 - 1}$, and $c = \sqrt{a^2 + b^2}$. By direct calculation, show that $c/a = e$ and that $b^2/c = d$.

Review Problem Set

In problems 1 through 10, find the equation of the circle that satisfies the conditions given.

1 $(-3, 5)$ and $(7, -3)$ are the endpoints of the diameter.

2 Center at $(4, -1)$ and containing the point $(-1, 3)$.

3 Center at $(4, 3)$ and containing the origin.

4 Radius is 8, center in quadrant I, and tangent to both axes.

5 Center is on the x axis and containing the points $(0, 4)$ and $(6, 8)$.

6 Tangent to the y axis and containing the points $(16, 12)$ and $(2, -2)$.

7 Tangent to both axes and containing the point $(18, -25)$.

8 Center at $(-2, 4)$ and tangent to the line $x - y - 6 = 0$.

9 Containing the points $(1, 2)$, $(3, 1)$, and $(-3, -1)$.

10 Containing the points $(-4, -3)$, $(-1, -7)$, and $(0, 0)$.

11 Find the equations of the tangent and normal lines to the following circles at the points given.
  (a) $x^2 + y^2 = 25$ at $(-3, -4)$ (b) $(x - 4)^2 + (y - 1)^2 = 9$ at $(7, 1)$

12 Find the dimensions of the rectangle of maximum area that can be cut from a semicircle of radius 4.

In problems 13 through 18, find the equation of the ellipse that satisfies the conditions given.

13 Center at $(0,0)$, one vertex at $(5,0)$, and one focus at $(3,0)$.

14 Center at $(0,0)$, containing the points $(4,3)$ and $(6,2)$, and with major axis on the x axis.

15 Major axis 16, minor axis 8, center at the origin, and vertices on the y axis.

16 Major axis 20, minor axis 12, center at the origin, and vertices on the x axis.

17 Vertices at $(0,0)$, $(10,0)$ and foci at $(1,0)$ and $(9,0)$.

18 Foci at $(3,-2)$ and $(9,-2)$, minor axis 8.

In problems 19 through 23, find the coordinates of the center, the vertices, and the foci of each ellipse. Also find the eccentricity and the equations of the directrices.

19 $\dfrac{x^2}{8} + \dfrac{y^2}{12} = 1$ **20** $144x^2 + 169y^2 = 24{,}336$

21 $9x^2 + 25y^2 + 18x - 50y - 191 = 0$ **22** $3x^2 + 4y^2 - 28x - 16y + 48 = 0$

23 $9x^2 + 4y^2 + 72x - 48y + 144 = 0$

24 Find the equations of the tangent and normal to the ellipse $x^2 + 3y^2 = 21$ at the point $(3,-2)$.

25 At what point(s) on the ellipse $16x^2 + 9y^2 = 400$ does y decrease at the same rate that x increases?

26 The lower base of an isosceles trapezoid is the major axis of an ellipse; the ends of the upper base are points on the ellipse. Show that the length of the upper base of the trapezoid of maximum area is half the length of the lower base.

27 Reduce each of the following equations to an equation of an ellipse in a "simpler" form by using a suitable translation.
(a) $16x^2 + y^2 - 32x + 4y - 44 = 0$ (b) $9x^2 + 4y^2 + 36x - 24y - 252 = 0$

28 A point P moves so that the product of the slopes of the line segments $\overline{PQ}$ and $\overline{PR}$, where $Q = (3,-2)$ and $R = (-2,1)$, is -6. Find the equation of the curve traced out by P and sketch it.

29 An arch in the form of a semiellipse has a span of 150 feet and a maximum height of 45 feet. There are two vertical supports equidistant from each other and the end of the arch. Find their heights.

In problems 30 through 34, find the equation of the hyperbola that satisfies the conditions given.

30 Vertices at $(3,-6)$, $(3,6)$ and foci at $(3,-10)$ and $(3,10)$.

31 Vertices at $(-2,3)$, $(6,3)$ and one focus at $(7,3)$.

32 Containing the point $(1,1)$ and with equations of asymptotes $y = -2x$ and $y = 2x$.

33 Center at $(0,0)$, transverse axis on the y axis, length of latus rectum 36, and distance between its foci 24.

34 Center at $(0,0)$, a focus at $(8,0)$, and a vertex at $(6,0)$.

In problems 35 through 39, find the coordinates of the center, the vertices, and the foci of each hyperbola. Also find the eccentricity and the equations of the asymptotes. Sketch the graph.

35 $x^2 - 9y^2 = 72$ **36** $y^2 - 9x^2 = 54$ **37** $x^2 - 4y^2 + 4x + 24y - 48 = 0$

38 $16x^2 - 9y^2 - 96x = 0$ **39** $4y^2 - x^2 - 24y + 2x + 34 = 0$

40 Find the equations of the tangents to the hyperbola $x^2 - y^2 + 16 = 0$ that are perpendicular to the line $5x + 3y - 15 = 0$.

41 Find the equations of the tangent and normal lines to the hyperbola $x^2 - 8y^2 = 1$ at the point $(3, 1)$.

42 Find the points on the hyperbola $x^2 - y^2 - 16 = 0$ that are nearest to the point $(0, 6)$.

43 A point moves on the hyperbola $x^2 - 4y^2 = 20$ in such a way that x is increasing at the rate of 3 units per second. Find the rate at which y is changing when the moving point passes through $(6, -2)$.

44 Two points are 2000 feet apart. At one of these points the report of a cannon is heard 1 second later than at the other. By means of the definition of a hyperbola, show that the cannon is somewhere on a certain hyperbola, and write its equation after making a suitable choice of axes. (Consider the velocity of sound to be 1100 ft/sec.)

In problems 45 through 48, find the equation of the parabola satisfying the given conditions. Sketch the graph.

45 Vertex at $(0, 0)$, focus on the x axis, and containing the point $(-2, 6)$.

46 Containing the points $(-2, 1)$, $(1, 2)$, and $(-1, 3)$ and whose axis is parallel to the x axis.

47 Vertex at $(2, 3)$, containing the point $(4, 5)$, and whose axis is parallel to the y axis.

48 Focus at $(6, -2)$ and whose directrix is the line $x - 2 = 0$.

49 Find the equation of the circle containing the vertex and the ends of the latus rectum of the parabola $y^2 = 8x$.

50 Find the equations of the tangent and normal lines to the following parabolas at the point given.
(a) $y^2 + 5x = 0$ at $(-5, 5)$ (b) $4y = 8 + 16x - x^2$ at $(1, \frac{23}{4})$

51 Find the dimensions of the rectangle of maximum area that can be inscribed in the segment of the parabola $x^2 = 24y$ cut off by the line $y = 6$.

52 Show that the line from the focus of the parabola $y^2 = 24x$ to the point where the tangent to the parabola at the point $(24, 24)$ cuts the y axis is perpendicular to the tangent.

53 At what point on the graph of the parabola $y = 4 + 2x - x^2$ is the tangent line parallel to the line $2x + y - 6 = 0$?

54 A parabola whose axis is parallel to the y axis contains the origin and is tangent to the line whose equation is $x - 2y = 8$ at the point $(20, 6)$. What is the equation of the parabola? Sketch the graph.

55 The cable of a suspension bridge assumes the shape of a parabola if the weight of the suspended roadbed (together with that of the cable) is uniformly distributed horizontally. Suppose that the towers of such a bridge are 240 feet apart and 60 feet high and that the lowest point of the cable is 20 feet above the roadway. Find the vertical distance from the roadway to the cables at intervals of 20 feet.

56 We know that the eccentricity $e = c/a < 1$ for the ellipse $x^2/a^2 + y^2/b^2 = 1$, where $c^2 = a^2 - b^2$. Now, as e comes closer and closer to zero in value, $c/a = e$ approaches zero; hence, c^2/a^2 approaches zero, so that a approaches b in value. (Why?) What is the shape of an ellipse in which a and b are close in value? What if $a = b$? Does this give an indication of how to define a conic with eccentricity $e = 0$? Use sketches and examples to answer these questions.

57 Suppose that e_1 is the eccentricity of the hyperbola $x^2/a^2 - y^2/b^2 = 1$ and e_2 is the eccentricity of the hyperbola $y^2/b^2 - x^2/a^2 = 1$. Prove that $1/e_1^2 + 1/e_2^2 = 1$.

5

ANTIDIFFERENTIATION, DIFFERENTIAL EQUATIONS, AND AREA

In the foregoing chapters we have seen that differentiation has applications to problems in areas ranging from geometry and physics to business and economics. In the present chapter we introduce *antidifferentiation,* which reverses the procedure of differentiation and allows us to find all functions that have a given function as their derivative. We show how antidifferentiation permits us to solve simple *differential equations* and we relate antidifferentiation to the problem of calculating the area of regions in the xy plane. However, before going into these matters, we first introduce the idea of *differentials,* since *differential notation* will be useful in connection with antidifferentiation.

1 Differentials

In Section 2.1 of Chapter 2 we introduced the Leibniz notation dy/dx for the derivative; however, we were careful to point out that dy/dx should not be regarded as being a fraction until the "numerator" dy and the "denominator" dx are given separate meanings. Shortly, we provide such meanings for the so-called "differentials" dx and dy.

Leibniz himself regarded dx and dy as being "infinitesimals," that is, quantities that, although they are nonzero, are smaller in magnitude than any finite quantity. He imagined that, in the limit, Δx and Δy somehow become "infinitesimal quantities" dx and dy, respectively, so that the difference quotient $\Delta y/\Delta x$ becomes the derivative dy/dx. Leibniz's point of view has persisted and, even today, some mathematicians and most engineers and scientists prefer to think of dx and dy as "infinitesimals."

Part of Leibniz's concept can be salvaged by regarding dy/dx as being an honest ratio, if not of "infinitesimals," then of "differentials" dy and dx, and by rewriting the equation $dy/dx = f'(x)$ as $dy = f'(x)\, dx$. The latter equation will serve as a *definition* of the differential dy, provided that an appropriate meaning can be given to the differential dx. This is accomplished by the following definition.

DEFINITION 1   **Differentials**

Let f be a function and let x and y be variables so related that $y = f(x)$. Then the *differential dx* is a quantity that can take on (or be assigned) any value in $\mathbb{R}$. If x is any number in the domain of f for which $f'(x)$ exists, then the *differential dy* is defined by

$$dy = f'(x)\, dx.$$

EXAMPLE   If $y = f(x) = 3x^2 - 2x + 1$, find dy.

SOLUTION
Here $f'(x) = 6x - 2$, so $dy = (6x - 2)\, dx$.

Notice the distinction between the differential dx of the independent variable x and the differential dy of the dependent variable y. Whereas dx can be given any value whatsoever, dy depends for its value on x and dx (not to mention f). For instance, in the example above, when $x = 1$ and $dx = 0.002$, we have $dy = (6 - 2)(0.002) = 0.008$.

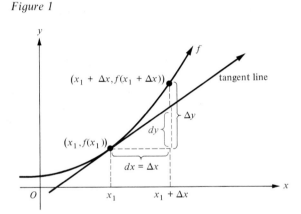

Figure 1

Figure 1 gives dy a geometric interpretation and compares it with Δy. Here, we suppose that f is differentiable at x_1, we set $dx = \Delta x$, and we show Δx as an increase in the value of x from x_1 to $x_1 + \Delta x$. Thus, Δy is the corresponding increase in the value of y from the value $f(x_1)$ to the value $f(x_1 + \Delta x)$ as determined by moving up along the graph of f. However, since $f'(x_1)$ is the slope of the tangent line to the graph of f at $(x_1, f(x_1))$, it follows that $dy = f'(x_1)\, dx$ gives the corresponding increase in the value of y as determined by moving up along the *tangent line*.

From Figure 1 it is clear that dy is a good approximation to Δy, provided that $dx = \Delta x$ and that Δx is sufficiently small. A precise statement and rigorous proof of the fact that

$$dy \approx \Delta y \quad \text{if} \quad dx = \Delta x \quad \text{and} \quad \Delta x \text{ is "small"}$$

can be given using the linear approximation theorem in Section 7 of Chapter 2 (Problem 32). p.127

EXAMPLES 1 Let $y = f(x) = 3x^2 - 2x + 4$ and put $x_1 = 1$, $dx = \Delta x = 0.02$. (a) Calculate $\Delta y = f(x_1 + \Delta x) - f(x_1)$ exactly, (b) find an estimate for Δy using $dy = f'(x_1)\, dx$, and (c) determine the error $\Delta y - dy$ involved in this estimate.

SOLUTION
(a) $f(x_1) = f(1) = 3(1)^2 - 2(1) + 4 = 5$ and $f(x_1 + \Delta x) = f(1.02) = 3(1.02)^2 - 2(1.02) + 4 = 5.0812$; hence, $\Delta y = 5.0812 - 5 = 0.0812$.
(b) Since $f'(x) = 6x - 2$, it follows that
$dy = f'(x_1)\, dx = f'(1)\, \Delta x = [6(1) - 2](0.02) = 0.08$.
(c) The error is given by $\Delta y - dy = 0.0812 - 0.08 = 0.0012$.

2 Use differentials to estimate $\sqrt{35}$.

SOLUTION

Let $y = \sqrt{x}$ and let Δy be the change in the value of y caused by decreasing the value of x from 36 to 35. Put $dx = \Delta x = 35 - 36 = -1$. Thus, $\sqrt{35} = \sqrt{36} + \Delta y = 6 + \Delta y$. Since $\dfrac{dy}{dx} = \dfrac{d}{dx}\sqrt{x} = \dfrac{1}{2\sqrt{x}}$, it follows that, when $x = 36$, we have $dy/dx = 1/(2\sqrt{36}) = \frac{1}{12}$. Therefore,

$$\Delta y \approx dy = \frac{dy}{dx}\,dx = \frac{dy}{dx}\,\Delta x = \left(\frac{1}{12}\right)(-1) = \frac{-1}{12},$$

so $\sqrt{35} = 6 + \Delta y \approx 6 - \frac{1}{12} \approx 5.9167$.

3 The radius of a spherical steel ball is measured to be 1.5 centimeters and it is known that the error involved in this measurement does not exceed 0.1 centimeter. The volume V of the ball is calculated from its measured radius using the standard formula $V = \frac{4}{3}\pi r^3$. Estimate the possible error in the calculated volume.

SOLUTION

The true value of the radius is $1.5 + \Delta r$, where Δr is the measurement error. We are given that $|\Delta r| \leq 0.1$. The true value of the volume is $\frac{4}{3}\pi(1.5 + \Delta r)^3$, while the value of the volume calculated from the measurement of the radius is $\frac{4}{3}\pi(1.5)^3$. The difference $\Delta V = \frac{4}{3}\pi(1.5 + \Delta r)^3 - \frac{4}{3}\pi(1.5)^3$ represents the error in the calculated volume. We put $dr = \Delta r$ and estimate ΔV by dV as follows: Note that

$$\frac{dV}{dr} = \frac{d}{dr}\left(\frac{4}{3}\pi r^3\right) = 4\pi r^2.$$

Consequently,

$$\Delta V \approx dV = \frac{dV}{dr}\,dr = \frac{dV}{dr}\,\Delta r = 4\pi r^2\,\Delta r = 4\pi(1.5)^2\,\Delta r = 9\pi\,\Delta r.$$

Therefore, $|\Delta V| \approx |9\pi\,\Delta r| = 9\pi|\Delta r| \leq 9\pi(0.1) = 0.9\pi$, so the possible error is bounded in absolute value by about $0.9\pi \approx 2.8$ cubic centimeters.

4 Use differentials to find the approximate volume of a right circular cylindrical shell (Figure 2) of height 6 inches whose inner radius is 2 inches and whose thickness is $\frac{1}{10}$ inch.

SOLUTION

The volume of a right circular cylinder is its height times the area of its base. If V denotes the volume of a (solid) cylinder of height 6 inches and radius r, then $V = 6\pi r^2$. Figure 2 shows a cylinder of height 6 inches and radius 2 inches inside a larger cylinder of height 6 inches and radius $r + \Delta r = 2 + \frac{1}{10} = 2.1$ inches. The difference ΔV in the volumes of these two cylinders is the required volume of the cylindrical shell. We put $dr = \Delta r = \frac{1}{10}$ and use the approximation

$$\Delta V \approx dV = \frac{dV}{dr}\,dr = \frac{d}{dr}\left(6\pi r^2\right)\,dr = 12\pi r\,dr = 12\pi(2)(\tfrac{1}{10}) = \tfrac{12}{5}\pi.$$

Thus, the volume of the shell is approximately $\frac{12}{5}\pi \approx 7.5$ cubic inches.

Figure 2

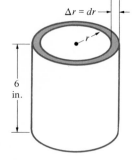

1.1 Formulas Involving Differentials

Since $dy = \dfrac{dy}{dx} dx$, it follows that any formula which gives the derivative $\dfrac{dy}{dx}$ can be converted into a formula which gives the differential dy simply by multiplying the derivative by the differential dx. For instance, if $y = u + v$, then $\dfrac{dy}{dx} = \dfrac{du}{dx} + \dfrac{dv}{dx}$, so $dy = \left(\dfrac{du}{dx} + \dfrac{dv}{dx}\right) dx = du + dv$; that is, $d(u + v) = du + dv$. In the following table, some of the standard formulas for derivatives have been converted into formulas for differentials. In these formulas u and v are variables that are assumed to be differentiable functions of x, and c represents a constant function or a constant number as the context indicates. Also, n represents a constant rational exponent.

	Derivatives		Differentials
1	$\dfrac{dc}{dx} = 0$	I	$dc = 0$
2	$\dfrac{d(cu)}{dx} = c\dfrac{du}{dx}$	II	$d(cu) = c\,du$
3	$\dfrac{d(u + v)}{dx} = \dfrac{du}{dx} + \dfrac{dv}{dx}$	III	$d(u + v) = du + dv$
4	$\dfrac{d(uv)}{dx} = u\dfrac{dv}{dx} + v\dfrac{du}{dx}$	IV	$d(uv) = u\,dv + v\,du$
5	$\dfrac{d\left(\dfrac{u}{v}\right)}{dx} = \dfrac{v\dfrac{du}{dx} - u\dfrac{dv}{dx}}{v^2}$	V	$d\left(\dfrac{u}{v}\right) = \dfrac{v\,du - u\,dv}{v^2}$
6	$\dfrac{d(u^n)}{dx} = nu^{n-1}\dfrac{du}{dx}$	VI	$d(u^n) = nu^{n-1}\,du$
7	$\dfrac{d(cu^n)}{dx} = ncu^{n-1}\dfrac{du}{dx}$	VII	$d(cu^n) = ncu^{n-1}\,du$
8	$\dfrac{d(cx^n)}{dx} = ncx^{n-1}$	VIII	$d(cx^n) = ncx^{n-1}\,dx$

EXAMPLES 1 Let $y = 47x^3 - 21x^2 + 3x^{-1}$. Find dy.

SOLUTION

$$dy = 141x^2\,dx - 42x\,dx - 3x^{-2}\,dx \quad \text{or} \quad dy = (141x^2 - 42x - 3x^{-2})\,dx.$$

2 Let $y = \sqrt{3 - x^5}$. Find dy.

SOLUTION

$$dy = \frac{d(3 - x^5)}{2\sqrt{3 - x^5}} = \frac{-5x^4}{2\sqrt{3 - x^5}}\,dx.$$

3 Let x and y be functions of a third variable t and suppose that $x^3 + 4x^2y + y^3 = 2$. Find the relationship between dx and dy.

SOLUTION

"Taking the differential" on both sides of the given equation, we obtain

$$d(x^3 + 4x^2y + y^3) = d(2) = 0 \quad \text{or} \quad d(x^3) + 4d(x^2y) + d(y^3) = 0.$$

Since $\quad d(x^3) = 3x^2\,dx, \qquad d(x^2 y) = x^2\,dy + y\,d(x^2) = x^2\,dy + y(2x\,dx), \qquad$ and $d(y^3) = 3y^2\,dy$, it follows that

$$3x^2\,dx + 4x^2\,dy + 8yx\,dx + 3y^2\,dy = 0 \quad \text{or} \quad (3x^2 + 8yx)\,dx + (4x^2 + 3y^2)\,dy = 0.$$

1.2 A Remark on Differentials and the Chain Rule

Suppose that u is a function of v and, in turn, v is a function of x. According to the chain rule, we have

$$\frac{du}{dx} = \frac{du}{dv}\frac{dv}{dx},$$

provided that the derivatives du/dv and dv/dx exist. We pointed out in Section 4 of Chapter 2 that the Leibniz notation makes the chain rule—a rather deep and important fact—*look* obvious. Now that du/dx, du/dv, and dv/dx are actual fractions, can we not relegate the chain rule to an algebraic triviality? The answer is still *no*, since the differential dv in du/dv is the differential of v *regarded as an independent variable* (upon which u depends), while the differential dv in dv/dx is the differential of v *regarded as a dependent variable* (depending, in fact, on x). Because of this distinction between the dv in du/dv and the dv in dv/dx, we cannot conclude that $\dfrac{du}{dx} = \dfrac{du}{dv}\dfrac{dv}{dx}$ on purely algebraic grounds. It is the chain rule that justifies the algebraic manipulation rather than vice versa.

In casual calculation with differentials, one does not bother to distinguish between differentials of dependent and independent variables. Miraculously, such carelessness rarely causes any difficulty. The "miracle," of course, is really the chain rule!

Problem Set 1

In problems 1 through 6, set dx equal to the given value of Δx and use the indicated value of x_1 to calculate (a) $\Delta y = f(x_1 + \Delta x) - f(x_1)$, (b) $dy = f'(x_1)\,dx$, and (c) $\Delta y - dy$.

1 f is defined by $y = 3x^2 + 1$, $x_1 = 1$, and $\Delta x = 0.1$.

2 f is defined by $y = -5x^2 + x$, $x_1 = 2$, and $\Delta x = 0.02$.

3 f is defined by $y = -2x^2 + 4x + 1$, $x_1 = 2$, and $\Delta x = 0.4$.

4 f is defined by $y = 2x^3 + 5$, $x_1 = -1$, and $\Delta x = 0.05$.

5 f is defined by $y = 9/\sqrt{x}$, $x_1 = 9$, and $\Delta x = -1$.

6 f is defined by $y = \dfrac{3}{x + 4}$, $x_1 = 3$, and $\Delta x = -2$.

In problems 7 through 14, find dy in terms of x and dx.

7 $y = 5 - 4x + x^3$

8 $y = \dfrac{3 - x^2}{\sqrt{x}}$

9 $y = \sqrt{9 - 3x^2}$

10 $y = x^4 \sqrt[5]{x^2 + 2}$

11 $y = \sqrt{\dfrac{x - 3}{x + 3}}$

12 $y = \dfrac{x}{\sqrt{x^2 + 5}}$

13 $y = \dfrac{\sqrt{3x + 1}}{x^2 + 7}$

14 $y = \dfrac{x^3}{\sqrt[3]{x + 1}}$

In problems 15 through 22, use differentials to approximate each expression.

15 $\sqrt{101}$

16 $\sqrt[3]{28}$

17 $\sqrt[4]{17}$

18 $\sqrt[3]{27.5}$

19 $\dfrac{1}{\sqrt{10}}$

20 $\dfrac{1}{\sqrt[5]{31}}$

21 $\sqrt{0.041}$

22 $\sqrt[3]{0.000063}$

In problems 23 through 31, suppose that x and y are functions of t which are related as shown. Find the relation between dx and dy by "taking the differential" on both sides of each equation.

23 $x^2 + y^2 = 36$

24 $9x^2 = 36 - 4y^2$

25 $9x^2 - 16y^2 = 144$

26 $x^{2/3} + y^{2/3} = 4$

27 $\sqrt[3]{x} = 2 - \sqrt[3]{y}$

28 $x^2 + xy + 3y^2 = 51$

29 $2x^3 + 5y^3 = 13 + 4xy^2 - xy$

30 $\sqrt{1 - x} + \sqrt{1 - y} = 9$

31 $x^3 + y^3 = \sqrt[3]{x + y}$

32 Assume that the function f is differentiable at the number x_1 and that $x_1 + \Delta x$ belongs to the domain of f. Put $dx = \Delta x$, $dy = f'(x_1)\,dx$. Use the linear approximation theorem (Section 7 of Chapter 2, Theorem 1 on page 127) to prove that dy is a good approximation to $\Delta y = f(x_1 + \Delta x) - f(x_1)$ provided that Δx is sufficiently small. (*Hint:* In the linear approximation theorem, put $x = x_1 + \Delta x = x_1 + dx$.)

33 Find the approximate volume of a spherical shell whose inner radius is 3 inches and whose thickness is $\frac{3}{32}$ inch.

34 The volume of a sphere of radius r is $V = \frac{4}{3}\pi r^3$. Notice that $dV/dr = 4\pi r^2$. The surface area of a sphere of radius r is $A = 4\pi r^2$. Is it just an "accident" that the same expression, $4\pi r^2$, shows up as dV/dr and as A?

35 The edge of a cube is measured to be 10 centimeters with a possible error of 0.02 centimeter. Use differentials to find an approximate upper bound for the error involved in calculating its volume to be $10^3 = 1000$ cubic centimeters.

36 The region between two concentric circles in the plane is called an *annulus*. Find:
(a) The exact area of an annulus with inner radius 5 inches and outer radius $\frac{81}{16}$ inches.
(b) An approximation to the exact area found in part (a) by using differentials.
(c) The error involved in the approximation above.

37 The altitude of a certain right circular cone is a constant equal to 2 meters. If the radius of its base is increased from 100 centimeters to 105 centimeters, find the approximate increase in the volume of the cone.

38 A particle moves along a straight line in accordance with the equation $s = \frac{1}{3}t^3 - 2t + 3$, where t is the elapsed time in seconds and s is the directed distance, measured in feet, from the origin to the particle. Find the approximate distance covered by the particle in the interval from $t = 2$ to $t = 2.1$ seconds.

39 Approximately how many cubic centimeters of chromium plate must be applied to coat the lateral surface of a cylindrical rod of radius 2.34 centimeters to a thickness of 0.01 centimeter, if the rod is 30 centimeters long?

40 The period of oscillation of a pendulum of length L units is given by $T = 2\pi\sqrt{L/g}$, where g is the acceleration of gravity in units of length per second per second and T is in seconds. Find the approximate percent that the pendulum in a grandfather clock should be lengthened if the clock gains 3 minutes in 24 hours.

41 The attractive force between unlike electrically charged particles is expressed by $F = k/x^2$, where x is the distance between the particles and k is a certain constant. If x is increased by 2 percent, find the approximate percent decrease in F.

42 The total cost in dollars of producing x toys is $C = \dfrac{x^3}{15{,}000} - \dfrac{3x^2}{100} + 11x + 75$, and each toy is sold at \$10.
   (a) Find the total profit P as a function of x.
   (b) Find dP in terms of x and dx.
   (c) When the production level changes from $x = 350$ to $x = 355$, what is the approximate change in P?

43 The law for adiabatic expansion of a certain gas is $PV^{1.7} = C$, where V is the volume of the gas, P is the pressure of the gas, and C is a constant. Derive the equation $\dfrac{dP}{P} + \dfrac{1.7\,dV}{V} = 0$.

2   Antiderivatives

In applied mathematics it often happens that we know the derivative of a function and would like to find the function itself. For instance, we might know the velocity ds/dt of a particle and be required to find its equation of motion $s = f(t)$, or we might wish to find the revenue function for a certain product when we know the marginal revenue. The solutions of such problems require us to "undo" the operation of differentiation; that is, we are required to *anti-differentiate*.

If f and g are functions such that $g' = f$, we say that g is an *antiderivative* of f. Thus, $g(x) = x^2$ is an antiderivative of $f(x)$, since $D_x(x^2) = 2x$. If C is any constant, then the function defined by $y = x^2 + C$ is also an antiderivative of f, since $D_x(x^2 + C) = 2x$. This is geometrically clear, since the graph of $y = x^2 + C$ is obtained by translating the graph of $y = x^2$ vertically by C units, and this does not change the slope of the tangent line for a given value of x (Figure 1).

In general, we define antiderivatives as follows.

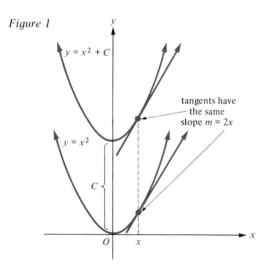

Figure 1

$y = x^2 + C$

$y = x^2$

tangents have the same slope $m = 2x$

C

DEFINITION 1   **Antiderivative**

A function g is called an *antiderivative* of a function f on a set of numbers I if $g'(x) = f(x)$ holds for every value of x in I. The procedure of finding anti-derivatives is called *antidifferentiation*.

If we assert that g is an antiderivative of f without explicitly mentioning the set I in Definition 1, then it is to be understood that I is the entire domain of f, so that $g'(x) = f(x)$ holds for all values of x in the domain of f.

EXAMPLE   Show that $g(x) = \dfrac{x+1}{x-1}$ is an antiderivative of $f(x) = \dfrac{-2}{(x-1)^2}$.

SOLUTION

$$g'(x) = \frac{(x-1)-(x+1)}{(x-1)^2} = \frac{-2}{(x-1)^2}.$$

Notice that, if g is an antiderivative of f on a set I, then so is $g + C$, where C is any constant. The reason is that, if $D_x g(x) = f(x)$, then $D_x[g(x) + C] = D_x g(x) + D_x C = f(x) + 0 = f(x)$. Therefore, as soon as we find one antiderivative g of a function f, we automatically have an infinite number of antiderivatives of f, namely, all functions of the form $g + C$, where C is any constant.

EXAMPLE   Given that $g(x) = \dfrac{x}{1+x}$ is an antiderivative of $f(x) = \dfrac{1}{(1+x)^2}$, find an infinite number of additional antiderivatives of f.

SOLUTION
Let C be any constant, and define $h = g + C$; that is,

$$h(x) = \frac{x}{1+x} + C = \frac{x + C + Cx}{1+x}.$$

Any such function h is an antiderivative of f.

Since the derivative of a constant function is the zero function, it follows that any constant function is an antiderivative of the zero function. The following important theorem shows, conversely, that the constant functions are the *only* antiderivatives of the zero function.

THEOREM 1   **Antidifferentiation of the Zero Function**
Let g be a function such that $g'(x) = 0$ holds for all values of x in some open interval I. Then g has a constant value on I.

PROOF
It is enough to prove that the value of g at a number a in I is the same as the value of g at every other number b in I. By the mean value theorem (Chapter 3, Section 1.2), there exists a number c between a and b such that

$$g(b) - g(a) = g'(c)(b - a) = 0(b - a) = 0.$$

Thus, $g(a) = g(b)$, and the theorem is proved.

The following theorem, which is a direct consequence of Theorem 1, tells us how to find *all* the antiderivatives of a function on an open interval, provided that we know one such antiderivative.

THEOREM 2   **Antidifferentiation on an Open Interval**
Suppose that g is an antiderivative of the function f on the open interval I. Then, a function h with domain I is an antiderivative of f on I if and only if $h = g + C$ for some constant C.

PROOF

If $h = g + C$, then $h' = g' = f$, so that h is an antiderivative of f on I. Conversely, suppose that h is an antiderivative of f on I. Then, the function $h - g$ satisfies $(h - g)' = h' - g' = f - f = 0$ on the open interval I. It follows from Theorem 1 that there exists a constant C such that $h - g = C$; that is, $h = g + C$.

EXAMPLE   Given that the function $g(x) = \frac{1}{2}x^2$ is an antiderivative of the function $f(x) = x$, find all antiderivatives of f.

SOLUTION

Here, the interval I is $\mathbb{R}$. By Theorem 2, the antiderivatives of f are all functions h of the form $h(x) = \frac{1}{2}x^2 + C$, where C is a constant.

2.1   Notation for Antiderivatives

Antiderivatives are traditionally written using a special symbolism which has some of the same advantages as the Leibniz notation for derivatives and which, in fact, was used by Leibniz himself. This symbolism can be understood by thinking of the differential dy as an "infinitesimal bit of y" and imagining that y is the "sum" of all these "infinitesimals." Leibniz used a stylized letter s, written $\int$, for such "summations," so that $y = \int dy$ would symbolize the idea that "y is the sum of all its own differentials." Johann Bernoulli, a contemporary of Leibniz, suggested that the process of assembling infinitesimals so as to get a whole or complete quantity as expressed by $y = \int dy$ should properly be called *integration* rather than summation. People accepted Bernoulli's suggestion; hence, the symbol $\int$ is referred to as the *integral sign*.

Now suppose that g is an antiderivative of f, so that $g' = f$. If we let $y = g(x)$, then $dy = g'(x)\,dx = f(x)\,dx$, so that

$$y = \int dy = \int f(x)\,dx; \quad \text{that is,} \quad g(x) = \int f(x)\,dx.$$

If C is any constant, then $g(x) + C$ is also an antiderivative of f; therefore, we make the following definition.

DEFINITION 2   **Integral Notation for Antiderivatives**

The notation $\int f(x)\,dx = g(x) + C$, where C denotes an arbitrary constant, means that the function g is an antiderivative of the function f, so that $g'(x) = f(x)$ holds for all values of x in the domain of f.

Naturally, if I is a set of numbers, then the assertion that $\int f(x)\,dx = g(x) + C$ on I (or *for x in I*) means that g is an antiderivative of f on I. In Definition 2 the constant C is called the *constant of integration*, the symbol $\int$ is called the *integral sign*, and the function f [or the expression $f(x)$] is called the *integrand* of the expression $\int f(x)\, dx$. One also says that $f(x)$ is *under the integral sign* in the expression $\int f(x)\, dx$. The procedure of calculating $\int f(x)\, dx$, that is, of finding $g(x) + C$, is called *indefinite integration*. The adjective "indefinite" is presumably used because the constant C can have any value and therefore is not definitely determined by the function f. Because of the arbitrary nature of the constant C, $\int f(x)\, dx$, which is called the *indefinite integral* of the function f, does not represent a particular quantity or function; hence, care should be taken in manipulating this expression.

In order to verify an assertion of the form $\int f(x)\, dx = g(x) + C$ [or $\int f(x)\, dx = g(x) + C$ on I], it is only necessary to check that $g'(x) = f(x)$ for all values of x in the domain of f (or for all values of x in I).

EXAMPLES   Verify the given equation.

1 $\int x^2\, dx = \tfrac{1}{3}x^3 + C$

SOLUTION

$$D_x(\tfrac{1}{3}x^3) = x^2.$$

2 $\int dx = x + C$

SOLUTION

$$\int dx = \int 1\, dx = x + C, \text{ since } D_x(x) = 1.$$

2.2 Basic Rules for Antidifferentiation

Since antidifferentiation (or indefinite integration) "undoes" differentiation, each rule or formula for differentiation should yield a corresponding rule for anti-differentiation when "read backwards." Some of these rules are as follows.

Basic Rules for Antidifferentiation

1 $D_x \int f(x)\, dx = f(x).$

2 $\int f'(x)\, dx = f(x) + C.$

3 $\int dx = x + C.$

4 *Power Rule*: If n is a rational number other than -1, then

$$\int x^n \, dx = \frac{x^{n+1}}{n+1} + C.$$

5 *Homogeneous Rule*: If a is a constant, then $\int af(x) \, dx = a \int f(x) \, dx.$

6 *Additive Rule*: $\int [f(x) + g(x)] \, dx = \int f(x) \, dx + \int g(x) \, dx.$

7 *Linear Rule*: $\int [a_1 f(x) + a_2 g(x)] \, dx = a_1 \int f(x) \, dx + a_2 \int g(x) \, dx$
if a_1 and a_2 are constants.

8 *General Linear Rule*: If $a_1, a_2, \ldots, a_m$ are constants, then

$$\int [a_1 f_1(x) + a_2 f_2(x) + \cdots + a_m f_m(x)] \, dx$$

$$= a_1 \int f_1(x) \, dx + a_2 \int f_2(x) \, dx + \cdots + a_m \int f_m(x) \, dx.$$

All eight of the basic rules work on any prescribed set of numbers I provided that the functions involved have antiderivatives on I.

EXAMPLE   Use the basic rules to evaluate $\int (5x + \sqrt{x} - 7) \, dx.$

SOLUTION

$$\int (5x + \sqrt{x} - 7) \, dx = \int 5x \, dx + \int \sqrt{x} \, dx + \int (-7) \, dx \qquad \text{(Rule 6)}$$

$$= 5 \int x \, dx + \int x^{1/2} \, dx - 7 \int dx \qquad \text{(Rule 5)}$$

$$= 5\left(\frac{x^{1+1}}{1+1} + C_1\right) + \left(\frac{x^{(1/2)+1}}{\frac{1}{2}+1} + C_2\right) - 7(x + C_3) \qquad \text{(Rule 4)}$$

$$= \tfrac{5}{2}x^2 + \tfrac{2}{3}x^{3/2} - 7x + 5C_1 + C_2 - 7C_3$$

$$= \tfrac{5}{2}x^2 + \tfrac{2}{3}x^{3/2} - 7x + C,$$

where $C = 5C_1 + C_2 - 7C_3$.

In practice, when the basic rules are used to antidifferentiate (that is, to evaluate indefinite integrals), the individual constants of integration that arise are immediately combined into one constant. Thus, the solution given above would ordinarily be condensed as follows:

$$\int (5x + \sqrt{x} - 7) \, dx = 5 \int x \, dx + \int x^{1/2} \, dx - 7 \int dx = \tfrac{5}{2}x^2 + \tfrac{2}{3}x^{3/2} - 7x + C.$$

EXAMPLES   Evaluate the given antiderivative.

1 $\int \dfrac{x^4 + 3x^2 + 5}{x^2} \, dx$

SOLUTION

$$\int \frac{x^4 + 3x^2 + 5}{x^2}\, dx = \int \left(\frac{x^4}{x^2} + 3\,\frac{x^2}{x^2} + \frac{5}{x^2} \right) dx = \int (x^2 + 3 + 5x^{-2})\, dx$$

$$= \int x^2\, dx + 3 \int dx + 5 \int x^{-2}\, dx$$

$$= \frac{x^3}{3} + 3x + 5\,\frac{x^{-1}}{-1} + C = \frac{x^3}{3} + 3x - \frac{5}{x} + C.$$

2 $\int (y\sqrt[3]{y} + 1)^2\, dy$

SOLUTION

$$\int (y\sqrt[3]{y} + 1)^2\, dy = \int (y^{4/3} + 1)^2\, dy = \int (y^{8/3} + 2y^{4/3} + 1)\, dy$$

$$= \int y^{8/3}\, dy + 2 \int y^{4/3}\, dy + \int dy$$

$$= \frac{y^{11/3}}{\frac{11}{3}} + \frac{2y^{7/3}}{\frac{7}{3}} + y + C = \frac{3y^{11/3}}{11} + \frac{6y^{7/3}}{7} + y + C.$$

2.3 Change of Variable (Substitution)

In order to evaluate $\int x(x^2 + 5)^{100}\, dx$ using only the basic rules, it would be necessary to expand $(x^2 + 5)^{100}$ by the binomial theorem, multiply through by x, and then antidifferentiate the resulting expression term by term using the general linear rule. The required calculation could be quite tedious, to say the least. Fortunately, there is a simpler way to proceed, namely, by "changing the variable" from x to a new variable $u = x^2 + 5$. Notice that, if $u = x^2 + 5$, then we have $du = 2x\, dx$, or, $x\, dx = \frac{1}{2}\, du$, and

$$\int x(x^2 + 5)^{100}\, dx = \int (x^2 + 5)^{100} x\, dx = \int u^{100}\,\frac{1}{2}\, du = \frac{1}{2} \int u^{100}\, du = \frac{1}{2}\,\frac{u^{101}}{101} + C.$$

Substitution of $u = x^2 + 5$ into the latter expression gives

$$\int x(x^2 + 5)^{100}\, dx = \frac{(x^2 + 5)^{101}}{202} + C.$$

The method illustrated here is called *change of variable* or *substitution*.

There is more going on than meets the eye in the calculation above, since the integral notation has compressed what could otherwise be a complicated argument (using the chain rule) into a few simple calculations. In the same way that the basic rules for antidifferentiation are obtained by "reading the basic rules for differentiation backwards," the method of substitution (or change of variable) is just the chain rule "read backwards" (Problem 48). This method of integration is carried out according to the following procedure.

Procedure for Evaluating $\int f(x)\,dx$ by Change of Variable (Substitution)

Step 1 Find a portion of the integrand $f(x)$ that is especially "prominent" in the sense that, if it were replaced by a single new variable, say u, then the integrand would be noticeably simpler. Set u equal to this portion. The resulting equation will have the form $u = g(x)$.

Step 2 Using the equation $u = g(x)$ obtained in Step 1, find the differential du. The resulting equation will have the form $du = g'(x)\,dx$.

Step 3 Using the two equations $u = g(x)$ and $du = g'(x)\,dx$ obtained in Steps 1 and 2, rewrite the entire integrand, *including dx*, in terms of u and du only.

Step 4 Evaluate the resulting indefinite integral in terms of u.

Step 5 Using the equation $u = g(x)$ of Step 1, rewrite the answer obtained in Step 4 in terms of the original variable x.

There is never a guarantee of success when the method of substitution is used—one can only try and see what happens. After the algebraic manipulations required in Step 3 are carried out, it may happen that the resulting integral (involving u) is more complicated than the original integral (involving x). However, if the procedure fails for one choice of u, it may still work for another—just try again.

EXAMPLES   Use the procedure for change of variable (substitution) and the basic rules to evaluate the given antiderivative (indefinite integral).

1 $\displaystyle\int \sqrt{7x + 2}\,dx$

SOLUTION
Let $u = 7x + 2$. Then $du = 7\,dx$. It follows that $dx = \frac{1}{7}\,du$ and

$$\int \sqrt{7x + 2}\,dx = \int \sqrt{u}\cdot\frac{1}{7}\,du = \frac{1}{7}\int u^{1/2}\,du$$

$$= \frac{1}{7}\frac{u^{3/2}}{\frac{3}{2}} + C = \frac{2}{21}\,(7x + 2)^{3/2} + C.$$

2 $\displaystyle\int \frac{x^2\,dx}{(x^3 + 4)^5}$

SOLUTION
Let $u = x^3 + 4$. Then $du = 3x^2\,dx$, so $x^2\,dx = \frac{1}{3}\,du$. Thus,

$$\int \frac{x^2\,dx}{(x^3 + 4)^5} = \int \frac{\frac{1}{3}\,du}{u^5} = \frac{1}{3}\int u^{-5}\,du = \frac{1}{3}\left(\frac{u^{-4}}{-4}\right) + C = -\frac{1}{12}\,(x^3 + 4)^{-4} + C.$$

3 $\displaystyle\int x^2\sqrt{3 - 2x}\,dx$

SOLUTION
Let $u = 3 - 2x$. Then $du = -2\,dx$, so $dx = -\frac{1}{2}\,du$. Solving the equation $u = 3 - 2x$ for x, we obtain $x = \dfrac{3 - u}{2}$. It follows that

$$x^2 = \left(\frac{3 - u}{2}\right)^2 = \frac{9 - 6u + u^2}{4}.$$

Thus,

$$\int x^2\sqrt{3-2x}\,dx = \int \frac{9-6u+u^2}{4}\sqrt{u}\left(-\frac{1}{2}\,du\right) = -\frac{1}{8}\int (9-6u+u^2)u^{1/2}\,du$$

$$= -\frac{1}{8}\int (9u^{1/2} - 6u^{3/2} + u^{5/2})\,du$$

$$= -\frac{1}{8}\left(9\,\frac{u^{3/2}}{\frac{3}{2}} - 6\,\frac{u^{5/2}}{\frac{5}{2}} + \frac{u^{7/2}}{\frac{7}{2}}\right) + C$$

$$= -\frac{3}{4}u^{3/2} + \frac{3}{10}u^{5/2} - \frac{1}{28}u^{7/2} + C$$

$$= -\frac{3}{4}(3-2x)^{3/2} + \frac{3}{10}(3-2x)^{5/2} - \frac{1}{28}(3-2x)^{7/2} + C.$$

4 $\displaystyle\int \frac{t\,dt}{\sqrt{t+5}}$

SOLUTION

Let $u = \sqrt{t+5}$, so that $du = \dfrac{dt}{2\sqrt{t+5}}$ and $\dfrac{dt}{\sqrt{t+5}} = 2\,du$. Also, $u^2 = t+5$, so $t = u^2 - 5$ and

$$\int \frac{t\,dt}{\sqrt{t+5}} = \int (u^2 - 5)2\,du = 2\int (u^2 - 5)\,du = \frac{2u^3}{3} - 10u + C$$

$$= \frac{2}{3}(\sqrt{t+5})^3 - 10\sqrt{t+5} + C.$$

Problem Set 2

In problems 1 and 2, verify that the function g is an antiderivative of the function f.

1 $f(x) = 12x^2 - 6x + 1$, $g(x) = 4x^3 - 3x^2 + x - 1$

2 $f(x) = (x-1)^3$, $g(x) = \frac{1}{4}x^4 - x^3 + \frac{3}{2}x^2 - x + 753$

In problems 3 through 14, use the basic rules for antidifferentiation to evaluate each indefinite integral.

3 $\displaystyle\int (3x^2 - 4x - 5)\,dx$

4 $\displaystyle\int (x^3 - 3x^2 + 2x - 4)\,dx$

5 $\displaystyle\int (2x^3 - 4x^2 - 5x + 6)\,dx$

6 $\displaystyle\int (2x^3 - 1)(x^2 + 5)\,dx$

7 $\displaystyle\int \frac{x^3 - 1}{x - 1}\,dx$

8 $\displaystyle\int (4t^2 + 3)^2\,dt$

9 $\displaystyle\int \left(t^2 + 3t + \frac{1}{t^2}\right)dt$

10 $\displaystyle\int \left(\frac{3}{x^2} + \frac{5}{x^4}\right)dx$

11 $\displaystyle\int \frac{25x^3 - 1}{\sqrt{x}}\,dx$

12 $\displaystyle\int \left(\sqrt{2x} + 2x\sqrt{x} + \frac{1}{\sqrt{x}}\right)dx$

13 $\displaystyle\int \frac{(\sqrt{x} - 1)^2}{\sqrt{x}}\,dx$

14 $\displaystyle\int \frac{t^3 + 2t^2 - 3}{\sqrt[3]{t}}\,dt$

In problems 15 through 40, find the antiderivative by using substitution and the basic rules for antidifferentiation. (In some cases a suitable substitution is suggested.)

15 $\int (4x + 3)^4 \, dx, u = 4x + 3$

16 $\int t(4t^2 + 7)^9 \, dt, u = 4t^2 + 7$

17 $\int x\sqrt{4x^2 + 15} \, dx, u = 4x^2 + 15$

18 $\int \dfrac{3x \, dx}{(4 - 3x^2)^8}, u = 4 - 3x^2$

19 $\int \dfrac{s \, ds}{\sqrt[3]{5s^2 + 16}}, u = 5s^2 + 16$

20 $\int \dfrac{(8t + 2) \, dt}{(4t^2 + 2t + 6)^{17}}, u = 4t^2 + 2t + 6$

21 $\int (1 - x^{3/2})^{5/3}\sqrt{x} \, dx, u = 1 - x^{3/2}$

22 $\int (x^2 - 6x + 9)^{11/3} \, dx, u = x - 3$

23 $\int \dfrac{x^2 \, dx}{(4x^3 + 1)^7}$

24 $\int \dfrac{x^2 + 1}{\sqrt{x^3 + 3x}} \, dx$

25 $\int (5t^2 + 1)\sqrt[4]{5t^3 + 3t - 2} \, dt$

26 $\int \dfrac{\sqrt[3]{1 + 1/(2t)}}{t^2} \, dt$

27 $\int \dfrac{2x^2 - 1}{(6x^3 - 9x + 1)^{3/2}} \, dx$

28 $\int \dfrac{\sqrt{1 + \sqrt{x}}}{\sqrt{x}} \, dx$

29 $\int \left(x + \dfrac{5}{x}\right)^{21}\left(\dfrac{x^2 - 5}{x^2}\right) dx$

30 $\int (49x^2 - 42x + 9)^{6/7} \, dx$

31 $\int x\sqrt{5 - x} \, dx$

32 $\int x^2\sqrt{1 + x} \, dx$

33 $\int \dfrac{t \, dt}{\sqrt{t + 1}}$

34 $\int \dfrac{y + 2}{\sqrt[3]{2 - y}} \, dy$

35 $\int \dfrac{2x \, dx}{(2 - x)^{2/3}}$

36 $\int (x + 2)^2\sqrt{1 + x} \, dx$

37 $\int \sqrt[3]{3x^2 + 5} \, x^3 \, dx$

38 $\int \sqrt[4]{x^3 + 1} \, x^5 \, dx$

39 $\int \dfrac{t^2 \, dt}{\sqrt{t + 4}}$

40 $\int \dfrac{y \, dy}{\sqrt{3 - y}}$

41 Evaluate $\int x^2\sqrt{5x - 1} \, dx$ by two methods and compare your answers:

    (a) Use the substitution $y = 5x - 1$.

    (b) Use the substitution $y = \sqrt{5x - 1}$.

42 If n is a positive integer, find $\int |x|^n \, dx$.

43 (a) Give an example to show that $\int f(x) \, dx \neq f(x)\int dx$.

    (b) Give an example to show that $\int f(x)g(x) \, dx \neq \left[\int f(x) \, dx\right]\left[\int g(x) \, dx\right]$.

    (c) Give an example to show that $\int \dfrac{f(x)}{g(x)} \, dx \neq \dfrac{\int f(x) \, dx}{\int g(x) \, dx}$.

44 Explain why an antiderivative of a polynomial function is again a polynomial function.

45 Given that f is a function with domain $(-1, \infty)$ such that $f(0) = 0$ and $f'(x) = 2/(1 + x)^2$, find f.

46 Suppose that $g'(x) = 1/(1 + x)^2$ holds for all values of x except for $x = -1$. Given that $\lim\limits_{x \to +\infty} g(x) = \lim\limits_{x \to -\infty} g(x) = 0$, find g.

47 Use Theorem 1 on antidifferentiation of the zero function to prove the following result: If two functions have the same derivative on an open interval, then they differ by a constant on this interval.

48 The following theorem is sometimes given as a justification of the method of change of variable: Suppose that $\int g(u)\, du = G(u) + C$ holds on a set I. Suppose, further, that the function h is differentiable at each number x in a set J. Assume, finally, that, for each number x in J, the number $h(x)$ belongs to I. Then, $\int g[h(x)]h'(x)\, dx = G[h(x)] + C$ holds on J. Prove this theorem.

49 Prove the additive rule for antidifferentiation.

50 Let f, g, and h be defined by the equations $f(x) = -2/x^3$, $g(x) = 1/x^2$, and

$$h(x) = \begin{cases} \dfrac{1 - 2x^2}{x^2} & \text{if } x < 0 \\[2mm] \dfrac{1}{x^2} & \text{if } x > 0, \end{cases}$$

respectively. Show that both g and h are antiderivatives of f but that there is no constant such that $h = g + C$. Does this contradict Theorem 2? Explain.

3 Simple Differential Equations and Their Solution

Relationships among the variables that arise in applied problems can often be expressed by equations such as $dy/dx = 2x$, or $dy = 2x\, dx$, which involve derivatives or differentials. Such an equation is called a *differential equation*.

Notice that $y = x^2$ is a *solution* of the differential equation $dy/dx = 2x$ in the sense that, if $y = x^2$, then the equation $dy/dx = 2x$ is satisfied. The simplest type of differential equation has the form $dy/dx = f(x)$, or, what is the same thing, $dy = f(x)\, dx$, where f is a given function. Evidently, $y = g(x)$ is a solution of $dy/dx = f(x)$ if and only if g is an antiderivative of f. Given any one solution $y = g(x)$ of $dy/dx = f(x)$, called a *particular solution*, we can write $y = g(x) + 7$ or $y = g(x) - \frac{967}{971}$ or, indeed, $y = g(x) + C$, where C is an arbitrary constant, to obtain other particular solutions.

Conversely, according to Theorem 2 of Section 2, any particular solution of $dy/dx = f(x)$ on an open interval I has the form $y = g(x) + C$ for a suitable value of the constant C. In this sense, $y = g(x) + C$ represents the *complete solution* of the differential equation $dy/dx = f(x)$ on the interval I, provided that g is an anti-derivative of f. Since $\int f(x)\, dx = g(x) + C$, this complete solution can be written as $y = \int f(x)\, dx$. For instance, the complete solution of the differential equation $dy/dx = 2x$ is given by $y = \int 2x\, dx$, that is, by $y = x^2 + C$.

The circumstances that give rise to differential equations often entail additional conditions called *constraints, side conditions, initial conditions,* or *boundary conditions* on the variables involved. These additional conditions can be used to single out the relevant particular solution from the complete solution.

EXAMPLES Find the complete solution of the given differential equation, and then find the particular solution satisfying the indicated initial condition.

1 $\dfrac{dy}{dx} = 6x^2 - \dfrac{1}{x^2} + 3; \; y = 10$ when $x = 1$.

SOLUTION

The complete solution is given by

$$y = \int \left(6x^2 - \frac{1}{x^2} + 3 \right) dx; \quad \text{that is,} \quad y = 2x^3 + \frac{1}{x} + 3x + C.$$

Putting $x = 1$ and $y = 10$ in the complete solution, we obtain $10 = 6 + C$, so that $C = 4$, and the desired particular solution is

$$y = 2x^3 + \frac{1}{x} + 3x + 4.$$

2 $dy = x\sqrt{x^2 + 5}\, dx; \; y = 8$ when $x = 2$.

SOLUTION

The complete solution is given by $y = \int x\sqrt{x^2 + 5}\, dx$. Making the change of variable $u = x^2 + 5$, so that $du = 2x\, dx$ and $x\, dx = \frac{1}{2}\, du$, we have

$$y = \int \frac{1}{2}\sqrt{u}\, du = \frac{1}{2}\left(\frac{u^{3/2}}{\frac{3}{2}} \right) + C = \frac{1}{3}(\sqrt{u})^3 + C = \frac{1}{3}(\sqrt{x^2 + 5})^3 + C.$$

Putting $x = 2$ and $y = 8$ in the latter equation, we obtain

$$8 = \tfrac{1}{3}(\sqrt{4 + 5})^3 + C = 9 + C,$$

so that $C = -1$. The desired particular solution is, therefore,

$$y = \tfrac{1}{3}(\sqrt{x^2 + 5})^3 - 1.$$

The differential equation $dy/dx = f(x)$ is said to be *separable* since it can be rewritten in the form $dy = f(x)\, dx$, in which the variables x and y are "separated" so that all expressions involving x are on the right and all expressions involving y are on the left. More generally, an equation that can be rewritten in the form

$$G(y)\, dy = F(x)\, dx,$$

where F and G are functions, is called *separable*. To find the so-called "*general solution*" of a separable differentiable equation, simply separate the variables, so that the equation takes the form $G(y)\, dy = F(x)\, dx$, and then take the indefinite integral on both sides. The two constants of integration corresponding to $\int G(y)\, dy$ and $\int F(x)\, dx$ can be combined, as usual, into a single constant C. If a side condition is given, it can be used to determine the value of C.

EXAMPLE   Solve the differential equation $y' = 4x^2 y^2$ with the side condition that $x = 1$ when $y = -1$.

SOLUTION
Here, y' is used as an abbreviation for dy/dx, so the given differential equation has the form $dy/dx = 4x^2 y^2$. Separating the variables, we have $(1/y^2)\,dy = 4x^2\,dx$. Thus, $\int (1/y^2)\,dy = \int 4x^2\,dx$, so

$$\frac{y^{-1}}{-1} + C_1 = \frac{4x^3}{3} + C_2 \quad \text{or} \quad \frac{1}{y} = -\frac{4x^3}{3} + C_0,$$

where $C_0 = C_1 - C_2$. The latter equation can be rewritten as

$$y = \frac{3}{3C_0 - 4x^3} \quad \text{or} \quad y = \frac{3}{C - 4x^3}, \qquad \text{where } C = 3C_0.$$

Letting $x = 1$ and $y = -1$, we find $C = 1$; hence, the desired particular solution is

$$y = \frac{3}{1 - 4x^3}.$$

In the preceding example, the general solution $y = \dfrac{3}{C - 4x^3}$ is not the complete solution of the differential equation $dy/dx = 4x^2 y^2$ since the constant function given by $y = 0$ is also a solution of $dy/dx = 4x^2 y^2$, yet it cannot be obtained by assigning a value to the constant C. A solution (such as $y = 0$ in the present example) which cannot be obtained directly from the general solution by assigning a value to the constant of integration is called a *singular solution.*

The general solution of a separable differential equation involving x and y may be an equation involving x and y which only implicitly determines y as a function of x. This situation arises in the following example.

EXAMPLE   Find the general solution of the differential equation $x\,dx + y\,dy = 0$.

SOLUTION
Separating the variables and antidifferentiating, we have

$$x\,dx = -y\,dy, \quad \text{so} \quad \int x\,dx = \int (-y)\,dy;$$

hence,

$$\frac{x^2}{2} + C_1 = -\frac{y^2}{2} + C_2 \quad \text{or} \quad \frac{x^2}{2} + \frac{y^2}{2} = C_2 - C_1.$$

Therefore, $x^2 + y^2 = 2(C_2 - C_1)$, that is, $x^2 + y^2 = C$, where we have put $C = 2(C_2 - C_1)$. Thus, the general solution of $x\,dx + y\,dy = 0$ is $x^2 + y^2 = C$.

3.1 Second-Order Differential Equations

Up to now, we have considered only "first-order" differential equations, that is, equations involving only first derivatives. By definition, the highest order of all the derivatives involved in a differential equation is called the *order* of the equation

For instance, $\dfrac{d^2y}{dx^2} + 2\dfrac{dy}{dx} = x$ is a second-order differential equation, while

$\dfrac{d^n y}{dx^n} = x^3 + 7$ is an nth-order differential equation.

Frequently, the general solution of a second-order differential equation can be obtained by two *successive* antidifferentiations and the resulting solution will involve two arbitrary constants that cannot be combined into one constant.

EXAMPLE   Find the general solution of the second-order differential equation $d^2y/dx^2 = -2x + 1$.

SOLUTION

We begin by rewriting the differential equation as $\dfrac{d}{dx}\left(\dfrac{dy}{dx}\right) = -2x + 1$, or

$d\left(\dfrac{dy}{dx}\right) = (-2x + 1)\,dx$. Therefore,

$$\frac{dy}{dx} = \int d\left(\frac{dy}{dx}\right) = \int (-2x + 1)\,dx, \quad \text{that is,} \quad \frac{dy}{dx} = -x^2 + x + C_1,$$

where the two constants of integration arising from $\int d(dy/dx)$ and $\int (-2x + 1)\,dx$ have been combined into the single constant C_1. This first-order differential equation can now be solved as usual by separating the variables and again antidifferentiating both sides, as follows:

$$dy = (-x^2 + x + C_1)\,dx,$$

so

$$\int dy = \int (-x^2 + x + C_1)\,dx = -\int x^2\,dx + \int x\,dx + C_1 \int dx$$

$$y = -\frac{x^3}{3} + \frac{x^2}{2} + C_1 x + C_2,$$

where the individual constants of integration arising from $\int dy$, $\int x^2\,dx$, $\int x\,dx$, and $\int dx$ have been combined into the single constant C_2. Thus, the general solution is

$$y = -\frac{x^3}{3} + \frac{x^2}{2} + C_1 x + C_2,$$

where C_1 and C_2 are arbitrary constants. Notice, here, that there is no way to combine the two constants C_1 and C_2 into a single constant since C_1 occurs as a multiplier of x.

Because the general solution of a second-order differential equation involves two arbitrary constants, *two* side conditions are required to determine these constants.

EXAMPLE   Suppose that y depends on x in such a way that $y'' = -2x + 1$. Assume further that $y = -1$ when $x = 0$ and that $y' = 1$ when $x = 0$. Find an explicit equation giving y in terms of x.

SOLUTION

Since y'' is an abbreviation for d^2y/dx^2, the given differential equation is the same as the equation $d^2y/dx^2 = -2x + 1$, whose general solution

$$y = -\frac{x^3}{3} + \frac{x^2}{2} + C_1 x + C_2$$

was found in the previous example. Substituting $x = 0$ and $y = -1$ from the first side condition into the latter equation, we obtain $-1 = C_2$. Therefore,

$$y = -\frac{x^3}{3} + \frac{x^2}{2} + C_1 x - 1.$$

Differentiating the latter equation on both sides, we obtain

$$y' = \frac{dy}{dx} = -x^2 + x + C_1.$$

Substituting $x = 0$ and $y' = 1$ from the second side condition into this equation, we find that $1 = C_1$. Hence,

$$y = -\frac{x^3}{3} + \frac{x^2}{2} + x - 1$$

is the desired solution.

3.2 Linear Motion

Differential equations are important in the study of the motion of physical objects (cars, projectiles, balls, planets, and so forth). Often we disregard the size, shape, and orientation of such objects and think of them as particles. Here we consider the motion of a particle P along a linear scale, which we call the s axis (Figure 1). The equation of motion of P, $s = f(t)$, gives the coordinate s of P in terms of the elapsed time t since some arbitrary (but fixed) initial instant. The instantaneous velocity v and the instantaneous acceleration a of P are given by the equations

$$v = \frac{ds}{dt} = f'(t) \quad \text{and} \quad a = \frac{dv}{dt} = \frac{d^2s}{dt^2} = f''(t).$$

(See Section 2 of Chapter 3.)

Figure 1

If v or a is a known function of t, subject to suitable initial conditions, it may be possible to solve the resulting differential equations for the law of motion $s = f(t)$. This is illustrated in the following examples.

EXAMPLES   Find the law of motion $s = f(t)$ from the information given.

1 $a = 2t - t^2$; $v = 0$ when $t = 0$, and $s = 0$ when $t = 0$.

SOLUTION

We have $dv/dt = a = 2t - t^2$, so $dv = (2t - t^2)\, dt$. A first antidifferentiation gives

$$v = \int dv = \int (2t - t^2)\, dt = t^2 - \frac{t^3}{3} + C_1.$$

Since $v = 0$ when $t = 0$, it follows that $C_1 = 0$. Also, $ds/dt = v = t^2 - t^3/3$, so $ds = (t^2 - t^3/3)\,dt$, and a second antidifferentiation gives

$$s = \int ds = \int \left(t^2 - \frac{t^3}{3} \right) dt = \frac{t^3}{3} - \frac{t^4}{12} + C_2.$$

Since $s = 0$ when $t = 0$, it follows that $C_2 = 0$ and the equation of motion is $s = t^3/3 - t^4/12$.

2 A car is braked to a stop with constant deceleration. The car stops 8 seconds after the brakes are applied and travels 200 feet during this time. Find the law of motion of the car during this 8-second interval. Also, find the acceleration and the speed of the car at the instant the brakes are first applied.

SOLUTION

Represent the car by a particle on the s axis moving (say) to the right. Start reckoning time t from the instant when the brakes are first applied and place the origin at the position of the car when $t = 0$. Here the acceleration a is constant and we have

$$s = 0 \qquad \text{when } t = 0,$$
$$s = 200 \qquad \text{when } t = 8,$$
$$v = 0 \qquad \text{when } t = 8.$$

Since $dv/dt = a$ and a is constant, then $dv = a\,dt$ and

$$v = \int dv = \int a\,dt = a \int dt = at + C_1.$$

Therefore, $ds/dt = v = at + C_1$, so $ds = (at + C_1)\,dt$ and

$$s = \int ds = \int (at + C_1)\,dt = \tfrac{1}{2}at^2 + C_1 t + C_2.$$

Since $s = 0$ when $t = 0$, it follows that $C_2 = 0$, so that $s = \tfrac{1}{2}at^2 + C_1 t$.

If we use the fact that $v = 0$ when $t = 8$ and the equation $v = at + C_1$, we find that $0 = 8a + C_1$, so that $C_1 = -8a$. Substituting $-8a$ for C_1 in the equation $s = \tfrac{1}{2}at^2 + C_1 t$, we obtain $s = \tfrac{1}{2}at^2 - 8at$. Now, $s = 200$ when $t = 8$, so $200 = \tfrac{1}{2}a(8)^2 - 8a(8) = -32a$; hence, $a = -200/32 = -25/4$ ft/sec^2. (The negative sign indicates *deceleration*.) Therefore, the law of motion is

$$s = \frac{1}{2}\left(\frac{-25}{4} \right)t^2 - 8\left(\frac{-25}{4} \right)t \quad \text{or} \quad s = 50t - \frac{25}{8}t^2.$$

The speed of the car at the instant when $t = 0$ is obtained by putting $t = 0$ in the equation $v = at + C_1$ to obtain

$$v = C_1 = -8a = -8\left(\frac{-25}{4} \right) = 50 \text{ ft/sec.}$$

An object falling near the surface of the earth experiences some force because of air resistance; however, in many situations this force is negligible, especially when the object has a high density, is more or less "streamlined," and has not attained a very high velocity. Neglecting air resistance, such an object falls with a constant acceleration g, called the *acceleration of gravity*. The value of g is approximately 32 ft/sec^2, or 980 cm/sec^2, or 9.8 m/sec^2.

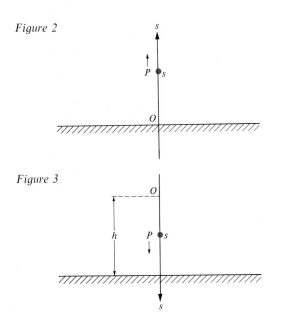

Figure 2

Figure 3

If such an object P is projected straight upward, it is usually convenient to take the s axis pointing straight upward with the origin at the surface of the earth (Figure 2). Then the acceleration a of P is negative (it will slow down, stop, then begin to fall back down); hence,

$$a = \frac{d^2s}{dt^2} = \frac{dv}{dt} = -g.$$

On the other hand, if the object P is dropped or thrown straight downward from an initial height h, it is usually convenient to take the s axis pointing downward with the origin h units above the surface of the earth (Figure 3). Then the acceleration a of P is positive; hence,

$$a = \frac{d^2s}{dt^2} = \frac{dv}{dt} = g.$$

EXAMPLE An iron ball is thrown vertically upward, starting 4 feet above the ground, with an initial velocity of 24 ft/sec. How many seconds will elapse before the ball strikes the ground?

SOLUTION
We set up the coordinate axis as in Figure 2. Here, $dv/dt = -g$, so $dv = -g\,dt$ and

$$v = \int dv = \int (-g)\,dt = -g\int dt = -gt + C_1.$$

When $t = 0$, $v = 24$ ft/sec, so $24 = (-g)(0) + C_1$ and $C_1 = 24$. Therefore, $v = -gt + 24$, that is, $ds/dt = -gt + 24$, or $ds = (-gt + 24)\,dt$. It follows that

$$s = \int ds = \int (-gt + 24)\,dt = -\tfrac{1}{2}gt^2 + 24t + C_2.$$

Since $s = 4$ when $t = 0$, it follows that $4 = -\tfrac{1}{2}g(0)^2 + 24(0) + C_2$, so $C_2 = 4$. Therefore, $s = -\tfrac{1}{2}gt^2 + 24t + 4$. Here $g = 32$ ft/sec^2, and so $s = -16t^2 + 24t + 4$. When the ball strikes the surface of the earth, $s = 0$, so that

$$0 = -16t^2 + 24t + 4 \quad \text{or} \quad 4t^2 - 6t - 1 = 0.$$

By the quadratic formula, $t = \dfrac{3 \pm \sqrt{13}}{4}$. When the ball lands, t is positive; hence, we reject the negative solution and conclude that

$$t = \frac{3 + \sqrt{13}}{4} \approx 1.65 \text{ sec.}$$

Problem Set 3

In problems 1 through 6, find the complete solution of each differential equation.

1 $\dfrac{dy}{dx} = 5x^4 + 3x^2 + 1$ **2** $\dfrac{dy}{dx} = 20x^3 - 6x^2 + 17$ **3** $\dfrac{dy}{dx} = \dfrac{6}{x^2} + 15x^2 + 10$

4 $y' = \dfrac{(x^2 - 4)^2}{2x^2}$ $\qquad\qquad$ **5** $\dfrac{dy}{dx} = \sqrt{7x^3}$ $\qquad\qquad$ **6** $dy = (5t + 12)^3 \, dt$

In problems 7 through 10, find the particular solution of the given differential equation that satisfies the initial condition indicated.

7 $\dfrac{dy}{dx} = 5 - 3x$; $y = 4$ when $x = 0$. $\qquad\qquad$ **8** $\dfrac{dy}{dx} = 3x^2 + x$; $y = -2$ when $x = 1$.

9 $\dfrac{dy}{dt} = t^3 + \dfrac{1}{t^2}$; $y = 1$ when $t = -2$. $\qquad\qquad$ **10** $y' = \sqrt{x} + 2$; $y = 5$ when $x = 4$.

In problems 11 through 18, find the general solution of each separable differential equation.

11 $y' = x(x^2 - 3)^4$ $\qquad\qquad\qquad\qquad$ **12** $(3x^2 + 2x + 1)^5 \, dy = (6x + 2) \, dx$

13 $\sqrt{x^3 + 7} \, dy = x^2 \, dx$ $\qquad\qquad\qquad$ **14** $\dfrac{ds}{dt} = (t + 1)^2 t^3$

15 $\sqrt{2x + 1} \, dy = y^2 \, dx$ $\qquad\qquad\qquad$ **16** $(y^2 - \sqrt{y}) \, dy = (x^2 + \sqrt{x}) \, dx$

17 $\dfrac{dy}{dx} = \dfrac{x\sqrt[3]{y^4 + 7}}{5y^3}$ $\qquad\qquad\qquad$ **18** $y\dfrac{dy}{dx} = x^3\sqrt{10y^2 + 1}$

In problems 19 through 22, find the particular solution of the given separable differentiable equation that satisfies the initial condition indicated.

19 $\dfrac{dx}{y} = \dfrac{dy}{2 - x^{3/2}}$; $y = 2$ when $x = 9$. $\qquad\qquad$ **20** $\dfrac{ds}{dt} = \dfrac{t^2}{\sqrt{t^3 + 1}}$; $s = \frac{1}{2}$ when $t = 2$.

21 $t^{2/3} \, dW = (1 - t^{1/3})^3 \, dt$; $W = -1$ when $t = 8$.

22 $y' = \sqrt{x^3 + x^2 - 2x + 8}\,(9x^2 + 6x - 6)$; $y = 0$ when $x = 1$.

In problems 23 through 30, find the general solution of each second-order differential equation.

23 $\dfrac{d^2y}{dx^2} = 3x^2 + 2x + 1$ $\quad$ **24** $y'' = (5x + 1)^4$ $\quad$ **25** $y'' = \sqrt[3]{4x + 5}$ $\quad$ **26** $S'' = \dfrac{5}{(t + 7)^3}$

27 $\dfrac{d^2y}{dx^2} = 2x^4 + 3$ $\quad$ **28** $y'' = (x + 1)^2$ $\quad$ **29** $\dfrac{d^2y}{dx^2} = 0$ $\quad$ **30** $D_x^2 y = 1$

In problems 31 through 36, find the particular solution of each differential equation that satisfies the side conditions given.

31 $\dfrac{d^2y}{dx^2} = 6x + 1$; $y = 2$ and $y' = 3$ when $x = 0$. $\qquad$ **32** $\dfrac{d^2y}{dx^2} = \sqrt{x}$; $y = 3$ and $y' = 2$ when $x = 9$.

33 $\dfrac{d^2y}{dx^2} = 2$; $y = 0$ when $x = 1$ and $y = 0$ when $x = -3$. $\qquad$ **34** $\dfrac{d^2y}{dx^2} = 3x^2$; $y = -1$ when $x = 0$ and $y = 9$ when $x = 2$.

35 $y'' = 3(2 + 5x)^2$; $y = 2$ and $y' = -1$ when $x = 1$. $\qquad$ **36** $\dfrac{d^2s}{dt^2} = \sqrt[4]{5t - 4}$; $s = 2$ and $s' = -3$ when $t = 4$.

37 The work W done in stretching a certain spring through s units satisfies the differential equation $dW/ds = 5s$. Find W in terms of s if $W = 0$ when $s = 0$.

38 The graph of $y = f(x)$ has a relative minimum at the point $(-4, 1)$. Find f if $f''(x) = \frac{1}{2}$ holds for all values of x in $\mathbb{R}$.

39 If k is a constant, find the general solution of the differential equation $d^2y/dx^2 = kx$.

40 Suppose that f is a twice-differentiable function on the open interval I such that $f''(x) = 0$ for all values of x in I. Give a *rigorous* proof that there exist constants A and B such that $f(x) = Ax + B$ for all values of x in I.

41 A particle moving along a straight line has the equation of motion $s = f(t)$, where t is in seconds and s is in feet. Its velocity v satisfies the equation $v = t^2 - 8t + 15$. If $s = 1$ when $t = 0$, find s when $t = 3$.

42 A particle, starting with an initial velocity of 25 meters per second, moves in a straight line through a resisting medium which decreases the velocity of the particle at a constant rate of 10 meters per second each second. How far will the particle travel before coming to rest?

43 The brakes on a certain car can stop the car in 200 feet from a speed of 55 miles/hr. Assume that, when the brakes are applied, the car has a constant negative acceleration.
  (a) How much time in seconds is required to bring the car to a stop from 55 miles/hr?
  (b) If the car is brought to a stop from 55 miles/hr, how far will it have moved by the time its speed is reduced to 25 miles/hr? (Recall that 1 mile is 5280 feet.)

44 From the top edge of a building 20 meters high, a stone is thrown vertically upward with an initial velocity of 30 meters per second.
  (a) In how many seconds will the stone strike the ground?
  (b) How high will the stone rise?
  (c) How fast will the stone be falling when it hits the ground?

45 A balloon is rising at the constant rate of 10 ft/sec, and is 100 feet from the ground at the instant when the aeronaut drops his binoculars.
  (a) How long will it take the binoculars to strike the ground?
  (b) With what speed will the binoculars strike the ground?

46 A projectile is fired vertically upward by a cannon with an initial velocity of v_0 ft/sec. At what speed will the projectile be moving when it returns and strikes the hapless cannoneer? (Neglect air resistance).

47 A balloon is rising vertically at the constant rate of 5 ft/sec and has reached an altitude of 26 feet at the instant when an assistant on the ground directly under the balloon attempts to toss the aeronaut's binoculars up to her. What is the minimum velocity with which the binoculars should be thrown straight upward if they are released from the assistant's hand at an altitude of 6 feet?

48 Suppose that a particle P moves along the s axis with a constant acceleration a. Let v_0 be the velocity of the particle when $t = 0$ and let s_0 be its coordinate when $t = 0$. Show that:
  (a) $v = at + v_0$.
  (b) $s = \frac{1}{2}at^2 + v_0t + s_0$.

49 A stone is dropped from a height of h feet with zero initial velocity and it hits the ground T seconds later. Show that $h = 16T^2$.

4 Applications of Differential Equations

In Section 3.2 we saw an application of differential equations to linear motion problems. Further simple applications to geometry, physics, and economics are given in the present section.

A first-order differential equation of the form $dy/dx = g(x)$, where g is a given function, can be interpreted geometrically as a condition on the slope dy/dx of the tangent line to the graph of $y = f(x)$, where f is an unknown function. The complete solution of the differential equation gives all functions f whose graphs satisfy this condition. An initial condition amounts to an additional requirement that the graph of f contain a specified point.

EXAMPLE   (a) Find all curves in the plane satisfying the condition that the slope of the tangent at each point is three times the abscissa of that point.

(b) Sketch graphs of several such curves on the same xy coordinate system.

(c) Find the equation of that particular curve satisfying the condition in part (a) which contains the point $(1, 2)$.

SOLUTION

(a) The differential equation expressing the condition on the slope of the tangent is $dy/dx = 3x$. The complete solution of this differential equation is

$$y = \int 3x \, dx = 3 \int x \, dx = 3 \cdot \frac{x^2}{2} + C.$$

Thus, the equation of a curve satisfying the given condition is of the form $y = 3x^2/2 + C$, where C is a constant. There is a different curve for each different value of C.

(b) The curves corresponding to $C = -2$, $C = 0$, $C = \frac{1}{2}$, $C = 2$, and $C = 4$ are sketched in Figure 1.

Figure 1

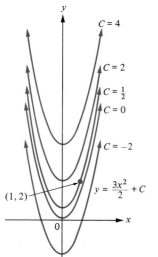

(c) Here we have to impose the side condition that $y = 2$ when $x = 1$. Setting $y = 2$ and $x = 1$ in the equation $y = \frac{3}{2}x^2 + C$ and solving for C, we find that $C = 2 - \frac{3}{2} = \frac{1}{2}$. The desired equation is, therefore, $y = \frac{3}{2}x^2 + \frac{1}{2}$.

4.1 Work Done by a Variable Force

Suppose that a constant force F, having a direction parallel to the s axis, acts on a particle P which moves along this axis from an original position with coordinate s_0 to a final position with coordinate s_1 (Figure 2). Then, by definition, the force does an amount of work W on the particle given by $W = F \cdot (s_1 - s_0)$. A positive (respectively, a negative) force is understood to act in the positive (respectively, the negative) direction along the s axis.

We now take up the problem of calculating the work done when the applied force is not necessarily constant, but still acts in a direction parallel to the s axis. We suppose that P starts at an initial position with coordinate s_0, and we denote by W the net work done by the variable force F in moving P from its original position to the position s (Figure 3). Since F may not be constant, we cannot calculate W simply by multiplying F by $s - s_0$.

In this case, suppose that P moves from s to $s + \Delta s$, causing the net work done by the force to change from W to $W + \Delta W$ (Figure 4). Here, ΔW is the work done by F in moving P from s to

Figure 2

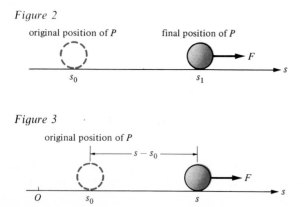

original position of P final position of P

s_0 s_1

Figure 3

original position of P

$s - s_0$

O s_0 s

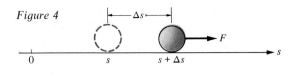

Figure 4

$s + \Delta s$. Although the force F might change as P is so moved, it should not change appreciably if Δs is very small. Thus, if F is the force acting on P at position s, we should have

$$\Delta W \approx F \, \Delta s \quad \text{or} \quad \frac{\Delta W}{\Delta s} \approx F.$$

As Δs approaches 0, this approximation should become better and better; hence, we should have

$$\frac{dW}{ds} = \lim_{\Delta s \to 0} \frac{\Delta W}{\Delta s} = F.$$

Therefore, the net work W done by the variable force F satisfies the differential equation $dW = F \, ds$ with the initial condition that $W = 0$ when $s = s_0$.

EXAMPLES **1** The force F acting on a particle P is given by $F = 1/s^2$, where s is the coordinate of P. How much work is done by this force in moving P from $s = 1$ to $s = 9$?

SOLUTION
Let W denote the net work done by the force in moving P from the point with coordinate 1 to the point with coordinate s. We want to find the value of W when $s = 9$. Since $dW = F \, ds = (1/s^2) \, ds$, it follows that $W = \int (1/s^2) \, ds = (-1)/s + C$. When $s = 1$, we have $W = 0$, so that $0 = (-1)/1 + C$; hence, $C = 1$ and $W = (-1)/s + 1$. When $s = 9$, we have $W = -\frac{1}{9} + 1 = \frac{8}{9}$ unit of work.

2 Consider the apparatus shown in Figure 5, in which a particle P attached to a stanchion by a perfectly elastic spring can slide without friction along the horizontal s axis. The particle starts at $s = 0$ and is pulled by a force F to a final position $s = b$. At the start, when $s = 0$, assume that the spring is relaxed and $F = 0$. Find the work done by F in pulling P from $s = 0$ to $s = b$.

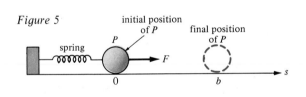

Figure 5

SOLUTION
Because the spring is perfectly elastic, *Hooke's law* requires that the force F be proportional to the displacement s; that is, $F = ks$, where k is a constant called the *spring constant*. (The stiffer the spring, the larger the value of k.) Let W be the net work done by F in pulling P from the origin to the position s, so that $dW = F \, ds = ks \, ds$. Antidifferentiation gives

$$W = \int ks \, ds = k \int s \, ds = k \frac{s^2}{2} + C.$$

When $s = 0$, we have $W = 0$, so that $0 = k(0^2/2) + C$, and it follows that $C = 0$ and $W = ks^2/2$. Thus, when $s = b$, we have $W = kb^2/2$.

3 A perfectly elastic spring is stretched from its relaxed position through 6 feet. When extended these 6 feet, the stretching force on the spring is 20 pounds. How much work is done?

SOLUTION
Reasoning as in Example 2, we have $F = ks$, so $k = F/s$. When $s = 6$ feet, then $F = 20$ pounds; hence, $k = \frac{20}{6} = \frac{10}{3}$ lb/ft. Using the result of Example 2, we have

$$W = k\frac{b^2}{2} = \frac{10}{3} \cdot \frac{6^2}{2} = 60 \text{ foot-pounds.}$$

4.2 Remarks on Setting Up Differential Equations

As has been mentioned, many engineers and physical scientists like to think of dx as an "infinitesimal bit of x" and they prefer to regard $\int dx$ as a "summation" of all the "infinitesimal bits of x" to give the quantity x; that is, apart from an additive constant of integration, $x = \int dx$. They persist in this point of view, in spite of its possible lack of "mathematical rigor," because it allows them to set up differential equations quickly and easily.

For instance, a physicist might argue that a force F acting through an "infinitesimal" distance ds should remain virtually constant and should result in the accomplishment of an "infinitesimal" amount of work dW given by $dW = F \, ds$. Thus, from the point of view of infinitesimals, the differential equation for work is found with a minimum of fuss. To be sure, the "infinitesimal point of view" can lead to invalid arguments; however, differential equations set up using this point of view generally turn out to be correct.

In the remainder of this book, we use arguments involving "infinitesimals" whenever it is convenient to do so. In every case, rigorous arguments can be made to justify the result.

The following example illustrates a typical physical application of the "infinitesimal point of view."

EXAMPLE   According to Newton's law of gravity, two particles with masses m_1 and m_2 grams which are separated by a distance of s centimeters attract each other with a force of $F = \gamma(m_1 m_2 / s^2)$ dynes, where γ is a constant given by $\gamma = 6.6732 \times 10^{-8}$ dyne cm^2/g^2. If 1000 grams of lead are distributed uniformly along the x axis between the origin and the point with x coordinate 2 centimeters, how much gravitational force does it exert on a 1-gram particle P situated on the x axis 1 centimeter to the left of the origin?

Figure 6

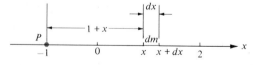

SOLUTION
Let F denote the gravitational force of attraction on P of the mass lying on the interval $[0, x]$ (Figure 6). The linear density of lead on the interval $[0, 2]$ is $\frac{1000}{2} = 500$ g/cm; hence, the "infinitesimal" mass dm of the portion of lead on the subinterval $[x, x + dx]$ is given by $dm = 500 \, dx$. This "infinitesimal" mass exerts an "infinitesimal" gravitational force dF on the unit mass P given by

$$dF = \gamma\frac{(1)(dm)}{(1 + x)^2} = \gamma\frac{500}{(1 + x)^2} \, dx,$$

since the distance between P and dm is $1 + x$ centimeters (Figure 6). Therefore,

$$F = \int dF = \int \gamma \frac{500}{(1 + x)^2}\, dx = 500\gamma \int \frac{1}{(1 + x)^2}\, dx.$$

Making the change of variable $u = 1 + x$, so that $du = dx$, we obtain

$$F = 500\gamma \int u^{-2}\, du = -500\gamma u^{-1} + C = C - \frac{500\gamma}{1 + x}.$$

Here we have the side condition that $F = 0$ when $x = 0$ (why?), so that $0 = C - \dfrac{500\gamma}{1 + 0}$, and therefore $C = 500\gamma$. It follows that $F = 500\gamma - \dfrac{500\gamma}{1 + x}$. Putting $x = 2$, we obtain

$$F = 500\gamma - \frac{500\gamma}{3} = \frac{1000\gamma}{3} = \frac{1000}{3}(6.6732 \times 10^{-8}) = 2.2244 \times 10^{-5} \text{ dyne.}$$

4.3   Applications to Economics

Here we give two very simple examples illustrating how the total cost or total revenue function can be found if the marginal cost or revenue function, respectively, is known. We use K for the constant of integration since C is used to represent the total cost. As before, R denotes the total revenue.

EXAMPLES **1** The marginal cost dC/dx for producing x electric alarm clocks is given by $dC/dx = 0.05 + 5000/x^2$ dollars per item. Find the total cost C as a function of x, given that $C = \$5500$ when $x = 1000$.

SOLUTION

$$C = \int dC = \int \left(0.05 + \frac{5000}{x^2}\right) dx = 0.05x - \frac{5000}{x} + K.$$

Putting $C = 5500$ and $x = 1000$ into the latter equation, we obtain $K = 5455$. Therefore, $C = 0.05x - (5000/x) + 5455$ dollars.

2 The marginal revenue for a digital watch is expressed by $dR/dx = 60{,}000 - 40{,}000(1 + x)^{-2}$ dollars per thousand watches, where x represents the demand in thousands of watches. Express the total sales revenue in terms of x, given that, for $x = 1$ (thousand watches), the total sales revenue is $\$38{,}000$. If the demand increases to $x = 4$ (thousand watches), what is the total sales revenue?

SOLUTION

$$R = \int [60{,}000 - 40{,}000(1 + x)^{-2}]\, dx = 60{,}000 \int dx - 40{,}000 \int (1 + x)^{-2}\, dx.$$

The change of variable $u = 1 + x$ in the second integral gives

$$\int (1 + x)^{-2}\, dx = \int u^{-2}\, du = -u^{-1} + K = -\frac{1}{1 + x} + K;$$

hence,

$$R = 60{,}000x + \frac{40{,}000}{1 + x} + K_1, \qquad \text{where } K_1 = -40{,}000K.$$

Putting $x = 1$ and $R = 38,000$ in the latter equation and solving for K_1, we find that $K_1 = -42,000$; hence, $R = 60,000x + \dfrac{40,000}{1 + x} - 42,000$. When $x = 4$, we have $R = 240,000 + 8000 - 42,000 = \$206,000$.

4.4 Newton's Law of Motion — Momentum and Energy

If a particle P of constant mass m moves along a linear scale because of an unopposed (possibly variable) force F (Figure 7), then Newton's (second) law of motion—force equals mass times acceleration—can be written

$$F = ma \quad \text{or} \quad F = m\frac{dv}{dt} \quad \text{or} \quad F = m\frac{d^2s}{dt^2}.$$

Figure 7

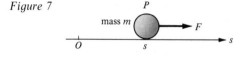

Thus, Newton's law of motion is really a second-order differential equation whose solution (subject to suitable initial value conditions) gives the equation of motion of the particle. To obtain this solution, two successive antidifferentiations are required. In order to carry out these antidifferentiations, we need to know F explicitly as a function of time t.

Even without knowing F explicitly as a function of t, it is possible to carry out the first of the two antidifferentiations, at least formally. This can be done either by multiplying by dt and integrating or by multiplying by ds and integrating. Multiplying $F = m(dv/dt)$ by dt and integrating, we have

$$\int F \, dt = \int m \, dv = m \int dv = mv + C.$$

The quantity mv, mass times velocity, is called the *linear momentum* of the particle P and plays an important role in dynamics. Thus, the idea of momentum arises naturally from Newton's law by a first antidifferentiation *with respect to time.*

Multiplying $F = m(dv/dt)$ by ds and integrating, we have

$$\int F \, ds = \int m \frac{dv}{dt} \, ds = m \int \frac{ds}{dt} \, dv = m \int v \, dv = m(\tfrac{1}{2}v^2) + C = \tfrac{1}{2}mv^2 + C.$$

The quantity $\tfrac{1}{2}mv^2$, one-half mass times the square of velocity, is called the *kinetic energy* of the particle P. Thus, the notion of kinetic energy arises naturally from Newton's law by a first antidifferentiation *with respect to distance.*

In Section 4.1 we obtained the differential equation $dW = F \, ds$ for the work W done by a variable force F. Thus, $W = \int dW = \int F \, ds$, so the equation $\int F \, ds = \tfrac{1}{2}mv^2 + C$ can be rewritten as $W = \tfrac{1}{2}mv^2 + C$. Therefore, *apart from an additive constant, the net work done by a variable (unopposed) force F acting on the particle* (Figure 7) *is equal at all times to the kinetic energy of the particle.*

The quantity V defined by $V = -W = -\int F \, ds$ is called the *potential energy* of the particle P. The *total energy* E of the particle is defined to be the sum of its kinetic and potential energy, so $E = \tfrac{1}{2}mv^2 + V$. Since $W = \tfrac{1}{2}mv^2 + C$ and $V = -W$, we see that $E = \tfrac{1}{2}mv^2 - W = -C$, which shows that *the total energy*

E of the particle P remains constant. This is a simple case of the famous *law of conservation of energy.*

An object of mass m near the surface of the earth is acted upon by a constant force F of gravity and, if allowed to fall, accelerates with a constant acceleration g. By Newton's law, $F = mg$. We refer to the force F as the *weight* of the object. In the British system of units, if the pound is taken as the unit of force (hence, in particular, of weight), then the unit of mass is called a *slug*. The weight of a 1-slug mass is therefore given by $F = mg = (1)(32) = 32$ pounds. The basic unit of energy (either potential or kinetic) as well as of work in the British system is the *foot-pound*. One foot-pound is the work done in raising a 1-pound weight through 1 foot.

In the cgs (centimeter–gram–second) system of units, the *gram* is the basic unit of mass. One thousand grams (that is, 1 *kilogram*) of mass weighs about 2.2 pounds near the surface of the earth. The basic unit of force in the cgs system is the *dyne*. One dyne is the amount of force required to accelerate a 1-gram mass by 1 centimeter per second per second. Since $g = 980$ cm/sec^2, a 1-gram mass near the surface of the earth will weigh 980 dynes (weight = mg). One pound is approximately 445,000 dynes. The unit of energy in the cgs system is the *dyne-centimeter*, also called the *erg*. One erg is the work done in raising a 1-gram mass through 1 centimeter. One foot-pound is approximately 13,560,000 ergs.

EXAMPLE   The force F in dynes acting on an electrically charged particle of mass m grams because of an electrostatic field is given by $F = s^{-2}$, where s is the coordinate of the particle and distances are measured in centimeters. The potential energy V of the particle satisfies $V = 1$ erg when $s = 1$. The particle starts at $s = 1$ with an initial velocity $v_0 = 0$. Find:

(a) V as a function of s.
(b) $\lim\limits_{s \to +\infty} V$.
(c) The velocity of the particle when it reaches the position with coordinate s.

SOLUTION

(a) $V = -\int F\, ds = -\int s^{-2}\, ds = s^{-1} + C$. Since $V = 1$ when $s = 1$, it follows that $C = 0$; hence, $V = s^{-1}$.

(b) $\lim\limits_{s \to +\infty} V = \lim\limits_{s \to +\infty} s^{-1} = 0$.

(c) The total energy $E = \frac{1}{2}mv^2 + V = \frac{1}{2}mv^2 + s^{-1}$ is constant. When $s = 1$, we have $v = 0$; hence, $E = \frac{1}{2}m(0)^2 + 1^{-1} = 1$. Therefore, $1 = \frac{1}{2}mv^2 + s^{-1}$ holds at all times. Solving the latter equation for v, we obtain $v = \pm\sqrt{2(s - 1)/ms}$. Since the force F acting on the particle is always positive and the initial velocity is 0, it follows that the velocity cannot be negative (why?); hence, $v = \sqrt{2(s - 1)/ms}$.

4.5   The Motion of a Projectile

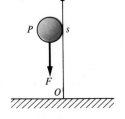

Figure 8

In the following examples, assume that a projectile P of mass m is fired vertically upward from the surface of the earth at time $t = 0$ with an initial velocity v_0 (Figure 8). Take the s axis pointing straight upward with its origin at the surface of the earth. Then the (constant) force F of gravity on the projectile is acting downward; hence, F is negative. The acceleration a of P is also negative; hence, $a = -g$ and $F = ma = -mg$. Choose the potential energy V so that $V = 0$ when $t = 0$, that is, when $s = 0$. Let h be the maximum height to which P climbs.

EXAMPLES **1** Find the equation of motion $s = f(t)$ of P.

SOLUTION

Since $dv/dt = a = -g$, we have $dv = -g\, dt$, so that $v = \int (-g)\, dt = -gt + C_1$. When $t = 0$, we have $v = v_0$; hence, $C_1 = v_0$ and so $v = -gt + v_0$. Therefore, $ds/dt = -gt + v_0$, or $ds = -gt\, dt + v_0\, dt$. Hence, $s = \int (-g)t\, dt + \int v_0\, dt = -\frac{1}{2}gt^2 + v_0 t + C_2$. Since $s = 0$ when $t = 0$, it follows that $C_2 = 0$, so that the equation of motion is $s = -\frac{1}{2}gt^2 + v_0 t$.

2 Find the velocity v in terms of s.

SOLUTION

From Example 1, $v = -gt + v_0$ and $s = -\frac{1}{2}gt^2 + v_0 t$. Eliminating t from these two equations, we find by simple algebra that $v^2 = v_0^2 - 2gs$; hence, $v = \pm\sqrt{v_0^2 - 2gs}$.

3 Find the maximum height h in terms of v_0.

SOLUTION

To maximize s, we set $ds/dt = v = 0$. In Example 2 we found that $v^2 = v_0^2 - 2gs$; so, setting $v = 0$ and $s = h$, we obtain $0 = v_0^2 - 2gh$, or $h = v_0^2/(2g)$.

4 Find the kinetic and the potential energy of P as functions of t.

SOLUTION

Using Example 1, we see that the kinetic energy is given by

$$\tfrac{1}{2}mv^2 = \tfrac{1}{2}m(-gt + v_0)^2 = \tfrac{1}{2}mg^2t^2 - mgv_0 t + \tfrac{1}{2}mv_0^2.$$

The potential energy is given by

$$V = -\int F\, ds = -\int (-mg)\, ds = mg \int ds = mgs + C.$$

When $t = 0$, then $s = 0$ and $V = 0$, so $0 = mg(0) + C$ and $C = 0$. Thus, $V = mgs$; hence, using Example 1 again, we see that

$$V = mgs = mg(-\tfrac{1}{2}gt^2 + v_0 t) = -\tfrac{1}{2}mg^2t^2 + mgv_0 t.$$

5 Find the kinetic and the potential energy of P as functions of the position coordinate s.

SOLUTION

By Example 2, $v^2 = v_0^2 - 2gs$; hence, the kinetic energy is given by $\tfrac{1}{2}mv^2 = \tfrac{1}{2}mv_0^2 - mgs$. By Example 4, the potential energy is given by $V = mgs$.

6 Find the (constant) total energy E of P and explain what happens to the kinetic and to the potential energy of P as it climbs from $s = 0$ to $s = h$.

SOLUTION

By Example 5, $E = \tfrac{1}{2}mv^2 + V = (\tfrac{1}{2}mv_0^2 - mgs) + mgs$, so $E = \tfrac{1}{2}mv_0^2$. When $s = 0$, all the energy is kinetic. As P climbs from $s = 0$ to $s = h$, there is a "trade-off" between kinetic and potential energy—the kinetic energy decreases and the potential energy increases. Finally, when $s = h = v_0^2/(2g)$ (Example 3), then all the energy is potential.

Problem Set 4

In problems 1 through 4, the slope dy/dx of the tangent line to the graph of a function f is given by a differential equation. Find the function f if the graph passes through the indicated point (a, b).

1 $\dfrac{dy}{dx} = 1 - 3x$; $(a, b) = (-1, 4)$

2 $\dfrac{dy}{dx} = x^2 + 1$; $(a, b) = (-3, 5)$

3 $\dfrac{dy}{dx} = \left(\dfrac{y}{x}\right)^2$; $(a, b) = (2, 1)$

4 $\dfrac{dy}{dx} = 2xy^2$; $(a, b) = (0, 1)$

In problems 5 through 8, the force F acting on a particle P moving along the s axis is given in terms of the coordinate s of P. Find the work done by F in moving P from $s = s_0$ to $s = s_1$.

5 $F = 2s$, s in feet, F in pounds, $s_0 = 1$, $s_1 = 5$.

6 $F = 400s\sqrt{1 + s^2}$, s in centimeters, F in dynes, $s_0 = 0$, $s_1 = 3$.

7 $F = \sqrt{s}$, s in centimeters, F in dynes, $s_0 = 0$, $s_1 = 8$.

8 $F = (1 + s)^{2/3}$, s in feet, F in pounds, $s_0 = -7$, $s_1 = 7$.

In problems 9 through 12, a perfectly elastic spring is stretched from its relaxed position through b units. When extended these b units, the stretching force on the spring is F_b units. The spring constant is k. From the given information, find the work done.

9 $b = 10$ feet, $F_b = 100$ pounds.

10 $b = 0.03$ centimeter, $F_b = 15{,}000$ dynes.

11 $k = 2500$ dynes/cm, $F_b = 10{,}000$ dynes.

12 $k = 32$ lb/in., $b = 6$ inches.

13 A perfectly elastic spring with spring constant $k = 150$ lb/in. is stretched through 6 inches. At the start of this stretching, the spring is not relaxed, and the force on the spring is 300 pounds. How much work is done in stretching the spring through the 6 inches?

14 A bucket containing sand is lifted, starting at ground level, at a constant speed of 2 ft/sec. The bucket weighs 3 pounds and, at the start, is filled with 70 pounds of sand. As the bucket is lifted, sand runs out of a hole in the bottom at the constant rate of 1 pound per second. How much work is done in lifting the bucket up to the height at which the last of the sand runs out?

15 If M grams of mass are distributed uniformly along the x axis between the origin and the point $x = a$ centimeters, where $a > 0$, find the gravitational force of attraction of this mass on a 1-gram particle P situated at the point $x = -b$, where $b > 0$. Use Newton's formula for the gravitational force F between two particles of mass m_1 and m_2 separated by the distance s, $F = \gamma(m_1 m_2/s^2)$, $\gamma = 6.6732 \times 10^{-8}$ dyne cm^2/g^2.

16 Two thin rods made of gold are each 1 meter long and each rod contains 250 grams of gold. These rods are placed end to end along the x axis so that they meet at the origin. Let F be the gravitational force of attraction of the right-hand rod on the portion of the left-hand rod between its left end and the point with coordinate x, $-100 \le x \le 0$. Find the differential equation for F and give the initial value condition.

17 The marginal cost C' of producing x wigs is given by the equation $C' = 12 - (8/\sqrt{x})$ dollars per wig.
  (a) Find the total cost C as a function of x given that the total cost of manufacturing 100 wigs is $1200.
  (b) If each wig sells for $21, find the total profit P as a function of x.

18 The marginal cost C' of producing x electric razors per month is given by $C' = Ax - B$, where A and B are positive constants. Each razor sells for \$$k$. Give a formula for the number of razors that should be manufactured to maximize the monthly profit in terms of A, B, and k.

19 The marginal cost dC/dx of producing x thousand cans of baby food is given by $dC/dx = 30x^{-2/3}$, where the production cost is in dollars. Given that 8000 cans can be produced for \$600, how much will it cost to produce 125,000 cans?

20 The total cost of manufacturing x thousand yoyos is given by $C = 600 + 60x$ dollars. The corresponding marginal revenue is $dR/dx = 400 - 8x$. Find the number of yoyos (in thousands) that will maximize the manufacturer's profit.

21 The cost of production of a newspaper in a small town is $C = 3.5x + 100$ dollars per month, where x is the number of subscriptions to the paper. The marginal revenue is given by $dR/dx = 13 - (x/40)$ dollars per subscriber per month, and $R = 0$ if $x = 0$.
(a) Find R as a function of x.
(b) Find the total profit P as a function of x.
(c) What value of x will maximize the total profit?
(d) If the value of x is that which maximizes the profit, what price does each subscriber pay per month for a subscription?

22 The Frostbite Frozen Foods Company can manufacture x thousand frozen macaroni and cheese dinners for $350x + 10,000$ dollars. The marginal revenue is $dR/dx = 515 - (x/2)$ dollars per thousand dinners.
(a) How many macaroni and cheese dinners should be manufactured to maximize the total revenue?
(b) How many dinners should be manufactured to maximize total profit?
(c) When total profit is maximized, what price does Frostbite Co. charge for a single macaroni dinner? (Assume that $R = 0$ when $x = 0$.)

Figure 9

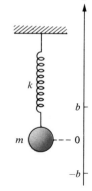

Problems 23 through 27 refer to Figure 9, in which a mass m is suspended by a perfectly elastic spring with spring constant k. The mass of the spring itself and air friction are to be neglected. A vertical s axis is set up so that, when the mass and spring are hanging in equilibrium, the s coordinate of m is zero. The mass m is lifted to the position with coordinate $s = b$ and released at time $t = 0$ with an initial velocity $v_0 = 0$. Naturally, the mass bobs up and down between b and $-b$.

23 By Hooke's law, the unbalanced force F exerted on the mass m by the spring is given by $F = -ks$. Explain the negative sign.

24 Show that the differential equation governing the motion of the mass m is $\dfrac{m}{k}\dfrac{d^2s}{dt^2} + s = 0$.

25 Given that the potential energy is zero when the mass passes through the origin, find a formula giving the potential energy V in terms of s.

26 (a) Find the (constant) total energy E of the system.
(b) Prove that $mv^2 + ks^2 = kb^2$ holds at all times.

27 Find the velocity v of the mass m at the instant when it passes through the position $s = 0$ for the first time after it is released. (*Hint:* At this instant, v is negative and all the energy is kinetic.)

28 A particle P is moving along the s axis under the influence of a force F.
(a) Explain why the potential energy V of P is only defined up to an additive constant.
(b) Explain why $F = -dV/ds$.

29 A particle P is moving along the positive s axis under the influence of a force F. The total energy E of P is zero and the potential energy is given by $V = -1/s$. Find the velocity $v = ds/dt$ of P in terms of s.

30 The particle P in problem 29 is at $s = 25$ centimeters when $t = 0$ and its velocity is negative when $t = 0$. Find the equation of motion of P.

31 Suppose that the particle P in problem 29 starts at $s = 25$ centimeters with a positive velocity when $t = 0$. Find the equation of motion of P and show that the velocity v of P is positive for all $t > 0$. Also find $\lim\limits_{t \to +\infty} v$.

32 Two particles P_1 and P_2 with masses m_1 and m_2, respectively, are moving on the s axis with variable velocities v_1 and v_2, respectively. A variable force of attraction exists between the two particles, so the force exerted on P_1 by P_2 is F_1 and the force exerted by P_1 on P_2 is F_2. Given that $-F_1 = F_2$ at all times, prove that the sum $m_1 v_1 + m_2 v_2$ of the linear momentum of P_1 and the linear momentum of P_2 remains constant during the motion.

33 A particle with unit positive electrical charge is fixed at the origin and a similarly charged particle P is brought in toward the origin starting from a point 10 units away from the origin. In physics, it is shown that the work W done in moving P against the force of repulsion caused by the similarly charged particle at O is independent of the path followed and depends only on the distance r from O to P (Figure 10). Furthermore, W satisfies the differential equation $dW/dr = -1/r^2$. Given that $W = 0$ when $r = 10$, find the equation that gives W in terms of r.

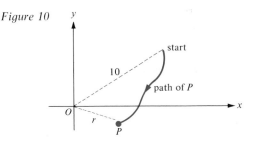

Figure 10

34 The graph of $y = f(x)$ contains the point $(1, 4)$ and the tangent line to this graph at the point $(1, 4)$ has slope 1. Find f if $f''(x) = 4x$ holds for all values of x in R.

5 Areas of Regions in the Plane by the Method of Slicing

Figure 1

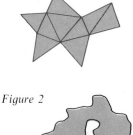

One of the most important applications of antidifferentiation is finding the areas of regions in the plane. If a region has a boundary consisting of a finite number of straight-line segments, it can be broken up into a finite number of nonoverlapping triangles (Figure 1) and its area can be found by summing up the areas of these triangles. However, if the region is bounded by curves (Figure 2), it may not be immediately clear how to calculate its area.

Before we attack the problem of calculating the area of a region such as the one shown in Figure 2, we should perhaps ask just what is *meant* by the "area" of such a figure. Unfortunately, it is not possible to provide an answer without getting into complications which we prefer to avoid here. Thus, we are going to assume that the reader understands what is meant by the "area of a region," if only in some rough-and-ready sense.

Figure 2

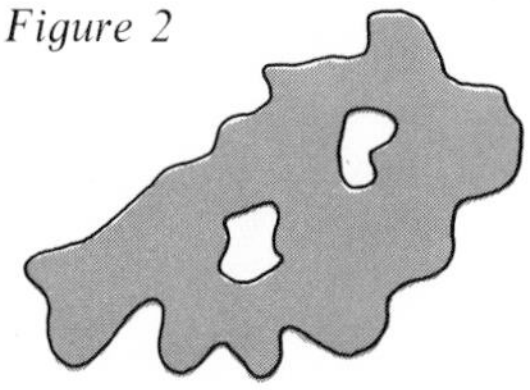

5.1 Areas by Slicing

The *method of slicing* is an effective technique for finding the areas of many planar regions. Given such a region, choose a convenient "reference" axis, say the s axis (Figure 3). At each point along this axis, construct a perpendicular line intersecting the region in a line segment of length l. Notice that l is a function of s. Suppose that the entire region lies between the perpendicular at $s = a$ and the perpendicular at $s = b$. Let A denote the area of the portion of the region between the perpendiculars at a and at s (Figure 3). Evidently, A is a function of s and $A = 0$ when $s = a$.

The "infinitesimal" point of view permits us to set up a differential equation for A. If s is increased by an "infinitesimal amount" ds, then A increases by a corresponding "infinitesimal amount" dA (Figure 4). Notice that dA is virtually the

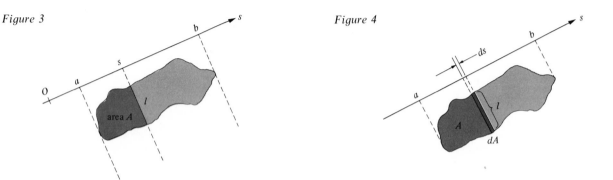

Figure 3

Figure 4

area of a small rectangle of length l and width ds; that is, $dA = l\,ds$. Therefore, A can be obtained by solving the differential equation $dA = l\,ds$ subject to the side condition $A = 0$ when $s = a$. The value of A when $s = b$ is the desired area.

EXAMPLES   Find the area of the given region by the method of slicing.

1 A triangle with base 5 meters and height 8 meters.

SOLUTION
In Figure 5, we take the reference s axis perpendicular to the base of the triangle with the origin at the level of the base, so that $dA = l\,ds$. From Figure 5 and similar triangles, $\dfrac{l}{8 - s} = \dfrac{5}{8}$; hence, $l = \frac{5}{8}(8 - s)$. The differential equation $dA = l\,ds$ can now be solved to obtain

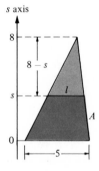

Figure 5

$$A = \int l\,ds = \int \tfrac{5}{8}(8 - s)\,ds = 5\int ds - \tfrac{5}{8}\int s\,ds = 5s - \tfrac{5}{16}s^2 + C.$$

Since $A = 0$ when $s = 0$, we have $C = 0$; hence, $A = 5s - \frac{5}{16}s^2$. The area of the entire triangle is obtained by putting $s = 8$ and calculating $A = (5)(8) - (\frac{5}{16})(8)^2 = 20$ square meters. (This, of course, corresponds to the result obtained by the usual formula "one-half the height times the base.")

2 The region in the xy plane bounded below by the parabola $y = x^2$ and above by the horizontal line $y = 4$.

SOLUTION
In Figure 6, we take the y axis as the reference axis for the method of slicing, so that $dA = l\,dy$. From Figure 6, $l = 2x$, where $x > 0$ and (x, y) lies on the graph of $y = x^2$. Therefore, $x = \sqrt{y}$, $l = 2\sqrt{y}$, and so

$$A = \int l\,dy = \int 2\sqrt{y}\,dy = 2\int y^{1/2}\,dy = 2(\tfrac{2}{3}y^{3/2}) + C = \tfrac{4}{3}y^{3/2} + C.$$

Figure 6

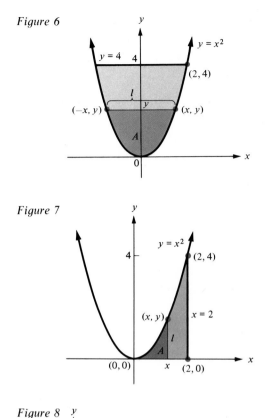

Figure 7

Figure 8

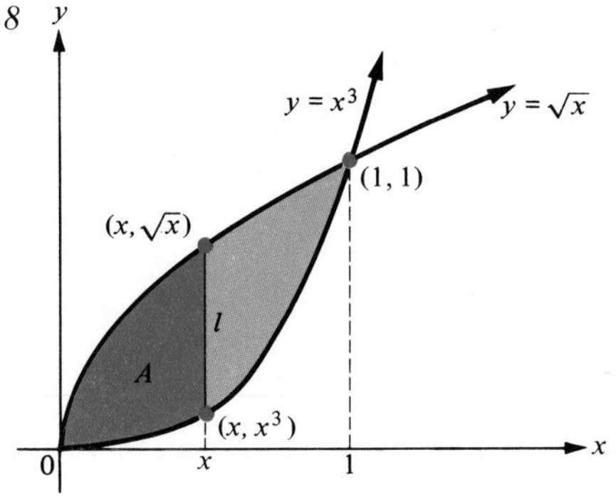

Since $A = 0$ when $y = 0$, it follows that $C = 0$ and $A = \frac{4}{3}y^{3/2}$. When $y = 4$, $A = \frac{4}{3}(4^{3/2}) = \frac{32}{3}$ square units.

3 The region in the first quadrant of the xy plane bounded above by the parabola $y = x^2$, below by the x axis, and on the right by the vertical line $x = 2$.

SOLUTION
We use the method of slicing, taking the x axis as the axis of reference. Here, the differential equation is $dA = l\,dx$, where, by Figure 7, $l = y = x^2$. Thus,

$$A = \int l\,dx = \int x^2\,dx = \frac{x^3}{3} + C.$$

Since $A = 0$ when $x = 0$, we have $C = 0$ and $A = x^3/3$. Hence, when $x = 2$, $A = \frac{8}{3}$ square units.

4 The region in the xy plane between the curve $y = \sqrt{x}$ and the curve $y = x^3$.

SOLUTION
The region in question is shown in Figure 8. If we take the x axis as the axis of reference for the method of slicing, then $dA = l\,dx$. Since $l = \sqrt{x} - x^3$, it follows that

$$A = \int l\,dx = \int (\sqrt{x} - x^3)\,dx = \int x^{1/2}\,dx - \int x^3\,dx$$

$$= \frac{2}{3}x^{3/2} - \frac{x^4}{4} + C.$$

When $x = 0$, $A = 0$, so $C = 0$. When $x = 1$,

$$A = \frac{2}{3}(1)^{3/2} - \frac{1}{4} = \frac{5}{12} \text{ square unit.}$$

Two remarks should be made in connection with the method of slicing. First, although our "derivation" of the differential equation $dA = l\,ds$ for area (Figure 4) might make it seem that an approximation is involved, nevertheless, it is possible to prove that the differential equation holds *exactly*, and, with the appropriate choice of the constant of integration, $A = \int l\,ds$ gives the area *exactly*. Second, the "reference axis" can be chosen arbitrarily—the answer will always be the same. Of course, adroit selection of the reference axis may simplify the actual details of the calculation.

Problem Set 5

In problems 1 through 5, find the area of the shaded region in Figures 9 through 13 by the method of slicing, using the indicated axis as the reference axis.

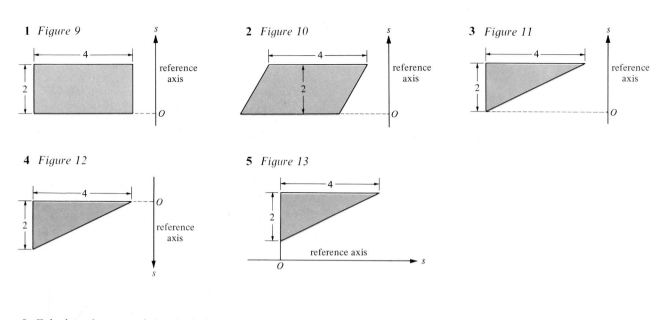

1 *Figure 9*

2 *Figure 10*

3 *Figure 11*

4 *Figure 12*

5 *Figure 13*

6 Calculate the area of the shaded region in Figure 14 by the method of slicing in two ways:
(a) Using the x axis as the reference axis.
(b) Using the y axis as the reference axis.

Figure 14

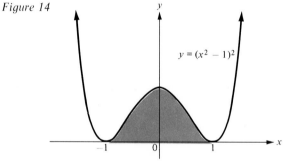

$y = (x^2 - 1)^2$

In problems 7 through 20, calculate the area of the given region by the method of slicing using the given axis as the reference axis. Sketch the region in the xy plane.

7 The region bounded above by $y = \sqrt{x - 2}$, on the left by $x = 2$, on the right by $x = 6$, and below by $y = 0$. Take the x axis as the reference axis.

8 Same as problem 7, but take the y axis as the reference axis.

9 The region between $y^2 = x$ and $y - x + 2 = 0$. Take the y axis as the reference axis.

10 The region between $y = x^2 - 6x + 8$ and $y = -x^2 + 4x - 3$. Take the x axis as the reference axis.

11 The triangular region bounded by the straight lines $y + 2x - 2 = 0$, $y - x - 5 = 0$, and $y = 7$. Take the y axis as the reference axis.

12 The region between $x = y^2 - 4$ and $x = 2 - y^2$. Take the y axis as the reference axis.

13 The region between $y^2 = 1 - x$ and $y = x + 5$. Take the y axis as the reference axis.

14 The region under the curve $y = x\sqrt{25 - x^2}$ between $x = -5$ and $x = 0$. Take the x axis as the reference axis.

15 The region between $y = 4 - x^2$ and $y = x^2 - 2$. Take the x axis as the reference axis.

16 The region between $y = x^2$ and $y = 2x$. Take the x axis as the reference axis.

17 The region between $y = x^2$ and $x = y^2$. Take the x axis as the reference axis.

18 The region between $y = x$, $y = 2 - x$, and $y = 0$. Take the y axis as the reference axis.

19 The region between $y = 4 - x^2$ and $y = -2$. Take the x axis as the reference axis.

20 The region between $y = x^4 + 1$ and $y = 17$. Take the y axis as the reference axis.

6 Area Under the Graph of a Function—The Definite Integral

It is traditional to refer to the region between the graph of a function f and the x axis as the region "under" the graph of f, in spite of the fact that, if the graph drops below the x axis, part of this region may actually lie above the graph. Thus, we speak of the shaded region in Figure 1 as the *region under the graph of f between $x = a$ and $x = b$.* Denote by A_1 the area of the portion of this region that lies above the x axis and by A_2 the area of the portion that lies below the x axis, so that $A_1 + A_2$ is the total area of the region. In many applications of calculus, it is appropriate to subtract the area A_2 from the area A_1 to form the quantity $A_1 - A_2$, which is called the *signed area* under the graph of f between $x = a$ and $x = b$.

The idea of signed area enables us to give a preliminary informal definition of the *definite integral*—one of the most useful and significant concepts in calculus. (In Section 2 of Chapter 6, we give a formal analytic definition of the definite integral.)

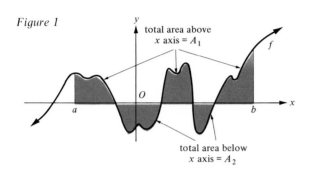

Figure 1

total area above
x axis $= A_1$

total area below
x axis $= A_2$

DEFINITION 1   **The Definite Integral (Preliminary Informal Version)**

Let f be a function defined at least on the closed interval $[a, b]$. Then the signed area under the graph of f between $x = a$ and $x = b$ is denoted by $\int_a^b f(x)\, dx$. Thus,

$$\int_a^b f(x)\, dx = A_1 - A_2 \text{ (Figure 1)}.$$

The expression $\int_a^b f(x)\, dx$ is called *the definite integral from a to b of $f(x)\, dx$;* the function f [or the expression $f(x)$] is called the *integrand;* and the interval $[a, b]$ is called the *interval of integration.* The numbers a and b are called the *lower* and the *upper limits of integration,* respectively. (This use of the word "limit" should not be confused with the limit of a function.)

We shall soon see just why the notation chosen for the definite integral is so similar to the notation $\int f(x)\, dx$ for the indefinite integral. For now, notice that the definite integral $\int_a^b f(x)\, dx$ is a definite number $A_1 - A_2$, while the indefinite integral $\int f(x)\, dx$ is a function $g(x) + C$ which involves an arbitrary constant C.

The following example shows how a simple definite integral can be found geometrically.

EXAMPLE   Let f be the function defined by $f(x) = 1 - x$. Use elementary geometry to find $\int_{-1}^{2} f(x)\, dx$.

Figure 2

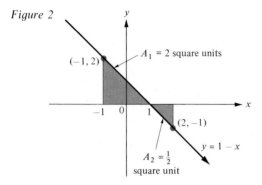

SOLUTION

In Figure 2 the area A_1 above the x axis is the area of a triangle with base 2 units and height 2 units; hence, $A_1 = \frac{1}{2}(2)(2) = 2$ square units. Similarly, the area A_2 below the x axis is the area of a triangle with base 1 unit and height 1 unit; hence, $A_2 = \frac{1}{2}(1)(1) = \frac{1}{2}$ square unit. Thus, by Definition 1,

$$\int_{-1}^{2} f(x)\,dx = \int_{-1}^{2} (1-x)\,dx = A_1 - A_2 = 2 - \tfrac{1}{2} = \tfrac{3}{2}.$$

If the graph of the function f is more complicated than the graph in Figure 2, then elementary geometry may not provide us with a numerical value for the definite integral $\int_{a}^{b} f(x)\,dx$. However, there is a theorem that gives us an analytic means for evaluating definite integrals—a theorem so basic and so important that it is called the *fundamental theorem of calculus.* Here we give a preliminary informal version of this theorem. (A formal version and a rigorous proof are given in Section 4.1 of Chapter 6.)

THEOREM 1   Fundamental Theorem of Calculus—Preliminary Informal Version

Suppose that f is a continuous function on the closed interval $[a, b]$ and that

$$\int f(x)\,dx = g(x) + C.$$

Then,

$$\int_{a}^{b} f(x)\,dx = g(b) - g(a).$$

Notice that the fundamental theorem of calculus connects the indefinite integral $\int f(x)\,dx$ and the definite integral $\int_{a}^{b} f(x)\,dx$; in fact, it allows us to calculate the definite integral if we know the indefinite integral. We cannot give a rigorous proof of this theorem here—for one thing, we have no formal definition of "area." However, we can make the fundamental theorem *plausible* by using the method of slicing as follows:

In Figure 3 let $a \le x \le b$ and denote the signed area under the graph of f between a and x by I. Note that I depends on x, that $I = 0$ when $x = a$, and that $I = \int_{a}^{b} f(x)\,dx$ when $x = b$. We proceed as before with the method of slicing, taking the x axis as the axis of reference. When the graph of f is above the x axis, the ordinate $f(x)$ is positive, and we have the differential equation $dI = f(x)\,dx$ as usual. However, when the graph drops below the x axis, the ordinate $f(x)$ becomes *negative*, so $f(x)\,dx$ becomes *negative.* But, since area below the x axis is to be *subtracted* when computing the signed area I, we see that dI is also negative when the graph drops below the x axis; hence, the differential equation $dI = f(x)\,dx$ continues to hold.

Figure 3

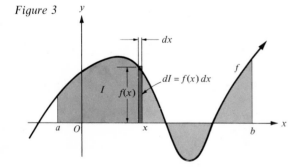

Antidifferentiating both sides of the differential equation $dI = f(x)\,dx$, we obtain $I = \int f(x)\,dx = g(x) + C$. From the initial value condition $I = 0$ when $x = a$, we have $0 = g(a) + C$, and it follows that $C = -g(a)$. Therefore, $I = g(x) - g(a)$. When $x = b$, $I = \int_a^b f(x)\,dx$; hence, $\int_a^b f(x)\,dx = g(b) - g(a)$. This argument should make the fundamental theorem of calculus plausible.

EXAMPLE   Use the fundamental theorem of calculus to find the area of the region under the graph of $f(x) = x^{2/3}$ between $x = 0$ and $x = 1$.

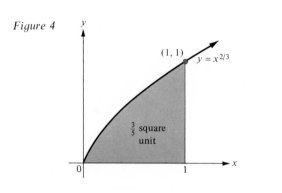

Figure 4

SOLUTION
In Figure 4 we see that the region under the graph of $f(x) = x^{2/3}$ between $x = 0$ and $x = 1$ *is entirely above the x axis.* Therefore, by Definition 1, its area is equal to $\int_0^1 x^{2/3}\,dx$. The indefinite integral $\int x^{2/3}\,dx$ is given by $\int x^{2/3}\,dx = \frac{3}{5}x^{5/3} + C$. Consequently, by the fundamental theorem of calculus,

$$\int_0^1 x^{2/3}\,dx = \tfrac{3}{5}(1)^{5/3} - \tfrac{3}{5}(0)^{5/3} = \tfrac{3}{5} \text{ square unit.}$$

The simple notational device introduced in the following definition makes the fundamental theorem easier to state and to use.

DEFINITION 2   **Special Notation**
If g is any function and if the numbers a and b belong to the domain of g, then the notation $g(x)\Big|_a^b$, read "$g(x)$ *evaluated between* $x = a$ *and* $x = b$," is defined by

$$g(x)\Big|_a^b = g(b) - g(a).$$

For instance, $x^2\Big|_1^2 = 2^2 - 1^2$; $f(x)\Big|_t^{t+\Delta t} = f(t + \Delta t) - f(t)$; and

$$(x^3 - 3x + 1)\Big|_{-1}^{1} = [1^3 - 3(1) + 1] - [(-1)^3 - 3(-1) + 1] = -4.$$

Using this special notation, we can now write the fundamental theorem of calculus as

$$\int_a^b f(x)\,dx = \left[\int f(x)\,dx\right]_a^b,$$

a form in which it is easy to remember and to use. Indeed, if $\int f(x)\,dx = g(x) + C$, then

$$\left[\int f(x)\,dx\right]_a^b = [g(x) + C]\Big|_a^b = [g(b) + C] - [g(a) + C] = g(b) - g(a) = \int_a^b f(x)\,dx.$$

Notice how the constant of integration C cancels out in the calculation above. Thus, in using the fundamental theorem to evaluate a definite integral, the constant of integration in the corresponding indefinite integral can safely be neglected.

EXAMPLES   Evaluate the given definite integral by using the fundamental theorem of calculus.

1 $\displaystyle\int_0^1 (x^2 + 1)\, dx$

SOLUTION

$$\int_0^1 (x^2 + 1)\, dx = \left[\int (x^2 + 1)\, dx \right]\Big|_0^1 = \left[\frac{x^3}{3} + x \right]\Big|_0^1 = \left(\frac{1^3}{3} + 1 \right) - \left(\frac{0^3}{3} + 0 \right) = \frac{4}{3}.$$

2 $\displaystyle\int_1^4 \frac{1 - x}{\sqrt{x}}\, dx$

SOLUTION

$$\int_1^4 \frac{1 - x}{\sqrt{x}}\, dx = \int_1^4 \left(\frac{1}{\sqrt{x}} - \frac{x}{\sqrt{x}} \right) dx = \int_1^4 (x^{-1/2} - x^{1/2})\, dx$$

$$= \left[\int (x^{-1/2} - x^{1/2})\, dx \right]\Big|_1^4 = \left[\frac{x^{1/2}}{\frac{1}{2}} - \frac{x^{3/2}}{\frac{3}{2}} \right]\Big|_1^4$$

$$= \left[2x^{1/2} - \frac{2}{3} x^{3/2} \right]\Big|_1^4$$

$$= \left[2(4)^{1/2} - \frac{2}{3}(4)^{3/2} \right] - \left[2(1)^{1/2} - \frac{2}{3}(1)^{3/2} \right]$$

$$= -\frac{4}{3} - \frac{4}{3} = -\frac{8}{3}.$$

The following examples illustrate the use of the fundamental theorem of calculus to calculate areas.

EXAMPLES   Use the fundamental theorem of calculus to find the area of the given region.

1 The region below the graph of $f(x) = \dfrac{x^2 - 2x + 8}{8}$ between $x = -2$ and $x = 4$.

SOLUTION
Since the region lies entirely above the x axis (Figure 5), its area A is given by

$$A = \int_{-2}^4 \frac{x^2 - 2x + 8}{8}\, dx = \left[\int \frac{x^2 - 2x + 8}{8}\, dx \right]\Big|_{-2}^4$$

$$= \left[\frac{x^3}{24} - \frac{x^2}{8} + x \right]\Big|_{-2}^4$$

$$= \left[\frac{4^3}{24} - \frac{4^2}{8} + 4 \right] - \left[\frac{(-2)^3}{24} - \frac{(-2)^2}{8} - 2 \right]$$

$$= \frac{15}{2} \text{ square units.}$$

Figure 5

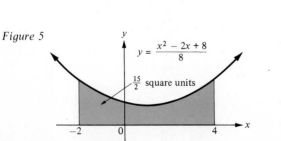

$$y = \frac{x^2 - 2x + 8}{8}$$

$\frac{15}{2}$ square units

Figure 6

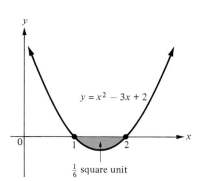

$y = x^2 - 3x + 2$

$\frac{1}{6}$ square unit

2 The region bounded below by the graph of

$$y = x^2 - 3x + 2$$

and bounded above by the x axis (Figure 6).

SOLUTION

By definition,

$$\int_1^2 (x^2 - 3x + 2)\, dx = A_1 - A_2,$$

where A_1 is the area above the x axis and A_2 is the area below the x axis between $x = 1$ and $x = 2$ (Figure 6). Since all the area is below the x axis, $A_1 = 0$ and A_2 is the desired area. Thus,

$$A_2 = -\int_1^2 (x^2 - 3x + 2)\, dx = -\left[\int (x^2 - 3x + 2)\, dx\right]\bigg|_1^2$$

$$= -\left[\frac{x^3}{3} - \frac{3x^2}{2} + 2x\right]\bigg|_1^2$$

$$= -\left[\left(\frac{2^3}{3} - \frac{3(2)^2}{2} + 2(2)\right) - \left(\frac{1^3}{3} - \frac{3(1)^2}{2} + 2(1)\right)\right]$$

$$= -\left(-\frac{1}{6}\right) = \frac{1}{6}\ \text{square unit.}$$

The situation illustrated in the example above arises quite frequently. Evidently, whenever the region between $x = a$ and $x = b$ bounded by a curve $y = f(x)$ and the x axis lies entirely below the x axis, then the area of this region is the *negative* of the definite integral $\int_a^b f(x)\, dx$.

A region may have geometric symmetries which can be exploited in calculating its area. In particular, the region under the graph of an even or odd function exhibits such symmetries.

EXAMPLE   Find the area of the region under the curve $f(x) = x\sqrt{4 - x^2}$ between $x = -\sqrt{2}$ and $x = \sqrt{2}$ (Figure 7).

Figure 7

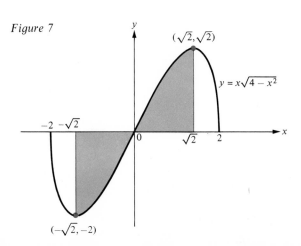

$(\sqrt{2}, \sqrt{2})$

$y = x\sqrt{4 - x^2}$

$-2\ -\sqrt{2}$

$\sqrt{2}$ 2

$(-\sqrt{2}, -2)$

SOLUTION

Here, we cannot find the area by evaluating $\int_{-\sqrt{2}}^{\sqrt{2}} x\sqrt{4 - x^2}\, dx$, since this definite integral has the value zero. (Why?) However, since f is an odd function, the shaded region in Figure 7 is symmetric about the origin, so the desired area is given by $2\int_0^{\sqrt{2}} x\sqrt{4 - x^2}\, dx$. Making the change of variable $u = 4 - x^2$, and noting that $du = -2x\, dx$, so that $x\, dx = -\frac{1}{2}\, du$, we have

$$\int x\sqrt{4 - x^2}\, dx = -\frac{1}{2}\int \sqrt{u}\, du = (-\tfrac{1}{2})(\tfrac{2}{3})u^{3/2} + C$$

$$= (-\tfrac{1}{3})(4 - x^2)^{3/2} + C.$$

Consequently, the shaded area A in Figure 7 is given by

$$A = 2 \int_0^{\sqrt{2}} x\sqrt{4 - x^2}\, dx = 2[(-\tfrac{1}{3})(4 - x^2)^{3/2}]\Big|_0^{\sqrt{2}}$$

$$= 2[(-\tfrac{1}{3})(4 - 2)^{3/2} - (-\tfrac{1}{3})(4 - 0)^{3/2}]$$

$$= 2\left(\frac{-\sqrt{8} + 8}{3}\right) = \frac{16 - 4\sqrt{2}}{3} \approx 3.45 \text{ square units.}$$

Problem Set 6

In problems 1 through 6, evaluate each definite integral by using *elementary geometry*—do not use the fundamental theorem of calculus.

1 $\int_0^5 x\, dx$

2 $\int_{-2}^1 2x\, dx$

3 $\int_{-1}^1 \sqrt{1 - x^2}\, dx$ (*Hint*: The area of a circle of radius r is πr^2.)

4 $\int_{-1}^1 f(x)\, dx$, where $f(x) = \begin{cases} \sqrt{1 - x^2} & \text{if } -1 \leq x \leq 0 \\ 1 - x & \text{if } 0 < x \leq 1 \end{cases}$

5 $\int_{-2}^2 |x|\, dx$

6 $\int_{-2}^2 \dfrac{x\, dx}{1 + x^2}$

In problems 7 through 14, write the area of the given region as a definite integral or as a sum or difference of definite integrals. Do not evaluate the integrals. Also sketch the graph.

7 The region under the curve $y = 3x - x^2$ between $x = 0$ and $x = 3$.

8 The region under the curve $y = 1/x$ between $x = 1$ and $x = 2.718$.

9 The region under the curve $y = 2x - 3$ between $x = 0$ and $x = \frac{3}{2}$.

10 The region under the curve $y = 2x - 3$ between $x = 0$ and $x = 5$.

11 The region between the curve $y = 2 - x^2$ and the x axis.

12 The region between the curve $y = x^2 - 1$ and the x axis.

13 The region under the curve $y = \frac{1}{3}x^3 - x$ between $x = -4$ and $x = 4$.

14 The region under the curve $y = 1 - x^{2/3}$ between $x = -8$ and $x = 8$.

In problems 15 through 20, use the fundamental theorem of calculus to evaluate each definite integral.

15 $\int_0^2 (3x^2 - 2x + 1)\, dx$

16 $\int_{-1}^0 (x^3 + x^2 + x - 1)\, dx$

17 $\int_0^{64} \dfrac{\sqrt{x}}{8}\, dx$

18 $\int_{-1/4}^{1/4} x\sqrt{1 - x^2}\, dx$

19 $\int_{-1}^2 \dfrac{x\, dx}{\sqrt{x^2 + 1}}$

20 $\int_1^2 x\sqrt{1 + x}\, dx$

In problems 21 through 30, set up an expression for the area of the given region in terms of definite integrals, and then use the fundamental theorem of calculus to evaluate the integrals and thus find the area of the region. Sketch the region in the xy plane.

21 The region under the curve $y = 10x - (x^2 + 24)$ between $x = 4$ and $x = 6$.

22 The region under the curve $y = 1 - x^2$ between $x = -1$ and $x = \frac{1}{2}$.

23 The region under the curve $y = x^2 - 3x$ between $x = 0$ and $x = 3$.

24 The region under the curve $y = -x^2$ between $x = 0$ and $x = 2$.

25 The region under the curve $y = x^4$ between $x = -2$ and $x = 1$.

26 The region under the curve $y = 2x^2 - 11x + 5$ between $x = 0$ and $x = 5$.

27 The region under the curve $y = -\frac{1}{3}x^3 + x^2$ between $x = -2$ and $x = 4$.

28 The region under the curve $y = x/(x^2 + 1)^2$ between $x = -2$ and $x = 2$.

29 The region under the curve $y = \frac{1}{3}x^3 - x$ between $x = -\sqrt{3}$ and $x = \sqrt{3}$.

30 The region given in problem 14 under the curve $y = 1 - x^{2/3}$ between $x = -8$ and $x = 8$.

In problems 31 through 39, find the area of each region.

31 The region bounded by $y = x^3$, $x = -2$, and $x = 2$.

32 The region bounded above by $y = x^2$, below by the
x axis, and on the left and right by $y = 2 - x^2$.

Figure 8

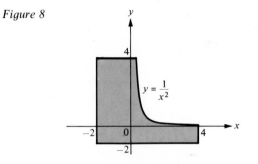

33 The region shown in Figure 8.

34 The region under the graph of the function f given by

$$f(x) = \begin{cases} 2x - 1 & \text{for } -3 \le x < 0 \\ x + 1 & \text{for } 0 \le x \le 4 \\ 5 & \text{for } x > 4 \end{cases}$$

between $x = -3$ and $x = 8$.

35 The region bounded above by $y = 1 - x^2$ and below
by $y = |x| - 1$.

36 The region between the two parallel lines $y = 2x + 8$ and $y = 2x + 3$ cut off by the
parabola $y = x^2$.

37 The quadrilateral region with vertices $(0, 2)$, $(1, 0)$, $(4, 1)$, and $(1, 3)$.

38 The region consisting of all points (x, y) that satisfy $|x| + |y| \le 1$.

39 The region bounded by $y = x + 2$, $y = 3x - 3$, $y = 1 - x$, and $2y + 3x + 6 = 0$.

40 Justify the following assertions geometrically:

(a) If f is a continuous even function and $a > 0$, then $\int_{-a}^{a} f(x)\, dx = 2 \int_{0}^{a} f(x)\, dx$.

(b) If f is a continuous odd function and $a > 0$, then $\int_{-a}^{a} f(x)\, dx = 0$.

(c) If f is a continuous function and $a < b < c$, then

$$\int_{a}^{c} f(x)\, dx = \int_{a}^{b} f(x)\, dx + \int_{b}^{c} f(x)\, dx.$$

Review Problem Set

In problems 1 through 4, put $dx = \Delta x$ and find Δy, dy, and $\Delta y - dy$.

1 $y = x^2 + 1$; $x = 2$, $\Delta x = 0.01$

2 $y = \sqrt{x}$; $x = 1$, $\Delta x = 0.23$

3 $y = x^3$; $x = 2$, $\Delta x = 0.02$

4 $y = 1/x$; $x = 2$, $\Delta x = -0.5$

In problems 5 through 8, find dy.

5 $y = x^3 - x$ **6** $y = \sqrt{4 - x^2}$ **7** $y = \dfrac{x^2 + 5}{2x + 1}$ **8** $x^3 + y^3 - 6xy = 2$

9 If x and y are functions of t and $x^4 + y^4 - 4xy = 13$, find the algebraic relationship between dx and dy. From this relationship, solve for dy/dx.

10 Given $y = x^3 - 3x + 2$ and $x = 4t^2 - 3$, put $dt = \Delta t = 0.2$, $t = 1$, and find the corresponding values of Δy and dy.

11 Use differentials to find the approximate area of a walk 1 meter wide around a city square which is 100 meters on each side not counting the walk.

12 A large spherical tank of inside radius r has a thin metal wall of thickness Δr. Use differentials to write a formula for the approximate volume of metal in the wall.

13 Derive an approximate formula for the volume of metal in a can having the shape of a cube of edge x, the thickness Δx of the metal being small.

14 Sketch the graph of the function defined by $y = 1/x$, show a typical value x_0 in the domain of this function, indicate a small change $\Delta x = dx$ in the value of x from x_0 to $x_0 + \Delta x$, and show the corresponding values of Δy and dy on the graph. Show analytically that Δy is approximately equal to $-\Delta x/x^2$. Use this fact to find a decimal approximation of $\frac{1}{102}$. Indicate how much error is involved in this approximation.

15 Use differentials to find an approximation to $\sqrt[5]{33}$.

16 The diameter of a bar in the shape of a right circular cylinder is measured as 4.2 centimeters with a possible error of 0.05 centimeter. Give a reasonable estimate (using differentials) for the possible error in the cross-sectional area computed from the measured diameter.

In problems 17 through 28, evaluate each antiderivative.

17 $\displaystyle\int (3x^4 + 4x^2 + 11)\, dx$ **18** $\displaystyle\int (4x^3 + 3x^2 - x + 91)\, dx$ **19** $\displaystyle\int 3t \sqrt[3]{t}\, dt$

20 $\displaystyle\int (1 + 2t)^5\, dt$ **21** $\displaystyle\int \sqrt[7]{3t + 9}\, dt$ **22** $\displaystyle\int x^2(x^3 - 1)^{40}\, dx$

23 $\displaystyle\int x^2(x^3 + 8)^{17}\, dx$ **24** $\displaystyle\int x(x^2 + 4)^{-1/3}\, dx$ **25** $\displaystyle\int \dfrac{x^7\, dx}{\sqrt[5]{x^8 + 13}}$

26 $\displaystyle\int \dfrac{(\sqrt{x} - 3)^{44}\, dx}{\sqrt{x}}$ **27** $\displaystyle\int x\sqrt{7 + x}\, dx$ **28** $\displaystyle\int \dfrac{3t\, dt}{\sqrt{t + 5}}$

29 Evaluate $\displaystyle\int x^5\sqrt{x^3 + 1}\, dx$, starting with the change of variable $u = x^3 + 1$, and using the fact that $x^5 = x^3 \cdot x^2$.

In problems 30 through 37, find the general solution of each differential equation.

30 $\dfrac{dy}{dx} = 2x + 1$ **31** $\dfrac{dy}{dx} = (x - 4)(3x - 2)$ **32** $\dfrac{dy}{dx} = \dfrac{1}{(3 - x)^2}$ **33** $\dfrac{dy}{dx} = \dfrac{1 + x}{\sqrt{x}}$

34 $\dfrac{dy}{dx} = (1 - x^{3/2})^{15}\sqrt{x}$ **35** $\dfrac{dy}{dx} = \dfrac{\sqrt{1 + \sqrt{x}}}{\sqrt{x}}$ **36** $\dfrac{d^2y}{dx^2} = 3 - 2x + 6x^2$ **37** $\dfrac{d^2y}{dx^2} = \dfrac{1}{(1 - x)^4}$

In problems 38 through 43, solve each differential equation subject to the side conditions given.

38 $\dfrac{dy}{dx} = 2x^3 + 2x + 1$; $y = 0$ when $x = 0$.   **39** $\dfrac{dy}{dx} = x^{-1/3}$; $y = 0$ when $x = 1$.

40 $\dfrac{dy}{dx} = \dfrac{x}{\sqrt{1-x^2}}$; $y = -1$ when $x = 0$.

41 $\dfrac{dy}{dx} = x^2(1 + x^3)^{10}$; $y = 2$ when $x = 0$.

42 $\dfrac{d^2 y}{dx^2} = x^3 + 1$; $y = 0$ and $y' = 1$ when $x = 0$.

43 $\dfrac{d^2 y}{dx^2} = \dfrac{1}{x^3}$; $y = 2$ and $y' = 1$ when $x = 1$.

44 Suppose that $G' = g$, $F' = f$, and that u is a solution of the differential equation $g[u(x)] \cdot u'(x) = f(x)$. Show that there is a constant C such that $G \circ u = F + C$.

45 Find a formula for $f(x)$ given that the graph of f contains the point $(2,6)$ and that the slope of the tangent to this graph at each point $(x, f(x))$ is $x\sqrt{x^2 + 5}$.

46 (a) Show that there is no function f satisfying the differential equation $f'(x) = 3x^2 + 1$ such that $f(0) = 0$ and $f(1) = 3$.
(b) Show that there is a function f satisfying the differential equation $f''(x) = 3x^2 + 1$ such that $f(0) = 0$ and $f(1) = 3$. Explain, contrasting part (a) with part (b).

47 Solve the differential equation $dy/dx = |x| + |x - 1| + |x - 2|$ subject to the initial value condition that $y = 1$ when $x = 0$. $\left(\text{Hint: } \int |x| \, dx = x\,\dfrac{|x|}{2} + C.\right)$

48 The force F acting on a particle P moving along the s axis is given by $F = \sqrt{1 + \sqrt{s}}$ dynes. Find the work done by this force as the particle moves from $s = 1$ centimeter to $s = 0$ centimeter.

49 A perfectly elastic spring is compressed 6 inches from its natural length because a weight of 2 tons is placed on it. How much work (in foot-pounds) is done?

50 What constant negative acceleration is required to bring a train to rest in 500 meters if it is initially going at 44 m/sec?

51 An athlete running the 100-yard dash maintains a constant acceleration for the first 32 yards and thereafter has zero acceleration. What must the acceleration be if the athlete is to run the race in 9.4 seconds?

52 A certain type of motorcar can be brought to a stop from a speed of 72 km/hr in 4 seconds. How long will it take to bring the car to a stop from a speed of 96 km/hr? Assume the same constant deceleration in both cases.

53 A stone is thrown straight down with an initial velocity of 96 feet per second from a bridge 256 feet above a river.
(a) How many seconds elapse before the stone hits the water?
(b) What will be the velocity of the stone when it strikes the water?

54 Given that the acceleration of gravity at the surface of the earth is 980 cm/sec², explain why a 1-gram mass weighs 980 dynes at the surface of the earth.

55 A 500-pound weight is suspended 150 feet below a windlass by a cable weighing 0.75 lb/ft. Neglecting friction, how many foot-pounds of work will be required to lift the weight through the 150 feet?

56 Let W be the work done in charging a capacitor having (constant) capacitance C with Q coulombs. If E is the voltage drop across the capacitor and I is the current in amperes flowing into the capacitor, then $dW = E \, dQ$, $Q = CE$, and $dQ = I \, dt$, where t denotes time in seconds. (a) Show that the instantaneous power dW/dt required to charge the capacitor is given by $dW/dt = EI$. (b) Assuming that $W = 0$ when $E = 0$, prove that $W = \frac{1}{2}CE^2$.

57 Suppose that m grams of mass are distributed uniformly along the s axis between $s = a$ and $s = b$ centimeters, where $0 < a < b$. An M-gram particle P is placed on the s axis at the origin. Find the net gravitational force exerted on P by the distributed mass, given that the universal constant of gravitation is $\gamma = 6.6732 \times 10^{-8}$ dyne cm²/g².

58 A particle moves along a curved path in the plane in such a way that its coordinates at time t are $(f(t), g(t))$. Assume that f and g are differentiable functions and that f' and g' are continuous. Write a differential equation for the distance s, measured along the curved path, traveled by the particle since the instant when $t = 0$.

59 A company manufacturing pantyhose expects a marginal revenue given by $dR/dx = 5 - (x/25,000)$ dollars per pair manufactured. Assume that $R = \$0$ when $x = 0$. Its marginal cost is given by $dC/dx = 3$ dollars per pair manufactured, where $C = \$200$ when $x = 0$.
(a) Find the total revenue expected if x pairs of pantyhose are manufactured.
(b) Find the total cost of manufacturing x pairs of pantyhose.
(c) How many pairs of pantyhose should be manufactured to maximize the total revenue?
(d) How many pairs of pantyhose should be manufactured to maximize the total profit?
(e) How much should the company charge per pair to maximize the profit?

60 The expected total revenue for the sale of x sweaters is given by $R = x(27 - x/1000)$ dollars, while the marginal cost is given by $dC/dx = 700/\sqrt{x}$ dollars per sweater. When $x = 0$, then $C = \$500$.
(a) Find the total cost as a function of x.
(b) Show that the maximum profit is achieved by selling 10,000 sweaters.
(c) What would be the price per sweater if 10,000 sweaters were sold?

In problems 61 through 64, find the area of each region by the method of slicing, taking the given axis as the reference axis.
61 The region bounded above by $y = 1$, on the left by $y = x^2$, on the right by $y = x$, and below by $y = 0$. Use the y axis for the reference axis.

62 The region in the second quadrant under the graph of $y = \sqrt{x + 1}$. Use the y axis for the reference axis.

63 The same region as in problem 62, but use the x axis as the reference axis.

64 The region under the graph of $y = 1 - |x|$ between $x = -1$ and $x = 1$. Use the y axis as the reference axis.

In problems 65 through 71, evaluate each definite integral using only elementary geometry —do not use the fundamental theorem of calculus.

65 $\int_{0}^{2} (10 + 3x)\, dx$

66 $\int_{-1}^{2} (x + 4)\, dx$

67 $\int_{-2}^{2} (2x - 3)\, dx$

68 $\int_{0}^{5} -\sqrt{25 - x^2}\, dx$

69 $\int_{-1}^{2} (5 - 2|x|)\, dx$

70 $\int_{-1}^{3} f(x)\, dx$, where $f(x) = \begin{cases} 2x + 2 & \text{for } x < 0 \\ \dfrac{6 - 2x}{3} & \text{for } x \geq 0 \end{cases}$

71 $\int_{a}^{b} dx$, where $a < b$

72 Use elementary geometry to evaluate $\int_{-a}^{a} \sqrt{a^2 - x^2}\, dx$, where $a > 0$.

73 Suppose that n is an odd integer and that a is a positive constant. Explain *geometrically* why $\int_{-a}^{a} x^n\, dx = 0$.

74 Evaluate $\int_{-43}^{43} \dfrac{x^{17}\, dx}{43 + x^4}$.

In problems 75 through 78, write the area of each region as a definite integral or as a sum or difference of definite integrals. Do not evaluate the integrals. Sketch the region.

75 The region under the curve $y = \dfrac{1}{1 + x}$ between $x = -3$ and $x = -2$.

76 The region under the curve $y = \dfrac{x^3}{1 + x^2}$ between $x = -1$ and $x = 1$.

77 The region under the curve $y = \dfrac{|x| - x}{|x| + 1}$ between $x = -2$ and $x = 2$.

78 The region under the curve $y = \sin x$ between $x = -2\pi$ and $x = 2\pi$.

In problems 79 through 84, use the fundamental theorem of calculus to evaluate each definite integral.

79 $\displaystyle\int_{-2}^{2} (4x^3 - 1)\, dx$

80 $\displaystyle\int_{1}^{4} \frac{(2x^3 + x^2 - 1)\, dx}{\sqrt{x}}$

81 $\displaystyle\int_{0}^{1} x^2 \sqrt{3 + x^3}\, dx$

82 $\displaystyle\int_{a}^{b} f'(x)\, dx$

83 $\displaystyle\int_{-8}^{8} (2 - x^{2/3})\, dx$

84 $\displaystyle\int_{-1/2}^{1} \frac{x^2\, dx}{(x^3 + 1)^2}$

6 THE DEFINITE OR RIEMANN INTEGRAL

In Section 6 of Chapter 5 we gave a preliminary definition of the definite integral and stated a preliminary version of the fundamental theorem of calculus. In the present chapter we give an indication of the manner in which definite integrals can be defined and handled formally and rigorously. We also give a careful treatment of the fundamental theorem of calculus and develop methods for approximating values of definite integrals.

1 The Sigma Notation for Sums

The formal definition of the definite integral involves the sum of many terms, and thus calls for some special notation. In mathematical symbolism, the capital Greek letter sigma, which is written $\sum$ and corresponds to the letter S, stands for the words "the sum of all terms of the form. . . ." For instance, rather than writing $1 + 2 + 3 + 4 + 5 + 6$, we can write $\sum k$, and $1^2 + 2^2 + 3^2 + 4^2 + 5^2 + 6^2$ can be written as $\sum k^2$, provided that we make the convention that k is to run through all integral values from 1 to 6. If we wish to include the range of values of k as part of the summation notation we write, for instance,

$$\sum_{1 \le k \le 6} \quad \text{or} \quad \sum_{k=1}^{k=6} \quad \text{or} \quad \sum_{k=1}^{6}.$$

Thus, $\sum_{k=1}^{6} k^2$ means "the sum of all terms of the form k^2 as k runs through the integers from 1 to 6."

EXAMPLES   Write out the given sum explicitly and then find its numerical value.

1 $\displaystyle\sum_{k=1}^{7} k^2$

SOLUTION

$$\sum_{k=1}^{7} k^2 = 1^2 + 2^2 + 3^2 + 4^2 + 5^2 + 6^2 + 7^2$$
$$= 1 + 4 + 9 + 16 + 25 + 36 + 49 = 140.$$

2 $\displaystyle\sum_{k=1}^{4} 2^k$

SOLUTION

$$\sum_{k=1}^{4} 2^k = 2^1 + 2^2 + 2^3 + 2^4 = 2 + 4 + 8 + 16 = 30.$$

3 $\displaystyle\sum_{k=1}^{n} (3^k - 3^{k-1})$

SOLUTION

$$\sum_{k=1}^{n} (3^k - 3^{k-1}) = (3^1 - 3^0) + (3^2 - 3^1) + (3^3 - 3^2) + \cdots + (3^n - 3^{n-1})$$
$$= -3^0 + (3^1 - 3^1) + (3^2 - 3^2) + \cdots + (3^{n-1} - 3^{n-1}) + 3^n$$
$$= 3^n - 3^0 = 3^n - 1.$$

In the examples above, the variable k, which runs from 1 to 7 in $\sum_{k=1}^{7} k^2$, from 1 to 4 in $\sum_{k=1}^{4} 2^k$, and from 1 to n in $\sum_{k=1}^{n} (3^k - 3^{k-1})$, is called the *summation index.* There is no particular reason to use k for the summation index—any letter of the alphabet will do. For instance,

$$\sum_{k=1}^{4} 2^k = \sum_{i=1}^{4} 2^i = \sum_{j=1}^{4} 2^j = 2^1 + 2^2 + 2^3 + 2^4.$$

Notice that if f is a function and if the integers from 1 to n belong to the domain of f, then

$$\sum_{k=1}^{n} f(k) = f(1) + f(2) + f(3) + \cdots + f(n).$$

For instance, if f is a constant function, say $f(x) = C$ for all values of x, then

$$\sum_{k=1}^{n} f(k) = f(1) + f(2) + f(3) + \cdots + f(n) = \underbrace{C + C + C + \cdots + C}_{n \text{ terms}} = nC.$$

The fact expressed by the equation above is often written simply as $\sum_{k=1}^{n} C = nC$.

Thus, $\sum_{k=1}^{7} 5 = (7)(5) = 35$, $\sum_{k=1}^{n} (-2) = -2n$, $\sum_{i=1}^{100} 1 = 100$, and so forth.

The sigma notation is especially handy for indicating the sum of the terms of a sequence of numbers. A sequence of numbers consisting of a first number a_1, a second number a_2, a third number a_3, and so on, is written as $a_1, a_2, a_3, \ldots$. Thus, the kth term in this sequence is a_k, and the sum of (say) the first six terms can be written

$$\sum_{k=1}^{6} a_k = a_1 + a_2 + a_3 + a_4 + a_5 + a_6.$$

Similarly, the sum of the second, third, and fourth terms of the sequence can be written

$$\sum_{k=2}^{4} a_k = a_2 + a_3 + a_4.$$

Sometimes, in dealing with sequences, it is convenient to start with a "zeroth term" a_0. For such a sequence the sum of the terms a_k, k running from 0 to n, can be expressed as

$$\sum_{k=0}^{n} a_k = a_0 + a_1 + a_2 + \cdots + a_n.$$

Some basic properties of summation, which are easy to establish either by inspection or by using the principle of mathematical induction, are assembled in the following list.

Basic Properties of Summation

Let a_0, a_1, a_2, ..., a_n and b_0, b_1, b_2, ..., b_n denote sequences of numbers and let A, B, and C be constant numbers. Then:

1 Constant Property: $\displaystyle\sum_{k=1}^{n} C = nC.$

2 Homogeneous Property: $\displaystyle\sum_{k=1}^{n} Ca_k = C \sum_{k=1}^{n} a_k.$

3 Additive Property: $\displaystyle\sum_{k=1}^{n} (a_k + b_k) = \sum_{k=1}^{n} a_k + \sum_{k=1}^{n} b_k.$

4 Linear Property: $\displaystyle\sum_{k=1}^{n} (Aa_k + Bb_k) = A \sum_{k=1}^{n} a_k + B \sum_{k=1}^{n} b_k.$

5 Generalized Triangle Inequality: $\displaystyle\left| \sum_{k=1}^{n} a_k \right| \leq \sum_{k=1}^{n} |a_k|.$

6 Sum of Arithmetic Sequence: $\displaystyle\sum_{k=0}^{n} (A + Ck) = (n + 1)\left(A + \frac{nC}{2}\right).$

7 Sum of Geometric Sequence: $\displaystyle\sum_{k=0}^{n} AC^k = A\left(\frac{1 - C^{n+1}}{1 - C}\right),$ if $C \neq 1.$

8 Sum of Successive Integers: $\displaystyle\sum_{k=1}^{n} k = \frac{n(n + 1)}{2}.$

9 Sum of Successive Squares: $\displaystyle\sum_{k=1}^{n} k^2 = \frac{n(n + 1)(2n + 1)}{6}.$

10 Sum of Successive Cubes: $\displaystyle\sum_{k=1}^{n} k^3 = \frac{n^2(n + 1)^2}{4}.$

11 Telescoping Property: $\displaystyle\sum_{k=1}^{n} (b_k - b_{k-1}) = b_n - b_0.$

EXAMPLES   Use the basic properties to evaluate the given sum.

1 $\displaystyle\sum_{k=1}^{20} (2k^2 - 3k + 1)$

SOLUTION

$$\sum_{k=1}^{20} (2k^2 - 3k + 1) = \sum_{k=1}^{20} 2k^2 + \sum_{k=1}^{20} (-3k) + \sum_{k=1}^{20} 1 \qquad \text{(Property 4)}$$

$$= 2\sum_{k=1}^{20} k^2 - 3\sum_{k=1}^{20} k + 20 \qquad \text{(Properties 2 and 1)}$$

$$= 2\frac{(20)(21)(41)}{6} - 3\frac{(20)(21)}{2} + 20 = 5130.$$

2 $\displaystyle\sum_{k=1}^{n} k^2(5k + 1)$

SOLUTION

$$\sum_{k=1}^{n} k^2(5k + 1) = \sum_{k=1}^{n} (5k^3 + k^2) = 5\sum_{k=1}^{n} k^3 + \sum_{k=1}^{n} k^2$$

$$= \tfrac{5}{4}n^2(n + 1)^2 + \frac{n(n + 1)(2n + 1)}{6}$$

$$= n(n + 1)\left[\frac{5n(n + 1)}{4} + \frac{2n + 1}{6}\right]$$

$$= n(n + 1)\frac{15n(n + 1) + 2(2n + 1)}{12}$$

$$= \frac{n(n + 1)(15n^2 + 19n + 2)}{12}.$$

3 $\displaystyle\sum_{k=0}^{n} \frac{1}{2^k}$

SOLUTION

$$\sum_{k=0}^{n} \frac{1}{2^k} = \sum_{k=0}^{n} (\tfrac{1}{2})^k = \frac{1 - (\tfrac{1}{2})^{n+1}}{1 - \tfrac{1}{2}} = 2 - \frac{1}{2^n} \qquad \text{(by Property 7 with } A = 1 \text{ and } C = \tfrac{1}{2}).$$

4 The sum of the first 100 odd integers.

SOLUTION
In Property 6, take $A = 1$, $C = 2$, $n = 99$. Then

$$\sum_{k=0}^{99} (1 + 2k) = 1 + 3 + 5 + 7 + \cdots + 199;$$

hence, the desired sum is given by

$$\sum_{k=0}^{99} (1 + 2k) = (100)\left[1 + \frac{(99)(2)}{2}\right] = 10{,}000.$$

1.1 The Area Under a Parabola

As an example of the use of the sigma notation for sums, we are going to calculate the area A under the parabola $y = x^2$ between $x = 0$ and $x = 1$ (Figure 1). Using the method developed in Section 6 of Chapter 5, we calculate this area as

$$A = \int_0^1 x^2\, dx = \frac{x^3}{3}\bigg|_0^1 = \frac{1^3}{3} - \frac{0^3}{3} = \frac{1}{3} \text{ square unit;}$$

however, we never really *proved* that this method works. At best, our arguments were merely plausible since they were based on the Leibnizian notion that a differential is an "infinitesimal." Now, using the sigma notation, we present a more conclusive argument that the desired area is, in fact, $\frac{1}{3}$ square unit.

In Figure 1, we have indicated a subdivision of the interval $[0, 1]$ into n equal subintervals: $\left[0, \frac{1}{n}\right]$, $\left[\frac{1}{n}, \frac{2}{n}\right]$, $\left[\frac{2}{n}, \frac{3}{n}\right]$, $\left[\frac{3}{n}, \frac{4}{n}\right], \ldots, \left[\frac{n-1}{n}, \frac{n}{n}\right]$. Notice that the kth subinterval is $\left[\frac{k-1}{n}, \frac{k}{n}\right]$. Above each subinterval we have formed a corresponding circumscribed rectangle. Evidently, the height of the kth circumscribed rectangle is $\left(\frac{k}{n}\right)^2$ and its area is $\frac{1}{n}\left(\frac{k}{n}\right)^2$. An estimate of the area A can now be obtained by adding the areas of the n circumscribed rectangles; that is, $A \approx \sum_{k=1}^{n} \frac{1}{n}\left(\frac{k}{n}\right)^2$. Since

$$\sum_{k=1}^{n} \frac{1}{n}\left(\frac{k}{n}\right)^2 = \sum_{k=1}^{n} \left(\frac{1}{n}\right)^3 k^2 = \left(\frac{1}{n}\right)^3 \sum_{k=1}^{n} k^2$$

$$= \frac{1}{n^3}\left[\frac{n(n+1)(2n+1)}{6}\right] = \frac{(n+1)(2n+1)}{6n^2}$$

by the formula for the summation of successive squares; it follows that

$$A \approx \frac{(n+1)(2n+1)}{6n^2} = \frac{1}{3} + \frac{1}{2n} + \frac{1}{6n^2}.$$

Moreover, since the approximating rectangles are circumscribed, $A \leq \frac{1}{3} + 1/(2n) + 1/(6n^2)$.

In Figure 2, we have again subdivided the interval $[0, 1]$ into n equal subintervals, but we have taken the inscribed rather than the circumscribed rectangles. Evidently, the height of the kth inscribed rectangle is $\left(\frac{k-1}{n}\right)^2$ and its area is $\frac{1}{n}\left(\frac{k-1}{n}\right)^2$. Again, we obtain an estimate of the area A by adding the areas of the n inscribed rectangles; that is,

$$A \approx \sum_{k=1}^{n} \frac{1}{n}\left(\frac{k-1}{n}\right)^2.$$

Figure 1

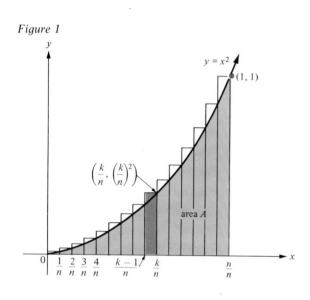

Figure 2

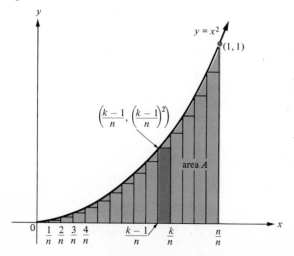

Now,

$$\sum_{k=1}^{n} \frac{1}{n}\left(\frac{k-1}{n}\right)^2 = \sum_{k=1}^{n}\left(\frac{1}{n}\right)^3(k-1)^2 = \left(\frac{1}{n}\right)^3 \sum_{k=1}^{n}(k^2 - 2k + 1)$$

$$= \frac{1}{n^3}\left[\frac{n(n+1)(2n+1)}{6} - \frac{2n(n+1)}{2} + n\right]$$

$$= \frac{2n^2 - 3n + 1}{6n^2},$$

so

$$A \approx \frac{2n^2 - 3n + 1}{6n^2} = \frac{1}{3} - \frac{1}{2n} + \frac{1}{6n^2}.$$

Moreover, since the approximating rectangles are inscribed, $\frac{1}{3} - \frac{1}{2n} + \frac{1}{6n^2} \le A$.

The above considerations show that, for every positive integer n,

$$\sum_{k=1}^{n} \frac{1}{n}\left(\frac{k-1}{n}\right)^2 = \frac{1}{3} - \frac{1}{2n} + \frac{1}{6n^2} \le A \le \frac{1}{3} + \frac{1}{2n} + \frac{1}{6n^2} = \sum_{k=1}^{n} \frac{1}{n}\left(\frac{k}{n}\right)^2.$$

As n becomes larger and larger, both $\frac{1}{3} - \frac{1}{2n} + \frac{1}{6n^2}$ and $\frac{1}{3} + \frac{1}{2n} + \frac{1}{6n^2}$ approach $\frac{1}{3}$ as a limit. Since the constant number A is "trapped" between two quantities, both of which can be made to come as close to $\frac{1}{3}$ as we please, it follows that A must be equal to $\frac{1}{3}$ (Problem 24).

Problem Set 1

In problems 1 through 8, write out each sum explicitly and then find its numerical value.

1 $\sum_{k=1}^{6}(2k+1)$

2 $\sum_{k=1}^{5} 7k^2$

3 $\sum_{i=0}^{6}(2i-1)^2$

4 $\sum_{j=3}^{7} \frac{j}{j-2}$

5 $\sum_{k=2}^{6} \frac{1}{k(k-1)}$

6 $\sum_{i=-1}^{3} 3^i$

7 $\sum_{j=0}^{3} \frac{1}{j^2+3}$

8 $\sum_{k=-1}^{3} \frac{k}{k+2}$

In problems 9 through 18, evaluate each sum by using the basic properties of summation.

9 $\sum_{k=1}^{50}(2k+3)$

10 $\sum_{i=1}^{30} i(i-1)$

11 $\sum_{k=1}^{100} 5^k$

12 $\sum_{k=0}^{100}(5^{k+1} - 5^k)$

13 $\sum_{k=1}^{n} k(k+1)$

14 $\sum_{k=1}^{100}\left(\frac{1}{k} - \frac{1}{k+1}\right)$

15 $\sum_{k=1}^{n-1} k^2$

16 $\sum_{k=1}^{n}(a_k - a_{k-1})$

17 $\sum_{k=1}^{n}(k-1)^2$

18 $\sum_{j=1}^{100} \frac{1}{10^j}$

19 Verify the generalized triangle inequality by using the principle of mathematical induction.

20 Verify the formula for the sum of successive integers by using the principle of mathematical induction.

21 Verify the formula for the sum of successive squares by using the principle of mathematical induction.

22 Verify the formula for the sum of a geometric sequence by completing the following argument. Let

$$S = \sum_{k=0}^{n} AC^k = A + AC + AC^2 + \cdots + AC^n \qquad (C \neq 1).$$

Then $SC = AC + AC^2 + \cdots + AC^n + AC^{n+1}$, so that $S - SC = A - AC^{n+1}$. (Now, solve the latter equation for S.)

23 Verify the formula for the sum of successive cubes by using the principle of mathematical induction.

24 Prove that, if A is a constant number such that $f(n) \leq A \leq g(n)$ holds for all positive integral values of n, where f and g are functions such that $\lim_{n \to +\infty} f(n) = \lim_{n \to +\infty} g(n) = L$, then $A = L$. (To say that $\lim_{n \to +\infty} f(n) = L$ means, by definition, that, for each positive number ε, there exists a positive integer N such that $|f(n) - L| < \varepsilon$ holds whenever $n \geq N$.)

25 Using the inequality derived in Section 1.1, estimate the area A under the parabola $y = x^2$ between $x = 0$ and $x = 1$ by taking n to be (a) $n = 1000$ and (b) $n = 10,000$.

2   The Definite (Riemann) Integral— Analytic Definition

The definition of $\int_a^b f(x)\,dx$ given in Section 6 of Chapter 5 was called a preliminary informal definition because of possible difficulty with the very meaning of the "area" involved. In Section 1.1 we successfully calculated the "area" under the graph of $y = x^2$ between $x = 0$ and $x = 1$ by using circumscribed and inscribed rectangles and the sigma notation for sums. This calculation suggests a way to overcome the difficulty inherent in our preliminary definition of $\int_a^b f(x)\,dx$. Thus, in this section we give a purely analytic definition of the definite integral based on the technique used in Section 1.1.

Suppose that f is a function defined (but not necessarily continuous) on a closed interval $[a, b]$. By a *partition* of the interval $[a, b]$, we mean a sequence of n subintervals $[x_0, x_1]$, $[x_1, x_2]$, $[x_2, x_3]$, ..., $[x_{n-1}, x_n]$ of $[a, b]$, where $x_0 = a$ and $x_n = b$ (Figure 1). The subinterval $[x_{k-1}, x_k]$ is called the kth *subinterval* in the partition and its length is denoted by $\Delta x_k = x_k - x_{k-1}$. Thus, the length of the first subinterval is $\Delta x_1 = x_1 - x_0$, the length of the second subinterval is $\Delta x_2 = x_2 - x_1$, and so on. There is no requirement that all subintervals in the partition have the same length. Denote the partition $[x_0, x_1]$, $[x_1, x_2]$, ..., $[x_{n-1}, x_n]$ by the symbol $\mathscr{P}$.

By the *norm* of the partition $\mathscr{P}$, in symbols $\|\mathscr{P}\|$, we mean the largest of the numbers Δx_1, Δx_2, Δx_3, ..., Δx_n representing the lengths of the n subintervals in $\mathscr{P}$. Note that two or more of the numbers Δx_1, Δx_2, Δx_3, ..., Δx_n may have the

Figure 1

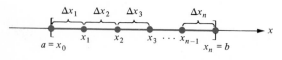

same largest value $\|\mathscr{P}\|$. In particular, if all the subintervals in the partition $\mathscr{P}$ happen to have the same length, then

$$\Delta x_1 = \Delta x_2 = \Delta x_3 = \cdots = \Delta x_n = \frac{b-a}{n},$$

so that $\|\mathscr{P}\| = \dfrac{b-a}{n}$ in this special case.

Figure 2

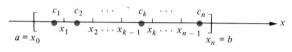

We now choose a number from each subinterval in the partition $\mathscr{P}$. Let c_1 denote the number chosen from the first subinterval $[x_0, x_1]$, let c_2 denote the number chosen from the second subinterval $[x_1, x_2]$, and so forth. Thus, c_k denotes the number chosen from the kth subinterval (Figure 2). The partition $\mathscr{P}$, together with the chosen numbers $c_1, c_2, \ldots, c_n$, is called an *augmented partition* and is denoted by $\mathscr{P}^*$.

Figure 3

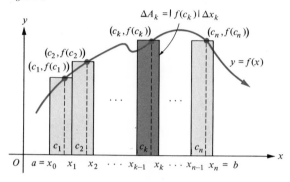

On each subinterval of the partition $\mathscr{P}$, we now construct a rectangle as in Figure 3. Notice that the kth rectangle has the kth subinterval $[x_{k-1}, x_k]$ of length Δx_k as its base and extends to the point $(c_k, f(c_k))$ on the graph of f. (If the graph of f is below the x axis at c_k, then the kth rectangle extends *downward*.) In any case, the height of the kth rectangle is $|f(c_k)|$ and its area ΔA_k is given by $\Delta A_k = |f(c_k)| \, \Delta x_k$. Therefore, $f(c_k) \, \Delta x_k = \pm \Delta A_k$, where the plus or the minus sign is used according to whether the kth rectangle extends upward or downward, respectively, from the x axis.

The sum $\displaystyle\sum_{k=1}^{n} (\pm \Delta A_k)$ of the signed areas $\pm \Delta A_k$ of the rectangles determined by the augmented partition $\mathscr{P}^*$ is called the *Riemann sum* corresponding to $\mathscr{P}^*$ for the function f. (This terminology is used in honor of the German mathematician Bernhard Riemann, who, during the nineteenth century, did some of the early definitive work on the problem of giving a precise mathematical formalization of the integral of Newton and Leibniz.) Since $\pm \Delta A_k = f(c_k) \, \Delta x_k$, this Riemann sum can also be written as $\displaystyle\sum_{k=1}^{n} f(c_k) \, \Delta x_k$.

The Riemann sum $\displaystyle\sum_{k=1}^{n} f(c_k) \, \Delta x_k = \sum_{k=1}^{n} (\pm \Delta A_k)$ should provide a reasonable approximation for what we provisionally defined as the definite integral $\displaystyle\int_a^b f(x) \, dx$ in Chapter 5. [Recall that, in calculating $\displaystyle\int_a^b f(x) \, dx$, the area below the x axis was to be subtracted, and notice that corresponding areas of approximating rectangles in the Riemann sum $\displaystyle\sum_{k=1}^{n} (\pm \Delta A_k)$ would also be subtracted in carrying out the summation.] If an exceedingly large number of very skinny rectangles are involved in the Riemann sum, the approximation should be quite accurate. Since the norm $\|\mathscr{P}\|$ gives the length of the base of the widest of these rectangles, we might expect that

$$\lim_{\|\mathscr{P}\| \to 0} \sum_{k=1}^{n} f(c_k) \, \Delta x_k = \int_a^b f(x) \, dx.$$

The latter equation suggests the following formal definition.

DEFINITION 1   **The Definite (Riemann) Integral—Analytic Version**

If the limit $\lim\limits_{\|\mathscr{P}\|\to 0} \sum\limits_{k=1}^{n} f(c_k)\,\Delta x_k$ exists, then the function f is said to be *integrable* on $[a,b]$ *in the sense of Riemann*. If f is integrable, then the *definite (Riemann) integral of f on the interval* $[a,b]$ is defined by

$$\int_a^b f(x)\,dx = \lim_{\|\mathscr{P}\|\to 0} \sum_{k=1}^{n} f(c_k)\,\Delta x_k.$$

The limit indicated in Definition 1 is to be understood in the following sense: To assert that a number I is the limit of the Riemann sum $\sum\limits_{k=1}^{n} f(c_k)\,\Delta x_k$ as the norm $\|\mathscr{P}\|$ approaches zero means that, for each positive number ε (no matter how small), there exists a positive number δ (depending on ε) such that

$$\left| \sum_{k=1}^{n} f(c_k)\,\Delta x_k - I \right| < \varepsilon$$

holds *for every augmented partition* $\mathscr{P}^*$ with $\|\mathscr{P}\| < \delta$.

EXAMPLE   Find the Riemann sum for the function f given by $f(x) = 1 + x^3$ on the interval $[-2,2]$ using the augmented partition $\mathscr{P}^*$ consisting of the eight subintervals $[-2, -\frac{3}{2}]$, $[-\frac{3}{2}, -1]$, $[-1, -\frac{1}{2}]$, $[-\frac{1}{2}, 0]$, $[0, \frac{1}{2}]$, $[\frac{1}{2}, 1]$, $[1, \frac{3}{2}]$, and $[\frac{3}{2}, 2]$ with $c_1 = -2$, $c_2 = -\frac{3}{2}$, $c_3 = -\frac{1}{2}$, $c_4 = 0$, $c_5 = \frac{1}{2}$, $c_6 = 1$, $c_7 = \frac{3}{2}$, and $c_8 = 2$. Sketch a graph showing the eight rectangles corresponding to this Riemann sum.

SOLUTION
For the given partition each of the eight subintervals has length $\frac{1}{2}$ unit; that is,

$$\Delta x_1 = \Delta x_2 = \Delta x_3 = \Delta x_4 = \Delta x_5 = \Delta x_6 = \Delta x_7 = \Delta x_8 = \tfrac{1}{2}.$$

Also,

$$f(c_1) = 1 + (-2)^3 = -7,$$
$$f(c_2) = 1 + (-\tfrac{3}{2})^3 = -\tfrac{19}{8},$$
$$f(c_3) = 1 + (-\tfrac{1}{2})^3 = \tfrac{7}{8},$$
$$f(c_4) = 1 + 0^3 = 1,$$
$$f(c_5) = 1 + (\tfrac{1}{2})^3 = \tfrac{9}{8},$$
$$f(c_6) = 1 + 1^3 = 2,$$
$$f(c_7) = 1 + (\tfrac{3}{2})^3 = \tfrac{35}{8},$$
$$f(c_8) = 1 + 2^3 = 9.$$

Therefore, the required Riemann sum is given by

$$\sum_{k=1}^{8} f(c_k)\,\Delta x_k = f(c_1)\,\Delta x_1 + f(c_2)\,\Delta x_2 + f(c_3)\,\Delta x_3 + f(c_4)\,\Delta x_4$$
$$+ f(c_5)\,\Delta x_5 + f(c_6)\,\Delta x_6 + f(c_7)\,\Delta x_7 + f(c_8)\,\Delta x_8$$
$$= (-7)(\tfrac{1}{2}) + (-\tfrac{19}{8})(\tfrac{1}{2}) + (\tfrac{7}{8})(\tfrac{1}{2}) + (1)(\tfrac{1}{2}) + (\tfrac{9}{8})(\tfrac{1}{2})$$
$$+ (2)(\tfrac{1}{2}) + (\tfrac{35}{8})(\tfrac{1}{2}) + (9)(\tfrac{1}{2}) = \tfrac{9}{2}.$$

Figure 4

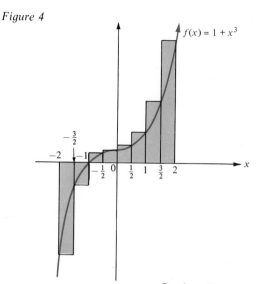

$f(x) = 1 + x^3$

Figure 4 shows the eight rectangles whose signed areas are the terms in the Riemann sum. Notice that the first two rectangles contribute negative terms to the Riemann sum.

From the example above and the definition of the Riemann integral we can conclude that

$$\int_{-2}^{2} (1 + x^3)\, dx \approx \tfrac{9}{2},$$

provided that the integral exists. For a better approximation, we would have to take an augmented partition with a smaller norm (and, hence, with more rectangles).

2.1   Existence of Riemann Integrals

Virtually every function encountered in practical scientific work is integrable in the sense of Riemann on any closed interval contained in its domain. This includes not only all continuous functions and all monotone (increasing or decreasing) functions, but all functions that are bounded and "piecewise continuous" or "piecewise monotone" as well. The proofs of the Riemann integrability of such functions can be found in more advanced textbooks on analysis; here, we content ourselves with precise statements of some of the existence theorems for the Riemann integrals.

THEOREM 1   Existence of the Riemann Integral of a Continuous Function
If f is a continuous function on the closed interval $[a,b]$, then f is Riemann-integrable on $[a,b]$.

Figure 5

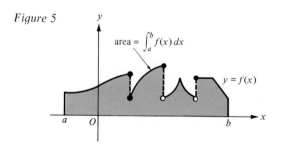

area $= \int_{a}^{b} f(x)\, dx$

$y = f(x)$

For instance, a polynomial function is Riemann-integrable on any closed interval since polynomial functions are continuous.

The function whose graph is shown in Figure 5 is not continuous, but it is *piecewise continuous* on the interval $[a,b]$ in the sense that $[a,b]$ can be partitioned into a finite number of subintervals such that f is continuous on each subinterval. The function f is also *bounded* on the interval $[a,b]$ in the sense that there are fixed numbers A and B such that $A \le f(x) \le B$ holds for all values of x in $[a,b]$. The integrability of any such function is assured by the following theorem.

THEOREM 2   Existence of the Riemann Integral of a Piecewise Continuous Bounded Function
If f is a bounded and piecewise continuous function on the closed interval $[a,b]$, then f is Riemann-integrable on $[a,b]$.

The function defined by

$$f(x) = \begin{cases} \dfrac{1}{x} & \text{for } x > 0 \\ 1 & \text{for } x \le 0 \end{cases}$$

Figure 6

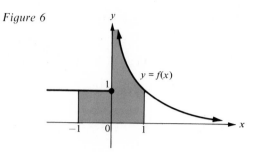

is piecewise continuous, but not bounded on the interval $[-1, 1]$ (Figure 6). A glance at Figure 6 suggests that great care should be taken in talking about the "area" of the region under the graph of an unbounded function. The following theorem gives further evidence for the necessity of such care.

THEOREM 3   **Boundedness of Riemann-Integrable Functions**
If f is defined and Riemann-integrable on $[a, b]$, then f is bounded on $[a, b]$.

For instance, the function f whose graph appears in Figure 6 is *not* Riemann-integrable on $[-1, 1]$, since it is unbounded on this interval.

The following theorem is very useful in deciding whether or not certain functions are Riemann-integrable.

THEOREM 4   **Change of a Function at Finitely Many Points**
If f is defined and Riemann-integrable on $[a, b]$ and if h is also defined on $[a, b]$ and satisfies $h(x) = f(x)$ for all but a finite number of values of x in $[a, b]$, then h is also Riemann-integrable on $[a, b]$ and $\int_a^b h(x)\, dx = \int_a^b f(x)\, dx$.

The use of Theorem 4 is illustrated by the following example.

EXAMPLE   Let the function h be defined by

$$h(x) = \begin{cases} 1 & \text{for } x \neq 0 \\ 0 & \text{for } x = 0. \end{cases}$$

Find $\int_{-1}^1 h(x)\, dx$, given that $\int_{-1}^1 dx = 2$.

SOLUTION
Let f be the constant function defined by $f(x) = 1$ for all values of x. Evidently, $h(x) = f(x)$ for all values of x on $[-1, 1]$ except for $x = 0$. Therefore, by Theorem 4, $\int_{-1}^1 h(x)\, dx$ exists and

$$\int_{-1}^1 h(x)\, dx = \int_{-1}^1 f(x)\, dx = \int_{-1}^1 1\, dx = \int_{-1}^1 dx = 2.$$

2.2 Calculation of Riemann Integrals by Direct Use of the Definition

The definite integral $\int_a^b f(x)\, dx$ of any Riemann-integrable function f can evidently be calculated by selecting any sequence of augmented partitions

$$\mathscr{P}_1^*, \mathscr{P}_2^*, \mathscr{P}_3^*, \ldots, \mathscr{P}_n^*, \ldots$$

such that $\lim\limits_{n \to +\infty} \|\mathscr{P}_n\| = 0$, calculating the Riemann sums corresponding to each augmented partition, and finding the limit as $n \to +\infty$ of the resulting sequence of Riemann sums. In doing this, it is often convenient to consider only partitions of $[a, b]$ consisting of n equal subintervals each of which has length $\Delta x = \dfrac{b - a}{n}$.

Furthermore, in augmenting these partitions, it is often helpful to select the numbers $c_1, c_2, c_3, \ldots, c_n$ so that all approximating rectangles are circumscribed (or inscribed).

EXAMPLES Evaluate the given Riemann integral directly by calculating a limit of Riemann sums. Use partitions consisting of subintervals of equal lengths and use inscribed or circumscribed rectangles as indicated.

1 $\displaystyle\int_0^2 x^3 \, dx$ (inscribed rectangles)

Figure 7

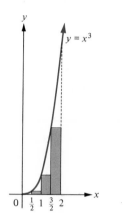

SOLUTION

Let $\mathscr{P}_n$ denote the partition of the interval $[0, 2]$ into n equal subintervals of length Δx, so that $\Delta x = \dfrac{2 - 0}{n} = \dfrac{2}{n}$. (For instance, Figure 7 shows the case $n = 4$ with four subintervals each of length $\frac{2}{4} = \frac{1}{2}$ unit.) The function f given by $f(x) = x^3$ is *increasing* on the interval $[0, 2]$; hence, *inscribed* rectangles are obtained by forming the augmented partition $\mathscr{P}_n^*$ for which $c_1, c_2, c_3, \ldots, c_n$ are the *left endpoints* of the corresponding intervals. Thus, $\mathscr{P}_n^*$ consists of the n subintervals

$$[0, \Delta x], [\Delta x, 2\Delta x], [2\Delta x, 3\Delta x], \ldots, [2 - \Delta x, 2]$$

with

$$c_1 = 0, \ c_2 = \Delta x, \ c_3 = 2\,\Delta x, \ \ldots, \ c_n = 2 - \Delta x.$$

Evidently,

$$c_k = (k - 1)\,\Delta x = (k - 1)\frac{2}{n} = \frac{2(k - 1)}{n}.$$

Consequently, the Riemann sum corresponding to $\mathscr{P}_n^*$ is given by

$$\sum_{k=1}^{n} f(c_k)\,\Delta x_k = \sum_{k=1}^{n} (c_k)^3\,\Delta x = \sum_{k=1}^{n} \left[\frac{2(k-1)}{n}\right]^3 \frac{2}{n} = \sum_{k=1}^{n} \frac{16}{n^4}(k-1)^3.$$

Using the basic properties of summation in Section 1, we have

$$\sum_{k=1}^{n} f(c_k)\,\Delta x_k = \sum_{k=1}^{n} \frac{16}{n^4}(k-1)^3 = \frac{16}{n^4}\sum_{k=1}^{n}(k-1)^3$$

$$= \frac{16}{n^4}\sum_{k=1}^{n}(k^3 - 3k^2 + 3k - 1) = \frac{16}{n^4}\left(\sum_{k=1}^{n}k^3 - 3\sum_{k=1}^{n}k^2 + 3\sum_{k=1}^{n}k - \sum_{k=1}^{n}1\right)$$

$$= \frac{16}{n^4}\left[\frac{n^2(n+1)^2}{4} - 3\frac{n(n+1)(2n+1)}{6} + 3\frac{n(n+1)}{2} - n\right]$$

$$= \frac{16}{n^4}\left[\frac{n^4 - 2n^3 + n^2}{4}\right] = 4\left(1 - \frac{2}{n} + \frac{1}{n^2}\right).$$

Therefore,

$$\int_0^2 x^3\, dx = \lim_{n \to +\infty} \sum_{k=1}^{n} f(c_k)\, \Delta x_k = \lim_{n \to +\infty} 4\left(1 - \frac{2}{n} + \frac{1}{n^2}\right) = 4.$$

2 $\displaystyle\int_{-2}^{0} x^2\, dx$ (circumscribed rectangles)

Figure 8

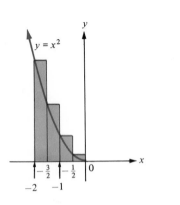

SOLUTION

Let $\mathscr{P}_n$ denote the partition of the interval $[-2,0]$ into n equal subintervals of length Δx, so that $\Delta x = \dfrac{0 - (-2)}{n} = \dfrac{2}{n}$. (For instance, Figure 8 shows the case $n = 4$ with four subintervals each of length $\frac{2}{4} = \frac{1}{2}$ unit.) The function f given by $f(x) = x^2$ is *decreasing* on the interval $[-2,0]$; hence, *circumscribed* rectangles are obtained by forming the augmented partition $\mathscr{P}_n^*$ for which $c_1, c_2, c_3, \ldots, c_n$ are the *left endpoints* of the corresponding intervals. Thus, $\mathscr{P}_n^*$ consists of the n subintervals

$$[-2, -2 + \Delta x], [-2 + \Delta x, -2 + 2\Delta x],$$
$$[-2 + 2\Delta x, -2 + 3\Delta x], \ldots, [-\Delta x, 0]$$

with

$$c_1 = -2,\ c_2 = -2 + \Delta x,\ c_3 = -2 + 2\,\Delta x, \ldots,\ c_n = -\Delta x.$$

Evidently,

$$c_k = -2 + (k - 1)\,\Delta x = -2 + (k - 1)\frac{2}{n} = \frac{2}{n}[k - (n + 1)],$$

so that

$$f(c_k) = (c_k)^2 = \frac{4}{n^2}[k - (n + 1)]^2 = \frac{4}{n^2}[k^2 - 2(n + 1)k + (n + 1)^2].$$

Consequently, the Riemann sum corresponding to $\mathscr{P}_n^*$ is given by

$$\sum_{k=1}^{n} f(c_k)\, \Delta x_k = \sum_{k=1}^{n} \frac{4}{n^2}[k^2 - 2(n + 1)k + (n + 1)^2]\frac{2}{n}$$

$$= \frac{8}{n^3}\left[\sum_{k=1}^{n} k^2 - 2(n + 1)\sum_{k=1}^{n} k + (n + 1)^2 \sum_{k=1}^{n} 1\right]$$

$$= \frac{8}{n^3}\left[\frac{n(n + 1)(2n + 1)}{6} - 2(n + 1)\frac{n(n + 1)}{2} + (n + 1)^2 n\right]$$

$$= \frac{8}{6}\left(\frac{n}{n}\right)\left(\frac{n + 1}{n}\right)\left(\frac{2n + 1}{n}\right) = \frac{4}{3}\left(1 + \frac{1}{n}\right)\left(2 + \frac{1}{n}\right).$$

Therefore,

$$\int_{-2}^{0} x^2\, dx = \lim_{n \to +\infty} \left[\tfrac{4}{3}\left(1 + \frac{1}{n}\right)\left(2 + \frac{1}{n}\right)\right] = \tfrac{8}{3}.$$

Problem Set 2

In problems 1 through 4, find the Riemann sum for each function on the prescribed interval using the indicated augmented partition. Also, sketch the graph of the function on the given interval showing the rectangles corresponding to the Riemann sum.

1 $f(x) = 3x + 1$ on $[0, 3]$. $\mathscr{P}^*$ consists of $[0, \frac{1}{2}]$, $[\frac{1}{2}, 1]$, $[1, \frac{3}{2}]$, $[\frac{3}{2}, 2]$, $[2, \frac{5}{2}]$, and $[\frac{5}{2}, 3]$ with $c_1 = \frac{1}{2}$, $c_2 = 1$, $c_3 = \frac{3}{2}$, $c_4 = 2$, $c_5 = \frac{5}{2}$, and $c_6 = 3$.

2 $f(x) = -2x^2$ on $[0, 3]$. $\mathscr{P}^*$ consists of $[0, \frac{1}{2}]$, $[\frac{1}{2}, 1]$, $[1, \frac{3}{2}]$, $[\frac{3}{2}, 2]$, $[2, \frac{5}{2}]$, and $[\frac{5}{2}, 3]$ with $c_1 = \frac{1}{4}$, $c_2 = \frac{3}{4}$, $c_3 = \frac{5}{4}$, $c_4 = \frac{7}{4}$, $c_5 = \frac{9}{4}$, and $c_6 = \frac{11}{4}$.

3 $f(x) = 1/x$ on $[1, 3]$. $\mathscr{P}^*$ consists of $[1, \frac{3}{2}]$, $[\frac{3}{2}, 2]$, $[2, \frac{5}{2}]$, and $[\frac{5}{2}, 3]$ with $c_1 = \frac{5}{4}$, $c_2 = \frac{7}{4}$, $c_3 = \frac{9}{4}$, and $c_4 = \frac{11}{4}$.

4 $f(x) = \dfrac{1}{2 + x}$ on $[-1, 2]$. $\mathscr{P}^*$ consists of $[-1, -\frac{1}{2}]$, $[-\frac{1}{2}, 0]$, $[0, \frac{1}{2}]$, $[\frac{1}{2}, 1]$, $[1, \frac{3}{2}]$, and $[\frac{3}{2}, 2]$ with $c_1 = -1$, $c_2 = -\frac{1}{2}$, $c_3 = 0$, $c_4 = \frac{1}{2}$, $c_5 = 1$, and $c_6 = \frac{3}{2}$.

In problems 5 through 12, indicate whether each Riemann integral exists and give a reason for your answer.

5 $\displaystyle\int_1^{1000} \frac{1}{x}\, dx$

6 $\displaystyle\int_0^1 \frac{1}{x}\, dx$

7 $\displaystyle\int_{-1}^1 |x|\, dx$

8 $\displaystyle\int_{-1}^1 \frac{x + 1}{\sqrt{x}}\, dx$

9 $\displaystyle\int_0^{\pi} \tan x\, dx$

10 $\displaystyle\int_1^{100} [\![x]\!]\, dx$

11 $\displaystyle\int_0^3 f(x)\, dx$, where $f(x) = \begin{cases} x & \text{for } 0 \le x < 1 \\ x - 1 & \text{for } 1 \le x < 2 \\ 1 - x^2 & \text{for } 2 \le x \le 3 \end{cases}$

12 $\displaystyle\int_0^2 f(x)\, dx$, where $f(x) = \begin{cases} 7 & \text{for } x \ne 1 \\ 2 & \text{for } x = 1 \end{cases}$

In problems 13 through 19, evaluate each Riemann integral directly by calculating a limit of Riemann sums. Use partitions consisting of subintervals of equal lengths and use inscribed or circumscribed rectangles as indicated.

13 $\displaystyle\int_0^3 2x\, dx$ (circumscribed rectangles)

14 $\displaystyle\int_0^3 2x\, dx$ (inscribed rectangles)

15 $\displaystyle\int_4^7 (2x - 6)\, dx$ (circumscribed rectangles)

16 $\displaystyle\int_1^3 (9 - x^2)\, dx$ (inscribed rectangles)

17 $\displaystyle\int_{-2}^{-1} (x^2 - x - 2)\, dx$ (inscribed rectangles)

18 $\displaystyle\int_0^2 (x^3 + 2)\, dx$ (circumscribed rectangles)

19 $\displaystyle\int_0^2 (x^3 + 2)\, dx$ (inscribed rectangles)

20 Define the function f by

$$f(x) = \begin{cases} 0 & \text{for } x \ne 0 \\ 1 & \text{for } x = 0. \end{cases}$$

Evaluate $\displaystyle\int_0^1 f(x)\, dx$ *directly from the definition.*

21 Use the formal analytic definition to evaluate $\displaystyle\int_1^2 7\, dx$ as a limit of Riemann sums. Interpret the result geometrically.

22 It can be shown that $\int_{1}^{2}(1/x)\,dx \approx 0.693$ (correct to three decimal places). Make an estimate of $\int_{1}^{2}(1/x)\,dx$ using a Riemann sum involving 10 circumscribed rectangles with equal bases.

3 Basic Properties of the Definite Integral

The basic properties of the definite (Riemann) integral can be derived deductively from the formal analytic definition given in Section 2. For the most part, we do not give such proofs; however, we state these properties as formal theorems and interpret them in terms of our intuitive idea of "area." We intend to use these properties in the next section to give a proof of the fundamental theorem of calculus; therefore, to avoid circular reasoning, we do not allow ourselves to use the preliminary version of the fundamental theorem in this section.

THEOREM 1   Integral of a Constant Function

Let f be the constant function defined by the equation $f(x) = K$, where K is a constant number. Then $\int_{a}^{b} f(x)\,dx = \int_{a}^{b} K\,dx = K(b-a)$.

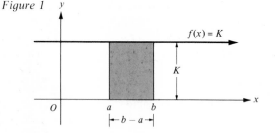

Figure 1

Geometrically, Theorem 1 simply asserts that a rectangle with width $b - a$ and height $|K|$ has an area of $|K|(b-a)$ square units (Figure 1). In particular,

$$\int_{a}^{b} dx = \int_{a}^{b} 1\,dx = b - a$$

and

$$\int_{a}^{b} 0\,dx = 0 \cdot (b-a) = 0.$$

EXAMPLE   Find $\int_{-2}^{33}(-7)\,dx$.

SOLUTION

By Theorem 1, $\int_{-2}^{33}(-7)\,dx = (-7)[33 - (-2)] = -245.$

THEOREM 2   Homogeneous Property

If f is a Riemann-integrable function on the interval $[a,b]$ and K is a constant number, then Kf is also Riemann-integrable on $[a,b]$ and $\int_{a}^{b} Kf(x)\,dx = K\int_{a}^{b} f(x)\,dx.$

Figure 2 shows the underlying geometric reason for the homogeneous property; namely, when f is multiplied by K, the approximating rectangles in the Riemann sum have their heights multiplied by K, and therefore their areas are multiplied by K.

Figure 2

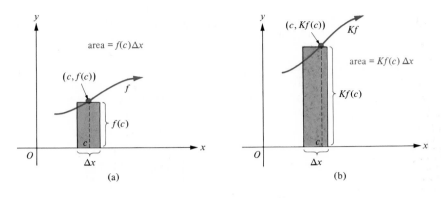

(a)

(b)

EXAMPLE   Given that $\int_1^{2.7} f(x)\,dx = -13$, find $\int_1^{2.7} 52f(x)\,dx$.

SOLUTION

By Theorem 2, $\int_1^{2.7} 52f(x)\,dx = 52\int_1^{2.7} f(x)\,dx = 52(-13) = -676$.

THEOREM 3   **Additive Property**
If f and g are Riemann-integrable functions on the interval $[a,b]$, then $f + g$ is also Riemann-integrable on $[a,b]$ and

$$\int_a^b [f(x) + g(x)]\,dx = \int_a^b f(x)\,dx + \int_a^b g(x)\,dx.$$

The underlying geometric reason for the additive property of the Riemann integral can be seen in Figure 3, which shows corresponding approximating rectangles in Riemann sums for f, g, and $f + g$. The area of such a rectangle for $f + g$ is $[f(c) + g(c)]\,\Delta x = f(c)\,\Delta x + g(c)\,\Delta x$; hence, its area is the sum of the areas of the corresponding rectangles for f and for g.

Figure 3

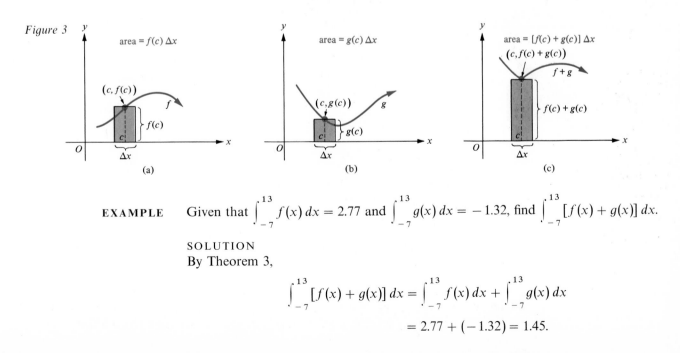

(a)

(b)

(c)

EXAMPLE   Given that $\int_{-7}^{13} f(x)\,dx = 2.77$ and $\int_{-7}^{13} g(x)\,dx = -1.32$, find $\int_{-7}^{13} [f(x) + g(x)]\,dx$.

SOLUTION
By Theorem 3,

$$\int_{-7}^{13} [f(x) + g(x)]\,dx = \int_{-7}^{13} f(x)\,dx + \int_{-7}^{13} g(x)\,dx$$

$$= 2.77 + (-1.32) = 1.45.$$

The homogeneous and additive properties of the Riemann integral can be combined to yield the linear property.

THEOREM 4   **Linear Property**

If f and g are Riemann-integrable functions on the interval $[a, b]$ and if A and B are constant numbers, then $Af + Bg$ is also Riemann-integrable on $[a, b]$ and

$$\int_a^b [Af(x) + Bg(x)]\, dx = A \int_a^b f(x)\, dx + B \int_a^b g(x)\, dx.$$

EXAMPLE   Given that $\int_2^3 x^4\, dx = \frac{211}{5}$ and $\int_2^3 x\, dx = \frac{5}{2}$, find $\int_2^3 (10x^4 + 16x)\, dx$.

SOLUTION

By Theorem 4,

$$\int_2^3 (10x^4 + 16x)\, dx = 10 \int_2^3 x^4\, dx + 16 \int_2^3 x\, dx = 10\left(\frac{211}{5}\right) + 16\left(\frac{5}{2}\right) = 462.$$

By using the principle of mathematical induction, the linear property can be extended to more than two functions. Thus, for example, we have

$$\int_0^2 (4x^3 - 3x^2 + 7x - 8)\, dx = 4 \int_0^2 x^3\, dx + (-3) \int_0^2 x^2\, dx + 7 \int_0^2 x\, dx + (-8) \int_0^2 dx$$

$$= 4 \int_0^2 x^3\, dx - 3 \int_0^2 x^2\, dx + 7 \int_0^2 x\, dx - 8 \int_0^2 dx.$$

THEOREM 5   **Positivity**

If f is a Riemann-integrable function on the interval $[a, b]$ and if $f(x) \geq 0$ for all values of x in $[a, b]$, then $\int_a^b f(x)\, dx \geq 0$.

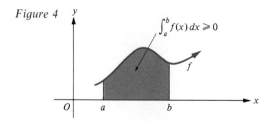

Figure 4

Theorem 5 is geometrically evident since the integral $\int_a^b f(x)\, dx$ is supposed to represent the algebraic difference between the area of the region under the graph of f that is above the x axis and the area of the region under the graph of f that is below the x axis. If $f(x) \geq 0$ for $a \leq x \leq b$, then none of the region lies below the x axis; hence, $\int_a^b f(x)\, dx$ cannot be negative (Figure 4).

EXAMPLE   Show that $\int_0^1 x^3\, dx \leq \int_0^1 x\, dx$.

SOLUTION

For $0 \leq x \leq 1$, $x^3 \leq x$ holds; hence, $x - x^3 \geq 0$. By Theorem 5, $\int_0^1 (x - x^3)\, dx \geq 0$; that is, $\int_0^1 x\, dx - \int_0^1 x^3\, dx \geq 0$. From the latter inequality, $\int_0^1 x\, dx \geq \int_0^1 x^3\, dx$; that is, $\int_0^1 x^3\, dx \leq \int_0^1 x\, dx$.

The argument given in the example above is quite general. Indeed, suppose that f and g are Riemann-integrable functions on the interval $[a, b]$ such that $f(x) \le g(x)$. Then, $g(x) - f(x) \ge 0$, so that, by Theorem 5, $\int_a^b [g(x) - f(x)] \, dx \ge 0$; that is, $\int_a^b g(x) \, dx - \int_a^b f(x) \, dx \ge 0$, or $\int_a^b f(x) \, dx \le \int_a^b g(x) \, dx$. Therefore, we have the following theorem.

THEOREM 6   **Comparison**

If f and g are Riemann-integrable functions on the interval $[a, b]$ and if $f(x) \le g(x)$ holds for all values of x in $[a, b]$, then $\int_a^b f(x) \, dx \le \int_a^b g(x) \, dx$.

Theorem 6 provides the basis for the following theorem.

THEOREM 7   **Absolute Value Property**

If f is Riemann-integrable on the interval $[a, b]$, then so is $|f|$ and

$$\left| \int_a^b f(x) \, dx \right| \le \int_a^b |f(x)| \, dx.$$

A rigorous proof of Theorem 7 would require an argument to show that $|f|$ is Riemann-integrable and is beyond the scope of this book. However, assuming that f and $|f|$ are Riemann-integrable functions on the interval $[a, b]$, we can derive the inequality in Theorem 7 as follows: Applying Theorem 6 to the inequality $-|f(x)| \le f(x) \le |f(x)|$, we obtain

$$\int_a^b -|f(x)| \, dx \le \int_a^b f(x) \, dx \le \int_a^b |f(x)| \, dx;$$

that is,

$$-\int_a^b |f(x)| \, dx \le \int_a^b f(x) \, dx \le \int_a^b |f(x)| \, dx.$$

It follows that

$$\left| \int_a^b f(x) \, dx \right| \le \int_a^b |f(x)| \, dx.$$

The following theorem expresses one of the most important features of the Riemann integral.

THEOREM 8   **Additivity with Respect to the Interval of Integration**

Let $a < b < c$ and suppose that the function f is Riemann-integrable on the interval $[a, b]$ as well as on the interval $[b, c]$. Then f is also Riemann-integrable on the interval $[a, c]$ and

$$\int_a^c f(x) \, dx = \int_a^b f(x) \, dx + \int_b^c f(x) \, dx.$$

Figure 5

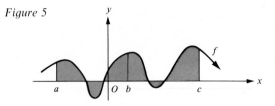

Geometrically, Theorem 8 simply means that the total signed area under the graph of $y = f(x)$ from $x = a$ to $x = c$ is the sum of the signed area from $x = a$ to $x = b$ and the signed area from $x = b$ to $x = c$ (Figure 5).

EXAMPLES 1 Given that $\int_{-1}^{2} f(x)\,dx = 7$ and that $\int_{2}^{3} f(x)\,dx = -5$, find $\int_{-1}^{3} f(x)\,dx$.

SOLUTION
By Theorem 8,

$$\int_{-1}^{3} f(x)\,dx = \int_{-1}^{2} f(x)\,dx + \int_{2}^{3} f(x)\,dx = 7 + (-5) = 2.$$

2 Given that $\int_{a}^{b} x\,dx = \tfrac{1}{2}(b^2 - a^2)$, use Theorem 8 to find $\int_{-3}^{2} |x|\,dx$.

SOLUTION
By Theorem 8, $\int_{-3}^{2} |x|\,dx = \int_{-3}^{0} |x|\,dx + \int_{0}^{2} |x|\,dx$. For $-3 \le x \le 0$, $|x| = -x$, while, for $0 \le x \le 2$, $|x| = x$. Therefore,

$$\int_{-3}^{2} |x|\,dx = \int_{-3}^{0} (-x)\,dx + \int_{0}^{2} x\,dx = -\int_{-3}^{0} x\,dx + \int_{0}^{2} x\,dx$$

$$= -[\tfrac{1}{2}(0^2 - (-3)^2)] + \tfrac{1}{2}(2^2 - 0^2) = \tfrac{9}{2} + 2 = \tfrac{13}{2}.$$

Theorem 8 is especially useful for integrating "piecewise-defined" functions, as the following example illustrates.

EXAMPLE   Define the function f by the equation

$$f(x) = \begin{cases} x^2 - 1 & \text{for } x < 0 \\ x - 1 & \text{for } 0 \le x < 1 \\ 3 & \text{for } x \ge 1 \end{cases}$$

(Figure 6). Assume that $\int_{a}^{b} x^2\,dx = \tfrac{1}{3}(b^3 - a^3)$ and that $\int_{a}^{b} x\,dx = \tfrac{1}{2}(b^2 - a^2)$. Use Theorem 8 to calculate $\int_{-2}^{4} f(x)\,dx$.

SOLUTION

$$\int_{-2}^{4} f(x)\,dx = \int_{-2}^{0} f(x)\,dx + \int_{0}^{1} f(x)\,dx + \int_{1}^{4} f(x)\,dx$$

$$= \int_{-2}^{0} (x^2 - 1)\,dx + \int_{0}^{1} (x - 1)\,dx + \int_{1}^{4} 3\,dx$$

$$= \int_{-2}^{0} x^2\,dx - \int_{-2}^{0} dx + \int_{0}^{1} x\,dx - \int_{0}^{1} dx + 3\int_{1}^{4} dx$$

$$= \tfrac{1}{3}[0^3 - (-2)^3] - [0 - (-2)] + \tfrac{1}{2}(1^2 - 0^2) - (1 - 0) + 3(4 - 1)$$

$$= \tfrac{8}{3} - 2 + \tfrac{1}{2} - 1 + 9 = \tfrac{55}{6}.$$

Figure 6

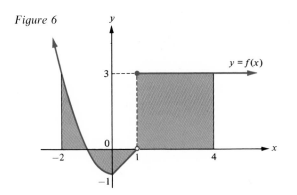

In Figure 6 we have $\int_0^1 f(x)\,dx = \int_0^1 (x-1)\,dx$ in spite of the fact that $f(1) \neq 1 - 1$. This is justified by Theorem 4 of Section 2 since $f(x) = x - 1$ for all values of x in $[0, 1]$ except for $x = 1$.

The following property of the Riemann integral plays an important role in our forthcoming proof of the fundamental theorem of calculus; hence, we give a careful proof of this property.

THEOREM 9

Mean Value Theorem for Integrals

Suppose that f is a continuous function on the interval $[a, b]$. Then, there exists a number c in $[a, b]$ such that

$$f(c) \cdot (b - a) = \int_a^b f(x)\,dx.$$

PROOF

In Chapter 3, Section 5, we learned that a continuous function f on a closed interval $[a, b]$ takes on a maximum value (say) B and a minimum value (say) A. Thus, $A \leq f(x) \leq B$ holds for all values of x in $[a, b]$. By Theorem 6, then, $\int_a^b A\,dx \leq \int_a^b f(x)\,dx \leq \int_a^b B\,dx$. By Theorem 1, $\int_a^b A\,dx = A(b-a)$ and $\int_a^b B\,dx = B(b-a)$; hence, $A(b-a) \leq \int_a^b f(x)\,dx \leq B(b-a)$. But $b - a > 0$, so the latter inequality can be rewritten as $A \leq \dfrac{1}{b-a}\int_a^b f(x)\,dx \leq B$. Now the intermediate value theorem for continuous functions (Chapter 3, Section 1.1, Theorem 1) asserts that f takes on every value between any two of its values. Thus, since A and B are two such values of f and since $\dfrac{1}{b-a}\int_a^b f(x)\,dx$ lies between these two values, there must exist a number c in $[a, b]$ such that $f(c) = \dfrac{1}{b-a}\int_a^b f(x)\,dx$; that is,

$$f(c)(b-a) = \int_a^b f(x)\,dx.$$

Figure 7

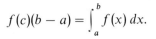

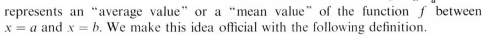

area under curve = area of rectangle

The condition $f(c)(b-a) = \int_a^b f(x)\,dx$ (Figure 7) means that the area under the curve $y = f(x)$ between $x = a$ and $x = b$ is the same as the area of a rectangle whose base is the interval $[a, b]$ and whose height is $f(c)$. Thus, if the curve $y = f(x)$ were "flattened out" between $x = a$ and $x = b$ so as to have constant height $f(c)$, then the area under the curve would stay the same. In this sense, the number $f(c) = \dfrac{1}{b-a}\int_a^b f(x)\,dx$ represents an "average value" or a "mean value" of the function f between $x = a$ and $x = b$. We make this idea official with the following definition.

DEFINITION 1   **Mean Value of a Function on an Interval**

Let f be a Riemann-integrable function on the interval $[a,b]$. Then the *mean value* of f on $[a,b]$ is given by

$$\frac{1}{b-a} \int_a^b f(x)\,dx.$$

The mean value theorem for integrals simply asserts that a continuous function f on an interval $[a,b]$ takes on its own mean value at some number c on this interval.

EXAMPLE   Given that $\int_a^b x^2\,dx = \frac{1}{3}(b^3 - a^3)$, find the mean value of the function f defined by $f(x) = x^2$ on the interval $[1,4]$, and find a value of c on this interval such that $f(c)$ gives this mean value.

SOLUTION
The desired mean value is given by

$$\frac{1}{4-1} \int_1^4 x^2\,dx = \frac{1}{3}[\frac{1}{3}(4^3 - 1^3)] = 7.$$

We need to find a value of c with $1 \le c \le 4$ such that $f(c) = c^2 = 7$. Evidently, $c = \sqrt{7} \approx 2.65$.

In the definition of the definite (Riemann) integral $\int_a^b f(x)\,dx$, it was assumed that $a < b$. For certain purposes, it is convenient to be able to deal with expressions such as $\int_a^b f(x)\,dx$ without worrying about whether or not $a < b$. This suggests that we should make a suitable definition of $\int_a^b f(x)\,dx$ for the case in which $a \ge b$. If $a = b$, it is natural to define $\int_a^b f(x)\,dx = \int_a^a f(x)\,dx = 0$ (why?). The clue to the proper way to define $\int_a^b f(x)\,dx$ when $a > b$ is Theorem 8, which asserts that $\int_a^c f(x)\,dx = \int_a^b f(x)\,dx + \int_b^c f(x)\,dx$. If we formally put $c = a$ in the latter equation and set $\int_a^c f(x)\,dx = \int_a^a f(x)\,dx = 0$, we obtain $0 = \int_a^b f(x)\,dx + \int_b^a f(x)\,dx$; that is, $\int_a^b f(x)\,dx = -\int_b^a f(x)\,dx$. These considerations lead us to the following definition.

DEFINITION 2   **The Definite Integral $\int_a^b f(x)\,dx$ for $a \ge b$**

(i) If f is any function and a is a number in the domain of f, then we define

$$\int_a^a f(x)\,dx = 0.$$

(ii) If $a > b$ and f is Riemann-integrable on $[b, a]$, then we define
$$\int_a^b f(x)\, dx = -\int_b^a f(x)\, dx.$$

It is important to notice that the properties of definite integrals expressed by Theorems 1 through 4 are still operative when the lower limit of integration is not smaller than the upper limit of integration. For instance, Theorem 1 asserts that $\int_a^b K\, dx = K(b - a)$, at least when $a < b$. When $a = b$, both sides of the equation become zero, so that the equation still holds. For $a > b$,
$$\int_a^b K\, dx = -\int_b^a K\, dx = -K(a - b) = K(b - a),$$

and the equation $\int_a^b K\, dx = K(b - a)$ continues to hold. The reader can check to see that the homogeneous, additive, and linear properties also remain operative when $a \geq b$ (Problems 56 and 57). Of course, Theorem 5 fails to hold unless $a \leq b$ and so do Theorems 6 and 7. (Why?) With minor modifications, Theorem 8 continues to hold even if the condition $a < b < c$ fails. Thus, we have the following theorem.

THEOREM 10   **General Additivity with Respect to the Interval of Integration**
Suppose that the function f is Riemann-integrable on a closed bounded interval I and that a, b, and c are three numbers in I. Then
$$\int_a^c f(x)\, dx = \int_a^b f(x)\, dx + \int_b^c f(x)\, dx.$$

To prove Theorem 10, one uses Theorem 8 and Definition 2 to check each of the six possible cases $a < b < c$, $a < c < b$, $b < a < c$, $b < c < a$, $c < a < b$, and $c < b < a$ as well as the three cases in which two of the numbers a, b, or c are equal. For instance, if $a < b < c$, then Theorem 10 is the same as Theorem 8. But, suppose that $a < c < b$. Then, by Theorem 8,
$$\int_a^b f(x)\, dx = \int_a^c f(x)\, dx + \int_c^b f(x)\, dx;$$

hence,
$$\int_a^c f(x)\, dx = \int_a^b f(x)\, dx - \int_c^b f(x)\, dx = \int_a^b f(x)\, dx + \int_b^c f(x)\, dx,$$

so the desired equation continues to hold. The reader should check the remaining cases (Problem 58).

As usual, there is no particular reason for using the letter x as the "variable of integration" in $\int_a^b f(x)\, dx$—any other letter would do. In fact, $\int_a^b f(x)\, dx$ is a number depending on a, b and the function f; it really has nothing to do with x at all. The reader should always keep in mind that the variable of integration in a *definite integral* is a "dummy variable," so that, for instance,
$$\int_0^1 x^2\, dx = \int_0^1 t^2\, dt = \int_0^1 s^2\, ds = \int_0^1 y^2\, dy = \tfrac{1}{3}.$$

Often it is necessary to consider a definite integral in which one or both of the limits of integration are variable quantities. For instance, if the upper limit of integration is a variable quantity (and the lower limit of integration is fixed), then the value of the integral will be a function of the upper limit. Since it is customary to use the symbol x for the independent variable whenever possible, such an integral might be written as $\int_a^x f(x)\,dx$; however, the "x" in the expression "$f(x)\,dx$" is the (dummy) variable of integration and should not be confused with the variable upper limit x. Therefore, in such situations, we usually write

$$\int_a^x f(t)\,dt \quad \text{or} \quad \int_a^x f(s)\,ds \quad \text{or} \quad \int_a^x f(w)\,dw,$$

and so forth.

Problem Set 3

In problems 1 through 16, use the basic properties of the definite integral to evaluate each expression. You may assume that $\int_a^b x\,dx = \frac{1}{2}(b^2 - a^2)$ and that $\int_a^b x^2\,dx = \frac{1}{3}(b^3 - a^3)$.

1 $\displaystyle\int_{-5}^{4} (7 + \pi)\,dx$

2 $\displaystyle\int_{2}^{7} (-dx)$

3 $\displaystyle\int_{0.5}^{0.75} (1 + \sqrt{2} - \sqrt{3})\,dx$

4 $\displaystyle\int_{0}^{\pi} \left(\sum_{k=1}^{100} k\right)dx$

5 $\displaystyle\int_{1}^{3} 5x\,dx$

6 $\displaystyle\int_{1}^{2} (x + x^2)\,dx$

7 $\displaystyle\int_{-2}^{3} (3x^2 - 2x + 1)\,dx$

8 $\displaystyle\int_{-1}^{1} (x + 1)^2\,dx$

9 $\displaystyle\int_{3/2}^{5/3} (2x - 3)(3x + 2)\,dx$

10 $\displaystyle\int_{-1}^{a} x\,dx + \int_{a}^{1} x\,dx$, where $-1 < a < 1$

11 $\displaystyle\int_{0}^{\pi} (2x - 1)\,dx + \int_{\pi}^{4} (2x - 1)\,dx$

12 $\displaystyle\int_{1}^{3} f(x)\,dx$, given that $\int_{1}^{4} f(x)\,dx = -1$ and $\int_{3}^{4} f(x)\,dx = 5$

13 $\displaystyle\int_{0}^{4} f(x)\,dx$, where $f(x) = \begin{cases} 2x^2 & \text{for } 0 \le x \le 2 \\ 4x & \text{for } 2 < x \le 4 \end{cases}$

14 $\displaystyle\int_{0}^{1} f(x)\,dx$, where $f(x) = \begin{cases} 0 & \text{for } x = 0 \\ x + 1 & \text{for } 0 < x < 1 \\ 0 & \text{for } x = 1 \end{cases}$ (*Hint:* Use Theorem 4 of Section 2.)

15 $\displaystyle\int_{-1}^{1} f(x)\,dx$, where $f(x) = \begin{cases} 1 & \text{for } x \ne 0 \\ 0 & \text{for } x = 0 \end{cases}$ (*Hint:* Use Theorem 4 of Section 2.)

16 $\displaystyle\int_{-2}^{3} f(x)\,dx$, where $f(x) = \begin{cases} 1 - x & \text{for } -2 \le x \le 0 \\ 1 + x & \text{for } 0 < x \le 3 \end{cases}$

In problems 17 through 20, use the positivity or the comparison theorem to decide whether or not each inequality holds without evaluating the integrals involved.

17 $\displaystyle\int_0^1 x\,dx \leq \int_0^1 dx$

18 $\displaystyle\int_1^2 x^2\,dx < \int_1^2 x\,dx$

19 $\displaystyle 0 \leq \int_0^1 \frac{dx}{1+x^2}$

20 $\displaystyle\int_0^1 x^5\,dx \leq \int_0^1 x^6\,dx$

21 If $0 < K \leq f(x)$ holds for all values of x in $[a,b]$, prove that $0 < \displaystyle\int_a^b f(x)\,dx$. [*Hint:* Use the comparison theorem and the fact that $\displaystyle\int_a^b K\,dx = K(b-a)$.]

22 Suppose that f is a continuous function such that $\displaystyle\int_a^b f(x)\,dx = 0$ for *every* closed interval $[a,b]$. Prove that f is the zero function.

23 Suppose that f and g are Riemann-integrable functions on $[a,b]$ such that $|f(x) - g(x)| \leq K$ for every number x in $[a,b]$, where K is a positive constant. Prove that

$$\left| \int_a^b f(x)\,dx - \int_a^b g(x)\,dx \right| \leq K \cdot (b-a).$$

(*Hint:* Use the absolute value property.)

24 If f and g are Riemann-integrable functions on $[a,b]$, prove that

$$\left| \int_a^b [f(x) + g(x)]\,dx \right| \leq \int_a^b |f(x)|\,dx + \int_a^b |g(x)|\,dx.$$

In problems 25 through 30, find the average (or mean) value M of the given function f on each interval. You may assume that

$$\int_a^b x\,dx = \tfrac{1}{2}(b^2 - a^2) \quad \text{and that} \quad \int_a^b x^2\,dx = \tfrac{1}{3}(b^3 - a^3).$$

25 $f(x) = x^2 + 1$ on $[1,4]$

26 $f(x) = |x|$ on $[-1,1]$

27 $f(x) = x^2 - 2x + 3$ on $[-1,5]$

28 $f(x) = \begin{cases} \dfrac{|x|}{x} & \text{if } x \neq 0 \\ 0 & \text{if } x = 0 \end{cases}$ on $[-1,1]$

29 $f(x) = Ax^2 + Bx + C$ on $[a,b]$

30 $f(x) = [\![x]\!]$ on $[-1,3]$

In problems 31 through 35, find a value of c on the interval $[a,b]$ such that $f(c)$ is the mean value of the function f on $[a,b]$. You may assume that $\displaystyle\int_a^b x\,dx = \tfrac{1}{2}(b^2 - a^2)$ and that $\displaystyle\int_a^b x^2\,dx = \tfrac{1}{3}(b^3 - a^3)$.

31 $f(x) = x^2 + 5$ on $[0,1]$

32 $f(x) = x^2$ on $[-a,a]$

33 $f(x) = Ax + B$ on $[a,b]$

34 $f(x) = (x-2)(x+3)$ on $[-3,2]$

35 $f(x) = |x|$ on $[-2,5]$

36 The force required to stretch a spring from its relaxed position through s units is given by $F = ks$, where k is the spring constant. Find the mean value of F if the spring is stretched from $s = a$ to $s = b$. Also find a value of c between a and b such that the value of F when $s = c$ is the same as this mean value.

37 Explain why $\int_a^b f(x)\,dx = -\int_b^a f(x)\,dx$ holds for *all* values of a and b, assuming that f is Riemann-integrable on the closed bounded interval I and that a and b belong to I.

38 Explain why it is reasonable to define $\int_a^a f(x)\,dx$ to be zero.

In problems 39 through 45, evaluate each integral using only the basic properties and Definition 2. You may assume that $\int_a^b x\,dx = \frac{1}{2}(b^2 - a^2)$ and $\int_a^b x^2\,dx = \frac{1}{3}(b^3 - a^3)$ hold when $a < b$.

39 $\displaystyle\int_1^0 (x+1)\,dx$

40 $\displaystyle\int_1^{-1} (x+1)(x-1)\,dx$

41 $\displaystyle\int_\pi^\pi (x^2 + 2x + 1)\,dx$

42 $\displaystyle\int_5^{-2} |x|\,dx$

43 $\displaystyle\int_a^b x\,dx$, where $a \geq b$

44 $\displaystyle\int_3^{-2} 2[\![x]\!]\,dx$

45 $\displaystyle\int_a^b x^2\,dx$, where $a \geq b$

In problems 46 through 55, use only the basic properties of the integral to justify each assertion.

46 $\displaystyle\int_0^3 (4 + 3x - x^2)\,dx \geq 0$

47 $\displaystyle\int_0^1 x^4\,dx \leq \int_0^1 x\,dx$

48 $\displaystyle\int_0^1 x^4\,dx \leq \int_0^1 t\,dt$

49 $\displaystyle\int_0^{1/2} y^2\,dy \geq \int_0^{1/2} x^6\,dx$

50 $\displaystyle\int_{-1000}^{1000} [x^{1776} + \sqrt{|x|^{1699}}]\,dx \geq 0$

51 $\displaystyle\int_5^1 \sqrt{x^2 + 1}\,dx = \int_0^1 \sqrt{x^2 + 1}\,dx - \int_0^5 \sqrt{x^2 + 1}\,dx$

52 $\displaystyle\int_4^{-1} \sqrt[3]{5x^2 + 3}\,dx - \int_4^2 \sqrt[3]{5x^2 + 3}\,dx + \int_{-1}^2 \sqrt[3]{5x^2 + 3}\,dx = 0$

53 $\displaystyle\int_3^4 \frac{dx}{1 + x^2} - \int_5^5 \frac{dx}{1 + x^2} = \int_6^4 \frac{dt}{1 + t^2} + \int_3^6 \frac{dy}{1 + y^2}$

54 $\displaystyle\int_a^b \frac{dx}{\sqrt{1 + x^2}} + \int_b^c \frac{dy}{\sqrt{1 + y^2}} + \int_c^a \frac{dz}{\sqrt{1 + z^2}} = 0$

55 $\displaystyle\frac{d}{dx}\int_a^x K\,dt = K$, where K is any constant.

56 Prove that if f is Riemann-integrable on the closed interval I, then the homogeneous property $\int_a^b Kf(x)\,dx = K\int_a^b f(x)\,dx$ holds even if $a \geq b$ for a, b in I, provided that K is constant.

57 Show that Theorem 4 continues to hold even if $a \geq b$.

58 Complete the proof of Theorem 10 by considering the remaining cases $b < a < c$, $b < c < a$, $c < a < b$, and $c < b < a$ as well as the cases in which two of the numbers a, b, or c are equal.

59 Prove that Theorem 9 continues to hold even if $a \geq b$; that is, show that if f is continuous on the closed interval between a and b, then there exists a number c on this interval such that $f(c) \cdot (b - a) = \int_a^b f(x)\,dx$.

4   The Fundamental Theorem of Calculus

In this section we give a proof of the fundamental theorem of calculus based on the analytic definition of the Riemann integral in Section 2 and the properties presented in Section 3. Essentially, the fundamental theorem of calculus asserts that the operations of differentiation and integration are inverses of each other— that is, differentiation "undoes" integration and vice versa.

Actually, the fundamental theorem of calculus consists of two parts. The first part says roughly that the derivative of an integral is the integrand, while the second part corresponds to the preliminary version presented in Section 6 of Chapter 5. Later in Section 4.1 we state the two parts of the fundamental theorem precisely and give rigorous proofs. Here we give an *informal* statement of the theorem and present some examples illustrating its use.

Fundamental Theorem of Calculus—Informal Version
Let f be a continuous function on an interval I, suppose that a and b are fixed numbers in I, and let x denote a variable number in I. Then:

First Part. $\dfrac{d}{dx} \displaystyle\int_a^x f(t)\, dt = f(x).$

Second Part.   If g is an antiderivative of f, so that $g'(x) = f(x)$ holds for all x in I,
then $\displaystyle\int_a^b f(x)\, dx = g(b) - g(a).$

EXAMPLES   Use the first part of the fundamental theorem of calculus to find the indicated derivative.

1 $\dfrac{d}{dx} \displaystyle\int_0^x (2t^2 - t + 1)\, dt$

SOLUTION
By the first part of the fundamental theorem with $f(t) = 2t^2 - t + 1$ and $a = 0$, we have

$$\frac{d}{dx} \int_0^x (2t^2 - t + 1)\, dt = 2x^2 - x + 1.$$

2 $D_x y$ if $y = \displaystyle\int_{-43}^x \dfrac{dt}{5 + t^4}$

SOLUTION

$$D_x y = \frac{1}{5 + x^4}.$$

The first part of the fundamental theorem of calculus asserts that the function g given by $g(x) = \displaystyle\int_a^x f(t)\, dt$ is an antiderivative of the continuous function f. From this, it follows that *every continuous function f has an antiderivative.*

It sometimes happens that one (or both) of the limits of integration of a Riemann integral are functions of x, and it is required to find the derivative of the integral. The technique for making such a calculation, which depends on the first part of the fundamental theorem of calculus and the chain rule, is shown in the following examples.

EXAMPLES   Find the indicated derivative.

1 $\dfrac{dy}{dx}$ if $y = \displaystyle\int_{3}^{x^2} (5t + 7)^{25}\, dt$.

SOLUTION

We put $u = x^2$, so that $y = \displaystyle\int_{3}^{u} (5t + 7)^{25}\, dt$. By the first part of the fundamental theorem of calculus,

$$\frac{dy}{du} = \frac{d}{du} \int_{3}^{u} (5t + 7)^{25}\, dt = (5u + 7)^{25} = (5x^2 + 7)^{25}.$$

Therefore, by the chain rule,

$$\frac{dy}{dx} = \frac{dy}{du}\frac{du}{dx} = (5x^2 + 7)^{25}(2x).$$

2 $D_x y$ if $y = \displaystyle\int_{-x}^{3x+2} \sqrt{1 + t^2}\, dt$.

SOLUTION

$$y = \int_{-x}^{0} \sqrt{1 + t^2}\, dt + \int_{0}^{3x+2} \sqrt{1 + t^2}\, dt = -\int_{0}^{-x} \sqrt{1 + t^2}\, dt + \int_{0}^{3x+2} \sqrt{1 + t^2}\, dt;$$

hence,

$$D_x y = -\sqrt{1 + (-x)^2}\,(-1) + \sqrt{1 + (3x + 2)^2}\,(3) = \sqrt{1 + x^2} + 3\sqrt{1 + (3x + 2)^2}.$$

The second part of the fundamental theorem of calculus has already been illustrated by numerous examples in Section 6 of Chapter 5, where we introduced the notation

$$g(x)\Big|_{a}^{b} = g(b) - g(a).$$

Using this notation, we wrote the second part of the fundamental theorem of calculus as

$$\int_{a}^{b} f(x)\, dx = \left[\int f(x)\, dx\right]\Bigg|_{a}^{b}$$

To refresh the reader's memory, we give some additional examples.

EXAMPLES   Use the second part of the fundamental theorem of calculus to evaluate the given integral.

1 $\displaystyle\int_{1}^{2} (4x^3 - 3x^2 + 2x + 3)\, dx$

SOLUTION

$$\int_{1}^{2} (4x^3 - 3x^2 + 2x + 3)\, dx = \left[\int (4x^3 - 3x^2 + 2x + 3)\, dx\right]\Bigg|_{1}^{2}$$

$$= \left(4\frac{x^4}{4} - 3\frac{x^3}{3} + 2\frac{x^2}{2} + 3x\right)\Bigg|_{1}^{2}$$

$$= (16 - 8 + 4 + 6) - (1 - 1 + 1 + 3) = 14.$$

2 $\displaystyle\int_1^2 \frac{t^2\,dt}{(t^3+2)^2}$

SOLUTION

We begin by evaluating the indefinite integral $\displaystyle\int \frac{t^2\,dt}{(t^3+2)^2}$ using the change of variable $u = t^3 + 2$, so that $du = 3t^2\,dt$, or $t^2\,dt = \frac{1}{3}\,du$. Therefore,

$$\int \frac{t^2\,dt}{(t^3+2)^2} = \int \frac{1}{3}\frac{du}{u^2} = \frac{1}{3}\int u^{-2}\,du = \frac{1}{3}\left(\frac{u^{-1}}{-1}\right) + C = \frac{-1}{3u} + C = \frac{-1}{3(t^3+2)} + C.$$

Hence,

$$\int_1^2 \frac{t^2\,dt}{(t^3+2)^2} = \frac{-1}{3(t^3+2)}\bigg|_1^2 = \frac{-1}{3(8+2)} - \frac{-1}{3(1+2)} = \frac{7}{90}.$$

The calculation in Example 2 above can be shortened by the trick of *changing the limits of integration in accordance with the change of variable*. Indeed, the limits of integration in the original definite integral refer to the variable t, a fact that is sometimes emphasized by writing

$$\int_{t=1}^{t=2} \frac{t^2\,dt}{(t^3+2)^2} \quad \text{rather than} \quad \int_1^2 \frac{t^2\,dt}{(t^3+2)^2}.$$

To make the change of variable $u = t^3 + 2$, notice that

when $t = 1$, $u = 1^3 + 2 = 3$ and   when $t = 2$, $u = 2^3 + 2 = 10$.

Thus, since $t^2\,dt = \frac{1}{3}\,du$, we have

$$\int_{t=1}^{t=2} \frac{t^2\,dt}{(t^3+2)^2} = \int_{u=3}^{u=10} \frac{1}{3}\frac{du}{u^2} = \frac{1}{3}\int_3^{10} u^{-2}\,du = \frac{1}{3}\left(\frac{-1}{u}\right)\bigg|_3^{10}$$

$$= \frac{1}{3}\left(\frac{-1}{10}\right) - \frac{1}{3}\left(\frac{-1}{3}\right) = \frac{7}{90}.$$

More generally, when changing the variable from (say) x to (say) $u = g(x)$ in a definite integral of the form $\displaystyle\int_a^b f[g(x)]g'(x)\,dx$, not only must we change the integrand just as we did for an indefinite integral, but we must also change the limits of integration so that the integral takes the form $\displaystyle\int_{g(a)}^{g(b)} f(u)\,du$.

EXAMPLES   Use a change of variable to evaluate the given integral.

1 $\displaystyle\int_0^1 x\sqrt{9-5x^2}\,dx$

SOLUTION
We make the change of variable $u = 9 - 5x^2$, noting that $du = -10x\,dx$, or $x\,dx = (-1/10)\,du$. Also, $u = 9$ when $x = 0$, and $u = 4$ when $x = 1$. Therefore,

$$\int_0^1 x\sqrt{9-5x^2}\,dx = \int_{x=0}^{x=1} \sqrt{9-5x^2}\,x\,dx = \int_{u=9}^{u=4} \sqrt{u}\left(\frac{-1}{10}\right)du = \frac{-1}{10}\int_9^4 \sqrt{u}\,du$$

$$= \frac{1}{10}\int_4^9 u^{1/2}\,du = \frac{1}{10}\frac{u^{3/2}}{3/2}\bigg|_4^9 = \frac{1}{15}9^{3/2} - \frac{1}{15}4^{3/2} = \frac{19}{15}.$$

2 $\displaystyle\int_{2}^{5} x\sqrt{x-1}\,dx$

SOLUTION
Let $u = \sqrt{x-1}$, so that $u^2 = x - 1$, $x = u^2 + 1$, and $dx = 2u\,du$. Because $u = \sqrt{x-1}$, we see that $u = 1$ when $x = 2$, and $u = 2$ when $x = 5$. Thus, we have

$$\int_{2}^{5} x\sqrt{x-1}\,dx = \int_{1}^{2} (u^2 + 1)u(2u\,du) = 2\int_{1}^{2} (u^4 + u^2)\,du = 2\left[\frac{u^5}{5} + \frac{u^3}{3}\right]\Bigg|_{1}^{2}$$

$$= 2\left[\left(\frac{32}{5} + \frac{8}{3}\right) - \left(\frac{1}{5} + \frac{1}{3}\right)\right] = \frac{256}{15}.$$

Sometimes it is useful to split the interval of integration into two (or more) subintervals before using the fundamental theorem of calculus.

EXAMPLE   Evaluate $\displaystyle\int_{0}^{2} |1 - x|\,dx$.

SOLUTION
For $x \leq 1$, we have $|1 - x| = 1 - x$, while, for $x \geq 1$, we have $|1 - x| = -(1 - x) = x - 1$. Thus,

$$\int_{0}^{2} |1 - x|\,dx = \int_{0}^{1} |1 - x|\,dx + \int_{1}^{2} |1 - x|\,dx = \int_{0}^{1} (1 - x)\,dx + \int_{1}^{2} (x - 1)\,dx$$

$$= \left(x - \frac{x^2}{2}\right)\Bigg|_{0}^{1} + \left(\frac{x^2}{2} - x\right)\Bigg|_{1}^{2} = \left[\frac{1}{2} - 0\right] + \left[0 - \left(-\frac{1}{2}\right)\right] = 1.$$

4.1 Proof of the Fundamental Theorem of Calculus

We now give a rigorous proof of the fundamental theorem of calculus; however, we state and prove the two parts as two separate theorems. Also, we state these two parts more precisely than we did previously.

THEOREM 1   **Fundamental Theorem of Calculus—First Part**
Let f be a continuous function on the closed interval $[b, c]$ and suppose that a is a fixed number in this interval. Define the function g with domain $[b, c]$ by

$$g(x) = \int_{a}^{x} f(t)\,dt$$

for x in $[b, c]$. Then g is differentiable on the open interval (b, c) and

$$g'(x) = f(x)$$

holds for all x in (b, c). Furthermore,

$$g'_{+}(b) = f(b) \quad \text{and} \quad g'_{-}(c) = f(c).$$

PROOF
Suppose that x belongs to the open interval (b, c) and that Δx is small enough

so that $x + \Delta x$ also belongs to (b, c). Then $g(x) = \int_a^x f(t)\, dt$ and $g(x + \Delta x) = \int_a^{x + \Delta x} f(t)\, dt$. It follows that

$$g(x + \Delta x) - g(x) = \int_a^{x + \Delta x} f(t)\, dt - \int_a^x f(t)\, dt = \int_a^{x + \Delta x} f(t)\, dt + \int_x^a f(t)\, dt$$

by Definition 2 of Section 3. Therefore, by Theorem 10 of Section 3,

$$g(x + \Delta x) - g(x) = \int_x^a f(t)\, dt + \int_a^{x + \Delta x} f(t)\, dt = \int_x^{x + \Delta x} f(t)\, dt.$$

Since f is continuous on the interval $[b, c]$, it is also continuous on the closed subinterval between x and $x + \Delta x$. By Theorem 9 of Section 3, it follows that there exists a number x^* on the closed interval between x and $x + \Delta x$ such that

$$\int_x^{x + \Delta x} f(t)\, dt = f(x^*)[(x + \Delta x) - x] = f(x^*)\, \Delta x.$$

Consequently,

$$g(x + \Delta x) - g(x) = f(x^*)\, \Delta x \quad \text{or} \quad \frac{g(x + \Delta x) - g(x)}{\Delta x} = f(x^*).$$

Since x^* lies between x and $x + \Delta x$, then x^* approaches x as Δx approaches zero. Therefore,

$$g'(x) = \lim_{\Delta x \to 0} \frac{g(x + \Delta x) - g(x)}{\Delta x} = \lim_{x^* \to x} f(x^*) = f(x),$$

where we have used the continuity of f in the last equation. This establishes the desired result for values of x on the open interval (b, c). The proof that one-sided derivatives of g at the endpoints b and c give the values of f at these endpoints is similar and is left to the reader as an exercise (Problem 64).

THEOREM 2   **Fundamental Theorem of Calculus—Second Part**
Let f be a continuous function on the closed interval $[a, b]$, and suppose that g is a continuous function on $[a, b]$ such that $g'(x) = f(x)$ holds for all values of x in the open interval (a, b). Then

$$\int_a^b f(x)\, dx = g(b) - g(a).$$

PROOF

Define a function F with domain $[a, b]$ by $F(x) = \int_a^x f(t)\, dt$ for x in $[a, b]$. By Theorem 1, $F'(x) = f(x)$ holds for all values of x in the open interval (a, b), while $F'_+(a) = f(a)$ and $F'_-(b) = f(b)$ hold at the endpoints. Since F is differentiable on (a, b), it is continuous on (a, b). Because F has right and left derivatives at a and b, respectively, F is continuous from the right at a and F is continuous from the left at b. It follows that F is continuous on the closed interval $[a, b]$. On the open interval (a, b), we have $F' = f = g'$; hence, $(F - g)' = 0$. It follows from Theorem 1, Chapter 5, Section 2, that $F(x) - g(x) = C$ holds for all values of x in the open interval (a, b), where C is a constant number.

Since F and g are continuous from the right at a, the equality $C = F(x) - g(x)$ for $a < x < b$ implies that

$$C = \lim_{x \to a^+} C = \lim_{x \to a^+} [F(x) - g(x)] = \lim_{x \to a^+} F(x) - \lim_{x \to a^+} g(x) = F(a) - g(a).$$

But, $F(a) = \int_a^a f(t)\, dt = 0$; hence, $C = -g(a)$. Thus, the equation $C = F(x) - g(x)$ for $a < x < b$ can be rewritten as $F(x) = g(x) - g(a)$. Because F and g are continuous from the left at b, the latter equality implies that

$$F(b) = \lim_{x \to b^-} F(x) = \lim_{x \to b^-} [g(x) - g(a)] = \lim_{x \to b^-} g(x) - \lim_{x \to b^-} g(a) = g(b) - g(a).$$

Since $F(b) = \int_a^b f(t)\, dt = \int_a^b f(x)\, dx$, we therefore have $\int_a^b f(x)\, dx = g(b) - g(a)$.

From now on, we shall often refer simply to "the fundamental theorem of calculus" and leave it to the reader to discern from the context whether we mean the first or the second part of this theorem. The fundamental theorem of calculus establishes a profound connection between differentiation and integration and allows us to convert facts about differentiation into facts about integration.

Problem Set 4

In problems 1 through 12, use the first part of the fundamental theorem of calculus to find dy/dx.

1 $y = \displaystyle\int_0^x (t^2 + 1)\, dt$

2 $y = \displaystyle\int_1^x (w^3 - 2w + 1)\, dw$

3 $y = \displaystyle\int_{-1}^x \dfrac{ds}{1 + s^2}$

4 $y = \displaystyle\int_0^x \dfrac{ds}{1 + s} + \int_2^x \dfrac{ds}{1 + s}$

5 $y = \displaystyle\int_0^x |v|\, dv$

6 $y = \displaystyle\int_{-1}^x f(t)\, dt$, where $f(t) = \begin{cases} t + 1 & \text{for } t \le -1 \\ (t+1)^3 & \text{for } t > -1 \end{cases}$

7 $y = \displaystyle\int_{-1}^x \sqrt{t^2 + 4}\, dt$

8 $y = \displaystyle\int_x^1 (t^3 - 3t + 1)^{10}\, dt$

9 $y = \displaystyle\int_x^1 (w^{10} + 3)^{25}\, dw$

10 $y = \displaystyle\int_x^4 \sqrt[3]{4s^2 + 7}\, ds$

11 $y = \displaystyle\int_x^0 \sqrt[3]{t^2 + 1}\, dt + \int_0^x \sqrt[3]{t^2 + 1}\, dt$

12 $y = \dfrac{d}{dx} \displaystyle\int_1^x \dfrac{1}{1 + t^2}\, dt$

In problems 13 through 20, use the first part of the fundamental theorem of calculus together with the chain rule to find dy/dx.

13 $y = \displaystyle\int_1^{3x} (5t^3 + 1)^7\, dt$

14 $y = \displaystyle\int_1^{5x+1} \dfrac{dt}{9 + t^2}$

15 $y = \displaystyle\int_1^{8x+2} (w - 3)^{15}\, dw$

16 $y = \displaystyle\int_1^{x-1} \sqrt{s^2 - 1}\, ds$

17 $y = \displaystyle\int_{-x}^0 \sqrt{t + 2}\, dt$

18 $y = \displaystyle\int_{x^2+1}^2 \sqrt[3]{u - 1}\, du$

19 $y = \displaystyle\int_x^{3x^2+2} \sqrt[4]{t^4 + 17}\, dt$

20 $y = \displaystyle\int_{x^3}^{x-x^2} \sqrt{t^3 + 1}\, dt$

21 Given that u and v are differentiable functions of x and that f is a continuous function, justify the equation

$$\frac{d}{dx} \int_u^v f(t)\, dt = f(v)\,\frac{dv}{dx} - f(u)\,\frac{du}{dx}.$$

22 Let f be a continuous function and let M be the mean value of f on the interval $[-x, x]$. Find dM/dx.

In problems 23 through 45, use the second part of the fundamental theorem of calculus to evaluate each integral.

23 $\int_2^3 (3x + 4)\, dx$

24 $\int_{-3}^{-1} (4 - 8x + 3x^2)\, dx$

25 $\int_1^5 (x^3 - 3x^2 + 1)\, dx$

26 $\int_1^3 (x - 1)(x^2 + x + 1)\, dx$

27 $\int_0^1 (x^2 + 2)^2\, dx$

28 $\int_1^5 \frac{x^4 - 16}{x^2 + 4}\, dx$

29 $\int_0^8 (2 - \sqrt[3]{t})^2\, dt$

30 $\int_1^{32} \frac{1 + \sqrt[5]{t^2}}{\sqrt[3]{t}}\, dt$

31 $\int_0^1 \frac{y^2\, dy}{(y^3 + 1)^5}$

32 $\int_0^1 (2x + 3)^{10}\, dx$

33 $\int_{-1}^1 \sqrt{1 - x}\, dx$

34 $\int_0^1 \sqrt{4 - 3x}\, dx$

35 $\int_{1/4}^3 \frac{dx}{\sqrt{1 + x}}$

36 $\int_0^2 \frac{x\, dx}{(4 + x^2)^{3/2}}$

37 $\int_0^2 x^2 \sqrt[3]{x^3 + 1}\, dx$

38 $\int_0^1 \frac{x + 3}{\sqrt{x^2 + 6x + 2}}\, dx$

39 $\int_7^{10} \frac{x\, dx}{\sqrt{x - 6}}$

40 $\int_{-1}^1 \sqrt{|t|} + t\, dt$

41 $\int_0^3 |3 - x^2|\, dx$

42 $\int_{-1}^3 \sqrt[3]{2(|x| - x)}\, dx$

43 $\int_0^3 y|2 - y|\, dy$

44 $\int_{-1}^3 [\![x]\!]x\, dx$

45 $\int_{-3}^5 f(x)\, dx$, where $f(x) = \begin{cases} (1 - x)^{3/2} & \text{for } x \le 0 \\ (x + 4)^{1/2} & \text{for } x > 0 \end{cases}$

46 Does the following computation violate the positivity property:
$$\int_{-1}^1 \frac{1}{x^2}\, dx = \left[-\frac{1}{x}\right]\Big|_{-1}^1 = -1 - 1 = -2?$$

47 (a) Show that $\int_0^x |t|\, dt = x|x|/2$ by considering the separate cases $x \le 0$ and $x > 0$.

(b) Use the result in part (a) and the first part of the fundamental theorem of calculus to prove that $\dfrac{d}{dx}\left[\dfrac{x|x|}{2}\right] = |x|$.

48 Use the method suggested by problem 47 to find an antiderivative of the function f, where $f(x) = |x|^n$.

In problems 49 through 52, find the mean value of each function on the interval given.

49 $f(x) = x^2 + 1$ on $[1, 4]$

50 $f(x) = x^3 - 1$ on $[1, 3]$

51 $f(x) = \sqrt{x}$ on $[1, 9]$

52 $f(x) = |x| + 1$ on $[a, b]$

53 Find $f''(x)$ if f is the function defined by
$$f(x) = \int_0^x \left[\int_0^t (u^2 + 7)\, du\right] dt.$$

54 Show that $\int_0^{-x} f(t)\, dt + \int_0^x f(-t)\, dt$ is a constant by taking the derivative with respect to x. (Assume that f is continuous.) What is the value of this constant?

In problems 55 through 58, sketch graphs of f and g.

55 $f(t) = \begin{cases} 1 - t^2 & \text{for } t \le 0 \\ 1 + t^2 & \text{for } t > 0 \end{cases}$, $g(x) = \int_{-1}^{x} f(t)\,dt$

56 $f(t) = \begin{cases} 1 - t^2 & \text{for } t \le 0 \\ t^2 & \text{for } t > 0 \end{cases}$, $g(x) = \int_{-1}^{x} f(t)\,dt$

57 $f(t) = [\![t]\!]$ for $-3 \le t \le 3$, $g(x) = \int_{0}^{x} f(t)\,dt$ for $-3 \le x \le 3$

58 $f(t) = \begin{cases} -2 & \text{for } t \le 0 \\ t & \text{for } 0 < t \le 5 \\ 2t & \text{for } t > 5 \end{cases}$, $g(x) = \int_{-1}^{x} f(t)\,dt$

59 Assume that f is a differentiable function and that f' is continuous. Explain why
$$D_x \int_{a}^{x} f(t)\,dt = \int_{a}^{x} D_t f(t)\,dt + f(a).$$

60 The operator of a trucking company wants to determine the optimal period of time T (in months) between overhauls of a truck. Let the rate of depreciation of the truck be given by $f(t)$, where t is the time in months since the last overhaul. If K represents the fixed cost of an overhaul, explain why T is the value of t that minimizes
$$g(t) = t^{-1}\left[K + \int_{0}^{t} f(x)\,dx\right].$$

61 Solve the following equations for c.

(a) $\int_{0}^{c} x^2\,dx = \int_{c}^{10} x^2\,dx$ (b) $\int_{c}^{c+1} x\,dx = 10$ (c) $\int_{0}^{4} (x - c)\sqrt{4 - x}\,dx = 0$

62 Find the maximum value of the function f defined by $f(x) = \int_{0}^{x} (|t| - |t - 1|)\,dt$ on the interval $[-1, 2]$.

63 The marginal cost of processing a certain grade of tuna fish is given by $200 - 30\sqrt{x}$, where x is the number of cans in thousands and the cost is in dollars. What would be the total increase in cost if production were increased from 4000 cans to 25,000 cans?

64 Complete the proof of Theorem 1 in Section 4.1 by attending to the details at the endpoints of the interval $[b, c]$.

5 Approximation of Definite Integrals—Simpson and Trapezoidal Rules

The fundamental theorem of calculus enables us to calculate the numerical value of a definite integral $\int_{a}^{b} f(x)\,dx$, provided that we can find an antiderivative g of the function f. However, in practical applications of calculus, it is sometimes necessary to evaluate a definite integral $\int_{a}^{b} f(x)\,dx$ for which it is difficult or impossible to find an antiderivative g of f such that $g(b)$ and $g(a)$ can be calculated explicitly. In these cases, numerical methods of approximation can be used to estimate the value of $\int_{a}^{b} f(x)\,dx$ within acceptable error bounds.

In this section we present some numerical methods for estimating the value of a definite integral $\int_{a}^{b} f(x)\,dx$ by formulas that use the values of $f(x)$ at only a finite number of points on the interval $[a, b]$. These methods involve only simple computations and thus lend themselves well to the use of hand calculators or computers.

5.1   Direct Use of the Analytic Definition

Perhaps the most obvious way to approximate the definite integral $\int_{a}^{b} f(x)\,dx$ is to use its very definition as a limit of Riemann sums. Thus, we can select a partition $\mathscr{P}$ with a small norm, augment $\mathscr{P}$ to obtain $\mathscr{P}^*$, evaluate the corresponding Riemann sum $\sum_{k=1}^{n} f(c_k)\,\Delta x_k$, and observe that

$$\sum_{k=1}^{n} f(c_k)\,\Delta x_k \approx \int_{a}^{b} f(x)\,dx.$$

Figure 1

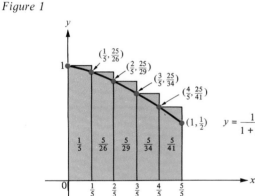

Figure 1 illustrates how this procedure, with $n = 5$, is used to estimate $\int_{0}^{1} \dfrac{1}{1 + x^2}\,dx$. The areas of the five rectangles are shown in Figure 1, and the sum of these areas, which is approximately 0.834, provides an estimate for the integral above. The true value of this integral, rounded off to four decimal places, is 0.7854, so the estimate $\int_{0}^{1} \dfrac{1}{1 + x^2}\,dx \approx 0.834$ is not particularly good. To obtain a reasonably accurate estimate by this method, we would have to partition the interval $[0, 1]$ into a very large number of subintervals.

5.2   The Trapezoidal Rule

Figure 1 makes it clear why the above estimation of $\int_{0}^{1} \dfrac{1}{1 + x^2}\,dx$ is so crude—the five rectangles overhang the region under the graph of $y = \dfrac{1}{1 + x^2}$, so that their total area is considerably larger than the desired area.

Figure 2

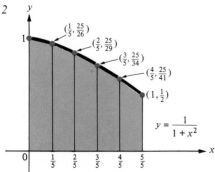

Evidently, a much more accurate estimate of $\int_{0}^{1} \dfrac{1}{1 + x^2}\,dx$ is obtained by adding the areas of the five trapezoids shown in Figure 2. Each of these trapezoids has the same width, namely, $\frac{1}{5}$ unit. Since the area of a trapezoid is given by one-half the sum of the lengths of its two parallel edges times the distance between these edges, the total area of the five trapezoids in Figure 2 is

$$\left(\frac{1+\frac{25}{26}}{2}\right)\frac{1}{5} + \left(\frac{\frac{25}{26}+\frac{25}{29}}{2}\right)\frac{1}{5} + \left(\frac{\frac{25}{29}+\frac{25}{34}}{2}\right)\frac{1}{5} + \left(\frac{\frac{25}{34}+\frac{25}{41}}{2}\right)\frac{1}{5} + \left(\frac{\frac{25}{41}+\frac{1}{2}}{2}\right)\frac{1}{5}$$

$$= \left(\frac{1}{2} + \frac{25}{26} + \frac{25}{29} + \frac{25}{34} + \frac{25}{41} + \frac{1}{4}\right)\frac{1}{5}.$$

The numerical value of this total area, rounded off to four decimal places, is 0.7837. Therefore, by this "trapezoidal method," $0.7837 \approx \int_0^1 \frac{1}{1+x^2}\,dx$. The trapezoidal rule for estimating definite integrals in general is given by the following theorem.

THEOREM 1   Trapezoidal Rule

Let the function f be defined and Riemann-integrable on the closed interval $[a, b]$. For each positive integer n, define

$$T_n = \left(\frac{y_0}{2} + y_1 + y_2 + \cdots + y_{n-1} + \frac{y_n}{2}\right)\Delta x,$$

where $\Delta x = \dfrac{b-a}{n}$ and $y_k = f(a + k\,\Delta x)$ for $k = 0, 1, 2, \ldots, n$. Then $T_n \approx \int_a^b f(x)\,dx$, the approximation becoming better and better as n increases in the sense that

$$\lim_{n \to +\infty} T_n = \int_a^b f(x)\,dx.$$

Figure 3

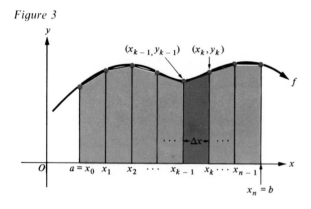

PROOF

Let $\mathscr{P}_n$ be the partition of $[a, b]$ consisting of n subintervals of equal length $\Delta x = \dfrac{b-a}{n}$. Thus, the subintervals in $\mathscr{P}_n$ are $[x_0, x_1]$, $[x_1, x_2]$, $[x_2, x_3]$, ..., $[x_{n-1}, x_n]$ where $x_0 = a$, $x_n = b$, and $x_k - x_{k-1} = \Delta x$ for $k = 1, 2, \ldots, n$ (Figure 3). Evidently, $x_1 = a + \Delta x$, $x_2 = a + 2\,\Delta x$, $x_3 = a + 3\,\Delta x, \ldots, x_k = a + k\,\Delta x, \ldots,$ and $x_n = a + n\,\Delta x = b$. Now, augment the partition $\mathscr{P}_n$ by choosing the points $c_1, c_2, c_3, \ldots, c_n$, where $c_1 = x_1, c_2 = x_2, c_3 = x_3, \ldots, c_k = x_k, \ldots,$ and $c_n = x_n$. Then

$$f(c_k) = f(x_k) = f(a + k\,\Delta x) = y_k$$

and the corresponding Riemann sum is given by

$$\sum_{k=1}^{n} f(c_k)\,\Delta x_k = \sum_{k=1}^{n} y_k\,\Delta x = \left[\sum_{k=1}^{n} y_k\right]\Delta x = [y_1 + y_2 + \cdots + y_n]\,\Delta x$$

$$= \left[\frac{y_0}{2} + y_1 + y_2 + \cdots + y_{n-1} + \frac{y_n}{2}\right]\Delta x - \frac{y_0}{2}\,\Delta x + \frac{y_n}{2}\,\Delta x$$

$$= T_n + \frac{y_n - y_0}{2}\,\Delta x = T_n + \frac{f(b) - f(a)}{2}\,\Delta x$$

$$= T_n + \frac{[f(b) - f(a)](b-a)}{2n}.$$

Therefore,

$$T_n = \sum_{k=1}^{n} f(c_k)\,\Delta x_k - \frac{[f(b) - f(a)](b-a)}{2n},$$

so that

$$\lim_{n \to +\infty} T_n = \lim_{n \to +\infty} \sum_{k=1}^{n} f(c_k)\, \Delta x_k - \lim_{n \to +\infty} \frac{[f(b) - f(a)](b - a)}{2n}$$

$$= \int_a^b f(x)\, dx - 0 = \int_a^b f(x)\, dx.$$

The theorem above admits a simple geometric explanation since the quantity T_n represents the total area of all the trapezoids shown in Figure 3 (Problem 14).

EXAMPLES   Use the trapezoidal rule, $T_n \approx \int_a^b f(x)\, dx$, with the indicated value of n to estimate the given definite integral.

1 $\displaystyle\int_0^1 \sqrt{1 + x^3}\, dx$, $n = 4$

SOLUTION

Here $[a, b] = [0, 1]$ and $n = 4$, so $\Delta x = \dfrac{b - a}{n} = \dfrac{1 - 0}{4} = \dfrac{1}{4}$. Also, $y_k = f(0 + k\,\Delta x) = \sqrt{1 + (k/4)^3}$ for $k = 0, 1, 2, 3, 4$; hence,

$$y_0 = \sqrt{1 + 0^3} = 1, \qquad y_1 = \sqrt{1 + (\tfrac{1}{4})^3} = \frac{\sqrt{65}}{8},$$

$$y_2 = \sqrt{1 + (\tfrac{2}{4})^2} = \sqrt{\tfrac{9}{8}}, \qquad y_3 = \sqrt{1 + (\tfrac{3}{4})^3} = \frac{\sqrt{91}}{8}, \qquad y_4 = \sqrt{1 + 1^3} = \sqrt{2}.$$

Therefore,

$$T_4 = \left(\frac{y_0}{2} + y_1 + y_2 + y_3 + \frac{y_4}{2}\right) \Delta x = \left(\frac{1}{2} + \frac{\sqrt{65}}{8} + \sqrt{\frac{9}{8}} + \frac{\sqrt{91}}{8} + \frac{\sqrt{2}}{2}\right)\left(\frac{1}{4}\right)$$

$$\approx (0.50 + 1.01 + 1.06 + 1.19 + 0.71)(0.25) = (4.47)(0.25) \approx 1.12.$$

Thus, $\displaystyle\int_0^1 \sqrt{1 + x^3}\, dx \approx 1.12$.

2 $\displaystyle\int_1^2 \frac{dx}{1 + x^2}$, $n = 5$

SOLUTION
Here

$$\Delta x = \frac{2 - 1}{5} = \frac{1}{5} \quad \text{and} \quad y_k = \frac{1}{1 + (1 + k/5)^2} = \frac{25}{25 + (5 + k)^2}$$

for $k = 0, 1, 2, 3, 4, 5$. Therefore,

$$T_5 = \left(\frac{1}{4} + \frac{25}{61} + \frac{25}{74} + \frac{25}{89} + \frac{25}{106} + \frac{25}{250}\right)\left(\frac{1}{5}\right)$$

$$\approx (0.2500 + 0.4098 + 0.3378 + 0.2809 + 0.2358 + 0.1000)(0.2) \approx 0.323;$$

hence, $\displaystyle\int_1^2 \frac{dx}{1 + x^2} \approx 0.323$.

The following theorem, whose proof can be found in more advanced textbooks, provides an upper bound for the error involved in using the trapezoidal rule.

THEOREM 2   **Error Bound for Trapezoidal Rule**

Suppose that f'' is defined and continuous on $[a, b]$ and that M is the maximum value of $|f''(x)|$ for x in $[a, b]$. Then, if T_n is the approximation to $\int_a^b f(x)\, dx$ given by the trapezoidal rule,

$$\left| T_n - \int_a^b f(x)\, dx \right| \leq M \frac{(b-a)^3}{12 n^2}.$$

EXAMPLE   Use the trapezoidal rule to estimate $\int_1^2 \frac{dx}{x}$ with $n = 6$, and use Theorem 2 to find a bound for the error of the estimate.

SOLUTION

We have $\Delta x = \dfrac{2-1}{6} = \dfrac{1}{6}$. Also, $y_k = \dfrac{1}{1 + k/6} = \dfrac{6}{6+k}$ for $k = 0, 1, 2, 3, 4, 5, 6$; hence,

$$y_0 = \tfrac{6}{6} = 1, \quad y_1 = \tfrac{6}{7}, \quad y_2 = \tfrac{6}{8}, \quad y_3 = \tfrac{6}{9}, \quad y_4 = \tfrac{6}{10}, \quad y_5 = \tfrac{6}{11}, \quad \text{and} \quad y_6 = \tfrac{6}{12}.$$

Therefore,

$$\begin{aligned}
T_6 &= \left(\tfrac{1}{2} + \tfrac{6}{7} + \tfrac{6}{8} + \tfrac{6}{9} + \tfrac{6}{10} + \tfrac{6}{11} + \tfrac{1}{4}\right)\left(\tfrac{1}{6}\right)\\
&\approx (0.5 + 0.8571 + 0.75 + 0.6667 + 0.6 + 0.5455 + 0.25)(0.1667)\\
&\approx 0.6950,
\end{aligned}$$

so that $\int_1^2 \frac{dx}{x} \approx 0.695$.

Here, $f(x) = 1/x$, $f'(x) = -1/x^2$, and $f''(x) = 2/x^3$. Since f'' is a decreasing function on $[1, 2]$, it follows that its maximum value on this interval is taken on at the left endpoint 1. Hence, in Theorem 2, $M = f''(1) = 2/1^3 = 2$. In Theorem 2 we also put $b = 2$, $a = 1$, and $n = 6$ to conclude that the error in the estimate above does not exceed $M \dfrac{(b-a)^3}{12 n^2} = (2)\dfrac{1^3}{432} = \dfrac{1}{216}$. Since $\tfrac{1}{216} < 0.005$, it follows that the estimation $\int_1^2 \frac{dx}{x} \approx 0.695$ is correct to at least two decimal places.

5.3 Simpson's Rule

A third method for approximating the value of a definite integral is known as *Simpson's rule* or the *parabolic rule* and is usually more efficient than either the direct use of the analytic definition or the trapezoidal rule. The method is based on the use of a number of adjacent regions having the shape shown in Figure 4 to approximate the area under the graph of f. (See Problem 29.) The result is the following theorem, whose proof is analogous to the proof of Theorem 1 and is therefore omitted.

Figure 4

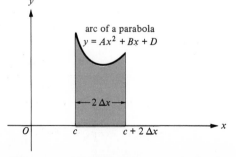

THEOREM 3   **Simpson's Parabolic Rule**

Let the function f be defined and Riemann-integrable on the closed interval $[a, b]$. For each positive integer n, define

$$S_{2n} = \frac{\Delta x}{3}(y_0 + 4y_1 + 2y_2 + 4y_3 + 2y_4 + \cdots + 2y_{2n-2} + 4y_{2n-1} + y_{2n}),$$

where $\Delta x = \dfrac{b - a}{2n}$ and $y_k = f(a + k\,\Delta x)$ for $k = 0, 1, 2, \ldots, 2n$. Then $S_{2n} \approx \displaystyle\int_a^b f(x)\,dx$, the approximation becoming better and better as n increases in the sense that

$$\lim_{n \to +\infty} S_{2n} = \int_a^b f(x)\,dx.$$

EXAMPLES   Use Simpson's parabolic rule, $S_{2n} \approx \displaystyle\int_a^b f(x)\,dx$, to estimate the given definite integral using the indicated value of n.

1 $\displaystyle\int_0^1 (1 + x)^{-1}\,dx$; $n = 4$

SOLUTION

Here, the interval $[0, 1]$ must be subdivided into $2n = 8$ parts, each of length $\Delta x = \dfrac{1 - 0}{8} = \dfrac{1}{8}$. Also,

$$y_0 = (1 + 0)^{-1} = 1, \qquad y_1 = (1 + \tfrac{1}{8})^{-1} = \tfrac{8}{9}, \qquad y_2 = (1 + \tfrac{2}{8})^{-1} = \tfrac{8}{10},$$
$$y_3 = (1 + \tfrac{3}{8})^{-1} = \tfrac{8}{11}, \qquad y_4 = (1 + \tfrac{4}{8})^{-1} = \tfrac{8}{12}, \qquad y_5 = (1 + \tfrac{5}{8})^{-1} = \tfrac{8}{13},$$
$$y_6 = (1 + \tfrac{6}{8})^{-1} = \tfrac{8}{14}, \qquad y_7 = (1 + \tfrac{7}{8})^{-1} = \tfrac{8}{15}, \qquad y_8 = (1 + \tfrac{8}{8})^{-1} = \tfrac{1}{2}.$$

Therefore,

$$S_{2n} = S_8 = \frac{\Delta x}{3}(y_0 + 4y_1 + 2y_2 + 4y_3 + 2y_4 + 4y_5 + 2y_6 + 4y_7 + y_8)$$

$$= \tfrac{1}{24}(1 + \tfrac{32}{9} + \tfrac{16}{10} + \tfrac{32}{11} + \tfrac{16}{12} + \tfrac{32}{13} + \tfrac{16}{14} + \tfrac{32}{15} + \tfrac{1}{2})$$

$$\approx \tfrac{1}{24}(1 + 3.5556 + 1.6000 + 2.9091 + 1.3333$$
$$+ 2.4615 + 1.1429 + 2.1333 + 0.5000)$$

$$= \tfrac{1}{24}(16.6357) \approx 0.6932.$$

Consequently, $\displaystyle\int_0^1 (1 + x)^{-1}\,dx \approx 0.6932$. Incidentally, the correct value, rounded off to five decimal places, is $\displaystyle\int_0^1 (1 + x)^{-1}\,dx \approx 0.69315$.

2 $\displaystyle\int_0^1 \sqrt{1 - x^4}\,dx$; $n = 3$

SOLUTION

Here

$$\Delta x = \frac{1 - 0}{6} = \frac{1}{6}, \qquad y_k = \sqrt{1 - \left(\frac{k}{6}\right)^4} = \sqrt{\frac{6^4 - k^4}{6^4}} = \sqrt{\frac{1296 - k^4}{1296}};$$

$k = 0, 1, 2, 3, 4, 5, 6$. Therefore,

$$S_{2n} = S_6 = \frac{1}{18}\left(1 + 4\sqrt{\frac{1295}{1296}} + 2\sqrt{\frac{1280}{1296}} + 4\sqrt{\frac{1215}{1296}} + 2\sqrt{\frac{1040}{1296}} + 4\sqrt{\frac{671}{1296}} + 0\right)$$

$$\approx \tfrac{1}{18}(1 + 3.9985 + 1.9876 + 3.8730 + 1.7916 + 2.8782)$$

$$= \tfrac{1}{18}(15.5289) \approx 0.8627 \approx \int_0^1 \sqrt{1 - x^4}\,dx.$$

The following theorem, whose proof can be found in more advanced textbooks, provides an upper bound for the error involved in using Simpson's parabolic rule.

THEOREM 4   **Error Bound for Simpson's Parabolic Rule**
Suppose that the fourth derivative $f^{(4)}(x)$ is defined and continuous on $[a, b]$ and that N is the maximum value of $|f^{(4)}(x)|$ for x in $[a, b]$. If S_{2n} is the approximation to $\int_a^b f(x)\,dx$ given by Simpson's parabolic rule, then

$$\left[S_{2n} - \int_a^b f(x)\,dx \right] \leq N \frac{(b-a)^5}{2880n^4}.$$

EXAMPLE   Use Simpson's parabolic rule to estimate $\int_1^2 \dfrac{dx}{x}$, making certain that the error of estimation does not exceed 0.001.

SOLUTION
Here, $f(x) = 1/x$, $f'(x) = -1/x^2$, $f''(x) = 2/x^3$, $f'''(x) = -6/x^4$, and $f^{(4)}(x) = 24/x^5$. Since $f^{(4)}$ is a decreasing function on $[1, 2]$, its maximum value on this interval is taken on at the left-hand endpoint 1. Hence, in Theorem 4, $N = |f^{(4)}(1)| = 24/1^5 = 24$ and $b - a = 2 - 1 = 1$, so the error of estimation cannot exceed

$$N \frac{(b-a)^5}{2880n^4} = (24) \frac{1}{2880n^4} = \frac{1}{120n^4}.$$

Therefore, we require that $1/(120n^4) \leq 0.001$; that is, $25/3 \leq n^4$. The smallest value of n satisfying the latter inequality is $n = 2$. Thus, we estimate $\int_1^2 dx/x$ using $S_{2n} = S_4$. We have $\Delta x = \dfrac{2-1}{4} = \dfrac{1}{4}$, so $y_0 = 1$, $y_1 = \frac{4}{5}$, $y_2 = \frac{4}{6}$, $y_3 = \frac{4}{7}$, and $y_4 = \frac{1}{2}$; hence,

$$S_{2n} = \tfrac{1}{12}\left(1 + \tfrac{16}{5} + \tfrac{8}{6} + \tfrac{16}{7} + \tfrac{1}{2}\right) = \tfrac{1747}{2520}.$$

It follows that $\int_1^2 dx/x \approx \tfrac{1747}{2520}$ with an error not exceeding $\tfrac{1}{1000}$. In fact, rounding off to five decimal places, $\tfrac{1747}{2520} \approx 0.69325$, while the correct value of $\int_1^2 dx/x$, rounded off to five decimal places, is 0.69315.

Theorem 4 has an interesting consequence; namely, *Simpson's parabolic rule gives the exact value of $\int_a^b f(x)\,dx$ if f is a polynomial function of degree not exceeding 3.* The reason for this is simply that, for such a polynomial function, $f^{(4)}$ is the zero function; hence, in Theorem 4, $N = 0$. Thus, taking $n = 1$ in Theorem 3, we obtain the following theorem.

THEOREM 5   **Prismoidal Formula**
Let f be any polynomial function whose degree does not exceed three. Then

$$\int_a^b f(x)\,dx = \frac{b-a}{6}\left[f(a) + 4f\left(\frac{a+b}{2}\right) + f(b) \right].$$

EXAMPLE   Use the prismoidal formula to evaluate $\int_0^2 (x^3 + 1)\, dx$.

SOLUTION

$$\int_0^2 (x^3 + 1)\, dx = \tfrac{2}{6}[(0^3 + 1) + 4(1^3 + 1) + (2^3 + 1)] = 6.$$

Problem Set 5

In problems 1 through 10, use the trapezoidal rule, $T_n \approx \int_a^b f(x)\, dx$, with the indicated value of n, to estimate each integral.

1 $\displaystyle\int_0^1 \frac{dx}{1 + x^2}; n = 4$

2 $\displaystyle\int_1^3 \frac{dx}{x}; n = 3$

3 $\displaystyle\int_2^8 \frac{dx}{1 + x}; n = 6$

4 $\displaystyle\int_0^3 \sqrt{9 - x^2}\, dx; n = 6$

5 $\displaystyle\int_0^1 \frac{dx}{\sqrt{1 + x^4}}; n = 5$

6 $\displaystyle\int_0^1 \frac{dx}{1 + x^3}; n = 4$

7 $\displaystyle\int_2^8 (4 + x^2)^{-1/3}\, dx; n = 6$

8 $\displaystyle\int_2^3 \sqrt{1 + x^2}\, dx; n = 7$

9 $\displaystyle\int_1^2 \frac{dx}{x\sqrt{1 + x}}; n = 5$

10 $\displaystyle\int_{\pi/2}^{\pi} \frac{\sin x}{x}\, dx; n = 4$

11 Use Theorem 2 to find an error bound for the estimate in problem 1.

12 Use Theorem 2 to find an error bound for the estimate in problem 2.

13 Suppose that $f(x) \geq 0$ for all x in $[a, b]$ and that the graph of f is concave downward on $[a, b]$. Explain geometrically why $T_n \leq \int_a^b f(x)\, dx$ in this case.

14 Show that the quantity T_n in Theorem 1 represents the total area of all the trapezoids in Figure 3.

In problems 15 through 22, use Simpson's parabolic rule, $S_{2n} \approx \int_a^b f(x)\, dx$, with the indicated value of n to estimate each integral.

15 $\displaystyle\int_{-1}^1 \frac{dx}{1 + x^2}; n = 2$

16 $\displaystyle\int_0^4 x^2\sqrt{x + 1}\, dx; n = 4$

17 $\displaystyle\int_0^8 \frac{dx}{x^3 + x + 1}; n = 4$

18 $\displaystyle\int_2^{10} \frac{dx}{1 + x^3}; n = 4$

19 $\displaystyle\int_0^2 x\sqrt{9 - x^3}\, dx; n = 3$

20 $\displaystyle\int_0^2 \frac{dx}{\sqrt{1 + x^2}}; n = 2$

21 $\displaystyle\int_0^2 \sqrt{1 + x^4}\, dx; n = 4$

22 $\displaystyle\int_0^2 \sqrt[3]{1 - x^2}\, dx; n = 4$

23 Find the smallest value of n for which the error involved in the estimation $S_{2n} \approx \int_1^2 dx/x$ does not exceed 0.0001. (Use Theorem 4.)

24 Use Simpson's parabolic rule, $S_{2n} \approx \int_a^b f(x)\, dx$, with $n = 1$ to estimate $\int_{2.5}^{2.7} dx/x$. Give an upper bound for the error of estimation.

25 On geometric grounds, $4\int_0^1 \sqrt{1 - x^2}\, dx = \pi$. Why? Using Simpson's parabolic rule, $S_{2n} \approx \int_0^1 \sqrt{1 - x^2}\, dx$, to estimate $\int_0^1 \sqrt{1 - x^2}\, dx$, give an estimate for π. Use $n = 2$.

26 Use the procedure of problem 25 to estimate π taking $n = 5$. Compare the result with the correct value of π, which is $3.14159\ldots$.

27 (a) Use the prismoidal formula to prove that

$$\int_{-a}^{a} (Ax^3 + Bx^2 + Cx + D)\, dx = \frac{a}{3}(2Ba^2 + 6D).$$

(b) Prove the formula in part (a) directly using the fundamental theorem of calculus.

28 Prove the prismoidal formula directly using the fundamental theorem of calculus.

29 Prove that, by a suitable choice of the three coefficients A, B, and D, the graph of $y = Ax^2 + Bx + D$ can be made to pass through any three points of the form (c, p), $(c + \Delta x, q)$, and $(c + 2\,\Delta x, r)$ (Figure 4).

30 A loaded freighter is anchored in still water. At water level, the boat is 200 feet long and for each $k = 0, 1, 2, \ldots, 20$ has breadth y_k at a distance $10k$ feet from the bow. Assume that $y_0 = 0$ and $y_{20} = 0$.
(a) Use Simpson's parabolic rule to write a formula giving the approximate area of the water-level section of the boat.
(b) Recalling an exploit of Archimedes, write a formula for the approximate number of tons of freight that should be removed to raise the level of the boat by 1 foot. (Assume that the water weighs 64 lb/ft^3.)

6   Areas of Planar Regions

Our preliminary definition of the definite integral in Section 6 of Chapter 5 depends on the idea of the area of a planar region, whereas the analytic definition given in Section 2 of the present chapter is formally independent of this idea. Because we have used the same notation, $\int_{a}^{b} f(x)\, dx$, for the definite integral in either sense, and since we have used the idea of the area under a graph to understand the analytic definition and to illustrate its basic properties, the reader may already be convinced that the two definitions are equivalent. Indeed they are; however, a rigorous proof of this important fact requires a precise definition of the area of a planar region—a definition that is too sophisticated to be given in this book. Therefore, we simply assume the following fact: Let f be a piecewise continuous and bounded function on the closed interval $[a, b]$. Then the definite (Riemann) integral defined analytically in Section 2 is numerically equal to the signed area under the graph of f between $x = a$ and $x = b$. Thus, in Figure 1, we have

$$\int_{a}^{b} f(x)\, dx = A_1 - A_2.$$

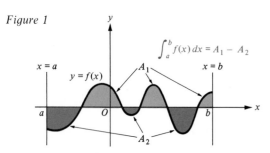

Figure 1

In Section 6 of Chapter 5 we gave several examples of the use of the definite integral to calculate the area under the graph of a function. The following additional examples will refresh the reader's memory.

EXAMPLES 1 Find the area A under the graph of the function $f(x) = \frac{1}{3}x^3$ between $x = -1$ and $x = 2$.

SOLUTION
A sketch of the graph of f (Figure 2) shows that it falls below the x axis on the interval $[-1, 0]$. We cannot find A simply by calculating $\int_{-1}^{2} \frac{1}{3}x^3\, dx$ since the area below the x axis provides a negative contribution to this integral. However, by splitting the interval $[-1, 2]$ into two subintervals, we obtain the desired area A as follows:

$$A = -\int_{-1}^{0} \tfrac{1}{3}x^3\, dx + \int_{0}^{2} \tfrac{1}{3}x^3\, dx = -\left[\tfrac{1}{12}x^4\right]\Big|_{-1}^{0} + \left[\tfrac{1}{12}x^4\right]\Big|_{0}^{2}$$

$$= \tfrac{1}{12} + \tfrac{16}{12} = \tfrac{17}{12} \text{ square units.}$$

2 Find the area between $x = -2$ and $x = 5$ under the graph of

$$f(x) = \begin{cases} 2 + \dfrac{x^3}{4} & \text{if } x < 0 \\ x^2 - x - 2 & \text{if } 0 \le x < 3 \\ 16 - 4x & \text{if } 3 \le x. \end{cases}$$

Also, find $\int_{-2}^{5} f(x)\, dx$.

SOLUTION
Figure 3 shows a sketch of the graph of f with the desired area A shaded.

Figure 2

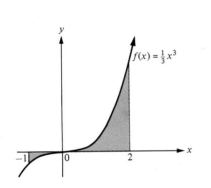

Figure 3

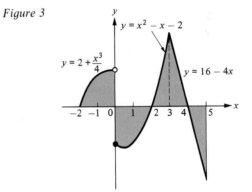

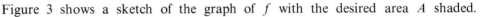

Thus,

$$A = \int_{-2}^{0} \left(2 + \frac{x^3}{4}\right) dx - \int_{0}^{2} (x^2 - x - 2)\, dx + \int_{2}^{3} (x^2 - x - 2)\, dx$$

$$+ \int_{3}^{4} (16 - 4x)\, dx - \int_{4}^{5} (16 - 4x)\, dx$$

$$= \left(2x + \frac{x^4}{16}\right)\Big|_{-2}^{0} - \left(\frac{x^3}{3} - \frac{x^2}{2} - 2x\right)\Big|_{0}^{2} + \left(\frac{x^3}{3} - \frac{x^2}{2} - 2x\right)\Big|_{2}^{3}$$

$$+ (16x - 2x^2)\Big|_{3}^{4} - (16x - 2x^2)\Big|_{4}^{5}$$

$$= 3 + \frac{10}{3} + \frac{11}{6} + 2 + 2 = \frac{73}{6} \text{ square units.}$$

Also,

$$\int_{-2}^{5} f(x)\,dx = \int_{-2}^{0} \left(2 + \frac{x^3}{4}\right) dx + \int_{0}^{3} (x^2 - x - 2)\,dx + \int_{3}^{5} (16 - 4x)\,dx$$

$$= \left(2x + \frac{x^4}{16}\right)\Bigg|_{-2}^{0} + \left(\frac{x^3}{3} - \frac{x^2}{2} - 2x\right)\Bigg|_{0}^{3} + (16x - 2x^2)\Bigg|_{3}^{5}$$

$$= 3 - \frac{3}{2} + 0 = \frac{3}{2}.$$

6.1   Areas by Slicing

The method of slicing, first introduced in Section 5 of Chapter 5, enables us to find areas of planar regions by solving a differential equation of the form $dA = l\,ds$. We now show how such areas can be calculated using the definite integral.

In the present section we consider only regions R in the plane that satisfy the following two conditions:

Figure 4

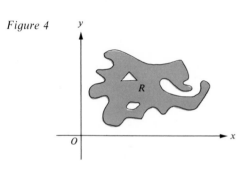

1   The boundary of R consists of a finite number of straight line segments or smooth arcs which can meet in a finite number of "corners" or "vertices."

2   The region R is *bounded* in the sense that there is an upper bound to the distances between the points of R.

A region R satisfying conditions 1 and 2 will be called an *admissible* region. The region R shown in Figure 4 is an example of an admissible region. Notice that such a region is permitted to have a finite number of "holes," provided that the boundaries of these "holes" satisfy condition 1. The boundaries of the "holes," if any, are regarded as part of the boundary of R.

We assume that any admissible region R in the plane has an area, $A(R)$ square units, associated with it. The following theorem shows how to calculate $A(R)$ in terms of a definite (Riemann) integral.

THEOREM 1   Areas by Slicing Using the Definite Integral

Let R be an admissible region and choose a convenient coordinate axis, called the s axis, as a "reference axis" (Figure 5). At each point along the reference axis, erect a perpendicular line and suppose that the region R is entirely contained between the two perpendiculars at the points with coordinates a and b, respectively. Let the perpendicular at the point with coordinate s intersect the region R in one or more line segments of total length $l(s)$. Then the area of the region R is given by $A(R) = \displaystyle\int_{a}^{b} l(s)\,ds.$

Figure 5

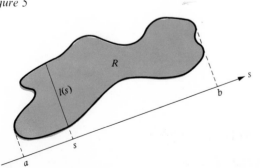

A rigorous proof of Theorem 1 is beyond the scope of this textbook; however, we give an informal argu-

Figure 6

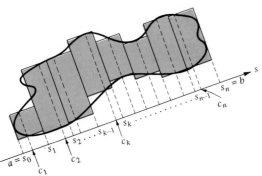

ment to indicate its plausibility. Figure 6 shows a partition of the reference axis consisting of n sub-intervals $[s_0, s_1], [s_1, s_2], \ldots, [s_{k-1}, s_k], \ldots, [s_{n-1}, s_n]$, where $a = s_0$ and $b = s_n$. We have augmented this partition by selecting numbers $c_1, c_2, c_3, \ldots, c_n$ from the successive subintervals. Above each subinterval $[s_{k-1}, s_k]$ we have constructed a rectangle of width $\Delta s_k = s_k - s_{k-1}$ and of length $l(c_k)$. The area of the kth rectangle is $l(c_k)\,\Delta s_k$. Evidently, the desired area $A(R)$ is approximated by the sum of the areas of all these rectangles, so that $A(R) \approx \sum_{k=1}^{n} l(c_k)\,\Delta s_k$. As the norm of the partition approaches zero and the rectangles become narrower and more numerous, the approximation obviously should improve. The sum $\sum_{k=1}^{n} l(c_k)\,\Delta s_k$ is a Riemann sum and its limit as the norm of the partition approaches zero is, by definition, $\int_a^b l(s)\,ds$. Therefore, the result

$$A(R) = \int_a^b l(s)\,ds \text{ seems geometrically reasonable.}$$

EXAMPLE   Find the area of the region R bounded by the graphs of the equations $y^2 = 2x$ and $y = x - 4$ (Figure 7).

SOLUTION

We use Theorem 1, taking the y axis as the reference axis. To determine the points of intersection of the two graphs, we solve the two equations $y^2 = 2x$ and $y = x - 4$ simultaneously. Substituting $y = x - 4$ into $y^2 = 2x$, we have $(x - 4)^2 = 2x$; that is, $x^2 - 8x + 16 = 2x$, or $x^2 - 10x + 16 = 0$. Factoring to solve the latter equation, we obtain $(x - 2)(x - 8) = 0$, so $x = 2$ or $x = 8$. When $x = 2$, then $y = x - 4 = 2 - 4 = -2$. When $x = 8$, then $y = x - 4 = 8 - 4 = 4$. Therefore, the two graphs meet at $(2, -2)$ and at $(8, 4)$.

Figure 7

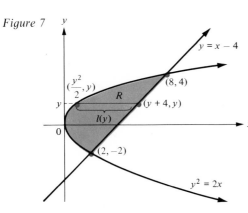

The point on the graph of $y^2 = 2x$ with ordinate y has abscissa $x = y^2/2$, while the point on the graph of $y = x - 4$ with ordinate y has abscissa $x = y + 4$ (Figure 7). Therefore,

$$l(y) = (y + 4) - \frac{y^2}{2} \quad \text{for} \quad -2 \le y \le 4.$$

By Theorem 1,

$$A(R) = \int_{-2}^{4} l(y)\,dy = \int_{-2}^{4} \left(y + 4 - \frac{y^2}{2}\right) dy = \left(\frac{y^2}{2} + 4y - \frac{y^3}{6}\right)\Bigg|_{-2}^{4}$$

$$= \frac{40}{3} - \left(-\frac{14}{3}\right) = 18 \text{ square units.}$$

6.2 Area Between Two Graphs

Theorem 1 can be used to prove the following theorem.

THEOREM 2   Area Between Two Graphs

Let f and g be continuous functions on the closed interval $[a, b]$. Then the area of the region R between the graph of f and the graph of g, to the right of $x = a$ and to the left of $x = b$ (Figure 8), is given by

$$A(R) = \int_a^b |f(x) - g(x)| \, dx.$$

Figure 8

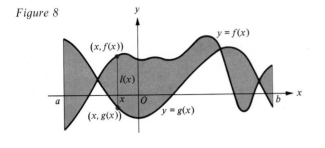

PROOF

In Figure 8, we take the x axis as the reference axis. Note that $l(x)$ is the distance between the point $(x, f(x))$ and the point $(x, g(x))$; hence, $l(x) = |f(x) - g(x)|$. Therefore, by Theorem 1,

$$A(R) = \int_a^b l(x) \, dx = \int_a^b |f(x) - g(x)| \, dx.$$

EXAMPLES   Use Theorem 2 to calculate the area of the given region R.

Figure 9

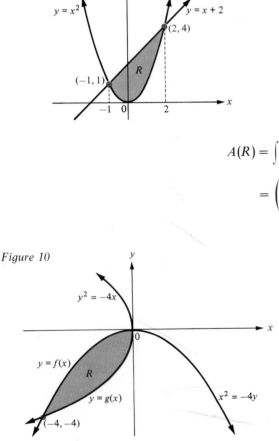

1 R is the region between the graph of $y = x + 2$ and the graph of $y = x^2$ (Figure 9).

SOLUTION

To determine the points of intersection of the two graphs, we solve the two equations $y = x + 2$ and $y = x^2$ simultaneously and obtain the points $(-1, 1)$ and $(2, 4)$. Evidently, the graph of $y = x + 2$ lies above the graph of $y = x^2$ between $x = -1$ and $x = 2$. By Theorem 2,

$$A(R) = \int_{-1}^{2} |(x + 2) - x^2| \, dx = \int_{-1}^{2} (x + 2 - x^2) \, dx$$

$$= \left(\frac{x^2}{2} + 2x - \frac{x^3}{3} \right) \Big|_{-1}^{2} = \frac{10}{3} - \left(-\frac{7}{6} \right) = \frac{9}{2} \text{ square units.}$$

2 The region R bounded by the graphs of the equations $y^2 = -4x$ and $x^2 = -4y$ (Figure 10).

Figure 10

SOLUTION

Solving the equations $y^2 = -4x$ and $x^2 = -4y$ simultaneously, we find that the two graphs intersect at the points $(-4, -4)$ and $(0, 0)$. In order to use Theorem 2, we must find the equations $y = f(x)$ and $y = g(x)$ of the upper and lower boundaries, respectively, of the region R.

The upper boundary of R is the portion of the graph of $x^2 = -4y$ between $x = -4$ and $x = 0$; hence, its equation can be written as $y = f(x)$, where $f(x) = -\frac{1}{4}x^2$ and $-4 \leq x \leq 0$.

The lower boundary of R is the portion of the graph of $y^2 = -4x$ between $x = -4$ and $x = 0$. On this lower boundary, y is negative or zero; hence, solving the equation $y^2 = -4x$ for y, we obtain $y = -\sqrt{-4x} = -2\sqrt{-x}$. Therefore, the equation of the lower boundary of R is $y = g(x)$, where $g(x) = -2\sqrt{-x}$ and $-4 \leq x \leq 0$.

Now, by Theorem 2,

$$A(R) = \int_{-4}^{0} |f(x) - g(x)|\, dx = \int_{-4}^{0} \left[\left(-\frac{1}{4}x^2 \right) - \left(-2\sqrt{-x} \right) \right] dx$$

$$= \int_{-4}^{0} \left(2\sqrt{-x} - \frac{1}{4}x^2 \right) dx = 2 \int_{-4}^{0} \sqrt{-x}\, dx - \frac{1}{4}\int_{-4}^{0} x^2\, dx$$

$$= 2 \int_{-4}^{0} \sqrt{-x}\, dx - \frac{1}{4}\left(\frac{x^3}{3} \right)\Big|_{-4}^{0} = 2 \int_{-4}^{0} \sqrt{-x}\, dx - \frac{16}{3}.$$

To evaluate the definite integral $\int_{-4}^{0} \sqrt{-x}\, dx$, we use the change of variable $u = -x$, so $du = -dx$ and

$$\int_{-4}^{0} \sqrt{-x}\, dx = \int_{4}^{0} \sqrt{u}\,(-du) = -\int_{4}^{0} u^{1/2}\, du = -\frac{2}{3}u^{3/2}\Big|_{4}^{0} = 0 - \left(-\frac{16}{3} \right) = \frac{16}{3}.$$

It follows that

$$A(R) = 2 \int_{-4}^{0} \sqrt{-x}\, dx - \frac{16}{3} = 2\left(\frac{16}{3} \right) - \frac{16}{3} = \frac{16}{3} \text{ square units.}$$

3 The region R bounded by the graphs of the equations $x = y^2 - 2y$ and $x = 2y - 3$ (Figure 11).

SOLUTION
Simultaneous solution of the two equations shows that the graphs intersect at the points $(-1, 1)$ and $(3, 3)$. We use Theorem 2 but with the roles of x and y interchanged. Thus,

Figure 11

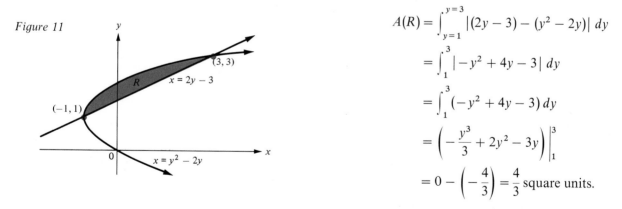

$$A(R) = \int_{y=1}^{y=3} |(2y - 3) - (y^2 - 2y)|\, dy$$

$$= \int_{1}^{3} |-y^2 + 4y - 3|\, dy$$

$$= \int_{1}^{3} (-y^2 + 4y - 3)\, dy$$

$$= \left(-\frac{y^3}{3} + 2y^2 - 3y \right)\Big|_{1}^{3}$$

$$= 0 - \left(-\frac{4}{3} \right) = \frac{4}{3} \text{ square units.}$$

Problem Set 6

In problems 1 through 10, find the area under the graph of each function between $x = a$ and $x = b$. Sketch the graph of the function.

1 $f(x) = 1 - x^2$; $a = -1$, $b = 1$

2 $g(x) = x^2 - 2$; $a = 0$, $b = 1$

3 $h(x) = x^3 - x$; $a = -1$, $b = 1$

4 $F(x) = x^2 - 9$; $a = -3$, $b = 3$

5 $G(x) = x^3$; $a = -2$, $b = 2$

6 $H(x) = x^2 - 6x + 5$; $a = 1$, $b = 3$

7 $f(x) = x^3 - 4x^2 + 3x$; $a = 0$, $b = 2$

8 $g(x) = x^3 - 6x^2 + 8x$; $a = 0$, $b = 4$

9 $f(x) = \frac{1}{3}(x - x^3)$; $a = -1$, $b = 2$

10 $f(x) = x^n$; $a = 0$, $b = 1$, where $n \geq 1$

In problems 11 through 28, (a) sketch the graphs of the two equations, (b) find the points of intersection of the two graphs, and (c) find the area of the region bounded by the two graphs.

11 $f(x) = x^2$ and $g(x) = 2x + \frac{5}{4}$

12 $y = \dfrac{x^2}{4}$ and $7x - 2y = 20$

13 $f(x) = -x^2 - 4$ and $g(x) = -8$

14 $y = \sqrt{x}$ and $y = x$

15 $y = x^2 - x$ and $y = x$

16 $f(x) = x^3$ and $g(x) = x$

17 $x = (y - 2)^2$ and $x = y$

18 $y^2 = 3x$ and $y = x$

19 $x = 6y^2 - 3$ and $x + 3y = 0$

20 $x = 4y^2 - 1$ and $8x - 6y + 3 = 0$

21 $f(x) = x^3$ and $g(x) = \sqrt[3]{x}$

22 $y = 2x^2 - x^3$ and $y = 2x - x^2$

23 $y = x^2$ and $y = x^4$

24 $y = |x|$ and $y = x^4$

25 $x = y^2 - 2$ and $x = 6 - y^2$

26 $x = y^2 - y$ and $x = y - y^2$

27 $f(x) = x|x|$ and $g(x) = x^3$

28 $f(x) = -x$ and $g(x) = 2x - 3x^2$

In problems 29 through 32, find the area of the region bounded by the graphs of the given equations.

29 $y = x + 6$, $y = \frac{1}{2}x^2$, $x = 1$, and $x = 4$

30 $y = x^3$, $y = 12 - x^2$, and $x = 0$

31 $y = 2 - x$, $y = x^2$, and above $y = \sqrt[3]{x}$

32 $y = x^3$, $y = 0$, and $y = 3x - 2$ (first quadrant)

In problems 33 through 38, sketch a graph of each function f, find the area under the graph of f between $x = a$ and $x = b$, and find $\displaystyle\int_a^b f(x)\,dx$.

33 $f(x) = \begin{cases} x^3 & \text{for } -2 \le x \le 1 \\ \sqrt{x} & \text{for } 1 < x \le 4 \\ 10 - 2x & \text{for } 4 < x \le 7 \\ 2x - 18 & \text{for } 7 < x \le 12 \end{cases}$; $a = -2, b = 12$

34 $f(x) = \begin{cases} -x - 3 & \text{for } -5 \le x < -2 \\ x^2 + 2x - 1 & \text{for } -2 \le x \le 1 \\ 2 & \text{for } 1 < x \le 4 \end{cases}$; $a = -5, b = 4$

35 $f(x) = \begin{cases} x^2 + 6x - 7 & \text{for } -7 \le x \le -6 \\ -x^2 - 4x + 5 & \text{for } -6 < x \le 0 \\ |x - 5| & \text{for } 0 < x \le 8 \end{cases}$; $a = -7, b = 8$

36 $f(x) = \begin{cases} x\sqrt{x^2 - 4} & \text{for } -3 \le x \le -2 \\ -x^2 & \text{for } -2 < x \le 0 \\ 3 - x & \text{for } 0 < x \le 4 \\ \sqrt{2x + 1} & \text{for } 4 < x \le 6 \end{cases}$; $a = -3, b = 6$

37 $f(x) = \begin{cases} x^3 - 1 & \text{for } -1 \le x < 0 \\ x^2 & \text{for } 0 \le x < 2 \\ 2[\![x]\!] & \text{for } 2 \le x < 4 \\ -\sqrt{x - 4} & \text{for } 4 \le x \le 8 \end{cases}$; $a = -1, b = 8$

38 $f(x) = \begin{cases} \dfrac{x}{(x^2 + 1)^2} & \text{for } -3 \le x < 0 \\ \sqrt[3]{2x} & \text{for } 0 \le x < 4 \\ x^2 - 12x + 32 & \text{for } 4 \le x \le 10 \\ 5 & \text{for } 10 < x \le 12 \end{cases}$; $a = -3, b = 12$

39 Use the method of slicing to find the area of the triangle whose vertices are the points $(1, 1)$, $(2, 4)$, and $(4, 3)$.

40 Use the method of slicing to find the area of the trapezoid whose vertices are the points $(1, 1)$, $(6, 1)$, $(2, 4)$, and $(5, 4)$.

Review Problem Set

In problems 1 through 4, write out each sum explicitly and then find its numerical value.

1 $\displaystyle\sum_{k=1}^{5}(5k+3)$
2 $\displaystyle\sum_{i=1}^{3}5(i+1)^2$
3 $\displaystyle\sum_{k=0}^{4}\frac{k}{k+1}$
4 $\displaystyle\sum_{i=0}^{3}\frac{1}{i^2+1}$

In problems 5 through 8, evaluate each sum by using the basic properties of summation.

5 $\displaystyle\sum_{k=1}^{n}k(2k-1)$
6 $\displaystyle\sum_{j=1}^{n}(6^{j+1}-6^j)$
7 $\displaystyle\sum_{j=0}^{n}(3^j+3^{j+1})$
8 $\displaystyle\sum_{k=0}^{n}(k+1)^3$

9 Find the sum of the first 1000 even integers, $2+4+6+8+\cdots+2000$.

10 Use the principle of mathematical induction to show that $\displaystyle\sum_{k=1}^{n}(2k-1)=n^2$.

11 Evaluate $\displaystyle\sum_{k=1}^{n}\frac{1}{k^2+k}\cdot\left(Hint:\frac{1}{k^2+k}=\frac{1}{k}-\frac{1}{k+1}\cdot\right)$

12 (a) Suppose that f is a Riemann-integrable function on the interval $[0,1]$. Use the analytic definition of the definite integral to express $\displaystyle\lim_{n\to+\infty}\frac{1}{n}\sum_{k=1}^{n}f\left(\frac{k}{n}\right)$ as a certain integral.

(b) Write $\displaystyle\lim_{n\to+\infty}\frac{\sum_{k=1}^{n}k^5}{n^6}$ as a definite integral.

13 Evaluate the Riemann sum $\displaystyle\sum_{k=1}^{n}f(c_k)\Delta x_k$ corresponding to the augmented partition $[0,\frac14]$, $[\frac14,\frac12]$, $[\frac12,\frac34]$, $[\frac34,1]$; $c_1=\frac18$, $c_2=\frac38$, $c_3=\frac58$, and $c_4=\frac78$ if f is given by $f(x)=x^2$. Interpret this Riemann sum as an approximation to a certain area.

In problems 14 through 16, find the Riemann sum corresponding to each augmented partition for the indicated function, sketch the graph of the function, and show the Riemann sum as a sum of areas of rectangles.

14 $f(x)=x^2+1$, $\mathscr{P}_4^*$ is the partition of $[1,4]$ into four equal subintervals with $c_1, c_2, c_3,$ and c_4 chosen at the midpoints of these subintervals.

15 $f(x)=x^3-x$, $\mathscr{P}_4^*$ is the partition of $[1,3]$ into four equal subintervals with $c_1, c_2, c_3,$ and c_4 chosen so that all approximating rectangles are inscribed.

16 Same as problem 15, except that all approximating rectangles are to be circumscribed.

In problems 17 through 22, indicate whether each Riemann integral exists and give a reason for your answer.

17 $\displaystyle\int_{1}^{100}\frac{[\![x]\!]}{x}\,dx$
18 $\displaystyle\int_{1}^{2}\frac{dx}{[\![x]\!]}$

19 $\displaystyle\int_{1}^{2}\sqrt{1-x^2}\,dx$
20 $\displaystyle\int_{0}^{1}\frac{dx}{1+x^2}$

21 $\displaystyle\int_{0}^{\pi/2}\frac{dx}{\cos x}$
22 $\displaystyle\int_{-1}^{1}f(x)\,dx$, where $f(x)=\begin{cases}x^2 & \text{for }x<0\\x^4 & \text{for }x\geq0\end{cases}$

In problems 23 through 26, evaluate each Riemann integral directly by calculating a limit of Riemann sums. Use partitions consisting of subintervals of equal lengths and augment these partitions so as to obtain inscribed or circumscribed rectangles as indicated.

23 $\int_1^4 (x^2 + 1)\, dx$ (circumscribed rectangles)

24 $\int_1^4 (x^2 + 1)\, dx$ (inscribed rectangles)

25 $\int_1^3 (x^3 + x)\, dx$ (circumscribed rectangles)

26 $\int_{-1}^0 (x^2 - 2x + 3)\, dx$ (inscribed rectangles)

27 Prove that, if K is a constant, then $\int_a^b K\, dx = K(b - a)$, by making direct use of the analytic definition of the definite (Riemann) integral.

28 Prove that $\int_a^b x\, dx = \frac{1}{2}(b^2 - a^2)$ by making direct use of the analytic definition of the definite (Riemann) integral.

In problems 29 through 33, assume that $\int_1^3 f(x)\, dx = 6$ and $\int_3^5 g(x)\, dx = 8$. Use the properties of definite integrals to evaluate each expression.

29 $\int_1^3 (-5)f(x)\, dx$

30 $\int_3^1 7f(x)\, dx$

31 $\int_1^5 h(x)\, dx$, where $h(x) = \begin{cases} 4f(x) & \text{for } 1 \le x < 3 \\ -2g(x) & \text{for } 3 \le x \le 5 \end{cases}$

32 $\int_1^3 F(x)\, dx$, where $F(x) = \begin{cases} 0 & \text{for } x = 1 \\ f(x) & \text{for } 1 < x < 3 \\ 52 & \text{for } x = 3 \end{cases}$

33 $\int_5^1 H(x)\, dx$, where $H(x) = \begin{cases} 1 + f(x) & \text{for } 1 \le x \le 3 \\ g(x) - 1 & \text{for } 3 < x \le 5 \end{cases}$

34 Suppose that f is a Riemann-integrable function on $[a, b]$ and that $|f(x)| \le K$ holds for all values of x in $[a, b]$, where K is a constant. Prove that $\left| \int_a^b f(x)\, dx \right| \le K \cdot |b - a|$.

35 Show that $\int_0^1 \frac{dx}{1 + x^2} \ge \int_0^1 \frac{dx}{1 + x}$.

36 Assume that f and g are continuous functions on $[a, b]$ and that K is a constant. Show that

$$2K \int_a^b f(x)g(x)\, dx \le \int_a^b [f(x)]^2\, dx + K^2 \int_a^b [g(x)]^2\, dx.$$

(*Hint*: $0 \le [f(x) - Kg(x)]^2$.)

37 Use problem 36 to derive the inequality

$$\int_a^b f(x)g(x)\, dx \le \frac{1}{2} \int_a^b [f(x)]^2\, dx + \frac{1}{2} \int_a^b [g(x)]^2\, dx.$$

38 Use problem 36 to derive the *Cauchy–Bunyakovski–Schwarz* inequality

$$\left[\int_a^b f(x)g(x)\, dx \right]^2 \le \int_a^b [f(x)]^2\, dx \int_a^b [g(x)]^2\, dx.$$

$$\left[\textit{Hint}: \text{ In problem 36, put } K = \frac{\int_a^b f(x)g(x)\, dx}{\int_a^b [g(x)]^2\, dx}. \right]$$

39 Assuming the fundamental theorem of calculus, find the mean value of the given function on the indicated interval.
(a) $f(x) = x^3$ on $[0, 2]$ (b) $f(x) = \sqrt{x}$ on $[0, 9]$
(c) $f(x) = x^n$ on $[-1, 1]$, where n is a positive integer

40 Assuming the fundamental theorem of calculus, find a number c on the interval $[a, b]$ such that $f(c)$ is the mean value of the given function f on $[a, b]$.

    (a) $f(x) = x|x|, [a, b] = [-0.5, 1]$

    (b) $f(x) = \dfrac{1}{x^2}, [a, b] = [1, 2]$

    (c) $f(x) = \sqrt{x}, [a, b] = [0, 1]$

In problems 41 through 48, use the first part of the fundamental theorem of calculus and the basic properties of the definite integral to find each derivative.

41 $D_x \displaystyle\int_3^x (4t + 1)^{300} \, dt$

42 $\dfrac{d}{dx} \displaystyle\int_2^x (3w^2 - 7)^{15} \, dw$

43 $g'(x)$, where $g(x) = \displaystyle\int_1^x (8t^{17} + 5t^2 - 13)^{40} \, dt$

44 $h''(t)$, where $h(t) = \displaystyle\int_0^t \sqrt{1 + x^{16}} \, dx$

45 $D_x \displaystyle\int_x^{1000} \dfrac{t^2 \, dt}{\sqrt{t^4 + 8}}$

46 $\dfrac{d}{dx} \displaystyle\int_x^0 |w| \, dw$

47 $g''(t)$, where $g(t) = \displaystyle\int_t^0 \sqrt{1 + x^2} \, dx + \int_0^t \sqrt{1 + w^2} \, dw$

48 $h'(t)$, where $h(t) = \displaystyle\int_t^0 \dfrac{dx}{1 + x^2} + \int_1^t \dfrac{dx}{1 + x^2}$

In problems 49 through 51, use the first part of the fundamental theorem of calculus together with the chain rule to find each derivative.

49 $D_x \displaystyle\int_1^{x^2} \dfrac{t^2 \, dt}{1 + t^2}$

50 $\dfrac{d}{dx} \displaystyle\int_{3x+1}^{x^2} \dfrac{t + \sqrt{t}}{t^3 + 5} \, dt$

51 $D_t \displaystyle\int_{4t+3}^{5t^2+t} (w^5 + 1)^{17} \, dw$

52 Assuming that all required derivatives and integrals exist:

    (a) Find $\dfrac{d}{dx} \left[\displaystyle\int_a^x f[g(t)]g'(t) \, dt - \int_{g(a)}^{g(x)} f(u) \, du \right]$.

    (b) Conclude that $\displaystyle\int_a^b f[g(t)]g'(t) \, dt = \int_{g(a)}^{g(b)} f(u) \, du$.

53 Find the maximum value of each function g on the interval indicated.

    (a) $g(x) = \displaystyle\int_0^x (t^2 - 4t + 4) \, dt$ on $[0, 4]$

    (b) $g(x) = \displaystyle\int_0^x (\sqrt{t} - t) \, dt$ on $[0, 1]$

54 Suppose that the function f is piecewise continuous and bounded on the interval $[a, b]$ and define g by $g(x) = \displaystyle\int_a^x f(t) \, dt$. Prove that g is continuous on $[a, b]$.

In problems 55 through 66, find the given definite integral by using the second part of the fundamental theorem of calculus and the basic properties of the definite integral, sketch a graph of the integrand, and interpret the integral as an area or as a difference of areas.

55 $\displaystyle\int_0^1 (2x + 3x^2) \, dx$

56 $\displaystyle\int_0^1 5(x - \sqrt{x})^2 \, dx$

57 $\displaystyle\int_{-1}^3 (x + |x|) \, dx$

58 $\displaystyle\int_{-1}^3 |x + 1| \, dx$

59 $\displaystyle\int_{-1}^1 \sqrt{x + 1} \, dx$

60 $\displaystyle\int_{-1}^4 x^2 |x| \, dx$

61 $\displaystyle\int_{-1}^3 \dfrac{2x \, dx}{(1 + x^2)^2}$

62 $\displaystyle\int_0^1 x^3 \sqrt{x^4 + 1} \, dx$

63 $\displaystyle\int_0^4 (|x - 1| + |x - 2|) \, dx$

64 $\displaystyle\int_0^2 \dfrac{x \, dx}{\sqrt{3x + 10}}$

65 $\displaystyle\int_0^3 f(x) \, dx$, where $f(x) = \begin{cases} 1 - x & \text{for } 0 \le x < 1 \\ x^2 - 1 & \text{for } 1 \le x < 2 \\ x + 1 & \text{for } 2 \le x \le 3 \end{cases}$

66 $\displaystyle\int_{-1}^2 g(x) \, dx$, where $g(x) = \begin{cases} -\sqrt{|x|} & \text{for } -1 \le x < 0 \\ \sqrt{x + 1} & \text{for } 0 \le x < 1 \\ x\sqrt{1 + x^2} & \text{for } 1 \le x < 2 \end{cases}$

In problems 67 through 74, indicate whether each statement is true or false. You may assume that all required derivatives and integrals exist, that functions appearing in denominators are nonzero, and so forth. If the statement is false, give a specific example to show it is false. (Such an example is called a *counterexample*.)

67 $\int_a^b f(x)g(x)\,dx = \int_a^b f(x)\,dx \int_a^b g(x)\,dx?$

68 $\int_a^b f(x)g(x)\,dx = f(x)\int_a^b g(x)\,dx + g(x)\int_a^b f(x)\,dx?$

69 $\int_a^b \dfrac{f(x)}{g(x)}\,dx = \dfrac{\int_a^b f(x)\,dx}{\int_a^b g(x)\,dx}?$

70 $\int_a^b f(x)g(x)\,dx = [f(x)h(x)]\Big|_a^b - \int_a^b h(x)f'(x)\,dx$, where $h'(x) = g(x)?$

71 $\int_{-100}^{100} (x^{775} + x^{776})\,dx > 0?$

72 $\int_a^x f(kt)\,dt = \dfrac{1}{k}\int_{ka}^{kx} f(t)\,dt$, where k is a constant?

73 $\left|\int_a^b g(x)\,dx\right| = \int_a^b |g(x)|\,dx?$

74 If f is a decreasing function, then the function g defined by $g(x) = \int_0^x f(t)\,dt$ is also decreasing.

In problems 75 through 78, use the trapezoidal rule, $T_n \approx \int_a^b f(x)\,dx$, with the indicated value of n to estimate each integral.

75 $\int_0^2 x\sqrt{16 - x^3}\,dx; n = 4$

76 $\int_1^2 \sqrt{4 + x^3}\,dx; n = 4$

77 $\int_0^{10} \sqrt[3]{125 + x^3}\,dx; n = 5$

78 $\int_4^8 \sqrt{64 - x^2}\,dx; n = 8$

In problems 79 through 82, use the Simpson parabolic rule, $S_{2n} \approx \int_a^b f(x)\,dx$, with the indicated value of n to estimate the given integral.

79 $\int_0^8 \dfrac{3x}{1 + x^3}\,dx; n = 4$

80 $\int_0^4 \sqrt{16 - x^2}\,dx; n = 2$

81 $\int_2^8 \dfrac{x\,dx}{\sqrt[3]{3 + x^3}}; n = 3$

82 $\int_0^5 \dfrac{x^3\,dx}{\sqrt{1 + x^3}}; n = 3$

In problems 83 through 94, (a) sketch the graphs of the equations, (b) find the points of intersection of these graphs, and (c) find the area of the region bounded by these graphs.

83 $y = \frac{1}{4}x^3$ and $y = x, x \geq 0$

84 $y = x^3$, the y axis, and $y = -27$

85 $y = 9 - x^2$ and $y = x^2$

86 $2y^2 + 9x = 36$ and $3x + 2y = 0$

87 $y = 2x^2$ and $y = x^2 + 2x + 3$

88 $y^2 = -4(x - 1)$ and $y^2 = -2(x - 2)$

89 $2x + 3y + 1 = 0$ and $x + 3 = (y - 1)^2$

90 $x^2y = x^2 - 1, y = 1, x = 1,$ and $x = 4$

91 $y = 4 - x^2$ and $y = 4 - 4x$

92 $x = \frac{1}{4}y^2 - 1$ and $y = 4x - 16$

93 $y^3 = 9x, y^2 = -3(x - 6),$ and $y = -3$

94 $y^2 = -16(x - 1)$ and $y^2 = \frac{16}{3}(x + 3)$

95 Find the area of the region bounded by the curve $y = -\frac{2}{27}x^3$ and the tangent to this curve at $(3, -2)$.

7

APPLICATIONS OF THE DEFINITE INTEGRAL

In Chapter 6 we have seen that definite integrals can be used to find the area of planar regions. Here we apply definite integrals to the problem of finding the volume of three-dimensional regions. In addition, we consider applications of definite integrals to the calculation of arc length, surface area, consumer's surplus, blood flow, work, and energy.

1 Volumes of Solids of Revolution

The techniques used in Section 6 of Chapter 6 can be modified to express the volume of a three-dimensional region S as an integral. We assume that any three-dimensional region S which has a "reasonable shape" has a definite volume $V(S)$ cubic units associated with it. We refer to such a region as an *admissible three-dimensional region* or simply as a *solid*. In particular, let us stipulate that any three-dimensional region having the following two properties is a solid:

1 The boundary of S consists of a finite number of smooth surfaces that can intersect in a finite number of edges. These edges, in turn, can intersect in a finite number of vertices.

2 S is *bounded* in the sense that there is an upper bound to the distances between the points of S.

For instance, a solid spherical ball, a solid right circular cone, a solid cube, or the solid region between two coaxial right circular cylinders would satisfy the above criteria (Figure 1).

A solid consisting of all points lying between an admissible planar region B_1 and a second admissible planar region B_2 obtained by parallel translation of B_1

Figure 1

Figure 2

base B_2

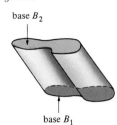

base B_1

is called a *solid cylinder with bases* B_1 *and* B_2 (Figure 2). All the line segments joining points on the base B_1 to corresponding points on the base B_2 are parallel to one another. If all these line segments are perpendicular to the bases, then the solid cylinder is called a solid *right* cylinder (Figure 3). The distance, measured perpendicularly, between the two bases of a solid cylinder is called its *height*. Henceforth, we assume that the volume of a solid right cylinder is its height times the area of one of its bases.

If one solid S_1 is contained in a similar but slightly larger solid S_2, then the three-dimensional region S_3 consisting of all points in S_2 that are not in S_1 is sometimes called a *shell* (Figure 4). Notice that $V(S_3) = V(S_2) - V(S_1)$; that is, the volume of the shell is the difference between the volumes of the larger and the smaller solids. For instance, in Figure 4, the volume of the right cylindrical shell is given by
$$V(S_3) = V(S_2) - V(S_1) = \pi r_2^2 h - \pi r_1^2 h.$$

Figure 3

Solid right cylinder

Figure 4

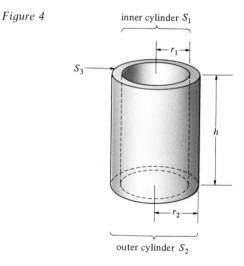

inner cylinder S_1

S_3

r_1

h

r_2

outer cylinder S_2

1.1 Solids of Revolution—Method of Circular Disks

In this section and the next we develop methods for finding the volume of solids called *solids of revolution*. Solids of revolution are formed as follows: Let R be an admissible planar region and let l be a straight line lying in the same plane as R, but touching R, if at all, only at boundary points of R (Figure 5a). The solid S swept out or generated when R is revolved about the line l as an axis is called a *solid of revolution* (Figure 5b).

Consider the special case in which R is the region under the graph of a continuous nonnegative function f between $x = a$ and $x = b$ (Figure 6a). Denote by S the solid of revolution generated by revolving R about the x axis (Figure 6b). Figure 6c shows an "infinitesimal" portion dV of the volume V of S consisting of a circular disk of "infinitesimal" thickness dx, perpendicular to the axis of revolution, and intersecting it at the point with coordinate x. Evidently, the radius r of this disk is given by $r = f(x)$.

Figure 5

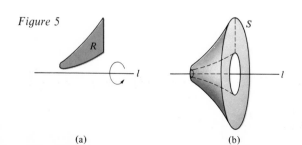

R

l

S

l

(a)   (b)

Figure 6

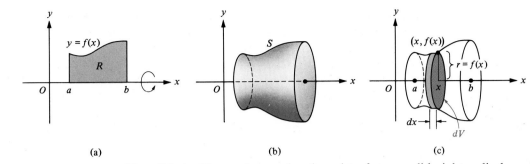

(a)   (b)   (c)

The disk in Figure 6c can be thought of as a solid right cylinder whose base is a circle of radius r and whose height is dx. The area of its base is πr^2; hence, its volume dV is given by

$$dV = \pi r^2\, dx = \pi[f(x)]^2\, dx.$$

The total volume V of the solid S should be obtained by "summing"—that is to say, integrating—all the "infinitesimal" volumes dV of such disks, as x runs from a to b. Therefore, we should have

$$V = \int_{x=a}^{x=b} dV = \int_a^b \pi[f(x)]^2\, dx = \pi \int_a^b [f(x)]^2\, dx.$$

The calculation of volumes by the formula above is called the *method of circular disks*. Although our derivation of the formula using Leibnizian infinitesimals may not be mathematically rigorous, the formula is correct and can be derived rigorously from the analytic definition of the definite integral.

EXAMPLES   Use the method of circular disks to find the volume V of the solid S generated by revolving the region R under the graph of the given function f on the indicated interval $[a,b]$ about the x axis. Sketch the graph of f and the solid S.

1 $f(x) = x^3$ on $[1,2]$

Figure 7

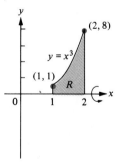

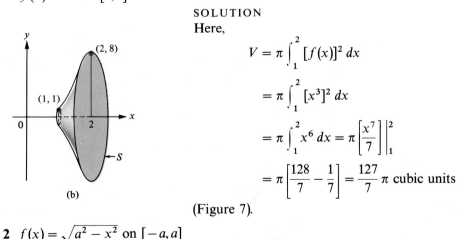

(a)   (b)

SOLUTION
Here,

$$V = \pi \int_1^2 [f(x)]^2\, dx$$

$$= \pi \int_1^2 [x^3]^2\, dx$$

$$= \pi \int_1^2 x^6\, dx = \pi \left[\frac{x^7}{7}\right]\bigg|_1^2$$

$$= \pi \left[\frac{128}{7} - \frac{1}{7}\right] = \frac{127}{7}\pi \text{ cubic units}$$

(Figure 7).

2 $f(x) = \sqrt{a^2 - x^2}$ on $[-a, a]$

SOLUTION
Here,

$$V = \pi \int_{-a}^a [f(x)]^2\, dx = \pi \int_{-a}^a [\sqrt{a^2 - x^2}]^2\, dx = \pi \int_{-a}^a (a^2 - x^2)\, dx = \pi \left[a^2 x - \frac{x^3}{3}\right]\bigg|_{-a}^a$$

$$= \pi \left[\left(a^3 - \frac{a^3}{3}\right) - \left(-a^3 + \frac{a^3}{3}\right)\right] = \pi \frac{4a^3}{3} = \frac{4}{3}\pi a^3.$$

Figure 8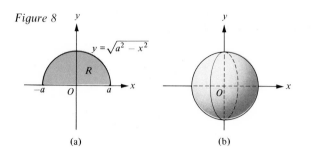

(a)   (b)

Notice that the graph of $f(x) = \sqrt{a^2 - x^2}$ on $[-a, a]$ is a semicircle and the corresponding solid of revolution is a sphere of radius a (Figure 8). Thus, by the method of circular disks, we have obtained the familiar formula $V = \frac{4}{3}\pi a^3$ for the volume of a sphere of radius a.

Of course, a planar region can be revolved about the y axis rather than the x axis, and again a solid of revolution is generated. For instance, suppose that R is a planar region bounded by the y axis, the horizontal lines $y = a$ and $y = b$, where $a < b$, and the graph of $x = g(y)$, where the function g is continuous and $g(y) \geq 0$ for $a \leq y \leq b$ (Figure 9a). Figure 9b shows the solid of revolution S generated by revolving R about the y axis. In Figure 9b, $dV = \pi r^2 \, dy = \pi[g(y)]^2 \, dy$; hence,

$$V = \int_{y=a}^{y=b} dV = \int_a^b \pi[g(y)]^2 \, dy = \pi \int_a^b [g(y)]^2 \, dy.$$

The use of the formula above to find the volume of S is still referred to as the *method of circular disks.*

Figure 9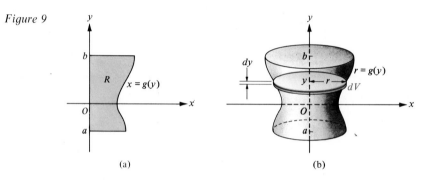

(a)   (b)

EXAMPLE   Find the volume of the solid S generated by revolving the region R bounded by the y axis, the line $y = 4$, and the graph of $y = x^2$ with $x \geq 0$ about the y axis. Use the method of circular disks and sketch both R and S.

SOLUTION
Solving the equation $y = x^2$ for x in terms of y and using the fact that $x \geq 0$, we have $x = \sqrt{y}$ (Figure 10). By the method of circular disks,

$$V = \pi \int_0^4 [\sqrt{y}]^2 \, dy = \pi \int_0^4 y \, dy = \pi \left[\frac{y^2}{2}\right]\Big|_0^4 = 8\pi \text{ cubic units.}$$

Figure 10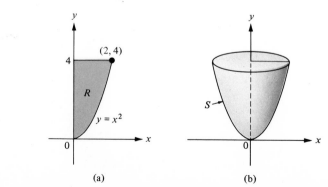

(a)   (b)

1.2 The Method of Circular Rings

Volumes of solids of revolution more general than those considered in Section 1.1 can be found by using the *method of circular rings*. This method works as follows: Suppose that f and g are nonnegative continuous functions on the interval $[a,b]$ such that $f(x) \geq g(x)$ holds for all values of x in $[a,b]$, and let R be the planar region bounded by the graphs of f and g between $x = a$ and $x = b$ (Figure 11a). Let S be the solid generated by revolving R about the x axis (Figure 11b). Here we

Figure 11

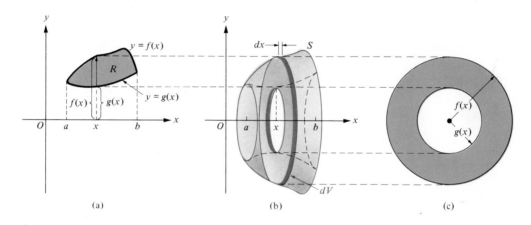

(a)   (b)   (c)

consider an "infinitesimal" portion dV of the volume V of S consisting of a circular ring or washer of "infinitesimal" thickness dx, perpendicular to the axis of revolution, and centered at the point with coordinate x. The base of this circular ring is the region between two concentric circles of radii $f(x)$ and $g(x)$ (Figure 11c); hence, the area of this base is $\pi[f(x)]^2 - \pi[g(x)]^2$ square units. It follows that

$$dV = \{\pi[f(x)]^2 - \pi[g(x)]^2\}\, dx,$$

so that

$$V = \int_{x=a}^{x=b} dV = \int_a^b \{\pi[f(x)]^2 - \pi[g(x)]^2\}\, dx = \pi \int_a^b \{[f(x)]^2 - [g(x)]^2\}\, dx.$$

EXAMPLE Using the method of circular rings, find the volume V of the solid S generated by revolving the region R about the x axis, where R is bounded by the curves $y = x^2$ and $y = x + 2$.

SOLUTION
The points of intersection of the two curves are $(2,4)$ and $(-1,1)$ (Figure 12a). By the method of circular rings (Figure 12b), we have

$$V = \pi \int_{-1}^2 [(x+2)^2 - (x^2)^2]\, dx = \pi \int_{-1}^2 (x^2 + 4x + 4 - x^4)\, dx$$

$$= \pi \left[\frac{x^3}{3} + 2x^2 + 4x - \frac{x^5}{5} \right]\Bigg|_{-1}^2 = \pi \left[\frac{184}{15} - \left(-\frac{32}{15} \right) \right]$$

$$= \frac{72\pi}{5} \text{ cubic units.}$$

Figure 12

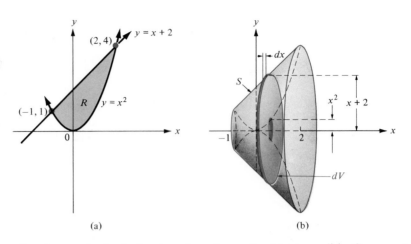

(a)   (b)

Naturally, the method of circular rings is applicable to solids S generated by revolving planar regions R about the y axis rather than the x axis. Thus, in Figure 13a, the planar region R is bounded on the right by the graph of $x = F(y)$

Figure 13

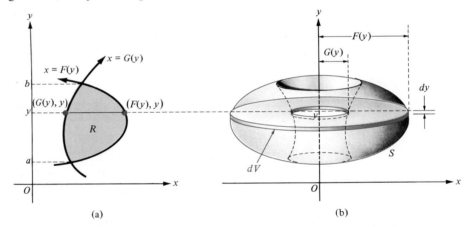

(a)   (b)

and on the left by the graph of $x = G(y)$, while these two graphs intersect at points with ordinates a and b. If R is revolved about the y axis, then a solid of revolution S is generated (Figure 13b). The circular ring perpendicular to the axis of revolution and centered at the point $(0, y)$ has "infinitesimal" volume dV given by $dV = \pi\{[F(y)]^2 - [G(y)]^2\}\,dy$; hence,

$$V = \int_{y=a}^{y=b} dV = \int_a^b \pi\{[F(y)]^2 - [G(y)]^2\}\,dy = \pi \int_a^b \{[F(y)]^2 - [G(y)]^2\}\,dy.$$

EXAMPLE   Use the method of circular rings to find the volume V of the solid of revolution S generated by revolving the region R about the y axis, where R is the planar region bounded on the right by the graph of $x = 2$, on the left by the graph of $y = x^3$, and below by the x axis. Sketch R and S.

SOLUTION
The region R and the solid S are shown in Figure 14a and b, respectively. Let $F(y) = 2$ and $G(y) = \sqrt[3]{y}$. By the method of circular rings,

$$V = \pi \int_0^8 \{[F(y)]^2 - [G(y)]^2\}\,dy$$

so

$$V = \pi \int_0^8 [4 - (\sqrt[3]{y})^2]\,dy = \pi \left[4y - \frac{3}{5}y^{5/3}\right]\Big|_0^8 = \pi\left(32 - \frac{96}{5}\right) = \frac{64\pi}{5} \text{ cubic units.}$$

Figure 14

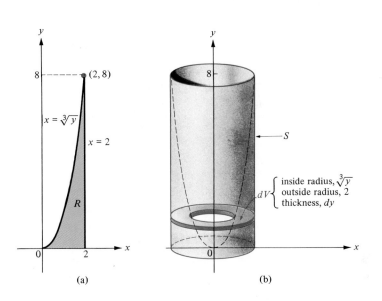

(a) (b)

The method of circular rings is also effective for solids generated by revolving planar regions about axes other than the x or the y axis. This is illustrated by the following examples.

EXAMPLES 1 Use the method of circular rings to find the volume of the solid S obtained by revolving the region R about the line $x = 6$, where R is bounded by the graphs of $y^2 = 4x$ and $x = 4$.

SOLUTION
From Figure 15, the inside radius of the circular ring at level y is 2 units and its outside radius is $6 - (y^2/4)$ units; hence, $dV = \pi[(6 - y^2/4)^2 - 2^2]\,dy$. Integrating, we obtain

$$V = \int_{y=-4}^{y=4} dV = \int_{-4}^{4} \pi\left[\left(6 - \frac{y^2}{4}\right)^2 - 2^2\right] dy$$

$$= \pi \int_{-4}^{4} \left(\frac{y^4}{16} - 3y^2 + 32\right) dy = \pi\left(\frac{y^5}{80} - y^3 + 32y\right)\Big|_{-4}^{4}$$

$$= \pi\left[\frac{384}{5} - \left(-\frac{384}{5}\right)\right] = \frac{768\pi}{5} \text{ cubic units.}$$

Figure 15

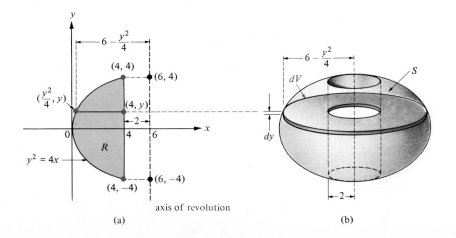

axis of revolution

(a) (b)

Figure 16

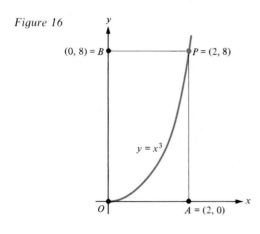

2 In Figure 16, the curve OP has the equation $y = x^3$. Find the volume of the solid of revolution generated by revolving the region
(a) OBP about the line $y = 8$.
(b) OAP about the line $x = 2$.
(c) OAP about the line $y = 8$.

SOLUTION

(a) When the region OBP is revolved about the axis $y = 8$, the "infinitesimal" rectangle of height $8 - x^3$ and width dx shown in Figure 17 sweeps out a circular *disk* of thickness dx and radius $8 - x^3$. Its volume is given by

$$dV = \pi(8 - x^3)^2\, dx = \pi(64 - 16x^3 + x^6)\, dx.$$

Therefore,

$$V = \int_{x=0}^{x=2} dV = \int_0^2 \pi(64 - 16x^3 + x^6)\, dx$$

$$= \pi\left(64x - 4x^4 + \frac{x^7}{7}\right)\Bigg|_0^2 = \frac{576\pi}{7} \text{ cubic units.}$$

Figure 17   Figure 18

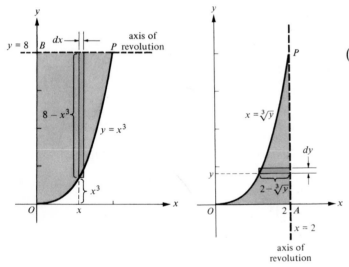

(b) When the region OAP is revolved about the axis $x = 2$, the "infinitesimal" rectangle of height dy and length $2 - \sqrt[3]{y}$ shown in Figure 18 sweeps out a circular *disk* of thickness dy and radius $2 - \sqrt[3]{y}$. Its volume is given by

$$dV = \pi(2 - \sqrt[3]{y})^2\, dy = \pi(4 - 4y^{1/3} + y^{2/3})\, dy.$$

Therefore,

$$V = \int_{y=0}^{y=8} dV = \int_0^8 \pi(4 - 4y^{1/3} + y^{2/3})\, dy$$

$$= \pi\left(4y - 3y^{4/3} + \frac{3}{5} y^{5/3}\right)\Bigg|_0^8 = \frac{16\pi}{5} \text{ cubic units.}$$

(c) When the region OAP is revolved about the axis $y = 8$, the "infinitesimal" rectangle of height x^3 and width dx shown in Figure 19 sweeps out a circular *ring* of thickness dx with inside radius $8 - x^3$, outside radius 8, and "infinitesimal" volume

$$dV = \pi[8^2 - (8 - x^3)^2]\, dx = \pi(16x^3 - x^6)\, dx.$$

Therefore,

$$V = \int_{x=0}^{x=2} dV = \int_0^2 \pi(16x^3 - x^6)\, dx$$

$$= \pi\left(4x^4 - \frac{x^7}{7}\right)\Bigg|_0^2 = \frac{320\pi}{7} \text{ cubic units.}$$

Figure 19

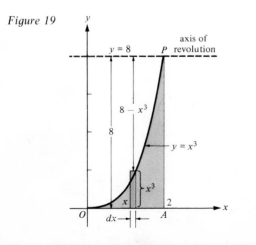

Problem Set 1

In problems 1 through 6, find the volume of the solid generated by revolving the region under the graph of each function over the indicated interval about the x axis.

1 $f(x) = 3x^2$; $[-1, 3]$

2 $g(x) = 3\sqrt{x}$; $[1, 4]$

3 $h(x) = \sqrt{9 - x^2}$; $[-1, 3]$

4 $G(x) = |x|$; $[-2, 1]$

5 $F(x) = \sqrt{2 + x^2}$; $[0, 2]$

6 $f(x) = |x| - x$; $[-3, 2]$

In problems 7 through 12, find the volume of the solid generated by revolving the region bounded by the graphs of the given equations about the y axis.

7 $y = x^3$, $y = 8$, and $x = 0$

8 $y^2 = x$, $y = 4$, and $x = 0$

9 $y^2 = 4x$, $y = 4$, and $x = 0$

10 $y = x^2 + 2$, $y = 4$, and $x = 0$ (first quadrant)

11 $y^2 = x^3$, $y = 8$, and $x = 0$

12 $y = 2x^3$, $y = 2$, and $x = 0$

In problems 13 through 25, find the volume of the solid generated by rotating the region bounded by the given curves about the indicated axis. Use the method of circular disks or the method of circular rings.

13 $y = x^2$ and $y = 2x$ about the x axis

14 $y = x^3$ and $y^2 = x$ about the x axis

15 $y = x^2$ and $y = x$ about the y axis

16 $y = x^2 + 4$ and $y = 2x^2$ about the y axis

17 $y = 12 - x^2$, $y = x$, and $x = 0$ (first quadrant) about the y axis

18 $y = 2x$, $y = x$, and $x + y = 6$ about the x axis

19 $y = x^3$, $x = 2$, and the x axis about the y axis

20 $y = 3x$, $y = x$, and $x + y = 8$ about the y axis

21 $y^2 = 4x + 16$ and the y axis about the y axis

22 $y = x^3$, $x = 0$, and $y = 8$ about the line $y = 8$

23 $y = x^2$ and $y^2 = x$ about the line $x = -1$

24 $y = x^2 - x$ and $y = 3 - x^2$ about the line $y = 4$

25 $y = 4x - x^2$ and $y = x$ about the line $x = 3$

26 A *torus* or *anchor ring* is a doughnut-shaped solid generated by revolving a circular region R about an axis in its plane that does not cut the region R. Find the volume V of such a torus if the radius of R is a and the distance from the center of R to the axis of revolution is b.

In problems 27 through 34, find the volume of the solid generated when the given region in Figure 20 is rotated about the axis indicated. In Figure 20, the curve OP has the equation $y = 3x^2$ for $0 \le x \le 2$.

27 OAP about the x axis

28 OBP about the x axis

29 OBP about the y axis

30 OAP about the y axis

31 OAP about the line $\overline{AP}$

32 OBP about the line $\overline{AP}$

33 OBP about the line $\overline{BP}$

34 OAP about the line $\overline{BP}$

Figure 20

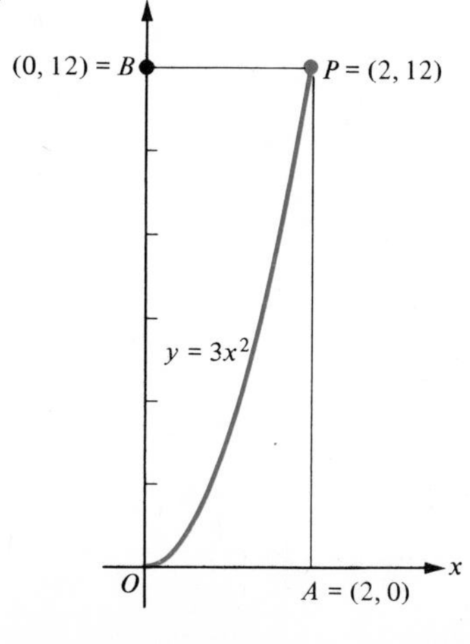

2   The Method of Cylindrical Shells

In this section we present an alternative method for finding the volume of a solid of revolution based on cylindrical shells rather than on circular disks or rings. We begin by considering the volume V of a solid right circular cylinder of fixed height h, but with variable radius x (Figure 1a). Here $V = \pi x^2 h$. If the radius x is increased by a small amount $\Delta x = dx$, then the volume V is increased by a corresponding amount ΔV (Figure 1b). Evidently, ΔV is the volume of a thin cylindrical shell of height h with inner radius x and thickness $\Delta x = dx$. Using the differential to approximate ΔV, we have $\Delta V \approx dV = d(\pi x^2 h) = 2\pi x h\,dx$. As Δx approaches zero, this approximation becomes better and better. Thus, a cylindrical shell of height h, inside radius x, and "infinitesimal" thickness dx has an "infinitesimal" volume $dV = 2\pi x h\,dx$.

Figure 1

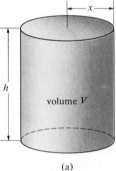
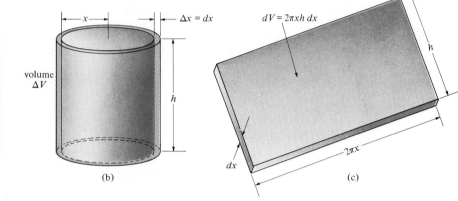

(a)   (b)   (c)

The formula $dV = 2\pi x h\,dx$ can be remembered by imagining that the cylindrical shell (Figure 1b) is cut vertically and "unwrapped" to form a rectangular slab of height h and thickness dx (Figure 1c). The length of this slab is approximately the inner circumference of the shell, $2\pi x$ units; hence, its volume is approximately $(2\pi x)(h)(dx) = 2\pi x h\,dx$ cubic units.

Now, let R be a region in the xy plane bounded above by the curve $y = f(x)$, below by the curve $y = g(x)$, on the left by $x = a$ and on the right by $x = b$, where $0 < a < b$ and $f(x) \geq g(x)$ for $a \leq x \leq b$ (Figure 2a). Let S be the solid of revolution generated by revolving R about the y axis (Figure 2b). Consider the rectangle with

Figure 2

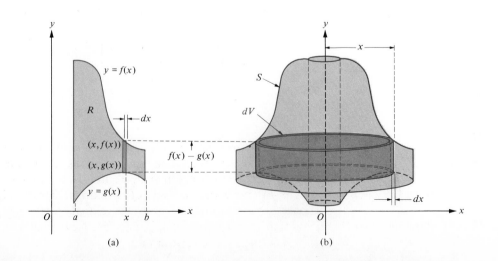

(a)   (b)

"infinitesimal" width dx and height $f(x) - g(x)$ situated above the point $(x, 0)$ as in Figure 2a. As R revolves about the y axis to generate S, this rectangle sweeps out an "infinitesimal" portion dV of the volume of the solid S having the shape of a cylindrical shell of height $f(x) - g(x)$, inside radius x, and "infinitesimal" thickness dx. Therefore, by the argument in the preceding paragraph,

$$dV = 2\pi x[f(x) - g(x)]\, dx.$$

Integrating the differential equation above, and noting that x runs from a to b, we obtain

$$V = \int_{x=a}^{x=b} dV = \int_a^b 2\pi x[f(x) - g(x)]\, dx = 2\pi \int_a^b x[f(x) - g(x)]\, dx.$$

The calculation of volumes by this formula (which can be rigorously justified) is called the *method of cylindrical shells*.

EXAMPLE   Let R be the planar region bounded by the graphs of $y = x^{3/2}$, $y = 1$, $x = 1$, and $x = 3$ (Figure 3a), and let S be the solid generated by revolving R about the y axis (Figure 3b). Use the method of cylindrical shells to find the volume V of S.

Figure 3

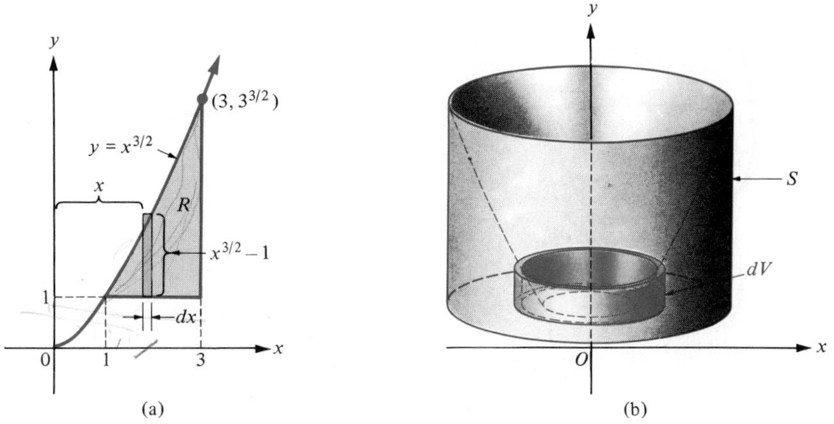

(a)   (b)

SOLUTION

$$V = 2\pi \int_1^3 x[x^{3/2} - 1]\, dx = 2\pi \int_1^3 (x^{5/2} - x)\, dx$$

$$= 2\pi \left[\frac{2}{7} x^{7/2} - \frac{x^2}{2} \right] \Big|_1^3 = 2\pi \left[\left(\frac{2}{7}(3)^{7/2} - \frac{3^2}{2} \right) - \left(\frac{2}{7}(1)^{7/2} - \frac{1^2}{2} \right) \right]$$

$$= 2\pi \left[\frac{54\sqrt{3} - 30}{7} \right] \approx 57.03 \text{ cubic units.}$$

With obvious modifications, the method of cylindrical shells is applicable to solids generated by revolving planar regions about the x axis or, indeed, any axis lying in the plane of the region. Figure 4 contrasts the method of circular rings (Figure 4a) with the method of cylindrical shells (Figure 4b). Notice that the circular rings are generated by "infinitesimal" rectangles *perpendicular* to the axis of revolution, while cylindrical shells are generated by "infinitesimal" rectangles *parallel* to the axis of revolution.

Figure 4

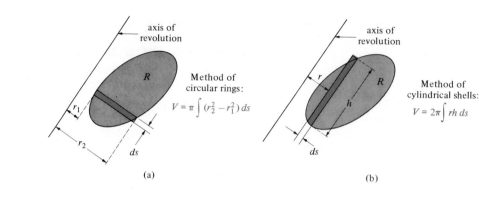

(a)   (b)

EXAMPLE   Let R be the planar region bounded by the graphs of $y = \sqrt{x}$, $y = 1$, and $x = 4$ (Figure 5). Use the method of cylindrical shells to find the volume V of the solid S generated by revolving R about the line $y = -2$.

Figure 5

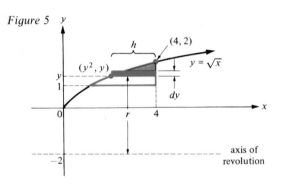

SOLUTION
When revolved about the line $y = -2$, the "in-finitesimal" rectangle of length h and width dy shown in Figure 5 generates a cylindrical shell of radius r, thickness dy, height h, and infinitesimal volume $dV = 2\pi rh\, dy$. From Figure 5 we see that

$$r = y + 2 \quad \text{and} \quad h = 4 - y^2;$$

hence,

$$dV = 2\pi rh\, dy = 2\pi(y + 2)(4 - y^2)\, dy$$
$$= 2\pi(-y^3 - 2y^2 + 4y + 8)\, dy.$$

Here, y runs from 1 to 2, so

$$V = \int_{y=1}^{y=2} dV = \int_1^2 2\pi(-y^3 - 2y^2 + 4y + 8)\, dy$$

$$= 2\pi\left(-\frac{y^4}{4} - \frac{2y^3}{3} + 2y^2 + 8y\right)\Bigg|_1^2 = 2\pi\left(\frac{44}{3} - \frac{109}{12}\right) = \frac{67\pi}{6} \text{ cubic units.}$$

Problem Set 2

In problems 1 through 4, use the method of cylindrical shells to find the volume V of the solid of revolution S generated by revolving each region R about the y axis.

1 R is bounded by the graphs of $y = x^3$, $y = x^2 + 1$, $x = 0$, and $x = 1$.

2 R is bounded by the graphs of $y = \sqrt[3]{x}$, $y = 1$, $x = 1$, and $x = 27$.

3 R is bounded by the graphs of $3x - 2y + 1 = 0$, $y = x$, $x = 1$, and $x = 3$.

4 R is bounded by the graphs of $y = \sqrt{1 - x^2}$, $y = -\sqrt{1 - x^2}$, and $x = 0$ and lies in the first and fourth quadrants.

In problems 5 through 8, use the method of cylindrical shells to find the volume V of the solid of revolution S generated by revolving each region R about the axis indicated.

5 R is bounded by the graphs of $y = x^3$, $y = 27$, and $x = 0$; about the x axis.

6 R is the same region as in problem 5, but the axis of revolution is $y = 27$.

7 R is bounded by the line $y = 16$ and the parabola $y = x^2$; about the x axis.

8 R is the same region as in problem 7, but the axis of revolution is $x = 20$.

In problems 9 through 12, use the method of cylindrical shells to work the indicated problem in Problem Set 1.

9 Problem 16 in Problem Set 1.   **10** Problem 18 in Problem Set 1.

11 Problem 23 in Problem Set 1.   **12** Problem 24 in Problem Set 1.

13 Let a be a positive constant. Use the method of cylindrical shells to find the volume of a sphere of radius a generated by revolving the region $R: x^2 + y^2 \le a^2$, $x \ge 0$ about the y axis.

14 Use the cylindrical shell method to show that the volume of a right circular cone is one-third of its height times the area of its base.

15 Use the method of cylindrical shells to find the volume V of the solid S generated by revolving the region R in the first quadrant above the graph of $y = x^2$ and below the graph of $y = x + 2$ about the y axis.

16 A solid S is generated by revolving the region R in the second quadrant above the graph of $y = -x^3$ and below the graph of $y = 3x^2$ about the line $y = -3$. Using the method of cylindrical shells, find the volume V of S.

17 Find the volume of the solid generated by revolving the right triangle with vertices at $(a, 0)$, $(b, 0)$, and (a, h) about the y axis. Assume that $0 < a < b$ and $h > 0$.

18 Let V be the volume of the solid of revolution generated by revolving a region R, which lies entirely to the right of the y axis, about the y axis. If b is a positive constant, show that the volume of the solid of revolution generated by revolving R about the axis $x = -b$ is $V + 2\pi b A$, where A is the area of R.

19 Find the volume V of the football-shaped solid generated by rotating the region R bounded above by the graph of $16x^2 + 64y^2 = 1024$ with $y \ge 0$ and bounded below by the x axis about the x axis.

20 A cylindrical hole is bored through the center of a sphere of unknown radius. However, the length of the hole is known to be L units. Show that the volume of the portion of the sphere that remains is equal to the volume of a sphere of diameter L.

3   Volumes by the Method of Slicing

In Section 5 of Chapter 5 and Section 6.1 of Chapter 6 we studied the method of slicing for determining the areas of admissible planar regions. In this section we use an analogous method, also called the *method of slicing,* for finding the volumes of solids. Actually, this method is just a generalization of the method of circular disks, or circular rings, presented in Section 1.

In order to calculate the volume V of a solid S by the method of slicing, select a convenient reference axis and let $A(s)$ be the area of the cross section

Figure 1

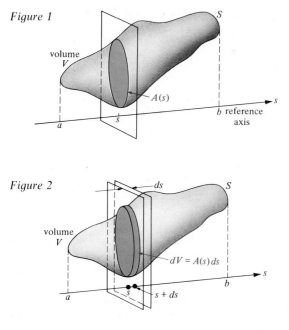

volume
V

$A(s)$

b reference
axis

s

a

s

Figure 2

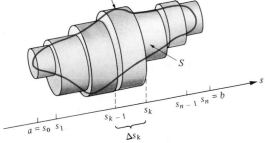

volume
V

ds S

$dV = A(s)\,ds$

a s $s + ds$ b s

Figure 3

volume of kth slab $\approx A(s_k)\Delta s_k$

S

s

$a = s_0$ s_1 s_{k-1} s_k s_{n-1} $s_n = b$

Δs_k

of S intercepted by the plane perpendicular to the reference axis at the point with coordinate s (Figure 1). Suppose that the entire solid S is contained between the plane at $s = a$ and the plane at $s = b$.

In Figure 2, let dV denote the "infinitesimal" volume of the portion of the solid S between the plane perpendicular to the reference axis at the point with coordinate s and the corresponding plane at the point with coordinate $s + ds$. Here ds represents an "infinitesimal" increase in the reference coordinate. Evidently, dV is the volume of an "infinitesimal" solid cylinder of height ds with base area $A(s)$; hence, $dV = A(s)\,ds$.

The total volume V of the solid S should be obtained by "summing"—that is, integrating—all the "infinitesimal" volumes dV as s runs from a to b. Therefore, we should have $V = \int_a^b A(s)\,ds$.

This argument using Leibnizian infinitesimals, although perhaps not mathematically rigorous, produces the correct result. A more conclusive argument, based directly on the analytic definition of the definite integral $\int_a^b A(s)\,ds$, can be made as follows.

We partition the reference axis into n subintervals $[s_0, s_1], [s_1, s_2], \ldots, [s_{k-1}, s_k], \ldots, [s_{n-1}, s_n]$ (Figure 3). Above each subinterval $[s_{k-1}, s_k]$ of this partition we have constructed a circumscribed cylindrical slab of thickness Δs_k. The cross-sectional area of such a slab is approximately $A(s_k)$, so its volume is approximately $A(s_k)\,\Delta s_k$. Therefore, $V \approx \sum_{k=0}^{n} A(s_k)\,\Delta s_k$. As the norm of the partition becomes smaller and smaller and the number of slabs increases while the slabs get thinner and thinner, the approximation $V \approx \sum_{k=0}^{n} A(s_k)\,\Delta s_k$ becomes better and better. In the limit, as the norm of the partition approaches zero, $\sum_{k=0}^{n} A(s_k)\,\Delta s_k$ approaches $\int_a^b A(s)\,ds$ by definition; therefore, we have $V = \int_a^b A(s)\,ds$.

EXAMPLES 1 Find the volume of a solid whose base is a circle of radius 2 if all cross sections perpendicular to a fixed diameter of the base are squares.

SOLUTION
Figure 4 shows the square cross section of the solid at a distance s units from the center O along the fixed diameter. The reference axis is taken to lie along this diameter. If x is the length of one side of the square, then, from Figure 4 and the Pythagorean theorem,

$$s^2 + \left(\frac{x}{2}\right)^2 = 2^2 \quad \text{or} \quad \frac{x^2}{4} = 4 - s^2.$$

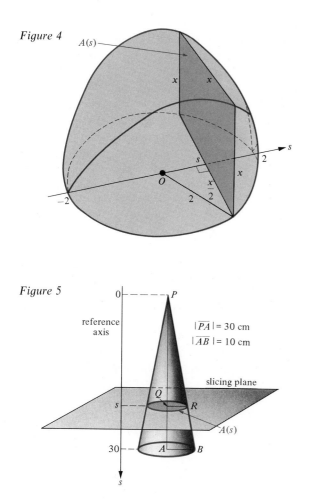

Figure 4

Therefore, the area of the square is

$$A(s) = x^2 = 4(4 - s^2) = 16 - 4s^2.$$

By the method of slicing, the required volume is

$$V = \int_{-2}^{2} A(s)\, ds = \int_{-2}^{2} (16 - 4s^2)\, ds = \left[16s - \tfrac{4}{3}s^3\right]\Big|_{-2}^{2}$$

$$= \tfrac{128}{3} \approx 42.67 \text{ cubic units.}$$

2 Find the volume of a solid right circular cone of height 30 centimeters if the base radius is 10 centimeters.

SOLUTION

We choose the reference axis (Figure 5) pointing downward. The cross section of the cone cut by the slicing plane at level s is a circle of radius $|\overline{QR}|$ and area $A(s) = \pi|\overline{QR}|^2$. We find $|\overline{QR}|$ in terms of s by similar triangles as follows: $|\overline{QR}|/|\overline{PQ}| = |\overline{AB}|/|\overline{PA}|$; that is, $|\overline{QR}|/s = \tfrac{10}{30}$, or $|\overline{QR}| = \tfrac{1}{3}s$. Consequently,

$$A(s) = \pi|\overline{QR}|^2 = \pi(\tfrac{1}{3}s)^2 = \pi s^2/9.$$

By the method of slicing, the volume of the cone is therefore given by

$$V = \int_0^{30} A(s)\, ds = \int_0^{30} \frac{\pi s^2}{9}\, ds$$

$$= \left[\frac{\pi s^3}{27}\right]\Big|_0^{30} = \frac{\pi}{27}(30)^3 = 1000\pi \text{ cubic centimeters.}$$

Figure 5

$|\overline{PA}| = 30$ cm
$|\overline{AB}| = 10$ cm

This coincides with the volume as calculated by the familiar formula—one-third the height times the area of the base.

3 Gasoline is stored in a spherical tank of radius $r = 10$ meters. How many cubic meters of gasoline are in the tank if the surface of the gasoline is 3 meters below the center of the tank?

SOLUTION

The cross section of the tank cut by the slicing plane at level s is a circle of radius $\overline{QP}$ and area $A(s) = \pi|\overline{QP}|^2$ (Figure 6). We find $|\overline{QP}|^2$ in terms of s by the Pythagorean theorem as follows: $|\overline{QP}|^2 + |\overline{QC}|^2 = |\overline{CP}|^2$; that is,

$$|\overline{QP}|^2 = |\overline{CP}|^2 - |\overline{QC}|^2$$
$$= r^2 - |s|^2 = r^2 - s^2.$$

Consequently,

$$A(s) = \pi(r^2 - s^2) = \pi(100 - s^2).$$

Notice that the surface of the gasoline is at level $s = -3$. Therefore, by the method of slicing, the

Figure 6

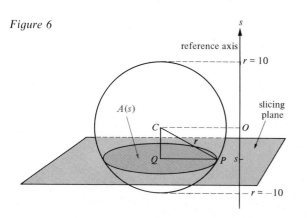

reference axis
$r = 10$
slicing plane
$r = -10$

volume of the portion of the sphere occupied by the gasoline is given by

$$V = \int_{-10}^{-3} A(s)\, ds = \int_{-10}^{-3} \pi(100 - s^2)\, ds = \pi\left(100s - \frac{s^3}{3}\right)\Bigg|_{-10}^{-3}$$

$$= \left[\pi\left(100(-3) - \frac{(-3)^3}{3}\right)\right] - \left[\pi\left(100(-10) - \frac{(-10)^3}{3}\right)\right]$$

$$= -291\pi + \frac{2000}{3}\pi = \frac{1127}{3}\pi \approx 1180.19 \text{ cubic meters.}$$

4   A solid right circular cylinder has a radius of 3 inches. A wedge is cut from this cylinder by a plane through a diameter of the base and inclined to the base at an angle of 30°. Find the volume of the wedge.

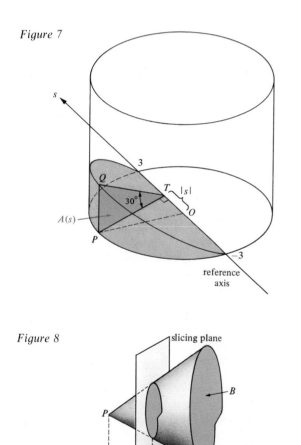

Figure 7

Figure 8

SOLUTION

Let the reference axis lie along the intersection of the plane and the base of the cylinder, with the origin O at the center of the base (Figure 7). The cutting plane for the method of slicing evidently cuts the wedge in a triangular cross section TPQ. We must find the area $A(s) = \frac{1}{2}|\overline{PQ}| \cdot |\overline{PT}|$ of TPQ in terms of the coordinate s of the point T. Triangle OTP is a right triangle and $|\overline{OT}| = |s|$, while $|\overline{OP}| = 3$; hence, $|\overline{OP}|^2 = |\overline{OT}|^2 + |\overline{PT}|^2$, so $|\overline{PT}|^2 = |\overline{OP}|^2 - |\overline{OT}|^2 = 9 - |s|^2 = 9 - s^2$. Referring to right triangle TPQ, we have $|\overline{PQ}| = |\overline{PT}| \tan 30°$. Therefore,

$$A(s) = \tfrac{1}{2}|\overline{PQ}| \cdot |\overline{PT}| = \tfrac{1}{2}|\overline{PT}| \tan 30° |\overline{PT}|$$
$$= \tfrac{1}{2}|\overline{PT}|^2 \tan 30°.$$

Since $|\overline{PT}|^2 = 9 - s^2$, $A(s) = \frac{1}{2}(9 - s^2)\tan 30° = \frac{1}{2}(9 - s^2)(\sqrt{3}/3)$. By the method of slicing,

$$V = \int_{-3}^{3} A(s)\, ds = \int_{-3}^{3} \frac{1}{2}(9 - s^2)\frac{\sqrt{3}}{3}\, ds$$

$$= \frac{\sqrt{3}}{6}\left[9s - \frac{s^3}{3}\right]\Bigg|_{-3}^{3} = \frac{\sqrt{3}}{6}[18 - (-18)]$$

$$= \frac{\sqrt{3}}{6}(36) = 6\sqrt{3} \approx 10.39 \text{ cubic inches.}$$

If B is an admissible planar region and P is a point not lying in the same plane as B, then the three-dimensional solid consisting of all points lying on straight line segments between P and points of B is called a *solid cone* with *vertex* P and *base* B (Figure 8). The perpendicular distance h between the vertex P and the base B is called the *height* of the cone.

If a reference axis is chosen perpendicular to the base B, then a slicing plane at a distance s from the vertex P will intercept the cone in a cross-sectional region which is similar to the base B. Furthermore, the linear dimensions of this cross section are proportional to its distance s from P. Since the area $A(s)$ of the cross section is proportional to the *square* of its linear dimensions, $A(s)$ is proportional to s^2. Thus, $A(s) = Ks^2$, where K is a constant. When $s = h$, then $A(s) = A(h) =$ the area of the base B. Therefore, $A(h) = Kh^2$; and so $K = A(h)/h^2$. Hence, $A(s) = \dfrac{A(h)}{h^2} s^2$. By the method of slicing,

$$V = \int_0^h A(s)\, ds = \int_0^h \frac{A(h)}{h^2} s^2\, ds = \frac{A(h)}{h^2} \int_0^h s^2\, ds$$

$$= \frac{A(h)}{h^2} \left[\frac{s^3}{3} \right]\Bigg|_0^h = \frac{A(h)}{h^2} \cdot \frac{h^3}{3} = \frac{A(h)}{3} h.$$

Hence, the volume of a solid cone is given by one-third of its height times the area of the base.

EXAMPLE Find the volume of a pyramid with a square base 7 meters on each side if the perpendicular distance from the vertex to the base is 12 meters.

Figure 9

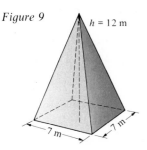
$h = 12\text{ m}$
7 m 7 m

SOLUTION
Such a pyramid is a solid cone with a square base (Figure 9); hence, by the result above, $V = \frac{1}{3}(12)(7)^2 = 196$ cubic meters.

Problem Set 3

1 A certain solid has a circular base of radius 3 inches. If cross sections perpendicular to one of the diameters of the base are squares, find the volume of the solid.

2 Find the volume of a solid whose base is a circle of radius 5 centimeters if all cross sections perpendicular to a fixed diameter of the base are equilateral triangles.

3 A monument is 30 meters high. A horizontal cross section x meters above the base is an equilateral triangle whose sides are $\dfrac{30 - x}{15}$ meters long. Find the volume of the monument.

4 A tower is 24 meters tall. A horizontal cross section of the tower x meters from its top is a square whose sides are $\frac{1}{13}(x + 1.5)$ meters long. Find the volume of the tower.

5 Find the volume of a spherical shell whose inside diameter is Y_1 units and whose outside diameter is Y_2 units.

6 The base of a solid is a planar region bounded by an ellipse with semimajor axis 4 units and semiminor axis 3 units. Every cross section perpendicular to the major axis of the ellipse is a semicircle. Find the volume of the solid.

7 The base of a solid lies in a plane and the height of the solid is 5 feet. Find the volume of the solid if the area of a cross section parallel to the base and s feet above the base is given by the equation (a) $A(s) = 3s^2 + 2$, (b) $A(s) = s^2 + s$.

8 Use the method of slicing to find the volume of a solid right circular cylinder of radius r and height h, taking the cross sections perpendicular to a fixed diameter of the base. You may assume that $\displaystyle\int_{-r}^{r} \sqrt{r^2 - s^2}\, ds = \frac{1}{2}\pi r^2$.

9 The Department of Public Works of East Mattoon intends to cut down a diseased elm tree 4 feet in diameter. They first cut out a wedge bounded below by a horizontal plane and bounded above by a plane that meets the horizontal plane along a diameter of the tree at a 45° angle. What is the volume of the wedge?

10 A solid right circular cylinder has a radius of r units. A wedge is cut from this cylinder by a plane through a diameter of the base circle and inclines to this base circle at an angle θ. Find a formula for the volume of this wedge.

Figure 10

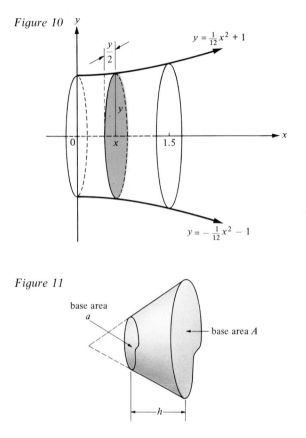

11 The plans for a wave-guide antenna are shown in Figure 10. Each cross section in a plane perpendicular to the central axis of the wave guide (the x axis) is an ellipse whose major diameter is twice as long as its minor diameter. The upper and lower boundaries of the vertical cross section through the central axis are parabolas $y = \frac{1}{12}x^2 + 1$ and $y = -\frac{1}{12}x^2 - 1$, respectively. Find the volume enclosed by the wave guide if it is 1.5 meters long. Assume that the area of an ellipse is given by π times the product of its semimajor and semiminor axes.

12 Show that the volume V of the frustum of a cone shown in Figure 11 is given by $V = \frac{h}{3}(A + \sqrt{Aa} + a)$.

Figure 11

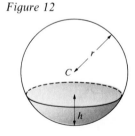

base area a

base area A

13 The base of a solid is the planar region bounded by the hyperbola $16x^2 - 9y^2 = 144$ and the line $x = 6$. Every cross section of the solid perpendicular to the x axis is an equilateral triangle. Find the volume of the solid.

14 Show that the volume of a solid cylinder (not necessarily a *right* cylinder) is given by its height times the area of one of its bases.

15 Find the volume of the spherical segment of one base shown in Figure 12 if the radius of the sphere is r units and the height of the segment is h units.

Figure 12

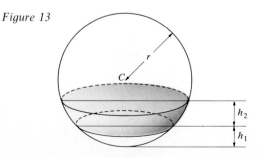

16 Find the volume of a solid whose cross section made by a plane perpendicular to a reference axis at the point with coordinate s has area given by

$$A(s) = \begin{cases} as^2 + bs + c & \text{for } 0 \le s \le h \\ 0 & \text{otherwise.} \end{cases}$$

Express this volume in terms of $A_0 = A(0)$, $A_1 = A(h/2)$, and $A_2 = A(h)$. (*Hint:* Use the prismoidal formula of Chapter 6, Section 5.3.)

17 A tent is made by stretching canvas from a circular base of radius a to a semicircular rib erected at right angles to the base and meeting the base at the ends of a diameter. Find the volume enclosed by the tent.

Figure 13

18 Find the volume of the spherical segment of two bases shown in Figure 13 if the sphere has radius r, the height of the segment is h_2, and the lower base of the segment is h_1 units from the bottom of the sphere.

19 A regular *octahedron* is a solid bounded by eight congruent equilateral triangles (Figure 14). Find the volume of an octahedron if each of its eight bounding triangles has sides of length *l*.

20 Find the volume of the spherical sector shown in Figure 15.

21 A tower is 60 meters high and every horizontal cross section is a square. The vertices of these squares lie in four congruent parabolas whose planes pass through the central axis of the tower and which open outward away from this central axis. Each of the four parabolas has its vertex in the upper square base of the tower and each of these parabolas has a horizontal axis. The diagonals of the upper and lower bases of the tower are 2 meters and 12 meters long, respectively. Find the volume enclosed by the tower.

22 Two right circular cylinders of radius *r* have central axes meeting at right angles. Find the volume *V* of the solid common to both cylinders.

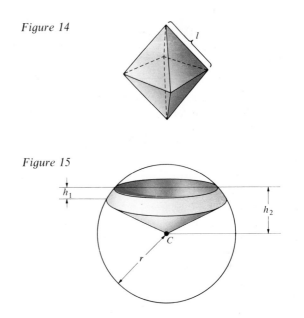

Figure 14

Figure 15

4   Arc Length and Surface Area

In the present section we give two further geometric applications of the definite integral: the calculation of arc length and the calculation of the area of a surface of revolution.

4.1   Arc Length of the Graph of a Function

If a straight piece of wire of length *s* is bent into a curve *C*, we understand that the curve *C* has *arc length s* (Figure 1). For instance, the arc length of a circle of diameter *D* is known to be given by $s = \pi D$. Here we do not attempt to give a formal definition of arc length, but we assume that the reader has an intuitive understanding of this concept.

Now, let *f* be a function with a continuous first derivative on some open interval *I* containing the closed interval $[a, b]$. We now give an informal derivation of a formula for the arc length of the portion of the graph of *f* between the points $(a, f(a))$ and $(b, f(b))$ (Figure 2). Let *s* denote the arc length of the portion of the graph of *f*

Figure 1

Figure 2

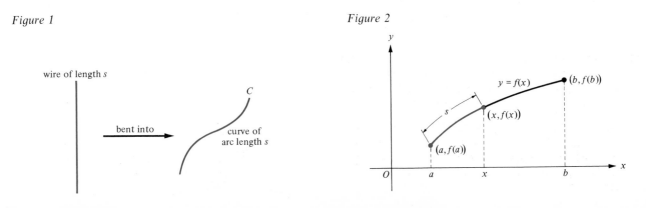

Figure 3

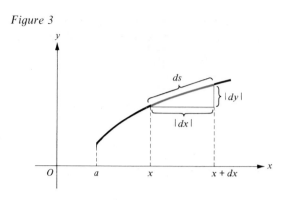

between the points $(a, f(a))$ and $(x, f(x))$, where $a \le x \le b$. If x is increased by an "infinitesimal" amount dx, then $y = f(x)$ will change by a corresponding "infinitesimal" amount dy and likewise s will increase by an "infinitesimal" amount ds (Figure 3). Notice the "infinitesimal right triangle" with legs $|dx|$ and $|dy|$ and with hypotenuse ds. By the Pythagorean theorem,

$$(ds)^2 = (dx)^2 + (dy)^2$$
$$= \left[1 + \left(\frac{dy}{dx}\right)^2\right](dx)^2;$$

hence,

$$ds = \sqrt{1 + \left(\frac{dy}{dx}\right)^2}\, dx = \sqrt{1 + [f'(x)]^2}\, dx.$$

Integration of the differential equation above gives the arc length of the graph between the point with abscissa a and the point with abscissa b; thus,

$$s = \int_{x=a}^{x=b} ds = \int_a^b \sqrt{1 + [f'(x)]^2}\, dx.$$

EXAMPLES 1 Find the arc length of the graph of the function f defined by $f(x) = x^{2/3} - 1$ between the points $(8, 3)$ and $(27, 8)$.

SOLUTION
Here $f'(x) = \frac{2}{3}x^{-1/3}$, so the desired arc length is given by

$$s = \int_8^{27} \sqrt{1 + [f'(x)]^2}\, dx = \int_8^{27} \sqrt{1 + \frac{4}{9}x^{-2/3}}\, dx$$

$$= \int_8^{27} \sqrt{\frac{9x^{2/3} + 4}{9x^{2/3}}}\, dx = \int_8^{27} \sqrt{9x^{2/3} + 4}\, \frac{dx}{3x^{1/3}}.$$

Making the change of variable $u = 9x^{2/3} + 4$ and noting that $du = 6x^{-1/3}\, dx$, $u = 40$ when $x = 8$, and $u = 85$ when $x = 27$, we have

$$s = \int_{x=8}^{x=27} \sqrt{9x^{2/3} + 4}\, \frac{dx}{3x^{1/3}} = \int_{u=40}^{u=85} \sqrt{u}\, \frac{du}{18}$$

$$= \left[\frac{u^{3/2}}{27}\right]\bigg|_{40}^{85} = \frac{85^{3/2} - 40^{3/2}}{27} \approx 19.65 \text{ units.}$$

2 Set up an integral representing the arc length of the portion of the parabola $y = x^2$ between the origin and the point $(2, 4)$. Then use Simpson's parabolic rule, $S_{2n} \approx \int_0^2 \sqrt{1 + [f'(x)]^2}\, dx$, with $n = 2$ to estimate this integral and thus the arc length.

SOLUTION
Here $dy/dx = 2x$, so the desired arc length is given by

$$s = \int_0^2 \sqrt{1 + (2x)^2}\, dx = \int_0^2 \sqrt{1 + 4x^2}\, dx.$$

To find the quantities needed for Simpson's parabolic rule with $n = 2$, we subdivide the closed interval $[0, 2]$ into $2n = 4$ equal subintervals by means of the points $x_k = k/2$ for $k = 0, 1, 2, 3, 4$. Now let $y_k = \sqrt{1 + 4(x_k)^2} = \sqrt{1 + k^2}$, so that $y_0 = 1$, $y_1 = \sqrt{2}$, $y_2 = \sqrt{5}$, $y_3 = \sqrt{10}$, and $y_4 = \sqrt{17}$. Then

$$S_{2n} = S_4 = \frac{\frac{1}{2}}{3}[y_0 + 4y_1 + 2y_2 + 4y_3 + y_4] = \frac{1}{6}(1 + 4\sqrt{2} + 2\sqrt{5} + 4\sqrt{10} + \sqrt{17}).$$

Evaluating S_4 and rounding off to two decimal places, we have $S_4 \approx 4.65$; hence, the desired arc length is approximately 4.65 units. (Incidently, the correct value of the desired arc length, rounded off to four decimal places, is 4.6468 units.)

If we express the equation of the curve between two points in the form $x = g(y)$, where g' is a continuous function in the closed interval $[c, d]$, the arc length of the graph of g between $(g(c), c)$ and $(g(d), d)$ is given by the formula

$$s = \int_c^d \sqrt{1 + [g'(y)]^2}\, dy = \int_c^d \sqrt{1 + \left(\frac{dx}{dy}\right)^2}\, dy.$$

EXAMPLE   Find the arc length of the graph of the equation $8x = y^4 + 2/y^2$ from $(\frac{3}{8}, 1)$ to $(\frac{33}{16}, 2)$.

SOLUTION
Here $x = \frac{1}{8}y^4 + \frac{1}{4}y^{-2}$, so

$$\frac{dx}{dy} = \frac{1}{2}y^3 - \frac{2}{4}y^{-3} = \frac{y^3}{2} - \frac{1}{2y^3}.$$

The desired arc length is given by

$$s = \int_1^2 \sqrt{1 + \left(\frac{dx}{dy}\right)^2}\, dy = \int_1^2 \sqrt{1 + \left(\frac{y^3}{2} - \frac{1}{2y^3}\right)^2}\, dy$$

$$= \int_1^2 \sqrt{1 + \frac{y^6}{4} - \frac{1}{2} + \frac{1}{4y^6}}\, dy = \int_1^2 \sqrt{\frac{y^6}{4} + \frac{1}{2} + \frac{1}{4y^6}}\, dy$$

$$= \int_1^2 \sqrt{\left(\frac{y^3}{2} + \frac{1}{2y^3}\right)^2}\, dy = \int_1^2 \left(\frac{y^3}{2} + \frac{1}{2y^3}\right) dy$$

$$= \left(\frac{y^4}{8} - \frac{1}{4y^2}\right)\Bigg|_1^2 = \frac{33}{16} \text{ units.}$$

4.2   Area of a Surface of Revolution

Once again we do not attempt to give a formal definition of the "surface area" of a solid S (Figure 4). Suffice it to say that, if the solid S has surface area A, then the same amount of paint required to apply a uniform thin coat to the surface of S would be required to apply a uniform thin coat to a flat surface of area A.

To find the lateral surface area A of the right circular cone shown in Figure 5a, we cut the cone along the dashed line and flatten it out to form a sector of a circle, as in Figure 5b. The slant height a of the cone is the radius of the sector and the arc length of the sector is the circumference $2\pi r$ of the base of the cone.

Figure 4

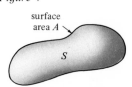

surface
area A

S

Figure 5

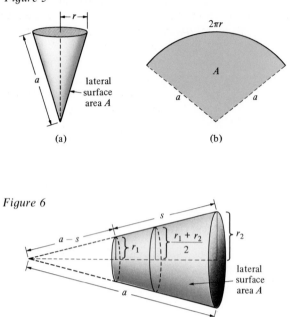

(a)   (b)

Figure 6

The sector in Figure 5b is part of a circle of area πa^2 with circumference $2\pi a$. The ratio of the area A of the sector to the area πa^2 of the circle is the same as the ratio of the arc length $2\pi r$ of the sector to the total arc length $2\pi a$ of the circle; that is,

$$\frac{A}{\pi a^2} = \frac{2\pi r}{2\pi a} \quad \text{or} \quad A = \pi ar.$$

Now, consider the frustum of a right circular cone shown in Figure 6. The lateral surface area A of this frustum is evidently the difference between the lateral surface area of the large cone with slant height a and base radius r_2 and the lateral surface area of the small cone with slant height $a - s$ and base radius r_1 (Figure 6). Thus, by the formula previously obtained for the lateral surface area of a cone,

$$A = \pi ar_2 - \pi(a - s)r_1 = \pi a(r_2 - r_1) + \pi sr_1.$$

By similar triangles, $\dfrac{a}{r_2} = \dfrac{a - s}{r_1}$, so

$$ar_1 = ar_2 - sr_2 \quad \text{or} \quad a = \frac{sr_2}{r_2 - r_1}.$$

Therefore,

$$A = \pi \frac{sr_2}{r_2 - r_1}(r_2 - r_1) + \pi sr_1 = \pi sr_2 + \pi sr_1 = \pi s(r_1 + r_2) = 2\pi \frac{r_1 + r_2}{2} s.$$

Notice that $\dfrac{r_1 + r_2}{2}$ is the radius of the cross section of the frustum midway between its two bases and $2\pi \dfrac{r_1 + r_2}{2}$ is the circumference of this midsection. Hence, the lateral surface area A of a frustum of a right circular cone is given by $A = 2\pi \dfrac{r_1 + r_2}{2} s$, the circumference of its midsection times its slant height.

We now consider the problem of finding the surface area A of the surface of revolution generated by revolving the portion of the graph of the nonnegative continuous function f between the lines $x = a$ and $x = b$ about the x axis (Figure 7). Let ds

Figure 7

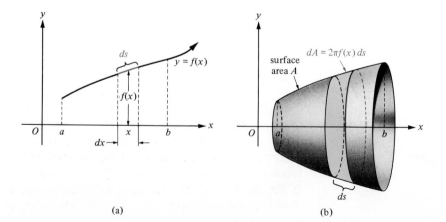

(a)   (b)

denote the "infinitesimal" arc length of the portion of the graph of f above the interval of "infinitesimal" length dx as shown in Figure 7a, and let x denote the coordinate of the center of this interval. When the "infinitesimal" arc of length ds is revolved about the x axis, it generates an "infinitesimal" frustum of a cone of slant height ds whose midsection has radius $f(x)$ (Figure 7b). The surface area of this "infinitesimal" frustum is $dA = 2\pi f(x)\, ds$. Assume that the function f has a continuous first derivative, so that $ds = \sqrt{1 + [f'(x)]^2}\, dx$; hence, $dA = 2\pi f(x)\sqrt{1 + [f'(x)]^2}\, dx$. The desired surface area A can now be obtained by integrating dA, so

$$A = \int_a^b 2\pi f(x)\sqrt{1 + [f'(x)]^2}\, dx \quad \text{or} \quad A = \int_{x=a}^{x=b} 2\pi f(x)\, ds,$$

where $ds = \sqrt{1 + [f'(x)]^2}\, dx$.

EXAMPLES **1** Let m be a positive constant. Find the area of the surface of revolution generated by revolving the graph of $f(x) = mx$ between $x = 0$ and $x = b$ about the x axis. Interpret the result geometrically.

SOLUTION
When the graph of f between $x = 0$ and $x = b$ is revolved about the x axis, it generates a right circular cone of height b with base radius mb (Figure 8). Here $f'(x) = m$, so the integral formula for surface area gives

$$A = \int_0^b 2\pi mx\sqrt{1 + m^2}\, dx = (2\pi\sqrt{1 + m^2})m \int_0^b x\, dx$$

$$= (2\pi\sqrt{1 + m^2})m\,\frac{b^2}{2} = \pi mb^2\sqrt{1 + m^2}.$$

Figure 8

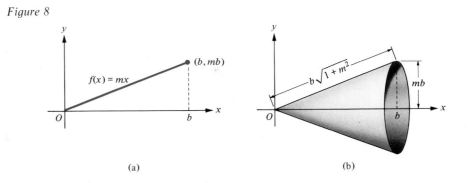

(a)   (b)

Since the slant height of the cone is the distance from the origin to the point (b, mb), the slant height is $\sqrt{b^2 + m^2 b^2} = b\sqrt{1 + m^2}$. Therefore, the formula $A = \pi mb^2\sqrt{1 + m^2} = \pi[b\sqrt{1 + m^2}]mb$ corresponds to the previously derived formula $A = \pi(\text{slant height})(\text{base radius})$ for the lateral surface area of a cone.

2 Find the area of the surface obtained by revolving the curve $y = \sqrt{x}$ between $x = 1$ and $x = 4$ about the x axis. Sketch the curve and the surface.

SOLUTION

The curve, part of a parabola, and the corresponding paraboloid of revolution are shown in Figure 9. Here $dy/dx = 1/(2\sqrt{x})$ and

$$A = 2\pi \int_1^4 \sqrt{x} \sqrt{1 + \left(\frac{dy}{dx}\right)^2}\, dx = 2\pi \int_1^4 \sqrt{x} \sqrt{1 + \frac{1}{4x}}\, dx = 2\pi \int_1^4 \sqrt{x + \frac{1}{4}}\, dx.$$

Figure 9

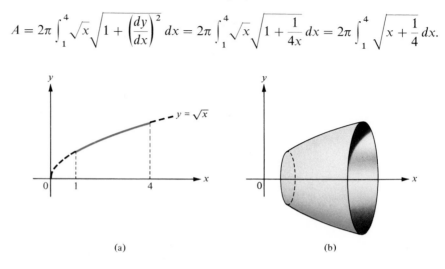

(a)   (b)

We make the change of variable $u = x + \frac{1}{4}$, noting that $du = dx$, so

$$A = 2\pi \int_{x=1}^{x=4} \sqrt{x + \frac{1}{4}}\, dx = 2\pi \int_{u=5/4}^{u=17/4} \sqrt{u}\, du = 2\pi \left[\frac{2}{3} u^{3/2}\right]\Bigg|_{5/4}^{17/4}$$

$$= \frac{4\pi}{3} \left[\left(\frac{17}{4}\right)^{3/2} - \left(\frac{5}{4}\right)^{3/2}\right] \approx 30.85 \text{ square units.}$$

3 Find the surface area of a sphere of radius r.

SOLUTION

The sphere of radius r is generated by revolving the semicircle whose equation is $y = \sqrt{r^2 - x^2}$, $-r \leq x \leq r$, about the x axis (Figure 10). Here, $dy/dx = -x/\sqrt{r^2 - x^2}$ is not defined on the closed interval $[-r, r]$, but only on the open

Figure 10

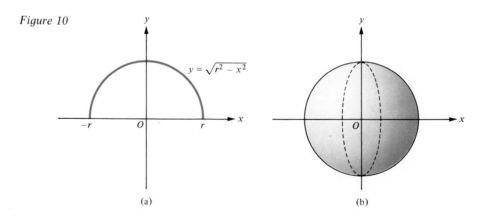

(a)   (b)

interval $(-r, r)$. Thus, we let ε be a small positive number, calculate the surface area of the portion of the sphere generated by revolving the part of the semicircle between $x = -r + \varepsilon$ and $x = r - \varepsilon$ about the x axis, and then take the limit as ε approaches zero. The desired area A is, therefore, given by

$$A = \lim_{\varepsilon \to 0^+} \int_{-r+\varepsilon}^{r-\varepsilon} 2\pi\sqrt{r^2 - x^2}\sqrt{1 + \left(\frac{dy}{dx}\right)^2}\, dx$$

$$= \lim_{\varepsilon \to 0^+} 2\pi \int_{-r+\varepsilon}^{r-\varepsilon} \sqrt{r^2 - x^2}\sqrt{1 + \frac{x^2}{r^2 - x^2}}\, dx$$

$$= \lim_{\varepsilon \to 0^+} 2\pi \int_{-r+\varepsilon}^{r-\varepsilon} r\, dx = \lim_{\varepsilon \to 0^+} 2\pi r x \Big|_{-r+\varepsilon}^{r-\varepsilon}$$

$$= \lim_{\varepsilon \to 0^+} 2\pi r[(r - \varepsilon) - (-r + \varepsilon)]$$

$$= \lim_{\varepsilon \to 0^+} 2\pi r(2r - 2\varepsilon) = 4\pi r^2.$$

This confirms the familiar formula $A = 4\pi r^2$ for the surface area of a sphere.

If the axis of revolution is the y axis, then the corresponding formula for the surface area of the surface of revolution is given by

$$A = \int_{y=c}^{y=d} 2\pi g(y)\, ds, \qquad \text{where } ds = \sqrt{1 + [g'(y)]^2}\, dy.$$

EXAMPLE   Find the area of the surface obtained by revolving the curve $x = \sqrt{y}$ between $y = 0$ and $y = 4$ about the y axis. Sketch the curve and the surface.

SOLUTION
The curve, part of a parabola, and the corresponding paraboloid of revolution are shown in Figure 11. Here,

$$A = 2\pi \int_{y=0}^{y=4} x\, ds = 2\pi \int_0^4 x\sqrt{1 + \left(\frac{dx}{dy}\right)^2}\, dy$$

$$= 2\pi \int_0^4 \sqrt{y}\sqrt{1 + \left(\frac{1}{2\sqrt{y}}\right)^2}\, dy = 2\pi \int_0^4 \sqrt{y + \frac{1}{4}}\, dy$$

$$= 2\pi \left[\frac{2}{3}\left(y + \frac{1}{4}\right)^{3/2}\right]\Big|_0^4 = \frac{4\pi}{3}\left[\left(\frac{17}{4}\right)^{3/2} - \left(\frac{1}{4}\right)^{3/2}\right]$$

$$\approx 36.18 \text{ square units.}$$

Figure 11

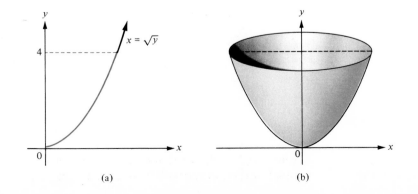

(a)   (b)

Problem Set 4

In problems 1 through 10, find the arc length of the graph of each equation between the indicated points.

1 $y = x^{3/2}$ from $(0,0)$ to $(4,8)$

2 $x = \frac{1}{4}y^{2/3}$ from $(0,0)$ to $(1,8)$

3 $y = mx + b$ from $(0,b)$ to $(a, ma + b)$

4 $y = \dfrac{x^3}{6} + \dfrac{1}{2x}$ from $(1, \frac{2}{3})$ to $(3, \frac{14}{3})$

5 $12xy = 4x^4 + 3$ from $(1, \frac{7}{12})$ to $(3, \frac{109}{12})$

6 $y = \displaystyle\int_4^x \sqrt{t - 1}\, dt$ from $(4,0)$ to $\left(9, \displaystyle\int_4^9 \sqrt{t - 1}\, dt\right)$

$\left(Hint\text{: Do not evaluate } \displaystyle\int_4^x \sqrt{t - 1}\, dt \text{—use the fundamental theorem of calculus.}\right)$

7 $y = \displaystyle\int_0^x t\sqrt{t^2 + 2}\, dt$ from $(0,0)$ to $\left(2, \displaystyle\int_0^2 t\sqrt{t^2 + 2}\, dt\right)$

8 $y = \displaystyle\int_1^x \sqrt{t^4 + t^2 - 1}\, dt$ from $(1,0)$ to $\left(3, \displaystyle\int_1^3 \sqrt{t^4 + t^2 - 1}\, dt\right)$

9 $x = \dfrac{y^3}{3} + \dfrac{1}{4y}$ from $(\frac{7}{12}, 1)$ to $(\frac{67}{24}, 2)$

10 $x = \dfrac{y^5}{5} + \dfrac{1}{12y^3}$ from $(\frac{17}{60}, 1)$ to $(\frac{3077}{480}, 2)$

In problems 11 through 13, set up an integral representing the arc length of each curve, and then use Simpson's parabolic rule with $n = 2$ (that is, with *four* subintervals) to estimate this arc length.

11 $y = 1/x$ from $(1, 1)$ to $(2, \frac{1}{2})$

12 $y = \displaystyle\int_1^x \dfrac{dt}{t}$ from $(1,0)$ to $\left(2, \displaystyle\int_1^2 \dfrac{dt}{t}\right)$

13 $y = x^3$ from $(1, 1)$ to $(2, 8)$

14 Let f be a function with a continuous first derivative and let Δs denote the arc length of the graph of f between the point $(a, f(a))$ and the point $(a + \Delta x, f(a + \Delta x))$. If Δl denotes the length of the straight line segment joining the point $(a, f(a))$ and the point $(a + \Delta x, f(a + \Delta x))$, prove that $\displaystyle\lim_{\Delta x \to 0} \dfrac{\Delta l}{\Delta s} = 1$.

15 Suppose that a particle P moves in the xy plane in such a way that, at time t, its x and y coordinates are given by $x = f(t)$ and $y = g(t)$, where f and g are functions with continuous first derivatives. Write an integral that gives the arc length of the path of P between the instant when $t = a$ and the instant when $t = b$, where $a < b$. $[Hint: (ds)^2 = (dx)^2 + (dy)^2.]$

16 Suppose that f is an invertible function with a continuous first derivative on the interval $[a,b]$. Make a *geometric* argument to show that the arc length of the graph of f between $(a, f(a))$ and $(b, f(b))$ is the same as the arc length of the graph of f^{-1} between $(f(a), a)$ and $(f(b), b)$.

17 Find a formula for the *total* surface area (lateral surface plus base) of a right circular cone of height h with base radius r.

18 (a) If the linear dimensions of a right circular cone are multiplied by a positive constant k, how is its total surface area affected? Why?
(b) Complete the following sentence: If the linear dimensions of a solid S are multiplied by a positive constant k, then the surface area of S is _____.

In problems 19 through 23, find the (lateral) surface area of the surface of revolution obtained by revolving each curve about the x axis.

19 $y = 3x + 2$ between $(0, 2)$ and $(3, 11)$

20 $y^2 = kx$, where k is a positive constant and $y \geq 0$, between $(a, \sqrt{ka})$ and $(b, \sqrt{kb})$, where $0 < a < b$

21 $y = x^3$ between $(0, 0)$ and $(2, 8)$

22 $y = \sqrt{2x - x^2}$ between $\left(\frac{1}{2}, \frac{\sqrt{3}}{2}\right)$ and $\left(\frac{3}{2}, \frac{\sqrt{3}}{2}\right)$

23 $y = \frac{\sqrt{2}}{4}x\sqrt{1 - x^2}$ between $(0, 0)$ and $\left(\frac{1}{2}, \frac{\sqrt{6}}{16}\right)$

In problems 24 through 27, find the (lateral) surface area of the surface of revolution obtained by revolving each curve about the y axis.

24 $y^2 = x^3$ between $(0, 0)$ and $(4, 8)$

25 $y = 4x^2$ between $(0, 0)$ and $(3, 36)$

26 $y^3 = 9x$ between $(0, 0)$ and $\left(\frac{8}{9}, 2\right)$

27 $8x = y^4 + 2/y^2$ between $\left(\frac{3}{8}, 1\right)$ and $\left(\frac{33}{16}, 2\right)$

28 Find the area of the surface generated when the graph of $y = (\sqrt{2}/4)x\sqrt{1 - x^2}$ between $x = 0$ and $x = 1$ is revolved about the x axis.

29 Use Simpson's parabolic rule with $n = 2$ to estimate the area of the surface of revolution generated by revolving the graph of $y = \frac{2}{3}\sqrt{9 - x^2}$ between $(0, 2)$ and $(2, \frac{2}{3}\sqrt{5})$ about the x axis.

30 Suppose that the function f is nonnegative and has a continuous derivative on $[a, b]$. Let S_0 denote the arc length of the graph of f between $(a, f(a))$ and $(b, f(b))$, and suppose that A_0 is the area of the surface of revolution generated when this graph is revolved about the x axis. Let k be a positive constant and define a function g by $g(x) = f(x) + k$. Express the area A of the surface generated by revolving the graph of g between $(a, g(a))$ and $(b, g(b))$ about the x axis in terms of S_0 and A_0.

31 One cubic centimeter of a certain substance is formed into n equal spheres.
(a) What is the total surface area of all these spheres?
(b) What is the limit as n grows larger and larger without bound of the total surface area in part (a)?
(c) Assuming that the rate at which this substance dissolves in a certain solvent is proportional to its surface area, explain why it dissolves more rapidly if it is first ground into a fine powder.

32 An artificial "organism" in the shape of a sphere with a semipermeable surface is to be suspended in a fluid having the same density, d grams per cubic centimeter, as the "organism." Nutrients diffuse through the surface from the fluid into the "organism" at the rate of k grams per second per square centimeter of surface and waste products diffuse into the surrounding fluid at the same rate. To sustain itself, the "organism" requires at least b grams of nutrients per second per gram of its own total weight. Find an upper bound R for the radius r of the "organism."

33 A sphere of radius r is inscribed in a right circular cylinder of radius r. Two planes perpendicular to the central axis of the cylinder cut off a spherical zone of area A on the surface of the sphere. Show that the same two planes cut off a region on the cylinder with the same surface area A.

5 Applications to Economics and the Life Sciences

We now present a few simple, but typical, applications of the definite integral to economics and the life sciences.

5.1 Consumer's Surplus

Suppose that the *demand equation* $x = g(y)$ expresses the number of units x of a certain commodity that are demanded when the market price is y dollars per unit. If this demand is met, the resulting *total revenue* R dollars to the producer (or producers) is given by $R = xy$. Economists often solve the equation $x = g(y)$ for y in terms of x, so the demand equation takes the equivalent form $y = f(x)$, where f is the inverse of the function g (Section 9 of Chapter 0).

Usually, fewer units of the commodity are demanded as the price per unit increases and, when the price reaches a sufficiently large value c dollars per unit, none of the commodity is demanded. Thus, we assume that $0 = g(c)$; that is, $c = f(0)$. Now suppose that the actual market price, y_0 dollars per unit, is less than c dollars per unit and that the corresponding demand, given by $x_0 = g(y_0)$, is positive. Clearly, those customers who are willing to pay more than y_0 dollars gain from the fact that the price is only y_0 dollars. Economists measure this gain, the *consumer's surplus*, by the integral

$$\text{consumer's surplus} = \int_{y_0}^{c} g(y)\, dy.$$

We now derive an alternative formula for consumer's surplus in terms of the function f. If we take the differential on both sides of the revenue equation $R = xy$, we obtain $dR = x\, dy + y\, dx$. It follows that

$$g(y)\, dy = x\, dy = dR - y\, dx = dR - f(x)\, dx.$$

Integrating both sides of the latter equation from $y = y_0$ to $y = c$, we obtain

$$\text{consumer's surplus} = \int_{y_0}^{c} g(y)\, dy = \int_{y=y_0}^{y=c} dR - \int_{y=y_0}^{y=c} f(x)\, dx$$

$$= R \Big|_{y=y_0}^{y=c} + \int_{y=c}^{y=y_0} f(x)\, dx = xy \Big|_{y=y_0}^{y=c} + \int_{x=0}^{x=x_0} f(x)\, dx$$

$$= (0c) - (x_0 y_0) + \int_{0}^{x_0} f(x)\, dx.$$

Therefore,

$$\text{consumer's surplus} = \int_{0}^{x_0} f(x)\, dx - x_0 y_0.$$

EXAMPLE   The demand equation for coal in a particular marketing area is given by $y = 50(600 - 10x - x^2)$, where x is the demand in thousands of tons and y is the price in dollars per thousand tons. If the current price is \$20,000 per thousand tons, find the consumer's surplus.

SOLUTION

Here $f(x) = 50(600 - 10x - x^2)$ and $y_0 = \$20{,}000$. Solving the quadratic equation $20{,}000 = 50(600 - 10x_0 - x_0^2)$ for x_0, and recalling that x_0 cannot be negative, we find that $x_0 = 10$. Therefore,

$$
\begin{aligned}
\text{consumer's surplus} &= \int_0^{x_0} f(x)\,dx - x_0 y_0 \\
&= \int_0^{10} 50(600 - 10x - x^2)\,dx - 10(20{,}000) \\
&= 50\left(600x - 5x^2 - \frac{x^3}{3}\right)\Bigg|_0^{10} - 200{,}000 \\
&= \frac{775{,}000}{3} - 200{,}000 \approx \$58{,}333.33.
\end{aligned}
$$

5.2   Production over a Period of Time

When a new production process is set in motion, the rate of production is often rather slow at first because of "bugs" in the production techniques, unfamiliarity with new methods, and so forth. As the "bugs" are worked out and the personnel become accustomed to the new methods, the rate of production ordinarily increases and, after a suitable period of time, approaches a steady value.

Thus, let x denote the number of units of a certain commodity produced in the first t units of working time by a new process, so the derivative dx/dt represents the rate of production at time t. Then, according to the fundamental theorem of calculus, the total number of units of the commodity produced during the interval of time from $t = a$ to $t = b$ is given by

$$
x\Bigg|_{t=a}^{t=b} = \int_{t=a}^{t=b} \frac{dx}{dt}\,dt.
$$

EXAMPLE   Solar Electric Company has set up a new production line to manufacture small wind-powered generators to be used as alternative energy sources. The rate of production t weeks after the start is given by $dx/dt = 300[1 - 400(t + 20)^{-2}]$ generators per week. How many generators are produced during the fifth week of operation?

SOLUTION

From the end of the fourth week, when $t = 4$, to the end of the fifth week, when $t = 5$, the number of generators produced is

$$
\int_4^5 300[1 - 400(t + 20)^{-2}]\,dt = 300\left(t + \frac{400}{t + 20}\right)\Bigg|_4^5 = 6300 - 6200
$$

$$
= 100 \text{ generators.}
$$

5.3   Pollution

The rate at which pollutants are introduced into a given ecosystem may vary with time as a consequence of a number of factors. For instance, the rate at which a factory dumps pollutants into a lake may increase as the level of production increases and as antipollution devices in the factory wear out and become less efficient. If we denote by x the number of units of pollutant accumulated in a given ecosystem after t units of time, the rate of pollution of the ecosystem is given by the derivative dx/dt. Consequently, the total number of units of pollutant accumulated in the ecosystem during the interval of time from $t = a$ to $t = b$ is given by

$$x \Big|_{t=a}^{t=b} = \int_{t=a}^{t=b} \frac{dx}{dt}\, dt.$$

EXAMPLE   Filters at the O.L. Factory, designed to remove sulfur dioxide from vented air, are replaced every 90 days. However, t days after the filters are changed, they allow sulfur dioxide to escape into the atmosphere at the rate of $25\sqrt{\dfrac{t}{10}}$ pounds per day. How many pounds of sulfur dioxide is introduced into the atmosphere over one 90-day interval?

SOLUTION
Making the change of variable $u = t/10$, we have

$$\int_{0}^{90} 25\sqrt{\frac{t}{10}}\, dt = 250 \int_{0}^{9} u^{1/2}\, du = 250\left(\frac{2}{3} u^{3/2}\right)\Big|_{0}^{9} = 4500 \text{ pounds.}$$

5.4   Blood Flow in the Circulatory System

Provided that certain factors (such as pressure and viscosity) are held within prescribed limits, blood will flow smoothly through a cylindrical blood vessel in such a way that the velocity v of flow increases continuously from a value close to zero at the wall of the vessel to a maximum value at its center. Figure 1 shows a cross section of the blood vessel perpendicular to its central axis and an "infinitesimal" circular ring of width dr at a distance r from the center. Assume that the velocity v of blood flow depends only on r; then the "infinitesimal" volume of blood dV flowing across the circular ring in unit time will be given by

Figure 1

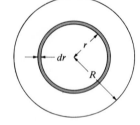

$$dV = v(2\pi r\, dr),$$

the product of the velocity of flow across the circular ring and the area of this ring. If R denotes the radius of the blood vessel, the total volume of blood passing the entire cross section in unit time will be given by

$$V = \int_{r=0}^{r=R} dV = 2\pi \int_{r=0}^{r=R} vr\, dr.$$

EXAMPLE   A cylindrical blood vessel has a radius $R = 0.25$ cm and blood is flowing through this vessel in such a way that its velocity r cm from the center is given by $v = 2.5 - 40r^2$ cm/sec. Find the rate of flow of blood through the vessel.

SOLUTION
The rate of flow, measured by the total volume of blood passing a cross section of the vessel in unit time, is given by

$$V = 2\pi \int_{r=0}^{r=R} vr\, dr = 2\pi \int_{0}^{0.25} (2.5 - 40r^2)r\, dr = 2\pi \int_{0}^{0.25} (2.5r - 40r^3)\, dr$$

$$= 2\pi \left(\frac{2.5r^2}{2} - \frac{40r^4}{4}\right)\Big|_{0}^{0.25} = \frac{5\pi}{64} \approx 0.25 \text{ cm}^3/\text{sec.}$$

Problem Set 5

1 Consider the demand equation $x = g(y)$, where $g(y) = 100(4 - \sqrt{4y})$.
   (a) Solve the equation $x = g(y)$ for y in terms of x, and thus rewrite the demand equation in the equivalent form $y = f(x)$.
   (b) Calculate the consumer's surplus when the price is $y_0 = \$1$ per unit using the definition of consumer's surplus as the integral $\int_{y_0}^{c} g(y)\,dy$.
   (c) Calculate the consumer's surplus using the alternative formula $\int_{0}^{x_0} f(x)\,dx - x_0 y_0$.

2 Sketch a graph showing an arbitrary demand curve $y = f(x)$ and a point (x_0, y_0) on the curve. Shade in an appropriate region on your figure whose area corresponds to the consumer's surplus when the price is y_0 dollars per unit.

3 On the basis of market surveys, a manufacturer determines that 5000 souvenirs will be sold in a certain resort area if the price is $1 per souvenir, but that, for every 5-cent increase in price per souvenir, 500 fewer souvenirs will be sold. Find the demand equation for the souvenirs and calculate the consumer's surplus if the souvenirs are priced at $1.25 each.

4 Suppose that the *supply equation* $x = g(y)$ gives the number of units of a certain commodity that manufacturers would be willing to put on the market at a selling price of y dollars per unit. Supposing that $0 = g(c)$ and that the current selling price is y_0 dollars per unit, economists define the *producer's surplus* to be the value of $\int_{c}^{y_0} g(y)\,dy$. If f is the inverse of g, find a formula for the producer's surplus in terms of f.

5 Heron Motors has set up an assembly line for their new steam-powered automobile and expects to be producing them at the rate of $30\sqrt{t}$ automobiles per week at the end of t weeks. How many automobiles do they expect to produce during the first 36 weeks of production?

6 Suppose that the rate of production of a new product is $A\left[1 - \left(\dfrac{k}{t + k}\right)^{p}\right]$ units per week at the end of t weeks, where A and k are positive constants and p is a constant greater than 1. Find a formula for the number of units produced during the nth week of production.

7 A factory is dumping pollutants into a lake at the rate of $t^{2/3}/600$ tons per week, where t is the time in weeks since the factory commenced operations. After 10 years of operation, how much pollutant has the factory dumped into the lake?

8 In problem 7, assume that natural processes can remove up to 0.015 ton of pollutant per week from the lake and that there was no pollution in the lake when the factory commenced operations 10 years ago. How many tons of pollutant have now accumulated in the lake?

9 A cylindrical blood vessel has a radius $R = 0.1$ cm and blood is flowing through this vessel with a velocity $v = 0.30 - 30r^2$ cm/sec at points r cm from the center. Find the rate of flow of the blood.

10 On the basis of hydrodynamical theory, as well as laboratory experiments, it is often assumed in medical research that blood, flowing smoothly through a cylindrical blood vessel of radius R, has a velocity given by $v = K(R^2 - r^2)$ at points r units from the center. (Here K is a constant depending on such factors as the viscosity of the blood and the blood pressure.) If this is so, find the rate of flow of the blood through the vessel.

6 Force, Work, and Energy

In Chapter 5 we used antidifferentiation to calculate the work done by a variable force and we related the notion of work to the concept of energy. Here we show how the definite integral can be used in calculating force, work, and energy.

6.1 Work Done by a Variable Force

In Section 4 of Chapter 5 we showed that if a particle P moves along the s axis under the influence of a possibly variable force F acting parallel to the s axis (Figure 1), the net work W done by F on P satisfies the differential equation $dW = F\, ds$. If a and b are coordinates of two points on the s axis, we can take the definite integral from $s = a$ to $s = b$ of both sides of the latter equation to obtain $\displaystyle\int_{s=a}^{s=b} dW = \int_{a}^{b} F\, ds$. Since $\displaystyle\int_{s=a}^{s=b} dW = W\Big|_{s=a}^{s=b}$ represents the difference between the value of W

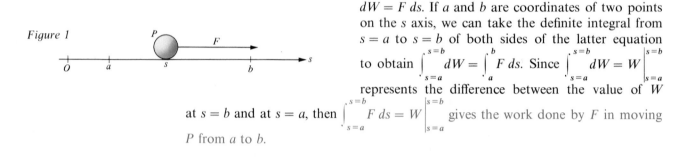

Figure 1

at $s = b$ and at $s = a$, then $\displaystyle\int_{s=a}^{s=b} F\, ds = W\Big|_{s=a}^{s=b}$ gives the work done by F in moving P from a to b.

EXAMPLES 1 A force F given by $F = s^3/3 + 1$ pounds acts on a particle P on the s axis and moves the particle from $s = 2$ feet to $s = 5$ feet. How much work is done?

SOLUTION
The work done is given by

$$\int_{2}^{5} F\, ds = \int_{2}^{5} \left(\frac{s^3}{3} + 1\right) ds = \left[\frac{s^4}{12} + s\right]\Big|_{2}^{5} = \frac{215}{4} = 53.75 \text{ foot-pounds.}$$

2 A spring has a natural length of 10 inches. If a force of 8 pounds is required to stretch the spring by 2 inches, how much work is done in stretching the spring by 6 inches?

SOLUTION
By Hooke's law, the force F on the spring is proportional to its displacement, $F = ks$. When $s = 2$, then $F = 8$; hence, $k = \frac{8}{2} = 4$ pounds per inch. Therefore, $F = 4s$, so the work done is given by

$$\int_{0}^{6} F\, ds = \int_{0}^{6} 4s\, ds = [2s^2]\Big|_{0}^{6} = 72 \text{ inch-pounds.}$$

6.2 Work Done in Pumping a Liquid

A container C contains a liquid weighing w units of force per cubic unit of volume (Figure 2). It is desired to pump some of this liquid up above the rim of the container to a certain height, and then discharge it. If we wish to calculate the work done by the pump, we establish a vertical s axis with its origin at the level up to which the liquid is to be pumped. For simplicity, let the positive direction on the s axis be downward.

Assume that the level of the liquid at the start of the pumping is $s = a$ and that its level at the end is $s = b$. Suppose that the cross-sectional area of the surface of the liquid at level s is $A(s)$ square units. The "infinitesimal" slab of liquid between level s and level $s + ds$ has a volume of $A(s)\,ds$ cubic units and weighs $wA(s)\,ds$ units of force (Figure 3). The "infinitesimal" work done in raising this slab through

Figure 2

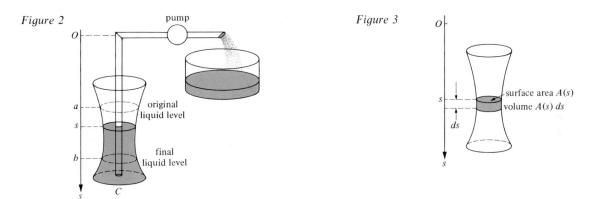

Figure 3

s units to the level of the origin is given by $dW = swA(s)\,ds$, so the total work required to pump the liquid from level $s = a$ to level $s = b$ is given by

$$\int_{s=a}^{s=b} dW = \int_a^b swA(s)\,ds = w\int_a^b sA(s)\,ds.$$

EXAMPLE   Water, which weighs 1000 kg/m³, fills a hemispherical reservoir of radius 5 meters. Water is pumped out of the reservoir up to a level of 6 meters above the rim until the surface of the remaining water is 4 meters below the rim of the reservoir. How much work is done?

SOLUTION

Figure 4

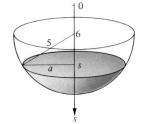

In Figure 4 we have chosen a vertical reference axis pointing straight down through the center of the reservoir with its origin 6 meters above the rim. When the upper surface of the water is at the position with coordinate s, its radius a satisfies $a^2 + (s - 6)^2 = 5^2$, by the Pythagorean theorem. Hence, the cross-sectional area $A(s)$ is given by $A(s) = \pi a^2 = \pi[25 - (s - 6)^2]$. At the start of the pumping, $s = 6$ meters, while at the end of the pumping, $s = 10$ meters. Hence, the work done is given by

$$w\int_6^{10} sA(s)\,ds = 1000\int_6^{10} s\pi[25 - (s - 6)^2]\,ds$$

$$= 1000\pi\int_6^{10}(-s^3 + 12s^2 - 11s)\,ds = 1000\pi\left[-\frac{s^4}{4} + 4s^3 - \frac{11s^2}{2}\right]\Bigg|_6^{10}$$

$$= 608{,}000\pi \text{ kilogram-meters.}$$

6.3 Compression or Expansion of a Gas

It requires work to compress a gas; conversely, when a gas expands, it does work upon its surroundings. Consider, for instance, a quantity of gas in a

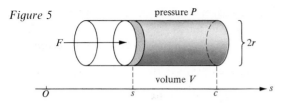

Figure 5

cylinder of radius r closed by a movable piston (Figure 5). Set up a reference axis parallel to the central axis of the cylinder, denote the coordinate of the piston by s, and let c be the coordinate of the end of the cylinder.

Denote the pressure of the gas by P units of force per unit area and let V represent the volume of the gas in cubic units. The force F on the piston is given by $F = \pi r^2 P$, the product of its cross-sectional area and the pressure of the gas. Also, $V = \pi r^2(c - s) = \pi r^2 c - \pi r^2 s$; hence, $dV = -\pi r^2\, ds$, or $ds = -(1/\pi r^2)\, dV$. The work done on the gas by F in moving the piston from s to $s + ds$ is given by $dW = F\, ds = [\pi r^2 P] \cdot [-(1/\pi r^2)\, dV] = -P\, dV$. Hence, if the gas is compressed from an initial volume V_0 to a final volume V_1, the total work done is given by

$$W = \int_{V=V_0}^{V=V_1} dW = \int_{V_0}^{V_1} (-P)\, dV \quad \text{or} \quad W = \int_{V_1}^{V_0} P\, dV.$$

Here, we have switched the limits of integration to remove the negative sign. Notice that, if the gas is being compressed, its volume decreases, so that $V_1 < V_0$ and $W = \int_{V_1}^{V_0} P\, dV$ is positive.

If the gas is expanding, it does just as much work on its surroundings as would be required to compress it back to its original volume. Hence, if a gas expands from an original volume V_0 to a final volume V_1, it does an amount of work given by $W = \int_{V_0}^{V_1} P\, dV$.

EXAMPLE   One cubic foot of air at an initial pressure of 50 pounds per square inch expands adiabatically (that is, without transfer of any heat energy) to a final volume of 3 cubic feet according to the adiabatic gas law $P = kV^{-1.4}$, where k is a constant. Find the work done.

SOLUTION
When $V = 1$, $P = 50$ lb/in.^{2} = $(50)(144)$ lb/ft^2. Therefore, $(50)(144) = k(1^{-1.4})$, so $k = (50)(144) = 7200$. Hence, $P = 7200V^{-1.4}$, $V_0 = 1$, $V_1 = 3$, and

$$W = \int_{V_0}^{V_1} P\, dV = \int_1^3 7200V^{-1.4}\, dV = \left[\frac{7200}{-0.4} V^{-0.4} \right]\Bigg|_1^3$$

$$= -18{,}000[3^{-0.4} - 1^{-0.4}] \approx 6400.9 \text{ foot-pounds.}$$

6.4 Energy

If a particle P of mass m is moving along the s axis as a consequence of a (possibly) variable unopposed force F, always acting in a direction parallel to the s axis, then the *kinetic energy* K of the particle has been defined in Section 4.4 of Chapter 5 by $K = \frac{1}{2}mv^2$, where $v = ds/dt$ is the velocity of P. This kinetic energy represents the capacity of the particle P to do work owing to its *motion*.

The *potential energy* of the particle P also represents the capacity of P to do work, not because it is moving, but because of where it is located (or how it got there). For instance, if the particle is connected to a spring and the spring is stretched, then work can be done by letting the spring return to its natural length. If the force F acting on P depends only on the position coordinate s of P, then the potential energy V of the particle at the point with coordinate s is given by $V = \int_{s}^{s_0} F \, ds$. Here, s_0 is understood to be the coordinate of an arbitrary, but fixed, reference point from which the potential energy is calculated.

Notice that $V = \int_{s}^{s_0} F \, ds$ is the work that F would do on P if P were moved from its position at s to the reference point at s_0. (Here we have not followed our usual convention of using different symbols for the dummy variable and the variable limit of integration.) We can also write $V = -\int_{s_0}^{s} F \, ds$, so that, by the fundamental theorem of calculus, $dV/ds = -F$, or $F = -dV/ds$.

The *total energy* E of the particle P is defined by $E = K + V$. As we saw in Chapter 5, $dE/dt = dK/dt + dV/dt = 0$, so E is a constant.

EXAMPLES **1** What velocity must a 10-gram projectile have in order to have the same kinetic energy as a 2000-kilogram car traveling at 24 km/hr?

SOLUTION
Let v be the velocity of the projectile in kilometers per hour. Setting the two kinetic energies equal, we have

$$\frac{1}{2}\left(\frac{10}{1000}\right)v^2 = \frac{1}{2}(2000)(24)^2,$$

so

$$v \approx 10{,}733.13 \text{ km/hr.}$$

Figure 6

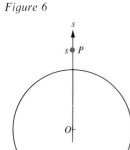

2 Establish an s axis, calibrated in centimeters, with origin at the center of the earth (Figure 6). A particle P of mass $m = 1$ gram on this axis at the point s experiences a gravitational force F given by $F = \dfrac{-3.99 \times 10^{20}}{s^2}$ dynes, $s \geq 6.38 \times 10^8$ centimeters (the radius of the earth). Choose the potential energy V of the particle so that $V = 0$ when $s = 6.38 \times 10^8$ centimeters.
(a) Find V in terms of s for $s \geq 6.38 \times 10^8$ centimeters.
(b) Find $\lim_{s \to +\infty} V$.
(c) With what initial velocity should P be shot upward from the surface of the earth so that it never returns? (Neglect air resistance.)

SOLUTION

(a) $V = -\displaystyle\int_{6.38 \times 10^8}^{s} F \, ds = -\int_{6.38 \times 10^8}^{s} \frac{-3.99 \times 10^{20}}{s^2}\, ds$

$$= \left[-\frac{3.99 \times 10^{20}}{s}\right]\Bigg|_{6.38 \times 10^8}^{s} = \left[-\frac{3.99 \times 10^{20}}{s}\right] - \left[-\frac{3.99 \times 10^{20}}{6.38 \times 10^8}\right]$$

$$\approx 6.25 \times 10^{11} - \frac{3.99 \times 10^{20}}{s} \text{ dyne-centimeters.}$$

(b) $\displaystyle\lim_{s \to +\infty} V \approx \lim_{s \to +\infty}\left[6.25 \times 10^{11} - \frac{3.99 \times 10^{20}}{s}\right] = 6.25 \times 10^{11}$ dyne-centimeters.

(c) Let v_0 be the required "escape velocity." The total energy E is given by $E = \frac{1}{2}mv^2 + V = \frac{1}{2}v^2 + V$. When $s = 6.38 \times 10^8$, $V = 0$, and $E = \frac{1}{2}v_0^2$. Since E is constant, $\frac{1}{2}v_0^2 = \frac{1}{2}v^2 + V$ holds at all times. As P moves away from the earth, its velocity v decreases; in fact, $\lim_{s \to +\infty} v = 0$. Taking the limit on both sides of the equation $\frac{1}{2}v_0^2 = \frac{1}{2}v^2 + V$ as s approaches $+\infty$, and using the result in part (b), we have $\frac{1}{2}v_0^2 \approx 0 + 6.25 \times 10^{11}$; hence,

$$v_0 \approx \sqrt{2 \times 6.25 \times 10^{11}} \approx 1.12 \times 10^6 \text{ cm/sec}$$

(approximately 7 miles/sec).

6.5 Force Caused by Fluid Pressure

If an admissible planar region R is exposed to a fluid under pressure, there is a resulting net force F on R. If the pressure of the fluid is constant over the region R, say with a value of P units of force per square unit of area, then $F = PA$, where A is the area of R. However, if P is not constant over R, integration is required to find F.

Here we consider a special case of the problem of calculating F, the situation in which the plane of the region R is vertical and R is submerged in an incompressible liquid of constant density. If such a liquid has w units of weight per cubic unit of its volume, then a column of this liquid h units high and with unit cross-sectional area has a volume of $h \cdot 1$ cubic units and a weight of $w \cdot h \cdot 1 = wh$ units (Figure 7). This weight causes a total force of wh units spread over the 1 square unit at the bottom of the column; hence, the pressure at the bottom of this column is given by $P = wh$. In words, then, the pressure at a point in an incompressible liquid of constant density is the product of its weight per unit volume and the distance of the point below the surface of the liquid.

Figure 7

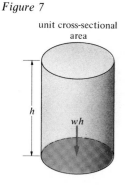

unit cross-sectional area

h

wh

Figure 8

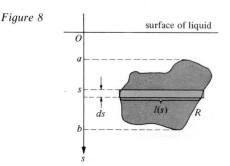

surface of liquid

O

a

s

ds

$l(s)$

R

b

s

Now, let R be an admissible planar region placed vertically beneath the surface of an incompressible liquid of constant weight per unit volume w (Figure 8). Establish a vertical s axis for reference, pointing downward, with its origin at the surface level of the liquid. Denote the total length of a horizontal cross section of R at level s units below the surface of the liquid by $l(s)$. The pressure P at depth s is given by ws units of force per square unit of area; hence, the "infinitesimal" force on the strip of height ds at depth s (Figure 8) is given by

$$dF = P\, dA = (ws)[l(s)\, ds] = wsl(s)\, ds.$$

Therefore, if R lies entirely between the horizontal lines $s = a$ and $s = b$, the total force on R is given by

$$F = \int_{s=a}^{s=b} dF = \int_a^b wsl(s)\, ds = w \int_a^b sl(s)\, ds.$$

EXAMPLE   Find the force exerted on the semicircular end of a trough if the trough is full of water and the semicircular end has a radius of 3 feet. Assume that the density of water is given by $w = 62.4$ pounds per cubic foot.

Figure 9

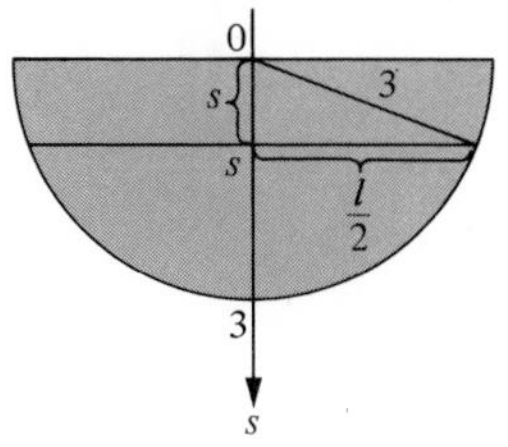

SOLUTION

From Figure 9 and the Pythagorean theorem, $(l/2)^2 + s^2 = 3^2$, or $l = 2\sqrt{9 - s^2}$. The desired force is given by

$$F = w \int_0^3 sl \, ds = 62.4 \int_0^3 2s\sqrt{9 - s^2} \, ds = 62.4\left[-\tfrac{2}{3}(9 - s^2)^{3/2}\right]\Big|_0^3$$

$$= (62.4)(18) = 1123.2 \text{ pounds.}$$

Problem Set 6

1 How much work is done in stretching a spring of natural length 12 inches from 15 inches to 18 inches if the final stretching force is 25 pounds?

2 A spring has a natural length of 30 centimeters. A force of 2000 kilograms is required to compress the spring by 3 centimeters. How much work is done in compressing the spring from its natural length to a length of 25 centimeters?

3 A 5-cent piece has a mass of about 5.2 grams.
(a) How many dynes does the 5-cent piece weigh?
(b) How many ergs of work are required to lift the 5-cent piece 1 meter high?

4 Is the work necessary to stretch a spring from 21 to 22 centimeters the same as the work necessary to stretch it from 22 to 23 centimeters? Explain.

5 Is the work necessary to lift a 5-cent piece from 21 to 22 centimeters above ground level the same as the work necessary to lift it from 22 to 23 centimeters? Explain.

6 An elastic spring with a spring constant of 17,640 dynes per centimeter is stretched 4 centimeters from its relaxed position. How much (potential) energy in ergs is stored in the stretched spring?

7 A reservoir is in the form of a hemisphere of radius 3 meters. If it is filled with salt water weighing 1032 kg/m^3, how much work (in kilogram-meters) is required to pump all the salt water out over the rim of the reservoir?

8 Water is pumped directly up from the surface of a lake into a water tower. The tank of this water tower is a vertical right circular cylinder of height 20 feet, with radius 5 feet, whose bottom is 60 feet above the surface of the lake. The pump is driven by a 1.5-horsepower motor. Neglecting friction, how long will it take to fill the tank with water? Assume that water weighs 62.4 lb/ft^3 and that 1 horsepower is 33,000 ft-lb/min.

9 A conical cistern is 6 meters across the top, 4.5 meters deep, and is filled to within 1.5 meters of the top with rainwater weighing 1 gm/cm^3. Find the work in joules done in pumping the water over the top to empty the tank. (One joule is 10^7 ergs.)

10 An elevator weighing 1 ton is lifted through 60 feet by winding its cable onto a winch. Neglecting friction, how much work is done if the cable weighs 10 lb/ft?

11 A water trough is 6 feet long and has a cross section consisting of an isosceles trapezoid with altitude 4 feet, upper base of 3 feet, and lower base of 2 feet. If the trough is filled with water weighing 62.4 lb/ft^3, how much work will be required to pump all this water up to a level 20 feet above the top of the trough?

12 An object immersed in water is buoyed up by a force equal to the weight of the water that the object displaces (Archimedes' principle). Find the work required to completely submerge a spherical float of negligible weight if its diameter is 30 centimeters. Assume that the water is in such a large reservoir that its level rises negligibly as we submerge the float.

13 The piston in a cylinder compresses a gas adiabatically from 50 cubic inches to 25 cubic inches. Assuming that $PV^{1.3} = 100$, where P is the pressure in pounds per square inch and V is the volume in cubic inches, how much work in *foot-pounds* is done on the gas?

14 Gas being compressed in a cylinder satisfies the equation $PV^\gamma = $ constant, where γ is a constant greater than 1, if the compression takes place adiabatically (that is, without heat passing into or out of the cylinder). If the compression takes place isothermally (that is, at constant temperature), then the gas satisfies the equation $PV = $ constant. If a gas is compressed from an initial volume V_0 to a final volume V_1, will more work be required for adiabatic or for isothermal compression?

15 Steam expands adiabatically according to the law $PV^{1.4} = $ constant. How much work is done if 0.5 cubic meter of steam at a pressure of 2000 kg/m² expands by 60 percent?

16 Two positively charged particles with charges of 1 statcoulomb each repel each other with a force equal to $1/r^2$ dynes, where r is the distance between the particles measured in centimeters. One particle is held fixed at the origin of the s axis and the other is moved directly toward the first from an initial point at $s_0 = 100$ centimeters. Find the work done in terms of the final position s_1 of the second particle.

17 The weight of a certain particle is $10^8/r^2$ pounds, where r is its distance in miles from the center of the earth. Find the work in *foot-pounds* required to lift the particle from the surface of the earth (where $r = 4000$ miles) to:
(a) 1000 miles above the surface.
(b) 10,000 miles above the surface.

18 Figure 10 shows an s axis with its origin at the center of a homogeneous solid sphere of radius r centimeters with constant density w grams per cubic centimeter. A particle P of mass m grams at the point with coordinate s on the axis will experience a gravitational force F given in dynes by

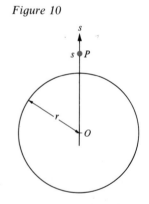

Figure 10

$$F = \begin{cases} \dfrac{-4\gamma m\pi r^3 w}{3s^2} & \text{for } s \geq r \\ -\tfrac{4}{3}\gamma m\pi ws & \text{for } 0 \leq s \leq r, \end{cases}$$

where $\gamma = 6.6732 \times 10^{-8}$ dyne cm²/g² is Newton's universal constant of gravitation.
(a) Find the potential energy V of P in terms of s if $\lim\limits_{s \to +\infty} V = 0$.
(b) Find the "escape velocity" for P starting on the surface of the sphere.

19 Given that the potential energy of a relaxed spring is zero, and that the spring constant is 40 lb/ft, how far must the spring be stretched so that its potential energy is 25 foot-pounds?

20 A 1-ton weight is dropped from a height of h feet and it hits the ground with the same kinetic energy as that possessed by a car weighing 1.5 tons and traveling 10 miles/hr. Find h.

21 A rectangular oil can is filled with oil weighing 0.93 gm/cm³. What is the force on one side of the can that is 20 cm wide and 40 cm high?

22 A water main in the shape of a horizontal cylinder 1.9 meters in radius is half filled with water. Find the force on the gate that closes the main.

23 Find the force on one face of a plank 18 feet long and 8 inches wide submerged vertically in water with its upper end at the surface.

24 What force will water exert on the lower half of a vertical ellipse whose semiaxes are 2 and 3 feet.
(a) When the major axis lies on the surface of the water?
(b) When the minor axis lies on the surface of the water?

25 What force must be withstood by a vertical dam 30 meters long and 6 meters deep?

26 A tank in the shape of a horizontal right circular cylinder of radius 5 centimeters is half filled with water and half filled with oil. The oil, which weighs 0.93 g/cm³, floats on top of the water, which weighs 1 g/cm³. Find the total force caused by these liquids on the circular end of the tank.

27 The face of a vertical dam has the shape of an isosceles trapezoid of altitude 16 feet with an upper base of 42 feet and a lower base of 30 feet. Find the total force exerted by the water on the dam when the water is 12 feet deep.

28 Show that, for a region R submerged as shown in Figure 11, the force F caused by a liquid weighing w units of weight per unit volume is given by $F = \frac{w}{2} \int_{a}^{b} [h(s)]^2\, ds.$

Figure 11

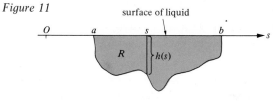

29 A horizontal tank in the shape of a right circular cylinder of radius 2 meters is sealed and water is forced in until it is exactly half full. The upper half of the tank is then subjected to an air pressure of one atmosphere, 10,333 kg/m². Find the total force on one end of the tank caused by air and water pressure.

Review Problem Set

In problems 1 through 8, find the volume of the solid generated by revolving the region bounded by the given curves about the indicated axis.

1 $y = \sqrt[3]{x}$, $y = 0$, and $x = 8$ about the x axis.

2 $y = x^2$, $y = 0$, and $x = 1$ about the y axis.

3 $x^2 + 4 = 4y$, $x = 0$, and $y = x$ about the x axis.

4 $y = x^3$, $x = 1$, and $y = 0$ about the line $x = -2$.

5 $y = x$, $x = 4$, and $y = 0$ about the line $y = -2$.

6 $x^2 = 4(1 - y)$ and $y = 0$ about the line $y = 3$.

7 $y^2 = 4x$ and $x = 4$ about the line $x = -2$.

8 The loop of $y^2 = x^4(x + 4)$ about the y axis.

9 Find the volume of the solid generated by revolving the region bounded by a circle of radius 3 centimeters about a line in its plane that is 7 centimeters from its center.

10 A solid is generated by revolving the region bounded by the graph of $y = f(x)$, the lines $x = 0$ and $x = a$, and the x axis about the x axis. Its volume for all values of a is given by $V = a^2 + 7a$. Find a formula for $f(x)$.

11 Show that the volume enclosed by the surface generated by revolving the semiellipse $x^2/a^2 + y^2/b^2 = 1$, $y \geq 0$, about the x axis is given by $V = \frac{4}{3}\pi ab^2$.

12 The region bounded by the y axis and the semiellipse $x^2/a^2 + y^2/b^2 = 1$, $x \geq 0$, is revolved about the y axis to generate a solid spheroid. Use the method of cylindrical shells to find the volume of this solid.

13 Find the volume generated when the region bounded by the hyperbola $16y^2 - 9x^2 = 144$ and the line $y = 6$ is revolved about the x axis.

14 Assume that f is a continuous invertible function on the closed interval $[a, b]$ and that $f(x) \geq 0$ for $a \leq x \leq b$. Prove that

$$\int_{a}^{b} [f(x)]^2\, dx = b[f(b)]^2 - a[f(a)]^2 - 2 \int_{f(a)}^{f(b)} y f^{-1}(y)\, dy.$$

(*Hint*: Find the volume of a certain solid of revolution in two different ways.)

15 A solid paraboloid of revolution is generated by revolving the region bounded by the graph of $y = b\sqrt{x/a}$, the x axis, and the line $x = a$ about the x axis.
(a) Sketch the paraboloid of revolution.
(b) Find the volume of the solid by the method of circular disks.
(c) Find the volume of the solid by the method of cylindrical shells.

16 A vessel has the form generated by revolving the graph of $y = \frac{1}{4}x^2$ about the y axis. If 4 cubic units of liquid leak out of the bottom of the vessel per minute, at what rate is the depth of liquid changing at the instant when the depth is 4 units?

17 Sketch the graph of the portion of the parabola $Ax^2 = B^2y$ lying in the first quadrant, where A and B are positive constants. Find the volume of the solid generated when the region bounded by this parabola, the y axis, and the line $y = A$ is revolved about the y axis. Compare the answer with the volume of a cylinder of height A and base radius B.

18 For each positive integer n, let U_n denote the volume of the solid generated by revolving the region under the graph of $y = x^n$ between $x = 0$ and $x = 1$ about the y axis. Let V_n be the volume of the solid generated by revolving the same region about the x axis. Evaluate:

(a) $\lim\limits_{n \to +\infty} U_n$
(b) $\lim\limits_{n \to +\infty} V_n$
(c) $\lim\limits_{n \to +\infty} \dfrac{U_n}{V_n}$

19 Find the volume of a solid whose cross section perpendicular to the x axis is a square x units on a side for $0 \le x \le 2$.

20 Find the volume of a solid whose cross section perpendicular to the x axis is an equilateral triangle $|x|$ units on a side for $-2 \le x \le 3$.

21 A spire is 26 feet high. A horizontal cross section x feet above the base is a rectangle whose long side is $\dfrac{26 - x}{13}$ feet long and whose short side is $\dfrac{26 - x}{20}$ feet long. Find the volume of the spire.

22 A *conoid* is a wedge-shaped solid whose lateral surface is generated by a straight line segment which moves so as always to be parallel to a fixed plane and which has one endpoint on a fixed circle and the other endpoint on a fixed straight line, both the circle and the straight line being perpendicular to the fixed plane (Figure 1). If the distance from the fixed straight line to the plane of the circle is h units and the circle has radius a units, find the volume enclosed by the conoid.

Figure 1

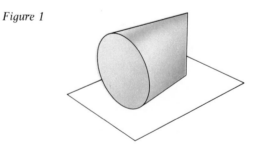

23 A solid is generated by a variable hexagon which moves so that its plane is always perpendicular to a given diameter of a fixed circle of radius a, the center of the hexagon lying in this diameter, and two opposite vertices lying on the circle. Find the volume of the solid.

24 Liquid in a reservoir evaporates at a rate (cubic units per unit time) proportional to the surface area of the liquid exposed to the air. Show that, no matter what the shape of the reservoir, the surface of the liquid drops at a constant rate.

25 A plane section passing through two opposite seams of a football is an ellipse whose equation is $x^2/49 + 4y^2/49 = 1$. Find the volume of the football if the leather is so stiff that every section perpendicular to the x axis is a square.

26 A horn-shaped solid is generated by a variable circle whose plane turns about a fixed line. The point of the circle nearest the line describes a quadrant AB of a circle of radius a, and the radius of the variable circle is $c\theta$, where θ denotes the angle between the variable plane and its initial position when it passes through A. Show that the volume of the horn is given by $V = \frac{1}{192}\pi^4 c^2 (8a + 3\pi c)$.

In problems 27 through 36, find the arc length of the graph of each equation between the points indicated.

27 $(y - 8)^2 = x^3$ from $(0, 8)$ to $(1, 9)$

28 $y^3 = 4x^2$ from $(4, 4)$ to $(32, 16)$

29 $y = \frac{2}{3}x^{3/2}$ from $(0, 0)$ to $(4, \frac{16}{3})$

30 $y = \frac{1}{8}[x^4 + (2/x^2)]$ from $(1, \frac{3}{8})$ to $(2, \frac{33}{16})$

31 $y = mx + b$ from $(x_0, mx_0 + b)$ to $(x_1, mx_1 + b)$

32 $x = \frac{y^5}{5} + \frac{1}{12y^3}$ from $(\frac{323}{480}, \frac{1}{2})$ to $(\frac{17}{60}, 1)$

33 $y = (x + 1)^{3/2} + 2$ from $(3, 10)$ to $(8, 29)$

34 $y = \int_0^x \sqrt{t^2 + 2t}\, dt$ from $(0, 0)$ to $\left(1, \int_0^1 \sqrt{t^2 + 2t}\, dt\right)$

35 $y = \int_1^x \sqrt{2t^4 + t^7 - 1}\, dt$ from $(1, 0)$ to $\left(2, \int_1^2 \sqrt{2t^2 + t^7 - 1}\, dt\right)$

36 $y = \sqrt{4 - x^2}$ from $(-2, 0)$ to $(2, 0)$

In problems 37 through 40, set up an integral representing the arc length of each curve, and use Simpson's parabolic rule with $n = 2$ (that is, with *four* subintervals) to estimate this arc length.

37 $y = 2\sqrt{x}$ from $(1, 2)$ to $(4, 4)$

38 $y = 2(1 + x^2)^{1/2}$ from $(0, 2)$ to $(1, 2\sqrt{2})$

39 $y = x^2$ from $(0, 0)$ to $(1, 1)$

40 $y = \int_0^x \frac{dt}{1 + t^2}$ from $(0, 0)$ to $\left(2, \int_0^2 \frac{dt}{1 + t^2}\right)$

41 Write an expression involving an integral for the total arc length of the ellipse $x^2/a^2 + y^2/b^2 = 1$, but do not attempt to evaluate the integral.

42 Suppose that the function f has a continuous first derivative on an open interval containing the closed interval $[0, 1]$ and that f is monotone decreasing on $[0, 1]$. Assume that $f(0) = 1$ and $f(1) = 0$. Write an integral for the amount of time T required for a particle P of mass m to slide frictionlessly down the graph of $y = f(x)$ from $(0, 1)$ to $(1, 0)$ under the influence of gravity.

In problems 43 through 52, find the (lateral) surface area of the surface of revolution obtained by revolving each curve about the axis given.

43 $y = 3\sqrt{x}$ from $(1, 3)$ to $(4, 6)$ about the x axis

44 $y = 3x^2$ from $(0, 0)$ to $(2, 12)$ about the y axis

45 $y = x^3$ from $(1, 1)$ to $(3, 27)$ about the x axis

46 $y = 4x^2$ from $(0, 0)$ to $(1, 4)$ about the y axis

47 $y = \frac{1}{3}x^3$ from $(0, 0)$ to $(3, 9)$ about the x axis

48 $y = x^4/4 + 1/(8x^2)$ from $(1, \frac{3}{8})$ to $(3, \frac{1459}{72})$ about the x axis

49 $y = \frac{1}{4}x^2$ from $(0, 0)$ to $(4, 4)$ about the y axis

50 $y^2 = x^3$ from $(1, 1)$ to $(4, 8)$ about the y axis

51 $y^2 = 9 - x$ from $(0, 3)$ to $(9, 0)$ about the x axis

52 $y = x^5/5 + 1/(12x^3)$ from $(1, \frac{17}{60})$ to $(2, \frac{3077}{480})$ about the x axis

53 Joe, Jamal, and Gus have set up a lemonade stand. The demand equation for lemonade at their stand is $y = -x^2/100 - 7x/20 + 30$, where y is their asking price in cents per glass and x is the number of glasses demanded. Calculate the consumer's surplus if the asking price is 15 cents per glass.

54 David, Dean, and Scott have grown 100 pumpkins, which they are planning to sell. They believe that they can sell all their pumpkins at a price of 75 cents each, but that each 25-cent increase in price per pumpkin will result in the sale of 20 fewer pumpkins. What is the consumer's surplus if they sell their pumpkins at the price that brings them the maximum revenue?

55 A former mathematics professor is going into business as an income tax consultant and expects to be receiving $100\sqrt{t}$ dollars per week in consulting fees t weeks after opening a new office. The fixed expenses for office rental, telephone, newspaper advertisements, and so forth, will amount to $200 per week. What is the consultant's net profit after 49 weeks in this new job?

56 A new employee at Solar Electric Company can solder $50 - 50/\sqrt{t+1}$ connections per hour at the end of t working hours. How many connections does the employee solder during the first 8 working hours?

57 Blood is flowing smoothly through a cylindrical artery of radius R in such a way that its velocity r units from the center is given by $v = K(R^2 - r^2)$ units per second. Here K is a constant depending on blood pressure, viscosity, and so forth. A drug is administered which increases the radius of the artery by 5 percent. Calculate the percentage increase in the rate of flow of blood through the artery.

58 The vertex of a conical tank points downward. If the tank is of height h units and the radius of its base is r units, how much work is done in pumping a liquid of density w mass units per unit volume over the rim of the tank if the tank is full at the start and empty at the end of the pumping?

59 Find the work done in pumping all the water out of a full cistern in the shape of a hemisphere of radius r feet surmounted by a right circular cylinder of radius r feet and height h feet.

60 A tank is made in the shape of a right circular cylinder surmounted by a frustum of a cone with a vertical cross section as shown in Figure 2. How much work is done in pumping the tank full of water *from the bottom*?

Figure 2

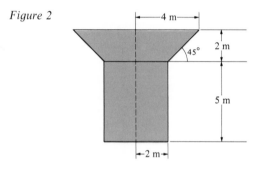

61 A cable that is 30 meters long and weighs 400 g/m hangs vertically from a winch. Find the work done in winding half of the cable onto the winch.

62 A cable that is l units long weighs w units of force per unit of length and hangs vertically from a winch. If a weight W hangs at the end of the cable, how much work is done in winding k units of the cable onto the winch?

63 How much work is done compressing 400 cubic feet of air at 15 pounds per square inch to a volume of 30 cubic feet, provided that the compression is adiabatic $(PV^{1.4} = \text{constant})$?

64 A mass m is falling freely near the earth's surface. At a certain instant, $t = 0$ seconds, the mass is s_0 units above the ground and is falling with a velocity v_0. Take the s axis positively upward with its origin at ground level. (Thus, v_0 is *negative* and the acceleration of gravity is $-g$.)
(a) Find the velocity of the mass in terms of its distance s above the ground.
(b) Find the increase in the kinetic energy of the mass between $s = s_0$ and $s = s_1$, where $0 \le s_1 \le s_0$.
(c) Find the potential energy V of the mass in terms of s if the potential energy is zero when $s = s_0$.
(d) Compute the decrease in the potential energy of the mass between $s = s_0$ and $s = s_1$, where $0 \le s_1 \le s_0$.
(e) Find the kinetic energy K and the potential energy V of the mass in terms of the time t.

65 The velocity of a 200-pound sled is $60 - 4t$ ft/sec at time t seconds. Find the change in its kinetic energy as well as the change in its potential energy between $t = 0$ and $t = 10$ seconds.

66 A rod of length b is spinning n times per second about one of its ends. The material of which the rod is composed has a density of $f(x)$ mass units per unit of length at a distance of x units from the stationary end. Find an integral representing the kinetic energy of the spinning rod.

67 A right triangle ABC is submerged in water so that edge AB lies on the surface and edge BC is vertical. Find the total force on the triangle caused by the pressure of the water if $|\overline{BC}| = 18$ feet and $|\overline{AB}| = 8$ feet.

68 A vertical masonry dam in the form of an isosceles trapezoid is 200 feet long at the surface of the water, 150 feet long at the bottom, and 60 feet high. Find the total force that it must withstand.

8

TRIGONOMETRIC FUNCTIONS AND THEIR INVERSES

Although trigonometric functions were introduced in Chapter 0 and used in a number of examples, we have used these functions sparingly in the previous chapters, mainly because we have not yet established their analytic properties. Thus, our purpose in the present chapter is to develop formulas for derivatives and integrals of the trigonometric functions and the inverse trigonometric functions so that these important functions can be used freely in subsequent chapters.

1 Limits and Continuity of the Trigonometric Functions

Recall that all angles are to be measured in radians unless we explicitly indicate otherwise. From Figure 1 it seems geometrically obvious that sine and cosine are continuous functions, since a small change in the angle t should produce only a small change in the position of the point $(\cos t, \sin t)$.

In this textbook, the very definition of the sine and cosine functions is geometric rather than analytic; hence, ultimately, all properties of the trigonometric functions must hinge on geometrical arguments. The property of continuity is no exception, and the argument above for the continuity of sine and cosine is probably as conclusive as any other argument that can be given at the level of rigor of this book. However, we give an alternative argument which uses the standard identities and the geometric fact that a chord of a circle is always shorter than the corresponding arc.

Figure 1

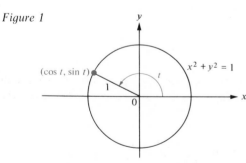

1.1 Continuity of the Trigonometric Functions

We begin by establishing the following theorem.

THEOREM 1 **Comparison of an Angle and Its Sine**
If $0 < t < \pi/2$, then $0 < \sin t < t$.

Figure 2

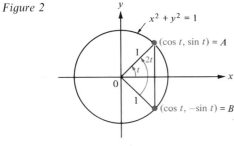

We submit the following geometric argument for the plausibility of Theorem 1. Consider the point A whose coordinates are $(\cos t, \sin t)$ in Figure 2. A lies in the first quadrant because $0 < t < \pi/2$. If $B = (\cos t, -\sin t)$, then the length of the chord $\overline{AB}$ is given by $|\overline{AB}| = 2 \sin t$. Since t is measured in radians and since the circle in Figure 2 has radius 1, the circular arc between A and B has length $2t$. The geometrically obvious fact that the chord $\overline{AB}$ has positive length and is shorter than the circular arc between A and B implies that $0 < 2 \sin t < 2t$, or $0 < \sin t < t$, as claimed by Theorem 1.

Since $\sin(-t) = -\sin t$, Theorem 1 implies that $0 < |\sin t| < |t|$ holds whenever $0 < |t| < \pi/2$. From this, it follows immediately that $\lim\limits_{t \to 0} \sin t = 0$. (Why?)

EXAMPLE Using the fact just established that $\lim\limits_{t \to 0} \sin t = 0$, prove that $\lim\limits_{t \to 0} \cos t = 1$.

SOLUTION
For $|t| \leq \pi/2$, $\cos t = \sqrt{1 - \sin^2 t}$. Hence,

$$\lim_{t \to 0} \cos t = \lim_{t \to 0} \sqrt{1 - \sin^2 t} = \sqrt{\lim_{t \to 0} (1 - \sin^2 t)}$$

$$= \sqrt{1 - \left(\lim_{t \to 0} \sin t\right)^2} = \sqrt{1 - 0^2} = \sqrt{1} = 1.$$

The facts that $\lim\limits_{t \to 0} \sin t = 0$ and $\lim\limits_{t \to 0} \cos t = 1$ can now be used to prove the following theorem.

THEOREM 2 **Continuity of the Trigonometric Functions**
All six trigonometric functions—sine, cosine, tangent, cotangent, secant, and cosecant—are continuous at each number in their domains.

PROOF
We begin by showing that the sine function is continuous. To do this, it is enough to prove that $\lim\limits_{\Delta t \to 0} \sin(t + \Delta t) = \sin t$. Using the addition formula for the sine function, we have

$$\lim_{\Delta t \to 0} \sin(t + \Delta t) = \lim_{\Delta t \to 0} [\sin t \cos \Delta t + \sin \Delta t \cos t]$$

$$= (\sin t)\left(\lim_{\Delta t \to 0} \cos \Delta t\right) + \left(\lim_{\Delta t \to 0} \sin \Delta t\right)(\cos t)$$

$$= (\sin t)(1) + (0)(\cos t) = \sin t.$$

Therefore, the sine function is continuous. To prove the continuity of the cosine function, we use the identity $\cos t = \sin (\pi/2 - t)$. Thus,

$$\lim_{\Delta t \to 0} \cos (t + \Delta t) = \lim_{\Delta t \to 0} \sin \left(\frac{\pi}{2} - t - \Delta t\right).$$

Since the sine function is continuous, we have

$$\lim_{\Delta t \to 0} \sin \left(\frac{\pi}{2} - t - \Delta t\right) = \sin \left[\lim_{\Delta t \to 0} \left(\frac{\pi}{2} - t - \Delta t\right)\right] = \sin \left(\frac{\pi}{2} - t\right) = \cos t.$$

It follows that $\lim_{\Delta t \to 0} \cos (t + \Delta t) = \cos t$; hence, the cosine is also a continuous function. Therefore, $\tan t = \sin t/\cos t$, $\cot t = \cos t/\sin t$, $\sec t = 1/\cos t$, and $\csc t = 1/\sin t$ are continuous at each number where they are defined.

1.2 Special Limits Involving Trigonometric Functions

The inequalities in the next theorem are useful for establishing certain special limits involving trigonometric functions.

THEOREM 3   **Fundamental Inequalities**

If $0 < t < \dfrac{\pi}{2}$, then $0 < \cos t < \dfrac{\sin t}{t} < \dfrac{1}{\cos t}$.

Again, we cannot give an analytic proof of this theorem because of our geometric definition of sine and cosine. However, we give the following geometric argument to make Theorem 3 plausible.

Since $0 < t < \pi/2$, the point $P = (\cos t, \sin t)$ lies in the first quadrant (Figure 3). In Figure 3, $\sin t = |\overline{PQ}|$, $\cos t = |\overline{OQ}|$ (why?), and, from the right triangle OAB,

Figure 3

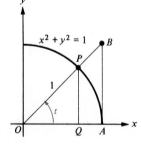

$$\tan t = \frac{|\overline{AB}|}{|\overline{OA}|} = \frac{|\overline{AB}|}{1} = |\overline{AB}|.$$

Since the circular sector AOP occupies a fraction $t/(2\pi)$ of the whole circle of radius 1, the area of the sector AOP is given by $(t/2\pi)(\pi \cdot 1^2) = t/2$. Evidently, the area $t/2$ of the circular sector AOP is larger than the area $\frac{1}{2}|\overline{PQ}||\overline{OQ}| = \frac{1}{2} \sin t \cos t$ of the triangle QOP, so $\frac{1}{2} \sin t \cos t < t/2$, or $\sin t \cos t < t$. Similarly, the area $t/2$ of the circular sector AOP is smaller than the area $\dfrac{1}{2}|\overline{AB}||\overline{OA}| = \dfrac{1}{2}(\tan t)(1) = \dfrac{1}{2} \cdot \dfrac{\sin t}{\cos t}$ of triangle OAB, so that $\dfrac{t}{2} < \dfrac{1}{2} \cdot \dfrac{\sin t}{\cos t}$, or $t < \dfrac{\sin t}{\cos t}$. Therefore, we have $0 < \sin t \cos t < t < \sin t/\cos t$. Dividing these inequalities by $\sin t$, we obtain $0 < \cos t < t/\sin t < 1/\cos t$; hence, when we take reciprocals, we obtain $0 < \cos t < (\sin t)/t < 1/\cos t$, as claimed by Theorem 3.

Now suppose that $0 < -t < \pi/2$. Then, by Theorem 3, we have $0 < \cos (-t) < \sin (-t)/(-t) < 1/\cos (-t)$. Since $\cos (-t) = \cos t$ and $\sin (-t) = -\sin t$, the latter inequalities can be written as $0 < \cos t < (\sin t)/t < 1/\cos t$. It follows that $0 < \cos t < (\sin t)/t < 1/\cos t$ holds whenever $0 < |t| < \pi/2$.

What happens to the ratio $(\sin t)/t$ as t approaches zero? Since this ratio is "squeezed" between the two quantities $\cos t$ and $1/\cos t$, and since $\lim_{t \to 0} \cos t = 1$ and

$$\lim_{t \to 0} \frac{1}{\cos t} = \frac{1}{\lim_{t \to 0} \cos t} = \frac{1}{1} = 1,$$

it follows that $(\sin t)/t$ must also approach 1 as t approaches zero. We record this important fact for future use in the next theorem.

THEOREM 4   **Limit of the Ratio $\dfrac{\sin t}{t}$**

$$\lim_{t \to 0} \frac{\sin t}{t} = 1.$$

Theorem 4 implies that $\sin t \approx t$ when $|t|$ is very small. For instance, consulting tables that give $\sin t$ rounded off to four decimal places, we see that $\sin 0.5 \approx 0.4794$, $\sin 0.1 \approx 0.0998$, $\sin 0.09 \approx 0.0899$, and $\sin 0.05 \approx 0.0500$. (Again, we stress that all angles are in *radians*.)

The following examples illustrate the use of Theorem 4 to calculate various special limits involving trigonometric functions.

EXAMPLES 1 Show that $\lim_{t \to 0} \dfrac{1 - \cos t}{t} = 0$.

SOLUTION

By the "half-angle formula," $\sin^2 \dfrac{t}{2} = \dfrac{1 - \cos t}{2}$; hence,

$$\frac{1 - \cos t}{t} = \frac{2 \sin^2 t/2}{t} = \frac{\sin^2 t/2}{t/2}.$$

Put $s = t/2$ and note that s approaches zero when t approaches zero. Thus,

$$\frac{1 - \cos t}{t} = \frac{\sin^2 s}{s},$$

so that

$$\lim_{t \to 0} \frac{1 - \cos t}{t} = \lim_{s \to 0} \frac{\sin^2 s}{s} = \lim_{s \to 0} \left[\sin s \, \frac{\sin s}{s} \right]$$

$$= \left[\lim_{s \to 0} \sin s \right] \left[\lim_{s \to 0} \frac{\sin s}{s} \right] = (0)(1) = 0.$$

2 Evaluate $\lim_{x \to 0} \dfrac{\sin 5x}{x}$.

SOLUTION

Put $t = 5x$ and note that t approaches zero when x approaches zero. Since $x = t/5$, it follows that

$$\lim_{x \to 0} \frac{\sin 5x}{x} = \lim_{t \to 0} \frac{\sin t}{t/5} = \lim_{t \to 0} \left[5 \, \frac{\sin t}{t} \right]$$

$$= 5 \lim_{t \to 0} \frac{\sin t}{t} = 5(1) = 5.$$

3 Evaluate $\displaystyle\lim_{x\to 0}\frac{\sin 7x}{\sin 9x}$.

SOLUTION

$$\lim_{x\to 0}\frac{\sin 7x}{\sin 9x}=\lim_{x\to 0}\frac{\dfrac{\sin 7x}{x}}{\dfrac{\sin 9x}{x}}=\lim_{x\to 0}\frac{7\left(\dfrac{\sin 7x}{7x}\right)}{9\left(\dfrac{\sin 9x}{9x}\right)}$$

$$=\frac{7}{9}\lim_{x\to 0}\frac{\dfrac{\sin 7x}{7x}}{\dfrac{\sin 9x}{9x}}=\frac{7}{9}\left[\frac{\displaystyle\lim_{x\to 0}\dfrac{\sin 7x}{7x}}{\displaystyle\lim_{x\to 0}\dfrac{\sin 9x}{9x}}\right].$$

Put $u=7x$ and $v=9x$. Then,

$$\lim_{x\to 0}\frac{\sin 7x}{\sin 9x}=\frac{7}{9}\left[\frac{\displaystyle\lim_{u\to 0}\dfrac{\sin u}{u}}{\displaystyle\lim_{v\to 0}\dfrac{\sin v}{v}}\right]=\frac{7}{9}\left(\frac{1}{1}\right)=\frac{7}{9}.$$

Problem Set 1

In problems 1 through 4, evaluate each limit.

1 $\displaystyle\lim_{x\to \pi/6}\frac{\cos x}{\sin x}$
2 $\displaystyle\lim_{x\to \pi}\frac{1}{\cos x}$
3 $\displaystyle\lim_{x\to \pi}\frac{\sin x}{x}$
4 $\displaystyle\lim_{x\to -\infty}\frac{\cos x}{x}$

In problems 5 through 12, determine (a) the domain of each function and (b) the points of this domain (if any) at which the function is discontinuous.

5 $f(x)=\tan x$
6 $f(x)=\sec \dfrac{x}{2}$
7 $g(x)=\csc x$

8 $h(x)=\cot x$
9 $f(x)=\dfrac{1-\sin x}{\cos x}$
10 $f(x)=\begin{cases}\dfrac{\sin 2x}{x} & \text{if } x\neq 0\\ 1 & \text{if } x=0\end{cases}$

11 $h(x)=\begin{cases}\tan x & \text{if } x\leq \pi/4\\ \sqrt{2}\,\sin x & \text{if } x>\pi/4\end{cases}$
12 $f(x)=\begin{cases}\sin \dfrac{1}{x} & \text{if } x\neq 0\\ 0 & \text{if } x=0\end{cases}$

In problems 13 through 24, evaluate each limit.

13 $\displaystyle\lim_{x\to 0}\frac{\sin 6x}{x}$
14 $\displaystyle\lim_{x\to 0}\frac{x}{\sin 3x}$
15 $\displaystyle\lim_{x\to 0}\frac{\sin 2x}{\sin 5x}$

16 $\displaystyle\lim_{t\to 0}\frac{1-\cos 2t}{\sin t}$
17 $\displaystyle\lim_{\theta\to 0}\frac{\sin^2 \theta}{\theta^2}$
18 $\displaystyle\lim_{x\to 0}\frac{\sin x-\cos x\sin x}{x^2}$

19 $\displaystyle\lim_{u\to 0}\frac{1-\cos^2 u}{u^2}$
20 $\displaystyle\lim_{x\to 3}\frac{x-3}{\sin (x-3)}$
21 $\displaystyle\lim_{x\to \pi}\frac{\cos x/2}{x/2-\pi/2}$

22 $\displaystyle\lim_{t\to 0}\frac{\tan 4t}{2t}$
23 $\displaystyle\lim_{\theta\to 0}\frac{\tan 2\theta}{\sin \theta}$
24 $\displaystyle\lim_{v\to \pi}\frac{1+\cos v}{(\pi-v)^2}$

25 Without using tables, estimate sin $1°$. (*Hint:* $1° = \pi/180$ radian.)

26 Let f be the function defined by $f(x) = \sin x$. Prove that $f'(0) = 1$.

27 (a) Show that $\lim\limits_{x \to 0} \dfrac{1 - \cos x}{x^2} = \dfrac{1}{2}$.

    (b) Use the result of part (a) to estimate $\cos 1°$ without using tables. (*Hint:* $1° = \pi/180$ radian.)

28 Use the results in problems 25 and 27 to estimate sin $31°$ without using tables. (*Hint:* $31° = 30° + 1° = \pi/6 + \pi/180$ radian.)

2   Derivatives of the Trigonometric Functions

In Section 1 we showed that $\lim\limits_{t \to 0} \dfrac{\sin t}{t} = 1$ and $\lim\limits_{t \to 0} \dfrac{1 - \cos t}{t} = 0$. These two special limits enable us to find the derivative of the sine function as in the following theorem.

THEOREM 1   **Derivative of the Sine Function**

The sine function is differentiable at each real number and the derivative of the sine function is the cosine function. In symbols,

$$D_x \sin x = \cos x \quad \text{or} \quad \frac{d}{dx} \sin x = \cos x.$$

PROOF

$D_x \sin x = \lim\limits_{\Delta x \to 0} \dfrac{\sin (x + \Delta x) - \sin x}{\Delta x}$. By the addition formula for sine, we have $\sin (x + \Delta x) = \sin x \cos \Delta x + \sin \Delta x \cos x$; hence,

$$D_x \sin x = \lim_{\Delta x \to 0} \frac{\sin x \cos \Delta x + \sin \Delta x \cos x - \sin x}{\Delta x}$$

$$= \lim_{\Delta x \to 0} \left[\frac{\sin x(\cos \Delta x - 1)}{\Delta x} + \frac{\sin \Delta x \cos x}{\Delta x} \right]$$

$$= \lim_{\Delta x \to 0} \left[(-\sin x)\frac{1 - \cos \Delta x}{\Delta x} + (\cos x) \frac{\sin \Delta x}{\Delta x} \right]$$

$$= (-\sin x) \lim_{\Delta x \to 0} \frac{1 - \cos \Delta x}{\Delta x} + (\cos x) \lim_{\Delta x \to 0} \frac{\sin \Delta x}{\Delta x}$$

$$= (-\sin x)(0) + (\cos x)(1) = \cos x.$$

More generally, if u is any differentiable function of x, then by Theorem 1 and the chain rule,

$$D_x \sin u = \cos u \, D_x u.$$

Using Theorem 1 together with the identity $\cos x = \sin (\pi/2 - x)$ and the chain rule, we now find the derivative of the cosine function.

THEOREM 2   **Derivative of the Cosine Function**

The cosine function is differentiable at each real number, and the derivative of the cosine function is the negative of the sine function. In symbols,

$$D_x \cos x = -\sin x \quad \text{or} \quad \frac{d}{dx} \cos x = -\sin x.$$

PROOF

$$D_x \cos x = D_x \sin (\pi/2 - x) = \cos (\pi/2 - x)D_x(\pi/2 - x) = (\sin x)(-1) = -\sin x.$$

Again, combining Theorem 2 with the chain rule, we have

$$D_x \cos u = -\sin u \, D_x u,$$

provided that u is a differentiable function of x.

EXAMPLES   Find the indicated derivative.

1 $D_x \sin (5x^2)$

SOLUTION

$$D_x \sin (5x^2) = \cos (5x^2)D_x(5x^2) = [\cos (5x^2)](10x) = 10x \cos (5x^2).$$

2 dy/dx, where $y = \cos \sqrt{4x^2 + 3}$.

SOLUTION

$$\frac{dy}{dx} = -[\sin \sqrt{4x^2 + 3}] \cdot \frac{d}{dx} \sqrt{4x^2 + 3}$$

$$= -[\sin \sqrt{4x^2 + 3}] \cdot \frac{8x}{2\sqrt{4x^2 + 3}} = \frac{-4x \sin \sqrt{4x^2 + 3}}{\sqrt{4x^2 + 3}}.$$

3 $f'(x)$, where $f(x) = 2 \sin^{3/2} (2x^2 + 7)$.

SOLUTION

$$f'(x) = D_x[2 \sin^{3/2} (2x^2 + 7)]$$
$$= D_x 2[\sin (2x^2 + 7)]^{3/2}$$
$$= (2)(\tfrac{3}{2})[\sin (2x^2 + 7)]^{(3/2)-1}D_x \sin (2x^2 + 7)$$
$$= 3[\sin (2x^2 + 7)]^{1/2}[\cos (2x^2 + 7)]D_x(2x^2 + 7)$$
$$= 3 \sin^{1/2} (2x^2 + 7)[\cos (2x^2 + 7)](4x)$$
$$= 12x \sin^{1/2} (2x^2 + 7) \cos (2x^2 + 7).$$

4 $D_x y$, where $x \cos y + y \sin x = 5$.

SOLUTION

By implicit differentiation,

$$\cos y + xD_x \cos y + (D_x y) \sin x + yD_x \sin x = 0,$$

or

$$\cos y - x \sin yD_x y + (D_x y) \sin x + y \cos x = 0,$$

so

$$D_x y = \frac{\cos y + y \cos x}{x \sin y - \sin x}.$$

The derivatives of the remaining trigonometric functions are now easily obtained using the derivatives of the sine and cosine functions together with the usual differentiation rules.

THEOREM 3   **Derivatives of Tan, Cot, Sec, and Csc**

(i) $D_x \tan x = \sec^2 x,$

(ii) $D_x \cot x = -\csc^2 x,$

(iii) $D_x \sec x = \sec x \tan x,$ and

(iv) $D_x \csc x = -\csc x \cot x$

hold for all values of x in the domains of the respective functions.

PROOF

We prove (i) and (iii), leaving (ii) and (iv) as exercises (Problem 64).

$$(i) \quad D_x \tan x = D_x \frac{\sin x}{\cos x} = \frac{\cos x D_x \sin x - \sin x D_x \cos x}{\cos^2 x}$$

$$= \frac{\cos x \cos x - \sin x(-\sin x)}{\cos^2 x} = \frac{\cos^2 x + \sin^2 x}{\cos^2 x}$$

$$= \frac{1}{\cos^2 x} = \sec^2 x.$$

$$(iii) \quad D_x \sec x = D_x \frac{1}{\cos x} = -\frac{1}{\cos^2 x} D_x \cos x = -\frac{1}{\cos^2 x}(-\sin x)$$

$$= \frac{1}{\cos x} \cdot \frac{\sin x}{\cos x} = \sec x \tan x.$$

The differentiation formulas established in Theorems 1, 2, and 3 can be incorporated with the chain rule to obtain the following:

(1) $D_x \sin u = \cos u D_x u,$

(2) $D_x \cos u = -\sin u D_x u,$

(3) $D_x \tan u = \sec^2 u D_x u,$

(4) $D_x \cot u = -\csc^2 u D_x u,$

(5) $D_x \sec u = \sec u \tan u D_x u,$ and

(6) $D_x \csc u = -\csc u \cot u D_x u,$

where u is a differentiable function of x.

EXAMPLES   **1** Find dy/dx if $y = \tan (5x^3 - 13).$

SOLUTION

$$\frac{dy}{dx} = \sec^2 (5x^3 - 13) \frac{d}{dx} (5x^3 - 13) = [\sec^2 (5x^3 - 13)] \cdot (15x^2)$$

$$= 15x^2 \sec^2 (5x^3 - 13).$$

2 Find $f'(t)$ if $f(t) = (3 + 2 \cot t)^4.$

SOLUTION

$$f'(t) = 4(3 + 2 \cot t)^3 D_t(3 + 2 \cot t)$$

$$= 4(3 + 2 \cot t)^3(-2 \csc^2 t) = -8 \csc^2 t (3 + 2 \cot t)^3.$$

3 Find $D_x y$ if $y = x^3 \sec^{5/3} x$.

SOLUTION

$$
\begin{aligned}
D_x y &= 3x^2 \sec^{5/3} x + x^3 (\tfrac{5}{3} \sec^{2/3} x D_x \sec x) \\
&= 3x^2 \sec^{5/3} x + \tfrac{5}{3} x^3 \sec^{2/3} x \sec x \tan x \\
&= x^2 \sec^{5/3} x \, (3 + \tfrac{5}{3} x \tan x).
\end{aligned}
$$

4 Find $D_x \left(\dfrac{\csc x}{x^2 + 1} \right)$.

SOLUTION

$$
\begin{aligned}
D_x \left(\frac{\csc x}{x^2 + 1} \right) &= \frac{(x^2 + 1)D_x(\csc x) - \csc x D_x(x^2 + 1)}{(x^2 + 1)^2} \\
&= \frac{-(x^2 + 1) \csc x \cot x - (\csc x)(2x)}{(x^2 + 1)^2} \\
&= -\csc x \left[\frac{(x^2 + 1) \cot x + 2x}{(x^2 + 1)^2} \right].
\end{aligned}
$$

5 Find dy/dx and $d^2 y/dx^2$ if $y = \tan^2 3x$.

SOLUTION

$dy/dx = 2 \tan 3x \sec^2 3x \cdot 3 = 6 \tan 3x \sec^2 3x$; hence,

$$
\begin{aligned}
\frac{d^2 y}{dx^2} &= 6 \left(\frac{d}{dx} \tan 3x \right) \sec^2 3x + 6 \tan 3x \left(\frac{d}{dx} \sec^2 3x \right) \\
&= 6(3 \sec^2 3x) \sec^2 3x + 6 \tan 3x \left(2 \sec 3x \frac{d}{dx} \sec 3x \right) \\
&= 18 \sec^4 3x + 12 \tan 3x \sec 3x(3 \sec 3x \tan 3x) \\
&= 18 \sec^4 3x + 36 \sec^2 3x \tan^2 3x.
\end{aligned}
$$

6 Let g be the function defined by the equation

$$
g(x) = \begin{cases} x^2 \sin \dfrac{1}{x} & \text{if } x \neq 0 \\ 0 & \text{if } x = 0. \end{cases}
$$

Find $g'(x)$.

SOLUTION
For $x \neq 0$,

$$
g'(x) = 2x \sin \frac{1}{x} + x^2 \cos \frac{1}{x} \left(-\frac{1}{x^2} \right) = 2x \sin \frac{1}{x} - \cos \frac{1}{x}.
$$

By definition,

$$
g'(0) = \lim_{\Delta x \to 0} \frac{g(0 + \Delta x) - g(0)}{\Delta x} = \lim_{\Delta x \to 0} \frac{g(\Delta x) - 0}{\Delta x}
$$

$$
= \lim_{\Delta x \to 0} \frac{(\Delta x)^2 \sin \dfrac{1}{\Delta x}}{\Delta x} = \lim_{\Delta x \to 0} \left(\Delta x \sin \frac{1}{\Delta x} \right).
$$

Since $\left|\sin \left(1/\Delta x\right)\right| \leq 1$, we have

$$0 < \left|\Delta x \sin \frac{1}{\Delta x}\right| = |\Delta x| \cdot \left|\sin \frac{1}{\Delta x}\right| \leq |\Delta x|;$$

hence, $\lim\limits_{\Delta x \to 0} \left[\Delta x \sin \left(1/\Delta x\right)\right] = 0$. It follows that $g'(0) = 0$.

Problem Set 2

In problems 1 through 36, differentiate each function.

1 $f(x) = 5 \sin 7x$

2 $f(x) = 8 \cos (3x + 5)$

3 $g(x) = 4 \sin 6x^2$

4 $g(t) = 3 \sin (5t^2 + t)$

5 $h(x) = \sin \sqrt{x}$

6 $H(s) = s^2 \sin s^3$

7 $g(t) = \sin^4 3t$

8 $g(x) = \cos^2 5x - \sin^2 5x$

9 $H(x) = \cos (\sin x)$

10 $f(t) = (1 - 2 \sin 3t)^{3/2}$

11 $f(x) = \sqrt{\cos 5x}$

12 $G(x) = \dfrac{4 - \cos 3x}{x^2}$

13 $H(x) = \dfrac{\sin x}{1 + \cos 5x}$

14 $g(x) = \dfrac{\sin x - x \cos x}{\cos x}$

15 $H(t) = \dfrac{27}{\sin 2t} + \dfrac{35}{\cos 2t}$

16 $g(r) = \tan 5r^4$

17 $g(t) = \cot (3t^5)$

18 $h(r) = \sec \sqrt[3]{r}$

19 $F(u) = \csc \sqrt{u^2 + 1}$

20 $g(s) = \cot \dfrac{7}{s}$

21 $h(x) = \sqrt{1 + \sec 5x}$

22 $g(t) = \tan \dfrac{t}{t + 2}$

23 $h(t) = \sec^2 7t - \tan^2 7t$

24 $g(x) = \csc^2 15x - \cot^2 15x$

25 $H(s) = \sec^4 13s - \tan^4 13s$

26 $g(x) = (\tan x + \sec x)^3$

27 $g(x) = x^3 \tan^5 2x$

28 $f(t) = \dfrac{\cot 3t}{t^2 + 1}$

29 $H(x) = \dfrac{2x}{1 + \sec 5x}$

30 $g(t) = \tan 3t \cot 3t$

31 $f(x) = \frac{1}{3}x^2 - \cot^3 2x$

32 $G(r) = \frac{3}{2}r^2 \csc^5 3r$

33 $g(t) = \dfrac{\sec^2 3t}{t^3}$

34 $f(\theta) = \left(\dfrac{\theta}{\tan \theta}\right)^3$

35 $f(x) = \sin (\tan 5x^2)$

36 $g(x) = \sec (\csc^2 7x)$

In problems 37 through 46, find dy/dx and d^2y/dx^2.

37 $y = 7 \cos 11x$

38 $y = -6 \sin (-2x + 5)$

39 $y = -4 \sec 5x$

40 $y = 2 \csc^2 7x$

41 $y = 5 \tan^3 4x$

42 $y = x \cot^2 3x$

43 $y = \dfrac{\csc 3x}{x}$

44 $y = \sqrt{1 + \sin 5x}$

45 $y = \sin \dfrac{x}{x + 1}$

46 $y = 3 \sec \dfrac{11}{x}$

47 Let f be the function defined by

$$f(x) = \begin{cases} x^4 \sin \dfrac{1}{x} & \text{if } x \neq 0 \\ 0 & \text{if } x = 0. \end{cases}$$

Find $f'(x)$.

48 Let f be the function defined in problem 47. Find $f''(x)$.

49 Does $\lim\limits_{x \to +\infty} \sin x$ exist? Why or why not?

50 Let f be a function such that $f(0) = f'(0) = 0$ and define the function g by the equation

$$g(x) = \begin{cases} f(x) \sin \dfrac{1}{x} & \text{for } x \neq 0 \\ 0 & \text{for } x = 0. \end{cases}$$

Find $g'(0)$.

In problems 51 through 56, use implicit differentiation to find $D_x y$.

51 $y = \sin(2x + y)$ **52** $x \cos y = (x + y)^2$ **53** $\tan xy + xy = 2$

54 $\tan^2 x + \tan^2 y = 4$ **55** $\sin^2 x + \cos^2 y = 1$ **56** $\sec(x + y) + \csc(x + y) = 5$

In problems 57 through 60, use the fundamental theorem of calculus, the chain rule, and the formulas for the derivatives of the trigonometric functions to find dy/dx.

57 $y = \displaystyle\int_1^{\sin x} \frac{dt}{9 + t^2}$ **58** $y = \displaystyle\int_3^{\cot 2x} \frac{dt}{1 + t^4}$ **59** $y = \displaystyle\int_0^{\sec x} (1 + t^2)^{300}\, dt$ **60** $y = \displaystyle\int_0^{\csc 5x} (5 + w^4)^{12}\, dw$

61 In differential form, $d \sin x = \cos x\, dx$. Write the expressions for the differentials of the remaining five trigonometric functions.

62 Use the identity $\sin(x + h) - \sin x = 2 \sin(h/2) \cos(x + h/2)$ to give an alternative proof for Theorem 1.

63 Show that $y = \sin x$ as well as $y = \cos x$ are solutions of the differential equation $d^2y/dx^2 + y = 0$.

64 Prove parts (ii) and (iv) of Theorem 3.

65 Show that $\dfrac{d^n \cos x}{dx^n} = \cos\left(x + \dfrac{n\pi}{2}\right)$ for any positive integer n.

66 Show that $\dfrac{d^n \sin x}{dx^n} = \sin\left(x + \dfrac{n\pi}{2}\right)$ for any positive integer n.

3 Applications of Derivatives of Trigonometric Functions

The differentiation formulas established in Section 2 are used in this section to help sketch graphs of functions, to solve maximum–minimum problems, and to solve related rate problems involving trigonometric functions.

EXAMPLES 1 Let $f(x) = \sin^2 x + 2 \cos x$ for $-\pi \leq x \leq \pi$. Find the maximum and minimum points on the graph of f, locate the intervals on which the graph is concave upward or downward, find the points of inflection, and sketch the graph.

SOLUTION

Here, $f'(x) = 2 \sin x \cos x - 2 \sin x = 2 \sin x(\cos x - 1)$. Using the fact that $\sin 2x = 2 \sin x \cos x$, we can also write $f'(x) = \sin 2x - 2 \sin x$; hence,

$$f''(x) = 2 \cos 2x - 2 \cos x = 2(\cos 2x - \cos x) = 2(2 \cos^2 x - 1 - \cos x)$$
$$= 2(2 \cos x + 1)(\cos x - 1).$$

(Because f is only defined on $[-\pi, \pi]$, the derivatives at $-\pi$ and at π must be regarded as being one-sided.)

Notice that $f''(x) = 0$ when $\cos x = -\frac{1}{2}$ and when $\cos x = 1$; that is (for $-\pi \le x \le \pi$), when $x = \pm 2\pi/3$ and when $x = 0$. Since f'' is continuous, it cannot change its algebraic sign on the intervals $[-\pi, -2\pi/3)$, $(-2\pi/3, 0)$, $(0, 2\pi/3)$, and $(2\pi/3, \pi]$. By evaluating f'' at (say) $-\pi$, $-\pi/2$, $\pi/2$, and π, respectively, we see that $f'' > 0$ on the first and fourth intervals, while $f'' < 0$ on the second and third intervals. Hence, the graph of f is concave upward on $[-\pi, -2\pi/3)$ and on $(2\pi/3, \pi]$, while it is concave downward on $(-2\pi/3, 0)$ and on $(0, 2\pi/3)$. Evidently, the only points of inflection are $(-2\pi/3, f(-2\pi/3))$ and $(2\pi/3, f(2\pi/3))$.

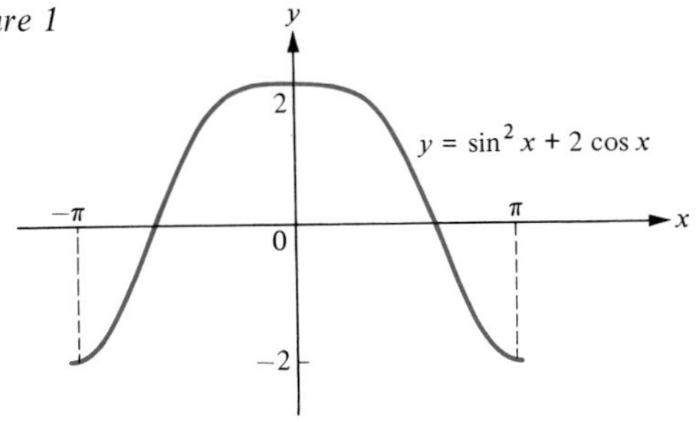

Now, $f'(x) = 0$ when $\sin x = 0$ and again when $\cos x = 1$, that is, when $x = \pm \pi$ or when $x = 0$. Since these are the only critical numbers and since $f(-\pi) = f(\pi) = -2 < 2 = f(0)$, it follows that f takes on its absolute minimum value at $x = \pm \pi$ and its absolute maximum value at $x = 0$. Note that $f(-x) = f(x)$, so the graph of f is symmetric about the y axis (Figure 1).

Figure 1

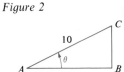

$y = \sin^2 x + 2\cos x$

2 Find the dimensions of the right triangle ABC with hypotenuse $|\overline{AC}| = 10$ inches if its area is a maximum.

SOLUTION

Figure 2

Let θ represent angle CAB (Figure 2), so that $\sin \theta = |\overline{BC}|/10$ and $\cos \theta = |\overline{AB}|/10$. If y denotes the area of the triangle, then

$$y = \frac{1}{2} |\overline{AB}| \cdot |\overline{BC}| = \frac{10^2}{2} \sin \theta \cos \theta = 50 \sin \theta \cos \theta.$$

Therefore,

$$\frac{dy}{d\theta} = 50 \cos \theta \cos \theta - 50 \sin \theta \sin \theta = 50(\cos^2 \theta - \sin^2 \theta).$$

When $\theta = \pi/4$, $\sin \theta = \cos \theta$, so $dy/d\theta = 0$ and the area is a maximum. Thus, we want $|\overline{AB}| = 10 \cos (\pi/4) = 5\sqrt{2}$ inches and $|\overline{BC}| = 10 \sin (\pi/4) = 5\sqrt{2}$ inches.

3 An airplane flies at a height of 9 kilometers in the direction of an observer on the ground at a speed of 800 km/hr. Find the rate of change of the angle of elevation of the plane from the observer at the instant when this angle is $\pi/3$ radians.

SOLUTION

Figure 3

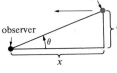

Denote the angle of elevation by θ and let x be the horizontal distance between the plane and the observer (Figure 3). Then $\cot \theta = x/9$ and $dx/dt = -800$ km/hr. Differentiating both sides of $\cot \theta = x/9$ with respect to t, we obtain

$$-\csc^2 \theta \frac{d\theta}{dt} = \frac{1}{9} \frac{dx}{dt} = -\frac{800}{9}.$$

At the instant when $\theta = \pi/3$, we have $\csc \theta = 2/\sqrt{3}$, so

$$-\frac{4}{3} \frac{d\theta}{dt} = -\frac{800}{9} \quad \text{or} \quad \frac{d\theta}{dt} = \frac{200}{3} \text{ radians per hour}$$

(about $63.66°$ per minute).

Figure 4

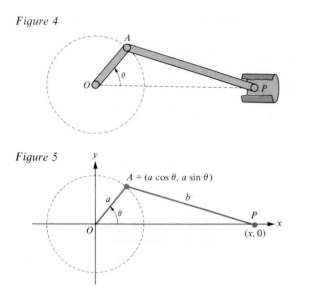

Figure 5

4 The crankshaft of an engine is turning at the constant rate of 25 revolutions/sec. Suppose that the arm $\overline{OA}$ is 2 inches long and the connecting rod $\overline{AP}$ is 8 inches long (Figure 4). At what rate is the piston P moving? What is this rate at the instant when $\theta = 3\pi/4$?

SOLUTION

Each revolution is 2π radians, so $d\theta/dt = (25)(2\pi) = 50\pi$ radians/sec. In Figure 5, we have set up an xy coordinate system with origin at O and with P at $(x, 0)$. For simplicity, let $a = |\overline{AO}| = 2$ and $b = |\overline{AP}| = 8$. We want to find dx/dt. By the distance formula,

$$b^2 = |\overline{AP}|^2 = (a\cos\theta - x)^2 + (a\sin\theta - 0)^2$$
$$= a^2\cos^2\theta - (2a\cos\theta)x + x^2 + a^2\sin^2\theta$$
$$= a^2(\cos^2\theta + \sin^2\theta) - (2a\cos\theta)x + x^2$$
$$= a^2(1) - (2a\cos\theta)x + x^2.$$

Therefore, $x^2 - (2a\cos\theta)x + (a^2 - b^2) = 0$. Using the quadratic formula to find x, we have

$$x = \frac{2a\cos\theta \pm \sqrt{4a^2\cos^2\theta - 4(a^2 - b^2)}}{2} = a\cos\theta \pm \sqrt{a^2\cos^2\theta - a^2 + b^2}$$

$$= a\cos\theta \pm \sqrt{b^2 - a^2(1 - \cos^2\theta)} = a\cos\theta \pm \sqrt{b^2 - a^2\sin^2\theta}.$$

Since P lies to the right of A, $x > a\cos\theta$; hence, we must use the plus sign in the solution above. Therefore, $x = a\cos\theta + \sqrt{b^2 - a^2\sin^2\theta}$. Differentiating the latter equation, we have

$$\frac{dx}{dt} = \left[-a\sin\theta - \frac{a^2\sin\theta\cos\theta}{\sqrt{b^2 - a^2\sin^2\theta}}\right]\frac{d\theta}{dt} = -a\sin\theta\left[1 + \frac{a\cos\theta}{\sqrt{b^2 - a^2\sin^2\theta}}\right]\frac{d\theta}{dt}.$$

Putting $a = 2$, $b = 8$, $\theta = 3\pi/4$, and $d\theta/dt = 50\pi$, we obtain

$$\frac{dx}{dt} = -2\left(\frac{\sqrt{2}}{2}\right)\left[1 + \frac{2(-\sqrt{2}/2)}{\sqrt{8^2 - (2)^2(\frac{1}{2})}}\right]50\pi$$

$$= -50\sqrt{2}\,\pi\left(1 - \sqrt{\frac{1}{31}}\right) \approx -182.25 \text{ inches per second.}$$

Problem Set 3

In problems 1 through 8, find the maximum and minimum points, locate the intervals in which the graph of the given function is concave upward or downward, find the points of inflection, and sketch the graph for $0 \le x \le 2\pi$.

1 $f(x) = \sin 2x$

2 $g(x) = \sin x + \cos x$

3 $f(x) = \frac{1}{3} + \frac{2}{3}\cos 2x$

4 $h(x) = \sin 2x + 2\cos x$

5 $g(x) = x + 2\cos(x/2)$

6 $f(x) = x - \tan x$

7 $f(x) = \sec x + \cos x$

8 $h(x) = 10\csc x - 5\cot x$

9 What is the altitude of the isosceles triangle with minimum area that can be circumscribed about a circle of radius 6 centimeters?

10 A steel ball bearing of radius a units is to fit into a hollow right circular cone so that the ball is barely but entirely inside the cone. Find the dimensions of the cone of smallest volume for which this is possible.

11 Find the dimensions of the right triangle ABC with hypotenuse $|\overline{AC}| = c$, a given constant, if its area is a maximum.

12 Figure 6 shows a first circle of radius a and part of a second circle of radius r whose center is on the first circle. Let l be the length of the arc of the second circle that lies inside the first circle. Show that l is a maximum when the angle θ satisfies the equation $\cot \theta = \theta$. (Here a is a constant, while θ and r can vary.)

Figure 6

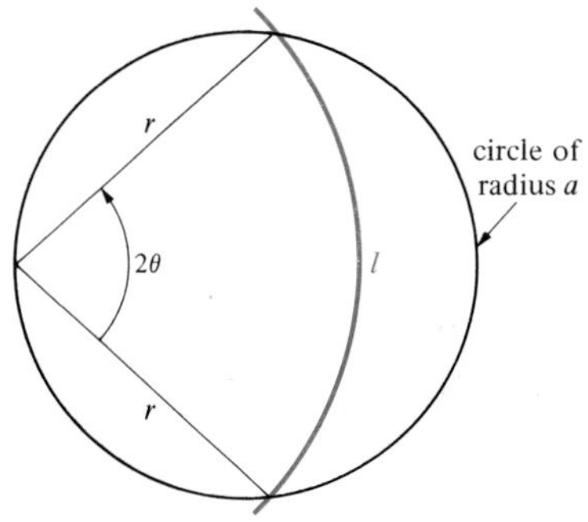

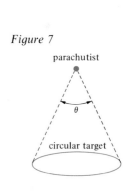

circle of radius a

13 The crankshaft of an engine is turning at the constant rate of $d\theta/dt$ radians/sec (Figure 4). Suppose that arm $\overline{OA}$ is a centimeters long and the connecting rod $\overline{AP}$ is b centimeters long. When $\theta = \pi/6$ radians, the piston is moving at r cm/sec. Find $d\theta/dt$ in terms of r, a, and b.

14 A spherical ball of radius r settles slowly into a full conical glass of water causing the water to overflow. The line segments connecting the vertex to the points on the upper rim each make an angle of θ radians with the central axis of the glass. If the conical glass has height a, find the value of r for which the amount of overflow is a maximum.

15 The range of a projectile is given by the formula $R = (v_0^2 \sin 2\theta)/g$, where v_0 is the muzzle velocity, g is the gravitational acceleration, and θ is the angle of elevation. Find the angle of elevation that gives the maximum range.

Figure 7

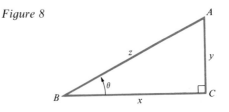

parachutist

circular target

16 A parachutist is descending at a constant (but unknown) velocity from an unknown height directly toward the center of a horizontal circular target of unknown radius (Figure 7). She carries a small optical device that measures the angle θ (in radians) subtended by the circular target and calculates the quantity T given by $T = \sin \theta/(d\theta/dt)$. Show that she will hit the target in T seconds.

17 An irrigation ditch is to have a cross section in the shape of an isosceles trapezoid whose bottom and sides are each 5 feet long. What should be the width across the top of the ditch if its carrying capacity is to be as large as possible?

In problems 18 through 22, consider the triangle ABC (Figure 8). Assume that θ is decreasing at the rate of $\frac{1}{30}$ radian/sec. Find the indicated rate of change using the given information.

Figure 8

18 dy/dt when $\theta = \pi/3$ given that x is a constant equal to 12 centimeters.

19 dz/dt when $\theta = \pi/4$ given that y is a constant equal to $10\sqrt{2}$ centimeters.

20 dx/dt when $y = 20$ centimeters given that z is a constant equal to 40 centimeters.

21 dz/dt when $y = x$ if x remains 1.6 kilometers at all times.

22 dz/dt when $x = 1$ foot and $z = 2$ feet if x and y are both changing and y is increasing at the rate of $\frac{2}{15}$ ft/sec.

23 A ladder 3 meters long leans against a house. The upper end slips down the wall at the rate of 1.5 m/sec. How fast is the ladder turning when it makes an angle of $\pi/6$ radian with the ground?

24 A weight is lifted by a rope which passes over a pulley and down to a truck which is moving at 5 ft/sec. If the pulley is 100 feet above the level of the truck, how fast is the weight rising when the rope from the pulley to the truck makes an angle of $\pi/4$ radian with the ground?

25 In an isosceles triangle, whose equal sides are each 3 centimeters long, the vertex angle increases at the rate of $\pi/90$ radian/sec. How fast is the area of the triangle increasing when the vertex angle is $\pi/3$ radians?

26 A man is walking along a straight sidewalk at the rate of 2 m/sec. A searchlight on the ground 12 meters from the sidewalk is kept trained on him. At what rate is the searchlight revolving when the man is 7 meters away from the point on the sidewalk nearest the light?

27 An airplane is flying horizontally at a speed of 400 miles per hour at an elevation of 10,000 feet toward an observer on the ground. Find the rate of change of the angle of elevation of the airplane from the observer when the angle is $\pi/4$.

28 The hands of a tower clock are 4.5 feet and 6 feet long. How fast are the tips of the hands approaching each other at four o'clock?

4　Integration of Trigonometric Functions

The formulas obtained in Section 2 for the derivatives of the trigonometric functions can be reversed to obtain the following formulas for indefinite integrals:

$$1 \quad \int \sin u \, du = -\cos u + C.$$

$$2 \quad \int \cos u \, du = \sin u + C.$$

$$3 \quad \int \sec^2 u \, du = \tan u + C.$$

$$4 \quad \int \csc^2 u \, du = -\cot u + C.$$

$$5 \quad \int \sec u \tan u \, du = \sec u + C.$$

$$6 \quad \int \csc u \cot u \, du = -\csc u + C.$$

Each of the formulas is easily verified by differentiation of the right-hand side. The following examples illustrate the use of these formulas, together with suitable substitutions, to evaluate integrals involving trigonometric functions.

EXAMPLES　Evaluate the given integral.

$$1 \quad \int \sin 9x \, dx$$

SOLUTION

Put $u = 9x$, so that $du = 9\,dx$ and $dx = \frac{1}{9}\,du$. Thus,

$$\int \sin 9x\,dx = \int \sin u\left(\frac{1}{9}\,du\right) = \frac{1}{9}\int \sin u\,du = -\frac{1}{9}\cos u + C = -\frac{\cos 9x}{9} + C.$$

2 $\displaystyle\int \frac{\cos\sqrt{x}}{\sqrt{x}}\,dx$

SOLUTION

Put $u = \sqrt{x}$, so that $du = dx/(2\sqrt{x})$, or $2\,du = dx/\sqrt{x}$. Thus,

$$\int \frac{\cos\sqrt{x}}{\sqrt{x}}\,dx = \int \cos u(2\,du) = 2\int \cos u\,du = 2\sin u + C = 2\sin\sqrt{x} + C.$$

3 $\displaystyle\int \sec 15x\,\tan 15x\,dx$

SOLUTION

Put $u = 15x$, so that $du = 15\,dx$ and $dx = du/15$. Thus,

$$\int \sec 15x\,\tan 15x\,dx = \int \sec u\,\tan u\,\frac{du}{15} = \frac{\sec u}{15} + C = \frac{\sec 15x}{15} + C.$$

4 $\displaystyle\int \cot 3x\,\csc^2 3x\,dx$

SOLUTION

Put $u = \cot 3x$, so that $du = -3\csc^2 3x\,dx$, or $\csc^2 3x\,dx = -\frac{1}{3}\,du$. Therefore,

$$\int \cot 3x\,\csc^2 3x\,dx = \int u\left(-\frac{1}{3}\,du\right) = -\frac{1}{3}\int u\,du$$

$$= -\frac{1}{3}\cdot\frac{u^2}{2} + C = -\frac{\cot^2 3x}{6} + C.$$

5 $\displaystyle\int \frac{\csc 7x\,\cot 7x\,dx}{(1 + \csc 7x)^4}$

SOLUTION

Put $u = 1 + \csc 7x$, so that $du = -7\csc 7x\,\cot 7x\,dx$, or $\csc 7x\,\cot 7x\,dx = -\frac{1}{7}\,du$. Thus,

$$\int \frac{\csc 7x\,\cot 7x\,dx}{(1 + \csc 7x)^4} = \int\left(-\frac{1}{7}\right)\frac{du}{u^4} = \frac{1}{21u^3} + C$$

$$= \frac{1}{21(1 + \csc 7x)^3} + C.$$

6 $\displaystyle\int x^2 \csc^2 5x^3\,dx$

SOLUTION

Let $u = 5x^3$, so that $du = 15x^2\,dx$, or $x^2\,dx = \frac{1}{15}\,du$. Therefore,

$$\int x^2 \csc^2 5x^3\,dx = \int \csc^2 u\left(\frac{1}{15}\,du\right) = \frac{1}{15}\int \csc^2 u\,du$$

$$= -\frac{1}{15}\cot u + C = -\frac{1}{15}\cot 5x^3 + C.$$

7 $\displaystyle\int_{\pi/36}^{\pi/9} \sec^2 (4x - \pi/9)\, dx$

SOLUTION
Put $u = 4x - \pi/9$, so that $du = 4\, dx$. Then, $u = 0$ when $x = \pi/36$, and $u = \pi/3$ when $x = \pi/9$, so that

$$\int_{\pi/36}^{\pi/9} \sec^2 \left(4x - \frac{\pi}{9}\right) dx = \int_0^{\pi/3} \sec^2 u\, \frac{du}{4} = \frac{\tan u}{4}\bigg|_0^{\pi/3}$$

$$= \frac{\tan \pi/3}{4} - \frac{\tan 0}{4} = \frac{\sqrt{3}}{4}.$$

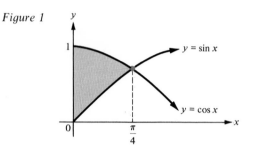

Figure 1

8 Find the area of the region between the graph of $y = \sin x$ and $y = \cos x$ on the interval $[0, \pi/4]$ (Figure 1).

SOLUTION
The area is given by

$$\int_0^{\pi/4} (\cos x - \sin x)\, dx = \int_0^{\pi/4} \cos x\, dx - \int_0^{\pi/4} \sin x\, dx$$

$$= [\sin x]\bigg|_0^{\pi/4} - [-\cos x]\bigg|_0^{\pi/4}$$

$$= \left[\sin \frac{\pi}{4} - \sin 0\right] - \left[-\cos \frac{\pi}{4} + \cos 0\right]$$

$$= \left[\frac{\sqrt{2}}{2} - 0\right] - \left[-\frac{\sqrt{2}}{2} + 1\right] = \sqrt{2} - 1 \approx 0.41 \text{ square unit.}$$

Problem Set 4

In problems 1 through 16, evaluate the integral.

1 $\displaystyle\int (2 \sin x + 3 \cos x)\, dx$
2 $\displaystyle\int (x + \sin 3x)\, dx$
3 $\displaystyle\int 2 \sin 35x\, dx$

4 $\displaystyle\int (7 \sin 5x + 3 \cos 7x)\, dx$
5 $\displaystyle\int 3 \cos (8x - 1)\, dx$
6 $\displaystyle\int 5 \cos (3x - 8)\, dx$

7 $\displaystyle\int \frac{dx}{\sin^2 3x}$ $\left(Hint: \dfrac{1}{\sin^2 u} = \csc^2 u.\right)$
8 $\displaystyle\int \frac{dx}{\cos^2 4x}$
9 $\displaystyle\int \sec^2 11x\, dx$

10 $\displaystyle\int -\csc^2 5x\, dx$
11 $\displaystyle\int \sec x(\tan x + \sec x)\, dx$
12 $\displaystyle\int \tan^2 3x\, dx$

13 $\displaystyle\int \sec (2x + 1) \tan (2x + 1)\, dx$
14 $\displaystyle\int \csc 10x \cot 10x\, dx$

15 $\displaystyle\int -\sec \frac{x}{5} \tan \frac{x}{5}\, dx$
16 $\displaystyle\int \csc 2x(\csc 2x + \cot 2x)\, dx$

In problems 17 through 28, evaluate the integral by using the given substitution or by finding an appropriate substitution for yourself.

17 $\int \cos x \cos (\sin x) \, dx; u = \sin x$

18 $\int \dfrac{\sin \sqrt{x + 1}}{\sqrt{x + 1}} \, dx; u = \sqrt{x + 1}$

19 $\int \dfrac{\sin x}{(2 + \cos x)^2} \, dx; u = 2 + \cos x$

20 $\int \cos 2x \sqrt{5 + \sin 2x} \, dx; u = 5 + \sin 2x$

21 $\int \dfrac{\sin x}{\cos^3 x} \, dx; u = \cos x$

22 $\int \dfrac{\cos 2x}{\sqrt[3]{\sin 2x}} \, dx; u = \sin 2x$

23 $\int \dfrac{\sec^2 x}{(3 + 2 \tan x)^3} \, dx$

24 $\int \cot^3 5x \csc^2 5x \, dx$

25 $\int \sec^2 3x \tan 3x \, dx$

26 $\int \csc^2 \dfrac{x}{2} \cot \dfrac{x}{2} \, dx$

27 $\int x^3 \sec 10x^4 \tan 10x^4 \, dx$

28 $\int \dfrac{\cot \sqrt{x} \csc \sqrt{x}}{\sqrt{x}} \, dx$

In problems 29 through 40, evaluate the definite integral.

29 $\int_0^{\pi/6} 2 \sin 3x \, dx$

30 $\int_0^{\pi/3} (2 + \cos 3x) \, dx$

31 $\int_0^1 \sec^2 \dfrac{\pi x}{4} \, dx$

32 $\int_{1/3}^{1/2} \csc^2 \pi x \, dx$

33 $\int_{\pi/2}^{2\pi/3} \sec^2 \dfrac{x}{2} \, dx$

34 $\int_0^1 \sec \dfrac{\pi x}{4} \tan \dfrac{\pi x}{4} \, dx$

35 $\int_{1/2}^1 \csc \dfrac{\pi x}{3} \cot \dfrac{\pi x}{3} \, dx$

36 $\int_{\pi/4}^{\pi/2} \csc^2 \left(\dfrac{x}{2} - \dfrac{\pi}{2} \right) dx$

37 $\int_0^{5\pi} \cos \dfrac{x + \pi}{2} \, dx$

38 $\int_{7\pi/12}^{11\pi/12} \dfrac{\cos 2x}{\sin^2 2x} \, dx$

39 $\int_0^{\pi/2} \sin^2 x \cos x \, dx$

40 $\int_0^{\pi/4} \cos^3 x \sin x \, dx$

In problems 41 through 44, find the area of the region bounded by the given curves.

41 One arch of $y = 3 \cos 2x$ and the x axis.

42 $y = \sin x$, $y = -3 \sin x$, $x = \pi/3$, and $x = \pi$.

43 $y = \tan x \sec^2 x$, the x axis, $x = \pi/6$, and $x = 0$.

44 $y = \tan^2 (\pi x/4)$, the x axis, and the lines $x = 0$ and $x = 1$.

45 Let R be the region in the first quadrant bounded by the curve $y = \sec (\pi x/2)$, the x axis, and the line $x = \frac{1}{2}$. Find the volume of the solid generated by revolving R about the x axis.

46 Evaluate $\displaystyle\int_{-\pi}^{\pi} \sin mx \cos nx \, dx$, where m and n are integers, by using the identity

$\sin a \cos b = \frac{1}{2} \sin (a + b) + \frac{1}{2} \sin (a - b)$.

47 (a) Evaluate $\int \sin x \cos x \, dx$ using the substitution $u = \sin x$.

(b) Evaluate $\int \sin x \cos x \, dx$ using the identity $\sin x \cos x = \frac{1}{2} \sin 2x$.

(c) Show that the answers to parts (a) and (b) are consistent with each other.

5   Inverse Trigonometric Functions

Recall from Chapter 0, Section 9, that the graph of the inverse f^{-1} of a function f is obtained by reflecting the graph of f across the line $y = x$

Figure 1

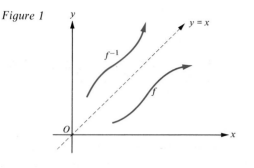

Figure 2

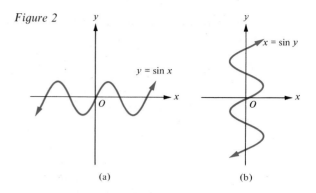

(a)   (b)

(Figure 1). However, in order for f to have an inverse function at all, the reflection of the graph of f across the line $y = x$ must not pass under itself.

The graph of the sine function forms a repeating pattern because the sine function is periodic (Figure 2a); hence, when this graph is reflected about the line $y = x$, the resulting graph repeatedly passes under itself (Figure 2b). Therefore, the sine function is not invertible.

Notice, however, that the sine function is monotone increasing on the interval $[-\pi/2, \pi/2]$ (Figure 3a); hence, *on this interval* it has an inverse (Figure 3b). The function whose graph appears in Figure 3b is called the "inverse sine function"

Figure 3

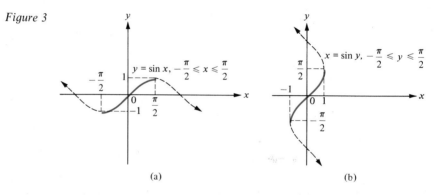

(a)   (b)

and is written $\sin^{-1}$. The terminology here is not quite correct since the sine function (defined on all of $\mathbb{R}$) has no inverse. Thus, $\sin^{-1}$ really is not the inverse of the sine function, but it is the inverse of the *portion* of the sine function whose graph lies between $x = -\pi/2$ and $x = \pi/2$ inclusive. Nearly everyone uses this terminology and notation, in spite of the fact that it could be misleading. Thus, we set forth the following definition.

DEFINITION 1   The Inverse Sine Function

Figure 4

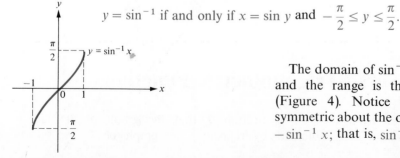

$y = \sin^{-1}$ if and only if $x = \sin y$ and $-\dfrac{\pi}{2} \leq y \leq \dfrac{\pi}{2}$.

The domain of $\sin^{-1}$ is the closed interval $[-1, 1]$ and the range is the closed interval $[-\pi/2, \pi/2]$ (Figure 4). Notice that the graph of $\sin^{-1}$ is symmetric about the origin. Therefore, $\sin^{-1}(-x) = -\sin^{-1} x$; that is, $\sin^{-1}$ is an odd function.

EXAMPLE   Find $\sin^{-1}\frac{1}{2}$ and $\sin^{-1}\left(-\sqrt{3}/2\right)$.

SOLUTION
To say that $y = \sin^{-1}\frac{1}{2}$ is to say that $\sin y = \frac{1}{2}$ and $-\pi/2 \le y \le \pi/2$. Since $\sin(\pi/6) = \frac{1}{2}$ and $-\pi/2 \le \pi/6 \le \pi/2$, it follows that $\pi/6 = \sin^{-1}\frac{1}{2}$. Similarly, $\sin^{-1}\left(-\sqrt{3}/2\right) = -\pi/3$.

Notice that, although $\sin^2 x$ means $(\sin x)^2$ and $\sin^{3/2} x$ means $(\sin x)^{3/2}$, $\sin^{-1} x$ *does not mean* $(\sin x)^{-1}$. In order to avoid any possible confusion between $\sin^{-1} x$ on the one hand, and $1/\sin x$ on the other, some people prefer to use the notation "arcsin x"—meaning "the arc (that is, the angle) whose sine is x"—rather than $\sin^{-1} x$.

The remaining five trigonometric functions—cosine, tangent, cotangent, secant, and cosecant—are also periodic; hence, they are not invertible on the interval $\mathbb{R}$. However, by restricting each of these functions to suitable intervals on which they are monotone, we can define corresponding inverses, just as we did for the sine function. Figure 5 (below and p. 400) shows the intervals usually selected for this purpose. (In the figure we have included the sine function for completeness.) The definitions of the six inverse trigonometric functions are as follows.

DEFINITION 2   **The Inverse Trigonometric Functions**

(i) $y = \sin^{-1} x$ if and only if $x = \sin y$ and $-\pi/2 \le y \le \pi/2$.

(ii) $y = \cos^{-1} x$ if and only if $x = \cos y$ and $0 \le y \le \pi$.

(iii) $y = \tan^{-1} x$ if and only if $x = \tan y$ and $-\pi/2 < y < \pi/2$.

(iv) $y = \cot^{-1} x$ if and only if $x = \cot y$ and $0 < y < \pi$.

Figure 5

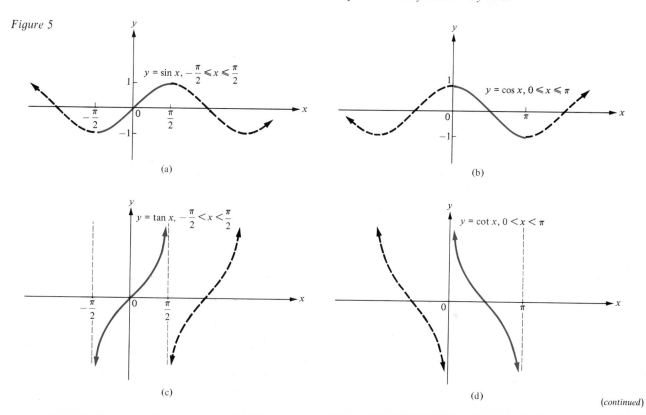

(a)

(b)

(c)

(d)

(continued)

Figure 5 (continued)

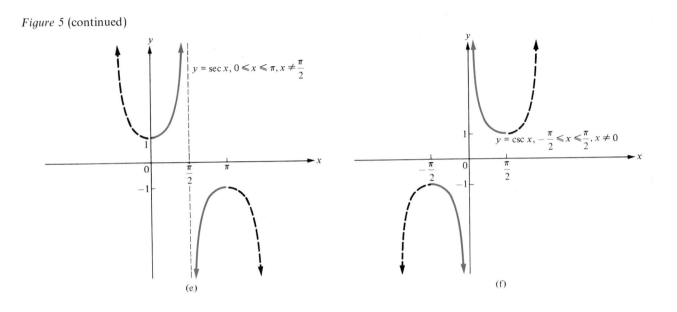

(e)   (f)

(v) $y = \sec^{-1} x$ if and only if $x = \sec y$, $y \neq \pi/2$, and $0 \leq y \leq \pi$.

(vi) $y = \csc^{-1} x$ if and only if $x = \csc y$, $y \neq 0$, and $-\pi/2 \leq y \leq \pi/2$.

In part (i) of Definition 2, we have repeated the definition of $\sin^{-1}$ for completeness. Just as $\sin^{-1} x$ is sometimes written as arcsin x, so also $\cos^{-1} x$ is sometimes written as arccos x, $\tan^{-1} x$ is sometimes written as arctan x, and so forth.

The graphs of the six inverse trigonometric functions are obtained by reflecting each of the graphs in Figure 5 about the line $y = x$. Figure 6 shows the resulting graphs and the following facts:

(1) The domain of $\sin^{-1}$ is $[-1, 1]$ and its range is $[-\pi/2, \pi/2]$.

(2) The domain of $\cos^{-1}$ is $[-1, 1]$ and its range is $[0, \pi]$.

(3) The domain of $\tan^{-1}$ is $\mathbb{R}$ and its range is $(-\pi/2, \pi/2)$.

(4) The domain of $\cot^{-1}$ is $\mathbb{R}$ and its range is $(0, \pi)$.

(5) The domain of $\sec^{-1}$ is $(-\infty, -1]$ together with $[1, \infty)$ and its range is $[0, \pi/2)$ together with $(\pi/2, \pi]$.

(6) The domain of $\csc^{-1}$ is $(-\infty, -1]$ together with $[1, \infty)$ and its range is $[-\pi/2, 0)$ together with $(0, \pi/2]$.

EXAMPLE   Find $\cos^{-1} \frac{1}{2}$, $\tan^{-1}(-\sqrt{3})$, $\cot^{-1} 1$, $\cot^{-1}(-\sqrt{3})$, $\sec^{-1} 2$, and $\csc^{-1}(-2\sqrt{3}/3)$.

SOLUTION
To say that $y = \cos^{-1} \frac{1}{2}$ is to say that $0 \leq y \leq \pi$ and $\cos y = \frac{1}{2}$. Evidently, $y = \pi/3$; hence, $\cos^{-1} \frac{1}{2} = \pi/3$. Similarly, $\tan^{-1}(-\sqrt{3}) = -\pi/3$, $\cot^{-1} 1 = \pi/4$, $\cot^{-1}(-\sqrt{3}) = 5\pi/6$, $\sec^{-1} 2 = \pi/3$, and $\csc^{-1}(-2\sqrt{3}/3) = -\pi/3$.

The following theorem is useful when dealing with inverse trigonometric functions.

Figure 6

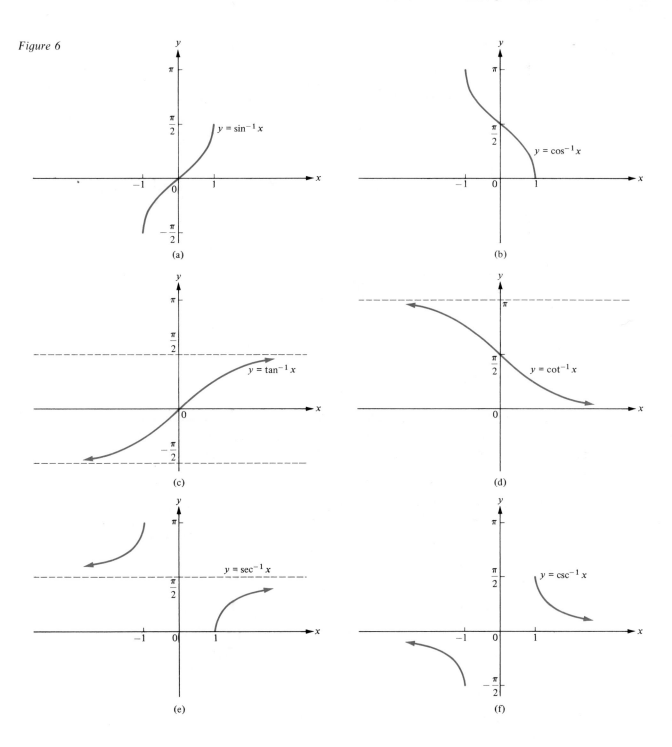

(a)

(b)

(c)

(d)

(e)

(f)

THEOREM 1   Basic Relations Between Inverse Trigonometric Functions

(i) $\cos^{-1} x = (\pi/2) - \sin^{-1} x$, $-1 \leq x \leq 1$.

(ii) $\cot^{-1} x = (\pi/2) - \tan^{-1} x$, for all x in $\mathbb{R}$.

(iii) $\sec^{-1} x = \cos^{-1} (1/x)$, $|x| \geq 1$.

(iv) $\csc^{-1} x = \sin^{-1} (1/x)$, $|x| \geq 1$.

PROOF

(i)   Suppose that $y = \cos^{-1} x$, so that $0 \leq y \leq \pi$ and $x = \cos y$. Since $0 \leq y \leq \pi$, we have $0 \geq -y \geq -\pi$; hence, $\pi/2 \geq (\pi/2) - y \geq -\pi/2$. Recall that $\cos y = \sin (\pi/2 - y)$. Therefore, $x = \sin (\pi/2 - y)$ with $-\pi/2 \leq (\pi/2) - y \leq \pi/2$; hence, $(\pi/2) - y = \sin^{-1} x$. It follows that $(\pi/2) - \cos^{-1} x = \sin^{-1} x$, or $\cos^{-1} x = (\pi/2) - \sin^{-1} x$. By a similar argument, we can prove (ii) (Problem 39).

(iii) Suppose that $y = \sec^{-1} x$, so that $0 \leq y \leq \pi$ and $\sec y = x$; that is, $1/\cos y = x$, or $\cos y = 1/x$. Since $0 \leq y \leq \pi$ and $\cos y = 1/x$, it follows that $y = \cos^{-1} (1/x)$; hence, $\sec^{-1} x = \cos^{-1} (1/x)$. By a similar argument, we can prove (iv) (Problem 40).

Since $\sin^{-1} x$ represents an angle between $-\pi/2$ and $\pi/2$, inclusive, whose sine is x, it follows that, for $-1 \leq x \leq 1$, $\sin (\sin^{-1} x) = x$. Also, for $-\pi/2 \leq y \leq \pi/2$, $\sin^{-1} (\sin y) = y$. Similarly, for $-1 \leq x \leq 1$, $\cos (\cos^{-1} x) = x$, while, for $0 \leq y \leq \pi$, $\cos^{-1} (\cos y) = y$. Analogous assertions can be made for $\tan^{-1}$, $\cot^{-1}$, $\sec^{-1}$, and $\csc^{-1}$ (Problem 37).

Further calculations with inverse trigonometric functions are illustrated in the following examples.

EXAMPLES 1 Find the exact value of $\sin [\tan^{-1} (-1)]$.

SOLUTION
We have $\tan^{-1} (-1) = -\pi/4$, so

$$\sin [\tan^{-1} (-1)] = \sin \left(-\frac{\pi}{4} \right) = -\frac{\sqrt{2}}{2}.$$

2 Prove geometrically that, for $-1 \leq x \leq 1$, $\cos (\sin^{-1} x) = \sqrt{1 - x^2}$.

SOLUTION
First suppose that $0 \leq x \leq 1$, so that $0 \leq \sin^{-1} x \leq \pi/2$. Let $\theta = \sin^{-1} x$, and construct a right triangle ABC with angle CAB equal to θ and with hypotenuse $\overline{AC}$ of length 1 unit (Figure 7). Thus,

Figure 7

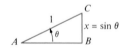

$$x = \sin \theta = \frac{|\overline{BC}|}{|\overline{AC}|} = |\overline{BC}|.$$

By the Pythagorean theorem, $|\overline{AB}|^2 + |\overline{BC}|^2 = |\overline{AC}|^2 = 1^2 = 1$, so that we have $|\overline{AB}|^2 = 1 - |\overline{BC}|^2 = 1 - x^2$, or $|\overline{AB}| = \sqrt{1 - x^2}$. Therefore,

$$\cos (\sin^{-1} x) = \cos \theta = \frac{|\overline{AB}|}{1} = \sqrt{1 - x^2}.$$

For $-1 \leq x < 0$, we have $0 < -x \leq 1$; hence,

$$\cos (\sin^{-1} x) = \cos \{ -[\sin^{-1} (-x)] \}$$
$$= \cos [\sin^{-1} (-x)] = \sqrt{1 - (-x)^2} = \sqrt{1 - x^2}.$$

Therefore, in any case, $\cos (\sin^{-1} x) = \sqrt{1 - x^2}$, provided that $-1 \leq x \leq 1$.

3 Simplify the expression $\cos (2 \sin^{-1} x)$, for $-1 \leq x \leq 1$.

SOLUTION
Let $y = \sin^{-1} x$. By the double-angle formula,

$$\cos (2 \sin^{-1} x) = \cos 2y = 2 \cos^2 y - 1 = 2[\cos (\sin^{-1} x)]^2 - 1.$$

Using the formula $\cos(\sin^{-1} x) = \sqrt{1 - x^2}$ from Example 2, we therefore have

$$\cos(2\sin^{-1} x) = 2(\sqrt{1 - x^2})^2 - 1 = 2(1 - x^2) - 1 = 1 - 2x^2.$$

4 Simplify $\tan(\tan^{-1} a + \tan^{-1} b)$, assuming that $ab \neq 1$.

SOLUTION

By the addition formula for tangent,

$$\tan(\tan^{-1} a + \tan^{-1} b) = \frac{\tan(\tan^{-1} a) + \tan(\tan^{-1} b)}{1 - \tan(\tan^{-1} a)\tan(\tan^{-1} b)} = \frac{a + b}{1 - ab}.$$

Problem Set 5

In problems 1 through 18, evaluate each expression.

1 $\sin^{-1} 1$

2 $\sin^{-1}\left(-\frac{\sqrt{3}}{2}\right)$

3 $\sin^{-1}\left(-\frac{\sqrt{2}}{2}\right)$

4 $\cos^{-1}\left(-\frac{1}{2}\right)$

5 $\cos^{-1} 0$

6 $\cos^{-1}\frac{\sqrt{3}}{2}$

7 $\cos^{-1} 1$

8 $\tan^{-1}(-1)$

9 $\tan^{-1}\left(-\frac{\sqrt{3}}{3}\right)$

10 $\tan^{-1}\sqrt{3}$

11 $\cot^{-1}(-1)$

12 $\cot^{-1}\sqrt{3}$

13 $\cot^{-1}\left(-\frac{\sqrt{3}}{3}\right)$

14 $\sec^{-1}\sqrt{2}$

15 $\sec^{-1}(-2)$

16 $\csc^{-1}\sqrt{2}$

17 $\csc^{-1} 2$

18 $\csc^{-1}(-\sqrt{2})$

In problems 19 through 28, evaluate each expression.

19 $\cos(\cos^{-1}\frac{3}{4})$

20 $\sin(\sin^{-1}\frac{2}{5})$

21 $\cos^{-1}\left(\cos\frac{\pi}{4}\right)$

22 $\tan^{-1}\left(\cot\frac{\pi}{4}\right)$

23 $\sec(\csc^{-1}\sqrt{2})$

24 $\sin^{-1}\left(\sin\frac{5\pi}{4}\right)$

25 $\cos^{-1}\left[\cos\left(-\frac{\pi}{3}\right)\right]$

26 $\cos[\tan^{-1}(-2)]$

27 $\tan[\sec^{-1}(-5)]$

28 $\csc[\cot^{-1}(-2)]$

29 Given that $y = \sin^{-1}\frac{2}{3}$, find the exact value of each of the following expressions: (a) $\cos y$, (b) $\tan y$, (c) $\cot y$, (d) $\sec y$, (e) $\csc y$.

30 Given that $y = \tan^{-1} x$, where $x \geq 0$, find the exact value of each of the following expressions: (a) $\sin y$, (b) $\cos y$, (c) $\cot y$, (d) $\sec y$, (e) $\csc y$.

In problems 31 through 36, simplify each expression.

31 $\sin(2\sin^{-1}\frac{4}{5})$

32 $\cos(2\sin^{-1} x)$, $-1 \leq x \leq 1$

33 $\tan[2\sec^{-1}(-\frac{5}{3})]$

34 $\sin(2\csc^{-1} x)$, $|x| \geq 1$

35 $\sin(\sin^{-1}\frac{3}{4} + \cos^{-1}\frac{1}{4})$

36 $\cos(\sin^{-1} x + \sin^{-1} y)$, $-1 \leq x \leq 1$, $-1 \leq y \leq 1$

37 Prove that:
(a) $\tan^{-1}(\tan y) = y$ for $-\pi/2 < y < \pi/2$.
(b) $\tan(\tan^{-1} x) = x$ for all values of x.
(c) $\cot^{-1}(\cot y) = y$ for $0 < y < \pi$.
(d) $\cot(\cot^{-1} x) = x$ for all values of x.
(e) $\sec^{-1}(\sec y) = y$ for $y \neq \pi/2$ and $0 \leq y \leq \pi$.
(f) $\sec(\sec^{-1} x) = x$ for $|x| \geq 1$.
(g) $\csc^{-1}(\csc y) = y$ for $y \neq 0$ and $-\pi/2 \leq y \leq \pi/2$.
(h) $\csc(\csc^{-1} x) = x$ for $|x| \geq 1$.

38 Prove that *all* solutions x of the equation $\sin x = y$, where $-1 \leq y \leq 1$, are given by $x = 2\pi n + \sin^{-1} y$ or $x = (2n + 1)\pi - \sin^{-1} y$, where $n = 0, \pm 1, \pm 2, \pm 3, \ldots$.

In problems 39 through 43, prove each assertion.

39 $\cot^{-1} x = \dfrac{\pi}{2} - \tan^{-1} x$

40 $\csc^{-1} x = \sin^{-1}\left(\dfrac{1}{x}\right), |x| \geq 1$

41 $\cos^{-1}(-x) = \pi - \cos^{-1} x$

42 $\tan^{-1}\left(-\dfrac{1}{x}\right) = \pi - \cot^{-1} x, x < 0$

43 $\tan\left(\frac{1}{2}\cos^{-1} x\right) = \sqrt{\dfrac{1 - x}{1 + x}}$

44 Find *all* solutions x of the equation $\tan x = y$ in terms of $\tan^{-1} y$.

45 Solve for x: $\tan^{-1}\frac{1}{3} + \tan^{-1}\frac{1}{2} = \sin^{-1} x$.

46 Solve for x: $\cot^{-1}\frac{1}{3} + \cot^{-1}\frac{1}{2} = \cos^{-1} x$.

47 Solve for x: $2\cot^{-1} x = \cot^{-1}\left(\dfrac{x^2 - 1}{2x}\right)$.

48 Use the graph of $y = \tan^{-1} x$ to decide whether $D_x(\tan^{-1} x)^2$ is positive or negative.

49 Make a geometric argument using graphs of $y = \tan^{-1} x$ and $y = \cot x$ to answer the following question: We know that $\tan^{-1} x$ does not mean $1/\tan x$, but is there any particular value of x for which the equation $\tan^{-1} x = 1/\tan x$ does hold?

50 Triangle ABC has a right angle at vertex C and side $\overline{AC}$ is a units long, where a is a constant. Let D denote the midpoint of side $\overline{CB}$ and define θ to be angle BDA. If $|\overline{BC}| = 2x$, express θ as a function of x.

Figure 8

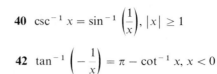

circle of radius r

$\longmapsto 2a \longmapsto$

51 A picture 0.6 meter high hangs on a wall so that its bottom is 0.3 meter above the eye level of an observer. If the observer stands x meters from the wall, show that the angle θ subtended by the picture is given by $\theta = \tan^{-1}(0.9/x) - \tan^{-1}(0.3/x)$.

52 Two parallel lines, each at a distance a from the center of a circle of radius r, cut off the shaded region R in Figure 8. Write a formula for the area of R.

6 Differentiation of Inverse Trigonometric Functions

The derivatives of the six inverse trigonometric functions are given by the following formulas, where u is a differentiable function of x.

$$1 \quad D_x \sin^{-1} u = \dfrac{D_x u}{\sqrt{1 - u^2}},$$

$$2 \quad D_x \cos^{-1} u = \dfrac{-D_x u}{\sqrt{1 - u^2}},$$

$$3 \quad D_x \tan^{-1} u = \dfrac{D_x u}{1 + u^2},$$

$$4 \quad D_x \cot^{-1} u = \dfrac{-D_x u}{1 + u^2},$$

$$5 \quad D_x \sec^{-1} u = \frac{D_x u}{|u|\sqrt{u^2 - 1}}, \text{ and}$$

$$6 \quad D_x \csc^{-1} u = \frac{-D_x u}{|u|\sqrt{u^2 - 1}}.$$

Rigorous proofs of these formulas are given in Section 6.1; however, they can be confirmed informally by implicit differentiation. For example, to confirm the formula for $D_x \sin^{-1} u$, suppose that u is a differentiable function of x and that $-1 < u < 1$. Put $y = \sin^{-1} u$, so that $\sin y = u$. Implicitly differentiating the latter equation with respect to x, we obtain $\cos y D_x y = D_x u$, or $D_x y = D_x u / \cos y$. Using the result of Example 2 in Section 5, we have

$$\cos y = \cos (\sin^{-1} u) = \sqrt{1 - u^2};$$

hence,

$$D_x \sin^{-1} u = D_x y = \frac{D_x u}{\cos y} = \frac{D_x u}{\sqrt{1 - u^2}}.$$

EXAMPLES   Differentiate the given function.

1 $f(x) = \sin^{-1} 2x$

SOLUTION

$$f'(x) = D_x \sin^{-1} 2x = \frac{D_x(2x)}{\sqrt{1 - (2x)^2}} = \frac{2}{\sqrt{1 - 4x^2}}.$$

2 $g(t) = \cos^{-1} t^4$

SOLUTION

$$g'(t) = D_t \cos^{-1} t^4 = \frac{-D_t t^4}{\sqrt{1 - (t^4)^2}} = \frac{-4t^3}{\sqrt{1 - t^8}}.$$

3 $F(x) = \tan^{-1} (x/5)$

SOLUTION

$$F'(x) = \frac{d}{dx} \tan^{-1} \frac{x}{5} = \frac{\frac{d}{dx}\left(\frac{x}{5}\right)}{1 + \left(\frac{x}{5}\right)^2} = \frac{\frac{1}{5}}{1 + \frac{x^2}{25}} = \frac{5}{25 + x^2}.$$

4 $G(t) = \cot^{-1} t^2$

SOLUTION

$$G'(t) = \frac{d}{dt} \cot^{-1} t^2 = \frac{-\frac{d}{dt}(t^2)}{1 + (t^2)^2} = \frac{-2t}{1 + t^4}.$$

5 $f(x) = \sec^{-1} (5x - 7)$

SOLUTION

$$f'(x) = D_x \sec^{-1} (5x - 7)$$

$$= \frac{D_x(5x - 7)}{|5x - 7|\sqrt{(5x - 7)^2 - 1}} = \frac{5}{|5x - 7|\sqrt{25x^2 - 70x + 48}}.$$

6 $h(t) = \csc^{-1} t^3$

SOLUTION

$$h'(t) = D_t \csc^{-1} t^3 = \frac{-D_t(t^3)}{|t^3|\sqrt{(t^3)^2 - 1}} = \frac{-3t^2}{|t|^3\sqrt{t^6 - 1}} = \frac{-3}{|t|\sqrt{t^6 - 1}}.$$

Further calculations involving derivatives of inverse trigonometric functions are illustrated by the following examples.

EXAMPLES 1 Given that $\tan^{-1} x + \tan^{-1} y = \pi/2$, find dy/dx.

SOLUTION
We use implicit differentiation. Differentiation of both sides of the given equation with respect to x yields

$$\frac{1}{1 + x^2} + \frac{dy/dx}{1 + y^2} = 0 \quad \text{or} \quad \frac{dy}{dx} = -\frac{1 + y^2}{1 + x^2}.$$

2 Let $f(x) = x \sin^{-1} x$ for $-1 < x < 1$. Find f' and f'' and sketch the graph of f.

SOLUTION
For $-1 < x < 1$ we have $f'(x) = \sin^{-1} x + x/\sqrt{1 - x^2}$ and

$$f''(x) = \frac{1}{\sqrt{1 - x^2}} + \frac{1}{(\sqrt{1 - x^2})^3} = \frac{2 - x^2}{(\sqrt{1 - x^2})^3}.$$

Evidently,

$$f(-x) = -x \sin^{-1}(-x) = -(-x \sin^{-1} x) = f(x);$$

Figure 1

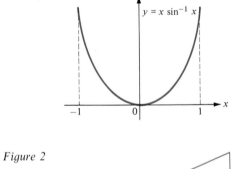

$y = x \sin^{-1} x$

hence, f is an even function. Also, $f'(0) = 0$ and $f''(0) = 2 > 0$, so that the graph of f shows a relative minimum at $(0,0)$. Since $f''(x) > 0$ for $-1 < x < 1$, the graph of f is concave upward (Figure 1).

Figure 2

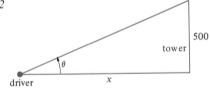

500

tower

θ

driver x

3 A high tower stands at the end of a level road. A truck driver approaches the tower at the rate of 50 miles/hr. The tower rises 500 feet above the level of the driver's eyes. How fast is the angle θ subtended by the tower increasing at the instant when the distance x between the driver and the base of the tower is 1200 feet (Figure 2)?

SOLUTION
Since $\tan \theta = 500/x$, then $\theta = \tan^{-1}(500/x)$. Therefore,

$$\frac{d\theta}{dt} = \frac{\dfrac{d}{dt}\left(\dfrac{500}{x}\right)}{1 + \left(\dfrac{500}{x}\right)^2} = \frac{-\left(\dfrac{500}{x^2}\right)\dfrac{dx}{dt}}{1 + \dfrac{500^2}{x^2}} = -\frac{500}{x^2 + 500^2}\frac{dx}{dt}.$$

Now, $dx/dt = -50$ miles/hr $= -(50 \times 5280)/3600$ ft/sec ≈ -73.33 ft/sec; hence,

$$\frac{d\theta}{dt} \approx -\frac{500}{1200^2 + 500^2}(-73.33) \approx 0.0217 \text{ radian/sec}.$$

6.1 Proofs of the Differentiation Formulas for the Inverse Trigonometric Functions

The proofs of Formulas 1 through 6 on pages 404–405 provide nice illustrations of the use of the inverse function theorem (Theorem 1, Section 5.1, Chapter 2).

THEOREM 1   **Derivatives of Sin^{-1} and Cos^{-1}**

If u is a function of x such that $-1 < u < 1$ and $D_x u$ exists, then

(i) $D_x \sin^{-1} u = \dfrac{D_x u}{\sqrt{1 - u^2}}.$

(ii) $D_x \cos^{-1} u = \dfrac{-D_x u}{\sqrt{1 - u^2}}.$

PROOF
Let f be the function defined by $f(x) = \sin x$ for $-\pi/2 < x < \pi/2$. Then $f'(x) = \cos x \neq 0$ for all values of x in the open interval $(-\pi/2, \pi/2)$. Therefore, by the inverse function theorem, f is invertible, f^{-1} is differentiable, and $(f^{-1})'(x) = 1/f'[f^{-1}(x)]$. But, $f^{-1} = \sin^{-1}$; hence, $D_x \sin^{-1} x = 1/\cos(\sin^{-1} x)$. Using the identity $\cos(\sin^{-1} x) = \sqrt{1 - x^2}$ (Example 2, Section 5, p. 402), we obtain $D_x \sin^{-1} x = 1/\sqrt{1 - x^2}$. Combining the latter equation with the chain rule, we obtain part (i) of the theorem. To prove part (ii), we use part (i) and calculate as follows:

$$D_x \cos^{-1} u = D_x\left(\frac{\pi}{2} - \sin^{-1} u\right) = -D_x \sin^{-1} u = \frac{-D_x u}{\sqrt{1 - u^2}}.$$

THEOREM 2   **Derivatives of Tan^{-1} and Cot^{-1}**

If u is a function of x such that $D_x u$ exists, then

(i) $D_x \tan^{-1} u = \dfrac{D_x u}{1 + u^2}.$

(ii) $D_x \cot^{-1} u = \dfrac{-D_x u}{1 + u^2}.$

PROOF
Let f be the function defined by $f(x) = \tan x$ for $-\pi/2 < x < \pi/2$. Then $f'(x) = \sec^2 x \neq 0$ for all values of x in the open interval $(-\pi/2, \pi/2)$. Therefore, by the inverse function theorem, f is invertible, f^{-1} is differentiable, and $(f^{-1})'(x) = 1/f'[f^{-1}(x)]$. But, $f^{-1} = \tan^{-1}$; hence, $D_x \tan^{-1} x = 1/\sec^2(\tan^{-1} x)$. From the identity $\sec^2 t = 1 + \tan^2 t$, we have $\sec^2(\tan^{-1} x) = 1 + \tan^2(\tan^{-1} x) = 1 + [\tan(\tan^{-1} x)]^2 = 1 + x^2$; hence, $D_x \tan^{-1} x = 1/(1 + x^2)$. Combining the latter equation with the chain rule, we obtain part (i) of the theorem. To prove part (ii), we use part (i) and calculate as follows:

$$D_x \cot^{-1} u = D_x\left(\frac{\pi}{2} - \tan^{-1} u\right) = -D_x \tan^{-1} u = \frac{-D_x u}{1 + u^2}.$$

THEOREM 3   **Derivatives of Sec^{-1} and Csc^{-1}**
If u is a function of x such that $|u| > 1$ and $D_x u$ exists, then

(i) $D_x \sec^{-1} u = \dfrac{D_x u}{|u|\sqrt{u^2 - 1}}.$

(ii) $D_x \csc^{-1} u = \dfrac{-D_x u}{|u|\sqrt{u^2 - 1}}.$

PROOF

We prove only part (i), leaving part (ii) as an exercise (Problem 22). Recall that $\sec^{-1} u = \cos^{-1}(1/u)$ [part (iii) of Theorem 1, Section 5, p. 401]; hence, by part (ii) of Theorem 1,

$$D_x \sec^{-1} u = D_x \cos^{-1}\left(\frac{1}{u}\right) = \frac{-D_x\left(\frac{1}{u}\right)}{\sqrt{1 - \left(\frac{1}{u}\right)^2}} = \frac{\left(\frac{1}{u^2}\right)D_x u}{\sqrt{1 - \frac{1}{u^2}}} = \frac{D_x u}{u^2\sqrt{\frac{u^2 - 1}{u^2}}}$$

$$= \frac{D_x u}{\frac{u^2}{\sqrt{u^2}}\sqrt{u^2 - 1}} = \frac{D_x u}{\frac{u^2}{|u|}\sqrt{u^2 - 1}} = \frac{D_x u}{|u|\sqrt{u^2 - 1}}.$$

Problem Set 6

In problems 1 through 21, differentiate each function.

1 $f(x) = \sin^{-1} 3x$

2 $g(x) = \cos^{-1} 7x$

3 $h(x) = \tan^{-1}\dfrac{x}{5}$

4 $H(x) = \cot^{-1}\dfrac{2x}{3}$

5 $G(t) = \sec^{-1} t^3$

6 $f(x) = \csc^{-1} x^2$

7 $f(t) = \cot^{-1}(t^2 + 3)$

8 $F(x) = \tan^{-1}\left(\dfrac{2x}{1 - x^2}\right)$

9 $g(x) = \csc^{-1}\dfrac{3}{2x}$

10 $f(r) = \tan^{-1}\left(\dfrac{r + 2}{1 - 2r}\right)$

11 $h(u) = \sec^{-1}\left(\dfrac{1}{\sqrt{1 - u^2}}\right)$

12 $f(x) = x\sqrt{4 - x^2} + 4\sin^{-1}\dfrac{x}{2}$

13 $f(s) = \sin^{-1}\dfrac{2}{s} + \cot^{-1}\dfrac{s}{2}$

14 $g(t) = t\cos^{-1} 2t - \tfrac{1}{2}\sqrt{1 - 4t^2}$

15 $G(r) = \sec^{-1} r + \csc^{-1} r$

16 $F(x) = \sec^{-1}\sqrt{x^2 + 9}$

17 $g(x) = x^2 \cos^{-1} 3x$

18 $h(t) = t(\sin^{-1} t)^3 - 3t$

19 $H(x) = \dfrac{1}{x^2}\tan^{-1}\dfrac{5}{x}$

20 $F(x) = \dfrac{\sec^{-1}\sqrt{x}}{x^2 + 1}$

21 $g(x) = \dfrac{\csc^{-1}(x^2 + 1)}{\sqrt{x^2 + 1}}$

22 Prove part (ii) of Theorem 3.

In problems 23 through 26, use implicit differentiation to find $D_x y$.

23 $x\sin^{-1} y = x + y$

24 $\cos^{-1} xy = \sin^{-1}(x + y)$

25 $\tan^{-1} x + \cot^{-1} y = \dfrac{\pi}{2}$

26 $\sec^{-1} x + \csc^{-1} y = \pi/2$

27 If $y = \int_0^{\sin^{-1} x} \dfrac{dt}{5 + t^4}$, find $\dfrac{dy}{dx}$.

28 If $y = \int_0^{\tan^{-1} 2x} (5 + u^2)^{20} \, du$, find $D_x y$.

29 If $y = \tan\left(2 \tan^{-1} \dfrac{x}{2}\right)$, show that $\dfrac{dy}{dx} = \dfrac{4(1 + y^2)}{4 + x^2}$.

30 Find $\dfrac{d}{dx}\left(\tan^{-1} \dfrac{2x}{\sqrt{1 - x^2}}\right)$.

31 Sketch the graph of $y = \sin^{-1}(\cos x) + \cos^{-1}(\sin x)$ in the interval $[0, 2\pi]$.

32 Find the acute angle between the tangents to the curves $y = \tan^{-1} x$ and $y = \cot^{-1} x$ at their point of intersection.

33 A picture 2.13 meters high is hung on a wall in such a way that its lower edge is 2.74 meters above the level of an observer's eye. How far from the wall should the observer stand in order to maximize the angle subtended by the picture?

34 Solve problem 33 for the case in which the picture is h units high and its lower edge is a units above the observer's eye level.

35 A ladder 15 meters long is leaning against a vertical wall of an office building. A window washer pulls the bottom of the ladder horizontally away from the building so that the top of the ladder slides down the wall at the rate of 2 meters per minute. How fast is the angle between the ladder and the ground changing when the bottom of the ladder is 6 meters from the wall?

36 A missile rises vertically from a point on the ground 10 miles from a radar station. If the missile is rising at the rate of 4000 ft/min at the instant when it is 2000 feet high, find the rate of change of the angle of elevation of the missile from the radar station at this instant.

37 A police officer in a patrol car is approaching an intersection at 80 ft/sec. When she is 210 feet from the intersection, a car crosses it, traveling at right angles to the police car's path at 60 ft/sec. If the officer trains her spotlight on the other car, how fast is the light beam turning 2 seconds later provided that both vehicles continue at their original rates?

Figure 3

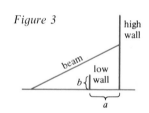

38 A high wall is to be braced by a beam which must pass over a lower wall b units high. If the distance between the low and the high walls is a units, what is the length of the shortest beam that can be used (Figure 3)?

7   Integrals Yielding Inverse Trigonometric Functions

Integration of the six inverse trigonometric functions involves the use of techniques that we have not yet introduced. However, we can obtain integration formulas that *yield* the inverse trigonometric functions simply by reversing the differentiation formulas obtained in Section 6. For instance, we have the following three formulas:

1 $\displaystyle\int \frac{du}{\sqrt{1 - u^2}} = \sin^{-1} u + C$, for $|u| < 1$.

2 $\displaystyle\int \frac{du}{1 + u^2} = \tan^{-1} u + C$, for u in $\mathbb{R}$.

3 $\displaystyle\int \frac{du}{u\sqrt{u^2 - 1}} = \sec^{-1} |u| + C$, for $|u| > 1$.

Formulas 1 and 2 follow directly from the formulas for the derivatives of $\sin^{-1} u$ and $\tan^{-1} u$. To establish Formula 3, recall that for $u \neq 0$, $\dfrac{d}{du} |u| = \dfrac{|u|}{u}$; hence,

$$\frac{d}{du} \sec^{-1} |u| = \frac{\dfrac{d}{du} |u|}{|u| \sqrt{|u|^2 - 1}} = \frac{\left(\dfrac{|u|}{u}\right)}{|u| \sqrt{u^2 - 1}} = \frac{1}{u \sqrt{u^2 - 1}}$$

holds for $|u| > 1$. It follows that

$$\int \frac{du}{u \sqrt{u^2 - 1}} = \sec^{-1} |u| + C \qquad \text{for } |u| > 1.$$

Formulas 1, 2, and 3 can easily be generalized as in the following theorem.

THEOREM 1 **Integrals Yielding Inverse Trigonometric Functions**

(i) $\displaystyle \int \frac{dx}{\sqrt{a^2 - x^2}} = \sin^{-1} \frac{x}{a} + C$ for $a > 0$ and $|x| < a$.

(ii) $\displaystyle \int \frac{dx}{a^2 + x^2} = \frac{1}{a} \tan^{-1} \frac{x}{a} + C.$

(iii) $\displaystyle \int \frac{dx}{x \sqrt{x^2 - a^2}} = \frac{1}{|a|} \sec^{-1} \left| \frac{x}{a} \right| + C$ for $a \neq 0$ and $|x| > |a|$.

PROOF
We prove (i) and leave the similar proofs of (ii) and (iii) as exercises (Problems 37 and 38). To prove (i), we make the substitution $x = au$, so that $dx = a\,du$, and use Formula 1. Thus,

$$\int \frac{dx}{\sqrt{a^2 - x^2}} = \int \frac{a\,du}{\sqrt{a^2 - a^2 u^2}} = \int \frac{a\,du}{\sqrt{a^2}\sqrt{1 - u^2}} = \int \frac{a\,du}{a\sqrt{1 - u^2}} = \int \frac{du}{\sqrt{1 - u^2}}$$

$$= \sin^{-1} u + C = \sin^{-1} \frac{x}{a} + C.$$

EXAMPLES **1** Evaluate $\displaystyle \int \frac{dx}{\sqrt{9 - x^2}}.$

SOLUTION
By part (i) of Theorem 1, $\displaystyle \int \frac{dx}{\sqrt{9 - x^2}} = \sin^{-1} \frac{x}{3} + C.$

2 Evaluate $\displaystyle \int \frac{dx}{\sqrt{1 - 16x^2}}.$

SOLUTION
Put $u = 4x$, so $u^2 = 16x^2$ and $dx = du/4$. Then,

$$\int \frac{dx}{\sqrt{1 - 16x^2}} = \int \frac{du}{4\sqrt{1 - u^2}} = \frac{1}{4} \sin^{-1} u + C = \frac{1}{4} \sin^{-1} 4x + C.$$

3 Find $\displaystyle\int_0^{5\sqrt{3}} \frac{dx}{25 + x^2}$.

SOLUTION
By part (ii) of Theorem 1,

$$\int_0^{5\sqrt{3}} \frac{dx}{25 + x^2} = \left[\frac{1}{5} \tan^{-1} \frac{x}{5} \right]\Big|_0^{5\sqrt{3}} = \frac{1}{5} \tan^{-1} \sqrt{3} - \frac{1}{5} \tan^{-1} 0$$

$$= \frac{1}{5}\left(\frac{\pi}{3}\right) - 0 = \frac{\pi}{15}.$$

4 Evaluate $\displaystyle\int_0^{\sqrt{3}} \frac{\tan^{-1} x}{1 + x^2}\, dx$.

SOLUTION

Put $u = \tan^{-1} x$ and note that $du = \dfrac{dx}{1 + x^2}$, $u = 0$ when $x = 0$, and $u = \pi/3$ when $x = \sqrt{3}$. Therefore,

$$\int_0^{\sqrt{3}} \frac{\tan^{-1} x}{1 + x^2}\, dx = \int_0^{\pi/3} u\, du = \left[\frac{u^2}{2}\right]\Big|_0^{\pi/3} = \frac{\pi^2}{18}.$$

5 Evaluate $\displaystyle\int_{3\sqrt{2}}^{6} \frac{dx}{x\sqrt{x^2 - 9}}$.

SOLUTION
By part (iii) of Theorem 1,

$$\int_{3\sqrt{2}}^{6} \frac{dx}{x\sqrt{x^2 - 9}} = \left[\frac{1}{3} \sec^{-1} \left|\frac{x}{3}\right| \right]\Big|_{3\sqrt{2}}^{6}$$

$$= \frac{1}{3} \sec^{-1} 2 - \frac{1}{3} \sec^{-1} \sqrt{2}$$

$$= \frac{1}{3}\left(\frac{\pi}{3}\right) - \frac{1}{3}\left(\frac{\pi}{4}\right) = \frac{\pi}{36}.$$

Problem Set 7

In problems 1 through 20, evaluate each integral.

1 $\displaystyle\int \frac{dx}{\sqrt{4 - x^2}}$

2 $\displaystyle\int \frac{dt}{\sqrt{16 - 9t^2}}$

3 $\displaystyle\int \frac{dx}{\sqrt{9 - 4x^2}}$

4 $\displaystyle\int \frac{dy}{\sqrt{25 - 11y^2}}$

5 $\displaystyle\int \frac{dt}{\sqrt{1 - 9t^2}}$

6 $\displaystyle\int \frac{dx}{x^2 + 9}$

7 $\displaystyle\int \frac{dy}{4 + 9y^2}$

8 $\displaystyle\int \frac{du}{9u^2 + 1}$

9 $\displaystyle\int \frac{dx}{4x^2 + 9}$

10 $\displaystyle\int \frac{dx}{x\sqrt{x^2 - 4}}$

11 $\displaystyle\int \frac{dt}{t\sqrt{16t^2 - 25}}$

12 $\displaystyle\int \frac{du}{u\sqrt{9u^2 - 100}}$

13 $\displaystyle\int \frac{4\, dx}{x\sqrt{x^2 - 16}}$

14 $\displaystyle\int_0^{1/2} \frac{dt}{\sqrt{1 - t^2}}$

15 $\displaystyle\int_{-3}^{3} \frac{dx}{\sqrt{12 - x^2}}$

16 $\displaystyle\int_0^{2} \frac{2\, du}{\sqrt{8 - u^2}}$

17 $\displaystyle\int_0^{3} \frac{dt}{3 + t^2}$

18 $\displaystyle\int_{-1}^{1} \frac{dx}{4 + x^2}$

19 $\displaystyle\int_{-2}^{-\sqrt{2}} \frac{dt}{t\sqrt{t^2 - 1}}$

20 $\displaystyle\int_{\sqrt{2}/2}^{1} \frac{du}{u\sqrt{4u^2 - 1}}$

In problems 21 through 30, evaluate each integral by substitution. (In some cases, an appropriate substitution is indicated.)

21 $\displaystyle\int \frac{\cos x}{\sqrt{36 - \sin^2 x}}\, dx, u = \sin x$

22 $\displaystyle\int \frac{\sec^2 t}{1 + \tan^2 t}\, dt$

23 $\displaystyle\int \frac{x}{4 + x^4}\, dx, u = x^2$

24 $\displaystyle\int \frac{dx}{7 + (3x - 1)^2}$

25 $\displaystyle\int \frac{\cos (x/2)}{1 + \sin^2 (x/2)}\, dx, u = \sin \frac{x}{2}$

26 $\displaystyle\int \frac{\sec^2 t\, dt}{\sqrt{1 - 9 \tan^2 t}}$

27 $\displaystyle\int_{\pi/6}^{\pi/3} \frac{\csc t \cot t}{1 + 9 \csc^2 t}\, dt, u = 3 \csc t$

28 $\displaystyle\int_{\sqrt{2}/2}^{\sqrt{3}/2} \frac{\cos^{-1} x}{\sqrt{1 - x^2}}\, dx$

29 $\displaystyle\int_1^8 \frac{dt}{t^{2/3}(1 + t^{2/3})}$

30 $\displaystyle\int_0^2 \frac{\cot^{-1} (x/2)}{4 + x^2}\, dx$

31 Evaluate $\displaystyle\int \frac{dx}{\sqrt{a^2 - b^2 x^2}}, b \neq 0.$

32 Evaluate $\displaystyle\int \frac{du}{4u^2 - 4u + 5}.$

33 If $0 < x < \pi/2$, calculate $\displaystyle\int_{\cos x}^{\sin x} du/\sqrt{1 - u^2}.$

34 Find the area of the region under the graph of $y = 3/(9 + x^2)$ between $x = 0$ and $x = \sqrt{3}$.

35 Find the volume of the solid generated when the region bounded by the graphs of $y = 1/\sqrt{1 + x^2}$, $x = 1$, $x = 0$, and $y = 0$ is rotated about the x axis.

36 Notice that $\pi = \displaystyle\int_0^{1/2} 6\, dx/\sqrt{1 - x^2}$. Use Simpson's parabolic rule with $n = 2$ (four subintervals) to estimate this definite integral and thus to estimate π.

37 Prove part (ii) of Theorem 1.   **38** Prove part (iii) of Theorem 1.

Review Problem Set

In problems 1 through 4, evaluate each limit.

1 $\displaystyle\lim_{t \to 0} \frac{\sin 13t}{t}$

2 $\displaystyle\lim_{x \to 0} \frac{x}{\sin 47x}$

3 $\displaystyle\lim_{u \to 0} \frac{\sin 19u}{\sin 7u}$

4 $\displaystyle\lim_{x \to 0^+} \frac{\sin x}{1 - \cos x}$

In problems 5 through 28, differentiate each function.

5 $f(x) = \sin (x + x^2)$

6 $g(t) = \sin t + 2 \sin t^2$

7 $h(t) = \sin (\cos t)$

8 $g(x) = \dfrac{\sin (\cos x)}{x}$

9 $H(x) = \cos (x + \sin x)$

10 $f(\theta) = \tan^2 \theta \tan \theta^2$

11 $F(u) = \cot^3 (\tan 5u)$

12 $h(x) = \dfrac{\cot^2 x}{1 + \sec x}$

13 $g(x) = \dfrac{\sec^3 5x}{x^3}$

14 $f(x) = (x^2 + 7) \csc^3 x$

15 $G(t) = \sqrt{\tan 17t}$

16 $H(x) = (1 + \csc x)^{5/2}$

17 $f(x) = \sqrt[5]{\cos^{-1} (x^3 + 1)}$

18 $f(t) = \sin^{-1} \sqrt{3t}$

19 $F(u) = u^2 \tan^{-1} u^4$

20 $f(x) = (\sec^{-1} x)^2$

21 $f(x) = (\cot^{-1} x^2)^4$

22 $g(\theta) = \theta^3 \csc^{-1} 5\theta$

23 $H(t) = \dfrac{\cot^{-1} (t^2 + 1)}{t^4}$

24 $f(x) = \dfrac{\sin^{-1} x^2}{\sqrt{x^2 - 1}}$

25 $f(x) = \displaystyle\int_0^{\cos x} (5 + t^4)^{26}\, dt$

26 $g(u) = \displaystyle\int_1^{\sin^{-1} u} (17 + x^2)^{34}\, dx$

27 $h(x) = \displaystyle\int_1^{\tan^{-1} x} \left(\frac{1 - t^2}{1 + t^2}\right)^{14} dt$

28 $f(x) = \displaystyle\int_1^{\tan x^2} \frac{du}{16 + u^6}$

In problems 29 through 32, find dy/dx by implicit differentiation.

29 $x \cos^{-1}(x + y) - \pi = y^2$

30 $y \tan x^2 - xy^4 = 15$

31 $3x \sin(x - y) + 107 = x^2 y$

32 $y \tan^{-1} x - x \tan^{-1} y = \pi/2$

In problems 33 through 58, evaluate each integral.

33 $\displaystyle\int 3 \sin 6x \, dx$

34 $\displaystyle\int \left(\frac{\sec(1/x)}{x} \right)^2 dx$

35 $\displaystyle\int \frac{\sec^2 x}{\sqrt[5]{\tan x}} dx$

36 $\displaystyle\int \frac{dx}{\cos^2(x/2)}$

37 $\displaystyle\int \frac{\sec 3x \tan 3x}{(\sec 3x + 8)^{10}} dx$

38 $\displaystyle\int \csc^2(\sin x) \cos x \, dx$

39 $\displaystyle\int \cos(\sec x) \sec x \tan x \, dx$

40 $\displaystyle\int x \sec^2 5x^2 \, dx$

41 $\displaystyle\int \frac{\csc^2 4x \, dx}{(7 + 3 \cot 4x)^8}$

42 $\displaystyle\int \frac{\cos x}{(1 - \sin x)^4} dx$

43 $\displaystyle\int \frac{\sec^2 x \, dx}{\sqrt{1 + \tan x}}$

44 $\displaystyle\int \frac{\csc^2 x}{\sqrt{5 + \cot x}} dx$

45 $\displaystyle\int \frac{a + b \cos x}{(ax + b \sin x)^7} dx$

46 $\displaystyle\int \frac{\sin 5x}{(\cos 5x + 13)^2} dx$

47 $\displaystyle\int \frac{\sin x \, dx}{4 + \cos^2 x}$

48 $\displaystyle\int \frac{dx}{\sin^2 x \sqrt{1 - \cot^2 x}}$

49 $\displaystyle\int \frac{x \, dx}{\sqrt{9 - 2x^4}}$

50 $\displaystyle\int \frac{x \sin^{-1} x^2}{\sqrt{1 - x^4}} dx$

51 $\displaystyle\int \frac{x \cos x^2 \, dx}{\sqrt{9 - \sin^2 x^2}}$

52 $\displaystyle\int \frac{\sec u \tan u \, du}{\sqrt{16 - \sec^2 u}}$

53 $\displaystyle\int \frac{\cot^{-1} v \, dv}{1 + v^2}$

54 $\displaystyle\int \frac{\tan^{-1}(x/2)}{4 + x^2} dx$

55 $\displaystyle\int_0^{\sqrt{\pi}} x \sin x^2 \, dx$

56 $\displaystyle\int_0^{\pi/4} \sin \theta \cos \theta \sec \theta \csc \theta \, d\theta$

57 $\displaystyle\int_0^5 \frac{dx}{25 + x^2}$

58 $\displaystyle\int_0^{\sqrt{3}} \frac{x \, dx}{9 + x^4}$

59 Find the values of x in the interval $(-2\pi, 2\pi)$ for which the graph of $f(x) = \cos x$ has a point of inflection.

60 If g is defined by $g(x) = a \cos kx + b \sin kx$, show that g satisfies the differential equation $g''(x) + k^2 g(x) = 0$.

61 A projectile fired from the foot of a $30°$ slope will strike the slope at a horizontal distance x given by the equation

$$ x = \frac{v_0^2}{g} \left[\sin 2\theta - \frac{1}{\sqrt{3}}(1 + \cos 2\theta) \right], $$

where v_0 is the muzzle velocity, θ is the angle of elevation of the gun, g is the acceleration of gravity, and air resistance is neglected (Figure 1). For what value of θ will the projectile reach the farthest distance up the slope?

Figure 1

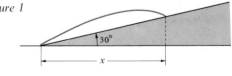

62 Let $f(x) = \cos 2x + 2 \cos x$ for $-\pi \le x \le \pi$. Find the extreme points of the graph of f and sketch this graph.

63 For what point (x, y) on the ellipse $x = 3 \cos \theta$, $y = 4 \sin \theta$, $0 \le \theta \le 2\pi$ is the sum $x + y$ of the coordinates a maximum?

64 A thin rod (whose weight can be ignored) is 40 centimeters long and passes through the center of a small heavy ball that is fixed on the rod at a distance of 10 centimeters from one end. The rod is placed in a smooth hollow hemisphere of radius 40 centimeters and the slightly rounded ends of the rod slide frictionlessly on the walls of the hemisphere. Find the angle that the rod makes with the horizontal when it finally comes to rest at equilibrium. (At equilibrium the center of mass of the rod and ball is as low as possible.)

65 A right circular cone whose generators make an angle θ with its axis is inscribed in a sphere of radius a. For what value of θ will the lateral area of the cone be greatest?

66 A gutter is to be made out of a long sheet of metal 30 centimeters wide by turning up strips 10 centimeters long on each side so that they make equal angles θ with the vertical (Figure 2). For what angle θ will the carrying capacity be maximum?

Figure 2

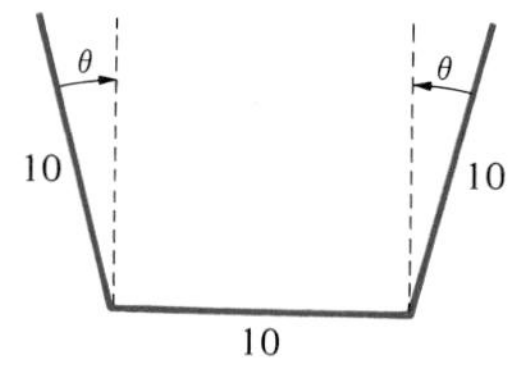

67 A ladder 25 feet long leans against a building. How far from the building should the foot of the ladder be placed to give maximum headroom under the ladder at a point 7 feet from the building?

68 From a point $\frac{1}{2}$ mile from a straight road, a searchlight is kept trained upon a car which travels along the road at a constant speed of 55 miles/hr. At what rate is the beam of light turning when the car is at the nearest point of the road?

69 A balloon rises from the ground 1000 feet from an observer and ascends vertically at the rate of 20 ft/sec. How fast is the angle of elevation changing at the observer's position at the instant when the balloon is 2000 feet from the ground?

70 Let $y = \sin^{-1}(a \sin t) + \sin^{-1}[a \sin(2 - t)]$, where a is a constant. Find the value of t that minimizes y.

71 An isosceles triangle has a 10-meter base. Its vertex is moving straight downward toward the base at the rate of 2 m/sec. How fast is the vertex angle changing at the instant when the vertex is 25 meters above the base?

72 A liquid nitrogen tank has the shape of a horizontal cylinder with a radius of 2 meters and a length of 10 meters. The tank is vented to allow the evaporating liquid nitrogen to escape. At the instant when the tank is half full, the surface level of the liquid nitrogen is dropping at the rate of 2 cm/hr. Find the rate of evaporation, in cubic meters per hour, at this instant.

73 Angle BAC of a right triangle is computed from measurement of the opposite side $y = |\overline{BC}|$ and the adjacent side $x = |\overline{AC}|$. (Angle ACB is the right angle.) Use differentials to approximate the possible error in the calculated value of angle BAC if the measurements of x and y are subject to at most 1 percent error.

74 A beam 30 feet long is to be carried horizontally around the corner of two passageways that intersect at right angles. One passageway is 10 feet wide. How narrow can the other passageway be if we assume no allowance for the thickness of the beam?

75 An isosceles triangle whose two equal sides are each 6 centimeters long has an included angle θ which is increasing at the constant rate of $1°$ per minute. How fast is the area of the triangle changing?

76 A mass m is suspended from the ceiling by a massless, perfectly elastic spring with spring constant k (units of force per unit of distance). A vertical s axis is established for reference with the origin at the equilibrium level of the mass (Figure 3). The mass is pulled down to level $s = -a$, where $a > 0$, and released. The differential equation of the motion is $\dfrac{d^2s}{dt^2} + \dfrac{k}{m}s = 0$. The initial conditions are $s = -a$ and $ds/dt = 0$ when $t = 0$. Find the equation of motion, show that the motion is periodic, and calculate the frequency of vibration of the mass. (*Hint:* See problem 60.)

Figure 3

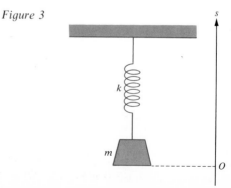

9 EXPONENTIAL, LOGARITHMIC, AND HYPERBOLIC FUNCTIONS

Algebraic and trigonometric functions, useful as they are, are not sufficient for the applications of mathematics to physics, chemistry, engineering, economics, and the life sciences. Thus, in this chapter, we add the exponential, logarithmic, and hyperbolic functions to our repertory.

All the functions that can be built up from algebraic, trigonometric, exponential, and logarithmic functions by addition, subtraction, multiplication, division, composition, and inversion are called *elementary* functions. Although nonelementary functions sometimes have to be used in applied mathematics, the elementary functions are sufficient for our purposes in this textbook.

1 The Natural Logarithm Function

Recall from precalculus courses that, if $x = b^y$, then y is called the *logarithm* of x to the *base* b, and we write $y = \log_b x$. A sketch of the graph of the function $f(x) = \log_b x$ (Figure 1) reveals that f is continuous on the interval $(0, \infty)$. Furthermore, the graph appears to have a tangent line at every point, so f is evidently differentiable.

Figure 1

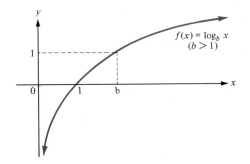

$f(x) = \log_b x$
$(b > 1)$

Assuming that the function $f(x) = \log_b x$ is differentiable, we can calculate its derivative as follows: Suppose that $b > 0$, $b \neq 1$, and $x > 0$. Then

$$f'(x) = \lim_{\Delta x \to 0} \frac{f(x + \Delta x) - f(x)}{\Delta x} = \lim_{\Delta x \to 0} \frac{\log_b (x + \Delta x) - \log_b x}{\Delta x}$$

$$= \lim_{\Delta x \to 0} \frac{1}{\Delta x} \log_b \frac{x + \Delta x}{x} = \lim_{\Delta x \to 0} \frac{1}{x} \cdot \frac{x}{\Delta x} \log_b \left(1 + \frac{\Delta x}{x}\right)$$

$$= \frac{1}{x} \lim_{\Delta x \to 0} \frac{x}{\Delta x} \log_b \left(1 + \frac{\Delta x}{x}\right) = \frac{1}{x} \lim_{\Delta x \to 0} u \log_b \left(1 + \frac{1}{u}\right),$$

where we have put $u = x/\Delta x$. As Δx approaches zero from the right, u approaches $+\infty$; hence,

$$f'(x) = \frac{1}{x} \lim_{u \to +\infty} u \log_b \left(1 + \frac{1}{u}\right) = \frac{1}{x} \lim_{u \to +\infty} \log_b \left(1 + \frac{1}{u}\right)^u$$

$$= \frac{1}{x} \log_b \lim_{u \to +\infty} \left(1 + \frac{1}{u}\right)^u.$$

If we provisionally define $e = \lim_{u \to +\infty} (1 + 1/u)^u$, assuming that the limit exists, then $f'(x) = (1/x) \log_b e$; that is, $D_x \log_b x = (1/x) \log_b e$.

The constant e given by $e = \lim_{u \to +\infty} (1 + 1/u)^u$ can be evaluated to as many decimal places as necessary by various methods—rounded off to 10 decimal places, $e \approx 2.7182818285$. The symbol e is used in honor of the Swiss mathematician Leonhard Euler (1707–1789), who was one of the first to recognize the importance of this number.

The formula $D_x \log_b x = (1/x) \log_b e$ becomes especially simple if we choose the base b to be Euler's constant e, for then $\log_b e = \log_e e = 1$, so that $D_x \log_e x = 1/x$. It follows that $\int \frac{1}{x} dx = \log_e x + C$ for $x > 0$; hence,

$$\int_1^x \frac{1}{t} dt = [\log_e t] \Big|_1^x = \log_e x - \log_e 1 = \log_e x - 0 = \log_e x.$$

Unfortunately, the discussion above suffers from a number of logical gaps arising from the fact that, whereas it is clear what is meant by b^y when y is an integer, or even a rational number, there could be some question about the meaning of b^y when y is not a rational number. For instance, what does $57^{\sqrt{2}}$ really mean? This difficulty can be avoided by *starting with* the equation $\log_e x = \int_1^x \frac{1}{t} dt$ as the *definition* of a logarithm, and then using logarithms to define b^y. This is exactly what we are going to do; however, to avoid the possibility of circular reasoning, we use the symbol "ln" rather than "$\log_e$" for the function so defined. These considerations lead us to the following definition.

DEFINITION 1   **The Natural Logarithm Function**

The *natural logarithm* function, denoted by ln, is defined by

$$\ln x = \int_1^x \frac{1}{t} dt \quad \text{for } x > 0.$$

Evidently, the domain of ln is the interval $(0, \infty)$ and, for $x > 1$, $\ln x$ can be interpreted geometrically as the area under the graph of $y = 1/t$, $t > 0$, between $t = 1$ and $t = x$ (Figure 2). From Definition 1, we have

$$\ln 1 = \int_1^1 \frac{1}{t} dt = 0.$$

Also, for $0 < x < 1$, ln x is the negative of the area under the graph of $y = 1/t$, $t > 0$, between $t = x$ and $t = 1$ (Figure 3).

Figure 2

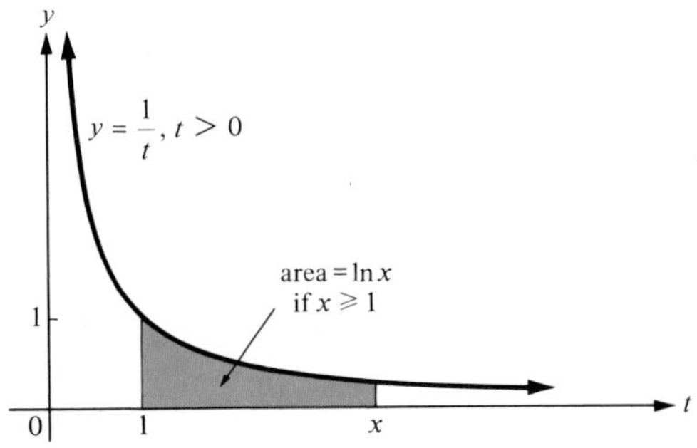

Figure 3

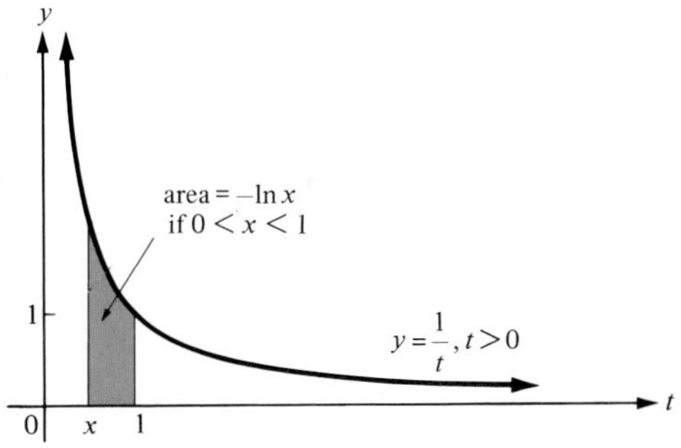

In summary, we have ln $x < 0$ if $0 < x < 1$, ln $x = 0$ if $x = 1$, and ln $x > 0$ if $x > 1$.

1.1   The Derivative of the Natural Logarithm

Since ln $x = \int_1^x \dfrac{1}{t}\, dt$ for $x > 0$, it follows from the fundamental theorem of calculus that D_x ln $x = \dfrac{1}{x}$. More generally, we have the following theorem.

THEOREM 1   **The Derivative of ln u**
If u is a differentiable function of x and $u > 0$, then

$$D_x \ln u = \frac{1}{u} D_x u.$$

PROOF
We already have $D_u \ln u = 1/u$ by the definition of ln and the fundamental theorem of calculus; hence, by the chain rule, $D_x \ln u = \dfrac{1}{u} D_x u.$

EXAMPLES 1   Given $y = \ln (5x + 7)$, find $D_x y$.

SOLUTION
Applying Theorem 1 with $u = 5x + 7$, we get

$$D_x y = \frac{1}{5x + 7} D_x(5x + 7) = \frac{5}{5x + 7}.$$

2   Find $D_x \ln \dfrac{x}{x + 3}$.

SOLUTION

$$D_x \ln \frac{x}{x + 3} = \frac{x + 3}{x} D_x\left(\frac{x}{x + 3}\right) = \left(\frac{x + 3}{x}\right)\left[\frac{3}{(x + 3)^2}\right] = \frac{3}{x(x + 3)}.$$

3   Given $f(x) = 5x \ln \sqrt{\cos x}$, find $f'(x)$.

SOLUTION

Using the product rule, we have

$$f'(x) = 5 \ln \sqrt{\cos x} + 5x \, \frac{1}{\sqrt{\cos x}} \cdot \frac{-\sin x}{2\sqrt{\cos x}}$$

$$= 5 \left[\ln \sqrt{\cos x} - \frac{x}{2} \tan x \right].$$

4 If $\ln(xy) + 3x + y = 5$, use implicit differentiation to find dy/dx.

SOLUTION

Here, we have

$$\frac{d}{dx}(\ln xy) + \frac{d(3x)}{dx} + \frac{dy}{dx} = \frac{d}{dx} 5,$$

so that $\dfrac{1}{xy}\left(y + x\,\dfrac{dy}{dx}\right) + 3 + \dfrac{dy}{dx} = 0$. Therefore,

$$\frac{1}{x} + \frac{1}{y}\frac{dy}{dx} + 3 + \frac{dy}{dx} = 0 \quad \text{or} \quad \frac{dy}{dx} = -\frac{3 + 1/x}{1 + 1/y} = -\frac{3xy + y}{xy + x}.$$

1.2 Integrals Yielding the Natural Logarithm

For $u \neq 0$, we have

$$D_u|u| = D_u\sqrt{u^2} = \frac{2u}{2\sqrt{u^2}} = \frac{u}{|u|};$$

hence, by Theorem 1,

$$D_u \ln|u| = \frac{1}{|u|}\,D_u|u| = \frac{1}{|u|} \cdot \frac{u}{|u|} = \frac{u}{|u|^2} = \frac{u}{u^2} = \frac{1}{u}.$$

Rewriting the latter equation in terms of antidifferentiation, we obtain the following theorem.

THEOREM 2   **Integration of $1/u$**

For $u \neq 0$, $\displaystyle\int \frac{1}{u}\,du = \ln|u| + C$.

EXAMPLES   Evaluate the given integral.

1 $\displaystyle\int \frac{x}{x^2 + 7}\,dx$

SOLUTION

We make the change of variable $u = x^2 + 7$, so that $du = 2x\,dx$ and $x\,dx = \frac{1}{2}\,du$. Then, since $x^2 + 7 > 0$, Theorem 2 gives

$$\int \frac{x}{x^2 + 7}\,dx = \frac{1}{2}\int \frac{1}{u}\,du = \frac{1}{2}\ln|u| + C = \frac{1}{2}\ln|x^2 + 7| + C = \frac{1}{2}\ln(x^2 + 7) + C.$$

2 $\displaystyle\int \frac{(\ln x)^2}{x}\,dx$

SOLUTION
Put $u = \ln x$, so that $du = dx/x$. Thus,

$$\int \frac{(\ln x)^2}{x}\,dx = \int u^2\,du = \frac{u^3}{3} + C = \frac{(\ln x)^3}{3} + C.$$

3 $\displaystyle\int \tan x\,dx$

SOLUTION
Noting that $\displaystyle\int \tan x\,dx = \int \frac{\sin x}{\cos x}\,dx$, we put $u = \cos x$, so that $du = -\sin x\,dx$. Thus,

$$\int \tan x\,dx = -\int \frac{du}{u} = -\ln|u| + C = -\ln|\cos x| + C.$$

4 $\displaystyle\int_{-9}^{-5} \frac{dx}{x+1}$

SOLUTION
Put $u = x + 1$, so that $du = dx$. When $x = -9$, $u = -8$, and when $x = -5$, $u = -4$. Thus,

$$\int_{-9}^{-5} \frac{dx}{x+1} = \int_{-8}^{-4} \frac{du}{u} = \left.(\ln|u|)\right|_{-8}^{-4} = \ln|-4| - \ln|-8| = \ln 4 - \ln 8.$$

5 $\displaystyle\int_{0}^{\pi/4} \frac{\sin 2x}{3 + 5\cos 2x}\,dx$

SOLUTION
Put $u = 3 + 5\cos 2x$, so that $du = -10\sin 2x\,dx$, or $\sin 2x\,dx = -\frac{1}{10}\,du$. When $x = 0$, $u = 8$, and when $x = \pi/4$, $u = 3$. Thus,

$$\int_{0}^{\pi/4} \frac{\sin 2x}{3 + 5\cos 2x}\,dx = \int_{8}^{3} \frac{-\frac{1}{10}\,du}{u} = -\frac{1}{10}\int_{8}^{3} \frac{du}{u} = -\frac{1}{10}\left.\ln|u|\right|_{8}^{3}$$

$$= -\frac{1}{10}(\ln 3 - \ln 8) = \frac{1}{10}(\ln 8 - \ln 3).$$

Problem Set 1

In problems 1 through 22, differentiate each function.

1 $f(x) = \ln(4x^2 + 1)$

2 $g(x) = \ln(\cos x^2)$

3 $f(x) = \sin(\ln x)$

4 $f(t) = \tan^{-1}(\ln t)$

5 $g(x) = x - \ln(\sin 6x)$

6 $H(x) = \ln(x + \cos x) - \tan^{-1} x$

7 $f(t) = \sin t \ln(t^2 + 7)$

8 $G(x) = \ln(4x + x^2 + 5)$

9 $F(u) = \ln(\ln u)$

10 $f(x) = \ln(\csc x - \cot x)$

11 $h(x) = x \ln \dfrac{\sec x}{5}$

12 $f(r) = \ln \sqrt[5]{1 + 5r^3}$

13 $g(v) = \ln(v^2\sqrt{v+1})$

14 $g(t) = \ln(t^3 \ln t^2)$

15 $h(x) = \ln(\cos^2 x)$

16 $g(r) = r^2 \csc (\ln r^2)$

17 $f(x) = \dfrac{1}{6} \ln \dfrac{8x^2}{4x^3 + 1}$

18 $g(x) = \ln \sqrt[3]{\dfrac{x}{x^2 + 1}}$

19 $h(t) = \dfrac{\ln t}{t^3 + 5}$

20 $H(x) = \sqrt{\ln \dfrac{x}{x + 2}}$

21 $g(x) = \dfrac{\ln (\tan^2 x)}{x^2}$

22 $f(x) = \dfrac{\ln [\cot (x/3)]}{x}$

In problems 23 through 26, find dy/dx by implicit differentiation.

23 $\ln \dfrac{x}{y} + \dfrac{y}{x} = 5$

24 $y = \ln |\sec x + \tan x| - \csc y$

25 $y \ln (\sin x) - xy^2 = 4$

26 $\ln y - \cos (x + y) = 2$

27 Find $D_x \displaystyle\int_1^{\ln x} \cos t^2 \, dt$ using the chain rule and the fundamental theorem of calculus.

28 Evaluate dy/dx if $y = \displaystyle\int_1^{\cos x} \ln (\tan t^4) \, dt$.

In problems 29 through 40, evaluate each integral by using a suitable substitution. (In some problems an appropriate substitution is suggested.)

29 $\displaystyle\int \dfrac{dx}{7 + 5x}, u = 7 + 5x$

30 $\displaystyle\int \dfrac{\sin x}{9 + \cos x} \, dx$

31 $\displaystyle\int \cot x \, dx, u = \sin x$

32 $\displaystyle\int \dfrac{dx}{(x + 2) \ln (x + 2)}$

33 $\displaystyle\int \dfrac{\sec^2 (\ln 4x)}{x} \, dx, u = \ln 4x$

34 $\displaystyle\int \dfrac{dx}{3\sqrt[3]{x^2} (1 + \sqrt[3]{x})}$

35 $\displaystyle\int \dfrac{4x \, dx}{x^2 + 7}$

36 $\displaystyle\int \dfrac{(\ln 5x)^2}{x} \, dx$

37 $\displaystyle\int \dfrac{\cos (\ln x)}{x} \, dx$

38 $\displaystyle\int \dfrac{\sec^2 x + \sec x \tan x}{\sec x + \tan x} \, dx$

39 $\displaystyle\int \dfrac{dx}{x\sqrt{1 - (\ln x)^2}}$

40 $\displaystyle\int \dfrac{dx}{x[1 + (\ln x)^2]}$

In problems 41 through 46, evaluate each definite integral.

41 $\displaystyle\int_{1/8}^{1/5} \dfrac{dx}{x}$

42 $\displaystyle\int_{0.01}^{10} \dfrac{dt}{t}$

43 $\displaystyle\int_1^{\sqrt{6}} \dfrac{x \, dx}{x^2 + 3}$

44 $\displaystyle\int_1^9 \dfrac{dx}{\sqrt{x} (1 + \sqrt{x})}$

45 $\displaystyle\int_1^4 \dfrac{\cos (\ln x) \, dx}{x}$

46 $\displaystyle\int_0^{\pi/2} \dfrac{\sin x}{2 - \cos x} \, dx$

47 A particle moves along the s axis in such a way that its velocity at time t seconds is $\dfrac{1}{t + 1}$ m/sec. If the particle is at the origin at $t = 0$, find the distance that it travels during the time interval from $t = 0$ to $t = 3$.

48 Find the arc length of the graph of $y = x^2/4 - (\ln x)/2$ between $(1, \frac{1}{4})$ and $(2, 1 - (\ln 2)/2)$.

2 Properties of the Natural Logarithm Function

Since the natural logarithm function is defined by $\ln x = \displaystyle\int_1^x \dfrac{dt}{t}$ for $x > 0$, we can use the properties of the integral to obtain an analytic derivation of the properties of ln. For instance, we have already shown that ln is a differentiable function on $(0, \infty)$ with $D_x \ln x = 1/x$, from which it follows that ln is a continuous function on $(0, \infty)$ (Theorem 1, Section 2.2, Chapter 2). The following theorem establishes one of the most important properties of ln.

THEOREM 1   **The Natural Logarithm of a Product**
If $a > 0$ and $b > 0$, then $\ln ab = \ln a + \ln b.$

PROOF
Using Theorem 10 in Section 3 of Chapter 6, we have

$$\ln ab = \int_1^{ab} \frac{dt}{t} = \int_1^a \frac{dt}{t} + \int_a^{ab} \frac{dt}{t} = \ln a + \int_a^{ab} \frac{dt}{t}.$$

In the latter integral, we make the substitution $t = au$, so that $dt = a\, du$. Also $u = 1$ when $t = a$, and $u = b$ when $t = ab$. Thus,

$$\int_a^{ab} \frac{dt}{t} = \int_1^b \frac{a\, du}{au} = \int_1^b \frac{du}{u} = \ln b.$$

It follows that $\ln ab = \ln a + \ln b$.

EXAMPLE   Given that $\ln 2 \approx 0.6931$ and $\ln 3 \approx 1.0986$, estimate $\ln 6$.

SOLUTION
By Theorem 1, we have
$$\ln 6 = \ln (2)(3) = \ln 2 + \ln 3 \approx 0.6931 + 1.0986 = 1.7917.$$

THEOREM 2   **The Natural Logarithm of a Quotient**

If $a > 0$ and $b > 0$, then $\ln \dfrac{a}{b} = \ln a - \ln b.$

PROOF
$a = (a/b)b$; hence, by Theorem 1, $\ln a = \ln [(a/b)b] = \ln (a/b) + \ln b$. Solving the latter equation for $\ln (a/b)$, we obtain $\ln (a/b) = \ln a - \ln b$.

EXAMPLES   **1**   Given that $\ln 4 \approx 1.3863$ and $\ln 3 \approx 1.0986$, estimate $\ln \frac{3}{4}$.

SOLUTION
By Theorem 2, we have
$$\ln \tfrac{3}{4} = \ln 3 - \ln 4 \approx 1.0986 - 1.3863 = -0.2877.$$

2   Show that, if $b > 0$, then $\ln (1/b) = -\ln b$.

SOLUTION
$$\ln \frac{1}{b} = \ln 1 - \ln b = 0 - \ln b = -\ln b.$$

Since $\log_b x^k = k \log_b x$, we might expect that $\ln x^k = k \ln x$. However, as we pointed out, there may be some question about the meaning of x^k when k is an irrational number. Thus, the following theorem establishes the desired identity only for the case in which k is rational.

THEOREM 3   **The Natural Logarithm of a Rational Power**
If $x > 0$ and k is a rational number, then $\ln x^k = k \ln x.$

PROOF

$$D_x \ln x^k = \frac{1}{x^k}\left(kx^{k-1}\right) = \frac{k}{x} \text{ and } D_x k \ln x = kD_x \ln x = \frac{k}{x}\,; \text{ hence,}$$

$$D_x(\ln x^k - k \ln x) = k/x - k/x = 0.$$

It follows from Theorem 1 in Section 2 of Chapter 5 that $\ln x^k - k \ln x = C$, where C is a constant. Putting $x = 1$ in the latter equation, we obtain $C = \ln 1^k - k \ln 1 = \ln 1 - k \ln 1 = 0$. Thus, $\ln x^k - k \ln x = 0$, so $\ln x^k = k \ln x$.

EXAMPLE   Given that $\ln 2 \approx 0.6931$, estimate $\ln 2048$.

SOLUTION
By Theorem 3, we have

$$\ln 2048 = \ln 2^{11} = (11) \ln 2 \approx (11)(0.6931) \approx 7.624.$$

Since $D_x \ln x = 1/x > 0$ for $x > 0$, it follows that $\ln$ is an increasing function. In fact, we have the following theorem.

THEOREM 4   **Monotonicity of the Natural Logarithm**
If $0 < a < b$, then $\ln a < \ln b$. Therefore, if $x > 0$, $y > 0$, and $\ln x = \ln y$, it follows that $x = y$.

PROOF
We have already seen that $\ln$ is an increasing function, so $0 < a < b$ implies that $\ln a < \ln b$. Now suppose that $x > 0$, $y > 0$, and $\ln x = \ln y$. Then it cannot be that $x < y$, since $x < y$ would imply that $\ln x < \ln y$, contradicting $\ln x = \ln y$. Similarly, it cannot be that $y < x$, since $y < x$ would imply that $\ln y < \ln x$, again contradicting $\ln y = \ln x$. The only remaining possibility is $x = y$.

EXAMPLE   Solve the equation $\ln (5 - 3x) + \ln (1 + x) = \ln 4$ for x.

SOLUTION
By Theorem 1, we can rewrite the given equation as

$$\ln 4 = \ln (5 - 3x) + \ln (1 + x) = \ln [(5 - 3x)(1 + x)].$$

Using Theorem 4, we see that the condition $\ln 4 = \ln [(5 - 3x)(1 + x)]$ implies that $4 = (5 - 3x)(1 + x)$, so $4 = 5 + 2x - 3x^2$, or $3x^2 - 2x - 1 = 0$. Factoring the latter equation, we have $(3x + 1)(x - 1) = 0$, so $x = -\frac{1}{3}$ or $x = 1$.

Notice that $\ln (5 - 3x)$ and $\ln (1 + x)$ are defined only when $5 - 3x > 0$ and $1 + x > 0$, so we must take the trouble to check that these inequalities hold for our alleged solutions $x = -\frac{1}{3}$ and $x = 1$. If $x = -\frac{1}{3}$, then $5 - 3x = 6 > 0$ and $1 + x = \frac{2}{3} > 0$; hence, $x = -\frac{1}{3}$ is indeed a solution. Similarly, if $x = 1$, then $5 - 3x = 2 > 0$ and $1 + x = 2 > 0$; hence, $x = 1$ is also a solution.

2.1   The Graph of the Natural Logarithm Function

In order to find the value of $\ln x$, it is necessary to evaluate $\int_1^x \frac{1}{t}\, dt$. Thus, for instance, to find $\ln 2$, we must evaluate $\int_1^2 \frac{1}{t}\, dt$. In Chapter 6, Section 5.3, page 324, we used Simpson's parabolic rule to estimate the latter integral, and thus obtained the approximation $\ln 2 \approx 0.69325$ with an error not exceeding 0.001. Therefore, we can be certain that $\ln 2$ lies between $0.69325 - 0.001$ and $0.69325 + 0.001$; that is, $0.69225 \le \ln 2 \le 0.69425$. Rounding off to two decimal places, we obtain $\ln 2 \approx 0.69$. It follows that

$$\ln 4 = \ln 2^2 = 2 \ln 2 \approx 2(0.69) = 1.38,$$
$$\ln 8 = \ln 2^3 = 3 \ln 2 \approx 3(0.69) = 2.07,$$
$$\ln \tfrac{1}{2} = -\ln 2 \approx -0.69,$$
$$\ln \tfrac{1}{4} = -\ln 4 \approx -1.38,$$
$$\ln \tfrac{1}{8} = -\ln 8 \approx -2.07, \text{ and so forth.}$$

From the considerations above, together with the fact that $\ln 1 = 0$, we can compile a rudimentary table of *approximate* values of the natural logarithm which is useful in plotting a graph of $y = \ln x$ (Table A). (A more extensive and more accurate table of values of the natural logarithm is given in Table II of the Appendix.) Table A also shows the corresponding values of $dy/dx = 1/x$. Since $dy/dx = 1/x > 0$, the graph of $y = \ln x$ is always rising as x increases, and since $\dfrac{d^2y}{dx^2} = \dfrac{d}{dx}\left(\dfrac{1}{x}\right) = -\dfrac{1}{x^2} < 0$, the graph is concave downward.

Table A

x	(approximate) $\ln x$	$D_x \ln x = \dfrac{1}{x}$
$\tfrac{1}{8}$	-2.07	8
$\tfrac{1}{4}$	-1.38	4
$\tfrac{1}{2}$	-0.69	2
1	0	1
2	0.69	$\tfrac{1}{2}$
4	1.38	$\tfrac{1}{4}$
8	2.07	$\tfrac{1}{8}$

Figure 1

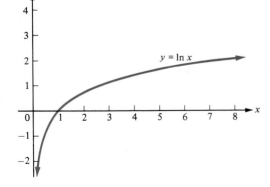

This information can now be used to sketch a graph of $y = \ln x$ (Figure 1). Unfortunately, this graph does not make it clear whether $y = \ln x$ becomes arbitrarily large as x increases or whether the graph has a horizontal asymptote. Notice, however, that $\ln 4 \approx 1.38$, so that $\ln 4 > 1$. It follows from Theorem 3 that, for any positive integer n, $\ln 4^n = n \ln 4 > (n)(1) = n$. The inequality $\ln 4^n > n$ shows that $\ln x$ can be made as large as we please by taking x large enough. For instance, to make $\ln x$ larger than 1000 it is only necessary to take x larger than 4^{1000}. It follows that $\lim\limits_{x \to +\infty} \ln x = +\infty$; hence, the graph of $y = \ln x$ climbs without bound and so it has no horizontal asymptote.

Since $\ln (1/t) = -\ln t$, it follows that

$$\lim_{x \to 0^+} \ln x = \lim_{t \to +\infty} \ln \frac{1}{t} = \lim_{t \to +\infty} (-\ln t) = -\lim_{t \to +\infty} \ln t = -\infty.$$

Therefore, ln x can be made smaller than any given number simply by taking x close enough to 0. These observations can be used to prove the following theorem.

THEOREM 5   **Range of the Natural Logarithm**
The range of ln is the entire real number system $\mathbb{R}$.

PROOF
We must prove that, given any real number y, there exists a positive real number x such that $y = \ln x$. Since $\lim\limits_{x \to +\infty} \ln x = +\infty$, there exists a positive real number b so large that $\ln b > y$. Since $\lim\limits_{x \to 0^+} \ln x = -\infty$, there exists a positive real number a so small that $\ln a < y$. Therefore, because $\ln a < y < \ln b$ and ln is a continuous function, the intermediate value theorem (Section 1.1 of Chapter 3) implies that there exists a value of x between a and b such that $\ln x = y$.

The following examples illustrate some of the uses of the properties of the natural logarithm.

EXAMPLES   **1** Air is compressed isothermally according to the equation $PV = C$, from an initial volume of $V_0 = 8$ cubic inches to a final volume of $V_1 = 1$ cubic inch. If the initial pressure was $P_0 = 15$ lb/in.^{2}, how much work is done?

SOLUTION
Here $C = P_0 V_0 = (15)(8) = 120$. The work done during the isothermal compression is given by

$$W = \int_1^8 P\, dV = \int_1^8 \frac{C}{V}\, dV = C \int_1^8 \frac{1}{V}\, dV = C \ln |V| \Big|_1^8$$

$$= C(\ln 8 - \ln 1) = C \ln 8 = 120 \ln 8 \text{ inch-pounds.}$$

From Table A on page 423, $\ln 8 \approx 2.07$; hence,

$$W \approx (120)(2.07) \approx 248 \text{ inch-pounds.}$$

2 Sketch the graph of the equation $y = \ln |x|$.

SOLUTION
Since $\ln |x| = \ln |-x|$, the required graph is symmetric about the y axis. For positive values of x, the required graph coincides with the graph of $y = \ln x$. Therefore, the graph of $y = \ln |x|$ consists of the graph of $y = \ln x$ together with the reflection of this graph across the y axis (Figure 2).

Figure 2

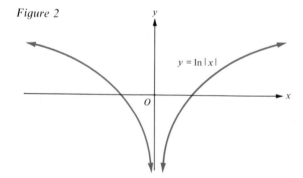

$y = \ln|x|$

3 Find the volume of the solid generated by revolving the region bounded by the graphs of $y = x^{-1/2}$, $x = 1$, $x = 4$, and $y = 0$ about the x axis.

SOLUTION
Using the method of circular disks, we have

$$V = \int_1^4 \pi(x^{-1/2})^2\, dx = \pi \int_1^4 \frac{1}{x}\, dx = \pi \left[\ln |x| \right]_1^4 = \pi[\ln |4| - \ln |1|] = \pi \ln 4.$$

From Table A, $\ln 4 \approx 1.38$; hence, $V \approx (3.14)(1.38) \approx 4.33$ cubic units.

Problem Set 2

In problems 1 through 6, assume only the following information and use the properties of the natural logarithm to estimate each quantity: $\ln 2 \approx 0.6931$, $\ln 3 \approx 1.0986$, $\ln 5 \approx 1.6094$.

1 $\ln 10$ **2** $\ln 0.6667$ **3** $\ln 100$

4 $\ln 66.67$ **5** $\ln 2.5$ **6** $\ln 25$

In problems 7 through 10, solve each equation for x.

7 $\ln (x - 1) + \ln (x - 2) = \ln 6$ **8** $\ln (x^2 - 4) + \ln (x - 2) = \ln 3$

9 $2 \ln (x - 2) = \ln x$ **10** $\ln (6 - x - x^2) - \ln (x + 3) = \ln (2 - x)$

In problems 11 through 17, sketch the graph of each function. Indicate the domain and the range of the function. Also indicate the extreme points and the inflection points of the graph and any horizontal or vertical asymptotes that may exist.

11 $g(x) = \ln (2 - x)$ **12** $h(x) = \ln (-x)$ **13** $F(x) = \ln |x + 1|$ **14** $G(x) = \ln \dfrac{4}{x}$

15 $H(x) = x \ln x$ **16** $f(x) = \dfrac{x}{\ln x}$ **17** $L(x) = x - \ln x$

18 Sketch the graph of the equation $x = \ln y$ for $y > 0$. Use implicit differentiation to find the slope of the graph at the point $(\ln 2, 2)$.

19 Find the equations of the tangent and normal lines to the curve $y = x^2 \ln x$ at the point $(2, 4 \ln 2)$.

20 Find the work done during the isothermal compression of a sample of gas from an initial pressure of P_0 and an initial volume of V_0 to a final pressure of P_1.

21 The speed of the signal in a submarine telegraph cable is proportional to $x^2 \ln (1/x)$, where x is the ratio of the radius of the core to the thickness of the winding. For what value of this ratio will the speed be maximum?

22 A product sells for \$25 per unit. Production cost C for x units is expressed by the equation $C = 250 + x(6 + 2 \ln x)$. Find the output level that yields maximum profit assuming that all manufactured units are sold.

23 Assuming that $a > 0$ and $b > 0$, find $D_x \left[\dfrac{1}{2\sqrt{ab}} \ln \dfrac{x\sqrt{a} - \sqrt{b}}{x\sqrt{a} + \sqrt{b}} \right]$ and simplify your answer.

24 A particle is moving along a straight line according to the equation of motion $s = \ln \dfrac{8t}{4t^2 + 5}$. Find the velocity ds/dt and the acceleration d^2s/dt^2.

In problems 25 through 28, find the area of the region under each curve.

25 $y = \dfrac{1}{x}$ between $x = -7$ and $x = -5$ **26** $y = \dfrac{4}{x - 1}$ between $x = 2$ and $x = 3$

27 $y = \dfrac{3}{x - 2}$ between $x = 3$ and $x = 4$ **28** $y = \dfrac{1}{2x - 1}$ between $x = 2$ and $x = 3$

In problems 29 through 31, find the volume of the solid generated by revolving the region bounded by the given curves about the x axis.

29 $(1 + x^2)y^2 = x$, $x = 1$, $x = 4$, $y = 0$ **30** $(x - 6)y^2 = x$, $x = 7$, $x = 10$, $y = 0$

31 $(x^2 + 2x)y^2 = x + 1$, $x = 1$, $x = 4$, $y = 0$

32 Given that $\ln 10 \approx 2.3025851$, use the differential to estimate $\ln 10.007$.

3   The Exponential Function

Since $D_x \ln x = 1/x$ for $x > 0$, we see that the natural logarithm function has a nonzero derivative on the open interval $(0, \infty)$; hence, according to the inverse function theorem (Theorem 1, Section 5.1, Chapter 2), ln is an invertible function on $(0, \infty)$. Later we shall prove that $\ln x$ is exactly the same as $\log_e x$, from which it will follow that the inverse of ln is the function $f(x) = e^x$. In anticipation of this result, we are going to refer to the inverse of ln as the *exponential function.* Thus, we set forth the following definition.

DEFINITION 1   **The Exponential Function**

The inverse of the natural logarithm function is called the *exponential function.* We denote the exponential function by exp.

Hence, by definition, $y = \exp x$ is equivalent to $x = \ln y$. For instance, from Table II in the Appendix, $1.1378 \approx \ln 3.12$; therefore, $\exp 1.1378 \approx 3.12$. Similarly, $\exp 0.9163 \approx 2.5$, since $0.9163 \approx \ln 2.5$. Thus, (approximate) values of $\exp x$ can be found simply by reading the table of values of the natural logarithm "backwards." For convenience, Table III in the Appendix gives (approximate) values of the exponential function directly. For instance, from the latter table, $\exp 1.6 \approx 4.9530$ and $\exp 2.9 \approx 18.174$.

Because the domain of ln is $(0, \infty)$, it follows that the range of exp is $(0, \infty)$. Likewise, because the range of ln is the set $\mathbb{R}$ of all real numbers, it follows that $\mathbb{R}$ is also the domain of exp. Since exp is the inverse of ln, the graph of exp is obtained by reflecting the graph of ln about the line $y = x$ (Figure 1).

From the graph of $y = \exp x$, we notice that $\exp x > 0$ for all values of x in $\mathbb{R}$. Also,

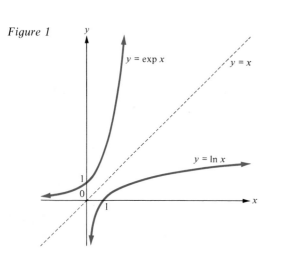

Figure 1

$$0 < \exp x < 1 \qquad \text{if } x < 0,$$

$$\exp x = 1 \qquad \text{if } x = 0, \text{ and}$$

$$\exp x > 1 \qquad \text{if } x > 0.$$

The fact that $\exp 0 = 1$ (which, later, will be seen to correspond to the fact that $e^0 = 1$) is used freely in what follows. Also, Figure 1 shows that

$$\lim_{x \to -\infty} \exp x = 0 \quad \text{and} \quad \lim_{x \to +\infty} \exp x = +\infty.$$

3.1   Properties of the Exponential Function

Because exp is the inverse of ln, we have

1 $\exp (\ln x) = x$ for $x > 0$.

2 $\ln (\exp x) = x$ for all values of x.

Relations 1 and 2 can be used to establish many of the basic properties of the exponential function. For instance, we have the following theorem.

THEOREM 1 **Basic Properties of the Exponential Function**
If x and y are any two real numbers and if k is a rational number, then

(i) $\exp(x + y) = (\exp x)(\exp y)$.

(ii) $\exp(kx) = (\exp x)^k$.

(iii) $\exp(x - y) = \dfrac{\exp x}{\exp y}$.

PROOF

(i)   Let $A = \exp x$, so that $x = \ln A$, and let $B = \exp y$, so that $y = \ln B$. By Theorem 1 of Section 2, $\ln A + \ln B = \ln(AB)$; hence, $x + y = \ln(AB)$. It follows that

$$\exp(x + y) = \exp[\ln(AB)] = AB = (\exp x)(\exp y).$$

(ii) Again put $A = \exp x$, so that $x = \ln A$. By Theorem 3 of Section 2, $kx = k\ln A = \ln A^k$; hence,

$$\exp(kx) = \exp(\ln A^k) = A^k = (\exp x)^k.$$

(iii) The proof of part (iii) is left as an exercise (Problem 61).

Now we replace our provisional definition of e by a formal definition based on the official definitions of ln and exp. The clue as to how this should be done comes from the anticipated result $\exp x = e^x$. Putting $x = 1$, we should have $\exp 1 = e$. This leads us to the following formal definition of the constant e.

DEFINITION 2 **The Constant e**
By definition, $e = \exp 1$; that is, e is the positive real number for which $\ln e = 1$.

Figure 2

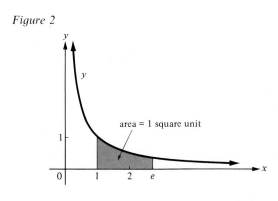

From Table II of the Appendix, it appears that the number e whose natural logarithm is 1 is given approximately by $e \approx 2.72$. Using more refined methods, it can be shown that e is an irrational number, $e = 2.71828182845904523536\ldots$. Graphically, of course, e is determined by the condition that the area $\displaystyle\int_1^e \frac{1}{x}\,dx$ between $x = 1$ and $x = e$ under the graph of $y = 1/x$ is exactly one square unit (Figure 2).

Combining the definition of e with part (ii) of Theorem 1, we obtain the following result.

THEOREM 2 **The Exponential of a Rational Number**
If k is a rational number, then $\exp k = e^k$.

PROOF
Put $x = 1$ in part (ii) of Theorem 1 to obtain $\exp k = (\exp 1)^k$. By Definition 2, $\exp 1 = e$; hence, $\exp k = e^k$, as desired.

By Theorem 2, $e^x = \exp x$ holds for all rational numbers x, so it is reasonable to ask whether it also holds when x is irrational. However, as we originally pointed out, if x is irrational, it is perhaps not clear just what is meant by e^x

anyway. To clarify matters once and for all, we simply *define* e^x to be equal to exp x when x is irrational.

DEFINITION 3

Definition of e^x for Irrational Values of x
If x is an irrational number, we define e^x as follows:

$$e^x = \exp x.$$

By Theorem 2 and Definition 3, $e^x = \exp x$ holds for all real numbers x, whether rational or irrational. Therefore, $\ln e^x = \ln (\exp x) = x$; that is, e^x is the number whose natural logarithm is x. Similarly, $e^{\ln x} = \exp (\ln x) = x$ holds for $x > 0$. In summary, we have:

1 $y = e^x$ if and only if $x = \ln y$.
2 $e^{\ln x} = x$ for $x > 0$.
3 $\ln e^x = x$ for all values of x.
4 $e^{x+y} = e^x e^y$ for all values of x and y.
5 $e^{kx} = (e^x)^k$ for all values of x and all rational numbers k.
6 $e^{x-y} = e^x/e^y$ for all values of x and y.

Of course, Properties 4, 5, and 6 are just parts (i), (ii), and (iii), respectively, of Theorem 1, rewritten using Definition 3.

EXAMPLE   Simplify each of the following expressions:

(a) $e^{\ln a}$ (b) $e^{-2+2\ln 3}$ (c) $\ln e^{5x}$ (d) $\dfrac{e^{\ln (x^2-9)}}{x-3}$

SOLUTION
(a) $e^{\ln a} = a.$
(b) $e^{-2+2\ln 3} = e^{-2}e^{2\ln 3} = (e^2)^{-1}(e^{\ln 3^2}) = (e^2)^{-1}(3^2) = 9/e^2.$
(c) $\ln e^{5x} = 5x.$
(d) $\dfrac{e^{\ln (x^2-9)}}{x-3} = \dfrac{x^2-9}{x-3} = x + 3$, for $x \neq 3.$

3.2   Differentiation of the Exponential Function

Henceforth, we can use either the notation e^x or the notation exp x for the inverse of $\ln$ evaluated at x. The inverse function theorem can now be invoked to obtain the derivative $D_x e^x$. The result—which is most remarkable—is given by the following theorem.

THEOREM 3

The Derivative of the Exponential Function
The exponential function is its own derivative, that is, $D_x e^x = e^x$.

A formal proof of Theorem 3 can be obtained using the inverse function theorem (Theorem 1, Section 5.1, Chapter 2). However, the following informal proof is perhaps more illuminating: Let $y = e^x$, so that $x = \ln y$. Differentiating both sides of the latter equation with respect to x, we obtain $1 = (1/y)D_x y$, that is, $D_x y = y$, or $D_x e^x = e^x$.

Combining the result of Theorem 3 with the chain rule, we see that, if u is a differentiable function of x, then

$$D_x e^u = e^u D_x u.$$

EXAMPLES **1** If $y = e^{x^2 + x}$, find $D_x y$.

SOLUTION

$$D_x y = D_x e^{x^2 + x} = e^{x^2 + x} D_x (x^2 + x) = e^{x^2 + x}(2x + 1).$$

2 Find $f'(x)$ if $f(x) = e^{\cos x}$.

SOLUTION

$$f'(x) = e^{\cos x} D_x \cos x = e^{\cos x}(-\sin x) = -\sin x \, e^{\cos x}.$$

3 If $xe^y = \tan^{-1} x$, use implicit differentiation to find dy/dx.

SOLUTION

$$\frac{d}{dx}(xe^y) = \frac{d}{dx}\tan^{-1} x, \quad \text{so} \quad e^y + xe^y \frac{dy}{dx} = \frac{1}{1 + x^2}.$$

Solving for dy/dx, we obtain

$$\frac{dy}{dx} = \frac{1}{(1 + x^2)xe^y} - \frac{e^y}{xe^y} = \frac{1}{(1 + x^2)\tan^{-1} x} - \frac{1}{x}.$$

4 The *normal probability density function* f is defined in probability theory by the equation

$$f(x) = \frac{1}{\sigma\sqrt{2\pi}} \exp\left[\frac{-(x - \mu)^2}{2\sigma^2}\right],$$

where σ and μ are certain constants called the *standard deviation* and the *mean*, respectively, and $\sigma > 0$. For what value of x is $f(x)$ a maximum and what is the maximum value of $f(x)$?

SOLUTION

$$f'(x) = \frac{1}{\sigma\sqrt{2\pi}}\left\{\exp\left[-\frac{(x - \mu)^2}{2\sigma^2}\right]\right\} D_x\left[-\frac{(x - \mu)^2}{2\sigma^2}\right]$$

$$= \frac{1}{\sigma\sqrt{2\pi}}\left\{\exp\left[-\frac{(x - \mu)^2}{2\sigma^2}\right]\right\}\left[-\frac{x - \mu}{\sigma^2}\right].$$

Since $\exp\left[-(x - \mu)^2/2\sigma^2\right]$ cannot be zero (recall that $\exp x > 0$ holds for all values of x), $f'(x) = 0$ when $-(x - \mu)/\sigma^2 = 0$, that is, when $x = \mu$. Since $f'(x) > 0$ for $x < \mu$ and $f'(x) < 0$ for $x > \mu$ (why?), it follows that f takes on a maximum value when $x = \mu$. This maximum value is given by $f(\mu) = \dfrac{1}{\sigma\sqrt{2\pi}} \exp 0 = \dfrac{1}{\sigma\sqrt{2\pi}}$.

3.3 Integration of e^u

Theorem 3 can now be used to obtain the following important integral formula.

THEOREM 4   **Integration of e^u**

$$\int e^u \, du = e^u + C.$$

PROOF
By Theorem 3, $D_u e^u = e^u$; that is, e^u is an antiderivative of e^u.

EXAMPLES   **1** Evaluate $\int e^{4x} \, dx$.

SOLUTION
Put $u = 4x$, so $du = 4 \, dx$ and $dx = \frac{1}{4} \, du$. Thus, by Theorem 4,

$$\int e^{4x} \, dx = \int e^u (\tfrac{1}{4} \, du) = \tfrac{1}{4} \int e^u \, du = \tfrac{1}{4} e^u + C = \tfrac{1}{4} e^{4x} + C.$$

2 Evaluate $\int x^2 e^{x^3} \, dx$.

SOLUTION
Put $u = x^3$, so $du = 3x^2 \, dx$, or $x^2 \, dx = \frac{1}{3} \, du$. Thus,

$$\int x^2 e^{x^3} \, dx = \tfrac{1}{3} \int e^u \, du = \tfrac{1}{3} e^u + C = \tfrac{1}{3} e^{x^3} + C.$$

3 Find $\int \dfrac{e^x \, dx}{\sqrt{1 - e^{2x}}}$.

SOLUTION
We put $u = e^x$, noting that $du = e^x \, dx$ and that $e^{2x} = (e^x)^2 = u^2$. Therefore,

$$\int \frac{e^x \, dx}{\sqrt{1 - e^{2x}}} = \int \frac{du}{\sqrt{1 - u^2}} = \sin^{-1} u + C = \sin^{-1} e^x + C.$$

4 Find the volume V of the solid generated by revolving the region bounded by the curves $y = e^{-4x}$, $x = 0$, $x = 2$, and $y = 0$ about the x axis.

SOLUTION

By the method of circular disks, $V = \int_0^2 \pi (e^{-4x})^2 \, dx = \pi \int_0^2 e^{-8x} \, dx$. We put $u = -8x$, so that $du = -8 \, dx$ and $dx = -\frac{1}{8} \, du$. When $x = 0$, $u = 0$, and when $x = 2$, $u = -16$; hence,

$$V = \pi \int_0^{-16} e^u \left(-\frac{1}{8} \, du \right) = -\frac{\pi}{8} \int_0^{-16} e^u \, du = -\frac{\pi}{8} e^u \Big|_0^{-16}$$

$$= -\frac{\pi}{8} (e^{-16} - 1) = \frac{\pi}{8} (1 - e^{-16}).$$

Since $e > 2$, $e^{-1} < 2^{-1}$, so

$$e^{-16} = (e^{-1})^{16} < (2^{-1})^{16} = 2^{-16} = \frac{1}{2^{16}} = \frac{1}{65,536}.$$

Thus, e^{-16} is very small, so $V \approx \pi/8 \approx 0.39$ cubic unit.

Problem Set 3

1 Simplify each expression.

(a) $e^{\ln 5}$ 　　　　　　(b) $e^{-3\ln 2}$ 　　　　　　(c) $e^{3+4\ln 2}$ 　　　　　　(d) $\ln e^{1/x}$

(e) $\ln e^{x-x^2}$ 　　　　(f) $e^{-\ln(1/x)}$ 　　　　(g) $\ln e^{x^2-4}$ 　　　(h) $e^{(\ln x^2)-4}$

(i) $\dfrac{e^{\ln(x^2-4)}}{x+2}$ 　　　(j) $e^{\ln x-3\ln y}$

2 Solve the following equations for x:

(a) $2e^x + 1 = e^{-x}$ 　　　　(b) $e^x + 20e^{-x} = 21$

In problems 3 through 24, find the derivative of each function.

3 $f(x) = e^{7x}$ 　　　　　　　　　**4** $g(t) = (e^{4t})^3$ 　　　　　　　　　**5** $g(x) = e^{\ln x^3}$

6 $f(u) = \exp(\sin u)$ 　　　　　**7** $f(x) = \cos(\exp x)$ 　　　　　**8** $g(x) = e^{-x}\cos 2x$

9 $f(t) = e^{-2t}\sin t$ 　　　　　**10** $g(r) = \tan^{-1}(\exp r)$ 　　　**11** $h(x) = e^{x^2+5\ln x}$

12 $F(x) = \exp\sqrt{4-x^2}$ 　　　**13** $f(t) = e^{t\ln t}$ 　　　　　　**14** $g(x) = e^{x^2}\cot 4x$

15 $h(x) = \sec^{-1}(e^{3x})$ 　　　**16** $f(x) = e^{\sqrt{x}}\ln\sqrt{x}$ 　　　**17** $G(s) = (1-e^{3s})^2$

18 $f(t) = \dfrac{e^{2t}}{e^t+1}$ 　　　　**19** $f(x) = \dfrac{1}{\sqrt{2\pi}}\exp\left(-\dfrac{x^2}{2}\right)$ 　　　**20** $f(t) = \ln\dfrac{e^t}{t+1}$

21 $f(x) = (8x^2 - 3x + 1)e^{-x}$ 　　　**22** $H(r) = \displaystyle\int_1^{e^r}\sqrt{1+t^2}\,dt$

23 $g(\theta) = \theta^3\sin(e^\theta)$ 　　　　　**24** $f(x) = e^{-3x}(3\cos x + \sin 2x)$

In problems 25 through 28, use implicit differentiation to find dy/dx.

25 $y(1+e^x) - xy^2 = 7$ 　　　　　**26** $e^y - \sin(x+y) = 2$

27 $x\sin y = e^{x+y}$ 　　　　　　　**28** $e^{-x}\ln y + e^y\ln x = 4$

29 Show that $y = e^{-3x}$ satisfies the differential equation $y'' + 2y' - 3y = 0$.

30 Show that $y = 5xe^{-4x}$ satisfies the differential equation $\dfrac{d^2y}{dx^2} + 8\dfrac{dy}{dx} + 16y = 0$.

In problems 31 through 48, evaluate each integral. (In some cases a suitable substitution is suggested.)

31 $\displaystyle\int e^{3x}\,dx$ 　　　　　**32** $\displaystyle\int e^{-7x}\,dx$ 　　　　**33** $\displaystyle\int e^{5x+3}\,dx,\ u = 5x+3$

34 $\displaystyle\int e^{-4x+5}\,dx$ 　　　**35** $\displaystyle\int xe^{5x^2}\,dx,\ u = 5x^2$ 　　**36** $\displaystyle\int e^{\sin x}\cos x\,dx$

37 $\displaystyle\int \frac{e^x\,dx}{1+e^{2x}},\ u = e^x$ 　**38** $\displaystyle\int \frac{e^{\sqrt[3]{x}}}{\sqrt[3]{x^2}}\,dx$ 　　　**39** $\displaystyle\int \frac{3e^x\,dx}{\sqrt{e^x+4}},\ u = e^x+4$

40 $\displaystyle\int \frac{5e^{-3x}}{(e^{-3x}+7)^8}\,dx$ 　**41** $\displaystyle\int e^{\cot x}\csc^2 x\,dx,\ u = \cot x$ 　**42** $\displaystyle\int e^{\sec x}\sec x\tan x\,dx$

43 $\displaystyle\int_0^1 e^{2x}\,dx$ 　　　　**44** $\displaystyle\int_0^{\ln 5} e^{-3x}\,dx$ 　　　**45** $\displaystyle\int_0^1 2x^2(e^{x^3}+1)\,dx$

46 $\displaystyle\int_1^2 (1+e^{-x})^2\,dx$ 　**47** $\displaystyle\int_0^{\pi/2} e^{\sin 2x}\cos 2x\,dx$ 　**48** $\displaystyle\int_0^1 \frac{3+e^{4x}}{e^{4x}}\,dx$

In problems 49 through 51, sketch the graph of each function and indicate the intervals where the function is increasing, decreasing, concave upward, and concave downward. Also, locate all extreme points, points of inflection, and asymptotes.

49 $f(x) = e^{-3x}$ **50** $f(x) = e^{-x^2}$ **51** $f(x) = xe^{-x}$

52 Find the minimum value of f if $f(x) = 3e^{4x} + 5e^{-5x}$.

53 The total potential audience for an advertising campaign numbers 10,000. Average revenue per response is $3 and the cost of the campaign is $300 per day, plus a fixed cost of $500. To maximize profit, find the number of days the campaign should continue if the number of responses y is given in terms of the number of campaign days t by $y = (1 - e^{-0.25t})(10,000)$.

54 Let $f(x) = e^x - 1 - x$. Show that $f'(x) \geq 0$ if $x \geq 0$ and that $f'(x) \leq 0$ if $x \leq 0$. Use this fact to prove that $e^x \geq 1 + x$ and that $e^{-x} \geq 1 - x$.

55 If the value of a certain piece of property at time t years is given by the equation $V = \$20,000 - \$10,000e^{-0.1t}$, find the rate of change of V with respect to t when $t = 5$ years.

56 If the x intercept of the tangent to the curve $y = e^{-x}$ at the variable point (p, q) is increasing at the constant rate of 5 units per second, find the rate at which the y intercept is changing when the x intercept is 10 units.

57 Atmospheric pressure at altitude h feet above sea level is given by $P = 15e^{-0.0004h}$ lb/in.^{2}. A jetliner is climbing through 10,000 feet at the rate of 1000 ft/min. Find the rate of change of external air pressure as measured by a gauge on board the plane.

58 Find the area under the curve $y = e^{2x} - x$ between $x = 1$ and $x = 5$.

59 Find the area bounded by the curves $y = e^{2x}$, $y = e^{3x}$, and $x = 1$.

60 Find the volume of the solid generated by revolving the region under the curve $y = e^{2x}$ between $x = 0$ and $x = 2$ about the x axis.

61 Finish the proof of Theorem 1 by demonstrating that $\exp(x - y) = \exp x / \exp y$.

62 Find the arc length of the graph of $y = \dfrac{e^x + e^{-x}}{2}$ from $(0, 1)$ to $\left(1, \dfrac{e + e^{-1}}{2}\right)$.

4 Exponential and Logarithmic Functions with Base Other Than e

In the present section we use the natural logarithm and the exponential function to define b^x and $\log_b x$ for values of the base b other than e. Of course, b^k is already defined when $b > 0$ and k is rational. In fact, we have the following theorem.

THEOREM 1 b^k **for** $b > 0$ **and** k **Rational**

If $b > 0$ and k is a rational number, then $b^k = e^{k \ln b}$.

PROOF
By part (ii) of Theorem 1 in Section 3.1, we have $[\exp(\ln b)]^k = \exp(k \ln b)$, that is, $(e^{\ln b})^k = e^{k \ln b}$. Since $e^{\ln b} = b$, it follows that $b^k = e^{k \ln b}$.

Theorem 1 gives us the key to the proper definition of b^x when x is irrational. Thus, we set forth the following definition.

DEFINITION 1   **Definition of b^x for Irrational Values of x**
If $b > 0$ and x is irrational, we define $b^x = e^{x \ln b}$.

As a consequence of Theorem 1 and Definition 1, $b^x = e^{x \ln b}$ holds for all real values of x, provided that $b > 0$.

EXAMPLE   Using Tables II and III in the Appendix, estimate $\pi^{\sqrt{7}}$.

SOLUTION

$$\pi^{\sqrt{7}} = e^{\sqrt{7} \ln \pi} \approx e^{2.65 \ln 3.14} \approx e^{(2.65)(1.14)} \approx e^{3.02} \approx 20.5.$$

4.1 Basic Properties of b^x

Using the equation $b^x = e^{x \ln b}$ and the known properties of the exponential function, we can now derive the basic properties of b^x.

THEOREM 2   **Laws of Exponents**
Let x and y be real numbers and suppose that a and b are positive real numbers. Then:

(i) $b^x b^y = b^{x+y}$.

(ii) $\dfrac{b^x}{b^y} = b^{x-y}$.

(iii) $(b^x)^y = b^{xy}$.

(iv) $(ab)^x = a^x b^x$.

(v) $\left(\dfrac{a}{b}\right)^x = \dfrac{a^x}{b^x}$.

(vi) $b^{-x} = \dfrac{1}{b^x}$.

(vii) $\ln b^x = x \ln b$.

PROOF
We prove (i), (iii), (iv), and (v) and leave (ii), (vi), and (vii) as exercises (Problems 51, 52, and 53).

(i) $b^x b^y = e^{x \ln b} e^{y \ln b} = e^{x \ln b + y \ln b} = e^{(x+y) \ln b} = b^{x+y}$.

(iii)   Since $\ln e^{x \ln b} = x \ln b$, it follows that $\ln b^x = x \ln b$. Therefore, $(b^x)^y = e^{y \ln b^x} = e^{yx \ln b} = b^{yx} = b^{xy}$.

(iv) $(ab)^x = e^{x \ln ab} = e^{x(\ln a + \ln b)} = e^{x \ln a + x \ln b} = e^{x \ln a} e^{x \ln b} = a^x b^x$.

(v) $\left(\dfrac{a}{b}\right)^x = e^{x \ln (a/b)} = e^{x(\ln a - \ln b)} = e^{x \ln a - x \ln b} = \dfrac{e^{x \ln a}}{e^{x \ln b}} = \dfrac{a^x}{b^x}$.

According to the power rule (Theorem 3, Section 5.2, Chapter 2, p. 118), $D_x x^k = k x^{k-1}$ holds for $x > 0$ and k a constant rational exponent. The following theorem generalizes the power rule to arbitrary (constant) exponents.

THEOREM 3   **General Power Rule**

Let c be a constant real number and suppose that u is a differentiable function of x. Then, for $u > 0$, $D_x u^c = cu^{c-1}D_x u$.

PROOF

$$D_x u^c = D_x e^{c \ln u} = e^{c \ln u}D_x(c \ln u) = u^c\left(c\,\frac{D_x u}{u}\right) = cu^{c-1}D_x u.$$

EXAMPLES   Differentiate the given function.

1 $f(x) = x^{-e} + e^{-x}$

SOLUTION
Using Theorem 3 to differentiate x^{-e}, we have

$$f'(x) = D_x(x^{-e}) + D_x(e^{-x}) = (-e)x^{-e-1} + e^{-x}D_x(-x) = -ex^{-e-1} - e^{-x}.$$

2 $g(x) = (x^3 + 1)^\pi.$

SOLUTION

$$g'(x) = D_x(x^3 + 1)^\pi = \pi(x^3 + 1)^{\pi-1}D_x(x^3 + 1) = \pi(x^3 + 1)^{\pi-1}(3x^2).$$

In applying the general power rule, it is important to realize that the *base is the variable* and the *exponent is constant*. **On the other hand,** if the *base is constant* and the *exponent is the variable,* we can find the derivative using the formula $b^x = e^{x \ln b}$. For instance, $D_x 2^x = D_x e^{x \ln 2} = e^{x \ln 2}D_x(x \ln 2) = 2^x(\ln 2)$. More generally, we have the following theorem.

THEOREM 4   **Derivative of b^x**

If b is a positive constant, then $D_x b^x = b^x \ln b$.

PROOF

$$D_x b^x = D_x e^{x \ln b} = e^{x \ln b}D_x(x \ln b) = b^x(\ln b) = b^x \ln b.$$

Of course, Theorem 4 can be combined with the chain rule to obtain the formula

$$D_x b^u = b^u \ln b \, D_x u.$$

EXAMPLES **1** If $y = 2^{x^2}$, find dy/dx.

SOLUTION

$$\frac{dy}{dx} = 2^{x^2}(\ln 2)(2x) = 2x2^{x^2} \ln 2 = x2^{x^2+1} \ln 2.$$

2 Let $f(x) = 3^{\tan x}$. Find $f'(x)$.

SOLUTION

$$f'(x) = 3^{\tan x} \ln 3 \sec^2 x.$$

3 Find $D_x(5^{x^3}e^{\sin x})$.

SOLUTION

$$D_x(5^{x^3}e^{\sin x}) = 5^{x^3} \ln 5(3x^2)e^{\sin x} + 5^{x^3}e^{\sin x} \cos x$$
$$= 5^{x^3}e^{\sin x}(3x^2 \ln 5 + \cos x).$$

The differentiation formula $D_x b^x = b^x \ln b$ yields a corresponding integral formula as follows:

$$\int b^x \, dx = \int \frac{b^x \ln b}{\ln b} \, dx = \frac{1}{\ln b} \int b^x \ln b \, dx = \frac{1}{\ln b} b^x + C.$$

Thus, replacing x by u, we have

$$\int b^u \, du = \frac{b^u}{\ln b} + C, \qquad \text{provided that } b > 0 \text{ and } b \neq 1.$$

EXAMPLES   Evaluate the given integral.

1 $\int 7^x \, dx$

SOLUTION

$$\int 7^x \, dx = \frac{7^x}{\ln 7} + C.$$

2 $\int 5^{\sin 2x} \cos 2x \, dx$

SOLUTION
Put $u = \sin 2x$, so that $du = 2 \cos 2x \, dx$. Thus,

$$\int 5^{\sin 2x} \cos 2x \, dx = \frac{1}{2} \int 5^u \, du = \frac{5^u}{2 \ln 5} + C = \frac{5^{\sin 2x}}{2 \ln 5} + C.$$

3 $\int_0^1 3^{-x} \, dx$

SOLUTION
Put $u = -x$, so that $du = -dx$ and

$$\int_0^1 3^{-x} \, dx = -\int_0^{-1} 3^u \, du = -\frac{3^u}{\ln 3}\Big|_0^{-1}$$
$$= \left(-\frac{3^{-1}}{\ln 3}\right) - \left(-\frac{3^0}{\ln 3}\right) = \frac{1 - \frac{1}{3}}{\ln 3} = \frac{2}{3 \ln 3}.$$

4.2 Logarithmic Differentiation

The properties of the natural logarithm (as developed in Section 2) are useful for finding the derivatives of functions involving products, quotients, and powers of other functions. The technique, called *logarithmic differentiation*, works as follows.

Procedure for Logarithmic Differentiation

To find the derivative of a function f, carry out the following steps:

Step 1 Write the equation $y = f(x)$.

Step 2 Take the natural logarithm on both sides of the equation.

Step 3 Differentiate the resulting equation implicitly with respect to x.

EXAMPLES Use logarithmic differentiation to find dy/dx.

1 $y = x^x$

SOLUTION

Taking the natural logarithm on both sides of the equation $y = x^x$, we obtain

$$\ln y = \ln x^x = x \ln x.$$

Differentiating the equation $\ln y = x \ln x$ on both sides with respect to x, we have

$$\frac{1}{y}\frac{dy}{dx} = x \frac{d}{dx}\ln x + \frac{dx}{dx}\ln x = x\frac{1}{x} + \ln x = 1 + \ln x.$$

Therefore,

$$\frac{dy}{dx} = y(1 + \ln x) = x^x(1 + \ln x).$$

2 $y = \dfrac{(x^2 + 5)(5x + 2)^{3/2}}{\sqrt[4]{(3x + 1)(x^3 + 2)}}$

SOLUTION

$$\ln y = \ln (x^2 + 5) + \ln (5x + 2)^{3/2} - \ln \sqrt[4]{(3x + 1)(x^3 + 2)}$$
$$= \ln (x^2 + 5) + \tfrac{3}{2} \ln (5x + 2) - \tfrac{1}{4}[\ln (3x + 1) + \ln (x^3 + 2)],$$

so that

$$\frac{1}{y}\frac{dy}{dx} = \frac{1}{x^2 + 5}(2x) + \left(\frac{3}{2}\right)\frac{1}{5x + 2}(5) - \frac{1}{4}\left[\frac{1}{3x + 1}(3) + \frac{1}{x^3 + 2}(3x^2)\right]$$

or

$$\frac{dy}{dx} = y\left[\frac{2x}{x^2 + 5} + \frac{15}{2(5x + 2)} - \frac{3}{4(3x + 1)} - \frac{3x^2}{4(x^3 + 2)}\right].$$

Therefore,

$$\frac{dy}{dx} = \frac{(x^2 + 5)(5x + 2)^{3/2}}{\sqrt[4]{(3x + 1)(x^3 + 2)}}\left[\frac{2x}{x^2 + 5} + \frac{15}{2(5x + 2)} - \frac{3}{4(3x + 1)} - \frac{3x^2}{4(x^3 + 2)}\right].$$

4.3 The Function $\log_a$

The base $e = 2.71828\ldots$ is preferred in all work with logarithms or exponents in calculus because of the relative simplicity of the resulting differentiation and integration formulas. However, other bases are used in some applications. For instance, for purely computational purposes, $\log_{10}$ is often advantageous because numbers are usually written to the base 10. Also, in information theory, $\log_2$ is used in reckoning with so-called "bits" of information.

Recall the definition of logarithm to the base a: If $a > 0$, $a \neq 1$ and $x > 0$, then we say that y is the *logarithm of x to the base a*, and we write $y = \log_a x$, to mean that $a^y = x$. Thus, $a^{\log_a x} = x$. For example,

$$\log_6 36 = 2 \qquad \text{since} \quad 6^2 = 36,$$

$$\log_{10} 0.0001 = -4 \qquad \text{since } 10^{-4} = 0.0001,$$

and so forth. In particular, $y = \log_e x$ means that $e^y = x$; that is, it means that $y = \ln x$. Therefore, $\ln x = \log_e x$. The following important theorem allows us to convert from logarithms in base a to logarithms in another base b.

THEOREM 5   **Conversion of Base for Logarithms**
If $a > 0$, $b > 0$, $a \neq 1$, $b \neq 1$, and $x > 0$, then

(i) $\log_b x = \log_a x \log_b a$.

In particular,

(ii) $\ln x = \log_a x \ln a$.

(iii) $\log_a x = \dfrac{\ln x}{\ln a}$.

PROOF

(i) $b^{\log_a x \log_b a} = (b^{\log_b a})^{\log_a x} = a^{\log_a x} = x$; hence, $\log_b x = \log_a x \log_b a$, by the definition of logarithm.

(ii)   By (i), $\log_e x = \log_a x \log_e a$. Since $\ln$ is the same as $\log_e$, it follows that $\ln x = \log_a x \ln a$.

(iii)   Follows immediately from part (ii).

Using Theorem 5 and the properties of the natural logarithm, it is a simple matter to verify the properties of $\log_a$, such as

$$\log_a (xy) = \log_a x + \log_a y, \qquad \log_a \left(\frac{x}{y}\right) = \log_a x - \log_a y, \qquad \log_a x^y = y \log_a x.$$

(See Problem 59.) Similarly, we have the following theorem.

THEOREM 6   **Derivative of Logarithmic Functions**
Let u be a differentiable function of x with $u > 0$ and suppose that $a > 0$ with $a \neq 1$. Then $D_x \log_a u = \dfrac{1}{u \ln a} D_x u$.

PROOF
By part (iii) of Theorem 5, $\log_a u = \ln u / \ln a$. Since $D_x \ln u = (1/u)D_x u$, it follows that $D_x \log_a u = \dfrac{D_x \ln u}{\ln a} = \dfrac{1}{u \ln a} D_x u$.

EXAMPLES   1   Find dy/dx if $y = \log_{10}(x^2 + 5)$.

SOLUTION
By Theorem 6,

$$\frac{dy}{dx} = \frac{1}{(x^2 + 5) \ln 10} \frac{d}{dx}(x^2 + 5) = \frac{2x}{(x^2 + 5) \ln 10}.$$

2 Find $D_x(\log_2 \sin x)$.

SOLUTION

$$D_x \log_2 \sin x = \frac{\cos x}{\sin x \ln 2} = \frac{\cot x}{\ln 2}.$$

4.4 The Exponential Function as a Limit

Our provisional definition of e as $\lim_{u \to +\infty} (1 + 1/u)^u$ was not really legitimate since, at that stage of the game, we did not have Definition 1 at our disposal, so the very meaning of $(1 + 1/u)^u$ when u is irrational was in doubt. Now we can establish the formula $e = \lim_{u \to +\infty} (1 + 1/u)^u$ on a rigorous basis.

THEOREM 7 e **as a Limit**

$$e = \lim_{u \to +\infty} \left(1 + \frac{1}{u}\right)^u = \lim_{u \to -\infty} \left(1 + \frac{1}{u}\right)^u.$$

PROOF
In the expression $(1 + 1/u)^u$, we put $\Delta x = 1/u$, so $u = 1/\Delta x$ and $(1 + 1/u)^u = (1 + \Delta x)^{1/\Delta x}$. Notice that $u \to +\infty$ as $\Delta x \to 0^+$ and that $u \to -\infty$ as $\Delta x \to 0^-$. Hence, the equations to be established can be written

$$e = \lim_{\Delta x \to 0^+} (1 + \Delta x)^{1/\Delta x} = \lim_{\Delta x \to 0^-} (1 + \Delta x)^{1/\Delta x},$$

or simply as the single equation $e = \lim_{\Delta x \to 0} (1 + \Delta x)^{1/\Delta x}$. To prove this, we use Definition 1 to write

$$(1 + \Delta x)^{1/\Delta x} = e^{(1/\Delta x) \ln (1 + \Delta x)} = \exp\left[\frac{1}{\Delta x} \ln (1 + \Delta x)\right].$$

The proof will be complete once we show that $\lim_{\Delta x \to 0} [(1/\Delta x) \ln (1 + \Delta x)] = 1$, for then

$$\lim_{\Delta x \to 0} (1 + \Delta x)^{1/\Delta x} = \exp\left\{\lim_{\Delta x \to 0} \left[\frac{1}{\Delta x} \ln (1 + \Delta x)\right]\right\} = \exp 1 = e$$

will follow from the continuity of the exponential function.

To prove that $\lim_{\Delta x \to 0} [(1/\Delta x) \ln (1 + \Delta x)] = 1$, put $f(x) = \ln x$ for $x > 0$, so that $f(1) = \ln 1 = 0$, $f'(x) = 1/x$, and $f'(1) = 1$. Thus,

$$\lim_{\Delta x \to 0} \left[\frac{1}{\Delta x} \ln (1 + \Delta x)\right] = \lim_{\Delta x \to 0} \frac{f(1 + \Delta x)}{\Delta x} = \lim_{\Delta x \to 0} \frac{f(1 + \Delta x) - f(1)}{\Delta x}$$

$$= f'(1) = 1,$$

as desired.

Using Theorem 7, we establish an interesting formula for e^a in the following theorem.

THEOREM 8   **The Exponential Function as a Limit**

$$e^a = \lim_{h \to +\infty} \left(1 + \frac{a}{h}\right)^h = \lim_{h \to -\infty} \left(1 + \frac{a}{h}\right)^h.$$

PROOF

The equations obviously hold when $a = 0$. Thus, we can assume that $a \neq 0$. First we consider the case in which $a > 0$. Put $u = h/a$, noting that $u \to +\infty$ as $h \to +\infty$. Therefore,

$$\lim_{h \to +\infty} \left(1 + \frac{a}{h}\right)^h = \lim_{h \to +\infty} \left(1 + \frac{a}{h}\right)^{(h/a)a} = \lim_{u \to +\infty} \left(1 + \frac{1}{u}\right)^{ua} = \lim_{u \to +\infty} \left[\left(1 + \frac{1}{u}\right)^u\right]^a.$$

We put $v = (1 + 1/u)^u$, noting that $\lim\limits_{u \to +\infty} v = e$ by Theorem 7. By Theorem 3, v^a is a differentiable function of v; hence, it is a continuous function of v. Therefore,

$$\lim_{h \to +\infty} \left(1 + \frac{a}{h}\right)^h = \lim_{u \to +\infty} v^a = \left(\lim_{u \to +\infty} v\right)^a = e^a.$$

Similar arguments take care of the cases in which $a < 0$ and in which $h \to -\infty$ (Problem 63).

As an example of Theorem 8, notice that $(1 + \frac{5}{1000})^{1000} \approx 146.58$, while $e^5 \approx 148.41$; hence, $(1 + \frac{5}{1000})^{1000}$ approximates e^5 to within 2 percent. Also, the following example indicates the usefulness of Theorem 8 in connection with practical calculation.

EXAMPLE   Telephone calls coming into a certain switchboard follow a "Poisson probability distribution," *averaging* c calls per minute. The theory of probability gives $P = \lim\limits_{n \to +\infty} c(1 - c/n)^{n-1}$ as the probability that exactly one call comes into the switchboard during any given 1-minute period. Find P if $c = 4$ calls per minute.

SOLUTION

$$P = \lim_{n \to +\infty} \frac{c\left(1 - \dfrac{c}{n}\right)^n}{\left(1 - \dfrac{c}{n}\right)} = \frac{c \lim\limits_{n \to +\infty} \left(1 + \dfrac{-c}{n}\right)^n}{\lim\limits_{n \to +\infty} \left(1 - \dfrac{c}{n}\right)} = \frac{ce^{-c}}{1} = ce^{-c}$$

by Theorem 8. For $c = 4$, we have $P = 4e^{-4} \approx 0.073 = \frac{73}{1000}$ and we could expect exactly one incoming call during approximately 73 of 1000 one-minute periods.

Problem Set 4

In problems 1 through 26, differentiate each function.

1 $f(x) = x^{-3\pi}$

2 $g(t) = t^{\pi - 2}$

3 $h(t) = (t^2 + 1)^{3e}$

4 $H(r) = (2r^2 + 7)^{-e^2}$

5 $g(x) = 3^{2x+1}$

6 $f(x) = 2^{7x^2}$

7 $f(t) = 6^{-5t}$

8 $h(x) = 2^{\tan x^2}$

9 $g(x) = 5^{\sec x}$

10 $H(x) = 3^{\cos x^2}$

11 $h(x) = (x^2 + 5)2^{-7x^2}$

12 $g(t) = \sin t \cdot 3^{5t^2}$

13 $f(x) = \dfrac{2^{x+1}}{x^2 + 5}$

14 $h(x) = 2^{5x} \cot x$

15 $g(x) = \log_2 x^2$

16 $h(t) = \log_3 (t^3 + 1)$

17 $f(x) = e^{\log_4 x}$

18 $F(x) = \log_5 (\sin x^2)$

19 $h(t) = \log_{10} \dfrac{2t}{1 + t}$

20 $f(t) = \log_5 (\ln \cos t)$

21 $F(u) = 3^{\tan u} \log_8 u$

22 $g(t) = \sqrt{\log_5 t}$

23 $f(x) = \dfrac{\log_3 (x^2 + 5)}{x + 2}$

24 $F(x) = \csc x \log_3 (x^4 + 1)$

25 $f(x) = \tan^{-1} (\log_3 x)$

26 $h(x) = \log_7 \dfrac{3^x}{1 + 5^x}$

In problems 27 through 38, use logarithmic differentiation to find dy/dx.

27 $y = x^{\sqrt{x}}$

28 $y = (\cos x)^x$

29 $y = (\sin x^2)^{3x}$

30 $y = (x + 1)^x$

31 $y = (x^2 + 4)^{\ln x}$

32 $y = (\sin x)^{\cos x}$

33 $y = (x^2 + 7)^2 (6x^3 + 1)^4$

34 $y = x^2 \sin x^3 \cos (3x + 7)$

35 $y = \dfrac{\sin x \sqrt[3]{1 + \cos x}}{\sqrt{\cos x}}$

36 $y = \dfrac{\tan^2 x}{\sqrt{1 - 4 \sec x}}$

37 $y = \dfrac{x^2 \sqrt[5]{x^2 + 7}}{\sqrt[4]{11x + 8}}$

38 $y = \sqrt{\dfrac{\sec x + \tan x}{\sec x - \tan x}}$

In problems 39 through 48, evaluate each integral. (In some cases an appropriate change of variable is suggested.)

39 $\displaystyle\int 3^{5x}\, dx, u = 5x$

40 $\displaystyle\int \dfrac{5^{\ln x^2}}{x}\, dx$

41 $\displaystyle\int 7^{x^4 + 4x^3}(x^3 + 3x^2)\, dx, u = x^4 + 4x^3$

42 $\displaystyle\int 3^{\tan x} \sec^2 x\, dx$

43 $\displaystyle\int \dfrac{2^{\ln (1/x)}}{x}\, dx, u = \ln x$

44 $\displaystyle\int 8^{\sec x} \sec x \tan x\, dx$

45 $\displaystyle\int 4^{\cot x} \csc^2 x\, dx$

46 $\displaystyle\int 2^{x \ln x}(1 + \ln x)\, dx$

47 $\displaystyle\int_0^1 5^{-2x}\, dx$

48 $\displaystyle\int_0^{\pi/2} 3^{\sin x} \cos x\, dx$

49 Which number is larger, e^π or π^e?

50 Find the maximum value of $(\ln x)/x$ for $x > 0$ and use the result to answer problem 49 without evaluating e^π or π^e.

51 Prove that $b^x/b^y = b^{x-y}$ for $b > 0$.

52 Prove that $b^{-x} = 1/b^x$ for $b > 0$.

53 Prove that $\ln b^x = x \ln b$ for $b > 0$.

54 Patients arriving at the emergency room of a certain hospital follow a so-called "Poisson probability distribution," averaging c patients per hour. According to the theory of probability,

$$P(k) = \lim_{n \to +\infty} \dfrac{n(n-1)(n-2)\cdots(n-k+1)}{k(k-1)(k-2)\cdots 1}\left(\dfrac{c}{n}\right)^k \left(1 - \dfrac{c}{n}\right)^{n-k}$$

gives the probability that *exactly* k patients will arrive at the emergency room during a given 1-hour period. Evaluate this limit and thus find $P(k)$.

55 For what value of x is $f(x) = x^{1/x}$, $x > 0$, a maximum?

56 Let g be defined by $g(x) = \frac{1}{2}(3^x + 3^{-x})$. Show that $g(c + d) + g(c - d) = 2g(c)g(d)$.

57 Prove that, if $a > 0$ and $b > 0$, then $\log_a b = 1/\log_b a$.

58 Use problem 57 and Theorem 6 to show that $D_x \log_a u = \dfrac{\log_a e}{u} D_x u$.

59 Prove that, if $a > 0$, $a \neq 1$, $x > 0$, and $y > 0$, then:

(i) $\log_a xy = \log_a x + \log_a y$, (ii) $\log_a \dfrac{x}{y} = \log_a x - \log_a y$, and (iii) $\log_a x^u = u \log_a x$.

60 Suppose that $a > 0$ and $a \neq 1$ is a given constant. Find $D_x \log_x a$ for $x > 0$, $x \neq 1$.

61 Find the volume of the solid generated by revolving the region under the graph of $y = 3^x$ between $x = 0$ and $x = 2$ about the x axis.

62 If y is a differentiable function of x, then the (point) *elasticity* of y with respect to x is defined to be the limit, $\lim\limits_{\Delta x \to 0} \dfrac{\Delta y/y}{\Delta x/x}$, of the proportional change in y to the proportional change in x. Show that, if x and y are positive, then the elasticity is given by dY/dX, where $Y = \log_b y$ and $X = \log_b x$.

63 Complete the proof of Theorem 8 by considering the cases in which $a < 0$ and in which $h \to -\infty$.

5   Hyperbolic Functions

Certain combinations of exponential functions, which are related to a hyperbola in somewhat the same way that trigonometric functions are related to a circle, prove to be important in applied mathematics. These functions are called the *hyperbolic functions* and their similarity to the trigonometric functions is emphasized by calling them *hyperbolic sine, hyperbolic cosine, hyperbolic tangent,* and so on. They are defined as follows.

DEFINITION 1   **The Hyperbolic Functions**

(i) $\sinh x = \dfrac{e^x - e^{-x}}{2}$.

(ii) $\cosh x = \dfrac{e^x + e^{-x}}{2}$.

(iii) $\tanh x = \dfrac{\sinh x}{\cosh x} = \dfrac{e^x - e^{-x}}{e^x + e^{-x}}$.

(iv) $\coth x = \dfrac{\cosh x}{\sinh x} = \dfrac{e^x + e^{-x}}{e^x - e^{-x}}$.

(v) $\operatorname{sech} x = \dfrac{1}{\cosh x} = \dfrac{2}{e^x + e^{-x}}$.

(vi) $\operatorname{csch} x = \dfrac{1}{\sinh x} = \dfrac{2}{e^x - e^{-x}}$.

The six hyperbolic functions satisfy identities that correspond to the standard trigonometric identities except for an occasional switch of plus and minus signs. For instance, we have the following identities:

1 $\cosh^2 x - \sinh^2 x = 1.$

2 $1 - \tanh^2 x = \operatorname{sech}^2 x.$

3 $\coth^2 x - 1 = \operatorname{csch}^2 x.$

4 $\sinh(s + t) = \sinh s \cosh t + \sinh t \cosh s$ and
   $\sinh(s - t) = \sinh s \cosh t - \sinh t \cosh s.$

5 $\cosh(s + t) = \cosh s \cosh t + \sinh s \sinh t$ and
   $\cosh(s - t) = \cosh s \cosh t - \sinh s \sinh t.$

6 $\sinh 2x = 2 \sinh x \cosh x.$

7 $\cosh 2x = \cosh^2 x + \sinh^2 x = 2\cosh^2 x - 1 = 2\sinh^2 x + 1.$

EXAMPLES   Prove the given identity.

1 $\cosh^2 x - \sinh^2 x = 1$

SOLUTION

$$\cosh^2 x - \sinh^2 x = \left(\frac{e^x + e^{-x}}{2}\right)^2 - \left(\frac{e^x - e^{-x}}{2}\right)^2$$

$$= \frac{e^{2x} + 2e^x e^{-x} + e^{-2x}}{4} - \frac{e^{2x} - 2e^x e^{-x} + e^{-2x}}{4}$$

$$= \frac{e^{2x} + 2 + e^{-2x} - e^{2x} + 2 - e^{-2x}}{4} = \frac{4}{4} = 1.$$

2 $1 - \tanh^2 x = \operatorname{sech}^2 x$

SOLUTION
We divide the equation $\cosh^2 x - \sinh^2 x = 1$ on both sides by $\cosh^2 x$ to obtain

$$1 - \frac{\sinh^2 x}{\cosh^2 x} = \frac{1}{\cosh^2 x} \quad \text{or} \quad 1 - \tanh^2 x = \operatorname{sech}^2 x.$$

The remaining identities for hyperbolic functions can be proved similarly (Problems 1 through 5). The trigonometric identity $\cos^2 t + \sin^2 t = 1$ implies that the point $(\cos t, \sin t)$ always belongs to the circle whose equation is $x^2 + y^2 = 1$. Likewise, the hyperbolic identity $\cosh^2 t - \sinh^2 t = 1$ implies that the point $(\cosh t, \sinh t)$ always belongs to the hyperbola whose equation is $x^2 - y^2 = 1$.
Like the sine function, the hyperbolic sine function is odd. Indeed,

$$\sinh(-x) = \frac{e^{(-x)} - e^{-(-x)}}{2} = -\frac{e^x - e^{-x}}{2} = -\sinh x;$$

hence, the graph of $y = \sinh x$ is symmetric about the origin. Just as $\sin 0 = 0$, so also $\sinh 0 = \frac{e^0 - e^{-0}}{2} = \frac{0}{2} = 0$. For large values of x, e^{-x} is very small, so $\sinh x = \frac{e^x - e^{-x}}{2} \approx \frac{e^x}{2}$; thus, for large values of x, the graph of $\sinh x$ is very close to the graph of $y = e^x/2$ (Figure 1). Unlike the sine function, $\sinh$ is not periodic—rather, it is monotone-increasing.

Figure 1

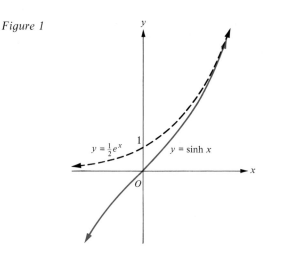

Like the cosine function, the hyperbolic cosine function is even. Indeed,

$$\cosh(-x) = \frac{e^{(-x)} + e^{-(-x)}}{2} = \frac{e^x + e^{-x}}{2} = \cosh x;$$

hence, the graph of $y = \cosh x$ is symmetric about the y axis. Just as $\cos 0 = 1$, so also $\cosh 0 = \frac{e^0 + e^{-0}}{2} = \frac{2}{2} = 1$. For large values of x, e^{-x} is very small, so $\cosh x = \frac{e^x + e^{-x}}{2} \approx \frac{e^x}{2}$; thus, for large values of x, the graph of $y = \cosh x$ is very close to the graph of $y = e^x/2$ (Figure 2). Again, cosh is not periodic, but is increasing on $(0, \infty)$ and decreasing on $(-\infty, 0)$.

Since sinh is odd and cosh is even, it is easy to see that $\tanh = \sinh/\cosh$ is an odd function; hence, the graph of $y = \tanh x$ is symmetric about the origin. For large values of x we have $\sinh x \approx e^x/2$ and $\cosh x \approx e^x/2$; therefore, $\tanh x = \sinh x/\cosh x \approx 1$. Indeed, $y = 1$ and $y = -1$ are horizontal asymptotes of the graph of $y = \tanh x$ (Figure 3).

Figure 2   Figure 3

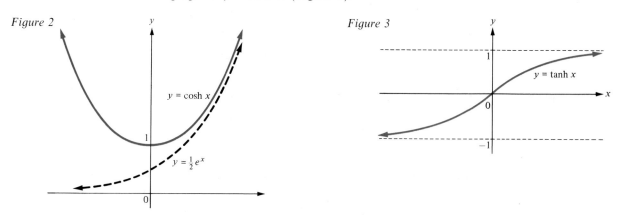

The graphs of $y = \coth x$, $y = \operatorname{sech} x$, and $y = \operatorname{csch} x$ can be sketched with relative ease using the information available in Figures 1 through 3. We leave it as an exercise for the reader to actually sketch these graphs (Problems 34, 35, and 36).

5.1 Differentiation of the Hyperbolic Functions

The differentiation formulas for the hyperbolic functions closely resemble those for the trigonometric functions except for some algebraic signs. In fact, since $D_x e^u = e^u D_x u$ and $D_x e^{-u} = -e^{-u} D_x u$, the following theorem is a direct consequence of Definition 1.

THEOREM 1   **Derivatives of the Hyperbolic Functions**
Let u be a differentiable function of x. Then:

(i) $D_x \sinh u = \cosh u D_x u$.

(ii) $D_x \cosh u = \sinh u D_x u$.

(iii) $D_x \tanh u = \operatorname{sech}^2 u D_x u$.

(iv) $D_x \coth u = -\operatorname{csch}^2 u D_x u.$

(v) $D_x \operatorname{sech} u = -\operatorname{sech} u \tanh u D_x u.$

(vi) $D_x \operatorname{csch} u = -\operatorname{csch} u \coth u D_x u.$

For instance, to prove (i), we calculate as follows:

$$D_x \sinh u = D_x \frac{e^u - e^{-u}}{2} = \frac{D_x e^u - D_x e^{-u}}{2} = \frac{e^u + e^{-u}}{2} D_x u = \cosh u D_x u.$$

Part (ii) is proved similarly. To prove part (iii), we use parts (i) and (ii) and calculate as follows:

$$D_x \tanh u = D_x \frac{\sinh u}{\cosh u} = \frac{\cosh u D_x \sinh u - \sinh u D_x \cosh u}{\cosh^2 u}$$

$$= \frac{\cosh^2 u - \sinh^2 u}{\cosh^2 u} D_x u = \frac{1}{\cosh^2 u} D_x u = \operatorname{sech}^2 u D_x u.$$

Proofs of parts (ii), (iv), (v), and (vi) are left as an exercise (Problem 33).

EXAMPLES 1 Find $D_x \sinh (5x + 2)$.

SOLUTION
By part (i) of Theorem 1,

$$D_x \sinh (5x + 2) = \cosh (5x + 2) D_x (5x + 2) = 5 \cosh (5x + 2).$$

2 If $y = \cosh (\ln x)$, find dy/dx.

SOLUTION
By part (ii) of Theorem 1,

$$\frac{dy}{dx} = \sinh (\ln x) \frac{d}{dx} \ln x = \frac{1}{x} \sinh (\ln x).$$

3 If $f(x) = x^2 \tanh x$, find $f'(x)$.

SOLUTION
Here, $f'(x) = 2x \tanh x + x^2 \operatorname{sech}^2 x.$

5.2 Integrals Yielding Hyperbolic Functions

The differentiation formulas in Section 5.1 can be inverted to yield the following integration formulas:

1 $\displaystyle\int \sinh u \, du = \cosh u + C.$

2 $\displaystyle\int \cosh u \, du = \sinh u + C.$

3 $\displaystyle\int \operatorname{sech}^2 u \, du = \tanh u + C.$

4 $\displaystyle\int \operatorname{csch}^2 u \, du = -\coth u + C.$

5 $\displaystyle\int \operatorname{sech} u \tanh u \, du = -\operatorname{sech} u + C.$

6 $\displaystyle\int \operatorname{csch} u \coth u \, du = -\operatorname{csch} u + C.$

EXAMPLES   Evaluate the given integral.

1 $\int \sinh 5x \, dx$

SOLUTION
Put $u = 5x$, so that $dx = \frac{1}{5} \, du$ and

$$\int \sinh 5x \, dx = \int \sinh u(\tfrac{1}{5} \, du) = \tfrac{1}{5} \int \sinh u \, du$$
$$= \tfrac{1}{5} \cosh u + C = \tfrac{1}{5} \cosh 5x + C.$$

2 $\int e^{\coth x} \operatorname{csch}^2 x \, dx$

SOLUTION
Put $u = \coth x$, so that $du = -\operatorname{csch}^2 x \, dx$ and

$$\int e^{\coth x} \operatorname{csch}^2 x \, dx = \int e^u(-du) = -e^u + C = -e^{\coth x} + C.$$

3 $\int_4^9 \dfrac{\coth \sqrt{x} \operatorname{csch} \sqrt{x}}{\sqrt{x}} \, dx$

SOLUTION
Put $u = \sqrt{x}$, so that $du = dx/(2\sqrt{x})$ and

$$\int_4^9 \frac{\coth \sqrt{x} \operatorname{csch} \sqrt{x}}{\sqrt{x}} \, dx = \int_2^3 \coth u \operatorname{csch} u(2 \, du)$$
$$= \left[-2 \operatorname{csch} u \right]\Big|_2^3 = 2 \operatorname{csch} 2 - 2 \operatorname{csch} 3.$$

Problem Set 5

In problems 1 through 14, prove each identity.

1 $\sinh (s \pm t) = \sinh s \cosh t \pm \sinh t \cosh s$

2 $\cosh (s \pm t) = \cosh s \cosh t \pm \sinh s \sinh t$

3 $\coth^2 x - 1 = \operatorname{csch}^2 x$

4 $\sinh 2x = 2 \sinh x \cosh x$

5 $\cosh 2x = \cosh^2 x + \sinh^2 x = 2 \cosh^2 x - 1 = 2 \sinh^2 x + 1$

6 $\tanh (s \pm t) = \dfrac{\tanh s \pm \tanh t}{1 \pm \tanh s \tanh t}$

7 $\sinh^2 s - \sinh^2 t = \sinh (s + t) \sinh (s - t)$

8 $\tanh 2t = \dfrac{2 \tanh t}{1 + \tanh^2 t}$

9 $\cosh \dfrac{t}{2} = \sqrt{\dfrac{\cosh t + 1}{2}}$

10 $(\cosh t + \sinh t)^3 = \cosh 3t + \sinh 3t$

11 $(\cosh t - \sinh t)^4 = \cosh 4t - \sinh 4t$

12 $\dfrac{e^{2t} - 1}{e^{2t} + 1} = \tanh t$

13 $\sinh (\ln x) = \dfrac{x^2 - 1}{2x}$

14 $e^x = \sinh x + \cosh x$

In problems 15 through 28, differentiate each function.

15 $f(x) = \sinh(3x^2 + 5)$

16 $g(x) = \cosh(\ln x)$

17 $f(t) = \ln(\sinh t^3)$

18 $f(u) = e^{2u} \tanh u$

19 $h(t) = \coth(e^{3t})$

20 $F(r) = \sin^{-1} r \tanh(3r + 5)$

21 $G(s) = \sin^{-1}(\operatorname{sech} s^2)$

22 $f(x) = \int_1^{\tanh x} \dfrac{dt}{1 + t^2}$

23 $g(x) = \int_1^{\sinh x^2} \dfrac{dt}{\sqrt{1 + t^2}}$

24 $G(x) = \dfrac{\operatorname{sech}(e^{2x})}{x^2 + 7}$

25 $f(x) = \dfrac{\tan^{-1}(\coth x)}{x^2 + 3}$

26 $g(x) = x^{\cosh x}$

27 $F(t) = \dfrac{e^{\cosh t}}{\sqrt{1 - t^2}}$

28 $f(x) = (\sinh x)^{5x}$

In problems 29 through 32, use implicit differentiation to find dy/dx.

29 $x^2 = \sinh y$

30 $\sinh x = \cosh y$

31 $\tanh^2 x - 2 \sinh y = \tanh y$

32 $\sin x = \sinh y$

33 Prove parts (ii), (iv), (v), and (vi) of Theorem 1.

34 Sketch the graph of $y = \coth x$.

35 Sketch the graph of $y = \operatorname{sech} x$.

36 Sketch the graph of $y = \operatorname{csch} x$.

In problems 37 through 48, evaluate each integral. (In some cases an appropriate substitution is suggested.)

37 $\displaystyle\int \cosh 7x \, dx, \ u = 7x$

38 $\displaystyle\int \sinh \dfrac{5x}{3} \, dx$

39 $\displaystyle\int \tanh 3x \operatorname{sech}^2 3x \, dx, \ u = \tanh 3x$

40 $\displaystyle\int \coth 5x \operatorname{csch}^2 5x \, dx$

41 $\displaystyle\int \dfrac{\operatorname{csch}^2 \sqrt{x}}{\sqrt{x}} \, dx, \ u = \sqrt{x}$

42 $\displaystyle\int \sinh^{10} 3x \cosh 3x \, dx, \ u = \sinh 3x$

43 $\displaystyle\int e^{\operatorname{sech} x} \operatorname{sech} x \tanh x \, dx, \ u = \operatorname{sech} x$

44 $\displaystyle\int \tanh x \operatorname{sech}^3 x \, dx$

45 $\displaystyle\int \dfrac{\sinh 5x}{\cosh^3 5x} \, dx$

46 $\displaystyle\int \dfrac{\operatorname{sech} x \tanh x}{1 + \operatorname{sech}^2 x} \, dx$

47 $\displaystyle\int_0^1 \cosh^3 x \sinh x \, dx$

48 $\displaystyle\int_0^2 \sinh^4 x \cosh x \, dx$

49 Evaluate $\displaystyle\int \cosh(\ln x) \, dx$. (*Hint:* Use the definition of cosh before integrating.)

50 Evaluate $\displaystyle\int \sinh(\ln 2x) \, dx$.

51 Simplify the equation $y = \tanh(\frac{1}{2} \ln x)$.

52 Find the area under the curve $y = \sinh x$ between $x = 0$ and $x = 1$.

53 Show that, if A, B, and k are constants, the function $y = A \sinh kx + B \cosh kx$ satisfies the differential equation $d^2y/dx^2 - k^2 y = 0$.

54 Given the function f defined by $f(x) = 9 \sinh x - 17 \cosh x$, find the maximum of f and find the intervals in which the graph of f is concave upward or downward.

55 When a flexible cord or chain is suspended from its ends, it hangs in a curve called a *catenary* whose equation is $y = a \cosh(x/a)$, a being a constant. Find the length of the catenary between $(-b, a \cosh(b/a))$ and $(b, a \cosh(b/a))$.

6   The Inverse Hyperbolic Functions

By analogy with the six inverse trigonometric functions, there are six inverse hyperbolic functions which we consider in the present section. Unlike the analogous trigonometric functions, sinh, tanh, coth, and csch are monotone, so there is no problem about the existence of $\sinh^{-1}$, $\tanh^{-1}$, $\coth^{-1}$, and csch^{-1}. The graphs of these four inverse hyperbolic functions are sketched in Figure 1. Of course, the graphs of $\sinh^{-1}$, $\tanh^{-1}$, $\coth^{-1}$, and csch^{-1} are obtained by reflecting the graphs of sinh, tanh, coth, and csch, respectively, about the line $y = x$.

Figure 1

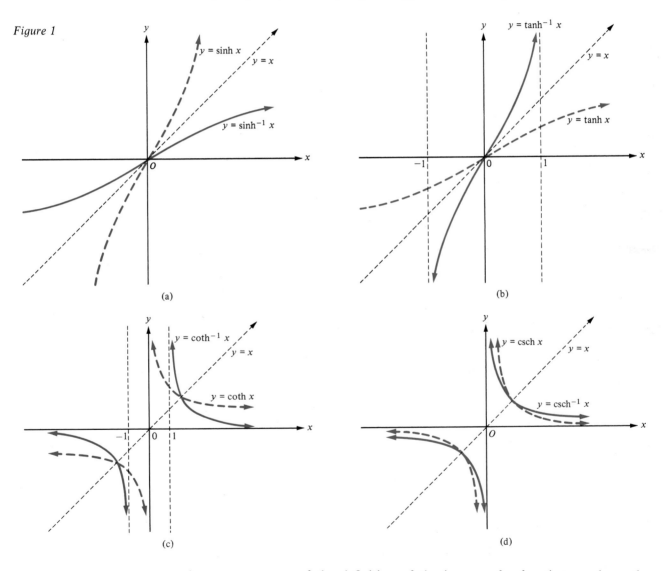

(a)

(b)

(c)

(d)

As a consequence of the definition of the inverse of a function, we have the following facts:

1 $y = \sinh^{-1} x$ if and only if $\sinh y = x$. **Therefore,** $\sinh(\sinh^{-1} x) = x$ and $\sinh^{-1}(\sinh y) = y$.

2 $y = \tanh^{-1} x$ if and only if $\tanh y = x$. **Therefore,** $\tanh(\tanh^{-1} x) = x$ and $\tanh^{-1}(\tanh y) = y$ hold if $|x| < 1$.

3 $y = \coth^{-1} x$ if and only if $\coth y = x$. Therefore, $\coth(\coth^{-1} x) = x$ and $\coth^{-1}(\coth y) = y$ hold if $|x| > 1$ and $y \neq 0$.

4 $y = \operatorname{csch}^{-1} x$ if and only if $\operatorname{csch} y = x$. Therefore, $\operatorname{csch}(\operatorname{csch}^{-1} x) = x$ and $\operatorname{csch}^{-1}(\operatorname{csch} y) = y$ hold if $x \neq 0$ and $y \neq 0$.

The hyperbolic cosine and its reciprocal, the hyperbolic secant, are not monotone functions; hence, they are not invertible. However, the portions of these functions whose graphs lie to the right of the y axis are monotone and consequently invertible (Figure 2). The inverses of these portions of cosh and sech are denoted by $\cosh^{-1}$ and sech^{-1}, respectively. Thus, we make the following definition.

Figure 2

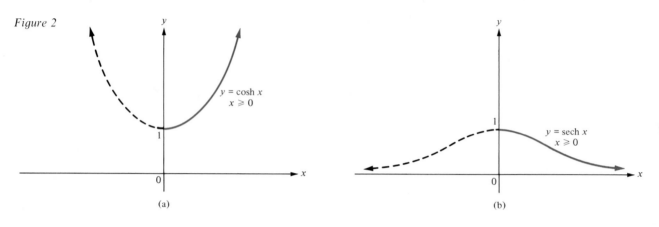

(a)   (b)

DEFINITION 1   **The Inverse Hyperbolic Cosine and Secant**

(i) The *inverse hyperbolic cosine* function, denoted by $\cosh^{-1}$, is defined by $y = \cosh^{-1} x$ if and only if $\cosh y = x$ and $y \geq 0$.

(ii) The *inverse hyperbolic secant* function, denoted by sech^{-1}, is defined by $y = \operatorname{sech}^{-1} x$ if and only if $\operatorname{sech} y = x$ and $y \geq 0$.

The graphs of $\cosh^{-1}$ and sech^{-1} are easily obtained from the graphs in Figure 2 by reflection about the line $y = x$ (Figure 3).

Figure 3

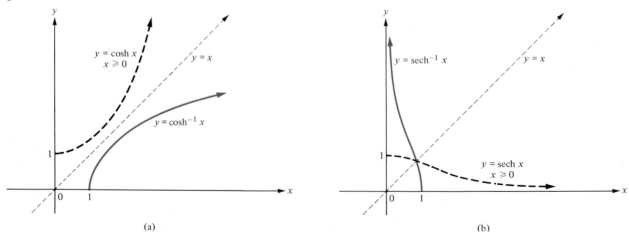

(a)   (b)

As a consequence of the definition of $\cosh^{-1}$ and of sech^{-1}, we have the following facts:

1 $\cosh(\cosh^{-1} x) = x$ and $\cosh^{-1}(\cosh y) = y$ hold if $x \geq 1$ and $y \geq 0$.

2 $\operatorname{sech}(\operatorname{sech}^{-1} x) = x$ and $\operatorname{sech}^{-1}(\operatorname{sech} y) = y$ hold if $0 < x \leq 1$ and $y \geq 0$.

The following theorem shows that all six of the inverse hyperbolic functions can be expressed in terms of the natural logarithm.

THEOREM 1   **Formulas for the Inverse Hyperbolic Functions**

(i) $\sinh^{-1} x = \ln(x + \sqrt{x^2 + 1})$ for any real number x.

(ii) $\cosh^{-1} x = \ln(x + \sqrt{x^2 - 1})$ for $x \geq 1$.

(iii) $\tanh^{-1} x = \frac{1}{2}\ln\dfrac{1 + x}{1 - x}$ for $|x| < 1$.

(iv) $\coth^{-1} x = \frac{1}{2}\ln\dfrac{x + 1}{x - 1}$ for $|x| > 1$.

(v) $\operatorname{sech}^{-1} x = \ln\left(\dfrac{1 + \sqrt{1 - x^2}}{x}\right)$ for $0 < x \leq 1$.

(vi) $\operatorname{csch}^{-1} x = \ln\left(\dfrac{1}{x} + \dfrac{\sqrt{1 + x^2}}{|x|}\right)$ for $x \neq 0$.

PROOF

We prove (i) and (iii) and leave the remaining formulas for the reader to check (Problems 2, 3, 4, and 5).

(i)   Let $y = \sinh^{-1} x$, so that $x = \sinh y = \dfrac{e^y - e^{-y}}{2}$. Thus, $2x = e^y - e^{-y}$.

Multiplying both sides of the latter equation by e^y, we obtain $2xe^y = (e^y)^2 - 1$, or $(e^y)^2 - 2x(e^y) - 1 = 0$. Solving this equation for e^y by the quadratic formula, we have

$$e^y = \frac{2x \pm \sqrt{4x^2 + 4}}{2} = x \pm \sqrt{x^2 + 1}.$$

Since e^y is always positive and $x - \sqrt{x^2 + 1}$ is always negative, the solution must be $e^y = x + \sqrt{x^2 + 1}$. Taking logarithms on both sides of the latter equation, we obtain $y = \ln(x + \sqrt{x^2 + 1})$. Therefore, we have $\sinh^{-1} x = \ln(x + \sqrt{x^2 + 1})$.

(iii) Let $y = \tanh^{-1} x$, so that $-1 < x < 1$ and

$$x = \tanh y = \frac{e^y - e^{-y}}{e^y + e^{-y}} = \frac{(e^y)^2 - 1}{(e^y)^2 + 1}.$$

Therefore, $x(e^y)^2 + x = (e^y)^2 - 1$, or $x + 1 = (1 - x)e^{2y}$. Hence, $e^{2y} = \dfrac{1 + x}{1 - x}$, so that $2y = \ln\dfrac{1 + x}{1 - x}$ and $\tanh^{-1} x = y = \dfrac{1}{2}\ln\dfrac{1 + x}{1 - x}$.

EXAMPLES   Find the approximate numerical value of the given quantity.

1 $\sinh^{-1} 4$

SOLUTION

$$\sinh^{-1} 4 = \ln(4 + \sqrt{4^2 + 1}) = \ln(4 + \sqrt{17}) \approx 2.0947.$$

2 $\tanh^{-1}\left(-\frac{1}{3}\right)$

SOLUTION

$$\tanh^{-1}\left(-\frac{1}{3}\right) = \frac{1}{2}\ln\frac{1-\frac{1}{3}}{1+\frac{1}{3}} = \frac{1}{2}\ln\frac{1}{2} \approx -0.3466.$$

6.1 Differentiation of the Inverse Hyperbolic Functions

The derivatives of the six inverse hyperbolic functions can now be obtained with ease, either by using the inverse function theorem or by using Theorem 1. The results are collected in the following theorem.

THEOREM 2   **Derivatives of the Inverse Hyperbolic Functions**
If u is a differentiable function of x, then we have:

(i) $D_x \sinh^{-1} u = \dfrac{1}{\sqrt{1+u^2}} D_x u.$

(ii) $D_x \cosh^{-1} u = \dfrac{1}{\sqrt{u^2-1}} D_x u$ if $u > 1.$

(iii) $D_x \tanh^{-1} u = \dfrac{1}{1-u^2} D_x u$ if $|u| < 1.$

(iv) $D_x \coth^{-1} u = \dfrac{1}{1-u^2} D_x u$ if $|u| > 1.$

(v) $D_x \operatorname{sech}^{-1} u = \dfrac{-1}{u\sqrt{1-u^2}} D_x u$ if $0 < u < 1.$

(vi) $D_x \operatorname{csch}^{-1} u = \dfrac{-1}{|u|\sqrt{1+u^2}} D_x u$ if $u \neq 0.$

PROOF
We prove (i) and (iii), leaving the remaining parts for the reader to check (Problems 22, 23, 24, and 25).

(i)   We elect to use the inverse function theorem and implicit differentiation rather than the formula $\sinh^{-1} x = \ln\left(x + \sqrt{x^2+1}\right)$. Since $D_x \sinh x = \cosh x > 0$, it follows that $\sinh^{-1}$ is a differentiable function by the inverse function theorem (Section 5.1, Chapter 2). Thus, let $y = \sinh^{-1} u$, so that $u = \sinh y$. Differentiation of both sides of the latter equation with respect to x gives $D_x u = \cosh y D_x y$ by the chain rule. Therefore, $D_x y = \dfrac{1}{\cosh y} D_x u$. Now, since $\cosh^2 y - \sinh^2 y = 1$, we have

$$\cosh y = \sqrt{1 + \sinh^2 y} = \sqrt{1 + u^2}$$

and it follows that

$$D_x y = \frac{1}{\sqrt{1+u^2}} D_x u \quad \text{or} \quad D_x \sinh^{-1} u = \frac{1}{\sqrt{1+u^2}} D_x u.$$

(iii) Although we could use the inverse function theorem and implicit differentiation as in part (i), we use the formula $\tanh^{-1} u = \frac{1}{2} \ln \frac{1+u}{1-u}$ from Theorem 1 instead. Thus,

$$D_x \tanh^{-1} u = \frac{1}{2} D_x \ln \frac{1+u}{1-u} = \frac{1}{2} \cdot \frac{1-u}{1+u} D_x\left(\frac{1+u}{1-u}\right) = \frac{1}{2} \cdot \frac{1-u}{1+u}\left(\frac{2D_x u}{(1-u)^2}\right)$$

$$= \frac{1}{(1+u)(1-u)} D_x u = \frac{1}{1-u^2} D_x u.$$

EXAMPLES 1 If $y = \tanh^{-1}(\sin 2x)$, find dy/dx.

SOLUTION

$$\frac{dy}{dx} = \frac{1}{1 - \sin^2 2x} D_x \sin 2x = \frac{2 \cos 2x}{\cos^2 2x} = 2 \sec 2x.$$

2 Figure 3a shows that the graph of $y = \cosh^{-1} x$ always lies below the straight line $y = x$ for $x \geq 1$. Prove this analytically.

SOLUTION
We must prove that $x - \cosh^{-1} x > 0$ holds for $x \geq 1$. Since $\cosh^{-1} 1 = 0$, the desired inequality holds for $x = 1$, so we need only consider the case $x > 1$. Define a function f by $f(x) = x - \cosh^{-1} x$ for $x > 1$. Then $f'(x) = 1 - \dfrac{1}{\sqrt{x^2 - 1}}$ and $f'(\sqrt{2}) = 0$. For $1 < x < \sqrt{2}$ we have $0 < x^2 - 1 < 1$, so that $f'(x) = 1 - \dfrac{1}{\sqrt{x^2 - 1}} < 0$. Similarly, for $x > \sqrt{2}$ we have $f'(x) > 0$, and it follows that $f(x)$ takes on its absolute minimum value at $x = \sqrt{2}$. Since this minimum value is $f(\sqrt{2}) = \sqrt{2} - \cosh^{-1}\sqrt{2} \approx 0.53 > 0$, $f(x) > 0$ must hold for all values of $x \geq 1$. Therefore, $x - \cosh^{-1} x > 0$ holds for all values of $x \geq 1$.

6.2 Integrals Yielding Inverse Hyperbolic Functions

The differentiation formulas obtained in Theorem 2 enable us to obtain some corresponding integration formulas as in the following theorem.

THEOREM 3   **Integrals Yielding Inverse Hyperbolic Functions**

(i) $\displaystyle\int \frac{dx}{\sqrt{a^2 + x^2}} = \sinh^{-1} \frac{x}{a} + C$ for $a > 0$.

(ii) $\displaystyle\int \frac{dx}{\sqrt{x^2 - a^2}} = \cosh^{-1} \frac{x}{a} + C$ for $a > 0$.

(iii) $\displaystyle\int \frac{dx}{a^2 - x^2} = \begin{cases} \dfrac{1}{a} \tanh^{-1} \dfrac{x}{a} + C & \text{for } |x| < a. \\[2mm] \dfrac{1}{a} \coth^{-1} \dfrac{x}{a} + C & \text{for } |x| > a > 0. \end{cases}$

(iv) $\displaystyle\int \frac{dx}{x\sqrt{a^2 - x^2}} = -\frac{1}{a}\operatorname{sech}^{-1}\frac{|x|}{a} + C$ for $0 < |x| < a$.

(v) $\displaystyle\int \frac{dx}{x\sqrt{a^2 + x^2}} = -\frac{1}{a}\operatorname{csch}^{-1}\frac{|x|}{a} + C$ for $a > 0$, $x \neq 0$.

PROOF

To verify each formula, we simply differentiate the right side using Theorem 2 and the fact that $D_x|x| = |x|/x$ for $x \neq 0$. The detailed verification is left as an exercise for the reader (Problem 41).

EXAMPLES   Evaluate the given integral.

1 $\displaystyle\int \frac{dx}{\sqrt{4 + x^2}}$

SOLUTION
By part (i) of Theorem 3,

$$\int \frac{dx}{\sqrt{4 + x^2}} = \sinh^{-1}\frac{x}{2} + C.$$

2 $\displaystyle\int \frac{dx}{x\sqrt{9 - x^2}}$

SOLUTION
By part (iv) of Theorem 3,

$$\int \frac{dx}{x\sqrt{9 - x^2}} = \frac{-1}{3}\operatorname{sech}^{-1}\frac{|x|}{3} + C \text{ for } 0 < |x| < 3.$$

3 $\displaystyle\int \frac{dx}{16 - x^2}$

SOLUTION
By part (iii) of Theorem 3,

$$\int \frac{dx}{16 - x^2} = \begin{cases} \dfrac{1}{4}\tanh^{-1}\dfrac{x}{4} + C & \text{for } |x| < 4. \\[2ex] \dfrac{1}{4}\coth^{-1}\dfrac{x}{4} + C & \text{for } |x| > 4. \end{cases}$$

Problem Set 6

1 Use Theorem 1 to find the approximate numerical value of
   (a) $\sinh^{-1} 3$,   (b) $\tanh^{-1}\left(-\frac{1}{3}\right)$,   (c) $\coth^{-1} 3$,   and (d) $\operatorname{sech}^{-1}\left(\frac{1}{3}\right)$.

2 Prove part (ii) of Theorem 1.   **3** Prove part (iv) of Theorem 1.

4 Prove part (v) of Theorem 1.   **5** Prove part (vi) of Theorem 1.

6 Prove
   (a) $\sinh^{-1}\sqrt{x^2 - 1} = \cosh^{-1} x$ if $x > 1$.   (b) $\cosh^{-1}\sqrt{x^2 + 1} = \sinh^{-1}|x|$.

In problems 7 through 20, differentiate each function.

7 $f(x) = \sinh^{-1} x^3$

8 $g(x) = \cosh^{-1} \dfrac{x}{3}$

9 $G(t) = \tanh^{-1} 5t$

10 $h(u) = \coth^{-1} u^2$

11 $g(r) = \coth^{-1} (5r + 2)$

12 $H(x) = \text{sech}^{-1} x^2$

13 $f(x) = \text{sech}^{-1} (\sin x)$

14 $G(t) = t \cosh^{-1} t - \sqrt{1 + t^2}$

15 $h(u) = u \tanh^{-1} (\ln u)$

16 $F(x) = \dfrac{\coth^{-1} x}{x^2 + 4}$

17 $g(x) = x \cosh^{-1} (e^x)$

18 $G(x) = x \coth^{-1} \sqrt{x^2 - 1}$

19 $F(x) = \dfrac{\text{csch}^{-1} x}{e^x + 3}$

20 $H(t) = \ln \sqrt{t^2 - 1} - t \tanh^{-1} t$

21 If $x^2 - y^2 = 1$ with $x > 0$, show that there is exactly one value of t for which $x = \cosh t$ and $y = \sinh t$.

22 Prove part (ii) of Theorem 2.

23 Prove part (iv) of Theorem 2.

24 Prove part (v) of Theorem 2.

25 Prove part (vi) of Theorem 2.

26 Verify that $\displaystyle\int \sinh^{-1} x \, dx = x \sinh^{-1} x - \sqrt{1 + x^2} + C$ by differentiation of the right side.

In problems 27 through 40, evaluate each integral.

27 $\displaystyle\int \dfrac{dx}{\sqrt{9 + x^2}}$

28 $\displaystyle\int \dfrac{dx}{\sqrt{1 + 4x^2}}$

29 $\displaystyle\int \dfrac{dx}{\sqrt{x^2 - 16}}$

30 $\displaystyle\int \dfrac{dx}{\sqrt{9x^2 - 1}}$

31 $\displaystyle\int \dfrac{dx}{25 - x^2}$

32 $\displaystyle\int \dfrac{dx}{1 - 9x^2}$

33 $\displaystyle\int \dfrac{dx}{x\sqrt{36 - x^2}}$

34 $\displaystyle\int \dfrac{dx}{x\sqrt{1 - 4x^2}}$

35 $\displaystyle\int \dfrac{dx}{x\sqrt{16 + x^2}}$

36 $\displaystyle\int \dfrac{dx}{x\sqrt{1 + 4x^2}}$

37 $\displaystyle\int_0^1 \dfrac{dx}{\sqrt{1 + x^2}}$

38 $\displaystyle\int_0^{0.5} \dfrac{dx}{1 - x^2}$

39 $\displaystyle\int_5^7 \dfrac{dx}{\sqrt{x^2 - 9}}$

40 $\displaystyle\int_{-1}^1 \dfrac{dx}{\sqrt{3 - x^2}}$

41 Prove Theorem 3.

42 Find the area of the region bounded by the curve $y = 16/\sqrt{16 + x^2}$ and the lines $x = 0$, $y = 0$, and $x = 3\sqrt{2}$.

43 Given that $y = \sinh^{-1} xy$, use implicit differentiation to find dy/dx.

7 Exponential Growth

In the next two sections we briefly consider some of the many applications of the exponential and logarithmic functions. Often these applications involve the solution of a separable differential equation containing a term of the form dy/y. Upon integration, this term contributes a term of the form $\ln |y|$ to the solution.

Applications of logarithms and exponentials range all the way from calculating the age of a fossilized bone by radioactive dating to calculating the probability of accidents occurring in a large industrial plant. Here we begin with applications to problems involving growth and decay.

7.1 The Natural Law of Growth

Many natural quantities change in time at an instantaneous rate depending on the value of the quantity itself. For instance, the rate at which the amount of a radioactive substance decreases because of radioactive disintegration depends on the amount of substance present. Similarly, the rate at which the amount of money in a savings account at compound interest increases depends on the amount of money in the account.

In general, if q represents the quantity of some substance at time t, then dq/dt represents the instantaneous rate of change of q. As we have mentioned, it often happens that dq/dt depends on q in some definite way; in fact, in many cases dq/dt turns out to be proportional to q, so that $dq/dt = kq$, where k is the constant of proportionality. If $k > 0$, then $dq/dt > 0$, so that q is increasing as time goes on, whereas if $k < 0$, then $dq/dt < 0$ and q decreases as time passes. The differential equation $dq/dt = kq$ applies to so many different natural phenomena that it is often called the *natural law of growth*. (Strictly speaking, when $k < 0$, it should be called the natural law of *decay*.)

The differential equation $dq/dt = kq$ is solved as follows.

THEOREM 1   **Solution of $dq/dt = kq$**

The general solution of the differential equation $\dfrac{dq}{dt} = kq$, where k is a constant, is $q = q_0 e^{kt}$, where q_0 is a constant equal to the value of q when $t = 0$.

PROOF
The differential equation $dq/dt = kq$ is separable, so it can be rewritten as $dq/q = k\,dt$. Integration of both sides of the latter equation yields $\ln |q| = kt + C$, where C is the constant of integration. Exponentiating both sides of the latter equation, we have $e^{\ln |q|} = e^{kt+C}$; that is, $|q| = e^{kt}e^C$. Since $|q| = \pm q$, we can write $q = (\pm e^C)e^{kt}$ and, because C is a constant, so is $\pm e^C$. If q_0 is the value of q when $t = 0$, then $q_0 = (\pm e^C)e^{(k)(0)} = (\pm e^C)e^0 = \pm e^C$; hence, $q = q_0 e^{kt}$.

Actually, it can be shown that $q = q_0 e^{kt}$ is the *complete* solution of the natural law of growth $dq/dt = kq$; that is, there are no singular solutions, so every solution (on an open interval) has the form $q = q_0 e^{kt}$ (Problem 8).

EXAMPLE   The population q of the United States in 1975 was approximately 220 million. Suppose that q grows at a rate proportional to itself, say $dq/dt = kq$, with $k = 0.017$ (that is, 1.7 percent per year). Calculate the value of q: (a) in 1990 and (b) in 2001.

SOLUTION
By Theorem 1, $q = q_0 e^{0.017t}$, where $q_0 = 220$ million. In 1990, $t = 1990 - 1975 = 15$ and

$$q = 220 e^{(0.017)(15)} \approx 284 \text{ million}.$$

In 2001, $t = 2001 - 1975 = 26$ and

$$q = 220 e^{(0.017)(26)} \approx 342 \text{ million}.$$

In many problems involving the natural law of growth, we are given data concerning the values of q at certain times, but we are not given the value of k. Under these circumstances, the following theorem is useful.

THEOREM 2   **Natural-Law-of-Growth Formulas**

Suppose that the quantity q satisfies the natural law of growth, so that $q = q_0 e^{kt}$. Assume that $q = q_1$ when $t = t_1$ and that $q = q_2$ when $t = t_2$, where $t_1 \neq t_2$. Then:

(i) $k = \dfrac{1}{t_2 - t_1} \ln \left(\dfrac{q_2}{q_1} \right).$

(ii) $q_0 = q_1 \left(\dfrac{q_1}{q_2} \right)^{t_1/(t_2-t_1)}$

(iii) $q_2 = q_0 \left(\dfrac{q_1}{q_0} \right)^{t_2/t_1}$ if $t_1 \neq 0.$

PROOF

We have $q_1 = q_0 e^{kt_1}$ and $q_2 = q_0 e^{kt_2}$. Taking natural logarithms on both sides of these two equations yields

$$\ln q_1 = \ln q_0 + kt_1 \quad \text{and} \quad \ln q_2 = \ln q_0 + kt_2.$$

Subtracting the first equation from the second, we obtain $\ln q_2 - \ln q_1 = kt_2 - kt_1$, so that

$$\ln \left(\frac{q_2}{q_1} \right) = k(t_2 - t_1) \quad \text{or} \quad k = \frac{1}{t_2 - t_1} \ln \left(\frac{q_2}{q_1} \right).$$

This proves (i). To prove (ii), we solve the equation $q_1 = q_0 e^{kt_1}$ for q_0 to obtain $q_0 = q_1 e^{-kt_1}$, then substitute the value just obtained for k. Thus,

$$q_0 = q_1 e^{-kt_1} = q_1 \exp \left[\frac{-t_1}{t_2 - t_1} \ln \left(\frac{q_2}{q_1} \right) \right] = q_1 \exp \left[\frac{t_1}{t_2 - t_1} \ln \left(\frac{q_2}{q_1} \right)^{-1} \right]$$

$$= q_1 \exp \left[\frac{t_1}{t_2 - t_1} \ln \left(\frac{q_1}{q_2} \right) \right] = q_1 \left(\frac{q_1}{q_2} \right)^{t_1/(t_2-t_1)},$$

by Definition 1 of Section 4. To prove (iii), we solve the equation in (ii) for q_2 as follows: From (ii), we obtain

$$\frac{q_0}{q_1} = \left(\frac{q_1}{q_2} \right)^{t_1/(t_2-t_1)} \quad \text{so that} \quad \left(\frac{q_0}{q_1} \right)^{(t_2-t_1)/t_1} = \frac{q_1}{q_2}$$

or

$$q_2 = q_1 \left(\frac{q_0}{q_1} \right)^{-(t_2-t_1)/t_1} = q_1 \left(\frac{q_1}{q_0} \right)^{(t_2-t_1)/t_1} = q_1 \left(\frac{q_1}{q_0} \right)^{(t_2/t_1)-1}$$

$$= q_1 \left(\frac{q_1}{q_0} \right)^{t_2/t_1} \left(\frac{q_1}{q_0} \right)^{-1} = q_1 \left(\frac{q_1}{q_0} \right)^{t_2/t_1} \left(\frac{q_0}{q_1} \right) = q_0 \left(\frac{q_1}{q_0} \right)^{t_2/t_1}.$$

EXAMPLES 1 The rate of decay of radium is proportional to the amount present at any time. If after 25 years a quantity of radium has decreased to 4.948 grams and if at the end of an additional 25 years it has decreased to 4.896 grams, how many grams were originally present?

SOLUTION

In Theorem 2, put $q_1 = 4.948$, $t_1 = 25$, $q_2 = 4.896$, and $t_2 = 50$. By part (ii) of Theorem 2,

$$q_0 = q_1 \left(\frac{q_1}{q_2}\right)^{t_1/(t_2 - t_1)} = 4.948\left(\frac{4.948}{4.896}\right)^{25/(50 - 25)}$$

$$= \frac{(4.948)^2}{4.896} \approx 5 \text{ grams.}$$

2 Bacteria grown in a culture increase at a rate proportional to the number of bacteria present. If there are 2500 bacteria present initially and if the number of bacteria triples in $\frac{1}{2}$ hour, how many bacteria are present after t hours? How many are present after 2 hours?

SOLUTION

In Theorem 2 we have $q_0 = 2500$ and $q_1 = 7500$ when $t_1 = \frac{1}{2}$ hour. By part (iii) of Theorem 2 with $t_2 = t$ and $q_2 = q$, at the end of t hours the number of bacteria present is given by

$$q = q_0 \left(\frac{q_1}{q_0}\right)^{t/t_1} = 2500\left(\frac{7500}{2500}\right)^{2t} = 2500(3)^{2t};$$

hence, $q = 2500(9^t)$. When $t = 2$ hours, $q = 2500(81) = 202{,}500$.

3 In Example 2, how long will it take before there are 1,000,000 bacteria present?

SOLUTION

Put $q = 1{,}000{,}000$ in the formula $q = 2500(9^t)$ and solve for t as follows:

$$1{,}000{,}000 = 2500(9^t), \qquad 9^t = 400;$$

hence,

$$\log_{10} 9^t = \log_{10} 400, \; t \log_{10} 9 = \log_{10} 400, \text{ and}$$

$$t = \frac{\log_{10} 400}{\log_{10} 9} \approx \frac{2.6021}{0.9542} \approx 2.73 \text{ hours.}$$

If a substance is decaying in accordance with the natural law of growth, then the amount q of the substance present at time t is given by $q = q_0 e^{kt}$, where $k < 0$. In this case, one usually rewrites the equation in the form $q = q_0 e^{-at}$ where $a = -k > 0$. Here the constant a gives a measure of how rapidly the substance is decaying. Another such measure, which is perhaps easier to comprehend, is the *half-life* of the substance, defined to be the length of time T during which exactly half of the substance decays. Thus, at time T the remaining amount of substance is $\frac{1}{2}q_0$, so $\frac{1}{2}q_0 = q_0 e^{-aT}$ or $e^{aT} = 2$. Taking natural logarithms on both sides of the latter equation, we obtain $aT = \ln 2$, or $T = (\ln 2)/a$.

EXAMPLES **1** Find a formula for the quantity q of a radioactive substance present after t years if the half-life of the substance is T years.

SOLUTION

Here we have $q = q_0 e^{-at}$, where $a = (\ln 2)/T$; hence,

$$q = q_0 e^{-(t/T)\ln 2} = q_0 2^{-t/T}.$$

2 If the half-life of polonium is 140 days, how long does it take for 2 grams of polonium to decay to 0.1 gram?

SOLUTION

By the previous example,

$$q = q_0 e^{-(t/T)\ln 2}, \frac{q}{q_0} = e^{-(t/T)\ln 2}, \ln \frac{q}{q_0} = -\frac{t}{T}\ln 2,$$

so that

$$t = -\frac{T}{\ln 2}\ln\frac{q}{q_0} = \frac{T}{\ln 2}\ln\left(\frac{q}{q_0}\right)^{-1} = \frac{T}{\ln 2}\ln\frac{q_0}{q}.$$

In the present case, $q_0 = 2$, $q = 0.1$, $T = 140$ days, and so

$$t = \frac{140}{\ln 2}\ln 20 \approx 605 \text{ days.}$$

Problem Set 7

1 The rate at which the population of a certain city increases is proportional to the population. If there were 125,000 people in the city in 1960 and 140,000 in 1975, what population can be predicted for the year 2000?

2 Given that $dy/dx = 2y$ and that $y = 10$ when $x = 0$, find the value of y when $x = 1$.

3 In the chemical processing of a certain mineral, the rate of change of the amount of mineral remaining is proportional to this amount. If, after 8 hours, 100 kilograms of mineral has been reduced to 70 kilograms, what quantity of the mineral remains after 24 hours?

4 The half-life of radium is 1656 years. If a sample of radium now weighs 50 milligrams, how much radium remains in the sample 100 years from now?

5 Bacteria grown in a certain culture increase at a rate proportional to the number of bacteria present. If there are 1000 bacteria initially and if the number of bacteria doubles in 15 minutes, how long will it take before there are 2,000,000 bacteria present?

6 True or false: A radioactive substance has a half-life of 1800 years; that is, half of it will have decayed after 1800 years. Therefore, after 3600 years, all of the substance will be gone.

7 At a certain instant 100 grams of a radioactive substance is present. After 4 years, 20 grams remain. (a) How much of the substance remains after 8 years? (b) What is the half-life of the substance?

8 Prove that the solution $q = q_0 e^{kt}$ of the differential equation $dq/dt = kq$ in Theorem 1 is actually the complete solution. Do this by assuming that $q = f(t)$ is a solution of $dq/dt = kq$ and that $q_0 = f(0)$, then calculate $D_t[f(t)e^{-kt}]$.

9 The number of bacteria in an unrefrigerated chicken salad triples in 6 hours. In how many hours is the number multiplied by a factor of 50?

10 Bacteria in a culture have a natural tendency to increase by 25 percent in 1 hour; however, a certain drug present in the culture kills 20 percent of the bacteria per hour. How much time is required for the number of bacteria to double under these conditions?

11 A mobile home initially costs $18,000. If it depreciates at a rate that is proportional to its value, and has a value of $15,000 after 2 years, what is its value after 10 years?

12 A cylindrical tank 2 feet high has a cross-sectional area of 9 square feet and is initially full of benzene. However, there is a leak in the bottom of the tank, and the benzene is running out at a rate proportional to its depth. Find the volume of benzene in the tank at the end of 2 days if the tank is half full at the end of 12 hours.

13 In Theorem 2, find a formula expressing t_2 in terms of q_0, q_1, q_2, and t_1.

14 Let p be a continuous function of t and let g be any antiderivative of p. Show that $q = Ce^{-g(t)}$ is a solution of the differential equation $dq/dt + p(t)q = 0$. Explain how this result generalizes Theorem 1.

15 The differential equation $dq/dt = Kb^t q$, where K and b are positive constants and $b < 1$, is sometimes used to describe the growth of a quantity q under restrictive conditions, for instance, the growth of an animal population with a restricted food supply. Solve this separable differential equation for q as a function of time t. (The solution is called a *Gompertz* function.)

16 Let P and Q be continuous functions of x. The differential equation $dy/dx + P(x)y = Q(x)$ is called a *first-order linear differential equation.* If g is any antiderivative of P, show that a solution of this differential equation is given by $y = e^{-g(x)} \int e^{g(x)} Q(x) \, dx$.

17 The current I flowing in an electrical circuit consisting of a constant inductance of L henrys, a constant resistance of R ohms, and a driving electromotive force of $E(t)$ volts at time t satisfies the first-order linear differential equation $\dfrac{dI}{dt} + \dfrac{R}{L} I = \dfrac{E(t)}{L}$.

   (a) Use the result of problem 16 to solve for I as a function of time t.
   (b) Assuming that E is constant and that $I = 0$ when $t = 0$, find $\lim\limits_{t \to +\infty} I$.

8 Other Applications of Logarithms and Exponentials

In Section 7 we considered applications to problems involving growth and decay. Here, we consider some other applications of logarithms and exponentials; specifically, applications to problems involving mixing, cooling, continuously compounded interest, and Poisson processes.

8.1 Mixing and Cooling

Problems involving the uniform mixing of various substances, as well as problems concerning the rate at which objects lose heat to their surroundings, often give rise to differential equations whose solutions involve logarithms or exponentials.

EXAMPLE   A tank initially holds 100 gallons of salt water at a concentration of $\frac{3}{10}$ pound of salt per gallon of water. Water containing $\frac{1}{10}$ pound of salt per gallon runs into the tank at the rate of 2 gal/min and the solution, kept uniform by stirring, runs out at the same rate.

   (a) Find an equation for the number of pounds q of salt in the tank at the end of t minutes and sketch the graph of this equation.
   (b) Find the concentration of salt in the tank at the end of 25 minutes.

SOLUTION

(a) The concentration of salt in the tank at time t is $q/100$ lb/gal. In the "infinitesimal" time interval dt, $2\,dt$ gallons of salt water carrying $\frac{1}{10}(2\,dt)$ pounds of salt run into the tank, while $2\,dt$ gallons of salt water carrying $(q/100)(2\,dt)$ pounds of salt flow out of the tank. Thus, during the time interval dt, the amount of salt in the tank changes by the "infinitesimal" amount

$$dq = \tfrac{1}{10}(2\,dt) - (q/100)(2\,dt) \text{ pounds.}$$

Therefore, q satisfies the separable differential equation

$$dq = \frac{20 - 2q}{100}\,dt$$

with the initial condition that $q = \left(\frac{3}{10}\right)(100) = 30$ pounds when $t = 0$. Separating variables and integrating, we obtain

$$\frac{dq}{20 - 2q} = \frac{dt}{100} \quad \text{so that} \quad -\frac{1}{2}\ln|20 - 2q| = \frac{t}{100} + C_1$$

or

$$\ln|2q - 20| = -\frac{t}{50} + C, \qquad \text{where } C = -2C_1.$$

Exponentiating both sides of the latter equation, we obtain

$$|2q - 20| = e^C e^{-t/50},$$

so that

$$2q - 20 = \pm e^C e^{-t/50}$$

or

$$q = 10 + Ke^{-t/50}, \text{ where } K = \pm e^C/2.$$

Since $q = 30$ when $t = 0$, it follows that $30 = 10 + Ke^0$; hence, $K = 20$. Thus, at the end of t minutes, there are q pounds of salt in the tank, where $q = 10 + 20e^{-t/50}$. Figure 1 shows the graph of this equation. Notice that q approaches 10 pounds as t approaches $+\infty$.

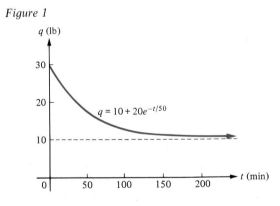

Figure 1

(b) At the end of 25 minutes, there are $10 + 20e^{-25/50} = 10 + 20e^{-1/2}$ pounds of salt in the tank, so the salt concentration is given by

$$\frac{10 + 20e^{-1/2}}{100} \approx 0.22 \text{ pound per gallon.}$$

Newton's law of cooling states that the instantaneous rate of change of temperature of a cooling body is proportional to the difference between the temperature of the body and that of the surrounding medium. Thus, if x is the temperature of the body at time t, then

$$\frac{dx}{dt} = k(x - a),$$

where k is the constant of proportionality and a is the temperature of the surrounding medium. Here we assume that a is constant. Note that $k < 0$. (Why?)

The differential equation $dx/dt = k(x - a)$ is similar to the natural law of growth $dq/dt = kq$; in fact, if we put $q = x - a$, then, assuming that a is constant, $dq/dt = dx/dt$. Consequently, Newton's law of cooling $dx/dt = k(x - a)$ is equivalent to the natural law of growth $dq/dt = kq$ if we put $q = x - a$. Therefore, the results obtained in Section 7.1 can be used to help solve problems about Newton's law of cooling. For instance, by Theorem 1 in Section 7.1, $q = q_0 e^{kt}$; that is, $x - a = q_0 e^{kt}$, or $x = a + q_0 e^{kt}$, where $q_0 = x_0 - a$ and x_0 is the value of x when $t = 0$.

EXAMPLE   An object is heated to $300°F$ and then allowed to cool in a room whose temperature is $80°F$. If, after 10 minutes, the temperature of the object is $250°$, what is its temperature after 20 minutes?

SOLUTION
Let x be the temperature at time t minutes and let $q = x - a$, where $a = 80°$. Thus, $q_0 = x_0 - a = 300° - 80° = 220°$. At time $t_1 = 10$ minutes, we have a temperature of $x_1 = 250°$ and a corresponding value $q_1 = x_1 - a = 250° - 80° = 170°$. By part (iii) of Theorem 2 in Section 7.1,

$$q_2 = q_0 \left(\frac{q_1}{q_0} \right)^{t_2/t_1}$$

gives the value of q at time t_2. Therefore, the value of x at time t_2 is given by

$$x_2 = a + q_2 = a + q_0 \left(\frac{q_1}{q_0} \right)^{t_2/t_1}.$$

To solve our current problem, we put $t_2 = 20$ minutes, so

$$x_2 = 80 + 220 \left(\frac{170}{220} \right)^{20/10} = 80 + 220 \left(\frac{17}{22} \right)^2 \approx 211.36°F.$$

8.2 Continuously Compounded Interest

Suppose that q_0 dollars is deposited in a savings account at a compound interest of k percent per year compounded n times per year—that is, compounded every $(1/n)$th of a year. For instance, if $n = 2$, the interest is compounded semi-annually; if $n = 4$, it is compounded quarterly; if $n = 52$, it is compounded weekly; and so forth.

At the end of the first $(1/n)$th part of the year, an interest of $(k/100)(1/n)q_0$ dollars is added to the principal and $\left(q_0 + \dfrac{k}{100n} q_0 \right)$ dollars is then in the account. At the end of the second $(1/n)$th part of the year, an interest of $\left(\dfrac{k}{100} \right)\left(\dfrac{1}{n} \right)\left(q_0 + \dfrac{k}{100n} q_0 \right)$ dollars is added to the principal and the account contains

$$\left(q_0 + \frac{k}{100n} q_0 \right) + \frac{k}{100n}\left(q_0 + \frac{k}{100n} q_0 \right) = q_0 \left(1 + \frac{k}{100n} \right)^2 \text{ dollars.}$$

At the end of the third $(1/n)$th part of the year, the amount in the account will be $q_0[1 + k/(100n)]^3$ dollars, and so on. Thus, at the end of the year the account will contain $q_0 \left(1 + \dfrac{k}{100n} \right)^n$ dollars.

EXAMPLE If $1000 is invested at a compound interest rate of 7 percent per year compounded daily, how much is the investment worth at the end of 1 year?

SOLUTION
Using the formula above with $k = 7$ and $n = 365$, we see that the amount of the investment at the end of 1 year is

$$\$1000\left(1 + \frac{7}{(100)(365)}\right)^{365} = \$1072.50.$$

If, in the example above, the interest were compounded hourly rather than daily, then the amount at the end of the 1-year period would be

$$\$1000\left[1 + \frac{7}{(100)(365)(24)}\right]^{(365)(24)} = \$1072.51.$$

Since q_0 dollars invested at k percent interest per year compounded every $(1/n)$th part of the year is worth $q_0[1 + k/(100n)]^n$ dollars at the end of the year, it seems reasonable to define

$$q = \lim_{n \to +\infty} q_0\left(1 + \frac{k}{100n}\right)^n$$

to be the value of the investment at the end of the year if the interest is compounded *continuously*. In the formula above, put $a = k/100$ and use Theorem 8 in Section 4.4 to obtain

$$q = \lim_{n \to +\infty} q_0\left(1 + \frac{a}{n}\right)^n = q_0 \lim_{n \to +\infty}\left(1 + \frac{a}{n}\right)^n = q_0 e^a.$$

Therefore, q_0 dollars invested at k percent interest per year compounded continuously is worth $q_0 e^{k/100}$ dollars at the end of 1 year. At the end of t years, it is worth $q_0 e^{kt/100}$ dollars.

EXAMPLE If $100,000 is invested at an interest rate of 6 percent per year compounded continuously, how much is the investment worth after 1 year? How much is it worth after t years? In how many years will the original amount double?

SOLUTION
After 1 year, the investment is worth $\$100,000 e^{7/100} = \$107,250.82$. After t years, the original investment is worth $\$100,000 e^{7t/100}$. Thus, the investment doubles when $e^{7t/100} = 2$; that is, when $\frac{7t}{100} = \ln 2$, or $t = \frac{100}{7} \ln 2 \approx 9.9$ years.

8.3 Poisson Processes

Many phenomena can be viewed as a number of separate and distinct "happenings" occurring relative to a continuous "background." For instance, if the continuous "background" is time, the happenings might be anything from radioactive disintegrations of individual uranium atoms to suicides in New York City. On the other hand, the continuous "background" might be the volume of water in the supply reservoir of a city and the "happenings" might be the existence of individual coliform bacteria distributed within this volume.

A phenomenon such as any of those described above is called a *Poisson process* (in honor of the French mathematician S. D. Poisson), provided that the following conditions are met:

1   If the variable t represents the continuous background, then the probability of one happening in a very small portion Δt of t is proportional to Δt.

2   The probability of two or more happenings in the same very small portion Δt of t is negligible.

3   If Δt_1 and Δt_2 are two very small nonoverlapping portions of t, then the occurrence or nonoccurrence of a happening in Δt_1 exerts no influence over the occurrence or nonoccurrence of a happening in Δt_2.

Under conditions 1, 2, and 3, it can be shown that the probability of exactly k happenings in t units of the continuous background is given by the Poisson formula

$$p_k = \frac{(ct)^k}{k!} e^{-ct},$$

where the constant c is the long-run average number of happenings per unit of background. (See Problem 54 in Problem Set 4 for an indication of how the Poisson formula is derived.)

EXAMPLES 1 Coliform bacteria are known to be present in the water supply reservoir of a certain city at the average rate of $c = 2$ bacteria per cubic inch of water. Assume that the presence of a coliform bacterium in a sample of this water is a Poisson process—that is, a happening is the occurrence of a single bacterium and the continuous background is the volume of water involved. If a sample of 9 cubic inches of water is drawn from the reservoir, what is the probability that the sample contains exactly 20 coliform bacteria?

SOLUTION
Here $t = 9$ cubic inches, $c = 2$ bacteria per cubic inch on the average, and $k = 20$ bacteria; hence, by the Poisson formula,

$$p_k = \frac{(ct)^k}{k!} e^{-ct} = \frac{18^{20}}{20!} e^{-18} \approx 0.08 \quad \text{or} \quad \text{about } 8\%.$$

2 In a certain calculus textbook there are 4000 exercises. Answers to all of these exercises are in the back of the book; however, 1 percent of the given answers are incorrect. Dolores works 10 exercises and then checks her answers against those in the back of the book. What is the probability that all 10 of the answers in the back of the book are correct?

SOLUTION
Even though the background (that is, answers in the back of the book) is not really continuous, there are so many answers (4000 of them) that we can safely neglect this discontinuity and use the Poisson formula. Here, of course, a happening is an erroneous answer. Since 1 percent of the 4000 answers are wrong, there are 40 incorrect answers and $c = 40/4000 = 0.01$ error per answer. For $t = 10$ answers and $k = 0$ errors, the Poisson formula gives

$$p_0 = \frac{[(0.01)(10)]^0}{0!} e^{-(0.01)(10)} = e^{-0.1} \approx 0.9 \quad \text{or} \quad \text{about } 90\%$$

probability of no error in the 10 answers.

Suppose that we have a Poisson process whose background is time, so that, in any short interval of time Δt, we may or may not observe a happening. Let c be the long-run average number of happenings per unit time interval. By the Poisson formula, the probability of exactly k happenings in a time interval of length t is given by $p_k = \dfrac{(ct)^k}{k!} e^{-ct}$.

Let us begin to observe this Poisson process and record the elapsed time T until the *first* happening. We denote by $P\{T > t\}$ the probability that the waiting time T until the first happening exceeds t. Thus, for example, if $P\{T > 2\} = 0.57$, where time is measured in hours, then in about 57 percent of the cases, we could expect to wait at least 2 hours before the first happening. Similarly, we denote by $P\{T \le t\}$ the probability that the waiting time T does not exceed t. Notice that if $P\{T > 2\} = 0.57$, then $P\{T \le 2\} = 0.43$. (Why?) More generally, $P\{T \le t\} = 1 - P\{T > t\}$.

Evidently, to assert that the waiting time T until the first happening exceeds t is equivalent to asserting that there are exactly zero happenings in the time interval t. Therefore, using the Poisson formula $p_k = \dfrac{(ct)^k}{k!} e^{-ct}$ with $k = 0$, we have

$$P\{T > t\} = p_0 = \frac{(ct)^0}{0!} e^{-ct} = e^{-ct} \quad \text{and} \quad P\{T \le t\} = 1 - e^{-ct},$$

where c is the average number of happenings per unit time.

EXAMPLES 1 Flameouts, occurring for a certain type of jet engine, have been observed to be approximately a Poisson process averaging one flameout every 1000 hours of operation. What is the probability that one of these engines operates for 1500 hours without a flameout?

SOLUTION
Here $c = \frac{1}{1000}$ flameout per hour; hence,

$$P\{T > 1500\} = e^{-(1/1000)(1500)} = e^{-3/2} \approx 0.22 \quad \text{or} \quad \text{about } 22\%.$$

2 Misprints occur in a certain book at the average rate of one per 100 pages. What is the probability that the first misprint occurs before page 76?

SOLUTION
Here $c = \frac{1}{100}$ misprint per page; hence,

$$P\{T \le 75\} = 1 - e^{-(1/100)(75)} \approx 1 - 0.47 \approx 0.53 = 53\%.$$

Problem Set 8

1 A tank initially contains 50 gallons of water into which 10 pounds of salt is dissolved. Pure water runs into the tank at the rate of 3 gal/min and is uniformly stirred into the solution. Meanwhile, the mixture runs out of the tank at the constant rate of 2 gal/min. After how long is only 2 pounds of dissolved salt left in the tank?

2 Solve problem 1 if the water running into the tank is not pure but contains $\frac{1}{30}$ pound of salt per gallon and this water runs into the tank at the rate of 2 gal/min.

3 Air in a chemical laboratory contains 1 percent hydrogen sulfide. An exhaust fan is turned on which removes air from the room at the rate of 500 ft^3/min. Meanwhile, fresh air is drawn into the room through cracks under doors, and so forth. If the volume of the room is 10,000 cubic feet, what is the concentration of hydrogen sulfide after 5 minutes?

4 Water containing A pounds of pollutant per gallon flows into a reservoir at R gal/min. The reservoir initially contained G gallons of water containing B pounds of pollutant. The polluted water is drawn out of the reservoir at the rate of r gal/min. Set up the differential equation for the number of pounds q of pollutant in the reservoir at the end of t minutes.

5 A baked potato at a temperature of 200°F is placed on Emily's dinner plate. She does not start eating the potato until 7 minutes have elapsed. After 4 minutes, the temperature of the potato has dropped to 175°F. How hot is the potato when Emily starts to eat it if the air temperature is 70°F?

6 A body, initially at the temperature 160°F, is allowed to cool in air. The body cools to 100°F after 50 minutes. After 100 minutes, it has cooled to 80°. What is the air temperature?

7 An iron ball at a temperature of 190°F is placed in a water bath at a constant temperature of 35°F. After 10 minutes, the ball has cooled to a temperature of 75°F. How many more minutes is required for the ball to cool to 50°F?

8 If \$2000 were invested at an interest rate of 6 percent per year compounded continuously, how much would the investment be worth after 10 years?

9 If \$5000 is invested at an interest rate of 7 percent compounded continuously, how long is required for the investment to triple in value?

10 A savings bank with \$28,000,000 in regular savings accounts is paying compound interest of $5\frac{1}{2}$ percent per year compounded quarterly. The bank is contemplating offering the same rate of interest, but compounding continuously rather than quarterly. How much more interest will the bank have to pay out per year if the new plan is implemented?

11 If \$1 invested at 6 percent per year compounded quarterly yields the same profit at the end of 1 year as \$1 invested at k percent per year compounded continuously, find the value of k.

12 Suppose that the number of deaths from automobile accidents averages 2 per day in a certain state. What is the probability of exactly 4 such deaths in a 2-day period?

13 A machine produces copper wire coated with enamel insulation. On the average, there are two imperfections in the insulation every 7000 feet of wire. What is the probability of exactly one imperfection in 5000 feet of wire?

14 A newsboy on a corner sells an average of 70 papers per hour. What is the probability that he sells exactly 70 papers in a particular hour? [Use the approximation $\log_{10}(70!) \approx 100.07841$.]

15 In problem 14, what is the probability that the newsboy sells (a) no papers, (b) exactly one paper, (c) exactly two papers, during a given 1-minute period of time?

16 In problem 14, what is the probability that the newsboy sells more than two papers during a given 1-minute interval of time?

17 What is the probability that the newsboy in problem 14, having just arrived on the corner, must wait more than 5 minutes before selling his first paper?

18 Air traffic controllers at a certain busy airport receive an average of seven requests to land each quarter of an hour. During a given quarter of an hour, what is the probability that they receive:
(a) Exactly six requests?   (b) Exactly seven requests?   (c) No requests?

19 Philo Empiric, who is accident-prone, injures himself seriously, on the average, once every 7 months. What is the probability that 1 year passes before Philo's next serious injury?

20 In problem 19, what is the probability that Philo's next serious injury occurs within 1 week?

21 In problem 4, show that if $R \neq r$, then

$$q = (B - AG)\left(1 + \frac{R - r}{G}t\right)^{-r/(R-r)} + AG + A(R - r)t$$

is a solution of the differential equation satisfying the initial value ⸳ ⸳ion. What is the solution if $R = r$?

22 Carry out the following steps to show that the function f defined by $f(x) = (1 + a/x)^x$ is increasing if a is a positive constant and $x > 0$:
(a) Use the mean value theorem to show that there is a number t with

$$\ln\left(1 + \frac{x}{a}\right) - \ln\left(\frac{x}{a}\right) = \frac{1}{t} \quad \text{and} \quad \frac{x}{a} \leq t \leq 1 + \frac{x}{a}.$$

(b) From part (a), show that $\ln(1 + a/x) \geq a/(a + x)$.
(c) Use part (b) to prove that $D_x(1 + a/x)^x \geq 0$.

23 Interpret the result that $f(x) = (1 + a/x)^x$ is increasing (problem 22) in terms of compound interest.

Review Problem Set

In problems 1 through 48, differentiate each function.

1 $f(x) = \ln(x^2 + 7)$

2 $f(t) = \ln(\cos 3t)$

3 $g(r) = \ln(r\sqrt{r + 2})$

4 $G(u) = \ln(u - 1)^3$

5 $g(x) = x^2 \tan^{-1}(\ln x)$

6 $h(x) = \sin(\ln x^2)$

7 $F(u) = \dfrac{(\ln u)^2}{u}$

8 $G(x) = \sqrt[7]{\ln x}$

9 $f(x) = \ln \dfrac{\sin x}{x}$

10 $H(x) = \ln(x + \sqrt{x^2 + 1})$

11 $g(x) = e^{-4x^3}$

12 $H(x) = -xe^{-x}$

13 $F(x) = \sin^{-1} e^{-2x}$

14 $g(u) = e^u \cot e^u$

15 $f(t) = \ln \dfrac{e^t + 2}{e^t - 2}$

16 $H(x) = \ln(e^{\sin x} + 5)$

17 $g(x) = \cos^{-1} x^2 - xe^{x^3}$

18 $g(x) = e^x \ln(\sin x)$

19 $f(x) = \tan e^x$

20 $f(x) = e^x(x^2 - 2x + 5)$

21 $f(x) = (3 - e^{4x})^4$

22 $g(x) = \dfrac{e^x - e^{-x}}{e^x + e^{-x}}$

23 $f(x) = \ln \dfrac{e^x + 2}{e^{-x} + 2}$

24 $h(z) = e^{3z} \cos^{-1} e^{2z}$

25 $F(x) = x^{4e}$

26 $g(x) = x^{-17\pi}$

27 $f(x) = 5^{\cos x}$

28 $g(t) = 3^{t^2 + 2}$

29 $f(x) = 7^{\sin x^2}$

30 $g(x) = (x^2 + 7)2^{-5x}$

31 $g(x) = 3^{5x} \cdot 2^{4x^2}$

32 $H(x) = e^{\cos x} \cdot 2^{4x}$

33 $f(t) = \dfrac{\log_7 t}{t}$

34 $g(u) = \log_3 \dfrac{u}{u + 7}$

35 $g(x) = \sqrt[4]{\log_{10} x}$

36 $f(x) = \sqrt[5]{\log_{10} \dfrac{1 + x}{1 - x}}$

37 $g(x) = \cosh e^{4x}$

38 $H(s) = \sinh(\sin^{-1} s)$

39 $g(t) = \operatorname{csch}(e^{-t})$

40 $F(x) = \dfrac{\tanh(\sin x)}{x}$

41 $f(x) = e^{\operatorname{sech} x^2}$

42 $g(u) = \sinh u^2 \tanh 3u$

43 $g(t) = \ln(\tanh t + \operatorname{sech} t)$

44 $f(x) = \tan^{-1}(\sinh x^3)$

45 $g(x) = \sinh^{-1}(3x + 1)$

46 $g(u) = \coth^{-1} e^{5u}$

47 $f(x) = \tanh^{-1} e^x$

48 $F(x) = \coth^{-1} e^{-x^2}$

49 If $(\ln x)/x = (\ln 2)/2$, does it necessarily follow that $x = 2$? Justify your answer. [*Hint*: Sketch a graph of $y = (\ln x)/x$.]

50 If $(\ln x)/x = (\ln \frac{1}{2})/\frac{1}{2}$, does it necessarily follow that $x = \frac{1}{2}$? Justify your answer.

In problems 51 through 60, use logarithmic differentiation to find dy/dx.

51 $y = (3x)^x$

52 $y = x^{x^3}$

53 $y = (\sin x)^{x^2}$

54 $y = (\cosh x)^{2x}$

55 $y = (\tanh^{-1} x)^{x^3}$

56 $y = x^{\cos^{-1} x}$

57 $y = \dfrac{\cos x \sqrt[3]{1 + \sin^2 x}}{\sin^5 x}$

58 $y = \dfrac{x^2 \sin x}{\sqrt{1 - 3 \tan x \cdot \sec 2x}}$

59 $y = \dfrac{x^2(x + 5)^3 \sin 2x}{\sec 3x}$

60 $y = \dfrac{x \cot x}{(x + 1)(x + 3)^2(x + 7)^4}$

61 Let a be a positive constant and define the function f by $f(x) = a^x$. Use the fact that $f'(0) = a^0 \ln a$ to prove that $\lim\limits_{h \to 0} \dfrac{a^h - 1}{h} = \ln a$.

62 Let $0 < a < b$, where a and b are constants, and define the function p by $p(t) = \displaystyle\int_a^b x^t \, dx$, with the understanding that t is to be held fixed while calculating $\displaystyle\int_a^b x^t \, dx$. Is the function p continuous at the number -1? (*Hint*: Use the result in problem 61.)

In problems 63 through 66, use implicit differentiation to find dy/dx.

63 $xe^{4y} + x \cos y = 2$

64 $y \cdot 2^x + xe^y = 3$

65 $\cosh (x - y) + \sinh (x + y) = 1$

66 $\log_{10} (x + y) + \log_{10} (x - y) = 2$

In problems 67 and 68, use the fundamental theorem of calculus and the chain rule to calculate $D_x y$.

67 $y = \displaystyle\int_1^{\ln x} \dfrac{dt}{5 + t^3}$

68 $y = \displaystyle\int_1^{\cosh x} \dfrac{dt}{3 + t^2}$

In problems 69 through 88, evaluate each integral.

69 $\displaystyle\int \dfrac{dx}{8 + 3x}$

70 $\displaystyle\int \dfrac{\sin (\ln x)}{x} \, dx$

71 $\displaystyle\int \dfrac{\ln x}{x} \, dx$

72 $\displaystyle\int xe^{x^2 - 4} \, dx$

73 $\displaystyle\int e^{\sqrt{x}} \dfrac{dx}{\sqrt{x}}$

74 $\displaystyle\int \dfrac{e^x \, dx}{\sqrt{1 - e^{2x}}}$

75 $\displaystyle\int \dfrac{e^x \, dx}{\cos^2 e^x}$

76 $\displaystyle\int \dfrac{\pi^{1/x}}{x^2} \, dx$

77 $\displaystyle\int 2^x \cdot 5^x \, dx$

78 $\displaystyle\int 3^{2x} \cos 3^{2x} \, dx$

79 $\displaystyle\int_1^4 \dfrac{1/x^2}{1 + 1/x} \, dx$

80 $\displaystyle\int_0^{1/5} e^{\cosh 5x} \sinh 5x \, dx$

81 $\displaystyle\int \text{csch}^2 x \coth x \, dx$

82 $\displaystyle\int \dfrac{(\sinh^{-1} x)^5}{\sqrt{1 + x^2}} \, dx$

83 $\displaystyle\int \dfrac{e^{\cosh^{-1} x}}{\sqrt{x^2 - 1}} \, dx$

84 $\displaystyle\int \dfrac{dx}{\sqrt{16 + x^2}}$

85 $\displaystyle\int \dfrac{dx}{\sqrt{x^2 - 1}}$

86 $\displaystyle\int \dfrac{dx}{\sqrt{16 - 9x^2}}$

87 $\displaystyle\int \dfrac{dx}{x\sqrt{16 - 4x^2}}$

88 $\displaystyle\int \dfrac{dx}{x\sqrt{16 + 4x^2}}$

89 Using the definition of the natural logarithm, prove that $t > \ln t$ holds for all positive values of t.

90 Using the result of problem 89, show that $e^t > t$, hence that $e^{nt} > t^n$ holds for all positive integers n, if $t > 0$. Conclude that $e^x/x^n > (1/n)^n$ holds for all positive integers n if $x > 0$. (*Hint:* Put $t = x/n$.)

91 Using the result of problem 90, prove:
(a) If $x > 0$, then $e^x/x > x/4$.   (b) $\lim_{x \to +\infty} e^x/x = +\infty$.
(c) If $x > 0$, then $e^x/x^2 > x/27$.   (d) $\lim_{x \to +\infty} e^x/x^2 = +\infty$.

92 Prove that, if n is a positive integer, then (a) $\lim_{x \to +\infty} e^x/x^n = +\infty$ and therefore that
(b) $\lim_{x \to +\infty} x^n e^{-x} = 0$.

93 Apply the mean value theorem to the expression $\ln(1 + x) - \ln 1$, and thus conclude that $\dfrac{x}{1 + x} < \ln(1 + x) < x$ holds for $x > 0$.

94 Show that $\lim_{x \to 0} \dfrac{\sinh x}{x} = 1$. (*Hint:* Use the result of problem 61.)

In problems 95 through 98, sketch a graph of the given function indicating all significant features such as extreme points, inflection points, and asymptotes.
95 $f(x) = x^2 e^{2x}$ **96** $g(x) = x - \cosh^{-1} x$

97 $h(x) = x^2 e^{-x}$ **98** $F(x) = \begin{cases} \dfrac{\sinh x}{x} & \text{if } x \neq 0 \\ 1 & \text{if } x = 0 \end{cases}$

99 Find the volume of the solid generated by revolving the region bounded by the curves $xy^2 = 1$, $y = 0$, $x = 1$, and $x = e^3$ about the x axis.

100 Find the area under the graph of $y = 4/\sqrt{4x^2 + 9}$ between $x = 0$ and $x = 2\sqrt{2}$.

101 A radioactive substance has a half-life of 2 hours. How long does it take it to decay to $\frac{1}{10}$ of its original mass?

102 At 2:00 P.M. a dish of cream that has been standing in a warm room contains 150,000 bacteria of a certain kind. At 4:00 P.M. there are 900,000 of these bacteria present. How many were present at 3:00 P.M.?

103 Let a, b, and K be constants with $a \neq b$. Verify that $y = Ce^{-bx} + \dfrac{K}{b - a}e^{-ax}$ is a solution of the differential equation $dy/dx + by = Ke^{-ax}$, where C is an arbitrary constant.

104 Radioactive substance A with half-life $(\ln 2)/a$ decomposes into radioactive substance B with half-life $(\ln 2)/b$, where $a \neq b$; that is, when an atom of substance A decomposes, it is transmuted into an atom of substance B. If the number of atoms of substance A at time t is given by $q = q_0 e^{-at}$, and if y denotes the number of atoms of substance B at this same instant, show that y satisfies the differential equation $dy/dt + by = aq_0 e^{-at}$.

105 In paleontology, radioactive dating is sometimes used to determine the age of plant and animal fossils. The age t is calculated in terms of the ratio y/q of radioactive "daughter" substance B to "mother" substance A in the fossil. In problem 104, assume that $y = 0$ when $t = 0$.
(a) Use problem 103 to derive the equation $y = \dfrac{aq_0}{a - b}(e^{-bt} - e^{-at})$ for the number of atoms of substance B present at time t.
(b) Show that $t = \dfrac{1}{a - b}\ln\left(1 + \dfrac{a - b}{a} \cdot \dfrac{y}{q}\right)$.

106 When a capacitor with capacitance C farads discharges through a resistor with resistance R ohms, the current I in amperes flowing through the resistor at time t satisfies the differential equation $dI/dt + I/RC = 0$.
(a) If $I = I_0$ when $t = 0$, find a formula giving I in terms of t.
(b) If $RC = \frac{1}{200}$ and if $I_0 = 40$ milliamperes, plot a graph of I as a function of t.

107 An industrial plant dumps 10 pounds of pollutant into a lake every minute. Fresh water from streams runs into the lake at the rate of 100 ft^3/min and mixes uniformly with the polluted water, and meanwhile the polluted water runs out of the lake into a river at the same rate. The plant commenced operations 10 years ago, at which time the lake was unpolluted. The lake now tests at $\frac{1}{20}$ pound of pollutant per cubic foot. How many cubic feet of water are there in the lake? (Take 10 years to be 5.2596×10^6 minutes.)

108 An amount q_0 dollars is invested at k percent per year compounded n times per year.
(a) Write a formula for the value q of the investment after t years.
(b) Holding q_0, t, and k constant, find $\lim_{n \to +\infty} q$.

109 Black coffee is poured into a cup at an initial temperature of 200°F. One minute later, the coffee has cooled to 160°F. If the room temperature is 80°F, find how many *additional* minutes will be required for the coffee to cool to 110°F.

110 A certain bank pays $5\frac{1}{2}$ percent interest per year on its regular savings accounts compounded daily. Find the approximate equivalent yearly rate of interest if the bank compounded interest just once a year.

111 You have just won first prize in the state lottery and you have your choice of the following:
(a) $30,000 will be placed in a savings account in your name, compounded continuously at the rate of 10 percent per year.
(b) $0.01 will be placed in a fund in your name and the amount will be doubled every 6 months over the next 12 years. Which plan should you choose?

112 Scott, who is just learning to ride his bicycle, falls, on the average, once every $\frac{1}{4}$ mile. What is the probability that Scott falls exactly twice during a 1-mile journey on his bicycle?

113 What is the probability that Scott, in problem 112, rides his bicycle more than $\frac{1}{2}$ mile before falling?

114 A machine is producing plastic fishing line in a continuous string which is wound on a large drum to be cut later into appropriate lengths. On the average, there are three weak spots along the line every 1200 yards. A fisherman buys 100 yards of this line. What is the probability that there are now weak spots in his line?

115 Sociologists sometimes use the following simplified model for the spread of a rumor in a population: The rate at which the rumor spreads is proportional to the number of contacts between those who have heard it and those who have not. Thus, if p denotes the proportion of the population who have heard the rumor at time t, so that $1 - p$ is the proportion who have not yet heard it, then $dp/dt = Kp(1 - p)$, where K is the constant of proportionality. Solve this differential equation for p as a function of t.
$$\left[Hint: \frac{1}{p(1 - p)} = \frac{1}{p} + \frac{1}{1 - p}. \right]$$

10 TECHNIQUES OF INTEGRATION

Our purpose in this chapter is to consolidate the methods of integration presented previously, introduce some new methods, and give a unified development of the most important basic techniques of integration. There are really only three general procedures for evaluating integrals:

1 *Substitution* or *change of variable.*
2 *Manipulation of the integrand* using algebraic or other identities so as to convert it into a more tractable form.
3 *Integration by parts.*

Although we have made extensive use of the first two methods in the preceding five chapters, we have yet to introduce a few special "tricks of the trade" such as *trigonometric substitution* and *partial fractions.* The third method, *integration by parts,* is also introduced in this chapter.

The three general procedures are often used in combination; however, none of these procedures—singly or in combination—is guaranteed to be effective under all circumstances. Frequently, the choice of the appropriate technique is a matter of trial and error guided by experience.

1 Integrals That Involve Products of Powers of Sines and Cosines

In this section we use the formulas developed in Section 4 of Chapter 8, in conjunction with appropriate trigonometric identities and changes of variable, to evaluate integrals involving products of powers of sines and cosines.

1.1 Integrals of the Form $\int \sin^m x \cos^n x\, dx$

In order to evaluate $\int \sin^m x \cos^n x\, dx$, where m and n are constant exponents, we consider separately the case in which at least one of the exponents is a positive odd integer and the case in which both of the exponents are nonnegative even integers.

Case 1: At least one of the exponents m, n is a positive odd integer.

In this case we use the identity $\sin^2 x + \cos^2 x = 1$ to rewrite the integrand either in the form $F(\cos x)\sin x$ or in the form $F(\sin x)\cos x$. In the former case, the substitution $u = \cos x$ is effective, while in the latter case, the substitution $u = \sin x$ works.

EXAMPLES   Evaluate the given integral.

1 $\int \sin^3 x\, dx$

SOLUTION

$$\int \sin^3 x\, dx = \int \sin^2 x \sin x\, dx = \int (1 - \cos^2 x) \sin x\, dx.$$

Making the substitution $u = \cos x$, so that $du = -\sin x\, dx$, we have

$$\int (1 - \cos^2 x) \sin x\, dx = \int (1 - u^2)(-du) = -\int (1 - u^2)\, du$$

$$= -\left(u - \frac{u^3}{3}\right) + C = -\cos x + \frac{1}{3}\cos^3 x + C.$$

2 $\int \sin^2 x \cos^5 x\, dx$

SOLUTION

$$\int \sin^2 x \cos^5 x\, dx = \int \sin^2 x \cos^4 x \cos x\, dx = \int \sin^2 x (\cos^2 x)^2 \cos x\, dx$$

$$= \int \sin^2 x (1 - \sin^2 x)^2 \cos x\, dx.$$

Making the substitution $u = \sin x$, so that $du = \cos x\, dx$, we have

$$\int \sin^2 x (1 - \sin^2 x)^2 \cos x\, dx = \int u^2 (1 - u^2)^2\, du = \int u^2 (1 - 2u^2 + u^4)\, du$$

$$= \int (u^2 - 2u^4 + u^6)\, du = \frac{1}{3}u^3 - \frac{2}{5}u^5 + \frac{1}{7}u^7 + C$$

$$= \frac{1}{3}\sin^3 x - \frac{2}{5}\sin^5 x + \frac{1}{7}\sin^7 x + C.$$

3 $\int \sin^5 kx \cos^3 kx \, dx$, k a constant, $k \neq 0$.

SOLUTION
Here, both exponents are odd positive integers. Naturally, we rewrite the factor $\cos^3 kx$ with the *smaller* exponent as $\cos^2 kx \cos kx$. Thus,

$$\int \sin^5 kx \cos^3 kx \, dx = \int \sin^5 kx \cos^2 kx \cos kx \, dx$$

$$= \int \sin^5 kx (1 - \sin^2 kx) \cos kx \, dx.$$

Put $u = \sin kx$, so that $du = k \cos kx \, dx$ and $\cos kx \, dx = \dfrac{1}{k} du$. Thus,

$$\int \sin^5 kx (1 - \sin^2 kx) \cos kx \, dx = \int u^5 (1 - u^2) \frac{1}{k} \, du = \frac{1}{k} \int (u^5 - u^7) \, du$$

$$= \frac{1}{k} \left(\frac{u^6}{6} - \frac{u^8}{8} \right) + C = \frac{1}{24k} (4u^6 - 3u^8) + C$$

$$= \frac{1}{24k} (4 \sin^6 kx - 3 \sin^8 kx) + C.$$

4 $\int \dfrac{\sin^3 x}{\sqrt{\cos x}} \, dx$

SOLUTION

$$\int \frac{\sin^3 x}{\sqrt{\cos x}} \, dx = \int \frac{\sin^2 x}{\sqrt{\cos x}} \sin x \, dx = \int \frac{1 - \cos^2 x}{\sqrt{\cos x}} \sin x \, dx.$$

Put $u = \cos x$, so that $du = -\sin x \, dx$ and

$$\int \frac{1 - \cos^2 x}{\sqrt{\cos x}} \sin x \, dx = -\int \frac{1 - u^2}{\sqrt{u}} \, du = \int (u^{3/2} - u^{-1/2}) \, du$$

$$= \frac{2}{5} u^{5/2} - 2u^{1/2} + C = \frac{2}{5} (\cos x)^{5/2} - 2\sqrt{\cos x} + C.$$

In the last example, notice that the technique works even when one of the powers of the sine or cosine is not an integer—just so the other power is an odd positive integer.

Case 2: Both of the exponents m, n are nonnegative even integers.

In this case the integral $\int \sin^m x \cos^n x \, dx$ can be evaluated by using the trigonometric identities

$$\sin^2 x = \frac{1}{2} (1 - \cos 2x) \quad \text{and} \quad \cos^2 x = \frac{1}{2} (1 + \cos 2x),$$

which are direct consequences of the double-angle formula

$$\cos 2x = 1 - 2 \sin^2 x = 2 \cos^2 x - 1.$$

EXAMPLES   In Examples 1 through 3, evaluate the given integral.

1 $\int \cos^2 kx \, dx$, where k is a constant, $k \neq 0$.

SOLUTION

$$\int \cos^2 kx \, dx = \int \frac{1}{2} (1 + \cos 2kx) \, dx = \frac{1}{2} \int dx + \frac{1}{2} \int \cos 2kx \, dx$$

$$= \frac{x}{2} + \frac{1}{2} \int \cos 2kx \, dx.$$

Put $u = 2kx$, so that $du = 2k \, dx$ and $dx = \dfrac{1}{2k} \, du$. Thus,

$$\int \cos 2kx \, dx = \int \cos u \left(\frac{1}{2k} \, du \right) = \frac{1}{2k} \int \cos u \, du = \frac{\sin u}{2k} + C = \frac{\sin 2kx}{2k} + C.$$

It follows that

$$\int \cos^2 kx \, dx = \frac{x}{2} + \frac{1}{2} \left(\frac{\sin 2kx}{2k} \right) + C = \frac{x}{2} + \frac{\sin 2kx}{4k} + C.$$

2 $\int \sin^4 x \, dx$

SOLUTION

$$\int \sin^4 x \, dx = \int (\sin^2 x)^2 \, dx = \int \left[\frac{1}{2} (1 - \cos 2x) \right]^2 dx$$

$$= \frac{1}{4} \int (1 - 2 \cos 2x + \cos^2 2x) \, dx$$

$$= \frac{1}{4} \int dx - \frac{1}{2} \int \cos 2x \, dx + \frac{1}{4} \int \cos^2 2x \, dx$$

$$= \frac{x}{4} - \frac{1}{2} \left(\frac{\sin 2x}{2} \right) + \frac{1}{4} \left(\frac{x}{2} + \frac{\sin 4x}{8} \right) + C,$$

where we used the substitution $u = 2x$ to evaluate the middle integral and we used the result of Example 1 with $k = 2$ to evaluate $\int \cos^2 2x \, dx$. Combining terms and simplifying, we have

$$\int \sin^4 x \, dx = \frac{3x}{8} - \frac{\sin 2x}{4} + \frac{\sin 4x}{32} + C.$$

3 $\int \sin^4 2x \cos^2 2x \, dx$

SOLUTION

$$\int \sin^4 2x \cos^2 2x \, dx = \int \left[\frac{1}{2} (1 - \cos 4x) \right]^2 \left[\frac{1}{2} (1 + \cos 4x) \right] dx$$

$$= \frac{1}{8} \int (1 - 2 \cos 4x + \cos^2 4x)(1 + \cos 4x) \, dx$$

$$= \frac{1}{8} \int (1 - \cos 4x - \cos^2 4x + \cos^3 4x) \, dx$$

$$= \frac{1}{8} \int dx - \frac{1}{8} \int \cos 4x \, dx - \frac{1}{8} \int \cos^2 4x \, dx + \frac{1}{8} \int \cos^3 4x \, dx$$

$$= \frac{x}{8} - \frac{\sin 4x}{32} - \frac{1}{8} \left(\frac{x}{2} + \frac{\sin 8x}{16} \right) + \frac{1}{8} \int \cos^2 4x \cos 4x \, dx$$

$$= \frac{x}{16} - \frac{\sin 4x}{32} - \frac{\sin 8x}{128} + \frac{1}{8} \int (1 - \sin^2 4x) \cos 4x \, dx.$$

To evaluate the remaining integral, put $u = \sin 4x$, so that $du = 4 \cos 4x \, dx$ and

$$\frac{1}{8} \int (1 - \sin^2 4x) \cos 4x \, dx = \frac{1}{8} \int (1 - u^2) \frac{du}{4} = \frac{1}{32} \left(u - \frac{u^3}{3} \right) + C$$

$$= \frac{\sin 4x}{32} - \frac{\sin^3 4x}{96} + C.$$

Therefore,

$$\int \sin^4 2x \cos^2 2x \, dx = \frac{x}{16} - \frac{\sin 4x}{32} - \frac{\sin 8x}{128} + \frac{\sin 4x}{32} - \frac{\sin^3 4x}{96} + C$$

$$= \frac{x}{16} - \frac{\sin 8x}{128} - \frac{\sin^3 4x}{96} + C.$$

4 Find the volume of the solid generated if the region under one arch of the curve $y = \sin x$ is revolved about the line $y = -2$.

SOLUTION
By the method of circular rings, the desired volume V is given by

$$V = \pi \int_0^\pi [(2 + \sin x)^2 - 2^2] \, dx = \pi \int_0^\pi (4 \sin x + \sin^2 x) \, dx$$

$$= 4\pi \int_0^\pi \sin x \, dx + \pi \int_0^\pi \sin^2 x \, dx = 4\pi(-\cos x) \Big|_0^\pi + \pi \int_0^\pi \frac{1}{2} (1 - \cos 2x) \, dx$$

$$= 4\pi(-\cos \pi + \cos 0) + \frac{\pi}{2} \int_0^\pi dx - \frac{\pi}{2} \int_0^\pi \cos 2x \, dx$$

$$= 8\pi + \frac{\pi}{2} (\pi) - \frac{\pi}{2} \left(\frac{\sin 2x}{2} \right) \Big|_0^\pi = 8\pi + \frac{\pi^2}{2} \text{ cubic units.}$$

1.2 Integrals Involving the Products
$\sin mx \cos nx$, $\sin mx \sin nx$, or $\cos mx \cos nx$

Integrals involving the products mentioned in the title above play an important role in the mathematical analysis of periodic phenomena ranging from ocean tides to "brain waves." Such integrals are easily handled by using the following trigonometric identities:

$$1 \quad \sin s \cos t = \frac{1}{2} \sin (s + t) + \frac{1}{2} \sin (s - t).$$

$$2 \quad \sin s \sin t = \frac{1}{2} \cos (s - t) - \frac{1}{2} \cos (s + t).$$

$$3 \quad \cos s \cos t = \frac{1}{2} \cos (s - t) + \frac{1}{2} \cos (s + t).$$

These identities are readily verified by using the addition formulas to expand the right sides (Problem 33). Their use in evaluating integrals of the type now under consideration is illustrated as follows.

EXAMPLE   Evaluate $\int \sin 3x \cos 4x \, dx$.

SOLUTION
By Identity 1,

$$\int \sin 3x \cos 4x \, dx = \int \left[\frac{1}{2} \sin 7x + \frac{1}{2} \sin (-x) \right] dx = \frac{1}{2} \int (\sin 7x - \sin x) \, dx$$

$$= -\frac{\cos 7x}{14} + \frac{\cos x}{2} + C.$$

Problem Set 1

In problems 1 through 16, use the identity $\cos^2 x + \sin^2 x = 1$ and appropriate substitutions to evaluate each integral.

1 $\int \cos^3 x \, dx$

2 $\int \sin^3 4x \, dx$

3 $\int \sin^5 2t \, dt$

4 $\int \cos^5 3v \, dv$

5 $\int \sin^7 2x \cos^3 2x \, dx$

6 $\int \cos^3 x \sin^3 x \, dx$

7 $\int \sin^2 x \cos^3 x \, dx$

8 $\int \sin^3 4x \cos^2 4x \, dx$

9 $\int \sin^5 x \cos^2 x \, dx$

10 $\int \sin^4 2x \cos^5 2x \, dx$

11 $\int \frac{\sin^3 x}{\cos^4 x} \, dx$

12 $\int \sqrt[3]{\sin^2 3x} \cos^5 3x \, dx$

13 $\int_{\pi/4}^{\pi/2} \frac{\cos^3 x}{\sqrt{\sin x}} \, dx$

14 $\int_0^{\pi/3} \sin^2 3x \cos^5 3x \, dx$

15 $\int_0^{1/2} \sqrt[4]{\sin \pi t} \cos^3 \pi t \, dt$

16 $\int_{1/4}^{1/2} \frac{\cos^5 \pi u}{\sin^2 \pi u} \, du$

In problems 17 through 24, use the identities $\cos^2 x = \frac{1}{2}(1 + \cos 2x)$ and $\sin^2 x = \frac{1}{2}(1 - \cos 2x)$ and appropriate substitutions to evaluate each integral.

17 $\displaystyle\int \sin^2 3x \, dx$

18 $\displaystyle\int \cos^2 \frac{x}{2} \, dx$

19 $\displaystyle\int \sin^2 \frac{t}{2} \, dt$

20 $\displaystyle\int \cos^4 2x \, dx$

21 $\displaystyle\int \sin^6 u \, du$

22 $\displaystyle\int \sin^2 \pi t \cos^2 \pi t \, dt$

23 $\displaystyle\int_0^{\pi/8} \sin^4 2x \cos^2 2x \, dx$

24 $\displaystyle\int_0^{\pi} \sin^8 x \, dx$

In problems 25 through 32, use Identities 1, 2, and 3 of Section 1.2 and appropriate substitutions to evaluate each integral.

25 $\displaystyle\int \sin 5x \cos 2x \, dx$

26 $\displaystyle\int \sin 4x \cos 2x \, dx$

27 $\displaystyle\int \cos 4x \cos 3x \, dx$

28 $\displaystyle\int \sin 3t \cos 5t \, dt$

29 $\displaystyle\int \sin 7u \sin 3u \, du$

30 $\displaystyle\int \cos 8v \cos 4v \, dv$

31 $\displaystyle\int_0^1 \sin 2\pi x \cos 3\pi x \, dx$

32 $\displaystyle\int_0^5 \cos \frac{2\pi x}{5} \cos \frac{7\pi x}{5} \, dx$

33 Verify Identities 1, 2, and 3 of Section 1.2.

In problems 34 through 36, let m and n denote positive integers and verify each formula.

34 $\displaystyle\int_{-\pi}^{\pi} \cos mx \cos nx \, dx = \begin{cases} 0 & \text{if } m \neq n \\ \pi & \text{if } m = n \end{cases}$

35 $\displaystyle\int_{-\pi}^{\pi} \sin mx \sin nx \, dx = \begin{cases} 0 & \text{if } m \neq n \\ \pi & \text{if } m = n \end{cases}$

36 $\displaystyle\int_{-\pi}^{\pi} \cos mx \sin nx \, dx = 0$

37 Find the volume of the solid generated by revolving one arch of the sine curve about the x axis.

38 Find the area under the curve $y = \cos^2 x$ between $x = 0$ and $x = 2\pi$.

39 If n is an odd positive integer, show that $\displaystyle\int_0^{\pi} \cos^n x \, dx = 0.$ $\left(Hint: \text{Let } y = x - \dfrac{\pi}{2}.\right)$

40 Show that the volume of the solid generated by revolving the region under the graph of $y = \sin x$ between $x = 0$ and $x = \pi$ about the y axis is four times the volume of the solid generated when the same region is revolved about the x axis. (*Hint:* When revolving about the y axis, use cylindrical shells and use the integration formula
$\displaystyle\int x \sin x \, dx = \sin x - x \cos x + C.$)

41 A particle is moving on the s axis according to the law of motion $s = f(t)$, where $v = ds/dt = \sin^2 \pi t$. If $s = 0$ when $t = 0$, find a formula for $f(t)$ and locate the particle when $t = 8$ seconds.

42 Suppose that $f(x) = \displaystyle\sum_{n=1}^{N} a_n \sin nx$, where N is a positive integer and $a_1, a_2, \ldots, a_N$ are constants. Show that $a_m = \dfrac{1}{\pi}\displaystyle\int_{-\pi}^{\pi} f(x) \sin mx \, dx$ holds for $m = 1, 2, \ldots, N$. (*Hint:* Use problem 35.)

43 If $n \geq 2$ and a is a nonzero constant, verify the reduction formula

$$\int \sin^n ax \, dx = -\frac{\sin^{n-1} ax \cos ax}{an} + \frac{n-1}{n}\int \sin^{n-2} ax \, dx.$$

(*Hint:* Differentiate the right side of the equation.)

44 Use the reduction formula of problem 43 to evaluate

(a) $\int \sin^2 ax \, dx$ (b) $\int \sin^3 ax \, dx$ (c) $\int \sin^4 ax \, dx$

In problems 45 through 50, suppose that $I_k = \int_0^{\pi/2} \sin^k x \, dx$ for $k = 1, 2, 3, 4, \ldots$ and let n denote a positive integer.

45 (a) Show that $I_1 = 1$, $I_2 = \pi/4$, $I_3 = \frac{2}{3}$, and $I_4 = 3\pi/16$.

(b) Use problem 43 to show that if $n \geq 2$, then $I_n = \dfrac{n-1}{n} I_{n-2}$.

46 If $k \geq 1$, show that

(a) $I_{2k} = \dfrac{1 \cdot 3 \cdot 5 \cdot 7 \cdots (2k-1)}{2 \cdot 4 \cdot 6 \cdot 8 \cdots (2k)} \cdot \dfrac{\pi}{2}$ and that (b) $I_{2k+1} = \dfrac{2 \cdot 4 \cdot 6 \cdot 8 \cdots (2k)}{3 \cdot 5 \cdot 7 \cdot 9 \cdots (2k+1)}$.

47 Use problem 46 to show that $\pi/2 = (2k+1)I_{2k+1} \cdot I_{2k}$.

48 Use problem 46 to show that $\pi/2 = 2kI_{2k-1} \cdot I_{2k}$.

49 Show that if $1 \leq k \leq n$, then $I_n \leq I_k$. [*Hint:* For $0 \leq x \leq \pi/2$, $0 \leq \sin x \leq 1$, so that $(\sin x)^n \leq (\sin x)^k$.]

50 Use problems 47, 48, and 49 to show that

$$\frac{1}{2k+1} \cdot \frac{\pi}{2I_{2k}} \leq I_{2k} \leq \frac{1}{2k} \cdot \frac{\pi}{2I_{2k}}.$$

51 Using problem 50 and part (a) of problem 46, show that

$$\frac{2}{\pi(2k+1)} \leq \left[\frac{1 \cdot 3 \cdot 5 \cdot 7 \cdots (2k-1)}{2 \cdot 4 \cdot 6 \cdot 8 \cdots (2k)} \right]^2 \leq \frac{1}{\pi k}.$$

52 Use problem 51 to prove Wallis's formula for $\pi/4$, namely,

$$\frac{\pi}{4} = \lim_{k \to +\infty} \frac{2 \cdot 4 \cdot 4 \cdot 6 \cdot 6 \cdot 8 \cdot 8 \cdots (2k) \cdot (2k)}{3 \cdot 3 \cdot 5 \cdot 5 \cdot 7 \cdot 7 \cdot 9 \cdots (2k-1)(2k+1)}.$$

2   Integrals That Involve Powers and Products of Trigonometric Functions Other Than Sine and Cosine

In this section we present some techniques for dealing with integrals involving the tangent, cotangent, secant, and cosecant functions. These techniques involve the use of trigonometric identities and change of variable, either separately or in combination.

For instance, to evaluate $\int \tan u \, du$, we write $\tan u$ as $\sin u/\cos u$ and make the change of variable $v = \cos u$. Thus, since $dv = -\sin u \, du$, we have

$$\int \tan u \, du = \int \frac{\sin u}{\cos u} \, du = \int \frac{-dv}{v} = -\ln |v| + C$$

$$= -\ln |\cos u| + C = \ln |\cos u|^{-1} + C$$

$$= \ln |\sec u| + C.$$

A similar maneuver handles the integral of the cotangent (Problem 42), and we have the following.

THEOREM 1   **Integral of Tangent and Cotangent**

(i) $\int \tan u \, du = -\ln |\cos u| + C = \ln |\sec u| + C.$

(ii) $\int \cot u \, du = \ln |\sin u| + C.$

The integration of the secant and cosecant functions requires a clever trick. To find $\int \sec u \, du$, we write

$$\int \sec u \, du = \int \frac{\sec u(\sec u + \tan u)}{\sec u + \tan u} \, du = \int \frac{\sec^2 u + \sec u \tan u}{\sec u + \tan u} \, du$$

and then notice that the numerator of the integrand is the derivative of the denominator. Thus, putting $v = \sec u + \tan u$, we have

$$dv = (\sec u \tan u + \sec^2 u) \, du,$$

and so

$$\int \sec u \, du = \int \frac{dv}{v} = \ln |v| + C = \ln |\sec u + \tan u| + C.$$

A similar trick—writing csc u as $\dfrac{\csc u(\csc u - \cot u)}{\csc u - \cot u}$ —effects the integration of the cosecant (Problem 43). Thus, we have the following theorem.

THEOREM 2   **Integral of Secant and Cosecant**

(i) $\int \sec u \, du = \ln |\sec u + \tan u| + C.$

(ii) $\int \csc u \, du = \ln |\csc u - \cot u| + C.$

EXAMPLES   Evaluate the given integral.

1 $\int \tan 4x \, dx$

SOLUTION
Put $u = 4x$, so that $du = 4 \, dx$. Using Theorem 1, we have

$$\int \tan 4x \, dx = \int \tan u \left(\frac{du}{4}\right) = \frac{1}{4} \int \tan u \, du$$

$$= \frac{1}{4} \ln |\sec u| + C = \frac{1}{4} \ln |\sec 4x| + C.$$

2 $\displaystyle\int \frac{dx}{\cos 5x}$

SOLUTION

Put $u = 5x$, so that $du = 5\, dx$. Using Theorem 2, we have

$$\int \frac{dx}{\cos 5x} = \int \sec 5x\, dx = \int \sec u \frac{du}{5} = \frac{1}{5}\int \sec u\, du$$

$$= \frac{1}{5}\ln|\sec u + \tan u| + C$$

$$= \frac{1}{5}\ln|\sec 5x + \tan 5x| + C.$$

2.1 Powers of Tangent, Cotangent, Secant, and Cosecant

In integrating positive integral powers or products of positive integral powers of tan, cot, sec, and csc, the identities

$$1 + \tan^2 u = \sec^2 u \quad \text{and} \quad 1 + \cot^2 u = \csc^2 u$$

can often be used to advantage. The following examples illustrate the techniques involved.

EXAMPLES   In Examples 1 through 4, evaluate the given integral.

1 $\displaystyle\int \cot^2 x\, dx$

SOLUTION

$$\int \cot^2 x\, dx = \int (\csc^2 x - 1)\, dx = \int \csc^2 x\, dx - \int dx = -\cot x - x + C.$$

2 $\displaystyle\int \tan^3 2x\, dx$

SOLUTION

We write $\tan^3 2x$ as $\tan 2x \tan^2 2x$, then use the identity $\tan^2 2x = \sec^2 2x - 1$ to obtain

$$\int \tan^3 2x\, dx = \int \tan 2x \tan^2 x\, dx = \int \tan 2x(\sec^2 2x - 1)\, dx$$

$$= \int (\tan 2x \sec^2 2x - \tan 2x)\, dx$$

$$= \int \tan 2x \sec^2 2x\, dx - \int \tan 2x\, dx.$$

To evaluate the integral $\int \tan 2x \sec^2 2x \, dx$, we make the substitution $u = \tan 2x$, so that $du = 2 \sec^2 2x \, dx$ and

$$\int \tan 2x \sec^2 2x \, dx = \int u \frac{du}{2} = \frac{1}{2} \int u \, du = \frac{u^2}{4} + C = \frac{\tan^2 2x}{4} + C.$$

The integral $\int \tan 2x \, dx$ is easily evaluated by the method already illustrated in Example 1, page 477. Therefore,

$$\int \tan^3 2x \, dx = \frac{\tan^2 2x}{4} - \frac{1}{2} \ln |\sec 2x| + C.$$

3 $\int \sec^4 3x \, dx$

SOLUTION

$$\int \sec^4 3x \, dx = \int \sec^2 3x \sec^2 3x \, dx = \int (1 + \tan^2 3x) \sec^2 3x \, dx.$$

Put $u = \tan 3x$, so that $du = 3 \sec^2 3x \, dx$, and

$$\int (1 + \tan^2 3x) \sec^2 3x \, dx = \int (1 + u^2) \frac{du}{3} = \frac{1}{3} \int du + \frac{1}{3} \int u^2 \, du$$

$$= \frac{u}{3} + \frac{u^3}{9} + C = \frac{\tan 3x}{3} + \frac{\tan^3 3x}{9} + C.$$

4 $\int \tan^3 x \sec^4 x \, dx$

SOLUTION

$$\int \tan^3 x \sec^4 x \, dx = \int \tan^3 x \sec^2 x \sec^2 x \, dx$$

$$= \int \tan^3 x (1 + \tan^2 x) \sec^2 x \, dx$$

$$= \int (\tan^3 x + \tan^5 x) \sec^2 x \, dx.$$

Put $u = \tan x$, so that $du = \sec^2 x \, dx$ and

$$\int \tan^3 x \sec^4 x \, dx = \int (u^3 + u^5) \, du = \frac{u^4}{4} + \frac{u^6}{6} + C$$

$$= \frac{1}{4} \tan^4 x + \frac{1}{6} \tan^6 x + C.$$

5 Find the area A under the graph of $f(x) = \sqrt{\csc x \cot^3 x}$ between $x = \pi/4$ and $x = \pi/2$.

SOLUTION

$$A = \int_{\pi/4}^{\pi/2} \sqrt{\csc x}\, \cot^3 x\, dx = \int_{\pi/4}^{\pi/2} \sqrt{\csc x}\,(\csc^2 x - 1)\cot x\, dx$$

$$= \int_{\pi/4}^{\pi/2} \frac{\sqrt{\csc x}}{\csc x}\,(\csc^2 x - 1)\csc x \cot x\, dx$$

$$= \int_{\pi/4}^{\pi/2} \frac{1}{\sqrt{\csc x}}\,(\csc^2 x - 1)\csc x \cot x\, dx.$$

We put $u = \csc x$, so that $du = -\csc x \cot x\, dx$. Then we have $u = \sqrt{2}$ when $x = \pi/4$ and $u = 1$ when $x = \pi/2$. Thus,

$$A = \int_{\sqrt{2}}^{1} \frac{1}{\sqrt{u}}\,(u^2 - 1)(-du) = \int_{\sqrt{2}}^{1} \left(\frac{1}{\sqrt{u}} - \frac{u^2}{\sqrt{u}}\right)du = \int_{\sqrt{2}}^{1} (u^{-1/2} - u^{3/2})\, du$$

$$= \left(2u^{1/2} - \frac{2}{5}u^{5/2}\right)\Big|_{\sqrt{2}}^{1} = \frac{8}{5} - \frac{6}{5}\sqrt[4]{2} \approx 0.173 \text{ square unit.}$$

In integrating expressions such as those appearing above, the following suggestions may prove useful:

1　If the integrand involves an even power of $\sec x$ (**respectively,** of $\csc x$), factor out $\sec^2 x$ (**respectively,** $\csc^2 x$) and try to write the remaining part of the integrand in terms of $\tan x$ (**respectively,** of $\cot x$). Then make the substitution $u = \tan x$ (**respectively,** $u = \cot x$).

2　If the integrand involves an odd power of $\tan x$ (**respectively,** of $\cot x$), try to factor out $\sec x \tan x$ (**respectively,** $\csc x \cot x$) and write the remaining part of the integrand in terms of $\sec x$ (**respectively,** of $\csc x$).

EXAMPLES　1 Evaluate $\int \cot^5 x\, \csc^3 x\, dx$.

SOLUTION
Since the integrand involves an odd power of $\cot x$, we try Suggestion 2. Thus,

$$\int \cot^5 x\, \csc^3 x\, dx = \int \cot^4 x\, \csc^2 x\,(\csc x \cot x)\, dx$$

$$= \int (\cot^2 x)^2\, \csc^2 x\,(\csc x \cot x)\, dx$$

$$= \int (\csc^2 x - 1)^2\, \csc^2 x\,(\csc x \cot x)\, dx.$$

Putting $u = \csc x$, so that $du = -\csc x \cot x\, dx$, we have

$$\int \cot^5 x\, \csc^3 x\, dx = \int (u^2 - 1)^2 u^2 (-du) = -\int (u^4 - 2u^2 + 1)u^2\, du$$

$$= -\int (u^6 - 2u^4 + u^2)\, du$$

$$= -\frac{1}{7}u^7 + \frac{2}{5}u^5 - \frac{1}{3}u^3 + C$$

$$= -\frac{1}{7}\csc^7 x + \frac{2}{5}\csc^5 x - \frac{1}{3}\csc^3 x + C.$$

2 Find the volume of the solid that is generated if the region under the curve $y = \cot 2x \csc^2 2x$ between $x = \pi/6$ and $x = \pi/3$ is revolved about the x axis.

SOLUTION

By the method of circular disks, the volume V is given by

$$V = \pi \int_{\pi/6}^{\pi/3} (\cot 2x \csc^2 2x)^2 \, dx = \pi \int_{\pi/6}^{\pi/3} \cot^2 2x \csc^4 2x \, dx.$$

Using Suggestion 1, we have

$$V = \pi \int_{\pi/6}^{\pi/3} \cot^2 2x \csc^2 2x \csc^2 2x \, dx$$

$$= \int_{\pi/6}^{\pi/3} \cot^2 2x \,(1 + \cot^2 2x) \csc^2 2x \, dx.$$

Thus, putting $u = \cot 2x$, so that $du = -2 \csc^2 2x \, dx$, $u = 1/\sqrt{3}$ when $x = \pi/6$, and $u = -1/\sqrt{3}$ when $x = \pi/3$, we have

$$V = \pi \int_{1/\sqrt{3}}^{-1/\sqrt{3}} u^2(1 + u^2)\left(\frac{du}{-2}\right) = \frac{\pi}{2} \int_{-1/\sqrt{3}}^{1/\sqrt{3}} (u^2 + u^4) \, du$$

$$= \frac{\pi}{2}\left(\frac{u^3}{3} + \frac{u^5}{5}\right)\Bigg|_{-1/\sqrt{3}}^{1/\sqrt{3}} = \frac{2\pi}{15\sqrt{3}} \approx 0.242 \text{ cubic unit.}$$

Unfortunately, the suggestions above are not particularly helpful for integrating odd powers of $\sec x$ and $\csc x$ such as $\int \sec^3 x \, dx$ or $\int \csc^3 x \, dx$. The latter two integrals can be obtained by the method to be introduced in Section 4—integration by parts. Even powers of the tangent or cotangent can be integrated by using the reduction formulas given in Problems 38 and 48.

Problem Set 2

In problems 1 through 4, evaluate each integral.

1 $\int \cot 4x \, dx$

2 $\int \tan \frac{x}{2} \, dx$

3 $\int \frac{dx}{\cos 3x}$

4 $\int \csc \frac{x}{5} \, dx$

In problems 5 through 30, evaluate the given integral by using the standard identities $1 + \tan^2 x = \sec^2 x$ and $1 + \cot^2 x = \csc^2 x$ and appropriate substitutions.

5 $\int \tan^2 \frac{2x}{3} \, dx$

6 $\int \cot^3 5x \, dx$

7 $\int \cot^4 4x \, dx$

8 $\int \tan^3 \frac{\pi t}{2} \, dt$

9 $\int \csc^4 3t \, dt$

10 $\int \sec^6 2x \, dx$

11 $\int \tan^4 2t \sec^2 2t \, dt$

12 $\int \cot^4 3x \csc^4 3x \, dx$

13 $\int \tan^3 5x \sec^5 5x \, dx$

14 $\displaystyle\int \cot^3 \frac{\pi x}{2} \csc^3 \frac{\pi x}{2}\, dx$

15 $\displaystyle\int (\tan 2x + \cot 2x)^2\, dx$

16 $\displaystyle\int (\sec 3x + \tan 3x)^2\, dx$

17 $\displaystyle\int \frac{\sec^4 t\, dt}{\sqrt{\tan t}}$

18 $\displaystyle\int \frac{\tan^3 3x\, dx}{\sqrt{\sec 3x}}$

19 $\displaystyle\int \tan^3 7x \sec^4 7x\, dx$

20 $\displaystyle\int \left(\frac{\tan x}{\cos x}\right)^4 dx$

21 $\displaystyle\int \cot 3x \csc^3 3x\, dx$

22 $\displaystyle\int \cot^{7/2} 2x \csc^4 2x\, dx$

23 $\displaystyle\int \tan^3 5x \sec 5x\, dx$

24 $\displaystyle\int \cot^3 \frac{x}{2} \csc^3 \frac{x}{2}\, dx$

25 $\displaystyle\int \tan^3 2x\sqrt{\sec 2x}\, dx$

26 $\displaystyle\int \sqrt{\tan 7x}\, \sec^4 7x\, dx$

27 $\displaystyle\int \tan^5 x \sec^7 x\, dx$

28 $\displaystyle\int \frac{\csc^4 2\pi x}{\cot^2 2\pi x}\, dx$

29 $\displaystyle\int \frac{\csc^2 8x}{\cot^4 8x}\, dx$

30 $\displaystyle\int \sec^3 2x \tan^5 2x\, dx$

In problems 31 through 34, evaluate each definite integral.

31 $\displaystyle\int_{\pi/6}^{\pi/9} \cot 3x\, dx$

32 $\displaystyle\int_{\pi/8}^{\pi/6} 5 \sec 2x\, dx$

33 $\displaystyle\int_{\pi/4}^{\pi/2} \cot^4 x \csc^4 x\, dx$

34 $\displaystyle\int_0^{\pi/4} \tan^5 x\, dx$

35 Find the volume of the solid generated when the region under the curve $y = \sec^2 x$ between $x = 0$ and $x = \pi/3$ is rotated about the x axis.

36 Find the area of the region under the curve $y = \sec x$ between $x = -\pi/3$ and $x = \pi/3$.

37 Find the area of the region under the curve $y = 5 \tan^2 x$ from $x = -\pi/4$ to $x = \pi/4$.

38 If $n \geq 2$ and a is a nonzero constant, verify the reduction formula

$$\int \tan^n ax\, dx = \frac{1}{a(n-1)} \tan^{n-1} ax - \int \tan^{n-2} ax\, dx.$$

(*Hint*: Differentiate the right side.)

39 Show that $\displaystyle\int \csc u\, du = \ln |\tan (u/2)| + C$ and reconcile this with part (ii) of Theorem 2.

40 Use the formula in problem 38 twice to find $\displaystyle\int \tan^6 x\, dx$.

41 Find the arc length of the curve $y = \ln (\sin x)$ from $x = \pi/4$ to $x = \pi/2$.

42 Prove part (ii) of Theorem 1.

43 Prove part (ii) of Theorem 2.

In problems 44 through 47, evaluate each integral.

44 $\displaystyle\int \frac{\csc^3 8x}{\cot^4 8x}\, dx$ (*Hint*: Rewrite the integrand in terms of sine and cosine.)

45 $\displaystyle\int \frac{dx}{\sec 3x}$

46 $\displaystyle\int \frac{1}{\sqrt{1 + 3 \sec^2 x}}\, dx, \quad -\frac{\pi}{2} < x < \frac{\pi}{2}$

47 $\displaystyle\int \frac{\tan^2 \theta}{\sec^5 \theta}\, d\theta$

48 If $n \geq 2$ and a is a nonzero constant, verify the reduction formula

$$\int \cot^n ax\, dx = -\frac{\cot^{n-1} ax}{a(n-1)} - \int \cot^{n-2} ax\, dx.$$

3 Integration by Trigonometric Substitution

So far we have integrated functions involving powers and products of trigonometric functions. In this section we use trigonometric substitution to cope with integrals involving expressions such as $\sqrt{a^2 - u^2}$, $\sqrt{a^2 + u^2}$, and $\sqrt{u^2 - a^2}$, where a is a positive constant. Also, we show how to handle the general forms $\sqrt{a + bu + cu^2}$ and $a + bu + cu^2$, where a, b, and c are constants.

The appropriate substitutions when the integrand involves $\sqrt{a^2 - u^2}$, $\sqrt{a^2 + u^2}$, or $\sqrt{u^2 - a^2}$ are suggested by the right triangles in Figure 1. In Figure 1a,

$$u = a \sin\theta \quad \text{and} \quad \sqrt{a^2 - u^2} = a\cos\theta;$$

in Figure 1b,

$$u = a \tan\theta \quad \text{and} \quad \sqrt{a^2 + u^2} = a\sec\theta;$$

and in Figure 1c,

$$u = a \sec\theta \quad \text{and} \quad \sqrt{u^2 - a^2} = a\tan\theta.$$

Figure 1

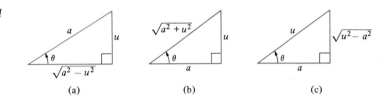

(a)   (b)   (c)

The substitutions suggested geometrically in Figure 1 can be carried out analytically, provided that a is a positive constant and u is positive. In fact, we have the following:

1 If the integrand involves $\sqrt{a^2 - u^2}$, where $0 < u < a$, put $u = a\sin\theta$ with $0 < \theta < \pi/2$, so that $du = a\cos\theta\,d\theta$ and

$$\sqrt{a^2 - u^2} = \sqrt{a^2 - a^2\sin^2\theta} = \sqrt{a^2(1 - \sin^2\theta)} = \sqrt{a^2\cos^2\theta} = a\cos\theta.$$

2 If the integrand involves $\sqrt{a^2 + u^2}$, where $u > 0$ and $a > 0$, put $u = a\tan\theta$ with $0 < \theta < \pi/2$, so that $du = a\sec^2\theta\,d\theta$ and

$$\sqrt{a^2 + u^2} = \sqrt{a^2 + a^2\tan^2\theta} = \sqrt{a^2(1 + \tan^2\theta)} = \sqrt{a^2\sec^2\theta} = a\sec\theta.$$

3 If the integrand involves $\sqrt{u^2 - a^2}$, where $u > a > 0$, put $u = a\sec\theta$ with $0 < \theta < \pi/2$, so that $du = a\sec\theta\tan\theta\,d\theta$ and

$$\sqrt{u^2 - a^2} = \sqrt{a^2\sec^2\theta - a^2} = \sqrt{a^2(\sec^2\theta - 1)} = \sqrt{a^2\tan^2\theta} = a\tan\theta.$$

After the integration has been carried out with respect to the variable θ, the answer can be written in terms of the original variable by referring to an appropriate right triangle.

This method is called *trigonometric substitution*.

EXAMPLES 1 Evaluate $\displaystyle\int \frac{x^2\,dx}{(4 - x^2)^{3/2}}$.

Figure 2

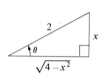

SOLUTION

Make the substitution $x = 2 \sin \theta$ (Figure 2), so that $dx = 2 \cos \theta \, d\theta$ and $(4 - x^2) = 4 - 4 \sin^2 \theta = 4 \cos^2 \theta$. Thus,

$$\int \frac{x^2 \, dx}{(4 - x^2)^{3/2}} = \int \frac{(2 \sin \theta)^2 (2 \cos \theta \, d\theta)}{(4 \cos^2 \theta)^{3/2}} = \int \frac{\sin^2 \theta}{\cos^2 \theta} \, d\theta$$

$$= \int \tan^2 \theta \, d\theta = \int (\sec^2 \theta - 1) \, d\theta = \tan \theta - \theta + C.$$

By Figure 2, $\tan \theta = \dfrac{x}{\sqrt{4 - x^2}}$ and $\theta = \sin^{-1} \dfrac{x}{2}$; hence,

$$\int \frac{x^2 \, dx}{(4 - x^2)^{3/2}} = \frac{x}{\sqrt{4 - x^2}} - \sin^{-1} \frac{x}{2} + C.$$

2 Evaluate $\displaystyle\int \frac{dx}{x^2 \sqrt{x^2 + 9}}$.

Figure 3

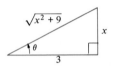

SOLUTION

Make the substitution $x = 3 \tan \theta$ (Figure 3), so that $dx = 3 \sec^2 \theta \, d\theta$ and $x^2 + 9 = 9 \tan^2 \theta + 9 = 9 \sec^2 \theta$. Thus,

$$\int \frac{dx}{x^2 \sqrt{x^2 + 9}} = \int \frac{3 \sec^2 \theta \, d\theta}{(9 \tan^2 \theta) \sqrt{9 \sec^2 \theta}} = \frac{1}{9} \int \frac{\sec \theta \, d\theta}{\tan^2 \theta}$$

$$= \frac{1}{9} \int \frac{1/\cos \theta}{\sin^2 \theta / \cos^2 \theta} \, d\theta = \frac{1}{9} \int \frac{\cos \theta \, d\theta}{\sin^2 \theta}.$$

In order to evaluate the latter integral, we put $v = \sin \theta$, so that $dv = \cos \theta \, d\theta$ and

$$\frac{1}{9} \int \frac{\cos \theta \, d\theta}{\sin^2 \theta} = \frac{1}{9} \int \frac{dv}{v^2} = -\frac{1}{9v} + C = -\frac{1}{9 \sin \theta} + C = -\frac{1}{9} \csc \theta + C.$$

From Figure 3, $\csc \theta = \dfrac{\sqrt{x^2 + 9}}{x}$; hence,

$$\int \frac{dx}{x^2 \sqrt{x^2 + 9}} = -\frac{\sqrt{x^2 + 9}}{9x} + C.$$

3 Evaluate $\displaystyle\int \frac{dt}{t^3 \sqrt{t^2 - 25}}$.

Figure 4

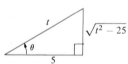

SOLUTION

Make the substitution $t = 5 \sec \theta$ (Figure 4), so that $dt = 5 \sec \theta \tan \theta \, d\theta$ and $t^2 - 25 = 25 \sec^2 \theta - 25 = 25 \tan^2 \theta$. Thus,

$$\int \frac{dt}{t^3 \sqrt{t^2 - 25}} = \int \frac{5 \sec \theta \tan \theta \, d\theta}{(125 \sec^3 \theta) \sqrt{25 \tan^2 \theta}}$$

$$= \frac{1}{125} \int \frac{d\theta}{\sec^2 \theta} = \frac{1}{125} \int \cos^2 \theta \, d\theta$$

$$= \frac{1}{125} \int \frac{1}{2} (1 + \cos 2\theta) \, d\theta = \frac{1}{250} \left(\theta + \frac{\sin 2\theta}{2} \right) + C$$

$$= \frac{1}{250} \left(\theta + \frac{2 \sin \theta \cos \theta}{2} \right) + C = \frac{1}{250} (\theta + \sin \theta \cos \theta) + C.$$

By Figure 4, $\theta = \sec^{-1} \dfrac{t}{5}$, $\sin\theta = \dfrac{\sqrt{t^2 - 25}}{t}$, and $\cos\theta = \dfrac{5}{t}$; hence

$$\int \frac{dt}{t^3\sqrt{t^2 - 25}} = \frac{1}{250}\left(\sec^{-1}\frac{t}{5} + \frac{5\sqrt{t^2 - 25}}{t^2}\right) + C.$$

4 Evaluate $\displaystyle\int_{2/3}^{2\sqrt{3}/3} \frac{du}{u\sqrt{9u^2 + 4}}$.

SOLUTION

Make the substitution $3u = 2\tan\theta$ or $u = \frac{2}{3}\tan\theta$, so that $du = \frac{2}{3}\sec^2\theta\,d\theta$ and $\sqrt{9u^2 + 4} = \sqrt{4\tan^2\theta + 4} = 2\sec\theta$. Here, $\theta = \tan^{-1}(3u/2)$, so that $\theta = \pi/4$ when $u = \frac{2}{3}$ and $\theta = \pi/3$ when $u = 2\sqrt{3}/3$. Therefore,

$$\int_{2/3}^{2\sqrt{3}/3} \frac{du}{u\sqrt{9u^2 + 4}} = \int_{\pi/4}^{\pi/3} \frac{\frac{2}{3}\sec^2\theta\,d\theta}{\frac{2}{3}\tan\theta \cdot 2\sec\theta} = \frac{1}{2}\int_{\pi/4}^{\pi/3} \frac{\sec\theta\,d\theta}{\tan\theta}$$

$$= \frac{1}{2}\int_{\pi/4}^{\pi/3} \frac{1/\cos\theta}{\sin\theta/\cos\theta}\,d\theta = \frac{1}{2}\int_{\pi/4}^{\pi/3} \frac{1}{\sin\theta}\,d\theta = \frac{1}{2}\int_{\pi/4}^{\pi/3} \csc\theta\,d\theta$$

$$= \frac{1}{2}\ln|\csc\theta - \cot\theta|\Big|_{\pi/4}^{\pi/3}$$

$$= \frac{1}{2}\left[\ln\left(\frac{2}{\sqrt{3}} - \frac{1}{\sqrt{3}}\right) - \ln(\sqrt{2} - 1)\right]$$

$$= \frac{1}{2}\ln\left(\frac{1}{\sqrt{6} - \sqrt{3}}\right) \approx 0.166.$$

3.1 Integrals Involving $ax^2 + bx + c$

By completing the square, the expression $ax^2 + bx + c$, where $a \neq 0$, can be rewritten as $a\left(x + \dfrac{b}{2a}\right)^2 + \left(\dfrac{4ac - b^2}{4a}\right)$. Thus, if the expression $ax^2 + bx + c$ occurs in an integrand, it is often useful to complete the square and then make the substitution $u = \sqrt{|a|}\left(x + \dfrac{b}{2a}\right)$. This procedure is illustrated by the following examples.

EXAMPLES   Evaluate the given integral.

1 $\displaystyle\int \frac{dx}{(5 - 4x - x^2)^{3/2}}$

SOLUTION

Completing the square, we have

$$5 - 4x - x^2 = 5 - (x^2 + 4x) = 5 + 4 - (x^2 + 4x + 4) = 9 - (x + 2)^2 = 9 - u^2,$$

where we have put $u = x + 2$. Thus,

$$\int \frac{dx}{(5 - 4x - x^2)^{3/2}} = \int \frac{du}{(9 - u^2)^{3/2}}.$$

Now, we make the trigonometric substitution $u = 3 \sin \theta$, so that $du = 3 \cos \theta \, d\theta$ and $9 - u^2 = 9 - 9 \sin^2 \theta = 9 \cos^2 \theta$. Thus,

$$\int \frac{du}{(9 - u^2)^{3/2}} = \int \frac{3 \cos \theta \, d\theta}{(9 \cos^2 \theta)^{3/2}} = \int \frac{3 \cos \theta \, d\theta}{27 \cos^3 \theta} = \frac{1}{9} \int \frac{d\theta}{\cos^2 \theta}$$

$$= \frac{1}{9} \int \sec^2 \theta \, d\theta = \frac{1}{9} \tan \theta + C.$$

Figure 5

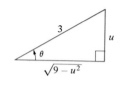

Referring to Figure 5, we find that $\tan \theta = \dfrac{u}{\sqrt{9 - u^2}}$; hence,

$$\int \frac{dx}{(5 - 4x - x^2)^{3/2}} = \frac{1}{9} \tan \theta + C = \frac{u}{9\sqrt{9 - u^2}} + C = \frac{x + 2}{9\sqrt{5 - 4x - x^2}} + C.$$

2 $\displaystyle\int \frac{x \, dx}{\sqrt{4x^2 + 8x + 5}}$

SOLUTION

Completing the square, we have

$$4x^2 + 8x + 5 = 4(x^2 + 2x) + 5 = 4(x^2 + 2x + 1) + 5 - 4$$
$$= 4(x + 1)^2 + 1 = 4u^2 + 1,$$

where we have put $u = x + 1$. Here, $x = u - 1$ and $dx = du$, so that

$$\int \frac{x \, dx}{\sqrt{4x^2 + 8x + 5}} = \int \frac{(u - 1) \, du}{\sqrt{4u^2 + 1}}.$$

Now, we make the trigonometric substitution $2u = \tan \theta$, so that $u = \frac{1}{2} \tan \theta$, $du = \frac{1}{2} \sec^2 \theta \, d\theta$, and $4u^2 + 1 = \tan^2 \theta + 1 = \sec^2 \theta$. Thus,

$$\int \frac{(u - 1) \, du}{\sqrt{4u^2 + 1}} = \int \frac{(\frac{1}{2} \tan \theta - 1)(\frac{1}{2} \sec^2 \theta) \, d\theta}{\sqrt{\sec^2 \theta}} = \frac{1}{4} \int (\tan \theta - 2) \sec \theta \, d\theta$$

$$= \frac{1}{4} \int (\sec \theta \tan \theta - 2 \sec \theta) \, d\theta$$

$$= \frac{1}{4} \sec \theta - \frac{1}{2} \ln |\sec \theta + \tan \theta| + C.$$

Here we have $\sec \theta = \sqrt{\tan^2 \theta + 1} = \sqrt{4u^2 + 1} = \sqrt{4x^2 + 8x + 5}$ and $\tan \theta = 2u = 2(x + 1) = 2x + 2$; hence,

$$\int \frac{x \, dx}{\sqrt{4x^2 + 8x + 5}} = \frac{1}{4} \sqrt{4x^2 + 8x + 5} - \frac{1}{2} \ln \left| \sqrt{4x^2 + 8x + 5} + 2x + 2 \right| + C.$$

In the examples above, we have not taken care to impose the conditions on the algebraic signs of the variable required for trigonometric substitution. For instance, in Example 2 we should have insisted that $u > 0$ (that is, $x + 1 > 0$ or $x > -1$) before making the substitution $2u = \tan \theta$. People often permit themselves the luxury of such carelessness, and their answers frequently are correct regardless of such negligence. (See Problems 44 through 46.) Fortunately, the answer—once obtained by any method, no matter how questionable—can and should be checked by differentiation (see Problem 23).

Problem Set 3

In problems 1 through 22, use an appropriate trigonometric substitution to evaluate each integral.

1 $\int \dfrac{dx}{x^2\sqrt{16-x^2}}$

2 $\int \dfrac{\sqrt{9-x^2}}{x^2}\,dx$

3 $\int \dfrac{dt}{t^4\sqrt{4-t^2}}$

4 $\int \dfrac{y^3\,dy}{\sqrt{4-y^2}}$

5 $\int \dfrac{x^2\,dx}{\sqrt{4-9x^2}}$

6 $\int \sqrt{9-2u^2}\,du$

7 $\int \dfrac{\sqrt{7-4t^2}}{t^4}\,dt$

8 $\int \dfrac{dx}{x^2(a^2+x^2)^{3/2}},\ a>0$

9 $\int \dfrac{t\,dt}{\sqrt{t^2-a^2}}$

10 $\int \dfrac{x^3\,dx}{\sqrt{x^2+4}}$

11 $\int \dfrac{dx}{x^2\sqrt{1+x^2}}$

12 $\int \dfrac{dx}{x^4\sqrt{4+x^2}}$

13 $\int \dfrac{dt}{t\sqrt{t^2+5}}$

14 $\int \dfrac{dx}{(x^2+9)^2}$

15 $\int \dfrac{7x^3\,dx}{(4x^2+9)^{3/2}}$

16 $\int \dfrac{dt}{\sqrt{16t^2+9}}$

17 $\int \dfrac{dx}{x^2\sqrt{x^2-4}}$

18 $\int \dfrac{dt}{t^4\sqrt{t^2-4}}$

19 $\int \dfrac{dt}{\sqrt{9t^2-4}}$

20 $\int \dfrac{\sqrt{y^2-9}}{y}\,dy$

21 $\int \dfrac{dx}{(4x^2-9)^{3/2}}$

22 $\int x^3\sqrt{x^2-1}\,dx$

23 Check your answers to problems 1, 13, and 19 by differentiation, carefully noting any required restrictions on the values of the variables involved.

24 For each of the given integrals, make a suitable change of variable so that trigonometric substitution can subsequently be employed.

(a) $\int \dfrac{\sec^2 t\,dt}{(\tan^2 t+9)^{3/2}}$

(b) $\int \dfrac{\sin v\,dv}{(25-\cos^2 v)^{3/2}}$

(c) $\int \dfrac{e^{-x}\,dx}{\sqrt{4+9e^{-2x}}}$

(d) $\int \dfrac{dt}{t[(\ln t)^2-4]^{3/2}}$

In problems 25 through 32, complete the square in the quadratic expression, then use trigonometric substitution to evaluate the integral.

25 $\int \dfrac{dt}{(5-4t-t^2)^{3/2}}$

26 $\int \dfrac{x\,dx}{\sqrt{2-x-x^2}}$

27 $\int \dfrac{dx}{\sqrt{-3+8x-4x^2}}$

28 $\int \dfrac{dt}{\sqrt{2t^2-6t+5}}$

29 $\int \dfrac{x\,dx}{\sqrt{1-x+3x^2}}$

30 $\int \sqrt{2-3x+x^2}\,dx$

31 $\int \dfrac{2t}{(t^2+3t+4)^2}\,dt$

32 $\int \dfrac{3}{(x^2+6x+1)^2}\,dx$

In problems 33 through 38, evaluate each definite integral.

33 $\displaystyle\int_{3/8}^{2/3} \dfrac{dx}{\sqrt{4x^2+1}}$

34 $\displaystyle\int_{3}^{4} \sqrt{25-t^2}\,dt$

35 $\displaystyle\int_{0}^{1} \dfrac{t^2\,dt}{(25-9t^2)^{3/2}}$

36 $\displaystyle\int_{0}^{1} \dfrac{x\,dx}{\sqrt{6-x-x^2}}$

37 $\displaystyle\int_{3}^{6} \dfrac{\sqrt{t^2-9}}{t}\,dt$

38 $\displaystyle\int_{\ln 2}^{\ln 3} \dfrac{dz}{e^z-e^{-z}}$

39 Find the area bounded by the curves $y = \dfrac{45}{\sqrt{16x^2 - 175}}$, $y = 0$, $x = 4$, and $x = 5$.

40 Find the area enclosed by the ellipse $\dfrac{x^2}{a^2} + \dfrac{y^2}{b^2} = 1$.

41 Find the volume of the solid generated when the region under the curve $y = \dfrac{1}{x^2 + 4}$ between $x = 0$ and $x = 3$ is revolved about the x axis.

42 Find the area of the region bounded by the hyperbola $x^2 - y^2 = 9$ and the line $y = 2x - 6$.

43 Find the volume of the solid generated when the region bounded by the curves $y = \dfrac{x}{(x^2 + 16)^{3/2}}$, $y = 0$, and $x = 4$ is revolved about the x axis.

44 Let $a > 0$ and suppose that it is required to integrate an expression involving $\sqrt{a^2 - u^2}$, where $|u| < a$. Since $|u/a| < 1$, we can make the substitution $\theta = \sin^{-1}(u/a)$, where $-\pi/2 < \theta < \pi/2$. Show that $u = a \sin \theta$ and that

(a) $\sqrt{a^2 - u^2} = a \cos \theta$

(b) $du = a \cos \theta \, d\theta$

(c) $\csc \theta = \dfrac{a}{u}$ if $u \neq 0$

(d) $\sec \theta = \dfrac{a}{\sqrt{a^2 - u^2}}$

(e) $\tan \theta = \dfrac{u}{\sqrt{a^2 - u^2}}$

(f) $\cot \theta = \dfrac{\sqrt{a^2 - u^2}}{u^2}$ if $u \neq 0$

45 Using the results in problem 44, explain why it is not really necessary to assume that u is positive when using the trigonometric substitution $u = a \sin \theta$.

46 Discuss the trigonometric substitution $u = a \tan \theta$ for the case in which u is not necessarily positive.

47 Evaluate $\displaystyle\int \dfrac{x \, dx}{\sqrt{x^2 - 1}}$ two ways:

(a) By trigonometric substitution.
(b) By using the substitution $u = x^2 - 1$.

48 A vertical wire of uniform density with total mass M and length l stands with its lower end on the x axis at the point $(a, 0)$. If γ denotes the constant of gravitation, calculate the horizontal component of the force of gravity exerted by the wire on a particle of mass m situated at the origin.

49 Solve the separable differential equation $x^2 \, dy - \sqrt{x^2 - 9} \, dx = 0$.

50 Find the surface area generated by revolving the arc of the curve $x = e^y$ between $(1, 0)$ and $(e^2, 2)$ about the y axis.

51 Find the arc length of the curve $y = \sin^{-1} e^x$ from $x = \ln \tfrac{1}{2}$ to $x = \ln \tfrac{3}{5}$.

4 Integration by Parts

Virtually every rule or technique for differentiation can be inverted to yield a corresponding rule or technique for integration. For instance, the technique of integration by change of variable is essentially the inversion of the chain rule for differentiation. In the present section we study the integration technique, called *integration by parts,* that results from an inversion of the product rule for differentiation.

4.1 Formula and Procedure for Integration by Parts

According to the product rule, $(f \cdot g)' = f' \cdot g + f \cdot g'$, the function $f \cdot g$ is an antiderivative of $f' \cdot g + f \cdot g'$; that is,

$$\int [f'(x)g(x) + f(x)g'(x)] \, dx = f(x)g(x) + C$$

or

$$\int f'(x)g(x) \, dx + \int f(x)g'(x) \, dx = f(x)g(x) + C.$$

The latter equation can be rewritten as

$$\int f(x)g'(x) \, dx = f(x)g(x) - \int g(x)f'(x) \, dx,$$

provided it is understood that the constant C is absorbed into the constant of integration corresponding to $\int g(x)f'(x) \, dx$.

The formula $\int f(x)g'(x) \, dx = f(x)g(x) - \int g(x)f'(x) \, dx$ just obtained is easier to remember and to use if it is written in terms of variables u and v defined by

$$u = f(x) \quad \text{and} \quad v = g(x).$$

Then $du = f'(x) \, dx$, $dv = g'(x) \, dx$, and, after substituting on both sides, the formula is rewritten

$$\int u \, dv = uv - \int v \, du.$$

The latter is called the formula for *integration by parts*. Its use is illustrated in the following example.

EXAMPLE Evaluate $\int x \sin x \, dx$.

SOLUTION
In order to use integration by parts, we put $u = x$ and $dv = \sin x \, dx$, so that $du = dx$ and $v = \int \sin x \, dx = -\cos x + C_0$. Thus,

$$\int \underset{u \quad dv}{\underline{x \, \sin x \, dx}} = \int u \, dv = uv - \int v \, du = x(-\cos x + C_0) - \int (-\cos x + C_0) \, dx$$

$$= -x \cos x + xC_0 + \int \cos x \, dx - \int C_0 \, dx$$

$$= -x \cos x + xC_0 + \sin x + C - C_0 x$$

$$= -x \cos x + \sin x + C.$$

In the example above, the constant of integration C_0 arising from the preliminary integration $v = \int \sin x \, dx$ cancels out in the end. In carrying out integration by parts, this always happens (Problem 42); hence, the constant arising from the preliminary integration $v = \int dv$ need not be written into the calculation. Thus, we would write

$$\int x \sin x \, dx = \int u \, dv = uv - \int v \, du = x(-\cos x) - \int (-\cos x) \, dx$$
$$= -x \cos x + \sin x + C.$$

More generally, we have the following procedure.

Procedure for Integration by Parts

To evaluate $\int F(x) \, dx$, carry out the following steps:

Step 1 Factor the integrand into two parts in some appropriate way, say $F(x) = F_1(x)F_2(x)$.

Step 2 Choose one of the two factors and set it equal to u. Then put dv equal to the remaining factor times dx. For definiteness, suppose that we have chosen to put $u = F_1(x)$, so that $dv = F_2(x) \, dx$.

Step 3 Calculate $du = F'_1(x) \, dx$ and calculate $v = \int F_2(x) \, dx$. Do not bother to write a constant of integration for the latter integral. We now have

$$u = F_1(x) \qquad dv = F_2(x) \, dx$$
$$du = F'_1(x) \, dx \qquad v = \int F_2(x) \, dx.$$

Step 4 Calculate $\int v \, du$. Again, do not bother to write a constant of integration for this integral.

Step 5 Write the solution

$$\int F(x) \, dx = \int \underbrace{F_1(x)}_{u}\underbrace{F_2(x) \, dx}_{dv} = \int u \, dv = uv - \int v \, du + C.$$

Notice that the constant of integration C is finally provided in the last step.

In Step 2, try to select u and dv so that the product of du and v—which must be integrated in Step 4—is as "simple" as possible. The proper factorization in Step 1 and choice of u and dv in Step 2 are often a matter of trial and error.

EXAMPLES   Evaluate the given integral using integration by parts.

1 $\int x \sec^2 x \, dx$

SOLUTION
The integrand is already factored, so we pass directly to the second step. There seem to be two possible choices:

Choice 1: $u = x$, $dv = \sec^2 x \, dx$, so that $du = dx$, $v = \int \sec^2 x \, dx = \tan x$.

Choice 2: $u = \sec^2 x$, $dv = x \, dx$, so that $du = 2 \sec^2 x \tan x$, $v = \int x \, dx = x^2/2$.

Evidently, $v \, du$ is "simpler" if we use Choice 1; hence, we follow Choice 1. Thus, in Step 3 we have

$$u = x \qquad dv = \sec^2 x \, dx$$
$$du = dx \qquad v = \tan x.$$

Carrying out Step 4, we have

$$\int v \, du = \int \tan x \, dx = -\ln |\cos x|.$$

Therefore, by Step 5,

$$\int x \sec^2 x \, dx = uv - \int v \, du = x \tan x + \ln |\cos x| + C.$$

2 $\int \ln x \, dx$

SOLUTION
Here it is not clear how to factor $\ln x$; however, we try the "trivial" factorization $\ln x = (\ln x)(1)$. Thus, put $u = \ln x$ and $dv = 1 \, dx = dx$, so that

$$u = \ln x \qquad dv = dx$$
$$du = \frac{1}{x} \, dx \qquad v = x.$$

Since $\int v \, du = \int x \left(\frac{1}{x} \, dx \right) = \int dx = x$, then

$$\int \ln x \, dx = uv - \int v \, du = x \ln x - x + C.$$

3 $\int x^3 \cos x^2 \, dx$

SOLUTION
Since $\cos x^2$ cannot be integrated in terms of "elementary" functions, we use the preliminary substitution $t = x^2$, so that $dt = 2x \, dx$. Thus,

$$\int x^3 \cos x^2 \, dx = \frac{1}{2} \int x^2 \cos x^2 (2x \, dx) = \frac{1}{2} \int t \cos t \, dt.$$

Now we can set

$$u = t \quad \text{and} \quad dv = \cos t \, dt,$$

so that

$$du = dt \quad \text{and} \quad v = \sin t.$$

Therefore,

$$\int x^3 \cos x^2 \, dx = \frac{1}{2} \int t \cos t \, dt = \frac{1}{2} \left(uv - \int v \, du \right)$$

$$= \frac{1}{2} \left(t \sin t - \int \sin t \, dt \right) = \frac{1}{2} \left(t \sin t + \cos t \right) + C$$

$$= \frac{1}{2} \left(x^2 \sin x^2 + \cos x^2 \right) + C.$$

In the formula for integration by parts, suppose that u and v are functions of (say) the variable x. Then uv, as well as the indefinite integrals $\int u \, dv$ and $\int v \, du$, depend on x. Consider the definite integral $\int_{x=a}^{x=b} u \, dv$, where we have written the limits of integration as $x = a$ to $x = b$ to emphasize that x (rather than u or v) is the variable in question. We have

$$\int_{x=a}^{x=b} u \, dv = \left(uv - \int v \, du \right) \Bigg|_{x=a}^{x=b} = (uv) \Bigg|_{x=a}^{x=b} - \int_{x=a}^{x=b} v \, du,$$

the formula for integration by parts of definite integrals.

EXAMPLE   Use integration by parts to find $\int_1^e x^2 \ln x \, dx$.

SOLUTION
Put $u = \ln x$ and $dv = x^2 \, dx$. Then $du = dx/x$, $v = x^3/3$, and

$$\int_1^e x^2 \ln x \, dx = \int_{x=1}^{x=e} u \, dv = (uv) \Bigg|_{x=1}^{x=e} - \int_{x=1}^{x=e} v \, du$$

$$= (\ln x)\left(\frac{x^3}{3} \right) \Bigg|_{x=1}^{x=e} - \int_{x=1}^{x=e} \frac{x^3}{3} \frac{dx}{x} = \left(\frac{x^3 \ln x}{3} \right) \Bigg|_{x=1}^{x=e} - \left(\frac{x^3}{9} \right) \Bigg|_{x=1}^{x=e}$$

$$= \frac{e^3}{3} - \frac{e^3}{9} + \frac{1}{9} = \frac{2e^3 + 1}{9}.$$

4.2 Repeated Integration by Parts

The formula $\int u \, dv = uv - \int v \, du$ for integration by parts converts the problem of evaluating $\int u \, dv$ into the problem of evaluating $\int v \, du$. By suitable choice of u and dv, we can often arrange it so that $\int v \, du$ is "simpler" than $\int u \, dv$; however, we may not be able to evaluate $\int v \, du$ directly. A second integration by parts might be necessary to evaluate $\int v \, du$. In fact, several successive integrations by parts could be required.

EXAMPLE   Evaluate $\int x^2 e^{2x} \, dx$.

SOLUTION

Put $u = x^2$ and $dv = e^{2x} \, dx$, so that $du = 2x \, dx$ and $v = \int e^{2x} \, dx = \frac{1}{2}e^{2x}$. Thus,

$$\int x^2 e^{2x} \, dx = uv - \int v \, du = \frac{1}{2}x^2 e^{2x} - \int x e^{2x} \, dx.$$

Although $\int x e^{2x} \, dx$ is "simpler" than the original $\int x^2 e^{2x} \, dx$, a second integration by parts is required to evaluate it. Thus, put $u_1 = x$ and $dv_1 = e^{2x} \, dx$, so that $du_1 = dx$ and $v_1 = \int e^{2x} \, dx = \frac{1}{2}e^{2x}$. Now,

$$\int x e^{2x} \, dx = u_1 v_1 - \int v_1 \, du_1 = \frac{1}{2}x e^{2x} - \int \frac{1}{2}e^{2x} \, dx = \frac{1}{2}x e^{2x} - \frac{1}{4}e^{2x} + C_1.$$

Hence,

$$\begin{aligned}
\int x^2 e^{2x} \, dx &= \frac{1}{2}x^2 e^{2x} - \int x e^{2x} \, dx \\
&= \frac{1}{2}x^2 e^{2x} - \left(\frac{1}{2}x e^{2x} - \frac{1}{4}e^{2x} + C_1\right) \\
&= \frac{1}{2}x^2 e^{2x} - \frac{1}{2}x e^{2x} + \frac{1}{4}e^{2x} + C,
\end{aligned}$$

where $C = -C_1$.

The calculations in the example above are depicted schematically in the following table:

Therefore,

$$\int \underbrace{x^2 \, e^{2x}}_{u \quad v' \, dx} \, dx = \frac{1}{2}x^2 e^{2x} - \frac{1}{2}x e^{2x} + \frac{1}{4}e^{2x} + C.$$

In general, repeated integration by parts of an integrand of the form uv', *where u is a polynomial*, can be accomplished by the *short* or *tabular method* as follows:

Step 1 Write the integrand as uv', where u is a polynomial.

Step 2 Make two parallel columns, labeling them the "u column" and the "v' column." The top entry in the u column is to be the polynomial u, while the top entry in the v' column is to be v'.

Step 3 Successive entries in the u column (reading downward) are obtained by repeated differentiation until arriving at 0. Corresponding successive entries in the v' column (reading downward) are obtained by repeated indefinite integration, omitting constants of integration at each stage, until the v' column is just as long as the u column.

Step 4 Multiply each nonzero entry in the u column by the *next entry below it* in the v' column, change the algebraic sign of *every other* product so obtained, and add the resulting terms. The desired integral is this sum plus an arbitrary constant of integration.

EXAMPLE Use the tabular method of repeated integration by parts to evaluate

$$\int (2x^4 - 8x^3)e^{-3x}\, dx.$$

SOLUTION

u		v'		
$2x^4 - 8x^3$		e^{-3x}		
$8x^3 - 24x^2$	*times*	$-\frac{1}{3}e^{-3x}$	(+)	$+(2x^4 - 8x^3)(-\frac{1}{3}e^{-3x})$
$24x^2 - 48x$	*times*	$\frac{1}{9}e^{-3x}$	(−)	$-(8x^3 - 24x^2)(\frac{1}{9}e^{-3x})$
$48x - 48$	*times*	$-\frac{1}{27}e^{-3x}$	(+)	$+(24x^2 - 48x)(-\frac{1}{27}e^{-3x})$
48	*times*	$\frac{1}{81}e^{-3x}$	(−)	$-(48x - 48)(\frac{1}{81}e^{-3x})$
0	*times*	$-\frac{1}{243}e^{-3x}$	(+)	$+48(-\frac{1}{243}e^{-3x})$.

Adding the terms in the last column and simplifying, we have

$$\int (2x^4 - 8x^3)e^{-3x}\, dx = \left(-\tfrac{2}{3}x^4 + \tfrac{16}{9}x^3 + \tfrac{16}{9}x^2 + \tfrac{32}{27}x + \tfrac{32}{81}\right)e^{-3x} + C.$$

Repeated integration by parts of integrands of the form $e^x \cos x$, $\sec^3 x$, $\csc^5 x$, and so on leads back to an integrand similar to the original one. When this happens, the resulting equation can often be solved for the unknown integral. This is illustrated by the following examples.

EXAMPLES 1 Evaluate $\int e^x \cos x\, dx$.

SOLUTION
An initial integration by parts with $u = e^x$, $dv = \cos x\, dx$, $du = e^x\, dx$, and $v = \sin x$ gives

(i) $\int e^x \cos x\, dx = e^x \sin x - \int e^x \sin x\, dx + C_1$.

A second integration by parts with $u_1 = e^x$, $dv_1 = \sin x\, dx$, $du_1 = e^x\, dx$, and $v_1 = -\cos x$ applied to $\int e^x \sin x\, dx$ gives

(ii) $\int e^x \sin x\, dx = -e^x \cos x + \int e^x \cos x\, dx + C_2$.

Substitution of (ii) into (i) yields

(iii) $\int e^x \cos x\, dx = e^x \sin x - \left(-e^x \cos x + \int e^x \cos x\, dx + C_2 \right) + C_1,$ or

(iv) $\int e^x \cos x\, dx = e^x \sin x + e^x \cos x - \int e^x \cos x\, dx + C_3,$

where we have put $C_3 = -C_2 + C_1$. Solving equation (iv) for $\int e^x \cos x\, dx$, we

obtain $2 \int e^x \cos x\, dx = e^x \sin x + e^x \cos x + C_3$, so that

(v) $\int e^x \cos x\, dx = \dfrac{e^x \sin x + e^x \cos x}{2} + C,$

where we have put $C = C_3/2$.

2 Evaluate $\int \sec^3 x\, dx.$

SOLUTION

We already noticed in Section 2 that the methods used there are not effective in evaluating the integral of $\sec^3 x$; however, we now use integration by parts to obtain such an integral. Factor $\sec^3 x$ as $\sec x \sec^2 x$ and put $u = \sec x$ and $dv = \sec^2 x\, dx$. Then $du = \sec x \tan x\, dx$ and $v = \tan x$, so that

$$\int \sec^3 x\, dx = \int \sec x \sec^2 x\, dx = \sec x \tan x - \int \sec x \tan^2 x\, dx + C_0.$$

Since

$$\int \sec x \tan^2 x\, dx = \int \sec x(\sec^2 x - 1)\, dx = \int \sec^3 x\, dx - \int \sec x\, dx,$$

it follows that

$$\int \sec^3 x\, dx = \sec x \tan x - \left(\int \sec^3 x\, dx - \int \sec x\, dx \right) + C_0$$

$$= \sec x \tan x - \int \sec^3 x\, dx + \int \sec x\, dx + C_0$$

$$= \sec x \tan x - \int \sec^3 x\, dx + \ln |\sec x + \tan x| + C_0.$$

From the latter equation,

$$2 \int \sec^3 x\, dx = \sec x \tan x + \ln |\sec x + \tan x| + C_0$$

or

$$\int \sec^3 x\, dx = \tfrac{1}{2} \sec x \tan x + \tfrac{1}{2} \ln |\sec x + \tan x| + C,$$

where we have put $C = \tfrac{1}{2}C_0$.

Repeated integration by parts leading back to an integrand similar to the original one can also be handled by a "short" or "tabular" method. See Problem 48 for an indication of how this is done.

Problem Set 4

In problems 1 through 14, use integration by parts to evaluate each integral.

1 $\int x \cos 2x \, dx$

2 $\int x \sin kx \, dx$

3 $\int xe^{3x} \, dx$

4 $\int xe^{-4x} \, dx$

5 $\int \ln 5x \, dx$

6 $\int x \ln 2x \, dx$

7 $\int \dfrac{x^3}{\sqrt{1-x^2}} \, dx$ (*Hint:* Let $u = x^2$.)

8 $\int x^3 \ln(x^2) \, dx$

9 $\int x \csc^2 x \, dx$

10 $\int \sin^{-1} 3x \, dx$

11 $\int t \sec t \tan t \, dt$

12 $\int \tan^{-1} x \, dx$

13 $\int \cos^{-1} x \, dx$

14 $\int \cosh^{-1} x \, dx$

In problems 15 through 22, use the short method of repeated integration by parts to evaluate each integral.

15 $\int x^2 \sin 3x \, dx$

16 $\int x^3 \sin 5x \, dx$

17 $\int x^4 \cos 2x \, dx$

18 $\int x^2 \sin^2 x \, dx$

19 $\int t^4 e^{-t} \, dt$

20 $\int (3x^2 - 2x + 1) \cos x \, dx$

21 $\int (x^5 - x^3 + x)e^{-x} \, dx$

22 $\int x^2 \sec^2 x \tan x \, dx$

In problems 23 through 28, use a suitable substitution to express the integral in a form so that integration by parts is applicable, and then evaluate the integral. (In some cases a suitable substitution is suggested.)

23 $\int x^3 e^{x^2} \, dx; \; x^2 = t$

24 $\int x^3 \sin 2x^2 \, dx$

25 $\int \sqrt{1+x^2} \, dx; \; x = \tan \theta$

26 $\int \dfrac{x^2}{\sqrt{x^2-1}}; \; x = \sec \theta$

27 $\int x^{11} \cos x^4 \, dx; \; x^4 = t$

28 $\int x^{3/2} \cos \sqrt{x} \, dx$

In problems 29 through 34, use repeated integration by parts to evaluate each integral.

29 $\int e^{-x} \cos 2x \, dx$

30 $\int e^{2x} \sin x \, dx$

31 $\int \csc^3 x \, dx$

32 $\int \sec^5 x \, dx$

33 $\int \sin x \sin 2x \, dx$

34 $\int e^{ax} \sin bx \, dx$

In problems 35 through 40, evaluate each definite integral.

35 $\int_0^{\pi/9} 4x^2 \sin 3x \, dx$

36 $\int_0^1 \dfrac{xe^x}{(1+x)^2} \, dx$

37 $\int_2^3 \sec^{-1} x \, dx$

38 $\int_{-1}^1 \cos^{-1} x \, dx$

39 $\int_0^{\pi/4} (5x^2 - 3x + 1) \sin x \, dx$

40 $\int_1^e \sin(\ln x) \, dx$

41 Use integration by parts to show that

$$\int_0^a x^2 f'''(x) \, dx = a^2 f''(a) - 2af'(a) + 2f(a) - 2f(0).$$

42 It is desired to evaluate $\int F_1(x)F_2(x) \, dx$ by parts. To this end, we put $u = F_1(x)$ and $dv = F_2(x) \, dx$. Then $du = F'_1(x) \, dx$ and $v = G(x) + C_0$, where G is an antiderivative of

F_2 and C_0 is the constant of integration. Show that C_0 always cancels in the calculation

of $\int F_1(x)F_2(x)\,dx$ by parts.

43 Use integration by parts to prove that

$$\int f(x)\,dx = xf(x) - \int xf'(x)\,dx.$$

44 Suppose that $g'' = f$. Evaluate $\int x^2 f(x)\,dx$.

45 Let $f(x) = xe^{-x}$ for $x \geq 0$. Suppose that $f(x)$ takes on its maximum value when $x = a$. Find the area of the region under the graph of f between $x = 0$ and $x = a$.

46 Find the volume of the solid generated if the region described in problem 45 is rotated about the x axis.

47 Use integration by parts to establish the reduction formula in problem 43 of Problem Set 1. Take $u = \sin^{n-1} ax$ and $dv = \sin ax\,dx$.

48 In the following two columns,

assume that the derivative of each term in the first column is the term below it and that (apart from a constant of integration) the indefinite integral of each term in the second column is the term below it. Prove that

$$\int u\,dv = uv - u_1 v_1 + u_2 v_2 - u_3 v_3 + \cdots \mp u_{n-1} v_{n-1} \pm \int u_n\,dv_n,$$

where the plus or minus sign is used on the last term (involving the integral) according to whether n is even or odd, respectively, while the opposite sign is used on the next-to-last term.

49 Use the result of problem 48 to justify the short method of repeated integration by parts.

50 Use integration by parts to establish the reduction formula

$$\int \sec^n kx\,dx = \frac{1}{k(n-1)} \cdot \frac{\sin kx}{\cos^{n-1} kx} + \frac{n-2}{n-1}\int \sec^{n-2} kx\,dx,$$

where $n \geq 2$ and $k \neq 0$.

5 Integration of Rational Functions by Partial Fractions—Linear Case

In Section 6.3 of Chapter 0 we defined a rational function to be a function h given by $h(x) = P(x)/Q(x)$, where $P(x)$ and $Q(x)$ are polynomials and $Q(x)$ is not the zero polynomial. As we shall see in this and the succeeding section, it is possible to find the integral of a rational function by expanding it into a sum of simple rational functions called *partial fractions*, and then finding the integrals of these partial fractions.

Before discussing the various possible cases, let us observe that if the degree of the numerator $P(x)$ of the fraction $P(x)/Q(x)$ is not smaller than the degree of the denominator $Q(x)$, then we can always perform a long division of $Q(x)$ into $P(x)$ to obtain a quotient $f(x)$ and a remainder $R(x)$. Thus,

$$Q(x)\overline{)P(x)} \quad \overset{f(x)}{} \quad \text{with remainder } R(x),$$

where $f(x)$ is a polynomial and where $R(x)$ is either the zero polynomial or else $R(x)$ is a polynomial of degree less than the degree of $Q(x)$. Then we have

$$h(x) = \frac{P(x)}{Q(x)} = f(x) + \frac{R(x)}{Q(x)},$$

so that

$$\int h(x)\,dx = \int f(x)\,dx + \int \frac{R(x)}{Q(x)}\,dx.$$

Since $f(x)$ is a polynomial, there is no difficulty in evaluating $\int f(x)\,dx$. Therefore, the problem of finding $\int h(x)\,dx$ reduces to the problem of finding $\int \frac{R(x)}{Q(x)}\,dx$, where the fraction $R(x)/Q(x)$ is *proper* in the sense that the degree of the denominator $Q(x)$ exceeds the degree of the numerator $R(x)$.

In view of the procedure explained above, we can concentrate on the problem of integrating proper rational functions. Of course, we also assume that all common factors in numerator and denominator have been canceled.

5.1   The Case in Which the Denominator Factors into Distinct Linear Factors

Notice that the two fractions $\dfrac{2}{x-2}$ and $\dfrac{3}{x+1}$ can be added to obtain

$$\frac{2}{x-2} + \frac{3}{x+1} = \frac{2(x+1) + 3(x-2)}{(x-2)(x+1)} = \frac{5x-4}{(x-2)(x+1)}.$$

It follows that

$$\int \frac{(5x-4)\,dx}{(x-2)(x+1)} = \int \frac{2\,dx}{x-2} + \int \frac{3\,dx}{x+1} = 2\ln|x-2| + 3\ln|x+1| + C.$$

Therefore, it is a simple matter to integrate $\dfrac{5x-4}{(x-2)(x+1)}$ *provided that we know it decomposes into the sum of* $\dfrac{2}{x-2}$ *and* $\dfrac{3}{x+1}$.

The method of partial fractions is just a systematic procedure for decomposing proper rational functions $P(x)/Q(x)$ into a sum of simple rational functions, each of which can be integrated by a standard method. The simple rational functions whose sum is $P(x)/Q(x)$ are called the *partial fractions* for $P(x)/Q(x)$. For instance, the equation

$$\frac{5x - 4}{(x - 2)(x + 1)} = \frac{2}{x - 2} + \frac{3}{x + 1}$$

shows the decomposition of the rational function on the left into the two partial fractions on the right.

Decomposition of a rational function into partial fractions is easiest to accomplish when the denominator factors into distinct (that is, different) linear factors. In this case, it is only necessary to provide a partial fraction of the form $\dfrac{\text{constant}}{ax + b}$ for each of the linear factors $ax + b$ in the denominator. The constant numerators of the partial fractions can be denoted by A, B, C, and so forth.

For instance, the fraction $\dfrac{5x - 4}{(x - 2)(x + 1)}$ is decomposed as follows:

$$\frac{5x - 4}{(x - 2)(x + 1)} = \frac{A}{x - 2} + \frac{B}{x + 1}.$$

Another example is provided by

$$\frac{27x^3 - 4x + 13}{(3x - 1)(x + 2)\left(\dfrac{x}{2} - 7\right)x} = \frac{C}{3x - 1} + \frac{D}{x + 2} + \frac{E}{\dfrac{x}{2} - 7} + \frac{F}{x}.$$

Here, of course, the problem is to determine the values of the constant numerators appearing in the partial fractions.

In the case under consideration there are two general methods for finding the values of the numerators of the partial fractions.

1 Equating Coefficients

We have

$$\frac{P(x)}{Q(x)} = (\text{a sum of partial fractions}),$$

where the numerators of the partial fractions are denoted by A, B, C, and so forth. Multiply this equation by the denominator $Q(x)$ on both sides to obtain

$$P(x) = Q(x) \, (\text{sum of partial fractions}),$$

and then collect like terms on the right side. Equate the coefficients of powers of x on the left side to the coefficients of corresponding powers of x on the right side, and thus obtain a system of equations involving the unknowns A, B, C, and so forth. Solve this system for the unknowns.

EXAMPLE   Find the partial fractions decomposition of $\dfrac{5x - 4}{(x - 2)(x + 1)}$ by equating coefficients.

SOLUTION
We have

$$\frac{5x - 4}{(x - 2)(x + 1)} = \frac{A}{x - 2} + \frac{B}{x + 1},$$

where the constants A and B must be determined. Multiplying both sides of the equation above by the denominator $(x-2)(x+1)$ yields

$$5x - 4 = A(x + 1) + B(x - 2) \quad \text{or} \quad 5x - 4 = (A + B)x + (A - 2B).$$

Equating coefficients of like powers of x on both sides of the latter identity gives

$$\begin{cases} 5 = A + B \\ -4 = A - 2B. \end{cases}$$

Solving (say, by elimination) the system of linear equations above, we find that $A = 2$ and $B = 3$. Therefore,

$$\frac{5x - 4}{(x - 2)(x + 1)} = \frac{2}{x - 2} + \frac{3}{x + 1}.$$

2 Substitution

Suppose that $\dfrac{A}{ax + b}$ is any one of the partial fractions in the decomposition of $P(x)/Q(x)$. Then $ax + b$ is one of the factors of $Q(x)$, so we can write $Q(x) = (ax + b)Q_1(x)$, where $Q_1(x)$ is the remaining part of the denominator. Thus,

$$\frac{P(x)}{(ax + b)Q_1(x)} = \frac{A}{ax + b} + \text{(other partial fractions)}.$$

Multiply both sides of this equation by $(ax + b)$ to obtain

$$\frac{P(x)}{Q_1(x)} = A + (ax + b)\,\text{(other partial fractions)}.$$

If we now put $x = -b/a$, then $ax + b = 0$ and the equation becomes

$$\frac{P(-b/a)}{Q_1(-b/a)} = A.$$

Thus, we obtain the numerical value of A. Repeat this procedure for each of the partial fractions.

EXAMPLE Find the partial fractions decomposition of $\dfrac{5x - 4}{(x - 2)(x + 1)}$ by substitution.

SOLUTION

Again $\dfrac{5x - 4}{(x - 2)(x + 1)} = \dfrac{A}{x - 2} + \dfrac{B}{x + 1}$, where the constants A and B must be determined. Multiply both sides of this equation by $x - 2$, and then substitute $x = 2$ to obtain

$$\frac{5x - 4}{x + 1} = A + (x - 2)\frac{B}{x + 1},$$

$$\frac{10 - 4}{2 + 1} = A + (0)\frac{B}{2 + 1}, \qquad \frac{6}{3} = A, \quad A = 2.$$

Similarly, to find B, multiply both sides of the equation

$$\frac{5x - 4}{(x - 2)(x + 1)} = \frac{A}{x - 2} + \frac{B}{x + 1}$$

by $x + 1$, and then substitute $x = -1$ to obtain $\dfrac{-5-4}{-1-2} = B$, $B = 3$. Therefore,

$$\frac{5x-4}{(x-2)(x+1)} = \frac{2}{x-2} + \frac{3}{x+1}.$$

The method of substitution can be shortened simply by temporarily "covering up" or "disregarding" appropriate portions of the equation and then substituting. For instance, to find the constant A in

$$\frac{5x-4}{(x-2)(x+1)} = \frac{A}{x-2} + \frac{B}{x+1},$$

we "cover up" or "disregard" everything on the right side of the equation except A and we "cover up" or "disregard" the factor in the denominator on the left side that corresponds to A. This produces the equation

$$\frac{5x-4}{(\cancel{x-2})(x+1)} = \frac{A}{\cancel{x-2}} + \frac{\cancel{B}}{\cancel{x+1}}$$

into which we substitute the value $x = 2$ to obtain

$$\frac{10-4}{2+1} = A \quad \text{or} \quad A = 2.$$

Similarly, to find B, we substitute $x = -1$ into

$$\frac{5x-4}{(x-2)(\cancel{x+1})} = \frac{\cancel{A}}{\cancel{x-2}} + \frac{B}{\cancel{x+1}}$$

to obtain

$$\frac{-5-4}{-1-2} = B \quad \text{or} \quad B = 3.$$

This method is called the *short method of substitution*.

EXAMPLES Evaluate the given integral.

1 $\displaystyle\int \frac{3x-5}{x^2-x-2}\,dx$

SOLUTION
$x^2 - x - 2 = (x-2)(x+1)$; hence,

$$\frac{3x-5}{x^2-x-2} = \frac{3x-5}{(x-2)(x+1)} = \frac{A}{x-2} + \frac{B}{x+1}.$$

By the short method of substitution $\dfrac{6-5}{2+1} = A$, so that $A = \tfrac{1}{3}$; and $\dfrac{-3-5}{-1-2} = B$, so that $B = \tfrac{8}{3}$. Therefore,

$$\int \frac{3x-5}{x^2-x-2}\,dx = \int \frac{\tfrac{1}{3}}{x-2}\,dx + \int \frac{\tfrac{8}{3}}{x+1}\,dx$$

$$= \tfrac{1}{3}\ln|x-2| + \tfrac{8}{3}\ln|x+1| + C$$

$$= \ln\left[|x-2|^{1/3}|x+1|^{8/3}\right] + C.$$

2 $\displaystyle\int \frac{5x^3 - 6x^2 - 68x - 16}{x^3 - 2x^2 - 8x}\,dx$

SOLUTION

The integrand is an improper fraction. By long division

$$
x^3 - 2x^2 - 8x\,\overline{\big)\,5x^3 - 6x^2 - 68x - 16}
$$

$$
\begin{array}{r}
5 \\
\underline{5x^3 - 10x^2 - 40x} \\
4x^2 - 28x - 16,
\end{array}
$$

so that

$$
\frac{5x^3 - 6x^2 - 68x - 16}{x^3 - 2x^2 - 8x} = 5 + \frac{4x^2 - 28x - 16}{x^3 - 2x^2 - 8x}.
$$

Since $x^3 - 2x^2 - 8x = x(x^2 - 2x - 8) = x(x - 4)(x + 2)$, we have

$$
\frac{4x^2 - 28x - 16}{x^3 - 2x^2 - 8x} = \frac{4x^2 - 28x - 16}{x(x - 4)(x + 2)} = \frac{A}{x} + \frac{B}{x - 4} + \frac{C}{x + 2}.
$$

By the short method of substitution, $\dfrac{(4)(0)^2 - (28)(0) - 16}{(0 - 4)(0 + 2)} = A$, so that $A = 2$;

$\dfrac{(4)(4)^2 - (28)(4) - 16}{4(4 + 2)} = B$, so that $B = -\tfrac{8}{3}$; and $\dfrac{(4)(-2)^2 - (28)(-2) - 16}{(-2)(-2 - 4)} = C$,

so that $C = \tfrac{14}{3}$. Therefore,

$$
\int \frac{4x^2 - 28x - 16}{x^3 - 2x^2 - 8x}\,dx = \int \frac{2}{x}\,dx + \int \frac{-\tfrac{8}{3}}{x - 4}\,dx + \int \frac{\tfrac{14}{3}}{x + 2}\,dx
$$

$$
= 2 \ln |x| - \tfrac{8}{3} \ln |x - 4| + \tfrac{14}{3} \ln |x + 2| + K
$$

$$
= \ln \left(\frac{x^2 |x + 2|^{14/3}}{|x - 4|^{8/3}} \right) + K,
$$

where we have written the constant of integration as K to avoid confusing it with the numerator of the partial fraction $\dfrac{C}{x + 2}$. Consequently,

$$
\int \frac{5x^3 - 6x^2 - 68x - 16}{x^3 - 2x^2 - 8x}\,dx = \int 5\,dx + \int \frac{4x^2 - 28x - 16}{x^3 - 2x^2 - 8x}\,dx
$$

$$
= 5x + \ln \left(\frac{x^2 |x + 2|^{14/3}}{|x - 4|^{8/3}} \right) + K.
$$

5.2 The Case in Which the Denominator Has Repeated Linear Factors

We now consider the case of a rational function $P(x)/Q(x)$ whose denominator $Q(x)$ factors completely into linear factors, not all of which are distinct. An example would be

$$
Q(x) = (3x - 2)(x - 1)(3x - 2)(x - 1)(x + 1)(x - 1),
$$

in which the factor $(3x - 2)$ appears twice and the factor $(x - 1)$ appears three times. The first step in handling such a case is to gather like factors and write them with the aid of exponents. In the example at hand, we have

$$Q(x) = (x - 1)^3(3x - 2)^2(x + 1).$$

If the denominator contains a factor of the form $(ax + b)^k$ where $k > 1$, it is necessary to provide k corresponding partial fractions of the form

$$\frac{A_1}{ax + b} + \frac{A_2}{(ax + b)^2} + \frac{A_3}{(ax + b)^3} + \cdots + \frac{A_k}{(ax + b)^k},$$

where $A_1, A_2, A_3, \ldots, A_k$ are constants that need to be determined. This must be done for *each* factor in the denominator. Nonrepeated factors are handled just as before.

For instance, the partial fractions decomposition for $\dfrac{3x^2 + 4x + 2}{x(x + 1)^2}$ has the form

$$\frac{3x^2 + 4x + 2}{x(x + 1)^2} = \frac{A}{x} + \frac{B_1}{x + 1} + \frac{B_2}{(x + 1)^2}.$$

Although the short method of substitution can no longer be used to determine *all* the unknown constants, it can still be used to find the constants, such as A in the example above, that correspond to nonrepeated factors. Indeed,

$$\frac{(3)(0)^2 + (4)(0) + 2}{(0 + 1)^2} = A,$$

so $A = 2$.

The short method of substitution is also effective for determining the constant numerators of the partial fractions involving the *highest powers* of the repeated factors. (Why?) In the example above, for instance,

$$\frac{(3)(-1)^2 + (4)(-1) + 2}{(-1)} = B_2,$$

so $B_2 = -1$. Unfortunately, the short method of substitution does not work for the constant numerators of the partial fractions involving remaining (lower) powers of the repeated factors, such as B_1 in the present example. These remaining constants can be found by the method of equating coefficients.

For instance, we now have

$$\frac{3x^2 + 4x + 2}{x(x + 1)^2} = \frac{2}{x} + \frac{B_1}{x + 1} - \frac{1}{(x + 1)^2}.$$

In order to find B_1, we multiply both sides of the equation by $x(x + 1)^2$ to obtain $3x^2 + 4x + 2 = 2(x + 1)^2 + B_1x(x + 1) - x$; that is,

$$3x^2 + 4x + 2 = (2 + B_1)x^2 + (3 + B_1)x + 2.$$

Equating coefficients of like powers, we have $3 = 2 + B_1$ and $4 = 3 + B_1$. Therefore, $B_1 = 1$, and so

$$\frac{3x^2 + 4x + 2}{x(x + 1)^2} = \frac{2}{x} + \frac{1}{x + 1} - \frac{1}{(x + 1)^2}.$$

Rational functions of the form $P(x)/Q(x)$ in which the denominator factors completely into linear factors can always be expressed in terms of partial fractions of the form $\dfrac{A}{(ax + b)^n}$. Since $\displaystyle\int \dfrac{A\,dx}{(ax + b)^n}$ can easily be evaluated using the change of variable $u = ax + b$, we see that $\displaystyle\int \dfrac{P(x)}{Q(x)}\,dx$ can always be evaluated. This is illustrated by the following examples.

EXAMPLES   Evaluate the given integral.

1 $\displaystyle\int \dfrac{3x^2 + 4x + 2}{x(x + 1)^2}\,dx$

SOLUTION
Using the decomposition above, we have

$$\int \frac{3x^2 + 4x + 2}{x(x + 1)^2}\,dx = \int \left[\frac{2}{x} + \frac{1}{x + 1} - \frac{1}{(x + 1)^2} \right] dx$$

$$= \int \frac{2\,dx}{x} + \int \frac{dx}{x + 1} - \int \frac{dx}{(x + 1)^2}$$

$$= 2 \ln |x| + \ln |x + 1| + \frac{1}{x + 1} + C$$

$$= \ln |x|^2 + \ln |x + 1| + \frac{1}{x + 1} + C$$

$$= \ln |x^2(x + 1)| + \frac{1}{x + 1} + C.$$

2 $\displaystyle\int_2^3 \dfrac{3x - 2}{x^3 - x^2}\,dx$

SOLUTION

$$\frac{3x - 2}{x^3 - x^2} = \frac{3x - 2}{x^2(x - 1)} = \frac{A_1}{x} + \frac{A_2}{x^2} + \frac{B}{x - 1}.$$

By the short method of substitution, $\dfrac{3 - 2}{1^2} = B$, so that $B = 1$; $\dfrac{0 - 2}{(0 - 1)} = A_2$, so that $A_2 = 2$. Thus, $\dfrac{3x - 2}{x^2(x - 1)} = \dfrac{A_1}{x} + \dfrac{2}{x^2} + \dfrac{1}{x - 1}$. Using the method of equating coefficients, we have

$$3x - 2 = A_1 x(x - 1) + 2(x - 1) + x^2 = (A_1 + 1)x^2 + (-A_1 + 2)x - 2;$$

hence, $A_1 + 1 = 0$ and $-A_1 + 2 = 3$, so that $A_1 = -1$. Therefore,

$$\int_2^3 \frac{3x - 2}{x^3 - x^2}\,dx = \int_2^3 \left(\frac{-1}{x} + \frac{2}{x^2} + \frac{1}{x - 1} \right) dx$$

$$= \left(-\ln |x| - \frac{2}{x} + \ln |x - 1| \right) \Big|_2^3 = \left(\ln \left| \frac{x - 1}{x} \right| - \frac{2}{x} \right) \Big|_2^3$$

$$= \left(\ln \frac{2}{3} - \frac{2}{3} \right) - \left(\ln \frac{1}{2} - 1 \right) = \ln \frac{4}{3} + \frac{1}{3}.$$

Problem Set 5

In problems 1 through 24, evaluate each integral.

1 $\int \dfrac{x+1}{x(x-2)}\,dx$

2 $\int \dfrac{x+1}{x^2-x-2}\,dx$

3 $\int \dfrac{x^3+5x^2-4x-20}{x^2+3x-10}\,dx$

4 $\int \dfrac{x\,dx}{(x-1)(x+1)(x+2)}$

5 $\int \dfrac{x^3+5x^2-x-22}{x^2+3x-10}\,dx$

6 $\int \dfrac{8x+7}{2x^2+3x+1}\,dx$

7 $\int \dfrac{2x+1}{x^3+x^2-2x}\,dx$

8 $\int \dfrac{x^3+2x^2-3x+1}{x^3+3x^2+2x}\,dx$

9 $\int \dfrac{dx}{x^3-x}$

10 $\int \dfrac{(t+7)\,dt}{(t+1)(t^2-4t+3)}$

11 $\int \dfrac{x^2\,dx}{x^2-x-6}$

12 $\int \dfrac{3z+1}{z(z^2-4)}\,dz$

13 $\int \dfrac{x^2\,dx}{x^2+x-6}$

14 $\int \dfrac{x^4+2x^3+1}{x^3-x^2-2x}\,dx$

15 $\int \dfrac{5x^2-7x+8}{x^3+3x^2-4x}\,dx$

16 $\int \dfrac{2x\,dx}{(x+2)(x^2-1)}$

17 $\int \dfrac{2z+3}{z^2(4z+1)}\,dz$

18 $\int \dfrac{x^2+1}{(x+3)(x^2+4x+4)}\,dx$

19 $\int \dfrac{x+3}{(x+1)^2(x+7)}\,dx$

20 $\int \dfrac{x+4}{(x^2+2x+1)(x-1)^2}\,dx$

21 $\int \dfrac{4x^2-7x+10}{(x+2)(3x-2)^2}\,dx$

22 $\int \dfrac{x^3-3x^2+5x-12}{(x-1)^2(x^2-3x-4)}\,dx$

23 $\int \dfrac{4z^2\,dz}{(z-1)^2(z^2-4z+3)}$

24 $\int \dfrac{(t+2)\,dt}{(t^2-1)(t+3)^2}$

In problems 25 through 30, evaluate each definite integral.

25 $\displaystyle\int_2^4 \dfrac{x\,dx}{(x+1)(x+2)}$

26 $\displaystyle\int_1^2 \dfrac{5t^2-3t+18}{t(9-t^2)}\,dt$

27 $\displaystyle\int_2^3 \dfrac{4t^5-3t^4-6t^3+4t^2+6t-1}{(t-1)(t^2-1)}\,dt$

28 $\displaystyle\int_1^2 \dfrac{x^5+3x^4-4x^3-x^2+11x+12}{x^2(x^2+4x+5)}\,dx$

29 $\displaystyle\int_3^5 \dfrac{x^2-2}{(x-2)^2}\,dx$

30 $\displaystyle\int_1^2 \dfrac{2z^3+1}{z(z+1)^2}\,dz$

31 Evaluate $\int \dfrac{ax+b}{cx+d}\,dx$ if $c \neq 0$. $\left(Hint: \dfrac{ax+b}{cx+d} = \dfrac{a}{c} + \dfrac{bc-ad}{c(cx+d)}.\right)$

32 Evaluate $\int \dfrac{dx}{(x-a)(x-b)}$ if $a \neq b$.

33 Evaluate $\int \dfrac{dx}{(x-a)(x-b)}$ if $a = b$.

34 Is it true that $\displaystyle\lim_{a \to b} \int \dfrac{dx}{(x-a)(x-b)} = \int \dfrac{dx}{(x-b)^2}$?

35 Evaluate $\int \dfrac{x+1}{x^2-x-6}\,dx$ in two ways:

(a) By completing the square in the denominator and substituting $u = x - \frac{1}{2}$.

(b) By factoring the denominator and using partial fractions.

36 Evaluate $\int \dfrac{x+c}{(x-a)^2}\,dx$.

37 Find the area of the region in the first quadrant bounded by the curve $(x+2)^2 y = 4 - x$.

38 Find the volume of the solid generated by revolving the region in problem 37 about the x axis.

39 In connection with the study of the production of ions by radiation it is necessary to deal with the integral $\int \dfrac{dx}{q-ax^2}$, where q and a are positive constants. Evaluate this integral.

40 The differential equation $dy/dt = ak[1 - (1 + b)y][1 - (1 - b)y]$ applies to the velocity of the reaction between ethyl alcohol and chloroacetic acid. Here, a, k, and b represent positive constants. Solve this differential equation.

41 The differential equation $dx/dt = k(a - x)^4$ applies to the velocity of the reaction between hydrobromic and bromic acids. Here a and k are positive constants. Solve this differential equation.

42 Find the arc length of the curve $y = \ln(1 - x^2)$ from $x = 0$ to $x = \frac{1}{3}$.

43 The marginal cost function of a certain product is given by $C'(x) = \dfrac{400x^2 + 1300x - 900}{x(x - 1)(x + 3)}$,

where x is the number of units produced. If it costs \$47 to produce 2 units, find a formula for C in terms of x. (Assume that $x \geq 2$.)

44 Explain why every rational function whose denominator factors completely into linear factors has an integral that can be expressed in terms of rational functions and logarithms (of absolute values). Assume—as is, in fact, the case—that such a rational function can always be decomposed into partial fractions.

45 Ecologists sometimes use the following simplified growth model for the population q of a sexually reproducing species: The rate of change of q is the birth rate minus the death rate. The birth rate depends on the frequency of contacts between males and females; hence, it is proportional to q^2. The death rate is proportional to q. Hence, $dq/dt = Bq^2 - Aq$, where B and A are constants of proportionality. Solve this differential equation by separating the variables and integrating.

6   Integration of Rational Functions by Partial Fractions—Quadratic Case

We now turn our attention to the problem of integrating a rational function $P(x)/Q(x)$ whose denominator $Q(x)$ cannot be factored completely into linear factors. It can be proved that any polynomial $Q(x)$ with real numbers as coefficients can be factored completely into a finite number of polynomials, each of which is either linear or quadratic. Furthermore, the quadratic factors (if any) can be assumed to be *irreducible*—that is, incapable of further factorization into linear polynomials. For instance,

$$x^5 + x^4 - x^3 - 3x^2 + 2 = (x - 1)(x^2 + 2x + 2)(x + 1)(x - 1),$$

as can be confirmed by direct multiplication of the factors on the right. The quadratic polynomial $x^2 + 2x + 2$ cannot be factored into linear factors (unless imaginary numbers are introduced); hence, it is irreducible.

It is easy to check whether a quadratic polynomial $ax^2 + bx + c$ is irreducible—it is irreducible if and only if its *discriminant*, $b^2 - 4ac$, is negative. For instance, $x^2 + 2x + 2$ is irreducible since its discriminant $2^2 - (4)(1)(2) = -4$ is negative.

As in Section 5, we can assume without loss of generality that $P(x)/Q(x)$ is a proper fraction and that all common factors have been canceled from its numerator and denominator.

6.1 The Case in Which the Denominator Involves Distinct Irreducible Quadratic Factors

Assume that the denominator $Q(x)$ of $P(x)/Q(x)$ factors into linear and irreducible quadratic polynomials, but that none of the quadratic factors are repeated. Again, we seek to decompose $P(x)/Q(x)$ into suitable partial fractions. This is accomplished as follows:

For each nonrepeated irreducible quadratic factor $ax^2 + bx + c$ in the denominator of $P(x)/Q(x)$, one must provide a corresponding partial fraction of the form $\dfrac{Ax + B}{ax^2 + bx + c}$ in order to obtain the decomposition of $P(x)/Q(x)$. Again, it is necessary to determine the numerical values of A and B, say by equating coefficients. Partial fractions corresponding to linear factors of the denominator must be introduced just as in Section 5.

EXAMPLES   Decompose the given rational fraction into partial fractions.

1 $\dfrac{8x^2 + 3x + 20}{x^3 + x^2 + 4x + 4}$

SOLUTION
Factoring the denominator, we obtain

$$x^3 + x^2 + 4x + 4 = x^2(x + 1) + 4(x + 1) = (x + 1)(x^2 + 4).$$

The quadratic factor $x^2 + 4$ is irreducible; hence we must provide a corresponding partial fraction of the form $\dfrac{Ax + B}{x^2 + 4}$. We must also introduce a partial fraction $\dfrac{C}{x + 1}$ corresponding to the linear factor $x + 1$. Consequently,

$$\frac{8x^2 + 3x + 20}{x^3 + x^2 + 4x + 4} = \frac{8x^2 + 3x + 20}{(x + 1)(x^2 + 4)} = \frac{Ax + B}{x^2 + 4} + \frac{C}{x + 1}.$$

We can evaluate the constant C by the short method of substitution just as in Section 5; thus, $\dfrac{(8)(-1)^2 + (3)(-1) + 20}{(-1)^2 + 4} = C$, so that $C = 5$. Therefore,

$$\frac{8x^2 + 3x + 20}{(x + 1)(x^2 + 4)} = \frac{Ax + B}{x^2 + 4} + \frac{5}{x + 1}.$$

Multiplying both sides of the latter equation by $(x + 1)(x^2 + 4)$ and collecting terms on the right, we have

$$8x^2 + 3x + 20 = (Ax + B)(x + 1) + 5(x^2 + 4)$$

or

$$8x^2 + 3x + 20 = (A + 5)x^2 + (A + B)x + (B + 20).$$

Equating coefficients of like powers in the latter equation, we have $8 = A + 5$, $3 = A + B$, and $20 = B + 20$. Therefore, $B = 0$, $A = 3$, and

$$\frac{8x^2 + 3x + 20}{x^3 + x^2 + 4x + 4} = \frac{8x^2 + 3x + 20}{(x + 1)(x^2 + 4)} = \frac{3x}{x^2 + 4} + \frac{5}{x + 1}.$$

2 $\dfrac{3x^3 + 11x - 16}{(x^2 + 1)(x^2 + 4x + 13)}$

SOLUTION

$$\frac{3x^3 + 11x - 16}{(x^2 + 1)(x^2 + 4x + 13)} = \frac{Ax + B}{x^2 + 1} + \frac{Cx + D}{x^2 + 4x + 13}.$$

Multiplying by $(x^2 + 1)(x^2 + 4x + 13)$, we have

$$3x^3 + 11x - 16 = (Ax + B)(x^2 + 4x + 13) + (Cx + D)(x^2 + 1)$$

or

$$3x^3 + 11x - 16 = (A + C)x^3 + (4A + B + D)x^2 + (13A + 4B + C)x + (13B + D).$$

Equating the coefficients, we have

$$\begin{cases} 3 = A + C \\ 0 = 4A + B + D \\ 11 = 13A + 4B + C \\ -16 = 13B + D. \end{cases}$$

Solving these simultaneous equations for A, B, C, and D (say by elimination), we obtain $A = 1$, $B = -1$, $C = 2$, and $D = -3$. Therefore,

$$\frac{3x^3 + 11x - 16}{(x^2 + 1)(x^2 + 4x + 13)} = \frac{x - 1}{x^2 + 1} + \frac{2x - 3}{x^2 + 4x + 13}.$$

As soon as a rational function has been decomposed into a sum of partial fractions, it can be integrated by integrating each of these partial fractions. We have already seen in Section 5 that there is no problem in integrating the partial fractions that correspond to the linear factors (repeated or not) in the denominator. The partial fractions corresponding to nonrepeated irreducible quadratic factors in the denominator have the form $\dfrac{Ax + B}{ax^2 + bx + c}$. The integral $\displaystyle\int \dfrac{Ax + B}{ax^2 + bx + c}\, dx$ can be handled by completing the square in the denominator (if necessary) as in Section 3.1.

EXAMPLES   Evaluate the given integral.

1 $\displaystyle\int \dfrac{8x^2 + 3x + 20}{(x + 1)(x^2 + 4)}\, dx$

SOLUTION

In the preceding set of examples (see Example 1, page 507) we saw that the partial fractions decomposition of the integrand is

$$\frac{8x^2 + 3x + 20}{(x + 1)(x^2 + 4)} = \frac{3x}{x^2 + 4} + \frac{5}{x + 1}.$$

Hence,

$$\int \frac{8x^2 + 3x + 20}{(x + 1)(x^2 + 4)}\, dx = 3\int \frac{x\, dx}{x^2 + 4} + 5\int \frac{dx}{x + 1}$$

$$= \tfrac{3}{2}\ln(x^2 + 4) + 5\ln|x + 1| + C.$$

2 $\displaystyle\int \frac{3x^3 + 11x - 16}{(x^2 + 1)(x^2 + 4x + 13)}\, dx$

SOLUTION

Using the partial fractions decomposition of the integrand already found in the last set of examples (see Example 2, page 508), we have

$$\int \frac{3x^3 + 11x - 16}{(x^2 + 1)(x^2 + 4x + 13)}\, dx$$

$$= \int \frac{x - 1}{x^2 + 1}\, dx + \int \frac{2x - 3}{x^2 + 4x + 13}\, dx$$

$$= \int \frac{x\, dx}{x^2 + 1} - \int \frac{dx}{x^2 + 1} + \int \frac{2x - 3}{x^2 + 4x + 13}\, dx$$

$$= \int \frac{x\, dx}{x^2 + 1} - \int \frac{dx}{x^2 + 1} + \int \frac{2x + 4}{x^2 + 4x + 13}\, dx - 7 \int \frac{dx}{(x + 2)^2 + 9}$$

$$= \frac{1}{2}\ln(x^2 + 1) - \tan^{-1} x + \ln(x^2 + 4x + 13) - \frac{7}{3}\tan^{-1}\left(\frac{x + 2}{3}\right) + C,$$

where $\displaystyle\int \frac{2x - 3}{x^2 + 4x + 13}\, dx$ has been found by adding and subtracting 7 in the numerator and splitting the integral into two integrals such that in the first integral the numerator is the derivative of the denominator. The last integral is evaluated by completing the square in the denominator and then substituting $u = x + 2$.

3 $\displaystyle\int \frac{3x^3 + 2x - 2}{x^2(x^2 + 2)}\, dx$

SOLUTION

Here we have $\displaystyle\frac{3x^3 + 2x - 2}{x^2(x^2 + 2)} = \frac{A}{x} + \frac{B}{x^2} + \frac{Cx + D}{x^2 + 2}$. By the short method of substitution, $B = \dfrac{-2}{0 + 2} = -1$. Putting $B = -1$ and clearing fractions, we have

$$3x^3 + 2x - 2 = Ax(x^2 + 2) - (x^2 + 2) + (Cx + D)x^2$$
$$= (A + C)x^3 + (D - 1)x^2 + 2Ax - 2.$$

Equating coefficients, we obtain $3 = A + C$, $0 = D - 1$, and $2 = 2A$; hence, $A = 1$, $C = 2$, and $D = 1$. Consequently,

$$\int \frac{3x^3 + 2x - 2}{x^2(x^2 + 2)}\, dx = \int \frac{dx}{x} - \int \frac{dx}{x^2} + \int \frac{2x + 1}{x^2 + 2}\, dx$$

$$= \ln|x| + \frac{1}{x} + \int \frac{2x\, dx}{x^2 + 2} + \int \frac{dx}{x^2 + 2}$$

$$= \ln|x| + \frac{1}{x} + \ln(x^2 + 2) + \frac{\sqrt{2}}{2}\tan^{-1}\frac{\sqrt{2}}{2}x + C.$$

6.2   The Case in Which the Denominator Involves Repeated Irreducible Quadratic Factors

We now consider the case in which the denominator, after being completely factored, involves *repeated* irreducible quadratic factors. After such repeated quadratic factors as well as all linear factors are gathered with the aid of exponents, the denominator is a product of factors having the form $(ax + b)^k$ or $(ax^2 + bx + c)^k$, where $k \geq 1$. The factors of the form $(ax + b)^k$ are handled just as in Section 5. The k partial fractions corresponding to $(ax^2 + bx + c)^k$ are those appearing in the expression

$$\frac{A_1 x + B_1}{ax^2 + bx + c} + \frac{A_2 x + B_2}{(ax^2 + bx + c)^2} + \cdots + \frac{A_k x + B_k}{(ax^2 + bx + c)^k}.$$

Again, A_1, B_1, A_2, B_2, ..., A_k, B_k must be determined, say by the method of equating coefficients.

EXAMPLES   Decompose the given rational fraction into partial fractions.

1 $\dfrac{x^3 + x + 2}{x(x^2 + 1)^2}$

SOLUTION

$$\frac{x^3 + x + 2}{x(x^2 + 1)^2} = \frac{A}{x} + \frac{Bx + C}{x^2 + 1} + \frac{Dx + E}{(x^2 + 1)^2}.$$

Here, the constant A can be found by the short method of substitution as $\dfrac{(0)^3 + 0 + 2}{(0^2 + 1)^2} = A$, or $A = 2$. In order to find B, C, D, and E, we put $A = 2$ and multiply both sides of the equation by $x(x^2 + 1)^2$ to clear fractions and obtain

$$x^3 + x + 2 = 2(x^2 + 1)^2 + (Bx + C)x(x^2 + 1) + (Dx + E)x$$
$$= (2x^4 + 4x^2 + 2) + (Bx^4 + Cx^3 + Bx^2 + Cx) + (Dx^2 + Ex)$$
$$= (2 + B)x^4 + Cx^3 + (4 + B + D)x^2 + (C + E)x + 2.$$

Equating coefficients of like powers of x on both sides of the equation above, we obtain

$$\begin{cases} 0 = 2 + B \\ 1 = C \\ 0 = 4 + B + D \\ 1 = C + E. \end{cases}$$

Solving these equations simultaneously, we have $B = -2$, $C = 1$, $D = -2$, and $E = 0$. Hence,

$$\frac{x^3 + x + 2}{x(x^2 + 1)^2} = \frac{2}{x} - \frac{2x - 1}{x^2 + 1} - \frac{2x}{(x^2 + 1)^2}.$$

2 $\dfrac{x^5 - 2x^4 + 2x^3 + x - 2}{x^2(x^2 + 1)^2}$

SOLUTION

$$\frac{x^5 - 2x^4 + 2x^3 + x - 2}{x^2(x^2 + 1)^2} = \frac{A}{x} + \frac{B}{x^2} + \frac{Cx + D}{x^2 + 1} + \frac{Ex + G}{(x^2 + 1)^2}.$$

Here we find that $B = -2$ by the short method of substitution. Thus,

$$\frac{x^5 - 2x^4 + 2x^3 + x - 2}{x^2(x^2 + 1)^2} = \frac{A}{x} - \frac{2}{x^2} + \frac{Cx + D}{x^2 + 1} + \frac{Ex + G}{(x^2 + 1)^2}.$$

Clearing fractions, we have

$$\begin{aligned}
x^5 - 2x^4 + 2x^3 + x - 2 &= Ax(x^2 + 1)^2 - 2(x^2 + 1)^2 \\
&\quad + (Cx + D)x^2(x^2 + 1) + (Ex + G)x^2 \\
&= (Ax^5 + 2Ax^3 + Ax) - (2x^4 + 4x^2 + 2) \\
&\quad + (Cx^5 + Dx^4 + Cx^3 + Dx^2) + (Ex^3 + Gx^2) \\
&= (A + C)x^5 + (D - 2)x^4 + (2A + C + E)x^3 \\
&\quad + (D + G - 4)x^2 + Ax - 2.
\end{aligned}$$

Equating coefficients, we have

$$\begin{cases}
1 = A + C \\
-2 = D - 2 \\
2 = 2A + C + E \\
0 = D + G - 4 \\
1 = A.
\end{cases}$$

Simultaneous solution of these equations yields $A = 1$, $C = 0$, $D = 0$, $E = 0$, and $G = 4$. Hence,

$$\frac{x^5 - 2x^4 + 2x^3 + x - 2}{x^2(x^2 + 1)^2} = \frac{1}{x} - \frac{2}{x^2} + \frac{4}{(x^2 + 1)^2}.$$

In order to integrate rational functions containing repeated irreducible quadratic factors $ax^2 + bx + c$ in the denominator, it is necessary to be able to integrate partial fractions of the form $\dfrac{Ax + B}{(ax^2 + bx + c)^k}$. After completing the square, if necessary, in the expression $ax^2 + bx + c$ and making the usual change of variable, the required integral can be brought into the form

$$\int \frac{Cu + D}{(au^2 + q)^k}\, du \quad \text{or} \quad C \int \frac{u\, du}{(au^2 + q)^k} + D \int \frac{du}{(au^2 + q)^k}.$$

The integral $\displaystyle\int \frac{u\, du}{(au^2 + q)^k}$ can be handled with ease by the substitution $t = au^2 + q$ (Problem 36). However, the integral $\displaystyle\int \frac{du}{(au^2 + q)^k}$ can present more of a challenge. Since the original quadratic polynomial $ax^2 + bx + c$ was irreducible, it can be shown (Problem 37) that $q/a > 0$. Thus, we can make the substitution $u = \sqrt{q/a}\,w$ and obtain

$$\int \frac{du}{(au^2 + q)^k} = \int \frac{\sqrt{\dfrac{q}{a}}\, dw}{(qw^2 + q)^k} = \frac{\sqrt{\dfrac{q}{a}}}{q^k} \int \frac{dw}{(w^2 + 1)^k}.$$

The required integral can be found provided that we can evaluate an integral of the form $\int \dfrac{dw}{(w^2 + 1)^k}$. The trigonometric substitution $w = \tan \theta$ converts the latter integral into the form $\int \cos^n \theta \, d\theta$, where $n = 2(k - 1)$ (Problem 38). By this or other means (Problems 39 and 40), $\int \dfrac{dw}{(w^2 + 1)^k}$ can always be evaluated.

EXAMPLES 1 Evaluate $\displaystyle\int \dfrac{x^3 + x + 2}{x(x^2 + 1)^2} \, dx$.

SOLUTION
Here we must make a decomposition of the integrand into partial fractions. By Example 1 in the preceding set (page 510),

$$\frac{x^3 + x + 2}{x(x^2 + 1)^2} = \frac{2}{x} - \frac{2x - 1}{x^2 + 1} - \frac{2x}{(x^2 + 1)^2};$$

hence,

$$\int \frac{x^3 + x + 2}{x(x^2 + 1)^2} \, dx = 2 \int \frac{dx}{x} - \int \frac{2x - 1}{x^2 + 1} \, dx - \int \frac{2x \, dx}{(x^2 + 1)^2}$$

$$= 2 \int \frac{dx}{x} - \int \frac{2x \, dx}{x^2 + 1} + \int \frac{dx}{x^2 + 1} - \int \frac{2x \, dx}{(x^2 + 1)^2}$$

$$= 2 \ln |x| - \ln (x^2 + 1) + \tan^{-1} x + \frac{1}{x^2 + 1} + C,$$

where the second and fourth integrals have been evaluated by using the substitution $t = x^2 + 1$.

2 Evaluate $\displaystyle\int \dfrac{x^5 - 2x^4 + 2x^3 + x - 2}{x^2(x^2 + 1)^2} \, dx$.

SOLUTION
By Example 2 in the preceding set (pages 510 and 511),

$$\frac{x^5 - 2x^4 + 2x^3 + x - 2}{x^2(x^2 + 1)^2} = \frac{1}{x} - \frac{2}{x^2} + \frac{4}{(x^2 + 1)^2};$$

hence,

$$\int \frac{x^5 - 2x^4 + 2x^3 + x - 2}{x^2(x^2 + 1)^2} \, dx = \int \frac{dx}{x} - 2 \int \frac{dx}{x^2} + 4 \int \frac{dx}{(x^2 + 1)^2}$$

$$= \ln |x| + \frac{2}{x} + 4 \left[\frac{\tan^{-1} x}{2} + \frac{x}{2(x^2 + 1)} \right] + C,$$

where we have used the trigonometric substitution $x = \tan \theta$ to evaluate $\int \dfrac{dx}{(x^2 + 1)^2}$. Thus, $dx = \sec^2 \theta \, d\theta$ and

$$\int \frac{dx}{(x^2 + 1)^2} = \int \frac{\sec^2 \theta \, d\theta}{(\tan^2 \theta + 1)^2} = \int \frac{\sec^2 \theta \, d\theta}{(\sec^2 \theta)^2} = \int \frac{d\theta}{\sec^2 \theta}$$

$$= \int \cos^2 \theta \, d\theta = \frac{\theta}{2} + \frac{\sin 2\theta}{4} = \frac{\theta}{2} + \frac{\sin \theta \cos \theta}{2} + C.$$

Using an appropriate right triangle, we have $\theta = \tan^{-1} x$, $\sin \theta = x/\sqrt{x^2 + 1}$, and $\cos \theta = 1/\sqrt{x^2 + 1}$, so that

$$\int \frac{dx}{(x^2 + 1)^2} = \frac{\tan^{-1} x}{2} + \frac{1}{2} \frac{x}{\sqrt{x^2 + 1}} \cdot \frac{1}{\sqrt{x^2 + 1}} + C$$

$$= \frac{\tan^{-1} x}{2} + \frac{x}{2(x^2 + 1)} + C.$$

Problem Set 6

In problems 1 through 22, evaluate each integral. (*Caution:* Some integrands may require a preliminary long division and some may be partial fractions to begin with.)

1 $\displaystyle\int \frac{dx}{(x - 1)(x^2 + 4)}$

2 $\displaystyle\int \frac{x^5 + 9x^3 + 1}{x^3 + 9x} dx$

3 $\displaystyle\int \frac{(x + 3)\, dx}{x(x^2 + 1)}$

4 $\displaystyle\int \frac{dy}{y^4 - 16}$

5 $\displaystyle\int \frac{2t^2 - t + 1}{t(t^2 + 25)} dt$

6 $\displaystyle\int \frac{x - 3}{2x^2 - 12x + 19} dx$

7 $\displaystyle\int \frac{x\, dx}{x^4 - 1}$

8 $\displaystyle\int \frac{x^3 + 2x^2 + 7x + 2}{x^2 + 2x + 5} dx$

9 $\displaystyle\int \frac{6x^2 - 8x - 1}{(x - 2)(2x^2 - 3x + 5)} dx$

10 $\displaystyle\int \frac{dy}{(y - 2)(y^2 + 4y + 5)}$

11 $\displaystyle\int \frac{16\, dx}{x(x^2 + 4)^2}$

12 $\displaystyle\int \frac{2x^3 + 9}{x^4 + x^3 + 12x^2} dx$

13 $\displaystyle\int \frac{5t^3 - 3t^2 + 2t - 1}{t^4 + 9t^2} dt$

14 $\displaystyle\int \frac{dx}{(x^2 + 1)^3}$

15 $\displaystyle\int \frac{x^3 + 4}{x^2(x^2 + 1)^2} dx$

16 $\displaystyle\int \frac{2y^2}{y^4 + y^3 + 12y^2} dy$

17 $\displaystyle\int \frac{x^5 + 4x^3 + 3x^2 - x + 2}{x^5 + 4x^3 + 4x} dx$

18 $\displaystyle\int \frac{4x^2\, dx}{(x - 1)^2(x^2 - x + 1)}$

19 $\displaystyle\int \frac{(2t + 2)\, dt}{t(t^2 + 2t + 2)^2}$

20 $\displaystyle\int \frac{x^2\, dx}{(x - 1)^2(x^2 - x + 1)}$

21 $\displaystyle\int \frac{dx}{x^3 + 3x^2 + 7x + 5}$

22 $\displaystyle\int \frac{3x^2 + 8x + 6}{x^3 + 4x^2 + 6x + 4} dx$

In problems 23 through 28, evaluate each definite integral.

23 $\displaystyle\int_0^3 \frac{t + 10}{(t + 1)(t^2 + 1)} dt$

24 $\displaystyle\int_0^1 \frac{dx}{8x^3 + 27}$

25 $\displaystyle\int_0^{1/2} \frac{8x\, dx}{(2x + 1)(4x^2 + 1)}$

26 $\displaystyle\int_1^2 \frac{4\, dx}{x^3 + 4x}$

27 $\displaystyle\int_1^2 \frac{1 - x^2}{x(x^2 + 1)} dx$

28 $\displaystyle\int_2^5 \frac{x^4 - x^3 + 2x^2 - x + 2}{(x - 1)(x^2 + 2)} dx$

In problems 29 through 32, find a change of variable that reduces each integrand to a rational function.

29 $\displaystyle\int \frac{\cos x\, dx}{\sin^3 x + \sin^2 x + 9 \sin x + 9}$

30 $\displaystyle\int \frac{dx}{x\sqrt{x} + x + 1}$

31 $\displaystyle\int \frac{3e^{2x} + 2e^x - 2}{e^{3x} - 1} dx$

32 $\displaystyle\int \frac{3e^{3x} + e^x + 3}{(e^{2x} + 1)^2} dx$

33 Find the partial fractions decomposition of $\dfrac{ax^3 + bx^2 + cx + d}{(x^2 + 1)^2}$.

34 Find a formula for $\displaystyle\int \frac{ax^3 + bx^2 + cx + d}{(x^2 + 1)^2} dx$.

35 Integrate $\displaystyle\int \frac{5x^4 + 6x^2 + 1}{x^5 + 2x^3 + x}\, dx$ in two ways:

(a) By using the substitution $u = x^5 + 2x^3 + x$.   (b) By using partial fractions.

36 Find a formula for $\displaystyle\int \frac{u\, du}{(au^2 + q)^k}$.

37 Suppose that $ax^2 + bx + c$ is irreducible, so that $b^2 - 4ac < 0$. Complete the square and make the appropriate change of variable so that $ax^2 + bx + c = au^2 + q$, then prove that $q/a > 0$.

38 Prove that the trigonometric substitution $w = \tan \theta$ converts the integral $\displaystyle\int \frac{dw}{(w^2 + 1)^k}$

into the form $\displaystyle\int \cos^n \theta\, d\theta$, where $n = 2(k - 1)$.

39 Prove the following reduction formula: For $k \geq 2$,

$$\int \frac{dw}{(w^2 + 1)^k} = \frac{1}{2k - 2} \cdot \frac{w}{(w^2 + 1)^{k-1}} + \frac{2k - 3}{2k - 2} \int \frac{dw}{(w^2 + 1)^{k-1}}.$$

40 Show that $\displaystyle\int \frac{dx}{(x^2 + 1)^k}$, where k is a positive integer, can always be expressed in terms of rational functions and the inverse tangent function. (*Hint:* Use problem 39.)

41 Use the reduction formula in problem 39 to evaluate

(a) $\displaystyle\int \frac{dw}{(w^2 + 1)^2}$ (b) $\displaystyle\int \frac{dw}{(w^2 + 1)^3}$

42 It is an algebraic fact that every rational function can be decomposed into partial fractions. Show, therefore, that the integral of any rational function can be expressed in terms of rational functions, inverse tangents, and logarithms (of absolute values).

43 An integral of the form $\displaystyle\int \frac{dx}{(a - bx)^{2/3}(c - x)}$, where a, b, and c are positive constants, has to be evaluated to determine the time required for a homogeneous sphere of iron to dissolve in an acid bath. Make the change of variable $\sqrt[3]{a - bx} = z$ and show that the integrand then becomes a rational function of z.

44 Find a formula for the integral in problem 43.

7 Integration by Special Substitutions

In this section we examine some special substitutions that can be effective when the integrand contains sines and cosines or when the integrand involves fractional powers of the variable of integration.

7.1 Integration of Functions Involving Fractional Powers

If the integrand involves an expression of the form $\sqrt[n]{x}$, then the substitution $z = \sqrt[n]{x}$ may be helpful. If the integrand involves both $\sqrt[n]{x}$ and $\sqrt[m]{x}$, then the substitution $u = \sqrt[p]{x}$, where $p = nm$, may prove effective. In the first case, we have

$x = z^n$, so that $dx = nz^{n-1} \, dz$; while in the second case, we have $x = u^p$, so that $dx = pu^{p-1} \, du$.

EXAMPLES   Evaluate the given integral.

1 $\displaystyle\int \frac{dx}{1 + \sqrt{x}}$

SOLUTION
Put $z = \sqrt{x}$, so that $x = z^2$, $dx = 2z \, dz$, and

$$\int \frac{dx}{1 + \sqrt{x}} = \int \frac{2z \, dz}{1 + z}.$$

Since $\dfrac{2z}{1 + z}$ is an improper fraction, we divide numerator by denominator to obtain

a quotient of 2 and a remainder of -2. Thus, $\dfrac{2z}{1 + z} = 2 - \dfrac{2}{1 + z}$. It follows that

$$\int \frac{dx}{1 + \sqrt{x}} = \int \frac{2z \, dz}{1 + z} = \int \left(2 - \frac{2}{1 + z}\right) dz = 2\int dz - 2\int \frac{dz}{1 + z}$$

$$= 2z - 2\ln|1 + z| + C = 2\sqrt{x} - 2\ln|1 + \sqrt{x}| + C.$$

2 $\displaystyle\int \frac{dx}{\sqrt{1 + \sqrt[3]{x}}}$

SOLUTION
We could put $z = \sqrt[3]{x}$, but, with some foresight, it seems better to try $z = \sqrt{1 + \sqrt[3]{x}}$. Then $z^2 = 1 + \sqrt[3]{x}$, $z^2 - 1 = \sqrt[3]{x}$, and so $x = (z^2 - 1)^3$. In particular, we have $dx = 3(z^2 - 1)^2(2z \, dz) = 6z(z^2 - 1)^2 \, dz$, so

$$\int \frac{dx}{\sqrt{1 + \sqrt[3]{x}}} = \int \frac{6z(z^2 - 1)^2 \, dz}{z} = 6\int (z^2 - 1)^2 \, dz$$

$$= 6\int (z^4 - 2z^2 + 1) \, dz = 6\left(\frac{z^5}{5} - \frac{2z^3}{3} + z\right) + C$$

$$= 6\left[\frac{(1 + \sqrt[3]{x})^{5/2}}{5} - \frac{2(1 + \sqrt[3]{x})^{3/2}}{3} + (1 + \sqrt[3]{x})^{1/2}\right] + C.$$

3 $\displaystyle\int \frac{dt}{\sqrt{t} - \sqrt[3]{t}}$

SOLUTION
Put $u = \sqrt[6]{t}$, so that $t = u^6$ and $dt = 6u^5 \, du$. Thus,

$$\int \frac{dt}{\sqrt{t} - \sqrt[3]{t}} = \int \frac{6u^5 \, du}{\sqrt{u^6} - \sqrt[3]{u^6}} = \int \frac{6u^5 \, du}{u^3 - u^2} = 6\int \frac{u^3 \, du}{u - 1}.$$

Since $\dfrac{u^3}{u - 1}$ is an improper fraction, we divide numerator by denominator to obtain

the quotient $u^2 + u + 1$ and the remainder 1. Hence,

$$\frac{u^3}{u-1} = u^2 + u + 1 + \frac{1}{u-1},$$

so that

$$\int \frac{dt}{\sqrt{t} - \sqrt[3]{t}} = 6 \int \left(u^2 + u + 1 + \frac{1}{u-1}\right) du$$

$$= 6\left(\frac{u^3}{3} + \frac{u^2}{2} + u + \ln|u-1|\right) + C$$

$$= 2\sqrt{t} + 3\sqrt[3]{t} + 6\sqrt[6]{t} + 6 \ln|\sqrt[6]{t} - 1| + C.$$

Naturally, if the integrand involves an expression of the form $\sqrt[n]{u}$, where u is a function of x, then the substitution $z = \sqrt[n]{u}$ suggests itself. Such a substitution was effective in Example 2 above. Further examples are given below.

EXAMPLES   Evaluate the given integral.

1 $\int_4^{12} x\sqrt{x-3}\, dx.$

SOLUTION
Put $z = \sqrt{x-3}$, so that $z^2 = x - 3$, $x = z^2 + 3$, $dx = 2z\, dz$, $z = 1$ when $x = 4$, and $z = 3$ when $x = 12$. Thus,

$$\int_4^{12} x\sqrt{x-3}\, dx = \int_1^3 (z^2 + 3)z(2z\, dz) = 2\int_1^3 (z^4 + 3z^2)\, dz$$

$$= \left(\frac{2}{5}z^5 + 2z^3\right)\Big|_1^3 = \frac{756}{5} - \frac{12}{5} = \frac{744}{5}.$$

2 $\int \frac{1 + x^2}{(3+x)^{1/3}}\, dx$

SOLUTION
Put $z = (3+x)^{1/3}$, so that $z^3 = 3 + x$, $x = z^3 - 3$, $dx = 3z^2\, dz$, and

$$\int \frac{1+x^2}{(3+x)^{1/3}}\, dx = \int \frac{1 + (z^3 - 3)^2}{z}(3z^2\, dz) = 3\int (z^7 - 6z^4 + 10z)\, dz$$

$$= \frac{3z^8}{8} - \frac{18z^5}{5} + 15z^2 + C$$

$$= \frac{3}{8}(3+x)^{8/3} - \frac{18}{5}(3+x)^{5/3} + 15(3+x)^{2/3} + C.$$

7.2 Integration of Rational Functions of Sine and Cosine

It turns out that the substitution $z = \tan(x/2)$ reduces any integrand that is a rational function of $\sin x$ and $\cos x$ to a rational function of z. The appropriate formulas are contained in the following theorem.

THEOREM 1   **Tangent-Half-Angle Substitution**
Suppose that $z = \tan (x/2)$. Then

(i) $\cos x = \dfrac{1 - z^2}{1 + z^2}$.

(ii) $\sin x = \dfrac{2z}{1 + z^2}$.

(iii) $dx = \dfrac{2\,dz}{1 + z^2}$.

PROOF
Let $z = \tan (x/2)$. Then, using the double-angle formula, we have

$$\cos x = 2 \cos^2 \frac{x}{2} - 1 = \frac{2}{\sec^2 (x/2)} - 1 = \frac{2}{1 + \tan^2 (x/2)} - 1$$

$$= \frac{2}{1 + z^2} - 1 = \frac{1 - z^2}{1 + z^2},$$

so that (i) holds. Also,

$$\sin x = 2 \sin \frac{x}{2} \cos \frac{x}{2} = 2 \frac{\sin (x/2)}{\cos (x/2)} \cos^2 \frac{x}{2} = \tan \frac{x}{2} \left(2 \cos^2 \frac{x}{2} \right)$$

$$= \tan \frac{x}{2} (\cos x + 1) = z \left(\frac{1 - z^2}{1 + z^2} + 1 \right) = \frac{2z}{1 + z^2},$$

so that (ii) holds. To prove (iii), note that $z = \tan (x/2)$ implies that

$$dz = \left(\sec^2 \frac{x}{2} \right) \frac{dx}{2} = \frac{1}{2} \left(1 + \tan^2 \frac{x}{2} \right) dx = \frac{1 + z^2}{2} dx;$$

hence, $dx = \dfrac{2\,dz}{1 + z^2}$, as desired.

The following examples illustrate the use of the tangent-half-angle substitution.

EXAMPLES   Use the tangent-half-angle substitution to evaluate the given integral.

1 $\displaystyle\int \frac{dx}{1 - \cos x}$

SOLUTION
Put $z = \tan \dfrac{x}{2}$, so that $\cos x = \dfrac{1 - z^2}{1 + z^2}$, $dx = \dfrac{2\,dz}{1 + z^2}$, and

$$\int \frac{dx}{1 - \cos x} = \int \frac{\dfrac{2\,dz}{1 + z^2}}{1 - \left(\dfrac{1 - z^2}{1 + z^2} \right)} = \int \frac{dz}{z^2} = -\frac{1}{z} + C = \frac{-1}{\tan \dfrac{x}{2}} + C$$

$$= -\cot \frac{x}{2} + C.$$

2 $\displaystyle\int \frac{dx}{\sin x - \cos x + 1}$

SOLUTION

Put $z = \tan (x/2)$, so that $\sin x = \dfrac{2z}{1 + z^2}$, $\cos x = \dfrac{1 - z^2}{1 + z^2}$, $dz = \dfrac{2\,dz}{1 + z^2}$, and

$$\int \frac{dx}{\sin x - \cos x + 1} = \int \frac{\dfrac{2\,dz}{1 + z^2}}{\dfrac{2z}{1 + z^2} - \dfrac{1 - z^2}{1 + z^2} + 1} = \int \frac{dz}{z^2 + z}$$

$$= \int \left(\frac{1}{z} - \frac{1}{z + 1} \right) dz = \ln |z| - \ln |z + 1| + C$$

$$= \ln \left| \frac{z}{z + 1} \right| + C = \ln \left| \frac{\tan (x/2)}{\tan (x/2) + 1} \right| + C.$$

Problem Set 7

In problems 1 through 8, use an appropriate substitution of the form $z = \sqrt[n]{x}$ and evaluate each integral.

1 $\displaystyle\int \frac{dx}{1 - \sqrt{x}}$

2 $\displaystyle\int \frac{dx}{4 + \sqrt{x}}$

3 $\displaystyle\int \frac{dx}{1 + \sqrt[3]{x}}$

4 $\displaystyle\int \frac{x\,dx}{1 - \sqrt[3]{x}}$

5 $\displaystyle\int \frac{x\,dx}{2 + \sqrt{x}}$

6 $\displaystyle\int \frac{2\sqrt{x}\,dx}{1 + \sqrt[3]{x}}$

7 $\displaystyle\int \frac{dx}{\sqrt[4]{x} + \sqrt{x}}$

8 $\displaystyle\int \frac{dx}{x^{1/2} - x^{3/4}}$

In problems 9 through 24, use an appropriate substitution of the form $z = \sqrt[n]{u}$, where u is a function of x, and evaluate each integral.

9 $\displaystyle\int x^3 \sqrt{2x^2 - 1}\,dx$

10 $\displaystyle\int x^5 \sqrt{5 - 2x^2}\,dx$

11 $\displaystyle\int x\sqrt[3]{3x + 1}\,dx$

12 $\displaystyle\int x^9 \sqrt{1 + 2x^5}\,dx$

13 $\displaystyle\int x^2 (4x + 1)^{3/2}\,dx$

14 $\displaystyle\int x(1 + x)^{2/3}\,dx$

15 $\displaystyle\int \frac{dx}{1 + \sqrt{x + 1}}$

16 $\displaystyle\int \frac{x\,dx}{\sqrt[4]{1 - x}}$

17 $\displaystyle\int \frac{x^3\,dx}{(2 - 3x^2)^{3/4}}$

18 $\displaystyle\int \sqrt{\frac{1 - x}{x}}\,dx$

19 $\displaystyle\int e^x \sqrt{1 - e^x}\,dx$

20 $\displaystyle\int e^{2x} \sqrt{1 + e^x}\,dx$

21 $\displaystyle\int \sin x \cos x \sqrt{1 + \sin x}\,dx$

22 $\displaystyle\int \frac{3e^{2x}}{1 + e^{-x}}\,dx$

23 $\displaystyle\int \frac{x}{(3x + 1)^2} \sqrt{\frac{1}{3x + 1}}\,dx$

24 $\displaystyle\int \frac{1}{(1 + x)^2} \sqrt[3]{\frac{1 - x}{1 + x}}\,dx$

In problems 25 through 34, use the tangent-half-angle substitution to evaluate each integral.

25 $\displaystyle\int \frac{dx}{3 + 5 \sin x}$

26 $\displaystyle\int \frac{\sin t}{1 + \cos t}\,dt$

27 $\displaystyle\int \frac{\cos x\,dx}{\sin x \cos x + \sin x}$

28 $\displaystyle\int \frac{dx}{\sin x + \sqrt{3}\,\cos x}$

29 $\displaystyle\int \frac{\sec t}{1 + \sin t}\,dt$

30 $\displaystyle\int \frac{du}{2\,\csc u - \sin u}$

31 $\displaystyle\int \frac{dt}{\sin t + \cos t}$

32 $\displaystyle\int \frac{dx}{\csc x - \cot x}$

33 $\displaystyle\int \frac{dx}{1 + \sin x + \cos x}$

34 $\displaystyle\int \frac{\sec \theta \csc \theta \; d\theta}{3 + 5 \cos \theta}$

In problems 35 through 38, use the substitution $x = 1/t$, $dx = -dt/t^2$ to simplify each integrand, then evaluate the integral.

35 $\displaystyle\int \frac{dx}{x\sqrt{1 + x^2}}$

36 $\displaystyle\int \frac{dx}{x^2\sqrt{x^2 + 2x}}$

37 $\displaystyle\int \frac{dx}{x\sqrt{3x^2 - 2x - 1}}$

38 $\displaystyle\int \frac{dx}{x\sqrt{x^2 + 4x - 1}}$

39 If $z = \tanh (x/2)$, show that

    (a) $\cosh x = \dfrac{1 + z^2}{1 - z^2}$, (b) $\sinh x = \dfrac{2z}{1 - z^2}$, and (c) $dx = \dfrac{2\,dz}{1 - z^2}$.

In problems 40 through 42, use the substitution suggested by problem 39 to evaluate each integral.

40 $\displaystyle\int \frac{dx}{1 - \sinh x}$

41 $\displaystyle\int \frac{dx}{\cosh x - \sinh x}$

42 $\displaystyle\int \frac{\tanh x}{1 + \cosh x}\,dx$

In problems 43 through 50, evaluate each integral.

43 $\displaystyle\int_{1}^{4} x\sqrt{x - 1}\,dx$

44 $\displaystyle\int_{3}^{11} x\sqrt{2x + 3}\,dx$

45 $\displaystyle\int_{1}^{4} \frac{4 - \sqrt{x}}{1 + x}\,dx$

46 $\displaystyle\int_{4}^{9} \frac{1 - \sqrt{x}}{1 + \sqrt{x}}\,dx$

47 $\displaystyle\int_{-3}^{-1} \frac{x^2\,dx}{\sqrt{1 - x}}$

48 $\displaystyle\int_{1}^{7/3} \frac{1 - \sqrt{3x + 2}}{1 + \sqrt{3x + 2}}\,dx$

49 $\displaystyle\int_{0}^{\pi/2} \frac{\cos x}{2 + \sin x}\,dx$

50 $\displaystyle\int_{1}^{2} \frac{\sqrt{t^4 + 1}}{t}\,dt$

51 Find the area of the region bounded by the curves $y = \dfrac{5x}{1 + \sqrt{x}}$, $y = 0$, and $x = 9$.

52 Find the volume generated by revolving the region bounded by the curves given by $y = x + \sqrt{x + 1}$ and $y = 0$ between $x = 0$ and $x = 8$ about the x axis.

Review Problem Set

In problems 1 through 90, evaluate each integral.

1 $\displaystyle\int \cos^3 2x\,dx$

2 $\displaystyle\int \sin^3 4x \cos^2 4x\,dx$

3 $\displaystyle\int \sin^3 3x \cos^3 3x\,dx$

4 $\displaystyle\int \sqrt{\cos x}\,\sin^5 x\,dx$

5 $\displaystyle\int \sin^3 (1 - 2x)\,dx$

6 $\displaystyle\int \sin^3 \frac{x}{2} \cos^{3/2} \frac{x}{2}\,dx$

7 $\displaystyle\int \sin^{-2/3} 5x \cos^3 5x\,dx$

8 $\displaystyle\int \sin^4 \frac{2x}{5} \cos^3 \frac{2x}{5}\,dx$

9 $\displaystyle\int \sin^2 (2 - 3x)\,dx$

10 $\displaystyle\int \cos^2 \frac{2x}{7}\,dx$

11 $\displaystyle\int (\sin x - \cos x)^2\,dx$

12 $\displaystyle\int \sin^2 (1 - 2x) \cos^2 (1 - 2x)\,dx$

14 $\int \sin^4 4x \cos^2 4x \, dx$

15 $\int \dfrac{\cos^3 (3t/2)}{\sqrt[3]{\sin (3t/2)}} \, dt$

17 $\int \sin 8x \sin 3x \, dx$

18 $\int \cos 13x \cos 2x \, dx$

1? $\int \sin x \sin 2x \sin 3x \, dx$

20 $\int \cos 3x \cos 5x \cos 9x \, dx$

21 $\int \tan^4 (2x - 1) \, dx$

22 $\int \cot^4 (2 - 3x) \, dx$

23 $\int x \tan^3 5x^2 \, dx$

24 $\int x^2 \cot^3 (5 - x^3) \, dx$

25 $\int (\sec t - \tan t)^2 \, dt$

26 $\int \dfrac{\cos (\tan x)}{\cos^2 x} \, dx$

27 $\int \dfrac{dx}{(1 - \sin x)^2}$

28 $\int \sqrt{1 + \cos x} \, dx$

29 $\int \sec^4 (1 + 2x) \, dx$

30 $\int \csc^4 (3 - 2x) \, dx$

31 $\int \tan^3 (2 + 3x) \sec^4 (2 + 3x) \, dx$

32 $\int \cot^3 (1 - x) \csc^4 (1 - x) \, dx$

33 $\int \dfrac{dx}{\sqrt{x^2 + 64}}$

34 $\int \dfrac{dx}{x^2 \sqrt{81 - x^2}}$

35 $\int \dfrac{dx}{(1 - x^2)^{5/2}}$

36 $\int \dfrac{x}{(4 - x^2)^2} \, dx$

37 $\int x \sqrt{x^2 - 4} \, dx$

38 $\int \sqrt{4 - 9x^2} \, dx$

39 $\int \dfrac{dx}{\sqrt{2x - x^2}}$

40 $\int \dfrac{dt}{\sqrt{1 + 2t - 2t^2}}$

41 $\int \dfrac{dx}{\sqrt{x^2 + 6x + 13}}$

42 $\int \dfrac{dx}{\sqrt{8 + 4x - 4x^2}}$

43 $\int x^2 e^{-7x} \, dx$

44 $\int \sqrt{x} \ln 2x \, dx$

45 $\int t^2 \sin^{-1} 2t \, dt$

46 $\int \ln (x^2 + 16) \, dx$

47 $\int (x + 2)e^{3x} \, dx$

48 $\int (x + 1)^2 e^{-x} \, dx$

49 $\int t^3 \cos 3t \, dt$

50 $\int \sin (\ln x) \, dx$

51 $\int \tan^{-1} \sqrt{x} \, dx$

52 $\int \dfrac{xe^{-x}}{(1 - x)^2} \, dx$

53 $\int e^{-7x} \cosh x \, dx$

54 $\int e^{2x} \sin^2 x \, dx$

55 $\int e^{3x} \cos^2 x \, dx$

56 $\int \sec^3 5x \, dx$

57 $\int x^{11} e^{-x^4} \, dx$

58 $\int x^5 \sin x^2 \, dx$

59 $\int x^3 \cos (-3x^2) \, dx$

60 $\int x^{17} \cos x^6 \, dx$

61 $\int \dfrac{3y^2 - y + 1}{(y^2 - y)(y + 1)} \, dy$

62 $\int \dfrac{2x + 1}{x(x + 1)(x + 2)} \, dx$

63 $\int \dfrac{3x^2 - x + 1}{x^3 - x^2} \, dx$

64 $\int \dfrac{dx}{x^3(1 + x)}$

65 $\int \dfrac{t^2 + 6t + 4}{t^4 + 5t^2 + 4} \, dt$

66 $\int \dfrac{x^2 - 4x - 4}{(x - 2)(x^2 + 9)} \, dx$

67 $\int \dfrac{3x + 1}{x^2(x^2 + 1)} \, dx$

68 $\int e^{\sin t} \left(\dfrac{t \cos^3 t - \sin t}{\cos^2 t} \right) dt$

69 $\int \dfrac{x \, dx}{\sqrt[4]{1 + 2x}}$

70 $\int \dfrac{dt}{\sqrt[4]{t} + 3}$

71 $\int \dfrac{\sqrt[5]{x^3} + \sqrt[6]{x}}{\sqrt{x}} \, dx$

72 $\int \dfrac{dy}{\sqrt{y} + y^{3/4}}$

73 $\displaystyle\int \frac{dt}{\sqrt{e^t + 1}}$

74 $\displaystyle\int \frac{\sqrt{x+1}}{\sqrt{x-1}}\, dx$

75 $\displaystyle\int \frac{dx}{\sqrt{4+\sqrt{x+1}}}$

76 $\displaystyle\int \frac{dy}{y \ln y\,(\ln y + 5)}$

77 $\displaystyle\int \ln \sqrt{x^2 + 3}\; dx$

78 $\displaystyle\int y \ln \sqrt[3]{5y + 2}\; dy$

79 $\displaystyle\int x\sqrt{1 - x^{2/3}}\; dx$

80 $\displaystyle\int \sqrt{\frac{1 + \sqrt{x}}{x}}\; dx$

81 $\displaystyle\int \cos \sqrt[3]{x}\; dx$

82 $\displaystyle\int \frac{e^{2y}}{\sqrt[4]{e^y + 3}}\, dy$

83 $\displaystyle\int \frac{dx}{10 + 11 \cos x}$

84 $\displaystyle\int \frac{\sin x}{8 + \cos x}\, dx$

85 $\displaystyle\int \frac{dy}{3 + 2 \sin y + \cos y}$

86 $\displaystyle\int \frac{\cot x}{\cot x + \csc x}\, dx$

87 $\displaystyle\int \frac{\sec x}{1 + \sin x}\, dx$

88 $\displaystyle\int \frac{dx}{3 - \cos x + 2 \sin x}$

89 $\displaystyle\int \frac{e^{4x}}{\sqrt[4]{e^{2x} + 1}}\, dx$

90 $\displaystyle\int \frac{dx}{a^2 \cos x + b^2 \sin x}$

In problems 91 through 122, evaluate each definite integral.

91 $\displaystyle\int_0^{\pi/4} \cos x \cos 5x\; dx$

92 $\displaystyle\int_0^{\pi/4} \sin^3 2t \cos^3 2t\; dt$

93 $\displaystyle\int_{\pi/12}^{\pi/8} \tan^3 2x\; dx$

94 $\displaystyle\int_0^1 x \tan^{-1} x\; dx$

95 $\displaystyle\int_1^2 (\ln t)^2\; dt$

96 $\displaystyle\int_0^{\pi/4} x^2 \sin 2x\; dx$

97 $\displaystyle\int_1^2 t^3 \ln t\; dt$

98 $\displaystyle\int_0^{\pi} \sin^3 x\; dx$

99 $\displaystyle\int_{-\pi/8}^{\pi/8} |\tan^3 2x|\; dx$

100 $\displaystyle\int_0^1 \cosh^4 x\; dx$

101 $\displaystyle\int_3^{3/2} \frac{(9 - x^2)^{3/2}}{x^2}\, dx$

102 $\displaystyle\int_0^{1/3} \frac{t\, dt}{\sqrt{1 - 9t^4}}$

103 $\displaystyle\int_5^{10} \frac{\sqrt{t^2 - 25}}{t}\, dt$

104 $\displaystyle\int_0^a x^2\sqrt{a^2 - x^2}\; dx$

105 $\displaystyle\int_0^{\pi} \sqrt{1 + \cos \frac{x}{3}}\; dx$

106 $\displaystyle\int_{\pi/4}^{\pi/2} \frac{\cot x\, dx}{1 - \cos x}$

107 $\displaystyle\int_3^5 \frac{t^2 - 1}{(t - 2)^2}\, dt$

108 $\displaystyle\int_1^2 \frac{5x^2 - 3x + 18}{x(9 - x^2)}\, dx$

109 $\displaystyle\int_0^1 \frac{x^2 + 3x + 1}{x^4 + 2x^2 + 1}\, dx$

110 $\displaystyle\int_0^1 \frac{t^5\, dt}{(t^2 + 1)^2}$

111 $\displaystyle\int_{1/2}^2 \frac{dx}{x\sqrt{5x^2 + 4x - 1}}$

112 $\displaystyle\int_0^{1/2} (2x - x^2)^{3/2}\; dx$

113 $\displaystyle\int_1^8 \frac{dx}{x + \sqrt[3]{x}}$

114 $\displaystyle\int_1^4 \frac{\sqrt{x} + 1}{\sqrt{x}\,(x + 1)}\, dx$

115 $\displaystyle\int_2^5 \frac{t\, dt}{(t - 1)^{3/2}}$

116 $\displaystyle\int_{-1}^8 \frac{dx}{\sqrt{1 + \sqrt{1 + x}}}$

117 $\displaystyle\int_{1/4}^{5/4} \frac{dt}{\sqrt{t + 1} - \sqrt{t}}$

118 $\displaystyle\int_{16}^{25} \frac{dy}{y - 2\sqrt{y} - 3}$

119 $\displaystyle\int_0^{\ln 4} \frac{dx}{\sqrt{e^{-2x} + 2e^{-x}}}$

120 $\displaystyle\int_0^a \sqrt{\sqrt[3]{a} + \sqrt[3]{x}}\; dx$

121 $\displaystyle\int_{\pi/4}^{\pi/8} \frac{dx}{\sin x + \tan x}$

122 $\displaystyle\int_{1/8}^1 \frac{x\, dx}{x + \sqrt[3]{x}}$

123 Determine constants A and B so that

$$\frac{c \sin \theta + d \cos \theta}{e \sin \theta + f \cos \theta} = A + B\,\frac{e \cos \theta - f \sin \theta}{e \sin \theta + f \cos \theta}.$$

Then obtain a formula for the integral $\displaystyle\int \frac{c \sin \theta + d \cos \theta}{e \sin \theta + f \cos \theta}\, d\theta.$

$$\frac{1}{\sqrt{1-t^2}} \quad \text{for } 0 < t < 1. \text{ Then show that } \int_0^x \frac{dt}{\sqrt{1-t^3}} < \sin^{-1} x,$$

$$\frac{dx}{\sqrt{x^2 - 2x + 5}} \quad \text{by using the substitution } \sqrt{x^2 - 2x + 5} + x = z.$$

126 If $f(x) = \int_1^x \dfrac{dt}{t + \sqrt{t^2 - 1}}$, $x \geq 1$, show that $\frac{1}{2} \ln x \leq f(x) \leq \ln x$.

127 Show that $\displaystyle\int_0^x e^{-y} y^2 \, dy = 2e^{-x}\left(e^x - 1 - x - \dfrac{x^2}{2}\right).$

128 If $g(x) = \displaystyle\int_1^x f\left(t + \dfrac{1}{t}\right)\left(\dfrac{dt}{t}\right)$, show that $g\left(\dfrac{1}{x}\right) = -g(x).$

129 Derive the reduction formula

$$\int x^m (\ln x)^n \, dx = \frac{x^{m+1}(\ln x)^n}{m+1} - \frac{n}{m+1} \int x^m (\ln x)^{n-1} \, dx, \qquad m \neq -1.$$

130 Use graphical considerations to prove that

$$\int_0^{2\pi} \sin^{2n} t \, dt = \int_0^{2\pi} \cos^{2n} t \, dt.$$

131 Find the area under the curve $y = \frac{1}{4}x^2 - \frac{1}{2} \ln x$ from $x = 1$ to $x = 4$.

132 Find the area under the curve $y = x^2 e^{-x}$ between $x = 0$ and $x = 1$.

133 Find the area bounded by one arch of $y = \sin^3 x$ from $x = 0$ to $x = \pi$.

134 Find the volume of the solid generated by revolving the region bounded by the curves $y = x \ln x$, $y = 0$, and $x = 4$ about the x axis.

135 Find the volume of the solid generated by revolving the region bounded by one arch of $y = \sin x$ and the line $y = 0$ about the line $y = -2$.

136 If the velocity v in feet per second of a particle that is moving along a straight line is expressed by the formula $v = \dfrac{t + 3}{t^3 + t}$, find the distance s in feet that the particle traveled from $t = 1$ second to $t = 3$ seconds.

137 Evaluate $\displaystyle\int_0^{\pi/3} \sqrt{1 + (dy/dx)^2} \, dx$, if $y = \ln (\cos x)$.

138 Find the arc length of the curve $y = \ln x$ from $x = 1$ to $x = \sqrt{3}$.

139 Find the arc length of the curve $y = \ln (\csc x)$ from $x = \pi/6$ to $x = \pi/2$.

140 Find the surface area generated by revolving the arc of the curve $y = e^x$ from $x = 0$ to $x = 1$ about the x axis.

11 POLAR COORDINATES AND ROTATION OF AXES

Until now we have specified the position of points in the plane by means of cartesian coordinates; however, in some situations it is more natural to use a different coordinate system. In this chapter, we study the *polar coordinate system,* the conversion of coordinates from cartesian to polar and vice versa, the graphs of polar equations, and area and arc length in polar coordinates. The chapter also includes *rotation of coordinate axes.*

1 Polar Coordinates

In order to establish a *polar coordinate system* in the plane, we choose a fixed point O called the *pole* and a fixed ray (or half-line) with endpoint O called the *polar axis* (Figure 1). An angle in the *standard position* is understood to have its vertex at the pole O and to have the polar axis as its initial side.

Figure 1

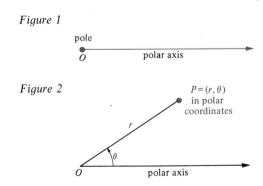

Now let P be any point in the plane and denote by r the distance between P and the pole O, so that $r = |\overline{OP}|$ (Figure 2). If $P \neq O$, then P lies on a uniquely determined ray with endpoint O, and this ray forms the terminal side of an angle in the standard position. We denote this angle, measured in degrees or radians (whichever is preferred), by θ and we refer to the ordered pair (r, θ) as the *polar coordinates* of the point P. (As usual, positive angles are measured counterclockwise.) Thus, in polar coordinates

$$P = (r, \theta).$$

Figure 2

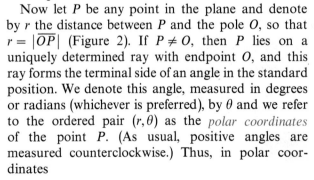

The polar coordinates (r, θ) locate the point P with respect to a "grid" formed by concentric circles with center O and rays emanating from O (Figure 3). The value of r locates P on a circle of radius r, the value of θ locates P on a ray which is the terminal side of the angle θ in standard position, and P itself

523

lies at the intersection of the circle and the ray. For instance, the point with polar coordinates $(r, \theta) = (4, 240°)$ is shown in Figure 3.

Figure 3

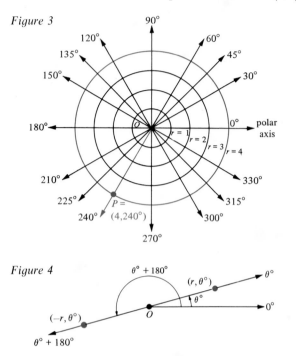

If $r = 0$ in the polar coordinate system, we understand that the point (r, θ) is at the pole O no matter what the angle θ may be. Also, it is convenient to allow r to be negative, making the convention that the point $(-r, \theta°)$ is located $|r|$ units from the pole, but on the ray opposite to the $\theta°$ ray, that is, on the ray $\theta° + 180°$ (Figure 4). Thus, $(-r, \theta°) = (r, \theta° + 180°)$, or $(-r, \theta) = (r, \theta + \pi)$ for θ in radians.

Unlike the cartesian coordinate system, a point P has many different representations in the polar coordinate system. Not only do we have as above, $(r, \theta°) = (-r, \theta° + 180°)$, but we also have $(r, \theta°) = (r, \theta° + 360°) = (r, \theta° - 360°)$, since $\pm 360°$ corresponds to a full revolution about the pole. Indeed, if n is any integer, we have $(r, \theta°) = (r, \theta° + 360° \cdot n)$, or, in radians, $(r, \theta) = (r, \theta + 2n\pi)$.

Figure 4

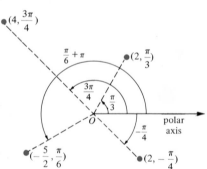

If we say "Plot the polar point (r, θ)," or "Plot the point (r, θ) in the polar coordinate system," we mean to draw a diagram showing the pole, the polar axis, and the point P whose *polar* coordinates are (r, θ).

EXAMPLES   1 Plot the points $(2, \pi/3)$, $(4, 3\pi/4)$, $(-5/2, \pi/6)$, and $(2, -\pi/4)$ in the polar coordinate system.

SOLUTION

To plot the polar point $(2, \pi/3)$, we construct an angle of $\pi/3$ radians (that is, $60°$) in the standard position and then locate the point 2 units from the pole on the terminal side of this angle (Figure 5). The remaining polar points are plotted similarly.

Figure 5

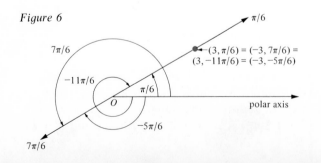

2 Plot the point $(3, \pi/6)$ in the polar coordinate system, and then give three other polar representations of the same point for which

(a) $r < 0$ and $0 \le \theta < 2\pi$.
(b) $r > 0$ and $-2\pi < \theta \le 0$.
(c) $r < 0$ and $-2\pi < \theta \le 0$.

SOLUTION

The point $(3, \pi/6)$ is 3 units from the pole and lies on the ray which is the terminal side of the angle $30° = \pi/6$ radians in the standard position (Figure 6). The same point can also be represented by

(a) $\left(-3, \dfrac{\pi}{6} + \pi\right) = \left(-3, \dfrac{7\pi}{6}\right)$.

(b) $\left(3, \dfrac{\pi}{6} - 2\pi\right) = \left(3, -\dfrac{11\pi}{6}\right)$.

(c) $\left(-3, \dfrac{7\pi}{6} - 2\pi\right) = \left(-3, -\dfrac{5\pi}{6}\right)$.

Figure 6

1.1 Conversion of Coordinates

At times it may be advantageous to convert from a cartesian representation to a polar representation, or vice versa. When making such a conversion, it is important to realize that the geometric points in the plane do not change—only the method by which they are assigned numerical "addresses" changes.

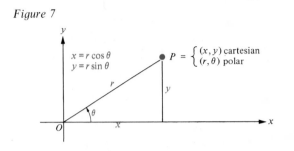

Figure 7

The usual arrangement is to take the pole for the polar coordinate system at the origin of the cartesian coordinate system and the polar axis along the positive x axis, so that the positive y axis is the polar ray $\theta = \pi/2$. If we consider the point P whose polar coordinates are (r, θ) with $r \geq 0$, it is clear that the cartesian coordinates (x, y) of P are given by $x = r \cos \theta$ and $y = r \sin \theta$ (Figure 7). This is certainly true if $r = 0$, while, if $r > 0$, it follows from $\cos \theta = x/r$ and $\sin \theta = y/r$.

Now, suppose that a point P has polar coordinates (r, θ) with $r < 0$ and we desire to find the cartesian coordinates (x, y) of P. Since $(r, \theta) = (-r, \theta + \pi)$ and $-r > 0$, it follows from the equations developed above that

$$x = (-r) \cos (\theta + \pi) = -r(\cos \theta \cos \pi - \sin \theta \sin \pi) = -r(-\cos \theta) = r \cos \theta,$$

$$y = (-r) \sin (\theta + \pi) = -r(\sin \theta \cos \pi + \cos \theta \sin \pi) = -r(-\sin \theta) = r \sin \theta.$$

Therefore, the equations

$$x = r \cos \theta \quad \text{and} \quad y = r \sin \theta$$

work in all possible cases to convert from the polar coordinates (r, θ) of a point P to the cartesian coordinates (x, y) of P.

From the latter equations, we have

$$x^2 + y^2 = r^2 \cos^2 \theta + r^2 \sin^2 \theta = r^2(\cos^2 \theta + \sin^2 \theta) = r^2,$$

so that

$$r = \pm\sqrt{x^2 + y^2}.$$

Also, if $x \neq 0$, we have $\dfrac{y}{x} = \dfrac{r \sin \theta}{r \cos \theta} = \dfrac{\sin \theta}{\cos \theta} = \tan \theta$, so that

$$\tan \theta = \frac{y}{x} \quad \text{for } x \neq 0.$$

The equations above do not determine r and θ uniquely simply because the point P whose cartesian coordinates are (x, y) has an unlimited number of different representations in the polar coordinate system. In finding the polar coordinates of P, one should pay attention to the quadrant in which P lies, since this will help to determine θ.

EXAMPLES 1 Convert the given polar coordinates to cartesian coordinates.
(a) $(4, 30°)$ (b) $(-2, 5\pi/6)$

SOLUTION
(a) $(x, y) = (4 \cos 30°, 4 \sin 30°) = (4 \cdot \sqrt{3}/2, 4 \cdot \tfrac{1}{2}) = (2\sqrt{3}, 2)$.

(b) $(x, y) = \left(-2 \cos \dfrac{5\pi}{6}, -2 \sin \dfrac{5\pi}{6}\right) = \left(-2 \cdot \left(-\dfrac{\sqrt{3}}{2}\right), -2 \cdot \dfrac{1}{2}\right) = (\sqrt{3}, -1)$.

2 Convert the given cartesian coordinates to polar coordinates with $r \geq 0$ and $-\pi < \theta \leq \pi$.
(a) $(2,2)$ (b) $(5, -5/\sqrt{3})$ (c) $(0, -7)$ (d) $(-3,3)$

SOLUTION
(a) $r = \sqrt{2^2 + 2^2} = \sqrt{8} = 2\sqrt{2}$ and $\tan \theta = \frac{2}{2} = 1$. Since the point lies in the first quadrant, it follows that $0 < \theta < \pi/2$; hence, $\theta = \pi/4$. The polar coordinates are $(2\sqrt{2}, \pi/4)$.

(b) $r = \sqrt{25 + \frac{25}{3}} = \frac{10}{\sqrt{3}}$ and $\tan \theta = \frac{(-5/\sqrt{3})}{5} = \frac{-1}{\sqrt{3}}$. Here the point lies in the fourth quadrant, so that $-\pi/2 < \theta < 0$; hence, $\theta = -\pi/6$. The polar coordinates are $(10/\sqrt{3}, -\pi/6)$.

(c) $r = \sqrt{0 + 49} = 7$. Since $x = 0$ and $y < 0$, the point lies on the negative y axis; hence, $\theta = -\pi/2$. The polar coordinates are $(7, -\pi/2)$.

(d) $r = \sqrt{9 + 9} = 3\sqrt{2}$ and $\tan \theta = 3/(-3) = -1$. Since the point lies in the second quadrant, it follows that $\pi/2 < \theta < \pi$; hence, $\theta = 3\pi/4$. The polar coordinates are $(3\sqrt{2}, 3\pi/4)$.

1.2 Graphs of Polar Equations

A *polar equation* is an equation relating the polar coordinates r and θ, such as $r = \theta^2$ or $r^2 = 9 \cos 2\theta$. Because a single point P in the plane has a multitude of different polar representations, it is necessary to define the graph of a polar equation with some care.

DEFINITION 1   **Graph of a Polar Equation**
The *graph* of a polar equation consists of all the points P in the plane that have at least one pair of polar coordinates (r, θ) satisfying the equation.

Thus, if none of the pairs of polar coordinates that represent P satisfy the polar equation, then P does not belong to the graph of this equation. However, in order for P to belong to this graph, it is not necessary for all its polar representations to satisfy the equation—any one will do.

EXAMPLE   Sketch the graph of the polar equation.

(a) $r = 4$ (b) $r^2 = 16$ (c) $\theta = \frac{\pi}{6}$ (d) $\theta^2 - \frac{4\pi}{3}\theta + \frac{7\pi^2}{36} = 0$

SOLUTION
(a) The graph of $r = 4$ is a circle of radius 4 with center at the pole (Figure 8). (The polar axis is drawn for reference, but it is not part of the graph.) Notice, for instance, that the point $P = (4, -\pi)$ belongs to the graph in spite of the fact that not all its representations, such as $(-4, 0)$ or $(-4, 2\pi)$, satisfy the equation $r = 4$.
(b) The equation $r^2 = 16$ is equivalent to $|r| = 4$ and its graph is the same as the graph of $r = 4$ (Figure 8).

Figure 8

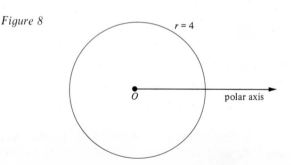

$r = 4$

O polar axis

(c) The graph of $\theta = \pi/6$ consists of the entire straight line through O making an angle of $\pi/6$ radians (30°) with the polar axis—not just the *ray*, as one might at first think (Figure 9). A point $P = (r, \pi/6 + \pi)$ qualifies for belonging to the graph of $\theta = \pi/6$ because it can be rewritten as $P = (-r, \pi/6)$.

(d) The equation is written in factored form as

$$\theta^2 - \frac{4\pi}{3}\theta + \frac{7\pi^2}{36} = \left(\theta - \frac{\pi}{6}\right)\left(\theta - \frac{7\pi}{6}\right) = 0,$$

so that

$$\theta - \frac{\pi}{6} = 0 \quad \text{or} \quad \theta - \frac{7\pi}{6} = 0.$$

Now, $\theta = 7\pi/6$ has the same graph as $\theta = \pi/6$. Therefore, the equation $\theta^2 - \frac{4\pi}{3}\theta + \frac{7\pi^2}{36} = 0$ has the same graph as $\theta = \pi/6$ (Figure 9). (Why?)

Figure 9

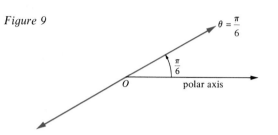

Sometimes it is easy to find the shape of a polar graph by converting the polar equation to a cartesian equation by means of the equations $x = r \cos \theta$ and $y = r \sin \theta$, and then sketching the cartesian graph as usual. Conversely, any cartesian equation can be converted into a corresponding polar equation simply by substituting $r \cos \theta$ for x and $r \sin \theta$ for y.

EXAMPLES 1 Find a cartesian equation corresponding to the polar equation $r = 4 \tan \theta \sec \theta$. Sketch the graph.

SOLUTION

We have $r = 4 \tan \theta \sec \theta = 4 \frac{\sin \theta}{\cos \theta} \cdot \frac{1}{\cos \theta}$, so that $r \cos^2 \theta = 4 \sin \theta$. Multiplication by r gives

$$r^2 \cos^2 \theta = 4r \sin \theta \quad \text{or} \quad (r \cos \theta)^2 = 4r \sin \theta.$$

We now use the conversion equations $x = r \cos \theta$ and $y = r \sin \theta$ to rewrite the polar equation in cartesian form $x^2 = 4y$. Therefore, the graph is a parabola (Figure 10).

Figure 10

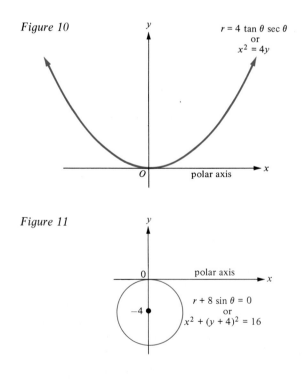

2 Find a polar equation corresponding to the cartesian equation $x^2 + (y + 4)^2 = 16$ and sketch the graph.

SOLUTION

The graph is a circle of radius 4 with center at the point with cartesian coordinates $(0, -4)$ (Figure 11). Rewriting the equation as

$$x^2 + y^2 + 8y + 16 = 16 \quad \text{or} \quad x^2 + y^2 + 8y = 0,$$

and substituting $x = r \cos \theta$, $y = r \sin \theta$, so that $x^2 + y^2 = r^2$, we obtain

$$r^2 + 8r \sin \theta = 0 \quad \text{or} \quad r + 8 \sin \theta = 0.$$

Figure 11

In simplifying the equation $r^2 + 8r \sin \theta = 0$ to $r + 8 \sin \theta = 0$ in the last example, we have divided by r; hence, we may have lost the solution $r = 0$. In this particular case, we have not lost the solution $r = 0$, since, if $r + 8 \sin \theta = 0$, then $r = 0$ when $\theta = 0$. More generally, when we multiply a polar equation by r, we *may* introduce an extraneous solution $r = 0$ that does not really belong, and, conversely, when we divide a polar equation by r, we *may* lose a solution $r = 0$ that really does belong. Each case should be checked!

1.3 Explicit Formulas for Conversion from Cartesian to Polar Coordinates

Although the conversion from polar to cartesian coordinates can always be obtained analytically using the equations $x = r \cos \theta$ and $y = r \sin \theta$, the equations $r = \pm\sqrt{x^2 + y^2}$ and $\tan \theta = y/x$ do not uniquely determine the polar coordinates in terms of the cartesian coordinates. In most cases, this slight difficulty is easily overcome by simple geometric considerations; however, in some cases (for instance, in programming a computer) it is necessary to have explicit formulas for r and θ in terms of x and y.

The following equations give the polar coordinates (r, θ), with $r \geq 0$ and $-\pi < \theta \leq \pi$, of the point with cartesian coordinates (x, y):

$$r = \sqrt{x^2 + y^2}$$

$$\theta = \begin{cases} \tan^{-1} \dfrac{y}{x} & \text{if } x > 0, \\[2mm] \dfrac{\pi}{2} & \text{if } x = 0 \text{ and } y > 0, \\[2mm] 0 & \text{if } x = 0 \text{ and } y = 0, \\[2mm] -\dfrac{\pi}{2} & \text{if } x = 0 \text{ and } y < 0, \\[2mm] \tan^{-1} \dfrac{y}{x} + \pi & \text{if } x < 0 \text{ and } y \geq 0, \\[2mm] \tan^{-1} \dfrac{y}{x} - \pi & \text{if } x < 0 \text{ and } y < 0. \end{cases}$$

The values of r and θ given by the equations above can be taken as *standard values*, in which case other polar representations of the same point can be found by adding $2n\pi$ to θ and leaving r alone, or by adding $(2n + 1)\pi$ to θ and changing the sign of r, where n is any integer. Many small electronic calculators are preprogrammed to calculate standard values of r and θ using the equations above.

Problem Set 1

In problems 1 through 6, plot each point in the polar coordinate system, and then give three other polar representations of the same point for which (a) $r < 0$ and $0 \leq \theta < 2\pi$, (b) $r > 0$ and $-2\pi < \theta \leq 0$, and (c) $r < 0$ and $-2\pi < \theta \leq 0$.

1 $(3, \pi/4)$ **2** $(6, 2\pi/3)$ **3** $(-2, \pi/6)$

4 $(3, 150°)$ **5** $(4, 180°)$ **6** $(4, 5\pi/4)$

In problems 7 through 12, convert the given polar coordinates to cartesian coordinates.

7 $(7, \pi/3)$ **8** $(0, \pi/3)$ **9** $(-2, \pi/4)$

10 $(6, 13\pi/6)$ **11** $(1, -\pi/3)$ **12** $(-5, 150°)$

In problems 13 through 18, convert the given cartesian coordinates to polar coordinates (r, θ) with $r \geq 0$ and $-\pi < \theta \leq \pi$.

13 $(7, 7)$ **14** $(1, -\sqrt{3})$ **15** $(-3, -3\sqrt{3})$

16 $(-5, 5)$ **17** $(0, 7)$ **18** $(-2, 0)$

In problems 19 and 20, rewrite the answers to problems 13 through 18 subject to the conditions given.

19 $r \geq 0, 0 \leq \theta < 2\pi$ **20** $r \leq 0, 0 \leq \theta < 2\pi$

In problems 21 and 22, plot the given point P in the polar coordinate system and then give five other polar representations for the same point.

21 $P = \left(-3, \frac{187}{6}\pi\right)$ **22** $P = \left(4, -\frac{6002}{3}\pi\right)$

In problems 23 through 30, sketch the graph of each polar equation.

23 $r = 1$ **24** $r^2 = 9$ **25** $\theta = \dfrac{\pi}{2}$ **26** $\theta^2 = \dfrac{25\pi^2}{36}$

27 $\theta = -\dfrac{\pi}{2}$ **28** $r = 2\cos\theta + 2\sin\theta$ **29** $r = 4\cos\theta$ **30** $r = \dfrac{1}{2\cos\theta - 3\sin\theta}$

In problems 31 through 34, convert each polar equation into a cartesian equation.

31 $r = 3\cos\theta$ **32** $r\cos 2\theta = 2$ **33** $r = \cos\theta + \sin\theta$ **34** $r = 5\theta$

In problems 35 through 38, convert each cartesian equation into a polar equation.

35 $x^2 + y^2 = 25$ **36** $xy = 12$ **37** $\dfrac{x^2}{4} + y^2 = 1$ **38** $y = 4x^3$

39 Show that the distance between the point (r_1, θ_1) and the point (r_2, θ_2) in the polar coordinate system is given by $\sqrt{r_1^2 - 2r_1 r_2 \cos(\theta_1 - \theta_2) + r_2^2}$.

40 Give an analytic proof—without reference to diagrams—to establish rules governing the conditions under which $(r_1, \theta_1) = (r_2, \theta_2)$ in polar coordinates. Use the fact that $(r_1, \theta_1) = (r_2, \theta_2)$ if and only if $r_1 \cos\theta_1 = r_2 \cos\theta_2$ and $r_1 \sin\theta_1 = r_2 \sin\theta_2$. The identities

$$\sin\theta_1 - \sin\theta_2 = 2\cos\tfrac{1}{2}(\theta_1 + \theta_2) \cdot \sin\tfrac{1}{2}(\theta_1 - \theta_2),$$

$$\cos\theta_1 - \cos\theta_2 = -2\sin\tfrac{1}{2}(\theta_1 + \theta_2) \cdot \sin\tfrac{1}{2}(\theta_1 - \theta_2),$$

$$\sin\theta_1 + \sin\theta_2 = 2\sin\tfrac{1}{2}(\theta_1 + \theta_2) \cdot \cos\tfrac{1}{2}(\theta_1 - \theta_2), \quad \text{and}$$

$$\cos\theta_1 + \cos\theta_2 = 2\cos\tfrac{1}{2}(\theta_1 + \theta_2) \cdot \cos\tfrac{1}{2}(\theta_1 - \theta_2)$$

may be used.

2 Sketching Polar Graphs

Although we managed to sketch certain polar graphs in Section 1.2 by converting the polar equation to cartesian form and then sketching the graph as usual, there are polar equations that are quite difficult to express in cartesian form. Thus, in this section we consider the problem of sketching the graph of a polar equation *directly*, without converting it into cartesian form.

To plot the graph of a polar equation, it is often useful to start with a fixed value of θ (say $\theta = 0$), then investigate the corresponding value (or values) of r as θ increases or decreases. It may be helpful to make a table of values of r corresponding to selected values of θ and to plot polar points (r, θ) corresponding to pairs of values in the table. If r depends continuously on θ, then a sketch of the graph is obtained simply by connecting these points with a continuous curve. With patience, virtually any continuous polar graph can be sketched in this way.

Polar graph sketching can often be expedited by the same techniques that are effective for sketching cartesian graphs. It is useful to find the intercepts of the graph with the polar axis and with a few of the special rays such as $\theta = \pm\pi/2$, $\theta = \pm\pi/4$, and so forth. Symmetries of the graph can be especially helpful. If the polar equation of the graph involves trigonometric functions, then the periodicity of these functions should be taken into consideration.

In this section we discuss and illustrate some of the more practical techniques for sketching polar graphs, including the use of calculus to find the direction of the tangent line to the graph.

We begin with an example in which a polar graph is sketched simply by plotting points and connecting them by a continuous curve.

EXAMPLE   Sketch the graph of $r = 1 + \dfrac{6}{\pi}\theta$ for $0 \le \theta \le 2\pi$.

SOLUTION
The following table shows some selected values of θ between 0 and 2π and the corresponding values of r:

θ	0	$\dfrac{\pi}{6}$	$\dfrac{\pi}{3}$	$\dfrac{\pi}{2}$	$\dfrac{2\pi}{3}$	$\dfrac{5\pi}{6}$	π	$\dfrac{7\pi}{6}$	$\dfrac{4\pi}{3}$	$\dfrac{3\pi}{2}$	$\dfrac{5\pi}{3}$	$\dfrac{11\pi}{6}$	2π
r	1	2	3	4	5	6	7	8	9	10	11	12	13

Plotting the polar points (r, θ) shown in the table and connecting them with a continuous curve, we obtain the desired graph (Figure 1).

Figure 1

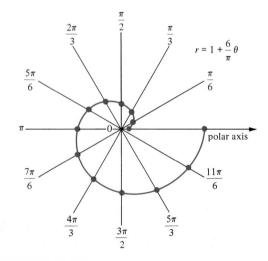

2.1 Symmetries of Polar Graphs

Symmetries of the graph of a polar equation can often be detected by making suitable replacements in the equation and checking to see whether the new equation is equivalent to the original one. The following table shows some replacements which, when they produce equations equivalent to the original one, imply the type of symmetry indicated:

Replace	An Equivalent Equation Implies
1. θ by $-\theta$ or 2. θ by $\pi - \theta$ and r by $-r$	Symmetry about the straight line obtained by extending the polar axis (Figure 2a)
3. θ by $\pi - \theta$ or 4. θ by $-\theta$ and r by $-r$	Symmetry about the straight line through the pole perpendicular to the polar axis (that is, about the $\pi/2$ axis) (Figure 2b)
5. θ by $\theta + \pi$ or 6. r by $-r$	Symmetry about the pole (Figure 2c)

Figure 2

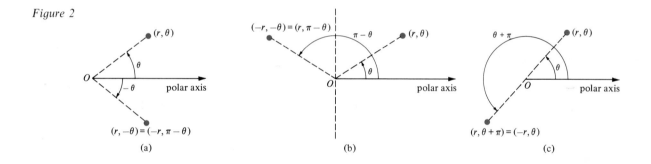

(a)   (b)   (c)

Because of the nonuniqueness of representations in polar coordinates, the conditions given above *may* fail even though the graph displays the indicated symmetry. Nevertheless, the given conditions are the ones most likely to be encountered in practice.

EXAMPLES   Sketch the graph of the given polar equation. Discuss the symmetry.

1 $r = 2(1 - \cos \theta)$

SOLUTION
Testing all the rules for symmetry, we observe that the graph is symmetric only about the polar axis; indeed, replacing θ by $-\theta$ yields

$$r = 2[1 - \cos(-\theta)],$$

which is equivalent to the original equation. We construct a table by giving the coordinates of some points on the graph corresponding to selected values of θ. Plotting these points, we sketch the upper half of the graph. The lower half

Figure 3

θ	r
0	0
$\dfrac{\pi}{6}$	$2 - \sqrt{3} \approx 0.27$
$\dfrac{\pi}{4}$	$2 - \sqrt{2} \approx 0.59$
$\dfrac{\pi}{3}$	1
$\dfrac{\pi}{2}$	2
$\dfrac{2\pi}{3}$	3
$\dfrac{5\pi}{6}$	$2 + \sqrt{3} \approx 3.73$
π	4

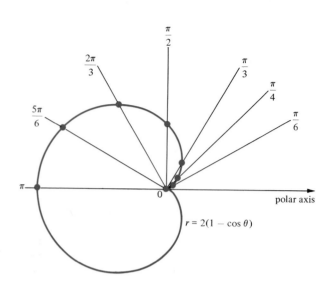

$$r = 2(1 - \cos \theta)$$

Figure 4

θ	r
$-\dfrac{\pi}{2}$	3
$-\dfrac{\pi}{3}$	0
$-\dfrac{\pi}{4}$	$-\dfrac{3\sqrt{2}}{2} \approx 2.12$
$-\dfrac{\pi}{6}$	-3
0	0
$\dfrac{\pi}{6}$	3
$\dfrac{\pi}{4}$	$\dfrac{3\sqrt{2}}{2} \approx 2.12$
$\dfrac{\pi}{3}$	0
$\dfrac{\pi}{2}$	-3

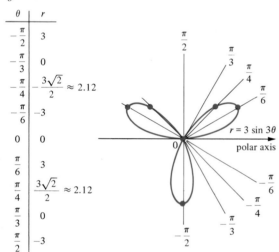

$r = 3 \sin 3\theta$

Figure 5

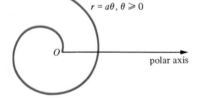

$r = a\theta, \theta \geqslant 0$

Archimedean spiral

Figure 6

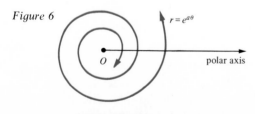

$r = e^{a\theta}$

logarithmic spiral

is then drawn using the symmetry about the polar axis (Figure 3). This graph is called a *cardioid*.

2 $r = 3 \sin 3\theta$

SOLUTION
Testing all the rules for symmetry, we find that the graph is symmetric only about the line $\theta = \pi/2$; indeed, replacing θ by $-\theta$ and r by $-r$ yields the equation

$$-r = 3 \sin 3(-\theta) = -3 \sin 3\theta$$

which is equivalent to the original equation. We construct a table giving the coordinates of some points on the graph. Plotting these points, we sketch part of the graph. The other part is obtained by using the symmetry about the vertical line $\theta = \pi/2$ (Figure 4).

The graph in Figure 4 is called a *three-leaved rose*. More generally, either one of the equations

$$r = a \sin k\theta \quad \text{or} \quad r = a \cos k\theta$$

gives an *N-leaved rose*, where

$$N = \begin{cases} k & \text{if } k \text{ is an odd integer} \\ 2k & \text{if } k \text{ is an even integer.} \end{cases}$$

We now give equations, names, and sketches of other common polar curves. Except for the spirals, each displays some type of symmetry. The curve $r = a\theta$ for $\theta \geq 0$ is called an *Archimedean spiral* (Figure 5). The curve in Figure 6 has the equation $r = e^{a\theta}$ and is called a *logarithmic spiral*.

Figure 7

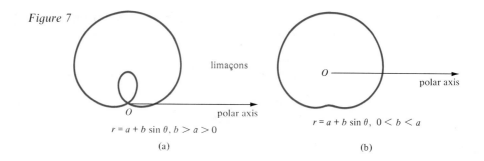

limaçons

$r = a + b \sin \theta$, $b > a > 0$

(a)

$r = a + b \sin \theta$, $0 < b < a$

(b)

We have already graphed one of the *cardioids* in Figure 3. More generally, the polar graph of $r = a(1 + \cos \theta)$ $r = a(1 - \cos \theta)$ or $r = a(1 + \sin \theta)$ or $r = a(1 - \sin \theta)$ forms a cardioid similar to the one shown in Figure 3, except that it is rotated about the pole through an angle of 90°, 180°, or 270°.

A polar graph of $r = a \pm b \cos \theta$ or $r = a \pm b \sin \theta$ forms a curve called a *limaçon*. If $b > a > 0$, the limaçon has a loop (Figure 7a), while if $0 < b < a$, it merely has an indentation (Figure 7b). Note that when $a = b$, the limaçon becomes a cardioid.

Figure 8

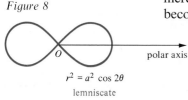

$r^2 = a^2 \cos 2\theta$

lemniscate

The graph of the polar equation

$$r^2 = a^2 \cos 2\theta \quad \text{or} \quad r^2 = a^2 \sin 2\theta$$

is called a *lemniscate* (Figure 8).

2.2 Direction of a Polar Graph

Now we derive formulas for the direction of a polar graph at a point, that is, for the direction of the tangent line to the polar graph at the point. Thus, consider a polar curve $r = f(\theta)$, where f is a differentiable function of θ, and let l denote the tangent line to the curve at the polar point $P = (r, \theta)$ (Figure 9). Let α be the angle measured from the polar axis to the tangent line l. Thus, since the polar axis coincides with the x axis, $\tan \alpha$ is the slope of the tangent line l in the cartesian xy coordinate system.

If $P = (x, y)$ in the cartesian coordinate system, then we have $dy/dx = \tan \alpha$ as usual. Using the formulas for conversion from polar to cartesian coordinates obtained in Section 1.1, we have

$$x = r \cos \theta = f(\theta) \cos \theta$$

and

$$y = r \sin \theta = f(\theta) \sin \theta;$$

hence, x and y can be regarded as functions of θ. By the chain rule, $\dfrac{dy}{d\theta} = \dfrac{dy}{dx}\dfrac{dx}{d\theta}$. Therefore, if $\dfrac{dx}{d\theta} \neq 0$, we have

$$\tan \alpha = \frac{dy}{dx} = \frac{\dfrac{dy}{d\theta}}{\dfrac{dx}{d\theta}} = \frac{\dfrac{d}{d\theta}(r \sin \theta)}{\dfrac{d}{d\theta}(r \cos \theta)} = \frac{\dfrac{dr}{d\theta}\sin \theta + r \cos \theta}{\dfrac{dr}{d\theta}\cos \theta - r \sin \theta}.$$

Figure 9

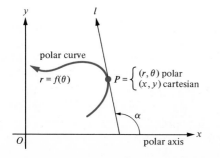

polar curve

$r = f(\theta)$

$P = \begin{cases} (r, \theta) \text{ polar} \\ (x, y) \text{ cartesian} \end{cases}$

α

polar axis

EXAMPLE   Find the slope of the tangent line to the three-leaved rose $r = 3 \sin 3\theta$ at the point $(3, \pi/6)$ (Figure 4).

SOLUTION
Here $dr/d\theta = 9 \cos 3\theta$, so

$$\tan \alpha = \frac{\dfrac{dr}{d\theta} \sin \theta + r \cos \theta}{\dfrac{dr}{d\theta} \cos \theta - r \sin \theta} = \frac{9 \cos 3\theta \sin \theta + 3 \sin 3\theta \cos \theta}{9 \cos 3\theta \cos \theta - 3 \sin 3\theta \sin \theta}.$$

Thus, putting $\theta = \pi/6$, we have

$$\tan \alpha = \frac{9 \cos \dfrac{\pi}{2} \sin \dfrac{\pi}{6} + 3 \sin \dfrac{\pi}{2} \cos \dfrac{\pi}{6}}{9 \cos \dfrac{\pi}{2} \cos \dfrac{\pi}{6} - 3 \sin \dfrac{\pi}{2} \sin \dfrac{\pi}{6}} = \frac{3\left(\dfrac{\sqrt{3}}{2}\right)}{-3\left(\dfrac{1}{2}\right)} = -\sqrt{3}.$$

In the formula for $\tan \alpha$ derived above, if the numerator $\dfrac{dr}{d\theta} \sin \theta + r \cos \theta$ is zero and the denominator $\dfrac{dr}{d\theta} \cos \theta - r \sin \theta$ is nonzero, then $\tan \alpha = 0$ and the tangent line l to the polar curve is *horizontal*. Likewise, if the denominator is zero and the numerator is nonzero, the tangent line l is *vertical*.

The case in which $r = 0$ in the formula for $\tan \alpha$ is of particular interest, since the formula then gives the slope of the tangent line at the pole. Specifically, we have

$$\tan \alpha = \frac{\dfrac{dr}{d\theta} \sin \theta}{\dfrac{dr}{d\theta} \cos \theta} = \frac{\sin \theta}{\cos \theta} = \tan \theta \qquad \text{when } r = 0.$$

Therefore, if the polar curve passes through the pole for a particular value of θ, the tangent line to the polar curve at the pole has slope equal to $\tan \theta$.

A polar graph may pass repeatedly through the pole and it may have different directions on each passage. (For instance, consider the three-leaved rose in Figure 4.) To find the direction of the curve on any particular passage, simply use the appropriate value of θ corresponding to that passage.

EXAMPLE   Find the slope of the tangent to the four-leaved rose $r = \cos 2\theta$ on that passage through the origin for which $\theta = 5\pi/4$ (Figure 10). Also discuss the motion of the point (r, θ) along the curve as θ increases from 0 to 2π.

SOLUTION
When $\theta = 0$, $r = 1$ and we start at the indicated point on the polar axis (Figure 10). As θ increases from 0 to $\pi/4$, $r = \cos 2\theta$ decreases from 1 to 0, and the curve makes its first passage through the pole. As θ continues to increase, the point (r, θ) moves along the four-leaved rose according to the arrows, passing through the pole for the second time when $\theta = 3\pi/4$, for the third time when $\theta = 5\pi/4$, and for the fourth time when $\theta = 7\pi/4$. When θ reaches 2π, the point (r, θ) returns to the starting point. On its third passage through the pole, when $\theta = 5\pi/4$, the slope of the tangent line is equal to $\tan(5\pi/4) = 1$.

Figure 10

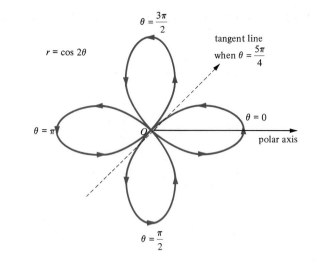

$\theta = \dfrac{3\pi}{2}$

tangent line

when $\theta = \dfrac{5\pi}{4}$

$r = \cos 2\theta$

$\theta = \pi$

O

$\theta = 0$

polar axis

$\theta = \dfrac{\pi}{2}$

In dealing with polar curves, it is often convenient to specify the direction of the tangent line l at a polar point $P = (r, \theta)$ by giving the angle ψ measured from the radial line through P to the tangent line l (Figure 11a). (The symbol ψ is the Greek letter *psi*.) Since the three vertex angles of triangle OQP must add up to

Figure 11

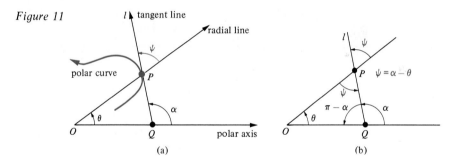

l tangent line

radial line

polar curve

ψ

P

θ

α

O Q polar axis

(a)

l

ψ

P $\psi = \alpha - \theta$

ψ

$\pi - \alpha$ α

θ

O Q

(b)

$180° = \pi$ radians, it follows that $\theta + (\pi - \alpha) + \psi = \pi$; that is, $\psi = \alpha - \theta$ (Figure 11b). Using the trigonometric identity for the tangent of the difference between two angles, the fact that $\psi = \alpha - \theta$, and the formula previously derived for $\tan \alpha$, we have

$$\tan \psi = \tan(\alpha - \theta) = \frac{\tan \alpha - \tan \theta}{1 + \tan \alpha \tan \theta}$$

$$= \frac{\dfrac{\dfrac{dr}{d\theta} \sin \theta + r \cos \theta}{\dfrac{dr}{d\theta} \cos \theta - r \sin \theta} - \tan \theta}{1 + \dfrac{\dfrac{dr}{d\theta} \sin \theta + r \cos \theta}{\dfrac{dr}{d\theta} \cos \theta - r \sin \theta} \tan \theta}.$$

Multiplying numerator and denominator of the fraction above by $\dfrac{dr}{d\theta} \cos \theta - r \sin \theta$, we have

$$\tan \psi = \frac{\dfrac{dr}{d\theta}\sin\theta + r\cos\theta - \left(\dfrac{dr}{d\theta}\cos\theta - r\sin\theta\right)\tan\theta}{\dfrac{dr}{d\theta}\cos\theta - r\sin\theta + \left(\dfrac{dr}{d\theta}\sin\theta + r\cos\theta\right)\tan\theta}$$

$$= \frac{\dfrac{dr}{d\theta}\sin\theta + r\cos\theta - \dfrac{dr}{d\theta}\cos\theta\tan\theta + r\sin\theta\tan\theta}{\dfrac{dr}{d\theta}\cos\theta - r\sin\theta + \dfrac{dr}{d\theta}\sin\theta\tan\theta + r\cos\theta\tan\theta}$$

$$= \frac{r\cos\theta + r\dfrac{\sin^2\theta}{\cos\theta}}{\dfrac{dr}{d\theta}\cos\theta + \dfrac{dr}{d\theta}\dfrac{\sin^2\theta}{\cos\theta}} = \frac{r\cos^2\theta + r\sin^2\theta}{\dfrac{dr}{d\theta}\cos^2\theta + \dfrac{dr}{d\theta}\sin^2\theta}$$

$$= \frac{r}{dr/d\theta}.$$

Thus, we obtain the very simple formula

$$\tan \psi = \frac{r}{dr/d\theta},$$

which explains why it is often advantageous to use the angle ψ to indicate the direction of the tangent line. (The above derivation has to be modified slightly if the point P is in different portions of the plane; however, the final result is always the same.)

EXAMPLE   Find $\tan \psi$ at the polar point $(2, \pi/2)$ on the cardioid $r = 2(1 - \cos\theta)$ (Figure 12).

Figure 12

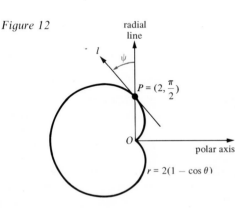

SOLUTION
Here we have $dr/d\theta = 2\sin\theta$, so that

$$\tan \psi = \frac{r}{dr/d\theta} = \frac{2(1 - \cos\theta)}{2\sin\theta}$$

$$= \csc\theta - \cot\theta.$$

Therefore, when $\theta = \pi/2$, we have

$$\tan \psi = \csc\frac{\pi}{2} - \cot\frac{\pi}{2} = 1.$$

Problem Set 2

In problems 1 through 18, test for symmetry with respect to the polar axis, the line $\theta = \pm\pi/2$, and the pole. Sketch the graph of the equation.

1 $r = 4\cos\theta$ **2** $r\sin\theta = 5$ **3** $r\cos\theta = 5$

4 $r = 2$ **5** $r = 2\sin\theta$ **6** $\theta = 3$

7 $r = 2\sin 3\theta$ (three-leaved rose)   **8** $r = 2\cos 2\theta$ (four-leaved rose)   **9** $r = 4\sin 2\theta$ (four-leaved rose)

10 $r = 2\sin 4\theta$ (eight-leaved rose)   **11** $r = 4(1 + \cos\theta)$ (cardioid)   **12** $r = 2(1 - \sin\theta)$ (cardioid)

13 $r = 3 - 2 \cos \theta$ (limaçon)

14 $r = 3 + 4 \sin \theta$ (limaçon)

15 $r = 1 - 2 \sin \theta$ (limaçon)

16 $r = 1/\theta, \theta > 0$ (reciprocal spiral)

17 $r^2 = 8 \cos 2\theta$ (lemniscate)

18 $9\theta = \ln r$ (logarithmic spiral)

In problems 19 through 22, find the slope of the tangent line to the graph of each polar equation at the given point.

19 $r = 3(1 + \cos \theta)$ at $(3, \pi/2)$.

20 $r = 2(1 - \sin \theta)$ at $(1, \pi/6)$.

21 $r = 8 \sin 2\theta$ at $(0, 0)$.

22 $r = \sec^2 \theta$ at $(4, \pi/3)$.

23 Find all points where the tangent to the limaçon $r = 4 + 3 \sin \theta$ is either horizontal or vertical.

24 Find the polar coordinates of the tips of the petals of the *N*-leaved rose $r = 2 \cos k\theta$ by finding the values of θ where r takes on its maximum values. (*Note:* Here it is not necessary to use calculus.)

In problems 25 through 29, find $\tan \psi$ at the indicated point (r, θ) on the graph of each polar curve.

25 $r = \sin \theta$ at $(\frac{1}{2}, \pi/6)$.

26 $r = \cos 2\theta$ at $(0, \pi/4)$.

27 $r = e^\theta$ at $(e^2, 2)$.

28 $r^2 = \csc 2\theta$ at $(1, \pi/6)$.

29 $r = 4 \sec \theta$ at $(4, 0)$.

30 If $b > a > 0$, find the slope of *both* tangents to the limaçon $r = a + b \sin \theta$ at the pole.

31 If $0 < b < a$, find the minimum value of r for the limaçon $r = a + b \sin \theta$.

32 Show that the polar graph of $r = \sin (\theta/3)$ is symmetric about the line $\theta = \pm \pi/2$ even though the tests for such symmetry given in Section 2.1 both fail. Sketch the graph.

33 Show that the polar graph of $r = f(\theta)$ is:
(a) Symmetric about the line $\theta = 0$ if f is an even function.
(b) Symmetric about the line $\theta = \pm \pi/2$ if f is an odd function.

34 In view of problem 32, the symmetry tests given in Section 2.1 may fail to detect actual symmetries. Develop a set of tests that cannot fail.

35 Find and sketch the graph of a polar curve such that, at each point of the curve, $\tan \psi = K$, where K is a nonzero constant.

3   Conics in Polar Form and Intersections of Polar Curves

In Chapter 4 we studied the cartesian equations of the conics. In this section we derive the equations for the conics in polar form. We also study the method for finding the points at which polar curves intersect.

3.1   Ellipses, Parabolas, and Hyperbolas in Polar Form

Recall from Section 5 of Chapter 4 that an ellipse, parabola, or hyperbola can be defined in terms of a point F called the *focus*, a line D called the *directrix*, and a positive number e called the *eccentricity*.

To find the polar equation for the conic with focus F, directrix D, and eccentricity e, place the focus F at the pole O and place the directrix D

Figure 1

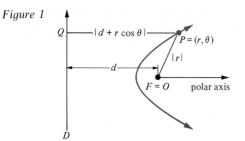

Figure 2

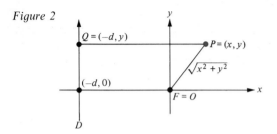

Figure 3

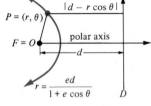

perpendicular to the polar axis and d units to the *left* of the pole, $d > 0$ (Figure 1).

Now consider an arbitrary point $P = (r, \theta)$ in the plane and let Q be the point at the foot of the perpendicular from P to the directrix D. Switching momentarily to cartesian coordinates, so that $P = (x, y)$, $Q = (-d, y)$, $x = r \cos \theta$, and $y = r \sin \theta$ (Figure 2), we see that

$$|\overline{PQ}| = |d + x| = |d + r \cos \theta|$$

and $|\overline{PF}| = \sqrt{x^2 + y^2} = |r|$. By definition, P belongs to the conic if and only if

$$\frac{|\overline{PF}|}{|\overline{PQ}|} = e, \quad \text{that is,} \quad \frac{|r|}{|d + r \cos \theta|} = e.$$

Therefore, the equation of the conic in polar form is

$$\left| \frac{r}{d + r \cos \theta} \right| = e \quad \text{or} \quad \frac{\pm r}{d + r \cos \theta} = e.$$

If $P = (r_1, \theta_1)$ satisfies the equation $\dfrac{-r}{d + r \cos \theta} = e$, then $P = (-r_1, \theta_1 + \pi)$ also satisfies the equation $\dfrac{r}{d + r \cos \theta} = e$; hence, we lose no points on the conic by writing its equation as $\dfrac{r}{d + r \cos \theta} = e$. Solving for r, we obtain

$$r = \frac{ed}{1 - e \cos \theta},$$

which is therefore the polar equation of the conic in Figure 1.

If the focus F remains at the pole, but the directrix D is placed to the *right* of the focus and perpendicular to the polar axis (Figure 3), then the polar equation of the conic becomes $r = \dfrac{ed}{1 + e \cos \theta}$. (See Problem 35.)

If the directrix D is placed parallel to the polar axis, with the focus F still at the pole, then the polar equation of the conic becomes

$$r = \frac{ed}{1 \pm e \sin \theta},$$

where the plus sign is used if the directrix is *above* the polar axis (Figure 4a) and the minus sign is used if the directrix is *below* the polar axis (Figure 4b). We leave the derivation as an exercise (Problem 36).

Figure 4

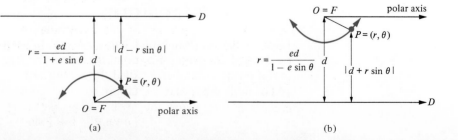

(a)   (b)

EXAMPLES   In Examples 1 and 2, find the eccentricity, the directrices, the foci, and the vertices of the given conic in polar coordinates. Identify the conic and sketch its graph.

1 $r = \dfrac{12}{3 - 2\cos\theta}$

SOLUTION
Dividing numerator and denominator of the given fraction by 3 gives

$$r = \frac{4}{1 - \frac{2}{3}\cos\theta} = \frac{\frac{2}{3}(6)}{1 - \frac{2}{3}\cos\theta},$$

which is the polar equation of a conic with focus at the pole, directrix perpendicular to the polar axis and $d = 6$ units to the left of the pole, and eccentricity $e = \frac{2}{3}$. Since $e < 1$, the conic is an ellipse. Using Case (i) of Theorem 1 in Section 5 of Chapter 4, we have

$$a = \frac{ed}{1 - e^2} = \frac{4}{1 - \frac{4}{9}} = \frac{36}{5}, \qquad b = \frac{ed}{\sqrt{1 - e^2}} = \frac{4}{\sqrt{\frac{5}{9}}} = \frac{12}{\sqrt{5}}, \qquad \text{and} \qquad c = ae = \frac{24}{5}.$$

Hence, the second focus is on the polar axis $2c = \frac{48}{5}$ units from the pole, and the second directrix is perpendicular to the polar axis and $2c + d = \frac{78}{5}$ units to the right of the pole. The vertices in cartesian coordinates are

$$V_1 = (c - a, 0) = (-\tfrac{12}{5}, 0), \quad V_2 = (c + a, 0) = (12, 0),$$
$$V_3 = (c, b) = (\tfrac{24}{5}, 12/\sqrt{5}), \quad \text{and} \quad V_4 = (c, -b) = (\tfrac{24}{5}, -12/\sqrt{5}).$$

Figure 5

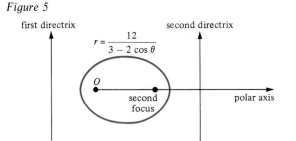

first directrix   second directrix

$r = \dfrac{12}{3 - 2\cos\theta}$

O

second focus   polar axis

In polar coordinates, therefore, the vertices are $V_1 = (\tfrac{12}{5}, \pi)$, $V_2 = (12, 0)$, $V_3 = (\tfrac{36}{5}, \tan^{-1}(\sqrt{5}/2))$, and $V_4 = (\tfrac{36}{5}, -\tan^{-1}(\sqrt{5}/2))$ (Figure 5).

2 $r = \dfrac{3}{2 + 2\cos\theta}$

SOLUTION
Divide the numerator and denominator by 2, so that

$r = \dfrac{\frac{3}{2}}{1 + \cos\theta}$, which is the polar equation of a conic with focus at the pole, directrix perpendicular to the polar axis and $d = \frac{3}{2}$ units to the right of the pole, and eccentricity $e = 1$. Since $e = 1$, the conic is a parabola and its vertex is on the polar axis midway between the focus and the directrix. Hence, the vertex is at $V = \left(\dfrac{d}{2}, 0\right) = (\tfrac{3}{4}, 0)$ in either cartesian or polar coordinates (Figure 6).

Figure 6

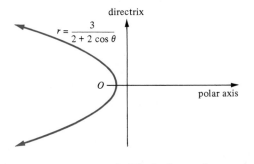

directrix

$r = \dfrac{3}{2 + 2\cos\theta}$

O

polar axis

3 Find the polar equation of the hyperbola of eccentricity $e = \frac{5}{4}$, one of whose foci is at the pole with the corresponding directrix parallel to the polar axis and $d = 3$ units above the pole.

SOLUTION

$$r = \frac{ed}{1 + e\sin\theta} = \frac{\frac{5}{4}(3)}{1 + \frac{5}{4}\sin\theta} = \frac{15}{4 + 5\sin\theta}.$$

3.2 Circles in Polar Form

By Problem 39 in Section 1, the distance from the point $P = (r, \theta)$ to the point $P_0 = (r_0, \theta_0)$ in the polar coordinate system is given by

$$|\overline{PP_0}| = \sqrt{r^2 - 2rr_0 \cos(\theta - \theta_0) + r_0^2}.$$

Therefore, the equation of the circle with radius R and center $P_0 = (r_0, \theta_0)$ is given by $|\overline{PP_0}|^2 = R^2$, or

$$r^2 - 2rr_0 \cos(\theta - \theta_0) + r_0^2 = R^2.$$

EXAMPLE   Find the polar equation of the circle of radius $R > 0$ whose center P_0 is R units above the pole on the ray $\theta = \pi/2$ (Figure 7).

Figure 7

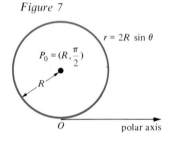

SOLUTION
Put $r_0 = R$, $\theta_0 = \pi/2$ in the general equation above to obtain

$$r^2 - 2rR \cos(\theta - \pi/2) + R^2 = R^2 \quad \text{or} \quad r^2 - 2rR \sin\theta = 0.$$

Thus, the equation of the circle is $r = 2R \sin\theta$. (Canceling the r does not lose the solution $r = 0$, since $r = 2R \sin\theta$ still gives $r = 0$ when $\theta = 0$.)

3.3 Intersection Points of Polar Curves

To find the intersection points of two curves in the cartesian coordinate system, one just solves the equations of the curves simultaneously. However, in the polar coordinate system, this procedure does not necessarily yield *all* intersection points, because an intersection point may have two *different* polar representations, one that satisfies the equation of the first curve while the other satisfies the equation of the second curve.

   In order to find *all* the common points P of the polar graphs of the equations

$$r = f(\theta) \quad \text{and} \quad r = g(\theta),$$

we must take account of the multiplicity of representations in polar coordinates by carrying out the following procedure:

Step 1   Check each polar graph separately to see whether it passes through the pole O. If both do, then O is an intersection point of the graphs.

Step 2   Solve, if possible, the simultaneous equations

$$\begin{cases} r = f(\theta) \\ r = g(\theta + 2n\pi) \end{cases}$$

for r, θ, and n, where $r \neq 0$ and n must be an *integer*. For each such solution, the point $P = (r, \theta)$ is an intersection point of the two graphs.

Step 3   Solve, if possible, the simultaneous equations

$$\begin{cases} r = f(\theta) \\ r = -g(\theta + (2n + 1)\pi) \end{cases}$$

for r, θ, and n, where $r \neq 0$ and n must be an *integer*. For each such solution, the point $P = (r, \theta)$ is an intersection point of the two graphs.

EXAMPLE   Find all points of intersection of the circle $r = -6 \cos \theta$ and the limaçon $r = 2 - 4 \cos \theta$.

SOLUTION
   Step 1: Since $O = (0, \pi/2)$ is on the circle and $O = (0, \pi/3)$ is on the limaçon, then O is a point of intersection.
   Step 2: We solve the simultaneous equations

$$\begin{cases} r = -6 \cos \theta \\ r = 2 - 4 \cos (\theta + 2n\pi), \; r \neq 0, \; n \text{ an integer.} \end{cases}$$

Noting that $\cos (\theta + 2n\pi) = \cos \theta$, we rewrite the equations above as

$$r = -6 \cos \theta = 2 - 4 \cos \theta.$$

Thus, $-2 \cos \theta = 2$, so that $\cos \theta = -1$, $r = 6$, and $\theta = \pi$. Thus, Step 2 yields the intersection point $P_1 = (6, \pi)$.
   Step 3: We solve the simultaneous equations

$$\begin{cases} r = -6 \cos \theta \\ r = -[2 - 4 \cos (\theta + (2n + 1)\pi)], \end{cases} \qquad r \neq 0, \quad n \text{ an integer.}$$

Figure 8

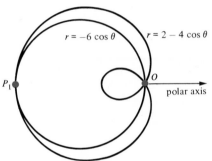

$r = -6 \cos \theta$ $r = 2 - 4 \cos \theta$

P_1 O polar axis

Since $\cos (\theta + 2n\pi + \pi) = \cos (\theta + \pi) = -\cos \theta$, we rewrite the equations above as

$$r = -6 \cos \theta = -(2 + 4 \cos \theta).$$

Thus, $2 \cos \theta = 2$, so that $\cos \theta = 1$, $r = -6$, and $\theta = 0$. Therefore, Step 3 yields the same intersection point $P_1 = (-6, 0) = (6, \pi)$ as Step 2. Hence, the circle and the limaçon intersect in the two points O and P_1 (Figure 8).

Problem Set 3

In problems 1 through 10, find (a) the eccentricity, (b) the directrix (directrices), (c) the focus (foci), and (d) the vertex (vertices) of the conic in the polar coordinate system. Also, identify the conic and sketch the graph.

1 $r = \dfrac{16}{5 - 3 \cos \theta}$

2 $r = \dfrac{16}{5 + 3 \cos \theta}$

3 $r = \dfrac{6}{1 - \cos \theta}$

4 $r = \dfrac{24}{5 - 7 \cos \theta}$

5 $r = \dfrac{10}{1 - \sin \theta}$

6 $r = \dfrac{6}{1 - 2 \sin \theta}$

7 $r = \dfrac{1}{1 + 2 \sin \theta}$

8 $r = \dfrac{6}{10 + 5 \sin \theta}$

9 $r = \dfrac{4}{1 + \cos \theta}$

10 $r = \csc^2 \theta - \csc \theta \cot \theta$

In problems 11 through 14, find the *polar* equation of the conic having one focus at the pole and satisfying the condition given. (In each case, the directrix given is the one corresponding to the focus at the pole.)

11 Eccentricity is $\frac{1}{3}$, directrix is horizontal and 5 units above the pole.

12 Contains the *cartesian* point $(5, 12)$ and has directrix $x = -4$.

13 Directrix 4 units to the right of the pole, second directrix 6 units to the right of the pole.

14 Directrix 4 units to the right of the pole, second directrix 6 units to the left of the pole.

15 If $0 < d$ and $0 < e < 1$, find the *polar* coordinates of the two foci, the center, and the four vertices of the ellipse $r = \dfrac{ed}{1 - e \cos \theta}$ in terms of d and e.

16 If $0 < d$ and $e > 1$, find the *polar* coordinates of the two foci, the center, and the two vertices of the hyperbola $r = \dfrac{ed}{1 - e \cos \theta}$ in terms of d and e. Also find the slopes of the two asymptotes.

17 Consider the ellipse $r = \dfrac{ed}{1 - e \cos \theta}$, where $0 < e < 1$ and $d = \dfrac{a}{e}$, $a > 0$, a constant.
  (a) As $e \to 0^+$, what happens to the directrix of the ellipse corresponding to the focus at the pole? What about the second directrix?
  (b) As $e \to 0^+$, what happens to the shape of the ellipse?

18 Show that $r = \dfrac{d}{2} \csc^2 \dfrac{\theta}{2}$ is the polar equation of a parabola.

In problems 19 through 22, find the polar equation of each circle.

19 Center at $(5, \pi/4)$, radius $R = 5$.

20 Center at $(R, \pi/4)$, radius R.

21 Tangent to polar axis at $(7, 0)$, center above polar axis, radius $R = 5$.

22 Tangent to polar axis at $(a, 0)$, $a > 0$, center R units above polar axis.

23 Find the polar equation of the straight line perpendicular to the polar axis and
  (a) d units to the right of the pole.   (b) d units to the left of the pole, $d > 0$.

24 Show that, if $C \neq 0$, the straight line whose cartesian equation is $Ax + By + C = 0$ has the polar equation

$$r = \frac{-C}{A \cos \theta + B \sin \theta}.$$

25 Find the polar equation of the straight line parallel to the polar axis and
  (a) d units above the pole.   (b) d units below the pole, $d > 0$.

26 Prove that if the straight line L contains the point $P_1 = (r_1, \theta_1)$ where $r_1 > 0$ and is perpendicular to the line segment $\overline{OP}_1$, then its polar equation is

$$r = \frac{r_1}{\cos \theta_1 \cos \theta + \sin \theta_1 \sin \theta}.$$

In problems 27 through 34, find all points of intersection of each pair of polar curves. Sketch graphs of the two curves showing these points.

27 $r = -3 \sin \theta$ and $r = 2 + \sin \theta$.

28 $r = \dfrac{2}{\sqrt{2} \cos \theta + \sqrt{2} \sin \theta}$ and $r = 2 \cos \left(\theta - \dfrac{\pi}{4} \right)$.

29 $r = 1$ and $r = 2 \cos 3\theta$.

30 $r = \theta$ for $\theta \geq 0$ and $r = -\theta$ for $\theta \geq 0$.

31 $r = \sin \theta$, $r = \sin 2\theta$.

32 $r = 1 + \cos \theta$, $r = (1 + \sqrt{2}) \cos \theta$.

33 $r = 2 \sin 3\theta$ and $r = -2/\sin \theta$.

34 $r = \theta$ and $r = 2\theta$.

35 Derive the polar equation of a conic with focus at the pole and directrix perpendicular to the polar axis and d units to the right of the pole.

36 Derive the polar equation of a conic with focus at the pole and directrix parallel to the polar axis.

37 Find the *polar* equations of the two directrices of the ellipse $r = \dfrac{ed}{1 - e\cos\theta}$, $0 < e < 1$, $d > 0$.

38 In problem 37, find the condition or conditions on e and d so that the left vertex of the ellipse lies on the inner loop of the limaçon $r = a - b\cos\theta$, $b > a > 0$.

39 Find all points of intersection of the hyperbola $r = \dfrac{2}{1 - 2\cos\theta}$ and the limaçon $r = 2 - 4\cos\theta$. Sketch the graphs.

4 Area and Arc Length in Polar Coordinates

Geometrically, the two fundamental problems of calculus are finding the slope of the tangent to a curve and determining the area of a region bounded by a curve. In Section 2 we attended to the first problem for polar curves; in the present section we see how to handle the second. We also develop a formula for arc length of polar curves.

4.1 Area of a Region in Polar Coordinates

Figure 1

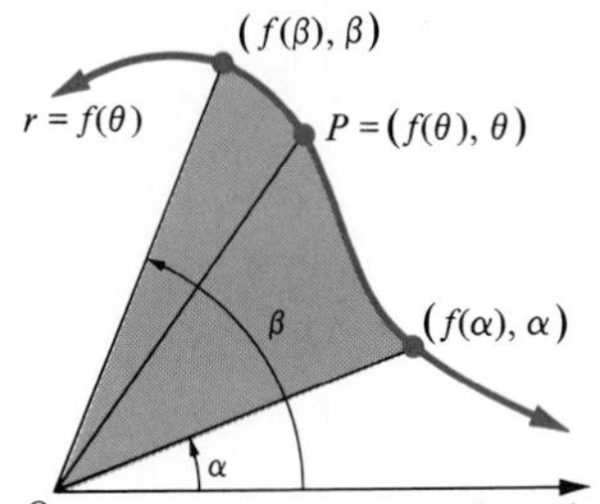

Consider the curve whose polar equation is $r = f(\theta)$, where f is a continuous function (Figure 1). As θ increases from $\theta = \alpha$ to $\theta = \beta$, the point $P = (f(\theta), \theta)$ moves along the polar curve from $(f(\alpha), \alpha)$ to $(f(\beta), \beta)$ and the radial line segment $\overline{OP}$ sweeps out a planar region. We refer to this region as the region *enclosed* by the polar curve between $\theta = \alpha$ and $\theta = \beta$. Below, we develop a formula for its area.

The simplest polar region is perhaps the circular sector enclosed by the circle of radius r between $\theta = \alpha$ and $\theta = \beta$ (Figure 2). Since the area of the circle is πr^2 and the sector occupies a fraction $\dfrac{\beta - \alpha}{2\pi}$ of the whole circle, the area A of the sector is given by

$$A = \pi r^2 \frac{\beta - \alpha}{2\pi} = \frac{1}{2} r^2 (\beta - \alpha).$$

More generally, even if the polar curve is not a circle, when the angle increases from θ to $\theta + d\theta$, the radial segment $\overline{OP}$ in Figure 1 will sweep out a small region which is virtually a sector of a circle of radius $|r| = |f(\theta)|$ (Figure 3). If $d\theta$ is

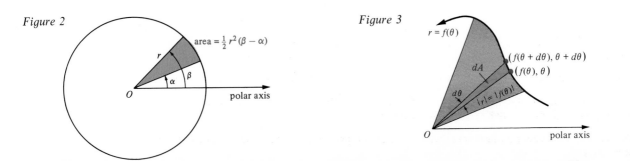

Figure 2

Figure 3

regarded as an "infinitesimal," then the "infinitesimal" area of this sector is given by

$$dA = \tfrac{1}{2}|r|^2 \, d\theta = \tfrac{1}{2}r^2 \, d\theta = \tfrac{1}{2}[f(\theta)]^2 \, d\theta.$$

If we "sum," that is, integrate the areas of all such "infinitesimal" sectors, running θ from α to β, then we obtain the required area A. Thus, we have

$$A = \int_\alpha^\beta dA = \int_\alpha^\beta \tfrac{1}{2}[f(\theta)]^2 \, d\theta = \tfrac{1}{2}\int_\alpha^\beta [f(\theta)]^2 \, d\theta.$$

This formula is also written as

$$A = \tfrac{1}{2}\int_\alpha^\beta r^2 \, d\theta.$$

EXAMPLES   1 Find the area of the "top half" of the region inside the cardioid $r = 3(1 + \cos\theta)$ (Figure 4).

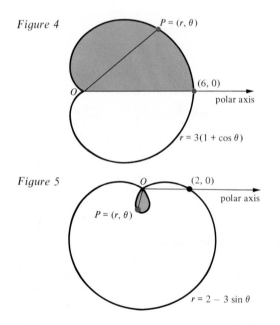

Figure 4

Figure 5

SOLUTION
As θ goes from 0 to π, the radial segment $\overline{OP}$ sweeps out the "top half" of the region inside the cardioid. Therefore, the area A of the region is given by

$$A = \frac{1}{2}\int_0^\pi r^2 \, d\theta = \frac{1}{2}\int_0^\pi [3(1 + \cos\theta)]^2 \, d\theta$$

$$= \frac{9}{2}\int_0^\pi (1 + 2\cos\theta + \cos^2\theta) \, d\theta$$

$$= \frac{9}{2}\left(\theta + 2\sin\theta + \frac{\theta}{2} + \frac{\sin 2\theta}{4}\right)\Bigg|_0^\pi$$

$$= \frac{27\pi}{4} \text{ square units.}$$

2 Find the area of the region enclosed by the inner loop of the limaçon $r = 2 - 3\sin\theta$ (Figure 5).

SOLUTION
When $\theta = 0$, $r = 2$ and the point $(r,\theta) = (2,0)$ starts on the polar axis. Now as θ begins to increase, $r = 2 - 3\sin\theta$ begins to decrease, and it reaches 0 when $\theta = \sin^{-1}\frac{2}{3} \approx 42°$. At this point, the radial line segment $\overline{OP}$ begins to sweep out the desired region. When θ reaches the value $\pi/2$, then $r = -1$ and the radial line segment $\overline{OP}$, which points straight downward, has swept out exactly half of the desired area. Therefore, the area A of the region is given by

$$A = 2\left[\frac{1}{2}\int_{\sin^{-1}(2/3)}^{\pi/2} (2 - 3\sin\theta)^2 \, d\theta\right]$$

$$= \int_{\sin^{-1}(2/3)}^{\pi/2} (4 - 12\sin\theta + 9\sin^2\theta) \, d\theta$$

$$= \left(4\theta + 12\cos\theta + \frac{9}{2}\theta - \frac{9}{4}\sin 2\theta\right)\Bigg|_{\sin^{-1}(2/3)}^{\pi/2}$$

$$= \frac{17\pi}{4} - \left[\frac{17}{2}\sin^{-1}\frac{2}{3} + 12\cos\left(\sin^{-1}\frac{2}{3}\right) - \frac{9}{4}\sin\left(2\sin^{-1}\frac{2}{3}\right)\right].$$

Since $\cos(\sin^{-1}\frac{2}{3}) = \sqrt{5}/3$ and $\sin(2\sin^{-1}\frac{2}{3}) = 2\sin(\sin^{-1}\frac{2}{3})\cos(\sin^{-1}\frac{2}{3}) = 4\sqrt{5}/9$, then

$$A = \frac{17\pi}{4} - \frac{17}{2}\sin^{-1}\frac{2}{3} - 3\sqrt{5} \approx 0.44 \text{ square unit.}$$

3 Find the area enclosed by the lemniscate $r^2 = 4\cos 2\theta$ (Figure 6).

Figure 6

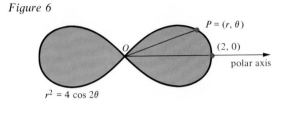

$r^2 = 4\cos 2\theta$

SOLUTION

Consider the portion of the lemniscate for which $r = 2\sqrt{\cos 2\theta}$. When $\theta = 0$, $r = 2$, and, as θ increases, the point $P = (r,\theta)$ moves to the left along the top of the lemniscate until it arrives at the pole O when $\theta = \pi/4$. Thus, as θ goes from 0 to $\pi/4$, the line segment $\overline{OP}$ sweeps out one fourth of the desired area. Therefore, the area A of the entire region is given by

$$A = 4\left(\frac{1}{2}\int_0^{\pi/4} r^2\,d\theta\right) = 2\int_0^{\pi/4} 4\cos 2\theta\,d\theta = 4\sin 2\theta\Big|_0^{\pi/4} = 4 \text{ square units.}$$

In using the formula $A = \frac{1}{2}\int_\alpha^\beta r^2\,d\theta$ to find the area of the region enclosed by a polar curve $r = f(\theta)$ between $\theta = \alpha$ and $\theta = \beta$, we must be certain that $\alpha \le \beta$ and that the radial line segment $\overline{OP}$, where $P = (f(\theta),\theta)$, sweeps *just once* across each point in the interior of the region. For instance, if we wished to find the *total* area inside the limaçon $r = 2 - 3\sin\theta$ (Figure 5), it would be incorrect to integrate from 0 to 2π. The reason is simply that, as θ goes from 0 to 2π, the radial segment $\overline{OP}$ sweeps *twice* across all the points within the inner loop, so the area of the inner loop gets counted twice.

Often we need to find the area of a planar region enclosed by the graphs of two polar equations such as $r = f(\theta)$ and $r = g(\theta)$ between two successive points of intersection P_1 and P_2, where $P_1 = (r_1,\alpha)$ and $P_2 = (r_2,\beta)$ (Figure 7). If the region enclosed by the curve $r = g(\theta)$ between P_1 and P_2 is contained in the region enclosed by the curve $r = f(\theta)$ between P_1 and P_2, then the desired area A is just the difference of the areas of the two regions and is given by

$$A = \frac{1}{2}\int_\alpha^\beta [f(\theta)]^2\,d\theta - \frac{1}{2}\int_\alpha^\beta [g(\theta)]^2\,d\theta = \frac{1}{2}\int_\alpha^\beta \left[[f(\theta)]^2 - [g(\theta)]^2\right]\,d\theta.$$

Figure 7

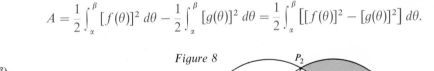

Figure 8

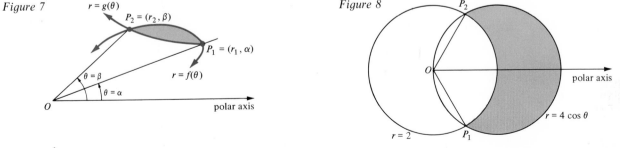

EXAMPLE   Find the area of the region inside the circle $r = 4\cos\theta$ but outside the circle $r = 2$ (Figure 8).

SOLUTION

The two circles intersect at $P_1 = (2, -\pi/3)$ and $P_2 = (2, \pi/3)$. The area A of the region outside the circle $r = 2$ and inside the circle $r = 4 \cos \theta$ between $\theta = -\pi/3$ and $\theta = \pi/3$ is given by

$$A = \frac{1}{2} \int_{-\pi/3}^{\pi/3} [(4 \cos \theta)^2 - 2^2] \, d\theta$$

$$= \frac{1}{2} \int_{-\pi/3}^{\pi/3} (16 \cos^2 \theta - 4) \, d\theta = \frac{1}{2} \int_{-\pi/3}^{\pi/3} \left[\frac{16}{2} (1 + \cos 2\theta) - 4 \right] d\theta$$

$$= \frac{1}{2} \int_{-\pi/3}^{\pi/3} (4 + 8 \cos 2\theta) \, d\theta = 2(\theta + \sin 2\theta) \Big|_{-\pi/3}^{\pi/3}$$

$$= 2 \left[\left(\frac{\pi}{3} + \frac{\sqrt{3}}{2} \right) - \left(-\frac{\pi}{3} - \frac{\sqrt{3}}{2} \right) \right] = \frac{4\pi}{3} + 2\sqrt{3} \text{ square units.}$$

4.2 Arc Length of a Polar Curve

For a curve in cartesian coordinates, $ds = \sqrt{(dx)^2 + (dy)^2}$ gives the differential of arc length (Figure 9). If we convert to polar coordinates, we have

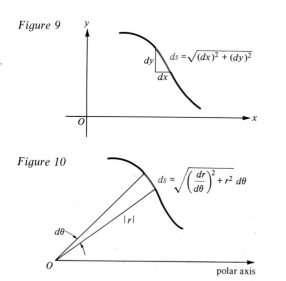

Figure 9

$$dx = d(r \cos \theta) = dr \cos \theta - r \sin \theta \, d\theta \quad \text{and}$$
$$dy = d(r \sin \theta) = dr \sin \theta + r \cos \theta \, d\theta.$$

Thus,

$$(dx)^2 + (dy)^2 = (dr)^2 \cos^2 \theta - 2r \cos \theta \sin \theta (dr)(d\theta)$$
$$+ r^2 \sin^2 \theta (d\theta)^2 + (dr)^2 \sin^2 \theta$$
$$+ 2r \cos \theta \sin \theta (dr)(d\theta)$$
$$+ r^2 \cos^2 \theta (d\theta)^2$$
$$= (dr)^2 (\cos^2 \theta + \sin^2 \theta)$$
$$+ r^2 (d\theta)^2 (\sin^2 \theta + \cos^2 \theta)$$
$$= (dr)^2 + r^2 (d\theta)^2,$$

so that

$$ds = \sqrt{(dx)^2 + (dy)^2} = \sqrt{(dr)^2 + r^2 (d\theta)^2}$$
$$= \sqrt{\left[\left(\frac{dr}{d\theta} \right)^2 + r^2 \right] (d\theta)^2}.$$

Figure 10

Therefore, in polar coordinates,

$$ds = \sqrt{\left(\frac{dr}{d\theta} \right)^2 + r^2} \, d\theta$$

(Figure 10).

The arc length of the portion of a polar curve $r = f(\theta)$ between $\theta = \alpha$ and $\theta = \beta$ is accordingly given by

$$s = \int_\alpha^\beta ds = \int_\alpha^\beta \sqrt{\left(\frac{dr}{d\theta} \right)^2 + r^2} \, d\theta = \int_\alpha^\beta \sqrt{[f'(\theta)]^2 + [f(\theta)]^2} \, d\theta,$$

provided that the derivative f' exists and is continuous on the interval $[\alpha, \beta]$.

EXAMPLE   Find the total arc length of the cardioid $r = 2(1 - \cos\theta)$ (Figure 11).

SOLUTION
Because the cardioid is symmetric about the polar axis and its extension, we have

Figure 11

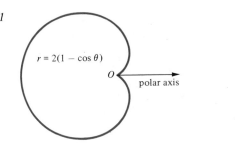

$r = 2(1 - \cos\theta)$

$$s = 2\int_0^\pi \sqrt{\left(\frac{dr}{d\theta}\right)^2 + r^2}\, d\theta$$

$$= 2\int_0^\pi \sqrt{(2\sin\theta)^2 + 4(1 - 2\cos\theta + \cos^2\theta)}\, d\theta$$

$$= 4\int_0^\pi \sqrt{\sin^2\theta + \cos^2\theta + 1 - 2\cos\theta}\, d\theta$$

$$= 4\int_0^\pi \sqrt{2 - 2\cos\theta}\, d\theta = 4\sqrt{2}\int_0^\pi \sqrt{1 - \cos\theta}\, d\theta.$$

Since $2\sin^2(\theta/2) = 1 - \cos\theta$, it follows that

$$s = 4\sqrt{2}\int_0^\pi \sqrt{2\sin^2\frac{\theta}{2}}\, d\theta = 4\sqrt{2}\int_0^\pi \sqrt{2}\sin\frac{\theta}{2}\, d\theta$$

$$= 8\left(-2\cos\frac{\theta}{2}\right)\Big|_0^\pi = 16 \text{ units.}$$

Problem Set 4

1-17, 25, 29

In problems 1 through 6, find the area enclosed by each polar curve between the indicated values of θ. Sketch the curve.

1 $r = 4\theta$ from $\theta = \pi/4$ to $\theta = 5\pi/4$.

2 $r = \sin^2(\theta/6)$ from $\theta = 0$ to $\theta = \pi$.

3 $r = 4/\theta$ from $\theta = \pi/4$ to $\theta = 5\pi/4$.

4 $r = 3\sin 3\theta$ from $\theta = 0$ to $\theta = \pi/3$.

5 $r = 3\csc\theta$ from $\theta = \pi/2$ to $\theta = 5\pi/6$.

6 $r = 2\sec^2\theta$ from $\theta = 0$ to $\theta = \pi/3$.

In problems 7 through 14, find the area enclosed by each polar curve. In each case sketch the curve, determine the appropriate limits of integration, and make certain that none of the area is being counted twice.

7 $r = 4\sin\theta$

8 $r = 2(1 + \cos\theta)$

9 $r = 2\sin 3\theta$

10 $r = 2\sqrt{|\cos\theta|}$

11 $r = 2 + \cos\theta$

12 $r = 1 + 2\cos\theta$

13 $r = 2 - 2\sin\theta$

14 $r = 3\cos\theta + 4\sin\theta$

In problems 15 through 19, find all points of intersection of the two curves, sketch the two curves, and find the area of each region described.

15 Inside $r = 6\sin\theta$ and outside $r = 3$.

16 Inside both $r = 1$ and $r = 1 + \cos\theta$.

17 Inside both $r = 3\cos\theta$ and $r = 1 + \cos\theta$.

18 Inside $r^2 = 8\cos 2\theta$ and outside $r = 2$.

19 Inside $r = 6\cos\theta$ and outside $r = 6(1 - \cos\theta)$.

20 Assuming that $b > a > 0$, find a formula for the area enclosed by the inner loop of the limaçon $r = a + b\sin\theta$.

21 Find the area of the shaded region enclosed by the graph of $r = \theta$ in Figure 12.

Figure 12

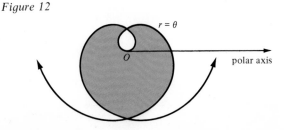

$r = \theta$

polar axis

In problems 22 through 29, find the arc length of each polar curve between the indicated values of θ.

22 $r = -2$ from $\theta = 0$ to $\theta = 2\pi$.

23 $r = 6 \sin \theta$ from $\theta = 0$ to $\theta = \pi$.

24 $r = 2(1 + \cos \theta)$ from $\theta = 0$ to $\theta = 2\pi$.

25 $r = 4\theta^2$ from $\theta = 0$ to $\theta = \frac{3}{2}$.

26 $r = 2 \sin^3 (\theta/3)$ from $\theta = 0$ to $\theta = 3\pi$.

27 $r = e^\theta$ from $\theta = 0$ to $\theta = 4\pi$.

28 $r = -7 \csc \theta$ from $\theta = \pi/4$ to $\theta = 3\pi/4$.

29 $r = 3 \cos \theta + 4 \sin \theta$ from $\theta = 0$ to $\theta = \pi/2$.

30 Criticize the following naive use of differentials: In Figure 10 ds is virtually the arc length of the portion of the circumference of a circle of radius $|r|$ cut off by the angle $d\theta$ radians; therefore, ds should be given by $|r| \, d\theta$ rather than by $\sqrt{(dr/d\theta)^2 + r^2} \, d\theta$.

31 Find the error in the following argument: Since the graph of the polar equation $r = 4 \cos \theta$ is a circle of radius 2, the circumference of this circle should be given by

$$s = \int_0^{2\pi} \sqrt{\left(\frac{dr}{d\theta}\right)^2 + r^2} \, d\theta = \int_0^{2\pi} \sqrt{16 \sin^2 \theta + 16 \cos^2 \theta} \, d\theta = \int_0^{2\pi} 4 \, d\theta = 8\pi.$$

(But the circumference of a circle of radius 2 should be only 4π.)

32 Show that, with suitable restrictions, the area of the surface generated by revolving the portion of the polar curve $r = f(\theta)$ between $\theta = \alpha$ and $\theta = \beta$ about

(a) the polar axis is $A = 2\pi \int_\alpha^\beta r \sin \theta \sqrt{(dr/d\theta)^2 + r^2} \, d\theta$, and

(b) the axis $\theta = \pi/2$ is $A = 2\pi \int_\alpha^\beta r \cos \theta \sqrt{(dr/d\theta)^2 + r^2} \, d\theta$.

In problems 33 through 36, use the formulas given in problem 32 to find the area of the surface generated by revolving each polar curve about the given axis.

33 $r = 2$ about the polar axis.

34 $r = 4$ about the axis $\theta = \pi/2$.

35 $r = 5 \cos \theta$ about the polar axis.

36 $r^2 = 4 \cos 2\theta$ about the axis $\theta = \pi/2$.

5   Rotation of Axes

In Chapter 4 we concentrated on conics whose axes of symmetry were parallel to the coordinate axes, and we found that the equations of such conics could be "simplified" and brought into standard form by a suitable translation of the cartesian coordinate axes. Notice that the translation equations

$$\begin{cases} \bar{x} = x - h \\ \bar{y} = y - k \end{cases}$$

are especially simple in the cartesian coordinate system, whereas the conversion of these equations to polar coordinates would introduce some complication. The cartesian coordinate system is naturally adapted to translation—the polar coordinate system is not.

On the other hand, the polar coordinate system is naturally adapted to *rotation* about the pole. Figure 1 shows an "old" polar axis, a "new" polar axis obtained by rotating the old polar axis

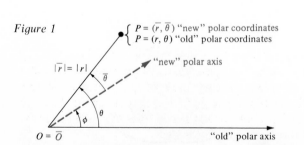

Figure 1

$P = (\bar{r}, \bar{\theta})$ "new" polar coordinates
$P = (r, \theta)$ "old" polar coordinates

"new" polar axis

$|\bar{r}| = |r|$

$\bar{\theta}$

ϕ θ

$O = \bar{O}$ "old" polar axis

counterclockwise about the pole through the angle ϕ, and a point P in the plane. Evidently, if $P = (r, \theta)$ in the old polar coordinate system, then we have $P = (r, \theta - \phi) = (\bar{r}, \bar{\theta})$ in the new polar coordinate system. Thus, for polar coordinates, we have the very simple rotation equations

$$\begin{cases} \bar{r} = r \\ \bar{\theta} = \theta - \phi. \end{cases}$$

which give the new polar coordinates $(\bar{r}, \bar{\theta})$ of the point P whose old polar coordinates are (r, θ).

EXAMPLE The equation of the circle of radius $a > 0$ with center at the point $(a, 0)$ on the old horizontal polar axis is $r = 2a \cos \theta$ (Figure 2). Find the equation of the same circle with respect to a new polar axis making the angle $\phi = \pi/2$ with the old polar axis.

SOLUTION
Using the polar rotation equations

Figure 2

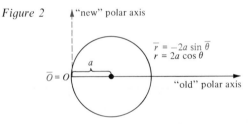

$\bar{r} = -2a \sin \bar{\theta}$
$r = 2a \cos \theta$

$\bar{O} = O$ "old" polar axis

$$\begin{cases} \bar{r} = r \\ \bar{\theta} = \theta - \phi \end{cases} \quad \text{or} \quad \begin{cases} r = \bar{r} \\ \theta = \bar{\theta} + \phi, \end{cases}$$

we rewrite $r = 2a \cos \theta$ as $\bar{r} = 2a \cos (\bar{\theta} + \phi)$. Putting $\phi = \pi/2$ and noticing that

$$\cos (\bar{\theta} + \phi) = \cos (\bar{\theta} + \pi/2) = -\sin \bar{\theta},$$

we see that the equation in the new polar coordinate system is $\bar{r} = -2a \sin \bar{\theta}$.

5.1 Rotation of Cartesian Axes

The rotation equations for *cartesian* coordinates are not quite so simple as the rotation equations for polar coordinates developed above. Nevertheless, it is often necessary to rotate cartesian coordinates, and we give the appropriate equations for such a rotation in the following theorem.

THEOREM 1 **Cartesian Rotation Equations**
Suppose that the new $\bar{x}\bar{y}$ coordinate system has the same origin as the old xy coordinate system, but that the new $\bar{x}$ axis is obtained by rotating the old x axis counterclockwise about the origin through an angle ϕ (Figure 3). Let the point P have old cartesian coordinates (x, y) and new cartesian coordinates $(\bar{x}, \bar{y})$. Then

Figure 3

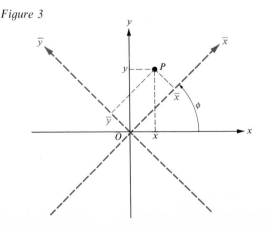

$$\begin{cases} x = \bar{x} \cos \phi - \bar{y} \sin \phi, \\ y = \bar{x} \sin \phi + \bar{y} \cos \phi. \end{cases}$$

The proof of Theorem 1 is easily accomplished by converting from cartesian coordinates to polar coordinates, rotating the polar coordinate system through the angle ϕ, and then converting back to cartesian coordinates; hence, it is left as an exercise for the reader (Problem 28).

EXAMPLES 1 The old xy coordinate system is rotated through $\pi/6$ radian to obtain a new $\bar{x}\bar{y}$ coordinate system (Figure 4). Find the old xy coordinates of the point P whose new $\bar{x}\bar{y}$ coordinates are $(2, 1)$.

SOLUTION
Using the cartesian rotation equations with $\phi = \pi/6$, we obtain

Figure 4

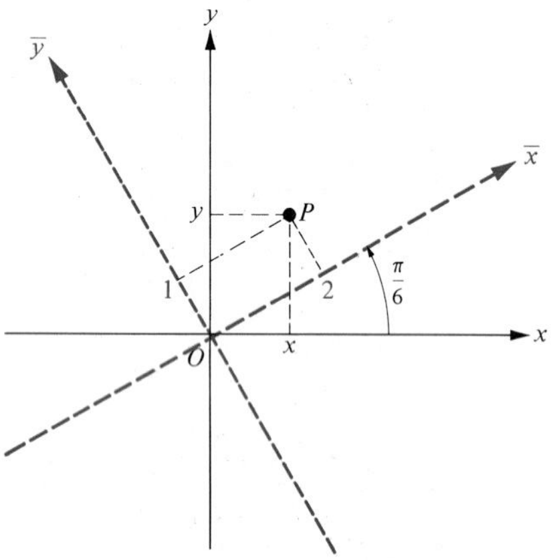

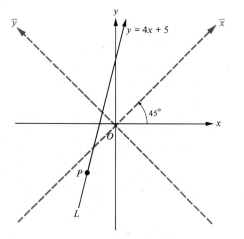

$$x = \bar{x} \cos \phi - \bar{y} \sin \phi$$

$$= 2\left(\frac{\sqrt{3}}{2}\right) - 1\left(\frac{1}{2}\right) = \sqrt{3} - \frac{1}{2}$$

and

$$y = \bar{x} \sin \phi + \bar{y} \cos \phi$$

$$= 2\left(\frac{1}{2}\right) + 1\left(\frac{\sqrt{3}}{2}\right) = 1 + \frac{\sqrt{3}}{2}.$$

Thus, the old xy coordinates of P are

$$\left(\sqrt{3} - \frac{1}{2}, 1 + \frac{\sqrt{3}}{2}\right).$$

2 A straight line L has the equation $L: y = 4x + 5$ in the old xy coordinate system. By rotation about the origin, a new $\bar{x}\bar{y}$ coordinate system is established whose axes make an angle of $45°$ with the corresponding xy axes (Figure 5). Find the equation of L in the new coordinate system.

Figure 5

SOLUTION
Since

$$x = \bar{x} \cos 45° - \bar{y} \sin 45° \quad \text{and}$$

$$y = \bar{x} \sin 45° + \bar{y} \cos 45°,$$

it follows that

$$x = \bar{x}\frac{\sqrt{2}}{2} - \bar{y}\frac{\sqrt{2}}{2} = \frac{\sqrt{2}}{2}(\bar{x} - \bar{y}) \quad \text{and}$$

$$y = \bar{x}\frac{\sqrt{2}}{2} + \bar{y}\frac{\sqrt{2}}{2} = \frac{\sqrt{2}}{2}(\bar{x} + \bar{y}).$$

A point P on the straight line L has old coordinates satisfying $y = 4x + 5$; hence, its new $\bar{x}\bar{y}$ coordinates will satisfy

$$\frac{\sqrt{2}}{2}(\bar{x} + \bar{y}) = 4\left[\frac{\sqrt{2}}{2}(\bar{x} - \bar{y})\right] + 5.$$

Simplifying the latter equation, we obtain $\bar{y} = \frac{3}{5}\bar{x} + \sqrt{2}$.

The cartesian rotation equations

$$\begin{cases} x = \bar{x} \cos \phi - \bar{y} \sin \phi, \\ y = \bar{x} \sin \phi + \bar{y} \cos \phi \end{cases}$$

can be solved for $\bar{x}$ and $\bar{y}$ in terms of x and y to obtain

$$\begin{cases} \bar{x} = x \cos \phi + y \sin \phi, \\ \bar{y} = -x \sin \phi + y \cos \phi \end{cases}$$

(see Problem 29). The latter two equations enable us to find the new $\bar{x}\bar{y}$ coordinates in terms of the old xy coordinates and the rotation angle ϕ.

EXAMPLE   The new $\bar{x}\bar{y}$ coordinate system is obtained by rotating the old xy coordinate system through $-30°$. Find the new $\bar{x}\bar{y}$ coordinates of the points whose old xy coordinates are given. (a) $(\sqrt{3}, 2)$ (b) $(-\sqrt{3}, 2)$

SOLUTION
Here we have

$$\bar{x} = x \cos(-30°) + y \sin(-30°) = x \frac{\sqrt{3}}{2} + y(-\tfrac{1}{2}) = \frac{\sqrt{3}\,x - y}{2} \quad \text{and}$$

$$\bar{y} = -x \sin(-30°) + y \cos(-30°) = -x(-\tfrac{1}{2}) + y\left(\frac{\sqrt{3}}{2}\right) = \frac{x + \sqrt{3}\,y}{2}.$$

Thus:

(a) $\bar{x} = \dfrac{\sqrt{3}\sqrt{3} - 2}{2} = \dfrac{1}{2}$ and $\bar{y} = \dfrac{\sqrt{3} + \sqrt{3}\,(2)}{2} = \dfrac{3\sqrt{3}}{2}$, so $(\bar{x}, \bar{y}) = \left(\dfrac{1}{2}, \dfrac{3\sqrt{3}}{2}\right)$.

(b) $\bar{x} = \dfrac{\sqrt{3}\,(-\sqrt{3}) - 2}{2} = -\dfrac{5}{2}$ and $\bar{y} = \dfrac{-\sqrt{3} + \sqrt{3}\,(2)}{2} = \dfrac{\sqrt{3}}{2}$,

so $(\bar{x}, \bar{y}) = \left(-\dfrac{5}{2}, \dfrac{\sqrt{3}}{2}\right)$.

Figure 6

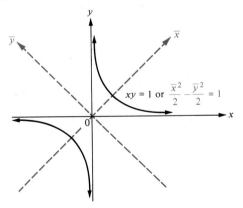

$xy = 1$ or $\dfrac{\bar{x}^2}{2} - \dfrac{\bar{y}^2}{2} = 1$

By a suitable rotation of coordinate axes, the equation of a curve can often be "simplified" or brought into recognizable form. For instance, consider the graph of the equation $xy = 1$ (Figure 6). This graph has two branches, one in the first quadrant and the other in the third quadrant. It looks suspiciously like a hyperbola whose transverse axis makes an angle of $45°$ with the x axis. To see if this is indeed the case, establish a new $\bar{x}\bar{y}$ coordinate system at the angle $\phi = 45°$ to the old xy coordinate system. Substituting

$$x = \bar{x} \cos 45° - \bar{y} \sin 45° = \frac{\sqrt{2}}{2}(\bar{x} - \bar{y}) \quad \text{and}$$

$$y = \bar{x} \sin 45° + \bar{y} \cos 45° = \frac{\sqrt{2}}{2}(\bar{x} + \bar{y})$$

into the equation $xy = 1$, we obtain

$$\frac{\sqrt{2}}{2}(\bar{x} - \bar{y}) \cdot \frac{\sqrt{2}}{2}(\bar{x} + \bar{y}) = 1 \quad \text{or} \quad \frac{\bar{x}^2}{2} - \frac{\bar{y}^2}{2} = 1.$$

Thus, just as we suspected, the curve in Figure 6 is a hyperbola, since its equation has the standard form in the new $\bar{x}\bar{y}$ coordinate system.

Problem Set 5

In problems 1 through 10, the "old" horizontal polar axis is rotated through the indicated angle ϕ to obtain a "new" polar axis. Find the new equation in terms of $\bar{r}$ and $\bar{\theta}$ of the polar curve whose old polar equation is given.

1 $r = 5;\ \phi = \pi/2$

2 $\theta = \pi/3;\ \phi = -\pi/6$

3 $r = 4\cos\theta;\ \phi = -\dfrac{\pi}{2}$

4 $r = \dfrac{1}{1 - \cos\theta};\ \phi = \pi$

5 $r = 3 + 5\sin\theta;\ \phi = \pi$

6 $r = 0;\ \phi = \pi/3$

7 $r = 3 - 5\sin\theta;\ \phi = -\dfrac{\pi}{2}$

8 $r = \dfrac{ed}{1 + e\cos\theta};\ \phi = \dfrac{\pi}{2}$

9 $r^2 = 25\cos 2\theta;\ \phi = \dfrac{\pi}{2}$

10 $r = \dfrac{-C}{A\cos\theta + B\sin\theta};\ \phi$ arbitrary

In problems 11 through 19, the old xy axes have been rotated through the angle ϕ to form a new $\bar{x}\bar{y}$ coordinate system. The point P has coordinates (x, y) in the old xy system and it has coordinates $(\bar{x}, \bar{y})$ in the new $\bar{x}\bar{y}$ system. Find the missing information.

11 $(x, y) = (4, -7),\ \phi = 90°,\ (\bar{x}, \bar{y}) = ?$

12 $(x, y) = (2, 0),\ (\bar{x}, \bar{y}) = (1, \sqrt{3}),\ \phi = ?$

13 $(\bar{x}, \bar{y}) = (-3, -3),\ \phi = \pi/3$ radians, $(x, y) = ?$

14 $(x, y) = (5\sqrt{2}, \sqrt{2}),\ \phi = 45°,\ (\bar{x}, \bar{y}) = ?$

15 $(\bar{x}, \bar{y}) = (-4, -2),\ \phi = 30°,\ (x, y) = ?$

16 $(\bar{x}, \bar{y}) = (-3\sqrt{2}, \sqrt{2}),\ \phi = 3\pi/4$ radians, $(x, y) = ?$

17 $(x, y) = (-4, 0),\ \phi = \pi$ radians, $(\bar{x}, \bar{y}) = ?$

18 $(x, y) = (1, -7),\ \phi = 240°,\ (\bar{x}, \bar{y}) = ?$

19 $(x, y) = (0, 8),\ \phi = 360°,\ (\bar{x}, \bar{y}) = ?$

20 Find an angle ϕ (if one exists) for which the rotation equations give each of the following:

(a) $\bar{x} = y$ and $\bar{y} = -x$ (b) $\bar{x} = -x$ and $\bar{y} = -y$ (c) $\bar{x} = y$ and $\bar{y} = x$ (d) $\bar{x} = -y$ and $\bar{y} = -x$

In problems 21 through 26, the xy axes are rotated 30° to form the $\bar{x}\bar{y}$ axes. Express each equation in the other coordinate system.

21 $y^2 = 3x$

22 $\bar{y} = 3\bar{x}$

23 $\bar{x}^2 + \bar{y}^2 = 1$

24 $5x - y = 4$

25 $x^2 + y^2 = 1$

26 $x^2 = 25$

27 "Simplify" the equation $x^2 - 2xy + y^2 - \sqrt{2}\,x - \sqrt{2}\,y = 0$ by rotating the coordinate system through the angle $\phi = -\pi/4$.

28 Derive the equations in Theorem 1 by converting to polar coordinates, rotating the polar axis, then reconverting to cartesian coordinates.

29 Solve the equations in Theorem 1 for $\bar{x}$ and $\bar{y}$ in terms of x and y.

6   General Second-Degree Equation and Rotation Invariants

At the end of the previous section we showed that the graph of the equation $xy = 1$ is a hyperbola; in fact, this equation was transformed into the standard equation of a hyperbola by a rotation of the coordinate axes through 45°. The equation $xy = 1$ is a special case of the *general second-degree equation in x and y,*

$$Ax^2 + Bxy + Cy^2 + Dx + Ey + F = 0.$$

In the general second-degree equation, the coefficients A, B, C, D, E, and F are understood to be constants. Notice that by taking $A = C = D = E = 0$, $B = 1$, and $F = -1$ in the general equation, we obtain $xy - 1 = 0$, or $xy = 1$. Similarly, by taking $A = 1/a^2$, $B = 0$, $C = 1/b^2$, $D = 0$, $E = 0$, and $F = -1$, we obtain $\frac{x^2}{a^2} + \frac{y^2}{b^2} = 1$, the standard form for the equation of an ellipse. Also, if we set $A = 1/a^2$, $B = 0$, $C = -1/b^2$, $D = 0$, $E = 0$, and $F = -1$, then the general second-degree equation becomes $\frac{x^2}{a^2} - \frac{y^2}{b^2} = 1$, which is the standard form for the equation of a hyperbola. Finally, by putting $A = 1$, $B = 0$, $C = 0$, $D = 0$, $E = -4p$, and $F = 0$, we obtain $x^2 = 4py$, the equation of a parabola.

Every conic section considered until now has had a cartesian equation which was a special case of the general second-degree equation

$$Ax^2 + Bxy + Cy^2 + Dx + Ey + F = 0;$$

however, except for the hyperbola $xy = 1$, the coefficient B has always been zero and so the "mixed" term Bxy did not show up. We now prove that the mixed term can always be removed from the general second-degree equation by a suitable rotation of the coordinate axes.

THEOREM 1   **Removal of the Mixed Term by Rotation**
If the "old" xy coordinate system is rotated about the origin through the angle ϕ to obtain a "new" $\bar{x}\bar{y}$ coordinate system, then the curve whose old equation with respect to the xy axes was $Ax^2 + Bxy + Cy^2 + Dx + Ey + F = 0$ will have a new equation $\bar{A}\bar{x}^2 + \bar{B}\bar{x}\bar{y} + \bar{C}\bar{y}^2 + \bar{D}\bar{x} + \bar{E}\bar{y} + \bar{F} = 0$ with respect to the $\bar{x}\bar{y}$ axes, where

$$\bar{A} = A \cos^2 \phi + B \cos \phi \sin \phi + C \sin^2 \phi,$$

$$\bar{B} = 2(C - A) \cos \phi \sin \phi + B(\cos^2 \phi - \sin^2 \phi),$$

$$\bar{C} = A \sin^2 \phi - B \cos \phi \sin \phi + C \cos^2 \phi,$$

$$\bar{D} = D \cos \phi + E \sin \phi,$$

$$\bar{E} = -D \sin \phi + E \cos \phi, \quad \text{and}$$

$$\bar{F} = F.$$

In particular, if $B \neq 0$ and if ϕ is chosen so that $0 < \phi < \pi/2$ and $\cot 2\phi = \dfrac{A - C}{B}$, then $\bar{B} = 0$ and the mixed term $\bar{B}\bar{x}\bar{y}$ will not appear in the new equation.

PROOF
We substitute $x = \bar{x} \cos \phi - \bar{y} \sin \phi$ and $y = \bar{x} \sin \phi + \bar{y} \cos \phi$ from the rotation equations into $Ax^2 + Bxy + Cy^2 + Dx + Ey + F = 0$ to obtain

$$A(\bar{x}^2 \cos^2 \phi - 2\bar{x}\bar{y} \cos \phi \sin \phi + \bar{y}^2 \sin^2 \phi) + B(\bar{x}^2 \cos \phi \sin \phi + \bar{x}\bar{y} \cos^2 \phi$$
$$- \bar{x}\bar{y} \sin^2 \phi - \bar{y}^2 \cos \phi \sin \phi) + C(\bar{x}^2 \sin^2 \phi + 2\bar{x}\bar{y} \cos \phi \sin \phi + \bar{y}^2 \cos^2 \phi)$$
$$+ D(\bar{x} \cos \phi - \bar{y} \sin \phi) + E(\bar{x} \sin \phi + \bar{y} \cos \phi) + F = 0.$$

Collecting similar terms in the latter equation, we obtain

$$\bar{A}\bar{x}^2 + \bar{B}\bar{x}\bar{y} + \bar{C}\bar{y}^2 + \bar{D}\bar{x} + \bar{E}\bar{y} + \bar{F} = 0,$$

where the coefficients $\bar{A}$, $\bar{B}$, $\bar{C}$, $\bar{D}$, $\bar{E}$, and $\bar{F}$ are as shown in the statement of the theorem. Since $\sin 2\phi = 2 \cos \phi \sin \phi$ and $\cos 2\phi = \cos^2 \phi - \sin^2 \phi$, then

$\bar{B} = 2(C - A) \cos \phi \sin \phi + B(\cos^2 \phi - \sin^2 \phi)$ can also be written in the form $\bar{B} = (C - A) \sin 2\phi + B \cos 2\phi$. Thus, if $B \neq 0$ and if ϕ is chosen so that $0 < \phi < \pi/2$ and $\cot 2\phi = \dfrac{A - C}{B}$, then $\dfrac{\cos 2\phi}{\sin 2\phi} = \dfrac{A - C}{B}$, so that $B \cos 2\phi = (A - C) \sin 2\phi$ and $\bar{B} = (C - A) \sin 2\phi + B \cos 2\phi = (C - A) \sin 2\phi + (A - C) \sin 2\phi = 0$.

In applying Theorem 1 to remove the mixed term from a general second-degree equation, it is not necessary to use the given formulas to obtain the new coefficients, since one can always substitute the rotation equations into the given second-degree equation and collect terms.

EXAMPLE   Rotate the coordinate axes to remove the mixed term from the equation $x^2 - 4xy + y^2 - 6 = 0$ and sketch the graph showing both the old xy and the new $\bar{x}\bar{y}$ coordinate system.

SOLUTION
The equation has the form $Ax^2 + Bxy + Cy^2 + Dx + Ey + F = 0$ with $A = 1$, $B = -4$, $C = 1$, $D = 0$, $E = 0$, and $F = -6$. According to Theorem 1, the mixed term can be removed by rotating the coordinate system through the angle ϕ, where

$$0 < \phi < \pi/2 \quad \text{and} \quad \cot 2\phi = \frac{A - C}{B} = \frac{1 - 1}{-4} = 0.$$ Hence, we take $2\phi = \pi/2$, so that $\phi = \pi/4$, $\sin \phi = \sqrt{2}/2$, and $\cos \phi = \sqrt{2}/2$. Then, by the rotation equations, $x = \bar{x} \cos \phi - \bar{y} \sin \phi = \dfrac{\sqrt{2}}{2}(\bar{x} - \bar{y})$ and $y = \bar{x} \sin \phi + \bar{y} \cos \phi = \dfrac{\sqrt{2}}{2}(\bar{x} + \bar{y})$. Substituting these expressions for x and y into the equation $x^2 - 4xy + y^2 - 6 = 0$, we have

$$\left[\frac{\sqrt{2}}{2}(\bar{x} - \bar{y})\right]^2 - 4\left[\frac{\sqrt{2}}{2}(\bar{x} - \bar{y})\right]\left[\frac{\sqrt{2}}{2}(\bar{x} + \bar{y})\right] + \left[\frac{\sqrt{2}}{2}(\bar{x} + \bar{y})\right]^2 - 6 = 0.$$

The latter equation simplifies to

Figure 1

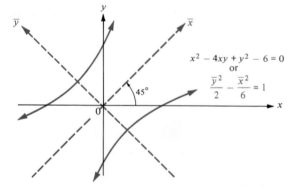

$$\frac{1}{2}(\bar{x} - \bar{y})^2 - 2(\bar{x} - \bar{y})(\bar{x} + \bar{y}) + \frac{1}{2}(\bar{x} + \bar{y})^2 = 6$$

or

$$\frac{1}{2}(\bar{x}^2 - 2\bar{x}\bar{y} + \bar{y}^2) - 2(\bar{x}^2 - \bar{y}^2)$$

$$+ \frac{1}{2}(\bar{x}^2 + 2\bar{x}\bar{y} + \bar{y}^2) = 6.$$

Collecting terms in the equation above, we get $-\bar{x}^2 + 3\bar{y}^2 = 6$, or $\dfrac{\bar{y}^2}{2} - \dfrac{\bar{x}^2}{6} = 1$, a hyperbola (Figure 1).

If $0 < \phi < \pi/2$, then the trigonometric identities

$$\cos 2\phi = \frac{\cot 2\phi}{\sqrt{\cot^2 2\phi + 1}}, \quad \cos \phi = \sqrt{\frac{1 + \cos 2\phi}{2}}, \quad \text{and} \quad \sin \phi = \sqrt{\frac{1 - \cos 2\phi}{2}}$$

permit us to find $\cos \phi$ and $\sin \phi$ algebraically in terms of the value of $\cot 2\phi$.

This is useful in applying Theorem 1 to remove the mixed term from a second-degree equation.

EXAMPLES Rotate the coordinate system to remove the mixed term and sketch the graph showing both the old xy and the new $\bar{x}\bar{y}$ axes.

1 $8x^2 - 4xy + 5y^2 = 144$

SOLUTION
Here $A = 8$, $B = -4$, $C = 5$, $D = 0$, $E = 0$, and $F = -144$; therefore,
$\cot 2\phi = \dfrac{A - C}{B} = \dfrac{8 - 5}{-4} = -\dfrac{3}{4}$. Thus,

$$\cos 2\phi = \frac{\cot 2\phi}{\sqrt{\cot^2 2\phi + 1}} = \frac{-\frac{3}{4}}{\sqrt{(-\frac{3}{4})^2 + 1}} = \frac{-\frac{3}{4}}{\sqrt{\frac{9}{16} + 1}} = \frac{-\frac{3}{4}}{\frac{5}{4}} = -\frac{3}{5},$$

so that

$$\cos \phi = \sqrt{\frac{1 + \cos 2\phi}{2}} = \sqrt{\frac{1 - \frac{3}{5}}{2}} = \sqrt{\frac{2}{10}} = \sqrt{\frac{1}{5}} = \frac{\sqrt{5}}{5},$$

and

$$\sin \phi = \sqrt{\frac{1 - \cos 2\phi}{2}} = \sqrt{\frac{1 + \frac{3}{5}}{2}} = \sqrt{\frac{8}{10}} = \sqrt{\frac{4}{5}} = \frac{2\sqrt{5}}{5}.$$

Substituting

$$x = \frac{\sqrt{5}}{5}\,\bar{x} - \frac{2\sqrt{5}}{5}\,\bar{y} \quad \text{and} \quad y = \frac{2\sqrt{5}}{5}\,\bar{x} + \frac{\sqrt{5}}{5}\,\bar{y}$$

into the given equation, we obtain

$$8\left(\frac{\sqrt{5}}{5}\,\bar{x} - \frac{2\sqrt{5}}{5}\,\bar{y}\right)^2 - 4\left(\frac{\sqrt{5}}{5}\,\bar{x} - \frac{2\sqrt{5}}{5}\,\bar{y}\right)\left(\frac{2\sqrt{5}}{5}\,\bar{x} + \frac{\sqrt{5}}{5}\,\bar{y}\right) + 5\left(\frac{2\sqrt{5}}{5}\,\bar{x} + \frac{\sqrt{5}}{5}\,\bar{y}\right)^2 = 144.$$

This equation simplifies to $4\bar{x}^2 + 9\bar{y}^2 = 144$, or $\dfrac{\bar{x}^2}{36} + \dfrac{\bar{y}^2}{16} = 1$, which is an ellipse. Since $\sin \phi = \dfrac{2\sqrt{5}}{5}$, it follows that $\phi \approx 63.43°$ (Figure 2).

Figure 2

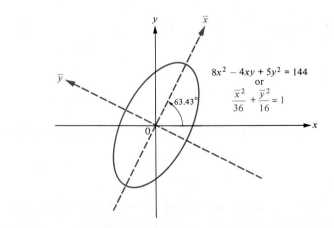

$$8x^2 - 4xy + 5y^2 = 144$$
$$\text{or}$$
$$\frac{\bar{x}^2}{36} + \frac{\bar{y}^2}{16} = 1$$

2 $16x^2 - 24xy + 9y^2 - 80x - 190y + 425 = 0$

SOLUTION
Here $A = 16$, $B = -24$, $C = 9$, $D = -80$, $E = -190$, and $F = 425$;

$$\cot 2\phi = \frac{A - C}{B} = -\frac{7}{24}, \qquad \cos 2\phi = \frac{-\frac{7}{24}}{\sqrt{\left(-\frac{7}{24}\right)^2 + 1}} = -\frac{7}{25},$$

$$\cos \phi = \sqrt{\frac{1 + \left(-\frac{7}{25}\right)}{2}} = \frac{3}{5}, \qquad \text{and} \qquad \sin \phi = \sqrt{\frac{1 - \left(-\frac{7}{25}\right)}{2}} = \frac{4}{5}.$$

Figure 3

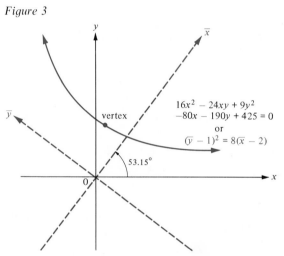

vertex

$16x^2 - 24xy + 9y^2$
$-80x - 190y + 425 = 0$
or
$(\bar{y} - 1)^2 = 8(\bar{x} - 2)$

53.15°

By direct substitution in the equation of Theorem 1, we have $\bar{A} = A(\frac{3}{5})^2 + B(\frac{3}{5})(\frac{4}{5}) + C(\frac{4}{5})^2 = 0$, $\bar{B} = 0$, $\bar{C} = A(\frac{4}{5})^2 - B(\frac{3}{5})(\frac{4}{5}) + C(\frac{3}{5})^2 = 25$, $\bar{D} = D(\frac{3}{5}) + E(\frac{4}{5}) = -200$, $\bar{E} = -D(\frac{4}{5}) + E(\frac{3}{5}) = -50$, $\bar{F} = F = 425$. Thus, the equation relative to the new $\bar{x}\bar{y}$ coordinate system is $25\bar{y}^2 - 200\bar{x} - 50\bar{y} + 425 = 0$; that is, $\bar{y}^2 - 8\bar{x} - 2\bar{y} + 17 = 0$. Now, we complete the square to get

$$\bar{y}^2 - 2\bar{y} + 1 = 8\bar{x} - 17 + 1 \quad \text{or} \quad (\bar{y} - 1)^2 = 8(\bar{x} - 2).$$

This is a parabola with vertex at $(2, 1)$ in the new coordinate system. Since $\sin \phi = \frac{4}{5}$, it follows that $\phi \approx 53.13°$ (Figure 3).

Although a rotation of the coordinate axes through the angle ϕ changes the equation $Ax^2 + Bxy + Cy^2 + Dx + Ey + F = 0$ into the equation

$$\bar{A}\bar{x}^2 + \bar{B}\bar{x}\bar{y} + \bar{C}\bar{y}^2 + \bar{D}\bar{x} + \bar{E}\bar{y} + \bar{F} = 0,$$

both of these equations have precisely the same graph. This graph, as well as any other entity that remains unchanged as a consequence of the rotation, is called a *rotation invariant*. An *algebraic rotation invariant*, in particular, is defined to be any expression involving the coefficients A, B, C, D, E, and F which remains unchanged as the axes are rotated. For instance, according to Theorem 1, $\bar{F} = F$; hence, F is an algebraic rotation invariant. The following theorem, whose proof is left as an exercise (Problem 14), gives two additional algebraic rotation invariants.

THEOREM 2   Rotation Invariants
When the coordinate axes are rotated through an angle ϕ, causing the second-degree equation $Ax^2 + Bxy + Cy^2 + Dx + Ey + F = 0$ to transform into $\bar{A}\bar{x}^2 + \bar{B}\bar{x}\bar{y} + \bar{C}\bar{y}^2 + \bar{D}\bar{x} + \bar{E}\bar{y} + \bar{F} = 0$ as in Theorem 1. then the quantities F, $A + C$, and $B^2 - 4AC$ are invariant in the sense that:

1 $F = \bar{F}$.
2 $A + C = \bar{A} + \bar{C}$.
3 $B^2 - 4AC = \bar{B}^2 - 4\bar{A}\bar{C}$.

The algebraic rotation invariants in Theorem 2 may be used as a check against numerical errors in performing a rotation of axes for a second-degree equation. For instance, to check the numerical calculation in Example 2, we have $A = 16$, $B = -24$, $C = 9$, $\bar{A} = 0$, $\bar{B} = 0$, and $\bar{C} = 25$, so that

$$A + C = \bar{A} + \bar{C} = 25 \quad \text{and} \quad B^2 - 4AC = \bar{B}^2 - 4\bar{A}\bar{C} = 0.$$

We now have the means to sketch the graph of any second-degree equation in x and y. If the equation contains a mixed term, then a suitable rotation of the coordinate axes will remove it. Thus, by completing the squares if necessary, we shall *usually* obtain the standard form for the equation of a circle, an ellipse, a parabola, or a hyperbola. However, there are some cases in which this does not happen (Problem 12). In these exceptional cases, the graph is called a degenerate conic. The only possible degenerate conics are (1) the whole xy plane, (2) the empty set, (3) a straight line, (4) two straight lines, and (5) a single point.

Assuming that the graph of a given second-degree equation in x and y is a nondegenerate conic, the conic can be identified as follows:

1 The conic is a circle or an ellipse if $B^2 - 4AC < 0$.

2 The conic is a parabola if $B^2 - 4AC = 0$.

3 The conic is a hyperbola if $B^2 - 4AC > 0$.

(See Problem 10.)

EXAMPLE Identify the nondegenerate conic $2x^2 - 12xy - 3y^2 = 84$.

SOLUTION
Here $A = 2$, $B = -12$, and $C = -3$; hence, $B^2 - 4AC = 168 > 0$. Therefore, the conic is a hyperbola.

Problem Set 6

In problems 1 through 9, determine (a) the angle ϕ of rotation that removes the mixed term $(0 < \phi < \pi/2)$, (b) x and y in terms of $\bar{x}$ and $\bar{y}$, and (c) the "new" equation in terms of $\bar{x}$ and $\bar{y}$. Then sketch the graph showing both the "old" xy and the "new" $\bar{x}\bar{y}$ coordinate systems.

1 $x^2 + 4xy - 2y^2 = 12$

2 $x^2 + 2xy + y^2 + x + y = 0$

3 $x^2 + 2xy + y^2 = 1$

4 $x^2 + 2xy + y^2 - 4\sqrt{2}x + 4\sqrt{2}y = 0$

5 $9x^2 - 24xy + 16y^2 = 144$

6 $2x^2 + 4\sqrt{3}xy - 2y^2 - 4 = 0$

7 $6x^2 - 6xy + 14y^2 = 45$

8 $17x^2 - 12xy + 8y^2 - 68x + 24y - 12 = 0$

9 $2x^2 + 6xy - 6y^2 + 2\sqrt{10}x + 3\sqrt{10}y - 16 = 0$

10 Suppose that $Ax^2 + Bxy + Cy^2 + Dx + Ey + F = 0$ is the equation of a nondegenerate conic. Show that the conic is:
(i) A circle or ellipse if $B^2 - 4AC < 0$.
(ii) A parabola if $B^2 - 4AC = 0$.
(iii) A hyperbola if $B^2 - 4AC > 0$.
Also show that in case (i) the conic is a circle if $A = C$ and $B = 0$.

11 Assuming the facts given in problem 10, identify the given conics without bothering to rotate the coordinate system. (Assume nondegeneracy.)
(a) $6x^2 - 4xy + 3y^2 + x + 11y + 10 = 0$
(b) $18x^2 + 12xy + 2y^2 + x - 3y - 4 = 0$
(c) $2x^2 + 2y^2 - 8x + 12y + 1 = 0$
(d) $x^2 - 3xy - 3y^2 + 21 = 0$

12 The following second-degree equations in x and y have graphs that are degenerate conics. In each case, verify that the graph is as described.
(a) $0x^2 + 0xy + 0y^2 + 0x + 0y + 0 = 0$ (the whole xy plane).
(b) $x^2 + y^2 + 1 = 0$ (the empty set).
(c) $2x^2 - 4xy + 2y^2 = 0$ (a straight line).
(d) $4x^2 - y^2 + 16x + 2y + 15 = 0$ (two intersecting straight lines).
(e) $x^2 - 2xy + y^2 - 18 = 0$ (two parallel straight lines).
(f) $x^2 + y^2 - 6x + 4y + 13 = 0$ (a single point).

13 Sketch the graph of $2x^2 + 3xy + 2y^2 = 4$.   **14** Using the formulas in Theorem 1, prove Theorem 2.

15 In Problems 1, 3, 5, 7, and 9, use the algebraic rotation invariants in Theorem 2 to check for a possible error in your calculation.

16 Assume that $AC > 0$. Show that the graph of $Ax^2 + Cy^2 + Dx + Ey + F = 0$ is an ellipse, a circle, a single point, or the empty set.

17 Derive the equations of the following conics from the given information. Then rotate the coordinate axes so that the new $\bar{x}$ axis contains the foci of the conic. Find the equation of the conic in the new $\bar{x}\bar{y}$ coordinate system.
(a) Foci at $F_1 = (3, 1)$ and $F_2 = (-3, -1)$; distance from the center to either vertex along the major axis is $a = 4$.
(b) Foci at $F_1 = (4, 3)$ and $F_2 = (-4, -3)$; distance from the center to either vertex along the transverse axis is $a = 3$.

18 Assume that $AC < 0$. Show that the graph of $Ax^2 + Cy^2 + Dx + Ey + F = 0$ is a hyperbola or a pair of intersecting straight lines.

19 Find a second-degree equation in x and y whose graph consists of a pair of straight lines intersecting at the origin and having slopes 3 and -3, respectively.

20 Show that the graph of the equation $Ax^2 + Dx + Ey + F = 0$, where $A \neq 0$, is a parabola, a pair of parallel straight lines, a single straight line, or the empty set.

21 Sketch the graph of $x^2 + 4xy + 4y^2 = 8$.

22 Show that $2A^2 + B^2 + 2C^2$ is a rotation invariant in the sense that $2A^2 + B^2 + 2C^2 = 2\bar{A}^2 + \bar{B}^2 + 2\bar{C}^2$.

23 Sketch the graph of $x^2 - 9xy + y^2 = 12$.

24 In Theorem 1 suppose that $0 < \phi < \pi/2$ and $\cot 2\phi = \dfrac{A - C}{B}$. Prove that

$$\bar{A} = \tfrac{1}{2}[A + C + S\sqrt{(A - C)^2 + B^2}] \quad \text{and} \quad \bar{C} = \tfrac{1}{2}[A + C - S\sqrt{(A - C)^2 + B^2}],$$

where

$$S = \frac{B}{|B|}.$$

Review Problem Set

In problems 1 through 6, plot each point in the polar coordinate system, and then give three other polar representations of the same point for which (a) $r < 0$ and $0 \leq \theta < 2\pi$, (b) $r > 0$ and $-2\pi < \theta \leq 0$, and (c) $r < 0$ and $-2\pi \leq \theta < 0$.

1 $(1, \pi/3)$ **2** $(2, 5\pi/6)$ **3** $(2, 7\pi/4)$

4 $(-1, -11\pi/3)$ **5** $(3, \pi)$ **6** (π, π)

In problems 7 through 12, convert the polar coordinates to cartesian coordinates.

7 $(0, 0)$ **8** $(17, 0)$ **9** $(17, \pi)$

10 $(-3, \pi/6)$ **11** $(11, 3\pi/4)$ **12** $(-3, -\pi/3)$

In problems 13 through 18, convert the cartesian coordinates to polar coordinates (r, θ) with $r \geq 0$ and $-\pi < \theta \leq \pi$.

13 $(17, 0)$ **14** $(2, 3)$ **15** $(2, -2\sqrt{3})$

16 $(-\sqrt{3}, -1)$ **17** $(-17, 17)$ **18** $(0, -1)$

In problems 19 through 22, convert each polar equation into a cartesian equation.

19 $r^2 \cos 2\theta = 1$

20 $r = \dfrac{1}{3 \cos \theta - 4 \sin \theta}$

21 $r = \sqrt{|\sec \theta|}$

22 $r = \dfrac{ed}{1 - e \cos \theta}, \; 0 < e < 1, \; d > 0$

In problems 23 through 26, convert each cartesian equation into a polar equation.

23 $y = 2x - 1$

24 $(x - 1)^2 + (y - 3)^2 = 4$

25 $y^2 = 4x$

26 $\dfrac{x^2}{a^2} + \dfrac{y^2}{b^2} = 1$

In problems 27 through 32, sketch the graph of each polar equation. Discuss whatever symmetry the graph may possess.

27 $r = 2 \cos 3\theta$

28 $r = 2 \cos (3\theta + \pi)$

29 $r = \dfrac{2}{\cos \theta + \sqrt{3} \sin \theta}$

30 $r = \frac{1}{2} + \sin \theta$

31 $r = \dfrac{3}{1 + \sin \theta}$

32 $r = 1 + \frac{1}{2} \sin \theta$

In problems 33 through 36, find the slope of the tangent line to the polar graph at the point indicated.

33 $r = \dfrac{1}{5 \cos \theta - 3 \sin \theta}$ at $(\frac{1}{5}, 0)$

34 $r = a \sin k\theta$, $a > 0$, k a positive integer at $(0, \pi/k)$

35 $r = 2 - 3 \cos \theta$ at $(2, \pi/2)$ **36** $r^2 = -9 \cos 2\theta$ at $(3, \pi/2)$

37 Find all points where the tangent to the limaçon $r = 3 - 4 \cos \theta$ is either horizontal or vertical.

38 In problems 35 and 36, find $\tan \psi$ at the indicated points.

39 Find $\tan \psi$ in terms of θ for the curve $r = \sin^3 \theta$.

40 Let $P \neq 0$ be a point of intersection of the two polar curves $r = f(\theta)$ and $r = g(\theta)$. If ψ_1 is the value of ψ for $r = f(\theta)$ at P and ψ_2 is the value of ψ for $r = g(\theta)$ at P, show that

$$\tan \phi = \frac{\tan \psi_1 - \tan \psi_2}{1 + \tan \psi_1 \tan \psi_2},$$

where ϕ is the angle between the tangent to $r = f(\theta)$ at P and the tangent to $r = g(\theta)$ at P.

In problems 41 through 46, identify the type of conic, give its eccentricity, and specify the position of the directrix corresponding to the focus at the pole.

41 $r = \dfrac{17}{1 - \cos \theta}$

42 $r = \dfrac{15}{3 + 5 \sin \theta}$

43 $r = \dfrac{10}{5 - 2 \sin \theta}$

44 $r = \dfrac{3}{2} \csc^2 \dfrac{\theta}{2}$ **45** $r = \dfrac{1}{\frac{1}{2} + \sin \theta}$ **46** $r = \dfrac{1}{1 + \cos(\theta + \pi/4)}$

47 Find the polar coordinates of all points of intersection of the limaçon $r = 1 - 2\cos\theta$ and the parabola $r = \dfrac{2}{1 - \cos\theta}$. Sketch the graphs showing these intersection points.

In problems 48 through 51, find the area enclosed by each polar curve between the indicated values of θ.

48 $r = -\theta$ from $\theta = 0$ to $\theta = \pi/2$.   **49** $r = 4\sin\theta$ from $\theta = 0$ to $\theta = \pi$.

50 $r = \dfrac{1}{\sin\theta + \cos\theta}$ from $\theta = 0$ to $\theta = \pi/6$.

51 $r = 1 - \cos\theta$ from $\theta = 0$ to $\theta = \pi$.

52 Find the area of the region inside the outer loop but outside the inner loop of the limaçon $r = a + b\sin\theta$, $b > a > 0$.

In problems 53 through 56, find the arc length of the portion of each polar curve between the indicated values of θ.

53 $r = e^{-3\theta}$ for θ between 0 and 2π.   **54** $r = 5\sin\theta$ for θ between 0 and π.

55 $r = \cos^2(\theta/2)$ for θ between 0 and π.   **56** $r = 1 - \cos\theta$ for θ between 0 and 2π.

57 Find the equation into which the polar equation $r^2 \sin 2\theta = 2$ is transformed by rotating the polar axis through $45°$.

58 Find the equation into which the cartesian equation $\sqrt{x} + \sqrt{y} = 2$ is transformed by rotation of the axes through an angle of $45°$ and elimination of radicals. Is the graph of this equation a parabola?

In problems 59 through 62, rotate the axes to transform each equation into an equation that has no mixed term. Sketch the graph showing both the old and the new coordinate axes.

59 $3x^2 + 4\sqrt{3}\,xy - y^2 = 15$ **60** $4x^2 + 4xy + 4y^2 = 24$

61 $\sqrt{3}\,x^2 + xy = 11$ **62** $y^2 + xy - 3x = 7$

In problems 63 through 67, use the invariant $B^2 - 4AC$ to identify each conic. You may assume that the conic is nondegenerate.

63 $13x^2 + 6\sqrt{3}\,xy + 7y^2 = 16$ **64** $x^2 + 6xy + y^2 + 8 = 0$ **65** $7x^2 + 2\sqrt{3}\,xy + 5y^2 + 1 = 0$

66 $x^2 - 3xy + 5y^2 = 16$ **67** $81x^2 - 18xy + 5x + y^2 = 41$

12

INDETERMINATE FORMS, IMPROPER INTEGRALS, AND TAYLOR'S FORMULA

Since the two basic notions of calculus—the derivative and the definite integral—are defined in terms of limits, it should come as no surprise that derivatives and integrals can often be used to help evaluate limits. In this chapter we see how certain limits, whose values are not obvious by inspection, can be found using derivatives according to special rules called *L'Hôpital's rules*. We also use limits to study the behavior of *improper integrals*. The chapter ends with an investigation of *Taylor's formula*, which can be used to approximate certain functions by polynomials.

1 The Indeterminate Form 0/0

From our earlier study of the properties of limits, we know that

$$\lim_{x \to a} \frac{f(x)}{g(x)} = \frac{\lim_{x \to a} f(x)}{\lim_{x \to a} g(x)}$$

holds, provided that $\lim_{x \to a} f(x)$ and $\lim_{x \to a} g(x)$ both exist and that $\lim_{x \to a} g(x)$ is not zero (Property 4, Section 2, Chapter 1, page 54). If $\lim_{x \to a} g(x) = 0$, this rule simply does not apply. But, notice that, if $\lim_{x \to a} g(x) = 0$ and if $\lim_{x \to a} \frac{f(x)}{g(x)}$ exists at all, then

$$\lim_{x \to a} f(x) = \lim_{x \to a} \left[\frac{f(x)}{g(x)} g(x) \right] = \left[\lim_{x \to a} \frac{f(x)}{g(x)} \right] \left[\lim_{x \to a} g(x) \right] = \left[\lim_{x \to a} \frac{f(x)}{g(x)} \right] \cdot 0 = 0.$$

Therefore, if the limit of a quotient exists and the limit of the denominator is zero, then the limit of the numerator must also be zero.

An expression of the form $\frac{f(x)}{g(x)}$ such that both $\lim_{x \to a} f(x) = 0$ and $\lim_{x \to a} g(x) = 0$ is said to have the *indeterminate form* 0/0 *at a*. [Here we are assuming that

$g(x) \neq 0$ for values of x near a but different from a, so that the fraction $f(x)/g(x)$ is actually defined for such values of x.]

For instance, the fraction $\dfrac{x^2 - x - 2}{x^2 + x - 6}$ has the indeterminate form $0/0$ at 2 since $\lim_{x \to 2} (x^2 - x - 2) = 0$ and $\lim_{x \to 2} (x^2 + x - 6) = 0$. In Chapter 1 we evaluated limits of such fractions by the "trick" of factoring numerator and denominator; thus,

$$\lim_{x \to 2} \frac{x^2 - x - 2}{x^2 + x - 6} = \lim_{x \to 2} \frac{(x + 1)(x - 2)}{(x + 3)(x - 2)} = \lim_{x \to 2} \frac{x + 1}{x + 3} = \frac{3}{5}.$$

Other examples, such as $\lim_{x \to 0} \dfrac{\sin x}{x}$ and $\lim_{x \to 0} \dfrac{1 - \cos x}{x}$ (both of which involve indeterminates of the form $0/0$ at 0) were not so amenable and had to be found by special techniques in Section 1.2 of Chapter 8.

In one of the first textbooks on calculus, the amateur French mathematician L'Hôpital (1661–1704) published a simple and elegant method for handling limits of indeterminate forms. Here, we give an informal statement of L'Hôpital's rule; later, in Section 1.1 we give a formal statement and proof.

L'Hôpital's Rule

Suppose that $\dfrac{f(x)}{g(x)}$ has the indeterminate form $0/0$ at a and that $\lim_{x \to a} \dfrac{f'(x)}{g'(x)}$ exists. Then $\lim_{x \to a} \dfrac{f(x)}{g(x)} = \lim_{x \to a} \dfrac{f'(x)}{g'(x)}$. $\left[\textit{Caution:} \text{ Notice that } \dfrac{f'(x)}{g'(x)} \text{ is } \textit{not} \text{ the derivative of } \dfrac{f(x)}{g(x)}. \right.$ In L'Hôpital's rule, the fraction $\dfrac{f'(x)}{g'(x)}$ is obtained by separately differentiating the numerator and denominator of the fraction $\left. \dfrac{f(x)}{g(x)}. \right]$

EXAMPLES Use L'Hôpital's rule to find the given limit.

1 $\lim\limits_{x \to 3} \dfrac{x^2 - 6x + 9}{x^2 - 7x + 12}$

SOLUTION
Both numerator and denominator approach zero as $x \to 3$, so the fraction has the indeterminate form $0/0$ at 3. By L'Hôpital's rule, we have

$$\lim_{x \to 3} \frac{x^2 - 6x + 9}{x^2 - 7x + 12} = \lim_{x \to 3} \frac{D_x(x^2 - 6x + 9)}{D_x(x^2 - 7x + 12)} = \lim_{x \to 3} \frac{2x - 6}{2x - 7} = \frac{0}{-1} = 0.$$

2 $\lim\limits_{x \to 0} \dfrac{e^x - e^{-x}}{\ln (x + 1)}$

SOLUTION
The fraction has the indeterminate form $0/0$ at 0, so that

$$\lim_{x \to 0} \frac{e^x - e^{-x}}{\ln (x + 1)} = \lim_{x \to 0} \frac{D_x(e^x - e^{-x})}{D_x \ln (x + 1)} = \lim_{x \to 0} \frac{e^x + e^{-x}}{1/(x + 1)}$$

$$= \lim_{x \to 0} (x + 1)(e^x + e^{-x}) = (0 + 1)(e^0 + e^0) = 2.$$

Sometimes an application of L'Hôpital's rule to an indeterminate form simply produces a new indeterminate form. When this happens, a second application of L'Hôpital's rule may be necessary; in fact, several successive applications of the rule might be required to remove the indeterminacy. (However, there are cases in which the indeterminacy stubbornly persists no matter how many times L'Hôpital's rule is applied—see Problem 36 for an example.)

EXAMPLE   Evaluate $\lim\limits_{x \to 0} \dfrac{x^2}{1 - \cos 2x}$.

SOLUTION
The fraction has the indeterminate form 0/0 at 0. Applying L'Hôpital's rule, we have

$$\lim_{x \to 0} \frac{x^2}{1 - \cos 2x} = \lim_{x \to 0} \frac{D_x x^2}{D_x(1 - \cos 2x)} = \lim_{x \to 0} \frac{2x}{2 \sin 2x} = \lim_{x \to 0} \frac{x}{\sin 2x}.$$

But, the fraction $x/\sin 2x$ still has the indeterminate form 0/0 at 0; hence, we use L'Hôpital's rule a second time to obtain

$$\lim_{x \to 0} \frac{x^2}{1 - \cos 2x} = \lim_{x \to 0} \frac{x}{\sin 2x} = \lim_{x \to 0} \frac{D_x x}{D_x \sin 2x} = \lim_{x \to 0} \frac{1}{2 \cos 2x} = \frac{1}{2}.$$

In using L'Hôpital's rule repeatedly as above, we must make certain that the rule remains applicable at each stage—in particular, we must make certain that the fraction under consideration really is indeterminate before each application of L'Hôpital's rule. A common type of error is illustrated by the following *incorrect* calculation:

$$\lim_{x \to 1} \frac{3x^2 - 2x - 1}{x^2 - x} = \lim_{x \to 1} \frac{6x - 2}{2x - 1} = \lim_{x \to 1} \frac{6}{2} = 3\,?$$

Here, the first step is correct, but the second is not! (Why?) In fact,

$$\lim_{x \to 1} \frac{3x^2 - 2x - 1}{x^2 - x} = \lim_{x \to 1} \frac{6x - 2}{2x - 1} = \frac{6 - 2}{2 - 1} = 4.$$

L'Hôpital's rule may be extended in various ways; for instance, it works for one-sided limits as $x \to a^+$ or as $x \to a^-$. As we prove in Section 1.1, it also works for limits at infinity; that is, for limits as $x \to +\infty$ or as $x \to -\infty$.

EXAMPLE   Evaluate $\lim\limits_{x \to +\infty} \dfrac{\sin (5/x)}{2/x}$.

SOLUTION
Here, $\lim\limits_{x \to +\infty} \sin \dfrac{5}{x} = \sin \left(\lim\limits_{x \to +\infty} \dfrac{5}{x} \right) = \sin 0 = 0$ and $\lim\limits_{x \to +\infty} \dfrac{2}{x} = 0$; hence, the fraction has the indeterminate form 0/0 at $+\infty$. Applying L'Hôpital's rule, we have

$$\lim_{x \to +\infty} \frac{\sin (5/x)}{2/x} = \lim_{x \to +\infty} \frac{D_x \sin (5/x)}{D_x(2/x)} = \lim_{x \to +\infty} \frac{(-5/x^2) \cos (5/x)}{-2/x^2} = \lim_{x \to +\infty} \frac{5}{2} \cos \frac{5}{x}$$

$$= \frac{5}{2} \cos \left(\lim_{x \to +\infty} \frac{5}{x} \right) = \frac{5}{2} \cos 0 = \frac{5}{2}.$$

1.1 Cauchy's Generalized Mean Value Theorem and a Proof of L'Hôpital's Rule

To prove L'Hôpital's rule, we need the following generalized version of the mean value theorem (Theorem 2, Section 1.2, Chapter 3, page 138), which is attributed to the French mathematician Augustin Louis Cauchy (1789–1857).

THEOREM 1   **Cauchy's Generalized Mean Value Theorem**
Let f and g be functions that are continuous on the closed interval $[a, b]$ and differentiable on the open interval (a, b), and suppose that $g'(x) \neq 0$ for all values of x in (a, b). Then there is a number c in the interval (a, b) such that

$$\frac{f(b) - f(a)}{g(b) - g(a)} = \frac{f'(c)}{g'(c)}.$$

PROOF
Notice that, if $g(a) = g(b)$, then by Rolle's theorem (Theorem 3, Section 1.3, Chapter 3, page 140), there is a number x in (a, b) such that $g'(x) = 0$, contrary to the hypothesis of the theorem. Therefore, $g(a) \neq g(b)$, so that $g(b) - g(a) \neq 0$. Define a function K by the equation

$$K(x) = [f(b) - f(a)]g(x) - [g(b) - g(a)]f(x) \qquad \text{for } x \text{ in } [a, b].$$

Evidently, K is continuous on $[a, b]$ since f and g are continuous on $[a, b]$. Likewise, since f and g are differentiable on (a, b), then so is K; furthermore,

$$K'(x) = [f(b) - f(a)]g'(x) - [g(b) - g(a)]f'(x) \qquad \text{for } x \text{ in } (a, b).$$

Now,

$$K(b) = [f(b) - f(a)]g(b) - [g(b) - g(a)]f(b) = f(b)g(a) - f(a)g(b),$$

$$K(a) = [f(b) - f(a)]g(a) - [g(b) - g(a)]f(a) = f(b)g(a) - f(a)g(b),$$

and it follows that $K(a) = K(b)$. Therefore, we can apply Rolle's theorem to the function K and conclude that there exists a number c in the open interval (a, b) such that $K'(c) = 0$; that is,

$$0 = [f(b) - f(a)]g'(c) - [g(b) - g(a)]f'(c).$$

We have seen that $g(b) - g(a) \neq 0$; hence, since $g'(c) \neq 0$ by hypothesis, the latter equation can be rewritten as $\dfrac{f(b) - f(a)}{g(b) - g(a)} = \dfrac{f'(c)}{g'(c)}$, and the theorem is proved.

Notice that, if $g(x) = x$, then $g'(x) = 1$ and the conclusion of Theorem 1 reduces to the conclusion of the original mean value theorem.

EXAMPLE   If $f(x) = x^3 + 12$ and $g(x) = x^2 - 2$, find a number c in the open interval $(0, 2)$ such that $\dfrac{f(2) - f(0)}{g(2) - g(0)} = \dfrac{f'(c)}{g'(c)}$.

SOLUTION
Theorem 1 shows that such a number c exists but does not tell us how to actually find it. To find c, we calculate as follows: $f(2) = 20$, $f(0) = 12$, $g(2) = 2$, $g(0) = -2$, $f'(c) = 3c^2$, and $g'(c) = 2c$. Therefore, the required condition is

$$\frac{20 - 12}{2 - (-2)} = \frac{3c^2}{2c} \qquad \text{or} \qquad \frac{8}{4} = \frac{3c}{2}.$$

Evidently, $c = \frac{4}{3}$.

We now state and prove L'Hôpital's rule for the case in which x approaches *a from the right*. An analogous result holds for the case in which x approaches *a from the left* (Problem 33), and the two results, taken together, provide the justification for our original informal statement of L'Hôpital's rule.

THEOREM 2   **L'Hôpital's Rule for the Indeterminate Form 0/0 as $x \rightarrow a^+$**

Let the functions f and g be defined and differentiable on an open interval (a, b) and suppose that $g(x) \neq 0$ for $a < x < b$. Assume that $\lim\limits_{x \to a^+} f(x) = 0$, $\lim\limits_{x \to a^+} g(x) = 0$, and $g'(x) \neq 0$ for $a < x < b$. Then, if $\lim\limits_{x \to a^+} \dfrac{f'(x)}{g'(x)}$ exists, so does $\lim\limits_{x \to a^+} \dfrac{f(x)}{g(x)}$ and

$$\lim_{x \to a^+} \frac{f(x)}{g(x)} = \lim_{x \to a^+} \frac{f'(x)}{g'(x)}.$$

PROOF

Define functions F and G as follows:

$$F(x) = \begin{cases} f(x) & \text{if } a < x < b \\ 0 & \text{if } x = a \end{cases} \qquad G(x) = \begin{cases} g(x) & \text{if } a < x < b \\ 0 & \text{if } x = a \end{cases}$$

for all values of x in $[a, b]$. Notice that F coincides with f on the open interval (a, b); hence, F is differentiable on (a, b) and $F'(x) = f'(x)$ for $a < x < b$. It follows that F is continuous on (a, b) and, since $\lim\limits_{x \to a^+} F(x) = \lim\limits_{x \to a^+} f(x) = 0 = F(a)$, F is actually continuous on $[a, b]$. Likewise, G is continuous on $[a, b]$ and differentiable on (a, b) with $G'(x) = g'(x)$ for $a < x < b$.

Now, choose any number x in the open interval (a, b) and notice that F and G are continuous on the closed interval $[a, x]$ and differentiable on the open interval (a, x). Furthermore, for any number t in (a, x), $G'(t) = g'(t) \neq 0$. Applying the generalized mean value theorem (Theorem 1) to the functions F and G on the interval $[a, x]$, we conclude that there is a number c in the open interval (a, x) such that

$$\frac{F(x) - F(a)}{G(x) - G(a)} = \frac{F'(c)}{G'(c)}; \quad \text{that is,} \quad \frac{f(x)}{g(x)} = \frac{f'(c)}{g'(c)}.$$

Notice that $a < c < x$ and that the value of c might depend on the choice of x. Evidently, as x approaches a from the right, then c—which is "trapped" between a and x—must also approach a from the right. Thus,

$$\lim_{x \to a^+} \frac{f(x)}{g(x)} = \lim_{x \to a^+} \frac{f'(c)}{g'(c)} = \lim_{c \to a^+} \frac{f'(c)}{g'(c)} = \lim_{x \to a^+} \frac{f'(x)}{g'(x)}.$$

The following theorem shows that L'Hôpital's rule is also effective for limits at infinity. We state and prove the theorem only for the case $x \to +\infty$ and leave the statement and proof of the analogous result for the case $x \to -\infty$ to the reader as an exercise (Problem 38).

THEOREM 3   **L'Hôpital's Rule at Infinity**

Let the functions f and g be defined and differentiable on an open interval of the form $(k, +\infty)$, where $k > 0$ and $g(x) \neq 0$ for $x > k$. Suppose that $\lim\limits_{x \to +\infty} f(x) = 0$, $\lim\limits_{x \to +\infty} g(x) = 0$, and $g'(x) \neq 0$ for $x > k$. Then, if $\lim\limits_{x \to +\infty} \dfrac{f'(x)}{g'(x)}$ exists, so does

$$\lim_{x \to +\infty} \frac{f(x)}{g(x)} \quad \text{and} \quad \lim_{x \to +\infty} \frac{f(x)}{g(x)} = \lim_{x \to +\infty} \frac{f'(x)}{g'(x)}.$$

PROOF

We put $t = 1/x$ for $x > k$, noting that $x = 1/t$, that $0 < t < 1/k$, and that $t \to 0^+$ as $x \to +\infty$. We define functions F and G on the interval $(0, 1/k)$ by the equations

$$F(t) = f\left(\frac{1}{t}\right) \quad \text{and} \quad G(t) = g\left(\frac{1}{t}\right) \quad \text{for } 0 < t < \frac{1}{k}.$$

Notice that $\lim_{t \to 0^+} F(t) = \lim_{t \to 0^+} f\left(\frac{1}{t}\right) = \lim_{x \to +\infty} f(x) = 0$. Similarly, $\lim_{t \to 0^+} G(t) = 0$. By the chain rule, F and G are differentiable on $(0, 1/k)$ and we have

$$F'(t) = (-t^{-2})f'\left(\frac{1}{t}\right) \quad \text{and} \quad G'(t) = (-t^{-2})g'\left(\frac{1}{t}\right) \quad \text{for } 0 < t < \frac{1}{k}.$$

Applying Theorem 2 to the functions F and G on the interval $(0, 1/k)$, we have

$$\lim_{t \to 0^+} \frac{F(t)}{G(t)} = \lim_{t \to 0^+} \frac{F'(t)}{G'(t)};$$

hence,

$$\lim_{x \to +\infty} \frac{f(x)}{g(x)} = \lim_{t \to 0^+} \frac{f\left(\frac{1}{t}\right)}{g\left(\frac{1}{t}\right)} = \lim_{t \to 0^+} \frac{F(t)}{G(t)} = \lim_{t \to 0^+} \frac{F'(t)}{G'(t)}$$

$$= \lim_{t \to 0^+} \frac{(-t^{-2})f'\left(\frac{1}{t}\right)}{(-t^{-2})g'\left(\frac{1}{t}\right)} = \lim_{t \to 0^+} \frac{f'\left(\frac{1}{t}\right)}{g'\left(\frac{1}{t}\right)} = \lim_{x \to +\infty} \frac{f'(x)}{g'(x)},$$

and the proof is complete.

Problem Set 1

In problems 1 through 22, use L'Hôpital's rule to evaluate each limit.

1 $\lim\limits_{x \to 0} \dfrac{x + \sin 2x}{x - \sin 2x}$

2 $\lim\limits_{x \to \pi/2} \dfrac{\sin x - 1}{\pi/2 - x}$

3 $\lim\limits_{x \to -2} \dfrac{2x^2 + 3x - 2}{3x^2 - x - 14}$

4 $\lim\limits_{x \to 1} \dfrac{x^3 - 3x^2 + 5x - 3}{x^2 + x - 2}$

5 $\lim\limits_{t \to 2} \dfrac{t^4 - 16}{t - 2}$

6 $\lim\limits_{x \to 0} \dfrac{\cos x - \cos 3x}{\sin x^2}$

7 $\lim\limits_{t \to \pi/2} \dfrac{\sin t - 1}{\cos t}$

8 $\lim\limits_{x \to 0} \dfrac{xe^{3x} - x}{1 - \cos 2x}$

9 $\lim\limits_{x \to 0} \dfrac{e^x - 1 - x}{\cos 2x - 1}$

10 $\lim\limits_{t \to 2} \dfrac{\sqrt{16t - t^4} - 2\sqrt[3]{4t}}{2 - \sqrt[4]{2t^3}}$

11 $\lim\limits_{x \to 7} \dfrac{\ln (x/7)}{7 - x}$

12 $\lim\limits_{x \to 0} \dfrac{x - \tan^{-1} x}{x - \sin^{-1} x}$

13 $\lim\limits_{x \to 1} \dfrac{\ln x - \sin (x - 1)}{(x - 1)^2}$

14 $\lim\limits_{t \to 0^+} \dfrac{\ln (e^t + 1) - \ln 2}{t^2}$

15 $\lim\limits_{t \to 0} \dfrac{e^t - e^{-t} - 2 \sin t}{4t^3}$

16 $\lim\limits_{y \to 0} \dfrac{y^2 - y \sin y}{e^y + e^{-y} - y^2 - 2}$

17 $\lim\limits_{x \to \pi/2} \dfrac{\ln (\sin x)}{(\pi - 2x)^2}$

18 $\lim\limits_{x \to 1} \dfrac{\ln x}{x - \sqrt{x}}$

19 $\lim\limits_{x \to +\infty} \dfrac{\sin (7/x)}{(5/x)}$

20 $\lim\limits_{x \to +\infty} \dfrac{\sin (3/x)}{\tan^{-1} (2/x)}$

21 $\lim\limits_{x \to +\infty} \dfrac{1 - \cos (2/x)}{\tan (3/x)}$

22 $\lim\limits_{x \to +\infty} \dfrac{\sin (1/x)}{\sin (2/x)}$

In problems 23 and 24, use L'Hôpital's rule and the fundamental theorem of calculus to evaluate each limit.

23 $\lim\limits_{x \to 0} \dfrac{\displaystyle\int_0^x 3\cos^4 7t \, dt}{\displaystyle\int_0^x e^{5t^2} \, dt}$

24 $\lim\limits_{x \to 0} \dfrac{\displaystyle\int_0^x e^{7t}(4t^3 + t^2 + 11) \, dt}{\displaystyle\int_0^x e^{7t}(-7t^3 + 6t + 8) \, dt}$

In problems 25 through 30, find a number c satisfying Cauchy's generalized mean value theorem for each pair of functions on the indicated interval.

25 $f(x) = 2x^3$ and $g(x) = 3x^2 - 1$ on $[0, 2]$.

26 $f(x) = \sin x$ and $g(x) = \cos x$ on $[0, \pi/4]$.

27 $f(x) = \ln x$ and $g(x) = 1/x$ on $[1, e]$.

28 $f(x) = \sin^{-1} x$ and $g(x) = x$ on $[0, 1]$.

29 $f(x) = \tan x$ and $g(x) = 4x/\pi$ on $[-\pi/4, \pi/4]$.

30 $f(x) = x^2(x^2 - 2)$ and $g(x) = x$ on $[-1, 1]$.

31 Find constants a and b such that

$$\lim_{x \to 0} \frac{1}{bx - \sin x} \int_0^x \frac{t^2 \, dt}{\sqrt{a + t}} = 1.$$

32 Show that

$$\lim_{x \to 1} \frac{nx^{n+1} - (n+1)x^n + 1}{(x-1)^2} = \frac{n(n+1)}{2}.$$

33 State and prove the analog of Theorem 2 for the case $x \to a^-$.

34 Explain why the calculation of a derivative directly from the definition always involves the evaluation of a limit of an indeterminate form 0/0.

35 If we had known about L'Hôpital's rule in Chapter 8, could we have used it to find $\lim\limits_{x \to 0} \dfrac{\sin x}{x}$?

36 Explain what happens when we try to find $\lim\limits_{t \to 0^+} \dfrac{e^{-1/t}}{t}$ by repeated applications of L'Hôpital's rule.

37 In problem 36, make the change of variable $t = 1/x$. Then use the result of problem 92(b) in the Review Problem Set for Chapter 9 to evaluate the limit.

38 State and prove the analog of Theorem 3 for the case $x \to -\infty$.

39 The equation $I = (E/R)(1 - e^{-Rt/L})$, where t is time in seconds, R is resistance in ohms, L is inductance in henrys, and I is current in amperes, arises in the theory of electrical circuits. If t, E, and L are held fixed, find the limit of I as R approaches zero.

40 Let A be an endpoint of a diameter of a fixed circle with center O and radius r. In Figure 1, $\overline{AQ}$ is tangent to the circle at point A and $|\overline{AQ}|$ is equal to the length of the arc AP. If B is the point of intersection of the line through Q and P with the line through A and O, find the limiting position of B as P approaches A.

Figure 1

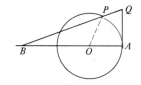

41 A weight hanging from a spring is caused to vibrate by a driving force so that its displacement at time t seconds is given by $y = \dfrac{A}{p^2 - w^2}(\sin wt - \sin pt)$, where A, p, and w are positive constants with $p \neq w$. Determine the limiting value of y as p approaches w holding A and t fixed.

42 For each positive integer n, let V_n denote the volume of the solid obtained by revolving the region under the graph of $f(x) = x^n$ between $x = 0$ and $x = 1$ about the y axis and let H_n be the volume of the solid obtained by revolving the same region about the x axis. Evaluate:

(a) $\lim\limits_{n \to +\infty} V_n$

(b) $\lim\limits_{n \to +\infty} H_n$

(c) $\lim\limits_{n \to +\infty} \dfrac{V_n}{H_n}$

2   Other Indeterminate Forms

In Section 1 we have seen that L'Hôpital's rule can often be used to handle limits of fractions having indeterminate form 0/0. In the present section we discuss additional indeterminate forms, ∞/∞, $\infty \cdot 0$, $\infty - \infty$, 0^0, ∞^0, and 1^∞.

2.1   The Indeterminate Form ∞/∞

By methods similar to those used in Section 1.1, it can be proved that L'Hôpital's rule remains effective when the numerator and denominator become infinite in absolute value, or, as we say, when the fraction has the *indeterminate form* ∞/∞. The facts are set forth in the following theorem.

THEOREM 1   **L'Hôpital's Rule for the Indeterminate Form ∞/∞**

Suppose that the functions f and g are defined and differentiable on an open interval I, except possibly at the number a in I. Moreover, assume that $\lim_{x \to a} |g(x)| = +\infty$. Then, if $g'(x) \neq 0$ for all values of x other than a in I and if $\lim_{x \to a} \dfrac{f'(x)}{g'(x)}$ exists, it follows that $\lim_{x \to a} \dfrac{f(x)}{g(x)}$ also exists and

$$\lim_{x \to a} \frac{f(x)}{g(x)} = \lim_{x \to a} \frac{f'(x)}{g'(x)}.$$

Notice that the condition $\lim_{x \to a} |f(x)| = +\infty$ is not required in Theorem 1; the rule expressed in this theorem is valid even when the fraction does not have the form ∞/∞ at a. However, the most interesting applications are those in which not only does $|g(x)|$ approach $+\infty$, but $|f(x)|$ also approaches $+\infty$ as x approaches a. The theorem is also correct when $x \to a^+$, $x \to a^-$, $x \to +\infty$, or $x \to -\infty$, provided that obvious changes are made in the hypotheses. Since rigorous proofs of this theorem and its alternative versions are a bit complicated, we shall be content here with illustrating these forms of L'Hôpital's rule as follows.

EXAMPLES   Evaluate the given limit.

1 $\lim_{x \to 0^+} \dfrac{1 - \ln x}{e^{1/x}}$

SOLUTION

The fraction has the indeterminate form ∞/∞ at 0. Applying Theorem 1, we have

$$\lim_{x \to 0^+} \frac{1 - \ln x}{e^{1/x}} = \lim_{x \to 0^+} \frac{D_x(1 - \ln x)}{D_x e^{1/x}} = \lim_{x \to 0^+} \frac{-1/x}{(-1/x^2)e^{1/x}} = \lim_{x \to 0^+} \frac{x}{e^{1/x}}$$

$$= \left(\lim_{x \to 0^+} x \right)\left(\lim_{x \to 0^+} \frac{1}{e^{1/x}} \right) = (0)(0) = 0.$$

2 $\lim_{x \to \pi/2^+} \dfrac{7 \tan x}{5 + \sec x}$

SOLUTION

The fraction has the indeterminate form ∞/∞ at $\pi/2$. By L'Hôpital's rule, we have

$$\lim_{x \to \pi/2^+} \frac{7 \tan x}{5 + \sec x} = \lim_{x \to \pi/2^+} \frac{7 \sec^2 x}{\sec x \tan x} = \lim_{x \to \pi/2^+} \frac{7 \sec x}{\tan x}$$

$$= \lim_{x \to \pi/2^+} \frac{7/\cos x}{\sin x/\cos x} = \lim_{x \to \pi/2^+} \frac{7}{\sin x} = 7.$$

Just as was the case for the indeterminate form 0/0, it may be necessary to apply L'Hôpital's rule several times in order to evaluate a limit of an indeterminate form ∞/∞. Also, it can be shown that L'Hôpital's rule still holds if the quotient $f'(x)/g'(x)$ approaches $+\infty$ or $-\infty$; that is,

$$\lim_{x \to a} \frac{f(x)}{g(x)} = \lim_{x \to a} \frac{f'(x)}{g'(x)} = \pm \infty.$$

This is illustrated by the following example.

EXAMPLE Evaluate $\displaystyle\lim_{x \to +\infty} \frac{e^x}{x^3}$.

SOLUTION

The fraction has the indeterminate form ∞/∞ at $+\infty$. Using L'Hôpital's rule three times, we have

$$\lim_{x \to +\infty} \frac{e^x}{x^3} = \lim_{x \to +\infty} \frac{e^x}{3x^2} = \lim_{x \to +\infty} \frac{e^x}{6x} = \lim_{x \to +\infty} \frac{e^x}{6} = +\infty.$$

2.2 Other Cases of Indeterminate Forms

We now consider indeterminate forms of the type $\infty \cdot 0$, $\infty - \infty$, 0^0, ∞^0, and 1^∞. These are handled by algebraic manipulation of the indeterminate form so as to reduce it to the type 0/0 or ∞/∞ so that L'Hôpital's rule can be used. In each case suitable examples are given to illustrate the technique.

1 The Case $\infty \cdot 0$

If $\displaystyle\lim_{x \to a} |f(x)| = +\infty$ and $\displaystyle\lim_{x \to a} g(x) = 0$, we say that the product $f(x)g(x)$ *has the indeterminate form* $\infty \cdot 0$ *at a*. To find $\displaystyle\lim_{x \to a} f(x)g(x)$ in this case, we can write $f(x)g(x)$ as $\dfrac{g(x)}{1/f(x)}$ or as $\dfrac{f(x)}{1/g(x)}$, thus obtaining the form 0/0 or the form ∞/∞, respectively, whichever is more convenient.

EXAMPLE Evaluate $\displaystyle\lim_{x \to 0^+} x^2 \ln x$.

SOLUTION

We write $x^2 \ln x = \dfrac{\ln x}{x^{-2}}$, noting that $\dfrac{\ln x}{x^{-2}}$ has the indeterminate form ∞/∞ at 0. By applying L'Hôpital's rule, we obtain

$$\lim_{x \to 0^+} x^2 \ln x = \lim_{x \to 0^+} \frac{\ln x}{x^{-2}} = \lim_{x \to 0^+} \frac{1/x}{-2x^{-3}} = \lim_{x \to 0^+} \left(-\tfrac{1}{2}\right)x^2 = 0.$$

2 The Case $\infty - \infty$

If $\lim_{x \to a} f(x) = +\infty$ and $\lim_{x \to a} g(x) = +\infty$, we say that the difference $f(x) - g(x)$ *has the indeterminate form* $\infty - \infty$ *at a.* By performing the indicated subtraction, the indeterminate form $\infty - \infty$ can usually be converted into the indeterminate form $0/0$. Otherwise, one can resort to the trick of rewriting $f(x) - g(x)$ as $\dfrac{1/g(x) - 1/f(x)}{1/[f(x)g(x)]}$, noting that the latter fraction has the indeterminate form $0/0$ at a.

EXAMPLE   Evaluate $\lim\limits_{x \to 0^+} \left(\csc x - \dfrac{1}{x} \right)$.

SOLUTION
The expression has the indeterminate form $\infty - \infty$ at $x = 0$; however, if we write $\csc x$ as $1/\sin x$ and subtract, we have

$$\lim_{x \to 0^+} \left(\csc x - \frac{1}{x} \right) = \lim_{x \to 0^+} \left(\frac{1}{\sin x} - \frac{1}{x} \right) = \lim_{x \to 0^+} \frac{x - \sin x}{x \sin x}.$$

The fraction has the indeterminate form $0/0$ at 0. Applying L'Hôpital's rule twice, we obtain

$$\lim_{x \to 0^+} \frac{x - \sin x}{x \sin x} = \lim_{x \to 0^+} \frac{1 - \cos x}{\sin x + x \cos x} = \lim_{x \to 0^+} \frac{\sin x}{2 \cos x - x \sin x}$$

$$= \frac{0}{2} = 0.$$

3 The Case 0^0, ∞^0, or 1^∞

If $f(x) > 0$, $\lim_{x \to a} f(x) = 0$, and $\lim_{x \to a} g(x) = 0$, we say that the expression $f(x)^{g(x)}$ *has the indeterminate form* 0^0 *at a.* Expressions with the indeterminate forms ∞^0 or 1^∞ are defined analogously. In order to evaluate $\lim\limits_{x \to a} f(x)^{g(x)}$ in such indeterminate cases, we carry out the following procedure:

Step 1   Calculate $\lim\limits_{x \to a} [g(x) \ln f(x)] = L$.

Step 2   Conclude that $\lim\limits_{x \to a} f(x)^{g(x)} = e^L$.

This procedure is justified by the observation that

$$\lim_{x \to a} f(x)^{g(x)} = \lim_{x \to a} e^{g(x) \ln f(x)} = e^L.$$

Notice that the product $g(x) \ln f(x)$ in Step 1 has the indeterminate form $\infty \cdot 0$ at a.

EXAMPLES   Evaluate the indicated limit.

1 $\lim\limits_{x \to \pi/2^-} \left(\dfrac{5\pi}{2} - 5x \right)^{\cos x}$

SOLUTION
The indeterminate form is 0^0, so we carry out the above procedure.

Step 1: We must evaluate $\lim\limits_{x\to\pi/2^-} [\cos x \ln (5\pi/2 - 5x)]$. Here the product has the indeterminate form $0 \cdot \infty$ at $\pi/2$; hence, we rewrite it as

$$\cos x \ln \left(\frac{5\pi}{2} - 5x\right) = \frac{\ln \left(\dfrac{5\pi}{2} - 5x\right)}{1/\cos x} = \frac{\ln \left(\dfrac{5\pi}{2} - 5x\right)}{\sec x}.$$

The resulting fraction has the indeterminate form ∞/∞ at $\pi/2$. Using L'Hôpital's rule twice, we have

$$L = \lim_{x\to\pi/2^-} \left[\cos x \ln \left(\frac{5\pi}{2} - 5x\right)\right] = \lim_{x\to\pi/2^-} \frac{\ln \left(\dfrac{5\pi}{2} - 5x\right)}{\sec x}$$

$$= \lim_{x\to\pi/2^-} \frac{\left[\dfrac{-5}{(5\pi/2) - 5x}\right]}{\sec x \tan x} = \lim_{x\to\pi/2^-} \frac{\left[\dfrac{-5}{(5\pi/2) - 5x}\right]}{\sin x/\cos^2 x}$$

$$= \lim_{x\to\pi/2^-} \frac{-5 \cos^2 x}{(\sin x)\left(\dfrac{5\pi}{2} - 5x\right)} = \lim_{x\to\pi/2^-} \frac{10 \cos x \sin x}{-5 \sin x + (\cos x)\left(\dfrac{5\pi}{2} - 5x\right)}$$

$$= \frac{0}{-5} = 0.$$

Step 2: $\lim\limits_{x\to\pi/2^-} \left(\dfrac{5\pi}{2} - 5x\right)^{\cos x} = e^L = e^0 = 1.$

2 $\lim\limits_{x\to 0^+} (\csc x)^{\sin x}$

SOLUTION
The indeterminate form is ∞^0. Since

$$\lim_{x\to 0^+} (\csc x)^{\sin x} = \lim_{x\to 0^+} \left(\frac{1}{\sin x}\right)^{\sin x},$$

we begin by making the change of variable $u = \sin x$. Notice that $u \to 0^+$ as $x \to 0^+$; hence,

$$\lim_{x\to 0^+} (\csc x)^{\sin x} = \lim_{u\to 0^+} \left(\frac{1}{u}\right)^u.$$

Now we use our procedure.

Step 1: $L = \lim\limits_{u\to 0^+} \left[u \ln \left(\dfrac{1}{u}\right)\right] = \lim\limits_{u\to 0^+} (-u \ln u) = \lim\limits_{u\to 0^+} \dfrac{-\ln u}{(1/u)}$

$$= \lim_{u\to 0^+} \frac{D_u(-\ln u)}{D_u(1/u)} = \lim_{u\to 0^+} \frac{-1/u}{-1/u^2} = \lim_{u\to 0^+} u = 0.$$

Step 2: $\lim\limits_{x\to 0^+} (\csc x)^{\sin x} = \lim\limits_{u\to 0^+} \left(\dfrac{1}{u}\right)^u = e^L = e^0 = 1.$

Problem Set 2

In problems 1 through 50, evaluate each limit.

1 $\displaystyle\lim_{x\to\pi/2}\frac{1+\sec x}{\tan x}$

2 $\displaystyle\lim_{x\to 1/2}\frac{\sec 3\pi x}{\tan 3\pi x}$

3 $\displaystyle\lim_{x\to+\infty}\frac{\ln\,(17+x)}{x}$

4 $\displaystyle\lim_{x\to 1^+}\frac{\ln\,(x-1)+\tan\,(\pi x/2)}{\cot\,(x-1)}$

5 $\displaystyle\lim_{x\to+\infty}\frac{e^x+1}{x^4+x^3}$

6 $\displaystyle\lim_{x\to+\infty}\frac{2^x}{x^3}$

7 $\displaystyle\lim_{x\to+\infty}\frac{2x^4}{e^{3x}}$

8 $\displaystyle\lim_{x\to+\infty}\frac{\ln\,(x+e^x)}{x}$

9 $\displaystyle\lim_{t\to+\infty}\frac{t\ln t}{(t+2)^2}$

10 $\displaystyle\lim_{x\to+\infty}\frac{x+e^{2x}}{\ln x+e^{2x}}$

11 $\displaystyle\lim_{x\to+\infty}xe^{-x}$

12 $\displaystyle\lim_{t\to 0}\sin 3t\cot 2t$

13 $\displaystyle\lim_{x\to 0^+}xe^{1/x}$

14 $\displaystyle\lim_{x\to+\infty}x\sin\frac{\pi}{x}$

15 $\displaystyle\lim_{x\to 0^+}x(\ln x)^2$

16 $\displaystyle\lim_{x\to\pi/2}\cos 3x\sec 5x$

17 $\displaystyle\lim_{x\to\pi/2}\tan x\tan 2x$

18 $\displaystyle\lim_{x\to 0}\csc x\sin^{-1}x$

19 $\displaystyle\lim_{x\to 1}\left(\frac{x}{x-1}-\frac{1}{\ln x}\right)$

20 $\displaystyle\lim_{x\to 1}\left(\frac{x}{\ln x}-\frac{1}{x\ln x}\right)$

21 $\displaystyle\lim_{x\to 0^+}(\csc x-\csc 2x)$

22 $\displaystyle\lim_{x\to+\infty}\left[\ln\,(x-2)-\ln\frac{x}{2}\right]$

23 $\displaystyle\lim_{x\to+\infty}(x^2-\sqrt{x^4+x^2+7})$

24 $\displaystyle\lim_{x\to\pi/2}\left(x\tan x-\frac{\pi}{2}\sec x\right)$

25 $\displaystyle\lim_{x\to 4}\left(\frac{7}{x^2-x-12}-\frac{1}{x-4}\right)$

26 $\displaystyle\lim_{x\to 1}\left(\frac{n}{x^n-1}-\frac{m}{x^m-1}\right)$

27 $\displaystyle\lim_{x\to 0^+}x^x$

28 $\displaystyle\lim_{x\to 0^+}(\sin x)^x$

29 $\displaystyle\lim_{x\to 0^+}x^{1/\ln x}$

30 $\displaystyle\lim_{x\to 0^+}x^{\sin x}$

31 $\displaystyle\lim_{x\to\pi/2}(\cos x)^{x-(\pi/2)}$

32 $\displaystyle\lim_{x\to\pi/4^-}\left(\frac{\pi}{4}-x\right)^{\cos 2x}$

33 $\displaystyle\lim_{x\to 0^+}(\cot x)^{x^2}$

34 $\displaystyle\lim_{x\to 0^+}(\cot x)^{\sin x}$

35 $\displaystyle\lim_{x\to+\infty}\left(\frac{x}{x-2}\right)^x$

36 $\displaystyle\lim_{x\to+\infty}(e^x+x)^{1/x}$

37 $\displaystyle\lim_{x\to+\infty}x^{1/x}$

38 $\displaystyle\lim_{x\to 0^+}(-\ln x)^x$

39 $\displaystyle\lim_{x\to 0}(1+\tan x)^{1/x}$

40 $\displaystyle\lim_{x\to+\infty}\left(1+\frac{3}{x}\right)^x$

41 $\displaystyle\lim_{x\to 0}(1+2x)^{3/x}$

42 $\displaystyle\lim_{x\to 0^+}(1+x)^{\ln x}$

43 $\displaystyle\lim_{x\to+\infty}\left(\cos\frac{2}{x}\right)^{x^2}$

44 $\displaystyle\lim_{x\to 0}(e^{x^2/2}\cos x)^{4/x^4}$

45 $\displaystyle\lim_{x\to 0}(1+x)^{\cot x}$

46 $\displaystyle\lim_{x\to 2^-}\left(1-\frac{x}{2}\right)^{\tan\pi x}$

47 $\displaystyle\lim_{x\to 0^-}(1+x)^{\ln|x|}$

48 $\displaystyle\lim_{x\to 0}(e^{2x}+2x)^{1/(4x)}$

49 $\displaystyle\lim_{x\to\pi/2}(\sin x)^{\sec x}$

50 $\displaystyle\lim_{x\to 1}x^{1/(1-x)}$

51 Suppose that the function f has the following properties:

$$\lim_{x\to+\infty}f(x)=\lim_{x\to+\infty}f'(x)=\lim_{x\to+\infty}f''(x)=+\infty\quad\text{and}\quad\lim_{x\to+\infty}\frac{xf'''(x)}{f''(x)}=1.\text{ Evaluate }\lim_{x\to+\infty}\frac{xf'(x)}{f(x)}.$$

52 Find a value of the constant c such that $\displaystyle\lim_{x\to+\infty}\left(\frac{x+c}{x-c}\right)^x=4$.

3 Improper Integrals with Infinite Limits

In Chapters 5 and 6 we obtained the areas of admissible regions in the plane by using definite integrals. Recall, however, that an admissible region must be *bounded*. If we wish to find areas of unbounded regions, we have to deal with "improper" integrals.

Consider, for instance, the region R under the graph of $y = 1/x^2$ to the right of $x = 1$ (Figure 1a). Notice that the region R extends indefinitely to the right,

Figure 1

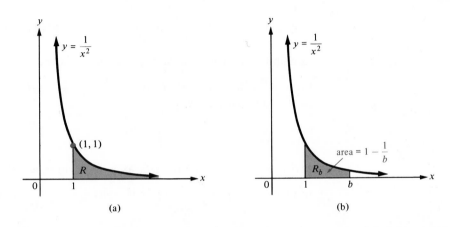

(a)   (b)

and therefore is unbounded. Offhand, it is perhaps not clear what is meant by the "area" of such an unbounded region. However, let R_b denote the bounded region under the graph of $y = 1/x^2$ between $x = 1$ and $x = b$ (Figure 1b). The area of R_b is given by

$$\int_1^b \frac{1}{x^2}\, dx = \frac{-1}{x}\bigg|_1^b = 1 - \frac{1}{b}.$$

For large values of b, the bounded region R_b might be considered to be a good approximation to the unbounded region R; in fact, one is tempted to write $R = \lim_{b \to +\infty} R_b$. Hence, one might expect that

$$\text{area of } R = \lim_{b \to +\infty} (\text{area of } R_b) = \lim_{b \to +\infty} \left(1 - \frac{1}{b}\right) = 1 \text{ square unit.}$$

Figure 2

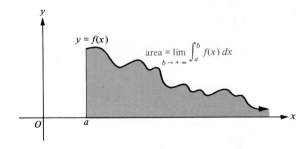

In general, if f is a function defined on an infinite interval of the form $[a, +\infty)$, and if $f(x) \geq 0$ holds for $x \geq a$, we define the area of the unbounded region under the graph of f and to the right of $x = a$ to be $\lim_{b \to +\infty} \int_a^b f(x)\, dx$ (Figure 2). Often we write this area simply as $\int_a^{+\infty} f(x)\, dx$. More generally, we make the following definition.

DEFINITION 1 **Improper Integrals with Infinite Upper Limits**

Let f be a function defined at least on the infinite interval $[a, +\infty)$. Suppose that f is Riemann-integrable on the closed interval $[a, b]$ for every value of b larger than a. Then we define

$$\int_a^{+\infty} f(x)\, dx = \lim_{b \to +\infty} \int_a^b f(x)\, dx,$$

provided that the limit exists as a finite number.

The expression $\int_a^{+\infty} f(x)\, dx$, which is often written as $\int_a^{\infty} f(x)\, dx$ for simplicity, is called an *improper integral* with an *infinite upper limit*. If $\lim\limits_{b \to +\infty} \int_a^b f(x)\, dx$ exists as a finite number, then we say that the improper integral $\int_a^{+\infty} f(x)\, dx$ is *convergent*. Otherwise, it is *divergent*. An *improper integral* with an *infinite lower limit* is defined in a similar way by

$$\int_{-\infty}^b f(x)\, dx = \lim_{a \to -\infty} \int_a^b f(x)\, dx,$$

provided that the limit exists as a finite number, in which case we say that the improper integral $\int_{-\infty}^b f(x)\, dx$ is *convergent*. Otherwise, it is *divergent*.

EXAMPLES   Evaluate the given improper integral (if it is convergent).

1 $\displaystyle\int_1^{\infty} \frac{dx}{x^3}$

SOLUTION
By Definition 1, we have

$$\int_1^{\infty} \frac{dx}{x^3} = \lim_{b \to +\infty} \int_1^b \frac{dx}{x^3} = \lim_{b \to +\infty} \left(\frac{-1}{2x^2} \Big|_1^b \right) = \lim_{b \to +\infty} \left(\frac{-1}{2b^2} + \frac{1}{2} \right) = \frac{1}{2}.$$

2 $\displaystyle\int_0^{\infty} \frac{dx}{1 + x^2}$

SOLUTION

$$\int_0^{\infty} \frac{dx}{1 + x^2} = \lim_{b \to +\infty} \int_0^b \frac{dx}{1 + x^2} = \lim_{b \to +\infty} (\tan^{-1} x) \Big|_0^b$$

$$= \lim_{b \to +\infty} (\tan^{-1} b - \tan^{-1} 0) = \lim_{b \to +\infty} \tan^{-1} b = \frac{\pi}{2}.$$

3 $\displaystyle\int_{-\infty}^3 \frac{dx}{(9 - x)^2}$

SOLUTION

$$\int_{-\infty}^3 \frac{dx}{(9 - x)^2} = \lim_{a \to -\infty} \int_a^3 \frac{dx}{(9 - x)^2} = \lim_{a \to -\infty} \left(\frac{1}{9 - x} \Big|_a^3 \right)$$

$$= \lim_{a \to -\infty} \left(\frac{1}{9 - 3} - \frac{1}{9 - a} \right) = \frac{1}{6}.$$

4 $\displaystyle\int_{-\infty}^{0} e^{-x}\, dx$

SOLUTION

$$\int_{-\infty}^{0} e^{-x}\, dx = \lim_{a \to -\infty} \int_{a}^{0} e^{-x}\, dx = \lim_{a \to -\infty} (-e^{-x}) \Big|_{a}^{0} = \lim_{a \to -\infty} (e^{-a} - e^{0}) = +\infty;$$

hence, $\displaystyle\int_{-\infty}^{0} e^{-x}\, dx$ is divergent.

Improper integrals can be used to find areas of unbounded regions as in the following examples.

EXAMPLES 1 Find the area A of the region in the first quadrant bounded by the curve $y = 2^{-x}$, the x axis, and the y axis (Figure 3).

SOLUTION
The area A is given by

$$A = \int_{0}^{\infty} 2^{-x}\, dx = \lim_{b \to +\infty} \int_{0}^{b} 2^{-x}\, dx = \lim_{b \to +\infty} \left(\frac{2^{-x}}{-\ln 2} \right) \Big|_{0}^{b}$$

$$= \lim_{b \to +\infty} \left(\frac{2^{-b}}{-\ln 2} - \frac{2^{-0}}{-\ln 2} \right) = \frac{1}{\ln 2} \text{ square units,}$$

since $\displaystyle\lim_{b \to +\infty} 2^{-b} = 0$.

Figure 3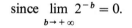

Figure 4

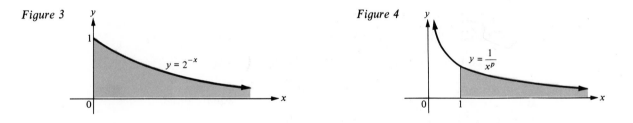

2 Suppose that $p > 1$. Find the area A of the region in the first quadrant bounded by the curve $y = 1/x^{p}$, the x axis, and the line $x = 1$ (Figure 4).

SOLUTION

$$A = \int_{1}^{\infty} \frac{dx}{x^{p}} = \lim_{b \to +\infty} \int_{1}^{b} \frac{dx}{x^{p}} = \lim_{b \to +\infty} \left(\frac{x^{1-p}}{1-p} \Big|_{1}^{b} \right)$$

$$= \lim_{b \to +\infty} \left(\frac{b^{1-p}}{1-p} - \frac{1^{1-p}}{1-p} \right) = 0 - \frac{1}{1-p} = \frac{1}{p-1} \text{ square units,}$$

since the fact that $p > 1$ implies that $\displaystyle\lim_{b \to +\infty} b^{1-p} = \lim_{b \to +\infty} \frac{1}{b^{p-1}} = 0$.

In Example 2, notice that, if $p < 1$, then $\displaystyle\lim_{b \to +\infty} b^{1-p} = +\infty$, so that $\displaystyle\int_{1}^{\infty} \frac{dx}{x^{p}}$ diverges and the unbounded region in Figure 4 has an infinite area. If $p = 1$, the integral also diverges (Problem 5); hence,

Figure 5

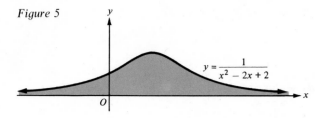

$y = \dfrac{1}{x^2 - 2x + 2}$

$$\int_1^\infty \frac{dx}{x^p} \begin{cases} \text{converges to } \dfrac{1}{p-1} & \text{if } p > 1, \\ \text{diverges} & \text{if } p \le 1. \end{cases}$$

Now consider the entire unbounded region under the curve $y = \dfrac{1}{x^2 - 2x + 2}$ (Figure 5). The portion of this region to the right of the y axis has an area A_1 given by

$$A_1 = \int_0^\infty \frac{dx}{x^2 - 2x + 2} = \lim_{b \to +\infty} \int_0^b \frac{dx}{x^2 - 2x + 2} = \lim_{b \to +\infty} \int_0^b \frac{dx}{(x-1)^2 + 1}$$

$$= \lim_{b \to +\infty} \left[\tan^{-1}(x-1) \Big|_0^b \right] = \lim_{b \to +\infty} \left[\tan^{-1}(b-1) - \tan^{-1}(-1) \right]$$

$$= \frac{\pi}{2} + \frac{\pi}{4} = \frac{3\pi}{4} \text{ square units.}$$

Likewise, the portion of the region in Figure 5 to the left of the y axis has an area A_2 given by

$$A_2 = \int_{-\infty}^0 \frac{dx}{x^2 - 2x + 2} = \lim_{a \to -\infty} \int_a^0 \frac{dx}{(x-1)^2 + 1}$$

$$= \lim_{a \to -\infty} \left[\tan^{-1}(x-1) \Big|_a^0 \right] = \lim_{a \to -\infty} \left[\tan^{-1}(-1) - \tan^{-1}(a-1) \right]$$

$$= -\frac{\pi}{4} + \frac{\pi}{2} = \frac{\pi}{4} \text{ square unit.}$$

Therefore, the total area under the curve is given by $A_1 + A_2 = \dfrac{3\pi}{4} + \dfrac{\pi}{4} = \pi$ square units. It would seem reasonable to write this area as

$$\int_{-\infty}^\infty \frac{dx}{x^2 - 2x + 2} = \pi \text{ square units.}$$

More generally, we make the following definition.

DEFINITION 2   **Improper Integrals with Both Limits Infinite**
Let the function f be defined for all real numbers and suppose that the improper integrals $\int_0^\infty f(x)\,dx$ and $\int_{-\infty}^0 f(x)\,dx$ both converge. Then, by definition,

$$\int_{-\infty}^\infty f(x)\,dx = \int_{-\infty}^0 f(x)\,dx + \int_0^\infty f(x)\,dx,$$

and we say that the improper integral $\int_{-\infty}^\infty f(x)\,dx$ is *convergent*.

If either of the improper integrals $\int_0^\infty f(x)\,dx$ or $\int_{-\infty}^0 f(x)\,dx$ is divergent, we say that the *improper integral* $\int_{-\infty}^\infty f(x)\,dx$ is *divergent*.

EXAMPLES Evaluate the given improper integral (if it is convergent).

1 $\int_{-\infty}^{\infty} \dfrac{x\,dx}{(x^2+1)^2}$

SOLUTION
Using the substitution $u = x^2 + 1$, we find that

$$\int \frac{x\,dx}{(x^2+1)^2} = \frac{1}{2}\int \frac{du}{u^2} = -\frac{1}{2u} + C = \frac{-1}{2(x^2+1)} + C.$$

Thus,

$$\int_0^\infty \frac{x\,dx}{(x^2+1)^2} = \lim_{b\to+\infty}\left[\frac{-1}{2(x^2+1)}\Big|_0^b\right]$$

$$= \lim_{b\to+\infty}\left[\frac{-1}{2(b^2+1)} - \frac{-1}{2(0^2+1)}\right] = \frac{1}{2}.$$

Similarly, $\displaystyle\int_{-\infty}^0 \frac{x\,dx}{(x^2+1)^2} = -\frac{1}{2}$. Therefore,

$$\int_{-\infty}^\infty \frac{x\,dx}{(x^2+1)^2} = \int_{-\infty}^0 \frac{x\,dx}{(x^2+1)^2} + \int_0^\infty \frac{x\,dx}{(x^2+1)^2} = -\frac{1}{2} + \frac{1}{2} = 0.$$

2 $\int_{-\infty}^{\infty} x\,dx$

SOLUTION

$$\int_{-\infty}^\infty x\,dx = \int_{-\infty}^0 x\,dx + \int_0^\infty x\,dx,$$

provided that the latter two improper integrals converge. But they do not converge; in particular,

$$\int_0^\infty x\,dx = \lim_{b\to+\infty}\int_0^b x\,dx = \lim_{b\to+\infty}\frac{b^2}{2} = +\infty.$$

Therefore, $\displaystyle\int_{-\infty}^\infty x\,dx$ diverges.

Problem Set 3

In problems 1 through 24, evaluate each improper integral (if it is convergent).

1 $\displaystyle\int_1^\infty \frac{dx}{x\sqrt{x}}$

2 $\displaystyle\int_1^\infty \frac{dx}{(4x+3)^2}$

3 $\displaystyle\int_3^\infty \frac{dx}{x^2+9}$

4 $\displaystyle\int_{-\infty}^0 \frac{dx}{x^2+16}$

5 $\displaystyle\int_1^\infty \frac{dx}{x}$

6 $\displaystyle\int_1^\infty \frac{x\,dx}{5x^2+3}$

7 $\displaystyle\int_0^\infty \frac{dx}{(x+1)(x+2)}$

8 $\displaystyle\int_2^\infty \frac{x\,dx}{(x+1)(x+2)}$

9 $\displaystyle\int_0^\infty 4e^{8x}\,dx$

10 $\displaystyle\int_0^\infty \frac{dx}{\sqrt[3]{e^x}}$

11 $\displaystyle\int_1^\infty \frac{x\,dx}{1+x^4}$

12 $\displaystyle\int_e^\infty \frac{dx}{x(\ln x)^2}$

13 $\displaystyle\int_e^\infty \frac{dx}{x\ln x}$

14 $\displaystyle\int_0^\infty e^{-x}\sin x\,dx$

15 $\displaystyle\int_{-\infty}^\infty \frac{dx}{1+x^2}$

16 $\displaystyle\int_{-\infty}^0 (e^t - e^{2t})\,dt$

17 $\displaystyle\int_{-\infty}^0 xe^x\,dx$

18 $\displaystyle\int_{-\infty}^\infty (x^2 + 2x + 2)^{-1}\,dx$

19 $\displaystyle\int_{-\infty}^\infty \frac{|x|\,dx}{1+x^4}$

20 $\displaystyle\int_{-\infty}^1 xe^{3x}\,dx$

21 $\displaystyle\int_0^\infty e^{-3x}\,dx$

22 $\displaystyle\int_{-\infty}^\infty \frac{e^x\,dx}{\cosh x}$

23 $\displaystyle\int_{-\infty}^\infty \operatorname{sech} x\,dx$

24 $\displaystyle\int_{-\infty}^\infty \frac{dx}{a^2 + x^2}$

25 Show that $\displaystyle\int_2^\infty \frac{dx}{x(\ln x)^p}$ converges to $\dfrac{(\ln 2)^{1-p}}{p-1}$ if $p > 1$.

26 Show that $\displaystyle\int_3^\infty \frac{dx}{x\ln x[\ln(\ln x)]^p}$ converges to $\dfrac{[\ln(\ln 3)]^{1-p}}{p-1}$ if $p > 1$.

27 For what value of n does the improper integral $\displaystyle\int_0^\infty \left(\frac{2}{x+1} - \frac{n}{x+3}\right) dx$ converge? Evaluate the integral for this value of n.

28 If $\displaystyle\int_{-\infty}^\infty f(x)\,dx$ is convergent, show that $\displaystyle\int_{-\infty}^\infty f(x)\,dx = \lim_{c\to+\infty} \int_{-c}^c f(x)\,dx$.

29 Show that $\displaystyle\lim_{c\to+\infty} \int_{-c}^c \sin x\,dx = 0$.

30 In view of problems 28 and 29, can we conclude that $\displaystyle\int_{-\infty}^\infty \sin x\,dx = 0$? Explain.

31 Find the area of the unbounded region under the curve $y = \dfrac{1}{e^x + e^{-x}}$.

32 Find the volume of the unbounded solid generated by revolving the region under the curve $y = \sqrt{x}\,e^{-x^2}$, $x \geq 0$, about the x axis.

33 Find the volume of the unbounded solid generated by revolving the region under the curve $y = 1/x$, $x \geq 1$, about the x axis.

34 Show that the surface area of the unbounded surface generated by revolving the curve $y = 1/x$, $x \geq 1$, about the x axis is infinite. [*Hint:* For $x \geq 1$, $(1/x)\sqrt{1 + (-1/x^2)^2} > 1/x$.]

35 Find the area of the unbounded region bounded by the two curves $y = 1/x^2$ and $y = e^{-2x}$ over the interval $[1, \infty)$.

36 Give an example of:
(a) An unbounded region in the plane with infinite area which, when rotated about the x axis, generates an unbounded solid with finite volume.
(b) An unbounded solid with finite volume which has infinite surface area.

37 If a business firm predicts a profit of $\$P(t)$ per year t years from now, then the *present value* of all future profit, when interest is compounded continuously at the rate of k per cent per year, is defined to be $\displaystyle\int_0^\infty e^{-kt/100}P(t)\,dt$. Calculate the present value of all future profit if $P(t) = At + B$, where A and B are nonnegative constants.

38 In problem 37, show that, for a constant yearly profit, present value is inversely proportional to the interest rate k.

4 Improper Integrals with Unbounded Integrands

The improper integrals considered in Section 3 enabled us to calculate the areas of unbounded regions in the xy plane that extend indefinitely to the right or to the left. In this section we consider unbounded regions that extend indefinitely upward or downward.

For instance, consider the unbounded region R under the curve $y = 1/\sqrt{x}$, to the right of the y axis and to the left of the vertical line $x = 9$ (Figure 1a). A good approximation to this unbounded region is afforded by the bounded region R_ε under the curve $y = 1/\sqrt{x}$ between $x = \varepsilon$ and $x = 9$, provided that ε is a small positive number (Figure 1b). As $\varepsilon \to 0^+$, R_ε becomes a better and better approximation to R,

Figure 1

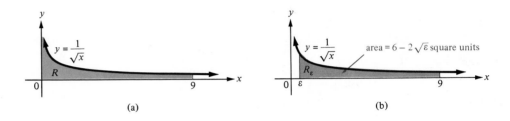

(a)   (b)

and one is tempted to write $R = \lim\limits_{\varepsilon \to 0^+} R_\varepsilon$. Hence, one might expect that

$$\text{area of } R = \lim_{\varepsilon \to 0^+} (\text{area of } R_\varepsilon) = \lim_{\varepsilon \to 0^+} \int_\varepsilon^9 \frac{dx}{\sqrt{x}} = \lim_{\varepsilon \to 0^+} \left(2\sqrt{x} \,\Big|_\varepsilon^9 \right)$$

$$= \lim_{\varepsilon \to 0^+} (6 - 2\sqrt{\varepsilon}) = 6 \text{ square units.}$$

In view of this calculation, we write $\displaystyle\int_0^9 \frac{dx}{\sqrt{x}} = 6$; however, it is important to realize

that, as a definite (Riemann) integral, $\displaystyle\int_0^9 \frac{dx}{\sqrt{x}}$ *fails to exist* because the integrand

$\dfrac{1}{\sqrt{x}}$ is undefined on $[0,9]$ and unbounded on $(0,9]$. Thus, $\displaystyle\int_0^9 \frac{dx}{\sqrt{x}}$ is called an

improper integral. More generally, we make the following definition.

DEFINITION 1   **Improper Integral at Lower Limit**

Suppose that the function f is defined on the half-open interval $(a,b]$ and is Riemann-integrable on every closed interval of the form $[a + \varepsilon, b]$ for $0 < \varepsilon < b - a$.

Then we define the *improper integral* $\displaystyle\int_a^b f(x)\,dx$ by

$$\int_a^b f(x)\,dx = \lim_{\varepsilon \to 0^+} \int_{a+\varepsilon}^b f(x)\,dx,$$

provided that the limit exists as a finite number.

If $\displaystyle\lim_{\varepsilon \to 0^+} \int_{a+\varepsilon}^b f(x)\,dx$ exists as a finite number, we say that the improper integral

$\displaystyle\int_a^b f(x)\,dx$ is *convergent*; otherwise, it is called *divergent*.

Definition 1 takes care of the case in which the definite integral $\int_a^b f(x)\,dx$ may fail to exist because $f(a)$ is undefined. A similar definition,

$$\int_a^b f(x)\,dx = \lim_{\varepsilon \to 0^+} \int_a^{b-\varepsilon} f(x)\,dx,$$

of an *improper integral at the upper limit* covers the case in which $f(b)$ is undefined.

To avoid confusing the improper integrals just defined with ordinary definite integrals, some authors use the notation $\int_{a^+}^b f(x)\,dx$ or $\int_a^{b^-} f(x)\,dx$. However, since we can always determine whether an integral is improper by examining the integrand, we need not be so careful about notation.

EXAMPLES   In Examples 1 through 4, evaluate the given improper integral (if it is convergent).

1 $\int_{3/2}^4 \dfrac{dx}{\left(x - \frac{3}{2}\right)^{2/5}}$

SOLUTION
The integral is improper because the integrand is undefined at the lower limit $\frac{3}{2}$. We begin by using the substitution $u = x - \frac{3}{2}$ to evaluate the corresponding indefinite integral. Thus,

$$\int \frac{dx}{\left(x - \frac{3}{2}\right)^{2/5}} = \int \frac{du}{u^{2/5}} = \tfrac{5}{3}u^{3/5} + C = \tfrac{5}{3}\left(x - \tfrac{3}{2}\right)^{3/5} + C.$$

Applying Definition 1, we have

$$\int_{3/2}^4 \frac{dx}{\left(x - \frac{3}{2}\right)^{2/5}} = \lim_{\varepsilon \to 0^+} \int_{(3/2)+\varepsilon}^4 \frac{dx}{\left(x - \frac{3}{2}\right)^{2/5}}$$

$$= \lim_{\varepsilon \to 0^+} \left[\tfrac{5}{3}\left(x - \tfrac{3}{2}\right)^{3/5} \Big|_{(3/2)+\varepsilon}^4 \right]$$

$$= \lim_{\varepsilon \to 0^+} \left[\tfrac{5}{3}\left(\tfrac{5}{2}\right)^{3/5} - \tfrac{5}{3}\varepsilon^{3/5} \right] = \tfrac{5}{3}\left(\tfrac{5}{2}\right)^{3/5}.$$

2 $\int_0^{\pi/2} \sec x\,dx$

SOLUTION
The integral is improper because the integrand is undefined at the upper limit $\pi/2$. Here we have

$$\int_0^{\pi/2} \sec x\,dx = \lim_{\varepsilon \to 0^+} \int_0^{(\pi/2)-\varepsilon} \sec x\,dx = \lim_{\varepsilon \to 0^+} \left[\ln\,(\sec x + \tan x)\Big|_0^{(\pi/2)-\varepsilon} \right]$$

$$= \lim_{\varepsilon \to 0^+} \left\{ \ln\left[\sec\left(\frac{\pi}{2} - \varepsilon\right) + \tan\left(\frac{\pi}{2} - \varepsilon\right) \right] - \ln\,(\sec 0 + \tan 0) \right\}$$

$$= +\infty,$$

because $\lim_{\varepsilon \to 0^+} \sec\,(\pi/2 - \varepsilon) = +\infty$ and $\lim_{\varepsilon \to 0^+} \tan\,(\pi/2 - \varepsilon) = +\infty$. Therefore, the improper integral is divergent.

3 $\displaystyle\int_0^{\pi/2} \frac{\cos x}{\sqrt{\sin x}}\,dx$

SOLUTION

Here the integral is improper at the lower limit 0. Making the substitution $u = \sin x$, we have

$$\int \frac{\cos x}{\sqrt{\sin x}}\,dx = \int u^{-1/2}\,du = 2u^{1/2} + C = 2\sqrt{\sin x} + C.$$

Therefore,

$$\int_0^{\pi/2} \frac{\cos x}{\sqrt{\sin x}}\,dx = \lim_{\varepsilon \to 0^+} \int_\varepsilon^{\pi/2} \frac{\cos x}{\sqrt{\sin x}}\,dx = \lim_{\varepsilon \to 0^+}\left(2\sqrt{\sin x}\,\Big|_\varepsilon^{\pi/2}\right)$$

$$= \lim_{\varepsilon \to 0^+}\left(2\sqrt{\sin(\pi/2)} - 2\sqrt{\sin \varepsilon}\right) = 2\sqrt{\sin(\pi/2)} = 2.$$

4 $\displaystyle\int_0^b \frac{dx}{x-b}$, where $b > 0$

SOLUTION

Here,

$$\int_0^b \frac{dx}{x-b} = \lim_{\varepsilon \to 0^+} \int_0^{b-\varepsilon} \frac{dx}{x-b} = \lim_{\varepsilon \to 0^+}\left[\left(\ln|x-b|\right)\Big|_0^{b-\varepsilon}\right]$$

$$= \lim_{\varepsilon \to 0^+}\left(\ln \varepsilon - \ln b\right) = -\infty;$$

hence, the improper integral is divergent.

5 Let R be the unbounded region under the curve $y = 1/\sqrt{4 - x^2}$ to the right of the y axis and to the left of the vertical line $x = 2$ (Figure 2). Determine whether the region R has a finite area and, if it does, find this area.

Figure 2

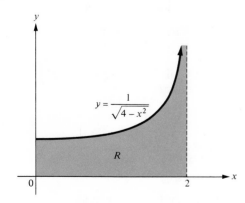

$y = \dfrac{1}{\sqrt{4-x^2}}$

R

SOLUTION

The area in question is given by the improper integral

$$\int_0^2 \frac{dx}{\sqrt{4-x^2}} = \lim_{\varepsilon \to 0^+} \int_0^{2-\varepsilon} \frac{dx}{\sqrt{4-x^2}}$$

$$= \lim_{\varepsilon \to 0^+}\left[\left(\sin^{-1}\frac{x}{2}\right)\Big|_0^{2-\varepsilon}\right]$$

$$= \lim_{\varepsilon \to 0^+}\left(\sin^{-1}\frac{2-\varepsilon}{2} - \sin^{-1} 0\right)$$

$$= \frac{\pi}{2} \text{ square units.}$$

If a function f is defined at every number on a closed interval $[a, b]$ except for a number c, with $a < c < b$, we also refer to $\displaystyle\int_a^b f(x)\,dx$ as an *improper integral.* Such an improper integral is said to *converge* provided that both of the improper integrals

$$\int_a^c f(x)\,dx \quad \text{and} \quad \int_c^b f(x)\,dx$$

converge. Otherwise, it is said to *diverge*. If the improper integral $\int_a^b f(x)\,dx$ converges, then its value is defined by

$$\int_a^b f(x)\,dx = \int_a^c f(x)\,dx + \int_c^b f(x)\,dx.$$

EXAMPLES   In Examples 1 and 2, evaluate the given improper integral (if it is convergent).

1 $\int_{-4}^1 \dfrac{dx}{\sqrt[3]{x+2}}$

SOLUTION

The integrand is undefined at the number -2 between the limits of integration; hence, the integral is improper. Here we have

$$\int_{-4}^1 \frac{dx}{\sqrt[3]{x+2}} = \int_{-4}^{-2} \frac{dx}{\sqrt[3]{x+2}} + \int_{-2}^{1} \frac{dx}{\sqrt[3]{x+2}}$$

$$= \lim_{\varepsilon \to 0^+} \int_{-4}^{-2-\varepsilon} \frac{dx}{\sqrt[3]{x+2}} + \lim_{\varepsilon \to 0^+} \int_{-2+\varepsilon}^{1} \frac{dx}{\sqrt[3]{x+2}}$$

$$= \lim_{\varepsilon \to 0^+} \left[\tfrac{3}{2}(x+2)^{2/3}\Big|_{-4}^{-2-\varepsilon} \right] + \lim_{\varepsilon \to 0^+} \left[\tfrac{3}{2}(x+2)^{2/3}\Big|_{-2+\varepsilon}^{1} \right]$$

$$= \lim_{\varepsilon \to 0^+} \left[\tfrac{3}{2}(-\varepsilon)^{2/3} - \tfrac{3}{2}(-2)^{2/3}\right] + \lim_{\varepsilon \to 0^+} \left[\tfrac{3}{2}(3^{2/3}) - \tfrac{3}{2}(\varepsilon^{2/3})\right]$$

$$= -\tfrac{3}{2}(2^{2/3}) + \tfrac{3}{2}(3^{2/3}) = \tfrac{3}{2}(3^{2/3} - 2^{2/3}).$$

2 $\int_0^\pi \tan x\,dx$

SOLUTION

The integrand is undefined at $x = \pi/2$, so that the improper integral is convergent only if both $\int_0^{\pi/2} \tan x\,dx$ and $\int_{\pi/2}^{\pi} \tan x\,dx$ are convergent. But

$$\int_0^{\pi/2} \tan x\,dx = \lim_{\varepsilon \to 0^+} \int_0^{(\pi/2)-\varepsilon} \tan x\,dx = \lim_{\varepsilon \to 0^+} \left[\ln\,(\sec x)\Big|_0^{(\pi/2)-\varepsilon} \right]$$

$$= \lim_{\varepsilon \to 0^+} \left[\ln \sec \left(\frac{\pi}{2} - \varepsilon \right) - \ln \sec 0 \right] = +\infty;$$

hence, $\int_0^{\pi/2} \tan x\,dx$ diverges. Therefore, $\int_0^{\pi} \tan x\,dx$ also diverges.

Figure 3

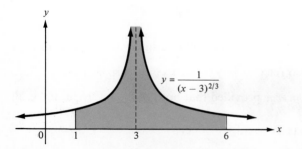

$y = \dfrac{1}{(x-3)^{2/3}}$

3 Let R be the unbounded region under the graph of $y = \dfrac{1}{(x-3)^{2/3}}$ between $x = 1$ and $x = 6$ (Figure 3). Determine whether the region R has a finite area and, if it does, find this area.

SOLUTION

We cannot simply evaluate $\int_1^6 \dfrac{dx}{(x-3)^{2/3}}$ as if it were an ordinary definite integral because the integrand is not defined at $x = 3$. However, if we split the region R into two portions, one to the left and the other to the right of the vertical line $x = 3$, then the areas of these two portions are given by the improper integrals

$$\int_{1}^{3} \frac{dx}{(x - 3)^{2/3}} = \lim_{\varepsilon \to 0^+} \int_{1}^{3-\varepsilon} \frac{dx}{(x - 3)^{2/3}} = \lim_{\varepsilon \to 0^+} \left[3(x - 3)^{1/3} \Big|_{1}^{3-\varepsilon} \right] = 3(2^{1/3}) \quad \text{and}$$

$$\int_{3}^{6} \frac{dx}{(x - 3)^{2/3}} = \lim_{\varepsilon \to 0^+} \int_{3+\varepsilon}^{6} \frac{dx}{(x - 3)^{2/3}} = \lim_{\varepsilon \to 0^+} \left[3(x - 3)^{1/3} \Big|_{3+\varepsilon}^{6} \right] = 3(3^{1/3}).$$

Therefore, the desired area A is actually given by the convergent *improper* integral

$$\int_{1}^{6} \frac{dx}{(x - 3)^{2/3}} = \int_{1}^{3} \frac{dx}{(x - 3)^{2/3}} + \int_{3}^{6} \frac{dx}{(x - 3)^{2/3}} = 3(2^{1/3}) + 3(3^{1/3}) \approx 8.11.$$

Hence, $A \approx 8.11$ square units.

Problem Set 4

In problems 1 through 24, evaluate each improper integral (if it is convergent.)

1 $\int_{0}^{4} \frac{dx}{\sqrt{x}}$

2 $\int_{0}^{9} \frac{dx}{x\sqrt{x}}$

3 $\int_{1}^{28} \frac{dx}{\sqrt[3]{x - 1}}$

4 $\int_{1}^{2} \frac{x \, dx}{\sqrt[3]{x - 1}}$

5 $\int_{0}^{1} \frac{\cos \sqrt[3]{x}}{\sqrt[3]{x^2}} \, dx$

6 $\int_{0}^{1} \frac{dx}{(1 + x)\sqrt{x}}$

7 $\int_{0}^{1} x \ln x \, dx$

8 $\int_{0}^{1} \frac{(\ln x)^2}{x} \, dx$

9 $\int_{0}^{4} \frac{dx}{\sqrt{16 - x^2}}$

10 $\int_{0}^{5} \frac{x \, dx}{\sqrt{25 - x^2}}$

11 $\int_{0}^{4} \frac{e^{-\sqrt{x}}}{\sqrt{x}} \, dx$

12 $\int_{0}^{\pi/2} \tan^5 x \sec^2 x \, dx$

13 $\int_{1/2}^{1} \frac{dt}{t(\ln t)^{2/7}}$

14 $\int_{0}^{1} \frac{1}{x^2} \sin \frac{1}{x} \, dx$

15 $\int_{-1}^{1} \frac{dx}{x^3}$

16 $\int_{1}^{3} \frac{x \, dx}{2 - x}$

17 $\int_{0}^{\pi} \frac{\sin x}{\sqrt[5]{\cos x}} \, dx$

18 $\int_{0}^{2} \frac{x \, dx}{(x - 1)^{2/3}}$

19 $\int_{-\pi}^{\pi} \frac{dt}{1 - \cos t}$

20 $\int_{-1}^{3} \frac{dx}{x\sqrt{x + 4}}$

21 $\int_{0}^{\pi/3} \frac{\sec^2 t}{\sqrt{1 - \tan t}} \, dt$

22 $\int_{e/3}^{2} \frac{dx}{x(\ln x)^3}$

23 $\int_{-1}^{1} \frac{e^x \, dx}{\sqrt[5]{e^x - 1}}$

24 $\int_{-1}^{1} \frac{e^{-1/x}}{x^2} \, dx$

25 Show that the integral $\int_{0}^{1} x^n \ln x \, dx$ converges and has the value $\frac{-1}{(n + 1)^2}$ if and only if $n > -1$.

26 (a) Show that $\lim_{\varepsilon \to 0^+} \left(\int_{-1}^{-\varepsilon} \frac{dx}{x} + \int_{\varepsilon}^{1} \frac{dx}{x} \right) = 0$.

(b) From part (a), can we conclude that $\int_{-1}^{1} \frac{dx}{x} = 0$? Explain.

27 Criticize the following calculation:

$$\int_{-1}^{1} \frac{dx}{x} = (\ln |x|) \Big|_{-1}^{1} = \ln |1| - \ln |-1| = 0.$$

28 Show that the improper integral $\int_0^1 \dfrac{dx}{1 - x^4}$ is divergent.

29 Find the area of the unbounded region under the curve $y = \dfrac{1}{\sqrt{x(2 - x)}}$ between $x = 0$ and $x = 2$.

30 Find the volume of the unbounded solid generated by revolving the region of problem 29 about the x axis.

31 Find the volume of the unbounded solid generated by revolving the region bounded by the curve $y = \dfrac{1}{x - 2}$ and the straight lines $x = 2$, $x = 4$, and $y = 0$ about the x axis.

32 Find the area of the unbounded region beneath the curve $y = \ln(1/x)$ between $x = 0$ and $x = 1$.

33 Find the volume of the unbounded solid generated by revolving the region beneath the curve $y = 1/x^p$ between $x = 0$ and $x = 1$ about the x axis. Assume that $p > 0$.

34 If the function f is defined and continuous on the closed interval $[a, b]$, show that

$$\int_a^b f(x)\, dx = \lim_{\varepsilon \to 0^+} \int_{a+\varepsilon}^b f(x)\, dx.$$

$$\left[Hint: \int_a^b f(x)\, dx = \int_a^{a+\varepsilon} f(x)\, dx + \int_{a+\varepsilon}^b f(x)\, dx. \right]$$

35 Find the volume of the unbounded solid generated by revolving the region described in problem 33 about the y axis.

36 Suppose that f is a continuous nonnegative function defined on the interval $[1, \infty)$ and that R is the unbounded region beneath the graph of f. Assume that the unbounded solid obtained by revolving the region R about the x axis has a finite volume. Is it necessarily true that the unbounded solid obtained by revolving the region R about the y axis has a finite volume? Explain.

5   Taylor's Formula

In Section 1.1 we introduced Cauchy's generalized mean value theorem and used it in establishing L'Hôpital's rule. In this section we present another extension of the mean value theorem which is usually associated with the English mathematician Brook Taylor (1685–1731). This remarkable theorem enables us to approximate various complicated functions by much simpler polynomial functions. The polynomial functions involved in Taylor's theorem are formed according to the following definition.

DEFINITION 1   **Taylor Polynomial**

Let f be a function having a derivative $f^{(n)}$ of order $n \geq 1$ on an open interval I and let a be a fixed number in I. Then the *nth-degree Taylor polynomial for the function f at the number a is the polynomial function P_n defined by*

$$P_n(x) = f(a) + \frac{f'(a)}{1!}(x - a) + \frac{f''(a)}{2!}(x - a)^2 + \frac{f'''(a)}{3!}(x - a)^3 + \cdots + \frac{f^{(n)}(a)}{n!}(x - a)^n.$$

Recall that, if n is a positive integer, then $n!$ is defined by

$$n! = n(n-1)(n-2) \cdots 3 \cdot 2 \cdot 1.$$

Also, by definition, $0! = 1$.

EXAMPLES 1 Find the fourth-degree Taylor polynomial P_4 for the function f defined by $f(x) = \sin x$ at the number $a = \pi/4$.

SOLUTION
Here we arrange the work as follows:

$$f(x) = \sin x, \qquad f\left(\frac{\pi}{4}\right) = \frac{\sqrt{2}}{2},$$

$$f'(x) = \cos x, \qquad f'\left(\frac{\pi}{4}\right) = \frac{\sqrt{2}}{2},$$

$$f''(x) = -\sin x, \qquad f''\left(\frac{\pi}{4}\right) = -\frac{\sqrt{2}}{2},$$

$$f'''(x) = -\cos x, \qquad f'''\left(\frac{\pi}{4}\right) = -\frac{\sqrt{2}}{2}, \quad \text{and}$$

$$f^{(4)}(x) = \sin x, \qquad f^{(4)}\left(\frac{\pi}{4}\right) = \frac{\sqrt{2}}{2}.$$

Hence,

$$P_4(x) = f\left(\frac{\pi}{4}\right) + \frac{f'\left(\frac{\pi}{4}\right)}{1!}\left(x - \frac{\pi}{4}\right) + \frac{f''\left(\frac{\pi}{4}\right)}{2!}\left(x - \frac{\pi}{4}\right)^2$$

$$+ \frac{f'''\left(\frac{\pi}{4}\right)}{3!}\left(x - \frac{\pi}{4}\right)^3 + \frac{f^{(4)}\left(\frac{\pi}{4}\right)}{4!}\left(x - \frac{\pi}{4}\right)^4$$

$$= \frac{\sqrt{2}}{2} + \frac{\sqrt{2}}{2}\left(x - \frac{\pi}{4}\right) - \frac{\sqrt{2}}{4}\left(x - \frac{\pi}{4}\right)^2 - \frac{\sqrt{2}}{12}\left(x - \frac{\pi}{4}\right)^3 + \frac{\sqrt{2}}{48}\left(x - \frac{\pi}{4}\right)^4.$$

2 Find the Taylor polynomial P_3 of degree 3 for the function f defined by $f(x) = \dfrac{1}{1+x}$ for $x > -1$ at the number $a = 0$.

SOLUTION

$$f(x) = (1+x)^{-1}, \qquad f(0) = 1,$$
$$f'(x) = -(1+x)^{-2}, \qquad f'(0) = -1,$$
$$f''(x) = 2(1+x)^{-3}, \qquad f''(0) = 2, \quad \text{and}$$
$$f'''(x) = -6(1+x)^{-4}, \qquad f'''(0) = -6.$$

Hence,

$$P_3(x) = f(0) + \frac{f'(0)}{1!}(x-0) + \frac{f''(0)}{2!}(x-0)^2 + \frac{f'''(0)}{3!}(x-0)^3$$

$$= 1 - x + x^2 - x^3.$$

The Taylor polynomial P_n for the function f at the number a always has the following property, whose proof is left as an exercise (Problem 24):

> The values of the successive derivatives of P_n at the number a—up to and including order n—are equal to the values of the corresponding successive derivatives of f at the number a.

Thus, $f'(a) = P_n'(a)$, $f''(a) = P_n''(a)$, $f'''(a) = P_n'''(a), \ldots$, and $f^{(n)}(a) = P_n^{(n)}(a)$. Also, of course, $f(a) = P_n(a)$. For instance, in Example 2, $P_3(x) = 1 - x + x^2 - x^3$, $P_3'(x) = -1 + 2x - 3x^2$, $P_3''(x) = 2 - 6x$, and $P_3'''(x) = -6$, so that

$$f(0) = 1 = P_3(0),$$
$$f'(0) = -1 = P_3'(0),$$
$$f''(0) = 2 = P_3''(0), \quad \text{and}$$
$$f'''(0) = -6 = P_3'''(0).$$

Perhaps the most important feature of the Taylor polynomial P_n for the function f at a is that the approximation

$$f(x) \approx P_n(x)$$

often is exceedingly accurate, especially if n is large and x is close to a. For instance, in Example 1 (rounding off to six decimal places), we have

$$\sin \frac{\pi}{6} = 0.500000 \approx P_4\left(\frac{\pi}{6}\right) = 0.500008.$$

In practical work, if we use the value $P_n(x)$ to estimate the value $f(x)$, it is necessary to be able to calculate a bound for the error of the estimate. This error, the difference between the true value $f(x)$ and the estimated value $P_n(x)$, is called the *Taylor remainder* $R_n(x)$. Thus, we make the following definition.

DEFINITION 2

Taylor Remainder

If P_n is the nth-degree Taylor polynomial for the function f at the number a, we define the corresponding *Taylor remainder* to be the function R_n given by

$$R_n(x) = f(x) - P_n(x).$$

Notice that $f(x) = P_n(x) + R_n(x)$, so the approximation $f(x) \approx P_n(x)$ will be accurate when $|R_n(x)|$ is small. The extended mean value theorem, referred to in the introduction to the present section, provides an effective means for estimating $|R_n(x)|$.

THEOREM 1

Extended Mean Value Theorem

Let n be a nonnegative integer and suppose that f is a function having a derivative $f^{(n+1)}$ of order $n + 1$ on an open interval I. Then, if a and b are any two distinct numbers in I, there exists a number c strictly between a and b such that

$$f(b) = f(a) + \frac{f'(a)}{1!}(b - a) + \frac{f''(a)}{2!}(b - a)^2 + \frac{f'''(a)}{3!}(b - a)^3 + \cdots + \frac{f^{(n)}(a)}{n!}(b - a)^n + r_n$$

where $r_n = \dfrac{f^{(n+1)}(c)}{(n + 1)!}(b - a)^{n+1}$.

PROOF

Define the number r_n by

$$r_n = f(b) - f(a) - \frac{f'(a)}{1!}(b-a) - \frac{f''(a)}{2!}(b-a)^2$$

$$- \frac{f'''(a)}{3!}(b-a)^3 - \cdots - \frac{f^{(n)}(a)}{n!}(b-a)^n.$$

Then,

$$f(b) = f(a) + \frac{f'(a)}{1!}(b-a) + \frac{f''(a)}{2!}(b-a)^2$$

$$+ \frac{f'''(a)}{3!}(b-a)^3 + \cdots + \frac{f^{(n)}(a)}{n!}(b-a)^n + r_n.$$

Define a new function g with domain I by the equation

$$g(x) = f(x) + \frac{f'(x)}{1!}(b-x) + \frac{f''(x)}{2!}(b-x)^2 + \frac{f'''(x)}{3!}(b-x)^3 + \cdots$$

$$+ \frac{f^{(n)}(x)}{n!}(b-x)^n + r_n \frac{(b-x)^{n+1}}{(b-a)^{n+1}}.$$

Then,

$$g'(x) = f'(x) + \left[-\frac{f'(x)}{1!} + \frac{f''(x)}{1!}(b-x) \right] + \left[-\frac{2f''(x)}{2!}(b-x) + \frac{f'''(x)}{2!}(b-x)^2 \right]$$

$$+ \left[-\frac{3f'''(x)}{3!}(b-x)^2 + \frac{f^{(4)}(x)}{3!}(b-x)^3 \right] + \cdots$$

$$+ \left[-\frac{nf^{(n)}(x)}{n!}(b-x)^{n-1} + \frac{f^{(n+1)}(x)}{n!}(b-x)^n \right] - \frac{(n+1)r_n(b-x)^n}{(b-a)^{n+1}}.$$

In the expression above, each set of brackets contains a term of the form $-\frac{kf^{(k)}(x)}{k!}(b-x)^{k-1}$. Since

$$\frac{k}{k!} = \frac{k}{k(k-1)(k-2)\cdots 1} = \frac{1}{(k-1)!},$$

this term can be rewritten as $-\frac{f^{(k)}(x)}{(k-1)!}(b-x)^{k-1}$, and thus cancels with the term $\frac{f^{(k)}(x)}{(k-1)!}(b-x)^{k-1}$ in the preceding set of brackets.

Performing all such cancellations, we find that all but the last two terms drop out, and we have

$$g'(x) = \frac{f^{(n+1)}(x)}{n!}(b-x)^n - \frac{(n+1)r_n(b-x)^n}{(b-a)^{n+1}}.$$

Going back to our original definition of g and putting $x = a$, we find that

$$g(a) = f(a) + \frac{f'(a)}{1!}(b-a) + \frac{f''(a)}{2!}(b-a)^2$$

$$+ \frac{f'''(a)}{3!}(b-a)^3 + \cdots + \frac{f^{(n)}(a)}{n!}(b-a)^n + r_n \cdot 1,$$

so that $g(a) = f(b)$. Similarly, putting $x = b$ in the definition of g, we obtain $g(b) = f(b)$; hence, $g(a) = g(b)$. Thus, applying Rolle's theorem to the function g on the closed interval from a to b, we conclude that there is a number c strictly between a and b such that $g'(c) = 0$. Using the formula above for $g'(x)$, we have

$$g'(c) = \frac{f^{(n+1)}(c)}{n!}(b-c)^n - \frac{(n+1)r_n(b-c)^n}{(b-a)^{n+1}} = 0$$

or

$$\frac{f^{(n+1)}(c)}{n!}(b-c)^n = \frac{(n+1)r_n(b-c)^n}{(b-a)^{n+1}}.$$

Since $b \neq c$, it follows that $b - c \neq 0$ and the common factor $(b-c)^n$ on both sides of the latter equation can be canceled to yield

$$\frac{f^{(n+1)}(c)}{n!} = \frac{(n+1)r_n}{(b-a)^{n+1}} \quad \text{or} \quad r_n = \frac{f^{(n+1)}(c)}{(n+1)n!}(b-a)^{n+1}.$$

Because $(n+1)n! = (n+1)!$ (why?), we have

$$r_n = \frac{f^{(n+1)}(c)}{(n+1)!}(b-a)^{n+1},$$

as desired.

Notice that if we put $n = 0$ in Theorem 1, we obtain $f(b) = f(a) + r_0$, where $r_0 = \frac{f'(c)}{1!}(b-a)^1$; that is, $f(b) - f(a) = (b-a)f'(c)$, where c is strictly between a and b. Therefore, for $n = 0$, the conclusion of Theorem 1 coincides with the conclusion of the mean value theorem.

As promised, we can now use Theorem 1 to obtain an expression for the Taylor remainder R_n. The appropriate theorem is nothing but a restatement of Theorem 1 as follows.

THEOREM 2 **Taylor's Formula with Lagrange Remainder**
Let f be a function having a derivative $f^{(n+1)}$ of order $n+1$ on an open interval I and let a be a fixed number in I. Denote the nth-degree Taylor polynomial for f at a and the corresponding Taylor remainder by P_n and R_n, respectively. Then, for any x in I, we have

$$f(x) = P_n(x) + R_n(x).$$

If $x \neq a$, there is a number c strictly between a and x such that

$$R_n(x) = \frac{f^{(n+1)}(c)}{(n+1)!}(x-a)^{n+1}.$$

PROOF
Put $b = x$ in Theorem 1, noting that

$$f(a) + \frac{f'(a)}{1!}(x-a) + \frac{f''(a)}{2!}(x-a)^2 + \frac{f'''(a)}{3!}(x-a)^3 + \cdots + \frac{f^{(n)}(a)}{n!}(x-a)^n = P_n(x),$$

and, therefore, that

$$R_n(x) = r_n = \frac{f^{(n+1)}(c)}{(n+1)!}(x-a)^{n+1}.$$

Theorems 1 and 2, which really make the same assertion, but with different notation, are both referred to as "Taylor's theorem" or "Taylor's formula." The expression

$$\frac{f^{(n+1)}(c)}{(n+1)!}(x-a)^{n+1}$$

for $R_n(x)$ given in Theorem 2 is called the *Lagrange form* for the Taylor remainder.

EXAMPLES 1 Find the fourth-degree Taylor polynomial P_4 and the corresponding Taylor remainder R_4 in Lagrange form for the function f defined by $f(x) = \dfrac{1}{x+2}$ for $x > -2$ at the number $a = 1$.

SOLUTION
Here, we have $f'(x) = -(x+2)^{-2}$, $f''(x) = 2(x+2)^{-3}$, $f'''(x) = -6(x+2)^{-4}$, $f^{(4)}(x) = 24(x+2)^{-5}$, and $f^{(5)}(x) = -120(x+2)^{-6}$. Hence, $f(1) = \frac{1}{3}$, $f'(1) = -\frac{1}{9}$, $f''(1) = \frac{2}{27}$, $f'''(1) = -\frac{6}{81}$, $f^{(4)}(1) = \frac{24}{243}$, and

$$P_4(x) = f(1) + \frac{f'(1)}{1!}(x-1) + \frac{f''(1)}{2!}(x-1)^2 + \frac{f'''(1)}{3!}(x-1)^3 + \frac{f^{(4)}(1)}{4!}(x-1)^4$$

$$= \tfrac{1}{3} - \tfrac{1}{9}(x-1) + \tfrac{1}{27}(x-1)^2 - \tfrac{1}{81}(x-1)^3 + \tfrac{1}{243}(x-1)^4.$$

In Lagrange form, the corresponding remainder is given by

$$R_4(x) = \frac{f^{(5)}(c)}{5!}(x-1)^5 = -\frac{120(x-1)^5}{5!(c+2)^6} = -\frac{(x-1)^5}{(c+2)^6},$$

where c is a number strictly between 1 and x.

2 Use the third-degree Taylor polynomial for the function $f(x) = \ln(1+x)$ at the number $a = 0$ to estimate $\ln 1.1$, and then use the Lagrange form of the remainder to place a bound on the error of this estimate.

SOLUTION
Here, we have $f'(x) = (1+x)^{-1}$, $f''(x) = -(1+x)^{-2}$, $f'''(x) = 2(1+x)^{-3}$, and $f^{(4)}(x) = -6(1+x)^{-4}$. Thus, $f(0) = 0$, $f'(0) = 1$, $f''(0) = -1$, $f'''(0) = 2$, so the third-degree Taylor polynomial at $a = 0$ is given by

$$P_3(x) = f(0) + \frac{f'(0)}{1!}(x-0) + \frac{f''(0)}{2!}(x-0)^2 + \frac{f'''(0)}{3!}(x-0)^3$$

$$= 0 + x - \tfrac{1}{2}x^2 + \tfrac{1}{3}x^3.$$

Putting $x = 0.1$ in the approximation $f(x) \approx P_3(x)$, we obtain the estimate

$$\ln 1.1 = f(0.1) \approx P_3(0.1) = 0.1 - \tfrac{1}{2}(0.1)^2 + \tfrac{1}{3}(0.1)^3 = \tfrac{143}{1500}$$

or

$$\ln 1.1 \approx 0.0953333\ldots.$$

The error involved in this estimation is given by the Lagrange remainder

$$f(0.1) - P_3(0.1) = R_3(0.1) = \frac{f^{(4)}(c)}{4!}(0.1 - 0)^4$$

$$= \frac{-6(1+c)^{-4}}{4!}10^{-4} = \frac{-1}{10^4(1+c)^4(4)},$$

where $0 < c < 0.1$. Since $c > 0$, it follows that

$$|R_3(0.1)| = \frac{1}{10^4(1+c)^4(4)} < \frac{1}{10^4(4)};$$

hence,

$$|R_3(0.1)| < \frac{1}{40,000} = 0.000025.$$

Notice that $R_3(0.1)$ is negative, so that the estimated value $\ln 1.1 \approx 0.0953333\ldots$ is actually a little larger than the true value of $\ln 1.1$. However, since the absolute value of the error cannot exceed 0.000025, the given estimate is correct to at least four decimal places, and we therefore write $\ln 1.1 \approx 0.0953$.

The following theorem can often be used to determine—in advance—the degree n of the Taylor polynomial required to guarantee that the absolute value of the error involved in the estimation $f(b) \approx P_n(b)$ does not exceed a specified bound.

THEOREM 3   **Error Bound for Taylor Polynomial Approximation**
Let f be a function having a derivative $f^{(n+1)}$ of order $n+1$ on an open interval I and let a and b be two distinct numbers in I. Suppose that M_n is a constant (depending only on n) and that $|f^{(n+1)}(c)| \le M_n$ holds for all values of c strictly between a and b. Then, if P_n is the nth-degree Taylor polynomial for f at a, the absolute value of the error involved in the estimate $f(b) \approx P_n(b)$ does not exceed
$$M_n \frac{|b-a|^{n+1}}{(n+1)!}.$$

PROOF
By Theorem 2, there is a number c strictly between a and b such that

$$R_n(b) = \frac{f^{(n+1)}(c)}{(n+1)!}(b-a)^{n+1}.$$

Therefore,

$$|f(b) - P_n(b)| = |R_n(b)| = \left| \frac{f^{(n+1)}(c)}{(n+1)!}(b-a)^{n+1} \right|$$

$$= |f^{(n+1)}(c)| \frac{|b-a|^{n+1}}{(n+1)!} \le M_n \frac{|b-a|^{n+1}}{(n+1)!}.$$

EXAMPLES 1 Estimate $\ln 0.99$, making certain that the error does not exceed 10^{-7} in absolute value.

SOLUTION
In Theorem 3, we take $f(x) = \ln x$, $a = 1$ and $b = 0.99$. (We have chosen $a = 1$ because it is near 0.99 and we know that $\ln 1 = 0$.) Here, $f'(x) = x^{-1}$, $f''(x) = -x^{-2}$, $f'''(x) = 2x^{-3}$, $f^{(4)}(x) = -6x^{-4}$, $\ldots$, $f^{(n)}(x) = (-1)^{n-1}(n-1)! \, x^{-n}$, and $f^{(n+1)}(x) = (-1)^n n! \, x^{-(n+1)}$. (If desired, this can be established rigorously by induction on n.) Hence, for $0.99 < c < 1$,

$$|f^{(n+1)}(c)| = |(-1)^n n! \, c^{-(n+1)}| = \frac{n!}{c^{n+1}} < \frac{n!}{(0.99)^{n+1}}$$

and we can take $M_n = n!/(0.99)^{n+1}$ in Theorem 3. Thus, the absolute value of the error involved in the estimation $\ln 0.99 = f(b) \approx P_n(b)$ does not exceed

$$M_n \frac{|0.99 - 1|^{n+1}}{(n+1)!} = \frac{n!}{(0.99)^{n+1}} \frac{(0.01)^{n+1}}{(n+1)!} = \frac{1}{(n+1)(99)^{n+1}}.$$

The smallest value of n for which $1/[(n+1)(99)^{n+1}] \le 10^{-7}$ can be found by trial and error to be $n = 3$; hence, $P_3(0.99)$ approximates $\ln 0.99$ with the desired accuracy. Here,

$$P_3(x) = f(1) + \frac{f'(1)}{1!}(x-1) + \frac{f''(1)}{2!}(x-1)^2 + \frac{f'''(1)}{3!}(x-1)^3$$

$$= 0 + (x-1) - \tfrac{1}{2}(x-1)^2 + \tfrac{1}{3}(x-1)^3,$$

so that

$$\ln 0.99 \approx P_3(0.99) = -0.01 - \tfrac{1}{2}(0.01)^2 - \tfrac{1}{3}(0.01)^3 = -0.0100503333\ldots.$$

Since the error involved in this estimation does not exceed $10^{-7} = 0.0000001$ in absolute value, we can be certain that the approximation $\ln 0.99 \approx -0.0100503$ is accurate to six decimal places. (The true value, rounded off to eight decimal places, is $\ln 0.99 \approx -0.01005034$.)

2 Estimate $\sin 40°$ with an error no more than 10^{-5} in absolute value.

SOLUTION

In Theorem 3 we take $f(x) = \sin x$, $a = \dfrac{\pi}{4} = 45°$ and $b = \dfrac{40\pi}{180} = \dfrac{2\pi}{9} = 40°$. (We have chosen $a = 45°$ because it is near $40°$ and its sine is well known.) Here, $f'(x) = \cos x$, $f''(x) = -\sin x$, $f'''(x) = -\cos x$, $f^{(4)}(x) = \sin x$, and so forth. Hence, $f^{(n+1)}(c)$ is one of $\pm\sin c$ or $\pm\cos c$, so that $|f^{(n+1)}(c)| \le 1$ and we can take $M_n = 1$ in Theorem 3. Since the absolute value of the error of estimation cannot exceed

$$M_n \frac{|b-a|^{n+1}}{(n+1)!} = (1)\frac{\left|\dfrac{2\pi}{9} - \dfrac{\pi}{4}\right|^{n+1}}{(n+1)!} = \frac{\left(\dfrac{\pi}{36}\right)^{n+1}}{(n+1)!},$$

we must choose n large enough so that $\dfrac{(\pi/36)^{n+1}}{(n+1)!} \le 10^{-5}$. By trial and error, the smallest such value is $n = 3$. Since

$$P_3(x) = \sin\frac{\pi}{4} + \frac{\cos(\pi/4)}{1!}\left(x - \frac{\pi}{4}\right) - \frac{\sin(\pi/4)}{2!}\left(x - \frac{\pi}{4}\right)^2 - \frac{\cos(\pi/4)}{3!}\left(x - \frac{\pi}{4}\right)^3,$$

it follows that

$$\sin 40° = \sin\frac{2\pi}{9} \approx P_3\left(\frac{2\pi}{9}\right) = \frac{\sqrt{2}}{2} + \frac{\sqrt{2}}{2}\left(-\frac{\pi}{36}\right) - \frac{\sqrt{2}}{4}\left(-\frac{\pi}{36}\right)^2 - \frac{\sqrt{2}}{12}\left(-\frac{\pi}{36}\right)^3$$

$$= 0.6427859\ldots$$

with an error no greater than $10^{-5} = 0.00001$ in absolute value. (The true value, rounded off to seven decimal places, is $\sin 40° \approx 0.6427876$.)

Problem Set 5

In problems 1 through 16, find the Taylor polynomial of degree n at the indicated number a for each function and write the corresponding Taylor remainder in the Lagrange form.

1 $f(x) = 1/x$, $a = 2$, $n = 6$.

2 $g(x) = \sqrt{x}$, $a = 4$, $n = 5$.

3 $f(x) = 1/\sqrt{x}$, $a = 100$, $n = 4$.

4 $f(x) = \sqrt[3]{x}$, $a = 1000$, $n = 4$

5 $g(x) = (x - 2)^{-2}$, $a = 3$, $n = 5$.

6 $f(x) = (1 - x)^{-1/2}$, $a = 0$, $n = 3$.

7 $f(x) = \sin x$, $a = 0$, $n = 6$.

8 $g(x) = \cos x$, $a = -\pi/3$, $n = 3$.

9 $g(x) = \tan x$, $a = \pi/4$, $n = 4$.

10 $f(x) = e^{2x}$, $a = 0$, $n = 5$.

11 $f(x) = xe^x$, $a = 1$, $n = 3$.

12 $f(x) = e^{-x^2}$, $a = 0$, $n = 3$.

13 $g(x) = 2^x$, $a = 1$, $n = 3$.

14 $f(x) = \ln x$, $a = 1$, $n = 4$.

15 $g(x) = \sinh x$, $a = 0$, $n = 4$.

16 $f(x) = \ln(\cos x)$, $a = \pi/3$, $n = 3$.

In problems 17 through 23, use an appropriate Taylor polynomial to approximate each function value with an error of no more than 10^{-5} in absolute value. In each case, write the answer rounded off to four decimal places.

17 $\sin 1$

18 $\cos 29°$

19 e (*Hint:* $e = e^1$.)

20 $e^{-1.1}$

21 $\ln(0.98)$

22 $\ln 17$ [*Hint:* Write $\ln 17 = \ln 16(1 + \frac{1}{16}) = 4 \ln 2 + \ln(1 + \frac{1}{16})$.]

23 $\sqrt{9.04}$

24 Use mathematical induction to prove that, if P_n is the nth-degree Taylor polynomial at a of the function f, then $f^{(k)}(a) = P_n^{(k)}(a)$ holds for $k = 1, 2, \ldots, n$.

25 Give a bound on the absolute value of the error involved in estimating $\sin 5°$ (that is, $\sin \pi/36$) by using the nth-degree Taylor polynomial for $\sin x$ at $a = 0$. $\left(\text{Use the fact that } \frac{\pi}{36} < \frac{1}{10}.\right)$

26 Give a bound on the absolute value of the error involved in estimating $\sqrt{1 + x}$ by $1 + \frac{1}{2}x$ for $|x| \le 0.1$.

27 Show that the error of the estimate $\cos x = 1 - x^2/2$ does not exceed $x^4/24$ in absolute value. [*Hint:* Notice that $1 - x^2/2 = P_3(x)$.]

28 (a) Show that the difference between the arc length s of an arc of a circle of fixed radius r and the length of the corresponding chord is given by $s - 2r \sin(s/2r)$.
   (b) Use a third-degree Taylor polynomial to approximate the quantity $s - 2r \sin(s/2r)$ and give a bound for the absolute value of the error involved in this approximation.

29 Use a third-degree Taylor polynomial to approximate the area of the smaller segment cut off by the chord in problem 28 and give a bound for the absolute value of the error involved in this approximation.

30 Use a third-degree Taylor polynomial approximation for $\sin x$ at $a = 0$ to find, approximately, a value of $x > 0$ for which $5 \sin x - 4x = 0$.

31 If A dollars are borrowed and $A + B$ dollars are repaid in n equal periodic installments, then the interest rate I per period, charged on the unpaid balance, satisfies the equation

$$(A + B)[1 - (1 + I)^{-n}] = AnI.$$

Using a second-degree Taylor approximation for $(1 + I)^{-n}$, find an approximate solution of the equation above for I in terms of A, B, and n, assuming that I is small.

32 If a flexible cable weighs w kilograms per meter, has a span of s meters between points of support at the same level, and has a tension of H kilograms at its lowest point, then it has a sag of f meters given by

$$f = \frac{H}{2w}\left[\exp\left(\frac{ws}{2H}\right) + \exp\left(\frac{-ws}{2H}\right) - 2\right].$$

Using an appropriate second-degree Taylor polynomial, and assuming that ws/H is small, obtain an approximate formula for f.

33 The tension T in the cable of problem 32 at either point of support is given by $T = \dfrac{H}{2}\left[\exp\left(\dfrac{ws}{2H}\right) + \exp\left(\dfrac{-ws}{2H}\right)\right]$. Assuming that ws/H is small, obtain an approximate formula for T using a suitable second-degree Taylor polynomial.

34 Let f be a polynomial function of degree n and let P_n be the nth-degree Taylor polynomial for f at a. Show that $f(x) = P_n(x)$ for all values of x.

35 (a) Show that, for $x \leq 0$, $e^x = 1 + x + \dfrac{x^2}{2} + \dfrac{x^3}{6} + \dfrac{x^4}{24} + R_4(x)$, where $\dfrac{x^5}{120} \leq R_4(x) \leq 0$.

(b) Replace x by $-t^2$ in part (a) to conclude that

$$e^{-t^2} = 1 - t^2 + \frac{t^4}{2} - \frac{t^6}{6} + \frac{t^8}{24} - r(t), \qquad \text{where } 0 \leq r(t) \leq \frac{t^{10}}{120}.$$

(c) If $b > 0$, use part (b) to show that

$$\int_0^b e^{-t^2}\, dt = b - \frac{b^3}{3} + \frac{b^5}{10} - \frac{b^7}{42} + \frac{b^9}{216} - \varepsilon, \qquad \text{where } 0 \leq \varepsilon \leq \frac{b^{11}}{1320}.$$

(d) Evaluate $\displaystyle\int_0^{3/4} e^{-t^2}\, dt$, rounded off to three decimal places.

36 Assume that f is a function with a continuous derivative $f^{(n+1)}$ of order $n + 1$ on an open interval I. Let a and b be distinct numbers in I and let P_n be the nth-degree Taylor polynomial for f at a. Use induction to prove that the value $R_n(b)$ of the corresponding Taylor remainder is given by

$$R_n(b) = \frac{1}{n!} \int_a^b (b - x)^n f^{(n+1)}(x)\, dx.$$

37 (a) If $f(x) = \dfrac{1}{1 - x}$, show that the Taylor polynomial P_n for f at $a = 0$ is given by
$P_n(x) = 1 + x + x^2 + x^3 + \cdots + x^n$.

(b) Show that the Taylor remainder corresponding to P_n is given by $R_n(x) = \dfrac{x^{n+1}}{1 - x}$.

38 Suppose that f is a function having a derivative of order n on the open interval I and that a is a number in I. Let P be a polynomial function of degree n such that $f(a) = P(a)$ and $f^{(k)}(a) = P^{(k)}(a)$ for $k = 1, 2, \ldots, n$. Prove that P is the nth-degree Taylor polynomial for f at a.

39 Let P_n be the nth-degree Taylor polynomial for f at a, where the function f has a derivative of order n on the open interval I and a belongs to I. Let the function g be defined on I by $g(x) = \displaystyle\int_a^x f(t)\, dt$ and let the function Q be defined on I by $Q(x) = \displaystyle\int_a^x P_n(x)\, dx$. Show that Q is the Taylor polynomial of degree $n + 1$ for g at a.

(*Hint*: Use the result of problem 38.)

40 Suppose that h is a function that has a derivative of order n on the open interval I and that the function g is defined by $g(x) = x^{n+1}h(x)$ for x in I.
(a) Prove that $g(0) = 0$ and that $g^{(k)}(0) = 0$ for $k = 1, 2, \ldots, n$.
(b) Prove that the nth-degree Taylor polynomial for g at $a = 0$ is the zero polynomial.

41 Assume that f is a function having a derivative of order n on the open interval I and that 0 belongs to I. Suppose that the function h also has a derivative of order n on I and that P is a polynomial of degree n such that

$$f(x) = P(x) + x^{n+1}h(x)$$

holds for all values of x in I. Prove that P is the nth-degree Taylor polynomial for f at 0. [*Hint*: Use part (a) of problem 40 and problem 38.]

42 (a) Prove that

$$\frac{1}{1+x^2} = 1 - x^2 + x^4 - x^6 + \cdots + (-1)^n x^{2n} + \frac{(-1)^{n+1} x^{2n+2}}{1+x^2}.$$

(*Hint*: Use problem 37.)

(b) If $f(x) = \dfrac{1}{1+x^2}$, show that the Taylor polynomial P_{2n} of degree $2n$ for f at $a = 0$

is given by $P_{2n}(x) = 1 - x^2 + x^4 - x^6 + \cdots + (-1)^n x^{2n}$ and show that the corresponding Taylor remainder is given by $R_{2n}(x) = \dfrac{(-1)^{n+1} x^{2n+2}}{1+x^2}$. [*Hint*: Use problem 41 and part (a).]

43 Use part (b) of problem 42, problem 39, and the fact that $\tan^{-1} x = \displaystyle\int_0^x \frac{dt}{1+t^2}$ to show that the Taylor polynomial of degree $2n+1$ for the inverse tangent function at $a = 0$ is given by

$$P_{2n+1}(x) = x - \frac{x^3}{3} + \frac{x^5}{5} - \frac{x^7}{7} + \cdots + (-1)^n \frac{x^{2n+1}}{2n+1}.$$

Review Problem Set

In problems 1 through 32, use L'Hôpital's rules to evaluate each limit (if it exists).

1 $\displaystyle\lim_{x\to 0} \frac{xe^x}{1-e^x}$

2 $\displaystyle\lim_{x\to 0} \frac{8^x - 2^x}{4x}$

3 $\displaystyle\lim_{x\to 0} \frac{\ln(\sec 2x)}{\ln(\sec x)}$

4 $\displaystyle\lim_{x\to 0} \frac{\cos 2x - \cos x}{\sin^2 x}$

5 $\displaystyle\lim_{x\to 1^+} \left(\frac{x}{\ln x} - \frac{1}{1-x} \right)$

6 $\displaystyle\lim_{x\to 0} \frac{e^x - 1}{x^2 - x}$

7 $\displaystyle\lim_{x\to 0^-} \frac{2 - 3e^{-x} + e^{-2x}}{2x^2}$

8 $\displaystyle\lim_{x\to 1} \frac{2x^3 + 5x^2 - 4x - 3}{x^3 + x^2 - 10x + 8}$

9 $\displaystyle\lim_{x\to 0} \frac{\sqrt{1-x} - \sqrt{1+x}}{x}$

10 $\displaystyle\lim_{x\to 1^+} \frac{1}{x-1} - \frac{1}{\sqrt{x-1}}$

11 $\displaystyle\lim_{x\to 0^+} x^3 (\ln x)^3$

12 $\displaystyle\lim_{x\to 0^+} \frac{\ln x}{\cot x}$

13 $\displaystyle\lim_{x\to 1^+} \frac{(\ln x)^2}{\sin(x-1)}$

14 $\displaystyle\lim_{x\to 0} \frac{\sqrt{1+\sin x} - \sqrt{1-\sin x}}{\tan x}$

15 $\displaystyle\lim_{x\to 1^+} x \sin \frac{a}{x}$

16 $\displaystyle\lim_{x\to 0} \csc x \sin(\tan x)$

17 $\displaystyle\lim_{x\to 1} \left(\frac{2}{x^2-1} - \frac{1}{x-1} \right)$

18 $\displaystyle\lim_{x\to +\infty} \frac{\sqrt[3]{1+x^6}}{1 - x + 2\sqrt{1 + x^2 + x^4}}$

19 $\displaystyle\lim_{x\to 0^+} \frac{\sin x}{x} \cdot \frac{\sin x}{x - \sin x}$

20 $\displaystyle\lim_{x\to +\infty} (\cosh x - \sinh x)$

21 $\displaystyle\lim_{x\to 1^-} x^{1/(1-x^2)}$

22 $\displaystyle\lim_{x\to 0^+} \left(\frac{\sin x}{x} \right)^{1/x^3}$

23 $\displaystyle\lim_{x\to +\infty} \left(1 + \frac{1}{x} \right)^{x^2}$

24 $\displaystyle\lim_{x\to 0} (1 + ax^2)^{a/x}$

25 $\displaystyle\lim_{x\to +\infty} (x^2 + 4)^{1/x}$

26 $\displaystyle\lim_{x\to 4^+} (x-4)^{x^2-16}$

27 $\displaystyle\lim_{x\to 0} (\cos x)^{1/x^2}$

28 $\lim\limits_{x \to 0} (1 + \sin x)^{\cot x}$

29 $\lim\limits_{x \to 0} [\ln (x + 1)]^x$

30 $\lim\limits_{x \to 0^+} (\tan^{-1} x)^{1/\ln x}$

31 $\lim\limits_{x \to 0^+} \left(\ln \dfrac{1}{x} \right)^x$

32 $\lim\limits_{x \to 0} (\sin^{-1} x)^x$

In problems 33 and 34, find all numbers c satisfying the conclusion of Cauchy's generalized mean value theorem for each of the functions f and g on the indicated interval $[a, b]$.

33 $f(x) = \sqrt{x + 9}$, $g(x) = \sqrt{x}$, and $[a, b] = [0, 16]$.

34 $f(x) = \sin x$, $g(x) = \cos x$, and $[a, b] = [\pi/6, \pi/3]$.

In problems 35 and 36, use the fundamental theorem of calculus and L'Hôpital's rule to find each limit.

35 $\lim\limits_{x \to +\infty} \dfrac{\displaystyle\int_0^x e^t(t^2 - t + 5) \, dt}{\displaystyle\int_0^x e^t(3t^2 + 7t + 1) \, dt}$

36 $\lim\limits_{x \to 0} \dfrac{\displaystyle\int_0^x (\cos^2 t + 5 \cos t^2) \, dt}{\displaystyle\int_0^x e^{-t^2} \, dt}$

37 Show that if α is a fixed positive number, then $\lim\limits_{x \to 0^+} x^\alpha \ln x = 0$.

38 Find constants a and b so that $\lim\limits_{t \to 0} \left(\dfrac{\sin 3t}{t^3} + \dfrac{a}{t^2} + b \right) = 0$.

In problems 39 through 56, evaluate each improper integral (if it is convergent).

39 $\displaystyle\int_1^\infty \dfrac{dx}{x\sqrt{2x^2 - 1}}$

40 $\displaystyle\int_1^\infty \dfrac{t \, dt}{(1 + t^2)^2}$

41 $\displaystyle\int_1^\infty \dfrac{e^{2/t^2} \, dt}{t^3}$

42 $\displaystyle\int_{-\infty}^0 (e^x - e^{2x}) \, dx$

43 $\displaystyle\int_1^\infty \dfrac{x^2 - 1}{x^4} \, dx$

44 $\displaystyle\int_2^\infty \dfrac{x \, dx}{(x^2 - 1)^{3/2}}$

45 $\displaystyle\int_e^\infty \dfrac{dx}{x(\ln x)^{7/2}}$

46 $\displaystyle\int_{-\infty}^0 \dfrac{e^x + 2x}{e^x + x^2} \, dx$

47 $\displaystyle\int_{-\infty}^1 xe^{3x} \, dx$

48 $\displaystyle\int_{-\infty}^\infty x^3 e^{-x} \, dx$

49 $\displaystyle\int_{-3}^1 \dfrac{dx}{x + 3}$

50 $\displaystyle\int_{-2}^6 \dfrac{dx}{\sqrt[3]{x + 2}}$

51 $\displaystyle\int_0^1 \dfrac{e^t \, dt}{\sqrt[3]{e^t - 1}}$

52 $\displaystyle\int_{-2}^2 \dfrac{dx}{\sqrt[5]{x + 1}}$

53 $\displaystyle\int_0^{3a} \dfrac{2x \, dx}{(x^2 - a^2)^{2/3}}$

54 $\displaystyle\int_a^{2a} \dfrac{x^2 \, dx}{\sqrt{x^2 - a^2}}$

55 $\displaystyle\int_0^\infty \dfrac{1}{x^2 + 9} \, dx$

56 $\displaystyle\int_0^\infty \sqrt{x} \, e^{-\sqrt{x}} \, dx$

57 Find the area of the unbounded region beneath the graph of $y = \dfrac{1}{x \ln x}$ and to the right of the line $x = e$.

58 Find the area of the unbounded region beneath the graph of $y = \dfrac{1}{x(x + 2)^2}$ and to the right of the line $x = 1$.

59 Let $f(x) = x^2 e^{-ax}$, where a is a positive constant. Find the volume of the unbounded solid generated by revolving the region in the first quadrant beneath the graph of f: (a) about the x axis, and (b) about the y axis.

60 The gravitational force of attraction F between two particles of masses m_1 and m_2 is given by $F = \gamma(m_1 m_2/s^2)$, where γ is a constant and s is the distance between the particles. Find the work done in moving the particle with mass m_2 along a straight line until it is "infinitely far away" from the other particle if the two particles are initially unit distance apart.

In problems 61 through 64, find the Taylor polynomial P_n at a for each function f and write the corresponding Taylor remainder in the Lagrange form.

61 $f(x) = \sin 2x$, $a = 0$, $n = 3$.

62 $f(x) = \dfrac{1}{(1 + x)^2}$, $a = 1$, $n = 3$.

63 $f(x) = e^{-x}$, $a = 0$, $n = 7$.

64 $f(x) = \cos 3x$, $a = \pi/6$, $n = 6$.

In problems 65 through 70, use an appropriate Taylor polynomial to estimate each quantity. In each case, round your answer off to five decimal places and be certain that $|R_n(b)| \le 5/10^6$.

65 $\sin 88°$

66 $\cos \dfrac{59\pi}{180}$

67 $\ln (1.5)$

68 $\sqrt[10]{e}$

69 $\sqrt{1.03}$

70 $\displaystyle\int_0^{1/2} \sin t^2 \, dt$

13

INFINITE SERIES

In Section 5 of Chapter 12 we discussed the use of Taylor polynomials to approximate the value of various functions and observed that the approximation often becomes more and more accurate as the degree of the Taylor polynomial is increased. For instance, the approximation

$$e^x \approx 1 + \frac{x}{1!} + \frac{x^2}{2!} + \frac{x^3}{3!} + \cdots + \frac{x^n}{n!}$$

becomes better and better as the number of terms on the right becomes larger and larger. This suggests that, in some appropriate sense, e^x should be given exactly by the "infinite sum" of all terms of the form $x^k/k!$; that is,

$$e^x = 1 + \frac{x}{1!} + \frac{x^2}{2!} + \frac{x^3}{3!} + \cdots + \frac{x^n}{n!} + \frac{x^{n+1}}{(n+1)!} + \cdots.$$

Such "infinite sums," which are called *infinite series*, are studied in this chapter. Most of the important functions considered in calculus can be represented as a "sum" of an infinite series in which the terms involve powers of the independent variable; such a series is called a *power series*.

In the chapter we consider the following questions: First, how do we assign meaning to the "sum" of an infinite series? Second, how can we tell whether a given infinite series has such a "sum" (that is, whether it is *convergent*)? Third, when and how can we represent a function by a power series? Fourth, when and how can we differentiate and integrate functions represented by power series?

To answer these questions and to develop a theory of infinite series, we begin by studying the closely related idea of *infinite sequences*.

1 Sequences

The word "sequence" is used in ordinary language to mean a succession of things arranged in a definite order. Here, we are concerned with sequences of numbers such as

$$1, \quad 3, \quad 5, \quad 7, \quad 9 \quad \text{or such as}$$
$$0, \quad 1, \quad 4, \quad 9, \quad 16, \quad 25, \quad 36, \quad 49, \quad 64, \quad \ldots.$$

The individual numbers that appear in a sequence are called the *terms* of the sequence. A sequence having only a finite number of terms (such as the sequence 1, 3, 5, 7, 9) is called a *finite sequence*. Notice that the sequence 0, 1, 4, 9, 16, 25, 36, 49, 64, ... (whose terms are the "perfect squares" arranged in ascending order) involves an infinite number of terms and is therefore an *infinite sequence*. Of course, we cannot write down *all* the terms of an infinite sequence; hence, we resort to the convention of writing the first few terms and then appending three dots to mean "and so forth."

Here, our concern is with infinite sequences only, so from now on we refer to an infinite sequence simply as a *sequence*. Such a sequence can be indicated by

$$a_1, a_2, a_3, \ldots, a_n, \ldots,$$

where a_1 is the first term, a_2 is the second term, a_3 is the third term, and so forth. The "general term," or the *n*th term, is here denoted by a_n. In order to specify the sequence, it is sufficient to provide a rule or a formula for the *n*th term a_n. For instance, the sequence whose *n*th term is given by the formula $a_n = 3n - 1$ has first term $a_1 = 3(1) - 1 = 2$, second term $a_2 = 3(2) - 1 = 5$, and so on. The resulting sequence is

$$2, 5, 8, 11, 14, 17, 20, \ldots, 3n - 1, \ldots.$$

It is important to realize that a sequence is more than a mere collection of numbers; indeed, the numbers in a sequence *appear in a definite order* and also *repetitions of these numbers are permitted*. For instance, the following are perfectly legitimate sequences:

$$1, -1, 1, -1, 1, -1, \ldots, (-1)^{n+1}, \ldots$$

and

$$0, 0, 0, 0, 0, 0, 0, \ldots, 0, \ldots.$$

Sometimes a listing of the first few terms of a sequence indicates beyond any reasonable doubt the rule or formula determining the general term. Examples are:

$$1, 2, 3, 4, 5, 6, \ldots \qquad (a_n = n)$$

$$2, 4, 6, 8, 10, 12, \ldots \qquad (a_n = 2n)$$

$$1, \frac{1}{2}, \frac{1}{3}, \frac{1}{4}, \frac{1}{5}, \frac{1}{6}, \ldots \qquad \left(a_n = \frac{1}{n}\right)$$

$$-1, \frac{1}{3}, -\frac{1}{5}, \frac{1}{7}, -\frac{1}{9}, \frac{1}{11}, \ldots \qquad \left(a_n = (-1)^n \frac{1}{2n - 1}\right).$$

However, it can often be very difficult—if not impossible—to determine the intended general rule from an examination of the numerical pattern formed by the first few terms. When there is the slightest doubt, the safe thing is to specify the general term explicitly. (In this connection, see Problem 40.)

In rigorous mathematical treatises, a *sequence* is defined to be a function f whose domain is the set of positive integers. Then $f(1)$ is called the *first term*, $f(2)$ the *second term*, and in general, $f(n)$ is called the *n*th *term* of the sequence f. From this point of view, the sequence

$$3, \frac{21}{4}, \frac{17}{3}, \frac{93}{16}, \ldots, 6 - \frac{3}{n^2}, \ldots$$

would be identified with the function f whose domain is the positive integers and which is defined by $f(n) = 6 - (3/n^2)$.

The definition of a sequence as a function not only has the advantage of technical precision, but it also permits many of the ideas previously developed for functions to be applied directly to sequences. The reader who is so inclined is encouraged to regard sequences as being functions; however, in this book, we deal with sequences more informally.

We use the notation $\{a_n\}$ as shorthand for the sequence whose nth term is a_n. For instance, $\left\{\dfrac{n}{3n+1}\right\}$ denotes the sequence

$$\frac{1}{4}, \frac{2}{7}, \frac{3}{10}, \frac{4}{13}, \ldots, \frac{n}{3n+1}, \ldots$$

Since $\dfrac{n}{3n+1} = \dfrac{1}{3+(1/n)}$ and, as n grows larger, $1/n$ grows smaller, it is clear that, as we go farther and farther out in this sequence, the terms come closer and closer to the value $\frac{1}{3}$. Thus, we write

$$\lim_{n \to +\infty} \frac{n}{3n+1} = \frac{1}{3}$$

and we say that the sequence $\left\{\dfrac{n}{3n+1}\right\}$ *converges to the limit* $\frac{1}{3}$.

More generally, we say that a sequence $\{a_n\}$ *converges to the limit L* in case $\lim_{n \to +\infty} a_n = L$ in the sense that the difference between a_n and L can be made as small as we please in absolute value, provided that n is made sufficiently large. The following definition expresses the idea of convergence of a sequence more formally.

DEFINITION 1   **Convergence and Divergence of a Sequence**

We write $\lim_{n \to +\infty} a_n = L$ and say that the sequence $\{a_n\}$ *converges to the limit L* provided that, for each positive number ε, there exists a positive integer N (possibly depending on ε) such that

$$|a_n - L| < \varepsilon \qquad \text{whenever } n \geq N.$$

A sequence that converges to a limit is called a *convergent* sequence, while a sequence that is not convergent is said to be *divergent*.

In dealing with an unfamiliar sequence, it is a good idea to write out the first few terms explicitly to gain some insight into its general behavior. However, care must be exercised, since it is the "tail end" and not the first few terms that determines the convergence or divergence of a sequence.

EXAMPLES   Determine whether the given sequence converges or diverges. If it converges, find its limit.

1 $\{10^{1-n}\}$

SOLUTION
Here the sequence is

$$1, \frac{1}{10}, \frac{1}{100}, \frac{1}{1000}, \frac{1}{10,000}, \ldots, \frac{1}{10^{n-1}}, \ldots$$

and it is clear that the terms are steadily becoming smaller and smaller. By choosing

n large enough, we can make $|1/10^{n-1} - 0| = 1/10^{n-1}$ as small as we please; hence, by Definition 1, the sequence converges to the limit 0.

2 $\{(-1)^n\}$

SOLUTION
This sequence,

$$-1, \ 1, \ -1, \ 1, \ -1, \ldots, (-1)^n, \ldots,$$

simply jumps back and forth forever between -1 and 1; hence it cannot approach a limit and is therefore divergent.

1.1 Properties of Limits of Sequences

The calculation of limits of convergent sequences is carried out in much the same way as the calculation of limits of functions. In fact, the two procedures are closely related, as is indicated by the following theorem.

THEOREM 1   **Convergence of Sequences and Functions**
Let the function f be defined on the interval $[1, \infty)$ and define the sequence $\{a_n\}$ by $a_n = f(n)$ for each positive integer n. Then, if $\lim\limits_{x \to +\infty} f(x) = L$, it follows that

$$\lim_{n \to +\infty} a_n = L.$$

The proof of Theorem 1 is left as an exercise (Problem 42).

EXAMPLE   Show that the sequence $\left\{ \dfrac{\ln n}{n} \right\}$ converges and find its limit.

SOLUTION
The function $\dfrac{\ln x}{x}$ is indeterminate of the form ∞/∞ at $+\infty$, so we can apply L'Hôpital's rule to obtain

$$\lim_{x \to +\infty} \frac{\ln x}{x} = \lim_{x \to +\infty} \frac{1/x}{1} = 0.$$

By Theorem 1, $\lim\limits_{n \to +\infty} \left(\dfrac{\ln n}{n} \right) = 0$; that is, the sequence $\left\{ \dfrac{\ln n}{n} \right\}$ converges to the limit 0.

The following properties of limits of sequences are analogous to the properties of limits of functions (see Chapter 1, Section 2); therefore, we simply assume them here and illustrate their use by examples.

Properties of Limits of Sequences

Suppose that the sequences $\{a_n\}$ and $\{b_n\}$ converge to the limits A and B, respectively, and that c is a constant. Then:

1 $\lim\limits_{n \to +\infty} c = c.$

2 $\displaystyle\lim_{n\to+\infty} ca_n = c \lim_{n\to+\infty} a_n = cA.$

3 $\displaystyle\lim_{n\to+\infty} (a_n + b_n) = \left(\lim_{n\to+\infty} a_n\right) + \left(\lim_{n\to+\infty} b_n\right) = A + B$ and

$\displaystyle\lim_{n\to+\infty} (a_n - b_n) = \left(\lim_{n\to+\infty} a_n\right) - \left(\lim_{n\to+\infty} b_n\right) = A - B.$

4 $\displaystyle\lim_{n\to+\infty} (a_n b_n) = \left(\lim_{n\to+\infty} a_n\right)\left(\lim_{n\to+\infty} b_n\right) = AB.$

5 If $b_n \neq 0$ for all positive integers n and if $B \neq 0$, then

$$\lim_{n\to+\infty} \frac{a_n}{b_n} = \frac{\displaystyle\lim_{n\to+\infty} a_n}{\displaystyle\lim_{n\to+\infty} b_n} = \frac{A}{B}.$$

6 $\displaystyle\lim_{n\to+\infty} \frac{c}{n^k} = 0$, if k is a positive constant.

7 If $|a| < 1$, then $\displaystyle\lim_{n\to+\infty} a^n = 0$. If $|a| > 1$, then $\{a^n\}$ diverges.

EXAMPLES Use Theorem 1 together with Properties 1 through 7 to determine the limit of the given sequence (provided that it converges).

1 $\left\{\dfrac{3n^2 + 7n + 11}{8n^2 - 5n + 3}\right\}$

SOLUTION
Dividing the numerator and the denominator of the given fraction by n^2 and applying Properties 1 through 6, we obtain

$$\lim_{n\to+\infty} \frac{3n^2 + 7n + 11}{8n^2 - 5n + 3} = \lim_{n\to+\infty} \frac{3 + \dfrac{7}{n} + \dfrac{11}{n^2}}{8 - \dfrac{5}{n} + \dfrac{3}{n^2}} = \frac{\displaystyle\lim_{n\to+\infty}\left(3 + \dfrac{7}{n} + \dfrac{11}{n^2}\right)}{\displaystyle\lim_{n\to+\infty}\left(8 - \dfrac{5}{n} + \dfrac{3}{n^2}\right)}$$

$$= \frac{\displaystyle\lim_{n\to+\infty} 3 + \lim_{n\to+\infty} \dfrac{7}{n} + \lim_{n\to+\infty} \dfrac{11}{n^2}}{\displaystyle\lim_{n\to+\infty} 8 - \lim_{n\to+\infty} \dfrac{5}{n} + \lim_{n\to+\infty} \dfrac{3}{n^2}}$$

$$= \frac{3 + 0 + 0}{8 - 0 + 0} = \frac{3}{8}.$$

2 $\left\{\dfrac{2^n}{3^{n+1}}\right\}$

SOLUTION

$$\lim_{n\to+\infty} \frac{2^n}{3^{n+1}} = \lim_{n\to+\infty} \frac{1}{3}\left(\frac{2}{3}\right)^n = \frac{1}{3} \lim_{n\to+\infty} \left(\frac{2}{3}\right)^n = \frac{1}{3}(0) = 0.$$

Here we used Property 7 to conclude that $\displaystyle\lim_{n\to+\infty} \left(\frac{2}{3}\right)^n = 0.$

3 $\left\{n \sin \dfrac{\pi}{2n}\right\}$

SOLUTION

$$\lim_{x \to +\infty} x \sin \frac{\pi}{2x} = \lim_{x \to +\infty} \frac{\pi}{2} \cdot \frac{\sin(\pi/2x)}{\pi/2x} = \frac{\pi}{2} \cdot \lim_{x \to +\infty} \frac{\sin(\pi/2x)}{\pi/2x}$$

$$= \frac{\pi}{2} \lim_{t \to 0^+} \frac{\sin t}{t} = \frac{\pi}{2}(1) = \frac{\pi}{2},$$

where we have put $t = \pi/2x$ and observed that $t \to 0^+$ as $x \to +\infty$. Therefore, by Theorem 1,

$$\lim_{n \to +\infty} n \sin \frac{\pi}{2n} = \frac{\pi}{2}.$$

4 $\left\{\dfrac{n^3 + 5n}{7n^2 + 1}\right\}$

SOLUTION
Here

$$\frac{n^3 + 5n}{7n^2 + 1} = \frac{\dfrac{1}{n^2}(n^3 + 5n)}{\dfrac{1}{n^2}(7n^2 + 1)} = \frac{n + \dfrac{5}{n}}{7 + \dfrac{1}{n^2}}.$$

As n grows larger and larger, the numerator $n + (5/n)$ grows larger and larger, while the denominator $7 + (1/n^2)$ approaches 7. Therefore, the fraction becomes large without bound as $n \to +\infty$; hence, the sequence diverges.

1.2 Monotonic and Bounded Sequences

Consider the sequence $\left\{\dfrac{2n}{5n + 3}\right\}$, and notice that its terms

$$\frac{2}{8}, \frac{4}{13}, \frac{6}{18}, \frac{8}{23}, \frac{10}{28}, \ldots, \frac{2n}{5n + 3}, \ldots$$

are steadily growing larger and larger. This can be seen algebraically by writing $\dfrac{2n}{5n + 3} = \dfrac{2}{5 + (3/n)}$ and noticing that, as n grows larger, $3/n$ gets smaller, so that the fraction grows larger.

More generally, we make the following definition.

DEFINITION 2   **Increasing and Decreasing Sequences**
A sequence $\{a_n\}$ is said to be *increasing* (respectively, *decreasing*) if $a_n \le a_{n+1}$ (respectively, $a_n \ge a_{n+1}$) holds for all positive integers n.

A sequence that is either increasing or decreasing is called *monotonic*; otherwise, it is called *nonmonotonic*.

EXAMPLES   Determine whether the given sequence is increasing, decreasing, or nonmonotonic.

1 $\left|\dfrac{2n+1}{3n-2}\right|$

SOLUTION

Here, $a_n = \dfrac{2n+1}{3n-2}$ and $a_{n+1} = \dfrac{2(n+1)+1}{3(n+1)-2} = \dfrac{2n+3}{3n+1}$; hence, for any n,

$$a_{n+1} - a_n = \frac{2n+3}{3n+1} - \frac{2n+1}{3n-2} = \frac{(3n-2)(2n+3) - (2n+1)(3n+1)}{(3n+1)(3n-2)}$$

$$= \frac{(6n^2+5n-6) - (6n^2+5n+1)}{(3n+1)(3n-2)} = \frac{-7}{(3n+1)(3n-2)}.$$

If n is a positive integer, then $3n+1>0$ and $3n-2>0$, so it follows that $a_{n+1} - a_n < 0$; that is, $a_n > a_{n+1}$. Therefore, the given sequence is decreasing.

2 $\left\{\sin \dfrac{n\pi}{2}\right\}$

SOLUTION
The sequence is

$$1, 0, -1, 0, 1, 0, -1, 0, \ldots, \sin \frac{n\pi}{2}, \ldots$$

and it repeats the pattern $1, 0, -1, 0$ again and again; hence, it is nonmonotonic.

3 $\left\{\dfrac{n+5}{n^2+6n+4}\right\}$

SOLUTION

Consider the function $f(x) = \dfrac{x+5}{x^2+6x+4}$ and notice that

$$f'(x) = -\frac{x^2+10x+26}{(x^2+6x+4)^2} = -\frac{(x+5)^2+1}{(x^2+6x+4)^2} < 0 \qquad \text{for } x \geq 1;$$

hence, f is a decreasing function on the interval $[1, \infty)$. In particular, $f(n) > f(n+1)$ holds for all positive integers n; that is, the given sequence is decreasing.

Consider the sequence

$$\frac{1}{2}, \frac{2}{3}, \frac{3}{4}, \frac{4}{5}, \ldots, \frac{n}{n+1}, \ldots$$

and notice that every term in this sequence is less than or equal to 1. Thus, we say that 1 is an *upper bound* for the sequence $\left\{\dfrac{n}{n+1}\right\}$. Similarly, every term in the sequence $\left\{\dfrac{n}{n+1}\right\}$ is greater than or equal to 0, so we say that 0 is a *lower bound* for the sequence. More generally, we make the following definition.

DEFINITION 3   **Bounded Sequences**
A number C (respectively, a number D) is called a *lower bound* (respectively, an *upper bound*) for the sequence $\{a_n\}$ if $C \leq a_n$ (respectively, $a_n \leq D$) holds for every positive integer n.

If the sequence $\{a_n\}$ has a lower bound (respectively, an upper bound), it is said to be *bounded from below* (respectively, *bounded from above*). A sequence is said to be *bounded* if it is bounded both from below and from above.

It is easy to show that a sequence $\{a_n\}$ is bounded if and only if there exists a positive constant M such that $|a_n| < M$ holds for all positive integers n (**Problem 48**). If a sequence converges, it must be bounded (**Problem 46**); however, a bounded sequence need not converge (**Problem 47**).

EXAMPLES   Determine whether the given sequence is bounded from above or from below.

1 $\left\{(-1)^n \dfrac{2n}{3n+1}\right\}$

SOLUTION

Since $0 \le \dfrac{2n}{3n+1} = \dfrac{2}{3+(1/n)} < \dfrac{2}{3}$, then $-\dfrac{2}{3} \le (-1)^n \dfrac{2n}{3n+1} \le \dfrac{2}{3}$ and the sequence is bounded both from above and from below.

2 $\left\{\dfrac{n!}{2^n}\right\}$

SOLUTION
Since all terms in the sequence

$$\frac{1}{2}, \frac{2}{4}, \frac{6}{8}, \frac{24}{16}, \frac{120}{32}, \ldots, \frac{n!}{2^n}, \ldots$$

are positive, 0 serves as a lower bound for $\{n!/2^n\}$. We must decide whether the sequence has an upper bound. Using a table of factorials and experimenting a bit with the given sequence, we find that the terms seem to be growing rather large, even for reasonably small values of n; for instance, the 10th term is 3543.75, while the 15th term is nearly 40,000,000. This leads us to *suspect* that the sequence may be unbounded from above.

To confirm that the sequence $\{n!/2^n\}$ has no upper bound, we must show that, given any positive number K—no matter how large—there is a term $n!/2^n$ far enough out in the sequence so that $n!/2^n > K$; that is,

$$\frac{1}{2} \cdot \frac{2}{2} \cdot \frac{3}{2} \cdot \frac{4}{2} \cdot \frac{5}{2} \cdot \frac{6}{2} \cdots \frac{n}{2} > K.$$

In the product on the left of the desired inequality, all the factors after the first three are greater than or equal to 2, and there are $n - 3$ such factors; hence,

$$\frac{n!}{2^n} = \left(\frac{1}{2} \cdot \frac{2}{2} \cdot \frac{3}{2}\right) \cdot \left(\frac{4}{2} \cdot \frac{5}{2} \cdot \frac{6}{2} \cdots \frac{n}{2}\right) \ge \left(\frac{3}{4}\right)(2^{n-3}).$$

Therefore, $\dfrac{n!}{2^n} > K$ certainly holds if $\left(\dfrac{3}{4}\right)(2^{n-3}) > K$; that is, if $2^{n-3} > \dfrac{4K}{3}$, or $(n-3)\ln 2 > \ln(4K/3)$. Thus, by choosing the positive integer n so that $n > 3 + \dfrac{\ln(4K/3)}{\ln 2}$, we can be certain that $\dfrac{n!}{2^n} > K$. Hence, the sequence $\left\{\dfrac{n!}{2^n}\right\}$ has no upper bound.

Imagine a sequence $a_1, a_2, a_3, \ldots, a_n, \ldots$ whose terms are growing steadily larger but which has an upper bound D. Thus,

$$a_1 \le a_2 \le a_3 \le \cdots \le a_n \le a_{n+1} \le \cdots \le D.$$

If we think of the terms of this sequence as corresponding to points on the number line (Figure 1), then we see an inevitable "pileup" of these points to the left of the point corresponding to D, and we easily persuade ourselves that the sequence must be converging to a limit somewhere in the interval $[a_1, D]$. In this instance our geometric intuition is vindicated by the following theorem, whose proof, unfortunately, is beyond the scope of this book.

Figure 1

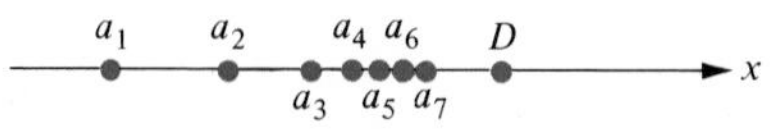

THEOREM 2 **Convergence of Bounded Monotonic Sequences**
Every increasing sequence that is bounded from above is convergent. Likewise, every decreasing sequence that is bounded from below is convergent.

EXAMPLES Use Theorem 2 to show that the given sequence is convergent.

1 $\left\{\dfrac{n}{e^n}\right\}$

SOLUTION

Consider the function $f(x) = \dfrac{x}{e^x}$ and notice that $f'(x) = \dfrac{e^x - xe^x}{(e^x)^2} = \dfrac{1 - x}{e^x} < 0$ for $x > 1$; hence, f is decreasing on $[1, \infty)$. It follows that $f(n) > f(n + 1)$ for all positive integers n; that is, the sequence $\{n/e^n\}$ is decreasing. Since all terms of the sequence are positive, it follows that it is bounded below by the number 0. Hence, by Theorem 2, the sequence converges.

2 $\left\{\dfrac{5^n}{3^n n!}\right\}$

SOLUTION
The first four terms of this sequence (rounded off to three decimal places) are 1.667, 1.389, 0.772, 0.322, ..., so the sequence may be decreasing. To prove that it really is decreasing, we must show that

$$\frac{5^n}{3^n n!} \geq \frac{5^{n+1}}{3^{n+1}(n+1)!};$$

that is,

$$\frac{(n+1)!}{n!} \geq \frac{(5^{n+1})(3^n)}{(5^n)(3^{n+1})} \quad \text{or} \quad n + 1 \geq \frac{5}{3}.$$

The latter inequality is obviously true for any positive integer, so that the sequence is indeed decreasing. All terms of the sequence are positive; hence, 0 is a lower bound. Since the sequence is decreasing and bounded below, it converges, by Theorem 2.

We bring this section to a close by showing how Definition 1 is used to make formal proofs of theorems on limits of sequences. For this purpose, the following theorem (which is intuitively quite clear) serves as a typical example.

THEOREM 3 **Convergent Monotonic Sequences**
The limit of a convergent increasing (respectively, decreasing) sequence is an upper bound (respectively, a lower bound) for the sequence.

PROOF

We prove only the part of the theorem pertaining to increasing sequences since the proof for decreasing sequences is quite similar and only requires reversing some inequalities (Problem 52). Thus, assume that $\{a_n\}$ is monotone increasing and that $\lim_{n \to +\infty} a_n = L$. We must prove that all the terms of the sequence are less than or equal to L. If this were not so, there would be at least one term, say a_q, with $L < a_q$. Thus, let $\varepsilon = a_q - L$, so that $\varepsilon > 0$. By Definition 1, there exists a positive integer N such that $|a_n - L| < \varepsilon$ holds whenever $n \geq N$.

Now, choose the integer n to be larger than both N and q. Since $q < n$, it follows that $a_q \leq a_n$, so that $L < a_q \leq a_n$ and $a_n - L > 0$. Consequently,

$$a_n - L = |a_n - L| < \varepsilon = a_q - L,$$

from which it follows that $a_n < a_q$, contrary to the fact that $a_q \leq a_n$. Therefore, the supposition that there is a term a_q with $L < a_q$ leads to a contradiction. It follows that no term such as a_q can exist, so that L is an upper bound for the sequence, and the theorem is proved.

Problem Set 1

In problems 1 through 4, write out the first six terms of each sequence. Also, write the 100th term.

1 $\{n^2 + 1\}$ **2** $\left\{\dfrac{(-1)^{n+1}}{n+1}\right\}$ **3** $\left\{\dfrac{n}{n^2 + 5}\right\}$ **4** $\left\{2 + \dfrac{1}{n}\right\}$

In problems 5 through 8, find an expression for the general term (nth term) of each sequence.

5 $1, \dfrac{3}{2}, 2, \dfrac{5}{2}, 3, \dfrac{7}{2}, \ldots$ **6** $1, 0, 1, 0, 1, 0, \ldots$ **7** $\dfrac{1}{2}, \dfrac{1}{3}, \dfrac{1}{4}, \dfrac{1}{5}, \dfrac{1}{6}, \ldots$ **8** $1, 9, 25, 49, 81, 121, \ldots$

In problems 9 through 26, determine whether each sequence converges or diverges. If it converges, find its limit.

9 $\left\{\dfrac{100}{n}\right\}$ **10** $\left\{\dfrac{n^2}{5n^2 + 1}\right\}$ **11** $\left\{\dfrac{n^3 - 5n}{7n^3 + 2n}\right\}$

12 $\left\{\dfrac{2n^2 + 1}{9n^2 + 5}\right\}$ **13** $\left\{\dfrac{5n^2}{3n + 1}\right\}$ **14** $\left\{\dfrac{(-1)^n}{10^n}\right\}$

15 $\left\{\dfrac{2n^2 + n}{n + 1} \sin \dfrac{\pi}{2n}\right\}$ **16** $\left\{\dfrac{e^n + e^{-n}}{e^n - e^{-n}}\right\}$ **17** $\left\{\dfrac{\ln(n+1)}{n+1}\right\}$

18 $\{1 + (\tfrac{1}{3})^n - (\tfrac{3}{4})^n\}$ **19** $\left\{\dfrac{\ln(1/n)}{\ln(n+4)}\right\}$ **20** $\{\ln(e^n + 2) - \ln(e^n + 1)\}$

21 $\left\{\dfrac{1}{\sqrt{n^2 + 1} - n}\right\}$ **22** $\{\ln(e^n + 2) - n\}$ **23** $\{n^{1/\sqrt{n}}\}$

24 $\{n^{1/n^2}\}$ **25** $\left\{\left(1 + \dfrac{1}{n}\right)^n\right\}$ **26** $\left\{\left(1 + \dfrac{5}{n}\right)^n\right\}$

In problems 27 through 38, determine whether each sequence is increasing, decreasing, or nonmonotonic, and also whether it is bounded from above or below. Indicate whether the sequence is convergent or divergent.

27 $\left\{\dfrac{2n+1}{3n+2}\right\}$ **28** $\{\sin n\pi\}$ **29** $\{3^n - n\}$

30 $\left\{\dfrac{3^n}{1 + 3^n}\right\}$ **31** $\{(-1)^{n^2}\}$ **32** $\left\{\dfrac{3n^4}{n + 3^n}\right\}$

33 $\left|\dfrac{(-1)^n n}{n+1}\right|$ **34** $\{\sqrt{n+4}-\sqrt{n+3}\}$ **35** $\left|1-\dfrac{2^n}{n}\right|$

36 $\left|\dfrac{n^n}{n!}\right|$ **37** $\left|\dfrac{\sin(n\pi/4)}{n}\right|$ **38** $\left|\dfrac{1\cdot 3\cdot 5\cdot 7\cdots(2n-1)}{n!}\right|$

39 Give an example to show that the sum $\{a_n+b_n\}$ of two unbounded sequences $\{a_n\}$ and $\{b_n\}$ can be a bounded sequence.

40 (a) Write out the first six terms of the sequence

$$\{n+(n-1)(n-2)(n-3)(n-4)(n-5)(n-6)\}.$$

   (b) What is the seventh term of the sequence in part (a)?
   (c) What can you conclude about determining the general term of a sequence from an examination of its first few terms?

41 Determine the convergence or divergence of the sequence $\{a^n\}$ in the following cases:
   (a) $a<-1$ (b) $a=-1$ (c) $-1<a<1$
   (d) $a=1$ (e) $a>1$

42 Prove Theorem 1 in Section 1.1.

43 Assume that a and b are constants with $a>1$ and $b>0$. Show that the sequence $\{n^b/a^n\}$ converges to the limit 0.

44 Suppose that $\{a_n\}$ and $\{b_n\}$ are two sequences whose corresponding terms agree from some point on; that is, suppose that there is a positive integer k such that $a_n=b_n$ holds for all $n\geq k$. If $\lim\limits_{n\to+\infty}a_n=L$, prove that $\lim\limits_{n\to+\infty}b_n=L$, too.

45 Suppose that the sequence $\{a_n\}$ has the property that $a_{n+1}=\frac{1}{2}(a_n+a_{n-1})$ for $n\geq2$ and that $a_1=1$, while $a_2=3$.
   (a) Write out the first eight terms of the sequence $\{a_n\}$.
   (b) Use mathematical induction to prove that $a_n=\dfrac{7}{3}+\dfrac{(-1)^n}{3\cdot 2^{n-3}}$.
   (c) Find $\lim\limits_{n\to+\infty}a_n$.

46 Assume that the sequence $\{a_n\}$ is convergent. Show that $\{a_n\}$ is bounded.

47 By means of a suitable example, show that a bounded sequence need not be convergent.

48 Show that the sequence $\{a_n\}$ is bounded if and only if there exists a positive constant M such that $|a_n|\leq M$ holds for all positive integers n.

49 Suppose that the sequence $\{a_n\}$ is convergent and satisfies the condition that $a_{n+1}=A+Ba_n$ for all positive integers n, where A and B are constants and $B\neq1$. Find $\lim\limits_{n\to+\infty}a_n$. (*Hint:* Take the limit as $n\to+\infty$ on both sides of the equation $a_{n+1}=A+Ba_n$.)

50 In advanced calculus, one proves Stirling's formula:

$$\sqrt{2n\pi}\left(\frac{n}{e}\right)^n<n!<\sqrt{2n\pi}\left(\frac{n}{e}\right)^n\left(1+\frac{1}{12n-1}\right).$$

Use Stirling's formula to prove that the sequence $\left\{\dfrac{n^n}{e^n n!}\right\}$ has an upper bound.

51 Let $|a|<1$. Show that:
   (a) $\{na^n\}$ converges to the limit 0. (b) $\{n^2a^n\}$ converges to the limit 0.

52 Prove the part of Theorem 3 that pertains to decreasing sequences.

2 Infinite Series

An indicated sum of all the terms of an infinite sequence $\{a_n\}$, such as

$$a_1 + a_2 + a_3 + \cdots + a_n + \cdots,$$

is called an *infinite series*, or simply a *series*. Using the "sigma notation" introduced in Section 1 of Chapter 6, we can write this series more compactly as $\sum_{k=1}^{\infty} a_k$. (The notation $\sum_{k=1}^{+\infty} a_k$ would perhaps be preferable, but most textbooks drop the sign on $+\infty$ in this case.) The numbers a_1, a_2, a_3, and so forth are called the *terms* of the series, and a_n is called the *nth term* or the *general term* of the series.

Although we cannot literally add an infinite number of terms, it is sometimes useful to assign a numerical value to an infinite series by means of a special definition and to refer to this value as being the "sum" of the series. This is accomplished by using the "partial sums" of the series.

The sum s_n of the first n terms of a series $\sum_{k=1}^{\infty} a_k$ is called the *nth partial sum* of the series; thus,

$$s_n = a_1 + a_2 + a_3 + \cdots + a_n = \sum_{k=1}^{n} a_k.$$

The sequence $\{s_n\}$ is referred to as the *sequence of partial sums* of the series. Notice that, for each positive integer n,

$$s_{n+1} = s_n + a_{n+1}. \qquad \text{(Why?)}$$

For instance, the first few partial sums of the infinite series

$$1 + \frac{1}{2} + \frac{1}{4} + \frac{1}{8} + \frac{1}{16} + \cdots + \frac{1}{2^{n-1}} + \cdots$$

are as follows:

First partial sum $= s_1 = 1$.

Second partial sum $= s_2 = 1 + \dfrac{1}{2} = 1.5$.

Third partial sum $= s_3 = 1 + \dfrac{1}{2} + \dfrac{1}{4} = 1.75$.

Fourth partial sum $= s_4 = 1 + \dfrac{1}{2} + \dfrac{1}{4} + \dfrac{1}{8} = 1.875$.

Fifth partial sum $= s_5 = 1 + \dfrac{1}{2} + \dfrac{1}{4} + \dfrac{1}{8} + \dfrac{1}{16} = 1.9375$.

We can continue in this way as long as we please. In the present case, we find that the sequence s_n of partial sums

$$1, \quad 1.5, \quad 1.75, \quad 1.875, \quad 1.9375, \ldots$$

seems to be approaching 2 as a limit; for example, the twenty-fifth partial sum is

$$s_{25} = 1 + \frac{1}{2} + \frac{1}{4} + \frac{1}{8} + \frac{1}{16} + \cdots + \frac{1}{2^{24}} \approx 1.99999998.$$

Here it is not difficult to verify that the sequence of partial sums really does converge to 2 as a limit (see Section 2.1); hence, it seems reasonable to *define* the "sum" of the infinite series to be 2 and to write

$$2 = 1 + \frac{1}{2} + \frac{1}{4} + \frac{1}{8} + \frac{1}{16} + \cdots + \frac{1}{2^{n-1}} + \cdots.$$

More generally, we make the following definition.

DEFINITION 1   **Convergence of an Infinite Series**

If the sequence $\{s_n\}$ of partial sums of the infinite series $\sum_{k=1}^{\infty} a_k$ converges to a limit $S = \lim_{n \to +\infty} s_n$, we say that the infinite series $\sum_{k=1}^{\infty} a_k$ *converges* and that its *sum* is S.

If the infinite series $\sum_{k=1}^{\infty}$ converges and its sum is S, we write

$$S = \sum_{k=1}^{\infty} a_k.$$

Of course, an infinite series that does not converge is said to *diverge*.

EXAMPLES   Write out the first five terms of the given series, and then write out the first five terms of the sequence of partial sums of the series. Find a formula for the nth partial sum of the series, determine whether the series converges or diverges, and, if it converges, find its sum S.

1 $\displaystyle\sum_{k=1}^{\infty} \frac{1}{k(k+1)}$

SOLUTION
Here,

$$\sum_{k=1}^{\infty} \frac{1}{k(k+1)} = \frac{1}{2} + \frac{1}{6} + \frac{1}{12} + \frac{1}{20} + \frac{1}{30} + \cdots.$$

The first five partial sums are accordingly given by

$$s_1 = a_1 \qquad = \frac{1}{2}$$

$$s_2 = s_1 + a_2 = \frac{1}{2} + \frac{1}{6} = \frac{2}{3}$$

$$s_3 = s_2 + a_3 = \frac{2}{3} + \frac{1}{12} = \frac{3}{4}$$

$$s_4 = s_3 + a_4 = \frac{3}{4} + \frac{1}{20} = \frac{4}{5}$$

$$s_5 = s_4 + a_5 = \frac{4}{5} + \frac{1}{30} = \frac{5}{6}.$$

These first five partial sums *suggest* that the nth partial sum might be given by the formula

$$s_n = s_{n-1} + a_n = \frac{n}{n+1}.$$

This can be confirmed by mathematical induction on n, but we present an alternative argument. By partial fractions

$$a_k = \frac{1}{k(k+1)} = \frac{1}{k} - \frac{1}{k+1};$$

hence,

$$s_n = \sum_{k=1}^{n} a_k = \sum_{k=1}^{n} \left(\frac{1}{k} - \frac{1}{k+1}\right)$$

$$= \left(\frac{1}{1} - \frac{1}{2}\right) + \left(\frac{1}{2} - \frac{1}{3}\right) + \left(\frac{1}{3} - \frac{1}{4}\right) + \cdots + \left(\frac{1}{n} - \frac{1}{n+1}\right)$$

$$= 1 - \frac{1}{2} + \frac{1}{2} - \frac{1}{3} + \frac{1}{3} - \frac{1}{4} + \cdots + \frac{1}{n} - \frac{1}{n+1}$$

$$= 1 - \frac{1}{n+1} = \frac{n}{n+1}.$$

Therefore, we have

$$S = \lim_{n \to +\infty} s_n = \lim_{n \to +\infty} \frac{n}{n+1} = \lim_{n \to +\infty} \frac{1}{1 + (1/n)} = 1;$$

hence, the series converges and $\sum_{k=1}^{\infty} \frac{1}{k(k+1)} = 1$.

2 $\sum_{k=1}^{\infty} k(k-1)$

SOLUTION
Here,

$$\sum_{k=1}^{\infty} k(k-1) = 0 + 2 + 6 + 12 + 20 + \cdots.$$

The first five partial sums are therefore given by

$$s_1 = a_1 \quad\quad = 0$$
$$s_2 = s_1 + a_2 = 0 \ + 2 \ = 2$$
$$s_3 = s_2 + a_3 = 2 \ + 6 \ = 8$$
$$s_4 = s_3 + a_4 = 8 \ + 12 = 20$$
$$s_5 = s_4 + a_5 = 20 + 20 = 40.$$

Using the formulas for the sums of successive integers and successive squares from Section 1 of Chapter 6, we find that

$$s_n = \sum_{k=1}^{n} k(k-1) = \sum_{k=1}^{n} (k^2 - k) = \sum_{k=1}^{n} k^2 - \sum_{k=1}^{n} k$$

$$= \frac{n(n+1)(2n+1)}{6} - \frac{n(n+1)}{2} = \frac{n(n^2-1)}{3}.$$

Therefore, we have

$$\lim_{n \to +\infty} s_n = \lim_{n \to +\infty} \frac{n(n^2-1)}{3} = +\infty.$$

Thus, the sequence of partial sums—hence also the given series—diverges.

The series $\sum\limits_{k=1}^{\infty} \dfrac{1}{k(k+1)}$ of Example 1, when rewritten in the form $\sum\limits_{k=1}^{\infty} \left(\dfrac{1}{k} - \dfrac{1}{k+1} \right)$, is called a *telescoping series* because of the cancellation that occurs in the calculation of its partial sums. More generally, if $\{b_n\}$ is a sequence, then a series of the form $\sum\limits_{k=1}^{\infty} (b_k - b_{k+1})$ is called a *telescoping series*. The nth partial sum is given by

$$s_n = \sum_{k=1}^{n} (b_k - b_{k+1}) = (b_1 - b_2) + (b_2 - b_3) + \cdots + (b_n - b_{n+1})$$
$$= b_1 - \cancel{b_2} + \cancel{b_2} - \cancel{b_3} + \cdots + \cancel{b_n} - b_{n+1}$$
$$= b_1 - b_{n+1}.$$

Therefore, if $\lim\limits_{n \to +\infty} b_{n+1}$ exists, say, $\lim\limits_{n \to +\infty} b_{n+1} = L$, then we have

$$\sum_{k=1}^{\infty} (b_k - b_{k+1}) = \lim_{n \to +\infty} s_n = \lim_{n \to +\infty} (b_1 - b_{n+1}) = b_1 - L.$$

EXAMPLE   Show that the series $\sum\limits_{k=1}^{\infty} \dfrac{3}{9k^2 + 3k - 2}$ converges and find its sum.

SOLUTION
By partial fractions,

$$\frac{3}{9k^2 + 3k - 2} = \frac{3}{(3k-1)(3k+2)} = \frac{1}{3k-1} - \frac{1}{3k+2} = \frac{1}{3k-1} - \frac{1}{3(k+1)-1};$$

hence,

$$\sum_{k=1}^{\infty} \frac{3}{9k^2 + 3k - 2} = \sum_{k=1}^{\infty} \left[\frac{1}{3k-1} - \frac{1}{3(k+1)-1} \right] = \frac{1}{3(1)-1} - \lim_{n \to +\infty} \frac{1}{3(n+1)-1}$$
$$= \frac{1}{2} - 0 = \frac{1}{2}.$$

In the study of infinite series, it is sometimes useful to be able to manufacture a series $\sum\limits_{k=1}^{\infty} a_n$ with a preassigned sequence $\{s_n\}$ of partial sums. In this connection, the equation

$$a_n = s_n - s_{n-1},$$

which must hold for integral values of n larger than 1, together with the fact that $a_1 = s_1$, provide the desired series.

EXAMPLE   Find an infinite series whose sequence of partial sums is $\left\{ \dfrac{3n}{2n+1} \right\}$, determine whether this series converges, and, if it converges, find its sum.

SOLUTION
Here we have

$$s_n = \frac{3n}{2n+1} \quad \text{and} \quad s_{n-1} = \frac{3(n-1)}{2(n-1)+1} = \frac{3n-3}{2n-1};$$

hence,

$$a_n = s_n - s_{n-1} = \frac{3n}{2n+1} - \frac{3n-3}{2n-1} = \frac{(2n-1)(3n) - (3n-3)(2n+1)}{(2n+1)(2n-1)} = \frac{3}{4n^2 - 1}$$

holds for $n > 1$. Here,

$$a_1 = s_1 = \frac{3(1)}{2(1)+1} = 1,$$

which happens to be the same as the value of $\dfrac{3}{4n^2-1}$ when $n = 1$. Thus, the desired series is given by

$$\sum_{k=1}^{\infty} a_k = \sum_{k=1}^{\infty} \frac{3}{4k^2-1}.$$

Since the nth partial sum of the series is s_n, it follows that

$$\sum_{k=1}^{\infty} \frac{3}{4k^2-1} = \lim_{n \to +\infty} s_n = \lim_{n \to +\infty} \frac{3n}{2n+1} = \lim_{n \to +\infty} \frac{3}{2+(1/n)} = \frac{3}{2}.$$

2.1 Geometric Series

By definition, a *geometric series* is a series of the form

$$\sum_{k=1}^{\infty} ar^{k-1} = a + ar + ar^2 + ar^3 + \cdots + ar^{n-1} + \cdots,$$

in which each term after the first is obtained by multiplying its immediate predecessor by a constant multiplier r. Since r is the ratio between any term (after the first) and its immediate predecessor, we refer to the series as a geometric series with *ratio r*.

Notice that a geometric series is completely specified by giving its initial term a and its ratio r. For instance, the geometric series with initial term $a = 1$ and ratio $r = \frac{1}{2}$ is

$$1 + \frac{1}{2} + \frac{1}{4} + \frac{1}{8} + \frac{1}{16} + \cdots + \frac{1}{2^{n-1}} + \cdots.$$

A negative ratio r produces alternating algebraic signs; for example, the geometric series

$$\frac{2}{3} - \frac{1}{2} + \frac{3}{8} - \frac{9}{32} + \frac{27}{128} - \frac{81}{512} + \cdots$$

has initial term $a = \frac{2}{3}$ and ratio $r = -\frac{3}{4}$.

By a clever trick, it is possible to obtain a simple formula for the nth partial sum s_n of a geometric series $\sum_{k=1}^{\infty} ar^{k-1}$. In fact, starting with

(i) $$s_n = a + ar + ar^2 + \cdots + ar^{n-1},$$

and multiplying through by r, we obtain

(ii) $$s_n r = ar + ar^2 + ar^3 + \cdots + ar^n.$$

Subtracting equation (ii) from equation (i), we have

$$s_n - s_n r = a - ar^n \quad \text{or} \quad s_n(1-r) = a(1-r^n).$$

Therefore,

$$s_n = a\left(\frac{1 - r^n}{1 - r}\right) \qquad \text{if } r \neq 1.$$

By Property 7 in Section 1, if $|r| < 1$, then $\lim_{n \to +\infty} r^n = 0$; hence,

$$\lim_{n \to +\infty} s_n = \lim_{n \to +\infty} a\left(\frac{1 - r^n}{1 - r}\right) = \frac{a}{1 - r} \qquad \text{if } |r| < 1.$$

On the other hand, if $|r| > 1$, then the sequence $\{r^n\}$ diverges, and it follows that the sequence $\{s_n\}$ also diverges. In the remaining case in which $|r| = 1$, so that $r = 1$ or $r = -1$, it is easy to see that the sequence of partial sums diverges (unless $a = 0$). Thus, we have the following theorem.

THEOREM 1 **Geometric Series**

The geometric series $\sum_{k=1}^{\infty} ar^{k-1}$ with initial term $a \neq 0$ and ratio r converges if and only if $|r| < 1$. If $|r| < 1$, then

$$\sum_{k=1}^{\infty} ar^{k-1} = \frac{a}{1 - r}.$$

EXAMPLES Determine whether the given geometric series converges or diverges and, if it converges, find its sum.

1 $\displaystyle\sum_{k=1}^{\infty} \frac{2}{3^{k-1}}$

SOLUTION
Since $\sum_{k=1}^{\infty} \frac{2}{3^{k-1}} = \sum_{k=1}^{\infty} 2\left(\frac{1}{3}\right)^{k-1}$, the given series is indeed geometric with ratio $r = \frac{1}{3}$ and initial term $a = 2$. Because $|r| = \frac{1}{3} < 1$, the series converges and its sum is given by

$$\sum_{k=1}^{\infty} \frac{2}{3^{k-1}} = \frac{a}{1 - r} = \frac{2}{1 - \frac{1}{3}} = 3.$$

2 $-1 + \dfrac{2}{3} - \dfrac{4}{9} + \dfrac{8}{27} - \dfrac{16}{81} + \cdots$

SOLUTION
Here $a = -1$ and the ratio is $r = -\frac{2}{3}$. [For instance, the ratio of the fourth term to the third term is $\left(\frac{8}{27}\right) \div \left(-\frac{4}{9}\right) = -\frac{2}{3}$.] Since $|r| = \frac{2}{3} < 1$, the series is convergent and

$$-1 + \frac{2}{3} - \frac{4}{9} + \frac{8}{27} - \frac{16}{81} + \cdots = \frac{a}{1 - r} = \frac{-1}{1 - (-\frac{2}{3})} = -\frac{3}{5}.$$

3 $\displaystyle\sum_{k=1}^{\infty} \left(\frac{3}{2}\right)^k$

SOLUTION
The series is geometric with initial term $a = \frac{3}{2}$ and ratio $r = \frac{3}{2}$. Since $|r| = \frac{3}{2} > 1$, the series is divergent.

2.2 Applications of Geometric Series

Geometric series arise quite naturally in many branches of mathematics, as the following examples illustrate.

EXAMPLES 1 The probability of making the point "8" in a game of craps—that is, the probability of rolling an 8 with two dice before rolling a 7—is given by

$$\frac{5}{36} + \left(\frac{5}{36}\right)\left(\frac{25}{36}\right) + \left(\frac{5}{36}\right)\left(\frac{25}{36}\right)^2 + \left(\frac{5}{36}\right)\left(\frac{25}{36}\right)^3 + \cdots.$$

Find this probability.

SOLUTION
The displayed series is geometric with initial term $a = \frac{5}{36}$ and ratio $r = \frac{25}{36}$. Its sum is accordingly given by

$$\frac{a}{1-r} = \frac{\frac{5}{36}}{1 - \frac{25}{36}} = \frac{5}{11}.$$

Therefore, the probability of making the point "8" is $\frac{5}{11}$.

2 A simple air pump is evacuating a container of volume V. The cylinder of the pump, with the piston at the top, has volume v and the total mass of air in the container at the outset is M. On the nth stroke of the pump, the mass of air removed from the container is

$$\frac{Mv}{V+v}\left(\frac{V}{V+v}\right)^{n-1}.$$

Assuming that the pump operates "forever," what is the total mass of the air removed from the container?

SOLUTION
The total mass removed is given by the sum of the infinite series

$$\frac{Mv}{V+v} + \frac{Mv}{V+v}\left(\frac{V}{V+v}\right) + \frac{Mv}{V+v}\left(\frac{V}{V+v}\right)^2 + \cdots + \frac{Mv}{V+v}\left(\frac{V}{V+v}\right)^{n-1} + \cdots$$

with initial term $a = \dfrac{Mv}{V+v}$ and ratio $r = \dfrac{V}{V+v}$. Its sum is

$$\frac{a}{1-r} = \frac{\dfrac{Mv}{V+v}}{1 - \left(\dfrac{V}{V+v}\right)} = M.$$

Thus, *all* the air is removed if the pump operates "forever." (Of course, no pump is perfect—valves leak, air leaks around the piston, and so forth—so our answer is only of theoretical interest.)

3 Express the infinite repeating decimal 1.267676767... as a ratio of whole numbers.

SOLUTION

$$1.267676767\ldots = 1.2 + 0.067 + 0.00067 + 0.0000067 + \cdots$$

$$= \frac{12}{10} + \left(\frac{67}{1000} + \frac{67}{100{,}000} + \frac{67}{10{,}000{,}000} + \cdots\right).$$

The geometric series in the parentheses has initial term $a = \frac{67}{1000}$ and ratio $r = \frac{1}{100}$; hence, it converges and its sum is given by

$$\frac{a}{1-r} = \frac{\frac{67}{1000}}{1 - \frac{1}{100}} = \frac{67}{990}.$$

Therefore,

$$1.267676767\ldots = \frac{12}{10} + \frac{67}{990} = \frac{251}{198}.$$

Problem Set 2

In problems 1 through 6, write out the first five terms of each series, and then write out the first five terms of the sequence $\{s_n\}$ of its partial sums. Find a "simple" formula for the nth partial sum s_n in terms of n, determine whether the series converges or diverges, and, if it converges, find its sum $S = \lim_{n \to +\infty} s_n$.

1 $\displaystyle\sum_{k=1}^{\infty} \frac{1}{(2k-1)(2k+1)}$ **2** $\displaystyle\sum_{k=1}^{\infty} \ln\left(1 - \frac{2}{2k+3}\right)$ **3** $\displaystyle\sum_{k=1}^{\infty} k(k+1)$

4 $\displaystyle\sum_{k=1}^{\infty} \frac{1}{k^2 + 2k}$ **5** $\displaystyle\sum_{k=1}^{\infty} \frac{2k+1}{k^2(k+1)^2}$ **6** $\displaystyle\sum_{k=1}^{\infty} \frac{3^{k-1}}{5^k}$

In problems 7 through 12, find an infinite series with the given sequence of partial sums, determine whether this series converges or diverges, and, if it converges, find its sum.

7 $\{s_n\} = \left\{\dfrac{n}{n+1}\right\}$ **8** $\{s_n\} = \left\{\dfrac{2n}{n+5}\right\}$ **9** $\{s_n\} = \left\{\dfrac{2n^2}{3n+5}\right\}$

10 $\{s_n\} = \{n\}$ **11** $\{s_n\} = \{1 - (-1)^n\}$ **12** $\{s_n\} = \left\{2 - \dfrac{1}{2^{n-1}}\right\}$

In problems 13 through 21, find the initial term a and the ratio r of each geometric series, determine whether the series converges, and, if it converges, find its sum.

13 $1 + \dfrac{2}{7} + \dfrac{4}{49} + \dfrac{8}{343} + \cdots$ **14** $\displaystyle\sum_{k=1}^{\infty} \left(\frac{9}{10}\right)^{k+1}$

15 $\displaystyle\sum_{k=1}^{\infty} \left(\frac{7}{6}\right)^k$ **16** $-\dfrac{5}{8} + \dfrac{25}{64} - \dfrac{125}{512} + \dfrac{625}{4096} - \cdots$

17 $1 - 1 + 1 - 1 + 1 - 1 + 1 - 1 + \cdots$ **18** $\displaystyle\sum_{k=1}^{\infty} e^{1-k}$

19 $\displaystyle\sum_{k=1}^{\infty} \frac{3^{k-1}}{4^{k+1}}$ **20** $0.9 + 0.09 + 0.009 + 0.0009 + \cdots$

21 $\displaystyle\sum_{k=1}^{\infty} 5^{-k}$

22 The series $1 - 1 + 1 - 1 + 1 - 1 + \cdots + (-1)^{n-1} + \cdots$ is geometric with ratio $r = -1$; hence, it diverges. Thus, the calculation $1 - 1 + 1 - 1 + 1 - 1 + \cdots + (-1)^{n-1} + \cdots = (1-1) + (1-1) + (1-1) + \cdots = 0 + 0 + 0 + \cdots = 0$ must be incorrect. What is wrong with this calculation?

In problems 23 through 26, express each repeating decimal as a ratio of whole numbers by using an appropriate geometric series.
23 $0.33333\ldots$ **24** $1.11111\ldots$ **25** $4.717171\ldots$ **26** $15.712712712\ldots$

27 Is it true that $\displaystyle\lim_{n \to +\infty} \sum_{k=1}^{n} a_k = \sum_{k=1}^{\infty} a_k$? Explain.

28 Find $\lim\limits_{n \to +\infty} \left(1 + \dfrac{1}{3^2} + \dfrac{1}{3^4} + \dfrac{1}{3^6} + \cdots + \dfrac{1}{3^{2n}} \right)$.

29 In a game of craps, the probability that the shooter wins (that is, rolls 7 or 11 on the first throw or rolls a number other than 2, 3, or 12, then, on a successive roll, repeats this number before rolling a 7) is given by the repeating decimal 0.4929292929.... Express this probability as a ratio of whole numbers.

30 A beaker originally contains 10 grams of salt dissolved in 1000 cubic centimeters of water. The following procedure is performed repeatedly: 250 cubic centimeters of salt water is poured out, replaced by 250 cubic centimeters of pure water, and the solution is thoroughly stirred.
   (a) After the procedure is repeated n times, how many grams of salt has been removed from the beaker?
   (b) If the procedure is repeated "infinitely often," how much salt remains in the beaker?

31 A rubber ball rebounds to 60 percent of the height from which it was dropped. If it is dropped from a height of 2 meters, how far does it travel before coming to rest?

32 Abner starts walking directly toward a brick wall d meters away at a constant speed of v meters per second. At the same instant, a fly departs from Abner's forehead flying directly toward the brick wall at a constant speed of V meters per second, where $V > v$. Upon arriving at the brick wall, the fly immediately turns about and flies back to Abner's forehead at the same speed, V meters per second. The fly continues to shuttle between Abner's forehead and the wall in this manner until Abner finally reaches the wall.
   (a) Show that, on the nth round trip from Abner's forehead to the wall and back, the fly covers a distance of $\dfrac{2Vd}{V + v} \left(\dfrac{V - v}{V + v} \right)^{n-1}$ meters.
   (b) Show that the fly requires $\dfrac{2d}{V + v} \left(\dfrac{V - v}{V + v} \right)^{n-1}$ seconds for the nth round trip.
   (c) By part (a), the total distance flown by the fly is given by $\sum\limits_{n=1}^{\infty} \dfrac{2Vd}{V + v} \left(\dfrac{V - v}{V + v} \right)^{n-1}$ meters. Find this distance by summing the series.
   (d) Using part (b), set up and sum a series to determine the total time required for Abner to reach the wall.
   (e) Determine the total distance flown by the fly without summing an infinite series.

33 Let $\sum\limits_{k=1}^{\infty} a_k$ be a given infinite series and let $\{s_n\}$ be its sequence of partial sums. Define the sequence $\{b_n\}$ by

$$b_n = \begin{cases} 0 & \text{if } n = 1 \\ -s_{n-1} & \text{if } n > 1 \end{cases}$$

for each integer $n \geq 1$. Show that the series $\sum\limits_{k=1}^{\infty} a_k$ is term-by-term exactly the same as the telescoping series $\sum\limits_{k=1}^{\infty} (b_k - b_{k+1})$. Thus, conclude that any infinite series can be rewritten as a telescoping series.

34 Show that

$$\sum_{k=1}^{n} (b_k - b_{k+2}) = (b_1 + b_2) - (b_{n+1} + b_{n+2}).$$

[Hint: $b_k - b_{k+2} = (b_k - b_{k+1}) + (b_{k+1} - b_{k+2})$.]

35 Using problem 34, show that if $\{b_n\}$ is a convergent sequence with $\lim\limits_{n \to +\infty} b_n = L$, then the series $\sum\limits_{k=1}^{\infty} (b_k - b_{k+2})$ is convergent and $\sum\limits_{k=1}^{\infty} (b_k - b_{k+2}) = b_1 + b_2 - 2L$.

3   Properties of Infinite Series

In Section 2 we were able to find the sum of certain infinite series by finding "nice" formulas for their partial sums. For instance, the nth partial sum of the telescoping series $\sum_{k=1}^{\infty} (b_k - b_{k+1})$ is simply $b_1 - b_{n+1}$, while the nth partial sum of the geometric series $\sum_{k=1}^{\infty} ar^{k-1}$ is just $a\left(\dfrac{1-r^n}{1-r}\right)$. Unfortunately, it is not always so easy to find tidy formulas for nth partial sums; hence, it is important to develop alternative methods for determining whether a given series converges or diverges and for dealing with its sum if it does converge. Some of these methods are consequences of the general properties of infinite series which we develop in this section.

The following theorem gives an important property of convergent series.

THEOREM 1   **Necessary Condition for Convergence**

If the infinite series $\sum_{k=1}^{\infty} a_k$ converges, then $\lim_{n \to +\infty} a_n = 0$.

PROOF

Let $\{s_n\}$ be the sequence of partial sums of the series $\sum_{k=1}^{\infty} a_k$. If $\sum_{k=1}^{\infty} a_k$ converges, then, by definition, the sequence $\{s_n\}$ converges and $\lim_{n \to +\infty} s_n = S = \sum_{k=1}^{\infty} a_k$. As $n \to +\infty$, then also $n - 1 \to +\infty$, so that $\lim_{n \to +\infty} s_{n-1} = S$. Notice that $a_n = s_n - s_{n-1}$; hence,

$$\lim_{n \to +\infty} a_n = \lim_{n \to +\infty} (s_n - s_{n-1}) = \lim_{n \to +\infty} s_n - \lim_{n \to +\infty} s_{n-1} = S - S = 0.$$

EXAMPLE   Given that $\sum_{k=1}^{\infty} \dfrac{2^k}{k!}$ converges, find $\lim_{n \to +\infty} \dfrac{2^n}{n!}$.

SOLUTION

By Theorem 1, because $\sum_{k=1}^{\infty} \dfrac{2^k}{k!}$ converges, $\lim_{n \to +\infty} \dfrac{2^n}{n!} = 0$.

Theorem 1 can be used to show that certain series *diverge*. Indeed, an immediate consequence of Theorem 1 is that, if it is false that $\lim_{n \to +\infty} a_n = 0$, then $\sum_{k=1}^{\infty} a_k$ cannot converge. We record this fact for future use as follows.

THEOREM 2   **Sufficient Condition for Divergence**

If $\lim_{n \to +\infty} a_n$ does not exist, or if $\lim_{n \to +\infty} a_n$ exists but is different from zero, then the series $\sum_{k=1}^{\infty} a_k$ is divergent.

EXAMPLES   Use Theorem 2 to show that the given series diverges.

1 $\sum_{k=1}^{\infty} \dfrac{k+1}{k}$

SOLUTION

Since $\lim\limits_{n \to +\infty} \dfrac{n+1}{n} = \lim\limits_{n \to +\infty} \left(1 + \dfrac{1}{n}\right) = 1 \neq 0$, it follows that $\sum\limits_{k=1}^{\infty} \dfrac{k+1}{k}$ diverges, by Theorem 2.

2 $\sum\limits_{k=1}^{\infty} (-1)^k$

SOLUTION

Here, $\lim\limits_{n \to +\infty} (-1)^n$ does not exist. By Theorem 2, therefore, $\sum\limits_{k=1}^{\infty} (-1)^k$ diverges.

(*Caution:* Do not misinterpret Theorem 1. It says that the general term of a convergent series must approach zero, but it emphatically does not assert the converse. *There are lots of divergent series whose general terms approach zero.*)

EXAMPLE   Consider the series $\sum\limits_{k=1}^{\infty} \ln \dfrac{k}{k+1}$, and notice that

$$\lim_{n \to +\infty} \ln \frac{n}{n+1} = \ln\left(\lim_{n \to +\infty} \frac{n}{n+1}\right) = \ln 1 = 0.$$

From this, can we conclude that the series $\sum\limits_{k=1}^{\infty} \ln \dfrac{k}{k+1}$ converges?

SOLUTION

No! Just because the general term approaches zero is no guarantee that the series converges. Here, in fact, $\sum\limits_{k=1}^{\infty} \ln \dfrac{k}{k+1}$ can be rewritten as $\sum\limits_{k=1}^{\infty} [\ln k - \ln(k+1)]$, and the latter telescoping series diverges since $\lim\limits_{n \to +\infty} \ln(n+1) = +\infty$.

Many properties of infinite series are analogs of corresponding properties of sequences. For instance, we have the following theorem.

THEOREM 3   Linear Properties of Series

(i) If $\sum\limits_{k=1}^{\infty} a_k$ and $\sum\limits_{k=1}^{\infty} b_k$ are convergent series, then $\sum\limits_{k=1}^{\infty} (a_k + b_k)$ and $\sum\limits_{k=1}^{\infty} (a_k - b_k)$ are also convergent and

$$\sum_{k=1}^{\infty} (a_k + b_k) = \sum_{k=1}^{\infty} a_k + \sum_{k=1}^{\infty} b_k,$$

while

$$\sum_{k=1}^{\infty} (a_k - b_k) = \sum_{k=1}^{\infty} a_k - \sum_{k=1}^{\infty} b_k.$$

(ii) If $\sum\limits_{k=1}^{\infty} a_k$ is a convergent series and c is a constant, then $\sum\limits_{k=1}^{\infty} ca_k$ is also convergent and $\sum\limits_{k=1}^{\infty} ca_k = c \sum\limits_{k=1}^{\infty} a_k$. If $\sum\limits_{k=1}^{\infty} a_k$ is a divergent series and c is a nonzero constant, then $\sum\limits_{k=1}^{\infty} ca_k$ is also divergent.

PROOF

We prove part (i) and leave part (ii) as an exercise (Problem 37). Thus, let $s_n = \sum_{k=1}^{n} a_k$ and $t_n = \sum_{k=1}^{n} b_k$ be the nth partial sums of the two given series. Then

$$s_n + t_n = \sum_{k=1}^{n} a_k + \sum_{k=1}^{n} b_k = \sum_{k=1}^{n} (a_k + b_k)$$

is the nth partial sum of the series $\sum_{k=1}^{\infty} (a_k + b_k)$. Since

$$\lim_{n \to +\infty} (s_n + t_n) = \lim_{n \to +\infty} s_n + \lim_{n \to +\infty} t_n = \sum_{k=1}^{\infty} a_k + \sum_{k=1}^{\infty} b_k,$$

$\sum_{k=1}^{\infty} (a_k + b_k)$ converges and its sum is given by $\sum_{k=1}^{\infty} (a_k + b_k) = \sum_{k=1}^{\infty} a_k + \sum_{k=1}^{\infty} b_k$. By a similar argument (Problem 35), $\sum_{k=1}^{\infty} (a_k - b_k) = \sum_{k=1}^{\infty} a_k - \sum_{k=1}^{\infty} b_k$.

EXAMPLE   Find the sum of the series $\sum_{k=1}^{\infty} \left(\dfrac{5}{2^{k-1}} + \dfrac{1}{3^{k-1}} \right)$.

SOLUTION

Notice that $\sum_{k=1}^{\infty} \dfrac{5}{2^{k-1}}$ is a geometric series with initial term $a = 5$ and ratio $= \dfrac{1}{2}$; hence it converges and

$$\sum_{k=1}^{\infty} \frac{5}{2^{k-1}} = \frac{a}{1-r} = \frac{5}{1 - \frac{1}{2}} = 10.$$

Similarly, $\sum_{k=1}^{\infty} \dfrac{1}{3^{k-1}}$ converges and

$$\sum_{k=1}^{\infty} \frac{1}{3^{k-1}} = \frac{1}{1 - \frac{1}{3}} = \frac{3}{2}.$$

It follows from part (i) of Theorem 3 that $\sum_{k=1}^{\infty} \left(\dfrac{5}{2^{k-1}} + \dfrac{1}{3^{k-1}} \right)$ converges and

$$\sum_{k=1}^{\infty} \left(\frac{5}{2^{k-1}} + \frac{1}{3^{k-1}} \right) = 10 + \frac{3}{2} = \frac{23}{2}.$$

The following useful theorem is an immediate consequence of part (i) of Theorem 3.

THEOREM 4   **Divergence of a Series of Sums**

If the series $\sum_{k=1}^{\infty} a_k$ converges and the series $\sum_{k=1}^{\infty} b_k$ diverges, then the series $\sum_{k=1}^{\infty} (a_k + b_k)$ diverges.

PROOF

Assume the contrary; that is, suppose that $\sum_{k=1}^{\infty}(a_k+b_k)$ converges. Since $\sum_{k=1}^{\infty}a_k$ converges, $\sum_{k=1}^{\infty}[(a_k+b_k)-a_k]=\sum_{k=1}^{\infty}b_k$ also converges, by part (i) of Theorem 3, contradicting the hypothesis that $\sum_{k=1}^{\infty}b_k$ diverges. Hence, the assumption that $\sum_{k=1}^{\infty}(a_k+b_k)$ converges must be wrong; that is, $\sum_{k=1}^{\infty}(a_k+b_k)$ must diverge.

EXAMPLE Determine whether the series $\sum_{k=1}^{\infty}\left(\ln\dfrac{k}{k+1}-\dfrac{1}{3^k}\right)$ converges or diverges.

SOLUTION

The series $\sum_{k=1}^{\infty}\ln\dfrac{k}{k+1}$ diverges. (Why?) However, the series $\sum_{k=1}^{\infty}\dfrac{-1}{3^k}$ is a geometric series with initial term $a=-\frac{1}{3}$ and ratio $r=\frac{1}{3}$; hence, it converges. Therefore, by Theorem 4, the series

$$\sum_{k=1}^{\infty}\left(\ln\frac{k}{k+1}-\frac{1}{3^k}\right)=\sum_{k=1}^{\infty}\left(\frac{-1}{3^k}+\ln\frac{k}{k+1}\right)$$

must be divergent.

Notice that even if *both* series $\sum_{k=1}^{\infty}a_k$ and $\sum_{k=1}^{\infty}b_k$ are divergent, the series $\sum_{k=1}^{\infty}(a_k+b_k)$ *may be convergent.* For instance, let $a_n=n$ and $b_n=-n$ for all positive integers n.

Calculations with infinite series are often simplified by various manipulations involving the summation index. Such maneuvers are usually obvious and virtually self-explanatory. For instance, it is not necessary to begin a series with $k=1$. Thus, we can write

$$\sum_{k=0}^{\infty}\frac{1}{2^k}=1+\frac{1}{2}+\frac{1}{4}+\frac{1}{8}+\cdots,$$

$$\sum_{k=2}^{\infty}\ln\frac{k-1}{k}=\ln\frac{1}{2}+\ln\frac{2}{3}+\ln\frac{3}{4}+\cdots,$$

and so forth. Also, there is no particular reason to use the symbol k for the summation index; indeed it is possible, and often desirable, to change the summation index in an infinite series in much the same way that variables are changed in integrals. For instance, in the series $\sum_{k=1}^{\infty}\dfrac{1}{2^{k-1}}$, let us put $j=k-1$, noting that $j=0$ when $k=1$. Then we obtain

$$\sum_{k=1}^{\infty}\frac{1}{2^{k-1}}=\sum_{j=0}^{\infty}\frac{1}{2^j}.$$

As the following theorem shows, the first few terms of an infinite series have no effect whatsoever on the convergence or divergence of the series—it is only the "tail end" of the series that matters as far as convergence or divergence is concerned.

THEOREM 5   **Removing the First M Terms from a Series**

If M is a fixed positive integer, then the series $\sum\limits_{k=1}^{\infty} a_k$ converges if and only if the series $\sum\limits_{k=M+1}^{\infty} a_k$ converges. Moreover, if these series converge, then

$$\sum_{k=1}^{\infty} a_k = \sum_{k=1}^{M} a_k + \sum_{k=M+1}^{\infty} a_k.$$

PROOF
For $n > M$, we have

$$\sum_{k=1}^{n} a_k = \sum_{k=1}^{M} a_k + \sum_{k=M+1}^{n} a_k.$$

Since $\sum\limits_{k=1}^{M} a_k$ is a constant, it follows that $\lim\limits_{n\to+\infty} \sum\limits_{k=1}^{n} a_k$ exists if and only if $\lim\limits_{n\to+\infty} \sum\limits_{k=M+1}^{n} a_k$ exists; that is, $\sum\limits_{k=1}^{\infty} a_k$ converges if and only if $\sum\limits_{k=M+1}^{\infty} a_k$ converges. Assuming that these limits do exist, and taking the limit on both sides of the above equation as $n \to +\infty$, we obtain

$$\sum_{k=1}^{\infty} a_k = \sum_{k=1}^{M} a_k + \sum_{k=M+1}^{\infty} a_k.$$

Let us illustrate Theorem 5 using the geometric series $\sum\limits_{k=1}^{\infty} ar^{k-1}$ with $|r| < 1$. Notice that

$$\sum_{k=M+1}^{\infty} ar^{k-1} = ar^M + ar^{M+1} + ar^{M+2} + \cdots$$

is also a geometric series with initial term ar^M, ratio r, and sum equal to $\dfrac{ar^M}{1-r}$. Also, $\sum\limits_{k=1}^{M} ar^{k-1}$ is just the Mth partial sum of $\sum\limits_{k=1}^{\infty} ar^{k-1}$; consequently, $\sum\limits_{k=1}^{M} ar^{k-1} = a\left(\dfrac{1-r^M}{1-r}\right)$. Thus, the equation of Theorem 5 becomes

$$\sum_{k=1}^{\infty} ar^{k-1} = \sum_{k=1}^{M} ar^{k-1} + \sum_{k=M+1}^{\infty} ar^{k-1}$$

or

$$\frac{a}{1-r} = a\left(\frac{1-r^M}{1-r}\right) + \frac{ar^M}{1-r},$$

an obvious algebraic identity.

By Theorem 2 in Section 1.2, if the sequence $\{s_n\}$ of partial sums of a series $\sum\limits_{k=1}^{\infty} a_k$ is monotonic and bounded, then the sequence $\{s_n\}$—hence also the series $\sum\limits_{k=1}^{\infty} a_k$—is convergent. In particular, we have the following theorem.

THEOREM 6   **Convergence of a Series of Nonnegative Terms Whose Partial Sums Are Bounded**

Let $\sum\limits_{k=1}^{\infty} a_k$ be an infinite series whose terms are all nonnegative (that is, $a_k \geq 0$ for all k). If the sequence $\{s_n\}$ of nth partial sums of $\sum\limits_{k=1}^{\infty} a_k$ is bounded above (that is, $s_n \leq M$ for all n, where M is a constant), then the series $\sum\limits_{k=1}^{\infty} a_k$ is convergent.

PROOF
Since $s_{n+1} = s_n + a_{n+1}$ (why?) and $a_{n+1} \geq 0$, then $s_{n+1} \geq s_n$ holds for all integers $n \geq 1$. Thus, $\{s_n\}$ is an increasing sequence that is bounded above. It follows that $\{s_n\}$ is convergent; hence, the series $\sum\limits_{k=1}^{\infty} a_k$ is convergent.

EXAMPLE   Use Theorem 6 to show that the series $\sum\limits_{k=1}^{\infty} \dfrac{k-1}{k \cdot 2^k}$ converges.

SOLUTION
Clearly, each term of the given series is nonnegative. By Theorem 6, the series converges if the sequence $\{s_n\}$ of partial sums $s_n = \sum\limits_{k=1}^{n} \dfrac{k-1}{k \cdot 2^k}$ is bounded. Notice that $\dfrac{k-1}{k} < 1$, so that

$$\frac{k-1}{k \cdot 2^k} = \frac{k-1}{k}\left(\frac{1}{2}\right)^k < \left(\frac{1}{2}\right)^k.$$

Therefore,

$$s_n = \sum_{k=1}^{n} \frac{k-1}{k \cdot 2^k} < \sum_{k=1}^{n} \left(\frac{1}{2}\right)^k = \frac{1}{2} \cdot \frac{1 - (\frac{1}{2})^n}{1 - \frac{1}{2}} < \frac{1}{2} \cdot \frac{1}{1 - \frac{1}{2}} = 1,$$

so $\{s_n\}$ is bounded above by $M = 1$ and $\sum\limits_{k=1}^{\infty} \dfrac{k-1}{k \cdot 2^k}$ converges.

Problem Set 3

In problems 1 through 8, show that each series diverges by showing that the general term does not approach zero.

1 $\sum\limits_{k=1}^{\infty} \dfrac{k}{5k+7}$

2 $\sum\limits_{k=1}^{\infty} \ln\left(\dfrac{5k}{12k+5}\right)$

3 $\sum\limits_{k=1}^{\infty} \dfrac{3k^2 + 5k}{7k^2 + 13k + 2}$

4 $\sum\limits_{k=1}^{\infty} \dfrac{e^k}{3e^k + 7}$

5 $\sum\limits_{k=1}^{\infty} \sin \dfrac{\pi k}{4}$

6 $\sum\limits_{k=1}^{\infty} \dfrac{k}{\cos k}$

7 $\sum\limits_{k=1}^{\infty} k \sin \dfrac{1}{k}$

8 $\sum\limits_{k=1}^{\infty} \dfrac{k!}{2^k}$

In problems 9 through 14, use the linear properties of series to find the sum of each series.

9 $\sum\limits_{k=1}^{\infty} \left[\left(\dfrac{1}{3}\right)^k + \left(\dfrac{1}{4}\right)^k\right]$

10 $\sum\limits_{k=1}^{\infty} \left[\left(\dfrac{1}{2}\right)^{k-1} - \left(-\dfrac{1}{3}\right)^{k+1}\right]$

11 $\sum\limits_{k=1}^{\infty} \left[\dfrac{1}{k(k+1)} - \left(\dfrac{3}{4}\right)^{k-1}\right]$

12 $\sum\limits_{k=0}^{\infty} \left[2\left(\dfrac{1}{3}\right)^k - 3\left(-\dfrac{1}{5}\right)^{k+1}\right]$

13 $\sum\limits_{k=1}^{\infty} \left(\dfrac{2^k + 3^k}{6^k} - \dfrac{1}{7^{k+1}}\right)$

14 $\sum\limits_{k=1}^{\infty} \left(\sin \dfrac{1}{k} + 2^{-k} - \sin \dfrac{1}{k+1}\right)$

15 Does the fact that $\lim_{n \to +\infty} \frac{1}{n} = 0$ guarantee the convergence of the series $\sum_{k=1}^{\infty} \frac{1}{k}$?

16 Given that $\sum_{k=1}^{\infty} \frac{c^k}{k!}$ converges for each value of the constant c, find $\lim_{n \to +\infty} \frac{c^n}{n!}$.

17 Given that $1 - \frac{1}{2} + \frac{1}{3} - \frac{1}{4} + \frac{1}{5} - \frac{1}{6} + \cdots = \ln 2$, find the sum of the series

$$-2 + 1 - \frac{2}{3} + \frac{2}{4} - \frac{2}{5} + \frac{2}{6} - \frac{2}{7} + \cdots.$$

18 Criticize the following calculation: Let $\sum_{k=1}^{\infty} (b_k - b_{k+1})$ be a convergent telescoping series. Then

$$\sum_{k=1}^{\infty} (b_k - b_{k+1}) = \sum_{k=1}^{\infty} b_k - \sum_{k=1}^{\infty} b_{k+1}$$
$$= (b_1 + b_2 + b_3 + \cdots) - (b_2 + b_3 + \cdots) = b_1?$$

19 Show that the series $\sum_{k=1}^{\infty} \left[\frac{1}{k(k+1)} - \ln \frac{k}{k+1} \right]$ diverges.

In problems 20 through 23, rewrite each series by changing the summation index from k to j as indicated.

20 $\sum_{k=1}^{\infty} ar^{k-1}; j = k - 1$

21 $\sum_{k=2}^{\infty} \frac{1}{k(k-1)}; j = k - 1$

22 $\sum_{k=M}^{\infty} a_k; j = k - M + 1$

23 $\sum_{k=1}^{\infty} a_k; j = k + M - 1$

24 Suppose that $\sum_{k=1}^{\infty} (b_k - b_{k+1})$ is a convergent telescoping series. By Theorem 5,

$$\sum_{k=1}^{\infty} (b_k - b_{k+1}) = \sum_{k=1}^{M} (b_k - b_{k+1}) + \sum_{k=M+1}^{\infty} (b_k - b_{k+1});$$

that is,

$$b_1 - \lim_{n \to +\infty} b_n = b_1 - b_{M+1} + \sum_{k=M+1}^{\infty} (b_k - b_{k+1}), \text{ or } \sum_{k=M+1}^{\infty} (b_k - b_{k+1}) = b_{M+1} - \lim_{n \to +\infty} b_n.$$

Verify the latter equation directly without using Theorem 5.

25 Use the fact that $\sum_{k=1}^{\infty} \frac{1}{k(k+1)} = 1$ and the fact that $\sum_{k=1}^{M} \frac{1}{k(k+1)} = 1 - \frac{1}{M+1}$ to find the sum of $\sum_{k=M+1}^{\infty} \frac{1}{k(k+1)}$.

26 (a) Use Theorem 5 to show that if two series agree, term by term, except possibly for the first M terms, then either they both converge or else they both diverge.
(b) Show that changing, deleting, or adding a single term cannot affect the convergence or divergence of a series.

27 Given that $e = \sum_{k=1}^{\infty} \frac{1}{(k-1)!}$, find the sum of the series $1 + \frac{1}{2!} + \frac{1}{3!} + \frac{1}{4!} + \cdots$.

28 If $\sum_{k=1}^{\infty} a_k$ is a convergent series all of whose terms are nonnegative, show that $\sum_{k=1}^{M} a_k \le \sum_{k=1}^{\infty} a_k$ holds for all positive integers M. (*Hint:* Use Theorem 3 in Section 1.)

In problems 29 through 34, all the series have nonnegative terms. In each case, establish the convergence of the series by proving directly that its sequence of partial sums is bounded above.

29 $\displaystyle\sum_{k=1}^{\infty} \frac{k}{(k+1)\cdot 3^k}$

30 $\displaystyle\sum_{k=1}^{\infty} \frac{(k-1)\ln 3}{4^{k-1}}$

31 $\displaystyle\sum_{k=0}^{\infty} \frac{4^{-k}k}{k^2+1}$

32 $\displaystyle\sum_{k=0}^{\infty} \frac{k}{5^k}$

33 $\displaystyle\sum_{k=1}^{\infty} \frac{1}{k^2}$ $\left[Hint: \dfrac{1}{k^2} \le \dfrac{1}{(k-1)k} \text{ for } k \ge 2.\right]$

34 $\displaystyle\sum_{k=1}^{\infty} \frac{1}{k!}$

35 Complete the proof of Theorem 3 by showing that if $\displaystyle\sum_{k=1}^{\infty} a_k$ and $\displaystyle\sum_{k=1}^{\infty} b_k$ are convergent, then so is $\displaystyle\sum_{k=1}^{\infty} (a_k - b_k)$ and

$$\sum_{k=1}^{\infty} (a_k - b_k) = \sum_{k=1}^{\infty} a_k - \sum_{k=1}^{\infty} b_k.$$

36 Prove that, if a series of nonnegative terms converges, then its sequence of partial sums must be bounded.

37 Prove part (ii) of Theorem 3.

4 Series of Nonnegative Terms

In Theorem 6 of Section 3 we showed that a series of nonnegative terms converges if its sequence of partial sums is bounded. In this section we present the *integral test* and the *comparison tests* for convergence or divergence of series whose terms are nonnegative. We begin with the integral test, which uses the convergence or divergence of an improper integral as a criterion for the convergence or divergence of the series.

4.1 The Integral Test

The integral test is based upon the comparison of the partial sums of a series of the form $\displaystyle\sum_{k=1}^{\infty} f(k)$ and certain areas below the graph of the function f. Geometrically, the basic idea is quite simple and is illustrated in Figure 1. In Figure 1a, the area under the graph of a continuous, decreasing, nonnegative function f between $x = 1$ and $x = n + 1$ is overestimated by the sum $f(1) + f(2) + f(3) + \cdots + f(n)$ of the areas of the shaded rectangles; that is,

$$\int_1^{n+1} f(x)\,dx \le f(1) + f(2) + f(3) + \cdots + f(n).$$

Similarly, in Figure 1b, the area under the graph of the same function f between $x = 1$ and $x = n$ is underestimated by the sum $f(2) + f(3) + f(4) + \cdots + f(n)$ of the shaded rectangles; that is,

$$f(2) + f(3) + f(4) + \cdots + f(n) \le \int_1^n f(x)\,dx.$$

Figure 1

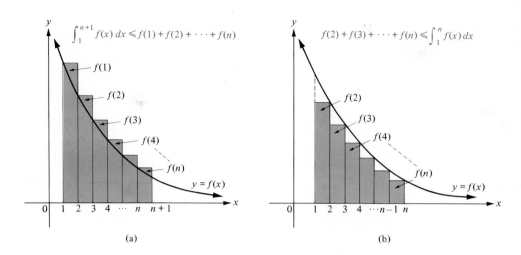

(a)   (b)

Adding $f(1)$ to both sides of the latter inequality, we obtain

$$f(1) + f(2) + f(3) + \cdots + f(n) \le f(1) + \int_1^n f(x)\,dx.$$

In summary, we have the following result. If f is a continuous, decreasing, nonnegative function defined at least on the closed interval $[1, n+1]$, where n is a positive integer, then

$$\int_1^{n+1} f(x)\,dx \le f(1) + f(2) + f(3) + \cdots + f(n) \le f(1) + \int_1^n f(x)\,dx.$$

(See Problems 41 and 42 for an analytic derivation of the preceding inequalities.) These inequalities are used to prove the following theorem.

THEOREM 1   **The Integral Test**
Suppose that the function f is continuous, decreasing, and nonnegative on the interval $[1, \infty)$.

(i) If the improper integral $\int_1^\infty f(x)\,dx$ converges, then the infinite series

$\displaystyle\sum_{k=1}^\infty f(k)$ converges.

(ii) If the improper integral $\int_1^\infty f(x)\,dx$ diverges, then the infinite series

$\displaystyle\sum_{k=1}^\infty f(k)$ diverges.

PROOF
(i) We have seen that the nth partial sum $s_n = f(1) + f(2) + \cdots + f(n)$ of the series $\displaystyle\sum_{k=1}^\infty f(k)$ satisfies $\int_1^{n+1} f(x)\,dx \le s_n \le f(1) + \int_1^n f(x)\,dx$. If $\int_1^\infty f(x)\,dx$ converges, then $s_n \le f(1) + \int_1^n f(x)\,dx \le f(1) + \int_1^\infty f(x)\,dx$, so that $\{s_n\}$ has the upper bound $M = f(1) + \int_1^\infty f(x)\,dx$ and, consequently, $\displaystyle\sum_{k=1}^\infty f(k)$ converges by Theorem 6 of Section 3.

(ii) If $\int_1^\infty f(x)\,dx$ diverges, then $\int_1^{n+1} f(x)\,dx$ grows large without bound as $n \to +\infty$; hence, since $\int_1^{n+1} f(x)\,dx \le s_n$, then s_n also grows large without bound as $n \to +\infty$. It follows in this case that $\{s_n\}$ diverges, and so $\sum_{k=1}^\infty f(k)$ diverges.

In the integral test, there is no necessity to start the infinite series at $k = 1$. For instance, to test for convergence or divergence of the series $\sum_{k=2}^\infty f(k)$, we would use the improper integral $\int_2^\infty f(x)\,dx$.

EXAMPLES   Use the integral test to determine whether the given series converges or diverges.

1 $\sum_{k=1}^\infty \dfrac{1}{k^2 + 1}$

SOLUTION

The function f defined by $f(x) = \dfrac{1}{x^2 + 1}$ is continuous, decreasing, and nonnegative on the interval $[1, \infty)$. Also,

$$\int_1^\infty \frac{1}{x^2 + 1}\,dx = \lim_{b \to +\infty} \int_1^b \frac{dx}{x^2 + 1} = \lim_{b \to +\infty} \left[(\tan^{-1} x) \Big|_1^b \right]$$

$$= \lim_{b \to +\infty} (\tan^{-1} b - \tan^{-1} 1) = \frac{\pi}{2} - \frac{\pi}{4} = \frac{\pi}{4}.$$

Thus, the improper integral $\int_1^\infty \dfrac{1}{x^2 + 1}\,dx$ converges and so the series $\sum_{k=1}^\infty \dfrac{1}{k^2 + 1}$ converges.

2 $\sum_{k=2}^\infty \dfrac{1}{k(\ln k)^{1/4}}$

SOLUTION

The function f defined by $f(x) = \dfrac{1}{x(\ln x)^{1/4}}$ is continuous, decreasing, and non-negative on the interval $[2, \infty)$. Here, by the change of variable $u = \ln x$, we have

$$\int \frac{1}{x(\ln x)^{1/4}}\,dx = \int u^{-1/4}\,du = \tfrac{4}{3}u^{3/4} + C = \tfrac{4}{3}(\ln x)^{3/4} + C;$$

so

$$\lim_{b \to +\infty} \int_2^b \frac{1}{x(\ln x)^{1/4}}\,dx = \lim_{b \to +\infty} [\tfrac{4}{3}(\ln b)^{3/4} - \tfrac{4}{3}(\ln 2)^{3/4}] = +\infty.$$

Thus, the improper integral $\int_2^\infty \dfrac{1}{x(\ln x)^{1/4}}\,dx$ diverges, and so the series $\sum_{k=2}^\infty \dfrac{1}{k(\ln k)^{1/4}}$ diverges.

The integral test makes it especially easy to study the convergence or divergence of a *p series*, which is by definition a series of the form $\sum_{k=1}^{\infty} \dfrac{1}{k^p}$, where p is a constant.

When $p = 1$, the p series becomes $\sum_{k=1}^{\infty} \dfrac{1}{k}$, or $1 + \dfrac{1}{2} + \dfrac{1}{3} + \dfrac{1}{4} + \dfrac{1}{5} + \cdots$, and is called the *harmonic series*.

THEOREM 2   **Convergence and Divergence of the p Series**

The p series $\sum_{k=1}^{\infty} \dfrac{1}{k^p}$ converges if $p > 1$ and diverges if $p \leq 1$. In particular, the harmonic series $\sum_{k=1}^{\infty} \dfrac{1}{k}$ diverges.

PROOF

If $p < 0$, then $\lim\limits_{n \to +\infty} \dfrac{1}{n^p} = +\infty$ (why?), so that $\sum_{k=1}^{\infty} \dfrac{1}{k^p}$ diverges, by Theorem 2 of Section 3. Thus, we can assume that $p \geq 0$. The function f defined by $f(x) = 1/x^p$ is continuous, decreasing, and nonnegative on $[1, \infty)$ and

$$\int_1^b \frac{1}{x^p}\,dx = \begin{cases} \dfrac{b^{1-p} - 1}{1 - p} & \text{if } p \neq 1 \\[2ex] \ln b & \text{if } p = 1. \end{cases}$$

Thus, for $p \leq 1$, $\lim\limits_{b \to +\infty} \displaystyle\int_1^b \dfrac{1}{x^p}\,dx = +\infty$, so that the improper integral $\displaystyle\int_1^{\infty} \dfrac{1}{x^p}\,dx$ diverges and so does the series $\sum_{k=1}^{\infty} \dfrac{1}{k^p}$. However, for $p > 1$,

$$\lim_{b \to +\infty} \int_1^b \frac{1}{x^p}\,dx = \lim_{b \to +\infty} \frac{b^{1-p} - 1}{1 - p} = \frac{1}{p - 1},$$

so that $\displaystyle\int_1^{\infty} \dfrac{1}{x^p}\,dx$—hence also $\sum_{k=1}^{\infty} \dfrac{1}{k^p}$—is convergent.

EXAMPLE   Test for convergence or divergence.

(a) $\sum_{k=1}^{\infty} \dfrac{1}{k^3}$ (b) $\sum_{k=1}^{\infty} \dfrac{1}{\sqrt[5]{k}}$

SOLUTION

(a) $\sum_{k=1}^{\infty} \dfrac{1}{k^3}$ is a p series with $p = 3 > 1$; hence, it converges.

(b) $\sum_{k=1}^{\infty} \dfrac{1}{\sqrt[5]{k}} = \sum_{k=1}^{\infty} \dfrac{1}{k^{1/5}}$ is a p series with $p = \frac{1}{5} < 1$; hence, it diverges.

The harmonic series $1 + \frac{1}{2} + \frac{1}{3} + \frac{1}{4} + \cdots$ is a particularly intriguing series, since it marks the boundary between the convergent and the divergent p series. Although its partial sums $s_n = 1 + \frac{1}{2} + \frac{1}{3} + \cdots + 1/n$ become large without bound as $n \to +\infty$, they do so rather slowly. To see this, consider the continuous, decreasing, nonnegative function f defined by $f(x) = 1/x$ for $x \geq 1$. For this function, the inequality

$$\int_1^{n+1} f(x)\, dx \le f(1) + f(2) + \cdots + f(n) \le f(1) + \int_1^n f(x)\, dx$$

(Figure 1) becomes

$$\ln(n+1) \le s_n \le 1 + \ln n.$$

If we put $n = 1{,}000{,}000$, we obtain

$$13.81 \le s_{1,000,000} \le 14.82,$$

so that the sum of the first million terms of the harmonic series is less than 15.

There is no "nice" formula for the sum of the p series $\sum\limits_{k=1}^{\infty} \dfrac{1}{k^p}$ with $p > 1$. The function ζ defined on $(1, \infty)$ by $\zeta(p) = \sum\limits_{k=1}^{\infty} \dfrac{1}{k^p}$ is called the *Riemann zeta function* and plays an important role in analytic number theory.

4.2 Comparison Tests

The most practical tests for convergence or divergence of infinite series are based on the idea of comparing a given series with a series that is known to converge or to diverge. Geometric series and p series are especially useful in such comparison tests. We begin with the following definition.

DEFINITION 1

Domination of Series

Let $\sum\limits_{k=1}^{\infty} a_k$ and $\sum\limits_{k=1}^{\infty} b_k$ be two series whose terms are nonnegative. We say that the series $\sum\limits_{k=1}^{\infty} b_k$ *dominates* the series $\sum\limits_{k=1}^{\infty} a_k$ if $a_k \le b_k$ holds for all positive integral values of k.

More generally, if there is a positive integer N such that $a_k \le b_k$ holds for all integers $k \ge N$, we say that the series $\sum\limits_{k=1}^{\infty} b_k$ *eventually dominates* the series $\sum\limits_{k=1}^{\infty} a_k$.

THEOREM 3

Direct Comparison Test

Let $\sum\limits_{k=1}^{\infty} a_k$ and $\sum\limits_{k=1}^{\infty} b_k$ be series all of whose terms are nonnegative, and suppose that $\sum\limits_{k=1}^{\infty} b_k$ dominates $\sum\limits_{k=1}^{\infty} a_k$ $\left(\text{or that } \sum\limits_{k=1}^{\infty} b_k \text{ eventually dominates } \sum\limits_{k=1}^{\infty} a_k\right)$.

   (i) If $\sum\limits_{k=1}^{\infty} b_k$ converges, then $\sum\limits_{k=1}^{\infty} a_k$ converges.

   (ii) If $\sum\limits_{k=1}^{\infty} a_k$ diverges, then $\sum\limits_{k=1}^{\infty} b_k$ diverges.

PROOF

We prove the theorem under the hypothesis that $\sum\limits_{k=1}^{\infty} b_k$ dominates $\sum\limits_{k=1}^{\infty} a_k$. Since the convergence or divergence of an infinite series is controlled by its "tail end," conclusions (i) and (ii) must still hold if $\sum_{k=1}^{\infty} b_k$ eventually dominates $\sum_{k=1}^{\infty} a_k$ (Problem 46).

(i) Assume that $\sum_{k=1}^{\infty} b_k$ converges to the sum B. Then, for any positive integer n,

$\sum_{k=1}^{n} b_k \leq \sum_{k=1}^{\infty} b_k = B$ (see Problem 28 of Problem Set 3). Since $a_k \leq b_k$ holds

for all positive integral values of k, $\sum_{k=1}^{n} a_k \leq \sum_{k=1}^{n} b_k \leq \sum_{k=1}^{\infty} b_k = B$; hence, the

sequence of partial sums of $\sum_{k=1}^{\infty} a_k$ is bounded above by B. It follows from

Theorem 6 of Section 3 that $\sum_{k=1}^{\infty} a_k$ is convergent.

(ii) Assume that $\sum_{k=1}^{\infty} a_k$ is divergent. Then, the partial sums $\sum_{k=1}^{n} a_k$ become large

without bound as $n \to +\infty$ (see Problem 47). Since $\sum_{k=1}^{n} a_k \leq \sum_{k=1}^{n} b_k$, it follows

that the partial sums $\sum_{k=1}^{n} b_k$ become large without bound as $n \to +\infty$; hence,

the series $\sum_{k=1}^{\infty} b_k$ cannot be convergent.

EXAMPLES   Use the direct comparison test to determine the convergence or divergence of the given series.

1 $\sum_{k=1}^{\infty} \dfrac{1}{7k^2 + 1}$

SOLUTION
Comparing the kth term of the given series with the kth term of the p series
$\sum_{k=1}^{\infty} \dfrac{1}{k^2}$, we have $\dfrac{1}{7k^2 + 1} \leq \dfrac{1}{k^2}$, since $7k^2 + 1 \geq 7k^2 \geq k^2$ for every positive integer
k. Therefore, the convergent p series $\sum_{k=1}^{\infty} \dfrac{1}{k^2}$ dominates the series $\sum_{k=1}^{\infty} \dfrac{1}{7k^2 + 1}$,
forcing the latter series to converge.

2 $\sum_{k=1}^{\infty} \dfrac{1}{\sin k + 5^k}$

SOLUTION

We are going to show that the series $\sum_{k=1}^{\infty} \dfrac{1}{\sin k + 5^k}$ is dominated by the convergent

geometric series $\sum_{k=1}^{\infty} \dfrac{1}{5^{k-1}}$ by showing that $\dfrac{1}{\sin k + 5^k} \leq \dfrac{1}{5^{k-1}}$ holds for $k \geq 1$.
The desired inequality is equivalent to $5^{k-1} \leq \sin k + 5^k$; that is,

$$-\sin k \leq 5^k - 5^{k-1} = 5^{k-1}(5 - 1) = 5^{k-1} \cdot 4.$$

The latter inequality is true, since

$$-\sin k \leq 1 \leq 4 \leq 5^{k-1} \cdot 4.$$

Hence, the given series converges.

3 $\displaystyle\sum_{k=1}^{\infty}\frac{1}{\sqrt{k+2}}$

SOLUTION

We are going to show that the given series dominates the series $\displaystyle\sum_{k=1}^{\infty}\frac{1}{2}\cdot\frac{1}{\sqrt{k}}$, which diverges because the p series $\displaystyle\sum_{k=1}^{\infty}\frac{1}{\sqrt{k}}$ diverges. Thus, we wish to prove that

$$\frac{1}{2}\cdot\frac{1}{\sqrt{k}}\le\frac{1}{\sqrt{k+2}};\quad\text{that is,}\quad\frac{1}{4k}\le\frac{1}{k+2}\quad\text{or}\quad k+2\le 4k\qquad\text{for }k\ge 1.$$

Since $k+2\le 4k$ is equivalent to $\dfrac{2}{3}\le k$, it follows that $\dfrac{1}{2}\cdot\dfrac{1}{\sqrt{k}}\le\dfrac{1}{\sqrt{k+2}}$ holds for $k\ge 1$ and the given series diverges.

4 $\displaystyle\sum_{k=2}^{\infty}\frac{1}{\ln k}$

SOLUTION

Because $0<\ln k<k$ for $k\ge 2$, we have $1/k<1/\ln k$, so that the series $\displaystyle\sum_{k=2}^{\infty}\frac{1}{\ln k}$ dominates the series $\displaystyle\sum_{k=2}^{\infty}\frac{1}{k}$. Since the harmonic series $\displaystyle\sum_{k=1}^{\infty}\frac{1}{k}$ diverges, so does the series $\displaystyle\sum_{k=2}^{\infty}\frac{1}{k}$, by Theorem 5 of Section 3. It follows that the given series diverges.

The choice of a suitable series with which to compare a given series is not always obvious and may require some trial and error; however, a geometric series or a constant multiple of a p series whose form is similar to that of the given series often works.

The following test is essentially another version of the direct comparison test, but it is sometimes easier to apply.

THEOREM 4   **Limit Comparison Test**

Let $\displaystyle\sum_{k=1}^{\infty}a_k$ be a series of nonnegative terms and suppose that $\displaystyle\sum_{k=1}^{\infty}b_k$ is a series of positive terms such that $\displaystyle\lim_{n\to+\infty}\frac{a_n}{b_n}=c$, where $c>0$. Then, either both series converge or else both series diverge.

PROOF

Since $\displaystyle\lim_{n\to+\infty}\frac{a_n}{b_n}=c$, it follows that, given any positive number ε, there exists a positive integer N such that

$$\left|\frac{a_n}{b_n}-c\right|<\varepsilon\quad\text{holds whenever}\quad n\ge N.$$

The condition $\left|\dfrac{a_n}{b_n}-c\right|<\varepsilon$ can be rewritten as $-\varepsilon<\dfrac{a_n}{b_n}-c<\varepsilon$, or as

$c - \varepsilon < \dfrac{a_n}{b_n} < c + \varepsilon.$ Putting $\varepsilon = \dfrac{c}{2},$ we see that $\dfrac{c}{2} < \dfrac{a_n}{b_n} < \dfrac{3c}{2}$ holds for all

integers $n \geq N$. Therefore, if $n \geq N$, it follows that $\dfrac{c}{2} b_n < a_n < \dfrac{3c}{2} b_n$; hence, the series

$\displaystyle\sum_{k=1}^{\infty} a_k$ eventually dominates the series $\displaystyle\sum_{k=1}^{\infty} \dfrac{c}{2} b_k$, while the series $\displaystyle\sum_{k=1}^{\infty} \dfrac{3c}{2} b_k$ eventually

dominates the series $\displaystyle\sum_{k=1}^{\infty} a_k$. Consequently, if the series $\displaystyle\sum_{k=1}^{\infty} b_k$ converges, then the series

$\displaystyle\sum_{k=2}^{\infty} \dfrac{3c}{2} b_k$ converges [part (ii) of Theorem 3 in Section 3], and so the series $\displaystyle\sum_{k=1}^{\infty} a_k$

converges by the direct comparison test. On the other hand, if the series $\displaystyle\sum_{k=1}^{\infty} b_k$

diverges, then the series $\displaystyle\sum_{k=1}^{\infty} \dfrac{c}{2} b_k$ diverges [part (ii) of Theorem 3 in Section 3 again],

and so the series $\displaystyle\sum_{k=1}^{\infty} a_k$ diverges, by the direct comparison test.

EXAMPLES Use the limit comparison test to determine whether the given series converges or diverges.

1 $\displaystyle\sum_{k=1}^{\infty} \dfrac{1}{\sqrt[4]{k^3 + 1}}$

SOLUTION

We use the divergent p series $\displaystyle\sum_{k=1}^{\infty} \dfrac{1}{\sqrt[4]{k^3}}$ for the limit comparison test. Let a_n be

the nth term of the given series and let b_n be the nth term of the series

$\displaystyle\sum_{k=1}^{\infty} \dfrac{1}{\sqrt[4]{k^3}}$. Then

$$\lim_{n \to +\infty} \frac{a_n}{b_n} = \lim_{n \to +\infty} \frac{1/\sqrt[4]{n^3 + 1}}{1/\sqrt[4]{n^3}} = \lim_{n \to +\infty} \frac{\sqrt[4]{n^3}}{\sqrt[4]{n^3 + 1}} = \lim_{n \to +\infty} \sqrt[4]{\frac{n^3}{n^3 + 1}}$$

$$= \lim_{n \to +\infty} \sqrt[4]{\frac{1}{1 + (1/n)^3}} = 1.$$

It follows from the limit comparison test that the given series diverges.

2 $\displaystyle\sum_{k=1}^{\infty} \dfrac{7k + 3}{(5k + 1) \cdot 3^k}$

SOLUTION

We use the convergent geometric series $\displaystyle\sum_{k=1}^{\infty} \dfrac{1}{3^k}$ for the limit comparison test. Thus,

if a_n is the nth term of the given series and b_n is the nth term of the series

$\displaystyle\sum_{k=1}^{\infty} \dfrac{1}{3^k}$, then

$$\lim_{n \to +\infty} \frac{a_n}{b_n} = \lim_{n \to +\infty} \frac{\left[\dfrac{7n + 3}{(5n + 1) \cdot 3^n} \right]}{1/3^n} = \lim_{n \to +\infty} \frac{7n + 3}{5n + 1} = \lim_{n \to +\infty} \frac{7 + (3/n)}{5 + (1/n)} = \frac{7}{5},$$

and so the given series converges by Theorem 4.

The following theorem can easily be proved by slightly modifying the proof of Theorem 4. Its proof is left as an exercise (Problem 48).

THEOREM 5   **Modified Limit Comparison Test**

Let $\sum\limits_{k=1}^{\infty} a_k$ be a series of nonnegative terms and suppose that $\sum\limits_{k=1}^{\infty} b_k$ is a series of positive terms.

(i) If $\lim\limits_{n \to +\infty} \dfrac{a_n}{b_n} = 0$ and $\sum\limits_{k=1}^{\infty} b_k$ converges, then $\sum\limits_{k=1}^{\infty} a_k$ converges.

(ii) If $\lim\limits_{n \to +\infty} \dfrac{a_n}{b_n} = +\infty$ and $\sum\limits_{k=1}^{\infty} b_k$ diverges, then $\sum\limits_{k=1}^{\infty} a_k$ diverges.

EXAMPLES   Determine whether the given series converges or diverges by using the modified limit comparison test.

1 $\displaystyle\sum_{k=1}^{\infty} \frac{\ln k}{k^4}$

SOLUTION

We use the convergent p series $\displaystyle\sum_{k=1}^{\infty} \frac{1}{k^3}$ for the modified limit comparison test. If a_n is the nth term of the given series and b_n is the nth term of the series $\displaystyle\sum_{k=1}^{\infty} \frac{1}{k^3}$, then

$$\lim_{n \to +\infty} \frac{a_n}{b_n} = \lim_{n \to +\infty} \frac{(\ln n)/n^4}{1/n^3} = \lim_{n \to +\infty} \frac{\ln n}{n} = \lim_{x \to +\infty} \frac{\ln x}{x}$$

$$= \lim_{x \to +\infty} \frac{1/x}{1} = \lim_{x \to +\infty} \frac{1}{x} = 0,$$

where we have used Theorem 1 in Section 1.1 and L'Hôpital's rule to evaluate the limit. By part (i) of Theorem 5, the given series converges.

2 $\displaystyle\sum_{k=1}^{\infty} \frac{1}{\sqrt{2k+1}}$

SOLUTION

We use the divergent harmonic series $\displaystyle\sum_{k=1}^{\infty} \frac{1}{k}$ for the modified limit comparison test. If a_n is the nth term of the given series and b_n is the nth term of the series $\displaystyle\sum_{k=1}^{\infty} \frac{1}{k}$, then

$$\lim_{n \to +\infty} \frac{a_n}{b_n} = \lim_{n \to +\infty} \frac{1/\sqrt{2n+1}}{1/n} = \lim_{n \to +\infty} \frac{n}{\sqrt{2n+1}}$$

$$= \lim_{n \to +\infty} \sqrt{n} \sqrt{\frac{n}{2n+1}} = \lim_{n \to +\infty} \sqrt{n} \sqrt{\frac{1}{2 + (1/n)}} = +\infty.$$

By part (ii) of Theorem 5, the given series diverges.

Problem Set 4

In problems 1 through 16, use the integral test to determine whether each series converges or diverges.

1 $\displaystyle\sum_{k=1}^{\infty} \frac{1}{k\sqrt[3]{k}}$

2 $\displaystyle\sum_{k=1}^{\infty} \frac{1}{k^2+4}$

3 $\displaystyle\sum_{k=1}^{\infty} \frac{3k^2}{k^3+16}$

4 $\displaystyle\sum_{k=1}^{\infty} \frac{1}{1+\sqrt{k}}$

5 $\displaystyle\sum_{n=1}^{\infty} \left(\frac{1000}{n}\right)^2$

6 $\displaystyle\sum_{m=1}^{\infty} e^{-m}$

7 $\displaystyle\sum_{k=2}^{\infty} \frac{\ln k}{k}$

8 $\displaystyle\sum_{k=2}^{\infty} \frac{1}{k \ln k}$

9 $\displaystyle\sum_{j=1}^{\infty} je^{-j}$

10 $\displaystyle\sum_{n=1}^{\infty} \coth n$

11 $\displaystyle\sum_{k=2}^{\infty} \frac{1}{k\sqrt{\ln k}}$

12 $\displaystyle\sum_{k=3}^{\infty} \frac{1}{k \ln k \ln (\ln k)}$

13 $\displaystyle\sum_{m=1}^{\infty} \frac{\tan^{-1} m}{1+m^2}$

14 $\displaystyle\sum_{r=1}^{\infty} \frac{r}{2^r}$

15 $\displaystyle\sum_{k=1}^{\infty} \frac{1}{(2k+1)(3k+1)}$

16 $\displaystyle\sum_{n=1}^{\infty} \frac{1}{n(n+1)(n+2)}$

In problems 17 through 30, use the direct comparison test with either a p series or a geometric series to determine whether each series converges or diverges.

17 $\displaystyle\sum_{k=1}^{\infty} \frac{k^2}{k^4+3k+1}$

18 $\displaystyle\sum_{k=1}^{\infty} \frac{k}{k^3+2k+7}$

19 $\displaystyle\sum_{k=1}^{\infty} \frac{1}{k \cdot 5^k}$

20 $\displaystyle\sum_{n=1}^{\infty} \frac{5}{(n+1)3^n}$

21 $\displaystyle\sum_{j=1}^{\infty} \frac{j+1}{(j+2) \cdot 7^j}$

22 $\displaystyle\sum_{r=1}^{\infty} \frac{5r}{\sqrt[3]{r^7+3}}$

23 $\displaystyle\sum_{k=1}^{\infty} \frac{8}{\sqrt[3]{k}+1}$

24 $\displaystyle\sum_{k=1}^{\infty} \frac{1}{4k+6}$

25 $\displaystyle\sum_{j=1}^{\infty} \frac{j^2}{j^3+4j+3}$

26 $\displaystyle\sum_{k=2}^{\infty} \frac{\ln k}{k}$

27 $\displaystyle\sum_{q=1}^{\infty} \frac{\sqrt{q}}{q+2}$

28 $\displaystyle\sum_{j=1}^{\infty} \frac{1+e^{-j}}{e^j}$

29 $\displaystyle\sum_{k=1}^{\infty} \frac{k+3}{k!}$

30 $\displaystyle\sum_{n=1}^{\infty} \frac{8}{3^n+n^2}$

In problems 31 through 40, use a limit comparison test with either a p series or a geometric series to determine whether each series converges or diverges.

31 $\displaystyle\sum_{k=1}^{\infty} \frac{1}{\sqrt[3]{k^2+5}}$

32 $\displaystyle\sum_{k=1}^{\infty} \frac{1}{3 \cdot 2^k+2}$

33 $\displaystyle\sum_{k=1}^{\infty} \frac{5k^2}{(k+1)(k+2)(k+3)(k+4)}$

34 $\displaystyle\sum_{j=1}^{\infty} \frac{1+e^j}{j+5^j}$

35 $\displaystyle\sum_{k=1}^{\infty} \frac{k^2}{1+k^3}$

36 $\displaystyle\sum_{j=1}^{\infty} \frac{1}{j\sqrt{2j^3+5}}$

37 $\displaystyle\sum_{j=1}^{\infty} \frac{1}{7^j-\cos j}$

38 $\displaystyle\sum_{k=2}^{\infty} \frac{\ln k}{k^2+4}$

39 $\displaystyle\sum_{k=1}^{\infty} \frac{\ln k}{k^2}$

40 $\displaystyle\sum_{j=1}^{\infty} \frac{j!}{(2j)!}$

41 Assume that f is a continuous and decreasing function on the interval $[k-1, k]$.
   (a) Use the mean value theorem for integrals (Theorem 9 in Section 3 of Chapter 6) to show that there exists a number c with $k-1 \le c \le k$ such that

$$\int_{k-1}^{k} f(x)\, dx = f(c).$$

   (b) Explain why $f(k) \le f(c) \le f(k-1)$.
   (c) Conclude that $f(k) \le \displaystyle\int_{k-1}^{k} f(x)\, dx \le f(k-1)$.

42 Assume that the function f is a continuous and decreasing function on the interval $[1, n+1]$, where n is a positive integer.
   (a) Use part (c) of problem 41 to show that $\displaystyle\sum_{k=2}^{n} f(k) \le \int_{1}^{n} f(x)\, dx$.

(b) Use part (c) of problem 41 to show that $\displaystyle\int_{1}^{n+1} f(x)\,dx \le \sum_{k=2}^{n+1} f(k-1)$.

(c) Conclude that $\displaystyle\int_{1}^{n+1} f(x)\,dx \le \sum_{k=1}^{n} f(k) \le f(1) + \int_{1}^{n} f(x)\,dx$.

43 Suppose that the function f is continuous, decreasing, and nonnegative on the interval $[1, \infty)$ and that the improper integral $\displaystyle\int_{1}^{\infty} f(x)\,dx$ converges. By the integral test, $\displaystyle\sum_{k=1}^{\infty} f(k)$ converges. Using part (c) of problem 42, show that

$$\int_{1}^{\infty} f(x)\,dx \le \sum_{k=1}^{\infty} f(k) \le f(1) + \int_{1}^{\infty} f(x)\,dx.$$

44 Use the result of problem 43 to prove that

$$\frac{\pi}{4} \le \sum_{k=1}^{\infty} \frac{1}{k^2 + 1} \le \frac{\pi}{4} + \frac{1}{2}.$$

45 Give an example to show that a series $\displaystyle\sum_{k=1}^{\infty} a_k$ with positive terms can be convergent and yet the series $\displaystyle\sum_{k=1}^{\infty} \sqrt{a_k}$ can be divergent.

46 Show that the conclusions of the direct comparison test (Theorem 3 of Section 4.2) still hold if $\displaystyle\sum_{k=1}^{\infty} b_k$ eventually dominates $\displaystyle\sum_{k=1}^{\infty} a_k$.

47 Suppose that the series $\displaystyle\sum_{k=1}^{\infty} a_k$ diverges and that its terms are nonnegative. Prove that the partial sums $\displaystyle s_n = \sum_{k=1}^{n} a_k$ become large without bound as $n \to +\infty$.

48 Prove the modified limit comparison test (Theorem 5).

49 Suppose that f is a continuous, decreasing, nonnegative function on the interval $[m, M]$, where m and M are positive integers and $m < M$. Prove that

$$f(M) + \int_{m}^{M} f(x)\,dx \le \sum_{k=m}^{M} f(k) \le f(m) + \int_{m}^{M} f(x)\,dx.$$

5   Series Whose Terms Change Sign

The tests developed in Section 4 allow us to handle series whose terms do not change sign. (If all terms are nonpositive, we just multiply by -1 to convert to a series whose terms are nonnegative.) In the present section we consider series whose terms change sign. The simplest such series is an *alternating series* whose terms alternate in sign, for instance, the series

$$\sum_{k=1}^{\infty} (-1)^{k+1} \frac{1}{k} = 1 - \frac{1}{2} + \frac{1}{3} - \frac{1}{4} + \cdots + (-1)^{n+1}\frac{1}{n} + \cdots,$$

which is called the *alternating harmonic series*. Notice that a geometric series with a negative ratio r, such as

$$\sum_{k=1}^{\infty} (-1)\left(-\frac{1}{2}\right)^{k-1} = -1 + \frac{1}{2} - \frac{1}{4} + \frac{1}{8} - \cdots + (-1)\left(\frac{-1}{2}\right)^{n-1} + \cdots,$$

is an alternating series.

The following theorem, which exhibits an important feature of an alternating series whose terms decrease in absolute value, will be used to prove a test for convergence of such a series.

THEOREM 1   **Alternating Series Whose Terms Decrease in Absolute Value**
Let $\{a_n\}$ be a decreasing sequence of positive terms. Then, the partial sums s_n of the alternating series

$$a_1 - a_2 + a_3 - a_4 + \cdots + (-1)^{n+1}a_n + \cdots$$

satisfy the following conditions:

(i) $0 \le s_2 \le s_4 \le s_6 \le s_8 \le \cdots$.

(ii) $s_1 \ge s_3 \ge s_5 \ge s_7 \ge s_9 \ge \cdots$.

(iii) If n is an even positive integer, then $s_{n+1} - s_n = a_{n+1}$.

(iv)   If n is an even positive integer, then $0 \le s_n \le s_{n+1} \le s_1$.

PROOF

(i)   If n is an *even* positive integer, then we can form the partial sum $s_n = a_1 - a_2 + a_3 - a_4 + \cdots + a_{n-1} - a_n$ and group the terms in pairs to obtain

$$s_n = (a_1 - a_2) + (a_3 - a_4) + \cdots + (a_{n-1} - a_n).$$

Since $a_1 \ge a_2 \ge a_3 \ge a_4 \ge \cdots$, it follows that each quantity enclosed in parentheses is nonnegative. The next *even* integer after n is $n + 2$, and we have $s_{n+2} = s_n + (a_{n+1} - a_{n+2}) \ge s_n$. It follows that

$$0 \le s_2 \le s_4 \le s_6 \le s_8 \le \cdots.$$

(ii)   Similarly, if m is an *odd* positive integer, we can write

$$s_m = a_1 - a_2 + a_3 - a_4 + \cdots - a_{m-1} + a_m$$
$$= a_1 - (a_2 - a_3) - (a_4 - a_5) - \cdots - (a_{m-1} - a_m),$$

where, again, each quantity enclosed in parentheses is nonnegative. The next *odd* integer after m is $m + 2$, and we have

$$s_{m+2} = s_m - a_{m+1} + a_{m+2} = s_m - (a_{m+1} - a_{m+2}) \le s_m;$$

hence,

$$s_1 \ge s_3 \ge s_5 \ge s_7 \ge s_9 \ge \cdots.$$

(iii) If n is even, then $n + 1$ is odd and $s_{n+1} = s_n + a_{n+1}$. Therefore,

$$s_{n+1} - s_n = a_{n+1}.$$

(iv)   Again, if n is even, then by (iii), $s_{n+1} - s_n = a_{n+1} \ge 0$; hence, $s_n \le s_{n+1}$. By (i), $0 \le s_n$, and by (ii), $s_{n+1} \le s_1$, so that

$$0 \le s_n \le s_{n+1} \le s_1.$$

In Theorem 1, if $\{a_n\}$ is a strictly decreasing sequence, so that $a_1 > a_2 > a_3 > \cdots$, then all inequalities appearing in the proof and in the conclusions can be made strict.

The following theorem, discovered by Leibniz, provides a useful test for convergence of alternating series.

THEOREM 2   **Leibniz's Alternating Series Test**

If $\{a_n\}$ is a decreasing sequence of positive terms with $\lim\limits_{n\to+\infty} a_n = 0$, then the alternating series

$$a_1 - a_2 + a_3 - a_4 + \cdots + (-1)^{n+1}a_n + \cdots$$

is convergent. Moreover, if S is its sum and if s_n is its nth partial sum, then

$$0 \le (-1)^n(S - s_n) \le a_{n+1}.$$

PROOF

Putting $n = 2j$ in part (iv) of Theorem 1, we see that $0 \le s_{2j} \le s_{2j+1} \le s_1$ holds for all positive integers j. By parts (i) and (ii) of Theorem 1, $\{s_{2j}\}$ is an increasing sequence bounded above by s_1 while $\{s_{2j+1}\}$ is a decreasing sequence bounded below by 0. By Theorem 2 of Section 1.2, it follows that both of the sequences $\{s_{2j}\}$ and $\{s_{2j+1}\}$ converge. By part (iii) of Theorem 1, $s_{2j+1} - s_{2j} = a_{2j+1}$; hence,

$$0 = \lim_{j\to+\infty} a_{2j+1} = \lim_{j\to+\infty}(s_{2j+1} - s_{2j}) = \lim_{j\to+\infty} s_{2j+1} - \lim_{j\to+\infty} s_{2j},$$

and it follows that $\lim\limits_{j\to+\infty} s_{2j+1} = \lim\limits_{j\to+\infty} s_{2j}$. Since the terms in the sequence $\{s_n\}$ whose indices are even and also the terms in this sequence whose indices are odd converge to the same limit, call it S, it is easy to see (Problem 43) that the entire sequence $\{s_n\}$ must converge to S. Therefore, $\sum\limits_{k=1}^{\infty}(-1)^{k+1}a_k$ converges and its sum is S.

To prove the second part of the theorem, notice that the sequence $s_2, s_4, s_6, s_8, \ldots$ is increasing and converges to S; hence, $s_n \le S$ holds for all even positive integers n by Theorem 3 in Section 1.2. Similarly, since the sequence $s_1, s_3, s_5, s_7, \ldots$ is decreasing and converges to S, then $S \le s_m$ holds for all odd positive integers m. If n is even, then $n+1$ is odd and so $s_n \le S \le s_{n+1}$. Subtracting s_n and noting that $s_{n+1} - s_n = a_{n+1}$, we have $0 \le S - s_n \le a_{n+1}$ when n is even. If m is odd, then $m+1$ is even and so $s_{m+1} \le S \le s_m$. Subtracting s_m and noting that $s_{m+1} - s_m = -a_{m+1}$, we have $-a_{m+1} \le S - s_m \le 0$ when m is odd. It follows that $0 \le (-1)^n(S - s_n) \le a_{n+1}$ holds for every positive integer n, whether even or odd, and the proof is complete.

In Theorem 2, if $\{a_n\}$ is strictly decreasing, so that $a_1 > a_2 > a_3 \cdots$, then the conclusion $0 \le (-1)^n(S - s_n) \le a_{n+1}$ can be sharpened to $0 < (-1)^n(S - s_n) < a_{n+1}$.

Notice that $S - s_n$ is the error involved in estimating the sum $S = \sum\limits_{k=1}^{\infty}(-1)^{k+1}a_k$ by the nth partial sum $s_n = \sum\limits_{k=1}^{n}(-1)^{k+1}a_k$. If n is even, then

$$0 \le (-1)^n(S - s_n) = S - s_n,$$

so that s_n *underestimates* S. If n is odd, then $0 \le (-1)^n(S - s_n) = -(S - s_n)$, so that s_n *overestimates* S. In any case, $|S - s_n| \le a_{n+1}$; hence, *the absolute value of the error of estimation does not exceed the absolute value of the first neglected term.*

For example, although the harmonic series $1 + \frac{1}{2} + \frac{1}{3} + \frac{1}{4} + \cdots$ diverges, it follows from Leibniz's theorem that the alternating harmonic series $1 - \frac{1}{2} + \frac{1}{3} - \frac{1}{4} + \cdots$ does converge. In fact, it can be shown that $1 - \frac{1}{2} + \frac{1}{3} - \frac{1}{4} + \cdots = \ln 2$. If we wish to estimate $\ln 2$ by a partial sum of the alternating harmonic series, then the error does not exceed the absolute value of the first neglected term. For instance, $\ln 2 \approx 1 - \frac{1}{2} + \frac{1}{3} - \frac{1}{4} + \frac{1}{5} - \frac{1}{6} + \frac{1}{7} - \frac{1}{8} + \frac{1}{9}$ with an error that does not exceed $\frac{1}{10}$. Moreover, since we have an *odd* number of terms in this estimate, we have *overestimated* $\ln 2$, so that $\ln 2$ is less than

$$1 - \tfrac{1}{2} + \tfrac{1}{3} - \tfrac{1}{4} + \tfrac{1}{5} - \tfrac{1}{6} + \tfrac{1}{7} - \tfrac{1}{8} + \tfrac{1}{9} = 0.7456\ldots \text{ (Actually, } \ln 2 = 0.6931\ldots\text{)}$$

EXAMPLES (a) Show that the given series is convergent, (b) find the partial sum s_4 of its first four terms, and (c) find a bound on the absolute value of the error involved in estimating its sum by s_4.

1 $\displaystyle\sum_{k=1}^{\infty} (-1)^{k+1}\,\frac{(k+3)}{k(k+2)}$

SOLUTION

(a) Let f be the function defined by $f(x) = \dfrac{x+3}{x(x+2)}$. Then

$$f'(x) = -\frac{x^2+6x+6}{x^2(x+2)^2} < 0 \qquad \text{for } x \ge 1;$$

hence, the function f is decreasing on $[1, \infty)$. It follows that the sequence $\{f(n)\}$—that is, the sequence $\left\{\dfrac{n+3}{n(n+2)}\right\}$—is decreasing. Since $\displaystyle\lim_{n \to +\infty} \frac{n+3}{n(n+2)} = 0$ (why?) and $\dfrac{n+3}{n(n+2)} \ge 0$, the given alternating series converges, by Theorem 2.

(b) $s_4 = \dfrac{4}{3} - \dfrac{5}{8} + \dfrac{6}{15} - \dfrac{7}{24} = \dfrac{49}{60}$.

(c) The absolute value of the error of the estimate

$$s_4 = \frac{49}{60} \approx \sum_{k=1}^{\infty} (-1)^{k+1}\,\frac{k+3}{k(k+2)}$$

does not exceed the fifth term, $\dfrac{5+3}{5(5+2)} = \dfrac{8}{35}$. Here s_4 involves an *even* number of terms, so that $\dfrac{49}{60}$ *under*estimates $\displaystyle\sum_{k=1}^{\infty} (-1)^{k+1}\,\frac{k+3}{k(k+2)}$.

2 $\displaystyle\sum_{k=1}^{\infty} \frac{(-1)^k}{k!}$

SOLUTION

(a) This series begins with a negative term, $-1/1!$, whereas Leibniz's theorem as stated above concerns alternating series which begin with a positive term. However, we can write

$$\sum_{k=1}^{\infty} \frac{(-1)^k}{k!} = -\sum_{k=1}^{\infty} \frac{(-1)^{k+1}}{k!}$$

and apply Leibniz's theorem to the series $\displaystyle\sum_{k=1}^{\infty} \frac{(-1)^{k+1}}{k!}$. Here, $\left\{\dfrac{1}{n!}\right\}$ is a decreasing sequence of nonnegative terms and $\displaystyle\lim_{n \to +\infty} \frac{1}{n!} = 0$, so that $\displaystyle\sum_{k=1}^{\infty} \frac{(-1)^{k+1}}{k!}$; hence also $\displaystyle\sum_{k=1}^{\infty} \frac{(-1)^k}{k!}$ converges.

(b) and (c) Here, $1 - \dfrac{1}{2} + \dfrac{1}{6} - \dfrac{1}{24} = \dfrac{15}{24}$ underestimates $\displaystyle\sum_{k=1}^{\infty} \frac{(-1)^{k+1}}{k!}$ with an error no more than $\dfrac{1}{5!} = \dfrac{1}{120}$, and therefore, $-\dfrac{15}{24}$ *over*estimates

$$-\sum_{k=1}^{\infty} \frac{(-1)^{k+1}}{k!} = \sum_{k=1}^{\infty} \frac{(-1)^k}{k!}$$

with an error whose absolute value does not exceed $\frac{1}{120}$.

5.1   Absolute and Conditional Convergence

Consider the alternating geometric series $1 - \frac{1}{2} + \frac{1}{4} - \frac{1}{8} + \frac{1}{16} - \frac{1}{32} + \cdots$ with ratio $r = -\frac{1}{2}$. Not only does this series converge, but also the corresponding series $|1| + \left|-\frac{1}{2}\right| + \left|\frac{1}{4}\right| + \left|-\frac{1}{8}\right| + \left|\frac{1}{16}\right| + \left|-\frac{1}{32}\right| + \cdots$ of absolute values; that is, the geometric series $1 + \frac{1}{2} + \frac{1}{4} + \frac{1}{8} + \frac{1}{16} + \frac{1}{32} + \cdots$ is also convergent. Such a series is called *absolutely convergent.*

On the other hand, consider the alternating harmonic series

$$1 - \tfrac{1}{2} + \tfrac{1}{3} - \tfrac{1}{4} + \tfrac{1}{5} - \tfrac{1}{6} + \cdots,$$

which converges by Leibniz's theorem. The corresponding series of absolute values is the harmonic series $1 + \frac{1}{2} + \frac{1}{3} + \frac{1}{4} + \frac{1}{5} + \frac{1}{6} + \cdots$, which diverges. Thus, the convergence of the alternating harmonic series actually depends upon the fact that its terms change sign. Such a series is called *conditionally convergent.* More generally, we make the following definition.

DEFINITION 1   Absolute and Conditional Convergence

(i) If the series $\displaystyle\sum_{k=1}^{\infty} |a_k|$ converges, we say that the series $\displaystyle\sum_{k=1}^{\infty} a_k$ is *absolutely convergent.*

(ii) If the series $\displaystyle\sum_{k=1}^{\infty} a_k$ is convergent, but the series $\displaystyle\sum_{k=1}^{\infty} |a_k|$ is divergent, we say that the series $\displaystyle\sum_{k=1}^{\infty} a_k$ is *conditionally convergent.*

EXAMPLES   Determine whether the given series is divergent, conditionally convergent, or absolutely convergent.

1 $\displaystyle\sum_{k=1}^{\infty} \frac{(-1)^k}{k^2 + 1}$

SOLUTION
Since the series

$$\sum_{k=1}^{\infty} \left| \frac{(-1)^k}{k^2 + 1} \right| = \sum_{k=1}^{\infty} \frac{1}{k^2 + 1}$$

converges by comparison with the convergent p series $\displaystyle\sum_{k=1}^{\infty} \frac{1}{k^2}$, it follows that $\displaystyle\sum_{k=1}^{\infty} \frac{(-1)^k}{k^2 + 1}$ is absolutely convergent.

2 $\displaystyle\sum_{k=1}^{\infty} (-1)^{k+1} \frac{k+1}{k+2}$

SOLUTION
Since $\displaystyle\lim_{n \to +\infty} (-1)^{n+1} \frac{n+1}{n+2}$ does not exist (why?), the given series is divergent (Section 3, Theorem 2).

3 $\displaystyle\sum_{k=2}^{\infty} \frac{(-1)^k}{\ln k}$

SOLUTION

The given series converges, by Leibniz's theorem. However, $\displaystyle\sum_{k=2}^{\infty}\left|\frac{(-1)^k}{\ln k}\right| = \sum_{k=2}^{\infty}\frac{1}{\ln k}$

diverges since $\dfrac{1}{\ln k} \geq \dfrac{1}{k}$ holds for $k \geq 2$ and $\displaystyle\sum_{k=2}^{\infty}\frac{1}{k}$ diverges. Hence, $\displaystyle\sum_{k=2}^{\infty}\frac{(-1)^k}{\ln k}$ is

conditionally convergent.

THEOREM 3   **Absolute Convergence Implies Convergence**

If a series $\displaystyle\sum_{k=1}^{\infty} a_k$ is absolutely convergent, then it is convergent.

PROOF

Assume that $\displaystyle\sum_{k=1}^{\infty} |a_k|$ is convergent. The inequalities $-|a_k| \leq a_k \leq |a_k|$ can be

rewritten $0 \leq a_k + |a_k| \leq 2|a_k|$. Now, $\displaystyle\sum_{k=1}^{\infty} 2|a_k|$ converges since $\displaystyle\sum_{k=1}^{\infty} |a_k|$ converges

[part (ii) of Theorem 3 in Section 3]; hence, by the comparison test, $\displaystyle\sum_{k=1}^{\infty} (a_k + |a_k|)$

converges. Therefore, $\displaystyle\sum_{k=1}^{\infty} [(a_k + |a_k|) - |a_k|] = \sum_{k=1}^{\infty} a_k$ converges [part (i) of Theorem

3 in Section 3].

EXAMPLE   Determine whether the series $\displaystyle\sum_{k=1}^{\infty} \frac{\sin k}{k^3 + 4}$ converges or diverges.

SOLUTION
Although the given series contains both positive and negative terms, it is *not* an
alternating series (why?). However, we have

$$\left|\frac{\sin k}{k^3 + 4}\right| = \frac{|\sin k|}{k^3 + 4} \leq \frac{1}{k^3}$$

for every positive integer k, so that $\displaystyle\sum_{k=1}^{\infty}\left|\frac{\sin k}{k^3 + 4}\right|$ is dominated by the convergent

p series $\displaystyle\sum_{k=1}^{\infty}\frac{1}{k^3}$. Therefore, the given series is absolutely convergent and hence

convergent by Theorem 3.

The following theorem gives one of the most practical tests for absolute
convergence.

THEOREM 4   **The Ratio Test**

Let $\displaystyle\sum_{k=1}^{\infty} a_k$ be a given series of nonzero terms.

(i) If $\displaystyle\lim_{n \to +\infty}\left|\frac{a_{n+1}}{a_n}\right| < 1$, then the series converges absolutely.

(ii) If $\displaystyle\lim_{n \to +\infty}\left|\frac{a_{n+1}}{a_n}\right| > 1$, or if $\displaystyle\lim_{n \to +\infty}\left|\frac{a_{n+1}}{a_n}\right| = +\infty$, then the series diverges.

(iii) If $\lim\limits_{n \to +\infty} \left| \dfrac{a_{n+1}}{a_n} \right| = 1$, then the test is inconclusive.

PROOF

(i) Suppose that $\lim\limits_{n \to +\infty} \left| \dfrac{a_{n+1}}{a_n} \right| = L < 1$. Choose and fix a number r with $L < r < 1$ $\left(\text{for instance, } r = \dfrac{L+1}{2} \right)$. Let $\varepsilon = r - L$, noting that $\varepsilon > 0$. Since $\lim\limits_{n \to +\infty} |a_{n+1}/a_n| = L$, there exists a positive integer N such that

$$\left| \left| \frac{a_{n+1}}{a_n} \right| - L \right| < \varepsilon$$

for all integers $n \geq N$; that is, $-\varepsilon < \left| \dfrac{a_{n+1}}{a_n} \right| - L < \varepsilon$, or

$$L - \varepsilon < \left| \frac{a_{n+1}}{a_n} \right| < L + \varepsilon$$

for all $n \geq N$. Since $L + \varepsilon = L + (r - L) = r$, then $|a_{n+1}/a_n| < r$, or $|a_{n+1}| < |a_n| r$ holds for $n \geq N$. Therefore,

$$|a_{N+1}| < |a_N| r,$$
$$|a_{N+2}| < |a_{N+1}| r < |a_N| r^2,$$
$$|a_{N+3}| < |a_{N+2}| r < |a_N| r^3,$$

and so forth. In fact,

$$|a_{N+j}| < |a_N| r^j$$

holds for all positive integers j. (For an inductive proof, see Problem 44.) Therefore, the geometric series $\sum\limits_{j=1}^{\infty} |a_N| r^j$ dominates the series $\sum\limits_{j=1}^{\infty} |a_{N+j}|$. Since $0 < r < 1$, the geometric series converges; hence,

$$\sum_{j=1}^{\infty} |a_{N+j}| = \sum_{k=N+1}^{\infty} |a_k|$$

converges by the direct comparison test. Therefore, by Theorem 5 of Section 3, $\sum\limits_{k=1}^{\infty} |a_k|$ converges; that is, $\sum\limits_{k=1}^{\infty} a_k$ is absolutely convergent.

(ii) Assume that $\lim\limits_{n \to +\infty} \left| \dfrac{a_{n+1}}{a_n} \right| = L > 1$. Choose and fix a number r with $1 < r < L$ and put $\varepsilon = L - r$. Thus, there exists a positive integer N such that $L - \varepsilon < \left| \dfrac{a_{n+1}}{a_n} \right| < L + \varepsilon$ holds for $n \geq N$. Hence, $1 < r = L - \varepsilon < \left| \dfrac{a_{n+1}}{a_n} \right| = \dfrac{|a_{n+1}|}{|a_n|}$, so that $|a_n| < |a_{n+1}|$ holds for $n \geq N$. Therefore,

$$|a_N| < |a_{N+1}| < |a_{N+2}| < |a_{N+3}| < \cdots,$$

so that $0 < |a_N| < |a_n|$ holds for all $n \geq N$. This shows that the condition

$\lim\limits_{n \to +\infty} a_n = 0$ cannot hold; hence, $\sum\limits_{k=1}^{\infty} a_k$ is divergent (Theorem 2 in Section 3).

Similarly, if $\lim\limits_{n \to +\infty} \left| \dfrac{a_{n+1}}{a_n} \right| = +\infty$, then there exists a positive integer N such that $1 < \left| \dfrac{a_{n+1}}{a_n} \right|$ holds for all $n \geq N$, and we can complete the argument just as above and see that $\sum\limits_{k=1}^{\infty} a_k$ diverges.

(iii) To see that the test really is inconclusive if $\lim\limits_{n \to +\infty} |a_{n+1}/a_n| = 1$, consider the following two series: (a) $\sum\limits_{k=1}^{\infty} 1/k^2$ and (b) $\sum\limits_{k=1}^{\infty} 1/k$. Series (a) is convergent, but

$$\lim_{n \to +\infty} \frac{1/(n+1)^2}{1/n^2} = \lim_{n \to +\infty} \frac{n^2}{n^2 + 2n + 1} = 1.$$

Series (b) is divergent, but

$$\lim_{n \to +\infty} \frac{1/(n+1)}{1/n} = \lim_{n \to +\infty} \frac{n}{n+1} = 1.$$

EXAMPLES   Use the ratio test to determine the convergence or divergence of the given series.

1 $\displaystyle\sum_{k=1}^{\infty} \frac{2^k}{7^k(k+1)}$

SOLUTION

Here, $a_n = \dfrac{2^n}{7^n(n+1)}$ and $a_{n+1} = \dfrac{2^{n+1}}{7^{n+1}(n+2)}$. Therefore,

$$\lim_{n \to +\infty} \left| \frac{a_{n+1}}{a_n} \right| = \lim_{n \to +\infty} \left(\frac{2^{n+1}}{7^{n+1}(n+2)} \cdot \frac{7^n(n+1)}{2^n} \right) = \lim_{n \to +\infty} \frac{2(n+1)}{7(n+2)}$$

$$= \lim_{n \to +\infty} \frac{2(1 + 1/n)}{7(1 + 2/n)} = \frac{2}{7} < 1,$$

so the series converges absolutely by the ratio test.

2 $\displaystyle\sum_{k=0}^{\infty} \frac{(-1)^{k+1}5^k}{k!}$

SOLUTION

Here, $a_n = \dfrac{(-1)^{n+1}5^n}{n!}$ and $a_{n+1} = \dfrac{(-1)^{n+2}5^{n+1}}{(n+1)!}$. Therefore,

$$\lim_{n \to +\infty} \left| \frac{a_{n+1}}{a_n} \right| = \lim_{n \to +\infty} \left(\frac{5^{n+1}}{(n+1)!} \cdot \frac{n!}{5^n} \right) = \lim_{n \to +\infty} \frac{5}{n+1} = 0 < 1,$$

so the series converges absolutely.

3 $\displaystyle\sum_{k=0}^{\infty} (-1)^k \frac{(4k)!}{(k!)^2}$

SOLUTION

Here, $a_n = \dfrac{(-1)^n (4n)!}{(n!)^2}$ and $a_{n+1} = \dfrac{(-1)^{n+1} [4(n+1)]!}{[(n+1)!]^2}$. Therefore,

$$\lim_{n \to +\infty} \left| \frac{a_{n+1}}{a_n} \right| = \lim_{n \to +\infty} \frac{n!\, n!\, (4n+4)!}{(n+1)!\,(n+1)!\,(4n)!}$$

$$= \lim_{n \to +\infty} \frac{(4n+4)(4n+3)(4n+2)(4n+1)}{(n+1)(n+1)} = +\infty;$$

hence, the given series diverges.

4 $\displaystyle\sum_{k=1}^{\infty} \frac{1 \cdot 3 \cdot 5 \cdots (2k-1)}{k!}$

SOLUTION
Here,

$$a_n = \frac{1 \cdot 3 \cdot 5 \cdots (2n-1)}{n!} \quad \text{and} \quad a_{n+1} = \frac{1 \cdot 3 \cdot 5 \cdots (2n-1)(2n+1)}{(n+1)!}.$$

Therefore,

$$\lim_{n \to +\infty} \left| \frac{a_{n+1}}{a_n} \right| = \lim_{n \to +\infty} \frac{(2n+1)n!}{(n+1)!} = \lim_{n \to +\infty} \frac{2n+1}{n+1} = 2 > 1,$$

so the given series diverges.

Another useful test for absolute convergence is given by the following theorem.

THEOREM 5   **The Root Test**

Let $\displaystyle\sum_{k=1}^{\infty} a_k$ be a given series.

(i)   If $\displaystyle\lim_{n \to +\infty} \sqrt[n]{|a_n|} < 1$, then the series converges absolutely.

(ii) If $\displaystyle\lim_{n \to +\infty} \sqrt[n]{|a_n|} > 1$, or if $\displaystyle\lim_{n \to +\infty} \sqrt[n]{|a_n|} = +\infty$, then the series diverges.

(iii) If $\displaystyle\lim_{n \to +\infty} \sqrt[n]{|a_n|} = 1$, then the test is inconclusive.

PROOF
The proof is very similar to the proof of the ratio test, so we only sketch it here and leave the details as an exercise (Problem 46). If $\displaystyle\lim_{n \to +\infty} \sqrt[n]{|a_n|} = L < 1$, choose r with $L < r < 1$. Then, for large enough values of n, $\sqrt[n]{|a_n|} < r$, or $|a_n| < r^n$; hence, $\displaystyle\sum_{k=1}^{\infty} |a_k|$ converges by comparison with $\displaystyle\sum_{k=1}^{\infty} r^k$. If the hypothesis of (ii) holds, then $\sqrt[n]{|a_n|} > 1$ holds for large enough values of n, so that $\displaystyle\lim_{n \to +\infty} a_n = 0$ cannot hold. For (iii), the same examples used in the proof of the ratio test are still effective.

EXAMPLE   Use the root test to decide whether the series $\sum_{k=1}^{\infty} \dfrac{(-1)^k}{[\ln (k+1)]^k}$ converges or diverges.

SOLUTION

Here $a_n = \dfrac{(-1)^n}{[\ln (n+1)]^n}$. Therefore,

$$\lim_{n \to +\infty} \sqrt[n]{|a_n|} = \lim_{n \to +\infty} \sqrt[n]{\left|\dfrac{(-1)^n}{[\ln (n+1)]^n}\right|} = \lim_{n \to +\infty} \dfrac{1}{\ln (n+1)} = 0 < 1,$$

so the given series converges absolutely by the root test, and thus converges.

Problem Set 5

In problems 1 through 14, determine whether the given alternating series converges or diverges. Use Leibniz's alternating series test to establish convergence whenever it applies.

1 $\sum_{k=1}^{\infty} \dfrac{(-1)^{k+1}}{k^2}$

2 $\sum_{k=1}^{\infty} \dfrac{(-1)^{k+1}}{(2k)!}$

3 $\sum_{k=1}^{\infty} \dfrac{(-1)^{k+1}k}{k^3+2}$

4 $\sum_{k=1}^{\infty} \dfrac{-\cos k\pi}{k^3}$

5 $\sum_{k=1}^{\infty} \dfrac{(-1)^k k}{\sqrt{k^5+7}}$ $\left[Hint\colon \text{First consider } \sum_{k=1}^{\infty} \dfrac{(-1)^{k+1}k}{\sqrt{k^5+7}}. \right]$

6 $\sum_{k=1}^{\infty} \dfrac{(-1)^{k+1}}{k^2-10k+26}$ $\left[Hint\colon \text{First consider } \sum_{k=5}^{\infty} \dfrac{(-1)^{k+1}}{k^2-10k+26}. \right]$

7 $\sum_{k=1}^{\infty} \dfrac{(-1)^k \sqrt{k}}{k+3}$ $\left[Hint\colon \text{First consider } \sum_{k=3}^{\infty} \dfrac{(-1)^k \sqrt{k}}{k+3}. \right]$

8 $\sum_{k=1}^{\infty} \ln k \cos k\pi$

9 $\sum_{k=1}^{\infty} (-1)^{k+1} \dfrac{k+1}{k+7}$

10 $\sum_{k=1}^{\infty} (-1)^k \dfrac{3k^2}{4k^2+1}$

11 $\sum_{k=0}^{\infty} \dfrac{(-1)^k}{\ln (k+2)}$

12 $\sum_{k=1}^{\infty} (-1)^{k+1} \dfrac{\ln (k+1)}{k\sqrt{k}}$

13 $\sum_{k=1}^{\infty} (-1)^{k+1} \sin \dfrac{\pi}{k}$

14 $\sum_{k=2}^{\infty} (-1)^{k+1} \dfrac{k}{\ln k}$

In problems 15 through 20, approximate the sum of each series by finding the partial sum of its first n terms for the indicated value of n. Also, give a bound on the absolute value of the error involved in this approximation and state whether the approximation overestimates or underestimates the true value.

15 $\sum_{k=1}^{\infty} \dfrac{(-1)^{k+1}}{3k-1}, n=5$

16 $\sum_{k=1}^{\infty} \dfrac{(-1)^{k+1}}{2^k}, n=100$

17 $\sum_{k=1}^{\infty} \dfrac{(-1)^{k+1}}{k^2}, n=4$

18 $\sum_{k=1}^{\infty} \dfrac{(-1)^k}{k^3+1}, n=4$

19 $\sum_{k=1}^{\infty} \dfrac{(-1)^k}{k \cdot 5^k}, n=3$

20 $\sum_{k=1}^{\infty} \dfrac{\sin (k+\frac{1}{2})\pi}{2k!}, n=3$

In problems 21 and 22, find the sum of each series with an error not exceeding 5×10^{-4} in absolute value and write your answer to three decimal places.

21 $\sum_{k=1}^{\infty} \dfrac{(-1)^{k+1}}{k \cdot 2^k}$

22 $\sum_{k=1}^{\infty} \dfrac{(-1)^k k}{(2k)!}$

In problems 23 through 28, apply the ratio test to determine whether each series converges absolutely or diverges.

23 $\sum_{k=1}^{\infty} \dfrac{(-1)^{k+1}5^k}{k \cdot 4^k}$

24 $\sum_{k=1}^{\infty} \dfrac{(-1)^{k+1}(k^3+1)}{k!}$

25 $\sum_{k=1}^{\infty} (-1)^{k+1} \dfrac{7^k}{(3k)!}$

26 $\sum_{k=1}^{\infty} \dfrac{(-1)^k (2k-1)!}{e^k}$

27 $\sum_{k=1}^{\infty} \dfrac{(-1)^{k+1}k^4}{(1.02)^k}$

28 $\sum_{k=1}^{\infty} (-1)^k \dfrac{1+e^k}{2^k}$

In problems 29 through 32, apply the root test to determine whether each series converges or diverges.

29 $\displaystyle\sum_{k=1}^{\infty} (-1)^{k+1} \left(\frac{k}{3k+1}\right)^k$

30 $\displaystyle\sum_{k=2}^{\infty} \frac{(-1)^k k^k}{(\ln k)^k}$

31 $\displaystyle\sum_{k=1}^{\infty} (\sqrt[k]{k} - 1)^k$

32 $\displaystyle\sum_{k=1}^{\infty} \frac{k^k}{(2k+1/k)^k}$

In problems 33 through 42, determine whether each series is divergent, conditionally convergent, or absolutely convergent. Use whatever tests or theorems seem most appropriate to justify your answer.

33 $\displaystyle\sum_{k=1}^{\infty} (-1)^k \frac{3^k}{k!}$

34 $\displaystyle\sum_{k=1}^{\infty} k \left(\frac{3}{5}\right)^k$

35 $\displaystyle\sum_{k=1}^{\infty} \frac{(-1)^{k+1}}{\ln(k+1)}$

36 $\displaystyle\sum_{k=1}^{\infty} \frac{(-1)^{k+1} k^2}{k^3 + 10}$

37 $\displaystyle\sum_{n=1}^{\infty} \frac{(-1)^{n+1} \ln n}{n}$

38 $\displaystyle\sum_{k=1}^{\infty} (-1)^k \frac{k!}{(2k+1)!}$

39 $\displaystyle\sum_{j=1}^{\infty} \frac{(-1)^j}{j^2 + 1}$

40 $\displaystyle\sum_{k=1}^{\infty} \frac{2 \cdot 4 \cdot 6 \cdots (2k)}{1 \cdot 4 \cdot 7 \cdots (3k-2)}$

41 $\displaystyle\sum_{k=1}^{\infty} \frac{k^k}{k!}$

42 $\displaystyle\sum_{k=1}^{\infty} \frac{(k!)^2}{(2k)!}$

43 Suppose that the sequence $s_2, s_4, s_6, s_8, \ldots$ converges to the limit S and that the sequence $s_1, s_3, s_5, s_7, \ldots$ converges to the same limit S. Prove: The sequence $s_1, s_2, s_3, s_4, s_5, s_6, \ldots$ converges to the limit S.

44 (a) If $|a_{n+1}| < |a_n| r$ holds for all integers $n \geq N$, where r is a positive constant, prove that $|a_{N+j}| < |a_N| r^j$ holds for all positive integers j. (Use mathematical induction.)
(b) If $|a_n| r < |a_{n+1}|$ holds for all integers $n \geq N$, where r is a positive constant, prove that $|a_N| r^j < |a_{N+j}|$ holds for all positive integers j.

45 Is it true that if the series $\displaystyle\sum_{k=1}^{\infty} a_k$ converges absolutely, then the series $\displaystyle\sum_{k=1}^{\infty} \frac{a_k^2}{1 + a_k^2}$ also converges? Why?

46 Fill in the details in the proof of the root test (Theorem 5).

47 If the series $\displaystyle\sum_{k=1}^{\infty} a_k$ converges absolutely, show that $\left| \displaystyle\sum_{k=1}^{\infty} a_k \right| \leq \displaystyle\sum_{k=1}^{\infty} |a_k|$.

6   Power Series

An infinite series of the form

$$\sum_{k=0}^{\infty} c_k(x-a)^k = c_0 + c_1(x-a) + c_2(x-a)^2 + c_3(x-a)^3 + \cdots$$

is called a *power series in x* or simply a *power series*. The constants $c_0, c_1, c_2, c_3, \ldots$ are called the *coefficients* of the power series and the constant a is called its *center*. A power series in x with center $a = 0$ has the form

$$\sum_{k=0}^{\infty} c_k x^k = c_0 + c_1 x + c_2 x^2 + c_3 x^3 + \cdots$$

and thus generalizes the idea of a polynomial in x.

In a power series $\displaystyle\sum_{k=0}^{\infty} c_k(x-a)^k$ we usually think of x as a quantity that can be varied at will. The series may converge for some values of x, but not for others. Naturally, when $x = a$, we understand that the series converges and that its sum is c_0. The following three examples show that the ratio test (Theorem 4 in Section 5) can be useful in determining the values of x for which a power series converges.

EXAMPLES   Find the values of x for which the given power series converges.

1 $\displaystyle\sum_{k=0}^{\infty} (-1)^k \frac{k}{3^k} x^k = 0 - \frac{1}{3} x + \frac{2}{9} x^2 - \frac{3}{27} x^3 + \cdots$

SOLUTION
Of course, the series converges for $x = 0$. For $x \neq 0$, we use the ratio test with

$$a_n = \frac{(-1)^n n x^n}{3^n} \quad \text{and} \quad a_{n+1} = \frac{(-1)^{n+1}(n+1)x^{n+1}}{3^{n+1}}.$$

Here,

$$\lim_{n\to+\infty} \left| \frac{a_{n+1}}{a_n} \right| = \lim_{n\to+\infty} \left| \frac{(-1)^{n+1}(n+1)x^{n+1}}{3^{n+1}} \cdot \frac{3^n}{(-1)^n n x^n} \right| = \lim_{n\to+\infty} \frac{n+1}{3n} |x|$$

$$= |x| \lim_{n\to+\infty} \frac{n+1}{3n} = \frac{|x|}{3},$$

so the series converges for $|x|/3 < 1$, that is, for $-3 < x < 3$. If $x < -3$ or if $x > 3$, then $|x|/3 > 1$ and the series diverges. When $|x| = 3$, we have $|a_n| = \left| (-1)^n \frac{n}{3^n} x^n \right| = \frac{n}{3^n} |x|^n = \frac{n}{3^n} 3^n = n$, so that $\lim_{n\to+\infty} a_n \neq 0$ and the series must diverge. Therefore, the series converges for values of x in the open interval $(-3, 3)$ and only for such values of x.

2 $\displaystyle\sum_{k=0}^{\infty} \frac{(x-5)^{2k}}{k!} = 1 + (x-5)^2 + \frac{1}{2}(x-5)^4 + \frac{1}{6}(x-5)^6 + \frac{1}{24}(x-5)^8 + \cdots$

SOLUTION
The series converges for $x = 5$. For $x \neq 5$, we use the ratio test with

$$a_n = \frac{(x-5)^{2n}}{n!} \quad \text{and} \quad a_{n+1} = \frac{(x-5)^{2(n+1)}}{(n+1)!} = \frac{(x-5)^{2n+2}}{(n+1)!}.$$

Here,

$$\lim_{n\to+\infty} \left| \frac{a_{n+1}}{a_n} \right| = \lim_{n\to+\infty} \left| \frac{(x-5)^{2n+2}}{(n+1)!} \cdot \frac{n!}{(x-5)^{2n}} \right| = \lim_{n\to+\infty} \frac{(x-5)^2}{n+1} = 0 < 1$$

for all values of x; hence, the series converges for all values of x.

3 $\displaystyle\sum_{k=0}^{\infty} (k!)(x+2)^k = 1 + (x+2) + 2(x+2)^2 + 6(x+2)^3 + 24(x+2)^4 + \cdots.$

SOLUTION
Here we have $a_n = (n!)(x+2)^n$ and $a_{n+1} = [(n+1)!](x+2)^{n+1}$. For $x \neq -2$, the ratio test gives

$$\lim_{n\to+\infty} \left| \frac{a_{n+1}}{a_n} \right| = \lim_{n\to+\infty} \left| \frac{[(n+1)!](x+2)^{n+1}}{(n!)(x+2)^n} \right| = \lim_{n\to+\infty} (n+1)|x+2| = +\infty;$$

hence, the series diverges. Of course, the series converges for $x = -2$, and so it converges only when $x = -2$.

The set I of all numbers x for which a power series $\sum_{k=0}^{\infty} c_k(x-a)^k$ converges is called its *interval of convergence*. In Example 1, $I=(-3,3)$, in Example 2, $I=(-\infty,\infty)$, and in Example 3, I is the "interval" containing the single number -2.

For any power series $\sum_{k=1}^{\infty} c_k(x-a)^k$, the interval of convergence I always has one of the following forms:

Case 1 I is a bounded interval with center a and with endpoints $a-R$ and $a+R$, where R is a positive real number.

Case 2 $I=(-\infty,\infty)$.

Case 3 I consists of the single number a.

In Case 1, we call the number R the *radius of convergence* of the power series. In Case 2, it is convenient to say that the radius of convergence of the power series is infinite and to write $R=+\infty$. Naturally, in Case 3, we say that the power series has radius of convergence zero and we write $R=0$. Examples 1 through 3 above illustrate these three possibilities.

Figure 1

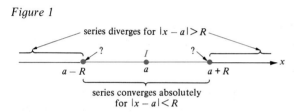

series diverges for $|x-a|>R$

series converges absolutely for $|x-a|<R$

In Case 1, the endpoints $a-R$ and $a+R$ of the interval of convergence I may or may not belong In Example 1, neither endpoint belongs to I, so that I is an open interval. In general, anything can happen—the series may diverge, converge conditionally, or converge absolutely at an endpoint of I. Thus, in Case 1, the interval of convergence can be any one of the four sets $I=[a-R,a+R]$, $I=[a-R,a+R)$, $I=(a-R,a+R]$, or $I=(a-R,a+R)$ (Figure 1). The power series always converges *absolutely* on the open interval $(a-R,a+R)$.

The following theorem provides an efficient means for finding the radius of convergence of a power series.

THEOREM 1 **Radius of Convergence of a Power Series**

Let $\sum_{k=0}^{\infty} c_k(x-a)^k$ be a power series with radius of convergence R. Suppose that

$$\lim_{n\to+\infty}\left|\frac{c_{n+1}}{c_n}\right|=L, \text{ where } L \text{ is either a nonnegative real number or } L=+\infty.$$

(i) If L is a positive real number, then $R=1/L$.

(ii) If $L=0$, then $R=+\infty$.

(iii) If $L=+\infty$, then $R=0$.

PROOF

We attend to part (i) here. Parts (ii) and (iii), which are handled similarly, are left as an exercise (Problem 30). Thus, suppose that $\lim_{n\to+\infty}\left|\frac{c_{n+1}}{c_n}\right|=L$, where L is a positive real number. We apply the ratio test to the infinite series $\sum_{k=0}^{\infty} c_k(x-a)^k$. Here, $a_n=c_n(x-a)^n$ and $a_{n+1}=c_{n+1}(x-a)^{n+1}$, so that

$$\lim_{n\to+\infty}\left|\frac{a_{n+1}}{a_n}\right|=\lim_{n\to+\infty}\left|\frac{c_{n+1}(x-a)^{n+1}}{c_n(x-a)^n}\right|=\lim_{n\to+\infty}\left|\frac{c_{n+1}}{c_n}\right|\cdot|x-a|=L|x-a|.$$

By the ratio test, the series converges absolutely for $L|x - a| < 1$ and it diverges for $L|x - a| > 1$; that is, it converges for $|x - a| < 1/L$ and it diverges for $|x - a| > 1/L$. Thus, $1/L = R$, the radius of convergence of the power series.

Several important comments need to be made in connection with Theorem 1:

1 Notice that the ratio c_{n+1}/c_n in Theorem 1 is a ratio of *coefficients*—not *terms*—of the power series. Do not confuse Theorem 1 with the original ratio test (Theorem 4 in Section 5), which involves the *terms* of a series.

2 Theorem 1 is easy to remember if, *for purposes of this theorem only*, we agree that $1/L = +\infty$ when $L = 0$ and that $1/L = 0$ when $L = +\infty$. Then, the theorem simply says that $R = 1/L$ in every case.

3 Theorem 1 may not apply in certain cases because the sequence $|c_{n+1}/c_n|$ has no real number as a limit and does not approach $+\infty$ as $n \to +\infty$. In such a case, the power series still has a definite radius of convergence R, but methods beyond the scope of this book may be required to find it.

4 Theorem 1 *does not* apply to a power series of the form

$$\sum_{k=0}^{\infty} c_k(x - a)^{kp} = c_0 + c_1(x - a)^p + c_2(x - a)^{2p} + c_3(x - a)^{3p} + \cdots,$$

where p is a constant integer greater than 1, since c_k is not the coefficient of the kth power of $x - a$. In such a case, the radius of convergence can often be found by applying the original ratio test (Theorem 4 in Section 5) directly to the *terms* of the series as in Example 2 on page 645.

5 Theorem 1 says nothing, one way or the other, about whether the power series converges at the *endpoints* of its interval of convergence. This has to be checked by substituting $x = a - R$ and $x = a + R$ in the power series and using the standard tests for convergence of a series.

6 Theorem 1 is still valid for a power series such as $\sum_{k=M}^{\infty} c_k(x - a)^k$, where M is a positive integer and the summation starts at $k = M$ rather than at $k = 0$. (Why?)

EXAMPLES Find the center a, the radius of convergence R, and the interval of convergence I of the given power series. Be sure to check the endpoints of I for divergence, absolute convergence, or conditional convergence.

1 $\sum_{k=1}^{\infty} \frac{1}{k} x^k$

SOLUTION
Here the center is $a = 0$ and $c_k = 1/k$. In Theorem 1, we put $c_n = 1/n$ and $c_{n+1} = 1/(n + 1)$, so that

$$\lim_{n \to +\infty} \left| \frac{c_{n+1}}{c_n} \right| = \lim_{n \to +\infty} \left| \frac{1/(n + 1)}{1/n} \right| = \lim_{n \to +\infty} \frac{n}{n + 1} = 1 = L;$$

hence, $R = 1/L = 1$. Therefore, the series converges absolutely for values of x in the open interval $(a - R, a + R) = (0 - 1, 0 + 1) = (-1, 1)$ and it diverges for values of x outside the closed interval $[-1, 1]$. Substituting $x = 1$ in the series, we obtain the harmonic series $\sum_{k=1}^{\infty} \frac{1}{k}$, which diverges. For $x = -1$, the series becomes the alternating harmonic series $\sum_{k=1}^{\infty} \frac{1}{k}(-1)^k$, which converges by Leibniz's test. Hence, we have divergence at the endpoint 1 and conditional convergence at the endpoint -1. The interval of convergence is $I = [-1, 1)$.

2 $\sum_{k=0}^{\infty} \frac{(x+3)^k}{3^k}$

SOLUTION
The power series is centered at $a = -3$. We have $c_n = 1/3^n$ and $c_{n+1} = 1/3^{n+1}$, so that

$$\lim_{n \to +\infty} \left| \frac{c_{n+1}}{c_n} \right| = \lim_{n \to +\infty} \left| \frac{1/3^{n+1}}{1/3^n} \right| = \lim_{n \to +\infty} \frac{1}{3} = \frac{1}{3} = L;$$

hence, $R = 1/L = 3$ by Theorem 1. Therefore, the series converges absolutely on the open interval $(a - R, a + R) = (-3 - 3, -3 + 3) = (-6, 0)$. When $x = -6$, the series becomes

$$\sum_{k=0}^{\infty} \frac{(-3)^k}{3^k} = 1 - 1 + 1 - 1 + \cdots,$$

which diverges because the general term does not approach zero. When $x = 0$, the series becomes

$$\sum_{k=0}^{\infty} \frac{3^k}{3^k} = 1 + 1 + 1 + \cdots,$$

which also diverges. Therefore, $I = (-6, 0)$.

3 $\sum_{k=0}^{\infty} \frac{(-1)^k}{k!}(x - 17)^k$

SOLUTION
The power series is centered at $a = 17$. We have

$$c_n = \frac{(-1)^n}{n!} \quad \text{and} \quad c_{n+1} = \frac{(-1)^{n+1}}{(n+1)!},$$

so that

$$\lim_{n \to +\infty} \left| \frac{c_{n+1}}{c_n} \right| = \lim_{n \to +\infty} \left| \frac{(-1)^{n+1}}{(n+1)!} \cdot \frac{n!}{(-1)^n} \right| = \lim_{n \to +\infty} \frac{1}{n+1} = 0 = L;$$

hence, $R = +\infty$ by part (ii) of Theorem 1. Therefore, $I = (-\infty, \infty)$.

4 $\sum_{k=1}^{\infty} k^k x^k$

SOLUTION
The power series is centered at $a = 0$. We have $c_n = n^n$ and $c_{n+1} = (n+1)^{n+1}$, so that

$$\lim_{n \to +\infty} \left| \frac{c_{n+1}}{c_n} \right| = \lim_{n \to +\infty} \left| \frac{(n+1)^{n+1}}{n^n} \right| = \lim_{n \to +\infty} \left[\left(\frac{n+1}{n} \right)^n (n+1) \right]$$

$$= \left[\lim_{n \to +\infty} \left(1 + \frac{1}{n} \right)^n \right] \cdot \left[\lim_{n \to +\infty} (n+1) \right] = e \left[\lim_{n \to +\infty} (n+1) \right] = +\infty.$$

Thus, $R = 0$ by part (iii) of Theorem 1, and so I consists of the single number 0.

5 $\displaystyle\sum_{k=1}^{\infty} \frac{3^k(x-4)^{2k}}{k^2}$

SOLUTION

The power series is centered at $a = 4$. Here we cannot use Theorem 1 since $3^k/k^2$ is not the coefficient of the kth power of $x - 4$. Thus, we resort to the original ratio test (Theorem 4 in Section 5). The nth *term* (not coefficient!) of the series is

$$a_n = \frac{3^n(x-4)^{2n}}{n^2}, \quad \text{so that} \quad a_{n+1} = \frac{3^{n+1}(x-4)^{2(n+1)}}{(n+1)^2} = \frac{3^{n+1}(x-4)^{2n+2}}{(n+1)^2}.$$

Therefore,

$$\lim_{n \to +\infty} \left| \frac{a_{n+1}}{a_n} \right| = \lim_{n \to +\infty} \left| \frac{3^{n+1}(x-4)^{2n+2}}{(n+1)^2} \cdot \frac{n^2}{3^n(x-4)^{2n}} \right|$$

$$= \lim_{n \to +\infty} 3\left(\frac{n}{n+1}\right)^2 |x-4|^2 = 3|x-4|^2.$$

It follows that the series converges absolutely when $3|x-4|^2 < 1$, that is, when $|x-4| < 1/\sqrt{3}$. It diverges when $3|x-4|^2 > 1$, that is, when $|x-4| > 1/\sqrt{3}$. Therefore, $R = \dfrac{1}{\sqrt{3}}$. When $x = 4 - \dfrac{1}{\sqrt{3}}$, the series becomes $\displaystyle\sum_{k=1}^{\infty} \frac{1}{k^2}$, which converges absolutely. (Why?) Similarly, when $x = 4 + \dfrac{1}{\sqrt{3}}$ the series becomes $\displaystyle\sum_{k=1}^{\infty} \frac{1}{k^2}$, which converges absolutely. It follows that the series converges absolutely on its entire interval of convergence $I = \left[4 - \dfrac{1}{\sqrt{3}}, \, 4 + \dfrac{1}{\sqrt{3}} \right]$.

Problem Set 6

In problems 1 through 25, find the center a, the radius of convergence R, and the interval of convergence I of the given power series. Be sure to check the endpoints of I for divergence, absolute convergence, or conditional convergence.

1 $\displaystyle\sum_{k=0}^{\infty} 7^k x^k$

2 $\displaystyle\sum_{k=0}^{\infty} \frac{x^{k+1}}{\sqrt{k+1}}$

3 $\displaystyle\sum_{k=0}^{\infty} \frac{x^k}{k!}$

4 $\displaystyle\sum_{k=0}^{\infty} \frac{(-1)^k(x-1)^k}{k+1}$

5 $\displaystyle\sum_{k=1}^{\infty} \frac{(-1)^{k+1}x^{2k-1}}{(2k-1)!}$

6 $\displaystyle\sum_{k=1}^{\infty} \frac{(-1)^{k+1}k(x+3)^{k-1}}{7^{k-1}}$

7 $\displaystyle\sum_{k=1}^{\infty} \frac{(x+2)^{k-1}}{k^2}$

8 $\displaystyle\sum_{k=1}^{\infty} \frac{(x+1)^k}{k\sqrt{k+1}}$

9 $\displaystyle\sum_{k=1}^{\infty} \frac{(x+5)^k}{(2k-1)(2k)}$

10 $\displaystyle\sum_{k=2}^{\infty} \frac{(x-1)^{2k-2}}{(2k-4)!}$

11 $\displaystyle\sum_{j=0}^{\infty} \frac{(-1)^j 2^j x^j}{(j+1)^3}$

12 $\displaystyle\sum_{j=0}^{\infty} \frac{\sqrt{j}\, x^j}{1 \cdot 3 \cdot 5 \cdots (2j+1)}$

13 $\displaystyle\sum_{n=0}^{\infty} \frac{(x-1)^n}{(n+2)!}$

14 $\displaystyle\sum_{k=0}^{\infty} \left[\frac{(-1)^k}{2k+1}\right]\left(\frac{x}{2}\right)^{2k}$

15 $\displaystyle\sum_{k=0}^{\infty} \frac{(1-x)^k}{(k+1)\cdot 3^k}$

16 $\displaystyle\sum_{k=0}^{\infty} \frac{(x+1)^{5k}}{(k+1)\cdot 5^k}$

17 $\displaystyle\sum_{k=1}^{\infty} \frac{1}{k}\left(\frac{x}{4}-1\right)^k$

18 $\displaystyle\sum_{n=0}^{\infty} \frac{1}{3n-1}\left(\frac{x}{3}+\frac{2}{3}\right)^n$

19 $\displaystyle\sum_{j=1}^{\infty} \frac{(3-x)^{j-1}}{\sqrt{j}}$

20 $\displaystyle\sum_{k=1}^{\infty} \frac{(-1)^k 2^k(x-5)^{2k}}{k^3}$

21 $\displaystyle\sum_{k=1}^{\infty} (5^k + 5^{-k})(x+1)^{3k-2}$

22 $\displaystyle\sum_{k=1}^{\infty} \frac{k!}{k^k} x^k$

23 $\displaystyle\sum_{k=1}^{\infty} (\tan^{-1} k)(x-1)^k$

24 $\displaystyle\sum_{k=1}^{\infty} \frac{(x-3)^{4k}}{\sqrt[k]{k}}$

25 $(x-8) + (x-8)^2 + 2!(x-8)^3 + 3!(x-8)^4 + 4!(x-8)^5 + \cdots$.

26 If R is the radius of convergence of the power series $\sum\limits_{k=0}^{\infty} c_k(x-a)^k$, $0 < R < +\infty$, and p is a positive integer, show that the radius of convergence of the power series $\sum\limits_{k=0}^{\infty} c_k(x-a)^{pk}$ is $\sqrt[p]{R}$.

27 If R is the radius of convergence of the power series $\sum\limits_{k=0}^{\infty} c_k(x-a)^k$, and p is a positive integer, find the radius of convergence of the power series $\sum\limits_{k=0}^{\infty} c_k(x-a)^{p+k}$.

28 Let $\sum\limits_{k=0}^{\infty} c_k(x-a)^k$ be a given power series.

    (a) If $\lim\limits_{n \to +\infty} \sqrt[n]{|c_n|} = +\infty$, prove that the radius of convergence of the power series is zero.

    (b) If $\lim\limits_{n \to +\infty} \sqrt[n]{|c_n|} = 0$, prove that the power series has an infinite radius of convergence.

    (c) If $\lim\limits_{n \to +\infty} \sqrt[n]{|c_n|} = L \neq 0$, prove that the radius of convergence of the power series is given by $R = 1/L$.

29 Suppose that b is a constant greater than 1. Find the radius of convergence and the interval of convergence of the power series $\sum\limits_{k=0}^{\infty} \dfrac{x^k}{1+b^k}$.

30 Complete the proof of Theorem 1 by considering parts (ii) and (iii).

31 If $a > b \geq 0$, find the radius of convergence of the power series $\sum\limits_{k=0}^{\infty} \dfrac{x^k}{a^k + b^k}$.

7   Continuity, Integration, and Differentiation of Power Series

In the present section we study functions of the form

$$f(x) = \sum_{k=0}^{\infty} c_k(x-a)^k,$$

where $\sum\limits_{k=0}^{\infty} c_k(x-a)^k$ is a given power series. Here it is understood that the domain of f is the interval of convergence of the power series.

Since a *finite* sum can be differentiated term by term, and since

$$f(x) = c_0 + c_1(x-a) + c_2(x-a)^2 + c_3(x-a)^3 + \cdots,$$

we might guess that the derivative $D_x f(x)$ can be obtained by differentiating term by term; that is,

$$D_x f(x) = D_x c_0 + D_x c_1(x-a) + D_x c_2(x-a)^2 + D_x c_3(x-a)^3 + \cdots$$
$$= 0 + c_1 + 2c_2(x-a) + 3c_3(x-a)^2 + \cdots.$$

Likewise, we might guess that the integral $\int f(x)\,dx$ can be obtained by integrating term by term; that is,

$$\int f(x)\,dx = \int c_0\,dx + \int c_1(x-a)\,dx + \int c_2(x-a)^2\,dx + \int c_3(x-a)^3\,dx + \cdots$$

$$= \left[c_0(x-a) + \frac{c_1}{2}(x-a)^2 + \frac{c_2}{3}(x-a)^3 + \frac{c_3}{4}(x-a)^4 + \cdots \right] + C.$$

Later in this section we see that such "calculations" are quite legitimate, provided that $|x-a| < R$, where R is the radius of convergence of the power series.

Notice that term-by-term differentiation or integration of a power series

$$\sum_{k=0}^{\infty} c_k(x-a)^k$$

produces a new power series,

$$\sum_{k=0}^{\infty} D_x[c_k(x-a)^k] = \sum_{k=1}^{\infty} kc_k(x-a)^{k-1}$$

or

$$\sum_{k=0}^{\infty} \int c_k(x-a)^k\,dx = \sum_{k=0}^{\infty} \frac{c_k}{k+1}(x-a)^{k+1},$$

respectively. (In the second equation, we have suppressed the constants of integration.)

If $f(x) = \sum_{k=0}^{\infty} c_k(x-a)^k$, where the power series has radius of convergence R, we shall establish the following properties of f and $\sum_{k=0}^{\infty} c_k(x-a)^k$:

Property I   The function f is continuous on the open interval $(a-R, a+R)$.

Property II   The three power series

$$\sum_{k=0}^{\infty} c_k(x-a)^k,$$

$$\sum_{k=0}^{\infty} D_x[c_k(x-a)^k] = \sum_{k=1}^{\infty} kc_k(x-a)^{k-1},$$

$$\sum_{k=0}^{\infty} \int c_k(x-a)^k\,dx = \sum_{k=0}^{\infty} \frac{c_k}{k+1}(x-a)^{k+1}$$

all have the same radius of convergence R.

Property III   For $|x-a| < R$,

$$f'(x) = D_x\left[\sum_{k=0}^{\infty} c_k(x-a)^k \right] = \sum_{k=1}^{\infty} kc_k(x-a)^{k-1}.$$

Property IV   For $|x-a| < R$,

$$\int f(x)\,dx = \int \left[\sum_{k=0}^{\infty} c_k(x-a)^k \right] dx = \sum_{k=0}^{\infty} \frac{c_k}{k+1}(x-a)^{k+1} + C.$$

Property V   For $|b-a| < R$,

$$\int_a^b f(x)\,dx = \int_a^b \left[\sum_{k=0}^{\infty} c_k(x-a)^k \right] dx = \sum_{k=0}^{\infty} \int_a^b c_k(x-a)^k\,dx$$

$$= \sum_{k=0}^{\infty} \frac{c_k}{k+1}(b-a)^{k+1}.$$

The proofs of Properties I through V are somewhat technical, so we prefer to postpone them until Section 7.1. Meanwhile, we illustrate these properties in the following examples.

EXAMPLES **1** Find $D_x(1 + 2x + 3x^2 + 4x^3 + \cdots)$.

SOLUTION

By Theorem 1 in Section 6, the radius of convergence of the power series $\sum_{k=0}^{\infty} (k + 1)x^k$ is $R = 1$; hence, by Properties II and III,

$$D_x(1 + 2x + 3x^2 + 4x^3 + \cdots) = 0 + 2 + 6x + 12x^2 + \cdots;$$

that is,

$$D_x\left[\sum_{k=0}^{\infty} (k + 1)x^k\right] = \sum_{k=1}^{\infty} k(k + 1)x^{k-1} \qquad \text{for } |x| < 1.$$

2 Find $\int (1 + 2x + 3x^2 + 4x^3 + \cdots)\, dx$ for $|x| < 1$.

SOLUTION

By Property IV,

$$\int (1 + 2x + 3x^2 + 4x^3 + \cdots)\, dx = (x + x^2 + x^3 + x^4 + \cdots) + C;$$

that is,

$$\int \left[\sum_{k=0}^{\infty} (k + 1)x^k\right] dx = \sum_{k=0}^{\infty} x^{k+1} + C \qquad \text{for } |x| < 1.$$

3 Use the formula $\sum_{k=0}^{\infty} x^k = \dfrac{1}{1 - x}$, which gives the sum of the geometric series $1 + x + x^2 + x^3 + \cdots$ for $|x| < 1$, to write the given expression as the sum of an infinite series:

(a) $\dfrac{1}{(1 - x)^2}$ (b) $\dfrac{1}{1 + x}$ (c) $\ln{(1 + x)}$ (d) $\dfrac{1}{1 + x^2}$ (e) $\tan^{-1} x$

(f) $\dfrac{1}{3 - x}$ (g) $\dfrac{2}{x^2 - 4x + 3}$

SOLUTION

(a) $\dfrac{1}{(1 - x)^2} = D_x\left(\dfrac{1}{1 - x}\right) = D_x\left(\sum_{k=0}^{\infty} x^k\right) = \sum_{k=1}^{\infty} kx^{k-1}$

$\qquad = 1 + 2x + 3x^2 + 4x^3 + \cdots \qquad \text{for } |x| < 1.$

(b) Replacing x by $-x$ in $\sum_{k=0}^{\infty} x^k = \dfrac{1}{1 - x}$, we obtain

$$\dfrac{1}{1 + x} = \sum_{k=0}^{\infty} (-x)^k = 1 - x + x^2 - x^3 + \cdots$$

for $|-x| < 1$, that is, for $|x| < 1$.

(c) Using the result of (b) and Property V, we have

$$\ln (1 + x) = \int_0^x \frac{dt}{1 + t} = \int_0^x (1 - t + t^2 - t^3 + \cdots)\, dt$$

$$= \int_0^x dt - \int_0^x t\, dt + \int_0^x t^2\, dt - \int_0^x t^3\, dt + \cdots$$

$$= x - \frac{x^2}{2} + \frac{x^3}{3} - \frac{x^4}{4} + \cdots$$

$$= \sum_{k=0}^{\infty} (-1)^k \frac{x^{k+1}}{k+1} \qquad \text{for } |x| < 1.$$

(d) Replacing x by $-x^2$ in $\displaystyle\sum_{k=0}^{\infty} x^k = \frac{1}{1-x}$, we obtain

$$\frac{1}{1+x^2} = \sum_{k=0}^{\infty} (-x^2)^k = 1 - x^2 + x^4 - x^6 + \cdots \qquad \text{for } |x| < 1.$$

(Note that we get convergence of the series when $|-x^2| < 1$; however, $|-x^2| = |x^2| = |x|^2$, and $|x|^2 < 1$ holds exactly when $|x| < 1$.)

(e) Using the result of (d) and Property V, we have

$$\tan^{-1} x = \int_0^x \frac{dt}{1+t^2} = \int_0^x (1 - t^2 + t^4 - t^6 + \cdots)\, dt$$

$$= \int_0^x dt - \int_0^x t^2\, dt + \int_0^x t^4\, dt - \int_0^x t^6\, dt + \cdots$$

$$= x - \frac{x^3}{3} + \frac{x^5}{5} - \frac{x^7}{7} + \cdots$$

$$= \sum_{k=0}^{\infty} (-1)^k \frac{x^{2k+1}}{2k+1} \qquad \text{for } |x| < 1.$$

(f) $\displaystyle \frac{1}{3-x} = \frac{1}{3} \cdot \frac{1}{1 - \left(\dfrac{x}{3}\right)} = \frac{1}{3} \sum_{k=0}^{\infty} \left(\frac{x}{3}\right)^k = \sum_{k=0}^{\infty} \frac{1}{3} \cdot \frac{x^k}{3^k} = \sum_{k=0}^{\infty} \frac{x^k}{3^{k+1}} \qquad \text{for } |x| < 3.$

(g) $\displaystyle \frac{2}{x^2 - 4x + 3} = \frac{2}{(x-1)(x-3)} = \frac{-1}{x-1} + \frac{1}{x-3} = \frac{1}{1-x} - \frac{1}{3-x}$ by partial fractions. Therefore, using the result of (f), we have

$$\frac{2}{x^2 - 4x + 3} = \frac{1}{1-x} - \frac{1}{3-x} = \sum_{k=0}^{\infty} x^k - \sum_{k=0}^{\infty} \frac{x^k}{3^{k+1}}$$

$$= \sum_{k=0}^{\infty} \left(x^k - \frac{x^k}{3^{k+1}} \right) = \sum_{k=0}^{\infty} \left(\frac{3^{k+1} - 1}{3^{k+1}} \right) x^k$$

$$= \frac{2}{3} + \frac{8}{9} x + \frac{26}{27} x^2 + \frac{80}{81} x^3 + \cdots \qquad \text{for } |x| < 1.$$

4 If $\displaystyle f(x) = \sum_{k=1}^{\infty} (-1)^{k+1} \frac{(x-2)^k}{\sqrt{k}}$ for $|x - 2| < 1$, find an infinite series representation for $f'(x)$.

SOLUTION

For $|x - 2| < 1$,

$$f'(x) = \sum_{k=1}^{\infty} (-1)^{k+1} \frac{k}{\sqrt{k}} (x - 2)^{k-1} = \sum_{k=1}^{\infty} (-1)^{k+1} \sqrt{k} (x - 2)^{k-1}.$$

5 If $f(t) = \sum_{k=0}^{\infty} (-1)^k \frac{(t + 3)^{2k}}{(2k)!}$, find an infinite series representation for $\int_{-3}^{x} f(t)\, dt$.

SOLUTION

By the ratio test, the power series $\sum_{k=0}^{\infty} (-1)^k \frac{(t + 3)^{2k}}{(2k)!}$ has radius of convergence $R = +\infty$; hence, $f(t)$ is defined for all values of t. Therefore, for every value of x, we have

$$\int_{-3}^{x} f(t)\, dt = \sum_{k=0}^{\infty} \int_{-3}^{x} (-1)^k \frac{(t + 3)^{2k}}{(2k)!}\, dt = \sum_{k=0}^{\infty} \frac{(-1)^k (x + 3)^{2k+1}}{(2k)!\,(2k + 1)}$$

$$= \sum_{k=0}^{\infty} \frac{(-1)^k (x + 3)^{2k+1}}{(2k + 1)!}$$

7.1   Proofs of the Properties of Power Series

We now set forth and prove the theorems required to justify Properties I through V. For simplicity, we consider only power series with center $a = 0$. The general case in which $a \neq 0$ can be handled simply by replacing x by $x - a$ in the following theorems.

Thus, let $\sum_{k=0}^{\infty} c_k x^k$ be a power series with radius of convergence R and define the function f by

$$f(x) = \sum_{k=0}^{\infty} c_k x^k.$$

For each value of x in the interval of convergence of $\sum_{k=0}^{\infty} c_k x^k$, we obtain an approximation

$$f(x) \approx \sum_{k=0}^{n} c_k x^k$$

by taking the nth partial sum of the series. The error

$$E_n(x) = f(x) - \sum_{k=0}^{n} c_k x^k = \sum_{k=0}^{\infty} c_k x^k - \sum_{k=0}^{n} c_k x^k$$

involved in this approximation can be written as

$$E_n(x) = \sum_{k=n+1}^{\infty} c_k x^k,$$

by Theorem 5 of Section 3.

The following important theorem shows that the error $E_n(x)$ can be made as small as we please for all values of x in a closed subinterval of $(-R, R)$, provided only that n is sufficiently large.

THEOREM 1

On Making the Error $E_n(x)$ Small

Let $E_n(x)$ be the error as defined above, and fix any number r with $0 < r < R$. Then, given any $\varepsilon > 0$, there exists a positive integer N such that $|E_n(x)| < \varepsilon$ provided that $n \geq N$ and $|x| \leq r$.

PROOF

Since $0 < r < R$, the series $\sum\limits_{k=0}^{\infty} c_k r^k$ converges absolutely; that is,

$$\lim_{n \to +\infty} \sum_{k=0}^{n} |c_k r^k| = \lim_{n \to +\infty} \sum_{k=0}^{n} |c_k| r^k = \sum_{k=0}^{\infty} |c_k| r^k$$

exists. It follows that

$$\lim_{n \to +\infty} \left(\sum_{k=0}^{\infty} |c_k| r^k - \sum_{k=0}^{n} |c_k| r^k \right) = 0;$$

that is, $\lim\limits_{n \to +\infty} \sum\limits_{k=n+1}^{\infty} |c_k| r^k = 0$; hence, there exists a positive integer N such that $\sum\limits_{k=n+1}^{\infty} |c_k| r^k < \varepsilon$, provided that $n \geq N$. Now, suppose that $n \geq N$ and $|x| \leq r$. Then, by Problem 47 in Problem Set 5,

$$|E_n(x)| = \left| \sum_{k=n+1}^{\infty} c_k x^k \right| \leq \sum_{k=n+1}^{\infty} |c_k x^k|$$

$$= \sum_{k=n+1}^{\infty} |c_k| \cdot |x|^k \leq \sum_{k=n+1}^{\infty} |c_k| r^k < \varepsilon.$$

THEOREM 2

Continuity of Power Series

If $R > 0$ is the radius of convergence of the power series $\sum\limits_{k=0}^{\infty} c_k x^k$, then the function f defined by $f(x) = \sum\limits_{k=0}^{\infty} c_k x^k$ is continuous on the open interval $(-R, R)$.

PROOF

Given a number b in $(-R, R)$ and a number $\varepsilon_0 > 0$, we must prove that $|f(x) - f(b)| < \varepsilon_0$ holds provided that x is sufficiently close to b. Noting that $|b| < R$, we choose and fix a number r with $0 \leq |b| < r < R$ and confine our attention to values of x close enough to b so that $|x| < r$. Putting $\varepsilon = \varepsilon_0/3$ in Theorem 1, we find a positive integer N so large that $|E_N(x)| < \varepsilon_0/3$ for all such values of x. In particular, $|E_N(b)| < \varepsilon_0/3$. Since the polynomial function P defined by $P(x) = \sum\limits_{k=0}^{N} c_k x^k$ is continuous, we can choose x close enough to b so that $|P(x) - P(b)| < \varepsilon_0/3$. Then we have

$$|f(x) - f(b)| = \left\| \left[\sum_{k=0}^{N} c_k x^k + E_N(x) \right] - \left[\sum_{k=0}^{N} c_k b^k + E_N(b) \right] \right\|$$

$$= |P(x) + E_N(x) - P(b) - E_N(b)|$$

$$= |P(x) - P(b) + E_N(x) + (-1)E_N(b)|$$

$$\leq |P(x) - P(b)| + |E_N(x)| + |(-1)E_N(b)|$$

$$< \frac{\varepsilon_0}{3} + \frac{\varepsilon_0}{3} + \frac{\varepsilon_0}{3} = \varepsilon_0,$$

as desired.

THEOREM 3

On Making the Integral of the Error Small

Let $E_n(x)$ be the error as defined above and let b be a fixed number with $|b| < R$. Then $\lim\limits_{n \to +\infty} \int_0^b E_n(x)\,dx = 0$.

PROOF

Suppose that $b \geq 0$. (The case $b < 0$ is handled similarly.) For each positive integer n, define the polynomial function P_n by $P_n(x) = \sum\limits_{k=0}^{n} c_k x^k$. For $|x| < R$, we have $E_n(x) = f(x) - P_n(x)$, where f is a continuous function by Theorem 2. Since P_n is continuous, it follows that the function E_n is continuous, so the integral $\int_0^b E_n(x)\,dx$ exists. Let $\varepsilon_0 > 0$ be given. We must find a positive integer N so large that $\left| \int_0^b E_n(x)\,dx \right| \leq \varepsilon_0$ provided that $n \geq N$. Since $\int_0^b E_n(x)\,dx = 0$ if $b = 0$, we can assume $b \neq 0$. Put $\varepsilon = \varepsilon_0/b$ and $r = b$ in Theorem 1. Then, for all values of x between 0 and b inclusive, $|E_n(x)| < \varepsilon$ if $n \geq N$. It follows that, for $n \geq N$,

$$\left| \int_0^b E_n(x)\,dx \right| \leq \int_0^b |E_n(x)|\,dx \leq \int_0^b \varepsilon\,dx = \varepsilon b = \varepsilon_0.$$

THEOREM 4

Term-by-Term Integration of a Power Series

If $R > 0$ is the radius of convergence of the power series $\sum\limits_{k=0}^{\infty} c_k x^k$ and if $|b| < R$, then

$$\int_0^b \left(\sum_{k=0}^{\infty} c_k x^k \right) dx = \sum_{k=0}^{\infty} \int_0^b c_k x^k\,dx.$$

PROOF

Let $f(x) = \sum\limits_{k=0}^{\infty} c_k x^k$ and $E_n(x) = f(x) - \sum\limits_{k=0}^{n} c_k x^k$, as on page 654. We must prove that $\lim\limits_{n \to +\infty} \sum\limits_{k=0}^{n} \int_0^b c_k x^k\,dx = \int_0^b f(x)\,dx$. We have

$$\lim_{n \to +\infty} \sum_{k=0}^{n} \int_0^b c_k x^k\,dx = \lim_{n \to +\infty} \int_0^b \left(\sum_{k=0}^{n} c_k x^k \right) dx = \lim_{n \to +\infty} \int_0^b [f(x) - E_n(x)]\,dx$$

$$= \lim_{n \to +\infty} \left[\int_0^b f(x)\,dx - \int_0^b E_n(x)\,dx \right]$$

$$= \int_0^b f(x)\,dx - \lim_{n \to +\infty} \int_0^b E_n(x)\,dx$$

$$= \int_0^b f(x)\,dx - 0 = \int_0^b f(x)\,dx \qquad \text{(Theorem 3)}.$$

THEOREM 5

Antidifferentiation of a Power Series

If $R > 0$ is the radius of convergence of the power series $\sum\limits_{k=0}^{\infty} c_k x^k$, then the power series $\sum\limits_{k=0}^{\infty} \dfrac{c_k}{k+1} x^{k+1}$ converges at least for $|x| < R$. Furthermore, for $|x| < R$,

$$\sum_{k=0}^{\infty} c_k x^k = D_x \sum_{k=0}^{\infty} \frac{c_k}{k+1} x^{k+1}.$$

PROOF

By Theorem 4,

$$\int_0^b \left(\sum_{k=0}^\infty c_k x^k \right) dx = \sum_{k=0}^\infty \int_0^b c_k x^k \, dx = \sum_{k=0}^\infty \frac{c_k}{k+1} b^{k+1} \qquad \text{for } |b| < R.$$

Rewriting the variable of integration as t rather than x, we have

$$\int_0^b \left(\sum_{k=0}^\infty c_k t^k \right) dt = \sum_{k=0}^\infty \frac{c_k}{k+1} b^{k+1} \qquad \text{for } |b| < R.$$

In the latter equation, we replace b by x to obtain

$$\int_0^x \left(\sum_{k=0}^\infty c_k t^k \right) dt = \sum_{k=0}^\infty \frac{c_k}{k+1} x^{k+1} \qquad \text{for } |x| < R.$$

It follows that the power series $\sum_{k=0}^\infty \dfrac{c_k}{k+1} x^{k+1}$ converges at least for $|x| < R$. Also, by the fundamental theorem of calculus,

$$\sum_{k=0}^\infty c_k x^k = D_x \int_0^x \left(\sum_{k=0}^\infty c_k t^k \right) dt = D_x \left(\sum_{k=0}^\infty \frac{c_k}{k+1} x^{k+1} \right) \qquad \text{for } |x| < R.$$

THEOREM 6 **Radius of Convergence of the Antidifferentiated Power Series**

If R is the radius of convergence of the power series $\sum_{k=0}^\infty c_k x^k$, then R is also the radius of convergence of the power series $\sum_{k=0}^\infty \dfrac{c_k}{k+1} x^{k+1}$.

PROOF

Let R_1 be the radius of convergence of $\sum_{k=0}^\infty \dfrac{c_k}{k+1} x^{k+1}$. By Theorem 5, $\sum_{k=0}^\infty \dfrac{c_k}{k+1} x^{k+1}$ converges at least for $|x| < R$; hence, $R \le R_1$. We are going to prove that $R = R_1$ by showing that the supposition $R < R_1$ leads to a contradiction. Thus, suppose that $R < R_1$ and choose numbers x_0 and x_1 such that $R < x_0 < x_1 < R_1$. Since $0 < x_1 < R_1$, then $\sum_{k=0}^\infty \dfrac{c_k}{k+1} x_1^{k+1}$ converges; hence, factoring out the constant x_1, we find that $\sum_{k=0}^\infty \dfrac{c_k}{k+1} x_1^k$ converges. It follows that $\lim_{n \to +\infty} \dfrac{c_n}{n+1} x_1^n = 0$. Therefore, the sequence $\left\{ \dfrac{c_n}{n+1} x_1^n \right\}$ is bounded; that is, there exists a constant M such that $\left| \dfrac{c_n}{n+1} x_1^n \right| \le M$ for all n.

Now, let $q = x_0/x_1$, and note that $0 < q < 1$. By the ratio test, the series $\sum_{k=0}^\infty M(k+1)q^k$ converges. For all values of n we have

$$|c_n x_0^n| = \left| \frac{c_n}{n+1} x_1^n (n+1) \frac{x_0^n}{x_1^n} \right| = \left| \frac{c_n}{n+1} x_1^n \right| \cdot (n+1) \left(\frac{x_0}{x_1} \right)^n \le M(n+1)q^n.$$

Therefore, the series $\sum_{k=0}^\infty |c_k x_0^k|$ converges by comparison with the series

$\sum\limits_{k=0}^{\infty} M(k+1)q^k$. Hence, $\sum\limits_{k=0}^{\infty} c_k x_0^k$ converges, and so $-R \le x_0 \le R$, contradicting the fact that $R < x_0$ and completing the proof.

THEOREM 7   **Term-by-Term Differentiation of a Power Series**

If R is the radius of convergence of the power series $\sum\limits_{k=0}^{\infty} b_k x^k$, then R is also the radius of convergence of the power series $\sum\limits_{k=1}^{\infty} k b_k x^{k-1}$ and

$$D_x\left(\sum_{k=0}^{\infty} b_k x^k\right) = \sum_{k=1}^{\infty} k b_k x^{k-1} \qquad \text{for } |x| < R.$$

PROOF
For each nonnegative integer k, define $c_k = (k+1)b_{k+1}$, and consider the power series

$$\sum_{k=0}^{\infty} c_k x^k = c_0 + c_1 x + c_2 x^2 + \cdots$$

$$= b_1 + 2b_2 x + 3b_3 x^2 + \cdots = \sum_{k=0}^{\infty} k b_k x^{k-1}.$$

By Theorem 6, the radius of convergence of $\sum\limits_{k=1}^{\infty} k b_k x^{k-1} = \sum\limits_{k=0}^{\infty} c_k x^k$ is the same as the radius of convergence of

$$\sum_{k=0}^{\infty} \frac{c_k}{k+1} x^{k+1} = \frac{c_0}{1} x + \frac{c_1}{2} x^2 + \frac{c_2}{3} x^3 + \cdots$$

$$= b_1 x + b_2 x^2 + b_3 x^3 + \cdots = \sum_{k=1}^{\infty} b_k x^k.$$

Therefore, the radius of convergence of $\sum\limits_{k=1}^{\infty} k b_k x^{k-1}$ is the same as the radius of convergence of $\sum\limits_{k=1}^{\infty} b_k x^k$, which, in turn, is the same as the radius of convergence R of the power series $\sum\limits_{k=0}^{\infty} b_k x^k$. Now, using Theorem 5 and the above equations, we have, for $|x| < R$,

$$D_x\left(\sum_{k=0}^{\infty} b_k x^k\right) = D_x\left(b_0 + \sum_{k=1}^{\infty} b_k x^k\right) = 0 + D_x \sum_{k=1}^{\infty} b_k x^k$$

$$= D_x \sum_{k=0}^{\infty} \frac{c_k}{k+1} x^{k+1} = \sum_{k=0}^{\infty} c_k x^k = \sum_{k=1}^{\infty} k b_k x^{k-1}.$$

Problem Set 7

In problems 1 through 15, use the formula $\sum\limits_{k=0}^{\infty} x^k = \dfrac{1}{1-x}$, $|x| < 1$, for the sum of the geometric series to obtain an infinite series representation for each expression. In each case specify the values of x for which the representation is correct.

1 $\dfrac{1}{1-x^4}$ **2** $\dfrac{x}{1-x^4}$ **3** $\dfrac{1}{1-4x}$ **4** $\dfrac{x^3}{(1-x^4)^2}$

5 $\dfrac{x}{1-x^2}$

6 $\displaystyle\int_0^x \dfrac{t\,dt}{1-t^2}$

7 $\dfrac{1}{2+x}$

8 $\dfrac{1+x^2}{(1-x^2)^2}$

9 $\ln(1-x)$

10 $\ln\dfrac{1+x}{1-x}$

11 $\displaystyle\int_0^x \ln(1-t)\,dt$

12 $\tanh^{-1}x$

13 $\displaystyle\int_0^x \tanh^{-1}t\,dt$

14 $\dfrac{1}{6-x-x^2}$

15 $\displaystyle\int_0^x \dfrac{dt}{6-t-t^2}$

16 Find the sum of the series $\displaystyle\sum_{k=1}^{\infty}\dfrac{k}{2^k}$. [*Hint:* Use the infinite series expansion of $\dfrac{1}{(1-x)^2}$ obtained in Example 3(a) on page 652.]

17 Find:

(a) $D_x\left(x-\dfrac{x^3}{3!}+\dfrac{x^5}{5!}-\dfrac{x^7}{7!}+\dfrac{x^9}{9!}-\dfrac{x^{11}}{11!}+\cdots\right)$.

(b) $D_x^2\left(x-\dfrac{x^3}{3!}+\dfrac{x^5}{5!}-\dfrac{x^7}{7!}+\dfrac{x^9}{9!}-\dfrac{x^{11}}{11!}+\cdots\right)$.

18 In probability theory it is necessary to evaluate $\displaystyle\sum_{k=1}^{\infty}kp(1-p)^{k-1}$, where $0\le p\le 1$, in order to find the mean value of a geometric random variable. Evaluate this infinite sum.

In problems 19 through 24, let the function f be defined by the given power series. Write a power series for $f'(x)$ and find its radius of convergence.

19 $f(x)=\displaystyle\sum_{k=0}^{\infty}k^2x^k$

20 $f(x)=\displaystyle\sum_{k=1}^{\infty}(-1)^{k+1}k^2(x-2)^k$

21 $f(x)=\displaystyle\sum_{k=0}^{\infty}\dfrac{x^k}{k!}$

22 $f(x)=\displaystyle\sum_{k=0}^{\infty}\dfrac{(-1)^{k+1}x^{2k+1}}{(2k+1)!}$

23 $f(x)=\displaystyle\sum_{k=0}^{\infty}2^{k/2}(x+1)^{2k}$

24 $f(x)=\displaystyle\sum_{k=1}^{\infty}\dfrac{(x-1)^{k^3}}{k^3}$

In problems 25 through 28, let the function f be defined by the given power series. Write a power series for $\displaystyle\int_0^x f(t)\,dt$ and find its radius of convergence.

25 $f(t)=\displaystyle\sum_{k=0}^{\infty}\dfrac{(-1)^k t^{2k}}{(2k)!}$

26 $f(t)=\displaystyle\sum_{k=0}^{\infty}\dfrac{t^k}{2^{k+1}}$

27 $f(t)=\displaystyle\sum_{k=0}^{\infty}\dfrac{t^{2k+1}}{(2k+1)!}$

28 $f(t)=\displaystyle\sum_{k=1}^{\infty}\dfrac{t^k}{k^3}$

29 Given $f(x)=\displaystyle\sum_{k=0}^{\infty}\dfrac{(-1)^k x^{2k}}{(2k)!}$, find

(a) $f(0)$ (b) $f'(0)$ (c) $f''(0)$ (d) $f'''(0)$

30 Use the identity $\pi/4=\tan^{-1}\frac{1}{7}+2\tan^{-1}\frac{1}{3}$ and the known power series expansion $\tan^{-1}x=x-x^3/3+x^5/5-x^7/7+\cdots$ of Example 3(e) on page 653 to approximate the value of π to four decimal places.

8 Taylor and Maclaurin Series

We have seen in Section 7 that a power series $\displaystyle\sum_{k=0}^{\infty}c_k(x-a)^k$ with a nonzero radius of convergence R defines a function f by $f(x)=\displaystyle\sum_{k=0}^{\infty}c_k(x-a)^k$. Thus, *starting with a power series*, we obtained a function f. In the present section, we study the reverse procedure—*starting with a function f*, we try to find a power series that converges to it; that is, we try to *expand* f into a power series. Although a power series expansion cannot *always* be obtained, most of the familiar functions in calculus can be represented as the sum of a convergent power series.

Suppose that f is a function that can be expanded into a power series, so that

$$f(x) = \sum_{k=0}^{\infty} c_k(x-a)^k \qquad \text{for } a - R < x < a + R.$$

Then, by Property III in Section 7, page 651, f is differentiable on the open interval $(a - R, a + R)$ and we have

$$f'(x) = \sum_{k=1}^{\infty} k c_k(x-a)^{k-1} \qquad \text{for } a - R < x < a + R.$$

Thus, not only is f differentiable, but the derivative f' of f can itself be expanded into a power series. Hence, we can apply Property III once again to obtain

$$f''(x) = \sum_{k=2}^{\infty} k(k-1) c_k(x-a)^{k-2} \qquad \text{for } a - R < x < a + R.$$

Continuing in this way, we have

$$f'''(x) = \sum_{k=3}^{\infty} k(k-1)(k-2) c_k(x-a)^{k-3},$$

$$f^{(4)}(x) = \sum_{k=4}^{\infty} k(k-1)(k-2)(k-3) c_k(x-a)^{k-4},$$

$$f^{(5)}(x) = \sum_{k=5}^{\infty} k(k-1)(k-2)(k-3)(k-4) c_k(x-a)^{k-5},$$

and so on, for $a - R < x < a + R$. The pattern here is obvious—evidently

$$f^{(n)}(x) = \sum_{k=n}^{\infty} k(k-1)(k-2) \cdots (k-n+1) c_k(x-a)^{k-n} \qquad \text{for } a - R < x < a + R.$$

(For a rigorous proof, see Problem 35.)

Putting $x = a$ in the formula just obtained, we find that all terms of the infinite series reduce to zero, except for the first term (for which $k = n$), since all the terms after the first contain a factor $(x - a)$. Therefore,

$$f^{(n)}(a) = n(n-1)(n-2) \cdots 3 \cdot 2 \cdot 1 \cdot c_n = (n!)c_n.$$

It follows that $c_n = \dfrac{f^{(n)}(a)}{n!}$; hence, the coefficients of the power series are given by the same formula as the coefficients of the Taylor polynomial for f (Definition 1, Section 5, Chapter 12, page 584).

These considerations lead us to the following definitions.

DEFINITION 1   **Infinitely Differentiable Function**
A function f defined on an open interval J is said to be *infinitely differentiable* on J if f has derivatives $f^{(n)}$ of all orders $n \geq 1$ on J.

DEFINITION 2   **Taylor Series**
Let the function f be infinitely differentiable on an open interval J and let a be a number in J. Then, the *Taylor series for f at a* is the power series

$$\sum_{k=0}^{\infty} c_k(x-a)^k \quad \text{where} \quad c_k = \frac{f^{(k)}(a)}{k!} \qquad \text{for } k = 0, 1, 2, 3, \ldots.$$

The Taylor series for f at $a = 0$ is called the *Maclaurin series* for f. Notice that there is no implication intended in Definition 2 that the Taylor series for f actually converges to f. Also notice the distinction between the nth Taylor *polynomial* for f at a,

$$P_n(x) = \sum_{k=0}^{n} \frac{f^{(k)}(a)}{k!} (x - a)^k,$$

and the Taylor *series* for f at a,

$$\sum_{k=0}^{\infty} \frac{f^{(k)}(a)}{k!} (x - a)^k.$$

The former is a *polynomial* of degree at most n, while the latter is an *infinite series*. In fact, $P_n(x)$ is the nth partial sum of the Taylor series.

EXAMPLES 1 Find the Taylor series for $f(x) = \sin x$ at $a = \pi/4$.

SOLUTION

We arrange our work as follows:

$$f(x) = \quad \sin x, \qquad f\left(\frac{\pi}{4}\right) = \quad \sin \frac{\pi}{4} = \frac{\sqrt{2}}{2},$$

$$f'(x) = \quad \cos x, \qquad f'\left(\frac{\pi}{4}\right) = \quad \cos \frac{\pi}{4} = \frac{\sqrt{2}}{2},$$

$$f''(x) = -\sin x, \qquad f''\left(\frac{\pi}{4}\right) = -\sin \frac{\pi}{4} = -\frac{\sqrt{2}}{2},$$

$$f'''(x) = -\cos x, \qquad f'''\left(\frac{\pi}{4}\right) = -\cos \frac{\pi}{4} = -\frac{\sqrt{2}}{2},$$

$$f^{(4)}(x) = \quad \sin x, \qquad f^{(4)}\left(\frac{\pi}{4}\right) = \quad \sin \frac{\pi}{4} = \frac{\sqrt{2}}{2},$$

and so forth. Therefore, the coefficients of the Taylor series are

$$c_k = \frac{f^{(k)}(\pi/4)}{k!} = \frac{\pm \sqrt{2}/2}{k!} = \pm \frac{\sqrt{2}}{2 \cdot k!},$$

where the plus and minus signs alternate *in pairs*. The Taylor series $\sum_{k=0}^{\infty} c_k \left(x - \frac{\pi}{4}\right)^k$ for $\sin x$ at $a = \pi/4$ is accordingly given by

$$\frac{\sqrt{2}}{2} + \frac{\sqrt{2}}{2}\left(x - \frac{\pi}{4}\right) - \frac{\sqrt{2}}{2 \cdot 2!}\left(x - \frac{\pi}{4}\right)^2 - \frac{\sqrt{2}}{2 \cdot 3!}\left(x - \frac{\pi}{4}\right)^3 + \frac{\sqrt{2}}{2 \cdot 4!}\left(x - \frac{\pi}{4}\right)^4 + \cdots.$$

2 Find the Maclaurin series for $f(x) = e^x$.

SOLUTION

The Maclaurin series for e^x is just the Taylor series for e^x at $a = 0$. The work is arranged as follows:

$$f(x) = e^x, \qquad f(0) = e^0 = 1,$$

$$f'(x) = e^x, \qquad f'(0) = e^0 = 1,$$

$$f''(x) = e^x, \qquad f''(0) = e^0 = 1,$$

$$f'''(x) = e^x, \qquad f'''(0) = e^0 = 1,$$

and so forth. Evidently, $f^{(k)}(0) = e^0 = 1$ holds for all $k = 0, 1, 2, 3, \dots$. Therefore, the coefficients of the Maclaurin series are given by

$$c_k = \frac{f^{(k)}(0)}{k!} = \frac{1}{k!}.$$

Hence, the Maclaurin series $\sum\limits_{k=0}^{\infty} c_k(x - 0)^k = \sum\limits_{k=0}^{\infty} c_k x^k$ for e^x is given by

$$1 + x + \frac{x^2}{2!} + \frac{x^3}{3!} + \frac{x^4}{4!} + \cdots + \frac{x^n}{n!} + \cdots = \sum_{k=0}^{\infty} \frac{x^k}{k!}.$$

As we have seen, if a function can be expanded into a power series, then the function must be infinitely differentiable and the power series must be its Taylor series. However, even if a function is infinitely differentiable, there is no automatic guarantee that it can be expanded into a power series at all! In other words, although an infinitely differentiable function has a Taylor series, this Taylor series need not converge to the function. (See Problem 34 for an example.) The following theorem, which is a consequence of Taylor's formula (Theorem 2, Section 5, Chapter 12, page 588), gives a condition under which the Taylor series for a function actually converges to the function.

THEOREM 1   **Expansion of a Function into Its Taylor Series**
Let the function f be infinitely differentiable on some open interval containing the number a. Suppose that there exists a positive number r and a positive constant M such that

$$\left| f^{(n)}(x) \right| \le M$$

holds for all values of x in the interval $(a - r, a + r)$ and all positive integers n. Then f can be expanded into a Taylor series; that is,

$$f(x) = \sum_{k=0}^{\infty} \frac{f^{(k)}(a)}{k!} (x - a)^k$$

holds for all values of x in the interval $(a - r, a + r)$.

PROOF
Let $P_n(x) = \sum\limits_{k=0}^{n} \frac{f^{(k)}(a)}{k!} (x - a)^k$ be the nth Taylor polynomial for f at a and let $R_n(x) = f(x) - P_n(x)$ be the corresponding Taylor remainder. Notice that $P_n(x)$ is just the nth partial sum of the Taylor series $\sum\limits_{k=0}^{\infty} \frac{f^{(k)}(a)}{k!} (x - a)^k$; hence, this Taylor series converges to $f(x)$ if and only if $f(x) = \lim\limits_{n \to +\infty} P_n(x)$. The latter condition is equivalent to $\lim\limits_{n \to +\infty} [f(x) - P_n(x)] = 0$; that is, it is equivalent to $\lim\limits_{n \to +\infty} R_n(x) = 0$.
    Now assume that $\left| f^{(n)}(x) \right| \le M$ holds for $a - r < x < a + r$ and for all positive integers n. By Taylor's formula with Lagrange remainder (Theorem 2, Section 5, Chapter 12, page 588), $R_n(x) = \frac{f^{(n+1)}(c)}{(n+1)!} (x - a)^{n+1}$ holds for some c between a and x. Hence, for x in $(a - r, a + r)$, we have

$$\left| R_n(x) \right| = \left| f^{(n+1)}(c) \right| \frac{|x - a|^{n+1}}{(n+1)!} \le M \frac{|x - a|^{n+1}}{(n+1)!}.$$

The infinite series $\sum_{k=0}^{\infty} \dfrac{|x-a|^{k+1}}{(k+1)!}$ converges by the ratio test, and therefore its general term approaches zero; that is, $\lim_{n \to +\infty} \dfrac{|x-a|^{n+1}}{(n+1)!} = 0$. Hence, since M is a constant, the inequality $|R_n(x)| \le M \dfrac{|x-a|^{n+1}}{(n+1)!}$ shows that $\lim_{n \to +\infty} R_n(x) = 0$ holds for all values of x in the interval $(a-r, a+r)$. Therefore, the Taylor series converges to $f(x)$, and the proof is complete.

The following examples not only illustrate the use of Theorem 1, but also provide power series expansions that are so important that they should be memorized for future use.

EXAMPLES   Justify the following power series expansions:

1 $e^x = 1 + x + \dfrac{x^2}{2!} + \dfrac{x^3}{3!} + \dfrac{x^4}{4!} + \cdots$ for all values of x.

SOLUTION
In Theorem 1, we let $f(x) = e^x$ and $a = 0$, noting that f is infinitely differentiable on $(-\infty, \infty)$ and that $f^{(n)}(x) = e^x$ holds for all values of $n \ge 1$. If r is any positive number, then, for x in the interval $(-r, r)$, we have

$$|f^{(n)}(x)| = |e^x| = e^x \le e^r,$$

since $x < r$. Thus, taking $M = e^r$ in Theorem 1, we conclude that

$$e^x = \sum_{k=0}^{\infty} \frac{f^{(k)}(0)}{k!} x^k = \sum_{k=0}^{\infty} \frac{e^0}{k!} x^k = \sum_{k=0}^{\infty} \frac{x^k}{k!} = 1 + x + \frac{x^2}{2!} + \frac{x^3}{3!} + \cdots$$

holds for all values of x between $-r$ and r. Since we can choose r as large as we please, then $e^x = 1 + x + x^2/2! + x^3/3! + \cdots$ holds for all values of x.

2 $\sin x = x - \dfrac{x^3}{3!} + \dfrac{x^5}{5!} - \dfrac{x^7}{7!} + \dfrac{x^9}{9!} - \dfrac{x^{11}}{11!} + \cdots$ for all values of x.

SOLUTION
Successive derivatives of $\sin x$ give only $\pm \sin x$ or $\pm \cos x$. In any case, $|D_x^n \sin x| \le 1$, so we can take $M = 1$ in Theorem 1 and r can be chosen as large as we please. Then

$$\sin x = \sin 0 + \frac{\cos 0}{1!} x - \frac{\sin 0}{2!} x^2 - \frac{\cos 0}{3!} x^3 + \frac{\sin 0}{4!} x^4 + \frac{\cos 0}{5!} x^5 - \cdots;$$

that is, $\sin x = x - \dfrac{x^3}{3!} + \dfrac{x^5}{5!} - \dfrac{x^7}{7!} + \cdots$ holds for all values of x.

3 $\cos x = 1 - \dfrac{x^2}{2!} + \dfrac{x^4}{4!} - \dfrac{x^6}{6!} + \cdots$ for all values of x.

SOLUTION
This expansion can be obtained in essentially the same way that the expansion for $\sin x$ was obtained in Example 2. However, it is more interesting to obtain the power series for the cosine by differentiating the power series for the sine as follows:

$$\cos x = D_x \sin x = D_x\left(x - \frac{x^3}{3!} + \frac{x^5}{5!} - \frac{x^7}{7!} + \cdots\right)$$

$$= 1 - \frac{3x^2}{3!} + \frac{5x^4}{5!} - \frac{7x^6}{7!} + \cdots = 1 - \frac{x^2}{2!} + \frac{x^4}{4!} - \frac{x^6}{6!} + \cdots.$$

Some elementary consequences of the expansions obtained above are shown in the following examples.

EXAMPLES **1** Find a power series expansion for $\dfrac{1 - \cos x}{x}$, $x \neq 0$.

SOLUTION

$$\cos x = 1 - \frac{x^2}{2!} + \frac{x^4}{4!} - \frac{x^6}{6!} + \cdots,$$

so that

$$1 - \cos x = \frac{x^2}{2!} - \frac{x^4}{4!} + \frac{x^6}{6!} - \frac{x^8}{8!} + \cdots \qquad \text{for all values of } x.$$

Hence, for $x \neq 0$,

$$\frac{1 - \cos x}{x} = \frac{1}{x}\left(\frac{x^2}{2!} - \frac{x^4}{4!} + \frac{x^6}{6!} - \frac{x^8}{8!} + \cdots\right)$$

$$= \frac{x}{2!} - \frac{x^3}{4!} + \frac{x^5}{6!} - \frac{x^7}{8!} + \cdots.$$

2 Find a power series expansion for $\displaystyle\int_0^x e^{t^2}\, dt$.

SOLUTION

We replace x by t^2 in the expansion $e^x = \displaystyle\sum_{k=0}^{\infty} \frac{x^k}{k!}$ to obtain $e^{t^2} = \displaystyle\sum_{k=0}^{\infty} \frac{t^{2k}}{k!}$. Hence,

$$\int_0^x e^{t^2}\, dt = \int_0^x \left(\sum_{k=0}^{\infty} \frac{t^{2k}}{k!}\right) dt = \sum_{k=0}^{\infty} \int_0^x \frac{t^{2k}\, dt}{k!} = \sum_{k=0}^{\infty} \frac{x^{2k+1}}{(2k+1)k!}.$$

3 Estimate $1/e$ with an error of no more than $5/10^4$.

SOLUTION

$$\frac{1}{e} = e^{-1} = \sum_{k=0}^{\infty} \frac{(-1)^k}{k!} = 1 - \frac{1}{1!} + \frac{1}{2!} - \frac{1}{3!} + \frac{1}{4!} - \frac{1}{5!} + \frac{1}{6!} - \frac{1}{7!} + \cdots.$$

Estimating $1/e$ by using only the first seven terms, it follows from Leibniz's theorem that we make an error not exceeding the absolute value of the first omitted term, $\dfrac{1}{7!} = \dfrac{1}{5040} < \dfrac{5}{10^4}$. Hence,

$$\frac{1}{e} \approx 1 - 1 + \frac{1}{2} - \frac{1}{6} + \frac{1}{24} - \frac{1}{120} + \frac{1}{720} = \frac{53}{144} = 0.368\ldots.$$

We now assemble some of the more important power series expansions obtained in this and the previous section.

$$1 \quad e^x = 1 + x + \frac{x^2}{2!} + \frac{x^3}{3!} + \frac{x^4}{4!} + \cdots = \sum_{k=0}^{\infty} \frac{x^k}{k!} \text{ for all } x.$$

$$2 \quad \sin x = x - \frac{x^3}{3!} + \frac{x^5}{5!} - \frac{x^7}{7!} + \cdots = \sum_{k=0}^{\infty} \frac{(-1)^k x^{2k+1}}{(2k+1)!} \text{ for all } x.$$

$$3 \quad \cos x = 1 - \frac{x^2}{2!} + \frac{x^4}{4!} - \frac{x^6}{6!} + \cdots = \sum_{k=0}^{\infty} \frac{(-1)^k x^{2k}}{(2k)!} \text{ for all } x.$$

$$4 \quad \ln(1+x) = x - \frac{x^2}{2} + \frac{x^3}{3} - \frac{x^4}{4} + \cdots = \sum_{k=0}^{\infty} \frac{(-1)^k x^{k+1}}{k+1} \text{ for } |x| < 1.$$

$$5 \quad \tan^{-1} x = x - \frac{x^3}{3} + \frac{x^5}{5} - \frac{x^7}{7} + \cdots = \sum_{k=0}^{\infty} \frac{(-1)^k x^{2k+1}}{2k+1} \text{ for } |x| < 1.$$

$$6 \quad \frac{1}{1-x} = 1 + x + x^2 + x^3 + x^4 + \cdots = \sum_{k=0}^{\infty} x^k \text{ for } |x| < 1.$$

$$7 \quad \frac{1}{1+x} = 1 - x + x^2 - x^3 + x^4 - \cdots = \sum_{k=0}^{\infty} (-1)^k x^k \text{ for } |x| < 1.$$

Of course, additional power series expansions can be obtained from 1 through 7 above by various substitutions. For instance, if $t \geq 0$, then

$$\cos \sqrt{t} = 1 - \frac{t}{2!} + \frac{t^2}{4!} - \frac{t^3}{6!} + \cdots$$

follows by substituting $x = \sqrt{t}$ in the above power series expansion for $\cos x$.

Problem Set 8

In problems 1 through 16, find the Taylor series for each function f at the indicated value of a. (In some cases it may be easiest to develop the desired Taylor series by starting with the known Taylor series expansion of a related function.)

1 $f(x) = \sin x$ at $a = \pi/6$.

2 $f(x) = \sqrt{x}$ at $a = 9$.

3 $f(x) = 1/x$ at $a = 2$.

4 $f(x) = \sqrt{x^3}$ at $a = 1$.

5 $f(x) = e^x$ at $a = 4$.

6 $f(x) = \cos x$ at $a = \pi/6$.

7 $f(x) = \sqrt{x-1}$ at $a = 2$.

8 $f(x) = \cos x$ at $a = \pi/3$.

9 $f(x) = \sinh x$ at $a = 0$.

10 $f(x) = \cosh x$ at $a = 0$.

11 $f(x) = e^{-x^2}$ at $a = 0$.

12 $f(x) = \ln \dfrac{1+x}{1-x}$ at $a = 0$.

13 $f(x) = \begin{cases} \dfrac{\sin x}{x} & \text{if } x \neq 0 \\ 1 & \text{if } x = 0 \end{cases}$ at $a = 0$.

14 $f(x) = \displaystyle\int_0^x \sin t^2 \, dt$ at $a = 0$.

15 $f(x) = \begin{cases} \dfrac{\tan^{-1} x}{x} & \text{if } x \neq 0 \\ 1 & \text{if } x = 0 \end{cases}$ at $a = 0$.

16 $f(x) = \begin{cases} \ln(1+x)^{1/x} & \text{if } x \neq 0 \\ 1 & \text{if } x = 0 \end{cases}$ at $a = 0$.

17 Find the Maclaurin series expansion for $\sin^2 x$. [*Hint:* $\sin^2 x = \frac{1}{2}(1 - \cos 2x)$.]

18 Find the Maclaurin series expansion for $\displaystyle\int_0^x e^{-t^2} \, dt$.

19 Estimate $e^{-0.02}$ with an error not exceeding $5/10^5$.

20 If f is a polynomial function of degree n and a is any constant, prove that
  (a) f can be represented in the form

$$f(x) = c_0 + c_1(x - a) + c_2(x - a)^2 + \cdots + c_n(x - a)^n.$$

  (b) The coefficients $c_0, c_1, \ldots, c_n$ in part (a) are uniquely determined by the requirement that

$$c_k = \frac{f^{(k)}(a)}{k!} \qquad \text{for } k = 0, 1, 2, \ldots, n.$$

21 If $f(x) = \tan^{-1} x$, find a formula for $f^{(n)}(0)$. (*Hint:* Use the known Maclaurin series expansion for $\tan^{-1} x$.)

22 Suppose that $\sum_{k=0}^{\infty} b_k(x - a)^k$ and $\sum_{k=0}^{\infty} c_k(x - a)^k$ are two power series with positive radii of convergence. If there exists $\varepsilon > 0$ such that

$$\sum_{k=0}^{\infty} b_k(x - a)^k = \sum_{k=0}^{\infty} c_k(x - a)^k$$

holds for all values of x with $|x - a| < \varepsilon$, prove that $b_k = c_k$ for all nonnegative integers k.

In problems 23 through 28, find the Maclaurin series expansion for each function and indicate the range of values of x for which the expansion is correct.

23 $f(x) = e^{-x}$ **24** $f(x) = \ln(1 + x^2)$ **25** $f(x) = \dfrac{1}{4 - x}$

26 $f(x) = \sinh x^2$ **27** $f(x) = 2^x$ **28** $f(x) = x \sin x$

In problems 29 through 33, use the Maclaurin series expansion $f(x) = \sum_{k=0}^{\infty} c_k x^k$ and the fact that $c_k = f^{(k)}(0)/k!$ to find the value of the indicated higher-order derivative.

29 $f^{(15)}(0)$, where $f(x) = x \sin x$.   **30** $f^{(16)}(0)$, where $f(x) = \cos x^2$.   **31** $f^{(17)}(0)$, where $f(x) = \displaystyle\int_0^x e^{-t^2}\, dt$.

32 $f^{(19)}(0)$, where $f(x) = xe^{-x}$.   **33** $f^{(20)}(0)$, where $f(x) = \ln(1 + x^2)$.

34 Let f be the function defined by

$$f(x) = \begin{cases} e^{-1/x^2} & \text{for } x \neq 0 \\ 0 & \text{for } x = 0. \end{cases}$$

  (a) Sketch the graph of f.
  (b) Find $f'(x)$, $f''(x)$, and $f'''(x)$ for $x \neq 0$.
  (c) Prove by induction that, for every positive integer n, there exists a polynomial function P (depending on n) such that $f^{(n)}(x) = P(1/x) \cdot f(x)$ holds for $x \neq 0$.
  (d) Using (c), prove that $\displaystyle\lim_{n \to +\infty} \frac{f^{(n)}(x)}{x} = 0$ for all positive integers n.
  (e) Show that f is infinitely differentiable on $(-\infty, \infty)$ and that $f^{(n)}(0) = 0$ for all positive integers n.
  (f) Show that f cannot be expanded into a power series about 0.

35 Suppose that the power series $\sum_{k=0}^{\infty} c_k(x - a)^k$ converges at least for x in the interval $(a - r, a + r)$, where $r > 0$. Define the function f by $f(x) = \sum_{k=0}^{\infty} c_k(x - a)^k$ for $a - r < x < a + r$. Prove by induction that, for any positive integer n,

$$f^{(n)}(x) = \sum_{k=n}^{\infty} k(k - 1)(k - 2) \cdots (k - n + 1)c_k(x - a)^{k-n},$$

for $a - r < x < a + r$.

9   The Binomial Series

We now set for ourselves the problem of finding the Maclaurin series expansion for the function f defined by $f(x) = (1 + x)^p$, where p is a constant and $1 + x > 0$. Since $f'(x) = p(1 + x)^{p-1}$, we have

$$(1 + x)f'(x) = pf(x).$$

Now, assume that f can be expanded into a Maclaurin series

$$f(x) = c_0 + c_1 x + c_2 x^2 + c_3 x^3 + \cdots.$$

Then,

$$f'(x) = c_1 + 2c_2 x + 3c_3 x^2 + \cdots,$$

so that

$$
\begin{aligned}
(1 + x)f'(x) &= f'(x) + xf'(x) \\
&= (c_1 + 2c_2 x + 3c_3 x^2 + \cdots) + (c_1 x + 2c_2 x^2 + 3c_3 x^3 + \cdots) \\
&= c_1 + (2c_2 + c_1)x + (3c_3 + 2c_2)x^2 + (4c_4 + 3c_3)x^3 + \cdots.
\end{aligned}
$$

The condition $(1 + x)f'(x) = pf(x)$ therefore becomes

$$c_1 + (2c_2 + c_1)x + (3c_3 + 2c_2)x^2 + (4c_4 + 3c_3)x^3 + \cdots$$
$$= pc_0 + pc_1 x + pc_2 x^2 + pc_3 x^3 + \cdots.$$

Equating coefficients of like powers of x in this equation (see Problem 22 in Section 8), we have

$$c_1 = pc_0, \quad 2c_2 + c_1 = pc_1, \quad 3c_3 + 2c_2 = pc_2, \quad 4c_4 + 3c_3 = pc_3,$$

and so forth. Evidently, $(n + 1)c_{n+1} + nc_n = pc_n$; that is, $c_{n+1} = \dfrac{p - n}{n + 1} c_n$ holds for all $n \geq 0$. Hence,

$$c_1 = pc_0, \quad c_2 = \frac{p - 1}{2} c_1 = \frac{(p - 1)p}{2} c_0, \quad c_3 = \frac{p - 2}{3} c_2 = \frac{(p - 2)(p - 1)p}{3 \cdot 2} c_0,$$

$$c_4 = \frac{p - 3}{4} c_3 = \frac{(p - 3)(p - 2)(p - 1)p}{4 \cdot 3 \cdot 2} c_0,$$

and so forth. Since $f(0) = (1 + 0)^p = 1^p = 1$, it follows that $c_0 = 1$.

The above calculations suggest the general formula

$$c_n = \frac{\overbrace{p(p - 1)(p - 2) \cdots (p - n + 1)}^{n \text{ factors}}}{n!}$$

for $n \geq 1$, and this is easily confirmed by mathematical induction (Problem 17). Thus, putting $c_0 = 1$, we obtain the power series

$$\sum_{k=0}^{\infty} c_k x^k = 1 + \sum_{k=1}^{\infty} \frac{p(p - 1)(p - 2) \cdots (p - k + 1)}{k!} x^k,$$

which is called a *binomial series*.

The reader should notice that we really have not proved that the equation

$$(1 + x)^p = \sum_{k=0}^{\infty} c_k x^k$$

holds. What we have shown is that, *if the function f can be expanded into a power series at all*, the above equation holds at least for values of x in an open interval around 0. However, we can now prove the following theorem.

THEOREM 1

Binomial Series Expansion

Let p be any constant other than 0 or a positive integer. Define

$$c_0 = 1 \quad \text{and} \quad c_n = \frac{p(p-1)(p-2)\cdots(p-n+1)}{n!}$$

for each positive integer n. Then

(i) $c_{n+1} = \dfrac{p-n}{n+1}\, c_n$ for $n \geq 0$.

(ii) The binomial series $\displaystyle\sum_{k=0}^{\infty} c_k x^k$ has radius of convergence $R = 1$.

(iii) For $|x| < 1$,

$$(1+x)^p = \sum_{k=0}^{\infty} c_k x^k = 1 + px + \frac{p(p-1)}{2!}x^2 + \frac{p(p-1)(p-2)}{3!}x^3 + \cdots.$$

PROOF

(i) None of the coefficients $c_0, c_1, c_2, \ldots$ is equal to zero since, if $c_n = 0$, one of the factors $p, p-1, p-2, \ldots, p-n+1$ in the numerator of the fraction defining c_n would have to be 0, so that p would be a nonnegative integer—contrary to the hypothesis. The relation $c_{n+1} = \dfrac{p-n}{n-1}\, c_n$ surely holds since it was originally used to generate the numbers $c_0, c_1, c_2, \ldots$. (See Problem 18 for direct verification.)

(ii) $\displaystyle\lim_{n \to +\infty} \left| \frac{c_{n+1}}{c_n} \right| = \lim_{n \to +\infty} \left| \frac{p-n}{n+1} \right| = \lim_{n \to +\infty} \frac{\left| \dfrac{p}{n} - 1 \right|}{1 + (1/n)} = 1$; hence, the radius of convergence of the power series is given by $R = 1/1 = 1$ (Theorem 1 of Section 6).

(iii) Define the function g by $g(x) = \displaystyle\sum_{k=0}^{\infty} c_k x^k$. The power series $\displaystyle\sum_{k=0}^{\infty} c_k x^k$ was set up in the first place so that

$$(1+x)D_x \sum_{k=0}^{\infty} c_k x^k = p \sum_{k=0}^{\infty} c_k x^k$$

for $|x| < R$; hence, $(1+x)g'(x) = pg(x)$ holds for $|x| < 1$. (For the details, see Problem 18.) Using the fact that $(1+x)g'(x) = pg(x)$ for $|x| < 1$, we calculate as follows:

$$D_x \frac{g(x)}{(1+x)^p} = \frac{(1+x)^p g'(x) - g(x)p(1+x)^{p-1}}{(1+x)^{2p}}$$

$$= \frac{(1+x)^{p-1}}{(1+x)^{2p}} [(1+x)g'(x) - pg(x)] = 0$$

for $|x| < 1$. It follows that $\dfrac{g(x)}{(1+x)^p}$ is a constant, say $\dfrac{g(x)}{(1+x)^p} = K$, for

$|x| < 1$. Putting $x = 0$, we find that $K = \dfrac{g(0)}{(1 + 0)^p} = g(0) = c_0 = 1$. There-fore,

$$g(x) = K(1 + x)^p = 1 \cdot (1 + x)^p = (1 + x)^p \qquad \text{for } |x| < 1;$$

that is,

$$(1 + x)^p = g(x) = \sum_{k=0}^{\infty} c_k x^k \text{ holds for } |x| < 1.$$

EXAMPLES 1 Find a power series expansion for $\sqrt[3]{1 + x}$, $|x| < 1$.

SOLUTION
Take $p = \frac{1}{3}$ in Theorem 1. Thus, $c_0 = 1$, $c_1 = \frac{1}{3}$, $c_2 = -\frac{1}{9}$, $c_3 = \frac{5}{81}$, ... In general,

$$c_n = \frac{\frac{1}{3}(\frac{1}{3} - 1)(\frac{1}{3} - 2) \cdots (\frac{1}{3} - n + 1)}{n!} = \frac{1(1 - 3)(1 - 6) \cdots [1 - 3(n - 1)]}{3^n n!}$$

$$= \frac{(-1)^{n+1} 2 \cdot 5 \cdot 8 \cdot 11 \cdots (3n - 4)}{3^n n!} \qquad \text{for } n \geq 2,$$

and therefore

$$\sqrt[3]{1 + x} = 1 + \frac{1}{3}x - \frac{1}{9}x^2 + \frac{5}{81}x^3 - \cdots$$

$$= 1 + \frac{1}{3} x + \sum_{k=2}^{\infty} \frac{(-1)^{k+1} 2 \cdot 5 \cdot 8 \cdot 11 \cdots (3k - 4)}{3^k k!} x^k, \qquad |x| < 1.$$

2 Use the first three terms of the expansion obtained in Example 1 to approximate $\sqrt[3]{28}$. Give a bound on the error involved.

SOLUTION
Naturally, we wish to use the fact that $\sqrt[3]{27} = 3$. Thus, we write

$$\sqrt[3]{28} = \sqrt[3]{27 + 1} = \sqrt[3]{27(1 + \tfrac{1}{27})} = \sqrt[3]{27} \sqrt[3]{1 + \tfrac{1}{27}} = 3\sqrt[3]{1 + \tfrac{1}{27}}.$$

Putting $x = \frac{1}{27}$ in Example 1, we obtain

$$\sqrt[3]{1 + \tfrac{1}{27}} \approx 1 + \tfrac{1}{3}(\tfrac{1}{27}) - \tfrac{1}{9}(\tfrac{1}{27})^2 = \tfrac{6641}{6561}$$

with an error no larger than $\frac{5}{81}(\frac{1}{27})^3$, since the series (apart from the first term) is alternating and Leibniz's theorem applies. Therefore,

$$\sqrt[3]{28} = 3\sqrt[3]{1 + \tfrac{1}{27}} \approx 3(\tfrac{6641}{6561}) = 3.036579\ldots.$$

This approximation *under*estimates $\sqrt[3]{28}$ with an error not exceeding $3 \cdot \dfrac{5}{81}\left(\dfrac{1}{27}\right)^3 < 0.0000095$. (The true value of $\sqrt[3]{28}$ is $3.03658897\ldots$.)

3 Find the Maclaurin series expansion for $\dfrac{1}{\sqrt{1 + x}}$ along with an upper bound for the absolute value of the error involved in approximating $\dfrac{1}{\sqrt{1 + x}}$ by the first n terms of the series.

SOLUTION
$\dfrac{1}{\sqrt{1 + x}} = (1 + x)^{-1/2}$, so that $p = -\frac{1}{2}$ in Theorem 1, and

$$c_n = \frac{1}{n!}\left(-\tfrac{1}{2}\right)\left(-\tfrac{1}{2} - 1\right)\left(-\tfrac{1}{2} - 2\right)\cdots\left(-\tfrac{1}{2} - n + 1\right) = \frac{(-1)^n}{2^n n!} \cdot 1 \cdot 3 \cdot 5 \cdot 7 \cdots (2n - 1).$$

The desired expansion is

$$\frac{1}{\sqrt{1 + x}} = 1 - \tfrac{1}{2}x + \tfrac{3}{8}x^2 - \tfrac{5}{16}x^3 + \cdots = 1 + \sum_{k=1}^{\infty} (-1)^k \frac{1 \cdot 3 \cdot 5 \cdot 7 \cdots (2k - 1)}{2^k k!} x^k$$

for $|x| < 1$. If $0 \le x < 1$, the series is alternating and its terms decrease in absolute value; hence, by Leibniz's theorem,

$$\frac{1}{\sqrt{1 + x}} \approx 1 + \sum_{k=1}^{n-1} (-1)^k \frac{1 \cdot 3 \cdot 5 \cdot 7 \cdots (2k - 1)}{2^k k!} x^k$$

with an error whose absolute value does not exceed $\dfrac{1 \cdot 3 \cdot 5 \cdot 7 \cdots (2n - 1)}{2^n n!} x^n$, the absolute value of the first omitted term. If $-1 < x < 0$, the series is no longer alternating and Leibniz's theorem is inapplicable. However, in this case we can use Taylor's theorem with the Lagrange form of the remainder to conclude that the absolute value of the error involved in the preceding estimation does not exceed

$$|R_{n-1}(x)| = \frac{|f^{(n)}(c)|}{n!}\,|x|^n, \qquad \text{where } x < c < 0 \text{ and } f(x) = (1 + x)^{-1/2}.$$

4 Use the first three terms of the expansion obtained in Example 3 above to approximate $1/\sqrt{15}$. Give a bound on the error involved.

SOLUTION

$$\frac{1}{\sqrt{15}} = (15)^{-1/2} = \left(16 \cdot \tfrac{15}{16}\right)^{-1/2} = 16^{-1/2}\left(\tfrac{15}{16}\right)^{-1/2} = \left(\tfrac{1}{4}\right)\left(1 - \tfrac{1}{16}\right)^{-1/2}$$

By Example 3, with $x = -\tfrac{1}{16}$,

$$\left(1 - \tfrac{1}{16}\right)^{-1/2} \approx 1 - \left(\tfrac{1}{2}\right)\left(-\tfrac{1}{16}\right) + \left(\tfrac{3}{8}\right)\left(-\tfrac{1}{16}\right)^2 = \tfrac{2115}{2048}$$

with an error whose absolute value does not exceed $\dfrac{|f^{(3)}(c)|}{3!} \cdot \left|-\tfrac{1}{16}\right|^3$, where $-\tfrac{1}{16} < c < 0$ and $f(x) = (1 + x)^{-1/2}$. Now,

$$f'(x) = \left(-\tfrac{1}{2}\right)(1 + x)^{-3/2}, \qquad f''(x) = \tfrac{3}{4}(1 + x)^{-5/2},$$

and

$$f'''(x) = -\tfrac{15}{8}(1 + x)^{-7/2},$$

so the absolute value of the error of our estimation does not exceed

$$\frac{\left|-\tfrac{15}{8}(1 + c)^{-7/2}\right|}{3!} \cdot \left|-\frac{1}{16}\right|^3 = \frac{5}{16^4(1 + c)^{7/2}}.$$

Since $-\tfrac{1}{16} < c < 0$, then $\tfrac{15}{16} < 1 + c < 1$, so that $\left(\tfrac{15}{16}\right)^{7/2} < (1 + c)^{7/2}$, $\dfrac{1}{(1 + c)^{7/2}} < \left(\tfrac{16}{15}\right)^{7/2}$ and the absolute value of the error does not exceed

$$\frac{5}{16^4(1 + c)^{7/2}} < \frac{5}{16^4}\left(\frac{16}{15}\right)^{7/2} < \frac{5}{16^4}\left(\frac{16}{15}\right)^{8/2} = \frac{5}{15^4}.$$

Therefore,

$$\frac{1}{\sqrt{15}} = \left(\frac{1}{4}\right)\left(1 - \frac{1}{16}\right)^{-1/2} \approx \left(\frac{1}{4}\right)\left(\frac{2115}{2048}\right) = 0.25817\ldots$$

with an error whose absolute value does not exceed $(\frac{1}{4})(5/15^4) < 0.00003$. Here all the terms in the binomial series are positive, so $0.25817\ldots$ must *under*estimate the true value of $1/\sqrt{15}$. (The true value of $1/\sqrt{15}$ is $0.25819\ldots$.)

5 Estimate $\displaystyle\int_0^{1/2} \sqrt[3]{1 + x^3}\, dx$ and place a bound on the error involved.

SOLUTION

By Example 1, for $|x| < 1$, we have

$$\sqrt[3]{1 + x} = 1 + \tfrac{1}{3}x - \tfrac{1}{9}x^2 + \tfrac{5}{81}x^3 - \cdots.$$

Replacing x by x^3, we obtain

$$\sqrt[3]{1 + x^3} = 1 + \tfrac{1}{3}x^3 - \tfrac{1}{9}x^6 + \tfrac{5}{81}x^9 - \cdots.$$

Therefore,

$$\int_0^{1/2} \sqrt[3]{1 + x^3}\, dx = \int_0^{1/2} 1\, dx + \int_0^{1/2} \tfrac{1}{3}x^3\, dx - \int_0^{1/2} \tfrac{1}{9}x^6\, dx + \int_0^{1/2} \tfrac{5}{81}x^9\, dx - \cdots$$

$$= \tfrac{1}{2} + (\tfrac{1}{3})(\tfrac{1}{4})(\tfrac{1}{2})^4 - (\tfrac{1}{9})(\tfrac{1}{7})(\tfrac{1}{2})^7 + (\tfrac{5}{81})(\tfrac{1}{10})(\tfrac{1}{2})^{10} - \cdots.$$

Apart from the first term, the latter series is alternating and the terms are decreasing in absolute value (see Problems 21 and 22). Hence, Leibniz's theorem applies, so that, using (say) the first three terms of the series, we have

$$\int_0^{1/2} \sqrt[3]{1 + x^3}\, dx \approx \tfrac{1}{2} + (\tfrac{1}{3})(\tfrac{1}{4})(\tfrac{1}{2})^4 - (\tfrac{1}{9})(\tfrac{1}{7})(\tfrac{1}{2})^7 = 0.505084\ldots$$

with an error whose absolute value does not exceed $(\tfrac{5}{81})(\tfrac{1}{10})(\tfrac{1}{2})^{10} < 0.000007$. Therefore, rounding off to four decimal places, we have $\displaystyle\int_0^{1/2} \sqrt[3]{1 + x^3}\, dx \approx 0.5051$.

Problem Set 9

In problems 1 through 8, use the binomial series expansion (Theorem 1) to find a Maclaurin series expansion for each expression. Specify the range of values of x for which the expansion is correct.

1 $\dfrac{1}{\sqrt[3]{1 + x}}$
     2 $\sqrt{1 + x^2}$
     3 $\dfrac{1}{\sqrt[3]{1 - x^2}}$
     4 $\sqrt[3]{27 + x}$

5 $\dfrac{1}{\sqrt{1 + x^3}}$
     6 $\dfrac{x}{\sqrt[3]{1 - x^2}}$
     7 $\dfrac{x}{(1 + 2x)^2}$
     8 $(9 + x)^{3/2}$

In problems 9 through 12, use the first three terms of an appropriate binomial series to estimate each number. Give an upper bound for the absolute value of the error.

9 $\sqrt{1.03}$
     10 $\sqrt[5]{33}$
     11 $\sqrt[4]{17}$
     12 $\dfrac{1}{\sqrt[3]{100}}$

In problems 13 through 16, estimate each quantity, rounded off to three decimal places. (Take sufficiently many terms so that the absolute value of the error does not exceed 5×10^{-4}.)

13 $\sqrt{101}$
     14 $\sqrt{99}$
     15 $\displaystyle\int_0^{2/3} \sqrt{1 + x^3}\, dx$
     16 $\displaystyle\int_0^{1/2} \dfrac{dx}{\sqrt{1 + x^3}}$

17 Given a sequence $c_0, c_1, c_2, c_3, c_4, \ldots$ such that $c_0 = 1$ and $c_{n+1} = \dfrac{p-n}{n+1} c_n$ for $n \geq 0$, where p is a constant, prove by mathematical induction that

$$c_n = \frac{1}{n!} p(p-1)(p-2) \cdots (p-n+1) \qquad \text{for } n \geq 1.$$

18 Let p be a given constant and define $c_0 = 1$ and

$$c_n = (1/n!)p(p-1)(p-2) \cdots (p-n+1) \qquad \text{for } n \geq 1.$$

(a) Prove that $c_{n+1} = \dfrac{p-n}{n+1} c_n$ for $n \geq 0$.

(b) Prove that $(1+x)D_x \displaystyle\sum_{k=0}^{\infty} c_k x^k = p \sum_{k=0}^{\infty} c_k x^k$, $|x| < 1$.

19 Compare the binomial series expansion of $(1+x)^{-1}$ with the geometric series expansion of the same expression.

20 If a is a positive constant and p is any constant, show that

$$(a+x)^p = a^p + pa^{p-1}x + \frac{p(p-1)}{2!} a^{p-2}x^2 + \frac{p(p-1)(p-2)}{3!} a^{p-3}x^3 + \cdots$$

$$= a^p + \sum_{k=1}^{\infty} \frac{p(p-1) \cdots (p-k+1)}{k!} a^{p-k} x^k$$

for $|x| < a$.

21 Let $(1+x)^p = \displaystyle\sum_{k=0}^{\infty} c_k x^k$ for $|x| < 1$ be the binomial series expansion. Show that, if $0 \leq x < 1$ and $n > p$, then $\displaystyle\sum_{k=n+1}^{\infty} c_k x^k$ is an alternating series. $\left(Hint: c_{n+1} = \dfrac{p-n}{n+1} c_n. \right)$

22 In problem 21, assume that $p > -1$ and prove that the terms in the series $\displaystyle\sum_{k=n+1}^{\infty} c_k x^k$ are decreasing in absolute value (so that Leibniz's theorem is applicable).

23 Exactly what happens in the binomial series expansion when the exponent p is a positive integer? Is the expansion still correct? For what values of x is it correct? Why?

24 From the binomial series expansion for $\dfrac{1}{\sqrt{1-x^2}}$ and the fact that $\sin^{-1} x = \displaystyle\int_0^x \dfrac{dt}{\sqrt{1-t^2}}$, find a power series expansion for $\sin^{-1} x$.

Review Problem Set

In problems 1 through 12, determine whether each sequence converges or diverges. If the sequence converges, find its limit.

1 $\left\{ \dfrac{n(n+1)}{3n^2 + 7n} \right\}$

2 $\left\{ \dfrac{\sin n}{n} \right\}$

3 $\left\{ \dfrac{\sqrt{n+1}}{\sqrt{3n+1}} \right\}$

4 $\left\{ \dfrac{7n^3 + 3n^2 - n^3(\frac{1}{2})^n}{3n^2 + n^2(\frac{3}{4})^n} \right\}$

5 $\left\{ \dfrac{1 + (-1)^n}{n} \right\}$

6 $\left\{ \left(50 + \dfrac{1}{n}\right)^2 \cdot \left(1 + \dfrac{n-1}{n^2}\right)^{50} \right\}$

7 $\left\{ \dfrac{\cos(n\pi/2)}{\sqrt{n}} \right\}$

8 $\{ n[1 + (-1)^n] \}$

9 $\{ n^2 + (-1)^n 2n \}$

10 $\left\{ \dfrac{1}{(n+1) + (-1)^n(1-n)} \right\}$

11 $\left\{ 1 - \dfrac{3^n}{n!} \right\}$

12 $\left\{ \dfrac{2^n n!}{(2n+1)!} \right\}$

In problems 13 through 16, indicate whether each sequence is increasing, decreasing, or nonmonotone.

13 $\{2^n\}$

14 $\left\{\dfrac{1}{2^n}\right\}$

15 $\left\{\dfrac{(-1)^n}{n}\right\}$

16 $\{(-1)^n\}$

17 Is the sequence $\left\{n - \dfrac{2^n}{n}\right\}$ monotone? Why?

18 For each positive integer n, let $a_n = \dfrac{1}{n} + \dfrac{1}{n+1} + \dfrac{1}{n+2} + \cdots + \dfrac{1}{2n}$.

  (a) Show that $\{a_n\}$ is a monotone sequence. Is it increasing or decreasing?
  (b) Does $\{a_n\}$ converge or diverge? Why?

19 Indicate whether the sequence $\left\{1 - \dfrac{1}{4} + \dfrac{1}{16} - \cdots + \dfrac{(-1)^{n-1}}{4^{n-1}}\right\}$ is bounded or unbounded; increasing, decreasing, or nonmonotone; convergent or divergent.

20 If $\{a_n\}$ and $\{b_n\}$ are convergent sequences and $a_n \leq b_n$ holds for all positive integers n, prove that $\lim\limits_{n \to +\infty} a_n \leq \lim\limits_{n \to +\infty} b_n$.

21 Explain carefully the distinction between a *sequence* and a *series*.

In problems 22 through 25, find the sum of each series by forcing the terms in the partial sums to telescope.

22 $\sum\limits_{k=1}^{\infty} \dfrac{k}{(k+1)(k+2)(k+3)}$

23 $\sum\limits_{k=1}^{\infty} \dfrac{\sqrt{k+1} - \sqrt{k}}{\sqrt{k^2 + k}}$

24 $\sum\limits_{k=1}^{\infty} \dfrac{4}{(2k-1)(2k+3)}$

25 $\sum\limits_{k=1}^{\infty} \left(\sin \dfrac{1}{k} - \sin \dfrac{1}{k+1}\right)$

26 If $\{b_n\}$ is a given sequence and p is a fixed positive integer, find a formula for the nth partial sum of the series $\sum\limits_{k=1}^{\infty} (b_k - b_{k+p})$. Assuming that $\lim\limits_{n \to +\infty} b_n = L$, find a formula for the sum of the series $\sum\limits_{k=1}^{\infty} (b_k - b_{k+p})$.

27 Find an infinite series whose nth partial sum is given by $s_n = \dfrac{3n}{2n+5}$. Determine whether the resulting series converges and, if it does, find its sum.

In problems 28 through 31, use the facts concerning geometric series to find the sum of each series.

28 $\sum\limits_{k=2}^{\infty} \left[5\left(\dfrac{1}{2}\right)^k + 3\left(\dfrac{1}{3}\right)^k\right]$

29 $\sum\limits_{k=1}^{\infty} \dfrac{3}{10^k}$

30 $\sum\limits_{k=1}^{\infty} 2\left(-\dfrac{1}{3}\right)^{k+7}$

31 $\sum\limits_{k=0}^{\infty} \left[2\left(\dfrac{1}{4}\right)^k + 7\left(\dfrac{1}{7}\right)^{k+1}\right]$

32 Assume that A and B are positive constants and that the sequence $\{a_n\}$ satisfies $|a_n| \leq AB^n$ for all positive integers n. Prove that the series $\sum\limits_{k=0}^{\infty} a_k x^k$ converges if $|x| < 1/B$.

33 Criticize the following argument: Since $\lim\limits_{n \to +\infty} \ln \dfrac{2n+1}{2n-1} = 0$, the series $\sum\limits_{k=1}^{\infty} \ln \dfrac{2k+1}{2k-1}$ must be convergent.

34 Make an informal argument to show that, if the series $\sum\limits_{k=1}^{\infty} b_k$ converges and if the sequence $\{a_n\}$ converges, then the series $\sum\limits_{k=1}^{\infty} a_k b_k$ must converge.

In problems 35 through 38, use the integral test to determine whether each series converges.

35 $\sum\limits_{k=2}^{\infty} \dfrac{1}{k(\ln k)^6}$

36 $\sum\limits_{k=1}^{\infty} \dfrac{k}{10 + k^2}$

37 $\sum\limits_{k=1}^{\infty} \dfrac{k^2}{e^k}$

38 $\sum\limits_{k=2}^{\infty} \dfrac{\ln k}{k}$

In problems 39 through 42, use the direct comparison test to determine the convergence of each series.

39 $\displaystyle\sum_{k=1}^{\infty} \frac{k^2}{k^2 + 2}\left(\frac{1}{3}\right)^k$ **40** $\displaystyle\sum_{k=0}^{\infty} \frac{1}{3 + k!}$ **41** $\displaystyle\sum_{k=1}^{\infty} \frac{1}{5k + 1}$ **42** $\displaystyle\sum_{k=1}^{\infty} \frac{1}{\sqrt{10k}}$

In problems 43 through 52, use any appropriate method to decide whether each series converges or diverges. If it converges, determine whether the convergence is absolute or conditional.

43 $\displaystyle\sum_{k=1}^{\infty} \frac{(-1)^k \sqrt{k}}{k + 10}$ **44** $\displaystyle\sum_{k=2}^{\infty} \frac{(-1)^k}{k^2 + (-1)^k}$ **45** $\displaystyle\sum_{k=1}^{\infty} \frac{k}{k+1}\left(\frac{1}{9}\right)^k$

46 $\displaystyle\sum_{k=1}^{\infty} \frac{(-1)^k}{\ln(1 + 1/k)}$ **47** $\displaystyle\sum_{k=1}^{\infty} \frac{1 + (-1)^k}{k}$ **48** $\displaystyle\sum_{k=1}^{\infty} \frac{(-1)^k}{\ln(e^k + e^{-k})}$

49 $\displaystyle\sum_{k=1}^{\infty} \frac{1 \cdot 3 \cdot 5 \cdots (2k - 1)}{3^k k!}$ **50** $\displaystyle\sum_{k=1}^{\infty} \left[\frac{2 \cdot 4 \cdot 6 \cdots (2k)}{1 \cdot 3 \cdot 5 \cdots (2k - 1)}\right]^2$

51 $\displaystyle\sum_{k=1}^{\infty} (-1)^{k+1} e^{-k^2}$ **52** $\displaystyle\sum_{k=2}^{\infty} \sin\left(\pi k + \frac{1}{\ln k}\right)$

53 Estimate the sum of the given series with an error not exceeding 5×10^{-4} in absolute value.

    (a) $\displaystyle\sum_{k=1}^{\infty} (-1)^{k+1} \frac{1}{k \cdot 2^k}$ (b) $\displaystyle\sum_{k=1}^{\infty} (-1)^{k+1} \frac{1}{(3k)^3}$

54 Give an example of a convergent series $\displaystyle\sum_{k=1}^{\infty} a_k$ of positive terms for which

    $\displaystyle\lim_{k \to +\infty} \frac{a_{k+1}}{a_k}$ does not exist.

In problems 55 through 64, find the center a, the radius R, and the interval I of convergence of the given power series.

55 $\displaystyle\sum_{k=1}^{\infty} \frac{(x - 1)^{2k}}{k \cdot 5^k}$ **56** $\displaystyle\sum_{k=0}^{\infty} \left(\sin\frac{\pi k}{2}\right) x^k$

57 $\displaystyle\sum_{k=0}^{\infty} (\cos \pi k)(x + 2)^k$ **58** $\displaystyle\sum_{k=1}^{\infty} 1 \cdot 3 \cdot 5 \cdot 7 \cdots (2k - 1) x^k$

59 $\displaystyle\sum_{k=1}^{\infty} \frac{1 \cdot 3 \cdot 5 \cdot 7 \cdots (2k - 1)}{2^{3k+1}} (x - 10)^k$ **60** $\displaystyle\sum_{k=0}^{\infty} 2^k (x + 4)^k$

61 $\displaystyle\sum_{k=0}^{\infty} (-1)^k \frac{10^k}{k!} (x + \pi)^k$ **62** $\displaystyle\sum_{k=1}^{\infty} \frac{1 \cdot 5 \cdot 9 \cdot 13 \cdots (4k - 3)}{2 \cdot 4 \cdot 6 \cdot 8 \cdots (2k)} (x + 6)^k$

63 $\displaystyle\sum_{k=0}^{\infty} \frac{(-1)^k 2^{2k+1}}{2k + 1} (x - 3)^{2k}$ **64** $\displaystyle\sum_{k=1}^{\infty} (1 + 2 + 3 + 4 + \cdots + k) x^{2k-1}$

In problems 65 and 66, find the Taylor series expansion for each function about the indicated center a. Give the range of values of x for which the expansion is correct.

65 $f(x) = \ln x$, $a = 1$.   **66** $f(x) = \sqrt{x}$, $a = 4$.

In problems 67 through 72, find the first four terms of the Taylor series for each function at the given value of a.

67 $f(x) = e^x$, $a = -1$.   **68** $f(x) = \tan x$, $a = \pi/4$.   **69** $f(x) = \sqrt{x}$, $a = 1$.

70 $f(x) = \ln(1/x)$, $a = 2$.   **71** $g(x) = \sin 2x$, $a = \pi/4$.   **72** $h(x) = \sec x$, $a = \pi/6$.

73 Show that the Taylor series in problem 67 actually converges to the function for all values of x by direct use of Theorem 1 in Section 8.

74 (a) Let f be the function defined by

$$f(x) = \begin{cases} \dfrac{e^x - 1}{x} & \text{for } x \neq 0 \\ 1 & \text{for } x = 0. \end{cases}$$

Find the Maclaurin series expansion for $f(x)$ and indicate the values of x for which it represents the function. Show that f is continuous.

(b) Find the Maclaurin series expansion for f'.

(c) Use the result of part (b) to find the sum of the series $\displaystyle\sum_{k=1}^{\infty} \frac{k}{(k+1)!}$.

75 Use power series to prove that

$$\int_0^x \tan^{-1} t \, dt = x \tan^{-1} x - \tfrac{1}{2} \ln (1 + x^2) \qquad \text{for } |x| < 1.$$

In problems 76 through 81, use the binomial series to find a power series expansion for each expression. In each case, specify the range of values of x for which the expansion works.

76 $\displaystyle\int_0^x \sqrt{1 + t^2} \, dt$

77 $\dfrac{1}{\sqrt{1 + x^2}}$

78 $D_x \sqrt[3]{1 + x^2}$

79 $(1 - 2x)^{2/3}$

80 $(16 + x^4)^{1/4}$

81 $\displaystyle\int_0^x \sqrt[3]{1 + t^3} \, dt$

82 Find a function f such that $f''(x) + af(x) = 0$, where a is a positive constant, $f(0) = 0$ and $f'(0) = \sqrt{a}$. (*Hint:* Expand f into a Maclaurin series with unknown coefficients. Then determine these coefficients.)

In problems 83 and 84, use the first three terms of an appropriate binomial series to estimate each number. Give an upper bound for the absolute value of the error.

83 $\sqrt[5]{30}$

84 $\displaystyle\int_0^{1/2} \sqrt[3]{1 + x^2} \, dx$

85 Find Maclaurin series expansions for
   (a) $\sin x + \cos x$
   (b) $\cos^2 x - \sin^2 x$
   (c) $\tan^{-1} x^3$
   (d) 10^x

86 Suppose that the power series $\displaystyle\sum_{k=0}^{\infty} c_k x^k$ has a positive radius of convergence R and that the function f defined by $f(x) = \displaystyle\sum_{k=0}^{\infty} c_k x^k$ for $|x| < R$ is an even function. Show that $c_k = 0$ for every odd positive integer k.

In problems 87 through 92, a function f is defined in terms of a power series. Write a formula for f in finite terms.

87 $f(x) = x - \dfrac{x^3}{3!} + \dfrac{x^5}{5!} - \dfrac{x^7}{7!} + \cdots$

88 $f(x) = \displaystyle\sum_{k=0}^{\infty} (-1)^k \dfrac{x^{2k}}{(2k)!}$

89 $f(x) = x - \dfrac{x^2}{3!} + \dfrac{x^4}{5!} - \dfrac{x^6}{7!} + \dfrac{x^8}{9!} - \cdots$

90 $f(x) = 1 + \sin x + \dfrac{\sin^2 x}{2!} + \dfrac{\sin^3 x}{3!} + \dfrac{\sin^4 x}{4!} + \cdots$

91 $f(x) = 1 + x \ln 2 + \dfrac{(\ln 2)^2}{2!} x^2 + \dfrac{(\ln 2)^3}{3!} x^3 + \dfrac{(\ln 2)^4}{4!} x^4 + \cdots$

92 $f(x) = \displaystyle\sum_{k=1}^{\infty} (-1)^k \dfrac{3^k + 1}{2^k k!} x^k$

14 VECTORS IN THE PLANE

Until now, we have dealt exclusively with quantities—such as length, area, volume, angle, mass, density, speed, time, temperature, and probability—that can be measured or represented by real numbers. Since real numbers can be represented by points on a number scale, such quantities are often called *scalars*.

On the other hand, mathematicians and scientists often must deal with quantities that cannot be described or represented by a single real number—quantities that have both magnitude and direction. Such quantities are called *vectors*, and include force, velocity, acceleration, displacement, momentum, electric field, magnetic field, gravitational field, angular velocity, and angular momentum. (We have previously treated some of these quantities, for instance, force, velocity, and acceleration, as if they were scalars—but only in connection with straight-line motion where the direction was understood.)

Actually, a vector is more than just a quantity with magnitude and direction, since it can interact with other vectors and with scalars in certain well-defined ways. Physicists define vectors to be quantities that transform in a particular manner under groups of symmetries, while mathematicians use an even more abstract definition of a vector as an element of a "linear space." However, for our purposes, a simple geometric definition is quite adequate.

1 Vectors and Vector Addition

By definition, a *vector* is a directed line segment; that is, in everyday language, an arrow. Figure 1 shows some vectors. Each vector has a *tail end* (also called an *initial point* and a *head end* (also called a *terminal point*) and is understood to be directed from its tail end to its head end. Reversing the arrow gives us a vector in the opposite direction.

It is essential to use some special notation for vectors so that they can be distinguished from scalars. In this book we use $\overline{A}$, $\overline{B}$, $\overline{C}$, and so forth to denote vectors.

If a particle is moved from the point P to the point Q, we say that the particle has undergone a *displacement* from P to Q (Figure 2). Such a displacement can be

Figure 1

Figure 2

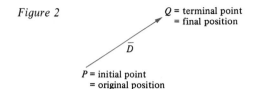

Q = terminal point
= final position

$\overline{D}$

P = initial point
= original position

Figure 3

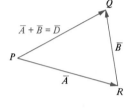

$\overline{A} + \overline{B} = \overline{D}$

Q

$\overline{B}$

P

$\overline{A}$

R

Figure 4

$\overrightarrow{P_1 Q_1} = \overrightarrow{P_2 Q_2}$

Q_1

$\overrightarrow{P_1 Q_1}$

P_1

Q_2

$\overrightarrow{P_2 Q_2}$

P_2

Figure 5

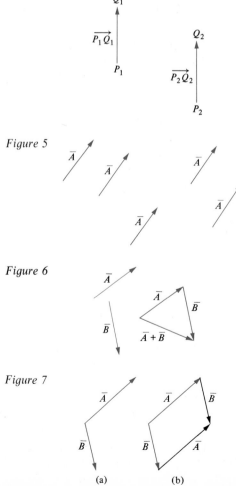

$\overline{A}$ $\overline{A}$ $\overline{A}$ $\overline{A}$ $\overline{A}$

Figure 6

$\overline{A}$ $\overline{A}$ $\overline{B}$

$\overline{B}$ $\overline{A} + \overline{B}$

Figure 7

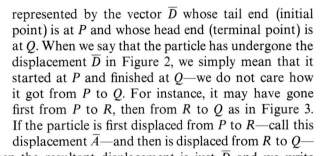

$\overline{A}$ $\overline{A}$ $\overline{B}$

$\overline{B}$ $\overline{B}$ $\overline{A}$

(a)   (b)

represented by the vector $\overline{D}$ whose tail end (initial point) is at P and whose head end (terminal point) is at Q. When we say that the particle has undergone the displacement $\overline{D}$ in Figure 2, we simply mean that it started at P and finished at Q—we do not care how it got from P to Q. For instance, it may have gone first from P to R, then from R to Q as in Figure 3. If the particle is first displaced from P to R—call this displacement $\overline{A}$—and then is displaced from R to Q— call this displacement $\overline{B}$—then the resultant displacement is just $\overline{D}$ and we write $\overline{A} + \overline{B} = \overline{D}$ (Figure 3). This illustrates the *head-to-tail rule* for adding vectors:

> If the head end of $\overline{A}$ coincides with the tail end of $\overline{B}$, then the vector $\overline{D}$ whose tail end is the tail end of $\overline{A}$ and whose head end is the head end of $\overline{B}$ is called the *sum* of $\overline{A}$ and $\overline{B}$ and written as $\overline{D} = \overline{A} + \overline{B}$.

A vector $\overline{D}$ whose tail end is at the point P and whose head end is at the point Q is written as $\overline{D} = \overrightarrow{PQ}$. With this notation, the head-to-tail rule can be written in the form

$$\overrightarrow{PQ} = \overrightarrow{PR} + \overrightarrow{RQ},$$

which is quite easy to remember.

If two particles, starting from different points P_1 and P_2, are moved due north through the same distance, it is traditional to say that they have undergone the *same* displacement (Figure 4). More generally, two vectors that are parallel, point in the same direction, and have the same length are usually regarded as being *equal*. Following this convention, the two vectors in Figure 4 are regarded as being equal, and we write $\overrightarrow{P_1 Q_1} = \overrightarrow{P_2 Q_2}$.

Actually, the equality convention for vectors is extremely useful, since it implies that a vector may be moved around freely without being changed—provided only that it is always kept parallel to its original position and that its direction and length remain unchanged. Thus, for instance, all the arrows in Figure 5 represent the same vector $\overline{A}$.

The fact that vectors can be moved around as specified above implies that any two vectors $\overline{A}$ and $\overline{B}$ can be brought into a head-to-tail position (Figure 6); hence, any two vectors have a sum.

We can also move any two vectors $\overline{A}$ and $\overline{B}$ into a tail-to-tail position (Figure 7a) so that they form two adjacent sides of a parallelogram (Figure 7b). We refer to this parallelogram as the parallelogram *spanned* by $\overline{A}$ and $\overline{B}$.

Notice that the diagonal vector whose tail end coincides with the common tail ends of $\overline{A}$ and $\overline{B}$ in the parallelogram spanned by $\overline{A}$ and $\overline{B}$ is the sum of $\overline{A}$ and $\overline{B}$, as can be seen by looking at the top half

Figure 8

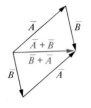

of the parallelogram in Figure 8. Glancing at the bottom half of this parallelogram, we see that the same diagonal vector is the sum of $\bar{B}$ and $\bar{A}$; that is, we see that

$$\bar{A} + \bar{B} = \bar{B} + \bar{A}.$$

This equation expresses the *commutative law of vector addition*. The technique of finding the sum of two vectors by forming the diagonal of a parallelogram is called the *parallelogram rule* for adding vectors.

Vectors also satisfy the *associative law of vector addition*

$$\bar{A} + (\bar{B} + \bar{C}) = (\bar{A} + \bar{B}) + \bar{C},$$

as can be seen by examining Figure 9. Because of the associative law, parentheses are not really needed when adding several vectors, and we can simply write $\bar{A} + \bar{B} + \bar{C}$, $\bar{A} + \bar{B} + \bar{C} + \bar{D}$, $\bar{A}_1 + \bar{A}_2 + \bar{A}_3 + \cdots + \bar{A}_n$, and so forth. Notice that these sums can be obtained geometrically by an obvious extension of the head-to-tail rule (Figure 10).

Figure 9

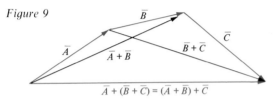

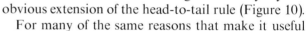

For many of the same reasons that make it useful to have a zero scalar, it is also convenient to introduce a *zero vector*, written $\bar{0}$. Intuitively, $\bar{0}$ can be thought of as an arrow that has shrunk to a single point. Alternatively, it can be regarded as a vector $\overrightarrow{QQ}$ that starts and ends at the same point Q. Just as nonzero vectors can be moved around subject to the aforementioned conditions, so also the zero vector

Figure 10

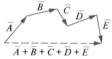

can be moved around and we have $\bar{0} = \overrightarrow{QQ} = \overrightarrow{RR} = \overrightarrow{SS} = \cdots$. Of course, the zero vector is the only vector whose length is 0 and it is the only vector whose direction is indeterminate. Notice that $\overrightarrow{PQ} + \overrightarrow{QQ} = \overrightarrow{PQ}$; that is, we have the law

$$\bar{A} + \bar{0} = \bar{A},$$

which, in algebraic parlance, says that $\bar{0}$ is the *additive identity* for vectors.

If we turn a vector $\bar{A}$ around by interchanging its tail and head ends, we obtain a vector in the opposite direction called the *negative* of $\bar{A}$ and written $-\bar{A}$ (Figure 11). Evidently, if $\bar{A} = \overrightarrow{PQ}$, then $-\bar{A} = \overrightarrow{QP}$. Notice that the sum $\overrightarrow{PQ} + \overrightarrow{QP} = \overrightarrow{PP} = \bar{0}$; that is, we have the law

Figure 11

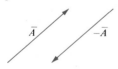

$$\bar{A} + (-\bar{A}) = \bar{0},$$

which expresses the fact that $-\bar{A}$ is the *additive inverse* of $\bar{A}$.

We now *define* the operation of vector subtraction by the equation

Figure 12

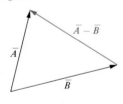

$$\bar{A} - \bar{B} = \bar{A} + (-\bar{B}).$$

Notice that $\bar{A} - \bar{B}$ is the solution $\bar{X}$ of the vector equation $\bar{X} + B = \bar{A}$ (Problem 19); that is, $\bar{A} - \bar{B}$ is the vector which, when added to $\bar{B}$, gives $\bar{A}$ (Figure 12). Thus, if $\bar{A}$ and $\bar{B}$ have the same tail end, as in Figure 12, then $\bar{A} - \bar{B}$ is the vector running between their head ends and *pointing toward the head end of $\bar{A}$*. We refer to this method of finding the vector difference $\bar{A} - \bar{B}$ geometrically as the *triangle rule for vector subtraction*.

EXAMPLE   Show geometrically that $-(\bar{A} - \bar{B}) = \bar{B} - \bar{A}$.

Figure 13

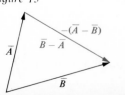

SOLUTION
Figure 12 shows the vector $\bar{A} - \bar{B}$. Its negative, $-(\bar{A} - \bar{B})$, is obtained by reversing its direction (Figure 13); however, the triangle rule for subtraction gives the same vector for $\bar{B} - \bar{A}$.

In addition to the identity $-(\bar{A} - \bar{B}) = \bar{B} - \bar{A}$, we have the identities $-(-\bar{A}) = \bar{A}$ and $-(\bar{A} + \bar{B}) = -\bar{A} - \bar{B}$, which are also easy to verify geometrically.

In summary, for vector addition we have the following four basic laws:

1 $\bar{A} + \bar{B} = \bar{B} + \bar{A}$ (Commutative Law)

2 $\bar{A} + (\bar{B} + \bar{C}) = (\bar{A} + \bar{B}) + \bar{C}$ (Associative Law)

3 $\bar{A} + \bar{0} = \bar{A}$ (Identity Law)

4 $\bar{A} + (-\bar{A}) = \bar{0}$ (Inverse Law).

Using these four basic laws, together with the definition of vector subtraction, $\bar{A} - \bar{B} = \bar{A} + (-\bar{B})$, one can develop all the identities for vector addition and subtraction in a purely deductive manner; however, we do not attempt to do this here. Although the algebra of vectors is pretty much like the algebra of scalars, one must take care not to confuse vectors with scalars; for instance, an equation such as $\bar{A} = 3$ is meaningless since the left side is a vector and the right side is a scalar.

Problem Set 1

In problems 1 through 14, copy the appropriate vectors from Figure 14 onto your paper, then find each vector by means of a suitable geometric construction.

Figure 14

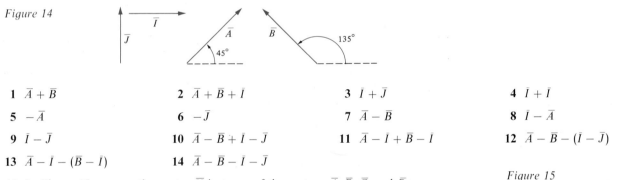

1 $\bar{A} + \bar{B}$ 2 $\bar{A} + \bar{B} + \bar{I}$ 3 $\bar{I} + \bar{J}$ 4 $\bar{I} + \bar{I}$

5 $-\bar{A}$ 6 $-\bar{J}$ 7 $\bar{A} - \bar{B}$ 8 $\bar{I} - \bar{A}$

9 $\bar{I} - \bar{J}$ 10 $\bar{A} - \bar{B} + \bar{I} - \bar{J}$ 11 $\bar{A} - \bar{I} + \bar{B} - \bar{I}$ 12 $\bar{A} - \bar{B} - (\bar{I} - \bar{J})$

13 $\bar{A} - \bar{I} - (\bar{B} - \bar{I})$ 14 $\bar{A} - \bar{B} - \bar{I} - \bar{J}$

15 In Figure 15, express the vector $\bar{X}$ in terms of the vectors $\bar{A}$, $\bar{B}$, $\bar{C}$, and $\bar{D}$.

16 One of the diagonals of the parallelogram spanned by two vectors $\bar{A}$ and $\bar{B}$ is $\bar{A} + \bar{B}$. Express the other diagonal in terms of $\bar{A}$ and $\bar{B}$.

In problems 17 and 18, solve each equation for the vector $\bar{X}$.

17 $\bar{A} + \bar{X} + \bar{B} = \bar{0}$ 18 $(\bar{A} - \bar{X}) - (\bar{B} - \bar{X}) = \bar{C} - \bar{X}$

19 Show by a diagram or otherwise that the law of transposition works for vectors; that is, $\bar{X} + \bar{B} = \bar{A}$ holds if and only if $\bar{X} = \bar{A} - \bar{B}$.

20 A car travels 30 miles due east, then 40 miles due north, then 30 miles due east again. Draw a scale diagram and represent by vectors the successive displacements of the car. Add these vectors using the head-to-tail rule and thus find the resultant displacement of the car.

21 Suppose that the four distinct points P, Q, R, and S all lie on the same straight line. If the vector $\overrightarrow{PQ}$ has the same direction as the vector $\overrightarrow{RS}$, what are the possible orders in which the points lie on the line? (For instance, S, Q, R, P—meaning first S, then Q, then R, and finally P—is one possibility.)

22 In all our deliberations above, we have been picturing vectors—that is, arrows—as lying in a given plane. Does anything change if the arrows are presumed to lie in space?

23 What is wrong with the following equations?
(a) $\bar{A} - \bar{B} = 5$ (b) $\bar{A} + 3 = \bar{B}$ (c) $\bar{A} + \bar{B} = 0$

24 Does $\bar{A} > \bar{0}$ make any sense?

Figure 15

2   Multiplication of Vectors by Scalars

If $\bar{A}$ is a vector, it seems reasonable on algebraic grounds to define $2\bar{A}$ by the equation $2\bar{A} = \bar{A} + \bar{A}$. Using the head-to-tail rule to add $\bar{A}$ to itself, we find that $2\bar{A}$ has the same direction as $\bar{A}$, but is twice as long (Figure 1). More generally, we make the following rules for multiplication of a vector $\bar{A}$ by a scalar s:

Figure 1

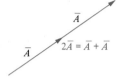

1   If $s > 0$, then $s\bar{A}$ is the vector in the same direction as $\bar{A}$, but s times as long.
2   If $s < 0$, then $s\bar{A}$ is the vector in the direction opposite to $\bar{A}$, but $|s|$ times as long.

Naturally, we understand that $s\bar{A} = \bar{0}$ if either $s = 0$ or $\bar{A} = \bar{0}$ (or both).

Figure 2 shows a vector $\bar{A}$ and some of its scalar multiples. Notice that $(-1)\bar{A} = -\bar{A}$. More generally,

Figure 2

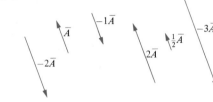

$$(-s)\bar{A} = s(-\bar{A})$$

holds for all scalars s and all vectors $\bar{A}$. (Why?)

Multiplication of a vector by a scalar $s > 1$ "stretches" the vector by a factor of s, while multiplication by a scalar s between 0 and 1 "shrinks" it by a factor of s. If all vectors appearing in a diagram (Figure 3a) are multiplied by the same positive scalar s, $s \neq 1$, then the resulting diagram is either a magnification (Figure 3b) or a compression (Figure 3c) of the original diagram, according to whether $s > 1$ or $s < 1$, respectively. In Figure 3 notice that $\bar{A} + \bar{B} = \bar{C}$ by the head-to-tail rule, and likewise, $s\bar{A} + s\bar{B} = s\bar{C}$. Substituting from

Figure 3

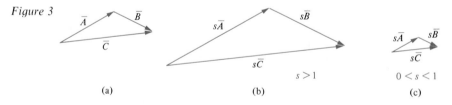

(a)   (b)   (c)

the first equation into the second, we obtain the important *distributive law* $s\bar{A} + s\bar{B} = s(\bar{A} + \bar{B})$, $s > 0$. Now suppose that $s > 0$, so that $-s < 0$. Using the distributive law above, we have

$$(-s)\bar{A} + (-s)\bar{B} = s(-\bar{A}) + s(-\bar{B}) = s[(-\bar{A}) + (-\bar{B})] = s[-(\bar{A} + \bar{B})]$$
$$= (-s)(\bar{A} + \bar{B});$$

hence, the distributive law works for negative scalars as well. Since it obviously works when $s = 0$, we have

$$s\bar{A} + s\bar{B} = s(\bar{A} + \bar{B})$$

for all scalars s and for all vectors $\bar{A}$ and $\bar{B}$.

There is a *second distributive law* involving two scalars and one vector,

$$s\bar{A} + t\bar{A} = (s + t)\bar{A}.$$

This is an obvious consequence of the rules for multiplication by scalars when both

s and t are nonnegative. If the reader checks the other possible cases, it will be clear that the equation holds for all scalars s and t.

In addition to the two distributive laws, there is an *associative* type law for scalar multiplication,

$$(st)\overline{A} = s(t\overline{A}),$$

which can be confirmed geometrically simply by considering the various possible cases according to whether s and t are positive, negative, or zero.

The basic laws for multiplication of vectors by scalars can now be summarized:

1   $(st)\overline{A} = s(t\overline{A})$ (Associative Law)

2   $s(\overline{A} + \overline{B}) = s\overline{A} + s\overline{B}$ (First Distributive Law)

3   $(s + t)\overline{A} = s\overline{A} + t\overline{A}$ (Second Distributive Law)

4   $1\overline{A} = \overline{A}$ (Multiplicative Identity Law).

These four laws, together with the four basic laws for vector addition (Section 1, page 679), can be taken as postulates and used to give a purely deductive development of the algebra of vectors. Although we do not attempt such a development, we hereby report that there are no real surprises, and the manipulative rules for vector algebra turn out to be largely the same as for the ordinary algebra of scalars. Some of the important rules are stated as follows:

5   $s(-\overline{A}) = (-s)\overline{A} = -(s\overline{A})$

6   $-\overline{A} = (-1)\overline{A}$

7   $0\overline{A} = \overline{0}$

8   $s\overline{0} = \overline{0}$

9   If $s\overline{A} = \overline{0}$, then $s = 0$ or $\overline{A} = \overline{0}$ (or both).

10   $s(\overline{A} - \overline{B}) = s\overline{A} - s\overline{B}$

11   $(s - t)\overline{A} = s\overline{A} - t\overline{A}$

If s is a nonzero scalar and $\overline{A}$ is a vector, the expression $(1/s)\overline{A}$ is often written as $\overline{A}/s$. Thus, when we speak of *dividing* the vector $\overline{A}$ by the nonzero scalar s, we simply mean to multiply $\overline{A}$ by $1/s$.

EXAMPLE   Let P, Q, and R be three points in the plane and let M be the midpoint of the line segment $\overrightarrow{PQ}$. Show that $\overrightarrow{RM} = \frac{1}{2}(\overrightarrow{RP} + \overrightarrow{RQ})$ (Figure 4).

SOLUTION
Evidently, $\overrightarrow{PM} = \frac{1}{2}\overrightarrow{PQ}$, since M is the midpoint of $\overrightarrow{PQ}$. Therefore,

$$
\begin{aligned}
\overrightarrow{RM} &= \overrightarrow{RP} + \overrightarrow{PM} && \text{(head-to-tail rule)} \\
&= \overrightarrow{RP} + \tfrac{1}{2}\overrightarrow{PQ} && \text{(substitution of equals)} \\
&= \overrightarrow{RP} + \tfrac{1}{2}(\overrightarrow{RQ} - \overrightarrow{RP}) && \text{(triangle rule for} \\
& && \qquad \text{subtraction)} \\
&= \overrightarrow{RP} + \tfrac{1}{2}\overrightarrow{RQ} - \tfrac{1}{2}\overrightarrow{RP} && \text{(rule 10)} \\
&= (1 - \tfrac{1}{2})\overrightarrow{RP} + \tfrac{1}{2}\overrightarrow{RQ} && \text{(rule 11)} \\
&= \tfrac{1}{2}\overrightarrow{RP} + \tfrac{1}{2}\overrightarrow{RQ} && \text{(scalar arithmetic)} \\
&= \tfrac{1}{2}(\overrightarrow{RP} + \overrightarrow{RQ}) && \text{(first distributive law).}
\end{aligned}
$$

Figure 4

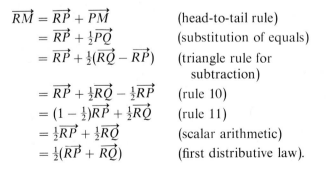

In order to show that two vectors $\overline{A}$ and $\overline{B}$ are *parallel*, one must show that one of the two vectors (either one) is a scalar multiple of the other one. Indeed, we take this as our official definition.

DEFINITION 1   **Parallelism of Vectors**

We say that the vectors $\bar{A}$ and $\bar{B}$ are *parallel* if there is a scalar s such that $\bar{A} = s\bar{B}$ or if there is a scalar t such that $\bar{B} = t\bar{A}$. Two vectors $\bar{A}$ and $\bar{B}$ that are not parallel are said to be *linearly independent*.

If $\bar{A} = s\bar{B}$, where $s > 0$, then we say that $\bar{A}$ and $\bar{B}$ are not only parallel, but that they have the *same direction*; however, if $\bar{A} = s\bar{B}$, where $s < 0$, then we say that $\bar{A}$ and $\bar{B}$ are parallel, but have *opposite directions*.

We notice three simple consequences of Definition 1. First, any vector is parallel to itself. (Why?) Second, the zero vector is parallel to every vector because, if $\bar{A}$ is any vector, then $\bar{0} = 0\bar{A}$. (This is not unreasonable since the zero vector has an indeterminate direction anyway.) Third, if two vectors are linearly independent, then neither of them can be zero, since the zero vector *is* parallel to every other vector.

The following theorem gives an important property of two linearly independent (that is, nonparallel) vectors.

THEOREM 1   **Linear Independence of Two Vectors**

If $\bar{A}$ and $\bar{B}$ are linearly independent vectors, then the only scalars s and t for which $s\bar{A} + t\bar{B} = \bar{0}$ are $s = 0$ and $t = 0$.

PROOF

Suppose that $s\bar{A} + t\bar{B} = \bar{0}$, but that at least one of the scalars s or t is nonzero. Then, if $s \neq 0$, we have $\frac{1}{s}(s\bar{A}) + \frac{1}{s}(t\bar{B}) = \frac{1}{s}\bar{0} = \bar{0}$, or $\bar{A} + \frac{t}{s}\bar{B} = \bar{0}$; that is, $\bar{A} = \left(-\frac{t}{s}\right)\bar{B}$. Similarly, if $t \neq 0$, we have $\bar{B} = (-s/t)\bar{A}$. In either case, the vectors A and B are parallel—hence, by definition, not linearly independent. Therefore, if $\bar{A}$ and $\bar{B}$ are linearly independent, the equation $s\bar{A} + t\bar{B} = \bar{0}$ can hold only if $s = t = 0$.

EXAMPLE   Suppose that $\bar{A}$ and $\bar{B}$ are linearly independent vectors and that

$$(3x - y)\bar{A} + (1 - x)\bar{B} = 7\bar{A} + y\bar{B}.$$

Find x and y.

SOLUTION

From the given equation, we have

$$(3x - y)\bar{A} - 7\bar{A} + (1 - x)\bar{B} - y\bar{B} = \bar{0}$$

or

$$(3x - y - 7)\bar{A} + (1 - x - y)\bar{B} = \bar{0}.$$

Since $\bar{A}$ and $\bar{B}$ are linearly independent, Theorem 1 implies that

$$\begin{cases} 3x - y - 7 = 0 \\ 1 - x - y = 0. \end{cases}$$

Solving these two simultaneous scalar equations, we find that $x = 2$ and $y = -1$.

An expression of the form $s\bar{A} + t\bar{B}$ is called a *linear combination* of the vectors $\bar{A}$ and $\bar{B}$ with *coefficients* s and t. To say that $\bar{A}$ and $\bar{B}$ are linearly independent is equivalent to saying that the only way to obtain $\bar{0}$ as a linear combination of $\bar{A}$ and $\bar{B}$ is to make both of the coefficients equal to zero (Problem 28).

2.1 Standard Basis Vectors

In dealing with vectors lying in a plane, it is often useful to choose two perpendicular vectors of length 1 and to write all other vectors in the plane as linear combinations of these two vectors. The two chosen vectors, called *basis vectors*, are usually denoted by i and j (Figure 5). In our diagrams, we usually take i to be a horizontal vector pointing to the right and we take j to be a vertical vector pointing upward. Although i and j can be moved around in the plane, as all vectors can, we usually draw them with a common tail end O. Naturally, if we have a cartesian or a polar coordinate system already established in the plane, we ordinarily take O to be the origin or the pole. The basis i, j is called the *standard basis*. Note that i and j are linearly independent.

Figure 5

To see that any vector $\bar{R}$ in the plane can be written as a linear combination of i and j, we argue as follows: Establish a cartesian coordinate system so that i lies along the positive x axis and j lies along the positive y axis. Move the given vector $\bar{R}$ so that its tail end is at the origin O and let its head end be the point $P = (x, y)$. Drop perpendiculars from P to the x and y axes, meeting these axes respectively at the points $S = (x, 0)$ and $T = (0, y)$ (Figure 6). If $x > 0$, then $x i$ is a vector in the same direction as i, but x times as long; hence, $x i = \overrightarrow{OS}$. If $x \leq 0$, we also have $x i = \overrightarrow{OS}$. (Why?) Similarly, $y j = \overrightarrow{OT}$. By the parallelogram rule,

$$\bar{R} = \overrightarrow{OP} = \overrightarrow{OS} + \overrightarrow{OT} = x i + y j;$$

Figure 6

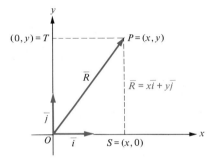

hence, $\bar{R}$ is expressed as a linear combination of i and j. The numbers x and y are called the (scalar) *components* of the vector $\bar{R}$ with respect to the standard i, j basis. Notice that these components are just the coordinates of the head end of $\bar{R}$ when its tail end is at the origin.

Now suppose that $\bar{A} = \overrightarrow{P_1 P_2}$, where $P_1 = (x_1, y_1)$ and $P_2 = (x_2, y_2)$ (Figure 7). Then $\overrightarrow{OP_1} = x_1 i + y_1 j$ and $\overrightarrow{OP_2} = x_2 i + y_2 j$; hence, by the triangle rule for subtraction,

Figure 7

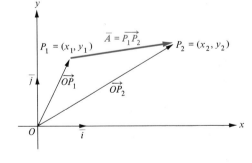

$$\bar{A} = \overrightarrow{P_1 P_2} = \overrightarrow{OP_2} - \overrightarrow{OP_1}$$
$$= (x_2 i + y_2 j) - (x_1 i + y_1 j)$$
$$= x_2 i + y_2 j - x_1 i - y_1 j$$
$$= x_2 i - x_1 i + y_2 j - y_1 j$$
$$= (x_2 - x_1) i + (y_2 - y_1) j.$$

In words, the scalar components of the vector $\overrightarrow{P_1 P_2}$ are the differences of the coordinates of P_2 and the corresponding coordinates of P_1.

EXAMPLES 1 If $P = (-2, 5)$ and $Q = (7, 3)$, write the vectors $\overrightarrow{OP}$ and $\overrightarrow{PQ}$ in component form.

SOLUTION
$$\overrightarrow{OP} = -2i + 5j \text{ and } \overrightarrow{PQ} = [7 - (-2)]i + [3 - 5]j = 9i - 2j.$$

2 Use vectors to find the coordinates of a point $\frac{3}{4}$ of the way from $(-5, -9)$ to $(7, 7)$.

SOLUTION

Put $P = (-5, -9)$ and $Q = (7, 7)$. Let $R = (x, y)$ be the desired point (Figure 8), so that $\overrightarrow{PR} = \frac{3}{4}\overrightarrow{PQ}$. Now, $\overrightarrow{OR} = \overrightarrow{OP} + \overrightarrow{PR} = \overrightarrow{OP} + \frac{3}{4}\overrightarrow{PQ}$. Since

$$\overrightarrow{OR} = x\bar{i} + y\bar{j}, \quad \overrightarrow{OP} = -5\bar{i} - 9\bar{j}, \quad \text{and}$$
$$\overrightarrow{PQ} = [7 - (-5)]\bar{i} + [7 - (-9)]\bar{j} = 12\bar{i} + 16\bar{j},$$

the equation $\overrightarrow{OR} = \overrightarrow{OP} + \frac{3}{4}\overrightarrow{PQ}$ can be rewritten as

$$x\bar{i} + y\bar{j} = -5\bar{i} - 9\bar{j} + \frac{3}{4}(12\bar{i} + 16\bar{j})$$
$$= -5\bar{i} - 9\bar{j} + 9\bar{i} + 12\bar{j}$$
$$= 4\bar{i} + 3\bar{j}.$$

Therefore,

$$(x - 4)\bar{i} + (y - 3)\bar{j} = \bar{0}.$$

Since $\bar{i}$ and $\bar{j}$ are linearly independent, the latter equation implies that $x - 4 = 0$ and $y - 3 = 0$. Therefore, $(x, y) = (4, 3)$.

3 Show that, if two vectors are equal, then their components are equal.

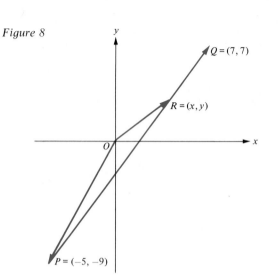

Figure 8

$Q = (7, 7)$

$R = (x, y)$

O

x

y

$P = (-5, -9)$

SOLUTION

If $x\bar{i} + y\bar{j} = a\bar{i} + b\bar{j}$, then $(x - a)\bar{i} + (y - b)\bar{j} = \bar{0}$. Since $\bar{i}$ and $\bar{j}$ are linearly independent, the latter equation implies that $x - a = 0$ and $y - b = 0$; that is, $x = a$ and $y = b$.

Calculation with vectors in component form is really very simple because of the fact that vectors can be added, subtracted, or multiplied by scalars in a "componentwise" manner. In fact, if $\bar{A} = a\bar{i} + b\bar{j}$ and $\bar{B} = c\bar{i} + d\bar{j}$, we have

1 $\bar{A} + \bar{B} = (a + c)\bar{i} + (b + d)\bar{j}$.
2 $\bar{A} - \bar{B} = (a - c)\bar{i} + (b - d)\bar{j}$.
3 $t\bar{A} = (ta)\bar{i} + (tb)\bar{j}$.

These facts are easily verified using the identities for vector addition and multiplication by scalars; for instance, to confirm the second equation, we calculate as follows:

$$\bar{A} - \bar{B} = \bar{A} + (-\bar{B}) = a\bar{i} + b\bar{j} - c\bar{i} - d\bar{j} = a\bar{i} - c\bar{i} + b\bar{j} - d\bar{j}$$
$$= (a - c)\bar{i} + (b - d)\bar{j}.$$

EXAMPLE   If $\bar{A} = 2\bar{i} + 17\bar{j}$ and $\bar{B} = 13\bar{i} - 3\bar{j}$, find (a) $\bar{A} + \bar{B}$, (b) $-7\bar{A}$, and (c) $-7\bar{A} - \bar{B}$ in component form.

SOLUTION

(a) $\bar{A} + \bar{B} = (2 + 13)\bar{i} + (17 - 3)\bar{j} = 15\bar{i} + 14\bar{j}$.
(b) $-7\bar{A} = -7(2\bar{i} + 17\bar{j}) = -14\bar{i} - 119\bar{j}$.
(c) $-7\bar{A} - \bar{B} = (-14\bar{i} - 119\bar{j}) - (13\bar{i} - 3\bar{j}) = (-14 - 13)\bar{i} + (-119 + 3)\bar{j}$
$$= -27\bar{i} - 116\bar{j}.$$

If we fix a standard basis $\bar{i}, \bar{j}$, then a vector $\bar{A} = x\bar{i} + y\bar{j}$ both determines and is determined by its scalar components x and y. For this reason, the vector

$\bar{A} = x\bar{i} + y\bar{j}$ is sometimes denoted by the ordered pair $\langle x, y \rangle$ of its components, and is written $\bar{A} = \langle x, y \rangle$.   Thus, in the previous example, we have $\bar{A} = \langle 2, 17 \rangle$ and $\bar{B} = \langle 13, -3 \rangle$; hence,

$$\bar{A} + \bar{B} = \langle 2 + 13, 17 - 3 \rangle = \langle 15, 14 \rangle, \quad -7\bar{A} = \langle -14, -119 \rangle,$$

$$-7\bar{A} - \bar{B} = \langle -14 - 13, -119 + 3 \rangle = \langle -27, -116 \rangle.$$

The reader is encouraged to use ordered-pair notation for vectors whenever it seems convenient.

Problem Set 2

In problems 1 through 5, copy the appropriate vectors from Figure 9 onto your paper, then find each vector by means of a suitable geometric construction.

Figure 9

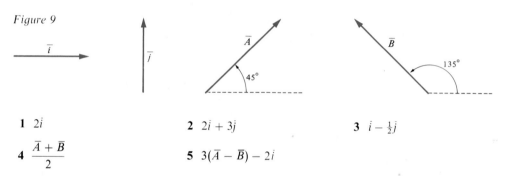

1 $2\bar{i}$

2 $2\bar{i} + 3\bar{j}$

3 $\bar{i} - \frac{1}{2}\bar{j}$

4 $\dfrac{\bar{A} + \bar{B}}{2}$

5 $3(\bar{A} - \bar{B}) - 2\bar{i}$

6 In Figure 10, $PQRS$ is a square and T is the midpoint of $\overrightarrow{SR}$. Express $\overrightarrow{PT}$ in terms of $\overrightarrow{PQ}$ and $\overrightarrow{QR}$.

Figure 10

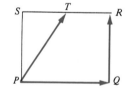

7 Let $\bar{A} = \overrightarrow{PQ}$, $\bar{B} = \overrightarrow{PR}$, and $\bar{C} = \overrightarrow{PS}$, where the point S is on the line segment $\overline{QR}$ and $\frac{3}{8}$ of the way from Q to R (Figure 11). Express $\bar{C}$ in terms of $\bar{A}$ and $\bar{B}$.

8 Generalize problem 7, replacing $\frac{3}{8}$ by an arbitrary fraction t, $0 < t < 1$.

In problems 9 and 10, let $\bar{A} = 4\bar{i} + 2\bar{j}$, $\bar{B} = -3\bar{i} + 4\bar{j}$, and $\bar{C} = -5\bar{i} + 7\bar{j}$.

9 Find (a) $\bar{A} + \bar{B}$; (b) $\bar{A} - \bar{B}$.

10 Find (a) $2\bar{A} + 3\bar{B}$; (b) $7\bar{A} - 5\bar{C}$.

In problems 11 and 12, let $\bar{A} = \langle 4, 2 \rangle$, $\bar{B} = \langle -3, 4 \rangle$, and $\bar{C} = \langle -5, 7 \rangle$.

11 Find (a) $5\bar{B} + 2\bar{C}$; (b) $4\bar{B} - \bar{C}$.

12 Find (a) $3\bar{A} + 4\bar{C}$; (b) $5\bar{A} - 2\bar{C}$.

Figure 11

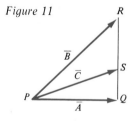

In problems 13 through 19, let $\bar{A} = 2\bar{i} + 7\bar{j}$, $\bar{B} = \bar{i} - 6\bar{j}$, and $\bar{C} = -5\bar{i} + 10\bar{j}$.

13 Find:

(a) $\bar{A} + \bar{B}$

(b) $\bar{A} - \bar{B}$

(c) $\dfrac{\bar{A} + \bar{B}}{2} - \dfrac{2}{5}\bar{C}$

(d) $s\bar{A} + t\bar{B} + u\bar{C}$

14 Show that $\bar{A}$ and $\bar{B}$ are linearly independent.

15 Find scalars s and t such that $s\bar{A} + t\bar{B} = \bar{C}$.

16 Show that all vectors in the plane can be obtained by forming linear combinations of $\bar{A}$ and $\bar{B}$.

17 Solve the vector equation $(3x + 4y - 12)\bar{A} + (3x - 8y)\bar{B} = \bar{0}$ for x and y.

18 Find a scalar s such that $s\bar{A} + \bar{B}$ is parallel to $\bar{C}$.

19 Solve the vector equation $(x - y)\bar{A} = (3x + 2y)\bar{B} - \bar{A}$ for x and y.

20 In Figure 12, let $\bar{A} = \overrightarrow{PQ}$, $\bar{B} = \overrightarrow{PR}$, $\overrightarrow{PS} = 5\bar{A}$, and $\overrightarrow{PT} = 3\bar{B}$. Express each of the following vectors in terms of $\bar{A}$ and $\bar{B}$:
(a) $\overrightarrow{QR}$ (b) $\overrightarrow{QS}$ (c) $\overrightarrow{ST}$ (d) $\overrightarrow{SR}$ (e) $\overrightarrow{QM}$ (f) $\overrightarrow{MS}$

21 Let $O = (0,0)$, $P = (3,6)$, $Q = (-1,3)$, $R = (-7,-1)$, and $S = (3,-6)$ be points in the xy plane. Find the (scalar) components of each of the following vectors with respect to the standard i, j basis.
(a) $\overrightarrow{OP}$ (b) $\overrightarrow{OQ}$ (c) $\overrightarrow{PQ}$ (d) $\overrightarrow{QP}$ (e) $\overrightarrow{PR}$
(f) $\overrightarrow{RS}$ (g) $\overrightarrow{PQ} + \overrightarrow{RS}$ (h) $3\overrightarrow{QR} - 5\overrightarrow{SP}$

Figure 12

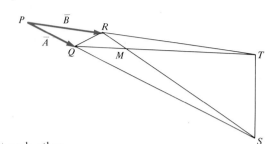

22 Let $ABCD$ be a quadrilateral and let P, Q, R, and S be the midpoints of $\overline{AB}$, $\overline{BC}$, $\overline{CD}$, and $\overline{DA}$, respectively. Use vectors to show that $PQRS$ is a parallelogram.

23 Use vectors to show that the diagonals of a parallelogram bisect each other.

24 In this section we have been picturing vectors as lying in a *plane*. Which considerations, if any, depend on this in an essential way?

25 Let P and Q be two distinct fixed points in the cartesian plane. If O is the origin and t is a variable scalar, let R be the point such that $\overrightarrow{OR} = \overrightarrow{OP} + t\overrightarrow{PQ}$. Describe the motion of the point R as t varies. Sketch a diagram.

26 If $\bar{A}$ and $\bar{B}$ are two fixed linearly independent vectors in the plane, give a geometric construction to show how to find scalars s and t so that $s\bar{A} + t\bar{B} = \bar{C}$, where $\bar{C}$ is an arbitrary given vector in the plane.

27 If $\bar{A} = u\bar{B}$, where u is a scalar, show that there are scalars s and t, *not both zero*, such that $s\bar{A} + t\bar{B} = \bar{0}$.

28 Show that two vectors $\bar{A}$ and $\bar{B}$ are linearly independent if and only if the only way to obtain $\bar{0}$ as a linear combination of $\bar{A}$ and $\bar{B}$ is to make both coefficients zero.

3   Dot Products, Lengths, and Angles

In Sections 1 and 2 we have seen that vectors can be added, subtracted, and multiplied by scalars. Vectors can also be multiplied by each other; in fact, there are several different products defined for vectors, each with its own special notation. One of the most useful is the "*dot product*" of two vectors $\bar{A}$ and $\bar{B}$, which is so called because of the notation "$\bar{A} \cdot \bar{B}$" traditionally used to denote it. In this section we define and study the dot product (also called the *inner product* or the *scalar product*) and calculate lengths of vectors, angles between vectors, and projections of vectors upon other vectors using dot products.

We begin with a geometric definition of the dot product.

DEFINITION 1   Dot Product of Vectors

Figure 1

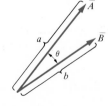

Let $\bar{A}$ and $\bar{B}$ be two vectors with lengths (magnitudes) a and b, respectively, and let θ be the angle between $\bar{A}$ and $\bar{B}$ (Figure 1). Then the *dot product* of $\bar{A}$ and $\bar{B}$ is defined by

$$\bar{A} \cdot \bar{B} = ab \cos \theta.$$

(If either $\bar{A}$ or $\bar{B}$ is the zero vector, then the angle θ is indeterminate; but, in this case $a = 0$ or $b = 0$, so that $\bar{A} \cdot \bar{B} = 0$.)

Notice that the dot product of two vectors is a *scalar*—not a vector. In order to measure the angle θ between two vectors $\bar{A}$ and $\bar{B}$, we can always move $\bar{A}$ and $\bar{B}$ so that they have a common tail end (as in Figure 1) and then use a protractor in the usual way. We express the angle either in degrees or in radians.

EXAMPLE If the length of $\bar{A}$ is 5 units and the length of $\bar{B}$ is 4 units, find $\bar{A} \cdot \bar{B}$ given that the angle θ between $\bar{A}$ and $\bar{B}$ is (a) $\theta = 0$, (b) $\theta = \dfrac{\pi}{2}$, and (c) $\theta = \dfrac{5\pi}{6}$.

SOLUTION
(a) $\bar{A} \cdot \bar{B} = (5)(4) \cos 0 = (5)(4)(1) = 20.$
(b) $\bar{A} \cdot \bar{B} = (5)(4) \cos (\pi/2) = (5)(4)(0) = 0.$
(c) $\bar{A} \cdot \bar{B} = (5)(4) \cos (5\pi/6) = (5)(4)(-\sqrt{3}/2) = -10\sqrt{3}.$

Now, suppose that $\bar{A}$ is a nonzero vector of length a. Since the angle between $\bar{A}$ and itself is zero, $\bar{A} \cdot \bar{A} = aa \cos 0 = a^2$; hence, $a = \sqrt{\bar{A} \cdot \bar{A}}$. On the other hand, if $\bar{A} = \bar{0}$, then $\bar{A} \cdot \bar{A} = 0$ and $\sqrt{\bar{A} \cdot \bar{A}}$ still gives the length of $\bar{A}$. We use the symbol $|\bar{A}|$ for the *length* or *magnitude* of a vector $\bar{A}$; hence, we have the formulas

$$\bar{A} \cdot \bar{A} = |\bar{A}|^2 \quad \text{and} \quad |\bar{A}| = \sqrt{\bar{A} \cdot \bar{A}}.$$

Using the notation $|\bar{A}|$ and $|\bar{B}|$ for the lengths of the vectors $\bar{A}$ and $\bar{B}$, we can now rewrite the formula for the dot product as

$$\bar{A} \cdot \bar{B} = |\bar{A}||\bar{B}| \cos \theta, \qquad \text{where } \theta \text{ is the angle between } \bar{A} \text{ and } \bar{B}.$$

EXAMPLE Let i and j be the standard basis vectors in the xy plane. Find $i \cdot i, j \cdot j$, and $i \cdot j$.

SOLUTION
Since $|i| = 1$, it follows that $i \cdot i = |i|^2 = 1^2 = 1$. Similarly, $j \cdot j = 1$. The angle between i and j is $\pi/2$ radians; hence, $i \cdot j = |i| \, |j| \cos (\pi/2) = (1)(1)(0) = 0.$

3.1 Properties of the Dot Product

Let $\bar{A}$ and $\bar{D}$ be nonzero vectors making an acute angle θ as in Figure 2. Then, perpendiculars dropped from the endpoints of $\bar{A}$ to the straight line through $\bar{D}$ cut off a segment $\overline{ST}$ of length

Figure 2

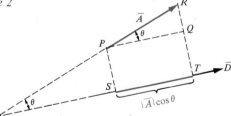

$$|\overline{ST}| = |\overline{PQ}| = |\overline{PR}| \cos \theta = |\bar{A}| \cos \theta.$$

The number $|\bar{A}| \cos \theta$ is called the *scalar component of $\bar{A}$ in the direction of $\bar{D}$* or the *scalar projection of $\bar{A}$ in the direction of $\bar{D}$*. If $\bar{A}$ and $\bar{D}$ make an obtuse angle θ, then $\pi/2 < \theta < \pi$, $\cos \theta < 0$, and so the scalar component of $\bar{A}$ in the direction of $\bar{D}$ is negative.

Since $\bar{A} \cdot \bar{D} = |\bar{A}||\bar{D}| \cos \theta = (|\bar{A}| \cos \theta)|\bar{D}|$, it follows that the dot product of two vectors is the scalar component of the first vector in the direction of the second vector times the length of the second vector. Also note that the scalar component of $\bar{A}$ in the direction of $\bar{D}$ is given by

$$\frac{\bar{A} \cdot \bar{D}}{|\bar{D}|}.$$

(Why?) A glance at Figure 3 shows that the sum of the scalar components of $\bar{A}$ and $\bar{B}$ in the direction of $\bar{D}$ is equal to the scalar component of $\bar{A} + \bar{B}$ in the direction of $\bar{D}$, so that

Figure 3

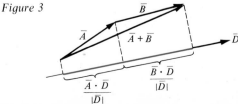

$$\frac{\bar{A} \cdot \bar{D}}{|\bar{D}|} + \frac{\bar{B} \cdot \bar{D}}{|\bar{D}|} = \frac{(\bar{A} + \bar{B}) \cdot \bar{D}}{|\bar{D}|};$$

that is, scalar components are *additive*. (The reader

should check appropriate diagrams to see that scalar components are additive in all cases, even when some of the angles involved are obtuse.) Multiplying the latter equation on both sides by $|\bar{D}|$, we obtain the important *distributive law* for the dot product,

$$(\bar{A} + \bar{B}) \cdot \bar{D} = \bar{A} \cdot \bar{D} + \bar{B} \cdot \bar{D}.$$

From the rule for the product of a vector by a scalar, we have $|s\bar{A}| = |s||\bar{A}|$ for all scalars s and all vectors $\bar{A}$. If θ is the angle between $\bar{A}$ and $\bar{B}$ and if s is a positive scalar, then θ is still the angle between $s\bar{A}$ and $\bar{B}$; hence,

$$(s\bar{A}) \cdot \bar{B} = |s\bar{A}||\bar{B}| \cos \theta = |s||\bar{A}||\bar{B}| \cos \theta = |s|(\bar{A} \cdot \bar{B}) = s(\bar{A} \cdot \bar{B}).$$

On the other hand, if s is a negative scalar, then $s\bar{A}$ makes an angle of $\pi - \theta$ with $\bar{B}$ (Figure 4) and we have

Figure 4

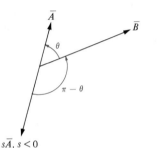

$s\bar{A}, s < 0$

$$(s\bar{A}) \cdot \bar{B} = |s\bar{A}||\bar{B}| \cos (\pi - \theta)$$
$$= |s||\bar{A}||\bar{B}|(-\cos \theta)$$
$$= (-s)|\bar{A}||\bar{B}|(-\cos \theta)$$
$$= s|\bar{A}||\bar{B}| \cos \theta$$
$$= s(\bar{A} \cdot \bar{B}).$$

Consequently, $(s\bar{A}) \cdot \bar{B} = s(\bar{A} \cdot \bar{B})$ holds for $s \neq 0$. Since this equation also holds when $s = 0$, we have the *homogeneous law* for the dot product,

$$(s\bar{A}) \cdot \bar{B} = s(\bar{A} \cdot \bar{B}).$$

Because of this law, we may simply write $s\bar{A} \cdot \bar{B}$, without parentheses.

We can now present the four basic laws governing the behavior of the dot product.

1 $\bar{A} \cdot \bar{B} = \bar{B} \cdot \bar{A}$ (Commutative Law)
2 $(\bar{A} + \bar{B}) \cdot \bar{D} = \bar{A} \cdot \bar{D} + \bar{B} \cdot \bar{D}$ (Distributive Law)
3 $(s\bar{A}) \cdot \bar{B} = s(\bar{A} \cdot \bar{B})$ (Homogeneous Law)
4 $\bar{A} \cdot \bar{A} \geq 0$ and $\bar{A} \cdot \bar{A} = 0$ only if $\bar{A} = \bar{0}$ (Positive Definite Law).

The commutative law follows from the fact that $|\bar{A}||\bar{B}| \cos \theta = |\bar{B}||\bar{A}| \cos \theta$, and the positive definite law is a consequence of the fact that $\bar{A} \cdot \bar{A} = |\bar{A}|^2$.

All the properties of the dot product and of lengths of vectors can be derived deductively from the four basic laws together with the identities developed in Sections 1 and 2. Some of these properties are as follows:

5 $\bar{D} \cdot (\bar{A} + \bar{B}) = \bar{D} \cdot \bar{A} + \bar{D} \cdot \bar{B}$
6 $(\bar{A} - \bar{B}) \cdot \bar{D} = \bar{A} \cdot \bar{D} - \bar{B} \cdot \bar{D}$
7 $\bar{D} \cdot (\bar{A} - \bar{B}) = \bar{D} \cdot \bar{A} - \bar{D} \cdot \bar{B}$
8 $(s\bar{A}) \cdot (t\bar{B}) = st(\bar{A} \cdot \bar{B})$
9 $(-\bar{A}) \cdot \bar{B} = -(\bar{A} \cdot \bar{B}) = \bar{A} \cdot (-\bar{B})$
10 $|\bar{A} \cdot \bar{B}| \leq |\bar{A}||\bar{B}|$ (Schwarz Inequality)
11 $|\bar{A} + \bar{B}| \leq |\bar{A}| + |\bar{B}|$ (Triangle Inequality)
12 $|\bar{A}| = |-\bar{A}|$
13 $|s\bar{A}| = |s||\bar{A}|$

EXAMPLES 1 Derive the identity $\bar{D} \cdot (\bar{A} - \bar{B}) = \bar{D} \cdot \bar{A} - \bar{D} \cdot \bar{B}$ using only the four basic laws for the dot product and vector algebra.

SOLUTION

$$
\begin{aligned}
\bar{D} \cdot (\bar{A} - \bar{B}) &= (\bar{A} - \bar{B}) \cdot \bar{D} && \text{(commutative law)} \\
&= [\bar{A} + (-\bar{B})] \cdot \bar{D} && \text{(definition of subtraction)} \\
&= \bar{A} \cdot \bar{D} + (-\bar{B}) \cdot \bar{D} && \text{(distributive law)} \\
&= \bar{A} \cdot \bar{D} + [(-1)\bar{B}] \cdot \bar{D} && \text{(vector algebra)} \\
&= \bar{A} \cdot \bar{D} + (-1)(\bar{B} \cdot \bar{D}) && \text{(homogeneous law)} \\
&= \bar{D} \cdot \bar{A} + (-1)(\bar{D} \cdot \bar{B}) && \text{(commutative law)} \\
&= \bar{D} \cdot \bar{A} - \bar{D} \cdot \bar{B} && \text{(scalar arithmetic).}
\end{aligned}
$$

2 Give a geometric interpretation of the triangle inequality, $|\bar{A} + \bar{B}| \le |\bar{A}| + |\bar{B}|$.

SOLUTION

Using the head-to-tail rule for adding vectors, we see that $\bar{A}$, $\bar{B}$, and $\bar{A} + \bar{B}$ form three sides of a triangle (Figure 5). Thus, the triangle rule says that the length $|\bar{A} + \bar{B}|$ of one side of a triangle cannot exceed the sum of the lengths $|\bar{A}| + |\bar{B}|$ of the other two sides.

Figure 5

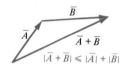

$|\bar{A} + \bar{B}| \le |\bar{A}| + |\bar{B}|$

3.2 Dot Products of Vectors in Component Form

The following theorem gives a useful formula for calculating the dot product of two vectors given in component form.

THEOREM 1 **Dot Products of Vectors in Component Form**

If $\bar{A} = a\dot{i} + b\dot{j}$ and $\bar{B} = c\dot{i} + d\dot{j}$, then $\bar{A} \cdot \bar{B} = ac + bd$.

PROOF
Using the distributive and homogeneous laws, we have

$$\bar{A} \cdot \bar{B} = (a\dot{i} + b\dot{j}) \cdot \bar{B} = (a\dot{i}) \cdot \bar{B} + (b\dot{j}) \cdot \bar{B} = a(\dot{i} \cdot \bar{B}) + b(\dot{j} \cdot \bar{B}).$$

In order to calculate $\dot{i} \cdot \bar{B}$, we observe that $\dot{i} \cdot \dot{i} = 1$ and $\dot{i} \cdot \dot{j} = 0$ (why?); hence,

$$\dot{i} \cdot \bar{B} = \dot{i} \cdot (c\dot{i} + d\dot{j}) = \dot{i} \cdot (c\dot{i}) + \dot{i} \cdot (d\dot{j}) = c(\dot{i} \cdot \dot{i}) + d(\dot{i} \cdot \dot{j}) = c.$$

Similarly,

$$\dot{j} \cdot \bar{B} = \dot{j} \cdot (c\dot{i} + d\dot{j}) = \dot{j} \cdot (c\dot{i}) + \dot{j} \cdot (d\dot{j}) = c(\dot{j} \cdot \dot{i}) + d(\dot{j} \cdot \dot{j}) = d,$$

since $\dot{j} \cdot \dot{i} = 0$ and $\dot{j} \cdot \dot{j} = 1$. (Why?) It follows that

$$\bar{A} \cdot \bar{B} = a(\dot{i} \cdot \bar{B}) + b(\dot{j} \cdot B) = ac + bd.$$

In words, Theorem 1 says that the dot product of two vectors is the sum of the products of their corresponding scalar components. Using the ordered-pair notation, we have

$$\langle a, b \rangle \cdot \langle c, d \rangle = ac + bd.$$

EXAMPLE Let $\bar{A} = 2\dot{i} + 3\dot{j}$ and $B = 4\dot{i} - 5\dot{j}$. Calculate (a) $\bar{A} \cdot \bar{B}$ and (b) $\bar{A} \cdot (2\bar{A} - 3\bar{B})$.

SOLUTION

Using Theorem 1, we have

(a) $\bar{A} \cdot \bar{B} = (2)(4) + (3)(-5) = -7.$

(b) $\bar{A} \cdot (2\bar{A} - 3\bar{B}) = (2\bar{i} + 3\bar{j}) \cdot (-8\bar{i} + 21\bar{j}) = (2)(-8) + (3)(21) = 47.$

An important consequence of Theorem 1 is the formula

$$|x\bar{i} + y\bar{j}| = \sqrt{x^2 + y^2},$$

which says that the length of a vector is the square root of the sum of the squares of its scalar components. Indeed, if $\bar{A} = x\bar{i} + y\bar{j}$, then by Theorem 1, $\bar{A} \cdot \bar{A} = x^2 + y^2$, and so $|\bar{A}| = \sqrt{\bar{A} \cdot \bar{A}} = \sqrt{x^2 + y^2}$. Notice that, if $\bar{A} \neq \bar{0}$, then

$$\left| \frac{\bar{A}}{|\bar{A}|} \right| = \left| \frac{1}{|\bar{A}|} \bar{A} \right| = \left| \frac{1}{|\bar{A}|} \right| |\bar{A}| = \frac{1}{|\bar{A}|} |\bar{A}| = 1;$$

hence, $\dfrac{\bar{A}}{|\bar{A}|}$ is a vector of unit length in the same direction as $\bar{A}$. A vector of unit length is called a *unit vector* and the procedure of dividing a nonzero vector by its own length to obtain a unit vector in the same direction is called *normalizing*.

EXAMPLES 1 If $\bar{A} = 3\bar{i} + 4\bar{j}$, find (a) $|\bar{A}|$ and (b) a unit vector in the same direction as $\bar{A}$.

SOLUTION

(a) $|\bar{A}| = \sqrt{3^2 + 4^2} = \sqrt{25} = 5$ units.

(b) Normalizing $\bar{A}$, we obtain $\dfrac{\bar{A}}{|\bar{A}|} = \dfrac{3\bar{i} + 4\bar{j}}{5} = \dfrac{3}{5}\bar{i} + \dfrac{4}{5}\bar{j}.$

2 If $\bar{A} = 3\bar{i} - 5\bar{j}$ and $\bar{D} = 4\bar{i} + 3\bar{j}$, find the scalar component of $\bar{A}$ in the direction of $\bar{D}$

SOLUTION

$$\frac{\bar{A} \cdot \bar{D}}{|\bar{D}|} = \frac{(3)(4) + (-5)(3)}{\sqrt{4^2 + 3^2}} = -\frac{3}{5}.$$

3.3 Applications of the Dot Product

If $\bar{A}$ and $\bar{B}$ are nonzero vectors, then the formula $\bar{A} \cdot \bar{B} = |\bar{A}||\bar{B}| \cos \theta$ can be rewritten in the form

$$\cos \theta = \frac{\bar{A} \cdot \bar{B}}{|\bar{A}||\bar{B}|}$$

and used to find the cosine of the angle θ between $\bar{A}$ and $\bar{B}$. Notice that $\theta = \pi/2$ if and only if $\bar{A} \cdot \bar{B} = 0$; that is, $\bar{A}$ and $\bar{B}$ are perpendicular if and only if $\bar{A} \cdot \bar{B} = 0$. (Some authors use the word "*orthogonal*" rather than the word "perpendicular.") Since the zero vector has an indeterminate direction, it is convenient to say, by definition, that $\bar{0}$ is perpendicular to every vector, even to itself.

EXAMPLES 1 If $|\bar{A}| = 5$, $|\bar{B}| = \sqrt{2}$, and $\bar{A} \cdot \bar{B} = 1$, find the angle θ between $\bar{A}$ and $\bar{B}$.

SOLUTION

$$\cos \theta = \frac{\bar{A} \cdot \bar{B}}{|\bar{A}||\bar{B}|} = \frac{1}{5\sqrt{2}}; \text{ hence, } \theta = \cos^{-1} \frac{1}{5\sqrt{2}} \approx 81.87°.$$

2 Find the angle θ between $\bar{A} = \langle 2,3 \rangle$ and $\bar{B} = \langle 3, -1 \rangle$.

SOLUTION

$$\cos\theta = \frac{\bar{A} \cdot \bar{B}}{|\bar{A}|\,|\bar{B}|} = \frac{(2)(3) + (3)(-1)}{\sqrt{2^2 + 3^2}\sqrt{3^2 + (-1)^2}} = \frac{3}{\sqrt{13}\sqrt{10}} = \frac{3}{\sqrt{130}};$$

hence, $\theta = \cos^{-1}(3/\sqrt{130}) \approx 74.74°$.

3 Find a value of the scalar t so that $\bar{A} = -6\bar{i} + 3\bar{j}$ and $\bar{B} = 4\bar{i} + t\bar{j}$ are perpendicular vectors.

SOLUTION

$$\bar{A} \cdot \bar{B} = (-6)(4) + 3t = 3t - 24.$$

Since $\bar{A}$ and $\bar{B}$ are perpendicular if and only if $\bar{A} \cdot \bar{B} = 0$, we require that $3t - 24 = 0$; that is, $t = 8$.

All the propositions of Euclidean geometry can be proved algebraically using vectors. The following examples indicate how this is done.

EXAMPLES 1 Use vectors to demonstrate the Pythagorean theorem.

SOLUTION

Consider the right triangle made up of the perpendicular vectors $\bar{A}$ and $\bar{B}$ with hypotenuse $\bar{A} + \bar{B}$ (Figure 6). We have

Figure 6

$$\begin{aligned} |\bar{A} + \bar{B}|^2 &= (\bar{A} + \bar{B}) \cdot (\bar{A} + \bar{B}) = \bar{A} \cdot (\bar{A} + \bar{B}) + \bar{B} \cdot (\bar{A} + \bar{B}) \\ &= \bar{A} \cdot \bar{A} + \bar{A} \cdot \bar{B} + \bar{B} \cdot \bar{A} + \bar{B} \cdot \bar{B} = \bar{A} \cdot \bar{A} + 2\bar{A} \cdot \bar{B} + \bar{B} \cdot \bar{B} \\ &= |\bar{A}|^2 + 2A \cdot B + |\bar{B}|^2. \end{aligned}$$

Since $\bar{A}$ and $\bar{B}$ are perpendicular, $\bar{A} \cdot \bar{B} = 0$; hence, $|\bar{A} + \bar{B}|^2 = |\bar{A}|^2 + |\bar{B}|^2$, which is the Pythagorean theorem.

Figure 7

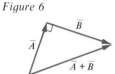

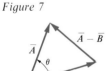

2 Use vectors to demonstrate the law of cosines.

SOLUTION

Consider the triangle in Figure 7. We have

$$\begin{aligned} |\bar{A} - \bar{B}|^2 &= (\bar{A} - \bar{B}) \cdot (\bar{A} - \bar{B}) = \bar{A} \cdot (\bar{A} - \bar{B}) - \bar{B} \cdot (\bar{A} - \bar{B}) \\ &= \bar{A} \cdot \bar{A} - \bar{A} \cdot \bar{B} - \bar{B} \cdot \bar{A} + \bar{B} \cdot \bar{B} = |\bar{A}|^2 - 2\bar{A} \cdot \bar{B} + |\bar{B}|^2 \\ &= |\bar{A}|^2 + |\bar{B}|^2 - 2|\bar{A}|\,|\bar{B}| \cos\theta, \end{aligned}$$

which is the law of cosines.

Figure 8

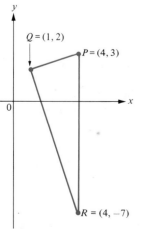

3 Show that the points $P = (4,3)$, $Q = (1,2)$, and $R = (4,-7)$ form the vertices of a right triangle.

SOLUTION

Figure 8 leads us to suspect that the angle PQR is a right angle. We propose to confirm this by showing that $(\overrightarrow{QP}) \cdot (\overrightarrow{QR}) = 0$. Here,

$$\begin{aligned} \overrightarrow{QP} &= (4-1)\bar{i} + (3-2)\bar{j} = 3\bar{i} + \bar{j}, \\ \overrightarrow{QR} &= (4-1)\bar{i} + (-7-2)\bar{j} = 3\bar{i} - 9\bar{j}; \end{aligned}$$

hence,

$$\begin{aligned} (\overrightarrow{QP}) \cdot (\overrightarrow{QR}) &= (3\bar{i} + \bar{j}) \cdot (3\bar{i} - 9\bar{j}) \\ &= (3)(3) + (1)(-9) = 0. \end{aligned}$$

Figure 9

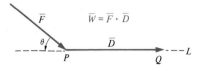

Dot products are useful in mechanics for calculating the work done by a force acting on a particle in a direction other than the direction of displacement. Indeed, suppose that a constant force, represented by the vector $\bar{F}$, acts on a particle, starting at the point P, and moves it along a straight line to the point Q. Then the work W done by $\bar{F}$ in producing the displacement $\bar{D} = \overrightarrow{PQ}$ is defined to be the product of the scalar component of $\bar{F}$ in the direction of $\bar{D}$ and the magnitude $|\bar{D}|$ of the displacement (Figure 9). Thus,

$$W = \frac{\bar{F} \cdot \bar{D}}{|\bar{D}|} |\bar{D}| = \bar{F} \cdot \bar{D}.$$

Notice that, when $\bar{F}$ and $\bar{D}$ have the same direction, $W = \bar{F} \cdot \bar{D} = |\bar{F}||\bar{D}|$ as usual. (Why?)

EXAMPLES 1 A person pushes on a lawnmower handle with a force of 35 pounds and moves the mower through a distance of 100 feet. How much work is done if the handle makes an angle of 30° with the ground (Figure 10)?

Figure 10

SOLUTION
For the force vector $\bar{F}$ we have $|\bar{F}| = 35$ pounds, while for the displacement vector $\bar{D}$ we have $|\bar{D}| = 100$ feet. Since the angle θ between $\bar{F}$ and $\bar{D}$ is $\theta = 30°$,

$$W = \bar{F} \cdot \bar{D} = |\bar{F}||\bar{D}| \cos \theta = (35)(100)\left(\frac{\sqrt{3}}{2}\right) = 1750\sqrt{3} \text{ foot-pounds.}$$

2 Find the work done by a force $\bar{F}$ in the xy plane whose magnitude is 10 units of force and which makes an angle of 60° with the positive x axis if $\bar{F}$ moves an object along a straight line from the origin O to the point $Q = (6, 4)$ (Figure 11).

SOLUTION
From Figure 11,

$$x = |\bar{F}| \cos 60° = (10)(\tfrac{1}{2}) = 5,$$

$$y = |\bar{F}| \sin 60° = (10)\left(\frac{\sqrt{3}}{2}\right) = 5\sqrt{3};$$

hence,

$$\bar{F} = x\bar{i} + y\bar{j} = 5\bar{i} + 5\sqrt{3}\,\bar{j}.$$

Figure 11

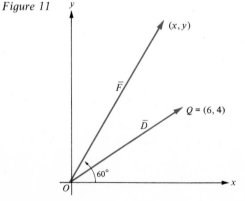

The displacement vector is $\bar{D} = 6\bar{i} + 4\bar{j}$; hence, the work W is

$$W = \bar{F} \cdot \bar{D} = (5\bar{i} + 5\sqrt{3}\,\bar{j}) \cdot (6\bar{i} + 4\bar{j})$$

$$= (5)(6) + (5\sqrt{3})(4)$$

$$= 30 + 20\sqrt{3} \text{ units of work.}$$

Problem Set 3

In problems 1 through 10, a and b denote the lengths (magnitudes) of vectors $\bar{A}$ and $\bar{B}$, respectively, while θ denotes the angle between $\bar{A}$ and $\bar{B}$.

1 If $a = 15$, $b = 2$, and $\theta = 60°$, find $\bar{A} \cdot \bar{B}$.

2 If $a = 7$, $b = 4$, and $\theta = 5\pi/6$, find $\bar{A} \cdot \bar{B}$.

3 If $a = 1$, $b = 3$, and $\theta = \pi/2$, find $\bar{A} \cdot \bar{B}$.

4 If $a = 1$, $b = 2$, and $\bar{A} \cdot \bar{B} = \sqrt{3}$, find θ.

5 If $a \neq 0$, $b \neq 0$, and $\bar{A} \cdot \bar{B} = 0$, find θ.

6 If $a = 2$, $b = 3$, and $\bar{A} \cdot \bar{B} = -6$, find θ.

7 If $\bar{A} \cdot \bar{A} = 25$, find a.

8 If $a = 10$, $\theta = \dfrac{\pi}{3}$, and $\bar{A} \cdot \bar{B} = 30$, find b.

9 If $a = 3$, $b = 4$, and $\theta = \pi$, find $\bar{A} \cdot \bar{B}$.

10 If $a = 2$ and $b = 3$, explain why $|\bar{A} \cdot \bar{B}| \leq 6$.

In problems 11 through 16, use the information to find the scalar component (scalar projection) of $\bar{A}$ in the direction of $\bar{D}$. Denote the angle between $\bar{A}$ and $\bar{D}$ by θ.

11 $|\bar{A}| = 10$, $\theta = 45°$.

12 $\bar{A} \cdot \bar{D} = 5$, $|\bar{D}| = 2$.

13 $|\bar{A}| = 3$, $\theta = \pi/3$.

14 $\bar{A} \cdot \bar{D} = -7$, $|\bar{D}| = \frac{1}{2}$.

15 $|\bar{A}| = 6$, $\theta = \pi/2$.

16 $|\bar{A}| = \frac{1}{2}$, $\theta = \pi$.

In problems 17 through 22, i, j represents the standard basis in the xy plane. Find (a) $\bar{A} \cdot \bar{B}$, (b) $|\bar{A}|$, (c) $|\bar{B}|$, (d) $|\bar{A} - \bar{B}|$, and (e) $\cos \theta$, where θ is the angle between $\bar{A}$ and $\bar{B}$. (f) Find a unit vector having the same direction as $\bar{A}$, as $\bar{B}$, and as $\bar{A} - \bar{B}$. (g) Find the scalar component (scalar projection) of $\bar{A}$ in the direction of $\bar{B}$.

17 $\bar{A} = i + j$ and $\bar{B} = i - j$

18 $\bar{A} = 3i + 2j$ and $\bar{B} = 4i - 5j$

19 $\bar{A} = 4i + j$ and $\bar{B} = i - 2j$

20 $\bar{A} = -\dfrac{i}{2} + \dfrac{j}{3}$, $\bar{B} = \dfrac{4i}{3} + \dfrac{7j}{2}$

21 $\bar{A} = 2i + 4j$, $\bar{B} = -3j$

22 $\bar{A} = \dfrac{i}{3}$, $\bar{B} = \dfrac{j}{5}$

In problems 23 through 28, let $\bar{A} = \langle -1, 3 \rangle$, $\bar{B} = \langle 5, 3 \rangle$, and $\bar{C} = \langle -2, -3 \rangle$. Find the value of each expression.

23 $\bar{A} \cdot (\bar{B} + \bar{C})$

24 $(2\bar{A}) \cdot (3\bar{B})$

25 $(-\bar{A}) \cdot (4\bar{B} - 2\bar{C})$

26 $\dfrac{\bar{A}}{|\bar{A}|} \cdot \dfrac{\bar{B}}{|\bar{B}|}$

27 $(\bar{A} \cdot \bar{C})\bar{B} - (\bar{A} \cdot \bar{B})\bar{C}$

28 $(i + j) \cdot (\bar{A} - \bar{B} + 2\bar{C})$

29 Let $\bar{A} = ti - 3j$ and $\bar{B} = 5i + 7j$, where t is a scalar. Find t so that $\bar{A}$ and $\bar{B}$ are perpendicular.

30 Let $\bar{D}$ be a nonzero vector.

   (a) The *vector projection* of $\bar{A}$ onto $\bar{D}$ is defined to be the vector $\bar{P} = \dfrac{\bar{A} \cdot \bar{D}}{\bar{D} \cdot \bar{D}} \bar{D}$. Find $|\bar{P}|$ and describe the direction of $\bar{P}$ in terms of the direction of $\bar{D}$.

   (b) Let $\bar{A} = 3i + 5j$ and $\bar{D} = -4i + 3j$. Find the vector projection $\bar{P}$ of $\bar{A}$ onto $\bar{D}$, and find $|\bar{P}|$.

31 Explain the geometric meaning of the following conditions:

   (a) $\bar{A} \cdot \bar{B} > 0$ (b) $\bar{A} \cdot \bar{B} = 0$ (c) $\bar{A} \cdot \bar{B} < 0$

32 If $\bar{A}$ is a vector in the cartesian plane and if i, j is the standard basis, prove that $\bar{A} = (\bar{A} \cdot i)i + (\bar{A} \cdot j)j$.

33 Use the properties of dot products to confirm the following identities:

   (a) $(\bar{A} + \bar{B}) \cdot (\bar{A} - \bar{B}) = |\bar{A}|^2 - |\bar{B}|^2$

   (b) $|\bar{A} + \bar{B}|^2 = |\bar{A}|^2 + 2\bar{A} \cdot \bar{B} + |\bar{B}|^2$

   (c) $|s\bar{A} + t\bar{B}|^2 = s^2|\bar{A}|^2 + 2st(\bar{A} \cdot \bar{B}) + t^2|\bar{B}|^2$

   (d) $|\bar{A} + \bar{B}|^2 + |\bar{A} - \bar{B}|^2 = 2|\bar{A}|^2 + 2|\bar{B}|^2$

   (e) $\bar{A} \cdot \bar{B} = \frac{1}{2}(|\bar{A} + \bar{B}|^2 - |\bar{A}|^2 - |\bar{B}|^2)$

34 Prove the converse of the Pythagorean theorem; that is, prove that, if the equation $|\bar{A}|^2 + |\bar{B}|^2 = |\bar{A} + \bar{B}|^2$ holds, then $\bar{A}$ is perpendicular to $\bar{B}$.

35 Suppose that $\bar{A}$ and $\bar{B}$ are perpendicular vectors. Write each of the following expressions in terms of $|\bar{A}|$ and $|\bar{B}|$.

(a) $|\bar{A} + \bar{B}|$ (b) $|2\bar{A} - 3\bar{B}|$ (c) $|\bar{A} - \bar{B}|$

36 Find a scalar t so that the vectors $\bar{A} = i + j$ and $\bar{B} = ti - j$ make an angle of $3\pi/4$ radians.

37 Use the dot product to show that the triangle with vertices $A = (2, 1)$, $B = (6, 3)$, and $C = (4, 7)$ is a right triangle.

38 Use the dot product to prove that the diagonals of a rhombus are perpendicular to each other.

39 Use the dot product to prove that the four points $P = (1, 2)$, $Q = (2, 3)$, $R = (1, 4)$, and $S = (0, 3)$ are vertices of a square.

40 Show, using vectors, that the medians of a triangle meet at a point two thirds of the distance from any vertex of the triangle to the midpoint of the opposite side.

41 A block weighing 15 pounds slides 6 feet down an incline making a $60°$ angle with the horizontal. Find the work done by the force of gravity.

42 How much work is done by the force of gravity in moving a particle all the way around a vertical triangle with vertices P, Q, and R, starting and ending at vertex P? Assume that the mass of the particle is m so that its weight is mg, g being the acceleration of gravity.

43 How much work is done by the constant force vector $\bar{F}$ in moving a particle first along the line segment from point P to point Q, then along the line segment from point Q to point R? How much work is done by $\bar{F}$ in moving the particle directly from P to R?

44 Let $\bar{A}$ and $\bar{B}$ be unit vectors that have initial points at the origin and that make angles α and β, respectively, with the positive x axis.

(a) Write $\bar{A}$ and $\bar{B}$ in component form.

(b) Compute $\bar{A} \cdot \bar{B}$ to derive a formula for $\cos(\alpha - \beta)$.

4 Equations in Vector Form

In this section we use the vector algebra developed in Sections 1 through 3 to write the equations of curves in *vector form*. This is accomplished by using the idea of the "position vector" of a point.

We begin by choosing and fixing a point O in the plane called the *origin*. If we already have a cartesian or polar coordinate system, we naturally take O to be the origin or the pole of the coordinate system. If P is any point, then the vector $\bar{R} = \overrightarrow{OP}$ is called the *position vector* of P (Figure 1). Notice that $|\bar{R}| = |\overrightarrow{OP}|$ gives the distance between the point P and the origin O.

Evidently, the point P determines the position vector $\bar{R} = \overrightarrow{OP}$ uniquely (since O is fixed beforehand). Conversely, the position vector $\bar{R}$ determines the point P uniquely; in fact, if the tail end of $\bar{R}$ is placed at the origin, then the vector $\bar{R}$ will point to P. For this reason, we ordinarily place all position vectors with their tail ends at the origin.

Suppose that $\bar{R}_1 = \overrightarrow{OP_1}$ and $\bar{R}_2 = \overrightarrow{OP_2}$ are the position vectors of the points P_1

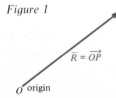

Figure 1

$\bar{R} = \overrightarrow{OP}$

O origin

Figure 2

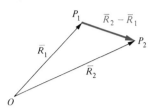

and P_2, respectively. Then $\bar{R}_2 - \bar{R}_1$ is the vector $\overrightarrow{P_1P_2}$ from P_1 to P_2 (Figure 2). We usually regard the difference $\bar{R}_2 - \bar{R}_1$ of two position vectors as representing a displacement from P_1 to P_2. Notice that $|\bar{R}_2 - \bar{R}_1|$, the magnitude of this displacement, is just the distance between the points P_1 and P_2.

If we have established a cartesian or polar coordinate system in the plane, then the point P with cartesian coordinates $P = (x, y)$ has the position vector

$$\bar{R} = \overrightarrow{OP} = x\bar{i} + y\bar{j} \qquad \text{(Figure 3a)},$$

while, if $P = (r, \theta)$ in polar coordinates, then

$$\bar{R} = \overrightarrow{OP} = (r \cos \theta)\bar{i} + (r \sin \theta)\bar{j} \qquad \text{(Figure 3b)}.$$

Figure 3

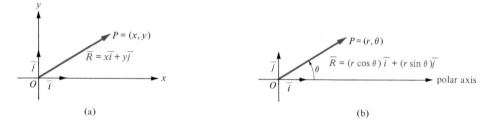

(a)

(b)

EXAMPLES 1 Find the position vectors $\bar{R}_1$ and $\bar{R}_2$ of the cartesian points $P_1 = (7, 13)$ and $P_2 = (-3, 1)$, respectively. Also find the displacement vector $\bar{R}_2 - \bar{R}_1$ and the distance between P_1 and P_2.

SOLUTION
The position vectors $\bar{R}_1$ and $\bar{R}_2$ are

$$\bar{R}_1 = \overrightarrow{OP_1} = 7\bar{i} + 13\bar{j} \quad \text{and} \quad \bar{R}_2 = \overrightarrow{OP_2} = -3\bar{i} + \bar{j}.$$

Hence, the displacement vector $\bar{R}_2 - \bar{R}_1$ is given by

$$\bar{R}_2 - \bar{R}_1 = (-3\bar{i} + \bar{j}) - (7\bar{i} + 13\bar{j}) = -10\bar{i} - 12\bar{j}$$

and the distance between P_1 and P_2 is

$$|\bar{R}_2 - \bar{R}_1| = \sqrt{(-10)^2 + (-12)^2} = \sqrt{244} \approx 15.62 \text{ units}.$$

2 Find the position vectors $\bar{R}_0$ and $\bar{R}_1$ of the polar points $P_0 = (2, \pi/3)$ and $P_1 = (-3, \pi/4)$, respectively.

SOLUTION
The position vectors $\bar{R}_0$ and $\bar{R}_1$ are given by

$$\bar{R}_0 = \overrightarrow{OP_0} = \left(2 \cos \frac{\pi}{3}\right)\bar{i} + \left(2 \sin \frac{\pi}{3}\right)\bar{j} = \bar{i} + \sqrt{3}\bar{j} \quad \text{and}$$

$$\bar{R}_1 = \overrightarrow{OP_1} = \left(-3 \cos \frac{\pi}{4}\right)\bar{i} + \left(-3 \sin \frac{\pi}{4}\right)\bar{j} = -\frac{3}{2}\sqrt{2}\bar{i} - \frac{3}{2}\sqrt{2}\bar{j}.$$

Figure 4

4.1 Graphs of Vector Equations

A continuously moving point P traces out a curve C (Figure 4). As P moves, its position vector $\bar{R} = \overrightarrow{OP}$ changes, generally in both length and direction. If the head end (terminal point) P of a variable position vector $\bar{R}$ traces out a curve C, then, for simplicity, we say that "$\bar{R}$ traces out the curve C."

Figure 5

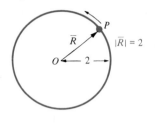

The idea of a position vector $\bar{R}$ varying and thus tracing out a curve leads naturally to the notion of the *vector equation* of the curve. For instance, if $\bar{R}$ is a variable position vector in the plane, then, as $\bar{R}$ varies subject to the condition $|\bar{R}| = 2$, it traces out a circle of radius 2 units with center at the origin (Figure 5). Therefore, we say that $|\bar{R}| = 2$ is the *vector equation* of this circle. Notice that the circle consists of all points P and only those points P whose position vector $\bar{R}$ satisfies the equation $|\bar{R}| = 2$.

More generally, we make the following definition.

DEFINITION 1   **Graph of a Vector Equation**
The *graph* of an equation involving the position vector $\bar{R}$ is the set of all points P, and only those points P, whose position vector $\bar{R}$ satisfies the equation.

EXAMPLES 1 Sketch a graph of the vector equation

$$|\bar{R} - \bar{R}_0| = a,$$

where all vectors lie in the same plane, a is a positive constant, and $\bar{R}_0$ is a constant position vector.

SOLUTION
Let $\bar{R} = \overrightarrow{OP}$ and $\bar{R}_0 = \overrightarrow{OP_0}$, so that $\bar{R} - \bar{R}_0 = \overrightarrow{P_0P}$ by the triangle rule for vector subtraction (Figure 6a). The condition $|\bar{R} - \bar{R}_0| = a$ means that the distance between the points P_0 and P is a units. Thus, as $\bar{R}$ varies subject to the condition $|\bar{R} - \bar{R}_0| = a$, the point P traces out a circle of radius a with center at the point P_0 (Figure 6b).

Figure 6

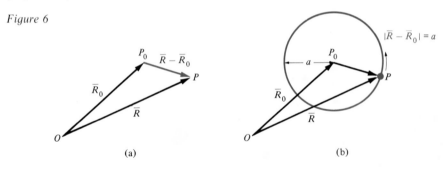

(a)

(b)

2 Find the vector equation for the ellipse whose foci are at the points F_1 and F_2 and whose semimajor axis is $a > 0$ (Figure 7).

SOLUTION
Recall from Section 2 of Chapter 4 that a point P belongs to the given ellipse if and only if

$$|\overline{F_1P}| + |\overline{F_2P}| = 2a.$$

Figure 7

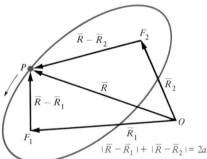

Thus, let $\bar{R}_1 = \overrightarrow{OF_1}$, $\bar{R}_2 = \overrightarrow{OF_2}$, and $\bar{R} = \overrightarrow{OP}$, so that

$$\overrightarrow{F_1P} = \overrightarrow{OP} - \overrightarrow{OF_1} = \bar{R} - \bar{R}_1$$

and

$$\overrightarrow{F_2P} = \overrightarrow{OP} - \overrightarrow{OF_2} = \bar{R} - \bar{R}_2.$$

Then the equation of the ellipse can be rewritten in vector form as

$$|\bar{R} - \bar{R}_1| + |\bar{R} - \bar{R}_2| = 2a.$$

If $\overline{N}$ is a fixed nonzero vector and $\overline{R}_0 = \overrightarrow{OP_0}$, then the condition $\overline{N} \cdot (\overline{R} - \overline{R}_0) = 0$ means that the vectors $\overline{N}$ and $\overline{R} - \overline{R}_0$ are perpendicular (Figure 8). As $\overline{R} = \overrightarrow{OP}$ varies, subject to the condition that $\overline{R} - \overline{R}_0$ is perpendicular to $\overline{N}$, it is geometrically clear that P traces out a straight line through P_0 perpendicular to $\overline{N}$. We say that the vector $\overline{N}$ is *normal* to this line. Therefore,

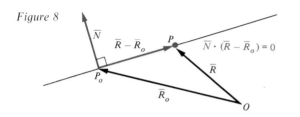

Figure 8

$$\overline{N} \cdot (\overline{R} - \overline{R}_0) = 0$$

is the vector equation of the straight line containing the point whose position vector is $\overline{R}_0$ and having $\overline{N}$ as a normal vector.

The equation $\overline{N} \cdot (\overline{R} - \overline{R}_0) = 0$ can be rewritten as $\overline{N} \cdot \overline{R} - \overline{N} \cdot \overline{R}_0 = 0$, or as $\overline{N} \cdot \overline{R} = \overline{N} \cdot \overline{R}_0$. Since $\overline{N} \cdot \overline{R}_0$ is a constant K, the equation can also be written as $\overline{N} \cdot \overline{R} = K$. We can convert the latter equation into cartesian scalar form simply by writing $\overline{R} = xi + yj$ and $\overline{N} = Ai + Bj$, so that $\overline{N} \cdot \overline{R} = Ax + By$ and the equation takes the form $Ax + By = K$. This equation can also be written as

$$Ax + By + C = 0,$$

where we have put $C = -K$. In particular, notice that $\overline{N} = Ai + Bj$ gives a normal vector to the straight line $Ax + By + C = 0$.

EXAMPLES 1 Find a normal vector to the line $y = 2x - 3$ and write the equation of this line in vector form.

SOLUTION
The given equation can be rewritten in the form $2x + (-1)y - 3 = 0$; hence, $\overline{N} = 2i + (-1)j$ is a normal vector to the given line. Let the position vector $\overline{R}$ be given by $\overline{R} = xi + yj$. Then $\overline{N} \cdot \overline{R} = (2i - j) \cdot (xi + yj) = 2x - y$, and the given equation can be rewritten in vector form as $\overline{N} \cdot \overline{R} = 3$.

2 Convert the vector equation $\overline{N} \cdot (\overline{R} - \overline{R}_0) = 0$ into cartesian scalar form if $\overline{N} = -i + 3j$ and $\overline{R}_0 = 7i - 2j$.

SOLUTION
With $\overline{R} = xi + yj$, we have

$$\overline{R} - \overline{R}_0 = (xi + yj) - (7i - 2j) = (x - 7)i + (y + 2)j;$$

hence,

$$\overline{N} \cdot (\overline{R} - \overline{R}_0) = (-i + 3j) \cdot [(x - 7)i + (y + 2)j] = -(x - 7) + 3(y + 2).$$

The condition $\overline{N} \cdot (\overline{R} - \overline{R}_0) = 0$ is therefore equivalent to

$$-(x - 7) + 3(y + 2) = 0 \quad \text{or} \quad 3y - x + 13 = 0.$$

4.2 Perpendicular Distance from a Point to a Line

We now use the vector equation of a straight line to derive a formula for the perpendicular distance from a point to a line.

THEOREM 1   **Distance from a Point to a Line**

Let P_1 be a point in the plane and let L be a straight line in the plane whose vector equation is $\bar{N} \cdot \bar{R} = K$. Let $\bar{R}_1$ be the position vector of P_1. Then the perpendicular distance d from P_1 to L is given by the equation

$$d = \frac{|\bar{N} \cdot \bar{R}_1 - K|}{|\bar{N}|}.$$

Figure 9

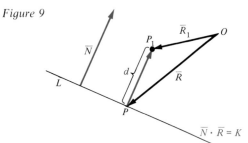

PROOF
Let P be the point at the foot of the perpendicular from P_1 to L (Figure 9). Note that the desired distance is $d = |\overrightarrow{PP_1}| = |\bar{R}_1 - \bar{R}|$. Since $\bar{R}_1 - \bar{R}$ is parallel to $\bar{N}$, there is a scalar t such that $\bar{R}_1 - \bar{R} = t\bar{N}$. Therefore,

$$d = |\bar{R}_1 - \bar{R}| = |t\bar{N}| = |t|\,|\bar{N}|.$$

Taking the dot product with $\bar{N}$ on both sides of the equation $\bar{R}_1 - \bar{R} = t\bar{N}$, and noting that $\bar{N} \cdot \bar{R} = K$, we obtain

$$\bar{N} \cdot \bar{R}_1 - K = \bar{N} \cdot (t\bar{N}) = t\bar{N} \cdot \bar{N} = t|\bar{N}|^2.$$

Solving the latter equation for t, we find that

$$t = \frac{\bar{N} \cdot \bar{R}_1 - K}{|\bar{N}|^2}.$$

Therefore,

$$d = |t|\,|\bar{N}| = \left|\frac{\bar{N} \cdot \bar{R}_1 - K}{|\bar{N}|^2}\right| |\bar{N}| = \frac{|\bar{N} \cdot \bar{R}_1 - K|}{|\bar{N}|}.$$

We now recast Theorem 1 in cartesian form with $P_1 = (x_1, y_1)$, $\bar{N} = A\bar{i} + B\bar{j}$, $\bar{R} = x\bar{i} + y\bar{j}$, $\bar{R}_1 = x_1\bar{i} + y_1\bar{j}$, and $K = -C$. Then $\bar{N} \cdot \bar{R}_1 - K = Ax_1 + By_1 + C$ and $|\bar{N}| = \sqrt{A^2 + B^2}$, so that the perpendicular distance d from the point (x_1, y_1) to the straight line $Ax + By + C = 0$ is

$$d = \frac{|Ax_1 + By_1 + C|}{\sqrt{A^2 + B^2}}.$$

EXAMPLE   Find the perpendicular distance from the point $P_1 = (3, 7)$ to the straight line $y = 2x - 5$.

SOLUTION
We rewrite the equation $y = 2x - 5$ in the form $2x - y - 5 = 0$, that is, in the form $Ax + By + C = 0$ with $A = 2$, $B = -1$, and $C = -5$. Here $P_1 = (x_1, y_1) = (3, 7)$, so the desired distance d is

$$d = \frac{|Ax_1 + By_1 + C|}{\sqrt{A^2 + B^2}} = \frac{|(2)(3) + (-1)(7) + (-5)|}{\sqrt{2^2 + (-1)^2}} = \frac{6}{\sqrt{5}} \approx 2.68 \text{ units.}$$

Problem Set 4

In problems 1 through 6, find the position vector $\bar{R}$ of each point P. Express $\bar{R}$ in terms of $\bar{i}$ and $\bar{j}$.

1 $P = (5, -3)$

2 $P = (0, 7)$

3 $P = O$

4 $P = (5, \pi/3)$ (polar coordinates)

5 $P = (-2, 3\pi/4)$ (polar coordinates)

6 $P = (a, a^2)$

In problems 7 through 10, find the vector $\bar{R}_2 - \bar{R}_1$ if $\bar{R}_1$ and $\bar{R}_2$ are the position vectors of P_1 and P_2, respectively.

7 $P_1 = (2, -1)$, $P_2 = (7, 2)$ **8** $P_1 = (a, a^2)$, $P_2 = (b, b^2)$

9 $P_1 = (1, \pi/4)$, $P_2 = (5, 5\pi/3)$ (polar coordinates)   **10** $P_1 = (x, f(x))$, $P_2 = (a, f(a))$

In problems 11 through 22, sketch a graph of each vector equation in the xy plane.

11 $|\bar{R}| = 3$ **12** $\bar{R} \cdot \bar{R} = 16$ **13** $|\bar{R} - (2i + 3j)| = 4$

14 $|\bar{R} + i - j| = 3$ **15** $|\bar{R} - i| + |\bar{R} + i| = 6$ **16** $\left| |\bar{R} - i| - |\bar{R} + i| \right| = 1$

17 $\bar{R} \cdot j = 0$ **18** $(i + j) \cdot \bar{R} = 0$ **19** $(2i + 3j) \cdot (\bar{R} - i) = 0$

20 $(i - 4j) \cdot \bar{R} = 1$ **21** $(2i + 3j) \cdot \bar{R} = (2i + 3j) \cdot j$ **22** $i \cdot \bar{R} = 2 + j \cdot \bar{R}$

In problems 23 through 28, find the vector equation of the curve described.

23 The circle of radius 9 with center at $(-3, 3)$.

24 The ellipse with semimajor axis 3 and foci at $(0, 1)$ and $(0, -1)$.

25 The straight line through the point $(-1, -5)$ and perpendicular to the vector $\bar{N} = 2i - 7j$.

26 The hyperbola with transverse axis 6 and foci at $(7, 7)$ and $(-2, -2)$.

27 The straight line whose cartesian scalar equation is $12x - 6y = 7$.

28 The parabola whose directrix has the equation $i \cdot \bar{R} = -2$ and whose focus is at the origin.

In problems 29 through 32, find a normal vector $\bar{N}$ to the straight line whose cartesian equation is given.

29 $2x - 17y + 2 = 0$ **30** $y = mx + b$ **31** $\dfrac{x}{2} + \dfrac{y}{3} = 17$ **32** $y - y_0 = m(x - x_0)$

In problems 33 through 37, find the perpendicular distance from the indicated point $P_1 = (x_1, y_1)$ to the straight line whose equation is given.

33 $P_1 = (1, 2)$; $3x - y = 4$ **34** $P_1 = (-7, 3)$; $y = \frac{1}{2}x + 3$ **35** $P_1 = (1, 2)$; $(i + j) \cdot (\bar{R} - 2i + j) = 0$

36 $P_1 = (0, 0)$; $Ax + By + C = 0$ **37** $P_1 = (4, 0)$; $y = 2x - 5$

38 Find a formula for the perpendicular distance from the origin to the line $\bar{N} \cdot \bar{R} = K$.

39 Find a formula for the perpendicular distance from the point with position vector $\bar{R}_1$ to the straight line $\bar{D} \cdot (\bar{R} - \bar{D}) = 0$, where $\bar{D}$ is a constant nonzero vector.

40 Find the vector equation of the parabola with focus at the origin whose directrix is the straight line $\bar{D} \cdot (\bar{R} - \bar{D}) = 0$, where $\bar{D}$ is a fixed nonzero vector.

41 Let $\bar{N} \neq \bar{0}$, $K \neq 0$, and suppose that L is the straight line whose vector equation is $\bar{N} \cdot \bar{R} = K$. Put $\bar{D} = (K/|\bar{N}|^2)\bar{N}$. Show that L also has the vector equation $\bar{D} \cdot (\bar{R} - \bar{D}) = 0$.

5   Parametric Equations

In Section 4 we studied equations in vector form. Often the variable position vector that traces out a curve is controlled by a variable scalar called a *parameter*. Equations involving parameters in an explicit way are called *parametric equations*.

As an indication of the way in which parameters are used, we begin by deriving a vector equation for a straight line L, not in terms of a normal vector $\bar{N}$ as in Section 4, but rather in terms of a *direction vector* $\bar{M}$ parallel to the line. Figure 1 shows a straight line L parallel to the nonzero vector $\overline{M}$ and containing the point P_0. Here, $\overline{R}_0$ is the position vector of P_0.

Evidently, the point P whose position vector is $\overline{R}$ belongs to the line L if and only if $\overrightarrow{P_0P}$ is parallel to $\overline{M}$, that is, if and only if $\overline{R} - \overline{R}_0$ is parallel to $\overline{M}$. By Definition 1 in Section 2, $\overline{R} - \overline{R}_0$ is parallel to $\overline{M}$ if and only if there is a scalar t such that $\overline{R} - \overline{R}_0 = t\overline{M}$; that is, $\overline{R} = \overline{R}_0 + t\overline{M}$. As the scalar t varies over the real numbers, the position vector $\overline{R} = \overline{R}_0 + t\overline{M}$ traces out the straight line L. Here t is a *parameter* whose value determines the position vector $\overline{R}$. The equation

$$\overline{R} = \overline{R}_0 + t\overline{M}$$

is therefore called a *vector parametric equation* for L.

Figure 1

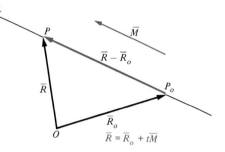

EXAMPLE   Let L be the straight line in the xy plane that contains the two points $P_0 = (6,2)$ and $P_1 = (3,5)$ (Figure 2).
(a) Find a vector $\overline{M}$ that is parallel to L.
(b) Find a vector parametric equation for L.

Figure 2

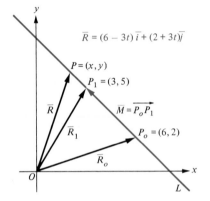

SOLUTION
Put

$$\overline{R} = \overrightarrow{OP} = x i + y j,$$
$$\overline{R}_0 = \overrightarrow{OP_0} = 6 i + 2 j,$$
$$\overline{R}_1 = \overrightarrow{OP_1} = 3 i + 5 j.$$

(a) $\overline{M} = \overrightarrow{P_0P_1} = \overline{R}_1 - \overline{R}_0 = -3 i + 3 j.$
(b) $\overline{R} = \overline{R}_0 + t\overline{M} = (6 i + 2 j) + t(-3 i + 3 j)$
      $\qquad = (6 - 3t) i + (2 + 3t) j.$
Hence, $\overline{R} = (6 - 3t) i + (2 + 3t) j$ is a vector parametric equation for L.

To obtain a vector parametric equation for a curve, it is necessary to select an appropriate quantity t to use as a parameter. The selected quantity t should determine the position of a point P on the curve and, as t is varied, P should trace out the entire curve. Sometimes a suitable parameter suggests itself on algebraic grounds, while other times geometric or physical quantities such as distances, angles, or time can be used.

When angles are used as parameters in the xy plane, it is important to keep the following fact in mind. If $\overline{A}$ is a vector in the xy plane, making an angle t with the positive x axis, and if $a = |\overline{A}|$, then

$$\overline{A} = (a \cos t) i + (a \sin t) j$$

Figure 3

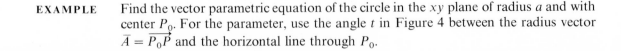

(Figure 3).

EXAMPLE   Find the vector parametric equation of the circle in the xy plane of radius a and with center P_0. For the parameter, use the angle t in Figure 4 between the radius vector $\overline{A} = \overrightarrow{P_0P}$ and the horizontal line through P_0.

Figure 4

$\bar{R} = \bar{R}_0 + (a \cos t)\bar{i} + (a \sin t)\bar{j}$

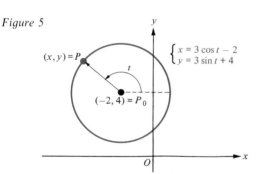

SOLUTION
Here $a = |\bar{A}|$, so that $\bar{A} = (a \cos t)\bar{i} + (a \sin t)\bar{j}$. We put $\bar{R} = \overrightarrow{OP}$, $\bar{R}_0 = \overrightarrow{OP_0}$ as in Figure 4. Then $\bar{R} = \bar{R}_0 + \bar{A}$, so that

$$\bar{R} = \bar{R}_0 + (a \cos t)\bar{i} + (a \sin t)\bar{j}.$$

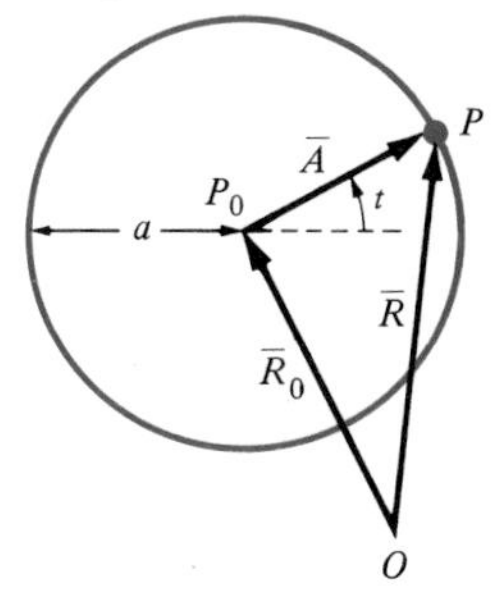

As the parameter t goes from 0 to 2π, the position vector $\bar{R}$ traces out the circle once. As a special case of this equation, notice that the vector parametric equation of the circle of radius a with center at O is

$$\bar{R} = (a \cos t)\bar{i} + (a \sin t)\bar{j}.$$

If we have a vector parametric equation giving the variable position vector $\bar{R} = x\bar{i} + y\bar{j} = \overrightarrow{OP}$ in terms of a parameter, then we can always find *two* equations giving the cartesian coordinates x and y of the point $P = (x, y)$ in terms of the parameter. These two equations are called the (cartesian) *scalar parametric* equations of the curve under consideration. We illustrate this procedure in the following examples.

EXAMPLES 1 Find scalar parametric equations for the circle of radius $a = 3$ units with center $P_0 = (-2, 4)$ (Figure 5).

Figure 5

SOLUTION
By the previous example, the vector parametric equation of the circle is

$$\bar{R} = \bar{R}_0 + (a \cos t)\bar{i} + (a \sin t)\bar{j}$$

or

$$x\bar{i} + y\bar{j} = -2\bar{i} + 4\bar{j} + (3 \cos t)\bar{i} + (3 \sin t)\bar{j}$$
$$= (3 \cos t - 2)\bar{i} + (3 \sin t + 4)\bar{j}.$$

Since equal vectors have equal components, it follows that

$$\begin{cases} x = 3 \cos t - 2 \\ y = 3 \sin t + 4. \end{cases}$$

The latter equations are the scalar parametric equations for the circle. As the parameter t goes from 0 to 2π, the point $P = (x, y)$ traces out the circle.

2 Find scalar parametric equations for the straight line:
(a) Containing the distinct points $P_0 = (x_0, y_0)$ and $P_1 = (x_1, y_1)$.
(b) Containing the points $P_0 = (-1, 7)$ and $P_1 = (5, 3)$.

SOLUTION
Let $\bar{R} = \overrightarrow{OP}$ denote the position vector of a point on the line and let $\bar{R}_0 = \overrightarrow{OP_0}$.
(a) The line is parallel to the vector $\bar{M} = \overrightarrow{P_0 P_1} = (x_1 - x_0)\bar{i} + (y_1 - y_0)\bar{j}$; hence, the vector parametric equation is $\bar{R} = \bar{R}_0 + t\bar{M}$, or

$$x\bar{i} + y\bar{j} = x_0\bar{i} + y_0\bar{j} + t[(x_1 - x_0)\bar{i} + (y_1 - y_0)\bar{j}]$$
$$= [x_0 + t(x_1 - x_0)]\bar{i} + [y_0 + t(y_1 - y_0)]\bar{j}.$$

The resulting scalar parametric equations are

$$\begin{cases} x = x_0 + t(x_1 - x_0) \\ y = y_0 + t(y_1 - y_0). \end{cases}$$

(b) Putting $x_0 = -1$, $y_0 = 7$, $x_1 = 5$, and $y_1 = 3$ in part (a), we obtain

$$\begin{cases} x = -1 + t(5 + 1) \\ y = 7 + t(3 - 7); \end{cases} \quad \text{that is,} \quad \begin{cases} x = 6t - 1 \\ y = 7 - 4t. \end{cases}$$

Parametric equations can provide simple descriptions of curves generated by physical motion in cases where it may be difficult to find a cartesian equation. One such curve, called a *cycloid*, is traced out by a tack P stuck on the periphery of a wheel of radius a as the wheel rolls without slipping along a straight line—say the x axis. The cycloid consists of a sequence of arches, one arch for each revolution of the wheel (Figure 6).

Figure 6

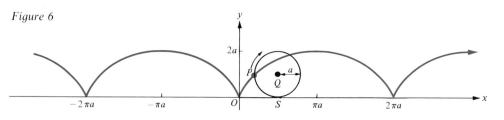

EXAMPLE   Find a vector parametric equation for the cycloid in Figure 6.

SOLUTION

Figure 7

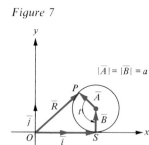

$|\bar{A}| = |\bar{B}| = a$

We use the angle t in Figure 7 as the parameter. Notice that t is just the angle through which the wheel has turned, starting with P at the origin O. We measure t in radians so that the arc length of the portion of the circle between P and S is ta units. Since this portion of the circle has rolled along the segment from O to S, it follows that $|\overrightarrow{OS}| = ta$; hence, $\overrightarrow{OS} = ta\hat{\imath}$. From Figure 7,

$$\bar{R} = \overrightarrow{OS} + \bar{B} + \bar{A} = ta\hat{\imath} + \bar{B} + \bar{A}.$$

Figure 8

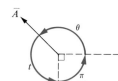

We must write the vectors $\bar{B}$ and $\bar{A}$ in terms of the parameter t. Since $\bar{B}$ has the same direction as $\hat{\jmath}$ and $|\bar{B}| = a$, it follows that $\bar{B} = a\hat{\jmath}$. Let θ be the angle between $\bar{A}$ and the positive x axis. By Figure 8, $\theta + t + \dfrac{\pi}{2} = 2\pi$, so that $\theta = (2\pi - t) - \dfrac{\pi}{2}$. It follows that

$$\cos \theta = \sin(2\pi - t) = -\sin t, \quad \sin \theta = -\cos(2\pi - t) = -\cos t,$$

and

$$\bar{A} = (a \cos \theta)\hat{\imath} + (a \sin \theta)\hat{\jmath}$$
$$= (-a \sin t)\hat{\imath} + (-a \cos t)\hat{\jmath}.$$

Since we have

$$\overrightarrow{OS} = ta\hat{\imath}, \quad \bar{B} = a\hat{\jmath}, \quad \text{and} \quad \bar{A} = (-a \sin t)\hat{\imath} + (-a \cos t)\hat{\jmath},$$

we can rewrite $\bar{R} = \overrightarrow{OS} + \bar{B} + \bar{A}$ as

$$\bar{R} = ta\hat{\imath} + a\hat{\jmath} - (a \sin t)\hat{\imath} - (a \cos t)\hat{\jmath},$$

or

$$\bar{R} = a(t - \sin t)\hat{\imath} + a(1 - \cos t)\hat{\jmath}.$$

5.1 Equations of Curves in the Plane

The equation of a curve in the plane may now be expressed in any one of the following forms:

1. *Scalar nonparametric form.* Here we have a single equation involving the cartesian x and y coordinates (or, perhaps, the polar r and θ coordinates), but not involving vectors in any explicit way. *Example:* $2x - 3y = 8$.

2. *Vector parametric form.* Here we have a single equation in which the only variable quantity is the position vector $\bar{R}$ of the point P that traces out the curve. *Example:* $\bar{N} \cdot \bar{R} = 8$, where $\bar{N} = 2\bar{i} - 3\bar{j}$.

3. *Vector parametric form.* Here we have a single equation in which the position vector $\bar{R}$ is expressed in terms of a scalar variable, say t, called a parameter. As the parameter t varies, the position vector $\bar{R}$ traces out the curve. *Example:* $\bar{R} = (3t + 1)\bar{i} + (2t - 2)\bar{j}$.

4. *Scalar parametric form.* Here we have two equations that give the cartesian x and y coordinates (or the polar r and θ coordinates) of a point P in terms of a third scalar variable, say t, called a parameter. As the parameter t varies, the point P traces out the curve. *Example:* $\begin{cases} x = 3t + 1 \\ y = 2t - 2. \end{cases}$

The following examples illustrate how we may change from one of the above forms to another.

EXAMPLES **1** Find the vector parametric form for the straight line

$$\begin{cases} x = 3t + 1 \\ y = 2t - 2. \end{cases}$$

SOLUTION
$\bar{R} = (3t + 1)\bar{i} + (2t - 2)\bar{j}$.

2 Find a vector parametric form for the equation of the parabola $y = x^2$ using $t = x$ as the parameter.

SOLUTION
In scalar parametric form, we have

$$\begin{cases} x = t \\ y = t^2; \end{cases}$$

hence, the vector parametric equation is $\bar{R} = t\bar{i} + t^2\bar{j}$.

3 Find a scalar nonparametric form for the curve $\bar{R} = 3t\bar{i} + t^2\bar{j}$ and sketch the graph.

SOLUTION
With $\bar{R} = x\bar{i} + y\bar{j}$, we have $x\bar{i} + y\bar{j} = 3t\bar{i} + t^2\bar{j}$; hence,

$$\begin{cases} x = 3t \\ y = t^2. \end{cases}$$

To eliminate the parameter t from the scalar parametric equations, we solve the first equation for t to obtain $t = x/3$ and substitute into the second equation, so that

$$y = \left(\frac{x}{3}\right)^2 \quad \text{or} \quad y = \frac{x^2}{9} \qquad \text{(Figure 9)}.$$

Figure 9

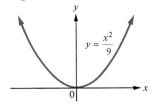

$y = \dfrac{x^2}{9}$

4 Find a scalar nonparametric form for the curve

$$\begin{cases} x = \dfrac{1}{t-1} \\ y = 3t + 1 \end{cases}$$

and sketch the graph.

SOLUTION

From the first equation, we have $t - 1 = 1/x$, or $t = 1 + 1/x$. Substitution into the second equation yields

$$y = 3\left(1 + \frac{1}{x}\right) + 1$$

or

$$y = 4 + \frac{3}{x} \qquad \text{(Figure 10)}.$$

Figure 10

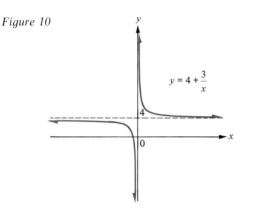

$y = 4 + \dfrac{3}{x}$

5 Find a scalar nonparametric form for $\bar{R} = (\cosh t)\bar{i} + (\sinh t)\bar{j}$.

SOLUTION

In scalar parametric form, we have

$$\begin{cases} x = \cosh t \\ y = \sinh t. \end{cases}$$

Here the parameter t is most easily eliminated by using the identity $\cosh^2 t - \sinh^2 t = 1$ to obtain $x^2 - y^2 = 1$.

Given the equation of a curve in scalar parametric form, one can always find the derivative dy/dx by eliminating the parameter and then proceeding as usual; however, it is possible to find dy/dx directly from the scalar parametric equations. Indeed, assuming the existence of the required derivatives, we have

$$\frac{dy}{dx} \cdot \frac{dx}{dt} = \frac{dy}{dt}$$

by the chain rule. Hence, if $dx/dt \neq 0$, then

$$\frac{dy}{dx} = \frac{dy/dt}{dx/dt}.$$

Similarly, if we put $y' = dy/dx$, then we can obtain the second derivative d^2y/dx^2 by using the chain rule again; hence,

$$\frac{d^2y}{dx^2} = \frac{dy'}{dx} = \frac{dy'/dt}{dx/dt}.$$

EXAMPLE   Given

$$\begin{cases} x = t^2 - 6 \\ y = t^3 + 5, \end{cases}$$

find dy/dx and d^2y/dx^2.

SOLUTION

$$y' = \frac{dy}{dx} = \frac{dy/dt}{dx/dt} = \frac{3t^2}{2t} = \frac{3}{2}t \qquad \text{for } t \neq 0$$

and

$$\frac{d^2y}{dx^2} = \frac{dy'}{dx} = \frac{dy'/dt}{dx/dt} = \frac{\frac{3}{2}}{2t} = \frac{3}{4t} \qquad \text{for } t \neq 0.$$

Problem Set 5

In problems 1 through 6, (a) find a vector $\overline{M}$ parallel to the straight line L containing the two given points P_0 and P_1; (b) write a vector parametric equation for L; (c) write scalar parametric equations for L; (d) eliminate the parameter from the equations in part (c), thus obtaining the scalar nonparametric equation of L; (e) find a normal vector $\overline{N}$ to L; (f) write a vector nonparametric equation for L; and (g) sketch a graph of L.

1 $P_0 = (1, 2), P_1 = (3, 4)$

2 $P_0 = (0, 0), P_1 = (-1, 3)$

3 $P_0 = (-\frac{3}{2}, \frac{5}{2}), P_1 = (\frac{2}{3}, -\frac{1}{3})$

4 $P_0 = (0, a), P_1 = (b, 0), a \neq 0, b \neq 0$

5 $P_0 = (\pi, e), P_1 = (-\sqrt{2}, \sqrt{3})$

6 $P_0 = (x_0, y_0), P_1 = (x_1, y_1)$

In problems 7 through 12, find a vector parametric equation for each curve.

7 The straight line through the point whose position vector is $\overline{R}_0 = 7i - 2j$ and which is parallel to the vector $\overline{M} = -i + 3j$.

8 The straight line through the point whose position vector is $\overline{R}_0 = -i/7 + j/3$ and which has $\overline{N} = 3i - 7j$ as a normal vector.

9 The circle with center at the point whose position vector is $\overline{R}_0 = 3i + 4j$ and whose radius is 4 units.

10 The parabola $y = (x - 1)^2$.

11 The cycloid generated by a point P on the circumference of a rolling circle of radius 5 units.

12 The ellipse whose polar equation is $r = \dfrac{4}{3 - \cos \theta}$.

In problems 13 through 20, (a) convert the vector parametric equation to scalar parametric form, (b) eliminate the parameter t and thus obtain a scalar nonparametric equation, and (c) sketch the graph.

13 $\overline{R} = (2 \cos t)i + (2 \sin t)j$ for $0 \leq t \leq 2\pi$.

14 $\overline{R} = (3 \cos t)i + (4 \sin t)j$ for $0 \leq t \leq 2\pi$.

15 $\overline{R} = (5 \cos 2t)i - (5 \sin 2t)j$ for $0 \leq t \leq \pi/2$.

16 $\overline{R} = t^2 i + tj$ for $-1 \leq t \leq 1$.

17 $\overline{R} = 4ti + (3t + 5)j$ for $-\infty < t < \infty$.

18 $\overline{R} = e^t i + e^{-t} j$ for $-\infty < t < \infty$.

19 $\overline{R} = \dfrac{1}{t - 2} i + (2t + 1)j$ for $0 \leq t < 2$.

20 $\overline{R} = (\sin t)i + (\sin t)j$ for $0 \leq t \leq \pi/2$.

In problems 21 through 28, (a) eliminate the parameter t, (b) write the equation in scalar nonparametric form, and (c) express dy/dx and d^2y/dx^2 as functions of t.

21 $\begin{cases} x = 3t + 1 \\ y = 2 - t \end{cases}$

22 $\begin{cases} x = \dfrac{1}{t^2 - 1} \\ y = 3t + 1 \end{cases}$

23 $\begin{cases} x = t^2 - 2 \\ y = 5t \end{cases}$

24 $\begin{cases} x = \dfrac{1}{(t - 2)^2} \\ y = -3t + 2 \end{cases}$

25 $\begin{cases} x = 2 - \dfrac{1}{t} \\ y = 2t + \dfrac{1}{t} \end{cases}$

26 $\begin{cases} x = 3 \cos^2 t \\ y = 4 \sin^2 t \end{cases}$

27 $\begin{cases} x = t^2 - 2t \\ y = t^3 - 3t \end{cases}$

28 $\begin{cases} x = 4 \sin^2 t \cos t \\ y = 4 \sin t \cos^2 t \end{cases}$

29 Show that the vector $\overline{M} = -Bi + Aj$ is parallel to the straight line whose equation is $Ax + By + C = 0$.

30 A point P is located on a spoke of a wheel of radius $a > 0$ at a distance $b > 0$ from the center. Derive vector and scalar parametric equations for the curve traced out by P as the wheel rolls without slipping along the x axis. Assume that, when $x = 0$, $P = (0, a - b)$.

31 Write a vector parametric equation for the curve whose cartesian equation is $y = f(x)$. Use $t = x$ as the parameter.

32 Write a vector parametric equation for the curve whose polar equation is $r = g(\theta)$. Use $t = \theta$ as the parameter.

6   Vector-Valued Functions of a Scalar

The vector parametric equations developed in Section 5 can be better understood in terms of the idea of a *vector-valued function*. Thus, a vector parametric equation has the form

$$\bar{R} = \bar{F}(t),$$

where the "independent variable" is the parameter t, $\bar{F}$ is the vector-valued function, and the "dependent variable" is the position vector $\bar{R}$. The following definition makes this idea more precise.

DEFINITION 1   **Vector-Valued Function**

A *vector-valued function* $\bar{F}$ assigns a unique vector $\bar{F}(t)$ to each real number t in its *domain*. The *range* of $\bar{F}$ is the set of all vectors of the form $\bar{F}(t)$ as t runs through the domain of $\bar{F}$.

EXAMPLE   Let $\bar{F}$ be the vector-valued function defined by the equation

$$\bar{F}(t) = (2 \cos t)\bar{i} + (2 \sin t)\bar{j}, \;\; 0 \le t \le 2\pi.$$

(a) What is the domain of $\bar{F}$? (b) What is the range of $\bar{F}$? (c) Find $\bar{F}(\pi)$.

SOLUTION
(a) The domain of $\bar{F}$ is the interval $[0, 2\pi]$.
(b) As t runs from 0 to 2π, $\bar{F}(t) = (2 \cos t)\bar{i} + (2 \sin t)\bar{j}$, regarded as a position vector, traces out a circle in the xy plane with radius 2 units and center at the origin. Therefore, the range of $\bar{F}$ is the set of all position vectors of points on this circle.
(c) $\bar{F}(\pi) = (2 \cos \pi)\bar{i} + (2 \sin \pi)\bar{j} = -2\bar{i}$.

A vector-valued function $\bar{F}$ whose range is contained in the xy plane can always be defined by an equation of the form

$$\bar{F}(t) = g(t)\bar{i} + h(t)\bar{j},$$

where g and h are ordinary real-valued functions called the *scalar component functions of* $\bar{F}$.

Figure 1

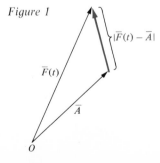

6.1   Limits of Vector-Valued Functions

Let $\bar{F}$ be a vector-valued function and let $\bar{A}$ be a fixed vector. The scalar $|\bar{F}(t) - \bar{A}|$ represents the distance between the head ends (terminal points) of $\bar{F}(t)$ and $\bar{A}$ when their tail ends (initial points) are at the origin (Figure 1). Evidently, $|\bar{F}(t) - \bar{A}|$ is close to zero when $\bar{F}(t)$ is close to $\bar{A}$, both in magnitude and direction. Thus, we make the following definition.

DEFINITION 2 **Limit of a Vector-Valued Function**

Suppose that the number c belongs to an open interval I and that every number in I, except possibly for c, belongs to the domain of the vector-valued function $\bar{F}$. If $\bar{A}$ is a vector such that $\lim_{t \to c} |\bar{F}(t) - \bar{A}| = 0$, then we say that $\bar{F}(t)$ *approaches* $\bar{A}$ *as a limit* when t approaches c, and we write $\lim_{t \to c} \bar{F}(t) = \bar{A}$.

Limits of vector-valued functions can be calculated "component-wise," as is shown by the following theorem.

THEOREM 1 **Limits of Vector-Valued Functions**

Suppose that $\bar{F}(t) = g(t)i + h(t)j$ and $\bar{A} = ai + bj$. Then $\lim_{t \to c} \bar{F}(t) = \bar{A}$ if and only if $\lim_{t \to c} g(t) = a$ and $\lim_{t \to c} h(t) = b$.

PROOF

$$|\bar{F}(t) - \bar{A}| = |[g(t) - a]i + [h(t) - b]j| = \sqrt{[g(t) - a]^2 + [h(t) - b]^2}.$$

Therefore, $|\bar{F}(t) - \bar{A}|$ approaches 0 if and only if both $g(t) - a$ and $h(t) - b$ approach 0; that is, if and only if $g(t)$ approaches a and $h(t)$ approaches b.

EXAMPLE Given $\bar{F}(t) = (t^2 + 1)i + (3t - 2)j$, find $\lim_{t \to 1} \bar{F}(t)$.

SOLUTION
By Theorem 1,

$$\lim_{t \to 1} \bar{F}(t) = \left[\lim_{t \to 1} (t^2 + 1)\right]i + \left[\lim_{t \to 1} (3t - 2)\right]j = 2i + j.$$

Each part of the following theorem can be confirmed by writing the vector-valued functions in terms of their scalar component functions and then applying Theorem 1 (Problems 31 through 33).

THEOREM 2 **Properties of Limits of Vector-Valued Functions**

Let $\bar{F}$ and $\bar{G}$ be vector-valued functions and let f be a real-valued function. Suppose that $\lim_{t \to c} \bar{F}(t) = \bar{A}$, $\lim_{t \to c} \bar{G}(t) = \bar{B}$, and $\lim_{t \to c} f(t) = k$. Then:

(i) $\lim_{t \to c} [\bar{F}(t) + \bar{G}(t)] = \bar{A} + \bar{B}$.

(ii) $\lim_{t \to c} [\bar{F}(t) - \bar{G}(t)] = \bar{A} - \bar{B}$.

(iii) $\lim_{t \to c} [f(t)\bar{F}(t)] = k\bar{A}$.

(iv) $\lim_{t \to c} [\bar{F}(t) \cdot \bar{G}(t)] = \bar{A} \cdot \bar{B}$.

(v) $\lim_{t \to c} |\bar{F}(t)| = |\bar{A}|$.

(vi) $\lim_{t \to c} \dfrac{\bar{F}(t)}{f(t)} = \dfrac{\bar{A}}{k}$ if $k \neq 0$.

EXAMPLE If $\lim_{t \to c} \bar{F}(t) = i$ and $\lim_{t \to c} \bar{G}(t) = j$, find $\lim_{t \to c} [\bar{F}(t) \cdot \bar{G}(t)]$.

SOLUTION
By part (iv) of Theorem 2, $\lim_{t \to c} [\bar{F}(t) \cdot \bar{G}(t)] = i \cdot j = 0$.

6.2 Continuity of Vector-Valued Functions

By direct analogy with the definition of continuity for real-valued functions, we make the following definition for vector-valued functions.

DEFINITION 3

Continuity of a Vector-Valued Function

A vector-valued function $\bar{F}$ is said to be *continuous* at the number c if (i) c belongs to the domain of $\bar{F}$, (ii) $\lim_{t \to c} \bar{F}(t)$ exists, and (iii) $\lim_{t \to c} \bar{F}(t) = \bar{F}(c)$.

A vector-valued function is said to be *continuous* provided that it is continuous at every number in its domain. One can also define right- and left-sided limits and right- and left-sided continuity for vector-valued functions in essentially the same way that these notions were defined for real-valued functions. There are no surprises, and everything works as expected (Problems 39 and 44).

Using Theorem 1, we easily establish the following result (Problem 34).

THEOREM 3

Continuity of Vector-Valued Functions

A vector-valued function $\bar{F}$ is continuous at a number c if and only if both of its scalar component functions are continuous at c.

EXAMPLE

Discuss the continuity of $\bar{F}(t) = \dfrac{1}{t - 2}\, i + (5t + 1)j$.

SOLUTION

The scalar component functions of $\bar{F}$ are $g(t) = \dfrac{1}{t - 2}$ and $h(t) = 5t + 1$, where g is continuous for every value of t except for $t = 2$ and h is continuous for every value of t. The domain of $\bar{F}$ consists of all real numbers different from 2, and $\bar{F}$ is continuous at every number t in its domain.

6.3 Derivatives of Vector-Valued Functions

Given a vector-valued function $\bar{F}$, we can form the difference quotient

$$\frac{\bar{F}(t + \Delta t) - \bar{F}(t)}{\Delta t} = \frac{1}{\Delta t}\left[\bar{F}(t + \Delta t) - \bar{F}(t)\right]$$

by direct analogy with the difference quotient of a real-valued function. Notice, however, that $(1/\Delta t)[\bar{F}(t + \Delta t) - \bar{F}(t)]$ is the product of a scalar $1/\Delta t$ and a vector $\bar{F}(t + \Delta t) - \bar{F}(t)$; hence, it is a *vector*. Pursuing the analogy with real-valued functions, we make the following definition.

DEFINITION 4

Derivative of a Vector-Valued Function

Let t be a number belonging to an open interval contained in the domain of a vector-valued function $\bar{F}$. We define the *derivative* of $\bar{F}$ at t, in symbols, $\bar{F}'(t)$, by

$$\bar{F}'(t) = \lim_{\Delta t \to 0} \frac{\bar{F}(t + \Delta t) - \bar{F}(t)}{\Delta t}, \qquad \text{provided that this limit exists.}$$

Using Definition 4 and Theorem 1, it is easy to prove the following theorem (Problem 40).

THEOREM 4   **Derivatives of Vector-Valued Functions**

Let $\bar{F}(t) = g(t)\bar{i} + h(t)\bar{j}$ and suppose that t belongs to an open interval contained in the domain of $\bar{F}$. Then:

(i) $\bar{F}$ has a derivative at t if and only if both of the scalar component functions g and h have derivatives at t.

(ii) If $g'(t)$ and $h'(t)$ both exist, then $\bar{F}'(t) = g'(t)\bar{i} + h'(t)\bar{j}$.

EXAMPLE   If $\bar{F}(t) = e^{2t}\bar{i} - (\sinh t)\bar{j}$, find $\bar{F}'(t)$.

SOLUTION
$$\bar{F}'(t) = (D_t e^{2t})\bar{i} + [D_t(-\sinh t)]\bar{j} = 2e^{2t}\bar{i} - (\cosh t)\bar{j}.$$

The notation and terminology used in connection with derivatives of vector-valued functions is the direct analog of that used for real-valued functions. For instance, if $\bar{F}$ is differentiable for all values of t in some open interval, then $\bar{F}'$, the *derived* vector-valued function, is defined on this interval and we can ask whether $\bar{F}'$ has a derivative. If it does, we denote this *second derivative* by F''. Third- and higher-order derivatives of $\bar{F}$ are treated similarly.

EXAMPLE   Let $\bar{F}(t) = (\cos t)\bar{i} + (\sin t)\bar{j}$. Find (a) $\bar{F}'(t)$, (b) $\bar{F}''(t)$, and (c) $\bar{F}'(t) \cdot \bar{F}''(t)$.

SOLUTION
(a) $\bar{F}'(t) = (-\sin t)\bar{i} + (\cos t)\bar{j}$.
(b) $\bar{F}''(t) = (-\cos t)\bar{i} - (\sin t)\bar{j}$.
(c) $\bar{F}'(t) \cdot \bar{F}''(t) = \sin t \cos t - \sin t \cos t = 0$.

The Leibniz differential notation is used in connection with vector-valued functions in much the same way that it is used for real-valued functions. For instance, if $\bar{F}$ is a differentiable vector-valued function and the variable vector $\bar{R}$ is defined by $\bar{R} = \bar{F}(t)$, we write

$$\frac{d\bar{R}}{dt} = \bar{F}'(t) \quad \text{and} \quad d\bar{R} = F'(t)\, dt.$$

Notice that the differential $d\bar{R}$ is a *vector* since it is a product of the scalar dt and the vector $\bar{F}'(t)$. Thus, we can rewrite Theorem 4 as follows. If $\bar{R} = u\bar{i} + v\bar{j}$, where the scalars u and v are differentiable functions of t, then

$$\frac{d\bar{R}}{dt} = \frac{d}{dt}(u\bar{i} + v\bar{j}) = \frac{du}{dt}\bar{i} + \frac{dv}{dt}\bar{j} \quad \text{or} \quad d\bar{R} = du\,\bar{i} + dv\,\bar{j}.$$

EXAMPLE   If $\bar{R} = t^2\bar{i} + (\tan t)\bar{j}$, find $\dfrac{d\bar{R}}{dt}$ and $d\bar{R}$.

SOLUTION
$$\frac{d\bar{R}}{dt} = 2t\bar{i} + (\sec^2 t)\bar{j} \quad \text{and} \quad d\bar{R} = (2t\,dt)\bar{i} + (\sec^2 t\,dt)\bar{j}.$$

Using the Leibniz notation and Theorem 4, we now establish the basic properties of derivatives of vector-valued functions.

THEOREM 5   **Properties of Derivatives of Vector-Valued Functions**

Let R and $\bar{S}$ be variable vectors that depend on the scalar parameter t and let the scalar w be a function of t. Assume that $\dfrac{d\bar{R}}{dt}, \dfrac{d\bar{S}}{dt}$, and $\dfrac{dw}{dt}$ exist. Then:

(i) $\dfrac{d}{dt}(\bar{R} + \bar{S}) = \dfrac{d\bar{R}}{dt} + \dfrac{d\bar{S}}{dt}.$

(ii) $\dfrac{d}{dt}(\bar{R} - \bar{S}) = \dfrac{d\bar{R}}{dt} - \dfrac{d\bar{S}}{dt}.$

(iii) $\dfrac{d}{dt}(w\bar{R}) = \dfrac{dw}{dt}\bar{R} + w\dfrac{d\bar{R}}{dt}.$

(iv) $\dfrac{d}{dt}(\bar{R} \cdot \bar{S}) = \dfrac{d\bar{R}}{dt} \cdot \bar{S} + \bar{R} \cdot \dfrac{d\bar{S}}{dt}.$

(v) $\dfrac{d}{dt}|\bar{R}| = \dfrac{\bar{R}}{|\bar{R}|} \cdot \dfrac{d\bar{R}}{dt}$ if $\bar{R} \neq \bar{0}.$

(vi) $\dfrac{d}{dt}\left(\dfrac{\bar{R}}{w}\right) = \dfrac{w(d\bar{R}/dt) - (dw/dt)\bar{R}}{-w^2}$ if $w \neq 0.$

PROOF

Parts (i) through (iv) can be proved by expanding the vectors in terms of their components and using Theorem 4. We leave parts (i), (ii), and (iii) as exercises (Problems 41 and 42), and attend here to parts (iv), (v), and (vi).

(iv) Let $\bar{R} = x\bar{i} + y\bar{j}$ and let $\bar{S} = u\bar{i} + v\bar{j}$. Then, $\bar{R} \cdot \bar{S} = xu + yv$ and

$$\frac{d}{dt}(\bar{R} \cdot \bar{S}) = \frac{d}{dt}(xu + yv) = \frac{dx}{dt}u + x\frac{du}{dt} + \frac{dy}{dt}v + y\frac{dv}{dt}$$

$$= \left(\frac{dx}{dt}u + \frac{dy}{dt}v\right) + \left(x\frac{du}{dt} + y\frac{dv}{dt}\right)$$

$$= \left(\frac{dx}{dt}\bar{i} + \frac{dy}{dt}\bar{j}\right) \cdot (u\bar{i} + v\bar{j}) + (x\bar{i} + y\bar{j}) \cdot \left(\frac{du}{dt}\bar{i} + \frac{dv}{dt}\bar{j}\right)$$

$$= \frac{d\bar{R}}{dt} \cdot \bar{S} + \bar{R} \cdot \frac{d\bar{S}}{dt}.$$

(v) Assume that $\bar{R} \neq \bar{0}$. By part (iv), we have

$$\frac{d}{dt}|\bar{R}| = \frac{d}{dt}\sqrt{\bar{R} \cdot \bar{R}} = \frac{1}{2\sqrt{\bar{R} \cdot \bar{R}}} \cdot \frac{d}{dt}(\bar{R} \cdot \bar{R}) = \frac{1}{2|\bar{R}|}\left(\frac{d\bar{R}}{dt} \cdot \bar{R} + \bar{R} \cdot \frac{d\bar{R}}{dt}\right)$$

$$= \frac{1}{2|\bar{R}|}\left(2\bar{R} \cdot \frac{d\bar{R}}{dt}\right) = \frac{\bar{R}}{|\bar{R}|} \cdot \frac{d\bar{R}}{dt}.$$

(vi) Assume that $w \neq 0$, By part (iii), we have

$$\frac{d}{dt}\left(\frac{\bar{R}}{w}\right) = \frac{d}{dt}(w^{-1}\bar{R}) = \left(\frac{d}{dt}w^{-1}\right)\bar{R} + w^{-1}\frac{d\bar{R}}{dt} = -w^{-2}\frac{dw}{dt}\bar{R} + w^{-1}\frac{d\bar{R}}{dt}$$

$$= \frac{w(d\bar{R}/dt) - (dw/dt)\bar{R}}{w^2}.$$

EXAMPLES 1 Let $\bar{R} = (5\sin 2t)\bar{i} + (5\cos 2t)\bar{j}$, $\bar{S} = e^{2t}\bar{i} + e^{-2t}\bar{j}$, and $w = e^{-5t}$. Find:

(a) $\dfrac{d}{dt}|\bar{R}|$ (b) $\dfrac{d}{dt}(\bar{R} \cdot \bar{S})$ (c) $\dfrac{d}{dt}(w\bar{S}).$

SOLUTION

By Theorem 5, we have

(a) $\dfrac{d}{dt}\,|\bar{R}| = \dfrac{\bar{R}}{|\bar{R}|}\cdot\dfrac{d\bar{R}}{dt} = \dfrac{(5\sin 2t)\bar{i} + (5\cos 2t)\bar{j}}{\sqrt{(5\sin 2t)^2 + (5\cos 2t)^2}}\cdot[(10\cos 2t)\bar{i} - (10\sin 2t)\bar{j}]$

$= \dfrac{50\sin 2t\cos 2t - 50\cos 2t\sin 2t}{5} = 0.$

(b) $\dfrac{d}{dt}\,(\bar{R}\cdot\bar{S}) = \dfrac{d\bar{R}}{dt}\cdot\bar{S} + \bar{R}\cdot\dfrac{d\bar{S}}{dt}$

$= [(10\cos 2t)\bar{i} - (10\sin 2t)\bar{j}]\cdot[e^{2t}\bar{i} + e^{-2t}\bar{j}]$

$\qquad + [(5\sin 2t)\bar{i} + (5\cos 2t)\bar{j}]\cdot[2e^{2t}\bar{i} - 2e^{-2t}\bar{j}]$

$= 10\cos 2t\,e^{2t} - 10\sin 2t\,e^{-2t} + 10\sin 2t\,e^{2t} - 10\cos 2t\,e^{-2t}$

$= 10(\cos 2t + \sin 2t)(e^{2t} - e^{-2t}).$

(c) $\dfrac{d}{dt}\,(w\bar{S}) = \dfrac{dw}{dt}\,\bar{S} + w\,\dfrac{d\bar{S}}{dt}$

$= -5e^{-5t}(e^{2t}\bar{i} + e^{-2t}\bar{j}) + e^{-5t}(2e^{2t}\bar{i} - 2e^{-2t}\bar{j})$

$= -5e^{-3t}\bar{i} - 5e^{-7t}\bar{j} + 2e^{-3t}\bar{i} - 2e^{-7t}\bar{j} = -3e^{-3t}\bar{i} - 7e^{-7t}\bar{j}.$

2 Assuming the existence of the required derivatives, find a formula for $\dfrac{d^2}{dt^2}\,(\bar{R}\cdot\bar{S})$.

SOLUTION

$$\dfrac{d^2}{dt^2}\,(\bar{R}\cdot\bar{S}) = \dfrac{d}{dt}\left[\dfrac{d}{dt}\,(\bar{R}\cdot\bar{S})\right] = \dfrac{d}{dt}\left[\dfrac{d\bar{R}}{dt}\cdot\bar{S} + \bar{R}\cdot\dfrac{d\bar{S}}{dt}\right]$$

$$= \dfrac{d}{dt}\left(\dfrac{d\bar{R}}{dt}\cdot\bar{S}\right) + \dfrac{d}{dt}\left(\bar{R}\cdot\dfrac{d\bar{S}}{dt}\right)$$

$$= \dfrac{d^2\bar{R}}{dt^2}\cdot\bar{S} + \dfrac{d\bar{R}}{dt}\cdot\dfrac{d\bar{S}}{dt} + \dfrac{d\bar{R}}{dt}\cdot\dfrac{d\bar{S}}{dt} + \bar{R}\cdot\dfrac{d^2\bar{S}}{dt^2}$$

$$= \dfrac{d^2\bar{R}}{dt^2}\cdot\bar{S} + 2\,\dfrac{d\bar{R}}{dt}\cdot\dfrac{d\bar{S}}{dt} + \bar{R}\cdot\dfrac{d^2\bar{S}}{dt^2}.$$

The chain rule also works for vector-valued functions, as the following theorem shows.

THEOREM 6   **Chain Rule for Vector-Valued Functions**

Let $\bar{R}$ be a differentiable vector-valued function of the scalar t and let t be a differentiable function of the scalar s. Then, regarding $\bar{R}$ as a function of s, we have

$$\dfrac{d\bar{R}}{ds} = \dfrac{d\bar{R}}{dt}\dfrac{dt}{ds}.$$

PROOF

Let $\bar{R} = u\bar{i} + v\bar{j}$, where the scalar components u and v are functions of t. By the usual chain rule for real-valued functions,

$$\dfrac{du}{ds} = \dfrac{du}{dt}\dfrac{dt}{ds} \quad\text{and}\quad \dfrac{dv}{ds} = \dfrac{dv}{dt}\dfrac{dt}{ds};$$

hence, by Theorem 4,

$$\frac{d\bar{R}}{ds} = \frac{du}{ds}\,i + \frac{dv}{ds}\,j = \frac{du}{dt}\frac{dt}{ds}\,i + \frac{dv}{dt}\frac{dt}{ds}\,j = \left(\frac{du}{dt}\,i + \frac{dv}{dt}\,j\right)\frac{dt}{ds} = \frac{d\bar{R}}{dt}\frac{dt}{ds}.$$

EXAMPLE   Given that $\bar{R} = \bar{F}(t)$, $\bar{F}'(t) = 2ti - e^{-t}j$, and $t = \sin\theta$, find (a) $\dfrac{d\bar{R}}{d\theta}$ and (b) $\dfrac{d^2\bar{R}}{d\theta^2}$.

SOLUTION
By Theorem 6,

(a) $\dfrac{d\bar{R}}{d\theta} = \dfrac{d\bar{R}}{dt}\dfrac{dt}{d\theta} = \bar{F}'(t)\dfrac{d}{d\theta}(\sin\theta) = (2ti - e^{-t}j)\cos\theta$

$= [(2\sin\theta)i - e^{-\sin\theta}j]\cos\theta$

$= (2\sin\theta\cos\theta)i - (e^{-\sin\theta}\cos\theta)j$

$= (\sin 2\theta)i - (e^{-\sin\theta}\cos\theta)j.$

(b) Using the result of part (a), we have

$$\frac{d^2\bar{R}}{d\theta^2} = (2\cos 2\theta)i + e^{-\sin\theta}(\sin\theta + \cos^2\theta)j.$$

Problem Set 6

In problems 1 through 6, (a) find the domain of $\bar{F}$, (b) find $\bar{F}(t_0)$ if t_0 belongs to the domain of $\bar{F}$, (c) find $\lim\limits_{t\to t_0}\bar{F}(t)$ if it exists, and (d) determine where $\bar{F}$ is continuous.

1 $\bar{F}(t) = (3t + 2)i + \dfrac{5}{t^2 - 1}j;\ t_0 = 2.$

2 $\bar{F}(t) = \sqrt{t - 1}\,i + \sqrt{5 - t}\,j;\ t_0 = 1.$

3 $\bar{F}(t) = \dfrac{t^2 - 5t + 6}{t - 3}i + \dfrac{t^2 + 7t - 30}{t - 3}j;\ t_0 = 3.$

4 $\bar{F}(t) = \dfrac{t^2 + 2t + 1}{t - 1}i + (\sin^{-1}t)j;\ t_0 = 1.$

5 $\bar{F}(t) = \dfrac{1}{t}\,i + (\sin 3t)j;\ t_0 = \dfrac{\pi}{6}.$

6 $\bar{F}(t) = \ln(t + 1)i + e^{-2t}j;\ t_0 = 0.$

In problems 7 through 12, find $\bar{F}'(t)$ and $\bar{F}''(t)$.

7 $\bar{F}(t) = (3t^2 - 1)i + (9t^4 + 5)j.$

8 $\bar{F}(t) = \ln(3 + t)i + (\sin t)j.$

9 $\bar{F}(t) = e^{3t}i + (\ln 2t)j.$

10 $\bar{F}(t) = (3\sec t)i + (4\tan t)j.$

11 $\bar{F}(t) = (5\cos t)i + (3\sin t)j.$

12 $\bar{F}(t) = (\tan^{-1}2t)i + e^{7t}j.$

In problems 13 through 16, find $d\bar{R}/dt$ and $d^2\bar{R}/dt^2$.

13 $\bar{R} = (\cos t^2)i + (\sin t^2)j$

14 $\bar{R} = te^{-2t}i + te^{2t}j$

15 $\bar{R} = \dfrac{1}{t}\,i + \dfrac{1}{t^2}\,j$

16 $\bar{R} = \dfrac{t - 2}{t + 2}\,i + \dfrac{t - 3}{t + 3}\,j$

In problems 17 through 20, find (a) $\bar{F}'(t)$, (b) $\bar{F}''(t)$, (c) $\bar{F}'(t)\cdot\bar{F}''(t)$, and (d) $\dfrac{d}{dt}|\bar{F}(t)|$.

17 $\bar{F}(t) = (e^t + 3)i + (e^t + 7)j$

18 $\bar{F}(t) = (3\cos 2t)i + (3\sin 2t)j$

19 $\bar{F}(t) = (4\sin 3t)i - (4\sin 3t)j$

20 $\bar{F}(t) = e^{-5t}i + e^{5t}j$

In problems 21 through 24, find (a) $\dfrac{d}{dt}(\bar{R}\cdot\bar{S})$ and (b) $\dfrac{d}{dt}(w\bar{R})$.

21 $\bar{R} = e^{2t}i + e^{-4t}j,\ \bar{S} = (\cos 2t)i + (\sin 2t)j,\ w = e^{-7t}.$

22 $\bar{R} = 5ti + t^2j,\ \bar{S} = (\sec t)i + (3\sin t)j,\ w = \cot t.$

23 $\bar{R} = (\ln t)i + \dfrac{1}{t - 1}\,j,\ \bar{S} = \dfrac{i}{t} + \dfrac{3i}{t^2},\ w = \cos 7t.$

24 $\bar{R} = \dfrac{i}{t^2 + 4} + \dfrac{2j}{1 - t^2},\ \bar{S} = (\ln t)i + t^2j,\ w = 5t^2 + 8.$

25 Let $\bar{F}(t) = (\sin t)\bar{i} + (\cos t)\bar{j}$, $\bar{G}(t) = e^t\bar{i} + e^{-t}\bar{j}$, and $t = s^3$. Use Theorems 5 and 6 to find:

(a) $\dfrac{d}{ds}\,[\bar{F}(t)]$ (b) $\dfrac{d^2}{ds^2}\,[\bar{G}(t)]$ (c) $\dfrac{d}{dt}\,[\,|\bar{F}(t)|^2\,]$ (d) $\dfrac{d^2}{dt^2}\,[\bar{F}(t)\cdot\bar{G}(t)]$

26 Show that, if $\dfrac{dw}{dt}$ exists and $\bar{A}$ is a constant vector, then $\dfrac{d}{dt}\,(w\bar{A}) = \dfrac{dw}{dt}\,\bar{A}$.

27 Show that, if $\dfrac{d\bar{R}}{dt}$ exists and c is a constant scalar, then $\dfrac{d}{dt}\,(c\bar{R}) = c\,\dfrac{d\bar{R}}{dt}$.

28 Suppose that $d\bar{R}/dt$ exists and that $|\bar{R}|$ is constant. Show that $\bar{R}$ and $d\bar{R}/dt$ are perpendicular.

29 Suppose that $\bar{R}$ is a variable vector that is always parallel to a fixed nonzero vector $\bar{A}$. If $d\bar{R}/dt$ exists, show that $d\bar{R}/dt$ is always parallel to the fixed vector $\bar{A}$.

30 If $\bar{R} \neq \bar{0}$ and $\dfrac{d\bar{R}}{dt}$ exists, find a formula for $\dfrac{d}{dt}\left(\dfrac{\bar{R}}{|\bar{R}|}\right)$.

31 Prove parts (i) and (ii) of Theorem 2.   32 Prove parts (iii) and (iv) of Theorem 2.

33 Prove parts (v) and (vi) of Theorem 2.   34 Prove Theorem 3.

35 Prove that the sum or difference of continuous vector-valued functions is again continuous.

36 Prove that the dot product of continuous vector-valued functions is a continuous real-valued function.

37 If $\bar{F}$ is a continuous vector-valued function, show that the real-valued function f defined by $f(t) = |\bar{F}(t)|$ is also continuous.

38 If $\bar{F}$ is a continuous vector-valued function and if f is a continuous real-valued function, both of which have the same domain, show that the vector-valued function $\bar{G}$ defined by $\bar{G}(t) = f(t)\bar{F}(t)$ is continuous.

39 Give an appropriate definition of the following one-sided limits:
(a) $\lim\limits_{t \to c^+} \bar{F}(t) = \bar{A}$ (b) $\lim\limits_{t \to c^-} \bar{F}(t) = \bar{A}$

40 Prove Theorem 4.   41 Prove parts (i) and (ii) of Theorem 5.

42 Prove part (iii) of Theorem 5.

43 Prove that a differentiable vector-valued function is continuous.

44 Give an appropriate definition of the following: The vector-valued function $\bar{F}$ is continuous on the closed interval $[a, b]$.

7 Velocity, Acceleration, and Arc Length

Figure 1

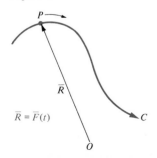

$\bar{R} = \bar{F}(t)$

A vector parametric equation $\bar{R} = \bar{F}(t)$ can be regarded as giving the position vector $\bar{R}$ of a moving point P at the time t (Figure 1). As the time t varies, the point P traces out a curve C. Even if the parameter t actually represents some quantity other than time, it may be useful to think of it—at least for the purposes of this section—as corresponding to the elapsed time since some arbitrary (but fixed) initial instant. This point of view enables us to study the vector-valued function $\bar{F}$ in terms of the intuitively appealing idea of physical motion.

For the remainder of this section, we assume that the moving point P traces out a smooth curve C in the plane and that the variable position vector $\bar{R}$ of P at

Figure 2

Figure 2

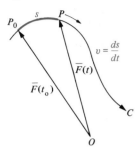

Figure 3

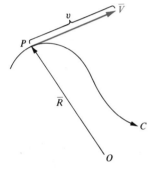

Figure 4

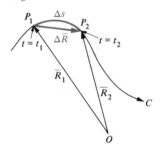

Figure 5

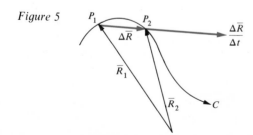

time t is given by $\bar{R} = \bar{F}(t)$. Furthermore, we assume that $\bar{F}$ has a first derivative $\bar{F}'$ and a second derivative $\bar{F}''$. We imagine that the moving point P is equipped with an odometer to measure the *distance s* that it has traveled along the path C, starting at P_0 when $t = t_0$, and that it is equipped with a speedometer to measure its *instantaneous speed* $v = \dfrac{ds}{dt}$ at any time t (Figure 2). Here s is just the arc length of the curve C between P_0 and P, while v is the instantaneous rate of change of s with respect to the parameter t.

The *velocity vector* of the moving point P is defined to be the vector $\bar{V}$ whose length is the speed of P, so that

$$v = |\bar{V}|,$$

and whose direction is parallel to the tangent line to the curve C at P (Figure 3). Here we understand that $\bar{V}$ points in the "instantaneous direction of motion" of P, so that, if the constraints causing P to follow the curve C were suddenly removed, then P would "fly off along the tangent line" in the direction of $\bar{V}$.

We are now going to give an informal argument to show that *the velocity vector of a moving point is the derivative with respect to time of the position vector of the point.* To begin the argument, let $\Delta t = t_2 - t_1$ denote a small positive interval of time and put

$$\bar{R}_1 = \overrightarrow{OP_1} = \bar{F}(t_1) \quad \text{and} \quad \bar{R}_2 = \overrightarrow{OP_2} = \bar{F}(t_2) = \bar{F}(t_1 + \Delta t)$$

(Figure 4). During the time interval Δt from t_1 to t_2, the point P moves along a small portion of the curve C between P_1 and P_2. Denote the arc length of this portion of C by Δs. In moving from P_1 to P_2, the point P undergoes the *displacement*

$$\Delta \bar{R} = \bar{R}_2 - \bar{R}_1 = \bar{F}(t_1 + \Delta t) - \bar{F}(t_1).$$

Evidently,

$$\Delta s \approx |\Delta \bar{R}|,$$

with better and better approximation as $\Delta t \to 0^+$. Dividing by Δt, we obtain

$$\frac{\Delta s}{\Delta t} \approx \frac{1}{\Delta t} |\Delta \bar{R}| = \left| \frac{\Delta \bar{R}}{\Delta t} \right|;$$

hence, taking the limit as $\Delta t \to 0^+$, we have

$$|\bar{V}| = v = \frac{ds}{dt} = \left| \frac{d\bar{R}}{dt} \right|.$$

Therefore, $d\bar{R}/dt$ has the same *length* as the velocity vector $\bar{V}$.

To finish the argument, we only need to show that $d\bar{R}/dt$ has the same *direction* as the velocity vector $\bar{V}$. If the time interval Δt is small, then $1/\Delta t$ is large; hence, the difference quotient

$$\frac{\Delta \bar{R}}{\Delta t} = \frac{\bar{F}(t_1 + \Delta t) - \bar{F}(t_1)}{\Delta t} = \frac{1}{\Delta t} \Delta \bar{R}$$

not only has the same direction as the displacement vector $\Delta \bar{R}$, but it is also considerably longer than $\Delta \bar{R}$ (Figure 5).

As $\Delta t \to 0^+$, the displacement vector $\Delta \bar{R}$ becomes

Figure 6

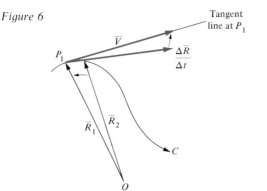

shorter and shorter; however, the difference quotient $\Delta \bar{R}/\Delta t$ approaches $d\bar{R}/dt$, which is not necessarily $\bar{0}$. At the same time, $\Delta \bar{R}/\Delta t$ pivots about the point P_1 and approaches the direction of the tangent line to the curve C at P_1 (Figure 6). Since $d\bar{R}/dt = \lim_{\Delta t \to 0} \Delta \bar{R}/\Delta t$, it follows that $\dfrac{d\bar{R}}{dt}$ is parallel to this tangent line; hence, $d\bar{R}/dt$ has the same *direction* as the velocity vector $\bar{V}$. Since $d\bar{R}/dt$ has the same length and direction as $\bar{V}$, we conclude that

$$\frac{d\bar{R}}{dt} = \bar{V}.$$

As the point P moves along the curve C, its velocity vector $\bar{V}$ can change in length, in direction, or both. The instantaneous rate of change of $\bar{V}$ with respect to time is a vector $d\bar{V}/dt$ called the *acceleration vector* of the moving point P and denoted by $\bar{A}$. Thus,

$$\bar{V} = \frac{d\bar{R}}{dt} = \bar{F}'(t) \quad \text{and} \quad \bar{A} = \frac{d\bar{V}}{dt} = \frac{d^2\bar{R}}{dt^2} = \bar{F}''(t).$$

EXAMPLES 1 A point P is moving on an elliptical path according to the equation $\bar{R} = (5 \cos 2t)\bar{i} + (3 \sin 2t)\bar{j}$. Distances are measured in centimeters and time in seconds. Find (a) the velocity vector $\bar{V}$, (b) the acceleration vector $\bar{A}$, and (c) the speed v at the instant t.

SOLUTION

(a) $\bar{V} = \dfrac{d\bar{R}}{dt} = (-10 \sin 2t)\bar{i} + (6 \cos 2t)\bar{j}.$

(b) $\bar{A} = \dfrac{d\bar{V}}{dt} = (-20 \cos 2t)\bar{i} + (-12 \sin 2t)\bar{j}.$

(c) $v = |\bar{V}| = \sqrt{100 \sin^2 2t + 36 \cos^2 2t}$ cm/sec.

2 A moving point P has position vector $\bar{R} = \dfrac{t^2}{2}\bar{i} - t^3\bar{j}$ at time t. Find (a) the velocity vector $\bar{V}$, (b) the acceleration vector $\bar{A}$, and (c) the speed v of the moving point at the instant when $t = 10$.

SOLUTION
(a) $\bar{V} = t\bar{i} - 3t^2\bar{j}$; hence, when $t = 10$, $\bar{V} = 10\bar{i} - 300\bar{j}$.
(b) $\bar{A} = \bar{i} - 6t\bar{j}$; hence, when $t = 10$, $\bar{A} = \bar{i} - 60\bar{j}$.
(c) $v = |\bar{V}| = \sqrt{t^2 + 9t^4}$; hence, when $t = 10$, $v = \sqrt{100(1 + 900)} = 10\sqrt{901}$.

Figure 7

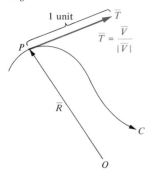

If the velocity vector $\bar{V}$ is not the zero vector, then we can normalize $\bar{V}$ by dividing it by its own length $|\bar{V}|$. The resulting vector $\bar{T}$ has the same direction as the velocity vector, but has *unit length* (Figure 7). Since $\bar{V}$ is parallel to the tangent line to C at P, so is $\bar{T}$; hence, we call $\bar{T}$ the *unit tangent vector* to the curve C at the point P. Evidently,

$$\bar{T} = \frac{\bar{V}}{|\bar{V}|} = \frac{\bar{V}}{v} = \frac{\dfrac{d\bar{R}}{dt}}{\left|\dfrac{d\bar{R}}{dt}\right|} = \frac{\dfrac{d\bar{R}}{dt}}{\dfrac{ds}{dt}}.$$

EXAMPLE   Find a unit tangent vector $\bar{T}$ to the curve C whose vector parametric equation is $\bar{R} = (t^2 - 4t)\bar{i} + (\frac{3}{4}t^4)\bar{j}$.

SOLUTION
Here we have

$$\frac{d\bar{R}}{dt} = (2t - 4)\bar{i} + (3t^3)\bar{j},$$

so that

$$\left|\frac{d\bar{R}}{dt}\right| = \sqrt{(2t - 4)^2 + (3t^3)^2} = \sqrt{9t^6 + 4t^2 - 16t + 16},$$

$$\bar{T} = \frac{\dfrac{d\bar{R}}{dt}}{\left|\dfrac{d\bar{R}}{dt}\right|} = \frac{(2t - 4)\bar{i} + (3t^3)\bar{j}}{\sqrt{9t^6 + 4t^2 - 16t + 16}}.$$

Since $ds/dt = v = |d\bar{R}/dt|$, we have the differential equation

$$ds = \left|\frac{d\bar{R}}{dt}\right| dt$$

for the arc length s of the curve whose vector parametric equation is $\bar{R} = \bar{F}(t)$. Integrating, we find that the arc length s between the point corresponding to $t = a$ and the point corresponding to $t = b$ is

Figure 8

$$s = \int_a^b \left|\frac{d\bar{R}}{dt}\right| dt \qquad \text{(Figure 8)}.$$

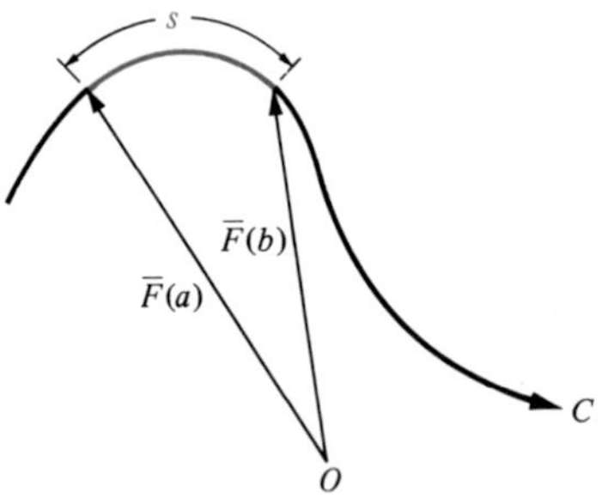

If $\bar{R} = \bar{F}(t) = g(t)\bar{i} + h(t)\bar{j}$, then

$$\left|\frac{d\bar{R}}{dt}\right| = |g'(t)\bar{i} + h'(t)\bar{j}|$$

$$= \sqrt{[g'(t)]^2 + [h'(t)]^2}$$

and

$$s = \int_a^b \sqrt{[g'(t)]^2 + [h'(t)]^2} \, dt.$$

EXAMPLES   Find the arc length s of the given curve between the indicated values of the parameter.

1 $\bar{R} = (3t^2)\bar{i} + (t^3 - 3t)\bar{j}$ between $t = 0$ and $t = 1$.

SOLUTION
Here, $d\bar{R}/dt = (6t)\bar{i} + (3t^2 - 3)\bar{j}$ and

$$\left|\frac{d\bar{R}}{dt}\right| = \sqrt{(6t)^2 + (3t^2 - 3)^2} = \sqrt{36t^2 + 9t^4 - 18t^2 + 9}$$

$$= \sqrt{9t^4 + 18t^2 + 9} = \sqrt{(3t^2 + 3)^2} = 3t^2 + 3.$$

Therefore,

$$s = \int_0^1 \left|\frac{d\bar{R}}{dt}\right| dt = \int_0^1 (3t^2 + 3) \, dt = (t^3 + 3t)\Big|_0^1 = 4 \text{ units}.$$

2 $\begin{cases} x = e^{-t} \cos t \\ y = e^{-t} \sin t \end{cases}$ between $t = 0$ and $t = \pi$.

SOLUTION

The given scalar parametric equations are equivalent to the vector parametric equation $\bar{R} = g(t)\bar{i} + h(t)\bar{j}$, where $g(t) = e^{-t}\cos t$ and $h(t) = e^{-t}\sin t$. Thus,

$$g'(t) = -e^{-t}(\cos t + \sin t), \quad h'(t) = -e^{-t}(\sin t - \cos t),$$

and

$$s = \int_0^\pi \sqrt{[g'(t)]^2 + [h'(t)]^2}\, dt$$

$$= \int_0^\pi \sqrt{e^{-2t}(\cos^2 t + 2\cos t \sin t + \sin^2 t) + e^{-2t}(\sin^2 t - 2\sin t \cos t + \cos^2 t)}\, dt$$

$$= \int_0^\pi e^{-t}\sqrt{2(\cos^2 t + \sin^2 t)}\, dt = \int_0^\pi \sqrt{2}\, e^{-t}\, dt$$

$$= \left(-\sqrt{2}\, e^{-t}\right)\Big|_0^\pi = \sqrt{2}(1 - e^{-\pi}) \text{ units.}$$

3 $y = f(x)$ between $x = a$ and $x = b$.

SOLUTION

With x as the parameter, the scalar equation $y = f(x)$ is equivalent to the vector parametric equation $\bar{R} = x\bar{i} + f(x)\bar{j}$. Here,

$$\frac{d\bar{R}}{dx} = \bar{i} + f'(x)\bar{j}, \quad \left|\frac{d\bar{R}}{dx}\right| = \sqrt{1^2 + [f'(x)]^2}, \quad \text{and } s = \int_a^b \sqrt{1 + [f'(x)]^2}\, dx,$$

which coincides with the formula for the arc length in Chapter 7, Section 4.1.

Problem Set 7

In problems 1 through 10, the vector equation of motion of a point P in the xy plane is given. Find (a) the velocity vector $\bar{V}$, (b) the acceleration vector $\bar{A}$, and (c) the speed v of the moving point at the time t.

1 $\bar{R} = (3t^2 - 1)\bar{i} + (2t + 5)\bar{j}$

2 $\bar{R} = (2\cos t)\bar{i} - (7\sin t)\bar{j}$

3 $\bar{R} = t^2\bar{i} + (\ln t)\bar{j}$

4 $\bar{R} = (t^2 + 1)\bar{i} + t^3\bar{j}$

5 $\bar{R} = 2(t - \sin t)\bar{i} + 2(1 - \cos t)\bar{j}$

6 $\bar{R} = (4\cos^3 t)\bar{i} + (4\sin^3 t)\bar{j}$

7 $\bar{R} = (\cos t + \sin t)\bar{i} + (\cos t - \sin t)\bar{j}$

8 $\bar{R} = (2\cot t)\bar{i} + (2\sin^2 t)\bar{j}$

9 $\bar{R} = (e^t \cos t)\bar{i} + (e^t \sin t)\bar{j}$

10 $\bar{R} = \phi(t)\bar{i} + f(\phi(t))\bar{j}$

In problems 11 through 16, the vector equation of motion of a point P in the xy plane is given. Find (a) the velocity vector $\bar{V}$, (b) the acceleration vector $\bar{A}$, and (c) the speed v of the moving point at the instant when t has the indicated value t_1.

11 $\bar{R} = (7t^2 - t)\bar{i} + (5t - 7)\bar{j}$; $t_1 = 2$.

12 $\bar{R} = (t\sin t)\bar{i} + (\cos t)\bar{j}$; $t_1 = \pi/2$.

13 $\bar{R} = 3(1 + \cos \pi t)\bar{i} + 4(1 + \sin \pi t)\bar{j}$; $t_1 = \frac{5}{4}$.

14 $\bar{R} = t\bar{i} + e^t\bar{j}$; $t_1 = 0$.

15 $\bar{R} = (\ln \sin t)\bar{i} + (\ln \cos t)\bar{j}$; $t_1 = \pi/6$.

16 $\bar{R} = (4\cos t^2)\bar{i} + (4\sin t^2)\bar{j}$; $t_1 = \sqrt{\pi/2}$.

In problems 17 through 24, find (a) the velocity vector $\bar{V}$, (b) the speed v at time t, and (c) the unit tangent vector $\bar{T}$.

17 $\bar{R} = (12t - 3)\bar{i} - (7t + 9)\bar{j}$.

18 $\bar{R} = (at + b)\bar{i} + (ct + d)\bar{j}$, where a, b, c, and d are constants and $a^2 + c^2 \neq 0$.

19 $\bar{R} = (2t - 1)\bar{i} + (3t^2 + 7)\bar{j}$.

20 $\bar{R} = \bar{R}_0 + t\bar{M}$, where $\bar{R}_0$ and $\bar{M}$ are constant vectors and $|\bar{M}| \neq 0$.

21 $\bar{R} = (-7 + \cos 2t)\bar{i} + (5 + \sin 2t)\bar{j}$.

22 $\bar{R} = (x_0 + a \cos t)\bar{i} + (y_0 + b \sin t)\bar{j}$, where $x_0, y_0, a,$ and b are constants, $a > 0$ and $b > 0$.

23 $\bar{R} = \dfrac{\cos t}{1 + \cos t}\bar{i} + \dfrac{\sin t}{1 + \cos t}\bar{j}$.   **24** $\bar{R} = \ln (t^2 + 1)\bar{i} + e^{2t}\bar{j}$.

In problems 25 through 36, find the the arc length s of each curve between the indicated values of the parameter.

25 $\bar{R} = (7t - 9)\bar{i} - (5t + 4)\bar{j}$; $t = 0$ to $t = 2$.   **26** $\bar{R} = (at + b)\bar{i} + (ct + d)\bar{j}$; $t = t_1$ to $t = t_2$.

27 $\bar{R} = e^{2t}\bar{i} + (\frac{1}{4}e^{4t} - t)\bar{j}$; $t = 0$ to $t = \frac{1}{2}$.   **28** $\bar{R} = t^2\bar{i} + t^3\bar{j}$; $t = 0$ to $t = 4$.

29 $\bar{R} = (3 \cos 2t)\bar{i} + (3 \sin 2t)\bar{j}$; $t = 0$ to $t = \pi$.   **30** $\begin{cases} x = 3 \cos t - \cos 3t \\ y = 3 \sin t - \sin 3t \end{cases}$; $t = 0$ to $t = \pi$.

31 $\begin{cases} x = e^{-t} \cos t \\ y = e^{-t} \sin t \end{cases}$; $t = 0$ to $t = 2\pi$.   **32** $\begin{cases} x = t - \sin t \\ y = 1 - \cos t \end{cases}$; $t = 0$ to $t = 2\pi$.

33 $\bar{R} = (\cos t + t \sin t)\bar{i} + (\sin t - t \cos t)\bar{j}$; $t = 0$ to $t = \pi/4$.   **34** $\bar{R} = t\bar{i} + t^2\bar{j}$; $t = 0$ to $t = a$.

35 $\bar{R} = (\ln \sqrt{1 + t^2})\bar{i} + (\tan^{-1} t)\bar{j}$; $t = 0$ to $t = 1$.   **36** $\bar{R} = \dfrac{t^2}{2}\bar{i} + \left|\dfrac{(6t + 9)^{3/2}}{9}\right|\bar{j}$; $t = 0$ to $t = 4$.

37 The curve C whose equation in polar coordinates is $r = f(\theta)$ can be expressed in parametric cartesian form as

$$\begin{cases} x = f(\theta) \cos \theta \\ y = f(\theta) \sin \theta, \end{cases}$$

using θ as parameter. Using the methods of this section, find a formula for the arc length of the portion of C between the point where $\theta = \theta_1$ and the point where $\theta = \theta_2$.

38 Find a formula for the unit tangent vector $\bar{T}$ to the curve C in problem 37.

39 The curve C whose cartesian equation is $y = f(x)$ can be expressed in vector parametric form as $\bar{R} = x\bar{i} + f(x)\bar{j}$, using x as the parameter. Find a formula for the unit tangent vector $\bar{T}$ to C.

40 If $\bar{R} = \bar{F}(s)$, where the parameter s is the arc length, find a formula for the unit tangent vector $\bar{T}$.

8   Normal Vectors and Curvature

In this section we continue the study, initiated in Section 7, of the curve C traced out by a moving point P in the plane according to the vector parametric equation

$$\bar{R} = \overrightarrow{OP} = \bar{F}(t)$$

(Figure 1). As in Section 7, the parameter t can be thought of as time, so that the vector derivatives

$$\bar{V} = \bar{F}'(t) \quad \text{and} \quad \bar{A} = \bar{F}''(t),$$

which we assume exist, represent the velocity and acceleration vectors of P, respectively, at the time t. We continue to denote by s the arc length traced out by P along

Figure 1

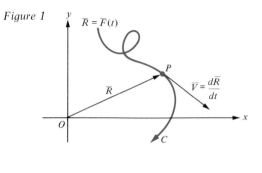

Figure 2

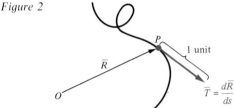

Figure 3

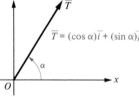

Figure 4

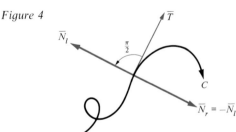

the curve C since some arbitrary (but fixed) initial time t_0, so that

$$v = \frac{ds}{dt} = |\overline{V}|$$

is the instantaneous speed of the moving point P.

For the remainder of this section we assume that the instantaneous speed of the moving point P is never zero, $v \neq 0$, so that we can form the unit tangent vector

$$\overline{T} = \frac{\overline{V}}{v}$$

to the curve C at the point P (Figure 2). Using the chain rule, we have

$$\frac{d\overline{R}}{ds} = \frac{d\overline{R}}{dt}\frac{dt}{ds} = \frac{d\overline{R}/dt}{ds/dt} = \frac{\overline{V}}{v} = \overline{T};$$

that is, the unit tangent vector $\overline{T}$ is the derivative $d\overline{R}/ds$ of the position vector with respect to *arc length*.

As the moving point P traces out the curve C, the unit tangent vector $\overline{T}$ at P can change its direction, but not its length (which is always 1 unit). In order to study the change in direction of $\overline{T}$, we denote by α the angle in radians from the positive x axis to $\overline{T}$ when the initial point of $\overline{T}$ is moved to the origin (Figure 3). Thus, since $|\overline{T}| = 1$, we have

$$\overline{T} = (\cos \alpha)\overline{i} + (\sin \alpha)\overline{j}$$

(see Figure 3, Section 5, page 700).

Any vector perpendicular to the unit tangent vector $\overline{T}$ at the point P on the curve C is called a *normal vector* to C at P. Of course, a normal vector whose length is 1 unit is called a *unit normal vector*. If you were to sit on the moving point P, facing in the direction of the unit tangent vector $\overline{T}$, as P traces out the curve C, you would find that one of the unit normal vectors, $\overline{N}_l$, is always on your left and the other unit normal vector, $\overline{N}_r$, is always on your right (Figure 4). Evidently,

$$\overline{N}_r = -\overline{N}_l.$$

Since $\overline{N}_l$ is obtained by rotating $\overline{T}$ *counterclockwise* through $\pi/2$ radians, it follows that $\overline{N}_l$ makes an angle of $\alpha + \pi/2$ radians with the positive x axis. Therefore, since $|\overline{N}_l| = 1$, we have

$$\overline{N}_l = \cos\left(\alpha + \frac{\pi}{2}\right)\overline{i} + \sin\left(\alpha + \frac{\pi}{2}\right)\overline{j}$$

$$= (-\sin \alpha)\overline{i} + (\cos \alpha)\overline{j};$$

hence,

$$\overline{N}_r = -\overline{N}_l = (\sin \alpha)\overline{i} - (\cos \alpha)\overline{j}.$$

If g and h are the component functions of $\overline{F}$, so that $\overline{R} = \overline{F}(t) = g(t)\overline{i} + h(t)\overline{j}$, then, as we have seen in Section 7,

$$\bar{T} = \frac{d\bar{R}/dt}{|d\bar{R}/dt|} = \frac{g'(t)}{\sqrt{[g'(t)]^2 + [h'(t)]^2}}\, i + \frac{h'(t)}{\sqrt{[g'(t)]^2 + [h'(t)]^2}}\, j.$$

Also, $\bar{T} = (\cos \alpha)i + (\sin \alpha)j$; hence, it follows that

$$\cos \alpha = \frac{g'(t)}{\sqrt{[g'(t)]^2 + [h'(t)]^2}} \quad \text{and} \quad \sin \alpha = \frac{h'(t)}{\sqrt{[g'(t)]^2 + [h'(t)]^2}}.$$

Therefore, since $\bar{N}_l = (-\sin \alpha)i + (\cos \alpha)j$, we have

$$\bar{N}_l = \frac{-h'(t)}{\sqrt{[g'(t)]^2 + [h'(t)]^2}}\, i + \frac{g'(t)}{\sqrt{[g'(t)]^2 + [h'(t)]^2}}\, j.$$

EXAMPLE Find the unit normal vector $\bar{N}_l$ to the curve $\bar{R} = ti + t^2 j$ at the point P whose position vector is $\bar{R}$.

SOLUTION
Here the component functions g and h are given by $g(t) = t$ and $h(t) = t^2$. Thus,

$$g'(t) = 1, \quad h'(t) = 2t, \quad \sqrt{[g'(t)]^2 + [h'(t)]^2} = \sqrt{1 + 4t^2},$$

and

$$\bar{N}_l = \frac{-2t}{\sqrt{1 + 4t^2}}\, i + \frac{1}{\sqrt{1 + 4t^2}}\, j.$$

When the curve C bends rather sharply, the unit tangent vector $\bar{T}$ changes its direction quite rapidly; that is, the rate of change $d\alpha/ds$ of the angle α (Figure 3) with respect to arc length s is large in absolute value. Consequently, the instantaneous rate $d\alpha/ds$ at which the tangent vector is turning, in radians per unit of arc length, is called the *curvature* of C at the point P and is traditionally denoted by the Greek letter κ (called "kappa"). Thus, by definition,

$$\kappa = \frac{d\alpha}{ds}.$$

Notice that $\kappa > 0$ if α is increasing as P moves to trace out the curve; that is, $\kappa > 0$ if the unit tangent vector turns *counterclockwise* as s is increased (Figure 5a). Similarly, $\kappa < 0$ if the unit tangent vector turns *clockwise* as s is increased (Figure 5b).

Figure 5

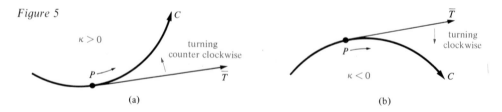

(a)   (b)

If we differentiate the equation $\bar{T} = (\cos \alpha)i + (\sin \alpha)j$ on both sides with respect to arc length s, we obtain

$$\frac{d\bar{T}}{ds} = \left(-\sin \alpha \frac{d\alpha}{ds}\right)i + \left(\cos \alpha \frac{d\alpha}{ds}\right)j = \frac{d\alpha}{ds}[(-\sin \alpha)i + (\cos \alpha)j];$$

hence,

$$\frac{d\bar{T}}{ds} = \kappa \bar{N}_l.$$

The equation $d\overline{T}/ds = \kappa\overline{N}_l$ is extremely important in the theory of curves in the plane. Since $\overline{N}_l$ is normal to the curve C and κ is a scalar, it implies that the vector $d\overline{T}/ds$ is always a normal vector to the curve C at the point P. Furthermore, since $|\overline{N}_l| = 1$, it follows that

$$\left|\frac{d\overline{T}}{ds}\right| = |\kappa\overline{N}_l| = |\kappa|\,|\overline{N}_l| = |\kappa|.$$

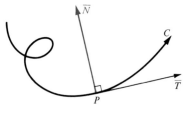

Thus, the vector $d\overline{T}/ds$ is long where C is rather sharply curved, and not so long where C bends slowly. Finally, it can be shown that $d\overline{T}/ds$ always points in the direction of concavity of the curve C (**Figure 6**) (see Problem 30).

The vector $d\overline{T}/ds$ is called the *curvature vector* of the curve C at the point P. If the curvature vector is nonzero, that is, if $|\kappa| = |d\overline{T}/ds| \neq 0$, then the vector $\overline{N}$ defined by

Figure 7

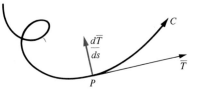

$$\overline{N} = \frac{\dfrac{d\overline{T}}{ds}}{\left|\dfrac{d\overline{T}}{ds}\right|} = \frac{1}{|\kappa|}\frac{d\overline{T}}{ds}$$

is called the *principal normal vector* to C at P (**Figure 7**). Notice that $\overline{N}$ is a unit normal vector to the curve C and points in the direction of concavity of C. (Why?) It follows that

$$\overline{N} = \overline{N}_l \quad \text{if } \kappa > 0, \qquad \text{while} \qquad \overline{N} = \overline{N}_r \quad \text{if } \kappa < 0.$$

Evidently, the equation

$$\frac{d\overline{T}}{ds} = |\kappa|\overline{N}$$

always holds, provided that $\kappa \neq 0$. If $\kappa = 0$, the principal normal vector is undefined.

The following important theorem relates the acceleration vector $\overline{A}$, the unit tangent vector $\overline{T}$, the speed v, the curvature vector $d\overline{T}/ds$, the principal normal vector $\overline{N}$, and the curvature κ.

THEOREM 1   **Resolution of the Acceleration Vector into Tangential and Normal Components**

(i) $\overline{A} = \dfrac{dv}{dt}\overline{T} + v^2\dfrac{d\overline{T}}{ds}.$

(ii) $\overline{A} = \dfrac{dv}{dt}\overline{T} + v^2|\kappa|\overline{N}$ if $\kappa \neq 0.$

PROOF
Since $\overline{T} = \overline{V}/v$, we have that $\overline{V} = v\overline{T}$. Differentiating both sides of the latter equation with respect to t, we obtain

$$\overline{A} = \frac{d\overline{V}}{dt} = \frac{dv}{dt}\overline{T} + v\frac{d\overline{T}}{dt}.$$

By the chain rule, $\dfrac{d\overline{T}}{dt} = \dfrac{ds}{dt}\dfrac{d\overline{T}}{ds} = v\dfrac{d\overline{T}}{ds}$; hence,

$$\overline{A} = \frac{dv}{dt}\,\overline{T} + v\left(v\frac{d\overline{T}}{ds}\right) = \frac{dv}{dt}\,\overline{T} + v^2\frac{d\overline{T}}{ds},$$

and part (i) is proved. To obtain part (ii), we substitute $d\overline{T}/ds = |\kappa|\overline{N}$ into part (i).

Theorem 1 expresses the acceleration vector $\overline{A}$ as a sum of its *tangential component* $\dfrac{dv}{dt}\,\overline{T}$ and its *normal component* $v^2\dfrac{d\overline{T}}{ds}$ (or $v^2|\kappa|\overline{N}$ if $\kappa \neq 0$) and has useful applications to mechanics (see Section 9). It also has the following important consequence.

THEOREM 2   Formula for Curvature

$$\kappa = \frac{\overline{A}\cdot\overline{N}_l}{v^2}.$$

PROOF
We have $d\overline{T}/ds = \kappa\overline{N}_l$; hence,

$$\frac{d\overline{T}}{ds}\cdot\overline{N}_l = \kappa\overline{N}_l\cdot\overline{N}_l = \kappa|\overline{N}_l|^2 = \kappa,$$

since $|\overline{N}_l| = 1$. Therefore, taking the dot product on both sides of the equation in part (i) of Theorem 1 with $\overline{N}_l$, we obtain

$$\overline{A}\cdot\overline{N}_l = \frac{dv}{dt}\,\overline{T}\cdot\overline{N}_l + v^2\frac{d\overline{T}}{ds}\cdot\overline{N}_l = \frac{dv}{dt}\,(0) + v^2\kappa = v^2\kappa,$$

since the vectors $\overline{T}$ and $\overline{N}_l$ are perpendicular. Solving the latter equation for κ, we obtain the formula $\kappa = (\overline{A}\cdot\overline{N}_l)/v^2$, as desired.

The formula for curvature in Theorem 2 can be rewritten in terms of the component functions g and h of $\overline{R}$ (Problem 22). Thus, if $\overline{R} = g(t)\overline{i} + h(t)\overline{j}$, then

$$\kappa = \frac{g'(t)h''(t) - g''(t)h'(t)}{\{[g'(t)]^2 + [h'(t)]^2\}^{3/2}}.$$

EXAMPLES   1 Find the curvature κ of $\overline{R} = t\overline{i} + t^2\overline{j}$.

SOLUTION
Here the component functions g and h are given by $g(t) = t$ and $h(t) = t^2$. Using the formula above and noting that

$$g'(t) = 1, \quad g''(t) = 0, \quad h'(t) = 2t, \quad \text{and } h''(t) = 2,$$

we obtain

$$\kappa = \frac{(1)(2) - (0)(2t)}{[1^2 + (2t)^2]^{3/2}} = \frac{2}{(1 + 4t^2)^{3/2}}.$$

2 Find the curvature κ and the principal normal vector $\overline{N}$ for the ellipse

$$\begin{cases} x = 2\cos t \\ y = 3\sin t. \end{cases}$$

SOLUTION

Here the component functions are given by $g(t) = 2\cos t$ and $h(t) = 3\sin t$. Since $g'(t) = -2\sin t$, $g''(t) = -2\cos t$, $h'(t) = 3\cos t$, and $h''(t) = -3\sin t$, we have

$$\kappa = \frac{g'(t)h''(t) - g''(t)h'(t)}{\{[g'(t)]^2 + [h'(t)]^2\}^{3/2}} = \frac{6\sin^2 t + 6\cos^2 t}{(4\sin^2 t + 9\cos^2 t)^{3/2}} = \frac{6}{(4\sin^2 t + 9\cos^2 t)^{3/2}}.$$

Since $\kappa > 0$,

$$\bar{N} = \bar{N}_l = \frac{-h'(t)\vec{i} + g'(t)\vec{j}}{\sqrt{[g'(t)]^2 + [h'(t)]^2}} = \frac{(-3\cos t)\vec{i} - (2\sin t)\vec{j}}{\sqrt{4\sin^2 t + 9\cos^2 t}}.$$

3 Find a formula for the curvature κ of the graph of $y = f(x)$, using x as the parameter.

SOLUTION

Writing the equation in vector parametric form, we have $\bar{R} = x\vec{i} + f(x)\vec{j}$; hence, the component functions g and h are given by $g(x) = x$ and $h(x) = f(x)$. Thus, replacing t by x in the formula for κ, we have $g'(x) = 1$, $g''(x) = 0$, $h'(x) = f'(x)$, and $h''(x) = f''(x)$; hence,

$$\kappa = \frac{g'(x)h''(x) - g''(x)h'(x)}{\{[g'(x)]^2 + [h'(x)]^2\}^{3/2}} = \frac{f''(x)}{\{1 + [f'(x)]^2\}^{3/2}}.$$

4 Find the curvature κ of the sine curve $y = \sin x$ at the point $(\pi/2, 1)$.

SOLUTION

Using the formula obtained in Example 3 with $f(x) = \sin x$, $f'(x) = \cos x$, and $f''(x) = -\sin x$, we have

$$\kappa = \frac{-\sin x}{(1 + \cos^2 x)^{3/2}}.$$

Thus, when $x = \frac{\pi}{2}$, $\kappa = -\frac{1}{1} = -1$.

For convenience, we now gather the various formulas pertaining to the curve C in the xy plane traced out by the variable position vector $\bar{R} = g(t)\vec{i} + h(t)\vec{j}$.

1 $\bar{V} = \dfrac{d\bar{R}}{dt} = g'(t)\vec{i} + h'(t)\vec{j}.$

2 $\bar{A} = \dfrac{d^2\bar{R}}{dt^2} = g''(t)\vec{i} + h''(t)\vec{j}.$

3 $v = \dfrac{ds}{dt} = \sqrt{[g'(t)]^2 + [h'(t)]^2}.$ *instantaneous velocity*

4 $\bar{T} = \dfrac{g'(t)\vec{i} + h'(t)\vec{j}}{\sqrt{[g'(t)]^2 + [h'(t)]^2}}.$ *unit tangent vector*

5 $\bar{N}_l = \dfrac{-h'(t)\vec{i} + g'(t)\vec{j}}{\sqrt{[g'(t)]^2 + [h'(t)]^2}}.$

6 $\bar{N}_r = \dfrac{h'(t)\vec{i} - g'(t)\vec{j}}{\sqrt{[g'(t)]^2 + [h'(t)]^2}}.$

7 $\kappa = \dfrac{g'(t)h''(t) - g''(t)h'(t)}{\{[g'(t)]^2 + [h'(t)]^2\}^{3/2}}.$

8 $\overline{N} = \begin{cases} \overline{N}_l & \text{if } \kappa > 0 \\ \overline{N}_r & \text{if } \kappa < 0. \end{cases}$

9 $s = \int_a^b \sqrt{[g'(t)]^2 + [h'(t)]^2}\, dt.$

10 $\dfrac{d\overline{R}}{ds} = \overline{T}.$

11 $\dfrac{d^2\overline{R}}{ds^2} = \dfrac{d\overline{T}}{ds} = \kappa\overline{N}_l = |\kappa|\overline{N}.$

12 $\dfrac{d\overline{N}}{ds} = -|\kappa|\overline{T}.$

13 $\overline{V} = v\overline{T}.$

14 $\overline{A} = \dfrac{dv}{dt}\overline{T} + v^2\dfrac{d\overline{T}}{ds}$

$= \dfrac{dv}{dt}\overline{T} + v^2|\kappa|\overline{N}$

$= \dfrac{dv}{dt}\overline{T} + v^2\kappa\overline{N}_l.$

15 $v = |\overline{V}|.$

16 $\overline{T} = \dfrac{\overline{V}}{v} = \dfrac{d\overline{R}}{ds}.$

17 $\overline{N}_r = -\overline{N}_l.$

18 $\overline{N} = \dfrac{\dfrac{d\overline{T}}{ds}}{\left|\dfrac{d\overline{T}}{ds}\right|} = \dfrac{\dfrac{d\overline{T}}{ds}}{|\kappa|}.$

19 $\kappa = \dfrac{\overline{A}\cdot\overline{N}_l}{v^2} = \dfrac{d\alpha}{ds}.$

20 $|\kappa| = \dfrac{\overline{A}\cdot\overline{N}}{v^2} = \left|\dfrac{d\overline{T}}{ds}\right| = \dfrac{\left|\dfrac{d\overline{T}}{dt}\right|}{v}.$

All the formulas above have either been obtained or follow easily from what has already been established, with the possible exception of Formula 12 and part of Formula 20. The proofs of Formulas 12 and 20 are left as exercises (Problems 32 and 33).

Problem Set 8

In problems 1 through 16, find (a) the unit tangent vector $\overline{T}$, (b) the unit normal vector $\overline{N}_l$, (c) the curvature κ, and (d) the principal unit normal vector $\overline{N}$ for each curve.

1 $\overline{R} = (7t - 4)\hat{i} + (9 - 3t)\hat{j}$

2 $\overline{R} = (at + b)\hat{i} + (ct + d)\hat{j}$

3 $\begin{cases} x = 3\cos t \\ y = 3\sin t \end{cases}$

4 $\bar{R} = (a \cos t)\bar{i} + (a \sin t)\bar{j}$, where a is a positive constant.

5 $\bar{R} = 3(1 + \cos \pi t)\bar{i} + 5(1 + \sin \pi t)\bar{j}$

6 $\begin{cases} x = t^2 - 2t \\ y = 1 - 7t \end{cases}$ **7** $\bar{R} = t\bar{i} + e^t\bar{j}$ **8** $\bar{R} = (2 \cos \theta)\bar{i} + (5 \sin \theta)\bar{j}$

9 $\begin{cases} x = 3t^2 - 1 \\ y = 2t^2 + 7 \end{cases}$ **10** $\bar{R} = t^2\bar{i} + (\ln \sec t)\bar{j}$ **11** $R = \dfrac{1}{t}\bar{i} + (t^2 + 1)\bar{j}$

12 $\begin{cases} x = e^t \\ y = e^{-t} \end{cases}$ **13** $\bar{R} = (\cos 2\theta)\bar{i} + (\sin \theta)\bar{j}$ **14** $\bar{R} = (\ln u)\bar{i} + e^{2u}\bar{j}$

15 $y = \ln x$, with x as the parameter. **16** $y = 4x^2$, with x as the parameter.

In problems 17 through 21, find (a) the unit tangent vector $\bar{T}$, (b) the unit normal vector $\bar{N}_t$, (c) the curvature κ, and (d) the principal unit normal vector $\bar{N}$ for each curve at the point where the parameter has the indicated value.

17 $\bar{R} = 3(1 - \cos t)\bar{i} + 3(1 - \sin t)\bar{j}$ when $t = \pi/6$.

18 $\bar{R} = (\sin t)\bar{i} + (\tan t)\bar{j}$ when $t = 5\pi/6$.

19 $\begin{cases} x = \ln (t + 3) \\ y = \dfrac{1}{4}t^4 \end{cases}$ when $t = 1$. **20** $\begin{cases} x = ue^u \\ y = ue^{-u} \end{cases}$ when $u = 0$. **21** $\bar{R} = \dfrac{u^2 + 1}{u^2 - 1}\bar{i} + \dfrac{1}{u^2}\bar{j}$ when $u = 2$.

22 Using Theorem 2, derive the formula for the curvature κ of the curve $\bar{R} = g(t)\bar{i} + h(t)\bar{j}$.

23 Consider the curve C whose equation in polar coordinates is $r = f(\theta)$. With θ as a parameter, the corresponding vector parametric equation is $\bar{R} = [f(\theta) \cos \theta]\bar{i} + [f(\theta) \sin \theta]\bar{j}$ and the component functions g and h are given by $g(\theta) = f(\theta) \cos \theta$ and $h(\theta) = f(\theta) \sin \theta$. Find formulas for (a) $\bar{T}$, (b) $\bar{N}_t$, (c) κ, and (d) $\bar{N}$.

In problems 24 through 28, use the results of problem 23 to find (a) $\bar{T}$, (b) $\bar{N}_t$, (c) κ, and (d) $\bar{N}$ for the curve C whose equation is given in polar coordinates.

24 $r = 1 - \cos \theta$ **25** $r = 1 + 2 \cos \theta$

26 $r = \dfrac{ed}{1 + e \sin \theta}, d > 0$ **27** $r = \theta$

28 Show that the principal normal vector for a circle always points toward the center of the circle.

29 Show that the absolute value of the curvature of a circle is a constant equal to the reciprocal of its radius.

30 Explain why the curvature vector $d\bar{T}/ds$, and therefore also the principal unit normal vector $\bar{N}$, always point in the direction of concavity of the curve. (*Hint:* For s small, $d\bar{T}/ds$ is approximated by $\Delta\bar{T}/\Delta s$.)

31 If we use x as the parameter, explain the geometric significance of the algebraic sign of the curvature κ of the graph of $y = f(x)$.

32 Prove that $\dfrac{d\bar{N}}{ds} = -|\kappa|\bar{T}$. [*Hint:* Begin by noticing that $\bar{N} = \dfrac{\kappa}{|\kappa|}\bar{N}_t$, where $\bar{N}_t = (-\sin \alpha)\bar{i} + (\cos \alpha)\bar{j}$.]

33 Prove that $|\kappa| = \bar{A} \cdot \bar{N}/v^2$.

34 Assume that $\bar{R} = \bar{F}(t)$ is the vector parametric equation of a curve with constant nonzero curvature κ. Prove that the curve is a circle (or a portion of a circle) by carrying out the following steps:
(a) Show that $\bar{R} + (1/|\kappa|)\bar{N}$ is a constant vector by proving that its derivative is zero. (Use problem 32.)
(b) Put $\bar{R}_0 = \bar{R} + (1/|\kappa|)\bar{N}$.
(c) Show that $\bar{R}$ satisfies an equation of the form $|\bar{R} - \bar{R}_0| = $ constant.

35 Assume that $\bar{R} = \bar{F}(t)$ is the vector parametric equation of a curve whose curvature is equal to zero at all points. Prove that the curve is a straight line (or a portion of a straight line). [*Hint:* First use the relation $|d\bar{T}/ds| = |\kappa|$ to prove that $\bar{T}$ is a constant. Then select a fixed vector $\bar{N} \neq \bar{0}$ such that $\bar{N} \cdot \bar{T} = 0$. Choose and fix a value of $\bar{R}$, say $\bar{R}_0$. Show that $\dfrac{d}{ds}[\bar{N} \cdot (\bar{R} - \bar{R}_0)] = 0$, conclude that $\bar{N} \cdot (\bar{R} - \bar{R}_0)$ is a constant, and then show that $\bar{N} \cdot (\bar{R} - \bar{R}_0) = 0$.]

9 Applications to Mechanics

In this section we use the ideas and techniques developed in Sections 7 and 8 to study the motion of a particle in the xy plane. Briefly, the idea here is to consider a curve that is traced out by a physical particle of mass m, rather than by a geometric point.

Figure 1

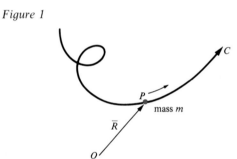

Thus, suppose that a particle P of mass m moves in the xy plane and traces out a curve C (Figure 1). The curve C is called the *path*, or the *trajectory*, or the *orbit* of the particle P. As before, we can locate the particle at any given time t by the position vector $\bar{R} = \overrightarrow{OP}$, and we have

$$\bar{R} = \bar{F}(t),$$

where we assume that $\bar{F}$ is a twice-differentiable vector-valued function. The equation $\bar{R} = \bar{F}(t)$ is called the *equation of motion* of the particle P.

By Formulas 1 and 13 in Section 8, pages 723 and 724, the velocity of the particle at any instant is the vector

$$\bar{V} = \frac{d\bar{R}}{dt} = v\bar{T},$$

where v is the instantaneous speed of the particle and $\bar{T}$ is the unit tangent vector to the path of the particle.

EXAMPLE A particle moves along the parabola $y^2 = 2x$ from left to right away from the vertex at a speed of 5 units per second. Find the velocity vector $\bar{V}$ as the particle moves through the point $(2, 2)$.

SOLUTION
Using y (not t) as a parameter, we can write a vector parametric equation of the parabola as

$$\bar{R} = \frac{y^2}{2}\,i + y\,j,$$

since $x = \dfrac{y^2}{2}$. Here we have

$$\frac{d\bar{R}}{dy} = yi + j \quad \text{and} \quad \left|\frac{d\bar{R}}{dy}\right| = \sqrt{y^2 + 1};$$

hence,

$$\bar{T} = \frac{\dfrac{d\bar{R}}{dy}}{\left|\dfrac{d\bar{R}}{dy}\right|} = \frac{yi + j}{\sqrt{y^2 + 1}}$$

is the unit tangent vector to the parabola at the point $(y^2/2, y)$. When $y = 2$, we have $\bar{T} = \dfrac{2i + j}{\sqrt{5}}$ and

$$\bar{V} = v\bar{T} = (5)\frac{2i + j}{\sqrt{5}} = \sqrt{5}(2i + j).$$

In the example above, we were not given the equation of motion of the particle—we were only given the equation of its *path* and the speed of the particle at a certain instant. Thus, we were not able to calculate $\bar{V}$ directly as $d\bar{R}/dt$ and had to rely instead on the formula $\bar{V} = v\bar{T}$.

Newton's second law of motion can now be expressed in the form

$$\bar{f} = m\bar{A},$$

where the vector $\bar{f}$ represents the force acting on the particle and

$$\bar{A} = \frac{d\bar{V}}{dt} = \frac{d^2\bar{R}}{dt^2}$$

is the acceleration of the particle. (Here we are assuming that the mass m is constant and we are neglecting relativistic effects.)

EXAMPLE Suppose that a projectile is fired at an angle of 60° with the level ground at an initial speed of 800 ft/sec. Assuming that the only force acting on the projectile is the force of gravity,

Figure 2

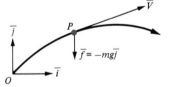

$$\bar{f} = -mgj,$$

and that the projectile is fired from the origin O at time $t = 0$, find (a) the velocity vector $\bar{V}$ and (b) the position vector $\bar{R}$ at time t (Figure 2).

SOLUTION
Let the equation of motion of the projectile be $\bar{R} = xi + yj$, where x and y are functions of time t. Then

$$\bar{A} = \frac{d^2\bar{R}}{dt^2} = \frac{d^2x}{dt^2} i + \frac{d^2y}{dt^2} j$$

and the equation $\bar{f} = m\bar{A}$ becomes

$$-mgj = m\frac{d^2x}{dt^2} i + m\frac{d^2y}{dt^2} j.$$

Equating scalar components gives the two scalar differential equations

$$\begin{cases} m\dfrac{d^2x}{dt^2} = 0 \\[2mm] m\dfrac{d^2y}{dt^2} = -mg; \end{cases} \quad \text{that is,} \quad \begin{cases} \dfrac{d^2x}{dt^2} = 0 \\[2mm] \dfrac{d^2y}{dt^2} = -g. \end{cases}$$

Integrating twice, we find that the solution of $d^2x/dt^2 = 0$ is

$$x = C_1 t + C_2,$$

where C_1 and C_2 are the constants of integration. Similarly, the differential equation $d^2y/dt^2 = -g$ has the solution

$$y = -\tfrac{1}{2}gt^2 + C_3 t + C_4.$$

Therefore, the equation of motion has the form

$$\bar{R} = x i + y j = (C_1 t + C_2) i + \left(-\frac{1}{2}gt^2 + C_3 t + C_4\right) j,$$

where the constants $C_1, C_2, C_3,$ and C_4 must still be determined. Differentiating the equation of motion with respect to t, we obtain

$$\bar{V} = \frac{d\bar{R}}{dt} = C_1 i + (-gt + C_3) j.$$

When $t = 0$, we have

$$\bar{V} = \bar{V}_0 = C_1 i + C_3 j.$$

Now, the initial velocity $\bar{V}_0$ has length 800 ft/sec and makes an angle of 60° with the positive x axis; hence,

$$\bar{V}_0 = (800\cos 60°) i + (800\sin 60°) j = 400 i + 400\sqrt{3}\, j.$$

Therefore,

$$C_1 i + C_3 j = 400 i + 400\sqrt{3}\, j,$$

so that $C_1 = 400$ and $C_3 = 400\sqrt{3}$. Hence,

(a) $\bar{V} = 400 i + (-gt + 400\sqrt{3}) j.$

The equation of motion now becomes

$$\bar{R} = (400t + C_2) i + (-\tfrac{1}{2}gt^2 + 400\sqrt{3}\,t + C_4) j.$$

Putting $t = 0$ and recalling that the projectile starts from the origin, we have $\bar{0} = C_2 i + C_4 j$; hence, $C_2 = 0$ and $C_4 = 0$. Therefore,

(b) $\bar{R} = (400t) i + (-\tfrac{1}{2}gt^2 + 400\sqrt{3}\,t) j.$

9.1 Tangential and Normal Components of Acceleration

By part (ii) of Theorem 1 in Section 8, we have

$$\bar{A} = \frac{dv}{dt}\,\bar{T} + |\kappa| v^2 \bar{N}, \qquad \text{if } \kappa \neq 0.$$

Geometrically, this equation says that the acceleration vector $\bar{A}$ can be resolved into a sum of two perpendicular vectors, $(dv/dt)\bar{T}$ and $|\kappa| v^2 \bar{N}$, the first of which is tangent to the trajectory and the second of which is normal to the trajectory (Figure 3). (If the curvature $\kappa = 0$, then the normal component of the acceleration is $\bar{0}$ and we have $\bar{A} = \dfrac{dv}{dt}\,\bar{T}$ in this case.) The normal component vector $|\kappa|v^2\bar{N}$ of the acceleration is also called the *centripetal acceleration.* It can be regarded as that part of the acceleration $\bar{A}$ caused by the change in *direction* of the velocity vector. If P moves along a straight line, then $\kappa = 0$, so that the centripetal acceleration is $\bar{0}$. On the other hand, if the speed v of the particle is constant, then $dv/dt = 0$, the tangential component of the acceleration is $\bar{0}$, and the acceleration is entirely centripetal.

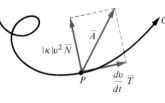

Figure 3

EXAMPLE A particle P is moving with a constant speed of 10 units per second in a counter-clockwise direction around the ellipse $x^2/4 + y^2/9 = 1$. Find the acceleration vector $\bar{A}$ at the moment when the particle passes through the vertex $(0, 3)$.

SOLUTION

We find $\bar{A}$ by using the formula $\bar{A} = \dfrac{dv}{dt}\,\bar{T} + |\kappa|v^2\bar{N}$. Since v is constant, we have $dv/dt = 0$ and $\bar{A} = |\kappa|v^2\bar{N}$. To find κ, we begin by implicitly differentiating the equation of the ellipse to obtain $\dfrac{2x}{4} + \dfrac{2y}{9}\dfrac{dy}{dx} = 0$, so that

$$\frac{dy}{dx} = -\frac{9x}{4y} \quad \text{and} \quad \frac{d^2y}{dx^2} = \frac{-36y + 36x\dfrac{dy}{dx}}{16y^2}.$$

Thus, when $x = 0$ and $y = 3$, we have

$$\frac{dy}{dx} = 0 \quad \text{and} \quad \frac{d^2y}{dx^2} = \frac{-(36)(3) + 0}{(16)(3)^2} = -\frac{3}{4}.$$

Using the formula obtained in Example 3, Section 8, page 723, we have

$$\kappa = \frac{d^2y/dx^2}{[1 + (dy/dx)^2]^{3/2}} = \frac{-\frac{3}{4}}{(1 + 0)^{3/2}} = -\frac{3}{4}.$$

When $(x, y) = (0, 3)$, the tangent vector is horizontal, so that the principal unit normal vector points straight *down* (in the direction of concavity of the ellipse at this point). Therefore, at the point $(0, 3)$, $\bar{N} = -\bar{j}$, and we have

$$\bar{A} = |\kappa|v^2\bar{N} = |-\tfrac{3}{4}|(10)^2(-\bar{j}) = -75\bar{j}.$$

9.2 Centripetal Force

If we combine Newton's law $\bar{f} = m\bar{A}$ with the equation $\bar{A} = \dfrac{dv}{dt}\,\bar{T} + |\kappa|v^2\bar{N}$, we obtain the equation

$$\bar{f} = m\frac{dv}{dt}\,\bar{T} + |\kappa|mv^2\bar{N};$$

that is, the force $\bar{f}$ acting on the particle can always be resolved into a sum of two perpendicular component vectors, a *tangential component vector* $m\dfrac{dv}{dt}\overline{T}$ of magnitude $m\dfrac{dv}{dt}$, and a *normal component vector* $|\kappa|mv^2\overline{N}$ of magnitude $|\kappa|mv^2$. The normal component vector $|\kappa|mv^2\overline{N}$ is called the *centripetal force vector* and its length $|\kappa|mv^2$ is called the *magnitude of the centripetal force* or sometimes simply the *centripetal force.*

EXAMPLE
A particle P is moving on a circle of radius r with center at the point P_0. Find the tangential and the normal component vectors of the force $\overline{f}$ acting on the particle in terms of the instantaneous speed v and its time derivative dv/dt. Discuss the case in which the speed v of the particle is a constant (Figure 4).

Figure 4

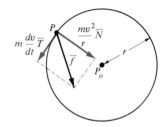

SOLUTION
By Problem 29 in Problem Set 8, the absolute value of the curvature of the circle is given by $|\kappa| = 1/r$. Thus,

$$\overline{f} = m\frac{dv}{dt}\overline{T} + \frac{mv^2}{r}\overline{N},$$

where the principal normal unit vector $\overline{N}$ points directly toward the center of the circle and the unit tangent vector $\overline{T}$ points in the instantaneous direction of motion of the particle. If the particle moves with constant speed v, then $dv/dt = 0$ and the force vector $\overline{f}$ coincides with the centripetal force vector $(mv^2/r)\overline{N}$. In this case the centripetal force is

$$|\overline{f}| = \left|\frac{mv^2}{r}\overline{N}\right| = \frac{mv^2}{r}.$$

The negative of the centripetal force vector in the example above, that is, the vector $-(mv^2/r)\overline{N}$, points directly *away* from the center of the circle and is called the *centrifugal force vector*. If a stone is whirled at the end of a string, the inertia of the stone causes it to pull away from the center of rotation, and this pull is represented by the centrifugal force vector $-(mv^2/r)\overline{N}$. The magnitude of the centrifugal force vector, mv^2/r, is often called the *centrifugal force* and it manifests itself as tension in the string. Notice that the centrifugal force and the centripetal force are numerically equal, but the corresponding force vectors point in opposite directions.

The formula mv^2/r for centripetal (or centrifugal) force gives the force in dynes when m is in grams, v is in centimeters per second, and r is in centimeters. To convert from dynes to kilograms of force, divide by 980,000. If m is expressed in pounds, v in feet per second, and r in feet, then mv^2/r will be the force in poundals. To convert from poundals to pounds of force, divide by 32.

EXAMPLES 1 A 100-gram mass is whirled around in a circle, three times per second, at the end of a string 70 centimeters long. Find the tension in the string in kilograms.

SOLUTION
The tension in the string is equal to the magnitude of the centrifugal force, mv^2/r. Since the mass travels three times around the circle each second, its speed v is given by

$$v = (2\pi r)(3) = 6\pi r = (6\pi)(70) = 420\pi \text{ cm/sec.}$$

Thus, the tension is given by

$$\frac{mv^2}{r} = \frac{(100)(420\pi)^2}{70} = 252{,}000\pi^2 \text{ dynes} \approx 2{,}487{,}140 \text{ dynes} \approx 2.54 \text{ kg.}$$

2 A 9-ton truck goes around a curve with a 100-foot radius at a speed of 15 miles per hour. How much centripetal force is required?

SOLUTION
Here $m = 9(2000) = 18{,}000$ pounds, $r = 100$ feet, and $v = 15(5280)/3600 = 22$ ft/sec. The centripetal force is directed toward the center of curvature and has magnitude

$$\frac{mv^2}{r} = \frac{(18{,}000)(22)^2}{100} = 87{,}120 \text{ poundals} = 2722.5 \text{ pounds.}$$

Problem Set 9

1 A particle moves along the upper branch of the hyperbola $y^2/4 - x^2/9 = 1$ from left to right, with a constant speed of 5 units of distance per second. Find (a) the velocity vector and (b) the acceleration vector at the instant when the particle passes through the vertex $(0, 2)$.

2 A particle moves along the parabola $y = x^2$ from left to right. As it passes through the point $(2, 4)$, v has the value 3 units per second and dv/dt has the value 7 units per second per second. Find (a) the velocity vector and (b) the acceleration vector at this point.

3 A particle moves along an ellipse according to the equations $\begin{cases} x = \cos 2\pi t \\ y = 3 \sin 2\pi t. \end{cases}$ Find the speed v and the rate of change of the speed dv/dt of the particle at time t.

4 A particle moves along a curve in the xy plane according to the equations $\begin{cases} x = g(t) \\ y = h(t). \end{cases}$ Derive the following formulas:

(a) $\bar{V} = \dfrac{dx}{dt}i + \dfrac{dy}{dt}j$

(b) $\dfrac{ds}{dt} = \sqrt{\left(\dfrac{dx}{dt}\right)^2 + \left(\dfrac{dy}{dt}\right)^2}$

(c) $\bar{A} = \dfrac{d^2x}{dt^2}i + \dfrac{d^2y}{dt^2}j$

(d) $\dfrac{dv}{dt} = \dfrac{\dfrac{dx}{dt}\dfrac{d^2x}{dt^2} + \dfrac{dy}{dt}\dfrac{d^2y}{dt^2}}{\sqrt{\left(\dfrac{dx}{dt}\right)^2 + \left(\dfrac{dy}{dt}\right)^2}}$

(e) $\dfrac{dv}{dt}\bar{T} = \dfrac{\dfrac{dx}{dt}\dfrac{d^2x}{dt^2} + \dfrac{dy}{dt}\dfrac{d^2y}{dt^2}}{(dx/dt)^2 + (dy/dt)^2}\left(\dfrac{dx}{dt}i + \dfrac{dy}{dt}j\right)$

(f) $|\kappa|v^2\bar{N} = \dfrac{1}{(dx/dt)^2 + (dy/dt)^2}\left[\dfrac{dy}{dt}\left(\dfrac{dy}{dt}\dfrac{d^2x}{dt^2} - \dfrac{dx}{dt}\dfrac{d^2y}{dt^2}\right)i + \dfrac{dx}{dt}\left(\dfrac{dx}{dt}\dfrac{d^2y}{dt^2} - \dfrac{dy}{dt}\dfrac{d^2x}{dt^2}\right)j\right]$

5 A player kicks a football at an angle of 30° with the level ground at a speed of 48 ft/sec. Assuming that the origin O of the xy coordinate system is placed at the point where the football is kicked, and neglecting air friction, find:
(a) The equation of motion of the football.
(b) The velocity vector $\bar{V}$ at time t.
(c) The total time of flight T of the football.
(d) The velocity vector at the point of impact.
(e) The horizontal distance between the origin and the point of impact.

6 A projectile is fired from the point (a, b) in the xy plane in a direction making an angle θ with the positive x axis at an initial speed of v_0 units of distance per second. Neglect air resistance.
(a) Derive the vector equation of motion of the projectile.
(b) Show that the projectile moves in a parabola.

7 A particle of mass 2 pounds is whirled in a horizontal circle of radius 2 feet. The particle makes 4 revolutions per second. Find the centripetal force on the particle.

8 A horizontal centrifuge rotates at 4300 revolutions per minute and has a radius of 3 inches. The centrifugal force developed by an object in the centrifuge is how many times the force of gravity on the object?

9 A jet plane is flying at a speed of 600 miles per hour in a horizontal circle. Find the radius of the circle in miles if the pilot feels a centrifugal force of "3 g's," that is, three times his own weight.

10 A particle of mass m is moving in a circle of radius r at a uniform speed. If the particle makes N revolutions per second, show that the centrifugal force is given by $4\pi^2 N^2 mr$.

11 A girl whirls an open pail of water in a vertical circle over her head. What is the minimum number of revolutions per second that will keep the water in the pail if the radius of the circle is 60 centimeters?

Review Problem Set

1 If A, B, C, and D are points in the xy plane such that $\overrightarrow{OB} - \overrightarrow{OA} = \overrightarrow{OC} - \overrightarrow{OD}$, show that $ABCD$ is a parallelogram.

2 Let $\bar{A}$ and $\bar{B}$ be position vectors of points P and Q, respectively, in the plane. Find the position vector of the point $\frac{4}{5}$ of the way from P to Q.

3 If $OABC$ is a parallelogram in the xy plane with A and C as opposite vertices, show that

$$\overrightarrow{OA} + \tfrac{1}{2}(\overrightarrow{OC} - \overrightarrow{OA}) = \tfrac{1}{2}\overrightarrow{OB}.$$

4 Explain the geometric significance of the condition $\bar{A} + \bar{B} + \bar{C} = 0$, where $\bar{A}$, $\bar{B}$, and $\bar{C}$ are vectors in the plane.

5 If u and v are scalars such that $u\bar{A} = v\bar{A}$, where $\bar{A}$ is a nonzero vector, is it true that $u = v$? Why?

6 Suppose that $ABCDEF$ is a regular hexagon with center at the origin O (Figure 1). Simplify the following expressions:
(a) $(\overrightarrow{AB} + \overrightarrow{OE}) + (\overrightarrow{AF} + \overrightarrow{BC}) + (\overrightarrow{AO} + \overrightarrow{CD}) + (\overrightarrow{ED} + \overrightarrow{AF})$.
(b) $(\overrightarrow{AD} - \overrightarrow{AF}) + (\overrightarrow{FE} - \overrightarrow{BA}) + (\overrightarrow{AO} - \overrightarrow{OF}) + (\overrightarrow{FD} - \overrightarrow{DB})$.

Figure 1

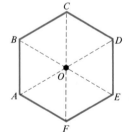

7 If s is a nonzero scalar and if $\bar{A}$ and $\bar{B}$ are vectors such that $s\bar{A} = s\bar{B}$, is it true that $\bar{A} = \bar{B}$? Why?

8 If i, j is the standard basis in the xy plane, write $\bar{A}$ as a linear combination of i and j and find $|\bar{A}|$, where $\bar{A}$ is the vector whose tail end is at $(2, -1)$ and whose head end is at $(-1, -2)$.

9 Let $\bar{A} = 2i - 3j$ and $\bar{B} = 4i + j$ be position vectors in the xy plane. Draw the vectors $\bar{A}$, $\bar{B}$, and $\bar{C} = \bar{A} + t\bar{B}$ on the same diagram for (a) $t = \frac{1}{2}$ and (b) $t = -2$.

10 Let $\bar{A} = 2i + j$ and $\bar{B} = i + 3j$ be position vectors in the xy plane. Draw the vectors $\bar{A}$, $\bar{B}$, and $\bar{C} = s\bar{A} + t\bar{B}$ on the same diagram for (a) $s = t = \frac{1}{2}$ and (b) $s = -1$ and $t = 2$.

11 Find the midpoint of the line segment containing the terminal points of the position vectors $\bar{A} = -i + 3j$ and $\bar{B} = 2i + 7j$.

12 Find the coordinates of the point that is $\frac{7}{10}$ of the way from $P = (-5, -9)$ to $Q = (7, 7)$.

13 Let $\bar{A} = 2i - j$ and $\bar{B} = 2i - 3j$. Find each of the following expressions:
(a) $5\bar{A}$ (b) $-4\bar{B}$ (c) $\bar{A} + \bar{B}$ (d) $\bar{A} - \bar{B}$ (e) $2\bar{A} + 3\bar{B}$
(f) $\bar{A} \cdot \bar{B}$ (g) $(2\bar{A} - 3\bar{B}) \cdot (2\bar{A} + 3\bar{B})$ (h) $|\bar{A}|$ (i) $|\bar{A} - \bar{B}|$ (j) $|2\bar{A}| + |3\bar{B}|$

14 Let $\bar{A} = \bar{i} + 2\bar{j}$, $\bar{B} = 2\bar{i} - 4\bar{j}$, and $\bar{C} = 3\bar{i} - 5\bar{j}$. Find scalars s and t such that $\bar{C} = s\bar{A} + t\bar{B}$.

15 Use dot products to find the three vertex angles of the triangle ABC where $A = (-2, -1)$, $B = (-1, 6)$, and $C = (2, 2)$.

16 Find two vectors $\bar{X}$ and $\bar{Y}$ in the plane such that both $\bar{X}$ and $\bar{Y}$ are perpendicular to the vector $\bar{A} = 2\bar{i} - 3\bar{j}$ and $|\bar{X}| = |\bar{Y}| = |\bar{A}|$.

17 Find the scalar components with respect to the standard $\bar{i}$, $\bar{j}$ basis of the vector obtained by rotating the vector $\bar{A} = x\bar{i} + y\bar{j}$ counterclockwise through $90°$.

18 Find the scalar components with respect to the standard $\bar{i}$, $\bar{j}$ basis of the vector obtained by rotating the vector $\bar{A} = x\bar{i} + y\bar{j}$ counterclockwise through θ radians.

19 Given that $|\bar{A} + \bar{B}|^2 - |\bar{A} - \bar{B}|^2 = 0$, show that $\bar{A}$ must be perpendicular to $\bar{B}$.

20 Let $\bar{A}$ and $\bar{B}$ be two given vectors in the plane such that $\bar{A} \cdot \bar{B} = 0$, $|\bar{A}| \neq 0$, and $|\bar{B}| \neq 0$. Show that, if $\bar{C}$ is any vector in the plane, then $\bar{C} = \dfrac{\bar{A} \cdot \bar{C}}{\bar{A} \cdot \bar{A}} \bar{A} + \dfrac{\bar{B} \cdot \bar{C}}{\bar{B} \cdot \bar{B}} \bar{B}$.

21 Find the scalar component of $\bar{A}$ in the direction of $\bar{B}$ if $\bar{A} = 2\bar{i} + 3\bar{j}$ and $\bar{B} = -3\bar{i} + 4\bar{j}$.

22 A manufacturer sells x sofas at a dollars per sofa and y chairs at b dollars per chair. The vector $\bar{D} = \langle x, y \rangle$ is called the *demand vector* while the vector $\bar{R} = \langle a, b \rangle$ is called the *revenue vector*. The *cost vector* is defined to be $\bar{C} = \langle c, d \rangle$, where c is the cost of manufacturing one sofa and d is the cost of manufacturing one chair. Give the economic interpretation of:
(a) $\bar{D} \cdot \bar{R}$ (b) $\bar{D} \cdot \bar{C}$ (c) $\bar{D} \cdot (\bar{R} - \bar{C})$

23 Draw two vectors $\bar{A}$ and $\bar{B}$ in the plane such that:
(a) $\bar{A} \cdot \bar{B} > 0$ (b) $\bar{A} \cdot \bar{B} = 0$ (c) $\bar{A} \cdot \bar{B} < 0$

24 Sketch the graph of the given vector nonparametric equation, where $\bar{A} = \bar{i} - 3\bar{j}$ and $\bar{B} = 2\bar{i} + \bar{j}$.
(a) $\bar{A} \cdot (\bar{R} - \bar{B}) = 0$ (b) $(\bar{R} - \bar{B}) \cdot (\bar{R} - \bar{B}) = 9$
(c) $\bar{A} \cdot \bar{R} = 8$ (d) $\bar{R} \cdot (\bar{R} - 2\bar{B}) = 0$

25 Find the equation of the straight line that contains the point $(3, 4)$ and has slope $-\frac{2}{5}$.
(a) In scalar nonparametric form.   (b) In vector nonparametric form.
(c) In vector parametric form.   (d) In scalar parametric form.

26 Rewrite equations (a) through (d) in problem 24 in vector parametric form.

27 Show that the following two vector parametric equations have the same graphs:
(a) $\bar{R} = (3 \sin t)\bar{i} + (3 \cos t)\bar{j}$ (b) $\bar{R} = (3 \cos 2t)\bar{i} + (3 \sin 2t)\bar{j}$

28 Let $\bar{F}(t) = \sqrt{t - 1}\,\bar{i} + \ln (2 - t)\bar{j}$.
(a) What is the domain of $\bar{F}$?   (b) Find $\lim\limits_{t \to 1.5} \bar{F}(t)$.   (c) Is $\bar{F}$ continuous? Why?

29 Find a vector parametric equation for the hyperbola $x^2 - y^2 = 1$. (*Hint:* Use hyperbolic functions.)

In problems 30 through 34, find (a) $\bar{F}'(t)$ and (b) $\bar{F}''(t)$.

30 $\bar{F}(t) = (t^2 + 7)\bar{i} + (t + t^3)\bar{j}$ **31** $\bar{F}(t) = (t - 1)^{-1}\bar{i} + 3(t^2 - 2)^{-1}\bar{j}$ **32** $\bar{F}(t) = (2 \cos 5t)\bar{i} + (5 \sin 5t)\bar{j}$

33 $\bar{F}(t) = e^{5t}\bar{i} + (e^{-3t} + 7)\bar{j}$ **34** $\bar{F}(t) = (\cos t^2)\bar{i} + (\cos^2 t)\bar{j}$

In problems 35 through 38, find (a) $\dfrac{d}{dt}|\bar{F}(t)|$ and (b) $\bar{F}'(t) \cdot \bar{F}''(t)$.

35 $\bar{F}(t) = (t - \sin t)\bar{i} + (t + \cos t)\bar{j}$ **36** $\bar{F}(t) = (e^{-t} \cos t)\bar{i} + (e^{-t} \sin t)\bar{j}$

37 $\bar{F}(t) = e^{4t}\bar{i} + e^{-4t}\bar{j}$ **38** $\bar{F}(t) = te^{-3t}\bar{i} + te^{3t}\bar{j}$

In problems 39 through 44, let $\bar{F}(t) = e^t i + (t^2 - 1)j$, $\bar{G}(t) = (\cos^2 t)i + (\sin^2 t)j$, and $h(t) = \ln(t + 1)$. Evaluate each expression.

39 $\dfrac{d}{dt}[\bar{F}(t) + \bar{G}(t)]$

40 $\dfrac{d}{dt}[\bar{F}(t) \cdot \bar{G}(t)]$

41 $\dfrac{d}{dt}|\bar{F}(t)|$

42 $\dfrac{d}{dt}[h(t)\bar{F}(t)]$

43 $\bar{F}'(t) \cdot \bar{F}''(t)$

44 $\dfrac{d}{dt}\left[\dfrac{\bar{F}(t)}{|\bar{F}(t)|}\right]$

45 Find the arc length of the curve whose vector equation is $\bar{R} = 3(\sin t - 1)i + 3(\cos t - 1)j$ between the point where $t = 0$ and the point where $t = 2$.

46 Find the arc length of the curve $\begin{cases} x = 4 \cos^3 t \\ y = 4 \sin^3 t \end{cases}$ between the point where $t = 0$ and the point where $t = 2\pi$.

47 The position vector of a particle at time t is given by $\bar{R} = ti + \left(\dfrac{t^3}{6} + \dfrac{1}{2t}\right)j$. Find the distance traveled by the particle during the interval of time from $t = 1$ to $t = 4$.

48 A ball is moving in accordance with the equations $\begin{cases} x = 32t \\ y = -16t^2 \end{cases}$ where t is the time in seconds. How far does the ball move along its path during the first 3 seconds?

In problems 49 through 52, find (a) the unit tangent vector $\bar{T}$, (b) the curvature κ of the given curve at the given value of t, and (c) the principal unit normal vector $\bar{N}$.

49 $\bar{R}(t) = (3 \cos t)i + (\sin t)j$ at $t = 0$

50 $\bar{R}(t) = (\tan 2t)i + (\cot 2t)j$ at $t = \dfrac{\pi}{8}$

51 $\begin{cases} x = t^2 \\ y = t^3 \end{cases}$ at $t = 1$

52 $\begin{cases} x = t \\ y = \ln \sec t \end{cases}$ at $t = \dfrac{\pi}{4}$

In problems 53 through 57, find (a) the velocity vector $\bar{V}$, (b) the acceleration vector $\bar{A}$, (c) the speed v, (d) the rate of change $\dfrac{dv}{dt}$ of the speed, (e) the tangential component vector of the acceleration, and (f) the normal component vector of the acceleration for a particle at time t moving according to the given equation of motion.

53 $\bar{R}(t) = (t^3 + 6)i + (2t^4 - 1)j$

54 $\bar{R}(t) = e^{2t}i + e^{-3t}j$

55 $\bar{R}(t) = (3 \cos 7t)i + (3 \sin 7t)j$

56 $\bar{R}(t) = (t \cos t)i + (t \sin t)j$

57 $\bar{R}(t) = \left(1 + \dfrac{t^2}{2}\right)i + \dfrac{t^3}{3}j$

58 A particle is moving along the branch of the hyperbola $xy = 1$ that lies in the first quadrant in such a way that its abscissa is increasing in time. At the instant when the particle passes through the point $(1, 1)$, its speed is given by $v = 2$ units per second and the time rate of change of the speed is given by $dv/dt = -3$ units per second per second. Find (a) the velocity vector $\bar{V}$ and (b) the acceleration vector $\bar{A}$ at this instant.

59 A pilot is pulling out of a vertical dive by following an arc of a circle of radius 1 mile. The speed of her plane is a constant 500 miles per hour. With how many times her usual weight is the pilot being pressed into her seat at the lowest point on the arc of the circle?

60 Let $\bar{R} = g(t)i + h(t)j$ be the equation of motion of a particle of mass m in the xy plane. (a) If $\alpha(t)$ denotes the area swept out by the variable position vector $\bar{R}$ as a function of the time t, show that

$$\alpha'(t) = \left|\frac{g(t)h'(t) - h(t)g'(t)}{2}\right|.$$

(b) If the force acting on the particle is always directed toward the origin, show that the position vector $\bar{R}$ sweeps through equal areas in equal times.

15

COORDINATE SYSTEMS AND VECTORS IN THREE-DIMENSIONAL SPACE

In this chapter we study three-dimensional geometry with the aid of coordinate systems and vectors in three-dimensional space. All the operations defined for vectors in the plane—sums, differences, multiplication by scalars, and dot products—carry over directly to vectors in three-dimensional space. In addition, we define the "cross product" of vectors in space and use vectors to obtain the equations of lines and planes in space. The chapter also includes vector-valued functions, curves in space, surfaces of revolution, and quadric surfaces.

1 Cartesian Coordinate Systems in Three-Dimensional Space

In order to assign cartesian coordinates to a point P in three-dimensional space, we begin by choosing an origin O as usual. We then set up a cartesian coordinate system in the horizontal plane passing through O, with the positive x axis pointing toward us and the positive y axis extending to our right (Figure 1). To find the cartesian coordinates of a point P in space, we drop a perpendicular from P to the xy plane, and we denote the foot of this perpendicular by $Q = (x, y)$. We define the *z coordinate* of P by $z = \pm |\overline{PQ}|$, where we use the plus sign if P is above the xy plane and the minus sign if P is below the xy plane. Of course, if P lies on the xy plane, then $P = Q$ and $z = 0$. Thus, z is the *directed distance* from P to the xy plane. The point P is understood to have *three* cartesian coordinates x, y, and z, and we write $P = (x, y, z)$.

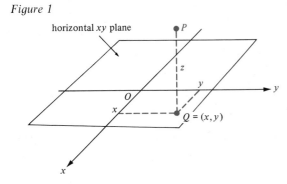

Figure 1

horizontal xy plane

In dealing with cartesian coordinates in three-dimensional space, it is customary to introduce a third coordinate axis, called the *z axis*, perpendicular to the xy plane, passing through the origin O, and directed upward (Figure 2). The z axis is equipped

Figure 2

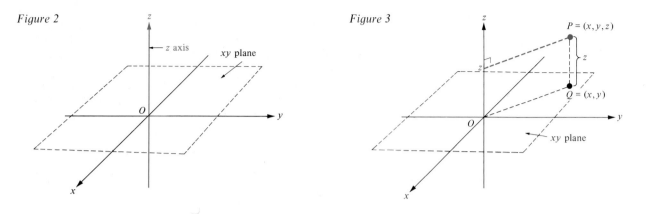

Figure 3

with its own number scale, just as the x and y axes are, and the unit of distance on the z axis is usually chosen to be the same as the unit of distance on the x and y axes.

If P is any point in space, then the z coordinate of P can be determined not only by measuring the directed distance from P to the xy plane, but also by dropping a perpendicular from P to the z axis (Figure 3). The scale value at the foot of this perpendicular is evidently the z coordinate of P. Similarly, the x and y coordinates of P can be found by dropping perpendiculars to the x and y axes, respectively (Figure 4). Since all three coordinates of P can be found by dropping perpendiculars to the three coordinate axes, it is not necessary to draw the xy plane at all when sketching three-dimensional diagrams—the three coordinate axes are sufficient.

Figure 4

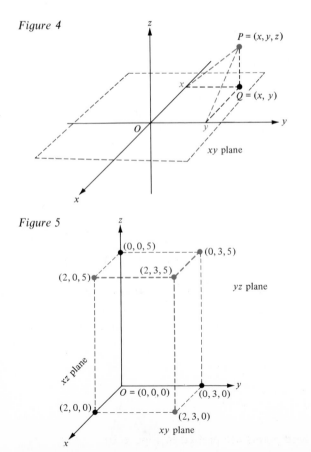

Figure 5

In the same way that the x and y axes determine the xy plane, the remaining pairs of axes determine planes, namely the *xz plane* and the *yz plane*. The xy, xz, and yz planes are called the three *coordinate planes*. To visualize these planes, we can think of the origin as a corner of a room with the floor as the xy plane, the wall to our left as the xz plane, and the wall facing us as the yz plane (Figure 5). The coordinates of a point P in space are then the directed distances x, y, and z from P to the yz, xz, and xy planes, respectively. Figure 5 shows the point $(2, 3, 5)$ and all the other points obtained by projecting it perpendicularly onto the coordinate planes and the coordinate axes.

The three coordinate planes divide the space into eight parts called *octants*. The octant in which all three coordinates are positive—that is, above the xy plane, to the right of the xz plane, and in front of the yz plane in Figure 5—is called the *first octant*.

The coordinate system described here is called a *right-handed, three-dimensional, cartesian* (or *rectangular*) *coordinate system*. If such a coordinate system has been set up, we refer to three-dimensional space as *xyz space*. (Incidentally, a "left-handed" cartesian coordinate system would have the x and y axes interchanged; however, here, and in what follows, we do not use "left-handed" coordinate systems.)

Figure 6

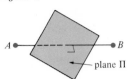

Two points A and B in space are said to be *symmetrically located with respect to a plane* Π if Π passes through the middle of the line segment $\overline{AB}$ and is perpendicular to this segment (Figure 6). For instance, the points $A = (x, y, z)$ and $B = (-x, y, z)$ are symmetrically located with respect to the yz plane (Figure 7). Similarly, the points $A = (x, y, z)$ and $C = (x, -y, z)$ are symmetrically located with respect to the xz plane, while the points $A = (x, y, z)$ and $D = (x, y, -z)$ are symmetrically located with respect to the xy plane.

Figure 7

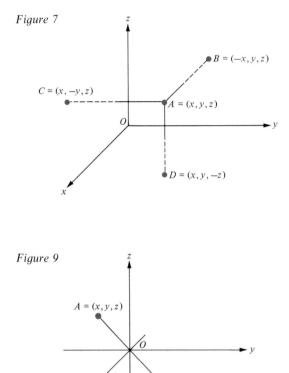

Figure 8

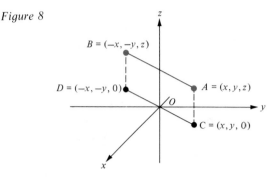

Notice in Figure 8 that the points $C = (x, y, 0)$ and $D = (-x, -y, 0)$ lie in the xy plane and are symmetrically located with respect to the origin O. If C and D are both raised through z units, we obtain the points $A = (x, y, z)$ and $B = (-x, -y, z)$, respectively. Evidently, $A = (x, y, z)$ and $B = (-x, -y, z)$ are symmetrically located with respect to the z axis.

Figure 9

Reasoning as above, we can see that the points $A = (x, y, z)$ and $E = (-x, y, -z)$ are symmetric with respect to the y axis, while the points $A = (x, y, z)$ and $F = (x, -y, -z)$ are symmetric with respect to the x axis. Finally, notice that the points $A = (x, y, z)$ and $G = (-x, -y, -z)$ are symmetric with respect to the origin O (Figure 9).

EXAMPLE   Plot the point $P = (3, 4, 5)$, and then give the coordinates of and plot the following points:

(a) The point Q symmetric to P with respect to the xy plane.

(b) The point R symmetric to P with respect to the y axis.

(c) The point S symmetric to P with respect to the origin.

(d) The point T which is 5 units directly below P.

Figure 10

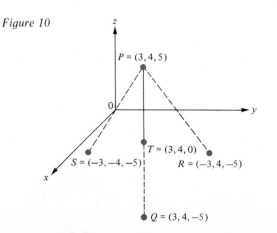

SOLUTION

(a) $Q = (3, 4, -5)$

(b) $R = (-3, 4, -5)$

(c) $S = (-3, -4, -5)$

(d) $T = (3, 4, 0)$

These points are plotted in Figure 10.

By the *graph* in *xyz* space of an equation (or of a set of simultaneous equations) involving one or more of the variables x, y, or z, we mean the set of all points $P = (x, y, z)$ whose coordinates x, y, and z satisfy the equation (or the equations). For instance, the graph of the equation $z = 0$ is the set of all points $P = (x, y, 0)$, and therefore consists of all points in the *xy* plane. Similarly, the graph of $x = 0$ is the *yz* plane and the graph of $y = 0$ is the *xz* plane.

EXAMPLE   Sketch the graph of $z = 0$ and the graph of $y = 1$ in *xyz* space. Show the intersection of these two graphs.

Figure 11

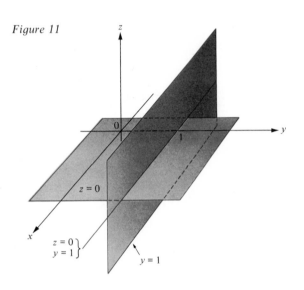

SOLUTION

The graph of $z = 0$ is the *xy* plane, part of which is shown in Figure 11. The graph of $y = 1$ is the set of all points whose y coordinate is 1; that is, all points one unit to the right of the *xz* plane. Thus, the graph of $y = 1$ is a plane parallel to the *xz* plane and 1 unit to the right of it (Figure 11). The two planes $z = 0$ and $y = 1$ evidently intersect in a straight line parallel to the x axis and 1 unit to the right of the origin in the *xy* plane.

Any two nonparallel planes, such as the planes $z = 0$ and $y = 1$ in Figure 11, intersect in a straight line. For instance, the equation $z = b$, where b is a constant, represents a plane perpendicular to the z axis and intersecting it at the point $(0, 0, b)$, while the equation $y = c$, where c is a constant, represents a plane perpendicular to the y axis and intersecting it at the point $(0, c, 0)$. (Why?). Thus, the intersection of the plane $z = b$ and the plane $y = c$ is the straight line parallel to the x axis and consisting of all points of the form $P = (x, c, b)$, where x runs through all real numbers. This straight line is the graph of the simultaneous equations $\begin{cases} z = b \\ y = c. \end{cases}$

EXAMPLES **1** Write a pair of simultaneous equations whose graph is the straight line perpendicular to the *xy* plane and which contains the point $(-3, 2, 1)$ (Figure 12).

Figure 12

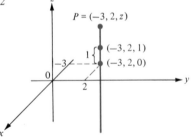

SOLUTION

The straight line in question contains the point $(-3, 2, 0)$ in the *xy* plane. Evidently, a point P is on this line if and only if the x coordinate of P is -3 and the y coordinate of P is 2—there is no restriction on the z coordinate of P. Therefore, the simultaneous equations $\begin{cases} x = -3 \\ y = 2 \end{cases}$ describe the given straight line.

2 Describe two planes whose intersection is the straight line in Figure 12.

SOLUTION

Evidently, the line $\begin{cases} x = -3 \\ y = 2 \end{cases}$ is the intersection of the plane $x = -3$ and the plane $y = 2$. The plane $x = -3$ is perpendicular to the x axis and contains the point $(-3, 0, 0)$. The plane $y = 2$ is perpendicular to the y axis and contains the point $(0, 2, 0)$.

3 Find the equation of the plane containing the point (a, b, c) and parallel to the yz plane.

SOLUTION

On the plane in question, all points have the same x coordinate, $x = a$. Conversely, any point P of the form $P = (a, y, z)$ belongs to the plane. Therefore, the equation of the plane is $x = a$.

Figure 13

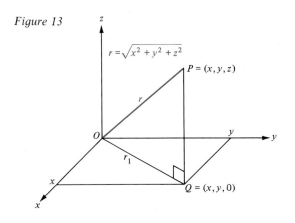

We now derive a formula for the distance r between the point $P = (x, y, z)$ and the origin O in xyz space. In Figure 13, let r_1 denote the distance in the xy plane between the point $Q = (x, y, 0)$ and the origin O. By the usual formula for the distance between two points in the xy plane, $r_1^2 = x^2 + y^2$. Applying the Pythagorean theorem to the triangle OQP, we have $r^2 = r_1^2 + |\overline{PQ}|^2 = x^2 + y^2 + z^2$; hence,

$$r = \sqrt{x^2 + y^2 + z^2}.$$

EXAMPLES **1** Find the distance r between the origin and the point $(2, 3, -1)$.

Figure 14

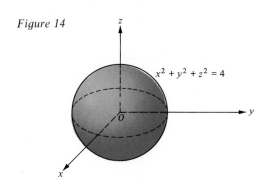

SOLUTION

From the formula above, we have

$$r = \sqrt{2^2 + 3^2 + (-1)^2} = \sqrt{14} \text{ units.}$$

2 Describe and sketch the graph in xyz space of the equation $x^2 + y^2 + z^2 = 4$.

SOLUTION

The point $P = (x, y, z)$ belongs to the graph if and only if $\sqrt{x^2 + y^2 + z^2} = 2$; that is, if and only if the distance from O to P is 2 units. Therefore, the graph is the surface of a sphere of radius 2 units with center at the origin O (Figure 14).

Problem Set 1

1 The points $(3, 5, 7)$, $(3, 0, 7)$, $(0, 0, 7)$, $(0, 5, 7)$, $(3, 5, 0)$, $(3, 0, 0)$, $(0, 0, 0)$, and $(0, 5, 0)$ form the vertices of a rectangular box in xyz space. Sketch this box and label the vertices.

2 A perpendicular is dropped from the point $P = (-1, 2, 5)$ to the yz plane. Find the coordinates of the point Q at the foot of this perpendicular.

In problems 3 through 6, plot the given point P in xyz space, and then give the coordinates of and plot the following points.
(a) The point Q symmetric to P with respect to the yz plane.
(b) The point R symmetric to P with respect to the z axis.
(c) The point S symmetric to P with respect to the origin.
(d) The point T that lies on the straight line through P parallel to the y axis but is 2 units farther to the right than P.

3 $P = (1, 2, 4)$ **4** $P = (3, 5, 0)$ **5** $P = (2, -1, -3)$ **6** $P = (\frac{5}{2}, -1, 0)$

7 Describe and sketch the set of points $P = (x, y, z)$ in xyz space satisfying both of the conditions $x = -1$ and $z = 2$.

8 Write the equation of the plane containing the three points $(-27, 14, 3)$, $(111, -75, 3)$, and $(666, 1720, 3)$.

9 Describe the set of points $P = (x, y, z)$ in xyz space satisfying the condition $z \geq 1$.

10 Give an equation or equations describing the x axis in xyz space.

In problems 11 through 22, sketch a graph in xyz space of each equation or equations.

11 $y = -2$

12 $\begin{cases} x = 3 \\ y = 4 \end{cases}$

13 $\begin{cases} x = -2 \\ y = -5 \end{cases}$

14 $x + y = 1$

15 $x^2 + y^2 + z^2 = 1$

16 $x^2 + y^2 = 1$

17 $\begin{cases} x = 1 \\ y = 2 \\ z = 3 \end{cases}$

18 $\begin{cases} x^2 + y^2 + z^2 = 4 \\ z = 1 \end{cases}$

19 $\begin{cases} y^2 = z \\ x = 0 \end{cases}$

20 $\begin{cases} y^2 = z \\ x = 2 \end{cases}$

21 $\begin{cases} x^2 + y^2 = 9 \\ z = 1 \end{cases}$

22 $\begin{cases} x = y \\ y = z \end{cases}$

23 A perpendicular is dropped from the point $P = (3, 5, 7)$ to the plane $y = -1$. Find the coordinates of the point Q at the foot of this perpendicular.

24 Given that the points $A = (x, y, z)$ and B are symmetrically located with respect to the plane $z = c$, find the coordinates of B.

In problems 25 through 34, write an equation or equations for the set of points described.

25 The set of points on the plane parallel to the xy plane and 4 units below it.

26 The set of points on the plane perpendicular to the y axis and passing through the point $(-1, -1, -1)$.

27 The set of points on the plane perpendicular to the x axis and containing the point $(27, \frac{1}{2}, -\pi)$.

28 The set of points on the straight line parallel to the z axis and containing the point $(-2, -7, -15)$.

29 The set of points on the straight line parallel to the y axis and containing the point $(5, \frac{16}{3}, -\frac{7}{2})$.

30 The set of points on the straight line containing the two points $(5, 0, 7)$ and $(5, -1, 7)$.

31 The set of points on the surface of a sphere of radius 5 units with its center at the origin.

32 The set of points on the plane containing the z axis and making a $45°$ angle with the xz plane and with the yz plane.

33 The set of points on a circle of radius 5 units, parallel to the yz plane, with center at the point $(7, 0, 0)$.

34 The set of points on the right circular cylinder of radius 5 units whose central axis coincides with the z axis.

35 Find the distance r between the origin and the given point P.
(a) $P = (-2, 1, 2)$
(b) $P = (8, 0, -6)$
(c) $P = (-4, -3, 0)$
(d) $P = (1, 4, 5)$

2   Vectors in Three-Dimensional Space

The concept of a vector and the algebra of vectors carry over almost without change to vectors in three-dimensional space. Thus, a vector in space is defined to be a directed line segment—that is, an arrow—and the directed line segment from P to Q is denoted by $\overrightarrow{PQ}$. Two vectors in space are regarded as being equal if they have the same length, are parallel, and point in the same direction. The definitions of sums and differences of vectors, the negative of a vector, the product of a scalar and a vector, and the dot product of two vectors all carry over verbatim to vectors in space.

The basic algebraic properties of vectors in space are exactly the same as the corresponding properties of vectors in the plane. The reader is encouraged to sketch diagrams illustrating the following.

Basic Algebraic Properties of Vectors in Space

Let $\bar{A}$, $\bar{B}$, and $\bar{C}$ be vectors in three-dimensional space and let s and t be scalars. Then:

1 $\bar{A} + \bar{B} = \bar{B} + \bar{A}$
2 $\bar{A} + (\bar{B} + \bar{C}) = (\bar{A} + \bar{B}) + \bar{C}$
3 $\bar{A} + \bar{0} = \bar{A}$
4 $\bar{A} + (-\bar{A}) = \bar{0}$
5 $(st)\bar{A} = s(t\bar{A})$
6 $s(\bar{A} + \bar{B}) = s\bar{A} + s\bar{B}$
7 $(s + t)\bar{A} = s\bar{A} + t\bar{A}$
8 $1\bar{A} = \bar{A}$

If just two vectors $\bar{A}$ and $\bar{B}$ in space are involved, they can always be placed so that they have the same tail end (initial point), and lie in the same plane (Figure 1). Notice that all linear combinations $s\bar{A} + t\bar{B}$ of $\bar{A}$ and $\bar{B}$ lie in the same plane. (Why?)

Three vectors $\bar{A}$, $\bar{B}$, and $\bar{C}$ in space are called *coplanar* (or *linearly dependent*) if, when they are placed so as to have a common tail end, they all lie in the same plane (Figure 2). It is easy to see that three vectors in space are coplanar if and

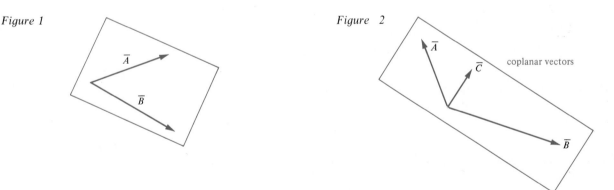

Figure 1

Figure 2

coplanar vectors

only if one of them can be expressed as a linear combination of the other two (Problem 50).

Three vectors $\bar{A}$, $\bar{B}$, and $\bar{C}$ are said to be *linearly independent* if they are not coplanar. The following theorem provides a handy test for the linear independence of three vectors. (For the proof, see Problem 52.)

THEOREM 1   **Linear Independence of Three Vectors**
Three vectors $\bar{A}$, $\bar{B}$, and $\bar{C}$ are linearly independent if and only if the following condition holds: The only scalars a, b, and c for which

$$a\bar{A} + b\bar{B} + c\bar{C} = \bar{0}$$

are $a = 0$, $b = 0$, and $c = 0$.

The unit vectors i, j, and $\bar{k}$ pointing in the positive directions along the x, y, and z axes, respectively, in xyz space are three noncoplanar—hence, linearly independent—vectors (Figure 3). We call i, j, and $\bar{k}$ the *standard basis vectors* in xyz space.

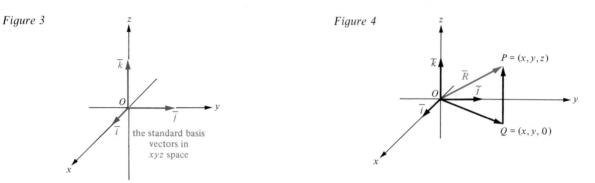

Figure 3

the standard basis vectors in xyz space

Figure 4

Just as every vector in the xy plane can be written as a linear combination of i and j, every vector in xyz space can be written as a linear combination of i, j, and $\bar{k}$. In fact, given a vector $\bar{R}$ in xyz space, we can move the tail end of $\bar{R}$ to the origin O, so that $\bar{R} = \overrightarrow{OP}$ is the *position vector* of the point $P = (x, y, z)$ (Figure 4). Let $Q = (x, y, 0)$ be the point at the foot of the perpendicular from P to the xy plane. Since $\overrightarrow{OQ}$ is the position vector of Q in the xy plane, it follows that $\overrightarrow{OQ} = xi + yj$. Because $\overrightarrow{QP}$ is parallel to $\bar{k}$, it is clear that $\overrightarrow{QP} = z\bar{k}$. Therefore,

$$\bar{R} = \overrightarrow{OQ} + \overrightarrow{QP} = xi + yj + z\bar{k}.$$

As before, x, y, and z are called the *(scalar) components* of $\bar{R}$ with respect to the standard i, j, $\bar{k}$ basis.

The sum $\bar{A} + \bar{B}$, the difference $\bar{A} - \bar{B}$, and the product $s\bar{A}$ are calculated "component by component" for vectors $\bar{A}$ and $\bar{B}$ in xyz space just as they are for vectors in the plane. Thus, if

$$\bar{A} = a_1 i + a_2 j + a_3 \bar{k} \quad \text{and} \quad \bar{B} = b_1 i + b_2 j + b_3 \bar{k},$$

then

1 $\bar{A} + \bar{B} = (a_1 + b_1)i + (a_2 + b_2)j + (a_3 + b_3)\bar{k}.$

2 $\bar{A} - \bar{B} = (a_1 - b_1)i + (a_2 - b_2)j + (a_3 - b_3)\bar{k}.$

3 $s\bar{A} = (sa_1)i + (sa_2)j + (sa_3)\bar{k}.$

EXAMPLE   Let $\bar{A} = 2i - 3j + \bar{k}$ and $\bar{B} = i + 2j + 5\bar{k}$. Find (a) $\bar{A} + \bar{B}$, (b) $\bar{A} - \bar{B}$, (c) $7\bar{A}$, and (d) $7\bar{A} - \bar{B}$.

SOLUTION
(a) $\bar{A} + \bar{B} = (2 + 1)i + (-3 + 2)j + (1 + 5)\bar{k} = 3i - j + 6\bar{k}.$
(b) $\bar{A} - \bar{B} = (2 - 1)i + (-3 - 2)j + (1 - 5)\bar{k} = i - 5j - 4\bar{k}.$
(c) $7\bar{A} = (7)(2)i + (7)(-3)j + (7)(1)\bar{k} = 14i - 21j + 7\bar{k}.$
(d) $7\bar{A} - \bar{B} = (14 - 1)i + (-21 - 2)j + (7 - 5)\bar{k} = 13i - 23j + 2\bar{k}.$

The three scalar components of a vector in xyz space are uniquely determined by the vector. To see this, assume that

$$x_1 i + y_1 j + z_1 \bar{k} = x_2 i + y_2 j + z_2 \bar{k}.$$

Then,

$$(x_1 - x_2)i + (y_1 - y_2)j + (z_1 - z_2)\bar{k} = \bar{0};$$

hence, since $\bar{i}$, $\bar{j}$, and $\bar{k}$ are linearly independent, it follows that $x_1 - x_2 = 0$, $y_1 - y_2 = 0$, and $z_1 - z_2 = 0$. Therefore, $x_1 = x_2$, $y_1 = y_2$, and $z_1 = z_2$.

Since a vector $\bar{R} = x\bar{i} + y\bar{j} + z\bar{k}$ both determines and is determined by its three scalar components x, y, and z, it is customary to represent $\bar{R}$ by the ordered triple $\langle x, y, z \rangle$ and to write $\bar{R} = \langle x, y, z \rangle$. Using this notation, we can write the solution to the last example as

(a) $\langle 2, -3, 1 \rangle + \langle 1, 2, 5 \rangle = \langle 3, -1, 6 \rangle$.

(b) $\langle 2, -3, 1 \rangle - \langle 1, 2, 5 \rangle = \langle 1, -5, -4 \rangle$.

(c) $7\langle 2, -3, 1 \rangle = \langle 14, -21, 7 \rangle$.

(d) $7\langle 2, -3, 1 \rangle - \langle 1, 2, 5 \rangle = \langle 13, -23, 2 \rangle$.

The reader is encouraged to use ordered triple notation for vectors in xyz space whenever it seems convenient.

The angle θ between two vectors $\bar{A}$ and $\bar{B}$ in three-dimensional space is measured by bringing the tail ends of the vectors together, so that the two vectors lie in the same plane (Figure 1), and then measuring the angle as usual. The dot product of $\bar{A}$ and $\bar{B}$ is defined, just as for vectors in the plane, by $\bar{A} \cdot \bar{B} = |\bar{A}| |\bar{B}| \cos \theta$, where $|\bar{A}|$ and $|\bar{B}|$ denote the lengths (magnitudes) of $\bar{A}$ and $\bar{B}$, respectively.

The basic properties of the dot product in three-dimensional space are established by using virtually the same arguments as for vectors in the plane.

Basic Properties of the Dot Product in Space

If $\bar{A}$, $\bar{B}$, and $\bar{C}$ are vectors in three-dimensional space and s is a scalar, then

1 $\bar{A} \cdot \bar{B} = \bar{B} \cdot \bar{A}$.

2 $(\bar{A} + \bar{B}) \cdot \bar{C} = \bar{A} \cdot \bar{C} + \bar{B} \cdot \bar{C}$.

3 $(s\bar{A}) \cdot \bar{B} = s(\bar{A} \cdot \bar{B}) = \bar{A} \cdot (s\bar{B})$.

4 $\bar{A} \cdot \bar{A} = |\bar{A}|^2 \geq 0$.

5 If $\bar{A} \cdot \bar{A} = 0$, then $\bar{A} = \bar{0}$.

6 $\bar{A} \cdot \bar{B} = 0$ if and only if $\bar{A}$ and $\bar{B}$ are perpendicular.

Furthermore, in xyz space, we have

7 $\bar{i} \cdot \bar{i} = \bar{j} \cdot \bar{j} = \bar{k} \cdot \bar{k} = 1$.

8 $\bar{i} \cdot \bar{j} = \bar{i} \cdot \bar{k} = \bar{j} \cdot \bar{k} = 0$.

Properties 7 and 8 follow from the facts that $\bar{i}$, $\bar{j}$, and $\bar{k}$ are unit vectors and that they are mutually perpendicular.

The following important theorem is the analog for xyz space of Theorem 1 in Section 3.2 of Chapter 14.

THEOREM 2   **Dot Product of Vectors in xyz Space**

Let $\bar{A} = a\bar{i} + b\bar{j} + c\bar{k}$ and $\bar{B} = x\bar{i} + y\bar{j} + z\bar{k}$. Then

$$\bar{A} \cdot \bar{B} = ax + by + cz.$$

The proof of Theorem 2 is accomplished simply by expanding the dot product $(a\bar{i} + b\bar{j} + c\bar{k}) \cdot (x\bar{i} + y\bar{j} + z\bar{k})$ and using Properties 7 and 8 (Problem 54). An immediate consequence of Theorem 2 is that, if $\bar{B} = x\bar{i} + y\bar{j} + z\bar{k}$, then

$\bar{B} \cdot \bar{B} = x^2 + y^2 + z^2$. Since $|\bar{B}| = \sqrt{\bar{B} \cdot \bar{B}}$ by Property 4, we have the following formula for the length of a vector in xyz space:

$$\text{If } \bar{B} = x\bar{i} + y\bar{j} + z\bar{k}, \text{ then } \bar{B} = \sqrt{x^2 + y^2 + z^2}.$$

EXAMPLE   Let $\bar{A} = 3\bar{i} - 2\bar{j} + \bar{k}$ and $\bar{B} = 2\bar{i} + 5\bar{j} - 2\bar{k}$. Find (a) $\bar{A} \cdot \bar{B}$, (b) $|\bar{A}|$, and (c) $|\bar{A} - \bar{B}|$.

SOLUTION
(a) $\bar{A} \cdot \bar{B} = (3)(2) + (-2)(5) + (1)(-2) = -6$.
(b) $|\bar{A}| = \sqrt{3^2 + (-2)^2 + 1^2} = \sqrt{14}$.
(c) $|\bar{A} - \bar{B}| = |\bar{i} - 7\bar{j} + 3\bar{k}| = \sqrt{1^2 + (-7)^2 + 3^2} = \sqrt{59}$.

Suppose that θ is the angle between the two vectors $\bar{A}$ and $\bar{B}$ in three-dimensional space. Then, since $\bar{A} \cdot \bar{B} = |\bar{A}||\bar{B}| \cos \theta$, we have

$$\cos \theta = \frac{\bar{A} \cdot \bar{B}}{|\bar{A}||\bar{B}|}.$$

This formula enables us to find the cosine of θ, and hence the angle θ, in terms of the dot product.

EXAMPLE   Find the angle θ between $\bar{A} = 2\bar{i} - 3\bar{j} + \bar{k}$ and $\bar{B} = \bar{i} + 2\bar{j} + 5\bar{k}$.

SOLUTION

$$\cos \theta = \frac{\bar{A} \cdot \bar{B}}{|\bar{A}||\bar{B}|} = \frac{(2)(1) + (-3)(2) + (1)(5)}{\sqrt{2^2 + (-3)^2 + 1^2}\sqrt{1^2 + 2^2 + 5^2}} = \frac{1}{2\sqrt{105}};$$

hence,

$$\theta = \cos^{-1} \frac{1}{2\sqrt{105}} \approx \cos^{-1} 0.0488 \approx 87.2°.$$

2.1 Distance Between Points in Space

We can now use vectors to derive a formula for the distance between two points $P = (x, y, z)$ and $Q = (a, b, c)$ in xyz space (Figure 5). Here,

$$\overrightarrow{OP} = x\bar{i} + y\bar{j} + z\bar{k}, \quad \overrightarrow{OQ} = a\bar{i} + b\bar{j} + c\bar{k},$$

and so

$$\overrightarrow{PQ} = \overrightarrow{OQ} - \overrightarrow{OP} = (a - x)\bar{i} + (b - y)\bar{j} + (c - z)\bar{k}.$$

Thus, the scalar components of the vector $\overrightarrow{PQ}$ in xyz space are the differences of the coordinates of Q and the corresponding coordinates of P. It follows that the distance between P and Q is given by

$$|\overrightarrow{PQ}| = |\overrightarrow{PQ}| = \sqrt{(a - x)^2 + (b - y)^2 + (c - z)^2}.$$

Figure 5

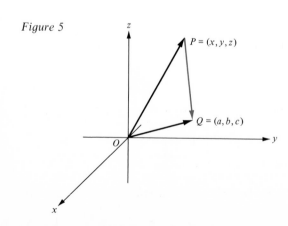

EXAMPLE If $P = (2, 3, -2)$ and $Q = (-1, 1, 5)$, find (a) the scalar components of $\overrightarrow{PQ}$ and (b) the distance between P and Q.

SOLUTION
(a) $\overrightarrow{PQ} = [(-1) - 2]\bar{i} + [1 - 3]\bar{j} + [5 - (-2)]\bar{k} = -3\bar{i} - 2\bar{j} + 7\bar{k}$.
(b) Using the distance formula above, we have

$$|\overrightarrow{PQ}| = \sqrt{[(-1) - 2]^2 + [1 - 3]^2 + [5 - (-2)]^2} = \sqrt{62} \text{ units.}$$

2.2 Direction Cosines of a Vector

Consider a nonzero vector $\bar{A}$ in xyz space. Move $\bar{A}$, if necessary, so that its tail end is the origin O, and let α, β, and γ be the angles between $\bar{A}$ and the positive x, y, and z axes, respectively (Figure 6). The angles α, β, and γ, which are the same as the angles between $\bar{A}$ and the standard $\bar{i}$, $\bar{j}$, and $\bar{k}$ basis vectors, respectively, are called the *direction angles* of the vector $\bar{A}$. The cosines of the three direction angles are called the *direction cosines* of the vector $\bar{A}$ and are given by the formulas

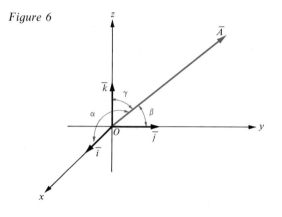

Figure 6

$$\cos \alpha = \frac{\bar{A} \cdot \bar{i}}{|\bar{A}|}, \quad \cos \beta = \frac{\bar{A} \cdot \bar{j}}{|\bar{A}|}, \quad \text{and} \quad \cos \gamma = \frac{\bar{A} \cdot \bar{k}}{|\bar{A}|}$$

(Problem 30).
Now, suppose that $\bar{A} = a\bar{i} + b\bar{j} + c\bar{k}$, so that

$$\bar{A} \cdot \bar{i} = a, \qquad \bar{A} \cdot \bar{j} = b, \qquad \bar{A} \cdot \bar{k} = c,$$

and

$$|\bar{A}| = \sqrt{a^2 + b^2 + c^2}.$$

Then,

$$\cos \alpha = \frac{a}{\sqrt{a^2 + b^2 + c^2}}, \qquad \cos \beta = \frac{b}{\sqrt{a^2 + b^2 + c^2}},$$

and

$$\cos \gamma = \frac{c}{\sqrt{a^2 + b^2 + c^2}}.$$

Notice that

$$(\cos \alpha)\bar{i} + (\cos \beta)\bar{j} + (\cos \gamma)\bar{k} = \frac{1}{\sqrt{a^2 + b^2 + c^2}} (a\bar{i} + b\bar{j} + c\bar{k}) = \frac{\bar{A}}{|\bar{A}|},$$

hence, the direction cosines of a nonzero vector $\bar{A}$ are the scalar components of the unit vector $\dfrac{\bar{A}}{|\bar{A}|}$ in the same direction as $\bar{A}$. Since $(\cos \alpha)\bar{i} + (\cos \beta)\bar{j} + (\cos \gamma)\bar{k}$ is a unit vector, it follows that

$$\cos^2 \alpha + \cos^2 \beta + \cos^2 \gamma = 1;$$

that is, the sum of the squares of the direction cosines of a nonzero vector is always equal to 1 (Problem 39).

EXAMPLES 1 Find the direction angles of $\bar{A} = 2i - j + 3\bar{k}$.

SOLUTION

$$\cos\alpha = \frac{2}{\sqrt{4+1+9}} = \frac{2}{\sqrt{14}}, \quad \cos\beta = \frac{-1}{\sqrt{14}}, \quad \text{and} \quad \cos\gamma = \frac{3}{\sqrt{14}};$$

hence,

$$\alpha = \cos^{-1}\frac{2}{\sqrt{14}} \approx 57.69°, \quad \beta = \cos^{-1}\frac{-1}{\sqrt{14}} \approx 105.50°,$$

and

$$\gamma = \cos^{-1}\frac{3}{\sqrt{14}} \approx 36.70°.$$

2 Let $P = (-1, 2, 3)$ and $Q = (-3, 3, 5)$. Find the direction cosines of the vector $\bar{A} = \overrightarrow{PQ}$ and verify that the sum of the squares of these direction cosines is equal to 1.

SOLUTION

$\bar{A} = \overrightarrow{PQ} = (-3+1)i + (3-2)j + (5-3)\bar{k} = -2i + j + 2\bar{k}$; hence,

$$|\bar{A}| = \sqrt{(-2)^2 + 1^2 + 2^2} = \sqrt{9} = 3 \quad \text{and} \quad \frac{\bar{A}}{|\bar{A}|} = -\frac{2}{3}i + \frac{1}{3}j + \frac{2}{3}\bar{k}.$$

Therefore, the direction cosines are $\cos\alpha = -\frac{2}{3}$, $\cos\beta = \frac{1}{3}$, and $\cos\gamma = \frac{2}{3}$, and

$$\cos^2\alpha + \cos^2\beta + \cos^2\gamma = \left(-\tfrac{2}{3}\right)^2 + \left(\tfrac{1}{3}\right)^2 + \left(\tfrac{2}{3}\right)^2 = \tfrac{4}{9} + \tfrac{1}{9} + \tfrac{4}{9} = 1.$$

Problem Set 2

In problems 1 through 25, suppose $\bar{A} = 3i + j - 4\bar{k}$, $\bar{B} = 4i + j - \bar{k}$, $\bar{C} = 2i - 5j + 4\bar{k}$, $\bar{D} = -2i + j + 7\bar{k}$, and $\bar{E} = 2i + j - \bar{k}$ are vectors in space. Evaluate each of the following.

1 $\bar{A} + \bar{B}$

2 $3\bar{A} + 2\bar{B}$

3 $3\bar{C} - 2\bar{D}$

4 $\bar{C} - \bar{B} + 2\bar{E}$

5 $4\bar{B} + 5\bar{E} - \bar{D} + 2\bar{A}$

6 $\frac{1}{3}(\bar{A} + \bar{B} - \bar{C} + \bar{D} - \bar{E})$

7 $|\bar{B}|\bar{A} - |\bar{A}|\bar{B}$

8 $\frac{\bar{B}}{|\bar{B}|} + 3\left|\frac{\bar{A}}{|\bar{A}|}\right|$

9 $|2\bar{A} + 5\bar{C}|$

10 $2|\bar{A}| + 5|\bar{C}|$

11 $\bar{A} \cdot \bar{B}$

12 $\bar{A} \cdot (-\bar{C}) + \bar{A} \cdot \bar{E}$

13 $\bar{A} \cdot (\bar{B} + \bar{C} - \bar{D} + \bar{E})$

14 $(\bar{A} \cdot \bar{B})\bar{C} - (\bar{A} \cdot \bar{C})\bar{B}$

15 $(2\bar{A} + \bar{E}) \cdot (\bar{A} - \bar{C})$

16 $(\bar{A} - \frac{1}{2}\bar{B} + \bar{C}) \cdot (\bar{E} - \frac{1}{3}\bar{D})$

17 $(\bar{C} \cdot \bar{B})\bar{E} - (\bar{C} \cdot \bar{A})\bar{B}$

18 $(\bar{A} - \bar{B}) \cdot (\bar{A} + \bar{B})$

19 The angle θ_1 between $\bar{A}$ and $\bar{B}$.

20 The angle θ_2 between $\bar{A}$ and $\bar{B} - \bar{A}$.

21 The angle θ_3 between $\bar{A} - \bar{B}$ and $\bar{A} + \bar{B}$.

22 The direction cosines of $\bar{A}$.

23 The direction cosines of $\bar{B}$.

24 The direction angles of $\bar{A} + \bar{B}$.

25 The direction angles of $\bar{A} - \bar{B}$.

26 Find a point $D = (x, y, z)$ such that $\overrightarrow{AB} = \overrightarrow{CD}$ if:
(a) $A = (1, 2, 3)$, $B = (7, 6, 5)$, $C = (4, 5, 6)$.
(b) $A = (0, -1, 7)$, $B = (16, -3, 5)$, $C = (-8, -1, -2)$.

In problems 27 through 29, find the distance between the given set of points P_1 and P_2.
27 $P_1 = (7, -1, 4)$; $P_2 = (8, 1, 6)$ **28** $P_1 = (1, 1, 5)$; $P_2 = (1, -2, 3)$ **29** $P_1 = (2, 3, -3)$; $P_2 = (2, 1, -2)$

30 Verify the formulas given on page 745 for the direction cosines of a nonzero vector $\bar{A}$.

In problems 31 through 33, find the direction cosines of the vector $\bar{A} = \overrightarrow{QP}$.
31 $P = (6, -1, -2)$ and $Q = (4, 9, 9)$ **32** $P = (3, 8, 1)$ and $Q = (1, 2, 10)$ **33** $P = (9, 1, -7)$ and $Q = (1, 0, -1)$

34 In problems 31, 32, and 33, write the vector $\overrightarrow{QP}$ in ordered triple notation $\langle x, y, z \rangle$.

In problems 35 through 38, determine whether each triple α, β, γ could possibly be direction angles of a vector $\bar{A}$.
35 $\alpha = 90°$, $\beta = 135°$, $\gamma = 45°$ **36** $\alpha = 2\pi/3$, $\beta = 3\pi/4$, $\gamma = \pi/3$

37 $\alpha = \pi/3$, $\beta = \pi/6$, $\gamma = \pi/4$ **38** $\alpha = 120°$, $\beta = 45°$, $\gamma = 60°$

39 Prove that the sum of the squares of the direction cosines of a nonzero vector is 1.

40 If $\alpha = \pi/3$ and $\beta = \pi/3$ are two direction angles of a vector $\bar{A}$, find the possible values of the third direction angle γ.

41 Show that a nonzero vector $\bar{A}$ in space is completely determined by its length l and its three direction cosines: $\cos \alpha$, $\cos \beta$, and $\cos \gamma$. [*Hint:* Show that the vector $\bar{A} = (l \cos \alpha)\bar{i} + (l \cos \beta)\bar{j} + (l \cos \gamma)\bar{k}$.]

42 Show that $\bar{A} = 7\bar{i} + 2\bar{j} - 10\bar{k}$, $\bar{B} = 4\bar{i} - 6\bar{j} + 5\bar{k}$, and $\bar{C} = 3\bar{i} - 2\bar{j} - 3\bar{k}$ are position vectors of the vertices of an isosceles triangle.

43 Use the dot product to determine whether the triangle whose vertices are the points $(1, 7, 2)$, $(0, 7, -2)$, and $(-1, 6, 1)$ is a right triangle.

44 Give a geometric interpretation of the "average" $(\bar{A} + \bar{B})/2$ of two vectors $\bar{A}$ and $\bar{B}$.

45 Let $\bar{A} = \overrightarrow{PQ}$ where $P = (-1, 2, -3)$ and $Q = (11, 11, 3)$. Find a point $S = (x, y, z)$ such that $\overrightarrow{PS} = 2\bar{A}$.

46 (a) Let $\bar{A}$ and $\bar{B}$ be vectors in three-dimensional space. Show that the *scalar projection* of $\bar{B}$ on $\bar{A}$ is given by $\dfrac{\bar{A} \cdot \bar{B}}{|\bar{A}|}$.
(b) Let $\bar{A} = 6\bar{i} - 3\bar{j} + 2\bar{k}$ and $\bar{B} = 2\bar{i} - \bar{j} + \bar{k}$. Find the scalar projection of $\bar{B}$ on $\bar{A}$.

47 If $\bar{A}$ is any vector in space, show that
$$\bar{A} = (\bar{A} \cdot \bar{i})\bar{i} + (\bar{A} \cdot \bar{j})\bar{j} + (\bar{A} \cdot \bar{k})\bar{k}.$$
(*Hint:* Begin by writing $\bar{A}$ as $\bar{A} = x\bar{i} + y\bar{j} + z\bar{k}$. Then take the dot product on both sides of the equation with $\bar{i}$, with $\bar{j}$, and with $\bar{k}$.)

48 Show that $\bar{D} = (\bar{B} \cdot \bar{B})\bar{A} - (\bar{A} \cdot \bar{B})\bar{B}$ is perpendicular to $\bar{B}$.

49 Are the vectors $\bar{A} = \langle 1, 1, 0 \rangle$, $\bar{B} = \langle 2, -1, 0 \rangle$, and $\bar{C} = \langle 1, 1, 1 \rangle$ coplanar? Explain.

50 Show that three vectors are coplanar if and only if one of them can be expressed as a linear combination of the other two.

51 Are the vectors $\bar{A} = 2\bar{i} - 3\bar{j} + 5\bar{k}$, $\bar{B} = 3\bar{i} + \bar{j} - 2\bar{k}$, and $\bar{C} = \bar{i} - 7\bar{j} + 12\bar{k}$ linearly independent? Explain.

52 Prove Theorem 1.

53 Suppose that $\bar{A}$, $\bar{B}$, and $\bar{C}$ are linearly independent vectors and that $x\bar{A} + y\bar{B} + z\bar{C} = a\bar{A} + b\bar{B} + c\bar{C}$. Prove that $x = a$, $y = b$, and $z = c$.

54 Prove Theorem 2.

55 If $\bar{A} = \langle x_1, y_1, z_1 \rangle$ and $\bar{B} = \langle x_2, y_2, z_2 \rangle$, find formulas for (a) $\bar{A} \cdot \bar{B}$, (b) $|\bar{A}|$, and (c) the direction cosines of $\bar{A}$.

56 Under what conditions is $|\bar{A} + \bar{B}| = |\bar{A}| + |\bar{B}|$ true?

3   Cross and Box Products of Vectors in Space

In this section we define and study a product that is available for vectors in three-dimensional space but not for vectors in the plane. This product is called the *cross product* (*outer product* or *vector product*). By combining the dot and cross products, we obtain the *box product* (or *determinant*) of three vectors in space.

The cross product of two vectors in space is defined to be a *vector* with a certain magnitude and direction. In order to specify the direction of the cross product of two vectors, we need the following concept: Three linearly independent vectors $\bar{A}$, $\bar{B}$, and $\bar{C}$ are said to form a *right-handed triple* in the order $\bar{A}$, $\bar{B}$, $\bar{C}$ if the thumb, index finger, and middle finger of the right hand (held in a "natural" position) can be aligned simultaneously with $\bar{A}$, $\bar{B}$, and $\bar{C}$, respectively (Figure 1). For instance, the standard basis vectors in the order $\bar{i}$, $\bar{j}$, $\bar{k}$ form a right-handed triple, but in the order $\bar{j}$, $\bar{i}$, $\bar{k}$ they do not. (Why?)

We can now give the definition of the cross product.

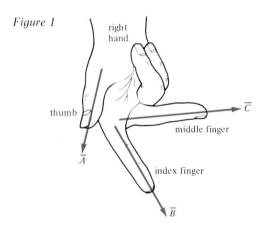

Figure 1

DEFINITION 1   Cross Product

Let $\bar{A}$ and $\bar{B}$ be two nonparallel vectors in three-dimensional space. Then the *cross product* of $\bar{A}$ and $\bar{B}$, denoted $\bar{A} \times \bar{B}$, is the vector whose length is numerically equal to the area of the parallelogram spanned by $\bar{A}$ and $\bar{B}$ and whose direction is perpendicular to both $\bar{A}$ and $\bar{B}$ in such a way that $\bar{A}$, $\bar{B}$, $\bar{A} \times \bar{B}$ is a right-handed triple (Figure 2).

Naturally, if $\bar{A}$ and $\bar{B}$ are parallel vectors, we define $\bar{A} \times \bar{B} = \bar{0}$.

For instance, the two perpendicular unit vectors $\bar{i}$ and $\bar{j}$ span a square of unit area; hence, $\bar{i} \times \bar{j}$ is a unit vector perpendicular to both $\bar{i}$ and $\bar{j}$ such that $\bar{i}, \bar{j}, \bar{i} \times \bar{j}$ is a right-handed triple (Figure 3). Therefore, $\bar{i} \times \bar{j} = \bar{k}$. Similarly, $\bar{j} \times \bar{k} = \bar{i}$ and $\bar{k} \times \bar{i} = \bar{j}$. Since $\bar{i}$ is parallel to itself, it follows that $\bar{i} \times \bar{i} = \bar{0}$. Similarly, $\bar{j} \times \bar{j} = \bar{0}$ and $\bar{k} \times \bar{k} = \bar{0}$.

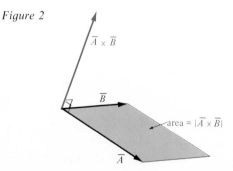

Figure 2

Figure 3

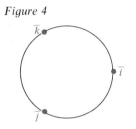

Figure 4

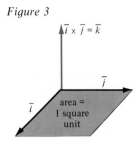

Figure 5

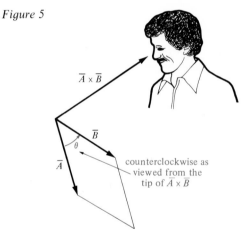

$\overline{A} \times \overline{B}$

$\overline{B}$

θ

$\overline{A}$

counterclockwise as viewed from the tip of $\overline{A} \times \overline{B}$

To find $\overline{j} \times \overline{i}$, we proceed just as above, but we must place our right thumb in the direction of $\overline{j}$ and our index finger in the direction of $\overline{i}$. Then, our middle finger points straight *downward*, so that $\overline{j} \times \overline{i} = -\overline{k}$. Similarly, $\overline{i} \times \overline{k} = -\overline{j}$ and $\overline{k} \times \overline{j} = -\overline{i}$.

In calculating cross products of the standard basis vectors, the following "memory device" may prove useful. Consider Figure 4 showing $\overline{i}$, $\overline{j}$, and $\overline{k}$ as symbols on the rim of a wheel. If, in naming two of these symbols successively, we go *clockwise* on the wheel, the cross product of the two named symbols is the remaining symbol; if we go *counterclockwise*, it is the negative of the remaining symbol. For instance, in moving from $\overline{k}$ to $\overline{j}$ we go counterclockwise; hence, $\overline{k} \times \overline{j} = -\overline{i}$.

The condition that $\overline{A}$, $\overline{B}$, $\overline{A} \times \overline{B}$ is a right-handed triple can be rephrased in various equivalent ways. For example, a right-handed screw turned through the smaller angle *from $\overline{A}$ to $\overline{B}$* advances in the direction of $\overline{A} \times \overline{B}$. Equivalently, if the parallelogram spanned by $\overline{A}$ and $\overline{B}$ is viewed from the tip of $\overline{A} \times \overline{B}$, then the angle θ from $\overline{A}$ to $\overline{B}$ is generated by a counterclockwise rotation (Figure 5). (Also, see Problem 35.)

Notice that the direction of the cross product is reversed when the two factors are interchanged; that is,

$$\overline{B} \times \overline{A} = -(\overline{A} \times \overline{B}).$$

Thus, *the cross product is not a commutative operation.* Consequently, in calculating with cross products, we must be very careful about the *order* in which the vectors are multiplied.

If s is a positive scalar, then $s\overline{A}$ has the same direction as $\overline{A}$ but is s times as long. It follows that the parallelogram spanned by $s\overline{A}$ and $\overline{B}$ has an area that is s times the area of the parallelogram spanned by $\overline{A}$ and $\overline{B}$; that is,

$$|(s\overline{A}) \times \overline{B}| = s|\overline{A} \times \overline{B}| \qquad \text{for } s > 0.$$

Obviously, the direction of $(s\overline{A}) \times \overline{B}$ is the same as the direction of $\overline{A} \times \overline{B}$ (why?), and it follows that

$$(s\overline{A}) \times \overline{B} = s(\overline{A} \times \overline{B}).$$

We leave it to the reader to make the simple geometric argument showing that the latter equation holds even when s is not positive (Problem 37). Similar considerations show that

$$\overline{A} \times (t\overline{B}) = t(\overline{A} \times \overline{B}).$$

Putting s or t equal to -1 gives

$$(-\overline{A}) \times B = -(\overline{A} \times \overline{B}) = \overline{A} \times (-\overline{B}).$$

3.1 The Box Product

The cross product $\overline{A} \times \overline{B}$ of vectors $\overline{A}$ and $\overline{B}$ is a vector and it is therefore possible to form its dot product $(\overline{A} \times \overline{B}) \cdot \overline{C}$ with a third vector $\overline{C}$. The scalar $(\overline{A} \times \overline{B}) \cdot \overline{C}$ is called the *box product* (*triple scalar product*, or *determinant*) of the vectors $\overline{A}$, $\overline{B}$, and $\bar{C}$. Suppose that $\bar{A}$, $\bar{B}$, $\bar{C}$ is a right-handed triple. Notice that $\bar{A}$, $\bar{B}$, and $\bar{C}$ form three adjacent edges of a "box" whose six faces are parallelograms and whose opposite pairs of faces are parallel and congruent (Figure 6). Such a box is called a *parallelepiped*, and its volume V is given by the area $|\bar{A} \times \bar{B}|$ of its base times its height h. If θ is the angle between $\bar{C}$ and $\bar{A} \times \bar{B}$, then $h = |\bar{C}| \cos \theta$ and by the definition of the dot product

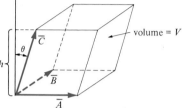

Figure 6

$$V = |\bar{A} \times \bar{B}| h = |\bar{A} \times \bar{B}||\bar{C}| \cos \theta = (\bar{A} \times \bar{B}) \cdot \bar{C}.$$

Thus, the box product $(\bar{A} \times \bar{B}) \cdot \bar{C}$ gives the volume of the box (Figure 6) spanned by $\bar{A}$, $\bar{B}$, and $\bar{C}$.

If $\bar{A}$, $\bar{B}$, and $\bar{C}$ are linearly independent vectors that do not form a right-handed triple $\bar{A}$, $\bar{B}$, and $\bar{C}$, we say that they form a *left-handed triple*. If $\bar{A}$, $\bar{B}$, $\bar{C}$ is a left-handed triple, it is easy to see that $\bar{B}$, $\bar{A}$, $\bar{C}$ is a right-handed triple (Problem 7); hence, if V is the volume of the box spanned by $\bar{B}$, $\bar{A}$, and $\bar{C}$, then

$$(\bar{A} \times \bar{B}) \cdot \bar{C} = [-(\bar{B} \times \bar{A})] \cdot \bar{C} = -[(\bar{B} \times \bar{A}) \cdot \bar{C}] = -V.$$

It follows from the considerations above that the volume V of the box (parallelepiped) spanned by the three linearly independent vectors $\bar{A}$, $\bar{B}$, and $\bar{C}$ is given by

$$V = \pm(\bar{A} \times \bar{B}) \cdot \bar{C},$$

where the plus or minus sign is used according to whether the triple $\bar{A}$, $\bar{B}$, $\bar{C}$ is right-handed or left-handed, respectively.

If $\bar{A}$, $\bar{B}$, $\bar{C}$ is a right-handed triple (as in Figure 6), then—as one can easily see—$\bar{B}$, $\bar{C}$, $\bar{A}$ is also a right-handed triple. Thus, since the volume V of the box spanned by the three vectors is the same regardless of which face is called the base, we must have

$$V = (\bar{B} \times \bar{C}) \cdot \bar{A} = (\bar{A} \times \bar{B}) \cdot \bar{C}.$$

Similarly, if $\bar{A}$, $\bar{B}$, $\bar{C}$ is a left-handed triple, we have

$$V = -(\bar{B} \times \bar{C}) \cdot \bar{A} = -(\bar{A} \times \bar{B}) \cdot \bar{C}.$$

In any case, then,

$$(\bar{B} \times \bar{C}) \cdot \bar{A} = (\bar{A} \times \bar{B}) \cdot \bar{C}.$$

Since $(\bar{B} \times \bar{C}) \cdot \bar{A} = \bar{A} \cdot (\bar{B} \times \bar{C})$, the latter equation can be rewritten as

$$\bar{A} \cdot (\bar{B} \times \bar{C}) = (\bar{A} \times \bar{B}) \cdot \bar{C};$$

that is, in the box product, *the dot and cross can be interchanged*, provided that the three vectors are kept in the same order.

3.2 The Distributive Law and the Cross-Product Formula

Although the preceding discussion describes the cross product $\bar{A} \times \bar{B}$ geometrically and relates it to the volume of a parallelepiped, it does not show how $\bar{A} \times \bar{B}$ can be found directly from $\bar{A}$ and $\bar{B}$ when $\bar{A}$ and $\bar{B}$ are given in component form. A "cross-product formula" that gives the components of $\bar{A} \times \bar{B}$ in terms of the components of $\bar{A}$ and of $\bar{B}$ can be derived on the basis of the following theorem.

THEOREM 1

Distributive Law for the Cross Product

For any three vectors $\bar{A}$, $\bar{B}$, and $\bar{C}$ in three-dimensional space,

$$\bar{A} \times (\bar{B} + \bar{C}) = (\bar{A} \times \bar{B}) + (\bar{A} \times \bar{C}).$$

PROOF

Let $\bar{D} = \bar{A} \times (\bar{B} + \bar{C}) - (\bar{A} \times \bar{B}) - (\bar{A} \times \bar{C})$, and note that it is sufficient to prove $\bar{D} = \bar{0}$. We do this by showing that $\bar{D} \cdot \bar{D} = |\bar{D}|^2 = 0$. We have

$$\bar{D} \cdot \bar{D} = \bar{D} \cdot [\bar{A} \times (\bar{B} + \bar{C}) - (\bar{A} \times \bar{B}) - (\bar{A} \times \bar{C})]$$
$$= \bar{D} \cdot [\bar{A} \times (\bar{B} + \bar{C})] - \bar{D} \cdot (\bar{A} \times \bar{B}) - \bar{D} \cdot (\bar{A} \times \bar{C}).$$

Interchanging the dot and cross in the three box products, we obtain

$$\bar{D} \cdot \bar{D} = (\bar{D} \times \bar{A}) \cdot (\bar{B} + \bar{C}) - (\bar{D} \times \bar{A}) \cdot \bar{B} - (\bar{D} \times \bar{A}) \cdot \bar{C}$$
$$= (\bar{D} \times \bar{A}) \cdot \bar{B} + (\bar{D} \times \bar{A}) \cdot \bar{C} - (\bar{D} \times \bar{A}) \cdot \bar{B} - (\bar{D} \times \bar{A}) \cdot \bar{C} = 0,$$

and the proof is complete.

Naturally, the distributive law for the cross product extends to three or more summands; for instance,

$$\bar{A} \times (\bar{B} + \bar{C} + \bar{D}) = [\bar{A} \times (\bar{B} + \bar{C})] + (\bar{A} \times \bar{D}) = (\bar{A} \times \bar{B}) + (\bar{A} \times \bar{C}) + (\bar{A} \times \bar{D}).$$

Also, the distributive law works "from the right"; that is,

$$(\bar{A} + \bar{B}) \times \bar{C} = -[\bar{C} \times (\bar{A} + \bar{B})] = -[(\bar{C} \times \bar{A}) + (\bar{C} \times \bar{B})]$$
$$= -(\bar{C} \times \bar{A}) - (\bar{C} \times \bar{B}) = (\bar{A} \times \bar{C}) + (\bar{B} \times \bar{C}).$$

THEOREM 2

The Cross-Product Formula

If $\bar{A} = a_1 i + a_2 j + a_3 \bar{k}$ and $\bar{B} = b_1 i + b_2 j + b_3 \bar{k}$, then

$$\bar{A} \times \bar{B} = (a_2 b_3 - a_3 b_2) i - (a_1 b_3 - a_3 b_1) j + (a_1 b_2 - a_2 b_1) \bar{k}.$$

PROOF

Using the distributive law for the cross product, we have

$$\bar{A} \times \bar{B} = \bar{A} \times (b_1 i + b_2 j + b_3 \bar{k}) = [\bar{A} \times (b_1 i)] + [\bar{A} \times (b_2 j)] + [\bar{A} \times (b_3 \bar{k})]$$
$$= b_1 (\bar{A} \times i) + b_2 (\bar{A} \times j) + b_3 (\bar{A} \times \bar{k}).$$

Here,

$$\bar{A} \times i = (a_1 i + a_2 j + a_3 \bar{k}) \times i = [(a_1 i) \times i] + [(a_2 j) \times i] + [(a_3 \bar{k}) \times i]$$
$$= a_1 (i \times i) + a_2 (j \times i) + a_3 (\bar{k} \times i) = a_1 \bar{0} + a_2 (-\bar{k}) + a_3 j$$
$$= a_3 j - a_2 \bar{k}.$$

Similar calculations yield

$$\bar{A} \times j = a_1 \bar{k} - a_3 i \quad \text{and} \quad \bar{A} \times \bar{k} = a_2 i - a_1 j.$$

Therefore, substituting into the original equation, we have

$$\bar{A} \times \bar{B} = b_1 (a_3 j - a_2 \bar{k}) + b_2 (a_1 \bar{k} - a_3 i) + b_3 (a_2 i - a_1 j)$$
$$= b_1 a_3 j - b_1 a_2 \bar{k} + b_2 a_1 \bar{k} - b_2 a_3 i + b_3 a_2 i - b_3 a_1 j$$
$$= (a_2 b_3 - a_3 b_2) i - (a_1 b_3 - a_3 b_1) j + (a_1 b_2 - a_2 b_1) \bar{k}.$$

The formula in Theorem 2 is not easily recalled as it stands; however, using *determinants*, it can be rewritten in a much simpler form. A *determinant* of order 2 is defined by

$$\begin{vmatrix} a & b \\ c & d \end{vmatrix} = ad - bc,$$

and a *determinant* of order 3 is defined by

$$\begin{vmatrix} a & b & c \\ x & y & z \\ u & v & w \end{vmatrix} = a \begin{vmatrix} y & z \\ v & w \end{vmatrix} - b \begin{vmatrix} x & z \\ u & w \end{vmatrix} + c \begin{vmatrix} x & y \\ u & v \end{vmatrix}.$$

In what follows, we often find it convenient to use determinant notation in which the first row contains vectors rather than scalars, so that, for instance,

$$\begin{vmatrix} i & j & k \\ x & y & z \\ u & v & w \end{vmatrix} = \begin{vmatrix} y & z \\ v & w \end{vmatrix} i - \begin{vmatrix} x & z \\ u & w \end{vmatrix} j + \begin{vmatrix} x & y \\ u & v \end{vmatrix} k$$

$$= (yw - zv)i - (xw - zu)j + (xv - yu)k.$$

Using determinant notation, we can now rewrite the cross-product formula in the following easily remembered form. If $\bar{A} = a_1 i + a_2 j + a_3 k$ and $\bar{B} = b_1 i + b_2 j + b_3 k$, then

$$\bar{A} \times \bar{B} = \begin{vmatrix} i & j & k \\ a_1 & a_2 & a_3 \\ b_1 & b_2 & b_3 \end{vmatrix}.$$

EXAMPLES 1 Evaluate (a) $\bar{A} \times \bar{B}$ and (b) $\bar{B} \times \bar{A}$ if $\bar{A} = 2i + 3j - 4k$ and $\bar{B} = 5i - 2j - k$.

SOLUTION

(a) $\bar{A} \times \bar{B} = \begin{vmatrix} i & j & k \\ 2 & 3 & -4 \\ 5 & -2 & -1 \end{vmatrix} = \begin{vmatrix} 3 & -4 \\ -2 & -1 \end{vmatrix} i - \begin{vmatrix} 2 & -4 \\ 5 & -1 \end{vmatrix} j + \begin{vmatrix} 2 & 3 \\ 5 & -2 \end{vmatrix} k$

$= [(3)(-1) - (-4)(-2)]i - [(2)(-1) - (-4)(5)]j + [(2)(-2) - (3)(5)]k$
$= -11i - 18j - 19k.$

(b) Since $\bar{B} \times \bar{A} = -(\bar{A} \times \bar{B})$, we have

$$\bar{B} \times \bar{A} = -(-11i - 18j - 19k) = 11i + 18j + 19k.$$

2 Evaluate $\bar{B} \times \bar{A}$ if $\bar{A} = \langle 0, 0, 1 \rangle$ and $\bar{B} = \langle 1, 1, 0 \rangle$.

SOLUTION

$$\bar{B} \times \bar{A} = \begin{vmatrix} i & j & k \\ 1 & 1 & 0 \\ 0 & 0 & 1 \end{vmatrix} = \begin{vmatrix} 1 & 0 \\ 0 & 1 \end{vmatrix} i - \begin{vmatrix} 1 & 0 \\ 0 & 1 \end{vmatrix} j + \begin{vmatrix} 1 & 1 \\ 0 & 0 \end{vmatrix} k$$

$$= i - j + (0)k = \langle 1, -1, 0 \rangle.$$

By combining the cross-product formula with the formula for calculating dot products, we obtain, in the following theorem, a formula for the box product. The proof is left as an exercise (Problem 42).

THEOREM 3   **The Box-Product Formula**

If $\bar{A} = a_1 i + a_2 j + a_3 k$, $\bar{B} = b_1 i + b_2 j + b_3 k$, and $\bar{C} = c_1 i + c_2 j + c_3 k$, then

$$(\bar{A} \times \bar{B}) \cdot \bar{C} = \begin{vmatrix} a_1 & a_2 & a_3 \\ b_1 & b_2 & b_3 \\ c_1 & c_2 & c_3 \end{vmatrix}.$$

EXAMPLES   Evaluate $(\bar{A} \times \bar{B}) \cdot \bar{C}$.

1 $\bar{A} = i + j$, $\bar{B} = 2i - 3j + k$, and $\bar{C} = j - 4k$.

SOLUTION

$$(\bar{A} \times \bar{B}) \cdot \bar{C} = \begin{vmatrix} 1 & 1 & 0 \\ 2 & -3 & 1 \\ 0 & 1 & -4 \end{vmatrix} = 1 \begin{vmatrix} -3 & 1 \\ 1 & -4 \end{vmatrix} - 1 \begin{vmatrix} 2 & 1 \\ 0 & -4 \end{vmatrix} + 0 \begin{vmatrix} 2 & -3 \\ 0 & 1 \end{vmatrix}$$

$$= 11 - (-8) + 0 = 19.$$

2 $\bar{A} = \langle 1, 0, -1 \rangle$, $\bar{B} = \langle 2, 1, 3 \rangle$, and $\bar{C} = \langle 5, -1, 0 \rangle$.

SOLUTION

$$(\bar{A} \times \bar{B}) \cdot \bar{C} = \begin{vmatrix} 1 & 0 & -1 \\ 2 & 1 & 3 \\ 5 & -1 & 0 \end{vmatrix} = 1 \begin{vmatrix} 1 & 3 \\ -1 & 0 \end{vmatrix} - 0 \begin{vmatrix} 2 & 3 \\ 5 & 0 \end{vmatrix} + (-1) \begin{vmatrix} 2 & 1 \\ 5 & -1 \end{vmatrix}$$

$$= 3 - 0 + 7 = 10.$$

Problem Set 3

In problems 1 through 6, determine whether each triple of vectors is right-handed or left-handed.

1 $\bar{i}, -\bar{j}, \bar{k}$ **2** $\bar{i}, -\bar{j}, -\bar{k}$ **3** $\bar{j}, \bar{k}, \bar{i}$

4 $\bar{j}, -\bar{i}, -\bar{k}$ **5** $-\bar{i}, -\bar{j}, -\bar{k}$ **6** $\bar{i} + \bar{j}, \bar{j} - \bar{i}, \bar{k}$

In problems 7 through 10, assume that $\bar{A}, \bar{B}, \bar{C}$ is a left-handed triple. Determine whether each triple is right- or left-handed.

7 $\bar{B}, \bar{A}, \bar{C}$ **8** $-\bar{A}, \bar{B}, -\bar{C}$ **9** $\bar{C}, \bar{A}, \bar{B}$ **10** $\bar{C}, \bar{B}, \bar{A}$

In problems 11 through 16, find each cross product by sketching an appropriate figure and using only geometric facts. (Do not use the "memory device" of Figure 4 or the cross-product formula.)

11 $\bar{i} \times \bar{k}$ **12** $(-\bar{i}) \times \bar{j}$ **13** $\bar{k} \times \bar{i}$

14 $(\bar{i} + \bar{j}) \times (\bar{j} - \bar{i})$ **15** $(-\bar{i}) \times (-\bar{j})$ **16** $(\bar{i} + \bar{j}) \times \bar{k}$

17 Using the "memory device" of Figure 4, fill out the remainder of the following cross-product multiplication table for the standard basis vectors.

$\times$	$\bar{i}$	$\bar{j}$	$\bar{k}$
$\bar{i}$	$\bar{0}$	$\bar{k}$	?
$\bar{j}$	?	?	$\bar{i}$
$\bar{k}$	?	?	?

18 Using the table in problem 17, calculate (a) $(\bar{i} \times \bar{j}) \times \bar{j}$ and (b) $\bar{i} \times (\bar{j} \times \bar{j})$. Does the associative law work for the cross product?

In problems 19 through 30, evaluate each cross product.

19 $\bar{A} \times \bar{B}$; $\bar{A} = 4\bar{k}$ and $\bar{B} = 6\bar{j}$ **20** $\bar{C} \times \bar{D}$; $\bar{C} = -2\bar{j}$ and $\bar{D} = 5\bar{k}$

21 $\bar{B} \times \bar{A}$; $\bar{A} = \bar{i}/3$ and $\bar{B} = 9\bar{k}$ **22** $\bar{E} \times \bar{F}$; $\bar{E} = \bar{k}/2$ and $\bar{F} = 3\bar{i}/5$

23 $\bar{A} \times \bar{B}$; $\bar{A} = \bar{i} + \bar{k}$ and $\bar{B} = 2\bar{k}$ **24** $\bar{D} \times \bar{E}$; $\bar{D} = \bar{i} - 2\bar{j}$ and $\bar{E} = 2\bar{k}$

25 $\bar{F} \times \bar{E}$; $\bar{E} = 3\bar{k}$ and $\bar{F} = 2\bar{i} + \bar{j}$ **26** $\bar{C} \times \bar{A}$; $\bar{A} = \bar{k} - 3\bar{j}$ and $\bar{C} = \bar{i} - 2\bar{j}$

27 $\bar{B} \times \bar{C}$; $\bar{B} = \bar{i} + \bar{j} + \bar{k}$ and $\bar{C} = 3\bar{i} - 2\bar{j} + \bar{k}$ **28** $\bar{F} \times \bar{G}$; $\bar{F} = 2\bar{i} - \bar{j} - 3\bar{k}$ and $\bar{G} = 5\bar{j} + 3\bar{k}$

29 $\bar{A} \times \bar{B}$; $\bar{A} = \langle 2, -3, 0 \rangle$ and $\bar{B} = \langle 5, -2, 7 \rangle$ **30** $\bar{B} \times \bar{A}$; $\bar{A} = \langle 1, -1, -2 \rangle$ and $\bar{B} = \langle 3, -4, 1 \rangle$

In problems 31 through 34, evaluate the box product $(\bar{A} \times \bar{B}) \cdot \bar{C}$.

31 $\bar{A} = i + 2j - 3\bar{k}, \bar{B} = 5i + 7j + 4\bar{k}, \bar{C} = 2i - 3j - 6\bar{k}$ **32** $\bar{A} = 7i - 5j + \bar{k}, \bar{B} = 2i - 4j + 6\bar{k}, \bar{C} = i + \bar{k}$

33 $\bar{A} = \langle 1, 1, 1 \rangle, \bar{B} = \langle -1, 3, 2 \rangle, \bar{C} = \langle 7, -9, 4 \rangle$ **34** $\bar{A} = \langle 1, -1, 0 \rangle, \bar{B} = \langle 0, -1, 1 \rangle, \bar{C} = \langle -1, 0, 1 \rangle$

35 If we grab the vector $\bar{A} \times \bar{B}$ with our right hand so that the fingers curl in the direction of the angle from $\bar{A}$ to $\bar{B}$ and our thumb lies along the shaft of the vector $\bar{A} \times \bar{B}$ (pointing away from our fist), will the thumb point in the same direction as the vector $\bar{A} \times \bar{B}$ or will it point in the opposite direction?

36 If $\bar{A}$ is perpendicular to $\bar{B}$, explain why $|\bar{A} \times \bar{B}| = |\bar{A}||\bar{B}|$.

37 Show that $(s\bar{A}) \times \bar{B} = s(\bar{A} \times \bar{B})$ even if the scalar s is not positive. (Make direct use of the definition of the cross product.)

38 Prove that $\bar{A} \times (t\bar{B}) = t(\bar{A} \times \bar{B})$.

39 Prove that $(\bar{A} - \bar{B}) \times \bar{C} = (\bar{A} \times \bar{C}) - (\bar{B} \times \bar{C})$. [*Hint*: $\bar{A} - \bar{B} = \bar{A} + (-1)\bar{B}$.]

40 Prove that $(\bar{A} - \bar{B}) \times (\bar{C} - \bar{D}) = (\bar{A} \times \bar{C}) - (\bar{A} \times \bar{D}) - (\bar{B} \times \bar{C}) + (\bar{B} \times \bar{D})$.

41 True or false: $(\bar{A} + \bar{B}) \times (\bar{A} - \bar{B}) = (\bar{A} \times \bar{A}) - (\bar{B} \times \bar{B})$?

42 Prove Theorem 3.

4 Algebraic Identities and Geometric Applications for the Cross and Box Products

In this section we obtain some algebraic identities that can be used to advantage in calculating with cross products and we present some simple geometric applications of cross and box products.

4.1 Identities Involving the Cross Product

Using Theorem 2 in Section 3, we can establish a number of useful identities involving the cross product. For instance, the identities in the following theorem can be verified by expanding both sides in terms of the scalar components of the three vectors (Problem 51).

THEOREM 1   **Triple Vector Product Identities**
For any three vectors $\bar{A}$, $\bar{B}$, and $\bar{C}$ in three-dimensional space:

(i) $\bar{A} \times (\bar{B} \times \bar{C}) = (\bar{A} \cdot \bar{C})\bar{B} - (\bar{A} \cdot \bar{B})\bar{C}$.

(ii) $(\bar{A} \times \bar{B}) \times \bar{C} = (\bar{A} \cdot \bar{C})\bar{B} - (\bar{B} \cdot \bar{C})\bar{A}$.

One immediate—and important—consequence of the triple vector product identities is that *the associative law does not hold for cross products*; that is, in general,

$$(\bar{A} \times \bar{B}) \times \bar{C} \neq \bar{A} \times (\bar{B} \times \bar{C}).$$

EXAMPLE   If $\bar{A} = 7i - j + 2\bar{k}$, $\bar{B} = -3i + 2j - \bar{k}$, and $\bar{C} = 5j - 3\bar{k}$, find $(\bar{A} \times \bar{B}) \times \bar{C}$ and $\bar{A} \times (\bar{B} \times \bar{C})$.

SOLUTION

$$(\bar{A} \times \bar{B}) \times \bar{C} = (\bar{A} \cdot \bar{C})\bar{B} - (\bar{B} \cdot \bar{C})\bar{A} = (-11)\bar{B} - 13\bar{A} = -58i - 9j - 15\bar{k},$$

$$\bar{A} \times (\bar{B} \times \bar{C}) = (\bar{A} \cdot \bar{C})\bar{B} - (\bar{A} \cdot \bar{B})\bar{C} = (-11)\bar{B} + 25\bar{C} = 33i + 103j - 64\bar{k}.$$

Using the triple vector product identities and the fact that the cross and dot can be exchanged in the box product, we can prove the following theorem.

THEOREM 2 **Lagrange's Identity**

For any four vectors $\bar{A}$, $\bar{B}$, $\bar{C}$, and $\bar{D}$ in three-dimensional space,

$$(\bar{A} \times \bar{B}) \cdot (\bar{C} \times \bar{D}) = \begin{vmatrix} \bar{A} \cdot \bar{C} & \bar{A} \cdot \bar{D} \\ \bar{B} \cdot \bar{C} & \bar{B} \cdot \bar{D} \end{vmatrix}.$$

PROOF

$$(\bar{A} \times \bar{B}) \cdot (\bar{C} \times \bar{D}) = \bar{A} \cdot [\bar{B} \times (\bar{C} \times \bar{D})] = \bar{A} \cdot [(\bar{B} \cdot \bar{D})\bar{C} - (\bar{B} \cdot \bar{C})\bar{D}]$$

$$= (\bar{B} \cdot \bar{D})(\bar{A} \cdot \bar{C}) - (\bar{B} \cdot \bar{C})(\bar{A} \cdot \bar{D}) = \begin{vmatrix} \bar{A} \cdot \bar{C} & \bar{A} \cdot \bar{D} \\ \bar{B} \cdot \bar{C} & \bar{B} \cdot \bar{D} \end{vmatrix}.$$

EXAMPLE Use Lagrange's identity to find a formula for $|\bar{A} \times \bar{B}|^2$.

SOLUTION

$$|\bar{A} \times \bar{B}|^2 = (\bar{A} \times \bar{B}) \cdot (\bar{A} \times \bar{B}) = \begin{vmatrix} \bar{A} \cdot \bar{A} & \bar{A} \cdot \bar{B} \\ \bar{B} \cdot \bar{A} & \bar{B} \cdot \bar{B} \end{vmatrix} = (\bar{A} \cdot \bar{A})(\bar{B} \cdot \bar{B}) - (\bar{A} \cdot \bar{B})^2$$

$$= |\bar{A}|^2 |\bar{B}|^2 - (\bar{A} \cdot \bar{B})^2.$$

From the identity obtained in the last example, we have the formula

$$|\bar{A} \times \bar{B}| = \sqrt{|\bar{A}|^2 |\bar{B}|^2 - (\bar{A} \cdot \bar{B})^2}$$

for the length of the cross product of $\bar{A}$ and $\bar{B}$. If θ is the angle between $\bar{A}$ and $\bar{B}$, then

$$|\bar{A} \times \bar{B}| = \sqrt{|\bar{A}|^2 |\bar{B}|^2 - (\bar{A} \cdot \bar{B})^2} = \sqrt{|\bar{A}|^2 |\bar{B}|^2 - |\bar{A}|^2 |\bar{B}|^2 \cos^2 \theta}$$

$$= |\bar{A}||\bar{B}|\sqrt{1 - \cos^2 \theta} = |\bar{A}||\bar{B}| \sin \theta.$$

4.2 Geometric Applications of the Cross and Box Products

Many of the geometric applications of the cross product follow directly from its definition.

EXAMPLE Find the area of the parallelogram spanned by the vectors $\bar{A} = i - 2j + 3\bar{k}$ and $\bar{B} = -i + j + 2\bar{k}$.

SOLUTION

The desired area is numerically equal to the length $|\bar{A} \times \bar{B}|$ of the cross product $\bar{A} \times \bar{B}$. Using the formula obtained in Section 4.1, we have

$$|\bar{A} \times \bar{B}| = \sqrt{|\bar{A}|^2 |\bar{B}|^2 - (\bar{A} \cdot \bar{B})^2} = \sqrt{(14)(6) - 3^2} = 5\sqrt{3} \text{ square units.}$$

Naturally, the box product is useful for computing the volume V of the parallelepiped spanned by the vectors $\bar{A}$, $\bar{B}$, and $\bar{C}$; in fact,

$$V = |(\bar{A} \times \bar{B}) \cdot \bar{C}|.$$

EXAMPLE   Find the volume V of the parallelepiped spanned by the vectors $\bar{A} = 3\bar{i} + \bar{j}$, $\bar{B} = \bar{i} + 2\bar{k}$, and $\bar{C} = \bar{i} + 2\bar{j} + 3\bar{k}$.

SOLUTION
Here,

$$(\bar{A} \times \bar{B}) \cdot \bar{C} = \begin{vmatrix} 3 & 1 & 0 \\ 1 & 0 & 2 \\ 1 & 2 & 3 \end{vmatrix} = 3 \begin{vmatrix} 0 & 2 \\ 2 & 3 \end{vmatrix} - 1 \begin{vmatrix} 1 & 2 \\ 1 & 3 \end{vmatrix} + 0 \begin{vmatrix} 1 & 0 \\ 1 & 2 \end{vmatrix}$$

$$= 3(-4) - 1(1) + 0 = -13;$$

hence, $V = |-13| = 13$ cubic units.

The algebraic sign of the box product can be used to determine whether a triple of vectors is right-handed or left-handed. In fact, $\bar{A}$, $\bar{B}$, $\bar{C}$ is a right-handed or left-handed triple according to whether $(\bar{A} \times \bar{B}) \cdot \bar{C}$ is positive or negative, respectively (Problem 60). Also, $(\bar{A} \times \bar{B}) \cdot \bar{C} = 0$ if and only if the vectors $\bar{A}$, $\bar{B}$, and $\bar{C}$ are coplanar—that is, linearly dependent (Problem 59).

EXAMPLES 1   Determine whether $\bar{A}$, $\bar{B}$, $\bar{C}$ is a right-handed or left-handed triple if $\bar{A} = -\bar{i} + 7\bar{j} + 2\bar{k}$, $\bar{B} = 2\bar{i} + 3\bar{j} + \bar{k}$, and $\bar{C} = -\bar{i} + 5\bar{j} + 2\bar{k}$.

SOLUTION

$$(\bar{A} \times \bar{B}) \cdot \bar{C} = \begin{vmatrix} -1 & 7 & 2 \\ 2 & 3 & 1 \\ -1 & 5 & 2 \end{vmatrix} = (-1) \begin{vmatrix} 3 & 1 \\ 5 & 2 \end{vmatrix} - 7 \begin{vmatrix} 2 & 1 \\ -1 & 2 \end{vmatrix} + 2 \begin{vmatrix} 2 & 3 \\ -1 & 5 \end{vmatrix} = -10 < 0;$$

hence, $\bar{A}$, $\bar{B}$, $\bar{C}$ is a left-handed triple.

2   Show that the three vectors $\bar{A} = \langle 1, -1, 1 \rangle$, $\bar{B} = \langle 2, 1, -1 \rangle$, and $\bar{C} = \langle 0, -1, 1 \rangle$ are coplanar (linearly dependent).

SOLUTION

$$(\bar{A} \times \bar{B}) \cdot \bar{C} = \begin{vmatrix} 1 & -1 & 1 \\ 2 & 1 & -1 \\ 0 & -1 & 1 \end{vmatrix} = 1 \begin{vmatrix} 1 & -1 \\ -1 & 1 \end{vmatrix} - (-1) \begin{vmatrix} 2 & -1 \\ 0 & 1 \end{vmatrix} + 1 \begin{vmatrix} 2 & 1 \\ 0 & -1 \end{vmatrix} = 0;$$

hence, $\bar{A}$, $\bar{B}$, and $\bar{C}$ are coplanar.

Many geometric objects can be built up from nonoverlapping triangles; hence, the following theorem, which gives a formula for the area of a triangle in terms of the coordinates of its vertices, can be quite useful.

THEOREM 3   **Area of a Triangle in Space**
Let PQR be a triangle in xyz space, where $P = (x_1, y_1, z_1)$, $Q = (x_2, y_2, z_2)$, and $R = (x_3, y_3, z_3)$. Then the area a of PQR is given by the formula

$$a = \frac{1}{2} |\overrightarrow{PQ} \times \overrightarrow{PR}|,$$

$$\overrightarrow{PQ} \times \overrightarrow{PR} = \begin{vmatrix} \bar{i} & \bar{j} & \bar{k} \\ x_2 - x_1 & y_2 - y_1 & z_2 - z_1 \\ x_3 - x_1 & y_3 - y_1 & z_3 - z_1 \end{vmatrix}.$$

Figure 1

PROOF
The area of triangle PQR is half the area of the parallelogram spanned by the vectors $\overrightarrow{PQ}$ and $\overrightarrow{PR}$ (Figure 1).

EXAMPLE Find the area a of the triangle whose vertices are $P = (2, 1, 5)$, $Q = (4, 0, 2)$, and $R = (-1, 0, -1)$.

SOLUTION
Here,

$$\vec{PQ} = (4 - 2)\vec{i} + (0 - 1)\vec{j} + (2 - 5)\vec{k} = 2\vec{i} - \vec{j} - 3\vec{k}, \quad \text{and}$$

$$\vec{PR} = (-1 - 2)\vec{i} + (0 - 1)\vec{j} + (-1 - 5)\vec{k} = -3\vec{i} - \vec{j} - 6\vec{k};$$

hence,

$$\vec{PQ} \times \vec{PR} = \begin{vmatrix} \vec{i} & \vec{j} & \vec{k} \\ 2 & -1 & -3 \\ -3 & -1 & -6 \end{vmatrix} = 3\vec{i} + 21\vec{j} - 5\vec{k}.$$

Therefore,

$$a = \frac{1}{2}|\vec{PQ} \times \vec{PR}| = \frac{1}{2}\sqrt{3^2 + (21)^2 + (-5)^2} = \frac{\sqrt{475}}{2} = \frac{5}{2}\sqrt{19} \text{ square units.}$$

Using Theorem 3, we can now derive an interesting formula for the area of a triangle in the xy plane.

THEOREM 4 **Area of a Triangle in the Plane**
Let PQR be a triangle in the xy plane with $P = (x_1, y_1)$, $Q = (x_2, y_2)$, and $R = (x_3, y_3)$. Then the area a of PQR is given by

$$a = \text{absolute value of } \frac{1}{2}\begin{vmatrix} x_1 & y_1 & 1 \\ x_2 & y_2 & 1 \\ x_3 & y_3 & 1 \end{vmatrix}.$$

PROOF
Regard the xy plane as being the xy coordinate plane in xyz space, so that $P = (x_1, y_1, 0)$, $Q = (x_2, y_2, 0)$, and $R = (x_3, y_3, 0)$. Then,

$$\vec{PQ} \times \vec{PR} = \begin{vmatrix} \vec{i} & \vec{j} & \vec{k} \\ x_2 - x_1 & y_2 - y_1 & 0 \\ x_3 - x_1 & y_3 - y_1 & 0 \end{vmatrix} = \begin{vmatrix} x_2 - x_1 & y_2 - y_1 \\ x_3 - x_1 & y_3 - y_1 \end{vmatrix}\vec{k},$$

so that

$$a = \frac{1}{2}|\vec{PQ} \times \vec{PR}| = \text{absolute value of } \frac{1}{2}\begin{vmatrix} x_2 - x_1 & y_2 - y_1 \\ x_3 - x_1 & y_3 - y_1 \end{vmatrix}$$

by Theorem 3. The reader can verify that

$$\begin{vmatrix} x_2 - x_1 & y_2 - y_1 \\ x_3 - x_1 & y_3 - y_1 \end{vmatrix} = \begin{vmatrix} x_1 & y_1 & 1 \\ x_2 & y_2 & 1 \\ x_3 & y_3 & 1 \end{vmatrix}$$

simply by expanding both determinants (Problem 53).

EXAMPLE Find the area of the triangle in the xy plane whose vertices are $P = (1, 2)$, $Q = (4, -3)$, and $R = (5, 0)$.

SOLUTION
Here,

$$\begin{vmatrix} 1 & 2 & 1 \\ 4 & -3 & 1 \\ 5 & 0 & 1 \end{vmatrix} = \begin{vmatrix} -3 & 1 \\ 0 & 1 \end{vmatrix} - 2\begin{vmatrix} 4 & 1 \\ 5 & 1 \end{vmatrix} + \begin{vmatrix} 4 & -3 \\ 5 & 0 \end{vmatrix} = 14,$$

so that $a = \frac{1}{2}(14) = 7$ square units.

Problem Set 4

In problems 1 through 30, let $\bar{A} = 2i - 3j + 5\bar{k}$, $\bar{B} = 3i + j - 2\bar{k}$, $\bar{C} = 2i - j$, $\bar{D} = -i + 4\bar{k}$, and $\bar{E} = 5\bar{k} - j$. Evaluate each expression using the identities for cross and box products whenever possible to simplify your calculations.

1 $(\bar{A} \times \bar{B}) \times \bar{C}$

2 $\bar{D} \times (\bar{C} \times \bar{E})$

3 $|\bar{E} \times \bar{A}|^2 + (\bar{A} \cdot \bar{E})^2$

4 $|\bar{B} \times \bar{D}|^2 - |\bar{B}|^2 |\bar{D}|^2$

5 $(\bar{A} \cdot \bar{C})\bar{B} - (\bar{A} \cdot \bar{B})\bar{C}$

6 $2\bar{B} \times [(-6\bar{D}) \times \bar{D}]$

7 $(3\bar{A}) \times (4\bar{B} + 3\bar{C})$

8 $\bar{A} \times (\bar{B} - \bar{A}) - (\bar{B} - \bar{A}) \times \bar{A}$

9 $\bar{A} \times (\bar{A} + \bar{E} - \bar{B})$

10 $\bar{A} \times (\bar{C} - 3\bar{B})$

11 $(\bar{A} \times \bar{C}) - 3(\bar{A} \times \bar{B})$

12 $\bar{B} \times (\bar{B} - 7\bar{C})$

13 $(\bar{A} \times \bar{B}) \cdot \bar{C}$

14 $\bar{B} \cdot (\bar{C} \times \bar{B})$

15 $\bar{A} \cdot (\bar{B} \times \bar{A})$

16 $\bar{B} \cdot (\bar{C} \times \bar{A})$

17 $(\bar{A} \times \bar{C}) \cdot \bar{B}$

18 $(\bar{A} + \bar{D}) \times (\bar{A} - \bar{D})$

19 $(2\bar{A} + 7\bar{D}) \times (4\bar{A} + 14\bar{D})$

20 $|(\bar{A} \times \bar{D}) - (\bar{D} \times \bar{A})|$

21 $|\bar{A} \times \bar{C}| - |\bar{C} \times \bar{A}|$

22 $|(2\bar{A} + \bar{B}) \times (\bar{A} - 2\bar{B})|$

23 $(\bar{A} \times \bar{C}) \cdot (\bar{C} \times \bar{E})$

24 $(\bar{C} \times \bar{E}) \cdot (\bar{B} \times \bar{D})$

25 $(\bar{A} \times \bar{B}) \cdot (\bar{A} \times \bar{C})$

26 $(\bar{A} \times \bar{B}) \cdot (\bar{A} \times \bar{B})$

27 $\bar{A} \times (\bar{B} \times \bar{A})$

28 $(\bar{A} \times \bar{B}) \times (\bar{B} \times \bar{A})$

29 $[\bar{B} \times (\bar{A} \times \bar{C})] - [(\bar{B} \times \bar{A}) \times \bar{C}]$

30 $(\bar{A} \times \bar{B}) \times (\bar{C} \times \bar{D})$

In problems 31 through 34, find the area of the parallelogram spanned by the given vectors $\bar{A}$ and $\bar{B}$.

31 $\bar{A} = i + j + \bar{k}$, $\bar{B} = i - j - \bar{k}$

32 $\bar{A} = 5i + 7j$, $\bar{B} = 3j - \bar{k}$

33 $\bar{A} = 2i + 5j - \bar{k}$, $\bar{B} = \dfrac{i}{2} - j + \dfrac{\bar{k}}{4}$

34 $\bar{A} = -i + j$, $\bar{B} = \dfrac{\bar{k}}{3}$

In problems 35 through 38, find the volume of the box (parallelepiped) spanned by the vectors $\bar{A}$, $\bar{B}$, and $\bar{C}$.

35 $\bar{A} = i + 2j$, $\bar{B} = 2i - j$, $\bar{C} = i + j + \bar{k}$

36 $\bar{A} = i$, $\bar{B} = j$, $\bar{C} = xi + yj + z\bar{k}$

37 $\bar{A} = \langle 1, -2, 1 \rangle$, $\bar{B} = \langle 3, -4, -1 \rangle$, $\bar{C} = \langle 4, -1, 3 \rangle$

38 $\bar{A} = \langle 1, 2, -1 \rangle$, $\bar{B} = \langle 0, 1, -1 \rangle$, $\bar{C} = \langle 3, 0, 1 \rangle$

In problems 39 through 42, (a) determine whether the three vectors $\bar{A}$, $\bar{B}$, and $\bar{C}$ are coplanar or linearly independent; and (b) if the vectors are linearly independent, determine whether the triple $\bar{A}$, $\bar{B}$, $\bar{C}$ is right-handed or left-handed.

39 $\bar{A} = \langle 1, 2, 3 \rangle$, $\bar{B} = \langle 2, -1, -1 \rangle$, $\bar{C} = \langle -2, 1, -1 \rangle$

40 $\bar{A} = i + j$, $\bar{B} = j + \bar{k}$, $\bar{C} = \bar{k} + i$

41 $\bar{A} = 2i - j + 7\bar{k}$, $\bar{B} = j + 3\bar{k}$, $\bar{C} = 2i - 3j + \bar{k}$

42 $\bar{A} = \langle 1, 1, -1 \rangle$, $\bar{B} = \langle 1, -1, 1 \rangle$, $\bar{C} = \langle -1, 1, 1 \rangle$

In problems 43 and 44, find the area of triangle PQR in xyz space.

43 $P = (1, 7, 2)$, $Q = (0, 7, -2)$, and $R = (-1, 6, 1)$

44 $P = (-1, 0, 1)$, $Q = (0, 1, 0)$, and $R = (1, 1, 1)$

In problems 45 and 46, find the area of triangle PQR in the xy plane.

45 $P = (1, -5)$, $Q = (-3, -4)$, and $R = (6, 2)$

46 $P = (-5, 7)$, $Q = (3, 6)$, and $R = (2, 1)$

47 Find a formula for the area of the triangle in space, two of whose adjacent edges are formed by the vectors $\bar{A}$ and $\bar{B}$ (Figure 2).

Figure 2

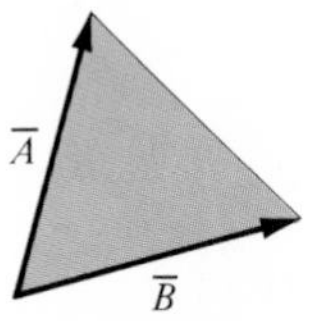

48 Show that the triangle in the xy plane two of whose adjacent edges are formed by the vectors $\bar{A} = x_1 i + y_1 j$ and $\bar{B} = x_2 i + y_2 j$ has an area given by the absolute value of

$$\frac{1}{2} \begin{vmatrix} x_1 & y_1 \\ x_2 & y_2 \end{vmatrix}.$$

49 Give a geometric interpretation of the determinant $\begin{vmatrix} x_1 & y_1 \\ x_2 & y_2 \end{vmatrix}$.

50 Let $P = (x_1, y_1)$, $Q = (x_2, y_2)$, $R = (x_3, y_3)$, and $S = (x_4, y_4)$ be the four vertices of a quadrilateral in the xy plane. Assume that the origin O is in the interior of $PQRS$ and that the vertices P, Q, R, and S occur in counterclockwise order around the diagonal. Show that the area a of the quadrilateral is given by

$$a = \frac{1}{2}\begin{vmatrix} x_1 & y_1 \\ x_2 & y_2 \end{vmatrix} + \frac{1}{2}\begin{vmatrix} x_2 & y_2 \\ x_3 & y_3 \end{vmatrix} + \frac{1}{2}\begin{vmatrix} x_3 & y_3 \\ x_4 & y_4 \end{vmatrix} + \frac{1}{2}\begin{vmatrix} x_4 & y_4 \\ x_1 & y_1 \end{vmatrix}.$$

51 Prove Theorem 1. (For simplicity, choose the x axis in the direction of the vector $\bar{C}$ and choose the y axis so that the vector $\bar{B}$ is contained in the xy plane.)

52 From part (i) of Theorem 1, it follows that the three vectors $\bar{B}$, $\bar{C}$, and $\bar{A} \times (\bar{B} \times \bar{C})$ are always coplanar. Explain *geometrically* why this should be so.

53 Complete the proof of Theorem 4.

54 Using Theorem 4, explain why the equation $\begin{vmatrix} x & y & 1 \\ x_1 & y_1 & 1 \\ x_2 & y_2 & 1 \end{vmatrix} = 0$ represents the straight line in the xy plane containing the two points (x_1, y_1) and (x_2, y_2).

55 Let $\bar{A} = \overrightarrow{OP}$, $\bar{B} = \overrightarrow{OQ}$, and $C = \overrightarrow{OR}$. Prove that the area of triangle PQR is given by $\frac{1}{2}|(\bar{A} \times \bar{B}) + (\bar{B} \times \bar{C}) + (\bar{C} \times \bar{A})|$.

56 Use the cross product to prove the law of sines; that is, for any triangle ABC with $a = |\overline{BC}|$, $b = |\overline{AC}|$, and $c = |\overline{AB}|$, $\dfrac{\sin A}{a} = \dfrac{\sin B}{b} = \dfrac{\sin C}{c}$.

57 Prove that $(\bar{A} - \bar{B}) \times (\bar{A} + \bar{B}) = 2(\bar{A} \times \bar{B})$.

58 Prove that $\bar{A} \times (\bar{B} \times \bar{C}) + \bar{B} \times (\bar{C} \times \bar{A}) + \bar{C} \times (\bar{A} \times \bar{B}) = \bar{0}$.

59 Show that $\bar{A}$, $\bar{B}$, and $\bar{C}$ are coplanar (linearly dependent) if and only if $(\bar{A} \times \bar{B}) \cdot \bar{C} = 0$.

60 If $\bar{A}$, $\bar{B}$, and $\bar{C}$ are linearly independent, show that $(\bar{A} \times \bar{B}) \cdot \bar{C}$ is positive or negative according to whether $\bar{A}$, $\bar{B}$, $\bar{C}$ forms a right-handed or a left-handed triple, respectively.

5 Equations of Lines and Planes in Space

In this section we make use of the vector algebra developed in Sections 2, 3, and 4 to derive vector equations for lines and planes in space. By considering the scalar components of the vectors in these equations, we also obtain cartesian (scalar) equations for lines and planes in space.

As before, $\bar{R} = \overrightarrow{OP}$ denotes the variable position vector of the point P, and the graph of an equation involving the vector $\bar{R}$ is the set of all points P (this time, in three-dimensional space) whose position vector $\bar{R}$ satisfies the equation.

5.1 Planes in Space

One of the simplest ways to specify a particular plane in space is to give a point P_0 contained in the plane and to give a nonzero vector $\bar{N}$ which is perpendicular to the plane. Figure 1 shows a portion of such a plane. The vector $\bar{N}$ is called a

Figure 1

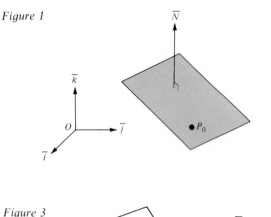

Figure 2

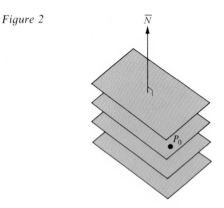

Figure 3

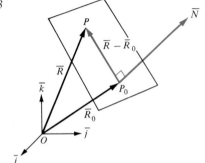

normal vector to the plane. Two planes with the same normal vector $\overline{N}$ are parallel to each other; hence, all the planes that have the same normal vector $\overline{N}$ form a "stack" of mutually parallel planes (Figure 2). Specification of the point P_0 chooses from this "stack" the one and only plane that contains the point P_0.

It is geometrically clear that a point P belongs to the plane containing the point P_0 and having the normal vector $\overline{N}$ if and only if the vector $\overrightarrow{P_0P}$ is perpendicular to $\overline{N}$; that is, $(\overrightarrow{P_0P}) \cdot \overline{N} = 0$ (Figure 3). If $\overline{R}$ is the position vector of P and $\overline{R}_0$ is the position vector of P_0, then $\overrightarrow{P_0P} = \overline{R} - \overline{R}_0$; hence, the condition that $\overrightarrow{P_0P}$ be perpendicular to $\overline{N}$ can be written

$$\overline{N} \cdot (\overline{R} - \overline{R}_0) = 0.$$

This is the *vector* equation of the plane with normal vector $\overline{N}$ and containing the point whose position vector is $\overline{R}_0$.

The equation $\overline{N} \cdot (\overline{R} - \overline{R}_0) = 0$ is easily converted into cartesian scalar form by putting

$$\overline{N} = a\overline{i} + b\overline{j} + c\overline{k}, \qquad \overline{R} = x\overline{i} + y\overline{j} + z\overline{k}, \qquad \text{and} \qquad \overline{R}_0 = x_0\overline{i} + y_0\overline{j} + z_0\overline{k}.$$

Then,

$$\overline{R} - \overline{R}_0 = (x - x_0)\overline{i} + (y - y_0)\overline{j} + (z - z_0)\overline{k},$$
$$\overline{N} \cdot (\overline{R} - \overline{R}_0) = a(x - x_0) + b(y - y_0) + c(z - z_0),$$

and the equation becomes

$$a(x - x_0) + b(y - y_0) + c(z - z_0) = 0.$$

EXAMPLE   Find the cartesian scalar equation of the plane containing the point $P_0 = (-3, 1, 7)$ and having $\overline{N} = 2\overline{i} + 3\overline{j} - \overline{k}$ as a normal vector.

SOLUTION
Here we have $x_0 = -3, y_0 = 1,$ and $z_0 = 7$. Also, $a = 2, b = 3,$ and $c = -1$. Therefore, the equation $a(x - x_0) + b(y - y_0) + c(z - z_0) = 0$ becomes

$$2(x + 3) + 3(y - 1) + (-1)(z - 7) = 0 \quad \text{or} \quad 2x + 3y - z + 10 = 0.$$

Given the cartesian scalar equation of a plane in xyz space,

$$a(x - x_0) + b(y - y_0) + c(z - z_0) = 0$$

we can always carry out the indicated multiplications and thus rewrite the equation in the form

$$ax + by + cz - (ax_0 + by_0 + cz_0) = 0 \quad \text{or} \quad ax + by + cz = D,$$

where D is a constant equal to $ax_0 + by_0 + cz_0$. In the scalar cartesian equation

$$ax + by + cz = D$$

for a plane in xyz space, the coefficients a, b, and c—which cannot all be zero—form the scalar components of a normal vector

$$\overline{N} = a\overline{i} + b\overline{j} + c\overline{k}.$$

As the constant D is assigned different values, we obtain different planes in space, all of which have the same normal vector, and therefore are mutually parallel. The value assigned to the constant D determines exactly one plane in this "stack" of mutually parallel planes.

EXAMPLE   Find the scalar cartesian equation of the plane that contains the point $(1, 2, 3)$ and is parallel to the plane $3x - y - 2z = 14$.

SOLUTION
The desired plane has an equation of the form $3x - y - 2z = D$, where the constant D has to be determined. Since the point $(1, 2, 3)$ belongs to the plane, we have $3(1) - 2 - 2(3) = D$, so $D = -5$. Therefore, the equation is

$$3x - y - 2z = -5 \quad \text{or} \quad 3x - y - 2z + 5 = 0.$$

Figure 4

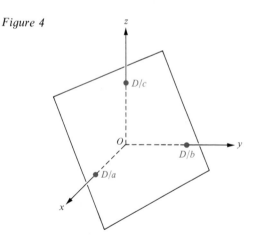

A plane $ax + by + cz = D$ that is not parallel to any of the three coordinate axes obviously must intersect each of these axes (Figure 4). The x, y, and z coordinates of the three points of intersection are called the x, y, and z *intercepts* of the plane, respectively. For instance, the point $(x, 0, 0)$ where the plane intersects the x axis satisfies $ax + b(0) + c(0) = D$, or $ax = D$. Hence, the x intercept is given by $x = D/a$. (Note that $a \neq 0$; otherwise, the plane is parallel to the x axis.) Similarly, the y and z intercepts are given by $y = D/b$ and $z = D/c$, respectively. Since there is exactly one plane containing three noncollinear points, the three intercepts determine the plane uniquely.

EXAMPLE   Consider the plane $-x + 2y + 3z = 6$. Find (a) a unit normal vector to the plane and (b) the intercepts of the plane. Then (c) sketch a portion of the plane.

SOLUTION
(a) We can take $\overline{N} = -\overline{i} + 2\overline{j} + 3\overline{k}$, so that $|\overline{N}| = \sqrt{(-1)^2 + 2^2 + 3^2} = \sqrt{14}$.
Therefore, a unit normal vector to the plane is given by

$$\frac{\overline{N}}{|\overline{N}|} = \frac{-\overline{i} + 2\overline{j} + 3\overline{k}}{\sqrt{14}}.$$

(b) The x, y, and z intercepts are given by $x = 6/(-1) = -6$, $y = \frac{6}{2} = 3$, and $z = \frac{6}{3} = 2$, respectively.
(c) The shaded triangle in Figure 5 is a portion of the plane.

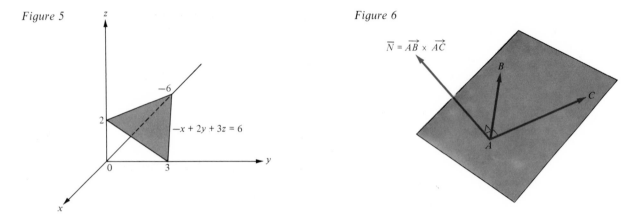

Figure 5

Figure 6

$\overline{N} = \overrightarrow{AB} \times \overrightarrow{AC}$

$-x + 2y + 3z = 6$

In general, to find the equation of the plane containing three given noncollinear points $A = (x_0, y_0, z_0)$, $B = (x_1, y_1, z_1)$, and $C = (x_2, y_2, z_2)$, simply notice that the vectors $\overrightarrow{AB}$ and $\overrightarrow{AC}$ are parallel to the plane; hence,

$$\overline{N} = \overrightarrow{AB} \times \overrightarrow{AC}$$

gives a normal vector to the plane (Figure 6). Since (x_0, y_0, z_0) lies on the plane, we can find either the vector equation or the scalar cartesian equation of the plane just as before.

EXAMPLES   Find the equation in scalar cartesian form of the plane satisfying the given conditions.

1 Containing the three points $A = (1, 1, -1)$, $B = (3, 3, 2)$, and $C = (3, -1, -2)$.

SOLUTION
Here,

$$\overrightarrow{AB} = (3 - 1)i + (3 - 1)j + (2 + 1)\overline{k} = 2i + 2j + 3\overline{k}, \quad \text{and}$$
$$\overrightarrow{AC} = (3 - 1)i + (-1 - 1)j + (-2 + 1)\overline{k} = 2i - 2j - \overline{k};$$

hence,

$$\overline{N} = \overrightarrow{AB} \times \overrightarrow{AC} = \begin{vmatrix} i & j & \overline{k} \\ 2 & 2 & 3 \\ 2 & -2 & -1 \end{vmatrix} = 4i + 8j - 8\overline{k}$$

gives a normal vector to the plane. Since the point $(1, 1, -1)$ belongs to the plane, the scalar cartesian equation is

$$4(x - 1) + 8(y - 1) - 8(z + 1) = 0.$$

This equation can also be written in the form

$$4x + 8y - 8z = 20 \quad \text{or} \quad x + 2y - 2z = 5.$$

2 Containing the points $(3, 2, 1)$ and $(-5, 1, 2)$, but not intersecting the y axis.

SOLUTION
Because the plane is parallel to the y axis, its equation has the form $ax + cz = D$. (Why?) Therefore, since the point $(3, 2, 1)$ satisfies the equation, so does the point $(3, 0, 1)$. Letting $A = (3, 2, 1)$, $B = (-5, 1, 2)$, and $C = (3, 0, 1)$, we have

$\vec{AB} = -8i - j + \bar{k}$ and $\vec{AC} = -2j$, so that a normal vector $\bar{N}$ to the plane is given by

$$\bar{N} = \vec{AB} \times \vec{AC} = \begin{vmatrix} i & j & \bar{k} \\ -8 & -1 & 1 \\ 0 & -2 & 0 \end{vmatrix} = 2\bar{i} + 16\bar{k}.$$

Therefore, the equation of the plane is

$$2(x - 3) + 0(y - 2) + 16(z - 1) = 0 \quad \text{or} \quad x + 8z = 11.$$

The equation of a plane can also be put into vector or scalar *parametric* form; since a plane is two-dimensional, *two* independent parameters are required. We do not take up this matter here.

5.2 Lines in Space

The equation of a straight line in space can be specified by giving a nonzero vector $\bar{M}$ parallel to the line and a point $P_0 = (x_0, y_0, z_0)$ on the line (Figure 7). The vector $\bar{M}$ is called a *direction vector* for the line. Two straight lines with the same direction vector are parallel to each other; hence all the straight lines that have the same direction vector $\bar{M}$ form a "bundle" of mutually parallel lines. Specification of the point P_0 chooses from this "bundle" the one and only line that contains the point P_0 (Figure 8).

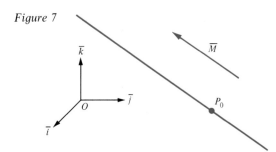

Figure 7

It is geometrically clear that a point P belongs to the line containing the point P_0 and having the direction vector $\bar{M}$ if and only if the vector $\vec{P_0P}$ is parallel to the direction vector $\bar{M}$ (Figure 9); that is, $\bar{M} \times \vec{P_0P} = \bar{0}$. (Recall that two vectors are parallel if and only if their cross product is the zero vector.) If $\bar{R}$ is the position vector of P and $\bar{R}_0$ is the position vector of P_0, then $\vec{P_0P} = \bar{R} - \bar{R}_0$; hence, the condition that $\bar{M}$ be parallel to $\vec{P_0P}$ can be written

$$\bar{M} \times (\bar{R} - \bar{R}_0) = \bar{0}.$$

This is the *vector* (nonparametric) equation of the straight line with direction vector $\bar{M}$ and containing the point whose position vector is $\bar{R}_0$.

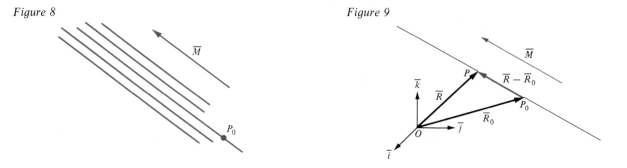

Figure 8

Figure 9

To convert the vector equation $\bar{M} \times (\bar{R} - \bar{R}_0) = \bar{0}$ to scalar cartesian form, let

$$\bar{M} = a\bar{i} + b\bar{j} + c\bar{k}, \qquad \bar{R} = x\bar{i} + y\bar{j} + z\bar{k}, \qquad \text{and} \qquad \bar{R}_0 = x_0\bar{i} + y_0\bar{j} + z_0\bar{k}.$$

Then $\bar{M} \times (\bar{R} - \bar{R}_0) = \bar{0}$ can be rewritten as

$$\begin{vmatrix} \bar{i} & \bar{j} & \bar{k} \\ a & b & c \\ x - x_0 & y - y_0 & z - z_0 \end{vmatrix} = 0\bar{i} + 0\bar{j} + 0\bar{k};$$

that is,

$$[b(z - z_0) - c(y - y_0)]\bar{i} - [a(z - z_0) - c(x - x_0)]\bar{j}$$
$$+ [a(y - y_0) - b(x - x_0)]\bar{k} = 0\bar{i} + 0\bar{j} + 0\bar{k}.$$

Equating the three scalar components, we obtain three simultaneous scalar equations

$$b(z - z_0) - c(y - y_0) = 0, \quad a(z - z_0) - c(x - x_0) = 0, \quad a(y - y_0) - b(x - x_0) = 0;$$

or

$$b(z - z_0) = c(y - y_0), \quad a(z - z_0) = c(x - x_0), \quad a(y - y_0) = b(x - x_0).$$

These three simultaneous equations give a cartesian scalar nonparametric form for the straight line.

If the coefficients a, b, and c in the equations above are nonzero, we can rewrite the equations in the form

$$\frac{z - z_0}{c} = \frac{y - y_0}{b}, \quad \frac{z - z_0}{c} = \frac{x - x_0}{a}, \quad \frac{y - y_0}{b} = \frac{x - x_0}{a};$$

or

$$\frac{x - x_0}{a} = \frac{y - y_0}{b} = \frac{z - z_0}{c}.$$

These simultaneous scalar nonparametric equations are referred to as the *symmetric equations* of the line. Here, (x_0, y_0, z_0) is a point on the line and the denominators a, b, and c are the scalar components of a vector $\bar{M}$ parallel to the line. If any one of the denominators in the symmetric equations of the line is zero, the understanding is that the corresponding numerator is also zero (Problem 36). For instance, if $a = 0$, but $b \neq 0$ and $c \neq 0$, the symmetric equations would be written

$$x - x_0 = 0, \quad \frac{y - y_0}{b} = \frac{z - z_0}{c}.$$

EXAMPLES 1 Find the symmetric equations of the straight line containing the point $P_0 = (x_0, y_0, z_0) = (3, 1, -1)$ and parallel to the vector $\bar{M} = 5\bar{i} - 7\bar{j} + 9\bar{k}$.

SOLUTION
The symmetric equations of the line are given by

$$\frac{x - 3}{5} = \frac{y - 1}{-7} = \frac{z + 1}{9}.$$

2 Write the equation of the line in Example 1 in vector nonparametric form.

SOLUTION
The equation of the line in vector nonparametric form is given by

$$\bar{M} \times (\bar{R} - \bar{R}_0) = \bar{0}, \quad \text{where} \quad \bar{M} = 5\bar{i} - 7\bar{j} + 9\bar{k} \quad \text{and} \quad \bar{R}_0 = 3\bar{i} + \bar{j} - \bar{k}.$$

Therefore, the equation of the line is

$$(5\bar{i} - 7\bar{j} + 9\bar{k}) \times [(x - 3)\bar{i} + (y - 1)\bar{j} + (z + 1)\bar{k}] = \bar{0}.$$

If $A = (x_0, y_0, z_0)$ and $B = (x_1, y_1, z_1)$ are two distinct points in three-dimensional space, there is one and only one straight line containing these points. Obviously,

$$\overline{M} = \overrightarrow{AB} = (x_1 - x_0)\overline{i} + (y_1 - y_0)\overline{j} + (z_1 - z_0)\overline{k}$$

is parallel to the line; hence, the equation of the line in vector form is

$$\overrightarrow{AB} \times (\overline{R} - \overline{R}_0) = \overline{0},$$

where $\overline{R}_0 = x_0\overline{i} + y_0\overline{j} + z_0\overline{k}$. Thus, in scalar symmetric form, the equations of the straight line through $A = (x_0, y_0, z_0)$ and $B = (x_1, y_1, z_1)$ are

$$\frac{x - x_0}{x_1 - x_0} = \frac{y - y_0}{y_1 - y_0} = \frac{z - z_0}{z_1 - z_0}.$$

We have seen that the equation $\overline{M} \times (\overline{R} - \overline{R}_0) = \overline{0}$ expresses the condition that $\overline{R} - \overline{R}_0$ is parallel to $\overline{M}$. This condition can be expressed equally well by the equation

$$\overline{R} - \overline{R}_0 = t\overline{M},$$

where t is a variable scalar. Solving the latter equation for the position vector $\overline{R}$, we obtain the *vector parametric equation*

$$\overline{R} = \overline{R}_0 + t\overline{M}$$

of the straight line parallel to the nonzero direction vector $\overline{M}$ and containing the point whose position vector is $\overline{R}_0$. As the parameter t varies, the position vector $\overline{R}$ traces out the straight line.

Letting $\overline{R} = x\overline{i} + y\overline{j} + z\overline{k}$, $\overline{R}_0 = x_0\overline{i} + y_0\overline{j} + z_0\overline{k}$, and $\overline{M} = a\overline{i} + b\overline{j} + c\overline{k}$ in the vector parametric equation $\overline{R} = \overline{R}_0 + t\overline{M}$, we have

$$x\overline{i} + y\overline{j} + z\overline{k} = (x_0\overline{i} + y_0\overline{j} + z_0\overline{k}) + t(a\overline{i} + b\overline{j} + c\overline{k})$$

or

$$x\overline{i} + y\overline{j} + z\overline{k} = (x_0 + at)\overline{i} + (y_0 + bt)\overline{j} + (z_0 + ct)\overline{k}.$$

Equating scalar components, we obtain the three *scalar parametric equations* of the straight line,

$$\begin{cases} x = x_0 + at \\ y = y_0 + bt \\ z = z_0 + ct. \end{cases}$$

As the parameter t varies, the point (x, y, z) given by these equations traces out the straight line. Notice that the point (x_0, y_0, z_0) corresponds to the parameter value $t = 0$.

EXAMPLES 1 Let $A = (-2, -1, -5)$ and $B = (2, 1, 5)$. Find (a) the equation in vector parametric form and (b) the equations in scalar parametric form for the straight line that contains the point $P_0 = (3, 2, 5)$ and is parallel to the straight line containing A and B.

SOLUTION
The vector

$$\overline{M} = \overrightarrow{AB} = (2 + 2)\overline{i} + (1 + 1)\overline{j} + (5 + 5)\overline{k} = 4\overline{i} + 2\overline{j} + 10\overline{k}$$

is parallel to the desired line and $\overline{R}_0 = 3\overline{i} + 2\overline{j} + 5\overline{k}$ is the position vector of the point P_0 on the line.

(a) $\overline{R} = \overline{R}_0 + t\overline{M} = 3\overline{i} + 2\overline{j} + 5\overline{k} + t(4\overline{i} + 2\overline{j} + 10\overline{k})$
$\quad = (3 + 4t)\overline{i} + (2 + 2t)\overline{j} + (5 + 10t)\overline{k}.$

(b) $\begin{cases} x = 3 + 4t \\ y = 2 + 2t \\ z = 5 + 10t. \end{cases}$

2 Find equations in scalar parametric form for the straight line containing the point $P_0 = (3, -2, 5)$ and perpendicular to the lines

$$\frac{x-1}{3} = \frac{y+2}{1} = \frac{z-3}{-2} \quad \text{and} \quad \frac{x+7}{1} = \frac{y-5}{2} = \frac{z}{1}.$$

SOLUTION

Here, vectors parallel to the indicated lines are given by

$$\overline{M}_1 = 3\overline{i} + \overline{j} - 2\overline{k} \quad \text{and} \quad \overline{M}_2 = \overline{i} + 2\overline{j} + \overline{k},$$

so that

$$\overline{M} = \overline{M}_1 \times \overline{M}_2 = \begin{vmatrix} \overline{i} & \overline{j} & \overline{k} \\ 3 & 1 & -2 \\ 1 & 2 & 1 \end{vmatrix} = 5\overline{i} - 5\overline{j} + 5\overline{k}$$

is perpendicular to both lines. Therefore, $\overline{M}$ is parallel to the desired line, and the scalar parametric equations for the line are

$$\begin{cases} x = 3 + 5t \\ y = -2 - 5t \\ z = 5 + 5t. \end{cases}$$

Problem Set 5

In problems 1 through 4, find the vector and the cartesian scalar equations of the plane containing the point P_0 and having the given normal vector $\overline{N}$.

1 $P_0 = (1, -1, 2)$ and $\overline{N} = \overline{i} + 2\overline{j} - 3\overline{k}$ **2** $P_0 = (1, 3, -1)$ and $\overline{N} = 2\overline{i} + \overline{j} - \overline{k}$

3 $P_0 = (0, 0, 0)$ and $\overline{N} = 5\overline{i} - 2\overline{j} + 10\overline{k}$ **4** $P_0 = (0, 0, 1)$ and $\overline{N} = \overline{j} + \overline{k}$

In problems 5 and 6, give the vector and the scalar equations of the plane containing the given points A, B, and C.

5 $A = (2, -1, 0)$, $B = (-3, -4, -5)$, and $C = (0, 8, 0)$ **6** $A = (2, 2, -2)$, $B = (4, 6, 4)$, and $C = (8, -1, 2)$

In problems 7 through 12, find (a) a unit normal vector to the plane and (b) the intercepts of the plane. Then (c) sketch a portion of the plane.

7 $2x + 3y + 6z = 12$ **8** $x - 4y + 8z = 8$

9 $\overline{N} \cdot (\overline{R} - \overline{R}_0) = 0$, where $\overline{N} = 12\overline{j} - 5\overline{k}$ and $\overline{R}_0 = 5\overline{j}$.   **10** $\overline{N} \cdot (\overline{R} - \overline{R}_0) = 0$, where $\overline{N} = \overline{k}$ and $\overline{R}_0 = \overline{i} + \overline{j} + 3\overline{k}$.

11 $5x = 3y + 4z$ **12** $3x = 4z + 12$

In problems 13 through 18, find the scalar cartesian equation of the plane satisfying the conditions given.

13 Containing the point $(-1, 3, 5)$ and parallel to the plane $6x - 3y - 2z + 9 = 0$.

14 Containing the origin and the points $P = (a, b, c)$ and $Q = (p, q, r)$.

15 Containing the points $(1, 2, 3)$ and $(-2, 1, 1)$, but not intersecting the x axis. (*Hint:* Begin by arguing that if the plane $ax + by + cz = D$ is parallel to the x axis, then $a = 0$.)

16 Containing the point $(4, -2, 1)$ and perpendicular to the straight line containing the two points $A = (2, -1, 2)$ and $B = (3, 2, -1)$.

17 Containing the point $(1, 1, 1)$ and perpendicular to the vector whose direction angles are $\pi/3$, $\pi/4$, and $\pi/3$.

18 Containing the point $(2, 3, 1)$ and parallel to the plane containing the origin and the points $P = (2, 0, -2)$ and $Q = (1, 1, 1)$.

In problems 19 through 24, give a geometric interpretation of the condition imposed on the plane whose equation is $ax + by + cz = D$ by the given equation or equations.

19 $D = 0$ **20** $a = 0$ **21** $a = 0$ and $b = 0$

22 $2a + 3b - 4c = 0$ **23** $b = 0$ **24** $b = 0$ and $c = 0$

In problems 25 through 34, find the equation of the straight line satisfying the given conditions in (a) vector nonparametric, (b) scalar symmetric, (c) vector parametric, and (d) scalar parametric form.

25 Containing the points $(1, 3, -2)$ and $(2, 2, 0)$.   **26** Containing the points $(-1, 3, 4)$ and $(4, 3, 9)$.

27 Containing the point $(3, 1, -4)$ and parallel to the vector $\overline{M} = 4\overline{i} - 2\overline{j} + 5\overline{k}$.

28 Containing the point $(1, 3, -2)$ and perpendicular to the plane $x - y + 2z = 5$.

29 Containing the point $(0, 0, 1)$ and parallel to the vector with direction angles $2\pi/3$, $\pi/3$, and $\pi/4$.

30 Containing the origin and perpendicular to the line $\dfrac{x - 2}{1} = \dfrac{y - 3}{2} = \dfrac{z}{7}$ at their intersection.

31 Containing the point $(0, 4, 5)$ and parallel to the line through the points $(1, 5, -2)$ and $(7, 7, 1)$.

32 Containing the origin and perpendicular to both of the vectors $-3\overline{i} + 2\overline{j} + \overline{k}$ and $6\overline{i} - 5\overline{j} + 2\overline{k}$.

33 Containing the point $(7, 1, -4)$ and parallel to the x axis.

34 Containing the origin and perpendicular to the two lines

$$\frac{x}{6} = \frac{y - 3}{5} = \frac{z - 4}{4} \quad \text{and} \quad x + 3 = \frac{y - 2}{4} = \frac{z - 5}{-2}.$$

35 Show that the points $(2, 0, 0)$, $(-2, -22, -10)$, and $(4, 11, 5)$ lie on a straight line and find the equation of this line in symmetric form.

36 Assuming that a, b, and c are nonzero, derive the symmetric form by eliminating the parameter t from the scalar parametric equations

$$\begin{cases} x = x_0 + at \\ y = y_0 + bt \\ z = z_0 + ct. \end{cases}$$

Exactly what happens if, say, $a = 0$, but b and c are not zero?

In problems 37 through 42, give a geometric interpretation of the condition imposed on the straight line $\begin{cases} x = x_0 + at \\ y = y_0 + bt \\ z = z_0 + ct \end{cases}$ by the given equation or equations.

37 $x_0 = y_0 = z_0 = 0$ **38** $a = 0$ **39** $a = 0$ and $b = 0$

40 $5a - 3b + 7c = 0$ **41** $b = 0$ **42** $b = 0$ and $c = 0$

6 The Geometry of Lines and Planes in Space

In Section 5 we developed equations in both vector and scalar forms for straight lines and planes in three-dimensional space. In the present section we obtain formulas for the distance from a point to a plane, the distance between two lines, the angle between two planes, the angle between two lines, and so forth.

6.1 The Distance from a Point to a Plane

We begin by proving the following theorem.

THEOREM 1   **The Distance from a Point to a Plane**

The distance d from the point P_1 in space whose position vector is $\bar{R}_1 = \overrightarrow{OP_1}$ to the plane whose vector equation is $\bar{N} \cdot (\bar{R} - \bar{R}_0) = 0$ is given by

$$d = \frac{|\bar{N} \cdot (\bar{R}_1 - \bar{R}_0)|}{|\bar{N}|}.$$

Figure 1

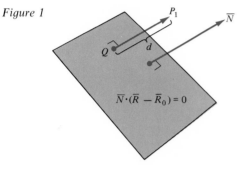

$\bar{N} \cdot (\bar{R} - \bar{R}_0) = 0$

PROOF

Let Q be the point at the foot of the perpendicular dropped from P_1 to the plane, so that $d = |\overrightarrow{QP_1}|$ (Figure 1). Since $\overrightarrow{QP_1}$ is parallel to the normal vector $\bar{N}$, there is a scalar u such that

$$\overrightarrow{QP_1} = u\bar{N}.$$

Let $\bar{R} = \overrightarrow{OQ}$ be the position vector of Q, noting that

$$\bar{N} \cdot (\bar{R} - \bar{R}_0) = 0$$

holds, since the point Q lies on the plane. Also,

$$\overrightarrow{OQ} + \overrightarrow{QP_1} = \overrightarrow{OP_1}; \quad \text{that is,} \quad \bar{R} + u\bar{N} = \bar{R}_1.$$

Therefore, subtracting $\bar{R}_0$ from both sides, we have

$$\bar{R} - \bar{R}_0 + u\bar{N} = \bar{R}_1 - \bar{R}_0.$$

Now, taking the dot product on both sides with $\bar{N}$ and using the fact that $\bar{N} \cdot (\bar{R} - \bar{R}_0) = 0$, we obtain

$$\bar{N} \cdot (u\bar{N}) = \bar{N} \cdot (\bar{R}_1 - \bar{R}_0) \quad \text{or} \quad u(\bar{N} \cdot \bar{N}) = \bar{N} \cdot (\bar{R}_1 - \bar{R}_0);$$

that is,

$$u|\bar{N}|^2 = \bar{N} \cdot (\bar{R}_1 - \bar{R}_0).$$

Taking absolute values on both sides, we get

$$|u||\bar{N}|^2 = |\bar{N} \cdot (\bar{R}_1 - \bar{R}_0)| \quad \text{or} \quad |u||\bar{N}| = \frac{|\bar{N} \cdot (\bar{R}_1 - \bar{R}_0)|}{|\bar{N}|}.$$

Since $d = |\overrightarrow{QP_1}| = |u\bar{N}| = |u||\bar{N}|$, it follows that

$$d = \frac{|\bar{N} \cdot (\bar{R}_1 - \bar{R}_0)|}{|\bar{N}|},$$

Figure 2

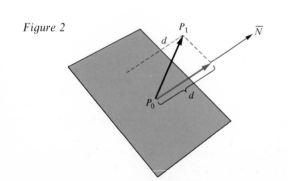

and the proof is complete.

Geometrically, Theorem 1 says that the distance d from P_1 to the plane is the absolute value of the scalar projection of $\bar{R}_1 - \bar{R}_0 = \overrightarrow{P_0P_1}$ in the direction of the normal vector $\bar{N}$ (Figure 2).

The distance formula in Theorem 1 is easy to convert to cartesian scalar form. Indeed, suppose that the cartesian scalar equation of the plane is $ax + by + cz = D$. Then,

$$\bar{N} = a\bar{i} + b\bar{j} + c\bar{k}, \qquad \bar{R} = x\bar{i} + y\bar{j} + z\bar{k}, \qquad \text{and} \qquad \bar{N} \cdot \bar{R}_0 = D.$$

If $P_1 = (x_1, y_1, z_1)$, so that $\bar{R}_1 = x_1\bar{i} + y_1\bar{j} + z_1\bar{k}$, then

$$\bar{N} \cdot (\bar{R}_1 - \bar{R}_0) = \bar{N} \cdot \bar{R}_1 - \bar{N} \cdot \bar{R}_0 = ax_1 + by_1 + cz_1 - D$$

and

$$|\bar{N}| = \sqrt{a^2 + b^2 + c^2}.$$

Therefore, substituting into the formula in Theorem 1, we see that

$$d = \frac{|ax_1 + by_1 + cz_1 - D|}{\sqrt{a^2 + b^2 + c^2}}$$

gives the distance from the point $P_1 = (x_1, y_1, z_1)$ to the plane $ax + by + cz = D$.

EXAMPLES 1 Find the distance d from the point $P_1 = (1, 3, -4)$ to the plane $2x + 2y - z = 6$.

SOLUTION
Here $a = 2$, $b = 2$, $c = -1$, $D = 6$, $x_1 = 1$, $y_1 = 3$, $z_1 = -4$, and the distance d is given by

$$d = \frac{|ax_1 + by_1 + cz_1 - D|}{\sqrt{a^2 + b^2 + c^2}} = \frac{|(2)(1) + (2)(3) + (-1)(-4) - 6|}{\sqrt{2^2 + 2^2 + (-1)^2}}$$

$$= \frac{6}{\sqrt{9}} = 2 \text{ units.}$$

2 Find the perpendicular distance between the two parallel planes $2x - y + 2z = -11$ and $2x - y + 2z + 2 = 0$.

SOLUTION
The required distance is the distance from *any* point on the plane $2x - y + 2z = -11$ to the plane $2x - y + 2z + 2 = 0$. For instance, the point $(-6, -1, 0)$ belongs to the plane $2x - y + 2z = -11$ and its distance from the plane $2x - y + 2z + 2 = 0$ is given by

$$d = \frac{|(2)(-6) + (-1)(-1) + (2)(0) + 2|}{\sqrt{2^2 + (-1)^2 + 2^2}} = \frac{|-9|}{3} = 3 \text{ units.}$$

6.2 The Angle Between Two Planes

Let $\bar{N}_0 \cdot (\bar{R} - \bar{R}_0) = 0$ and $\bar{N}_1 \cdot (\bar{R} - \bar{R}_1) = 0$ be the equations of two planes having normal vectors $\bar{N}_0$ and $\bar{N}_1$, respectively. We define *the angle α between the two planes* to be the smaller of the two angles θ and $\pi - \theta$, where θ is the angle between $\bar{N}_0$ and $\bar{N}_1$ (Figure 3). Since

Figure 3

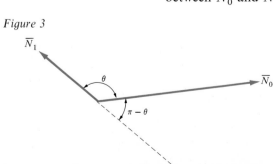

$$\cos \theta = \frac{\bar{N}_0 \cdot \bar{N}_1}{|\bar{N}_0| |\bar{N}_1|},$$

it is easy to check that the *smaller* of the angles θ and $\pi - \theta$ is given by

$$\alpha = \cos^{-1} \frac{|\bar{N}_0 \cdot \bar{N}_1|}{|\bar{N}_0| |\bar{N}_1|} \qquad \text{(Problem 45).}$$

Naturally, the two planes are said to be *perpendicular* if $\alpha = \pi/2$ and they are *parallel* if $\alpha = 0$.

EXAMPLES 1 Find the angle α between the two planes $3x - 2y + z = 5$ and $x + 2y - z = 17$.

SOLUTION
The normal vectors $\bar{N}_0$ and $\bar{N}_1$ to the two planes are $\bar{N}_0 = 3i - 2j + \bar{k}$ and $\bar{N}_1 = i + 2j - \bar{k}$. Therefore,

$$\alpha = \cos^{-1} \frac{|\bar{N}_0 \cdot \bar{N}_1|}{|\bar{N}_0||\bar{N}_1|} = \cos^{-1} \frac{|-2|}{\sqrt{14}\sqrt{6}} = \cos^{-1} \frac{1}{\sqrt{21}} \approx 77.4°.$$

2 Find the equation of the plane containing the points $P_0 = (1, 0, 3)$ and $P_1 = (0, 1, -2)$ that is perpendicular to the plane $2x + 3y + z + 4 = 0$.

SOLUTION
The point $P_0 = (1, 0, 3)$ belongs to the plane in question, so the desired equation has the form

$$\bar{N} \cdot (\bar{R} - \bar{R}_0) = 0,$$

where $\bar{R}_0 = \overrightarrow{OP_0} = i + 3\bar{k}$ and the normal vector $\bar{N}$ has yet to be determined. Since P_0 and P_1 both lie on the plane, $\overrightarrow{P_0P_1}$ is perpendicular to $\bar{N}$. The plane $2x + 3y + z + 4 = 0$ has normal vector $\bar{N}_1 = 2i + 3j + \bar{k}$; hence, $\bar{N}$ is perpendicular to $\bar{N}_1$, since the two planes are perpendicular. Thus, $\bar{N}$ will be perpendicular to both of the vectors $\overrightarrow{P_0P_1}$ and $\bar{N}_1$ provided that we take

$$\bar{N} = \overrightarrow{P_0P_1} \times \bar{N}_1$$

$$= (-i + j - 5\bar{k}) \times (2i + 3j + \bar{k}) = \begin{vmatrix} i & j & \bar{k} \\ -1 & 1 & -5 \\ 2 & 3 & 1 \end{vmatrix} = 16i - 9j - 5\bar{k}.$$

The desired equation is therefore

$$(16i - 9j - 5\bar{k}) \cdot [\bar{R} - (i + 3\bar{k})] = 0 \quad \text{or} \quad 16x - 9y - 5z - 1 = 0.$$

6.3 The Distance from a Point to a Straight Line

The following theorem provides a handy formula for the distance from a point to a straight line in space.

THEOREM 2

The Distance from a Point to a Straight Line
The distance d from the point P_1 whose position vector is $\bar{R}_1 = \overrightarrow{OP_1}$ to the straight line whose vector equation is $\bar{M} \times (\bar{R} - \bar{R}_0) = \bar{0}$ is given by

$$d = \frac{|\bar{M} \times (\bar{R}_1 - \bar{R}_0)|}{|\bar{M}|}.$$

Figure 4

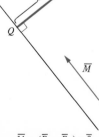

$$\bar{M} \times (\bar{R} - \bar{R}_0) = \bar{0}$$

PROOF
Let Q be the point at the foot of the perpendicular dropped from P_1 to the line, so that $d = |\overrightarrow{QP_1}|$ (Figure 4). Since $\overrightarrow{QP_1}$ is perpendicular to $\bar{M}$,

$$|\bar{M} \times \overrightarrow{QP_1}| = |\bar{M}||\overrightarrow{QP_1}| \sin 90° = |\bar{M}|d = d|\bar{M}|.$$

Let $\bar{R} = \overrightarrow{OQ}$ be the position vector of Q, noting that

$$\bar{M} \times (\bar{R} - \bar{R}_0) = \bar{0}$$

holds, since the point Q lies on the line. Also,

$$\overrightarrow{OQ} + \overrightarrow{QP_1} = \overrightarrow{OP_1}; \quad \text{that is,} \quad \bar{R} + \overrightarrow{QP_1} = \bar{R}_1.$$

Therefore, subtracting $\bar{R}_0$ from both sides, we have

$$\bar{R} - \bar{R}_0 + \overrightarrow{QP_1} = \bar{R}_1 - \bar{R}_0.$$

Now, taking the cross product on both sides with $\overline{M}$ and using the fact that $\overline{M} \times (\bar{R} - \bar{R}_0) = \bar{0}$, we obtain

$$\overline{M} \times \overrightarrow{QP_1} = \overline{M} \times (\bar{R}_1 - \bar{R}_0).$$

It follows that

$$|\overline{M} \times \overrightarrow{QP_1}| = |\overline{M} \times (\bar{R}_1 - \bar{R}_0)|; \quad \text{that is,} \quad d|\overline{M}| = |\overline{M} \times (\bar{R}_1 - \bar{R}_0)|.$$

Dividing the latter equation by $|\overline{M}|$, we obtain the desired formula for d.

EXAMPLE Find the distance d from the point $P_1 = (1, 7, -3)$ to the line

$$\frac{x - 5}{2} = \frac{y + 4}{-1} = \frac{z + 6}{2}.$$

SOLUTION
The direction vector $\overline{M}$ of the line is given by $\overline{M} = 2i - j + 2\bar{k}$ and $\bar{R}_0 = 5i - 4j - 6\bar{k}$ is the position vector of a point on the line. Thus, $\overline{M} \times (\bar{R} - \bar{R}_0) = \bar{0}$ is the vector equation of the line. Putting $\bar{R}_1 = \overrightarrow{OP_1}$, we have $\bar{R}_1 = i + 7j - 3\bar{k}$, and so $\bar{R}_1 - \bar{R}_0 = -4i + 11j + 3\bar{k}$.

$$\overline{M} \times (\bar{R}_1 - \bar{R}_0) = \begin{vmatrix} i & j & \bar{k} \\ 2 & -1 & 2 \\ -4 & 11 & 3 \end{vmatrix} = -25i - 14j + 18\bar{k},$$

$$|\overline{M} \times (\bar{R}_1 - \bar{R}_0)| = \sqrt{(-25)^2 + (-14)^2 + (18)^2} = \sqrt{1145}, \quad \text{and}$$

$$d = \frac{|\overline{M} \times (\bar{R}_1 - \bar{R}_0)|}{|\overline{M}|} = \frac{\sqrt{1145}}{\sqrt{2^2 + (-1)^2 + 2^2}} = \frac{1}{3}\sqrt{1145} \text{ units.}$$

6.4 The Distance Between Two Lines in Space

The distance between two *parallel* lines in space is easy to find using the formula in Theorem 2. Indeed, it is only necessary to select a point (any point will do!) on one line and then find the distance from that point to the other line. The distance between two nonparallel lines in space can be found using the following theorem, whose proof is similar to the proofs of Theorems 1 and 2, and is left as an exercise (Problem 46).

THEOREM 3 **The Distance Between Two Nonparallel Lines in Space**
Let

$$\overline{M}_0 \times (\bar{R} - \bar{R}_0) = \bar{0} \quad \text{and} \quad \overline{M}_1 \times (\bar{R} - \bar{R}_1) = \bar{0}$$

be the vector equations of two straight lines in space and suppose that $\overline{M}_0 \times \overline{M}_1 \neq \bar{0}$, so that the two lines are not parallel. Then, the distance d between the two lines is given by

$$d = \frac{|(\overline{M}_0 \times \overline{M}_1) \cdot (\bar{R}_1 - \bar{R}_0)|}{|\overline{M}_0 \times \overline{M}_1|}.$$

Figure 5

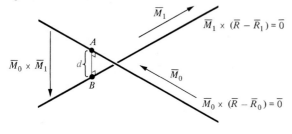

Notice that the distance d in Theorem 3 is the length of the *shortest* line segment $\overline{AB}$ between a point A on the first line and a point B on the second line (Figure 5). Evidently, the line segment $\overline{AB}$ is perpendicular to both lines. (Why?)

Since two straight lines intersect each other if and only if the distance between them is zero, Theorem 3 has the following consequence: The nonparallel straight lines

$$\overline{M}_0 \times (\overline{R} - \overline{R}_0) = \overline{0} \quad \text{and} \quad \overline{M}_1 \times (\overline{R} - \overline{R}_1) = \overline{0}$$

intersect each other in space if and only if $(\overline{M}_0 \times \overline{M}_1) \cdot (\overline{R}_1 - \overline{R}_0) = 0$.

EXAMPLES 1 Find the distance between the line $\dfrac{x + 2}{3} = y - 1 = \dfrac{z + 1}{-2}$ and the line $\dfrac{x - 3}{-1} = \dfrac{y - 1}{4} = z.$

SOLUTION
Here $\overline{M}_0 = 3i + j - 2\overline{k}, \overline{R}_0 = -2i + j - \overline{k}, \overline{M}_1 = -i + 4j + \overline{k}$, and $\overline{R}_1 = 3i + j + 0\overline{k}$; hence,

$$\overline{M}_0 \times \overline{M}_1 = \begin{vmatrix} i & j & \overline{k} \\ 3 & 1 & -2 \\ -1 & 4 & 1 \end{vmatrix} = 9i - j + 13\overline{k}, \qquad \overline{R}_1 - \overline{R}_0 = 5i + 0j + \overline{k}, \quad \text{and}$$

$$(\overline{M}_0 \times \overline{M}_1) \cdot (\overline{R}_1 - \overline{R}_0) = (9i - j + 13\overline{k}) \cdot (5i + 0j + \overline{k}) = 58.$$

Hence,

$$d = \frac{|(\overline{M}_0 \times \overline{M}_1) \cdot (\overline{R}_1 - \overline{R}_0)|}{|\overline{M}_0 \times \overline{M}_1|} = \frac{|58|}{\sqrt{9^2 + (-1)^2 + 13^2}} = \frac{58}{\sqrt{251}} \text{ units.}$$

2 (a) Show that the two straight lines

$$\frac{x + 9}{5} = \frac{y + 11}{7} = -z \quad \text{and} \quad x - 2 = \frac{y - 5}{2} = \frac{z - 1}{3}$$

meet in space.
(b) Find the coordinates of their point of intersection.

SOLUTION
(a) Direction vectors $\overline{M}_0$ and $\overline{M}_1$ and position vectors $\overline{R}_0$ and $\overline{R}_1$ of points on the two lines, respectively, are given by

$$\overline{M}_0 = 5i + 7j - \overline{k}, \qquad \overline{R}_0 = -9i - 11j,$$
$$\overline{M}_1 = i + 2j + 3\overline{k}, \qquad \overline{R}_1 = 2i + 5j + \overline{k}.$$

Here we have

$$\overline{M}_0 \times \overline{M}_1 = \begin{vmatrix} i & j & \overline{k} \\ 5 & 7 & -1 \\ 1 & 2 & 3 \end{vmatrix} = 23i - 16j + 3\overline{k},$$

so that

$$(\overline{M}_0 \times \overline{M}_1) \cdot (\overline{R}_1 - \overline{R}_0) = (23i - 16j + 3\overline{k}) \cdot (11i + 16j + \overline{k})$$
$$= 253 - 256 + 3 = 0;$$

hence, the two lines do meet in space.

(b) Let (x_0, y_0, z_0) denote the point of intersection. Then putting the common value of $\dfrac{x_0 + 9}{5}, \dfrac{y_0 + 11}{7}$, and $-z_0$ equal to t_0, we have

$$\frac{x_0 + 9}{5} = \frac{y_0 + 11}{7} = -z_0 = t_0.$$

Solving these equations for x_0, y_0, and z_0, we obtain

$$x_0 = -9 + 5t_0, \qquad y_0 = -11 + 7t_0, \qquad \text{and} \qquad z_0 = -t_0.$$

Similarly, putting the common value of $x_0 - 2$, $\dfrac{y_0 - 5}{2}$, and $\dfrac{z_0 - 1}{3}$ equal to u_0, we have

$$x_0 - 2 = \frac{y_0 - 5}{2} = \frac{z_0 - 1}{3} = u_0,$$

so that

$$x_0 = 2 + u_0, \qquad y_0 = 5 + 2u_0, \qquad \text{and} \qquad z_0 = 1 + 3u_0.$$

From the equations $x_0 = -9 + 5t_0$ and $x_0 = 2 + u_0$, we have

$$-9 + 5t_0 = 2 + u_0.$$

From the equations $y_0 = -11 + 7t_0$ and $y_0 = 5 + 2u_0$, we have

$$-11 + 7t_0 = 5 + 2u_0.$$

From the equations $z_0 = -t_0$ and $z_0 = 1 + 3u_0$, we have

$$-t_0 = 1 + 3u_0.$$

Thus, we have three equations in the two unknowns t_0 and u_0:

$$\begin{cases} -9 + 5t_0 = 2 + u_0 \\ -11 + 7t_0 = 5 + 2u_0 \\ -t_0 = 1 + 3u_0. \end{cases}$$

Solving, say, the second two equations simultaneously, we get

$$u_0 = -1 \quad \text{and} \quad t_0 = 2,$$

which also satisfies the first equation. Therefore,

$$x_0 = 2 + u_0 = 2 + (-1) = 1,$$
$$y_0 = 5 + 2u_0 = 5 - 2 = 3, \quad \text{and}$$
$$z_0 = 1 + 3u_0 = 1 - 3 = -2.$$

The point $(1, 3, -2)$ is the desired point of intersection.

6.5 The Angle Between Two Lines

Consider the two straight lines $\overline{M}_0 \times (\overline{R} - \overline{R}_0) = \overline{0}$ and $\overline{M}_1 \times (\overline{R} - \overline{R}_1) = \overline{0}$ having direction vectors $\overline{M}_0$ and $\overline{M}_1$, respectively. If θ is the angle between $\overline{M}_0$ and $\overline{M}_1$, we define *the angle α between the two lines* to be the smaller of the angles θ and $\pi - \theta$. Just as in Section 6.2, we have

$$\alpha = \cos^{-1} \frac{|\overline{M}_0 \cdot \overline{M}_1|}{|\overline{M}_0| |\overline{M}_1|}.$$

Notice that, by definition, we speak of the angle α between the two lines even if these lines do not meet in space. In particular, the two lines are said to be *perpendicular* if $\alpha = \pi/2$ and they are *parallel* if $\alpha = 0$.

EXAMPLE   Find the angle α between the two straight lines

$$\frac{x-5}{-7} = y + 3, \qquad z = 2, \qquad \text{and} \qquad \frac{x-3}{5} = \frac{y-5}{3} = \frac{z-7}{-2}.$$

SOLUTION
Here $\overline{M}_0 = -7i + j + 0\overline{k}$ and $\overline{M}_1 = 5i + 3j - 2\overline{k}$; hence,

$$\alpha = \cos^{-1}\frac{|\overline{M}_0 \cdot \overline{M}_1|}{|\overline{M}_0||\overline{M}_1|} = \cos^{-1}\frac{|-32|}{\sqrt{50}\sqrt{38}} = \cos^{-1}\frac{16}{5\sqrt{19}} \approx 42.77°.$$

6.6 Intersecting Planes

Two nonparallel planes $a_0x + b_0y + c_0z = D_0$ and $a_1x + b_1y + c_1z = D_1$ intersect in a straight line l (Figure 6). The normal vectors to the two planes are given by

$$\overline{N}_0 = a_0i + b_0j + c_0\overline{k} \quad \text{and} \quad \overline{N}_1 = a_1i + b_1j + c_1\overline{k},$$

Figure 6

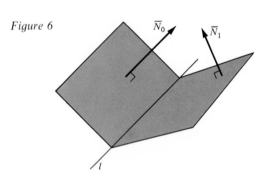

respectively. Since l is contained in both planes, it is perpendicular to both of the normal vectors $\overline{N}_0$ and $\overline{N}_1$; hence, l is parallel to the vector $\overline{M} = \overline{N}_0 \times \overline{N}_1$. The equation of l, in vector form, is therefore

$$\overline{M} \times (\overline{R} - \overline{R}_0) = \overline{0},$$

where $\overline{R}_0$ is the position vector of any particular point P_0 on l. To find such a point P_0, we just find any simultaneous solution (x_0, y_0, z_0) of the two algebraic equations

$$\begin{cases} a_0x + b_0y + c_0z = D_0 \\ a_1x + b_1y + c_1z = D_1. \end{cases}$$

This can be accomplished with ease by setting one of the three variables x, y, or z equal to 0 and solving the resulting pair of equations for the remaining two variables.

EXAMPLE   Find the equations in symmetric form of the straight line in which the planes $3x + 2y + z = 4$ and $x - 3y + 5z = 7$ intersect.

SOLUTION
The normal vectors to the two planes are given by $\overline{N}_0 = 3i + 2j + \overline{k}$ and $\overline{N}_1 = i - 3j + 5\overline{k}$; hence, a direction vector $\overline{M}$ for the line of intersection is given by

$$\overline{M} = \overline{N}_0 \times \overline{N}_1 = \begin{vmatrix} i & j & \overline{k} \\ 3 & 2 & 1 \\ 1 & -3 & 5 \end{vmatrix} = 13i - 14j - 11\overline{k}.$$

To find a point (x_0, y_0, z_0) common to both planes, we put $y = y_0 = 0$ in the simultaneous equations

$$\begin{cases} 3x + 2y + z = 4 \\ x - 3y + 5z = 7, \end{cases}$$

and then solve the resulting equations

$$\begin{cases} 3x + z = 4 \\ x + 5z = 7 \end{cases}$$

to obtain $x = x_0 = \frac{13}{14}$ and $z = z_0 = \frac{17}{14}$. The point $(x_0, y_0, z_0) = \left(\frac{13}{14}, 0, \frac{17}{14}\right)$ belongs to both planes; hence, it belongs to the line in which they intersect. The equations in symmetric form for this line are accordingly

$$\frac{x - \frac{13}{14}}{13} = \frac{y - 0}{-14} = \frac{z - \frac{17}{14}}{-11}.$$

Problem Set 6

In problems 1 through 6, find the distance from the given point to the indicated plane.

1 $(3, 2, -1)$; $7x - 6y + 6z = 8$ **2** $(0, 0, 0)$; $2x + y - 2z + 7 = 0$

3 $(-2, 8, -3)$; $9x - y - 4z = 0$ **4** $(-1, 2, -1)$; $x + y = 6$

5 $(0, -3, 0)$; $2x - 7y + z - 1 = 0$ **6** $(4, -4, 1)$; $2x + y - z = 3$

In problems 7 and 8, find the perpendicular distance between the two given parallel planes.

7 $8x + y - 4z + 6 = 0$; $8x + y - 4z = 24$ **8** $2x + y - z = 12$; $8x + 4y - 4z = 7$

In problems 9 through 12, find the angle α between the two given planes.

9 $x - 2y + 3z = 4$; $-2x + 2y - z = 3$ **10** $3x - z = 7$; $2y + 4z = 5$

11 $4x - 4y + 7z = 31$; $3x - 2y - 6z = 11$ **12** $2x + 3y + 5z = 17$; $x = 4$

13 Find a number c so that the plane $(c + 1)x - y + (2 - c)z = 5$ is perpendicular to the plane $2x + 6y - z + 3 = 0$.

14 Find the equation of the plane parallel to the two planes $8x - y + 3z = 12$ and $8x - y + 3z - 17 = 0$, but lying midway between them.

In problems 15 through 17, find the equation in scalar cartesian form for the plane satisfying the conditions given.

15 Containing the points $(-1, 3, 2)$ and $(-2, 0, 1)$, and perpendicular to the plane $3x - 2y + z = 15$.

16 Containing the origin, perpendicular to the plane $14x + 2y + 11 = 0$, and having a normal vector making a $45°$ angle with the vector $\bar{k}$.

17 Containing the point $(2, 0, 1)$ and perpendicular to the two planes $2x - 4y - z = 7$ and $x - y + z = 1$.

18 Show that the distance d between the two parallel planes $ax + by + cz = D_1$ and $ax + by + cz = D_2$ is given by

$$d = \frac{|D_1 - D_2|}{\sqrt{a^2 + b^2 + c^2}}.$$

19 Find a formula for the distance from the origin to the plane $ax + by + cz = D$.

20 Show that, if the point P_1 with position vector $\bar{R}_1$ is on the same side of the plane $\bar{N} \cdot (\bar{R} - \bar{R}_0) = 0$ as the normal vector $\bar{N}$ (when the tail end of $\bar{N}$ is on the plane), then $\dfrac{\bar{N} \cdot (\bar{R}_1 - \bar{R}_0)}{|\bar{N}|}$ is positive.

In problems 21 through 24, find the distance d from the given point to the indicated line.

21 $(1, 2, 3)$; $\dfrac{x + 1}{2} = \dfrac{y - 3}{-1} = \dfrac{z - 4}{3}$ **22** $(0, 0, 0)$; $x = \dfrac{y - 1}{2}$ and $z = 4$

23 $(-1, -1, -1);\ x - 2 = \dfrac{y-2}{3} = \dfrac{z-2}{-3}$

24 $(0,0,0);\ \dfrac{x - x_0}{a} = \dfrac{y - y_0}{b} = \dfrac{z - z_0}{c}$

In problems 25 through 28, find the distance d between the two given straight lines.

25 $\dfrac{x-1}{2} = \dfrac{y-3}{-2} = \dfrac{z+4}{-1}$ and $\dfrac{x}{2} = \dfrac{y+1}{3} = \dfrac{z+2}{6}$

26 $\dfrac{x - x_0}{a} = \dfrac{y - y_0}{b} = \dfrac{z - z_0}{c}$ and the x axis

27 $\dfrac{x+2}{4} = \dfrac{y-2}{-4} = \dfrac{z}{7}$ and the y axis

28 $\begin{cases} x = 4 - 4t \\ y = -4 + 4t \\ z = 7 - 7t \end{cases}$ and $\begin{cases} x = 2t \\ y = 2t \\ z = t \end{cases}$

In problems 29 and 30, determine whether or not the two given straight lines meet in space. If they do meet, find their point of intersection.

29 $\dfrac{x-5}{2} = \dfrac{y-3}{-1} = \dfrac{z-1}{-3}$ and $\dfrac{x-11}{7} = \dfrac{y}{8} = \dfrac{z+8}{-2}$

30 $\dfrac{x-3}{5} = \dfrac{y-5}{3} = z$ and $\dfrac{x-7}{6} = \dfrac{y+6}{-7} = \dfrac{z-2}{3}$

In problems 31 and 32, find the angle α between the two given lines.

31 $\dfrac{x}{2} = \dfrac{y}{2} = \dfrac{z}{1}$ and $\dfrac{x-3}{5} = \dfrac{y-2}{4} = \dfrac{z+1}{-3}$

32 $\dfrac{x-1}{6} = \dfrac{y-2}{1} = \dfrac{z-3}{10}$ and $\dfrac{x}{-2} = \dfrac{y+1}{3} = \dfrac{z}{-6}$

33 Consider the two straight lines $\dfrac{x}{1} = \dfrac{y+3}{2} = \dfrac{z+1}{3}$ and $\dfrac{x-3}{2} = \dfrac{y}{1} = \dfrac{z-1}{-1}$.

    (a) Find the angle α between the two lines.
    (b) Show that the two lines intersect in space.
    (c) Find the point of intersection of the two lines.

34 Find a necessary and sufficient condition for the straight line $\begin{cases} x = x_0 + at \\ y = y_0 + bt \\ z = z_0 + ct \end{cases}$ to intersect the x axis.

35 A certain plane contains the two points $(1,0,1)$ and $(1,1,0)$, but does not contain any of the points on the line $\dfrac{x}{1} = \dfrac{y}{1} = \dfrac{z-1}{1}$. Find the scalar cartesian equation of the plane.

In problems 36 through 38, determine whether the two given planes are parallel or not. If they are not parallel, find the equations in symmetric scalar form of their line of intersection.

36 $\begin{cases} 2x - 3y + 4z = 5 \\ 6x - 9y + 8z = 1 \end{cases}$

37 $\begin{cases} 3x + y - 2z = 7 \\ 6x - 5y - 4z = 7 \end{cases}$

38 $\begin{cases} x + 2y + 3z = 1 \\ -3x - 6y - 9z = 2 \end{cases}$

39 Find the equation of the plane that contains the points $P_1 = (1, 0, -1)$ and $P_2 = (-1, 2, 1)$ and is parallel to the line of intersection of the planes $3x + y - 2z = 6$ and $4x - y + 3z = 0$.

40 Show that the two straight lines

$$\begin{cases} x = -1 + 2t \\ y = 1 + t \\ z = -1 - 2t \end{cases} \quad \text{and} \quad \begin{cases} x = -7 + 3t \\ y = -3 + 2t \\ z = 1 - t \end{cases}$$

    meet in space and find the equation in cartesian scalar form of the plane containing them.

41 Find the equation of the plane that contains the straight line $\dfrac{x-2}{3} = \dfrac{y}{-1} = \dfrac{z+3}{2}$ but does not intersect the straight line $\begin{cases} x = 4t - 1 \\ y = 2t + 2 \\ z = 3t. \end{cases}$

42 Let l_1 be the line of intersection of the two planes $\begin{cases} x + y - 3z = 0 \\ 2x + 3y - 8z = 1 \end{cases}$ and let l_2 be the line of intersection of the two planes $\begin{cases} 3x - y - z = 3 \\ x + y - 3z = 5. \end{cases}$

    (a) Show that l_1 and l_2 are parallel lines.
    (b) Find the scalar cartesian equation of the plane containing l_1 and l_2.

43 Find a point on the line $x = y = z$ that is equidistant from the points $(3, 0, 5)$ and $(1, -1, 4)$.

44 (a) Give a reasonable definition of the angle α between the line $\overline{M}_0 \times (\overline{R} - \overline{R}_0) = \overline{0}$ and the plane $\overline{N}_1 \cdot (\overline{R} - \overline{R}_1) = 0$.

(b) Show that the angle α of part (a) is given by

$$\alpha = \sin^{-1} \frac{|\overline{M}_0 \cdot \overline{N}_1|}{|\overline{M}_0||\overline{N}_1|}.$$

(c) Find the angle between the line $\dfrac{x}{2} = \dfrac{y-1}{-2} = \dfrac{z-3}{2}$ and the plane $2x + 2y - z = 6$.

(d) Find the angle between the line $\dfrac{x-1}{3} = \dfrac{y}{1}$, $z = 0$, and the plane $x + 2y = 7$.

45 If θ is the angle between the nonzero vectors $\overline{N}_0$ and $\overline{N}_1$, show that $\alpha = \cos^{-1} \dfrac{|\overline{N}_0 \cdot \overline{N}_1|}{|\overline{N}_0||\overline{N}_1|}$ always gives the smaller of the two angles θ and $\pi - \theta$.

Figure 7

46 Prove Theorem 3.

47 A *tetrahedron* is a pyramid with four vertices P_0, P_1, P_2, and P_3 and four triangular faces (Figure 7). Its volume V is one-third of the distance from P_0 to the base $P_1P_2P_3$ times the area of the base. Find a formula for the volume V in terms of the coordinates of the vertices, $P_0 = (x_0, y_0, z_0)$, $P_1 = (x_1, y_1, z_1)$, $P_2 = (x_2, y_2, z_2)$, and $P_3 = (x_3, y_3, z_3)$.

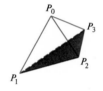

7   Vector-Valued Functions and Curves in Space

The concept of a vector-valued function and the results obtained in Sections 6 through 9 of Chapter 14 can be extended to vectors in three-dimensional space. Indeed, all the definitions in these sections were so formulated as to be applicable not only to vectors in the xy plane, but to vectors in space as well. Furthermore, all of the facts about continuity and differentiability of vector-valued functions carry over immediately to xyz space, provided that the vectors are written in terms of *three* scalar components. For instance, if $\overline{F}$ is a vector-valued function of the scalar t, then

$$\overline{F}(t) = u(t)\overline{i} + v(t)\overline{j} + w(t)\overline{k},$$

where u, v, and w are the three scalar component functions of $\overline{F}$; furthermore, $\overline{F}$ is differentiable if and only if u, v, and w are differentiable. If u, v, and w are differentiable, then

$$\overline{F}'(t) = u'(t)\overline{i} + v'(t)\overline{j} + w'(t)\overline{k},$$

and so forth.

The facts concerning the cross product of two vector-valued functions are exactly as might be expected. Indeed, if $\overline{F}$ and $\overline{G}$ are vector-valued functions, then:

1 If $\lim\limits_{t \to c} \overline{F}(t)$ and $\lim\limits_{t \to c} \overline{G}(t)$ both exist, it follows that $\lim\limits_{t \to c} [\overline{F}(t) \times \overline{G}(t)]$ exists and

$$\lim_{t \to c} [\overline{F}(t) \times \overline{G}(t)] = \left[\lim_{t \to c} \overline{F}(t)\right] \times \left[\lim_{t \to c} \overline{G}(t)\right].$$

2 If $\overline{F}$ and $\overline{G}$ are continuous, then so is the function $\overline{H}$ defined by $\overline{H}(t) = \overline{F}(t) \times \overline{G}(t)$.

3 If $\bar{F}$ and $\bar{G}$ are differentiable and $\bar{H}$ is defined by $\bar{H}(t) = \bar{F}(t) \times \bar{G}(t)$, then $\bar{H}$ is differentiable and

$$\bar{H}'(t) = \bar{F}'(t) \times \bar{G}(t) + \bar{F}(t) \times \bar{G}'(t).$$

These facts are easy to verify by considering the scalar component functions of $\bar{F}$ and $\bar{G}$ (Problems 34, 35, and 36). Using Leibniz notation, we can rewrite the equation for the derivative of the cross product as

$$\frac{d}{dt}(\bar{F} \times \bar{G}) = \frac{d\bar{F}}{dt} \times \bar{G} + \bar{F} \times \frac{d\bar{G}}{dt}.$$

EXAMPLE   Assuming that $\bar{F}$, $\bar{G}$, and $\bar{H}$ are differentiable vector-valued functions of t, find a formula for the derivative of the box product $(\bar{F} \times \bar{G}) \cdot \bar{H}$.

SOLUTION

$$\frac{d}{dt}[(\bar{F} \times \bar{G}) \cdot \bar{H}] = \left[\frac{d}{dt}(\bar{F} \times \bar{G})\right] \cdot \bar{H} + (\bar{F} \times \bar{G}) \cdot \frac{d\bar{H}}{dt}$$

$$= \left[\frac{d\bar{F}}{dt} \times \bar{G} + \bar{F} \times \frac{d\bar{G}}{dt}\right] \cdot \bar{H} + (\bar{F} \times \bar{G}) \cdot \frac{d\bar{H}}{dt}$$

$$= \left(\frac{d\bar{F}}{dt} \times \bar{G}\right) \cdot \bar{H} + \left(\bar{F} \times \frac{d\bar{G}}{dt}\right) \cdot \bar{H} + (\bar{F} \times \bar{G}) \cdot \frac{d\bar{H}}{dt}.$$

7.1 Curves and Motion in Space

If $\bar{F}$ is a three-dimensional vector-valued function, then a vector equation of the form $\bar{R} = \bar{F}(t)$ can be regarded as specifying a variable position vector $\bar{R}$ in terms of a parameter t. As t varies, $\bar{R}$ traces out a curve in space (Figure 1). For instance, if $\bar{M}$ is a nonzero constant vector, then the position vector $\bar{R}$ given by

$$\bar{R} = \bar{R}_0 + t\bar{M}$$

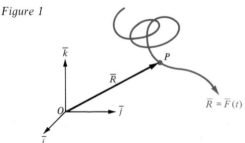

Figure 1

$\bar{R} = \bar{F}(t)$

traces out a straight line in space, as was seen in Section 5.

We can always regard $\bar{R} = \bar{F}(t)$ as giving the position vector of a moving particle P at time t. By the same considerations as those in Section 7 of Chapter 14, a particle P moving in space according to the *equation of motion*

$$\bar{R} = \bar{F}(t) = x\bar{i} + y\bar{j} + z\bar{k},$$

where x, y, and z are functions of time t, has a *velocity vector* $\bar{V}$ given by

$$\bar{V} = \frac{d\bar{R}}{dt} = \frac{dx}{dt}\bar{i} + \frac{dy}{dt}\bar{j} + \frac{dz}{dt}\bar{k} = \bar{F}'(t)$$

and an *acceleration vector* $\bar{A}$ given by

$$\bar{A} = \frac{d\bar{V}}{dt} = \frac{d^2\bar{R}}{dt^2} = \frac{d^2x}{dt^2}\bar{i} + \frac{d^2y}{dt^2}\bar{j} + \frac{d^2z}{dt^2}\bar{k} = \bar{F}''(t).$$

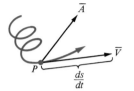

Figure 2

Reasoning exactly as in two dimensions, we find that the velocity vector $\overline{V}$ is always tangent to the path traced out by P. Also, the length $|\overline{V}|$ of the velocity vector gives the instantaneous *speed* v of P; that is,

$$v = \frac{ds}{dt} = |\overline{V}| = \left|\frac{d\overline{R}}{dt}\right| = \sqrt{\left(\frac{dx}{dt}\right)^2 + \left(\frac{dy}{dt}\right)^2 + \left(\frac{dz}{dt}\right)^2},$$

where s is the arc length measured along the path of P (Figure 2).

EXAMPLE A particle P is moving in space according to the equation of motion $\overline{R} = (4 \cos t)\overline{i} + (4 \sin t)\overline{j} + t^2\overline{k}$. At the instant when $t = \pi/2$, find (a) the velocity vector $\overline{V}$, (b) the speed v, and (c) the acceleration vector $\overline{A}$.

SOLUTION
At any time t we have $\overline{V} = d\overline{R}/dt = (-4 \sin t)\overline{i} + (4 \cos t)\overline{j} + 2t\overline{k}$, and $\overline{A} = d\overline{V}/dt = -(4 \cos t)\overline{i} - (4 \sin t)\overline{j} + 2\overline{k}$. Therefore, at the instant $t = \pi/2$,

(a) $\overline{V} = \left(-4 \sin \frac{\pi}{2}\right)\overline{i} + \left(4 \cos \frac{\pi}{2}\right)\overline{j} + 2\frac{\pi}{2}\overline{k} = -4\overline{i} + \pi\overline{k}.$

(b) $v = |\overline{V}| = \sqrt{16 + \pi^2}.$

(c) $\overline{A} = \left(-4 \cos \frac{\pi}{2}\right)\overline{i} - 4\left(\sin \frac{\pi}{2}\right)\overline{j} + 2\overline{k} = -4\overline{j} + 2\overline{k}.$

7.2 Tangent Vector and Arc Length

Let $\overline{R} = \overline{F}(t)$ be the vector parametric equation of a curve in space. Then, provided that $|d\overline{R}/dt| \neq 0$, a unit tangent vector $\overline{T}$ to the curve is given by

$$\overline{T} = \frac{d\overline{R}/dt}{|d\overline{R}/dt|} = \frac{d\overline{R}/dt}{ds/dt} = \frac{\overline{V}}{v}.$$

Notice that

$$\overline{V} = \frac{ds}{dt}\overline{T} = v\overline{T},$$

so that the velocity vector is obtained by multiplying the unit tangent vector by the speed. Henceforth, we assume that $|d\overline{R}/dt| \neq 0$, so that $\overline{T}$ is defined as above.

Since $\dfrac{ds}{dt} = \left|\dfrac{d\overline{R}}{dt}\right| = \sqrt{\left(\dfrac{dx}{dt}\right)^2 + \left(\dfrac{dy}{dt}\right)^2 + \left(\dfrac{dz}{dt}\right)^2}$, the arc length of the portion of the curve $\overline{R} = \overline{F}(t)$ between the points corresponding to $t = a$ and $t = b$, respectively, is given by

$$s = \int_a^b \left|\frac{d\overline{R}}{dt}\right| dt = \int_a^b \sqrt{\left(\frac{dx}{dt}\right)^2 + \left(\frac{dy}{dt}\right)^2 + \left(\frac{dz}{dt}\right)^2} \, dt.$$

EXAMPLE Let C be the curve whose vector parametric equation is $\overline{R} = \sqrt{6}\,t^2\overline{i} + 3t\overline{j} + \frac{4}{3}t^3\overline{k}$. Find (a) the unit tangent vector $\overline{T}$ to C when $t = 1$ and (b) the arc length s of C between the points where $t = 0$ and where $t = 1$.

SOLUTION

Here $d\bar{R}/dt = 2\sqrt{6}\,t\bar{i} + 3\bar{j} + 4t^2\bar{k}$, so that

$$\left|\frac{d\bar{R}}{dt}\right| = \sqrt{24t^2 + 9 + 16t^4} = \sqrt{(4t^2 + 3)^2} = 4t^2 + 3, \quad \text{and}$$

$$\bar{T} = \frac{d\bar{R}/dt}{|d\bar{R}/dt|} = \frac{2\sqrt{6}\,t\bar{i} + 3\bar{j} + 4t^2\bar{k}}{4t^2 + 3} \quad \text{for any value of } t.$$

(a) For $t = 1$, we have $\bar{T} = \dfrac{2\sqrt{6}\,\bar{i} + 3\bar{j} + 4\bar{k}}{7}$.

(b) $s = \displaystyle\int_0^1 \left|\frac{d\bar{R}}{dt}\right| dt = \int_0^1 (4t^2 + 3)\, dt = \frac{13}{3}$ units.

7.3 Normal Vector and Curvature

In what follows we denote by s the arc length measured along the curve C between a fixed starting point P_0 and the point P whose position vector is $\bar{R} = \bar{F}(t)$ (Figure 3). Then, by the chain rule,

$$\frac{d\bar{R}}{ds} = \frac{d\bar{R}}{dt}\frac{dt}{ds} = \frac{d\bar{R}/dt}{ds/dt} = \bar{T}.$$

Figure 3

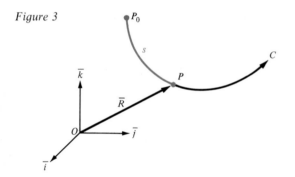

The vector $d\bar{T}/ds$ is still called the *curvature vector*, just as it was in two dimensions. In three-dimensional space, we do not attempt to give the *curvature* κ an algebraic sign, and we simply define $\kappa = \left|\dfrac{d\bar{T}}{ds}\right|$, so that $\kappa \geq 0$.

Since $\bar{T}$ is a unit vector, $\bar{T} \cdot \bar{T} = 1$. Differentiating with respect to s, we obtain

$$\frac{d}{ds}(\bar{T} \cdot \bar{T}) = \frac{d\bar{T}}{ds}\cdot\bar{T} + \bar{T}\cdot\frac{d\bar{T}}{ds} = 2\bar{T}\cdot\frac{d\bar{T}}{ds} = 0; \quad \text{that is,}$$

$$\bar{T}\cdot\frac{d\bar{T}}{ds} = 0.$$

Therefore, the curvature vector $d\bar{T}/ds$ is always perpendicular to the tangent vector $\bar{T}$. If $\kappa \neq 0$, then the vector $\bar{N}$ defined by

$$\bar{N} = \frac{d\bar{T}/ds}{|d\bar{T}/ds|} = \frac{d\bar{T}/ds}{\kappa}$$

Figure 4

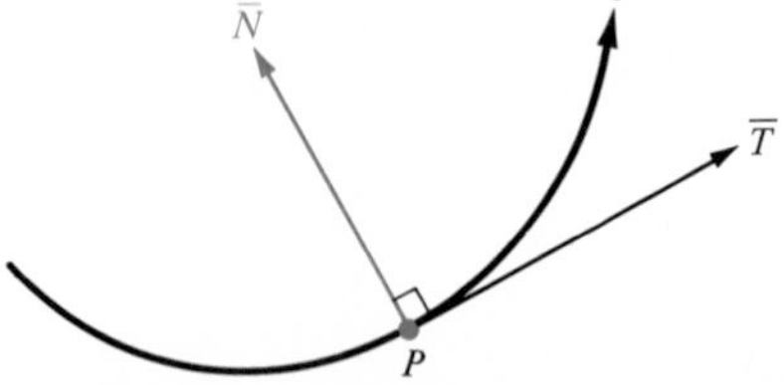

is consequently a unit vector perpendicular to the tangent vector $\bar{T}$. Just as in two dimensions, we refer to the vector $\bar{N}$ as the *principal unit normal vector* to the curve C (Figure 4). Notice that

$$\frac{d\bar{T}}{ds} = \kappa\bar{N}.$$

If we differentiate both sides of the equation $\bar{V} = v\bar{T}$ with respect to t, we obtain

$$\bar{A} = \frac{d\bar{V}}{dt} = \frac{dv}{dt}\bar{T} + v\frac{d\bar{T}}{dt} = \frac{dv}{dt}\bar{T} + v\frac{ds}{dt}\frac{d\bar{T}}{ds} \quad \text{or} \quad \bar{A} = \frac{dv}{dt}\bar{T} + v^2\kappa\bar{N};$$

hence, even in three-dimensional space, the acceleration vector $\bar{A}$ decomposes into the sum of a *tangential component vector* $\dfrac{dv}{dt}\bar{T}$ and a *normal component vector* $v^2\kappa\bar{N}$.

Taking the cross product with $\bar{V}$ on both sides of the equation $\bar{A} = \dfrac{dv}{dt}\bar{T} + v^2\kappa\bar{N}$, and using the fact that $\bar{V} = v\bar{T}$, we obtain

$$\bar{V} \times \bar{A} = \frac{dv}{dt}(\bar{V} \times \bar{T}) + v^2\kappa(\bar{V} \times \bar{N}) = \frac{dv}{dt}v(\bar{T} \times \bar{T}) + v^2\kappa v(\bar{T} \times \bar{N})$$

$$= \bar{0} + v^3\kappa(\bar{T} \times \bar{N}) = v^3\kappa(\bar{T} \times \bar{N}).$$

In what follows, we make extensive use of the equation

$$\bar{V} \times \bar{A} = v^3\kappa(\bar{T} \times \bar{N}).$$

Since $\bar{T}$ and $\bar{N}$ are perpendicular unit vectors, it follows that $|\bar{T} \times \bar{N}| = |\bar{T}||\bar{N}| = 1$; hence, by the equation above,

$$|\bar{V} \times \bar{A}| = |v^3\kappa(\bar{T} \times \bar{N})| = v^3\kappa|\bar{T} \times \bar{N}| = v^3\kappa.$$

Solving for κ, we obtain

$$\kappa = \frac{|\bar{V} \times \bar{A}|}{v^3}.$$

Because $v\bar{T} = \bar{V}$, the equation $\bar{V} \times \bar{A} = v^3\kappa(\bar{T} \times \bar{N})$ can be rewritten as $\bar{V} \times \bar{A} = v^2\kappa(\bar{V} \times \bar{N})$. Taking the cross product with $\bar{V}$, we obtain

$$(\bar{V} \times \bar{A}) \times \bar{V} = v^2\kappa(\bar{V} \times \bar{N}) \times \bar{V} = v^2\kappa[(\bar{V} \cdot \bar{V})\bar{N} - (\bar{N} \cdot \bar{V})\bar{V}],$$

where we have used the triple vector product identity (Theorem 1 in Section 4.1) to expand $(\bar{V} \times \bar{N}) \times \bar{V}$. Since $\bar{V} \cdot \bar{V} = |\bar{V}|^2 = v^2$ and $\bar{N} \cdot \bar{V} = \bar{N} \cdot (v\bar{T}) = v\bar{N} \cdot \bar{T} = v(0) = 0$, the equation above can be rewritten as

$$(\bar{V} \times \bar{A}) \times \bar{V} = v^2\kappa(v^2\bar{N} - 0\bar{V}) = v^4\kappa\bar{N}.$$

Thus,

$$\bar{N} = \frac{(\bar{V} \times \bar{A}) \times \bar{V}}{v^4\kappa}.$$

The equations $\kappa = \dfrac{|\bar{V} \times \bar{A}|}{v^3}$ and $\bar{N} = \dfrac{(\bar{V} \times \bar{A}) \times \bar{V}}{v^4k}$ allow us to find the curvature and the principal unit normal vector directly in terms of the original parameter t.

EXAMPLE   Find $\bar{T}$, $\bar{N}$, and κ if $\bar{R} = (2t^2 + 1)\bar{i} + (t^2 - 2t)\bar{j} + (t^2 - 1)\bar{k}$.

SOLUTION

Here, $\bar{V} = \dfrac{d\bar{R}}{dt} = 4t\bar{i} + (2t - 2)\bar{j} + 2t\bar{k}$, so that

$$v = |\bar{V}| = \sqrt{16t^2 + (2t - 2)^2 + 4t^2} = 2\sqrt{6t^2 - 2t + 1}, \text{ and}$$

$$\bar{A} = \frac{d\bar{V}}{dt} = 4\bar{i} + 2\bar{j} + 2\bar{k}.$$

Thus,

$$\bar{T} = \frac{\bar{V}}{v} = \frac{2t\bar{i} + (t - 1)\bar{j} + t\bar{k}}{\sqrt{6t^2 - 2t + 1}},$$

$$\bar{V} \times \bar{A} = \begin{vmatrix} i & j & \bar{k} \\ 4t & 2t-2 & 2t \\ 4 & 2 & 2 \end{vmatrix} = -4\bar{i} + 0\bar{j} + 8\bar{k} = -4(\bar{i} - 2\bar{k}),$$

$$(\bar{V} \times \bar{A}) \times \bar{V} = \begin{vmatrix} i & j & \bar{k} \\ -4 & 0 & 8 \\ 4t & 2t-2 & 2t \end{vmatrix} = (-16t + 16)\bar{i} + 40t\bar{j} + (-8t + 8)\bar{k},$$

$$\kappa = \frac{|\bar{V} \times \bar{A}|}{v^3} = \frac{4|\bar{i} - 2\bar{k}|}{8(6t^2 - 2t + 1)^{3/2}} = \frac{\sqrt{5}}{2(6t^2 - 2t + 1)^{3/2}},$$

$$\bar{N} = \frac{(\bar{V} \times \bar{A}) \times \bar{V}}{v^4 \kappa} = \frac{(2 - 2t)\bar{i} + 5t\bar{j} + (1 - t)\bar{k}}{\sqrt{5}\sqrt{6t^2 - 2t + 1}}.$$

7.4 Binormal Vector and Torsion

The vector $\bar{T} \times \bar{N}$ that appears in the equation

$$\bar{V} \times \bar{A} = v^3 \kappa (\bar{T} \times \bar{N})$$

is called the *unit binormal vector* to the curve considered in Section 7.3 and is symbolized by $\bar{B}$. Thus, by definition,

$$\bar{B} = \bar{T} \times \bar{N};$$

that is,

$$\bar{B} = \frac{\bar{V} \times \bar{A}}{v^3 \kappa} = \frac{\bar{V} \times \bar{A}}{|\bar{V} \times \bar{A}|}.$$

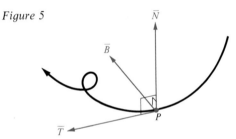

Figure 5

Note that $\bar{B}$ is a unit vector and it is perpendicular to both $\bar{T}$ and $\bar{N}$; furthermore, $\bar{T}, \bar{N}, \bar{B}$ is a right-handed triple, just as $\bar{i}, \bar{j}, \bar{k}$ is. However, $\bar{T}, \bar{N}$, and $\bar{B}$ are variable vectors, moving along with the point P as it traces out the curve (Figure 5).

Just as $\bar{i}, \bar{j}, \bar{k}$ forms a basis, so does any particular triple $\bar{T}, \bar{N}, \bar{B}$; hence, any vector $\bar{X}$ can be expressed as a linear combination of $\bar{T}, \bar{N}$, and $\bar{B}$ as follows:

$$\bar{X} = (\bar{X} \cdot \bar{T})\bar{T} + (\bar{X} \cdot \bar{N})\bar{N} + (\bar{X} \cdot \bar{B})\bar{B}$$

(see Problem 47 in Section 2). In particular,

$$\frac{d\bar{N}}{ds} = \left(\frac{d\bar{N}}{ds} \cdot \bar{T}\right)\bar{T} + \left(\frac{d\bar{N}}{ds} \cdot \bar{N}\right)\bar{N} + \left(\frac{d\bar{N}}{ds} \cdot \bar{B}\right)\bar{B}.$$

Differentiation of both sides of the equation $\bar{N} \cdot \bar{T} = 0$ with respect to s gives $(d\bar{N}/ds) \cdot \bar{T} + \bar{N} \cdot (d\bar{T}/ds) = 0$, so that

$$\frac{d\bar{N}}{ds} \cdot \bar{T} = -\bar{N} \cdot \frac{d\bar{T}}{ds} = -\bar{N} \cdot (\kappa \bar{N}) = -\kappa \bar{N} \cdot \bar{N} = -\kappa(1) = -\kappa.$$

Similarly, differentiation of both sides of the equation $\bar{N} \cdot \bar{N} = 1$ gives $2\dfrac{d\bar{N}}{ds} \cdot \bar{N} = 0$, so that $\dfrac{d\bar{N}}{ds} \cdot \bar{N} = 0$. Therefore, substituting into the equation above for $\dfrac{d\bar{N}}{ds}$, we obtain

$$\frac{d\bar{N}}{ds} = -\kappa \bar{T} + \left(\frac{d\bar{N}}{ds} \cdot \bar{B}\right)\bar{B}.$$

The scalar $(d\overline{N}/ds) \cdot \overline{B}$ in the latter equation is called the *torsion* of the curve and is traditionally denoted by the Greek latter τ (called "tau"). Thus, by definition,

$$\tau = \frac{d\overline{N}}{ds} \cdot \overline{B},$$

and the equation $d\overline{N}/ds = -\kappa\overline{T} + (d\overline{N}/ds \cdot \overline{B})\overline{B}$ can be written

$$\frac{d\overline{N}}{ds} = -\kappa\overline{T} + \tau\overline{B}.$$

The torsion τ is a measure of how rapidly the curve is "twisting" in space. The three equations

$$\frac{d\overline{T}}{ds} = \kappa\overline{N}, \quad \frac{d\overline{N}}{ds} = -\kappa\overline{T} + \tau\overline{B}, \quad \text{and} \quad \frac{d\overline{B}}{ds} = -\tau\overline{N}$$

are called the *Frenet formulas*. (For a proof of the third equation, see Problem 40.)

If we differentiate both sides of the identity $\overline{A} = \dfrac{dv}{dt}\overline{T} + v^2\kappa\overline{N}$ with respect to arc length s and use the identities $d\overline{T}/ds = \kappa\overline{N}$ and $d\overline{N}/ds = -\kappa\overline{T} + \tau\overline{B}$, we obtain

$$\frac{d\overline{A}}{ds} = \frac{d(dv/dt)}{ds}\overline{T} + \frac{dv}{dt}\frac{d\overline{T}}{ds} + \frac{d(v^2\kappa)}{ds}\overline{N} + v^2\kappa\frac{d\overline{N}}{ds}$$

$$= \frac{d(dv/dt)}{ds}\overline{T} + \frac{dv}{dt}\kappa\overline{N} + \frac{d(v^2\kappa)}{ds}\overline{N} + v^2\kappa(-\kappa\overline{T}) + v^2\kappa(\tau\overline{B}).$$

Taking the dot product on both sides of the latter equation with $\overline{B}$ and using the facts that $\overline{T} \cdot \overline{B} = \overline{N} \cdot \overline{B} = 0$ and $\overline{B} \cdot \overline{B} = 1$, we obtain $\dfrac{d\overline{A}}{ds} \cdot \overline{B} = v^2\kappa\tau$. Since

$v\dfrac{d\overline{A}}{ds} = \dfrac{ds}{dt}\dfrac{d\overline{A}}{ds} = \dfrac{d\overline{A}}{dt}$, we can multiply the above equation by v to obtain

$$\left(v\frac{d\overline{A}}{ds}\right) \cdot \overline{B} = v^3\kappa\tau \quad \text{or} \quad \frac{d\overline{A}}{dt} \cdot \overline{B} = v^3\kappa\tau.$$

Figure 6

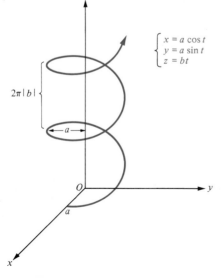

$$\begin{cases} x = a\cos t \\ y = a\sin t \\ z = bt \end{cases}$$

Since $\overline{B} = \dfrac{\overline{V} \times \overline{A}}{v^3\kappa}$, we have $\dfrac{d\overline{A}}{dt} \cdot \left[\dfrac{\overline{V} \times \overline{A}}{v^3\kappa}\right] = v^3\kappa\tau$, or

$$\tau = \frac{(d\overline{A}/dt) \cdot (\overline{V} \times \overline{A})}{v^6\kappa^2}.$$

This formula gives the torsion τ in terms of quantities that can be found directly in terms of the parameter t.

The following example involves a curve of the form $\overline{R} = (a\cos t)\overline{i} + (a\sin t)\overline{j} + bt\overline{k}$, where a and b are constants and $a > 0$. In scalar parametric form, $x = a\cos t$, $y = a\sin t$, and $z = bt$. As the parameter t increases, the point $(x, y, 0) = (a\cos t, a\sin t, 0)$ in the xy plane traces out a circle of radius a. Meanwhile, the z coordinate $z = bt$ changes uniformly, so that the point (x, y, z) traces out a *circular spiral* or *helix* (Figure 6). When t increases by 2π, x and y return to their previous values while z increases by $2\pi b$. We say that the spiral has *radius* a and *pitch* $2\pi|b|$.

EXAMPLE   If $\bar{R} = (2 \cos t)\bar{i} + (2 \sin t)\bar{j} + 3t\bar{k}$, find $\bar{V}$, v, $\bar{A}$, $\bar{T}$, $\bar{V} \times \bar{A}$, κ, $\bar{N}$, $\bar{B}$, $d\bar{A}/dt$, and τ.

SOLUTION

$$\bar{V} = \frac{d\bar{R}}{dt} = (-2 \sin t)\bar{i} + (2 \cos t)\bar{j} + 3\bar{k},$$

$$v = |\bar{V}| = \sqrt{4 \sin^2 t + 4 \cos^2 t + 9} = \sqrt{13},$$

$$\bar{A} = \frac{d\bar{V}}{dt} = (-2 \cos t)\bar{i} - (2 \sin t)\bar{j},$$

so that

$$\bar{T} = \frac{1}{v} \bar{V} = \frac{1}{\sqrt{13}} ((-2 \sin t)\bar{i} + (2 \cos t)\bar{j} + 3\bar{k}),$$

$$\bar{V} \times \bar{A} = \begin{vmatrix} \bar{i} & \bar{j} & \bar{k} \\ -2 \sin t & 2 \cos t & 3 \\ -2 \cos t & -2 \sin t & 0 \end{vmatrix} = (6 \sin t)\bar{i} - (6 \cos t)\bar{j} + 4\bar{k}.$$

Thus,

$$\kappa = \frac{|\bar{V} \times \bar{A}|}{v^3} = \frac{\sqrt{36 \sin^2 t + 36 \cos^2 t + 16}}{(13)^{3/2}} = \frac{\sqrt{52}}{13\sqrt{13}} = \frac{2}{13},$$

$$\bar{N} = \frac{(\bar{V} \times \bar{A}) \times \bar{V}}{v^4 \kappa} = \frac{1}{(13)^2 \cdot \frac{2}{13}} \begin{vmatrix} \bar{i} & \bar{j} & \bar{k} \\ 6 \sin t & -6 \cos t & 4 \\ -2 \sin t & 2 \cos t & 3 \end{vmatrix} = (-\cos t)\bar{i} - (\sin t)\bar{j},$$

and

$$\bar{B} = \frac{\bar{V} \times \bar{A}}{|\bar{V} \times \bar{A}|} = \frac{(6 \sin t)\bar{i} - (6 \cos t)\bar{j} + 4k}{\sqrt{36 \sin^2 t + 36 \cos^2 t + 16}} = \frac{(3 \sin t)\bar{i} - (3 \cos t)\bar{j} + 2\bar{k}}{\sqrt{13}}.$$

Also,

$$\frac{d\bar{A}}{dt} = (2 \sin t)\bar{i} - (2 \cos t)\bar{j},$$

so that

$$\tau = \frac{1}{v^6 \kappa^2} \frac{d\bar{A}}{dt} \cdot (\bar{V} \times \bar{A}) = \frac{1}{52} ((2 \sin t)\bar{i} - (2 \cos t)\bar{j}) \cdot ((6 \sin t)\bar{i} - (6 \cos t)\bar{j} + 4\bar{k})$$

$$= \tfrac{12}{52} = \tfrac{3}{13}.$$

7.5 Summary of Formulas

We now present a summary of the basic formulas in connection with a particle P moving according to the law of motion

$$\bar{R} = \bar{F}(t)$$

in three-dimensional space. Here $\bar{R} = \overrightarrow{OP}$, denominators are assumed to be nonzero, and functions are assumed to have as many successive derivatives as may be required.

1 $\bar{V} = \dfrac{d\bar{R}}{dt}$

2 $v = |\bar{V}| = \dfrac{ds}{dt}$

$$3 \quad s = \int_a^b v \, dt$$

$$4 \quad \overline{V} = v\overline{T}$$

$$5 \quad \overline{A} = \frac{d\overline{V}}{dt} = \frac{d^2\overline{R}}{dt^2}$$

$$6 \quad \overline{A} = \frac{dv}{dt}\overline{T} + \kappa v^2 \overline{N}$$

$$7 \quad \kappa = \frac{|\overline{V} \times \overline{A}|}{v^3}$$

$$8 \quad \overline{N} = \frac{(\overline{V} \times \overline{A}) \times \overline{V}}{v^4 \kappa}$$

$$9 \quad \overline{B} = \frac{\overline{V} \times \overline{A}}{v^3 \kappa} = \frac{\overline{V} \times \overline{A}}{|\overline{V} \times \overline{A}|}$$

$$10 \quad \tau = \frac{1}{v^6 \kappa^2} \frac{d\overline{A}}{dt} \cdot (\overline{V} \times \overline{A})$$

$$11 \quad \overline{T} = \frac{d\overline{R}}{ds}$$

$$12 \quad \kappa = \left| \frac{d\overline{T}}{ds} \right|$$

$$13 \quad \overline{N} = \frac{1}{\kappa} \frac{d\overline{T}}{ds}$$

$$14 \quad \overline{B} = \overline{T} \times \overline{N}$$

$$15 \quad \tau = \frac{d\overline{N}}{ds} \cdot \overline{B}$$

$$16 \quad \tau = \frac{1}{\kappa^2} \left(\frac{d\overline{R}}{ds} \times \frac{d^2\overline{R}}{ds^2} \right) \cdot \frac{d^3\overline{R}}{ds^3}$$

$$17 \quad \frac{d\overline{T}}{ds} = \kappa \overline{N}$$

$$18 \quad \frac{d\overline{N}}{ds} = -\kappa \overline{T} + \tau \overline{B} \quad \left. \right\} \text{ Frenet formulas}$$

$$19 \quad \frac{d\overline{B}}{ds} = -\tau \overline{N}$$

All of these formulas have been proved except for 16 and 19, which are left as exercises (Problems 39 and 40).

Problem Set 7

In problems 1 through 6, (a) find $\lim_{t \to t_0} \overline{F}(t)$, (b) determine where $\overline{F}$ is continuous, (c) find $\overline{F}'(t_0)$, and (d) find $\overline{F}''(t_0)$.

1 $\overline{F}(t) = 5t^2 i + (3t + 1)j + (3 - t^2)\overline{k}$, $t_0 = 1$

2 $\overline{F}(t) = \sqrt{t}\, i + e^{-3t}j + t\overline{k}$, $t_0 = 4$

3 $\overline{F}(t) = (1 + t)^2 i + (\cos t)j + \ln(1 - t)\overline{k}$, $t_0 = 0$

4 $\overline{F}(t) = (2 + \sin t)i + (\cos t)j + t\overline{k}$, $t_0 = \dfrac{\pi}{2}$

5 $\overline{F}(t) = (\cos 2t)i + (\sin 2t)j + t^3\overline{k}$, $t_0 = \dfrac{\pi}{4}$

6 $\overline{F}(t) = e^{-3t}i + e^{3t}j + t\overline{k}$, $t_0 = 0$

In problems 7 through 14, let $\bar{F}(t) = 5t i + (t^2 - 2)j + t^3\bar{k}$, $\bar{G}(t) = t i + (1/t)j + t^2\bar{k}$, and $h(t) = 3t$.

7 Calculate $h(t)\bar{F}(t)$, and then find $\dfrac{d}{dt}[h(t)\bar{F}(t)]$.

8 Find $\dfrac{d}{dt}[h(t)\bar{F}(t)]$ by using the formula

$$\frac{d}{dt}[h(t)\bar{F}(t)] = \frac{dh(t)}{dt}\bar{F}(t) + h(t)\frac{d}{dt}[\bar{F}(t)].$$

9 Calculate $\bar{F}(t) \cdot \bar{G}(t)$, and then find $\dfrac{d}{dt}[\bar{F}(t) \cdot \bar{G}(t)]$.

10 Find $\dfrac{d}{dt}[\bar{F}(t) \cdot \bar{G}(t)]$ by using the formula

$$\frac{d}{dt}[\bar{F}(t) \cdot \bar{G}(t)] = \left[\frac{d}{dt}\bar{F}(t)\right] \cdot \bar{G}(t) + \bar{F}(t) \cdot \left[\frac{d}{dt}\bar{G}(t)\right].$$

11 Calculate $\bar{F}(t) \times \bar{G}(t)$, and then find $\dfrac{d}{dt}[\bar{F}(t) \times \bar{G}(t)]$.

12 Find $\dfrac{d}{dt}[\bar{F}(t) \times \bar{G}(t)]$ by using the formula

$$\frac{d}{dt}[\bar{F}(t) \times \bar{G}(t)] = \left[\frac{d}{dt}\bar{F}(t)\right] \times \bar{G}(t) + \bar{F}(t) \times \left[\frac{d}{dt}\bar{G}(t)\right].$$

13 Find $\dfrac{d}{dt}|\bar{F}(t)|$.   **14** Find $\dfrac{d}{dt}\bar{F}(t^3)$.

In problems 15 through 20, find (a) the velocity vector $\bar{V}$, (b) the speed v, and (c) the acceleration vector $\bar{A}$ at the time t of the particle whose equation of motion is given.

15 $\bar{R} = (2\cos 2t)i + (3\sin 2t)j + t\bar{k}$ **16** $\bar{R} = (2 + t)i + 3tj + (5 - 3t)\bar{k}$

17 $\bar{R} = e^{-t}i + 2e^t j + t^3\bar{k}$ **18** $\bar{R} = (\ln\cos t)i + (\sin t)j + (\cos t)\bar{k}$

19 $\bar{R} = (t\sin t)i + (t\cos t)j + t^2\bar{k}$ **20** $\bar{R} = (e^{-t}\cos t)i + (e^{-t}\sin t)j + e^{-t}\bar{k}$

In problems 21 through 26, find the arc length of the portion of the given curve between the point where $t = a$ and the point where $t = b$.

21 $\bar{R} = (3\cos t)i + (4\cos t)j + (5\sin t)\bar{k}$; $a = 0$, $b = 2$ **22** $\bar{R} = (\cos 3t)i + (\sin 3t)j + 5t\bar{k}$; $a = 0$, $b = 2\pi$

23 $\bar{R} = (e^{-t}\cos t)i + (e^{-t}\sin t)j + e^{-t}\bar{k}$; $a = 0$, $b = \pi$ **24** $\bar{R} = (\cos t)i + (\ln\cos t)j + (\sin t)\bar{k}$; $a = 0$, $b = \pi/4$

25 $\bar{R} = (t^3/3 - 1/t)i + (t^3/3 - 7/t)j + 2t\bar{k}$; $a = 1$, $b = 4$ **26** $\bar{R} = (2t^2 + 3)i + (3 - 2t^2)j + 4t\bar{k}$; $a = 0$, $b = 2$

In problems 27 through 32, find (a) $\bar{V}$, (b) v, (c) $\bar{A}$, (d) $\bar{T}$, (e) $\bar{V} \times \bar{A}$, (f) κ, (g) $\bar{N}$, (h) $\bar{B}$, (i) $d\bar{A}/dt$, and (j) τ for each curve in space.

27 $\bar{R} = (2\cos t)i + (3\sin t)j + t\bar{k}$ **28** $\bar{R} = (t^2 - 3)i + (t^2 + 7)j + (t^2 + t)\bar{k}$

29 $\bar{R} = (e^{2t} + e^{-2t})i + (e^{2t} - e^{-2t})j + 5t\bar{k}$ **30** $\bar{R} = e^t i + 2te^t j + 3e^t\bar{k}$

31 $\bar{R} = t i + \dfrac{1}{\sqrt{2}}t^2 j + \dfrac{1}{3}t^3\bar{k}$ **32** $\bar{R} = \dfrac{\cos t}{4}i + \dfrac{\ln\cos t}{4}j + \dfrac{\sin t}{4}\bar{k}$

33 Find $\bar{T}$, $\bar{N}$, $\bar{B}$, κ, and τ for the circular spiral $\begin{cases} x = a\cos t \\ y = a\sin t \\ z = bt. \end{cases}$

34 Prove that if $\lim\limits_{t \to c}\bar{F}(t)$ and $\lim\limits_{t \to c}\bar{G}(t)$ both exist, then $\lim\limits_{t \to c}[\bar{F}(t) \times \bar{G}(t)]$ exists and equals

$$\left[\lim\limits_{t \to c}\bar{F}(t)\right] \times \left[\lim\limits_{t \to c}\bar{G}(t)\right].$$

35 If $\bar{F}$ and $\bar{G}$ are continuous vector-valued functions, show that $\bar{F} \times \bar{G}$ is also continuous.

36 Suppose that $\bar{F}$ and $\bar{G}$ are differentiable vector-valued functions and that $\bar{H}$ is defined by $\bar{H}(t) = \bar{F}(t) \times \bar{G}(t)$. Prove that $\bar{H}$ is differentiable and that

$$\frac{d\bar{H}(t)}{dt} = \left[\frac{d\bar{F}(t)}{dt}\right] \times \bar{G}(t) + \bar{F}(t) \times \left[\frac{d\bar{G}(t)}{dt}\right].$$

37 Evaluate $\dfrac{d^2(\bar{F} \times \bar{G})}{dt^2}$.

38 Suppose that the curve $\bar{R} = \bar{F}(t)$ in space has curvature $\kappa = 0$ at all points.
(a) Show that $\bar{T}$ is a constant vector.
(b) Show that $\bar{T} \times \bar{R}$ is a constant vector.
(c) Show that the curve is a portion of a straight line.

39 Prove formula 16 on page 785. (*Hint:* Use formula 10 and take $t = s$.)

40 Prove formula 19 on page 785.

$$\left[Hint: \frac{d\bar{B}}{ds} = \left(\frac{d\bar{B}}{ds} \cdot \bar{T}\right)\bar{T} + \left(\frac{d\bar{B}}{ds} \cdot \bar{N}\right)\bar{N} + \left(\frac{d\bar{B}}{ds} \cdot \bar{B}\right)\bar{B}.\right]$$

41 Suppose that the curve $\bar{R} = \bar{F}(t)$ in space lies in the plane $\bar{N}_0 \cdot (\bar{R} - \bar{R}_0) = 0$. Prove that the torsion τ of the curve is identically zero. (*Caution:* Here $\bar{N}_0$ is the *constant* normal vector *to the plane*—do not confuse it with $\bar{N}$, the variable principal normal vector *to the curve*.)

42 Suppose that the curve $\bar{R} = \bar{F}(t)$ in space has torsion $\tau = 0$ at all points. Prove that the curve lies in a plane.

43 Prove that $\tau = \pm \left|\dfrac{d\bar{B}}{ds}\right|$.

In Section 5 we found that a plane in xyz space has the cartesian scalar equation $ax + by + cz = D$. In general, an equation involving x, y, and z has a two-dimensional surface in xyz space as its graph. (The planes considered in Section 5 are just special surfaces.) In this section and the next, we consider some other types of surfaces and their equations.

8.1 Spheres

If $P = (x, y, z)$ and $P_0 = (x_0, y_0, z_0)$, then, as we saw in Section 2.1, the distance between P and P_0 is given by

$$|\overline{PP_0}| = \sqrt{(x - x_0)^2 + (y - y_0)^2 + (z - z_0)^2}.$$

Therefore, the equation

where r is a positive constant, means that the distance between P and P_0 is r units;

hence, this equation represents the surface of a sphere of radius r with center at the point $P_0 = (x_0, y_0, z_0)$ (Figure 1). In particular, if $P_0 = (0,0,0)$, then the equation of the sphere becomes

$$x^2 + y^2 + z^2 = r^2.$$

Figure 1

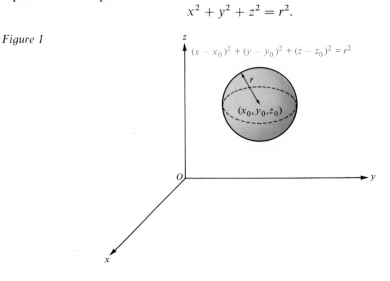

$(x - x_0)^2 + (y - y_0)^2 + (z - z_0)^2 = r^2$

r

(x_0, y_0, z_0)

EXAMPLES   1 Find the equation of the sphere with radius 5 units and center at $(-1, 2, 4)$.

SOLUTION
The equation of the sphere is $(x + 1)^2 + (y - 2)^2 + (z - 4)^2 = 25$.

2 Find the radius and the center of the sphere whose equation is

$$x^2 + y^2 + z^2 - 6x + 2y + 4z - 11 = 0.$$

SOLUTION
Regrouping terms and completing the squares, we obtain

$$(x^2 - 6x + 9) + (y^2 + 2y + 1) + (z^2 + 4z + 4) = 9 + 1 + 4 + 11$$

or

$$(x - 3)^2 + (y + 1)^2 + (z + 2)^2 = 25,$$

which represents a sphere of radius 5 with center $(3, -1, -2)$.

8.2 Cylinders

A straight line L in space which moves so as to remain parallel to a fixed straight line L_0 while intersecting a fixed curve C is said to sweep out a *cylindrical surface*—or just a *cylinder* for short (Figure 2). Any particular position of L is called a *generator* of the cylinder and the curve C is called the *base curve*. If the base curve C is a circle and all the generators are perpendicular to the plane of the circle, then the surface is called a *right circular cylinder*. A cylinder whose base curve is an ellipse, parabola, or hyperbola is naturally called an *elliptic*, *parabolic*, or *hyperbolic* cylinder, respectively.

Figure 2

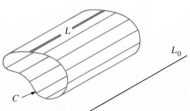

Figure 3

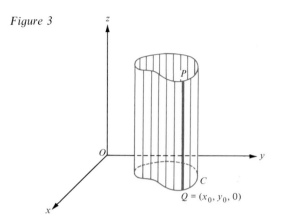

$Q = (x_0, y_0, 0)$

In what follows we consider only cylinders whose generators are parallel to one of the coordinate axes and whose base curve lies in the coordinate plane perpendicular to that axis. For instance, the cylinder in Figure 3 has generators parallel to the z axis; hence, if the point $Q = (x_0, y_0, 0)$ is on the base curve C, then all points of the form $P = (x_0, y_0, z)$ also belong to the cylinder. It follows that the equation of this cylinder cannot place any restriction on z whatsoever—that is, either z is not present in the equation, or else it can be removed from the equation by algebraic manipulation.

To sketch the graph of a cylindrical surface whose equation is missing one of the variables x, y, or z, we simply sketch the graph of the equation in the coordinate plane involving the variables that are present, thus obtaining the base curve C. The surface is then swept out by a straight line perpendicular to this coordinate plane and intersecting the base curve.

EXAMPLES The graph of the given equation is a cylinder.
(a) Identify the coordinate axis that is parallel to the generators.
(b) Specify the coordinate plane containing the base curve.
(c) Describe the base curve.
(d) Sketch the cylinder.

1 $y = e^z$

SOLUTION
(a) The variable x is missing; hence, the generators are parallel to the x axis.
(b) The base curve is contained in the yz plane.
(c) The base curve is an exponential curve.
(d) The graph appears in Figure 4.

2 $z = \sin x$

SOLUTION
(a) The variable y is missing; hence, the generators are parallel to the y axis.
(b) The base curve is contained in the xz plane.
(c) The base curve is a sine curve.
(d) The graph appears in Figure 5.

Figure 4   *Figure 5*

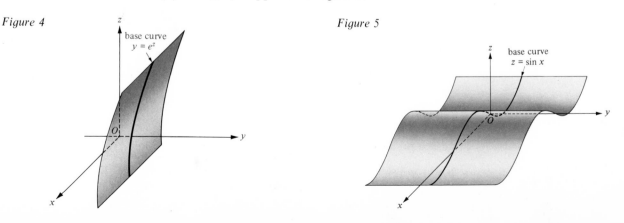

base curve
$y = e^z$

base curve
$z = \sin x$

8.3 Surfaces of Revolution

A *surface of revolution* is defined to be a surface generated by revolving a plane curve C about a straight line L lying in the same plane as the curve. The line L is called the *axis of revolution.* For instance, a sphere is the surface of revolution generated by revolving a circle C about an axis L passing through its center. Notice that a surface of revolution intersects a plane perpendicular to the axis of revolution in a circle or a single point, provided that it intersects the plane at all.

Figure 6

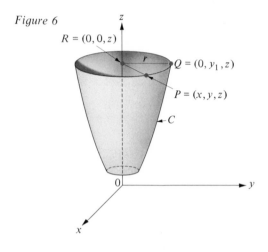

Figure 6 shows a portion of a surface of revolution obtained by revolving the curve C in the yz plane about the z axis. In order to find the equation of this surface, consider a general point $P = (x, y, z)$ on the surface, noting that P is obtained by revolving some point Q on the original curve C about the z axis as in Figure 6. Perpendiculars dropped from P and Q to the z axis meet the z axis at the same point $R = (0, 0, z)$. Notice that P, Q, and R have the same z coordinate. Let $r = |\overline{PR}| = |\overline{QR}|$. Since Q lies in the yz plane, its x coordinate is zero. Let $Q = (0, y_1, z)$, noting that $|y_1| = r$.

Now,

$$y_1^2 = r^2 = |\overline{PR}|^2 = (x - 0)^2 + (y - 0)^2 + (z - z)^2$$
$$= x^2 + y^2;$$

hence, $y_1 = \pm\sqrt{x^2 + y^2}$. We conclude that, if $P = (x, y, z)$ belongs to the surface of revolution, then either $Q = (0, \sqrt{x^2 + y^2}, z)$ or $Q = (0, -\sqrt{x^2 + y^2}, z)$ belongs to the generating curve. Conversely, if $Q = (0, y_1, z)$ is a given point on the generating curve C, then any point $P = (x, y, z)$ such that $\sqrt{x^2 + y^2} = |y_1|$ lies on a horizontal circle of radius $r = |y_1|$ with center at $R = (0, 0, z)$; hence, such a point P belongs to the surface of revolution. We conclude that $P = (x, y, z)$ belongs to the surface of revolution if and only if $Q = (0, \pm\sqrt{x^2 + y^2}, z)$ belongs to the generating curve. In other words, the equation of the surface generated by revolving the curve C in the yz plane about the z axis is obtained by replacing the variable y in the equation of C by the expression $\pm\sqrt{x^2 + y^2}$.

EXAMPLE   The parabola $z = y^2$ in the yz plane is revolved about the z axis to form a surface of revolution. Find the equation of this surface (Figure 7).

SOLUTION
Replacing y in the equation $z = y^2$ by $\pm\sqrt{x^2 + y^2}$, we obtain

Figure 7

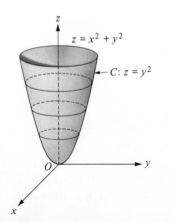

$$z = (\pm\sqrt{x^2 + y^2})^2 \quad \text{or} \quad z = x^2 + y^2.$$

The surface in Figure 7 is called a *paraboloid of revolution.* Similarly, a surface of revolution obtained by revolving a hyperbola or an ellipse about one of its own axes of symmetry is called a *hyperboloid of revolution* or an *ellipsoid of revolution*, respectively.

The preceding considerations apply to curves lying in any coordinate plane, and we have the following general rule for finding the equation of a surface of revolution:

1 List the three variables in such an order that the first variable represents the axis about which the generating curve C is revolved and the first two variables represent the plane in which C lies.

2 In the equation for C, replace the second listed variable by plus or minus the square root of the sum of the squares of the second and third listed variables.

EXAMPLE   Find the equation of the surface of revolution generated by revolving the curve $y^2 = 3x$ in the xy coordinate plane about the x axis. Sketch a graph of the surface.

Figure 8

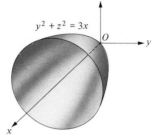

$y^2 + z^2 = 3x$

SOLUTION
Following step 1 of the rule above, we list the three variables in the order x, y, z. Carrying out step 2, we replace y in the equation $y^2 = 3x$ by $\pm\sqrt{y^2 + z^2}$ to obtain $(\pm\sqrt{y^2 + z^2})^2 = 3x$, or $y^2 + z^2 = 3x$ (Figure 8).

Notice that an equation represents a surface of revolution about one of the coordinate axes if it contains the variables corresponding to the other two coordinate axes only in combination as the sum of their squares. For instance, the equation $y^2 = 4(z^2 + x^2)$ contains the two variables x and z only in the combination $z^2 + x^2$ and so represents a surface of revolution obtained by revolving a curve about the y axis.

If a surface is known to be a surface of revolution about one of the coordinate axes, a generating curve C can be found by intersecting the surface with either one of the coordinate planes containing that coordinate axis. Of course, in order to find the equation of the intersection of a surface with the xy, xz, or yz coordinate plane, one simply sets z, y, or x, respectively, equal to 0 in the equation of the surface.

EXAMPLE   For the surface of revolution $x^2 - y^2 + z^2 = 1$:
(a) Find the axis of revolution.
(b) Find a generating curve in either one of the coordinate planes containing the axis of revolution.
(c) Sketch the graph.

SOLUTION
(a) The equation can be rewritten as $(x^2 + z^2) - y^2 = 1$, and thus contains the variables x and z only in the combination $x^2 + z^2$. It is therefore a surface of revolution about the y axis.

Figure 9

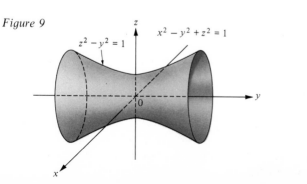

$z^2 - y^2 = 1$ $x^2 - y^2 + z^2 = 1$

(b) We can find a generating curve by intersecting the surface either with the xy plane or with the yz plane. The intersection with the yz plane is found by setting $x = 0$ in the equation $x^2 - y^2 + z^2 = 1$ to obtain $z^2 - y^2 = 1$, the equation of a hyperbola.

(c) The surface, which is sketched in Figure 9, is generated by revolving the hyperbola $z^2 - y^2 = 1$ about the y axis, and thus is a hyperboloid of revolution.

Problem Set 8

In problems 1 through 4, find an equation of the sphere satisfying the conditions given.
 1 Center at $(2, -1, 3)$ and radius 4 units.
 2 Its diameter is the line segment joining the points $(8, -1, 7)$ and $(-2, -5, 5)$.
 3 Center at $(4, 1, 3)$ and containing the point $(2, -1, 2)$.
 4 Containing the points $(3, 0, 4)$, $(3, 4, 0)$, $(9, 0, \sqrt{7})$, and $(-3, 0, 4)$.

In problems 5 through 9, find the center and radius of the sphere.
 5 $x^2 + y^2 + z^2 - 2x + 4z - 4 = 0$ **6** $x^2 + y^2 + z^2 + 2x + 14y - 10z + 74 = 0$

 7 $2x^2 + 2y^2 + 2z^2 + 4x - 4y - 14 = 0$ **8** $x^2 + y^2 + z^2 - 8x + 4y + 10z + 9 = 0$

 9 $x^2 + y^2 + z^2 - 6x + 2y - 4z - 19 = 0$

10 Find the set of points that are equidistant from the points $(1, 3, -1)$ and $(2, -1, 2)$.

In problems 11 through 22, (a) identify the coordinate axis to which the generators of the cylinder are parallel, (b) specify the coordinate plane containing the base curve, (c) describe the base curve, and (d) sketch the cylinder.

11 $y = z^2$ **12** $xy = 3$ **13** $yz = 1$ **14** $y^2 + z^2 = 4$

15 $|x| = z$ **16** $x^2 + y^2 = 4x$ **17** $y = e^x$ **18** $3x + 2y = 6$

19 $y^2 = 4z$ **20** $z = \sin 2x$ **21** $x^2 = z^2$ **22** $|y| + |z| = 1$

In problems 23 through 30, find the equation of the surface that results when the plane graph of the given equation is revolved about the indicated axis. Sketch the surface.

23 $y = x$ in the xy plane about the x axis.   **24** $y = 3x + 2$ in the xy plane about the y axis.

25 $y = \ln x$ in the xy plane about the x axis.   **26** $4x^2 - 9z^2 = 5$ in the xz plane about the z axis.

27 $z = x^2$ in the xz plane about the z axis.   **28** $\dfrac{y^2}{9} + \dfrac{z^2}{4} = 1$ in the yz plane about the z axis.

29 $6y^2 + 6z^2 = 7$ in the yz plane about the y axis.   **30** $x^2 - 4z^2 = 4$ in the xz plane about the x axis.

In problems 31 through 36, each equation represents a surface of revolution about one of the coordinate axes. Identify this axis, find the equation of a generating curve in the indicated coordinate plane, and sketch the graph.

31 $x^2 + y^2 + z = 3$; xz plane   **32** $9z^2 = x^2 + y^2$; xz plane   **33** $x^2 - 9y^2 - 9z^2 = 18$; xy plane

34 $4 - y = 3(x^2 + z^2)$; yz plane   **35** $x^2 + y^2 = \sin^2 z$; xz plane   **36** $9x^2 + 9y^2 + 4z^2 = 36$; yz plane

37 If $A^2 + B^2 + C^2 > 4D$, show that the equation $x^2 + y^2 + z^2 + Ax + By + Cz + D = 0$ can be rewritten in the form $(x - x_0)^2 + (y - y_0)^2 + (z - z_0)^2 = r^2$. (*Hint:* Complete the squares.)

9   Quadric Surfaces

In Chapter 11 we showed that, apart from certain degenerate cases, the graph in the xy plane of a second-degree equation

$$Ax^2 + Bxy + Cy^2 + Dx + Ey + F = 0$$

is a circle, an ellipse, a parabola, or a hyperbola; in any case, it is a conic section. A second-degree equation in the *three* variables x, y, and z has the form

$$Ax^2 + By^2 + Cz^2 + Dxy + Exz + Fyz + Gx + Hy + Iz + K = 0,$$

and the graph of such an equation in *xyz* space is called a *quadric surface*. By rotating and translating the coordinate system in three-dimensional space, it is possible to bring the second-degree equation into certain standard forms, and thus to classify the quadric surfaces.

In the present section we do not attempt a detailed study and classification of quadric surfaces. Rather, our intention is simply to familiarize the reader with the general shapes and with the standard equations of the common quadric surfaces.

9.1 Techniques for Studying Surfaces

We begin by mentioning, very briefly, some of the basic techniques for visualizing or graphing surfaces in space. These involve (1) *cross sections*, (2) *traces* and *intercepts*, and (3) *symmetries.*

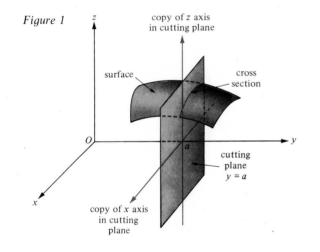

Figure 1

copy of *z* axis
in cutting plane

z

surface

cross
section

O

y

a

cutting
plane
y = a

x

copy of *x* axis
in cutting
plane

Cross Sections

The simplest (but in many cases the most effective) technique for visualizing, recognizing, or graphing surfaces in space is to find the *cross sections* of the surface formed by intersecting the surface with planes—especially by planes perpendicular to the coordinate axes or perpendicular to the axes of symmetry of the surface. For example, Figure 1 shows the cross section cut from a surface by a plane $y = a$ perpendicular to the *y* axis. The equation of the cross section *in the cutting plane* $y = a$ can be found simply by putting $y = a$ in the equation of the surface. The resulting equation, which will involve only *x* and *z*, is the equation of the cross section relative to copies of the *x* and *z* axes in the cutting plane as shown in Figure 1. Exactly the same idea applies to cross sections cut by planes perpendicular to the other coordinate axes.

Traces and Intercepts

The cross sections cut from a surface by the coordinate planes themselves are called the *traces* of the surface. For instance, the *yz trace* of the surface (that is, the cross section cut from the surface by the *yz* plane) is obtained by putting $x = 0$ in the equation of the surface. The *xy trace* and the *xz trace* are defined similarly.

The *x*, *y*, and *z intercepts* of the surface are defined to be the points (if any) in which the *x*, *y*, and *z* axes, respectively, intersect the surface. For instance, to find the *x* intercept, we set $y = 0$ and $z = 0$ in the equation of the surface.

Symmetries

Surfaces often display symmetries with respect to points, lines, or planes. Of course, a surface is said to be *symmetric* with respect to a point, line, or plane provided that, whenever a point *P* lies on the surface, so does the point *Q*, which is symmetrically located relative to the point, line, or plane. For instance, if an equation equivalent to the original equation of the surface is obtained when y is replaced by $-y$, then the surface is symmetric with respect to the xz plane. Symmetry with respect to the z axis can be tested by replacing *both* x and y by $-x$ and $-y$, respectively, and checking whether the resulting equation is equivalent to the original one. Similar tests are easily discovered for symmetries with respect to the other coordinate planes or coordinate axes as well as for symmetry with respect to the origin (Problems 25 through 27).

9.2 Central Quadric Surfaces

The graph in xyz space of an equation of the form

$$\pm \frac{x^2}{a^2} \pm \frac{y^2}{b^2} \pm \frac{z^2}{c^2} = 1,$$

where a, b, and c are positive constants and not all three algebraic signs are negative, is called a *central quadric* surface. Since only even powers of the variables appear, a central quadric surface is symmetric with respect to all three coordinate planes, all three coordinate axes, and the origin. The central quadric surfaces are classified as follows:

1 If all three algebraic signs are positive, then the surface is called an *ellipsoid.*

2 If two algebraic signs are positive and one is negative, then the surface is called a *hyperboloid of one sheet.*

3 If one algebraic sign is positive and the other two are negative, then the surface is called a *hyperboloid of two sheets.*

We now discuss each of the central quadrics briefly.

The Ellipsoid

The x, y, and z intercepts of the ellipsoid $\dfrac{x^2}{a^2} + \dfrac{y^2}{b^2} + \dfrac{z^2}{c^2} = 1$ are $(\pm a, 0, 0)$, $(0, \pm b, 0)$, and $(0, 0, \pm c)$, respectively, and the traces on the coordinate planes are the ellipses (or circles) $x^2/a^2 + y^2/b^2 = 1$, $x^2/a^2 + z^2/c^2 = 1$, and $y^2/b^2 + z^2/c^2 = 1$. In fact, all cross sections cut by planes perpendicular to the coordinate axes are ellipses (or circles) (Problem 29). The graph is sketched in Figure 2.

Figure 2

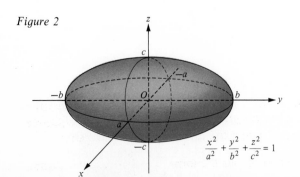

The Hyperboloid of One Sheet

Suppose, for definiteness, that the term involving z^2 carries the negative sign, so that the equation has the form $\dfrac{x^2}{a^2} + \dfrac{y^2}{b^2} - \dfrac{z^2}{c^2} = 1$. The x and y intercepts are $(\pm a, 0, 0)$ and $(0, \pm b, 0)$, but there is no z intercept since the equation $-z^2/c^2 = 1$ cannot be satisfied by any real number z. The traces on the coordinate planes are as follows:

Figure 3

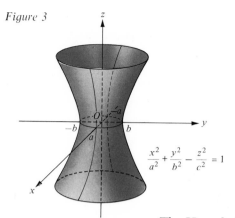

$$\frac{x^2}{a^2} + \frac{y^2}{b^2} - \frac{z^2}{c^2} = 1$$

1 The xy trace is the ellipse (or circle)
   $x^2/a^2 + y^2/b^2 = 1$.
2 The xz trace is the hyperbola
   $x^2/a^2 - z^2/c^2 = 1$.
3 The yz trace is the hyperbola
   $y^2/b^2 - z^2/c^2 = 1$.

Indeed, all cross sections cut by planes perpendicular to the x or y axes are hyperbolas or pairs of straight lines, while all cross sections cut by horizontal planes are ellipses (Problem 30). A portion of the surface is sketched in Figure 3.

The Hyperboloid of Two Sheets

Suppose, for definiteness, that the terms involving y^2 and z^2 carry the negative signs, so that the equation has the form $\dfrac{x^2}{a^2} - \dfrac{y^2}{b^2} - \dfrac{z^2}{c^2} = 1$. The x intercepts are $(\pm a, 0, 0)$, but there are no y and no z intercepts. (Why?) The traces on the coordinate planes are as follows:

1 The xy trace is the hyperbola $x^2/a^2 - y^2/b^2 = 1$.
2 The xz trace is the hyperbola $x^2/a^2 - z^2/c^2 = 1$.
3 There is no yz trace.

Figure 4

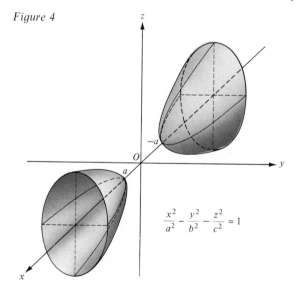

$$\frac{x^2}{a^2} - \frac{y^2}{b^2} - \frac{z^2}{c^2} = 1$$

Although the hyperboloid of two sheets has no trace on the yz plane, it does have cross sections cut by certain planes $x = k$ parallel to the yz plane. Indeed, the cross section cut by the plane $x = k$ has the equation

$$\frac{k^2}{a^2} - \frac{y^2}{b^2} - \frac{z^2}{c^2} = 1 \quad \text{or} \quad \frac{y^2}{b^2} + \frac{z^2}{c^2} = \frac{k^2}{a^2} - 1,$$

provided that $|k| > a$. The latter equation can be rewritten as

$$\frac{y^2}{b^2(k^2/a^2 - 1)} + \frac{z^2}{c^2(k^2/a^2 - 1)} = 1;$$

therefore, the cross sections cut by planes perpendicular to the x axis and at least a units from the origin are ellipses. A portion of the surface, which consists of two separated "parts" or "sheets," is shown in Figure 4.

9.3 Elliptic Cones

The graph in xyz space of an equation of the form

$$\pm \frac{x^2}{a^2} \pm \frac{y^2}{b^2} \pm \frac{z^2}{c^2} = 0,$$

where a, b, and c are positive constants and not all three algebraic signs are the same, is called an *elliptic cone*. By multiplying by -1 if necessary, we can arrange that two of the algebraic signs are positive and the remaining one is negative.

Suppose, for definiteness, that the equation has the form $\dfrac{x^2}{a^2} + \dfrac{y^2}{b^2} - \dfrac{z^2}{c^2} = 0$. Again, since only even powers of the variables occur, the elliptic cone is symmetric with respect to all coordinate planes, all coordinate axes, and the origin. If two of the variables are put equal to zero in the equation, then the third variable must also be zero; hence, the only x, y, or z intercept of the elliptic cone is the origin. The traces on the coordinate planes are as follows:

Figure 5

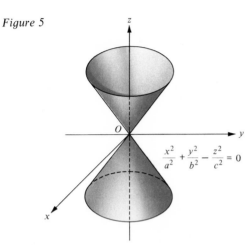

$$\frac{x^2}{a^2} + \frac{y^2}{b^2} - \frac{z^2}{c^2} = 0$$

1 The xy trace is $x^2/a^2 + y^2/b^2 = 0$, or $(x, y) = (0,0)$, just one point at the origin.
2 The xz trace is $x^2/a^2 - z^2/c^2 = 0$, or $z = \pm(c/a)x$, a pair of intersecting straight lines.
3 The yz trace is $y^2/b^2 - z^2/c^2 = 0$, or $z = \pm(c/b)x$, a pair of intersecting straight lines.

Cross sections cut by planes perpendicular to the x or y axis that do not pass through the origin are hyperbolas, while cross sections cut by horizontal planes which do not pass through the origin are ellipses (or circles) (Problem 31). A portion of the elliptic cone is shown in Figure 5. Notice that, if $a = b$, the elliptic cone becomes an ordinary right circular cone.

9.4 Elliptic and Hyperbolic Paraboloids

We now consider the graph in xyz space of an equation having one of the forms

$$\pm\frac{x^2}{a^2} \pm \frac{y^2}{b^2} = z \quad \text{or} \quad \pm\frac{y^2}{b^2} \pm \frac{z^2}{c^2} = x \quad \text{or} \quad \pm\frac{z^2}{c^2} \pm \frac{x^2}{a^2} = y,$$

where a, b, and c are positive constants. If both terms on the left carry the same algebraic sign, then the graph of any one of these equations is called an *elliptic paraboloid*. On the other hand, if the terms on the left carry opposite algebraic signs, then the graph of any one of the equations is called a *hyperbolic paraboloid*. We discuss these two cases briefly, considering only the first equation, $\pm x^2/a^2 \pm y^2/b^2 = z$. The other equations are handled similarly.

The Elliptic Paraboloid

For definiteness, we assume that the coefficients of x^2 and y^2 are both positive, so that the equation can be written in the form

$$\frac{x^2}{a^2} + \frac{y^2}{b^2} = z,$$

Figure 6

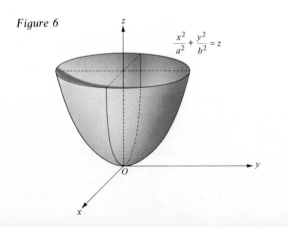

$$\frac{x^2}{a^2} + \frac{y^2}{b^2} = z$$

where a, $b > 0$. Evidently, this surface intersects the xy plane only at the origin and otherwise lies above the xy plane. It is symmetric with respect to the xz plane, the yz plane, and the z axis, because only even powers of x and y occur. The traces in the xz and yz planes are the parabolas $z = x^2/a^2$ and $z = y^2/b^2$, respectively. The cross sections cut by horizontal planes above the origin are ellipses (or circles) (Problem 32). A sketch of a portion of the elliptic paraboloid appears in Figure 6. Notice that, in the special case in which $a = b$, the elliptic paraboloid becomes a paraboloid of revolution (about the z axis).

The Hyperbolic Paraboloid

For definiteness, we assume that the coefficient of y^2 is positive while the coefficient of x^2 is negative, so that the equation can be written in the form

$$\frac{y^2}{b^2} - \frac{x^2}{a^2} = z,$$

where $a, b > 0$. The coordinate axes intersect this surface only at the origin and the surface is symmetric with respect to the xz plane, the yz plane, and the z axis. The traces are as follows:

1. The xy trace, $y^2/b^2 - x^2/a^2 = 0$, or $y = \pm(b/a)x$, is a pair of intersecting straight lines.
2. The xz trace, $-x^2/a^2 = z$, is a parabola opening downward.
3. The yz trace, $y^2/b^2 = z$, is a parabola opening upward.

Figure 7

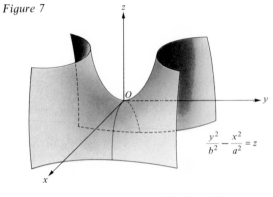

$$\frac{y^2}{b^2} - \frac{x^2}{a^2} = z$$

Moreover, cross sections cut by horizontal planes above the origin are hyperbolas with transverse axes parallel to the y axis, while cross sections cut by horizontal planes below the origin are hyperbolas with transverse axes parallel to the x axis (Problem 33). All cross sections cut by planes perpendicular to the x axis or y axis are parabolas opening upward or downward, respectively (Problem 34). A sketch of a portion of the hyperbolic paraboloid appears in Figure 7. Near the origin, the hyperbolic paraboloid has the shape of a saddle.

9.5 Examples of Quadric Surfaces

The surfaces discussed in Sections 9.2 through 9.4, together with cylinders whose base curves are conic sections, exhaust all possibilities for quadric surfaces with the exception of certain degenerate cases which we do not consider here. In the following examples we illustrate the technique of identifying and graphing quadrics whose equations are in the standard form.

EXAMPLES For the given quadric surface, (a) find the intercepts, (b) discuss the symmetry, (c) find the cross sections perpendicular to the coordinate axes, (d) find the traces, (e) identify the surface, and (f) sketch the graph.

1 $x^2 - 9y^2 + z^2 = 81$

SOLUTION
Dividing by 81, we see that the equation has the form $x^2/a^2 - y^2/b^2 + z^2/c^2 = 1$, where $a = 9$, $b = 3$, and $c = 9$.
(a) The x intercepts are $(\pm 9, 0, 0)$. There are no y intercepts. The z intercepts are $(0, 0, \pm 9)$.
(b) All variables are squared, so the surface is symmetric with respect to all three coordinate planes, all three coordinate axes, and the origin.
(c) The intersection with the plane $x = k$ is the curve $z^2 - 9y^2 = 81 - k^2$, which is a hyperbola except when $k = \pm 9$. When $k = \pm 9$, the intersection with the plane $x = k$ is the pair of intersecting straight lines $z = \pm 3y$. The intersection with the plane $y = k$ is the circle $x^2 + z^2 = 81 + 9k^2$ of radius $\sqrt{81 + 9k^2}$. The intersection with the plane $z = k$ is the curve $x^2 - 9y^2 = 81 - k^2$, which is a

Figure 8

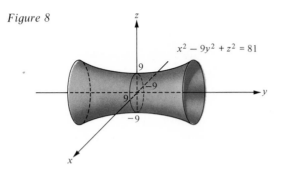

$$x^2 - 9y^2 + z^2 = 81$$

hyperbola except when $k = \pm 9$, in which case it is a pair of intersecting straight lines $x = \pm 3y$.

(d) The traces are found by putting $k = 0$ in (c). The yz trace is the hyperbola $z^2 - 9y^2 = 81$. The xz trace is the circle $x^2 + z^2 = 81$. The xy trace is the hyperbola $x^2 - 9y^2 = 81$.

(e) The surface is a hyperboloid of one sheet—in fact, a hyperboloid of revolution about the y axis.

(f) The graph is sketched in Figure 8.

2 $-9x^2 - 16y^2 + z^2 = 144$

SOLUTION
The equation has the form $-x^2/a^2 - y^2/b^2 + z^2/c^2 = 1$ with $a = 4$, $b = 3$, and $c = 12$.

(a) There are no x intercepts. There are no y intercepts. The z intercepts are $(0, 0, \pm 12)$.

Figure 9

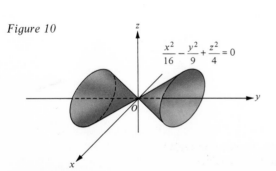

$$-9x^2 - 16y^2 + z^2 = 144$$

(b) All variables are squared, so that the surface is symmetric with respect to all three coordinate planes, all three coordinate axes, and the origin.

(c) The intersection with the plane $x = k$ is the hyperbola $z^2 - 16y^2 = 144 + 9k^2$. The intersection with the plane $y = k$ is the hyperbola $z^2 - 9x^2 = 144 + 16k^2$. For $|k| > 12$, the intersection with the plane $z = k$ is the ellipse $9x^2 + 16y^2 = k^2 - 144$.

(d) Putting $k = 0$ in (c), we find that the yz and xz traces are the hyperbolas $z^2 - 16y^2 = 144$ and $z^2 - 9x^2 = 144$, respectively, while there is no xy trace.

(e) The surface is a hyperboloid of two sheets.

(f) The graph is sketched in Figure 9.

3 $\dfrac{x^2}{16} - \dfrac{y^2}{9} + \dfrac{z^2}{4} = 0$

SOLUTION
(a) The only intercept along any coordinate axis is the origin $(0, 0, 0)$.

(b) The surface is symmetric with respect to all three coordinate planes, all three coordinate axes, and the origin.

(c) For $k \neq 0$, the intersection with the plane $x = k$ is the hyperbola $y^2/9 - z^2/4 = k^2/16$. For $k \neq 0$, the intersection with plane $y = k$ is the ellipse $x^2/16 + z^2/4 = k^2/9$. For $k \neq 0$, the intersection with the plane $z = k$ is the hyperbola $y^2/9 - x^2/16 = k^2/4$.

(d) The yz trace consists of two intersecting straight lines, $y = \pm\frac{3}{2}z$. The xz trace consists just of the origin $(0, 0, 0)$. The xy trace consists of two intersecting straight lines, $y = \pm\frac{3}{4}x$.

(e) The surface is an elliptic cone.

(f) The graph is sketched in Figure 10.

Figure 10

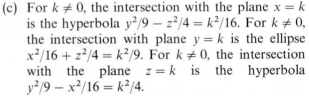

$$\frac{x^2}{16} - \frac{y^2}{9} + \frac{z^2}{4} = 0$$

4 $9y^2 - 25z^2 = x$

SOLUTION

The equation has the form $y^2/b^2 - z^2/c^2 = x$, where $b = \frac{1}{3}$ and $c = \frac{1}{5}$.

(a) The only intercept along any coordinate axis is the origin $(0,0,0)$.

(b) The squared variables are y and z, so the surface is symmetric with respect to the xy plane, the xz plane, and the x axis.

(c) For $k < 0$, the intersection with the plane $x = k$ is the hyperbola $25z^2 - 9y^2 = -k$ whose transverse axis is parallel to the z axis. For $k > 0$, the intersection with the plane $x = k$ is the hyperbola $9y^2 - 25z^2 = k$ whose transverse axis is parallel to the y axis. The intersection with the plane $y = k$ is the parabola $x = 9k^2 - 25z^2$, opening in the direction of the negative x axis. The intersection with the plane $z = k$ is the parabola $x = 9y^2 - 25k^2$, opening in the direction of the positive x axis.

(d) The yz trace is the pair of intersecting straight lines $y = \pm\frac{5}{3}z$. The xz trace is the parabola $x = -25z^2$. The xy trace is the parabola $x = 9y^2$.

(e) The surface is a hyperbolic paraboloid.

(f) The graph appears in Figure 11.

Figure 11

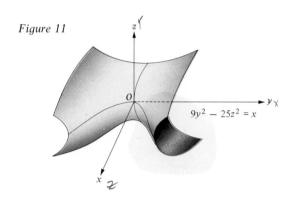

$9y^2 - 25z^2 = x$

Problem Set 9

In problems 1 through 8, identify the intersection of each quadric surface with the plane indicated.

1 $2x^2 + 3y^2 + z^2 = 6$; $x = 1$

2 $\dfrac{x^2}{9} - \dfrac{y^2}{4} + \dfrac{z^2}{25} = 1$; $z = 4$

3 $z^2 - \dfrac{y^2}{9} - \dfrac{x^2}{16} = 1$; $y = 2$

4 $3x^2 + 4y^2 = z$; $x = 2$

5 $25x^2 + 4y^2 - 100z^2 = 0$; $z = -1$

6 $3x^2 - 4z^2 + 5y - z - 2 = 0$; $z = 4$

7 $4x^2 - 16y^2 = z$; $x = \dfrac{3}{2}$

8 $\dfrac{x^2}{4} - \dfrac{y^2}{9} + \dfrac{z^2}{25} - x + y + xy = 4$; $x = 5$

In problems 9 through 24, (a) find the intercepts, (b) discuss the symmetry, (c) find the cross sections perpendicular to the coordinate axes, (d) find the traces, (e) identify the quadric surface, and (f) sketch a graph.

9 $x^2 + 3y^2 + 2z^2 = 6$

10 $144x^2 + 9y^2 + 16z^2 = 144$

11 $4x^2 - 9y^2 + 9z^2 = 36$

12 $x^2 + 2xy + y^2 = 1$

13 $x^2 - 4y^2 - 4z^2 = 4$

14 $z^2 = 1 - 2y + y^2$

15 $y^2 - 9x^2 - 9z^2 = 9$

16 $x^2 + y^2 + z^2 - 2y + 2z = 0$

17 $x^2 + 5y^2 - 8z^2 = 0$

18 $-x^2 - 25y^2 - 25z^2 = 25$

19 $3z = x^2 + y^2$

20 $4y = \dfrac{x^2}{9} + y^2$

21 $\dfrac{x^2}{16} - \dfrac{y^2}{9} = 3z$

22 $4y^2 - 9z^2 - 18x = 0$

23 $x^2 = y + z^2$

24 $\dfrac{x^2}{16} - \dfrac{y^2}{9} = 1$

25 Show that the graph of an equation is symmetric with respect to the xy plane provided that when z is replaced by $-z$ in the equation, an equivalent equation is obtained.

26 Find the conditions for symmetry of a graph with respect to (a) the yz plane, (b) the x axis, and (c) the y axis.

27 Find the condition for the symmetry of a graph with respect to the origin.

28 Discuss the symmetry of the graph of the equation $2xy + 3xz - 4yz = 24$.

29 Find all cross sections cut from the ellipsoid $x^2/a^2 + y^2/b^2 + z^2/c^2 = 1$ by (a) the planes $x = k$, (b) the planes $y = k$, and (c) the planes $z = k$.

30 Find all cross sections cut from the hyperboloid of one sheet $x^2/a^2 + y^2/b^2 - z^2/c^2 = 1$ by planes perpendicular to the coordinate axes.

31 Find all cross sections cut from the elliptic cone $x^2/a^2 + y^2/b^2 - z^2/c^2 = 0$ by planes perpendicular to the coordinate axes.

32 Find all cross sections cut from the elliptic paraboloid $x^2/a^2 + y^2/b^2 = z$ by the planes perpendicular to the coordinate axes.

33 Find all cross sections cut by planes $z = k$ from the hyperbolic paraboloid $y^2/b^2 - x^2/a^2 = z$. Discuss the cases $k < 0$, $k = 0$, and $k > 0$ separately.

34 Find all cross sections cut by planes perpendicular to the x or y axes from the hyperbolic paraboloid $y^2/b^2 - x^2/a^2 = z$.

35 Write an equation that describes the surface consisting of all points $P = (x, y, z)$ such that the distance from P to the point $(0, 0, -1)$ is the same as the distance from P to the plane $z = 1$. Identify this surface.

36 Prove that, if (x_0, y_0, z_0) is any point on a hyperboloid of one sheet, there are two straight lines in space, passing through (x_0, y_0, z_0), both of which lie on the hyperboloid.

37 Find an equation that describes the surface consisting of all points $P = (x, y, z)$ whose distance from the y axis is $\frac{2}{3}$ of its distance from P to the xz plane.

38 Prove that, if (x_0, y_0, z_0) is any point on a hyperbolic paraboloid, there are two straight lines in space, passing through (x_0, y_0, z_0), both of which lie on the hyperbolic paraboloid.

39 A central quadric surface $Ax^2 + By^2 + Cz^2 + K = 0$ contains the points $(3, -2, -1)$, $(0, 1, -3)$, and $(3, 0, 2)$. Find the equation of the surface and identify the surface.

40 A quadric surface $Ax^2 + By^2 + Cz = 0$ contains the points $(1, 0, 1)$ and $(0, 2, 1)$. Find the equation of the surface and identify the surface.

10   Cylindrical and Spherical Coordinates

In Chapter 11 we found that some problems in the plane were easier to formulate and to solve if polar coordinates, rather than cartesian coordinates, were used. Similarly, there are situations in which problems in three-dimensional space become more tractable if noncartesian coordinate systems are introduced. Two of the most important noncartesian coordinate systems in space, the *cylindrical* and the *spherical*, are discussed in the present section.

10.1 Cylindrical Coordinates

In Section 1 we obtained the cartesian coordinates of a point P in three-dimensional space by dropping a perpendicular $\overline{PQ}$ to the horizontal plane through the origin O and using the cartesian coordinates of Q in this plane together with the directed

Figure 1

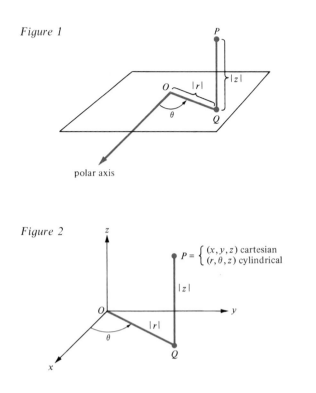

polar axis

Figure 2

distance $z = \pm |\overline{PQ}|$. The *cylindrical coordinate system* is quite similar, except that polar coordinates are used for the point Q in the horizontal plane (Figure 1). Thus, the *cylindrical coordinates* of the point P in Figure 1 are (r, θ, z).

Figure 2 shows the point P with respect to the cartesian coordinate system and with respect to the cylindrical coordinate system. If the polar coordinates of Q (with the x axis as the polar axis) are (r, θ), then the cartesian coordinates of Q are $x = r \cos \theta$ and $y = r \sin \theta$; hence, the cartesian coordinates of P are given by the equations

$$\begin{cases} x = r \cos \theta \\ y = r \sin \theta \\ z = z. \end{cases}$$

Since the point Q at the foot of the perpendicular from P to the xy plane has an unlimited number of different representations in the polar coordinate system, it follows that P has an unlimited number of different representations in the cylindrical coordinate system. For instance, if $P = (r, \theta, z)$, then also $P = (-r, \theta + \pi, z)$. In any case, if $P = (x, y, z)$ in cartesian coordinates, then the cylindrical coordinates (r, θ, z) of P must satisfy

$$r = \pm \sqrt{x^2 + y^2}$$

and, provided that $x \neq 0$,

$$\tan \theta = \frac{y}{x}.$$

If $x = 0$, then $\theta = \pi/2$ when $y > 0$, and $\theta = 3\pi/2$ when $y < 0$.

EXAMPLES 1 Find cartesian coordinates of the point P whose cylindrical coordinates are $(5, -\pi/3, 3)$ and plot the point P showing both coordinate systems.

SOLUTION
Here,

$$x = r \cos \theta = 5 \cos \left(-\frac{\pi}{3}\right) = \frac{5}{2},$$

$$y = 5 \sin \left(-\frac{\pi}{3}\right) = -\frac{5\sqrt{3}}{2}, \quad \text{and}$$

$$z = 3,$$

so that the point P has cartesian coordinates $(\frac{5}{2}, -5\sqrt{3}/2, 3)$ (Figure 3).

Figure 3

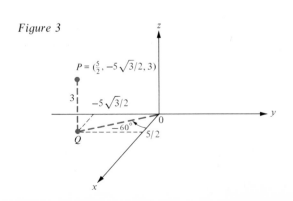

$P = (\frac{5}{2}, -5\sqrt{3}/2, 3)$

2 Find cylindrical coordinates of the point P whose cartesian coordinates are $(-2, 2\sqrt{3}, 4)$ and plot the point P showing both coordinate systems.

Figure 4

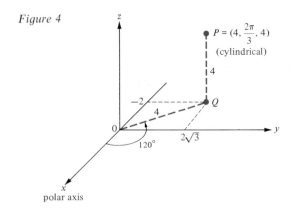

Figure 5

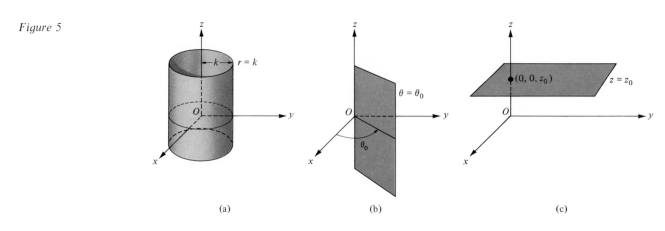

(a)   (b)   (c)

Figure 6

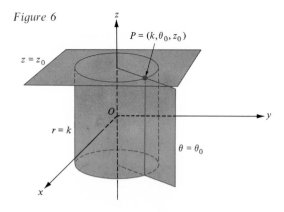

SOLUTION

Here, $r = \pm\sqrt{(-2)^2 + (2\sqrt{3})^2} = \pm\sqrt{16} = \pm 4$, and we take $r = 4$. Since $x = -2 < 0$ and $y = 2\sqrt{3} > 0$, it follows that θ is a second quadrant angle with $\tan\theta = \dfrac{y}{x} = -\dfrac{2\sqrt{3}}{2} = -\sqrt{3}$; hence, we take $\theta = 120° = 2\pi/3$ radians. Thus, the cylindrical coordinates of P are $(4, 2\pi/3, 4)$ (Figure 4).

In cylindrical coordinates, the graph of the equation $r = k$ is a circular cylinder of radius $|k|$ with the z axis as its central axis (Figure 5a). Likewise, the condition $\theta = \theta_0$ describes a plane through the z axis making an angle θ_0 with the positive x axis (Figure 5b), while the equation $z = z_0$ represents a horizontal plane intersecting the z axis at the point $(0, 0, z_0)$ (Figure 5c). The point P with cylindrical coordinates $P = (k, \theta_0, z_0)$ is a point at which the circular cylinder $r = k$, the plane $\theta = \theta_0$, and the plane $z = z_0$ intersect (Figure 6).

Since the equation of a circular cylinder is exceptionally simple in cylindrical coordinates, such coordinates are naturally adapted to the solutions of problems involving such cylinders. More generally, the equation of a surface of revolution about the z axis is usually simpler in cylindrical coordinates than in cartesian coordinates.

EXAMPLES **1** Write the equation in cylindrical coordinates of the right circular cylinder of radius 17 having the z axis as its central axis.

SOLUTION

The equation of the right circular cylinder is $r = 17$.

2 Find the equation in cylindrical coordinates for the paraboloid of revolution whose cartesian equation is $x^2 + y^2 = z$.

SOLUTION

Since $r^2 = x^2 + y^2$, the desired equation is $r^2 = z$.

3 Find the equation in cartesian coordinates of the surface whose equation in cylindrical coordinates is $z = 3r$, identify the surface, and sketch its graph.

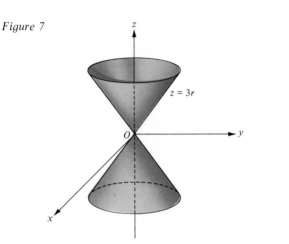

Figure 7

$z = 3r$

SOLUTION

Squaring both sides of the equation, we obtain $z^2 = 9r^2$, or $z^2 = 9(x^2 + y^2)$. The latter equation can be rewritten as $x^2 + y^2 - (z^2/9) = 0$, which represents an elliptic cone. In fact, since the coefficients of x^2 and y^2 are the same, the graph is a circular cone (Figure 7).

A curve in space can often be expressed parametrically by giving the cylindrical coordinates (r, θ, z) of a point P on the curve in terms of a parameter t. Thus, if

$$\begin{cases} r = f(t) \\ \theta = g(t) \\ z = h(t), \end{cases}$$

where f, g, and h are continuous functions, then the point $P = (r, \theta, z)$ traces out the curve as t varies. If f', g', and h' exist and are continuous, then from the equations

$$x = r \cos \theta, \qquad y = r \sin \theta, \qquad \text{and} \qquad z = z,$$

we obtain

$$\frac{dx}{dt} = \frac{dr}{dt} \cos \theta - r \sin \theta \frac{d\theta}{dt} \qquad \text{and} \qquad \frac{dy}{dt} = \frac{dr}{dt} \sin \theta + r \cos \theta \frac{d\theta}{dt}.$$

Thus,

$$\left(\frac{ds}{dt}\right)^2 = \left(\frac{dx}{dt}\right)^2 + \left(\frac{dy}{dt}\right)^2 + \left(\frac{dz}{dt}\right)^2$$

$$= \left(\frac{dr}{dt} \cos \theta - r \sin \theta \frac{d\theta}{dt}\right)^2 + \left(\frac{dr}{dt} \sin \theta + r \cos \theta \frac{d\theta}{dt}\right)^2 + \left(\frac{dz}{dt}\right)^2$$

$$= \left(\frac{dr}{dt}\right)^2 \cos^2 \theta - 2r \frac{dr}{dt} \frac{d\theta}{dt} \cos \theta \sin \theta + r^2 \sin^2 \theta \left(\frac{d\theta}{dt}\right)^2 + \left(\frac{dr}{dt}\right)^2 \sin^2 \theta$$

$$+ 2r \frac{dr}{dt} \frac{d\theta}{dt} \cos \theta \sin \theta + r^2 \cos^2 \theta \left(\frac{d\theta}{dt}\right)^2 + \left(\frac{dz}{dt}\right)^2$$

$$= \left(\frac{dr}{dt}\right)^2 (\cos^2 \theta + \sin^2 \theta) + r^2 \left(\frac{d\theta}{dt}\right)^2 (\cos^2 \theta + \sin^2 \theta) + \left(\frac{dz}{dt}\right)^2$$

$$= \left(\frac{dr}{dt}\right)^2 + r^2 \left(\frac{d\theta}{dt}\right)^2 + \left(\frac{dz}{dt}\right)^2.$$

It follows that the arc length of the curve between the point where the parameter has the value $t = a$ and the point where it has the value $t = b$ is given by

$$s = \int_a^b \sqrt{\left(\frac{dr}{dt}\right)^2 + r^2 \left(\frac{d\theta}{dt}\right)^2 + \left(\frac{dz}{dt}\right)^2} \, dt.$$

EXAMPLE   Find the arc length of the curve

$$\begin{cases} r = 5 \\ \theta = 2\pi t \\ z = 3t \end{cases}$$

between the point where $t = 0$ and the point where $t = 1$.

SOLUTION

Here $dr/dt = 0$, $d\theta/dt = 2\pi$, $dz/dt = 3$, and

$$s = \int_0^1 \sqrt{0^2 + 5^2(2\pi)^2 + 3^2}\, dt = \int_0^1 \sqrt{100\pi^2 + 9}\, dt$$

$$= \sqrt{100\pi^2 + 9} \int_0^1 dt = \sqrt{100\pi^2 + 9} \approx 31.56 \text{ units.}$$

10.2 Spherical Coordinates

In the *spherical coordinate system*, the angle θ has exactly the same meaning as it does in the cylindrical coordinate system. Thus, θ locates the point P on a plane containing the z axis and making the angle θ with the positive x axis (Figure 8). The distance between P and the origin O is denoted by the Greek letter ρ (called "rho"); thus, $\rho = |\overline{OP}|$. Finally, the angle from the positive z axis to the line segment $\overline{OP}$ is denoted by the Greek letter ϕ (called "phi"). The spherical coordinates of the point P are customarily written in the order (ρ, θ, ϕ) and they are usually chosen so that

$$\rho \geq 0, \qquad 0 \leq \theta < 2\pi, \qquad \text{and} \qquad 0 \leq \phi \leq \pi.$$

A point P with spherical coordinates $(\rho_0, \theta_0, \phi_0)$ is ρ_0 units from the origin; hence, it is located on a sphere of radius ρ_0 with center at O (Figure 9). The z axis intersects this sphere at the "north and south poles" and the xy plane intersects it in the "equator." Semicircles cut by half-planes passing through the north and south poles are called *meridians* and the meridian that intersects the positive x axis is called the *prime meridian*. The angles θ_0 and ϕ_0 locate P_0 on the surface of the sphere as shown in Figure 9. The spherical coordinate θ_0, which is called the *longitude* of P_0, measures the angle between the prime meridian and the meridian passing through P_0. (On the surface of the Earth, the meridian passing through Greenwich, England, is designated as the prime meridian.)

Circles cut on the surface $\rho = \rho_0$ by planes perpendicular to the z axis (hence, parallel to the equatorial plane) are called *parallels* and the angle measured from the equator to a parallel is called the *latitude* of that parallel (or of any point on the parallel) (Figure 10). Notice that the point P_0 with spherical coordinates $(\rho_0, \theta_0, \phi_0)$ has latitude $(\pi/2) - \phi_0$; in other words, the angle ϕ_0 is the complement of the latitude of P_0. For this reason, ϕ_0 is called the

Figure 8

Figure 9

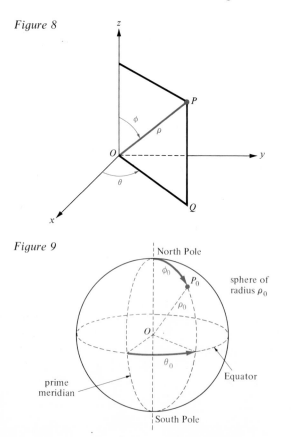

colatitude of P_0. For instance, the latitude of Boston, Massachusetts, is approximately 42.4°, so its colatitude is approximately 47.6°.

In Figure 11, suppose that the point P has cartesian coordinates (x, y, z) and spherical coordinates (ρ, θ, ϕ). The line segment $\overline{PQ}$ is parallel to the z axis, so that angle OPQ is equal to ϕ. Since OQP is a right triangle,

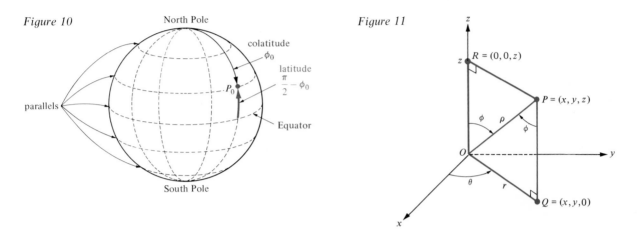

Figure 10

Figure 11

$$\sin \phi = \frac{|\overline{OQ}|}{|\overline{OP}|} = \frac{r}{\rho}; \quad \text{hence,} \quad r = \rho \sin \phi.$$

Therefore, we can rewrite the equations

$$x = r \cos \theta \quad \text{and} \quad y = r \sin \theta$$

as

$$x = \rho \sin \phi \cos \theta \quad \text{and} \quad y = \rho \sin \phi \sin \theta.$$

Triangle ORP in Figure 11 is a right triangle, so that

$$\cos \phi = \frac{|\overline{OR}|}{|\overline{OP}|} = \frac{z}{\rho}; \quad \text{hence,} \quad z = \rho \cos \phi.$$

The equations just derived,

$$\begin{cases} x = \rho \sin \phi \cos \theta \\ y = \rho \sin \phi \sin \theta \\ z = \rho \cos \phi, \end{cases}$$

give the cartesian coordinates of the point P whose spherical coordinates are (ρ, θ, ϕ). The argument just given applies if P lies above the xy plane and (Problem 38) the same equations are effective even when P lies on or below the xy plane. In some applications of spherical coordinates, the conditions $\rho \geq 0$, $0 \leq \theta < 2\pi$, and $0 \leq \phi \leq \pi$ are dropped and (ρ, θ, ϕ) is understood to locate P in spherical coordinates if x, y, and z are given by the equations above.

Since ρ is the distance between $P = (x, y, z)$ and the origin O,

$$\rho^2 = x^2 + y^2 + z^2.$$

Also, if $x \neq 0$, then

$$\frac{y}{x} = \frac{\sin \theta}{\cos \theta} = \tan \theta.$$

Finally, if $\rho \neq 0$, then

$$\frac{z}{\rho} = \cos \phi.$$

Therefore, if we choose the spherical coordinates (ρ, θ, ϕ) so that $\rho \geq 0$, $0 \leq \theta < 2\pi$, and $0 \leq \phi \leq \pi$, we have the formulas

$$\rho = \sqrt{x^2 + y^2 + z^2}, \qquad \tan \theta = \frac{y}{x} \quad \text{for } x \neq 0,$$

and

$$\phi = \cos^{-1} \frac{z}{\rho} = \cos^{-1} \frac{z}{\sqrt{x^2 + y^2 + z^2}} \quad \text{for } \rho \neq 0.$$

EXAMPLES 1 Find cartesian coordinates of the point P whose spherical coordinates are $(2, \pi/3, 2\pi/3)$ and plot the point P showing both coordinate systems.

Figure 12

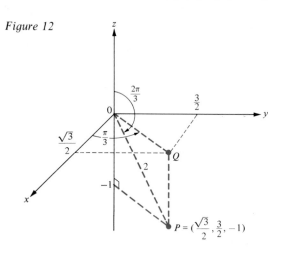

SOLUTION

Here we have $\rho = 2$, $\theta = \pi/3$, and $\phi = 2\pi/3$, so that

$$x = \rho \sin \phi \cos \theta = 2 \sin \frac{2\pi}{3} \cos \frac{\pi}{3}$$

$$= 2 \left(\frac{\sqrt{3}}{2} \right) \left(\frac{1}{2} \right) = \frac{\sqrt{3}}{2},$$

$$y = \rho \sin \phi \sin \theta = 2 \sin \frac{2\pi}{3} \sin \frac{\pi}{3}$$

$$= 2 \left(\frac{\sqrt{3}}{2} \right) \left(\frac{\sqrt{3}}{2} \right) = \frac{3}{2}, \quad \text{and}$$

$$z = \rho \cos \phi = 2 \cos \frac{2\pi}{3} = 2 \left(-\frac{1}{2} \right) = -1.$$

Thus, the cartesian coordinates of P are $(\sqrt{3}/2, \frac{3}{2}, -1)$ (Figure 12).

2 Find spherical coordinates of the point P whose cartesian coordinates are $(\sqrt{3}, -1, 2)$ and plot the point P showing both coordinate systems.

SOLUTION

Since $\tan \theta = y/x$, we have

$$\theta = \tan^{-1} \frac{-1}{\sqrt{3}} = -\frac{\pi}{6} = -30°.$$

If we require $0 \leq \theta < 2\pi$, we can take

$$\theta = 2\pi + (-\pi/6) = 11\pi/6 = 330°.$$

Here,

$$\rho = \sqrt{(\sqrt{3})^2 + (-1)^2 + 2^2} = \sqrt{8} = 2\sqrt{2},$$

and

$$\phi = \cos^{-1} \frac{z}{\rho} = \cos^{-1} \frac{2}{2\sqrt{2}} = \cos^{-1} \frac{\sqrt{2}}{2} = \frac{\pi}{4} = 45°.$$

Therefore, the spherical coordinates of P are $(\sqrt{8}, 11\pi/6, \pi/4)$ (Figure 13).

Figure 13

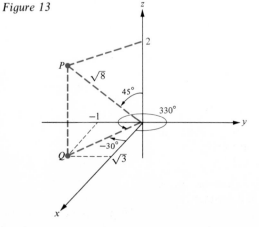

3 Convert the equation $\phi = \pi/4$ in spherical coordinates into (a) cartesian coordinates and (b) cylindrical coordinates.

SOLUTION

(a) $x = \rho \sin \phi \cos \theta = \left(\sin \dfrac{\pi}{4}\right)\rho \cos \theta = \dfrac{\sqrt{2}}{2}\rho \cos \theta,$

$y = \rho \sin \phi \sin \theta = \left(\sin \dfrac{\pi}{4}\right)\rho \sin \theta = \dfrac{\sqrt{2}}{2}\rho \sin \theta,$ and

$z = \rho \cos \phi = \rho \cos \dfrac{\pi}{4} = \dfrac{\sqrt{2}}{2}\rho.$

Since $z = (\sqrt{2}/2)\rho$, we can rewrite the first two equations as $x = z \cos \theta$ and $y = z \sin \theta$. From the latter two equations, we have

$$x^2 + y^2 = z^2 \cos^2 \theta + z^2 \sin^2 \theta = z^2(\cos^2 \theta + \sin^2 \theta) = z^2.$$

Thus, the equation in cartesian coordinates is $x^2 + y^2 = z^2$; hence the graph of the equation is a right circular cone.

(b) In cylindrical coordinates $r^2 = x^2 + y^2$, so the equation obtained in part (a) can be written as $r^2 = z^2$.

4 Rewrite the equation of the paraboloid $x^2 + y^2 = z$ in spherical coordinates.

SOLUTION

Here we have

$$x^2 + y^2 = (\rho \sin \phi \cos \theta)^2 + (\rho \sin \phi \sin \theta)^2$$
$$= \rho^2 \sin^2 \phi(\cos^2 \theta + \sin^2 \theta) = \rho^2 \sin^2 \phi.$$

Thus, since $z = \rho \cos \phi$, the equation $x^2 + y^2 = z$ becomes

$$\rho^2 \sin^2 \phi = \rho \cos \phi.$$

We lose no points on the graph by canceling ρ to obtain $\rho \sin^2 \phi = \cos \phi$ since the latter equation is still satisfied when $\rho = 0$ and $\phi = \pi/2$. Therefore, in spherical coordinates, the paraboloid has the equation $\rho \sin^2 \phi = \cos \phi$.

Problem Set 10

In problems 1 through 4, find the cartesian coordinates of the point whose cylindrical coordinates are given and plot the point.

1 $(4, \pi/3, 1)$ **2** $(3, \pi/2, 4)$ **3** $(5, \pi/6, -2)$ **4** $(2, 2, 2)$

In problems 5 through 8, find cylindrical coordinates (r, θ, z) with $r \geq 0$ and $0 \leq \theta < 2\pi$ for the point whose cartesian coordinates are given and plot the point.

5 $(4, 0, 1)$ **6** $(-2\sqrt{3}, -6, 0)$ **7** $(-3\sqrt{3}, 3, 6)$ **8** $(1, 1, -1)$

In problems 9 through 12, find the cartesian coordinates of the point whose spherical coordinates are given and plot each point.

9 $(2, \pi/6, \pi/3)$ **10** $(7, \pi/2, \pi)$ **11** $(12, 5\pi/6, 2\pi/3)$ **12** (π, π, π)

In problems 13 through 16, find spherical coordinates (ρ, θ, ϕ) with $\rho \geq 0$, $0 \leq \theta < 2\pi$, and $0 \leq \phi \leq \pi$ for the points whose cartesian coordinates are given and plot each point.

13 $(0, -1, 0)$ **14** $(0, 0, 5)$ **15** $(1, 2, -3)$ **16** $(0, 0, 0)$

In problems 17 through 26, convert each equation into an equivalent equation (a) in cylindrical coordinates and (b) in spherical coordinates. Sketch a graph of the surface.

17 $z = 2(x^2 + y^2)$ **18** $x^2 + y^2 - 4x = 0$ **19** $x = 2$ **20** $y = \sqrt{3}x$

21 $x^2 + y^2 = 5z^2$ **22** $x^2 + y^2 + (z-1)^2 = 1$ **23** $x^2 + y^2 = 25$

24 $x + 2y = 0$ **25** $x^2 + y^2 - z^2 = 1$ **26** $x^2 + y^2 + z^2 - 6z = 0$

In problems 27 through 32, each equation is written in cylindrical coordinates. Convert each equation into an equivalent equation in (a) cartesian coordinates and (b) spherical coordinates. Sketch a graph of the surface.

27 $z = r^2$ **28** $\theta = \dfrac{3\pi}{4}$ **29** $\dfrac{r^2}{9} + \dfrac{z^2}{4} = 1$

30 $z = \frac{1}{2}r^2 \sin 2\theta$ **31** $r = 4 \cos \theta$ **32** $r \cos \theta + 3r \sin \theta + 2z = 6$

In problems 33 through 37, each equation is written in spherical coordinates. Express each equation as an equivalent equation (a) in cartesian coordinates and (b) in cylindrical coordinates. Sketch a graph of the surface.

33 $\rho = 2$ **34** $\theta = \pi/3$ **35** $\rho \sin \phi = 3$

36 $\rho = 2 \cos \phi$ **37** $\phi = \pi/3$

38 Show that the equations given on page 805 for converting from spherical to cartesian coordinates are correct even when the point P lies on or below the xy plane.

39 Convert $\rho = \sin 2\phi$ to cylindrical coordinates.

40 Consider the curve in space whose equation is given parametrically in spherical coordinates by

$$\begin{cases} \rho = f(t) \\ \theta = g(t) \\ \phi = h(t), \end{cases}$$

where f', g', and h' exist and are continuous.

(a) Prove that $\left(\dfrac{dx}{dt}\right)^2 + \left(\dfrac{dy}{dt}\right)^2 + \left(\dfrac{dz}{dt}\right)^2 = \left(\dfrac{d\rho}{dt}\right)^2 + \rho^2 \sin^2 \phi \left(\dfrac{d\theta}{dt}\right)^2 + \rho^2 \left(\dfrac{d\phi}{dt}\right)^2.$

(b) Show that the arc length of the curve between the point where $t = a$ and the point where $t = b$ is given by

$$s = \int_a^b \sqrt{\left(\dfrac{d\rho}{dt}\right)^2 + \rho^2 \sin^2 \phi \left(\dfrac{d\theta}{dt}\right)^2 + \rho^2 \left(\dfrac{d\phi}{dt}\right)^2}\; dt.$$

41 Find the arc length of the curve whose equation is given parametrically in cylindrical coordinates by

$$\begin{cases} r = \dfrac{\sqrt{2}}{2} t \\ \theta = \sqrt{2}\, t \\ z = \dfrac{\sqrt{2}}{2} t \end{cases}$$

between $t = 0$ and $t = \sqrt{2}\,\pi$.

42 A cartesian coordinate system is established with the origin at the center of the Earth, the positive x axis passing through the point where the equator intersects the prime meridian, and the positive z axis passing through the North Pole. Taking the radius of the Earth to be 3959 miles, find:
(a) The cartesian coordinates of New York if the latitude of New York is 40.7° north of the equator and the longitude of New York is 74.02° west of the prime meridian.
(b) The cartesian coordinates of San Francisco if the latitude of San Francisco is 37.81° north of the equator and the longitude of San Francisco is 122.4° west of the prime meridian.

(c) The angle between the position vector of New York and the position vector of San Francisco.

(d) The great circle distance between New York and San Francisco.

Review Problem Set

In problems 1 through 6, for the given point P, find another point Q which is symmetric to P with respect to the given plane or axis.

1 $P = (2, -1, -3)$; yz plane

2 $P = (5, 6, 3)$; x axis

3 $P = (-3, 2, -1)$; xy plane

4 $P = (-1, -2, 3)$; z axis

5 $P = (3, -1, -5)$; y axis

6 $P = (1, 2, 5)$; xz plane

7 Find the distance from the point $P = (3, -1, 6)$ to (a) the origin, (b) the x axis, (c) the xy plane, (d) the point $(2, -3, 7)$, and (e) the point Q which is symmetric to P with respect to the origin.

8 Use the distance formula to show that the triangle with vertices $P = (1, 3, 3)$, $Q = (2, 2, 1)$, and $R = (3, 4, 2)$ is equilateral. Also, find the coordinates of the point where the medians of the triangle intersect.

9 Use vectors to determine whether the points $P = (-5, -10, 9)$, $Q = (-1, -5, 5)$, and $R = (11, 10, -9)$ are collinear (that is, lie on the same straight line).

10 Use vectors to determine whether the quadrilateral with vertices $P = (3, 2, 5)$, $Q = (1, 1, 1)$, $R = (4, 0, 3)$, and $S = (6, 1, 7)$ is a parallelogram.

11 A vector $\bar{R}$ makes an acute angle θ with the x axis, an angle $\pi/4$ with the y axis, and an angle $\pi/3$ with the z axis. Find θ.

12 If $P = (1, 2, 3)$ and $Q = (2, 5, 7)$, find (a) the scalar components of the vector $\bar{A} = \overrightarrow{PQ}$ and (b) the direction cosines of $\bar{A}$.

In problems 13 through 30, evaluate each of the following using

$$\bar{A} = i - 2j + 3\bar{k}, \qquad \bar{B} = 3i + 2j + \bar{k}, \qquad \bar{C} = 2i + 3j,$$
$$\bar{D} = -3i - 5j + 6\bar{k}, \qquad \bar{E} = i + 3j - 2\bar{k}, \qquad \bar{F} = 3i + 5j + 6\bar{k}.$$

13 $\bar{A} - 3\bar{B} + \bar{C}$

14 $|2\bar{C} - \bar{F}|$

15 $|\bar{A} - 3\bar{B} + \bar{C}|$

16 $(\bar{E} \cdot \bar{F})\bar{C} - (\bar{C} \cdot \bar{F})\bar{A}$

17 $(3\bar{A}) \cdot (\bar{F} + \bar{E})$

18 $\bar{A} \times \bar{D}$

19 $\bar{E} \cdot (\bar{D} \times \bar{E})$

20 $\bar{B} \cdot (\bar{C} \times \bar{E})$

21 $\bar{A} \times \bar{B}$

22 $\bar{A} \times (\bar{B} \times \bar{D})$

23 $(\bar{A} + \bar{B}) \times (\bar{A} - \bar{B})$

24 $(\bar{A} \times \bar{B}) \times \bar{D}$

25 $\bar{A} \cdot (\bar{B} \times \bar{C})$

26 $(3\bar{A} - \bar{D}) \times (\bar{E} - 2\bar{C})$

27 The angle between $\bar{C}$ and $\bar{D}$.

28 The scalar component of $\bar{C}$ in the direction of $\bar{F}$.

29 The volume of the parallelepiped whose edges are the vectors $\bar{A}$, $\bar{B}$, and $\bar{C}$.

30 $(\bar{A} \times \bar{B}) \cdot (\bar{C} \times \bar{D})$

31 Do the vectors $\bar{A} = i - 2j + 3\bar{k}$, $\bar{B} = 3i + 2j + \bar{k}$, and $\bar{C} = 2i + 3j$ form a right-handed triple $\bar{A}$, $\bar{B}$, $\bar{C}$? Why?

32 Find the area of the triangle whose vertices are $A = (2, 0, -1)$, $B = (5, 3, 3)$, and $C = (-1, 1, 2)$.

In problems 33 through 40, find the equation in scalar form of the plane satisfying the conditions given.

33 Containing the point $P = (1, 2, 3)$ and perpendicular to the radius vector $\overrightarrow{OP}$ at the point P.

34 Containing the three points $(1, 7, 2)$, $(3, 5, 1)$, and $(6, 3, -1)$.

35 Containing the point $P = (2, 0, -1)$ and perpendicular to the line joining the points $Q = (3, 4, 4)$ and $R = (-1, 2, 1)$.

36 At a distance of 4 units from the origin and perpendicular to the line joining the points $P = (-2, 3, 1)$ and $Q = (-5, 1, -5)$.

37 Containing the point $(4, 3, 1)$ and parallel to the plane $x + 3z = 8$.

38 Containing the points $(1, 3, 1)$ and $(4, 6, -2)$ and perpendicular to the plane $x + y - z = 3$.

39 Containing the point $(2, -1, 3)$ and perpendicular to the line $x = 3t + 5$, $y = 8t - 4$, $z = -7t + 16$.

40 Containing the point $(1, 2, 10)$ and the line $\dfrac{x - 1}{5} = \dfrac{y - 1}{2} = \dfrac{z - 6}{-7}$.

In problems 41 through 48, find the equations of the line (a) in scalar parametric form and (b) in symmetric form for which the given conditions hold.

41 Containing the points $(5, 6, -4)$ and $(2, -1, 1)$.

42 Containing the point $(-1, 2, 3)$ and perpendicular to the plane $-2x + 3y - z = -1$.

43 Containing the point $(2, -3, -2)$ and parallel to the line $\dfrac{x - 1}{3} = \dfrac{y + 7}{-2} = \dfrac{z}{7}$.

44 Containing the origin, perpendicular to the line of intersection of the two planes $x = y - 5$ and $z = 2y - 3$, and meeting the line of intersection of the two planes $y = 2x + 1$ and $z = x + 2$.

45 Formed by the intersection of the two planes $2x + y - z = 1$ and $x - y + 3z = 10$.

46 Containing the point $(3, 6, 4)$, parallel to the plane $x - 3y + 5z - 6 = 0$, and intersecting the z axis.

47 Containing the point $(2, 1, 4)$ and perpendicular to both the x and the y axes.

48 Containing the point $(2, -1, 4)$ and perpendicular to the lines $x = 5t + 1$, $y = -3t$, $z = t - 2$ and $x = 2t - 1$, $y = t + 1$, $z = -t$.

49 Find the distance from the point $(2, -3, 4)$ to the plane $2x - 2y + z = 5$.

50 Find the distance between the two parallel planes $x - 2y + 2z = 5$ and $x - 2y + 2z = 17$.

51 Show that the lines $\dfrac{x - 100}{99} = \dfrac{y + 94}{-97} = \dfrac{z - 51}{50}$ and $\dfrac{x - 102}{-101} = \dfrac{y + 96}{99} = \dfrac{z - 52}{-51}$ intersect and find the equation of the plane containing the two lines.

52 Given the equation $\dfrac{x - x_0}{a} = \dfrac{y - y_0}{b} = \dfrac{z - z_0}{c}$ of a straight line, explain how you would go about finding several points on this line.

53 Find the distance between the two lines $x = y + 3 = \dfrac{z - 2}{2}$ and $\dfrac{x - 3}{-1} = \dfrac{y + 1}{2} = z - 1$.

54 Let P be a variable point on a first straight line in space and let Q be a variable point on a second straight line in space. If $\overline{M}_1$ and $\overline{M}_2$ are direction vectors for the first and second lines, respectively, show that the quantity $(\overline{M}_1 \times \overline{M}_2) \cdot \overrightarrow{PQ}$ remains constant as P and Q vary.

55 Show that the straight lines $\dfrac{x - x_0}{a_0} = \dfrac{y - y_0}{b_0} = \dfrac{z - z_0}{c_0}$ and $\dfrac{x - x_1}{a_1} = \dfrac{y - y_1}{b_1} = \dfrac{z - z_1}{c_1}$ meet in space if and only if

$$\begin{vmatrix} a_0 & b_0 & c_0 \\ a_1 & b_1 & c_1 \\ x_1 - x_0 & y_1 - y_0 & z_1 - z_0 \end{vmatrix} = 0.$$

56 Let $\bar{R}_1$ be the position vector for a point P_1 in space

(a) Show that $\bar{R}_1 - \dfrac{\bar{N} \cdot (\bar{R}_1 - \bar{R}_0)}{|\bar{N}|^2} \bar{N}$ is the position vector of the point in the plane $\bar{N} \cdot (\bar{R} - \bar{R}_0) = 0$ that is closest to P_1.

(b) Show that $\bar{R}_0 + \dfrac{\bar{M} \cdot (\bar{R}_1 - \bar{R}_0)}{|\bar{M}|^2} \bar{M}$ is the position vector of the point on the line $\bar{M} \times (\bar{R} - \bar{R}_0) = \bar{0}$ that is closest to P_1.

57 Suppose that $\bar{A}$ is a known nonzero vector, that $\bar{X}$ is an unknown vector, but that the scalar $a = \bar{A} \cdot \bar{X}$ and vector $\bar{B} = \bar{A} \times \bar{X}$ are both known. Show that $\bar{X}$ is then determined by the equation

$$\bar{X} = \frac{a}{|\bar{A}|^2} \bar{A} - \frac{\bar{A} \times \bar{B}}{|\bar{A}|^2}.$$

(*Hint*: Start by expanding $\bar{A} \times \bar{B}$.)

58 If $\bar{A}$ is a nonzero vector and $\bar{A} \times \bar{X} = \bar{A} \times \bar{Y}$, can we "cancel" and conclude that $\bar{X} = \bar{Y}$? Explain.

In problems 59 and 60, find $\bar{F}'(t)$, $\bar{F}''(t)$, and $\dfrac{d}{dt}|\bar{F}(t)|$.

59 $\bar{F}(t) = e^{2t^2}\bar{i} - e^{-2t^2}\bar{j} + t\bar{k}$

60 $\bar{F}(t) = \tan\left(t + \dfrac{\pi}{2}\right)\bar{i} + (\tan t)\bar{j} + \tan\left(t - \dfrac{\pi}{2}\right)\bar{k}$

In problems 61 and 62, assume that a particle P moves according to the given equation of motion. Find (a) the velocity vector $\bar{V}$, (b) the acceleration vector $\bar{A}$, (c) the speed v, (d) the unit tangent vector $\bar{T}$, (e) the unit normal vector $\bar{N}$, (f) the unit binormal vector $\bar{B}$, and (g) the distance traveled by the particle along its path from the instant $t = 0$ to the instant $t = 1$.

61 $\bar{R} = (3\cos 2\pi t)\bar{i} + (3\sin 2\pi t)\bar{j} + 2t\bar{k}$

62 $\bar{R} = (e^{4t}\cos t)\bar{i} + (e^{4t}\sin t)\bar{j} + e^{4t}\bar{k}$

In problems 63 through 66, find (a) $\bar{V}$, (b) v, (c) $\bar{A}$, (d) $\bar{T}$, (e) $\bar{V} \times \bar{A}$, (f) κ, (g) $\bar{N}$, (h) $\bar{B}$, (i) $d\bar{A}/dt$, and (j) τ for each curve in xyz space.

63 $\bar{R} = t\bar{i} + t^2\bar{j} + t^3\bar{k}$

64 $\bar{R} = at\bar{i} + bt^2\bar{j} + ct^3\bar{k}$, where a, b, and c are constants.

65 $\bar{R} = (\sin t \cos t)\bar{i} + (\sin^2 t)\bar{j} + (\cos t)\bar{k}$

66 $\bar{R} = (t\sin t)\bar{i} + t(\cos t)\bar{j} + t\bar{k}$

67 A particle moves according to an equation of motion $\bar{R} = \bar{F}(t)$. Expand and simplify the following expressions:

(a) $\dfrac{d}{dt}(\bar{R} \cdot \bar{V})$

(b) $\dfrac{d}{dt}(\bar{R} \times \bar{V})$

(c) $\dfrac{d}{dt}(\bar{V} \cdot \bar{A})$

(d) $\dfrac{d}{dt}(\bar{V} \times \bar{A})$

(e) $\dfrac{d\bar{T}}{ds} \cdot \dfrac{d\bar{B}}{ds}$

(f) $\left(\dfrac{d\bar{R}}{ds} \times \dfrac{d^2\bar{R}}{ds^2}\right) \cdot \dfrac{d^3\bar{R}}{ds^3}$

68 (a) Sketch the curve $z = \cos(\pi y)$ in the yz plane.

(b) Sketch the cylinder with generators parallel to the x axis having the curve in part (a) as its base curve.

(c) Write the equation of the cylinder in part (b).

69 Find the equation of the surface of revolution generated by rotating the curve $z = 2(x - 3)^2$ in the xz plane about the x axis.

70 Find the equation of the *torus* (doughnut-shaped surface) that results when the circle $y^2 + (x - a)^2 = r^2$ $(a > r > 0)$ in the xy plane is revolved about the y axis. Sketch the surface.

71 Find the generating curve in the xz plane and the axis of revolution of the surface of revolution $y^2 + z^2 = e^{-2x}$. Sketch the surface.

72 Find the equation of the surface of revolution generated by revolving the cardioid $r = 1 - \cos\theta$ in the xy plane about the x axis.

In problems 73 through 78, identify and sketch a graph for each quadric surface.

73 $x^2 + 4y^2 + 4z^2 = 16$ **74** $x^2 + 4y^2 + 4z^2 = 16x$ **75** $x^2 + z^2 = 4 + y$

76 $9x^2 + z^2 = y$ **77** $y^2 - z^2 + 9x^2 = 1$ **78** $y^2 - z^2 + 9x^2 = 0$

79 A quadric surface has an equation of the form $Ax^2 + By^2 + Cz = 0$. Identify the surface if it contains the points $(3, 5, 8)$ and $(4, -2, -6)$.

80 A quadric surface has an equation of the form $Ax^2 + By^2 + Cz^2 = 1$ and contains the points $(2, -1, 1)$, $(-3, 0, 0)$, and $(1, -1, -2)$. Identify the surface.

81 Describe and sketch the surface whose equation in cylindrical coordinates is (a) $r = 2$; (b) $\theta = \pi/6$; (c) $r = \sin\theta$.

82 Describe and sketch the surface whose equation in spherical coordinates is $\rho = 5\cos\phi$.

83 Find the equation in cylindrical coordinates of the surface obtained by revolving the curve $z = f(x)$ in the xz plane about the z axis.

84 Find the arc length of the curve whose parametric equations in spherical coordinates are $\rho = t^2$, $\theta = \pi/6$, $\phi = t$ as t varies over the interval $[0, \sqrt{5}]$.

85 Convert the equation $\rho^2 \sin 2\phi = 4$ in spherical coordinates into cartesian coordinates and sketch the graph of the surface.

86 Find the great circle distance between Honolulu, with latitude $21.31°$N, longitude $157.87°$W, and Chicago, with latitude $41.83°$N, longitude $87.62°$W. Assume that the radius of the Earth is 3959 miles.

16 FUNCTIONS OF SEVERAL VARIABLES AND PARTIAL DERIVATIVES

In the preceding chapters we have dealt exclusively with functions of a single real variable; however, there are many practical situations in which functions depend on *several* variables. For instance, the frequency of a tuned circuit depends on its capacitance, its inductance, and its resistance; the pressure of a gas depends on its temperature and its volume; the demand for a commodity may depend not only on its price but also on the prices of related commodities, on income level, and on time; a person's income tax liability depends not only on income but also on several itemized deductions and on the number of dependents; and so forth.

In this chapter we study functions of more than one variable, we see that the concepts of limits and continuity are applicable to such functions, and we investigate their "partial" derivatives. Chain rules are also developed for functions of several variables. The chapter also includes a study of directional derivatives, tangent planes and normal lines to surfaces, and maxima and minima of functions of several variables.

1 Functions of Several Variables

A right circular cylinder, closed at the top and bottom, with base radius r and height h, has total surface area S given by

$$S = 2\pi rh + 2\pi r^2.$$

Here we say that the (dependent) variable S is a function of the two (independent) variables r and h, and we write

$$S = f(r, h).$$

For example, if $r = 11$ cm and $h = 5$ cm, then

$$S = f(11, 5) = 2\pi(11)(5) + 2\pi(11)^2 = 352\pi \text{ cm}^2.$$

Proceeding somewhat more formally, we make the following definition.

DEFINITION 1   **Function of Two Variables**

A *real-valued function f of two real variables* is a rule that assigns a unique real number z to each ordered pair (x, y) of real numbers in a certain set D, called the *domain* of the function f. If the rule f assigns the number z to the ordered pair (x, y) in D, then we write $z = f(x, y)$.

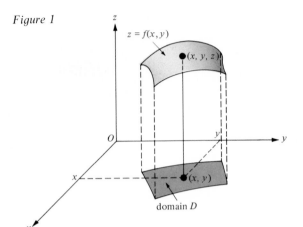

Figure 1

In the equation $z = f(x, y)$, we call z the *dependent* variable and we refer to x and y as the *independent* variables. The set of all possible values of z that can be obtained by applying the rule f to ordered pairs (x, y) in D is called the *range* of the function f.

We define the *graph* of a function f of two variables to be the set of all points (x, y, z) in cartesian three-dimensional space such that (x, y) belongs to the domain D of f and $z = f(x, y)$. The domain D can be pictured as a set of points in the xy plane and the graph of f as a surface whose perpendicular projection on the xy plane is D (Figure 1). [In Figure 1, the point shown as (x, y) in D is really $(x, y, 0)$; however, the third coordinate has been purposely omitted.] Notice that as the point (x, y) varies in D, the corresponding point $(x, y, z) = (x, y, f(x, y))$ varies over the surface.

EXAMPLES   Sketch the graph of the given function of two variables.

1 The function f whose domain D is the circular disk consisting of all points (x, y) with $x^2 + y^2 \leq 1$ and which is defined by the equation

$$f(x, y) = \sqrt{1 - x^2 - y^2}.$$

SOLUTION
A point (x, y, z) belongs to the graph of f if and only if $z = f(x, y)$; that is, $z = \sqrt{1 - x^2 - y^2}$. The condition $z = \sqrt{1 - x^2 - y^2}$ is equivalent to the two conditions $z \geq 0$ and $x^2 + y^2 + z^2 = 1$. Thus, the graph consists of the portion of the sphere $x^2 + y^2 + z^2 = 1$ lying on or above the xy plane (Figure 2).

2 The function f whose domain D is the entire xy plane and which is defined by the equation $f(x, y) = 1 - x - (y/2)$.

SOLUTION
The point (x, y, z) belongs to the graph of f if and only if $z = 1 - x - (y/2)$; that is, $2x + y + 2z = 2$. Therefore, the graph of f is a plane with intercepts $(1, 0, 0)$, $(0, 2, 0)$, and $(0, 0, 1)$. A portion of this plane, showing its traces with the xy, xz, and yz planes, appears in Figure 3.

Although it requires more persistence to sketch graphs of functions of two variables than it does to sketch graphs of functions of single variables, the basic idea is much the same and the techniques studied in Section 9 of Chapter 15— cross sections, intercepts, traces, symmetries, and so forth—can be quite helpful.

Functions of three or more variables are defined by an obvious extension of Definition 1 as follows.

Figure 2

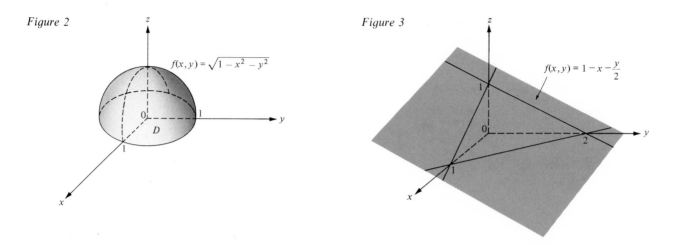

$f(x, y) = \sqrt{1 - x^2 - y^2}$

Figure 3

$f(x, y) = 1 - x - \dfrac{y}{2}$

DEFINITION 2   **Function of Several Variables**

A *real-valued function* f *of* n *real variables* is a rule that assigns a unique real number w to each ordered n-tuple $(x_1, x_2, x_3, \ldots, x_n)$ of real numbers in a certain set D, called the *domain* of the function f. If the rule f assigns the number w to the ordered n-tuple $(x_1, x_2, x_3, \ldots, x_n)$, then we write $w = f(x_1, x_2, x_3, \ldots, x_n)$.

In the equation $w = f(x_1, x_2, x_3, \ldots, x_n)$, we call w the *dependent* variable and we refer to x_1, x_2, x_3, $\ldots$, x_n as the *independent* variables. The set of all possible values of w that can be obtained by applying the rule f to ordered n-tuples $(x_1, x_2, x_3, \ldots, x_n)$ in D is called the *range* of the function f. In case $n = 2$, $w = f(x_1, x_2)$ is usually written $z = f(x, y)$ as in Definition 1. In case $n = 3$, $w = f(x_1, x_2, x_3)$ is usually written $w = f(x, y, z)$.

EXAMPLES **1** If f is defined by $f(x, y) = 3x + 2y$ for all values of x and y, find (a) $f(1, 2)$ and (b) $f(\sin t, \cos t)$.

SOLUTION
(a) $f(1, 2) = (3)(1) + (2)(2) = 7$.
(b) $f(\sin t, \cos t) = 3 \sin t + 2 \cos t$.

2 If $g(x, y, z) = \dfrac{xy}{x^2 + y^2 - z}$ for all values of x, y, and z except those that make the denominator equal to zero, find (a) $g(2, 3, 7)$ and (b) $g(\sin t, \cos t, 0)$.

SOLUTION

(a) $g(2, 3, 7) = \dfrac{(2)(3)}{2^2 + 3^2 - 7} = 1$.

(b) $g(\sin t, \cos t, 0) = \dfrac{\sin t \cos t}{\sin^2 t + \cos^2 t - 0} = \sin t \cos t$.

3 If $f(x_1, x_2, x_3, \ldots, x_n) = x_1^2 + x_2^2 + x_3^2 + \cdots + x_n^2$ for all integral values of x_1, x_2, x_3, $\ldots$, x_n, find $f(1, 2, 3, \ldots, n)$.

SOLUTION
Using the formula for the sum of successive squares (Chapter 6, Section 1), we have

$$f(1, 2, 3, \ldots, n) = 1^2 + 2^2 + 3^2 + \cdots + n^2 = \sum_{k=1}^{n} k^2 = \frac{n(n+1)(2n+1)}{6}.$$

If a function f of several variables is defined by an equation or a formula, then (unless we make a stipulation to the contrary) the domain of f is understood to be the set of all n-tuples of independent variables for which the equation or formula makes sense.

EXAMPLES 1 Find and sketch the domain of $f(x, y) = \dfrac{\sqrt{4 - x^2 - y^2}}{y}$.

SOLUTION

The domain of f consists of all ordered pairs (x, y) for which $x^2 + y^2 \le 4$ and $y \neq 0$. This is the set of all points not on the x axis which are either on the circle $x^2 + y^2 = 4$ or in the interior region bounded by the circle (Figure 4).

Figure 4

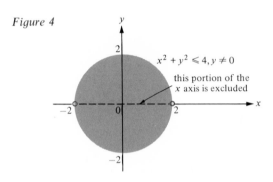

2 Find the domain of $f(x, y, z) = \dfrac{\sin^{-1} z}{x + y}$.

SOLUTION

Since $\sin^{-1} z$ is defined only when $|z| \le 1$, the domain of f consists of all ordered triples (x, y, z) such that $x + y \neq 0$ and $|z| \le 1$.

1.1 Scalar Fields

We have seen that a function f of two independent variables can be thought of in terms of its graph, which is a surface in xyz space. There is a second way of picturing such a function, which, for some purposes, is even more suggestive than using its graph; namely, the function f is regarded as a *scalar field* on a two-dimensional domain D as follows: The domain D of f is visualized as a set of points (x, y) lying in a certain region in the xy plane and each point (x, y) in this region is assigned a corresponding scalar $f(x, y)$ by the function f (Figure 5). The scalar value $f(x, y)$ corresponding to the point (x, y) in D is shown in Figure 5 on a "flag" attached to the point. As the point (x, y) is moved around in the region D, the "flag" moves with it and the number $f(x, y)$ on the "flag" changes.

Figure 5

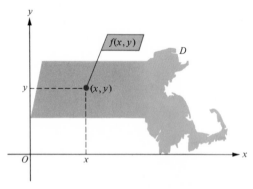

The scalar $f(x, y)$ assigned to the point (x, y) might represent, for instance, the temperature at (x, y), or the atmospheric pressure at (x, y), or the wind speed at (x, y), or the intensity of the magnetic field at (x, y), and so forth.

EXAMPLE   In Figure 6, suppose that $f(x, y)$ gives the temperature in degrees F at the point with cartesian coordinates (x, y), where x and y are measured in miles. Let $f(x, y) = 80 - (x/20) - (y/25)$.
(a) Find the temperature at the point $(60, 75)$.
(b) Find the equation of the curve along which the temperature has a constant value of $70°F$.
(c) Sketch the curve in part (b).

Figure 6

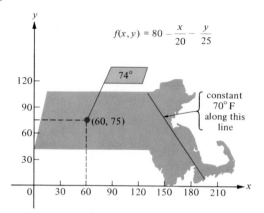

$$f(x,y) = 80 - \frac{x}{20} - \frac{y}{25}$$

SOLUTION

(a) $f(60, 75) = 80 - \dfrac{60}{20} - \dfrac{75}{25} = 74°\text{F}.$

(b) The equation is $f(x, y) = 70$; that is,

$$80 - \frac{x}{20} - \frac{y}{25} = 70,$$

or $5x + 4y = 1000$.

(c) The curve $5x + 4y = 1000$ is a straight line (Figure 6).

A curve along which a scalar field has constant value (such as the curve along which the temperature field in the previous example maintains the constant value 70°F) is called a *level curve* of the field or of the function f that describes the field. The equation of the level curve along which the function f assumes the constant value k is

$$f(x, y) = k.$$

The level curves for various scalar fields are customarily given special names depending on the nature of the field—*isotherms* for the level curves of a temperature field, *equipotential lines* for the level curves of an electric potential field, and so forth.

Suppose that the function f gives the height $z = f(x, y)$ of a certain surface S above the xy plane at the point (x, y). (Then S is the graph of the function f.) The intersection of the surface S with a horizontal plane $z = k$ produces a curve C consisting of all points on the surface which are k units above the xy plane (Figure 7). The perpendicular projection of the curve C on the xy plane gives a level curve for the function f. Such a level curve, whose equation in the xy plane is

$$f(x, y) = k,$$

is called a *contour line* for the surface S. By plotting a number of different contour lines, each labeled with its own value of k, we obtain a *contour map* of the surface S (Figure 8). Such a contour map enables us to visualize the surface as if we were

Figure 7

Figure 8

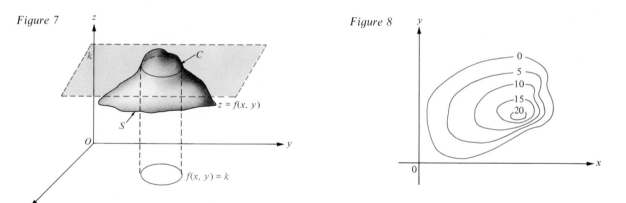

looking down upon it from above and seeing its intersections with horizontal cutting planes at various heights. If these heights are made to differ by equal amounts, then a crowding of successive contour lines indicates a relatively steep part of the surface.

EXAMPLE   Let S be the surface $z = x^2 - y^2 + 20$ for $x \geq 0$. Plot the contour lines for this surface corresponding to $z = 0$, $z = 10$, $z = 20$, $z = 30$, and $z = 40$.

SOLUTION
For $z = 0$ we obtain $0 = x^2 - y^2 + 20$, or $y^2 - x^2 = 20$, the equation of a hyperbola with a vertical transverse axis. Since $x \geq 0$, we get only the portion of this hyperbola lying in the first and fourth quadrants as the contour line corresponding to $z = 0$. For $z = 10$, we obtain $10 = x^2 - y^2 + 20$, or $y^2 - x^2 = 10$, another hyperbola. For $z = 20$, the equation is $20 = x^2 - y^2 + 20$, or $x = \pm y$, two straight lines meeting at the origin. Continuing in this way, we obtain the desired contour map (Figure 9).

Figure 9

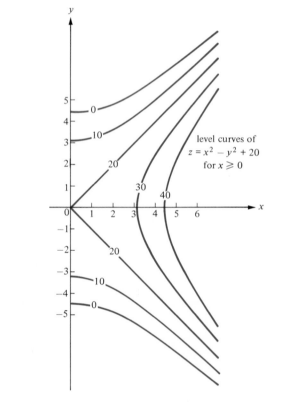

level curves of $z = x^2 - y^2 + 20$ for $x \geq 0$

Problem Set 1

In problems 1 through 6, find the domain of each function of two variables and sketch the graph of the function.

1 $f(x, y) = x + y$

2 $f(x, y) = 2x - 3y + 6$

3 $f(x, y) = \sqrt{4 - x^2 - y^2}$

4 $f(x, y) = x^2 + y^2$

5 $f(x, y) = 3 + \sqrt{9 - x^2 - y^2}$

6 $f(x, y) = \sqrt{1 + x^2 + y^2}$

In problems 7 through 20, evaluate each expression, using the functions f, g, and h defined by $f(x, y) = 5x^2 + 7xy$, $g(x, y) = \sqrt{xy}$, and $h(x, y, z) = \dfrac{2xy + z}{x^2 + y^2 - z^2}$.

7 $f(3, -4)$

8 $f(2, 0)g(2, 0)$

9 $g(k, k)$

10 $g(a^2, b^2)$

11 $h(1, 2, 3)$

12 $h(x, z, y)$

13 $f(\sqrt{a}, b)$

14 $g(\sin \theta, 2 \cos \theta)$

15 $h(\sin t, \cos t, 0)$

16 $\dfrac{f(x, y)}{[g(x, y)]^2}$

17 $f(x, y) + g(x, y)$

18 $f(x + h, y) - f(x, y)$

19 $h(x, y, 0)$

20 $\dfrac{g(x, y + k) - g(x, y)}{k}$

In problems 21 through 24, (a) specify the domain of the function and (b) evaluate $f(x, y)$ for the given values of x and y.

21 $f(x, y) = \sqrt{x + y - 4}$; $x = -4$, $y = 16$.

22 $f(x, y) = x\sqrt{1 - x^2 - y^2}$; $x = y = \frac{1}{4}$.

23 $f(x, y) = \dfrac{4x^2 - y^2}{2x - y}$; $x = 4$, $y = -1$.

24 $f(x, y) = \sin^{-1}(2x + y)$; $x = 1$, $y = -1$.

25 If $f(x_1, x_2, x_3, \ldots, x_n) = x_1 + x_2 + x_3 + \cdots + x_n$, evaluate and simplify $f(1, 2, 3, \ldots, n)$.

26 Find a function f such that $f(x, y)$ gives the surface area of a right circular cone with a closed base if x is the radius of the base and y is the height of the cone.

In problems 27 through 32, plot a contour map of the graph of $z = f(x, y)$ showing the contour lines corresponding to the given values of z.

27 $f(x, y) = 3x + 2y - 1$; $z = -1$, $z = 0$, $z = 1$, $z = 2$.

28 $f(x, y) = \sqrt{9 - x^2 - y^2}$; $z = 0$, $z = 1$, $z = 2$, $z = 3$.

29 $f(x, y) = \sqrt{1 - (x^2/4) - (y^2/9)}$; $z = 0$, $z = 1$, $z = \frac{1}{2}$.

30 $f(x, y) = xy$; $z = 0$, $z = 1$, $z = 2$, $z = 3$.

31 $f(x, y) = x + y^2$; $z = -1$, $z = 0$, $z = 1$, $z = 2$.

32 $f(x, y) = \dfrac{2x}{x^2 + y^2}$; $z = 0$, $z = 1$, $z = 2$, $z = 3$.

33 What does it mean if the isotherms of a temperature field tend to crowd together near a certain point $P_0 = (x_0, y_0)$?

34 A person is driving at 50 miles per hour from left to right along the straight line $y = 75$ in Figure 6. How rapidly does this person feel the temperature dropping, in degrees F per hour, if the temperature at the point (x, y) is given by $f(x, y) = 80 - (x/20) - (y/25)$?

35 Given a map showing the isotherms of a temperature field, explain how one would plot a path, starting at a point P_0, along which the temperature would increase most rapidly.

36 Look up the following words in a dictionary and explain how each of them refers to the level curves of a certain type of scalar field:
(a) isobar
(b) isocheim
(c) isoclinal line
(d) isodynamic line

2   Limits and Continuity

The idea of a limit extends easily to functions of two or more variables. For instance, to assert that $f(x, y)$ approaches the limit L as (x, y) approaches (x_0, y_0) means that the number $f(x, y)$ can be made to come as close to the number L as we wish by choosing the point (x, y) to be sufficiently close to the point (x_0, y_0), provided that $(x, y) \neq (x_0, y_0)$. The notation is

$$\lim_{(x, y) \to (x_0, y_0)} f(x, y) = L \quad \text{or} \quad \lim_{\substack{x \to x_0 \\ y \to y_0}} f(x, y) = L.$$

As a specific example, notice that

$$\lim_{(x, y) \to (0, 0)} \frac{1}{\sqrt{4 - x^2 - y^2}} = \frac{1}{\sqrt{4}} = \frac{1}{2},$$

since the quantity $1/\sqrt{4 - x^2 - y^2}$ comes closer and closer to $\frac{1}{2}$ as the point (x, y) comes closer and closer to $(0, 0)$. In Section 2.1 we give a formal definition of a limit of a function of two variables; however, for now, we prefer to proceed informally.

By arguing in much the same way as in Section 7 of Chapter 1, it can be shown that all the properties of limits of functions of single variables extend to functions of several variables; for instance, the limit of a sum, difference, product, or quotient is the sum, difference, product, or quotient of the limits, respectively, provided that these limits exist and that zeros do not appear in denominators. Also, we have

$$\lim_{\substack{x \to x_0 \\ y \to y_0}} f(x) = \lim_{x \to x_0} f(x) \quad \text{and} \quad \lim_{\substack{x \to x_0 \\ y \to y_0}} g(y) = \lim_{y \to y_0} g(y),$$

provided that $\lim\limits_{x \to x_0} f(x)$ and $\lim\limits_{y \to y_0} g(y)$ exist.

EXAMPLES   Evaluate the limit.

1 $\lim\limits_{\substack{x \to -1 \\ y \to 2}} \left(5x^2 y + 2xy - \dfrac{3y^2}{x + y} \right)$

SOLUTION
Applying the limit properties for sums and products, we have

$$\lim_{\substack{x \to -1 \\ y \to 2}} \left(5x^2 y + 2xy - \frac{3y^2}{x + y} \right)$$

$$= \lim_{\substack{x \to -1 \\ y \to 2}} 5x^2 y + \lim_{\substack{x \to -1 \\ y \to 2}} 2xy - \lim_{\substack{x \to -1 \\ y \to 2}} \frac{3y^2}{x + y}$$

$$= 5 \lim_{\substack{x \to -1 \\ y \to 2}} x^2 \lim_{\substack{x \to -1 \\ y \to 2}} y + 2 \lim_{\substack{x \to -1 \\ y \to 2}} x \lim_{\substack{x \to -1 \\ y \to 2}} y - \frac{3 \lim\limits_{\substack{x \to -1 \\ y \to 2}} y^2}{\lim\limits_{\substack{x \to -1 \\ y \to 2}} x + \lim\limits_{\substack{x \to -1 \\ y \to 2}} y}$$

$$= 5(-1)^2(2) + 2(-1)(2) - \frac{3(2)^2}{-1 + 2} = -6.$$

2 $\lim\limits_{(x,y) \to (0,0)} \left[e^{\sin (5x^2 + y)} + \cos (3xy) \right]$

SOLUTION

$$\lim_{(x,y) \to (0,0)} \left[e^{\sin (5x^2 + y)} + \cos (3xy) \right] = e^{\sin [5(0)^2 + 0]} + \cos [3(0)(0)]$$
$$= e^0 + \cos 0 = 2.$$

In Chapter 1 we observed that $\lim\limits_{x \to a} f(x)$ exists if and only if both one-sided limits, $\lim\limits_{x \to a^+} f(x)$ and $\lim\limits_{x \to a^-} f(x)$, exist and are the same. In dealing with a limit of a function f of two variables, say $\lim\limits_{(x,y) \to (a,b)} f(x, y)$, we can allow the point (x, y) to approach the point (a, b) not only from the right or from the left, but also from above, below, or indeed, any direction (Figure 1a). We can even allow (x, y) to approach (a, b) along a curved path (Figure 1b). To say that $\lim\limits_{(x,y) \to (a,b)} f(x, y) = L$ is to require that, as (x, y) approaches (a, b) *along any path whatsoever*, $f(x, y)$ approaches the same limit L. Thus, a convenient way to show that a particular limit $\lim\limits_{(x,y)\to(a,b)} f(x,y)$ *does not exist* is to show that $f(x,y)$ can be made to approach two different limits as (x,y) approaches (a,b) along two different paths.

Figure 1

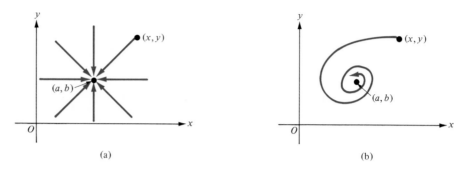

(a)   (b)

EXAMPLES 1 Let f be the function defined by $f(x,y) = \dfrac{2x^2y}{3x^2 + 3y^2}$.

(a) Evaluate the limit of $f(x,y)$ as (x,y) approaches $(0,0)$ along each of the following paths: (i) the x axis, (ii) the y axis, (iii) the line $y = x$, (iv) the parabola $y = x^2$.

(b) Does $\lim\limits_{(x,y)\to(0,0)} f(x,y)$ exist? If so, what is its value?

SOLUTION

(a) (i) On the x axis, $y = 0$ and $f(x,y) = f(x,0) = \dfrac{2x^2(0)}{3x^2 + 0} = 0$ for $x \neq 0$.

Therefore, $\lim\limits_{x\to 0} f(x,0) = \lim\limits_{x\to 0} 0 = 0$.

(ii) On the y axis, $x = 0$ and $f(x,y) = f(0,y) = \dfrac{2(0)^2y}{0 + 3y^2} = 0$ for $y \neq 0$.

Therefore, $\lim\limits_{y\to 0} f(0,y) = \lim\limits_{y\to 0} 0 = 0$.

(iii) On the line $y = x$, $f(x,y) = f(x,x) = \dfrac{2x^2x}{3x^2 + 3x^2} = \dfrac{2x^3}{6x^2} = \dfrac{x}{3}$ for $x \neq 0$.

Therefore, $\lim\limits_{x\to 0} f(x,x) = \lim\limits_{x\to 0} \dfrac{x}{3} = 0$.

(iv) On the parabola $y = x^2$, $f(x,y) = f(x,x^2) = \dfrac{2x^2x^2}{3x^2 + 3x^4} = \dfrac{2x^2}{3 + 3x^2}$ for

$x \neq 0$. Therefore, $\lim\limits_{x\to 0} f(x,x^2) = \lim\limits_{x\to 0} \dfrac{2x^2}{3 + 3x^2} = 0$.

(b) Along all the paths in part (a), the limit is the same, 0. Although there might be some *other path* along which the limit is not 0, the evidence leads us to *suspect* that the limit is 0. Our suspicion is confirmed as follows: For $x^2 + y^2 \neq 0$,

$x^2 \le x^2 + y^2$, so $x^2|y| \le (x^2 + y^2)|y|$; hence, $\dfrac{x^2|y|}{x^2 + y^2} \le |y|$. It follows that

$$|f(x,y)| = \left| \frac{2x^2y}{3x^2 + 3y^2} \right| = \left(\frac{2}{3}\right) \frac{x^2|y|}{x^2 + y^2} \le \frac{2}{3}|y|.$$

Therefore, as (x,y) approaches $(0,0)$, $f(x,y)$ must approach 0, since $|f(x,y)|$ is no larger than $\frac{2}{3}|y|$, which approaches 0. It follows that

$$\lim_{(x,y)\to(0,0)} \frac{2x^2y}{3x^2 + 3y^2} = 0.$$

2 Follow the directions in Example 1 for the function f defined by

$$f(x, y) = \frac{x^2 - y^2}{x^2 + y^2}.$$

SOLUTION

(a) (i)   On the x axis, $y = 0$ and $f(x, y) = f(x, 0) = \dfrac{x^2 - 0}{x^2 + 0} = 1$ for $x \neq 0$. Therefore,

$$\lim_{x \to 0} f(x, 0) = \lim_{x \to 0} 1 = 1.$$

(ii)   On the y axis, $x = 0$ and $f(x, y) = f(0, y) = \dfrac{0 - y^2}{0 + y^2} = -1$ for $y \neq 0$.

Therefore, $\lim\limits_{y \to 0} f(0, y) = \lim\limits_{y \to 0} (-1) = -1.$

(iii)   On the line $y = x$, $f(x, y) = f(x, x) = \dfrac{x^2 - x^2}{x^2 + x^2} = 0$ for $x \neq 0$. Therefore,

$$\lim_{x \to 0} f(x, x) = \lim_{x \to 0} 0 = 0.$$

(iv)   On the parabola $y = x^2$, $f(x, y) = f(x, x^2) = \dfrac{x^2 - x^4}{x^2 + x^4} = \dfrac{1 - x^2}{1 + x^2}$ for $x \neq 0$.

Therefore, $\lim\limits_{x \to 0} f(x, x^2) = \lim\limits_{x \to 0} \dfrac{1 - x^2}{1 + x^2} = 1.$

(b) Since the limits in (i) and (ii) are different, the limit $\lim\limits_{(x,y) \to (0,0)} f(x, y)$ does not exist.

The following example shows that $f(x, y)$ can approach the same limit as (x, y) approaches (a, b) along *every* straight line through (a, b), and yet $\lim\limits_{(x,y) \to (a,b)} f(x, y)$ can fail to exist.

EXAMPLE   Let f be the function defined by

$$f(x, y) = \begin{cases} \left(x + \dfrac{1}{x}\right) y^2 & \text{if } x \neq 0 \\[2mm] 0 & \text{if } x = 0. \end{cases}$$

(a) Calculate the limit of $f(x, y)$ as (x, y) approaches $(0, 0)$ along the line $y = mx$.
(b) Calculate the limit of $f(x, y)$ as (x, y) approaches $(0, 0)$ along the parabola $x = y^2$.
(c) Does $\lim\limits_{(x,y) \to (0,0)} f(x, y)$ exist?

SOLUTION

(a) On the straight line $y = mx$, $f(x, y) = f(x, mx) = \left(x + \dfrac{1}{x}\right)(mx)^2$ for $x \neq 0$, so

that $\lim\limits_{x \to 0} f(x, mx) = \lim\limits_{x \to 0} (x + 1/x)(mx)^2 = \lim\limits_{x \to 0} (m^2 x^3 + m^2 x) = 0.$

(b) Along the parabola $x = y^2$, $f(x, y) = f(y^2, y) = \left(y^2 + \dfrac{1}{y^2}\right) y^2$ for $y \neq 0$, so that

$$\lim_{y \to 0} f(y^2, y) = \lim_{y \to 0} \left(y^2 + \frac{1}{y^2}\right) y^2 = \lim_{y \to 0} (y^4 + 1) = 1.$$

(c) Since the limits in parts (a) and (b) are different, $\lim\limits_{(x,y) \to (0,0)} f(x, y)$ does not exist.

The definitions and the properties of limits are easily extended to functions of three or more variables. For instance, if the number $f(x, y, z)$ can be made to come as close to the number L as we wish by taking the point (x, y, z) sufficiently close to the point (x_0, y_0, z_0), but different from this point, then we write

$$\lim_{(x,y,z)\to(x_0,y_0,z_0)} f(x,y,z) = L \quad \text{or} \quad \lim_{\substack{x\to x_0 \\ y\to y_0 \\ z\to z_0}} f(x,y,z) = L.$$

2.1 The Formal Definition of Limit for Functions of Two Variables

The formal definition of the limit of a function of two variables is patterned on the definition in Section 1.1 of Chapter 1 of the limit of a function of a single variable.

DEFINITION 1

Limit of a Function of Two Variables

Let f be a function of two variables and let (x_0, y_0) be a point in the xy plane. Assume that there exists a circular disk with center at (x_0, y_0) and with positive radius such that every point within the disk, except possibly for the center (x_0, y_0), belongs to the domain of f. Then we say that *the limit as (x, y) approaches (x_0, y_0) of $f(x, y)$ is the number L*, and we write

$$\lim_{(x,y)\to(x_0,y_0)} f(x,y) = L \quad \left(\text{or } \lim_{\substack{x\to x_0 \\ y\to y_0}} f(x,y) = L \right),$$

provided that, for each positive number ε, there is a positive number δ such that $|f(x,y) - L| < \varepsilon$ holds whenever $(x,y) \neq (x_0, y_0)$ and the distance between (x, y) and (x_0, y_0) is less than δ.

The condition given in this definition can be written as follows: For every $\varepsilon > 0$, there exists $\delta > 0$ such that

$$0 < (x - x_0)^2 + (y - y_0)^2 < \delta^2 \quad \text{implies that} \quad |f(x,y) - L| < \varepsilon.$$

EXAMPLE   Show that $\lim\limits_{\substack{x\to 1 \\ y\to 2}} (3x + 2y) = 7$ by direct application of Definition 1.

SOLUTION

Let $\varepsilon > 0$ be given. We must find $\delta > 0$ such that $|3x + 2y - 7| < \varepsilon$ holds whenever $0 < (x - 1)^2 + (y - 2)^2 < \delta^2$. Now,

$$|3x + 2y - 7| = |3x - 3 + 2y - 4| \leq |3x - 3| + |2y - 4|$$
$$\leq |3(x - 1)| + |2(y - 2)| \leq 3|x - 1| + 2|y - 2|;$$

hence, if $3|x - 1| < \varepsilon/2$ and $2|y - 2| < \varepsilon/2$, then

$$|3x + 2y - 7| \leq 3|x - 1| + 2|y - 2| < \frac{\varepsilon}{2} + \frac{\varepsilon}{2} = \varepsilon.$$

The condition $3|x - 1| < \varepsilon/2$ is equivalent to $9(x - 1)^2 < \varepsilon^2/4$, or to $(x - 1)^2 < \varepsilon^2/36$, while the condition $2|y - 2| < \varepsilon/2$ is equivalent to $4(y - 2)^2 < \varepsilon^2/4$, or to $(y - 2)^2 < \varepsilon^2/16$. Therefore, if $(x - 1)^2 < \varepsilon^2/36$ and $(y - 2)^2 < \varepsilon^2/16$, we shall have $|3x + 2y - 7| < \varepsilon$. Thus, choose $\delta = \varepsilon/6$ and note that if

$$0 < (x - 1)^2 + (y - 2)^2 < \delta^2 = \varepsilon^2/36,$$

then

$$(x - 1)^2 \leq (x - 1)^2 + (y - 2)^2 < \frac{\varepsilon^2}{36} \quad \text{and} \quad (y - 2)^2 \leq (x - 1)^2 + (y - 2)^2 < \frac{\varepsilon^2}{36} < \frac{\varepsilon^2}{16};$$

hence, $(x - 1)^2 < \varepsilon^2/36$ and $(y - 2)^2 < \varepsilon^2/16$, so that $|3x + 2y - 7| < \varepsilon$.

2.2 Continuity

The definition of continuity for functions of a single variable can easily be extended to functions of several variables. For example, we make the following definition for functions of two variables.

DEFINITION 2

Continuity of a Function of Two Variables

Suppose that f is a function of two variables and that the point (x_0, y_0) is the center of a circular disk of positive radius contained in the domain of f. We say that f is *continuous* at (x_0, y_0) if

(i) $\lim\limits_{(x,y) \to (x_0,y_0)} f(x, y)$ exists and

(ii) $\lim\limits_{(x,y) \to (x_0,y_0)} f(x, y) = f(x_0, y_0)$.

EXAMPLES   Decide whether or not the given function is continuous at the indicated point.

1 $f(x, y) = 3x^2 + 2xy$ at $(-1, 3)$.

SOLUTION
Since, by the properties of limits,

$$\lim\limits_{(x,y) \to (-1,3)} (3x^2 + 2xy) = 3(-1)^2 + 2(-1)(3) = -3 = f(-1, 3),$$

it follows that f is continuous at $(-1, 3)$.

2 $f(x, y) = \begin{cases} \dfrac{4xy}{x^2 + y^2} & \text{if } (x, y) \neq (0,0) \\ 0 & \text{if } (x, y) = (0,0) \end{cases}$ at $(0,0)$.

SOLUTION
Along the x axis, we have

$$f(x, y) = f(x, 0) = \begin{cases} 0 & \text{if } x \neq 0 \\ 0 & \text{if } x = 0, \end{cases} \quad \text{so that } \lim\limits_{x \to 0} f(x, 0) = 0.$$

Along the line $y = x$, we have

$$f(x, y) = f(x, x) = \begin{cases} 2 & \text{if } x \neq 0 \\ 0 & \text{if } x = 0, \end{cases} \quad \text{so that } \lim\limits_{x \to 0} f(x, x) = 2.$$

Thus, $\lim\limits_{(x,y) \to (0,0)} f(x, y)$ does not exist, and so f is not continuous at $(0,0)$.

Figure 2

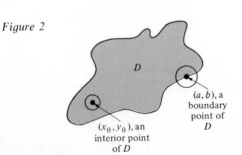

(a, b), a boundary point of D

(x_0, y_0), an interior point of D

Let D be a set of points in the xy plane. A point (x_0, y_0) is called an *interior point* of D if there is a circular disk of positive radius with (x_0, y_0) as its center, which is contained in D (Figure 2). On the other hand, a point (a, b) is called a *boundary point* of D if every circular disk of positive radius with (a, b) as its center contains at least one point that belongs to D and at least one point that does not belong to D (Figure 2). There is no requirement that a boundary point (a, b) of D belong to the set D. A set D is said to be *open* if it contains none of its own boundary points, while a set D is said to be *closed* if it contains all of its own boundary points. (This terminology is suggested by the ideas of open and closed intervals.)

Notice that Definition 2 applies only to the continuity of a function at an interior point of its domain. A modified definition of continuity, which we do not set forth here, is required for continuity of a function at a boundary point of its domain. Naturally, a function is said to be *continuous* if it is continuous at every point in its domain.

In order to show that a set D of points in the plane is open, it is enough to show that every point in D is an interior point of D (Problem 38).

EXAMPLE   Show that the domain of the function $f(x, y) = \dfrac{xy}{x - y}$ is an open set, illustrate this domain with a sketch, and show that f is continuous.

Figure 3

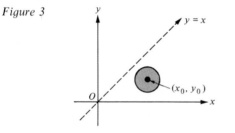

SOLUTION
The domain of f consists of all points in the xy plane except those lying on the line $y = x$ (Figure 3). Suppose that (x_0, y_0) is a point in the domain D of f, so that (x_0, y_0) is not on the line $y = x$. If d is the distance from (x_0, y_0) to the line $y = x$, then any circular disk of radius r with $0 < r < d$ is contained in D; hence, (x_0, y_0) lies in the interior of D. It follows that the domain D is open. Since, for $x_0 \neq y_0$,

$$\lim_{(x,y)\to(x_0,y_0)} \frac{xy}{x - y} = \frac{x_0 y_0}{x_0 - y_0} = f(x_0, y_0),$$

it follows that f is continuous at every point (x_0, y_0) of its domain. Therefore, f is a continuous function.

Functions of two variables have many of the same properties with regard to continuity as do functions of single variables. Among these are the following.

Continuity Properties for Functions of Two Variables

Suppose that (x_0, y_0) is an interior point of the domains of the functions f and g of two variables and assume that f and g are continuous at (x_0, y_0). Then:

1 $h(x, y) = f(x, y) + g(x, y)$ is continuous at (x_0, y_0).
2 $k(x, y) = f(x, y) - g(x, y)$ is continuous at (x_0, y_0).
3 $p(x, y) = f(x, y)g(x, y)$ is continuous at (x_0, y_0).

4 If $g(x_0, y_0) \neq 0$, then $q(x, y) = \dfrac{f(x, y)}{g(x, y)}$ is continuous at (x_0, y_0).

5 If w is a function of a single variable that is defined and continuous at the number $f(x_0, y_0)$ and if (x_0, y_0) is an interior point of the domain of $v(x, y) = w[f(x, y)]$, then v is continuous at (x_0, y_0).

EXAMPLE   What are the interior points of the domain of the function u defined by $u(x, y) = \sqrt{1 - x^2 - y^2} \ln (x + y)$? Is u continuous at such points? Illustrate with a sketch of the domain of u.

SOLUTION
The expression $\sqrt{1 - x^2 - y^2}$ is defined only for $1 - x^2 - y^2 \geq 0$, that is, for

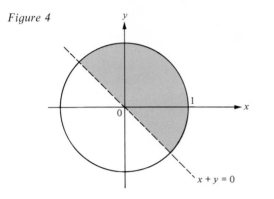

Figure 4

$x^2 + y^2 \leq 1$. The condition $x^2 + y^2 \leq 1$ holds exactly when (x, y) is on or inside the circle $x^2 + y^2 = 1$ of radius 1 with center at $(0, 0)$. The expression $\ln (x + y)$ is defined only for $x + y > 0$, that is, only when the point (x, y) lies above the straight line whose equation is $x + y = 0$. The domain of u is therefore the shaded region shown in Figure 4 inside or on the circle $x^2 + y^2 = 1$ but above the line $x + y = 0$. The interior points of this domain are just those points not lying on the rim of the circle (nor on the line $x + y = 0$). Since the square root function is continuous, the function f defined by $f(x, y) = \sqrt{1 - x^2 - y^2}$ is continuous inside the circle $x^2 + y^2 = 1$, by Property 5 above. Similarly, the function g defined by $g(x, y) = \ln (x + y)$ is continuous above the line $x + y = 0$. By Property 3, the function $u(x, y) = f(x, y)g(x, y)$ is continuous at all interior points of its domain.

The definitions and properties of continuous functions are easily extended to functions of three or more variables. Naturally, every polynomial function in several variables is continuous. Also, a ratio of such polynomial functions—that is, a *rational* function of several variables—has an open domain and is continuous at every point of its domain. For instance, the function f defined by

$$f(x, y, z) = \frac{5x^{17}y^{12}z^7 - 7xy^5z + x^2y^2z - x + 12y + 1984}{25x^3y^3z^2 - 37x + 15y - z + 33}$$

is continuous at every point (x, y, z) for which the denominator is not zero. Roughly speaking, any function built up in a "reasonable way" from continuous functions is continuous at every interior point of its domain.

Problem Set 2

In problems 1 through 8, evaluate each limit.

1 $\lim\limits_{(x,y)\to(1,-2)} (5x^2 + 3xy)$

2 $\lim\limits_{\substack{x\to\pi/4 \\ y\to\pi}} (\sin 2x + \sin 2y)$

3 $\lim\limits_{(x,y)\to(0,0)} (e^{\cos (3x+y)} + \sec 5xy)$

4 $\lim\limits_{(x,y)\to(0,0)} \dfrac{5}{\ln (x + 3y + e)}$

5 $\lim\limits_{\substack{x\to 0 \\ y\to 0}} \dfrac{\cos x + \cos y}{e^{-x} + 3e^y}$

6 $\lim\limits_{(x,y)\to(-1,2)} [\![7x + \tfrac{1}{3}y^3]\!]$

7 $\lim\limits_{\substack{x\to -1 \\ y\to 2 \\ z\to 0}} \dfrac{xyz - x + y - z}{x^2 + y^2 + z^2 - 4}$

8 $\lim\limits_{(x,y,z)\to(0,0,0)} z \sin \left(\dfrac{1}{\sqrt{x^2 + y^2 + z^2}}\right)$

In problems 9 through 14, (a) evaluate the limit of $f(x, y)$ as (x, y) approaches $(0, 0)$ along each of the indicated paths (i), (ii), (iii), and (iv), and (b) determine $\lim\limits_{(x,y)\to(0,0)} f(x, y)$, if it exists.

9 $f(x, y) = \dfrac{5xy^2}{x^2 + y^2}$ as (x, y) approaches $(0, 0)$ (i) along the x axis, (ii) along the y axis, (iii) along the line $y = 5x$, (iv) along the parabola $y = x^2$.

10 $f(x, y) = \dfrac{\sin (x^3 + y^3)}{x^3 + y^3}$ as (x, y) approaches $(0, 0)$ (i) along the x axis, (ii) along the y axis, (iii) along the line $y = 10x$, (iv) along the curve $y = x^4$.

11 $f(x, y) = \dfrac{xy}{x^2 + y^2}$ as (x, y) approaches $(0, 0)$ (i) along the x axis, (ii) along the y axis, (iii) along the line $y = x$, (iv) along the line $y = mx$.

12 $f(x, y) = \dfrac{3x^4 y^4}{(x^4 + y^2)^3}$ as (x, y) approaches $(0, 0)$ (i) along the x axis, (ii) along the y axis, (iii) along the line $y = x$, (iv) along the curve $y = -x^4$.

13 $f(x, y) = \begin{cases} (x + y) \sin \dfrac{1}{y} & \text{if } y \neq 0 \\ 0 & \text{if } y = 0 \end{cases}$ as (x, y) approaches $(0, 0)$ (i) along the x axis, (ii) along the y axis, (iii) along the curve $y = x^3$, (iv) along the curve $y = -x^5$.

14 $f(x, y) = \begin{cases} \dfrac{1}{y} \sin (xy) & \text{if } y \neq 0 \\ x & \text{if } y = 0 \end{cases}$ as (x, y) approaches $(0, 0)$ (i) along the x axis, (ii) along the line $y = 2x$, (iii) along the parabola $y = x^2$, (iv) along the curve $y = 5x^3$.

In problems 15 and 16, prove each assertion by making direct use of the definition of limit; that is, show that for each given $\varepsilon > 0$, there is a $\delta > 0$ so that the conditions of the definition hold.

15 $\lim\limits_{(x,y) \to (2,1)} (5x + 3y) = 13$

16 $\lim\limits_{\substack{x \to -4 \\ y \to 2}} (-x + 2y) = 8$

In problems 17 through 24, decide whether or not each function is continuous at the point indicated. Justify your answer.

17 $f(x, y) = \sqrt{25 - x^2 - y^2}$ at $(-3, 4)$.

18 $f(x, y) = e^{-xy} \ln (x^2 - 2y + 7)$ at $(0, 0)$.

19 $g(x, y) = \begin{cases} \dfrac{xy}{y - 2x} & \text{if } y \neq 2x \\ 1 & \text{if } y = 2x \end{cases}$ at $(1, 2)$.

20 $f(x, y, z) = \dfrac{1}{x^2 + y^2 + z^2}$ at $(0, 0, 1)$.

21 $g(x, y) = \dfrac{\sin x}{\sin y}$ at $\left(\pi, \dfrac{\pi}{2}\right)$.

22 $h(x, y) = \dfrac{xy}{1 + e^x}$ at $(0, 0)$.

23 $f(x, y) = \begin{cases} (x + y) \sin \dfrac{1}{x} & \text{if } x \neq 0 \\ 0 & \text{if } x = 0 \end{cases}$ at $(0, 0)$.

24 $f(x, y) = \begin{cases} \dfrac{7x^2 y}{x^2 + y^2} & \text{if } (x, y) \neq (0, 0) \\ 0 & \text{if } (x, y) = (0, 0) \end{cases}$ at $(0, 0)$.

In problems 25 through 35, (a) sketch a diagram showing the domain of each function in the xy plane, (b) specify which points of the domain are interior points, and (c) determine those interior points of the domain—if any—at which the function is discontinuous.

25 $f(x, y) = \sqrt{xy}$

26 $g(x, y) = \sin^{-1} (2x + y)$

27 $h(x, y) = \dfrac{4x^2 - y^2}{2x - y}$

28 $F(x, y) = \ln (xy - 2)$

29 $f(x, y) = \dfrac{x}{y^2 - 1}$

30 $H(x, y) = \ln (x^2 + y^2)$

31 $f(x, y) = \dfrac{xy}{\sqrt{9 - x^2 - y^2}}$

32 $G(x, y) = \sin^{-1} \left(\dfrac{y}{x}\right)$

33 $g(x, y) = \begin{cases} \dfrac{x - y}{x + y} & \text{if } (x, y) \neq (0, 0) \\ 0 & \text{if } (x, y) = (0, 0) \end{cases}$

34 $h(x, y) = \begin{cases} 4x^2 + 9y^2 & \text{if } 4x^2 + 9y^2 \leq 1 \\ \dfrac{1}{4x^2 + 9y^2} & \text{if } 4x^2 + 9y^2 > 1 \end{cases}$

35 $F(x, y) = \begin{cases} x^2 + y^2 & \text{if } x^2 + y^2 \geq 4 \\ 0 & \text{if } x^2 + y^2 < 4 \end{cases}$

36 Prove that the limit of the sum is the sum of the limits for functions of two variables. Use the precise definition of limits in terms of ε and δ and give a clear statement of whatever assumptions you need to make.

37 Write out a precise definition, in terms of ε and δ, for the statement

$$\lim_{(x,y,z)\to(x_0,y_0,z_0)} f(x,y,z) = L.$$

38 Prove that a set of points D in the xy plane is an open set if and only if every point in D is an interior point of D.

3   Partial Derivatives

The techniques, rules, and formulas developed in Chapter 2 for differentiating functions of a single variable can be extended to functions of two or more variables, provided that we hold all but one of the independent variables constant and then differentiate with respect to the remaining variable.

For instance, consider the function f of two variables given by

$$f(x,y) = x^2 + 3xy - 4y^2.$$

Let us agree, temporarily, to hold the second variable y constant and to differentiate with respect to the first variable x. Then, since y is constant,

$$\frac{d}{dx}(3xy) = 3y\frac{d}{dx}(x) = 3y \quad \text{and} \quad \frac{d}{dx}(-4y^2) = 0;$$

hence,

$$\frac{d}{dx}f(x,y) = \frac{d}{dx}(x^2) + \frac{d}{dx}(3xy) + \frac{d}{dx}(-4y^2) = 2x + 3y + 0 = 2x + 3y.$$

In order to emphasize that only x is being allowed to vary, so that y is held fixed while the derivative is calculated, it is traditional to modify the symbol $\frac{d}{dx}$ and write $\frac{\partial}{\partial x}$ instead. (The symbol ∂ is called the "round d.") Thus, we write the equation above as

$$\frac{\partial}{\partial x}f(x,y) = \frac{\partial}{\partial x}(x^2 + 3xy - 4y^2) = 2x + 3y.$$

The derivative calculated with respect to x while temporarily holding y constant is called the *partial derivative with respect to x*, and $\frac{\partial}{\partial x}$ is called the *partial derivative operator* with respect to x. Similarly, if we wish to hold the variable x fixed and differentiate with respect to y, we use the symbol $\frac{\partial}{\partial y}$. Thus, for the function f defined by $f(x,y) = x^2 + 3xy - 4y^2$, we have

$$\frac{\partial}{\partial y}f(x,y) = \frac{\partial}{\partial y}(x^2 + 3xy - 4y^2) = \frac{\partial}{\partial y}(x^2) + \frac{\partial}{\partial y}(3xy) + \frac{\partial}{\partial y}(-4y^2)$$

$$= 0 + 3x - 8y = 3x - 8y.$$

More formally, we have the following definition.

DEFINITION 1   **Partial Derivatives of Functions of Two Variables**

If f is a function of two variables and (x,y) is a point in the domain of f, then the *partial derivatives* $\dfrac{\partial f(x, y)}{\partial x}$ and $\dfrac{\partial f(x, y)}{\partial y}$ of f at (x, y) with respect to the first and second variables are defined by

$$\frac{\partial f(x, y)}{\partial x} = \lim_{\Delta x \to 0} \frac{f(x + \Delta x, y) - f(x, y)}{\Delta x}$$

and

$$\frac{\partial f(x, y)}{\partial y} = \lim_{\Delta y \to 0} \frac{f(x, y + \Delta y) - f(x, y)}{\Delta y},$$

provided that the limits exist. The procedure of finding partial derivatives is called *partial differentiation.*

It is convenient to have a notation for partial derivatives that is analogous to the notation $f'(x)$ for functions of one variable. Thus, if $z = f(x, y)$ we often write $f_1(x, y)$ or $f_x(x, y)$ rather than $\dfrac{\partial z}{\partial x}$ or $\dfrac{\partial}{\partial x} f(x, y)$ for the partial derivative of f with respect to x. The subscript 1 (respectively, the subscript x) denotes partial differentiation with respect to the first variable (or, with respect to x). The operator notation Df for ordinary derivatives can likewise be adapted to partial derivatives, and we have

$$\frac{\partial z}{\partial x} = \frac{\partial}{\partial x} f(x, y) = f_1(x, y) = f_x(x, y) = D_1 f(x, y) = D_x f(x, y).$$

Similarly, for the partial derivative with respect to y, we write

$$\frac{\partial z}{\partial y} = \frac{\partial}{\partial y} f(x, y) = f_2(x, y) = f_y(x, y) = D_2 f(x, y) = D_y f(x, y).$$

EXAMPLE Use Definition 1 directly to find $\partial z/\partial x$ and $\partial z/\partial y$ if $z = f(x, y) = 5x^2 - 7xy + 2y^2$.

SOLUTION

$$\frac{\partial z}{\partial x} = \lim_{\Delta x \to 0} \frac{f(x + \Delta x, y) - f(x, y)}{\Delta x}$$

$$= \lim_{\Delta x \to 0} \frac{[5(x + \Delta x)^2 - 7(x + \Delta x)y + 2y^2] - (5x^2 - 7xy + 2y^2)}{\Delta x}$$

$$= \lim_{\Delta x \to 0} \frac{5[x^2 + 2x\,\Delta x + (\Delta x)^2] - 7xy - 7y\,\Delta x + 2y^2 - 5x^2 + 7xy - 2y^2}{\Delta x}$$

$$= \lim_{\Delta x \to 0} \frac{10x\,\Delta x + 5(\Delta x)^2 - 7y\,\Delta x}{\Delta x} = \lim_{\Delta x \to 0} (10x - 7y + 5\,\Delta x)$$

$$= 10x - 7y.$$

$$\frac{\partial z}{\partial y} = \lim_{\Delta y \to 0} \frac{f(x, y + \Delta y) - f(x, y)}{\Delta y}$$

$$= \lim_{\Delta y \to 0} \frac{[5x^2 - 7x(y + \Delta y) + 2(y + \Delta y)^2] - (5x^2 - 7xy + 2y^2)}{\Delta y}$$

$$= \lim_{\Delta y \to 0} \frac{5x^2 - 7xy - 7x\,\Delta y + 2[y^2 + 2y\,\Delta y + (\Delta y)^2] - 5x^2 + 7xy - 2y^2}{\Delta y}$$

$$= \lim_{\Delta y \to 0} \frac{-7x\,\Delta y + 4y\,\Delta y + 2(\Delta y)^2}{\Delta y} = \lim_{\Delta y \to 0} (-7x + 4y + 2\,\Delta y)$$

$$= -7x + 4y.$$

Partial derivatives of functions of more than two variables are defined by the following obvious extension of Definition 1.

DEFINITION 2   **Partial Derivatives of Functions of n Variables**
Let f be a function of n variables and suppose that $(x_1, x_2, \ldots, x_k, \ldots, x_n)$ belongs to the domain of f. If $1 \le k \le n$, then the *partial derivative of f with respect to the kth variable x_k* is denoted by f_k and defined by

$$f_k(x_1, x_2, \ldots, x_k, \ldots, x_n)$$
$$= \lim_{\Delta x_k \to 0} \frac{f(x_1, x_2, \ldots, x_k + \Delta x_k, \ldots, x_n) - f(x_1, x_2, \ldots, x_k, \ldots, x_n)}{\Delta x_k},$$

provided that the limit exists.

If $w = f(x_1, x_2, \ldots, x_k, \ldots, x_n)$, then we also use the following notations for the partial derivative of f with respect to the kth variable x_k:

$$\frac{\partial w}{\partial x_k} = \frac{\partial}{\partial x_k} f(x_1, x_2, \ldots, x_k, \ldots, x_n) = f_{x_k}(x_1, x_2, \ldots, x_k, \ldots, x_n)$$
$$= D_k f(x_1, x_2, \ldots, x_k, \ldots, x_n) = D_{x_k} f(x_1, x_2, \ldots, x_k, \ldots, x_n).$$

We use the words "partial derivative" to refer both to the function f_k and to the value $f_k(x_1, x_2, \ldots, x_k, \ldots, x_n)$ of this function—one can always tell from the context what is intended.

In case $n = 3$ and the independent variables x_1, x_2, and x_3 in Definition 2 are written as x, y, and z, respectively, we have

$$f_1(x, y, z) = f_x(x, y, z) = \lim_{\Delta x \to 0} \frac{f(x + \Delta x, y, z) - f(x, y, z)}{\Delta x},$$

$$f_2(x, y, z) = f_y(x, y, z) = \lim_{\Delta y \to 0} \frac{f(x, y + \Delta y, z) - f(x, y, z)}{\Delta y},$$

$$f_3(x, y, z) = f_z(x, y, z) = \lim_{\Delta z \to 0} \frac{f(x, y, z + \Delta z) - f(x, y, z)}{\Delta z}.$$

3.1 Techniques for Calculating Partial Derivatives

Partial derivatives can be calculated by using the same techniques that are effective for ordinary derivatives, except that *all independent variables other than the variable with respect to which the differentiation takes place must temporarily be regarded as constants.*

EXAMPLES 1 If $w = xy^2z^3$, find $\partial w/\partial y$.

SOLUTION
Holding x and z constant, and differentiating with respect to y, we obtain

$$\frac{\partial w}{\partial y} = \frac{\partial}{\partial y}(xy^2z^3) = xz^3 \frac{\partial}{\partial y}(y^2) = xz^3(2y) = 2xyz^3.$$

2 If $f(x, y) = \dfrac{x + y}{x^2 + y^2}$, find $f_1(x, y)$ and $f_2(x, y)$.

SOLUTION
Using the quotient rule, holding y fixed, and differentiating with respect to x, we have

$$f_1(x, y) = \frac{\partial}{\partial x}\left(\frac{x + y}{x^2 + y^2}\right) = \frac{(x^2 + y^2)\dfrac{\partial}{\partial x}(x + y) - (x + y)\dfrac{\partial}{\partial x}(x^2 + y^2)}{(x^2 + y^2)^2}$$

$$= \frac{(x^2 + y^2)(1 + 0) - (x + y)(2x + 0)}{(x^2 + y^2)^2} = \frac{y^2 - 2xy - x^2}{(x^2 + y^2)^2}.$$

Similarly,

$$f_2(x, y) = \frac{\partial}{\partial y}\left(\frac{x + y}{x^2 + y^2}\right) = \frac{(x^2 + y^2)\dfrac{\partial}{\partial y}(x + y) - (x + y)\dfrac{\partial}{\partial y}(x^2 + y^2)}{(x^2 + y^2)^2}$$

$$= \frac{(x^2 + y^2)(0 + 1) - (x + y)(0 + 2y)}{(x^2 + y^2)^2} = \frac{x^2 - 2xy - y^2}{(x^2 + y^2)^2}.$$

3 If $f(x, y, z) = e^{xy^2}z^3$, find $f_z(x, y, z)$.

SOLUTION
Holding x and y constant, and differentiating with respect to z, we have

$$f_z(x, y, z) = \frac{\partial}{\partial z}(e^{xy^2}z^3) = e^{xy^2}\frac{\partial}{\partial z}(z^3) = 3e^{xy^2}z^2.$$

There are many versions of the *chain rule* applicable to partial derivatives, the simplest one being virtually a direct transcription of the ordinary chain rule for functions of a single variable. Thus, let g be a function of more than one variable— say two variables for definiteness. If $w = f(v)$ and $v = g(x, y)$, so that $w = f[g(x, y)]$, then, holding y constant and using the usual chain rule, we have

$$\frac{\partial w}{\partial x} = f'[g(x, y)]g_x(x, y) = f'(v)\frac{\partial v}{\partial x};$$

that is,

$$\frac{\partial w}{\partial x} = \frac{dw}{dv}\frac{\partial v}{\partial x},$$

provided that the derivatives dw/dv and $\partial v/\partial x$ exist. Similarly, holding x constant and using the usual chain rule, we have

$$\frac{\partial w}{\partial y} = f'[g(x, y)]g_y(x, y) = f'(v)\frac{\partial v}{\partial y};$$

that is,

$$\frac{\partial w}{\partial y} = \frac{dw}{dv}\frac{\partial v}{\partial y},$$

provided that the derivatives dw/dv and $\partial v/\partial y$ exist.

EXAMPLE   If $w = \sqrt{1 - x^2 - y^2}$, find $\partial w/\partial x$ and $\partial w/\partial y$.

SOLUTION
Put $v = 1 - x^2 - y^2$, so that $w = \sqrt{v}$. By the chain rule,

$$\frac{\partial w}{\partial x} = \frac{dw}{dv}\frac{\partial v}{\partial x} = \frac{1}{2\sqrt{v}}\frac{\partial}{\partial x}(1 - x^2 - y^2) = \frac{1}{2\sqrt{v}}(-2x) = \frac{-x}{\sqrt{1 - x^2 - y^2}},$$

$$\frac{\partial w}{\partial y} = \frac{dw}{dv}\frac{\partial v}{\partial y} = \frac{1}{2\sqrt{v}}\frac{\partial}{\partial y}(1 - x^2 - y^2) = \frac{1}{2\sqrt{v}}(-2y) = \frac{-y}{\sqrt{1 - x^2 - y^2}}.$$

In using the version of the chain rule given above, one proceeds in much the same way as in Section 4 of Chapter 2, except that one multiplies the *ordinary* derivative of the "outside function" by the appropriate *partial* derivative of the "inside function." For instance, if $f(x, y) = \tan(x^2 - y^2)$, then the "outside function" is the tangent function, while the "inside function" is the function g of two variables given by $g(x, y) = x^2 - y^2$. Thus,

$$f_x(x, y) = \sec^2(x^2 - y^2)\frac{\partial}{\partial x}(x^2 - y^2) = [\sec^2(x^2 - y^2)](2x)$$

$$= 2x\sec^2(x^2 - y^2) \quad\text{and}$$

$$f_y(x, y) = \sec^2(x^2 - y^2)\frac{\partial}{\partial y}(x^2 - y^2) = [\sec^2(x^2 - y^2)](-2y)$$

$$= -2y\sec^2(x^2 - y^2).$$

EXAMPLES   Find $\partial w/\partial x$ and $\partial w/\partial y$.

1 $w = e^{x/y}$

SOLUTION

$$\frac{\partial w}{\partial x} = e^{x/y}\frac{\partial}{\partial x}\left(\frac{x}{y}\right) = e^{x/y}\left(\frac{1}{y}\right) = \frac{e^{x/y}}{y},$$

$$\frac{\partial w}{\partial y} = e^{x/y}\frac{\partial}{\partial y}\left(\frac{x}{y}\right) = e^{x/y}\left(-\frac{x}{y^2}\right) = \frac{-xe^{x/y}}{y^2}.$$

2 $w = \displaystyle\int_1^{3x^2 - 5y^3} e^{t^2}\, dt$

SOLUTION

Put $u = 3x^2 - 5y^3$, so that $w = \displaystyle\int_1^u e^{t^2}\, dt$. By the fundamental theorem of calculus, $dw/du = e^{u^2}$; hence, by the chain rule,

$$\frac{\partial w}{\partial x} = \frac{dw}{du}\frac{\partial u}{\partial x} = e^{u^2}\frac{\partial}{\partial x}(3x^2 - 5y^3) = e^{u^2}(6x) = 6xe^{(3x^2 - 5y^3)^2},$$

$$\frac{\partial w}{\partial y} = \frac{dw}{du}\frac{\partial u}{\partial y} = e^{u^2}\frac{\partial}{\partial y}(3x^2 - 5y^3) = e^{u^2}(-15y^2) = -15y^2 e^{(3x^2 - 5y^3)^2}.$$

Problem Set 3

In problems 1 through 6, find each partial derivative by direct application of the formal definition.

1 $f_x(x, y)$, where $f(x, y) = 8x - 2y + 13$.

2 $\dfrac{\partial}{\partial x} f(x, y)$, where $f(x, y) = 3x^2 + 5xy + 7y^3$.

3 $f_1(-1, 2)$, where $f(x, y) = -7x^2 + 8xy^2$.

4 $\dfrac{\partial}{\partial z} f(x, y, z)$, where $f(x, y, z) = 2xy^2 - 7xz + 3xyz^2$.

5 $f_2(1, -1)$, where $f(x, y) = 5xy^3 + 6x^2 + 11$.

6 $\dfrac{\partial z}{\partial y}$ at the point $(3, -5)$, where $z = -7x^3y + xy + 3$.

In problems 7 through 20, find each partial derivative by treating all but one of the independent variables as constants and applying the rules of ordinary differentiation.

7 $\dfrac{\partial}{\partial x} f(x, y)$, where $f(x, y) = 7x^2 + 5x^2y + 2$.

8 $\dfrac{\partial}{\partial x} (x^2 \sin y)$.

9 $h_x(x, y)$, where $h(x, y) = \sin x \cos 7y$.

10 $f_2(x, y)$, where $f(x, y) = e^{-2x} \tan y$.

11 $\dfrac{\partial w}{\partial x}$, where $w = \dfrac{x^2 + y^2}{y^2 - x^2}$.

12 $D_x(x \sin y - y \ln x)$.

13 $f_1(r, \theta)$, where $f(r, \theta) = r^2 \cos 7\theta$.

14 $g_2(\theta, \phi)$, where $g(\theta, \phi) = \sin 2\theta \cos \phi$.

15 $f_1(x, y)$, where $f(x, y) = \displaystyle\int_y^x e^{-t^3} dt$.

16 $g_2(x, y)$, where $g(x, y) = \displaystyle\int_y^x e^{-t^2} dt$.

17 $f_z(x, y, z)$, where $f(x, y, z) = 6xyz + 3x^2y + 7z$.

18 $\dfrac{\partial}{\partial y} g(x, y, z)$, where $g(x, y, z) = 3x^2y^3z^4 - 4xyz^3$.

19 $\dfrac{\partial w}{\partial x}$, where $w = xy^2 + yz^2 + x^2y$.

20 $\dfrac{\partial w}{\partial z}$, where $w = x \cos z + y \sin x + xe^z$.

In problems 21 through 34, find each partial derivative by using the chain rule of Section 3.1.

21 $\dfrac{\partial w}{\partial x}$, where $w = \sqrt{u}$ and $u = 3x^2 + y^2$.

22 $\dfrac{\partial}{\partial x} g(x, y)$, where $g(x, y) = \sin (xy)^2$.

23 $\dfrac{\partial w}{\partial x}$, where $w = \ln u$ and $u = 7x^2 + 4y^3$.

24 $\dfrac{\partial}{\partial y} \ln (x^2/y)$.

25 $h_1(x, y)$, where $h(x, y) = \tan^{-1}(xy)$.

26 $D_1 f(x, y)$, where $f(x, y) = e^{x^2 + y^2}$.

27 $\dfrac{\partial w}{\partial y}$, where $w = \displaystyle\int_1^u e^{\sin t} dt$ and $u = x^2 - 5y$.

28 $\dfrac{\partial}{\partial y} \displaystyle\int_2^{\sin (x+y)} e^{t^3} dt$.

29 $f_1(x, y)$, where $f(x, y) = \displaystyle\int_{x^2 - y^2}^{\pi} e^{-t^2} dt$.

30 $f_z(x, y, z)$, where $f(x, y, z) = x^2/\sqrt{y^2 + z^2}$.

31 $g_3(x, y, z)$, where $g(x, y, z) = xz^2e^{xy} \cos (yz)$.

32 $\dfrac{\partial}{\partial y} [z \sin (xz) \cos (xy)]$.

33 $\dfrac{\partial w}{\partial x}$, where $w = (x^2 + y^2 + z^2)^{-3/2}$.

34 $\dfrac{\partial w}{\partial s}$, where $w = e^{-t} \sin (s + u)$.

35 Given that $w = x^3y^2 - 2xy^4 + 3x^2y^3$, verify that $x \dfrac{\partial w}{\partial x} + y \dfrac{\partial w}{\partial y} = 5w$.

36 Given that $w = (ax + by + cz)^n$, where a, b, and c are constants, verify that

$$x \frac{\partial w}{\partial x} + y \frac{\partial w}{\partial y} + z \frac{\partial w}{\partial z} = nw.$$

37 Given that $w = t^2 + \tan (te^{1/s})$, verify that $s^2 \dfrac{\partial w}{\partial s} + t \dfrac{\partial w}{\partial t} = 2t^2$.

38 Let f be a differentiable function and let $w = f(\frac{5}{2}s^2 - \frac{7}{2}t^2)$. Verify that $7t \dfrac{\partial w}{\partial s} + 5s \dfrac{\partial w}{\partial t} = 0$.

39 Given that $w = x^2 \sin \dfrac{y}{z} + y^2 \ln \dfrac{z}{x} + z^2 e^{x/y}$, verify that $x \dfrac{\partial w}{\partial x} + y \dfrac{\partial w}{\partial y} + z \dfrac{\partial w}{\partial z} = 2w$.

40 Given that $w = e^{x/y} + e^{y/z} + e^{z/x}$, verify that $x \dfrac{\partial w}{\partial x} + y \dfrac{\partial w}{\partial y} + z \dfrac{\partial w}{\partial z} = 0$.

41 Given that $w = \ln (x^3 + 5x^2y + 6xy^2 + 7y^3)$, verify that $x \dfrac{\partial w}{\partial x} + y \dfrac{\partial w}{\partial y} = 3$.

4   Elementary Applications of Partial Derivatives

Although most of the applications of partial derivatives depend on properties yet to be developed (such as the chain rules in Section 6), some of them are just adaptations of techniques that are already familiar for ordinary derivatives. We consider some of these elementary applications in this section.

4.1 Geometric Interpretation of Partial Derivatives

Suppose that f is a function of two variables and that f has partial derivatives f_1 and f_2. The graph of f is a surface with the equation $z = f(x, y)$ (Figure 1). Let $z_0 = f(x_0, y_0)$, so that $P = (x_0, y_0, z_0)$ is a point on this surface. The plane $y = y_0$ cuts a cross section APB from the surface, while the plane $x = x_0$ cuts a cross section CPD from the surface. As a point moves along the curve APB, its x and z coordinates vary according to the equation $z = f(x, y_0)$, while its y coordinate remains constant with $y = y_0$. The slope of the tangent line to APB at any point is the rate at which the z coordinate changes with respect to x; hence, this slope is given by $\partial z/\partial x = f_1(x, y_0)$. In particular, $f_1(x_0, y_0)$ represents the slope of the tangent line to APB at the point P. Similarly, $f_2(x_0, y_0)$ represents the slope of the tangent line to CPD at the point P. Thus, in Figure 1, we have

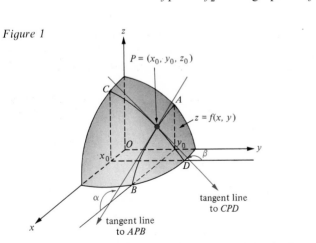

Figure 1

$$\tan \alpha = f_1(x_0, y_0) = f_x(x_0, y_0) = \frac{\partial z}{\partial x} \text{ evaluated at } (x_0, y_0)$$

and

$$\tan \beta = f_2(x_0, y_0) = f_y(x_0, y_0) = \frac{\partial z}{\partial y} \text{ evaluated at } (x_0, y_0).$$

EXAMPLE   Find the slope of the tangent line to the cross section cut from the surface $z = 4x^2y - xy^3$ by the plane $y = 2$ at the point $P = (3, 2, 48)$.

SOLUTION
Here we hold y constant and find $\partial z/\partial x$. We have

$$\frac{\partial z}{\partial x} = \frac{\partial}{\partial x}(4x^2y) - \frac{\partial}{\partial x}(xy^3) = 8xy - y^3,$$

so that, when $x = 3$ and $y = 2$,

$$\frac{\partial z}{\partial x} = 8(3)(2) - 2^3 = 40.$$

4.2 Rates of Change

Suppose that a variable quantity, say w, depends upon a number of other quantities, say $x_1, x_2, \ldots, x_n$. If all but one of the independent variables $x_1, x_2, \ldots, x_n$ are held constant, then w depends only on the remaining variable, and the instantaneous rate of change of w with respect to this variable is given as usual by a derivative—in this case, the partial derivative of w with respect to the variable in question.

EXAMPLE   The volume V of a right circular cone is given by

$$V = \frac{\pi}{24} y^2 \sqrt{4s^2 - y^2},$$

where s is the slant height and y is the diameter of the base.
(a) Find the instantaneous rate of change of the volume with respect to the slant height if the diameter remains constant with the value $y = 16$ inches while the slant height s varies. Evaluate this rate of change at the instant when $s = 10$ inches.
(b) Suppose that the slant height remains constant with the value $s = 10$ inches. Assuming that the diameter y is variable, find the rate of change of the volume with respect to the diameter at the instant when $y = 16$ inches.

SOLUTION
(a) $\dfrac{\partial V}{\partial s} = \dfrac{\partial}{\partial s}\left(\dfrac{\pi}{24} y^2 \sqrt{4s^2 - y^2}\right) = \dfrac{\pi}{24} y^2 \dfrac{\partial}{\partial s}\sqrt{4s^2 - y^2}$

$$= \frac{\pi}{24} y^2 \frac{\frac{\partial}{\partial s}(4s^2 - y^2)}{2\sqrt{4s^2 - y^2}} = \frac{\pi}{24} y^2 \frac{8s}{2\sqrt{4s^2 - y^2}} = \frac{\pi y^2 s}{6\sqrt{4s^2 - y^2}}.$$

When $s = 10$ and $y = 16$,

$$\frac{\partial V}{\partial s} = \frac{\pi(16)^2(10)}{6\sqrt{4(10)^2 - 16^2}} = \frac{320\pi}{9} \approx 111.70 \text{ cubic inches per inch.}$$

(b) $\dfrac{\partial V}{\partial y} = \dfrac{\partial}{\partial y}\left(\dfrac{\pi}{24}\, y^2 \sqrt{4s^2 - y^2}\right)$

$= \dfrac{\pi}{24}\left[\dfrac{\partial}{\partial y}(y^2)\right]\sqrt{4s^2 - y^2} + \dfrac{\pi}{24}\, y^2 \dfrac{\partial}{\partial y}\sqrt{4s^2 - y^2}$

$= \dfrac{\pi}{24}(2y)\sqrt{4s^2 - y^2} + \dfrac{\pi}{24}\, y^2 \dfrac{\dfrac{\partial}{\partial y}(4s^2 - y^2)}{2\sqrt{4s^2 - y^2}}$

$= \dfrac{\pi y \sqrt{4s^2 - y^2}}{12} + \dfrac{\pi}{24}\, y^2 \dfrac{-2y}{2\sqrt{4s^2 - y^2}}$

$= \dfrac{\pi y \sqrt{4s^2 - y^2}}{12} - \dfrac{\pi y^3}{24\sqrt{4s^2 - y^2}}.$

When $s = 10$ and $y = 16$,

$$\dfrac{\partial V}{\partial y} = \dfrac{\pi(16)\sqrt{4(10)^2 - 16^2}}{12} - \dfrac{\pi(16)^3}{24\sqrt{4(10)^2 - 16^2}}$$

$$= 16\pi - \dfrac{128\pi}{9} = \dfrac{16\pi}{9} \approx 5.59 \text{ cubic inches per inch.}$$

Problem Set 4

In problems 1 through 8, find the slope of the tangent line to the cross section cut from the given surface by the given plane at the indicated point.

1 The surface $z = 3x - 5y + 7$ and the plane $y = 2$ at the point $(1, 2, 0)$.

2 The surface $z = \sqrt{4 - x^2 - y^2}$ and the plane $x = 1$ at the point $(1, 1, \sqrt{2})$.

3 The surface $z = \sqrt{31 - 2x^2 - 3y^2}$ and the plane $y = 2$ at the point $(3, 2, 1)$.

4 The surface $x^2 + y^2 + z^2 = 14$ and the plane $x = 1$ at the point $(1, 3, 2)$.

5 The surface $z = e^{-x^2} \sin 3y$ and the plane $x = 1$ at the point $(1, 0, 0)$.

6 The surface $6x^2 + 9y^2 + 4z^2 = 61$ and the plane $y = -1$ at the point $(1, -1, 2)$.

7 The surface $z = \dfrac{2xy}{x^2 + y^2}$ and the plane $y = 4$ at the point $(3, 4, \frac{24}{25})$.

8 The surface $z = e^{x/y} - e^{y/x}$ and the plane $y = 2$ at the point $(2, 2, 0)$.

9 The lateral surface area A of a right circular cone of height h and with base radius r is given by $A = \pi r \sqrt{h^2 + r^2}$.
(a) If r is held fixed, with $r = 3$ inches, while h is varied, find the rate of change of A with respect to h at the instant when $h = 7$ inches.
(b) If h is held fixed, with $h = 7$ inches, while r is varied, find the rate of change of A with respect to r at the instant when $r = 3$ inches.

10 If two sides x and y of a triangle are held fixed while the angle θ between them is varied, find the rate of change of the third side z with respect to θ. (Use the law of cosines.)

11 Clairaut's formula for the weight W in dynes of a 1-gram mass at latitude L degrees and at height h centimeters above sea level is

$$W = 980.6056 - 2.5028 \cos \frac{\pi L}{90} - \frac{h}{3,000,000}.$$

(a) If L is held constant at $40°$N while h is varied, find the rate of change of W with respect to h at the instant when the mass is 6 kilometers above sea level.

(b) If the mass remains at a constant altitude of 6 kilometers above sea level while its latitude L is varied, find the rate of change of W per degree of latitude at the instant when $L = 40°$N.

12 A circular hill has a central vertical section in the form of a curve whose equation is $z = 10 - (x^2/160)$, where the units are in meters. The top is being cut down in horizontal layers at the constant rate of 100 cubic meters per day. How fast is the area of horizontal cross section increasing when the top has been cut down a vertical distance of 4 meters?

13 The resistance R ohms of an electrical circuit is given by the formula $R = E/I$, where I is the current in amperes and E is the electromotive force in volts. Calculate $\partial R/\partial I$ and $\partial R/\partial E$ when $I = 15$ amperes and $E = 110$ volts and give an interpretation of these two partial derivatives in terms of rates of change.

14 A mortgage company computes the maximum size of a 25-year mortgage that it is willing to give on a one-family house by using the formula

$$M = 3x + 0.025y^2,$$

where x is the size of the down payment in dollars and y is the applicant's monthly salary in dollars. Find $\partial M/\partial x$ and $\partial M/\partial y$ and interpret their meanings in terms of rate of change.

5 Linear Approximation and Differentiable Functions

In Section 4 we saw that the partial derivatives $f_1(x_0, y_0)$ and $f_2(x_0, y_0)$ involve only cross sections cut from the surface $z = f(x, y)$ by the two perpendicular planes $y = y_0$ and $x = x_0$; hence, in general, these two partial derivatives tell us very little about the shape of the surface "in between" the two cross sections. For this reason it is not appropriate to call a function of two (or more) variables "differentiable" just because its partial derivatives exist. The key to the proper definition of "differentiability" for a function of more than one variable is the important idea of *linear approximation*.

5.1 Linear Approximation

We begin by formulating a linear approximation procedure for a function f of two variables. Suppose that (x_0, y_0) is an interior point of the domain of f and that the two partial derivatives $f_1(x_0, y_0)$ and $f_2(x_0, y_0)$ exist. Then, by analogy with the procedure introduced in Section 7 of Chapter 2 for functions of a single variable, we have the following.

Linear Approximation Procedure

If the point (x, y) is near the point (x_0, y_0), then

$$f(x, y) \approx f(x_0, y_0) + f_1(x_0, y_0)(x - x_0) + f_2(x_0, y_0)(y - y_0).$$

Of course, the *error* involved in the linear approximation procedure is given by

$$E(x, y) = f(x, y) - f(x_0, y_0) - f_1(x_0, y_0)(x - x_0) - f_2(x_0, y_0)(y - y_0).$$

For most functions encountered in practical applications of calculus, the linear approximation procedure is quite accurate—that is, the absolute value of the error, $|E(x, y)|$, is small—when the point (x, y) is near the point (x_0, y_0).

We now define $\Delta x = x - x_0$ and $\Delta y = y - y_0$, so that the linear approximation procedure can be rewritten as

$$f(x_0 + \Delta x, y_0 + \Delta y) \approx f(x_0, y_0) + f_1(x_0, y_0)\, \Delta x + f_2(x_0, y_0)\, \Delta y.$$

The requirement that the point (x, y) is near the point (x_0, y_0) is evidently equivalent to the requirement that $|\Delta x|$ and $|\Delta y|$ are both small.

EXAMPLE   If $f(x, y) = \sqrt{x^2 + y^2 + 2}$, use the linear approximation procedure to estimate $f(3.01, 4.96)$.

SOLUTION
In the linear approximation

$$f(x_0 + \Delta x, y_0 + \Delta y) \approx f(x_0, y_0) + f_1(x_0, y_0)\, \Delta x + f_2(x_0, y_0)\, \Delta y,$$

we put $x_0 = 3$, $y_0 = 5$, $\Delta x = 0.01$, and $\Delta y = -0.04$, to obtain

$$f(3.01, 4.96) \approx \sqrt{3^2 + 5^2 + 2} + f_1(3, 5)(0.01) + f_2(3, 5)(-0.04).$$

Here,

$$f_1(x, y) = \frac{\partial}{\partial x} \sqrt{x^2 + y^2 + 2} = \frac{x}{\sqrt{x^2 + y^2 + 2}},$$

$$f_2(x, y) = \frac{\partial}{\partial y} \sqrt{x^2 + y^2 + 2} = \frac{y}{\sqrt{x^2 + y^2 + 2}},$$

so that

$$f_1(3, 5) = \frac{3}{\sqrt{36}} = \frac{1}{2} \quad \text{and} \quad f_2(3, 5) = \frac{5}{\sqrt{36}} = \frac{5}{6}.$$

Thus, $f(3.01, 4.96) \approx \sqrt{36} + \frac{1}{2}(0.01) + \frac{5}{6}(-0.04)$; that is, $f(3.01, 4.96) \approx 5.97166\ldots$. [The true value of $f(3.01, 4.96)$, correct to five decimal places, is 5.97174.]

5.2 Differentiable Functions of Two Variables

Roughly speaking, a function of two variables is said to be *differentiable* if the error involved in the linear approximation procedure is small in absolute value. More formally, suppose that (x_0, y_0) is an interior point of the domain of f and that the two partial derivatives $f_1(x_0, y_0)$ and $f_2(x_0, y_0)$ exist. Then we make the following definition.

DEFINITION 1   **Differentiable Function of Two Variables**
We say that f is *differentiable* at (x_0, y_0) if the error involved in the linear

approximation procedure has the form

$$E(x, y) = (x - x_0)\varepsilon_1(x, y) + (y - y_0)\varepsilon_2(x, y),$$

where

$$\lim_{\substack{x \to x_0 \\ y \to y_0}} \varepsilon_1(x, y) = \varepsilon_1(x_0, y_0) = 0 \quad \text{and} \quad \lim_{\substack{x \to x_0 \\ y \to y_0}} \varepsilon_2(x, y) = \varepsilon_2(x_0, y_0) = 0.$$

If the domain of f is an open set, then f is called a *differentiable* function provided that it is differentiable at every point in its domain.

Putting $\Delta x = x - x_0$ and $\Delta y = y - y_0$ in Definition 1, and abbreviating $E(x, y)$, $\varepsilon_1(x, y)$, and $\varepsilon_2(x, y)$ as E, ε_1, and ε_2, respectively, we have

$$E = \Delta x \varepsilon_1 + \Delta y \varepsilon_2, \qquad \text{where } \lim_{\substack{\Delta x \to 0 \\ \Delta y \to 0}} \varepsilon_1 = 0 \quad \text{and} \quad \lim_{\substack{\Delta x \to 0 \\ \Delta y \to 0}} \varepsilon_2 = 0.$$

Thus, for a differentiable function, the error E involved in the linear approximation approaches zero very rapidly as Δx and Δy approach zero. This is analogous to the condition on the linear approximation error for functions of a single variable in Theorem 1, Section 7, Chapter 2, page 127.

Unfortunately, the mere existence of the partial derivatives $f_1(x_0, y_0)$ and $f_2(x_0, y_0)$ does not guarantee that f is differentiable at (x_0, y_0) (see Problem 34). A condition that does ensure the differentiability of f is provided by the following definition.

DEFINITION 2   **Continuous Differentiability**

Let f be a function of two variables and let U be an open set of points contained in the domain of f. We say that f is *continuously differentiable on* U if the partial derivatives $f_1(x, y)$ and $f_2(x, y)$ exist at every point (x, y) in U and the functions f_1 and f_2 are continuous on U.

If the domain D of f is an open set and f is continuously differentiable on D, then we simply say that f is *continuously differentiable*.

In Section 5.5 we prove the following facts:

1 If f is differentiable at (x_0, y_0), then f is continuous at (x_0, y_0).

2 If f is continuously differentiable on an open set U, then f is differentiable at each point (x_0, y_0) in U.

For now, we assume these two facts.

EXAMPLES 1 Is the function $f(x, y) = e^{3x + 4y}$ differentiable?

SOLUTION
The domain of f is the entire xy plane, so it is an open set. (Why?) Here,

$$f_1(x, y) = \frac{\partial}{\partial x}\left(e^{3x+4y}\right) = e^{3x+4y}\,\frac{\partial}{\partial x}\,(3x + 4y) = 3e^{3x+4y},$$

$$f_2(x, y) = \frac{\partial}{\partial y}\left(e^{3x+4y}\right) = e^{3x+4y}\,\frac{\partial}{\partial y}\,(3x + 4y) = 4e^{3x+4y},$$

so that f_1 and f_2 are continuous on the entire xy plane; that is, f is continuously differentiable. It follows that f is differentiable at each point in its domain; hence, f is differentiable.

2 Is the function $f(x, y) = |xy|$ differentiable: (a) at $(1, 1)$, (b) at $(0, 1)$?

SOLUTION

(a) Inside a circular disk of radius, say, $\frac{1}{2}$ with center at $(1, 1)$, both x and y are positive, so that $f(x, y) = |xy| = xy$. Thus, inside this disk, $f_1(x, y) = \dfrac{\partial}{\partial x}(xy) = y$ and $f_2(x, y) = \dfrac{\partial}{\partial y}(xy) = x$; hence, f is continuously differentiable. It follows that f is differentiable at each point in the disk. In particular, it is differentiable at the center point $(1, 1)$ of the disk.

(b) Since $f(x, 1) = |x|$ and since the absolute value function (of one variable) is not differentiable at 0, it follows that the partial derivative $f_1(0, 1)$ does not exist. However, by definition, a differentiable function must have both of its partial derivatives. Consequently, f cannot be differentiable at $(0, 1)$.

It can be shown that the sum, difference, product, and quotient of continuously differentiable functions are continuously differentiable; hence, they are differentiable at each point of their domains. In particular, every polynomial function in two variables is continuously differentiable, and so is every rational function in two variables.

The ideas and facts presented above can be summarized *informally* as follows: A function with continuous partial derivatives is continuously differentiable. A continuously differentiable function is differentiable. A differentiable function is continuous. Finally, the linear approximation procedure, applied to a differentiable function, yields a small error.

5.3 The Total Differential

Suppose that f is a function of two variables and let

$$z = f(x, y).$$

If x and y are changed by small amounts Δx and Δy, respectively, then z changes by an amount Δz given by

$$\Delta z = f(x + \Delta x, y + \Delta y) - f(x, y).$$

Assuming that f is differentiable at (x, y), we know that the error involved in the linear approximation

$$f(x + \Delta x, y + \Delta y) \approx f(x, y) + f_1(x, y)\,\Delta x + f_2(x, y)\,\Delta y$$

will be small, and it follows that we can approximate Δz as

$$\Delta z \approx f_1(x, y)\,\Delta x + f_2(x, y)\,\Delta y.$$

Using the alternative notation $\partial z/\partial x$ and $\partial z/\partial y$ for the partial derivatives $f_1(x, y)$ and $f_2(x, y)$, we can write the approximation as

$$\Delta z \approx \frac{\partial z}{\partial x}\,\Delta x + \frac{\partial z}{\partial y}\,\Delta y.$$

By analogy with the notation used for functions of one variable in Section 1 of Chapter 5, the changes Δx and Δy in the two independent variables x and y are sometimes called the *differentials* of these variables and written as dx and dy, respectively. Thus, if dx and dy are small, then the change Δz in the value of z caused by changing x to $x + dx$ and changing y to $y + dy$ is approximated as

$$\Delta z \approx \frac{\partial z}{\partial x}\, dx + \frac{\partial z}{\partial y}\, dy.$$

Pursuing the analogy with functions of a single variable one step further, we define the *total differential dz* of the dependent variable z by

$$dz = \frac{\partial z}{\partial x}\, dx + \frac{\partial z}{\partial y}\, dy.$$

Thus, if dx and dy are small, then $\Delta z \approx dz$. Since $z = f(x, y)$, we also write dz as df, so that

$$df = f_1(x, y)\, dx + f_2(x, y)\, dy.$$

EXAMPLES **1** If $f(x, y) = 3x^3 y^2 - 2xy^3 + xy - 1$, find the total differential df.

SOLUTION
Here,

$$f_1(x, y) = 9x^2 y^2 - 2y^3 + y \quad \text{and} \quad f_2(x, y) = 6x^3 y - 6xy^2 + x;$$

hence,

$$df = (9x^2 y^2 - 2y^3 + y)\, dx + (6x^3 y - 6xy^2 + x)\, dy.$$

2 If $z = \dfrac{x - y}{x + y}$, find the total differential dz.

SOLUTION
Using the quotient rule, we have

$$\frac{\partial z}{\partial x} = \frac{2y}{(x + y)^2} \quad \text{and} \quad \frac{\partial z}{\partial y} = \frac{-2x}{(x + y)^2};$$

hence,

$$dz = \frac{\partial z}{\partial x}\, dx + \frac{\partial z}{\partial y}\, dy = \frac{2y\, dx}{(x + y)^2} + \frac{-2x\, dy}{(x + y)^2} = \frac{2y\, dx - 2x\, dy}{(x + y)^2}.$$

3 The volume V of a right circular cone of height h and base radius r is given by $V = \frac{1}{3}\pi r^2 h$. If the height is increased from 5 cm to 5.01 cm while the base radius is decreased from 4 cm to 3.98 cm, find an approximation to the change ΔV in the volume.

SOLUTION
We use the total differential dV to approximate ΔV. Here,

$$dV = \frac{\partial V}{\partial h}\, dh + \frac{\partial V}{\partial r}\, dr = \frac{\partial}{\partial h}\left(\frac{1}{3}\pi r^2 h\right) dh + \frac{\partial}{\partial r}\left(\frac{1}{3}\pi r^2 h\right) dr$$

$$= \frac{1}{3}\pi r^2\, dh + \frac{2}{3}\pi r h\, dr.$$

When $h = 5$, $dh = 0.01$, $r = 4$, and $dr = -0.02$, we have

$$\Delta V \approx dV = \frac{1}{3}\pi (4)^2 (0.01) + \frac{2}{3}\pi (4)(5)(-0.02);$$

that is,

$$\Delta V \approx -0.6702 \text{ cm}^3.$$

(The true value of ΔV, correct to five decimal places, is -0.66978.)

5.4 Functions of Three or More Variables

The notions of differentiability, continuous differentiability, and the total differential extend in an obvious way to functions of three or more variables. For instance, if f is a function of three variables, then a point (x_0, y_0, z_0) is called an *interior point* of the domain D of f if there is a positive number r such that all points (x, y, z) within a sphere of radius r with center at (x_0, y_0, z_0) belong to the domain D; that is, if $(x - x_0)^2 + (y - y_0)^2 + (z - z_0)^2 < r^2$, then (x, y, z) belongs to D. Similarly, a point (a, b, c) is called a *boundary point* of D if every sphere with positive radius r and center (a, b, c) contains a point that belongs to D and a point that does not. We say that D is *open* if it contains none of its own boundary points, that is, if it consists entirely of interior points. Similarly, D is called *closed* if it contains all its boundary points.

Now assume that (x_0, y_0, z_0) is an interior point of the domain D of f and that the three partial derivatives

$$A = f_1(x_0, y_0, z_0), \qquad B = f_2(x_0, y_0, z_0), \qquad C = f_3(x_0, y_0, z_0)$$

exist. Let $E(x, y, z)$ denote the error involved in the linear approximation

$$f(x, y, z) \approx f(x_0, y_0, z_0) + A(x - x_0) + B(y - y_0) + C(z - z_0).$$

Then we say that f is *differentiable* at (x_0, y_0, z_0) if there exist three functions ε_1, ε_2, and ε_3 such that

$$\varepsilon_1(x_0, y_0, z_0) = \varepsilon_2(x_0, y_0, z_0) = \varepsilon_3(x_0, y_0, z_0) = 0,$$

$$\lim_{\substack{x \to x_0 \\ y \to y_0 \\ z \to z_0}} \varepsilon_1(x, y, z) = \lim_{\substack{x \to x_0 \\ y \to y_0 \\ z \to z_0}} \varepsilon_2(x, y, z) = \lim_{\substack{x \to x_0 \\ y \to y_0 \\ z \to z_0}} \varepsilon_3(x, y, z) = 0, \quad \text{and}$$

$$E(x, y, z) = (x - x_0)\varepsilon_1(x, y, z) + (y - y_0)\varepsilon_2(x, y, z) + (z - z_0)\varepsilon_3(x, y, z).$$

If the domain D of f is open, we say that f is *continuously differentiable* if the partial derivatives f_1, f_2, and f_3 are defined and continuous on D. By arguments similar to those that will be given in Section 5.5 for functions of two variables, it can be shown that a continuously differentiable function of three variables is differentiable at each point of its domain and that a differentiable function of three variables is automatically continuous. Again, sums, products, differences, and quotients of continuously differentiable functions of three variables are continuously differentiable. Any polynomial function in three variables is continuously differentiable, as is any rational function in three variables.

If $w = f(x, y, z)$, where f is differentiable at an interior point (x, y, z) of its domain, we define the *total differential dw*, or *df*, at (x, y, z) by

$$dw = df = \frac{\partial w}{\partial x}\, dx + \frac{\partial w}{\partial y}\, dy + \frac{\partial w}{\partial z}\, dz = f_1(x, y, z)\, dx + f_2(x, y, z)\, dy + f_3(x, y, z)\, dz,$$

where dx, dy, and dz, the *differentials* of the independent variables, can be assigned arbitrary values. The total differential dw provides an approximation to the change Δw in the dependent variable w caused by changing x, y, and z by small amounts dx, dy, and dz, respectively.

EXAMPLES 1 If $w = x^2 y^3 z^4$, find the total differential dw.

SOLUTION

$$dw = \frac{\partial w}{\partial x}\,dx + \frac{\partial w}{\partial y}\,dy + \frac{\partial w}{\partial z}\,dz = 2xy^3z^4\,dx + 3x^2y^2z^4\,dy + 4x^2y^3z^3\,dz.$$

2 Three resistors, x ohms, y ohms, and z ohms, are connected in parallel to give a net resistance of w ohms, where $w = \dfrac{xyz}{xy + xz + yz}$. Each resistor is rated at 300 ohms, but is subject to 1 percent error. What is the approximate maximum possible error and the approximate maximum possible percent error in the value of w?

SOLUTION
We have

$$\Delta w \approx dw = \frac{\partial w}{\partial x}\,dx + \frac{\partial w}{\partial y}\,dy + \frac{\partial w}{\partial z}\,dz,$$

where dx, dy, and dz represent the errors in the values of the three resistors. Here $|dx|$, $|dy|$, and $|dz|$ do not exceed 3 ohms (1 percent of 300 ohms). We have

$$\frac{\partial w}{\partial x} = \frac{(xy + xz + yz)\dfrac{\partial}{\partial x}(xyz) - xyz\dfrac{\partial}{\partial x}(xy + xz + yz)}{(xy + xz + yz)^2}$$

$$= \frac{xy^2z + xyz^2 + y^2z^2 - xy^2z - xyz^2}{(xy + xz + yz)^2} = \frac{y^2z^2}{(xy + xz + yz)^2},$$

so that, when $x = y = z = 300$, $\partial w/\partial x = \frac{1}{9}$. Similar calculations yield $\partial w/\partial y = \frac{1}{9}$ and $\partial w/\partial z = \frac{1}{9}$ when $x = y = z = 300$. Thus,

$$dw = \frac{1}{9}\,dx + \frac{1}{9}\,dy + \frac{1}{9}\,dz$$

$$|dw| = \frac{1}{9}|dx + dy + dz| \le \frac{1}{9}(|dx| + |dy| + |dz|).$$

Since $|dx|$, $|dy|$, and $|dz|$ cannot exceed 3, it follows that

$$|dx| + |dy| + |dz| \le 3 + 3 + 3 = 9;$$

hence,

$$|dw| \le \frac{1}{9}(9) = 1 \text{ ohm}.$$

The approximate maximum possible error is 1 ohm. Since $w = 100$ ohms when $x = y = z = 300$ ohms, the approximate maximum possible percent error is $\frac{1}{100} \times 100$ percent $= 1$ percent.

5.5 Theorems on Differentiable Functions of Two Variables

Here, as promised, we prove the two facts about differentiable functions referred to in Section 5.2.

THEOREM 1

Continuity of a Differentiable Function

Let f be a function of two variables and let (x_0, y_0) be an interior point of the domain of f. Then, if f is differentiable at (x_0, y_0), it follows that f is continuous at (x_0, y_0).

PROOF

We have

$$f(x, y) = f(x_0, y_0) + a(x - x_0) + b(y - y_0) + E(x, y),$$

where $a = f_1(x_0, y_0)$, $b = f_2(x_0, y_0)$, and

$$E(x, y) = (x - x_0)\varepsilon_1(x, y) + (y - y_0)\varepsilon_2(x, y)$$

with

$$\lim_{\substack{x \to x_0 \\ y \to y_0}} E(x, y) = \lim_{\substack{x \to x_0 \\ y \to y_0}} (x - x_0) \lim_{\substack{x \to x_0 \\ y \to y_0}} \varepsilon_1(x, y) + \lim_{\substack{x \to x_0 \\ y \to y_0}} (y - y_0) \lim_{\substack{x \to x_0 \\ y \to y_0}} \varepsilon_2(x, y)$$

$$= (0)(0) + (0)(0) = 0;$$

hence,

$$\lim_{\substack{x \to x_0 \\ y \to y_0}} f(x, y) = f(x_0, y_0) + a \lim_{\substack{x \to x_0 \\ y \to y_0}} (x - x_0) + b \lim_{\substack{x \to x_0 \\ y \to y_0}} (y - y_0) + \lim_{\substack{x \to x_0 \\ y \to y_0}} E(x, y)$$

$$= f(x_0, y_0) + a(0) + b(0) + 0 = f(x_0, y_0).$$

Therefore, f is continuous at (x_0, y_0).

THEOREM 2

A Continuously Differentiable Function Is Differentiable

Let U be an open set contained in the domain of the function f of two variables. If f is continuously differentiable on U, then f is differentiable at each point (x_0, y_0) in U.

PROOF

Let (x_0, y_0) be a point in U. Since U is open, there is a circular disk of positive radius r with center (x_0, y_0) such that all points (x, y) within this disk are contained in U. From now on, we deal only with such points (x, y).

Define:

$$1 \quad \varepsilon_1(x, y) = \begin{cases} \dfrac{f(x, y_0) - f(x_0, y_0)}{x - x_0} - f_1(x_0, y_0) & \text{if } x \neq x_0 \\ 0 & \text{if } x = x_0. \end{cases}$$

$$2 \quad \varepsilon_2(x, y) = \begin{cases} \dfrac{f(x, y) - f(x, y_0)}{y - y_0} - f_2(x_0, y_0) & \text{if } y \neq y_0 \\ 0 & \text{if } y = y_0. \end{cases}$$

Since $\displaystyle\lim_{x \to x_0} \dfrac{f(x, y_0) - f(x_0, y_0)}{x - x_0} = f_1(x_0, y_0)$, it follows that

$$3 \quad \lim_{\substack{x \to x_0 \\ y \to y_0}} \varepsilon_1(x, y) = 0.$$

Now choose and temporarily fix a point (x, y) with $y \neq y_0$ and define a function ϕ of one variable by $\phi(t) = f(x, t)$. By the mean value theorem (Theorem 2, Section 1.2, Chapter 3, page 138), there exists a number c between y and y_0 such that $\phi'(c) = \dfrac{\phi(y) - \phi(y_0)}{y - y_0}$; that is, $f_2(x, c) = \dfrac{f(x, y) - f(x, y_0)}{y - y_0}$. Notice here that the value of c may depend on our original choice of the point (x, y); however, since c is between y and y_0, c approaches y_0 as y approaches y_0. From equation 2, we have

$$4 \quad \varepsilon_2(x, y) = f_2(x, c) - f_2(x_0, y_0) \text{ for } y \neq y_0.$$

Since f_2 is continuous and $c \to y_0$ as $y \to y_0$, it follows that

$$5 \quad \lim_{\substack{x \to x_0 \\ y \to y_0}} \varepsilon_2(x, y) = 0.$$

Let $E(x, y)$ be the error involved in the linear approximation

$$f(x, y) \approx f(x_0, y_0) + f_1(x_0, y_0)(x - x_0) + f_2(x_0, y_0)(y - y_0),$$

so that

$$6 \quad E(x, y) = f(x, y) - f(x_0, y_0) - f_1(x_0, y_0)(x - x_0) - f_2(x_0, y_0)(y - y_0).$$

By equations 1 and 2, we have

$$7 \quad E(x, y) = (x - x_0)\varepsilon_1(x, y) + (y - y_0)\varepsilon_2(x, y).$$

$$8 \quad \varepsilon_1(x_0, y_0) = \varepsilon_2(x_0, y_0) = 0.$$

By equations 3, 5, 7, 8, and Definition 1, f is differentiable at (x_0, y_0).

Problem Set 5

In problems 1 through 5, use the linear approximation procedure to estimate $f(x_0 + \Delta x, y_0 + \Delta y)$ for each function f and the indicated values of x_0, y_0, Δx, and Δy.

1 $f(x, y) = x^3 - 5x^2 + 6xy$; $x_0 = 1$, $y_0 = -2$, $\Delta x = 0.07$, $\Delta y = 0.02$.

2 $f(x, y) = x^3 - 2xy + 3y^3$; $x_0 = 2$, $y_0 = 1$, $\Delta x = 0.03$, $\Delta y = -0.01$.

3 $f(x, y) = x \ln y - y \ln x$; $x_0 = 1$, $y_0 = 1$, $\Delta x = 0.01$, $\Delta y = 0.02$.

4 $f(x, y) = x\sqrt{x - y}$; $x_0 = 6$, $y_0 = 2$, $\Delta x = 0.25$, $\Delta y = 0.25$.

5 $f(x, y) = y \tan^{-1}(xy)$; $x_0 = 2$, $y_0 = \frac{1}{2}$, $\Delta x = -\frac{1}{15}$, $\Delta y = -\frac{1}{10}$.

6 Find (a) the true value $f(x_0 + \Delta x, y_0 + \Delta y)$ and (b) the error involved in the linear approximation in problems 1, 3, and 5.

In problems 7 through 12, decide whether or not each function is differentiable at the indicated point. Give a reason for your answer.

7 $f(x, y) = xe^{-y}$ at (x, y).

8 $f(x, y) = |xy^2|$ at $(0, 1)$.

9 $f(x, y) = \dfrac{3xy}{x^3 + y^3}$ at $(1, 2)$.

10 $f(x, y) = \begin{cases} 1/(xy) & \text{if } x \neq 0 \text{ and } y \neq 0 \\ 1 & \text{if either } x = 0 \text{ or } y = 0 \end{cases}$ at $(0, 0)$.

11 $f(x, y) = \dfrac{xy^2}{x + y} \cos e^{x^2 + y^2}$ at $(1, 1)$.

12 $f(x, y, z) = |xyz|$ at $(1, 1, 1)$.

In problems 13 through 18, find each total differential.

13 df if $f(x, y) = 5x^3 + 4x^2y - 2y^3$.

14 dz if $z = \sqrt{x^2 + y^2}$.

15 dT if $T = PV/R$, where R is constant and P and V are variables.

16 df if $f(u, v) = \sin^{-1}(u/v)$.

17 df if $f(x, y, z) = xy^2 - 2zx^2 + 3xyz^2$.

18 dw if $w = xe^{yz} + ye^{zx}$.

In problems 19 through 26, use total differentials to make each required approximation.

19 The power P consumed in an electrical resistor is given by $P = E^2/R$ watts, where E is the electromotive force in volts and R is the resistance in ohms. If, at a given moment, $E = 100$ volts and $R = 5$ ohms, approximately how will the power change if E is decreased by 2 volts and R is decreased by 0.3 ohm?

20 Find the approximate area of a right triangle if the length of the long leg is 14.9 centimeters and the length of the hypotenuse is 17.1 centimeters.

21 The length l and the period T of a simple pendulum are connected by the equation $T = 2\pi\sqrt{l/g}$. If l is calculated for $T = 1$ second and $g = 32$ feet per second per second, approximate the error in l if T is really 1.02 seconds and g is really 32.01 feet per second per second. Also find the approximate percentage error.

22 The altitude and diameter of a right circular cylinder are 10 centimeters and 6 centimeters, respectively. If a measurement of the diameter produces a figure 4 percent too large, approximately what percent error in the measurement of the altitude will counteract the resulting error in the computed volume?

23 The dimensions of a rectangular box are 5, 6, and 8 inches. If each dimension is increased by 0.01 inch, what is the approximate resulting change in the volume?

24 The acceleration of gravity as determined by an Atwood machine is given by the formula $g = 82s/t^2$. Suppose that the true values of s and t are $s = 48$ cm and $t = 2$ seconds. Approximate the maximum possible error in the calculation of g if the measurements of s and t are subject to no more than 1 percent error.

25 Two electrical resistors r_1 and r_2 are connected in parallel, so that the net resistance R satisfies the equation $1/R = 1/r_1 + 1/r_2$. Suppose that $r_1 = 30$ ohms and $r_2 = 50$ ohms originally, but that r_1 increases by 0.03 ohm while r_2 decreases by 0.05 ohm. Approximate the resulting change in R.

26 A toy manufacturer makes two kinds of toys, x thousand toys of the first kind and y thousand toys of the second kind. They sell for $80 - 2x$ and $60 - 2y$ dollars per toy, respectively. The total sales revenue, in thousands of dollars, is given by the function $f(x, y) = 80x - 2x^2 + 60y - 2y^2$. Both toys are currently selling for \$20 per toy. Approximate the change in sales revenue in dollars if the price of the first kind of toy is increased by 50 cents and the price of the second kind of toy is increased by 70 cents.

27 Suppose that (x_0, y_0) is an interior point of the domain of f and that there are constants a and b such that
$$f(x, y) \approx f(x_0, y_0) + a(x - x_0) + b(y - y_0)$$
with an error $E(x, y) = f(x, y) - f(x_0, y_0) - a(x - x_0) - b(y - y_0)$ that satisfies the conditions in Definition 1. Prove that $a = f_1(x_0, y_0)$ and $b = f_2(x_0, y_0)$; hence, conclude that f is differentiable at (x_0, y_0).

28 Suppose that ϕ is a function of two variables and that (x_0, y_0) is an interior point of the domain of ϕ such that
$$\lim_{(x, y)\to(x_0, y_0)} \phi(x, y) = \phi(x_0, y_0) = 0.$$

Define two functions ε_1 and ε_2 with the same domain as ϕ by

$$\varepsilon_1(x, y) = \begin{cases} \dfrac{|x - x_0|}{x - x_0} \cdot \dfrac{\sqrt{(x - x_0)^2 + (y - y_0)^2}}{|x - x_0| + |y - y_0|} \phi(x, y) & \text{if } x \neq x_0 \\ 0 & \text{if } x = x_0 \end{cases}$$

and

$$\varepsilon_2(x, y) = \begin{cases} \dfrac{|y - y_0|}{y - y_0} \cdot \dfrac{\sqrt{(x - x_0)^2 + (y - y_0)^2}}{|x - x_0| + |y - y_0|} \phi(x, y) & \text{if } y \neq y_0 \\ 0 & \text{if } y = y_0. \end{cases}$$

Prove:

(a) $\sqrt{(x - x_0)^2 + (y - y_0)^2}\,\phi(x, y) = (x - x_0)\varepsilon_1(x, y) + (y - y_0)\varepsilon_2(x, y)$ for all points (x, y) in the domain of ϕ.

(b) $\displaystyle\lim_{(x,y)\to(x_0,y_0)} \varepsilon_1(x, y) = \varepsilon_1(x_0, y_0) = 0.$

(c) $\displaystyle\lim_{(x,y)\to(x_0,y_0)} \varepsilon_2(x, y) = \varepsilon_2(x_0, y_0) = 0.$

29 Let f be defined by

$$f(x, y) = \begin{cases} \dfrac{x^2 y^2}{x^2 + y^2} & \text{if } (x, y) \neq (0, 0) \\ 0 & \text{if } (x, y) = (0, 0). \end{cases}$$

(a) Find $f_1(x, y)$ and $f_2(x, y)$ for $(x, y) \neq (0, 0)$.
(b) Find $f_1(0, 0)$ and $f_2(0, 0)$ by direct calculation using the definitions of f_1 and f_2.
(c) Show that f is continuously differentiable.
(d) Explain why f is differentiable at $(0, 0)$.

30 Let (x_0, y_0) be an interior point of the domain of f and suppose that there exist constants a and b and a function ϕ with the same domain as f such that

$$f(x, y) = f(x_0, y_0) + a(x - x_0) + b(y - y_0) + \sqrt{(x - x_0)^2 + (y - y_0)^2}\,\phi(x, y)$$

holds for all (x, y) in the domain of f, where $\displaystyle\lim_{\substack{x\to x_0 \\ y\to y_0}} \phi(x, y) = \phi(x_0, y_0) = 0$. Use the results of problems 27 and 28 to prove that f is differentiable at (x_0, y_0).

31 Define ε_1 and ε_2 by

$$\varepsilon_1(x, y) = \begin{cases} \dfrac{x|y|}{|x| + |y|} & \text{if } (x, y) \neq (0, 0) \\ 0 & \text{if } (x, y) = (0, 0) \end{cases}$$

$$\varepsilon_2(x, y) = \begin{cases} \dfrac{y|x|}{|x| + |y|} & \text{if } (x, y) \neq (0, 0) \\ 0 & \text{if } (x, y) = (0, 0). \end{cases}$$

(a) Prove that $\displaystyle\lim_{\substack{x\to 0 \\ y\to 0}} \varepsilon_1(x, y) = \varepsilon_1(0, 0) = 0.$
(b) Prove that $\displaystyle\lim_{\substack{x\to 0 \\ y\to 0}} \varepsilon_2(x, y) = \varepsilon_2(0, 0) = 0.$
(c) Prove that $|xy| = x\varepsilon_1(x, y) + y\varepsilon_2(x, y).$
(d) Use (a), (b), and (c) to show that the function f defined by $f(x, y) = |xy|$ is differentiable at $(0, 0)$.

32 Suppose that (x_0, y_0) is an interior point of the domain of f. Prove that f is differentiable at (x_0, y_0) if and only if there exists a function ϕ satisfying the conditions of problem 30.

33 Prove that $f(x, y) = \sqrt{x^2 + y^2}$ is not differentiable at $(0, 0)$.

34 Let

$$f(x, y) = \begin{cases} \dfrac{xy}{x^2 + y^2} & \text{if } (x, y) \neq (0, 0) \\ 0 & \text{if } (x, y) = (0, 0). \end{cases}$$

Prove that both of the partial derivatives $f_1(0, 0)$ and $f_2(0, 0)$ exist but that f is not differentiable at $(0, 0)$. [*Hint:* Show that f is not continuous at $(0, 0)$ and use Theorem 1.]

6   The Chain Rules

In Section 3.1 we gave a version of the chain rule for partial derivatives which was an immediate extension of the chain rule for functions of a single variable. In this section we discuss some additional versions of the chain rule for partial derivatives that are not just restatements of the old chain rule.

The simplest such chain rule is suggested by the notation for total differentials introduced in Section 5.3. Indeed, suppose that z is a function of the two variables x and y, say $z = f(x, y)$, while x and y, in turn, are functions of another variable t, so that $x = g(t)$ and $y = h(t)$. Then z becomes a function of the single variable t; that is, $z = f(g(t), h(t))$. Since

$$dz = \frac{\partial z}{\partial x} dx + \frac{\partial z}{\partial y} dy,$$

we might expect that

$$\frac{dz}{dt} = \frac{\partial z}{\partial x} \frac{dx}{dt} + \frac{\partial z}{\partial y} \frac{dy}{dt}.$$

This version of the chain rule is, in fact, correct, provided that f, g, and h are differentiable functions. Indeed, let Δt denote a small change in t and let Δx, Δy, and Δz be the resulting changes in the variables x, y, and z, respectively. Since f is differentiable, we have

$$\Delta z = \frac{\partial z}{\partial x} \Delta x + \frac{\partial z}{\partial y} \Delta y + \varepsilon_1 \Delta x + \varepsilon_2 \Delta y,$$

where $\varepsilon_1 \Delta x + \varepsilon_2 \Delta y$ is the error involved in the linear approximation

$$\Delta z \approx \frac{\partial z}{\partial x} \Delta x + \frac{\partial z}{\partial y} \Delta y \quad \text{and}$$

$$\lim_{\substack{\Delta x \to 0 \\ \Delta y \to 0}} \varepsilon_1 = \lim_{\substack{\Delta x \to 0 \\ \Delta y \to 0}} \varepsilon_2 = 0.$$

Dividing by Δt, we have

$$\frac{\Delta z}{\Delta t} = \frac{\partial z}{\partial x} \frac{\Delta x}{\Delta t} + \frac{\partial z}{\partial y} \frac{\Delta y}{\Delta t} + \varepsilon_1 \frac{\Delta x}{\Delta t} + \varepsilon_2 \frac{\Delta y}{\Delta t}.$$

Taking the limit on both sides as $\Delta t \to 0$, and noting that $\Delta x \to 0$ and $\Delta y \to 0$, so that $\varepsilon_1 \to 0$ and $\varepsilon_2 \to 0$ as $\Delta t \to 0$, we obtain

$$\frac{dz}{dt} = \frac{\partial z}{\partial x} \frac{dx}{dt} + \frac{\partial z}{\partial y} \frac{dy}{dt} + (0)\frac{dx}{dt} + (0)\frac{dy}{dt} = \frac{\partial z}{\partial x} \frac{dx}{dt} + \frac{\partial z}{\partial y} \frac{dy}{dt},$$

as claimed.

More formally, we have the following theorem.

THEOREM 1   **First Chain Rule**

Let f be a function of two variables and let g and h be functions of a single variable. Assume that (x_0, y_0) is an interior point of the domain of f and that f is differentiable at (x_0, y_0). Suppose that $x_0 = g(t_0)$, that $y_0 = h(t_0)$, and that both g and h are differentiable at t_0. Define the function by $F(t) = f(g(t), h(t))$. Then F is differentiable at t_0 and

$$F'(t_0) = f_1(x_0, y_0)g'(t_0) + f_2(x_0, y_0)h'(t_0).$$

EXAMPLES 1 If $z = \sqrt{x^2 + y^2}$, $x = 2t + 1$, and $y = t^3$, use the first chain rule to find dz/dt.

SOLUTION

$$\frac{dz}{dt} = \frac{\partial z}{\partial x}\frac{dx}{dt} + \frac{\partial z}{\partial y}\frac{dy}{dt} = \frac{x}{\sqrt{x^2 + y^2}}(2) + \frac{y}{\sqrt{x^2 + y^2}}(3t^2)$$

$$= \frac{2x + 3t^2 y}{\sqrt{x^2 + y^2}} = \frac{2(2t + 1) + 3t^2(t^3)}{\sqrt{(2t + 1)^2 + (t^3)^2}} = \frac{3t^5 + 4t + 2}{\sqrt{t^6 + 4t^2 + 4t + 1}}.$$

2 If $F(t) = f(g(t), h(t))$, where $f(x, y) = e^{xy}$, $g(t) = \cos t$, and $h(t) = \sin t$, find $F'(t)$.

SOLUTION
By Theorem 1,

$$F'(t) = f_1(g(t), h(t)) g'(t) + f_2(g(t), h(t)) h'(t).$$

Here, $f_1(x, y) = y e^{xy}$, $f_2(x, y) = x e^{xy}$, $g'(t) = -\sin t$, and $h'(t) = \cos t$. Therefore,

$$F'(t) = \sin t \, e^{\cos t \sin t}(-\sin t) + \cos t \, e^{\cos t \sin t}(\cos t)$$

$$= (\cos^2 t - \sin^2 t)e^{\cos t \sin t} = \cos 2t \, e^{\cos t \sin t}.$$

3 The resistance R in ohms of a circuit is given by $R = E/I$, where I is the current in amperes and E is the electromotive force in volts. At a certain instant when $E = 120$ volts and $I = 15$ amperes, E is increasing at a rate of $\frac{1}{10}$ volt per second and I is decreasing at the rate of $\frac{1}{20}$ ampere per second. Find the instantaneous rate of change of R.

SOLUTION
Denoting time in seconds by the variable t, we have

$$\frac{dR}{dt} = \frac{\partial R}{\partial E}\frac{dE}{dt} + \frac{\partial R}{\partial I}\frac{dI}{dt},$$

by the first chain rule. Here,

$$\frac{\partial R}{\partial E} = \frac{\partial}{\partial E}\left(\frac{E}{I}\right) = \frac{1}{I} \quad \text{and} \quad \frac{\partial R}{\partial I} = \frac{\partial}{\partial I}\left(\frac{E}{I}\right) = \frac{-E}{I^2}.$$

Therefore,

$$\frac{dR}{dt} = \frac{1}{I}\frac{dE}{dt} - \frac{E}{I^2}\frac{dI}{dt}.$$

When $E = 120$, $I = 15$, $dE/dt = \frac{1}{10}$, and $dI/dt = -\frac{1}{20}$, we have

$$\frac{dR}{dt} = \left(\frac{1}{15}\right)\left(\frac{1}{10}\right) - \left(\frac{120}{225}\right)\left(-\frac{1}{20}\right) = \frac{1}{30} \text{ ohm per second.}$$

The first chain rule (Theorem 1) extends in an obvious way to functions of more than two variables. In fact, if w is a function of n variables $x_1, x_2, \ldots, x_n$ and each of these n variables is, in turn, a function of a single variable t, then

$$\frac{dw}{dt} = \frac{\partial w}{\partial x_1}\frac{dx_1}{dt} + \frac{\partial w}{\partial x_2}\frac{dx_2}{dt} + \cdots + \frac{\partial w}{\partial x_n}\frac{dx_n}{dt},$$

provided that the function giving w in terms of $x_1, x_2, \ldots, x_n$ is differentiable and that the derivatives $\dfrac{dx_1}{dt}, \dfrac{dx_2}{dt}, \ldots, \dfrac{dx_n}{dt}$ exist.

EXAMPLE   Let $w = \ln \dfrac{x^2 y^2}{4z^3}$, $x = e^t$, $y = \sec t$, and $z = \cot t$. Use the first chain rule to find $\dfrac{dw}{dt}$.

SOLUTION

Here, $w = 2 \ln x + 2 \ln y - 3 \ln z - \ln 4$, so that

$$\frac{\partial w}{\partial x} = \frac{2}{x} = \frac{2}{e^t}, \qquad \frac{\partial w}{\partial y} = \frac{2}{y} = \frac{2}{\sec t}, \qquad \frac{\partial w}{\partial z} = -\frac{3}{z} = \frac{-3}{\cot t}.$$

Therefore,

$$\frac{dw}{dt} = \frac{\partial w}{\partial x}\frac{dx}{dt} + \frac{\partial w}{\partial y}\frac{dy}{dt} + \frac{\partial w}{\partial z}\frac{dz}{dt}$$

$$= \frac{2}{e^t} e^t + \frac{2}{\sec t} \sec t \tan t + \frac{-3}{\cot t}(-\csc^2 t) = 2 + 2 \tan t + 3 \sec t \csc t.$$

Now consider the case in which a dependent variable z is a function of the two variables x and y, say

$$z = f(x, y),$$

while x and y, in turn, are functions of the two variables u and v, so that

$$x = g(u, v), \qquad y = h(u, v).$$

Then z becomes a function of u and v, namely

$$z = f(g(u, v), h(u, v)).$$

Suppose that we temporarily hold the variable v constant and ask for the (partial) derivative of z with respect to u. Using Theorem 1, we get

$$\frac{\partial z}{\partial u} = \frac{\partial z}{\partial x}\frac{\partial x}{\partial u} + \frac{\partial z}{\partial y}\frac{\partial y}{\partial u},$$

provided that f is differentiable and that the partial derivatives $\partial x/\partial u$ and $\partial y/\partial u$ exist. Similarly, if f is differentiable and the partial derivatives $\partial x/\partial v$ and $\partial y/\partial v$ both exist, then

$$\frac{\partial z}{\partial v} = \frac{\partial z}{\partial x}\frac{\partial x}{\partial v} + \frac{\partial z}{\partial y}\frac{\partial y}{\partial v}.$$

The following theorem expresses the preceding facts more precisely using subscript notation for partial derivatives.

THEOREM 2   **Second Chain Rule**

Let f, g, and h be functions of two variables, let (x_0, y_0) be an interior point of the domain of f, and suppose that f is differentiable at (x_0, y_0). Let $x_0 = g(u_0, v_0)$, $y_0 = h(u_0, v_0)$, and suppose that the partial derivatives $g_1(u_0, v_0)$, $g_2(u_0, v_0)$, $h_1(u_0, v_0)$, and $h_2(u_0, v_0)$ exist. Define the function F by

$$F(u, v) = f(g(u, v), h(u, v)).$$

Then F has partial derivatives $F_1(u_0, v_0)$ and $F_2(u_0, v_0)$ and

$$F_1(u_0, v_0) = f_1(x_0, y_0)g_1(u_0, v_0) + f_2(x_0, y_0)h_1(u_0, v_0),$$

$$F_2(u_0, v_0) = f_1(x_0, y_0)g_2(u_0, v_0) + f_2(x_0, y_0)h_2(u_0, v_0).$$

If (u_0, v_0) is an interior point of the domains of both g and h, and if both g and h are differentiable at (u_0, v_0), then it can be shown that the function F of Theorem 2 is actually differentiable at (u_0, v_0).

EXAMPLES Use Theorem 2 to work the following problems.

1 Given that $z = x^2 - y^2$, $x = u \cos v$, and $y = v \sin u$, find $\partial z/\partial u$ and $\partial z/\partial v$.

SOLUTION

$$\frac{\partial z}{\partial x} = 2x, \qquad \frac{\partial z}{\partial y} = -2y, \qquad \frac{\partial x}{\partial u} = \cos v, \qquad \frac{\partial x}{\partial v} = -u \sin v,$$

$$\frac{\partial y}{\partial u} = v \cos u, \qquad \frac{\partial y}{\partial v} = \sin u.$$

Applying the second chain rule, we obtain

$$\frac{\partial z}{\partial u} = \frac{\partial z}{\partial x}\frac{\partial x}{\partial u} + \frac{\partial z}{\partial y}\frac{\partial y}{\partial u} = 2x \cos v - 2yv \cos u$$

$$= 2(u \cos v) \cos v - 2(v \sin u)v \cos u = 2u \cos^2 v - v^2 \sin 2u,$$

$$\frac{\partial z}{\partial v} = \frac{\partial z}{\partial x}\frac{\partial x}{\partial v} + \frac{\partial z}{\partial y}\frac{\partial y}{\partial v} = 2x(-u \sin v) - 2y \sin u$$

$$= 2(u \cos v)(-u \sin v) - 2(v \sin u) \sin u = -u^2 \sin 2v - 2v \sin^2 u.$$

2 Let f be differentiable at $(3, 1)$ and suppose that $f_1(3, 1) = 2$ and $f_2(3, 1) = -5$. If $V = f(2x + 3y, e^x)$, find $\partial V/\partial x$ and $\partial V/\partial y$ when $x = 0$ and $y = 1$.

SOLUTION
Let $s = 2x + 3y$ and $t = e^x$, so that $V = f(s, t)$ and

$$\frac{\partial V}{\partial x} = \frac{\partial V}{\partial s}\frac{\partial s}{\partial x} + \frac{\partial V}{\partial t}\frac{\partial t}{\partial x} = \frac{\partial V}{\partial s}(2) + \frac{\partial V}{\partial t}e^x = 2f_1(s, t) + e^x f_2(s, t).$$

When $x = 0$ and $y = 1$, we have $s = 3$, $t = 1$, and

$$\frac{\partial V}{\partial x} = 2f_1(3, 1) + e^0 f_2(3, 1) = (2)(2) + (1)(-5) = -1.$$

Similarly,

$$\frac{\partial V}{\partial y} = \frac{\partial V}{\partial s}\frac{\partial s}{\partial y} + \frac{\partial V}{\partial t}\frac{\partial t}{\partial y} = \frac{\partial V}{\partial s}(3) + \frac{\partial V}{\partial t}(0) = 3f_1(s, t).$$

Thus, when $x = 0$ and $y = 1$, we have $s = 3$, $t = 1$, and

$$\frac{\partial V}{\partial y} = 3f_1(3, 1) = (3)(2) = 6.$$

The second chain rule given in Theorem 2 admits a natural extension to functions of more than two variables. In fact, if w is a differentiable function of m variables $y_1, y_2, \ldots, y_m$, and if each of these variables, in turn, is a function of n variables $x_1, x_2, \ldots, x_n$, then

$$\frac{\partial w}{\partial x_j} = \frac{\partial w}{\partial y_1}\frac{\partial y_1}{\partial x_j} + \frac{\partial w}{\partial y_2}\frac{\partial y_2}{\partial x_j} + \cdots + \frac{\partial w}{\partial y_m}\frac{\partial y_m}{\partial x_j}$$

holds for each $j = 1, 2, \ldots, n$, provided that the partial derivatives $\dfrac{\partial y_1}{\partial x_j}, \dfrac{\partial y_2}{\partial x_j}, \ldots, \dfrac{\partial y_m}{\partial x_j}$ exist. The equations above can be written more compactly as

$$\frac{\partial w}{\partial x_j} = \sum_{k=1}^{m} \frac{\partial w}{\partial y_k} \frac{\partial y_k}{\partial x_j}, \qquad \text{for } j = 1, 2, \ldots, n.$$

For example, if $w = f(x, y, z)$, $x = g(s, t, u)$, $y = h(s, t, u)$, and $z = p(s, t, u)$, and if f is differentiable, then

$$\frac{\partial w}{\partial s} = \frac{\partial w}{\partial x} \frac{\partial x}{\partial s} + \frac{\partial w}{\partial y} \frac{\partial y}{\partial s} + \frac{\partial w}{\partial z} \frac{\partial z}{\partial s},$$

$$\frac{\partial w}{\partial t} = \frac{\partial w}{\partial x} \frac{\partial x}{\partial t} + \frac{\partial w}{\partial y} \frac{\partial y}{\partial t} + \frac{\partial w}{\partial z} \frac{\partial z}{\partial t},$$

$$\frac{\partial w}{\partial u} = \frac{\partial w}{\partial x} \frac{\partial x}{\partial u} + \frac{\partial w}{\partial y} \frac{\partial y}{\partial u} + \frac{\partial w}{\partial z} \frac{\partial z}{\partial u},$$

provided that all the partial derivatives of x, y, and z with respect to s, t, and u exist.

EXAMPLES **1** Let $w = xy^2 + yz^2 + zx^2$, $x = r \cos\theta \sin\phi$, $y = r \sin\theta \sin\phi$, and $z = r\cos\phi$. Find $\partial w/\partial r$, $\partial w/\partial\theta$, and $\partial w/\partial\phi$.

SOLUTION
Using the chain rule, we have

$$\frac{\partial w}{\partial r} = \frac{\partial w}{\partial x}\frac{\partial x}{\partial r} + \frac{\partial w}{\partial y}\frac{\partial y}{\partial r} + \frac{\partial w}{\partial z}\frac{\partial z}{\partial r}$$

$$= (y^2 + 2xz)\cos\theta\sin\phi + (2xy + z^2)\sin\theta\sin\phi + (2yz + x^2)\cos\phi,$$

$$\frac{\partial w}{\partial\theta} = \frac{\partial w}{\partial x}\frac{\partial x}{\partial\theta} + \frac{\partial w}{\partial y}\frac{\partial y}{\partial\theta} + \frac{\partial w}{\partial z}\frac{\partial z}{\partial\theta}$$

$$= (y^2 + 2xz)(-r\sin\theta\sin\phi) + (2xy + z^2)(r\cos\theta\sin\phi) + (2yz + x^2)(0)$$

$$= -(y^2 + 2xz)r\sin\theta\sin\phi + (2xy + z^2)r\cos\theta\sin\phi, \quad \text{and}$$

$$\frac{\partial w}{\partial\phi} = \frac{\partial w}{\partial x}\frac{\partial x}{\partial\phi} + \frac{\partial w}{\partial y}\frac{\partial y}{\partial\phi} + \frac{\partial w}{\partial z}\frac{\partial z}{\partial\phi}$$

$$= (y^2 + 2xz)r\cos\theta\cos\phi + (2xy + z^2)r\sin\theta\cos\phi + (2yz + x^2)(-r\sin\phi).$$

2 Suppose that f is a differentiable function at $(0,0,0)$ and that $f_1(0,0,0) = 3$, $f_2(0,0,0) = 7$, and $f_3(0,0,0) = -2$. If the function g is defined by the equation $g(x, y) = f(x^2 - y^2, 4x - 4y, 5x - 5)$, find $g_1(1, 1)$ and $g_2(1, 1)$.

SOLUTION
Let $u = x^2 - y^2$, $v = 4x - 4y$, and $w = 5x - 5$, and put $z = f(u, v, w)$. Then $z = f(x^2 - y^2, 4x - 4y, 5x - 5) = g(x, y)$, so

$$g_1(x, y) = \frac{\partial z}{\partial x} = \frac{\partial z}{\partial u}\frac{\partial u}{\partial x} + \frac{\partial z}{\partial v}\frac{\partial v}{\partial x} + \frac{\partial z}{\partial w}\frac{\partial w}{\partial x}$$

$$= f_1(u, v, w)(2x) + f_2(u, v, w)(4) + f_3(u, v, w)(5).$$

When $x = 1$ and $y = 1$, we have $u = 0$, $v = 0$, $w = 0$ and

$$g_1(1,1) = 2f_1(0,0,0) + 4f_2(0,0,0) + 5f_3(0,0,0)$$
$$= 2(3) + 4(7) + 5(-2) = 24.$$

Likewise,

$$g_2(x,y) = \frac{\partial z}{\partial y} = \frac{\partial z}{\partial u}\frac{\partial u}{\partial y} + \frac{\partial z}{\partial v}\frac{\partial v}{\partial y} + \frac{\partial z}{\partial w}\frac{\partial w}{\partial y}$$
$$= f_1(u,v,w)(-2y) + f_2(u,v,w)(-4) + f_3(u,v,w)(0)$$
$$= -2yf_1(u,v,w) - 4f_2(u,v,w),$$

and so

$$g_2(1,1) = -2f_1(0,0,0) - 4f_2(0,0,0) = -2(3) - 4(7) = -34.$$

6.1 Implicit Differentiation

The procedure of implicit differentiation, originally discussed in Section 8 of Chapter 3, can be formulated with more precision and can be generalized by using partial derivatives.

For instance, given an equation involving the variables x and y, we can transpose all terms to the left of the equality sign so that the equation takes the form

$$f(x,y) = 0,$$

where f is a function of two variables. This equation is said to define y *implicitly* as a function g of x if

$$f(x,g(x)) = 0$$

holds for all values of x in the domain of g. Assuming that both f and g are differentiable, then by Theorem 1 we can differentiate both sides of the equation $f(x,g(x)) = 0$ with respect to x and obtain

$$f_1(x,g(x))\frac{dx}{dx} + f_2(x,g(x))\frac{d}{dx}g(x) = 0$$

or

$$f_1(x,y) + f_2(x,y)\frac{dy}{dx} = 0,$$

where $y = g(x)$. If $f_2(x,y) \neq 0$, we can solve the latter equation for dy/dx, and thus obtain

$$\frac{dy}{dx} = -\frac{f_1(x,y)}{f_2(x,y)}.$$

EXAMPLE Suppose that y is a function of x given implicitly by the equation $x^3y^2 + 3xy^2 + 5x^4 = 2y + 7$. Find the value of dy/dx when $x = 1$ and $y = 1$.

SOLUTION
By transposing all terms to the left, we put the equation into the form $f(x,y) = 0$, where $f(x,y) = x^3y^2 + 3xy^2 + 5x^4 - 2y - 7$. Here,

$$f_1(x,y) = 3x^2y^2 + 3y^2 + 20x^3 \quad \text{and} \quad f_2(x,y) = 2x^3y + 6xy - 2;$$

hence,

$$\frac{dy}{dx} = -\frac{f_1(x, y)}{f_2(x, y)} = -\frac{3x^2y^2 + 3y^2 + 20x^3}{2x^3y + 6xy - 2}.$$

Therefore, when $x = 1$ and $y = 1$,

$$\frac{dy}{dx} = -\frac{3 + 3 + 20}{2 + 6 - 2} = -\frac{13}{3}.$$

More generally, given an equation of the form

$$f(x, y, z) = 0$$

involving three variables, it may be possible to solve for one of the variables, say y, in terms of the other two variables x and z. If this solution has the form

$$y = g(x, z),$$

then

$$f(x, g(x, z), z) = 0$$

holds for all points (x, z) in the domain of the function g. Again, we say that the equation $f(x, y, z) = 0$ defines y *implicitly* as a function g of x and z. Assuming that the functions f and g are differentiable, we can take the partial derivatives with respect to x and also with respect to z on both sides of the equation $f(x, y, z) = 0$ to obtain

$$f_1(x, y, z)\frac{\partial x}{\partial x} + f_2(x, y, z)\frac{\partial y}{\partial x} + f_3(x, y, z)\frac{\partial z}{\partial x} = 0 \quad \text{and}$$

$$f_1(x, y, z)\frac{\partial x}{\partial z} + f_2(x, y, z)\frac{\partial y}{\partial z} + f_3(x, y, z)\frac{\partial z}{\partial z} = 0.$$

Since x and z are independent variables, we have $\partial z/\partial x = 0$, $\partial x/\partial z = 0$, $\partial x/\partial x = 1$, and $\partial z/\partial z = 1$. Therefore, we can rewrite the equations above as

$$f_2(x, y, z)\frac{\partial y}{\partial x} = -f_1(x, y, z) \quad \text{and} \quad f_2(x, y, z)\frac{\partial y}{\partial z} = -f_3(x, y, z).$$

Hence, if $f_2(x, y, z) \neq 0$, we can solve for $\partial y/\partial x$ and $\partial y/\partial z$ to obtain

$$\frac{\partial y}{\partial x} = -\frac{f_1(x, y, z)}{f_2(x, y, z)} \quad \text{and} \quad \frac{\partial y}{\partial z} = -\frac{f_3(x, y, z)}{f_2(x, y, z)}.$$

EXAMPLE   Suppose that y is a function of x and z given implicitly by the equation $7x^3y - 4xyz^3 + x^2y^3z^2 - z - 14 = 0$. Find $\partial y/\partial x$ and $\partial y/\partial z$ when $x = 1$, $z = 0$, and $y = 2$.

SOLUTION
The equation has the form $f(x, y, z) = 0$, where

$$f(x, y, z) = 7x^3y - 4xyz^3 + x^2y^3z^2 - z - 14.$$

Here,

$$f_1(x, y, z) = 21x^2y - 4yz^3 + 2xy^3z^2,$$

$$f_2(x, y, z) = 7x^3 - 4xz^3 + 3x^2y^2z^2,$$

$$f_3(x, y, z) = -12xyz^2 + 2x^2y^3z - 1.$$

Thus,

$$\frac{\partial y}{\partial x} = -\frac{21x^2y - 4yz^3 + 2xy^3z^2}{7x^3 - 4xz^3 + 3x^2y^2z^2} \quad \text{and} \quad \frac{\partial y}{\partial z} = -\frac{-12xyz^2 + 2x^2y^3z - 1}{7x^3 - 4xz^3 + 3x^2y^2z^2}.$$

Putting $x = 1$, $z = 0$, and $y = 2$, we obtain

$$\frac{\partial y}{\partial x} = -\frac{42}{7} = -6 \quad \text{and} \quad \frac{\partial y}{\partial z} = -\frac{-1}{7} = \frac{1}{7}.$$

The considerations above can be generalized in an obvious way to equations involving more than three variables (Problem 36). They can even be extended to simultaneous systems of such equations (Problem 42).

Problem Set 6

In problems 1 through 10, use the version of the chain rule given in Theorem 1 (or generalizations thereof) to find each derivative.

1 $\dfrac{dz}{dt}$, where $z = x^3y^2 - 3xy + y^2$, $x = 2t$, and $y = 6t^2$.

2 $\dfrac{dw}{dt}$, where $w = e^{x^2y}$, $x = \sin t$, and $y = \cos t$.

3 $\dfrac{dw}{dx}$, where $w = u \sin v + \cos (u - v)$, $u = x^2$, and $v = x^3$.

4 $\dfrac{dw}{d\theta}$, where $w = \sqrt{u^2 - v^2}$, $u = \sin \theta$, and $v = \cos \theta$.

5 $F'(t)$, where $F(t) = f(g(t), h(t))$, $f(x, y) = \sin (x + y) + \sin (x - y)$, $g(t) = 3t$, and $h(t) = t^3$.

6 $G'(t)$, where $G(t) = g(f(t), h(t))$, $g(u, v) = \ln (u^3 + v^3)$, $f(t) = e^{3t}$, and $h(t) = e^{-7t}$.

7 $F'(t)$, where $F(t) = f(g(t), h(t))$, $f(u, v) = \frac{1}{2}(e^{u/v} - e^{-u/v})$, $g(t) = \sinh t$, and $h(t) = t$.

8 $H'(u)$, where $H(u) = q(f(u), g(u), h(u))$, $q(x, y, z) = x^2y + xz^2 - yz^2 + xyz$, $f(u) = e^u$, $g(u) = e^{-u}$, and $h(u) = \cosh u$.

9 $\dfrac{dw}{dt}$, where $w = \ln \dfrac{x^3y^2}{5z}$, $x = 7t$, $y = \sec t$, and $z = \cot t$.

10 $F'(t)$, where $F(t) = f(g(t), h(t), p(t))$, $f(x, y, z) = \tan^{-1}(xyz)$, $g(t) = t^2$, $h(t) = t^3$, and $p(t) = t^4$.

In problems 11 through 24, use the version of the chain rule given in Theorem 2 (or generalizations thereof) to find each partial derivative.

11 $\dfrac{\partial z}{\partial u}$ and $\dfrac{\partial z}{\partial v}$, where $z = 3x^2 - 4y^2$, $x = uv$, and $y = \cos u + \sin v$.

12 $\dfrac{\partial w}{\partial r}$ and $\dfrac{\partial w}{\partial s}$, where $w = 4x^2 + 5xy - 2y^3$, $x = 3r + 5s$, and $y = 7r^2s$.

13 $\dfrac{\partial z}{\partial u}$ and $\dfrac{\partial z}{\partial v}$, where $z = 4x^3 - 3x^2y^2$, $x = u \cos v$, and $y = v \sin u$.

14 $\dfrac{\partial w}{\partial x}$ and $\dfrac{\partial w}{\partial y}$, where $w = u^2 - uv + 5v^2$, $u = x \cos 2y$, and $v = x \sin 2y$.

15 $\dfrac{\partial w}{\partial x}$ and $\dfrac{\partial w}{\partial y}$, where $w = \ln{(u^2 + v^2)}$, $u = x^2 + y^2$, and $v = 2x^2 + 3xy$.

16 $\dfrac{\partial z}{\partial x}$ and $\dfrac{\partial z}{\partial y}$, where $z = e^{s/t}$, $s = xy^2$, and $t = 5x + 2y^3$.

17 $\dfrac{\partial u}{\partial r}$ and $\dfrac{\partial u}{\partial s}$, where $u = \cosh{(3x + 7y)}$, $x = r^2 e^{-s}$, and $y = re^{3s}$.

18 $F_1(r, s)$ and $F_2(r, s)$, where $F(r, s) = f(g(r, s), h(r, s))$, $f(x, y) = \tan^{-1}{(y/x)}$, $g(r, s) = 2r + s$, and $h(r, s) = r^2 s$.

19 $F_1(u, v)$ and $F_2(u, v)$, where $F(u, v) = f(g(u, v), h(u, v))$, $f(x, y) = e^{xy^2}$, $g(u, v) = u^2 v$, and $h(u, v) = uv^2$.

20 $\dfrac{\partial w}{\partial r}$ and $\dfrac{\partial w}{\partial s}$, where $w = 6xyz^2$, $x = rs$, $y = 2r + s$, and $z = 3r^2 - s$.

21 $\dfrac{\partial w}{\partial u}$ and $\dfrac{\partial w}{\partial v}$, where $w = 2x^2 + 3y^2 + z^2$, $x = u\cos{v}$, $y = u\sin{v}$, and $z = uv$.

22 $F_1(r, s)$ and $F_2(r, s)$, where $F(r, s) = q(f(r, s), g(r, s), h(r, s))$, $q(x, y, z) = xy^2 + yz^2$, $f(r, s) = r\cosh{s}$, $g(r, s) = r\sinh{s}$, and $h(r, s) = re^s$.

23 $\dfrac{\partial w}{\partial \rho}$, $\dfrac{\partial w}{\partial \theta}$, and $\dfrac{\partial w}{\partial \phi}$, where $w = x^2 + y^2 - z^2$, $x = \rho\sin{\phi}\cos{\theta}$, $y = \rho\sin{\phi}\sin{\theta}$, and $z = \rho\cos{\phi}$.

24 $\dfrac{\partial w}{\partial r}$ and $\dfrac{\partial w}{\partial s}$, where $w = \displaystyle\int_x^y e^{t^2}\, dt$, $x = rs^4$, and $y = r^4 s$.

25 Given a differentiable function f of three variables such that $f_1(0, 0, 0) = 4$, $f_2(0, 0, 0) = 3$, and $f_3(0, 0, 0) = 5$, let $g(r, s) = f(r - s, r^2 - 1, 3s - 3)$, find $g_1(1, 1)$ and $g_2(1, 1)$.

26 Let g and h be differentiable functions of two variables and define $f(r, s) = [g(r, s)]^{h(r, s)}$. Assume that $g(1, 2) = 2$, $h(1, 2) = -2$, $g_1(1, 2) = -1$, $g_2(1, 2) = 3$, $h_1(1, 2) = 5$, and $h_2(1, 2) = 0$. Find
(a) $f(1, 2)$ (b) $f_1(1, 2)$ (c) $f_2(1, 2)$

In problems 27 through 30, assume that y is given implicitly as a differentiable function g of x by the given equation $f(x, y) = 0$.

(a) Use the result $\dfrac{dy}{dx} = -\dfrac{f_1(x, y)}{f_2(x, y)}$ to find $\dfrac{dy}{dx}$.

(b) Find the slope of the tangent line to the graph of $y = g(x)$ at the given point (x, y).

27 $6x^2 - 12xy + 4y^2 + 2 = 0$; $(1, 1)$.

28 $(x^2 - y^2)^2 - x^2 y^2 - 55 = 0$; $(3, 1)$.

29 $\sin{(x - y)} + \cos{(x + y)} = 0$; $(\pi/4, \pi/4)$.

30 $\tan^{-1}{(y/x)} + 3e^{2x - 2y} - 3 = \pi/4$; $(1, 1)$.

In problems 31 and 32, assume that y is given implicitly as a differentiable function of the remaining variables in each equation. Find the value of the indicated partial derivatives when the variables have the values given.

31 If $xy^2 z - 3x^2 yz + \dfrac{2xz}{y} - z^2 = 0$, find $\dfrac{\partial y}{\partial x}$ and $\dfrac{\partial y}{\partial z}$ when $x = 1$, $z = -1$, and $y = 2$.

32 If $\dfrac{\sin{(y - x)}}{z^2} = \dfrac{\cos{(x + y)}}{w^2 - 3}$, find $\dfrac{\partial y}{\partial x}$, $\dfrac{\partial y}{\partial z}$, and $\dfrac{\partial y}{\partial w}$ when $x = \dfrac{\pi}{4}$, $y = \dfrac{\pi}{4}$, $z = 1$, and $w = -2$.

33 If f is a differentiable function of two variables and $w = f(ay - x, x - ay)$, where a is a constant, prove that

$$a \frac{\partial w}{\partial x} + \frac{\partial w}{\partial y} = 0.$$

34 If f is a differentiable function of two variables and if a and b are constants, show that $w = f(x + az, y + bz)$ provides a solution of the partial differential equation

$$\frac{\partial w}{\partial z} = a \frac{\partial w}{\partial x} + b \frac{\partial w}{\partial y}.$$

35 Let $w = f(x, y)$, where f is a differentiable function. If $x = r \cos \theta$ and $y = r \sin \theta$, show that

$$\left(\frac{\partial w}{\partial x}\right)^2 + \left(\frac{\partial w}{\partial y}\right)^2 = \left(\frac{\partial w}{\partial r}\right)^2 + \frac{1}{r^2}\left(\frac{\partial w}{\partial \theta}\right)^2.$$

36 Suppose that f is a differentiable function of n variables and that it is possible to solve the equation $f(x_1, x_2, \ldots, x_k, \ldots, x_n) = 0$ for the kth variable x_k as a differentiable function of the remaining variables. Prove that $\dfrac{\partial x_k}{\partial x_i} = -\dfrac{f_i(x_1, x_2, \ldots, x_k, \ldots, x_n)}{f_k(x_1, x_2, \ldots, x_k, \ldots, x_n)}$ holds for $1 \le i \le n$, $i \ne k$, provided that $f_k(x_1, x_2, \ldots, x_k, \ldots, x_n) \ne 0$.

37 Let f be a differentiable function of three variables and let $w = f(x - y, y - z, z - x)$. Show that w satisfies the partial differential equation $\dfrac{\partial w}{\partial x} + \dfrac{\partial w}{\partial y} + \dfrac{\partial w}{\partial z} = 0$.

38 The volume V of a right circular cone is given by $V = (\pi r^2/3)\sqrt{s^2 - r^2}$, where r is the radius of the base and s is the slant height. At a certain instant when $r = 4$ centimeters and $s = 10$ centimeters, r is decreasing at the rate of 2 centimeters per minute and s is increasing at the rate of 3 centimeters per minute. Find the rate of change of V at this instant.

39 At a certain instant, the legs of a right triangle have lengths 2 feet and 4 feet and they are increasing at the rates of 1 foot per minute and 2 feet per minute, respectively. (a) How fast is the area of the triangle changing? (b) How fast is the perimeter of the triangle changing?

40 A motorcyclist starts at point A and travels toward point B at 25 miles per hour on a straight-line trail $\overline{AB}$ which is 56 miles long. At the same time, a second motorcyclist leaves B in a direction which makes an angle of $60°$ with $\overline{AB}$ and travels at 30 miles per hour. How fast is the straight-line distance between the two riders changing at the end of 1 hour?

41 The equation of a perfect gas is $PV = kT$, where T is the temperature, P is the pressure, V is the volume, and k is a constant. (The temperature is given in degrees Kelvin, where $0°$K is equivalent to $-273°$C.) At a certain instant, a sample of gas is under a pressure of 2×10^6 dynes/cm^2, its volume is 5000 cm^3, and its temperature is $300°$K. If the pressure is increasing at the rate of 1.5×10^5 dynes/cm^2 per minute and the volume is decreasing at the rate of 750 cm^3 per minute, find the rate at which the temperature is changing.

42 Suppose that f and g are differentiable functions of three variables and that the simultaneous equations $\begin{cases} f(x, y, z) = 0 \\ g(x, y, z) = 0 \end{cases}$ can be solved for x and y in terms of z. Assuming that x and y are differentiable functions of z and that the determinant $\begin{vmatrix} f_1 & f_2 \\ g_1 & g_2 \end{vmatrix}$ is not zero, show that

$$\frac{dx}{dz} = -\frac{\begin{vmatrix} f_3 & f_2 \\ g_3 & g_2 \end{vmatrix}}{\begin{vmatrix} f_1 & f_2 \\ g_1 & g_2 \end{vmatrix}} \quad \text{and} \quad \frac{dy}{dz} = -\frac{\begin{vmatrix} f_1 & f_3 \\ g_1 & g_3 \end{vmatrix}}{\begin{vmatrix} f_1 & f_2 \\ g_1 & g_2 \end{vmatrix}}.$$

7   Directional Derivatives, Gradients, Normal Lines, and Tangent Planes

In this section we study the *directional derivative* and the closely related idea of the *gradient* of a scalar field. The section also includes a discussion of the *normal line* and the *tangent plane* to a surface in space.

7.1   Directional Derivative and Gradient in the Plane

Consider a scalar field in the xy plane described by a differentiable function f of two variables. Thus, if $z = f(x, y)$, then z is the value of the scalar field at the point $P = (x, y)$. Let L denote a straight line in the xy plane. As P moves along L, z may change, and it makes sense to ask for the rate of change dz/ds of z with respect to distance s measured along L (Figure 1).

In order to find dz/ds, we introduce a unit vector $\bar{u} = a\bar{i} + b\bar{j}$ parallel to L and in the direction of motion of P along L (Figure 2). If $P = (x, y)$ is s units from a fixed point $P_0 = (x_0, y_0)$ on L, then $\overrightarrow{P_0P} = s\bar{u}$; that is,

$$(x - x_0)\bar{i} + (y - y_0)\bar{j} = as\bar{i} + bs\bar{j}.$$

Equating components, we have $x - x_0 = as$ and $y - y_0 = bs$; that is, $x = x_0 + as$ and $y = y_0 + bs$. Therefore,

$$\frac{dx}{ds} = a \quad \text{and} \quad \frac{dy}{ds} = b,$$

and it follows from the chain rule that

$$\frac{dz}{ds} = \frac{\partial z}{\partial x}\frac{dx}{ds} + \frac{\partial z}{\partial y}\frac{dy}{ds} = \frac{\partial z}{\partial x}a + \frac{\partial z}{\partial y}b.$$

Figure 1

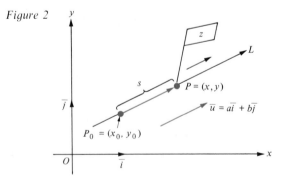

Figure 2

The derivative dz/ds, which is the rate of change of the scalar field z with respect to distance measured in the direction of the unit vector $\bar{u}$, is called the *directional derivative* of z (or the *directional derivative* of the function f) *in the direction of* $\bar{u}$ and is written as $D_{\bar{u}}z$ (or as $D_{\bar{u}}f$). Thus, we have

$$D_{\bar{u}}z = \frac{\partial z}{\partial x}a + \frac{\partial z}{\partial y}b \quad \text{or} \quad D_{\bar{u}}f(x, y) = f_1(x, y)a + f_2(x, y)b,$$

where

$$\bar{u} = a\vec{i} + b\vec{j}.$$

In particular, if $\bar{u}$ is the unit vector that makes the angle θ with the positive x axis, then $\bar{u} = (\cos \theta)\vec{i} + (\sin \theta)\vec{j}$ and

$$D_{\bar{u}}z = \frac{\partial z}{\partial x} \cos \theta + \frac{\partial z}{\partial y} \sin \theta \quad \text{or} \quad D_{\bar{u}}f(x, y) = f_1(x, y) \cos \theta + f_2(x, y) \sin \theta.$$

EXAMPLE   A temperature field in the xy plane is given by

$$z = 60 + \left(\frac{x}{20}\right)^2 + \left(\frac{y}{25}\right)^2,$$

where z is the temperature in degrees F at the point (x, y) and where distances are measured in miles. How rapidly is the temperature changing in degrees F per mile as we move from left to right through the point $(60, 75)$ along a straight line L making an angle of $30°$ with the positive x axis?

SOLUTION
A unit vector $\bar{u}$ parallel to L and in the direction of motion along L is given by $\bar{u} = (\cos 30°)\vec{i} + (\sin 30°)\vec{j} = (\sqrt{3}/2)\vec{i} + (1/2)\vec{j}$. Hence,

$$D_{\bar{u}}z = \frac{\partial z}{\partial x}\left(\frac{\sqrt{3}}{2}\right) + \frac{\partial z}{\partial y}\left(\frac{1}{2}\right) = \left[2\left(\frac{x}{20}\right)\right]\left(\frac{\sqrt{3}}{2}\right) + \left[2\left(\frac{y}{25}\right)\right]\left(\frac{1}{2}\right) = \frac{\sqrt{3}\,x}{20} + \frac{y}{25}.$$

Putting $x = 60$ and $y = 75$, we find that the rate of change of z as we move through the point $(60, 75)$ in the direction of $\bar{u}$ is

$$\frac{\sqrt{3}(60)}{20} + \frac{75}{25} = 3\sqrt{3} + 3 \approx 8.2 \text{ degrees F per mile.}$$

The vector $\vec{i}$ makes the angle $\theta = 0$ with the positive x axis; hence,

$$D_{\vec{i}}z = \frac{\partial z}{\partial x} \cos 0 + \frac{\partial z}{\partial y} \sin 0 = \frac{\partial z}{\partial x}.$$

Similarly, the vector $\vec{j}$ makes the angle $\theta = \pi/2$; hence,

$$D_{\vec{j}}z = \frac{\partial z}{\partial x} \cos \frac{\pi}{2} + \frac{\partial z}{\partial y} \sin \frac{\pi}{2} = \frac{\partial z}{\partial y}.$$

Therefore, the directional derivatives of z in the directions of the positive x and y axes are just the partial derivatives of z with respect to x and y, respectively.
The directional derivative $D_{\bar{u}}z$ can be expressed as a dot product

$$D_{\bar{u}}z = \frac{\partial z}{\partial x} a + \frac{\partial z}{\partial y} b = a\frac{\partial z}{\partial x} + b\frac{\partial z}{\partial y} = (a\vec{i} + b\vec{j}) \cdot \left(\frac{\partial z}{\partial x}\vec{i} + \frac{\partial z}{\partial y}\vec{j}\right) = \bar{u} \cdot \left(\frac{\partial z}{\partial x}\vec{i} + \frac{\partial z}{\partial y}\vec{j}\right).$$

The vector $\dfrac{\partial z}{\partial x}\vec{i} + \dfrac{\partial z}{\partial y}\vec{j}$ whose scalar components are the partial derivatives of z with respect to x and y is called the *gradient* of the scalar field z (or of the function f) and is written as ∇z (or as ∇f). The symbol ∇, an inverted Greek delta, is called "del" or "nabla." Thus, we have

$$\nabla z = \frac{\partial z}{\partial x}\vec{i} + \frac{\partial z}{\partial y}\vec{j} \quad \text{or} \quad \nabla f(x, y) = f_1(x, y)\vec{i} + f_2(x, y)\vec{j},$$

and we can write the directional derivative as

$$D_{\bar{u}}z = \bar{u} \cdot \nabla z \quad \text{or} \quad D_{\bar{u}}f(x,y) = \bar{u} \cdot \nabla f(x,y).$$

In words, the directional derivative of a scalar field in a given direction is the dot product of a unit vector in this direction and the gradient of the scalar field.

EXAMPLES 1 If $z = 4x^2 - 5xy^2$, find (a) ∇z, (b) the value of ∇z at the point $(2, -3)$, and (c) the directional derivative $D_{\bar{u}}z$ at the point $(2, -3)$ in the direction of the unit vector $\bar{u} = (\cos \pi/3)\bar{i} + (\sin \pi/3)\bar{j}$.

SOLUTION

(a) $\nabla z = \dfrac{\partial z}{\partial x}\bar{i} + \dfrac{\partial z}{\partial y}\bar{j} = (8x - 5y^2)\bar{i} + (-10xy)\bar{j}.$

(b) At the point $(2, -3)$,

$$\nabla z = [8(2) - 5(-3)^2]\bar{i} + [-10(2)(-3)]\bar{j} = -29\bar{i} + 60\bar{j}.$$

(c) At the point $(2, -3)$,

$$D_{\bar{u}}z = \bar{u} \cdot \nabla z = \left[\left(\cos\frac{\pi}{3}\right)\bar{i} + \left(\sin\frac{\pi}{3}\right)\bar{j}\right] \cdot (-29\bar{i} + 60\bar{j})$$

$$= -29\cos\frac{\pi}{3} + 60\sin\frac{\pi}{3} = (-29)\left(\frac{1}{2}\right) + (60)\left(\frac{\sqrt{3}}{2}\right) = \frac{60\sqrt{3} - 29}{2}.$$

2 If $f(x,y) = 4x^2 + xy + 9y^2$, find (a) $\nabla f(1,2)$ and (b) $D_{\bar{u}}f(1,2)$, where $\bar{u}$ is a unit vector in the direction of the vector $\bar{v} = 4\bar{i} - 3\bar{j}$.

SOLUTION
We obtain $\bar{u}$ by normalizing $\bar{v}$; thus,

$$\bar{u} = \frac{\bar{v}}{|\bar{v}|} = \frac{4\bar{i} - 3\bar{j}}{\sqrt{4^2 + (-3)^2}} = \frac{4\bar{i} - 3\bar{j}}{\sqrt{25}} = \frac{4}{5}\bar{i} - \frac{3}{5}\bar{j}.$$

(a) Here, $f_1(x,y) = 8x + y$ and $f_2(x,y) = x + 18y$, so that

$$\nabla f(1,2) = f_1(1,2)\bar{i} + f_2(1,2)\bar{j} = 10\bar{i} + 37\bar{j}.$$

(b) $D_{\bar{u}}f(1,2) = \bar{u} \cdot \nabla f(1,2) = \left(\dfrac{4}{5}\bar{i} - \dfrac{3}{5}\bar{j}\right) \cdot (10\bar{i} + 37\bar{j}) = \dfrac{40}{5} - \dfrac{111}{5} = -\dfrac{71}{5}.$

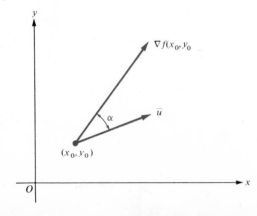

Figure 3

Notice that, if we fix a point (x_0, y_0) in the xy plane, then the directional derivative

$$D_{\bar{u}}f(x_0, y_0) = \bar{u} \cdot \nabla f(x_0, y_0)$$

depends only on our choice of the unit vector $\bar{u}$, since the gradient vector $\nabla f(x_0, y_0)$ is fixed. If α denotes the angle between $\bar{u}$ and $\nabla f(x_0, y_0)$ (Figure 3), then, by the definition of the dot product,

$$\bar{u} \cdot \nabla f(x_0, y_0) = |\bar{u}||\nabla f(x_0, y_0)| \cos \alpha.$$

Since $|\bar{u}| = 1$, it follows that

$$D_{\bar{u}}f(x_0, y_0) = |\nabla f(x_0, y_0)| \cos \alpha.$$

As we vary the angle α in the latter formula, we obtain the value of the directional derivative in various directions at the point (x_0, y_0). Taking $\alpha = \pi/2$, we have $\cos \alpha = 0$, so that $D_{\bar{u}}f(x_0, y_0) = 0$. Therefore, we have the following fact:

1 The directional derivative is zero when taken in a direction perpendicular to the gradient.

Since $\cos \alpha$ assumes its maximum value, namely 1, when $\alpha = 0$, we also obtain the following fact:

2 The directional derivative assumes its maximum value when taken in the direction of the gradient, and this maximum value is $|\nabla f(x_0, y_0)|$.

In other words, the gradient of a scalar field, evaluated at a point P, is a vector that points in the direction in which the scalar field increases most rapidly, while the length of the gradient vector is numerically equal to the instantaneous rate of increase of the field per unit distance in this direction as we move through the point P.

For instance, if we are at a given point in a temperature field and wish to "warm up" most rapidly, we should move in the direction of the gradient vector at this point. On the other hand, if we move perpendicular to the gradient, the instantaneous rate of change of temperature is zero and we find ourselves moving in the direction of the isotherm through the given point. Moving in the direction opposite to the gradient (that is, in the direction of the negative of the gradient), we "cool off" most rapidly.

EXAMPLES 1 Find (a) the maximum value of the directional derivative and (b) a unit vector $\bar{u}$ in the direction for which this maximum is attained for $f(x, y) = 2x^2y + xe^{y^2}$ at the point $(1, 0)$.

SOLUTION
Here,

$$\nabla f(x, y) = f_1(x, y)\mathbf{i} + f_2(x, y)\mathbf{j} = (4xy + e^{y^2})\mathbf{i} + (2x^2 + 2xye^{y^2})\mathbf{j},$$

so that

$$\nabla f(1, 0) = \mathbf{i} + 2\mathbf{j}.$$

Therefore:
(a) The maximum directional derivative at $(1, 0)$ is

$$|\nabla f(1, 0)| = |\mathbf{i} + 2\mathbf{j}| = \sqrt{1^2 + 2^2} = \sqrt{5}.$$

(b) The maximum directional derivative at $(1, 0)$ is attained in the direction of $\nabla f(1, 0) = \mathbf{i} + 2\mathbf{j}$. A unit vector $\bar{u}$ in the same direction is

$$\bar{u} = \frac{\nabla f(1, 0)}{|\nabla f(1, 0)|} = \frac{\mathbf{i} + 2\mathbf{j}}{\sqrt{5}}.$$

2 The temperature T in degrees C at the point (x, y) on a heated metal plate is given by $T = \dfrac{300}{x^2 + y^2 + 3}$, where x and y are measured in centimeters.
(a) In what direction should one move from the point $(-4, 3)$ in order that T should increase most rapidly?
(b) How rapidly does T increase as one moves through the point $(-4, 3)$ in the direction found in part (a)?

SOLUTION
(a) $\nabla T = \dfrac{\partial T}{\partial x}\mathbf{i} + \dfrac{\partial T}{\partial y}\mathbf{j} = \dfrac{-600x}{(x^2 + y^2 + 3)^2}\mathbf{i} + \dfrac{-600y}{(x^2 + y^2 + 3)^2}\mathbf{j}$; hence, when $x = -4$

and $y = 3$, we have

$$\nabla T = \frac{2400}{784} i - \frac{1800}{784} j \quad \text{and}$$

$$|\nabla T| = \sqrt{\left(\frac{2400}{784}\right)^2 + \left(-\frac{1800}{784}\right)^2} = \frac{3000}{784} = \frac{375}{98}.$$

Therefore, in order to maximize $D_{\bar{u}}T$ at $(-4, 3)$, we choose $\bar{u}$ to be a unit vector in the direction of ∇T; that is, we take

$$\bar{u} = \frac{\nabla T}{|\nabla T|} = \frac{4}{5} i - \frac{3}{5} j.$$

(b) As we move through the point $(-4, 3)$ in the direction of $\bar{u}$, the instantaneous rate of change of T with respect to distance is given by

$$D_{\bar{u}}T = |\nabla T| = \frac{375}{98} \approx 3.83 \text{ degrees C per centimeter.}$$

7.2 Normal Vectors to Level Curves in the Plane

Consider a scalar field in the plane given by $z = f(x, y)$, where f is a differentiable function. Recall from Section 1.1 that a curve in the plane along which z has a constant value, say k, has the equation

$$f(x, y) = k$$

Figure 4

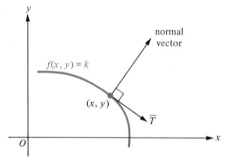

and is called a level curve of the field (Figure 4). Assume that the level curve $f(x, y) = k$ has a unit tangent vector

$$\overline{T} = \frac{dx}{ds} i + \frac{dy}{ds} j,$$

where s is arc length measured along the curve. Differentiating both sides of the equation $f(x, y) = k$ with respect to s by using the chain rule, we obtain

$$f_1(x, y) \frac{dx}{ds} + f_2(x, y) \frac{dy}{ds} = 0;$$

that is,

$$(\nabla f) \cdot \overline{T} = 0.$$

It follows that the gradient vector at a point P in a scalar field is normal to the level curve of the field passing through P—provided, of course, that there *is* a smooth level curve through P.

Since $\nabla f(x_0, y_0) = f_1(x_0, y_0) i + f_2(x_0, y_0) j$ is normal to the tangent line to the level curve of the scalar field $z = f(x, y)$ at the point (x_0, y_0), it follows that the equation of the tangent line is

$$f_1(x_0, y_0)(x - x_0) + f_2(x_0, y_0)(y - y_0) = 0.$$

EXAMPLE   Find a normal vector and the equation of the tangent line to the curve $2x^2 - 4xy^3 + y^5 = 1$ at the point $(2, 1)$.

SOLUTION
The curve can be regarded as the level curve $f(x, y) = 1$ of the scalar field $z = f(x, y)$, where $f(x, y) = 2x^2 - 4xy^3 + y^5$. Here

$$f_1(x, y) = 4x - 4y^3, \qquad f_2(x, y) = -12xy^2 + 5y^4,$$

and the gradient of f at $(2, 1)$ is given by

$$\nabla f(2, 1) = f_1(2, 1)\vec{i} + f_2(2, 1)\vec{j} = 4\vec{i} - 19\vec{j}.$$

Thus, $4\vec{i} - 19\vec{j}$ is normal to the curve at $(2, 1)$. Also, the equation of the tangent line to the curve at $(2, 1)$ is

$$4(x - 2) - 19(y - 1) = 0 \quad \text{or} \quad 4x - 19y + 11 = 0.$$

7.3 Directional Derivative and Gradient in Space

Just as a function of two variables can be regarded as specifying a scalar field in the plane, a function f of three variables can be thought of as describing a *scalar field in xyz space*; that is, we can think of f as assigning a scalar w, given by

$$w = f(x, y, z),$$

to each point (x, y, z) in its domain (Figure 5). Examples are temperature fields, pressure fields, density fields, electrical potential fields, and so forth.

All the ideas and techniques introduced above for scalar fields in the xy plane extend naturally to scalar fields in xyz space. For example, if $w = f(x, y, z)$, where f is a differentiable function, we define the *gradient* of w (or of f) by

Figure 5

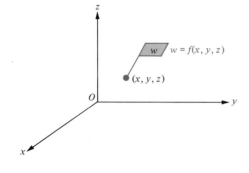

$$\nabla w = \frac{\partial w}{\partial x}\vec{i} + \frac{\partial w}{\partial y}\vec{j} + \frac{\partial w}{\partial z}\vec{k}$$

or

$$\nabla f(x, y, z) = f_1(x, y, z)\vec{i} + f_2(x, y, z)\vec{j} + f_3(x, y, z)\vec{k}.$$

If $\vec{u}$ is a unit vector in xyz space, it is easy to show (Problem 42) that the rate of change of the scalar field w with respect to distance measured in the direction of $\vec{u}$ is given by the *directional derivative*

$$D_{\vec{u}}w = \vec{u} \cdot \nabla w \qquad D_{\vec{u}}f(x, y, z) = \vec{u} \cdot \nabla f(x, y, z).$$

EXAMPLES 1   If $f(x, y, z) = 3x^2 + 8y^2 - 5z^2$, find the directional derivative of f at $(1, -1, 2)$ in the direction of the vector $\vec{v} = 2\vec{i} - 6\vec{j} + 3\vec{k}$.

SOLUTION

Here, $\bar{v}$ is not a unit vector; however, a unit vector in the direction of $\bar{v}$ is given by $\bar{u} = \bar{v}/|\bar{v}| = \frac{2}{7}i - \frac{6}{7}j + \frac{3}{7}k$. We have

$$\nabla f(x, y, z) = f_1(x, y, z)i + f_2(x, y, z)j + f_3(x, y, z)k$$
$$= 6xi + 16yj - 10zk,$$

so that

$$\nabla f(1, -1, 2) = 6(1)i + 16(-1)j - 10(2)k = 6i - 16j - 20k \quad \text{and}$$

$$D_{\bar{u}}f(1, -1, 2) = \bar{u} \cdot \nabla f(1, -1, 2) = \frac{2}{7}(6) - \frac{6}{7}(-16) + \frac{3}{7}(-20) = \frac{48}{7}.$$

2 The electrical potential V in volts at the point $P = (x, y, z)$ in xyz space is given by $V = 100(x^2 + y^2 + z^2)^{-1/2}$, where x, y, and z are measured in centimeters. How rapidly is the potential V changing at the instant when we move through the point $P_0 = (2, 1, -2)$ in the direction of the point $P_1 = (4, 3, 0)$?

SOLUTION

The rate of change of V, in volts per centimeter, is given by $D_{\bar{u}}V$, evaluated at $P_0 = (2, 1, -2)$ in the direction of the unit vector

$$\bar{u} = \frac{\overrightarrow{P_0P_1}}{|\overrightarrow{P_0P_1}|} = \frac{1}{\sqrt{3}}(i + j + k).$$

Here,

$$\frac{\partial V}{\partial x} = -100x(x^2 + y^2 + z^2)^{-3/2}, \qquad \frac{\partial V}{\partial y} = -100y(x^2 + y^2 + z^2)^{-3/2}, \quad \text{and}$$

$$\frac{\partial V}{\partial z} = -100z(x^2 + y^2 + z^2)^{-3/2};$$

hence,

$$\nabla V = \frac{\partial V}{\partial x}i + \frac{\partial V}{\partial y}j + \frac{\partial V}{\partial z}k = -100(x^2 + y^2 + z^2)^{-3/2}(xi + yj + zk).$$

Putting $x = 2$, $y = 1$, and $z = -2$, we find that the gradient at $P_0 = (2, 1, -2)$ is

$$\nabla V = -\frac{100}{27}(2i + j - 2k);$$

hence, the directional derivative at P_0 in the direction of $\bar{u}$ is

$$D_{\bar{u}}V = \bar{u} \cdot \nabla V = \left(\frac{1}{\sqrt{3}}\right)\left(-\frac{100}{27}\right)(i + j + k) \cdot (2i + j - 2k)$$

$$= \frac{-100}{27\sqrt{3}}(2 + 1 - 2) = \frac{-100}{27\sqrt{3}} \approx -2.14 \text{ volts per centimeter.}$$

Just as for scalar fields in the xy plane, the gradient of a scalar field in xyz space points in the direction for which the directional derivative attains its maximum and its length is numerically equal to this maximum directional derivative (Problem 44).

EXAMPLE Let $f(x, y, z) = xy + xz + yz + xyz$. Find (a) the maximum value of the directional derivative of f at $(8, -1, 4)$ and (b) a unit vector in the direction in which this maximum directional derivative is attained.

SOLUTION

Here,

$$\nabla f(x, y, z) = f_1(x, y, z)\bar{i} + f_2(x, y, z)\bar{j} + f_3(x, y, z)\bar{k}$$
$$= (y + z + yz)\bar{i} + (x + z + xz)\bar{j} + (x + y + xy)\bar{k}.$$

(a) The maximum value of the directional derivative at $(8, -1, 4)$ is

$$|\nabla f(8, -1, 4)| = |-\bar{i} + 44\bar{j} - \bar{k}| = \sqrt{(-1)^2 + 44^2 + (-1)^2} = \sqrt{1938}.$$

(b) The required vector, a unit vector in the direction of $\nabla f(8, -1, 4)$, is given by

$$\frac{\nabla f(8, -1, 4)}{|\nabla f(8, -1, 4)|} = \frac{-\bar{i} + 44\bar{j} - \bar{k}}{\sqrt{1938}}.$$

7.4 Level Surfaces, Normal Lines, and Tangent Planes

Let f be a differentiable function of three variables. If k is a constant belonging to the range of f, then the graph in xyz space of the equation

$$f(x, y, z) = k$$

is called a *level surface* for f [or for the scalar field $w = f(x, y, z)$ determined by f]. Suppose that (x_0, y_0, z_0) is a point on this level surface, so that $f(x_0, y_0, z_0) = k$, and assume that $\nabla f(x_0, y_0, z_0) \neq \bar{0}$. Then, by analogy with the considerations in Section 7.2, we define the *normal line* to the level surface at the point (x_0, y_0, z_0) to be the straight line containing the point (x_0, y_0, z_0) and parallel to the gradient vector $\nabla f(x_0, y_0, z_0)$ (Figure 6). Thus, in scalar symmetric form, the equation of the normal line to the level surface $f(x, y, z) = k$ at the point (x_0, y_0, z_0) is

$$\frac{x - x_0}{f_1(x_0, y_0, z_0)} = \frac{y - y_0}{f_2(x_0, y_0, z_0)} = \frac{z - z_0}{f_3(x_0, y_0, z_0)}. \qquad \text{(Why?)}$$

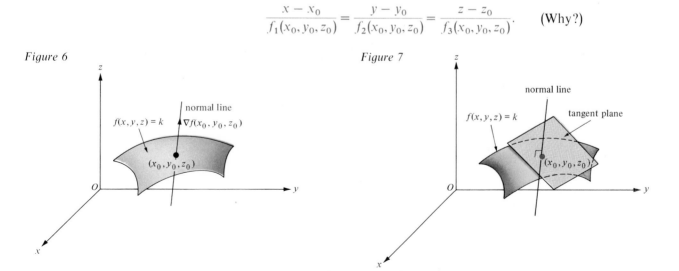

Figure 6

Figure 7

The plane containing the point (x_0, y_0, z_0) and perpendicular to the gradient vector $\nabla f(x_0, y_0, z_0)$ is called the *tangent plane* to the level surface $f(x, y, z) = k$ at the point (x_0, y_0, z_0) (Figure 7). In scalar form, the equation of the tangent plane to the surface $f(x, y, z) = k$ at the point (x_0, y_0, z_0) is

$$f_1(x_0, y_0, z_0)(x - x_0) + f_2(x_0, y_0, z_0)(y - y_0) + f_3(x_0, y_0, z_0)(z - z_0) = 0.$$

(Why?)

Figure 8

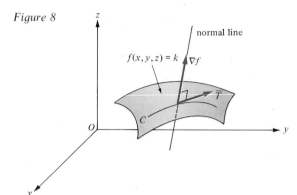

Now, let C be a curve lying on the surface $f(x, y, z) = k$, so that the coordinates x, y, and z of any point $P = (x, y, z)$ on C satisfy the equation

$$f(x, y, z) = k.$$

Using the chain rule, we differentiate both sides of the latter equation with respect to arc length s measured along the curve C to obtain

$$\frac{\partial f}{\partial x}\frac{dx}{ds} + \frac{\partial f}{\partial y}\frac{dy}{ds} + \frac{\partial f}{\partial z}\frac{dz}{ds} = 0;$$

that is, $\nabla f \cdot \overline{T} = 0$, where

$$\overline{T} = \frac{dx}{ds}\bar{i} + \frac{dy}{ds}\bar{j} + \frac{dz}{ds}\bar{k}.$$

Therefore, the unit tangent vector $\overline{T}$ to a curve C on the surface $f(x, y, z) = k$ is perpendicular to the gradient vector ∇f (Figure 8). It follows that the tangent plane to the surface $f(x, y, z) = k$ at a point P contains the tangent vectors at P to all (smooth) curves on the surface that pass through P.

EXAMPLES   Find the equations of (a) the tangent plane and (b) the normal line to the given surface at the indicated point.

1 The sphere $x^2 + y^2 + z^2 = 14$ at the point $(-1, 3, 2)$.

SOLUTION
Here we put $f(x, y, z) = x^2 + y^2 + z^2$ so that the equation of the sphere is $f(x, y, z) = 14$. We have

$$f_1(x, y, z) = 2x, \quad f_2(x, y, z) = 2y, \quad \text{and} \quad f_3(x, y, z) = 2z;$$

hence,

$$f_1(-1, 3, 2) = -2, \quad f_2(-1, 3, 2) = 6, \quad \text{and} \quad f_3(-1, 3, 2) = 4.$$

(a) The equation of the tangent plane at $(-1, 3, 2)$ is

$$f_1(-1, 3, 2)(x + 1) + f_2(-1, 3, 2)(y - 3) + f_3(-1, 3, 2)(z - 2) = 0;$$

that is,

$$-2(x + 1) + 6(y - 3) + 4(z - 2) = 0 \quad \text{or} \quad -2x + 6y + 4z = 28.$$

(b) The equation of the normal line at $(-1, 3, 2)$ is

$$\frac{x + 1}{f_1(-1, 3, 2)} = \frac{y - 3}{f_2(-1, 3, 2)} = \frac{z - 2}{f_3(-1, 3, 2)};$$

that is,

$$\frac{x + 1}{-2} = \frac{y - 3}{6} = \frac{z - 2}{4} \quad \text{or} \quad \frac{x + 1}{-1} = \frac{y - 3}{3} = \frac{z - 2}{2}.$$

2 The graph of $z = g(x, y)$ at the point $(x_0, y_0, g(x_0, y_0))$, where g is a differentiable function of two variables.

SOLUTION
Here we put $f(x, y, z) = g(x, y) - z$, so that the graph of $z = g(x, y)$ is the level surface $f(x, y, z) = 0$ of f. We have

$$f_1(x, y, z) = \frac{\partial}{\partial x}[g(x, y) - z] = \frac{\partial}{\partial x}g(x, y) = g_1(x, y),$$

$$f_2(x, y, z) = \frac{\partial}{\partial y}[g(x, y) - z] = \frac{\partial}{\partial y}g(x, y) = g_2(x, y),$$

$$f_3(x, y, z) = \frac{\partial}{\partial z}[g(x, y) - z] = \frac{\partial}{\partial z}(-z) = -1.$$

Therefore,

$$f_1(x_0, y_0, g(x_0, y_0)) = g_1(x_0, y_0), \qquad f_2(x_0, y_0, g(x_0, y_0)) = g_2(x_0, y_0),$$

$$f_3(x_0, y_0, g(x_0, y_0)) = -1.$$

(a) The equation of the tangent plane at $(x_0, y_0, g(x_0, y_0))$ is

$$g_1(x_0, y_0)(x - x_0) + g_2(x_0, y_0)(y - y_0) + (-1)(z - g(x_0, y_0)) = 0,$$

or

$$z = g(x_0, y_0) + g_1(x_0, y_0)(x - x_0) + g_2(x_0, y_0)(y - y_0).$$

(b) The equation of the normal line at $(x_0, y_0, g(x_0, y_0))$ is

$$\frac{x - x_0}{g_1(x_0, y_0)} = \frac{y - y_0}{g_2(x_0, y_0)} = \frac{z - z_0}{-1}.$$

3 The graph of $g(x, y) = 3x^4 y - 7x^3 y - x^2 + y + 1$ at the point $(1, 2, -6)$.

SOLUTION
We use the results in Example 2. Here we have

$$g_1(x, y) = 12x^3 y - 21x^2 y - 2x \quad \text{and} \quad g_2(x, y) = 3x^4 - 7x^3 + 1;$$

hence,

$$g_1(1, 2) = -20 \quad \text{and} \quad g_2(1, 2) = -3.$$

(a) The tangent plane has the equation

$$z = g(1, 2) + g_1(1, 2)(x - 1) + g_2(1, 2)(y - 2) = -6 - 20(x - 1) - 3(y - 2)$$

or

$$z = 20 - 20x - 3y.$$

(b) The normal line has the equation

$$\frac{x - 1}{g_1(1, 2)} = \frac{y - 2}{g_2(1, 2)} = \frac{z - g(1, 2)}{-1} \quad \text{or} \quad \frac{x - 1}{-20} = \frac{y - 2}{-3} = \frac{z + 6}{-1}.$$

Problem Set 7

In problems 1 through 4, find (a) the gradient ∇z of each scalar field, (b) the value of ∇z at the point (x_0, y_0), and (c) the directional derivative $D_{\bar{u}}z$ at (x_0, y_0) in the direction of the unit vector $\bar{u}$.

1 $z = 7x - 3y + 4$, $(x_0, y_0) = (1, 1)$, $\bar{u} = \frac{\sqrt{3}}{2}i + \frac{1}{2}j$.

2 $z = xy$, $(x_0, y_0) = (2, -1)$, $\bar{u} = \frac{\sqrt{2}}{2}i + \frac{\sqrt{2}}{2}j$.

3 $z = 2x^2 + 3y^2 - 1$, $(x_0, y_0) = (0,0)$, $\bar{u} = (\cos \theta)\bar{i} + (\sin \theta)\bar{j}$, $\theta = \pi/3$.

4 $z = e^{xy}$, $(x_0, y_0) = (1,1)$, $\bar{u} = (\cos \theta)\bar{i} + (\sin \theta)\bar{j}$, $\theta = -\pi/4$.

In problems 5 through 8, find (a) $\nabla f(x_0, y_0)$ and (b) the value of the directional derivative of f at (x_0, y_0) in the direction indicated.

5 $f(x, y) = x^2 y + 2xy^2$, $(x_0, y_0) = (1,2)$, in the direction of the unit vector $\bar{u} = \dfrac{1}{2}\bar{i} + \dfrac{\sqrt{3}}{2}\bar{j}$.

6 $f(x, y) = \tan^{-1}(y/x)$, $(x_0, y_0) = (-2,1)$, in the direction from $(-2,1)$ to $(-6,-2)$.

7 $f(x, y) = \dfrac{1}{x^2 + y^2}$, $(x_0, y_0) = (3,2)$, in the direction of the unit vector $\bar{u} = \dfrac{5}{13}\bar{i} + \dfrac{12}{13}\bar{j}$.

8 $f(x, y) = \ln \sqrt{x^2 + y^2}$, $(x_0, y_0) = (3,-1)$, in the direction of the vector $\bar{v} = 4\bar{i} + 3\bar{j}$. (*Caution:* $\bar{v}$ is not a unit vector.)

In problems 9 and 10, a scalar field $z = f(x, y)$ is given in the xy plane. Find the rate of change of this scalar field as we move from left to right through the given point (x_0, y_0) along a straight line making the indicated angle θ with the positive x axis.

9 $z = 3x^2 - xy + 3y$, $(x_0, y_0) = (3,1)$, $\theta = \pi/4$.

10 $z = e^{-y^2} \cos x$, $(x_0, y_0) = (\pi,1)$, $\theta = \pi/6$.

In problems 11 through 14, find (a) the maximum value of the directional derivative and (b) a unit vector $\bar{u}$ in the direction of the maximum directional derivative for each function at the point indicated.

11 $f(x, y) = x^2 - 7xy + 4y^2$ at $(1,-1)$.

12 $g(x, y) = (x + y - 2)^2 + (3x - y - 6)^2$ at $(1,1)$.

13 $h(x, y) = x^2 - y^2 - \sin y$ at $(1, \pi/2)$.

14 $F(x, y) = e^{-5x} \sin 5y$ at $(0, \pi/20)$.

15 The temperature T at the point (x, y) of a heated circular metal plate with center at the origin is given by $T = \dfrac{400}{2 + x^2 + y^2}$, where T is measured in degrees C and x and y are measured in centimeters. (a) In what direction should one move through the point $(1,1)$ in order that T should increase most rapidly? (b) How rapidly does T increase as one moves through the point $(1,1)$ in the direction found in part (a)?

16 Let f be a differentiable function of two variables such that $f_1(-1,2) = 2$ and $f_2(-1,2) = -3$. Find the directional derivative $D_{\bar{u}} f(-1,2)$ if $\bar{u} = \dfrac{\sqrt{2}}{2}\bar{i} - \dfrac{\sqrt{2}}{2}\bar{j}$.

In problems 17 and 18, (a) find a normal vector and (b) find the equation of the tangent line to each curve at the point indicated.

17 $x^2 + y^2 = 2$ at $(1,1)$.   **18** $2x^2 + y^2 - 9 = 0$ at $(2,1)$.

In problems 19 through 22, (a) find $\nabla f(x_0, y_0, z_0)$ and (b) find the directional derivative $D_{\bar{u}} f(x_0, y_0, z_0)$ for the given function f, the point (x_0, y_0, z_0), and the unit vector $\bar{u}$.

19 $f(x, y, z) = x^2 y + 3yz^2$, $(x_0, y_0, z_0) = (1,-1,1)$, $\bar{u} = \dfrac{1}{3}\bar{i} - \dfrac{2}{3}\bar{j} + \dfrac{2}{3}\bar{k}$.

20 $f(x, y, z) = \ln(x^2 + y^2 + z^2)$, $(x_0, y_0, z_0) = (1,1,1)$, $\bar{u} = -\dfrac{2}{3}\bar{i} + \dfrac{1}{3}\bar{j} + \dfrac{2}{3}\bar{k}$.

21 $f(x, y, z) = z - e^x \sin y$, $(x_0, y_0, z_0) = \left(\ln 3, \dfrac{3\pi}{2}, -3\right)$, $\bar{u} = \dfrac{2}{7}\bar{i} + \dfrac{3}{7}\bar{j} + \dfrac{6}{7}\bar{k}$.

22 $f(x, y, z) = e^{-y} \sin x + \dfrac{1}{3} e^{-3y} \sin 3x + z^2$, $(x_0, y_0, z_0) = \left(\dfrac{\pi}{3}, 0, 1\right)$,

$\bar{u} = \left(\cos \dfrac{2\pi}{3}\right)\bar{i} + \left(\cos \dfrac{\pi}{4}\right)\bar{j} + \left(\cos \dfrac{\pi}{3}\right)\bar{k}$.

In problems 23 through 26, find (a) the maximum value of the directional derivative and (b) a unit vector $\bar{u}$ in the direction in which this maximum directional derivative is attained for each function at the point indicated.

23 $f(x, y, z) = (x^2 + y^2 + z^2)^{-1}$ at $(1, 2, -3)$.

24 $h(x, y, z) = (x + y)^2 + (y + z)^2 + (x + z)^2$ at $(2, -1, 2)$.

25 $g(x, y, z) = e^x \cos(yz)$ at $(1, 0, \pi)$.

26 $f(x, y, z) = \dfrac{x}{x^2 + y^2} + \dfrac{y}{x^2 + z^2}$ at $(3, 1, 1)$.

In problems 27 through 36, find the equations of (a) the tangent plane and (b) the normal line to each surface at the point indicated.

27 $x^2 + 2y^2 + 3z^2 = 6$ at $(1, 1, 1)$.

28 $x^2 - 2y^2 + z^2 = 11$ at $(2, 1, 3)$.

29 $xyz = 6$ at $(1, 2, 3)$.

30 $8x - y^2 = 0$ at $(2, 4, 7)$.

31 $x^3 + y^3 - 6xy + z = 0$ at $(2, 2, 8)$.

32 $\sin(xz) = e^{xy}$ at $(1, 0, \pi/2)$.

33 $\cos(xy) + \sin(yz) = 0$ at $(1, \pi/6, -2)$.

34 $\ln(xy) + \sin(yz) = 2$ at $(e, 1, \pi/2)$.

35 $\sqrt{x} + \sqrt{y} + \sqrt{z} = 6$ at $(9, 4, 1)$.

36 $\tan^{-1}(y/x) - \ln(xyz) = \pi/4$ at $(1, 1, 1)$.

In problems 37 through 40, find (a) the equation of the tangent plane and (b) the equation of the normal line to the graph of each function g of two variables at the point indicated.

37 $g(x, y) = \sqrt{9 - x^2 - y^2}$ at $(-1, 2, 2)$.

38 $g(x, y) = x \sin(\pi y/2)$ at $(0, 0, 0)$.

39 $g(x, y) = x^y$ at $(1, 1, 1)$.

40 $g(x, y) = \ln \cos \sqrt{x^2 + y^2}$ at $(0, 0, 0)$.

41 Assuming that f and g are differentiable functions of three variables and that a and b are constants, show that $\nabla(af + bg) = a\nabla f + b\nabla g$.

42 Assuming that f is a differentiable function of three variables, prove that the rate of change of the scalar field $w = f(x, y, z)$ with respect to distance measured in the direction of the unit vector $\bar{u}$ is given by $\bar{u} \cdot \nabla f$.

43 Suppose that f is a differentiable function of a single variable and that $r = \sqrt{x^2 + y^2 + z^2}$. Verify the formula

$$\nabla f(r) = f'(r) \frac{x\bar{i} + y\bar{j} + z\bar{k}}{r}.$$

44 Prove that the gradient of a scalar field in xyz space points in the direction for which the directional derivative attains its maximum and its length is numerically equal to this maximum directional derivative.

45 Assume that f is a function of two variables which is differentiable at (x_0, y_0). Give a geometric interpretation of the differential df in terms of dx, dy, and the tangent plane to the graph of f at $(x_0, y_0, f(x_0, y_0))$.

46 Suppose that a point P is moving along a smooth curve C in a scalar field w. Assuming that the function that gives the scalar field w is differentiable, that $\bar{T}$ is the unit tangent vector to C, and that P is moving with the speed ds/dt, prove that the instantaneous rate of change of w with respect to time as measured at the point P is given by

$$\frac{ds}{dt} \bar{T} \cdot \nabla w = \frac{ds}{dt} D_{\bar{T}} w.$$

In problems 47 through 50, use the result of problem 46.

47 The temperature T at any point (x, y) of a rectangular plate lying in the xy plane is $T = x \sin 2y$. The point $P = (x, y)$ is moving clockwise around a circle of radius 1 unit with center at the origin at a constant speed of 2 units of arc length per second. How fast is the temperature at the point P changing at the instant when $(x, y) = \left(\frac{1}{2}, \frac{\sqrt{3}}{2}\right)$?

48 Suppose that the surface $z = \dfrac{x^3 - 2y^2}{5280}$ represents an uneven terrain and that a group of tourists is congregated at the origin. Here, x, y, and z are measured in miles. A Moslem tourist sets out for Mecca, going directly east along the positive direction of the x axis. If he travels at a constant speed of 3 miles per hour, how steep is his climb, in feet per minute, at the end of 1 hour?

49 The pressure P, in dynes per square centimeter, of a certain gas at the point (x, y, z) in space is given by $P = 10{,}000e^{-(x^2 + y^2 + z^2)}$, where x, y, and z are measured in centimeters. A variable point $Q = (x, y, z)$ is moving along a certain curve C in space. At the instant when $Q = (1, -2, 1)$, its speed is given by $ds/dt = 50$ cm/sec. If the unit tangent vector $\overline{T}$ to C at $(1, -2, 1)$ is given by $\overline{T} = \left(\cos \dfrac{2\pi}{3} \right)\overline{i} + \left(\cos \dfrac{\pi}{4} \right)\overline{j} + \left(\cos \dfrac{\pi}{3} \right)\overline{k}$, find the instantaneous time rate of change of pressure at Q just as Q passes through $(1, -2, 1)$.

50 The electrical potential V in volts at the point (x, y, z) in space caused by placing a charge q at the origin is given by $V = q/r$, where $r = \sqrt{x^2 + y^2 + z^2}$. (Assume that distances are measured in centimeters.) Let C be a curve in space having a unit tangent vector $\overline{T}$ at the point P. If P moves along C in the direction of $\overline{T}$ at a constant speed of v cm/sec, write an expression for the instantaneous time rate of change of electrical potential at P.

51 Suppose that two surfaces in xyz space, say $f(x, y, z) = k_1$ and $g(x, y, z) = k_2$, intersect in a smooth curve C (Figure 9). Assuming that f and g are differentiable and that (x_0, y_0, z_0) is a point on C, show that the vector

$$\nabla f(x_0, y_0, z_0) \times \nabla g(x_0, y_0, z_0)$$

is tangent to the curve C at the point (x_0, y_0, z_0).

In problems 52 through 55, use the result of problem 51 to find a tangent vector at the given point to the curve C in which the two surfaces intersect.

52 $x^2 - y^2 - z^2 + 12 = 0$ and $3x^2 + y^2 + z = 4$ at $(1, 2, -3)$.

53 $xz + 2x + 4z = 5$ and $4xy + 3y + 6z = 56$ at $(2, 5, \frac{1}{6})$.

54 $x^2 + \dfrac{y^2}{4} - \dfrac{z^2}{9} = 1$ and $x^2 + y^2 + z^2 = 14$ at $(-1, 2, 3)$.

55 $x^2 - 2xz + y^2 z = 1$ and $3xy + 2yz = -6$ at $(1, -2, 0)$.

Figure 9

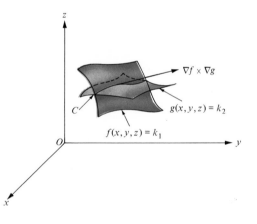

8   Higher-Order Partial Derivatives

In studying functions of a single variable, we found it useful to consider not only the first derivative, but higher-order derivatives as well. Similarly, in studying functions of several variables, it is useful to consider *higher-order partial derivatives*. Thus, consider a function f of two variables having partial derivatives f_1 and f_2, so that

$$f_1(x, y) = f_x(x, y) = \frac{\partial}{\partial x} f(x, y) \quad \text{and} \quad f_2(x, y) = f_y(x, y) = \frac{\partial}{\partial y} f(x, y).$$

The functions f_1 and f_2 are functions of two variables, and they may themselves have partial derivatives. For instance, if $f(x, y) = 3x^2y^3 + 6xy^2$, then

$$f_1(x, y) = f_x(x, y) = \frac{\partial}{\partial x} (3x^2y^3 + 6xy^2) = 6xy^3 + 6y^2,$$

$$f_2(x, y) = f_y(x, y) = \frac{\partial}{\partial y} (3x^2y^3 + 6xy^2) = 9x^2y^2 + 12xy.$$

Therefore,

$$\frac{\partial}{\partial x} f_1(x, y) = \frac{\partial}{\partial x} \left[\frac{\partial}{\partial x} f(x, y) \right] = \frac{\partial}{\partial x} (6xy^3 + 6y^2) = 6y^3,$$

$$\frac{\partial}{\partial y} f_1(x, y) = \frac{\partial}{\partial y} \left[\frac{\partial}{\partial x} f(x, y) \right] = \frac{\partial}{\partial y} (6xy^3 + 6y^2) = 18xy^2 + 12y,$$

$$\frac{\partial}{\partial x} f_2(x, y) = \frac{\partial}{\partial x} \left[\frac{\partial}{\partial y} f(x, y) \right] = \frac{\partial}{\partial x} (9x^2y^2 + 12xy) = 18xy^2 + 12y,$$

$$\frac{\partial}{\partial y} f_2(x, y) = \frac{\partial}{\partial y} \left[\frac{\partial}{\partial y} f(x, y) \right] = \frac{\partial}{\partial y} (9x^2y^2 + 12xy) = 18x^2y + 12x.$$

The four partial derivatives of partial derivatives found above are called the *second-order partial derivatives* of the original function f. Naturally, we can denote the partial derivatives of the function f_1 with respect to the first and second variables as $(f_1)_1$ and $(f_1)_2$, respectively; however, for simplicity, we omit the parentheses and write these second-order partial derivatives as f_{11} and f_{12}, respectively. Likewise, we write f_{xx} and f_{xy} rather than $(f_x)_x$ and $(f_x)_y$, respectively. For instance,

$$f_{12} = f_{xy} = \frac{\partial}{\partial y} \left(\frac{\partial f}{\partial x} \right).$$

The symbolism $\frac{\partial}{\partial y} \left(\frac{\partial f}{\partial x} \right)$ is also abbreviated to $\frac{\partial^2 f}{\partial y \, \partial x}$, much in the same way that $\frac{d^2 y}{dx^2}$ is used as an abbreviation for the ordinary second derivative. Similarly, we write $\frac{\partial^2 f}{\partial x^2}$ for the second-order partial derivative $\frac{\partial}{\partial x} \left(\frac{\partial f}{\partial x} \right)$, and so forth.

Summarizing the above, the four second-order partial derivatives of f can be written as follows:

$$f_{11} = f_{xx} = \frac{\partial^2 f}{\partial x^2} = \frac{\partial}{\partial x} \left(\frac{\partial f}{\partial x} \right),$$

$$f_{12} = f_{xy} = \frac{\partial^2 f}{\partial y \, \partial x} = \frac{\partial}{\partial y} \left(\frac{\partial f}{\partial x} \right),$$

$$f_{21} = f_{yx} = \frac{\partial^2 f}{\partial x \, \partial y} = \frac{\partial}{\partial x} \left(\frac{\partial f}{\partial y} \right),$$

$$f_{22} = f_{yy} = \frac{\partial^2 f}{\partial y^2} = \frac{\partial}{\partial y} \left(\frac{\partial f}{\partial y} \right).$$

In the subscript notation, $f_{12} = f_{xy}$ indicates an initial partial differentiation with respect to the first variable x followed by a partial differentiation with respect to the second variable y. However, the symbolism $\dfrac{\partial^2 f}{\partial x\, \partial y} = \dfrac{\partial}{\partial x}\left(\dfrac{\partial f}{\partial y}\right)$ indicates an initial partial differentiation with respect to y followed by a partial differentiation with respect to x. In the subscript notation, the order of the subscripts from *left to right* shows the order of partial differentiation, while in the notation $\dfrac{\partial^2 f}{\partial x\, \partial y}$, we must read the "denominator" from *right to left*.

EXAMPLES 1 If $f(x,y) = 7x^2 - 13xy + 18y^2$, find f_1, f_2, f_{11}, f_{12}, f_{21}, and f_{22}.

SOLUTION
Here,

$$f_1(x,y) = \frac{\partial}{\partial x}\left(7x^2 - 13xy + 18y^2\right) = 14x - 13y,$$

$$f_2(x,y) = \frac{\partial}{\partial y}\left(7x^2 - 13xy + 18y^2\right) = -13x + 36y.$$

Hence,

$$f_{11}(x,y) = \frac{\partial}{\partial x}\left(f_1(x,y)\right) = \frac{\partial}{\partial x}\left(14x - 13y\right) = 14,$$

$$f_{12}(x,y) = \frac{\partial}{\partial y}\left(f_1(x,y)\right) = \frac{\partial}{\partial y}\left(14x - 13y\right) = -13,$$

$$f_{21}(x,y) = \frac{\partial}{\partial x}\left(f_2(x,y)\right) = \frac{\partial}{\partial x}\left(-13x + 36y\right) = -13,$$

$$f_{22}(x,y) = \frac{\partial}{\partial y}\left(f_2(x,y)\right) = \frac{\partial}{\partial y}\left(-13x + 36y\right) = 36.$$

2 If $f(x,y) = 2e^{2x}\cos y$, find f_x, f_y, f_{xx}, f_{xy}, f_{yx}, and f_{yy}.

SOLUTION
Here,

$$f_x(x,y) = \frac{\partial}{\partial x}\left(2e^{2x}\cos y\right) = 4e^{2x}\cos y,$$

$$f_y(x,y) = \frac{\partial}{\partial y}\left(2e^{2x}\cos y\right) = -2e^{2x}\sin y.$$

Hence,

$$f_{xx}(x,y) = \frac{\partial}{\partial x}\left(f_x(x,y)\right) = \frac{\partial}{\partial x}\left(4e^{2x}\cos y\right) = 8e^{2x}\cos y,$$

$$f_{xy}(x,y) = \frac{\partial}{\partial y}\left(f_x(x,y)\right) = \frac{\partial}{\partial y}\left(4e^{2x}\cos y\right) = -4e^{2x}\sin y,$$

$$f_{yx}(x,y) = \frac{\partial}{\partial x}\left(f_y(x,y)\right) = \frac{\partial}{\partial x}\left(-2e^{2x}\sin y\right) = -4e^{2x}\sin y,$$

$$f_{yy}(x,y) = \frac{\partial}{\partial y}\left(f_y(x,y)\right) = \frac{\partial}{\partial y}\left(-2e^{2x}\sin y\right) = -2e^{2x}\cos y.$$

3 Let n, m, and k be constants and suppose that $f(x, y) = kx^n y^m$. Find $\dfrac{\partial f}{\partial x}$, $\dfrac{\partial f}{\partial y}$, $\dfrac{\partial^2 f}{\partial x^2}$, $\dfrac{\partial^2 f}{\partial y\, \partial x}$, $\dfrac{\partial^2 f}{\partial x\, \partial y}$, and $\dfrac{\partial^2 f}{\partial y^2}$.

SOLUTION
Here,

$$\frac{\partial f}{\partial x} = \frac{\partial}{\partial x}\left(kx^n y^m\right) = knx^{n-1}y^m \quad \text{and} \quad \frac{\partial f}{\partial y} = \frac{\partial}{\partial y}\left(kx^n y^m\right) = kmx^n y^{m-1}.$$

Hence,

$$\frac{\partial^2 f}{\partial x^2} = \frac{\partial}{\partial x}\left(\frac{\partial f}{\partial x}\right) = \frac{\partial}{\partial x}\left(knx^{n-1}y^m\right) = kn(n-1)x^{n-2}y^m,$$

$$\frac{\partial^2 f}{\partial y\, \partial x} = \frac{\partial}{\partial y}\left(\frac{\partial f}{\partial x}\right) = \frac{\partial}{\partial y}\left(knx^{n-1}y^m\right) = knmx^{n-1}y^{m-1},$$

$$\frac{\partial^2 f}{\partial x\, \partial y} = \frac{\partial}{\partial x}\left(\frac{\partial f}{\partial y}\right) = \frac{\partial}{\partial x}\left(kmx^n y^{m-1}\right) = kmnx^{n-1}y^{m-1},$$

$$\frac{\partial^2 f}{\partial y^2} = \frac{\partial}{\partial y}\left(\frac{\partial f}{\partial y}\right) = \frac{\partial}{\partial y}\left(kmx^n y^{m-1}\right) = km(m-1)x^n y^{m-2}.$$

The second-order partial derivatives $\dfrac{\partial^2 f}{\partial y\, \partial x}$ and $\dfrac{\partial^2 f}{\partial x\, \partial y}$ are called the *mixed* second-order partial derivatives of f. In all three examples above, notice that the mixed partial derivatives are the same. By Example 3, the mixed partial derivatives of any term in a polynomial function of two variables are the same; hence, $\dfrac{\partial^2 f}{\partial y\, \partial x} = \dfrac{\partial^2 f}{\partial x\, \partial y}$ holds for any polynomial function f of two variables. As a matter of fact, the two mixed partial derivatives are equal for a wide class of functions, as the following theorem shows.

THEOREM 1   **Equality of Mixed Second-Order Partial Derivatives**
Let f be a function of two variables and suppose that U is an open set of points contained in the domain of f. Then, if both of the mixed partial derivatives f_{12} and f_{21} exist and are continuous on U, it follows that $f_{12}(x, y) = f_{21}(x, y)$ for all points (x, y) in U.

The proof of this theorem, which depends on the mean value theorem for functions of one variable, is routinely presented in advanced calculus courses and is not given here.

The notation used for partial derivatives of order higher than 2 is almost self-explanatory. Thus,

$$\frac{\partial^3 f}{\partial x\, \partial y^2} = \frac{\partial}{\partial x}\left(\frac{\partial^2 f}{\partial y^2}\right) = f_{221} = f_{yyx},$$

$$\frac{\partial^5 f}{\partial y^2\, \partial x\, \partial y^2} = \frac{\partial^2}{\partial y^2}\left(\frac{\partial^3 f}{\partial x\, \partial y^2}\right) = f_{22122} = f_{yyxyy},$$

and so forth. Theorem 1 on the equality of mixed partial derivatives extends to such higher-order cases. For instance,

$$\frac{\partial^3 f}{\partial x\, \partial y^2} = \frac{\partial^3 f}{\partial y\, \partial x\, \partial y} = \frac{\partial^3 f}{\partial y^2\, \partial x}$$

holds on a set such as U in the theorem, provided that all three mixed partial derivatives exist and are continuous on U. Generally speaking, the order in which successive partial derivatives are taken when forming higher-order partial derivatives is irrelevant, provided that all of the partial derivatives in question exist and are continuous.

EXAMPLE   Let $f(x, y) = e^{xy} + \sin(x + y)$. Find (a) f_{xxy}, (b) f_{yyx}, and (c) f_{xxyxx}.

SOLUTION
(a) $f_x(x, y) = ye^{xy} + \cos(x + y)$, and

$$f_{xx}(x, y) = \frac{\partial}{\partial x}[ye^{xy} + \cos(x + y)] = y^2 e^{xy} - \sin(x + y),$$

so that

$$f_{xxy}(x, y) = \frac{\partial}{\partial y}[y^2 e^{xy} - \sin(x + y)] = 2ye^{xy} + xy^2 e^{xy} - \cos(x + y).$$

(b) $f_y(x, y) = xe^{xy} + \cos(x + y)$, and

$$f_{yy}(x, y) = \frac{\partial}{\partial y}[xe^{xy} + \cos(x + y)] = x^2 e^{xy} - \sin(x + y),$$

so that

$$f_{yyx}(x, y) = \frac{\partial}{\partial x}[x^2 e^{xy} - \sin(x + y)] = 2xe^{xy} + yx^2 e^{xy} - \cos(x + y).$$

(c) Since $f_{xxy}(x, y) = 2ye^{xy} + xy^2 e^{xy} - \cos(x + y)$, we have

$$f_{xxyx}(x, y) = \frac{\partial}{\partial x}[2ye^{xy} + xy^2 e^{xy} - \cos(x + y)]$$
$$= 2y^2 e^{xy} + y^2 e^{xy} + xy^3 e^{xy} + \sin(x + y)$$
$$= 3y^2 e^{xy} + xy^3 e^{xy} + \sin(x + y),$$

so that

$$f_{xxyxx}(x, y) = \frac{\partial}{\partial x}[3y^2 e^{xy} + xy^3 e^{xy} + \sin(x + y)]$$
$$= 3y^3 e^{xy} + y^3 e^{xy} + xy^4 e^{xy} + \cos(x + y)$$
$$= 4y^3 e^{xy} + xy^4 e^{xy} + \cos(x + y).$$

Naturally, the concept of higher-order partial derivatives extends immediately to functions of more than two variables. The notation is again self-explanatory. For instance, if $w = f(x, y, z)$, then

$$\frac{\partial^3 w}{\partial x\, \partial y\, \partial z} = \frac{\partial}{\partial x}\left[\frac{\partial}{\partial y}\left(\frac{\partial w}{\partial z}\right)\right] = f_{zyx} = f_{321},$$

$$\frac{\partial^4 w}{\partial x\, \partial z^2\, \partial y} = \frac{\partial}{\partial x}\left[\frac{\partial^2}{\partial z^2}\left(\frac{\partial w}{\partial y}\right)\right] = f_{yzzx} = f_{2331},$$

and so forth. Furthermore, the conditions for the equality of mixed partial derivatives remain effective.

EXAMPLE   Let $w = x^4y^2z + \sin xy$. Verify by direct calculation that

$$\frac{\partial^3 w}{\partial x\,\partial y\,\partial z} = \frac{\partial^3 w}{\partial z\,\partial y\,\partial x}.$$

SOLUTION
Here,

$$\frac{\partial w}{\partial z} = x^4y^2 \quad \text{and} \quad \frac{\partial^2 w}{\partial y\,\partial z} = 2x^4y,$$

so that $\dfrac{\partial^3 w}{\partial x\,\partial y\,\partial z} = 8x^3y$. On the other hand,

$$\frac{\partial w}{\partial x} = 4x^3y^2z + y\cos xy \quad \text{and} \quad \frac{\partial^2 w}{\partial y\,\partial x} = 8x^3yz + \cos xy - xy\sin xy,$$

so that $\dfrac{\partial^3 w}{\partial z\,\partial y\,\partial x} = 8x^3y$. Therefore,

$$\frac{\partial^3 w}{\partial x\,\partial y\,\partial z} = 8x^3y = \frac{\partial^3 w}{\partial z\,\partial y\,\partial x}.$$

Higher-order partial derivatives are usually calculated by successive differentiation, as in the example above. At each stage of such a calculation, the standard differentiation rules, including the chain rules, are effective. The use of the chain rules in this connection is illustrated by the following examples.

EXAMPLES 1 Let $w = u^2v^3$, where $u = x^2 + 3y^2$ and $v = 2x^2 - y^2$. Find $\dfrac{\partial^2 w}{\partial x\,\partial y}$.

SOLUTION
By the chain rule,

$$\frac{\partial w}{\partial y} = \frac{\partial w}{\partial u}\frac{\partial u}{\partial y} + \frac{\partial w}{\partial v}\frac{\partial v}{\partial y} = (2uv^3)(6y) + (3u^2v^2)(-2y) = (12uv^3 - 6u^2v^2)y.$$

Therefore,

$$\frac{\partial^2 w}{\partial x\,\partial y} = \frac{\partial}{\partial x}\left[(12uv^3 - 6u^2v^2)\right]y = \left[12\frac{\partial}{\partial x}(uv^3) - 6\frac{\partial}{\partial x}(u^2v^2)\right]y$$

$$= \left[12\left(\frac{\partial u}{\partial x}v^3 + 3uv^2\frac{\partial v}{\partial x}\right) - 6\left(2u\frac{\partial u}{\partial x}v^2 + 2u^2v\frac{\partial v}{\partial x}\right)y\right]$$

$$= [12(2xv^3 + 12uv^2x) - 6(4uxv^2 + 8u^2vx)]y$$

$$= 24(v^3 + 5uv^2 - 2u^2v)xy.$$

2 Suppose that z is a function of x and y and that both x and y are functions of t. Assuming the existence and continuity of the required derivatives, find $\dfrac{d^2z}{dt^2}$.

SOLUTION
By the chain rule,

$$\frac{dz}{dt} = \frac{\partial z}{\partial x}\frac{dx}{dt} + \frac{\partial z}{\partial y}\frac{dy}{dt}.$$

Differentiating again with respect to t, we obtain

$$\frac{d^2z}{dt^2} = \left[\frac{d}{dt}\left(\frac{\partial z}{\partial x}\right)\right]\frac{dx}{dt} + \frac{\partial z}{\partial x}\left[\frac{d}{dt}\left(\frac{dx}{dt}\right)\right] + \left[\frac{d}{dt}\left(\frac{\partial z}{\partial y}\right)\right]\frac{dy}{dt} + \frac{\partial z}{\partial y}\left[\frac{d}{dt}\left(\frac{dy}{dt}\right)\right]$$

$$= \left[\frac{d}{dt}\left(\frac{\partial z}{\partial x}\right)\right]\frac{dx}{dt} + \frac{\partial z}{\partial x}\frac{d^2x}{dt^2} + \left[\frac{d}{dt}\left(\frac{\partial z}{\partial y}\right)\right]\frac{dy}{dt} + \frac{\partial z}{\partial y}\frac{d^2y}{dt^2}.$$

The quantities remaining in the brackets can be found by additional applications of the chain rule. Thus,

$$\frac{d}{dt}\left(\frac{\partial z}{\partial x}\right) = \left[\frac{\partial}{\partial x}\left(\frac{\partial z}{\partial x}\right)\right]\frac{dx}{dt} + \left[\frac{\partial}{\partial y}\left(\frac{\partial z}{\partial x}\right)\right]\frac{dy}{dt} = \frac{\partial^2 z}{\partial x^2}\frac{dx}{dt} + \frac{\partial^2 z}{\partial y\,\partial x}\frac{dy}{dt},$$

$$\frac{d}{dt}\left(\frac{\partial z}{\partial y}\right) = \left[\frac{\partial}{\partial x}\left(\frac{\partial z}{\partial y}\right)\right]\frac{dx}{dt} + \left[\frac{\partial}{\partial y}\left(\frac{\partial z}{\partial y}\right)\right]\frac{dy}{dt} = \frac{\partial^2 z}{\partial x\,\partial y}\frac{dx}{dt} + \frac{\partial^2 z}{\partial y^2}\frac{dy}{dt}.$$

Substituting from the latter equations into the former one, and using the fact that $\dfrac{\partial^2 z}{\partial y\,\partial x} = \dfrac{\partial^2 z}{\partial x\,\partial y}$, we have

$$\frac{d^2z}{dt^2} = \left[\frac{\partial^2 z}{\partial x^2}\frac{dx}{dt} + \frac{\partial^2 z}{\partial x\,\partial y}\frac{dy}{dt}\right]\frac{dx}{dt} + \frac{\partial z}{\partial x}\frac{d^2x}{dt^2} + \left[\frac{\partial^2 z}{\partial x\,\partial y}\frac{dx}{dt} + \frac{\partial^2 z}{\partial y^2}\frac{dy}{dt}\right]\frac{dy}{dt} + \frac{\partial z}{\partial y}\frac{d^2y}{dt^2};$$

that is,

$$\frac{d^2z}{dt^2} = \frac{\partial^2 z}{\partial x^2}\left(\frac{dx}{dt}\right)^2 + 2\frac{\partial^2 z}{\partial x\,\partial y}\frac{dx}{dt}\frac{dy}{dt} + \frac{\partial^2 z}{\partial y^2}\left(\frac{dy}{dt}\right)^2 + \frac{\partial z}{\partial x}\frac{d^2x}{dt^2} + \frac{\partial z}{\partial y}\frac{d^2y}{dt^2}.$$

Problem Set 8

In problems 1 through 12, (a) find $\dfrac{\partial^2 f}{\partial x^2}$, (b) find $\dfrac{\partial^2 f}{\partial y^2}$, (c) find $\dfrac{\partial^2 f}{\partial y\,\partial x}$, (d) find $\dfrac{\partial^2 f}{\partial x\,\partial y}$, and

(e) verify that $\dfrac{\partial^2 f}{\partial y\,\partial x} = \dfrac{\partial^2 f}{\partial x\,\partial y}$.

1 $f(x, y) = 6x^2 + 7xy + 5y^2$ 2 $f(x, y) = 4x^2 y^3 - x^4 y + x$ 3 $f(x, y) = x \cos y - y^2$

4 $f(x, y) = \tan^{-1}\left(\dfrac{y}{x}\right)$ 5 $f(x, y) = (x^2 + y^2)^{3/2}$ 6 $f(x, y) = \dfrac{x + y}{x - y}$

7 $f(x, y) = y \cos x - xe^{2y}$ 8 $f(x, y) = e^{\sqrt{x^2 + y^2}}$ 9 $f(x, y) = \sin(x + 2y)$

10 $f(x, y) = e^{3x^2} - 2y^3$ 11 $f(x, y) = 5x \cosh 2y$ 12 $f(x, y) = \ln \sin \sqrt{x^2 + y^2}$

In problems 13 through 20, find each partial derivative.

13 $f(x, y, z) = xy^2 z + 3x^2 e^y$; (a) $f_{xy}(x, y, z)$, (b) $f_{yz}(x, y, z)$.   14 $g(x, y, z) = \cos(xyz^2)$; (a) $g_{xz}(x, y, z)$, (b) $g_{yz}(x, y, z)$.

15 $h(x, y, z) = xe^y + ze^x + e^{-3z}$; (a) $h_{112}(x, y, z)$, (b) $h_{213}(x, y, z)$.

16 $f(x, y, z) = x^2 e^{-2y} \sin z$; (a) $f_{113}(1, -1, \pi/4)$, (b) $f_{2231}(1, -1, \pi/4)$.

17 $f(x, y, z) = e^{xyz}$; (a) $f_{xyzx}(1, 2, 3)$, (b) $f_{xxyz}(1, 2, 3)$.

18 $f(s, r, t) = \ln(5r + 8s - 6t^2)$; (a) $f_{21}(r, s, t)$, (b) $f_{312}(r, s, t)$.

19 $h(x, y, z) = \sin x^2 y - z^2$; (a) $\dfrac{\partial^3 h}{\partial x\,\partial y\,\partial z}$, (b) $\dfrac{\partial^4 h}{\partial x\,\partial y\,\partial z^2}$.   20 $f(x, y, z) = y \ln(x - \csc z)$, (a) $\dfrac{\partial^2 f}{\partial y\,\partial z}$, (b) $\dfrac{\partial^2 f}{\partial x\,\partial z}$.

21 If $w = (Ax^2 + By^2)^3$, where A and B are constants, verify that $\dfrac{\partial^3 w}{\partial x^2\,\partial y} = \dfrac{\partial^3 w}{\partial y\,\partial x^2}$ by direct calculation.

22 If $w = \ln(x - y) + \tan(x + y)$, show that $\dfrac{\partial^2 w}{\partial x^2} = \dfrac{\partial^2 w}{\partial y^2}$.

In problems 23 through 26, show that each function f satisfies *Laplace's partial differential equation in two dimensions*, namely, $\dfrac{\partial^2 f}{\partial x^2} + \dfrac{\partial^2 f}{\partial y^2} = 0$.

23 $f(x, y) = e^x \sin y + e^y \sin x$ **24** $f(x, y) = e^{2x}(\cos^2 y - \sin^2 y)$

25 $f(x, y) = \ln(x^2 + y^2)$ **26** $f(x, y) = \tan^{-1}\left(\dfrac{y}{x}\right)$

In problems 27 and 28, show that each function f satisfies *Laplace's partial differential equation in three dimensions*, namely, $\dfrac{\partial^2 f}{\partial x^2} + \dfrac{\partial^2 f}{\partial y^2} + \dfrac{\partial^2 f}{\partial z^2} = 0$.

27 $f(x, y, z) = 3x^2 - 2y^2 + 5xy + 8xz - z^2$ **28** $f(x, y, z) = (x^2 + y^2 + z^2)^{-1/2}$

29 Show that $w = e^{-y^2} \cos x$ satisfies the partial differential equation

$$\frac{\partial^2 w}{\partial y^2} - 2 \frac{\partial^2 w}{\partial x^2} = 4y^2 w.$$

30 Assuming that the functions f and g are twice differentiable, show that, if

$$w = f(x + y) + g(x - y), \quad \text{then} \quad \frac{\partial^2 w}{\partial x^2} = \frac{\partial^2 w}{\partial y^2}.$$

31 If $w = Ax^2 + Bxy + Cy^2 + Dx + Ey + F$, determine the conditions on the constants A, B, C, D, E, and F so that w satisfies Laplace's partial differential equation in two dimensions, namely, $\dfrac{\partial^2 w}{\partial x^2} + \dfrac{\partial^2 w}{\partial y^2} = 0$.

32 A function f of two variables is said to be *homogeneous of degree n* if $f(tx, ty) = t^n f(x, y)$ holds for all values of x, y, and t.
 (a) Give an example of a function that is homogeneous of degree 1.
 (b) Give an example of a function that is homogeneous of degree 2.
 (c) Suppose that f is differentiable and that it is homogeneous of degree n. Prove that

$$xf_1(x, y) + yf_2(x, y) = nf(x, y).$$

33 Let $w = f(u, v)$, where $u = x + y + z$ and $v = 2x + y - z$. Assuming the existence and continuity of the required partial derivatives of f, find a formula for $\dfrac{\partial^2 w}{\partial x^2} + \dfrac{\partial^2 w}{\partial y^2} + \dfrac{\partial^2 w}{\partial z^2}$.

34 If f is a twice-differentiable function and c is a constant, show that $w = f(x - ct)$ is a solution of the *wave equation*, $\dfrac{\partial^2 w}{\partial t^2} = c^2 \dfrac{\partial^2 w}{\partial x^2}$. Also, show that $w = f(x + ct)$ is a solution of this equation.

35 If a, b, c, and k are constants, show that $w = (a \cos cx + b \sin cx)e^{-kc^2 t}$ is a solution of the *heat equation* $\dfrac{\partial w}{\partial t} = k \dfrac{\partial^2 w}{\partial x^2}$.

36 Suppose that $w = f(x, y, z)$ and let $x = \rho \sin \phi \cos \theta$, $y = \rho \sin \phi \sin \theta$, and $z = \rho \cos \phi$. Thus, if $(\rho, \theta, \phi) = (4, \pi/3, \pi/6)$, then $(x, y, z) = (1, \sqrt{3}, 2\sqrt{3})$. Suppose that f has continuous partial derivatives and that

$$f_x(1, \sqrt{3}, 2\sqrt{3}) = \sqrt{3}, \qquad f_y(1, \sqrt{3}, 2\sqrt{3}) = 2, \qquad f_z(1, \sqrt{3}, 2\sqrt{3}) = 1,$$
$$f_{xx}(1, \sqrt{3}, 2\sqrt{3}) = 4, \qquad f_{xy}(1, \sqrt{3}, 2\sqrt{3}) = -1, \qquad f_{xz}(1, \sqrt{3}, 2\sqrt{3}) = 2,$$
$$f_{yy}(1, \sqrt{3}, 2\sqrt{3}) = 4, \qquad f_{yz}(1, \sqrt{3}, 2\sqrt{3}) = -\sqrt{3}, \qquad f_{zz}(1, \sqrt{3}, 2\sqrt{3}) = -1.$$

Find:

(a) $\dfrac{\partial w}{\partial \phi}$ at $(4, \pi/3, \pi/6)$.   (b) $\dfrac{\partial^2 w}{\partial \rho \, \partial \phi}$ at $(4, \pi/3, \pi/6)$.

37 Let $w = f(u)$ where $u = g(x, y)$. Assuming that the required derivatives exist and are continuous, express $\dfrac{\partial^2 w}{\partial x^2} + \dfrac{\partial^2 w}{\partial y^2}$ in terms of f', f'', and partial derivatives of g.

38 Suppose that z is a function of u and v and that both u and v are functions of x and y. Assuming the existence and continuity of the required partial derivatives, find a formula for

(a) $\dfrac{\partial^2 z}{\partial x^2}$

(b) $\dfrac{\partial^2 z}{\partial x \, \partial y}$

39 Let $w = f(r)$, where $r = \sqrt{x^2 + y^2}$. Show that

$$\frac{\partial^2 w}{\partial x^2} + \frac{\partial^2 w}{\partial y^2} = \frac{d^2 w}{dr^2} + \frac{1}{r}\frac{dw}{dr}.$$

40 Suppose that $f(x, y) = 0$ defines y implicitly as a function of f. Assuming the existence and continuity of the required derivatives and that $\partial f / \partial y \neq 0$, show that

$$\frac{d^2 y}{dx^2} = -\frac{\dfrac{\partial^2 f}{\partial x^2}\left(\dfrac{\partial f}{\partial y}\right)^2 - 2\dfrac{\partial f}{\partial x}\cdot\dfrac{\partial f}{\partial y}\cdot\dfrac{\partial^2 f}{\partial x \, \partial y} + \dfrac{\partial^2 f}{\partial y^2}\left(\dfrac{\partial f}{\partial x}\right)^2}{\left(\dfrac{\partial f}{\partial y}\right)^3}.$$

41 If u and v are functions of x and y, and if u and v satisfy the *Cauchy–Riemann equations*, namely, $\dfrac{\partial u}{\partial x} = \dfrac{\partial v}{\partial y}$ and $\dfrac{\partial u}{\partial y} = -\dfrac{\partial v}{\partial x}$, show that $\dfrac{\partial^2 u}{\partial x^2} + \dfrac{\partial^2 u}{\partial y^2} = 0$ and that $\dfrac{\partial^2 v}{\partial x^2} + \dfrac{\partial^2 v}{\partial y^2} = 0$. Assume the existence and continuity of the required partial derivatives.

9 Extrema for Functions of More Than One Variable

In Chapter 3 we saw that the derivative is an indispensable tool for solving problems involving extrema of functions of single variables. In this section we study the problem of finding maximum and minimum values of functions of more than one variable, and we see that partial derivatives are especially useful in this connection.

The basic concepts, such as relative extrema, absolute extrema, critical points, and so forth, for functions of several variables are the natural analogs of the corresponding concepts for functions of one variable. For simplicity, we formulate these concepts for functions of two variables; it should be obvious how they can be extended to functions of more than two variables.

DEFINITION 1   Relative Extrema

(i) A function f of two variables has a *relative maximum value* $f(a,b)$ at the point (a,b) if there is a circular disk with positive radius and with center at (a,b) such that, if (x,y) is a point in this disk, then (x,y) is in the domain of f and $f(x,y) \leq f(a,b)$.

(ii) Similarly, a function f of two variables has a *relative minimum value* $f(a,b)$ at the point (a,b) if there is a circular disk with positive radius and with center at (a,b) such that, if (x,y) is a point in this disk, then (x,y) is in the domain of f and $f(x,y) \geq f(a,b)$.

(iii) A relative maximum or minimum value of a function is called a *relative extreme value* or a *relative extremum* of the function.

Figure 1

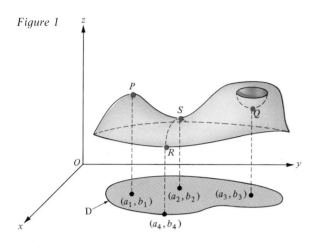

Consider, for instance, the surface in Figure 1 as the graph of a function f of two variables. The point P at the apex of the "hill" represents a relative maximum value of f, since $f(a_1, b_1)$—the height of P above the xy plane—is larger than all *nearby* values of $f(x, y)$. Similarly, the point Q at the bottom of the "hollow" corresponds to a relative minimum, $f(a_3, b_3)$.

The point S in the "pass" is neither a relative maximum nor a relative minimum; indeed, we *increase* our height above the xy plane as we move from S up the "hill" toward P, whereas we *decrease* our height above the xy plane as we move from S down toward the point R on the "rim" of the surface. Incidentally, R does *not* represent a relative extremum of f because there is no circular disk with positive radius and with center at (a_4, b_4) that is contained in the domain D of f. Notice that, by definition, a relative extremum of a function f can occur only at interior points of the domain D of f—never at a boundary point of D.

Sometimes it is possible to locate the relative extrema of a function f of more than one variable by simple algebraic considerations or by sketching the graph of f. This is illustrated by the following example.

EXAMPLE   Let the function f be defined by $f(x, y) = 1 - x^2 - y^2$. Find all relative extrema of f.

Figure 2

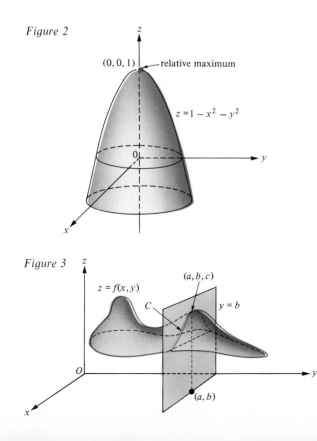

SOLUTION
Here, $f(x, y) = 1 - (x^2 + y^2)$ and $x^2 + y^2 \geq 0$, so that $f(x, y) \leq 1$ for all values of x and y. Since $f(0, 0) = 1$, f attains a relative maximum value of 1 at $(0, 0)$. The graph of f, a paraboloid of revolution about the z axis, clearly shows that f has no other relative extrema (Figure 2).

Suppose that f is a function of two variables which has a relative maximum $c = f(a, b)$ at the point (a, b) (Figure 3). Let C be the cross section cut from the graph of f by the plane $y = b$, and notice that the curve C attains a relative maximum at the point (a, b). Therefore, if C is a smooth curve, it must have a horizontal tangent line at (a, b, c), by Theorem 1 of Section 4 in Chapter 3. Since the slope of the tangent line to C at (a, b, c) is given by the partial derivative $f_1(a, b)$, then $f_1(a, b) = 0$.

By cutting the surface in Figure 3 with the plane $x = a$ and reasoning similarly, we can also conclude that, if $f_2(a, b)$ exists, then $f_2(a, b) = 0$. Likewise, if f has a relative minimum at (a, b) and if $f_1(a, b)$ and $f_2(a, b)$ exist, then $f_1(a, b) = 0$ and $f_2(a, b) = 0$. Thus, we have the following theorem.

Figure 3

THEOREM 1

Necessary Condition for Relative Extrema

Let (a, b) be an interior point of the domain of a function f and suppose that the two partial derivatives $f_1(a, b)$ and $f_2(a, b)$ exist. Then, if f has a relative extremum at (a, b), it is necessary that both $f_1(a, b) = 0$ and $f_2(a, b) = 0$.

Notice that the condition $f_1(a, b) = f_2(a, b) = 0$ in Theorem 1 can also be expressed as $\nabla f(a, b) = \bar{0}$, since $\nabla f(a, b) = f_1(a, b)\bar{i} + f_2(a, b)\bar{j}$. Therefore, at a point where a function attains a relative extremum, its gradient must either fail to exist, or be the zero vector. Thus, we make the following definition.

DEFINITION 2

Critical Point

Let f be a function of two variables. A point (a, b) in the domain of f such that either $\nabla f(a, b)$ fails to exist, or else $\nabla f(a, b) = \bar{0}$, is called a *critical point of f*. A critical point of f which is in the interior of the domain of f is called an *interior critical point* of the domain of f.

Now, Theorem 1 can be restated thus: If a function f has a relative extremum at a point (a, b), then (a, b) must be an interior critical point of the domain of f.

Therefore, to locate all the relative extrema of f, one begins by finding all the interior critical points of its domain. Some of these critical points may correspond to relative extrema; however, some of them may occur at "saddle points," that is, critical points where the function has neither a relative maximum nor a relative minimum. The point S on the surface in Figure 1 corresponds to a saddle point; thus, the interior critical point (a_2, b_2) in Figure 1 would not give a relative extremum for the function. (Notice that the graph of the function has the shape of a saddle near the point S in Figure 1.)

In order to sort out the interior critical points of a function and determine which of them correspond to relative maxima, relative minima, or saddle points, the following theorem (whose proof is best left to a course in advanced calculus) may be helpful.

THEOREM 2

Second Derivative Test

Let (a, b) be an interior point of the domain of a function f and suppose that the first and second partial derivatives of f exist and are continuous on some circular disk with (a, b) as its center and contained in the domain of f. Assume that (a, b) is a critical point of f, so that $f_1(a, b) = f_2(a, b) = 0$. Let

$$\Delta = \begin{vmatrix} f_{11}(a, b) & f_{12}(a, b) \\ f_{12}(a, b) & f_{22}(a, b) \end{vmatrix} = f_{11}(a, b)f_{22}(a, b) - [f_{12}(a, b)]^2.$$

Then:

(i) If $\Delta > 0$ and $f_{11}(a, b) + f_{22}(a, b) < 0$, then f has a relative maximum at (a, b).

(ii) If $\Delta > 0$ and $f_{11}(a, b) + f_{22}(a, b) > 0$, then f has a relative minimum at (a, b).

(iii) If $\Delta < 0$, then f has a saddle point at (a, b).

(iv) If $\Delta = 0$, then the test is inconclusive and other methods must be used.

EXAMPLES

Find all of the interior critical points of the domain of the given function, and then apply the second derivative test to decide (if possible) at which of these points relative maxima, relative minima, or saddle points occur.

1 $f(x, y) = x^2 + y^2 - 2x + 4y + 2$

SOLUTION

Here,

$$f_1(x,y) = 2x - 2, \qquad f_2(x,y) = 2y + 4,$$
$$f_{11}(x,y) = 2, \qquad f_{12}(x,y) = f_{21}(x,y) = 0, \qquad f_{22}(x,y) = 2.$$

To find all critical points, we must solve the simultaneous equations

$$\begin{cases} f_1(x,y) = 0 \\ f_2(x,y) = 0; \end{cases} \quad \text{that is,} \quad \begin{cases} 2x - 2 = 0 \\ 2y + 4 = 0. \end{cases}$$

In this case, the only solution is $x = 1$, $y = -2$; hence, $(1, -2)$ is the only critical point of f. Calculating the quantity Δ of Theorem 2 at the critical point $(1, -2)$, we obtain

$$\Delta = \begin{vmatrix} f_{11}(1, -2) & f_{12}(1, -2) \\ f_{12}(1, -2) & f_{22}(1, -2) \end{vmatrix} = \begin{vmatrix} 2 & 0 \\ 0 & 2 \end{vmatrix} = 4 - 0 = 4 > 0.$$

Also, $f_{11}(1, -2) + f_{22}(1, -2) = 2 + 2 = 4 > 0$. Therefore, by Theorem 2, f has a relative minimum at $(1, -2)$.

2 $f(x,y) = 12xy - 4x^2y - 3xy^2$

SOLUTION

Here,

$$f_1(x,y) = 12y - 8xy - 3y^2 = (12 - 8x - 3y)y,$$
$$f_2(x,y) = 12x - 4x^2 - 6xy = (12 - 4x - 6y)x,$$
$$f_{11}(x,y) = -8y,$$
$$f_{12}(x,y) = f_{21}(x,y) = 12 - 8x - 6y, \quad \text{and}$$
$$f_{22}(x,y) = -6x.$$

To find all critical points (x,y), we must find all solutions of the simultaneous equations

$$\begin{cases} (12 - 8x - 3y)y = 0 \\ (12 - 4x - 6y)x = 0. \end{cases}$$

Obviously, $x = 0$, $y = 0$ gives a solution, so that $(0,0)$ is a critical point. If $x \neq 0$ and $y = 0$, the equations become

$$\begin{cases} 0 = 0 \\ 12 - 4x = 0, \end{cases}$$

so that $x = 3$, $y = 0$ gives a solution, and $(3,0)$ is a critical point. If $x = 0$ and $y \neq 0$, the equations become

$$\begin{cases} 12 - 3y = 0 \\ 0 = 0, \end{cases}$$

giving the critical point $(0,4)$. Finally, if $x \neq 0$ and $y \neq 0$, the simultaneous equations become

$$\begin{cases} 12 - 8x - 3y = 0 \\ 12 - 4x - 6y = 0. \end{cases}$$

Solving the latter equations as usual (say, by elimination of variables), we obtain $x = 1$, $y = \frac{4}{3}$; hence, $(1, \frac{4}{3})$ is also a critical point.

The critical points of f are therefore $(0,0)$, $(3,0)$, $(0,4)$, and $(1,\frac{4}{3})$. Evaluating the quantity Δ at each of these critical points, we have:

1 At $(0,0)$, $\Delta = f_{11}(0,0)f_{22}(0,0) - [f_{12}(0,0)]^2 = 0 - (12)^2 = -144$.
2 At $(3,0)$, $\Delta = f_{11}(3,0)f_{22}(3,0) - [f_{12}(3,0)]^2 = 0 - (-12)^2 = -144$.
3 At $(0,4)$, $\Delta = f_{11}(0,4)f_{22}(0,4) - [f_{12}(0,4)]^2 = 0 - (-12)^2 = -144$.
4 At $(1,\frac{4}{3})$, $\Delta = f_{11}(1,\frac{4}{3})f_{22}(1,\frac{4}{3}) - [f_{12}(1,\frac{4}{3})]^2 = 64 - (-4)^2 = 48$.

Since $\Delta < 0$ at the first three critical points $(0,0)$, $(3,0)$, and $(0,4)$, these are saddle points. At $(1,\frac{4}{3})$, we have $\Delta > 0$ and $f_{11}(1,\frac{4}{3}) + f_{22}(1,\frac{4}{3}) = -\frac{32}{3} - 6 < 0$; hence, the function has a relative maximum at $(1,\frac{4}{3})$.

3 $f(x,y) = x^4 + y^4$

SOLUTION
Here,
$$f_1(x,y) = 4x^3, \qquad f_2(x,y) = 4y^3,$$
$$f_{11}(x,y) = 12x^2, \qquad f_{12}(x,y) = f_{21}(x,y) = 0, \qquad f_{22}(x,y) = 12y^2.$$

The only solution of the simultaneous equations
$$\begin{cases} 4x^3 = 0 \\ 4y^3 = 0 \end{cases}$$

is $x = 0$, $y = 0$; hence, $(0,0)$ is the only critical point of f. Thus,
$$\Delta = f_{11}(0,0)f_{22}(0,0) - [f_{12}(0,0)]^2 = 0,$$

and therefore the second derivative test is inconclusive at $(0,0)$. However, since $f(0,0) = 0$ and since $f(x,y) = x^4 + y^4 > 0$ when $(x,y) \neq (0,0)$, it is clear on purely algebraic grounds that f attains a relative minimum at $(0,0)$.

9.1 Absolute Extrema

In contrast to the notion of a relative extremum given in Definition 1, we now introduce the idea of an absolute extremum.

DEFINITION 3   **Absolute Extrema**
A function f of two variables has an *absolute maximum value* $f(a,b)$ at the point (a,b) of its domain D if $f(x,y) \leq f(a,b)$ holds for every point (x,y) in D. Similarly, f has an *absolute minimum value* $f(c,d)$ at the point (c,d) of its domain D if $f(x,y) \geq f(c,d)$ holds for every point (x,y) in D. An absolute maximum or minimum value of f is called an *absolute extremum* of f.

In Figure 2, the function f given by $f(x,y) = 1 - x^2 - y^2$ actually attains not only a relative maximum, but an absolute maximum at $(0,0)$. However, in Figure 1, the point P does not represent an absolute maximum, since the surface climbs even higher than P as we move to the right of the saddle point S.

It is intuitively clear that a continuous and bounded surface must have a highest point and a lowest point (provided that the "rim" of the surface is considered part of the surface). For instance, the point R on the surface in Figure 1 appears to be a lowest point, while a highest point would be found somewhere along the upper edge of the "hollow" surrounding the point Q. The following theorem confirms our intuition about these matters; however, its proof is best left to courses in advanced calculus or analysis.

THEOREM 3   **Existence of Absolute Extrema**

Let f be a continuous function of two variables whose domain D not only is bounded, but contains all of its own boundary points. Then f attains an absolute maximum value and an absolute minimum value.

Notice that an absolute extremum which occurs at an interior point of the domain D of a function f is automatically a relative extremum of f. (Why?) Consequently, an absolute extremum of f which is not a relative extremum must occur at a boundary point of D (that is, at a point of D which is not an interior point). Hence, in order to locate the absolute extrema of f, one first finds all the relative extrema and then compares the largest and the smallest of these with the values of f around the boundary of D. The technique is quite similar to that used in Chapter 3 to find absolute extrema of functions of one variable and is illustrated by the following example.

EXAMPLE   A circular metal plate of radius 1 meter is placed with its center at the origin in the xy plane and heated so that its temperature at the point (x, y) is given by $T = 64(3x^2 - 2xy + 3y^2 + 2y + 5)$ degrees C, where x and y are measured in meters. Find the highest and the lowest temperatures on the plate.

SOLUTION
Since

$$\frac{\partial T}{\partial x} = 64(6x - 2y) \quad \text{and} \quad \frac{\partial T}{\partial y} = 64(-2x + 6y + 2),$$

the critical points inside the circular disk are found by solving the simultaneous equations

$$\begin{cases} 64(6x - 2y) = 0 \\ 64(-2x + 6y + 2) = 0. \end{cases}$$

The only solution, $x = -\frac{1}{8}$, $y = -\frac{3}{8}$, gives exactly one internal critical point on the plate, namely $(-\frac{1}{8}, -\frac{3}{8})$. For the second derivative test at $(-\frac{1}{8}, -\frac{3}{8})$, we need the quantities

$$\frac{\partial^2 T}{\partial x^2} = (64)(6), \quad \frac{\partial^2 T}{\partial x\, \partial y} = (64)(-2), \quad \text{and} \quad \frac{\partial^2 T}{\partial y^2} = (64)(6).$$

Thus, at the critical point,

$$\Delta = \frac{\partial^2 T}{\partial x^2} \frac{\partial^2 T}{\partial y^2} - \left(\frac{\partial^2 T}{\partial x\, \partial y} \right)^2 = (384)(384) - (-128)^2 = 131{,}072 > 0,$$

and $\dfrac{\partial^2 T}{\partial x^2} + \dfrac{\partial^2 T}{\partial y^2} = 384 + 384 = 768 > 0$; hence, there is a relative minimum temperature of

$$64[3(-\tfrac{1}{8})^2 - 2(-\tfrac{1}{8})(-\tfrac{3}{8}) + 3(-\tfrac{3}{8})^2 + 2(-\tfrac{3}{8}) + 5] = 296 \text{ degrees C}$$

at the point $(-\frac{1}{8}, -\frac{3}{8})$.

Now, we must examine the values of T on the boundary $x^2 + y^2 = 1$ of the circular plate. If we put

$$\begin{cases} x = \cos\theta \\ y = \sin\theta, \end{cases}$$

then, as θ varies from 0 to 2π, the point (x, y) traverses the boundary of the plate. The temperature at the point corresponding to θ is given by

$$T = 64(3\cos^2\theta - 2\cos\theta\sin\theta + 3\sin^2\theta + 2\sin\theta + 5)$$
$$= 64(3 - 2\cos\theta\sin\theta + 2\sin\theta + 5)$$
$$= 64(8 - 2\cos\theta\sin\theta + 2\sin\theta)$$
$$= 128(4 - \cos\theta\sin\theta + \sin\theta).$$

Thus, on the boundary of the plate,

$$\frac{dT}{d\theta} = 128(\sin^2\theta - \cos^2\theta + \cos\theta) = 128(1 - 2\cos^2\theta + \cos\theta).$$

The critical values of θ on the boundary are the solutions of

$$1 - 2\cos^2\theta + \cos\theta = 0 \quad \text{or} \quad 2\cos^2\theta - \cos\theta - 1 = 0;$$

that is,

$$(2\cos\theta + 1)(\cos\theta - 1) = 0.$$

Thus, $\cos\theta = -\frac{1}{2}$ or $\cos\theta = 1$, so the critical values of θ are

$$\theta = \frac{2\pi}{3}, \qquad \theta = \frac{4\pi}{3}, \qquad \theta = 0.$$

When $\theta = 2\pi/3$, we have

$$T = 128\left(4 - \cos\frac{2\pi}{3}\sin\frac{2\pi}{3} + \sin\frac{2\pi}{3}\right) = 32(16 + 3\sqrt{3}) \approx 678.28 \text{ degrees C.}$$

When $\theta = 4\pi/3$, then

$$T = 128\left(4 - \cos\frac{4\pi}{3}\sin\frac{4\pi}{3} + \sin\frac{4\pi}{3}\right) = 32(16 - 3\sqrt{3}) \approx 345.72 \text{ degrees C.}$$

When $\theta = 0$, we have

$$T = 128(4 - \cos 0 \sin 0 + \sin 0) = 512 \text{ degrees C.}$$

On the boundary of the plate, the maximum temperature is therefore $32(16 + 3\sqrt{3}) \approx 678.28$ degrees C and the minimum temperature is $32(16 - 3\sqrt{3}) \approx 345.72$ degrees C. The relative minimum of 296 degrees C on the interior of the plate is smaller than the minimum on the boundary; hence, it is the absolute minimum temperature on the plate. Therefore, the absolute minimum temperature on the plate is 296 degrees C and the absolute maximum temperature on the plate is $32(16 + 3\sqrt{3}) \approx 678.28$ degrees C.

In applied problems, it is usually the *absolute* extrema that are desired, and a rigorous treatment ordinarily requires an exhaustive analysis, as illustrated in the example above. However, in many cases, people rely on their physical or geometric intuition and carry out an informal analysis involving only an examination of interior critical points. Such an informal procedure is illustrated in the following example.

EXAMPLE   A long piece of tin 12 inches wide is to be made into a trough by bending up strips of equal width along the edges so that the strips make equal angles with the horizontal. How wide should these strips be and what angle should they make with the horizontal so that the trough has a maximum carrying capacity?

Figure 4

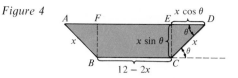

SOLUTION

Figure 4 shows a cross section of the trough. Here x denotes the width of the strips and θ denotes their angle with the horizontal. We must maximize the area z of the cross section. Since the triangles ABF and ECD each have areas of $\frac{1}{2}(x \cos \theta)(x \sin \theta)$ square inches and the rectangle $BCEF$ has area $x \sin \theta(12 - 2x)$ square inches, it follows that

$$z = x^2 \cos \theta \sin \theta + x \sin \theta \, (12 - 2x) \text{ square inches.}$$

Here,

$$\frac{\partial z}{\partial x} = \sin \theta \, [x(2 \cos \theta - 4) + 12],$$

$$\frac{\partial z}{\partial \theta} = x[x(2 \cos^2 \theta - 2 \cos \theta - 1) + 12 \cos \theta].$$

Hence, critical points will correspond to solutions of the simultaneous equations

$$\begin{cases} x(2 \cos \theta - 4) + 12 = 0 \\ x(2 \cos^2 \theta - 2 \cos \theta - 1) + 12 \cos \theta = 0. \end{cases}$$

Solving the first equation for $\cos \theta$, we obtain

$$\cos \theta = 2 - \frac{6}{x}.$$

Substituting this value of $\cos \theta$ into the second equation, we have

$$x \left[2\left(4 - \frac{24}{x} + \frac{36}{x^2} \right) - \left(4 - \frac{12}{x} \right) - 1 \right] + 24 - \frac{72}{x} = 0 \quad \text{or} \quad 3x - 12 = 0.$$

Hence, $x = 4$ and $\cos \theta = 2 - \frac{6}{4} = \frac{1}{2}$. The critical point, hence (we assume) the desired solution, is therefore given by

$$x = 4 \text{ inches} \quad \text{and} \quad \theta = 60°.$$

The solution $x = 4$ inches, $\theta = 60°$, in the problem above seems so reasonable on geometric grounds that we have not even invoked the second derivative test for a *relative* maximum. (The skeptical reader is invited to do so.) Of course, if great importance were somehow associated with this problem—for instance, if we were going into the trough business—we would not only test for a relative maximum, but we would go on to check that our solution represents the *absolute* maximum.

The facts concerning maxima and minima for functions of three or more variables are analogous to the facts set forth above for functions of two variables, except that the second derivative test becomes more complicated as the number of independent variables increases. Again, these matters are best left to advanced calculus.

Problem Set 9

In problems 1 through 18, find all critical points of each function, and then test each critical point to see whether it corresponds to a relative maximum, a relative minimum, or a saddle point.

1 $f(x, y) = x^2 + (y - 1)^2$

2 $f(x, y) = x^2 + 4xy - y^2 - 8x - 6y$

3 $f(x, y) = (x - y + 1)^2$

4 $g(x, y) = x^3 - 3xy^2 + y^3$

5 $f(x, y) = x^2 + y^2 + \dfrac{2}{xy}$

6 $f(x, y) = x \sin y$

7 $g(x, y) = xy - x - y$

8 $h(x, y) = e^x + e^y - e^{x+y}$

9 $f(x, y) = x^4 + y^4 + 4x + 4y$

10 $f(x, y) = xy^2 + x^2 + y$

11 $F(x, y) = x^3 y + 3x + y$

12 $G(x, y) = \sin x + \sin y + \cos(x + y)$

13 $H(x, y) = x^3 + y^3 - 3x$

14 $f(x, y) = xy(12 - 4x - 3y)$

15 $F(x, y) = x^2 + y^2 + (3x + 4y - 26)^2$

16 $f(x, y, z) = e^{-(x^2 + y^2 + z^2)}$

17 $f(x, y) = x^4 + y^4 + 32x - 4y + 52$

18 $f(x, y, z) = xy + xz$

In problems 19 through 22, find the absolute maximum and the absolute minimum of each function (if they exist).

19 $f(x, y) = \sqrt{1 - x^2 - y^2}$ for $x^2 + y^2 \leq 1$.

20 $f(x, y) = 3x^2 + 2xy + 4y^2 + 2x - 3y + 1$ for $0 \leq x \leq 1$ and $0 \leq y \leq 1$.

21 $f(x, y) = xy + 12(x + y) - (x + y)^2$ for $0 \leq x \leq 12$ and $0 \leq y \leq 12$.

22 $f(x, y) = 5x - 2y + 7$ for (x, y) on or inside of the ellipse

$$\begin{cases} x = 3 \cos \theta \\ y = 4 \sin \theta, \end{cases} \quad 0 \leq \theta \leq 2\pi.$$

23 A rectangular box is to have a volume of 20 cubic feet. The material used for the sides costs 1 dollar per square foot, the material used for the bottom costs 2 dollars per square foot, and the material used for the top costs 3 dollars per square foot. What are the dimensions of the cheapest box?

24 A manufacturer produces two grades of alloy in quantities of x and y tons, respectively. If the total cost of production is expressed by the function $C(x, y) = x^2 + 100x + y^2 - xy$ and the total revenue is given by the function $R(x, y) = 100x - x^2 + 2000y + xy$, find the production level that maximizes the profit.

25 The cost of paying for the inspection of an assembly line operation depends upon the number of inspections x and y at each of two sites according to the function $C(x, y) = x^2 + y^2 + xy - 20x - 25y + 1500$. How many inspections should be made at each site in order to minimize the cost?

26 One end of a house is to consist of solar panels in the shape of a rectangle surmounted by an isosceles triangle and is to have a perimeter p feet, where p is a given constant. If the house is to be constructed so as to collect a maximum amount of solar energy, show that the slope of the roof must be $1/\sqrt{3}$.

27 The temperature T in degrees centigrade at each point in the region $x^2 + y^2 \leq 1$ is given by $T = 16x^2 + 24x + 40y^2$. Find the temperatures at the hottest and coldest points in the region.

28 Suppose that the function f is continuous and that its domain D is bounded and contains all its own boundary points. Suppose also that f never takes on any negative values and that $f(x, y) = 0$ whenever (x, y) is on the boundary of D. If f has exactly one critical point in the interior of D, show that it takes on its absolute maximum at this point.

29 Find three nonnegative numbers x, y, and z such that $x + y + z = 2001$ and xyz is as large as possible. Show carefully that your answer actually represents an absolute maximum.

30 The electric potential V at each point (x, y) in the region $0 \leq x \leq 1$ and $0 \leq y \leq 1$ is given by $V = 48xy - 32x^3 - 24y^2$. Find the maximum and minimum potential in this region.

31 Explain why an absolute extremum that occurs at an interior point of the domain D of a function f must be a relative extremum of f.

32 A certain state plans to supplement its revenue by selling weekly lottery tickets. Opinion polls show that a potential 1,000,000 tickets will be purchased per week at $1 per ticket, but that 130,000 fewer tickets will be purchased per week for every 25-cent increase in the price per ticket. Fixed costs such as printing and distributing tickets, paying salaries of lottery officials, and advertising are expected to run $140,000 per week. Regardless of the price per ticket, it is estimated that each additional dollar (over the basic allotment provided for in fixed costs) spent in advertising per week will result in the sale of one additional ticket per week. The state, by law, must return one third of its weekly revenue from ticket sales as prizes to the purchasers. How much should the state charge for a lottery ticket in order to maximize its profits, and what maximum weekly profit from the lottery can it expect?

10   Lagrange Multipliers

In Section 9 we presented a routine for locating the extrema of functions of more than one variable. In this section we study the method of *Lagrange multipliers* for solving extremum problems that involve *constraints*.

A typical *constrained extremum problem* requires us to find the extrema of a function f of several variables when these variables are not independent, but satisfy one or more given conditions called *constraints*. The constraints are ordinarily specified by equations, called *constraint equations*, involving the variables in question. For instance, consider the following example.

EXAMPLE   Maximize the value of the function f given by

$$f(x, y, z) = xyz$$

subject to the constraint

$$g(x, y, z) = 42,$$

where g is the function defined by $g(x, y, z) = x + y + z$.

SOLUTION
Here the constraint equation $g(x, y, z) = 42$, or $x + y + z = 42$, can be solved for z in terms of x and y to obtain

$$z = 42 - x - y.$$

Therefore, the quantity to be maximized becomes

$$f(x, y, z) = f(x, y, 42 - x - y) = xy(42 - x - y).$$

If we let F be the function of x and y defined by

$$F(x, y) = xy(42 - x - y),$$

then our problem is simply to maximize $F(x, y)$, and we proceed as in Section 9. Thus,

$$\frac{\partial F}{\partial x} = y(42 - x - y) + xy(-1) = y(42 - 2x - y) \quad \text{and}$$

$$\frac{\partial F}{\partial y} = x(42 - x - y) + xy(-1) = x(42 - x - 2y),$$

so that the critical points of F are given by the solutions of the simultaneous equations

$$\begin{cases} y(42 - 2x - y) = 0 \\ x(42 - x - 2y) = 0. \end{cases}$$

Solving these equations, we obtain the critical points $(0,0)$, $(0,42)$, $(42,0)$, and $(14,14)$. The second derivative test shows a relative maximum only at $(14,14)$. Actually, $x = 14$, $y = 14$ gives an absolute maximum value of

$$F(14,14) = (14)(14)(42 - 14 - 14) = (14)^3.$$

The technique used to solve the constrained extremum problem above can be generalized as follows: Suppose that we require the extrema of

$$f(x, y, z)$$

subject to the constraint

$$g(x, y, z) = k,$$

where k is a constant and the functions f and g are differentiable. Assume that the constraint equation $g(x, y, z) = k$ can be solved for (say) z as a function of x and y, so that

$$z = h(x, y), \qquad \text{where } g(x, y, h(x, y)) = k.$$

The quantity $f(x, y, z)$ whose extrema are sought can now be written as

$$f(x, y, z) = f(x, y, h(x, y)).$$

If we define the function F of two variables by

$$F(x, y) = f(x, y, h(x, y)),$$

then our problem is simply to find the extrema of F, and we proceed as follows.

Assuming that the function h has partial derivatives, we can apply the chain rule to the equation $F(x, y) = f(x, y, h(x, y))$ to obtain

$$\frac{\partial F}{\partial x} = \frac{\partial f}{\partial x}\frac{\partial x}{\partial x} + \frac{\partial f}{\partial y}\frac{\partial y}{\partial x} + \frac{\partial f}{\partial z}\frac{\partial h}{\partial x} = \frac{\partial f}{\partial x} + \frac{\partial f}{\partial z}\frac{\partial h}{\partial x} \quad \text{and}$$

$$\frac{\partial F}{\partial y} = \frac{\partial f}{\partial x}\frac{\partial x}{\partial y} + \frac{\partial f}{\partial y}\frac{\partial y}{\partial y} + \frac{\partial f}{\partial z}\frac{\partial h}{\partial y} = \frac{\partial f}{\partial y} + \frac{\partial f}{\partial z}\frac{\partial h}{\partial y}.$$

Thus, the critical points of F are the solutions of the simultaneous equations

$$\begin{cases} \dfrac{\partial f}{\partial x} + \dfrac{\partial f}{\partial z}\dfrac{\partial h}{\partial x} = 0 \\[2mm] \dfrac{\partial f}{\partial y} + \dfrac{\partial f}{\partial z}\dfrac{\partial h}{\partial y} = 0. \end{cases}$$

Differentiating both sides of the equation $g(x, y, h(x, y)) = k$ with respect to x and with respect to y using the chain rule, we obtain a second pair of simultaneous equations:

$$\begin{cases} \dfrac{\partial g}{\partial x}\dfrac{\partial x}{\partial x} + \dfrac{\partial g}{\partial y}\dfrac{\partial y}{\partial x} + \dfrac{\partial g}{\partial z}\dfrac{\partial h}{\partial x} = 0 \\[2mm] \dfrac{\partial g}{\partial x}\dfrac{\partial x}{\partial y} + \dfrac{\partial g}{\partial y}\dfrac{\partial y}{\partial y} + \dfrac{\partial g}{\partial z}\dfrac{\partial h}{\partial y} = 0 \end{cases} \quad \text{or} \quad \begin{cases} \dfrac{\partial g}{\partial x} + \dfrac{\partial g}{\partial z}\dfrac{\partial h}{\partial x} = 0 \\[2mm] \dfrac{\partial g}{\partial y} + \dfrac{\partial g}{\partial z}\dfrac{\partial h}{\partial y} = 0. \end{cases}$$

The desired critical points of F are therefore to be found among the solutions of the four simultaneous equations

$$
\begin{cases}
\dfrac{\partial f}{\partial x} + \dfrac{\partial f}{\partial z}\dfrac{\partial h}{\partial x} = 0 \\[2mm]
\dfrac{\partial f}{\partial y} + \dfrac{\partial f}{\partial z}\dfrac{\partial h}{\partial y} = 0 \\[2mm]
\dfrac{\partial g}{\partial x} + \dfrac{\partial g}{\partial z}\dfrac{\partial h}{\partial x} = 0 \\[2mm]
\dfrac{\partial g}{\partial y} + \dfrac{\partial g}{\partial z}\dfrac{\partial h}{\partial y} = 0.
\end{cases}
$$

Now, assume that $\dfrac{\partial g}{\partial z} \neq 0$ and define λ to be the ratio

$$
\lambda = -\frac{\partial f / \partial z}{\partial g / \partial z}
$$

(λ is the Greek letter "lambda"). Solving the third equation in the latter set for $\partial h / \partial x$, we obtain

$$
\frac{\partial h}{\partial x} = -\frac{\partial g / \partial x}{\partial g / \partial z};
$$

hence, substituting this value of $\partial h / \partial x$ into the first of the simultaneous equations, we have

$$
\frac{\partial f}{\partial x} + \frac{\partial f}{\partial z}\left(-\frac{\partial g / \partial x}{\partial g / \partial z}\right) = 0 \quad \text{or} \quad \text{since } \lambda = -\frac{\partial f / \partial z}{\partial g / \partial z},
$$

$$
\frac{\partial f}{\partial x} + \lambda\frac{\partial g}{\partial x} = 0.
$$

Likewise, solving the fourth equation in the set above for $\partial h / \partial y$ and substituting into the second equation in the set, we obtain

$$
\frac{\partial f}{\partial y} + \lambda\frac{\partial g}{\partial y} = 0.
$$

From these calculations, we conclude that the critical points of F are among the solutions of the simultaneous equations

$$
\begin{cases}
\dfrac{\partial f}{\partial x} + \lambda\dfrac{\partial g}{\partial x} = 0 \\[2mm]
\dfrac{\partial f}{\partial y} + \lambda\dfrac{\partial g}{\partial y} = 0 \\[2mm]
\dfrac{\partial f}{\partial z} + \lambda\dfrac{\partial g}{\partial z} = 0.
\end{cases}
$$

(The third equation comes directly from the definition of λ.) Notice that these three simultaneous equations can be written as the single vector equation

$$
\nabla f + \lambda \nabla g = \overline{0} \quad \text{or} \quad \nabla f = -\lambda \nabla g.
$$

The argument above shows that, at a point where $f(x, y, z)$ attains a relative extremum, subject to the constraint $g(x, y, z) = k$, the gradient of f must be parallel to the gradient of g.

We can now formulate the main theorem of this section.

THEOREM 1   **Lagrange's Method of Multipliers**
Assume that the functions f and g are defined and have continuous partial derivatives on the subset D of xyz space, where D consists entirely of interior points. Suppose that, at each point (x, y, z) in D, at least one of the three partial derivatives $g_1(x, y, z)$, $g_2(x, y, z)$, and $g_3(x, y, z)$ is different from zero. Then the points (x, y, z) in D at which f attains relative extrema subject to the constraint

$$g(x, y, z) = k,$$

where k is a constant, can be found as follows:
Form the function u defined by

$$u(x, y, z) = f(x, y, z) + \lambda g(x, y, z)$$

for (x, y, z) in D, where λ (which is called the *Lagrange multiplier*) represents a constant yet to be determined. Then solve the simultaneous equations

$$\begin{cases} \dfrac{\partial u}{\partial x} = 0 \\[2mm] \dfrac{\partial u}{\partial y} = 0 \\[2mm] \dfrac{\partial u}{\partial z} = 0 \\[2mm] g(x, y, z) = k \end{cases}$$

for x, y, z, and λ. Several solutions may be obtained, but the desired points (x, y, z), where f attains its extrema subject to the constraint, are among these solutions.

INDICATION OF THE PROOF
A rigorous proof requires techniques beyond the scope of this book; however, we give an indication of why the theorem is true. If $u(x, y, z) = f(x, y, z) + \lambda g(x, y, z)$, then

$$\nabla u = \nabla f + \lambda \nabla g;$$

hence, the simultaneous equations $\partial u/\partial x = 0$, $\partial u/\partial y = 0$, and $\partial u/\partial z = 0$ are equivalent to $\nabla u = \bar{0}$, that is, to

$$\nabla f = -\lambda \nabla g.$$

However, the latter equation means that ∇f is parallel to ∇g, a condition that we previously found must hold at a point (x, y, z) where $f(x, y, z)$ attains a relative extremum subject to the constraint $g(x, y, z) = k$.

EXAMPLES   Use the Lagrange method of multipliers to work the given constrained extremum problem.

1 Find the extrema of the function f given by $f(x, y, z) = x^2 + y^2 + z^2$ subject to the constraint $x^2 + 2y^2 - z^2 = 1$.

SOLUTION
We let g be the function defined by $g(x, y, z) = x^2 + 2y^2 - z^2$ and form the quantity

$$u = f(x, y, z) + \lambda g(x, y, z);$$

that is,

$$u = x^2 + y^2 + z^2 + \lambda(x^2 + 2y^2 - z^2)$$

as in Theorem 1. Here,

$$\frac{\partial u}{\partial x} = 2x + 2\lambda x = 2(1 + \lambda)x,$$

$$\frac{\partial u}{\partial y} = 2y + 4\lambda y = 2(1 + 2\lambda)y, \quad \text{and}$$

$$\frac{\partial u}{\partial z} = 2z - 2\lambda z = 2(1 - \lambda)z.$$

After dividing by 2, we can write the simultaneous equations $\partial u/\partial x = 0$, $\partial u/\partial y = 0$, and $\partial u/\partial z = 0$ as $(1 + \lambda)x = 0$, $(1 + 2\lambda)y = 0$, and $(1 - \lambda)z = 0$. Appending the equation of constraint, we have

$$\begin{cases} (1 + \lambda)x = 0 \\ (1 + 2\lambda)y = 0 \\ (1 - \lambda)z = 0 \\ x^2 + 2y^2 - z^2 = 1. \end{cases}$$

Notice that $x = y = z = 0$ is not a solution of these simultaneous equations; hence, at least one of x, y, or z must be nonzero, and it follows that $1 + \lambda = 0$, or else $1 + 2\lambda = 0$, or else $1 - \lambda = 0$; that is, $\lambda = -1$, or else $\lambda = -\frac{1}{2}$, or else $\lambda = 1$. For $\lambda = -1$, the equations become

$$\begin{cases} 0 \quad\;\; = 0 \\ -y = 0 \\ 2z \quad = 0 \\ x^2 + 2y^2 - z^2 = 1, \end{cases}$$

so that $y = 0$, $z = 0$, and $x^2 = 1$. Thus, when $\lambda = -1$, we obtain the solutions $(1, 0, 0)$ and $(-1, 0, 0)$.

Similarly, when $\lambda = -\frac{1}{2}$, we get $x = 0$, $z = 0$, and $2y^2 = 1$; that is, we obtain the solutions $(0, \sqrt{2}/2, 0)$ and $(0, -\sqrt{2}/2, 0)$. Finally, when $\lambda = 1$, we get $x = 0$, $y = 0$, and $-z^2 = 1$. Since there are no real numbers z for which $-z^2 = 1$, there are no solutions corresponding to $\lambda = 1$. Thus, the critical points are $(\pm 1, 0, 0)$ and $(0, \pm\sqrt{2}/2, 0)$. At $(\pm 1, 0, 0)$, we have

$$f(\pm 1, 0, 0) = (\pm 1)^2 + 0^2 + 0^2 = 1,$$

and at $(0, \pm\sqrt{2}/2, 0)$, we have

$$f(0, \pm\sqrt{2}/2, 0) = 0^2 + (\pm\sqrt{2}/2)^2 + 0^2 = \tfrac{1}{2}.$$

Since we have given no "second derivative test" for constrained extrema problems, we must use algebraic or geometric means to decide whether the critical points obtained above correspond to relative maxima, relative minima, or saddle points.

Actually, $x^2 + 2y^2 - z^2 = 1$ is the equation of a hyperboloid of one sheet (Figure 1) and $f(x, y, z) = x^2 + y^2 + z^2$ gives the square of the distance from the origin to the

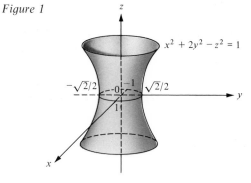

Figure 1

$x^2 + 2y^2 - z^2 = 1$

point (x, y, z); hence, it is geometrically evident that $(0, \pm\sqrt{2}/2, 0)$ are points where $f(x, y, z)$ takes on its absolute minimum value, namely $f(0, \pm\sqrt{2}/2, 0) = \frac{1}{2}$. Evidently, f has no absolute maximum, and $(\pm 1, 0, 0)$ are saddle points.

2 Find the dimensions of the rectangular box of largest volume which can be inscribed in the ellipsoid $(x^2/9) + (y^2/4) + z^2 = 1$, assuming that the edges of the box are parallel to the coordinate axes.

SOLUTION

Let (x, y, z) be the corner of the box in the first octant, so that the dimensions of the box are $2x$, $2y$, and $2z$ and its volume is given by

$$V = (2x)(2y)(2z) = 8xyz.$$

Thus, we must maximize V subject to the constraint

$$\frac{x^2}{9} + \frac{y^2}{4} + z^2 = 1.$$

We form the quantity

$$u = 8xyz + \lambda\left(\frac{x^2}{9} + \frac{y^2}{4} + z^2\right),$$

noting that

$$\frac{\partial u}{\partial x} = 8yz + \frac{2\lambda x}{9},$$

$$\frac{\partial u}{\partial y} = 8xz + \frac{2\lambda y}{4},$$

$$\frac{\partial u}{\partial z} = 8xy + 2\lambda z.$$

Setting these partial derivatives equal to zero and appending the equation of constraint, we have

$$\begin{cases} 8yz + \dfrac{2}{9}\lambda x = 0 \\[2mm] 8xz + \dfrac{2}{4}\lambda y = 0 \\[2mm] 8xy + 2\lambda z = 0 \\[2mm] \dfrac{x^2}{9} + \dfrac{y^2}{4} + z^2 = 1. \end{cases}$$

Multiplying the first three equations by $x/2$, $y/2$, and $z/2$, respectively, we have

$$4xyz + \lambda\frac{x^2}{9} = 0,$$

$$4xyz + \lambda\frac{y^2}{4} = 0,$$

$$4xyz + \lambda z^2 = 0.$$

Adding these three equations and using the fact that $(x^2/9) + (y^2/4) + z^2 = 1$, we obtain

$$12xyz + \lambda = 0; \quad \text{that is,} \quad \lambda = -12xyz.$$

Substituting this value of λ into the first three of the original simultaneous equations and simplifying yields

$$\begin{cases} yz(3 - x^2) = 0 \\ xz(4 - 3y^2) = 0 \\ xy(1 - 3z^2) = 0. \end{cases}$$

For a maximum volume, it is clear that x, y, and z must be positive; hence, we can cancel the factors yz, xz, and xy from the equations above to obtain

$$3 - x^2 = 0, \quad 4 - 3y^2 = 0, \quad \text{and} \quad 1 - 3z^2 = 0.$$

Therefore,

$$x = \sqrt{3}, \quad y = \frac{2}{\sqrt{3}}, \quad \text{and} \quad z = \frac{1}{\sqrt{3}}.$$

3 Find all critical points of the function f given by $f(x, y) = 3x^2 - 2xy + 5y^2$ subject to the constraint that $x^2 + 2y^2 = 6$.

SOLUTION

Here, f is a function of only two variables, but the method is still the same (Problem 18). We form the quantity

$$u = 3x^2 - 2xy + 5y^2 + \lambda(x^2 + 2y^2),$$

so that

$$\frac{\partial u}{\partial x} = 6x - 2y + 2\lambda x = 2[(3 + \lambda)x - y], \quad \text{and}$$

$$\frac{\partial u}{\partial y} = -2x + 10y + 4\lambda y = 2[-x + (5 + 2\lambda)y].$$

Thus, the desired critical points are found by solving the simultaneous equations

$$\begin{cases} (3 + \lambda)x - y = 0 \\ -x + (5 + 2\lambda)y = 0 \\ x^2 + 2y^2 = 6. \end{cases}$$

Multiplying the second equation by $3 + \lambda$ and adding it to the first, we cancel the x term and obtain

$$(2\lambda^2 + 11\lambda + 14)y = 0.$$

Similarly, multiplying the first equation by $5 + 2\lambda$ and adding it to the second, we cancel the y term and obtain

$$(2\lambda^2 + 11\lambda + 14)x = 0.$$

Since the equation of constraint, $x^2 + 2y^2 = 6$, must hold, we cannot have both x and y equal to zero; therefore, the coefficient $(2\lambda^2 + 11\lambda + 14)$ in the latter two equations must vanish, and we have

$$2\lambda^2 + 11\lambda + 14 = 0; \quad \text{that is,} \quad (2\lambda + 7)(\lambda + 2) = 0.$$

It follows that

$$\lambda = -\frac{7}{2} \quad \text{or} \quad \lambda = -2.$$

Putting $\lambda = -\frac{7}{2}$ in the first of the original simultaneous equations, $(3 + \lambda)x - y = 0$, we find that

$$y = -\frac{x}{2} \quad \text{when} \quad \lambda = -\frac{7}{2}.$$

Substituting $y = -x/2$ into the constraint equation $x^2 + 2y^2 = 6$, we obtain

$$x^2 + \frac{x^2}{2} = 6, \quad \text{so that} \quad x = \pm 2.$$

When $x = 2$, $y = -2/2 = -1$; when $x = -2$, $y = -(-2/2) = 1$. Similarly, putting $\lambda = -2$ in the equation $(3 + \lambda)x - y = 0$, we find that

$$y = x \quad \text{when} \quad \lambda = -2.$$

Substituting $y = x$ into the constraint equation $x^2 + 2y^2 = 6$, we obtain

$$x^2 + 2x^2 = 6, \quad \text{so that} \quad x = \pm\sqrt{2}.$$

When $x = \sqrt{2}$, $y = \sqrt{2}$; when $x = -\sqrt{2}$, $y = -\sqrt{2}$. Therefore, the desired critical points are $(2, -1)$, $(-2, 1)$, $(\sqrt{2}, \sqrt{2})$, and $(-\sqrt{2}, -\sqrt{2})$.

The method of Lagrange multipliers is effective when there is more than one constraint; however, more than one multiplier must then be used. Although we give an indication of this technique in Problem 20, the details are best left to an advanced course.

Problem Set 10

In problems 1 through 9, use the Lagrange multiplier method to find all critical points of each function f subject to the indicated constraint.

1 $f(x, y) = x^2 - y^2 - y$ with the constraint $x^2 + y^2 = 1$.

2 $f(x, y) = 3x^2 + 2\sqrt{2}\,xy + 4y^2$ with the constraint $x^2 + y^2 = 9$.

3 $f(x, y) = x^2 + y^2$ with the constraint $5x^2 + 6xy + 5y^2 = 8$.

4 $f(x, y) = x^2 + y^2$ with the constraint $2x^2 + y^2 = 1$.

5 $f(x, y) = x + y^2$ with the constraint $2x^2 + y^2 = 1$.

6 $f(x, y, z) = x^2 + y^2 + z^2$ with the constraint $x^2 + \frac{y^2}{4} + \frac{z^2}{9} = 1$.

7 $f(x, y, z) = xyz$ with the constraint $x^2 + \frac{y^2}{12} + \frac{z^2}{3} = 1$.

8 $f(x, y, z) = x^2 + y^2 + z^2$ with the constraint $x + 3y - 2z = 4$.

9 $f(x, y, z) = 3x^2 + 2y^2 + 4z^2$ with the constraint $2x + 4y - 6z = -5$.

10 Use the Lagrange multiplier method to find a point on the plane $3x - 4y + z = 2$ which is closest to the origin.

11 The cost of audit for tax purposes of a certain organization depends on the number of audits x and y at each of two headquarters according to the formula $C(x, y) = 2x^2 + xy + y^2 + 100$. How many audits should be made at each of the headquarters in order to minimize cost if the total number of audits must be 16?

12 A sheet metal container is to be made of a right circular cylinder with equal right circular conical caps on the ends. Show that, for a fixed volume, the total surface area is the smallest when the length of the cylinder is the same as the altitude of each cone and the diameter of the cylinder is $\sqrt{5}$ times its length.

13 The Postal Service specifies that the sum of the length and the perimeter of a cross section of a rectangular box accepted for parcel post must not exceed 100 inches. What are the dimensions of such a box that enable you to send the most by parcel post?

14 A manufacturer wants to make a rectangular box to hold a fixed volume V cubic units of a product. The material used for the sides costs $a per square unit, the material used for the bottom costs $b per square unit, and the material used for the top costs $c per square unit. What are the dimensions of the cheapest box?

15 The base of a rectangular box, open at the top, costs half as much per square foot as the sides. Find the dimensions of the box of largest volume that can be made for a fixed cost if the height of the box is to be 2 units.

16 Show that the maximum and minimum values of $Lx^2 + 2Mxy + Ny^2$, subject to the constraint $Ex^2 + 2Fxy + Gy^2 = 1$, where L, M, N, E, F, and G are constants and $EG - F^2 > 0$, are the two roots of the quadratic equation

$$(EG - F^2)t^2 - (LG - 2MF + NE)t + (LN - M^2) = 0.$$

17 Suppose that we wish to find the critical values of the function f of two variables subject to the constraint $g(x, y) = k$. Assuming that f and g have continuous partial derivatives, that $g_2(x, y) \neq 0$, and that the equation $g(x, y) = k$ can be solved for y in terms of x, show that ∇f and ∇g are parallel at a critical point.

18 Using the result of problem 17 as a guide, give a careful statement of a "Lagrange's method of multipliers theorem" for functions of two variables.

19 In Figure 2, assume that the curve $g(x, y) = k$ lies in a scalar field given by $w = f(x, y)$. At the point (a, b) on the curve the normal vector is $\nabla g(a, b)$. Suppose that $\nabla f(a, b)$ is not parallel to $\nabla g(a, b)$. In what direction should one move through (a, b) along the curve in order to (a) increase the value of w, (b) decrease the value of w? Thus explain geometrically why, at an extremum of f subject to the constraint $g(x, y) = k$, ∇f must be parallel to ∇g.

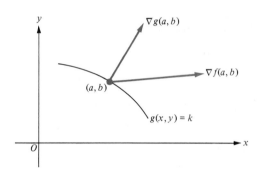

Figure 2

20 Suppose that we wish to find extrema of $f(x, y, z)$ subject to *two* contraints, namely $g(x, y, z) = k$ and $G(x, y, z) = K$. Show that, at such an extremum, there must exist two constants λ and μ (the Lagrange multipliers!) such that

$$\nabla f + \lambda \nabla g + \mu \nabla G = \bar{0}.$$

Assume that f and g have continuous partial derivatives and that the two simultaneous constraint equations can be solved for two of the variables x, y, and z in terms of the third.

Review Problem Set

In problems 1 through 4, find the domain of each function, sketch the domain, and identify the interior points of the domain.

1 $f(x, y) = \sqrt{64 - x^2 - y^2}$

2 $g(x, y) = \ln(x^2 + y^2 - 4)$

3 $g(x, y) = \dfrac{x^2y - x + 1}{1 - 2x - y}$

4 $f(x, y) = \dfrac{\sqrt{9 - x^2 - y^2}}{3y}$

5 Let f be a function of four variables defined by $f(x, y, z, w) = 3xyzw^2$. Find:
(a) $f(1, -1, 2, 1)$
(b) $f(3, 2, -1, 2)$
(c) $f(a, b^2, c, d^3)$
(d) the domain of f

In problems 6 through 8, evaluate the limit, if it exists.

6 $\displaystyle \lim_{(x,y)\to(0,0)} \frac{x^2 - y^2}{1 + x^2 + y^2}$

7 $\displaystyle \lim_{(x,y)\to(0,0)} \frac{6xy}{x^2 + y^2}$

8 $\displaystyle \lim_{(x,y)\to(0,0)} \frac{(1 + y^2)\sin x}{x}$

In problems 9 and 10, show that each function is discontinuous at the point $(0,0)$.

9 $f(x,y) = \begin{cases} \dfrac{x}{x - y} & \text{for } x \neq y \\ 0 & \text{for } x = y \end{cases}$

10 $f(x,y) = \begin{cases} \ln (x^2 + y^2) & \text{for } (x,y) \neq (0,0) \\ 0 & \text{for } (x,y) = (0,0) \end{cases}$

In problems 11 through 24, find the first partial derivatives of each function.

11 $f(x,y) = x^3 + 7xy^2 - 8y^3$

12 $g(x,y) = \dfrac{y^2}{y - x}$

13 $f(x,y) = y^3 e^x + x^3 e^y$

14 $f(x,y) = x^4 y - \sin (xy^4)$

15 $g(x,y) = \ln (x^3 - y^3)$

16 $h(x,y) = \displaystyle\int_x^y e^{-7t^2}\, dt$

17 $g(x,y) = \ln (x + \sqrt[3]{x^3 + y^3})$

18 $f(x,y) = \sin^{-1}\sqrt{1 - x^2 y^2}$

19 $g(x,y) = e^{-7x} \tan (x + y)$

20 $f(r,\theta,z) = zr^2 \sin \theta$

21 $g(x,y,z) = x^3 + y^3 + z^3 - 13x^2 y^2 z^2$

22 $h(x,y,z) = \tan \left(\dfrac{x}{y} + \dfrac{y}{z} + \dfrac{z}{x} \right)$

23 $f(x,y,z) = z \coth (xy)$

24 $g(x,y,z) = (\ln y)^x + \cos z$

25 Let $w = \dfrac{e^{x+y}}{e^x + e^y}$. Show that $\dfrac{\partial w}{\partial x} + \dfrac{\partial w}{\partial y} = w$.

26 If $w = x^2 y + y^2 z + xz^2$, show that $\dfrac{\partial w}{\partial x} + \dfrac{\partial w}{\partial y} + \dfrac{\partial w}{\partial z} = (x + y + z)^2$.

27 Suppose that $w = \cos (5x - 8y + 7z)$ and let $\bar{A} = \dfrac{\partial w}{\partial x}\, i + \dfrac{\partial w}{\partial y}\, j + \dfrac{\partial w}{\partial z}\, \bar{k}$. Show that $\bar{A}$ is perpendicular to the vector $\bar{B} = 10i + j - 6\bar{k}$.

28 Find $(2xi + yj + 2z\bar{k}) \cdot \nabla f$ if $f(x,y,z) = x \sin \dfrac{y^2}{z} - y^2 \tan \dfrac{z}{y^2}$.

29 Let $f(x,y,z) = \tan (y^2 e^{1/x})$. Verify that $(x^2 i + \tfrac{1}{2} y j) \cdot \nabla f = 0$.

30 At the point $(2, \sqrt{5}, 4)$ on the sphere $x^2 + y^2 + z^2 = 25$ a tangent line is drawn parallel to the xz plane. Find the angle between this tangent line and the xy plane.

31 Find the equation of (a) the normal line and (b) the tangent plane to the surface $xyz = 8$ at the point $(2,2,2)$.

32 Two sides and the included angle of a triangle are given by x, y, and θ, respectively. Find the rate of change of its area with respect to each of these quantities when the other two are held constant.

33 Let $f(x,y) = x^2 y$. (a) Find df. (b) Use the result of part (a) to find an approximation for $(5.04)^2(2.98)$.

34 Find the linear approximation for the function f defined by $f(x,y) = \dfrac{xy - 1}{xy + 1}$ near the point $(x_0, y_0) = (\tfrac{1}{2}, \tfrac{1}{4})$.

35 A solid metal cylinder with radius 4.02 centimeters and height 8.03 centimeters is cut down to a cylinder of radius 4 centimeters and height 8 centimeters. Approximately how much metal is removed?

36 A fence 4 feet high runs parallel to the wall of a building and 3 feet from it. A person standing at a window in the building looks directly over the fence at a point P on the ground. The person's eyes are 8 feet above the ground. If the fence were 3 inches higher and 4 inches farther from the building, approximately how much farther would the new point P be (a) from the building and (b) from the person's eyes?

37 If $r = \sqrt{x^2 + y^2 + z^2}$ and $\bar{R} = x\bar{i} + y\bar{j} + z\bar{k}$, show that (a) $\bar{R} \cdot \nabla r = r$ and $\bar{R} \cdot \nabla(r^2) = 2r^2$.

38 Suppose that $y = r \sin \theta$. If erroneous values are used for r and θ, explain why the approximate error in the resulting value of y should not exceed $|r \cos \theta| \, |\Delta\theta| + |\sin \theta| \, |\Delta r|$, where $\Delta\theta$ and Δr are the errors in θ and r, respectively, provided that the errors are small.

39 Compute $d\rho$ if $\rho = e^{\theta/2} \sin (\theta - \phi)$, $\theta = 0$, $\phi = \pi/2$, $d\theta = 0.2$, and $d\phi = -0.2$.

In problems 40 through 46, use the chain rule to find the indicated derivative.

40 $w = x^3 + 5xy - y^3$, $x = r^2 + s$, $y = r - s^2$; $\dfrac{\partial w}{\partial r}$ and $\dfrac{\partial w}{\partial s}$.

41 $w = 2x^4 - 3x^2y^2 + y^4$, $x = 3u + v$, $y = u - 2v$; $\dfrac{\partial w}{\partial u}$ and $\dfrac{\partial w}{\partial v}$.

42 $f(u, v) = \cos (u + v)$, $g(x, y) = x^2 + y^2$, $h(x, y) = x^2 - y^2$, $F(x, y) = f(g(x, y), h(x, y))$; F_1 and F_2.

43 $w = \displaystyle\int_x^y e^{t^4} \, dt$, $x = 2r + s$, $y = r - 3s$; $\dfrac{\partial w}{\partial r}$ and $\dfrac{\partial w}{\partial s}$.

44 $u = f(x - y, x + y)$; $\dfrac{\partial u}{\partial x}$ and $\dfrac{\partial u}{\partial y}$.

45 $f(u, v) = \cos (uv)$, $g(x) = \sqrt[3]{x}$, $h(x) = \sqrt[4]{x}$, $F(x) = f(g(x), h(x))$; F'.

46 $w = \tan^{-1} (uv)$, $u = e^x$, $v = e^{-5x}$; $\dfrac{dw}{dx}$.

47 Let f be the function defined by $f(x, y) = e^{-xy}$, and let g and h be functions such that $g(3) = 5$, $g'(3) = -2$, $h(3) = 4$, and $h'(3) = 7$. If $K(t) = f(g(t), h(t))$, find $K'(3)$.

48 If f is a differentiable function and $w = f\left(\dfrac{y - x}{xy}, \dfrac{z - x}{xy}\right)$, find the total differential dw.

49 Suppose that $w = f(x, y)$, $x = r \cos \theta$, and $y = r \sin \theta$. If $f_x(0, 1) = 2$ and $f_y(0, 1) = -3$, find:

(a) $\dfrac{\partial w}{\partial r}$ when $(r, \theta) = \left(1, \dfrac{\pi}{2}\right)$.

(b) $\dfrac{\partial w}{\partial \theta}$ when $(r, \theta) = \left(1, \dfrac{\pi}{2}\right)$.

50 If f is a differentiable function of two variables such that $f(1, 1) = 1$, $f_1(1, 1) = a$, and $f_2(1, 1) = b$, and if $g(x) = f(x, f(x, f(x, x)))$, find (a) $g(1)$ and (b) $g'(1)$.

51 If f is the function defined by $f(x, y) = x \sin y$, find (a) $f_1(y, x)$ and (b) $f_2(y, x)$.

52 If $w = 4x^3 + 3x^2y - 3zy^2$, find the directional derivative of w in the direction of the vector $\bar{u} = \dfrac{\sqrt{3}}{2\sqrt{2}} i + \dfrac{1}{2\sqrt{2}} j - \dfrac{1}{\sqrt{2}} k$ at the point $(1, -1, 3)$.

53 (a) Find the directional derivative of $w = x^2y$ at the point $(1, -3)$ in the direction of the unit vector $\bar{u} = \cos \dfrac{5\pi}{6} i + \sin \dfrac{5\pi}{6} j$.

(b) What is the maximum value of the directional derivative of $w = x^2y$ at $(1, -3)$ and in what direction is it attained?

54 Find the angle θ such that the directional derivative of $w = x^2 + \frac{1}{4}y^2$ at the point $(1, 2)$ is a maximum in the direction of the vector $\bar{u} = (\cos \theta)i + (\sin \theta)j$ and find this maximum value.

55 Find $\dfrac{dy}{dx}$ if $y \sin x - x \cos y = 0$.

56 If f is a differentiable function and $f(x, y) = 0$, find the value of dy/dx when $x = 3$ and $y = -7$ if $f_1(3, -7) = 2$ and $f_2(3, -7) = 5$.

57 Sand is being poured on a conical pile at the rate of 4 cubic feet per minute. At a given instant, the pile is 6 feet in diameter and 5 feet high. At what rate is the height increasing at this instant if the diameter is increasing at the rate of 2 inches per minute?

58 A piece of copper in the form of a rectangular parallelepiped has edges of lengths 2, 4, and 8 centimeters. Because of heating, the edges are increasing at 0.001 centimeter per minute. At what rate is the volume changing?

In problems 59 through 62, find the equation of the tangent plane and the normal line to each surface at the point indicated.

59 $z^3 + y^3 + x^3 - 3xyz = 8$; $(3,3,2)$.

60 $\dfrac{x^2}{16} + z = \dfrac{y^2}{9}$; $(15, \frac{75}{4}, 25)$.

61 $z^3 + 3xz - 2y = 0$; $(1,7,2)$.

62 $5x^2 + 4y^2 + 2z^2 = 17$; $(-1,1,2)$.

63 Let $w = e^{xy}$, $x = s^2 + 2st$, and $y = 2st + t^2$. Find (a) $\dfrac{\partial^2 w}{\partial s^2}$ and (b) $\dfrac{\partial^2 w}{\partial t^2}$.

64 Let $w = f(u,v)$, $u = g(x,y)$, and $v = h(x,y)$. Assume that f, g, and h have continuous second partial derivatives and that the conditions $\dfrac{\partial u}{\partial x} = \dfrac{\partial v}{\partial y}$ and $\dfrac{\partial u}{\partial y} = -\dfrac{\partial v}{\partial x}$ hold. Show that

$$\frac{\partial^2 w}{\partial x^2} + \frac{\partial^2 w}{\partial y^2} = \left[\left(\frac{\partial u}{\partial x}\right)^2 + \left(\frac{\partial v}{\partial x}\right)^2\right] \cdot \left(\frac{\partial^2 w}{\partial u^2} + \frac{\partial^2 w}{\partial v^2}\right).$$

In problems 65 through 68, find the relative maxima and minima for each function.

65 $f(x,y) = x^2 - y^2 + 2x - 4y + 3$

66 $g(x,y) = x^3 - 4y^2$

67 $h(x,y) = xy(3 - x - y)$

68 $F(x,y) = 3x^2 + 2y^2 + 3xy - 66x - 58y + 1600$

In problems 69 through 71, use the method of Lagrange multipliers to find the critical points of each function subject to the indicated constraint. In each case, determine whether the critical point corresponds to a relative (or absolute) maximum or minimum or whether it is a saddle point.

69 $f(x,y) = x^2 + y^2$ with the constraint $x + y - 1 = 0$.

70 $f(x,y,z) = x^2 + y^2 + z^2$ with the contraint $ax + by + cz + d = 0$.

71 $f(x,y) = x + y$ with the constraint $x^2 + y^2 = 1$.

72 A rectangular sheet metal tank is open at the top and is filled with 9 cubic meters of liquid. Find the dimensions of the tank so that the metal surface in contact with the liquid is minimal.

73 Show that a rectangular box (with top) made out of S square units of material has maximum volume when it is a cube.

74 There will be N lottery tickets sold at a price of $\$p$ per ticket provided that $\$A$ is spent for publicity. Every $\$x$ increase in the price per ticket will result in Bx fewer tickets being sold. Apart from publicity costs, a fixed overhead of $\$K$ is required to operate the lottery. Total prize money distributed to winners of the lottery will be $100k$ percent of the revenue from ticket sales. It is estimated that each additional dollar spent for publicity will result in C additional tickets being sold, regardless of the price per ticket. Find (a) the price per ticket and (b) the expenditure for publicity that will maximize the profit from the lottery. Also find (c) the number of tickets sold when the maximum profit is realized and (d) the maximum profit. (Your answers will be in terms of the constants p, A, B, K, k, and C.)

75 A manufacturer produces razors and razor blades at a cost of $\$0.60$ per razor and $\$0.30$ per dozen for the blades. If she charges x cents per razor and y cents per dozen for the blades, she will sell $\dfrac{6 \times 10^8}{x^2 y}$ razors and $\dfrac{48 \times 10^7}{xy^2}$ dozen blades per day. What should she charge for razors and for blades so as to maximize her profit?

17 MULTIPLE INTEGRATION

In Chapter 6 we introduced and studied the definite (Riemann) integral for a function of a single variable. In this chapter we extend the notion of a definite integral to functions of two or more variables in a natural and useful way to obtain *multiple integrals*. We find that most of the multiple integrals encountered in elementary applications to geometry and the physical sciences can be evaluated in terms of *iterated integrals*—that is, repeated definite integrals—in the cartesian, polar, cylindrical, or spherical coordinate system. The chapter also includes *line integrals, surface integrals, Green's theorem, Stokes' theorem,* and *the divergence theorem of Gauss.*

1 Iterated Integrals

In Section 3 of Chapter 16 we calculated partial derivatives of functions of several variables by regarding all but one of the independent variables as being constant and differentiating with respect to the remaining variable. Likewise, it is possible to take an indefinite *integral* of such a function with respect to one of its variables, while temporarily regarding the remaining variables as being held constant. For instance,

$$\int x^2 y^3 \, dx = y^3 \int x^2 \, dx = y^3 \frac{x^3}{3} + C$$

and

$$\int x^2 y^3 \, dy = x^2 \int y^3 \, dy = x^2 \frac{y^4}{4} + K.$$

Notice how the variable of integration is clearly indicated by the differential dx or dy under the integral sign.

In the above calculation of $\int x^2 y^3 \, dx$, we temporarily held y constant; however, different fixed values of y could require different values of the constant of

899

integration C. The possible dependence of C on y can be indicated by writing $C(y)$ rather than C; that is, we can regard the "constant" of integration as a function of y and write

$$\int x^2 y^3 \, dx = \frac{x^3 y^3}{3} + C(y).$$

Similarly, when integrating with respect to y, we should write

$$\int x^2 y^3 \, dy = \frac{x^2 y^4}{4} + K(x).$$

The integrals above are just the analogs for indefinite integration of partial derivatives for differentiation, and they could be called "partial integrals." However, we prefer to call them integrals *with respect to x* or *with respect to y*.

EXAMPLE   If $f(x, y) = x \cos y$, find $\int f(x, y) \, dx$ and $\int f(x, y) \, dy$.

SOLUTION

$$\int f(x, y) \, dx = \int x \cos y \, dx = \cos y \int x \, dx = \frac{x^2}{2} \cos y + C(y),$$

$$\int f(x, y) \, dy = \int x \cos y \, dy = x \int \cos y \, dy = x \sin y + K(x).$$

Now suppose that f is a function of two variables such that, for each *fixed* value of y, $f(x, y)$ is an integrable function of x. Then, for each fixed value of y, we can form the definite integral

$$\int_a^b f(x, y) \, dx.$$

Furthermore, for different fixed values of y, we can use different limits of integration a and b; that is, a and b can depend on y. Such dependence can be indicated by the usual function notation, and the integral becomes

$$\int_{a(y)}^{b(y)} f(x, y) \, dx.$$

EXAMPLE   Evaluate $\int_{\ln y}^{y^2} y e^{xy} \, dx$.

SOLUTION
Holding y temporarily constant and integrating with respect to x, we obtain

$$\int y e^{xy} \, dx = \frac{y e^{xy}}{y} + C(y) = e^{xy} + C(y).$$

Therefore,

$$\int_{\ln y}^{y^2} y e^{xy} \, dx = [e^{xy} + C(y)]\Big|_{\ln y}^{y^2} = [e^{y^2 y} + C(y)] - [e^{(\ln y)y} + C(y)]$$

$$= e^{y^3} - e^{y \ln y} + C(y) - C(y) = e^{y^3} - y^y.$$

In the preceding example, notice that the "constant" of integration, $C(y)$, cancels out as usual during the definite integration. Therefore, when dealing with *definite* integrals, there is no necessity to write the "constant" of integration at all.

Also, observe that the integration in the preceding example takes place *with respect to x*; hence, the limits of integration must be *substituted for x* after performing the indefinite integration. To emphasize this, one can write

$$\int_{x=a(y)}^{x=b(y)} f(x, y)\, dx = \left[\int f(x, y)\, dx\right]\Bigg|_{x=a(y)}^{x=b(y)} \quad \text{and} \quad \int_{y=g(x)}^{y=h(x)} f(x, y)\, dy = \left[\int f(x, y)\, dy\right]\Bigg|_{y=g(x)}^{y=h(x)}.$$

Notice that the quantity $\displaystyle\int_{y=g(x)}^{y=h(x)} f(x, y)\, dy$ depends only on x, while the quantity $\displaystyle\int_{x=a(y)}^{x=b(y)} f(x, y)\, dx$ depends only on y. Consequently, we can define functions F and G of the single variables x and y, respectively, by the equations

$$F(x) = \int_{y=g(x)}^{y=h(x)} f(x, y)\, dy \quad \text{and} \quad G(y) = \int_{x=a(y)}^{x=b(y)} f(x, y)\, dx.$$

In many cases, the functions F and G are themselves integrable, and we can write

$$\int_{x=c}^{x=d} F(x)\, dx = \int_{x=c}^{x=d} \left[\int_{y=g(x)}^{y=h(x)} f(x, y)\, dy\right] dx, \quad \text{and}$$

$$\int_{y=c}^{y=d} G(y)\, dy = \int_{y=c}^{y=d} \left[\int_{x=a(y)}^{x=b(y)} f(x, y)\, dx\right] dy.$$

The latter integrals are called *iterated* (or *repeated*) integrals, and are customarily written without the brackets and with the simpler notation for the limits of integration, thus:

$$\int_{c}^{d} \int_{g(x)}^{h(x)} f(x, y)\, dy\, dx = \int_{x=c}^{x=d} \left[\int_{y=g(x)}^{y=h(x)} f(x, y)\, dy\right] dx, \quad \text{and}$$

$$\int_{c}^{d} \int_{a(y)}^{b(y)} f(x, y)\, dx\, dy = \int_{y=c}^{y=d} \left[\int_{x=a(y)}^{x=b(y)} f(x, y)\, dx\right] dy.$$

Notice that, in order to evaluate

$$\int_{c}^{d} \int_{g(x)}^{h(x)} f(x, y)\, dy\, dx,$$

we first integrate $f(x, y)$ with respect to y, holding x fixed. The limits of integration, $g(x)$ and $h(x)$, depend on this fixed value of x and so does the resulting quantity,

$$\int_{g(x)}^{h(x)} f(x, y)\, dy.$$

Then we integrate the latter quantity with respect to x (now regarding x as a variable) between the constant limits of integration c and d.

On the other hand, the iterated integral

$$\int_{c}^{d} \int_{a(y)}^{b(y)} f(x, y)\, dx\, dy$$

involves a first integration of $f(x, y)$ with respect to x, holding y fixed, between the limits of integration $a(y)$ and $b(y)$, followed by an integration of the resulting quantity with respect to y between the constant limits of integration c and d.

The two successive integrations required to evaluate an iterated integral are carried out in the order in which the differentials (dx and dy in the integrals above) appear, *reading from left to right*. However, the corresponding limits of integration are associated with the integral signs reading from "inside out," that is, *from right to left.*

EXAMPLES   Evaluate the given iterated integrals.

1 $\displaystyle\int_{-1}^{2}\int_{0}^{2} x^2 y^3 \, dy \, dx$

SOLUTION

$$\int_{-1}^{2}\int_{0}^{2} x^2 y^3 \, dy \, dx = \int_{x=-1}^{x=2}\left[\int_{y=0}^{y=2} x^2 y^3 \, dy\right] dx = \int_{x=-1}^{x=2}\left[\frac{x^2 y^4}{4}\bigg|_{y=0}^{y=2}\right] dx$$

$$= \int_{x=-1}^{x=2}\left[\frac{16x^2}{4} - 0\right] dx = \frac{4}{3}x^3\bigg|_{-1}^{2} = \left(\frac{32}{3}\right) - \left(-\frac{4}{3}\right) = 12.$$

2 $\displaystyle\int_{0}^{4}\int_{0}^{3x/2} \sqrt{16 - x^2} \, dy \, dx$

SOLUTION

$$\int_{0}^{4}\int_{0}^{3x/2} \sqrt{16 - x^2} \, dy \, dx = \int_{x=0}^{x=4}\left[\int_{y=0}^{y=3x/2} \sqrt{16 - x^2} \, dy\right] dx$$

$$= \int_{x=0}^{x=4}\left[\sqrt{16 - x^2}\int_{y=0}^{y=3x/2} dy\right] dx$$

$$= \int_{x=0}^{x=4}\sqrt{16 - x^2}\left[y\bigg|_{y=0}^{y=3x/2}\right] dx$$

$$= \int_{0}^{4}\sqrt{16 - x^2}\left(\frac{3x}{2}\right) dx.$$

The latter integral can be evaluated using the substitution $u = 16 - x^2$, so that $du = -2x \, dx$, $x \, dx = -\frac{1}{2} du$, and $(3x/2) \, dx = -\frac{3}{4} du$. Since $u = 16$ when $x = 0$ and $u = 0$ when $x = 4$, we have

$$\int_{0}^{4}\sqrt{16 - x^2}\left(\frac{3x}{2}\right) dx = \int_{16}^{0}\sqrt{u}\left(-\frac{3}{4}\right) du = \frac{3}{4}\int_{0}^{16}\sqrt{u} \, du = \frac{3}{4}\left[\frac{2}{3}u^{3/2}\bigg|_{0}^{16}\right] = 32.$$

3 $\displaystyle\int_{0}^{\pi}\int_{0}^{y^2} \sin\frac{x}{y} \, dx \, dy$

SOLUTION

$$\int_{0}^{\pi}\int_{0}^{y^2} \sin\frac{x}{y} \, dx \, dy = \int_{0}^{\pi}\left[\int_{0}^{y^2} \sin\frac{x}{y} \, dx\right] dy = \int_{0}^{\pi}\left[-y\cos\frac{x}{y}\bigg|_{0}^{y^2}\right] dy$$

$$= \int_{0}^{\pi}\left(-y\cos\frac{y^2}{y} + y\cos\frac{0}{y}\right) dy = \int_{0}^{\pi}(y - y\cos y) \, dy$$

$$= \int_{0}^{\pi} y \, dy - \int_{0}^{\pi} y\cos y \, dy = \frac{y^2}{2}\bigg|_{0}^{\pi} - (y\sin y + \cos y)\bigg|_{0}^{\pi}$$

$$= \frac{\pi^2}{2} - (\pi\sin\pi + \cos\pi) + (0\sin 0 + \cos 0) = \frac{\pi^2}{2} + 2.$$

(The integral $\displaystyle\int_{0}^{\pi} y\cos y \, dy$ was evaluated using integration by parts.)

Problem Set 1

In problems 1 through 4, carry out each integration by holding all variables, other than the variable of integration, temporarily constant. Be certain to write the "constant" of integration as a function of the variables that were fixed during the integration.

1 $\int \sin(xy)\,dx$

2 $\int \dfrac{dy}{x^2 + y^2}$

3 $\int \left(x\sqrt{y} - \dfrac{y}{x} + \dfrac{1}{x\sqrt{y}} \right) dy$

4 $\int \sqrt{y^2 - x^2}\,dx$

In problems 5 and 6, evaluate each integral.

5 $\displaystyle\int_{x=\pi/2}^{x=y^3} y^2 \sin x\,dx$

6 $\displaystyle\int_{y=2}^{y=\ln x} y e^{xy}\,dy$

In problems 7 through 28, evaluate each iterated integral.

7 $\displaystyle\int_0^1 \int_0^2 x^4 y^2\,dy\,dx$

8 $\displaystyle\int_0^2 \int_0^3 y^3\,dy\,dx$

9 $\displaystyle\int_0^3 \int_0^1 x e^{xy}\,dy\,dx$

10 $\displaystyle\int_0^4 \int_0^1 y e^{xy}\,dx\,dy$

11 $\displaystyle\int_0^{\pi/2} \int_0^1 xy \cos(xy^2)\,dy\,dx$

12 $\displaystyle\int_0^{\pi} \int_0^{\pi/2} \sin x \cos y\,dx\,dy$

13 $\displaystyle\int_{-3}^5 \int_0^1 \dfrac{y^2}{1 + x^2}\,dx\,dy$

14 $\displaystyle\int_0^1 \int_1^2 u e^v\,dv\,du$

15 $\displaystyle\int_0^2 \int_0^{\sqrt{y}} x^3\,dx\,dy$

16 $\displaystyle\int_0^1 \int_0^{x^4} \sin(\pi x^5)\,dy\,dx$

17 $\displaystyle\int_1^5 \int_{\sqrt[3]{x}}^x \dfrac{1}{x}\,dy\,dx$

18 $\displaystyle\int_1^4 \int_{x/3}^{x/27} \sqrt[3]{\dfrac{x}{3y}}\,dy\,dx$

19 $\displaystyle\int_0^{\pi/3} \int_0^{\sin x} \dfrac{dy\,dx}{\sqrt{1 - y^2}}$

20 $\displaystyle\int_1^4 \int_0^t s^2 \ln t^4\,ds\,dt$

21 $\displaystyle\int_0^1 \int_{4u}^{6u} e^{u+v}\,dv\,du$

22 $\displaystyle\int_0^{\sqrt{\ln 2}} \int_0^y xy^5 e^{x^2 y^2}\,dx\,dy$

23 $\displaystyle\int_0^{\pi/2} \int_0^{2\cos\theta} \phi \cos\theta\,d\phi\,d\theta$

24 $\displaystyle\int_0^{\pi/6} \int_0^{\sec y \tan y} x^3 \cos^4 y\,dx\,dy$

25 $\displaystyle\int_1^2 \int_0^{x^2} e^{y/x}\,dy\,dx$

26 $\displaystyle\int_0^{\pi} \int_0^{3\cos\phi} \theta \sin\phi\,d\theta\,d\phi$

27 $\displaystyle\int_0^{\pi/4} \int_0^{\sin x} 4e^{-y} \cos x\,dy\,dx$

28 $\displaystyle\int_0^{\pi/2} \int_4^{4+3\cos t} s\,ds\,dt$

29 Let a, b, c, and d be constants and let f, F, and ϕ be functions of two variables such that $\dfrac{\partial}{\partial x} F(x, y) = f(x, y)$ and $\dfrac{\partial}{\partial y} \phi(x, y) = F(x, y)$. Show that

$$\int_a^b \int_c^d f(x, y)\,dx\,dy = \phi(d, b) - \phi(d, a) - \phi(c, b) + \phi(c, a).$$

30 With the notation of problem 29, and making the necessary assumptions about the existence and continuity of the required partial derivatives, show that

$$\int_a^b \int_c^d f(x, y)\,dx\,dy = \int_c^d \int_a^b f(x, y)\,dy\,dx.$$

2   The Double Integral

In Section 2 of Chapter 6 we defined the definite (Riemann) integral of a function f over a closed interval $[a,b]$ as a limit of Riemann sums. This definition was suggested by the problem of calculating the area under the graph of f between $x = a$ and $x = b$. Now, if f is a function of *two* variables and R is a region in the xy plane which is contained in the domain of f, we can formulate an analogous problem in three-dimensional space by asking for the volume V shown in Figure 1. Thus, if $f(x,y) \geq 0$ for (x,y) in R, we are asking for the volume of the solid that is bounded above by the graph of f, below by the region R, and laterally by the cylinder over the boundary of R whose generators are parallel to the z axis. We speak of this solid as "the solid below the graph of f and above the region R."

In this section we attack the problem of finding the volume V in the same way that we attacked the analogous two-dimensional problem in Chapter 6; namely, by setting up better and better approximations to V (called *Riemann sums*), and obtaining V as a limit of such approximations. This limit is called the *double integral of f over the region R* and is written as

$$\iint\limits_{R} f(x,y)\, dx\, dy.$$

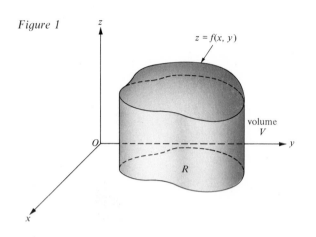

Figure 1

The problem of defining the double integral in all generality is best left to more advanced courses. For our purposes we assume that R is an admissible two-dimensional region (Chapter 6, Section 6.1), that R contains all of its own boundary points, and that f is a continuous function on R. We partition R into small rectangles in much the same way that we partitioned the interval of integration into small subintervals in Chapter 6; however, we consider only "regular" partitions in which all the small rectangles are congruent.

Since R is an admissible region, it is bounded, and can therefore be enclosed in a rectangle $a \leq x \leq b$, $c \leq y \leq d$ in the xy plane (Figure 2). Given a position integer n, we partition this rectangle into n^2 congruent subrectangles as follows:

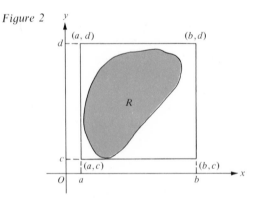

Figure 2

1 Divide the interval $[a,b]$ into n subintervals of equal length $\Delta x = \dfrac{b-a}{n}$ by means of the points

$$x_0 = a,\ x_1 = x_0 + \Delta x,\ x_2 = x_1 + \Delta x,\ \ldots,\ x_n = x_{n-1} + \Delta x = b.$$

2 Divide the interval $[c,d]$ into n subintervals of equal length $\Delta y = \dfrac{d-c}{n}$ by means of the points

$$y_0 = c,\ y_1 = y_0 + \Delta y,\ y_2 = y_1 + \Delta y,\ \ldots,\ y_n = y_{n-1} + \Delta y = d.$$

Figure 3

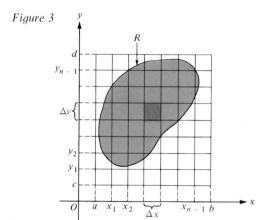

3 Within the rectangle $a \leq x \leq b$, $c \leq y \leq d$, form a grid consisting of vertical line segments $x = x_0$, $x = x_1$, $x = x_2$, ..., $x = x_n$ and horizontal line segments $y = y_0$, $y = y_1$, $y = y_2$, ..., $y = y_n$ (Figure 3). This grid divides the rectangle into n^2 congruent subrectangles, each of which has area $\Delta x \, \Delta y$. One of these subrectangles is shown shaded in Figure 3.

We call this decomposition of the rectangle $a \leq x \leq b$, $c \leq y \leq d$ into n^2 congruent subrectangles a *regular partition* and we refer to each of the n^2 subrectangles as a *cell* of the partition. Some of these cells may be contained in the region R, some of them may lie outside, and some of them may extend across part of the boundary of R. We now discard all cells that do not touch the region R and number the remaining cells (that do touch R) in some convenient manner, calling them (say) ΔR_1, ΔR_2, ΔR_3, ..., ΔR_m. Of course, each of these cells has an area $\Delta x \, \Delta y$ and, taken together, they contain the region R and approximate its shape and its area (Figure 4). As n increases, the grid becomes finer and the approximation improves.

Figure 4

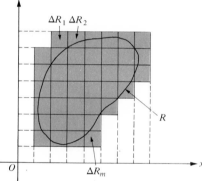

By analogy with the notion of an augmented partition considered in Chapter 6, we now choose a point inside each of the cells ΔR_1, ΔR_2, ..., ΔR_m, making certain that each chosen point belongs to the region R. For definiteness, denote the point chosen from the kth cell ΔR_k by (x_k^*, y_k^*) for $k = 1, 2, ..., m$.

Now, consider the solid below the graph of f and above the cell ΔR_k (Figure 5), noting that this solid is approximately a rectangular parallelepiped with base ΔR_k of area $\Delta x \, \Delta y$ and with height $f(x_k^*, y_k^*)$. Its volume is approximately

$$f(x_k^*, y_k^*) \, \Delta x \, \Delta y.$$

Adding the approximate volumes corresponding to each cell ΔR_1, ΔR_2, ..., ΔR_m, we obtain the approximation

$$V \approx \sum_{k=1}^{m} f(x_k^*, y_k^*) \, \Delta x \, \Delta y$$

Figure 5

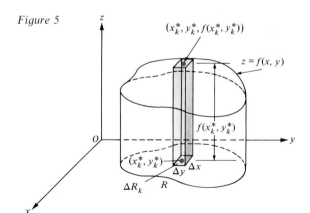

for the total volume below the graph of f and above the region R.

By analogy with the terminology introduced in Chapter 6, the sum

$$\sum_{k=1}^{m} f(x_k^*, y_k^*) \, \Delta x \, \Delta y$$

is called a *Riemann sum* corresponding to the given regular augmented partition. The limit of such Riemann sums, as the partition becomes finer and finer (in the present case, as $n \to +\infty$), is called the *double integral of f over R* and written as $\iint\limits_R f(x, y)\, dx\, dy$. Thus, by definition,

$$V = \iint\limits_R f(x, y)\, dx\, dy = \lim_{n \to +\infty} \sum_{k=1}^{m} f(x_k^*, y_k^*)\, \Delta x\, \Delta y,$$

provided that the limit exists.

EXAMPLE   Approximate the double integral $\iint\limits_R x^2 y^4\, dx\, dy$, where R is the region inside the circle $x^2 + y^2 = 1$. Use the regular partition of the rectangle $-1 \le x \le 1$, $-1 \le y \le 1$ into four congruent cells and use the midpoints of the cells to augment the partition (Figure 6).

Figure 6

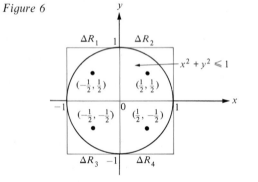

SOLUTION
Each of the four cells has dimensions $\Delta x = 1$ and $\Delta y = 1$. Here, $f(x, y) = x^2 y^4$,

$$(x_1^*, y_1^*) = \left(-\frac{1}{2}, \frac{1}{2}\right), \qquad (x_2^*, y_2^*) = \left(\frac{1}{2}, \frac{1}{2}\right),$$

$$(x_3^*, y_3^*) = \left(-\frac{1}{2}, -\frac{1}{2}\right), \qquad (x_4^*, y_4^*) = \left(\frac{1}{2}, -\frac{1}{2}\right),$$

and the Riemann sum

$$\sum_{k=1}^{4} f(x_k^*, y_k^*)\, \Delta x\, \Delta y = \sum_{k=1}^{4} (x_k^*)^2 (y_k^*)^4 (1)(1)$$

is given by

$$\left(-\frac{1}{2}\right)^2 \left(\frac{1}{2}\right)^4 + \left(\frac{1}{2}\right)^2 \left(\frac{1}{2}\right)^4 + \left(-\frac{1}{2}\right)^2 \left(-\frac{1}{2}\right)^4 + \left(\frac{1}{2}\right)^2 \left(-\frac{1}{2}\right)^4 = \frac{1}{16}.$$

Therefore,

$$\iint\limits_R x^2 y^4\, dx\, dy \approx \sum_{k=1}^{4} (x_k^*)^2 (y_k^*)^4 = \frac{1}{16}.$$

Now consider a region R in the xy plane and let f be the constant function defined by $f(x, y) = 1$ for all values of x and y. The graph of f is the horizontal plane $z = 1$. In this case, we write the double integral $\iint\limits_R f(x, y)\, dx\, dy$ simply as $\iint\limits_R dx\, dy$. Here, $\iint\limits_R dx\, dy$ represents the volume of the cylinder of height $h = 1$ with base R (Figure 7). If A is the area of the region R, then the cylinder has volume $V = hA = 1 \cdot A = A$, so that

$$V = A = \iint\limits_R dx\, dy.$$

Therefore, $\iint\limits_R dx\, dy$ is numerically equal to the area of the region R.

Figure 7

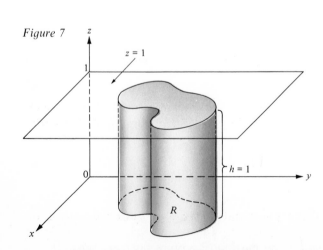

EXAMPLE   Let R be the interior of the triangle in the plane whose vertices are $A = (0,0)$, $B = (1,2)$, and $C = (5,14)$. Find $\iint_R dx\, dy$.

SOLUTION

By Theorem 4 in Section 4.2 of Chapter 15, the area of triangle ABC is the absolute value of

$$\frac{1}{2}\begin{vmatrix} 0 & 0 & 1 \\ 1 & 2 & 1 \\ 5 & 14 & 1 \end{vmatrix} = \frac{1}{2}\begin{vmatrix} 1 & 2 \\ 5 & 14 \end{vmatrix} = \frac{1}{2}(14 - 10) = 2.$$

Therefore, $\iint_R dx\, dy = 2$.

If a function f has nonnegative values over the region R, then $\iint_R f(x,y)\, dx\, dy$ can be interpreted as the volume V of the solid below the graph of f and above the region R. Often V can be found by the methods presented in Chapter 7 (slicing, circular disks, cylindrical shells, and so forth), and thus $\iint_R f(x,y)\, dx\, dy$ can be evaluated.

EXAMPLES 1 Let R be the region inside the circle $x^2 + y^2 \le 4$ and let f be defined by $f(x,y) = \sqrt{4 - x^2 - y^2}$. Evaluate $\iint_R f(x,y)\, dx\, dy$.

Figure 8

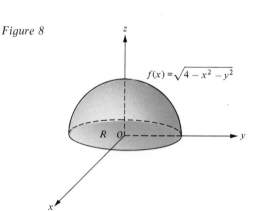

SOLUTION

The graph of f is a hemisphere of radius $r = 2$ units and the region R forms the base of this hemisphere. The solid above R and below the graph of f is therefore a hemispherical solid of radius $r = 2$ units (Figure 8); hence, its volume is

$$\frac{1}{2}\left(\frac{4}{3}\pi r^3\right) = \frac{16\pi}{3} \text{ cubic units.}$$

Therefore,

$$\iint_R \sqrt{4 - x^2 - y^2}\, dx\, dy = \frac{16\pi}{3}.$$

2 Let R be the rectangular region consisting of all points (x,y) such that $0 \le x \le 1$ and $0 \le y \le 2$. Define the function f by $f(x,y) = 2 + y - x$. Calculate $\iint_R f(x,y)\, dx\, dy$.

SOLUTION

The graph of f is the plane $z = 2 + y - x$ and $\iint_R f(x,y)\, dx\, dy$ is the volume of the solid below this plane and above the rectangle R (Figure 9). We determine this volume by the method of slicing, using the x axis as our reference axis, and taking cross sections perpendicular to the reference axis as in Figure 9. The cross section x units from the origin is a trapezoid with vertices $(x, 0, 0)$, $(x, 2, 0)$, $(x, 2, 4 - x)$, and $(x, 0, 2 - x)$. The two parallel bases of this trapezoid have lengths $2 - x$ units and $4 - x$ units and the distance between these bases is 2 units; hence, the area of the trapezoid is

$$A(x) = 2\,\frac{(2 - x) + (4 - x)}{2} = 6 - 2x \text{ square units.}$$

The volume V of the solid is therefore given by

$$V = \int_0^1 A(x)\,dx = \int_0^1 (6 - 2x)\,dx = (6x - x^2)\Big|_0^1$$

$$= 5 \text{ cubic units.}$$

Consequently,

$$\iint_R (2 + y - x)\,dx\,dy = 5.$$

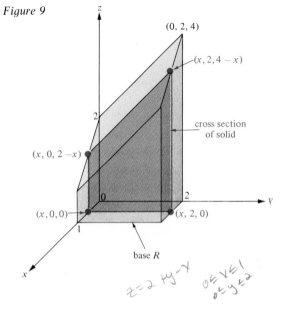

Figure 9

2.1 Basic Properties of the Double Integral

In the discussion above, we have considered only double integrals $\iint_R f(x, y)\,dx\,dy$ for which $f(x, y) \geq 0$ holds for all points (x, y) in R. If the latter condition does not hold, we can still form Riemann sums corresponding to regular augmented partitions and we can still ask for the limit of such Riemann sums as the partitions become finer and finer. Thus, we define the *double integral*,

$$\iint_R f(x, y)\,dx\,dy = \lim_{n \to +\infty} \sum_{k=1}^m f(x_k^*, y_k^*)\,\Delta x\,\Delta y,$$

just as before, for *any* continuous function f defined over an admissible region R, provided that the limit exists.

If the condition $f(x, y) \geq 0$ fails to hold for some points (x, y) in the region R, then part of the graph of f falls below the xy plane. In this case, the region R can be decomposed into two subregions R_1 and R_2 so that $f(x, y) \geq 0$ for (x, y) in R_1 and $f(x, y) \leq 0$ for (x, y) in R_2. The double integral $\iint_R f(x, y)\,dx\,dy$ can then be interpreted as the difference $V_1 - V_2$ between the volume V_1 of the solid below the graph of f and above R_1 and the volume V_2 of the solid above the graph of f and below R_2. [This is perfectly analogous to the interpretation of a Riemann integral $\int_a^b f(x)\,dx$ as a difference of two areas.]

In advanced calculus, a general definition is given of the double integral $\iint_R f(x, y)\,dx\,dy$, similar to our definition, but allowing partitions in which the cells are not all congruent (that is, nonregular partitions). Although the region R is usually required to have a "reasonably nice" shape (for instance, to be admissible in our sense), the function f is not required to be continuous. The question of whether the Riemann sums have a limit as the partitions become finer and finer has to be handled with considerable finesse. If such a limit exists, then one says that f is a (*Riemann*) *integrable function of two variables over the region* R. The following basic properties of the double integral are established in advanced calculus.

Basic Properties of the Double Integral

1 (*Existence*) If f is continuous on the admissible region R, then f is (Riemann) integrable over R; that is, $\iint\limits_{R} f(x, y)\, dx\, dy$ exists.

2 (*Interpretation as an Area*) If c is a constant and R is an admissible region of area A, then $\iint\limits_{R} c\, dx\, dy = cA$. In particular, $\iint\limits_{R} dx\, dy = A$.

3 (*Homogeneous Property*) If f is a (Riemann) integrable function on the admissible region R and K is a constant, then Kf is also (Riemann) integrable on R and $\iint\limits_{R} Kf(x, y)\, dx\, dy = K \iint\limits_{R} f(x, y)\, dx\, dy.$

4 (*Additive Property*) If f and g are (Riemann) integrable functions on the admissible region R, then $f + g$ is also (Riemann) integrable on R and

$$\iint\limits_{R} [f(x, y) + g(x, y)]\, dx\, dy = \iint\limits_{R} f(x, y)\, dx\, dy + \iint\limits_{R} g(x, y)\, dx\, dy.$$

5 (*Linear Property*) If f and g are (Riemann) integrable functions on the admissible region R and if A and B are constants, then $Af \pm Bg$ is also (Riemann) integrable on R and

$$\iint\limits_{R} [Af(x, y) \pm Bg(x, y)]\, dx\, dy = A \iint\limits_{R} f(x, y)\, dx\, dy \pm B \iint\limits_{R} g(x, y)\, dx\, dy.$$

6 (*Positivity*) If f is (Riemann) integrable on the admissible region R and if $f(x, y) \geq 0$ for all points (x, y) in R, then $\iint\limits_{R} f(x, y)\, dx\, dy \geq 0$.

7 (*Comparison*) If f and g are (Riemann) integrable functions on the admissible region R, and if $f(x, y) \leq g(x, y)$ holds for all points (x, y) in R, then $\iint\limits_{R} f(x, y)\, dx\, dy \leq \iint\limits_{R} g(x, y)\, dx\, dy.$

8 (*Additivity with Respect to the Region of Integration*) Let R be an admissible region and suppose that R can be decomposed into two non-overlapping admissible regions R_1 and R_2. (*Note*: The regions can share common boundary points.) If f is (Riemann) integrable on both R_1 and R_2, then f is (Riemann) integrable on R and

$$\iint\limits_{R} f(x, y)\, dx\, dy = \iint\limits_{R_1} f(x, y)\, dx\, dy + \iint\limits_{R_2} f(x, y)\, dx\, dy.$$

EXAMPLE Let R be the rectangle $0 \leq x \leq 1$, $0 \leq y \leq 1$, let R_1 be the portion of R above or on the diagonal $y = x$, and let R_2 be the portion of R below or on the diagonal

$y = x$. Suppose that

$$\iint\limits_{R_1} f(x, y)\, dx\, dy = 3, \qquad \iint\limits_{R_1} g(x, y)\, dx\, dy = -2,$$

$$\iint\limits_{R_2} f(x, y)\, dx\, dy = 5, \qquad \iint\limits_{R_2} g(x, y)\, dx\, dy = 1.$$

Find (a) $\iint\limits_{R} f(x, y)\, dx\, dy$, (b) $\iint\limits_{R} g(x, y)\, dx\, dy$, and (c) $\iint\limits_{R} [4 f(x, y) - 3g(x, y)]\, dx\, dy$.

SOLUTION
(a) By Property 8,

$$\iint\limits_{R} f(x, y)\, dx\, dy = \iint\limits_{R_1} f(x, y)\, dx\, dy + \iint\limits_{R_2} f(x, y)\, dx\, dy = 3 + 5 = 8.$$

(b) By Property 8,

$$\iint\limits_{R} g(x, y)\, dx\, dy = \iint\limits_{R_1} g(x, y)\, dx\, dy + \iint\limits_{R_2} g(x, y)\, dx\, dy = -2 + 1 = -1.$$

(c) Using parts (a) and (b) and Property 5, we have

$$\iint\limits_{R} [4 f(x, y) - 3g(x, y)]\, dx\, dy = 4 \iint\limits_{R} f(x, y)\, dx\, dy - 3 \iint\limits_{R} g(x, y)\, dx\, dy$$

$$= (4)(8) - (3)(-1) = 35.$$

Problem Set 2

In Problems 1 through 4, approximate the given double integral over the indicated rectangular region R. Use the regular partition of R into n^2 cells for the given value of n, and use the midpoints of the cells to augment the partition.

1 $\iint\limits_{R} xy\, dx\, dy$; $R: 0 \le x \le 2$ and $0 \le y \le 4$; $n = 2$

2 $\iint\limits_{R} (|x| + |y|)\, dx\, dy$; $R: -6 \le x \le 0$ and $-1 \le y \le 2$; $n = 3$

3 $\iint\limits_{R} (3x + 7y)\, dx\, dy$; $R: 0 \le x \le 3$ and $2 \le y \le 5$; $n = 3$

4 $\iint\limits_{R} (x - y^2)\, dx\, dy$; $R: -2 \le x \le 1$ and $0 \le y \le 2$; $n = 2$

5 Approximate the double integral $\iint\limits_{R} 4xy\, dx\, dy$, where R is the region inside the circle $x^2 + y^2 = 1$. Use the regular partition of the rectangle $-1 \le x \le 1$, $-1 \le y \le 1$ into four congruent cells and use the midpoints of the cells to augment the partition.

6 Approximate the double integral $\iint\limits_{R} 5x^2 y\, dx\, dy$, where R is the region in the first quadrant inside the circle $x^2 + y^2 = 1$. Use the regular partition of the rectangle $0 \le x \le 1$, $0 \le y \le 1$ into nine congruent cells and use the point in each cell that is nearest to the origin to augment the partition.

7 Approximate the double integral $\iint\limits_R x^3 y^2 \, dx \, dy$, where R is the region $0 \le x \le 3$, $0 \le y \le 9 - x^2$. Use the regular partition of the rectangle $0 \le x \le 3$, $0 \le y \le 9$ into nine congruent cells. Augment the partition by choosing, *for each cell not discarded*, the lower left corner.

8 Approximate the double integral $\iint\limits_R (3x^2 - 2y) \, dx \, dy$, where R is the region $-1 \le x \le 1$, $0 \le y \le 1 - x^2$. Use the regular partition of the rectangle $-1 \le x \le 1$, $0 \le y \le 1$ into nine congruent cells. Augment the partition by choosing, for each cell not discarded, the point nearest the origin.

In problems 9 through 20, interpret each double integral as a volume or as an area. Evaluate this volume or area, and thus the integral, by whatever method seems appropriate.

9 $\iint\limits_R \sqrt{25 - x^2 - y^2} \, dx \, dy$; $R: x^2 + y^2 \le 25$

10 $\iint\limits_R \sqrt{9 - x^2 - y^2} \, dx \, dy$; $R: x^2 + y^2 \le 9, x \ge 0, y \ge 0$

11 $\iint\limits_R (4 + \sqrt{2 - x^2 - y^2}) \, dx \, dy$; $R: x^2 + y^2 \le 2$

12 $\iint\limits_R (x - y + 1) \, dx \, dy$; $R: 0 \le x \le 1, 0 \le y \le 1$

13 $\iint\limits_R (5 + 2x + 3y) \, dx \, dy$; $R: 0 \le x \le 1, 0 \le y \le 2$

14 $\iint\limits_R (1 - x - y) \, dx \, dy$; $R: 0 \le y \le 1 - x, x \ge 0$

15 $\iint\limits_R dx \, dy$; $R: x^2 + y^2 \le r^2$

16 $\iint\limits_R dx \, dy$; $R: 0 \le y \le 1 - x, x \ge 0$

17 $\iint\limits_R dx \, dy$; $R:$ the interior of the triangle with vertices (x_1, y_1), (x_2, y_2), (x_3, y_3).

18 $\iint\limits_R (1 - x^2 - y^2) \, dx \, dy$; $R: x^2 + y^2 \le 1$

19 $\iint\limits_R (1 - \sqrt{x^2 + y^2}) \, dx \, dy$; $R: x^2 + y^2 \le 1$

20 $\iint\limits_R (5 - \sqrt{x^2 + y^2}) \, dx \, dy$; $R: x^2 + y^2 \le 1$

In problems 21 through 28, let R be the circular disk $x^2 + y^2 \le 1$, let R_1 be the upper half of R, and let R_2 be the lower half of R. Assume that $\iint\limits_{R_1} f(x, y) \, dx \, dy = 7$, $\iint\limits_{R_2} f(x, y) \, dx \, dy = -5$, $\iint\limits_{R_1} g(x, y) \, dx \, dy = -2$, and $\iint\limits_{R_2} g(x, y) \, dx \, dy = 4$. Evaluate the given double integral, using Properties 1 through 8 in Section 2.1.

21 $\iint\limits_R f(x, y) \, dx \, dy$

22 $\iint\limits_{R_1} [g(x, y) - f(x, y)] \, dx \, dy$

23 $\iint\limits_R g(x, y) \, dx \, dy$

24 $\iint\limits_{R_2} [1 - f(x, y) + g(x, y)] \, dx \, dy$

25 $\iint\limits_R [4g(x, y) - 6f(x, y)] \, dx \, dy$

26 $\iint\limits_R [3f(x, y) - 5g(x, y) + 9] \, dx \, dy$

27 $\iint\limits_R 8 \, dx \, dy$

28 $\iint\limits_R [f(x, y) - \sqrt{1 - x^2 - y^2}] \, dx \, dy$

29 Give a geometric explanation of Property 6 of Section 2.1.

30 Prove Property 7 of Section 2.1 using Properties 5 and 6.

31 Give a geometric explanation of Property 8 of Section 2.1, assuming that the integrand is nonnegative.

32 Suppose that M is the maximum value of the continuous function f and that m is the minimum value of f on the admissible region R. Using the properties of double integrals given in Section 2.1, prove:

(a) $\left| \iint\limits_{R} f(x,y)\, dx\, dy \right| \le \iint\limits_{R} |f(x,y)|\, dx\, dy.$

(b) $m \le \dfrac{\iint\limits_{R} f(x,y)\, dx\, dy}{\iint\limits_{R} dx\, dy} \le M,$ if $\iint\limits_{R} dx\, dy \neq 0.$

33 Suppose that f is a continuous function on the admissible region R and that (a,b) is a point in R. Let R_1 be an admissible subregion of R such that the point (a,b) belongs to the interior of R_1. Denote by A_1 the area of R_1 and denote by δ_1 the diameter of R_1 (that is, the maximum distance between any two points in R_1). Make informal arguments to show that:

(a) $\iint\limits_{R_1} f(x,y)\, dx\, dy \approx f(a,b)A_1$ if δ_1 is small.

(b) $\lim\limits_{\delta_1 \to 0} \dfrac{\iint\limits_{R_1} f(x,y)\, dx\, dy}{A_1} = f(a,b).$

3 Evaluation of Double Integrals by Iteration

In Section 1 we considered the iterated integral $\displaystyle\int_a^b \int_{g(y)}^{h(y)} f(x,y)\, dx\, dy$ of a function f of two variables, while in Section 2 we introduced the double integral $\displaystyle\iint\limits_{R} f(x,y)\, dx\, dy$ of f over a region R in the xy plane. These two kinds of integrals are defined in completely different ways and it is important not to confuse them. However, as we see in this section, it is sometimes possible to convert a double integral into an equivalent iterated integral and vice versa.

Consider, for instance, the double integral

$$\iint\limits_{R} f(x,y)\, dx\, dy,$$

where R is the specially shaped region shown in Figure 1, f is continuous on R, and $f(x,y) \ge 0$ for (x,y) in R. Notice that R is bounded below by the straight line $y = a$, above by the straight line $y = b$, on the left by the graph of the equation $x = g(y)$, and on the right by the graph of the equation $x = h(y)$. We assume that g and h are

Figure 1

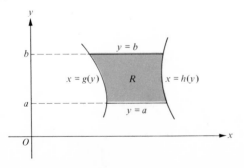

Figure 2

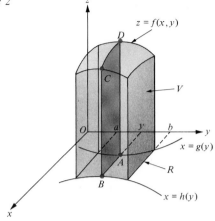

continuous functions defined on $[a, b]$ and that $g(y) \leq h(y)$ for $a \leq y \leq b$.

Since $f(x, y) \geq 0$ for (x, y) in R, the double integral $\iint_R f(x, y)\, dx\, dy$ can be interpreted as the volume V of the solid under the graph of f and above the region R (Figure 2). We propose to find the volume V by the method of slicing, using the y axis as our reference axis. Figure 2 shows the cross section $ABCD$ cut from the solid by the plane perpendicular to the y axis and y units from the origin. If we denote the area of $ABCD$ by $F(y)$, then, by the method of slicing,

$$\iint_R f(x, y)\, dx\, dy = V = \int_a^b F(y)\, dy.$$

Figure 3

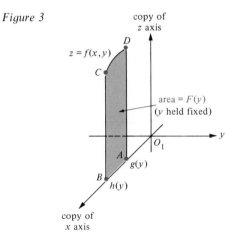

We are now going to find a formula for the cross-sectional area $F(y)$. To this end, we temporarily fix a value of y between a and b and set up parallel copies of the x and z axes in the same plane as the cross section $ABCD$ (Figure 3). The equation of the curve DC is $z = f(x, y)$. In Figure 2, the points A and B lie on the curves $x = g(y)$ and $x = h(y)$, respectively; hence, in Figure 3, the points A and B have x coordinates $g(y)$ and $h(y)$, respectively.

Figure 4 is obtained from Figure 3 by rotating the xz plane about the z axis so that the x axis extends to our right as usual. From Figure 4, it is now clear that the desired area $F(y)$ is just the area under the curve $z = f(x, y)$ between $x = g(y)$ and $x = h(y)$ (y being held fixed). Hence,

$$F(y) = \int_{g(y)}^{h(y)} f(x, y)\, dx.$$

It follows that

$$\iint_R f(x, y)\, dx\, dy = \int_a^b F(y)\, dy$$

$$= \int_a^b \left[\int_{g(y)}^{h(y)} f(x, y)\, dx \right] dy;$$

Figure 4

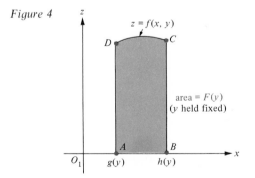

that is,

$$\iint_R f(x, y)\, dx\, dy = \int_a^b \int_{g(y)}^{h(y)} f(x, y)\, dx\, dy.$$

The following example illustrates the use of the latter equation for evaluating double integrals.

EXAMPLE   Let R be the region inside the trapezoid whose vertices are $(2,2)$, $(4,2)$, $(5,4)$, and $(1,4)$ (Figure 5). Evaluate $\iint\limits_R 8xy \, dx \, dy$ by converting it to an iterated integral.

SOLUTION

The equation of the straight line through the points $(2,2)$ and $(1,4)$ is $y = 6 - 2x$, or $x = \dfrac{6-y}{2}$. Similarly, the equation of the straight line through the points $(4,2)$ and $(5,4)$ is $y = 2x - 6$, or $x = \dfrac{6+y}{2}$. Using the result obtained above, we have

$$\iint\limits_R 8xy \, dx \, dy = \int_2^4 \int_{(6-y)/2}^{(6+y)/2} 8xy \, dx \, dy$$

$$= \int_2^4 \left[8y \frac{x^2}{2} \Big|_{(6-y)/2}^{(6+y)/2} \right] dy$$

$$= \int_2^4 4y \left[\left(\frac{6+y}{2}\right)^2 - \left(\frac{6-y}{2}\right)^2 \right] dy$$

$$= \int_2^4 24y^2 \, dy = \frac{24y^3}{3} \Big|_2^4 = 448.$$

Figure 5

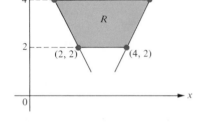

In our derivation of the equation for converting a double integral into an iterated integral, we used the method of slicing and took cross sections perpendicular to the y axis. We can also use the method of slicing with cross sections perpendicular to the x axis provided that the region R has the following shape: R is bounded on the left by the vertical line $x = a$, on the right by the vertical line $x = b$, above by the graph of an equation $y = h(x)$, and below by the graph of an equation $y = g(x)$, where g and h are continuous on $[a, b]$ and $g(x) \le h(x)$ for $a \le x \le b$ (Figure 6).

Figure 6

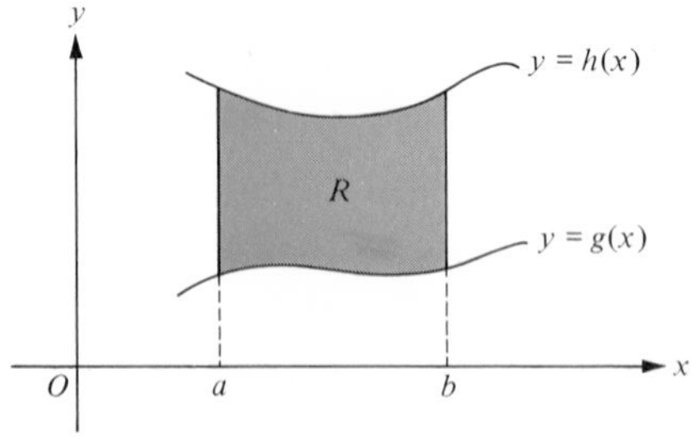

A region R bounded below and above by continuous curves $y = g(x)$ and $y = h(x)$, respectively, and bounded on the left and right by vertical straight lines $x = a$ and $x = b$, respectively, is called a *type* I *region* (Figure 6). On the other hand, a region R bounded on the left and on the right by continuous curves $x = g(y)$ and $x = h(y)$, respectively, and bounded below and above by horizontal lines $y = a$ and $y = b$, respectively, is called a *type* II *region* (Figure 1). These considerations indicate that *a double integral of a continuous function f over a region R of type* I *or of type* II *can be converted into an iterated integral.* In more advanced courses, this is proved rigorously, even for the case in which the function f takes on negative values.

Thus, we have the following method, called the *method of iteration,* for evaluating double integrals over special regions.

The Method of Iteration

Suppose that R is either a type I or a type II region in the plane and that the function f is continuous on R. In order to evaluate the double integral $\iint\limits_R f(x, y)\, dx\, dy$ by the method of iteration, proceed as follows:

Figure 7

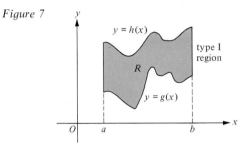

Case 1 If the region R is of type I, find the equations of the continuous curves $y = g(x)$ and $y = h(x)$ bounding R below and above, respectively. Also, find the constants a and b for which the vertical lines $x = a$ and $x = b$ bound R on the left and on the right, respectively (Figure 7). Then

$$\iint\limits_R f(x, y)\, dx\, dy = \int_{x=a}^{x=b} \left[\int_{y=g(x)}^{y=h(x)} f(x, y)\, dy \right] dx$$

$$= \int_a^b \int_{g(x)}^{h(x)} f(x, y)\, dy\, dx.$$

Figure 8

Case 2 If the region R is of type II, find the equations of the continuous curves $x = g(y)$ and $x = h(y)$ bounding R on the left and on the right, respectively. Also, find the constants a and b for which the horizontal lines $y = a$ and $y = b$ bound R below and above, respectively (Figure 8). Then

$$\iint\limits_R f(x, y)\, dx\, dy = \int_{y=a}^{y=b} \left[\int_{x=g(y)}^{x=h(y)} f(x, y)\, dx \right] dy = \int_a^b \int_{g(y)}^{h(y)} f(x, y)\, dx\, dy.$$

EXAMPLES   Evaluate the given double integral by the method of iteration.

1 $\iint\limits_R x \cos xy\, dx\, dy$; $R: 1 \le x \le 2$ and $\pi/2 \le y \le 2\pi/x$.

Figure 9

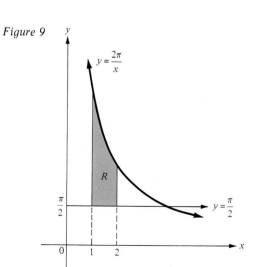

SOLUTION
The region is evidently of type I (Figure 9); hence,

$$\iint\limits_R x \cos xy\, dx\, dy = \int_{x=1}^{x=2} \left(\int_{y=\pi/2}^{y=2\pi/x} x \cos xy\, dy \right) dx$$

$$= \int_{x=1}^{x=2} \left(\sin xy \, \Big|_{y=\pi/2}^{y=2\pi/x} \right) dx$$

$$= \int_1^2 \left(\sin 2\pi - \sin \frac{\pi}{2} x \right) dx$$

$$= \int_1^2 \left(-\sin \frac{\pi x}{2} \right) dx = \frac{2}{\pi} \cos \frac{\pi x}{2} \, \Big|_1^2$$

$$= \frac{2}{\pi} \cos \pi - \frac{2}{\pi} \cos \frac{\pi}{2} = -\frac{2}{\pi}.$$

2 $\iint\limits_{R} (x + y)\, dx\, dy$ where R is the region in the first quadrant above the curve $y = x^2$ and below the curve $y = \sqrt{x}$.

SOLUTION

In this case the region R is both of type I and type II (Figure 10). Regarding R as a type II region, it is bounded on the left by the curve $x = y^2$, on the right by the curve $x = \sqrt{y}$, below by the line $y = 0$, and above by the line $y = 1$. Therefore,

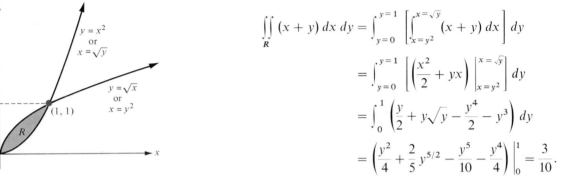

Figure 10

$$\iint\limits_{R} (x + y)\, dx\, dy = \int_{y=0}^{y=1} \left[\int_{x=y^2}^{x=\sqrt{y}} (x + y)\, dx \right] dy$$

$$= \int_{y=0}^{y=1} \left[\left(\frac{x^2}{2} + yx \right) \Bigg|_{x=y^2}^{x=\sqrt{y}} \right] dy$$

$$= \int_{0}^{1} \left(\frac{y}{2} + y\sqrt{y} - \frac{y^4}{2} - y^3 \right) dy$$

$$= \left(\frac{y^2}{4} + \frac{2}{5} y^{5/2} - \frac{y^5}{10} - \frac{y^4}{4} \right) \Bigg|_{0}^{1} = \frac{3}{10}.$$

The second example above illustrates two important facts about the method of iteration. First, the horizontal straight lines bounding a type II region below and above (and, likewise, the vertical straight lines bounding a type I region on the left and on the right) are permitted to touch the region R in single points, rather than along line segments. **Second**, there are regions that are both of type I and type II.

Since the region R in Figure 10 is of type I, we can also iterate the integral of Example 2 as follows:

$$\iint\limits_{R} (x + y)\, dx\, dy = \int_{x=0}^{x=1} \left[\int_{y=x^2}^{y=\sqrt{x}} (x + y)\, dy \right] dx = \frac{3}{10}.$$

Additional examples of regions that are both of type I and type II appear in Figure 11. A double integral over any such region can be iterated in two different ways, resulting in two iterated integrals with opposite orders of integration, but with the same value. These two iterated integrals are said to be obtained from each other by *reversing the order of integration*. A reversal of the order of integration often converts a complicated iterated integral into a simpler one.

Figure 11

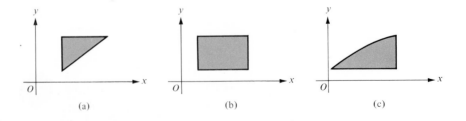

(a)   (b)   (c)

EXAMPLES   Reverse the order of integration, and then evaluate the resulting integral.

1 $\displaystyle\int_{0}^{1} \int_{x}^{1} x \sin y^3 \, dy\, dx$

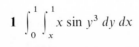

SOLUTION
The given iterated integral is equivalent to the double integral

$$\iint\limits_{R} x \sin y^3 \, dx \, dy$$

over the type I region R determined by the inequalities $0 \le x \le 1$ and $x \le y \le 1$ (Figure 12). Since R is also of type II, we have

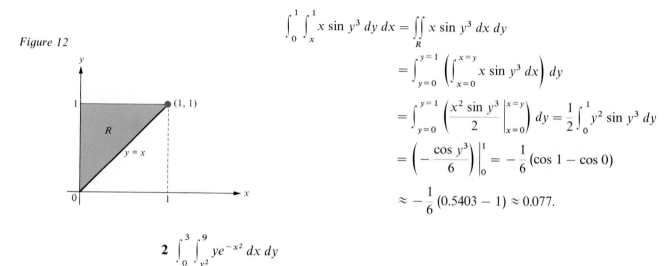

Figure 12

$$\int_0^1 \int_x^1 x \sin y^3 \, dy \, dx = \iint\limits_{R} x \sin y^3 \, dx \, dy$$

$$= \int_{y=0}^{y=1} \left(\int_{x=0}^{x=y} x \sin y^3 \, dx \right) dy$$

$$= \int_{y=0}^{y=1} \left(\frac{x^2 \sin y^3}{2} \Big|_{x=0}^{x=y} \right) dy = \frac{1}{2} \int_0^1 y^2 \sin y^3 \, dy$$

$$= \left(-\frac{\cos y^3}{6} \right) \Big|_0^1 = -\frac{1}{6} (\cos 1 - \cos 0)$$

$$\approx -\frac{1}{6} (0.5403 - 1) \approx 0.077.$$

2 $\displaystyle\int_0^3 \int_{y^2}^9 y e^{-x^2} \, dx \, dy$

SOLUTION
The given iterated integral is equivalent to the double integral $\iint\limits_{R} y e^{-x^2} \, dx \, dy$ over the type II region R determined by the inequalities $0 \le y \le 3$ and $y^2 \le x \le 9$ (Figure 13). Since R is also of type I, we have

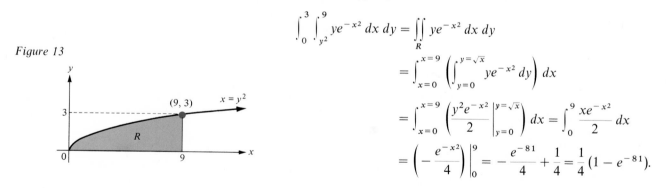

Figure 13

$$\int_0^3 \int_{y^2}^9 y e^{-x^2} \, dx \, dy = \iint\limits_{R} y e^{-x^2} \, dx \, dy$$

$$= \int_{x=0}^{x=9} \left(\int_{y=0}^{y=\sqrt{x}} y e^{-x^2} \, dy \right) dx$$

$$= \int_{x=0}^{x=9} \left(\frac{y^2 e^{-x^2}}{2} \Big|_{y=0}^{y=\sqrt{x}} \right) dx = \int_0^9 \frac{x e^{-x^2}}{2} \, dx$$

$$= \left(-\frac{e^{-x^2}}{4} \right) \Big|_0^9 = -\frac{e^{-81}}{4} + \frac{1}{4} = \frac{1}{4} (1 - e^{-81}).$$

Although there are regions that are neither of type I nor of type II, it is usually possible to cut such a region into nonoverlapping subregions, each of which is of type I or of type II. The double integral of a function over the large region can then be evaluated by integrating the function over each subregion and adding the resulting values.

EXAMPLE   Evaluate $\iint\limits_{R} (2x - y) \, dx \, dy$ over the region R of Figure 14.

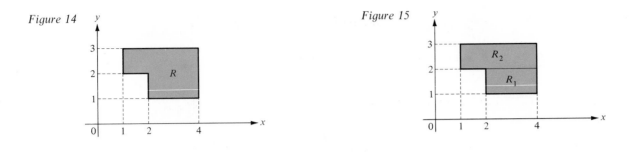

Figure 14

Figure 15

SOLUTION

Although R is neither of type I nor of type II, we can decompose it into two nonoverlapping regions R_1: $2 \leq x \leq 4$, $1 \leq y \leq 2$ and R_2: $1 \leq x \leq 4$, $2 \leq y \leq 3$ (Figure 15). Here,

$$\iint\limits_{R_1} (2x - y)\, dx\, dy = \int_2^4 \int_1^2 (2x - y)\, dy\, dx = \int_2^4 \left[\left(2xy - \frac{y^2}{2}\right) \Big|_{y=1}^{y=2} \right] dx$$

$$= \int_2^4 \left(2x - \frac{3}{2}\right) dx = \left(x^2 - \frac{3}{2}x\right) \Big|_2^4 = 9 \quad \text{and}$$

$$\iint\limits_{R_2} (2x - y)\, dx\, dy = \int_1^4 \int_2^3 (2x - y)\, dy\, dx = \int_1^4 \left[\left(2xy - \frac{y^2}{2}\right) \Big|_{y=2}^{y=3} \right] dx$$

$$= \int_1^4 \left(2x - \frac{5}{2}\right) dx = \left(x^2 - \frac{5}{2}x\right) \Big|_1^4 = \frac{15}{2}.$$

Therefore,

$$\iint\limits_{R} (2x - y)\, dx\, dy = \iint\limits_{R_1} (2x - y)\, dx\, dy + \iint\limits_{R_2} (2x - y)\, dx\, dy$$

$$= 9 + \frac{15}{2} = \frac{33}{2}.$$

Problem Set 3

In problems 1 through 14, (a) sketch the region R, (b) decide whether R is of type I or type II (or both), and (c) evaluate each double integral by using the method of iteration.

1 $\iint\limits_{R} x \sin(xy)\, dx\, dy$; $R: 0 \leq x \leq \pi, 0 \leq y \leq 1$

2 $\iint\limits_{R} \dfrac{x e^{x/y}}{y^2}\, dx\, dy$; $R: 0 \leq x \leq 1, 1 \leq y \leq 2$

3 $\iint\limits_{R} x \sin y\, dx\, dy$; $R: 0 \leq x \leq \pi, 0 \leq y \leq x$

4 $\iint\limits_{R} (x + y + 2)\, dx\, dy$; $R: 0 \leq x \leq 2, -\sqrt{x} \leq y \leq \sqrt{x}$

5 $\iint\limits_{R} (2x - 3y)\, dx\, dy$; $R: x^2 + y^2 \leq 1$

6 $\iint\limits_{R} e^{x+y}\, dx\, dy$; $R: |x| + |y| \leq 1$

7 $\iint\limits_{R} y\, dx\, dy$; $R: 0 \leq x \leq \pi, 0 \leq y \leq \sin x$

8 $\iint\limits_{R} y \sin^{-1} x\, dx\, dy$; $R: 0 \leq x \leq \frac{1}{2}, -1 \leq y \leq 1$

9 $\iint\limits_{R} x \, dx \, dy$; R: the region in the first quadrant bounded by $x = y - 2$, $y = x^2$, and $x = 0$.

10 $\iint\limits_{R} (6x + 5y) \, dx \, dy$; R: the region between the curves $y = 2\sqrt{x}$ and $y = 2x^3$.

11 $\iint\limits_{R} (7xy - 2y^2) \, dx \, dy$; R: the region in the first quadrant bounded by $y = x$, $y = \dfrac{x}{3}$, $x = 3$.

12 $\iint\limits_{R} \sqrt{1 + x^2} \, dx \, dy$; R: the inside of the triangle whose vertices are $(0,0)$, $(0,1)$, and $(1,1)$.

13 $\iint\limits_{R} \dfrac{1}{1 + y^2} \, dx \, dy$; R: the region in the first quadrant bounded by $y = x$, $y = 0$, and $x = 1$.

14 $\iint\limits_{R} x \, dx \, dy$; R: the smaller region cut from the circle $x^2 + y^2 \le 9$ by the straight line $x + y = 3$.

In problems 15 through 23, evaluate each iterated integral by reversing the order of integration. In each case, sketch an appropriate region in the xy plane over which the double integral corresponding to the given iterated integral is evaluated.

15 $\displaystyle\int_0^1 \int_y^1 e^{-3x^2} \, dx \, dy$

16 $\displaystyle\int_1^e \int_0^{\ln x} y \, dy \, dx$

17 $\displaystyle\int_0^2 \int_{3y}^6 \sin \frac{\pi x^2}{6} \, dx \, dy$

18 $\displaystyle\int_0^{\sqrt{2}/2} \int_0^{\sin^{-1} x} x \, dy \, dx$

19 $\displaystyle\int_0^1 \int_y^{\sqrt{y}} \frac{\sin x}{x} \, dx \, dy$

20 $\displaystyle\int_0^1 \int_{\sqrt{y}}^1 \sqrt{1 - x^3} \, dx \, dy$

21 $\displaystyle\int_0^2 \int_{x^3}^8 e^{x/\sqrt[3]{y}} \, dy \, dx$

22 $\displaystyle\int_0^1 \int_{-1}^{-\sqrt{y}} \sqrt[5]{x^3 + 1} \, dx \, dy$

23 $\displaystyle\int_0^4 \int_{\sqrt{y}}^2 \frac{y \, dx \, dy}{\sqrt{1 + x^5}}$

24 Evaluate $\iint\limits_{R} (2x - 3y^2) \, dx \, dy$ over the region R inside the triangle with vertices $(0,0)$, $(2,0)$, and $(-1, -1)$.

25 Evaluate $\iint\limits_{R} (1 - x + y) \, dx \, dy$ over the region R shown in Figure 16.

26 Suppose that R is the region bounded by the lines $y = 1$, $y = x + 6$, and the parabola $y = x^2$ (Figure 17). Evaluate $\iint\limits_{R} xy \, dx \, dy$ over the region R by:

(a) Dividing R into type II regions.
(b) Dividing R into type I regions.

Figure 16

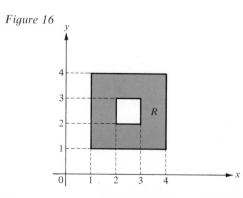

Figure 17

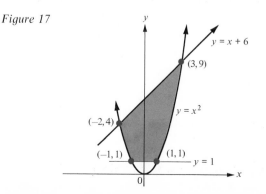

4   Elementary Applications of Double Integrals

The double integral, as defined in Section 2, was suggested by the problem of finding the volume under a surface; hence, it should not be surprising that double integrals find numerous applications in situations where volumes must be calculated. In this section we consider not only applications of double integrals to the calculation of volumes and areas, but also to problems involving *density, center of mass, centroids,* and *moments.*

4.1 Volumes by Double Integration

Using the double integral, we can express the volume V under the graph of a continuous nonnegative function f over a given region R (as we did in Section 2) by $V = \iint\limits_{R} f(x, y)\, dx\, dy$. By the method of iteration (Section 3), we can evaluate the double integral, and thus the volume V.

EXAMPLE   Let R be the region in the xy plane bounded above by the parabola $y = 4 - x^2$ and bounded below by the x axis (Figure 1). Find the volume V under the graph of $f(x, y) = x + 2y + 3$ and above the region R.

Figure 1

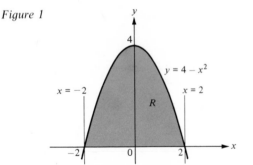

SOLUTION
The region R is of type I, since it is bounded on the left by the vertical straight line $x = -2$, on the right by the vertical straight line $x = 2$, above by the parabola $y = 4 - x^2$, and below by the straight line $y = 0$. Therefore, by the method of iteration

$$V = \iint\limits_{R} (x + 2y + 3)\, dx\, dy = \int_{-2}^{2} \left[\int_{0}^{4-x^2} (x + 2y + 3)\, dy \right] dx$$

$$= \int_{-2}^{2} \left[(xy + y^2 + 3y) \Big|_{0}^{4-x^2} \right] dx = \int_{-2}^{2} [x(4 - x^2) + (4 - x^2)^2 + 3(4 - x^2)]\, dx$$

$$= \int_{-2}^{2} (x^4 - x^3 - 11x^2 + 4x + 28)\, dx = \left(\frac{x^5}{5} - \frac{x^4}{4} - \frac{11x^3}{3} + 2x^2 + 28x \right) \Big|_{-2}^{2}$$

$$= \frac{992}{15} \text{ cubic units.}$$

4.2 Areas by Double Integration

As we noticed in Section 2, the area of a region R in the xy plane is given by $A = \iint\limits_{R} dx\, dy.$

EXAMPLE   Find the area A of the region R bounded by the curves $y = x^2$ and $y = x$.

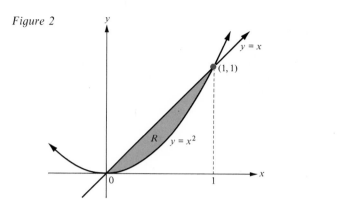

Figure 2

SOLUTION

The region (Figure 2) is type I; hence, the area is given by

$$A = \iint\limits_{R} dx \, dy = \int_{x=0}^{x=1} \left[\int_{y=x^2}^{y=x} dy \right] dx$$

$$= \int_{x=0}^{x=1} \left[y \Big|_{x^2}^{x} \right] dx = \int_{0}^{1} (x - x^2) \, dx$$

$$= \left(\frac{x^2}{2} - \frac{x^3}{3} \right) \Big|_{0}^{1} = \frac{1}{6} \text{ square unit.}$$

4.3 Density and Double Integrals

Consider a quantity such as mass or electric charge distributed in a continuous, but perhaps nonuniform, manner over a portion of the xy plane. We refer to a function σ of two variables (σ is the small Greek letter "sigma") as a *density function* for this two-dimensional distribution if, for every admissible region R in the xy plane,

$\iint\limits_{R} \sigma(x, y) \, dx \, dy$ gives the amount of the quantity contained in R.

EXAMPLE   Electric charge is distributed over the triangular region R of Figure 3 so that the charge density at any point (x, y) in R is given by

$$\sigma(x, y) = (x - x^2)(y - y^2) \text{ coulombs/cm}^2.$$

Find the total amount of electric charge on the region R.

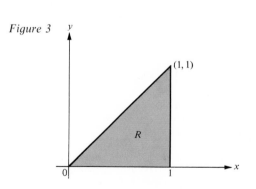

Figure 3

SOLUTION

The total charge on R is given by

$$\iint\limits_{R} \sigma(x, y) \, dx \, dy = \int_{0}^{1} \int_{0}^{x} (x - x^2)(y - y^2) \, dy \, dx$$

$$= \int_{0}^{1} (x - x^2) \left(\frac{y^2}{2} - \frac{y^3}{3} \right) \Big|_{0}^{x} dx$$

$$= \int_{0}^{1} \left(\frac{x^3}{2} - \frac{5x^4}{6} + \frac{x^5}{3} \right) dx$$

$$= \left(\frac{x^4}{8} - \frac{x^5}{6} + \frac{x^6}{18} \right) \Big|_{0}^{1} = \frac{1}{72} \text{ coulomb.}$$

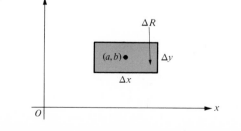

Figure 4

Now, suppose that a quantity is distributed over a region R in the xy plane and that σ is its density function. Choose a point (a, b) in the region R and consider the small rectangular region ΔR in Figure 4 with center at (a, b), with dimensions Δx and Δy, and with area $\Delta A = \Delta x \, \Delta y$. If Δq represents the amount of the quantity contained in the region ΔR, then

$$\Delta q = \iint_{\Delta R} \sigma(x, y)\, dx\, dy.$$

If σ is continuous and both Δx and Δy are very small, then the value of $\sigma(x, y)$ is close to the value of $\sigma(a, b)$ for all points (x, y) within the rectangle ΔR; that is,

$$\sigma(x, y) \approx \sigma(a, b) \qquad \text{for all points } (x, y) \text{ in } \Delta R.$$

Therefore, it seems plausible that

$$\Delta q = \iint_{\Delta R} \sigma(x, y)\, dx\, dy \approx \iint_{\Delta R} \sigma(a, b)\, dx\, dy = \sigma(a, b) \iint_{\Delta R} dx\, dy,$$

since $\sigma(a, b)$ is a constant. Notice that $\iint_{\Delta R} dx\, dy = \Delta A = \Delta x\, \Delta y$; hence,

$$\Delta q \approx \sigma(a, b)\, \Delta A = \sigma(a, b)\, \Delta x\, \Delta y.$$

Therefore,

$$\sigma(a, b) \approx \frac{\Delta q}{\Delta A},$$

presumably with better and better approximation as Δx and Δy become smaller and smaller. Consequently, the density $\sigma(a, b)$ at the point (a, b) can be interpreted as the limiting value of the amount of the quantity per unit area in a small region ΔR around the point (a, b) as ΔR "shrinks to zero." In other words,

$$\sigma(a, b) = \frac{dq}{dA} \quad \text{or} \quad dq = \sigma(a, b)\, dx\, dy.$$

We can regard the latter formula as giving the "infinitesimal" amount dq of the quantity contained in an "infinitesimal" rectangle of dimensions dx and dy with center at the point (a, b).

4.4 Moments and Center of Mass

Figure 5

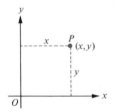

Figure 5

Suppose that a particle P of mass m is situated at the point (x, y) in the xy plane (Figure 5). Then the product mx, the mass m of the particle times its signed distance x from the y axis, is called the *moment of P about the y axis*. Similarly, the product my is called the *moment of P about the x axis*.

Now, suppose that a total mass m is continuously distributed over an admissible planar region R, say in the form of a thin sheet of material. Such a thin sheet is called a *lamina* (Figure 6). Let σ be the density function for this mass distribution.

If (x, y) is a point in R, consider the "infinitesimal" rectangle of dimensions dx and dy with center at (x, y). The mass contained in this "infinitesimal" rectangle is given by $dm = \sigma(x, y)\, dx\, dy$ and its signed distance from the x axis is y units; hence, its moment about the x axis is given by $(dm)y = \sigma(x, y)y\, dx\, dy$. The total amount of all the mass in the lamina is obtained by "summing"; that is, by *integrating* all the "infinitesimal" moments. Consequently, the

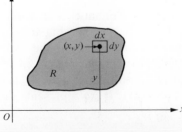

Figure 6

moment M_x of the lamina about the x axis is given by

$$M_x = \iint\limits_R \sigma(x,y)y \, dx \, dy.$$

Similarly, the moment M_y of the lamina about the y axis is given by

$$M_y = \iint\limits_R \sigma(x,y)x \, dx \, dy.$$

Also, the total mass m of the lamina is given by

$$m = \iint\limits_R \sigma(x,y) \, dx \, dy.$$

By definition, the coordinates $\bar{x}$ and $\bar{y}$ of the *center of mass* of the lamina are given by the equations

$$\bar{x} = \frac{M_y}{m} = \frac{\displaystyle\iint\limits_R \sigma(x,y)x \, dx \, dy}{\displaystyle\iint\limits_R \sigma(x,y) \, dx \, dy} \quad \text{and} \quad \bar{y} = \frac{M_x}{m} = \frac{\displaystyle\iint\limits_R \sigma(x,y)y \, dx \, dy}{\displaystyle\iint\limits_R \sigma(x,y) \, dx \, dy}.$$

Evidently, $m\bar{x} = M_y$ and $m\bar{y} = M_x$; that is, if all of the mass m of the lamina were concentrated in a particle P at the center of mass, then the moments of P about the x and y axes would be the same as the moments of the whole lamina about the x and y axes, respectively. In physics it is shown that a horizontal lamina balances perfectly on a sharp point placed at its center of mass.

EXAMPLE   A lamina R is bounded above by the graph of $y = \sqrt[3]{x}$, below by the x axis, and on the right by the vertical line $x = 8$. The mass density of the lamina at the point (x,y) is given by $\sigma(x,y) = kx$, where k is a positive constant. Find:

(a) The total mass m of the lamina.
(b) The moments M_x and M_y.
(c) The center of mass $(\bar{x}, \bar{y})$.

Figure 7

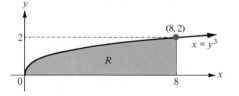

SOLUTION
The region R occupied by the lamina is both type I and type II (Figure 7). We regard it as type II in iterating our double integrals.

(a) $m = \displaystyle\iint\limits_R \sigma(x,y) \, dx \, dy = \int_0^2 \int_{y^3}^8 kx \, dx \, dy = \int_0^2 \left[\frac{kx^2}{2} \Big|_{y^3}^8 \right] dy$

$= k \int_0^2 \left(32 - \frac{y^6}{2} \right) dy = k\left(32y - \frac{y^7}{14} \right) \Big|_0^2 = \frac{384k}{7}$ mass units.

(b) $M_x = \displaystyle\iint\limits_R \sigma(x,y)y \, dx \, dy = \int_0^2 \int_{y^3}^8 kxy \, dx \, dy = k \int_0^2 \left[\frac{yx^2}{2} \Big|_{y^3}^8 \right] dy$

$= k \int_0^2 \left(32y - \frac{y^7}{2} \right) dy = k\left(16y^2 - \frac{y^8}{16} \right) \Big|_0^2 = 48k$ and

$M_y = \displaystyle\iint\limits_R \sigma(x,y)x \, dx \, dy = \int_0^2 \int_{y^3}^8 kx^2 \, dx \, dy = k \int_0^2 \left[\frac{x^3}{3} \Big|_{y^3}^8 \right] dy$

$= k \int_0^2 \left(\frac{512}{3} - \frac{y^9}{3} \right) dy = k\left(\frac{512}{3} y - \frac{y^{10}}{30} \right) \Big|_0^2 = \frac{1536k}{5}.$

(c) $\bar{x} = \dfrac{M_y}{m} = \dfrac{1536k}{5} \cdot \dfrac{7}{384k} = \dfrac{28}{5}$ and $\bar{y} = \dfrac{M_x}{m} = 48k \cdot \dfrac{7}{384k} = \dfrac{7}{8}$.

Hence, the center of mass is given by $(\bar{x}, \bar{y}) = \left(\frac{28}{5}, \frac{7}{8}\right)$.

4.5 Centroids

A distribution of mass (or any other quantity) whose density function σ is constant on a region R is said to be *uniform* or *homogeneous* on R. If a quantity is distributed uniformly on R, then the amount of this quantity in any subregion R_1 of R is proportional to the area of R_1 (Problem 31).

   The *centroid* of a planar region R is defined to be the center of mass of a *uniform* mass distribution on R. If the density of such a uniform distribution is given by $\sigma(x, y) = k$ for all (x, y) in R, where k is a constant, then the coordinates $(\bar{x}, \bar{y})$ of the centroid of R are given by

$$\bar{x} = \frac{\iint\limits_{R} \sigma(x, y)x \, dx \, dy}{\iint\limits_{R} \sigma(x, y) \, dx \, dy} = \frac{k\iint\limits_{R} x \, dx \, dy}{k\iint\limits_{R} dx \, dy} = \frac{\iint\limits_{R} x \, dx \, dy}{\iint\limits_{R} dx \, dy},$$

$$\bar{y} = \frac{\iint\limits_{R} \sigma(x, y)y \, dx \, dy}{\iint\limits_{R} \sigma(x, y) \, dx \, dy} = \frac{k\iint\limits_{R} y \, dx \, dy}{k\iint\limits_{R} dx \, dy} = \frac{\iint\limits_{R} y \, dx \, dy}{\iint\limits_{R} dx \, dy}.$$

Since $A = \iint\limits_{R} dx \, dy$ is the area of the region R, we can also write

$$\bar{x} = \frac{1}{A}\iint\limits_{R} x \, dx \, dy \quad \text{and} \quad \bar{y} = \frac{1}{A}\iint\limits_{R} y \, dx \, dy.$$

EXAMPLE   Find the centroid of the region R bounded by $y = x + 2$ and $y = x^2$.

SOLUTION
The region R is of type I (Figure 8), and we have

$$A = \int_{-1}^{2} \int_{x^2}^{x+2} dy \, dx = \int_{-1}^{2} (x + 2 - x^2) \, dx$$

$$= \left(\frac{x^2}{2} + 2x - \frac{x^3}{3}\right)\Bigg|_{-1}^{2} = \frac{9}{2} \text{ square units.}$$

Therefore,

$$\bar{x} = \frac{1}{A}\iint\limits_{R} x \, dx \, dy = \frac{2}{9}\int_{-1}^{2} \int_{x^2}^{x+2} x \, dy \, dx$$

$$= \frac{2}{9}\int_{-1}^{2} [x(x + 2) - x^3] \, dx = \frac{2}{9}\left[\frac{x^3}{3} + x^2 - \frac{x^4}{4}\right]\Bigg|_{-1}^{2}$$

$$= \frac{2}{9}\left(\frac{8}{3} - \frac{5}{12}\right) = \frac{1}{2} \quad \text{and}$$

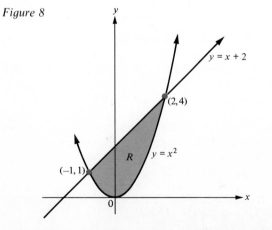

Figure 8

y

$y = x + 2$

$(2, 4)$

R $y = x^2$

$(-1, 1)$

0

x

$$\bar{y} = \frac{1}{A} \iint_R y \, dx \, dy = \frac{2}{9} \int_{-1}^{2} \int_{x^2}^{x+2} y \, dy \, dx = \frac{2}{9} \int_{-1}^{2} \frac{1}{2} \left[(x+2)^2 - x^4 \right] dx$$

$$= \frac{1}{9} \left[\frac{x^3}{3} + 2x^2 + 4x - \frac{x^5}{5} \right]_{-1}^{2} = \frac{1}{9} \left[\frac{184}{15} - \left(-\frac{32}{15} \right) \right] = \frac{8}{5}.$$

Consequently, the centroid of R is $(\bar{x}, \bar{y}) = (\frac{1}{2}, \frac{8}{5})$.

The centroid of a planar region is a purely geometric notion and is independent of the physical concept of mass. Indeed, if $(\bar{x}, \bar{y})$ is the centroid of a planar region R, then $\bar{x}$ should be thought of as the "average" x coordinate of points in R and $\bar{y}$ should be regarded as the "average" y coordinate of points in R. However, as can be confirmed by experiment, if a thin sheet of metal of uniform density is cut in the shape of R, it will balance perfectly on a sharp point placed at the centroid of R (Figure 9). Thus, if a planar region R has an axis of symmetry, then the centroid of R must lie on this axis. Also, the position of the centroid of a region is independent of the choice of the coordinate system; hence, in finding the centroid of a planar region, the coordinate axes can be chosen for convenience of calculation.

Figure 9

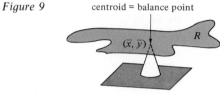
centroid = balance point

The following theorem provides an interesting connection between centroids and volumes of solids of revolution.

THEOREM 1

Theorem of Pappus for Volumes of Solids of Revolution
Let R be an admissible planar region lying in the same plane as a straight line L and entirely on one side of L. Let r be the distance from the centroid of R to the line L, and denote the area of R by A. Then, the volume V of the solid of revolution generated by revolving R about the line L is given by $V = 2\pi r A$.

Figure 10

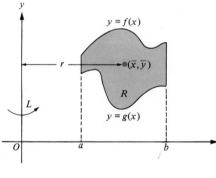

For an indication of a proof of Pappus' theorem, consider the case in which R is a type I region in the first quadrant (Figure 10) and L is the y axis. Notice that $r = \bar{x}$, so we must prove that $V = 2\pi \bar{x} A$. By definition of $\bar{x}$, we have

$$\bar{x} A = \iint_R x \, dx \, dy = \int_a^b \int_{g(x)}^{f(x)} x \, dy \, dx$$

$$= \int_a^b x[f(x) - g(x)] \, dx.$$

Therefore, using the method of cylindrical shells (Section 2 of Chapter 7), we obtain

$$V = 2\pi \int_a^b x[f(x) - g(x)] \, dx = 2\pi \bar{x} A, \qquad \text{as desired.}$$

EXAMPLE   A circular disk R of radius a units is revolved about an axis L which is in the same plane as R and r units from the center of R, $r > a$. Find the volume V of the doughnut-shaped solid (torus) thus generated.

SOLUTION
By symmetry, the centroid of R is its center. Since the area of R is given by $A = \pi a^2$, Pappus' theorem gives $V = 2\pi r A = 2\pi^2 r a^2$.

4.6 Moment of Inertia

Consider a force $\bar{F}$ acting at a point P in a rigid body and let $\overline{AB}$ be an axis not parallel to $\bar{F}$ and not passing through the point of application P. Let O be the point at the foot of a perpendicular dropped from P to the axis $\overline{AB}$ (Figure 11). The force $\bar{F}$ tends to cause the body to turn about the axis $\overline{AB}$; in fact, it produces a definite angular acceleration, α radians per second per second, about this axis. If we denote by F_p the absolute value of the (scalar) component of $\bar{F}$ in the direction perpendicular to the plane containing $\overline{AB}$ and $\overline{OP}$, then the quantity L defined by

$$L = F_p|\overline{OP}|$$

Figure 11

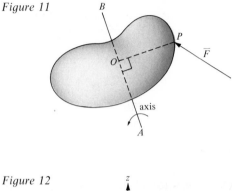

is called the magnitude of the *torque* about the axis $\overline{AB}$ caused by the application of the force $\bar{F}$ at the point P.

It is shown in elementary mechanics that the magnitude of the torque is proportional to the angular acceleration α; that is,

$$L = I_{\overline{AB}}\alpha,$$

where the constant of proportionality $I_{\overline{AB}}$, which is called the *moment of inertia of the body about the axis* $\overline{AB}$, depends only on the axis $\overline{AB}$ and the distribution of mass in the body.

Figure 12

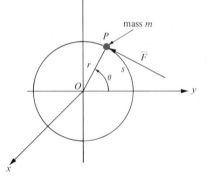

In Figure 12, a particle P of mass m is attached to the origin O by a rigid massless rod of length $r = |\overline{OP}|$ and made to move in a circle in the yz plane by a force $\bar{F}$ lying in the yz plane and perpendicular to $\overline{OP}$. If θ denotes the angle in radians between the y axis and $\overline{OP}$, then, by definition, $\alpha = d^2\theta/dt^2$ gives the angular acceleration of P about the x axis. Here it is easy to show that

$$|\bar{F}| = mr\frac{d^2\theta}{dt^2}$$

(Problem 48); hence, multiplying by r and noting that $L = F_p|\overline{OP}| = |\bar{F}|r$ and $\alpha = d^2\theta/dt^2$, we obtain

$$L = mr^2\alpha.$$

Thus, the moment of inertia of a particle P about an axis (in this case, the x axis) is given by

$$I = mr^2,$$

where m is the mass of the particle and r is its distance from the axis.

Using the result above, we can now tackle the problem of finding the moment of inertia of a lamina, occupying a region R in the plane, about an axis lying in this plane—say the x axis for definiteness. Suppose that the lamina has density $\sigma(x, y)$ at the point (x, y) and consider the "infinitesimal" rectangle of dimensions dx and dy with center at (x, y) (Figure 13).

Figure 13

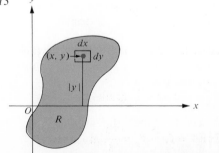

The mass in this "infinitesimal" rectangle is

$$dm = \sigma(x, y)\, dx\, dy$$

and its distance from the x axis is $|y|$ units; hence, its moment of inertia about the x axis is

$$dI_x = |y|^2\, dm = \sigma(x, y)y^2\, dx\, dy.$$

The total moment of inertia I_x of the lamina about the x axis is obtained by "summing," that is, by *integrating* all of the "infinitesimal" quantities $\sigma(x, y)y^2\, dx\, dy$. Thus,

$$I_x = \iint\limits_{R} \sigma(x, y)y^2\, dx\, dy.$$

Reasoning in a similar way, we find that the moment of inertia I_y of the lamina about the y axis is given by

$$I_y = \iint\limits_{R} \sigma(x, y)x^2\, dx\, dy.$$

More generally, the moment of inertia I_l about the straight line whose equation is $l: ax + by + c = 0$ is given by

$$I_l = \iint\limits_{R} \sigma(x, y)\frac{(ax + by + c)^2}{a^2 + b^2}\, dx\, dy$$

(Problem 46).

We can also ask for the moment of inertia of the lamina about an axis perpendicular to the plane of the lamina. If this axis passes through the origin, the result, called the *polar moment of inertia* I_o, is given by

$$I_o = \iint\limits_{R} \sigma(x, y)(x^2 + y^2)\, dx\, dy = I_x + I_y$$

(Problem 50).

EXAMPLES 1 Find the moments of inertia I_x, I_y, and I_o of a square lamina whose sides are 2 centimeters long and parallel to the x and y axes and whose center is at the origin. Assume that the lamina is homogeneous (that is, its mass is distributed uniformly) and that its total mass is 8 grams.

SOLUTION
The area of the lamina is 4 square centimeters. Since it is homogeneous, its mass density is a constant, $\frac{8}{4}$ grams per square centimeter. Thus, $\sigma(x, y) = 2$ for all points (x, y) within the lamina. Here

$$I_x = \iint\limits_{R} \sigma(x, y)y^2\, dx\, dy = 2 \iint\limits_{R} y^2\, dx\, dy = 2 \int_{-1}^{1} \int_{-1}^{1} y^2\, dx\, dy$$

$$= 2 \int_{-1}^{1} 2y^2\, dy = \left(\frac{4y^3}{3}\right)\Big|_{-1}^{1} = \frac{8}{3} \text{ gram cm}^2,$$

$$I_y = \iint\limits_{R} \sigma(x, y)x^2\, dx\, dy = 2 \iint\limits_{R} x^2\, dx\, dy = 2 \int_{-1}^{1} \int_{-1}^{1} x^2\, dx\, dy$$

$$= \frac{2}{3} \int_{-1}^{1} \left[x^3\Big|_{-1}^{1}\right] dy = \frac{2}{3} \int_{-1}^{1} 2\, dy = \frac{8}{3} \text{ gram cm}^2, \quad \text{and}$$

$$I_o = \iint\limits_{R} \sigma(x, y)(x^2 + y^2)\, dx\, dy = I_x + I_y = \frac{16}{3} \text{ gram cm}^2.$$

2 Find the moment of inertia I_x of a lamina occupying the region $R: 0 \le x \le 1$, $0 \le y \le \sqrt{1 - x^2}$ if the mass density at the point (x, y) is given by $\sigma(x, y) = 3y^3$ grams per square centimeter.

SOLUTION

$$I_x = \iint_R \sigma(x, y)y^2 \, dx \, dy = \int_0^1 \int_0^{\sqrt{1-x^2}} (3y^3)y^2 \, dy \, dx = \int_0^1 \left[\frac{y^6}{2} \Big|_0^{\sqrt{1-x^2}} \right] dx$$

$$= \frac{1}{2} \int_0^1 (1 - x^2)^3 \, dx = \frac{1}{2} \int_0^1 (1 - 3x^2 + 3x^4 - x^6) \, dx = \frac{1}{2} \left(x - x^3 + \frac{3x^5}{5} - \frac{x^7}{7} \right) \Big|_0^1$$

$$= \frac{8}{35} \text{ gram cm}^2.$$

Problem Set 4

In problems 1 through 8, use double integration to find the area of the region R in the xy plane bounded by the given curves.

1 $y = x^2$ and $y = 5x$.

2 $xy = 3$ and $x + y = 4$.

3 $y = x$ and $y = 3x - x^2$.

4 $y = \sin x$, $y = \cos x$, $x = 0$, and $x = \pi/4$.

5 $y = \sqrt[3]{x}$, $y = 0$, and $x = -8$.

6 $y = xe^{-x}$, $y = x$, and $x = 2$.

7 $y = \cos x$, $y = 0$, $x = -\pi/2$, and $x = \pi/2$.

8 $y = \cosh x$, $y = \sinh x$, $x = -1$, and $x = 1$.

In problems 9 through 12, use double integration to find the volume under the graph of each function f and above the indicated region R.

9 $f(x, y) = 12 + y + x^2$; $R: 0 \le x \le 1$ and $x^2 \le y \le \sqrt{x}$

10 $f(x, y) = Ax + By + C$; $R: a \le x \le b$ and $c \le y \le d$; $Ax + By + C \ge 0$ on R

11 $f(x, y) = xy$; $R: x^2 + y^2 \le 4$, $x \ge 0$, $y \ge 0$

12 $f(x, y) = x^2 + 9y^2$; R: the region inside the triangle whose vertices are $(0, 0, 0)$, $(0, 1, 0)$, and $(1, 1, 0)$.

In problems 13 through 18, use double integration to find the volume of each solid.

13 The solid below the paraboloid $z = 4 - x^2 - y^2$, above the plane $z = 0$, and bounded laterally by the planes $x = 0$, $y = 0$, $x = 1$, and $y = 1$.

14 The solid bounded above by the plane $z + y = 2$, below by the plane $z = 0$, and laterally by the right circular cylinder $x^2 + y^2 = 4$.

15 The solid in the first octant under the plane $x + y + z = 6$ and inside the parabolic cylinder $y = 4 - x^2$.

16 The solid in the first octant bounded by the graphs of $z = e^{x+y}$, $y = \ln x$, $x = 2$, $y = 0$, and $z = 0$.

17 The solid in the first octant bounded by the cylinder $x^2 + z = 1$, the plane $x + y = 1$, and the coordinate planes.

18 The solid bounded above by the plane $z = x + 2y + 2$, below by the surface $z = -\sqrt{2x - x^2 - y^2}$, and laterally by the cylinder $x^2 + y^2 = 2x$.

In problems 19 through 22, mass is distributed over the given region R in the indicated manner. Find the total mass m on the region R.

19 $R: 0 \le y \le 4$ and $y^2 - 2y \le x \le 2y$, with a constant density of 6 grams per square centimeter.

20 $R: 0 \le y \le 2$ and $\sqrt{3 - y} \le x \le 3 - y$, with a density function σ given by $\sigma(x, y) = 2x$ slugs per square foot.

21 R: the triangle with vertices $(0,0)$, $(4,0)$, and $(4,4)$, with a density function σ given by $\sigma(x,y) = y^2$ kilograms per square meter.

22 R: the region bounded by the curves $y = x - 1$ and $y = 1 - x^2$, with a density function given by $\sigma(x,y) = x^2 + y^2$ pounds of mass per square yard.

23 Electrical charge is distributed over the region R bounded by the parabolas $y^2 = x$ and $x^2 = y$ with a charge density σ given by $\sigma(x,y) = x^2 + 4y^2$ coulombs per square centimeter. Find the total electrical charge on the region R.

24 One coulomb of electrical charge is distributed uniformly over the region R bounded by the parabola $y = x^2$ and the straight line $y = x + 2$. Find the (constant) value of the charge density if distances are measured in centimeters.

In problems 25 through 30, assume that a lamina with mass density function σ occupies the region R in the xy plane. Find the total mass m and the coordinates $(\bar{x}, \bar{y})$ of the center of mass of the distribution. Assume that mass is measured in grams and distance in centimeters.

25 R: $0 \leq x \leq 3$, $0 \leq y \leq 3 - x$; $\sigma(x,y) = x^2 + y^2$.   **26** R: $2 \leq x \leq 3$, $2 \leq y \leq 5$; $\sigma(x,y) = xy$.

27 R: the triangle with vertices $(0,0)$, $(3,0)$, and $(3,5)$; $\sigma(x,y) = x$.

28 R: $-3 \leq x \leq 3$, $-3 \leq y \leq \sqrt{9 - x^2}$; $\sigma(x,y) = y + 3$.

29 R: the region in the first quadrant bounded by the curves $y = x + x^2$, $y = 0$, $x = 0$, and $x = 2$; $\sigma(x,y) = \dfrac{y}{1+x}$.

30 R: the region in the first quadrant bounded by the curves $y = e^x$, $x = 0$, $y = 0$, and $x = 2$; $\sigma(x,y)$ is proportional to the distance of the point (x,y) from the x axis and $\sigma(0,1) = 1$.

31 Suppose that mass is distributed uniformly on a planar region R. Show that the mass on an admissible subregion R_1 of R is proportional to the area of R_1.

In problems 32 through 36, find the centroid $(\bar{x}, \bar{y})$ of the region R bounded by the given curves.

32 $y = \sqrt{25 - x^2}$ and $xy = 12$ **33** $y = x^2$ and $y = 4x - x^2$

34 $y = \sin x$, $y = x$, and $x = \dfrac{\pi}{2}$ **35** $y = \dfrac{x^2}{4}$, $xy = 2$, and $3y = 2x + 4$, $x \geq 1$, $y \geq 1$

36 $y^2 = x^3$, $x + y = 2$, and $x = 4$

In problems 37 through 40, use the theorem of Pappus to find the volume of the solid generated by revolving the region bounded by the given curves about the indicated axis.

37 $y = x^2$ and $y = 2x + 3$ about the x axis.   **38** $y = x^3$, $x + y = 2$, and $y = 0$ about the line $y = -2$.

39 $2x + y = 2$, $x = 0$, and $y = 0$ about the y axis.   **40** $x = \sqrt{4 - y}$, $x = 0$, and $y = 0$ about the line $x = -2$.

In problems 41 through 44, find the moment of inertia I_x and the moment of inertia I_y about the x and y axes, respectively, of a homogeneous lamina of total mass 1 gram occupying each region R. Assume that all distances are measured in centimeters.

41 R: $0 \leq x \leq 1$, $x^4 \leq y \leq x^2$.   **42** R: the region bounded by $y = x^3/4$ and $y = |x|$.

43 R: the region bounded by $xy = 4$ and $2x + y = 6$.   **44** R: the region bounded by $y = e^x$, $y = e$, and $x = 0$.

45 Find the moments of inertia I_x, I_y, and I_o for a lamina occupying the region R: $0 \leq x \leq 2$, $0 \leq y \leq 2 - x$ if the mass density function σ is given by $\sigma(x,y) = x + 2y$ grams per square centimeter.

46 A lamina occupies the admissible region R and has mass density function σ. Show that the moment of inertia of the lamina about the straight line l: $ax + by + c = 0$ is given by

$$I_l = \iint\limits_R \sigma(x,y)\,\frac{(ax + by + c)^2}{a^2 + b^2}\,dx\,dy.$$

47 A lamina occupies the region $R\colon -\pi/2 \leq x \leq \pi/2,\ 0 \leq y \leq \cos x$ and has mass density function σ given by $\sigma(x,y) = y$ grams per square centimeter. Find its moment of inertia I_x about the x axis.

48 In Figure 12, let $\bar{R}$ be the variable position vector of P, so that $\bar{R} = (r \cos \theta)\bar{j} + (r \sin \theta)\bar{k}$. Here, $s = r\theta$. Show that:
(a) The tangential component vector of the acceleration is given by $r\alpha \bar{T}$, where $\alpha = d^2\theta/dt^2$.
(b) The normal component vector of the acceleration is given by $-(d\theta/dt)^2\bar{R}$.
(c) Since $\bar{F}$ is the tangential force acting on P, use part (a) to obtain the result $|\bar{F}| = mr\alpha$.
(d) Calculate the normal force acting on P and explain what provides this force.

49 Find the polar moment of inertia I_o of a ring-shaped lamina occupying the region $\frac{1}{4} \leq x^2 + y^2 \leq 1$ if the mass density function σ is given by $\sigma(x,y) = (x^2 + y^2)^{-1}$ grams per square centimeter.

50 Derive the formula for the moment of inertia of a lamina occupying an admissible region R in the xy plane about an axis perpendicular to the plane of the lamina and passing through the point with xy coordinates (a,b). Assume that σ is the mass density function for the lamina. In particular, show that $I_o = I_x + I_y$.

5 Double Integrals in Polar Coordinates

Often the region R over which a double integral is to be evaluated is more easily described by polar coordinates than by cartesian coordinates. For instance, the region R in Figure 1 is easily described in polar coordinates by the conditions $r_0 \leq r \leq r_1$ and $\theta_0 \leq \theta \leq \theta_1$; however, its description in cartesian coordinates is considerably more complicated. In this section we set forth a method for converting a double integral in cartesian coordinates to an equivalent iterated integral expressed in polar coordinates. The technique is analogous to a change of variable for the ordinary definite integral.

The clue to the appropriate method for changing from cartesian to polar coordinates can be found in Figure 2, which shows an "infinitesimal" portion dA of the area of the region R in Figure 1 corresponding to "infinitesimal" changes dr in r and $d\theta$ in θ. Evidently, dA is virtually the area of a rectangle of dimensions $r\, d\theta$ and dr, so that

$$dA = (r\, d\theta)\, dr = r\, dr\, d\theta.$$

Thus, whereas in cartesian coordinates the area of an "infinitesimal" rectangle of dimensions dx and dy is given by $dA = dx\, dy$, the analogous "infinitesimal" area in polar coordinates is given by $dA = r\, dr\, d\theta$.

In view of the argument above, it seems plausible that the integral

$$\iint\limits_{R} f(x,y)\, dx\, dy = \iint\limits_{R} f(x,y)\, dA$$

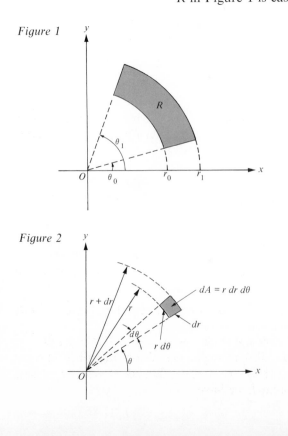

Figure 1

Figure 2

can be converted into an equivalent integral in polar coordinates by putting $x = r \cos \theta$, $y = r \sin \theta$, and $dA = r\, dr\, d\theta$. The following theorem shows exactly how such a conversion to polar coordinates is accomplished.

THEOREM 1 **Change to Polar Coordinates in a Double Integral**

Suppose that the function f is continuous on the region R in the xy plane consisting of all points of the form $(x, y) = (r \cos \theta, r \sin \theta)$, where $0 \le r_0 \le r \le r_1$ and $\theta_0 \le \theta \le \theta_1$ with $0 < \theta_1 - \theta_0 \le 2\pi$ (Figure 1). Then

$$\iint\limits_{R} f(x, y)\, dx\, dy = \int_{\theta = \theta_0}^{\theta = \theta_1} \left[\int_{r = r_0}^{r = r_1} f(r \cos \theta, r \sin \theta) r\, dr \right] d\theta$$

$$= \int_{\theta_0}^{\theta_1} \int_{r_0}^{r_1} f(r \cos \theta, r \sin \theta)\, r\, dr\, d\theta.$$

EXAMPLE By changing to polar coordinates as in Theorem 1, evaluate $\iint\limits_{R} e^{x^2 + y^2}\, dx\, dy$, where R is the region in the first quadrant inside the circle $x^2 + y^2 = 4$ and outside the circle $x^2 + y^2 = 1$ (Figure 3).

SOLUTION

The region R is described in polar coordinates by $1 \le r \le 2$ and $0 \le \theta \le \pi/2$; hence, by Theorem 1,

Figure 3

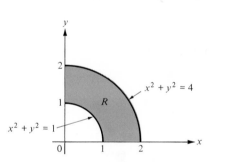

$$\iint\limits_{R} e^{x^2 + y^2}\, dx\, dy = \int_{\theta = 0}^{\theta = \pi/2} \left[\int_{r = 1}^{r = 2} e^{r^2} r\, dr \right] d\theta$$

$$= \int_{\theta = 0}^{\theta = \pi/2} \left[\frac{1}{2} e^{r^2} \Big|_{r = 1}^{r = 2} \right] d\theta$$

$$= \int_{0}^{\pi/2} \left(\frac{1}{2} e^4 - \frac{1}{2} e \right) d\theta$$

$$= \left(\frac{1}{2} e^4 - \frac{1}{2} e \right) \left(\theta \Big|_{0}^{\pi/2} \right) = \frac{\pi e}{4} (e^3 - 1).$$

Sometimes it is useful to rewrite a given iterated integral as an equivalent double integral, and then to evaluate the double integral by changing to polar coordinates. The following example illustrates the technique.

EXAMPLE Evaluate the iterated integral $\int_{-3}^{3} \int_{0}^{\sqrt{9 - x^2}} (2x + y)\, dy\, dx$ by switching to polar coordinates.

SOLUTION

The iterated integral $\int_{-3}^{3} \int_{0}^{\sqrt{9 - x^2}} (2x + y)\, dy\, dx$ is

Figure 4

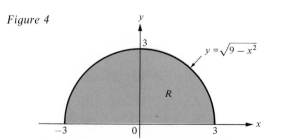

equivalent to the double integral $\iint\limits_{R} (2x + y)\, dx\, dy$

over the region R: $-3 \le x \le 3$, $0 \le y \le \sqrt{9 - x^2}$ (Figure 4). This region can also be described in polar coordinates by $0 \le r \le 3$ and $0 \le \theta \le \pi$. Using Theorem 1, we have

$$\int_{-3}^{3} \int_{0}^{\sqrt{9-x^2}} (2x + y)\, dy\, dx = \iint_{R} (2x + y)\, dx\, dy = \int_{0}^{\pi} \int_{0}^{3} (2r \cos \theta + r \sin \theta) r\, dr\, d\theta$$

$$= \int_{0}^{\pi} \left[\int_{0}^{3} (2 \cos \theta + \sin \theta) r^2\, dr \right] d\theta$$

$$= \int_{0}^{\pi} \left[(2 \cos \theta + \sin \theta) \frac{r^3}{3} \Big|_{0}^{3} \right] d\theta$$

$$= 9 \int_{0}^{\pi} (2 \cos \theta + \sin \theta)\, d\theta$$

$$= 9[2 \sin \theta - \cos \theta] \Big|_{0}^{\pi} = 18.$$

Figure 5

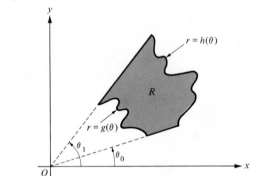

Theorem 1 can be generalized in a number of useful ways. For instance, consider the region R in the xy plane consisting of all points whose polar coordinates satisfy the conditions

$$\theta_0 \le \theta \le \theta_1 \quad \text{and} \quad g(\theta) \le r \le h(\theta),$$

where $0 < \theta_1 - \theta_0 \le 2\pi$ and g and h are continuous functions defined on the closed interval $[\theta_0, \theta_1]$ such that $0 \le g(\theta) \le h(\theta)$ holds for all values of θ in $[\theta_0, \theta_1]$ (Figure 5). Then, if f is a continuous function of two variables defined on the region R, we have

$$\iint_{R} f(x, y)\, dx\, dy = \int_{\theta=\theta_0}^{\theta=\theta_1} \left[\int_{r=g(\theta)}^{r=h(\theta)} f(r \cos \theta, r \sin \theta) r\, dr \right] d\theta$$

$$= \int_{\theta_0}^{\theta_1} \int_{g(\theta)}^{h(\theta)} f(r \cos \theta, r \sin \theta) r\, dr\, d\theta.$$

EXAMPLES   Use the preceding formula to work the following:

1 Evaluate $\displaystyle\iint_{R} x\, dx\, dy$ over the region R consisting of all points whose polar coordinates satisfy the conditions $0 \le \theta \le \pi/4$ and $2 \cos \theta \le r \le 2$ (Figure 6).

SOLUTION

$$\iint_{R} x\, dx\, dy = \int_{\theta=0}^{\theta=\pi/4} \left[\int_{r=2 \cos \theta}^{r=2} r \cos \theta\, r\, dr \right] d\theta = \int_{0}^{\pi/4} \cos \theta \left[\frac{r^3}{3} \Big|_{2 \cos \theta}^{2} \right] d\theta$$

Figure 6

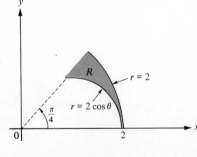

$$= \int_{0}^{\pi/4} \left(\frac{8}{3} \cos \theta - \frac{8 \cos^4 \theta}{3} \right) d\theta = \frac{8}{3} \int_{0}^{\pi/4} \cos \theta\, d\theta - \frac{8}{3} \int_{0}^{\pi/4} \cos^4 \theta\, d\theta$$

$$= \frac{8}{3} \int_{0}^{\pi/4} \cos \theta\, d\theta - \frac{1}{3} \int_{0}^{\pi/4} (3 + 4 \cos 2\theta + \cos 4\theta)\, d\theta$$

$$= \frac{8}{3} \sin \theta \Big|_{0}^{\pi/4} - \frac{1}{3} \left(3\theta + 2 \sin 2\theta + \frac{1}{4} \sin 4\theta \right) \Big|_{0}^{\pi/4}$$

$$= \frac{16\sqrt{2} - 3\pi - 8}{12}.$$

2 Find the area enclosed by the portion of the cardioid $r = 2(1 + \cos\theta)$ lying in the fourth and first quadrants (Figure 7).

Figure 7

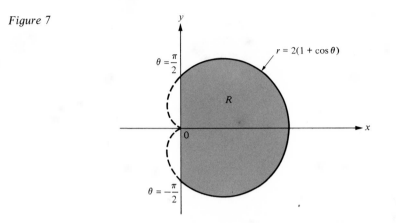

$r = 2(1 + \cos\theta)$

$\theta = \dfrac{\pi}{2}$

R

$\theta = -\dfrac{\pi}{2}$

SOLUTION

The region R whose area is desired can be described in polar coordinates by the conditions $-\pi/2 \le \theta \le \pi/2$ and $0 \le r \le 2(1 + \cos\theta)$. Therefore, the area A of R is given by

$$A = \iint_R dx\, dy = \int_{\theta=-\pi/2}^{\theta=\pi/2} \left[\int_{r=0}^{r=2(1+\cos\theta)} r\, dr \right] d\theta = \int_{-\pi/2}^{\pi/2} \left[\frac{r^2}{2}\Big|_0^{2(1+\cos\theta)} \right] d\theta$$

$$= \int_{-\pi/2}^{\pi/2} 2(1 + \cos\theta)^2\, d\theta = \int_{-\pi/2}^{\pi/2} 2(1 + 2\cos\theta + \cos^2\theta)\, d\theta$$

$$= 2\left(\theta + 2\sin\theta + \frac{\theta}{2} + \frac{\sin 2\theta}{4} \right)\Bigg|_{-\pi/2}^{\pi/2} = 3\pi + 8 \text{ square units.}$$

3 Find the volume of the solid in the first octant bounded by the cone $z = r$ and the cylinder $r = 4\sin\theta$.

SOLUTION

The cone $z = r$ has the cartesian equation $z = \sqrt{x^2 + y^2}$; hence, the required volume V is given by $V = \iint_R \sqrt{x^2 + y^2}\, dx\, dy$, where R is the region consisting of all points whose polar coordinates satisfy $0 \le \theta \le \pi/2$ and $0 \le r \le 4\sin\theta$ (Figure 8). Hence,

$$V = \iint_R \sqrt{x^2 + y^2}\, dx\, dy = \int_0^{\pi/2} \int_0^{4\sin\theta} r \cdot r\, dr\, d\theta = \int_0^{\pi/2} \left[\frac{r^3}{3}\Big|_0^{4\sin\theta} \right] d\theta$$

$$= \int_0^{\pi/2} \frac{64}{3} \sin^3\theta\, d\theta = \frac{64}{3} \int_0^{\pi/2} \sin\theta(1 - \cos^2\theta)\, d\theta.$$

Thus, making the change of variable $u = \cos\theta$, we obtain

$$V = \frac{64}{3} \int_1^0 (1 - u^2)(-du) = \frac{64}{3} \int_0^1 (1 - u^2)\, du$$

$$= \frac{64}{3}\left(u - \frac{u^3}{3} \right)\Bigg|_0^1 = \frac{128}{9} \text{ cubic units.}$$

Figure 8

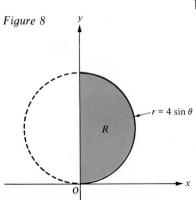

$r = 4\sin\theta$

R

4 Find the centroid of the region R inside the circle $r = 4 \sin \theta$ and outside the circle $r = 2$ (Figure 9).

SOLUTION

Solving for the points of intersection of the circles, we obtain $(2, \pi/6)$ and $(2, 5\pi/6)$ (in polar coordinates). Thus, the region R can be described in polar coordinates by the conditions $\pi/6 \le \theta \le 5\pi/6$ and $2 \le r \le 4 \sin \theta$. By symmetry, the centroid of R lies on the y axis; hence, $x = 0$ and it is only necessary to find $\bar{y}$. The area A of R is given by

$$A = \iint_R dx\, dy = \int_{\pi/6}^{5\pi/6} \int_{2}^{4\sin\theta} r\, dr\, d\theta$$

$$= \frac{1}{2} \int_{\pi/6}^{5\pi/6} \left[r^2 \Big|_{2}^{4\sin\theta} \right] d\theta$$

$$= \int_{\pi/6}^{5\pi/6} (8 \sin^2 \theta - 2)\, d\theta$$

$$= \int_{\pi/6}^{5\pi/6} \left[8\left(\frac{1 - \cos 2\theta}{2} \right) - 2 \right] d\theta$$

$$= \int_{\pi/6}^{5\pi/6} (2 - 4 \cos 2\theta)\, d\theta$$

$$= (2\theta - 2 \sin 2\theta) \Big|_{\pi/6}^{5\pi/6}$$

$$= \frac{4\pi}{3} + 2\sqrt{3} \text{ square units.}$$

Figure 9

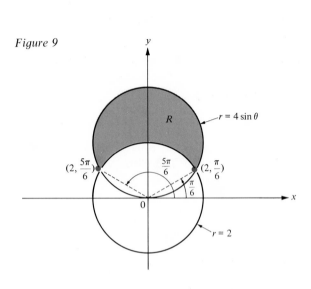

Therefore,

$$\bar{y} = \frac{1}{A} \iint_R y\, dx\, dy = \frac{1}{A} \int_{\pi/6}^{5\pi/6} \int_{2}^{4\sin\theta} r \sin \theta\, r\, dr\, d\theta = \frac{1}{A} \int_{\pi/6}^{5\pi/6} \left[\sin \theta \left(\frac{r^3}{3} \right) \Big|_{2}^{4\sin\theta} \right] d\theta$$

$$= \frac{1}{A} \int_{\pi/6}^{5\pi/6} \frac{1}{3} (64 \sin^4 \theta - 8 \sin \theta)\, d\theta$$

$$= \frac{1}{3A} \int_{\pi/6}^{5\pi/6} \left[64\left(\frac{3 - 4 \cos 2\theta + \cos 4\theta}{8} \right) - 8 \sin \theta \right] d\theta$$

$$= \frac{8}{3A} \int_{\pi/6}^{5\pi/6} (3 - 4 \cos 2\theta + \cos 4\theta - \sin \theta)\, d\theta$$

$$= \frac{8}{3A} \left(3\theta - 2 \sin 2\theta + \frac{\sin 4\theta}{4} + \cos \theta \right) \Big|_{\pi/6}^{5\pi/6}$$

$$= \frac{8}{3A} \left[\left(\frac{5\pi}{2} + \frac{3\sqrt{3}}{8} \right) - \left(\frac{\pi}{2} - \frac{3\sqrt{3}}{8} \right) \right] = \frac{1}{3A} (16\pi + 6\sqrt{3})$$

$$= \frac{16\pi + 6\sqrt{3}}{4\pi + 6\sqrt{3}} = \frac{8\pi + 3\sqrt{3}}{2\pi + 3\sqrt{3}}.$$

Thus, the centroid of R is $(\bar{x}, \bar{y}) = \left(0, \dfrac{8\pi + 3\sqrt{3}}{2\pi + 3\sqrt{3}} \right)$.

Problem Set 5

In problems 1 through 8, use polar coordinates to evaluate each double integral over the region indicated.

1 $\iint\limits_{R} \sqrt{4 - x^2 - y^2} \, dx \, dy;\ R: x^2 + y^2 \leq 4,\ x \geq 0 \text{ and } y \geq 0.$

2 $\iint\limits_{R} \frac{1}{(x^2 + y^2)^3} \, dx \, dy;\ R: 4 \leq x^2 + y^2 \leq 9.$

3 $\iint\limits_{R} \sqrt{x^2 + y^2} \, dx \, dy;\ R: 1 \leq x^2 + y^2 \leq 2,\ 0 \leq y \leq \sqrt{3}\,x.$

4 $\iint\limits_{R} \frac{dx \, dy}{1 + x^2 + y^2};\ R: x^2 + y^2 \leq 1.$

5 $\iint\limits_{R} (x - y) \, dx \, dy;\ R: x^2 + y^2 \leq 9,\ x \geq 0,\ y \geq 0.$

6 $\iint\limits_{R} 2x \, dx \, dy;\ R: 0 \leq \theta \leq \pi/4,\ 0 \leq r \leq 2 \sin \theta.$

7 $\iint\limits_{R} 3xy \, dx \, dy;\ R: 0 \leq \theta \leq \pi/2,\ 0 \leq r \leq 2.$

8 $\iint\limits_{R} dx \, dy;\ R: -\pi/2 \leq \theta \leq \pi/2,\ 3 \leq r \leq 3(1 + \cos \theta).$

In problems 9 through 16, evaluate each iterated integral by switching to polar coordinates.

9 $\int_{-3}^{3} \int_{-\sqrt{9-x^2}}^{\sqrt{9-x^2}} e^{-x^2 - y^2} \, dy \, dx$

10 $\int_{0}^{2} \int_{0}^{\sqrt{4-x^2}} \sqrt{x^2 + y^2} \, dy \, dx$

11 $\int_{0}^{a} \int_{0}^{\sqrt{a^2 - y^2}} (x^2 + y^2)^{3/2} \, dx \, dy$

12 $\int_{0}^{2} \int_{0}^{\sqrt{4-x^2}} \frac{dy \, dx}{4 + \sqrt{x^2 + y^2}}$

13 $\int_{0}^{3/\sqrt{2}} \int_{y}^{\sqrt{9-y^2}} x \, dx \, dy$

14 $\int_{0}^{a/\sqrt{2}} \int_{-\sqrt{a^2-x^2}}^{-x} y \, dy \, dx$

15 $\int_{-2}^{0} \int_{-\sqrt{4-y^2}}^{\sqrt{4-y^2}} x^2 \, dx \, dy$

16 $\int_{0}^{1} \int_{y}^{\sqrt{y}} \sqrt{x^2 + y^2} \, dx \, dy$

In problems 17 through 22, find the area of each region R by setting up and evaluating a suitable double integral.

17 $R: 0 \leq \theta \leq \pi,\ 0 \leq r \leq 4(1 - \cos \theta).$

18 $R: 0 \leq \theta \leq \pi/4,\ 0 \leq r \leq 3 \cos 2\theta.$

19 $R: 0 \leq \theta \leq \pi,\ 1 \leq r \leq 1 + \sin \theta.$

20 $R:$ the region enclosed by the lemniscate $r^2 = 2a^2 \cos 2\theta.$

21 $R:$ the region inside the circle $r = 2\sqrt{3} \sin \theta$ and outside the circle $r = 3.$

22 $R:$ the region inside the circle $r = 1$ and outside the cardioid $r = 1 - \cos \theta.$

23 Find the volume of the solid in the first octant bounded by the paraboloid $z = 1 - r^2$ and the cylinder $r = 1.$

24 Find the volume of the solid bounded above and below by the sphere $r^2 + z^2 = 4$ and bounded laterally by the cylinder $r = 1.$

25 Find the volume of the solid in the first octant bounded above by the plane $z = r \sin \theta$ and bounded laterally by the coordinate planes and by the cylinder $r = 2 \sin \theta.$

26 Find the volume of the solid in the first octant bounded by the paraboloid $z = \frac{1}{4}r^2$ and the planes $r = 2 \sec \theta,\ \theta = 0,\ \theta = \pi/4,$ and $z = 0.$

In problems 27 through 30, find the centroid of each region $R.$

27 $R:$ the region inside the circle $r = 2a \cos \theta$ and outside the circle $r = a,$ where a is a positive constant.

28 $R:$ the region in the first quadrant that lies inside the curve $r = 2 \sin 2\theta$ and outside the curve $r = \sqrt{3}.$

29 $R:$ the circular sector $-\theta_0 \leq \theta \leq \theta_0,\ 0 \leq r \leq r_0.$

30 $R: -\pi/4 \leq \theta \leq \pi/4,\ 0 \leq r \leq \cos 2\theta.$

31 Find the mass and the center of mass of a lamina contained in the region R: $x^2 + y^2 \leq 1,\ x \geq 0,\ y \geq 0$ if the density function σ is given by $\sigma(x, y) = \sqrt{x^2 + y^2}.$ Assume that mass is measured in kilograms and distance in meters.

32 Find the moment of inertia about the x axis, I_x, about the y axis, I_y, and about the origin, I_o, of a homogeneous lamina in the shape of one leaf of the four-leaved rose $r = a \cos 2\theta$. Use the leaf that cuts the positive x axis and assume that the total mass of the lamina is m grams.

33 Find the moment of inertia about the y axis, I_y, of a homogeneous lamina of total mass m grams in the shape of a lemniscate $r^2 = 2a^2 \cos 2\theta$. Assume that distances are measured in centimeters.

34 A circular lamina of radius r_1 is centered at the origin. Its total mass is m grams and its density at a point r units from the origin is proportional to r^n, where n is a constant. Find its moment of inertia I_x about the x axis. Assume that distances are measured in centimeters.

35 Let R be the region consisting of all points whose polar coordinates satisfy $\theta_0 \le \theta \le \theta_1$ and $r_0 \le r \le r_1$. Assume that F and G are continuous functions defined on the intervals $[\theta_0, \theta_1]$ and $[r_0, r_1]$, respectively. If (x, y) is a point in R whose polar coordinates are r and θ, define $f(x, y) = F(\theta)G(r)$. Prove that

$$\iint\limits_{R} f(x, y) \, dx \, dy = \left[\int_{\theta_0}^{\theta_1} F(\theta) \, d\theta \right] \cdot \left[\int_{r_0}^{r_1} G(r) r \, dr \right].$$

36 The improper integral $\int_{-\infty}^{\infty} e^{-x^2} \, dx$ is important in the theory of probability. Its exact value can be found by using polar coordinates and a clever trick. Write $A = \int_{-\infty}^{\infty} e^{-x^2} \, dx$, so that

$$A^2 = \left[\int_{-\infty}^{\infty} e^{-x^2} \, dx \right] \cdot \left[\int_{-\infty}^{\infty} e^{-y^2} \, dy \right].$$

(a) Show that $A^2 = \iint\limits_{R} e^{-(x^2 + y^2)} \, dx \, dy$, where R is the entire xy plane.

(b) Convert the (improper) double integral in part (a) into an equivalent integral in polar coordinates.

(c) Conclude that $\int_{-\infty}^{\infty} e^{-x^2} \, dx = \sqrt{\pi}$.

6   Triple Integrals

Triple integrals over solids in xyz space are defined by an obvious analogy with the definition of double integrals over regions in the xy plane. The details of the definition are best left to advanced calculus, so here we simply outline the main ideas.

Given a solid region S in three-dimensional space, such as a rectangular box, a cube, a pyramid, a ball, an ellipsoid, and so forth, and given a function f of three variables defined at each point (x, y, z) in S, we define the *triple integral*

$$\iiint\limits_{S} f(x, y, z) \, dx \, dy \, dz$$

(if it exists) as follows.

First we enclose the solid S in a rectangular box B with edges parallel to the coordinate axes (Figure 1). The box B is now partitioned into a large number of smaller boxes by intersecting it with planes parallel to the coordinate planes (Figure 2). These smaller boxes are called the *cells* of the partition. All cells of

Figure 1

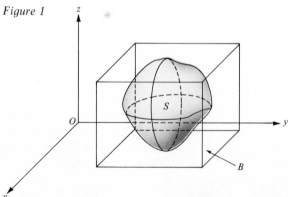

Figure 2

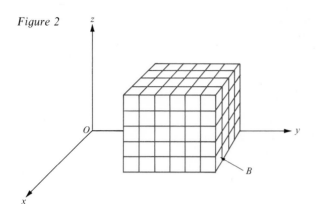

the partition that do not touch the solid region S are now discarded. The remaining cells, which, taken together, contain the solid S and roughly approximate its shape, are now numbered in some convenient way, and called (say) ΔS_1, ΔS_2, ..., ΔS_m. The maximum diagonal measurement of all these cells is called the *norm* of the partition and denoted by η (the Greek letter "eta").

Points are now chosen, one from each cell ΔS_1, ΔS_2, ..., ΔS_m, in such a way that each chosen point belongs to S, and the point chosen from the kth cell is denoted by (x_k^*, y_k^*, z_k^*) for $k = 1, 2, ..., m$. The partition, together with the chosen points, is called an *augmented partition.*

Corresponding to each augmented partition we can form a *Riemann sum*

$$\sum_{k=1}^{m} f(x_k^*, y_k^*, z_k^*)\, \Delta V_k,$$

where ΔV_k is the volume of the kth cell ΔS_k. We can now define the triple integral to be the limit (if it exists) of such Riemann sums as the number of cells becomes larger and larger in such a way that the norm η approaches zero; in symbols,

$$\iiint_S f(x, y, z)\, dx\, dy\, dz = \lim_{\eta \to 0} \sum_{k=1}^{m} f(x_k^*, y_k^*, z_k^*)\, \Delta V_k.$$

If the triple integral $\iiint_S f(x, y, z)\, dx\, dy\, dz$ exists (that is, if the above limit exists), then the function f is said to be (*Riemann*)-*integrable* on the solid S. In advanced calculus, it is shown that if S is an admissible three-dimensional region (Chapter 7, Section 1) which contains all of its own boundary, and if f is continuous on S, then f is (Riemann)-integrable on S. Thus, triple integrals satisfy the analog of Property 1 (the Existence Property) for double integrals set forth in Section 2.1. As a matter of fact, the obvious analogs of Properties 1 through 8 of double integrals given in Section 2.1 hold for triple integrals (Problem 24). For instance, the analog of Property 2 is the following: If c is a constant and S is an admissible three-dimensional region of volume V, then $\iiint_S c\, dx\, dy\, dz = cV$. In particular, $V = \iiint_S dx\, dy\, dz$.

It can be shown that a triple integral of a continuous function over an appropriately shaped solid can be reduced to an equivalent iterated integral—here, however, the iterated integral involves a *double* integral. In fact, we have the following.

Procedure for Evaluation of Triple Integrals by Iteration

Let R be an admissible region in the xy plane which contains all of its own boundary, and suppose that g and h are continuous functions defined on R and satisfying $g(x, y) \leq h(x, y)$ for all points (x, y) in R. Let S be the solid consisting of all points (x, y, z) satisfying the conditions that (x, y) belongs to R and

$$g(x, y) \leq z \leq h(x, y)$$

(Figure 3). Then, if f is a continuous function defined on S,

$$\iiint\limits_{S} f(x, y, z)\, dx\, dy\, dz = \iint\limits_{R} \left[\int_{z=g(x,y)}^{z=h(x,y)} f(x, y, z)\, dz \right] dx\, dy.$$

Figure 3

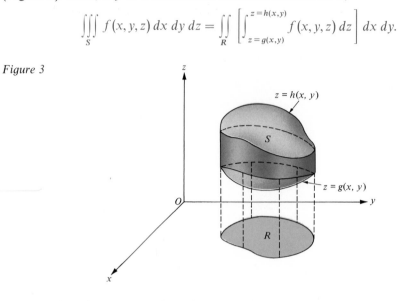

The solid S in the above iteration procedure is evidently bounded above by the surface $z = h(x, y)$, below by the surface $z = g(x, y)$, and laterally by the cylinder over the boundary of R with generators parallel to the z axis. The "inside" integral

$$\int_{z=g(x,y)}^{z=h(x,y)} f(x, y, z)\, dz$$

is, of course, calculated while temporarily holding x and y constant. After it is calculated, the "outside" double integral can be evaluated using the methods given in Sections 2, 3, and 5.

EXAMPLES 1 Evaluate $\iiint\limits_{S} (x + y + z)\, dx\, dy\, dz$, where S is the solid bounded above by the plane $z = 2 - x - y$, below by the plane $z = 0$, and laterally by the cylinder over the boundary of the triangular region R: $0 \le x \le 1$, $0 \le y \le 1 - x$ (Figure 4).

Figure 4

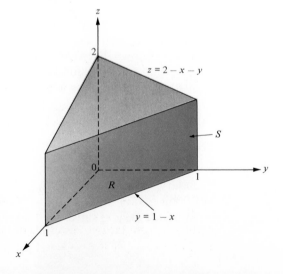

SOLUTION
We use the iteration procedure with $g(x, y) = 0$, $h(x, y) = 2 - x - y$, and R as described. Thus,

$$\iiint_S (x + y + z) \, dx \, dy \, dz = \iint_R \left[\int_{z=0}^{z=2-x-y} (x + y + z) \, dz \right] dx \, dy$$

$$= \iint_R \left[\left(xz + yz + \frac{z^2}{2} \right) \Big|_0^{2-x-y} \right] dx \, dy$$

$$= \iint_R \left[x(2 - x - y) + y(2 - x - y) + \frac{(2 - x - y)^2}{2} \right] dx \, dy$$

$$= \iint_R \left(2 - \frac{x^2}{2} - xy - \frac{y^2}{2} \right) dx \, dy$$

$$= \int_{x=0}^{x=1} \left[\int_{y=0}^{y=1-x} \left(2 - \frac{x^2}{2} - xy - \frac{y^2}{2} \right) dy \right] dx$$

$$= \int_{x=0}^{x=1} \left[\left(2y - \frac{x^2 y}{2} - \frac{xy^2}{2} - \frac{y^3}{6} \right) \Big|_{y=0}^{y=1-x} \right] dx$$

$$= \int_0^1 \left(\frac{11}{6} - 2x + \frac{x^3}{6} \right) dx$$

$$= \left(\frac{11}{6} x - x^2 + \frac{x^4}{24} \right) \Big|_0^1 = \frac{7}{8}.$$

2 Find the volume of the solid S bounded by $z + x^2 = 9$, $y + z = 4$, $y = 0$, and $y = 4$.

SOLUTION
The surface $z + x^2 = 9$ is a parabolic cylinder, opening downward, with generators parallel to the y axis; hence, it must form the upper boundary surface of S. The surface $y + z = 4$ is a plane which cuts the plane $z = 0$ in the straight line $y = 4$, and forms the lower boundary surface of S (Figure 5). The desired volume V is given by

$$V = \iiint_S dx \, dy \, dz.$$

To evaluate the triple integral by iteration, we must determine the region R obtained by projecting the solid S perpendicularly onto the xy plane. Obviously, the straight lines $y = 0$ and $y = 4$ provide two of the boundaries of R. In order to determine the remainder of the boundary of R, we notice that the upper and lower surfaces bounding the solid S meet in space in a curve consisting of all points (x, y, z) satisfying the simultaneous equations

$$\begin{cases} z + x^2 = 9 \\ y + z = 4. \end{cases}$$

The remaining boundary of R is obtained by projecting this curve perpendicularly on the xy plane, and this can be accomplished algebraically by eliminating the variable z from the two simultaneous equations. Thus, $y + 9 - x^2 = 4$ or $y = x^2 - 5$, so that $x = \pm\sqrt{y + 5}$.

Figure 5

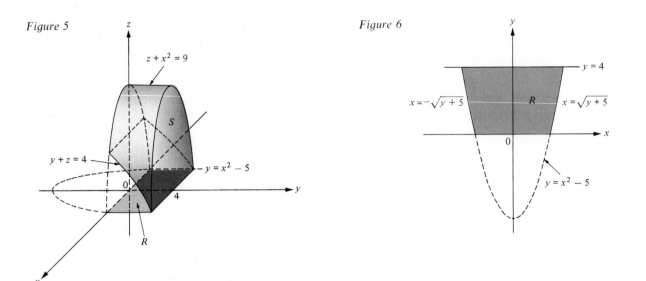

Figure 6

The region R is evidently of type II, and is described by the inequalities $0 \le y \le 4$ and $-\sqrt{y + 5} \le x \le \sqrt{y + 5}$ (Figure 6). Thus,

$$V = \iiint_S dx\, dy\, dz = \iint_R \left[\int_{z=4-y}^{z=9-x^2} dz \right] dx\, dy = \iint_R (5 - x^2 + y)\, dx\, dy$$

$$= \int_{y=0}^{y=4} \left[\int_{x=-\sqrt{y+5}}^{x=\sqrt{y+5}} (5 - x^2 + y)\, dx \right] dy = \int_{y=0}^{y=4} \left[\left(5x - \frac{x^3}{3} + xy \right) \Big|_{x=-\sqrt{y+5}}^{x=\sqrt{y+5}} \right] dy$$

$$= \int_0^4 \left[10\sqrt{y + 5} - \frac{2}{3}(y + 5)^{3/2} + 2y\sqrt{y + 5} \right] dy$$

$$= \left[\frac{20}{3}(y + 5)^{3/2} - \frac{4}{15}(y + 5)^{5/2} + \frac{4}{5}(y + 5)^{5/2} - \frac{20}{3}(y + 5)^{3/2} \right] \Big|_0^4$$

$$= \left[\frac{8}{15}(y + 5)^{5/2} \right] \Big|_0^4 = \frac{8}{15}(243 - 25\sqrt{5}) \approx 99.79 \text{ cubic units.}$$

(The integral $\int 2y\sqrt{y + 5}\, dy$ was found by using the substitution $u = y + 5$, so that $y = u - 5$ and $dy = du$.)

3 Find the volume of the solid S bounded by the elliptic paraboloids $z = 18 - x^2 - y^2$ and $z = x^2 + 5y^2$.

SOLUTION
The surface $z = 18 - x^2 - y^2$ is a paraboloid of revolution (about the z axis) opening downward; hence, it forms the upper boundary surface of the solid S. The elliptic paraboloid $z = x^2 + 5y^2$ opens upward and forms the lower boundary surface of S (Figure 7). Eliminating z from the simultaneous equations

$$\begin{cases} z = 18 - x^2 - y^2 \\ z = x^2 + 5y^2, \end{cases}$$

Figure 7

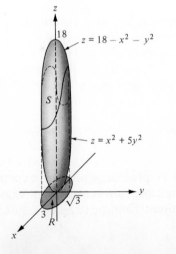

we obtain

$$x^2 + 5y^2 = 18 - x^2 - y^2 \quad \text{or} \quad \frac{x^2}{9} + \frac{y^2}{3} = 1,$$

the equation of the boundary of the region R in the xy plane.
Treating R as a type II region described by the inequalities

$$R: -\sqrt{3} \le y \le \sqrt{3}, \quad -\sqrt{9 - 3y^2} \le x \le \sqrt{9 - 3y^2},$$

and iterating the triple integral for the volume V of S, we obtain

$$V = \iiint_S dx\, dy\, dz = \iint_R \left[\int_{x^2 + 5y^2}^{18 - x^2 - y^2} dz \right] dx\, dy$$

$$= \iint_R (18 - 2x^2 - 6y^2)\, dx\, dy = \int_{-\sqrt{3}}^{\sqrt{3}} \int_{-\sqrt{9 - 3y^2}}^{\sqrt{9 - 3y^2}} [(18 - 6y^2) - 2x^2]\, dx\, dy$$

$$= \int_{-\sqrt{3}}^{\sqrt{3}} \left[2(18 - 6y^2)(9 - 3y^2)^{1/2} - \frac{4}{3}(9 - 3y^2)^{3/2} \right] dy$$

$$= \int_{-\sqrt{3}}^{\sqrt{3}} \left[4(9 - 3y^2)^{3/2} - \frac{4}{3}(9 - 3y^2)^{3/2} \right] dy$$

$$= \frac{8}{3} \int_{-\sqrt{3}}^{\sqrt{3}} (9 - 3y^2)^{3/2}\, dy = \frac{16}{3} \int_0^{\sqrt{3}} (9 - 3y^2)^{3/2}\, dy.$$

Making the change of variable $y = \sqrt{3} \sin \theta$, $0 \le \theta \le \pi/2$, we therefore have

$$V = 144\sqrt{3} \int_0^{\pi/2} \cos^4 \theta\, d\theta = 144\sqrt{3} \left(\frac{3\theta}{8} + \frac{\sin 2\theta}{4} + \frac{\sin 4\theta}{32} \right) \Big|_0^{\pi/2}$$

$$= 27\pi\sqrt{3} \text{ cubic units.}$$

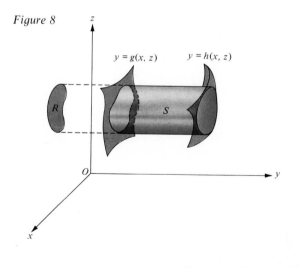

Figure 8

In the iteration procedure for triple integrals, we can change the roles of the variables x, y, and z. For instance, in Figure 8 we have a solid S bounded on the left and right by surfaces $y = g(x, z)$ and $y = h(x, z)$, respectively, and bounded laterally by the cylinder over the boundary of the region R with generators parallel to the y axis. Here, the admissible region R is contained in the xz plane, and we have

$$\iiint_S f(x, y, z)\, dx\, dy\, dz = \iint_R \left[\int_{y = g(x,z)}^{y = h(x,z)} f(x, y, z)\, dy \right] dx\, dz,$$

provided that f is a continuous function on S. A similar result obtains if S is bounded behind by $x = g(y, z)$, in front by $x = h(y, z)$, and laterally by a cylinder with generators parallel to the x axis.

Summarizing, then, there are really three cases of the iteration theorem for triple integrals of continuous functions over solid regions, depending on the shape of the solid:

Case 1 R is an admissible region in the xy plane and the solid S consists of all points (x, y, z) such that (x, y) is in R and $g(x, y) \le z \le h(x, y)$. Then

$$\iiint\limits_{S} f(x, y, z)\, dx\, dy\, dz = \iint\limits_{R} \left[\int_{z=g(x,y)}^{z=h(x,y)} f(x, y, z)\, dz \right] dx\, dy.$$

Case 2 R is an admissible region in the xz plane and the solid S consists of all points (x, y, z) such that (x, z) is in R and $g(x, z) \le y \le h(x, z)$. Then

$$\iiint\limits_{S} f(x, y, z)\, dx\, dy\, dz = \iint\limits_{R} \left[\int_{y=g(x,z)}^{y=h(x,z)} f(x, y, z)\, dy \right] dx\, dz.$$

Case 3 R is an admissible region in the yz plane and the solid S consists of all points (x, y, z) such that (y, z) is in R and $g(y, z) \le x \le h(y, z)$. Then

$$\iiint\limits_{S} f(x, y, z)\, dx\, dy\, dz = \iint\limits_{R} \left[\int_{x=g(y,z)}^{x=h(y,z)} f(x, y, z)\, dx \right] dy\, dz.$$

EXAMPLE   Evaluate $\iiint\limits_{S} 3z\, dx\, dy\, dz$ if S is the solid bounded by $x = 0$, $y = 0$, $z = 1$, and $x + y + z = 2$.

Figure 9

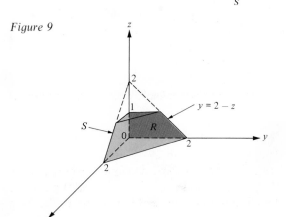

SOLUTION

The surface $x + y + z = 2$ is a plane with x, y, and z intercepts all equal to 2, and it forms the front boundary surface of S. The remaining boundary surfaces are the coordinate planes and the plane $z = 1$ (Figure 9). We treat S as in Case 3, with the region R in the yz plane described by

$$R: 0 \le z \le 1, \quad 0 \le y \le 2 - z$$

and the solid S described by

$$S: (y, z) \text{ in } R \quad \text{and} \quad 0 \le x \le 2 - y - z.$$

Thus,

$$\iiint\limits_{S} 3z\, dx\, dy\, dz = \iint\limits_{R} \left[\int_{x=0}^{x=2-y-z} 3z\, dx \right] dy\, dz = \iint\limits_{R} 3z(2 - y - z)\, dy\, dz$$

$$= \int_{z=0}^{z=1} \left[\int_{y=0}^{y=2-z} 3z(2 - y - z)\, dy \right] dz$$

$$= \int_{z=0}^{z=1} \left[\left(6yz - \frac{3}{2} y^2 z - 3yz^2 \right) \Big|_{y=0}^{y=2-z} \right] dz$$

$$= \int_{0}^{1} \left(\frac{3}{2} z^3 - 6z^2 + 6z \right) dz = \left(\frac{3z^4}{8} - 2z^3 + 3z^2 \right) \Big|_{0}^{1} = \frac{11}{8}.$$

6.1 Threefold Iterated Integrals

The integral in the previous example can be written as

$$\iint\limits_{R} \left[\int_{x=0}^{x=2-y-z} 3z\, dx \right] dy\, dz = \int_{z=0}^{z=1} \left\{ \int_{y=0}^{y=2-z} \left[\int_{x=0}^{x=2-y-z} 3z\, dx \right] dy \right\} dz,$$

and, in the latter form, it is called a *threefold iterated integral.* Unless confusion threatens, the brackets and the detailed information about the limits of integration are usually omitted, and the threefold iterated integral is simply written as

$$\int_0^1 \int_0^{2-z} \int_0^{2-y-z} 3z \, dx \, dy \, dz.$$

The order of integration is determined by the order of the differentials, reading from left to right, just as for the twice-iterated integrals considered in Section 1.

EXAMPLE   Evaluate the threefold iterated integral $\displaystyle\int_0^{\pi/2} \int_0^1 \int_0^{x^2} x \cos y \, dz \, dx \, dy.$

SOLUTION

$$\int_0^{\pi/2} \int_0^1 \int_0^{x^2} x \cos y \, dz \, dx \, dy = \int_0^{\pi/2} \left\{ \int_0^1 \left[\int_0^{x^2} x \cos y \, dz \right] dx \right\} dy$$

$$= \int_0^{\pi/2} \left\{ \int_0^1 \left[(x \cos y)z \Big|_0^{x^2} \right] dx \right\} dy$$

$$= \int_0^{\pi/2} \left\{ \int_0^1 x^3 \cos y \, dx \right\} dy = \int_0^{\pi/2} \left\{ \frac{x^4}{4} \cos y \Big|_0^1 \right\} dy$$

$$= \int_0^{\pi/2} \frac{\cos y}{4} \, dy = \frac{\sin y}{4} \Big|_0^{\pi/2} = \frac{1}{4}.$$

The evaluation of a triple integral by iteration always leads to an iterated integral of the form

$$\iint_R \left[\int_{w=g(u,v)}^{w=h(u,v)} f(u,v,w) \, dw \right] du \, dv,$$

where u, v, and w represent the variables x, y, and z in some order, and where R is an admissible region in the uv plane. If the planar region R is of type I or of type II, the latter integral can be rewritten as a threefold iterated integral having either the form

$$\int_{u=a}^{u=b} \int_{v=G(u)}^{v=H(u)} \int_{w=g(u,v)}^{w=h(u,v)} f(u,v,w) \, dw \, dv \, du,$$

or the form

$$\int_{v=a}^{v=b} \int_{u=G(v)}^{u=H(v)} \int_{w=g(u,v)}^{w=h(u,v)} f(u,v,w) \, dw \, du \, dv.$$

Problem Set 6

In problems 1 through 6, evaluate each threefold iterated integral.

1 $\displaystyle\int_0^1 \int_0^{3x} \int_0^{2x+y} y \, dz \, dy \, dx$

2 $\displaystyle\int_0^2 \int_0^{2y} \int_1^{y+3z} 5x \, dx \, dz \, dy$

3 $\displaystyle\int_0^1 \int_0^x \int_0^z (x+z) \, dy \, dz \, dx$

4 $\displaystyle\int_0^2 \int_0^{\sqrt{4-y^2}} \int_0^y xz \, dz \, dx \, dy$

5 $\displaystyle\int_0^{2\pi} \int_0^{\pi} \int_1^2 z^4 \sin y \, dz \, dy \, dx$

6 $\displaystyle\int_0^{\pi/2} \int_0^{2 \sin y} \int_0^{2-(z^2/2)} z \, dx \, dz \, dy$

In problems 7 and 8, (a) evaluate each threefold iterated integral, (b) rewrite the integral as an iterated integral of the form $\displaystyle\iint_R \left[\int_{g(u,v)}^{h(u,v)} f(u,v,w) \, dw \right] du \, dv$, where u, v, and w are the variables x, y, and z in some order, and (c) rewrite the integral as a triple integral.

7 $\displaystyle\int_0^1 \int_0^{\sqrt{1-z^2}} \int_0^{\sqrt{1-z^2}} xyz\, dy\, dx\, dz$ **8** $\displaystyle\int_0^\pi \int_0^y \int_z^{z+y} \sin(x+y)\, dx\, dz\, dy$

In problems 9 through 16, sketch the solid S and evaluate each triple integral.

9 $\displaystyle\iiint_S (3x+2y)\, dx\, dy\, dz$, where S is the solid bounded above by the plane $z=4$, below by the plane $z=0$, and laterally by the cylinder with generators parallel to the z axis over the boundary of the square region $R: -1 \le x \le 1,\ -1 \le y \le 3$.

10 $\displaystyle\iiint_S 3xy\, dx\, dy\, dz$, where S is the solid in the first octant bounded by the coordinate planes and the plane $x+y+z=6$.

11 $\displaystyle\iiint_S \sqrt{x^2+y^2}\, dx\, dy\, dz$, where S is the solid determined by the conditions $x^2+y^2 \le 1$ and $0 \le z \le \sqrt{x^2+y^2}$.

12 $\displaystyle\iiint_S xy^2z^3\, dx\, dy\, dz$, where S is the solid determined by the conditions $0 \le x \le 1, 0 \le y \le x$, and $0 \le z \le xy$.

13 $\displaystyle\iiint_S x^2\, dx\, dy\, dz$, where $S: x^2+y^2 \le 4$ and $0 \le z \le 5$.

14 $\displaystyle\iiint_S z\sin(x+y)\, dx\, dy\, dz$, where S is the solid bounded by the planes $z=x-y,\ z=x+y$, $y=x$, and $x=1$.

15 $\displaystyle\iiint_S x^2y^2\, dx\, dy\, dz$, where S is the solid bounded by the planes $z=0,\ z=1,\ x+y=0$, $x+y=1,\ x-y=0$, and $x-y=1$.

16 $\displaystyle\iiint_S x\, dx\, dy\, dz$, where $S: x^2+y^2 \le 1, 0 \le z \le x^2+y^2$.

In problems 17 through 21, use triple integration to find the volume of each solid S.

17 S is the solid bounded above by $z=y$, below by $z=0$, and laterally by the cylinder $y=1-x^2$.

18 S is the solid in the first octant bounded by the cylinder $y=4-x^2$ and the planes $z=x, y=0$, and $z=0$.

19 S is bounded by the surfaces $z=8-x^2-y^2$ and $z=x^2+3y^2$.

20 S is the solid in the first octant bounded by the coordinate planes and the planes $\dfrac{x}{a}+\dfrac{y}{b}+\dfrac{z}{c}=1$ and $z=k$, where a, b, c, and k are positive constants and $k \le c$.

21 S is the solid bounded by the planes $x+z=1, y=x, y=0$, and $z=0$.

22 Express the triple integral $\displaystyle\iiint_S f(x,y,z)\, dx\, dy\, dz$ as a threefold iterated integral in six different ways (that is, with six different orders of integration) if the solid S is given by:
(a) S is bounded by $x+y+z=1$ and the coordinate planes.
(b) S is bounded by $z=1-x^2-y^2$ and $z=0$.

23 Describe a solid S whose volume is given by

$$\int_0^1 \int_0^{1-z} \int_0^y dx\, dy\, dz.$$

24 State the analogs, for triple integrals, of Properties 3 through 8 of double integrals given in Section 2.1.

25 Describe a solid S whose volume is given by $\displaystyle\int_0^2 \int_{x^2}^4 \int_3^6 dz\, dy\, dx$.

26 Suppose that f is a continuous function defined on all of xyz space and that

$$\iiint_S f(x,y,z)\, dx\, dy\, dz = 0$$

for every solid S. Prove that $f(x,y,z) = 0$ for all points (x,y,z).

7 The Triple Integral in Cylindrical and Spherical Coordinates

In Section 5 we found that conversion to polar coordinates can render certain double integrals easier to evaluate. Similarly, as we show in this section, conversion to cylindrical or to spherical coordinates can be advantageous in evaluating triple integrals.

7.1 Conversion to Cylindrical Coordinates

A triple integral can be converted to cylindrical coordinates according to the following procedure.

Figure 1

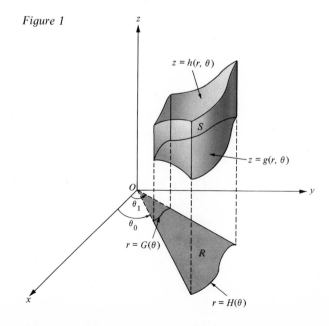

Procedure for Conversion of a Triple Integral to Cylindrical Coordinates

Let θ_0 and θ_1 be constants such that $0 < \theta_1 - \theta_0 \le 2\pi$, and suppose that G and H are continuous functions such that $0 \le G(\theta) \le H(\theta)$ holds for all values of θ in $[\theta_0, \theta_1]$. Let g and h be continuous functions such that $g(r,\theta) \le h(r,\theta)$ holds for all values of r and θ with $\theta_0 \le \theta \le \theta_1$ and $G(\theta) \le r \le H(\theta)$. Denote by S the solid consisting of all points whose cylindrical coordinates (r,θ,z) satisfy the conditions

$$\theta_0 \le \theta \le \theta_1, \quad G(\theta) \le r \le H(\theta), \quad g(r,\theta) \le z \le h(r,\theta)$$

(Figure 1). Then, if f is a continuous function defined for all points (x,y,z) in the solid S,

$$\iiint_S f(x,y,z)\, dx\, dy\, dz$$

$$= \int_{\theta_0}^{\theta_1} \int_{G(\theta)}^{H(\theta)} \int_{g(r,\theta)}^{h(r,\theta)} f(r\cos\theta, r\sin\theta, z)\, r\, dz\, dr\, d\theta.$$

In order to see why this procedure works, denote by R the region in the xy plane consisting of all points whose polar coordinates (r, θ) satisfy $\theta_0 \leq \theta \leq \theta_1$ and $G(\theta) \leq r \leq H(\theta)$. Also, let $z = a(x, y)$ and $z = b(x, y)$ be the equations of the lower and upper boundary surfaces of S, respectively, written in cartesian rather than cylindrical coordinates. Then, by the iteration procedure of Section 6,

$$\iiint_S f(x, y, z) \, dx \, dy \, dz = \iint_R \left[\int_{z=a(x,y)}^{z=b(x,y)} f(x, y, z) \, dz \right] dx \, dy.$$

Now, if the double integral over the region R in the latter equation is converted into polar coordinates as in Section 5, the result is the formula shown in the procedure above (Problem 28).

EXAMPLES 1 Express the integral $\int_0^2 \int_0^{\sqrt{4-x^2}} \int_0^6 \sqrt{x^2 + y^2} \, dz \, dy \, dx$ as an equivalent threefold iterated integral in cylindrical coordinates, and then evaluate the integral obtained.

Figure 2

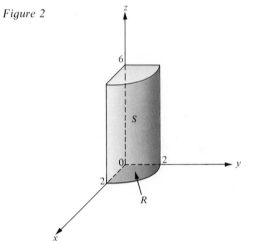

SOLUTION

The given threefold iterated integral is equivalent to the triple integral $\iiint_S \sqrt{x^2 + y^2} \, dx \, dy \, dz$, where

$$S: 0 \leq x \leq 2, \quad 0 \leq y \leq \sqrt{4 - x^2}, \quad 0 \leq z \leq 6$$

(Figure 2). The region

$$R: 0 \leq x \leq 2, \quad 0 \leq y \leq \sqrt{4 - x^2}$$

is the portion of the circular disk $x^2 + y^2 \leq 4$ lying in the first quadrant; hence, its description in polar coordinates is

$$R: 0 \leq \theta \leq \frac{\pi}{2}, \quad 0 \leq r \leq 2.$$

Therefore, using the procedure for conversion of a triple integral to cylindrical coordinates, we have

$$\iiint_S \sqrt{x^2 + y^2} \, dz \, dy \, dx = \int_0^{\pi/2} \int_0^2 \int_0^6 \sqrt{(r \cos \theta)^2 + (r \sin \theta)^2} \, r \, dz \, dr \, d\theta$$

$$= \int_0^{\pi/2} \int_0^2 \int_0^6 \sqrt{r^2} \, r \, dz \, dr \, d\theta = \int_0^{\pi/2} \int_0^2 \left(r^2 z \Big|_0^6 \right) dr \, d\theta$$

$$= \int_0^{\pi/2} \int_0^2 6r^2 \, dr \, d\theta = \int_0^{\pi/2} \left(2r^3 \Big|_0^2 \right) d\theta$$

$$= \int_0^{\pi/2} 16 \, d\theta = 16\theta \Big|_0^{\pi/2} = 8\pi.$$

2 Use cylindrical coordinates to find the volume of a right circular cone whose base radius is a units and whose altitude is h units (Figure 3).

SOLUTION

If we denote the solid cone by S and place its base R on the xy plane and its vertex on the z axis at the point $(0, 0, h)$, then its volume is given by

$$V = \iiint_S dx \, dy \, dz.$$

Figure 3

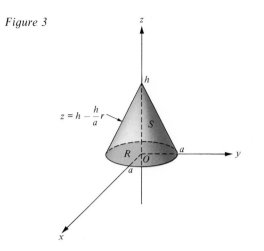

The upper boundary surface of S is formed by revolving the straight line in the yz plane whose equation is $(y/a) + (z/h) = 1$ about the z axis; hence, the equation in cartesian coordinates of the upper boundary surface is

$$\frac{\pm\sqrt{x^2 + y^2}}{a} + \frac{z}{h} = 1 \quad \text{or} \quad z = h \mp \frac{h}{a}\sqrt{x^2 + y^2}$$

(Chapter 15, Section 8.3). Since $z \leq h$ holds for the portion of the surface that actually bounds S, we must use the minus sign in the above equation, so that

$$z = h - \frac{h}{a}\sqrt{x^2 + y^2}.$$

Converting the latter equation to cylindrical coordinates, using $r = \sqrt{x^2 + y^2}$, we obtain

$$z = h - \frac{h}{a}r$$

for the equation of the upper boundary surface of S.

Since the base R of the solid cone S can be described in polar coordinates by $R: 0 \leq \theta \leq 2\pi,\ 0 \leq r \leq a$, it follows that

$$V = \iiint\limits_{S} dx\, dy\, dz = \int_{0}^{2\pi} \int_{0}^{a} \int_{0}^{h-(h/a)r} r\, dz\, dr\, d\theta = \int_{0}^{2\pi} \int_{0}^{a} \left(h - \frac{h}{a}r\right) r\, dr\, d\theta$$

$$= \int_{0}^{2\pi} \left(\frac{ha^2}{2} - \frac{ha^3}{3a}\right) d\theta = 2\pi\left(\frac{ha^2}{2} - \frac{ha^2}{3}\right) = \frac{h}{3}\pi a^2 \text{ cubic units.}$$

3 Find the volume of the solid S bounded by the paraboloid $z = 1 - (x^2 + y^2)$ and the plane $z = 0$ (Figure 4).

SOLUTION
In cylindrical coordinates, the equation of the upper boundary surface of S is $z = 1 - r^2$. This paraboloid cuts the plane $z = 0$ in a circle of radius 1, the inside of which comprises the base R of S. Thus, the desired volume V is given by

Figure 4

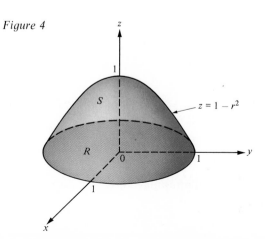

$$V = \iiint\limits_{S} dx\, dy\, dz = \int_{0}^{2\pi} \int_{0}^{1} \int_{0}^{1-r^2} r\, dz\, dr\, d\theta$$

$$= \int_{0}^{2\pi} \int_{0}^{1} r(1 - r^2)\, dr\, d\theta$$

$$= \int_{0}^{2\pi} \left[\left(\frac{r^2}{2} - \frac{r^4}{4}\right)\Big|_{0}^{1}\right] d\theta$$

$$= 2\pi\left(\frac{1}{2} - \frac{1}{4}\right) = \frac{\pi}{2} \text{ cubic units.}$$

As the examples above show, conversion to cylindrical coordinates is especially useful when the solid S is symmetric about the z axis.

7.2 Conversion to Spherical Coordinates

A triple integral can be converted to spherical coordinates according to the following procedure.

Procedure for Conversion of a Triple Integral to Spherical Coordinates

Let θ_0, θ_1, ϕ_0, ϕ_1, ρ_0, and ρ_1 be constants such that $0 < \theta_1 - \theta_0 \le 2\pi$ and $0 \le \rho_0 < \rho_1$. Suppose that the solid S consists of all points whose spherical coordinates (ρ, θ, ϕ) satisfy the conditions

$$\rho_0 \le \rho \le \rho_1, \qquad \theta_0 \le \theta \le \theta_1, \qquad \phi_0 \le \phi \le \phi_1$$

(Figure 5). Then, if f is a continuous function defined for all points (x, y, z) in the solid S,

$$\iiint\limits_S f(x, y, z)\, dx\, dy\, dz$$

$$= \int_{\phi_0}^{\phi_1} \int_{\theta_0}^{\theta_1} \int_{\rho_0}^{\rho_1} f(\rho \sin \phi \cos \theta,\, \rho \sin \phi \sin \theta,\, \rho \cos \phi)\, \rho^2 \sin \phi\, d\rho\, d\theta\, d\phi.$$

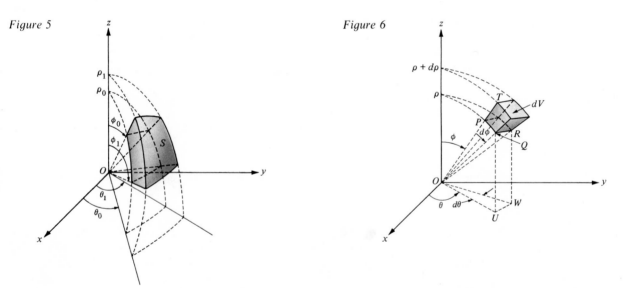

Figure 5

Figure 6

In order to see why this procedure works, consider Figure 6, which shows an "infinitesimal" portion dV of the volume V of the solid S in Figure 5 corresponding to "infinitesimal" changes in ρ, θ, and ϕ of amounts $d\rho$, $d\theta$, and $d\phi$, respectively. Thus, in spherical coordinates,

$$P = (\rho, \theta, \phi),$$
$$Q = (\rho, \theta, \phi + d\phi),$$
$$R = (\rho, \theta + d\theta, \phi + d\phi),$$
$$T = (\rho + d\rho, \theta, \phi).$$

Evidently, dV is virtually the volume of an "infinitesimal" box with dimensions $|\overline{PQ}|$, $|\overline{QR}|$, and $|\overline{PT}|$. Clearly,

$$|\overline{PT}| = d\rho.$$

Since P and Q lie on a circle of radius $|\overline{OP}| = |\overline{OQ}| = \rho$, and since the arc PQ subtends the angle $d\phi$, we have

$$|\overline{PQ}| \approx \rho \, d\phi.$$

Because $d\phi$ is "infinitesimal," angle UOQ is virtually the same as $(\pi/2) - \phi$; hence, considering the right triangle OUQ, we have

$$|\overline{OU}| \approx |\overline{OQ}| \cos\left(\frac{\pi}{2} - \phi\right) = \rho \sin \phi.$$

Note that $|\overline{QR}| = |\overline{UW}|$ and that U and W lie on a circle of radius $|\overline{OU}| \approx \rho \sin \phi$. Thus, since the arc UW subtends the angle $d\theta$, we have

$$|\overline{QR}| = |\overline{UW}| \approx |\overline{OU}| \, d\theta \approx \rho \sin \phi \, d\theta;$$

hence,

$$dV = |\overline{PQ}||\overline{QR}||\overline{PT}| = (\rho \, d\phi)(\rho \sin \phi \, d\theta)(d\rho) = \rho^2 \sin \phi \, d\rho \, d\theta \, d\phi.$$

Thus, whereas in cartesian coordinates the volume of an "infinitesimal" box of dimensions dx, dy, and dz is given by $dV = dx \, dy \, dz$, the analogous "infinitesimal" volume in spherical coordinates is given by $dV = \rho^2 \sin \phi \, d\rho \, d\theta \, d\phi$. These considerations should make the formula for conversion to spherical coordinates appear plausible.

EXAMPLES 1 Express the threefold iterated integral $\displaystyle\int_0^3 \int_0^{\sqrt{9-x^2}} \int_0^{\sqrt{9-x^2-y^2}} (x^2 + y^2 + z^2)^3 \, dz \, dy \, dx$ as an equivalent threefold iterated integral in spherical coordinates, and then evaluate the integral obtained.

Figure 7

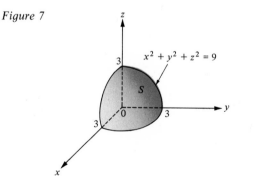

SOLUTION

The given threefold iterated integral is equivalent to the triple integral

$$\iiint_S (x^2 + y^2 + z^2)^3 \, dx \, dy \, dz,$$

where S is bounded by the first octant portion of the sphere $x^2 + y^2 + z^2 = 9$ and the three coordinate planes (Figure 7). Switching to spherical coordinates, and noting that S is described by the conditions

$$S: 0 \le \phi \le \frac{\pi}{2}, \quad 0 \le \theta \le \frac{\pi}{2}, \quad 0 \le \rho \le 3,$$

we have

$$\iiint_S (x^2 + y^2 + z^2)^3 \, dx \, dy \, dz = \int_0^{\pi/2} \int_0^{\pi/2} \int_0^3 (\rho^2)^3 \rho^2 \sin \phi \, d\rho \, d\theta \, d\phi$$

$$= \int_0^{\pi/2} \int_0^{\pi/2} \frac{\rho^9}{9} \left(\sin \phi \, \Big|_0^3\right) d\theta \, d\phi$$

$$= 2187 \int_0^{\pi/2} \left(\theta \sin \phi \, \Big|_0^{\pi/2}\right) d\phi = \frac{2187\pi}{2} \int_0^{\pi/2} \sin \phi \, d\phi$$

$$= \frac{2187\pi}{2} \left(-\cos \phi \, \Big|_0^{\pi/2}\right) = \frac{2187\pi}{2}.$$

2 Use spherical coordinates to find the volume V of a sphere of radius a.

SOLUTION

The solid sphere S of radius a with center at the origin is described in spherical coordinates by the conditions

$$S: 0 \le \phi \le \pi, \quad 0 \le \theta \le 2\pi, \quad 0 \le \rho \le a.$$

Therefore,

$$V = \iiint_S dx\, dy\, dz = \int_0^\pi \int_0^{2\pi} \int_0^a \rho^2 \sin \phi \, d\rho \, d\theta \, d\phi$$

$$= \int_0^\pi \int_0^{2\pi} \left(\frac{\rho^3 \sin \phi}{3} \Big|_0^a \right) d\theta \, d\phi = \int_0^\pi \int_0^{2\pi} \frac{a^3 \sin \phi}{3} \, d\theta \, d\phi$$

$$= \int_0^\pi \frac{2\pi a^3 \sin \phi}{3} \, d\phi = \frac{-2\pi a^3 \cos \phi}{3} \Big|_0^\pi$$

$$= \frac{-2\pi a^3(-1)}{3} - \frac{-2\pi a^3(1)}{3} = \frac{4}{3}\pi a^3 \text{ cubic units.}$$

3 Find the volume V of the solid S in the first octant bounded by the sphere $\rho = 4$, the coordinate planes, the cone $\phi = \pi/6$, and the cone $\phi = \pi/3$ (Figure 8).

SOLUTION

$$V = \iiint_S dx\, dy\, dz = \int_{\pi/6}^{\pi/3} \int_0^{\pi/2} \int_0^4 \rho^2 \sin \phi \, d\rho \, d\theta \, d\phi$$

$$= \int_{\pi/6}^{\pi/3} \int_0^{\pi/2} \frac{4^3}{3} \sin \phi \, d\theta \, d\phi = \int_{\pi/6}^{\pi/3} \frac{64}{3} (\sin \phi)\left(\frac{\pi}{2}\right) d\phi$$

$$= \frac{32\pi}{3} (-\cos \phi) \Big|_{\pi/6}^{\pi/3} = \frac{-32\pi}{3} \left(\frac{\cos \pi}{3} - \frac{\cos \pi}{6} \right)$$

$$= \frac{16\pi}{3} (\sqrt{3} - 1) \text{ cubic units.}$$

Figure 8

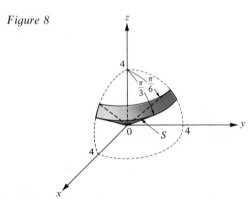

Problem Set 7

In problems 1 through 4, evaluate each threefold iterated integral.

1 $\displaystyle \int_0^\pi \int_0^{3\cos\theta} \int_0^3 r \sin \theta \, dz \, dr \, d\theta$

2 $\displaystyle \int_0^{\pi/2} \int_0^{\sin\theta} \int_0^{r^2} r \cos \theta \, dz \, dr \, d\theta$

3 $\displaystyle \int_0^{\pi/6} \int_0^{2\pi} \int_0^4 \rho^2 \sin \phi \, d\rho \, d\theta \, d\phi$

4 $\displaystyle \int_0^{\pi/2} \int_{\pi/4}^{\phi} \int_0^{2\csc\theta} \rho^3 \sin^2 \theta \sin \phi \, d\rho \, d\theta \, d\phi$

In problems 5 and 6, (a) rewrite each threefold iterated integral in cartesian coordinates as an equivalent triple integral over an appropriate solid S and sketch the solid, (b) rewrite the triple integral obtained in (a) as an equivalent threefold iterated integral in cylindrical coordinates, and (c) evaluate the integral obtained in part (b).

5 $\displaystyle \int_0^5 \int_0^{\sqrt{25-x^2}} \int_0^6 \frac{dz\, dy\, dx}{\sqrt{x^2+y^2}}$

6 $\displaystyle \int_0^2 \int_0^{\sqrt{4-x^2}} \int_0^{\frac{x^2+y^2}{2}} \frac{z\, dz\, dy\, dx}{\sqrt{x^2+y^2}}$

In problems 7 and 8, (a) rewrite each threefold iterated integral in cartesian coordinates as an equivalent triple integral over an appropriate solid S and sketch the solid, (b) rewrite the triple integral obtained in part (a) as an equivalent threefold iterated integral in spherical coordinates, and (c) evaluate the integral obtained in part (b).

7 $\int_0^1 \int_0^{\sqrt{1-x^2}} \int_0^{\sqrt{1-x^2-y^2}} \dfrac{dz\,dy\,dx}{1+x^2+y^2+z^2}$

8 $\int_0^3 \int_0^{\sqrt{9-x^2}} \int_0^{\sqrt{9-x^2-y^2}} xz\,dz\,dy\,dx$

In problems 9 through 12, convert to cylindrical coordinates and evaluate the integral.

9 $\iiint_S \sqrt{x^2+y^2}\,dx\,dy\,dz$, where S is the solid in the first octant bounded by the coordinate planes, the plane $z=4$, and the cylinder $x^2+y^2=25$.

10 $\iiint_S (x^2+y^2)^{3/2}\,dx\,dy\,dz$, where S is the solid bounded above by the paraboloid of revolution $z=\frac{1}{2}(x^2+y^2)$, below by the xy plane, and laterally by the cylinder $x^2+y^2=4$.

11 $\iiint_S \dfrac{dx\,dy\,dz}{\sqrt{x^2+y^2}}$, where S is the solid bounded above by the plane $z=4$, below by the plane $z=1$, and laterally by the cylinder $x^2+y^2=16$.

12 $\iiint_S dx\,dy\,dz$, where S is the cylindrical shell of height $h>0$ with base on the xy plane between the cylinders $x^2+y^2=a^2$ and $x^2+y^2=b^2$, where $b>a>0$.

In problems 13 through 16, convert to spherical coordinates and evaluate the integral.

13 $\iiint_S \sqrt{x^2+y^2+z^2}\,dx\,dy\,dz$, where S is the region bounded by the sphere of radius 3 with center at the origin.

14 $\iiint_S dx\,dy\,dz$, where S is the spherical shell determined by the condition

$$0 < a \le x^2+y^2+z^2 \le b.$$

15 $\iiint_S (x^2+y^2+z^2)^{3/2}\,dx\,dy\,dz$, where S is the solid in the first octant bounded by the sphere $x^2+y^2+z^2=25$, the cone $z=\sqrt{x^2+y^2}$, and the cone $z=2\sqrt{x^2+y^2}$.

16 $\iiint_S \sin\sqrt{x^2+y^2+z^2}\,dx\,dy\,dz$, where S is the solid bounded above by the sphere $x^2+y^2+z^2=49$ and below by the cone $z=\sqrt{x^2+y^2}$.

In problems 17 through 22, use an appropriate integration in cylindrical coordinates to find the volume V of each solid S.

17 S is the solid in the first octant bounded by the cylinder $x^2+y^2=9$, the plane $z=y$, and the coordinate planes.

18 S is the solid bounded by the plane $z=x$ and the paraboloid of revolution $z=x^2+y^2$.

19 S is the solid in the first octant bounded by the coordinate planes and the surfaces $x^2+y^2=z$ and $x^2+y^2=2y$.

20 S is the solid bounded above by the cone $z^2=x^2+y^2$, below by the plane $z=0$, and laterally by the cylinder $x^2+y^2=4x$.

21 S is the solid bounded above by the sphere $x^2+y^2+z^2=8$ and below by the paraboloid $x^2+y^2=2z$.

22 S is the solid bounded above by the sphere $x^2+y^2+(z-\frac{1}{2})^2=\frac{1}{4}$ and below by the cone $z=\sqrt{x^2+y^2}$.

In problems 23 through 26, use an appropriate integration in spherical coordinates to find the volume V of each solid S.

23 $S: 0 \le \phi \le \pi/4, \ \pi/6 \le \theta \le \pi/3, \ 2 \le \rho \le 4.$

24 S is bounded laterally by the sphere $x^2 + y^2 + z^2 = a^2$ and lies between the lower and upper nappes of the right circular cone $z^2 = b^2(x^2 + y^2)$, where a and b are positive constants.

25 S is bounded above by the sphere $x^2 + y^2 + z^2 = 9$ and below by the cone $z = 2\sqrt{x^2 + y^2}$.

26 S is the solid consisting of all points whose spherical coordinates satisfy $0 \le \theta \le 2\pi$, $0 \le \phi \le \pi$, and $0 \le \rho \le 2(1 - \cos \phi)$.

27 Find the volume of each of the two parts into which a solid sphere of radius $a > 0$ is cut by a plane passing k units from its center, where $0 < k < a$.

28 Complete the argument making the procedure for conversion of a triple integral to cylindrical coordinates plausible by actually iterating the double integral

$$\iint\limits_{R} \left[\int_{z=a(x,y)}^{z=b(x,y)} f(x, y, z)\, dz \right] dx\, dy$$

in polar coordinates as suggested in the text. What assumptions do you need to make?

29 Use the method of Example 3 in Section 7.2 to work Problem 20 in Problem Set 3 of Chapter 7.

30 Suppose that S is a solid consisting of all points whose spherical coordinates satisfy the conditions $\theta_0 \le \theta \le \theta_1$, $G(\theta) \le \phi \le H(\theta)$, and $g(\phi, \theta) \le \rho \le h(\phi, \theta)$, where G, H, g, and h are continuous functions. If f is a continuous function defined on S, find a threefold iterated integral in spherical coordinates equivalent to $\iiint\limits_{S} f(x, y, z)\, dx\, dy\, dz$.

31 The "hypervolume" of a "four-dimensional hypersphere" of radius a is given by $2\iiint\limits_{S} \sqrt{a^2 - x^2 - y^2 - z^2}\, dx\, dy\, dz$, where S is the ordinary solid sphere of radius a in xyz space. Evaluate this "hypervolume."

8   Elementary Applications of Triple Integrals

In Section 4 we applied the double integral to the solution of problems involving density, moments, centers of mass, centroids, and moments of inertia. In this section we present similar applications of triple integrals.

8.1 Density and Triple Integrals

Consider a quantity, such as mass or electric charge, which is distributed in a continuous, but perhaps nonuniform, manner over a portion of three-dimensional xyz space. By analogy with the two-dimensional case considered in Section 4.3, a function σ of three variables is called a *density function* for this three-dimensional distribution if, for every solid three-dimensional region S, the triple integral

$$\iiint\limits_{S} \sigma(x, y, z)\, dx\, dy\, dz$$

exists and gives the amount of the quantity contained in the region S.

EXAMPLE   Mass is distributed throughout the interior of a sphere of radius one meter in such a way that the density at a point P is inversely proportional to the distance of P from the center, and at the surface of the sphere the density is 1 gram per cubic centimeter. Find the total mass in kilograms inside the sphere.

SOLUTION

We take S to be the solid spherical ball of radius 100 centimeters with center at the origin. The density at the point (x, y, z) is given by

$$\sigma(x, y, z) = \frac{k}{\sqrt{x^2 + y^2 + z^2}} \text{ grams/cm}^3,$$

where k is the constant of proportionality. On the surface of the sphere, $\sigma(x, y, z) = 1$ gram/cm^3 and $\sqrt{x^2 + y^2 + z^2} = 100$; hence, $k = 100$ and

$$\sigma(x, y, z) = \frac{100}{\sqrt{x^2 + y^2 + z^2}}.$$

The total mass in S is given by

$$m = \iiint_S \sigma(x, y, z)\, dx\, dy\, dz = \iiint_S \frac{100\, dx\, dy\, dz}{\sqrt{x^2 + y^2 + z^2}}.$$

Using spherical coordinates, we have

$$m = \int_0^\pi \int_0^{2\pi} \int_0^{100} \frac{100}{\rho}\, \rho^2 \sin\phi\, d\rho\, d\theta\, d\phi = \int_0^\pi \int_0^{2\pi} \int_0^{100} 100\, \rho \sin\phi\, d\rho\, d\theta\, d\phi$$

$$= \int_0^\pi \int_0^{2\pi} \left(\frac{100\rho^2}{2} \sin\phi \, \bigg|_0^{100} \right) d\theta\, d\phi = \frac{(100)^3}{2} \int_0^\pi \int_0^{2\pi} \sin\phi\, d\theta\, d\phi$$

$$= 2\pi \frac{(100)^3}{2} \int_0^\pi \sin\phi\, d\phi = (100)^3 \pi \left(-\cos\phi \, \bigg|_0^\pi \right) = 2\pi(100)^3 \text{ grams}$$

$$= \frac{2\pi(100)^3}{1000} \text{ kilograms} = 2000\pi \text{ kilograms}.$$

The value of the density $\sigma(a, b, c)$ of a three-dimensional distribution at the point (a, b, c) can be interpreted as the limiting value of the amount of the quantity per unit volume in a small three-dimensional region ΔS around the point (a, b, c) as ΔS "shrinks to zero" in the sense that the maximum distance between any two points in ΔS approaches zero (Problem 29). It follows that the "infinitesimal" amount dq of the quantity contained within an "infinitesimal" three-dimensional region of volume dV around the point (x, y, z) is given by

$$dq = \sigma(x, y, z)\, dV.$$

8.2   Moments and Center of Mass

Consider a continuous distribution of mass over a solid S with mass density function σ. The total mass of S is given by

$$m = \iiint_S \sigma(x, y, z)\, dx\, dy\, dz.$$

By analogy with the definitions given for two-dimensional distributions in Section 4.4, we define the *moments* of the distribution *about the xy plane, the xz plane, and the yz plane* by

$$M_{xy} = \iiint_S z\sigma(x,y,z)\,dx\,dy\,dz,$$

$$M_{xz} = \iiint_S y\sigma(x,y,z)\,dx\,dy\,dz, \quad \text{and}$$

$$M_{yz} = \iiint_S x\sigma(x,y,z)\,dx\,dy\,dz,$$

respectively. Similarly, the point

$$(\bar{x}, \bar{y}, \bar{z}) = \left(\frac{M_{yz}}{m}, \frac{M_{xz}}{m}, \frac{M_{xy}}{m}\right)$$

is called the *center of mass* of the distribution.

EXAMPLE   Find the mass m and the center of mass $(\bar{x}, \bar{y}, \bar{z})$ of a solid S in the first octant bounded by the coordinate planes, the plane $z = 2$, and the cylinder $x^2 + y^2 = 9$ if its density at a point P is proportional to the distance of P from the z axis.

SOLUTION
The density at the point (x,y,z) is given by $\sigma(x,y,z) = k\sqrt{x^2 + y^2}$, where k is the constant of proportionality. We have

$$m = \iiint_S \sigma(x,y,z)\,dx\,dy\,dz = \iiint_S k\sqrt{x^2 + y^2}\,dx\,dy\,dz.$$

Using cylindrical coordinates, we have

$$m = \int_0^{\pi/2}\int_0^3\int_0^2 (kr)r\,dz\,dr\,d\theta = \int_0^{\pi/2}\int_0^3 2kr^2\,dr\,d\theta$$

$$= \int_0^{\pi/2}\left(\frac{2kr^3}{3}\Big|_0^3\right)d\theta = 18k\int_0^{\pi/2}d\theta = 9\pi k \text{ mass units.}$$

Here,

$$M_{xy} = \iiint_S z\sigma(x,y,z)\,dx\,dy\,dz = \iiint_S zk\sqrt{x^2 + y^2}\,dx\,dy\,dz$$

$$= \int_0^{\pi/2}\int_0^3\int_0^2 (zkr)r\,dz\,dr\,d\theta = \int_0^{\pi/2}\int_0^3\left(\frac{z^2kr^2}{2}\Big|_0^2\right)dr\,d\theta$$

$$= \int_0^{\pi/2}\int_0^3 2kr^2\,dr\,d\theta = 9\pi k, \text{ just as above.}$$

Also,

$$M_{xz} = \iiint_S y\sigma(x,y,z)\,dx\,dy\,dz = \iiint_S yk\sqrt{x^2 + y^2}\,dx\,dy\,dz$$

$$= \int_0^{\pi/2}\int_0^3\int_0^2 [(r\sin\theta)kr]r\,dz\,dr\,d\theta = \int_0^{\pi/2}\int_0^3 2kr^3\sin\theta\,dr\,d\theta$$

$$= \int_0^{\pi/2}\left(\frac{kr^4\sin\theta}{2}\Big|_0^3\right)d\theta = \frac{81k}{2}\int_0^{\pi/2}\sin\theta\,d\theta = \frac{81k}{2}(-\cos\theta)\Big|_0^{\pi/2} = \frac{81k}{2}.$$

Similarly,

$$M_{yz} = \iiint\limits_{S} x\sigma(x, y, z)\, dx\, dy\, dz = \iiint\limits_{S} xk\sqrt{x^2 + y^2}\, dx\, dy\, dz$$

$$= \int_{0}^{\pi/2} \int_{0}^{3} \int_{0}^{2} [(r\cos\theta)kr]r\, dz\, dr\, d\theta = \frac{81k}{2}.$$

Therefore,

$$\bar{x} = \frac{M_{yz}}{m} = \frac{81k/2}{9\pi k} = \frac{9}{2\pi},$$

$$\bar{y} = \frac{M_{xz}}{m} = \frac{81k/2}{9\pi k} = \frac{9}{2\pi}, \quad \text{and}$$

$$\bar{z} = \frac{M_{xy}}{m} = \frac{9\pi k}{9\pi k} = 1.$$

Hence, the center of mass of the distribution is

$$(\bar{x}, \bar{y}, \bar{z}) = \left(\frac{9}{2\pi}, \frac{9}{2\pi}, 1 \right).$$

8.3 Centroids

The center of mass of a uniform distribution (that is, a distribution with constant mass density) over a solid S is called the *centroid* of S. Since the volume of S is given by

$$V = \iiint\limits_{S} dx\, dy\, dz,$$

it is easy to derive the following formulas for the centroid $(\bar{x}, \bar{y}, \bar{z})$ of S (Problem 31):

$$\bar{x} = \frac{1}{V} \iiint\limits_{S} x\, dx\, dy\, dz, \qquad \bar{y} = \frac{1}{V} \iiint\limits_{S} y\, dx\, dy\, dz, \qquad \bar{z} = \frac{1}{V} \iiint\limits_{S} z\, dx\, dy\, dz.$$

EXAMPLE   Find the centroid $(\bar{x}, \bar{y}, \bar{z})$ of the solid S in the first octant bounded by the coordinate planes and the plane $\dfrac{x}{a} + \dfrac{y}{b} + \dfrac{z}{c} = 1$, where a, b, and c are positive constants.

SOLUTION
Here,

$$V = \iiint\limits_{S} dx\, dy\, dz = \int_{0}^{a} \int_{0}^{b-(b/a)x} \int_{0}^{c-(c/a)x-(c/b)y} dz\, dy\, dx$$

$$= \int_{0}^{a} \int_{0}^{b-(b/a)x} \left(c - \frac{c}{a}x - \frac{c}{b}y \right) dy\, dx = \int_{0}^{a} \int_{0}^{b-(b/a)x} \left[\left(c - \frac{c}{a}x \right) - \frac{c}{b}y \right] dy\, dx$$

$$= \int_{0}^{a} \left[\left(c - \frac{c}{a}x \right)\left(b - \frac{b}{a}x \right) - \frac{c}{2b}\left(b - \frac{b}{a}x \right)^2 \right] dx = \int_{0}^{a} \left(\frac{bc}{2} - \frac{bc}{a}x + \frac{bc}{2a^2}x^2 \right) dx$$

$$= \left(\frac{bc}{2}x - \frac{bc}{2a}x^2 + \frac{bc}{6a^2}x^3 \right)\Bigg|_{0}^{a} = \frac{abc}{6}.$$

Therefore,

$$\bar{x} = \frac{1}{V}\iiint_{S} x\, dx\, dy\, dz = \frac{6}{abc}\int_{0}^{a}\int_{0}^{b-(b/a)x}\int_{0}^{c-(c/a)x-(c/b)y} x\, dz\, dy\, dx$$

$$= \frac{6}{abc}\int_{0}^{a}\int_{0}^{b-(b/a)x}\left(cx - \frac{c}{a}x^2 - \frac{c}{b}xy\right) dy\, dx = \frac{6}{abc}\int_{0}^{a}\left(\frac{bc}{2}x - \frac{bc}{a}x^2 + \frac{bc}{2a^2}x^3\right) dx$$

$$= \frac{6}{abc}\left(\frac{bcx^2}{4} - \frac{bcx^3}{3a} + \frac{bcx^4}{8a^2}\right)\Bigg|_{0}^{a} = \frac{a}{4}.$$

Analogously,

$$\bar{y} = \frac{1}{V}\iiint_{S} y\, dx\, dy\, dz = \frac{6}{abc}\int_{0}^{a}\int_{0}^{b-(b/a)x}\int_{0}^{c-(c/a)x-(c/b)y} y\, dz\, dy\, dx = \frac{b}{4} \quad \text{and}$$

$$\bar{z} = \frac{1}{V}\iiint_{S} z\, dx\, dy\, dz = \frac{6}{abc}\int_{0}^{a}\int_{0}^{b-(b/a)x}\int_{0}^{c-(c/a)x-(c/b)y} z\, dz\, dy\, dx = \frac{c}{4}.$$

Hence, the centroid is $(\bar{x}, \bar{y}, \bar{z}) = \left(\dfrac{a}{4}, \dfrac{b}{4}, \dfrac{c}{4}\right)$.

8.4 Moment of Inertia

Again consider a continuous distribution of mass over a solid S with mass density function σ. Let A be a fixed straight line in space and consider an "infinitesimal" portion dm of the mass m of S in the "infinitesimal" box of dimensions dx, dy, and dz around the point (x, y, z) (Figure 1). Here,

Figure 1

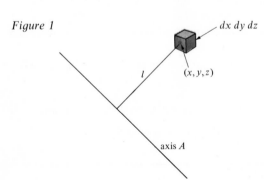

$$dm = \sigma(x, y, z)\, dx\, dy\, dz$$

and the moment of inertia dI_A of dm about the axis A is given by

$$dI_A = l^2\, dm = l^2\sigma(x, y, z)\, dx\, dy\, dz,$$

where l is the distance from (x, y, z) to the line A. Integrating over the solid S, we obtain the formula

$$I_A = \iiint_{S} l^2\sigma(x, y, z)\, dx\, dy\, dz$$

for the moment of inertia of S about the axis A.

In particular, the moments of inertia I_x, I_y, and I_z of S about the x, y, and z axes are given by

$$I_x = \iiint_{S} (y^2 + z^2)\sigma(x, y, z)\, dx\, dy\, dz,$$

$$I_y = \iiint_{S} (x^2 + z^2)\sigma(x, y, z)\, dx\, dy\, dz, \quad \text{and}$$

$$I_z = \iiint_{S} (x^2 + y^2)\sigma(x, y, z)\, dx\, dy\, dz.$$

EXAMPLES **1** Find the moment of inertia with respect to the central axis of a homogeneous solid right circular cylinder S of radius a, height h, and total mass m.

SOLUTION

We position S with its central axis along the positive z axis and its lower base on the xy plane. The volume of S is given by

$$V = \iiint\limits_S dx\, dy\, dz = \pi a^2 h.$$

Therefore, since the total mass m is distributed uniformly,

$$\sigma(x, y, z) = \frac{m}{\pi a^2 h} \text{ mass units per unit volume}$$

for every point (x, y, z) in S. It follows that

$$I_z = \iiint\limits_S (x^2 + y^2)\frac{m}{\pi a^2 h}\, dx\, dy\, dz = \frac{m}{\pi a^2 h}\iiint\limits_S (x^2 + y^2)\, dx\, dy\, dz.$$

Switching to cylindrical coordinates, we have

$$I_z = \frac{m}{\pi a^2 h}\int_0^{2\pi}\int_0^a\int_0^h (r^2)r\, dz\, dr\, d\theta$$

$$= \frac{m}{\pi a^2 h}\int_0^{2\pi}\int_0^a hr^3\, dr\, d\theta = \frac{m}{\pi a^2}\int_0^{2\pi}\frac{a^4}{4}\, d\theta$$

$$= \frac{m}{\pi a^2}\frac{2\pi a^4}{4} = \frac{ma^2}{2}.$$

2 Find the moment of inertia I_z of the spherical shell S: $10 \le \sqrt{x^2 + y^2 + z^2} \le 11$ if its mass density at a point P is inversely proportional to the square of the distance from P to the origin.

SOLUTION

Here, $\sigma(x, y, z) = \dfrac{k}{x^2 + y^2 + z^2}$, where k is the constant of proportionality; hence,

$$I_z = \iiint\limits_S (x^2 + y^2)\frac{k}{x^2 + y^2 + z^2}\, dx\, dy\, dz.$$

Switching to spherical coordinates, we have

$$I_z = \int_0^{\pi}\int_0^{2\pi}\int_{10}^{11}\left(\rho^2 \sin^2 \phi\, \frac{k}{\rho^2}\right)\rho^2 \sin \phi\, d\rho\, d\theta\, d\phi$$

$$= \int_0^{\pi}\int_0^{2\pi}\left(k\frac{\rho^3 \sin^3 \phi}{3}\Big|_{10}^{11}\right) d\theta\, d\phi = \frac{331k}{3}\int_0^{\pi}\int_0^{2\pi}\sin^3 \phi\, d\theta\, d\phi$$

$$= \frac{662\pi k}{3}\int_0^{\pi}\sin^3 \phi\, d\phi = \frac{662\pi k}{3}\int_0^{\pi}\sin \phi(1 - \cos^2 \phi)\, d\phi$$

$$= \frac{662\pi k}{3}\left(-\cos \phi + \frac{1}{3}\cos^3 \phi\right)\Big|_0^{\pi} = \frac{2648\pi k}{9}.$$

3 The spherical shell S in Example 2 has a total mass of 18 grams. Assuming that distances are measured in centimeters, find its moment of inertia I_z.

SOLUTION
The total mass $m = 18$ grams is given by

$$18 = m = \iiint_{S} \sigma(x, y, z) \, dx \, dy \, dz = \iiint_{S} \frac{k}{x^2 + y^2 + z^2} \, dx \, dy \, dz$$

$$= k \int_{0}^{\pi} \int_{0}^{2\pi} \int_{10}^{11} \frac{1}{\rho^2} \rho^2 \sin \phi \, d\rho \, d\theta \, d\phi$$

$$= k \int_{0}^{\pi} \int_{0}^{2\pi} \sin \phi \, d\theta \, d\phi = 2\pi k \int_{0}^{\pi} \sin \phi \, d\phi$$

$$= 2\pi k (-\cos \phi) \Big|_{0}^{\pi} = 4\pi k.$$

Therefore, $k = 18/(4\pi) = 9/(2\pi)$. Hence, by the result of Example 2,

$$I_z = \frac{2648\pi k}{9} = \frac{2648\pi}{9} \cdot \frac{9}{2\pi} = 1324 \text{ grams cm}^2.$$

Problem Set 8

In problems 1 through 4, find the mass m of the given solid S with the indicated mass density function σ.

1 $S: x^2 + y^2 \le 16, 0 \le z \le 10; \; \sigma(x, y, z) = \dfrac{\sqrt{x^2 + y^2}}{40}$

2 $S: 0 \le x \le 1, 0 \le y \le 1 - x^2, 0 \le z \le x^2 + y^2; \; \sigma(x, y, z) = x + 1$

3 $S: x^2 + y^2 \le 9, 0 \le z \le 9 - x^2 - y^2; \; \sigma(x, y, z) = z$

4 $S: x^2 + y^2 \le a^2, 0 \le z \le h - \dfrac{h}{a}\sqrt{x^2 + y^2}; \; h > 0, \; a > 0; \; \sigma(x, y, z) = \sqrt{x^2 + y^2}$

5 One kilogram of mass is distributed throughout a spherical ball in such a way that the density at a point is proportional to the distance of the point from the surface of the sphere. If the sphere is 10 centimeters in diameter, find the density in grams per cubic centimeter at the center of the sphere.

6 One coulomb of electrical charge is distributed throughout a solid right circular cone of height 10 centimeters and base diameter 4 centimeters in such a way that the charge density is proportional to the square of the distance from the vertex of the cone. Find the charge density at the center of the base of the cone.

In problems 7 through 10, find the coordinates $(\bar{x}, \bar{y}, \bar{z})$ of the center of mass of the given solid S with the indicated mass density function.

7 $S: 0 \le x \le 1, 0 \le y \le 1, 0 \le z \le 1; \; \sigma(x, y, z) = 3 - x - y - z$

8 $S: x^2 + y^2 \le a^2, \dfrac{h}{a}\sqrt{x^2 + y^2} \le z \le h; \; h > 0, \; a > 0; \; \sigma(x, y, z) = x^2 + y^2 + z^2$

9 $S: x^2 + y^2 + z^2 \le 100, \; \sigma(x, y, z) = (x^2 + y^2 + z^2)^n; \; n$ a positive constant.

10 $S: x^2 + y^2 + z^2 \le a^2, x \ge 0, y \ge 0, z \ge 0; \; a > 0. \; \sigma(x, y, z) = (x^2 + y^2 + z^2)^n; \; n$ a positive constant.

In problems 11 through 14, find the coordinates of the centroid of each solid S.

11 $S: x^2 + y^2 \le 1, x \ge 0, y \ge 0, 0 \le z \le xy$ **12** $S: x \ge 0, y \ge 0, 0 \le z \le 5 - x - y$

13 $S: a^2 \leq x^2 + y^2 + z^2 \leq b^2$, $z \geq 0$; where a and b are constants and $0 < a < b$.

14 $S: x^2 + y^2 \leq a^2$, $\dfrac{h}{a}\sqrt{x^2 + y^2} \leq z \leq h$; where h and a are positive constants.

In problems 15 through 18, find the designated moment for the given solid S with the indicated density function σ.

15 M_{xy}, $S: 0 \leq x \leq 1, 0 \leq y \leq 1 - x, 0 \leq z \leq 1$; $\sigma(x, y, z) = 3xy$

16 M_{xz}; $S: 0 \leq x \leq 1, 0 \leq y \leq 1, 0 \leq z \leq xy$; $\sigma(x, y, z) = 2(x + y)$

17 M_{yz}; $S: 0 \leq y \leq 4, 0 \leq x \leq \dfrac{y}{2}, 0 \leq z \leq \sqrt{8x}$; $\sigma(x, y, z) = 2$

18 M_{xz}; S is the solid in the first octant bounded by $x + z = 1$, $x = y$, $x = 0$, $y = 0$, and $z = 0$; $\sigma(x, y, z) = 5y$.

In problems 19 through 22, find the designated moment of inertia for the given solid S with the indicated density function σ.

19 I_z; $S: 0 \leq x \leq 1, 0 \leq y \leq 1 - x, 0 \leq z \leq 5$; $\sigma(x, y, z) = 3$.

20 I_z; S is the first octant solid bounded by $x + z = 1$, $y + z = 1$, and the coordinate planes; $\sigma(x, y, z) = z$.

21 I_x; $S: 0 \leq x \leq 1, 0 \leq y \leq 1, 0 \leq z \leq 1$; $\sigma(x, y, z) = 8z$

22 I_y; S is the first octant solid bounded by $x^2 + z^2 = 1$, $y = x$, $y = 0$, $z = 0$; $\sigma(x, y, z) = 2z$.

23 Find the moment of inertia about the central axis of a homogeneous right circular cylindrical shell of total mass m with inner radius a, outer radius b, and height h.

24 Find the moment of inertia of a homogeneous solid right circular cylinder about an axis perpendicular to the central axis and passing through the center of gravity of the solid cylinder. Assume that the height of the cylinder is h, that its radius is a, and that its mass is m.

25 Find the moment of inertia of a homogeneous solid spherical ball of radius a about an axis through the center of the ball. Let the ball have total mass m.

26 Find the moment of inertia of a homogeneous solid rectangular parallelepiped with sides a, b, and c about an axis passing through the center of gravity and parallel to the side of length c. Assume that the total mass is m.

27 Find the moment of inertia of a homogeneous spherical shell of inner radius a and outer radius b about an axis passing through its center. Assume that the total mass of the shell is m.

28 Let A be an axis passing through the center of gravity of a homogeneous solid S of total mass m and let B be an axis parallel to A. If h denotes the distance between the axes A and B, prove that

$$I_B = I_A + mh^2.$$

This result is known as the *parallel axis theorem*.

29 Give an informal argument, similar to the one given in Section 4.3 for two-dimensional distributions, to show that the density of a distribution at a point is the limit of the amount of the quantity per unit volume in a small region about the point as the region "shrinks to zero."

30 Combine the results of problems 24 and 28 to determine the moment of inertia of a homogeneous solid right circular cylinder about an axis passing through the center of one of the bases and perpendicular to the central axis.

31 Derive the formulas given in Section 8.3 for the centroid of a solid S.

32 The temperature of a hollow spherical shell with inner radius a and outer radius b is inversely proportional to the distance from the center and has the value $T_0 > 0$ on the inner surface. The quantity of heat required to raise any portion of the shell from one uniform temperature to another is proportional jointly to the volume of this portion and to the rise in temperature. Assume that C units of heat are required to raise the temperature of 1 cubic unit of the shell by 1 degree. How much heat will the shell give out if it is cooled to a uniform temperature of $0°$?

9   Line Integrals and Green's Theorem

In Section 2 we introduced double integrals over two-dimensional regions R, while in Section 6 we introduced triple integrals over three-dimensional solids S. In the present section we round out the picture by considering integrals over one-dimensional curves. It is customary to refer to a curve over which an integral is to be taken as a *line* (not necessarily a straight line!), and to call an integral over such a curve a *line integral*. Later in this section we present an important theorem, called *Green's theorem*, which relates line integrals in the plane and double integrals.

9.1 Line Integrals

In more advanced courses, line integrals are defined in terms of limits of Riemann sums in a manner similar to the definition of the definite integral. For our purposes, it is simpler to adopt the following definition, which is equivalent to the official definition for the kinds of curves and functions that we consider.

DEFINITION 1   **Line Integral in the Plane**
Let C be a curve in the xy plane with the parametric equations

$$C: \begin{cases} x = f(t) \\ y = g(t), \end{cases} \quad a \le t \le b,$$

where f and g have continuous first derivatives. Suppose that P and Q are continuous functions of two variables whose domains contain the curve C. Then the *line integral* $\int_C P(x, y)\, dx + Q(x, y)\, dy$ is defined by

$$\int_C P(x, y)\, dx + Q(x, y)\, dy = \int_a^b [P(f(t), g(t))f'(t)\, dt + Q(f(t), g(t))g'(t)\, dt].$$

Thus, to evaluate the line integral $\int_C P(x, y)\, dx + Q(x, y)\, dy$, we simply make the substitutions $x = f(t)$, $dx = f'(t)\, dt$, $y = g(t)$, and $dy = g'(t)\, dt$, and then integrate from $t = a$ to $t = b$.

EXAMPLE   Evaluate the line integral $\int_C (x^2 + 3y)\, dx + (y^2 + 2x)\, dy$ if

$$C: \begin{cases} x = t \\ y = t^2 + 1, \end{cases} \quad 0 \le t \le 1.$$

SOLUTION

Making the substitutions $x = t$, $dx = dt$, $y = t^2 + 1$, and $dy = 2t\,dt$, we have

$$\int_C (x^2 + 3y)\,dx + (y^2 + 2x)\,dy = \int_0^1 [t^2 + 3(t^2 + 1)]\,dt + [(t^2 + 1)^2 + 2t]2t\,dt$$

$$= \int_0^1 [(4t^2 + 3) + (t^4 + 2t^2 + 2t + 1)(2t)]\,dt$$

$$= \int_0^1 (2t^5 + 4t^3 + 8t^2 + 2t + 3)\,dt$$

$$= \left(\frac{t^6}{3} + t^4 + \frac{8}{3}t^3 + t^2 + 3t\right)\Bigg|_0^1 = 8.$$

Although Definition 1 makes it appear that the line integral $\int_C P\,dx + Q\,dy$ depends on the choice of the parameter t, it can be shown that the value of the line integral is unaffected by the choice of parameter, as long as the direction along the geometric curve C corresponding to increasing values of the parameter is kept the same. Thus, line integrals are actually taken over *directed* or *oriented* curves; that is, over curves for which one of the two endpoints is understood to be the starting or initial point, while the other endpoint is the ending or terminal point. In dealing with line integrals, the curve obtained from a given curve C by reversing its direction is usually denoted by $-C$ (Figure 1).

Figure 1

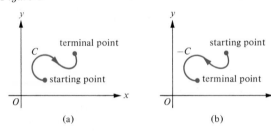

(a)   (b)

EXAMPLES   Evaluate the given line integral.

1 $\int_C (x + y)\,dx + (y - x)\,dy$ if C is the straight-line segment from $(1, 1)$ to $(4, 2)$.

SOLUTION

Using the methods of Section 5 of Chapter 14, we obtain the scalar parametric equations

$$C: \begin{cases} x = 1 + 3t \\ y = 1 + t, \end{cases} \quad 0 \leq t \leq 1$$

for the line segment from $(1, 1)$ to $(4, 2)$. Notice that $(x, y) = (1, 1)$ when $t = 0$, and $(x, y) = (4, 2)$ when $t = 1$, so that $(1, 1)$ is the initial point and $(4, 2)$ is the terminal point of C. Making the substitutions $x = 1 + 3t$, $dx = 3\,dt$, $y = 1 + t$, and $dy = dt$, we have

$$\int_C (x + y)\,dx + (y - x)\,dy = \int_0^1 [(1 + 3t) + (1 + t)]3\,dt + [(1 + t) - (1 + 3t)]\,dt$$

$$= \int_0^1 (10t + 6)\,dt = (5t^2 + 6t)\Bigg|_0^1 = 11.$$

2 $\int_C (x + 2y)\,dy$ if C is the arc of the parabola $x = y^2$ from $(1, -1)$ to $(9, -3)$.

SOLUTION

Here the line integral has the form $\int_C P(x, y)\, dx + Q(x, y)\, dy$ with $P(x, y) = 0$ and $Q(x, y) = x + 2y$. The arc of the parabola is described parametrically by

$$C: \begin{cases} x = t^2 \\ y = -t, \end{cases} \quad 1 \le t \le 3$$

in such a way that, as t goes from 1 to 3, (x, y) goes from $(1, -1)$ to $(9, -3)$. Making the substitutions $x = t^2$, $y = -t$, and $dy = -dt$, we have

$$\int_C (x + 2y)\, dy = \int_1^3 (t^2 - 2t)(-dt) = \int_1^3 (2t - t^2)\, dt = \left(t^2 - \frac{t^3}{3} \right) \bigg|_1^3 = -\frac{2}{3}.$$

Definition 1 extends in an obvious way to line integrals over curves in three-dimensional space. Indeed, suppose that C is such a curve, say with scalar parametric equations

$$C: \begin{cases} x = f(t) \\ y = g(t), \\ z = h(t) \end{cases} \quad a \le t \le b,$$

where f, g, and h have continuous first derivatives. Then, if M, N, and P are continuous functions of three variables whose domains contain the curve C, the line integral

$$\int_C M(x, y, z)\, dx + N(x, y, z)\, dy + P(x, y, z)\, dz$$

is evaluated by making the substitutions $x = f(t)$, $dx = f'(t)\, dt$, $y = g(t)$, $dy = g'(t)\, dt$, $z = h(t)$, and $dz = h'(t)\, dt$, and then integrating from a to b.

EXAMPLE   Evaluate $\int_C yz\, dx + xz\, dy + xy\, dz$ if

$$C: \begin{cases} x = t \\ y = t^2 \\ z = t^3, \end{cases} \quad -1 \le t \le 1.$$

SOLUTION

Making the substitutions $x = t$, $dx = dt$, $y = t^2$, $dy = 2t\, dt$, $z = t^3$, and $dz = 3t^2\, dt$, we have

$$\int_C yz\, dx + xz\, dy + xy\, dz = \int_{-1}^1 t^2 t^3\, dt + t t^3 (2t\, dt) + t t^2 (3t^2\, dt)$$

$$= \int_{-1}^1 (t^5 + 2t^5 + 3t^5)\, dt = \int_{-1}^1 6t^5\, dt = t^6 \bigg|_{-1}^1 = 0.$$

9.2 Vector Notation and Work

Consider the line integral $\int_C P(x, y)\, dx + Q(x, y)\, dy$, where the curve C is given by the parametric equations

$$C: \begin{cases} x = f(t) \\ y = g(t), \end{cases} \quad a \le t \le b.$$

If we let the vector

$$\bar{R} = x\bar{i} + y\bar{j} = f(t)\bar{i} + g(t)\bar{j}, \qquad a \le t \le b,$$

denote the variable position vector of a point (x, y) on C, then

$$\frac{d\bar{R}}{dt} = \frac{dx}{dt}\bar{i} + \frac{dy}{dt}\bar{j}, \quad \text{or in differential form,} \quad d\bar{R} = dx\,\bar{i} + dy\,\bar{j}.$$

Now, putting

$$\bar{F} = P(x, y)\bar{i} + Q(x, y)\bar{j},$$

we have

$$\bar{F} \cdot d\bar{R} = P(x, y)\,dx + Q(x, y)\,dy,$$

so that

$$\int_C P(x, y)\,dx + Q(x, y)\,dy = \int_C \bar{F} \cdot d\bar{R}.$$

Similarly, if C is a curve in three-dimensional space swept out by a variable position vector $\bar{R}$, then

$$\int_C M(x, y, z)\,dx + N(x, y, z)\,dy + P(x, y, z)\,dz = \int_C \bar{F} \cdot d\bar{R},$$

where

$$\bar{F} = M(x, y, z)\bar{i} + N(x, y, z)\bar{j} + P(x, y, z)\bar{k} \quad \text{and} \quad d\bar{R} = dx\,\bar{i} + dy\,\bar{j} + dz\,\bar{k}.$$

The vector notation $\int_C \bar{F} \cdot d\bar{R}$ for the line integral not only has the advantage of brevity, but it also suggests an important physical interpretation of the line integral. Suppose that $\bar{F}$ represents a variable force acting on a particle P which is moving along the curve C. If $\bar{R}$ is the variable position vector of P, then we can regard $d\bar{R}$ as representing an "infinitesimal" displacement of the particle, so that $\bar{F} \cdot d\bar{R}$ represents the work done on the particle by the force $\bar{F}$ during this displacement (Figure 2). Summing up—that is, integrating—all of these "infinitesimal" bits of work, we obtain $\int_C \bar{F} \cdot d\bar{R}$. Therefore, the line integral $\int_C \bar{F} \cdot d\bar{R}$ represents the net work done by the force $\bar{F}$ in moving a particle along the curve C from its initial point to its terminal point.

Figure 2

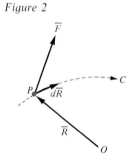

EXAMPLE The variable force $\bar{F} = (3x - 4y)\bar{i} + (4x + 2y)\bar{j}$ moves a particle along the curve

$$C: \begin{cases} x = 4t + 1 \\ y = 3t^2, \end{cases} \qquad 0 \le t \le 2, \quad \text{from } (1, 0) \text{ to } (9, 12).$$

Find the work done if distances are measured in centimeters and force is measured in dynes.

SOLUTION

The work is given by the line integral $\int_C \bar{F} \cdot d\bar{R}$, where $\bar{R} = x\bar{i} + y\bar{j}$, so that $d\bar{R} = dx\bar{i} + dy\bar{j}$. Thus,

$$\int_C \bar{F} \cdot d\bar{R} = \int_C (3x - 4y)\,dx + (4x + 2y)\,dy.$$

Making the substitutions $x = 4t + 1$, $dx = 4\,dt$, $y = 3t^2$, and $dy = 6t\,dt$, we have

$$\int_C \bar{F} \cdot d\bar{R} = \int_0^2 [3(4t + 1) - 4(3t^2)](4\,dt) + [4(4t + 1) + 2(3t^2)](6t\,dt)$$

$$= \int_0^2 (36t^3 + 48t^2 + 72t + 12)\,dt = (9t^4 + 16t^3 + 36t^2 + 12t)\Big|_0^2$$

$$= 440 \text{ ergs.}$$

9.3 Properties of Line Integrals

If C is a curve that is formed by joining successive curves $C_1, C_2, \ldots, C_n$, then we write

$$C = C_1 + C_2 + \cdots + C_n.$$

Figure 3 illustrates the case $n = 5$. If the variable vector $\bar{F}$ is continuous on each of the curves $C_1, C_2, \ldots, C_n$ and $C = C_1 + C_2 + \cdots + C_n$, then it can be shown that

$$\int_C \bar{F} \cdot d\bar{R} = \int_{C_1} \bar{F} \cdot d\bar{R} + \int_{C_2} \bar{F} \cdot d\bar{R} + \cdots + \int_{C_n} \bar{F} \cdot d\bar{R}.$$

Figure 3 $C = C_1 + C_2 + C_3 + C_4 + C_5$

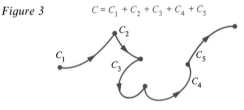

Figure 4

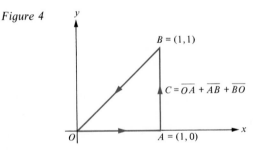

EXAMPLE   Let $A = (1, 0)$, $B = (1, 1)$. Calculate

$$\int_C (x^2 - y)\,dx + (x + y^2)\,dy$$

if $C = \overline{OA} + \overline{AB} + \overline{BO}$; that is, C is the perimeter of the triangle OAB taken in a counterclockwise direction (Figure 4).

SOLUTION
The line segments $\overline{OA}$, $\overline{AB}$, and $\overline{BO}$ are given parametrically by

$$\overline{OA}: \begin{cases} x = t \\ y = 0, \end{cases} \qquad \overline{AB}: \begin{cases} x = 1 \\ y = t, \end{cases} \qquad \overline{BO}: \begin{cases} x = 1 - t \\ y = 1 - t, \end{cases}$$

where $0 \le t \le 1$ in each case. Thus,

$$\int_{\overline{OA}} (x^2 - y)\,dx + (x + y^2)\,dy = \int_0^1 (t^2 - 0)\,dt + (t + 0)(0)\,dt = \int_0^1 t^2\,dt = \frac{t^3}{3}\Big|_0^1 = \frac{1}{3},$$

$$\int_{\overline{AB}} (x^2 - y)\,dx + (x + y^2)\,dy = \int_0^1 (1 - t)(0)\,dt + (1 + t^2)\,dt = \int_0^1 (1 + t^2)\,dt$$

$$= \left(t + \frac{t^3}{3}\right)\Big|_0^1 = \frac{4}{3}, \quad \text{and}$$

$$\int_{\overline{BO}} (x^2 - y)\, dx + (x + y^2)\, dy = \int_0^1 [(1-t)^2 - (1-t)](-dt) + [(1-t) + (1-t)^2](-dt)$$

$$= \int_0^1 (-2t^2 + 4t - 2)\, dt$$

$$= \left(-2\frac{t^3}{3} + 2t^2 - 2t\right)\Bigg|_0^1 = -\frac{2}{3}.$$

Therefore,

$$\int_C (x^2 - y)\, dx + (x + y^2)\, dy = \frac{1}{3} + \frac{4}{3} - \frac{2}{3} = 1.$$

A change in the direction (orientation) of the curve over which a line integral is taken results in a change in the algebraic sign of the integral; that is,

$$\int_{-C} \overline{F} \cdot d\overline{R} = -\int_C \overline{F} \cdot d\overline{R}.$$

(For an indication of the proof, see Problem 24.) For instance, in the preceding example, we found that

$$\int_{\overline{OA}} (x^2 - y)\, dx + (x + y^2)\, dy = \frac{1}{3},$$

and, since $-(\overline{OA}) = \overline{AO}$, it follows that

$$\int_{\overline{AO}} (x^2 - y)\, dx + (x + y^2)\, dy = -\frac{1}{3}.$$

Of course, line integrals, like ordinary integrals, are additive and homogeneous with respect to the integrand, so that

$$\int_C (\overline{F} + \overline{G}) \cdot d\overline{R} = \int_C \overline{F} \cdot d\overline{R} + \int_C \overline{G} \cdot d\overline{R}, \qquad \int_C (\overline{F} - \overline{G}) \cdot d\overline{R} = \int_C \overline{F} \cdot d\overline{R} - \int_C \overline{G} \cdot d\overline{R},$$

and

$$\int_C (a\overline{F}) \cdot d\overline{R} = a \int_C \overline{F} \cdot d\overline{R}$$

for any two continuous vector functions $\overline{F}$ and $\overline{G}$ and any constant scalar a. In particular, if M, N, and P are continuous functions of x, y, and z, then

$$\int_C M\, dx + N\, dy + P\, dz = \int_C M\, dx + \int_C N\, dy + \int_C P\, dz.$$

9.4 Green's Theorem

A remarkable theorem connecting a line integral around a closed curve in the plane with a double integral over the region enclosed by this curve is traditionally associated with the English mathematician George Green (1793–1841), although the result was actually obtained earlier by the German mathematician Karl Friedrich Gauss (1777–1855). Here we give a somewhat informal statement of the theorem. Later in Section 9.5 we give an argument to make the result plausible.

By a *closed curve*, we mean a curve whose initial point is the same as its terminal point. A *simple* closed curve is one that does not intersect itself, except at its common initial and terminal point.

THEOREM 1   Green's Theorem

Let C be a piecewise smooth, simple, closed curve in the xy plane and suppose that C forms the boundary of a two-dimensional region R. Assume that C is directed so that it turns counterclockwise around R **(Figure 5)**. Suppose that P and Q are continuous functions of two variables having continuous partial derivatives $\partial Q/\partial x$ and $\partial P/\partial y$ on R and its boundary C. Then

$$\int_C P(x,y)\,dx + Q(x,y)\,dy = \iint_R \left(\frac{\partial Q}{\partial x} - \frac{\partial P}{\partial y}\right)dx\,dy.$$

Figure 5

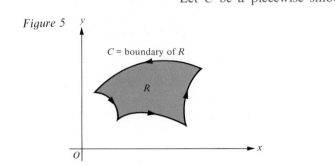

C = boundary of R

R

EXAMPLES   Use Green's theorem to evaluate the given line integral around the indicated simple closed curve C.

1 $\displaystyle\int_C (x^2 - y)\,dx + (x + y^2)\,dy$, where C is the perimeter of the triangle OAB taken in a counterclockwise sense, with $A = (1,0)$ and $B = (1,1)$ (Figure 4).

SOLUTION
Here $P(x,y) = x^2 - y$, $Q(x,y) = x + y^2$, and R is the triangular region enclosed by OAB. Thus,

$$\frac{\partial Q}{\partial x} = \frac{\partial}{\partial x}(x + y^2) = 1 \quad \text{and} \quad \frac{\partial P}{\partial y} = \frac{\partial}{\partial y}(x^2 - y) = -1.$$

Hence, by Green's theorem,

$$\int_C (x^2 - y)\,dx + (x + y^2)\,dy = \iint_R \left(\frac{\partial Q}{\partial x} - \frac{\partial P}{\partial y}\right)dx\,dy = \iint_R [1 - (-1)]\,dx\,dy$$

$$= \iint_R 2\,dx\,dy = 2\iint_R dx\,dy = 2\left(\frac{1}{2}\right) = 1,$$

since the area of the triangular region R is $\frac{1}{2}$ square unit. (This result is consistent with our previous solution of the example in Section 9.3.)

2 $\displaystyle\int_C (y + 3x)\,dx + (2y - x)\,dy$, where C is the circle $x^2 + y^2 = 9$ with the counterclockwise orientation.

SOLUTION
Here the region R enclosed by C is a circular disk of radius 3 and area $\iint_R dx\,dy = \pi 3^2 = 9\pi$ square units. Thus,

$$\int_C (y + 3x)\,dx + (2y - x)\,dy = \iint_R \left[\frac{\partial}{\partial x}(2y - x) - \frac{\partial}{\partial y}(y + 3x)\right]dx\,dy$$

$$= \iint_R [(-1) - 1]\,dx\,dy = -2\iint_R dx\,dy$$

$$= -2(9\pi) = -18\pi.$$

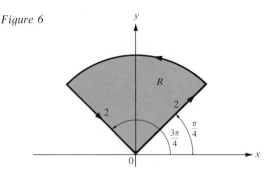

Figure 6

3 Let R be the region bounded by the three curves whose polar equations are $\theta = \pi/4$, $r = 2$, and $\theta = 3\pi/4$ (Figure 6), and let C be the boundary of R taken in a counterclockwise sense. Evaluate

$$\int_C xy\, dx + x^2\, dy.$$

SOLUTION

Using Green's theorem and then using polar coordinates to evaluate the resulting double integral, we have

$$\int_C xy\, dx + x^2\, dy = \iint_R \left[\frac{\partial}{\partial x}(x^2) - \frac{\partial}{\partial y}(xy)\right] dx\, dy = \iint_R (2x - x)\, dx\, dy$$

$$= \iint_R x\, dx\, dy = \int_{\pi/4}^{3\pi/4} \int_0^2 (r\cos\theta)r\, dr\, d\theta$$

$$= \int_{\pi/4}^{3\pi/4} \left[\frac{r^3}{3}\cos\theta \,\Big|_0^2\right] d\theta = \int_{\pi/4}^{3\pi/4} \frac{8}{3}\cos\theta\, d\theta$$

$$= \frac{8}{3}\sin\theta \,\Big|_{\pi/4}^{3\pi/4} = \frac{8}{3}\left(\sin\frac{3\pi}{4} - \sin\frac{\pi}{4}\right) = 0.$$

4 $\int_C \overline{F} \cdot d\overline{R}$, where $\overline{F} = (2xy - x^2)\overline{i} + (x + y^2)\overline{j}$ and C is the boundary, taken counterclockwise, of the region R bounded by the parabola $y = x^2$ and the straight line $y = x$.

SOLUTION

$$\int_C \overline{F} \cdot d\overline{R} = \iint_R \left[\frac{\partial}{\partial x}(x + y^2) - \frac{\partial}{\partial y}(2xy - x^2)\right] dx\, dy = \iint_R (1 - 2x)\, dx\, dy$$

$$= \int_0^1 \int_{x^2}^x (1 - 2x)\, dy\, dx = \int_0^1 \left[(1 - 2x)y \,\Big|_{x^2}^x\right] dx$$

$$= \int_0^1 (1 - 2x)(x - x^2)\, dx = \int_0^1 (2x^3 - 3x^2 + x)\, dx = \left(\frac{x^4}{2} - x^3 + \frac{x^2}{2}\right)\Big|_0^1 = 0.$$

9.5 Informal Proof of Green's Theorem

We now give an informal proof of Green's theorem for the special case in which the region R enclosed by the simple closed curve C is both of type I and of type II. The result for more general regions that can be decomposed into nonoverlapping subregions of both types then follows easily from this special case.

In Figure 7, the simple closed curve C bounds the region R and is directed in the counterclockwise sense. Here, C is the sum of the directed arcs ATB and BSA, and it is also the sum of the directed arcs SAT and TBS; that is,

$$C = ATB + BSA \quad \text{and} \quad C = SAT + TBS.$$

Suppose that the directed arcs ATB, $-(BSA)$, $-(SAT)$, and TBS are described as follows:

Figure 7

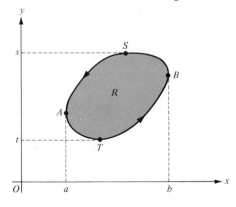

$$ATB: \quad y = g(x), \qquad a \le x \le b,$$
$$-(BSA): \quad y = h(x), \qquad a \le x \le b,$$
$$-(SAT): \quad x = G(y), \qquad t \le y \le s,$$
$$TBS: \quad x = H(y), \qquad t \le y \le s.$$

Now, assume that P and Q are continuous functions and that $\partial Q/\partial x$ and $\partial P/\partial y$ are defined and continuous on the region R and on its boundary curve C. We have

$$\iint\limits_{R} \frac{\partial Q}{\partial x}\, dx\, dy = \int_{t}^{s} \left[\int_{G(y)}^{H(y)} \frac{\partial Q}{\partial x}\, dx \right] dy = \int_{t}^{s} \left[Q(x,y) \Big|_{G(y)}^{H(y)} \right] dy$$

$$= \int_{t}^{s} [Q(H(y), y) - Q(G(y), y)]\, dy = \int_{t}^{s} Q(H(y), y)\, dy - \int_{t}^{s} Q(G(y), y)\, dy$$

$$= \int_{TBS} Q(x,y)\, dy - \int_{-(SAT)} Q(x,y)\, dy$$

$$= \int_{TBS} Q(x,y)\, dy + \int_{SAT} Q(x,y)\, dy = \int_{TBS+SAT} Q(x,y)\, dy$$

$$= \int_{C} Q(x,y)\, dy.$$

Similarly,

$$-\iint\limits_{R} \frac{\partial P}{\partial y}\, dx\, dy = -\int_{a}^{b} \left[\int_{g(x)}^{h(x)} \frac{\partial P}{\partial y}\, dy \right] dx = -\int_{a}^{b} \left[P(x,y) \Big|_{g(x)}^{h(x)} \right] dx$$

$$= -\int_{a}^{b} [P(x, h(x)) - P(x, g(x))]\, dx$$

$$= \int_{a}^{b} P(x, g(x))\, dx - \int_{a}^{b} P(x, h(x))\, dx$$

$$= \int_{ATB} P(x,y)\, dx - \int_{-(BSA)} P(x,y)\, dx$$

$$= \int_{ATB} P(x,y)\, dx + \int_{BSA} P(x,y)\, dx = \int_{ATB+BSA} P(x,y)\, dx$$

$$= \int_{C} P(x,y)\, dx.$$

Combining the two results, we obtain

$$\int_{C} P(x,y)\, dx + Q(x,y)\, dy = \int_{C} P(x,y)\, dx + \int_{C} Q(x,y)\, dy$$

$$= -\iint\limits_{R} \frac{\partial P}{\partial y}\, dx\, dy + \iint\limits_{R} \frac{\partial Q}{\partial x}\, dx\, dy = \iint\limits_{R} \left(\frac{\partial Q}{\partial x} - \frac{\partial P}{\partial y} \right) dx\, dy,$$

and our argument is complete.

Problem Set 9

In problems 1 through 10, evaluate each line integral using Definition 1.

1 $\int_C (3x^2 - 6y)\,dx + (3x + 2y)\,dy$, where $C: \begin{cases} x = t \\ y = t^2, \end{cases}$ $0 \leq t \leq 1$.

2 $\int_C y^2\,dx + x^2\,dy$, where $C: \begin{cases} x = t^2 \\ y = t + 1, \end{cases}$ $0 \leq t \leq 1$.

3 $\int_C x\,dy - y\,dx$, where $C: \begin{cases} x = 2 + \cos t \\ y = \sin t, \end{cases}$ $0 \leq t \leq 2\pi$.

4 $\int_C 2xy^3\,dx + 3x^2y^2\,dy$, where C is the portion of the parabola $y = x^2$ from $(0,0)$ to $(1,1)$.

5 $\int_C \bar{F} \cdot d\bar{R}$, where $\bar{F} = (x^2 + y^2)\bar{i} + 2xy\bar{j}$ and C is the portion of the parabola $y = x^2$ from $(0,0)$ to $(1,1)$.

6 $\int_C \bar{F} \cdot d\bar{R}$, where $\bar{F} = (\cos y)\bar{i} + (\sin x)\bar{j}$ and C is the triangle with vertices $(1,0)$, $(1,1)$, and $(0,0)$ taken in the counterclockwise sense.

7 $\int_C \bar{F} \cdot d\bar{R}$, where $\bar{F} = xy\bar{i} + x^2\bar{j}$ and C is the circle of radius 2 with center at the origin taken in the counterclockwise sense.

8 $\int_C (\nabla f) \cdot d\bar{R}$, where $f(x, y) = x^2 - 2xy$ and C is the path from $(0,0)$ to $(2,1)$, then from $(2,1)$ to $(2,3)$, then from $(2,3)$ to $(1,3)$.

9 $\int_C (y + z)\,dx + (2xz)\,dy + (x + z)\,dz$, where $C: \begin{cases} x = 2t + 1 \\ y = 3t - 1 \\ z = -t + 2, \end{cases}$ $0 \leq t \leq 2$.

10 $\int_C \bar{F} \cdot d\bar{R}$, where $\bar{F} = 2zy\bar{i} + x^2\bar{j} + (x + y)\bar{k}$, $C = \overline{OA} + \overline{AB} + \overline{BO}$, $A = (1,2,3)$, and $B = (2,0,-1)$.

In problems 11 through 18, use Green's theorem to evaluate each line integral. In each case, assume that the closed curve C is taken in the counterclockwise sense.

11 $\int_C (x^2 - xy^3)\,dx + (y^2 - 2xy)\,dy$, where C is the square with vertices at $(0,0)$, $(3,0)$, $(3,3)$, and $(0,3)$.

12 $\int_C (x^2 + 3y^2)\,dx$, where C is the square with vertices at $(-2,-2)$, $(2,-2)$, $(2,2)$, and $(-2,2)$.

13 $\int_C y\,dx + x^3\sqrt{4 - y^2}\,dy$, where C is the circle $x^2 + y^2 = 4$.

14 $\int_C x\cos y\,dx - y\sin x\,dy$, where C is the square with vertices at $(0,0)$, $(1,0)$, $(1,1)$, and $(0,1)$.

15 $\int_C \bar{F} \cdot d\bar{R}$, where $\bar{F} = (x^2 + y^2)\bar{i} + 3xy^2\bar{j}$ and C is the circle $x^2 + y^2 = 9$.

16 $\int_C \bar{F} \cdot d\bar{R}$, where $\bar{F} = (3x^2 - 8y^2)\bar{i} + (4y - 6xy)\bar{j}$ and C encloses the region $R: x \geq 0$, $y \geq 0$, $x + y \leq 2$.

17 $\int_C e^y\,dx + xe^y\,dy$, where C encloses the region $R: 0 \leq x \leq 1$, $\sin^{-1} x \leq y \leq \pi/2$.

18 $\int_C (x^3 - x^2 y)\, dx + xy^2\, dy$, where C encloses the region R bounded by $y = x^2$ and $x = y^2$.

19 Find the work done in moving a particle under the action of a force $\bar{F} = 3x^2 \bar{i} + (2xy - y^2)\bar{j}$ dynes along the straight line segment from $(0,0)$ to $(2,5)$. Assume that distances are measured in centimeters.

20 The gravitational force $\bar{F}$ acting on a particle of mass m near the earth's surface is (approximately) given by $\bar{F} = -mg\bar{j}$, where $\bar{j}$ is a unit vector pointing straight upward and g is the (constant) acceleration of gravity. Show that the work done by $\bar{F}$ on the particle as it moves in a vertical plane from height h_1 to height h_2 *along any path* depends only on h_1 and h_2.

21 Find the work done in moving a particle under the action of a force $\bar{F} = (2x + 3y)\bar{i} + xy\bar{j}$ dynes from $(0,0)$ to $(1,1)$:
(a) Along the straight-line segment from $(0,0)$ to $(1,1)$.
(b) Along the parabola $y = x^2$.
(c) Along the arc of the circle $x^2 + (y-1)^2 = 1$ from $(0,0)$ to $(1,1)$.
Assume that distances are measured in centimeters.

22 Suppose that f is a function with continuous first partial derivatives in a region R and that C is a smooth curve in R, directed from the initial point (x_0, y_0) to the terminal point (x_1, y_1). Prove that $\int_C (\nabla f) \cdot d\bar{R} = f(x_1, y_1) - f(x_0, y_0)$.

23 The line integral $\int_C \bar{F} \cdot d\bar{R}$ is sometimes written as $\int_C (\bar{F} \cdot \bar{T})\, ds$, where $\bar{T}$ denotes the unit tangent vector to C and s is the arc length measured along C. Explain why this notation is reasonable.

24 Let C be a curve with the parametric equations $C: \begin{cases} x = f(t) \\ y = g(t), \end{cases}$ $a \leq t \leq b$, where f and g have continuous first derivatives. Suppose that P and Q are continuous functions of two variables whose domains contain the curve C. Define functions F and G on $[a,b]$ by $F(t) = f(a + b - t)$ and $G(t) = g(a + b - t)$.
(a) Show that $\begin{cases} x = F(t) \\ y = G(t), \end{cases}$ $a \leq t \leq b$ are parametric equations for $-C$.
(b) Use the result of part (a) to prove that

$$\int_{-C} P(x, y)\, dx + Q(x, y)\, dy = -\int_C P(x, y)\, dx + Q(x, y)\, dy.$$

25 If $\bar{F} = \frac{1}{2}(-y\bar{i} + x\bar{j})$, show that $\int_C \bar{F} \cdot d\bar{R}$ gives the area of the region R enclosed by the simple closed curve C. (Use Green's theorem.)

26 Let $\bar{T}$ denote the unit tangent vector to the oriented curve C and let $\bar{N}_r$ denote the normal vector to C obtained by rotating $\bar{T}$ to the right through $90°$ (Chapter 14, Section 8). If $\bar{V} = f(x, y)\bar{i} + g(x, y)\bar{j}$, then the *flux* of $\bar{V}$ across C in the direction of $\bar{N}_r$ is defined to be $\int_C (\bar{V} \cdot \bar{N}_r)\, ds$. The scalar field $\dfrac{\partial f}{\partial x} + \dfrac{\partial g}{\partial y}$ is called the *divergence* of $\bar{V}$. Prove the following theorem, called the *divergence theorem* (in two dimensions): The outward flux of $\bar{V}$ across a simple closed curve C is the double integral of the divergence of $\bar{V}$ over the region R enclosed by C. (State carefully whatever assumptions about continuity, and so forth, that you need.)

27 Assume that C is a simple closed curve enclosing a region R and that f is a function of two variables, defined on R and its boundary C, which has continuous second-order partial derivatives. Use Green's theorem to show that $\int_C (\nabla f) \cdot d\bar{R} = 0$.

28 Suppose that a fluid is flowing in a thin sheet over the xy plane and that the velocity of a particle of this fluid at the point (x, y) is given by $\bar{V} = f(x, y)\bar{i} + g(x, y)\bar{j}$. Assume that the density of the fluid at the point (x, y) is given by $\sigma(x, y)$ mass units per unit of area. Show that the flux of $\sigma(x, y)\bar{V}$ across a curve C in the direction of the normal $\bar{N}_r$ (problem 26) gives the mass of fluid, per unit time, flowing across C in the direction of $\bar{N}_r$.

29 Criticize the following argument: Let $P(x, y) = -\dfrac{y}{x^2 + y^2}$ and let $Q(x, y) = \dfrac{x}{x^2 + y^2}$.

Then $\dfrac{\partial Q}{\partial x} - \dfrac{\partial P}{\partial y} = 0$, so that

P & Q not continuous on C, particularly at (0,0)

$$\int_C \frac{-y}{x^2 + y^2}\,dx + \frac{x}{x^2 + y^2}\,dy = \iint_R \left(\frac{\partial Q}{\partial x} - \frac{\partial P}{\partial y}\right)dx\,dy = 0,$$

where C is the circle $C: \begin{cases} x = \cos t \\ y = \sin t, \end{cases}$ $0 \leq t \leq 2\pi$ and R is the disk bounded by C.

(*Note*: Something must be wrong, because direct calculation of the line integral gives 2π, not 0.)

10 Surface Area and Surface Integrals

In Section 9 we considered line integrals, that is, integrals over curves. In this section, we study the analogous notion of *surface integrals*, that is, integrals over surfaces. A special case of the surface integral provides a technique for finding the area of a surface. The section also includes a discussion of the *flux* of a vector field through a surface.

10.1 Area of a Surface Described by Vector Parametric Equations

Since we previously used the letter S to denote solids in Section 6, we use the Greek letter Σ to denote surfaces in the present section. A surface Σ in xyz space can be described by a variable position vector $\bar{R}$ whose endpoint P traces out the surface (Figure 1a). If one wishes to express $\bar{R}$ parametrically, it is necessary to use *two* independent parameters, since Σ is "two-dimensional." Thus, if the two parameters are denoted by u and v, the vector parametric equation of Σ can be expressed as

$$\bar{R} = f(u, v)\bar{i} + g(u, v)\bar{j} + h(u, v)\bar{k}.$$

Here we assume that the functions f, g, and h are continuously differentiable and defined on an admissible region D in the uv plane (Figure 1b). As the point (u, v) moves through the region D, the position vector $\bar{R}$ traces out the surface Σ.

It is not necessary to denote the surface parameters by u and v; they could just as easily be denoted by θ and ϕ, or by x and y, or by s and t, and so forth. For

Figure 1

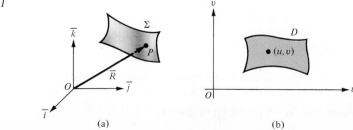

(a)   (b)

instance, the vector parametric equation of a sphere of radius a can be written in terms of the longitude θ and colatitude ϕ as

$$\bar{R} = (a \sin \phi \cos \theta)\bar{i} + (a \sin \phi \sin \theta)\bar{j} + (a \cos \phi)\bar{k},$$

where (θ, ϕ) ranges over the rectangular region

$$D: 0 \leq \theta \leq 2\pi, \quad 0 \leq \phi \leq \pi$$

in the $\theta\phi$ plane.

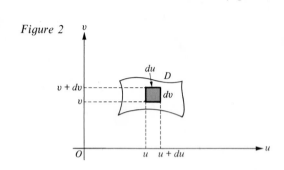

Figure 2

Now, let Σ be a surface described vector parametrically by

$$\bar{R} = f(u, v)\bar{i} + g(u, v)\bar{j} + h(u, v)\bar{k},$$

where (u, v) ranges over the region D in the uv plane. Consider the "infinitesimal" rectangle of dimensions du and dv with vertices (u, v), $(u + du, v)$, $(u + du, v + dv)$, and $(u, v + dv)$ within the region D (Figure 2). Let the four vertices, in the order just given, correspond to the four points P, Q, V, and W, respectively, on the surface Σ (Figure 3), so that

$$P = (f(u, v), g(u, v), h(u, v)),$$
$$Q = (f(u + du, v), g(u + du, v), h(u + du, v)),$$
$$V = (f(u + du, v + dv), g(u + du, v + dv), h(u + du, v + dv)), \quad \text{and}$$
$$W = (f(u, v + dv), g(u, v + dv), h(u, v + dv)).$$

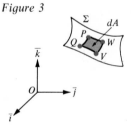

Figure 3

The "infinitesimal" region $PQVW$ on the surface Σ has area dA which is virtually the same as the area of the parallelogram spanned by the vectors $\overrightarrow{PQ}$ and $\overrightarrow{PW}$, so that

$$dA = |\overrightarrow{PQ} \times \overrightarrow{PW}|.$$

Since du represents an "infinitesimal" change in u,

$$f(u + du, v) - f(u, v) = f_1(u, v)\, du$$
$$g(u + du, v) - g(u, v) = g_1(u, v)\, du, \quad \text{and}$$
$$h(u + du, v) - h(u, v) = h_1(u, v)\, du.$$

Therefore,

$$\overrightarrow{PQ} = [f(u + du, v) - f(u, v)]\bar{i} + [g(u + du, v) - g(u, v)]\bar{j} + [h(u + du, v) - h(u, v)]\bar{k}$$
$$= [f_1(u, v)\, du]\bar{i} + [g_1(u, v)\, du]\bar{j} + [h_1(u, v)\, du]\bar{k}$$
$$= \left(\frac{\partial f}{\partial u}\, du\right)\bar{i} + \left(\frac{\partial g}{\partial u}\, du\right)\bar{j} + \left(\frac{\partial h}{\partial u}\, du\right)\bar{k} = \left(\frac{\partial f}{\partial u}\bar{i} + \frac{\partial g}{\partial u}\bar{j} + \frac{\partial h}{\partial u}\bar{k}\right) du$$
$$= \frac{\partial \bar{R}}{\partial u}\, du,$$

where we have used the obvious symbolism $\partial\bar{R}/\partial u$ for the vector obtained by partially differentiating $\bar{R}$, component by component, with respect to u. Similarly, we have

$$\overrightarrow{PW} = [f(u, v + dv) - f(u, v)]\bar{i} + [g(u, v + dv) - g(u, v)]\bar{j} + [h(u, v + dv) - h(u, v)]\bar{k}$$
$$= [f_2(u, v)\, dv]\bar{i} + [g_2(u, v)\, dv]\bar{j} + [h_2(u, v)\, dv]\bar{k}$$
$$= \left(\frac{\partial f}{\partial v}\, dv\right)\bar{i} + \left(\frac{\partial g}{\partial v}\, dv\right)\bar{j} + \left(\frac{\partial h}{\partial v}\, dv\right)\bar{k} = \frac{\partial \bar{R}}{\partial v}\, dv.$$

If we substitute the expressions obtained above for $\overrightarrow{PQ}$ and $\overrightarrow{PW}$ into the equation $dA = |\overrightarrow{PQ} \times \overrightarrow{PW}|$, we obtain

$$dA = \left| \frac{\partial \bar{R}}{\partial u} du \times \frac{\partial \bar{R}}{\partial v} dv \right| \quad \text{or} \quad dA = \left| \frac{\partial \bar{R}}{\partial u} \times \frac{\partial \bar{R}}{\partial v} \right| du \, dv.$$

Using the identity $|\bar{A} \times \bar{B}| = \sqrt{|\bar{A}|^2 |\bar{B}|^2 - (\bar{A} \cdot \bar{B})^2}$ obtained in Section 4.1 of Chapter 15, page 755, we can rewrite the formula for dA as

$$dA = \sqrt{\left| \frac{\partial \bar{R}}{\partial u} \right|^2 \left| \frac{\partial \bar{R}}{\partial v} \right|^2 - \left(\frac{\partial \bar{R}}{\partial u} \cdot \frac{\partial \bar{R}}{\partial v} \right)^2} \, du \, dv.$$

Naturally, the total area A of the surface Σ can be obtained by integrating dA; thus,

$$A = \iint_D dA = \iint_D \left| \frac{\partial \bar{R}}{\partial u} \times \frac{\partial \bar{R}}{\partial v} \right| du \, dv = \iint_D \sqrt{\left| \frac{\partial \bar{R}}{\partial u} \right|^2 \left| \frac{\partial \bar{R}}{\partial v} \right|^2 - \left(\frac{\partial \bar{R}}{\partial u} \cdot \frac{\partial \bar{R}}{\partial v} \right)^2} \, du \, dv.$$

EXAMPLES 1 Use the formula above to find the area of a sphere of radius a.

SOLUTION

The sphere is represented by the vector parametric equation

$$\bar{R} = (a \sin \phi \cos \theta)\bar{i} + (a \sin \phi \sin \theta)\bar{j} + (a \cos \phi)\bar{k},$$

where (θ, ϕ) ranges over the rectangular region

$$D: 0 \le \theta \le 2\pi, \quad 0 \le \phi \le \pi$$

in the $\theta\phi$ plane. Here,

$$\frac{\partial \bar{R}}{\partial \theta} = (-a \sin \phi \sin \theta)\bar{i} + (a \sin \phi \cos \theta)\bar{j} + 0\bar{k},$$

$$\frac{\partial \bar{R}}{\partial \phi} = (a \cos \phi \cos \theta)\bar{i} + (a \cos \phi \sin \theta)\bar{j} - (a \sin \phi)\bar{k};$$

hence,

$$\left| \frac{\partial \bar{R}}{\partial \theta} \right|^2 = a^2 \sin^2 \phi \sin^2 \theta + a^2 \sin^2 \phi \cos^2 \theta = a^2 \sin^2 \phi(\sin^2 \theta + \cos^2 \theta)$$

$$= a^2 \sin^2 \phi,$$

$$\left| \frac{\partial \bar{R}}{\partial \phi} \right|^2 = a^2 \cos^2 \phi \cos^2 \theta + a^2 \cos^2 \phi \sin^2 \theta + a^2 \sin^2 \phi$$

$$= a^2 \cos^2 \phi(\cos^2 \theta + \sin^2 \theta) + a^2 \sin^2 \phi = a^2 \cos^2 \phi + a^2 \sin^2 \phi$$

$$= a^2(\cos^2 \phi + \sin^2 \phi) = a^2, \quad \text{and}$$

$$\frac{\partial \bar{R}}{\partial \theta} \cdot \frac{\partial \bar{R}}{\partial \phi} = -a^2 \sin \phi \cos \phi \sin \theta \cos \theta + a^2 \sin \phi \cos \phi \sin \theta \cos \theta + 0 = 0.$$

Therefore,

$$\sqrt{\left| \frac{\partial \bar{R}}{\partial \theta} \right|^2 \left| \frac{\partial \bar{R}}{\partial \phi} \right|^2 - \left(\frac{\partial \bar{R}}{\partial \theta} \cdot \frac{\partial \bar{R}}{\partial \phi} \right)^2} = \sqrt{(a^2 \sin^2 \phi)a^2 - 0} = a^2 \sin \phi;$$

and it follows that, for the sphere of radius a with longitude and colatitude as parameters,

$$dA = a^2 \sin \phi \, d\theta \, d\phi.$$

Consequently, the surface area of the sphere is given by

$$A = \iint_D dA = \iint_D a^2 \sin\phi \; d\theta \; d\phi = a^2 \iint_D \sin\phi \; d\theta \; d\phi = a^2 \int_0^\pi \int_0^{2\pi} \sin\phi \; d\theta \; d\phi$$

$$= a^2 \int_0^\pi \left[\sin\phi \int_0^{2\pi} d\theta \right] d\phi = a^2 \int_0^\pi \left[\sin\phi \left(\theta \Big|_0^{2\pi} \right) \right] d\phi = a^2 \int_0^\pi 2\pi \sin\phi \; d\phi$$

$$= 2\pi a^2 (-\cos\phi) \Big|_0^\pi = -2\pi a^2 [-1 - 1] = 4\pi a^2.$$

2 Use the integral for surface area to rederive the formula for the area of a surface of revolution originally derived in Section 4.2 of Chapter 7.

SOLUTION
Let the surface Σ be generated by revolving the graph of $y = f(x)$, $a \leq x \leq b$, about the x axis, where f has a continuous first derivative and $f(x) \geq 0$ for $a \leq x \leq b$ (Figure 4). A point P on the surface Σ can be located by specifying its x coordinate and the angle α through which P has turned from its original position Q on the curve $y = f(x)$. Thus, we take x and α as our parameters. In Figure 4, notice that

Figure 4

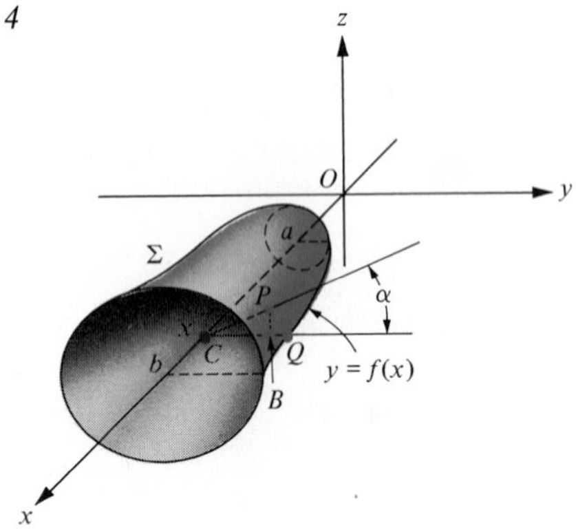

$$|\overline{CP}| = |\overline{CQ}| = f(x).$$

Consideration of the right triangle CBP shows that the y and z coordinates of P are given by

$$y = |\overline{CP}| \cos\alpha = f(x) \cos\alpha$$

and

$$z = |\overline{CP}| \sin\alpha = f(x) \sin\alpha.$$

Therefore, the position vector $\overline{R}$ of P is given by

$$\overline{R} = x\overline{i} + y\overline{j} + z\overline{k} = x\overline{i} + f(x)(\cos\alpha)\overline{j} + f(x)(\sin\alpha)\overline{k},$$

and, as (x, α) ranges over the rectangular region

$$D: a \leq x \leq b, 0 \leq \alpha \leq 2\pi$$

in the $x\alpha$ plane, the vector $\overline{R}$ traces out the surface Σ. Here,

$$\frac{\partial \overline{R}}{\partial x} = \overline{i} + f'(x)(\cos\alpha)\overline{j} + f'(x)(\sin\alpha)\overline{k},$$

$$\frac{\partial \overline{R}}{\partial \alpha} = 0\overline{i} - f(x)(\sin\alpha)\overline{j} + f(x)(\cos\alpha)\overline{k}.$$

Therefore,

$$\left| \frac{\partial \overline{R}}{\partial x} \right|^2 = 1^2 + [f'(x)]^2 \cos^2\alpha + [f'(x)]^2 \sin^2\alpha = 1 + [f'(x)]^2,$$

$$\left| \frac{\partial \overline{R}}{\partial \alpha} \right|^2 = 0^2 + [f(x)]^2 \sin^2\alpha + [f(x)]^2 \cos^2\alpha = [f(x)]^2, \quad \text{and}$$

$$\frac{\partial \overline{R}}{\partial x} \cdot \frac{\partial \overline{R}}{\partial \alpha} = 0 - f'(x)f(x) \cos\alpha \sin\alpha + f'(x)f(x) \cos\alpha \sin\alpha = 0.$$

Hence,

$$dA = \sqrt{\left|\frac{\partial \overline{R}}{\partial x}\right|^2 \left|\frac{\partial \overline{R}}{\partial \alpha}\right|^2 - \left(\frac{\partial \overline{R}}{\partial x} \cdot \frac{\partial \overline{R}}{\partial \alpha}\right)^2} \, dx \, d\alpha = \sqrt{\{1 + [f'(x)]^2\}[f(x)]^2 - 0} \, dx \, d\alpha$$

$$= f(x)\sqrt{1 + [f'(x)]^2} \, dx \, d\alpha.$$

It follows that

$$A = \iint_D dA = \iint_D f(x)\sqrt{1 + [f'(x)]^2} \, dx \, d\alpha = \int_a^b \int_0^{2\pi} f(x)\sqrt{1 + [f'(x)]^2} \, d\alpha \, dx$$

$$= \int_a^b \left[f(x)\sqrt{1 + [f'(x)]^2} \, \alpha \Big|_0^{2\pi} \right] dx = 2\pi \int_a^b f(x)\sqrt{1 + [f'(x)]^2} \, dx.$$

10.2   Area of the Graph of a Function of Two Variables

Let f be a continuously differentiable function of two variables defined over an admissible region D in the xy plane and denote the graph of f by Σ (Figure 5). Using x and y as parameters, we find that the vector parametric equation of Σ is

$$\overline{R} = x\overline{i} + y\overline{j} + f(x, y)\overline{k},$$

where (x, y) ranges over the region D. Here,

$$\frac{\partial \overline{R}}{\partial x} = \overline{i} + 0\overline{j} + f_1(x, y)\overline{k} \quad \text{and} \quad \frac{\partial \overline{R}}{\partial y} = 0\overline{i} + \overline{j} + f_2(x, y)\overline{k};$$

hence,

$$\left|\frac{\partial \overline{R}}{\partial x}\right|^2 = 1^2 + 0^2 + [f_1(x, y)]^2,$$

$$\left|\frac{\partial \overline{R}}{\partial y}\right|^2 = 0^2 + 1^2 + [f_2(x, y)]^2, \quad \text{and}$$

$$\frac{\partial \overline{R}}{\partial x} \cdot \frac{\partial \overline{R}}{\partial y} = 0 + 0 + f_1(x, y)f_2(x, y).$$

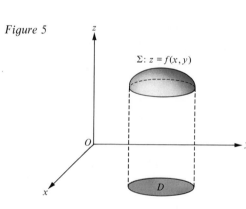

Figure 5

$\Sigma: z = f(x, y)$

It follows that

$$dA = \sqrt{\left|\frac{\partial \overline{R}}{\partial x}\right|^2 \left|\frac{\partial \overline{R}}{\partial y}\right|^2 - \left(\frac{\partial \overline{R}}{\partial x} \cdot \frac{\partial \overline{R}}{\partial y}\right)^2} \, dx \, dy$$

$$= \sqrt{\{1 + [f_1(x, y)]^2\}\{1 + [f_2(x, y)]^2\} - [f_1(x, y)f_2(x, y)]^2} \, dx \, dy$$

$$= \sqrt{1 + [f_1(x, y)]^2 + [f_2(x, y)]^2} \, dx \, dy.$$

Therefore, the area A of the portion of the graph of $z = f(x, y)$ lying above the region D in the xy plane is given by

$$A = \iint_D \sqrt{1 + [f_1(x, y)]^2 + [f_2(x, y)]^2} \, dx \, dy = \iint_D \sqrt{1 + \left(\frac{\partial z}{\partial x}\right)^2 + \left(\frac{\partial z}{\partial y}\right)^2} \, dx \, dy.$$

EXAMPLES 1   Find the area of the portion of the surface $z = x^2 + y^2$ that lies over the region $D: x^2 + y^2 \leq 1$.

SOLUTION

$$A = \iint_D \sqrt{1 + \left(\frac{\partial z}{\partial x}\right)^2 + \left(\frac{\partial z}{\partial y}\right)^2}\, dx\, dy = \iint_D \sqrt{1 + (2x)^2 + (2y)^2}\, dx\, dy$$

$$= \iint_D \sqrt{4x^2 + 4y^2 + 1}\, dx\, dy.$$

Converting the latter double integral to polar coordinates, we obtain

$$A = \int_0^{2\pi} \int_0^1 \sqrt{4r^2 \cos^2 \theta + 4r^2 \sin^2 \theta + 1}\; r\, dr\, d\theta = \int_0^{2\pi} \int_0^1 \sqrt{4r^2 + 1}\; r\, dr\, d\theta.$$

Making the change of variable $u = 4r^2 + 1$ in the "inside" integral, we have

$$A = \int_0^{2\pi} \left[\int_1^5 \sqrt{u}\,\frac{du}{8}\right] d\theta = \int_0^{2\pi} \left[\frac{u^{3/2}}{12}\Big|_1^5\right] d\theta = \int_0^{2\pi} \frac{\sqrt{125} - 1}{12}\, d\theta$$

$$= 2\pi \frac{5\sqrt{5} - 1}{12} \approx 5.33 \text{ square units.}$$

2 Find the area of the portion of the plane $2x + 3y + z = 6$ that is cut off by the three coordinate planes.

SOLUTION

The equation of the plane can be written in the form $z = 6 - 2x - 3y$, so that

$$\frac{\partial z}{\partial x} = -2 \quad \text{and} \quad \frac{\partial z}{\partial y} = -3.$$

The plane $z = 6 - 2x - 3y$ intersects the plane $z = 0$ in the straight line $2x + 3y = 6$; hence, the portion of the plane cut off by the three coordinate planes lies above the triangular region D in the xy plane bounded by the x axis, the y axis, and the line $2x + 3y = 6$ (Figure 6). The area of D is given by

Figure 6

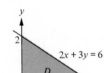

$$\iint_D dx\, dy = \frac{1}{2}(2)(3) = 3 \text{ square units.}$$

Thus, the area A of the portion of the plane $z = 6 - 2x - 3y$ lying above D is given by

$$A = \iint_D \sqrt{1 + \left(\frac{\partial z}{\partial x}\right)^2 + \left(\frac{\partial z}{\partial y}\right)^2}\, dx\, dy = \iint_D \sqrt{1 + (-2)^2 + (-3)^2}\, dx\, dy$$

$$= \sqrt{14} \iint_D dx\, dy = \sqrt{14}(3) = 3\sqrt{14} \approx 11.22 \text{ square units.}$$

10.3 The Surface Integral

Suppose that Σ is a surface described vector parametrically by

$$\Sigma: \overline{R} = f(u, v)\overline{i} + g(u, v)\overline{j} + h(u, v)\overline{k}$$

for (u, v) in an admissible region D in the uv plane. We assume, as in Section 10.1, that the scalar component functions f, g, and h are continuously differentiable, so that the differential dA of surface area is given by

$$dA = \left|\frac{\partial \overline{R}}{\partial u} \times \frac{\partial \overline{R}}{\partial v}\right| du\, dv = \sqrt{\left|\frac{\partial \overline{R}}{\partial u}\right|^2 \left|\frac{\partial \overline{R}}{\partial v}\right|^2 - \left(\frac{\partial \overline{R}}{\partial u} \cdot \frac{\partial \overline{R}}{\partial v}\right)^2}\, du\, dv.$$

If F is a function of three variables, defined and continuous on a subset of xyz space that contains the surface Σ, then the *surface integral* of F over Σ, denoted symbolically by $\iint_{\Sigma} F \, dA$, is defined by

$$\iint_{\Sigma} F \, dA = \iint_{D} F(f(u,v), g(u,v), h(u,v)) \sqrt{\left|\frac{\partial \overline{R}}{\partial u}\right|^2 \left|\frac{\partial \overline{R}}{\partial v}\right|^2 - \left(\frac{\partial \overline{R}}{\partial u} \cdot \frac{\partial \overline{R}}{\partial v}\right)^2} \, du \, dv.$$

EXAMPLE   Let Σ denote the surface of a sphere of radius a in xyz space with center at the origin. Evaluate the surface integral

$$\iint_{\Sigma} (x^2 + z) \, dA.$$

SOLUTION

Using longitude θ and colatitude ϕ as parameters, we can write the vector parametric equation of the sphere as

$$\overline{R} = (a \sin \phi \cos \theta)\overline{i} + (a \sin \phi \sin \theta)\overline{j} + (a \cos \phi)\overline{k},$$

where (θ, ϕ) ranges over the rectangular region

$$D : 0 \le \theta \le 2\pi, \quad 0 \le \phi \le \pi$$

in the $\theta\phi$ plane. In Example 1 in Section 10.1, we found that

$$dA = a^2 \sin \phi \, d\theta \, d\phi.$$

Therefore,

$$\iint_{\Sigma} (x^2 + z) \, dA = \iint_{D} [(a \sin \phi \cos \theta)^2 + (a \cos \phi)]a^2 \sin \phi \, d\theta \, d\phi$$

$$= \iint_{D} (a^4 \sin^3 \phi \cos^2 \theta + a^3 \sin \phi \cos \phi) \, d\theta \, d\phi$$

$$= \int_{0}^{\pi} \left[\int_{0}^{2\pi} \left(a^4 \sin^3 \phi \frac{1 + \cos 2\theta}{2} + a^3 \sin \phi \cos \phi \right) d\theta \right] d\phi$$

$$= \int_{0}^{\pi} \left[a^4 \sin^3 \phi \left(\frac{\theta}{2} + \frac{\sin 2\theta}{4} \right) + (a^3 \sin \phi \cos \phi)\theta \Big|_{0}^{2\pi} \right] d\phi$$

$$= \int_{0}^{\pi} (a^4 \pi \sin^3 \phi + 2a^3 \pi \sin \phi \cos \phi) \, d\phi$$

$$= a^4 \pi \int_{0}^{\pi} \sin^3 \phi \, d\phi + 2a^3 \pi \int_{0}^{\pi} \sin \phi \cos \phi \, d\phi$$

$$= a^4 \pi \int_{0}^{\pi} (1 - \cos^2 \phi) \sin \phi \, d\phi + 2a^3 \pi \left(\frac{\sin^2 \phi}{2} \Big|_{0}^{\pi} \right)$$

$$= a^4 \pi \int_{0}^{\pi} \sin \phi \, d\phi + a^4 \pi \int_{0}^{\pi} \cos^2 \phi(-\sin \phi) \, d\phi + 0$$

$$= a^4 \pi \left(-\cos \phi \Big|_{0}^{\pi} \right) + a^4 \pi \left(\frac{\cos^3 \phi}{3} \Big|_{0}^{\pi} \right)$$

$$= 2a^4 \pi - \frac{2}{3} a^4 \pi = \frac{4}{3} a^4 \pi.$$

10.4 The Flux of a Vector Field Through a Surface

Just as a scalar field assigns a scalar to each point in a three-dimensional region S, a *vector field* assigns a vector

$$\bar{F} = M(x, y, z)\bar{i} + N(x, y, z)\bar{j} + P(x, y, z)\bar{k}$$

Figure 7

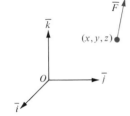

to each point (x, y, z) in S (Figure 7). As the point (x, y, z) moves around in S, the corresponding vector $\bar{F}$ may change, both in magnitude and direction. For instance, if a fluid is flowing through the three-dimensional region S, the vector $\bar{F}$ might represent the velocity of a particle of the fluid at the point (x, y, z). In what follows, we routinely assume that the scalar component functions M, N, and P of the vector field $\bar{F}$ are continuously differentiable.

Now suppose that Σ is a surface in xyz space and that $\bar{N}$ denotes a unit normal vector to Σ at the point (x, y, z) (Figure 8). We assume that, as the point (x, y, z) moves around on the surface Σ, the unit normal vector $\bar{N}$ varies in a continuous manner. Suppose that Σ is contained in the three-dimensional region S on which the vector field $\bar{F}$ is defined. For definiteness, we visualize $\bar{F}$ as the velocity field of a flowing fluid. Consider an "infinitesimal" region of area dA at a point on the surface Σ and let $\bar{N}$ be the unit normal to the surface at this point (Figure 9). If we let one unit of time go by, the fluid flowing through dA will fill an "infinitesimal" solid cylinder of height $h = \bar{F} \cdot \bar{N}$ and with volume

Figure 8

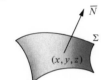

$$dV = h\, dA = \bar{F} \cdot \bar{N}\, dA.$$

Figure 9

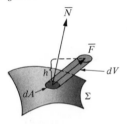

The "infinitesimal" volume dV of fluid flowing through the "infinitesimal" region dA in unit time is called the *flux* through dA. Integrating dV over the whole surface Σ, we obtain the total volume of fluid flowing through Σ in unit time; hence, we define the *flux of the vector field $\bar{F}$ through the surface Σ* to be the surface integral

$$\iint_{\Sigma} dV = \iint_{\Sigma} \bar{F} \cdot \bar{N}\, dA.$$

EXAMPLE   Let Σ denote the portion of the plane $z = x + 2y + 1$ that lies above the region D: $0 \le x \le 1$, $0 \le y \le 2$ in the xy plane. Find the flux of the vector field $\bar{F} = x^3\bar{i} + xy\bar{j} + z\bar{k}$ through the surface Σ in the direction of the normal $\bar{N}$ that makes an acute angle with the z axis.

SOLUTION
Writing the equation of the plane in the form

$$x + 2y - z + 1 = 0,$$

we find that the vector $\bar{N}_1 = \bar{i} + 2\bar{j} - \bar{k}$ is normal to the plane. So is the vector $\bar{N}_2 = -\bar{N}_1 = -\bar{i} - 2\bar{j} + \bar{k}$. Since $\bar{N}_2$ has a positive $\bar{k}$ component, it follows that $\bar{N}_2$ makes an acute angle with the positive z axis. (Why?) Normalizing $\bar{N}_2$, we obtain the desired unit normal vector

$$\bar{N} = \frac{\bar{N}_2}{|\bar{N}_2|} = \frac{-\bar{i} - 2\bar{j} + \bar{k}}{\sqrt{(-1)^2 + (-2)^2 + 1^2}} = \frac{-\bar{i} - 2\bar{j} + \bar{k}}{\sqrt{6}}.$$

Therefore, $\bar{F} \cdot \bar{N} = \dfrac{-x^3 - 2xy + z}{\sqrt{6}}$; hence the flux through Σ is given by

$$\iint_\Sigma \overline{F} \cdot \overline{N} \, dA = \iint_\Sigma \frac{-x^3 - 2xy + z}{\sqrt{6}} \, dA = \iint_D \frac{-x^3 - 2xy + z}{\sqrt{6}} \sqrt{1 + \left(\frac{\partial z}{\partial x}\right)^2 + \left(\frac{\partial z}{\partial y}\right)^2} \, dx \, dy$$

$$= \iint_D \frac{-x^3 - 2xy + x + 2y + 1}{\sqrt{6}} \sqrt{1 + (1)^2 + (2)^2} \, dx \, dy$$

$$= \int_0^2 \int_0^1 (-x^3 - 2xy + x + 2y + 1) \, dx \, dy$$

$$= \int_0^2 \left[\left(\frac{-x^4}{4} - x^2 y + \frac{x^2}{2} + 2xy + x \right) \Big|_0^1 \right] dy = \int_0^2 \left(y + \frac{5}{4} \right) dy$$

$$= \left(\frac{y^2}{2} + \frac{5}{4} y \right) \Big|_0^2 = \frac{9}{2}.$$

Problem Set 10

In problems 1 through 10, find the area A of each surface Σ.

1 Σ is the portion of the plane $x + y + z = 5$ that lies above the circular region $D: x^2 + y^2 \le 9$.

2 Σ is the portion of the cylinder $y^2 + z^2 = 16$ that lies above the triangular region $D: 0 \le x \le 2, 0 \le y \le 2 - x$.

3 Σ is the portion of the plane $3x + 2y + z = 7$ that is cut off by the three coordinate planes.

4 Σ is the portion of the parabolic cylinder $z^2 = 8x$ that lies above the region $D: 0 \le x \le 1$, $0 \le y \le \frac{1}{4}x^2$.

5 Σ is the portion of the cylinder $x^2 + z^2 = 9$ that lies above the rectangular region D with vertices $(0,0)$, $(1,0)$, $(1,2)$, and $(0,2)$.

6 Σ is the portion of the sphere $x^2 + y^2 + z^2 = 36$ that lies above the circular region $D: x^2 + y^2 \le 9$.

7 Σ is the portion of the cylinder $y^2 + z^2 = 4$ inside the cylinder $x^2 = 2y + 4$ and above the plane $z = 0$.

8 Σ is the portion of the cone $z = \sqrt{x^2 + y^2}$ inside the cylinder $x^2 + y^2 = 6y$.

9 Σ is generated by revolving the arc of $y^2 = 4x$ from $(0,0)$ to $(3, 2\sqrt{3})$ about the x axis.

10 Σ is generated by revolving the arc of $x = e^y$ from $(1,0)$ to $(e, 1)$ about the y axis.

11 Let D be an admissible region of area A_0 in the xy plane and let Σ be the portion of the plane $ax + by + cz + d = 0$ consisting of all points (x, y, z) on the plane for which (x, y) is in the region D. Assuming that $c > 0$, show that the area A of Σ is given by

$$A = \frac{A_0}{C} \sqrt{a^2 + b^2 + c^2}.$$

12 Let a and b be positive constants with $a > b$ and suppose that the surface Σ is described vector parametrically by

$$\overline{R} = (a + b \sin u)(\cos v)\overline{i} + (a + b \sin u)(\sin v)\overline{j} + (b \cos u)\overline{k}$$

for (u, v) in the rectangular region $D: 0 \le u \le 2\pi, 0 \le v \le 2\pi$ in the uv plane. Find the area A of the surface Σ.

13 In problem 11, show that $A_0 = A \cos \alpha$, where α is the angle between the plane containing D and the plane containing Σ.

14 Let $\bar{w}$ be a fixed vector of unit length in xyz space such that $\bar{w} \cdot \bar{k} \neq 0$ and let C be a curve with total arc length L in the xy plane having the scalar parametric equations

$$C: \begin{cases} x = f(s) \\ y = g(s), \end{cases} \quad 0 \leq s \leq L,$$

where the parameter s is arc length. Let Σ be the surface described by the vector parametric equation

$$\bar{R} = f(s)\bar{i} + g(s)\bar{j} + t\bar{w},$$

where the parameters s and t satisfy $0 \leq s \leq L$ and $a \leq t \leq b$. Show that the area of Σ is given by $A = L(b - a)$.

15 If c is a constant, what is the value of the surface integral $\iint\limits_{\Sigma} c \, dA$?

16 Let Σ be the portion of the right circular cylinder $x^2 + y^2 = 1$ lying between the plane $z = 0$ and the plane $z = 1$. Evaluate the surface integral $\iint\limits_{\Sigma} (x + y + z) \, dA$. (*Hint:* Use cylindrical coordinates.)

17 Evaluate $\iint\limits_{\Sigma} (y + z^2) \, dA$, where Σ is the sphere of radius 1 with center at the origin.

18 Let Σ be the right circular cone (with open base) whose vertex is at the origin and whose base is a circle of radius b which is parallel to the xy plane and h units above it. Evaluate $\iint\limits_{\Sigma} xyz \, dA$.

19 Let Σ be the portion of the plane $z = x + 4y + 5$ that lies above the region $D: 0 \leq x \leq 1$, $0 \leq y \leq 1$ in the xy plane. Evaluate $\iint\limits_{\Sigma} (2xy - z) \, dA$.

20 Suppose that the surface Σ is an admissible region in the xy plane and that F is a function of three variables, defined and continuous on a subset of xyz space that contains Σ. Show that, in this case, the surface integral $\iint\limits_{\Sigma} F \, dA$ is the same as the double integral $\iint\limits_{\Sigma} F(x, y, 0) \, dx \, dy$.

In problems 21 through 25, find the flux of each vector field $\bar{F}$ through the indicated surface Σ in the direction of the given unit normal vector $\bar{N}$.

21 $\bar{F} = yz\bar{i} + xz\bar{j} + xy\bar{k}$, Σ is the portion of the plane $z = x + y + 1$ that lies above the region $D: 0 \leq x \leq 1, 0 \leq y \leq 1$ in the xy plane, $\bar{N}$ is the unit normal to Σ that makes an acute angle with the positive z axis.

22 $\bar{F} = x\bar{i} + y\bar{j} + z\bar{k}$, Σ is the sphere of radius 1 with center at the origin, $\bar{N}$ is the unit normal to Σ that points away from the center.

23 $\bar{F} = x^2\bar{i} + xy\bar{j} + z\bar{k}$, Σ is the triangle with vertices $(1, 0, 0)$, $(0, 2, 0)$, and $(0, 0, 3)$, $\bar{N}$ is the unit normal to Σ that makes an acute angle with the positive x axis.

24 $\bar{F} = x^2\bar{i} + y^2\bar{j} + z^2\bar{k}$, Σ is the entire surface of the tetrahedron with vertices $(1, 0, 0)$, $(0, 1, 0)$, $(0, 0, 1)$, and $(0, 0, 0)$, $\bar{N}$ is the unit normal that points away from the interior of the tetrahedron. (*Hint:* Integrate over each of the four triangular faces and add the resulting numbers.)

25 $\bar{F} = \dfrac{1}{x}i + \dfrac{1}{y}j + \dfrac{1}{z}\bar{k}$, Σ is the portion of the right circular cylinder $x^2 + y^2 = 3$ that lies between the plane $z = 1$ and the plane $z = 2$, $\bar{N}$ is the unit normal to Σ that points away from the z axis.

26 If $\bar{R}$ is a variable position vector sweeping out a surface Σ and if u and v are the parameters, show that the vector $\dfrac{\partial \bar{R}}{\partial u} \times \dfrac{\partial \bar{R}}{\partial v}$ is normal to Σ and show that the flux of a vector field $\bar{F}$ through Σ in the direction of this normal is given by $\displaystyle\iint_{\Sigma} \bar{F} \cdot \left(\dfrac{\partial \bar{R}}{\partial u} \times \dfrac{\partial \bar{R}}{\partial v} \right) du\, dv$.

11 The Divergence Theorem and Stokes' Theorem

The *divergence theorem* and *Stokes' theorem* are generalizations to three-dimensional space of Green's theorem in the plane (Section 9.4). Before we present these theorems, we introduce the ideas of the *divergence* and the *curl* of a vector field.

11.1 Divergence and Curl

In dealing with scalar and vector fields it is useful to introduce the "symbolic vector" ∇ defined by

$$\nabla = i\,\frac{\partial}{\partial x} + j\,\frac{\partial}{\partial y} + \bar{k}\,\frac{\partial}{\partial z}.$$

Of course, ∇ is not really a vector at all, but merely an incomplete symbol that makes no sense unless it is "multiplied" by a scalar field or a vector field. For instance, if $w = f(x, y, z)$ is a scalar field, then

$$\nabla w = i\,\frac{\partial w}{\partial x} + j\,\frac{\partial w}{\partial y} + \bar{k}\,\frac{\partial w}{\partial z} = \frac{\partial w}{\partial x}\,i + \frac{\partial w}{\partial y}\,j + \frac{\partial w}{\partial z}\,\bar{k} = \text{the gradient of } w,$$

which explains why the notation ∇w is used for the gradient of w.

If $\bar{F} = M(x, y, z)i + N(x, y, z)j + P(x, y, z)\bar{k}$ is a vector field, then we can form either the "dot product" $\nabla \cdot \bar{F}$ or the "cross product" $\nabla \times \bar{F}$ of the "symbolic vector" ∇ with $\bar{F}$. The "dot product" $\nabla \cdot \bar{F}$ is called the *divergence* of $\bar{F}$, abbreviated div $\bar{F}$, while the "cross product" $\nabla \times \bar{F}$ is called the *curl* or *rotation* of $\bar{F}$, abbreviated curl $\bar{F}$ or rot $\bar{F}$. Thus, we have

$$\text{div } \bar{F} = \nabla \cdot \bar{F} = \left(i\,\frac{\partial}{\partial x} + j\,\frac{\partial}{\partial y} + \bar{k}\,\frac{\partial}{\partial z} \right) \cdot (Mi + Nj + P\bar{k}) = \frac{\partial M}{\partial x} + \frac{\partial N}{\partial y} + \frac{\partial P}{\partial z}$$

and

$$\text{curl } \bar{F} = \text{rot } \bar{F} = \nabla \times \bar{F} = \begin{vmatrix} i & j & \bar{k} \\[4pt] \dfrac{\partial}{\partial x} & \dfrac{\partial}{\partial y} & \dfrac{\partial}{\partial z} \\[6pt] M & N & P \end{vmatrix}$$

$$= \left(\frac{\partial P}{\partial y} - \frac{\partial N}{\partial z} \right) i + \left(\frac{\partial M}{\partial z} - \frac{\partial P}{\partial x} \right) j + \left(\frac{\partial N}{\partial x} - \frac{\partial M}{\partial y} \right) \bar{k}.$$

If the vector field $\bar{F}$ is thought of as the velocity field of a flowing fluid, then $\nabla \cdot \bar{F}$ and $\nabla \times \bar{F}$ have interesting physical interpretations. The scalar $\nabla \cdot \bar{F}$ is a measure of the tendency of the velocity vectors to "diverge" from one another. For instance, if the fluid is flowing with a constant velocity (Figure 1a), then the velocity vectors are parallel to one another, and the divergence is zero; however, near a "source" of fluid (Figure 1b) the divergence would be rather large. On the other hand, if the

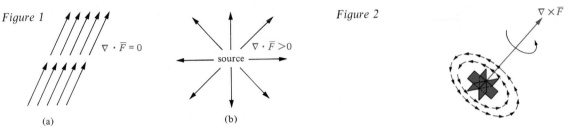

Figure 1

$\nabla \cdot \bar{F} = 0$

$\nabla \cdot \bar{F} > 0$

source

(a)

(b)

Figure 2

$\nabla \times \bar{F}$

tail end of the vector $\nabla \times \bar{F}$ were equipped with small vanes (Figure 2), then the flowing fluid would cause the vanes to rotate with an angular speed proportional to $|\nabla \times \bar{F}|$.

EXAMPLES 1 Let $\bar{F} = 2xy\bar{i} + 4yz\bar{j} - xz\bar{k}$. Find (a) $\nabla \cdot \bar{F}$ and (b) $\nabla \times \bar{F}$.

SOLUTION

(a) $\nabla \cdot \bar{F} = \dfrac{\partial}{\partial x}(2xy) + \dfrac{\partial}{\partial y}(4yz) + \dfrac{\partial}{\partial z}(-xz) = 2y + 4z - x.$

(b) $\nabla \times \bar{F} = \begin{vmatrix} \bar{i} & \bar{j} & \bar{k} \\ \dfrac{\partial}{\partial x} & \dfrac{\partial}{\partial y} & \dfrac{\partial}{\partial z} \\ 2xy & 4yz & -xz \end{vmatrix}$

$= \left[\dfrac{\partial}{\partial y}(-xz) - \dfrac{\partial}{\partial z}(4yz) \right]\bar{i} + \left[\dfrac{\partial}{\partial z}(2xy) - \dfrac{\partial}{\partial x}(-xz) \right]\bar{j}$

$+ \left[\dfrac{\partial}{\partial x}(4yz) - \dfrac{\partial}{\partial y}(2xy) \right]\bar{k}$

$= (-4y)\bar{i} + [-(-z)]\bar{j} + (-2x)\bar{k} = -4y\bar{i} + z\bar{j} - 2x\bar{k}.$

2 If $\bar{A}$ is a constant vector and $\bar{R} = x\bar{i} + y\bar{j} + z\bar{k}$, find (a) div $\bar{R}$ and (b) curl $(\bar{A} \times \bar{R})$.

SOLUTION

(a) div $\bar{R} = \dfrac{\partial}{\partial x}x + \dfrac{\partial}{\partial y}y + \dfrac{\partial}{\partial z}z = 1 + 1 + 1 = 3.$

(b) Let $\bar{A} = a\bar{i} + b\bar{j} + c\bar{k}$. Then,

$$\bar{A} \times \bar{R} = \begin{vmatrix} \bar{i} & \bar{j} & \bar{k} \\ a & b & c \\ x & y & z \end{vmatrix} = (bz - cy)\bar{i} + (cx - az)\bar{j} + (ay - bx)\bar{k};$$

hence,

$$\text{curl }(\bar{A} \times \bar{R}) = \begin{vmatrix} \bar{i} & \bar{j} & \bar{k} \\ \dfrac{\partial}{\partial x} & \dfrac{\partial}{\partial y} & \dfrac{\partial}{\partial z} \\ bz - cy & cx - az & ay - bx \end{vmatrix} = 2a\bar{i} + 2b\bar{j} + 2c\bar{k} = 2\bar{A}.$$

11.2 The Divergence Theorem

The *divergence theorem*, also called *Gauss's theorem* in honor of the renowned German mathematician Karl Friedrich Gauss (1777–1855), effects a profound connection between the divergence and the flux of a vector field, and represents a generalization to three-dimensional space of Green's theorem in the plane (see Problem 26 in Problem Set 9). The following is a somewhat informal statement of the theorem.

THEOREM 1   **The Divergence Theorem of Gauss**

Let S be a closed and bounded region in xyz space whose boundary is a piecewise smooth surface Σ. Suppose that $\overline{F}$ is a vector field defined on an open set U containing S, and assume that the scalar component functions of $\overline{F}$ are continuously differentiable on U. Let $\overline{N}$ denote the outward-pointing unit normal vector to the surface Σ. Then

$$\iiint_S \nabla \cdot \overline{F} \, dx \, dy \, dz = \iint_\Sigma \overline{F} \cdot \overline{N} \, dA.$$

In words, the integral over a solid S of the divergence of a vector field is the flux of the field through the boundary of the solid. In particular, if the integrand of a triple integral can be expressed as the divergence of a vector field, then the value of the integral depends only on the vectors on the surface that encloses the volume! Unfortunately, the proof of the divergence theorem is beyond the scope of this book.

EXAMPLES 1   Let $\overline{F} = (2x - z)\overline{i} + x^2 y \overline{j} + xz^2 \overline{k}$ and suppose that S is the solid cube bounded by the planes $x = 0$, $x = 1$, $y = 0$, $y = 1$, $z = 0$, and $z = 1$ (Figure 3). If Σ denotes the surface of S, use the divergence theorem to evaluate $\iint_\Sigma \overline{F} \cdot \overline{N} \, dA$, where $\overline{N}$ is the outward-pointing unit normal vector to Σ.

SOLUTION
By the divergence theorem,

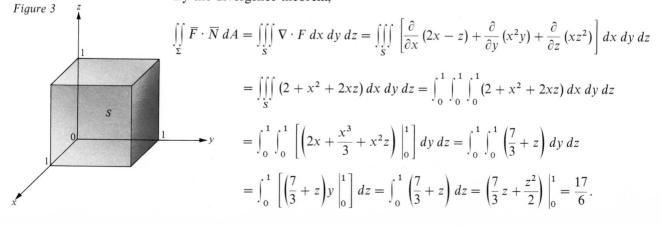

Figure 3

$$\iint_\Sigma \overline{F} \cdot \overline{N} \, dA = \iiint_S \nabla \cdot F \, dx \, dy \, dz = \iiint_S \left[\frac{\partial}{\partial x}(2x - z) + \frac{\partial}{\partial y}(x^2 y) + \frac{\partial}{\partial z}(xz^2) \right] dx \, dy \, dz$$

$$= \iiint_S (2 + x^2 + 2xz) \, dx \, dy \, dz = \int_0^1 \int_0^1 \int_0^1 (2 + x^2 + 2xz) \, dx \, dy \, dz$$

$$= \int_0^1 \int_0^1 \left[\left(2x + \frac{x^3}{3} + x^2 z \right) \Big|_0^1 \right] dy \, dz = \int_0^1 \int_0^1 \left(\frac{7}{3} + z \right) dy \, dz$$

$$= \int_0^1 \left[\left(\frac{7}{3} + z \right) y \Big|_0^1 \right] dz = \int_0^1 \left(\frac{7}{3} + z \right) dz = \left(\frac{7}{3} z + \frac{z^2}{2} \right) \Big|_0^1 = \frac{17}{6}.$$

Figure 4

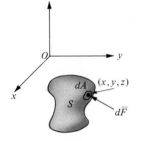

2 Use the divergence theorem to prove Archimedes' principle: The buoyant force on a solid S immersed in a fluid of constant density is equal to the weight of the displaced fluid.

SOLUTION

Let the xyz coordinate system be placed so that the solid S lies below the xy plane and the z axis points straight upward as usual (Figure 4). Since the pressure P of the fluid varies linearly with the depth, the pressure at the point (x, y, z) on the surface Σ of S is given by

$$P = a - \delta z,$$

where a is a constant and δ is the weight of a unit volume of the fluid. The "infinitesimal" force $d\bar{F}$ caused by this pressure on an "infinitesimal" area dA has magnitude

$$|d\bar{F}| = \text{pressure times area} = (a - \delta z)\, dA$$

and is directed *inward* along the normal to the surface. Thus, if $\bar{N}$ is the *outward-pointing* unit normal to Σ, then

$$d\bar{F} = |d\bar{F}|(-\bar{N}) = -(a - \delta z)\bar{N}\, dA = (\delta z - a)\bar{N}\, dA.$$

The vertical component of $d\bar{F}$ represents the magnitude of the "infinitesimal" buoyant force dF_b acting on dA; hence,

$$dF_b = \bar{k} \cdot d\bar{F} = \bar{k} \cdot [(\delta z - a)\bar{N}\, dA] = (\delta z - a)\bar{k} \cdot \bar{N}\, dA.$$

The magnitude of the net buoyant force on the entire surface Σ is obtained by summing—that is, integrating—dF_b over Σ; hence,

$$F_b = \iint_{\Sigma} (\delta z - a)\bar{k} \cdot \bar{N}\, dA.$$

Here we have

$$\nabla \cdot (\delta z - a)\bar{k} = \frac{\partial}{\partial z}(\delta z - a) = \delta.$$

Therefore, by the divergence theorem,

$$F_b = \iint_{\Sigma} (\delta z - a)\bar{k} \cdot \bar{N}\, dA = \iiint_{S} \nabla \cdot (\delta z - a)\bar{k}\, dx\, dy\, dz = \iiint_{S} \delta\, dx\, dy\, dz$$

$$= \delta \iiint_{S} dx\, dy\, dz = \delta V,$$

where V is the volume of S. Since δ is the weight of a unit volume of the fluid, it follows that δV is the weight of the displaced fluid.

3 Let D be a closed admissible region in the xy plane and suppose that g and h are continuous functions defined on D and satisfying $g(x,y) \le h(x,y)$ for all points (x,y) in D. Let S be the solid consisting of all points (x,y,z) satisfying the conditions that (x,y) belongs to D and $g(x,y) \le z \le h(x,y)$ (Figure 5). Suppose that f is a function of three variables that is continuously differentiable on some open set U containing S, and let $\overline{F}$ be the vector field defined by $\overline{F} = f(x,y,z)\overline{k}$. Verify the divergence theorem for the vector field $\overline{F}$ and the solid S.

Figure 5

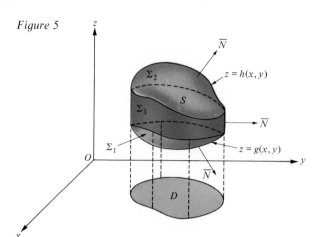

SOLUTION

The surface Σ of S consists of the three parts

Σ_1: the graph of $z = g(x,y)$ for (x,y) in D,

Σ_2: the graph of $z = h(x,y)$ for (x,y) in D,

Σ_3: the portion of the cylinder over the boundary of D, with vertical generators, cut off by Σ_1 and Σ_2.

Let $\overline{N}$ denote the outward-pointing unit normal vector to the surface Σ. The vector

$$g_1(x,y)\overline{i} + g_2(x,y)\overline{j} - \overline{k}$$

is normal to Σ_1 (Section 7.4, Chapter 16, page 867) and points outward from the solid S, since its $\overline{k}$ component is negative; hence, on Σ_1, we have

$$\overline{N} = \frac{g_1(x,y)\overline{i} + g_2(x,y)\overline{j} - \overline{k}}{\sqrt{[g_1(x,y)]^2 + [g_2(x,y)]^2 + 1}}.$$

Also, on Σ_1, the differential of surface area is given by

$$dA = \sqrt{[g_1(x,y)]^2 + [g_2(x,y)]^2 + 1}\, dx\, dy$$

(Section 10.2). Therefore, on Σ_1,

$$\overline{N}\, dA = [g_1(x,y)\overline{i} + g_2(x,y)\overline{j} - \overline{k}]\, dx\, dy \quad \text{and}$$

$$\overline{F} \cdot \overline{N}\, dA = f(x,y,z)\overline{k} \cdot [g_1(x,y)\overline{i} + g_2(x,y)\overline{j} - \overline{k}]\, dx\, dy = -f(x,y,z)\, dx\, dy$$

$$= -f(x,y,g(x,y))\, dx\, dy.$$

Consequently,

$$\iint_{\Sigma_1} \overline{F} \cdot \overline{N}\, dA = -\iint_D f(x,y,g(x,y))\, dx\, dy.$$

Reasoning in a similar way, but noting that the $\overline{k}$ component of $\overline{N}$ on Σ_2 is positive, we find that

$$\iint_{\Sigma_2} \overline{F} \cdot \overline{N}\, dA = \iint_D f(x,y,h(x,y))\, dx\, dy.$$

Since $\overline{N}$ is parallel to the xy plane on the cylinder Σ_3, it follows that

$$\overline{F} \cdot \overline{N} = f(x,y,z)\overline{k} \cdot \overline{N} = f(x,y,z)(0) = 0 \text{ on } \Sigma_3;$$

hence,

$$\iint_{\Sigma_3} \overline{F} \cdot \overline{N}\, dA = 0.$$

Therefore,

$$\iint_{\Sigma} \overline{F} \cdot \overline{N}\, dA = \iint_{\Sigma_1} \overline{F} \cdot \overline{N}\, dA + \iint_{\Sigma_2} \overline{F} \cdot \overline{N}\, dA + \iint_{\Sigma_3} \overline{F} \cdot \overline{N}\, dA$$

$$= -\iint_{D} f(x, y, g(x, y))\, dx\, dy + \iint_{D} f(x, y, h(x, y))\, dx\, dy + 0$$

$$= \iint_{D} [f(x, y, h(x, y)) - f(x, y, g(x, y))]\, dx\, dy$$

$$= \iint_{D} \left[f(x, y, z) \Big|_{z=g(x,y)}^{z=h(x,y)} \right] dx\, dy = \iint_{D} \left[\int_{g(x,y)}^{h(x,y)} \frac{\partial f}{\partial z}\, dz \right] dx\, dy$$

$$= \iiint_{S} \frac{\partial f}{\partial z}\, dx\, dy\, dz = \iiint_{S} \nabla \cdot \overline{F}\, dx\, dy\, dz,$$

in conformity with the divergence theorem.

11.3 Stokes' Theorem

We now turn to another generalization of Green's theorem, which is attributed to the Irish mathematical physicist Sir George G. Stokes (1819–1903). Roughly speaking, Stokes' theorem says that the flux of the curl of a vector field $\overline{F}$ through a surface Σ is equal to the line integral of the tangential component of $\overline{F}$ around the boundary of Σ.

To be more precise, suppose that Σ is a smooth surface and that $\overline{N}$ denotes a unit normal vector to Σ which varies continuously as we move around on the surface (Figure 6). Furthermore, we assume that the boundary of Σ consists of a single closed curve C in xyz space. Imagine standing on the boundary curve C with your head pointing in the direction of the normal vector $\overline{N}$ and with the surface Σ on your left. Now, if you walk forward along C, you will, by definition, move in the *positive* direction around the boundary. If we wish to describe C, or a part of C, parametrically, we always choose the parameter t so that, when t increases, we move along C in the positive direction. With this understanding, we can now give an informal statement of Stokes' theorem.

Figure 6

positive direction
around the boundary

THEOREM 2   **Stokes' Theorem**

Let $\overline{F}$ be a vector field whose component functions are continuously differentiable on an open set U containing the surface Σ and its boundary curve C. Then

$$\int_{C} \overline{F} \cdot d\overline{R} = \iint_{\Sigma} (\nabla \times \overline{F}) \cdot \overline{N}\, dA.$$

The line integral $\displaystyle\int_{C} \overline{F} \cdot d\overline{R}$ around the *closed* curve C is called the *circulation* of the vector field $\overline{F}$ around C. In particular, if $\overline{F}$ represents a force field, then the circulation of $\overline{F}$ around C is the total work done by the force $\overline{F}$ in carrying a particle once around the closed curve C. Thus, Stokes' theorem says that the circulation of a vector field around the boundary of a surface in xyz space is equal to the flux of the curl of the field through the surface. Unfortunately, the proof of Stokes' theorem is beyond the scope of this book.

EXAMPLE   Let Σ be the portion of the paraboloid of revolution

$$z = \frac{1}{2}(x^2 + y^2)$$

cut off by the plane $z = 2$ and lying below this plane, and let $\overline{N}$ be the unit normal vector to Σ that makes an acute angle with the positive z axis. If C is the boundary curve of Σ and $\overline{F}$ is the vector field

$$\overline{F} = 3y\overline{i} - xz\overline{j} + yz^2\overline{k},$$

verify Stokes' theorem for $\overline{F}$ and Σ.

SOLUTION

Proceeding as in Example 3 in Section 11.2, we find that

$$\overline{N}\,dA = -\left(\frac{\partial z}{\partial x}\overline{i} + \frac{\partial z}{\partial y}\overline{j} - \overline{k}\right)dx\,dy = (-x\overline{i} - y\overline{j} + \overline{k})\,dx\,dy.$$

Here,

$$\nabla \times \overline{F} = \begin{vmatrix} \overline{i} & \overline{j} & \overline{k} \\ \dfrac{\partial}{\partial x} & \dfrac{\partial}{\partial y} & \dfrac{\partial}{\partial z} \\ 3y & -xz & yz^2 \end{vmatrix} = (z^2 + x)\overline{i} + 0\overline{j} + (-z - 3)\overline{k};$$

hence,

$$(\nabla \times \overline{F}) \cdot \overline{N}\,dA = [(z^2 + x)(-x) + 0(-y) + (-z - 3)(1)]\,dx\,dy$$
$$= (-z^2 x - x^2 - z - 3)\,dx\,dy.$$

Notice that the surface Σ lies over the circular region

$$D: x^2 + y^2 \leq 4$$

in the xy plane. Therefore,

$$\iint_{\Sigma} (\nabla \times \overline{F}) \cdot \overline{N}\,dA = \iint_{\Sigma} (-z^2 x - x^2 - z - 3)\,dx\,dy$$

$$= \iint_{D} \left[-\frac{1}{4}(x^2 + y^2)^2 x - x^2 - \frac{1}{2}(x^2 + y^2) - 3\right] dx\,dy.$$

Switching to polar coordinates, we find that the flux through Σ of the curl of $\overline{F}$ is given by

$$\iint_{\Sigma} (\nabla \times \overline{F}) \cdot \overline{N}\,dA = \int_0^{2\pi} \int_0^2 \left(-\frac{1}{4}r^4 r \cos\theta - r^2 \cos^2\theta - \frac{1}{2}r^2 - 3\right) r\,dr\,d\theta$$

$$= \int_0^{2\pi} \int_0^2 \left(-\frac{r^6}{4}\cos\theta - r^3 \cos^2\theta - \frac{r^3}{2} - 3r\right) dr\,d\theta$$

$$= \int_0^{2\pi} \left[\left(-\frac{r^7}{28}\cos\theta - \frac{r^4}{4}\cos^2\theta - \frac{r^4}{8} - \frac{3r^2}{2}\right)\Big|_0^2\right] d\theta$$

$$= \int_0^{2\pi} \left(-\frac{32}{7}\cos\theta - 4\cos^2\theta - 8\right) d\theta$$

$$= \int_0^{2\pi} \left[-\frac{32}{7}\cos\theta - 2(1 + \cos 2\theta) - 8\right] d\theta$$

$$= \int_0^{2\pi} \left(-\frac{32}{7}\cos\theta - 2\cos 2\theta - 10\right) d\theta$$

$$= \left(-\frac{32}{7}\sin\theta - \sin 2\theta - 10\theta\right)\Big|_0^{2\pi} = -20\pi.$$

Now we calculate the circulation $\int_C \bar{F} \cdot d\bar{R}$. Notice that C is described vector parametrically by

$$\bar{R} = (2 \cos t)\bar{i} + (2 \sin t)\bar{j} + 2\bar{k}, \qquad 0 \le t \le 2\pi,$$

so that

$$d\bar{R} = [(-2 \sin t)\bar{i} + (2 \cos t)\bar{j}]\, dt$$

and

$$\begin{aligned}\bar{F} \cdot d\bar{R} &= [(3y)(-2 \sin t) + (-xz)(2 \cos t)]\, dt \\ &= [3(2 \sin t)(-2 \sin t) + (-2 \cos t)(2)(2 \cos t)]\, dt \\ &= (-12 \sin^2 t - 8 \cos^2 t)\, dt.\end{aligned}$$

Therefore,

$$\int_C \bar{F} \cdot d\bar{R} = \int_0^{2\pi} (-12 \sin^2 t - 8 \cos^2 t)\, dt = -\int_0^{2\pi} (4 \sin^2 t + 8 \sin^2 t + 8 \cos^2 t)\, dt$$

$$= -\int_0^{2\pi} (4 \sin^2 t + 8)\, dt = -\int_0^{2\pi} [2(1 - \cos 2t) + 8]\, dt$$

$$= -\int_0^{2\pi} (10 - 2 \cos 2t)\, dt = -(10t - \sin 2t)\Big|_0^{2\pi} = -20\pi,$$

in conformity with Stokes' theorem.

Problem Set 11

In problems 1 through 4, find (a) $\nabla \cdot \bar{F}$ and (b) $\nabla \times \bar{F}$.

1 $\bar{F} = xy^2\bar{i} - x^2\bar{j} + (x + y)\bar{k}$

2 $\bar{F} = (z^2 - x)\bar{i} - xy\bar{j} + 3z\bar{k}$

3 $\bar{F} = 3xyz^2\bar{i} + (5x^2y + z)\bar{j} + 2y^2z^3\bar{k}$

4 $\bar{F} = x(\cos y)\bar{i} + 3x^2(\sin y)\bar{j} + xz^2\bar{k}$

In problems 5 through 8, use the divergence theorem to evaluate the flux $\iint_\Sigma \bar{F} \cdot \bar{N}\, dA$ through the boundary Σ of the indicated solid S of the given vector field $\bar{F}$. Here, $\bar{N}$ denotes the outward-pointing unit normal vector to Σ.

5 $\bar{F} = (2xy + z)\bar{i} + y^2\bar{j} - (x + 3y)\bar{k}$; S is the solid bounded by the planes $2x + 2y + z = 6$, $x = 0$, $y = 0$, and $z = 0$.

6 $\bar{F} = x^2\bar{i} + y^2\bar{j} + z^2\bar{k}$; S is the solid cube bounded by the planes $x = 0$, $x = 1$, $y = 0$, $y = 1$, $z = 0$, and $z = 1$.

7 $\bar{F} = yz\bar{i} + xz\bar{j} + xy\bar{k}$; S is the solid sphere $x^2 + y^2 + z^2 \le 1$.

8 $\bar{F} = (z^2 - x)\bar{i} - xy\bar{j} + 3z\bar{k}$; S is the solid bounded above by the parabolic cylinder $z = 9 - y^2$, in back by the plane $x = 0$, in front by the plane $x = 4$, and below by the xy plane.

In problems 9 through 12, use Stokes' theorem to evaluate the flux $\iint_\Sigma (\nabla \times \bar{F}) \cdot \bar{N}\, dA$ of the curl of each vector field $\bar{F}$ through the indicated surface Σ in the direction of the given unit normal vector $\bar{N}$.

9 $\bar{F} = y^2\bar{i} + xy\bar{j} + xz\bar{k}$; Σ is the hemisphere $x^2 + y^2 + z^2 = 9$, $z \ge 0$; $\bar{N}$ has a nonnegative $\bar{k}$ component.

10 $\bar{F} = y\bar{i} + z\bar{j} + x\bar{k}$; Σ is the portion of the paraboloid of revolution $z = 1 - x^2 - y^2$ for which $z \geq 0$; $\bar{N}$ has a nonnegative $\bar{k}$ component.

11 $\bar{F} = (z + y)\bar{i} + (z + x)\bar{j} + (x + y)\bar{k}$; Σ is the triangle with vertices $(1,0,0)$, $(0,1,0)$, and $(0,0,1)$; $\bar{N}$ is the unit normal vector whose components are all positive.

12 $\bar{F} = \bar{R}/r^3$, $\bar{R} = x\bar{i} + y\bar{j} + z\bar{k}$, $r = |\bar{R}|$; Σ is the portion of the ellipsoid $\dfrac{x^2}{4} + \dfrac{y^2}{4} + \dfrac{z^2}{9} = 1$

for which $z \leq 0$; $\bar{N}$ is the unit normal vector whose $\bar{k}$ component is nonnegative.

In problems 13 and 14, verify the divergence theorem by direct calculation for the given vector field $\bar{F}$ and the indicated solid S.

13 $\bar{F} = x\bar{i} + y\bar{j} + z\bar{k}$; S is the solid cube bounded by the planes $x = 0$, $x = 1$, $y = 0$, $y = 1$, $z = 0$, and $z = 1$.

14 $\bar{F} = g(x, y, z)\bar{j}$, where g is a continuously differentiable function defined on an open set U containing the solid sphere $x^2 + y^2 + z^2 \leq 1$.

In problems 15 and 16, verify Stokes' theorem for the given vector field $\bar{F}$, the indicated surface Σ, and the given unit normal vector $\bar{N}$ to Σ.

15 $\bar{F} = (2y + z)\bar{i} + (x - z)\bar{j} + (y - x)\bar{k}$; Σ is the triangle cut from the plane $x + y + z = 1$ by the coordinate planes; $\bar{N}$ is the unit normal vector whose components are all positive.

16 $\bar{F} = 2y\bar{i} - x\bar{j} + z\bar{k}$; Σ is the hemisphere $x^2 + y^2 + z^2 = 4$, $z \geq 0$; $\bar{N}$ is the unit normal vector whose $\bar{k}$ component is nonnegative.

17 Prove the divergence theorem for the special case in which S is the solid cube bounded by the planes $x = 0$, $x = 1$, $y = 0$, $y = 1$, $z = 0$, and $z = 1$.

18 Let $\bar{F} = f(x, y, z)\bar{i} + g(x, y, z)\bar{j} + h(x, y, z)\bar{k}$. Explain why the surface integral $\displaystyle\iint_{\Sigma} \bar{F} \cdot \bar{N}\, dA$ is

sometimes written as

$$\iint_{\Sigma} f(x, y, z)\, dy\, dz + g(x, y, z)\, dz\, dx + h(x, y, z)\, dx\, dy.$$

19 Assuming the existence and continuity of the required partial derivatives, prove that (a) the curl of the gradient of a scalar field is zero and (b) the divergence of the curl of a vector field is zero.

20 Show that the flux of the curl of a vector field through the surface Σ of a solid S is zero. State carefully what assumptions you need to make.

21 Suppose that Σ is an admissible region in the xy plane whose boundary C forms a piecewise smooth simple closed curve. Show that Stokes' theorem for Σ is essentially just Green's theorem in the plane.

22 Show that a solid S immersed in a fluid of constant density experiences no net horizontal force because of the fluid pressure.

Review Problem Set

In problems 1 through 8, evaluate each iterated integral.

1 $\displaystyle\int_0^1 \int_0^y xy^2 \, dx \, dy$

2 $\displaystyle\int_0^4 \int_0^{\sqrt{x}} y\sqrt{x + y^2}\, dy\, dx$

3 $\displaystyle\int_1^2 \int_y^{y^2} (x + 2y)\, dx\, dy$

4 $\displaystyle\int_0^2 \int_0^{\sqrt{4 - x^2}} (x + y)\, dy\, dx$

5 $\displaystyle\int_1^2 \int_1^{x^2} x^3 y e^{y^2}\, dy\, dx$

6 $\displaystyle\int_3^5 \int_1^y \frac{y}{x \ln y}\, dx\, dy$

7 $\displaystyle\int_3^5 \int_{\pi/6}^{\pi/3} r^2 \sin\theta\, d\theta\, dr$

8 $\displaystyle\int_0^{\pi} \int_0^{2\cos\theta} r \sin\theta\, dr\, d\theta$

In problems 9 through 12, replace each iterated integral by an equivalent integral with the order of integration reversed, and then evaluate the integral obtained. Sketch the appropriate region.

9 $\displaystyle\int_0^4 \int_{4-(x/2)}^{\sqrt{4-x}} y \, dy \, dx$

10 $\displaystyle\int_0^3 \int_x^{4x-x^2} dy \, dx$

11 $\displaystyle\int_0^6 \int_{(2/3)(6-y)}^{(1/9)(36-y^2)} y^2 x \, dx \, dy$

12 $\displaystyle\int_0^6 \int_{y^3/18}^{2y} xy^2 \, dx \, dy$

In problems 13 through 16, evaluate each double integral over the region indicated.

13 $\displaystyle\iint_R (\sqrt{y} + x - 3xy^2)\, dx \, dy;\; R.\; 0 \le x \le 1,\, 1 \le y \le 3$

14 $\displaystyle\iint_R \sin(x+y)\, dx \, dy;\; R: 0 \le x \le \pi/2,\, 0 \le y \le \pi/2$

15 $\displaystyle\iint_R e^{x^2}\, dx \, dy;\; R: 0 \le x \le 2,\, 0 \le y \le x$

16 $\displaystyle\iint_R \frac{ds\, dt}{4-s};\; R: 2 \le s \le 3,\, 0 \le t \le s$

In problems 17 through 20, use double integration to find the area of the region R bounded by the given pair of curves.

17 $y = 4x - x^2$ and $y = x$.

18 $y^2 = 4x$ and $2x - y = 4$.

19 $4y^2 = x^3$ and $x = y$.

20 $y^2 = 4x$ and $x = 12 + 2y - y^2$.

In problems 21 through 26, express each iterated integral as an equivalent iterated integral in polar coordinates, and then evaluate the integral.

21 $\displaystyle\int_0^{10} \int_0^{\sqrt{100-x^2}} \sqrt{x^2+y^2}\, dy \, dx$

22 $\displaystyle\int_0^4 \int_0^x (x^2+y^2)^{3/2}\, dy \, dx$

23 $\displaystyle\int_0^2 \int_0^{\sqrt{4-y^2}} (1 - x^2 - y^2)^2\, dx \, dy$

24 $\displaystyle\int_0^2 \int_0^{\sqrt{24-y^2}} \sqrt{x^2+y^2}\, dx \, dy$

25 $\displaystyle\int_{-1}^1 \int_{-\sqrt{1-x^2}}^{\sqrt{1-x^2}} xy \, dy \, dx$

26 $\displaystyle\int_{-3}^3 \int_0^{\sqrt{9-y^2}} \frac{y}{\sqrt{x^2+y^2}}\, dx \, dy$

In problems 27 through 30, find the volume V of the solid under the graph of the given function f and above the indicated region R in the xy plane.

27 $f(x,y) = 4 - x^2;\; R: 0 \le y \le 2,\, y^2/2 \le x \le 2$

28 $f(x,y) = 2 - x;\; R: x^2 + y^2 \le 4$

29 $f(x,y) = 8 - x - y;\; R:$ the triangular region bounded by $x + y = 8$, $x + 2y = 8$, and $x = 0$.

30 $f(x,y) = x^2 + y^2;\; R:$ the region bounded by the cardioid whose polar equation is $r = 1 - \sin\theta$.

31 The volume V under the hyperbolic paraboloid $z = xy$ and above a region R in the xy plane is given by

$$V = \int_0^1 \int_0^y xy \, dx \, dy + \int_1^2 \int_0^{2-y} xy \, dx \, dy.$$

Sketch the region R in the xy plane, express V as a double integral in which the order of integration is reversed, and evaluate V.

32 If $b > a > 0$, show that $\displaystyle\int_0^\infty \frac{e^{-ax} - e^{-bx}}{x} \, dx = \ln\frac{b}{a}$ by using the fact that $\displaystyle\int_a^b e^{-xy} \, dy = \frac{e^{-ax} - e^{-bx}}{x}$, forming a suitable double integral, and reversing the order of integration.

In problems 33 through 36, use double integration to find the centroid of the region R.

33 $R: 0 \le x \le \pi/4,\, 0 \le y \le \sec^2 x$

34 $R: 0 \le y \le \ln 4,\, e^y \le x \le 4$

35 $R: 0 \le \theta \le \pi/3,\, 0 \le r \le \sin 2\theta$

36 $R:$ The triangular region with vertices $(0,0)$, $(b,0)$, and (c,h), where b, c, and h are positive constants.

37 Find the moments of inertia I_x, I_y, and I_o of a homogeneous lamina of mass m occupying the triangular region R of problem 36.

38 A lamina has the shape of a circular disk $x^2 + y^2 \le a^2$ and its density at the point (x,y) is given by $\sigma(x,y) = k(x^2 + y^2)^{3/2}$. Here, a and k are positive constants. Find I_x, I_y, and I_o.

In problems 39 through 43, evaluate each threefold iterated integral.

39 $\displaystyle\int_0^1 \int_0^{x^2} \int_0^{xy^2} xy^2z^3 \, dz \, dy \, dx$

40 $\displaystyle\int_0^1 \int_{-x}^{x} \int_0^{x+z} e^{x+y+z} \, dy \, dz \, dx$

41 $\displaystyle\int_0^1 \int_0^{1+x^2} \int_{3x^2+y}^{4-x^2} dz \, dy \, dx$

42 $\displaystyle\int_0^1 \int_0^{\sqrt{3}z} \int_0^{\sqrt{3(z^2+y^2)}} xyz\sqrt{x^2+y^2+z^2} \, dx \, dy \, dz$

43 $\displaystyle\int_0^\pi \int_0^{\pi/2} \int_0^{2a\cos\phi} \rho^2 \sin\phi \, d\rho \, d\phi \, d\theta$

44 Rewrite the threefold iterated integral $\displaystyle\int_0^1 \int_0^{2\sqrt{1-z}} \int_0^{\sqrt{1-z}} f(x, y, z) \, dy \, dx \, dz$ as an equivalent triple integral over a solid S and sketch S. Then rewrite the integral as an equivalent threefold iterated integral with as many different orders of integration as possible.

In problems 45 and 46, sketch the solid S and evaluate the triple integral by iteration.

45 $\displaystyle\iiint_S y \, dx \, dy \, dz$, where S is the solid bounded above by the plane $3x + 2y + z = 6$, below by the plane $z = 0$, and laterally by the cylinder with generators parallel to the z axis over the boundary of the region $R: 0 \le x \le 1, 0 \le y \le 2 - 2x$.

46 $\displaystyle\iiint_S xz \, dx \, dy \, dz$, where S is the solid bounded by the planes $x = 0$, $y = 0$, $z = 0$, $z = 1$, $x = 2 - 2z$, and $y = 3 - 3z$.

In problems 47 and 48, use triple integration to find the volume of each solid S.

47 S is the solid bounded above by the parabolic cylinder $x^2 + z = 4$, below by the plane $x + z = 2$, on the left by the plane $y = 0$, and on the right by the plane $y = 3$.

48 S is the solid in the first octant bounded by the circular cylinder $x^2 + z^2 = 9$ and the planes $y = 2x$, $y = 0$, and $z = 0$.

In problems 49 and 50, express each threefold iterated integral as an equivalent threefold iterated integral in cylindrical coordinates, and then evaluate it.

49 $\displaystyle\int_0^1 \int_0^{\sqrt{1-x^2}} \int_0^{\sqrt{1-x^2-y^2}} z^2 \, dz \, dy \, dx$

50 $\displaystyle\int_0^2 \int_0^{\sqrt{2x-x^2}} \int_0^{9-x^2-y^2} \sqrt{x^2+y^2} \, dz \, dy \, dx$

In problems 51 and 52, express each threefold iterated integral as an equivalent threefold iterated integral in spherical coordinates, and then evaluate it.

51 $\displaystyle\int_0^1 \int_0^{\sqrt{1-y^2}} \int_{\sqrt{3(x^2+y^2)}}^{\sqrt{4-x^2-y^2}} \sqrt{x^2+y^2+z^2} \, dz \, dx \, dy$

52 $\displaystyle\int_0^{\sqrt{2}/2} \int_y^{\sqrt{1-y^2}} \int_{-\sqrt{8(x^2+y^2)}}^{\sqrt{9-x^2-y^2}} f'[(x^2+y^2+z^2)^{3/2}] \, dz \, dx \, dy$

53 Use cylindrical coordinates to evaluate $\displaystyle\iiint_S z\sqrt{x^2+y^2} \, dx \, dy \, dz$, where S is the half of the solid right circular cone with vertex at $(0, 0, h)$ and base $x^2 + y^2 \le a^2$ lying to the right of the plane $y = 0$.

54 Use cylindrical coordinates to find the volume of the solid bounded above by the plane $z = ax + by + c$, below by the xy plane, and laterally by the cylinder $r = k\cos\theta$. [Assume that $k > 0$ and that $ax + by + c \ge 0$ for (x, y) inside the circle whose polar equation is $r = k\cos\theta$.]

In problems 55 through 58, find the coordinates $(\bar{x}, \bar{y}, \bar{z})$ of the centroid of each solid S.

55 S is bounded above by the plane $z = 4$ and below by the paraboloid of revolution $3z = r^2$.

56 S is bounded above by the sphere $x^2 + y^2 + z^2 = a^2$, below by the xy plane, and laterally by the cylinder $\left(x - \dfrac{a}{2}\right)^2 + y^2 = \left(\dfrac{a}{2}\right)^2$.

57 $S: 0 \le y \le 1, 0 \le x \le \sqrt{1-y^2}, \sqrt{3(x^2+y^2)} \le z \le \sqrt{4-x^2-y^2}$

58 S is bounded above by the paraboloid of revolution $x = 5 - y^2 - z^2$, below by the xy plane, behind by the yz plane, on the left by the xz plane, and on the right by the plane $y = 1$.

In problems 59 through 62, find the moment of inertia I_z about the z axis of a homogeneous solid S of total mass m as described.

59 S is a solid right circular cone of height h with vertex at the origin, central axis along the positive z axis, and base radius a.

60 S is bounded by the coordinate planes, the planes $x = 3$ and $z = 1$, and the paraboloid $y = 10 - x^2 - z^2$.

61 $S: (x - a)^2 + y^2 + z^2 \leq a^2$ **62** $S: r \leq 2 \cos \theta, 0 \leq \theta \leq \pi, 0 \leq z \leq \sqrt{4 - r^2}$

In problems 63 through 66, evaluate the line integral directly.

63 $\displaystyle\int_C (x + y)\, dx + (x + y^2)\, dy; \ C: \begin{cases} x = t + 1 \\ y = t^2, \end{cases} \quad 0 \leq t \leq 1$

64 $\displaystyle\int_C \bar{F} \cdot d\bar{R}; \ \bar{F} = ye^{xy}\bar{i} + xe^{xy}\bar{j}; \ C: \begin{cases} x = t^3 \\ y = 1 - t^6, \end{cases} \quad -1 \leq t \leq 1$

65 $\displaystyle\int_C x^2\, dx; \ C: \begin{cases} x = t \\ y = t^2, \end{cases} \quad 0 \leq t \leq 1.$

66 $\displaystyle\int_C (ax + by)\, dx + (cx + ky)\, dy; \ C$ is the straight line segment from (x_0, y_0) to (x_1, y_1).

In problems 67 through 70, use Green's theorem to evaluate each line integral.

67 $\displaystyle\int_C x^2 y\, dx + y^3\, dy; \ C$ is the boundary, taken in the counterclockwise sense, of the region R bounded by the curves $y^3 = x^2$ and $y = x$.

68 $\displaystyle\int_C (x^2 + y)\, dx + (x - y^2)\, dy; \ C$ is the counterclockwise boundary of the region R bounded by the curves $y = 2x^2$ and $y = 4x$.

69 $\displaystyle\int_C y\, dx - x\, dy; \ C$ is the counterclockwise boundary of the triangle with vertices $(0, 0)$, $(b, 0)$, and (c, h), where b, c, and h are positive constants.

70 $\displaystyle\int_C \bar{F} \cdot d\bar{R}; \ \bar{F}(x, y) = f(x)\bar{i} + g(y)\bar{j}; \ C$ is any simple closed curve; f and g are continuous functions.

71 Find the area of the portion of the spherical surface $x^2 + y^2 + z^2 = 9$ that lies outside the parabolic cylinder $z + y^2 + x = 9$.

72 Set up an integral which gives the area of the portion of the surface $f(x, y, z) = k$ that lies above the region R in the xy plane. Make whatever assumptions you need concerning continuity, differentiability, and so forth.

73 Two congruent cylinders are tangent to each other externally along a diameter of a sphere whose radius is twice that of the cylinders. Find the area of the portion of the surface of the sphere interior to the cylinders.

74 Suppose that the region R in the xy plane is bounded by the simple closed curve C. Write formulas for the coordinates of the centroid $(\bar{x}, \bar{y})$ of R that involve only line integrals around the curve C.

75 Evaluate $\displaystyle\iint_\Sigma (x^2 + y^2)\, dA$, where Σ is the portion of the surface of the cone $z^2 = x^2 + y^2$ between $z = 0$ and $z = 1$.

76 A thin homogeneous lamina of mass m has the shape of the hemisphere $x^2 + y^2 + z^2 = a^2$. Set up a surface integral which gives the moment of inertia I_y of this lamina about the y axis, and then evaluate this integral.

In problems 77 and 78, find (a) $\nabla \cdot \overline{F}$ and (b) $\nabla \times \overline{F}$ for each vector field $\overline{F}$.

77 $\overline{F} = (x^2 + yz)\overline{i} + (y^2 + xz)\overline{j} + (z^2 + xy)\overline{k}$ **78** $\overline{F} = (y \cos z)\overline{i} + (z \cos x)\overline{j} + (x \cos y)\overline{k}$

In problems 79 and 80, use the divergence theorem to calculate the flux $\displaystyle\iint_\Sigma \overline{F} \cdot \overline{N} \, dA$ of the given vector field $\overline{F}$ through the surface Σ of the indicated solid S in the direction of the outward-pointing unit normal vector $\overline{N}$.

79 $\overline{F} = x\overline{i} + y\overline{j} - z\overline{k}$; S is the solid right circular cylinder bounded above by $z = 2$, below by $z = 1$, and laterally by $x^2 + y^2 = 1$.

80 $\overline{F} = x^2\overline{i} + 2y^2\overline{j} + 3z^2\overline{k}$; S is the solid sphere $x^2 + y^2 + z^2 \leq 1$.

In problems 81 and 82, use Stokes' theorem to calculate the flux $\displaystyle\iint_\Sigma (\nabla \times \overline{F}) \cdot \overline{N} \, dA$ of the curl of the given vector field $\overline{F}$ through the indicated surface Σ in the direction of the given unit normal vector $\overline{N}$.

81 $\overline{F} = y\overline{i} + z\overline{j} + x\overline{k}$; Σ is the hemisphere $x^2 + y^2 + z^2 = 1$, $z \geq 0$; $\overline{N}$ is the unit normal vector whose $\overline{k}$ component is nonnegative.

82 $\overline{F} = (y + 3x)\overline{i} + (2y - x)\overline{j} + (xy^2 + z^3)\overline{k}$; Σ is the portion of the paraboloid of revolution $z = 1 - x^2 - y^2$ for which $z \geq 0$; $\overline{N}$ is the unit normal vector whose $\overline{k}$ component is non-negative.

83 Let $\overline{R} = x\overline{i} + y\overline{j} + z\overline{k}$. Show that the flux of $\overline{R}$ through the surface Σ of any solid S is three times the volume of S.

84 Let $\overline{R} = x\overline{i} + y\overline{j} + z\overline{k}$ and let $r = |\overline{R}|$. If $\overline{F} = r^2\overline{R}$, verify the divergence theorem for the solid sphere S of radius 1 with center at the origin.

APPENDICES
TABLES

TABLE I Trigonometric Functions

Degrees	Radians	Sin	Tan	Cot	Cos		
0	0	0	0	—	1.000	1.5708	90
1	0.0175	0.0175	0.0175	57.290	0.9998	1.5533	89
2	0.0349	0.0349	0.0349	28.636	0.9994	1.5359	88
3	0.0524	0.0523	0.0523	19.081	0.9986	1.5184	87
4	0.0698	0.0698	0.0699	14.301	0.9976	1.5010	86
5	0.0873	0.0872	0.0875	11.430	0.9962	1.4835	85
6	0.1047	0.1045	0.1051	9.5144	0.9945	1.4661	84
7	0.1222	0.1219	0.1228	8.1443	0.9925	1.4486	83
8	0.1396	0.1392	0.1405	7.1154	0.9903	1.4312	82
9	0.1571	0.1564	0.1584	6.3138	0.9877	1.4137	81
10	0.1745	0.1736	0.1763	5.6713	0.9848	1.3963	80
11	0.1920	0.1908	0.1944	5.1446	0.9816	1.3788	79
12	0.2094	0.2079	0.2126	4.7046	0.9781	1.3614	78
13	0.2269	0.2250	0.2309	4.3315	0.9744	1.3439	77
14	0.2443	0.2419	0.2493	4.0108	0.9703	1.3265	76
15	0.2618	0.2588	0.2679	3.7321	0.9659	1.3090	75
16	0.2793	0.2756	0.2867	3.4874	0.9613	1.2915	74
17	0.2967	0.2924	0.3057	3.2709	0.9563	1.2741	73
18	0.3142	0.3090	0.3249	3.0777	0.9511	1.2566	72
19	0.3316	0.3256	0.3443	2.9042	0.9455	1.2392	71
20	0.3491	0.3420	0.3640	2.7475	0.9397	1.2217	70
21	0.3665	0.3584	0.3839	2.6051	0.9336	1.2043	69
22	0.3840	0.3746	0.4040	2.4751	0.9272	1.1868	68
23	0.4014	0.3907	0.4245	2.3559	0.9205	1.1694	67
24	0.4189	0.4067	0.4452	2.2460	0.9135	1.1519	66
25	0.4363	0.4226	0.4663	2.1445	0.9063	1.1345	65
26	0.4538	0.4384	0.4877	2.0503	0.8988	1.1170	64
27	0.4712	0.4540	0.5095	1.9626	0.8910	1.0996	63
28	0.4887	0.4695	0.5317	1.8807	0.8829	1.0821	62
29	0.5061	0.4848	0.5543	1.8040	0.8746	1.0647	61
30	0.5236	0.5000	0.5774	1.7321	0.8660	1.0472	60
31	0.5411	0.5150	0.6009	1.6643	0.8572	1.0297	59
32	0.5585	0.5299	0.6249	1.6003	0.8480	1.0123	58
33	0.5760	0.5446	0.6494	1.5399	0.8387	0.9948	57
34	0.5934	0.5592	0.6745	1.4826	0.8290	0.9774	56
35	0.6109	0.5736	0.7002	1.4281	0.8192	0.9599	55
36	0.6283	0.5878	0.7265	1.3764	0.8090	0.9425	54
37	0.6458	0.6018	0.7536	1.3270	0.7986	0.9250	53
38	0.6632	0.6157	0.7813	1.2799	0.7880	0.9076	52
39	0.6807	0.6293	0.8098	1.2349	0.7771	0.8901	51
40	0.6981	0.6428	0.8391	1.1918	0.7660	0.8727	50
41	0.7156	0.6561	0.8693	1.1504	0.7547	0.8552	49
42	0.7330	0.6691	0.9004	1.1106	0.7431	0.8378	48
43	0.7505	0.6820	0.9325	1.0724	0.7314	0.8203	47
44	0.7679	0.6947	0.9657	1.0355	0.7193	0.8029	46
45	0.7854	0.7071	1.0000	1.0000	0.7071	0.7854	45
		Cos	Cot	Tan	Sin	Radians	Degrees

TABLE II   Natural Logarithms, ln *t*

t	0.00	0.01	0.02	0.03	0.04	0.05	0.06	0.07	0.08	0.09
1.0	0.0000	0.0100	0.0198	0.0296	0.0392	0.0488	0.0583	0.0677	0.0770	0.0862
1.1	0.0953	0.1044	0.1133	0.1222	0.1310	0.1398	0.1484	0.1570	0.1655	0.1740
1.2	0.1823	0.1906	0.1989	0.2070	0.2151	0.2231	0.2311	0.2390	0.2469	0.2546
1.3	0.2624	0.2700	0.2776	0.2852	0.2927	0.3001	0.3075	0.3148	0.3221	0.3293
1.4	0.3365	0.3436	0.3507	0.3577	0.3646	0.3716	0.3784	0.3853	0.3920	0.3988
1.5	0.4055	0.4121	0.4187	0.4253	0.4318	0.4383	0.4447	0.4511	0.4574	0.4637
1.6	0.4700	0.4762	0.4824	0.4886	0.4947	0.5008	0.5068	0.5128	0.5188	0.5247
1.7	0.5306	0.5365	0.5423	0.5481	0.5539	0.5596	0.5653	0.5710	0.5766	0.5822
1.8	0.5878	0.5933	0.5988	0.6043	0.6098	0.6152	0.6206	0.6259	0.6313	0.6366
1.9	0.6419	0.6471	0.6523	0.6575	0.6627	0.6678	0.6729	0.6780	0.6831	0.6881
2.0	0.6931	0.6981	0.7031	0.7080	0.7130	0.7178	0.7227	0.7275	0.7324	0.7372
2.1	0.7419	0.7467	0.7514	0.7561	0.7608	0.7655	0.7701	0.7747	0.7793	0.7839
2.2	0.7885	0.7930	0.7975	0.8020	0.8065	0.8109	0.8154	0.8198	0.8242	0.8286
2.3	0.8329	0.8372	0.8416	0.8459	0.8502	0.8544	0.8587	0.8629	0.8671	0.8713
2.4	0.8755	0.8796	0.8838	0.8879	0.8920	0.8961	0.9002	0.9042	0.9083	0.9123
2.5	0.9163	0.9203	0.9243	0.9282	0.9322	0.9361	0.9400	0.9439	0.9478	0.9517
2.6	0.9555	0.9594	0.9632	0.9670	0.9708	0.9746	0.9783	0.9821	0.9858	0.9895
2.7	0.9933	0.9969	1.0006	1.0043	1.0080	1.0116	1.0152	1.0188	1.0225	1.0260
2.8	1.0296	1.0332	1.0367	1.0403	1.0438	1.0473	1.0508	1.0543	1.0578	1.0613
2.9	1.0647	1.0682	1.0716	1.0750	1.0784	1.0818	1.0852	1.0886	1.0919	1.0953
3.0	1.0986	1.1019	1.1053	1.1086	1.1119	1.1151	1.1184	1.1217	1.1249	1.1282
3.1	1.1314	1.1346	1.1378	1.1410	1.1442	1.1474	1.1506	1.1537	1.1569	1.1600
3.2	1.1632	1.1663	1.1694	1.1725	1.1756	1.1787	1.1817	1.1848	1.1878	1.1909
3.3	1.1939	1.1970	1.2000	1.2030	1.2060	1.2090	1.2119	1.2149	1.2179	1.2208
3.4	1.2238	1.2267	1.2296	1.2326	1.2355	1.2384	1.2413	1.2442	1.2470	1.2499
3.5	1.2528	1.2556	1.2585	1.2613	1.2641	1.2669	1.2698	1.2726	1.2754	1.2782
3.6	1.2809	1.2837	1.2865	1.2892	1.2920	1.2947	1.2975	1.3002	1.3029	1.3056
3.7	1.3083	1.3110	1.3137	1.3164	1.3191	1.3218	1.3244	1.3271	1.3297	1.3324
3.8	1.3350	1.3376	1.3403	1.3429	1.3455	1.3481	1.3507	1.3533	1.3558	1.3584
3.9	1.3610	1.3635	1.3661	1.3686	1.3712	1.3737	1.3762	1.3788	1.3813	1.3838
4.0	1.3863	1.3888	1.3913	1.3938	1.3962	1.3987	1.4012	1.4036	1.4061	1.4085
4.1	1.4110	1.4134	1.4159	1.4183	1.4207	1.4231	1.4255	1.4279	1.4303	1.4327
4.2	1.4351	1.4375	1.4398	1.4422	1.4446	1.4469	1.4493	1.4516	1.4540	1.4563
4.3	1.4586	1.4609	1.4633	1.4656	1.4679	1.4702	1.4725	1.4748	1.4770	1.4793
4.4	1.4816	1.4839	1.4861	1.4884	1.4907	1.4929	1.4952	1.4974	1.4996	1.5019
4.5	1.5041	1.5063	1.5085	1.5107	1.5129	1.5151	1.5173	1.5195	1.5217	1.5239
4.6	1.5261	1.5282	1.5304	1.5326	1.5347	1.5369	1.5390	1.5412	1.5433	1.5454
4.7	1.5476	1.5497	1.5518	1.5539	1.5560	1.5581	1.5602	1.5623	1.5644	1.5665
4.8	1.5686	1.5707	1.5728	1.5748	1.5769	1.5790	1.5810	1.5831	1.5851	1.5872
4.9	1.5892	1.5913	1.5933	1.5953	1.5974	1.5994	1.6014	1.6034	1.6054	1.6074
5.0	1.6094	1.6114	1.6134	1.6154	1.6174	1.6194	1.6214	1.6233	1.6253	1.6273
5.1	1.6292	1.6312	1.6332	1.6351	1.6371	1.6390	1.6409	1.6429	1.6448	1.6467
5.2	1.6487	1.6506	1.6525	1.6544	1.6563	1.6582	1.6601	1.6620	1.6639	1.6658
5.3	1.6677	1.6696	1.6715	1.6734	1.6752	1.6771	1.6790	1.6808	1.6827	1.6845
5.4	1.6864	1.6882	1.6901	1.6919	1.6938	1.6956	1.6974	1.6993	1.7011	1.7029
5.5	1.7047	1.7066	1.7084	1.7102	1.7120	1.7138	1.7156	1.7174	1.7192	1.7210
5.6	1.7228	1.7246	1.7263	1.7281	1.7299	1.7317	1.7334	1.7352	1.7370	1.7387
5.7	1.7405	1.7422	1.7440	1.7457	1.7475	1.7492	1.7509	1.7527	1.7544	1.7561
5.8	1.7579	1.7596	1.7613	1.7630	1.7647	1.7664	1.7682	1.7699	1.7716	1.7733
5.9	1.7750	1.7766	1.7783	1.7800	1.7817	1.7834	1.7851	1.7867	1.7884	1.7901
6.0	1.7918	1.7934	1.7951	1.7967	1.7984	1.8001	1.8017	1.8034	1.8050	1.8066
6.1	1.8083	1.8099	1.8116	1.8132	1.8148	1.8165	1.8181	1.8197	1.8213	1.8229
6.2	1.8245	1.8262	1.8278	1.8294	1.8310	1.8326	1.8342	1.8358	1.8374	1.8390
6.3	1.8406	1.8421	1.8437	1.8453	1.8469	1.8485	1.8500	1.8516	1.8532	1.8547
6.4	1.8563	1.8579	1.8594	1.8610	1.8625	1.8641	1.8656	1.8672	1.8687	1.8703

TABLE II   Natural Logarithms, ln *t* (*continued*)

t	0.00	0.01	0.02	0.03	0.04	0.05	0.06	0.07	0.08	0.09
6.5	1.8718	1.8733	1.8749	1.8764	1.8779	1.8795	1.8810	1.8825	1.8840	1.8856
6.6	1.8871	1.8886	1.8901	1.8916	1.8931	1.8946	1.8961	1.8976	1.8991	1.9006
6.7	1.9021	1.9036	1.9051	1.9066	1.9081	1.9095	1.9110	1.9125	1.9140	1.9155
6.8	1.9169	1.9184	1.9199	1.9213	1.9228	1.9242	1.9257	1.9272	1.9286	1.9301
6.9	1.9315	1.9330	1.9344	1.9359	1.9373	1.9387	1.9402	1.9416	1.9430	1.9445
7.0	1.9459	1.9473	1.9488	1.9502	1.9516	1.9530	1.9544	1.9559	1.9573	1.9587
7.1	1.9601	1.9615	1.9629	1.9643	1.9657	1.9671	1.9685	1.9699	1.9713	1.9727
7.2	1.9741	1.9755	1.9769	1.9782	1.9796	1.9810	1.9824	1.9838	1.9851	1.9865
7.3	1.9879	1.9892	1.9906	1.9920	1.9933	1.9947	1.9961	1.9974	1.9988	2.0001
7.4	2.0015	2.0028	2.0042	2.0055	2.0069	2.0082	2.0096	2.0109	2.0122	2.0136
7.5	2.0149	2.0162	2.0176	2.0189	2.0202	2.0215	2.0229	2.0242	2.0255	2.0268
7.6	2.0282	2.0295	2.0308	2.0321	2.0334	2.0347	2.0360	2.0373	2.0386	2.0399
7.7	2.0412	2.0425	2.0438	2.0451	2.0464	2.0477	2.0490	2.0503	2.0516	2.0528
7.8	2.0541	2.0554	2.0567	2.0580	2.0592	2.0605	2.0618	2.0631	2.0643	2.0665
7.9	2.0669	2.0681	2.0694	2.0707	2.0719	2.0732	2.0744	2.0757	2.0769	2.0782
8.0	2.0794	2.0807	2.0819	2.0832	2.0844	2.0857	2.0869	2.0882	2.0894	2.0906
8.1	2.0919	2.0931	2.0943	2.0956	2.0968	2.0980	2.0992	2.1005	2.1017	2.1029
8.2	2.1041	2.1054	2.1066	2.1078	2.1090	2.1102	2.1114	2.1126	2.1138	2.1150
8.3	2.1163	2.1175	2.1187	2.1199	2.1211	2.1223	2.1235	2.1247	2.1258	2.1270
8.4	2.1282	2.1294	2.1306	2.1318	2.1330	2.1342	2.1353	2.1365	2.1377	2.1389
8.5	2.1401	2.1412	2.1424	2.1436	2.1448	2.1459	2.1471	2.1483	2.1494	2.1506
8.6	2.1518	2.1529	2.1541	2.1552	2.1564	2.1576	2.1587	2.1599	2.1610	2.1622
8.7	2.1633	2.1645	2.1656	2.1668	2.1679	2.1691	2.1702	2.1713	2.1725	2.1736
8.8	2.1748	2.1759	2.1770	2.1782	2.1793	2.1804	2.1815	2.1827	2.1838	2.1849
8.9	2.1861	2.1872	2.1883	2.1894	2.1905	2.1917	2.1928	2.1939	2.1950	2.1961
9.0	2.1972	2.1983	2.1994	2.2006	2.2017	2.2028	2.2039	2.2050	2.2061	2.2072
9.1	2.2083	2.2094	2.2105	2.2116	2.2127	2.2138	2.2148	2.2159	2.2170	2.2181
9.2	2.2192	2.2203	2.2214	2.2225	2.2235	2.2246	2.2257	2.2268	2.2279	2.2289
9.3	2.2300	2.2311	2.2322	2.2332	2.2343	2.2354	2.2364	2.2375	2.2386	2.2396
9.4	2.2407	2.2418	2.2428	2.2439	2.2450	2.2460	2.2471	2.2481	2.2492	2.2502
9.5	2.2513	2.2523	2.2534	2.2544	2.2555	2.2565	2.2576	2.2586	2.2597	2.2607
9.6	2.2618	2.2628	2.2638	2.2649	2.2659	2.2670	2.2680	2.2690	2.2701	2.2711
9.7	2.2721	2.2732	2.2742	2.2752	2.2762	2.2773	2.2783	2.2793	2.2803	2.2814
9.8	2.2824	2.2834	2.2844	2.2854	2.2865	2.2875	2.2885	2.2895	2.2905	2.2915
9.9	2.2925	2.2935	2.2946	2.2956	2.2966	2.2976	2.2986	2.2996	2.3006	2.3016

TABLE III   Exponential Functions

x	e^x	e^{-x}	x	e^x	e^{-x}
0.00	1.0000	1.0000	3.0	20.086	0.0498
0.05	1.0513	0.9512	3.1	22.198	0.0450
0.10	1.1052	0.9048	3.2	24.533	0.0408
0.15	1.1618	0.8607	3.3	27.113	0.0369
0.20	1.2214	0.8187	3.4	29.964	0.0334
0.25	1.2840	0.7788	3.5	33.115	0.0302
0.30	1.3499	0.7408	3.6	36.598	0.0273
0.35	1.4191	0.7047	3.7	40.447	0.0247
0.40	1.4918	0.6703	3.8	44.701	0.0224
0.45	1.5683	0.6376	3.9	49.402	0.0202
0.50	1.6487	0.6065	4.0	54.598	0.0183
0.55	1.7333	0.5769	4.1	60.340	0.0166
0.60	1.8221	0.5488	4.2	66.686	0.0150
0.65	1.9155	0.5220	4.3	73.700	0.0136
0.70	2.0138	0.4966	4.4	81.451	0.0123
0.75	2.1170	0.4724	4.5	90.017	0.0111
0.80	2.2255	0.4493	4.6	99.484	0.0101
0.85	2.3396	0.4274	4.7	109.95	0.0091
0.90	2.4596	0.4066	4.8	121.51	0.0082
0.95	2.5857	0.3867	4.9	134.29	0.0074
1.0	2.7183	0.3679	5.0	148.41	0.0067
1.1	3.0042	0.3329	5.1	164.02	0.0061
1.2	3.3201	0.3012	5.2	181.27	0.0055
1.3	3.6693	0.2725	5.3	200.34	0.0050
1.4	4.0552	0.2466	5.4	221.41	0.0045
1.5	4.4817	0.2231	5.5	244.69	0.0041
1.6	4.9530	0.2019	5.6	270.43	0.0037
1.7	5.4739	0.1827	5.7	298.87	0.0033
1.8	6.0496	0.1653	5.8	330.30	0.0030
1.9	6.6859	0.1496	5.9	365.04	0.0027
2.0	7.3891	0.1353	6.0	403.43	0.0025
2.1	8.1662	0.1225	6.5	665.14	0.0015
2.2	9.0250	0.1108	7.0	1096.6	0.0009
2.3	9.9742	0.1003	7.5	1808.0	0.0006
2.4	11.023	0.0907	8.0	2981.0	0.0003
2.5	12.182	0.0821	8.5	4914.8	0.0002
2.6	13.464	0.0743	9.0	8103.1	0.0001
2.7	14.880	0.0672	9.5	13,360	0.00007
2.8	16.445	0.0608	10.0	22,026	0.00004
2.9	18.174	0.0550			

TABLE IV   Hyperbolic Functions

x	$\sinh x$	$\cosh x$	$\tanh x$	x	$\sinh x$	$\cosh x$	$\tanh x$
0.0	0.00000	1.0000	0.00000	3.0	10.018	10.068	0.99505
0.1	0.10017	1.0050	0.09967	3.1	11.076	11.122	0.99595
0.2	0.20134	1.0201	0.19738	3.2	12.246	12.287	0.99668
0.3	0.30452	1.0453	0.29131	3.3	13.538	13.575	0.99728
0.4	0.41075	1.0811	0.37995	3.4	14.965	14.999	0.99777
0.5	0.52110	1.1276	0.46212	3.5	16.543	16.573	0.99818
0.6	0.63665	1.1855	0.53705	3.6	18.285	18.313	0.99851
0.7	0.75858	1.2552	0.60437	3.7	20.211	20.236	0.99878
0.8	0.88811	1.3374	0.66404	3.8	22.339	22.362	0.99900
0.9	1.0265	1.4331	0.71630	3.9	24.691	24.711	0.99918
1.0	1.1752	1.5431	0.76159	4.0	27.290	27.308	0.99933
1.1	1.3356	1.6685	0.80050	4.1	30.162	30.178	0.99945
1.2	1.5095	1.8107	0.83365	4.2	33.336	33.351	0.99955
1.3	1.6984	1.9709	0.86172	4.3	36.843	36.857	0.99963
1.4	1.9043	2.1509	0.88535	4.4	40.719	40.732	0.99970
1.5	2.1293	2.3524	0.90515	4.5	45.003	45.014	0.99975
1.6	2.3756	2.5775	0.92167	4.6	49.737	49.747	0.99980
1.7	2.6456	2.8283	0.93541	4.7	54.969	54.978	0.99983
1.8	2.9422	3.1075	0.94681	4.8	60.751	60.759	0.99986
1.9	3.2682	3.4177	0.95624	4.9	67.141	67.149	0.99989
2.0	3.6269	3.7622	0.96403	5.0	74.203	74.210	0.99991
2.1	4.0219	4.1443	0.97045	5.1	82.008	82.014	0.99993
2.2	4.4571	4.5679	0.97574	5.2	90.633	90.639	0.99994
2.3	4.9370	5.0372	0.98010	5.3	100.17	100.17	0.99995
2.4	5.4662	5.5569	0.98367	5.4	110.70	110.71	0.99996
2.5	6.0502	6.1323	0.98661	5.5	122.34	122.35	0.99997
2.6	6.6947	6.7690	0.98903	5.6	135.21	135.22	0.99997
2.7	7.4063	7.4735	0.99101	5.7	149.43	149.44	0.99998
2.8	8.1919	8.2527	0.99263	5.8	165.15	165.15	0.99998
2.9	9.0596	9.1146	0.99396	5.9	182.52	182.52	0.99998

TABLE V   Common Logarithms, $\log_{10} x$

x	0.00	0.01	0.02	0.03	0.04	0.05	0.06	0.07	0.08	0.09
1.0	.0000	.0043	.0086	.0128	.0170	.0212	.0253	.0294	.0334	.0374
1.1	.0414	.0453	.0492	.0531	.0569	.0607	.0645	.0682	.0719	.0755
1.2	.0792	.0828	.0864	.0899	.0934	.0969	.1004	.1038	.1072	.1106
1.3	.1139	.1173	.1206	.1239	.1271	.1303	.1335	.1367	.1399	.1430
1.4	.1461	.1492	.1523	.1553	.1584	.1614	.1644	.1673	.1703	.1732
1.5	.1761	.1790	.1818	.1847	.1875	.1903	.1931	.1959	.1987	.2014
1.6	.2041	.2068	.2095	.2122	.2148	.2175	.2201	.2227	.2253	.2279
1.7	.2304	.2330	.2355	.2380	.2405	.2430	.2455	.2480	.2504	.2529
1.8	.2553	.2577	.2601	.2625	.2648	.2672	.2695	.2718	.2742	.2765
1.9	.2788	.2810	.2833	.2856	.2878	.2900	.2923	.2945	.2967	.2989
2.0	.3010	.3032	.3054	.3075	.3096	.3118	.3139	.3160	.3181	.3201
2.1	.3222	.3243	.3263	.3284	.3304	.3324	.3345	.3365	.3385	.3404
2.2	.3424	.3444	.3464	.3483	.3502	.3522	.3541	.3560	.3579	.3598
2.3	.3617	.3636	.3655	.3674	.3692	.3711	.3729	.3747	.3766	.3784
2.4	.3802	.3820	.3838	.3856	.3874	.3892	.3909	.3927	.3945	.3962
2.5	.3979	.3997	.4014	.4031	.4048	.4065	.4082	.4099	.4116	.4133
2.6	.4150	.4166	.4183	.4200	.4216	.4232	.4249	.4265	.4281	.4298
2.7	.4314	.4330	.4346	.4362	.4378	.4393	.4409	.4425	.4440	.4456
2.8	.4472	.4487	.4502	.4518	.4533	.4548	.4564	.4579	.4594	.4609
2.9	.4624	.4639	.4654	.4669	.4683	.4698	.4713	.4728	.4742	.4757
3.0	.4771	.4786	.4800	.4814	.4829	.4843	.4857	.4871	.4886	.4900
3.1	.4914	.4928	.4942	.4955	.4969	.4983	.4997	.5011	.5024	.5038
3.2	.5051	.5065	.5079	.5092	.5105	.5119	.5132	.5145	.5159	.5172
3.3	.5185	.5198	.5211	.5224	.5237	.5250	.5263	.5276	.5289	.5302
3.4	.5315	.5328	.5340	.5353	.5366	.5378	.5391	.5403	.5416	.5428
3.5	.5441	.5453	.5465	.5478	.5490	.5502	.5514	.5527	.5539	.5551
3.6	.5563	.5575	.5587	.5599	.5611	.5623	.5635	.5647	.5658	.5670
3.7	.5682	.5694	.5705	.5717	.5729	.5740	.5752	.5763	.5775	.5786
3.8	.5798	.5809	.5821	.5832	.5843	.5855	.5866	.5877	.5888	.5899
3.9	.5911	.5922	.5933	.5944	.5955	.5966	.5977	.5988	.5999	.6010
4.0	.6021	.6031	.6042	.6053	.6064	.6075	.6085	.6096	.6107	.6117
4.1	.6128	.6138	.6149	.6160	.6170	.6180	.6191	.6201	.6212	.6222
4.2	.6232	.6243	.6253	.6263	.6274	.6284	.6294	.6304	.6314	.6325
4.3	.6335	.6345	.6355	.6365	.6375	.6385	.6395	.6405	.6415	.6425
4.4	.6435	.6444	.6454	.6464	.6474	.6484	.6493	.6503	.6513	.6522
4.5	.6532	.6542	.6551	.6561	.6571	.6580	.6590	.6599	.6609	.6618
4.6	.6628	.6637	.6646	.6656	.6665	.6675	.6684	.6693	.6702	.6712
4.7	.6721	.6730	.6739	.6749	.6758	.6767	.6776	.6785	.6794	.6803
4.8	.6812	.6821	.6830	.6839	.6848	.6857	.6866	.6875	.6884	.6893
4.9	.6902	.6911	.6920	.6928	.6937	.6946	.6955	.6964	.6972	.6981
5.0	.6990	.6998	.7007	.7016	.7024	.7033	.7042	.7050	.7059	.7067
5.1	.7076	.7084	.7093	.7101	.7110	.7118	.7126	.7135	.7143	.7152
5.2	.7160	.7168	.7177	.7185	.7193	.7202	.7210	.7218	.7226	.7235
5.3	.7243	.7251	.7259	.7267	.7275	.7284	.7292	.7300	.7308	.7316
5.4	.7324	.7332	.7340	.7348	.7356	.7364	.7372	.7380	.7388	.7396
5.5	.7404	.7412	.7419	.7427	.7435	.7443	.7451	.7459	.7466	.7474
5.6	.7482	.7490	.7497	.7505	.7513	.7520	.7528	.7536	.7543	.7551
5.7	.7559	.7566	.7574	.7582	.7589	.7597	.7604	.7612	.7619	.7627
5.8	.7634	.7642	.7649	.7657	.7664	.7672	.7679	.7686	.7694	.7701
5.9	.7709	.7716	.7723	.7731	.7738	.7745	.7752	.7760	.7767	.7774
6.0	.7782	.7789	.7796	.7803	.7810	.7818	.7825	.7832	.7839	.7846
6.1	.7853	.7860	.7868	.7875	.7882	.7889	.7896	.7903	.7910	.7917
6.2	.7924	.7931	.7938	.7945	.7952	.7959	.7966	.7973	.7980	.7987
6.3	.7993	.8000	.8007	.8014	.8021	.8028	.8035	.8041	.8048	.8055
6.4	.8062	.8069	.8075	.8082	.8089	.8096	.8102	.8109	.8116	.8122

TABLE V **Common Logarithms, $\log_{10} x$** (*continued*)

x	0.00	0.01	0.02	0.03	0.04	0.05	0.06	0.07	0.08	0.09
6.5	.8129	.8136	.8142	.8149	.8156	.8162	.8169	.8176	.8182	.8189
6.6	.8195	.8202	.8209	.8215	.8222	.8228	.8235	.8241	.8248	.8254
6.7	.8261	.8267	.8274	.8280	.8287	.8293	.8299	.8306	.8312	.8319
6.8	.8325	.8331	.8338	.8344	.8351	.8357	.8363	.8370	.8376	.8382
6.9	.8388	.8395	.8401	.8407	.8414	.8420	.8426	.8432	.8439	.8445
7.0	.8451	.8457	.8463	.8470	.8476	.8482	.8488	.8494	.8500	.8506
7.1	.8513	.8519	.8525	.8531	.8537	.8543	.8549	.8555	.8561	.8567
7.2	.8573	.8579	.8585	.8591	.8597	.8603	.8609	.8615	.8621	.8627
7.3	.8633	.8639	.8645	.8651	.8657	.8663	.8669	.8675	.8681	.8686
7.4	.8692	.8698	.8704	.8710	.8716	.8722	.8727	.8733	.8739	.8745
7.5	.8751	.8756	.8762	.8768	.8774	.8779	.8785	.8791	.8797	.8802
7.6	.8808	.8814	.8820	.8825	.8831	.8837	.8842	.8848	.8854	.8859
7.7	.8865	.8871	.8876	.8882	.8887	.8893	.8899	.8904	.8910	.8915
7.8	.8921	.8927	.8932	.8938	.8943	.8949	.8954	.8960	.8965	.8971
7.9	.8976	.8982	.8987	.8993	.8998	.9004	.9009	.9015	.9020	.9025
8.0	.9031	.9036	.9042	.9047	.9053	.9058	.9063	.9069	.9074	.9079
8.1	.9085	.9090	.9096	.9101	.9106	.9112	.9117	.9122	.9128	.9133
8.2	.9138	.9143	.9149	.9154	.9159	.9165	.9170	.9175	.9180	.9186
8.3	.9191	.9196	.9201	.9206	.9212	.9217	.9222	.9227	.9232	.9238
8.4	.9243	.9248	.9253	.9258	.9263	.9269	.9274	.9279	.9284	.9289
8.5	.9294	.9299	.9304	.9309	.9315	.9320	.9325	.9330	.9335	.9340
8.6	.9345	.9350	.9355	.9360	.9365	.9370	.9375	.9380	.9385	.9390
8.7	.9395	.9400	.9405	.9410	.9415	.9420	.9425	.9430	.9435	.9440
8.8	.9445	.9450	.9455	.9460	.9465	.9469	.9474	.9479	.9484	.9489
8.9	.9494	.9499	.9504	.9509	.9513	.9518	.9523	.9528	.9533	.9538
9.0	.9542	.9547	.9552	.9557	.9562	.9566	.9571	.9567	.9581	.9586
9.1	.9590	.9595	.9600	.9605	.9609	.9614	.9619	.9624	.9628	.9633
9.2	.9638	.9643	.9647	.9652	.9657	.9661	.9666	.9671	.9675	.9680
9.3	.9685	.9689	.9694	.9699	.9703	.9708	.9713	.9717	.9722	.9727
9.4	.9731	.9736	.9741	.9745	.9750	.9754	.9759	.9763	.9768	.9773
9.5	.9777	.9782	.9786	.9791	.9795	.9800	.9805	.9809	.9814	.9818
9.6	.9823	.9827	.9832	.9836	.9841	.9845	.9850	.9854	.9859	.9863
9.7	.9868	.9872	.9877	.9881	.9886	.9890	.9894	.9899	.9903	.9908
9.8	.9912	.9917	.9921	.9926	.9930	.9934	.9939	.9943	.9948	.9952
9.9	.9956	.9961	.9965	.9969	.9974	.9978	.9983	.9987	.9991	.9996

TABLE VI   Powers and Roots

Number	Square	Square Root	Cube	Cube Root	Number	Square	Square Root	Cube	Cube Root
1	1	1.000	1	1.000	51	2,601	7.141	132,651	3.708
2	4	1.414	8	1.260	52	2,704	7.211	140,608	3.733
3	9	1.732	27	1.442	53	2,809	7.280	148,877	3.756
4	16	2.000	64	1.587	54	2,916	7.348	157,464	3.780
5	25	2.236	125	1.710	55	3,025	7.416	166,375	3.803
6	36	2.449	216	1.817	56	3,136	7.483	175,616	3.826
7	49	2.646	343	1.913	57	3,249	7.550	185,193	3.849
8	64	2.828	512	2.000	58	3,364	7.616	195,112	3.871
9	81	3.000	729	2.080	59	3,481	7.681	205,379	3.893
10	100	3.162	1,000	2.154	60	3,600	7.746	216,000	3.915
11	121	3.317	1 331	2.224	61	3,721	7.810	226,981	3.936
12	144	3.464	1,728	2.289	62	3,844	7.874	238,328	3.958
13	169	3.606	2,197	2.351	63	3,969	7.937	250,047	3.979
14	196	3.742	2,744	2.410	64	4,096	8.000	262,144	4.000
15	225	3.873	3,375	2.466	65	4,225	8.062	274,625	4.021
16	256	4.000	4,096	2.520	66	4,356	8.124	287,496	4.041
17	289	4.123	4,913	2.571	67	4,489	8.185	300,763	4.062
18	324	4.243	5,832	2.621	68	4,624	8.246	314,432	4.082
19	361	4.359	6,859	2.668	69	4,761	8.307	328,509	4.102
20	400	4.472	8,000	2.714	70	4,900	8.367	343,000	4.121
21	441	4.583	9,261	2.759	71	5,041	8.426	357,911	4.141
22	484	4.690	10,648	2.802	72	5,184	8.485	373,248	4.160
23	529	4.796	12,167	2.844	73	5,329	8.544	389,017	4.179
24	576	4.899	13,824	2.884	74	5,476	8.602	405,224	4.198
25	625	5.000	15,625	2.924	75	5,625	8.660	421,875	4.217
26	676	5.099	17,576	2.962	76	5,776	8.718	438,976	4.236
27	729	5.196	19,683	3.000	77	5,929	8.775	456,533	4.254
28	784	5.292	21,952	3.037	78	6,084	8.832	474,552	4.273
29	841	5.385	24,389	3.072	79	6,241	8.888	493,039	4.291
30	900	5.477	27,000	3.107	80	6,400	8.944	512,000	4.309
31	961	5.568	29,791	3.141	81	6,561	9.000	531,441	4.327
32	1,024	5.657	32,768	3.175	82	6,724	9.055	551,368	4.344
33	1,089	5.745	35,937	3.208	83	6,889	9.110	571,787	4.362
34	1,156	5.831	39,304	3.240	84	7,056	9.165	592,704	4.380
35	1,225	5.916	42,875	3.271	85	7,225	9.220	614,125	4.397
36	1,296	6.000	46,656	3.302	86	7,396	9.274	636,056	4.414
37	1,369	6.083	50,653	3.332	87	7,569	9.327	658,503	4.431
38	1,444	6.164	54,872	3.362	88	7,744	9.381	681,472	4.448
39	1,521	6.245	59,319	3.391	89	7,921	9.434	704,969	4.465
40	1,600	6.325	64,000	3.420	90	8,100	9.487	729,000	4.481
41	1,681	6.403	68,921	3.448	91	8,281	9.539	753,571	4.498
42	1,764	6.481	74,088	3.476	92	8,464	9.592	778,688	4.514
43	1,849	6.557	79,507	3.503	93	8,649	9.644	804,357	4.531
44	1,936	6.633	85,184	3.530	94	8,836	9.695	830,584	4.547
45	2,025	6.708	91,125	3.557	95	9,025	9.747	857,375	4.563
46	2,116	6.782	97,336	3.583	96	9,216	9.798	884,736	4.579
47	2,209	6.856	103,823	3.609	97	9,409	9.849	912,673	4.595
48	2,304	6.928	110,592	3.634	98	9,604	9.899	941,192	4.610
49	2,401	7.000	117,649	3.659	99	9,801	9.950	970,299	4.626
50	2,500	7.071	125,000	3.684	100	10,000	10.000	1,000,000	4.642

ANSWERS TO SELECTED PROBLEMS

Chapter 0

Problem Set 1 page 3
1 (a) True; (b) true; (c) false; (d) true; (e) true   **3** $(3)(233) < (28)(25)$ **5** (a) $x < 0$;
(b) $x > 0$; (c) $x = 0$ **9** $2.646 > \sqrt{7}$ **11** No

Problem Set 2 page 8

1 (a)

(b)

(c)

(d)

(e)

(f)

3 $(-\infty, 3)$ **5** $(-2, 1]$ **7** $\left(-\dfrac{7}{4}, 1\right]$ **9** $(-\infty, -2]$ and $(1, \infty)$ **11** $(-\infty, -3)$

and $(3, \infty)$ **13** $(-1, 2)$ **15** $\left(-\infty, -\dfrac{2}{3}\right]$ and $[5, \infty)$ **17** $(-\infty, 0]$ and $(3, \infty)$

19 $[-1, 0)$ **21** $(-\infty, -4)$ and $\left[-\dfrac{8}{7}, 1\right)$ **23** $x = 5$ or $x = 1$ **25** $x = -2$ or $x = \dfrac{3}{2}$

27 $x = \dfrac{1}{2}$ or $x = -\dfrac{3}{4}$ **29** $(2, 3)$ **31** $\left(-\infty, -\dfrac{7}{3}\right)$ and $(-1, \infty)$ **33** $\left[-4, -\dfrac{6}{5}\right]$

35 $\left[\dfrac{1}{7}, \dfrac{1}{3}\right]$ **37** $\left(\dfrac{1}{1+k}, \dfrac{1}{1-k}\right)$ **39** $|x - y| \le |x - 2| + |2 - y|$

41 $|y + 2| \le |x + 2| + |y - x|$ **43** (a) $4x < 132$; (b) $x < 33$

Problem Set 3 page 11
1 (a) $N = (3, 2)$; (b) $R = (-3, 2)$; (c) $S = (-3, -2)$ **3** $\sqrt{13}$ **5** 4   **7** 10   **9** Side
lengths are 4, 6, and $\sqrt{52}$.   **11** Side lengths are 4, 5, and $\sqrt{41}$.   **13** Both distances are
$\sqrt{(x_1 - x_2)^2 + (y_1 - y_2)^2}$.   **15** Yes   **17** No   **19** Not isosceles

Problem Set 4 page 16

1 $-\dfrac{5}{3}$ **3** $\dfrac{1}{2}$ **5** $\dfrac{5}{11}$ **7** $y = 2x - 6$ **9** $y = \dfrac{x+5}{4}$ **11** $y = \dfrac{13}{3}x - 18$

13 $y = 6x - 16$ **15** $y = 20x + 138{,}000;\ 146{,}000$ **17** (a) $\left(\dfrac{15}{2}, 2\right)$; (b) $(2,5)$; (c) $(2,2)$;

(d) $\left(3, \dfrac{5}{2}\right)$ **19** (a) 2; (b) -3; (c) $\dfrac{3}{2}$; (d) $-\dfrac{b}{m}$ **21** Parallel   **23** Perpendicular

25 $\left(\dfrac{1}{7}, -\dfrac{2}{7}\right)$ **27** $(0,0)$ **29** Slope $\overline{AB}$ = slope $\overline{CD} = \dfrac{1}{6}$; slope $\overline{BC}$ = slope $\overline{AD} = \dfrac{5}{3}$.

31 (a) $d = 1$; (b) $k = -\dfrac{10}{3}$ **33** The lines have slopes $-\dfrac{A}{B}$ and $\dfrac{B}{A}$, respectively.

Problem Set 5 page 23

1 Domain is $\mathbb{R}$; range is $\mathbb{R}$.   **3** Domain is $\mathbb{R}$; range is $[0, \infty)$.   **5** Domain is $[-1, 1]$;
range is $[0, 1]$.   **7** Domain is $\left(-\infty, \dfrac{2}{3}\right)$ together with $\left(\dfrac{2}{3}, \infty\right)$; range is $(-\infty, 4)$
together with $(4, \infty)$.   **9** Domain is $\mathbb{R}$; range is $\{-3, -1, 2\}$.   **11** Domain is $\mathbb{R}$;

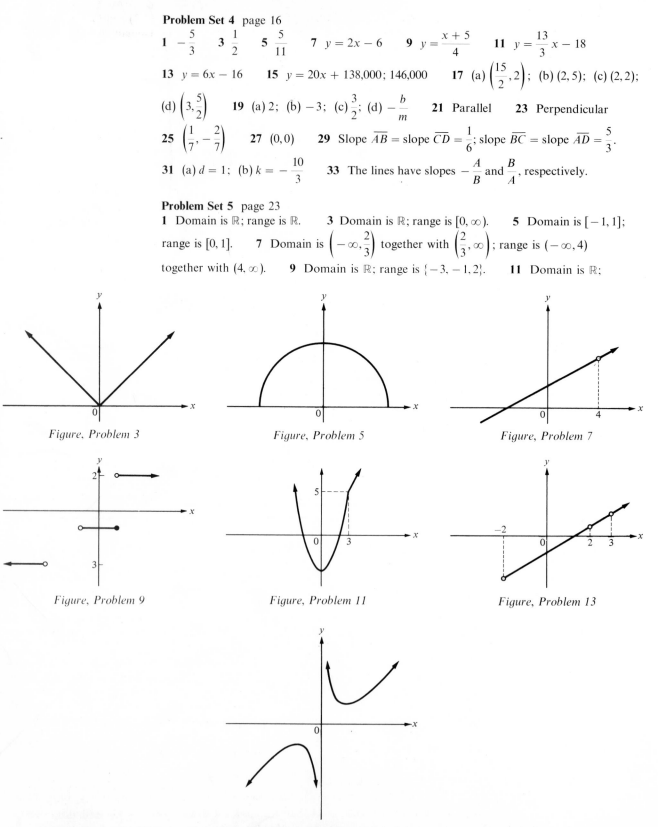

Figure, Problem 3

Figure, Problem 5

Figure, Problem 7

Figure, Problem 9

Figure, Problem 11

Figure, Problem 13

Figure, Problem 15

range is $[-4, \infty)$.   **13** Domain is $(-\infty, -2)$ together with $(2,3)$ and $(3, \infty)$; range is $(-\infty, -3)$ together with $(-3, 1)$, $(1, 2)$, and $(2, \infty)$.   **15** (a) Domain is $(-\infty, 0)$ together with $(0, \infty)$; (b) range is $(-\infty, -2]$ together with $[2, \infty)$; (d) $(-1, -2)$, $(1, 2)$, and $\left(-2, -\dfrac{5}{2}\right)$ lie on the graph of f.   **17** Domain is $\mathbb{R}$; range is $\left[-\dfrac{25}{4}, \infty\right)$. (a) -6; (b) -6; (c) 0;

(d) 0; (e) $a^2 - 3a - 4$; (f) $(a+b)^2 - 3(a+b) - 4$; (g) $(a-b)^2 - 3(a-b) - 4$;

(h) $x_0^2 - 3x_0 - 4$ **19** Domain is $\left(-\infty, -\dfrac{7}{3}\right)$ together with $\left(-\dfrac{7}{3}, \infty\right)$; (a) $-\dfrac{3}{17}$;

(b) $-\dfrac{5}{11}$; (c) $\dfrac{a-6}{3a+21}$; (d) $\dfrac{4-2a}{12+7a}$; (e) $\dfrac{a}{3a+13}$; (f) $\dfrac{a^2-2}{3a^2+7}$; (g) $\left(\dfrac{a-2}{3a+7}\right)^2$; (h) $\dfrac{x_0-2}{3x_0+7}$

21 Domain is $\left[-\dfrac{5}{3}, \infty\right)$. (a) 2; (b) 3; (c) $\sqrt{6}$; (d) $\sqrt{2}$; (e) $\sqrt{3a^2+5}$; (f) $3a+5$;

(g) $\sqrt{6x+8}$; (h) $\sqrt{3x_0 + 3h + 5}$ **23** Domain is $\left(-\infty, -\dfrac{1}{2}\right)$ together with $\left(-\dfrac{1}{2}, \infty\right)$.

(a) 1; (b) 2; (c) $\dfrac{3a-1}{1+2a}$; (d) $\dfrac{3a^2-1}{1+2a^2}$; (e) $\dfrac{-3a^2-1}{1-2a^2}$; (f) -1; (g) $\dfrac{3x-1}{1+2x}$

25 (a) $(a+1)(a+2)(a+3)(a+4)$; (b) $(a+2)(a+3)(a+4)(a+5)$

27 Yes; $r = \sqrt{\dfrac{A}{\pi}}$ **29** $V = \begin{cases} 216T & \text{for } 0 \le T < \dfrac{1}{12} \\ \\ 18 & \text{for } T \ge \dfrac{1}{12} \end{cases}$ **31** $p = 2x + \dfrac{50}{x}$ **33** 0

35 -8 **37** $\dfrac{-2}{x_0(x_0 + h)}$ **39** Functions are (b) and (c).   **41** True

Problem Set 6 page 29
1 Even   **3** Neither   **5** Odd   **7** Neither   **9** Neither   **11** Polynomial; degree 2; coefficients 6, -3, and -8 **13** Polynomial; degree 3; coefficients -1, 1, -5, and 6   **15** Polynomial; degree 4; coefficients $\sqrt{2}$, -5^{-1}, 0, 0, and 20   **17** Polynomial; no degree; coefficient 0   **19** 2 is a number, not a function.   **21** $y = -\dfrac{2}{5}x + \dfrac{29}{5}$

23 $f(x) = mx$, $m \ne 0$ **25** $x = -\dfrac{b}{m}$ is a root of $f(x) = mx + b$.   **27** Rational; domain is $(-\infty, 1)$ together with $(1, \infty)$.   **29** Rational; domain is $\mathbb{R}$.   **31** Not rational.
33 (a) $\operatorname{sgn}(-2) = -1$, $\operatorname{sgn}(-3) = -1$, $\operatorname{sgn}(0) = 0$, $\operatorname{sgn}(2) = 1$, $\operatorname{sgn}(3) = 1$, $\operatorname{sgn}(151) = 1$; (e) domain is $\mathbb{R}$, range is $\{-1, 0, 1\}$; (g) graph not one connected piece.   **35** Domain is $\mathbb{R}$, range is $[0, \infty)$.   **37** Domain is $\mathbb{R}$, range is $[-1, 1]$.   **39** Domain is $\mathbb{R}$, range is all integers.   **43** Domain is $\mathbb{R}$, range is all nonnegative integers.   **45** Domain is $\mathbb{R}$, range is $\left\{\dfrac{1}{3}\right\}$.

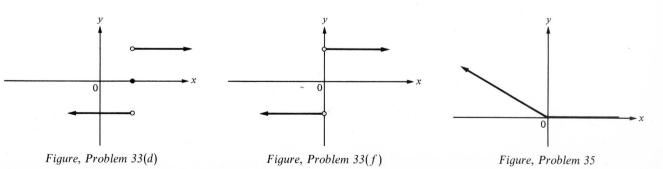

Figure, Problem 33(d)   Figure, Problem 33(f)   Figure, Problem 35

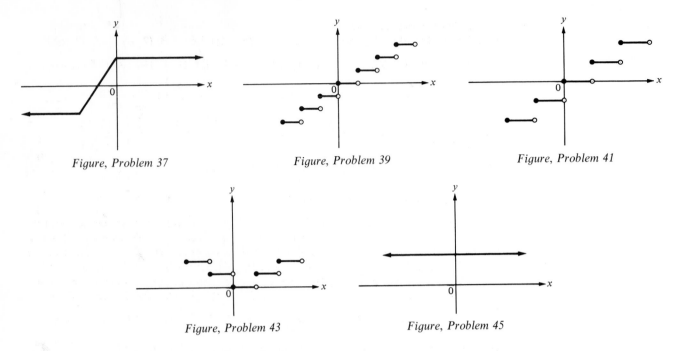

Figure, Problem 37

Figure, Problem 39

Figure, Problem 41

Figure, Problem 43

Figure, Problem 45

Problem Set 7 page 34

1 (a) (i) $-\dfrac{\pi}{4}$, (ii) $\dfrac{35\pi}{36}$, (iii) $-\dfrac{5\pi}{3}$; (b) (i) $40°$; (ii) $-157.5°$, (iii) $1290°$ **3** (a) $-\dfrac{3}{5}$; (b) $\dfrac{3}{4}$;

(c) $\dfrac{4}{3}$; (d) $-\dfrac{5}{4}$; (e) $-\dfrac{5}{3}$ **5** (a) $\sin^2 t$; (b) $2\cos t$; (c) $\csc^2 t$; (d) $\cot^2 t$; (e) $-\cos t$

7 (a) $\dfrac{\sqrt{2}}{4}(1-\sqrt{3})$, $\dfrac{\sqrt{2}}{4}(1+\sqrt{3})$; (b) $\dfrac{\sqrt{3}-1}{\sqrt{3}+1}$ **9** (a) $\sec^2 t$; (b) $\cot 2t$;

(c) $4\cos^4 t - 3\cos^2 t$; (d) $\cot t$; (e) $\cos s$ **11** (a) $\dfrac{33}{65}$; (b) $-\dfrac{16}{65}$; (c) $\dfrac{56}{33}$

Problem Set 8 page 38

5 (a) $5, -1, 6, \dfrac{2}{3}$; (b) $\operatorname{dom}(f \pm g) = \operatorname{dom} f \cdot g = \left[-1, \dfrac{3}{2}\right]$, $\operatorname{dom}\left(\dfrac{f}{g}\right) = \left[-1, -\dfrac{1}{2}\right)$ together

with $\left(-\dfrac{1}{2}, \dfrac{3}{2}\right]$.   **7** $\sqrt{x} + x^2 + 4$, $\sqrt{x} - x^2 - 4$, $\sqrt{x}(x^2 + 4)$, $\dfrac{\sqrt{x}}{x^2 + 4}$, and all domains are

$[0, \infty)$.   **9** $\sqrt{x-3} + \dfrac{1}{x}$, $\sqrt{x-3} - \dfrac{1}{x}$, $\sqrt{x-3}\left(\dfrac{1}{x}\right)$, $x\sqrt{x-3}$, and all domains are $[3, \infty)$.

11 $(f+g)(-x) = f(-x) + g(-x) = f(x) + g(x) = (f+g)(x)$, etc.

13 $(\sin x)(x^2 - \cos x)$ **15** $\tan x$ **17** (a) $c + x$; (b) $c - x$; (c) cx; (d) $\dfrac{c}{x}$; (e) $\dfrac{x}{c}$

21 (a) $\sin x^2$; (b) $\sin^2 x$; (c) x^4; (d) $(\sin x + \cos x)^2$; (e) $\tan^2 x$; (f) $\tan(\cot x)$;

(g) $\sin(\cos^2 x)$; (h) $\sin(\cos^2 x)$ **23** $k = \dfrac{3}{2}$ **25** For example, $f(x) = 1 - x$; but, if

$f(x) = \dfrac{1}{x}$, then 0 is not in the domain of f.   **27** $(f \circ g)(x) = (g \circ f)(x) = x$

Problem Set 9 page 43

1 $f^{-1}(x) = \dfrac{x + 13}{7}$ **3** $f^{-1}(x) = \sqrt{x + 3}$ **5** $f^{-1}(x) = \dfrac{2x - 3}{3x - 2}$ **7** $(f \circ g)(x) =$

$(g \circ f)(x) = x$ **11** No

Review Problem Set page 45
1 (a) False; (b) false; (c) true; (d) true; (e) false; (f) false   **3** $(2, \infty)$ **5** $[3, \infty)$

7 $(-5, 4)$ **9** No solutions   **11** $\left(\frac{1}{2}, 6\right)$ **13** $[-5, -2)$ **15** $a = -4, b = -2,$

$c = -3, d = 1$ or $a = 0, b = 2, c = -5, d = -3$ **17** (a) 11; (b) $\frac{1}{|x|} \cdot \frac{1}{|x+1|}$; (c) $\frac{1}{221}$;

(d) $\frac{1}{|x|} \cdot \frac{1}{|x+1|}$ **19** 4 or -1 **21** $x = -2$ **23** $x = \frac{4}{3}$ **25** $\left[-\frac{11}{2}, \frac{1}{2}\right]$

27 $\left(-\infty, \frac{1}{2}\right)$ **29** $\left(-\infty, \frac{3}{4}\right)$ together with $(1, \infty)$ **31** $\left(-\infty, -\frac{7}{2}\right]$ together with $\left[\frac{1}{2}, \infty\right)$

35 $\overline{AB}$, $\overline{BC}$, and $\overline{AC}$ all have the same slope, $\frac{1}{2}$.   **37** $y = -12x + 15$ **39** $y = -7$

41 $x + 3y - 1 = 0$ **43** Domain is $(-\infty, -1]$ together with $[1, \infty)$; symmetric about y axis.   **45** Domain is $\mathbb{R}$, not symmetric about y axis or origin.   **47** Domain is $(-\infty, 1)$ together with $(1, \infty)$, not symmetric about y axis or origin.   **49** No—domain is not $\mathbb{R}$.
51 (a) 4; (b) 4; (c) 36; (d) 36; (e) $64t^2$; (f) $36x^2$; (g) $4(x^2 + 2xh + h^2)$; (h) $8xh + 4h^2$;
(i) $4a$; (j) $2|a|$ **53** Domain is $\mathbb{R}$, f is even.   **55** Domain is $\mathbb{R}$, h is neither even nor odd.   **57** $\cos x$ **59** $-\csc t$ **61** $-\cos 2t$ **63** (a) 106; (b) 154; (c) -32;

(d) 323; (e) 2; (f) $5x^2 + 10x + 13$; (g) $\frac{5x + 7}{2(5x^2 - 1)}$; (h) $(5x^2 - 1)^2$; (i) $10x + 5h$; (j) 5

65 $\sqrt[4]{x}$ and $\sqrt[4]{x}$ **67** $9x^2 + 30x + 24$ and $3x^2 + 6x + 4$ **75** (a) $f^{-1}(x) = \frac{x + 19}{7}$;

(b) $g^{-1}(x) = -\sqrt[3]{\frac{x}{7}}$; (c) $h^{-1}(x) = \frac{13}{x}$

Chapter 1

Problem Set 1 page 53

1 12   **3** 8   **5** 0   **7** -1 **9** $\frac{3}{2}$ **11** 0   **13** (a) Yes; (b) yes; (c) yes; (d) yes;

(e) yes; (f) no   **15** $\delta = 0.0025$ **17** $\delta = 0.01$ **19** $\delta = 0.2$ **21** Take $\delta = \frac{\varepsilon}{2}$.

23 Take $\delta = \frac{\varepsilon}{4}$.   **25** Take any $\delta > 0$.

Problem Set 2 page 57

1 7   **3** 12   **5** $2\sqrt{3}$ **7** 7   **9** $\frac{7}{8}$ **11** 10   **13** $\frac{5}{16}$ **15** $\frac{3}{2}$ **17** -1

19 16   **21** $-\frac{1}{2}$ **23** 6   **25** 0   **27** $\frac{1}{2}$

Problem Set 3 page 62
1 (b) $\lim_{x \to 3^-} f(x) = 8$, $\lim_{x \to 3^+} f(x) = 6$; (c) does not exist; (d) discontinuous
3 (b) $\lim_{x \to 1^-} g(x) = 4$, $\lim_{x \to 1^+} g(x) = 2$; (c) does not exist; (d) discontinuous
5 (b) $\lim_{x \to 5^-} H(x) = 0$, $\lim_{x \to 5^+} H(x) = 0$; (c) $\lim_{x \to 5} H(x) = 0$; (d) discontinuous
7 (b) $\lim_{x \to 1^-} f(x) = 1$, $\lim_{x \to 1^+} f(x) = 1$; (c) $\lim_{x \to 1} f(x) = 1$; (d) continuous
9 (b) $\lim_{x \to 3^-} F(x) = 6$, $\lim_{x \to 3^+} F(x) = 6$; (c) $\lim_{x \to 3} F(x) = 6$; (d) discontinuous
11 (b) $\lim_{x \to 1/2^-} S(x) = 5$, $\lim_{x \to 1/2^+} S(x) = 5$; (c) $\lim_{x \to 1/2} S(x) = 5$; (d) continuous

13 (b) $\lim\limits_{x \to -1^-} f(x) = -4$, $\lim\limits_{x \to -1^+} f(x) = -4$; (c) $\lim\limits_{x \to -1} f(x) = -4$; (d) discontinuous

15 Continuous everywhere   **17** Continuous everywhere   **19** Continuous everywhere
21 Continuous for all a except $a = 0$ **23** (a) Most functions that we use are continuous;
(b) let $f(x) = \begin{cases} 0 & \text{if } x = 0 \\ 1 & \text{if } x \neq 0 \end{cases}$; then $\lim\limits_{x \to 0} f(x) = 1 \neq f(0)$.   **25** $\lim\limits_{x \to 2^+} \sqrt{x - 2} = 0$, but $\sqrt{x - 2}$
is undefined for $x < 2$.

Problem Set 4 page 65
1 Continuous at every number a **3** Continuous at every number a **5** Continuous
at every number except at 0   **7** Continuous at every number except at 1
9 Continuous everywhere except at -1 and 1   **11** (a), (b), (c), (d), and (e) are continuous
at 1.   **13** Continuous on $[-2, 2]$ and $(-2, 2)$ **15** Discontinuous on all indicated
intervals   **17** Continuous on $\left[-\dfrac{1}{2}, 0\right]$ and $\left(-1, -\dfrac{3}{4}\right)$ **19** Continuous on $\left(-\dfrac{3}{2}, \dfrac{3}{2}\right)$,
$\left(\dfrac{3}{2}, \infty\right)$, and $\left(\dfrac{5}{2}, \infty\right)$ **21** (b) E is continuous at each number in its domain except for a.

Problem Set 5 page 72
1 $+\infty$ **3** $+\infty$ **5** $-\infty$ **7** -1 **9** $-\infty$ **11** 6   **13** $\dfrac{5}{8}$ **15** 0
17 $\dfrac{8}{\sqrt[4]{3}}$ **19** $+\infty$ **21** $+\infty$ **23** (a) Given any positive number M, there exists a
positive number N such that $f(x) > M$ holds whenever $x > N$.

Problem Set 6 page 76
1 Horizontal asymptote: $y = \dfrac{7}{2}$; vertical asymptote: $x = \dfrac{5}{2}$ **3** Horizontal asymptote:
$y = -\dfrac{2}{5}$; vertical asymptote: $x = -\dfrac{3}{5}$ **5** Horizontal asymptotes: $y = -\dfrac{3}{\sqrt{2}}$ and $y = \dfrac{3}{\sqrt{2}}$
7 Horizontal asymptotes: $y = -2$ and $y = 2$ **9** Vertical asymptote: $x = 0$
13 Horizontal asymptote: $y = \dfrac{A}{a}$

Review Problem Set page 81
1 $\delta = 0.005$ **3** $\delta = 0.0004$ **5** $\delta = 0.002$ **7** 15   **9** 6   **11** 27   **13** 151
15 $\dfrac{3}{2}$ **17** -1 **19** $\dfrac{\sqrt{3}}{3}$ **21** 2   **23** $\dfrac{1}{6}$ **25** $-\infty$ **27** $+\infty$
29 $\lim\limits_{x \to 3/2^-} f(x) = \lim\limits_{x \to 3/2^+} f(x) = \lim\limits_{x \to 3/2} f(x) = 0$ **31** $\lim\limits_{x \to 2^-} g(x) = \lim\limits_{x \to 2^+} g(x) = \lim\limits_{x \to 2} g(x) = 4$
33 $+\infty$ **35** $\dfrac{2}{3}$ **37** $\dfrac{3}{7}$ **39** (a) Choose $|x - a| < \dfrac{\varepsilon}{3}$; (b) must choose $|x - a| < \dfrac{1}{300}$;
no, since $0.1 \nleq \dfrac{1}{300}$.   **41** Continuous at 3   **43** Continuous at 1   **45** Discontinuous
at multiples of $\dfrac{1}{4}$ **47** (b) $f(x) = -\dfrac{1}{2}x + |x| - |x - 1| + \dfrac{1}{2}|x - 2|$
49 (a) Continuous on $\left(-1, \dfrac{1}{2}\right)$; (b) continuous on all intervals   **51** Horizontal
asymptote: $y = 0$; vertical asymptote: $x = -1$

Chapter 2

Problem Set 1 page 90

1 $r \approx 54.55$ mph   **3** (a) 7.5; (b) 7   **5** (a) 30 ft/sec; (b) 24 ft/sec   **7** (a) 8 ft/sec;

(b) 7 ft/sec   **9** 0   **11** 2   **13** 2   **15** $\dfrac{1}{4}$ **17** $-\dfrac{1}{3}$ **19** (a) 80 ft/sec;

(b) 160 ft/sec   **21** $5\sqrt{3}$ square cm/cm of edge of length   **23** (a) -0.16 pound/in.^{2}/in.^{3};

(b) -0.2 pound/in.^{2}/in.^{3} **25** $\displaystyle\lim_{h \to 0^+} \dfrac{1}{\sqrt{h}} = +\infty$

Problem Set 2 page 96

1 $2x + 4$ **3** $6x^2 - 4$ **5** $-\dfrac{2}{x^2}$ **7** $\dfrac{-3}{(t-1)^2}$ **9** $\dfrac{1}{2\sqrt{v-1}}$ **11** $\dfrac{-2}{(x+1)^2}$

13 4   **15** $-\dfrac{14}{25}$ **17** $-\dfrac{1}{9}$ **19** $-\dfrac{4}{25}$ **23** $32t + 30$ **25** (b) Continuous at 3;

(c) differentiable at 3   **27** (b) Continuous at 3; (c) not differentiable at 3   **29** (b)
Continuous at 2; (c) not differentiable at 2   **31** (b) Continuous at 0; (c) differentiable
at 0   **33** (a) If a graph has a tangent line at a point, it cannot "jump" at that point;

(b) no, consider $f(x) = |x|$ at $x = 0$.   **35** (a) Not differentiable at $\dfrac{1}{3}$; (b) not differentiable

at 1

Problem Set 3 page 107

1 $5x^4 - 9x^2$ **3** $5x^9 + x^4$ **5** $8t^7 - 14t^6 + 3$ **7** $\dfrac{-6}{x^3} - \dfrac{4}{x^2}$ **9** $-\dfrac{25}{y^6} + \dfrac{25}{y^2}$

11 $-6x^{-3} + 7x^{-2}$ **13** $-\dfrac{2}{5x^2} + \dfrac{2\sqrt{2}}{3x^3}$ **15** $15x^4 - 2x$

17 $5x^4 + 12x^3 - 27x^2 - 54x$ **19** $24y^2 - 8y + 14$ **21** $\dfrac{16}{x^2} + 4x - 3x^2$

23 $-\dfrac{10}{x^6} - \dfrac{18}{x^4} - \dfrac{1}{x^2} + 3$ **25** $\dfrac{-23}{(3x-1)^2}$ **27** $\dfrac{-7x^2 + 6x + 5}{(x^2 - 3x + 2)^2}$ **29** $\dfrac{-20t}{(t^2 - 1)^2}$

31 $\dfrac{3x^2 + 12x + 37}{(x+2)^2}$ **33** (a) 4; (b) $-\dfrac{3}{16}$; (c) -9; (d) -2; (e) $-\dfrac{1}{18}$; (f) $\dfrac{64}{81}$

35 (a) $8x^3 - 3x^2 - 24x + 1$; (b) $54x^2 + 66x - 28$; (c) $9x^2 - 20x + 6 - 3x^{-2}$;
(d) $12x(2x^2 + 7)^2$ **37** (a) -1; (b) 5; (c) -5; (d) -2; (e) 16; (f) -4 **39** (a) 16;

(b) $-\dfrac{3}{49}$ **41** -4 **43** 7.5 ft/sec   **45** We must *first* calculate $f'(x) = 4x + 3$, *then*

substitute $x = 2$ to get $f'(2) = 11$.   **47** $c = \dfrac{1}{m+1}$, $n = m + 1$, $m \neq -1$

51 (a) $D_x(x \cdot 1) = 1$, but $(D_x x)(D_x 1) = (1)(0) = 0$; (b) $D_x\left(\dfrac{x}{1}\right) = 1$, but $\dfrac{D_x x}{D_x 1}$ is undefined,

since $D_x 1 = 0$.

Problem Set 4 page 113

1 $\dfrac{2x + 1}{2\sqrt{x^2 + x + 1}}$ **3** $\dfrac{-20x^3}{(x^4 + 1)^6}$ **5** $-20(5 - 2x)^9$ **7** $\dfrac{-20}{(4y + 1)^6}$

9 $45u^2(u^3 + 2)^{14}$ **11** $-7(5x^4 - 4x + 1)(x^5 - 2x^2 + x + 1)^{-8}$

13 $(3x^2 + 7)(5 - 3x)^2(-63x^2 + 60x - 63)$ **15** $\left(3x + \dfrac{1}{x}\right)(6x - 1)^4\left(126x - 6 + \dfrac{18}{x} + \dfrac{2}{x^2}\right)$

17 $\dfrac{(28y + 38)(2y - 1)^3}{(7y + 3)^3}$ **19** $4\dfrac{(x^2 + x)^3}{(1 - 2x)^5}(1 + 2x - 2x^2)$ **21** $\dfrac{-3(3x + 1)^2(3x + 2)}{x^7}$

23 $\dfrac{14\left(7t + \dfrac{1}{t}\right)^6\left(-7t^3 - 2t + 7 - \dfrac{1}{t^2}\right)}{(t^3 + 2)^8}$ **25** $\dfrac{-1}{2x\sqrt{x}}$ **27** $\dfrac{x + 1}{\sqrt{x^2 + 2x - 1}}$

29 $\dfrac{2t^3 - t}{\sqrt{t^4 - t^2 + \sqrt{3}}}$ **31** $2x - 2 - \dfrac{3}{2}\sqrt{x}$ **33** $\dfrac{(x + 1)^4(6x^2 + x + 10)}{\sqrt{x^2 + 2}}$ **37** $\dfrac{7}{2}$

41 (a) $5\cos 5x$; (b) $-8\sin(8x - 1)$; (c) $3\sec^2 3t$; (d) $-9\csc^2 9x$;

(e) $2\sec(2t + 9)\tan(2t + 9)$; (f) $-15\csc(15x - 2)\cot(15x - 2)$; (g) $3\sin^2\theta\cos\theta$;

(h) $\dfrac{\cos\sqrt{\theta}}{2\sqrt{\theta}}$; (i) $\dfrac{\pi}{90}\sin\dfrac{2\pi x}{360}\cos\dfrac{2\pi x}{360}$; (j) $\dfrac{-\sin\theta}{2\sqrt{\cos\theta}}$; (k) 0; (l) $2\cos 2\theta$ **45** (a) $\dfrac{5t^2 + 3t + 1}{2\sqrt{t}}$;

(b) $\dfrac{1 - t - 3t^2}{2\sqrt{t}(1 + t + t^2)^2}$ **47** $-\dfrac{1}{27}$

Problem Set 5 page 120

1 $\dfrac{1}{5}$ **3** $\dfrac{1}{4}$ **5** $-\dfrac{5}{9}$ **7** $-3^{3/2}$ **9** $\dfrac{\sqrt[5]{x}}{5x}$ **11** $-16x^{-13/9}$ **13** $\dfrac{2}{3}(1 - t)^{-5/3}$

15 $-18s(9 - s^2)^{-1/2}(9 + s^2)^{-3/2}$ **17** $-\dfrac{1}{2}x^{-3/2} - \dfrac{1}{3}x^{-4/3} - \dfrac{1}{4}x^{-5/4}$

19 $\dfrac{1}{5}(t^3 - t^{1/4})^{-4/5}\left(3t^2 - \dfrac{1}{4}t^{-3/4}\right)$ **21** $\dfrac{\sqrt[10]{\dfrac{x}{x + 1}}}{10x(x + 1)}$

23 $\dfrac{1 - 2x}{4}(1 + x)^{-7/4}(2x - 1)^{-1/2}$ **25** $\dfrac{(9t + 33)(t + 2)^{-3/4}(t + 5)^{-4/5}}{20}$

27 $\dfrac{1}{5}(\sin t)^{-4/5}\cos t$ **29** $-\dfrac{3}{4}\cos^{-1/4}x\sin x$ **31** (a), (b), and (c) $\dfrac{1}{4}x^{-3/4}$ **33** $2|x|$

35 $\dfrac{1}{2}$ **37** 7   **39** $\sqrt{2}$ **41** (b) and (c) $\dfrac{-1}{(3x - 2)^2}$ **43** (a) $x = \dfrac{1 + \sqrt{8y - 7}}{4}$;

(b) $f^{-1}(y) = \dfrac{1 + \sqrt{8y - 7}}{4}$; (c) and (d) $\dfrac{1}{\sqrt{8y - 7}}$ **45** $\dfrac{2}{3}$

Problem Set 6 page 124

1 Tangent line: $y = 8x - 15$; normal line: $x + 8y - 10 = 0$ **3** Tangent line: $y = 3x$; normal line: $x + 3y - 10 = 0$ **5** Tangent line: $x - 12y + 16 = 0$; normal line: $12x + y = 98$ **7** Tangent line: $y = 3$; normal line: $x = 0$ **9** Tangent line: $y = x + 1$; normal line: $x + y = 1$ **11** x intercept is -3, y intercept is $\dfrac{3}{2}$.   **13** $(8, 72)$; tangent line: $y = 16x - 56$ **15** $\left(-\dfrac{5}{6}, \dfrac{47}{12}\right)$ **17** $(1, 0)$; tangent line: $y = -x + 1$ **19** $(1, 0)$; normal lines: $y = \dfrac{1 - x}{2}$ or $y = \dfrac{-1 - x}{2}$ **21** $(5, 30)$, $(-1, 6)$; tangent lines: $y = 10x - 20$ or $y = -2x + 4$

Problem Set 7 page 129

1 3.01   **3** 6.0083   **5** 0.485   **7** 0.28   **9** (No. 1) estimate: 3.01; true: 3.00998...; (No. 3) estimate: 6.0083; true: 6.008327...; (No. 5) estimate: 0.485; true: 0.48543...; (No. 7) estimate: 0.28; true: 0.2849   **11** $x^3 - 3x - 2 = 9(x - 2) + \varepsilon(x)(x - 2)$, $\varepsilon(x) = x^2 + 2x - 8$ **13** $2x^3 = 54x - 108 + \varepsilon(x)(x - 3)$, $\varepsilon(x) = (x - 3)(2x + 12)$

15 $\dfrac{5}{x} = 2 - \dfrac{x}{5} + \varepsilon(x)(x-5), \varepsilon(x) = \dfrac{x-5}{5x}$ **19** (a) $m = 3, c = -2$; (b) $\lim\limits_{x\to 1}(x^2 + x - 2) = 0$;
(c) $m = f'(1)$, where $f(x) = x^3$

Review Problem Set page 130
1 (a) 120 ft/sec; (b) 136 ft/sec when $t = 2$, 104 ft/sec when $t = 3$; (c) 625 ft   **3** (a) 40.2 in.^{2}/in.; (b) 0.3216 in.^{2}/deg   **5** Tangent line: $y = 4x - 14$; normal line: $4y + x = 12$
7 Tangent line: $72x - 16y = 81$; normal line: $144y + 32x = 291$ **9** Tangent line: $y = kx + 1 - k$; y intercept: $1 - k$ **11** (a) Continuous at 2; (b) $f'_-(2) = -3$, $f'_+(2) = 4$; (c) not differentiable at 2   **13** (a) Continuous at -3; (b) $f'_-(-3) = -1$, $f'_+(-3) = 0$; (c) not differentiable at -3 **15** (a) $\lim\limits_{\Delta x\to 0}\dfrac{(x+\Delta x)^{2/3} - x^{2/3}}{\Delta x}$; (b) $\dfrac{1}{3}$ **17** (a) No;

(b) no   **19** (b) At 0 and at -1 **21** (a) $\dfrac{89}{30}$; (b) $\dfrac{91}{30}$; (c) $\dfrac{44}{3}$; (d) $\dfrac{46}{75}$; (e) $-\dfrac{23}{50}$; (f) $\dfrac{44}{15}$;

(g) $\dfrac{46}{675}$ **23** $\dfrac{x}{\sqrt{x^2 + 12}}$ **25** $\dfrac{3}{2\sqrt{3x - 11}}$ **27** $\dfrac{4x\sqrt{1 + x^3} + 3x^2}{4\sqrt{1 + x^3}\sqrt{x^2 + \sqrt{1 + x^2}}}$

29 $\dfrac{22x}{3(2 + 3x^2)^{2/3}(3 - x^2)^{4/3}}$ **31** $\dfrac{6t^2 + 2t - 15}{(2t^2 + 5)^2}$ **33** $\dfrac{(u^2 + 7)(-9u^2 + 8u - 7)}{2\sqrt{1 - u}}$

35 $\dfrac{z(z^2 + 3)^3(37 - 9z^2)}{\sqrt{5 - z^2}}$ **37** $\dfrac{2(-17t^2 - 4t + 10)}{3(4t^2 - 3t + 2)^{1/3}(t^2 - 5t)^{5/3}}$ **39** $\dfrac{ad - bc}{(cy + d)^2}$

41 $\dfrac{x(1 - 2x)^2}{2\sqrt{1 - 5x^3}}(130x^4 - 35x^3 - 20x + 4)$

43 $\dfrac{1}{2}[x + (x + x^{1/2})^{1/2}]^{-1/2}\left[1 + \dfrac{1}{2}(x + x^{1/2})^{-1/2}\left(1 + \dfrac{1}{2}x^{-1/2}\right)\right]$

45 $\dfrac{ad - bc}{2\sqrt{(ax + b)(cx + d)^3}}$ **47** $\dfrac{2}{5}$ **49** Slope: $-\dfrac{3}{4}$; tangent line: $4y + 3x = 20$
51 Approximately 4.0208   **53** 0, 1, 2, and 3   **55** Slopes are both n
59 $5f(x)[g(x)]^4g'(x) + f'(x)[g(x)]^5$ **61** $\dfrac{3g(x)[xf(x)]^2[xf'(x) + f(x)] - [xf(x)]^3g'(x)}{[g(x)]^2}$

65 Let $f(x) = (1 + x^2)^3$; then $f'(2) = 300$ **67** Yes   **71** (a) $\dfrac{1}{14}$; (b) $\dfrac{1}{14}$

73 $\dfrac{1}{3x^2 - 6x + 1}$ **75** $x(1 + E)$

Chapter 3

Problem Set 1 page 141
1 $f(1.4) = -0.04 < 0$, $f(1.5) = 0.25 > 0$ **3** $f(1) = -1 < 0$, $f(2) = 11 > 0$ **5** f is not continuous at 2   **7** $c = \dfrac{2}{3}\sqrt{3}$ **9** $c = 1$ **11** $c = \pm\dfrac{\sqrt{2}}{2}$ **13** $c = 3$

15 $c = \sqrt{\dfrac{13}{3}}$ **17** f is not differentiable at 0   **19** f is not differentiable at 1

21 (a) Not differentiable on (a,b); (b) not differentiable on (a,b); (c) not defined on $[a,b]$; (d) discontinuous at a; (e) not continuous on $[a,b]$ **25** $c = \dfrac{3}{2}$ **27** $c = 1 \pm\dfrac{2}{3}\sqrt{3}$

29 $c = \dfrac{4}{9}$ **31** Conclusion: the mean value theorem can be deduced from Rolle's theorem.

33 $c = \dfrac{a + b}{2}$

Problem Set 2 page 147

1 $v = 3t^2 + 4t$, $a = 6t + 4$ **3** $v = 10t + \dfrac{2}{\sqrt{4t-3}}$, $a = 10 - \dfrac{4}{(\sqrt{4t-3})^3}$ **5** $v =$

$\dfrac{25}{4}t^{3/2} + t^{1/2}$, $a = \dfrac{75}{8}t^{1/2} + \dfrac{1}{2}t^{-1/2}$ **7** (No. 1) $v = 7$ ft/sec, $a = 10$ ft/sec^2; (No. 3) $v = 12$

meters/sec, $a = 6$ meters/sec^2; (No. 5) $v = \dfrac{29}{4}$ km/hr, $a = \dfrac{79}{8}$ km/hr^2 **9** $f'(x) = 15x^2 + 4$,

$f''(x) = 30x$ **11** $f'(t) = 35t^4 - 46t + 1$, $f''(t) = 140t^3 - 46$ **13** $G'(x) = 6x^5$,

$G''(x) = 30x^4$ **15** $g'(t) = \dfrac{7}{2}t^{5/2} - 5$, $g''(t) = \dfrac{35}{4}t^{3/2}$ **17** $f'(x) = 2x + 3x^{-4}$, $f''(x) =$

$2 - 12x^{-5}$ **19** $f'(u) = 4(2-u)^{-2}$, $f''(u) = 8(2-u)^{-3}$ **21** $f'(t) = t(t^2+1)^{-1/2}$,

$f''(t) = (t^2+1)^{-3/2}$ **23** $F'(r) = 1 - r^{-1/2}$, $F''(r) = \dfrac{1}{2}r^{-3/2}$ **25** 0   **27** $\dfrac{736}{3}$

29 (a) $x(x^2-1)^{-1/2}$; (b) $-(x^2-1)^{-3/2}$; (c) $3x(x^2-1)^{-5/2}$ **31** $\dfrac{11}{3}$ **33** (a) When

$t = 2$; (b) a is never zero; (c) a is never zero   **35** $\dfrac{d^2s}{dt^2} = a = \dfrac{dv}{dt} = \dfrac{ds}{dt}\dfrac{dv}{ds} = v\dfrac{dv}{ds}$

37 $(-1)^n n!\, x^{-(n+1)}$ **39** $-\sin x$

Problem Set 3 page 156

1 Increasing: $(-\infty, -2]$ and $[2, \infty)$, decreasing: $[-2, 2]$ **3** Increasing: $(-\infty, 0)$ and

$[\sqrt[3]{6}, \infty)$, decreasing: $(0, \sqrt[3]{6}]$ **5** Increasing: $[4, \infty)$, decreasing: $(0, 4]$ **7** Increasing:

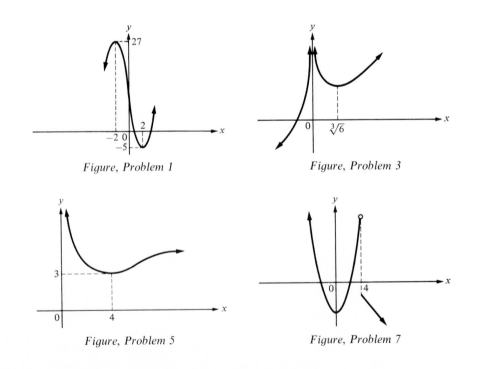

Figure, Problem 1   Figure, Problem 3

Figure, Problem 5   Figure, Problem 7

$[0, 4)$, decreasing: $(-\infty, 0]$ and $[4, \infty)$ **9** Concave downward: $\left(-\infty, \dfrac{1}{12}\right)$, point of

inflection: $\left(\dfrac{1}{12}, \dfrac{611}{432}\right)$; concave upward: $\left(\dfrac{1}{12}, \infty\right)$ **11** Concave upward on $(-\infty, \infty)$, no

point of inflection   **13** Concave downward on $(-\infty, 0)$ and $(0, \infty)$, no point of inflection

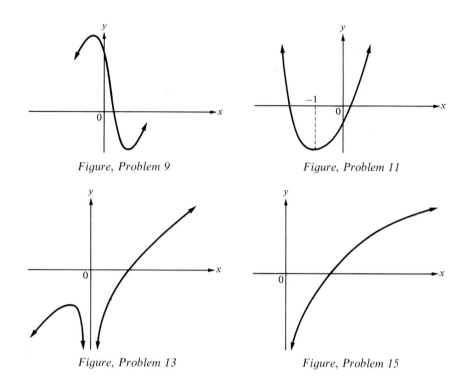

Figure, Problem 9

Figure, Problem 11

Figure, Problem 13

Figure, Problem 15

15 Always concave downward for $x > 0$ **17** (a) (i) Increasing on $[a,b]$ and $[c,d]$, (ii) decreasing on $[b,c]$ and $[d,e]$, (iii) never concave upward, (iv) never concave downward, (v) no inflection points; (b) (i) increasing on $[b,c]$, $[c,d]$, and $[d,e]$, (ii) not decreasing on any displayed subinterval, (iii) concave upward on (c,d), (iv) concave downward on (a,b), (b,c), and (d,e), (v) points of inflection at $(c, f(c))$ and $(d, f(d))$; (c) (i) increasing on $[a,b]$ and $[d,e]$, (ii) decreasing on $[b,c]$, (iii) concave upward on (c,d), (iv) concave downward on (a,b), (b,c), and (d,e), (v) points of inflection at $(c, f(c))$ and $(d, f(d))$; (d) (i) increasing on $[c,e]$, (ii) decreasing on $[a,b]$ and $[b,c]$, (iii) concave upward on (b,c) and (c,d), (iv) concave downward on (a,b) and (d,e), (v) points of inflection at $(b, f(b))$ and $(d, f(d))$ **19** (a) Increasing on $(-\infty, 1]$ and $[3, \infty)$; (b) decreasing on $[1,3]$; (c) concave upward on $(2, \infty)$; (d) concave downward on $(-\infty, 2)$; (e) point of inflection at $(2,3)$ **21** (a) Increasing on $[0, \infty)$; (b) decreasing on $(-\infty, 0]$; (c) concave upward on $(-\infty, \infty)$; (d) never concave downward; (e) no inflection points   **23** (a) Increasing on $(-\infty, -\sqrt{6}]$ and $[0, \sqrt{6}]$; (b) decreasing on $[-\sqrt{6}, 0]$ and $[\sqrt{6}, \infty)$; (c) concave upward on $(-\sqrt{2}, \sqrt{2})$; (d) concave downward on $(-\infty, -\sqrt{2})$ and $(\sqrt{2}, \infty)$; (e) points of inflection at $(-\sqrt{2}, 20)$ and $(\sqrt{2}, 20)$ **25** (a) Increasing on $(-\infty, -1]$ and $[1, \infty)$; (b) decreasing on $[-1, 0)$ and $(0, 1]$; (c) concave upward on $(0, \infty)$; (d) concave downward on $(-\infty, 0)$; (e) no points of

inflection   **27** (a) Increasing on $\left[0, \dfrac{6}{5}\right]$; (b) decreasing on $(-\infty, 0]$ and $\left[\dfrac{6}{5}, \infty\right)$;

(c) concave upward on $\left(-\infty, -\dfrac{3}{5}\right)$; (d) concave downward on $\left(-\dfrac{3}{5}, 0\right)$ and $(0, \infty)$;

(e) point of inflection at $\left(-\dfrac{3}{5}, \dfrac{18}{5} \sqrt[3]{\dfrac{9}{25}}\right)$ **29** (a) Increasing on $[2, \infty)$; (b) decreasing on

$(-\infty, 2]$; (c) never concave upward; (d) concave downward on $(-\infty, 2)$ and $(2, \infty)$; (e) no points of inflection   **31** (a) Increasing on $[-1, 1]$; (b) decreasing on $(-\infty, -1]$ and $[1, \infty)$; (c) concave upward on $(-\sqrt{3}, 0)$ and $(\sqrt{3}, \infty)$; (d) concave downward on

$(-\infty, -\sqrt{3})$ and $(0, \sqrt{3})$; (e) points of inflection at $\left(-\sqrt{3}, -\dfrac{\sqrt{3}}{2}\right)$, $(0,0)$, and $\left(\sqrt{3}, \dfrac{\sqrt{3}}{2}\right)$

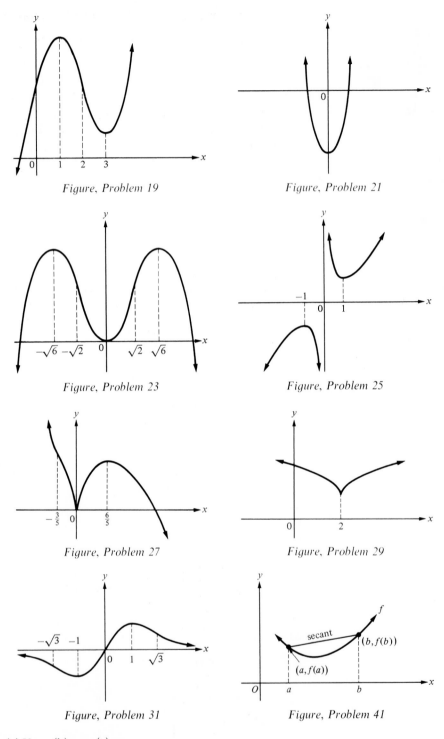

Figure, Problem 19

Figure, Problem 21

Figure, Problem 23

Figure, Problem 25

Figure, Problem 27

Figure, Problem 29

Figure, Problem 31

Figure, Problem 41

37 (a) Yes; (b) no; (c) yes

Problem Set 4 page 164

1 (a) Critical number at 3; (b) relative maximum at 3   **3** (a) Critical numbers at $-\dfrac{1}{3}$
and 1; (b) relative maximum at $-\dfrac{1}{3}$, relative minimum at 1   **5** (a) Critical number at 1;

(b) relative minimum at 1   **7** (a) Critical numbers at 0 and 1; (b) relative maximum at 1

9 (a) Critical number at 1; (b) relative maximum at 1   **11** (a) Critical number at $\frac{5}{2}$;

(b) $f''\left(\frac{5}{2}\right) = 2 > 0$; hence, relative minimum at $\frac{5}{2}$ **13** (a) Critical numbers at $-1 \pm \sqrt{2}$;

(b) $f''(-1 - \sqrt{2}) = -6\sqrt{2} < 0$; hence, relative maximum at $-1 - \sqrt{2}$.

$f''(-1 + \sqrt{2}) = 6\sqrt{2} > 0$; hence, relative minimum at $-1 + \sqrt{2}$ **15** (a) Critical

number at 1; (b) $f''(1) = 2 > 0$; hence, relative minimum at 1   **17** (a) Critical numbers

$\pm \sqrt[4]{5}$; (b) $f''(\pm \sqrt[4]{5}) = 8 > 0$; hence, relative minima at $\pm \sqrt[4]{5}$ **19** Relative maximum

at 1; relative minimum at 2

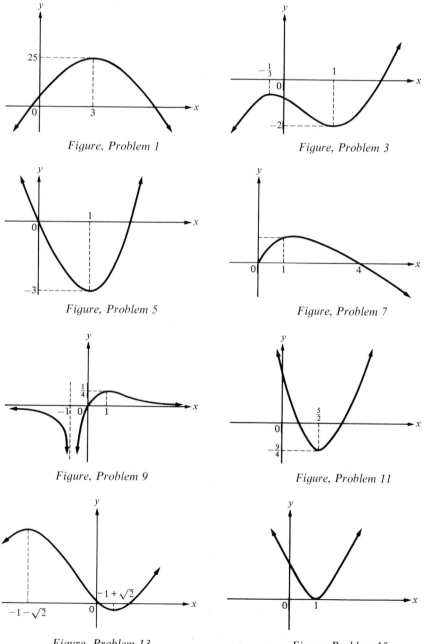

Figure, Problem 1

Figure, Problem 3

Figure, Problem 5

Figure, Problem 7

Figure, Problem 9

Figure, Problem 11

Figure, Problem 13

Figure, Problem 15

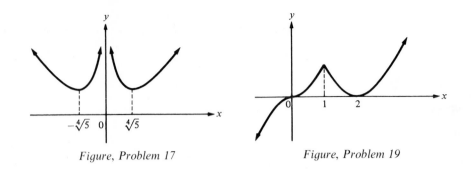

Figure, Problem 17   *Figure, Problem 19*

21 $a = -3$, $b = 5$ **23** $p = 12$, $q = 3$

Problem Set 5 page 169

1 Absolute maximum of 2 at -1, absolute minimum of -4 at 2   **3** Absolute maximum of 4 at 1, absolute minimum of 0 at -1 **5** Absolute maximum of 2 at 0, absolute minimum of 0 at -2 and at 2   **7** Absolute maximum of 15 at 5, absolute minimum of -4 at -6 **9** No absolute maximum, no absolute minimum   **11** Absolute maximum

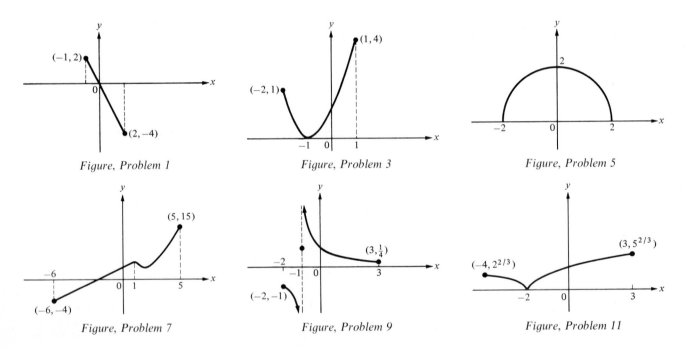

Figure, Problem 1   *Figure, Problem 3*   *Figure, Problem 5*

Figure, Problem 7   *Figure, Problem 9*   *Figure, Problem 11*

of $5^{2/3}$ at 3, absolute minimum of 0 at -2 **13** (a) Decreasing on $(-\infty, 2]$ and $[1, \infty)$, increasing on $[-2, 1]$; (b) concave upward on $\left(-\infty, -\frac{1}{2}\right)$, concave downward on $\left(-\frac{1}{2}, \infty\right)$; (c) relative maximum at 1, relative minimum at -2; (d) no absolute extrema; (e) point of inflection at $\left(-\frac{1}{2}, \frac{7}{2}\right)$ **15** (a) Increasing on $\left[-\frac{\sqrt{2}}{2}, 0\right]$ and $\left[\frac{\sqrt{2}}{2}, \infty\right)$, decreasing on $\left(-\infty, -\frac{\sqrt{2}}{2}\right]$ and $\left[0, \frac{\sqrt{2}}{2}\right]$; (b) concave upward on $\left(-\infty, -\frac{\sqrt{6}}{6}\right)$ and $\left(\frac{\sqrt{6}}{6}, \infty\right)$, concave downward on $\left(-\frac{\sqrt{6}}{6}, \frac{\sqrt{6}}{6}\right)$; (c) relative minimum at $\pm\frac{\sqrt{2}}{2}$, relative maximum at 0;   (d) no absolute maximum, absolute minimum of $\dfrac{7}{4}$ at $\pm\dfrac{\sqrt{2}}{2}$;   (e) points of inflection at $\left(-\dfrac{\sqrt{6}}{6},\dfrac{67}{36}\right)$ and $\left(\dfrac{\sqrt{6}}{6},\dfrac{67}{36}\right)$ **17** (a) Increasing on $(-\infty,0]$ and $[4,\infty)$, decreasing on $[0,4]$;   (b) concave upward on $(3,\infty)$, concave downward on $(-\infty,3)$;   (c) relative maximum at 0, relative minimum at 4;   (d) no absolute extrema;   (e) point of inflection at $(3,-648)$ **19** (a) Increasing on $[-3,3]$, decreasing on $(-\infty,-3]$ and $[3,\infty)$;   (b) concave downward on $(-\infty,-\sqrt{27})$ and $(0,\sqrt{27})$, concave upward on $(-\sqrt{27},0)$ and $(\sqrt{27},\infty)$;   (c) relative maximum at 3, relative minimum at -3;   (d) absolute maximum of $\dfrac{1}{2}$ at 3, absolute minimum of $-\dfrac{1}{2}$ at -3;   (e) points of inflection at $\left(-\sqrt{27},-\dfrac{\sqrt{27}}{12}\right)$, $(0,0)$,

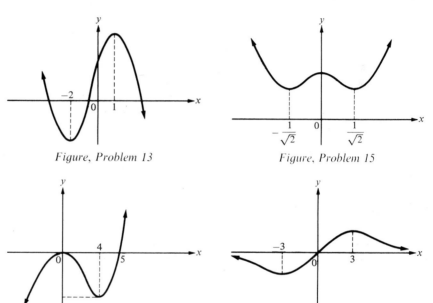

Figure, Problem 13   Figure, Problem 15

Figure, Problem 17   Figure, Problem 19

and $\left(\sqrt{27},\dfrac{\sqrt{27}}{12}\right)$ **21** (a) Increasing on $[-3,-1]$ and $[1,\infty)$, decreasing on $(-\infty,-3]$ and $[-1,1]$;   (b) concave upward on $\left(-\infty,\dfrac{-3-2\sqrt{3}}{3}\right)$ and $\left(\dfrac{-3+2\sqrt{3}}{3},\infty\right)$, concave downward on $\left(\dfrac{-3-2\sqrt{3}}{3},\dfrac{-3+2\sqrt{3}}{3}\right)$;   (c) relative maximum at -1, relative minima at -3 and 1;   (d) absolute minimum of 0 at -3 and 1, no absolute maximum;   (e) points of inflection $\left(\dfrac{-3-2\sqrt{3}}{3},\dfrac{64}{9}\right)$ and $\left(\dfrac{-3+2\sqrt{3}}{3},\dfrac{64}{9}\right)$ **23** (a) Decreasing on $(-\infty,1]$, increasing on $[1,\infty)$;   (b) concave downward on $(-\infty,1)$ and $(1,\infty)$;   (c) relative minimum at 1;   (d) absolute minimum of 2 at 1, no absolute maximum;   (e) no inflection points
25 (a) Increasing on $[-1-\sqrt{2},-1+\sqrt{2}]$, decreasing on $(-\infty,-1-\sqrt{2}]$ and $[-1+\sqrt{2},\infty)$;   (b) concave upward on $(-3,-\sqrt{3})$ and $(\sqrt{3},\infty)$, concave downward on $(-\infty,-3)$ and $(-\sqrt{3},\sqrt{3})$;   (c) relative maximum at $-1+\sqrt{2}$, relative minimum at $-1-\sqrt{2}$;   (d) absolute maximum of $\dfrac{\sqrt{2}}{4+2\sqrt{2}}$ at $-1+\sqrt{2}$, absolute minimum of $-\dfrac{\sqrt{2}}{4+2\sqrt{2}}$ at $-1-\sqrt{2}$;   (e) points of inflection at $(-3,-1)$, $\left(-\sqrt{3},\dfrac{1-\sqrt{3}}{8-4\sqrt{3}}\right)$, and $\left(\sqrt{3},\dfrac{1+\sqrt{3}}{8+4\sqrt{3}}\right)$

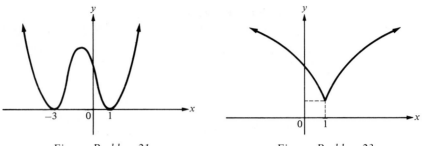

Figure, Problem 21   Figure, Problem 23

27 (a) Decreasing on $\left(-\infty, \frac{1}{2}\right)$, increasing on $\left[\frac{1}{2}, \frac{2}{3}\right)$ and $\left(\frac{2}{3}, \infty\right)$; (b) concave upward

on $\left(-\infty, \frac{2}{3}\right)$ and $(1, \infty)$, concave downward on $\left(\frac{2}{3}, 1\right)$; (c) relative minimum at $\frac{1}{2}$;

(d) no absolute extrema; (e) point of inflection at $(1,0)$ **29** (a) Increasing on

$(-\infty, -1]$ and $\left[-\frac{1}{7}, \infty\right)$, decreasing on $\left[-1, -\frac{1}{7}\right]$; (b) concave upward on

$((-2 - 3\sqrt{2})/14, 0)$ and $((-2 + 3\sqrt{2})/14, \infty)$, concave downward on $(-\infty, (-2 - 3\sqrt{2})/14)$

and $(0, (-2 + 3\sqrt{2})/14)$; (c) relative maximum at -1, relative minimum at $-\frac{1}{7}$; (d) no

absolute extrema; (e) points of inflection at $((-2 - 3\sqrt{2})/14, f((-2 - 3\sqrt{2})/14)) \approx$

$(-0.45, -0.23)$, $(0,0)$, and $((-2 + 3\sqrt{2})/14, f((-2 + 3\sqrt{2})/14)) \approx (0.16, 0.73)$

31 (a) Decreasing on $(-\infty, 0]$, increasing on $[0, \infty)$; (b) concave upward on $(-\sqrt{14}, \sqrt{14})$,

concave downward on $(-\infty, -\sqrt{14})$ and $(\sqrt{14}, \infty)$; (c) relative minimum at 0;

(d) absolute minimum of $\frac{1}{2}$ at 0; (e) points of inflection at $(-\sqrt{14}, f(-\sqrt{14})) \approx$

$(-3.74, 3.54)$ and $(\sqrt{14}, f(\sqrt{14})) \approx (3.74, 3.54)$

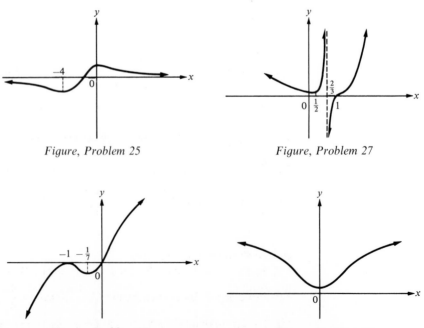

Figure, Problem 25   Figure, Problem 27

Figure, Problem 29   Figure, Problem 31

Problem Set 6 page 174

1 10 by 10   **3** 12 by 12   **5** 375 by 375   **7** 300 by 450   **9** 20   **11** 5 by 10

13 20   **15** $h = 10\sqrt[3]{\dfrac{18}{\pi}}, r = \dfrac{h}{\sqrt{2}}$ **17** $h = \dfrac{8}{\sqrt{3}}, r = \dfrac{h}{2}$ **19** $h = 16, r = 4\sqrt{2}$

Problem Set 7 page 181

1 $|\overline{PT}| = \dfrac{9}{4}$ miles   **3** $\dfrac{3}{4}$ mile   **5** $\dfrac{100}{\sqrt{13}}$ miles   **7** $m = 1$ **9** $15\sqrt{2}$ meters

11 $\dfrac{16{,}000P}{9\sqrt{3}\,EI}$ **13** (a) 80; (b) $x = 30$ **15** 200 tons   **17** 30   **19** 4   **21** 5 years

Problem Set 8 page 187

1 $-\dfrac{9x}{4y}$ **3** $\dfrac{y^2 - 2xy - 2x}{x^2 - 2xy}$ **5** $\dfrac{3y - 2x}{2y - 3x}$ **7** $-\sqrt[3]{\dfrac{y}{x}}$ **9** $-\dfrac{8x^3 y\sqrt{xy} + y}{2x^4\sqrt{xy} + x}$

11 $-\dfrac{2y\sqrt{x} + y\sqrt{y}}{2x\sqrt{y} + x\sqrt{x}}$ **13** $-\dfrac{2(1 + y)\sqrt{1 + x} + y\sqrt{1 + y}}{2(1 + x)\sqrt{1 + y} + x\sqrt{1 + x}}$ **15** $\dfrac{\sqrt{x + y} + \sqrt{x - y}}{\sqrt{x + y} - \sqrt{x - y}}$

17 $\dfrac{y}{x}$ **19** $\dfrac{y}{x}$ **21** $y = -\dfrac{1}{2}x + 4$ and $y = 2x - 1$ **23** $y = -x + 5$ and $y = x + 1$

27 $-\dfrac{32x}{y^5}$ **29** $\dfrac{-5x}{6x + 5y}$ **31** $\dfrac{2y - 1}{2x}$ **33** $y = \dfrac{5x - 6}{4}$ **35** $y = \dfrac{-x + \sqrt{x^2 + 4x}}{2}$

and $y = \dfrac{-x - \sqrt{x^2 + 4x}}{2}$ **37** $y = \dfrac{x + 1}{2 - 3x}$ **39** $y = \sqrt[4]{x}$ and $y = -\sqrt[4]{x}$

41 $y = 1 + \sqrt[3]{3x + 2}$ **43** Every point on the graph of f is on the graph of $\dfrac{x^2}{16} + \dfrac{y^2}{9} = 1.$

45 (a) $\dfrac{-3x}{4\sqrt{16 - x^2}}$; (b) $\dfrac{3x}{4\sqrt{16 - x^2}}$; (c) $-\dfrac{9x}{16y}$ **47** Slope $= -2$ **49** $\dfrac{-rh}{2r^2 + h^2}$

Problem Set 9 page 192

1 $-\dfrac{3}{4\pi} \approx -0.24$ **3** $-0.01224\pi \approx -0.038$ **5** (a) $-12{,}000 \text{ cm}^3/\text{min}$;

(b) $-2400 \text{ cm}^2/\text{min}$ **7** $\dfrac{1350}{\sqrt{2089}} \approx 29.54 \text{ mph}$ **9** (a) $\dfrac{18}{11}$ m/sec; (b) $\dfrac{40}{11}$ m/sec

11 $(0.7)\sqrt{3} \approx 1.21 \text{ m/sec}$ **13** $-(4\pi)^{-1/3}(1.2)^{-2/3}(0.17) \approx -0.06 \text{ m/min}$ **15** $\dfrac{5}{8}$ in./min

17 $-\dfrac{268}{1750\pi} \approx -0.049 \text{ ft/min}$ **19** 480 in.^{3}/min   **21** $\dfrac{1}{18\pi} \approx 0.02 \text{ ft/min}$

23 $\dfrac{8}{5\pi} \approx 0.51 \text{ in./min}$ **25** $\dfrac{1}{0.93} \approx 1.08 \text{ degrees C/min}$ **27** $\dfrac{100}{3}$ in.^{3}/sec

29 $\dfrac{3}{400} \approx \$0.01/\text{bushel/day}$ **31** $\dfrac{3613}{9100} \approx 0.397 \text{ cm}$

Review Problem Set page 195

1 (a) $f(1.7) < 0, f(1.8) > 0$; (b) $f(1) < 0, f(2) > 0$; (c) $f(0) > 0, f(1) < 0$ (the root

between 0 and 2 lies between 0 and 1)   **3** (a) $c = \dfrac{9}{4}$; (b) $c = 1$; (c) $c = \dfrac{4}{3}$; (d) $c =$

$4(\sqrt{2} - 1)$; (e) $c = \sqrt{2}, c = \dfrac{1}{2}$ **5** (a) $c = -3$; (b) $c = \dfrac{1}{\sqrt{3}}$; (c) $c = 0$; (d) $c =$

$\dfrac{27 - \sqrt{579}}{6} \approx 0.49$; (e) $c = -\dfrac{2}{\sqrt{3}} \approx -1.1$ **7** $\sqrt{1 - \sqrt{x}} = \dfrac{1}{2\sqrt{c}}(1 - x), 0 < x < c < 1$

11 (a) -16; (b) 31; (c) -7; (d) -10; (e) 132; (f) $-\dfrac{127}{128}$ **15** (a) $f'(x) = 3x|x|$,

$f''(x) = 6|x|$; (b) $f'(x) = 3x^2$ if $x \le 1$ and $f'(x) = 3$ if $x > 1$; $f''(x) = 6x$ if $x < 1$, $f''(x) = 0$
if $x > 1$, $f''(1)$ does not exist   **17** (a) Increasing on $(-\infty, -2]$ and $[0, \infty)$, decreasing on
$[-2, 0]$; (b) increasing on $(-\infty, -3]$ and $[-1, \infty)$, decreasing on $[-3, -1]$; (c) increasing

on $\left(-\infty, \dfrac{4}{7}\right]$ and $[4, \infty)$, decreasing on $\left[\dfrac{4}{7}, 4\right]$; (d) increasing on $\left(-\infty, -\sqrt[4]{\dfrac{4}{3}}\right]$ and

$\left[\sqrt[4]{\dfrac{4}{3}}, \infty\right)$, decreasing on $\left[-\sqrt[4]{\dfrac{4}{3}}, 0\right)$ and $\left(0, \sqrt[4]{\dfrac{4}{3}}\right]$; (e) increasing on $(-\infty, 0]$ and on $(0, \infty)$;

(f) decreasing on $(-\infty, 0)$, $[0, 1]$, and $(2, \infty)$, increasing on $[1, 2]$ **19** (a) $v = 12t - 6t^2$,
$a = 12 - 12t$; (b) $v = 24t^2 - 96t + 72$, $a = 48t - 96$; (c) $v = 128t - 64t^3$, $a = 128 - 192t^2$
23 Concave downward on $(-\infty, 2)$, concave upward on $(2, \infty)$ **25** Concave
downward on $(-\infty, -3)$ and $(0, 3)$, concave upward on $(-3, 0)$ and $(3, \infty)$ **27** Concave

upward on $\left(-\infty, \dfrac{1}{3}\right)$, concave downward on $\left(\dfrac{1}{3}, 1\right)$ and $(1, \infty)$ **29** (a) $\left(\dfrac{2}{3}, \dfrac{167}{27}\right)$;

(b) $(-1, -5)$ and $(1, 5)$; (c) $\left(-\dfrac{2}{3}, \dfrac{23}{27}\right)$; (d) $(-3, 18)$; (e) $\left(-\dfrac{1}{\sqrt{3}}, \dfrac{3}{4}\right)$ and $\left(\dfrac{1}{\sqrt{3}}, \dfrac{3}{4}\right)$;

(f) $\left(6, \dfrac{4}{27}\right)$; (g) $\left(-2 - \sqrt{3}, (-1 - \sqrt{3})/(8 + 4\sqrt{3})\right)$, $\left(-2 + \sqrt{3}, (-1 + \sqrt{3})/(8 - 4\sqrt{3})\right)$,

and $(1, 1)$ **35** (a) 1 and 2; (b) relative maximum at 1, relative minimum at 2
37 (a) 1 and -2; (b) relative minimum at -2 **39** (a) 0; (b) relative maximum at 0
41 (a) 0; (b) relative minimum at 0   **43** Absolute maximum of 5 at 1 and 4, absolute
minimum of 1 at 0 and 3   **45** Absolute minimum of -8 at -1, absolute maximum of

1 at 2   **47** No absolute maximum or minimum   **49** (a) Decreasing on $\left(-\infty, \dfrac{3}{2}\right)$,

increasing on $\left[\dfrac{3}{2}, \infty\right)$; (b) concave upward on $(-\infty, 0)$ and $(1, \infty)$, concave downward on

$(0, 1)$; (c) absolute minimum of $-\dfrac{11}{16}$ at $\dfrac{3}{2}$; (d) inflection points at $(0, 1)$ and $(1, 0)$ **51** (a)

Decreasing on $\left(-\infty, -\sqrt{\dfrac{2}{3}}\right]$ and $\left[\sqrt{\dfrac{2}{3}}, \infty\right)$, increasing on $\left[-\sqrt{\dfrac{2}{3}}, \sqrt{\dfrac{2}{3}}\right]$; (b) concave upward

on $(-\infty, 0)$, concave downward on $(0, \infty)$; (c) relative minimum at $-\sqrt{\dfrac{2}{3}}$, relative maximum

at $\sqrt{\dfrac{2}{3}}$; (d) inflection point at $(0, 5)$ **53** (a) Decreasing on $(-\infty, -1 - \sqrt{2}]$ and

$[-1 + \sqrt{2}, \infty)$, increasing on $[-1 - \sqrt{2}, -1 + \sqrt{2}]$; (b) concave upward on
$(-2 - \sqrt{3}, -2 + \sqrt{3})$ and $(1, \infty)$, concave downward on $(-\infty, -2 - \sqrt{3})$ and
$(-2 + \sqrt{3}, 1)$; (c) relative minimum at $-1 - \sqrt{2}$, relative maximum at $-1 + \sqrt{2}$;
(d) inflection points are $\left(-2 - \sqrt{3}, (-1 - \sqrt{3})/(8 + 4\sqrt{3})\right)$,
$\left(-2 + \sqrt{3}, (-1 + \sqrt{3})/(8 - 4\sqrt{3})\right)$, and $(1, 1)$ **59** 3 units   **63** $\dfrac{h}{r} = 2$ **65** $\dfrac{8}{3}$ and $\dfrac{4}{3}$

69 (a) A square $5\sqrt{2}$ by $5\sqrt{2}$; (b) base $r\sqrt{2}$, height $\dfrac{r}{\sqrt{2}}$; (c) a square 5 by 5; (d) base 6,

height 8   **71** (a) 2; (b) $\dfrac{5}{2}$; (c) 0; (d) 5   **73** $\dfrac{32\pi a^3}{81}$ **75** $|\overline{BC}| = \dfrac{5}{4}$ miles   **77** $\dfrac{2\pi a^3}{9\sqrt{3}}$

81 $D_x y = -\dfrac{3x^2}{2y^2}$, $D_x^2 y = -\dfrac{3x}{2y^5}$ **83** $D_x y = \dfrac{1}{12y^2}\left(\dfrac{4}{\sqrt{x + 1}} - 5x\right)$,

$$D_x^2 y = -\frac{1}{72y^5}\left(\frac{4}{\sqrt{x+1}} - 5x\right)^2 - \frac{1}{12y^2}\left[\frac{2}{(x+1)^{3/2}} + 5\right]$$ **85** Tangent: $2x + 5y = 16$,

normal: $5x - 2y = 11$ **87** Tangent: $y + 2x = 6$, normal: $2y - x = 2$ **89** (a) $-\dfrac{4}{\sqrt{15}}$;

(b) -3 **91** $\dfrac{268}{5}$ mph   **93** $\dfrac{49}{\sqrt{274}}$ ft/sec

Chapter 4

Problem Set 1   page 207
1 $x^2 + (y - 2)^2 = 9$ **3** $(x - 3)^2 + (y - 4)^2 = 25$ **5** $(x - 1)^2 + y^2 = 16$
7 $(h, k) = (-1, 2)$, $r = 3$ **9** $(h, k) = (-1, -2)$, $r = 1$ **11** $(h, k) = \left(-1, \frac{1}{2}\right)$, $r = 1$

13 $(h, k) = \left(-\frac{1}{2}, \frac{1}{2}\right)$, $r = \frac{1}{2}$ **15** $(x - 2)^2 + (y - 10)^2 = 106$ **17** $(x + 4)^2 + y^2 = 17$
and $(x - 4)^2 + y^2 = 17$ **19** $(x - 1)^2 + (y + 7)^2 = 49$ **21** $(x - 5)^2 + (y - 4)^2 = 4$ and
$(x - 1)^2 + (y - 4)^2 = 4$ **23** (a) $(1, -2)$;   (b) $(-1, -1)$;   (c) $(4, -5)$;   (d) $(-2, -4)$;
(e) $(6, 3)$;   (f) $(7, -2)$ **25** (a) $(3, -2)$;   (b) $(6, 0)$;   (c) $(0, 2)$;   (d) $(3 + \sqrt{2}, -4)$;
(e) $(3, -2 + \pi)$;   (f) $(0, 0)$ **27** $h = -\frac{7}{6}$, $k = \frac{5}{6}$ **29** Tangent: $x - 2y = 8$, normal:
$2x + y = 1$ **31** Tangent: $y = 2$, normal: $x = 3$ **33** (a) $(x + 2)^2 + (y - 4)^2 = 20$
35 Symmetric about the origin and any straight line through O **37** Yes, distance
between centers is less than sum of radii

Problem Set 2   page 214

	a	b	Foci	Vertices
1	4	2	$(-2\sqrt{2}, 0), (2\sqrt{2}, 0)$	$(-4, 0), (4, 0), (0, 2), (0, -2)$
3	4	2	$(0, -2\sqrt{2}), (0, 2\sqrt{2})$	$(-2, 0), (2, 0), (0, 4), (0, -4)$
5	4	1	$(-\sqrt{15}, 0), (\sqrt{15}, 0)$	$(-4, 0), (4, 0), (0, 1), (0, -1)$
7	$\frac{2}{3}$	$\frac{1}{3}$	$\left(-\frac{1}{\sqrt{3}}, 0\right), \left(\frac{1}{\sqrt{3}}, 0\right)$	$\left(-\frac{2}{3}, 0\right), \left(\frac{2}{3}, 0\right), \left(0, \frac{1}{3}\right), \left(0, -\frac{1}{3}\right)$

9 $\dfrac{x^2}{25} + \dfrac{y^2}{9} = 1$ **11** $\dfrac{x^2}{36} + \dfrac{y^2}{64} = 1$ **13** $\dfrac{x^2}{16} + \dfrac{7y^2}{64} = 1$

	Center	Vertices	Foci
15	$(1, -2)$	$(-2, -2), (4, -2), (1, 0), (1, -4)$	$(1 - \sqrt{5}, -2), (1 + \sqrt{5}, -2)$
17	$(-3, 0)$	$(-3, -6), (-3, 6), (-6, 0), (0, 0)$	$(-3, -3\sqrt{3}), (-3, 3\sqrt{3})$
19	$(-3, 0)$	$(-3 - \sqrt{2}, 0), (-3 + \sqrt{2}, 0),$ $(-3, 1), (-3, -1)$	$(-4, 0), (-2, 0)$
21	$(-5, 3)$	$(-5 - \sqrt{10}, 3), (-5 + \sqrt{10}, 3),$ $(-5, 5), (-5, 1)$	$(-5 - \sqrt{6}, 3), (-5 + \sqrt{6}, 3)$

23 $h = -1$, $k = 1$ **25** $h = 2$, $k = 3$ **27** $\dfrac{(x + 2)^2}{25} + \dfrac{(y - 1)^2}{16} = 1$

29 $\dfrac{(x - 1)^2}{4} + \dfrac{(y + 2)^2}{9} = 1$ **31** Tangent: $4y + x = 25$, normal: $4x - y = 32$

33 Tangent: $x + 6y = 15$, normal: $6x - y = 16$ **35** $\left(x_0\left(1 - \dfrac{b^2}{a^2}\right), 0\right)$ **37** (b) $\dfrac{9}{2}$ units

41 $\dfrac{x^2}{16} + \dfrac{y^2}{7} = 1$ **43** $2ab$ **45** $l = 2a + 2c = 60 + 40\sqrt{2}, 2c = 40\sqrt{2}$

47 $\dfrac{x^2}{50^2} + \dfrac{y^2}{25^2} = 1$ **51** It approaches a circle of radius p with center $(0, p)$.

Problem Set 3 page 221

1 Vertex $(0,0)$, focus $(1,0)$, directrix $x = -1$, latus rectum 4   **3** Vertex $(0,0)$, focus $\left(0, \dfrac{1}{4}\right)$, directrix $y = -\dfrac{1}{4}$, latus rectum 1   **5** Vertex $(0,0)$, focus $\left(0, -\dfrac{9}{4}\right)$, directrix $y = \dfrac{9}{4}$, latus rectum 9   **7** $y = \dfrac{1}{12} x^2$

	Vertex	Focus	Directrix	Latus Rectum
9	$(-3,2)$	$(-1,2)$	$x = -5$	8
11	$(4,-7)$	$(4,-4)$	$y = -10$	12
13	$(-3,4)$	$\left(-\dfrac{3}{2},4\right)$	$x = -\dfrac{9}{2}$	6
15	$(3,-1)$	$(3,1)$	$y = -3$	8

17 $(y-2)^2 = -4(x-5)$ **19** $(y+5)^2 = 32(x+6)$ **21** $(y+1)^2 = 8\left(x + \dfrac{1}{2}\right)$

23 $\bar{x} = x + \dfrac{1}{2}, \bar{y} = y + 1, \bar{y}^2 = 8\bar{x}$ **25** $\bar{x} = x + 2, \bar{y} = y - 3, \bar{x}^2 = 5\bar{y}^2$ **27** Tangent: $x + y = -2$, normal: $x - y = 6$ **29** Tangent: $y - x = 3$, normal: $x + y = -9$

31 Tangent: $x = \dfrac{9}{2}$, normal: $y = 1$ **33** $(1,5)$ **35** $(-1,5)$ **37** Height: 8 units, base: 4 units   **39** Out from center, lengths are 4, 10, 28, 58 meters.   **41** (a) $f(a) = \dfrac{a^2}{4p} + 2p$; (b) $\lim\limits_{a \to 0} f(a) = 2p$ **43** $-\dfrac{6}{\sqrt{2}}$ units/sec   **47** (a) Any two parabolas are similar in the sense that the "larger" is obtained from the "smaller" by magnification.

Problem Set 4 page 228

	Vertices	Foci	Asymptotes
1	$(-3,0), (3,0)$	$(-\sqrt{13},0), (\sqrt{13},0)$	$y = \pm \dfrac{2}{3}x$
3	$(0,-4), (0,4)$	$(0,-2\sqrt{5}), (0,2\sqrt{5})$	$y = \pm 2x$
5	$(-4,0), (4,0)$	$(-2\sqrt{5},0), (2\sqrt{5},0)$	$y = \pm \dfrac{1}{2}x$
7	$(0,-\sqrt{10}), (0,\sqrt{10})$	$(0,-\sqrt{46}), (0,\sqrt{46})$	$y = \pm \dfrac{\sqrt{10}}{6}x$
9	$\dfrac{x^2}{16} - \dfrac{y^2}{20} = 1$ **11** $\dfrac{x^2}{16} - \dfrac{y^2}{25} = 1$		

	Center	Vertices	Foci	Asymptotes
13	$(1, -2)$	$(-2, -2), (4, -2)$	$(1 + \sqrt{13}, -2),$ $(1 - \sqrt{13}, -2)$	$y + 2 = \pm\frac{2}{3}(x - 1)$
15	$(-2, -1)$	$(-2, 3), (-2, -5)$	$(-2, -1 + \sqrt{41}),$ $(-2, -1 - \sqrt{41})$	$y + 1 = \pm\frac{4}{5}(x + 2)$
17	$(2, -1)$	$(0, -1), (4, -1)$	$(2 - \sqrt{5}, -1),$ $(2 + \sqrt{5}, -1)$	$y + 1 = \pm\frac{1}{2}(x - 2)$
19	$(-4, -2)$	$(-4, 1), (-4, -5)$	$(-4, -2 + \sqrt{34}),$ $(-4, -2 - \sqrt{34})$	$y + 2 = \pm\frac{3}{5}(x + 4)$

21 (a) $\dfrac{(x - 4)^2}{1} - \dfrac{(y + 1)^2}{8} = 1$; (b) $\dfrac{(x + 2)^2}{4} - \dfrac{4(y - 3)^2}{9} = 1$;

(c) $\dfrac{(y - 3)^2}{25} - \dfrac{(x - 2)^2}{11} = 1$ **23** Tangent: $4y + 5x = -9$, normal: $5y - 4x = 40$

25 Tangent: $y = -2x$, normal: $y = \dfrac{1}{2}x$ **27** (a) $\dfrac{4x^2}{3} - \dfrac{y^2}{3} = 1$;

(b) $\dfrac{(y - 2)^2}{3} - \dfrac{4\left(x - \dfrac{1}{2}\right)^2}{3} = 1$; (c) $\dfrac{(x + 1)^2}{9} - \dfrac{(y - 4)^2}{9} = 1$ **29** $\bar{x} = x + 3, \bar{y} = y - 1$

31 $\bar{x} = x - 2, \bar{y} = y + 1$ **33** 8 units/sec   **35** $\dfrac{3\sqrt{10}}{2}$ units   **37** $\lim\limits_{x_0 \to +\infty} m = \dfrac{b}{a}$

39 $\dfrac{x^2}{a^2} - \dfrac{y^2}{a^2} = 1$ **41** It approaches the intersecting asymptotes.

Problem Set 5 page 234

1 $\dfrac{(x - 2)^2}{9} + \dfrac{y^2}{5} = 1$ **3** $8(x + 2) = y^2$ **5** $\dfrac{(x + 4)^2}{4} - \dfrac{y^2}{12} = 1$ **7** $e = \dfrac{5}{3}, D_1: x =$

$-\dfrac{9}{5}, D_2: x = \dfrac{9}{5}$ **9** $e = \dfrac{\sqrt{7}}{3}, D_1: x = -\dfrac{9}{\sqrt{7}}, D_2: x = \dfrac{9}{\sqrt{7}}$ **11** $e = \dfrac{4}{5}, D_1: y = -\dfrac{25}{2},$

$D_2: y = \dfrac{25}{2}$ **13** $e = 2\sqrt{\dfrac{2}{5}}, D_1: x = -\dfrac{5\sqrt{2}}{4}, D_2: x = \dfrac{5\sqrt{2}}{4}$ **15** $e = 1, D: y = -4$

17 (a) $(3, 2)$; (b) $(3 + 4\sqrt{3}, 2), (3 - 4\sqrt{3}, 2), (3, 4), (3, 0)$; (c) $D_1: x = 3 + \dfrac{24}{\sqrt{11}}, D_2: x =$

$3 - \dfrac{24}{\sqrt{11}}$; (d) $\dfrac{1}{2}\sqrt{\dfrac{11}{3}}$ **19** (a) $\dfrac{1}{2}$; (b) $2b = a\sqrt{3}, \dfrac{a^2}{c} = 2a$ **23** $D_1: y = k - \dfrac{a^2}{c}, D_2: y =$

$k + \dfrac{a^2}{c}$, where $c = \sqrt{a^2 - b^2}$ **25** $D_1: x = -\dfrac{a^2 - k}{\sqrt{a^2 - b^2}}, D_2: x = \dfrac{a^2 - k}{\sqrt{a^2 - b^2}}$

27 Ellipse gets "small" and becomes more circular.

Review Problem Set page 235

1 $(x - 2)^2 + (y - 1)^2 = 41$ **3** $(x - 4)^2 + (y - 3)^2 = 25$ **5** $(x - 7)^2 + y^2 = 65$
7 $(x - 13)^2 + (y + 13)^2 = 169$ or $(x - 73)^2 + (y + 73)^2 = 5329$
9 $x^2 + y^2 - x + 3y - 10 = 0$ **11** (a) Tangent: $4y + 3x + 25 = 0$, normal: $3y - 4x = 0$;

(b) tangent: $x = 7$, normal: $y = 1$ **13** $\dfrac{x^2}{25} + \dfrac{y^2}{16} = 1$ **15** $\dfrac{x^2}{16} + \dfrac{y^2}{64} = 1$

17 $\dfrac{(x-5)^2}{25} + \dfrac{y^2}{9} = 1$

	Center	Vertices	Foci	e	Directrices
19	$(0,0)$	$(0, 2\sqrt{3}),\ (0, -2\sqrt{3}),$ $(2\sqrt{2}, 0),\ (-2\sqrt{2}, 0)$	$(0, 2),\ (0, -2)$	$\dfrac{\sqrt{3}}{3}$	$y = \pm 6$
21	$(-1, 1)$	$(4, 1),\ (-6, 1),$ $(-1, 4),\ (-1, -2)$	$(3, 1),\ (-5, 1)$	$\dfrac{4}{5}$	$x = -1 \pm \dfrac{25}{4}$
23	$(-4, 6)$	$(-4, 12),\ (-4, 0),$ $(-8, 6),\ (0, 6)$	$(-4, 6 \pm 2\sqrt{5})$	$\dfrac{\sqrt{5}}{3}$	$y = \dfrac{30 \pm 18\sqrt{5}}{5}$

25 $\left(3, \dfrac{16}{3}\right)$ and $\left(-3, -\dfrac{16}{3}\right)$ **27** (a) $x = \bar{x} + 1,\ y = \bar{y} - 2,\ \dfrac{\bar{x}^2}{4} + \dfrac{\bar{y}^2}{64} = 1$;   (b) $x = \bar{x} - 2,$

$y = \bar{y} + 3,\ \dfrac{\bar{x}^2}{36} + \dfrac{\bar{y}^2}{81} = 1$ **29** $30\sqrt{2}$ ft   **31** $\dfrac{(x-2)^2}{16} - \dfrac{(y-3)^2}{9} = 1$ **33** $\dfrac{y^2}{36} - \dfrac{x^2}{108} = 1$

	Center	Vertices	Foci	e	Asymptotes
35	$(0, 0)$	$(6\sqrt{2}, 0),\ (-6\sqrt{2}, 0)$	$(4\sqrt{5}, 0),\ (-4\sqrt{5}, 0)$	$\dfrac{\sqrt{10}}{3}$	$y = \pm \dfrac{1}{3}x$
37	$(-2, 3)$	$(2, 3),\ (-6, 3)$	$(-2 + 2\sqrt{5}, 3),$ $(-2 - 2\sqrt{5}, 3)$	$\dfrac{\sqrt{5}}{2}$	$y - 3 = \pm \dfrac{1}{2}(x + 2)$
39	$(1, 3)$	$\left(1, \dfrac{7}{2}\right),\ \left(1, \dfrac{5}{2}\right)$	$\left(1, -3 - \dfrac{\sqrt{5}}{2}\right),$ $\left(1, -3 + \dfrac{\sqrt{5}}{2}\right)$	$\sqrt{5}$	$y - 3 = \pm \dfrac{1}{2}(x - 1)$

41 Tangent: $8y - 3x + 1 = 0$, normal: $3y + 8x - 27 = 0$ **43** $-\dfrac{9}{4}$ units/sec

45 $y^2 = -18x$ **47** $y - 3 = \dfrac{1}{2}(x - 2)^2$ **49** $x^2 + y^2 = 10x$ **51** $8\sqrt{3}$ by 4

53 $(2, 4)$ **55** $20, \dfrac{190}{9}, \dfrac{220}{9}, 30, \dfrac{340}{9}, \dfrac{430}{9}, 60$

Chapter 5

Problem Set 1 page 242

1 (a) 0.63; (b) 0.6; (c) 0.03   **3** (a) -1.92; (b) -1.6; (c) -0.32 **5** (a) 0.18; (b) $\dfrac{1}{6}$;

(c) 0.015   **7** $(-4 + 3x^2)\,dx$ **9** $\dfrac{-3x}{\sqrt{9 - 3x^2}}\,dx$ **11** $3(x + 3)^{-5/2}(x - 3)^{1/2}\,dx$

13 $\dfrac{-9x^2 - 4x + 21}{2\sqrt{3x + 1}(x^2 + 7)^2}\,dx$ **15** 10.05   **17** 2.03   **19** 0.31   **21** 0.2025

23 $x\,dx + y\,dy = 0$ **25** $9x\,dx - 16y\,dy = 0$ **27** $x^{-2/3}\,dx + y^{-2/3}\,dy = 0$

29 $(6x^2 - 4y^2 + y)\,dx + (15y^2 - 8xy + x)\,dy = 0$

31 $[9x^2(x + y)^{2/3} - 1]\,dx + [9y^2(x + y)^{2/3} - 1]\,dy = 0$ **33** 10.6 in.^{3} **35** 6 cm^3

37 0.21 m^3 **39** 4.41 cm^3 **41** -4%

Problem Set 2 page 251

1 $g'(x) = 12x^2 - 6x + 1 = f(x)$ **3** $x^3 - 2x^2 - 5x + C$

5 $\dfrac{x^4}{2} - \dfrac{4x^3}{3} - \dfrac{5x^2}{2} + 6x + C$ **7** $\dfrac{x^3}{3} + \dfrac{x^2}{2} + x + C$ **9** $\dfrac{t^3}{3} + \dfrac{3t^2}{2} - \dfrac{1}{t} + C$

11 $\dfrac{50}{7} x^{7/2} - 2x^{1/2} + C$ **13** $\dfrac{2}{3} x^{3/2} - 2x + 2x^{1/2} + C$ **15** $\dfrac{(4x + 3)^5}{20} + C$

17 $\dfrac{(4x^2 + 15)^{3/2}}{12} + C$ **19** $\dfrac{3}{20} (5s^2 + 16)^{2/3} + C$ **21** $-\dfrac{1}{4} (1 - x^{3/2})^{8/3} + C$

23 $\dfrac{-1}{72(4x^3 + 1)^6} + C$ **25** $\dfrac{4}{15} (5t^3 + 3t - 2)^{5/4} + C$ **27** $\dfrac{-2}{9\sqrt{6x^3 - 9x + 1}} + C$

29 $\dfrac{1}{22} \left(x + \dfrac{5}{x}\right)^{22} + C$ **31** $\dfrac{2}{5} (5 - x)^{5/2} - \dfrac{10}{3} (5 - x)^{3/2} + C$

33 $\dfrac{2}{3} (t + 1)^{3/2} - 2(t + 1)^{1/2} + C$ **35** $\dfrac{3}{2} (2 - x)^{4/3} - 12(2 - x)^{1/3} + C$

37 $\dfrac{1}{42} (3x^2 + 5)^{7/3} - \dfrac{5}{24} (3x^2 + 5)^{4/3} + C$

39 $\dfrac{2}{5} (t + 4)^{5/2} - \dfrac{16}{3} (t + 4)^{3/2} + 32(t + 4)^{1/2} + C$

41 $\dfrac{2}{125} \left[\dfrac{1}{7} (5x - 1)^{7/2} + \dfrac{2}{5} (5x - 1)^{5/2} + \dfrac{1}{3} (5x - 1)^{3/2}\right] + C$

43 Let $f(x) = g(x) = x$.   **45** $f(x) = \dfrac{2x}{1 + x}$

Problem Set 3 page 259

1 $y = x^5 + x^3 + x + C$ **3** $y = -\dfrac{6}{x} + 5x^3 + 10x + C$ **5** $y = \dfrac{2\sqrt{7}}{5} x^{5/2} + C$

7 $y = 5x - \dfrac{3}{2} x^2 + 4$ **9** $y = \dfrac{t^4}{4} - \dfrac{1}{t} - \dfrac{7}{2}$ **11** $y = \dfrac{(x^2 - 3)^5}{10} + C$

13 $y = \dfrac{2}{3} \sqrt{x^3 + 7} + C$ **15** $y = \dfrac{-1}{\sqrt{2x + 1} + C}$ **17** $15(y^4 + 7)^{2/3} = 4x^2 + C$

19 $20x - 4x^{5/2} = 5y^2 - 812$ **21** $W = \dfrac{-3(1 - t^{1/3})^4 - 1}{4}$ **23** $y =$

$\dfrac{x^4}{4} + \dfrac{x^3}{3} + \dfrac{x^2}{2} + C_1 x + C_2$ **25** $y = \dfrac{9}{448} (4x + 5)^{7/3} + C_1 x + C_2$ **27** $y =$

$\dfrac{1}{15} x^6 + \dfrac{3}{2} x^2 + C_1 x + C_2$ **29** $y = C_1 x + C_2$ **31** $y = x^3 + \dfrac{x^2}{2} + 3x + 2$

33 $y = x^2 + 2x - 3$ **35** $y = \dfrac{25}{4} x^4 + 10x^3 + 6x^2 - 68x + \dfrac{191}{4}$ **37** $W = \dfrac{5s^2}{2}$

39 $y = \dfrac{kx^3}{6} + C_1 x + C_2$ **41** 19 ft   **43** (a) 4.96 sec; (b) 158.68 ft   **45** (a) 2.83 sec;

(b) 80.62 ft/sec   **47** $v_0 \geq 5 + 16\sqrt{5} \approx 40.78$ ft/sec

Problem Set 4 page 269

1 $f(x) = x - \dfrac{3}{2} x^2 + \dfrac{13}{2}$ **3** $f(x) = \dfrac{2x}{2 + x}$ **5** 24 ft-lb   **7** $\dfrac{32}{3} \sqrt{2}$ ergs

9 500 ft-lb   **11** 20,000 ergs   **13** 2400 in.-lb   **15** $\dfrac{\gamma M}{b(a+b)}$ **17** (a) $C =$

$12x - 16\sqrt{x} + 160$; (b) $P = 9x + 16\sqrt{x} - 160$ **19** \$3300   **21** (a) $R = 13x - \dfrac{x^2}{80}$;

(b) $P = 9.5x - \dfrac{x^2}{80} - 100$; (c) \$8.25/month   **23** When s is negative, F pulls *upward*.

25 $V = \dfrac{ks^2}{2}$ **27** $v = -b\sqrt{\dfrac{k}{m}}$ **29** $v = \pm\sqrt{\dfrac{2}{ms}}$ **31** $s = \left(125 + \dfrac{3}{2}\sqrt{\dfrac{2}{m}}\,t\right)^{2/3}$,

$\lim\limits_{t \to +\infty} v = 0$ **33** $W = \dfrac{1}{r} - \dfrac{1}{10}$

Problem Set 5 page 273

1 8   **3** 4   **5** 4   **7** $\dfrac{16}{3}$ **9** $\dfrac{9}{2}$ **11** 6.75   **13** $\dfrac{125}{6}$ **15** $8\sqrt{3}$ **17** $\dfrac{1}{3}$

19 $8\sqrt{6}$

Problem Set 6 page 280

1 $\dfrac{25}{2}$ **3** $\dfrac{\pi}{2}$ **5** 4   **7** $\displaystyle\int_0^3 (3x - x^2)\,dx$ **9** $-\displaystyle\int_0^{3/2} (2x - 3)\,dx$

11 $\displaystyle\int_{-\sqrt{2}}^{\sqrt{2}} (2 - x^2)\,dx$ **13** $2\left[\displaystyle\int_{\sqrt{3}}^{4} \left(\dfrac{1}{3}x^3 - x\right)dx - \int_0^{\sqrt{3}} \left(\dfrac{1}{3}x^3 - x\right)dx\right]$ **15** 6

17 $\dfrac{128}{3}$ **19** $\sqrt{5} - \sqrt{2}$ **21** $\dfrac{4}{3}$ **23** $\dfrac{9}{2}$ **25** $\dfrac{33}{5}$ **27** $\dfrac{17}{2}$ **29** $\dfrac{3}{2}$ **31** 8

33 $\dfrac{95}{4}$ **35** $\dfrac{7}{3}$ **37** 6   **39** $\dfrac{27}{4}$

Review Problem Set page 281

1 $\Delta y = 0.0401$, $dy = 0.04$, $\Delta y - dy = 0.0001$ **3** $\Delta y = 0.242408$, $dy = 0.24$,

$\Delta y - dy = 0.002408$ **5** $dy = (3x^2 - 1)\,dx$ **7** $dy = \dfrac{2x^2 + 2x - 10}{(2x + 1)^2}\,dx$

9 $\dfrac{dy}{dx} = \dfrac{y - x^3}{y^3 - x}$ **11** 200 m^2 **13** $3x^2\,\Delta x$ cubic units   **15** 2.0125

17 $\dfrac{3x^5}{5} + \dfrac{4x^3}{3} + 11x + C$ **19** $\dfrac{9}{7}t^{7/3} + C$ **21** $\dfrac{7}{24}(3t + 9)^{8/7} + C$

23 $\dfrac{(x^3 + 8)^{18}}{54} + C$ **25** $\dfrac{5}{32}(x^8 + 13)^{4/5} + C$ **27** $\dfrac{2}{5}(7 + x)^{5/2} - \dfrac{14}{3}(7 + x)^{3/2} + C$

29 $\dfrac{2}{15}(x^3 + 1)^{5/2} - \dfrac{2}{9}(x^3 + 1)^{3/2} + C$ **31** $x^3 - 7x^2 + 8x + C$ **33** $2x^{1/2} + \dfrac{2}{3}x^{3/2} + C$

35 $\dfrac{4}{3}(1 + \sqrt{x})^{3/2} + C$ **37** $\dfrac{(1 - x)^{-2}}{6} + C_1 x + C_2$ **39** $\dfrac{3}{2}x^{2/3} - \dfrac{3}{2}$

41 $\dfrac{1}{33}[(1 + x^3)^{11} + 65]$ **43** $\dfrac{1}{2x} + \dfrac{3}{2}x$ **45** $\dfrac{1}{3}(x^2 + 5)^{3/2} - 3$

47 $\dfrac{x|x|}{2} + (x - 1)\dfrac{|x - 1|}{2} + (x - 2)\dfrac{|x - 2|}{2} + \dfrac{7}{2}$ **49** 1000 ft-lb   **51** 3.08 yards/sec^2

53 (a) 2 sec; (b) 160 ft/sec   **55** 83,437.5 ft-lb   **57** $\dfrac{\gamma Mm}{ab}$ dynes   **59** (a) $R =$

$5x - \dfrac{x^2}{50,000}$; (b) $C = 3x + 200$; (c) 125,000; (d) 50,000; (e) \$4   **61** $\dfrac{7}{6}$ **63** $\dfrac{2}{3}$

65 26   **67** -12 **69** 10   **71** $b - a$ **73** $\dfrac{\pi}{2}$ **75** $-\displaystyle\int_{-3}^{-2}(1 + x)^{-1}\,dx$

77 $\displaystyle\int_{-2}^{0}\dfrac{-2x}{1 - x}\,dx$ **79** -4 **81** $\dfrac{16 - 6\sqrt{3}}{9}$ **83** $-\dfrac{32}{5}$

Chapter 6

Problem Set 1 page 291

1 48   **3** 287   **5** $\dfrac{5}{6}$ **7** $\dfrac{17}{21}$ **9** 2700   **11** $\dfrac{5^{101} - 5}{4}$ **13** $\dfrac{n(n + 1)(n + 2)}{3}$

15 $\dfrac{(n - 1)n(2n - 1)}{6}$ **17** $\dfrac{(n - 1)n(2n - 1)}{6}$ **25** (a) $\dfrac{1}{3} - \dfrac{2999}{6,000,000} \le A \le \dfrac{1}{3} + \dfrac{3001}{6,000,000}$;

(b) $\dfrac{1}{3} - \dfrac{29,999}{600,000,000} \le A \le \dfrac{1}{3} + \dfrac{30,001}{600,000,000}$

Problem Set 2 page 299

1 $\dfrac{75}{4}$ **3** $\dfrac{3776}{3465}$ **5** Exists   **7** Exists   **9** Does not exist   **11** Exists   **13** 9

15 15   **17** $\dfrac{11}{6}$ **19** 8   **21** 7

Problem Set 3, page 308

1 $63 + 9\pi$ **3** $(1 + \sqrt{2} - \sqrt{3})(0.25)$ **5** 20   **7** 35   **9** $\dfrac{41}{216}$ **11** 12

13 $\dfrac{88}{3}$ **15** 2   **17** Yes   **19** Holds   **25** 8   **27** 6

29 $\dfrac{A}{3}(b^2 + ab + a^2) + \dfrac{B}{2}(b + a) + C$ **31** $\dfrac{\sqrt{3}}{3}$ **33** $\dfrac{a + b}{2}$ **35** $\dfrac{29}{14}$ **39** $-\dfrac{3}{2}$

41 0   **43** $\dfrac{1}{2}(b^2 - a^2)$ **45** $\dfrac{1}{3}(b^3 - a^3)$

Problem Set 4, page 316

1 $x^2 + 1$ **3** $\dfrac{1}{1 + x^2}$ **5** $|x|$ **7** $\sqrt{x^2 + 4}$ **9** $-(x^{10} + 3)^{25}$ **11** 0

13 $3(135x^3 + 1)^7$ **15** $8(8x - 1)^{15}$ **17** $\sqrt{-x + 2}$

19 $6x\sqrt[4]{(3x^2 + 2)^4 + 17} - \sqrt[4]{x^4 + 17}$ **23** $\dfrac{23}{2}$ **25** 36   **27** $\dfrac{83}{15}$ **29** $\dfrac{16}{5}$

31 $\dfrac{5}{64}$ **33** $\dfrac{4}{3}\sqrt{2}$ **35** $4 - \sqrt{5}$ **37** $\dfrac{1}{4}(9^{4/3} - 1)$ **39** $\dfrac{50}{3}$ **41** $4\sqrt{3}$ **43** $\dfrac{8}{3}$

45 $\dfrac{376}{15}$ **49** 8   **51** $\dfrac{13}{6}$ **53** $x^2 + 7$ **55** $g(x) = \begin{cases} x - \dfrac{x^3}{3} + \dfrac{2}{3} & \text{if } x \le 0 \\[2ex] x + \dfrac{x^3}{3} + \dfrac{2}{3} & \text{if } x > 0 \end{cases}$

$$57 \quad g(x) = \begin{cases} -3x - 3 & \text{if } -3 \le x < -2 \\ -2x - 1 & \text{if } -2 \le x < -1 \\ -x & \text{if } -1 \le x < 0 \\ 0 & \text{if } 0 \le x < 1 \\ x - 1 & \text{if } 1 \le x < 2 \\ 2x - 3 & \text{if } 2 \le x \le 3 \end{cases}$$

61 (a) $c = 5\sqrt[3]{4}$; (b) $c = \dfrac{19}{2}$; (c) $c = \dfrac{8}{5}$

63 $1860

Problem Set 5 page 325

1 0.783   **3** 1.107   **5** 0.925   **7** 2.050   **9** 0.448   **11** $\dfrac{1}{96}$ **13** Each trapezoid is contained in the region under the curve.   **15** 1.567   **17** 0.909   **19** 4.671   **21** 3.653   **23** $n = 4$ **25** 3.08

Problem Set 6 page 331

1 $\dfrac{4}{3}$ **3** $\dfrac{1}{2}$ **5** 8   **7** $\dfrac{3}{2}$ **9** $\dfrac{11}{12}$ **11** (b) $\left(-\dfrac{1}{2}, \dfrac{1}{4}\right)$ and $\left(\dfrac{5}{2}, \dfrac{25}{4}\right)$; (c) $\dfrac{9}{2}$

13 (b) $(2, -8)$ and $(-2, -8)$; (c) $\dfrac{32}{3}$ **15** (b) $(0, 0)$ and $(2, 2)$; (c) $\dfrac{4}{3}$ **17** (b) $(4, 4)$ and $(1, 1)$; (c) $\dfrac{9}{2}$ **19** (b) $\left(-\dfrac{3}{2}, \dfrac{1}{2}\right)$ and $(3, -1)$; (c) $\dfrac{27}{8}$ **21** (b) $(-1, -1)$, $(0, 0)$, and $(1, 1)$; (c) 1   **23** (b) $(-1, 1)$, $(0, 0)$, and $(1, 1)$; (c) $\dfrac{4}{15}$ **25** (b) $(2, 2)$ and $(2, -2)$; (c) $\dfrac{64}{3}$

27 (b) $(-1, -1)$, $(0, 0)$, and $(1, 1)$; (c) $\dfrac{1}{6}$ **29** 15   **31** $\dfrac{49}{12}$ **33** $\dfrac{323}{12}, \dfrac{35}{12}$

35 $\dfrac{344}{6}, \dfrac{130}{3}$ **37** $\dfrac{77}{4}, \dfrac{73}{12}$ **39** $\dfrac{7}{2}$

Review Problem Set page 333

1 90   **3** $\dfrac{163}{60}$ **5** $\dfrac{n(n + 1)(4n - 1)}{6}$ **7** $2(3^{n+1} - 1)$ **9** 1,001,000   **11** $\dfrac{n}{n + 1}$

13 $\dfrac{21}{64}$ **15** $\dfrac{21}{2}$ **17** Exists   **19** Does not exist   **21** Does not exist   **23** 24

25 24   **29** -30 **31** 8   **33** -14 **39** (a) 2; (b) 2; (c) $\dfrac{1}{n + 1}$ if n is even, 0 if n is odd   **41** $(4x + 1)^{300}$ **43** $(8x^{17} + 5x^2 - 13)^{40}$ **45** $-\dfrac{x^2}{\sqrt{x^4 + 8}}$ **47** 0

49 $\dfrac{2x^5}{1 + x^4}$ **51** $[(5t^2 + t)^5 + 1]^{17}(10t + 1) - 4[(4t + 3)^5 + 1]^{17}$ **53** (a) $\dfrac{16}{3}$; (b) $\dfrac{1}{6}$

55 2   **57** 9   **59** $\dfrac{4}{3}\sqrt{2}$ **61** $\dfrac{2}{5}$ **63** 9   **65** $\dfrac{16}{3}$ **67** False   **69** False

71 True   **73** False   **75** 7.012   **77** 68.268   **79** 3.307   **81** 5.892

83 (b) $(0, 0)$ and $(2, 2)$; (c) 1   **85** (b) $\left(\dfrac{3}{\sqrt{2}}, \dfrac{9}{2}\right)$ and $\left(-\dfrac{3}{\sqrt{2}}, \dfrac{9}{2}\right)$; (c) $18\sqrt{2}$

87 (b) $(3, 18)$ and $(-1, 2)$; (c) $\dfrac{32}{3}$ **89** (b) $(1, -1)$ and $\left(-\dfrac{11}{4}, \dfrac{3}{2}\right)$; (c) $\dfrac{125}{48}$

91 (b) $(0, 4)$ and $(4, -12)$; (c) $\dfrac{32}{3}$ **93** (b) $(3, 3)$, $(-3, -3)$, and $(3, -3)$; (c) 30   **95** $\dfrac{81}{2}$

Chapter 7

Problem Set 1 page 345

1 $\dfrac{2196}{5}\pi$ cubic inches   **3** $\dfrac{80\pi}{3}$ cubic units   **5** $\dfrac{20\pi}{3}$ cubic units   **7** $\dfrac{96}{5}\pi$ cubic units

9 $\dfrac{64}{5}\pi$ cubic units   **11** $\dfrac{384\pi}{7}$ cubic units   **13** $\dfrac{64}{15}\pi$ cubic units   **15** $\dfrac{\pi}{6}$ cubic unit

17 $\dfrac{99}{2}\pi$ cubic units   **19** $\dfrac{64}{5}\pi$ cubic units   **21** $\dfrac{1024}{15}\pi$ cubic units   **23** $\dfrac{29}{30}\pi$ cubic

units   **25** $\dfrac{27}{2}\pi$ cubic units   **27** $\dfrac{288}{5}\pi$ cubic units   **29** 24π cubic units

31 8π cubic units   **33** $\dfrac{768}{5}\pi$ cubic units

Problem Set 2 page 348

1 $\dfrac{11}{10}\pi$ **3** $\dfrac{38\pi}{3}$ **5** $\dfrac{13,122}{7}\pi$ **7** $\dfrac{8192}{5}\pi$ **9** 8π **11** $\dfrac{29}{30}\pi$ **13** $\dfrac{4}{3}\pi a^3$

15 $\dfrac{16}{3}\pi$ **17** $\dfrac{\pi h}{3}(b-a)(b+2a)$ **19** $\dfrac{512}{3}\pi$

Problem Set 3 page 353

1 144 in.^{3} **3** $10\sqrt{3}$ in.^{3} **5** $\dfrac{\pi}{6}(y_2^3 - y_1^3)$ **7** (a) 135 ft^3; (b) $\dfrac{325}{6}$ ft^3 **9** $\dfrac{16}{3}$ ft^3

11 $\dfrac{4347}{5120}\pi$ **13** $64\sqrt{3}$ **15** $\pi h^2\left(r - \dfrac{h}{3}\right)$ **17** $\dfrac{4}{3}a^3$ **19** $\dfrac{\sqrt{2}}{3}l^3$ **21** 280

Problem Set 4 page 362

1 $\dfrac{1}{27}(40^{3/2} - 8)$ **3** $a\sqrt{1 + m^2}$ **5** $\dfrac{53}{6}$ **7** $\dfrac{14}{3}$ **9** $\dfrac{59}{24}$ **11** 1.13 **13** 7.08

15 $\displaystyle\int_a^b \sqrt{[f'(t)]^2 + [g'(t)]^2}\, dt$ **17** $\pi r(r + \sqrt{h^2 + r^2})$ **19** $39\pi\sqrt{10}$ **21** $\dfrac{\pi}{27}(145^{3/2} - 1)$

23 $\dfrac{11\pi}{128}$ **25** $\dfrac{\pi}{96}(577^{3/2} - 1)$ **27** $\dfrac{1179}{256}\pi$ **29** 24.06 **31** (a) $\sqrt[3]{36n\pi}$; (b) $+\infty$

Problem Set 5 page 367

1 (a) $y = \dfrac{1}{40,000}(x - 400)^2$; (b) \$266.67; (c) \$266.67   **3** $y = \dfrac{15,000 - x}{100}$; \$312.50

5 $4,320$ **7** 33.63 tons   **9** 0.00075 cm^3/sec

Problem Set 6 page 373

1 $\dfrac{225}{4}$ in.-lb   **3** (a) 5096 dynes; (b) $509,600$ ergs   **5** Yes—force is constant.

7 $20,898\pi$ kg-meters   **9** 2.77088×10^5 joules   **11** $81,868.8$ ft-lb   **13** 1.99 ft-lb

15 428.47 kg-meters   **17** (a) $26,400,000$ ft-lb; (b) $94,285,714$ ft-lb   **19** $\dfrac{\sqrt{5}}{2}$ ft

21 14.88 kg   **23** 6739.2 lb   **25** $540,000$ kg   **27** $148,262.4$ lb   **29** $135,181.64$ kg

Review Problem Set page 375

1 $\dfrac{96\pi}{5}$ **3** $\dfrac{16}{15}\pi$ **5** $\dfrac{160}{3}\pi$ **7** $\dfrac{2816}{15}\pi$ **9** $126\pi^2$ cm^3 **13** $144\pi\sqrt{3}$

15 (b) $\dfrac{\pi b^2 a}{2}$; (c) $\dfrac{\pi b^2 a}{2}$ **17** $\dfrac{\pi B^2 A}{2}$; cylinder has twice the volume.   **19** $\dfrac{8}{3}$

21 $\dfrac{338}{15}$ ft^3 **23** $2\sqrt{3}\,a^3$ **25** $\dfrac{686}{3}$ **27** $\dfrac{8}{27}\left(\dfrac{13}{8}\sqrt{13}-1\right)$ **29** $\dfrac{2}{3}(5\sqrt{5}-1)$

31 $|x_1 - x_0|\sqrt{m^2+1}$ **33** $\dfrac{5}{27}(17\sqrt{85}-16\sqrt{10})$ **35** $\dfrac{2}{9}(10\sqrt{10}-3\sqrt{3})$ **37** 3.35

39 1.48   **41** $\displaystyle\lim_{h\to a^-}4\int_0^h\sqrt{1+\dfrac{b^2x^2}{a^2(a^2-x^2)}}\,dx$ **43** $\dfrac{\pi}{2}(125-13\sqrt{13})$

45 $\dfrac{10\pi}{27}(73\sqrt{730}-\sqrt{10})$ **47** $\dfrac{\pi}{9}(82\sqrt{82}-1)$ **49** $\dfrac{8\pi}{3}(5\sqrt{5}-1)$

51 $\dfrac{\pi}{6}(37\sqrt{37}-1)$ **53** \$2.14   **55** \$13,066.67   **57** 21.55%

59 $62.4\pi\left(\dfrac{h^2r^2}{2}+\dfrac{2hr^3}{3}+\dfrac{r^4}{4}\right)$ ft-lb   **61** 135 kg-m   **63** 3,927,363.64 ft-lb

65 $\Delta K = -10{,}000$ ft-lb; $\Delta V = 10{,}000$ ft-lb   **67** 26,956.8 lb

Chapter 8

Problem Set 1 page 384

1 $\sqrt{3}$ **3** 0   **5** (a) All real numbers except $\dfrac{n\pi}{2}$, $n = \pm 1, \pm 3, \pm 5, \pm 7, \ldots$; (b) none

7 (a) All real numbers except 0 and $n\pi$, $n = \pm 1, \pm 3, \pm 5, \pm 7, \ldots$; (b) none   **9** (a) All real numbers except $\dfrac{n\pi}{2}$, $n = \pm 1, \pm 3, \pm 5, \pm 7, \ldots$; (b) none   **11** (a) All real numbers;

(b) none   **13** 6   **15** $\dfrac{2}{5}$ **17** 1   **19** 1   **21** -1 **23** 2

25 $\sin 1° \approx 0.01745$ **27** (b) $\cos 1° \approx 0.9998$

Problem Set 2 page 389

1 $35\cos 7x$ **3** $48x\cos 6x^2$ **5** $\dfrac{\cos\sqrt{x}}{2\sqrt{x}}$ **7** $12\sin^3 3t\cos 3t$

9 $-\cos x\sin(\sin x)$ **11** $\dfrac{-5\sin 5x}{2\sqrt{\cos 5x}}$ **13** $\dfrac{\cos x + \cos 5x\cos x + 5\sin x\sin 5x}{(1+\cos 5x)^2}$

15 $70\sec 2t\tan 2t - 54\csc 2t\cot 2t$ **17** $-15t^4\csc^2(3t^5)$

19 $\dfrac{-u\csc\sqrt{u^2+1}\cot\sqrt{u^2+1}}{\sqrt{u^2+1}}$ **21** $\dfrac{5\sec 5x\tan 5x}{2\sqrt{1+\sec 5x}}$ **23** 0

25 $52\sec^2 13s\tan 13s$ **27** $x^2\tan^4 2x(10x\sec^2 2x + 3\tan 2x)$

29 $\dfrac{2\sec 5x - 10x\sec 5x\tan 5x + 2}{(1+\sec 5x)^2}$ **31** $\dfrac{2}{3}x + 6\cot^2 2x\csc^2 2x$

33 $\dfrac{3\sec^2 3t(2t\tan 3t - 1)}{t^4}$ **35** $10x\sec^2 5x^2\cos(\tan 5x^2)$ **37** $-77\sin 11x$,

$-847\cos 11x$ **39** $-20\sec 5x\tan 5x$, $-100\sec 5x(\sec^2 5x + \tan^2 5x)$
41 $60\tan^2 4x\sec^2 4x$, $480\tan 4x\sec^2 4x(\tan^2 4x + \sec^2 4x)$

43 $\dfrac{-\csc 3x(1+3x\cot 3x)}{x^2}$, $\dfrac{\csc 3x(9x^2\csc^2 3x + 9x^2\cot^2 3x + 6x\cot 3x + 2)}{x^3}$

45 $\dfrac{1}{(x+1)^2}\cos\left(\dfrac{x}{x+1}\right)$, $\dfrac{-1}{(x+1)^3}\left[\dfrac{1}{x+1}\sin\left(\dfrac{x}{x+1}\right) + 2\cos\left(\dfrac{x}{x+1}\right)\right]$ **47** For $x \neq 0$,

$f'(x) = 4x^3 \sin \dfrac{1}{x} - x^2 \cos \dfrac{1}{x};\ f'(0) = 0.$ **49** Does not exist

51 $\dfrac{2 \cos (2x + y)}{1 - \cos (2x + y)}$ **53** $-\dfrac{y}{x}$ **55** $\dfrac{\sin x \cos x}{\sin y \cos y}$ **57** $\dfrac{\cos x}{9 + (\sin x)^2}$

59 $\sec x \tan x\, (1 + \sec^2 x)^{300}$

Problem Set 3 page 392

1 Maximum at $x = \dfrac{\pi}{4}$ and $x = \dfrac{5\pi}{4}$; minimum at $x = \dfrac{3\pi}{4}$ and $x = \dfrac{7\pi}{4}$; concave upward on $\left(\dfrac{\pi}{2}, \pi\right)$ and $\left(\dfrac{3\pi}{2}, 2\pi\right)$; concave downward on $\left(0, \dfrac{\pi}{2}\right)$ and $\left(\pi, \dfrac{3\pi}{2}\right)$; inflection points at $\left(\dfrac{\pi}{2}, 0\right)$, $(\pi, 0)$ and $\left(\dfrac{3\pi}{2}, 0\right)$ **3** Maximum at $x = 0$, $x = \pi$ and $x = 2\pi$; minimum at $x = \dfrac{\pi}{2}$ and $x = \dfrac{3\pi}{2}$; concave upward on $\left(\dfrac{\pi}{4}, \dfrac{3\pi}{4}\right)$ and $\left(\dfrac{5\pi}{4}, \dfrac{7\pi}{4}\right)$; concave downward on $\left(0, \dfrac{\pi}{4}\right)$ and $\left(\dfrac{3\pi}{4}, \dfrac{5\pi}{4}\right)$; inflection points at $\left(\dfrac{\pi}{4}, \dfrac{1}{3}\right)$, $\left(\dfrac{3\pi}{4}, \dfrac{1}{3}\right)$, $\left(\dfrac{5\pi}{4}, \dfrac{1}{3}\right)$, and $\left(\dfrac{7\pi}{4}, \dfrac{1}{3}\right)$ **5** Maximum at $x = 2\pi$; minimum at $x = 0$; concave upward on $(\pi, 2\pi)$ and concave downward on $(0, \pi)$; inflection point at (π, π) **7** Maximum at $x = \pi$; concave upward on $\left(0, \dfrac{\pi}{2}\right)$ and $\left(\dfrac{3\pi}{2}, 2\pi\right)$; concave downward on $\left(\dfrac{\pi}{2}, \dfrac{3\pi}{2}\right)$; no inflection points   **9** 18 cm   **11** $\dfrac{c\sqrt{2}}{2}$, $\dfrac{c\sqrt{2}}{2}$, and c **13** $\dfrac{-4r\sqrt{b^2 - \dfrac{a^2}{4}}}{2a\sqrt{b^2 - \dfrac{a^2}{4}} + \sqrt{3}\, a^2}$ **15** $\dfrac{\pi}{4}$ **17** 10 ft   **19** $\dfrac{2}{3}$ **21** $\dfrac{-4\sqrt{2}}{75}$

23 $\dfrac{\sqrt{3}}{3}$ radian/sec   **25** $\dfrac{\pi}{40}$ cm²/sec   **27** $\dfrac{528}{5}$ radians/hr

Problem Set 4 page 396

1 $-2 \cos x + 3 \sin x + C$ **3** $-\dfrac{2}{35} \cos 35x + C$ **5** $\dfrac{3}{8} \sin (8x - 1) + C$

7 $\dfrac{-\cot 3x}{3} + C$ **9** $\dfrac{1}{11} \tan 11x + C$ **11** $\sec x + \tan x + C$ **13** $\dfrac{1}{2} \sec (2x + 1) + C$

15 $-5 \sec \dfrac{x}{5} + C$ **17** $\sin (\sin x) + C$ **19** $\dfrac{1}{2 + \cos x} + C$ **21** $\dfrac{1}{2 \cos^2 x} + C$

23 $\dfrac{-1}{4(3 + 2 \tan x)^2} + C$ **25** $\dfrac{1}{6} \sec^2 3x + C$ **27** $\dfrac{1}{40} \sec 10x^4 + C$ **29** $\dfrac{2}{3}$ **31** $\dfrac{4}{\pi}$

33 $2(\sqrt{3} - 1)$ **35** $\dfrac{2}{\pi}(3 - \sqrt{3})$ **37** -2 **39** $\dfrac{1}{3}$ **41** 3   **43** $\dfrac{1}{6}$ **45** 2

47 (a) $\dfrac{\sin^2 x}{2} + C$; (b) $-\dfrac{1}{4} \cos 2x + C$; (c) answers are the same, since $\sin^2 x = \dfrac{1 - \cos 2x}{2}$.

Problem Set 5 page 403

1 $\dfrac{\pi}{2}$ **3** $-\dfrac{\pi}{4}$ **5** $\dfrac{\pi}{2}$ **7** 0   **9** $-\dfrac{\pi}{6}$ **11** $\dfrac{3\pi}{4}$ **13** $\dfrac{2\pi}{3}$ **15** $\dfrac{2\pi}{3}$ **17** $\dfrac{\pi}{6}$

19 $\dfrac{3}{4}$ **21** $\dfrac{\pi}{4}$ **23** $\sqrt{2}$ **25** $\dfrac{\pi}{3}$ **27** $-2\sqrt{6}$ **29** (a) $\dfrac{\sqrt{5}}{3}$; (b) $\dfrac{2\sqrt{5}}{5}$;

(c) $\dfrac{\sqrt{5}}{2}$; (d) $\dfrac{3\sqrt{5}}{5}$; (e) $\dfrac{3}{2}$ **31** $\dfrac{24}{25}$ **33** $\dfrac{24}{7}$ **35** $\dfrac{3+\sqrt{105}}{16}$ **45** $\dfrac{\sqrt{2}}{2}$ **47** 1

49 Yes, $x \approx 0.92839486$

Problem Set 6 page 408

1 $\dfrac{3}{\sqrt{1-9x^2}}$ **3** $\dfrac{5}{25+x^2}$ **5** $\dfrac{3}{|t|\sqrt{t^6-1}}$ **7** $\dfrac{-2t}{t^4+6t^2+10}$ **9** $\dfrac{2}{\sqrt{9-4x^2}}$

11 $\dfrac{u}{|u|\sqrt{1-u^2}}$ **13** $\dfrac{-2}{|s|\sqrt{s^2-4}}-\dfrac{2}{s^2+4}$ **15** 0   **17** $\dfrac{-3x^2}{\sqrt{1-9x^2}}+2x\cos^{-1}3x$

19 $\dfrac{-5}{x^2(x^2+25)}-\dfrac{2}{x^3}\tan^{-1}\dfrac{5}{x}$ **21** $\dfrac{-2x-x\sqrt{x^4+2x^2}\csc^{-1}(x^2+1)}{(x^2+1)^{3/2}\sqrt{x^4+2x^2}}$

23 $\dfrac{\sqrt{1-y^2}(1-\sin^{-1}y)}{x-\sqrt{1-y^2}}$ **25** $\dfrac{1+y^2}{1+x^2}$ **27** $\dfrac{1}{[5+(\sin^{-1}x)^4]\sqrt{1-x^2}}$

31 $y=\pi-2x$ for $0\le x\le\dfrac{\pi}{2}$; $y=0$ for $\dfrac{\pi}{2}\le x\le\pi$; $y=2x-2\pi$ for $\pi\le x\le\dfrac{3\pi}{2}$; $y=0$ for

$\dfrac{3\pi}{2}\le x\le 2\pi$ **33** $\sqrt{13.3438}$ meters   **35** $-\dfrac{1}{3}$ radian/min   **37** $\dfrac{126}{169}$ radian/sec

Problem Set 7 page 411

1 $\sin^{-1}\dfrac{x}{2}+C$ **3** $\dfrac{1}{2}\sin^{-1}\dfrac{2x}{3}+C$ **5** $\dfrac{1}{3}\sin^{-1}3t+C$ **7** $\dfrac{1}{6}\tan^{-1}\dfrac{3y}{2}+C$

9 $\dfrac{1}{6}\tan^{-1}\dfrac{2}{3}x+C$ **11** $\dfrac{1}{5}\sec^{-1}\left|\dfrac{4t}{5}\right|+C$ **13** $\sec^{-1}\left|\dfrac{x}{4}\right|+C$ **15** $\dfrac{2\pi}{3}$ **17** $\dfrac{\pi\sqrt{3}}{9}$

19 $-\dfrac{\pi}{12}$ **21** $\sin^{-1}\left(\dfrac{\sin x}{6}\right)+C$ **23** $\dfrac{1}{4}\tan^{-1}\left(\dfrac{x^2}{2}\right)+C$ **25** $2\tan^{-1}\left(\sin\dfrac{x}{2}\right)+C$

27 $\dfrac{1}{3}(\tan^{-1}6-\tan^{-1}2\sqrt{3})$ **29** $3\tan^{-1}2-\dfrac{3\pi}{4}$ **31** $\dfrac{1}{b}\sin^{-1}\left(\dfrac{bx}{a}\right)+C$ **33** $\dfrac{\pi}{2}$

35 $\dfrac{\pi^2}{4}$ cubic units

Review Problem Set page 412

1 13   **3** $\dfrac{19}{7}$ **5** $(1+2x)\cos(x+x^2)$ **7** $(-\sin t)\cos(\cos t)$

9 $[-\sin(x+\sin x)](1+\cos x)$ **11** $-15\cot^2(\tan 5u)\csc^2(\tan 5u)\sec^2 5u$

13 $\dfrac{3\sec^3 5x(5x\tan 5x-1)}{x^4}$ **15** $\dfrac{17\sec^2 17t}{2\sqrt{\tan 17t}}$ **17** $\dfrac{-3x^2}{5[\cos^{-1}(x^3+1)]^{4/5}\sqrt{-x^6-2x^3}}$

19 $\dfrac{4u^5}{1+u^8}+2u\tan^{-1}u^4$ **21** $\dfrac{-8x(\cot^{-1}x^2)^3}{1+x^4}$ **23** $\dfrac{-2}{t^3(t^4+2t^2+2)}-\dfrac{4\cot^{-1}(t^2+1)}{t^5}$

25 $(5+\cos^4 x)^{26}(-\sin x)$ **27** $\left[\dfrac{1-(\tan^{-1}x)^2}{1+(\tan^{-1}x)^2}\right]^{14}\cdot\dfrac{1}{1+x^2}$

29 $\dfrac{\sqrt{1-(x+y)^2}\cos^{-1}(x+y)-x}{x+2y\sqrt{1-(x+y)^2}}$ **31** $\dfrac{3x\cos(x-y)+3\sin(x-y)-2xy}{x^2+3x\cos(x-y)}$

33 $-\dfrac{1}{2}\cos 6x+C$ **35** $\dfrac{5}{4}(\tan x)^{4/5}+C$ **37** $\dfrac{-1}{27(\sec^3 3x+8)^9}+C$

39 $\sin(\sec x)+C$ **41** $\dfrac{1}{84(7+3\cot 4x)^7}+C$ **43** $2\sqrt{1+\tan x}+C$

45 $\dfrac{-1}{6(ax + b \sin x)^6} + C$ **47** $-\dfrac{1}{2} \tan^{-1}\left(\dfrac{\cos x}{2}\right) + C$ **49** $\dfrac{1}{2\sqrt{2}} \sin^{-1}\left(\dfrac{\sqrt{2}\,x^2}{3}\right) + C$

51 $\dfrac{1}{2} \sin^{-1}\left(\dfrac{\sin x^2}{3}\right) + C$ **53** $\dfrac{-(\cot^{-1} v)^2}{2} + C$ **55** 1 **57** $\dfrac{\pi}{20}$ **59** $-\dfrac{3\pi}{2}, -\dfrac{\pi}{2},$

$\dfrac{\pi}{2}, \dfrac{3\pi}{2}$ **61** $\dfrac{\pi}{3}$ **63** $\left(\dfrac{9}{5}, \dfrac{16}{5}\right)$ **65** $\cos^{-1}\sqrt{\dfrac{2}{3}}$ **67** $25\sqrt[3]{\dfrac{7}{25}}$ **69** $\dfrac{1}{250}$ radian/sec

71 $\dfrac{2}{65}$ radian/sec   **73** $\dfrac{0.02xy}{x^2 + y^2}$ **75** $\dfrac{\pi \cos \theta}{10}$ cm^2/min

Chapter 9

Problem Set 1 page 419

1 $\dfrac{8x}{4x^2 + 1}$ **3** $\dfrac{1}{x} \cos (\ln x)$ **5** $1 - 6 \cot 6x$ **7** $(\sin t)\left(\dfrac{2t}{t^2 + 7}\right) + (\cos t) \ln (t^2 + 7)$

9 $\dfrac{1}{u \ln u}$ **11** $\ln\left(\dfrac{\sec x}{5}\right) + x \tan x$ **13** $\dfrac{5v + 4}{2v(v + 1)}$ **15** $-2 \tan x$ **17** $\dfrac{1 - 2x^3}{3x(4x^3 + 1)}$

19 $\dfrac{t^3 + 5 - 3t^3 \ln t}{t(t^3 + 5)^2}$ **21** $\dfrac{2x \sec^2 x - 2 \tan x \ln (\tan^2 x)}{x^3 \tan x}$ **23** $\dfrac{y}{x}$

25 $\dfrac{y^2 - y \cot x}{\ln (\sin x) - 2xy}$ **27** $\dfrac{\cos (\ln x)^2}{x}$ **29** $\dfrac{1}{5} \ln |7 + 5x| + C$ **31** $\ln |\sin x| + C$

33 $\tan (\ln 4x) + C$ **35** $2 \ln (x^2 + 7) + C$ **37** $\sin (\ln x) + C$ **39** $\sin^{-1} (\ln x) + C$

41 $\ln \dfrac{1}{5} - \ln \dfrac{1}{8}$ **43** $\dfrac{1}{2} (\ln 9 - \ln 4)$ **45** $\sin (\ln 4)$ **47** $\ln 4$

Problem Set 2 page 425

1 2.3025   **3** 4.6050   **5** 0.9163   **7** $x = 4$ **9** $x = 4$ **11** Domain $(-\infty, 2)$; range $\mathbb{R}$; $x = 2$ is a vertical asymptote; no maximum or minimum; no inflection points   **13** Domain $(-\infty, -1)$ and $(-1, \infty)$; range $\mathbb{R}$; $x = -1$ is a vertical asymptote; no maximum or minimum; no inflection points   **15** Domain $(0, \infty)$; range $(-a, \infty)$, where $\ln a = -1$, $a \approx 0.37$; no asymptotes; absolute minimum of $-a$ at $x = a$; no maximum; no inflection points   **17** Domain $(0, \infty)$; range $[1, \infty)$; y axis is a vertical asymptote; absolute minimum of 1 at $x = 1$; no maximum; no inflection points   **19** Tangent line:

$y - 4 \ln 2 = (2 + 4 \ln 2)(x - 2)$ and normal line: $y - 4 \ln 2 = \dfrac{-1}{2 + 4 \ln 2} (x - 2)$

21 When $\ln x = -\dfrac{1}{2}$, so that $x \approx 0.61$ **23** $\dfrac{1}{ax^2 - b}$ **25** $\ln 7 - \ln 5$ **27** $3 \ln 2$

29 $\dfrac{\pi}{2} (\ln 17 - \ln 2)$ **31** $\dfrac{3\pi}{2} \ln 2$

Problem Set 3 page 431

1 (a) 5; (b) $\dfrac{1}{8}$; (c) $16e^3$; (d) $\dfrac{1}{x}$; (e) $x - x^2$; (f) x; (g) $x^2 - 4$; (h) $\dfrac{x^2}{e^4}$; (i) $x - 2$; (j) $\dfrac{x}{y^3}$

3 $7e^{7x}$ **5** $3x^2$ **7** $(-\exp x) \sin (\exp x)$ **9** $e^{-2t} (\cos t - 2 \sin t)$

11 $e^{x^2 + 5 \ln x}\left(2x + \dfrac{5}{x}\right)$ **13** $(1 + \ln t)e^{t \ln t}$ **15** $\dfrac{3}{\sqrt{e^{6x-1}}}$ **17** $6e^{3s}(e^{3s} - 1)$

19 $\dfrac{-x}{\sqrt{2\pi}} \exp\left(-\dfrac{x^2}{2}\right)$ **21** $e^{-x}(-8x^2 + 19x - 4)$ **23** $\theta^2[\theta e^\theta \cos (e^\theta) + 3 \sin (e^\theta)]$

25 $\dfrac{y(y - e^x)}{1 + e^x - 2xy}$ **27** $\dfrac{e^{x+y} - \sin y}{x \cos y - e^{x+y}}$ **31** $\dfrac{e^{3x}}{3} + C$ **33** $\dfrac{1}{5} e^{5x+3} + C$

35 $\dfrac{1}{10} e^{5x^2} + C$ **37** $\tan^{-1} e^x + C$ **39** $6\sqrt{e^x + 4} + C$ **41** $-e^{\cot x} + C$

43 $\dfrac{e^2 - 1}{2}$ **45** $\dfrac{2}{3} e$ **47** 0 **49** Decreasing on $\mathbb{R}$; concave upward on $\mathbb{R}$; $y = 0$ is a horizontal asymptote; no maximum or minimum; no points of inflection.   **51** Increasing on $(-\infty, 1]$; decreasing on $[1, \infty)$; concave downward on $(-\infty, 2)$; concave upward on $(2, \infty)$; absolute maximum of $\dfrac{1}{e}$ at $x = 1$; inflection point at $\left(2, \dfrac{2}{e^2}\right)$; x axis is a horizontal asymptote.   **53** $4 \ln 25$ days   **55** $\dfrac{1000}{e^{0.5}}$ **57** $-\dfrac{6}{e^4} \dfrac{\text{lb/in.}^2}{\text{min}}$ **59** $\dfrac{e^3}{3} - \dfrac{e^2}{2} + \dfrac{1}{6}$

Problem Set 4 page 439

1 $-3\pi x^{-3\pi - 1}$ **3** $6et(t^2 + 1)^{3e-1}$ **5** $3^{2x+1}(\ln 3)(2)$ **7** $-5(\ln 6)(6^{-5t})$

9 $5^{\sec x} (\ln 5)(\sec x \tan x)$ **11** $2^{-7x^2}[2x - 14x (\ln 2)(x^2 + 5)]$

13 $\dfrac{2^{x+1}[(x^2 + 5) \ln 2 - 2x]}{(x^2 + 5)^2}$ **15** $\dfrac{2}{x \ln 2}$ **17** $\dfrac{e^{\log_4 x}}{x \ln 4}$ **19** $\dfrac{1}{t(\ln 10)(1 + t)}$

21 $3^{\tan u}\left[(\ln 3)(\sec^2 u) \log_8 u + \dfrac{1}{u \ln 8}\right]$ **23** $\dfrac{2x(x + 2) - \log_3 (x^2 + 5)(\ln 3)(x^2 + 5)}{(\ln 3)(x^2 + 5)(x + 2)^2}$

25 $\dfrac{1}{x(\ln 3)[1 + (\log_3 x)^2]}$ **27** $x^{\sqrt{x}}\left(\dfrac{2 + \ln x}{2\sqrt{x}}\right)$ **29** $(\sin x^2)^{3x}[3 \ln (\sin x^2) + 6x^2 \cot x^2]$

31 $(x^2 + 4)^{\ln x}\left[(\ln x)\left(\dfrac{2x}{x^2 + 4}\right) + \dfrac{\ln (x^2 + 4)}{x}\right]$ **33** $(6x^3 + 1)^3(x^2 + 7)(4x)(24x^3 + 126x + 1)$

35 $\dfrac{(\sin x)\sqrt[3]{1 + \cos x}}{\sqrt{\cos x}}\left[\cot x - \dfrac{\sin x}{3(1 + \cos x)} + \dfrac{1}{2} \tan x\right]$

37 $\dfrac{x^2 \sqrt[5]{x^2 + 7}}{\sqrt[4]{11x + 8}}\left[\dfrac{2}{x} + \dfrac{2x}{5(x^2 + 7)} - \dfrac{11}{4(11x + 8)}\right]$ **39** $\dfrac{3^{5x}}{5 \ln 3} + C$ **41** $\dfrac{7^{x^4 + 4x^3}}{4 \ln 7} + C$

43 $\dfrac{-2^{-\ln x}}{\ln 2} + C$ **45** $\dfrac{-4^{\cot x}}{\ln 4} + C$ **47** $\dfrac{12}{25 \ln 5}$ **49** $e^\pi > \pi^e$ **55** $x = e$ **61** $\dfrac{40\pi}{\ln 3}$

Problem Set 5 page 445

15 $6x \cosh (3x^2 + 5)$ **17** $3t^2 \coth t^3$ **19** $-3e^{3t} \operatorname{csch}^2 (e^{3t})$ **21** $-2s \operatorname{sech} s^2$

23 $2x$ **25** $\dfrac{-(\operatorname{csch}^2 x)(x^2 + 3) - 2x[\tan^{-1} (\coth x)](1 + \coth^2 x)}{(1 + \coth^2 x)(x^2 + 3)^2}$

27 $\dfrac{(1 - t^2)e^{\cosh t} \sinh t + te^{\cosh t}}{(1 - t^2)^{3/2}}$ **29** $2x \operatorname{sech} y$ **31** $\dfrac{2 \tanh x \operatorname{sech}^2 x}{\operatorname{sech}^2 y + 2 \cosh y}$

37 $\dfrac{\sinh 7x}{7} + C$ **39** $\dfrac{\tanh^2 3x}{6} + C$ **41** $-2 \coth \sqrt{x} + C$ **43** $-e^{\operatorname{sech} x} + C$

45 $-\dfrac{1}{10} \operatorname{sech}^2 5x + C$ **47** $\dfrac{\cosh^4 1 - 1}{4}$ **49** $\dfrac{x^2}{4} + \dfrac{1}{2} \ln |x| + C$ **51** $\dfrac{x - 1}{x + 1}$

55 $2a \sinh \dfrac{b}{a}$

Problem Set 6 page 452

1 (a) $\ln (3 + \sqrt{10}) \approx 1.8184$; (b) $\dfrac{1}{2} \ln \dfrac{1}{2} \approx -0.3466$; (c) $\dfrac{1}{2} \ln 2 \approx 0.3466$; (d) $\ln (3 + \sqrt{8}) \approx$

1.7627   **7** $\dfrac{3x^2}{\sqrt{1+x^6}}$ **9** $\dfrac{5}{1-25t^2}$ **11** $\dfrac{-5}{25r^2+20r+3}$ **13** $\dfrac{-1}{\sin x}\left(\dfrac{\cos x}{|\cos x|}\right)$

15 $\dfrac{1}{1-(\ln u)^2}+\tanh^{-1}(\ln u)$ **17** $\dfrac{xe^x}{\sqrt{e^{2x}-1}}+\cosh^{-1}e^x$

19 $-\dfrac{e^x+3+|x|\sqrt{1+x^2}(e^x)\operatorname{csch}^{-1}x}{|x|\sqrt{1+x^2}(e^x+3)^2}$ **27** $\sinh^{-1}\dfrac{x}{3}+C$ **29** $\cosh^{-1}\dfrac{x}{4}+C$

31 $\dfrac{1}{5}\tanh^{-1}\dfrac{x}{5}+C$ for $|x|<5$; $\dfrac{1}{5}\coth^{-1}\dfrac{x}{5}+C$ for $|x|>5$ **33** $-\dfrac{1}{6}\operatorname{sech}^{-1}\dfrac{|x|}{6}+C$

35 $-\dfrac{1}{4}\operatorname{csch}^{-1}\dfrac{|x|}{4}+C$ **37** $\ln(1+\sqrt{2})$ **39** $\ln\left(\dfrac{7+\sqrt{40}}{9}\right)$ **43** $\dfrac{y}{\sqrt{1+x^2y^2}-x}$

Problem Set 7 page 457
1 169,106   **3** 34.3 kg   **5** 164.49 min   **7** (a) 4 gm; (b) 1.72 years   **9** 21.37 hr

11 \$7,233.80   **13** $t_2=\dfrac{t_1\ln(q_2/q_0)}{\ln(q_1/q_0)}$

Problem Set 8 page 463
1 61.80 min   **3** 0.78%   **5** 159.5°F   **7** 7.24 more minutes   **9** 15.69 years
11 5.96   **13** 34.24%   **15** (a) 31.14%; (b) 36.33%; (c) 21.19%   **17** 0.29%
19 18.01%   **21** If $R=r$, $q=(B-AG)e^{-Rt/G}+AG$.   **23** The value of an investment at the end of 1 year will be greater if the interest is compounded more often during the year.

Review Problem Set page 465
1 $\dfrac{2x}{x^2+7}$ **3** $\dfrac{3r+4}{2r(r+2)}$ **5** $\dfrac{x}{1+(\ln x)^2}+2x\tan^{-1}(\ln x)$ **7** $\dfrac{(\ln u)(2-\ln u)}{u^2}$

9 $\cot x-\dfrac{1}{x}$ **11** $-12x^2e^{-4x^3}$ **13** $\dfrac{-2e^{-2x}}{\sqrt{1-e^{-4x}}}$ **15** $\dfrac{-4e^t}{e^{2t}-4}$

17 $\dfrac{-2x}{\sqrt{1-x^4}}-e^{x^3}(3x^3+1)$ **19** $e^x\sec^2 e^x$ **21** $-16e^{4x}(3-e^{4x})^3$

23 $\dfrac{2(e^x+e^{-x}+1)}{(e^x+2)(e^{-x}+2)}$ **25** $4ex^{4e-1}$ **27** $-(\ln 5)(\sin x)5^{\cos x}$

29 $(\ln 7)(\cos x^2)(2x)7^{\sin x^2}$ **31** $2^{4x^2}3^{5x}(5\ln 3+8x\ln 2)$ **33** $\dfrac{1-\ln t}{t^2\ln 7}$

35 $\dfrac{1}{4x(\ln 10)\sqrt[4]{(\log_{10}x)^3}}$ **37** $4e^{4x}\sinh e^{4x}$ **39** $e^{-t}\operatorname{csch}(e^{-t})\coth(e^{-t})$

41 $-2xe^{\operatorname{sech}x^2}\operatorname{sech}(x^2)\tanh(x^2)$ **43** $\dfrac{\operatorname{sech}t(\operatorname{sech}t-\tanh t)}{\tanh t+\operatorname{sech}t}$ **45** $\dfrac{3}{\sqrt{9x^2+6x+2}}$

47 $\dfrac{e^x}{1-e^{2x}}$ **49** No, $\dfrac{\ln 4}{4}=\dfrac{\ln 2}{2}$ **51** $(3x)^x(\ln 3x+1)$

53 $(\sin x)^{x^2}[x^2\cot x+2x\ln(\sin x)]$

55 $(\tanh^{-1}x)^{x^3}\left[\dfrac{x^3}{(\tanh^{-1}x)(1-x^2)}+3x^2\ln(\tanh^{-1}x)\right]$

57 $\dfrac{\cos x\sqrt[3]{1+\sin^2 x}}{\sin^5 x}\left[-\tan x+\dfrac{\sin 2x}{3(1+\sin^2 x)}-5\cot x\right]$

59 $\dfrac{x^2(x+5)^3\sin 2x}{\sec 3x}\left(\dfrac{2}{x}+\dfrac{3}{x+5}+2\cot 2x-3\tan 3x\right)$ **63** $\dfrac{2}{x^2(\sin y-4e^{4y})}$

65 $\dfrac{\sinh(x-y)+\cosh(x+y)}{\sinh(x-y)-\cosh(x+y)}$ **67** $\dfrac{1}{x[5+(\ln x)^3]}$ **69** $\dfrac{1}{3}\ln|8+3x|+C$

71 $\dfrac{(\ln x)^2}{2}+C$ **73** $2e^{\sqrt{x}}+C$ **75** $\tan(e^x)+C$ **77** $\dfrac{10^x}{\ln 10}+C$ **79** $\ln\dfrac{8}{5}$

81 $\dfrac{-\coth^2 x}{2}+C$ **83** $e^{\cosh^{-1}x}+C$ or $x+\sqrt{x^2-1}+C$ **85** $\cosh^{-1}x+C$

87 $-\dfrac{1}{4}\operatorname{sech}^{-1}\dfrac{|x|}{2}+C$ **95** Relative maximum of $\dfrac{1}{e^2}$ at $x=-1$; relative minimum of 0

at $x=0$; increasing on $(-\infty,-1]$ and on $[0,\infty)$; decreasing on $[-1,0]$; concave upward

on $\left(-\infty,\dfrac{-2-\sqrt{2}}{2}\right)$ and on $\left(\dfrac{-2+\sqrt{2}}{2},\infty\right)$; concave downward on

$\left(\dfrac{-2-\sqrt{2}}{2},\dfrac{-2+\sqrt{2}}{2}\right)$; inflection points at $\left(\dfrac{-2-\sqrt{2}}{2},f\left(\dfrac{-2-\sqrt{2}}{2}\right)\right)$ and

$\left(\dfrac{-2+\sqrt{2}}{2},f\left(\dfrac{-2+\sqrt{2}}{2}\right)\right)$; x axis is a horizontal asymptote.   **97** Relative minimum of

0 at $x=0$; relative maximum of $\dfrac{4}{e^2}$ at $x=2$; concave upward on $(-\infty,2-\sqrt{2})$ and on

$(2+\sqrt{2},\infty)$; concave downward on $(2-\sqrt{2},2+\sqrt{2})$; inflection points at

$(2-\sqrt{2},h(2-\sqrt{2}))$ and $(2+\sqrt{2},h(2+\sqrt{2}))$; x axis is a horizontal asymptote.   **99** 3π
101 6.64 hr   **107** 7.588×10^8 ft^3 **109** 2.42 additional minutes   **111** (a) At end of
12 years you have \$99,603.51; (b) at end of 12 years you have \$167,772.16.   **113** 14%

115 $p=\dfrac{Ae^{kt}}{1+Ae^{kt}}$, A a constant

Chapter 10

Problem Set 1 page 474

1 $\sin x-\dfrac{\sin^3 x}{3}+C$ **3** $-\dfrac{1}{2}\left(\cos 2t-\dfrac{2}{3}\cos^3 2t+\dfrac{1}{5}\cos^5 2t\right)+C$

5 $\dfrac{\sin^8 2x}{16}-\dfrac{\sin^{10}2x}{20}+C$ **7** $\dfrac{\sin^3 x}{3}-\dfrac{\sin^5 x}{5}+C$ **9** $\dfrac{2\cos^5 x}{5}-\dfrac{\cos^3 x}{3}-\dfrac{\cos^7 x}{7}+C$

11 $\dfrac{1}{3\cos^3 x}-\dfrac{1}{\cos x}+C$ **13** $\dfrac{8}{5}-2\sqrt{\dfrac{\sqrt{2}}{2}}+\dfrac{2}{5}\left(\dfrac{\sqrt{2}}{2}\right)^{5/2}$ **15** $\dfrac{32}{65\pi}$

17 $\dfrac{6x-\sin 6x}{2}+C$ **19** $\dfrac{t-\sin t}{2}+C$ **21** $\dfrac{5}{16}u-\dfrac{1}{4}\sin 2u+\dfrac{3}{64}\sin 4u+\dfrac{\sin^3 2u}{48}+C$

23 $\dfrac{3\pi-4}{384}$ **25** $-\dfrac{1}{14}\cos 7x-\dfrac{1}{6}\cos 3x+C$ **27** $\dfrac{1}{2}\sin x+\dfrac{1}{14}\sin 7x+C$

29 $\dfrac{1}{8}\sin 4u-\dfrac{1}{20}\sin 10u+C$ **31** $-\dfrac{4}{5\pi}$ **37** $\dfrac{\pi^2}{2}$ cubic units

41 $s=f(t)=\dfrac{2\pi t-\sin 2\pi t}{4\pi}$; when $t=8$, $s=4$

Problem Set 2 page 481

1 $\dfrac{1}{4}\ln|\sin 4x|+C$ **3** $\dfrac{1}{3}\ln|\sec 3x+\tan 3x|+C$ **5** $-\dfrac{3}{2}\tan\dfrac{2x}{3}-x+C$

7 $-\dfrac{\cot^3 4x}{12}+\dfrac{1}{4}\cot 4x+x+C$ **9** $-\dfrac{1}{3}\left(\cot 3t+\dfrac{\cot^3 3t}{3}\right)+C$ **11** $\dfrac{1}{10}\tan^5 2t+C$

13 $\dfrac{\sec^7 5x}{35}-\dfrac{\sec^5 5x}{25}+C$ **15** $\dfrac{1}{2}\tan 2x-\dfrac{1}{2}\cot 2x+C$ **17** $\dfrac{2}{5}\tan^{5/2}t+2\tan^{1/2}t+C$

19 $\dfrac{\tan^4 7x}{28} + \dfrac{\tan^6 7x}{42} + C$ **21** $\dfrac{-\csc^3 3x}{9} + C$ **23** $\dfrac{\sec^3 5x}{15} - \dfrac{\sec 5x}{5} + C$

25 $\dfrac{1}{5}\sec^{5/2} 2x - \sec^{1/2} 2x + C$ **27** $\dfrac{\sec^{11} x}{11} - \dfrac{2}{9}\sec^9 x + \dfrac{\sec^7 x}{7} + C$

29 $\dfrac{1}{24\cot^3 8x} + C$ **31** $\dfrac{1}{3}\ln\dfrac{\sqrt{3}}{2}$ **33** $\dfrac{12}{35}$ **35** $2\pi\sqrt{3}$ cubic units   **37** $5\left(\dfrac{4-\pi}{2}\right)$

square units   **41** $-\ln(\sqrt{2}-1)$ units   **45** $\dfrac{1}{3}\sin 3x + C$ **47** $\dfrac{\sin^3\theta}{3} - \dfrac{\sin^5\theta}{5} + C$

Problem Set 3 page 487

1 $\dfrac{-\sqrt{16-x^2}}{16x} + C$ **3** $-\dfrac{1}{16}\left[\dfrac{1}{3}\left(\dfrac{\sqrt{4-t^2}}{t}\right)^3 + \dfrac{\sqrt{4-t^2}}{t}\right] + C$

5 $\dfrac{2}{27}\sin^{-1}\dfrac{3x}{2} - 3x\dfrac{\sqrt{4-9x^2}}{4} + C$ **7** $\dfrac{-(\sqrt{7-4t^2})^3}{21t^3} + C$ **9** $\sqrt{t^2-a^2} + C$

11 $\dfrac{-\sqrt{x^2+1}}{x} + C$ **13** $\dfrac{1}{\sqrt{5}}\ln\left|\dfrac{\sqrt{t^2+5}-\sqrt{5}}{t}\right| + C$ **15** $\dfrac{7}{8}\left(\dfrac{2x^2+9}{\sqrt{4x^2+9}}\right) + C$

17 $\dfrac{\sqrt{x^2-4}}{4x} + C$ **19** $\dfrac{1}{3}\ln\left|\dfrac{3t+\sqrt{9t^2-4}}{2}\right| + C$ **21** $-\dfrac{x}{9\sqrt{4x^2-9}} + C$

25 $\dfrac{2+t}{9\sqrt{5-4t-t^2}} + C$ **27** $\dfrac{\sin^{-1}(2x-2)}{2} + C$

29 $\dfrac{1}{3}\sqrt{3x^2-x+1} + \dfrac{\sqrt{3}}{18}\ln\left|\sqrt{3x^2-x+1} + \sqrt{3}\,x - \dfrac{\sqrt{3}}{6}\right| + C$

31 $\dfrac{4\left(t+\dfrac{3}{2}\right)^2}{7(t^2+3t+4)} - \dfrac{12\sqrt{7}}{49}\tan^{-1}\left(\dfrac{2t+3}{\sqrt{7}}\right) - \dfrac{6t+9}{7(t^2+3t+4)} + C$ **33** $\dfrac{1}{2}\ln\dfrac{3}{2}$

35 $\dfrac{1}{27}\left(\dfrac{3}{4} - \sin^{-1}\dfrac{3}{5}\right)$ **37** $3\sqrt{3} - \pi$ **39** $\dfrac{45}{4}\ln\dfrac{7}{5}$ square units

41 $\dfrac{\pi}{16}\left(\tan^{-1}\dfrac{3}{2} + \dfrac{6}{13}\right)$ cubic units   **43** $\dfrac{\pi^2}{2048}$ cubic unit   **47** (a) and (b): $\sqrt{x^2-1} + C$

49 $y = \ln\left|\dfrac{x+\sqrt{x^2-9}}{3}\right| - \dfrac{\sqrt{x^2-9}}{x} + C$ **51** $\ln\left(\dfrac{1}{3}\right) - \ln(2-\sqrt{3})$ units

Problem Set 4 page 496

1 $\dfrac{1}{2}x\sin 2x + \dfrac{1}{4}\cos 2x + C$ **3** $\dfrac{1}{3}xe^{3x} - \dfrac{1}{9}e^{3x} + C$ **5** $x\ln 5x - x + C$

7 $-x^2\sqrt{1-x^2} - \dfrac{2}{3}(1-x^2)^{3/2} + C$ **9** $-x\cot x + \ln|\sin x| + C$

11 $t\sec t - \ln|\sec t + \tan t| + C$ **13** $x\cos^{-1} x - \sqrt{1-x^2} + C$

15 $\dfrac{-x^2}{3}\cos 3x + \dfrac{2}{9}x\sin 3x + \dfrac{2}{27}\cos 3x + C$

17 $\dfrac{x^4}{2}\sin 2x + x^3\cos 2x - \dfrac{3x^2}{2}\sin 2x - \dfrac{3x}{2}\cos 2x + \dfrac{3}{4}\sin 2x + C$

19 $-e^{-t}(t^4 + 4t^3 + 12t^2 + 24t + 24) + C$

21 $-e^{-x}(x^5 + 5x^4 + 19x^3 + 57x^2 + 115x + 115) + C$ **23** $\dfrac{e^{x^2}}{2}(x^2-1) + C$

25 $\dfrac{x\sqrt{1+x^2}}{2} + \dfrac{1}{2}\ln\left|\sqrt{1+x^2}+x\right| + C$ **27** $\dfrac{1}{4}\left(x^8\sin x^4 + 2x^4\cos x^4 - 2\sin x^4\right) + C$

29 $\dfrac{e^{-x}}{5}\left(2\sin 2x - \cos 2x\right) + C$ **31** $-\dfrac{1}{2}\csc x\cot x + \dfrac{1}{2}\ln\left|\csc x - \cot x\right| + C$

33 $-\dfrac{2}{3}\sin x\cos 2x + \dfrac{1}{3}\cos x\sin 2x + C$ **35** $12\pi\sqrt{3} - 2\pi^2 - \dfrac{36}{243}$

37 $3\sec^{-1} 3 - \dfrac{2\pi}{3} + \ln\dfrac{2+\sqrt{3}}{3+\sqrt{8}}$ **39** $\sqrt{2}\left(3 + \dfrac{13\pi}{8} - \dfrac{5\pi^2}{32}\right) - 9$ **45** $1 - \dfrac{2}{e}$ square units

Problem Set 5 page 505

1 $\ln\left(\dfrac{|x-2|^{3/2}}{|x|^{1/2}}\right) + C$ **3** $\dfrac{x^2}{2} + 2x + C$ **5** $\dfrac{x^2}{2} + 2x + \ln\left|(x+5)^{17/7}(x-2)^{4/7}\right| + C$

7 $\ln\left(\dfrac{|x-1|}{\sqrt{|x(x+2)|}}\right) + C$ **9** $\ln\left(\dfrac{\sqrt{|x^2-1|}}{|x|}\right) + C$ **11** $x + \ln\left|\dfrac{(x-3)^{9/5}}{(x+2)^{4/5}}\right| + C$

13 $x + \ln\left|\dfrac{(x-2)^{4/5}}{(x+3)^{9/5}}\right| + C$ **15** $\ln\left|\dfrac{(x+4)^{29/5}(x-1)^{6/5}}{x^2}\right| + C$

17 $-\dfrac{3}{z} + \ln\left(\dfrac{4z+1}{z}\right)^{10} + C$ **19** $\ln\left|\dfrac{x+1}{x+7}\right|^{1/9} - \dfrac{1}{3(x+1)} + C$

21 $\ln\left(\dfrac{|x+2|^{5/8}}{|3x-2|^{13/72}}\right) - \dfrac{8}{9(3x-2)} + C$ **23** $\ln\left|\dfrac{z-3}{z-1}\right|^{9/2} + \dfrac{1}{(z-1)^2} + \dfrac{5}{z-1} + C$

25 $\ln\dfrac{27}{20}$ **27** $\dfrac{167}{6} + \ln\dfrac{3}{2}$ **29** $\dfrac{10+12\ln 3}{3}$ **31** $\dfrac{ax}{c} + \dfrac{bc-ad}{c^2}\ln|cx+d| + K$

33 $-\dfrac{1}{x-a} + C$ **35** (a) and (b): $\ln\left|(x-3)^{4/5}(x+2)^{1/5}\right| + C$ **37** $2 - \ln 3$ square unit

39 $\dfrac{1}{2\sqrt{aq}}\ln\left|\dfrac{q-ax^2}{a}\right| + C$ **41** $x = a - \dfrac{1}{\sqrt[3]{3(kt+c)}}$

43 $C(x) = 100\ln\left(\dfrac{5x^3(x-1)^2}{8(x+3)}\right) + 47$ **45** $q = \dfrac{A}{B - Ce^{At}}$, where $C = B - \dfrac{A}{q_0}$

Problem Set 6 page 513

1 $\dfrac{1}{5}\ln|x-1| - \dfrac{1}{10}\ln(x^2+4) - \dfrac{1}{10}\tan^{-1}\dfrac{x}{2} + C$ **3** $\ln\dfrac{|x^3|}{(x^2+1)^{3/2}} + \tan^{-1}x + C$

5 $\ln\left|t^{1/25}(t^2+25)^{49/50}\right| - \dfrac{1}{5}\tan^{-1}\dfrac{t}{5} + C$ **7** $\ln\left(\dfrac{|x^2-1|}{x^2+1}\right)^{1/4} + C$

9 $\ln\left[|x-2|(2x^2-3x+5)^2\right] + \dfrac{6}{\sqrt{31}}\tan^{-1}\left[\dfrac{4}{\sqrt{31}}\left(x-\dfrac{3}{4}\right)\right] + C$

11 $\ln\left|\dfrac{x}{\sqrt{x^2+4}}\right| + \dfrac{2}{x^2+4} + C$ **13** $\ln\left|t^{2/9}(t^2+9)^{43/18}\right| + \dfrac{1}{9t} - \dfrac{26}{27}\tan^{-1}\dfrac{t}{3} + C$

15 $-\dfrac{4}{x} - 6\tan^{-1}x - \dfrac{1}{2(x^2+1)} - \dfrac{2x}{x^2+1} + C$

17 $x + \ln\left(\dfrac{|x|^{1/2}}{(x^2+2)^{1/4}}\right) - \dfrac{1}{x^2+2} - \dfrac{5\sqrt{2}}{8}\tan^{-1}\left(\dfrac{x}{\sqrt{2}}\right) - \dfrac{5x}{4(x^2+2)} + C$

19 $\ln\left(\dfrac{\sqrt{|t|}}{(t^2+2t+2)^{1/4}}\right) + \dfrac{t+2}{2(t^2+2t+2)} + C$ **21** $\dfrac{1}{8}\ln\left(\dfrac{x^2+2x+1}{x^2+2x+5}\right) + C$

23 $\frac{9}{2}\ln 4 - \frac{9}{4}\ln 10 + \frac{11}{2}\tan^{-1} 3$ **25** $\frac{\pi - 2\ln 2}{4}$ **27** $\ln\frac{4}{5}$ **29** Put $t = \sin x$.

31 Put $u = e^x$.   **33** $\frac{ax + b}{x^2 + 1} + \frac{(c - a)x + (d - b)}{(x^2 + 1)^2}$ **35** (a) and (b):

$\ln|x^5 + 2x^3 + x| + C$ **41** (a) $\frac{w}{2(w^2 + 1)} + \frac{1}{2}\tan^{-1} w + C$;

(b) $\frac{w}{4(w^2 + 1)^2} + \frac{3}{4}\left[\frac{w}{2(w^2 + 1)} + \frac{1}{2}\tan^{-1} w\right] + C$

Problem Set 7   page 518

1 $-2\sqrt{x} - \ln(1 - \sqrt{x})^2 + C$ **3** $\frac{3}{2}x^{2/3} - 3x^{1/3} + 3\ln|1 + \sqrt[3]{x}| + C$

5 $\frac{2}{3}x^{3/2} - 2x + 8\sqrt{x} - 16\ln|\sqrt{x} + 2| + C$ **7** $2\sqrt{x} - 4\sqrt[4]{x} + 4\ln(1 + \sqrt[4]{x}) + C$

9 $\frac{(2x^2 - 1)^{5/2}}{20} + \frac{(2x^2 - 1)^{3/2}}{12} + C$ **11** $\frac{(3x + 1)^{7/3}}{21} - \frac{(3x + 1)^{4/3}}{12} + C$

13 $\frac{(4x + 1)^{9/2}}{288} - \frac{(4x + 1)^{7/2}}{112} + \frac{(4x + 1)^{5/2}}{160} + C$ **15** $2\sqrt{x + 1} - 2\ln(1 + \sqrt{x + 1}) + C$

17 $\frac{-4\sqrt[4]{2 - 3x^2}}{9} + \frac{2(2 - 3x^2)^{5/4}}{45} + C$ **19** $-\frac{2}{3}(1 - e^x)^{3/2} + C$

21 $\frac{2(1 + \sin x)^{5/2}}{5} - \frac{2(1 + \sin x)^{3/2}}{3} + C$ **23** $-\frac{2}{27}\sqrt{\frac{1}{3x + 1}}\left(\frac{9x + 2}{3x + 1}\right) + C$

25 $\frac{1}{4}\ln\left|\dfrac{3\tan\frac{x}{2} + 1}{\tan\frac{x}{2} + 3}\right| + C$ **27** $\frac{1}{2}\left(\ln\left|\tan\frac{x}{2}\right| - \frac{1}{2}\tan^2\frac{x}{2}\right) + C$

29 $\ln\left(\dfrac{\sqrt{\left|1 + \tan\frac{t}{2}\right|}}{\sqrt{\left|1 - \tan\frac{t}{2}\right|}}\right) + \dfrac{1}{1 + \tan\frac{t}{2}} - \dfrac{1}{\left(1 + \tan\frac{t}{2}\right)^2} + C$ **31** $\frac{1}{\sqrt{2}}\ln\left|\dfrac{\tan\frac{t}{2} - 1 + \sqrt{2}}{\tan\frac{t}{2} - 1 - \sqrt{2}}\right| + C$

33 $\ln\left|\tan\frac{x}{2} + 1\right| + C$ **35** $-\text{csch}^{-1} x + C$ **37** $-\sin^{-1}\left(\frac{1 + x}{2x}\right) + C$

41 $\dfrac{2}{1 - \tanh\frac{x}{2}} + C$ **43** $\frac{28\sqrt{3}}{5}$ **45** $2\left(-1 + \tan^{-1} 2 - \frac{\pi}{4} + 2\ln\frac{5}{2}\right)$

47 $\frac{92 - 14\sqrt{2}}{15}$ **49** $\ln\frac{3}{2}$ **51** $75 - 10\ln 4$ square units

Review Problem Set   page 519

1 $\frac{\sin 2x}{2} - \frac{\sin^3 2x}{6} + C$ **3** $\frac{1}{3}\left(\frac{\sin^4 3x}{4} - \frac{\sin^6 3x}{6}\right) + C$

5 $\frac{1}{2}\left[\cos(1 - 2x) - \frac{\cos^3(1 - 2x)}{3}\right] + C$ **7** $\frac{3}{5}\sin^{1/3} 5x - \frac{3}{35}\sin^{7/3} 5x + C$

9 $\frac{1}{12}\sin(4 - 6x) + \frac{x}{2} + C$ **11** $x + \frac{\cos 2x}{2} + C$ **13** $\frac{1}{4}\left(\frac{x}{2} - \frac{\sin 24x}{48}\right) + C$

15 $\sin^{2/3}\frac{3t}{2} - \frac{1}{4}\sin^{8/3}\frac{3t}{2} + C$ **17** $\frac{\sin 5x}{10} - \frac{\sin 11x}{22} + C$

19 $-\dfrac{1}{16}\cos 4x - \dfrac{1}{8}\cos 2x - \dfrac{1}{12}\sin^2 3x + C$

21 $\dfrac{1}{6}\tan^3 (2x-1) - \dfrac{1}{2}\tan (2x-1) + x + C$

23 $\dfrac{1}{20}\tan^2 5x^2 - \dfrac{1}{10}\ln |\sec 5x^2| + C$ **25** $2\tan t - t - 2\sec t + C$

27 $\dfrac{2}{3}\tan^3 x + \tan x + \dfrac{2}{3}\sec^3 x + C$ **29** $\dfrac{1}{2}\left[\dfrac{\tan^3 (1+2x)}{3} + \tan (1+2x)\right] + C$

31 $\dfrac{1}{3}\left[\dfrac{\tan^6 (2+3x)}{6} + \dfrac{\tan^4 (2+3x)}{4}\right] + C$ **33** $\sinh^{-1}\dfrac{x}{8} + C$

35 $\dfrac{1}{3}\left(\dfrac{x}{\sqrt{1-x^2}}\right)^3 + \dfrac{x}{\sqrt{1-x^2}} + C$ **37** $\dfrac{1}{3}(x^2-4)^{3/2} + C$ **39** $\sin^{-1} (x-1) + C$

41 $\sinh^{-1}\left(\dfrac{x+3}{2}\right) + C$ **43** $-\dfrac{x^2}{7}e^{-7x} - \dfrac{2x}{49}e^{-7x} - \dfrac{2}{343}e^{-7x} + C$

45 $\dfrac{t^3}{3}\sin^{-1} 2t - \dfrac{1}{24}\left[\dfrac{(1-4t^2)^{3/2}}{3} - (1-4t^2)^{1/2}\right] + C$ **47** $\dfrac{1}{3}(x+2)e^{3x} - \dfrac{1}{9}e^{3x} + C$

49 $\dfrac{1}{3}t^3 \sin 3t + \dfrac{1}{3}t^2 \cos 3t - \dfrac{2t}{9}\sin 3t - \dfrac{2}{27}\cos 3t + C$

51 $x\tan^{-1}\sqrt{x} - \sqrt{x} + \tan^{-1}\sqrt{x} + C$ **53** $-\dfrac{e^{-7x}}{48}(7\cosh x + \sinh x) + C$

55 $e^{3x}\left(\dfrac{1}{6} + \dfrac{3}{26}\cos 2x + \dfrac{1}{13}\sin 2x\right) + C$ **57** $\dfrac{-e^{-x^4}}{4}(x^8 + 2x^4 + 2) + C$

59 $\dfrac{x^2 \sin 3x^2}{6} + \dfrac{\cos 3x^2}{18} + C$ **61** $\ln\left(\dfrac{|y-1|^{3/2}|y+1|^{5/2}}{|y|}\right) + C$

63 $\dfrac{1}{x} + 3\ln |x-1| + C$ **65** $\ln\left(\dfrac{t^2+1}{t^2+4}\right) + \tan^{-1} t + C$

67 $\ln\left(\dfrac{|x|}{\sqrt{x^2+1}}\right)^3 - \dfrac{1}{x} - \tan^{-1} x + C$ **69** $\dfrac{(1+2x)^{7/4}}{7} - \dfrac{(1+2x)^{3/4}}{3} + C$

71 $\dfrac{10}{11}x^{11/10} + \dfrac{3}{2}x^{2/3} + C$ **73** $\ln\left|\dfrac{\sqrt{e^t+1}-1}{\sqrt{e^t+1}+1}\right| + C$

75 $\dfrac{4}{3}(4 + \sqrt{x+1})^{3/2} - 16\sqrt{4 + \sqrt{x+1}} + C$

77 $x\ln\sqrt{x^2+3} - x + \sqrt{3}\,\tan^{-1}\left(\dfrac{x}{\sqrt{3}}\right) + C$

79 $-\dfrac{3}{7}(1-x^{2/3})^{7/2} + \dfrac{6}{5}(1-x^{2/3})^{5/2} - (1-x^{2/3})^{3/2} + C$

81 $3x^{2/3}\sin\sqrt[3]{x} + 6\sqrt[3]{x}\cos\sqrt[3]{x} - 6\sin\sqrt[3]{x} + C$ **83** $\dfrac{1}{\sqrt{21}}\ln\left|\dfrac{\sqrt{21} + \tan\dfrac{x}{2}}{\sqrt{21} - \tan\dfrac{x}{2}}\right| + C$

85 $\tan^{-1}\left(\tan\dfrac{y}{2} + 1\right) + C$ **87** $\ln\left|\dfrac{1+\tan\dfrac{x}{2}}{1-\tan\dfrac{x}{2}}\right|^{1/2} + \dfrac{\tan\dfrac{x}{2}}{\left(1+\tan\dfrac{x}{2}\right)^2} + C$

89 $\dfrac{2}{7}(e^{2x}+1)^{7/4} - \dfrac{2}{3}(e^{2x}+1)^{3/4} + C$ **91** $-\dfrac{1}{12}$ **93** $\dfrac{1}{6} + \dfrac{1}{4}\ln\dfrac{2}{3}$

95 $2(\ln 2)^2 - 4 \ln 2 + 2$ **97** $4 \ln 2 - \dfrac{15}{16}$ **99** $\dfrac{1}{2} - \ln \sqrt{2}$ **101** $\dfrac{9\pi}{2} - \dfrac{81\sqrt{3}}{8}$

103 $5\left(\sqrt{3} - \dfrac{\pi}{3}\right)$ **105** $3\sqrt{2}$ **107** $4 + 4 \ln 3$ **109** $\dfrac{\pi + 3}{4}$ **111** $\dfrac{\pi}{6}$

113 $\dfrac{3}{2} \ln \dfrac{5}{2}$ **115** 3 **117** $\dfrac{13}{6}$ **119** $3 - \sqrt{3}$

121 $\dfrac{1}{2} \ln \left(\dfrac{\tan \dfrac{\pi}{16}}{\tan \dfrac{\pi}{8}} \right) - \dfrac{1}{4}\left(\tan^2 \dfrac{\pi}{16} - \tan^2 \dfrac{\pi}{8}\right)$ **123** $A = \dfrac{ce + df}{e^2 + f^2}, \; B = \dfrac{de - cf}{e^2 + f^2};$

$\left(\dfrac{ce + df}{e^2 + f^2}\right)\theta + \left(\dfrac{de - cf}{e^2 + f^2}\right) \ln |e \sin \theta + f \cos \theta| + C$

125 $\dfrac{1}{\sqrt{5}} \ln \left| \dfrac{x + \sqrt{x^2 - 2x + 5} - \sqrt{5}}{x + \sqrt{x^2 - 2x + 5} + \sqrt{5}} \right| + C$ **131** $\dfrac{27}{4} - 2 \ln 4$ square units

133 $\dfrac{4}{3}$ square units   **135** $\dfrac{\pi}{2}(9\pi + 16)$ cubic units   **137** $\ln (2 + \sqrt{3})$

139 $\ln \left(\dfrac{1}{2 - \sqrt{3}}\right)$ units

Chapter 11

Problem Set 1 page 528

1 (a) $\left(-3, \dfrac{5\pi}{4}\right)$; (b) $\left(3, -\dfrac{7\pi}{4}\right)$; (c) $\left(-3, -\dfrac{3\pi}{4}\right)$ **3** (a) Already in this form;

(b) $\left(2, -\dfrac{5\pi}{6}\right)$; (c) $\left(-2, -\dfrac{11\pi}{6}\right)$ **5** (a) $(-4, 360°)$; (b) $(4, -180°)$; (c) $(-4, 0°)$

7 $x = \dfrac{7}{2}, y = \dfrac{7\sqrt{3}}{2}$ **9** $x = y = -\sqrt{2}$ **11** $x = \dfrac{1}{2}, y = -\dfrac{\sqrt{3}}{2}$ **13** $r = 7\sqrt{2}, \theta = \dfrac{\pi}{4}$

15 $r = 6, \theta = -\dfrac{2\pi}{3}$ **17** $r = 7, \theta = \dfrac{\pi}{2}$ **19** (13) Same; (14) $r = 2, \theta = \dfrac{5\pi}{3}$; (15) $r = 6,$

$\theta = \dfrac{4\pi}{3}$; (16) same; (17) same; (18) same   **21** $\left(-3, \dfrac{7\pi}{6}\right), \left(3, \dfrac{\pi}{6}\right), \left(3, -\dfrac{11\pi}{6}\right), \left(3, \dfrac{13\pi}{6}\right),$

$\left(-3, -\dfrac{5\pi}{6}\right)$ **23** Circle of radius 1, center at pole   **25** Straight line through pole, perpendicular to polar axis   **27** Same as problem 25   **29** Circle of radius 2 passing through the origin with center on polar axis   **31** $x^2 - 3x + y^2 = 0$
33 $x^2 - x + y^2 - y = 0$ **35** $r = 5$ **37** $r^2(1 + 3 \sin^2 \theta) = 4$

Problem Set 2 page 536
1 Symmetric with respect to (w.r.t.) polar axis (circle)   **3** Symmetric w.r.t. polar axis
(vertical straight line)   **5** Symmetric w.r.t. $\theta = \pm \dfrac{\pi}{2}$ (circle)   **7** Symmetric w.r.t.
$\theta = \pm \dfrac{\pi}{2}$ **9** Symmetric w.r.t. polar axis, $\theta = \pm \dfrac{\pi}{2}$, and the pole   **11** Symmetric w.r.t.
polar axis   **13** Symmetric w.r.t. polar axis   **15** Symmetric w.r.t. $\theta = \pm \dfrac{\pi}{2}$
17 Symmetric w.r.t. polar axis, $\theta = \pm \dfrac{\pi}{2}$, and pole   **19** 1

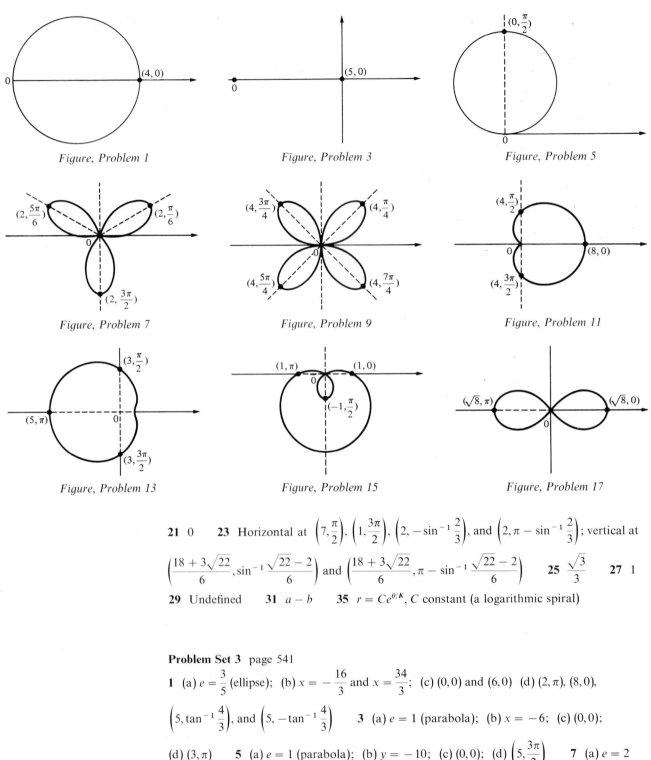

Figure, Problem 1   *Figure, Problem 3*   *Figure, Problem 5*

Figure, Problem 7   *Figure, Problem 9*   *Figure, Problem 11*

Figure, Problem 13   *Figure, Problem 15*   *Figure, Problem 17*

21 0   **23** Horizontal at $\left(7,\frac{\pi}{2}\right)$, $\left(1,\frac{3\pi}{2}\right)$, $\left(2,-\sin^{-1}\frac{2}{3}\right)$, and $\left(2,\pi-\sin^{-1}\frac{2}{3}\right)$; vertical at

$\left(\dfrac{18+3\sqrt{22}}{6},\sin^{-1}\dfrac{\sqrt{22}-2}{6}\right)$ and $\left(\dfrac{18+3\sqrt{22}}{6},\pi-\sin^{-1}\dfrac{\sqrt{22}-2}{6}\right)$ **25** $\dfrac{\sqrt{3}}{3}$ **27** 1

29 Undefined   **31** $a-b$ **35** $r=Ce^{\theta/K}$, C constant (a logarithmic spiral)

Problem Set 3 page 541

1 (a) $e=\dfrac{3}{5}$ (ellipse); (b) $x=-\dfrac{16}{3}$ and $x=\dfrac{34}{3}$; (c) $(0,0)$ and $(6,0)$ (d) $(2,\pi)$, $(8,0)$,

$\left(5,\tan^{-1}\dfrac{4}{3}\right)$, and $\left(5,-\tan^{-1}\dfrac{4}{3}\right)$ **3** (a) $e=1$ (parabola); (b) $x=-6$; (c) $(0,0)$;

(d) $(3,\pi)$ **5** (a) $e=1$ (parabola); (b) $y=-10$; (c) $(0,0)$; (d) $\left(5,\dfrac{3\pi}{2}\right)$ **7** (a) $e=2$

(hyperbola); (b) $y=\dfrac{1}{2}$ and $y=\dfrac{5}{6}$; (c) $(0,0)$ and $\left(\dfrac{4}{3},\dfrac{\pi}{2}\right)$; (d) $\left(\dfrac{1}{3},\dfrac{\pi}{2}\right)$ and $\left(1,\dfrac{\pi}{2}\right)$

9 (a) $e=1$ (parabola); (b) $x=4$; (c) $(0,0)$; (d) $(2,0)$ **11** $r=\dfrac{5}{3+\sin\theta}$

13 $r = \dfrac{4\sqrt{5}}{1 + \sqrt{5}\cos\theta}$ **15** $F_1 = (0,0),\ F_2 = \left(\dfrac{2e^2 d}{1 - e^2}, 0\right),\ C = \left(\dfrac{e^2 d}{1 - e^2}, 0\right),$

$V_1 = \left(\dfrac{ed}{1 + e}, \pi\right),\ V_2 = \left(\dfrac{ed}{1 - e}, 0\right),\ V_3 = \left(\dfrac{ed}{1 - e^2}, \tan^{-1}\dfrac{\sqrt{1 - e^2}}{e}\right),$

$V_4 = \left(\dfrac{ed}{1 - e^2}, -\tan^{-1}\dfrac{\sqrt{1 - e^2}}{e}\right)$ **17** (a) Directrices move outward and approach $+\infty$

and $-\infty$; (b) ellipse approaches the circle $r = a$.   **19** $r = 5\sqrt{2}(\cos\theta + \sin\theta)$

21 $r^2 - 2r(7\cos\theta + 5\sin\theta) + 49 = 0$ **23** (a) $r = \dfrac{d}{\cos\theta}$; (b) $r = -\dfrac{d}{\cos\theta}$

25 (a) $r = \dfrac{d}{\sin\theta}$; (b) $r = -\dfrac{d}{\sin\theta}$ **27** $\left(\dfrac{3}{2}, \dfrac{11\pi}{6}\right)$ and $\left(\dfrac{3}{2}, \dfrac{7\pi}{6}\right)$ **29** $(1, \theta)$ with $\theta =$

$\dfrac{\pi}{9}, \dfrac{5\pi}{9}, \dfrac{7\pi}{9}, \dfrac{11\pi}{9}, \dfrac{13\pi}{9}$, and $\dfrac{17\pi}{9}$ **31** $\left(\dfrac{\sqrt{3}}{2}, \dfrac{\pi}{3}\right), \left(\dfrac{\sqrt{3}}{2}, \dfrac{2\pi}{3}\right)$, and $(0,0)$ **33** $\left(2, \dfrac{3\pi}{2}\right)$

37 $r = -\dfrac{d}{\cos\theta}$ and $r = \dfrac{e^2 d + d}{(1 - e^2)\cos\theta}$ **39** $\left(2, \dfrac{\pi}{2}\right), \left(2, -\dfrac{\pi}{2}\right), (2, \pi), \left(-2 - 2\sqrt{2}, \dfrac{\pi}{4}\right),$

$\left(-2 - 2\sqrt{2}, \dfrac{5\pi}{4}\right), \left(-2 + 2\sqrt{2}, \dfrac{3\pi}{4}\right)$, and $\left(-2 + 2\sqrt{2}, \dfrac{5\pi}{4}\right)$

Problem Set 4 page 547

1 $\dfrac{31\pi^3}{6}$ **3** $\dfrac{128}{5\pi}$ **5** $\dfrac{9\sqrt{3}}{2}$ **7** 4π **9** π **11** $\dfrac{9\pi}{2}$ **13** 6π **15** $\dfrac{6\pi + 9\sqrt{3}}{2}$

17 $\dfrac{5\pi}{4}$ **19** $12(3\sqrt{3} - \pi)$ **21** $\dfrac{25\pi^3}{24}$ **23** 6π **25** $\dfrac{61}{6}$ **27** $\sqrt{2}(e^{4\pi} - 1)$

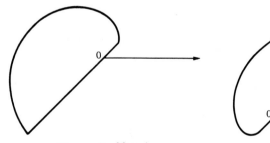

Figure, Problem 1

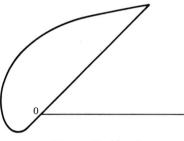

Figure, Problem 3

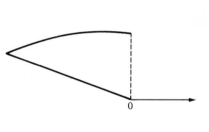

Figure, Problem 5

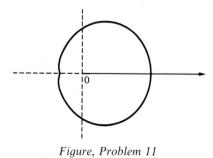

Figure, Problem 11

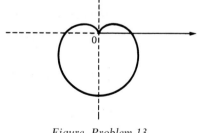

Figure, Problem 13

29 $\dfrac{5\pi}{2}$ **31** Limits of integration should be from 0 to π.   **33** 16π **35** 25π

Problem Set 5 page 552

1 $\bar{r}=5$ **3** $\bar{r}=4\sin\bar{\theta}$ **5** $\bar{r}=3-5\sin\bar{\theta}$ **7** $\bar{r}=3+5\cos\bar{\theta}$ **9** $\bar{r}=$

$-25\cos2\bar{\theta}$ **11** $(-7,-4)$ **13** $\left(\dfrac{3\sqrt{3}-3}{2},-\dfrac{3\sqrt{3}+3}{2}\right)$ **15** $(1-2\sqrt{3},-2-\sqrt{3})$

17 $(4,0)$ **19** $(0,8)$ **21** $\bar{x}^2+2\sqrt{3}\,\bar{x}\bar{y}+3\bar{y}^2-6\sqrt{3}\,\bar{x}+6\bar{y}=0$ **23** $x^2+y^2=1$

25 $\bar{x}^2+\bar{y}^2=1$ **27** $\bar{y}=\bar{x}^2$ **29** $\bar{x}=x\cos\phi+y\sin\phi,\ \bar{y}=-x\sin\phi+y\cos\phi$

Problem Set 6 page 557

1 (a) $\phi=\sin^{-1}\dfrac{\sqrt{5}}{5}\approx26.57°$; (b) $x=\dfrac{\sqrt{5}}{5}(2\bar{x}-\bar{y}),\ y=\dfrac{\sqrt{5}}{5}(\bar{x}+2\bar{y})$; (c) $2\bar{x}^2-3\bar{y}^2=12$

3 (a) $\phi=45°$; (b) $x=\dfrac{\sqrt{2}}{2}(\bar{x}-\bar{y}),\ y=\dfrac{\sqrt{2}}{2}(\bar{x}+\bar{y})$; (c) $2\bar{x}^2=1$ **5** (a) $\phi=$

$\sin^{-1}\dfrac{3}{5}\approx36.87°$; (b) $x=\dfrac{1}{5}(4\bar{x}-3\bar{y}),\ y=\dfrac{1}{5}(3\bar{x}+4\bar{y})$; (c) $25\bar{y}^2=144$

7 (a) $\phi=\sin^{-1}\dfrac{\sqrt{10}}{10}\approx18.43°$; (b) $x=\dfrac{\sqrt{10}}{10}(3\bar{x}-\bar{y}),\ y=\dfrac{\sqrt{10}}{10}(\bar{x}+3\bar{y})$; (c) $\bar{x}^2+3\bar{y}^2=9$

9 (a) $\phi=\sin^{-1}\dfrac{\sqrt{10}}{10}\approx18.43°$; (b) $x=\dfrac{\sqrt{10}}{10}(3\bar{x}-\bar{y}),\ y=\dfrac{\sqrt{10}}{10}(\bar{x}+3\bar{y})$;

(c) $3\bar{x}^2-7\bar{y}^2+9\bar{x}+7\bar{y}-16=0$ **11** (a) Ellipse; (b) parabola; (c) circle;

(d) hyperbola   **13** $\phi=45°,\dfrac{7}{2}\bar{x}^2+\dfrac{1}{2}\bar{y}^2=4$ (ellipse)   **17** (a) $\phi=\sin^{-1}\dfrac{\sqrt{10}}{10}\approx18.43°$,

$3\bar{x}^2+8\bar{y}^2=48$; (b) $\phi=\sin^{-1}\dfrac{3}{5}\approx36.87°,\ 16\bar{x}^2-9\bar{y}^2=144$ **19** $y^2-9x^2=0$

21 $\phi=\sin^{-1}\dfrac{2\sqrt{5}}{5}\approx63.43°,\ 5\bar{x}^2=8$ (two parallel straight lines)   **23** $\phi=45°$,

$-\dfrac{7}{2}\bar{x}^2+\dfrac{11}{2}\bar{y}^2=12$ (hyperbola)

Review Problem Set page 558

1 (a) $\left(-1,\dfrac{4\pi}{3}\right)$; (b) $\left(1,-\dfrac{5\pi}{3}\right)$; (c) $\left(-1,-\dfrac{2\pi}{3}\right)$ **3** (a) $\left(-2,\dfrac{3\pi}{4}\right)$; (b) $\left(2,-\dfrac{\pi}{4}\right)$;

(c) $\left(-2,-\dfrac{5\pi}{4}\right)$ **5** (a) $(-3,0)$; (b) $(3,-\pi)$; (c) $(-3,-2\pi)$ **7** $(0,0)$ **9** $(-17,0)$

11 $\left(\dfrac{-11\sqrt{2}}{2},\dfrac{11\sqrt{2}}{2}\right)$ **13** $(17,0)$ **15** $\left(4,-\dfrac{\pi}{3}\right)$ **17** $\left(17\sqrt{2},\dfrac{3\pi}{4}\right)$

19 $x^2-y^2=1$ **21** $y^2=\dfrac{1-x^4}{x^2}$ **23** $r=\dfrac{1}{2\cos\theta-1}$ **25** $r=4\cot\theta\csc\theta$

27 Three-leaved rose, symmetric about polar axis   **29** Straight line, no symmetry

31 Parabola, symmetric about $\theta=\dfrac{\pi}{2}$ **33** $\dfrac{5}{3}$ **35** $-\dfrac{3}{2}$ **37** Horizontal at

$\left(3-\dfrac{3+\sqrt{137}}{4},\cos^{-1}\dfrac{3+\sqrt{137}}{16}\right),\left(3-\dfrac{3-\sqrt{137}}{4},\cos^{-1}\dfrac{3-\sqrt{137}}{16}\right),$

$\left(3-\dfrac{3+\sqrt{137}}{4},2\pi-\cos^{-1}\dfrac{3+\sqrt{137}}{16}\right),$ and $\left(3-\dfrac{3-\sqrt{137}}{4},2\pi-\cos^{-1}\dfrac{3-\sqrt{137}}{16}\right)$;

vertical at $(-1,0),(7,\pi),\left(\dfrac{3}{2},\cos^{-1}\dfrac{3}{8}\right)$, and $\left(\dfrac{3}{2},2\pi-\cos^{-1}\dfrac{3}{8}\right)$ **39** $\dfrac{1}{3}\tan\theta$

41 Parabola, directrix $x = -17$ **43** Ellipse, $e = \frac{2}{5}$, directrix $y = -5$ **45** Hyp

$e = 2$, directrix $y = 1$ **47** $\left(\frac{\sqrt{17}-1}{2}, \cos^{-1} \frac{3-\sqrt{17}}{4} \right)$, $(\cdot, \pi)$, and

$\left(\frac{\sqrt{17}-1}{2}, 2\pi - \cos^{-1} \frac{3-\sqrt{17}}{4} \right)$ **49** 4π square units   **51** $\frac{5\pi}{4}$ square units

53 $\frac{(1 - e^{-6\pi})\sqrt{10}}{3}$ units   **55** 2 units   **57** $\bar{r}^2 \cos 2\bar{\theta} = 2$ **59** $5\bar{x}^2 - 3\bar{y}^2 = 15$

(hyperbola)   **61** $\left(\frac{\sqrt{3}+2}{2} \right)\bar{x}^2 - \left(\frac{2-\sqrt{3}}{2} \right)\bar{y}^2 = 11$ (hyperbola)   **63** Ellipse

65 Ellipse   **67** Parabola

Chapter 12

Problem Set 1 page 566

1 -3 **3** $\frac{5}{13}$ **5** 32   **7** 0   **9** $-\frac{1}{4}$ **11** $-\frac{1}{7}$ **13** $-\frac{1}{2}$ **15** $\frac{1}{6}$ **17** $-\frac{1}{8}$

19 $\frac{7}{5}$ **21** 0   **23** 3   **25** $\frac{4}{3}$ **27** $\frac{e}{e-1}$ **29** $\sec^{-1} \frac{2}{\sqrt{\pi}}$ **31** $a = 4, b = 1$

35 No, since we did not know $D_x \sin x$.   **37** 0   **39** $\frac{ET}{L}$ **41** $\frac{-At \cos wt}{2w}$

Problem Set 2 page 572

1 1   **3** 0   **5** $+\infty$ **7** 0   **9** 0   **11** 0   **13** $+\infty$ **15** 0   **17** -2

19 $\frac{1}{2}$ **21** $+\infty$ **23** $-\frac{1}{2}$ **25** $-\frac{1}{7}$ **27** 1   **29** e **31** 1   **33** 1   **35** e^2

37 1   **39** e **41** e^6 **43** e^{-2} **45** e **47** 1   **49** 1   **51** 3

Problem Set 3 page 577

1 2   **3** $\frac{\pi}{12}$ **5** Divergent   **7** $\ln 2$ **9** Divergent   **11** $\frac{\pi}{8}$ **13** Divergent

15 $\pi/4$ **17** -1 **19** $\frac{\pi}{2}$ **21** $\frac{1}{3}$ **23** π **27** $n = 2, 2 \ln 3$ **31** $\frac{\pi}{2}$ **33** π

35 $1 - \frac{1}{2e^2}$ **37** $(100)^2 A k^{-2} + 100 B k^{-1}$

Problem Set 4 page 583

1 4   **3** $\frac{27}{2}$ **5** $3 \sin 1$ **7** Divergent   **9** $\frac{\pi}{2}$ **11** $2 - \frac{2}{e^2}$ **13** $\frac{7}{5}(\ln 2)^{5/7}$

15 Diverges   **17** 0   **19** Diverges   **21** $-2\sqrt{1 - \sqrt{3}} + 2$

23 $\frac{5}{4}\left[\sqrt[5]{(e-1)^4} - \sqrt[5]{\left(\frac{1}{e} - 1 \right)^4} \right]$ **27** The fundamental theorem of calculus does not apply

to improper integrals.   **29** π square units   **31** Infinite   **33** $\frac{\pi}{1-2p}$ for $0 < p < \frac{1}{2}$,

otherwise, infinite   **35** $\frac{2\pi}{2-p}$ for $0 < p < 2$, infinite for $p \geq 2$

Problem Set 5 page 592

1 $\dfrac{1}{2} - \dfrac{(x-2)}{4} + \dfrac{(x-2)^2}{8} - \dfrac{(x-2)^3}{16} + \dfrac{(x-2)^4}{32} - \dfrac{(x-2)^5}{64} + \dfrac{(x-2)^6}{128}; C^{-8}(x-2)^7$

3 $\dfrac{1}{10} - \dfrac{(x-100)}{2(10^3)} + \dfrac{3(x-100)^2}{8(10^5)} - \dfrac{5(x-100)^3}{16(10^7)} + \dfrac{35(x-100)^4}{128(10^9)}; -\dfrac{63}{256} C^{-11/2}(x-100)^5$

5 $1 - 2(x-3) + 3(x-3)^2 - 4(x-3)^3 + 5(x-3)^4 - 6(x-3)^5; \dfrac{7(x-3)^6}{(c-2)^8}$

7 $x - \dfrac{x^3}{3!} + \dfrac{x^5}{5!}; \dfrac{-\cos c}{7!} x^7$ **9** $1 + 2\left(x - \dfrac{\pi}{4}\right) + 2\left(x - \dfrac{\pi}{4}\right)^2 + \dfrac{8}{3}\left(x - \dfrac{\pi}{4}\right)^3 + \dfrac{10}{3}\left(x - \dfrac{\pi}{4}\right)^4;$

$\dfrac{16 \sec^2 c \tan^4 c + 88 \sec^4 c \tan^2 c + 16 \sec^6 c}{5!}\left(x - \dfrac{\pi}{4}\right)^5$

11 $e + 2e(x-1) + \dfrac{3e}{2}(x-1)^2 + \dfrac{2e}{3}(x-1)^3; \dfrac{(c+4)e^c}{4!}(x-1)^4$

13 $2 + 2(\ln 2)(x-1) + (\ln 2)^2(x-1)^2 + \dfrac{(\ln 2)^3}{3}(x-1)^3; \dfrac{(\ln 2)^4 2^c (x-1)^4}{4!}$

15 $x + \dfrac{x^3}{3!}; \dfrac{\cosh c}{5!} x^5$ **17** 0.8415   **19** 2.7183   **21** -0.0202 **23** 3.0067

25 $\dfrac{1}{10^{n+1}(n+1)!}$ **29** $\dfrac{1}{2} r^2 s - \dfrac{1}{2} rs + \dfrac{1}{12r} s^3; \dfrac{s^5}{240r^3}$ **31** $I \approx \dfrac{2B}{(n+1)(A+B)}$

33 $T \approx \dfrac{w^2 s^2}{8H} + H$ **35** (d) 0.630

Review Problem Set page 594

1 -1 **3** 4   **5** $+\infty$ **7** $-\infty$ **9** -1 **11** 0   **13** 0   **15** a **17** $-\dfrac{1}{2}$

19 $+\infty$ **21** $e^{-1/2}$ **23** $+\infty$ **25** 1   **27** $e^{-1/2}$ **29** 1   **31** 1   **33** 3

35 $\dfrac{1}{3}$ **39** $\dfrac{\pi}{4}$ **41** $\dfrac{e^2-1}{4}$ **43** $\dfrac{2}{3}$ **45** $\dfrac{2}{5}$ **47** $\dfrac{2e^3}{9}$ **49** Diverges

51 $\dfrac{3}{2}(e-1)^{2/3}$ **53** $9\sqrt[3]{a^2}$ **55** $\dfrac{\pi}{6}$ **57** Infinite   **59** (a) $\dfrac{3\pi}{4a^5}$; (b) $\dfrac{12\pi}{a^4}$

61 $2x - \dfrac{8x^3}{3!}; \dfrac{2 \sin 2c}{3} x^4$ **63** $1 - x + \dfrac{x^2}{2!} - \dfrac{x^3}{3!} + \dfrac{x^4}{4!} - \dfrac{x^5}{5!} + \dfrac{x^6}{6!} - \dfrac{x^7}{7!}; \dfrac{e^{-c}}{8!} x^8$

65 0.99939   **67** 0.40547   **69** 1.01489

Chapter 13

Problem Set 1 page 606

1 2, 5, 10, 17, 26, 37; 10,001   **3** $\dfrac{1}{6}, \dfrac{2}{9}, \dfrac{3}{14}, \dfrac{4}{21}, \dfrac{5}{30}, \dfrac{6}{41}; \dfrac{100}{10,005}$ **5** $\dfrac{n+1}{2}$ **7** $\dfrac{1}{n+1}$

9 0   **11** $\dfrac{1}{7}$ **13** Diverges   **15** π **17** 0   **19** -1 **21** Diverges   **23** 1

25 e **27** Increasing, bounded, convergent   **29** Increasing, bounded below but not above, divergent   **31** Nonmonotonic, bounded, divergent   **33** Nonmonotonic, bounded, divergent   **35** Decreasing, bounded above but not below, divergent   **37** Nonmonotonic, bounded, convergent   **39** $a_n = n, b_n = -n$ **41** (a) Diverges;

(b) diverges; (c) converges; (d) converges; (e) diverges   **45** (a) 1, 3, 2, $\dfrac{5}{2}, \dfrac{9}{4}, \dfrac{19}{8}, \dfrac{37}{16}, \dfrac{75}{32}$;

(c) $\dfrac{7}{3}$ **47** $\left\{\dfrac{1}{2} + \dfrac{(-1)^{n+1}}{2}\right\}$ **49** $\dfrac{A}{1-B}$

Problem Set 2 page 615

1 $\dfrac{1}{3}+\dfrac{1}{15}+\dfrac{1}{35}+\dfrac{1}{63}+\dfrac{1}{99}+\cdots$; $\dfrac{1}{3},\dfrac{2}{5},\dfrac{3}{7},\dfrac{4}{9},\dfrac{5}{11},\ldots$; $s_n=\dfrac{n}{2n+1}$; converges to $\dfrac{1}{2}$

3 $2+6+12+20+30+\cdots$; $2,8,20,40,70,\ldots$; $s_n=\dfrac{n(n+1)(n+2)}{3}$; diverges

5 $\dfrac{3}{4}+\dfrac{5}{36}+\dfrac{7}{144}+\dfrac{9}{400}+\dfrac{11}{900}+\cdots$; $\dfrac{3}{4},\dfrac{8}{9},\dfrac{15}{16},\dfrac{24}{25},\dfrac{35}{36},\ldots$; $s_n=1-\dfrac{1}{(n+1)^2}$; converges to 1

7 $\displaystyle\sum_{k=1}^{\infty}\dfrac{1}{k(k+1)}=1$ **9** $\displaystyle\sum_{k=1}^{\infty}\dfrac{2(3k^2+7k-5)}{(3k+2)(3k+5)}$, diverges   **11** $\displaystyle\sum_{k=1}^{\infty}2(-1)^{k+1}$, diverges

13 $a=1, r=\dfrac{2}{7}$, converges to $\dfrac{7}{5}$ **15** $a=\dfrac{7}{6}, r=\dfrac{7}{6}$, diverges   **17** $a=1, r=-1$,

diverges   **19** $a=\dfrac{1}{16}, r=\dfrac{3}{4}$, converges to $\dfrac{1}{4}$ **21** $a=\dfrac{1}{5}, r=\dfrac{1}{5}$, converges to $\dfrac{1}{4}$ **23** $\dfrac{1}{3}$

25 $\dfrac{467}{99}$ **27** Yes, if it converges   **29** $\dfrac{244}{495}$ **31** 8 meters

Problem Set 3 page 622

9 $\dfrac{5}{6}$ **11** -3 **13** $\dfrac{31}{21}$ **15** No   **17** $-2\ln 2$ **21** $\displaystyle\sum_{j=1}^{\infty}\dfrac{1}{j(j+1)}$

23 $\displaystyle\sum_{j=M}^{\infty}a_{j-M+1}$ **25** $\dfrac{1}{M+1}$ **27** $e-1$

Problem Set 4 page 633

1 Converges   **3** Diverges   **5** Converges   **7** Diverges   **9** Converges
11 Diverges   **13** Converges   **15** Converges   **17** Converges by comparison with

$\displaystyle\sum_{k=1}^{\infty}\dfrac{1}{k^2}$ **19** Converges by comparison with $\displaystyle\sum_{k=1}^{\infty}\dfrac{1}{5^k}$ **21** Converges by comparison with

$\displaystyle\sum_{j=1}^{\infty}\dfrac{1}{7^j}$ **23** Diverges by comparison with $\displaystyle\sum_{k=1}^{\infty}\dfrac{1}{k^{1/3}}$ **25** Diverges by comparison with

$\displaystyle\sum_{j=1}^{\infty}\dfrac{1}{5j}$ **27** Diverges by comparison with $\displaystyle\sum_{q=1}^{\infty}\dfrac{1}{3q^{1/2}}$ **29** Converges by comparison with

$\displaystyle\sum_{k=1}^{\infty}\dfrac{10}{2^k}$ **31** Diverges $\left(\text{use }\displaystyle\sum_{k=1}^{\infty}\dfrac{1}{k^{2/3}}\right)$ **33** Converges $\left(\text{use }\displaystyle\sum_{k=1}^{\infty}\dfrac{1}{k^2}\right)$ **35** Diverges

$\left(\text{use }\displaystyle\sum_{k=1}^{\infty}\dfrac{1}{k}\right)$ **37** Converges $\left(\text{use }\displaystyle\sum_{k=1}^{\infty}\dfrac{1}{7^k}\right)$ **39** Converges $\left(\text{use }\displaystyle\sum_{k=1}^{\infty}\dfrac{1}{k^{3/2}}\right)$ **45** Let

$a_k=\dfrac{1}{k^2}$.

Problem Set 5 page 643

1 Converges   **3** Converges   **5** Converges   **7** Converges   **9** Diverges

11 Converges   **13** Converges   **15** $\dfrac{1249}{3080}$ overestimates with error $<\dfrac{1}{17}$

17 $\dfrac{115}{144}$ underestimates with error $<\dfrac{1}{25}$ **19** $-\dfrac{137}{750}$ underestimates with error $<\dfrac{1}{2500}$

21 0.406   **23** Diverges   **25** Converges absolutely   **27** Converges absolutely
29 Converges absolutely   **31** Converges   **33** Converges absolutely
35 Converges conditionally   **37** Converges conditionally   **39** Converges
absolutely   **41** Diverges   **45** True

Problem Set 6 page 649

1 $a = 0, R = \dfrac{1}{7}, I = \left(-\dfrac{1}{7}, \dfrac{1}{7}\right)$ **3** $a = 0, R = +\infty, I = (-\infty, \infty)$ **5** $a = 0, R = +\infty,$

$I = (-\infty, \infty)$ **7** $a = -2, R = 1, I = [-3, -1]$ **9** $a = -5, R = 1, I = [-6, -4]$

11 $a = 0, R = \dfrac{1}{2}, I = \left[-\dfrac{1}{2}, \dfrac{1}{2}\right]$ **13** $a = 1, R = +\infty, I = (-\infty, \infty)$ **15** $a = 1, R = 3,$

$I = (-2, 4]$ **17** $a = 4, R = 4, I = [0, 8)$ **19** $a = 3, R = 1, I = (2, 4]$ **21** $a = -1,$

$R = \dfrac{1}{\sqrt[3]{5}}, I = \left(-1 - \dfrac{1}{\sqrt[3]{5}}, -1 + \dfrac{1}{\sqrt[3]{5}}\right)$ **23** $a = 1, R = 1, I = (0, 2)$ **25** $a = 8, R = 0,$

$I = \{8\}$ **27** R **29** $R = b; I = (-b, b)$ **31** $R = a; I = (-a, a)$

Problem Set 7 page 658

1 $\displaystyle\sum_{k=0}^{\infty} x^{4k}, |x| < 1$ **3** $\displaystyle\sum_{k=0}^{\infty} (4x)^k, |x| < \dfrac{1}{4}$ **5** $\displaystyle\sum_{k=0}^{\infty} x^{2k+1}, |x| < 1$

7 $\displaystyle\sum_{k=0}^{\infty} \dfrac{(-1)^k x^k}{2^{k+1}}, |x| < 2$ **9** $-\displaystyle\sum_{k=0}^{\infty} \dfrac{x^{k+1}}{k+1}, |x| < 1$ **11** $-\displaystyle\sum_{k=0}^{\infty} \dfrac{x^{k+2}}{(k+1)(k+2)}, |x| < 1$

13 $\displaystyle\sum_{k=0}^{\infty} \dfrac{x^{2k+2}}{(2k+1)(2k+2)}, |x| < 1$ **15** $\displaystyle\sum_{k=0}^{\infty} \dfrac{[(-1)^k 2^{k+1} + 3^{k+1}]}{5(k+1)6^{k+1}} x^{k+1}, |x| < 2$

17 (a) $1 - \dfrac{x^2}{2!} + \dfrac{x^4}{4!} - \dfrac{x^6}{6!} + \cdots$; (b) $-x + \dfrac{x^3}{3!} - \dfrac{x^5}{5!} + \dfrac{x^7}{7!} - \dfrac{x^9}{9!} + \cdots$ **19** $\displaystyle\sum_{k=1}^{\infty} k^3 x^{k-1}, R = 1$

21 $\displaystyle\sum_{k=0}^{\infty} \dfrac{x^k}{k!}, R = +\infty$ **23** $\displaystyle\sum_{k=1}^{\infty} k \cdot 2^{(k+2)/2}(x+1)^{2k-1}, R = 2^{-1/4}$

25 $\displaystyle\sum_{k=0}^{\infty} \dfrac{(-1)^k x^{2k+1}}{(2k+1)!}, R = +\infty$ **27** $\displaystyle\sum_{k=0}^{\infty} \dfrac{x^{2k+2}}{(2k+2)!}, R = +\infty$ **29** (a) 1; (b) 0;

(c) -1; (d) 0

Problem Set 8 page 665

1 $\dfrac{1}{2} + \dfrac{\sqrt{3}}{2}\left(x - \dfrac{\pi}{6}\right) - \dfrac{1}{2(2!)}\left(x - \dfrac{\pi}{6}\right)^2 - \dfrac{\sqrt{3}}{2(3!)}\left(x - \dfrac{\pi}{6}\right)^3 + \dfrac{1}{2(4!)}\left(x - \dfrac{\pi}{6}\right)^4 + \cdots$

3 $\displaystyle\sum_{k=0}^{\infty} (-1)^k \dfrac{(x-2)^k}{2^{k+1}}$ **5** $\displaystyle\sum_{k=0}^{\infty} \dfrac{e^4(x-4)^k}{k!}$

7 $1 + \dfrac{1}{2}(x - 2) - \dfrac{1}{8}(x - 2)^2 + \displaystyle\sum_{k=3}^{\infty} \dfrac{(-1)^{k+1} 1 \cdot 3 \cdots (2k-3)}{k! \, 2^k}(x - 2)^k$

9 $\displaystyle\sum_{k=0}^{\infty} \dfrac{x^{2k+1}}{(2k+1)!}$ **11** $\displaystyle\sum_{k=0}^{\infty} (-1)^k \dfrac{x^{2k}}{k!}$ **13** $\displaystyle\sum_{k=0}^{\infty} \dfrac{(-1)^k x^{2k}}{(2k+1)!}$ **15** $\displaystyle\sum_{k=0}^{\infty} \dfrac{(-1)^k x^{2k}}{2k+1}$

17 $\displaystyle\sum_{k=1}^{\infty} \dfrac{(-1)^{k+1} 2^{2k-1} x^{2k}}{(2k)!}$ **19** 0.9802   **21** $f^{(n)}(0) = 0$ if n is even; $f^{(n)}(0) =$

$(-1)^{(n+3)/2}(n-1)!$ if n is odd   **23** $\displaystyle\sum_{k=0}^{\infty} \dfrac{(-1)^k x^k}{k!}$ for all x **25** $\displaystyle\sum_{k=0}^{\infty} \dfrac{x^k}{4^{k+1}}, |x| < 4$

27 $\displaystyle\sum_{k=0}^{\infty} \dfrac{(\ln 2)^k x^k}{k!}$ for all x **29** 0   **31** $\dfrac{16!}{8!}$ **33** $(-2)19!$

Problem Set 9 page 671

1 $1 + \displaystyle\sum_{k=1}^{\infty} \dfrac{(-1)^k [1 \cdot 4 \cdot 7 \cdots (3k-2)]}{3^k k!} x^k$ for $|x| < 1$

3 $1 + \displaystyle\sum_{k=1}^{\infty} \frac{1 \cdot 4 \cdot 7 \cdots (3k - 2)}{3^k k!} x^{2k}$ for $|x| < 1$ **5** $1 + \displaystyle\sum_{k=1}^{\infty} \frac{(-1)^k 1 \cdot 3 \cdot 5 \cdots (2k - 1)}{2^k k!} x^{3k}$ for

$|x| < 1$ **7** $\displaystyle\sum_{k=1}^{\infty} (-1)^k 2^{k-1} k x^k$ for $|x| < \frac{1}{2}$ **9** 1.0148875, $|\text{error}| \leq 1.7 \times 10^{-6}$

11 2.030518, $|\text{error}| \leq 2.7 \times 10^{-5}$ **13** 10.050 **15** 0.690 **19** They are the same.
23 The series becomes a finite sum, correct for all values of x.

Review Problem Set page 672

1 Converges; limit is $\dfrac{1}{3}$ **3** Converges; limit is $\sqrt{\dfrac{1}{3}}$ **5** Converges; limit is 0

7 Converges; limit is 0   **9** Diverges   **11** Converges; limit is 1   **13** Increasing
15 Nonmonotone   **17** No; it first increases and then decreases.   **19** Bounded;

nonmonotonic; convergent; limit is $\dfrac{4}{5}$ **23** 1   **25** $\sin 1$ **27** $\displaystyle\sum_{k=1}^{\infty} \frac{15}{(2k + 3)(2k + 5)}$;

converges; sum is $\dfrac{3}{2}$ **29** $\dfrac{1}{3}$ **31** $\dfrac{23}{6}$ **33** $\displaystyle\lim_{n \to +\infty} a_n = 0$ does not guarantee that $\displaystyle\sum_{k=1}^{\infty} a_k$ is

convergent.   **35** Converges   **37** Converges   **39** Converges   **41** Diverges
43 Conditionally convergent   **45** Absolutely convergent   **47** Diverges
49 Absolutely convergent   **51** Absolutely convergent   **53** (a) 0.4058; (b) 0.0332
55 $a = 1$; $R = \sqrt{5}$; $I = [1 - \sqrt{5}, 1 + \sqrt{5}]$ **57** $a = -2$; $R = 1$; $I = (-3, -1)$
59 $a = 10$; $R = 0$; $I = \{10\}$ **61** $a = -\pi$; $R = +\infty$; $I = (-\infty, \infty)$ **63** $a = 3$;

$R = \dfrac{1}{2}$; $I = \left[\dfrac{5}{2}, \dfrac{7}{2}\right]$ **65** $\displaystyle\sum_{k=1}^{\infty} (-1)^{k+1} \frac{(x - 1)^k}{k}$ for $0 < x < 2$

67 $\dfrac{1}{e} + \dfrac{1}{e}(x + 1) + \dfrac{1}{2!\,e}(x + 1)^2 + \dfrac{1}{3!\,e}(x + 1)^3$

69 $1 + \dfrac{1}{2}(x - 1) - \dfrac{1}{2!\,2^2}(x - 1)^2 + \dfrac{3}{3!\,2^3}(x - 1)^3$ **71** $1 + 0 - \dfrac{4}{2!}\left(x - \dfrac{\pi}{4}\right)^2 + 0$

77 $1 + \displaystyle\sum_{k=1}^{\infty} \frac{(-1)^k 1 \cdot 3 \cdot 5 \cdot 7 \cdots (2k - 1)}{2^k k!} x^{2k}$ for $|x| < 1$

79 $1 - \dfrac{4x}{3} - \displaystyle\sum_{k=1}^{\infty} \frac{2^{k+2} 1 \cdot 4 \cdot 7 \cdots (3k - 2)}{3^{k+1}(k + 1)!} x^{k+1}$ for $|x| < \dfrac{1}{2}$

81 $x + \dfrac{x^4}{12} + \displaystyle\sum_{k=2}^{\infty} \frac{(-1)^{k+1} 2 \cdot 5 \cdot 8 \cdots (3k - 4)}{(3k + 1) 3^k k!} x^{3k+1}$ for $|x| < 1$ **83** 1.974375

85 (a) $1 + x - \dfrac{x^2}{2!} - \dfrac{x^3}{3!} + \dfrac{x^4}{4!} + \dfrac{x^5}{5!} - \dfrac{x^6}{6!} - \dfrac{x^7}{7!} + \cdots$ for all x;

(b) $1 - 2x^2 + \dfrac{4^2 x^4}{4!} - \dfrac{4^3 x^6}{6!} + \dfrac{4^4 x^8}{8!} - \dfrac{4^5 x^{10}}{10!} + \cdots$ for all x;

(c) $x^3 - \dfrac{x^9}{3} + \dfrac{x^{15}}{5} - \dfrac{x^{21}}{7} + \dfrac{x^{27}}{9} - \dfrac{x^{33}}{11} + \cdots$ for $|x| < 1$;

(d) $1 + (\ln 10)x + \dfrac{(\ln 10)^2}{2!} x^2 + \dfrac{(\ln 10)^3}{3!} x^3 + \cdots$ for all x

87 $\sin x$ **89** $x - 1 + \dfrac{\sin x}{x}$ **91** 2^x

Chapter 14

Problem Set 1 page 679
15 $\bar{X} = \bar{A} + \bar{B} + \bar{C} - \bar{D}$ **17** $\bar{X} = -(\bar{A} + \bar{B})$ **21** $PRSQ$; $PRQS$; $PQRS$; $RSPQ$;
$RPSQ$; $RPQS$; $QSRP$; $SQRP$; $SRQP$; $QPSR$; $QSPR$; $RPQS$ **23** (a) $\bar{A} - \bar{B}$ is a vector, 5
is a scalar; (b) we cannot add a vector $\bar{A}$ and a scalar 3; (c) $\bar{A} + \bar{B}$ is a vector and 0 is a
scalar.

Problem Set 2 page 685

7 $\dfrac{5}{8}\bar{A}+\dfrac{3}{8}\bar{B}$ **9** (a) $i+6j$; (b) $7i-2j$ **11** (a) $\langle-25,34\rangle$; (b) $\langle-7,9\rangle$

13 (a) $3i+j$; (b) $i+13j$; (c) $\dfrac{7}{2}(i-j)$; (d) $(2s+t-5u)i+(7s-6t+10u)j$

15 $s=-\dfrac{20}{19}$ and $t=-\dfrac{55}{19}$ **17** $x=\dfrac{8}{3}$ and $y=1$ **19** $x=-\dfrac{2}{5}$ and $y=\dfrac{3}{5}$

21 (a) $3i+6j$; (b) $-i+3j$; (c) $-4i-3j$; (d) $4i+3j$; (e) $-10i-7j$; (f) $10i-5j$;
(g) $6i-8j$; (h) $-18i-72j$ **25** R moves along the straight line through P and Q.

Problem Set 3 page 693

1 15   **3** 0   **5** $\dfrac{\pi}{2}$ **7** 5   **9** -12 **11** $5\sqrt{2}$ **13** $\dfrac{3}{2}$ **15** 0   **17** (a) 0;

(b) $\sqrt{2}$; (c) $\sqrt{2}$; (d) 2; (e) 0; (f) $\dfrac{i+j}{\sqrt{2}},\dfrac{i-j}{\sqrt{2}},j$; (g) 0   **19** (a) 2; (b) $\sqrt{17}$; (c) $\sqrt{5}$;

(d) $3\sqrt{2}$; (e) $\dfrac{2}{\sqrt{85}}$; (f) $\dfrac{4i+j}{\sqrt{17}},\dfrac{i-2j}{\sqrt{5}},\dfrac{i+j}{\sqrt{2}}$; (g) $\dfrac{2}{\sqrt{5}}$ **21** (a) -12; (b) $2\sqrt{5}$; (c) 3;

(d) $\sqrt{53}$; (e) $\dfrac{-2}{\sqrt{5}}$; (f) $\dfrac{i+2j}{\sqrt{5}},-j,\dfrac{2i+7j}{\sqrt{53}}$; (g) -4 **23** -3 **25** -30

27 $\langle-27,-9\rangle$ **29** $\dfrac{21}{5}$ **31** (a) $\bar{A}$ and $\bar{B}$ make an acute angle; (b) $\bar{A}$ and $\bar{B}$ are
perpendicular; (c) $\bar{A}$ and $\bar{B}$ make an obtuse angle.   **35** (a) $\sqrt{|\bar{A}|^2+|\bar{B}|^2}$;
(b) $\sqrt{4|\bar{A}|^2+9|\bar{B}|^2}$; (c) $\sqrt{|\bar{A}|^2+|\bar{B}|^2}$ **37** $\overrightarrow{AB}\cdot\overrightarrow{BC}=0$ **41** $45\sqrt{3}$ ft-lb
43 $\bar{F}\cdot\overrightarrow{PR}$; work done is the same either way.

Problem Set 4 page 698

1 $5i-3j$ **3** $0i+0j$ **5** $\sqrt{2}i-\sqrt{2}j$ **7** $5i+3j$

9 $\left(\dfrac{5-\sqrt{2}}{2}\right)i+\left(\dfrac{-5\sqrt{3}-\sqrt{2}}{2}\right)j$ **11** Circle, $x^2+y^2=9$ **13** Circle,

$(x-2)^2+(y-3)^2=16$ **15** Ellipse, foci at $(1,0)$ and $(-1,0)$, semimajor axis 3
17 The x axis   **19** Straight line, $2x+3y=2$ **21** Straight line, $2x+3y=3$
23 $|\bar{R}-(-3i+3j)|=9$ **25** $(2i-7j)\cdot(\bar{R}+i+5j)=0$ **27** $(12i-6j)\cdot\bar{R}=7$

29 $2i-17j$ **31** $\dfrac{i}{2}+\dfrac{j}{3}$ **33** $\dfrac{3}{\sqrt{10}}$ **35** $\sqrt{2}$ **37** $\dfrac{3}{\sqrt{5}}$ **39** $\dfrac{|\bar{D}\cdot(\bar{R}_1-\bar{D})|}{|\bar{D}|}$

Problem Set 5 page 705

1 (a) $\bar{M}=2i+2j$; (b) $\bar{R}=(1+2t)i+(2+2t)j$; (c) $x=1+2t,\ y=2+2t$;

(d) $x-y+1=0$; (e) $\bar{N}=i-j$; (f) $(i-j)\cdot\bar{R}=-1$ **3** (a) $\bar{M}=\left(\dfrac{13}{6}\right)i-\left(\dfrac{17}{6}\right)j$;

(b) $\bar{R}=\left(-\dfrac{3}{2}+\dfrac{13}{6}t\right)i+\left(\dfrac{5}{2}-\dfrac{17}{6}t\right)j$; (c) $x=-\dfrac{3}{2}+\dfrac{13}{6}t,\ y=\dfrac{5}{2}-\dfrac{17}{6}t$; (d) $17x+13y=7$;

(e) $\bar{N}=17i+13j$; (f) $(17i+13j)\cdot\bar{R}=7$ **5** (a) $\bar{M}=(-\sqrt{2}-\pi)i+(\sqrt{3}-e)j$;

(b) $\bar{R}=[\pi+t(-\sqrt{2}-\pi)]i+[e+t(\sqrt{3}-e)]j$; (c) $x=\pi-(\sqrt{2}+\pi)t,\ y=e+(\sqrt{3}-e)t$;

(d) $(\sqrt{3}-e)x+(\sqrt{2}+\pi)y=\sqrt{3}\pi+\sqrt{2}e$; (e) $\bar{N}=(\sqrt{3}-e)i+(\sqrt{2}+\pi)j$;

(f) $[(\sqrt{3}-e)i+(\sqrt{2}+\pi)j]\cdot\bar{R}=\sqrt{3}\pi+\sqrt{2}e$ **7** $\bar{R}=(7-t)i+(3t-2)j$

9 $\bar{R}=(3+4\cos t)i+(4+4\sin t)j$ **11** $\bar{R}=5(t-\sin t)i+5(1-\cos t)j$
13 (a) $x=2\cos t,\ y=2\sin t$; (b) $x^2+y^2=4$ **15** (a) $x=5\cos 2t,\ y=-5\sin 2t$;

(b) $x^2 + y^2 = 25$ **17** (a) $x = 4t, y = 3t + 5$; (b) $y = \dfrac{3}{4}x + 5$ **19** (a) $x = (t - 2)^{-1}$,

$y = 2t + 1$; (b) $y = \dfrac{2}{x} + 5, x \le -\dfrac{1}{2}$ **21** (b) $x + 3y = 7$; (c) $\dfrac{dy}{dx} = -\dfrac{1}{3}, \dfrac{d^2y}{dx^2} = 0$

23 (b) $y^2 = 25(x + 2)$; (c) $\dfrac{dy}{dx} = \dfrac{5}{2t}, \dfrac{d^2y}{dx^2} = \dfrac{-5}{4t^3}$ **25** (b) $y = \dfrac{x^2 - 4x + 6}{2 - x}$;

(c) $\dfrac{dy}{dx} = 2t^2 - 1, \dfrac{d^2y}{dx^2} = 4t^3$ **27** (b) $y = 3x + 1 \pm (1 + x)^{3/2}$; (c) $\dfrac{dy}{dx} = \dfrac{3}{2}(t + 1)$,

$\dfrac{d^2y}{dx^2} = \dfrac{3}{4(t - 1)}$ **31** $\bar{R} = t\bar{i} + f(t)\bar{j}$

Problem Set 6   page 712

1 (a) All real numbers except -1 and 1; (b) $8\bar{i} + \dfrac{5}{3}\bar{j}$; (c) $8\bar{i} + \dfrac{5}{3}\bar{j}$; (d) all real numbers
except -1 and 1 **3** (a) All real numbers except 3; (b) not defined; (c) $\bar{i} + 13\bar{j}$;
(d) all real numbers except 3 **5** (a) All real numbers except 0; (b) $\dfrac{6}{\pi}\bar{i} + \bar{j}$; (c) $\dfrac{6}{\pi}\bar{i} + \bar{j}$;
(d) all real numbers except 3 **7** $6t\bar{i} + 36t^3\bar{j}$; $6\bar{i} + 108t^2\bar{j}$ **9** $3e^{3t}\bar{i} + \dfrac{1}{t}\bar{j}$; $9e^{3t}\bar{i} - \dfrac{1}{t^2}\bar{j}$

11 $(-5 \sin t)\bar{i} + (3 \cos t)\bar{j}$; $(-5 \cos t)\bar{i} - (3 \sin t)\bar{j}$ **13** $(-2t \sin t^2)\bar{i} + (2t \cos t^2)\bar{j}$;

$(-2 \sin t^2 - 4t^2 \cos t^2)\bar{i} + (2 \cos t^2 - 4t^2 \sin t^2)\bar{j}$ **15** $-\dfrac{1}{t^2}\bar{i} - \dfrac{2}{t^3}\bar{j}; \dfrac{2}{t^3}\bar{i} + \dfrac{6}{t^4}\bar{j}$

17 (a) $e^t\bar{i} + e^t\bar{j}$; (b) $e^t\bar{i} + e^t\bar{j}$; (c) $2e^{2t}$; (d) $\dfrac{2e^{2t} + 10e^t}{\sqrt{(e^t + 3)^2 + (e^t + 7)^2}}$

19 (a) $(12 \cos 3t)\bar{i} - (12 \cos 3t)\bar{j}$; (b) $(-36 \sin 3t)\bar{i} + (36 \sin 3t)\bar{j}$; (c) $-432 \sin 6t$;

(d) $(12\sqrt{2} \cos 3t)\dfrac{\sin 3t}{|\sin 3t|}$ **21** (a) $2e^{2t}(\cos 2t - \sin 2t) + 2e^{-4t}(\cos 2t - 2 \sin 2t)$;

(b) $-5e^{-5t} - 11e^{-11t}$ **23** (a) $\dfrac{1}{t^2} - \dfrac{3}{t^2(t - 1)^2} - \dfrac{\ln t}{t^2} - \dfrac{6}{t^3(t - 1)}$;

(b) $\left(\dfrac{\cos 7t}{t} - 7 \sin 7t \ln t\right)\bar{i} - \left[\dfrac{7 \sin 7t}{t - 1} + \dfrac{\cos 7t}{(t - 1)^2}\right]\bar{j}$ **25** (a) $(3s^2 \cos t)\bar{i} - (3s^2 \sin t)\bar{j}$;

(b) $(6s + 9s^4)e^t\bar{i} + (9s^4 - 6s)e^{-t}\bar{j}$; (c) 0; (d) $2e^t \cos t + 2e^{-t} \sin t$ **39** (a) $\lim_{t \to c^+} \bar{F}(t) = \bar{A}$ if
and only if $\lim_{t \to c^+} |\bar{F}(t) - \bar{A}| = 0$; (b) $\lim_{t \to c^-} \bar{F}(t) = \bar{A}$ if and only if $\lim_{t \to c^-} |\bar{F}(t) - \bar{A}| = 0$

Problem Set 7   page 717

1 (a) $6t\bar{i} + 2\bar{j}$; (b) $6\bar{i}$; (c) $2\sqrt{9t^2 + 1}$ **3** (a) $2t\bar{i} + \dfrac{1}{t}\bar{j}$; (b) $2\bar{i} - \dfrac{1}{t^2}\bar{j}$; (c) $\dfrac{1}{|t|}\sqrt{4t^4 + 1}$

5 (a) $2(1 - \cos t)\bar{i} + (2 \sin t)\bar{j}$; (b) $(2 \sin t)\bar{i} + (2 \cos t)\bar{j}$; (c) $2\sqrt{2(1 - \cos t)}$
7 (a) $(-\sin t + \cos t)\bar{i} + (-\sin t - \cos t)\bar{j}$; (b) $(-\cos t - \sin t)\bar{i} + (-\cos t + \sin t)\bar{j}$;
(c) $\sqrt{2}$ **9** (a) $e^t(\cos t - \sin t)\bar{i} + e^t(\cos t + \sin t)\bar{j}$; (b) $(-2e^t \sin t)\bar{i} + (2e^t \cos t)\bar{j}$; (c) $\sqrt{2} e^t$

11 (a) $27\bar{i} + 5\bar{j}$; (b) $14\bar{i}$; (c) $\sqrt{754}$ **13** (a) $\dfrac{3\pi\sqrt{2}}{2}\bar{i} - 2\pi\sqrt{2}\bar{j}$; (b) $\dfrac{3\pi^2\sqrt{2}}{2}\bar{i} + 2\pi^2\sqrt{2}\bar{j}$;

(c) $\dfrac{5\pi\sqrt{2}}{2}$ **15** (a) $\sqrt{3}\bar{i} - \dfrac{\sqrt{3}}{3}\bar{j}$; (b) $-4\bar{i} - \dfrac{4}{3}\bar{j}$; (c) $\sqrt{\dfrac{10}{3}}$ **17** (a) $12\bar{i} - 7\bar{j}$; (b) $\sqrt{193}$;

(c) $\dfrac{12\bar{i} - 7\bar{j}}{\sqrt{193}}$ **19** (a) $2\bar{i} + 6t\bar{j}$; (b) $2\sqrt{1 + 9t^2}$; (c) $\dfrac{\bar{i} + 3t\bar{j}}{\sqrt{1 + 9t^2}}$

21 (a) $(-2 \sin 2t)\vec{i} + (2 \cos 2t)\vec{j}$; (b) 2; (c) $(-\sin 2t)\vec{i} + (\cos 2t)\vec{j}$

23 (a) $\dfrac{-\sin t}{(1 + \cos t)^2}\vec{i} + \dfrac{1}{1 + \cos t}\vec{j}$; (b) $\sqrt{2}(1 + \cos t)^{-3/2}$;

(c) $\dfrac{-\sin t}{\sqrt{2}(1 + \cos t)^{1/2}}\vec{i} + \dfrac{1}{\sqrt{2}(1 + \cos t)^{-1/2}}\vec{j}$ **25** $2\sqrt{74}$ **27** $\dfrac{1}{4}(e^2 + 1)$ **29** 6π

31 $\sqrt{2} - \sqrt{2}e^{-2\pi}$ **33** $\dfrac{\pi^2}{32}$ **35** $\ln(\sqrt{2} + 1)$ **37** $\displaystyle\int_{\theta_1}^{\theta_2} \sqrt{\left(\dfrac{dr}{d\theta}\right)^2 + r^2}\, d\theta$

39 $\dfrac{\vec{i} + f'(x)\vec{j}}{\sqrt{1 + [f'(x)]^2}}$

Problem Set 8 page 724

1 (a) $\dfrac{7\vec{i} - 3\vec{j}}{\sqrt{58}}$; (b) $\dfrac{3\vec{i} + 7\vec{j}}{\sqrt{58}}$; (c) 0; (d) not defined   **3** (a) $(-\sin t)\vec{i} + (\cos t)\vec{j}$;

(b) $(-\cos t)\vec{i} - (\sin t)\vec{j}$; (c) $\dfrac{1}{3}$; (d) $(-\cos t)\vec{i} - (\sin t)\vec{j}$ **5** (a) $\dfrac{(-3 \sin \pi t)\vec{i} + (5 \cos \pi t)\vec{j}}{\sqrt{9 \sin^2 \pi t + 25 \cos^2 \pi t}}$;

(b) $\dfrac{(-5 \cos \pi t)\vec{i} - (3 \sin \pi t)\vec{j}}{\sqrt{9 \sin^2 \pi t + 25 \cos^2 \pi t}}$; (c) $\dfrac{15}{(9 \sin^2 \pi t + 25 \cos^2 \pi t)^{3/2}}$; (d) $\overline{N} = \overline{N}_l$

7 (a) $\dfrac{\vec{i} + e^t\vec{j}}{\sqrt{1 + e^{2t}}}$; (b) $\dfrac{-e^t\vec{i} + \vec{j}}{\sqrt{1 + e^{2t}}}$; (c) $\dfrac{e^t}{(1 + e^{2t})^{3/2}}$; (d) $\overline{N} = \overline{N}_l$ **9** (a) $\dfrac{t}{|t|}\left(\dfrac{3\vec{i} + 2\vec{j}}{\sqrt{13}}\right)$;

(b) $\dfrac{t}{|t|}\left(\dfrac{-2\vec{i} + 3\vec{j}}{\sqrt{13}}\right)$; (c) $\kappa = 0$; (d) not defined   **11** (a) $\dfrac{-\vec{i} + 2t^3\vec{j}}{\sqrt{1 + 4t^6}}$; (b) $\dfrac{-2t^3\vec{i} - \vec{j}}{\sqrt{1 + 4t^6}}$;

(c) $\dfrac{-6t^4}{(1 + 4t^6)^{3/2}}$; (d) $\overline{N} = -\overline{N}_l$ **13** (a) $\dfrac{(-2 \sin 2\theta)\vec{i} + (\cos \theta)\vec{j}}{\sqrt{4 \sin^2 2\theta + \cos^2 \theta}}$;

(b) $\dfrac{(-\cos \theta)\vec{i} - (2 \sin 2\theta)\vec{j}}{\sqrt{4 \sin^2 2\theta + \cos^2 \theta}}$; (c) $4(16 \sin^2 \theta + 1)^{-3/2}$; (d) $\overline{N} = \overline{N}_l$ **15** (a) $\dfrac{x\vec{i} + \vec{j}}{\sqrt{x^2 + 1}}$;

(b) $\dfrac{-\vec{i} + x\vec{j}}{\sqrt{x^2 + 1}}$; (c) $-x(x^2 + 1)^{-3/2}$; (d) $\overline{N} = -\overline{N}_l$ **17** (a) $\dfrac{\vec{i} - \sqrt{3}\vec{j}}{2}$; (b) $\dfrac{\sqrt{3}\vec{i} + \vec{j}}{2}$; (c) $\dfrac{1}{3}$;

(d) $\overline{N} = \overline{N}_l$ **19** (a) $\dfrac{\vec{i} + 4\vec{j}}{\sqrt{17}}$; (b) $\dfrac{-4\vec{i} + \vec{j}}{\sqrt{17}}$; (c) $\dfrac{52}{17^{3/2}}$; (d) $\overline{N} = \overline{N}_l$ **21** (a) $\dfrac{-32\vec{i} - 9\vec{j}}{\sqrt{1105}}$;

(b) $\dfrac{9\vec{i} - 32\vec{j}}{\sqrt{1105}}$; (c) $\dfrac{6912}{1105^{3/2}}$; (d) $\overline{N} = \overline{N}_l$

23 (a) $\overline{T} = \dfrac{\left(\dfrac{dr}{d\theta}\cos\theta - r\sin\theta\right)\vec{i} + \left(\dfrac{dr}{d\theta}\sin\theta + r\cos\theta\right)\vec{j}}{\sqrt{\left(\dfrac{dr}{d\theta}\right)^2 + r^2}}$;

(b) $N_l = \dfrac{-\left(\dfrac{dr}{d\theta}\sin\theta + r\sin\theta\right)\vec{i} + \left(\dfrac{dr}{d\theta}\cos\theta - r\cos\theta\right)\vec{j}}{\sqrt{\left(\dfrac{dr}{d\theta}\right)^2 + r^2}}$; (c) $\kappa = \dfrac{r^2 + 2\left(\dfrac{dr}{d\theta}\right)^2 - r\dfrac{d^2r}{d\theta^2}}{\sqrt{\left(\dfrac{dr}{d\theta}\right)^2 + r^2}}$;

(d) $\overline{N}_l = \begin{cases} \overline{N} & \text{if } \kappa > 0 \\ -\overline{N} & \text{if } \kappa < 0 \end{cases}$ **25** (a) $\dfrac{(-2 \sin 2\theta - \sin \theta)\vec{i} + (2 \cos 2\theta + \cos \theta)\vec{j}}{\sqrt{5 + 4 \cos \theta}}$;

(b) $-\dfrac{(2 \cos 2\theta + \cos \theta)\vec{i} + (2 \sin 2\theta + \sin \theta)\vec{j}}{\sqrt{5 + 4 \cos \theta}}$; (c) $\dfrac{9 + 6 \cos \theta}{(5 + 4 \cos \theta)^{3/2}}$; (d) $\overline{N} = \overline{N}_l$

27 (a) $\dfrac{(\cos\theta - \theta\sin\theta)\vec{i} + (\sin\theta + \theta\cos\theta)\vec{j}}{\sqrt{1+\theta^2}}$; (b) $\dfrac{(-\sin\theta - \theta\cos\theta)\vec{i} + (\cos\theta - \theta\sin\theta)\vec{j}}{\sqrt{1+\theta^2}}$;

(c) $\dfrac{\theta^2 + 2}{(\theta^2+1)^{3/2}}$; (d) $\vec{N} = \vec{N}_l$ **31** If $\kappa > 0$ (respectively, $\kappa < 0$), curve is concave upward (respectively, downward).

Problem Set 9   page 731

1 (a) $\vec{V} = 5\vec{i}$; (b) $\vec{A} = \dfrac{50}{9}\vec{j}$ **3** $v = 2\pi\sqrt{\sin^2 2\pi t + 9\cos^2 2\pi t}$ **5** (a) $\vec{R} =$

$24\sqrt{3}\,t\vec{i} + (24t - 16t^2)\vec{j}$; (b) $\vec{V} = 24\sqrt{3}\vec{i} + (24 - 32t)\vec{j}$; (c) $\dfrac{3}{2}$ sec; (d) $24\sqrt{3}\vec{i} - 24\vec{j}$;

(e) $36\sqrt{3}$ ft   **7** $8\pi^2$ lb   **9** $\dfrac{55}{36}$ miles   **11** $\dfrac{7}{\pi\sqrt{12}}$ rev/sec

Review Problem Set   page 732

5 Yes   **7** Yes   **11** $\dfrac{\vec{i}}{2} + 5\vec{j}$ **13** (a) $10\vec{i} - 5\vec{j}$; (b) $-8\vec{i} + 12\vec{j}$; (c) $4\vec{i} - 4\vec{j}$; (d) $2\vec{j}$;

(e) $10\vec{i} - 11\vec{j}$; (f) 7; (g) -97; (h) $\sqrt{5}$; (i) 2; (j) $2\sqrt{5} + 3\sqrt{13}$ **15** $\dfrac{\pi}{4}, \dfrac{\pi}{4}, \dfrac{\pi}{2}$

17 $-y\vec{i} + x\vec{j}$ **21** $\dfrac{6}{5}$ **25** (a) $2x + 5y = 26$; (b) $(2\vec{i} + 5\vec{j})\cdot(\vec{R} - 3\vec{i} - 4\vec{j}) = 0$;

(c) $\vec{R} = (3 + 5t)\vec{i} + (4 - 2t)\vec{j}$; (d) $x = 3 + 5t,\ y = 4 - 2t$ **29** $\vec{R} = (\cosh t)\vec{i} + (\sinh t)\vec{j}$

31 (a) $-(t-1)^{-2}\vec{i} - 6t(t^2 - 2)^{-2}\vec{j}$; (b) $2(t-1)^{-3}\vec{i} + [-6(t^2 - 2)^{-2} + 24t^2(t^2 - 2)^{-3}]\vec{j}$

33 (a) $5e^{5t}\vec{i} - 3e^{-3t}\vec{j}$; (b) $25e^{5t}\vec{i} + 9e^{-3t}\vec{j}$

35 (a) $\dfrac{2t - \sin t + \cos t - t\cos t - t\sin t}{\sqrt{2t^2 - 2t\sin t + 2t\cos t + 1}}$; (b) $\sin t - \cos t$ **37** (a) $\dfrac{4(e^{8t} - e^{-8t})}{\sqrt{e^{8t} + e^{-8t}}}$;

(b) $64(e^{8t} - e^{-8t})$ **39** $(e^t - \sin 2t)\vec{i} + (2t + \sin 2t)\vec{j}$ **41** $\dfrac{e^{2t} + 2t^3 - 2t}{\sqrt{e^{2t} + (t^2 - 1)^2}}$

43 $e^{2t} + 4t$ **45** 6   **47** $\dfrac{87}{8}$ **49** (a) $\vec{j}$; (b) 3; (c) $-\vec{i}$ **51** (a) $\dfrac{2\vec{i} + 3\vec{j}}{\sqrt{13}}$; (b) $\dfrac{6}{13^{3/2}}$;

(c) $\dfrac{-3\vec{i} + 2\vec{j}}{\sqrt{13}}$ **53** (a) $3t^2\vec{i} + 8t^3\vec{j}$; (b) $6t\vec{i} + 24t^2\vec{j}$; (c) $t^2\sqrt{9 + 64t^2}$; (d) $\dfrac{t(18 + 192t^3)}{\sqrt{9 + 64t^2}}$;

(e) $\dfrac{(54t + 576t^3)\vec{i} + (144t^2 + 1536t^4)\vec{j}}{9 + 64t^2}$; (f) $\dfrac{-192t^3\vec{i} + 72t^2\vec{j}}{9 + 64t^2}$

55 (a) $(-21\sin 7t)\vec{i} + (21\cos 7t)\vec{j}$; (b) $(-147\cos 7t)\vec{i} + (-147\sin 7t)\vec{j}$; (c) 21; (d) 0;

(e) 0; (f) $(-147\cos 7t)\vec{i} + (-147\sin 7t)\vec{j}$ **57** (a) $t\vec{i} + t^2\vec{j}$; (b) $\vec{i} + 2t\vec{j}$; (c) $\sqrt{t^2 + t^4}$;

(d) $\dfrac{t + 2t^3}{\sqrt{t^2 + t^4}}$; (e) $\dfrac{(1 + 2t^2)\vec{i} + (t + 2t^3)\vec{j}}{1 + t^2}$; (f) $\dfrac{-t^2\vec{i} + t\vec{j}}{1 + t^2}$ **59** $\dfrac{1807}{432}$

Chapter 15

Problem Set 1   page 739

3 (a) $(-1, 2, 4)$; (b) $(-1, -2, 4)$; (c) $(-1, -2, -4)$; (d) $(1, 4, 4)$ **5** (a) $(-2, -1, -3)$;
(b) $(-2, 1, -3)$; (c) $(-2, 1, 3)$; (d) $(2, 1, -3)$ **7** A straight line parallel to the y axis and containing $(-1, 0, 2)$ **9** All points on or above the horizontal plane $z = 1$

23 $(3, -1, 7)$ **25** $z = -4$ **27** $x = 27$ **29** $x = 5, z = -\dfrac{7}{2}$

31 $x^2 + y^2 + z^2 = 25$ **33** $x = 7, y^2 + z^2 = 25$ **35** (a) 3; (b) 10; (c) 5; (d) $\sqrt{42}$

Problem Set 2 page 746

1 $7i + 2j - 5\bar{k}$ **3** $10i - 17j - 2\bar{k}$ **5** $34i + 10j - 24\bar{k}$
7 $(3\sqrt{18} - 4\sqrt{26})i + (\sqrt{18} - \sqrt{26})j + (\sqrt{26} - 4\sqrt{18})\bar{k}$ **9** $\sqrt{929}$ **11** 17   **13** 46

15 98   **17** $58i + 14j - 14\bar{k}$ **19** $\cos^{-1}\left(\dfrac{17}{6\sqrt{13}}\right)$ **21** $\cos^{-1}\left(\dfrac{4}{\sqrt{195}}\right)$ **23** $\dfrac{4}{\sqrt{18}}$,

$\dfrac{1}{\sqrt{18}}, -\dfrac{1}{\sqrt{18}}$ **25** $\cos^{-1}\left(\dfrac{-1}{\sqrt{10}}\right), 90°, \cos^{-1}\left(\dfrac{-3}{\sqrt{10}}\right)$ **27** 3   **29** $\sqrt{5}$ **31** $\dfrac{2}{15}, -\dfrac{2}{3}, -\dfrac{11}{15}$

33 $\dfrac{8}{\sqrt{101}}, \dfrac{1}{\sqrt{101}}, \dfrac{-6}{\sqrt{101}}$ **35** Yes   **37** No   **43** Yes   **45** $(23, 20, 9)$ **49** No

51 No   **55** (a) $x_1 x_2 + y_1 y_2 + z_1 z_2$; (b) $\sqrt{(x_1)^2 + (y_1)^2 + (z_1)^2}$;

(c) $\dfrac{x_1}{\sqrt{(x_1)^2 + (y_1)^2 + (z_1)^2}}, \dfrac{y_1}{\sqrt{(x_1)^2 + (y_1)^2 + (z_1)^2}}, \dfrac{z_1}{\sqrt{(x_1)^2 + (y_1)^2 + (z_1)^2}}$

Problem Set 3 page 753

1 Left   **3** Right   **5** Left   **7** Right   **9** Left   **11** $-j$ **13** j **15** $\bar{k}$
17 $i \times \bar{k} = -j, j \times i = -\bar{k}, j \times j = \bar{0}, \bar{k} \times i = j, \bar{k} \times j = -i, \bar{k} \times \bar{k} = \bar{0}$ **19** $-24i$
21 $-3j$ **23** $-2j$ **25** $3i - 6j$ **27** $3i + 2j - 5\bar{k}$ **29** $-21i - 14j + 11\bar{k}$
31 133   **33** 36   **35** Same direction   **41** False

Problem Set 4 page 758

1 $11i + 22j - 39\bar{k}$ **3** 988   **5** $35i - 14\bar{k}$ **7** $57i + 318j + 168\bar{k}$
9 $-11i - 29j - 13\bar{k}$ **11** $2i - 47j - 29\bar{k}$ **13** -17 **15** 0   **17** 17   **19** 0
21 0   **23** -133 **25** 239   **27** $128i + 17j - 41\bar{k}$ **29** $35i - 14\bar{k}$ **31** $2\sqrt{2}$

33 $\dfrac{\sqrt{341}}{4}$ **35** 5   **37** 26   **39** (a) Linearly independent; (b) right-handed

41 Coplanar   **43** $\dfrac{1}{2}\sqrt{66}$ **45** $\dfrac{33}{2}$ **47** $\dfrac{1}{2}|\bar{A} \times \bar{B}|$

Problem Set 5 page 766

1 $(i + 2j - 3\bar{k}) \cdot [(x - 1)i + (y + 1)j + (z - 2)\bar{k}] = 0, x + 2y - 3z + 7 = 0$
3 $(5i - 2j + 10\bar{k}) \cdot (xi + yj + z\bar{k}) = 0, 5x - 2y + 10z = 0$
5 $(45i + 10j - 51\bar{k}) \cdot [xi + (y - 8)j + z\bar{k}] = 0, 45x + 10y - 51z = 80$

7 (a) $\dfrac{2i}{7} + \dfrac{3j}{7} + \dfrac{6\bar{k}}{7}$; (b) 6, 4, 2   **9** (a) $\dfrac{12}{13}j - \dfrac{5}{13}\bar{k}$; (b) None, 5, -12

11 (a) $\dfrac{i}{\sqrt{2}} - \dfrac{3j}{5\sqrt{2}} - \dfrac{4\bar{k}}{5\sqrt{2}}$; (b) 0, 0, 0   **13** $6x - 3y - 2z + 25 = 0$ **15** $2y - z = 1$

17 $x + \sqrt{2}\,y + z = 2 + \sqrt{2}$ **19** $(0, 0, 0)$ is on the plane
21 Parallel to the xy plane   **23** Parallel to the y axis
25 (a) $(i - j + 2\bar{k}) \times [(x - 1)i + (y - 3)j + (z + 2)\bar{k}] = \bar{0}$; (b) $\dfrac{x - 1}{1} = \dfrac{y - 3}{-1} = \dfrac{z + 2}{2}$;

(c) $\bar{R} = (1 + t)i + (3 - t)j + (-2 + 2t)\bar{k}$; (d) $x = 1 + t, y = 3 - t, z = -2 + 2t$
27 (a) $\bar{M} \times (\bar{R} - \bar{R}_0) = \bar{0}, \bar{M} = 4i - 2j + 5\bar{k}, \bar{R}_0 = 3i + j - 4\bar{k}$; (b) $\dfrac{x - 3}{4} = \dfrac{y - 1}{-2} = \dfrac{z + 4}{5}$;

(c) $\bar{R} = \bar{R}_0 + t\bar{M}$; (d) $x = 3 + 4t, y = 1 - 2t, z = -4 + 5t$ **29** (a) $\bar{M} \times (\bar{R} - \bar{R}_0) = \bar{0}$,

$\bar{M} = -\dfrac{1}{2}i + \dfrac{1}{2}j + \dfrac{\sqrt{2}}{2}\bar{k}, \bar{R}_0 = \bar{k}$; (b) $\dfrac{x}{-1} = \dfrac{y}{1} = \dfrac{z - 1}{\sqrt{2}}$; (c) $\bar{R} = \bar{R}_0 + t\bar{M}$; (d) $x = -t$,

$y = t, z = 1 + \sqrt{2}\,t$ **31** (a) $\bar{M} \times (\bar{R} - \bar{R}_0) = 0, M = 6\bar{i} + 2\bar{j} + 3\bar{k}, \bar{R}_0 = 4\bar{j} + 5\bar{k}$;

(b) $\dfrac{x}{6} = \dfrac{y-4}{2} = \dfrac{z-5}{3}$;   (c) $\bar{R} = \bar{R}_0 + t\bar{M}$;   (d) $x = 6t, y = 4 + 2t, z = 5 + 3t$

33 (a) $\bar{M} \times (\bar{R} - \bar{R}_0) = \bar{0}, \bar{M} = \bar{i}, \bar{R}_0 = 7\bar{i} + \bar{j} - 4\bar{k}$;   (b) $y - 1 = 0, z + 4 = 0$, no restriction

on x;   (c) $\bar{R} = \bar{R}_0 + t\bar{M}$;   (d) $x = 7 + t, y = 1, z = -4$ **35** $\dfrac{x-2}{2} = \dfrac{y}{11} = \dfrac{z}{5}$

37 Contains $(0,0,0)$ **39** Parallel to z axis   **41** Parallel to xz plane

Problem Set 6 page 775

1 $\dfrac{5}{11}$ **3** $\sqrt{2}$ **5** $\dfrac{10\sqrt{6}}{9}$ **7** $\dfrac{10}{3}$ **9** $\cos^{-1}\left(\dfrac{3}{\sqrt{14}}\right)$ **11** $\cos^{-1}\dfrac{22}{63}$ **13** 2

15 $5x + 2y - 11z + 21 = 0$ **17** $5x + 3y - 2z = 8$ **19** $\dfrac{|D|}{\sqrt{a^2 + b^2 + c^2}}$ **21** $\dfrac{2\sqrt{70}}{7}$

23 $\dfrac{6\sqrt{14}}{\sqrt{19}}$ **25** $\dfrac{85}{\sqrt{377}}$ **27** $\dfrac{14}{\sqrt{65}}$ **29** Meet at $(11, 0, -8)$ **31** $45°$

33 (a) $\cos^{-1}\left(\dfrac{1}{2\sqrt{21}}\right)$;   (c) $(1, -1, 2)$ **35** $2x - y - z = 1$ **37** Line of intersection:

$\dfrac{x}{2} = \dfrac{z+3}{3}, y = 1$ **39** $5x - 3y + 8z + 3 = 0$ **41** $7x + y - 10z = 44$ **43** $(2, 2, 2)$

47 Absolute value of $\dfrac{1}{6}\begin{vmatrix} x_1 - x_2 & y_1 - y_2 & z_1 - z_2 \\ x_3 - x_2 & y_3 - y_2 & z_3 - z_2 \\ x_0 - x_1 & y_0 - y_1 & z_0 - z_1 \end{vmatrix}$

Problem Set 7 page 785

1 (a) $5\bar{i} + 4\bar{j} + 2\bar{k}$;   (b) all real numbers;   (c) $10\bar{i} + 3\bar{j} - 2\bar{k}$;   (d) $10\bar{i} - 2\bar{k}$ **3** (a) $\bar{i} + \bar{j}$;

(b) $(-\infty, 1)$;   (c) $2\bar{i} - \bar{k}$;   (d) $2\bar{i} - \bar{j} - \bar{k}$ **5** (a) $\bar{j} + \dfrac{\pi^3}{64}\bar{k}$;   (b) all real numbers;

(c) $-2\bar{i} + \dfrac{3\pi^2}{16}\bar{k}$;   (d) $-4\bar{j} + \dfrac{3\pi}{2}\bar{k}$ **7** $30t\bar{i} + (9t^2 - 6)\bar{j} + 12t^3\bar{k}$

9 $5t^4 + 10t + 1 + \dfrac{2}{t^2}$ **11** $(4t^3 - 2t)\bar{i} + (4t^3 - 15t^2)\bar{j} + (2 - 3t^2)\bar{k}$

13 $\dfrac{25t + 2t(t^2 - 2) + 3t^5}{\sqrt{25t^2 + (t^2 - 2)^2 + t^6}}$ **15** (a) $(-4 \sin 2t)\bar{i} + (6 \cos 2t)\bar{j} + \bar{k}$;

(b) $\sqrt{16 \sin^2 2t + 36 \cos^2 2t + 1}$;   (c) $(-8 \cos 2t)\bar{i} - (12 \sin 2t)\bar{j}$
17 (a) $-e^{-t}\bar{i} + 2e^t\bar{j} + 3t^2\bar{k}$;   (b) $\sqrt{e^{-2t} + 4e^{2t} + 9t^4}$;   (c) $e^{-t}\bar{i} + 2e^t\bar{j} + 6t\bar{k}$
19 (a) $(\sin t + t \cos t)\bar{i} + (\cos t - t \sin t)\bar{j} + 2t\bar{k}$;   (b) $\sqrt{5t^2 + 1}$;
(c) $(2 \cos t - t \sin t)\bar{i} + (-2 \sin t - t \cos t)\bar{j} + 2k$ **21** 10 **23** $\sqrt{3}(1 - e^{-\pi})$

25 $\dfrac{99\sqrt{2}}{4}$ **27** (a) $(-2 \sin t)\bar{i} + (3 \cos t)\bar{j} + \bar{k}$;   (b) $\sqrt{4 \sin^2 t + 9 \cos^2 t + 1}$;

(c) $(-2 \cos t)\bar{i} - (3 \sin t)\bar{j}$;   (d) $\dfrac{-2 \sin t\,\bar{i} + 3 \cos t\,\bar{j} + \bar{k}}{\sqrt{4 \sin^2 t + 9 \cos^2 t + 1}}$;   (e) $(3 \sin t)\bar{i} - (2 \cos t)\bar{j} + 6\bar{k}$;

(f) $\dfrac{\sqrt{9 \sin^2 t + 4 \cos^2 t + 36}}{(4 \sin^2 t + 9 \cos^2 t + 1)^{3/2}}$;   (g) $\dfrac{(-4 \cos t)\bar{i} - (3 \sin t)\bar{j} + (\sin t \cos t)\bar{k}}{\sqrt{16 \cos^2 t + 9 \sin^2 t + \sin^2 t \cos^2 t}}$;

(h) $\dfrac{(3 \sin t)\bar{i} - (2 \cos t)\bar{j} + 6\bar{k}}{\sqrt{9 \sin^2 t + 4 \cos^2 t + 36}}$;   (i) $(2 \sin t)\bar{i} - (3 \cos t)\bar{j}$;   (j) $\dfrac{6}{9 \sin^2 t + 4 \cos^2 t + 36}$

29 (a) $(2e^{2t} - 2e^{-2t})\bar{i} + (2e^{2t} + 2e^{-2t})\bar{j} + 5\bar{k}$; (b) $\sqrt{8e^{4t} + 8e^{-4t} + 25}$;

(c) $(4e^{2t} + 4e^{-2t})\bar{i} + (4e^{2t} - 4e^{-2t})\bar{j}$; (d) $\dfrac{(2e^{2t} - 2e^{-2t})\bar{i} + (2e^{2t} + 2e^{-2t})\bar{j} + 5\bar{k}}{\sqrt{8e^{4t} + 8e^{-4t} + 25}}$;

(e) $20(e^{-2t} - e^{2t})\bar{i} + 20(e^{2t} + e^{-2t})\bar{j} - 32\bar{k}$; (f) $\kappa = \dfrac{4\sqrt{50(e^{4t} + e^{-4t}) + 64}}{[8(e^{4t} + e^{-4t}) + 25]^{3/2}}$;

(g) $\dfrac{41(e^{2t} + e^{-2t})\bar{i} + 9(e^{2t} - e^{-2t})\bar{j} - 20(e^{4t} - e^{-4t})\bar{k}}{\sqrt{[50(e^{4t} + e^{-4t}) + 64](8e^{4t} + 8e^{-4t} + 25)}}$;

(h) $\dfrac{5(e^{-2t} - e^{2t})\bar{i} + 5(e^{2t} + e^{-2t})\bar{j} - 8\bar{k}}{\sqrt{50(e^{4t} + e^{-4t}) + 64}}$; (i) $8(e^{2t} - e^{-2t})\bar{i} + 8(e^{2t} + e^{-2t})\bar{j}$;

(j) $\dfrac{20}{25(e^{4t} + e^{-4t}) + 32}$ **31** (a) $\bar{i} + 2t\bar{j} + t^2\bar{k}$; (b) $t^2 + 1$; (c) $\sqrt{2}\bar{j} + 2t\bar{k}$;

(d) $\dfrac{\bar{i} + \sqrt{2}\,t\bar{j} + t^2\bar{k}}{t^2 + 1}$; (e) $\sqrt{2}\,t^2\bar{i} - 2t\bar{j} + \sqrt{2}\,\bar{k}$; (f) $\dfrac{\sqrt{2}}{(t^2 + 1)^2}$; (g) $\dfrac{-\sqrt{2}\,t\bar{i} - (t^2 - 1)\bar{j} + \sqrt{2}\,t\bar{k}}{t^2 + 1}$;

(h) $\dfrac{t^2\bar{i} - \sqrt{2}\,t\bar{j} + \bar{k}}{t^2 + 1}$; (i) $2\bar{k}$; (j) $\dfrac{\sqrt{2}}{(t^2 + 1)}$ **33** $\bar{T} = \dfrac{(-a\sin t)\bar{i} + (a\cos t)\bar{j} + b\bar{k}}{\sqrt{a^2 + b^2}}$,

$\bar{N} = (-\cos t)\bar{i} - (\sin t)\bar{j}$, $\bar{B} = \dfrac{(b\sin t)\bar{i} - (b\cos t)\bar{j} + a\bar{k}}{\sqrt{a^2 + b^2}}$, $\kappa = \dfrac{a}{a^2 + b^2}$, $\tau = \dfrac{b}{a^2 + b^2}$

37 $\dfrac{d^2\bar{F}}{dt^2} \times \bar{G} + 2\dfrac{d\bar{F}}{dt} \times \dfrac{d\bar{G}}{dt} + \bar{F} \times \dfrac{d^2\bar{G}}{dt^2}$

Problem Set 8 page 792
1 $(x - 2)^2 + (y + 1)^2 + (z - 3)^2 = 16$ **3** $(x - 4)^2 + (y - 1)^2 + (z - 3)^2 = 9$
5 $(1, 0, -2), R = 3$ **7** $(-1, 1, 0), R = 3$ **9** $(3, -1, 2), R = \sqrt{33}$ **11** (a) x axis;
(b) yz plane; (c) parabola   **13** (a) x axis; (b) yz plane; (c) hyperbola   **15** (a) y axis;
(b) xz plane; (c) two half-lines meeting at O **17** (a) z axis; (b) xy plane;
(c) exponential curve   **19** (a) x axis; (b) yz plane; (c) parabola   **21** (a) y axis;
(b) xz plane; (c) two straight lines intersecting at O **23** $y^2 + z^2 = x^2$
25 $y^2 + z^2 = (\ln x)^2$ **27** $z = x^2 + y^2$ **29** $6x^2 + 6y^2 + 6z^2 = 7$
31 z axis, $x^2 + z = 3$ **33** x axis, $x^2 - 9y^2 = 18$ **35** z axis, $x^2 = \sin^2 z$

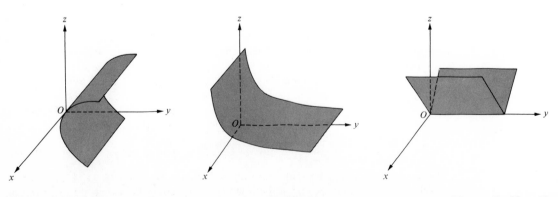

Figure, Problem 11   Figure, Problem 13   Figure, Problem 15

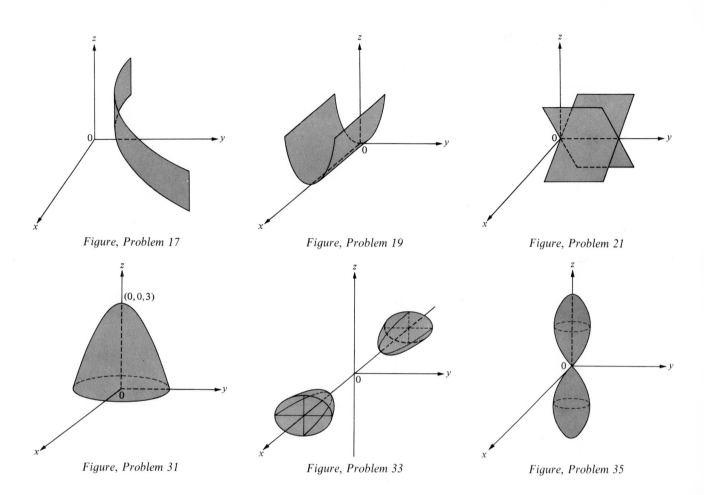

Figure, Problem 17   Figure, Problem 19   Figure, Problem 21

Figure, Problem 31   Figure, Problem 33   Figure, Problem 35

Problem Set 9 page 799

1 $3y^2 + z^2 = 4$, ellipse   **3** $z^2 - \dfrac{x^2}{16} = \dfrac{13}{9}$, hyperbola   **5** $25x^2 + 4y^2 = 100$, ellipse

7 $9 - 16y^2 = z$, parabola   **9** (a) $(\pm\sqrt{6}, 0, 0)$, $(0, \pm\sqrt{2}, 0)$, $(0, 0, \pm\sqrt{3})$; (b) all symmetries; (c) the ellipses $3y^2 + 2z^2 = 6 - k^2$, $x^2 + 2z^2 = 6 - 3k^2$, $x^2 + 3y^2 = 6 - 2k^2$; (d) the ellipses $3y^2 + 2z^2 = 6$, $x^2 + 2z^2 = 6$; $x^2 + 3y^2 = 6$; (e) ellipsoid   **11** (a) $(\pm 3, 0, 0)$, no y intercept, $(0, 0, \pm 2)$; (b) all symmetries; (c) the hyperbola $9y^2 - 9z^2 = 4k^2 - 36$ (pair of straight lines for $k = \pm 3$), the ellipse $4x^2 + 9z^2 = 36 + 9k^2$, the hyperbola $4x^2 - 9y^2 = 36 - 9k^2$ (pair of straight lines for $k = \pm 2$); (d) hyperbola $9z^2 - 9y^2 = 36$, ellipse $4x^2 + 9z^2 = 36$, hyperbola $4x^2 - 9y^2 = 36$; (e) hyperboloid of one sheet
13 (a) $(\pm 2, 0, 0)$, no y or z intercepts; (b) all symmetries; (c) ellipse $4y^2 + 4z^2 = k^2 - 4$, hyperbola $x^2 - 4z^2 = 4 + 4k^2$, hyperbola $x^2 - 4y^2 = 4 + 4k^2$; (d) hyperbola $x^2 - 4z^2 = 4$, hyperbola $x^2 - 4y^2 = 4$, no yz trace; (e) hyperboloid of two sheets   **15** (a) No x intercept, $(0, \pm 3, 0)$, no z intercept; (b) all symmetries; (c) hyperbola $y^2 - 9z^2 = 9 + 9k^2$, ellipse $9x^2 + 9z^2 = k^2 - 9$, hyperbola $y^2 - 9x^2 = 9 + 9k^2$; (d) hyperbola $y^2 - 9z^2 = 9$, hyperbola $y^2 - 9x^2 = 9$, no xz trace; (e) hyperboloid of two sheets   **17** (a) $(0, 0, 0)$; (b) all symmetries; (c) hyperbola $8z^2 - 5y^2 = k^2$, hyperbola $8z^2 - x^2 = 5k^2$, ellipse $x^2 + 5y^2 = 8k^2$; (d) two lines $z = \pm\sqrt{\dfrac{5}{8}}\, y$, two lines $z = \pm\sqrt{\dfrac{1}{8}}\, x$, point $(0, 0)$; (e) elliptic cone   **19** (a) $(0, 0, 0)$; (b) symmetry with respect to xz plane, yz plane, and z axis; (c) parabola $3z - k^2 = y^2$, parabola $3z - k^2 = x^2$, circle $3k = x^2 + y^2$; (d) parabola $3z = y^2$, parabola $3z = x^2$, point $(0, 0)$; (e) elliptic paraboloid (paraboloid of revolution)

21 (a) $(0,0,0)$; (b) symmetry with respect to xz plane, yz plane, and z axis; (c) parabola

$\dfrac{k^2}{16} - 3z = \dfrac{y^2}{9}$, parabola $\dfrac{x^2}{16} = 3z + \dfrac{k^2}{9}$, hyperbola $\dfrac{x^2}{16} - \dfrac{y^2}{9} = 3k$; (d) parabola $-3z = \dfrac{y^2}{9}$,

parabola $\dfrac{x^2}{16} = 3z$, two lines $y = \pm\dfrac{3}{4}x$; (e) hyperbolic paraboloid   **23** (a) $(0,0,0)$;

(b) symmetry with respect to xy plane, yz plane, and y axis; (c) parabola $k^2 = y + z^2$,
hyperbola $x^2 - z^2 = k$, parabola $x^2 = y + k^2$; (d) parabola $y = -z^2$, two lines $x = \pm z$,
parabola $x^2 = y$; (e) hyperbolic paraboloid   **27** An equivalent equation is obtained when
x, y, and z are replaced by $-x$, $-y$, and $-z$, respectively.   **29** (a) Ellipse

$\dfrac{y^2}{b^2} + \dfrac{z^2}{c^2} = 1 - \dfrac{k^2}{a^2}$ for $|k| < a$, points $(\pm a, 0, 0)$ for $k = \pm a$; (b) ellipse $\dfrac{x^2}{a^2} + \dfrac{z^2}{c^2} = 1 - \dfrac{k^2}{b^2}$ for

$|k| < b$, points $(0, \pm b, 0)$ for $k = \pm b$; (c) ellipse $\dfrac{x^2}{a^2} + \dfrac{y^2}{b^2} = 1 - \dfrac{k^2}{c^2}$ for $|k| < c$, points

$(0, 0, \pm c)$ for $k = \pm c$ **31** With $x = k \neq 0$, the hyperbola $\dfrac{z^2}{c^2} - \dfrac{y^2}{b^2} = \dfrac{k^2}{a^2}$; with $x = 0$, the

pair of intersecting lines $z = \pm\dfrac{c}{b}y$; with $y = k \neq 0$, the hyperbola $\dfrac{z^2}{c^2} - \dfrac{x^2}{a^2} = \dfrac{k^2}{b^2}$; with $y = 0$,

the pair of intersecting lines $z = \pm\dfrac{c}{a}x$; with $z = k \neq 0$, the ellipse $\dfrac{x^2}{a^2} + \dfrac{y^2}{b^2} = \dfrac{k^2}{c^2}$; with $z = 0$,

the point $(0,0,0)$ **33** With $z = k$, $k < 0$, the hyperbola $\dfrac{x^2}{a^2} - \dfrac{y^2}{b^2} = -k$; with $z = 0$, the

pair of intersecting lines $y = \pm\dfrac{b}{a}x$; with $z = k$, $k > 0$, the hyperbola $\dfrac{y^2}{b^2} - \dfrac{x^2}{a^2} = k$

35 $x^2 + y^2 = -4z$, a paraboloid of revolution about the negative z axis
37 $9x^2 + 9z^2 = 4y^2$, a circular cone   **39** $23x^2 + 27y^2 + 36z^2 = 351$, an ellipsoid

Problem Set 10 page 807

1 $(2, 2\sqrt{3}, 1)$ **3** $\left(\dfrac{5\sqrt{3}}{2}, \dfrac{5}{2}, -2\right)$ **5** $(4, 0, 1)$ **7** $\left(6, \dfrac{5\pi}{6}, 6\right)$ **9** $\left(\dfrac{3}{2}, \dfrac{\sqrt{3}}{2}, 1\right)$

11 $(-9, 3\sqrt{3}, -6)$ **13** $\left(1, \dfrac{3\pi}{2}, \dfrac{\pi}{2}\right)$ **15** $\left(\sqrt{14}, \tan^{-1} 2, \cos^{-1}\left(\dfrac{-3}{\sqrt{14}}\right)\right)$

17 (a) $z = 2r^2$; (b) $2\rho \sin^2 \phi = \cos \phi$ **19** (a) $r \cos \theta = 2$; (b) $\rho \sin \phi \cos \theta = 2$
21 (a) $r^2 = 5z^2$; (b) $\tan \phi = \sqrt{5}$ **23** (a) $r = 5$; (b) $\rho \sin \phi = 5$
25 (a) $r^2 - z^2 = 1$; (b) $\rho^2 \cos 2\phi = -1$ **27** (a) $z = x^2 + y^2$; (b) $\cos \phi = \rho \sin^2 \phi$

29 (a) $\dfrac{x^2}{9} + \dfrac{y^2}{9} + \dfrac{z^2}{4} = 1$; (b) $\rho^2(4 \sin^2 \phi + 9 \cos^2 \phi) = 36$ **31** (a) $x^2 + y^2 = 4x$;

(b) $\rho \sin \phi = 4 \cos \theta$ **33** (a) $x^2 + y^2 + z^2 = 4$; (b) $r^2 + z^2 = 4$ **35** (a) $x^2 + y^2 = 9$;
(b) $r = 3$ **37** (a) $3z^2 = x^2 + y^2$; (b) $3z^2 = r^2$ **39** $(r^2 + z^2)^{3/2} = 2rz$

41 $\dfrac{1}{2}\sqrt{2}\pi\sqrt{1 + 2\pi^2} + \dfrac{1}{2}\ln\left(\sqrt{1 + 2\pi^2} + \sqrt{2}\pi\right)$

Review Problem Set page 809
1 $(-2, -1, -3)$ **3** $(-3, 2, 1)$ **5** $(-3, -1, 5)$ **7** (a) $\sqrt{46}$; (b) $\sqrt{37}$; (c) 6;

(d) $\sqrt{6}$; (e) $2\sqrt{46}$ **9** Not collinear   **11** $\dfrac{\pi}{3}$ **13** $-6i - 5j$ **15** $\sqrt{61}$ **17** 0

19 0   **21** $-8i + 8j + 8k$ **23** $16i - 16j + 16k$ **25** 8   **27** $\cos^{-1}\left(\dfrac{-21}{\sqrt{910}}\right)$ **29** 8

31 Yes   **33** $x + 2y + 3z = 14$ **35** $4x + 2y + 3z = 5$ **37** $x + 3z = 7$

39 $3x + 8y - 7z + 23 = 0$ **41** (a) $x = 5 + 3t, y = 6 + 7t, z = -4 - 5t$; (b) $\dfrac{x-5}{3} =$

$\dfrac{y-6}{7} = \dfrac{z+4}{-5}$ **43** (a) $x = 2 + 3t, y = -3 - 2t, z = -2 + 7t$; (b) $\dfrac{x-2}{3} = \dfrac{y+3}{-2} = \dfrac{z+2}{7}$

45 (a) $x = 3 + 2t, y = -4 - 7t, z = 1 - 3t$; (b) $\dfrac{x-3}{2} = \dfrac{y+4}{-7} = \dfrac{z-1}{-3}$ **47** (a) $x = 2$,

$y = 1, z = 4 + t$; (b) $x = 2, y = 1, z$ arbitrary   **49** 3   **51** $3x + y - 4z = 2$ **53** $\dfrac{6}{\sqrt{3}}$

59 $4t(e^{2t^2}i + e^{-2t^2}j) + \bar{k}, 4e^{2t^2}(1 + 4t^2)i + 4e^{-2t^2}(1 - 4t^2)j, \dfrac{4te^{4t^2} - 4te^{-4t^2} + t}{\sqrt{e^{4t^2} + e^{-4t^2} + t^2}}$

61 (a) $(-6\pi \sin 2\pi\, t)i + (6\pi \cos 2\pi\, t)j + 2\bar{k}$; (b) $(-12\pi^2 \cos 2\pi t)i - (12\pi^2 \sin 2\pi t)j$;

(c) $2\sqrt{9\pi^2 + 1}$; (d) $\overline{T} = \dfrac{(-3\pi \sin 2\pi t)i + (3\pi \cos 2\pi t)j + \bar{k}}{\sqrt{9\pi^2 + 1}}$; (e) $(-\cos 2\pi t)i - (\sin 2\pi t)j$;

(f) $\dfrac{(\sin 2\pi t)i - (\cos 2\pi t)j + 3\pi\bar{k}}{\sqrt{9\pi^2 + 1}}$; (g) $2\sqrt{9\pi^2 + 1}$ **63** (a) $i + 2tj + 3t^2\bar{k}$; (b) $\sqrt{1 + 4t^2 + 9t^4}$;

(c) $2j + 6t\bar{k}$; (d) $\dfrac{i + 2tj + 3t^2\bar{k}}{\sqrt{1 + 4t^2 + 9t^4}}$; (e) $6t^2i - 6tj + 2\bar{k}$; (f) $\dfrac{\sqrt{36t^4 + 36t^2 + 4}}{(1 + 4t^2 + 9t^4)^{3/2}}$;

(g) $\dfrac{(-18t^3 - 4t)i + (2 - 18t^4)j + (12t^3 + 6t)\bar{k}}{\sqrt{1 + 4t^2 + 9t^4}\sqrt{36t^4 + 36t^2 + 4}}$; (h) $\dfrac{6t^2i - 6tj + 2\bar{k}}{\sqrt{36t^4 + 36t^2 + 4}}$; (i) $6\bar{k}$;

(j) $\dfrac{3}{9t^4 + 9t^2 + 1}$ **65** (a) $(\cos^2 t - \sin^2 t)i + (2 \sin t \cos t)j - (\sin t)\bar{k}$; (b) $\sqrt{1 + \sin^2 t}$;

(c) $(-4 \sin t \cos t)i + 2(\cos^2 t - \sin^2 t)j - (\cos t)\bar{k}$;

(d) $\dfrac{(\cos^2 t - \sin^2 t)i + (2 \sin t \cos t)j - (\sin t)\bar{k}}{\sqrt{1 + \sin^2 t}}$;

(e) $(-2 \sin^3 t)i + (\cos^3 t + 3 \sin^2 t \cos t)j + 2\bar{k}$; (f) $\dfrac{\sqrt{5 + 3 \sin^2 t}}{(1 + \sin^2 t)^{3/2}}$;

(g) $\dfrac{-\sin t \cos t(5 + 2 \sin^2 t)i + [4 - 2(1 + \sin^2 t)^2]j - (\cos t)\bar{k}}{\sqrt{1 + \sin^2 t}\sqrt{5 + 3 \sin^2 t}}$;

(h) $\dfrac{(-2 \sin^3 t)i + (\cos^3 t + 3 \sin^2 t \cos t)j + 2\bar{k}}{\sqrt{5 + 3 \sin^2 t}}$;

(i) $4(\sin^2 t - \cos^2 t)i - (8 \sin t \cos t)j + (\sin t)\bar{k}$;

(j) $\dfrac{-6 \sin t}{5 + 3 \sin^2 t}$ **67** (a) $\overline{R} \cdot \overline{A} + v^2$; (b) $\overline{R} \times \overline{A}$; (c) $|\overline{A}|^2 + \overline{V} \cdot \dfrac{d\overline{A}}{dt}$; (d) $\overline{V} \times \dfrac{d\overline{A}}{dt}$;

(e) $-\kappa\tau$; (f) $\kappa^2\tau$ **69** $z^2 + y^2 = 4(x - 3)^4$ **71** x axis, $z^2 = e^{-2x}$ **73** Ellipsoid
75 Paraboloid of revolution   **77** Hyperboloid of one sheet   **79** Hyperbolic
paraboloid   **81** (a) Right circular cylinder of radius 2; (b) plane making an angle of $\dfrac{\pi}{6}$

with the xz plane; (c) right circular cylinder of radius $\dfrac{1}{2}$ whose central axis is parallel to the

z axis and contains $\left(0, \dfrac{1}{2}, 0\right)$ **83** $z = f(\pm r)$ **85** $z = \dfrac{\pm 2}{\sqrt{x^2 + y^2}}$

Chapter 16

Problem Set 1 page 818
1 The entire xy plane   **3** All ordered pairs (x, y) such that $x^2 + y^2 \leq 4$ **5** All

ordered pairs (x, y) such that $x^2 + y^2 \leq 9$ **7** -39 **9** $|k|$ **11** $-\dfrac{7}{4}$

13 $5a + 7\sqrt{ab}$ **15** $\sin 2t$ **17** $5x^2 + 7xy + \sqrt{xy}$ **19** $\dfrac{2xy}{x^2 + y^2}$ **21** (a) All ordered pairs (x, y) such that $x + y \geq 4$; (b) $2\sqrt{2}$ **23** (a) All ordered pairs (x, y) such that $y \neq 2x$; (b) 7   **25** $\dfrac{n(n + 1)}{2}$ **33** Temperature changes rapidly near P

Problem Set 2 page 826

1 -1 **3** $e + 1$ **5** $\dfrac{1}{2}$ **7** 3   **9** (a) (i) 0, (ii) 0, (iii) 0, (iv) 0; (b) 0

11 (a) (i) 0, (ii) 0, (iii) $\dfrac{1}{2}$, (iv) $\dfrac{m}{1 + m^2}$; (b) does not exist   **13** (a) (i) 0, (ii) 0, (iii) 0, (iv) 0;

(b) 0   **15** $\delta \leq \dfrac{\varepsilon}{10}$ **17** Continuous   **19** Discontinuous   **21** Continuous

23 Continuous   **25** (a) All points (x, y) such that $x \geq 0$, $y \geq 0$ or $x \leq 0$, $y \leq 0$; (b) $x > 0$, $y > 0$ or $x < 0$, $y < 0$; (c) none   **27** (a) All points (x, y) such that $y \neq 2x$; (b) all points of the domain; (c) none   **29** (a) All points (x, y) such that $y \neq \pm 1$; (b) all points of the domain; (c) none   **31** (a) All points (x, y) such that $x^2 + y^2 < 9$; (b) all points of the domain; (c) none   **33** (a) All points $(x, y) \neq (0, 0)$ such that $y \neq -x$ together with $(0, 0)$; (b) all points of the domain except $(0, 0)$; (c) none   **35** (a) The entire plane; (b) all points; (c) discontinuities on $x^2 + y^2 = 4$

Problem Set 3 page 833

1 8   **3** 46   **5** 15   **7** $14x + 10xy$ **9** $\cos 7y \cos x$ **11** $\dfrac{4xy^2}{(y^2 - x^2)^2}$

13 $2r \cos 7\theta$ **15** e^{-x^3} **17** $6xy + 7$ **19** $2xy + y^2$ **21** $3x(3x^2 + y^2)^{-1/2}$

23 $\dfrac{14x}{7x^2 + 4y^3}$ **25** $\dfrac{y}{1 + x^2 y^2}$ **27** $-5e^{\sin (x^2 - 5y)}$ **29** $-2xe^{y^2 - x^2}$

31 $xze^{xy}(-yz \sin yz + 2 \cos yz)$ **33** $-3x(x^2 + y^2 + z^2)^{-5/2}$

Problem Set 4 page 836

1 3   **3** -6 **5** $\dfrac{3}{e}$ **7** $\dfrac{56}{625}$ **9** (a) $\dfrac{21\pi}{\sqrt{58}}$; (b) $\dfrac{67\pi}{\sqrt{58}}$ **11** (a) $-\dfrac{1}{3{,}000{,}000}$;

(b) $\dfrac{2.5028\pi}{90} \sin \dfrac{4\pi}{9}$ **13** $\dfrac{-22}{45}, \dfrac{1}{15}$

Problem Set 5 page 845

1 -17.21 **3** 0.01   **5** $\dfrac{12\pi - 7}{120}$ **7** Differentiable   **9** Differentiable

11 Differentiable   **13** $(15x^2 + 8xy)\,dx + (4x^2 - 6y^2)\,dy$ **15** $\dfrac{V}{R}\,dP + \dfrac{P}{R}\,dV$

17 $(y^2 - 4xz + 3yz^2)\,dx + (2xy + 3xz^2)\,dy + (6xyz - 2x^2)\,dz$ **19** 40 watts

21 $\dfrac{1.29}{4\pi^2}, 4\%$ **23** 1.18   **25** $\dfrac{3}{640} \approx 0.0047$ **29** (a) $\dfrac{2xy^4}{(x^2 + y^2)^2}, \dfrac{2x^4 y}{(x^2 + y^2)^2}$; (b) 0, 0

Problem Set 6 page 855

1 $2(3x^2 y^2 - 3y) + 12t(2x^3 y - 3x + 2y)$
3 $2x[\sin v - \sin (u - v)] + 3x^2[u \cos v + \sin (u - v)]$
5 $3[\cos (x + y) + \cos (x - y)] + 3t^2[\cos (x + y) - \cos (x - y)]$
7 $\dfrac{1}{2v}(e^{u/v} + e^{-u/v}) \cosh t - \dfrac{u}{2v^2}(e^{u/v} + e^{-u/v})$ **9** $\dfrac{21}{x} + \dfrac{2 \sec t \tan t}{y} + \dfrac{\csc^2 t}{z}$

11 $6xv + 8y \sin u$; $6xu - 8y \cos v$ **13** $(12x^2 - 6xy^2) \cos v - 6x^2yv \cos u$;

$(12x^2 - 6xy^2)(-u \sin v) - 6x^2y \sin u$ **15** $\dfrac{4ux + 8vx + 6vy}{u^2 + v^2}$; $\dfrac{4uy + 6vx}{u^2 + v^2}$

17 $6re^{-s} \sinh (3x + 7y) + 7e^{3s} \sinh (3x + 7y)$;
$-3r^2e^{-s} \sinh (3x + 7y) + 21re^{3s} \sinh (3x + 7y)$

19 $2uvy^2e^{xy^2} + 2v^2xye^{xy^2}$; $u^2y^2e^{xy^2} + 4uvxye^{xy^2}$ **21** $4x \cos v + 6y \sin v + 2zv$;
$-4xu \sin v + 6yu \cos v + 2zu$ **23** $2x \sin \phi \cos \theta + 2y \sin \phi \sin \theta - 2z \cos \phi$;
$-2x\rho \sin \phi \sin \theta + 2y\rho \sin \phi \cos \theta = 0$; $2x\rho \cos \phi \cos \theta + 2y\rho \cos \phi \sin \theta + 2z\rho \sin \phi$

25 10, 11   **27** (a) $\dfrac{3y - 3x}{2y - 3x}$; (b) 0   **29** (a) $\dfrac{\cos (x - y) - \sin (x + y)}{\cos (x - y) + \sin (x + y)}$; (b) 0

31 14, 2   **39** (a) 4 ft²/min; (b) $(3 + \sqrt{5})$ ft/min   **41** $-22.5°$K/min

Problem Set 7 page 867

1 (a) $7i - 3j$; (b) $7i - 3j$; (c) $\dfrac{7\sqrt{3} - 3}{2}$ **3** (a) $4xi + 6yj$; (b) $0i + 0j$; (c) 0

5 (a) $12i + 9j$; (b) $\dfrac{12 + 9\sqrt{3}}{2}$ **7** (a) $-\dfrac{6}{169}i - \dfrac{4}{169}j$; (b) $-\dfrac{6}{169}$ **9** $\dfrac{17\sqrt{2}}{2}$

11 (a) $\sqrt{306}$; (b) $\dfrac{3\sqrt{34}}{34}i - \dfrac{5\sqrt{34}}{34}j$ **13** (a) $\sqrt{4 + \pi^2}$; (b) $\dfrac{2}{\sqrt{4 + \pi^2}}i - \dfrac{\pi}{\sqrt{4 + \pi^2}}j$

15 (a) $-\dfrac{\sqrt{2}}{2}i - \dfrac{\sqrt{2}}{2}j$; (b) $50\sqrt{2}$°C/cm   **17** (a) $2i + 2j$; (b) $x + y - 2 = 0$

19 (a) $-2i + 4j - 6\bar{k}$; (b) $-\dfrac{22}{3}$ **21** (a) $3i + \bar{k}$; (b) $\dfrac{12}{7}$ **23** (a) $\dfrac{\sqrt{14}}{98}$;

(b) $-\dfrac{\sqrt{14}}{14}i - \dfrac{\sqrt{14}}{7}j + \dfrac{3\sqrt{14}}{14}\bar{k}$ **25** (a) e; (b) i **27** (a) $x + 2y + 3z = 6$;

(b) $\dfrac{x - 1}{2} = \dfrac{y - 1}{4} = \dfrac{z - 1}{6}$ **29** (a) $6x + 3y + 2z = 18$; (b) $\dfrac{x - 1}{6} = \dfrac{y - 2}{3} = \dfrac{z - 3}{2}$

31 (a) $z = 8$; (b) $x = 2, y = 2, z = t$ **33** (a) $-\dfrac{\pi}{12}x - \dfrac{3}{2}y + \dfrac{\pi}{12}z = -\dfrac{\pi}{2}$;

(b) $\dfrac{x - 1}{-\dfrac{\pi}{12}} = \dfrac{y - \dfrac{\pi}{6}}{-\dfrac{3}{2}} = \dfrac{z + 2}{\dfrac{\pi}{12}}$ **35** (a) $2x + 3y + 6z = 36$; (b) $\dfrac{x - 9}{2} = \dfrac{y - 4}{3} = \dfrac{z - 1}{6}$

37 (a) $x - 2y - 2z = -9$; (b) $\dfrac{x + 1}{1} = \dfrac{y - 2}{-2} = \dfrac{z - 2}{-2}$ **39** (a) $z = x$; (b) $\dfrac{x - 1}{1} = \dfrac{z - 1}{-1}$,

$y = 0$ **47** $-\cos \sqrt{3} + \sqrt{3} \sin \sqrt{3}$ **49** $1{,}000{,}000\sqrt{2}\, e^{-6}$ **53** $-66i + 107j + \dfrac{143}{6}k$

55 $-6i - 4j + 6\bar{k}$

Problem Set 8 page 876

1 (a) 12; (b) 10; (c) 7; (d) 7   **3** (a) 0; (b) $-x \cos y - 2$; (c) $-\sin y$; (d) $-\sin y$
5 (a) $(6x^2 + 3y^2)(x^2 + y^2)^{-1/2}$; (b) $(3x^2 + 6y^2)(x^2 + y^2)^{-1/2}$; (c) $3xy(x^2 + y^2)^{-1/2}$;
(d) $3xy(x^2 + y^2)^{-1/2}$ **7** (a) $-y \cos x$; (b) $-4xe^{2y}$; (c) $-\sin x - 2e^{2y}$;
(d) $-\sin x - 2e^{2y}$ **9** (a) $-\sin (x + 2y)$; (b) $-4 \sin (x + 2y)$; (c) $-2 \sin (x + 2y)$;
(d) $-2 \sin (x + 2y)$ **11** (a) 0; (b) $20x \cosh 2y$; (c) $10 \sinh 2y$; (d) $10 \sinh 2y$
13 (a) $2yz + 6xe^y$; (b) $2xy$ **15** (a) 0; (b) 0   **17** (a) $420e^6$; (b) $420e^6$ **19** (a) 0;

(b) 0   **31** $A = -C$ **33** $3\dfrac{\partial^2 f}{\partial u^2} + 4\dfrac{\partial^2 f}{\partial v\, \partial u} + 6\dfrac{\partial^2 f}{\partial v^2}$

37 $f''(u)\left[\left(\dfrac{\partial g}{\partial x}\right)^2 + \left(\dfrac{\partial g}{\partial y}\right)^2\right] + f'(u)\left(\dfrac{\partial^2 g}{\partial x^2} + \dfrac{\partial^2 g}{\partial y^2}\right)$

Problem Set 9 page 885

1 Relative minimum at $(0, 1)$ **3** Relative minimum at (x, y) such that $y = x + 1$
5 Relative minimum at $(1, 1)$ and $(-1, -1)$ **7** Saddle point at $(1, 1)$
9 Relative minimum at $(-1, -1)$ **11** Saddle point at $(-1, -1)$ **13** Critical points $(1, 0)$ and $(-1, 0)$; relative minimum at $(1, 0)$ and relative maximum at $(-1, 0)$
15 Relative minimum at $(3, 4)$ **17** Relative minimum at $(-2, 1)$ **19** Absolute maximum at $(0, 0)$ is 1; absolute minimum on the boundary of $x^2 + y^2 = 1$ is 0.
21 Absolute maximum at $(4, 4)$ is 48; absolute minimum at $(12, 12)$ is -144.   **23** 2, 2, 5

25 $x = 5, y = 10$ **27** $-9°$ at $\left(-\dfrac{3}{4}, 0\right)$, $46°$ at $\left(\dfrac{1}{2}, \pm\dfrac{\sqrt{3}}{2}\right)$ **29** Each is 667.

Problem Set 10 page 894

1 $(0, -1), (0, 1), \left(-\dfrac{\sqrt{15}}{4}, -\dfrac{1}{4}\right), \left(\dfrac{\sqrt{15}}{4}, \dfrac{1}{4}\right)$ **3** $\left(\dfrac{\sqrt{2}}{2}, \dfrac{\sqrt{2}}{2}\right), \left(-\dfrac{\sqrt{2}}{2}, -\dfrac{\sqrt{2}}{2}\right), (-\sqrt{2}, \sqrt{2}),$

$(\sqrt{2}, -\sqrt{2})$ **5** $\left(-\dfrac{\sqrt{2}}{2}, 0\right), \left(\dfrac{\sqrt{2}}{2}, 0\right), \left(\dfrac{1}{4}, -\dfrac{\sqrt{14}}{4}\right), \left(\dfrac{1}{4}, \dfrac{\sqrt{14}}{4}\right)$ **7** $(\pm 1, 0, 0),$

$(0, \pm 2\sqrt{3}, 0), (0, 0, \pm\sqrt{3}), \left(\pm\dfrac{\sqrt{3}}{3}, \pm 2, \pm 1\right)$ **9** $\left(-\dfrac{2}{11}, -\dfrac{6}{11}, \dfrac{9}{22}\right)$ **11** $x = 4, y = 12$

13 $\dfrac{100}{3}$ in., $\dfrac{50}{3}$ in., $\dfrac{50}{3}$ in.   **15** Length = width   **19** (a) In the direction so that the tangent to the curve makes an acute angle with $\nabla g(a, b)$; (b) the opposite direction to that in part (a)

Review Problem Set page 895

1 Domain: all points (x, y) such that $x^2 + y^2 \le 64$; interior points are all points (x, y) such that $x^2 + y^2 < 64$.   **3** Domain: all points (x, y) such that $2x + y \ne 1$; interior points are all points in the domain.   **5** (a) -6; (b) -72; (c) $3ab^2cd^6$; (d) the entire $xyzw$ space

7 Does not exist   **11** $\dfrac{\partial f}{\partial x} = 3x^2 + 7y^2, \dfrac{\partial f}{\partial y} = 14xy - 24y^2$ **13** $\dfrac{\partial f}{\partial x} = y^3 e^x + 3x^2 e^y,$

$\dfrac{\partial f}{\partial y} = 3y^2 e^x + x^3 e^y$ **15** $\dfrac{\partial g}{\partial x} = \dfrac{3x^2}{x^3 - y^3}; \dfrac{\partial g}{\partial y} = \dfrac{-3y^2}{x^3 - y^3}$

17 $\dfrac{\partial g}{\partial x} = \dfrac{(x^3 + y^3)^{2/3} + x^2}{x(x^3 + y^3)^{2/3} + (x^3 + y^3)}, \dfrac{\partial g}{\partial y} = \dfrac{y^2}{x(x^3 + y^3)^{2/3} + (x^3 + y^3)}$

19 $\dfrac{\partial g}{\partial x} = e^{-7x}[\sec^2 (x + y) - 7 \tan (x + y)]; \dfrac{\partial g}{\partial y} = e^{-7x} \sec^2 (x + y)$ **21** $\dfrac{\partial g}{\partial x} = 3x^2 - 26xy^2z^2;$

$\dfrac{\partial g}{\partial y} = 3y^2 - 26x^2yz^2; \dfrac{\partial g}{\partial z} = 3z^2 - 26x^2y^2z$ **23** $\dfrac{\partial f}{\partial x} = -yz \operatorname{csch}^2 (xy); \dfrac{\partial f}{\partial y} = -xz \operatorname{csch}^2 (xy);$

$\dfrac{\partial f}{\partial z} = \coth (xy)$ **31** (a) $\dfrac{x - 2}{1} = \dfrac{y - 2}{1} = \dfrac{z - 2}{1}$; (b) $x + y + z = 6$

33 (a) $2xy\, dx + x^2\, dy$; (b) 75.70   **35** 1.76π cm^3 **39** -0.1
41 $24x^3 - 18xy^2 - 6x^2y + 4y^3; 8x^3 - 6xy^2 + 12x^2y - 8y^3$ **43** $-2e^{x^4} + e^{y^4};$

$-e^{x^4} - 3e^{y^4}$ **45** $-\dfrac{1}{3} vx^{-2/3} \sin (uv) - \dfrac{1}{4} ux^{-3/4} \sin (uv)$ **47** $-27e^{-20}$ **49** (a) -3;

(b) -2 **51** (a) $\sin x$; (b) $y \cos x$ **53** (a) $\dfrac{6\sqrt{3} + 1}{2}$; (b) $\sqrt{37}$; $\dfrac{-6\sqrt{37}}{37} i + \dfrac{\sqrt{37}}{37} j$

55 $\dfrac{\cos y - y \cos x}{\sin x + x \sin y}$ **57** $\dfrac{48 - 10\pi}{3\pi}$ in./min   **59** $3x + 3y - 5z = 8;$

$\dfrac{x - 3}{3} = \dfrac{y - 3}{3} = \dfrac{z - 2}{-5}$ **61** $6x - 2y + 15z = 22; \dfrac{x - 1}{6} = \dfrac{y - 7}{-2} = \dfrac{z - 2}{15}$

63 (a) $e^{xy}[12st + 10t^2 + (6s^2t + 10st^2 + 2t^3)^2]$; (b) $e^{xy}[10s^2 + 12st + (2s^3 + 10s^2t + 6st^2)^2]$
65 No relative minimum or maximum; saddle point at $(-1, -2)$ **67** Relative maximum at $(1, 1)$; saddle points at $(0, 0)$, $(0, 3)$, and $(3, 0)$ **69** Relative minimum at $\left(\dfrac{1}{2}, \dfrac{1}{2}\right)$ **71** Maximum of $\left(\dfrac{\sqrt{2}}{2}, \dfrac{\sqrt{2}}{2}\right)$; minimum at $\left(\dfrac{-\sqrt{2}}{2}, \dfrac{-\sqrt{2}}{2}\right)$ **75** Razor \$1.00; blades \$0.40

Chapter 17

Problem Set 1 page 903

1 $-\dfrac{1}{y}\cos xy + C(y)$ **3** $\dfrac{2}{3}xy^{2/3} - \dfrac{y^2}{2x} + \dfrac{2\sqrt{y}}{x} + K(x)$ **5** $-y^2\cos y^3$ **7** $\dfrac{8}{15}$

9 $e^3 - 4$ **11** $\dfrac{1}{2}$ **13** $\dfrac{38\pi}{3}$ **15** $\dfrac{2}{3}$ **17** $7 - 3\sqrt[3]{5}$ **19** $\dfrac{\pi^2}{18}$ **21** $\dfrac{e^7}{7} - \dfrac{e^5}{5} + \dfrac{2}{35}$

23 $\dfrac{4}{3}$ **25** $\dfrac{2e^2 - 3}{2}$ **27** $4e^{-\sqrt{2}/2} + 2\sqrt{2} - 4$

Problem Set 2 page 910

1 16   **3** 261   **5** 0   **7** 351   **9** $\dfrac{250\pi}{3}$ **11** $\dfrac{24\pi + 4\sqrt{2}\,\pi}{3}$ **13** 18   **15** πr^2

17 $\dfrac{1}{2}|x_1y_2 - x_1y_3 - x_2y_1 + x_2y_3 + x_3y_1 - x_3y_2|$ **19** $\dfrac{\pi}{3}$ **21** 2   **23** 2   **25** -4

27 8π

Problem Set 3 page 918

1 π **3** $\dfrac{\pi^2 + 4}{4}$ **5** 0   **7** $\dfrac{\pi}{4}$ **9** $\dfrac{8}{3}$ **11** 50   **13** $\dfrac{\pi}{4} - \ln\sqrt{2}$

15 $\dfrac{1}{6}(1 - e^{-3})$ **17** 0   **19** $1 - \sin 1$ **21** $12(e - 1)$ **23** $\dfrac{1}{5}(\sqrt{33} - 1)$ **25** 8

Problem Set 4 page 928

1 $\dfrac{125}{6}$ square units   **3** $\dfrac{4}{3}$ square units   **5** 12 square units   **7** 2 square units

9 $\dfrac{593}{140}$ cubic units   **11** 2 cubic units   **13** $\dfrac{10}{3}$ cubic units   **15** $\dfrac{292}{15}$ cubic units

17 $\dfrac{5}{12}$ cubic unit   **19** 64 gm   **21** $\dfrac{64}{3}$ kg   **23** $\dfrac{3}{7}$ coulomb   **25** $\dfrac{27}{2}$ gm; $\left(\dfrac{6}{5}, \dfrac{6}{5}\right)$

27 15 gm; $\left(\dfrac{9}{4}, \dfrac{15}{8}\right)$ **29** $\dfrac{10}{3}$ gm; $\left(\dfrac{39}{25}, \dfrac{206}{75}\right)$ **33** $(1, 2)$ **35** $\left(\dfrac{21}{13 - 6\ln 2}, \dfrac{102}{65 - 30\ln 2}\right)$

37 $\dfrac{1088\pi}{15}$ cubic units   **39** $\dfrac{2\pi}{3}$ cubic units   **41** $\dfrac{15}{91}$ gm-cm^2, $\dfrac{3}{7}$ gm-cm^2

43 $\dfrac{1}{3 - 4\ln 2}$ gm-cm^2; $\dfrac{1}{6 - 8\ln 2}$ gm-cm^2 **45** $\dfrac{56}{15}$ gm-cm^2; $\dfrac{8}{3}$ gm-cm^2; $\dfrac{32}{5}$ gm-cm^2

47 $\dfrac{3}{32}\pi$ gm-cm^2 **49** $\dfrac{3\pi}{4}$ gm-cm^2

Problem Set 5 page 935

1 $\dfrac{4\pi}{3}$ **3** $\dfrac{\pi}{9}(2\sqrt{2} - 1)$ **5** 0   **7** 6   **9** $\pi(1 - e^{-9})$ **11** $\dfrac{\pi a^5}{10}$ **13** $\dfrac{9\sqrt{2}}{2}$

15 $\dfrac{2\pi}{3}$ **17** 12π square units   **19** $\dfrac{\pi + 8}{4}$ square units   **21** $\dfrac{3\sqrt{3} - \pi}{2}$ square units

23 $\dfrac{\pi}{8}$ cubic unit   **25** $\dfrac{\pi}{2}$ cubic units   **27** $\left(\dfrac{(8\pi + 3\sqrt{3})a}{4\pi + 6\sqrt{3}}, 0\right)$ **29** $\left(\dfrac{2r_0 \sin \theta_0}{3\theta_0}, 0\right)$

31 $\left(\dfrac{3}{2\pi}, \dfrac{3}{2\pi}\right)$ **33** $\dfrac{a^2 m(3\pi + 8)}{24}$ gm-cm^2

Problem Set 6   page 943

1 $\dfrac{9}{2}$ **3** $\dfrac{5}{24}$ **5** $\dfrac{124\pi}{5}$ **7** $\dfrac{1}{24}$ **9** 64   **11** $\dfrac{\pi}{2}$ **13** 20π **15** $\dfrac{1}{180}$

17 $\dfrac{8}{15}$ cubic unit   **19** $8\pi\sqrt{2}$ cubic units   **21** $\dfrac{1}{6}$ cubic unit

23   **25**

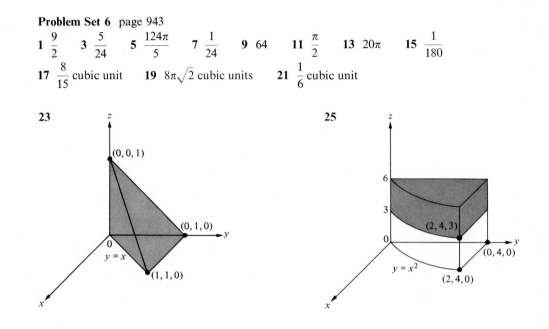

Problem Set 7   page 950

1 9   **3** $\dfrac{64\pi(2 - \sqrt{3})}{3}$ **5** 15π **7** $\dfrac{4\pi - \pi^2}{8}$ **9** $\dfrac{250\pi}{3}$ **11** 24π **13** 81π

15 $\dfrac{3125\pi(4\sqrt{5} - 5\sqrt{2})}{24}$ **17** 9 cubic units   **19** $\dfrac{3\pi}{2}$ cubic units   **21** $\dfrac{4\pi}{3}(8\sqrt{2} - 7)$

cubic units   **23** $\dfrac{14\pi}{9}(2 - \sqrt{2})$ cubic units   **25** $\dfrac{18\sqrt{5}\,\pi(\sqrt{5} - 2)}{5}$ cubic units

27 Top part: $\dfrac{\pi}{3}(2a^3 - 3a^2 k + k^3)$ cubic units; bottom part: $\dfrac{\pi}{3}(2a^3 + 3a^2 k - k^3)$ cubic units

29 $\dfrac{2\pi r^2 h_1}{3}$ cubic units   **31** $\dfrac{a^4 \pi^2}{2}$ (units)^4

Problem Set 8   page 958

1 $\dfrac{32\pi}{3}$ **3** $\dfrac{243\pi}{2}$ **5** $\dfrac{3}{\pi}$ gm/cm^2 **7** $\left(\dfrac{4}{9}, \dfrac{4}{9}, \dfrac{4}{9}\right)$ **9** $(0,0,0)$ **11** $\left(\dfrac{8}{15}, \dfrac{8}{15}, \dfrac{\pi}{24}\right)$

13 $\left(0, 0, \dfrac{3(b^4 - a^4)}{8(b^3 - a^3)}\right)$ **15** $\dfrac{1}{16}$ **17** $\dfrac{512}{35}$ **19** $\dfrac{5}{2}$ **21** $\dfrac{10}{3}$ **23** $\dfrac{m(a^2 + b^2)}{2}$

25 $\dfrac{2ma^2}{5}$ **27** $\dfrac{2m(b^5 - a^5)}{5(b^3 - a^3)}$

Problem Set 9 page 969

1 2   **3** 2π **5** $\dfrac{4}{3}$ **7** 0   **9** 32   **11** $\dfrac{189}{2}$ **13** $\dfrac{1024-60\pi}{15}$ **15** $\dfrac{243\pi}{4}$

17 0   **19** $-\dfrac{1}{3}$ erg   **21** (a) $\dfrac{17}{6}$ ergs (b) $\dfrac{12}{5}$ ergs (c) $\dfrac{22-3\pi}{6}$ ergs   **29** P and Q are discontinuous at $(0,0)$.

Problem Set 10 page 979

1 $9\sqrt{3}\,\pi$ square units   **3** $\dfrac{49\sqrt{14}}{12}$ square units   **5** $6\sin^{-1}\left(\dfrac{1}{3}\right)$ square units   **7** $16\sqrt{2}$ square units   **9** $\dfrac{56\pi}{3}$ square units   **17** $\dfrac{4\pi}{3}$ **19** $-21\sqrt{2}$ **21** $-\dfrac{23}{12}$ **23** $\dfrac{7}{4}$

25 $\dfrac{4\pi}{3}$

Problem Set 11 page 988

1 (a) y^2; (b) $i-j-2(x+xy)\overline{k}$ **3** (a) $3yz^2+5x^2+6y^2z^2$;
(b) $(4yz^3-1)i+6xyz j+(10xy-3xz^2)\overline{k}$ **5** 27   **7** 0   **9** 0   **11** 0   **13** 3 in both cases   **15** $\dfrac{3}{2}$ in both cases

Review Problem Set page 989

1 $\dfrac{1}{10}$ **3** $\dfrac{143}{30}$ **5** $\dfrac{e^{16}}{8}-2e$ **7** $\dfrac{49}{3}(\sqrt{3}-1)$ **9** $\dfrac{4}{3}$ **11** $\dfrac{2592}{35}$ **13** $\dfrac{6\sqrt{3}-38}{3}$

15 $\dfrac{1}{2}(e^4-1)$ **17** $\dfrac{9}{2}$ square units   **19** $\dfrac{8}{5}$ square units   **21** $\dfrac{500\pi}{3}$ **23** $\dfrac{7\pi}{3}$ **25** 0

27 $\dfrac{128}{21}$ cubic units   **29** $\dfrac{64}{3}$ cubic units   **31** $\dfrac{1}{3}$ cubic unit   **33** $\left(\dfrac{\pi}{4}+\ln\dfrac{\sqrt{2}}{2},\dfrac{2}{3}\right)$

35 $\left(\dfrac{466}{35(8\pi+3\sqrt{3})},\dfrac{234\sqrt{3}}{35(8\pi+3\sqrt{3})}\right)$ **37** $\dfrac{mh^2}{6};\dfrac{m}{6}(c^2+cb+b^2);\dfrac{m}{6}(h^2+c^2+cb+b^2)$

39 $\dfrac{1}{1232}$ **41** $\dfrac{34}{15}$ **43** $\dfrac{2a^3\pi}{3}$ **45** $\dfrac{13}{6}$ **47** $\dfrac{27}{2}$ cubic units   **49** $\dfrac{\pi}{30}$ **51** $\pi(2-\sqrt{3})$

53 $\dfrac{\pi a^3 h^2}{60}$ **55** $\left(0,0,\dfrac{8}{3}\right)$ **57** $\left(\dfrac{2\pi-3\sqrt{3}}{4\pi(2-\sqrt{3})},\dfrac{2\pi-3\sqrt{3}}{4\pi(2-\sqrt{3})},\dfrac{3}{8(2-\sqrt{3})}\right)$ **59** $\dfrac{3ma^2}{10}$

61 $\dfrac{7a^2m}{5}$ **63** $\dfrac{23}{6}$ **65** $\dfrac{1}{3}$ **67** $-\dfrac{1}{44}$ **69** $-bh$ **71** 80π **73** $4a^2(\pi-2)$,

$a=$ radius   **75** $\dfrac{\sqrt{2}\,\pi}{2}$ **77** (a) $2x+2y+2z$; (b) 0   **79** π **81** $-\pi$

INDEX

Numbers in italic type indicate pages on which related problems will be found.

Calculus (continued from front cover)

Rational Forms Involving $a + bu$

1 $\displaystyle \int \frac{du}{a+bu} = \frac{1}{b} \ln |a+bu| + C$

6 $\displaystyle \int \frac{u\,du}{(a+bu)^3} = -\frac{1}{b^2}\left[\frac{1}{a+bu} - \frac{a}{2(a+bu)^2}\right] + C$

2 $\displaystyle \int \frac{u\,du}{a+bu} = \frac{1}{b^2}\left[a+bu - a\ln|a+bu|\right] + C$

7 $\displaystyle \int \frac{du}{u(a+bu)} = -\frac{1}{a}\ln\left|\frac{a+bu}{u}\right| + C$

3 $\displaystyle \int \frac{u^2\,du}{a+bu} = \frac{1}{b^3}\left[\frac{1}{2}(a+bu)^2 - 2a(a+bu)\right.$
$\displaystyle \left. + a^2\ln|a+bu|\,\right] + C$

8 $\displaystyle \int \frac{du}{u^2(a+bu)} = -\frac{1}{au} + \frac{b}{a^2}\ln\left|\frac{a+bu}{u}\right| + C$

9 $\displaystyle \int \frac{du}{u(a+bu)^2} = \frac{1}{a(a+bu)} - \frac{1}{a}\ln\left|\frac{a+bu}{u}\right| + C$

4 $\displaystyle \int \frac{u\,du}{(a+bu)^2} = \frac{1}{b^2}\left[\frac{a}{a+bu} + \ln|a+bu|\right] + C$

5 $\displaystyle \int \frac{u^2\,du}{(a+bu)^2} = \frac{1}{b^3}\left[a+bu - \frac{a^2}{a+bu} - 2a\ln|a+bu|\right] + C$

Forms Involving $\sqrt{a + bu}$

1 $\displaystyle \int u\sqrt{a+bu}\,du = -\frac{2(2a-3bu)(a+bu)^{3/2}}{15b^2} + C$

5 $\displaystyle \int \frac{u^2\,du}{\sqrt{a+bu}} = \frac{2(8a^2 - 4abu + 3b^2u^2)\sqrt{a+bu}}{15b^3} + C$

2 $\displaystyle \int u^2\sqrt{a+bu}\,du$
$\displaystyle = \frac{2(8a^2 - 12abu + 15b^2u^2)(a+bu)^{3/2}}{105b^3} + C$

6 $\displaystyle \int \frac{u^n\,du}{\sqrt{a+bu}} = \frac{2u^n\sqrt{a+bu}}{b(2n+1)} - \frac{2an}{b(2n+1)}\int \frac{u^{n-1}\,du}{\sqrt{a+bu}}$

3 $\displaystyle \int u^n\sqrt{a+bu}\,du = \frac{2u^n(a+bu)^{3/2}}{b(2n+3)}$
$\displaystyle - \frac{2an}{b(2n+3)}\cdot\int u^{n-1}\sqrt{a+bu}\,du$

7 $\displaystyle \int \frac{du}{u\sqrt{a+bu}} = \frac{1}{\sqrt{a}}\ln\left|\frac{\sqrt{a+bu}-\sqrt{a}}{\sqrt{a+bu}+\sqrt{a}}\right| + C \quad (a>0)$

4 $\displaystyle \int \frac{u\,du}{\sqrt{a+bu}} = -\frac{2(2a-bu)\sqrt{a+bu}}{3b^2} + C$

8 $\displaystyle \int \frac{\sqrt{a+bu}}{u}\,du = 2\sqrt{a+bu} + a\int \frac{du}{u\sqrt{a+bu}}$

9 $\displaystyle \int \frac{\sqrt{a+bu}}{u^n}\,du = -\frac{(a+bu)^{3/2}}{a(n-1)u^{n-1}} - \frac{b(2n-5)}{2a(n-1)}\int \frac{\sqrt{a+bu}}{u^{n-1}}\,du \quad (n \neq 1)$

Forms Involving $\sqrt{a^2 + u^2}$

1 $\displaystyle \int \sqrt{a^2+u^2}\,du = \frac{u}{2}\sqrt{a^2+u^2} + \frac{a^2}{2}\ln\left|u+\sqrt{a^2+u^2}\right| + C$

2 $\displaystyle \int u^2\sqrt{a^2+u^2}\,du = \frac{u}{8}(a^2+2u^2)\sqrt{a^2+u^2} - \frac{a^4}{8}\ln\left|u+\sqrt{a^2+u^2}\right| + C$

3 $\displaystyle \int \frac{\sqrt{a^2+u^2}}{u}\,du = \sqrt{a^2+u^2} - a\ln\left|\frac{a+\sqrt{a^2+u^2}}{u}\right| + C$

4 $\displaystyle \int \frac{du}{\sqrt{a^2+u^2}} = \ln\left|u+\sqrt{a^2+u^2}\right| + C$

6 $\displaystyle \int \frac{\sqrt{a^2+u^2}}{u^2}\,du = -\frac{\sqrt{a^2+u^2}}{u} + \ln\left|u+\sqrt{a^2+u^2}\right| + C$

5 $\displaystyle \int \frac{du}{(a^2+u^2)^{3/2}} = \frac{u}{a^2\sqrt{a^2+u^2}} + C$

7 $\displaystyle \int \frac{du}{u^2\sqrt{a^2+u^2}} = -\frac{\sqrt{a^2+u^2}}{a^2u} + C$